W9-BUR-899

the science of zoology

the science of zoology

PAUL B. WEISZ

Professor of Biology
Brown University
Providence, Rhode Island

McGRAW-HILL BOOK COMPANY

New York • St. Louis • San Francisco • Toronto • London • Sydney

COLOR PHOTO CREDITS

Plate I: *A,C,D*, Eric V. Gravé; *B*, Douglas P. Wilson.

Plate II: *A–F*, Eric V. Gravé.

Plate III: *A–E*, Eric V. Gravé.

Plate IV: *A–F*, Eric V. Gravé.

Plate V: *A,C*, R. H. Noailles; *B*, General Biological Supply House, Inc.; *D*, courtesy Marineland of the Pacific.

Plate VI: *A,C*, General Biological Supply House, Inc.; *B*, W. M. Stephens; *D*, courtesy Marineland of the Pacific.

Plate VII: *A,B*, R. H. Noailles; *C,D*, courtesy Dr. Jonathan P. Green, Brown University; *E*, General Biological Supply House, Inc.

Plate VIII: *A*, R. H. Noailles; *B*, Douglas P. Wilson; *C,D*, General Biological Supply House, Inc.

Plate IX: *A,B*, General Biological Supply House, Inc.; *C*, courtesy Dr. Jonathan P. Green, Brown University; *D*, courtesy Dr. Richard J. Goss, Brown University.

Plate X: *A,C,D,E*, courtesy Dr. Jonathan P. Green, Brown University; *B*, R. H. Noailles.

Plate XI: *A*, courtesy Marineland of the Pacific; *B,C*, R. H. Noailles; *D,E*, General Biological Supply House, Inc.

Plate XII: *A,C,D*, General Biological Supply House, Inc.; *B*, courtesy Dr. Jonathan P. Green, Brown University.

Plate XIII: *A*, Eric V. Gravé; *B,C,D*, courtesy Dr. Jonathan P. Green, Brown University; *E*, courtesy Dr. Grover C. Stephens, University of California, Irvine.

Plate XIV: *A*, R. H. Noailles; *B*, General Biological Supply House, Inc.; *C*, W. M. Stephens; *D*, courtesy Dr. Jonathan P. Green, Brown University.

Plate XV: *A*, courtesy Dr. Richard J. Goss, Brown University; *B*, courtesy Marineland of the Pacific; *C,D,E*, courtesy Dr. Jonathan P. Green, Brown University.

Plate XVI: *A*, R. H. Noailles; *B,D*, courtesy Dr. Jonathan P. Green, Brown University; *C*, General Biological Supply House, Inc.; *E*, W. M. Stephens.

the science of zoology

Library of Congress Catalog Card Number 65-21577
69107

1234567890 RN 7321069876

Picture editor, Gabriele Wunderlich.

PHOTO CREDITS

Unit and part credits

Unit I. W. Suschitsky.
Unit II. R. H. Noailles.
Part 1. Paul Popper, Ltd.
Part 2. Dr. Keith R. Porter, Harvard University.
Part 3. Paul Popper, Ltd.
Part 4. R. H. Noailles.
Part 5. American Museum of Natural History.
Part 6. Buffalo Museum of Science.
Part 7. Natural History Photographic Agency.
Part 8. Ward's Natural Science Establishment, Inc.
Part 9. R. H. Noailles.
Part 10. Paul Popper, Ltd.

Chapter credits

3.3. Courtesy Dr. Norman E. Williams, University of Iowa.
3.4. Placoderm, American Museum of Natural History; trout, R. H. Noailles.
3.6. Courtesy Dr. Clifford Grobstein, Stanford University.
3.10. Douglas P. Wilson.
3.11. U.S. Fish and Wildlife Service.
3.14. *A*, American Museum of Natural History; *B*, U.S. Department of Agriculture; *C*, Eastman Kodak Company.
3.16. *A*,*B*,*C*, Paul Popper, Ltd.; *D*, Ward's Natural Science Establishment, Inc.
4.26. Eric V. Gravé.
4.27. Dr. Keith R. Porter, Harvard University.
4.28. Dr. Keith R. Porter, Harvard University.
4.29. *A*, Dr. Keith R. Porter, Harvard University; *B*, Dr. H. Fernandez-Moran, University of Chicago, and J. Cell Biol., vol. 22, 1964.
4.30. Dr. Keith R. Porter, Harvard University.
4.31. Courtesy Dr. W. G. Whaley, University of Texas.
4.32. Courtesy Dr. W. G. Whaley, University of Texas.
4.34. *A*, General Biological Supply House, Inc.; *B*, John L. Schumacher.
4.35. *B*, courtesy Dr. Norman E. Williams, University of Iowa; *C*, Dr. Keith R. Porter, Harvard University; *D*, courtesy Dr. Dorothy R. Pitelka, University of California, Berkeley.
6.1. *A*, Ward's Natural Science Establishment, Inc.; *B*, Dr. Keith R. Porter, Harvard University.
6.7. American Museum of Natural History.
6.28. General Biological Supply House, Inc.
6.29. General Biological Supply House, Inc.
6.30. General Biological Supply House, Inc.
6.31. General Biological Supply House, Inc.
6.32. General Biological Supply House, Inc.
6.34. Courtesy Dr. Herman B. Chase, Brown University.
7.2. © Walt Disney Productions.
7.4. *B*, Ward's Natural Science Establishment, Inc.; *C*,*D*, American Museum of Natural History.
7.5. *A*, Lynwood M. Chace.
7.7. Courtesy Dr. G. W. Corner and Carnegie Institution of Washington.
7.9. *A*, R. H. Noailles; *B*,*C*, U.S. Fish and Wildlife Service; *D*, American Museum of Natural History.
7.19. General Biological Supply House, Inc.
7.20. *B*, Eric V. Gravé.
7.21. *A*, General Biological Supply House, Inc.; *B*,*C*, Dr. B. J. Serber; *D*, Ward's Natural Science Establishment, Inc.
7.22. *A*, Ward's Natural Science Establishment, Inc.; *B*, R. H. Noailles.
7.24. M. C. Noailles.
7.25. M. C. Noailles.
8.3. *A*, courtesy Dr. Mac E. Hadley, Brown University; *B*, courtesy Dr. W. Montagna, Oregon Regional Primate Research Center.
8.4. *A*, Leonard L. Rue; *B*, R. H. Noailles.
8.5. Courtesy Dr. W. Montagna, Oregon Regional Primate Research Center.
8.7. *A*, Douglas P. Wilson; *B*, General Biological Supply House, Inc.
8.8. *B*, American Museum of Natural History; *C*, courtesy Dr. William Montagna, Oregon Regional Primate Research Center.
8.9. Ward's Natural Science Establishment, Inc.
8.11. M. C. Noailles.
8.12. *A*, M. C. Noailles; *B*, Ward's Natural Science Establishment, Inc.
8.17. General Biological Supply House, Inc.
8.18. M. C. Noailles.
8.19. Ward's Natural Science Establishment, Inc.
8.22. Ward's Natural Science Establishment, Inc.

8.23. Courtesy Dr. Mac V. Edds, Brown University.
8.25. *C*, Ward's Natural Science Establishment, Inc.
8.26. *F,G*, Photographic Department, Rhode Island Hospital.
8.28. Ward's Natural Science Establishment, Inc.
8.37. *A*, Ward's Natural Science Establishment, Inc.; *B*, General Biological Supply House, Inc.
8.45. Ward's Natural Science Establishment, Inc.
8.48. Paul Popper, Ltd.
8.50. Ward's Natural Science Establishment, Inc.
9.3. *B*, Dr. B. J. Serber; *C*, General Biological Supply House, Inc.
9.6. Photographic Department, Rhode Island Hospital.
9.11. General Biological Supply House, Inc.
9.12. Courtesy Dr. Robert Brenner, Oregon Regional Primate Research Center.
9.18. Ward's Natural Science Establishment, Inc.
9.21. Courtesy Dr. Elizabeth Leduc, Brown University.
9.22. *A,D*, Dr. B. J. Serber; *B,C*, M. C. Noailles; *E*, General Biological Supply House, Inc.
9.24. M. C. Noailles.
9.28. R. H. Noailles.
9.31. *A*, model designed by Dr. J. F. Mueller, photo by Ward's Natural Science Establishment, Inc.; *C*, Bell Telephone Laboratories, Inc.
10.2. Douglas P. Wilson.
10.3. Douglas P. Wilson.
10.4. Courtesy Dr. Charles Thornton, Michigan State University.
10.5. Ward's Natural Science Establishment, Inc.
10.11. R. H. Noailles.
10.17. *A,B*, General Biological Supply House, Inc.
10.25. Ward's Natural Science Establishment, Inc.
10.26. Ward's Natural Science Establishment, Inc.
10.29. Courtesy Carnegie Institution of Washington.
10.33. Dr. A. Gesell, from fig. 10, "The Embryology of Behavior," Harper and Brothers.
11.2. Courtesy Dr. Clifford Grobstein, and 13th Growth Symposium, Princeton University Press, 1954.
11.5. R. H. Noailles.
11.10. American Museum of Natural History.
11.12. Courtesy Dr. Roberts Rugh, from "Experimental Embryology," Burgess Publishing Company.
11.16. General Biological Supply House, Inc.
11.19. Eric V. Gravé.
11.20. Top, courtesy Dr. Roberts Rugh, from "Experimental Embryology," Harcourt, Brace & World, Inc.; bottom left and right, General Biological Supply House, Inc.; bottom middle, M. C. Noailles.
11.24. *B,C*, Douglas P. Wilson.
11.25. *A*, M. C. Noailles; *B*, Douglas P. Wilson.
11.28. L. E. Perkins, Natural History Photographic Agency.
12.1. American Museum of Natural History.
12.5. *A*, M. C. Noailles.
12.25. R. H. Noailles.
13.10. Genetics Laboratory, New York Zoological Society.
13.11. *A*, Paul Popper, Ltd.; *B*, New York Zoological Society.
13.12. *A,B,C*, American Museum of Natural History; *D*, Chicago Natural History Museum.
14.7. Courtesy Drs. R. M. Herriott and J. L. Barlow, J. Gen. Physiol., vol. 36, p. 17.
14.8. Courtesy Dr. A. G. Smith, J. Bacteriol., vol. 59, 1950, and American Society of Microbiology.
15.1. American Museum of Natural History.
15.2. *A*, American Museum of Natural History; *B*, Chicago Natural History Museum.
15.4. *A,B*, Chicago Natural History Museum; *C*, American Museum of Natural History.
15.5. American Museum of Natural History.
15.6. Chicago Natural History Museum.
15.8. Chicago Natural History Museum.
15.10. American Museum of Natural History.
15.12. Courtesy Peabody Museum, Yale University.
15.13. American Museum of Natural History.
15.14. Chicago Natural History Museum.
15.17. American Museum of Natural History.
15.18. American Museum of Natural History.
15.19. Chicago Natural History Museum.
15.20. Chicago Natural History Museum.
15.22. American Museum of Natural History.
15.23. American Museum of Natural History.
15.25. American Museum of Natural History.
15.26. *A*, Paul Popper, Ltd.; *B*, Lynwood M. Chace.
15.27. W. Suschitzky.
15.31. American Museum of Natural History.
15.32. American Museum of Natural History.
15.33. Chicago Natural History Museum.
15.34. American Museum of Natural History.
16.1. Standard Oil Company, N.J.
16.2. New York Zoological Society.
16.3. New York Zoological Society.
16.4. American Museum of Natural History.
16.7. M. Woodbridge Williams.
16.15. U.S. Department of Agriculture.
16.17. U.S. Department of Agriculture.

16.18. U.S. Department of Agriculture.
16.19. *A,B*, © Walt Disney Productions; *C*, Lynwood M. Chace.
16.20. R. H. Noailles.
16.21. *A,B,C*, U.S. Department of Agriculture; *D*, Buffalo Museum of Science.
16.22. *A*, American Museum of Natural History; *B*, Australian News and Information Bureau.
16.23. Leonard L. Rue.
16.24. R. H. Noailles.
17.8. *A*, R. H. Noailles; *B*, Paul Popper, Ltd.; *C*, S. Dalton, Natural History Photographic Agency.
17.10. E. W. Teale.
17.11. *A*, J. Carel and Larousse Publishing Company; *B*, R. H. Noailles.
17.12. *A*, New York Zoological Society; *B*, Paul Popper, Ltd.
17.13. Courtesy Dr. R. W. G. Wyckoff, from "Electron Microscopy," Interscience Publishers, Inc., 1949.
17.14. Armed Forces Institute of Pathology.
17.15. General Biological Supply House, Inc.
17.16. R. H. Noailles.
17.17. Courtesy Dr. R. W. G. Wyckoff, from "Electron Microscopy," Interscience Publishers, Inc., 1949.
17.18. E. W. Teale.
18.7. U.S. Department of Agriculture.
18.14. M. Woodbridge Williams.
18.17. R. H. Noailles.
18.18. American Museum of Natural History.
18.19. American Museum of Natural History.
18.20. R. H. Noailles.
18.21. © Walt Disney Productions.
18.22. *A*, South African Tourist Corporation; *B*, British Overseas Airways Corporation.
18.23. National Park Service.
18.24. *A*, Lynwood M. Chace; *B*, © Walt Disney Productions.
18.25. *A*, Leonard L. Rue; *B*, W. Suschitzky.
18.26. *A*, W. Suschitzky; *B*, © Walt Disney Productions.
18.27. *A*, Paul Popper, Ltd.; *B*, Australian News and Information Bureau.
19.11. Natural History Photographic Agency.
19.13. R. H. Noailles.
19.22. *B,C*, Eric V. Gravé.
19.28. *C*, Courtesy Dr. Maria A. Rudzinska and J. Gerontology, vol. 16, p. 213, 1961.
20.2. *A*, M. Woodbridge Williams; *B,C,D*, American Museum of Natural History.
20.4. General Biological Supply House, Inc.
21.8. *A*, R. H. Noailles; *B*, Douglas P. Wilson.
21.9. *A*, M. Woodbridge Williams; *B*, M. C. Noailles.
21.10. Douglas P. Wilson.
21.12. Ward's Natural Science Establishment, Inc.
21.14. *D*, Douglas P. Wilson.
21.18. Douglas P. Wilson.
21.20. *A*, Douglas P. Wilson; *B*, R. H. Noailles.
21.21. M. Woodbridge Williams.
21.22. R. H. Noailles.
21.23. *A*, Douglas P. Wilson; *C*, Natural History Photographic Agency.
21.25. Douglas P. Wilson.
22.10. *A*, General Biological Supply House, Inc.; *B*, Ward's Natural Science Establishment, Inc.
22.11. *A*, Douglas P. Wilson; *B*, General Biological Supply House, Inc.
22.13. Ward's Natural Science Establishment, Inc.
22.14. *B*, Ward's Natural Science Establishment, Inc.
22.17. Ward's Natural Science Establishment, Inc.
22.19. Douglas P. Wilson.
23.1. *A*, M. C. Noailles; *B*, American Museum of Natural History.
23.12. Ward's Natural Science Establishment, Inc.
24.4. Douglas P. Wilson.
24.6. *A*, Douglas P. Wilson; *B*, R. H. Noailles.
24.11. General Biological Supply House, Inc.
24.13. Dr. Richard A. Boolootian.
25.1. Natural History Photographic Agency.
25.5. American Museum of Natural History.
25.8. Douglas P. Wilson.
25.10. R. H. Noailles.
25.12. *A*, Douglas P. Wilson; *B*, R. H. Noailles.
25.13. Douglas P. Wilson.
25.15. *A*, R. H. Noailles; *B*, M. Woodbridge Williams.
25.16. M. Woodbridge Williams.
25.17. *A*, R. H. Noailles; *B*, Lynwood M. Chace.
25.20. American Museum of Natural History.
25.27. *A*, U.S. Fish and Wildlife Service; *B,C*, R. H. Noailles; *D*, Paul Popper, Ltd.
25.32. American Museum of Natural History.
25.34. *A*, R. H. Noailles; *B*, Natural History Photographic Agency.
26.5. R. H. Noailles.
26.6. Douglas P. Wilson.
26.8. Douglas P. Wilson.
26.10. *A*, M. Woodbridge Williams; *B*, R. H. Noailles.

26.11. American Museum of Natural History.
26.12. American Museum of Natural History.
26.15. Lynwood M. Chace.
26.17. *B*, Harold V. Green.
26.19. Australian News and Information Bureau.
27.1. U.S. Department of Agriculture.
27.7. Leonard L. Rue.
27.10. R. H. Noailles.
27.12. *A*, R. H. Noailles; *B*, Harold V. Green.
27.14. *A*, Lynwood M. Chace; *B*, S. Dalton, Natural History Photographic Agency.
27.16. *A*, Standard Oil Company, N.J.; *B*, S. Dalton, Natural History Photographic Agency.
27.17. *A*, R. H. Noailles; *B*, Lynwood M. Chace; *C*, General Biological Supply House, Inc.
27.18. Leonard L. Rue.
28.11. Eric V. Gravé.
28.12. *A*,*B*, M. Woodbridge Williams; *C*, American Museum of Natural History.
28.16. R. H. Noailles.
28.20. *A*, Ward's Natural Science Establishment, Inc.; *B*, Natural History Photographic Agency; *C*, J. Focht, Leonard L. Rue Enterprises; *D*, Eric V. Gravé.
28.21. *A*, R. H. Noailles; *B*, U.S. Department of Agriculture; *C*, Leonard L. Rue.
28.22. *A*, Lynwood M. Chace; *B*, M. C. Noailles; *C*,*E*, R. H. Noailles; *D*, S. Dalton, Natural History Photographic Agency.
28.23. *A*, Lynwood M. Chace; *B*, S. Dalton, Natural History Photographic Agency; *C*, U.S. Department of Agriculture.
28.24. *A*, R. H. Noailles; *B*, S. Dalton, Natural History Photographic Agency.
28.25. R. H. Noailles.
28.27. Lynwood M. Chace.
28.33. R. H. Noailles.
29.1. Douglas P. Wilson.
29.7. American Museum of Natural History.
29.11. Douglas P. Wilson.
29.17. *A*, Paul Popper, Ltd.
29.22. Douglas P. Wilson.
29.23. *B*, Douglas P. Wilson; *C*, M. Woodbridge Williams; *D*, General Biological Supply House, Inc.
29.25. Natural History Photographic Agency.
29.27. *B*, American Museum of Natural History.
29.29. Douglas P. Wilson.
29.30. M. Woodbridge Williams.
29.31. R. H. Noailles.
29.33. Douglas P. Wilson.
29.34. *A*, R. H. Noailles; *B*, Paul A. Knipping.
29.36. G. Cripps and Harold V. Green.
30.2. American Museum of Natural History.
30.6. *A*, Douglas P. Wilson; *B*, Natural History Photographic Agency; *C*, R. H. Noailles.
30.8. General Biological Supply House, Inc.
30.9. Douglas P. Wilson.
30.13. Douglas P. Wilson.
31.5. Courtesy Dr. W. Montagna, Oregon Regional Primate Research Center.
31.13. American Museum of Natural History.
31.16. R. H. Noailles.
31.17. Natural History Photographic Agency.
31.18. Canadian Pacific Railroad Company.
31.20. *C*, R. H. Noailles.
31.21. R. H. Noailles.
31.22. R. H. Noailles.
32.5. L. E. Perkins, Natural History Photographic Agency.
32.6. R. H. Noailles.
32.7. Lynwood M. Chace.
32.9. M. Woodbridge Williams.
32.11. Paul Popper, Ltd.
32.12. Ward's Natural Science Establishment, Inc.
32.13. © Walt Disney Productions.
32.14. Australian News and Information Bureau.
32.16. Leonard L. Rue.
32.17. American Museum of Natural History.
32.19. Ward's Natural Science Establishment, Inc.
32.22. Australian News and Information Bureau.
32.23. *A*, Lynwood M. Chace; *B*, J. Van Wormer.
32.24. *A*, Paul Popper, Ltd.; *B*, Australian News and Information Bureau.
32.25. Australian News and Information Bureau.
32.26. Australian News and Information Bureau.
32.28. *A*, W. Suschitzky; *B*, R. H. Noailles; *C*, Paul Popper, Ltd.
32.30. Marineland, Florida.
32.32. *A*, S. Dalton, Natural History Photographic Agency; *B*, W. Suschitzky; *C*, Standard Oil Company, N.J.

preface

At the present stage of biological knowledge, a college-level program in general zoology should consist of more than the traditional phylum-by-phylum presentation of descriptive gross morphology; particularly if, as is often the case, such a presentation tends to become equated with all of "zoology." Both the comparative maturity of the zoological sciences and the proved learning potential of the present-day student virtually demand that a meaningful course of study encompass the *whole* animal; gross morphology, while important, should comprise only part of the course. The student should also be given the opportunity to be intellectually excited by the biochemical, genic, and general molecular basis of animal life; the development of animals and the fascinating problems posed by eggs, embryos, and larvae; the main trends of current evolutionary and phylogenetic thought; and the reproductive, behavioral, ecological, physiological, and other attributes of animal nature. In other words, a significant discourse in zoology should examine not only the groups but also the fundamental *principles* underlying all animal life.

Accordingly, this book is divided into two broad units. The first, consisting of 6 three-chapter parts, emphasizes principles. In a sense it follows a "systems approach," though such a label here would embrace the systems of all levels of animal organization. The introduction lays the scientific, chemical, and specifically zoological foundations for all subsequent discussions. Then follows a thorough examination of the animal cell, beginning with the chemical, physical, and biological organization of cells and continuing with the structural and functional details of their molecular "systems." Based on this background, the next part outlines the ways in which cells form higher-level components of the animal body. It sketches the gross designs of various body forms, and describes the interrelations between taxonomy, embryology, and adult anatomy. A section on organ systems covers the structure, function, and development of these systems in both invertebrates and vertebrates. The fourth part focuses on the continuity of animals in space and time. Reproduction, development, and heredity are the chief topics, and the sections on development include a fairly detailed introduction to the patterns and processes of animal embryology. The fifth part concentrates on historical zoology, and covers evolutionary principles, paleontological data, and a critical survey of major phylogenetic hypotheses. Woven into this context are accounts of the origins of life, of animals, of the main animal groups, and of mammals and man specifically. The last part in this unit deals with animal "systems" on the most comprehensive level of organization: the ecosystem and the biosphere. The properties of species, populations, and symbiotic associations are discussed at length, as are the principal aspects of habitats, community life, zoogeography, and the environment as a whole. Taken together, the entire first unit thus represents a step-by-step analysis of the common foundations of animal life, on all levels from the molecular to the global.

The second unit contains 14 chapters organized into four parts, and constitutes a systematic examination of animal groups. Protozoa, sponges, and the radiate animals are the subject of the first part, acoelomates and pseudocoelomates of the second, and the coelomate animals of the following two parts. *All* groups, major as well as so-called "minor" ones, are treated in considerable detail. In most instances the characteristics of phyla are dealt with as far down as the ordinal rank. Because of the conceptual preparation provided in the first unit, the second can cover the subject in depth. For example, patterns of embryonic and larval development can be described specifically for each group; pertinent ecological and paleontological data can be introduced where warranted; and phylogenetic considerations already outlined generally for all groups can be probed as they apply directly to each group. Jointly, therefore, the two units should offer an adequate framework for an understanding of the whole animal.

It is by no means necessary that the segments of the book be studied consecutively in the order given, or that all segments be studied with equal emphasis. For example, it is eminently feasible to base a course primarily on Unit 1, with Unit 2 being used largely in a supplementary role. Or Unit 2 might provide the broad base of the course, with Unit 1 serving collaterally. Or if both units are to receive roughly equivalent attention, they can be programmed either consecutively or concurrently, Unit 2 perhaps being coordinated with laboratory work. Whatever basic pattern of procedure may fit a given course best, it is probably desirable that topic sequences within the pattern match those of the text in at least broad outline. Some shifting of topics is certainly possible however. For example ecological or evolutionary zoology might well be taken up at almost any point of a course. Also, with the possible exception of key chapters or sections, other particular portions of the book may be omitted from formal study if the instructor so specifies. In short, the text is designed to make available a maximum amount of material, and the instructor can decide how much of this material to use and how it should be arranged to fit his course.

Auxiliary study aids have been given considerable attention. Of the well over 800 illustrations, the photographs have been chosen carefully for pertinence, and the line drawings have been newly prepared. To each chapter are appended lists of review questions and collateral readings, the latter selected from both the popular and the technical literature. An extensive glossary (with pronunciation guide) and a complete index will be found at the end of the book. Available in addition is an Instructor's Manual for teachers, written specifically for use with the text.

A number of colleagues and friends have contributed in vital ways to the ultimate completion of this work. Dr. Grover C. Stephens, of the University of California at Irvine, and Dr. William C. Grant, Jr., of Williams College, Mass., have reviewed the whole manuscript generally and different portions of it in detail. Their constructive criticism has been exceedingly welcome and helpful, and I thank them for it. Mr. Russell Peterson has executed the drawings from sketches with a devotion and an artistic skill greater than I could have hoped for, and there can be little question that his work has added immeasurably to such positive values as this book may have. Miss Gabriele Wunderlich has labored diligently in procuring photographs, and numerous professional friends and organizations have supplied the photographs and have permitted them to be used. For all these efforts I extend my sincere and lasting appreciation. I also wish to record a special tribute to the members of my family, whose patience during my long preoccupation with this undertaking demands my most grateful recognition.

Paul B. Weisz

contents

UNIT 1 the heritage of animals

UNIT 

the heritage of animals

Despite their numerous and often obvious differences, animals of all kinds have far more in common than might at first be suspected. For example, all are alike in that they occupy specific parts of the environment; in that they have specific past histories; in that they possess body structures which reflect both their present places in the environment and their historical past; and, above all, in that very similar events occur within them which collectively make them *living* creatures. In effect, it is only in their more superficial traits that animals differ to greater or lesser degree. In most basic and fundamental respects all animals are, and indeed must be, very much alike.

All such common features represent an animal heritage, a set of life-maintaining properties inherited from nonanimal ancestors, i.e., the earliest living things on earth. The first unit of this book, encompassing roughly the first half of the whole volume, is devoted to a study of this common heritage of animals. The objective will be to examine in substantial detail the structural parts and the functional processes which underlie all life generally and animal life specifically. The second unit will then deal with specific animals as such, i.e., the actual embodiments of the ancestral heritage.

the study of animals

PART 1

The investigation of animal life is the concern of the science of zoology. As a science, zoology is interrelated closely with all other natural sciences, chemistry in particular. Therefore, a serious study of zoology may properly begin with three separate, equally pertinent questions. First, what is a science? Secondly, what areas of chemistry and other natural sciences are important toward an understanding of zoology? And thirdly, what is an animal—what, indeed, is life?

Answers to these questions are outlined in the three chapters of this introductory part.

CHAPTER 1

science and zoology

Our current civilization is so thoroughly permeated with science that, for many, the label "scientific" has become the highest badge of merit, the hallmark of progress, the dominant theme of the age of atoms and space. No human endeavor, so it is often claimed, can really be worthwhile or of basic significance unless it has a scientific foundation. Moreover, advertisements loudly proclaim the "scientific" nature of consumer goods, and their "scientifically proved" high quality is attested to by "scientific" experts. Human relations too are supposed to be "scientific" nowadays. Conversation and debate have become "scientific" discussions; and in a field such as sports, if one is a good athlete, he is a "scientific" athlete.

There are even those who claim to take their religion "scientifically" and those who stoutly maintain that literature, painting, and other artistic pursuits are reducible to "science," really. And then there are those who believe that science will eventually solve "everything" and that, if only the world were run more "scientifically," it would be a much better place.

Yet in contrast to this widespread confidence in things and activities which claim to be, and in a few cases actually are, scientific, large segments of society doubt and mistrust scientists as persons. To many, the scientist is somehow queer and "different." He is held to be naive and more or less uninformed outside his specialty. He is pictured as a cold, godless calculating machine living in a strange, illusory world of his own.

Many circumstances in our civilization conspire to foster such false, stereotyped notions about science and scientists. However, no one who wishes to consider himself properly educated can afford to know about the meaning of science only what popular misconceptions and "common knowledge" may have taught him. Especially is this true for one who is about to pursue studies in a modern science such as zoology.

What then *is* the actual meaning of science? How did truly scientific undertakings develop, and how does science "work"? What can it do and, more especially, what can it not do? How does science differ from other forms of activity, and what place does it have in the scheme of modern culture?

THE ORIGIN OF SCIENCE

Science began in the distant past, long before human history was being recorded. Its mother was tribal magic.

The same mother also gave birth to religion, and, probably even earlier, to art. Thus science, religion, and art have always been blood brothers. Their methods differ, but their aim is the same: to understand and interpret the universe and its workings and, from this, to promote the material and spiritual welfare of man where possible.

This was also the function of tribal magic. For long ages, magic was the rallying point of society, the central institution in which were concentrated the accumulated wisdom and experience of the day. The execution of magical procedures was in the hands of specially trained individuals, the medicine men and their equivalents. These were the forerunners of the scientists and the clergymen of today. How did science and religion grow out of magic? We may illustrate by means of an example.

Several thousand years ago, it was generally believed that magical rites were necessary to make wheat grow from planted seeds. In this particular instance, the rites took one of two forms. Either man intensified his sexual activity, in a solemn spring festival celebrated communally in the fields, or he abstained completely from sexual activity during the planting period.

The first procedure was an instance of imitative magic. The reasoning was that, since sowing seeds is like producing pregnancy in a woman, man could demonstrate to the soil what was wanted and so induce it to imitate man and be fertile. The second procedure, an instance of contagious magic, grew out of the assumption that only a limited amount of reproductive potency was available to living things. Consequently, if man did not use up his potency, that much more would become available for the soil. Depending on the tribe, the time, and the locality, either imitative or contagious magic might have been used to attain the same end, namely, to make the earth fruitful.

The fundamental weakness of magic was, of course, that it was unreliable. Sometimes it worked, and sometimes it did not work. Bad soil, bad grain, bad weather, and insect pests often must have defeated the best magic. In time, man must have realized that magical rites actually played no role in wheat growth, whereas soil conditions, grain quality, and good weather played very important roles. This was a momentous discovery—and a scientific one.

Magic became science when man accidentally found, or began to look for, situations which could be predictably controlled without magical rituals. In many situations where magic seemed to work successfully most of the time, man discovered an underlying scientific principle.

Yet there remained very many situations where magic did not work and where scientific principles could not be found. For example, in spite of good soil and good weather, wheat might not have grown because of virus or fungus infections. Such contingencies remained completely beyond man's understanding up to very modern times, and early man could only conclude that unseeable, uncontrollable "somethings" occasionally defeated his efforts. These somethings became spirits and gods. And unless prayers and sacrificial offerings maintained the good will of the gods, their wrath would undo human enterprise. Thus magical rituals evolved into primitively religious ones.

At this stage, medicine men ceased to be magicians and instead assumed the dual role of priest and scientist. Every personal or communal undertaking required both scientific and religious action: science, to put to use what was known; religion, to protect against possible failure by inducing the unknown to work on man's side.

In time, the "two-way" medicine man disappeared and made way for the specialized scientist and priest. In both religion and science, shades of the old magic lingered on for long periods. The religions still retain a high magical content today, and the sciences only recently dissociated from magic-derived pseudosciences such as alchemy, astrology, and the occult arts.

Throughout the early development of science and religion, emphasis was largely on practical matters. Science was primitively technological, and religion too was largely "applied," i.e., designed to deal with the concrete practical issues of the day. Man was preoccupied mainly with procuring food, shelter, and clothing, and science and religion served these necessities. Later, as a result of technological successes, more time became available for contemplation and cultural development, and this is when researchers and theorists appeared alongside the technologists, and theologians alongside the clergymen.

THE FORMS OF SCIENCE

Today there are three types of scientists carrying on two kinds of science.

One kind of scientist may be symbolized as a man who sits by the river on nice afternoons and who whittles away at a stick and wonders about things. Strange as it may seem to some, the most powerful science stems from such whittlers. Whereas most people who just sit manage merely to be lazy, a few boil quietly with rare powers and make the wheels of the world go round. Thinker-scientists of this sort usually are not too well known by the general public, unless their thoughts prove to be of outstanding importance. Newton, Einstein, Darwin, and Freud are among the best known.

A second kind of scientist is the serious young man in the white coat, reading the dials of monster machines while lights flash and buzzers purr softly. This picture symbolizes the technician, the lab man, the trained expert who tests, experiments, and works out the implications of what the whittler has been thinking.

The third kind of scientist is a relatively new phenomenon. He goes to an office, dictates to secretaries, and spends a good part of his time in conferences or in handling contracts, budgets, and personnel. This is the businessman-scientist, who gets and allocates the funds which buy time and privacy for the whittler and machines for the lab man.

Note, however, that every scientist worthy of the name actually is a complex mixture of philosopher, technician, and businessman all rolled into one, and none is a "pure" type. But the relative emphasis varies greatly in different scientists.

Whatever type mixture he may be, a scientist works either in basic research, often called *pure science*, or in technology, often called *applied science*.

Basic research is done primarily to further man's understanding of nature. Possible practical applications of the findings are here completely disregarded. Scientists in this field are more frequently of the philosopher–lab-man type than in technology. They may be found principally in university laboratories and research institutes and, in lesser numbers, in industry and government. They have little to show for their efforts beyond the written accounts of their work; hence it is comparatively hard for them to convince nonscientists that they are doing anything essential. However, government and every enlightened industry today either support independent research or conduct such research. And the public is beginning to realize that pure science is the soil from which applied science must develop.

Technology is concerned primarily with applying the results of pure science to practical uses. No lesser inventiveness and genius are required in this field than in basic research, though here the genius is more of a commercial and less of a philosophical nature. Physicians, engineers, crime detectives, drug manufacturers, agricultural scientists—all are technologists. They have services and tangible products to sell; hence the public recognizes their worth rather readily.

Here again, note that no scientist is pure researcher or pure technologist. Mixtures are in evidence once more, with emphasis one way or the other. Moreover, technology is as much the fertilizer of basic research as the other way round. As new theories suggest new ways of applying them, so new ideas for doing things suggest further advances in research. Thus, in most research today, pure and applied science work hand in hand. Many conclusions of pure science cannot be tested before the technologist thinks up the means of testing. Conversely, before the technologist can produce desirable new products, years of basic research may first be required. In so far as every basic researcher must use equipment, however modest, he is also a technologist; and insofar as every technologist must understand how and why his products work, he is also a basic researcher.

It follows that any science shrivels whenever either of its two branches ceases to be effective. If for every dollar spent on science an immediate, tangible return is expected, and if the budding scientist is prevented from being a whittler by the necessity of producing something salable, then basic research will be in danger of drying up. And when that happens, technology too will become obsolete before long.

THE PROCEDURE OF SCIENCE

Everything that is science ultimately has its basis in the *scientific method*. Both the powers and the limitations of science are defined by this method. And wherever the scientific method cannot be applied, there cannot be science.

Taken singly, most of the steps of the scientific method involve commonplace procedures carried out daily by every person. Taken together they amount to the most powerful tool man has devised to know and to control nature.

Observation

All science begins with *observation*, the first step of the scientific method.

At once this delimits the scientific domain; something that cannot be observed cannot be investigated by science. However, observation need not be direct. Atomic nuclei and magnetism, for example, cannot be perceived directly through our sense organs, but their effects can be observed with instruments. Similarly, mind cannot be observed directly, but its effects can be, as expressed, for example, in behavior.

For reasons which will become clear presently, it is necessary, furthermore, that an observation be *repeatable*, actually or potentially. Anyone who doubts that objects fall back to the ground after they have been thrown into the air can convince himself of it by repeating the observation. One-time events on earth are outside science.

Correct observation is a most difficult art acquired only after long experience and many errors. Everybody observes, with eyes, ears, touch, and all other senses, but few observe correctly. Lawyers experienced with witnesses, artists who teach students to draw objects in plain view, and scientists who try to see nature all can testify to this.

This difficulty of observation lies largely in unsuspected bias. People forever see what they *want* to see or what they think they *ought* to see. It is extremely hard to rid oneself of such unconscious prejudice and to see just what is actually there, no more and no less. Past experience, "common knowledge," and often teachers can be subtle obstacles to correct observation, and even experienced scientists may not always avoid them. That is why a scientific observation is not taken at face value until several scientists have repeated the observation independently and have reported the same thing. That is also a major reason why one-time, unrepeatable events normally cannot be science.

A scientific piece of work is only as good as the original observation. Observational errors persist into everything that follows, and the effort may be defeated before it has properly begun.

Problem

After an observation has been made, the second step of the scientific method is to define a *problem*. In other words, one asks a question about the observation. How does so and so come about? What is it that makes such and such happen in this or that fashion? Question asking additionally distinguishes the scientist from the layman; everybody makes observations, but not everybody shows further curiosity.

More significantly, not everyone sees that there may actually be a problem connected with an observation. During thousands of years, even curious people simply took it for granted that a detached, unsupported object falls to the ground. It took genius to ask, "How come?" and few problems, indeed, have ever turned out to be more profound.

Thus scientists take nothing for granted, and they ask questions, even at the risk of irritating others. Question askers are notorious for getting themselves into trouble, and so it has always been with scientists. But they have to continue to ask questions if they are to remain scientists. And society has to expect annoying questions if it wishes to have science.

Anyone can ask questions. However, good questioning, like good observing, is a high art. To be valuable scientifically, a question must be *relevant*, and it must be *testable*. The difficulty is that it is often very hard or impossible to tell in advance whether a question is relevant or irrelevant, testable or untestable. If a man collapses on the street and passers-by want to help him, it may or many not be irrelevant to ask when he had his last meal. Without experience one cannot decide on the relevance of this question, and a wrong procedure might be followed.

As to the testability of questions, it is clear that proper testing techniques must be available, actually or potentially. This cannot always be guaranteed. For example, Einstein's fame rests, in part, on his showing that it is impossible to test whether or not the earth moves through an "ether," an assumption held for many decades. All questions about an ether therefore become nonscientific, and we must reformulate associated problems until they become testable. Einstein did this, and he came up with relativity.

In general, science does best with "How?" or "What?" questions. "Why?" questions are more troublesome. Some of them can be rephrased to ask "How?" or "What?" But others such as "Why does the universe exist?" fall into the untestable category. These are outside the domain of science.

Hypothesis

Having asked a proper question, the scientist proceeds to the third step of the scientific method. This involves the seemingly quite unscientific procedure of guessing. One guesses what the answer to the question might conceivably be. Scientists call this postulating a *hypothesis*.

Hypothesizing distinguishes the scientist still further from the layman. For while many people observe and ask questions, most stop there. Some do wonder about likely answers, and scientists are among these.

Of course, a given question may have thousands of possible answers but only one right answer. Chances are therefore excellent that a random guess will be wrong. The scientist will not know whether his guess was or was not correct until he has completed the fourth step of the scientific method, *experimentation*. It is the function of every experiment to test the validity of a scientific guess.

If experimentation shows that the first guess was wrong, the scientist then must formulate a new hypothesis and once more test for validity by performing new experiments. Clearly, the guessing and guess-testing might go on for years, and a right answer might never be found. This happens.

But here again, artistry, genius, and experience usually provide shortcuts. There are good guesses and bad ones, and the skilled scientist is generally able to decide at the outset that, of a multitude of possible answers, so and so many are unlikely answers. His knowledge of the field, his past experience, and the experience of others working on related problems normally allow him to reduce the many possibilities to a few likelihoods.

This is also the place where hunches, intuitions, and lucky accidents aid science enormously. In one famous case, so the story has it, the German chemist Kekulé went to bed one night after a fairly alcoholic party and dreamed of six monkeys chasing one another in a circle, the tail of one held in the teeth of the other. Practically our whole chemical industry is based on that dream, for it told the sleeping scientist what the long-sought structure of benzene was—as we now know, six carbon atoms "chasing" one another in a circle. And benzene is the fundamental parent substance for thousands of chemical products.

The ideal situation for which the scientist generally strives is to reduce his problem to just two distinct alternative possibilities, one of which, when tested by experiment, may then be answered with a clear "yes," the other with a clear "no." It is exceedingly difficult to streamline problems in this way, and with many it cannot be done. Very often the answer obtained is "maybe." However, if a clear yes or no does emerge, scientists speak of an elegant piece of work, and such performances often are milestones in science.

Experiment

Experimentation is the fourth step in the scientific method. At this point, science and nonscience finally and completely part company.

Most people observe, ask questions, and also guess at answers. But the layman stops here: "My answer is so logical, so reasonable, and it sounds so 'right' that it must be correct." The listener considers the argument, finds that it is indeed logical and reasonable, and is convinced. He then goes out and in his turn converts others. Before long, the whole world rejoices that it has the answer.

Now the small, kill-joy voice of the scientist is heard in the background: "Where is the evidence?" Under such conditions in history, it has often been easier and more convenient to eradicate the scientist than to eradicate an emotionally fixed public opinion. But doing away with the scientist does not alter the fact that answers without evidence are at best unsupported opinions, at worst wishful thinking and fanatical illusions. Experimentation can provide the necessary evidence, and whosoever then experiments after guessing at answers becomes truly "scientific" in his approach, be he a professional scientist or not.

On the other hand, experiments do not guarantee a scientific conclusion. For there is ample room within experimentation and in succeeding steps to become unscientific again.

Experimentation is by far the hardest part of scientific procedure. There are no rules to follow; each experiment is a case unto itself. Knowledge and experience usually help technically, but to design the experiment, to decide on the means by which a hypothesis might best be tested, that separates the genius from the dilettante. The following example will illustrate the point:

Suppose you observe that a chemical substance *X*, which has accidentally spilled into a culture dish full of certain disease-causing bacteria, kills all the bacteria in that dish. Problem: Can drug *X* be used to protect human beings against these disease-causing bacteria? Hypothesis: yes. Experiment: You go to a hospital and find a patient with that particular bacterial disease and inject some of the drug into the patient.

Possible result 1: Two days later the patient is well. Conclusion: hypothesis confirmed. You proceed to market the drug at high prices. Shortly afterward, users of the drug die by the dozens, and you are tried and convicted for homicide.

Possible result 2: Two days later the patient is dead. Conclusion: The drug is worthless, and you abandon your project. A year later a colleague of yours is awarded the Nobel prize for having discovered a drug *X* which cures a certain bacterial disease in man—the same drug and the same disease in which you had been interested.

In this example, the so-called experiment was not an experiment at all.

First, no allowance was made for the possibility that people of different age, sex, eating habits, prior medical history, hereditary background, etc., might react differently to the same drug. Obviously, one would have to test the drug on many categories of carefully preselected patients, and there would have to be many patients in each such category. Besides, one would make the tests first on mice, or guinea pigs, or monkeys.

Secondly, the quantity of drug to be used was not determined. Clearly, a full range of dosages would have to be tested for each different category of patient. We tacitly assume, moreover, that the drug is a pure substance, i.e., that it does not contain traces of other chemicals which might obscure, or interfere with, the results. If impurities are suspected, whole sets of separate experiments would have to be made.

Thirdly, and most importantly, no account was taken of the possibility that your patient might have become well, or have died, in any case, even without your injecting the drug. What is needed here is *experimental control;* for every group of patients injected with drug solution, a precisely equal group must be injected with plain solution, without the drug. Then, by comparing results in the control and the experimental groups, one can determine whether or not the recovery or death of patients is really attributable to the drug.

Note that every experiment requires at least two parallel tests or sets of tests identical in all respects except one. Of these parallel tests, one is the control series, and it provides a standard of reference for assessing the results of the experimental series. In drug experiments on people, not fewer than about 100,000 to 200,000 test cases, half of them controls, half of them experimentals, would be considered adequate. It should be easy to see why a single test on a single test case may give completely erroneous conclusions. Many repetitions of the same test, under as nearly identical conditions as possible, and a full set of control tests for each set of the experimental tests—these are always prerequisite for any good experiment.

While an actual drug-testing program would be laborious, expensive, and time-consuming, the design of the experiment is nevertheless extremely simple. There are few steps to be gone through, and it is fairly clear what these steps must be. But there are many experiments in which the tests themselves may not take more than an hour or two, whereas thinking up appropriate, foolproof plans for the tests may have taken several years.

And despite a most ingenious design and a most careful execution, the result may still not be a clear yes or no. In a drug-testing experiment, for example, it is virtually certain that not 100 per cent of the experimental, drug-injected group will recover or that 100 per cent of the untreated control will remain sick.

The actual results might be something like 70 per cent recovery in the experimentals and something like 20 per cent recovery in the controls. The experimentals here show that 30 per cent of the patients with that particular disease do not recover despite treatment, and the controls show that 20 per cent of the patients get well even without treatment. Moreover, if 70 out of every 100 experimental patients recover, then 20 out of these 70 were not actually helped by the drug, since, from the control data, they would have recovered even without treatment. Hence the drug is effective in only 70 per cent minus 20 per cent, or 50 per cent, of the cases.

Medically, this may be a major accomplishment, for having the drug is obviously better than not having it. But scientifically, one is confronted with an equivocal "maybe" result. It will probably lead to new research based on the new observation that some people respond to the drug and some do not and to the new problem of why and what can be done about it.

The result of any experiment represents *evidence*. That is, the original guess in answer to a problem is confirmed as correct or is invalidated. If invalidated, a new hypothesis, with new experiments, must be thought up. This is repeated until a hypothesis may be hit upon which can be supported with confirmatory experimental evidence.

As with legal evidence, scientific evidence can be strong and convincing, or merely suggestive, or poor. In any case, nothing has been proved. Depending on the strength of the evidence, one merely has a basis for regarding the original hypothesis with a certain degree of confidence.

Our new drug, for example, may be just what we claim it to be when we use it in this country. In another part of the world it might not work at all or it might work better. All we can confidently say is that our evidence is based on so and so many experiments with American patients, American bacteria, and American drugs and that under specified hospital conditions, with proper allowance for unspotted errors, the drug has an effectiveness of 50 per cent. Experimental results are never better or broader than the experiments themselves.

This is where many who have been properly scientific up to this point become unscientific. Their claims exceed the evidence; they mistake their partial answer for the whole answer; they contend that they have proof for a fact, whereas all they actually have is some evidence for a hypothesis. There is always room for more and better evidence, or for new contradictory evidence, or indeed for better hypotheses.

Theory

Experimental evidence is the basis for the fifth and final step in the scientific method, the formulation of a *theory*.

When a hypothesis has been supported by really convincing evidence, best obtained in many different laboratories and by many independent researchers, and when the total accumulated evidence is unquestionably reliable within carefully specified limits, then a theory may be proposed.

In our drug example, after substantial corroborating evidence has also been obtained from many other test localities, an acceptable theory would be the statement that "in such and such a bacterial disease, drug *X* is effective in 50 per cent of the cases."

This statement is considerably broader than the experiments on which it is based. Theories always are. The statement implies, for example, that drug *X*, regardless of who manufactures it, will be 50 per cent effective anywhere in the world, under any conditions, and can be used also for animals other than man.

Direct evidence for these extended implications does not exist. But inasmuch as drug *X* is already known to work within certain limits, the theory expresses the belief, the *probability*, that it may also work within certain wider limits.

To that extent every good theory has predictive value. It prophesies certain results. In contrast to nonscientific prophecies, scientific ones always have a substantial body of evidence to back them up. Moreover, the scientific prophecy does not say that something will certainly happen, but says only that something is likely to happen with a stated degree of probability.

A few theories have proved to be so universally valid and to have such a high degree of probability that they are spoken of as *natural laws*. For example, no exception has ever been found to the observation that an apple, if disconnected from a tree and not otherwise supported, will fall to the ground. A law of gravitation is based on such observations.

Yet even laws do not pronounce certainties. For all practical purposes, it may well be irrational to assume that some day an apple will rise from a tree, yet there simply is no evidence that can absolutely guarantee the future. Evidence can be used only to estimate probabilities.

Most theories actually have rather brief life spans. For example, if, in chickens, our drug *X* should be found to perform not with 50 per cent but with 80 per cent efficiency, then our original theory becomes untenable and obsolete. And the exception to the theory becomes a new observation, beginning a new cycle of scientific procedure.

Thus new research might show that chickens contain a natural booster substance in their blood which materially bolsters the action of the drug. This might lead to isolation, identification, and mass production of the booster substance, hence to worldwide improvement in curing the bacterial disease. And we would also have a new theory of drug action, based on the new evidence.

Thus science is never finished. One theory predicts, holds up well for a time, exceptions are found, and a new, more inclusive theory takes over—for a while. We may note in passing that old theories do not become incorrect but merely become obsolete. Development of a new airplane does not mean that earlier planes can no longer fly. New theories, like new airplanes, merely range farther and serve more efficiently than earlier ones, but the latter still serve for their original purposes. Science is steady progression, not sudden revolution.

Clearly, knowledge of the scientific method does not by itself make a good scientist, any more than knowledge of English grammar alone makes a Shakespeare. At the same time, the demands of the scientific method should make it evident that scientists cannot be the cold, inhuman precision machines they are so often, and so erroneously, pictured to be. Scientists are essentially artists, and they require a sensitivity of eye and of mind as great as that of any master painter, and an imagination and keen inventiveness as powerful as that of any master poet.

THE LIMITATIONS OF SCIENCE

Observing, problem-posing, hypothesizing, experimenting, and theorizing—this sequence of procedural steps is both the beginning and the end of science. To determine what science means in wider contexts, we must examine what scientific method implies and, more especially, what it does not imply.

The Scientific Domain

First, scientific method defines the domain of science: *Anything to which the scientific method can be applied, now or in the future, is or will be science; anything to which the method cannot be applied is not science.*

This helps to clarify many a controversial issue. For example, does science have something to say about the concept of God? To determine this, we must find out if we can apply the scientific method.

Inasmuch as the whole universe and everything in it may be argued to be God's work, one may also argue that He is observable. It is possible, furthermore, to pose any number of problems, such as "Does He exist; is the universe indeed His doing?" and "Is He present everywhere and in everything?" One can also hypothesize; some might say "yes," some might say "no."

Can we design an experiment about God? For it to be reliable, we would need experimental control, i.e., two otherwise identical situations, one with God and one without. Now, what we wish to test is the hypothesis that God exists and is universal, i.e., that He is everywhere. Being a hypothesis thus far, this could be right or wrong.

If right, He would exist and exist everywhere; hence He would be present in every test we could possibly make. Thus we would never be able to devise a situation in which God is not present. But we need such a situation in order to have a controlled experiment.

But if the hypothesis is wrong, He would not exist, hence would be absent from every test we could possibly make. Therefore, we would never be able to devise a situation in which God *is* present. Yet we would need such a situation for a controlled experiment.

Right or wrong, our hypothesis is untestable either way, since we cannot run a controlled experiment. Hence we cannot apply the scientific method. The point is that the concept of God is outside the domain of science, and science cannot legitimately say anything about Him. He cannot be tested by science, because its method is inapplicable.

It should be carefully noted that this is a far cry from saying "Science disproves God," or "scientists must be godless; their method demands it." Nothing of the sort. Science specifically leaves anyone perfectly free to believe in any god whatsoever or in none. Many first-rate scientists are priests; many others are agnostics.

Science commits you to nothing more, and to nothing less, than adherence to scientific method.

Such adherence, it may be noted, is a matter of faith, just as belief in God or confidence in the telephone directory is a matter of faith. Whatever other faiths they may or may not hold, all scientists certainly have strong faith in the scientific method. So do those laymen who feel that having electric lights and not having bubonic plague are good things.

The Scientific Aim

A second consequence of the scientific method is that it defines the aim and purpose of science: *The objective of science is to make and to use theories.*

Many would say that the objective of science is to discover truth, to find out facts. We must be very careful here about the meaning of words. "Truth" is popularly used in two senses. It may indicate a temporary correctness, as in saying, "It is true that my hair is brown." Or it may indicate an absolute, eternal correctness, as in saying, "In plane geometry, the sum of the angles in a triangle is 180°."

From the earlier discussion on the nature of scientific method, it should be clear that science cannot deal with truth of the absolute variety. Something absolute is finished, known completely, once and for all. But science is never finished. Its method is unable to determine the absolute. Besides, once something is already known absolutely, there is no further requirement for science, since nothing further needs to be found out. Science can only adduce evidence for temporary truths, and another term for "temporary truth" is "theory." Because the word "truth," if not laboriously qualified, is ambiguous, scientists try not to use it at all.

The words "fact" and "proof" have a similar drawback. Both may indicate either something absolute or something temporary. If absolute, they are not science; if temporary, we have the less ambiguous word "evidence." Thus, science is content to find evidence for theories, and it leaves truths, proofs, and facts to others.

Speaking of words, "theorizing" is often popularly taken to mean "just talk and speculation." Consider, however, how successfully theorizing builds bridges!

Science and Values

A third important implication of the scientific method is that *it does not make value judgments or moral decisions.*

It is the user of scientific results who may place valuations on them. But the results by themselves do not carry built-in values. And nowhere in the scientific method is there a value-revealing step.

The consequences of this are vast. For example, the science which produced the atomic bomb and penicillin cannot, of itself, tell whether these products are good things or bad things. Every man must determine that for himself as best he can. The scientist who discusses the moral aspects of nuclear weapons can make weightier statements than a layman only in so far as he may know more about what damage such weapons may or may not do. This will certainly influence his opinions. But whatever opinion he gives, it will be a purely personal evaluation made as a citizen, and any other scientist—or layman—who is equally well informed about the capacities of the weapons may conceivably disagree completely. Human values are involved here; science is not.

In all other types of evaluations as well, science is silent and noncommittal. Beauty, love, evil, happiness, virtue, justice, liberty, property, financial worth—all these are human values which science cannot peg. To be sure, love, for example, might well be a subject of scientific research, and research might show much about what love is and how it works. But such research could never discover that love is wonderful, an evaluation clear to anyone who has done a certain amount of nonscientific research.

It also follows that it would be folly to strive for a strictly "scientific" way of life or to expect strictly "scientific" government. Certainly the role of science might profitably be enlarged in areas of personal and public life where science can make a legitimate contribution. But a completely scientific civilization, adhering strictly to the rules of the scientific method, could never tell, for example, whether it is right or wrong to commit murder, or whether it is good or bad to love one's neighbor. Science cannot and does not give such answers. However, this does not imply that science does away with morals. It merely implies that science cannot determine whether or not one ought to have moral standards, or what particular set of moral standards one ought to live by.

The Scientific Philosophy

A fourth and most important consequence of the scientific method is that it determines the philosophical foundation on which scientific pursuits must be based.

Inasmuch as the domain of science is the whole material universe, science must inquire into the nature of the forces which govern the universe and all happenings in it. What makes given events in the universe take place? What determines which event out of several possible ones will occur? And what controls or guides the course of any event to a particular conclusion?

Questions of this kind seek to discover the "prime mover" of the universe. As such they are actually philosophical questions of concern not only in science but in all other areas of human thought as well. Depending on how man answers such questions, he will adopt a particular philosophy of nature and this philosophy will then guide him in his various undertakings. Scientific man too must try to find answers, and we already know the framework within which the scientific answers must be given: to be useful in science, any statement about the universe or its parts must be consistent with the procedure of the scientific method. Therefore, if a given philosophy of nature can be verified wholly or even partly through experimental analysis, it will be valuable scientifically. But a philosophy which cannot be so verified will be without value in science, even though it may well be valuable in other areas of human thought.

Vitalism versus mechanism. In the course of history, two major answers have been proposed regarding the governing forces of the universe. These answers are incorporated in two systems of philosophy called *vitalism* and *mechanism.*

Vitalism is the doctrine of the supernatural. It holds, essentially, that the universe and all happenings in it are controlled by supernatural powers. Such powers have been variously called gods, spirits, or simply "vital forces." Their influence is held to determine the nature and guide the behavior of atoms, planets, stars, living things, and indeed all components of the universe. Clearly, most religious philosophies are vitalistic ones.

Whatever value a vitalistic philosophy might have elsewhere, it cannot have value in science. This is because the supernatural is by definition beyond reach of the natural. Inasmuch as the scientific method is a wholly natural procedure, it cannot be used for an investigation of the supernatural. We have already noted earlier, for example, that science cannot prove or disprove anything about God. Any other vitalistic conception is similarly untestable by experiment and is therefore unusable as a scientific philosophy of nature.

A philosophy which is usable in science is that of mechanism. In the mechanistic view, the prime mover of the universe is a set of natural laws, i.e., the laws of physics and chemistry. Experiments carried out in the course of several centuries have shown what some of these laws are, and any happening in the universe is held to be governed by the laws. The foundation of mechanism is therefore natural rather than supernatural and is amenable to experimental analysis.

On the basis of the total experimental experience, the mechanistic philosophy holds that if all physical and chemical phenomena in the universe can be accounted for, no other phenomena will remain. Therefore, the controlling agent of the material in the universe must reside within the material itself. Moreover, it must consist of physical and chemical events only. As a further consequence, the particular course of any happening must be guided automatically, by the way in which the natural laws permit physical and chemical events to occur within given materials. Note that living materials are included here; life too must be a result of physical and chemical events only. The course of life must be automatically self-determined by the physical and chemical events occurring within living matter.

Clearly, these differences between vitalism and mechanism point up a conceptual conflict between religion and science. But note that the conflict is not necessarily irreconcilable. To bridge the conceptual gap between the two philosophies, one might ask how the natural laws of the universe came into being to begin with. A possible answer is that they were created by God. In this view, the universe ran vitalistically up to the time that natural laws were created and ran mechanistically thereafter. The mechanist must then admit the existence of a supernatural Creator at the beginning of time (even though he has no scientific basis for either affirming or denying this; mechanism cannot, by definition, tell anything about a time at which natural laws might not have been in operation). Correspondingly, the vitalist must admit that any direct influence of God over the universe must have ceased once His natural laws were in operation. These laws would run the universe adequately, and further supernatural control would therefore not be necessary (or demonstrable, so long as the natural laws continued to operate without change).

Thus it is not necessarily illogical to hold both scientific and religious philosophies at the same time. However, it is decidedly illogical to try to use vitalistic ideas as explanations of scientific problems. Correct science does demand that supernatural concepts be kept out of natural events, i.e., those which can be investigated by means of the scientific method. However much a vitalist he might be in his nonscientific thinking, man in his scientific thinking must be a mechanist. And if he is not, he ceases to be scientific.

Many people, some scientists included, actually find it exceedingly difficult to keep vitalism out of science. Living events, undoubtedly the most complex of all known events in the universe, have in the past been particularly subject to attempts at vitalistic interpretation. How, it has been asked, can the beauty of a flower ever be understood simply as a series of physical and chemical events? How can an egg, transforming itself into a baby, be nothing more than a "mechanism" like a clock? And how can a man, who thinks and experiences visions of God, be conceivably regarded as nothing more than a piece of "machinery"? Mechanism *must* be inadequate as an explanation of life, it has been argued, and only something supernatural superimposed on the machine, some vital force, is likely to account for the fire of life.

In such replacements of mechanistic with mystical thought, the connotations of words often play a supporting role. For example, the words "mechanism" and "machine" usually bring to mind images of crude iron engines or clockworks. Such analogies tend to reinforce the suspicion of vitalists that those who regard living things as mere machinery must be simple-minded indeed. Consider, however, that the machines of today also include electronic computers which can learn, translate languages, compose music, play chess, make decisions, and improve their performance of such activities as they gather experience. In addition, theoretical knowledge now available would permit us to build a machine which could heal itself when injured and which could feed, sense, reproduce, and even evolve. Clearly, the term "mechanism" is not at all limited to crude, stupidly "mechanical" engines. And there is certainly nothing inherently simple-minded or reprehensible in the idea that living things are exquisitely complicated chemical mechanisms, some of which even have the capacity to think and to have visions of God.

On the contrary, if it could be shown that such a mechanistic view is at all justified, it would represent an enormous advance in our understanding of nature. In all the centuries of recorded history, vitalism in its various forms has hardly progressed beyond the mere initial assertion that living things are animated by supernatural forces. Just how such forces are presumed to do the animating has not been explained, nor have programs of inquiry been offered to find explanations. Actually, such inquiries are ruled out by definition, since natural man can never hope to fathom the supernatural. In the face of this closed door, mechanism provides the only way out for the curious. But is it justifiable to regard living things as pure mechanisms, even complicated chemical ones?

Notwithstanding the doubts expressed by some, a mechanistic interpretation of life is entirely justifiable and interjection of touches of vitalism is entirely unjustifiable. Science today can account for living properties in purely mechanistic terms. Moreover, zoologists are well on their way to being able to create a truly living entity "in the test tube," solely by means of physical and chemical procedures obeying known natural laws. We shall discuss some of the requirements for such laboratory creation in the course of this book. Evidently, vitalistic "aids" to explain the mechanistic universe are not only unjustifiable but also unnecessary.

It may be noted in this connection that, historically, vitalism has tended to fill the gaps left by incomplete scientific knowledge. Early man was a complete vitalist, who for want of better knowledge regarded even inanimate objects as "animated" by supernatural spirits. As scientific insight later increased, progressively more of the universe ceased to be in the domain of the supernatural. Thus it happened repeatedly that phenomena originally thought to be supernatural were later shown to be explainable naturally. So it has been with living phenomena as well. And those today who may still be prompted to fill gaps in scientific knowledge with vitalism must be prepared to have red faces tomorrow. Incidentally, it might also be pointed out in passing that even confirmed vitalists find it prudent on occasion to become ardent believers in mechanism, whether they realize it or not. For example, few vitalists hesitate to accept the mechanistic administrations of a physician at the first signs of disorder in their "machinery."

We conclude that a mechanistic view of nature is one

component of the philosophical attitude required in science. A second component may now be considered.

Teleology versus causalism. Even a casual observer must be impressed by the apparent nonrandomness of natural events. Every part of nature seems to follow a plan, and there is a distinct directedness to any given process. Living processes provide excellent instances of this. For example, developing eggs behave as if they knew exactly what the plan of the adult is to be. A chicken egg soon develops into an embryo with two wings and two legs, as if there existed a blueprint which specified that an adult chicken should have two wings and two legs. Moreover, since virtually all chicken eggs undergo the same course of growth, the impression of plan in development becomes reinforced strongly; one is led to conclude that the various parts of a chicken are there not just by random coincidence. Similarly, an earthworm which has been decapitated grows a new head, as if there were a plan which specified that every earthworm should have a head—not another tail and not two heads either, but one head.

All known natural processes, living or otherwise, thus start at given beginnings and proceed to particular endpoints. This observation poses a philosophical problem: how is a starting condition directed toward a given terminal condition; how does a starting point appear to "know" what the endpoint is to be?

It will be noted that such questions have to do with a specific aspect of the more general problem of the controlling agents of the universe. We should expect, therefore, that two sets of answers would be available, one vitalistic and the other mechanistic. This is the case. In view of the discussion in the preceding section, a book on science such as this could properly disregard the vitalistic answers as inadmissible from the outset and proceed at once with an outline of the mechanistic position. It is nevertheless advisable to examine both positions, partly because such a procedure adds to an understanding of the nature and limitations of science, partly because it is important to be able to recognize vitalistic answers if and when they occur (as they occasionally still do) in what is supposed to be scientific thought.

According to vitalistic doctrines, natural events *appear* to be planned because they *are* planned. A supernatural "divine plan" is held to fix the fate of every part of the universe, and all events in nature, past, present, and future, are programmed in this plan. All nature is therefore directed toward a preordained goal, namely, the fulfillment of the divine plan. As a consequence, nothing happens by chance, but everything happens on purpose.

Being a vitalistic, experimentally untestable conception, the notion of purpose in natural events has no place in science. Does the universe exist for a purpose? Does man live for a purpose? You cannot hope for an answer from science, for science is not designed to tackle such questions. Moreover, if you already hold certain beliefs in these areas, you cannot expect science either to prove or to disprove them for you.

Yet many arguments have been attempted to show purpose from science. For example, it has been maintained by some that the whole purpose of the evolution of living things was to produce man. Here the evidence supporting the theory of evolution is invoked to prove that man was the predetermined goal from the very beginning.

This implies several things besides the conceit that man is the finest product of creation. It implies, for example, that nothing could ever come after man, for he is supposed to be the last word in living magnificence. As a matter of record, man is sorely plagued by an army of parasites which cannot live anywhere except inside people. And it is clear that you cannot have a man-requiring parasite before you have a man.

Many human parasites did evolve after man. Thus, the purpose argument would at best show that the whole purpose of evolution was to produce those living organisms which cause influenza, diphtheria, gonorrhea, and syphilis. This even the most ardent arguer for purpose would probably not care to maintain.

If one is so inclined, he is of course perfectly free to believe that man is the pinnacle of it all. Then the rest of the universe with its billions of suns, including the living worlds which probably circle some of them, presumably are merely immense and fancy scenery for the microscopic stage on which man struts about. One may believe this, to be sure, but one cannot maintain that such beliefs are justified by evidence from science.

The essential point is that any purpose-implying argument, in this or in any other issue, stands on quicksand the moment science is invoked as a witness; for to say such and such is the goal, the ultimate purpose, is to state a belief and not a body of evidence adduced through the scientific method. Nowhere does this method include any purpose-revealing step.

The form of argumentation which takes recourse to purposes and supernatural planning is generally called *teleology*. In one system of teleology, the preordained plan exists outside natural objects, in an external Deity, for example. In another system, the plan resides within objects themselves. According to this view, a starting condition of an event proceeds toward a specific end condition because the starting object has built into it actual foreknowledge of what the end condition is to be. For example, the egg develops toward the goal of the adult because the egg knows what the adult state is to be. Similarly, evolution has occurred as it has because the participating starting chemicals had foreknowledge that the end should be man. Clearly, this and all other forms of teleology "explain" an end state by simply asserting it given at the beginning. And in thereby putting the future into the past, the effect before the cause, teleology negates time.

The scientifically useful alternative to teleology is called *causalism.* It has its foundations in mechanistic philosophy. Causalism denies foreknowledge of terminal states, preordination, purposes, goals, and fixed fates. It holds that natural events take place *sequentially.* Events occur only as other events permit them to occur, not as preordained goals or purposes make them occur. End states are consequences, not foregone conclusions, of beginning states. A headless earthworm regenerates a new head because conditions within the headless worm are such that only a head—*one* head—can develop. It becomes the task of the zoologist to find out what these conditions are and to see if, by changing the conditions, two heads or another tail could not be produced. Because scientists actually can obtain different end states after changing the conditions of initial states, the idea of predetermined goals loses all validity in scientific thought.

Care must therefore be taken in scientific endeavors not to fall unwittingly into the teleological trap. Consider often-heard statements such as: "the *purpose* of the heart is to pump blood"; "the ancestors of birds evolved wings *so that* they could fly"; "eggs have yolk *in order* to provide food for development." The last statement, for example, implies that eggs can "foresee" the nutritional problem in development and that food will be required; therefore, they proceed to store up some. In effect, eggs are given human mentality. The teleologist is always anthropocentric; i.e., he implies that the natural events he discusses have minds like his. Substitute "and" for every "so that" or "in order to," and "function" for every "purpose," in biological statements and they become properly nonteleological.

Clearly then, science in its present state of development must operate within carefully specified, self-imposed limits. The basic philosophical attitude must be mechanistic and causalistic, and we note that the results obtained through science are inherently without truth, without value, and without purpose.

But it is precisely because science is limited in this fashion that it advances. After centuries of earnest deliberation, mankind still does not agree on what truth is, values still change with the times and with places, and purposes remain as unfathomed as ever. On such shifting sands it has proved difficult to build a knowledge of nature. What little of nature we really know and are likely to know in the future stands on the bedrock of science and its powerful tool, the scientific method.

THE LANGUAGE OF SCIENCE

Science as a Whole

Fundamentally, science is a *language,* a system of communication. Religion, art, politics, English, and French are among other such languages. Like them, science enables man to travel into new countries of the mind and to understand and be understood in such countries. Like other languages, science too has its grammar—the scientific method; its authors and its literature—the scientists and their written work; and its various dialects or forms of expression—physics, chemistry, zoology, etc.

Indeed, science is one of the few truly universal languages, understood all over the globe. Art, religion, and politics are also universal. But each of these languages has several forms, so that Baptists and Hindus, for example, have little in common either religiously, artistically, or politically. Science, however, has the same single form everywhere, and Baptists and Hindus do speak the same scientific language.

It should be clear that no one language is "truer" or "righter" than any other. There are only different languages, each serving its function in its own domain. Many an idea is an idiom of a specific language and is best expressed in that language. For example, the German "Kindergarten" has been imported as is into English, and the American "baseball" has gone into the world without change. Likewise, one cannot discuss morality in the language of science, or thermodynamics in the language of religion, or artistic beauty in the language of politics; to the extent that each system of communication has specific idioms, there is no overlap or interchangeability among the systems.

On the other hand, many ideas can be expressed equally well in several languages. The English "water," the Latin "aqua," and the scientific "H_2O" are entirely equivalent, and no one of these is truer or righter than the others. They are merely different. Similarly, in one language man was created by God; in another man is a result of chance reactions among chemicals and of evolution. Again, neither the scientific nor the religious interpretation is the truer. If the theologian argues that everything was made by God, including scientists who think that man is the result of chance chemical reactions, then the scientist will argue back that chance chemical reactions created men with brains, including those theological brains which can conceive of a God who made everything. The impasse is permanent, and within their own systems of communication the scientist and the theologian are equally right. Many, of course, assume without warrant that it is the compelling duty of science to prove or disprove religious matters, and of religion, to prove or to disprove scientific matters.

The point is that there is no single "correct" formulation of any idea which spans various languages. There are only different formulations, and in given circumstances one or the other may be more useful, more satisfying, or more effective. Clearly, he who is adept in more than one language will be able to travel that much more widely and will be able to feel at ease in the company of more than one set of ideas.

We are, it appears, forever committed to multiple standards, according to the different systems of communication we use. But we have been in such a state all along, in many different ways. Thus, the color red means one thing politically, something else in a fall landscape, and is judged by a third standard in the fashion world. Or consider the different worth of the same dime to a child, to you, and to the United States Treasury. To be multilingual in his interpretation of the world has been the unique heritage of man from the beginning. Different proportions of the various languages may be mixed into the outlook of different individuals, but science, religion, art, politics, spoken language, all these and many more besides are always needed to make a full life.

Zoology

Within the language of science, one important dialect is *biology,* which permits travel in the domain of *living things.* Within biology in turn, a major subdialect is *zoology,* which permits travel in the domain of the *animal world.* Man probably was a zoologist very early during his history. His own body in health and disease; the phenomena of birth, growth, and death; and the other animals which gave him food and clothing undoubtedly were matters of serious concern to even the first of his kind. The motives were sheer necessity and the requirements of survival. These same motives still prompt the same zoological studies today; animal husbandry, fisheries, veterinary science, wildlife management, medicine, and fields allied to them are among important branches of modern applied zoology. In addition, zoology today is strongly experimental and pure research is done extensively all over the world. Some of this research promotes zoological technology; all of it increases our understanding of how animals are constructed and how they operate.

Over the decades, the frontiers of zoological investigations have been pushed into smaller and smaller realms. Some 100 to 150 years ago, when modern zoology began, the chief interest was the whole animal, how it lived, where it could be found, and how it was related to other whole living things. Such studies have been carried on ever since, but, in addition, techniques gradually became available for the investigation of progressively smaller parts of the whole, their structures, their functions, and their relationships to one another. Thus it happened that during the past few decades the frontiers of zoology were pushed down to the chemical level. And while research with larger living units continues as before, the newest zoology attempts to interpret living operations in terms of the chemicals out of which animals are constructed.

Zoology here merges with chemistry. Today there are already many signs that the next frontier will be the atoms which in turn compose the chemicals, and zoology tomorrow will undoubtedly merge with atomic physics. Such a trend is quite natural; for ultimately, animals are atomic things. Penultimately they are chemical things, and only on a large scale are they recognizable as whole animals. In the last analysis, therefore, zoology must attempt to show how atoms, and chemicals made out of atoms, are put together to form, on the one hand, something like a rock or a piece of metal and, on the other, something like a cat or a dog or a man.

This book is an outline of how successful the attempt has been thus far.

REVIEW QUESTIONS

1. What are the aims and the limitations of science? Review fully. In what sense is science a language, and how does it differ from other, similar languages?

2. What characterizes the different present-day forms of science and the different specializations of scientists?

3. Review the steps of the scientific method and discuss the nature of each of these steps. Define "controlled experiment."

4. How would you show by controlled experiment:
   a. Whether or not temperature affects the rate of animal growth?
   b. Whether or not houseflies can perceive differently colored objects?
   c. Whether or not earthworms use up some of the soil they live in?

5. Suppose that it were found in question 4*a* that, at an environmental temperature of 28°C, the growth of fertilized frog eggs into tadpoles occurs roughly twice as fast as at 18°C. What kinds of theories could such evidence suggest?

6. What are the historical and the modern relations of science and religion? Which of the ideas you have previously held about science should you now, after studying this chapter, regard as popular misconceptions?

7. Can you think of observations or problems which so far have not been investigated scientifically? Try to determine in each case whether or not such investigation is inherently possible. Why is mathematics not considered to be a science?

8. Describe the philosophical foundations of science. Define mechanism and causalism and contrast these systems of thought with those of vitalism and teleology. Can conceptual conflicts between science and religion be reconciled?

9. Consider the legal phrases "Do you swear to tell the truth and nothing but the truth?" and "Is it not a fact that on the night of . . . ?" If phrases of this sort were to be used in a strictly scientific context, how should they properly be formulated?

10. Zoology is called one of the *natural sciences,* all of which deal with the composition, properties, and behavior of matter in the universe. Which other sciences are customarily regarded as belonging to this category, and what distinguishes them from one another and from zoology? What are *social sciences?* Do they too operate by the scientific method?

COLLATERAL READINGS

Those who wish to read more on the general nature of science may find any of the following books and articles particularly instructive:

Arber, A.: "The Mind and the Eye," Cambridge, Cambridge, 1954.

Baker, J. R.: "The Scientific Life," Macmillan, New York, 1943.

Beveridge, W. I. B.: "The Art of Scientific Investigation," Norton, New York, 1957.

Bronowski, J.: "Science and Human Values," Harper Torchbooks 505, Harper & Row, New York, 1959.

Butterfield, H.: The Scientific Revolution, *Sci. American,* Sept., 1960.

Conant, J. B.: "Modern Science and Modern Man," Columbia, New York, 1952.

———: "Science and Common Sense," Yale, New Haven, Conn., 1951.

———: "On Understanding Science," Yale, New Haven, Conn., 1947.

Mausner, B. and J. Mausner: A Study of the Anti-Scientific Attitude, *Sci. American,* Feb., 1955.

Russell, B.: "The Scientific Outlook," Norton, New York, 1931.

Sullivan, J.: "The Limitations of Science," Mentor 35, or Viking, New York, 1933.

Terman, L. M.: Are Scientists Different? *Sci. American,* Jan., 1955.

Excellent accounts of various historical aspects of science may be found in the following:

Dampier, W. C.: "A History of Science," 3d ed., Macmillan, New York, 1942.

Singer, C.: "A History of Biology," rev. ed., Schuman, New York, 1950.

chemical foundations

As already noted, one of the most fruitful and significant advances in zoology during the last century has been the firm recognition that *all animals consist entirely of chemicals.* Moreover, it is now also clear that before there were living creatures of any kind on earth, there were only chemicals; living things originated out of chemicals (cf. Chap. 14). Chemicals in turn are composed of atoms. Thus, the story of life, animal life included, is largely a story of atoms and of chemicals; and it should not be surprising that physics and chemistry today are among the important background sciences to zoology. Indeed, much of modern zoology simply *is* physics or chemistry or both, and very many professional zoologists are good physicists or chemists. You too will have to understand certain of these background sciences, and this chapter is designed to provide the beginnings of such understanding.

CHEMICAL SUBSTANCES

The universe is made up of 92 different basic kinds of materials called chemical *elements*. Iron, silver, gold, copper, and aluminum are some familiar examples of elements. Some others, most of them present also in living matter, are listed in Table 1. Man has learned to create artificially several other elements in addition to the 92 kinds found in nature. Plutonium is an example of these man-made elements. Each element consists of unimaginably tiny particles called *atoms*. An atom may be said to be the very smallest complete unit of an element. For example, a gold atom is the basic unit of the element gold.

Each element is given a chemical symbol, often the first or the first two letters of its English or Latin name. For example, the symbol for hydrogen is H, that for carbon is C, and that for silicon is Si (see also Table 1). To represent one atom of an element, one simply writes the appropriate symbol. For example, the letter H stands for one atom of hydrogen. If more than one atom is to be indicated, the appropriate number is put before the atomic symbol. For example, 5 H stands for five separate hydrogen atoms.

Under specific conditions of temperature, pressure, and

TABLE 1. Some common chemical elements

Element	Symbol	Common valences	Common oxidation states in compounds
hydrogen	H	1	+1
sodium	Na	1	+1
potassium	K	1	+1
chlorine	Cl	1	−1
iodine	I	1	−1
calcium	Ca	2	+2
magnesium	Mg	2	+2
sulfur	S	2, 4	−2, +6
oxygen	O	2	−2
copper	Cu	1, 2	+1, +2
iron	Fe	2, 3	+2, +3
carbon	C	4	+4, −4
silicon	Si	4	+4
aluminum	Al	3	+3
nitrogen	N	3	−3, +3, +5
phosphorus	P	3	−3, +5

concentration, most atoms are able to attach to and to remain linked to certain other atoms. Such combinations of two or more atoms are called *compounds*. As we shall see below, the atoms of a compound are held together by specific bonding forces referred to as chemical *bonds*.

Each compound has a particular chemical name and a particular formula, both name and formula reflecting the kinds and numbers of atoms in the compound. For example, table salt is technically the compound "sodium chloride," the name indicating the presence of sodium and chlorine. The formula NaCl also shows the quantitative ratio of these components: one sodium atom is bonded to one chlorine atom. Water is technically the compound "hydrogen oxide" with the formula H_2O, which indicates the presence of two hydrogen atoms for every one of oxygen. Note generally that the number of like atoms in a compound is indicated as a subscript. For example, iron oxide (Fe_2O_3) contains two iron atoms for every three oxygen atoms. A more complex compound is a calcium phosphate, with the formula $Ca_3(PO_4)_2$. This is a shorthand notation for the following combination of atoms: three calcium atoms are bonded to two subcombinations, each of the latter consisting of one phosphorus and four oxygen atoms. Evidently, 13 atoms altogether form one unit of the compound calcium phosphate.

If more than one unit of a compound is to be written in symbols, the appropriate number is put before the formula. For example, H_2O stands for a single unit of the compound water and 5 H_2O stands for five such units.

How do atoms form chemical bonds between them? In other words, how are compounds produced? To answer this, we must consider the internal structure of atoms.

Atoms

The atoms of all elements are constructed out of components collectively known as *elementary particles*. Three types of elementary particles will concern us most: *protons, neutrons*, and *electrons*. Protons and neutrons occur in the center of an atom, where they form an *atomic nucleus*. Electrons are outside of such a nucleus.

A proton has mass, or "weight." This mass is the same for all protons, and it is given the arbitrary unit value 1. A mass of 1 also characterizes a neutron, which is consequently just as heavy as a proton. By contrast, the mass of an electron is very much less than 1, so much less, indeed, that its weight is practically negligible. Therefore, the total mass of a whole atom is concentrated almost entirely in its nucleus.

The mass of an atomic nucleus, i.e., the number of protons and neutrons present, determines the *atomic weight*. For example, the simplest type of atom is that of hydrogen. Its nucleus consists of a single proton, and there is a single electron on the outside; neutrons are absent. Since the nucleus therefore has a mass of 1, the atomic weight of hydrogen is said to be 1. By contrast, the most complex of the naturally occurring types of atoms is the atom of uranium. Its nucleus contains 92 protons as well as 146 neutrons. Therefore, the atomic weight of uranium is 238 (Fig. 2.1). The atomic weights of all other elements range between 1 and 238, according to the specific number of protons and neutrons in the atomic nuclei.

In addition to their mass, the elementary particles also have certain electrical properties. As is suggested by their name, neutrons are electrically neutral. Protons are electrically positive; more specifically, each proton carries one unit of positive electric charge. Electrons are electronegative, each carrying one unit of negative charge.

In each normal atom, the number of protons is exactly equal to the number of electrons. As noted, for example, a hydrogen atom consists of one proton (positively charged) and one electron (negatively charged). In a uranium atom there are 92 protons in the nucleus (see below), and there are also 92 electrons on the outside. All other atoms similarly display such numerical equality of the positive-charge carriers in the nucleus and the negative-charge carriers on the outside. Hence, each atom considered as a whole is electrically neutral.

Note, incidentally, that the number of electrons (or protons)

2.1. The atomic structure of hydrogen and uranium. The atomic nucleus of hydrogen contains a single proton; that of uranium, 92 protons and 146 neutrons.

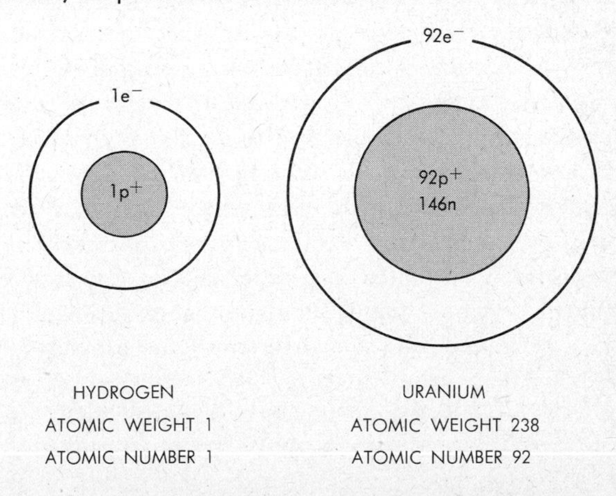

in an atom defines an *atomic number*. Hydrogen has atomic number 1, uranium, atomic number 92. It should be clear that if both the atomic number and the atomic weight of an atom are known, the composition of that atom is also known. For example, if the atomic number is 26 and the atomic weight is 56, then the atom contains 26 electrons on the outside and there must be 26 protons plus 30 neutrons in the nucleus. This happens to be the actual composition of an atom of iron. In certain contexts, rapid reference to either or both the atomic weight and atomic number of an element may be desirable. The appropriate figures are then indicated as a superscript and a subscript to the chemical symbol. For example, iron (Fe) may be symbolized as Fe^{56} or as $_{26}Fe^{56}$.

The electrons of an atom move in exceedingly rapid orbits around the atomic nucleus. An atom in effect resembles a miniature solar system. The nucleus is comparable to the central sun, and the electrons are comparable to the planets. Just as gravitational forces maintain the planets in orbit around the sun, so also do forces of electric attraction keep the negatively charged electrons in atomic orbits around the positively charged nucleus. Moreover, just as planetary orbits are located at various distances from the sun, so also are electron orbits spaced out from the atomic nucleus. Indeed, electrons can travel *only* at a number of fixed distances from the nucleus. The paths of the electrons at these distances may be said to mark out specific "shells," one outside the other.

Each such shell can hold only a fixed maximum number of electrons. The first shell, closest to the atomic nucleus, can hold a maximum of two electrons; the second shell, a maximum of eight electrons. Known maximums also characterize all other shells. In the case of a hydrogen atom, the single electron normally orbits in the first shell. Inasmuch as this shell could hold two electrons, hydrogen is said to have an *incomplete* or *open* shell. An atom of helium ($_2He^4$) possesses two electrons, both orbiting in the first shell. In this instance the shell holds the maximum possible number of electrons, and it is said to be *complete* or *closed*. In an atom of oxygen ($_8O^{16}$) eight orbital electrons are present. Two of these fill the first shell and the remaining six occupy the second. Since the second shell could hold eight electrons, this shell of oxygen is open. In atoms generally, electrons fill the orbital shells from the innermost outward. Thus, depending on the particular number of electrons present in a given atom, the outermost shell is either complete or to greater or lesser extent incomplete (Fig. 2.2).

It can be shown that an atom is electronically and chemically most stable when all of its electron shells are complete. A helium atom, possessing just the two electrons necessary to complete the first shell, is electronically entirely stable. Usually it is also quite inert chemically; i.e., it normally does not react with other atoms. This holds similarly for the atoms of a few other elements, among them neon, argon, krypton, xenon, and radon. All of these contain just enough electrons to complete all orbital shells present. For example, neon possesses 10 electrons, of which two fill the first shell and eight the second. Argon, analogously, possesses three closed shells; krypton, four; xenon, five; and radon, six. Elements of this kind are known as *inert gases*.

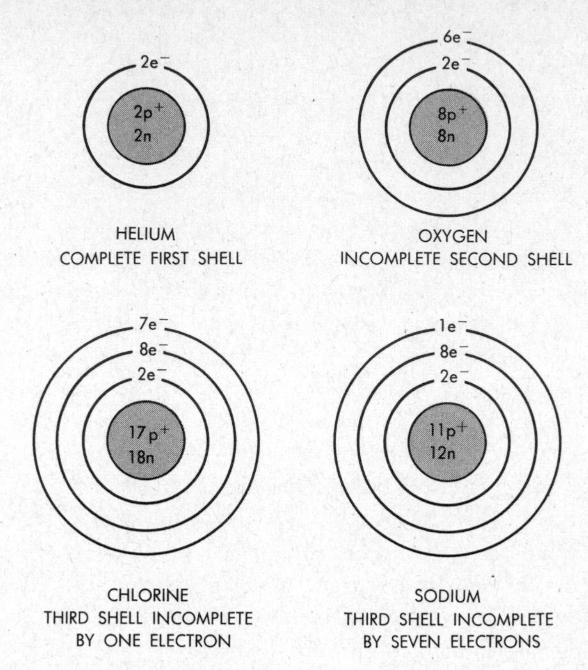

2.2. The electron shells of various atoms. Helium, $_2He^4$, has a complete first shell of two electrons. In oxygen, $_8O^{16}$, the second shell is incomplete by two electrons. Chlorine, $_{17}Cl^{35}$, possesses a nearly complete third shell, and sodium, $_{11}Na^{23}$, a nearly empty third shell.

In the atoms of all other elements, the outermost shells of electrons are incomplete, and such atoms are electronically relatively unstable. They reveal this comparative instability by being reactive chemically. In other words, if appropriate kinds and appropriate numbers of such atoms are brought into mutual contact, their incomplete outer electron shells may make them undergo a *chemical reaction*. The result of such a reaction is the formation of chemical bonds between the atoms; i.e., a chemical compound is produced. Note that *the chemical properties of atoms are determined by their outermost electron shells.*

Different kinds of atoms form bonds and compounds in different ways. The following sections outline the principal alternatives.

Ions

Every atom has a tendency to complete its outer electron shell and so to become as electronically stable as possible. This tendency to acquire complete outer shells is exhibited more or less forcefully by different kinds of atoms, and it constitutes the underlying cause for chemical interactions among atoms.

How can an originally incomplete electron shell become complete? Consider an atom of chlorine ($_{17}Cl^{35}$). Of the 17 orbital electrons, 2 form a complete first shell, 8 a complete second shell, and the remaining 7 an incomplete third shell (cf. Fig. 2.2). Like the second shell, the third similarly can hold a maximum of 8 electrons. Evidently, the chlorine atom is just one electron short of having a complete outer shell. If the atom could in some way *gain* one more electron, it would satisfy its very strong tendency for electronic completeness and greatest stability.

Consider now an atom of sodium ($_{11}Na^{23}$). Of the 11 electrons here present, 2 form a complete first shell, 8 a complete second shell, and the remaining 1 a highly incomplete third shell (cf. Fig. 2.2). If this atom were to *lose* the single electron in the third shell, its second shell would then in effect become the outermost shell. Inasmuch as this second shell is complete, the atom would have satisfied its tendency for completeness and would be stable.

It appears therefore that chlorine is unstable because it has one electron too few and that sodium is unstable because it has one electron too many. In view of this, could not both atoms become stable simultaneously if they transferred one electron from one atom to the other—if chlorine were to gain the one electron that sodium were to lose? This can indeed happen under appropriate conditions. When it does, it represents an example of one major class of chemical reactions: an *electron-transfer reaction* (Fig. 2.3).

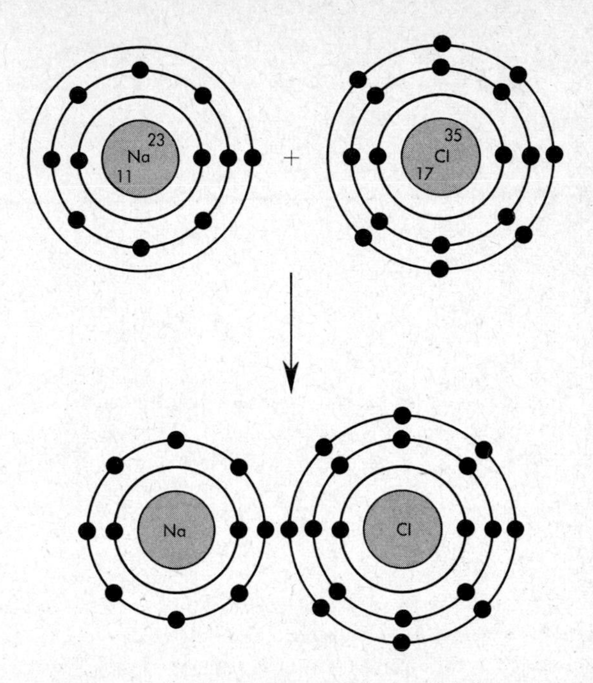

2.3. Electron-transfer reactions. In a reaction between one atom of sodium and one of chlorine, the single electron in the third shell of sodium becomes transferred to the third shell of chlorine. As a result, sodium now has a complete outer (second) shell and chlorine has a complete outer (third) shell of eight electrons. In this joined form, the sodium and chlorine atoms constitute the compound sodium chloride.

More than two atoms may participate in such a reaction and more than one electron may be transferred. For example, consider the interaction of magnesium and fluorine. Magnesium possesses two electrons in its incomplete third shell; if it were to lose these two, it would become stable. Fluorine possesses seven electrons in its nearly complete second shell; if it were to gain one more electron, its second shell would contain a full set of eight. Magnesium and fluorine may now interact by electron transfer. However, magnesium must lose two electrons, yet fluorine need gain only one. To make the transaction balance, therefore, each magnesium atom would have to interact with two fluorine atoms. This is how the reaction actually occurs (Fig. 2.4). In other words, if a magnesium-fluorine reaction is to achieve electronic stability for all participating atoms, then three atoms must interact and two electrons must be transferred. This reaction illustrates the general principle that a reaction can occur only if all participants achieve electronic stability. Different reactions therefore require the interaction of different numbers of given atoms and the transfer of different numbers of electrons.

In electron transfer among two or more atoms, those atoms which lose electrons may be called *electron donors* and those which gain them, *electron acceptors*. What determines

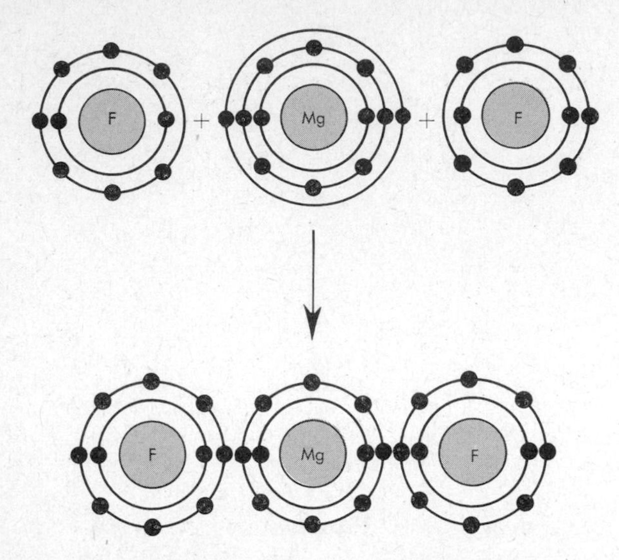

2.4. Electron-transfer reactions. A magnesium atom, with two electrons in its third shell, may lose these two electrons by reacting with two fluorine atoms, each of which requires one more electron for a complete second shell. The result of such an electron transfer is the compound magnesium fluoride, in which each of the three participating atoms now possesses a complete outer shell.

whether an atom is an electron donor or an electron acceptor? For example, could not fluorine become stable by losing its seven outer electrons instead of gaining an additional one? The answer is no, since it is exceedingly difficult to dislodge as many as seven electrons from an atom. Recall that electrons are negatively charged and are attracted to the positively charged protons in the atomic nucleus. Seven electrons are actually attracted very strongly, and they cannot be removed readily in one batch. Indeed, the nucleus exerts a sufficiently strong attracting force to capture and hold on to an additional electron from another atom. The situation is quite similar for chlorine and in general for all atoms in which the outermost shell is almost complete to begin with. Such atoms normally act as electron acceptors in transfer reactions.

Conversely, could not magnesium become stable by gaining six more electrons instead of losing the two in its outer shell? Here again the answer is no. In a shell capable of holding eight electrons, as few as two electrons are not attracted very strongly to the electropositive nucleus. Moreover, the attracting force of such a nucleus is not great enough to capture six additional electrons. Thus, magnesium normally acts as an electron donor. This holds also for sodium, for example, and generally for all atoms in which the outer shell is nearly empty to begin with.

The presence of few outer electrons therefore tends to identify an electron donor and the presence of many outer electrons, an electron acceptor. We may note that electron donors are commonly known as *metals,* electron acceptors as *nonmetals.* Sodium and magnesium are metals; fluorine and chlorine are nonmetals. Electron-transfer reactions commonly occur between metals and nonmetals.

Because of the negative charges of electrons, electron transfers have important electrical consequences. Consider again the transfer reaction between sodium and chlorine. Before the reaction, the sodium atom is electrically neutral; i.e., its total of 11 electrons is counterbalanced exactly by the 11 positively charged protons in the nucleus. During the reaction, one unit of negative charge, in the form of an electron, is lost from sodium. After the reaction, therefore, the sodium atom must be positively charged, for now there are only 10 electrons but the 11 protons are still present. Hence through the loss of one electron, sodium exhibits one unit of positive charge:

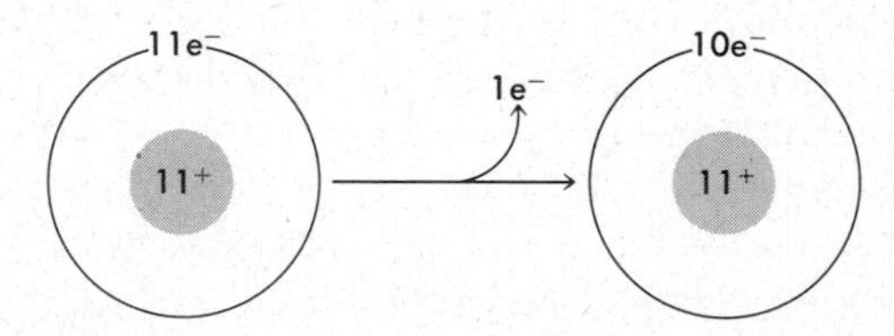

If we indicate a unit of positive charge by a superscript plus sign, we may also write

$$Na \longrightarrow e^- + Na^+$$

Analogously, chlorine is electrically neutral at the outset. During the reaction, it acquires one additional unit of negative charge in the form of an electron. After the reaction it must therefore be negatively charged:

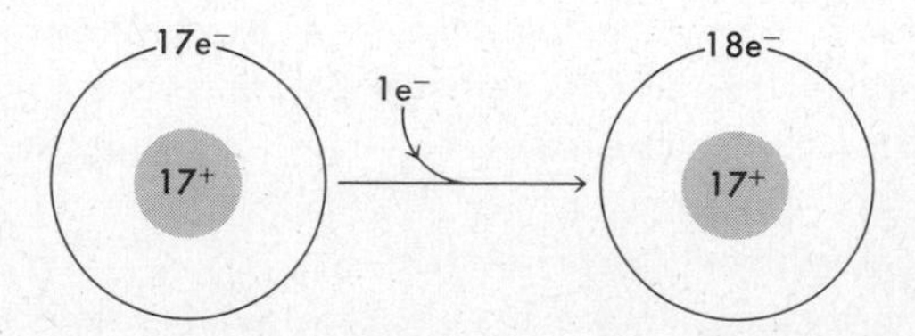

If we indicate a unit of negative charge by a superscript minus sign, we may write

$$Cl + e^- \longrightarrow Cl^-$$

We may now symbolize the sodium-chlorine reaction as a whole by writing

$$Na + Cl \longrightarrow Na^+ + Cl^-$$

or

$$Na + Cl \longrightarrow Na^+Cl^-$$

The equation implies that one electron has been transferred from sodium to chlorine, and it shows that the two atoms have acquired opposite unit charges as a result.

Similarly, in the reaction of magnesium and fluorine, the magnesium atom loses two electrons and then exhibits two units of positive charge. The two fluorine atoms accept the two electrons and so acquire negative charges. Symbolically,

$$Mg + 2\,F \longrightarrow Mg^{++} + 2\,F^-$$

or

$$Mg + 2\,F \longrightarrow Mg^{++}F_2{}^-$$

Atoms or groups of atoms carrying electric charges are known as *ions*. The symbols Na^+ and Mg^{++} stand for sodium ion and magnesium ion, respectively; the symbols F^- and Cl^- similarly stand for fluoride ion and chloride ion, respectively. Electron-transfer reactions may also be referred to as *ionic reactions*.

In such reactions, note that the total number of positive charges carried by one group of ions equals the total number of negative charges carried by the other group. Substances with opposite electric charges are attracted to each other, and we should therefore expect that positive and negative ions exert mutual electric attraction. This is the case; the force of attraction actually binds oppositely charged ions together. Any two ions so coupled are in effect united by a chemical bond, and the bonded group of ions represents a compound, an *ionic compound*. For example, in the reaction

$$Na + Cl \longrightarrow Na^+Cl^-$$

the endproduct is the ionic compound sodium chloride. It contains one bond between the sodium ion and the chloride ion, produced by the electric attraction these two ions have for each other.

The chemical bonds formed in electron-transfer reactions are called *ionic bonds*. Note that every electron transferred establishes one ionic bond. Thus the magnesium ion Mg^{++}, resulting from the transfer of two electrons, forms two ionic bonds with other ions.

The number of bonds an ion forms with others indicates the *valence* or, more specifically, the *electrovalence* of the ion. The sodium ion is said to have an electrovalence of 1, since it can form one ionic bond by the transfer of one electron. It is also said to have an *oxidation state* of $+1$, since it carries one unit of positive charge. Analogously, the magnesium ion has an electrovalence of 2 and an oxidation state of $+2$, having formed two ionic bonds by the transfer of two electrons and thus carrying two positive charges. The fluoride ion, and similarly the chloride ion, has an electrovalence of 1 and an oxidation state of -1; it is negatively charged and has formed one bond by acquiring one electron. Generally, metal ions have positive and nonmetal ions negative oxidation states (see Table 1). Note that, since oxidation states indicate the numbers and signs of charges, whole neutral atoms have oxidation states of zero. However, whole atoms are characterized by valence numbers, which indicate how many electrons can be transferred in ionic reactions.

In writing ionic compounds symbolically, it is not always necessary to indicate the electric charges of the ions. For example, instead of writing Na^+Cl^-, one may also write Na—Cl, the dash here representing the ionic bond. Or one may write, even more simply, NaCl. Similarly, the ionic compound magnesium fluoride, $Mg^{++}F_2{}^-$, may also be depicted simply as MgF_2. Even though such shorthand notations do not indicate the ionic nature of the components, ions are nevertheless present and ionic bonds unite them.

Molecules

Atoms may become electronically stable not only by transferring electrons but also by *sharing* electrons. For example, consider again a chlorine atom. As noted, it possesses seven outer electrons, but it requires eight for a complete shell. If appropriate electron donors such as sodium atoms are in the vicinity, the eighth electron may be gained by ionic reaction, as we have seen. However, suppose that appropriate electron donors are not available and that only chlorine atoms are present. Under such circumstances, a chlorine atom may complete its outer shell by reacting with another chlorine atom. We know that a chlorine atom can attract one additional electron rather strongly. Therefore, if two chlorine atoms come into contact, each will attempt to capture an electron from the other atom. But since each atom holds on strongly to its own electrons, an actual transfer cannot take place. Instead, a mutual "tug of war" will continue, each atom holding its own electrons and at the same time trying to pull one electron away from the other. The net result is a mutual attraction which will keep the two atoms in contact. Moreover, the atoms will *share* one pair of electrons: the sphere of influence of each atom will include seven outer electrons of its own plus one which is attracted from the other atom. Both atoms then behave as if they actually possessed eight outer electrons each, and this suffices to establish their electronic stability (Fig. 2.5).

Electron-sharing may also occur among hydrogen atoms, for example. Hydrogen possesses one orbital electron, and in the absence of electron acceptors, pairs of hydrogen atoms may share their electrons:

$$H\cdot + H\cdot \longrightarrow H{:}H$$

Through such pooling of the two electrons, the electrons no longer "belong" to either atom but belong equally to both. In effect, therefore, each of the two atoms attracts the two electrons necessary to complete its orbital shell.

More than one pair of electrons may be involved in a sharing

2.5. Electron-sharing. Two chlorine atoms are shown, only their seven outer electrons being indicated. These two atoms may share one pair of electrons, and so each atom may acquire a complete outer shell of eight electrons. The resulting compound is a molecule of chlorine, Cl_2.

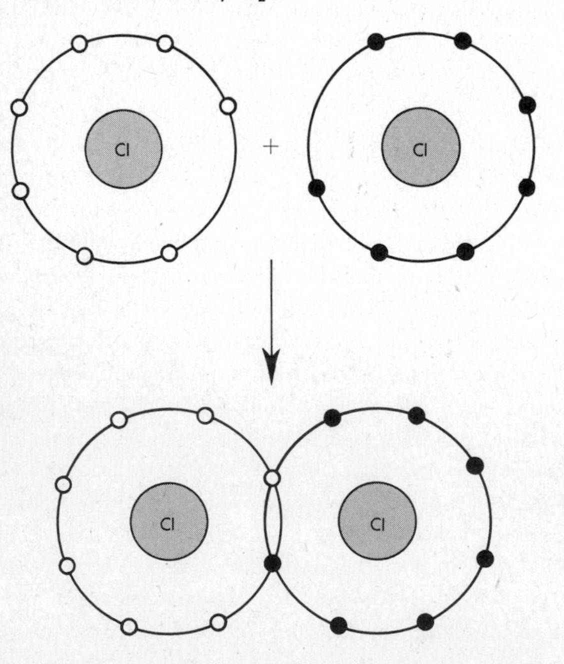

process. For example, oxygen possesses six outer electrons and requires two more for a complete shell. Completion may be achieved if two oxygen atoms share *two* pairs of electrons. If we indicate only the outer electrons, we may write

$$\underset{\cdot\cdot}{\ddot{O}}{:} + {:}\underset{\cdot\cdot}{\ddot{O}} \longrightarrow \underset{\cdot\cdot}{\ddot{O}}{::}\underset{\cdot\cdot}{\ddot{O}}$$

The two pairs of electrons shown between the two O's in the endproduct are the shared pairs. Each O atom now attracts eight electrons, the required number for a complete shell.

Electron-sharing may also occur among more than two atoms, and the atoms may be of different kinds. For example, one oxygen atom may share two of its six outer electrons with two hydrogen atoms. All three atoms may so acquire complete outer shells, i.e., oxygen a shell of eight electrons and each hydrogen a shell of two electrons:

$$\cdot H + \cdot H + \cdot\underset{\cdot\cdot}{\ddot{O}}\cdot \longrightarrow H{:}\underset{\cdot\cdot}{\ddot{O}}{:}H$$

We may conclude generally that electron-sharing characterizes a second major class of chemical reactions. In some instances, a given atom may participate either in electron-sharing or in electron-transfer processes, depending on what other kinds of atoms are available for reaction. In other cases, an atom participates almost exclusively in one kind of reaction only. Thus, sodium or magnesium almost always transfers electrons. On the other hand, atoms such as oxygen, nitrogen, or carbon almost always share electrons. Carbon, for example, possesses four outer electrons and needs eight for a complete shell. Here it is just as easy (or difficult) to gain four as to lose four, and the carbon atom instead shares all four of its outer electrons. In reacting with four hydrogen atoms, for example.

$$\cdot\underset{\cdot}{\dot{C}}\cdot + 4\,H\cdot \longrightarrow H{:}\overset{H}{\underset{H}{\overset{\cdot\cdot}{\underset{\cdot\cdot}{C}}}}{:}H$$

Or, carbon may also react with two oxygen atoms:

$${:}C{:} + 2\,{:}\underset{\cdot}{\dot{O}}{:} \longrightarrow {:}\underset{\cdot}{\dot{O}}{::}C{::}\underset{\cdot}{\dot{O}}{:}$$

As the electron distributions here indicate, each of the participating atoms again acquires a complete outer shell.

Through electron-sharing, two or more atoms become united into compounds. Compounds of this type are called *molecules*, and the reactions which produce them are called *molecular reactions*. In a molecule, the shared electrons represent the chemical bonds which hold the atoms together. Each shared electron pair represents one chemical bond, and we may note that bonds of this type are known as *covalent bonds.*

Since the formation of covalent bonds does not involve actual transfers of electrons, the participating atoms remain whole, electrically neutral atoms. The number of electrons an atom shares indicates its valence, more specifically, its *covalence.* Hydrogen exhibits a covalence of 1 when it shares its electron. Oxygen has a covalence of 2, nitrogen usually of 3, and carbon usually of 4. Put another way, oxygen can form two covalent bonds, nitrogen three, and carbon four (see Table 1). The atoms of a molecule may also be assigned oxidation states, according to how strongly they attract shared electron pairs. In $\begin{matrix} & H & \\ H\!: & \ddot{C} & :\!H \\ & \ddot{H} & \end{matrix}$, for example, the carbon atom attracts electron pairs more strongly than the hydrogen atoms, hence carbon is here said to have an oxidation state of -4, each H atom, an oxidation state of $+1$. Conversely, in $\ddot{O}::C::\ddot{O}$ it is the oxygen atoms which attract electron pairs more strongly, and each oxygen atom therefore is assigned an oxidation state of -2, whereas the carbon atom has an oxidation state of $+4$. Free, uncombined atoms again have oxidation states of zero, and their covalence numbers indicate how many bonds they can form by sharing.

In shorthand symbolization of molecules, bonds may be indicated either by pairs of dots representing electron pairs or by dashes. Alternatively, bonds need not be indicated at all. For example:

Cl:Cl	or	Cl—Cl	or	Cl_2	*chlorine molecule*
H:H		H—H		H_2	*hydrogen molecule*
O::O		O=O		O_2	*oxygen molecule*
:N:::N:		N≡N		N_2	*nitrogen molecule*
H:O:H		H—O—H		H_2O	*water molecule*
$\begin{matrix} & H & \\ H\!: & \ddot{C} & :\!H \\ & \ddot{H} & \end{matrix}$		$\begin{matrix} & H & \\ & \vert & \\ H— & C & —H \\ & \vert & \\ & H & \end{matrix}$		CH_4	*methane molecule*
O::C::O		O=C=O		CO_2	*carbon dioxide molecule*

A condensed shorthand formula obviously does not show whether a given compound is ionic or molecular. Only prior knowledge makes clear that a compound such as MgF_2 is ionic and a compound such as CO_2 is molecular. Where the distinction is important, ionic charges must be shown in one case and dot pairs or dashes in the other.

There are a great many compounds which are ionic but in which some or all of the ions consist of several atoms united by covalent bonds. For example, consider sulfuric acid:

$$H_2SO_4 = H_2^{++}SO_4^{=} = H^+ + H^+ + \left[\begin{matrix} & :\ddot{O}: & \\ :\ddot{O}: & \ddot{S} & :\ddot{O}: \\ & :\ddot{O}: & \end{matrix}\right]^{=}$$

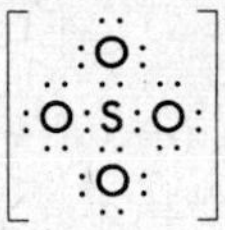

sulfuric acid — *hydrogen ions* — *sulfate ion*

The sulfate ion is a moleculelike complex in which one sulfur atom and four oxygen atoms share their outer electrons. Both sulfur and oxygen possess six outer electrons each and require eight for complete shells. One sulfur and four oxygen atoms together therefore possess 5×6, or a total of 30, outer electrons. However, 32 are needed in the arrangement shown above to give each of the five atoms a complete outer shell. The two missing electrons are provided by hydrogen atoms, which serve as electron donors. The result is the formation of a compound consisting of two hydrogen ions and one sulfate ion, the latter carrying two negative charges.

CHEMICAL CHANGE

With certain exceptions unimportant in the present context, free atoms today do not exist naturally on the surface of the earth. As we shall see in Chap. 14, atoms probably were free at one time, just after the origin of the earth. Later, atoms which could form compounds did so. Ever since, the earth has been very largely a collection of ionic and molecular compounds.

Compounds and Reactions

The chemical properties of a compound are determined by the *arrangement,* the *numbers,* and the *types* of atoms present. Two molecules, for example, may contain the same set of atoms; but if these are arranged differently, the molecules will have different properties. Thus the molecules

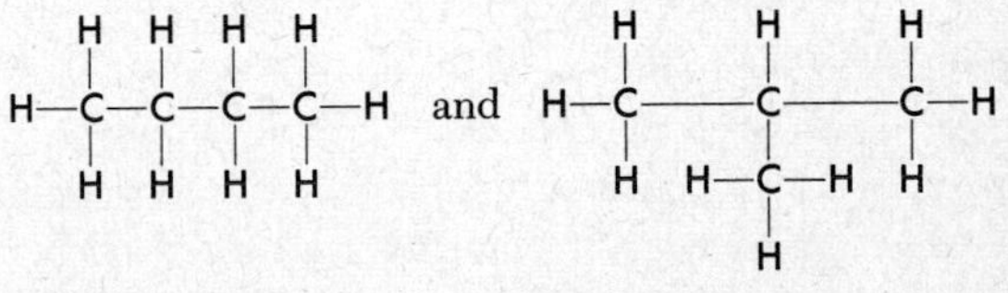

```
  H  H  H  H              H       H       H
  |  |  |  |              |       |       |
H—C——C——C——C—H   and   H—C———————C———————C—H
  |  |  |  |              |       |       |
  H  H  H  H              H    H—C—H      H
                                  |
                                  H
```

contain identical atoms, and both molecules may be written as C_4H_{10}. But since their atoms are bonded in different patterns, they are in fact different kinds of molecules with different properties. As we shall see, differences in the bonding patterns of otherwise similar molecules are particularly significant in the chemistry of living matter.

The numbers and types of atoms in a compound determine its size and mass. A molecule composed of but a few atoms of low atomic weight will obviously be smaller and lighter than a molecule composed of many atoms of high atomic weight. Of the compounds present in living matter, most consist of atoms of relatively low atomic weights. Compounds containing low-weight atoms, such as hydrogen, carbon, nitrogen, and oxygen, are particularly abundant. But although the atomic weights are low, the *molecular weights* can be exceedingly high; certain molecules in living matter are of very large size and contain hundreds and thousands of atoms each. Here it is the huge number of atoms, not their individual weights, which endows such compounds with high molecular weights.

As a direct consequence of its particular atomic composition and pattern of sructure, every compound has a greater or lesser *energy content.* As we have seen, forces of mutual electric attraction between atoms or ions produce the chemical bonds of a compound. These bonding forces, which hold atoms or ions together with a certain tenacity or strength, are said to represent *chemical energy* or *bond energy*. The greater the attracting force between two atoms or ions, the greater is the bond energy. The general concept of energy is roughly equivalent in meaning to work potential, or the capacity to do work. Bond energy may be defined as the amount of work necessary to break a chemical bond. Two bonded atoms or ions will become disunited only if some external force pushes them apart and so breaks the bond. Such forcible separation requires work, or energy, and the amount of energy needed clearly must be at least great enough to overcome the attraction between the two atoms or ions. In other words, the energy required to break a bond equals the bond energy. Correspondingly, the total chemical energy of a compound may be defined as the energy required to break all the bonds in the compound.

Once a given bond is broken, two atoms or ions will no longer be joined and will then be free to form new bonds. For example, depending on conditions, the two atoms or ions might rejoin each other and re-form the same bond which united them originally. Or they might be attracted independently to other appropriate atoms or ions and form new bonds with them. In either event, the *potential* for bond formation now exists. Evidently, when a bond is broken, the original bond energy does not disappear. Rather, it continues to exist in potential form, namely, as the potential of separated atoms or ions to make new bonds. We may conclude therefore that, by virtue of its bond energies, every compound represents a "package" of stored chemical energy. If a given amount of work is performed on the compound, the energy package can be opened. Some or all of the chemical energy can thereby be brought "out of storage" and can become available for the formation of new packages, i.e., new compounds.

The clear implication is that compounds are not absolutely stable or permanent structures. On the contrary, if they are subjected to the impact of appropriate amounts of external energy, they may undergo chemical reactions and become different compounds as a result. In the course of such a reaction, changes occur in the numbers or the types or the arrangements of the atoms of the participating compounds. Depending on the manner in which the structure of compounds becomes changed, four general categories of reactions may be distinguished.

First, two or more compounds may add together and form a single larger compound. This is a *synthesis* reaction. For example,

$$\underset{\text{carbon dioxide}}{CO_2} + \underset{\text{water}}{H_2O} \longrightarrow \underset{\text{carbonic acid}}{H_2CO_3}$$

Or, generally,

$$A + B \longrightarrow AB$$

Secondly, a given compound may break up into two or more smaller ones. This is a *decomposition* reaction, the reverse of synthesis. For example,

$$H_2CO_3 \longrightarrow H_2O + CO_2$$

Or, generally,

$$AB \longrightarrow A + B$$

Thirdly, one or more of the atoms or ions of one compound may trade places with one or more of the atoms or ions of another compound. This is an *exchange* reaction. For example,

$$H^+Cl^- + Na^+OH^- \longrightarrow H^+OH^-(= H_2O) + Na^+Cl^-$$

Or, generally,

$$AB + CD \longrightarrow AD + BC$$

Lastly, the numbers and types of atoms in a compound may remain the same, but the bonding pattern of the atoms changes. This is a *rearrangement* reaction. For example,

$$\begin{array}{ccccccccc} & H & H & H & H & & & & \\ & | & | & | & | & & & & \\ H- & C- & C- & C- & C & -H & & & \\ & | & | & | & | & & & & \\ & H & H & H & H & & & & \end{array} \longrightarrow \begin{array}{ccccccc} & H & & H & & H & \\ & | & & | & & | & \\ H- & C & - & C & - & C & -H \\ & | & & | & & | & \\ & H & H- & C & -H & H & \\ & & & | & & & \\ & & & H & & & \end{array}$$

Or, generally,

$$A \longrightarrow B$$

Note that in every equation illustrating a reaction, as above, the *total* numbers and types of atoms to the left of the arrow equal exactly the totals to the right of the arrow; in the reaction as a whole, atoms are neither gained nor lost. It is important to make sure that, whenever reactions are written out symbolically, the equations balance in this fashion.

Ionic Dissociation

Virtually all chemical reactions of biological interest take place in a water medium, i.e., in *aqueous solution.* When put into water, molecular compounds may dissolve. If they do, they then exist as whole molecules. In addition given molecular compounds may also *dissociate* to greater or lesser extent, that is, break up into free individual ions. For example,

$$\underset{\text{acetic acid}}{CH_3COOH} \xrightarrow{H_2O} \underset{\text{acetate ion}}{CH_3COO^-} + \underset{\text{hydrogen ion}}{H^+}$$

$$\underset{\text{ammonium hydroxide}}{NH_4OH} \xrightarrow{H_2O} \underset{\text{ammonium ion}}{NH_4^+} + \underset{\text{hydroxyl ion}}{OH^-}$$

Inasmuch as dissociation produces equal amounts of positive and negative electric charges, solutions containing dissociated compounds remain electrically neutral. However, the presence of free ions permits passage of electric currents through such solutions. Dissociated compounds are therefore also called *electrolytes,* and undissociated compounds are called *nonelectrolytes.*

Note that, in the first equation above, acetic acid dissociates in such a way that hydrogen ions are formed. This is what actually makes acetic acid an acid; any compound which dissociates to yield *hydrogen ions* is called an *acid.*

Analogously, any compound which dissociates to yield *hydroxyl ions,* OH^-, as in the second equation above, is a *base* or an *alkali.*

A compound resulting from the chemical interaction of an acid and a base is a *salt.* Sodium chloride (NaCl) is a salt because it is formed by the interaction of hydrochloric acid (HCl) and sodium hydroxide (NaOH):

$$\underset{\text{acid}}{HCl} + \underset{\text{base}}{NaOH} \longrightarrow \underset{\text{salt}}{NaCl} + H_2O$$

Every ion-forming compound is either an acid or a base or a salt. We may distinguish between "strong" acids and "weak" acids, for example, according to the *extent* to which an acid is dissociated. In a strong acid such as HCl, for example, all or virtually all of the molecules of the compound are dissociated into ions. By contrast, in a weak acid such as CH_3COOH, only a few ion pairs are formed. The situation is analogous for bases and salts. In other words, water breaks up potentially ion-producing compounds to different degrees. In the case of HCl, virtually all covalent bonds are broken and ion pairs form instead; in the case of CH_3COOH, only some bonds are broken. Consequently, an aqueous solution of acetic acid, for example, will contain partly the whole CH_3COOH molecules and partly the free acetate and hydrogen ions (Fig. 2.6).

It is often important to determine the acid or alkaline "strength" of a solution of compounds, i.e., the degree to which the compounds are dissociated. This can be done with appropriate electrical apparatus by measuring the relative number of free H^+ and OH^- ions present in the solution; for the more of these ions are found, the more the acids and bases present are dissociated. The result is expressed as a number called the pH of the solution. Mathematically, pH has been defined arbitrarily by the equation

$$pH = \log \frac{1}{[H^+]}$$

where $[H^+]$ indicates how many grams of H^+ ions are present

2.6. In a strong acid, many ionic compounds are dissociated into ions (many single white and black particles); in a weak acid, very few of the ionic compounds are dissociated (many joined pairs of white and black particles).

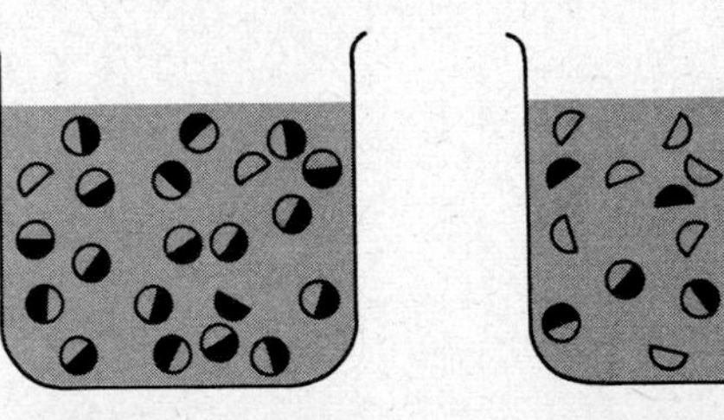

in 1 liter (1) of solution. For example, pure water dissociates to a very slight degree:

$$H_2O \longrightarrow H^+ + OH^-$$

Measurement shows that, in a liter of pure water, 0.0000001 gram (g) of H^+ ions is present. The pH of pure water therefore is

$$pH_{water} = \log \frac{1}{0.0000001} = 7$$

Since water contains as many H^+ ions as OH^- ions, it is neither acid nor basic but *chemically neutral.* We may note that any solution will be chemically neutral and will have a pH of 7 if its net H^+ ion concentration is as in pure water.

Suppose that a given mixture of dissolved compounds contains so much acid that, in a liter of the mixture, there is 0.1 g of H^+ ions present, i.e., 1 million times as many as in pure water. Then the pH of that mixture would be

$$pH = \log \frac{1}{0.1} = 1$$

We may note in general that the *lower* the pH of a solution below 7, the *more acid* it is; i.e., the more H^+ ions are present relative to OH^- ions. Analogously, the higher the pH above 7, the *more alkaline* is a solution, i.e., the fewer H^+ ions are present relative to OH^- ions (Fig. 2.7).

Acidities encountered in living animal matter usually do not go below pH 0, and maximal alkalinities are at pH 14. Indeed, inasmuch as it contains a mixture of variously dissociated acids, bases, and salts, animal material generally has a pH very near neutrality. For example, the pH of human blood is generally 7.3. Distinctly higher or lower pH levels do occur, however (e.g., in stomach cavities, which contain characteristically quite acid regions). Living animal matter does not tolerate significant variations of its normal acid-base balance, and its pH must remain within fairly narrow limits. If these limits should be exceeded, major chemical and physical disturbances would result which could be lethal.

2.7. The pH scale.

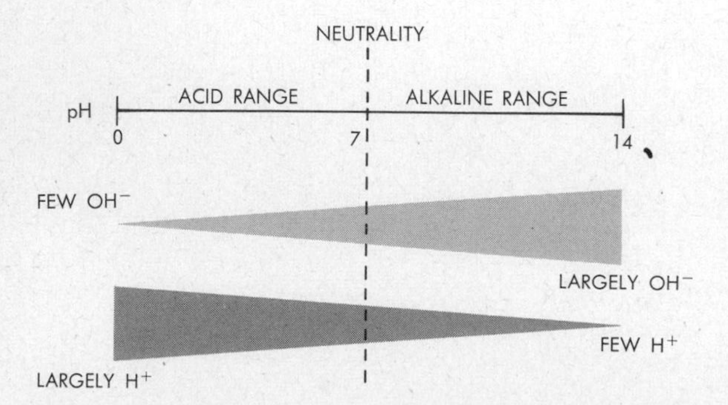

We shall find later that many normal processes within animals yield small amounts of excess acids or bases. But such small additions produce only negligible alterations of the pH. This is largely because living animal matter is *buffered;* i.e., it is protected to some extent against pH change.

For example, suppose we consider a solution of sodium bicarbonate ($NaHCO_3$), a salt normally present in animal matter. This salt exists as dissociated sodium ions (Na^+) and bicarbonate ions (HCO_3^-). If now a little hydrochloric acid (HCl) is added, we should expect the solution to become more acid. Actually, however, the pH change will be rather slight. This is because the hydrogen ions formed by dissociation of hydrochloric acid have a chance to react with the bicarbonate ions (HCO_3^-). The result is the formation of carbonic acid (H_2CO_3):

$$\begin{array}{l} NaHCO_3 \longrightarrow Na^+ + HCO_3^- \\ HCl \longrightarrow H^+ + Cl^- \\ \hline H^+ + HCO_3^- \longrightarrow H_2CO_3 \end{array}$$

Carbonic acid is a *weak* acid, i.e., the compound does not dissociate to any great extent. But the solution above at first contains very many H^+ and HCO_3^- ions, far more than a carbonic acid solution can actually hold. Therefore, H^+ and HCO_3^- will bond together into the whole compound H_2CO_3, as above, until the amounts of the free ions are reduced appropriately. In effect, the free H^+ ions from the added HCl are being "taken out of circulation," and the HCl consequently will not be able to change the pH appreciably.

We say that the presence of HCO_3^- ions *buffers* the solution, i.e., protects it from major pH change if a little acid is added. Analogous buffering effects against added bases are produced by a number of positively charged ions. Inasmuch as living matter contains complex mixtures of various ionic compounds, it is buffered by virtue of its composition. The bicarbonate ion and also phosphate ions ($PO_4^{\equiv}$) are among the most important biological buffers. To be sure, if a living animal system or any buffered solution is flooded with large quantities of additional acids or bases, then buffer protection will become insufficient and pH will change.

Energy Changes

Regardless of the nature of the participating compounds, a chemical reaction will proceed until all components present have attained a condition of greatest *stability*. In this respect chemical processes are like all other processes in the universe; i.e., they obey the *second law of thermodynamics*, one of the most fundamental laws of nature: *any closed system tends toward a state of greatest stability*. A closed system is one to which nothing is added and from which nothing is removed. Interacting chemicals often represent such a system, and in such a case the second law applies to them fully.

The stability of chemical systems depends primarily on two factors, both involving the energy of compounds. One factor is *enthalpy*, symbolized by the letter H; it denotes the *total energy content* of a chemical system. Compounds are most stable if H is at a minimum. It can be shown that, of all possible states of a system, a state of minimum energy content is the most likely; and it is because of this greatest likelihood, or greatest statistical probability, that minimum energy states are the most stable. Chemical reactions will therefore tend to proceed in such a way that, at the end, the total energy content of all participants will be least. For example, assume that in the generalized reaction

$$A + B \longrightarrow C + D$$

the total energy of all the bonds in A and B together is greater than the total energy of all the bonds in C and D together. In other words, more potential energy is available in the bonds of the starting materials than is needed to form the bonds of the endproducts. If all other conditions are suitable, such a reaction can occur readily because $C + D$, containing less total energy, is more stable than $A + B$. The energy difference will be lost from the reaction system into the environment, usually in the form of heat. We may symbolize the energy differential as ΔH, or *total energy change*. And since ΔH passes *from* the reaction system *to* the environment, it is given a negative sign. We may then write:

$$A + B \longrightarrow C + D - \Delta H$$

Energy-yielding reactions of this sort are said to be *exergonic* or *exothermic* (Fig. 2.8).

Conversely, assume that the total energy of $A + B$ is less than that of $C + D$. In this case $A + B$ is more stable, and the second law stipulates that, by itself, no process proceeds from a more stable to a less stable state. However, the reaction could be made to occur if energy were supplied *from* the environment *to* the reaction system, in an amount at least equal to ΔH. Then the total energy of $A + B$, together with ΔH from the environment, would actually suffice to form $C + D$.

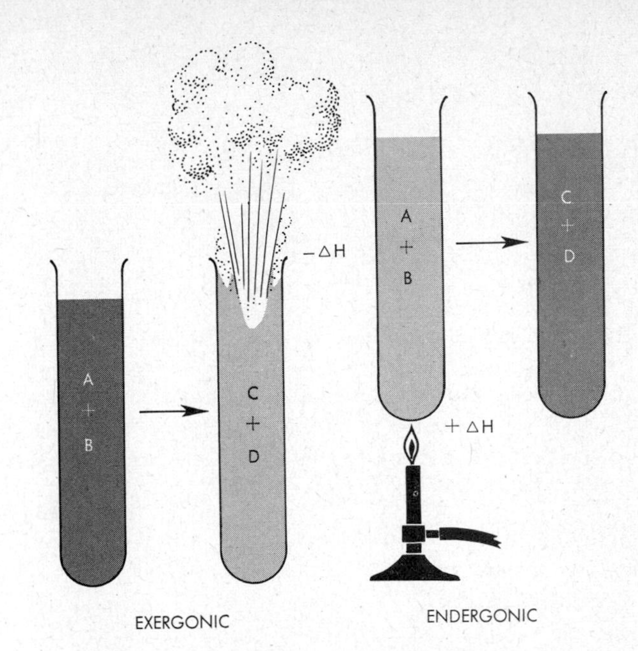

2.8. Left, exergonic reaction. The mixture $A + B$ contains more total energy than $C + D$; the energy difference, or total energy change ($-\Delta H$), escapes the reacting system. **Right, endergonic reaction.** The mixture $A + B$ contains less energy than $C + D$. Hence, if a reaction is to take place, the energy difference, $+\Delta H$, must be supplied to the reacting system from an external source.

We would then write

$$A + B \longrightarrow C + D + \Delta H$$

Energy-requiring reactions of this sort are said to be *endergonic* or *endothermic*. (Note that, by standard convention, a $-\Delta H$ on the right of an equation indicates energy *loss* from the system, a $+\Delta H$, energy *gain* by the system; the reacting system, not the environment, is the point of reference; cf. Fig. 2.8).

The second factor determining the relative stability of a chemical system is *entropy*, symbolized as S; it denotes the *energy distribution* in a chemical system. A system is most stable if S is at a maximum, i.e., if energy is distributed as uniformly or randomly as possible. It can be shown that, of all possible energy distributions in a system, the most random or unordered is the most likely; and it is because of this greatest statistical probability that states with the highest entropy are the most stable (Fig. 2.9). For example, consider again the reaction $A + B \longrightarrow C + D$, and assume that ΔH is zero, i.e., the total energy content is the same before and after the reaction. Assume however, that A is very rich in energy, that

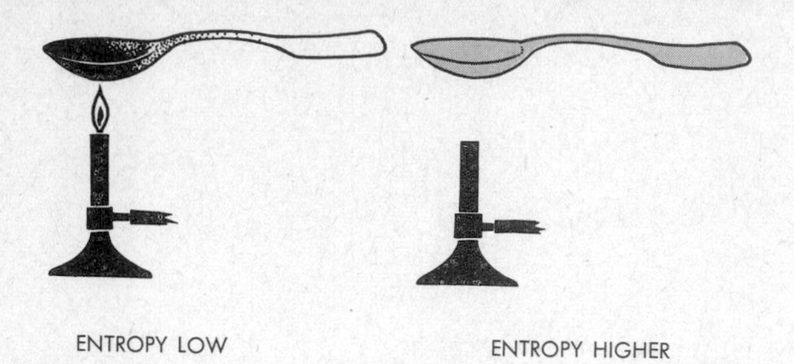

2.9. **Left,** heating a spoon at one end produces a nonuniform and comparatively unstable energy distribution, hence a state of low entropy. **Right,** if such a system is then left to itself (and assuming heat is not lost to the environment), the energy will become distributed more evenly throughout the spoon. Such a state is more stable and has a higher entropy.

B is very poor in energy, and that C and D each contain intermediate amounts of energy. Under such conditions, the energy distribution in the $A + B$ state is more uneven, or less random, than in the $C + D$ state; the latter will therefore be the more stable. If other conditions are suitable, the reaction may consequently occur and it will increase the entropy of the system. If we symbolize the net entropy change as ΔS, we may write

$$A + B \longrightarrow C + D + \Delta S$$

Consider now the reaction

$$A + B + C \longrightarrow D$$

and again assume that ΔH is zero. On the left side the total energy is distributed among three particles, but on the right all the energy is concentrated in one particle. Since, everything else being equal, the $A + B + C$ state has a more scattered, hence more random, energy distribution, this state will be more stable and will have a greater entropy than the D state. The reaction will therefore not be possible if the system is left to itself. However, by supplying energy from the environment, we could conceivably lower the entropy of $A + B + C$ sufficiently to make the reaction to D possible. We may write

$$A + B + C \longrightarrow D - \Delta S$$

Note that entropy is a directly temperature-dependent factor. For example, if equal amounts of water under otherwise equivalent conditions exist in the form of ice, liquid water, and steam, then ice will have the lowest and steam the highest entropy; water molecules are most scattered and disordered in the form of steam. In general, entropy tends to increase with rising temperature. For this reason, entropy changes are usually symbolized as a mathematical product of ΔS and T, where T is the absolute temperature at which a reaction occurs.

The overall stability change during a reaction thus is a function of two variables, ΔH and $T\Delta S$. By conceptual definition, these two are of opposite algebraic sign, since stability changes as enthalpy increases and entropy decreases or as enthalpy decreases and entropy increases. This circumstance finds expression in the formulation

$$\Delta F = \Delta H - T\Delta S$$

an important equation representing a symbolic statement of the second law. The term ΔF denotes overall change in system stability and is generally referred to as the *free-energy change*. If ΔF in a given reaction is numerically negative, the system has become more stable; and if it is positive, the system has become less stable. We may readily see why this should be so. If we disregard the entropy term for the moment, then ΔF will be negative if ΔH is negative, since in that case $-\Delta F = -\Delta H$; and we already know that a negative ΔH signifies an exergonic reaction system, one which has lost energy to the environment and thus has become more stable. If we now disregard the enthalpy term, then ΔF will be negative if ΔS is positive, since in that case $-\Delta F = -(+T\Delta S)$; but a positive ΔS means that the system has acquired a more randomized energy distribution and is therefore more stable. Conversely, if ΔF in the above equation is numerically positive, then the reaction system has become less stable; it must have gained total energy from the environment and/or the energy distribution of the system must have become less random.

Thus, the algebraic sign of ΔF indicates whether or not a given reaction possible on paper is likely to occur by itself in actuality; reactions with a positive ΔF lead to less stability and therefore require a supply of external energy if they are to take place. Note also that the ΔF is a measure of the theoretical amount of *usable* work one may obtain from a reacting system. For example, if a piece of wood is burned in a fireplace, the total energy content of the original wood minus the energy content of all the ashes and other combustion products represents ΔH, the total energy change. This total is not usable as such to perform work, e.g., to heat a room; for in the process of burning, the compact piece of wood has become dispersed into scattered ashes and into gases escaping into the air. Such scattering represents an increase in entropy, i.e., a loss of some of the potential energy of the original wood, dissipated in the scattering process. The amount of this lost energy is equivalent to the amount of energy we would have to spend to reassemble the dispersed combustion products back in one place. This quantity, equal to $T\Delta S$, is therefore not available to heat the room, and the net amount which does remain available is $\Delta H - T\Delta S$, or ΔF.

Changes in free energy, enthalpy, and entropy are expressed as *heat equivalents*, the units of heat being *calories*. A calory (cal) is defined as the amount of heat (or equivalent quantities of other energy forms) required to raise the temperature of 1 gram (g) of water by 1°C. For example, if 1 g of sugar at room temperature is decomposed in the presence of oxygen into water and carbon dioxide, the ΔF of this reaction can be shown to be $-3{,}810$ cal. The reaction evidently is feasible in practice, for it leads to an increase in stability. The ΔH here can be calculated to be $-3{,}738$ cal.; i.e., the reaction is exergonic, or energy-yielding; and $T\Delta S$ is $+72$ cal.; i.e., the reaction also increases the entropy. Thus, $-3{,}810 = (-3{,}738) - (+72)$.

We may note generally that a negative ΔF is largely characteristic of *decomposition* reactions, a positive ΔF, of *synthesis* reactions. Under suitable conditions, some of the energy produced by a decomposition reaction may be made to perform useful work. For example, it may drive an engine, as in the decomposition of fuels by burning, or it may simply provide useful heat as such, or, indeed, it may make possible energy-requiring synthesis reactions. In living systems, actually, decomposition reactions are *coupled* intimately to synthesis reactions; as we shall see, those with negative ΔFs yield the energy needed to sustain those with positive ΔFs. It is this circumstance which makes the sign and value of ΔFs of considerable zoological significance.

Whatever the ΔF, most reactions normally do not start by themselves. For example, if wood is simply exposed to the oxygen of air, nothing happens. To start a burning process, and in general to start any reaction, compounds must be sufficiently *activated*, or made ready to react. More specifically, certain amounts of starting energy, or *activation energy*, must be supplied from an external source. The basic reason for such a requirement is that, in any reaction, at least one existing chemical bond must be broken and at least one new bond must be created; the very meaning of "reaction" ultimately signifies "change in bonding pattern." But we already know that it takes energy to break or alter chemical bonds. Consequently, if a reaction is to start at all, the participating reactants must make direct *contact;* molecules and ions must actually collide with one another, and with sufficient intensity to break and remake chemical bonds between atoms. A certain amount of external energy, or activation energy, must therefore be supplied to produce enough initial contact collisions for a reaction to begin. Various kinds of external energy may serve to activate reactions. Heat is the most common kind, and, as is well known, many reactions can be started simply by applying heat to given compounds. Other types of energies suitable in particular cases include, for example, light, electricity, X rays, or mechanical energy in the form of pressure.

How do such energies produce their effect? The action of heat provides the general answer. Heat is a consequence of motion. All atoms, ions, and molecules, regardless of whether they are in a gas, a liquid, or a solid, vibrate uninterruptedly in random back-and-forth movements. We feel these movements as heat, and we measure them as temperature. A high temperature, for example, means that chemical units are in violent motion. Conversely, at -273°C, the theoretical absolute zero of temperature, heat is by definition entirely absent and all chemical units are stationary. However, every known natural or experimentally produced material always contains at least some heat; and a given temperature is always proportional to the amount of heat motion, or *thermal agitation*, of the chemical units present. Applying heat to a reaction system is therefore equivalent to intensifying the thermal agitation of its chemical units. As a result, the units will collide with one another more frequently, and at some critical collision rate specific for each type of reaction, interaction of the chemical units may begin.

The meaning of activation energy may be illustrated by considering a stone lying near the edge of a cliff. If the stone would fall over the edge, its descent into the valley would yield kinetic energy which could be used to perform work. But some external agency is first required to move the stone over the cliff edge to begin with; and this external agency is equivalent to activation energy. Assume that, in Fig. 2.10, point A represents the degree of stability (or free energy, F_A) of a chemical system before a reaction has started, and position B, the degree of stability (F_B) at the end of the reaction. The net stability change, or free-energy change ΔF, therefore may be represented as the difference in the levels of A and B. The lower position of B signifies that, as with a falling stone, stability is greater at the end of the reaction, and that the ΔF therefore is negative. But note that the stability curve from A to B passes through point C, located at a higher level than either A or B. This means that the stability of the system first *decreases* from A to C before it increases from C to B. The

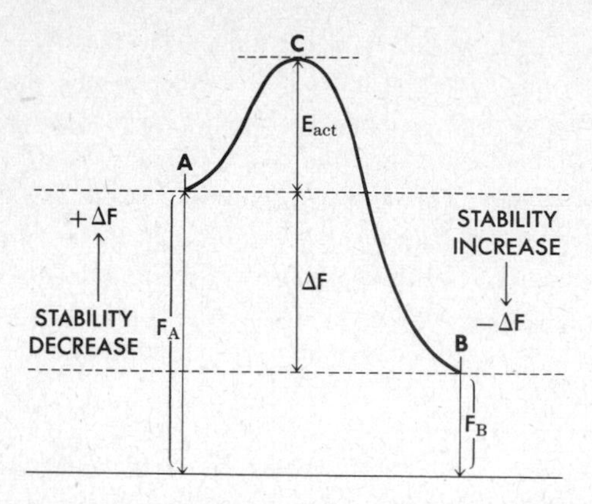

2.10. Free energy change and activation. The assumption in the diagram is that A is more stable than B (i.e., the free energy F_A of A is greater than the free energy F_B of B). If a reaction now proceeds from A to B, activation energy E_{act} must be supplied from the outside and the energy yield (free energy change) is then E_{act} plus ΔF, or $-\Delta F$ net. But if a reaction proceeds from B to A, the external energy supply must amount to ΔF plus E_{act}, and only E_{act} is then gained back. Hence the free energy change here is $+\ \Delta F$.

amount of this decrease from A to C represents the activation energy, E_{act} which must be supplied from the outside to bring the reaction system "over the cliff." Subsequently, just as a stone falls on its own, the reaction proceeds by itself; and in the process it not only "pays back" the energy expended for activation but also yields net free energy usable for work, in the amount of ΔF.

If we now consider the reverse reaction, from B to A, we note that, to provide adequate activation, we must supply external energy to the extent of ΔF plus E_{act}. Here only E_{act} is paid back when the reaction subsequently "falls" from C to A, and our net energy expenditure therefore is ΔF, which in this case is positive. In terms of our analogy, a great deal more work must be done to move a stone up from a valley over a cliff edge than in the opposite direction. Different kinds of reactions require different amounts of activation energy. In some cases, the activation produced by ordinary room temperature may suffice to allow a reaction to start. For example, water and metallic sodium react "spontaneously," i.e., at room temperature. But mix water and fat at room temperature and nothing happens. In this case, an appreciable reaction will occur only if more activation energy is supplied than is provided by the heat of the ordinary environment.

Inasmuch as heat produces collisions among reactants, it should follow that, after adequate activation, a reaction will proceed the faster the more heat is supplied to it. This is so. The higher the temperature the more often will the reactants collide and the greater will be the reaction speeds. It has been found that every temperature increase of 10°C approximately doubles to triples the speed of reactions. A *temperature coefficient,* conventionally designated by the letter Q, expresses how many times a reaction is speeded up by any stated increase of temperature. Thus, $Q_{10} = 2$ to 3 for chemical reactions generally. This statement reads: a 10° rise of temperature accelerates reactions two to three times. The implications for reactions in living systems are important. For example, the chemical reactions in a housefly will occur two or three times as fast on a day which is 10°C warmer than another.

It will have become clear that, of the three aspects of energy discussed, viz., bond energy, activation energy, and changes in free energy, bond energy is the most fundamental. Because compounds contain bond energies, which must be overcome or altered to start a reaction, activation energy is needed. And because different compounds contain different amounts and distributions of bond energies, a chemical process will be accompanied by free-energy changes, i.e., either will require the environment to provide energy or will release energy to the environment.

Catalysis

Most chemical reactions of zoological interest require fairly high activation energies—so high, indeed, that most living processes should require environments far hotter than room temperature. This is evidently not the case; if living animal matter were heated substantially above room temperature it would quickly be killed. It remains true, nevertheless, that at the comparatively low temperatures at which animal matter normally exists, the thermal agitation produced by the environment is insufficient to activate many reactions. How then are living processes possible? How are sufficient molecular collisions brought about without additional heat? The answer lies in *catalysis:* the acceleration of reactions by means of *catalysts* rather than heat.

Catalysts of various kinds are well known and widely used in the nonliving world. The special catalysts which occur in the living world are called *enzymes.* These substances are *proteins,* complex compounds about which much will be said later. For the present, we need note only that virtually every one of the thousands of chemical reactions in animal matter is speeded up enormously by a particular enzyme protein. Without enzymes the reactions could not occur fast enough

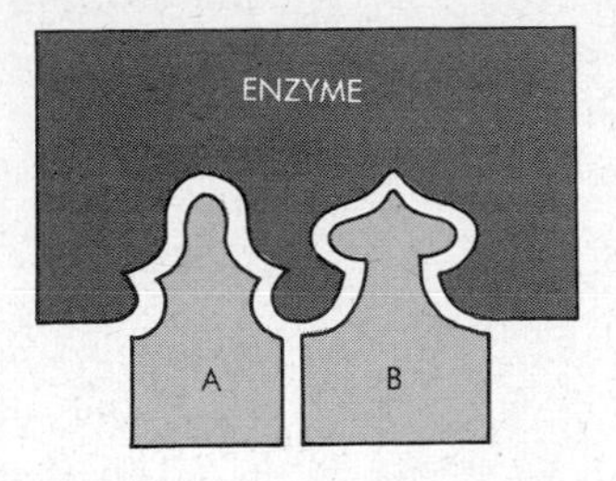

2.11. The surfaces of molecules *A* and *B* fit into the surface of the enzyme. Reaction between *A* and *B* is thus speeded up, for contact between *A* and *B* does now not depend on chance collision.

at ordinary temperatures to sustain life. Enzymes therefore represent a supplement to thermal agitation, a device through which reactions requiring high temperatures in test tubes can occur at low temperatures in animals. We may also say that the effect of enzymes is equivalent to a lowering of the activation-energy requirements, thus permitting reactions to take place at lower temperatures than would be possible otherwise.

How does an enzyme work? Best available evidence indicates that, like catalysts of all kinds, an enzyme combines temporarily with the reacting compounds. Mutual contact of these compounds is then no longer a matter of chance collision but a matter of certainty; hence reactions are faster.

The protein nature of enzymes is essential to this reaction-accelerating effect. Protein molecules are huge, and an almost unlimited number of different kinds of proteins exists. Accordingly, proteins have distinct molecular surfaces and the geometries of these surfaces differ as the internal structure of the proteins differs. The nature of the surface appears to be the key to enzyme action. Consider the reaction

fat + water ⟶ fatty acids + glycerin

Fat has a given unique surface geometry, and so does water. Enzymatic acceleration of this reaction may now occur if the surfaces of both fat and water happen to fit closely into the surface of a particular protein molecule. In other words, if the reacting molecules can become attached to a suitably shaped surface of an enzyme, then these molecules will be so close to each other that they may react chemically (Fig. 2.11). The enzyme itself remains almost passive here. It may provide a uniquely structured "platform," or *template*, on which particular molecules may become trapped. Such trapping brings reacting molecules into contact far faster than chance collisions at that temperature; hence reactions are accelerated. When they are held by the enzyme, fat and water interact and become fatty acid and glycerin, and these endproducts then disengage from the enzyme surface.

In enzyme-accelerated reactions, it is customary to speak of reacting molecules such as fat and water as the *substrates*. When substrate molecules are attached to an enzyme, the whole is referred to as an *enzyme-substrate complex*. Formation of such complexes may be thought of as a "lock-and-key" process. Only particularly shaped keys fit into particularly shaped locks. Just so, only certain types of molecules will establish a close fit with a given type of enzyme protein. For example, the enzyme in Fig. 2.12 may be effective in reactions involving substrates *a* or *b*, but not in those involving *c*.

Because enzyme proteins are huge molecules compared to most substrates, the whole surface of an enzyme is probably not required to promote a given reaction; only limited surface regions, called *active sites*, need be involved. Thus, even if other parts of enzyme molecules become altered chemically or physically, the enzymes may still be effective so long as their active sites remain intact. Until recently, it has been generally believed that enzymes were rigid molecular structures, templates of fixed shapes, and that enzyme activity depended on this permanence of configuration. Newer research indicates, however, that enzyme molecules might actually be flexible in a physical sense and that the structure of a particular substrate might *induce* the enzyme to bend or mold itself over the substrate. Such an "induced fit" hypothesis is consistent with the observation that only the active sites, not the whole enzyme molecules, need to retain permanent configurations. Moreover, the hypothesis may account in part for the long-established observation that many enzymes operate adequately only in the presence of

2.12. Reactants *A* and *B* fit partially into the surface of the enzyme but reactant *C* does not. Hence the enzyme may speed up reactions involving *A* and *B* but not those involving *C*.

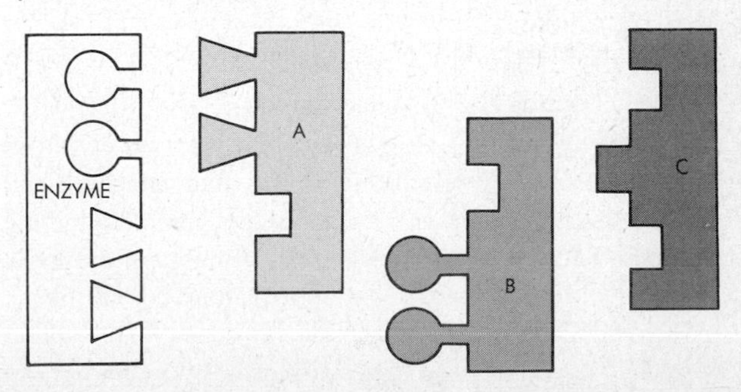

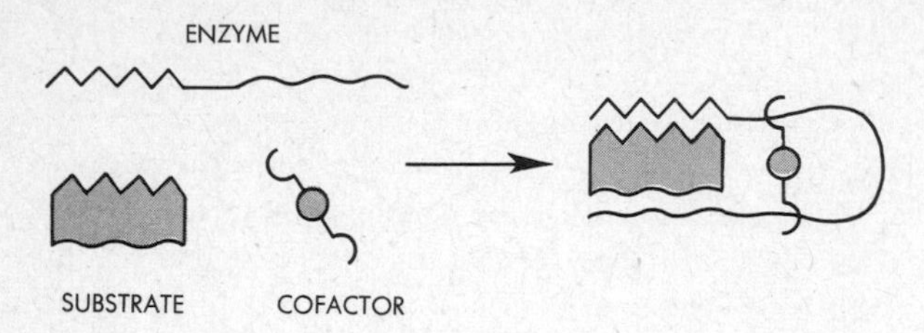

2.13. Diagrammatic representation of an enzyme with an active site at each end, and how a cofactor might aid in producing an induced fit between the active sites of the enzyme and the substrate.

certain *cofactors*. These are of various kinds and include for example, metal ions such as Mg^{++}. It is possible that cofactors are agents which, by virtue of their own chemical or physical properties or both, aid in molding an enzyme (or its substrates) into the shape required for a proper fit between enzyme and substrate (Fig. 2.13).

Differences in the surface configuration of different enzymes or their active sites undoubtedly account for the phenomenon of *enzyme specificity;* a given type of enzyme normally accelerates only one particular type of reaction. For example, the enzyme in the fat reaction above, called *lipase,* is specific and catalyzes only that particular type of reaction. In animal matter, there are actually almost as many different kinds of enzymes as there are different kinds of reactions. This specificity of enzymes is an important corollary of the more general phenomenon of *protein specificity,* about which more will be said in Chap. 4. Because of protein specificity, some proteins are enzymes to begin with and some are not. If a protein happens to have surface regions into which some other molecules could fit, then that protein could function as an enzyme in reactions involving those molecules.

Several other characteristics of enzymatic reactions may now be noted. If we assume the general reaction $A + B \longrightarrow C + D$ to be accelerated by an enzyme E, we may write

$$E + A + B \longrightarrow E(A \cdot B) \longrightarrow E(C \cdot D) \longrightarrow C + D + E$$

Note here that the enzyme molecule reappears unchanged at the end of the reaction, free to combine with a new set of starting substrates. We conclude again that enzymes, and catalysts in general, are not themselves affected by the reactions. Because of this, very small amounts of enzymes, used over and over, can catalyze large quantities of substrates.

Note further that a given enzyme can speed up a reaction in *either* direction; the reaction fat + water $\longrightarrow$ fatty acids + glycerin is accelerated by the same enzyme, namely lipase, that speeds up the reverse reaction. This is understandable if we keep in mind that enzymes are primarily passive reaction platforms. Thus, like heat, enzymes only influence reaction *speeds,* not the direction of a reaction.

Many substances not normally present in animals may, if introduced into living animal matter, combine with given enzymes and so act as poisons. For by attaching to enzymes, such poisons prevent the normal substances from combining with the enzymes. Since enzymes are present only in rather small amounts, small quantities of such poisons, or *enzyme inhibitors,* may block given normal reactions completely. This may be fatal. On the other hand, enzyme inhibition also accounts for the beneficial effect of many drugs (Fig. 2.14).

Enzymes may be classified according to the kind of substrate they affect. For example, any enzyme accelerating reactions of compounds called carbohydrates is referred to as a *carbohydrase.* Analogously, *proteinase* and *lipase* are enzymes that catalyze reactions of proteins and of fatty substances (= lipids), respectively. A suffix *-ase* always signifies that the substance in question is an enzyme. Note, however, that names of enzymes need not necessarily end

2.14. The principle of enzyme inhibition. The enzyme *E* normally speeds up reactions involving reactants *A* and *B*. But if an inhibitor molecule *I*, which fits into the surface of *E*, is supplied from the outside, then *I* may combine with *E* preferentially, thus preventing the normal reaction of *E*, *A*, and *B*. Poisons may be what they are because of such inhibiting effects. On the other hand, if the reaction of *E* with *A* and *B* gives an abnormal, disease-producing result, then a beneficial chemical may be introduced which acts like *I* and so prevents the disease.

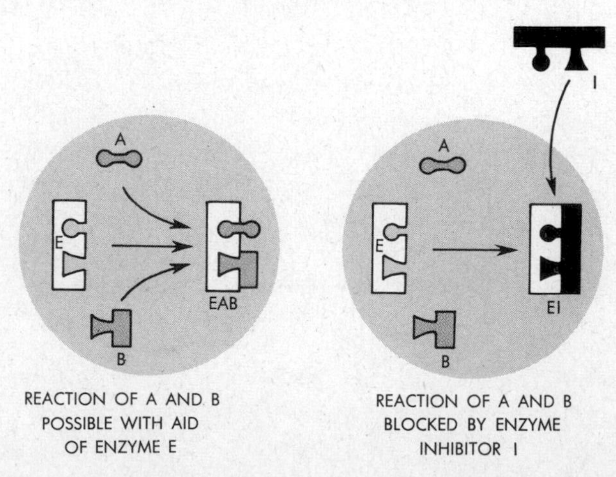

in *-ase*. Enzymes may also be classified according to the nature of the reaction they catalyze. For example, one distinguishes "splitting" and "synthesizing" enzymes, "transferring" enzymes (*transferases*), "rearranging" enzymes (*mutases*), and numerous others. In writing an enzymatic reaction symbolically, the name of the enzyme is conventionally put over the reaction arrow. Thus,

$$\text{fat} + \text{water} \xrightarrow{\textit{lipase}} \text{fatty acids} + \text{glycerin}$$

We have found that, like thermal agitation, enzymes increase the rate of contact among compounds, and so they affect the rate of reactions. But thermal agitation and enzymes are not the only two conditions to do so. A third is the *concentration* of the compounds present.

Equilibrium

Every chemical reaction has three basic attributes: it takes place at a certain *rate*, it proceeds in a certain *direction*, and it has a certain *duration*. What determines these attributes specifically for any given reaction?

As noted above, the rate of a reaction is determined by the environmental temperature and by catalysts. But it should be readily apparent that reaction rates will depend also on the concentration of the reacting compounds present; for if the temperature and the enzymes remain constant, then the greater the concentration of the starting compounds, the more frequently will contacts among the compounds become possible and the faster the reaction will therefore be. We may say that, other factors being equal, *the rate of a reaction is proportional to the concentrations of the participating compounds*. This is sometimes referred to as the *law of mass action* (Fig. 2.15).

This law actually relates not only to the rate, but by implication also to the direction and the duration of reactions. All chemical reactions are reversible in principle; if they can occur in one direction, they can also occur in the opposite direction, at least in theory. In practice, the conditions which permit a reaction to proceed in one direction usually preclude any appreciable simultaneous reverse reaction. However, assume that in the generalized reversible reaction system

$$A + B \rightleftharpoons C + D$$

conditions are such that they do permit a reaction in either direction. Assume also that the system at first contains compounds A and B only:

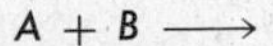

$$A + B \longrightarrow$$

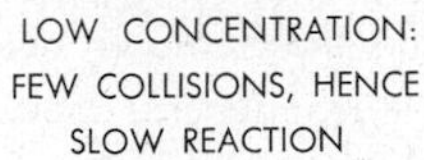

HIGH CONCENTRATION:
MANY COLLISIONS, HENCE
FAST REACTION

2.15. The principle of mass action. At left, few molecules are present and the concentration is low. Collisions are therefore infrequent, hence reaction speed is also low. At right, at the same temperature as at left, many molecules are present, and the concentration is high. Collisions at right are therefore frequent, hence reaction speed is high.

The rate of the ensuing reaction will be proportional to the concentrations of A and B, according to the mass-action law. If we symbolize these concentrations as $[A]$ and $[B]$, we may write

$$\text{forward rate} = k_{A,B}[A] \cdot [B]$$

where $k_{A,B}$ is a proportionality constant characteristic of the compounds A and B and of the particular conditions under which this reaction occurs. As A and B are transformed progressively into C and D, the concentrations of A and B will fall and the forward reaction rate will therefore decrease as well. On the other hand, as C and D accumulate, their concentrations—originally zero—will rise, and a reverse reaction will now take place at a gradually increasing rate. For this reverse reaction we may write

$$\text{reverse rate} = k_{C,D}[C] \cdot [D]$$

where $k_{C,D}$ again is a characteristic proportionality constant.

Eventually, a point will be reached at which the decreasing forward rate becomes just equal to the increasing reverse rate. This is the *equilibrium* point of the reaction system. At this point, A, B, C, and D will all be present in certain particular concentrations, and since the reaction rates in both directions now are equal, these concentrations will not change subsequently. For the equilibrium condition we may therefore write

$$k_{A,B}[A] \cdot [B] = k_{C,D}[C] \cdot [D]$$

or

$$\frac{k_{A,B}}{k_{C,D}} = \frac{[C] \cdot [D]}{[A] \cdot [B]}$$

Since both $k_{A,B}$ and $k_{C,D}$ are constants, the expression $k_{A,B}/k_{C,D}$ is also a constant. It is called the *equilibrium constant* and is symbolized as K. Each reaction, taking place under specified conditions (particularly temperature and pressure), has its own characteristic K value; it is an index of the concentrations of the participating compounds at equilibrium.

Note however, that, depending on the nature of a given reaction system, an equilibrium state may or may not be reached in practice. For example, if a reaction has the form

$$A + B \rightleftharpoons C + D \qquad \Delta F = 0$$

then an equilibrium may in fact be attained; for a zero ΔF indicates that both sides of the system are equally stable and that free energy is neither lost nor gained during a reaction in either direction. Since the forward and reverse reactions are therefore equally possible, the concentration product $[A] \cdot [B]$ will at equilibrium actually equal the product $[C] \cdot [D]$. In such a case, $K = 1$, and after an equilibrium has been attained, further *net* changes will not take place; the net reaction will in effect have stopped.

If, however,

$$A + B \longrightarrow C + D - \Delta F$$

then $C + D$ represents the more stable side. If the negative numerical magnitude of the ΔF is very great, then the forward reaction may proceed beyond the equilibrium state, and all of A and B might be converted into C and D. In that case a reverse reaction might not occur at all, and the forward reaction may stop only when A and B are no longer present. Such a system is in practice without equilibrium. But if the ΔF is only slightly negative, i.e., close to zero, then a comparatively larger forward reaction may be accompanied by a comparatively smaller reverse reaction, and an equilibrium point might be attained. At that point the concentration product $[C] \cdot [D]$ will be greater than the product $[A] \cdot [B]$, and K will be greater than 1. If

$$A + B \longrightarrow C + D + \Delta F$$

then, again depending on the (positive) magnitude of the ΔF, an equilibrium may or may not be attained in practice. If it is, $[A] \cdot [B]$ will be greater than $[C] \cdot [D]$, and $K < 1$.

Large free energy changes are not the only factors preventing attainment of equilibrium states. Even when $\Delta F = 0$, for example, if one of the reaction products is an escaping gas or an insoluble precipitate, then the loss of such compounds from a reaction mixture clearly will make a reverse reaction impossible. Here again the forward reaction will proceed beyond the equilibrium point and go to completion, i.e., to maximum yield. In general, therefore, we may note that equilibria imply *closed systems:* only if both energy (ΔF) and matter (reactants) are neither gained nor lost by a system can an equilibrium be attained and maintained. In practice, such closed conditions are seldom, if ever, realized completely. Yet even an analysis of the theoretical equilibrium state for any given reaction system is nevertheless of considerable significance. As we have seen, such an analysis gives information about the direction, rate, and duration of a reaction, and it interrelates concentrations and energy changes. Indeed it gives a substantially complete picture of most of the main characteristics of a reacting system. If we wanted to symbolize a reaction to such a degree of completeness, we would write

$$[A] + [B] \underset{k_{C,D}}{\overset{k_{A,B}}{\underset{\longleftarrow}{\overset{enzyme}{\longrightarrow}}}} [C] + [D] \qquad \Delta F \lesseqgtr 0$$

We may now conclude with the following summary. If chemical reactions occur in water, the participating substances will be more or less completely dissociated acids, bases, and salts as well as whole, nondissociated molecular compounds. Reactions will be possible if sufficient activation energy is made available. If it is, the chemical changes can be in the nature of synthesis, decomposition, exchange, and rearrangement. In the course of the reactions, the energy contents and energy distributions of the compounds will change, and the resulting free-energy changes will affect the directions, rates, and durations of the reactions. These system properties will also be influenced by other variously related and unrelated factors, including environmental temperature (and pressure), protein catalysts called enzymes, and reactant concentrations. Equilibria may or may not be attained, but in any event the endproducts will differ from the starting materials in the numbers, types, and bonding arrangements of the atoms present.

This account of chemical events suffices for a preliminary understanding of the life-producing events within animals, which, as already pointed out, are fundamentally chemical in nature.

REVIEW QUESTIONS

1. Define element, atom, compound, ion, molecule, chemical energy, chemical bond, valence.
2. What is an electrovalent bond? How is such a bond formed? Explain in terms of atomic structure. What is a covalent bond? How is such a bond formed? Again explain in terms of atomic structure.
3. In what kinds of chemical reactions may compounds participate? Give specific examples.
4. Consider the following equation:

$$Ca(OH)_2 + 2\ HCl \longrightarrow CaCl_2 + 2\ H_2O$$

a. Identify the different atoms by name and determine the valence of each.
b. Rewrite the equation to show the bonds, ionic, molecular, or both, within each compound.
c. Is the equation balanced?
d. Is this an exchange, synthesis, decomposition, or rearrangement reaction?

5. Define dissociation, electrolyte, acid, base, salt. Is H_2SO_4 an acid, a base, or a salt? How does sodium sulfate (Na_2SO_4) dissociate? The magnesium ion is Mg^{++} and the nitrate ion is NO_3^-; write the formulas for magnesium hydroxide, nitric acid, and magnesium nitrate.
6. What does the pH of a solution indicate? What would you expect the pH of a solution of NaCl to be? Of HCl? Of NaOH? Compared with pure water, how many more or fewer grams of H^+ ions will there be in a liter solution of (*a*) pH 5 and (*b*) pH 10? The pH of human blood is 7.3; what is the actual H^+ ion concentration per liter?
7. What is the zoological significance of dissociation and pH? What are buffers?
8. Define enthalpy, entropy, free energy change. What does the algebraic sign of a ΔF tell about a reaction? What is activation energy? How do energy changes differ for exergonic and endergonic reaction?
9. What is a catalyst? What is an enzyme and how does it work? What is an active site of an enzyme? Why is a carbohydrase ineffective in accelerating the reaction glycerin + fatty acids $\longrightarrow$ fat + water? What kind of enzyme does such a reaction require? Review the general operational characteristics of enzymes in living reactions.
10. State the law of mass action. What is the equilibrium point of a reaction system? What is an equilibrium constant? Under what conditions is a reaction reversible? Irreversible? How are ΔFs and equilibrium constants correlated? Show how the rate, direction, and amount of a chemical reaction are determined.

COLLATERAL READINGS

Substantial additional information on topics dealt with in this chapter may be found in almost any modern introductory college text of chemistry. The student desiring such information is urged to consult such texts, available in all college and most public libraries. The works listed below represent sample selections. Many other, equivalent books may be similarly adequate.

Daniels, F., and R. A. Alberty: "Physical Chemistry," Wiley, New York, 1955.

Edsall, J. T., and J. Wyman: "Biophysical Chemistry," Academic, New York, 1958.

Maron, S. H., and C. F. Prutton: "Principles of Physical Chemistry," 3d ed., Macmillan, New York, 1958.

Noller, C. R.: "Textbook of Organic Chemistry," Saunders, Philadelphia, 1958.

Pauling, L.: "College Chemistry," Freeman, San Francisco, 1950.

Sienko, M. J., and R. A. Plane: "Chemistry," 2d ed., McGraw-Hill, New York, 1961.

CHAPTER 

life, organism, animal

An animal is a particular kind of *organism,* or individual creature. Animal organisms share with all other kinds of organisms the property of being or having been *alive.* Thus if we wish to determine what an animal is, as is proper at the start of these zoological studies, we shall also have to determine what an organism is and what life is.

Surely the most obvious difference between something living and something nonliving is that the first *does* certain things the second does not do. Indeed, we may say that the essence of "living" lies in characteristic activities, or processes, or *functions.*

"Nonliving" may mean either "dead" or "inanimate," terms which are not equivalent. If a chicken does not perform its characteristic living functions it is dead, but then it is still distinguishable readily from an inanimate object such as a stone. Chickens, either living or dead, are organisms; stones are not. And we may note that all organisms are put together in such a way that the functions of life are or once were actually possible. Inanimate things, by contrast, are constructed in a way which makes living functions inherently impossible. Accordingly, the essence of "organism" lies in characteristic building materials and building patterns, or *structures.*

A "living organism," therefore, is what it is by virtue of its functions, which endow it with the property of life, and its structures, which make possible the execution of the life-sustaining functions. Animals, constituting a particular subgroup of living organisms, may be regarded as creatures in which at least some of the functions and structures are characteristically different from those in other subgroups of living organisms.

What, first, are the specific functions which endow all organisms with the potential of life?

THE NATURE OF LIFE

One of the principal life-sustaining activities of organisms is *nutrition,* a process which provides the raw materials for maintenance of life. All living matter has an unceasing requirement for such raw materials, for the very act of living continuously uses up two basic commodities: energy and matter. In this respect a living organism is like a mechanical engine or indeed like any other action-performing system in the universe. Energy is needed to power the system, to make the parts operate, to keep activity going—in short, to maintain function. And matter is needed to replace parts, to repair breakdowns, to continue the system intact and able to function—in short, to maintain structure. Therefore, by its very nature as an action-performing unit, a "living" organism can remain alive only if it continuously expends energy and matter. These commodities must be replenished from the outside at least as fast as they are used up inside; and this replenishment function is nutrition.

The nutritional raw materials are *nutrients,* including foods and required substances such as water and salts. They are available in the general environment of the earth, partly in the physical environment of air, water, and land, partly in the biological environment of other organisms, living or dead. The role of nutrition is to transfer appropriate kinds and amounts of nutrients from the external environment into the living organism. Note generally that, as nutrition permits continuation of life, so life also permits continuation of nutrition; nourished living matter must already preexist if its further nutrition is to be possible.

As just noted, one of the primary roles of nutrients within a living organism is to supply energy. Nutrients are chemicals, and as such they contain chemical energy (cf. Chap 2). All living matter runs on the chemical energy obtained from nutrients. The basic pattern here is that nutrient chemicals come to participate in exergonic decomposition reactions, of the general type described in the preceding chapter. Such reactions result in negative free-energy changes, $-\Delta F$s, and the usable energy so made available by the reactions sustains living activities. In this respect living systems are in principle quite similar to many familiar machines. In a gasoline or a steam engine, for example, fuel is burned, that is, decomposed exergonically. The burning process releases energy, $-\Delta F$, and this energy then drives the motor. In the living "motor," nutrients likewise function as fuels, and indeed nutrients and engine fuels belong to the same families of chemical substances. Secondly, nutrients are decomposed exergonically in a way which is actually a form of burning. And thirdly, the energy obtained through nutrient decomposition drives the living "machine."

In living organisms, the process of energy procurement through decomposition of nutrients is called *respiration*. In most cases respiration requires environmental oxygen as an accessory ingredient, and the collection of oxygen by the organism constitutes *breathing*. Respiration is a second major activity of living matter; it is the basic power-generating process which maintains *all* living functions. Note that this includes nutrition and even respiration itself. Nutritional activities can be sustained only with the aid of respiratory energy, but respiration in turn depends on the fuel-providing process of nutrition. Moreover, respiration is itself an activity of life and thus must be sustained by respiration. In other words, energy made available by previous respiration is required to make further respiration possible.

The second primary role of nutrients is to serve as construction materials. The whole structure of the living organism must be built from and kept intact with nutrient "bricks." Thus, the chemical stuff which forms living matter is fundamentally the same as that which forms nutrients. This consideration leads to an interesting inference. If nutrients and living matter are basically equivalent and if nutrients are also respiratory fuels, it should follow that living matter should be able to use *itself* as fuel. This is indeed the case; all living matter is inherently self-decomposing and self-consuming. In this respect a living organism may be likened to an engine which is built out of steel and in which steel is also the fuel. As such an engine runs, it burns up not only fuel supplied from the outside but also its own substance; the motor cannot tell the difference between external fuel and internal structural parts, because both are fundamentally the same. Such an engine would be quite unstable structurally and in fact would burn itself up very quickly. A living organism is similarly unstable structurally, but unlike a machine it counteracts this instability with the aid of nutrients. New structural parts are manufactured continually out of nutrients and the new parts replace those which burn away. Put another way, the structural damage resulting from the unceasing respiratory self-consumption is offset by an equally unceasing self-repair.

Respiration actually is not the only circumstance necessitating the use of nutrients as construction materials. For example, living matter frequently sustains structural damage through injuries from accidents and disease. Parts of the living structure also rub off, evaporate, and dissipate in other ways. Clearly, the structural wear and tear resulting unavoidably from the very activities of living has many forms and causes, and uninterrupted reconstruction must offset this wear and tear if the living structure is to persist. We note that living matter is actually never the same from moment to moment. As wear and tear and reconstruction occur side by side, the substance of living matter always "turns over"; although the fundamental structural pattern remains the same, every bit of the building material is replaced sooner or later. Moreover, if new building materials accumulate faster than old building materials wear away, the living organism will grow. Growth is a characteristic outcome of the use of nutrients in the construction of living matter.

The processes by which nutrients are fashioned into new structural parts of organisms may be referred to collectively as *synthesis* activities. They represent a third basic function of all living things. Like other functions of life, synthesis requires energy, ΔF, and respiration must provide it: the $-\Delta F$s resulting from respiration of some of the nutrients serve as the vital sources of the $+\Delta F$s required in synthesis reactions involving other nutrients. Thus, respiration is both the main cause making synthesis necessary and the main means making it pos-

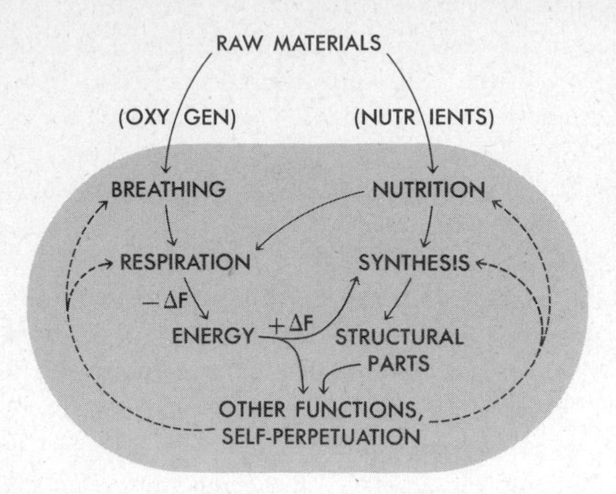

3.1. The pattern and interrelations of the main processes of metabolism and the main results of metabolism.

sible. In its own turn, synthesis maintains the structural apparatus required for respiration, for nutrition, and indeed for all other life functions as well.

The three functions of nutrition, respiration, and synthesis together constitute a broad living activity known as *metabolism* (Fig. 3.1). Taken as a whole, metabolism may be said to run the machinery of life. By running, the machinery may then carry out continued metabolism. As we have seen, a system which nourishes, respires, and synthesizes is capable of undertaking more nutrition, more respiration, and more synthesis.

However, being capable of continued activity is not the same as actually continuing the activity. Actual continuation becomes possible only if the activity is controlled. In this respect, living matter is again like an engine. It is not enough that an engine can obtain fuel, can generate energy by burning fuel, and can be repaired. Continuous operation demands that the rates of fuel supply, of burning, and of repair are finely geared to one another, and that if one rate changes for some external or internal reason all other rates change appropriately. In other words, the various operations of the engine must remain harmonized internally and must be adjusted and readjusted in line with events that may occur externally. Continuous operation, in short, requires control. Just so, continuation of metabolism in living matter depends on control. Metabolism as such is not equivalent to "life," but controlled metabolism in a general sense is.

The necessary control is provided by *self-perpetuation,* a broadly inclusive set of processes (Fig. 3.2). Self-perpetuation ensures that the metabolizing machinery does continue to run indefinitely, without outside help, and despite internal and external happenings which might otherwise stop its operation. We may also note that, in carrying out this controlling role, self-perpetuation uses up respiratory energy and the products of synthesis. Controlled metabolism thus is as much a result of self-perpetuation as a prerequisite.

The most direct and immediate regulation of metabolism is brought about by self-perpetuative processes we may collectively call *steady-state controls.* Fundamentally, such controls permit a living organism to receive information from within itself and from the external environment and to act on this information in a self-preserving manner. The information is received in the form of *stimuli* and the self-preserving actions are *responses.* For example, with the aid of energy and building materials, steady-state controls may cause the organ-

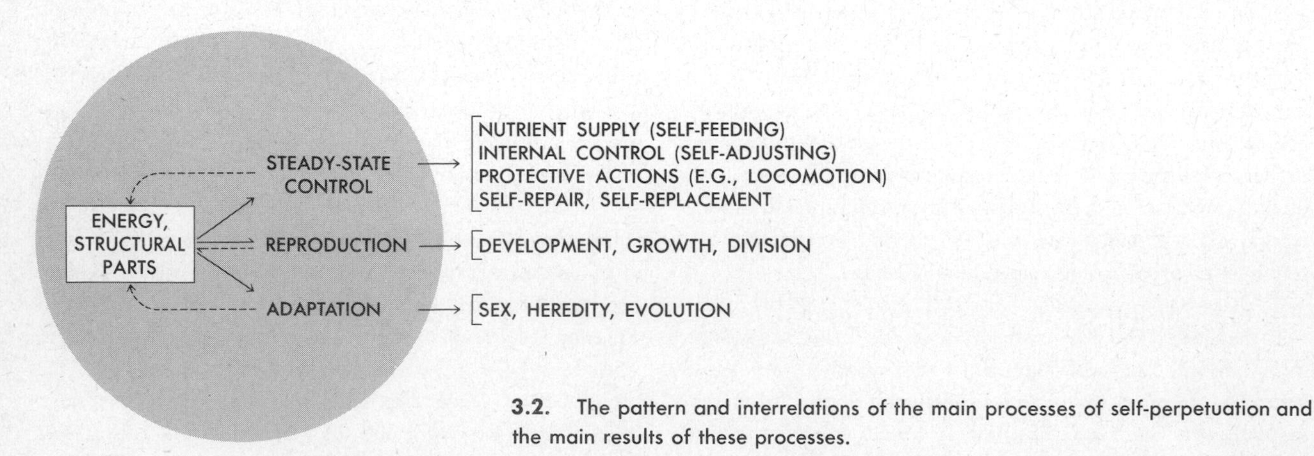

3.2. The pattern and interrelations of the main processes of self-perpetuation and the main results of these processes.

ism to procure fresh nutrients when past supplies are used up; may adjust respiration and synthesis in rate and amount according to given requirements; may channel the energy of respiration into protective responses like movement; may direct the repair of damaged parts or the construction of additional parts, as in growth. Taken together, such controls therefore preserve optimum operating conditions within living matter; they maintain a steady state. In this state a living organism may then remain intact and functioning as long as inherently possible.

But span of existence is invariably limited. Indeed, we know that individual life rarely lasts longer than a century and in most cases actually ceases within a single year. Death is a built-in characteristic of living matter because, like all other parts of the organism, those which maintain steady states are themselves subject to breakdown, to respiratory or accidental destruction, to wear and tear generally. When some of its controls become inoperative for any such reason, the organism suffers disease. We may define disease as any structural or functional breakdown of steady-state controls, one form of a temporary unsteady state. Other, still intact controls may then initiate self-repair. In time, however, so many controls break down simultaneously that too few remain intact to effect repairs. The organism then is in an irreversibly unsteady state and it must die. In this regard the organism again resembles a machine; even the most carefully serviced apparatus eventually becomes scrap, and the destructive impact of the environment ultimately can never be denied.

But unlike a machine, living matter here outwits the environment. For before it dies, the organism may have brought into play a second major self-perpetuative function, namely, *reproduction*. With the help of energy and raw materials, the living organism has enlarged, and such growth in size subsequently permits subdivision and growth in numbers. Reproduction in a sense anticipates and compensates for unavoidable individual death. Through reproduction over successive generations, the tradition of life may then be inherited and carried on indefinitely (Fig. 3.3).

Reproduction implies a still poorly understood capacity of rejuvenation. The material out of which the offspring is made is part of the parent, hence is really just as old as the rest of the parent. Yet the one lives and the other dies. Evidently, there is a profound distinction between "old" and "aged." Reproduction also implies the capacity of development, for the offspring is almost always not only smaller than the parent but also less nearly complete in form and function.

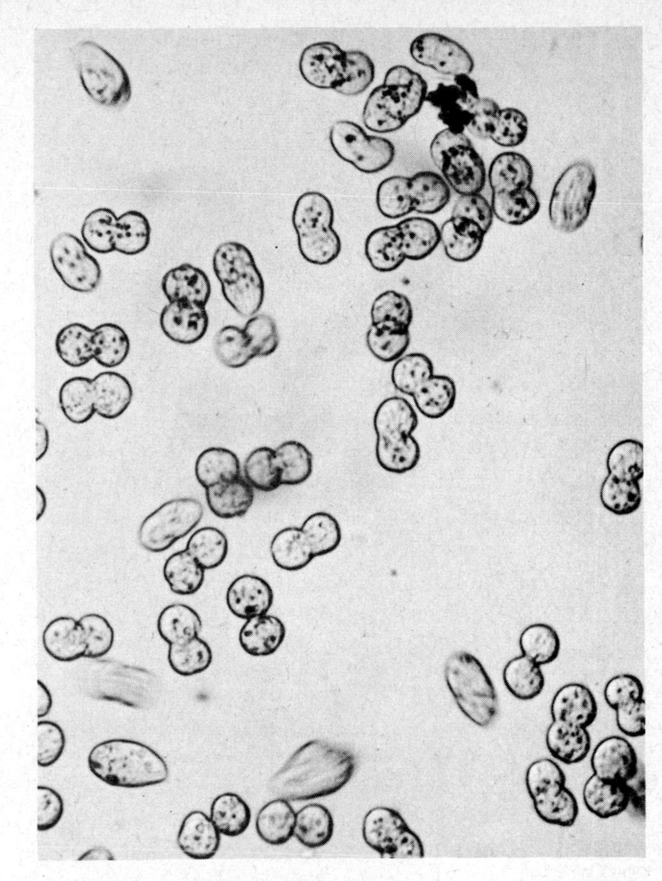

3.3. Reproduction: growth in size followed by subdivision and growth in numbers. The living ciliate protozoa *Tetrahymena* shown here have grown in size for a period of time and are now in various stages of reproduction by subdivision. Repeated at intervals, reproduction may maintain the living succession indefinitely.

Reproduction leads us to consider a third major self-perpetuative function, that of *adaptation*. As generation succeeds generation, long-term environmental changes are likely to have their effect on the living succession. In the course of thousands and millions of years, for example, climates may become altered profoundly; ice ages may come and go; mountains, oceans, vast tracts of land may appear and disappear. Moreover, living organisms themselves may in time alter the nature of a locality in major ways. Consequently, two related organisms many reproductive generations apart could find themselves in greatly different environments. And whereas the steady-state controls of the ancestor may have coped effectively with the early environment, these same controls, if inherited unchanged by the descendant, could be overpowered rapidly by the new environment. In the course of reproductive successions, therefore, organisms must change with the environment if they are to persist. They actually do change through adaptation. As we shall see later, this self-perpetuative function itself consists of several subfunctions, namely, *sex*, *heredity*, and *evolution* (Fig. 3.4).

We may note, therefore, that self-perpetuation as a whole encompasses three major control activities (cf. Fig. 3.2). Steady-state controls maintain optimum operating conditions within individual organisms as long as possible. Reproduction ensures a continuing succession of individual organisms. And adaptation molds and alters the members of this succession in step with the slowly changing nature of the environment. Self-perpetuation so adds the time dimension to metabolism; regardless of how the environment may change in time, self-perpetuation virtually guarantees the continuation of metabolism. Metabolism in turn makes possible uninterrupted self-perpetuation, and the system so able to metabolize and self-perpetuate can persist indefinitely; it becomes a "living" system.

3.4. An illustration of the process of adaptation, or change *with* the environment. The upper figure is a drawing of the placoderm *Dinichthys,* a type of fish long extinct but quite common some 300 million years ago. Fishes of the placoderm group were the ancestors of modern fish of which one, the speckled brook trout *Salvelinus,* is shown in the lower figure. In this evolutionary history, evidently, as in most others, descendants changed as their physical and biological environment changed and remained adapted to this changing environment.

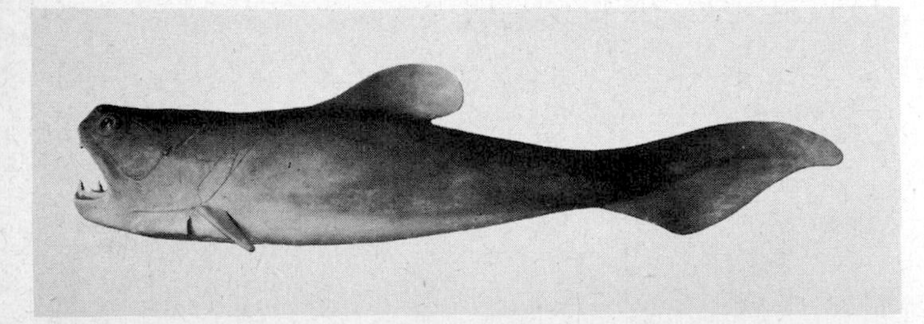

To define, then, the fundamental meaning of "living," we may say that *any structure which metabolizes and self-perpetuates is alive.* And we may add that the metabolic functions of nutrition, respiration, and synthesis make possible and are themselves made possible by the self-perpetuative functions of steady-state control, reproduction, and adaptation.

A first implication of this definition is that, by their very nature, living systems collectively are a highly permanent form of matter, perhaps the most permanent in the universe. They are certainly the most enduring on earth. Every inanimate or dead object on earth sooner or later decomposes and crumbles to dust under the impact of the environment. But every living object metabolizes and self-perpetuates and so may avoid such a fate. We come to realize that living matter, though soft and weak to the touch, is actually far more durable than the strongest steel, far more permanent than the hardest granite. Oceans, mountains, even whole continents have come and gone several times during the last 2 billion years, but living matter has persisted indestructibly during that time and, indeed, has become progressively more abundant.

A second implication is that any structure which does not satisfy the above definition in every particular is either inanimate or dead. Life must cease if even one of the fundamental functions of metabolism or self-perpetuation ceases; all these functions must be carried out simultaneously and interrelatedly. This criterion of life offers an instructive contrast to the operation of modern machines, many of which perform some of the functions also occurring in living organisms. As noted, for example, a machine may take on "nourishment" in the form of fuel and raw materials. The fuel may be "respired" to provide operating energy, and, with it, the raw materials may then be "synthesized" into nuts, bolts, and other structural components out of which such a machine might be built. If any one of these processes should stop, the machine would cease to operate even though it is still whole and intact.

Evidently, machines may carry out activities fully equivalent to those of metabolism; and metabolism, therefore, cannot be the special distinguishing feature of living nature. That distinguishing feature must lie, rather, in self-perpetuation. However, note that, like living systems, many "automated" machines have ingenious steady-state controls built into them. For example, such controls may make a machine automatically self-"feeding" and self-adjusting. But no machine is as yet self-protecting, self-repairing, or self-healing to any major extent, and no machine certainly is self-growing. On the other hand, it is known today how, theoretically, such a fully self-controlled, self-preserving machine could be built. If it is ever built, it will have steady-state controls conceivably quite as effective as the ones which have been standard equipment in living organisms for 2 billion years.

Machines resemble organisms further in that they "die" after a period of time, that is, enter so unsteady a state that they are beyond repair. But whereas living matter may reproduce before death, machines may not. It is in this capacity of reproduction that living systems differ most critically from inanimate systems. No machine self-reproduces, self-rejuvenates, or self-develops. However, it may be noted once again that the theoretical knowledge of how to build such a machine now exists. A device of this kind would metabolize, maintain steady states, and eventually "die" but, before that, would reproduce. It would almost be living. If it had the additional capacity of adaptation, it would be fully living. And here too the theoretical know-how is already available. On paper we may now design machines which could carry out "sexual" processes of a sort, which could pass on hereditary information to their self-reproduced "offspring," and which could "evolve" and change their properties in the course of many "generations."

Today, of course, such fully self-perpetuating machines do not—yet—exist in actuality. Living matter certainly may counteract the disruptive and destructive effect of the environment far more efficiently than any machine. It is primarily this which puts living objects into one category of matter and machines and all other inanimate objects into another. Nonliving objects are what they are because they cannot perform all the functions of metabolism and self-perpetuation. But if the time ever comes when machines will be able to carry out all these functions, then the essential distinction between "living" and "machine" will have disappeared.

This consideration brings us to a third implication of the definition above: the property of life basically does not depend on a particular substance. Any substance, of whatever composition, will be "living" provided that it metabolizes and self-perpetuates. It happens that only one such substance is now known. We call it "living matter," or often also *protoplasm*. It has certain clearcut characteristics of composition and makeup, as will become apparent below; and it is molded into organisms. But if someday we should be able to build a fully metabolizing and self-perpetuating system out of nuts, bolts, and wires, then it too will have to be regarded as being truly alive. Similarly, if someday we should encounter on another planet out in space a metabolizing and self-perpetuating entity made up of hitherto completely unknown materials, it also will have to be considered living. It may not be "life as we know it," that is, life based on the earthly variety of protoplasm, but in any case it will be truly living if it metabolizes and self-perpetuates.

We may conclude, then, that an object is defined as living or nonliving on the basis of its functional properties, not its structural properties. On the other hand, structural properties do determine whether an object is an "organism" or something inanimate. Linguistically as well as biologically, the root of "organism" is *organization*, a characteristic *structural order*. We shall examine this order in the following section.

THE NATURE OF ORGANISM

Levels of Organization

The smallest structural units of all matter, living matter included, are subatomic particles, e.g., the electrons, protons, and neutrons discussed in the previous chapter. The next larger units are atoms, each of which consists of subatomic particles. As we have seen, atoms in turn form still more complex combinations referred to as chemical compounds, and the latter are variously joined together into even more elaborate units we may call complexes of compounds. We may regard these units as successively higher *levels of organization of matter*. Such levels constitute a hierarchy in which any given level contains all lower levels as components. Moreover, any given level is also a component of all higher levels. For example, atoms contain subatomic particles as components, and atoms are themselves components of chemical compounds (Fig. 3.5).

All structural levels up to and including that of complexes of compounds are encountered both in the nonliving and in the living world. For example, two familiar chemical compounds found both in living and nonliving matter are water and table salt. Examples of complexes of compounds in the nonliving world are sea water and rock, each of which is composed of several compounds in combination (water and table salt among them). Note that differences in mere physical bulk do not necessarily indicate differences in level of organization. Thus, a pebble and a mountain range have the same organizational level, i.e., that of complexes of compounds. In living matter, complexes of compounds often occur as microscopic and submicroscopic bodies called *organelles*. We shall identify some of them and their functions in later chapters.

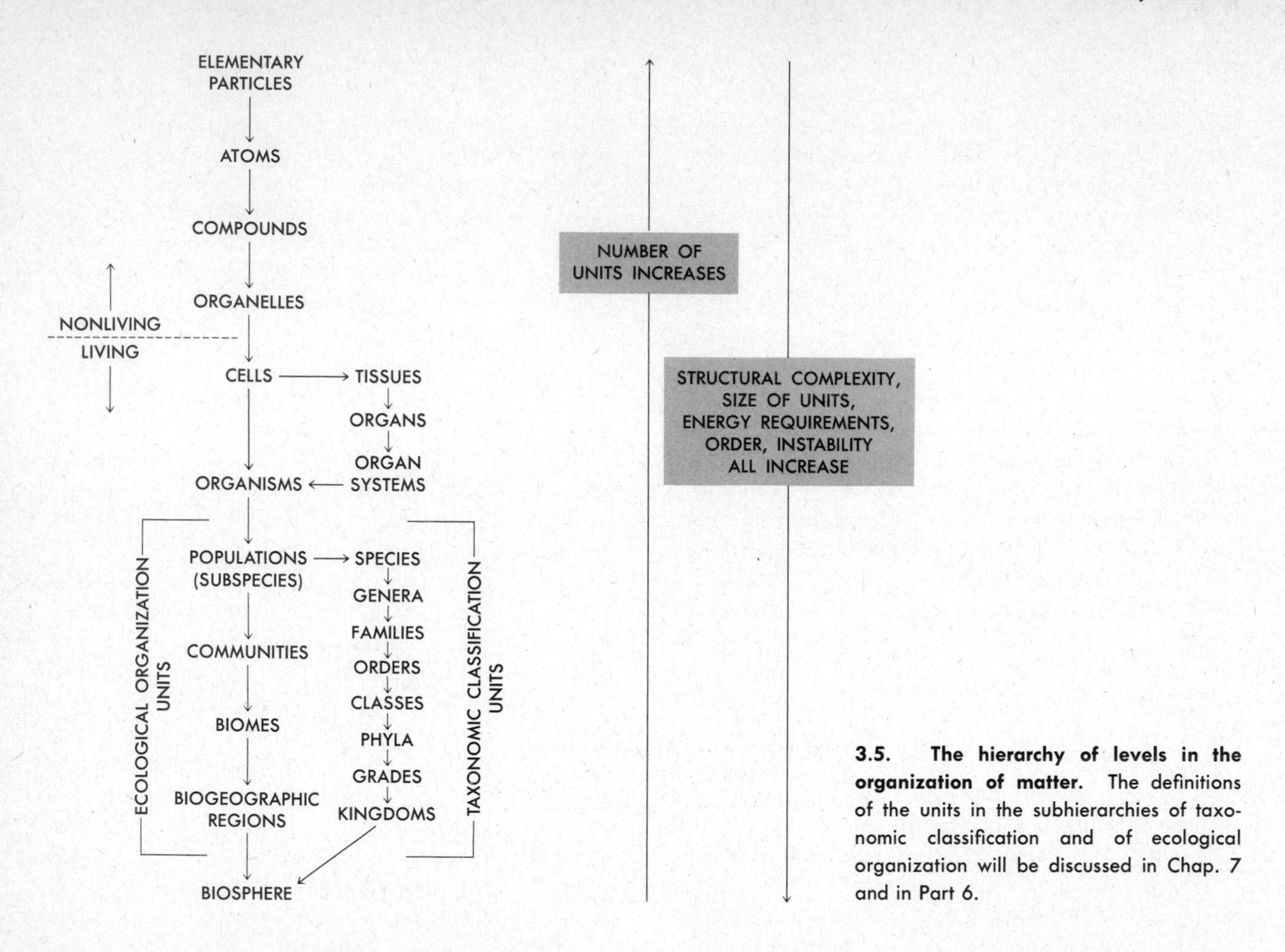

3.5. The hierarchy of levels in the organization of matter. The definitions of the units in the subhierarchies of taxonomic classification and of ecological organization will be discussed in Chap. 7 and in Part 6.

Even in their most elaborate and complicated forms—as organelles, for example—complexes of compounds as such cannot qualify as living units. To reach the level of life, we must go to the next higher structural level, namely, the level of *cells*. A cell is a specific combination of organelles, a usually microscopic bit of matter organized just complexly enough to contain all the necessary apparatus for the performance of metabolism and self-perpetuation. A cell in effect represents the least elaborate known structure that can be fully alive. It follows that a living organism must consist of at least one cell. Indeed, *unicellular* organisms probably constitute the majority of living creatures on earth. All other organisms are *multicellular,* each composed from two up to hundreds of trillions of joined cells.

Several distinct levels of organization may be recognized within multicellular organisms. The simplest multicellular types consist of aggregations of comparatively small numbers of cells. If all such cells are more or less alike, the organism is often referred to as a cellular *colony*. If two or more different groups of cells are present, each such group may be called a *tissue*. Structurally more complex organisms contain not only several tissues, but some of the tissues may also be joined further into one or more units called *organs*. The most complex organisms contain not only many tissues and organs, but groups of organs may also form one or more structural combinations known as *organ systems*. In terms of structural levels, therefore, living organisms exhibit at least five degrees of complexity: the single-celled form, the colonial form, the organism with tissues, the type with organs, and the type with organ systems.

Beyond organisms of all kinds, several still higher levels of life may be distinguished. A few individual organisms of one kind together may make up a *family*. Groups of families of one kind may form a *society*. Groups of families, societies, or simply large numbers of organisms of one kind make up a geographically localized *population*. All populations of the same kind together form a *species*. Different species aggregate into a local *community*. And the sum of all local communities represents the whole living world (cf. Fig. 3.5).

This hierarchial organization of matter into levels permits us to formulate a structural definition of life and nonlife. As we have seen, up to and including the level of the complex of compounds, matter is nonliving. At all higher levels matter is living, provided that, at each such level, metabolic and self-perpetuative functions are carried out. To be living, a society, for example, must metabolize and self-perpetuate on its own level, as well as on every subordinate level down to that of the cell.

Moreover, as life is organized by levels, so is death. Structural death occurs when one level is disrupted or decomposed into the next lower. For example, if a tissue is disaggregated into separate cells, the tissue ceases to exist. Structural death of this sort always entails functional death also, that is, disruption of the metabolic and self-perpetuative processes of the affected level. But note that disruption of one level need not necessarily mean disruption of lower levels. If a tissue is decomposed into cells, the cells may carry on as individuals (Fig. 3.6); if a family is disrupted, the member organisms still may survive on their own. On the other hand, death of one level always does entail death of higher levels. If many or all of its tissues are destroyed, the whole organ will be destroyed; if many or all of its families are dismembered, a society may cease to exist. In general, the situation is comparable to a pyramid of cards. Removal of a top card need not affect the rest of the pyramid, but removal of a bottom card usually topples the whole structure. We recognize that neither life nor death is a singular state but is organized and structured into levels.

Note, incidentally, that the hierarchy of levels provides a rough outline of the past history of matter. There is good reason to suspect that the universe as a whole began in the form of subatomic particles. These then became aggregated into atoms and formed galaxies, stars, and planets. As will be shown in Chap. 14, the atoms of planets subsequently gave rise to chemical compounds and complexes of compounds. On earth, some of the complexes of compounds eventually produced living matter in the form of cells, and unicellular types later gave rise to multicellular types. Among the latter, colonial types arose first, forms with organ systems last. Considered historically, therefore, matter appears to have become organized progressively, level by level. The presently existing hierarchy is the direct result, and it still gives mute, built-in testimony of its own historical development.

Several characteristics inherent in the structural hierarchy are of significance. For example, each level of organization includes fewer units than any lower level. There are fewer communities than species, fewer cells than organelles; and there is only one living world, but there are uncountable numbers of subatomic particles. It should be clear also that each level is structurally more complex than lower ones; a

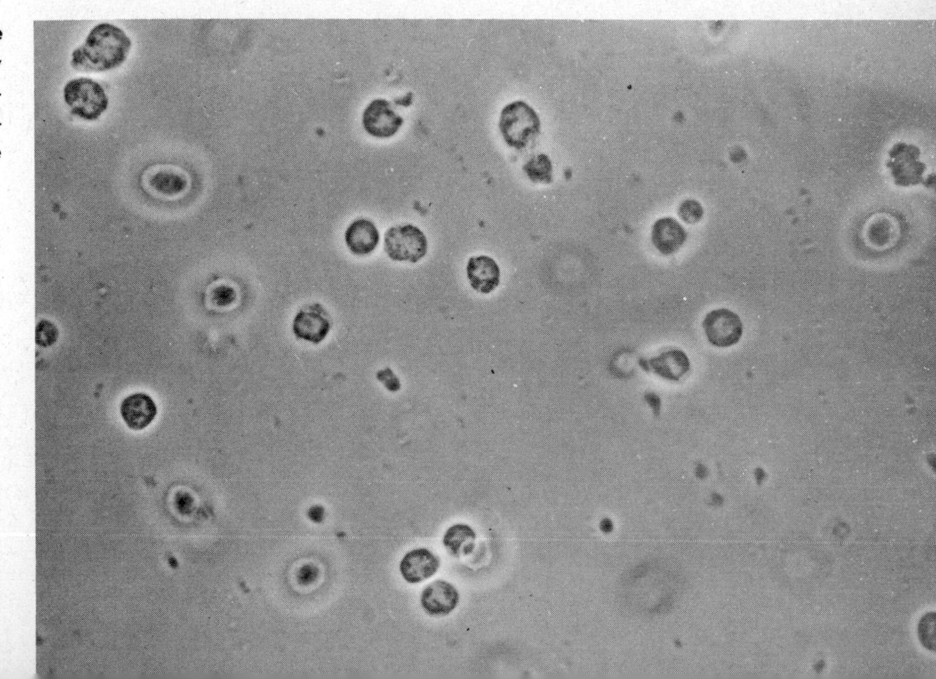

3.6. Disaggregated tissue cells of a mouse embryo, cultured in nutrient solution. Originally these cells were part of a compact tissue. Disaggregation destroyed the tissue level of organization but did not destroy the cellular level; the individual cells shown here remain alive.

given level combines the complexities of all lower levels, and it has an additional complexity of its own. Moreover, we may note that a jump from one organizational level to the next can be achieved only at the expense of energy. It takes energy to build atoms into chemical compounds, and it took energy to create cells out of chemical complexes. Similarly, energy is needed to produce tissues out of cells, societies out of families, or any other living level out of lower ones. Indeed, once a higher living level has been created, energy must continue to be expended thereafter to maintain that level. For example, if the energy supply to the cell, the organ, or the organism is stopped, death and decomposition soon follow and reversion to lower levels occurs. Similarly, maintenance of a family or a society requires work over and above that needed to maintain the organization of subordinate units.

This requirement is an expression of the second law of thermodynamics, already encountered in Chap. 2: if left to itself, any closed system tends toward a state of greatest stability. "Randomness," "disorder," and "probability" are equivalent to this meaning of stability. When we say that a system has a higher level of organization, we also say that the system exhibits a high degree of order, that it is nonrandom. The second law tells us that such a system is unstable and improbable and that if we leave it to itself it will eventually become disordered and therefore more stable. Living systems are the most ordered, unstable, and improbable systems known. If they are to avoid the fate predicted by the second law, they cannot be closed systems and a price must be paid. That price is external energy—energy to push the order up, against the constant tug to tear the order down.

But this price of energy is well worth paying, for each higher level exhibits new and useful properties over and above those found at lower levels. For example, a cell exhibits the property of life in addition to the various properties of the organelles which compose it; a multicellular organism with organs such as eyes and brains can see and distinguish objects, whereas cells taken singly at best can only sense the presence or absence of light. In general, the basic new property attained at each higher level may be described as united, integrated function: nonaggregated structure means independent function and, by extension, *competition;* aggregated structure means joint function and, by extension, *cooperation*. Atoms, for example, may remain structurally independent, and they may then be in functional competition for other suitable atoms with which they might aggregate. Once they do aggregate into a compound, they have lost structural independence and cannot but function unitedly, as a single cooperative unit. Similarly, cells may remain dependent structurally, and they may compete for space and raw materials. But if they aggregate into a tissue, they surrender their independence and become a cooperative, integrated unit.

This generalization applies at every other organizational level as well (Fig. 3.7). The results on the human level are very familiar. Men may be independent and competing, or they

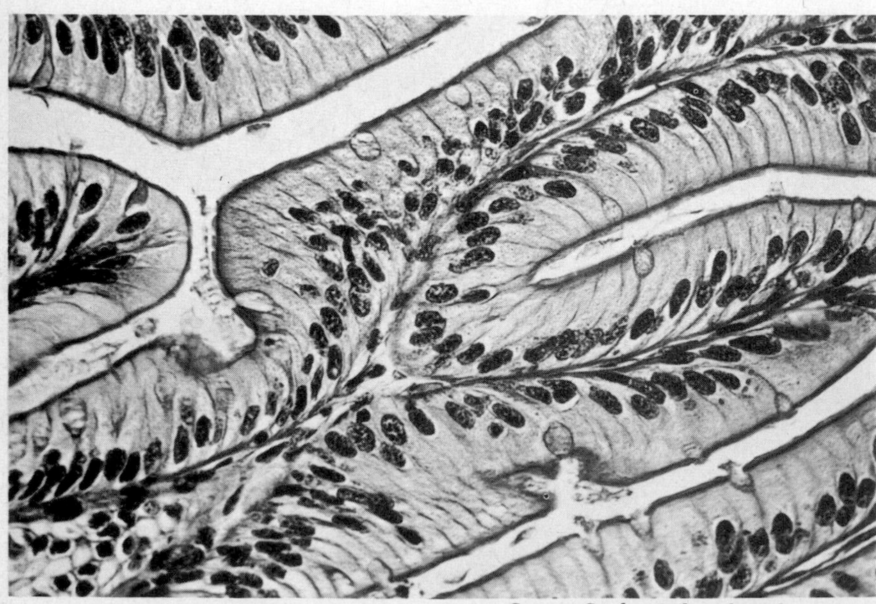

Courtesy Carolina Biological Supply Company.

3.7. Photomicrographic section through the small intestine of an amphibian, showing a portion of the wall. Several different types of tissues may be seen; they cooperate to produce and maintain this organ. In so doing, the tissues surrender much of their freedom and independence of action.

may give up a measure of independence, form families and societies, and start cooperating. Note here that sociological laws governing human society are based on and are reflections of the more fundamental laws governing the organization of all matter, from atoms to the whole living universe.

Note also that competition and cooperation are not in any basic sense willful, deliberate, planned, or thought out; atoms or cells neither think nor have political or economic motives. Structural units of any sort simply function as their internal makeup dictates. And the automatic result of such functioning among independent units may be competition; among aggregated units, cooperation. To be sure, human beings may decide to compete or to cooperate, but this merely channels, reinforces, makes conscious, and is superimposed on what they would necessarily do in any event.

Specialization

The fundamental advantage of cooperation is *operational efficiency:* the cooperating aggregate is more efficient in performing the functions of life than its subordinated components separately and competitively. For example, a given number of nonaggregated cells must expend more energy and materials to survive than if that same number of cells were integrated into a tissue. Similarly for all other organizational levels.

One underlying reason for this difference is that, in the aggregate, duplication of effort may be avoided. For example, in a set of nonaggregated cells, every cell is exposed to the environment on all sides and must therefore expend energy and materials on all sides to cope with the impact of the environment. However, if the same cells are aggregated into a compact tissue, only the outermost cells are in direct contact with the environment and inner cells then need not channel their resources into protective activities.

In addition to avoiding duplication of effort, aggregation also permits continuity of effort. Such continuity is not always possible in nonaggregated units. We may illustrate the general principle by contrasting unicellular and multicellular organisms, for example. A unicellular organism must necessarily carry out all survival functions within its one cell. In many instances, however, the performance of even one of these functions requires most or all of the capacities of the cell. In many cases, for example, the entire cell surface is designed to serve as gateway for entering nutrients and departing wastes. The entire substance of the cell functions to distribute materials within it. And all parts of the cell may be required directly in locomotion or in feeding, for example (Fig. 3.8).

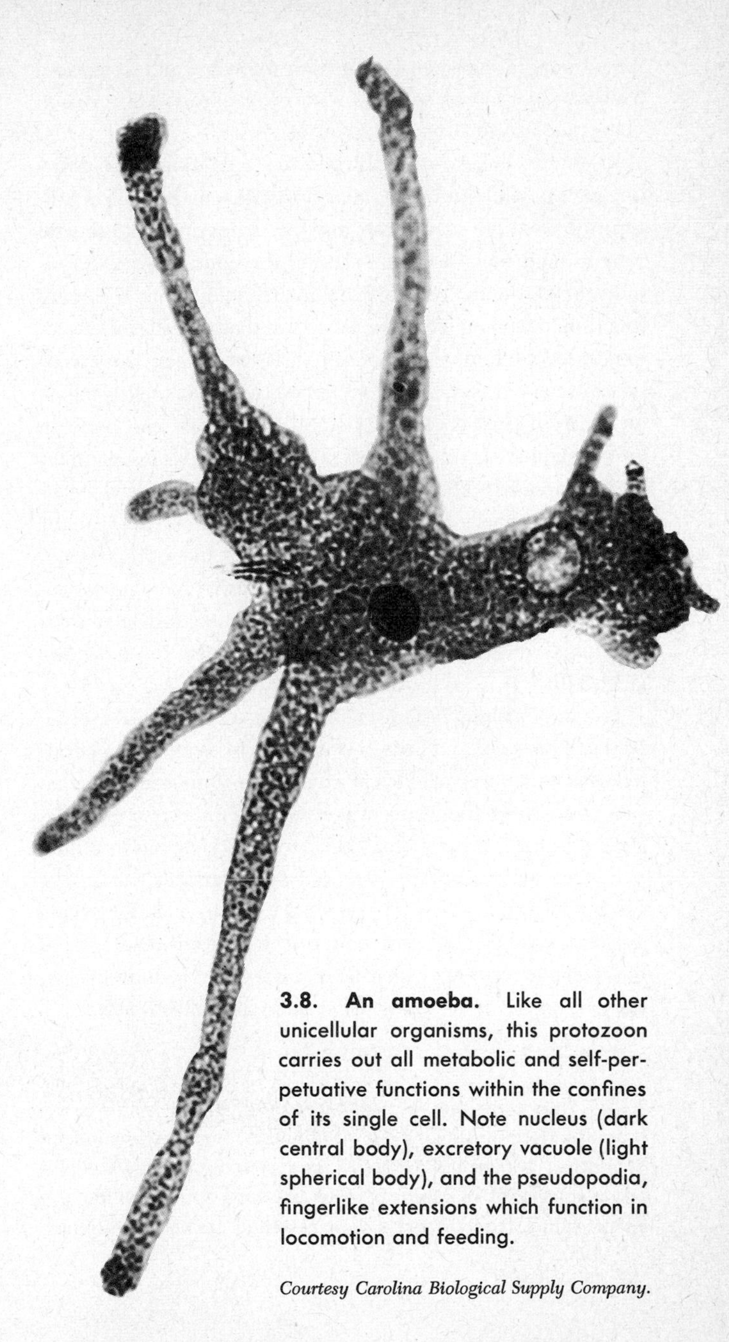

3.8. An amoeba. Like all other unicellular organisms, this protozoon carries out all metabolic and self-perpetuative functions within the confines of its single cell. Note nucleus (dark central body), excretory vacuole (light spherical body), and the pseudopodia, fingerlike extensions which function in locomotion and feeding.

Courtesy Carolina Biological Supply Company.

Very often, therefore, two such functions cannot be performed at the same time. Wherever locomotion and feeding each necessitate action by the whole cell surface, performance of one of these functions more or less precludes the simultaneous performance of the other. We shall find, moreover, that reproduction too involves the operational equipment of the whole cell, and in a unicellular organism this usually necessitates temporary suspension of both feeding and locomotion. Mutual exclusion of some functions by others is a common occurrence in all unicellular forms.

In multicellular forms, by contrast, continuity of effort becomes possible through *division of labor.* In such an organism, the total job of survival may be divided up into several subjobs, and each of these becomes the continuous responsibility of particular cells only. For example, some cells may function in feeding, continuously so, and other cells may function in locomotion, again continuously so. Indeed, division of labor in many cases is so pronounced that given cells are permanently limited in functional capacity; they can perform only certain jobs and no others. Thus, nerve cells can conduct nerve impulses only and are quite unable to reproduce or move. Muscle cells can move by contracting, but they cannot conduct nerve impulses; and normally they do not reproduce. Most cells in many multicellular organisms have analogous limitations. Each group of cells is more or less restricted in its functional versatility and exhibits a particular *specialization* (Fig. 3.9).

In a multicellular organism, therefore, an individual specialized cell does not and indeed cannot perform all the functions necessary for survival. This is why, when some cells are separated away from the whole organism, as in injury, for example, such cells must usually die. The specialized cell has lost independence mainly because it is not very versatile, because it can do only some of the jobs necessary for survival. The whole job of survival can be carried out only by the entire integrated multicellular system, which does possess the required versatility by virtue of its many differently specialized cells.

3.9. The principle of specialization. The single-celled organism (left) must carry out all required functions (symbolized by letters) within the limits of one cell. In the multicellular organism (right), by contrast, each cell may specialize to carry out a single function only, with resulting gains in efficiency.

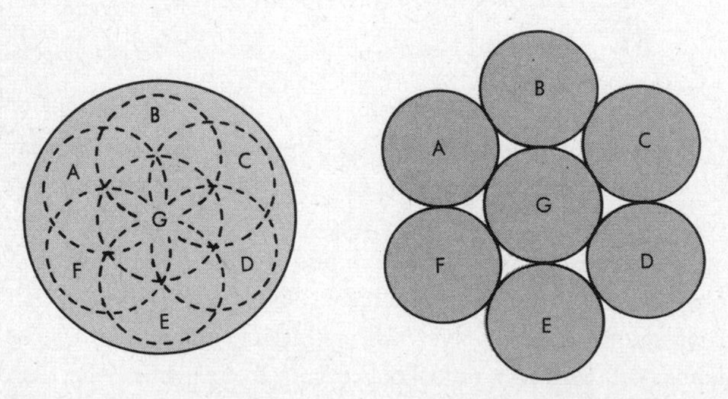

Specialization makes possible not only division of labor but also increased effectiveness of labor. For example, all unicellular organisms are sensitive to environmental stimuli. But such organisms, which must perform all survival functions within their single cells, are not particularly specialized for any one of these functions. Consequently, although they are sensitive to stimuli, the degree and range of sensitivity are quite modest. By contrast, many multicellular organisms possess highly specialized sensory cells. Such cells can be exceedingly sensitive and may respond to even very weak stimuli. Moreover, there may be several kinds of sensory cells, some specialized specifically for light stimuli, others for sound stimuli, still others for mechanical stimuli, etc. In short, the degree as well as the range of sensitivity can become enormously greater in cells which can specialize than in cells which cannot. This holds analogously for all other functions.

We may now understand the fundamental advantage of higher organizational levels generally and of multicellularity specifically. First, multicellularity makes possible division of labor, which avoids duplication of effort and permits continuity of effort. Secondly, division of labor leads to specialization, which permits any given effort to become highly effective. The overall result is an enormous saving of energy and materials, hence cheaper operation, and an enormous gain in efficiency. Basically, this gain of cheaper yet more efficient operation is what has favored more and more aggregation in matter generally and in living matter particularly; and this is why living history has produced multicellular organisms, equipped successively with tissues, organs, and organ systems, rather than only bigger and better unicellular organisms.

Note, incidentally, that loss of functional versatility in a specialized cell is never total. A cell cannot be so completely specialized that it performs just a single function. Certain irreducible "housekeeping" functions must be carried out by every living cell of a multicellular organism. Every such cell must absorb nutrients, must respire and synthesize, and must maintain steady states relative to its immediate environment. These metabolic and self-perpetuative functions cannot be specialized. Performed continuously and simultaneously in every cell, they are the bedrock of cellular survival. Specialization only affects additional functions, and the fewer of such additional functions a cell performs, the more specialized it is. Analogously, a cell can never be completely unspecialized and be so versatile functionally that it could survive under any or all conditions. All cells, even the most independent, still depend on very *particular* environments, for example. Cells therefore may only be more or less highly specialized; and within limits, the relative degree of functional versatility is an inverse measure of the relative degree of specialization.

Most multicellular organisms actually consist of cells which exhibit widely different degrees of specialization. In view of the earlier discussion about comparative efficiencies, would it not be most efficient if a multicellular organism consisted exclusively of highly specialized cells? Probably not, because certain functions need not be performed continuously. For example, it would be quite wasteful to maintain a permanent set of specialized scar-tissue cells—the organism might never sustain an injury. Analogously, it is a decided advantage that, in many multicellular organisms, reproductive structures become fully developed only during seasons when reproduction actually occurs. Thus, the actual design of the multicellular system permits the greatest possible economy of energy and materials. The most critical and continuously required functions are the responsibility of sets of permanently specialized cells. But less critical functions, and those required intermittently or only under unusual circumstances, are carried out by initially more versatile, less highly specialized cells.

In summary, the foregoing has shown that living units are organized structural aggregates at or above the cellular level of complexity, carrying out the functions of metabolism and self-perpetuation at every one of these levels. Energy must be expended to maintain the structural hierarchy, and the gain is greater division of labor and more specialization at successively higher levels. As a result of specialization, the versatility and independence of subordinate units is reduced, but the efficiency of maintaining life increases and the comparative cost in materials and energy decreases.

It follows also that a whole organism is more than the sum of its parts, for the organism exhibits properties over and above those of the totality of its components. The difference is a result of specialized organization, or pattern of arrangement. An organism is a collection of parts plus organization. It is not an accident that the words "organism" and "organization" have the same linguistic root.

It follows further that the specializations of one level determine the specializations of higher levels. If the cells composing a tissue are specialized as muscle cells, then the whole tissue will be specialized correspondingly for contraction and movement. If the organs of an organ system include teeth, stomach, and intestine, then, since the organs perform nutritional functions, the whole organ system will be specialized for nutrition. Every organism as a whole therefore is itself specialized, in line with the particular specializations of its subordinate parts. For example, every organism is able to live in a particular environment only and is able to pursue only a particular way of life. A fish must lead an aquatic existence, an earthworm cannot do without soil, and man too is specialized in his own way. He requires a terrestrial environment of particular properties, a social environment of variously specialized human beings, a community of wheat, cattle, and other food organisms. In effect, the specializations of his body allow him to pursue no other but a characteristically human mode of life. And by being specialized, every organism in effect is a dependent, necessarily cooperating unit of a higher living level: the population, the whole species, the community of several species. These higher-level units are specialized in their own turn, according to the specializations of their members.

We recognize, then, that although all organisms are alike in general characteristics, they differ in specific characteristics because of specialization. Functionally, all organisms pursue life identically through metabolism and self-perpetuation; structurally, all organisms are composed of cells. But in each organism both the functions and the structures are in some respects specialized, and the specializations differ in different cases.

What, then, are the characteristic specializations of animals?

THE NATURE OF ANIMALS

Altogether some two dozen major animal groups can be distinguished; a few of the more familiar ones are listed in Table 2, which may provide a preliminary background for discussion. Note that each major group constitutes a *phylum,* with a technical and usually also a general name. Each phylum is defined and distinguished from others by specific identifying criteria, a subject we shall deal with in detail in Unit 2. Note also that it is often useful to categorize animals as *vertebrates* and *invertebrates,* the latter group encompassing all animals without a vertebral column, i.e., all animals except the vertebrates.

Most of the organisms we traditionally call "animals" have two basic traits in common, one structural and one functional. Structurally, the animal body usually exhibits an *elaborate multicellular organization;* it is characterized in the vast majority of cases by the presence of organs and organ systems. Functionally, the typical mode of animal nutrition is *heterotrophism.* This means that food cannot be manufactured within the organism (as by photosynthesis in plants, for

TABLE 2. A few of the phyla of animals and some of their representative members

Phylum name	General name and chief characteristic	Representative animals
Protozoa	protozoans, predominantly one-celled types	amoebas, paramecia, malarial parasites
Porifera	sponges, or animals with water-pores	chalk, glass, and horny sponges
Cnidaria	coelenterates, or animals with digestive sacs	hydras, jellyfishes, corals, sea anemones, sea fans
Platyhelminthes	flatworms	planarians, flukes, tapeworms
Aschelminthes	sac worms	rotifers, roundworms, hair worms
Mollusca	mollusks, or soft-bodied animals	snails, clams, squids, octopuses
Annelida	segmented worms	clam worms, earthworms, leeches
Arthropoda	arthropods, or jointed-legged animals	crustacea: copepods, barnacles, water fleas, shrimps, crabs, lobsters, crayfishes
		insects, scorpions, spiders, ticks, mites, centipedes, millipedes
Echinodermata	echinoderms, or spiny-skinned animals	starfishes, sea urchins, sea cucumbers, brittle stars, sea lilies
Chordata	chordates, or notochord-possessing animals	tunicates, amphioxus
		vertebrates: jawless fishes (lampreys), bony fishes (herring, tuna, etc.), cartilage fishes (sharks), amphibia (salamanders, newts, toads, frogs), reptiles (turtles, lizards, snakes, alligators), birds, mammals

example), but must be acquired in some way from already preexisting organisms, living or dead.

Taken separately, neither of these two traits is unique to animals or necessarily characteristic of all animals. Much depends on how the concept "animal" is defined, a problem we shall discuss separately in a later context (Chap. 14). Here we need note only that certain organisms exhibit both of the two traits together, and that such organisms constitute the majority of what are commonly called "animals." In this majority, the two traits form the very essence of animal nature; they can be shown to condition virtually every other aspect of animal design and way of life.

Most importantly, the combined traits of elaborate body structure and heterotrophism virtually necessitate the further trait of *motility*, i.e., the capacity of active movement. Many nonanimal organisms also are heterotrophic (e.g., most bacteria, all fungi), yet most of these are incapable of active movement. However, the body of such organisms is not structured elaborately and indeed is comparatively small and is in many cases unicellular and microscopic. As a result, the correspondingly small amounts of food required can usually be obtained at the spot where the organism happens to be located, or where wind, water currents, and other such agencies of passive dispersal may have chanced to carry the organism. In animals, on the other hand, movement by passive means would be quite inadequate. Being complexly structured and on the whole therefore bulkier, animals require commensurately larger quantities of food; and random passive dispersion by environmental agencies in most cases is not likely to carry animals to adequate amounts or kinds of food organisms. Active, self-powered motion consequently becomes a necessary and indeed a rather familiar characteristic of animal design.

In the majority of animals the motion-producing equipment is a *muscular system*, composed of contractile organs (muscles) which move body parts such as tentacles, legs, wings, or creeping surfaces. In many animals, including in part those with muscular systems, motion is also made possible by *cilia* and *flagella*, tiny hairlike projections from the surfaces of certain cells, and by *pseudopodia*, flowing fingerlike extensions from amoeboid cells (cf. Fig. 3.8). Regardless of the specific means of motion, animals employ their motor capacity in two major ways to obtain food. In the more common case, an animal carries out some form of *locomotion;* i.e., it propels its whole body toward the location of a likely food source. In the second case, the animal remains stationary, or *sessile*, and lets the food source move toward it. All sessile animals are aquatic; they employ their motion-producing equipment either to create water currents which carry food organisms to them or to trap food organisms which happen to pass close by (Fig. 3.10).

Once the capacity of motion is given, it may serve not only in the search for food but also secondarily in other vital activities. As is well known, for example, locomotion aids animals significantly in protecting them from potential environmental dangers, both biological ones occasioned by other organisms and physical ones such as climatic changes. The latter affect all animals daily and seasonally, and animals respond primarily by moving, either to sheltering sites nearby or to more suitable territory farther away. Locomotion also plays an important role in mate selection and in reproduction generally. Yet in virtually all cases, the most consistent and continual locomotor effort made by animals probably is the search for food.

Before an animal can effectively move toward food or vice versa, it is clearly essential for the animal to recognize that, at such and such a spot in the environment, an object is located which is or appears to be a usable food. Moreover, after such identification and localization have been accomplished, it is equally essential to control the ensuing motion, i.e., to set and adjust course and speed and to determine when movement is to begin and to terminate. What is evidently needed is a complete guidance apparatus. Most animals actually possess it in the form of a *nervous system*. In it, sense organs of various kinds permit recognition of environmental detail, impulse signals via nerves produce control over motion, and a brain or brainlike organ correlates and coordinates, i.e., it fits a given set of recognitions to an appropriate set of motions. In exercising these functions, nervous systems actually must regulate more than muscular activity as such; for if muscles are to function at all, any structure or process which contributes to the maintenance of proper operating conditions within muscles requires regulation as well. It happens that virtually all internal components of animals play at least some role in maintaining the fitness of muscles. Correspondingly, nervous systems coordinate almost all internal operations of the body, and in effect the systems so become major controllers of steady states.

Indeed, steady-state control is essential not only in achieving proper motion but also in performing any other function. The general importance of such control is underscored by the existence, in most animals, of chemical coordinating systems in addition to the neural ones. In different animal groups the chemical regulators differ greatly in form and specific activity, and in many instances chemical control is exercised by components of other organ systems (e.g., blood, kidneys, gills). In some groups, notably in insects and vertebrates, various processes of chemical control are governed by specialized hormone-producing *endocrine systems*. These, and internal chemical coordinators in general, operate in conjunction with neural coordinators; and we may note that both basically serve in making the overt actions of an animal dovetail sensibly with

3.10. Motion and feeding. Animals obtain food either by active locomotion or by trapping small moving food organisms, as among the sessile featherduster worms illustrated here. These worms, named *Sabella* and related distantly to earthworms, live in attached tubular housings and project their feathery food-trapping crowns from the open ends of the tubes. Cf. also Fig. 26.10.

the requirements imposed by given environmental situations (Fig. 3.11). Motion is a major one of such overt actions. Evidently, inasmuch as internal coordination is a prerequisite to effective motion and inasmuch as motility is itself necessitated by heterotrophism, we may regard the presence of nervous and other steady-state controllers as a direct design consequence of the mode of animal nutrition.

3.11. A flounder against three different backgrounds. Note how the animal may adapt its skin pigmentation pattern to the color pattern of the environment. In such cases of adjustable camouflage, information about the environment is communicated to the skin via eyes, nerves, and hormones. Pigment cells in the skin respond to the information by contracting or expanding, thus altering body coloration.

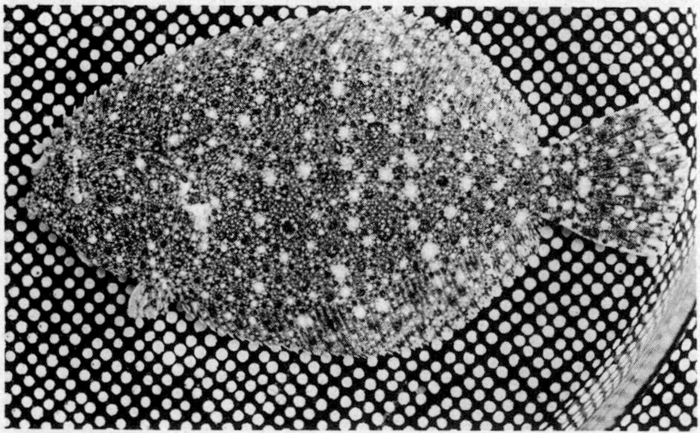

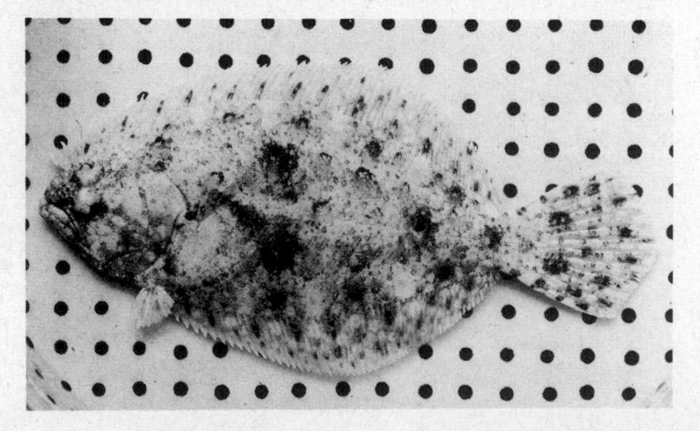

The requirement of motility has a significant effect on the architecture of animals. Motion must take place in water or air, and movement clearly will be most efficient if the external medium offers the least possible resistance. Unlike a tree, therefore, which is constructed in a ramified shape for maximum exposure to light, air, and soil, an animal is built as compactly as is feasible, for minimum surface exposure. Indeed, most motile animals also tend to be *bilaterally symmetrical* and *elongated* in the direction of motion, a shape which aids in reducing resistance to movement and provides stable mechanical balance between mirror-image right and left sides. Moreover, since one end of an elongated animal necessarily enters new environments first, that end will serve best as the place for the chief sense organs and nerve centers and for the food-catching apparatus. The leading part of the body so becomes a *head*. At the same time, elimination products of all kinds are best released at the hind end, where they do not impede forward progression. A general build of this sort is actually standard and nearly universal among moving animals (Fig. 3.12).

By contrast, sessile animals, and also many of the slow and sluggish types, face their environment more or less equally from all sides, like plants, and their architecture reflects this. They are or tend to be *radially symmetrical*, and a distinct head is usually not present (e.g., corals, starfish). Also, sense organs and other components of nervous systems tend to be greatly reduced (an observation which underscores clearly that the primary function of such systems in all animals, man included, is to control movement).

The physical problem of locomotion undoubtedly contributes to the preponderance of aquatic types among animals. Actually only relatively few groups are terrestrial: some snails, some worms, arthropods such as insects and spiders, and vertebrates such as reptiles, birds, and mammals. All other animal groups are primarily aquatic, and this is probably correlated with the requirement of locomotion. Propulsion can be accomplished with less effort in water than on land or in air; friction is greatest on land, support against gravity is least in the air. In water, however, an animal is buoyed up and comparatively little energy need be expanded for locomotion. Moreover, food is abundant in water and a water environment is biologically beneficial in other respects as well.

Once animal and food source are near each other, the

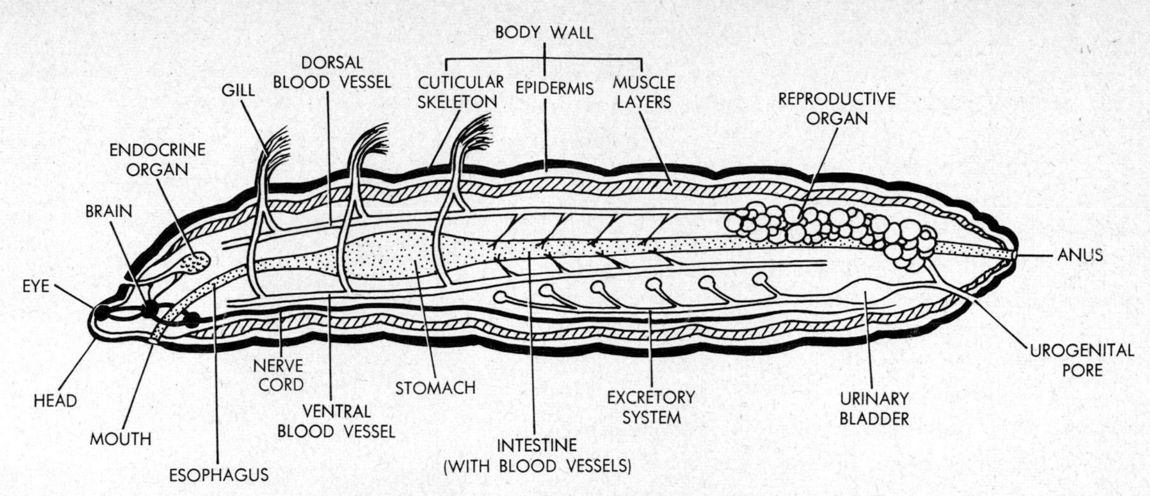

3.12. Diagrammatic representation of a moving animal. This is a hypothetical animal, showing the general position of the organ systems. Note that the integumentary and nervous systems are at or near the surface, that the alimentary, breathing, excretory, and reproductive systems communicate with the surface, and that the circulatory and muscular systems range throughout the body. If present, the endocrine and skeletal systems often range throughout the body as well, and the skeletal components may have exterior or interior positions. The endocrine system may consist of several organs which may or may not interconnect structurally.

animal must then make actual use of food. Here again the condition of heterotrophism imposes characteristically animal traits. Two basic "strategies" of handling foods are encountered. In one, the animal devours or eats food in bulk lots. Bulk feeding is a form of heterotrophic nutrition called *holotrophism*, and the eating process itself constitutes *ingestion*. It is accomplished by means of mouth openings and is generally facilitated by various accessory ingestive structures (e.g., teeth, tentacles and other body appendages, various pincer and clamplike devices in or near the mouth). However, food organisms ingested in whole or in part usually are not immediately usable in the bulk form in which they are eaten. Consequently, *digestion* of food into manageably molecular components becomes a necessary further activity. It is carried out in a digestive tract, typically tubular, leading away from the mouth opening. And since bulk food as eaten normally contains a variety of indigestible or otherwise unusable materials, the latter must eventually be eliminated, or *egested*. Most animals possess a separate egestive opening, the anus; in others a single opening serves both as mouth and anus. Taken together, the processes of ingestion, digestion, and egestion constitute the function of *alimentation*, and all structures associated with this function form an *alimentary system*.

The second principal method of food handling is *parasitism*, another form of heterotrophic nutrition. Here the food source does not enter the animal; instead, the animal enters the food source (Fig. 3.13). It should be added, however, that the food source in such cases is always a living organism; if an infected organism dies, the parasite must find another living host or it must die itself. Parasites obtain food from hosts by siphoning usable nutrient chemicals directly from the host's body. Parasitic animals are believed to have evolved from nonparasitic, holotrophic ancestors, and many still possess complete alimentary systems. But in many other parasites such systems are reduced, and in some cases they are no longer present at all. Food molecules then enter the parasite directly through its skin.

The necessity of alimentation has a major effect on the basic

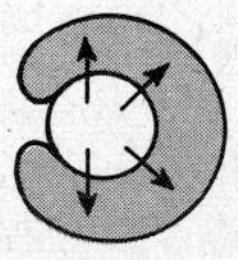

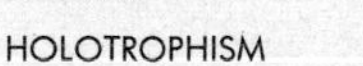

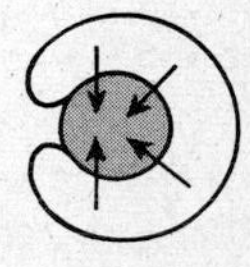

3.13. The food-obtaining animals (dark areas) have the food source within them in holotrophism, but are themselves within the host in parasitism.

animal way of life. Because the supply of appropriate food organisms is limited, animals tend to compete openly for whatever foods are available. As a result, some animals become predators, some become prey. Numerous adaptations in structure and function aid either the hunter or the hunted: brute strength, speed, body colors and shapes which protect by camouflage (cf. Fig. 3.11), colors and sounds which warn or lure, and many others. Significantly also, predator and prey are often characterized by distinct modes of *behavior*, as expressed, for example, through furtiveness, stealth, cunning, aggressiveness, timidity, and other manifestations of "mentality" and "personality."

In addition to channeling animal life into patterns of offense and defense, the condition of competitive heterotrophism is correlated with specializations in animal eating habits. Thus, *herbivores* are specialized to eat only plant foods; *carnivores* subsist only on other animals; and *omnivores* may eat both animal and plant foods, living or dead. In view of its abundance, plant food is easily come by. As a result, more herbivorous animal types exist than any other. Also, since plants do not put up a fight before being eaten, herbivores generally are of more or less gentle disposition and are more adept in defense than in offense.

A plant diet presents its own special problems. The large quantities of cellulose present make plant tissues tough and difficult to tear. Animals must therefore scrape or grind or chew plant foods or suck the juices out of them. Correspondingly, rasping, crushing, grinding, and sucking structures are particularly common among the ingestive devices of herbivores. Such animals also tend to possess comparatively long digestive tracts, which offer more surface and more time for the digestion of plant foods. A pound of fresh plant material consists largely of water and cellulose and of correspondingly less usable food. Accordingly, herbivores generally eat more, and more often, than other animal types.

Carnivores are specialized to overcome not only herbivores but also smaller carnivores and omnivores. Speed, strength, and varied prey-killing equipment in mouth or body appendage are familiar adaptations to a carnivorous way of life. Note, however, that virtually none of the carnivores kills wantonly, but only when hungry or threatened; it is to the carnivore's advantage to live amid a thriving population of herbivores. Animal tissue is softer than that of plants and tears fairly easily. Correspondingly, sharply pointed ingestive aids adapted for tearing are predominant among carnivores. Absence of cellulose in animal foods also tends to reduce chewing time and makes for easier digestion and relatively shorter

3.14. Skeletal types. A. The calcareous exoskeleton of a snail. **B.** The horny exoskeleton of a stag beetle. **C.** X-ray photograph of a girl, showing the human endoskeleton.

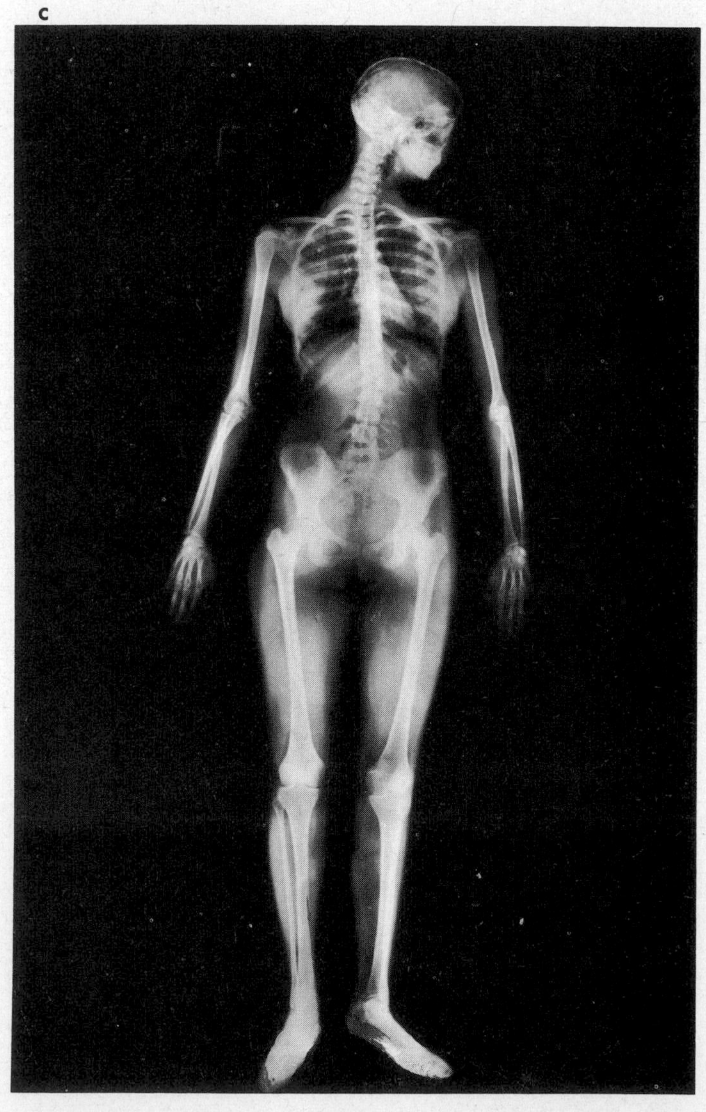

alimentary tracts. Nutritive values per pound of animal tissues are greater than in plant foods. Therefore, carnivores generally eat fewer, smaller meals.

Omnivores subsist on whatever nourishment they can find or catch. Many omnivores wait on the scene of battle between carnivore and herbivore, to scavenge among the remains. Others live on minute plant and animal debris in soil or water. The food-trapping and other alimentary structures of omnivores usually combine herbivorous and carnivorous features, as might be expected.

Numerous important characteristics of animal design result from the bulkiness and compactness of the animal body. Since animal cells possess comparatively little inherent rigidity, a bulky collection of cells cannot readily maintain shape and is likely to sag into a formless mass under the influence of gravity. Animals therefore require antigravity supports, and they actually possess them in the form of *muscular* and particularly *skeletal systems*. That muscles function not only in motion but also in support is well illustrated in animals such as earthworms, which do not possess a skeleton. The same muscles which move such animals also contribute to holding them together and to maintaining their shapes. Moreover, even an animal with a skeleton would sag into a formless mass if muscles did not maintain a taut, firm organization. On the other hand, that skeletons function not only in support but also in locomotion is also clear. A large, heavy animal could neither hold its shape nor propel itself forward by muscles alone, without rigid supports.

Animal skeletons are either calcium-containing calcareous supports or silicon-containing silicaceous supports or variously composed horny supports. The skeletons are organized either as *exoskeletons* or as *endoskeletons*. In an exoskeleton, the supporting material is on the outside of the animal and envelops it partly or wholly. Mollusks and arthropods are among groups with such exoskeletons. In an endoskeleton, the supports are internal and soft tissues are draped over them. The main animal groups characterized by this type of skeleton are the echinoderms and the chordates, the latter including the cartilage- and bone-possessing vertebrates. An endoskeleton permits an animal to become far larger than does an exoskeleton. With increasing body size, an exterior skeletal envelope rapidly becomes inadequate to support deep-lying tissues. Interior supports, however, can buttress all parts of even a large animal. It is not an accident, therefore, that the largest animals are the vertebrates, and that animals with exoskeletons or without skeletons of any kind are comparatively small (Fig. 3.14).

The bulky construction of animals also creates problems of internal logistics. For example, after food is eaten and digested, the usable nutrients must be distributed to all parts of the animal body. If the distances between the alimentary system and the farthest body parts range over only a few cell layers, as is the case in several groups of relatively simply constructed animals, then internal food distribution can be achieved adequately by ordinary physical diffusion (cf. Chap. 4). But if food substances must traverse very many cell layers before reaching distant body parts, as in the majority of animals, then distribution by diffusion no longer suffices. Some sort of internal transport system becomes essential, and the *circulatory systems* of animals are designed to meet this requirement (Fig. 3.15). In such networks of vessels, the transport vehicle of food is blood, more specifically, the water component of blood, in which nutrients are carried in solution. One or more muscular pumping organs, or

3.15. If an animal is small and its surface is thin and permeable **(A)**, then breathing and excretion may occur directly through the body surface and the internal transport of materials may be accomplished without special structures. But if an animal is larger and its surface is thick and impermeable **(B)**, then the breathing and excretory surfaces usually are parts of specialized, interiorized organ systems and internal transport of materials is accomplished by distinct circulatory systems.

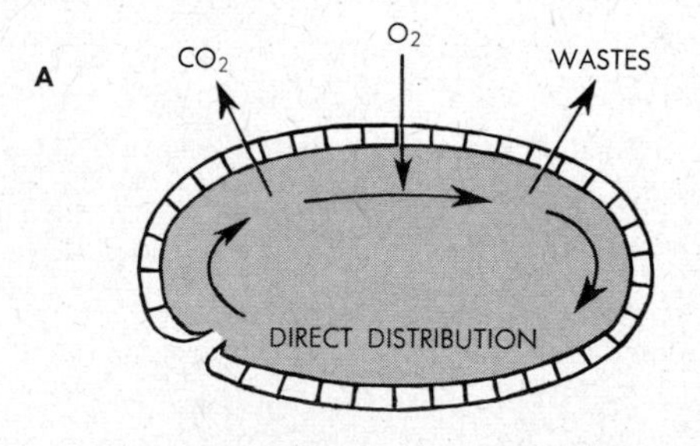

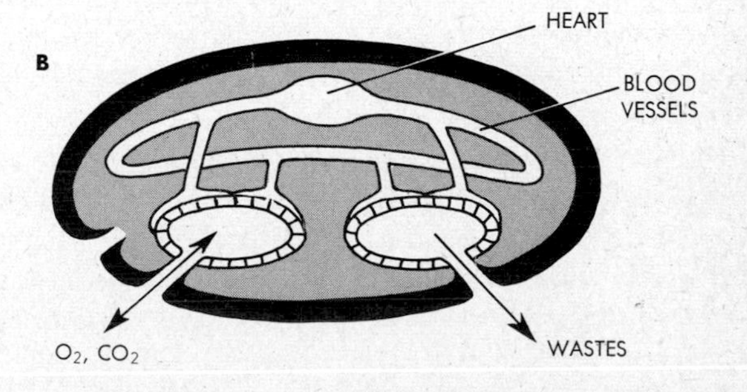

3.16. These stages in the life cycle of amphibia symbolize the main stages in the sexual development of animals generally. **A,** eggs; **B,** embryos (within jelly coats); **C,** larvae; **D,** adult. The transition from embryo to larva is achieved by hatching, that from larva to adult, by metamorphosis.

hearts, maintain a circulation of blood throughout the body. Blood is not red in all animals, but is blue or green in some and colorless in many others. Where blood is pigmented, the pigment functions specially in transport of oxygen, not food, a circumstance which points up another problem of internal logistics.

Because of the compact design of an animal, most of the cells are not in immediate contact with the external environment. Yet all cells require environmental oxygen for respiration, and every cell also must release to the environment any waste substance resulting from metabolism. In several groups of simply constructed animals, where the thickness of the body does not exceed a few cell layers, direct diffusion suffices adequately to supply all cells with oxygen and to rid them of metabolic wastes. In the majority of animals, however, the structural complexity is such that at best only the cells within diffusion distance from the body surface could exchange materials with the environment directly. Transport channels are therefore required, and in the majority of animals the circulatory system serves as the principal link between the environment and the interior of the body; blood serves as traffic vehicle to and from the body surface.

In some animals (e.g., earthworms, frogs), the entire body surface is adapted to absorb oxygen from the environment into the blood vessels of the skin, and the entire skin, similarly, may release waste substances from the blood into the environment. In such cases the skin, which represents the principal organ of the body-covering *integumentary system,* must remain relatively thin and permeable and can afford only limited protection. In most animals, however, the integumentary system is elaborated more complexly and is composed of skin, skin glands, and a large variety of different skin-covering structures (scales, hair, horn). Such a system protects, supports, and is relatively impermeable to the environmental medium. Animals so covered may exchange materials with the environment only at restricted areas, where surface thinness and permeability are preserved and where the blood supply is particularly abundant. For protection, such thin and sensitive areas are frequently tucked well into the body, away from the general body surface yet still in direct communication with the environment. These areas represent organs which make up two systems, the *breathing* and the *excretory systems* (cf. Fig. 3.15). Gills and lungs are the principal types of oxygen collectors, but these organs also contribute importantly in waste excretion. Serving primarily in excretion are kidneys and other, functionally equivalent types of organs.

Bulk and complex organization affect yet another aspect of animal design, namely, the pattern and process of reproduction. Like other organisms, animals too possess *reproductive systems* which manufacture reproductive cells, principally sperms and eggs. After such single cells fuse pairwise in a sexual process, the resulting fertilized egg, or *zygote,* still is a single cell. But the adult animal is complexly multicellular. Indeed it is so elaborately multicellular that successive cell divisions by themselves are quite insufficient to transform a unicellular zygote into the intricately organized adult. In addition to mere increase in cell number, therefore, groups of cells must be fashioned into structurally and functionally distinct organs, and these must come to interrelate in highly specific ways, both in location and operation. In other words, a specialized, lengthy course of development must take place. Animal development actually does occur in a uniquely characteristic manner, two major distinguishing features being the *embryo* and, typically, the *larva*. The embryonic phase of development starts with the zygote and usually terminates in a process of *hatching*. The ensuing larval phase then continues to *metamorphosis*, or transformation into the adult. Both embryo and larva are generally rather unlike the adult in structure or function; both developmental phases may be regarded as specifically animal devices designed to provide the necessary time, and the means, for the production of a complexly structured adult out of a single cell (Fig. 3.16).

We find, therefore, that if we take the basic animal attributes of heterotrophism and high level of structural complexity as a starting point, we may deduce all other principal facets of animal nature as contingent requirements. As we have seen, the two basic attributes at once necessitate the presence of nervous, muscular, skeletal, alimentary, circulatory, excretory, breathing, and integumentary systems. And if to these we add a reproductive system and in some cases also an endocrine system, we have a complete list of all the major architectural ingredients composing an animal. Implied also is a good deal about how an animal moves, behaves, feeds, develops, copes with its environment—in short, pursues life (Fig. 3.17). Moreover, we have learned in broad outline how the structural ingredients of the animal must be put together to form a sensibly functioning whole. As suggested in Fig. 3.12, some of the organ systems must be in surface positions in whole or in part (e.g., integumentary, nervous); others may lie deep but must at least communicate with the surface (e.g., alimentary, breathing, excretory, and also reproductive); and still others must range over and through the whole body (e.g., skeletal, circulatory, nervous, endocrine).

Based on such a preliminary sketch, the fundamental anatomy of a motile, elongated animal may readily be envisaged to resemble a complex tube having a triple-layered wall. The outermost layer of the tube is the body wall, the most conspicuous component here being the integumentary system. The innermost layer, which encloses the open channel through the tube, is represented principally by the alimentary system. And the bulky middle layer contains all other organs and systems. Such a triple-layered picture of animal architecture is actually more than a rough analogy; for at an early stage of development, most animal embryos consists of just three layers, one inside the other and each originally not more than one cell thick. From the outside inward, these so-called

3.17. The structural design and the functional attributes of animals are direct or indirect consequences of the requirements of motion and compactness, factors which in turn are predicated on the basic conditions of heterotrophism and elaborate organization.

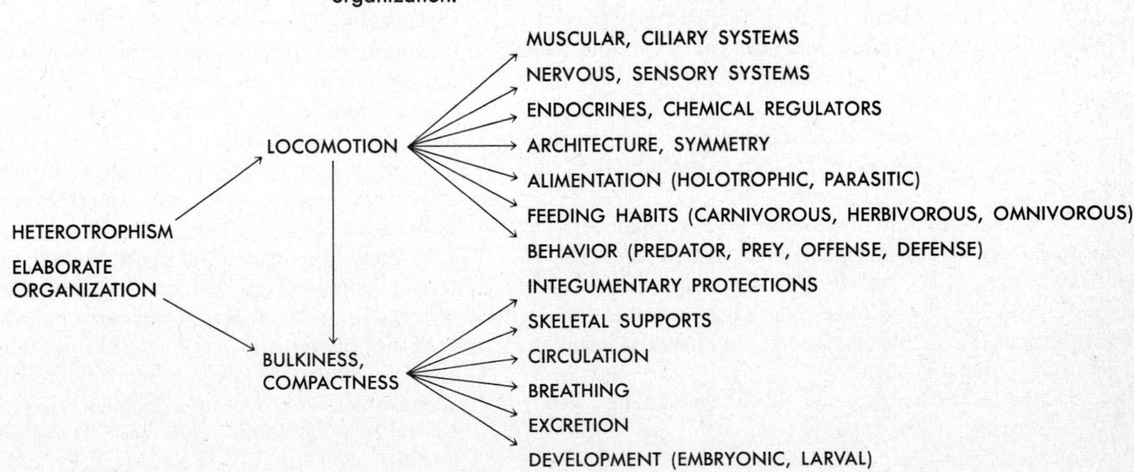

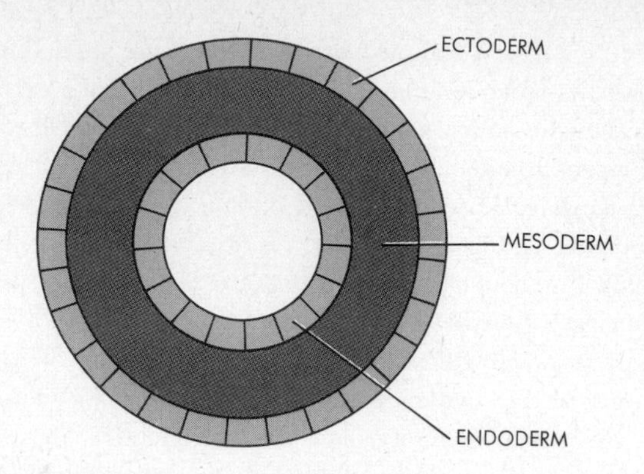

3.18. Diagram showing the three germ layers of an animal embryo. These layers will give rise to the triple-layered "tube" of the adult body, as in Fig. 3.12. Ectoderm generally forms epidermis and nervous system, endoderm forms the alimentary system, and mesoderm develops into muscular, circulatory, and reproductive systems. The other systems arise from different germ layers in different animal groups.

primary germ layers are known as the *ectoderm*, the *mesoderm*, and the *endoderm* (Fig. 3.18). Subsequently they each proliferate greatly and give rise to the triple-layered adult "tube." Ectoderm forms, for example, the integumentary and the nervous systems. Mesoderm develops in part into the muscular, circulatory, and reproductive systems. Endoderm produces the alimentary system. Other systems arise from different germ layers in different animal groups, as will be shown in later chapters.

In the resulting adult, all the structural parts are necessary and just sufficient to maintain the life-sustaining functions of metabolism and self-perpetuation. We may note here that, although given organs and systems are often more directly concerned with some functions than with others, all systems nevertheless play at least some role in all functions. Indeed, all functions also contribute importantly to the maintenance of all systems. For example, although the alimentary system serves primarily in nutrition, it also serves indirectly in all other functions of metabolism and contributes additionally to all processes of self-perpetuation, including even evolution: without a properly nourished reproductive system there can be no reproduction, hence also no heredity and no evolution. Analogously, it is easy to verify that every body part must provide functional support for every other part, and that only through such interdependent activities can all of metabolism and self-perpetuation take place.

Underlying these considerations is the firm recognition that all animals are composed of cells, the basic units of life both in structure and function. We know further that all cells are in turn composed of atoms and chemicals, and that life is a result of particular reactions among particular cellular chemicals. Accordingly, we must now direct our attention to the cellular foundations of life before we can appreciate the larger-scale aspects of animal life.

REVIEW QUESTIONS

1. What is metabolism? Self-perpetuation? What are the principal component functions of each of these, and what specific roles do these functions play in the maintenance of life?
2. What are the fundamental differences between inanimate, dead, and living systems? Discuss carefully and fully. Define living.
3. Review the hierarchy of levels in the organization of matter, and discuss how living matter is characterized in terms of levels.
4. Review the relation of levels of organization to energy, to aggregation, to complexity, to competition and cooperation, and to operational efficiency.
5. Define cell (*a*) structurally and (*b*) functionally. Define organelle, tissue, organ, organ system, organism.
6. In terms of cellular specializations, how does a cell of a single-celled organism differ from a cell of a multicellular organism? Cite examples of specialization on the tissue, organ, organism, and species levels of organization.
7. What are heterotrophism, holotrophism, parasitism, alimentation, hatching, metamorphosis, primary germ layers, integuments, hormones?
8. What basic traits identify animals generally? In what architectural respects are moving animals generally different from sessile ones?
9. Show in some detail how the basic traits of animals determine and influence the characteristics of all other aspects of animal nature. As far as you can, contrast such typically animal attributes with those generally considered to be characteristic of plants. Do you think that animals can be defined adequately as motile, heterotrophic organisms and plants, contrastingly, as sessile, photosynthetic organisms? (Defer a final answer until after you have studied Chap. 14, section on animal origins).

10. Name the organ systems characteristic of animals generally and describe the fundamental functions of each. Show how each system contributes specifically to metabolism and self-perpetuation. In man, which familiar organs belong to each of the organ systems? What are some of the specific functions of such organs?

COLLATERAL READINGS

Most of the topics dealing with animals specifically are taken up again in greater detail in later chapters; appropriate readings will be suggested in such contexts. In connection with the other topics introduced above, the following articles are recommended:

Brown, G. S., and D. P. Campbell: Control Systems, *Sci. American,* Sept., 1952.

Kemeny, J. G.: Man Viewed as a Machine, *Sci. American,* Apr., 1955.

King, G.: What Is Information? *Sci. American,* Sept., 1952.

Nagel, E.: Self-regulation, *Sci. American,* Sept., 1952.

Parker, G. H.: Criteria of Life, *Amer. Scientist,* vol. 41, 1953.

Penrose, L. S.: Self-reproducing Machines, *Sci. American,* July, 1959.

Walter, W. G.: An Imitation of Life, *Sci. American,* May, 1950.

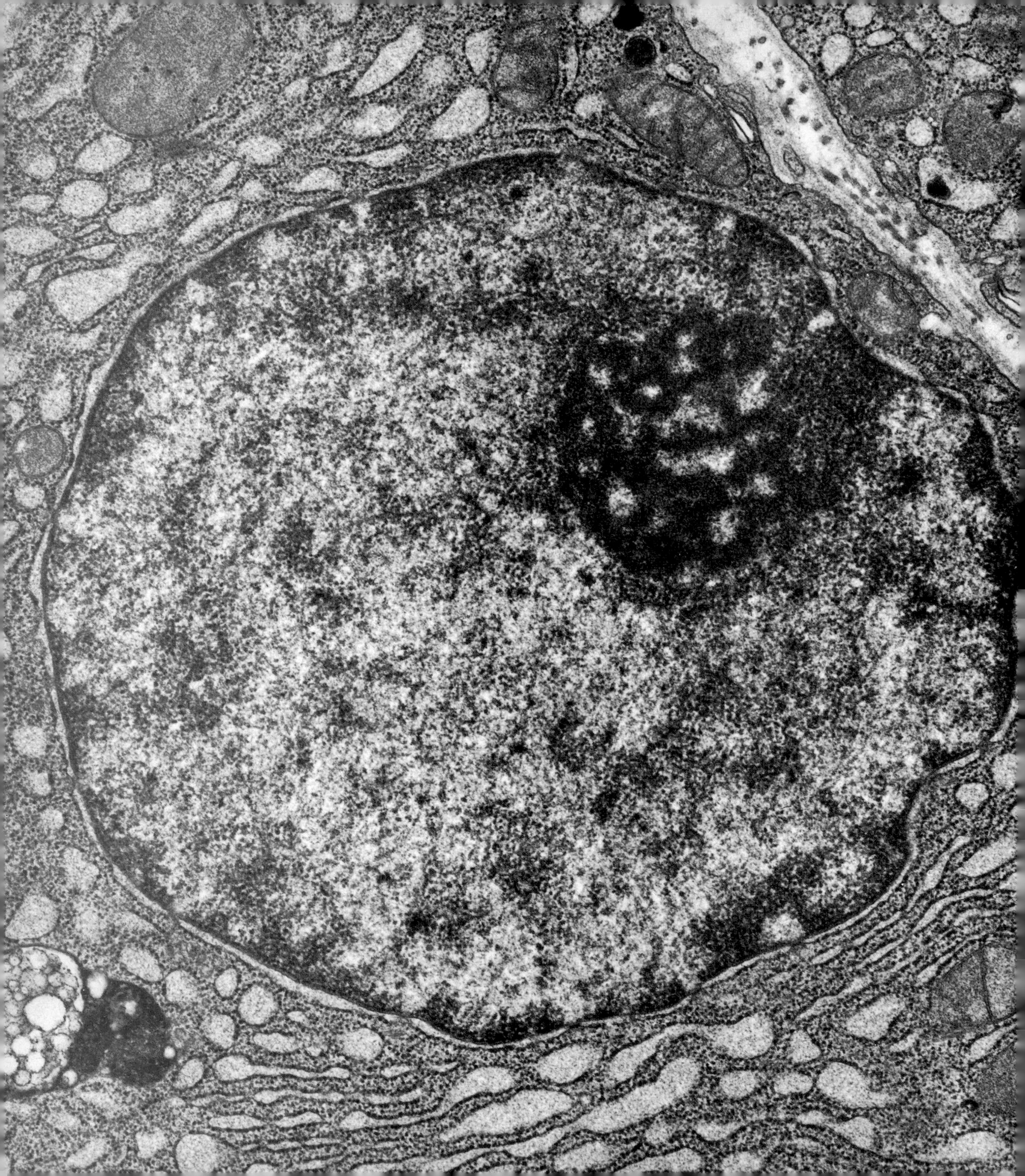

the cells of animals

PART 2

A searching discussion of animal cells must necessarily deal with cellular structure, or *organization*, and with cellular function, or *operation*. Concerning cell structure, we must examine the specific chemical compounds which compose a cell; the physical properties which these compounds confer on a cell; and the ways in which the compounds are joined together into higher levels of intracellular organization. Concerning cell function, we must show how, by virtue of its specific organization, a cell is able to metabolize and self-perpetuate, and how it actually performs each of the processes included within these activities.

In short, to the extent that we now understand it, the discussion must analyze how a cell can be alive. The three chapters of this part are devoted to such an analysis.

CHAPTER 

cellular organization

Different kinds of animal cells differ vastly in almost all their characteristics; it is a commonplace though nonetheless important zoological generalization that no one kind of cell is ever exactly like any other. Moreover, no one cell is ever exactly the same from moment to moment, for the substance of a living cell is not a static, passive material. New materials enter a cell continuously; wastes and manufactured products leave continuously; and substances in the cell interior are continuously transformed chemically and redistributed physically. As a result, a living cell is in persistent internal turmoil. To the human observer a sponge or a bone, for example, may appear to be a rather placid, inactive structure. But if the internal cellular components of a living sponge or bone could be seen, they would all be noted to be in constant, violent motion, colliding with one another and interacting and changing. Consequently, the parts of every animal change continuously, and so indeed does every animal as a whole.

However, despite such differences between cells and changes within cells, all animal cells nevertheless share certain very basic features; and it is this circumstance, actually, which makes the cell the universal minimum unit of all animal life. Representing the basic heritage passed on by the very first cells on earth, the common features of animal cells are partly *chemical*, partly *physical*, and partly *biological*. We shall examine each of these three aspects of cellular organization in this chapter.

CHEMICAL ORGANIZATION

Regardless of where, when, or how we investigate the structure of an animal cell, we ultimately find it to consist entirely of chemical compounds. And regardless of what particular function of a cell we examine, that function is ultimately always based on the properties of the cellular compounds.

Four of the most widely distributed chemical elements on earth make up approximately 95 per cent of the weight of cellular animal matter: oxygen, about 62 per cent; carbon, about 20 per cent; hydrogen, about 10 per cent; and nitrogen, about 3 per cent. Some 30 other elements contribute the remaining 5 per cent of the weight. The elements listed in Table 3 occur in almost all types of animal cells. Trace amounts of others are found only in particular types, and still other elements may become incorporated into animal matter accidentally, along with nutrient materials. All these elements are present in the ocean; animals have originated in water, as we shall see, and animal cells largely reflect the composition and content of the sea.

TABLE 3. The relative abundance of chemical elements in animal matter

Element	Symbol	Weight, per cent
oxygen	O	62
carbon	C	20
hydrogen	H	10
nitrogen	N	3
calcium	Ca	2.50
phosphorus	P	1.14
chlorine	Cl	0.16
sulfur	S	0.14
potassium	K	0.11
sodium	Na	0.10
magnesium	Mg	0.07
iodine	I	0.014
iron	Fe	0.010
		99.244
trace elements		0.756
		100.00

Most of the elements occur in the form of compounds. Cells consist of two great classes of compounds: mineral, or *inorganic*, compounds; and variously complex, carbon-containing, *organic* compounds.

The Inorganic Components

Directly or indirectly, all inorganic compounds in cells are of mineral origin; i.e., they are derived in finished form from the external physical environment. *Water,* the most abundant cellular mineral, is present in amounts ranging from 5 to 90 or more per cent. For example, the cellular water content of tooth enamel is about 5 per cent; of bone, about 25 per cent if without marrow and about 40 per cent if with marrow; of muscle, 75 per cent; of brain or milk, 80 to 90 per cent; and of jellyfish, 90 to 95 or more per cent. As a general average, cellular matter is about 65 to 75 per cent water.

Mineral solids constitute the other inorganic components of cells. Such solids are present in amounts ranging from about 1 to 5 per cent, on an average. A considerable fraction of the minerals may exist in the form of hard bulk deposits, either as crystals within cells or as secreted precipitates on the outside of cells. Such deposits are often silicon- or calcium-containing substances. For example, certain protozoa are protected externally with shells of glasslike silica; the hard part of bone is largely a deposit of calcium phosphate, secreted in layers around individual bone-forming cells; clam shells consist of calcium carbonate, secreted to the exterior by sheets of cells.

All other cellular minerals are in solution, either free or combined with organic compounds. These inorganic constituents exist largely in the form of *ions*. The most abundant positively charged inorganic ions are H^+, hydrogen ions; Ca^{++}, calcium ions; Na^+, sodium ions; K^+, potassium ions; and Mg^{++}, magnesium ions. Abundant negatively charged mineral constituents include OH^-, hydroxyl ions; $CO_3^=$, carbonate ions; HCO_3^-, bicarbonate ions; $PO_4^{\equiv}$, phosphate ions; Cl^-, chloride ions; and $SO_4^=$, sulfate ions. Note that the mineral constituents of cells are also major constituents of the ocean and of rocks and ores. This is not a coincidence, for rocks are dissolved by water, water finds its way into the ocean and into soil, and animals ultimately draw their mineral supplies from these sources.

The Organic Components

Organic substances may be defined as all those compounds of carbon in which the principal bonds are carbon-to-carbon and carbon-to-hydrogen links. Thus, carbon dioxide (CO_2), carbonate ions ($CO_3^=$), and compounds derived from them are not organic, since the carbon is bonded primarily to oxygen. But methane $\left(\begin{array}{c} \text{H} \\ | \\ \text{H—C—H} \\ | \\ \text{H} \end{array}, \text{ or } CH_4\right)$, with its carbon-hydrogen bonds, is considered to be an organic compound. Most of the organic compounds we shall be concerned with contain not only carbon-to-hydrogen but also numerous carbon-to-carbon bonds.

In this respect carbon is a rather unusual element. The atoms of most other elements may similarly bond to atoms of like kind, but the number of atoms so bondable is usually quite limited. For example, a hydrogen atom may link to one other hydrogen atom at most (H—H, H_2); an oxygen atom may link to two others at most (O—O with both bonded to O, O_3, ozone); and sulfur may form molecules of 8 atoms, S_8. But a carbon atom, having a covalence of 4 and therefore a greater bonding potential than the atoms of most other elements, may combine with a practically unlimited number of other carbon atoms. More or less long *chains* of carbon atoms may form in this way:

—C—C—C—C—. Such chains represent parts of organic molecules in which various other atoms or groups of atoms are attached to the carbon atoms. For example, hydrogen may be joined to all free bonds:

```
    H  H  H  H
    |  |  |  |
 H—C—C—C—C—H
    |  |  |  |
    H  H  H  H
```

Alternatively, one or more types of atoms other than hydrogen may be bonded to such chains.

Chains are by no means the only possible kinds of carbon-to-carbon combinations. If we imagine that one end of a carbon chain becomes connected to the other end, then a carbon *ring* will be the result. Benzene is one of such ring-containing compounds:

```
        H
        |
        C
      //  \
 H—C       C—H
     |      ||
 H—C       C—H
      \\  /
        C
        |
        H
```

Many additional types of configurations exist. For example, carbon chains can be branched, rings and chains can become joined to one another, and any of these "patterns in carbon" can extend into three as well as two dimensions. Such carbon structures form molecular "skeletons," as it were, and the other atoms bonded to the carbons may be thought of as the "flesh" on the skeletons.

No other element even approaches the self-bonding versatility displayed by carbon. Clearly, carbon-to-carbon combinations introduce the possibility of tremendous complexity as well as variety into molecular structure. Actually, carbon-containing substances display more complexity and more variety than all other chemicals put together. And it is this self-bonding capacity and the resulting diversity of carbon molecules which is at the very root of life; as we now know, only the largest, most complex kinds of molecules, such as are actually encountered among organic compounds, can have the properties necessary to form a "living" entity.

Cells contain hundreds of different categories of organic constituents. Of these, four broad categories in particular are found in all types of cells, and they form the organic basis of living animal matter. These four are

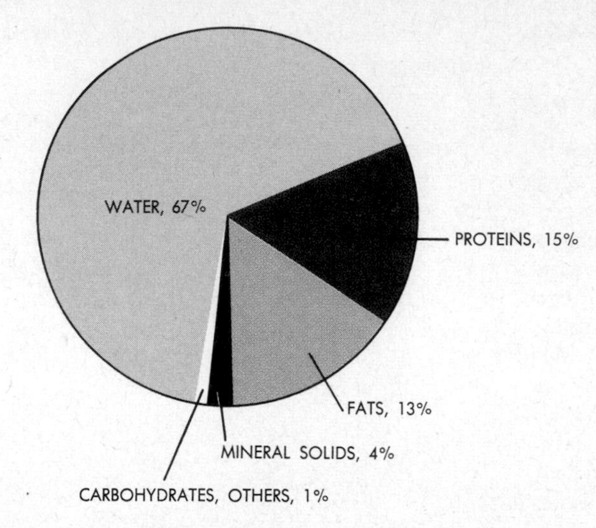

4.1. The average overall composition of animal cells.

1. carbohydrates and derivatives
2. fats and derivatives
3. proteins and derivatives
4. nucleotides and derivatives

Like mineral compounds, some of these organic substances may contribute to the formation of hard parts. For example, the various kinds of horny materials present in many exoskeletons and also in structures such as claws and hoofs are predominantly organic. More generally, however, organic materials are dissolved or suspended in cellular water, some in ionized, some in nonionized form. Their relative abundance varies considerably for different types of cells and for different types of animals. An animal such as man contains about 15 per cent protein, about 15 per cent fat, and other organic components to the extent of about 1 per cent (Fig. 4.1). Evidently, the inorganic matter (mainly water) far outweighs the organic, and this is the case in animals generally.

Carbohydrates

Organic compounds in this group are so called because they consist of carbon, hydrogen, and oxygen, the last two in a 2:1 ratio, as in water. The general atomic composition usually corresponds to the formula $C_x(H_2O)_y$, where x and y are whole numbers.

If x and y are low numbers, from 3 to about 7, then the formula describes the composition of the most common carbohydrates, the *monosaccharides,* or simple sugars. In these, the carbon atoms may be envisaged to form a chain, to which H and O atoms are variously attached. On the basis of the numbers of carbons present, different classes of monosaccharides may be distinguished: C_3 sugars are *trioses;* C_4 sugars, *tetroses,* C_5 sugars, *pentoses,* C_6 sugars, *hexoses;* and C_7 sugars, *heptoses*. The ending *-ose* always identifies a sugar. On the basis of how the H and O atoms are attached to the carbon chains, two basic series of sugars can be distinguished (Fig. 4.2). In one, called the *aldose* series, one end of the sugar molecule always consists of an aldehyde group, H—C=O; the other end has the form H_2C—OH (also written as CH_2OH); and the intermediate carbons all carry H and O in the pattern H—C—OH. In the other series, called the *ketose* series, both ends of the sugar have the form H_2C—OH; the second carbon near one end is a *ketone,* C=O; and all other carbons again

4.2. Monosaccharide series. Aldose sugars are identified by an aldehyde (—CHO) group in the 1 position. There are two aldotrioses, called D-glyceraldehyde and L-glyceraldehyde respectively. They differ in the alignment of the atomic groupings attached to carbon 2. Each of these sugars is related to a series of aldoses with higher carbon numbers. The configurations and names in the D-series are shown below. The L-series corresponds to the D-series and the names are likewise similar, the two series being distinguished by the prefixes D- and L-. Single examples of sugars in the L-series are shown. What are their names? Ketose sugars are identified by a ketone (—C=O) group in the 2 position. D- and L-series corresponding to those of aldose sugars are known here too. Individual examples are depicted in the right column. Cf. also Fig. 4.3.

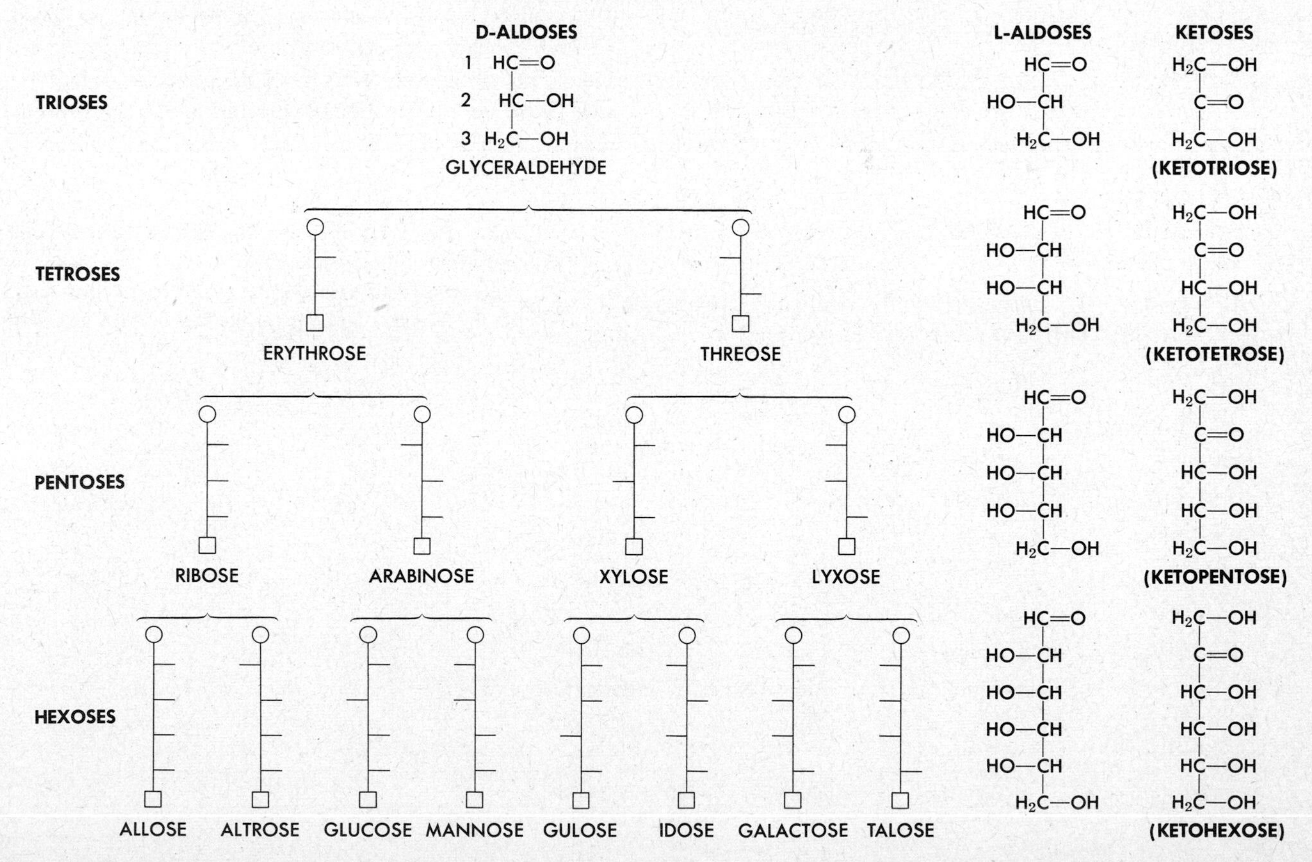

carry H and O in the pattern H—C—OH. Thus, aldoses and ketoses differ principally in that a —C=O group is terminal in one case, subterminal in the other.

On the basis of both class and series, given sugars may be named *aldotrioses* or *ketotrioses*, *aldotetroses* or *ketotetroses*, etc. Within each such category, sugars are further distinguished according to the left-right alignment of the H—C—OH groups relative to one another and the rest of the molecule. In an aldotetrose, for example, the two H—C—OH groups present may have four possible alignments:

```
H—C—O—H          H—C—O—H
H—C—O—H        H—O—C—H

H—O—C—H        H—O—C—H
    H—C—O—H    H—O—C—H
```

Thus there exist four distinct variant forms, or *isomers*, of aldotetroses, each with its own specific properties and name. Note again that, even where the same atoms are present, compounds have different properties if the atoms are bonded together in different configurations. In aldopentoses eight different configurations are possible, in aldohexoses 16 isomeric forms exist, etc. Analogous variations occur among the ketoses.

We shall have occasion to deal with only very few of these specific sugars. Reference will be made in several contexts to *ribose*, one of the aldopentoses, and to *ribulose*, one of the corresponding ketopentoses (the syllable *-ul-* in the name of a sugar always identifies it as a ketose). Frequently mentioned will be one of the aldohexoses, possibly the most common sugar of all, i.e., *glucose*. A corresponding ketohexose, which could or should be called "gluculose" but is named *fructose* instead, will also be referred to occasionally. The structure of these sugars is given in Fig. 4.3. Note that glucose (and other sugars as well) may be depicted in several different ways (Fig. 4.4).

Two or more monosaccharides of similar or identical type may become joined together end to end, forming chainlike larger molecules. If two monosaccharides become so joined, a "double sugar" or *disaccharide* results. For example, a combination of two glucose units forms the disaccharide *maltose*, malt sugar; a combination of glucose and fructose forms *sucrose*, the cane or beet sugar used familiarly as a sweetening

```
  HC=O          H2C—OH          HC=O         H2C—OH
  HC—OH          C=O            HC—OH         C=O
  H2C—OH        H2C—OH       HO—CH         HO—CH
GLYCERALDEHYDE  DIHYDROXYACETONE  HC—OH       HC—OH
                                HC—OH         HC—OH
                                H2C—OH        H2C—OH
                                GLUCOSE       FRUCTOSE

  HC=O          H2C—OH
  HC—OH          C=O
  HC—OH         HC—OH
  HC—OH         HC—OH
  H2C—OH        H2C—OH
  RIBOSE        RIBULOSE
                                 HC=O         H2C—OH
                                 HC—OH         C=O
  HC=O          H2C—OH        HO—CH        HO—CH
  HC—OH          C=O            HC—OH         HC—OH
HO—CH        HO—CH           HC—OH         HC—OH
  HC—OH         HC—OH           HC—OH         HC—OH
  H2C—OH        H2C—OH          H2C—OH        H2C—OH
  XYLOSE        XYLULOSE      SEDOHEPTOSE   SEDOHEPTULOSE
```

4.3. Five pairs of sugars, the members of each pair representing corresponding aldoses and ketoses. Note the structural similarities except for the groupings on the 1 and 2 carbons (at the top of each formula).

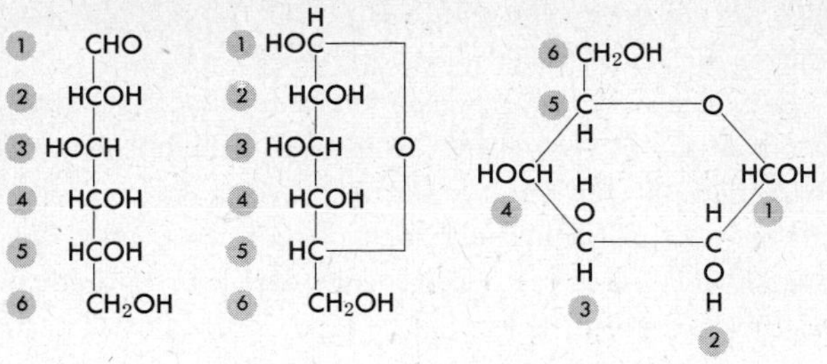

4.4. Different ways of writing the structural formula of glucose. Note that the structure may be depicted either as a chain or a ring.

agent. Both of these disaccharides have the atomic formula $C_{12}H_{22}O_{11}$, and their formation is described by the same equation:

$$2\ C_6H_{12}O_6 \longrightarrow H_2O + C_{12}H_{22}O_{11}$$

If many more than two monosaccharide units are joined together, "multiple sugars" or *polysaccharides* are the result. The general chemical process in which large molecules are built up from smaller units of like type is known as *polymerization.* Polysaccharides are said to be *polymers* of simple sugars. A polymer consisting of some hundreds or thousands of glucose units forms *glycogen,* an animal polysaccharide of considerable importance. Another polysaccharide is *cellulose,* a polymer of up to 2,000 glucose units, which is as rare among animals as it is common among plants. In many of these complex carbohydrates, the chains of monosaccharides are branched at given points. In a glycogen chain, for example, the majority of glucose units are joined via *1,4-linkages:* carbon 4 of one glucose unit links to carbon 1 of the adjacent unit, resulting in a linear array. However, in places along such an array, *1,6-linkages* occur as well; not only carbon 4 but also carbon 6 of a given unit links to carbon 1 of adjacent units. Variously forked chains are formed in this manner (Fig. 4.5).

As a group, carbohydrates function in animal cells in two general capacities: they are structural building blocks of the cellular substance, and they are energy-rich molecules suitable as fuels in respiration. Carbohydrates obtained from other organisms consequently are important animal foods. Glycogen specifically represents the chief form in which carbohydrates are stored in animals, and glucose is the chief form in which carbohydrates are transported, both from cell to adjacent cell and over greater distances via blood.

Lipids

Fats and their derivatives are known collectively as *lipids.* The principal lipids are the *fatty acids.* Like the sugars, these are composed of C, H, and O, the carbon atoms being

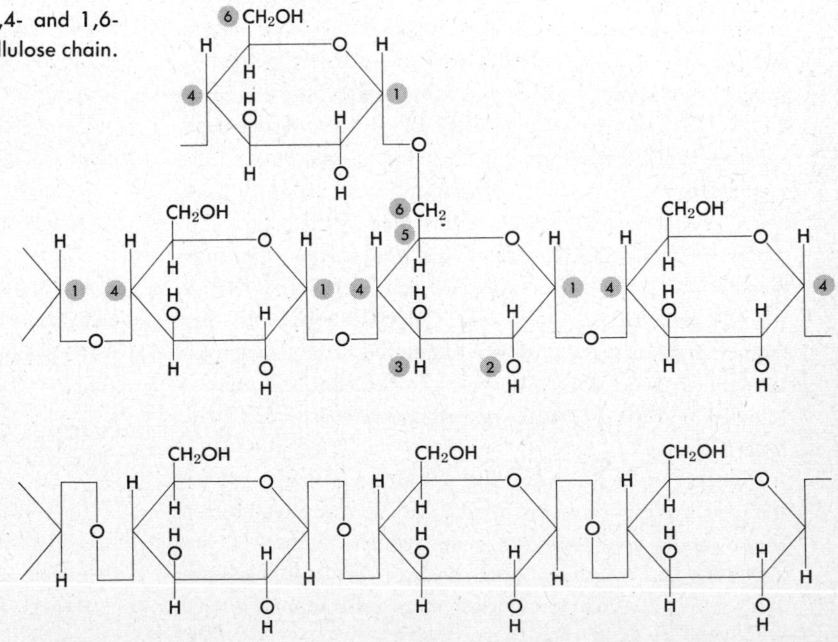

4.5. Top, portion of glycogen chain showing 1,4- and 1,6-linkages of the glucose units. **Bottom,** portion of cellulose chain.

STRUCTURE	COMPOSITION	NAME
H—COOH	CH_2O_2	FORMIC
α CH_3—COOH	$C_2H_4O_2$	ACETIC
β α CH_3—CH_2—COOH	$C_3H_6O_2$	PROPIONIC
γ β α CH_3—CH_2—CH_2—COOH	$C_4H_8O_2$	BUTYRIC
$CH_3(CH_2)_4COOH$	$C_6H_{12}O_2$	CAPROIC
$CH_3(CH_2)_6COOH$	$C_8H_{14}O_2$	CAPRYLIC
$CH_3(CH_2)_{14}COOH$	$C_{16}H_{32}O_2$	PALMITIC
$CH_3(CH_2)_{16}COOH$	$C_{18}H_{36}O_2$	STEARIC

4.6. Partial series of fatty acids with increasing molecular complexity. Note that carbon positions are identified by Greek letters, starting at the carbon next to the carboxyl group.

arranged as chains of various lengths (Fig. 4.6). All such chains carry a terminal *carboxyl* group, $-\overset{O}{\overset{\|}{C}}-OH$, which confers acid properties on a fatty acid; the H of the carboxyl group may dissociate: $-COOH \longrightarrow -COO^- + H^+$.

The simplest fatty acid is *formic acid*, HCOOH. This compound occurs occasionally in cellular excretions such as sweat and urine and also plays a protective role in some ants, where it may be squirted out as an irritant spray against potential enemies. A series of increasingly complex fatty acids is formed by successive addition of $-CH_2-$ groups to HCOOH. For example, addition of one such group produces *acetic acid*, CH_3COOH, the active ingredient of vinegar. Acetic acid serves a vital function in all cells, being an important intermediate in the respiratory decomposition of foods. Beyond acetic acid, fatty acids may be represented by the general formula $CH_3(CH_2)_nCOOH$, where n is any number other than zero. If $n = 1$, the acid is *propionic acid*, CH_3CH_2COOH; if $n = 2$, the result is *butyric acid*, $CH_3CH_2CH_2COOH$. The carbon atoms of fatty acids are numbered in sequence from the carboxyl end; the carbon next to the carboxyl group is referred to as the α-carbon; the carbon adjacent to that is the β-carbon, etc.

In most fatty acids of animals, n is an even number. As we shall see, this is because the cellular fatty acids are synthesized from acetic acid building units which themselves are 2-carbon, i.e., even-numbered, chains. Very common fatty acids in this category, present as components of most animal fats, are *palmitic acid*, $CH_3(CH_2)_{14}COOH$ (or $C_{16}H_{32}O_2$), and *stearic acid*, $CH_3(CH_2)_{16}COOH$ (or $C_{18}H_{36}O_2$).

Fatty acids like these are said to be *saturated;* all available bonds of the carbon chains are filled with hydrogen atoms. This contrasts with *unsaturated* fatty acids, characterized by the presence of one or more double bonds in a carbon chain. Such double bonds can become single bonds by the addition of more hydrogen:

$$-\underset{H}{\overset{}{C}}=\underset{H}{C}- \xrightarrow{+H} -\underset{H}{\overset{H}{C}}-\underset{H}{\overset{H}{C}}-$$

A widely occurring example of an unsaturated fatty acid is *oleic acid*, $CH_3(CH_2)_7CH{=}CH(CH_2)_7COOH$ (or $C_{18}H_{34}O_2$). If the double bond of this acid is converted to a single bond by addition of 2 H, then the result is the saturated compound $CH_3(CH_2)_{16}COOH$, i.e., stearic acid, as above. If more than one double bond is present, the acid is said to be *polyunsaturated*. In many of such cases, the double bonds alternate regularly with single bonds: —CH=CH—CH=CH—. Such acids are said to have *conjugated* double bonds.

In a fat molecule, three molecules of fatty acids are joined to one molecule of glycerin. The latter, a chainlike C_3 compound, serves as a sort of carrier for the fatty acids (Fig. 4.7). The physical nature of a fat is determined by the chain lengths and the degrees of saturation of the fatty acids present. Fats containing fatty acids that are short-chained, unsaturated, or both tend to be volatile or oily liquids. For example, oleic acid is oily, and fats containing oleic acid tend to be similarly liquid. By contrast, fats containing long-chained and saturated fatty acids tend to be hard tallow. This is the case, for example, in tristearin, a common animal fat containing three stearic acids per fat molecule.

Related to fats are various compounds in which, instead of three fatty acid units, only two are attached to glycerin. The

4.7. The formation of a fat. Three molecules of fatty acid combine with one molecule of glycerin, resulting in three water molecules and one fat molecule.

$$\begin{array}{lcl} CH_3(CH_2)_n-\overset{O}{\overset{\|}{C}}-O-H & HO-CH_2 & CH_3(CH_2)_n-\overset{O}{\overset{\|}{C}}-O-CH_2 \\ CH_3(CH_2)_n-\overset{O}{\overset{\|}{C}}-O-H \;+ & HO-CH \xrightarrow{-3\,H_2O} & CH_3(CH_2)_n-\overset{O}{\overset{\|}{C}}-O-CH \\ CH_3(CH_2)_n-\overset{O}{\overset{\|}{C}}-O-H & HO-CH_2 & CH_3(CH_2)_n-\overset{O}{\overset{\|}{C}}-O-CH_2 \end{array}$$

3 FATTY ACIDS GLYCERIN FAT

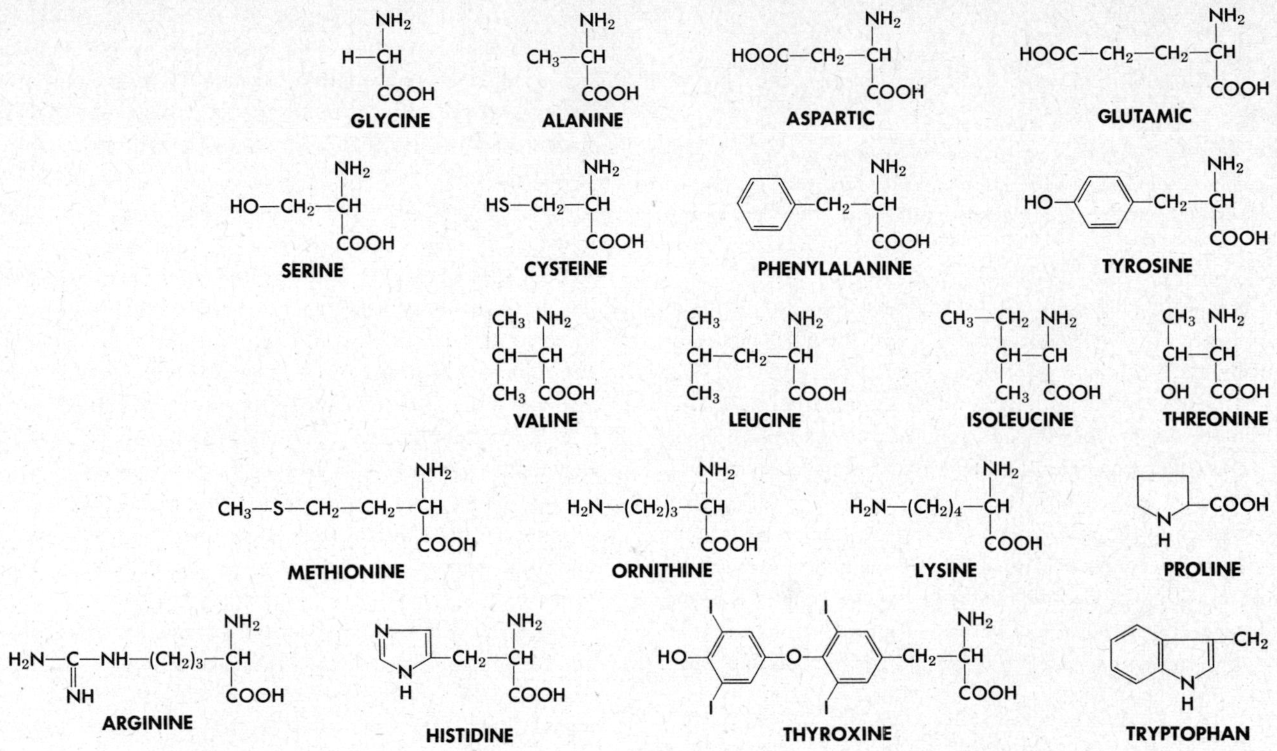

GLYCINE ALANINE ASPARTIC GLUTAMIC

SERINE CYSTEINE PHENYLALANINE TYROSINE

VALINE LEUCINE ISOLEUCINE THREONINE

METHIONINE ORNITHINE LYSINE PROLINE

ARGININE HISTIDINE THYROXINE TRYPTOPHAN

4.8. The structure of some amino acids.

third carbon of glycerin then holds some other atomic grouping. Substances of this type include lecithin, cephalin, and sphingomyelin, all present in small amounts in most accumulations of animal fats, and the last two present particularly in nerve tissue. Also related are compounds in which fatty acids are joined, not to glycerin, but to some other carrier molecule. Most waxes are of this type.

Lipids generally and fats specifically represent the chief food-storage compounds of animals. Like carbohydrates, lipids also play significant roles as structural components of cells; e.g., they are present in cellular membranes, where they probably contribute to controlling the traffic of materials into and out of cells. Moreover, lipids are even richer sources of respiratory energy than carbohydrates.

Proteins

These compounds are polymers of molecular units known as *amino acids*. The general structure of an amino acid may be represented as

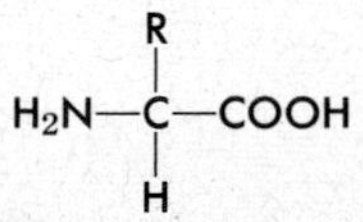

$$H_2N-\underset{\displaystyle H}{\overset{\displaystyle R}{\overset{|}{\underset{|}{C}}}}-COOH$$

where $-NH_2$ is an *amino* group, $-COOH$ a carboxyl group, and $-R$ an atomic grouping which may vary in structure considerably. For example, the simplest amino acid is *glycine*, where $R = H$. If $R = OH$, the amino acid is *serine*, and if $R = CH_3$, the acid is called *alanine*. Many other amino acids are characterized by comparatively more complex R— groups, some being carbon chains both straight and branched, others being rings (Fig. 4.8). Altogether, some 70 different amino acids are known, but only about 20 to 24 are actually encountered in animals.

Hundreds and even thousands of amino acid units may be joined together in a single protein molecule. Whenever molecules attain exceedingly large sizes, they are often referred to as *macromolecules*. Some of the carbohydrates (e.g., glycogen, cellulose) approach the dimensions of macromolecules. Most proteins are clearly macromolecules; among them, indeed, are

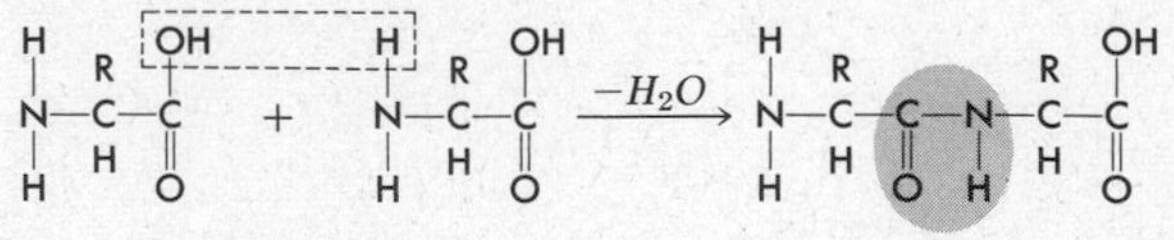

4.9. **The formation of a peptide bond.** Two amino acids combine with loss of water, resulting in a dipeptide. The peptide link in the dipeptide is indicated within the gray area above.

some of the largest chemical structures known. As such, proteins are associated most intimately with the phenomenon we call "life."

Adjacent amino acids in a protein are united in such a way that the amino group of one acid links to the carboxyl group of its neighbor; the bond is formed by the removal of one molecule of water (Fig. 4.9). The resulting grouping —NH—CO— is known as the *peptide bond,* and if two amino acids are so joined the result is a *dipeptide.* If many amino acids are polymerized via peptide bonds, the whole complex is a *polypeptide.* Such chains are the basis of protein structure.

Chemically, polypeptides may vary in practically unlimited fashion:

1. they may contain any or all of the two dozen or so naturally occurring *types* of amino acids.
2. they may contain almost any *number* of each of these types of amino acids.
3. the specific *sequence* in which given numbers and types of amino acids are joined into a chain can vary almost without restriction.

In other words, amino acid units may be envisaged to represent an "alphabet" of some 20+ "letters," and an astronomically large number of different polypeptide "sentences" may be constructed from this alphabet. Correspondingly, the possible number of chemically different proteins is likewise astronomical. Indeed, no two animals or organisms of any kind actually possess the same types of proteins.

A polypeptide chain with its particular sequence of amino acid units and peptide bonds forms what is known as the *primary structure* of a protein. A *secondary structure,* superimposed on the primary structure, is also recognized. More specifically, a polypeptide chain has the physical form of a longitudinally twisted ribbon. That is, if a line were drawn through all the R— portions of the consecutive amino acids present, that line would mark out a spiral. Such spirals are in some cases "right-handed" (α*-helix*), in others "left-handed" (β*-helix*). In either configuration, the spiral has quite uniform geometric characteristics of all kinds of polypeptide chains. For example, there are on the average 3.7 amino acid units per turn of the spiral. Accordingly, a polypeptide chain of some 18 amino acid units forms a helix with five complete turns (Fig. 4.10).

The configuration of the helix is maintained by so-called *hydrogen bonds.* A hydrogen bond is formed, for example, when a hydrogen atom is shared between an oxygen atom of one amino acid and a nitrogen atom of another amino acid. More specifically, the hydrogen atom in the —NH group of one amino acid unit may become bonded to the oxygen in the —C=O group of another amino acid unit. The H then shared by O and N links the two amino acid units together (Fig. 4.11). Such bonds generally arise between amino acid units spaced one spiral-turn apart, i.e., between acids which, on the average, are 3.7 units distant from each other. The helix and its hydrogen bonds represent the secondary structure of a protein.

Long coils of this sort may remain extended and threadlike, and the protein molecule then is said to be *fibrous.* In many cases, however, such coils may be looped and twisted and folded back on themselves, in an infinite variety of ways. Protein molecules then are *globular,* i.e., balled together somewhat like entangled twine. Such loops and bends give a protein a *tertiary structure.* Where present, a tertiary configuration is held together chiefly by three types of bonds. One is again the hydrogen bond, which in this case links together more or less distant portions of a polypeptide chain. Another is an *ionic bond,* formed when the carboxyl group of one amino acid dissociates and the resulting H^+ ion becomes attached to the amino nitrogen of another amino acid (cf. Fig. 4.11). The third type of link is the *disulfide bond,* which arises between sulfur-containing amino acids. Sulfur is present most

4.10. **The α helix.** Shown here is a portion of a polypeptide chain, illustrating the primary structure of a protein molecule. If a line is drawn to connect all —R— fractions of the consecutive amino acid units, such a line marks a spiral called an α helix.

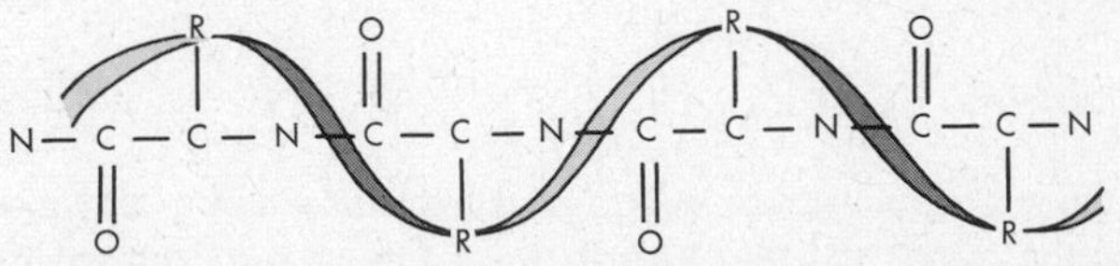

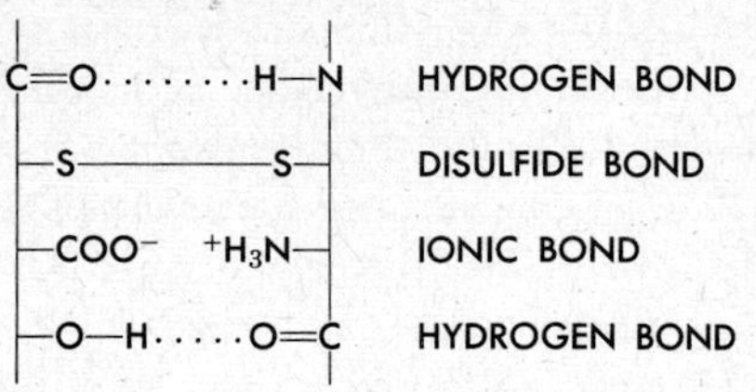

```
|                 |
C=O·······H—N      HYDROGEN BOND
|                 |
—S————————S—       DISULFIDE BOND
|                 |
—COO⁻    ⁺H3N—     IONIC BOND
|                 |
—O—H·····O=C       HYDROGEN BOND
|                 |
```

4.11. Two separate polypeptide chains or two segments of a single chain (vertical lines) may be held together by bonds such as shown above. In a hydrogen bond, an H atom is held in common by two side groups of the polypeptide chains. Disulfide bridges are formed by S-containing amino acid units, e.g., cysteine. Ionic links between charged side groups of polypeptide chains hold together by electric attraction. Hydrogen and ionic bonds may arise also in various different ways, and several other bond types between polypeptide chains, not shown here, are known as well.

often in the form of —SH groups, and the principal —SH-carrying amino acid is *cysteine.* If, for example, two distant cysteine units in a folded polypeptide come to lie close to each other, then their —SH groups may link together and form a "disulfide bridge," —S—S—. Such bridges contribute importantly toward maintenance of the tertiary structure (cf. Fig. 4.11).

Some proteins consist of not only one but of several separate polypeptide chains bonded to one another, often in the form of a bundle. Proteins of this type are said to possess a *quaternary structure.* Bundles of polypeptides are held together largely by the bond types already referred to: ionic bonds, hydrogen bonds, and disulfide bonds. For example, one of the hormones of the pancreas, *insulin,* is a protein consisting of two parallel polypeptide chains, held together by two disulfide bonds (Fig. 4.12). The composition of this protein was determined in 1954, the first time that the exact structure of any protein was established. One of the polypeptides of insulin has been shown to consist of 30 amino acid units, the other of 21. With only 51 units altogether, insulin is probably one of the smallest proteins. Most others are far more complex, and their exact structure is correspondingly less well known as yet. In the very large protein of the mammalian red blood cell, *hemoglobin,* it is now known that four polypeptide chains are linked, held together mainly by ionic bonds. Two of these chains appear to be identical to each other and to be arranged as α-spirals, the two others analogously identical to each other but arranged as β-spirals. Several other functionally important proteins have been found to be made up of four polypeptides.

It should be clear that proteins will differ not only according to their primary structure but also according to their secondary, tertiary, and quaternary structures. Thus, two proteins might have identical amino acid sequences, but if, for example, the folding pattern of the chains differs, the two molecules will have different properties and will in effect be different proteins. Many of the physical and biological properties of proteins actually depend on a specific secondary or

4.12. The structure of the insulin molecule. The molecule consists of two polypeptide chains held together by two disulfide bridges. The numbers represent amino acid units the names of which are listed below the diagram. Note that the disulfide bridges are formed by the amino acid cysteine. A bonded pair of cysteine units, as in a disulfide bridge below, represents the amino acid cysteine.

```
         NH2 S—————————S              NH2       NH2        NH2
          |  |         |               |         |          |
1—7—5—12—12—16—16—2—3—5—16—3—6—10—12—6—12—11—10—16—11
                |                                 |
                S                                 S
                |                                /
                S                               S
                |                               |
9—5—11—12—14—6—16—1—3—14—6—5—12—2—6—10—6—5—16—1—12—13—1—9—9—10—4—8—15—2
    |   |
   NH2 NH2
```

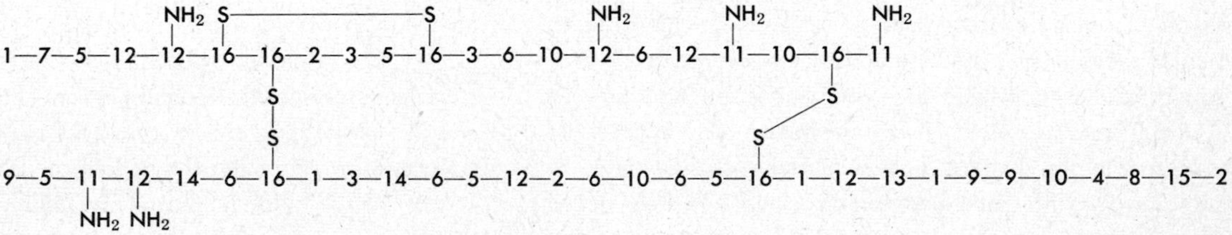

1 GLYCINE	5 VALINE	9 PHENYLALANINE	13 ARGININE
2 ALANINE	6 LEUCINE	10 TYROSINE	14 HISTIDINE
3 SERINE	7 ISOLEUCINE	11 ASPARTIC ACID	15 LYSINE
4 THREONINE	8 PROLINE	12 GLUTAMIC ACID	16 CYSTEINE

tertiary structure. For example, globular proteins are often soluble in water because, being globular in shape, the molecules may disperse freely.

The hydrogen bonds, and also the disulfide and ionic bonds which maintain a globular configuration of the molecules, are comparatively much weaker than peptide bonds. As a result, hydrogen bonds may be disrupted readily by changes in the physical and chemical environment of a protein. Excessive heat, pressure, electricity, heavy metals, increased acidity, and many other conditions may disrupt hydrogen bonds. When the bonds are so disrupted, the globular configuration of a protein will no longer hold together as before, and the molecule will lose its specific secondary or tertiary or quaternary structure. The originally highly folded protein may now stretch out and become a straight, fibrous protein; and a large collection of fibrous molecules may pile together like a log jam. Such piled-up molecules often cannot disperse freely in water, and in many cases fibrous proteins actually are water-insoluble. Changes in the physical configuration of proteins as above are called *denaturation*. If the environmental effect is mild and of brief duration, denaturation may be temporary and the protein may subsequently revert to its original *native* state. But if the environmental effect is drastic and persisting, then denaturation becomes permanent and irreversible, and the protein will be *coagulated*. For example, the protein of egg white, albumen, is globular (and water-soluble) in the raw native state, but becomes fibrous (and water-insoluble) in the cooked, coagulated state (like boiled egg white). A biological property a protein may have in the native state usually is lost after denaturation. This is a major reason why undue heat, or virtually any undue environmental change, kills cells.

Because proteins thus may vary in as many as four aspects of structure, they differ considerably in this respect from carbohydrates or fats. Even a highly complex carbohydrate, for example, is the same whether we obtain it from mushrooms or mangoes, from mice or from men. A given lipid, similarly, is the same lipid regardless of where we find it. Not so for proteins, however; these compounds can, and do, vary so much that, as noted, no two organisms contain precisely the same types. Even twin animals have slightly different proteins. The differences between proteins can be shown to be the greater the more unrelated two animals are evolutionally. We say that proteins have a high degree of *specificity:* the proteins of a given living unit have a unit-"specific" character; i.e., they are unique for that unit.

Protein specificity has major well-known consequences. For example, transfer of protein from one organism into the cells of another amounts to the introduction of foreign bodies, and disease may result. Thus, the proteins of plant pollen may produce allergy in animals such as mammals. Blood of one animal mixed with blood of another, if not of compatible type, may produce protein shock and death. Bacteria, partly because their proteins differ from those of other organisms, may produce many diseases if they infect given hosts. And portions of one animal, when grafted onto another animal, normally do not heal into place because the two sets of proteins differ.

To some extent, normally far less so than carbohydrates or fats, proteins are used in cells as foodstuffs. But by virtue of their special characteristics, proteins serve primarily in two far more important cellular roles. First, they represent the vital construction materials out of which the basic framework of cells is built. Proteins form the essential molecular "scaffolding," as it were, around which the carbohydrates, fats, minerals, and other cellular components are organized. Far more so than these other constituents, only the proteins include building "bricks" of the required size and diversity to make possible the construction of something so elaborate that it can have the properties of life. Indeed, structural differences among animals, and differences among the parts within an animal, are due primarily to the differences in the protein building materials present. As might be expected, insoluble fibrous proteins are particularly well suited to serve as cellular scaffolding. Good examples of such *structural proteins* in cells are *myosin,* the characteristic protein of muscle; *keratin,* the characteristic protein of hair and skin in mammals; and *collagen,* the fiber-forming protein produced by cells in bone, cartilage, tendons, and many other tissues of many animals.

Secondly, proteins play crucial functional roles in cells. For example, proteins are physical *carriers* of other, functionally important smaller molecules. A good illustration of this phenomenon is provided by *hemoglobin,* which is composed of two parts; a protein portion, *globin,* appears to be mainly a carrier vehicle for the second portion, *heme,* the compound actually performing the principal function of hemoglobin, namely, ferrying oxygen throughout the body (cf. Fig. 4.21). But by far the most significant functional role of proteins is that they serve as reaction-catalyzing *enzymes*. Life depends on enzymatic acceleration of reactions, and "living" therefore means protein-dependency. Enzyme proteins share the chemical and physical characteristics of proteins generally. Thus, as proteins are specific, so enzymes are specific: by virtue of its

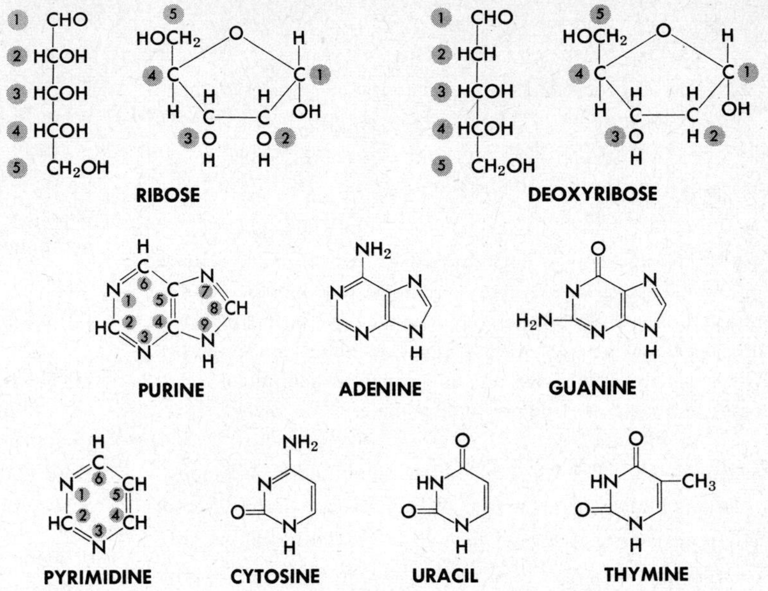

4.13. The structure of ribose, deoxyribose, and the purine and pyrimidine nitrogen bases.

particular structural attributes, a given enzyme can catalyze only one particular type of reaction. Also, enzymes are just as sensitive to environmental influences as proteins generally. Hence, if denaturation of enzyme proteins alters or destroys the native configuration of the active sites, specific enzyme activities will be lost. Note that enzymes, and most *functional proteins* of cells generally, tend to have a globular tertiary structure.

Clearly then, both for structural and functional reasons, cellular life would be unimaginable without molecular agents such as proteins. On the other hand, even if we grant the presence of proteins and all the other cellular constituents already described, a cell could not yet be alive; the molecular components discussed thus far only endow a cell with the potential of having a structure (proteins and other constituents), the potential of performing functions (enzymes), and the potential of possessing usable foods (carbohydrates and fats). The cell has not yet been equipped molecularly to make these potentials actual: how to use the foods, what actual structure to develop, and which functions to carry out. These all-important capacities emerge from the organic compounds to be dealt with next.

Nucleotides and Derivatives

A nucleotide is a molecular complex of three united subunits: a *phosphate group,* a *pentose sugar,* and a *nitrogen base.* Phosphate groups (often referred to as "organic" phosphates) are derivatives of phosphoric acid, H_3PO_4, an inorganic mineral substance. If this formula is rewritten as H—O—H_2PO_3, then —O—H_2PO_3 represents the phosphate group of present concern. We shall symbolize the —H_2PO_3 portion simply as P and write —O—P for organic phosphate.

The pentose sugar in a nucleotide is one of two kinds, viz., *ribose* or *deoxyribose.* The latter contains one O atom less than the former: at carbon 2 of deoxyribose is present a H—C—H group instead of the usual H—C—OH group, as in ribose. In other respects the two sugars are alike (Fig. 4.13).

The nitrogen base of a nucleotide is one of a series of ring compounds, the rings containing nitrogen as well as carbon. In *pyrimidines,* a single ring is present; a double ring characterizes *purines* (cf. Fig. 4.13). Pyrimidines include three variants of significance, viz. thymine, cytosine, and uracil. Among purines are two important types, viz., adenine and guanine.

If one nitrogen-base unit is joined with one pentose unit, the complex is referred to as a *nucleoside.* For example, an adenine-ribose combination forms a nucleoside called *adenosine;* others are listed in Fig. 4.14. If now a nucleoside is joined at the sugar end to a phosphate group, such a complex forms

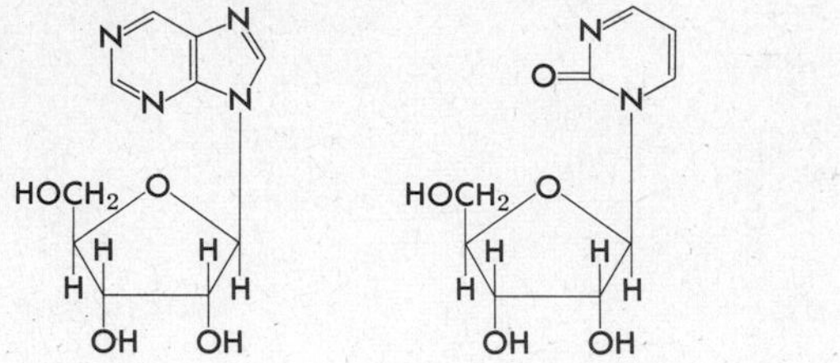

4.14. Nucleosides. Such molecules are formed when a purine base such as adenine or guanine (top left) is joined to ribose (bottom left), or when a pyrimidine base such as cytosine, uracil, or thymine (top right) is joined to ribose (bottom right). The resulting nucleosides are adenosine or guanosine in one case, and cytidine, uridine, or thymidine in the other.

a *nucleotide*. For example, the combination adenine—ribose—O—P is a nucleotide. It has at least three different but equivalent names: *adenosine monophosphate*, or AMP for short; or *adenylic acid;* or *adenine ribotide*. So far as is known, nucleotides in animals occur in two distinct series, depending on whether the pentose component is ribose or deoxyribose. Each series includes four specific kinds of nucleotides which differ in their N-base components (Fig. 4.15). Note that uracil occurs only in the ribose series, thymine only in the deoxyribose series; adenine, guanine, and cytosine occur in both series.

Nucleotides are building blocks in larger molecular complexes which serve three crucial functions in cells: some are *energy carriers;* others are *coenzymes;* and still others are components of *genetic systems*.

Energy carriers. Nucleotides have the property of being able to link up, at their phosphate ends, with one or two additional phosphate groups, in serial fashion. For example, if to adenosine monophosphate (AMP) is added one more organic phosphate, *adenosine diphosphate*, or ADP is formed; and if a third phosphate is added to ADP, *adenosine triphosphate*, or ATP results:

adenine—ribose—O—P	*AMP*
adenine—ribose—O—P—O∼P	*ADP*
adenine—ribose—O—P—O∼P—O∼P	*ATP*

The wavy symbol in the —O∼P links signifies the presence

4.15. Nucleotides. Such molecular complexes are formed when nucleosides are combined with phosphate groups (—O—PO_3H_2, in exchange for the —O—H in position 5 of the pentose sugar). Two nucleotide series are of importance, ribotides and deoxyribotides, each with four members. The structures of the ribotide UMP and the deoxyribotide TMP are depicted below; write out the structures of related members in each series from the information in this and earlier figures.

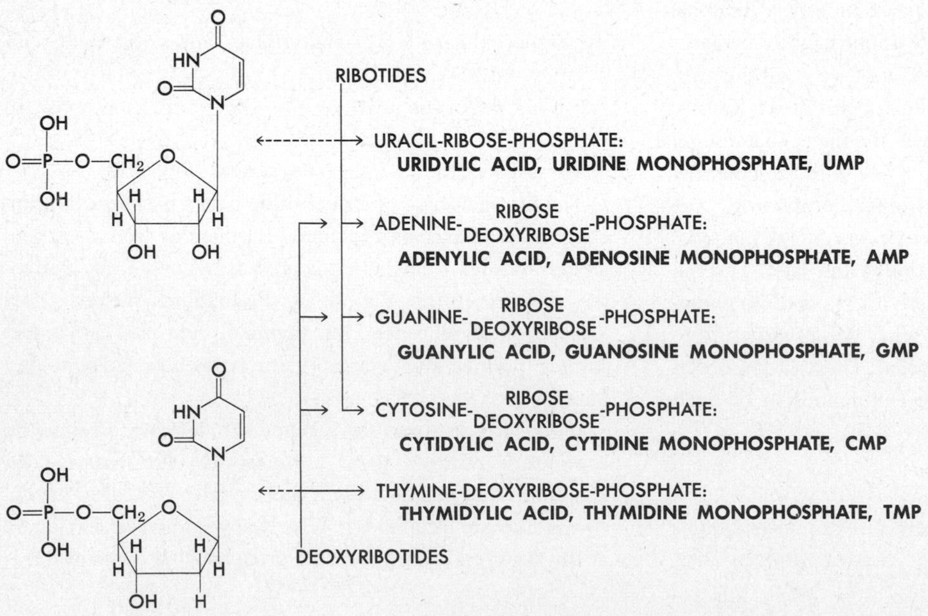

RIBOTIDES

URACIL-RIBOSE-PHOSPHATE:
URIDYLIC ACID, URIDINE MONOPHOSPHATE, UMP

ADENINE-RIBOSE/DEOXYRIBOSE-PHOSPHATE:
ADENYLIC ACID, ADENOSINE MONOPHOSPHATE, AMP

GUANINE-RIBOSE/DEOXYRIBOSE-PHOSPHATE:
GUANYLIC ACID, GUANOSINE MONOPHOSPHATE, GMP

CYTOSINE-RIBOSE/DEOXYRIBOSE-PHOSPHATE:
CYTIDYLIC ACID, CYTIDINE MONOPHOSPHATE, CMP

THYMINE-DEOXYRIBOSE-PHOSPHATE:
THYMIDYLIC ACID, THYMIDINE MONOPHOSPHATE, TMP

DEOXYRIBOTIDES

ADP + —O—P + RESPIRATORY FUEL ENERGY ⟶ ATP

ATP ⟶ USABLE ENERGY + —O—P + ADP

↓

OTHER CELLULAR ACTIVITIES

4.16. ATP produced during respiration is used subsequently to supply the energy necessary for all cellular activities.

of a so-called *high-energy bond.* The significance of such bonds will become clearer in Chap. 5. Here we need note only that bonds between certain atoms, notably the —O∼P— bonds, require particularly large amounts of energy to be formed; they also release correspondingly large amounts of energy when they are broken. Thus, to convert AMP to ADP and ADP to ATP requires not only additional phosphate groups but also large energy inputs. Such energy is derived in cells from respiratory fuels, and respiration actually functions primarily to create high-energy bonds in ATP. This latter compound may be said to be the significant energy-rich end-product of respiration. ATP subsequently supplies all parts of a cell with energy usable in other activities: ATP is split again to ADP and —O—P, and in this breaking of high-energy phosphate bonds large packets of energy become available to a cell (Fig. 4.16).

In this sense ATP is an *energy carrier.* Indeed, it happens to be the most abundant and most universal of such carriers. Others, playing a more limited energy-carrying role, are derivatives of some of the other nucleotides, e.g., as outlined in Fig. 4.17, guanosine di- and triphosphates (GDP, GTP), uridine di- and triphosphates (UDP, UTP), cytidine di- and triphosphates (CDP, CTP), and thymidine di- and triphosphates (TDP, TTP).

4.17. Apart from ATP, the nucleotide derivatives below occasionally play energy-carrying roles in cells. Note that the high-energy bonds are as in ATP.

GUANINE—RIBOSE—O—P—O∼P—O∼P
GUANOSINE TRIPHOSPHATE, GTP

URACIL—RIBOSE—O—P—O∼P—O∼P
URIDINE TRIPHOSPHATE, UTP

CYTOSINE—RIBOSE—O—P—O∼P—O∼P
CYTIDINE TRIPHOSPHATE, CTP

THYMINE—DEOXYRIBOSE—O—P—O∼P—O∼P
THYMIDINE TRIPHOSPHATE, TTP

Coenzymes. A coenzyme is a carrier molecule which functions in conjunction with a particular enzyme. It happens often in a metabolic process that a group of atoms is removed from one compound and is transferred to another. In such cases a specific enzyme catalyzes the removal, but a specific coenzyme must also be present to carry out the transfer. The coenzyme temporarily joins to, or accepts, the removed group of atoms and may subsequently "hand" it off to another acceptor compound. The majority of coenzymes happen to be chemical derivatives of nucleotides (Fig. 4.18).

More specifically, in most coenzymes the nitrogen-base portion of nucleotides is substituted by another chemical unit. This unit itself is usually a derivative of a particular vitamin. For example, one of the B vitamins in animal diets is riboflavin (B_2). This is a compound consisting of a ribose portion and, attached to it, a *flavin* portion, the latter being a complex triple-ring structure. In cells, a phosphate group becomes linked to riboflavin, resulting in a nucleotidelike complex

flavin—ribose—O—P

This compound is known as *flavin mononucleotide,* or *FMN* for short. If now FMN joins to AMP, a dinucleotide is formed:

flavin—ribose—O—P—O—P—O—ribose—adenine

This combination is known as *flavin adenine dinucleotide,* or FAD. Both FMN and FAD, and FAD particularly, function as coenzymes in cells. Their specific role is to serve as hydrogen carriers. As will be shown in the next chapter, one of the important subprocesses in respiration is the transfer of hydrogen from one compound to another. Being a light gas, hydrogen does not "stay put" and thus probably would not transfer properly by itself. The presence of hydrogen carriers such as FMN and FAD then becomes highly advantageous. In these coenzymes it is the flavin portion of the molecule which provides the specific place for temporary hydrogen attachment (cf. Fig. 4.18).

Two other hydrogen-carrying coenzymes are constructed from derivatives of *nicotinic acid,* or *niacin,* another B vitamin. A molecule of niacin possesses a so-called *pyridine* ring, and a derivative of such a ring (nicotinic amide) participates in the formation of the following complexes:

nicotinamide—ribose—O—P—O—P—O—ribose—adenine

nicotinamide—ribose—O—P—O—P—O—ribose—adenine
(with —O—P attached to the second ribose)

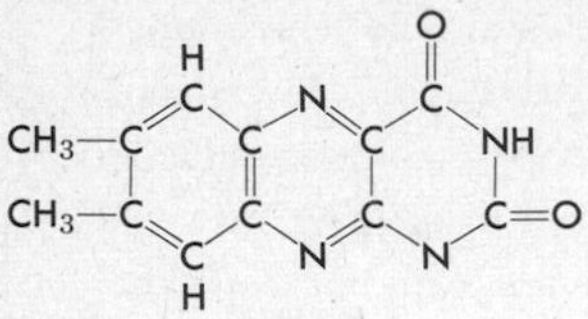

FLAVIN

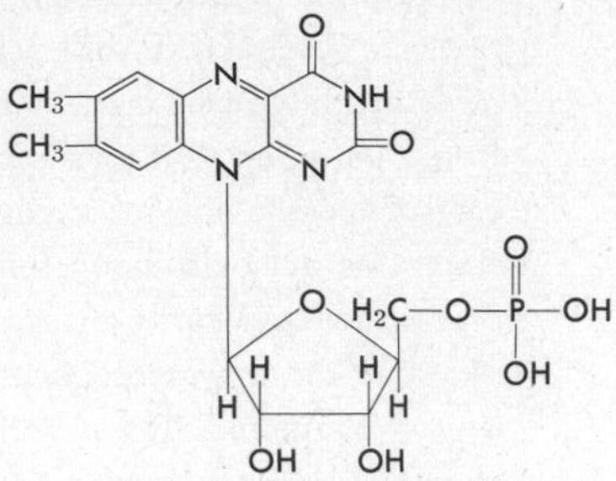

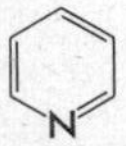

PYRIDINE

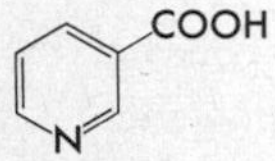

NICOTINIC ACID

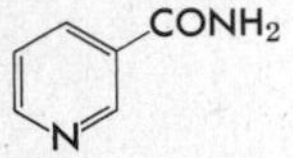

NICOTINAMIDE

FLAVIN—RIBOSE—PHOSPHATE,
RIBOFLAVIN—PHOSPHATE,
FLAVIN MONOPHOSPHATE, FMN

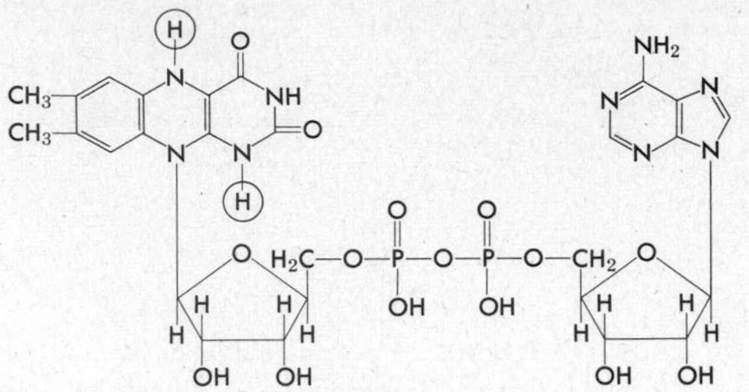

FAD
FLAVIN—RIBOSE—P—P—RIBOSE—ADENINE
FMN AMP

FLAVIN ADENINE DINUCLEOTIDE, FAD

NICOTINAMIDE—RIBOSE—P—P—RIBOSE—ADENINE: **diphosphopyridine nucleotide, DPN**

NICOTINAMIDE—RIBOSE—P—P—RIBOSE(—P)—ADENINE: **triphosphopyridine nucleotide, TPN**

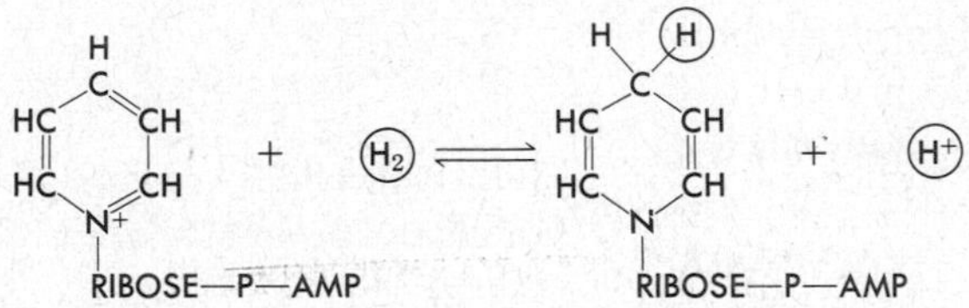

RIBOSE—P—AMP + H_2 ⇌ RIBOSE—P—AMP + H^+

DPN **DPN · H**

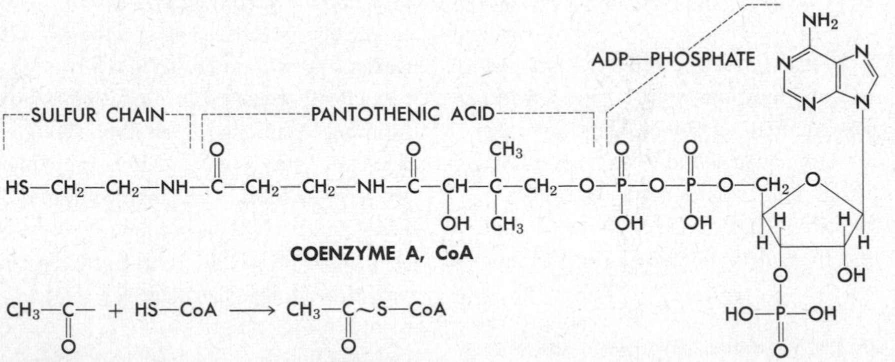

ADP—PHOSPHATE

SULFUR CHAIN PANTOTHENIC ACID

$HS{-}CH_2{-}CH_2{-}NH{-}C(=O){-}CH_2{-}CH_2{-}NH{-}C(=O){-}CH(OH){-}C(CH_3)_2{-}CH_2{-}O{-}P(=O)(OH){-}O{-}P(=O)(OH){-}O{-}CH_2$ (ribose–adenine, with $HO{-}P(=O){-}OH$ on the ribose)

COENZYME A, CoA

$CH_3{-}C(=O){-} + HS{-}CoA \longrightarrow CH_3{-}C(=O){\sim}S{-}CoA$

ACETYL GROUP **CoA** **ACETYL CoA**

4.18. Coenzymes. Among the flavin derivatives, flavin joined to ribose is riboflavin, and the latter joined to phosphate becomes FMN. This coenzyme joined to AMP as shown yields FAD. The places where FAD carries hydrogen are indicated by encircled H atoms. FMN carries H at corresponding locations. Among the pyridine derivatives, similarly, the H-carrying locations in DPN are shown by encircled H atoms; TPN functions analogously as H carrier. The structural components of coenzyme A are indicated above, as is the way in which this substance functions in carrying an acetyl group. Note the high-energy —C∼S— bond in acetyl CoA.

The first is referred to as *diphosphopyridine nucleotide*, or DPN for short; and the second, as *triphosphopyridine nucleotide*, or TPN for short. DPN is now also called, alternatively, *nicotinamide-adenine-dinucleotide*, or NAD for short; and an alternative name for TPN is *nicotinamide-adenine-dinucleotide-phosphate*, or NADP for short. For convenience, we shall here refer to these coenzymes as DPN and TPN. In both, it is the pyridine (nicotinamide) part of the molecule which serves specifically as the H carrier (cf. Fig. 4.18).

One of the coenzymes which carries not hydrogen but another specific group of atoms is *coenzyme A*, or CoA. A molecule of this compound consists of three parts, viz., a phosphate derivative of ADP, a B vitamin known as *pantothenic acid*, and a chain of nitrogen, carbon, and sulfur atoms to which hydrogen is attached (cf. Fig. 4.18). CoA serves as a carrier of a molecular group called *acetyl*, a 2-carbon combination of importance in respiration and synthesis. An acetyl group becomes attached to the sulfur end of CoA, and the product, *acetyl CoA*, is held together by an —S∼C— bond. Like the —O∼P bond discussed earlier, the —S∼C— bond too is a high-energy bond. Thus, CoA may be considered to be both an acetyl carrier and an energy carrier.

Several coenzymes which are not nucleotide derivatives will be encountered in later contexts.

Genetic systems. If any single entity could qualify as "the secret of life," that entity would unquestionably have to be the *nucleic acids*. To be sure, inasmuch as we can actually make such an identification today, it is really no longer possible to speak of any "secret." Nucleic acids are *polynucleotides*, extended chainlike polymers of up to thousands of nucleotide units. Nucleic acids thus are macromolecules (and it is because of this, as we shall see, that proteins themselves are macromolecules).

Nucleic acids are of two types, according to whether the nucleotides composing them are of the ribose series or the deoxyribose series. A polymer consisting of ribose nucleotides is called a *ribose nucleic acid*, or RNA for short; and a polymer of deoxyribotides is a *deoxyribose nucleic acid*, DNA for short. In either type, nucleotide units are linked so that the sugar component of one unit bonds to the phosphate component of the next. If we symbolize nitrogen bases as N, sugars as S, and phosphates as P, we may write, for a combination of four nucleotides,

```
—S—P—S—P—S—P—S—P—
 |    |    |    |
 N    N    N    N
```

In other words, the sugar and phosphate components form an extended molecular thread from which nitrogen bases project as side chains.

In the case of RNA, the particular types, numbers, and sequences of the four possible kinds of nucleotide units can vary in an infinite variety of ways. A given segment of a long RNA molecule might, for example, read as follows:

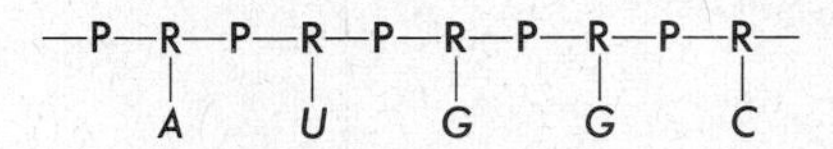

where R stands for ribose and *A, U, G, C*, for adenine, uracil, guanine, and cytosine, respectively. Evidently, RNA molecules differ according to their different N-base sequences—and that, as we shall see, is the key to their importance. The four possible N bases may be regarded as a four-letter "alphabet" out of which, just as with amino acids in proteins, any number of "sentences" may be constructed. Indeed, we shall find that the protein "sentences" precisely match and are in fact determined by the RNA "sentences."

Carrying this analogy one step further, we may say that the original "author" of the sentences is not RNA itself, but DNA. This type of nucleic acid appears to be a long *double* chain of polynucleotides:

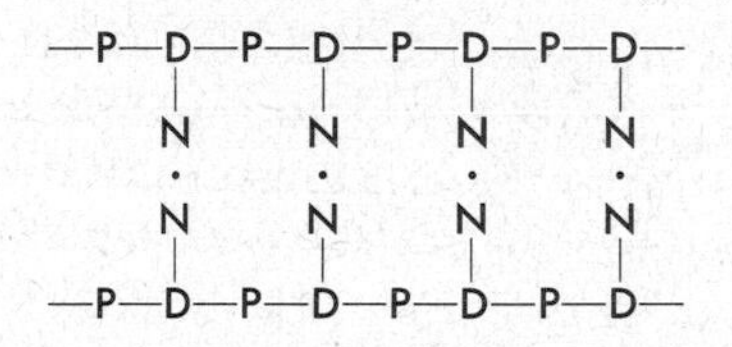

D here stands for deoxyribose, and N and P for nitrogen base and phosphate, respectively, as before. N bases are held together pair-wise by hydrogen bonds, which thus maintain the double-stranded structure of DNA. Moreover, there are only four different ways in which N bases may be paired, as outlined in Fig. 4.19. Note that *adenine is always paired with thymine, guanine always with cytosine*. The chemical properties of these purines and pyrimidines, and the space available between the parallel —P—D— chains, are such that only combinations shown in Fig. 4.19 can be formed.

But there is apparently no limit to the number of times each of these combinations can occur in a long double chain. Nor, apparently, are there restrictions as to their sequence. Thus $A \cdot T$, $T \cdot A$, $G \cdot C$, and $C \cdot G$ may again be regarded as an alphabet of four symbols, and sentences of any length may be constructed by using these symbols as often as desired and in any order. Evidently, the possible number of compositionally different DNAs is practically unlimited.

—D—P—D—P—D—P—D—P—
A G T C
T C A G
—D—P—D—P—D—P—D—P—

THYMINE CYTOSINE
ADENINE GUANINE

4.19. Top, the Watson-Crick model of DNA structure. *P*, phosphate; *D*, deoxyribose; *A, T, G, C*, purines and pyrimidines. A *P—D—A* unit represents one of the nucleotides. In this *—P—D—P—D—* double chain, four kinds of purine-pyrimidine pairs are possible, namely, *A · T, T · A, G · C*, and *C · G*. Each of the four may occur very many times, and the sequence of the pairs may vary in unlimited fashion. **Bottom, the hydrogen bonding between *T · A* and *C · G* pairs.**

A final structural characteristic of DNA is that its double chain is not straight but spiraled into a helix (Fig. 4.20). The DNA structure as outlined here is designated as the *Watson-Crick model*, after the investigators who proposed it.

Functionally, DNA exhibits three properties which make it the universal key to life. First, as will be shown in greater detail in Chap. 6, the specific sequence of nitrogen-base pairs in DNA represents *coded information*, which provides the cell with instructions on how to manufacture specific proteins. These coded instructions first are "transcribed" into matching nitrogen-base sequences within RNA, and the instructions in such RNA subsequently are translated into particular sequences of amino acid units within polypeptide chains and proteins. By so controlling protein manufacture, DNA ultimately controls the entire structural and functional makeup of every cell.

Secondly, under appropriate conditions (such as actually exist within cells), DNA has the property of being *self-replicating* or self-duplicating; i.e., DNA is a *reproducing* molecule. That a mere chemical should be able to multiply itself may perhaps be astounding, but this is nevertheless a known, unique property of DNA. Without doubt, this property is a direct result of the atomic complexity and the specific organization of the molecule. The reproduction of DNA is at the root of all reproduction—cellular, animal, species, as well as communal; in a fundamental sense even the reproduction of a whole animal is, after all, a reproduction of "chemicals." Through its reproduction, moreover, DNA is also the key to *heredity*.

Thirdly, DNA has the property of being *mutable*, i.e., under certain conditions a given sequence of nitrogen-base pairs may become altered slightly. Such alterations then are stable and persist into succeeding molecular generations of DNA. As the coded information of DNA becomes changed, however, the structural and functional traits of a cell become changed correspondingly. Through changes in its cells a whole animal and its progeny may thus become changed in the course of successive generations; this is equivalent to *evolution*.

In short, DNA is the material which forms *genes*, the ultimate controllers of all living operations, short-range or long-range. Together with RNA, DNA represents the substance of the *genetic systems* which direct metabolism and self-perpetuation and which so are the basis of life. In Chap. 6 we shall examine how the structure of the nucleic acids actually permits genes to function in such crucial ways.

Other Constituents

Carbohydrates, lipids, proteins, and nitrogen-base derivatives such as nucleic acids and nucleotide phosphates form the organic bulk of living matter. However, hundreds of other kinds of organic substances exist in cells. Although such

4.20. A DNA double chain is spiraled as shown in this diagram. The two spirals symbolize the *—P—D—P—D—* chains, and the connections between the spirals represent the purine-pyrimidine pairs.

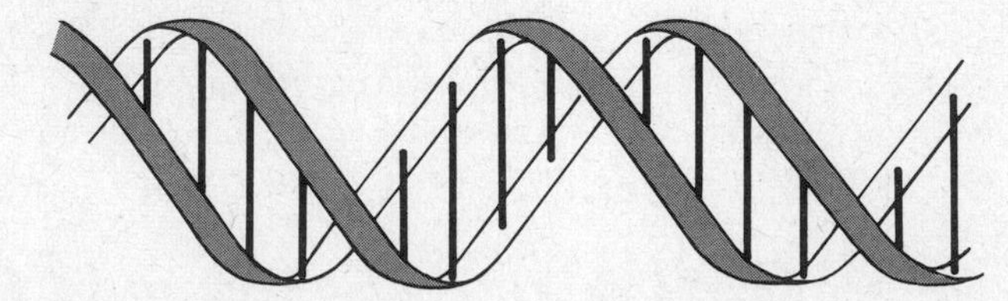

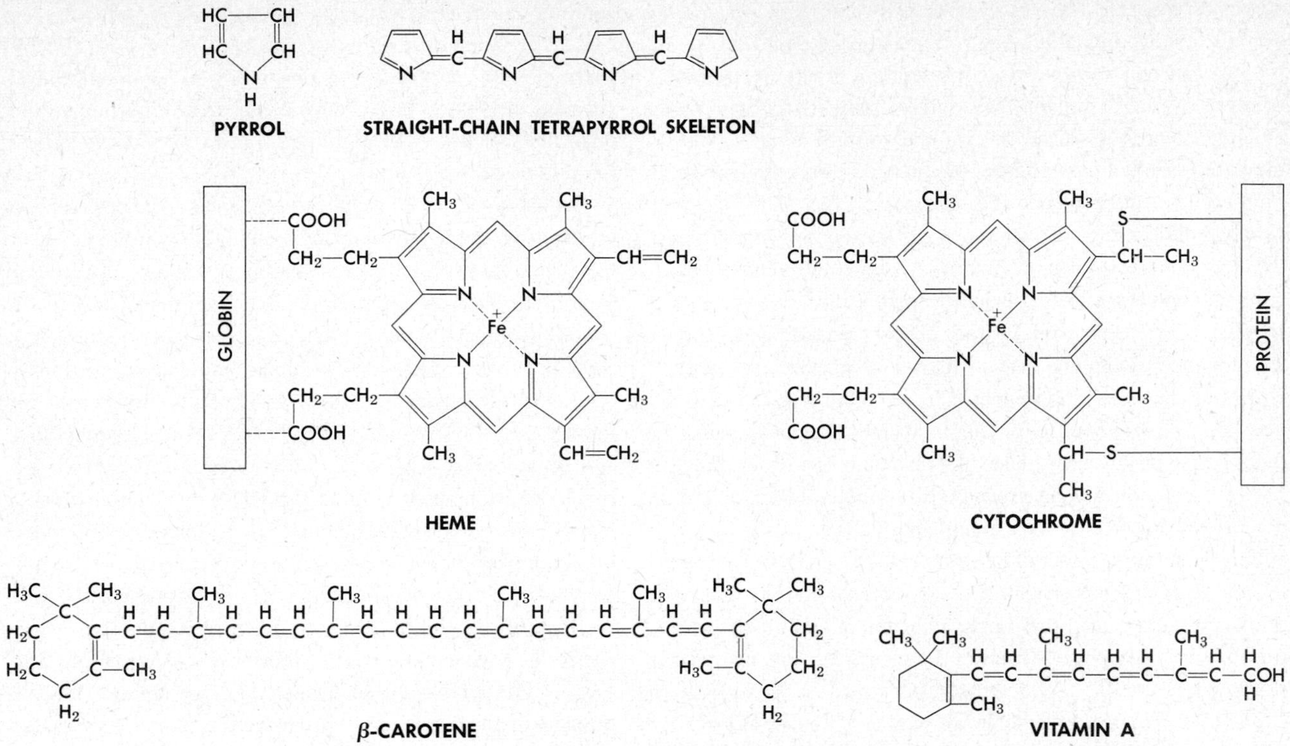

4.21. The structure of pyrrol and carotenoid derivatives. Note the basic similarity of heme and cytochrome, the chief difference being the nature and mode of attachment of the proteins. Note also that, in carotene and vitamin A, conjugated double bonds are present.

substances are often present in very small quantities only, they may nevertheless be of extreme importance in the maintenance of life. Some of these constituents are not related chemically to the four main categories above. Others are derivatives of one of the four groups, and still others are combinations of two or more of the basic four.

Few of these organic constituents occur universally in all types of animal cells. More usually, they are special components of particular cell types only, serving special functions. Of such special materials found in various animals, the following categories are of considerable importance structurally, functionally, or both (Fig. 4.21).

The *tetrapyrrols* are a group of pigmented compounds serving a variety of functions. A so-called pyrrol molecule contains a skeleton of five atoms, namely, carbon and nitrogen atoms; and the five are arranged as a ring. Four such pyrrol rings joined together form a tetrapyrrol. In some cases, the four joined pyrrol rings form a straight chain. Tetrapyrrols of this type include red, yellow, and other varieties of pigments found, for example, in the shells of robin and other bird eggs, and in mammalian feces and urine. In other tetrapyrrols, the four pyrrol rings are joined to form a larger ring in turn, and in the center of this larger ring is usually present a single atom of a metal. Major pigments of this type are the red *cytochromes*, in which the metal atom is iron. Cytochromes serve as hydrogen-carrying coenzymes, one of the few groups of such coenzymes that are not nucleotide derivatives. Structurally quite related to the cytochromes are the *hemes* which, as already noted, are joined to protein carriers and function as oxygen-transporting compounds in the bloods of many animals (cf. Fig. 4.21). It may be pointed out in passing that *chlorophyll*, the green pigment of photosynthetic organisms, is also a ring-like tetrapyrrol, the central metal atom here being magnesium.

Another group of widely occurring animal pigments are the

carotenes, responsible for the yellow and cream-yellow colors of, for example, animal fat, egg yolk, milk, butter, and cheese. In all of these, carotenes are present in greater or lesser amounts. Named after the carrot where they are particularly abundant, carotenes have the general formula $C_{40}H_{56}$, and they are essentially long chains of carbon atoms with carbon rings attached at both ends of the chains (cf. Fig. 4.21). When such a chain is split at the middle in the presence of water, the result is two molecules of vitamin A ($C_{20}H_{30}O$). Carotenes generally and vitamin A specifically play a direct role in the chemistry of vision and in the proper development and maintenance of exposed surface tissues, as in skin layers and linings of the breathing system. Closely related to the carotenes are the *xanthophyll* pigments, widely present in plants (where they produce the yellow and brown colors of autumn), but less common among animals. Nevertheless, xanthophyll derivatives produce the greens and reds on the body surfaces and in the interior tissues of lobsters, crayfish, and many other crustacea, as well as the yellows in egg yolks and in the feathers of canaries and some other birds. The general composition of the basic xanthophyll structure is given by the formula $C_{40}H_{56}O_2$.

Most conspicuous in animals are pigments known as *melanins*. They are responsible for all yellow-brown, brown, and particularly black colors, both on animal body surfaces and in interior tissues (as also in the "inks" squirted out by excited squids and octopuses). Melanin is a chemical derivative of the amino acid *tyrosine*. Specialized pigment cells produce melanin, which accumulates in granules within such cells. If only a few melanin granules are present, the cell appears to be yellowish or brownish in color; a black color is produced by dense masses of granules. Among other tyrosine derivatives are various red and purple pigments found among certain mollusks, as well as two vertebrate hormones, i.e., *thyroxine*, secreted by the thyroid gland, and *adrenaline*, secreted by the adrenal medulla and by certain nerve endings (Fig. 4.22).

Mention may be made of a derivative of glucose, *glucosamine*, in which the —OH group on carbon 2 of glucose is replaced by the amino group $—NH_2$. A high polymer of glucosamine forms the polysaccharidelike compound *chitin*, which is a hard, horny substance encountered very widely in the exoskeleton of invertebrate animals such as arthropods. Another important glucose derivative is *ascorbic acid*, more often called vitamin C (cf. Fig. 4.22).

Like amino acids and sugars, lipids too give rise to special derivatives. Notable among these are, for example, the *sterols*, complex molecules built on a quadruple-ring structure (cf. Fig. 4.22). One of the parent sterols is *cholesterol*, and chemically closely related to it is one group of vitamins, the D vitamins, as well as two groups of vertebrate hormones, the male and female sex hormones and the adrenocortical hormones.

The listing of special cellular compounds could be extended enormously, but a general principle may be glimpsed even at this point: almost the whole vast array of organic constituents in animal cells is related to or derived from only a half-dozen or so fundamental types of compounds; and among these, the principal types are sugars, fatty acids, amino acids, and nucleotides. Nature apparently builds with but a limited number of fundamental construction units, yet the possible combinations and variations among these units are practically unlimited. It should also be kept in mind that these diverse constituents of a cell are not "just there," randomly dissolved or suspended in water like the ingredients of a soup. Instead, as already emphasized earlier, the constituents must be organized if they are to form something living. Indeed, superimposed on the chemical organization within the molecules is a physical organization among the molecules. In large measure, such a physical organization results from the mixed macromolecular and micromolecular nature of the chemical units; as some of the constituents do and others do not dissolve in the water medium of a cell, the whole cell acquires very distinct physical characteristics. We shall examine these in the following section.

PHYSICAL ORGANIZATION

Any system composed of particles contained in another medium can be classified as belonging to one of three categories, depending on the size of the particles. If the particles are small enough to dissolve in the medium, then the system is a true *solution*. (In a water solution crystals can readily form, and such a system is therefore also called a *crystalloid*.) If the particles are large, e.g., the size of grains of soil, they soon settle out by gravity at the bottom of a container. Such a system is a coarse *suspension*. But if the particles are of intermediate size, they neither form a solution nor settle out. Such a system is a *colloid*.

The particles in a colloid range in diameter from $1/_{1,000,000}$ to $1/_{10,000}$ millimeter (mm). The larger figure corresponds very nearly to the limit of vision under a good microscope. In bio-

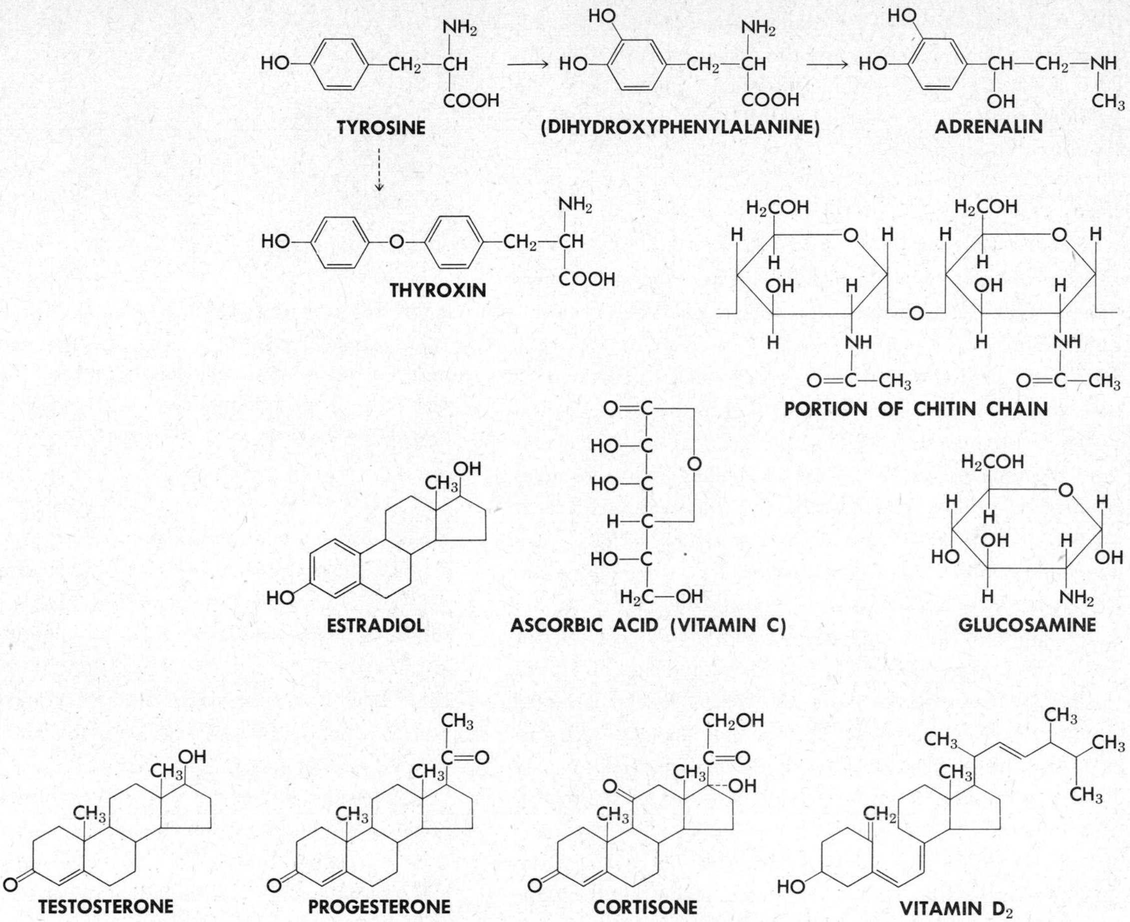

4.22. Derivatives of the amino acid tyrosine include hormones such as thyroxine and adrenaline, and pigments such as melanin (formed from dihydroxyphenyl alanine). Glucose derivatives include vitamin C and glucosamine, the latter being a structural unit in the horny skeletal material chitin. Estradiol is one of the estrogens, the female sex hormones of vertebrates. This compound illustrates the basic four-ring steroid structure. Other steroids include the male sex hormone testosterone, the female pregnancy hormone progesterone, the adrenal cortical hormone cortisone, and the D vitamins, as well as cholesterol (not shown).

logical practice, the unit 1/1,000 mm = 1 micron (μ) is frequently used. Hence the colloidal range is from 1/1,000 to 1/10 μ.

Any diphasic system is a colloid if one of the two components consists of particles of appropriate size. Eight general types of colloid systems are possible: a gas within either a solid or a liquid; a liquid within either a liquid, a solid, or a gas; and a solid within either a liquid, a solid, or a gas. Liquids within liquids are called *emulsions*. Among common colloidal systems are milk and mayonnaise (colloidal fat and protein in water), fog (colloidal water in air), cigarette smoke (colloidal ash in air), cheese (colloidal air in fat-protein), and ruby glass (colloidal gold in a solid).

The cell substance is partly a true solution, partly a colloidal system. Water is the medium in which many materials are dissolved, and it is also the *liquid phase* in which many insoluble materials of colloidal size are dispersed. This colloidal *dispersed phase* includes, for example, macromolecular solids such as the fibrous proteins and the nucleic acids, and liquids such as the oily fats. Therefore, in so far as animal matter is a colloidal system, it is of both the solid-within-liquid and the liquid-within-liquid type.

Cellular Colloids

What properties of an animal cell result from its colloidal nature? What, first, prevents the colloidal particles from settling out?

As noted earlier, the molecules of a liquid are under continuous thermal agitation; the more intense the agitation, the higher the temperature. When the liquid freezes, molecular motion is reduced sharply. Above the boiling point, molecules move so rapidly that many escape; i.e., the liquid vaporizes at great rate. If dispersed particles are present in a liquid, they are buffeted and bombarded constantly by the molecules of the liquid. Very large particles are unaffected by these tiny forces, and they fall straight to the bottom of a container. But smaller bodies of colloidal size may be pushed back and forth, up and down. Gravitational pull may thereby be counteracted partly or wholly, and the particles thus may be kept suspended. This random movement of small particles, called *Brownian motion,* is easily demonstrable under the microscope.

Brownian movement aids in keeping colloidal particles from settling out, but they cannot remain suspended by this force alone. They stay dispersed mainly because most types of colloids are ionized and thus carry *electric charges*. All colloidal particles of a given system are either electropositive or electronegative, the ions with opposite charges being noncolloidal and fully dissolved in the medium. The system as a whole is therefore electrically neutral, but the colloidal components carry like charges. And since like charges repel, the colloidal particles are kept apart. If the charge is neutralized by electricity of opposite type, e.g., by the addition of oppositely charged ions, which bond to the colloid particles and thereby reduce the degree of dissociation, then the colloid particles often do settle out (Fig. 4.23).

Cellular colloids undergo reversible *sol-gel transformations,* also called *phase reversals*. If large numbers of colloidal particles are added to the system or, alternatively, if water is gradually withdrawn, the particles are brought closer together and come into contact with one another eventually. Rod-shaped particles then pile up like a log jam; round or irregular particles interlock in intricate ways. In effect, the original dispersed phase now is a continuous spongelike network which holds water within its meshes, in discontinuous droplets. This is the *gel* state of a colloid. The quasi-solid, pliable aspect of animal substance, as in skin, or of protein colloids generally, as in Jell-O and gelatin, is due to the gel condition. We may understand, therefore, how even systems like jellyfish, which contain as much as 90 per cent or more water, can maintain definite form and shape.

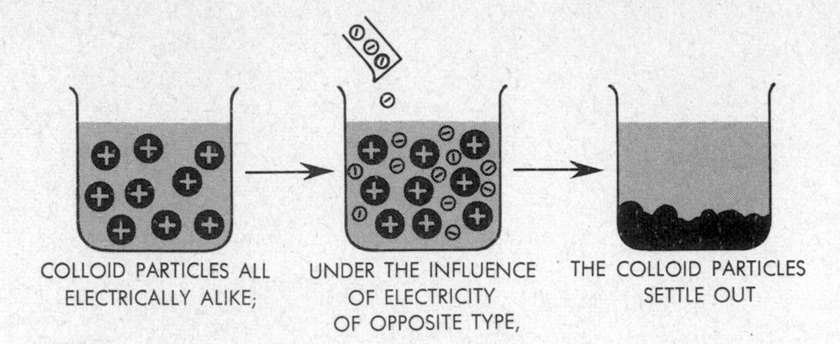

4.23. Colloid particles carry similar electric charges—in this illustration, positive ones (left). These charges make the particles repel one another and thus keep them suspended. If electricity of the opposite type is added (middle), the colloid charges are neutralized and the particles may settle out (right).

Conversely, addition of water to a colloidal system or removal of dispersed particles results in greater fluidity, the *sol* state of a colloid. In cells, sol and gel states alternate normally and repeatedly with local variations of particle concentrations.

Increased temperature may convert a gel into a sol; at higher temperature, colloidal particles in a gel become more agitated and the gelled meshwork is disrupted (e.g., liquefaction of Jell-O by heating). Many other physical and chemical influences, such as low or high pH or pressure, affect sol-gel conditions. For example, cream, a sol, when churned (i.e., when put under pressure), yields butter, a gel. Butter in turn can be creamed, i.e., returned to the sol state.

All colloids *age*. The particles in a young, freshly formed colloidal system are enveloped by layers of *bound water;* water molecules are held against the particle surfaces by electrochemical attraction. It is largely because of these forces that water within a gel does not "run out" through the gel meshes. With time, however, the binding capacity of the particles decreases, and some of the water does run out. The colloid "sets," i.e., contracts and gels progressively; examples are exudation of water from long-standing milk curd, custard, mustard. Such aging of colloids may be a factor contributing to the aging of animal systems.

Migratory movements occur in colloids, and also in true solutions, as a direct result of the thermal agitation of the particles. If ions, molecules, or colloidal particles are enevenly distributed, more collisions take place in more concentrated regions. For example, if a particle in the circle in Fig. 4.24 is displaced by thermal agitation or by Brownian bombardment toward a region of higher concentration, it will soon be stopped in its track by collision with other particles. But if it is displaced away from a high concentration, its movement will not be interrupted so soon, since neighboring particles are farther apart. On an average, therefore, a greater number of particles is displaced into more dilute regions then into more concentrated ones. In time, particles throughout the system will become distributed evenly. This equalization resulting from migration of particles is called *diffusion.*

Diffusion plays an important role in animal cells. For example, it happens often inside a cell that particles are unevenly distributed. Diffusion will then tend to equalize the distribution. Evidently, this is one way through which materials in cells can migrate about.

An important property of animal matter resulting from its colloidal makeup is that, as the following will show, it tends to form membranes.

Membranes and Permeability

The boundary between a colloidal system and a different medium (air, water, solid surfaces, or another colloid of different type) is called an *interface.* The molecules there are usually subjected to complex physical forces which act on and from both sides of the interface. The result is that the molecules at the interface pack together tightly and become oriented in parallel, in layers, or both; an interfacial membrane forms (e.g., the "skins" on puddings, custards, boiled milk). Complex molecular skins are present also on the surfaces of animal cells, where they are called *plasma membranes.* If the plasma membrane on the surface of a cell is punctured, a new membrane develops over the opening within seconds, before appreciable amounts of the interior can flow out.

4.24. Diffusion. In the initial state, particles are distributed unevenly (left). A given particle (for example, the circled one) will therefore have more freedom of movement in the direction of lower concentrations. An even particle distribution eventually results, as in the end state shown at right.

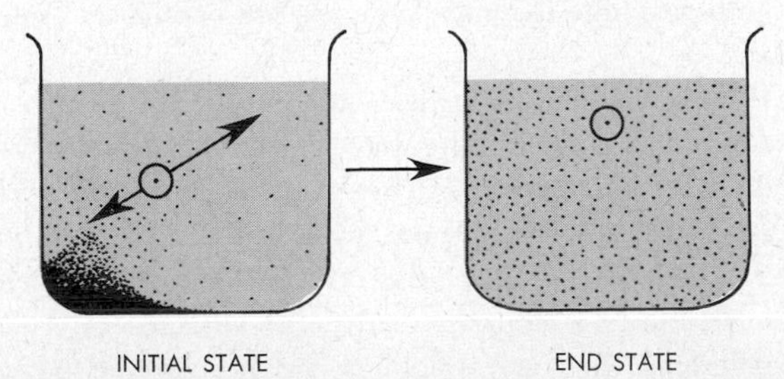

Plasma membranes are the gateways through which the molecular traffic into and out of cells must pass. How do materials get through such membranes?

Plasma membranes have different *permeability* to different substances. Most membranes are completely permeable to water; i.e., water molecules can pass through freely in either direction. As for other materials, organic or inorganic, there is no rule by which their passage potential can be determined beforehand. In general, three classes of materials can be distinguished: those that can pass through a membrane in either direction; those that can pass in one direction but not in the other; and those that cannot penetrate at all. These categories vary considerably for different kinds of membranes.

In the past, traffic through living membranes has been compared with traffic through nonliving ones like cellophane. Such nonliving membranes let water or small ions through, but not proteins, for example. Particle penetration here can be explained rather readily in terms of diffusion. Ions, for example, would strike the barrier; most of them would bounce off, but some would pass through *pores* in the membrane. If the ion concentration were greater on one side of the membrane than on the other, more ions on an average would migrate into the dilute side, thus equalizing concentrations.

However, a hypothesis postulating diffusion through pores is generally inadequate for living membranes. If cellular membranes were indeed passive, inert films with holes like cellophane, then it should not matter if such a membrane were poisoned; being nonliving, it could not be affected by a poison. But experiments actually show that the activity of cellular membranes is stopped or severely impaired by poisons, indicating that such membranes are not simply passive films. Moreover, if living membranes actually contained small holes, then the size of a particle should determine whether or not it could pass through such holes. However, particle size is often of little importance. For example, under certain conditions large protein molecules may pass through a given membrane readily, whereas very small molecules sometimes may not. Again, the molecules of the three sugars glucose, fructose, and galactose, all $C_6H_{12}O_6$, have the same size, yet they are passed through living membranes at substantially different rates.

Clearly, membranes are highly selective; i.e., they act as if they "knew" which substances to transmit and which to reject. Moreover, it is now known that active, energy-consuming work is often done by a living membrane in transmitting materials and that complex chemical reactions take place in the process. Therefore, rather than visualizing a passive membrane with small holes, we are led to consider plasma films as dynamic structures in which entering or leaving particles are actively "handed" across from one side to the other. The precise means by which cells accomplish such *active transport* across their boundaries is as yet understood for only a few types of materials (cf. Chap. 5).

Accordingly, if we encounter a situation where materials other than water pass through a living membrane, we will be quite wrong if we simply say offhandedly that this can be explained "by diffusion." Diffusion does play some role in most cases, but energy-requiring active transport by the membrane may play an equally important role.

Membranes also account for a final physical property we must discuss.

Osmosis

When a membrane separates one colloid or solution from another or from a different kind of medium, it often happens that the membrane is permeable to some of the particles present on either side of the membrane and is impermeable to others. If, initially, the transmissible particles are unequally concentrated on the two sides, diffusion will equalize the concentration, as we have seen. What happens when nontransmissible particles are unequally concentrated? For simplicity, let us assume that transmissible substances are not present at all and that we deal only with water containing nontransmissible particles. What events occur in such a system? Consult Fig. 4.25:

1. In the initial state, relatively more water molecules are in contact with the membrane *X* on the *A* side than on the *B* side, since fewer of the solid particles occupy membrane space on the *A* surface than on the *B* surface.

2. Therefore, more water molecules, on an average, are transmitted through the membrane from *A* to *B* than from *B* to *A*.

3. As a result, the water content decreases in *A* and increases in *B*. Particles in *A* become crowded into a smaller and smaller volume, and more and more of them therefore take up membrane space on the *A* surface. On the *B* side, the increasing water content permits the spreading of the particles into progressively larger volumes, thus reducing particle concentration along the *B* surface of the membrane.

4. A stage will be reached at which the number of particles along the *A* surface equals that along the *B* surface. From then on, the number of water molecules transmitted from *A* to *B* equals the number transmitted from *B* to *A*. Thereafter, no further *net* shift of water occurs.

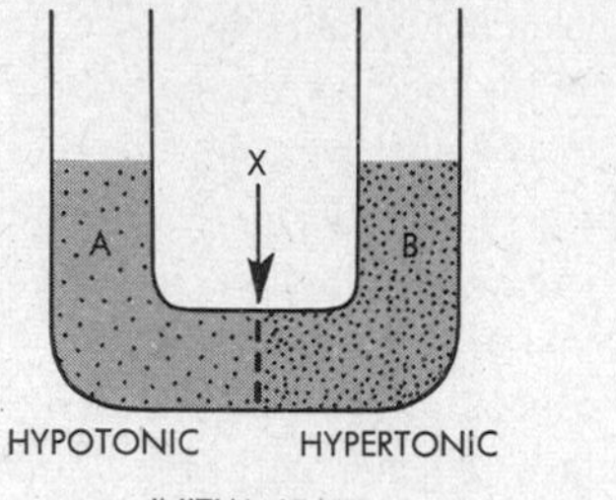

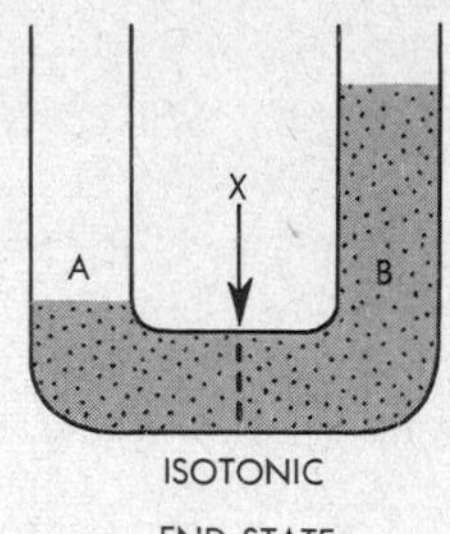

4.25. Osmosis. In the initial state, because *A* is less concentrated than *B*, water will move from *A* into *B*. This eventually leads to the isotonic end state, where concentrations in *A* and *B* are equal. From this point on, no further net migration of water occurs (i.e., just as much water moves from *A* into *B* as from *B* into *A*.) A semipermeable membrane is represented by *X*.

This movement of *water* is called *osmosis*.

Note that the extent of osmosis depends on the *concentration differential*, the relative numbers of particles in A and B. Actually, it makes no difference whether the particles involved are transmissible or not. If, as normally happens in cells, both kinds are present, transmissible particles on the more concentrated side will eventually diffuse over to the other side until concentrations are equal; but before this equality is reached, transmissible particles still exert a temporary osmotic effect. Nontransmissible particles, not being able to penetrate the membrane, exert a permanent osmotic effect.

If the difference in particle number is great enough (for example, if *A* contains pure water only and *B* contains water and a large number of particles), then the *A* side may dehydrate completely and collapse, while the *B* side might burst and so collapse also. Since this does not normally happen in animal systems, their parts, clearly, must be in general osmotic equilibrium. The concentration of particles must be the same on both sides of membranes, or as often stated, the two sides must be *isotonic* to each other. If the initial concentration on an arbitrary *A* side is lower than on a *B* side (as in Fig. 4.25), then the *A* side loses water. The *B* medium here is said to be *hypertonic* to *A*, that is, initially more highly concentrated than the *A* medium. The *B* side, having the higher initial concentration of particles, gains water. The *A* medium is said to be *hypotonic* to the *B* medium.

Note that the net effect of osmosis is to pull water into the region of higher concentration, i.e., from the hypotonic to the hypertonic side. The process will continue until the two sides are isotonic. And note that osmosis will occur whenever certain particles cannot or do not pass through a membrane. Then nothing migrates through the membrane except water (plus any particles present which can diffuse through).

Like diffusion, osmosis plays an important role in animal cells. It is one agency by which water is distributed and redistributed across membranes. As in diffusion also, care must be taken in explaining given membrane phenomena simply as "osmosis." Sometimes the event in question actually is osmosis, but many times it is not. In this connection, it is particularly poor practice simply to dismiss given events along membranes unthinkingly as "diffusion and osmosis."

We may conclude generally that, in its physical organization, the substance of animal cells is a mixed colloidal system bounded by variously permeable membranes, undergoing localized sol-gel transformations, and being kept in constant internal motion by molecular bombardments, by diffusion displacements, and by osmotic forces. As a result of these properties, the living substance is subjected to a continuous physical flux equally as profound as the chemical flux. Indeed, physical changes initiate chemical ones, and vice versa; and from any small-scale point of view, cellular matter is therefore never the same from moment to moment.

Superimposed on and ultimately resulting from its chemical and physical organization, the cell substance also displays a highly characteristic biological organization; the existence of such an organization is a third necessary requirement if a cell is to be alive. The next section focuses on this biological design of animal cells.

BIOLOGICAL ORGANIZATION

The generalization that all animals, and living organisms generally, consist of cells and cell products is known as the *cell theory*. Principal credit for its formulation is usually given to the German biologists Schleiden and Schwann, whose work was published in 1838. But the French biologist Dutrochet had made substantially the same generalization as early as 1824. The cell theory rapidly became one of the fundamental cornerstones of modern zoology and, with minor qualifications, it still has that status today.

Cells came to be recognized early as the "atomic" units of living matter, structurally as well as functionally. In 1831, the English biologist Robert Brown discovered the presence of nuclei within cells, and in 1839 the Bohemian biologist Purkinje coined the general term "protoplasm" for the living substance out of which cells are made. Virchow in 1855 concluded that *"omnis cellula e cellula"*—new living cells can arise only by reproduction of preexisting living cells. This important recognition of the continuity of life, and thus of the direct derivation of all cells from ancient cellular ancestors, introduced the notion of history into the study of cells. Ever since, the study of cells has revolved around three interrelated problems: cell *structure*, cell *function*, and short- and long-range cell *development*. The first of these concerns us here particularly.

The Basic Design

Examination of living or killed animal cells under various kinds of microscopes shows that cell sizes vary considerably, ranging in diameter from about 2μ to as much as several millimeters and more. However, the order of size of the vast majority of animal cells is remarkably uniform; a diameter of 5 to 15μ is fairly characteristic generally. We surmise that, notwithstanding the exceptions, cells can be neither much smaller nor much larger than a certain norm. Too small a size presumably would not provide enough room to accommodate the necessary parts, and too large a size would increase the maintenance problem and at the same time reduce the efficiency of compact operation. Note here that as a cell increases in size, its surface enlarges as the *square* of its radius, and that the available surface area determines the amount of nutrient uptake and waste elimination possible. However, cell volume increases with the *cube* of its radius, and the volume determines how much mass a cell must keep alive. Hence if a cell enlarges unduly, its mass would soon outrun the food-procuring capacity of its surface, and cell growth then would have to cease. The size limits of cells actually are such that available surfaces can adequately service the living volumes they delimit.

The two fundamental subdivisions of animal cells are the

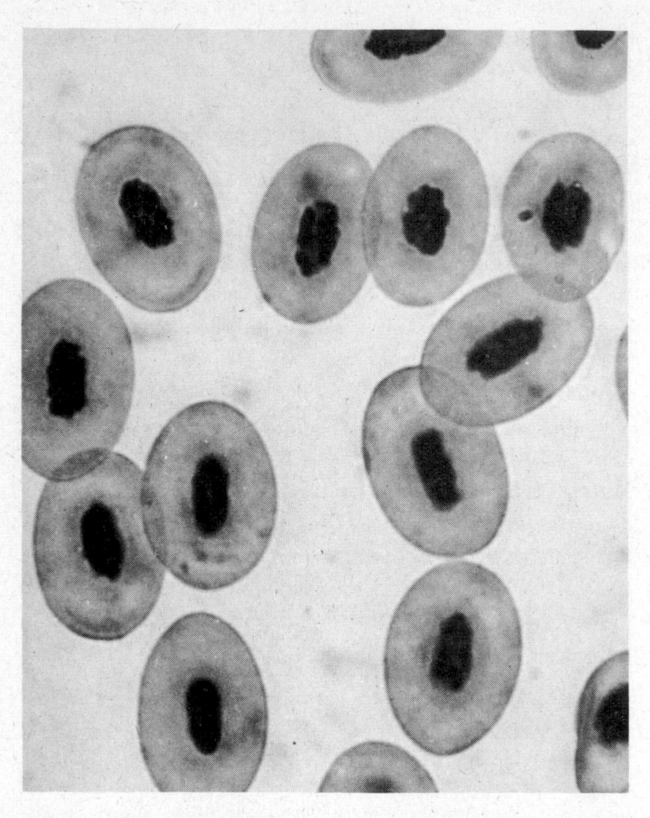

4.26. The general structure of cells is illustrated in this photo of red blood cells of a frog. Note darkly stained nucleus, external cell membrane, and the clear cytoplasm between nucleus and cell membrane.

nucleus and the living substance surrounding the nucleus, called the *cytoplasm*. The nucleus is bounded by a *nuclear membrane*, the cytoplasm by a *cell membrane*, also called *plasma membrane* (Fig. 4.26).

Animal cells usually contain a single nucleus each, but there are many exceptions. For example, mature mammalian red blood cells are without nuclei altogether (and are therefore referred to as "corpuscles" rather than as complete "cells"). Conversely, the cells of various animal tissues often contain more than one nucleus. A *binucleate* or *multinucleate* condition of this sort is particularly common among the cells of various protozoa.

There are exceptions too concerning the individuality of cells, normally maintained by the cell membrane. In certain cases, e.g., in the embryos of many animals, cell membranes at first form boundaries between individual cells. But at a later stage of development these boundaries dissolve, the result being a fused, continuous living mass with nuclei dispersed through it. Such a structure, in which cellular individuality has been lost, is called a *syncytium*. In certain animals the entire adult body is syncytial, in others only given tissues or organs exhibit such a structure.

Despite variations in the number of nuclei or the occasional loss of the structural discreteness of cells, the fundamentally cellular character of animal matter is undeniable even in such exceptional cases. And in all other cases the cellular character is unequivocal, for there we deal with distinct bits of living matter, each bounded by a plasma membrane and containing one nucleus.

Nucleus and Cytoplasm

A nucleus typically consists of three kinds of components: the more or less gel-like nuclear sap, or *nucleoplasm*, in which are suspended the *chromosomes* and one or more *nucleoli* (Fig. 4.27). The chromosomes are the principal nuclear organelles.

4.27. Electron micrograph of a cell nucleus (from the pancreas of a guinea pig). The whole spherical structure is the nucleus; cytoplasm is outside it. Note the nuclear bounding membrane. Within the nucleus, the large dark patch is the nucleolus, and the dark speckle elsewhere is the gene-containing chromatin.

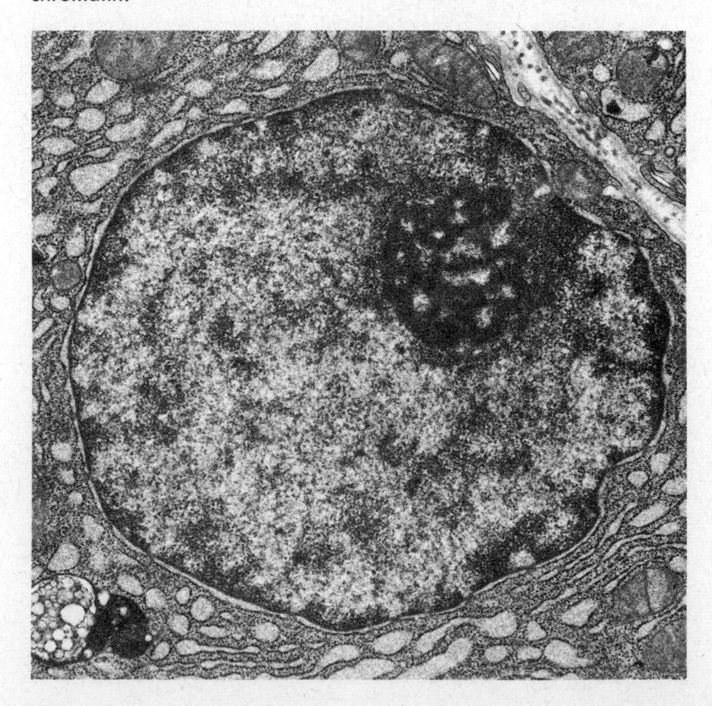

Indeed, a nucleus as a whole may be regarded primarily as a protective housing for these slender, threadlike bodies. Chemically, chromosomes consist largely of protein and of nucleic acids, intimately associated into complexes called *nucleoproteins*. DNA is the principal nucleic acid of the nucleoproteins, but RNA is also present. Functionally, chromosomes are the carriers of the genes which, as noted previously, are the ultimate controllers of cellular processes.

Chromosomes are conspicuous only during cell reproduction, when they become thickly coated with additional nucleoprotein. At other times such coats are absent, and chromosomes then are very fine filaments not easily identifiable within the nuclear sap. The exact number of chromosomes within each cell nucleus is an important species-specific trait. For example, cells of human beings contain 46 chromosomes each. Analogously, the cells of every other animal type have their own characteristic chromosome number. A cell rarely contains more than in the order of 100 chromosomes. Therefore, since there are at least 1.5 million different species of animals, many species share the same chromosome number. To be sure, possession of the same numbers of chromosomes does not mean possession of the same kinds.

A nucleolus ("little nucleus") is a spherical body which also consists largely of nucleoprotein. But the only type of nucleic acid present here is RNA. As we shall see in a later chapter, nucleoli are derivatives of chromosomes, and they appear to play an important role in the control of protein synthesis within cells. Given cell types contain a fixed number of nucleoli per nucleus.

The whole nucleus is separated from the surrounding cytoplasm by the nuclear membrane. This structure, like most other living membranes, is constructed mainly of proteins and lipids. It governs the vital traffic of materials between cytoplasm and nucleus. Examination with the electron microscope shows that the nuclear membrane is actually a double layer pierced by tiny pores (Fig. 4.28).

If the nucleus, by virtue of its genes, is the control center of cellular activities, then the cytoplasm is the executive center. In it the directives of the nucleus are carried out. But it should be emphasized at once that such a functional distinction between nucleus and cytoplasm should not be taken too rigorously. Although the nucleus primarily controls, it also executes many directives of the cytoplasm; and although the cytoplasm primarily executes, it also influences many nuclear processes. A vital reciprocal interdependence binds nucleus and cytoplasm, and experiment has repeatedly shown that one cannot long survive without the other.

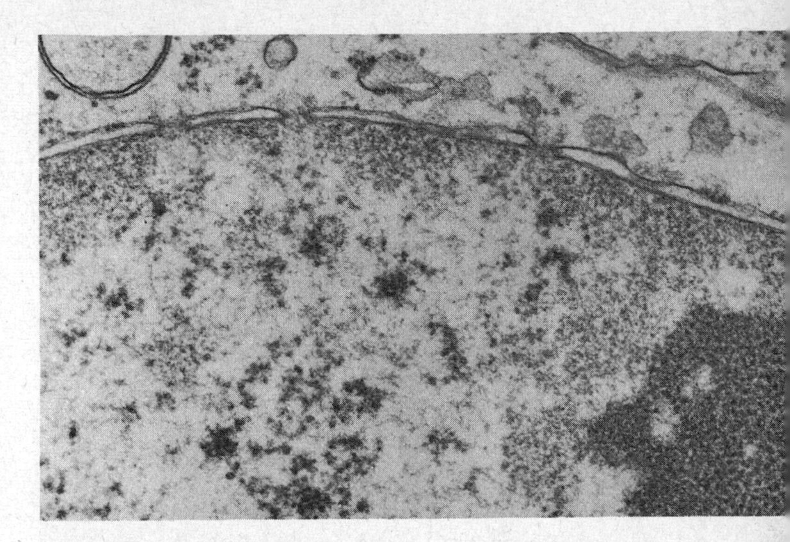

4.28. Electron-micrographic close-up of a nuclear membrane (from a cell in the spinal cord of a bat). Nuclear substance is at bottom of photo, cytoplasmic substance at top. Note the double-layered condition of the nuclear membrane and the pores in this membrane.

Cytoplasm consists of a semifluid *ground substance*, which is in a sol or a gel state at different times and in different cellular regions, and in which are suspended large numbers of several kinds of organelles. Such organelles may be shaped into granules, rodlets, filaments, or droplets. Each of these may have various sizes and chemical compositions and may have a variety of functions. Particular cell types often possess unique organelles not found elsewhere. The following organelles are widespread among many or all animal cell types.

Mitochondria. Found universally in all animal cells, these organelles are constructed predominantly out of lipids and proteins. Their principal functional constituents are *respiratory enzymes* and coenzymes, i.e., the compounds required in energy-producing reactions. Mitochondria are the chief chemical "factories" in which cellular respiration is carried out. Under the light microscope, mitochondria appear as short rods or thin filaments averaging 0.5 to 2 μ in length. The electron microscope shows that the surface of a mitochondrion consists of two fine membranes (Fig. 4.29). The inner one is greatly folded, the folds projecting into the interior of the mitochondrion. These folds, known as *mitochondrial cristae*, are believed to be the specific locations where respiratory reactions take place.

A

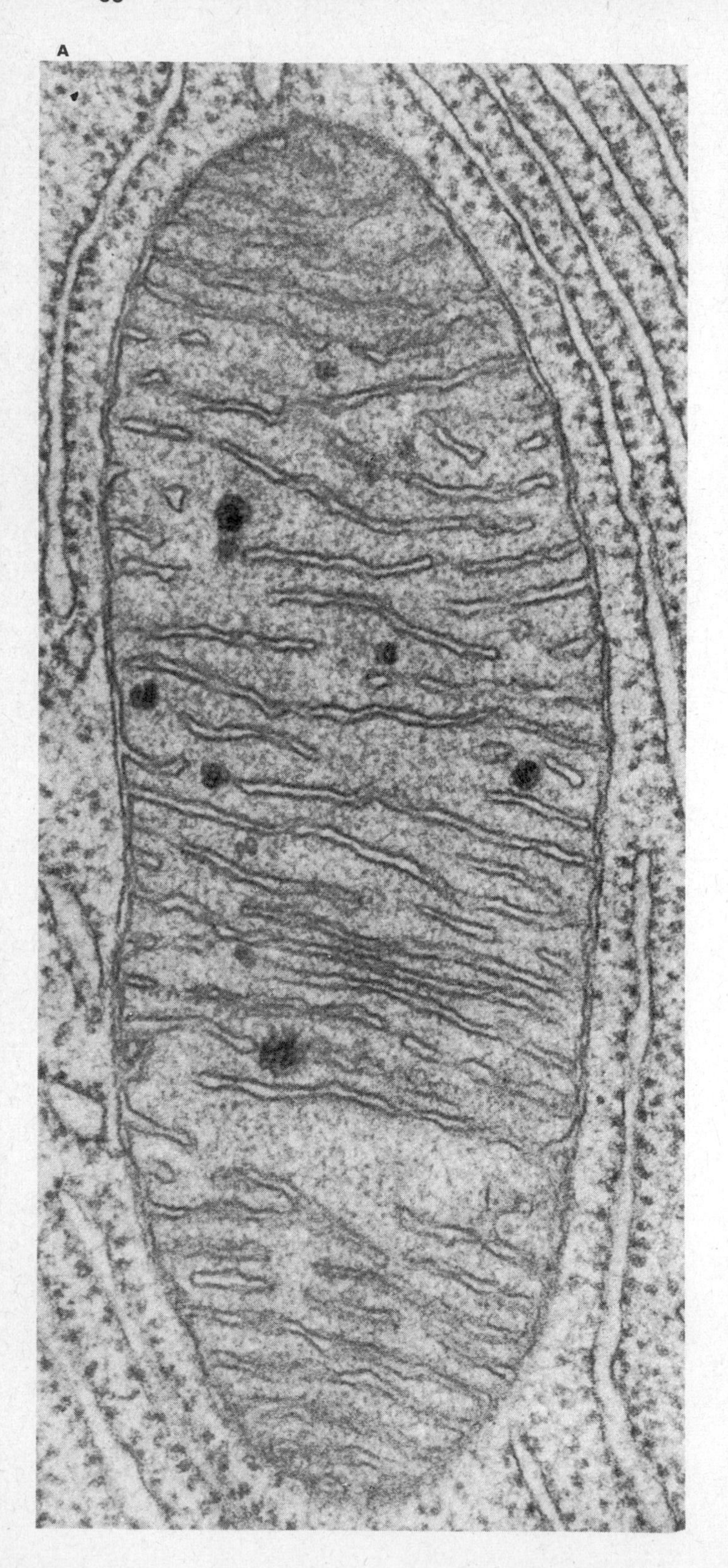

B

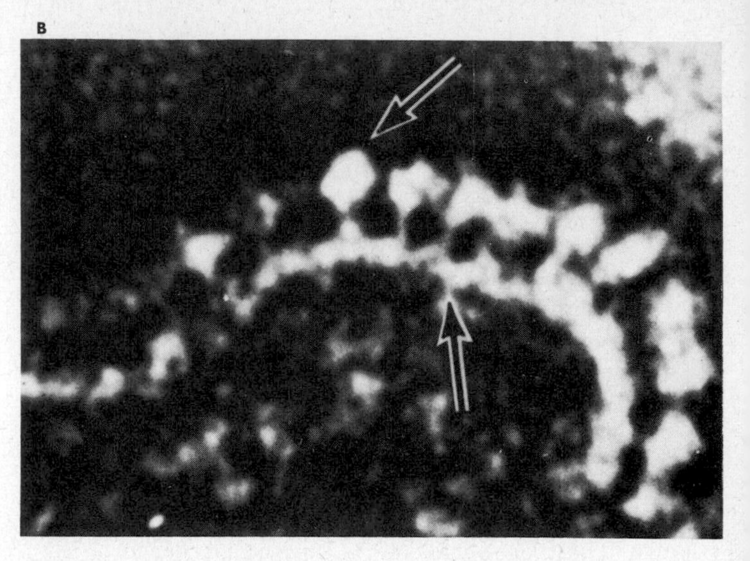

4.29. **A. Electron-micrographic section through a mitochondrion** (of a mammalian pancreatic cell). Note the double-layered exterior boundary and, in places along it, the infolding of the inner boundary membrane, forming the internal partitions called cristae. The dark spots in the interior of the mitochondrion are calcium-rich granules. **B. Highly enlarged portion of a mitochondrial crista** (in beef heart muscle), showing array of stalked particles attached to the cristal membrane. Arrows point to head piece and base piece of such a particle. Particles of this kind may contain the actual enzymatic apparatus for respiratory reactions.

Ribosomes. These organelles are exceedingly tiny granules, visible under the electron microscope (Fig. 4.30). Ribosomes contain RNA (hence the "ribo-" portion of their name) and the enzymes required in the manufacture of proteins; ribosomes appear to be the principal "factories" in which protein synthesis is carried out.

Golgi bodies. Depending partly on the way in which cells are prepared for microscopic study, Golgi bodies appear variously as complexes of droplets or as stacks of thin platelike layers or as mixtures of these and other configurations (Fig. 4.31). These organelles apparently function in the manufacture of cellular secretion products, for Golgi bodies are particularly conspicuous in actively secreting cells. For example, whenever gland cells are producing their characteristic secretions, the Golgi bodies of such cells become very prominent.

Centrioles. In animal cells a single small granule is located just outside the cell nucleus. As will be shown later, such centrioles function in cell reproduction.

Apart from organelles just listed, animal cytoplasm generally contains additional *granules* and fluid-filled droplets called *vacuoles,* which are bounded by membranes. Such cytoplasmic granules and vacuoles perform a large variety of functions. They may be vehicles transporting raw materials from the cell surface to interior processing centers (e.g., *food vacuoles*) or finished products in the opposite direction (e.g., *secretion granules*); they may be places of storage (e.g., *glycogen granules, fat vacuoles, water vacuoles, pigment granules*); they may be vehicles transporting waste materials to points of elimination (e.g., *excretory vacuoles*); or they may be special processing centers themselves.

In addition to all these, cytoplasm may or may not contain a variety of long, thin *fibrils* made predominantly out of protein (e.g., *contractile myofibrils, conducting neurofibrils*). Various other inclusions, unique to given cell types and serving unique functions, may also be present. In general, every function a cell performs, common or not, is based on a particular structure in which the machinery for that function is housed.

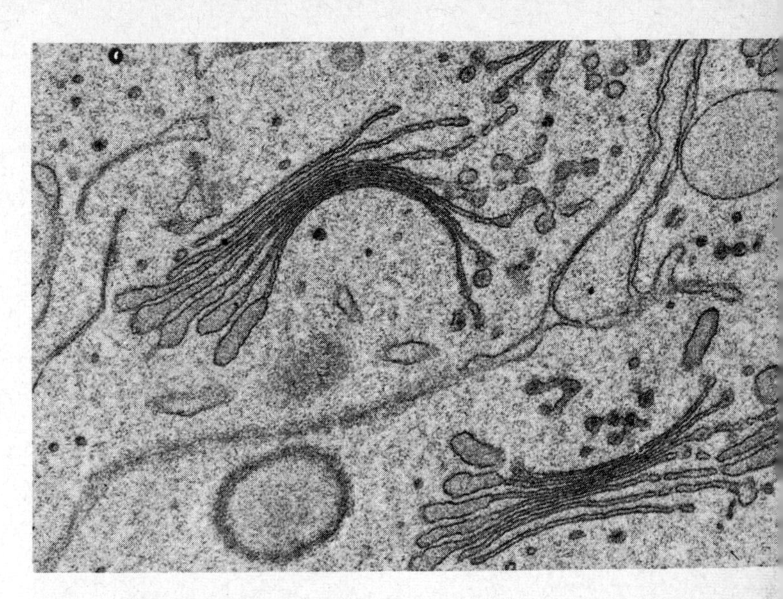

4.31. Electron micrograph of portion of cytoplasm showing two Golgi bodies. Each such body consists of stacks of parallel lamellae.

4.30. Electron micrograph of ribosomes (of a mammalian pancreatic cell). The ribosomes are the small dark granules. The double-layered membranes also visible here are portions of the endoplasmic reticulum, illustrated further in Fig. 4.32.

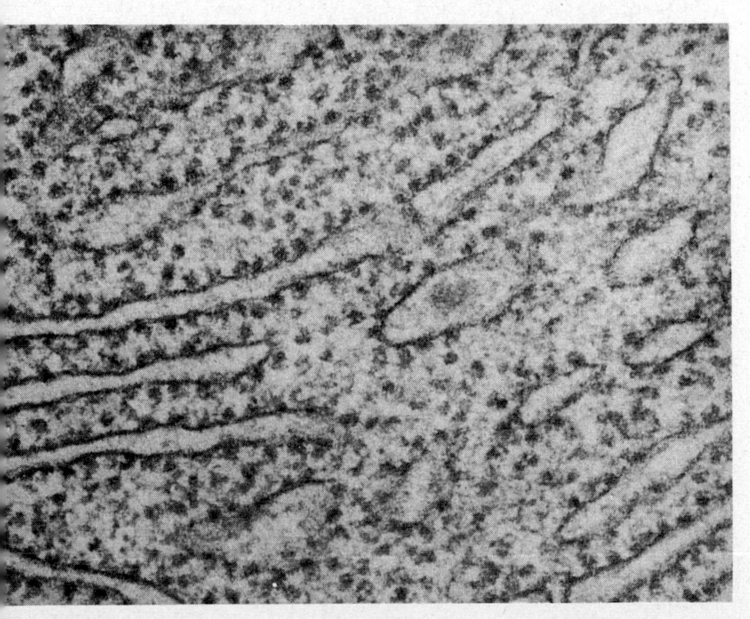

Cytoplasm as a whole is normally in motion. Irregular eddying and streaming occur at some times; and at others the substance of a cell is subjected to cyclical currents, a movement known as *cyclosis.* The organelles, nucleus included, are swept along passively in these streams. The specific cause of such motions is unknown, but there is little doubt that they are a reflection of the uninterrupted chemical and physical changes which take place on the molecular level. Whatever the specific causes may be, the apparently random movements might give the impression that nothing is fixed within a cell and that cytoplasm is simply a collection of loose particulate bodies suspended in "soup."

But this impression is erroneous, as examination under the electron microscope shows. The ground substance of cytoplasm, which under the light microscope does appear to be a fluid, structureless soup, actually turns out to be highly structured and organized. A network of exceedingly fine membranes can be shown to traverse the cytoplasm from plasma membrane to nuclear membrane and in many cases also from cell to cell. This network is known as the *endoplasmic reticulum* (Fig. 4.32). Linked to it are the nucleus, the mitochondria, the ribosomes, and all the other cytoplasmic organ-

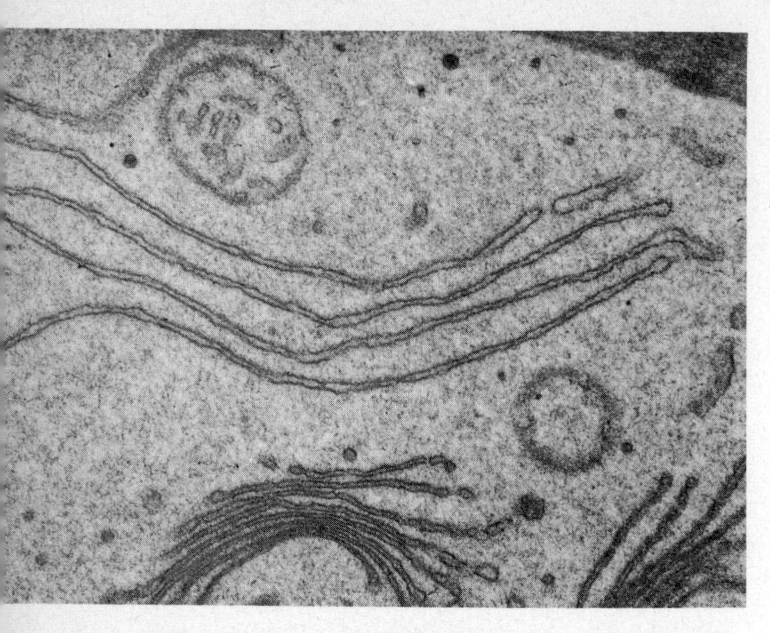

4.32. Electron micrograph of portions of the endoplasmic reticulum. The latter consists of an array of double membranes. A cross section through a mitochondrion is seen above the reticulum and a portion of a Golgi body below it.

elles. Thus, little is really "loose" in a cell. When cytoplasm as a whole streams and moves, the endoplasmic reticulum streams and moves too and the other organelles are carried along, still held to the ultramicroscopic network. Evidently, even though the contents of a cell may shift position and the cell as a whole may be deformable, an orderly structural integration of the interior persists nevertheless. Indeed, this is essential if cellular functions are to be orderly and integrated.

The Cell Surface

Every cell as a whole is bounded by a *cell* (or *plasma*) *membrane*. Composed predominantly of protein and lipid substances, this important structure is far more than a passive outer skin. As already noted, it is an active, highly selective, semipermeable membrane which regulates the entry and exit of materials into and out of a cell. The membrane therefore plays a critical role in all cell functions, since, directly or indirectly, every cell function necessitates *absorption* of materials from the exterior, *excretion* of materials from the interior, or both.

Unlike plant cells, most animal cells are without exterior cell walls. In many cases, the surfaces of such naked animal cells are fairly smooth, and in compact tissues the cells press against one another and assume more or less irregular, somewhat angular shapes. In certain packed tissues, however, the cell contours are far from smooth. Instead, numerous fingerlike extensions may project from one cell and interlock with similar extensions from neighboring cells. Such protrusions usually are not fixed structures; they may slowly retract, re-form again, and the whole cell so may change its contours in the course of time. Moreover, some of the protrusions of one cell occasionally may nip off a protrusion of an adjacent cell. The first cell thus acquires an internal vacuole, filled with a droplet of cytoplasm derived from an adjacent cell. This process is one of the means by which material may be transferred from cell to cell. Also, a cell may occasionally form a deepening surface depression which eventually nips off on the inside as a fluid vacuole. Through such fluid engulfment, or *pinocytosis*, a cell may transfer liquid droplets into its interior (Fig. 4.33).

All such transfers may be regarded as specialized forms of the more general phenomenon of *amoeboid motion*, in which transient, shifting fingerlike extensions, or *pseudopodia*, are again formed at any point on the cell surface (cf. Fig. 3.8). The protrusion-retraction process cannot be fully explained as yet, but it appears to involve sol-gel transformations in the layer of cytoplasm located just under the cell membrane. Some protozoa, particularly amoeboid protozoa, move and feed by means of pseudopods. In feeding, pseudopods flow all around a bit of food, which thus comes to be engulfed and forms a food vacuole in the cell interior. Many kinds of nonprotozoan

4.33. Diagrams of portions of cell surfaces to indicate variants of amoeboid engulfment. Series at top, pinocytosis, the cellular engulfment of fluid droplets. Series at bottom, phagocytosis, the cellular engulfment of solid particles.

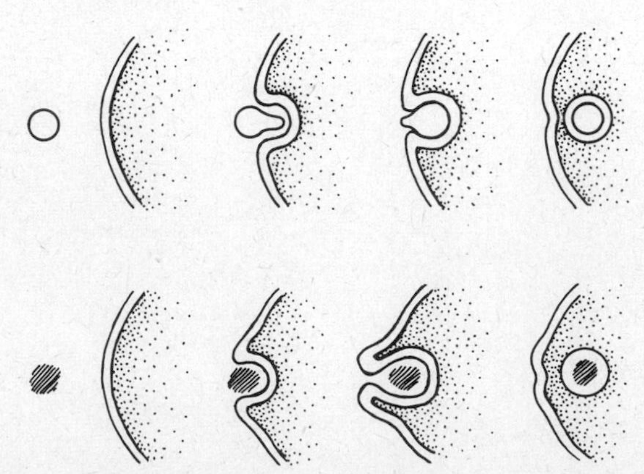

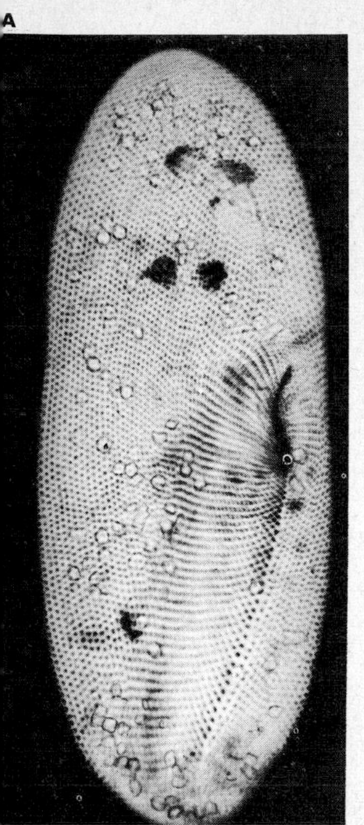

4.34. Pellicles and cuticles. A. The organic chitinlike pellicle of the ciliate protozoon *Paramecium*. The dark dots are tiny pores through which the cilia project. **B.** The chitin cuticle of the head of a carpenter ant, illustrating the high degree of structural elaboration possible in arthropod exoskeletons generally.

cells are similarly capable of amoeboid movement, e.g., many types of eggs, which engulf sperm cells in fertilization, and several categories of blood cells, which engulf foreign bodies, bacteria, and other potentially harmful bodies in blood. When one cell "swallows" such bodies or solid particles generally in amoeboid fashion, the term *phagocytosis* is often applied.

Some animal cells, particularly those exposed to the external environment, are not naked but are enveloped partially or wholly with wall-like *cuticles* or *pellicles*. For example, a protective coat of secreted chitin is found on the skin cells of insects and arthropods generally. Numerous other invertebrate animals secrete horny protein coats on their outer cells. Analogously, the surface cells of most vertebrates secrete external protective coats made of the protein keratin. In many other cases, surface cells secrete films of mucus, composed of polysaccharide derivatives. In tunicates, surface deposits of cellulose are present. Many mollusks manufacture *shells* over the surface cells, usually composed of lime. Shells of lime, glass (silica), or horny organic materials also characterize numerous protozoa (Fig. 4.34).

Many animal cell types are equipped with specialized locomotor organelles on the cell surface. Among these are, for example, *flagella*, long slender, threadlike projections from cells (Fig. 4.35). The base of a flagellum is anchored in the cell cytoplasm, on a distinct granule known as a *kinetosome* or *blepharoplast*. In some cases, a threadlike fibril, called a *rhizoplast*, connects the kinetosome with the centriole located just outside the nucleus. The kinetosome appears to control the motion of the flagellum. If a flagellum has a smooth external surface, it is said to be of the *whiplash* type; if it possesses exceedingly fine side branches set on the main stem like the bristles of a brush (*mastigonemes*), it is said to be of the *tinsel* type. A cell usually bears one flagellum, but in some cases many more are present, all anchored in the same kinetosome or complex of kinetosomes. Numerous cell types possess shorter variants of flagella called *cilia*. These are usually present in very large numbers and may cover all or major portions of a cell like tiny bristles. Each cilium has its own, separate kinetosome complex at its base, and cilia and flagella may be distinguished on this basis.

Internally, all flagella and cilia have the same structure. The electron microscope reveals a flagellum or cilium to be a bundle of eleven exceedingly fine fibrils. Two of the fibrils are central, and nine are arranged in a ring around the central two. All the fibrils connect with the kinetosome at the base. How motion is actually produced is as yet understood only poorly (cf. Fig. 4.35).

Flagella are the locomotor structures in numerous protozoa and in most sperm cells of animals generally. In the flagellate protozoa, the flagellum is at the anterior end of a cell and its sinuous beat pulls the cell behind it. In many sperms, on the other hand, the flagellum is at the posterior end and its beat pushes the cell forward. Flagella are also encountered in stationary animal cells. For example, many of the interior cells of sponges are flagellate, the flagella being used here to create food-bearing water currents (cf. Chap. 20). Analogously, the principal excretory cells of many invertebrate animals, so-called *flame cells*, bear flagella which are used to produce currents in waste-laden body fluids (cf. Chap. 9). Greatly modified flagellate cells exist in the retinas of mammalian eyes, in the form of rod and cone cells.

Cilia are even more widely distributed. They occur in a major group of protozoa, and most animal embryos, many types of larvae, and numerous groups of adults are ciliated externally. In such cases the surface cilia serve in locomotion, in creation of feeding currents, or both. Most animals also possess ciliated cells in the interior of the body, e.g., in the pressure-sensitive cells of certain sense organs, in the lining tissues of breathing, alimentary, and reproductive channels, and generally in any other location where movement of air, water, or solid materials must take place. Numerous modifica-

4.35. Cilia and flagella. **A.** Types of flagella and their insertion in cells. **B.** The ciliate protozoon *Tetrahymena* stained to show the cilia and their arrangement on the body surface. **C.** Electron-micrographic section through two regions of the flagellum of a mouse sperm, showing the two central and nine peripheral filaments characteristic of flagellar structure. **D.** Electron micrograph of a portion of the tinsel-type flagellum of a unicellular flagellate organism, showing the mastigonemes projecting laterally from the main shaft of the organelle. Cf. also Fig. 19.4.

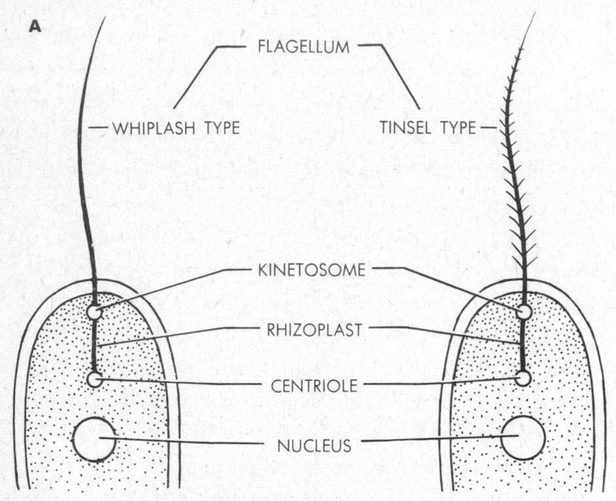

B C

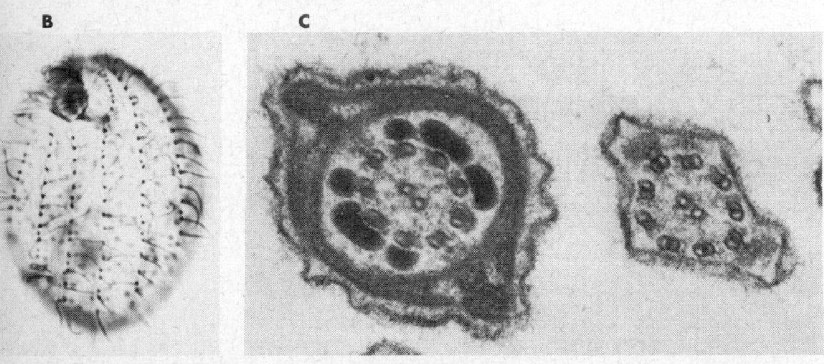

D

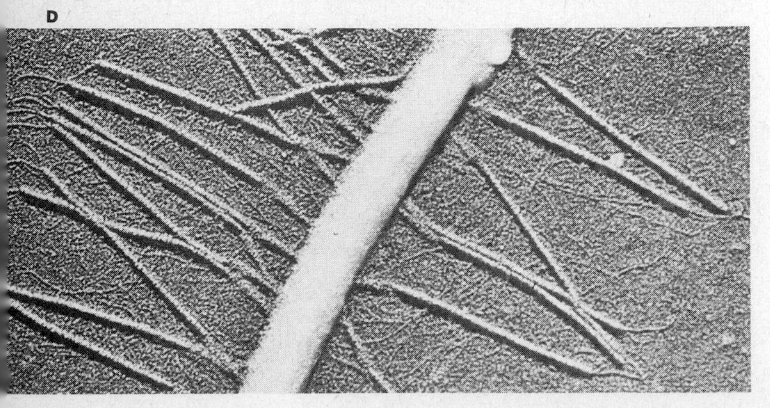

tions of cilia are encountered as well. For example, cilia are permanently stationary in some cases and enormously increase the surface area of a cell. Or, adjacent cilia may be fused together and form stiff, strong *cirri* which may be used as tiny walking legs, as in certain protozoa. Or again, fused cilia may form tiny veil-like sheets, such as *undulating membranes*, which also may serve as specialized locomotor organelles (cf. Chap. 19).

We clearly note that, in any cell, certain of the components of the nucleus, the cytoplasm, or the surface may be associated directly with well-circumscribed cell functions. Respiration and protein synthesis, for example, are distinct functions performed in distinct cytoplasmic structures; see Table 4 for a summary of such correlations. But many cell functions cannot be localized so neatly. For example, cellular reproduction or amoeboid movement requires the cooperative activity of many or all of the cell components present. Functions of this kind cannot be referred to any particular part of a cell; they must be referred to the cell as a whole.

TABLE 4. Some structural components of animal cells and their correlated functions

Structure	Function
nucleus	
chromosomes	gene carriers, ultimate control of cell activities
nucleolus	auxiliary to protein synthesis
nuclear membrane	traffic control to and from cytoplasm
cytoplasm	
mitochondria	site of respiration
ribosomes	site of protein synthesis
Golgi bodies	site of specific secretion synthesis
endoplasmic reticulum	secretion channels, connection between cell parts, attachment surfaces
centrioles	auxiliary to cell division
kinetosomes	anchor and control of flagella, cilia
myofibrils	contraction
neurofibrils	conduction
granules, vacuoles	transport, storage, processing centers
surface	
plasma membrane	traffic control to and from cell
cuticles, pellicles	support, protection, water-proofing
cilia, flagella	locomotion, current creation, feeding
pseudopodia	locomotion, feeding, phagocytosis

Note also that, whereas many organelles are bulky enough to be visible under the microscope, even more are not visible; individual molecules in a cell "function" no less than larger molecular aggregates. Be it a single dissolved molecule or a whole group of large suspended organelles, each cellular structure performs a function, and as the structures differ among cells, so do the functions.

By virtue of its chemical, physical, and biological organization, an animal cell has the capacity to be alive. Clearly, however, proper organization alone is not enough to make the capacity a reality; if the parts remain inert, no matter how perfectly they may be organized, the whole still remains dead. To exhibit life, the cellular organization must operate, i.e., it must perform the functions of metabolism and self-perpetuation. The next two chapters provide an outline of these fundamental cellular operations.

REVIEW QUESTIONS

1. What are inorganic compounds? Organic compounds? What principal classes of each occur in animal matter and in what relative amounts? Which of these substances are electrolytes and which are nonelectrolytes?
2. Review the chemical composition and molecular structure of carbohydrates. What are monosaccharides, disaccharides, and polysaccharides? Give examples of each. What are aldoses, ketoses, pentoses, hexoses? Write out a number of isomers of glucose. What is the difference between 1,4- and 1,6-linkages in polysaccharides?
3. Review the composition and structure of lipids. What are saturated and unsaturated fatty acids? In what kinds of reaction may carbohydrates and fats participate and what general roles do they play in living matter?
4. What are proteins and how are they constructed? In what ways do proteins differ from carbohydrates and fats? Discuss fully and carefully. Review the structure of amino acids, show how these compounds differ, and illustrate the formation of a peptide bond.
5. What is the primary, secondary, tertiary, and quaternary structure of proteins, and what is protein specificity? How is a coagulated protein different from a native or a denatured protein? Review the general biological roles of proteins.
6. Distinguish between nitrogen bases, nucleosides, nucleotides, ribotides, and deoxyribotides. Give examples of each. What are adenosine phosphates and coenzymes? What roles do such compounds play in animal cells?
7. What is the chemical composition and molecular structure of nucleic acids? In chemical terms, what are DNA and RNA? How are nucleotides related to DNA and RNA? What different kinds of nucleotides occur in nucleic acids? What are the general functions of such acids?
8. What main classes of pigments occur in animal cells? How do these substances differ chemically? What are some of their functions? What derivatives of carbohydrates and fats are common in animals?
9. What is a colloidal system? How does such a system differ from a solution? What kinds of colloidal systems are possible, and what kinds are found in animal matter? Review the properties of colloidal systems.
10. Define diffusion and show how and under what conditions this process will occur. What is the biological significance of diffusion?
11. How and where do plasma membranes form? What are the characteristics of such membranes? What roles do they play in cellular processes?
12. Define osmosis. Show how and under what conditions this process will occur. Distinguish carefully between osmosis and diffusion. Cite examples of biological situations characterized by isotonicity, hypertonicity, and hypotonicity.
13. What are the structural subdivisions of cells? What are the main components of each of these subdivisions, where are they found, and what functions do they carry out?
14. List cytoplasmic inclusions encountered in all cell types and inclusions found only in certain cell types. What is cyclosis?
15. What structures are found on the surfaces of various cell types? Which of these structures are primarily protective? What do they protect against? What are the functions of other surface organelles? Describe the structure of such organelles.

COLLATERAL READINGS

The articles cited below are good semipopular accounts dealing with a variety of cellular constituents.

Brachet, J.: The Living Cell, *Sci. American*, Sept., 1961.

Bushwell, A. M., and W. H. Rodebush: Water, *Sci. American*, Apr., 1956.

Crick, F. H. C.: Nucleic Acids, *Sci. American*, Sept., 1957.

Doty, P.: Proteins, *Sci. American*, Sept., 1957.

Fieser, L. F.: Steroids, *Sci. American*, Jan., 1955.

Frieden, E.: The Enzyme-Substrate Complex, *Sci. American*, Aug., 1959.

Fruton, J. S.: Proteins, *Sci. American*, June, 1950.

Kendrew, J. C.: The 3-Dimensional Structure of a Protein Molecule, *Sci. American*, Dec., 1961.

Linderstrom-Lang, K. U.: How Is a Protein Made? *Sci. American*, Sept., 1953.

Pauling, L., R. B. Corey, and R. Hayward: The Structure of Protein Molecules, *Sci. American*, July, 1954.

Robertson, J. D.: The Membrane of the Living Cell, *Sci. American*, Apr., 1962.

Satir, P.: Cilia, *Sci. American*, Feb., 1961.

Schmitt, F. O.: Giant Molecules in Cells and Tissues, *Sci. American*, Sept., 1957.

Solomon, A. K.: Pores in the Cell Membrane, *Sci. American*, Dec., 1960.

Stein, W. H., and S. Moore: The Chemical Structure of Proteins, *Sci. American*, Feb., 1961.

Vallee, B. L.: The Function of Trace Elements in Biology, *Sci. Monthly*, vol. 72, 1951.

The following paperbacks may be consulted for further background information on cell chemistry and structure:

Loewy, A. G., and P. Siekevitz: "Cell Structure and Function," Modern Biology Series, Holt, New York, 1963.

Swanson, C. P.: "The Cell," Foundations of Modern Biology Series, Prentice-Hall, Englewood Cliffs, N. J., 1960.

The following compilation contains abstracts from original research papers in major areas of biology; the cell is among the topics treated:

Gabriel, M. L., and S. Fogel (eds.): "Great Experiments in Biology," Prentice-Hall, Englewood Cliffs, N.J., 1955.

The following texts include accounts of the composition and the properties of cells:

Bloom, W., and D. W. Fawcett: "Textbook of Histology," Saunders, Philadelphia, 1962.

DeRobertis, E. D. P., W. W. Nowinski, and F. A. Saez: "General Cytology," 3d ed., Saunders, Philadelphia, 1960.

Giese, A. C.: "Cell Physiology," 2d ed., Saunders, Philadelphia, 1962.

Wilson, E. B.: "The Cell in Development and Heredity," Macmillan, New York, 1928, reprinted, 1953.

CHAPTER 5

cellular operations: nutrition, respiration

As shown in Chap. 3, the function of metabolism includes the subfunction of *nutrition*, which supplies raw materials; *respiration*, which procures usable energy from some of the raw materials; and *synthesis*, which, with the aid of the remaining raw materials and with respiratory energy, produces new living matter.

Note that the metabolism of a cell is not generally equivalent to the metabolism of a whole animal. Thus, nutrition on the level of the whole animal involves alimentation and circulation of molecular foods via the body fluids, whereas cellular nutrition only involves transfer of molecular foods from the body fluids into an individual cell. Respiration on the whole-animal level includes breathing and internal circulation of gases, whereas cellular respiration is a chemical process of decomposition of fuels and harvesting of energy. Synthesis on the whole-animal level includes, for example, development of an entire bony skeletal system, whereas cellular synthesis only comprises the manufacturing processes occurring within any one individual bone-forming cell.

In other words, whole-animal metabolism is a prerequisite for certain phases of cellular metabolism. At the same time, cellular metabolism is also a prerequisite for certain phases of whole-animal metabolism. The whole animal must live if its cells are to be alive, and the individual cells must live if the whole animal is to be alive. In this chapter we are concerned only with cellular metabolism, cellular respiration particularly; the processes of whole-animal metabolism will be discussed later, e.g., in Chap. 9.

THE PATTERN

Raw Materials: Cell Nutrition

With or without the aid of oxygen, and with organic nutrients as fuels, animal cells convert the chemical energy present within fuel molecules into energy usable metabolically. Such respiratory conversions basically depend on a form of burning, more specifically, exergonic decomposition reactions. What kinds of substances are the actual fuels in cells? The answer is, any organic compound that contains bond energies, i.e., in effect *any* organic constituent of cells: carbohydrates, fats, proteins, nucleotides, their various derivatives, vitamins, other special compounds, and indeed all the innumerable substances which together make up a cell. Like a fire, respiration is no respecter of materials. Anything that can be decomposed will be decomposed, and in cells this is the very substance of cells itself. Respiration does not distinguish between the expendable and the nonexpendable. For example, an amino acid which is an important structural member of the framework of a cell or is part of an enzyme may be respired away just as readily as an amino acid which has just been obtained by a cell as an external food.

However, if a fire is fed much of one fuel but little of another, more of the first is likely to be burned. Indeed, under normal conditions, an animal cell receives a steady enough supply of external foods to make *them* the primary fuels rather than the structural parts of a cell. Also, some kinds of materials

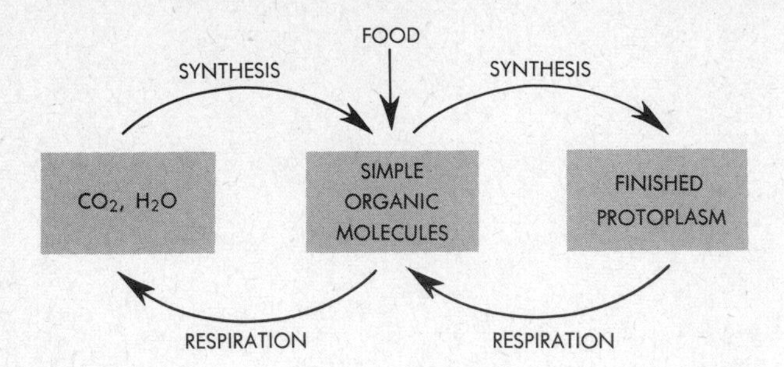

5.1. Destruction of cellular materials by respiration is offset by simultaneous construction of cellular materials by synthesis. A more detailed representation of this concept is given in Fig. 6.10.

decompose more readily than others, and some are more accessible to the decomposing apparatus than others. On this basis, foods, carbohydrates and fats in particular, are again favored as fuels, and the finished components of a cell tend to be spared. Yet the sparing is relative only. The formed parts of a cell *are* decomposed gradually, including even those which make up the decomposing apparatus itself.

But if a cell itself decomposes, how can it remain intact and functioning? Only by continuous construction of new living components, offsetting the continuous destruction through respiration. Note that these two processes go on side by side, at all times: destructive energy metabolism and constructive synthesis metabolism. One is in balance with the other, and foods serve both as fuel for the one and as building materials for the other. We say that the components of a cell are continuously "turned over"; i.e., existing parts are continuously replaced by new ones. The animal substance, we note, is never quite the same from instant to instant (Fig. 5.1).

Cellular nutrition takes place through the surface of a cell and may be characterized as a process of *absorption.* Nutrients, or *metabolites,* are absorbed from the cellular environment, which may be either the external physical world or adjacent cells or body fluids and blood. In almost all cases, the metabolites are individual ions or molecules, and absorption is accomplished by at least three kinds of subprocesses: water is absorbed in part by *osmosis;* inorganic and organic compounds dissolved in water are absorbed in part by *diffusion;* and dissolved compounds are absorbed additionally by energy-consuming *active transport* (Fig. 5.2).

The role of diffusion and of osmosis has already been pointed out (Chap. 4). If the concentration of dissolved materials is greater within a cell than in its surroundings, then some water will actually be absorbed osmotically; and if the concentration is greater in the surroundings, then some dissolved materials will be absorbed by diffusion. Regardless of concentrations, however, cells also absorb by active transport, very often against prevailing osmotic or diffusion gradients. As noted earlier, the precise mechanisms of active transport are still poorly understood. But in the case of at least one important metabolite, glucose, some of the details of absorption are known.

It is now believed that cells may not take up glucose as such (except to the extent that purely physical diffusion may occur). Instead, glucose absorption may be contingent on *phosphorylation,* i.e., chemical addition of a phosphate (—O—P) group (Fig. 5.3). The result is phosphorylated glucose, or glucose-6-phosphate, i.e., the —O—P group becomes attached at carbon 6 of glucose and there replaces a hydroxyl (—O—H) group. (Note, for later reference, that —O—P groups normally exchange places with hydroxyl groups.)

The phosphorylation reaction requires energy, which is obtained from a molecule of ATP. The latter is both the phosphate donor and the energy donor; as the terminal third phosphate group of ATP is split off and added to glucose, the high-energy bond of this third phosphate breaks and makes its energy available for the reaction. Note that, inasmuch as ATP is the endproduct of respiration, some of this endproduct must be used up to make cellular nutrition, hence respiration itself, possible to begin with. The reaction also requires a specific enzyme, *hexokinase* (the name "kinase" always referring to enzymes catalyzing reactions in which ATP participates). In vertebrates, moreover, glucose uptake into cells is additionally facilitated by the pancreatic hormone *insulin.* We may

5.2. Cellular nutrition. As indicated, a cell may obtain external raw materials by active transport regardless of concentration differentials between the environment and the cellular interior. If the exterior concentration is lower than that of the interior, water may be obtained by osmosis; and in the opposite case, raw materials may enter a cell by diffusion.

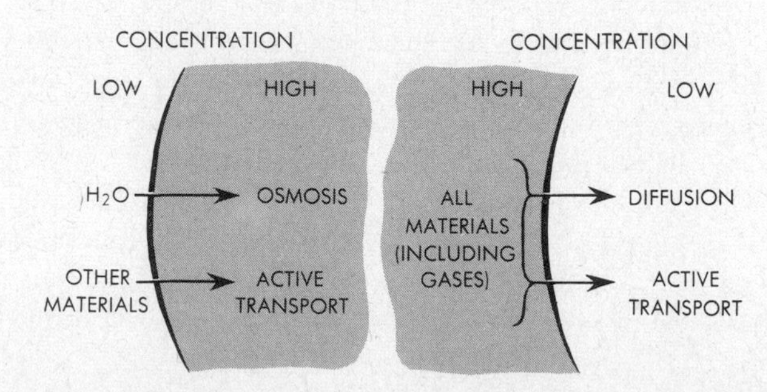

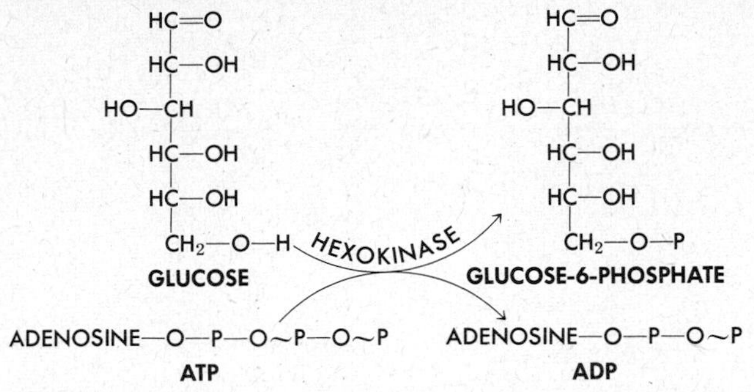

HC=O
HC—OH
HO—CH
HC—OH
HC—OH
CH₂—O—H
GLUCOSE

HEXOKINASE

HC=O
HC—OH
HO—CH
HC—OH
HC—OH
CH₂—O—P
GLUCOSE-6-PHOSPHATE

ADENOSINE—O—P—O~P—O~P
ATP

ADENOSINE—O—P—O~P
ADP

5.3. The phosphorylation reaction. Glucose reacts with ATP in the presence of the enzyme hexokinase, resulting in phosphorylated glucose and ADP. Note that ATP serves both as energy and phosphate donor. The —O—H at position 6 of glucose is replaced by —O~P of ATP, the high-energy bond becoming a low-energy bond in the process. (P stands for $—PO_3H_2$ here and in subsequent illustrations.)

understand, therefore, why insulin deficiency, as in diabetes, leads to impairment of cell function: without the hormone, glucose cannot be well absorbed into cells; and while the cells then starve, the sugar accumulates uselessly in blood and is eventually excreted in urine.

The phosphorylation reaction is believed to occur at, or in, the cell surface, and the carbohydrate then present within a cell therefore is glucose-6-phosphate. Various cell-surface reactions may also take place in the absorption of metabolites other than glucose, but in most of these instances the details are still obscure. Once a metabolite of any kind has been absorbed into a cell, intracellular distribution is achieved mainly by diffusion and cyclosis. Through these processes, every part of a cell comes to have access to all nutrients the cell may have acquired only at specific points of its surface. In the course of such internal distribution, some of the organic nutrients reach the mitochondria, the principal organelles in which respiration then occurs.

Apart from nutrient molecules, the only other respiratory raw material an animal cell requires is oxygen. This gas enters a cell by diffusion. As the cell continuously uses oxygen, the concentration of the gas within the cell always tends to be lower than in the surroundings. Consequently, the pressure gradient, or *tension gradient*, points into the cell, and gas diffusion suffices as the supply mechanism. At the same time, respiration continuously liberates CO_2, and the concentration of this byproduct therefore tends to be greater within a cell than in the surroundings. Diffusion then accomplishes the removal of CO_2 from the cell. We may note that the exchange of respiratory gases is one of the few instances of traffic through cell surfaces where diffusion actually appears to be the sole driving force.

So supplied with nutrients and oxygen, a cell may now respire.

Nutrient Oxidation

The decomposition reactions of respiration are a series of successive energy-yielding, exergonic processes characterized by an overall $-\Delta F$; the endproducts are more stable than the starting materials. It is this thermodynamic circumstance which ultimately "drives" all respiratory events.

Inasmuch as the energy yielded by respiration stems principally from the chemical bonds formed by carbon atoms, we may profitably examine the comparative stabilities of various carbon bonds. The least stable, hence most energy-rich, carbon combinations generally are the *hydrocarbon* groups, i.e., atomic groupings containing only carbon and hydrogen. These occur in organic molecules in forms such as CH_4, $—CH_3$, $—CH_2—$, $=CH—$. The most stable carbon combination on the other hand is CO_2, O=C=O, an *anhydride* (i.e., a hydrogen-free grouping). In general, therefore, we may predict that usable respiratory energy will result from conversions of hydrocarbons to anhydrides, i.e., from the replacement of H atoms by O atoms bonded to carbon.

Such conversions, and particularly the removal of hydrogen from carbon, or *dehydrogenations*, represent instances of a general category of chemical processes called *oxidation-reduction* reactions, or *redox* reactions. Every such reaction may be considered to consist of two *half-reactions*, one being an oxidation, the other a reduction (Fig. 5.4). In the case of dehydrogenations, for example, removal of H from a *hydrogen donor* compound represents the oxidizing half-reaction. The reducing half-reaction follows when the removed hydrogen is attached to another compound, a *hydrogen acceptor*. Dehydrogenation is oxidation, hydrogenation is reduction; the one can occur only if the other occurs as well. The energy relations are such that, at the end of both half-reactions, the endproducts are more stable than the starting materials, and the overall reaction has a $-\Delta F$. For example, in the overall reaction

$$CH_4 + 2\,O_2 \longrightarrow CO_2 + 2\,H_2O \qquad -\Delta F$$

the carbon of methane is oxidized to CO_2, but at the same time oxygen is also reduced to H_2O. Methane is the hydrogen

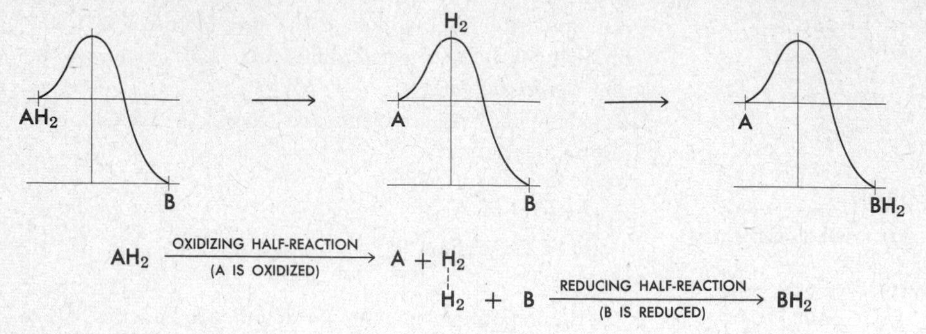

5.4. The general pattern of an oxidation-reduction (redox) process, as illustrated by a dehydrogenation. In the oxidizing half-reaction, AH_2 is the hydrogen donor, and energy is expended in the separation of H_2 from *A*. In the subsequent reducing half-reaction, *B* is the hydrogen acceptor, and energy is gained as *B* combines with H_2. This energy gain is greater than the earlier energy expenditure, hence BH_2 is stabler than AH_2 (and the ΔF will be negative, as per Fig. 2.10). The net overall redox reaction above is $AH_2 + B \longrightarrow A + BH_2$.

donor, oxygen the hydrogen acceptor. To symbolize the two half-reactions separately, we may write:

$$CH_4 + O_2 \longrightarrow CO_2 + 2\,H_2 \qquad +\Delta F,$$
oxidation of carbon, half-reaction

$$2\,H_2 + O_2 \longrightarrow 2\,H_2O \qquad -\Delta F,$$
reduction of oxygen, half-reaction

$$\underline{CH_4 + 2\,O_2 \longrightarrow CO_2 + 2\,H_2O \qquad -\Delta F,\ \text{net,}}$$
overall reaction

The oxidizing half-reaction requires a certain amount of energy, but the reducing half-reaction yields a greater amount. Hence the net ΔF of the overall reaction is negative; and CO_2 and water are more stable than methane and oxygen.

We may therefore speak of different *oxidation states,* or *oxidation levels,* of given carbon groupings. Thus, the carbon in methane has a lower oxidation state (but a higher reduction state) than that in CO_2 (in CH_4, the state is -4, in CO_2, it is $+4$; cf. Chap. 2). For convenience, we may refer to the whole reaction simply as an "oxidation." This is actually common practice in respiration reactions, which are often termed "biological oxidations." However, we must keep in mind that, notwithstanding the incomplete name, every oxidation implies and is accompanied by a reduction. Indeed, it is the reducing half, not the oxidizing half, of any "oxidation" that is the actual source of the energy yield.

In respiration, the transformation of hydrocarbon to anhydride does not occur in a single oxidation-reduction process, as just shown for methane. Instead, respiration takes place through a series of consecutive redox reactions, each resulting in a successively higher oxidation level. One H atom at a time is removed from a carbon atom, and one O atom at a time is added. At each step, therefore, the relative H:O ratio decreases. For example, if methane were to be oxidized in steps, the first step would be removal of one H atom from the carbon of methane and the addition of one O atom:

$$\begin{array}{c} H \\ | \\ H{-}C{-}H \\ | \\ H \end{array} \xrightarrow{+O} \begin{array}{c} H \\ | \\ H{-}C{-}O{-}H \\ | \\ H \end{array}$$

methane *methyl alcohol*

Note here that H need not be taken away altogether but need only be separated from its direct bonding to carbon. Oxygen achieves such a separation, and the resulting methyl alcohol is stabler than the original methane. Also, note that the oxidation state of carbon has changed from -4 to -2. As shown in the first equation of Fig. 5.5, the actual chemical process by which an O atom is interposed between C and H includes an oxidizing half-reaction in which H_2O is added and H_2 is removed. Subsequent combination of the removed H_2 with an appropriate hydrogen acceptor (oxygen in Fig. 5.5) then completes the overall reaction.

The next higher oxidation state of the carbon of methyl alcohol is attained if two H atoms are removed from the molecule entirely. The result, shown in Fig. 5.5, is *formaldehyde.* Carbon here has an oxidation level of zero, and we note that the H:O ratio now is even smaller; it was 4:0 in methane and 4:1 in methyl alcohol, and it is 2:1 in formaldehyde. A still higher oxidation level may be attained if an additional O atom

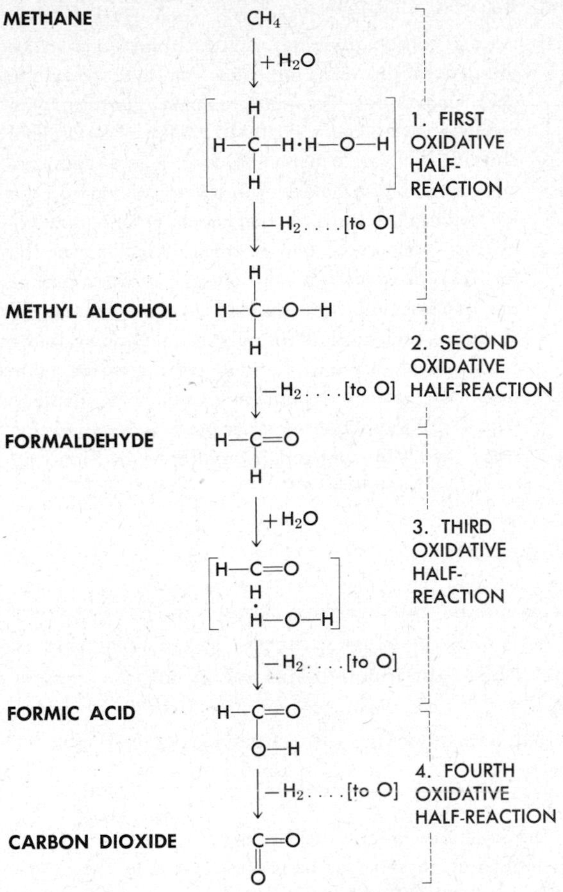

METHANE CH_4

$+H_2O$

$[H{-}C(H)(H){-}H \cdot H{-}O{-}H]$ — 1. FIRST OXIDATIVE HALF-REACTION

$-H_2 \ldots$ [to O]

METHYL ALCOHOL $H{-}C(H)(H){-}O{-}H$ — 2. SECOND OXIDATIVE HALF-REACTION

$-H_2 \ldots$ [to O]

FORMALDEHYDE $H{-}C(H){=}O$

$+H_2O$

$[H{-}C{=}O \;\; H \cdot H{-}O{-}H]$ — 3. THIRD OXIDATIVE HALF-REACTION

$-H_2 \ldots$ [to O]

FORMIC ACID $H{-}C(O{-}H){=}O$ — 4. FOURTH OXIDATIVE HALF-REACTION

$-H_2 \ldots$ [to O]

CARBON DIOXIDE $O{=}C{=}O$

$$CH_4 + 2\,H_2O \longrightarrow CO_2 + 4\,H_2 \quad \text{(SUM OF OXIDATIONS, } +\Delta F)$$

$$4\,H_2 + 2\,O_2 \longrightarrow 4\,H_2O \quad \text{(SUM OF REDUCTIONS, } -\Delta F)$$

$$CH_4 + 2\,O_2 \longrightarrow CO_2 + 2\,H_2O \quad \text{OVERALL REACTION, NET } -\Delta F$$

5.5. The stepwise oxidation of methane. Only the four oxidizing half-reactions are shown separately; at each step an H_2 emerges and combines with oxygen, a process representing the reducing half-reaction of that step. The sum of all oxidizing and reducing half-reactions is indicated at the bottom, as is also the overall reaction. Note that the latter is identical with the 1-step oxidation of methane discussed in the text. Note also that more energy is gained during the reduction than is expended during the oxidation, hence the net ΔF is negative; i.e., CO_2 is more stable than CH_4.

is bonded to the carbon of formaldehyde. This again takes the form of simultaneous addition of H_2O and removal of H_2 (Fig. 5.5, third equation). The endproduct, HCOOH, is *formic acid.* In it, carbon has an oxidation state of $+2$, and the H:O ratio now is 1:1. A final removal of the two H atoms remaining in formic acid leads to the highest attainable oxidation state of carbon ($+4$), i.e., the acid anhydride carbon dioxide (Fig. 5.5, fourth equation). The sum of all four oxidative half-reactions from methane to carbon dioxide is shown in Fig. 5.5, and this figure also indicates the overall result if oxygen is supplied as the hydrogen acceptor for all four reducing half-reactions.

We end up with the same final equation and the same net energy yield as in a single-step oxidation of methane, but here the oxidation has been a four-step process. Such stepwise oxidation occurs generally in the respiration of all foods. Indeed, although methane itself is not a food, the oxidation steps for actual foods are nevertheless the same as those outlined for methane: *the carbon groupings of food molecules are transformed successively from hydrocarbon to alcohol to aldehyde to acid to anhydride* (Fig. 5.6). Moreover, the oxidation states of carbon change successively from -4 to -2 to 0 to $+2$ to $+4$.

We may ask at this point why biological oxidations proceed stepwise at all, and why food molecules should not be completely oxidized to CO_2 in single steps. The answer is that a single-step oxidation would release the entire energy yield as a single large "packet." Most of this energy would then be wasted as heat. As we shall see shortly, cells are not equipped to make metabolic (i.e., chemical) use of single energy quantities which exceed certain magnitudes. Indeed, too intense an evolution of heat would kill a cell. Stepwise oxidations, on the other hand, make energy available in smaller, successive packets, and each of these can become metabolically useful. Moreover, heat wastage is thereby reduced and heat death is avoided.

Certain food molecules possess carbon groupings at various different oxidation levels. For example, one of the carbons of a fatty acid is at the acid level, the others are at the hydrocarbon level; a sugar such as glucose contains one carbon at the aldehyde level, five others at the alcohol level (Fig. 5.7). Oxidations of individual carbon groups here will proceed stepwise from any given starting level until the anhydride stage is attained. Correspondingly, the energy yields will differ according to how many oxidation steps may still occur for each carbon group.

How is oxidation actually accomplished in foods?

Hydrogen Transfer

If fuel dehydrogenation is to take place—and note from Fig. 5.6 that each oxidation step from hydrocarbon to CO_2

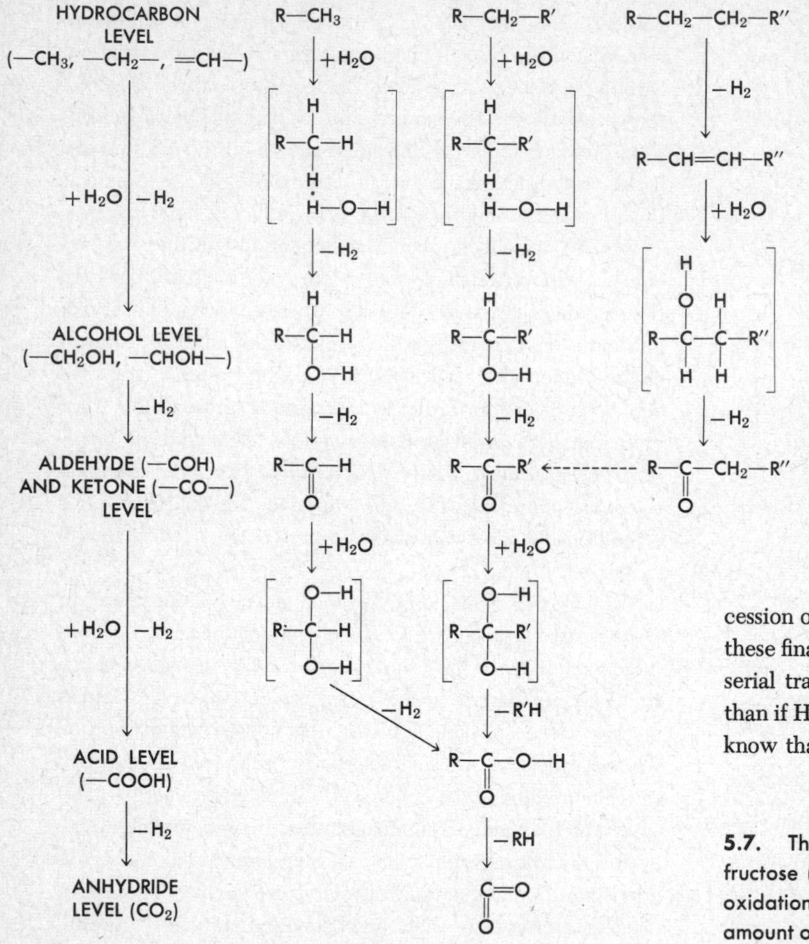

HYDROCARBON LEVEL ($—CH_3$, $—CH_2—$, $=CH—$)

$+H_2O$ | $-H_2$

ALCOHOL LEVEL ($—CH_2OH$, $—CHOH—$)

$-H_2$

ALDEHYDE (—COH) AND KETONE (—CO—) LEVEL

$+H_2O$ | $-H_2$

ACID LEVEL (—COOH)

$-H_2$

ANHYDRIDE LEVEL (CO_2)

5.6. The general pattern of the successive oxidation steps of carbon groupings. Column at far left indicates the general oxidative sequence from hydrocarbon to acid anhydride. The next column illustrates the changes taking place in the oxidation of a terminal hydrocarbon, i.e., a methyl group (—*R* here representing the rest of the molecule). The third column from the left outlines the corresponding steps for a nonterminal hydrocarbon group, and so does the column on the far right, in which two adjacent hydrocarbon groups are nonterminal. Note that, in the latter cases, the oxidative sequences differ only in detail down to the ketone level and become identical thereafter. Note also that, as in the stepwise oxidation of methane (Fig. 5.5), all sequences again involve four reaction steps, two characterized by hydration and all four by dehydrogenation.

actually does require the removal of a hydrogen pair—then two conditions must be fulfilled. First, a specific enzyme must catalyze the reaction. Such enzymes are known as *dehydrogenases.* Each particular food compound requires its own specific dehydrogenase for each specific oxidation step. Secondly, since hydrogen removal as such represents only an oxidizing half-reaction, the process can proceed only if a reducing half-reaction takes place as well. In other words, an appropriate hydrogen acceptor must be present. The ultimate hydrogen acceptor in cells is oxygen. This is why animals must obtain the gas by breathing and must supply it to all cells as a respiratory raw material. Water then becomes one of the final endproducts of respiration (Fig. 5.8).

However, cellular fuels do not release hydrogen to oxygen directly. Instead, H from fuel is first passed along a whole succession of intermediate hydrogen carriers, and only the last of these finally yields hydrogen to oxygen. The advantage of such serial transfers is that more usable energy may be obtained than if H and O were allowed to combine directly. We already know that, as a reducing half-reaction, a combination of hy-

5.7. The structures of fatty acid (top), glucose (middle), and fructose (bottom), to serve as examples indicating the various oxidation levels of different carbon groups in a molecule. The amount of further oxidation still possible for each group depends on its particular oxidation level in the intact molecule.

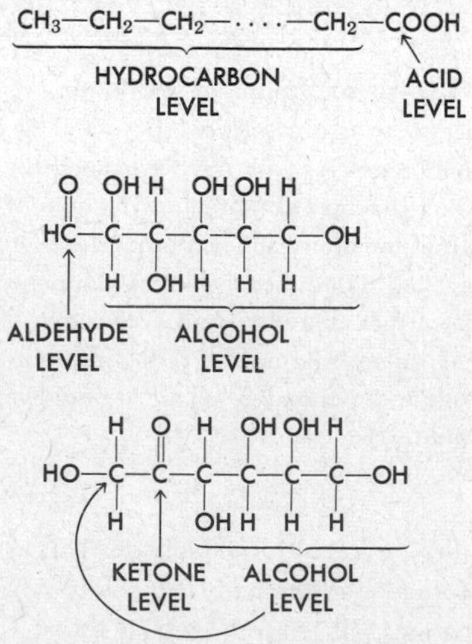

$CH_3—CH_2—CH_2—\cdots—CH_2—COOH$

HYDROCARBON LEVEL

ACID LEVEL

ALDEHYDE LEVEL

ALCOHOL LEVEL

KETONE LEVEL

ALCOHOL LEVEL

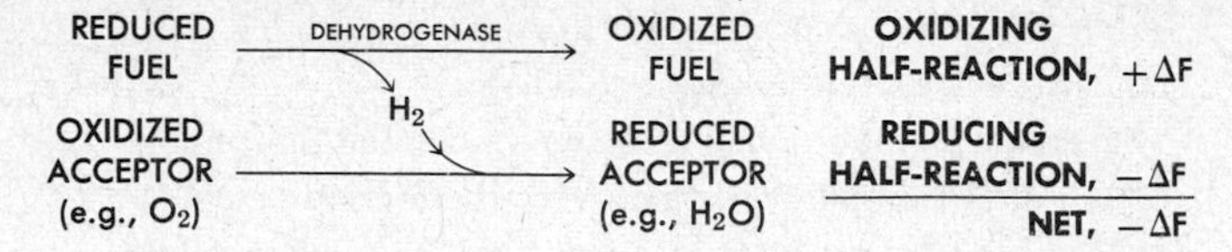

5.8. The general pattern of hydrogen transfer and the energy yield in respiration.

drogen and oxygen is an energy-yielding process ($-\Delta F$, above). That this is so can also be demonstrated readily in the test tube: when mixed in the right proportions, hydrogen and oxygen combine explosively. We should not conclude, however, that a similarly direct combination in cells would lead to explosion of cells; the quantities of gases involved there at any moment would probably be far too small to cause damage. The important conclusion is, rather, that in a direct combination of the two gases any energy released would appear as a single large packet, which would dissipate almost entirely as heat. Thus, H transfer to oxygen undoubtedly takes place in stepwise fashion for exactly the same reasons that fuel oxidation to CO_2 occurs in small steps; smaller, more numerous energy packets become available for metabolic use.

Indeed, like fuel oxidation to CO_2, each step in a serial hydrogen transfer is itself a complete redox process. The general nature of such serial redox reactions is indicated in Fig. 5.9. Note that, just as $A \cdot H_2$ in this figure is more stable than reduced fuel, so $B \cdot H_2$ is more stable than $A \cdot H_2$. Note

5.9. If *A* and *B* represent the first two carriers in a hydrogen transfer chain, then the successive transfers of H_2 from food to *A* to *B* may be symbolized as above. Note that each step of transfer is thermodynamically possible because it is characterized by a negative *net* ΔF.

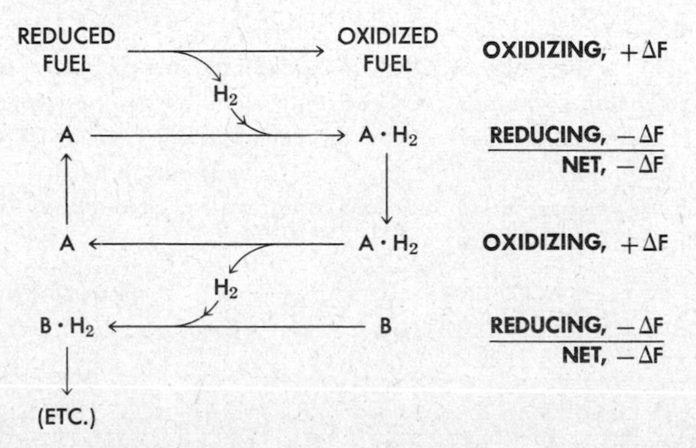

also that the intermediate H carriers are alternately reduced and oxidized; i.e., they operate cyclically, first accepting H_2, then passing it on, then accepting new H_2, etc.

The actual intermediate H carriers are some of the coenzymes already encountered in Chap. 4. The sequence in which such coenzymes actually transfer hydrogen is shown in Fig. 5.10. In this sequence, "Q" is an incompletely known coenzyme, and the cytochromes consist of a family of at least four slightly different variants which operate in succession. They are named, in order, cytochromes c_1, c, a, and a_3.

Note that, after H removal from food, either DPN or TPN may accept hydrogen. This is not a free choice, however, but depends on the nature of specific oxidation steps in specific foods. In certain particular cases, H transfers to DPN, in others to TPN. In the mitochondria of a cell, where most respiratory oxidations take place, it is usually DPN which serves as H carrier after food.

Some of the coenzymes, notably FAD, carry hydrogen in the form of whole atoms. DPN and the cytochromes, however, in part carry *electrons* rather than H atoms. In such cases H atoms dissociate, into H^+ and e^-. The resulting electron combines with the coenzyme, and the H ion remains free in the reaction medium. Subsequently the coenzyme either may transfer the electrons to another coenzyme, or may release the electron back to H^+, in which case a reconstituted whole H atom moves on to the next carrier. In the specific case of DPN, this coenzyme normally exists in the form of DPN^+; i.e., the molecule lacks one electron and is therefore electropositive (Fig. 5.11). When it accepts a hydrogen pair from fuel, one H atom dissociates and the resulting electron neutralizes DPN^+ to DPN; the second H atom remains whole and attaches to DPN. Thus, $DPN \cdot H$ represents the reduced state and an H^+ remains available in the reaction medium. This H^+ is used in the later reoxidation of the coenzyme, when H_2 is released again.

In the case of the cytochromes, both H atoms of a pair dissociate, and each resulting electron is carried by one coenzyme molecule:

$$H_2 \longrightarrow 2\,H^+ \longrightarrow 2\,H^+$$
$$\downarrow 2\,e^-$$
$$2\ \text{cyt.} \longrightarrow 2\ \text{cyt.}\ e^-$$

The active H-carrying component of a cytochrome molecule is the iron in the center of its tetrapyrrol ring. This iron normally exists in the form of Fe^{+++} (ferric ion). When it accepts

FOOD
H_2 → DPN or TPN → H_2 → FAD → H_2 → Q → H_2 → CYTOCHROMES → H_2 → OXYGEN → WATER

5.10. The basic sequence of coenzymes in respiratory hydrogen transfer from foods, leading to the ultimate production of water. In this series, Q stands for a carrier which is still known only poorly; it is sometimes referred to as "coenzyme Q" and it probably belongs to the chemical category of quinone compounds. A more detailed illustration of the steps in hydrogen transfer is given in Fig. 5.13.

an electron, therefore, it becomes Fe^{++} (ferrous ion); i.e., its electropositivity is reduced by one charge-unit:

$$Fe^{+++} + e^- \longrightarrow Fe^{++}$$

The pattern of hydrogen transport between two successive cytochromes is illustrated in Fig. 5.12. Note here especially that, like H transfer itself, electron transfers likewise are redox reactions. Electron loss is oxidation, electron gain, reduction. This holds not only for iron but quite generally. For example, the dissociation $H \longrightarrow H^+ + e^-$ involves electron loss by H, and H is therefore oxidized. Indeed, it is because H atoms contain electrons that H transfers are redox processes to begin with; every H removal or addition implies, and includes, removal or addition of an electron. The energy changes in redox reactions ultimately always result from actual or "hidden" electron transfers, and it is this which is the key to any oxidation or reduction. The cytochromes evidently undergo redox changes in the most basic form, whereas in foods and in coenzymes like FAD the essential redox agent, i.e., the electron, remains hidden within H atoms. We may conclude generally that the real significance of any hydrogen transfer is electron transfer.

The last cytochrome in the series, a_3, also known as *cytochrome oxidase,* releases electrons back to H^+, which has remained in the reaction medium; whole H atoms then combine with oxygen, forming water. The entire sequence of H transport is summarized in Fig. 5.13.

Aerobic and Anaerobic Transfer

Because H transport as above requires atmospheric oxygen as the ultimate hydrogen acceptor, this pattern of transfer is said to define an *aerobic* (with air) form of respiration. It is the standard, universal form among animals.

Several observations may be made concerning this aerobic pattern of H transport. First, as everywhere else in metabolism, each of the transport reactions must be catalyzed by a specific enzyme. We may note also that vitamins C, E, and K are known to be required in the reactions, although the precise role of these compounds is still obscure.

Secondly, it is clear that relatively small quantities of the hydrogen carriers suffice to transport comparatively large quantities of hydrogen; each carrier molecule functions cyclically and may be used repeatedly.

Thirdly, if any one of the reactions is stopped, the whole transport system becomes inoperative, and the energy it normally supplies cannot be obtained. Reaction blocks may occur in a number of ways. For example, *inhibitor* substances of various kinds may interfere specifically with given transport reactions. Thus, potassium cyanide specifically inhibits the cytochrome system, and this is why cyanide is such a violent poison. Another form of reaction block is produced if one of the carriers is in deficient supply. For example, inasmuch as riboflavin (vitamin B_2) is a structural part of the FAD molecule, a consistently riboflavin-deficient diet would soon impair the reactions in which FAD participates.

Any such reaction barrier introduced into the transport sequence will act like a roadblock and will lead to an accumulation of hydrogen back of the barrier. For example, if the cytochrome system is blocked, $FAD \cdot H_2$ cannot get rid of its hydrogen. All available FAD will then soon hold H_2 to capacity, and none will be free to accept more H_2 from DPN. Therefore, DPN cannot get rid of its own hydrogen, and free DPN will no longer become available to accept more hydrogen from fresh fuel. Respiration will be effectively stopped.

To be sure, cyanide poisoning and vitamin-B deficiencies

5.11. The role of DPN in accepting and subsequently releasing hydrogen. Of each H_2 yielded by food, one whole H atom and an electron from the other H atom join with DPN^+, resulting in neutral $DPN \cdot H$. An H^+ ion remains in the medium. In H release, these reactions are reversed. Note that TPN functions analogously.

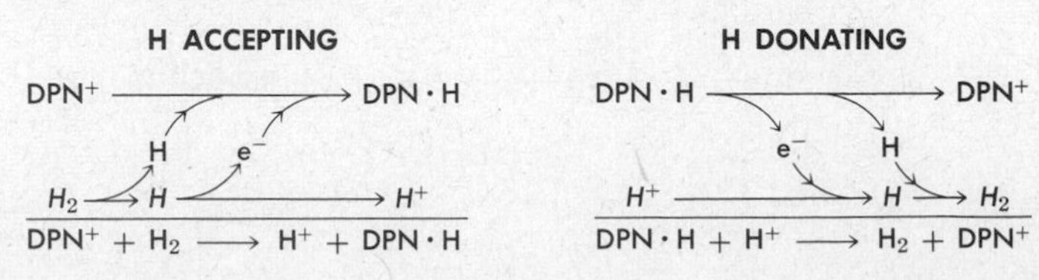

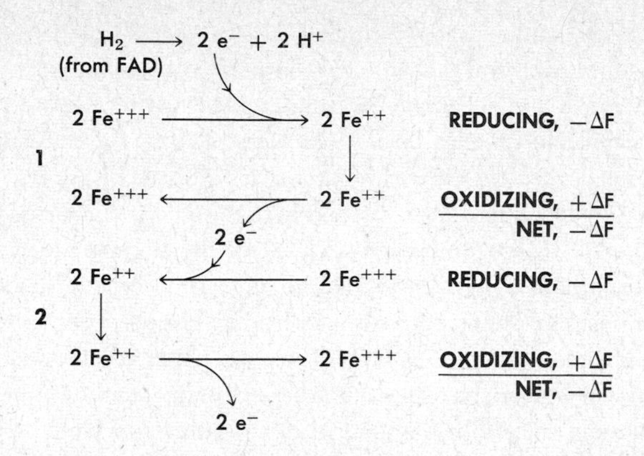

5.12. The pattern of H transfer between the iron components of two successive cytochromes (1 and 2). Each time a (negative) electron is gained, the positive charge on iron decreases; and each time an electron is lost, the positive charge increases. Note particularly that, like transport of whole H atoms, electron transport is likewise a redox process; electron gain is a reduction, electron loss, an oxidation. Thus, electron transport is characterized thermodynamically by a negative net ΔF.

are not particularly common hazards in the life of an animal, and most animals would probably survive quite well even without special protective adaptations against such contingencies. But there is one hazard affecting H transport which all animals do have to cope with quite frequently: unavailability of enough atmospheric oxygen. Lack of oxygen is a reaction barrier of the same sort as cyanide poisoning or vitamin deficiencies. The consequence is a damming up of hydrogen all the way back to DPN, with the further consequence that respiration as a whole becomes blocked.

Whenever oxygen supplies are inadequate or whenever hydrogen transport to oxygen is otherwise blocked, animals may respire in a way which does not require oxygen. This is *anaerobic respiration*, or *fermentation*, probably a more ancient form of energy production than the aerobic kind (cf. Chap. 14). Under conditions of oxygen deficiency, this anaerobic type of respiration may become a substitute or a subsidiary source of energy.

The principle of anaerobic hydrogen transport is relatively simple: with the path from DPN to FAD and oxygen blocked, another path, from DPN to another hydrogen acceptor, must be used. Such an alternative path is provided by *pyruvic acid*, $CH_3COCOOH$, one of the compounds normally formed in the course of carbohydrate oxidation (see below). If oxygen is amply available, pyruvic acid is merely one of the intermediate steps in the decomposition of carbohydrates. In other words, it is a fuel which, in the presence of oxygen, may be oxidized further to CO_2. But pyruvic acid has the property of reacting readily with hydrogen. And if DPN cannot use its normal hydrogen outlet to FAD, pyruvic acid is used instead. The acid then ceases to be a fuel and becomes a hydrogen carrier.

When pyruvic acid reacts with hydrogen, the result in animals is the formation of *lactic acid:*

$$DPN \cdot H + H^+ \longrightarrow DPN^+$$
$$\downarrow H_2$$
$$\underset{\textit{pyruvic acid}}{CH_3COCOOH} \longrightarrow \underset{\textit{lactic acid}}{CH_3CHOHCOOH}$$

This reaction in effect completes anaerobic respiration; lactic acid accumulates in a cell and eventually diffuses into the cellular surroundings. However, the energy gained

5.13. Summary of hydrogen transfer via coenzymes to oxygen. Note (1) that the carrier Q is probably interposed within the cytochrome transfer sequence, as indicated; (2) that cytochromes c_1, c, and *a* actually function individually and in succession, though they are shown below to operate as a group as a space-saving convenience; (3) that cytochrome a_3 is cytochrome oxidase; and (4) that all cytochromes transport electrons, not whole H atoms, that DPN transports partly electrons, partly H atoms, and that FAD, Q, and O transport whole H atoms only.

H_2 | DPN^+ | $FAD \cdot H_2$ | 2 cyt. b^{+++} | $Q \cdot H_2$ | 2 cyt. c_1, c, a^{+++} | $2H^+$ + 2 cyt. a_3^{++} | O

$DPN \cdot H$ + H^+ | FAD | 2 cyt. b^{++} + $2H^+$ | Q | 2 cyt. c_1, c, a^{++} + $2H^+$ | 2 cyt. a_3^{+++} | H_2O

through fermentation by itself is insufficient in most cases to sustain the life of an animal cell. Because oxygen is unavailable, the energy normally obtained through H transfer to oxygen cannot be realized. Moreover, anaerobic fuel oxidation has stopped at the pyruvic acid stage, and the potential energy still contained in this acid therefore remains untapped and locked in lactic acid. As we shall see below, fermentation actually yields only about 5 per cent of the energy obtainable by aerobic respiration. This is too little to maintain animal life; as is well known, complete absence of oxygen leads to death within minutes. (Animals generally cannot store oxygen. At any given moment only enough of the gas is in the body to suffice for immediate needs.)

But even though fermentation alone cannot sustain life, it may supplement the aerobic energy gains. Whenever energy demands are high, as during intensive muscular activity, the oxygen supply to the cells may become insufficient despite faster breathing. Under such conditions of *oxygen debt,* fermentation may proceed in parallel with aerobic respiration and provide a little extra energy. Lactic acid then accumulates, particularly in the muscles. This increase in concentration appears to be associated specifically with fatigue. Eventually, fatigue becomes so great that the animal must cease its intensive activity. During an ensuing rest period, faster breathing at first continues. The oxygen debt is thereby being repaid, and the extra oxygen permits the complete oxidation of the accumulated lactic acid. However, note that mammalian muscles do not use lactic acid as fuel directly. Instead, the acid diffuses from muscles into the blood, is carried by blood into the liver, and here the acid is converted to glucose. This sugar then passes back via the blood to the muscles, where it is transformed to glycogen. Finally, under aerobic conditions, muscles may oxidize glycogen to pyruvic acid and, ultimately, to CO_2. Thus, the potential energy still present in lactic acid need not be lost permanently to a fermenting cell. With the gradual disappearance of lactic acid, fatigue decreases and the breathing rate slows back to normal (Fig. 5.14).

5.14. If the path of H_2 to oxygen is blocked, pyruvic acid cannot be respired to CO_2 (dashed lines). Fermentation, or anaerobic respiration, then may take place, a process in which pyruvic acid accepts H_2 and becomes lactic acid. If produced in muscle, this acid passes via blood to the liver, becomes glucose there, and eventually returns via blood to muscle where it becomes glycogen. After the block to oxygen is no longer present, such glycogen may be respired completely to CO_2 (via pyruvic acid, which in that case is not required as a hydrogen acceptor).

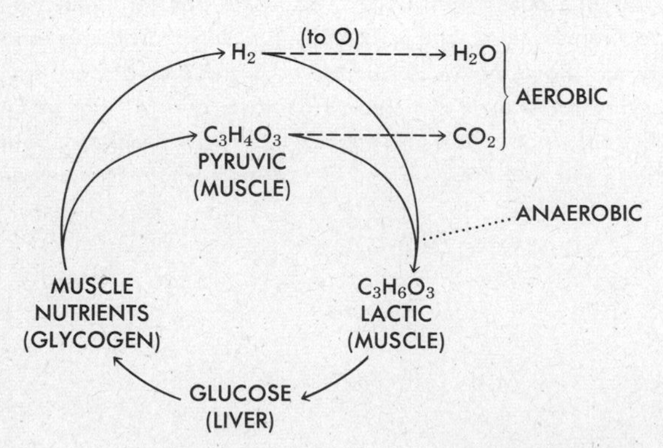

Up to this point, we have found that respiration consists of stepwise fuel oxidation accomplished through dehydrogenation and of hydrogen transport either to pyruvic acid or, more usually, to oxygen. The latter process as well as fuel oxidation itself constitutes the sources of the respiratory energy yields. However, we have so far considered energy yields only in general terms. How are such yields made useful, and how much of the yield can actually become useful?

Energy Transfer

Biological oxidations take place without appreciable increases in cellular temperatures, because the energy yields do not accumulate primarily as heat. Instead, they are harvested largely in the form of high-energy bonds within ATP:

$$\text{—O—P} \xrightarrow{\textit{respiratory energy}} \text{—O}\sim\text{P}$$

$$\underset{ADP}{\text{adenosine—O—P—O}\sim\text{P}} \longrightarrow \underset{ATP}{\text{adenosine—O—P—O}\sim\text{P—O}\sim\text{P}}$$

This scheme clearly implies, that, before —O ~ P groups can be transferred to ADP to form ATP, such groups must first be created. They are indeed created in respiration. Two general events take place. First, in a preparatory process, a fuel molecule is *phosphorylated,* i.e., a phosphate group, —O—P, is attached to it. Secondly, the phosphorylated fuel is oxidized, i.e., dehydrogenation occurs. We already know that this process yields energy, but inasmuch as an —O—P group is now present on the fuel molecule, the energy is not released as heat. Instead, the energy becomes redistributed within the fuel molecule and becomes incorporated directly into the

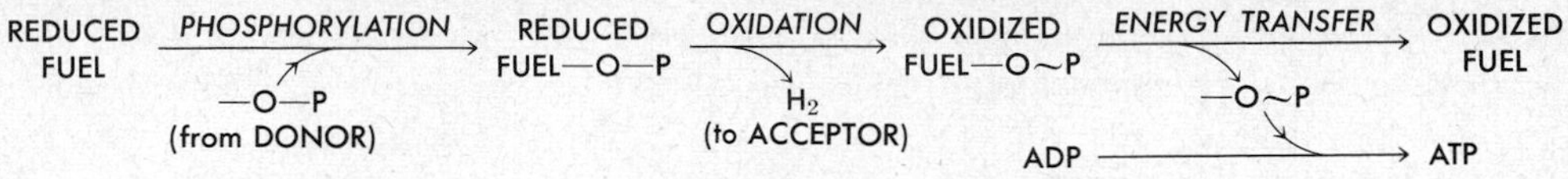

REDUCED FUEL —PHOSPHORYLATION→ REDUCED FUEL—O—P —OXIDATION→ OXIDIZED FUEL—O~P —ENERGY TRANSFER→ OXIDIZED FUEL

(—O—P from DONOR) (H_2 to ACCEPTOR) (—O~P; ADP → ATP)

5.15. The general pattern of high-energy bond creation by phosphorylation and oxidation, and of transfer of such bonds to ADP.

phosphate bond, without being "released" at all. The properties of this bond happen to be such that it can hold extra energy, more than is needed merely to join P to O. The phosphate, —O—P, so becomes a high-energy phosphate, —O ~ P. The latter is then transferred to ADP (Fig. 5.15).

Consider, for example, an actual aldehyde-to-acid oxidation:

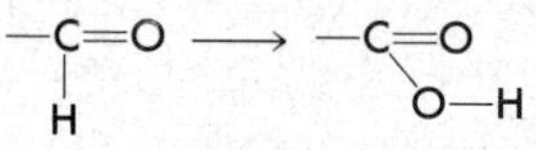

—C(H)=O ⟶ —C(=O)—O—H

Under biological conditions as just noted, the first event is a preliminary phosphorylation of the aldehyde. The phosphate donor here can be inorganic phosphoric acid, H—O—P:

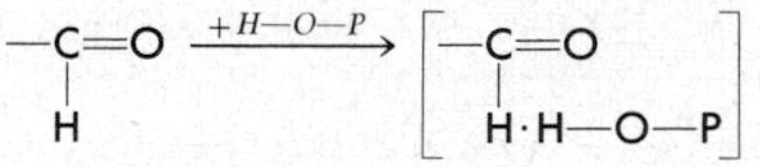

—C(H)=O —(+H—O—P)→ [—C(=O)H·H—O—P] (1)

The next step (which takes place almost simultaneously) is oxidation by hydrogen removal; the energy yield of this process appears as a high-energy bond:

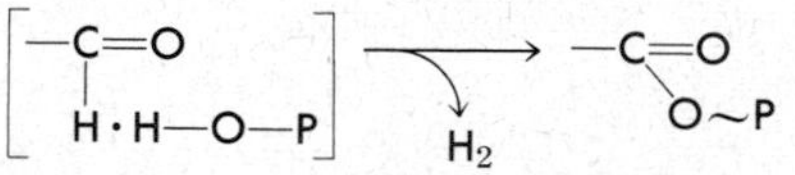

[—C(=O)H·H—O—P] ⟶ (H_2) —C(=O)—O~P (2)

The resulting hydrogen may be accepted by DPN, for example. The final step is transfer of the high-energy phosphate to ADP, in exchange for an —O—H group:

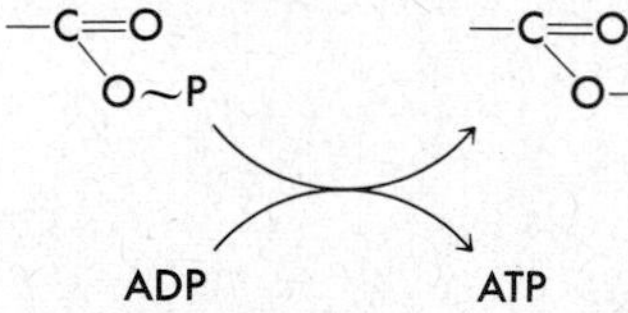

—C(=O)—O~P + ADP ⟶ —C(=O)—O—H + ATP (3)

The original aldehyde thus has become transforme... ..to an acid, —COOH, and ADP has become ATP. As we shall see shortly, this particular series of reactions occurs normally at one point in carbohydrate respiration. But we may note that the pattern is analogous in principle for any other biological oxidation.

In the example above, the phosphate donor for the preparatory phosphorylation has been phosphoric acid. This is relatively unusual, and in most cases the phosphate donor actually is ATP. We have already found earlier, for example, that glucose uptake into a cell is accompanied by phosphorylation and that ATP is the phosphorylating agent. In this and all analogous instances, the terminal high-energy phosphate of ATP is split off (and ATP thus becomes ADP), but an ordinary low-energy phosphate is added on to the fuel molecule:

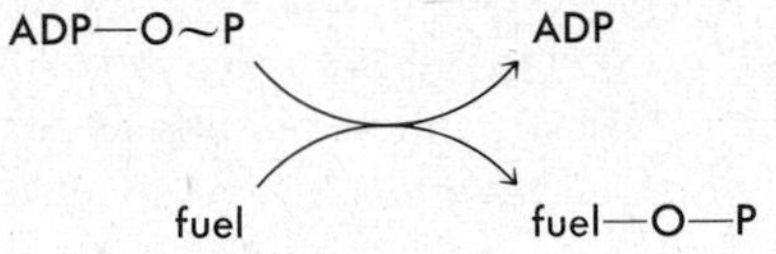

ADP—O~P + fuel ⟶ ADP + fuel—O—P $-\Delta F$, overall

In other words, ATP supplies not only phosphate, but also more than enough energy to produce a phosphorylated compound. Any energy excess here dissipates unavoidably as heat. Many phosphorylations can be achieved only by ATP, even though not all the energy supplied by ATP is being used. Note, furthermore, that ATP is the primary endproduct of respiration but that respiration cannot proceed unless ATP is available for preliminary phosphorylations; some of the endproduct is needed at the starting point, and prior respiration thus becomes a necessary condition for further respiration.

How much ATP can actually be formed in respiration? The quantity varies with the nature of the fuel, i.e., the number of possible oxidation steps. Moreover, it is important to realize that every single oxidation step does not necessarily yield a large enough amount of energy sufficient for the formation of ATP. It can be shown that the conversion of ADP to ATP requires on the average 7000 calories per gram-molecular weight (mole) of ADP. In other words, 7000 calories represent a kind of minimum energy packet in metabolism, and any oxidation which yields less than that will not result in ATP formation. Many oxidations actually yield more; in fuel oxidations, for example, energy yields range from about 5000 to about 11,000 calories, with an average of about 8000 calories, per mole of fuel oxidized. Most of such oxidations may result in ATP formation (and any energy excess dissipates as heat). On the other hand, several oxidation steps in hydrogen transport to oxygen produce less than the minimum energy amounts required for ATP formation. It can be shown that, for every pair of hydrogen atoms transferred from DPN to oxygen, a net total of three ATP molecules is formed. The steps at which these arise are indicated in Fig. 5.16, which also outlines the incompletely known mechanism of ATP creation at these steps.

Fig. 5.17 may be regarded as a comprehensive summary of the general pattern of respiration. The process always includes three principal phases: (1) *fuel decomposition,* which takes the form of phosphorylation and oxidation and which results in the endproduct CO_2; (2) *hydrogen transfer* from fuel to oxygen (or to pyruvic acid), which results in the endproduct H_2O (or lactic acid); and (3) *energy transfer* from fuel to ADP, which results in the endproduct ATP. The first two phases are the sources of usable energy. The second and third phases are the same for all kinds of fuels, and respiratory events differ principally according to the different specific fuels used as raw material. We now proceed to examine some of these actual reaction pathways by which given fuels are oxidized.

THE PROCESS

In the course of being oxidized progressively, certain carbon groupings of a fuel molecule eventually become CO_2, and this gas escapes into the environment. Thus, fuel molecules containing given numbers of carbon atoms at the start lose their carbons one at a time, until sooner or later the entire molecule will have been converted into C_1 fragments, or CO_2. If we follow this decomposition sequence backward, the next-to-last stage in fuel breakdown should be a 2-carbon C_2 fragment. This is the case; every fuel molecule sooner or later appears as a 2-carbon fragment, which in fact is always the same regardless of what the starting fuel may have been. In a last step, the C_2 fragment is then decomposed into two C_1 fragments, CO_2. The common C_2 stage for all fuels in respiration is a derivative of acetic acid called *acetyl,* $CH_3C{=}O$. It does not exist as such by itself but is attached to a carrier coenzyme, viz., the sulfur-containing coenzyme A, or CoA (cf. Chap. 4). The resulting complex is acetyl CoA, which may be symbolized as $CH_3CO{\sim}S{-}CoA$. The acetyl portion of this complex subsequently yields 2 CO_2.

The manner in which the acetyl CoA stage is reached differs for different types of fuels. For example, many carbohydrates are first broken up into 3-carbon compounds. Complex carbohydrates often are built up from 3-carbon units, and their carbon numbers then are whole multiples of 3. This holds, for example, for glucose and all other 6-carbon sugars, for 12-carbon disaccharides, and for polysaccharides such as glycogen. As we shall see, when any of these are used as respiratory fuels, the original 3-carbon units reappear in the course of breakdown. Many other organic substances, glycerin, for example, are 3-carbon molecules to begin with. All such C_3 compounds are eventually converted to *pyruvic acid* ($C_3H_4O_3$). This acid is the common representative of the 3-carbon stage in respiration. Pyruvic acid subsequently loses one carbon in the form of CO_2 and becomes acetyl CoA.

Fatty acids and related molecules consist of long, even-numbered carbon chains. These do not break up into 3-carbon units but become 2-carbon units directly. Other fuels are 2-

5.16. One respiratory source of ATP is fuel oxidation itself; another is H transport to oxygen, which yields high-energy phosphates in three places of the carrier sequence, as indicated. This second energy source thus yields 3 ATP for every H_2 transferred to oxygen.

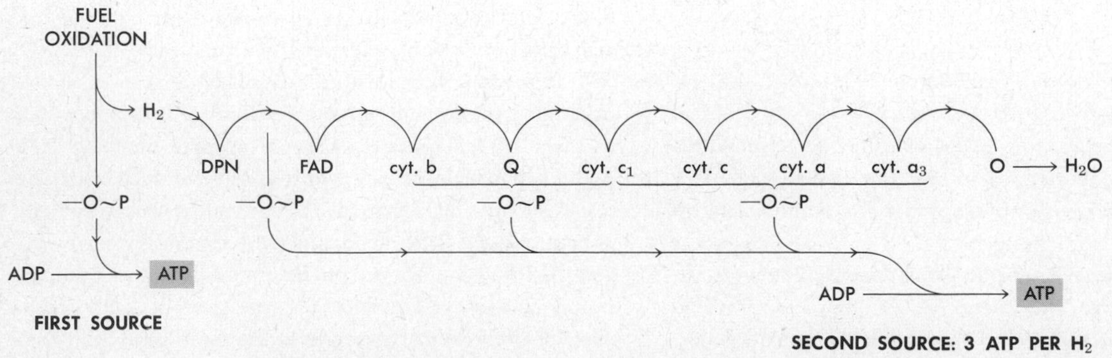

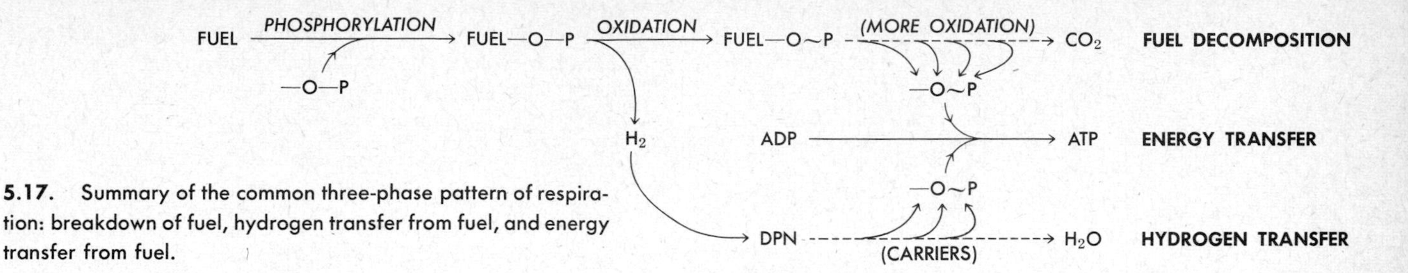

5.17. Summary of the common three-phase pattern of respiration: breakdown of fuel, hydrogen transfer from fuel, and energy transfer from fuel.

carbon molecules to begin with, and all such C_2 compounds eventually appear as acetyl CoA. Amino acids break down partly to pyruvic acid (which subsequently becomes acetyl CoA), partly to acetyl CoA directly. This holds also for many other organic substances which may happen to be used as fuel.

Thus, the overall pattern of aerobic fuel combustion may be likened to a tree with branches or to a river with tributaries (Fig. 5.18). A broad main channel is represented by the sequence pyruvic acid ⟶ acetyl CoA ⟶ carbon dioxide. Numerous side channels lead into this sequence, some funneling into the 3-carbon pyruvic acid step, others into the 2-carbon acetyl CoA step. The side channels themselves may be long or short, and each may have smaller side channels of its own. In the end, the flow from the entire system drains out as 1-carbon carbon dioxide.

Formation of C_3: Pyruvic Acid

The starting point of this sequence, often referred to as *glycolysis,* may be considered to be glucose-6-phosphate. Free glucose becomes glucose-6-phosphate with the phosphorylating aid of ATP, as noted earlier (cf. Fig. 5.3). Glycogen, a main fuel in many animal cell types, is phosphorylated by inorganic phosphates. Glycogen thereby breaks up into numerous molecules of glucose-1-phosphate, and the latter are then converted enzymatically to glucose-6-phosphate (Fig. 5.19).

The subsequent glycolytic fate of glucose-6-phosphate is charted in Fig. 5.20. Note here that glucose-6-phosphate is first transformed into fructose-6-phosphate, a stage also reached directly from free fructose. Phosphorylation (by ATP) of fructose-6-phosphate at the 1 position then results in fructose-1,6-diphosphate. The latter subsequently splits between carbons 3 and 4, yielding two slightly different C_3 molecules. These two become rearranged into two identical molecules, glyceraldehyde phosphate (or phosphoglyceraldehyde, or PGAL). Up to this point, all steps have been in the nature of preparatory reactions. Although the alcoholic carbons at positions 3 and 4 of the original glucose are now aldehyde carbons at position 1 of PGAL, ATP-forming oxidations have not yet taken place.

5.18. Some of the main pathways in the aerobic combustion of fuels. Pyruvic acid, acetyl CoA, and carbon dioxide form a main sequence which other pathways join, like branches of a tree.

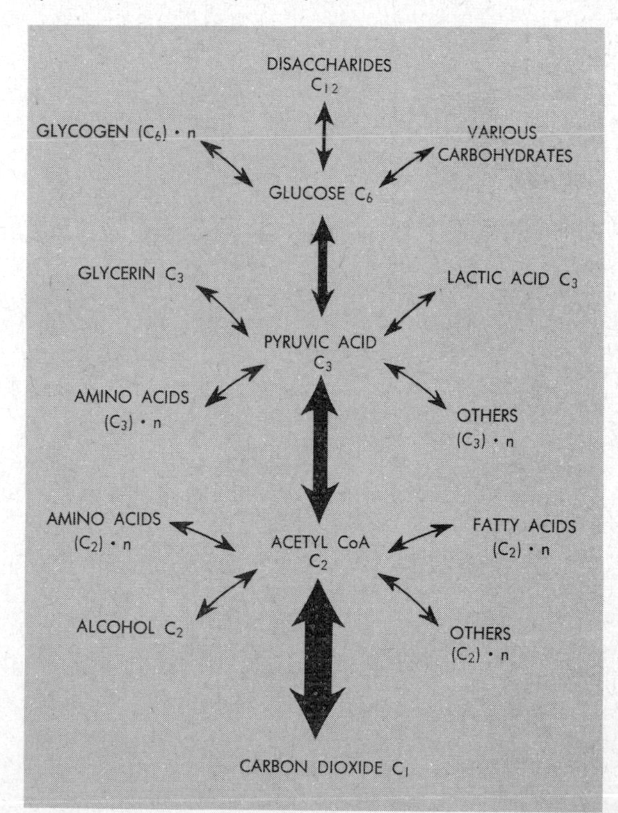

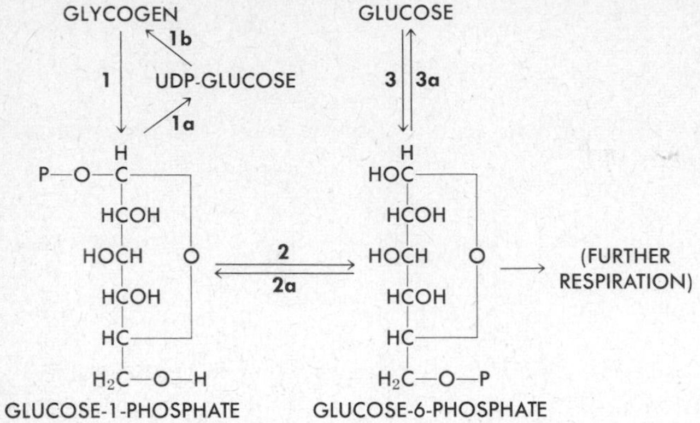

GLYCOGEN
1b
1 UDP-GLUCOSE
1a
GLUCOSE-1-PHOSPHATE
2
2a
GLUCOSE
3 3a
GLUCOSE-6-PHOSPHATE
(FURTHER RESPIRATION)

5.19. The reversible conversions of glycogen and glucose to glucose-6-phosphate. Numbers refer to the enzymes and energy sources required for each reaction, as follows: 1. glycogen phosphorylase; 1a. uridine triphosphate (UTP); 1b. transferase; 2. glucophosphokinase; 2a. glucophosphomutase; 3. ATP, hexokinase; 3a. phosphatase. Note that, in these particular instances, the forward reactions (*1, 2, 3,*) are catalyzed by enzymes differing from those of the reverse reactions (*1a, b, 2a, 3a*). Note also that each glycogen molecule produces many molecules of glucose-1-phosphate.

The next series of steps does produce ATP. PGAL is first phosphorylated at position 1, the phosphate donor being inorganic phosphoric acid (step 5 in Fig. 5.20). Simultaneously, dehydrogenation occurs in the presence of DPN^+, and a high-energy bond is thereby created at position 1 (1,3-diphosphoglyceric acid). The high-energy phosphate is then transferred to ADP, resulting in ATP. Events here clearly correspond to those in the example used earlier, in the section on energy transfer. The compound so formed in 3-phosphoglyceric acid (PGA), and we note that the *aldehyde* carbon of PGAL has been oxidized to the *acid* carbon of PGA.

In a following sequence, the alcoholic carbon at position 2 of PGA becomes oxidized. As a preliminary to this, the phosphate group at position 3 of PGA is shifted enzymatically to position 2, resulting in 2-phosphoglyceric acid. Next (step 8, Fig. 5.20), *oxidative dehydration* takes place: H is removed from position 2 and —OH is removed simultaneously from position 3. In other words, a hydrogen pair is extracted along with an O atom, and since oxygen is a perfect carrier for hydrogen, DPN^+ is here not required. The result is water and phosphoenolpyruvic acid, which contains a high-energy phosphate at position 2. When the latter is then transferred to ADP, the endproducts are ATP and pyruvic acid. This acid contains a ketone group at position 2; i.e., the *alcohol* carbon of PGA has been oxidized to a *ketone* carbon in pyruvic acid. We already know from Fig. 5.6 that ketones have essentially the same oxidation level as aldehydes, i.e., one level higher than alcohols.

The formation of pyruvic acid terminates glycolysis; a quantitative summary of the whole sequence is given in Fig. 5.21. If free glucose is considered to be the original raw material, we note that four phosphorylations have occurred, two by H—O—P and two by ATP. Each of the four added phosphates eventually becomes a high-energy phosphate; and of the four ATP then formed, two "pay back" for the two expended at the start of the sequence, while two represent the net gain. The fate of the atoms in glucose may be described by the equation

$$C_6H_{12}O_6 \longrightarrow 2\ C_3H_4O_3 + 2\ H_2$$

Thus the net loss of atoms from glucose amounts to 2 H_2, and these are held by DPN.

If respiration occurs under anaerobic conditions, pyruvic acid must now serve as the final hydrogen acceptor. Carbohydrate oxidation in this case stops with the formation of lactic acid, and the two ATP gained represent the net energy yield of the entire process.

But if conditions are aerobic, two desirable consequences supervene. First, the 2 H_2 held by DPN may be passed on to oxygen. As noted earlier, this transfer yields three additional ATP molecules per H_2, or 6 ATP total. Secondly, since pyruvic acid need not serve as a hydrogen carrier, it may be oxidized further. Most of the energy of the original glucose is actually still untapped. As we shall see, complete combustion of pyruvic acid will supply many more ATP molecules than have formed thus far.

Formation of C_2: Acetyl CoA

As pointed out above, the 2-carbon acetyl stage of respiration arises principally in two ways, viz., via compounds which first form pyruvic acid and via compounds such as fatty acids, which decompose to C_2 fragments directly.

Pyruvic acid $\longrightarrow$ *acetyl CoA.* The transformation of pyruvic acid to acetyl CoA may be described as an *oxidative decarboxylation;* it is oxidative inasmuch as hydrogen is removed, and it is a decarboxylation inasmuch as CO_2 is removed. The source of the CO_2 is the —COOH group at position 1 of pyruvic acid. This is a group which is originally alcoholic in glucose, then becomes aldehydic in PGAL, next becomes acidic in PGA, and which is still acidic in pyruvic acid (cf. Fig. 5.20). Now the group is oxidized by dehydrogenation to its highest level, i.e., the anhydride CO_2.

Conversion of —COOH to CO_2 requires the participation

5.20. Glycolysis. In the nine steps of this sequence, the four in the upper row include rearrangements and a preliminary phosphorylation. In the second row, steps 5 and 6 constitute a first oxidation and energy transfer, i.e., carbon 1 of glyceraldehyde-3-phosphate is simultaneously phosphorylated and oxidized. That carbon thereby changes from an *aldehyde* level to an *acid* level (in 3-phosphoglyceric acid). In this product of reaction 6, carbon 2 is at an *alcohol* level; in the remaining reactions this carbon is phosphorylated and then oxidized to the *ketone* level (in pyruvic acid). Thus, of the original glucose, carbon positions 2, 3, 4, and 5 are each oxidized once during glycolysis. The following listing indicates the enzymes participating specifically at each step and the basic actions of these enzymes. 1. phosphohexose isomerase, catalyzes rearrangement; 2. phosphofructokinase, adds —O—P in 1 position; 3. aldolase, splits chain between carbons 3 and 4; 4. phosphotriose isomerase, makes top compound same as one at bottom; 5. glyceraldehyde dehydrogenase, removes H_2 after phosphoric acid addition; 6. phosphoglyceric kinase, transfers —O∼P to ADP; 7. phosphoglyceromutase, switches phosphate from 3 to 2 position; 8. enolase, catalyzes dehydration with —O∼P formation; 9. pyruvic kinase, transfers —O∼P to ADP.

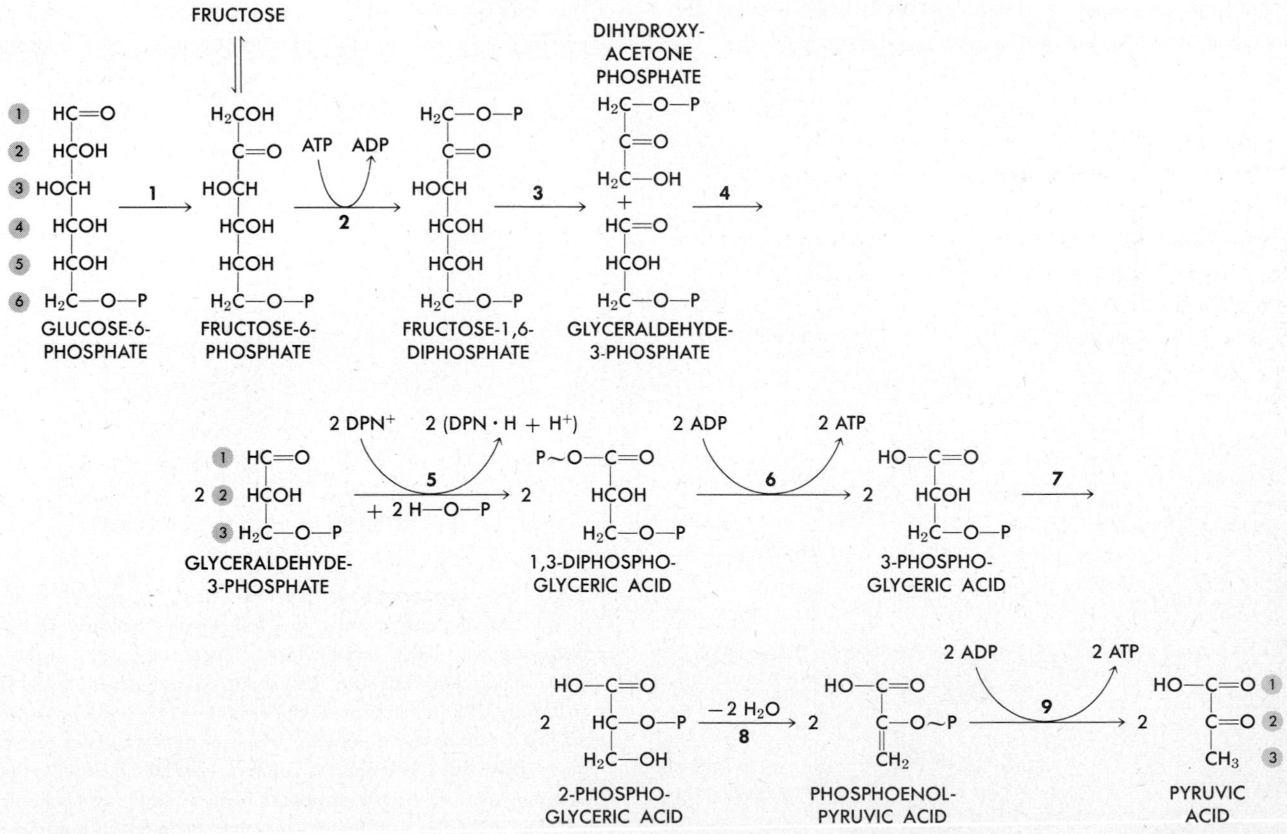

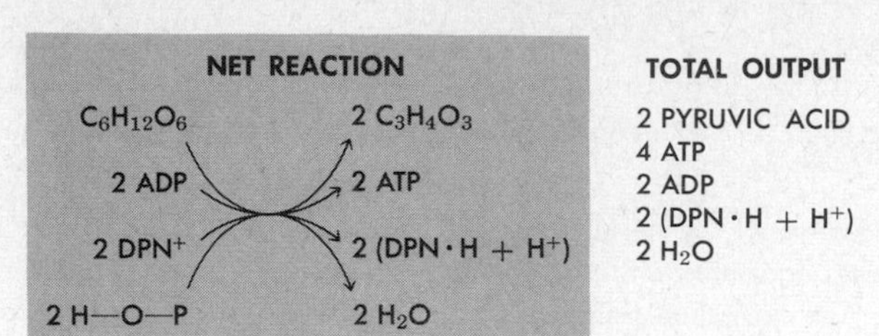

5.21. Quantitative summary of glycolysis. It is assumed here that glucose is the original starting fuel, i.e., one ATP must be expended and one ADP gained to convert glucose to glucose-6-phosphate. From that stage on, all inputs and outputs can be verified from Fig. 5.20. The summary *net* reaction then is as indicated in the center panel.

of a specific enzyme, *carboxylase,* as well as a specific coenzyme, *cocarboxylase.* The latter is a derivative of *thiamine,* or vitamin B_1. Thiamine reacts in cells with ATP to yield *thiamine pyrophosphate,* or cocarboxylase:

$$\text{thiamine} + \text{ATP} \longrightarrow \text{thiamine—O}{\sim}\text{P—O}{\sim}\text{P} + \text{AMP}$$

The coenzyme functions as a temporary CO_2 carrier after carboxylase catalyzes the extraction of CO_2 from —COOH (Fig. 5.22, reaction 1). We may note here, for later reference, that pyruvic acid is one of several kinds of *α-keto acids;* as in fatty acids, the α position is the carbon next to the acid group, and in pyruvic acid this α carbon (alternatively, position 2) is a ketone. Other α-keto acids similarly are of the type R—COCOOH, where R— may be one of a large variety of different groupings. All α-keto acids readily lose CO_2 from the acid group, and in such cases carboxylase, cocarboxylase, and Mg^{++} are likewise required. Moreover, the subsequent reactions of the resulting aldehydes also are quite similar to those of *acetaldehyde* formed in reaction 1, Fig. 5.22.

Acetaldehyde next is dehydrogenated to acetyl, CH_3CO—. This reaction specifically requires the presence of *coenzyme A,* CoA—S—H, which forms a sulfur-carbon link with acetyl. Like a —O—P— bond, the —S—C— bond similarly may incorporate excess energy and may so become a high-energy bond. This happens during the dehydrogenation of acetaldehyde (Fig. 5.22, reaction 2a). The hydrogen acceptor required

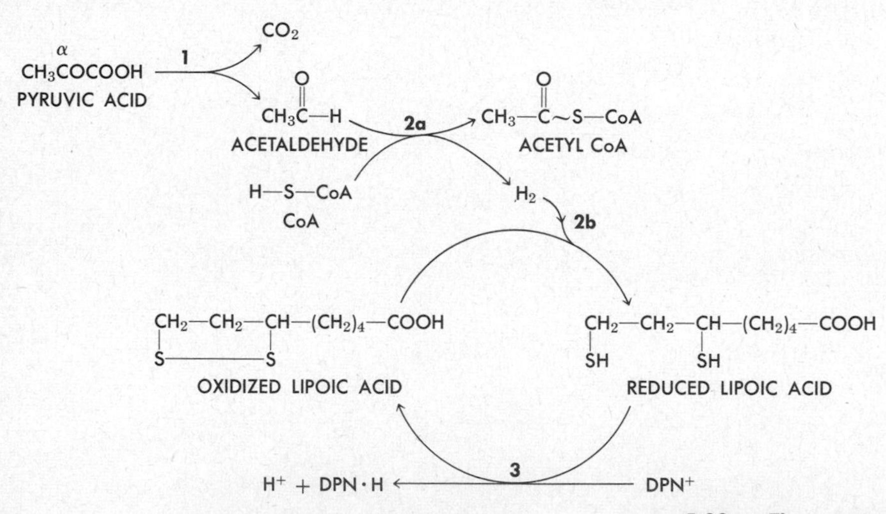

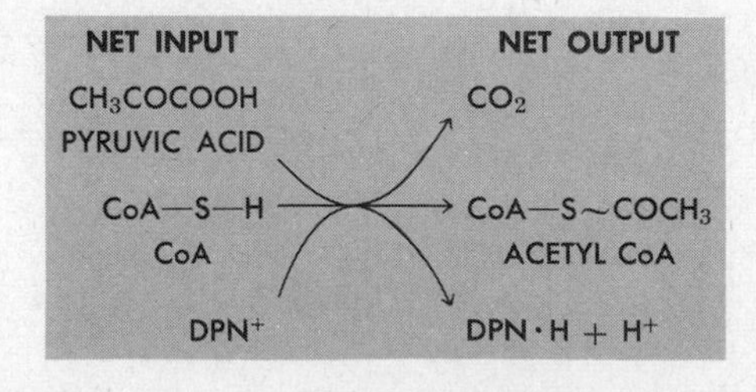

5.22. The conversion of pyruvic acid to acetyl CoA. The numbered steps identify the following reactions: 1. decarboxylation (with carboxylase, cocarboxylase, Mg^{++}); 2a. oxidizing half-reaction; 2b. reducing half-reaction; 3. H transfer to DPN. In this process the ketone (α) carbon of pyruvic acid first becomes an aldehyde, which is then oxidized and a high-energy —C~S— bond is formed. Lipoic acid accepts the hydrogen released and then passes it on to DPN; hence lipoic acid does not appear in the net reaction summary in the lower panel.

in conjunction with this oxidizing reaction is *lipoic acid,* a derivative of an 8-carbon fatty acid which contains a disulfide bond; when this acid accepts hydrogen and becomes reduced in the process, the —S—S— bond splits and forms two —SH groups (Fig. 5.22, reaction 2b). Reduced lipoic acid subsequently yields H_2 to DPN^+, resulting in reoxidized lipoic acid (Fig. 5.22, reaction 3). The hydrogen pair now held by DPN eventually transfers to oxygen, yielding three ATP molecules per H_2 (or 6 ATP if the two pyruvic acid molecules formed from glucose are converted to acetyl CoA). The fate of the atoms in pyruvic acid is given by the statement

$$CH_3COCOOH + CoA{-}S{-}H \longrightarrow CH_3CO{\sim}S{-}CoA + CO_2 + H_2$$

Fatty acids ⟶ *acetyl CoA.* The decomposition of fatty acids is known as *β-oxidation;* i.e., the second, or β, carbon of the acid undergoes oxidative changes. The result is the splitting off of successive 2-carbon fragments from a fatty acid chain until only a last 2-carbon fragment remains.

Fig. 5.23 outlines the five reactions included in β-oxidation. In the first, *activation,* a fatty acid molecule is linked terminally with CoA. ATP here provides the necessary energy by being split between its first and second phosphate groups, yielding AMP and a separate inorganic double phosphate (usually abbreviated as PP_i). The activated fatty acid next undergoes a first *dehydrogenation,* one H being removed from each of the α and β carbons. An unsaturated double bond, —CH═CH—, is so created; we know from Fig. 5.6 that such unsaturated hydrocarbons represent a higher oxidation level than the saturated ones. The specific hydrogen carrier in this reaction is FAD; i.e., DPN is bypassed completely in this particular H transfer to oxygen.

The next reaction is a *hydration,* which resolves the unsaturated double bond and produces an alcoholic group on the β-carbon. This group is subsequently oxidized to a ketone, DPN being the hydrogen acceptor (Fig. 5.23, reaction 4). This is the *β-oxidation* from which the whole sequence derives its name. The resulting compound lastly reacts with another molecule of CoA, yielding two fragments. One is acetyl CoA; the other is an activated fatty acid which is shorter by two carbons than the activation complex formed in reaction 1, at the start of the whole sequence. The shorter complex may now be β-oxidized in its own turn, and consecutive acetyl CoA molecules so may be cut off.

What is the energy gain? From the summary of β-oxidation in Fig. 5.23, we find that transfer of H_2 from FAD to oxygen

R—CH_2—CH_2(β)—CH_2(α)—C(═O)—OH

1. H—S—CoA, ATP → AMP + PP_i

R—CH_2—CH_2—CH_2—C(═O)~S—CoA

2. FAD → FAD · H_2

R—CH_2—CH(β)═CH(α)—C(═O)~S—CoA

3. H_2O

R—CH_2—C(H)(OH)—C(H)(H)—C(═O)~S—CoA

4. DPN^+ → DPN · H + H^+

R—CH_2—C(═O)(β)—CH_2(α)—C(═O)~S—CoA

5. H—S—CoA

R—CH_2—C(═O)(β)~S—CoA + CH_3—C(═O)(α)~S—CoA

FATTY ACID LESS C_2 | ACETYL CoA

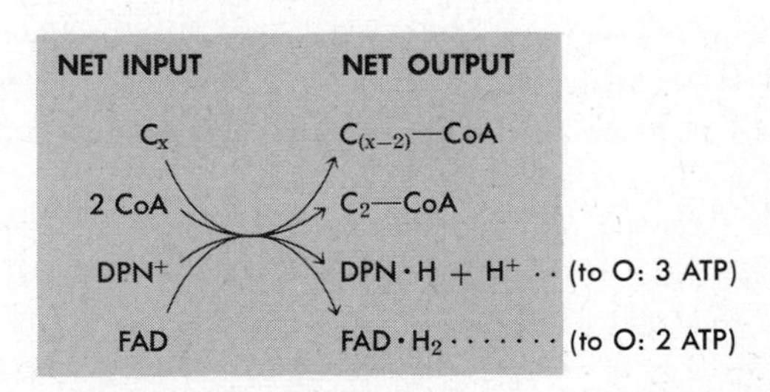

5.23. β-Oxidation of fatty acids. In step 1, a fatty acid is activated by combination with CoA. Step 2 is dehydrogenation, FAD here being the H acceptor. Step 3 is a hydration, and step 4 represents the β-oxidation proper: the *alcoholic* carbon becomes a *ketone.* In step 5 the 2-carbon fragment acetyl is split off, resulting in acetyl CoA and a fatty acid shorter by two carbons than the original one. The summary shows that 5 ATP are gained through H transport to oxygen. C_x here stands for a fatty acid with *x* numbers of C atoms.

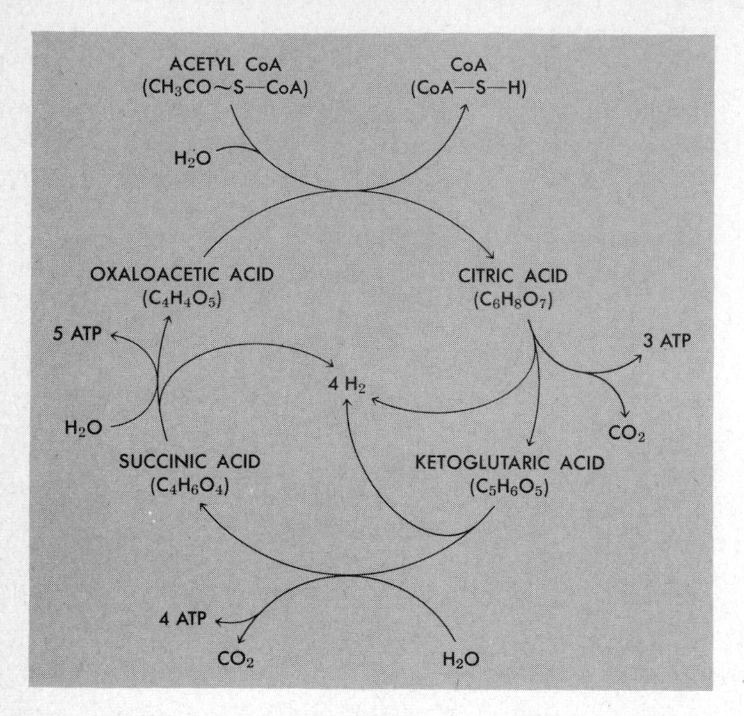

5.24. A simple version of the citric acid cycle. For greater detail see Fig. 5.25.

yields 2 ATP (not 3, since the DPN step is bypassed), and analogous transfer from DPN yields 3 ATP. Thus there is a gain of 5 ATP per one molecule of acetyl CoA formed. If, for example, we assume an actual starting fuel to be stearic acid, a C_{18} fatty acid very common in animal fats, then β-oxidation of this acid may occur successively eight times, yielding acetyl CoA each time and leaving a ninth acetyl CoA as a remainder. At 5 ATP per β-oxidation, the yield is therefore $5 \times 8 = 40$ ATP, minus 1 ATP expended for the original activation of the free stearic acid starting molecule. Therefore, one C_{18} fatty acid yields a net of 39 ATP and 9 acetyl CoA.

By way of comparison, one C_6 carbohydrate yields a net total of only 14 ATP and 2 acetyl CoA (cf. previous section: 8 ATP from glycolysis, 6 from acetyl CoA formation). One C_{18} molecule might be expected to yield just as much energy as three C_6 molecules. However, we shall soon see that, after all the acetyl CoA molecules are finally oxidized, an 18-carbons-long fatty acid actually produces *more* than three times as much—almost four times as much—ATP than a 6-carbons-long carbohydrate; fatty acids evidently are a richer source of usable energy than carbohydrates.

Formation of C_1: Carbon Dioxide

Once fuel decomposition has reached the acetyl CoA stage, all original distinctions between different carbohydrate and lipid fuels have already disappeared; acetyl CoA is the common 2-carbon stage, which subsequently oxidizes to CO_2 according to a single reaction pathway. As we shall see, amino acids also funnel into this pathway.

Citric acid cycle. The breakdown of the C_2 acetyl fragment takes the form of a *cycle* of reactions. The C_2 fragment first becomes attached to a C_4 molecule normally present in the mitochondria of a cell, yielding a C_6 molecule. The latter then loses one CO_2 and becomes a C_5 molecule; and in its turn the C_5 molecule subsequently loses another CO_2, resulting in a C_4 molecule. This last rearranges into the same C_4 compound which started the cycle. The net result is the conversion of acetyl into 2 CO_2 and the production of 12 ATP. The whole sequence is known as the *citric acid cycle,* after one of the participating (C_6) compounds, or as the *Krebs cycle,* after its discoverer. A simple version of the cycle is given in Fig. 5.24, a more detailed one in Fig. 5.25.

It will be seen from Fig. 5.25 that the cycle consists of nine consecutive steps. In a first step, acetyl CoA reacts with the C_4 compound *oxaloacetic acid* (an α-keto acid), yielding free CoA and *citric acid.* The —C~S— bond of acetyl CoA provides the energy for joining acetyl to oxaloacetic acid. Citric acid next reorganizes by two steps into *isocitric acid.* The net effect of this reorganization is a shift of the alcohol group at position 4 in citric acid to position 5 in isocitric acid. An oxidation now occurs at this position, yielding the ketone-containing *oxalosuccinic acid.* DPN is the hydrogen acceptor, and note that the oxidation has not affected either of the carbons contributed by the acetyl group at the start of the cycle.

Oxalosuccinic acid is subsequently decarboxylated at position 3, yielding *α-ketoglutaric acid.* This C_5 compound now undergoes an oxidative decarboxylation quite similar to that already discussed for pyruvic acid. The details are outlined in Fig. 5.26. The figure shows that α-ketoglutaric acid enters a series of subreactions in which it loses CO_2 and becomes an aldehyde, joins with CoA and loses H_2 to lipoic acid, and ends up as succinyl CoA. The latter contains a —C~S— bond, and the energy of this bond is transferred first into GTP, then into ATP. The net result of the whole sequence is the transformation of α-ketoglutaric acid to *succinic acid,* a C_4 compound.

Succinic acid now participates in three consecutive reactions (Fig. 5.25) which resemble the β-oxidation of a fatty acid. First, dehydrogenation produces an unsaturated double bond, FAD here being the hydrogen acceptor. Next, the resulting *fumaric acid* is hydrated to *malic acid,* which carries an alcoholic group at a subterminal carbon (position 4). Finally, this alcoholic group is dehydrogenated, DPN now being the H acceptor. This process results in the regeneration of *oxaloacetic acid.*

What is the overall tally? We may note, first, that the carbon atoms fed into the cycle as acetyl are not transformed immediately to CO_2. After a first turn of the cycle, these carbon atoms appear in the following turn at positions 5 and 6 of citric acid. Thus, an acetyl group fed into the cycle in any given turn appears as CO_2 byproduct only in later turns. Secondly, we may note that the 2 CO_2 eventually formed in any one cycle-turn derive from carbons 3 and 6 of citric acid; hence as one turn follows another, different carbon atoms become shifted successively into given positions within citric acid. Thirdly, carbon 5 of citric acid is oxidized twice (steps 4 and 6), carbon 4 is oxidized once (step 9). Carbon 5 is thus brought from an alcohol via a ketone to an acid level, and carbon 4, from an alcohol to a ketone level. Fourth, the cycle uses up four molecules of water (steps 1, 3, 6, and 8), but yields 1 H_2O at step 2.

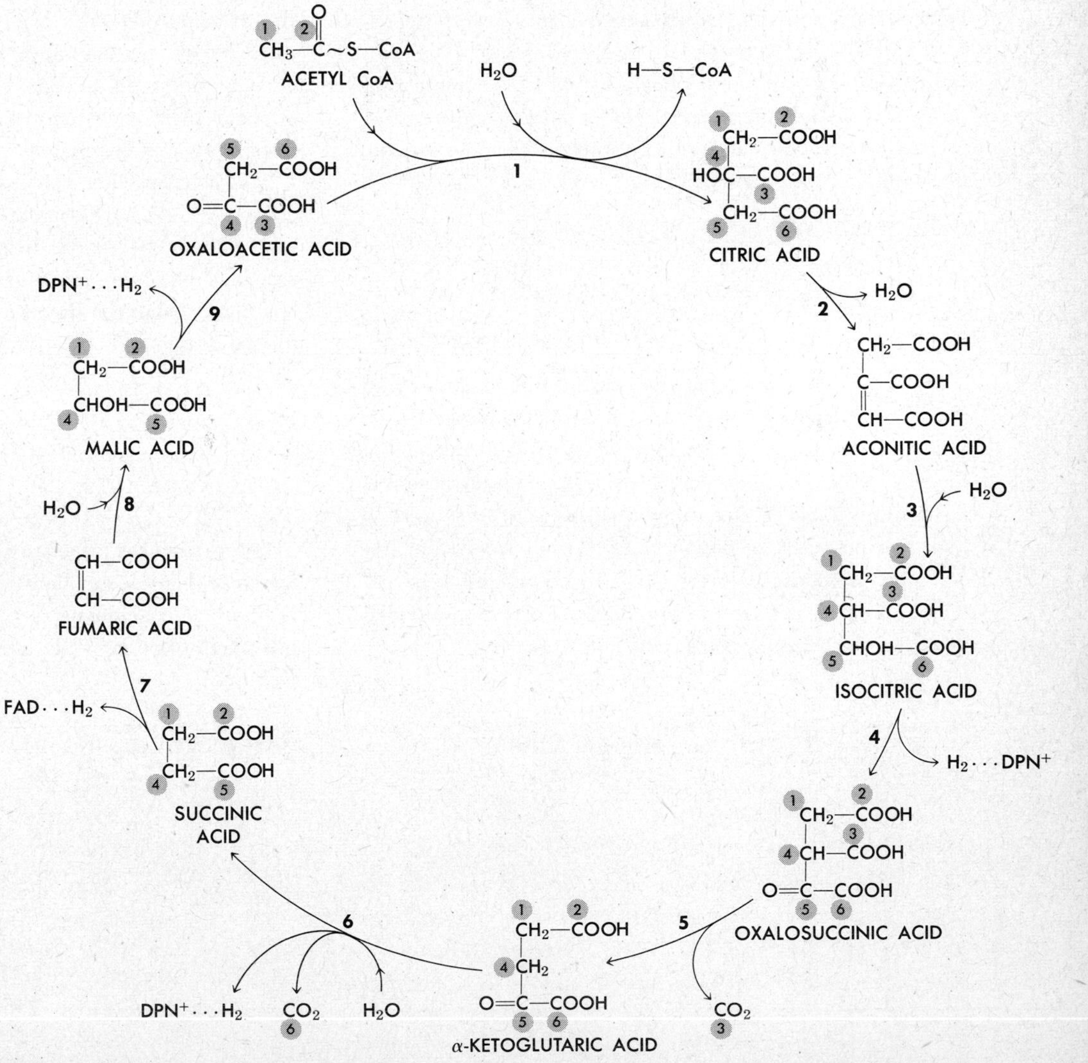

5.25. The citric acid cycle. Small numbers next to carbon atoms identify positions during a *first* turn of the cycle. Thus, during step 9, carbons 1, 2, and 4 of malic acid become carbons 1, 2, and 4 of a second-turn oxaloacetic acid and citric acid; carbon 5 of malic acid becomes carbon 3 of the second-turn citric acid; and the two carbons of the second-turn acetyl CoA assume positions 5 and 6 of the second-turn citric acid. —COO— groups identify the source of the CO_2 released in each turn of the cycle. The enzymes required at each step of the cycle are as follows: 1. condensing enzyme; 2., 3. aconitase; 4. isocitric dehydrogenase; 5. oxalosuccinic decarboxylase; 6. see Fig. 5.26; 7. succinic dehydrogenase; 8. fumarase; 9. malic dehydrogenase.

$$\underset{\alpha\text{-KETOGLUTARIC ACID}}{\mathrm{CH_2{-}COOH,\ CH_2,\ O{=}C{-}COOH}} \xrightarrow[1]{-CO_2} \mathrm{CH_2{-}COOH,\ CH{-}COOH,\ O{=}C{-}H}$$

Step 2: $\mathrm{CoA{-}S{-}H}$ in, $\mathrm{H_2}$ out → $\mathrm{CH_2{-}COOH,\ CH_2,\ O{=}C{\sim}S{-}CoA}$ SUCCINYL CoA; $\mathrm{H_2}$: LIPOIC ACID → LIPOIC ACID · $\mathrm{H_2}$; $\mathrm{DPN^+}$ → $\mathrm{DPN \cdot H + H^+}$

Step 3: SUCCINYL CoA → $\mathrm{CH_2{-}COOH,\ CH_2,\ O{=}C{-}OH}$ SUCCINIC ACID + $\mathrm{CoA{-}S{-}H}$; GDP → GTP; ADP → ATP

5.26. The transformation of α-ketoglutaric acid to succinic acid. The first step is a decarboxylation, the second a dehydrogenation (lipoic acid and DPN here being successive hydrogen acceptors), and the third step represents an energy transfer via GTP to ADP. Steps 1 and 2 are entirely similar to corresponding ones in acetyl CoA formation (Fig. 5.22).

Fifth, of the four hydrogen pairs removed in the cycle (steps 4, 6, 7, and 9), three pairs pass from DPN to oxygen, one pair from FAD to oxygen. These transfers yield 4 H_2O as well as 11 ATP; and one additional ATP arises from the conversion of succinyl CoA to succinic acid. The whole cycle may therefore be summarized as follows:

$$\underset{acetyl\ CoA}{\mathrm{CH_3COSCoA}} + 3\,\mathrm{H_2O} \longrightarrow 2\,\mathrm{CO_2} + 4\,\mathrm{H_2} + \mathrm{CoASH}$$

$\mathrm{H_2}$ → to oxygen: 4 H_2O, 12 ATP

If we now assume that the starting fuel is glucose and that it is respired completely under aerobic conditions, then this molecule can be shown to yield 38 ATP, as outlined in Fig. 5.27. This net total contrasts sharply with an anaerobic yield of only 2 ATP. If stearic acid is assumed to be the starting fuel (and note that oxidation of fatty acids always requires aerobic conditions), then the gain can be verified to be 147 ATP molecules, as shown in Fig. 5.28. An 18-carbon equivalent of carbohydrates yields only 114 ATP.

Amino-acid oxidation. If a cell obtains adequate supplies of carbohydrate and lipid fuels, proteins tend to be spared from decomposition. Yet to some extent even proteins are broken up, and their constituent amino acids then may become respiratory raw materials.

Amino acids enter oxidative pathways after their amino groups, $-NH_2$, have been removed. Such *deaminations* may take place in two ways. In one, *oxidative deamination*, an ammonia molecule, NH_3, is extracted from an amino acid. The reaction is catalyzed by an *oxidase*, i.e., an enzyme specific for the deamination of a particular type of amino acid:

$$\underset{amino\ acid}{\mathrm{NH_2{-}C(R)(H){-}COOH}} + \mathrm{O} \xrightarrow[oxidase]{R\text{-}specific} \underset{ammonia}{\mathrm{NH_3}} + \underset{\alpha\text{-}keto\ acid}{\mathrm{O{=}C(R){-}COOH}}$$

This process takes place particularly in liver cells, where any amino acid excess supplied via eaten food is deaminated. The resulting free ammonia is a toxic substance (because of a potentially very high pH after reaction with water), and is

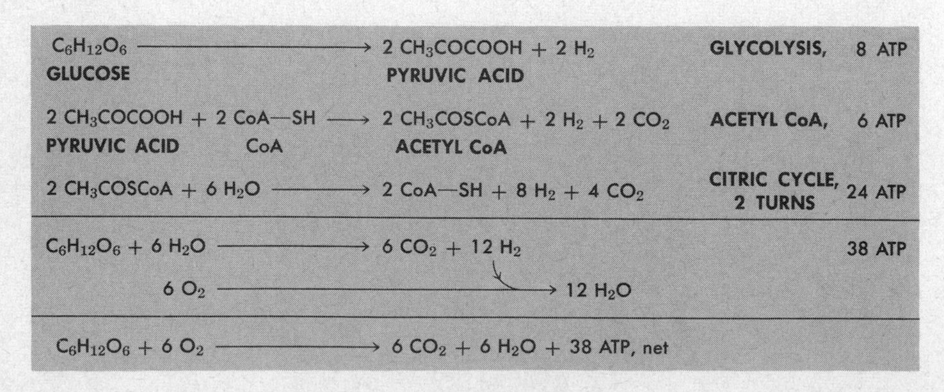

Reaction	Stage	Yield
$\mathrm{C_6H_{12}O_6}$ (GLUCOSE) → $2\,\mathrm{CH_3COCOOH} + 2\,\mathrm{H_2}$ (PYRUVIC ACID)	GLYCOLYSIS,	8 ATP
$2\,\mathrm{CH_3COCOOH} + 2\,\mathrm{CoA{-}SH}$ (PYRUVIC ACID, CoA) → $2\,\mathrm{CH_3COSCoA} + 2\,\mathrm{H_2} + 2\,\mathrm{CO_2}$ (ACETYL CoA)	ACETYL CoA,	6 ATP
$2\,\mathrm{CH_3COSCoA} + 6\,\mathrm{H_2O}$ → $2\,\mathrm{CoA{-}SH} + 8\,\mathrm{H_2} + 4\,\mathrm{CO_2}$	CITRIC CYCLE, 2 TURNS	24 ATP
$\mathrm{C_6H_{12}O_6} + 6\,\mathrm{H_2O}$ → $6\,\mathrm{CO_2} + 12\,\mathrm{H_2}$; $6\,\mathrm{O_2}$ (+ $12\,\mathrm{H_2}$) → $12\,\mathrm{H_2O}$		38 ATP
$\mathrm{C_6H_{12}O_6} + 6\,\mathrm{O_2}$ → $6\,\mathrm{CO_2} + 6\,\mathrm{H_2O}$ + 38 ATP, net		

5.27. Quantitative summary of glucose respiration. The last equation represents the net input and output.

$CH_3(CH_2)_{16}COOH + 8\,H_2O + 9\,CoASH \longrightarrow 9\,CH_3COSCoA + 16\,H_2 + H_2O$	39 ATP	8 β-OXIDATIONS
$9\,CH_3COSCoA + 27\,H_2O \longrightarrow 9\,CoASH + 36\,H_2 + 18\,CO_2$	108 ATP	CITRIC CYCLE, 9 TURNS
$CH_3(CH_2)_{16}COOH + 34\,H_2O \longrightarrow 18\,CO_2 + 52\,H_2$	147 ATP	
$26\,O_2 \longrightarrow 52\,H_2O$ (with the $52\,H_2$)		
$CH_3(CH_2)_{16}COOH + 26\,O_2 \longrightarrow 18\,CO_2 + 18\,H_2O + 147$ ATP, net		

5.28. Quantitative summary of the respiration of stearic acid. The last equation represents the net input and output.

ultimately excreted by the animal. Excretion occurs either in the form of ammonia as such (as in most aquatic types), or, after NH_3 has undergone various chemical transformations, in the form of *uric acid* (e.g., insects, birds) or *urea* (e.g., mammals, some reptiles, amphibia). We shall pursue this topic further in Chap. 9.

The other product of deamination is an α-keto acid which, depending on the nature of the R— group, is respired either like a carbohydrate or like a lipid. For example, if alanine is the original amino acid, then deamination yields pyruvic acid:

$$NH_2-\underset{H}{\overset{CH_3}{|}}{C}-COOH \xrightarrow[-NH_3]{+O} O{=}\overset{CH_3}{\overset{|}{C}}-COOH$$

alanine → *pyruvic acid*

This and analogous types of amino acids are said to be *glucogenic,* i.e., the α-keto acids resulting from deamination have chemical affinity to carbohydrates and are respired as such. However, if the original amino acid is *leucine*, for example, then deamination produces an α-keto acid which resembles a fatty acid rather closely:

$$CH_3-\underset{CH_3}{CH}-CH_2-\underset{NH_2}{CH}-COOH \xrightarrow[-NH_3]{+O} CH_3-\underset{CH_3}{CH}-CH_2-\underset{O}{\overset{}{\underset{\|}{C}}}-COOH$$

Such amino acids are said to be *ketogenic;* i.e., after deamination they have affinity to lipids. Their respiration actually yields acetyl CoA molecules directly. A few amino acids are both glucogenic and ketogenic at the same time; that is, their deamination products oxidize to 3-carbon as well as 2-carbon fragments.

From the generalized deamination reaction above, we may note that for each amino acid there exists a structurally corresponding α-keto acid. Conversely, if a given α-keto acid adds an amino group, a corresponding amino acid will be formed. These considerations lead us to the second method by which amino acids may lose their $—NH_2$ groups and enter respiratory pathways. On theoretical grounds, removal of an amino group could occur if a suitable amino acceptor were available; and in practice, α-keto acids actually represent excellently suited acceptors. Thus the $—NH_2$ group of an amino acid A may be transferred to a keto acid B, resulting in keto acid A and amino acid B:

$$NH_2-\underset{H}{\overset{A}{\overset{|}{C}}}-COOH + O{=}\overset{B}{\overset{|}{C}}-COOH \longrightarrow O{=}\overset{A}{\overset{|}{C}}-COOH + NH_2-\underset{H}{\overset{B}{\overset{|}{C}}}-COOH$$

amino acid A + *keto acid B* → *keto acid A* + *amino acid B*

This equation symbolizes a so-called *transamination* reaction. Its occurrence depends on the participation of a specific *transaminase,* and of a coenzyme which serves as temporary $—NH_2$ carrier and transferrer. This coenzyme is a derivative of *pyridoxine,* or vitamin B_6.

Unlike oxidative deaminations, transaminations occur in all types of cells. The reactions are significant for two reasons. First, they permit almost any given amino acid to be transformed into almost any other; the only requirement is availability of the appropriate α-keto acid. Transaminations actually are of considerable importance in the synthesis of different amino acids and proteins. Secondly, the transamination of certain particular amino acids yields corresponding α-keto acids which happen to be normal participants in glycolysis and in the citric acid cycle. More specifically, if alanine transaminates with some other α-keto acid, alanine becomes pyruvic acid, the α-keto acid corresponding to alanine. Analogously, if glutamic acid participates in a transamination, this amino acid becomes α-ketoglutaric acid. And if aspartic acid is transaminated, the corresponding α-keto acid formed is oxaloacetic acid (Fig. 5.29). These reactions, particularly the glutamic-ketoglutaric transformation, provide important direct pathways for the respiration of amino acids and of nitrogenous compounds generally.

Hexose monophosphate shunt. Where glycogen is used as a primary carbohydrate fuel, as in muscle, the respiratory pathway leads to CO_2 almost entirely via glycolysis and the citric acid cycle. In many other tissues, however (e.g., brain, liver), the primary carbohydrate fuel may be—and in brain must be—glucose, not glycogen. In such cases an important alternative respiratory path exists which involves neither glycolysis nor the Krebs cycle, nor indeed even the mitochondria. Known as the *hexose shunt,* this oxidation process occurs in the free cellular cytoplasm. In such cells it may account for anywhere from 10 to 90 per cent of all carbohydrate respiration, with an average at about 30 per cent.

The pattern of this cyclical pathway is illustrated in Fig. 5.30. As shown, the starting fuel is glucose-6-phosphate. Six molecules of it first are dehydrogenated twice in succession and decarboxylated once. The hydrogen acceptors in these reactions are TPN molecules. Six CO_2 form as byproduct and six molecules of ribulose-phosphate then remain. The latter undergo a complex series of rearrangements in which C_4, C_5, and C_7 sugars are intermediates at certain points. The ultimate result is the appearance of five molecules of glucose-6-phosphate, one less than present at the start. Thus the net effect of the cycle is the conversion of one out of six glucose molecules to 6 CO_2.

The chemical tally evidently is the same as that in glucose respiration via glycolysis and citric acid cycle. But the energy tally differs somewhat: twelve pairs of hydrogen atoms are eventually transferred to oxygen, yielding 36 ATP. Of these, 1 ATP must "pay back" for the ATP used up in converting one molecule of free glucose to glucose-6-phosphate. Thus the net yield is 35 ATP, which compares favorably with the 38 ATP obtainable via glycolysis. We may note here that TPN accepts hydrogen in the free cytoplasm, whereas H transfer to oxygen takes place within the mitochondria via DPN. A "transhydrogenation" reaction serves as a bridge between the free cytoplasm and the mitochondria; TPN · $[H_2]$ reacts with DPN across the mitochondrial boundary, resulting in TPN and DPN · $[H_2]$. The latter then transfers hydrogen to oxygen within the mitochondria.

Apart from representing an alternative process of glucose oxidation, the hexose shunt is significant also because it generates reduced TPN, required in fat synthesis, as we shall see, and because it generates sugars other than hexoses. We already know for example, that the C_5 sugars in particular are required raw materials in nucleotide synthesis.

A comprehensive summary of all major respiratory pathways is given in Fig. 5.31. These various metabolic events take place exceedingly rapidly within cells. It has been estimated, for example, that a glucose molecule may be oxidized completely within a single second. Considering the number of reactions, reactants, enzymes, carriers, and the like, such speed is truly impressive. In vertebrates, moreover, respiratory rates are greatly influenced by the thyroid hormone thyroxine.

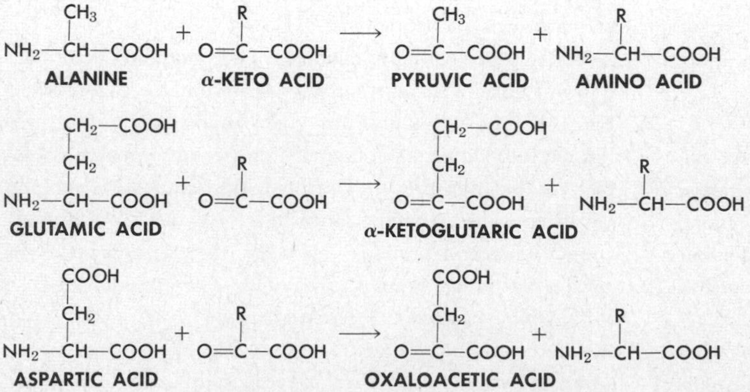

$$NH_2{-}CH(CH_3){-}COOH + O{=}C(R){-}COOH \longrightarrow O{=}C(CH_3){-}COOH + NH_2{-}CH(R){-}COOH$$

ALANINE + α-KETO ACID → PYRUVIC ACID + AMINO ACID

$$NH_2{-}CH(CH_2{-}CH_2{-}COOH){-}COOH + O{=}C(R){-}COOH \longrightarrow O{=}C(CH_2{-}CH_2{-}COOH){-}COOH + NH_2{-}CH(R){-}COOH$$

GLUTAMIC ACID → α-KETOGLUTARIC ACID

$$NH_2{-}CH(CH_2{-}COOH){-}COOH + O{=}C(R){-}COOH \longrightarrow O{=}C(CH_2{-}COOH){-}COOH + NH_2{-}CH(R){-}COOH$$

ASPARTIC ACID → OXALOACETIC ACID

5.29. Three transamination reactions of importance in amino acid respiration. The α-keto acids resulting from transamination happen to be participants in glycolysis and the citric acid cycle, hence through the reactions above amino acids may enter the same final respiratory pathways as carbohydrates and fatty acids. Various transaminases catalyze the reactions shown.

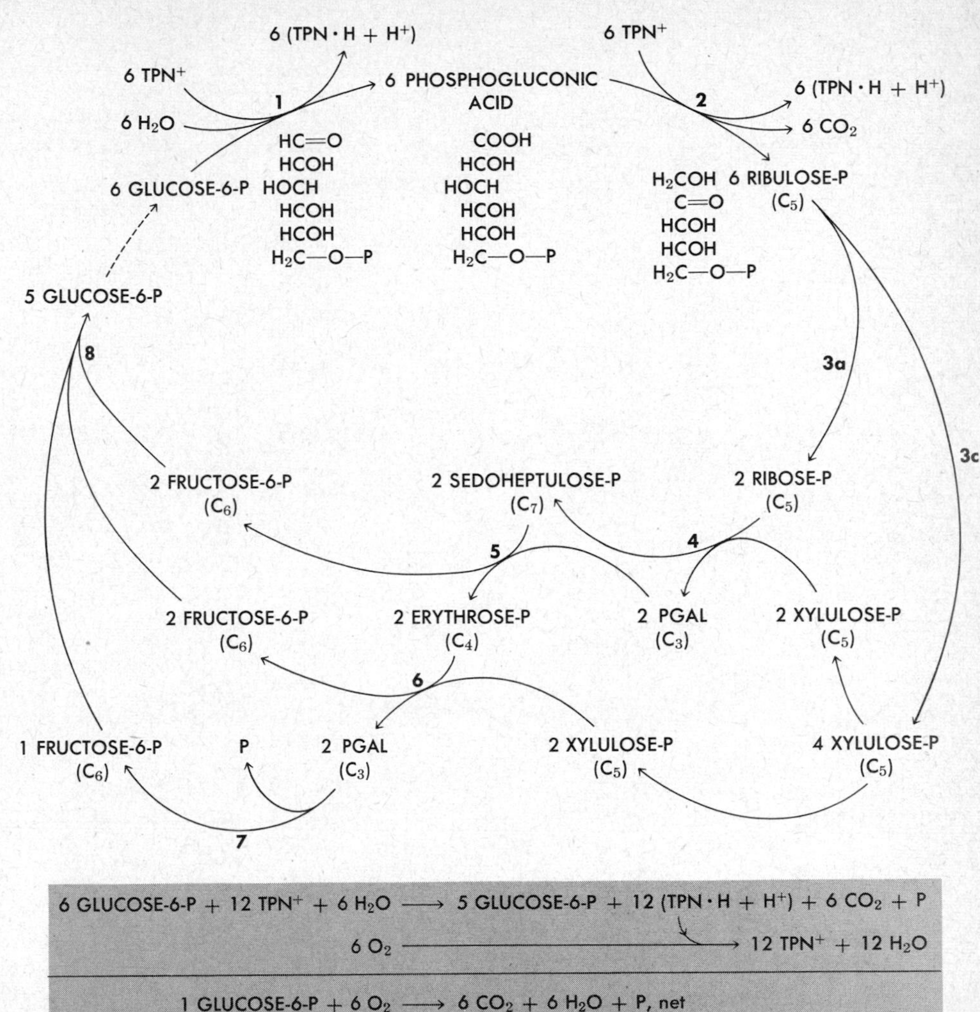

5.30. The hexose shunt. The formulas for the principal participants in reactions 1 and 2 are shown in the figure; formulas of the other components will be found in earlier parts of this and the previous chapter. Phosphates are indicated by the letter P; PGAL represents glyceraldehyde phosphate; the sources of H_2 and CO_2 in reactions 1 and 2 are indicated by atoms printed in boldface. Reaction 1 is oxidative in that it changes carbon position 1 from an aldehyde to an acid; reaction 2 is oxidative in that it converts carbon position 3 of phosphogluconic acid to the ketone of ribulose. Reactions 3 to 8 collectively regenerate glucose-6-phosphate, the total carbon numbers here remaining the same: six C_5 compounds (ribulose-P) become five C_6 compounds (glucose-P). The summary at the bottom shows that, net, one glucose-P molecule is converted into CO_2 and water.

This hormone accelerates respiration in proportion to its concentration. How this effect is achieved and what particular reactions are influenced are still more or less completely unknown. Most animals are not vertebrates, however, and their respiration is not under thyroxine control. Nevertheless, respiratory breakdowns still occur extremely rapidly. Very efficient enzyme action is probably one condition which makes speed possible. Another condition undoubtedly is the close, ordered proximity of all required ingredients in the submicroscopic recesses of the mitochondria. Just as a well-arranged industrial assembly line turns out products at a great rate, so do the even better-arranged mitochondria.

The fate and function of their chief product, ATP, is the subject of the next section.

THE ENERGY REQUIREMENT

How much energy must be expended by an animal cell to maintain life? The answer here varies, according to the varying intensities of cellular activity. But while a cell lives, its activities are never zero. Accordingly, if life is to continue, at least a basic minimum quantity of energy is required under all conditions. Every such energy requirement must be met by respiration; demand must be balanced by supply.

Thus far, we have measured the supply of energy in terms of ATP molecules. Utilization, on the other hand, is measured in terms of mechanical work, chemical work, and many other forms of activity. Evidently, before comparative statements can be made about supply and demand, a common yardstick

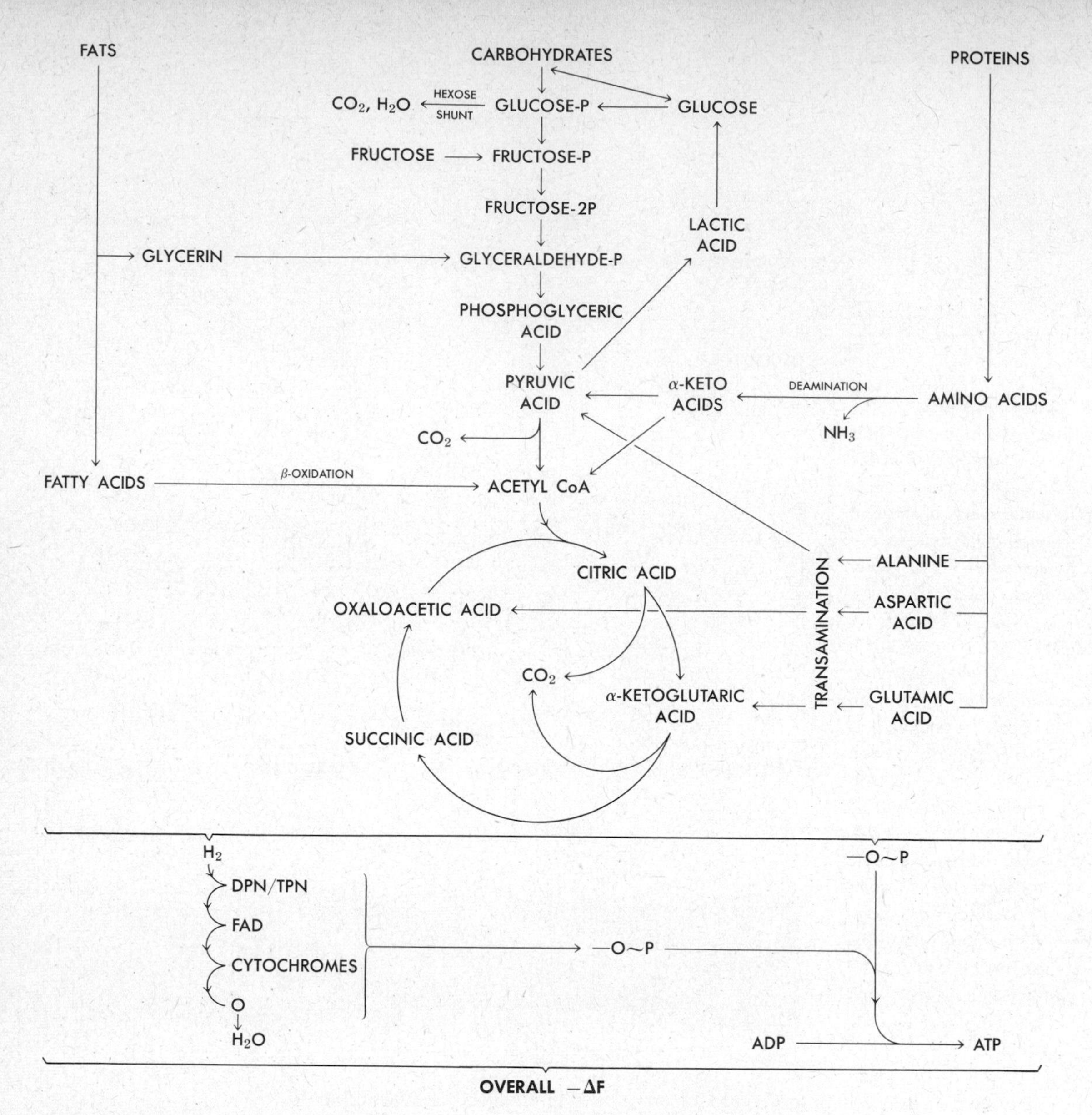

5.31. Generalized summary of the major metabolic pathways in respiration. Note that most of the reactions are theoretically reversible. Actual reversal would require a supply of external energy, to counteract the normal negative ΔFs of these reactions. Certain particular reactions are not reversible even if the ΔF is made positive, as will be shown in the following chapter and also in Chap. 9.

should be available. Such a yardstick is heat. All other forms of energy can be converted into heat, and any energy quantity may therefore be expressed as a *heat equivalent.* Two or more of these equivalents may then be compared directly. The unit of measurement here is the calorie.

What is the energy content of a metabolic fuel expressed in heat equivalents? To determine this, one simply burns the fuel in a laboratory furnace and measures the total heat given off. By such means, it is found that, for example, 1 g of glucose or of carbohydrates generally contains about 3800 cal. A gram of protein liberates roughly the same amount of heat, namely, 4000 cal, and a gram of fat yields about 9000 cal.

These figures suggest a reason why fats are the preferred animal storage foods, and why animal metabolism is highly fat-oriented generally: a given quantity of calories stored in the form of fat weighs less than half than if it were stored in the form of carbohydrate or protein. Fat storage therefore makes for less bulk, which is adaptively important in a motile animal. Yet even smaller quantities of fat still concentrate enough energy in them to provide sufficient power for motion. In this connection it is interesting to note that clams, for example, which move very little, store their food largely as carbohydrates, like rooted plants, whereas those plant seeds which are designed for passive motion through air store foods largely as lipids, like motile animals. Moreover, it should be clear that every 9000-cal surplus in the diet leads to the deposition of one additional gram of storage fat; and although extra weight of any kind is not advantageous, one extra gram (of fat) is preferable to two extra grams (of carbohydrates).

From such caloric values, one may also calculate how much energy is potentially available in given quantities of fuel. In cells, does all this energy become actually available in the form of ATP? In other words, how efficient is respiration? How much of the energy content of fuels can be harvested as useful chemical energy? We know from the preceding section that cellular combustion of, for example, one molecule of glucose yields 38 molecules of ATP. One gram molecular weight (mole) of glucose (180 g) should therefore yield 38 moles of ATP. It can be shown that if 1 mole of glucose is oxidized in a furnace, an energy potential of 686,000 cal (i.e., 180×3800) becomes available. At an average of 7000 cal for the formation of 1 mole of ATP, biological oxidation of glucose actually yields 38×7000, or 266,000 cal/mole in the form of ATP. Thus, the efficiency of aerobic glucose respiration is 266,000/686,000, or roughly 40 per cent. Sixty per cent of the potential energy dissipates unavoidably as heat. In the case of fat respiration, the efficiency level is quite comparable. For example, 1 mole of tristearin contains 890 g and thus makes available an energy potential of 890×9000, or roughly 8 million cal. Tristearin contains 3 C_{18} fatty acids and 1 C_3 carbohydrate, and complete biological oxidation therefore yields 3×147 plus 1×19, or 460 ATP. At 7000 cal per mole of ATP formed, the actual energy yield then is about 3.2 million cal. Hence the efficiency is 3.2/8, or 40 per cent.

How far do such ATP yields go toward support of life? To answer this, we must know the rate of energy expenditure of an animal under specified conditions.

To assess such expenditures, could one not simply determine the energy content of all the food eaten by an animal during a stated period? No, because all this food is normally not used toward energy production. An indeterminable fraction may be stored; another fraction may be used in synthesis rather than in respiration; and some food may also be eliminated unused. Moreover, an animal very often eats more food or less food than actually needed. The amount of energy potentially supplied by food thus is not a reliable measure of actual energy requirements.

A much better measure is *oxygen consumption.* Atmospheric oxygen is not stored; it is used specifically in respiration only; and it is taken into an animal in amounts geared precisely to actual requirements. Moreover, one can easily determine how much fuel may be burned with the aid of a given quantity of oxygen. For example, 1 liter of oxygen will support complete oxidation of 1.25 g of glucose.

Therefore, to determine the energy requirement of an animal it is necessary to specify, first, the activity of the animal and environmental conditions such as temperature, humidity, and other physical factors; secondly, the period of time during which the energy requirement is to be measured; and thirdly, the amount of oxygen consumed during this period.

If one measures not only oxygen consumption but at the same time also carbon dioxide output, then it is possible to determine what kinds of foods an animal burns to meet its energy requirement. The ratio of CO_2 released to O_2 consumed, known as the *respiratory quotient* (RQ), is quite characteristic for each of the main food classes. For example, when glucose is respired aerobically, the quantitative relation between the two respiratory gases is given by the statement:

input	*output*
$C_6H_{12}O_6$	$6\ H_2O$
$6\ O_2$	$6\ CO_2$

In other words, for every six molecules of oxygen consumed, six molecules of CO_2 are obtained. The respiratory quotient therefore is $CO_2/O_2 = 1.0$. Such an RQ is characteristic for carbohydrates generally. Accordingly, whenever measurement shows that RQ = 1.0, this indicates that the animal respires carbohydrates.

Different RQs are obtained for other food materials. For example, complete respiration of the fat tristearin is described by

input	*output*
$2\ C_{57}H_{110}O_6$	$110\ H_2O$
$163\ O_2$	$114\ CO_2$

Here RQ = 114/163 = 0.7, which is characteristic for fats

generally. Thus, when a measured RQ is about 0.7, the organism undoubtedly respires fats. In analogous manner, it can be shown that an RQ of about 0.8 characterizes the combustion of proteins; and if a mixture of carbohydrates, fats, and proteins is used as fuel, the RQ will usually fall somewhere between 0.8 and 0.9.

Most actual measurements of energy requirements have been made on man, but the same procedures apply in principle to any animal. The conditions chosen are often those of *basal* metabolism, i.e., when body activity is reduced to a minimum. The test animal is at complete physical and mental rest, as during quiet sleep, and the digestive system is empty. Oxygen consumption and CO_2 output are then measured over a given period of time. Under such conditions, the energy expended by the animal represents its *basal metabolic rate, BMR* for short. It indicates the energy necessary just to remain alive during complete rest or sleep: the energy required to maintain minimum breathing and heartbeat, minimum activity of brain, liver, kidneys, and all other vital organs, and minimum respiration and other chemical activities in all cells.

Tests have shown that under basal conditions a human adult consumes on an average about 14 liters of oxygen per hour. Since 1 liter burns 1.25 g of glucose and since 1 g of glucose yields 3800 cal, the energy expenditure will be $14 \times 1.25 \times 3800$, or 66,500 cal. In other words, at an efficiency of 40 per cent, some 18 g, or less than 1 ounce, of glucose will supply just enough energy to keep an adult alive during 1 hr of sleep.

BMR values vary widely. A growing animal expends more energy per pound of tissue than a nongrowing adult. A male metabolizes slightly more intensely than a female. If the temperature of the environment is low, more energy is expended toward maintenance of body temperature. A small animal possesses a large skin area in proportion to its volume, and it uses more energy to offset the greater heat loss through surface radiation and evaporation. Because of such variables, actual BMR determinations are quite complicated in practice and require control and measurement of numerous factors.

BMR varies not only with age, sex, weight, size, season, and climate but also with state of health. During disease, BMR values may become abnormally low or high, and this may sometimes be a clue to the nature of the disease. An abnormal BMR usually indicates that the utilization of foods is somehow defective. This is the case in diabetes, for example, where insulin production is inadequate and glucose utilization is impaired. Or respiration within cells could be impaired as a result of vitamin or thyroid deficiencies.

Conditions are not basal when the body is active. Energy requirements then are greater in proportion to the intensity or the amount of activity. It can be shown that a moderately active human adult requires about 3 million cal every 24 hr. Thus about ¾ lb of fat (333 g) would satisfy such an energy requirement. However, it should be clear that such an intake would not represent an adequate diet. Additional food is needed for cellular synthesis, and this, as well as respiration itself, requires a wide variety of foods. The caloric value of a diet is only one aspect of adequate nutrition.

The energy expended daily by an animal sustains both the physical and the chemical activities of its cells. What are these activities and how does the energy of ATP support them?

REVIEW QUESTIONS

1. How does cellular nutrition take place? What is active transport? Describe the uptake of glucose and oxygen by cells.
2. Compare and contrast a fire with respiration. What do they have in common? What is different? Which materials are fuels in respiration? What general types of events occur in respiration?
3. What are redox reactions? Half-reactions? What energy relations exist during redox reactions? Show how progressive changes of oxidation level may occur during the transformation of hydrocarbons to anhydrides.
4. What is dehydrogenation? Where does it occur, and what role does it play in respiration? Under what conditions does it take place? How is hydrogen transferred to oxygen? Describe the sequence of carriers and the specific role of each during aerobic hydrogen transport. What role does electron transport play?
5. Distinguish between aerobic and anaerobic respiration. Under what conditions does either take place? How and where may aerobic respiration become blocked? How is lactic acid formed? How and under what conditions is lactic acid respired? What is an oxygen debt and how is it paid?
6. Describe the role of adenosine phosphates in respiration. What is a high-energy bond? How and where are such bonds created in fuels? During H transfer? How much energy is required for ATP formation?
7. Review the sequence of events in glycolysis. Which

steps are oxidative, and what changes in oxidation level take place? How much energy is obtained? Which classes of nutrients pass through a pyruvic acid stage in respiration?

8. Review the conversion of pyruvic acid into acetyl CoA. What are the functions of carboxylase, cocarboxylase, lipoic acid, and coenzyme A? How much energy is gained and where? What classes of nutrients pass through an acetyl CoA stage in respiration?

9. Describe the process of β-oxidation. Which steps are oxidative? How much energy is gained, and where?

10. Describe the general sequence of events in the citric acid cycle. Which steps are oxidative, and what changes in oxidation level take place? What is the total input and output of the cycle? How much ATP is gained and through what steps?

11. Distinguish between oxidative deamination and transamination. What are glucogenic and ketogenic amino acids? Show how amino acids are respired.

12. Review the sequence of events in the hexose monophosphate shunt. What respiratory role does this process play, and where does it occur? How does the energy gain compare with that of glycolysis?

13. How efficient is respiration? How much potential energy does each of the main classes of nutrients contain? What is a respiratory quotient and how is it determined? What does it indicate? What is basal metabolism? How is it determined and what does it indicate?

14. Review and summarize the overall fate of one molecule of glucose during complete respiratory combustion. What is the total net input and what is the total net output? What happens to the individual atoms of glucose? What is the total ATP gain, and how much is gained during each of the main steps of breakdown?

15. Where in cells does respiration occur? What factors probably contribute to the speed of respiration? Inasmuch as respiratory reactions are reversible, how does it happen that energy continues to be produced?

COLLATERAL READINGS

The following are popularly written articles dealing with various aspects of respiration:

Green, D. E.: Enzymes in Teams, *Sci. American*, Sept., 1949.

———: The Metabolism of Fats, *Sci. American*, Jan., 1954.

———: Biological Oxidation, *Sci. American*, July, 1958.

———: The Mitochondrion, *Sci. American*, Jan., 1964.

Holter, H.: How Things Get Into Cells, *Sci. American*, Sept., 1961.

Lehninger, A. L.: Energy Transformation in the Cell, *Sci. American*, May, 1960.

———: How Cells Transform Energy, *Sci. American*, Sept., 1961.

Roberts, J. D.: Organic Chemical Reactions, *Sci. American*, Nov., 1957.

Siekevitz, P.: Powerhouse of the Cell, *Sci. American*, July, 1957.

Solomon, A. K.: Pores in the Cell Membrane, *Sci. American*, Dec., 1960.

———: Pumps in The Living Cell, *Sci. American*, Aug., 1962.

Stumpf, P. K.: ATP, *Sci. American*, Apr., 1953.

Accounts of the chemistry of respiration are included in the following paperbacks and books:

Baldwin, E. B.: "Dynamic Aspects of Biochemistry," 3d ed., Cambridge, New York, 1957.

Gerard, R. W.: "Unresting Cells," chaps. 5–7, Harper & Row, New York, 1949.

Harrison, K.: "A Guide Book to Biochemistry," Cambridge, New York, 1959.

Loewy, A. G., and P. Siekevitz: "Cell Structure and Function," Modern Biology Series, Holt, New York, 1963.

McElroy, W. D.: "Cellular Physiology and Biochemistry," Foundations of Modern Biology Series, Prentice-Hall, Englewood Cliffs, N.J., 1961.

CHAPTER 6

cellular operations: synthesis, self-perpetuation

Energy must be expended in all processes which maintain the metabolism and self-perpetuation of a cell. Such processes include physical as well as chemical ones. The most important physical roles of energy are to produce heat, to some extent also to produce light and electricity, and above all, to produce movement of cells and cell parts. The chief chemical roles of energy are maintenance of respiration itself and, most particularly, maintenance of activities associated with the synthesis of new cellular components. Such components must be manufactured to offset the respiratory decomposition and the wear and tear of existing ones, to make possible cellular repairs after injury, to maintain growth, and to permit reproduction. In all these processes of synthesis, energy is one requirement, structural building blocks in the form of nutrients are another. Under the heading of energy utilization, therefore, two major subtopics are the *physical uses* and *chemical uses* of energy.

PHYSICAL ROLES OF ATP

Probably the most abundant physical use of ATP is made in mechanical cell functions. Of these, the most readily discernible are those which produce movement, in the form of either locomotion of whole animals or internal motion of parts of animals. We shall first examine the basis of one major type of movement, namely, the contraction of specialized animal muscle cells.

Muscular Movement

The characteristic activity of muscle cells ranks among the most important activities carried out by animal cells generally, for few animal functions exist that do not include muscular contraction. Moreover, muscles are quantitatively the most abundant tissue of an animal, particularly a vertebrate. A proportionately large amount of all available energy must be expended to keep muscles contracting. Even during "inactive" periods like sleep, for example, the muscular system maintains not only posture and shape but also vital functions such as breathing, heartbeat, and blood pressure. Mainly because of muscular movement, the energy requirements of animals are far greater pound for pound than of any other kind of organism.

The contractile units. The functional units of all kinds of muscles are long, thin, intracellular filaments called *myofibrils*. Each muscle cell or muscle fiber contains many such myofibrils aligned in parallel and extending in the same direction as the long axis of the whole cell or fiber (Fig. 6.1). In striated muscles (cf. Chap. 8), the myofibrils exhibit alternate dark and light crossbands (*A bands* and *I bands*, respectively) visible under the microscope. The A and I bands in turn display finer cross-markings within them. When such a muscle contracts, only the I bands become shorter. The total contraction is the sum of all the individual contractions of the I bands.

The electron microscope shows that each myofibril is actually a bundle of many long, ultrathin, parallel filaments. These are composed principally of five kinds of materials: water, inorganic ions, ATP, and two proteins called *actin* and *myosin*. Together, these form the basic contraction apparatus.

That this is so has been demonstrated dramatically by experiment. With appropriate procedures, actin and myosin can be extracted from muscle, and it can be shown that neither actin nor myosin alone is able to contract. But by mixing actin and myosin together, artificial fibers of *actomyosin* can be made. To these fibers may be added water, inorganic ions, and ATP. When this is done, it is found that as soon as ATP reaches an actomyosin fiber, the latter contracts violently. Such contracting fibers may lift up to 1,000 times their own weight, just as a living muscle may do. And it is also found that, in a contracted fiber, ATP is no longer present but low-energy phosphates are present instead.

Experiments of this sort provide clues as to how contraction might be brought about in a living muscle. The process is far from being fully understood, but some of the main events are known. Muscle activity is at least a two-step cycle involving alternate *contraction* and *extension*. Energy is used up at some point or points in such a cycle. One view is that the energy makes possible the contraction of a muscle, like compressing a spring. Subsequent extension then is thought to be essentially an automatic recoil, like releasing a compressed spring. According to an alternative view, energy must be expended to extend a muscle, as in stretching a rubber band. Contraction would then be automatic, like releasing an extended rubber band. A good deal of evidence appears to favor

6.1. The structure of skeletal muscle. Whole muscle fibers are shown in **A.** Note the cross striations, the faintly visible internal longitudinal myofibrils, and the nuclei, which appear as dark patches. **B.** Electron micrograph of portions of two horizontal myofibrils (in the tail muscles of a frog tadpole), separated by a layer of cytoplasmic material (endoplasmic reticulum). Note that each myofibril in turn consists of bundles of still finer filaments. The predominantly dark A bands and predominantly light I bands are labeled. Cf. also Fig. 8.18.

A

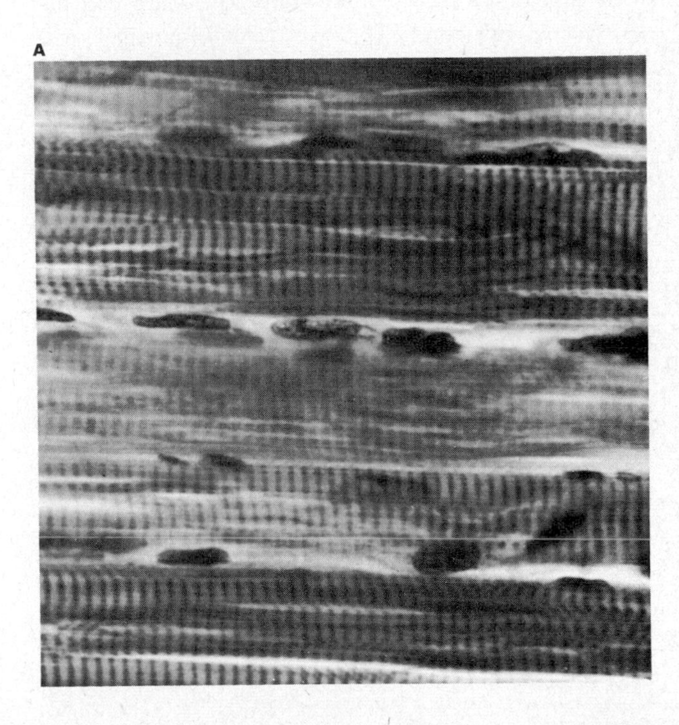

B

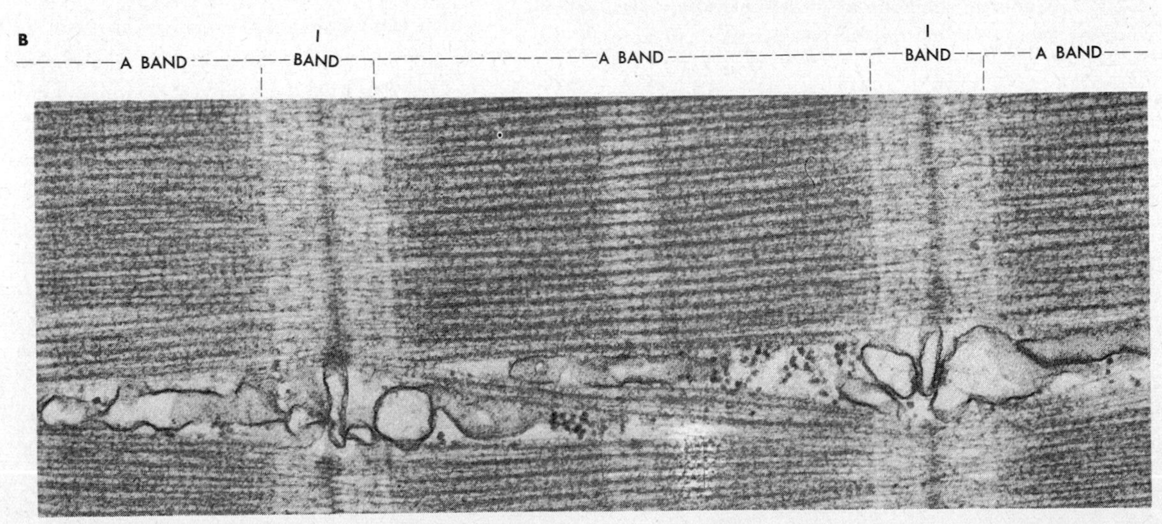

this second hypothesis, but the first cannot be ruled out; indeed there are indications that muscle may require energy for both contraction and extension. Muscle is shorter and thicker when contracted, longer and thinner when extended.

The energy cycle. The energy donor in muscle activity is ATP which, together with actin and myosin, forms an actomyosin-ATP complex. The ATP here appears to be not only the energy donor but also a necessary *structural* part of the contraction apparatus; ATP makes actomyosin supple and elastic and able to contract. When ATP disappears or separates from actomyosin—for example, in extreme fatigue or during rigor mortis after death—muscle becomes rigid and stiff.

During a contraction-extension cycle, the ATP of actomyosin-ATP yields up its energy. To prepare a muscle for a new contraction-extension cycle, new energy must be supplied from the outside. Respiration is the ultimate source of this energy, but it is not the immediate source. Fast though oxidation of muscle glycogen is, it is far too slow to supply the ATP required by an active muscle. A glycogen molecule in muscle may decompose within a second, but in that second the wing muscle of an insect may contract up to 100 times and use up energy far faster than could be supplied directly by fuel oxidation.

Unlike most other cells, muscles are able to store large amounts of energy. Cells other than muscles usually may store energy only in the form of ATP. The amounts of ADP available for conversion to ATP are limited, and in such cells energy utilization can occur only as rapidly as ATP can be formed and re-formed by respiration. But some cells, most notably muscles, are alternately highly active or virtually inactive. During periods of intense activity, therefore, more ATP may be required in a muscle than respiration is able to create. Conversely, during rest, fuel oxidation in muscles may produce more —O ∼ P than the entire ADP/ATP system can hold. Muscles and a few other animal tissues are able to cope with such excess supplies or demands. They possess a device which can store high-energy phosphate bonds beyond the storage capacity of ATP.

This device operates through one of several compounds. One of them is *creatine*, a nitrogen-containing organic substance found in the muscles of most vertebrates and some invertebrates. Another is *arginine*, an amino acid which, apart from its other functions, plays the same role in the muscles of many invertebrates as creatine plays in other animals. We may describe this role as follows:

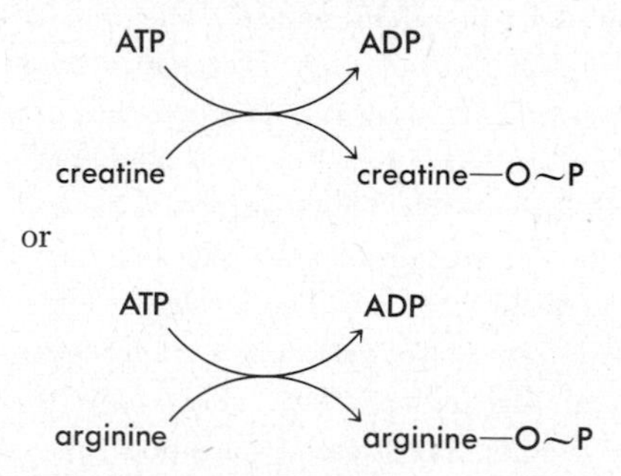

In other words, creatine or arginine may accept —O ∼ P from ATP and so become creatine-phosphate and arginine-phosphate, respectively. These last two compounds as well as others functioning in like manner are referred to collectively as *phosphagens.* Conversely, a phosphagen may donate —O ∼ P to ADP and so revert to compounds such as creatine or arginine.

The adaptive value of these reactions in muscle is clear. If, during rest, fuels supply more —O ∼ P than can be harvested as ATP, then the reactions above proceed to the right; ATP unloads —O ∼ P into phosphagen and becomes ADP. This ADP is now free to collect more —O ∼ P from fuel. Phosphagen stores accumulate in this manner in far greater quantities than ATP could accumulate. When a muscle subsequently becomes active, the energy of the ATP in actomyosin-ATP is used up, as noted earlier. If then the muscle is to be reenergized, new actomyosin-ATP must be formed. The immediate energy sources for such "recharging" of muscles are the phosphagens. They transfer their —O ∼ P and later

6.2. The energy relations in muscle activity. Respiration supplies energy for muscles via the phosphagen stores.

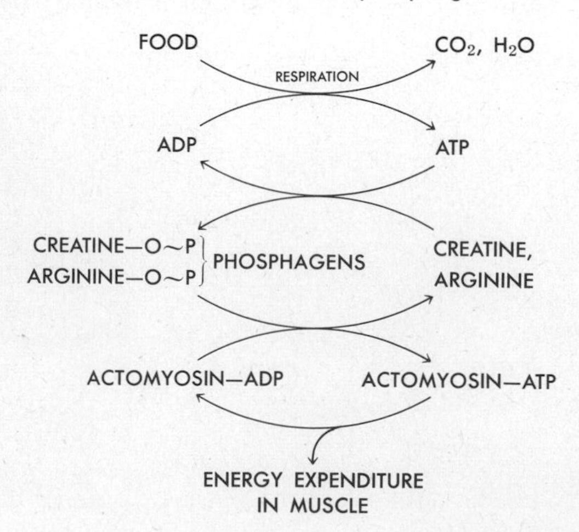

they are themselves reenergized by respiration. Thus, respiratory ATP slowly and continuously replenishes the phosphagen stores, and these ample stores rapidly and repeatedly re-create actomyosin-ATP while a muscle is active. Figure 6.2 summarizes these energy relations.

Muscular activity clearly can continue only as long as the energy stores of phosphagen last. If these stores become exhausted, actomyosin-ATP cannot be regenerated and muscle becomes fatigued. As noted in Chap. 5, fatigue is associated with comparative oxygen lack, fermentation, and lactic acid accumulation. Muscle *can* contract in the complete absence of oxygen so long as fermentation alone can maintain the phosphagen stores and so long as lactic acid concentrations are not excessive. Indeed, muscle normally probably respires anaerobically as well as aerobically, and any lactic acid formed can be carried off by blood as fast as it appears. Lactic acid tends to accumulate only during intense activity, and increasing fatigue then brings the activity to a halt sooner or later. Thereafter, aerobic combustion of muscle glycogen continues at a rapid pace, and depleted phosphagen stores are replenished. At the same time, lactic acid slowly diffuses into the blood, becomes liver glycogen, and returns to muscle and other tissues as blood glucose.

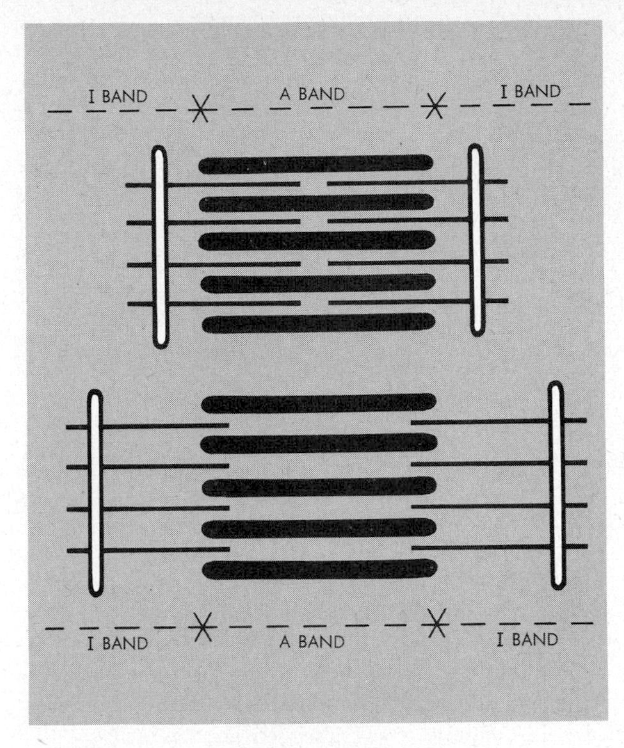

6.3. The presumed sliding action of actin during muscle activity. Myosin, black bars; actin, solid lines. Top, contracted state; bottom, extended state. The *A* band and the portions of the adjacent *I* bands of a muscle fiber shown here are intended to represent a working unit repeated horizontally many times in a whole fiber. Cf. Fig. 6.1.

The action cycle. What is the detailed mechanism of muscle contraction and extension? This aspect of muscle activity is just beginning to be clarified, and complete answers are not yet available. It is generally believed that, in an actomyosin-ATP complex, the actin and myosin components are arranged as parallel fibrils which may be joined side by side, perhaps by temporary chemical cross-linkages. The myosin fibrils are thought to extend lengthwise through the A bands, contributing to the dark appearance of these bands. The actin fibrils may extend through an I band and overlap partly at each end with the ends of myosin fibrils. Contraction and extension of such units is thought to be brought about by a sliding of actin fibrils past stationary myosin fibrils (Fig. 6.3).

At least three groups of data must be taken into account in any hypothesis designed to explain how such sliding movements would be initiated and controlled. First, it is known that the specific trigger for the contraction of a muscle is a nerve impulse. Secondly, it is known that inorganic ions, notably Mg^{++} and Ca^{++}, are associated with or attached to the actomyosin-ATP complex. By virtue of such positive charges, the complex may possess an electric potential over its surface; one of the known effects of nerve impulses is to bring about a reduction of electric potentials (cf. Chap. 8). Thirdly, it is known that muscles, like several other tissues, contain ATPases, i.e., enzymes which promote the conversion of ATP into ADP. The specific ATPase of muscle either is identical with myosin itself or is so closely linked with myosin that available techniques are unable to separate the two. Evidently, the actomyosin-ATP complex possesses not only built-in potential energy in the form of ATP but also the necessary built-in enzyme which may make this energy available. The action of the enzyme is believed to remain inhibited in some unknown way prior to arrival of a nerve impulse; directly or indirectly, the impulse appears to be the necessary stimulus for ATPase activation.

The following events might then take place during muscle action (Fig. 6.4). Initially, the actin and myosin components of actomyosin-ATP would be stretched apart, and the muscle would be extended. Such a condition might be maintained by the electric charges; since like charges repel one another, any contraction of the actomyosin-ATP complex would be pre-

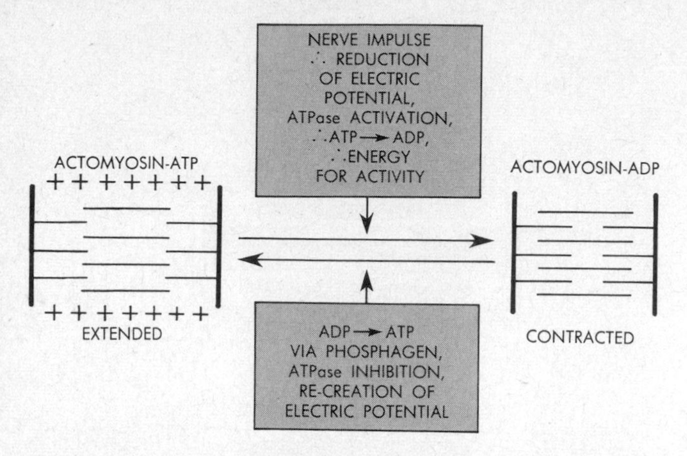

6.4. Summary of probable events during a unit cycle of muscle activity. Known events correlated with contraction are listed in the box above the reversible arrows; events correlated with extension, in the box below the arrows.

vented by electric repulsion of the actin and myosin. However, when a nerve impulse arrives, it may reduce the electric potentials and thus might remove the obstacle to contraction. The nerve impulse would also activate the ATPase, ATP would be split, and the potential energy of actomyosin-ATP would become actual. As noted, it is still not quite clear whether this energy brings about the actual contraction of the actomyosin complex or a reextension after the complex has contracted. In either event, the energy might promote the formation of new cross-linkages between actin and myosin. After the energy is spent, new potential energy is supplied by phosphagen. Muscle extension and recovery must also be accompanied by an inhibition of ATPase activity, by rebuilding of electric potentials, and by a sliding apart of the actin and myosin components.

Whatever the actual details of the fundamental action cycle in muscle may prove to be, cycles of this sort clearly take place fast enough to propel a cheetah, for example, at speeds of 50 mph; and they are powerful enough to permit many animals to lift objects weighing more than the animals themselves.

Other ATP Functions

Nonmuscular movement. All animal motion, muscular or otherwise, appears to be ATP-dependent. However, it is virtually unknown how the chemical energy of ATP is translated into the mechanical energy of motion. In the case of flagella and cilia, the locomotor apparatus is at least identifiable. Some evidence suggests that the beat of a flagellum or cilium might be produced by alternate contraction and relaxation of ultrafine protein filaments. If so, a machinery somewhat like that in muscles may conceivably be involved. In the case of amoeboid motion, distinct cell structures specialized to produce movement do not appear to exist. The machinery for locomotion here undoubtedly resides diffusely in all or most parts of the cell cytoplasm. Our understanding of these locomotor processes is at present limited to the elementary and rather unenlightening observation that if energy is unavailable to a cell, propulsion cannot occur.

Not all motion is locomotion, and some of these nonlocomotor movements within cells occur universally. For example, all cells move nutrients through their boundaries, both in absorption and secretion. We have already spoken of the requirement of respiratory energy in some of these processes, e.g., in the active transport of water and minerals and in phosphorylation during glucose uptake. All cells move compounds also within their substance, partly through diffusion, partly through cyclosis. The role of ATP is less clear here, but that it plays some role, even if very indirectly, seems almost certain. For example, localized heat production by ATP may create convection currents which might contribute to cyclotic streaming of cytoplasm. Moreover, cyclosis stops if respiration stops. Among other intracellular movements are the precise migrations of chromosomes during cell division (see below). The mechanism of these motions is again unknown. Some preliminary evidence suggests that contractile protein filaments energized by ATP might play a role here just as in flagellary and muscular motion.

In addition to such intracellular movements, groups of cells and indeed whole tissues and organs undergo numerous types of motions associated with growth, development, and the maintenance of steady states. We shall discuss some of these movements in later contexts and note here only that all of them are undoubtedly ATP-dependent too; if the ATP supply of an animal is stopped, the various movements also stop.

But the energizing of mechanical cell functions is not the only physical role of ATP. Production of heat, of light, and of electricity is an additional household task of many a cell type, and ATP is again the energy donor.

Heat production. One source of internal heat has already been referred to in the last chapter; if the high-energy phosphate of ATP is used in low-energy phosphorylations, then any excess energy of —O ~ P becomes heat. Another, and most important, heat source is food which, as we have seen, yields about 40 per cent of its energy as ATP but about 60 per cent directly as heat. Still another internal heat source is ATP-energized movement; friction of moving parts generates heat. Moreover, ATP is not used with 100 per cent efficiency in the production of movement. Conversion of the chemical energy of ATP into the mechanical energy of motion is accompanied by a loss of energy; this energy dissipates into the substance of a cell in the form of heat. Added to whatever heat is supplied by the external environment, food- and ATP-derived heat maintains the temperature of an animal and offsets heat lost to the environment by evaporation and radiation, creates tiny convection currents within cells and so assists in diffusion and cyclosis, and, above all, provides adequate operating temperatures for enzymes and all other functional parts of cells.

In birds and mammals, heat production is balanced dynamically against heat loss and a constant body temperature is thereby maintained. These animals are said to be "warm"-blooded, or *homoiothermic*. In all other animals, proper internal operating temperatures are maintained ultimately by the external environment, and the internal temperature of such animals by and large matches that of the external. These animals are said to be *poikilothermic* (the term "cold"-blooded often used is rather inappropriate since the blood may be either hot or cold, depending on external temperatures, and since many of these animals do not even possess blood). If the environment is either too cold or too hot, poikilothermic forms cannot survive. Within these extremes, however, ATP and food combustion generally create internal heat, which to some extent counteracts low environmental temperatures; just as the cooling effect of evaporation may reduce internal heat, and this to some extent counteracts high environmental temperatures.

Clearly then, heat is an essential requirement of every animal, and if food and ATP served in no other function than

6.5. The principal sources and functions of heat in animals.

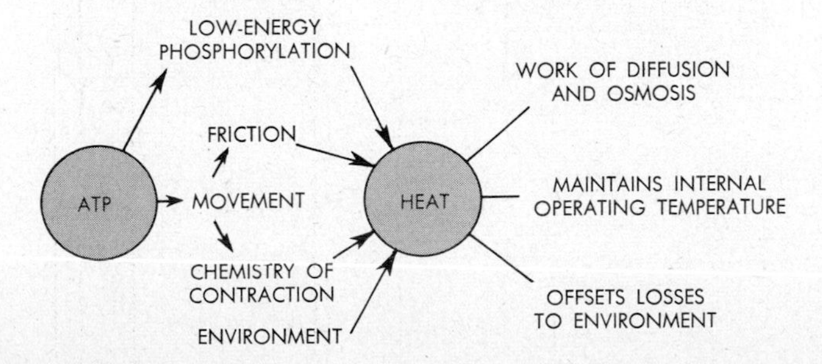

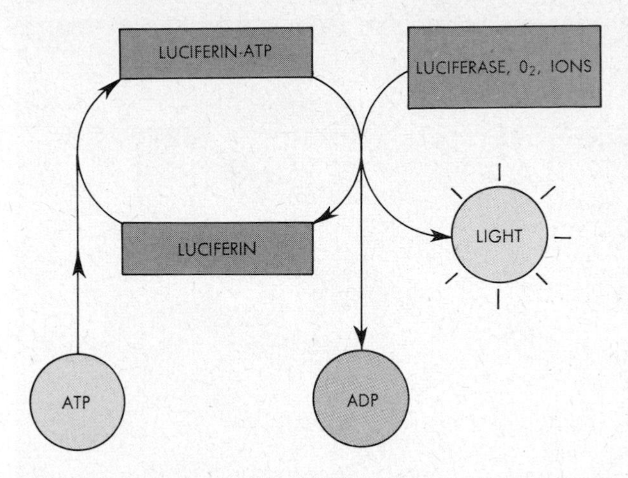

6.6. The general pattern of light production in bioluminescent animals.

heat production, they would still be among the most vitally necessary components of cells (Fig. 6.5).

Bioluminescence. "Living" light is emitted by virtually all major animal groups; most phyla include marine or terrestrial representatives which are bioluminescent. The capacity to produce light has developed independently several times during evolution, yet the essentials of the light-generating mechanism appear to be alike in all cases.

This mechanism consists of at least six components: water, inorganic ions, oxygen, ATP, and two groups of substances called, respectively, *luciferin* and *luciferase*. These last differ in composition in different species. Luciferin and the enzyme protein luciferase are the principal light-generating elements. They can be extracted from light-producing cells, and they are nonluminous on their own. If ATP is added to luciferin, a luciferin-ATP complex is formed. If, in the presence of ions and oxygen, a solution of luciferase is now added, the mixture emits light. At the same time, oxygen is used up and ATP becomes ADP. If, after the light disappears, more oxygen and more ATP are added, light is generated again. Light production evidently is an oxygen-requiring, ATP-dependent process (Fig. 6.6).

In bioluminescent animals, light emission depends on nervous stimulation of specialized cells in light-producing organs (Fig. 6.7). The light emitted by different animals may be of any wavelength in the visible spectrum; i.e., to the human eye it may be red, yellow, green, or blue. Little or no nonvisible radiation is generated. The actual wavelength of

6.7. Representation of a school of deep-sea squids, showing pattern of bioluminescent organs on each animal.

the emission is probably determined by the particular chemical makeup of luciferin. In some cases, two or more kinds of luciferin may occur in a single animal, and such an animal then may light up in several colors. In all cases, the available energy is spent very efficiently, for little heat is lost during light production. Hence the frequent designation of living light as "cold" light. Also, the unit intensity of the light is remarkably great. It compares favorably with that of modern fluorescent lamps.

Bioelectricity. Bioelectricity is a byproduct of all cellular processes in which ions play a part. In other words, electricity is as common throughout the living world as table salt. However, certain eels and rays are highly specialized in their capacity to produce electricity. These fish possess *electric organs* composed mainly of modified muscles. The component cells are disk-shaped and noncontractile, and they are piled into stacks. Assemblies of this sort have an appearance and a function reminiscent of storage batteries connected in series.

The details of operation here are understood less well than those of light production. However, it is known that the generation of electricity depends on ATP and a substance called *acetylcholine.* This chemical will be encountered again later, for it functions widely as a key agent in the transmission of nerve impulses. It also functions in the generation of bioelectricity. This event is apparently accompanied by a splitting of acetylcholine into separate acetyl and choline fractions. The two are then recombined into acetylcholine, with energy from ATP (Fig. 6.8).

As in light production, the efficiency of energy utilization is remarkably great. So also is the intensity of the electricity generated. An electric eel may deliver a shock of up to 400 volts, enough to kill another fish or to jolt a man severely or to light up a row of electric bulbs wired to a tank into which such an eel is put. Nervous stimulation of the electric organ triggers the production of electricity.

It is still unknown just how the chemical energy of ATP is actually converted into light energy or electric energy. But that ATP is the key is clearly established, and this versatile compound emerges as the source of all forms of living physical energy, usual or unusual (Fig. 6.9). Indeed, ATP is even more versatile, for it is also the source of all living chemical energy.

CHEMICAL ROLES OF ATP: SYNTHESIS

The energizing of *synthesis* reactions, i.e., those which are endergonic and have a net $+\Delta F$, represents the chief chemical role of ATP. A cyclical interrelation is therefore in evidence. On the one hand, breakdown of organic compounds leads to a net buildup of ATP through respiration. On the other hand, breakdown of ATP leads either to physical activity as discussed above or to a net buildup of organic compounds through chemical synthesis. Figure 6.10 outlines this basic cycle of energy and materials which governs the overall metabolism of all cells.

6.8. The general pattern of the production of bioelectricity.

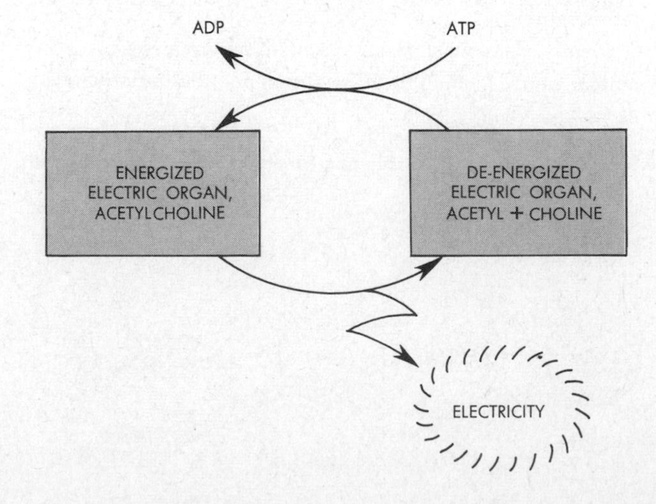

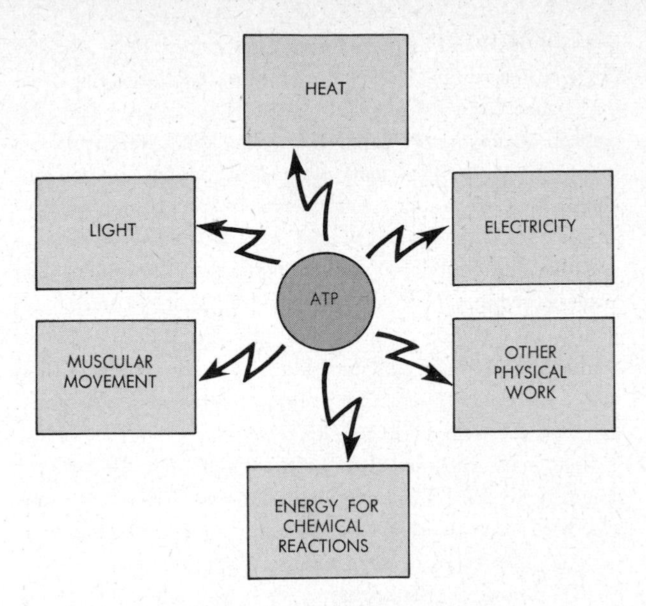

6.9. General summary of the functions of ATP.

Synthesis of cellular components and breakdown occur simultaneously, all the time. As already noted in the last chapter, breakdown may affect any cellular constituent regardless of composition or age. A protein just synthesized through long reaction sequences and at great expense of energy is as likely to be destroyed as a glucose molecule already present for days. A certain percentage of all cellular constituents is decomposed every second. Which constituents actually make up this percentage is largely a matter of chance.

Such randomness applies also to synthesis. Regardless of the source of materials, a certain percentage of available molecular components is synthesized every second into finished cell substance. If synthesis and breakdown are exactly balanced, the net characteristics of a cell may remain unchanged. But continuous turnover of energy and materials occurs nevertheless, and every brick in the building is sooner or later replaced by a new one. Thus the house always remains "fresh."

Synthesis and breakdown cannot sustain each other in a self-contained, self-sufficient cycle, even when the two processes are exactly balanced; for energy dissipates irretrievably through physical activities and through heat losses in chemical reactions, and materials dissipate through elimination, evaporation, and friction. Just to maintain a steady state, therefore, a cell must be supplied continuously with energy and raw materials. Very often, moreover, the rate of supply must exceed the rate required for mere maintenance, for net synthesis may exceed net breakdown. This is the case, for example, in growth, in repair after injury, and in cells which manufacture secretion products.

Two broad classes of synthesis reactions may be distinguished: *maintenance synthesis,* in which the reaction products stay within the producing cell and usually contribute to the survival of that cell, and *export synthesis,* in which the reaction products leave the producing cell and often contribute to the survival of other cells.

Patterns of Synthesis

The overall function of maintenance synthesis is the manufacture of all those cellular constituents which a cell does not obtain directly as prefabricated nutrients or secretions from other cells. Such missing constituents include most of the critically necessary compounds for cellular survival: nucleic acids, structural and enzymatic proteins, polysaccharides, fats, and numerous other groups of complex organic substances.

Many of these compounds are manufactured by reactions which are virtually the reverse of respiratory breakdown. For example, if read in reverse, the summary of respiratory pathways in Fig. 5.31 provides a general outline of polysaccharide and fat synthesis. Indeed, the same enzymes, coenzymes, and other reaction aids often function at the same steps in the

6.10. The fundamental metabolic balance of cellular energy and materials. A less comprehensive version is given in Fig. 5.1.

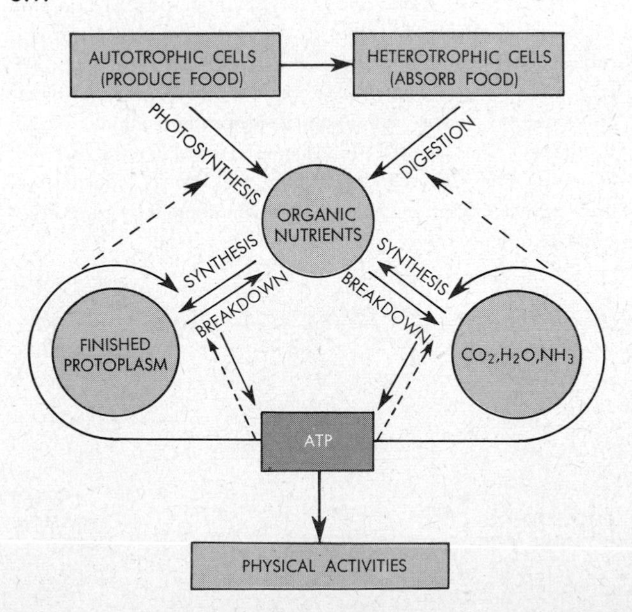

reversed sequences. ATP is expended here rather than gained, and this imposes a $+\Delta F$ and a reversed direction on the reactions. We may note, however, that animals are unable to reverse the final respiratory CO_2 release. Only so-called *autotrophic* cells can convert inorganic carbon in the form of CO_2 into organic carbon in the form of, for example, carbohydrates. Autotrophic cell types are largely the green photosynthesizing cells of plants and only these, in other words, are able to manufacture their own food. Animal cells are heterotrophic, we already know, and they depend on already preexisting foods, i.e., organic nutrient molecules. However, once organic compounds are given, animal cells can add, or "fix," CO_2 and thereby increase the carbon numbers of organic compounds. For example, phosphoenolpyruvic acid, a C_3 intermediate in glycolysis, may fix CO_2 and transform into oxaloacetic acid, a C_4 intermediate of the citric acid cycle. Analogously, pyruvic acid (C_3) may by CO_2 fixation convert into malic acid (C_4). The B vitamin *biotin* participates in such CO_2 fixations as a carrier coenzyme (Fig. 6.11).

In animal synthesis acetyl CoA often represents a fundamental starting compound. From it may be formed, for example, fatty acids of any length. The process is almost exactly the reverse of β-oxidation, the principal difference being the

6.11. CO_2 fixation in animal cells. The first reaction is reversible, the second apparently is not. Note that the first reaction provides a bypass of most of the citric acid cycle, or, in the reverse direction, a bypass of oxaloacetic acid and acetyl CoA. The second reaction analogously bypasses the acetyl CoA step of the citric acid cycle. Note also that the vitamin *biotin* is a required participant and that IDP is the —O~P acceptor. IDP is inosine diphosphate, inosine being a combination of hypoxanthine and ribose. Hypoxanthine in turn is a purine, related to adenine. Thus, the IDP-ITP system functions like the ADP-ATP system. Other CO_2-fixing reactions are known, and through them particular carbon chains may be lengthened by one carbon atom at a time.

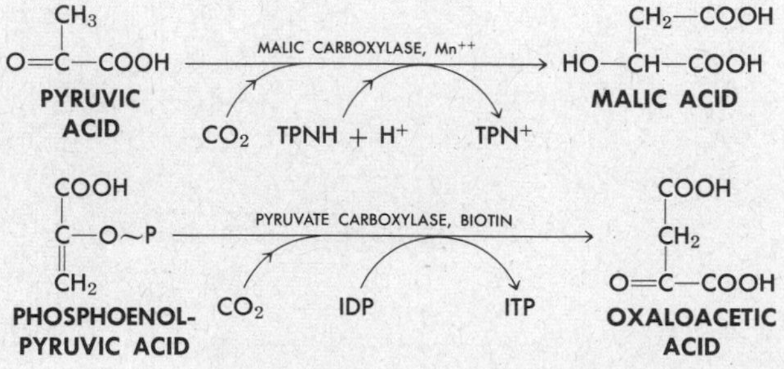

CH_3—C(=O)—COOH (PYRUVIC ACID) + CO_2 + TPNH + H^+ —MALIC CARBOXYLASE, Mn^{++}→ HO—CH(CH_2—COOH)—COOH (MALIC ACID) + TPN^+

COOH—C(—O~P)=CH_2 (PHOSPHOENOL-PYRUVIC ACID) + CO_2 + IDP —PYRUVATE CARBOXYLASE, BIOTIN→ COOH—CH_2—C(=O)—COOH (OXALOACETIC ACID) + ITP

last step. Here the hydrogen donor is reduced TPN, whereas in the corresponding step of β-oxidation (reaction 2, Fig. 5.23) the H acceptor is oxidized DPN. Acetyl CoA is also a participant in many other syntheses in which C_2 fragments are required, e.g., in the formation of C_3 compounds and polysaccharides. Pyruvic acid may serve as precursor in polysaccharide synthesis, and so also may glucose. Free glucose may be phosphorylated to form glucose-1-phosphate. The latter either may become glucose-6-phosphate or may react with UTP (uridine triphosphate) to form a UDP-glucose complex. Enzymatic polymerization of such complexes then yields glycogen. These and various related synthesis pathways are summarized in Fig. 6.12.

Cells may synthesize some of the more than 20 types of amino acids by transamination reactions. Alpha-keto acids may here serve as $—NH_2$ acceptors, and other amino acids or various other types of nitrogenous compounds may serve as $—NH_2$ donors. Notwithstanding such transaminations, however, animal cells are unable to convert carbohydrate or lipid raw materials into the R— groups of about 10 so-called "*essential*" kinds of amino acids. These R— groups must be supplied in fully prefabricated form via food and, ultimately, from plants. Indeed, plants also must be the ultimate source of several other categories of substances. Among these, as already pointed out, is primary organic carbon formed by plant photosynthesis. Another such category is usable nitrogen, e.g., $—NH_2$ groups. Plants are able to produce usable amino nitrogen from mineral nitrates (NO_3^- ions), but animals cannot utilize such inorganic sources. They also cannot make metabolic use of aerial nitrogen (cf. Chap. 8). Thus, they depend on one another and ultimately on plants for their usable nitrogen supply. In addition to providing essential amino acids, moreover, plants must be the source of several "essential" fatty acids, i.e., acids which animal cells require but cannot synthesize on their own. Finally, plants must supply most vitamins. Animals vary greatly in their vitamin-synthesizing capacities, but almost no animal is able to produce all required vitamins in required amounts (Fig. 6.13).

Such differences between plants and animals and among animals undoubtedly are a result of mutation and evolution. In all probability, the ancestors of animals—as of plants—were able to synthesize all needed cellular compounds on their own. In the course of time, random mutations must have led to the loss of various synthesizing abilities in different organisms. If the affected organism was photosynthetic, it must have become extinct, for it could not have obtained the

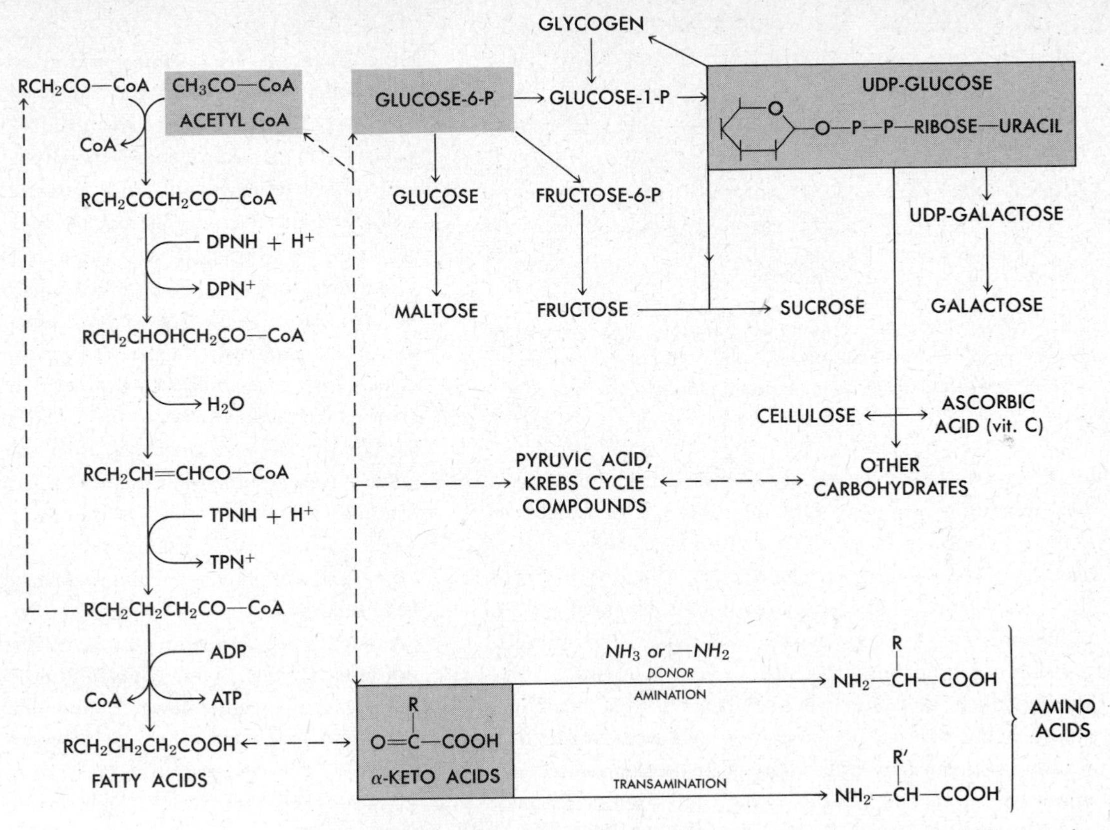

6.12. Some pathways in cellular synthesis. Compounds in shaded rectangles represent major precursor substances in synthesis. Solid-line arrows indicate synthesis sequences, broken-line arrows, interrelations already discussed in Chap. 5. Fatty acid synthesis is outlined in column at left. Reactions here are the reverse of β-oxidation until the TPN step. The source of TPN · H + H^+ is usually the hexose shunt cycle (cf. Fig. 5.30). Repeated additions of acetyl CoA to a given fatty acid (as suggested by the broken-line arrow) result in longer fatty acids, a 2-carbon fragment being linked on each time. In the upper right, above, some of the pathways of carbohydrate synthesis are summarized; the details of the reactions here have already been discussed (in the reverse direction) in Chap. 5 (e.g., cf. Fig. 5.31). In the lower right, analogously, pathways of amino acid synthesis are summarized. Note here that the particular —*R* groups which form part of the "essential" amino acids of animals must be supplied in finished form via food.

missing ingredients in any other way; and all plants surviving today still must synthesize all needed ingredients on their own. But if the affected organism was an animal, it could survive readily, for, as a heterotroph, it could obtain any missing compound from plants or another animal, by way of food. That mutations may indeed destroy cellular capacities of synthesizing certain compounds can be demonstrated experimentally.

Many maintenance syntheses in a cell do not follow any of the reaction pathways referred to thus far. Included here particularly are the reactions leading to the formation of nucleic acids and proteins, to be examined separately in the

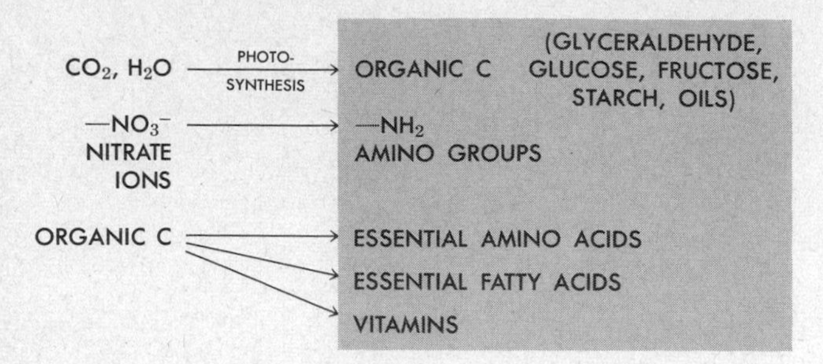

6.13. Five classes of compounds that green plants can synthesize and that the cells of most animals cannot synthesize. Animals consequently depend on green plants for these products.

next section. Also included are thousands of other, special manufacturing processes. Most of these we shall not have occasion to deal with at all. One, the biosynthesis of uracil- and cytosine-containing nucleotides, is outlined in Fig. 6.14, to serve as an example of special syntheses generally and also to illustrate one of the specific reaction sequences required as a necessary preliminary to nucleic acid synthesis.

Collectively, all maintenance syntheses of a cell function in counteracting normal decomposition and wear and tear, in replacing cellular parts after injury (as in healing and regeneration), and in making possible cellular growth, development, and reproduction. Additionally, depending on its particular specialization, a cell may also carry out export syntheses, as noted. Every cell actually is an exporting cell to some extent; for at the very least it exports metabolic wastes such as CO_2 and often water. The term *excretion* is generally used to refer to exported wastes, although what is waste in one cell may often be an essential metabolite in another cell (e.g., H_2O). Specially synthesized products which are exported from cells and are clearly not wastes are given the general designation *secretion*.

Substances exported from cells may have a variety of roles: *digestive* (e.g., enzymes poured into the gut); *excretory* (e.g., urea formation by liver cells); *regulative* (e.g., hormones secreted by given animal cells); *supportive* (e.g., secretion of bone or shell substance); *reproductive* (e.g., aromatic scents secreted by animals); or variously *protective* (e.g., secretion of irritants and poisons). Indeed there are few functions in any animal that do not require cellular exports of some sort.

Single cells and more particularly groups of cells specialized for the manufacture of given secretions are known as *glands*. Animal glands are broadly of two types. So-called *endocrine* glands are ductless, and they secrete into the blood. *Hormones* are the characteristic products of endocrine glands. Other secretions are manufactured in *exocrine* glands, which empty their products into free spaces or into ducts. Among glands of this type are digestive glands (e.g., liver, pancreas, salivary glands), skin glands (e.g., sweat glands, various oil- and wax-secreting glands), and numerous glands associated with the reproductive, circulatory, and other systems. We shall refer to a variety of such glands in later contexts.

We note that an animal cell obtains required ingredients in three general ways. Some components, like water, mineral ions, and organic materials such as vitamins and certain amino acids, must come in already synthesized final form from the external environment. Some others, like hormones, must come as fully synthesized secretions from other cells within the animal. And all other cellular materials must be synthesized from basic food compounds within the cell itself (Fig. 6.15). This total multitude of chemicals in a cell, built up at the expense of ATP, then maintains and perpetuates the body of a cell. But it must not be imagined that newly synthesized compounds just happen to arrange themselves into new living substance. If the proteins, fats, and other components were merely mixed together in water, the result would be a complex but lifeless soup. As has long been appreciated, *omnis cellula e cellula*—all cells arise from preexisting cells; all life arises from preexisting life. New cellular constituents become living matter only if older living matter provides the framework; the house may be added to and its parts may be replaced or modified, but an altogether new house cannot be built. That apparently occurred only once during the history of the earth, i.e., when living matter arose originally.

The most important single type of synthesis occurring in any cell is protein manufacture. As we shall see, it is associated intimately with nucleic acid synthesis. Proteins form the enzymes required for everyone of the hundreds of metabolic and self-perpetuative processes in a cell, and proteins are the principal architectural constituents of the cell. Thus, the whole structural and functional nature of a cell is determined and maintained by its proteins, and protein synthesis clearly demands extended attention.

Genes and Protein Synthesis

The pattern. The biosynthesis of proteins is bound to differ in at least one fundamental respect from the manufacture of many other cellular compounds: a cell cannot use just any newly made proteins, but only *specific* proteins. These must be largely different in structure from proteins formed in other types of cells or other animals; for only if newly formed proteins are exactly like those present earlier can a cell maintain its own special characteristics. Such a specificity problem does not arise with most other kinds of compounds. A glycogen molecule, for example, being composed of identical glucose units, will be almost automatically like any other glycogen

6.14. The pathway of cellular pyrimidine synthesis. The enzyme required specifically in step 2 is *aspartic transcarbamylase*. In this reaction, the atoms in carbamyl-P and aspartic acid emerge as phosphoric acid (H—O—P), and the NH_2—CO— part of carbamyl-P becomes linked to aspartic acid (where the dotted line is drawn in ureidosuccinic acid). In step 3, ring closure occurs by removal of H_2O. Also, a dehydrogenase in conjunction with DPN^+ catalyzes H_2 removal from the —CH_2—CH— group of ureidosuccinic acid. The resulting orotic acid is the key compound in the synthesis of pyrimidine nucleotides. The ribose-phosphate parts of a nucleotide are contributed by the compound phosphoribosyl pyrophosphate, and through steps 4 and 5 the uracil derivative UMP is formed (the source of the byproduct CO_2 here being the —COOH group of orotic acid). Through reactions outlined in steps 6 through 9, other uracil and also cytosine and thymine derivatives may be manufactured (i.e., all the principal pyrimidine nucleotides in cells). Note that TMP manufacture requires the vitamin *biotin*, an agent essential in the transfer of a methyl (—CH_3) group from a donor molecule to the TMP-producing reaction. See Chap. 4 for the structure of nucleotides not depicted fully below.

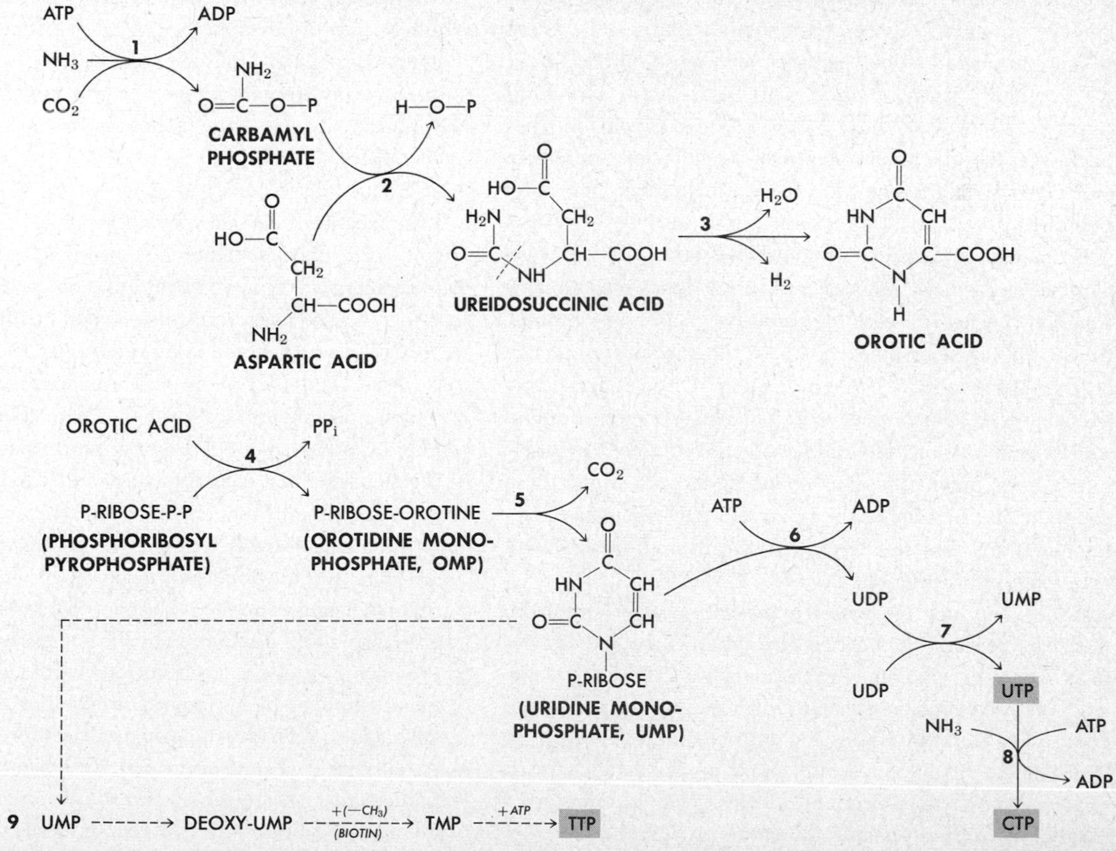

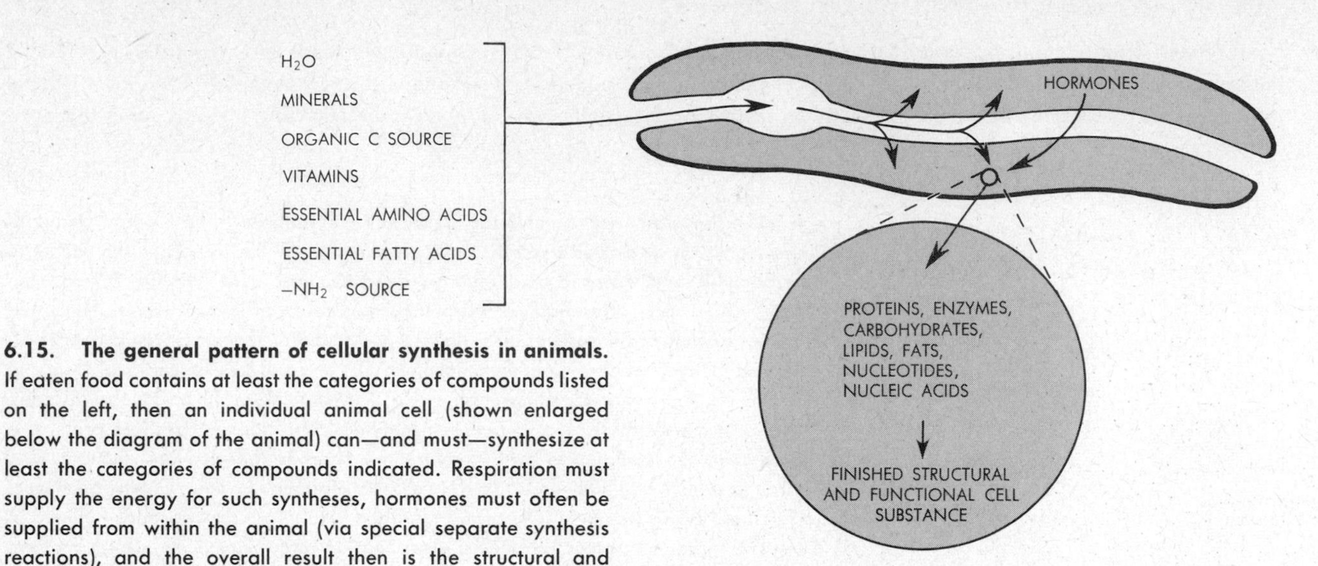

6.15. The general pattern of cellular synthesis in animals. If eaten food contains at least the categories of compounds listed on the left, then an individual animal cell (shown enlarged below the diagram of the animal) can—and must—synthesize at least the categories of compounds indicated. Respiration must supply the energy for such syntheses, hormones must often be supplied from within the animal (via special separate synthesis reactions), and the overall result then is the structural and functional maintenance of the cell.

molecule containing the same number of glucose units. But a protein molecule is composed of more than 20 different kinds of amino acid units, and random linking together here is likely to make one polypeptide chain quite different from every other. Clearly, protein synthesis requires *specificity control;* a "blueprint" must be available, to provide instructions regarding the precise sequence in which given numbers and types of amino acids are to be polymerized into polypeptides.

Ultimately, such specificity control is exercised by the *genes* of a cell, the genetic DNA of the chromosomes. *The primary function of genes is to control specificities in protein synthesis.* As already noted briefly in Chap. 4, the particular sequence of the nitrogen-base pairs in genetic DNA represents a chemical *code,* which specifies a particular sequence of amino acid polymerization. The different genes in a cell carry different codes, and a cell can manufacture proteins only as these codes dictate. In so far as the genes themselves remain stable, therefore, any new sets of proteins synthesized in a cell will match the preexisting sets precisely.

The "blueprints" are housed in the chromosomes within the nucleus, but the "factories" where proteins are actually put together are the ribosomes, in the cytoplasm. Evidently, some kind of instruction-transmitting device must exist. It does exist, in the form of RNA. The chromosomes are known to manufacture RNA, and in the course of this manufacturing process the chemical code of DNA becomes transferred exactly, or *transcribed,* into the structure of RNA. Chromosomal RNA thus becomes just as specific as the genetic DNA. The specific RNA molecules then leave the chromosomes and diffuse into the cytoplasm (possibly after temporary storage in the nucleoli), and they eventually reach the ribosomes. Here amino acids are joined together into proteins in accordance with the genetic code supplied by the RNA molecules (Fig. 6.16).

We may therefore regard genes essentially as passive information carriers. All that they do, or allow to be done to them, is to have their specific code-information copied by RNA molecules, which serve as information carriers in their turn. Genes consequently may be likened to important original "texts" carefully stored and preserved in the "library" of the nucleus. There they are available as permanent, authoritative "master documents" from which expendable duplicate copies may be prepared. RNA passing into the cytoplasm actually is expendable and comparatively short-lived. Very soon after it has exercised its function as code carrier, it is destroyed and respired away; new RNA from the nucleus must then become available in the cytoplasm if repeated protein synthesis is to occur. The genes on the other hand persist and are protected from respiratory destruction by the nuclear boundary. Clearly, presence of a distinct cell nucleus with a distinct boundary membrane may be an important evolutionary adaptation, an advantageous device which might permit preservation of gene stability. We may also appreciate the advantage of having the permanent message center and the manufacturing center at different locations within a cell. If both were at the same location in the nucleus, the manufacturing center probably would be too distant and isolated from the energy sources and the raw material supplies; and if both were at the same location in the cytoplasm, the message center would be subject to rapid respiratory destruction.

Inasmuch as the RNA manufactured in the chromosomes carries chemical messages to the ribosomes, it is called *messenger* RNA, or *m*RNA. Two additional types exist in a cell. One is *ribosomal* RNA, or *r*RNA; it is a normal structural component of the ribosomes. Conceivably, *r*RNA may play a role in ensuring the proper attachment of *m*RNA to the ribosomes. It is known that *m*RNA, a chainlike polymer of nucleotides and thus shaped like a filament, must preserve its extended, filamentous form if it is to function properly. *m*RNA is believed to become draped over the surface of one or more ribosomes, like a thread becoming draped over one or more spherical objects; and it is possible that *r*RNA in the ribosomes supplies a matching surface which may force or otherwise direct the proper linear deposition of *m*RNA (Fig. 6.17).

The third type of RNA is *transfer* RNA, or *t*RNA. It functions as *amino acid carrier.* A cell possesses more than 20 different kinds of *t*RNA, just as many as there are different kinds of amino acids. When a particular kind of amino acid enters a cell as food and is to be used in protein synthesis, a specific corresponding kind of *t*RNA becomes attached to the amino acid and carries it to the ribosomes. Here the *t*RNA "delivers"

6.16. The general pattern of protein synthesis. The genetic code in the chromosomal DNA of a cell is transcribed into RNA, and from there the code is further transcribed into specific amino acid sequences in proteins manufactured in the ribosomes of the cytoplasm.

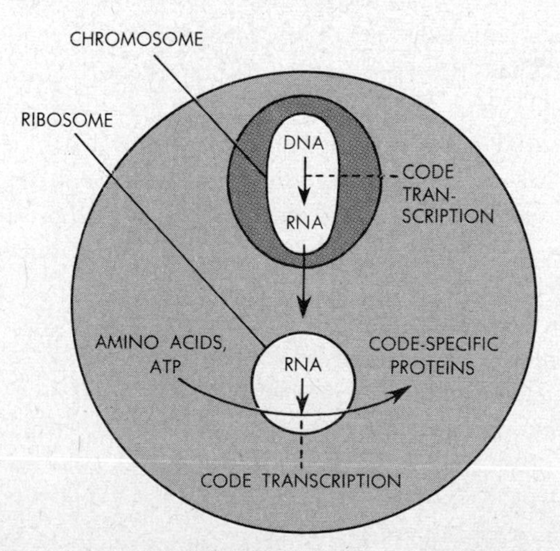

its amino acid at a particular place along the *m*RNA filament already present, and other *t*RNA carriers similarly deliver their amino acids at other specific places along the *m*RNA. Large numbers of amino acids so become lined up along *m*RNA in a particular sequence. Through a mechanism to be described presently, the specific nature of this sequence has been determined by the code within the *m*RNA. The "correctly" stationed amino acids then become joined to one another by enzymatic formation of peptide bonds, and a polypeptide chain with a gene-determined specificity so results (cf. Fig. 6.17).

Code transcription: DNA $\longrightarrow$ *m*RNA. We already know that, in line with the Watson-Crick model of DNA structure, DNA is a spiraled double chain of polynucleotides (cf. Figs. 4.19, 4.20). In it, the genetic code is based on a four-letter alphabet of nitrogen-base pairs, viz., $A \cdot T$, $T \cdot A$, $G \cdot C$, $C \cdot G$, and the code consists of a specific succession of such pairs. If the code of a given segment of a DNA chain is to be transcribed, a first requirement appears to be an unspiraling of this segment and an at least temporary "unzipping" of the double chain into two separate single chains. The exact mechanisms through which these events might occur are still uncertain. For purposes of illustration, let us assume that a double DNA chain containing six nitrogen-base pairs does unzip into two single chains (Fig. 6.18, part 1). Let us also assume that the nitrogen-base sequence *CAATGA* of one of the single chains is to be transcribed into RNA.

The manufacture of RNA requires adequate supplies of raw materials. We know that RNA contains four types of ribonucleotides, viz., adenine-ribose-phosphate (adenylic acid, AMP), guanine-ribose-phosphate (guanylic acid, GMP), cytosine-ribose-phosphate (cytidylic acid, CMP), and uracil-ribose-phosphate (uridylic acid, UMP, which substitutes for TMP of DNA). It has been found that the actual cellular raw materials for RNA synthesis are not these monophosphates as such but their diphosphate derivatives: ADP, GDP, CDP, and UDP. Each of the latter carries a high-energy bond between the first and second phosphate groups.

The first nitrogen base in the DNA sequence to be transcribed (Fig. 6.18, part 1) is noted to be *C*, cytosine. We know that such a base may, by hydrogen bonding, join specifically to the nitrogen base *G*, guanine. If therefore the available raw materials in a chromosome include GDP ($G—R—P$ in Fig. 6.18, part 2) this molecule may become bonded to the *C* of DNA. Quite analogously, the *A*s of DNA may each bond to a UDP raw material, the *T* may bond to an ADP, and the

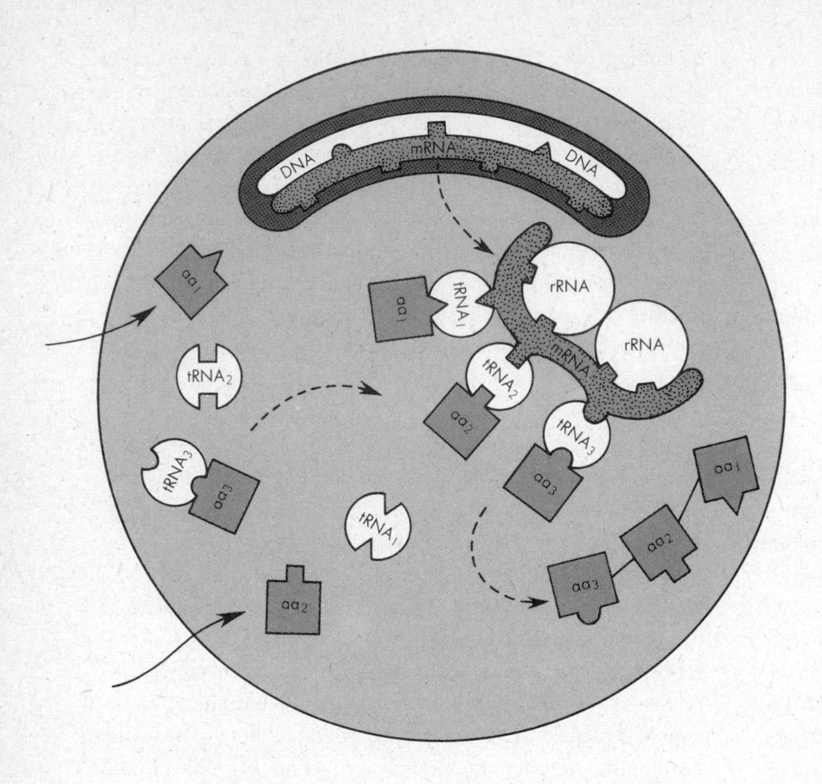

6.17. The pattern of protein synthesis. Specific (messenger) *m*RNA manufactured in chromosomes becomes attached to ribosomes, the (ribosomal) *r*RNA of the latter probably aiding in achieving the attachment. Amino acids (*aa*) entering a cell as food become joined to specific (transfer) *t*RNA molecules, and the *t*RNA-*aa* complexes then attach at specific, code-determined sites along *m*RNA. The amino acids subsequently link together through peptide bonding, and finished, code-specific polypeptide chains ultimately become free.

G may bond to a CDP (Fig. 6.18, part 2). In other words, the DNA sequence serves as a *template*, or mold, along which specific molecules of raw material may become attached in specific sequence. To form RNA, it is then necessary only to link the raw materials together into a polynucleotide chain. In cells this is accomplished with the aid of the high-energy bonds already built into the raw materials and with the aid of an enzyme, *RNA polymerase:*

$$(\text{N-base—ribose—O—P—O}{\sim}\text{P})_x \xrightarrow{\textit{RNA polymerase}} (\text{—O—P})_x + (\text{N-base—ribose—O—P})_x$$

In this equation, x refers to the number of ribonucleotide units linked together. In our example sequence, therefore, the result will be as in Fig. 6.18, part 3.

The finished RNA chain represents *m*RNA, which separates from the DNA template and eventually reaches a ribosome. Note that the specific DNA code is imprinted in *m*RNA in the form of *corresponding* nitrogen bases, in somewhat the same way that a photographic negative shows light objects as dark areas or that a plaster cast shows elevated objects as depressions. Such "inverted," negative codes in *m*RNA represent the actual working blueprints for protein synthesis.

*Code transcription: m*RNA ⟶ *protein.* How does the code in *m*RNA specify a given amino acid sequence—just what does the genetic code actually say?

We know the code must somehow "spell out" in chemical terms an identification of 20-plus different amino acids and that it must do so with a four-letter alphabet. Assuming that nature is as terse as it can be, how are 20 or more different identifying "words" constructed out of four letters, such that each word contains as few letters as possible? If the code consisted of one-letter words, a four-letter alphabet *A-B-C-D* would allow only four different identifications: *A, B, C, D.* If the code were made up of two-letter words, there could be 4^2, or 16 different letter combinations: *AA, AB, AC, AD, BB, BA, BC,* etc. Yet 16 combinations are still too few to specify 20+ words. However, if the code contained three-letter words, there could be 4^3, or 64 different letter combinations, more than enough to spell out 20+ words.

On the basis of such reasoning, it has been hypothesized and later actually confirmed that the genetic code "names" each amino acid by a sequence of three "letters," i.e., three adjacent nitrogen bases in DNA and *m*RNA. In such a *triplet code,* 20+ triplets would be "meaningful" and spell out amino acid identities; the remainder of the 64 possible triplets would be "meaningless."

Which N-base triplets identify which amino acid? The answer has been obtained through ingenious experiments. It has been possible to extract RNA polymerase from bacteria and to add to this enzyme, in the test tube, known mixtures of ADP, GDP, CDP, and UDP. From such reaction systems could be obtained artificial RNA molecules which contained known sequences of N-bases. To the RNA could then be added a mixture of known amino acids, and artificial polypeptides could so be synthesized in the test tube. Analysis of such polypeptides shows which amino acids they contain and in what sequence, and this information could be related directly to the N-base sequence in the RNA. In the first experiment of this kind, for example, only UDP was used as raw material for artificial RNA synthesis. The resulting RNA molecules then were "poly-*U*," i.e., they consisted entirely of a sequence of uracil-containing ribonucleotides. After mixed amino acids were added, it was found that the polypeptide formed in the test tube consisted entirely of the amino acid *phenyl alanine*. From this it could be concluded that the triplet code for phenyl alanine must be *UUU*.

6.18. Code transcription from DNA to *m*RNA. *A, T, G, C, U,* purine and pyrimidine bases; *R,* ribose; *P,* phosphate. In a first step, the DNA double chain unzips. The assumption is that the N-base sequence of the lower single chain is to be transcribed. In the second step, the nucleotide diphosphate raw materials ADP, GDP, CDP, and UDP become hydrogen-bonded to appropriate N-bases along the DNA chain. Formation of a linked ribose-phosphate chain, as in the third step, then yields finished *m*RNA, with an N-base sequence specifically determined by that of DNA.

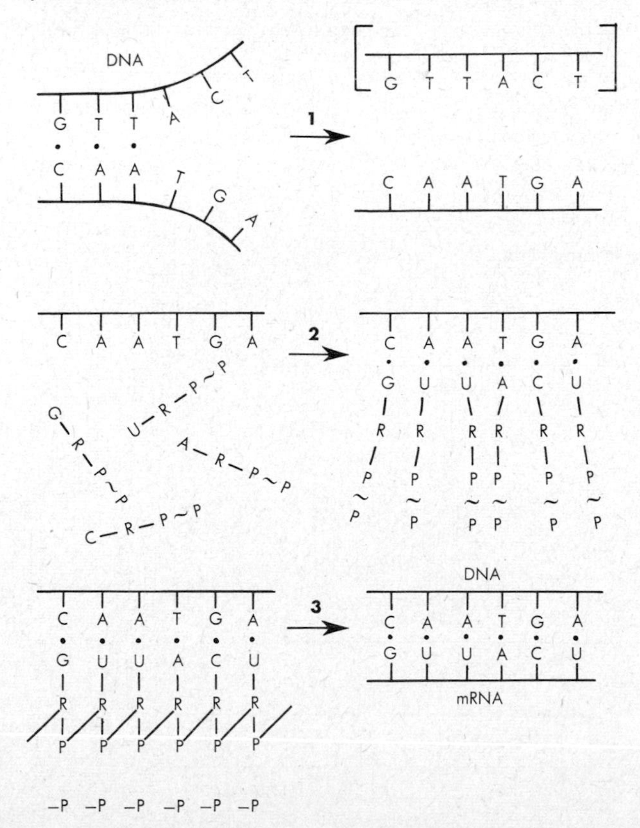

TABLE 5. Tentative triplet codes for amino acids in *m*RNA (*A*, adenine; *U*, uracil; *G*, guanine; *C*, cytosine)

Amino acid	Code triplet
alanine	*CCG, CGU*
arginine	*CCG*
aspartic acid	*GAU, CAA*
asparagine	*CAU, AAU, AAC*
cysteine	*GUU*
glutamic acid	*GAU, GAA*
glutamine	*AAC, AGG*
glycine	*GGU*
histidine	*CAU, CAC*
isoleucine	*AUU*
leucine	*AUU, GUU, CUU*
lysine	*AAU, AAA*
methionine	*GAU*
phenylalanine	*UUU*
proline	*CCU, CCC*
serine	*CUU, CCU*
threonine	*AAC, ACC*
tryptophane	*GGU*
tyrosine	*AUU*
valine	*GUU*

Later work along related lines has led to the identification of the code triplets for all 20+ amino acids (Table 5). It should be noted here that, although the N-bases for each meaningful triplet are now known, the exact sequence of the three bases within each triplet is still not clear in all cases. For example, we know that the code triplet for the amino acid aspartic acid contains *G*, *A*, and *U*, but we do not yet know whether the triplet reads *GAU* or *AGU* or *GUA* or as some other sequence of these three bases. Thus, although the code triplets for aspartic acid, glutamic acid, and methionine all contain *A*, *G*, and *U*, the three triplets undoubtedly differ in their internal base sequences. Moreover, it has been found that certain amino acids are coded by more than a single triplet. Leucine, for example, is coded as *AUU*, *GUU*, and *CUU*. Because of

R | R—enz. | R
NH₂—CH—C—OH → (1) → NH₂—CH—C—O—P—ADENOSINE → (2) → NH₂—CH—C—O—tRNA
O | O (AMP) | O

enz. + ATP → (1) → PPᵢ
tRNA → (2) → enz. + AMP

6.19. Amino acid activation. In reaction 1, the energy of ATP serves to link a specific enzyme and AMP to amino acid. In reaction 2, a *t*RNA specific for that amino acid and that enzyme replaces the AMP, and the enzyme disengages as well. The amino-acid-*t*RNA complex then migrates to *m*RNA.

such multiple codings, the genetic code is said to be somewhat *degenerate*. There are therefore more than 20+ meaningful triplets in actuality, and fewer meaningless ones than might be assumed theoretically. A limited degree of degeneracy is of considerable adaptive advantage, for change of one code letter—by mutation, for example—still may preserve the meaning of the whole triplet. Thus, if *AUU* were to mutate to *GUU*, the specification of leucine could remain intact nevertheless.

In our sample *m*RNA above, (Fig. 6.18), the six-base sequence *GUUACU* thus consists of two consecutive triplets, which could represent the code for the amino acid combination valine-histidine. How does such an *m*RNA code in the ribosomes control the formation of a valine-histidine portion in a polypeptide? To answer this, we must turn our attention to *t*RNA.

The joining of amino acids with their specific *t*RNA carriers in the free cytoplasm is referred to as *amino acid activation.* This process requires the participation of ATP as energy donor and of an enzyme, specific for each given amino acid, which links the amino acid with its specific *t*RNA carrier (Fig. 6.19). Each *t*RNA is a comparatively short ribonucleotide polymer in which both ends of the molecule play a critical role. At one end, which serves as "carrier" end, all *t*RNA types appear to possess the same N-base triplet, viz., *ACC,* and it is at this end

6.20. The structure of transfer RNA. At the *ACC* carrier end, the amino acid is linked to carbon 2 of ribose. The phosphate joins adjacent nucleotides between carbon 3 of one ribose and carbon 5 of the next, as shown. Behind the *ACC* carrier triplet, *t*RNA possesses a series of other nucleotides. The terminal triplet forms a recognition end, which is *CAA* in the case of valine, *UGA* in the case of histidine. It is this recognition end which specifically attaches to an appropriate N-base along *m*RNA on the ribosomes.

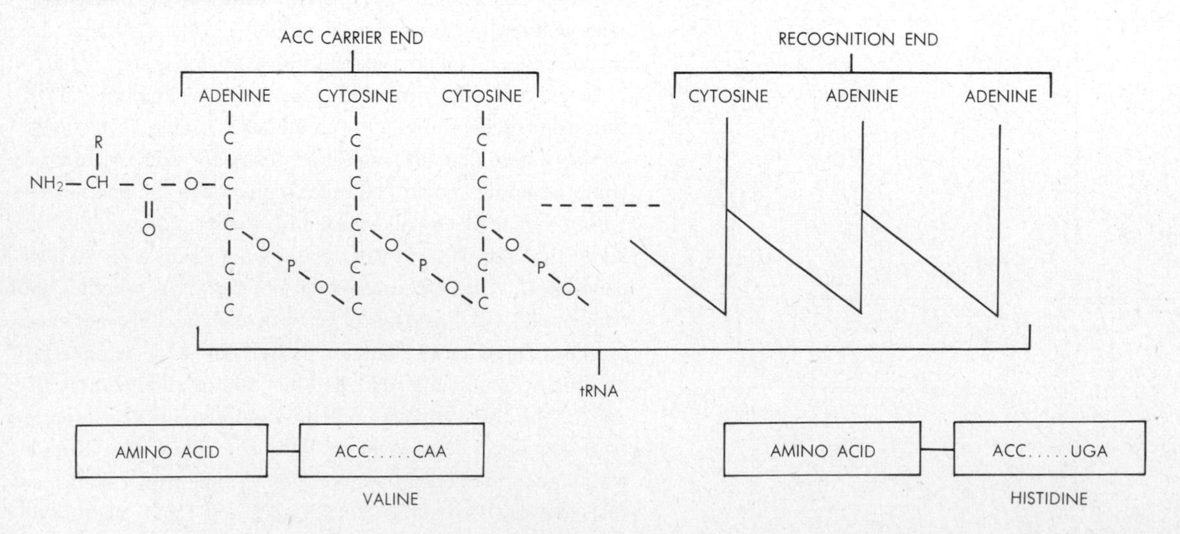

that *t*RNA is joined to its amino acid (Fig. 6.20). Thus the carrier end of *t*RNA appears to be nonspecific, and all amino acids become linked in the same way. However, the other end of *t*RNA *is* specific for a given amino acid. At this "recognition" end is present an N-base triplet which spells out a positive (i.e., DNA-like) code for a particular amino acid. In the case of valine, for example, the *t*RNA specific for valine would carry the terminal recognition triplet *CAA*, i.e., the inverse of *GUU*; and in the case of histidine, analogously, the recognition triplet would be *UGA*, the inverse of *ACU* (cf. Fig. 6.20). It is the specific enzyme in the activation reaction which "recognizes" and promotes interaction of a given amino acid and its corresponding *t*RNA; the enzyme thus must ensure that a particular acid becomes linked to the "correct" type of *t*RNA.

When a *t*RNA carrier then arrives at a ribosome, the positive recognition triplet of *t*RNA will be able to bond only to a corresponding negative code triplet along *m*RNA. Thus, as in our example, a valine-carrying *t*RNA with the recognition triplet *CAA* will be able to bond to a *GUU* triplet along *m*RNA; and a *t*RNA with the triplet *UGA* may bond to a *ACU* triplet of *m*RNA. In this way, amino acids become stationed along *m*RNA in a code-determined sequence (Fig. 6.21). The final link-up of amino acids into polypeptides may then be accomplished by formation of peptide bonds between adjacent amino acids. This process requires ATP and specific enzymes, among them various *peptidases* and *proteinases*. The reactions may take the form of dehydrations (cf. Fig. 6.21). The finished polypeptide disengages from its RNA connections, and the *t*RNA and *m*RNA molecules on the ribosomes soon decompose. The polypeptide formed represents the primary structural element of a protein molecule built according to gene-determined instructions.

Figure 6.17 may now be consulted again for a summary representation of the whole pattern of protein synthesis. In full accordance with this scheme, and with the use of extracts of *m*RNA, *t*RNA, and ribosomes from bacterial cells, it has actually been possible to synthesize in the test tube several artificial protein enzymes which are otherwise manufactured only in living cells. But note that, although we can now explain the formation of a specific primary structure of a protein, we have as yet little precise knowledge of how specific secondary, tertiary, or quaternary structures are determined. If two proteins differed, for example, in tertiary structure but not in primary structure, then the same *m*RNA should provide the code for both. Conceivably, the higher structural specificities of proteins may be governed by the various specific proteinases which promote amino acid polymerization, and perhaps by other reaction participants as well.

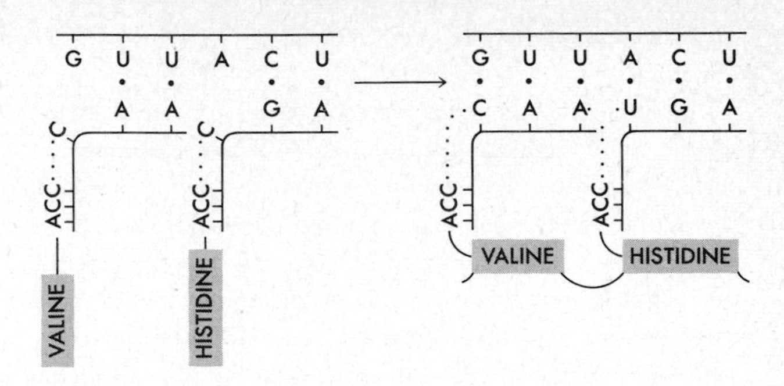

$$NH_2\text{—}RCH\text{—}\underset{\underset{O}{\|}}{C}\text{—}OH + H\text{—}\underset{\underset{H}{|}}{N}\text{—}R'CH\text{—}COOH \xrightarrow[H_2O]{enz.,\ ATP}$$

$$NH_2\text{—}RCH\text{—}\underset{\underset{O}{\|}}{C}\text{—}\underset{\underset{H}{|}}{N}\text{—}R'CH\text{—}COOH$$

PEPTIDE BOND

6.21. Top, diagrammatic representation of link-up between *m*RNA and recognition ends of amino-acid-carrying *t*RNA. Bottom, reaction pattern in link-up between two adjacent *t*RNA-carried amino acids, resulting in peptide bonding (and formation of polypeptide chains).

Newly formed proteins become part of the structural and functional makeup of a cell. Thus, by virtue of their particular specificities, some proteins might become incorporated into various fibrils, membranes, mitochondria, chromosomes, ribosomes, or indeed any other cellular organelle. Other proteins, again by virtue of their particular specificities, might come to function as specific enzymes and thus determine what kinds of reactions can take place in a cell. It has been estimated that there are in the order of 2,000 to 10,000 genes per chromosome, and it is known that there may be up to several dozen different chromosomes in a cell. Each of the many genes present controls the manufacture of a different kind of protein, and the totality of the proteins formed then maintains the nature of the cell.

Through its metabolism generally and its synthesis activities specifically, a cell is able to self-perpetuate. Indeed, self-perpetuative processes can be shown to be linked particularly closely with synthesis reactions. Actually this is to be expected, for self-perpetuation is achieved primarily through control of synthesis—most especially, synthesis of nucleic acids and proteins. We shall examine this fundamental interconnection in the following section.

CELLULAR SELF-PERPETUATION

Self-perpetuation includes the processes of steady-state control, reproduction, and adaptation. As in the case of metabolism, the functions of *cellular* self-perpetuation are not necessarily of the same type as those of whole-animal self-perpetuation. To cite just one obvious difference, we recall that adaptation on the level of the whole animal includes sex, heredity, and evolution. However, apart from reproductive cells, any other cell within an animal is directly concerned neither with sex nor with evolution (although indirectly it contributes to both). Thus, if we limit ourselves here strictly to cellular self-perpetuation, we may say that steady-state control means maintenance of optimal cellular operating conditions, notwithstanding frequent and often rapid changes in the surrounding body fluids or in neighboring cells; that reproduction means cell division, and growth in cell numbers; and that adaptation means long-range cell maintenance, which reduces essentially to faithful transmission of cellular characteristics through successive cell generations, i.e., to cellular heredity.

How are these various processes related to and governed by synthesis reactions?

DNA and Operons: Steady States

It should be readily apparent that, in any animal, the reaction patterns taking place within each of its cells must change repeatedly, both in time and in space. In time, an animal develops from egg via embryo and larva to the adult stage, and in the course of this developmental history given cell strains *differentiate,* i.e., they acquire characteristic specializations progressively and gradually. In space, different groups of cells within a developing animal specialize in different ways; hence in an adult even two adjacent cells may be quite dissimilar in structure and function. Moreover, the activities of any cell are likely to require moment-to-moment adjustment, in response to changing conditions in the surrounding body fluids or in neighboring cells.

All such changes in and among cells are expressions of changing patterns of cellular *chemical reactions.* We already know, however, that reaction patterns are determined by enzyme contents, and that enzyme contents in turn are determined by patterns of protein synthesis. Evidently, cells differ in their protein-synthesizing activities at different times and in different places. Yet at all times and all places, all cells of an animal contain identical genes; each cell has inherited a like set of genes through an ancestry which traces back to the original set of genes present in the egg. It follows that, although all cells possess the potential of making all the proteins characteristic of an animal, all cells do not actually manufacture all these proteins; different groups of genes must be active in different cells and at different times. In other words, the rate of protein synthesis in each cell must be adjustable; the functioning of genes must somehow be capable of being "turned on and off."

Inasmuch as gene functioning means enzyme synthesis, hence reaction control, we are dealing here with a fundamental aspect of cellular steady-state control. We may say that, ultimately, *cells maintain steady states by controlling and adjusting the rates of protein synthesis.* Control and adjustment of cellular reactions then follows, via presence or absence of particular enzymes.

An answer to the question of how genes may be turned on or off has been suggested by the chemical phenomena of *repression* and *induction.* It has long been known that, in certain metabolic reaction sequences within cells, the endproduct of the sequence often tends to inhibit some earlier point of the sequence. For example, in a sequential transformation of compound *A* into compound Z, the endproduct Z, once formed in given amounts, might inhibit some reaction such as $B \longrightarrow C$:

$$A \xrightarrow{enz.\ 1} B \xrightarrow{enz.\ 2} C \xrightarrow{enz.\ 3} \cdots \xrightarrow{enx.\ x} Z$$

Such "endproduct inhibitions," or repressions (symbolized by the transverse double bar above), prevent more endproduct from being formed. The evidence indicates that repressions are brought about by interference with the enzymes which catalyze given reaction steps. To cite a specific example, it has been shown in the synthesis of pyrimidine-containing nucleotides that the endproducts CMP or UMP repress the reaction step in which carbamyl phosphate becomes linked to aspartic acid (reaction 2, Fig. 6.14); and the repression appears to be a specific result of an inhibition of the enzyme aspartic transcarbamylase.

Conversely, for some sequential reactions in cells it has been found that addition of excess amounts of reactants leads to a rapid formation of excess amounts of specific enzymes. Thus, in a sequential transformation of *A* into Z, an excess of *A* may lead to an increase in the amount of enzyme which converts *A* into *B*, hence to a more rapid formation of Z:

$$A \xrightarrow{enz.\ 1} B \xrightarrow{enz.\ 2} C \xrightarrow{enz.\ 3} \cdots \xrightarrow{enz.\ x} Z$$

A classic instance of such "adaptive enzyme formation," or induction, is the effect of the C_6 sugar-derivative *galactoside* which, when added to a cell in excess amounts, brings about the appearance of large quantities of the corresponding enzyme *galactosidase*.

Inasmuch as enzymes are proteins, both repressions and inductions must somehow relate to the mechanism of protein synthesis or, more specifically, to the *rate control* of protein synthesis. A recent hypothesis, based on considerable evidence, shows how repressions and inductions might actually operate through a switch mechanism which turns gene activity on and off.

This so-called *operon hypothesis* (Fig. 6.22) postulates the existence of two types of genes, *regulator genes* (*RG*) and *structural genes* (*SG*). Regulator genes are thought to control the manufacture of protein products *R*, which function specifically by affecting the activity of structural genes. The latter are genes which control protein synthesis generally; i.e., they transcribe their codes via *m*RNA and ribosomes into cellular proteins. Such proteins then function in part as enzymes and promote metabolic reactions, e.g., sequential transformation of *A* into Z. All the structural genes which control the successive steps of a given reaction sequence are thought to be located close to one another on a chromosome. Such a chromosome region is also postulated to contain an *operator gene* (*Op*), which must be active if the nearby structural genes are to be active. The whole region, including operator and associated structural genes, is said to form an *operon;* it represents the section of a chromosome which

6.22. The postulated structure of an operon. *SG*, structural gene; *OP*, operator gene; *RG*, regulator gene (all three of these types located in a chromosome); *R*, protein product formed by *RG* and influencing *OP* in ways to be shown in next two illustrations. *A* to *Z* represents a reaction sequence in which a succession of specifically required enzymes (*enz*) is manufactured under the control of *SG* and *m*RNA.

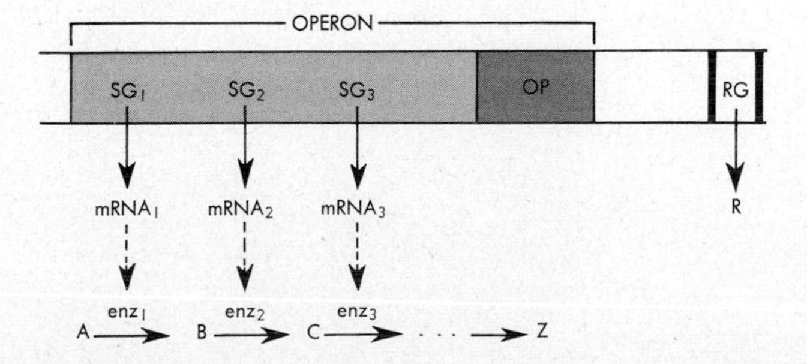

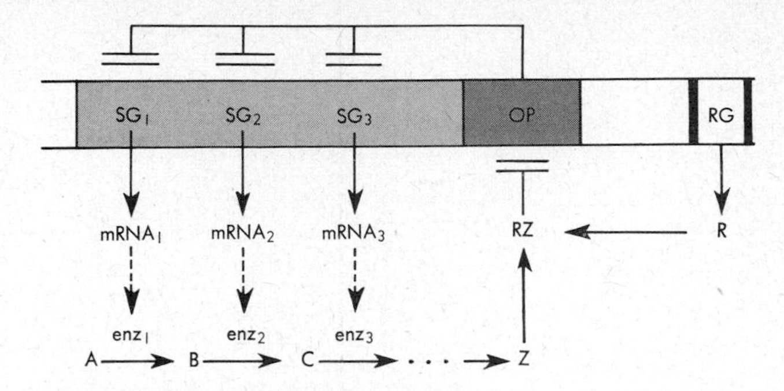

6.23. Operon functioning: repression. *R* here by itself is without effect on *OP*, but if *R* combines with reaction endproduct *Z*, then the complex *RZ* inhibits the operator gene *OP* (transverse double bar denotes inhibition). As a result, *OP* also prevents the *SG*s from functioning, enzymes will not be produced, and the reaction sequence *A* to *Z* will cease to take place. Thus, the endproduct *Z* represses the continuation of its own manufacture.

controls all the reaction steps in the transformation of a raw material *A* into an endproduct Z.

In a repression (Fig. 6.23), the product *R* of the regulator gene is assumed not to affect the operator gene *Op*, and the latter is then active. This permits the structural genes to be active as well, resulting ultimately in the formation of endproduct Z. However, Z now is believed to combine with *R*, and the complex *RZ* then "covers," or attaches to, the operator *Op*, thereby inhibiting it. The structural genes thus would become inactive also, specific transcription would cease, and the reaction sequence *A* to Z would soon be halted. Endproduct Z then would no longer be formed. The repression would last as long as Z is present above critical amounts, after which *Op* would again become uncovered and formation of Z could recur temporarily.

In an induction (Fig. 6.24), the product *R* of the regulator gene would be an inhibitor of *Op*, and endproduct Z thus could not be manufactured. But if raw material *A* were introduced into the cell, *A* would combine with *R*, and the complex *RA* would abolish the inhibition of *Op*. The structural genes could then become active, and *A* could be transformed into Z. After all of *A* is used up, *Op* would again become inhibited, and the enzymes (*enz. 1*, *enz. 2*) would no longer be synthesized. We note that induction is equivalent to removal of a repression, i.e., it is a *derepression*.

Clearly, an operon control mechanism saves a cell considerable amounts of energy and materials; enzyme proteins are

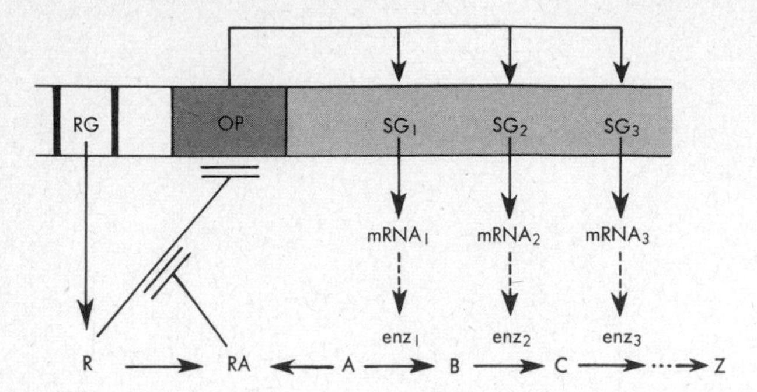

6.24. Operon functioning: induction. *R* by itself inhibits *OP*, hence the *SG*s do not operate and *Z* is not formed. But if starting material *A* is introduced into the system, *R* combines with *A* and the complex *RA* inhibits the inhibition exerted by *R* on *OP*; i.e., *RA* in effect permits *OP* to function again. The *SG*s then will likewise function, enzymes will be formed, and *A* can be converted to *Z*. Thus, *A* removes a repression, or it derepresses, *OP*, hence it promotes its own conversion to *Z*.

synthesized only when they are actually needed, i.e., when given raw materials are actually available and when given endproducts are not already present in excessive quantities. Quite as importantly, an operon mechanism permits a cell to be responsive to its environment: compounds entering a cell as food can be the specific stimuli for their own utilization, via enzyme induction; and compounds accumulating as finished products can be the specific stimuli halting their own manufacture, via enzyme repression. As a result, a cell may control its metabolic activities in accordance with the conditions of the moment, and the cell so may exercise steady-state control. Moreover, an operon mechanism permits certain reaction pathways to be inactivated and certain alternative ones to become operational instead (Fig. 6.25). Such switch-on, switch-off alternatives are important not only in steady-state control generally but may also account for the observed developmental specialization of a cell in one of several possible directions, a phenomenon we shall refer to again in Chap. 11.

DNA Synthesis: Reproduction and Mitosis

If genes ultimately regulate the production of all other cellular components, what regulates the production of genes themselves? Genes govern their own synthesis; each existing gene controls the manufacture of new genes exactly like it. DNA is therefore said to be "self"-duplicating.

New-formation of DNA has many features in common with code transcription from DNA to *m*RNA. Thus, as outlined in Fig. 6.26, the DNA double chain first unwinds and unzips into two single chains. Each single chain then links to itself, via hydrogen bonds, appropriate nucleotide raw materials. The latter in this case are the deoxyribose derivatives ADP, GDP, CDP, and TDP. After such raw materials have become attached to appropriate places along single DNA chains, an enzyme, *DNA polymerase,* links the raw materials together into a new DNA chain. Note that the action of this enzyme is quite comparable to that of RNA polymerase. Note also that DNA synthesis is equivalent to DNA *reproduction:* one DNA double chain gives rise to two identical DNA double chains. In each "daughter" DNA, one of the two single chains has preexisted in the "parent" DNA, the other single chain has been newly manufactured. In this manner, the genetic code is transmitted faithfully from one DNA "generation" to the next.

That old DNA actually becomes part of new DNA can be demonstrated experimentally. A cell about to divide can be supplied with radio-labeled raw materials needed in the manufacture of DNA. After cell division, the chromosomes of both offspring cells then contain radio-labeled DNA. This

6.25. Alternative pathway switching. Assume that A_1, A_2, A_3 and B_1, B_2, B_3 are two reaction pathways requiring the enzymes a', a'' and b', b'' respectively. Assume also that A_2 has the property of inhibiting b', and B_2, of inhibiting a'. If then the *A* sequence is operative, it will inhibit the *B* sequence; and if the *B* sequence is operative, it will inhibit the *A* sequence. Also, if presence of *A* induces the enzymes a' and a'' through an operon mechanism, then the *B* sequence will be automatically inhibited through the effect of A_2 on b'. By such means, two or more separate metabolic pathways may exert significant control over one another and thus over the kinds and amounts of materials synthesized in a cell.

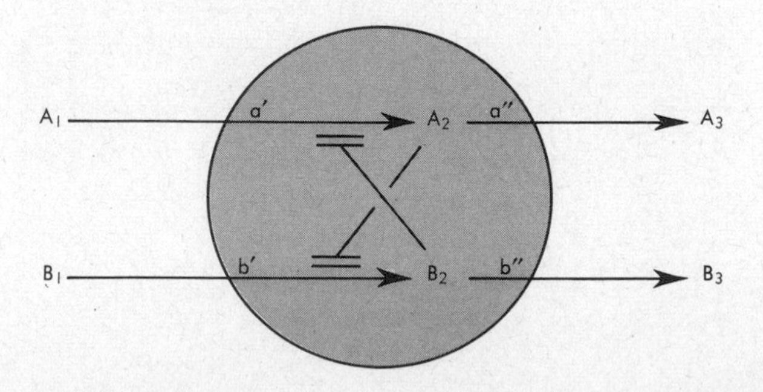

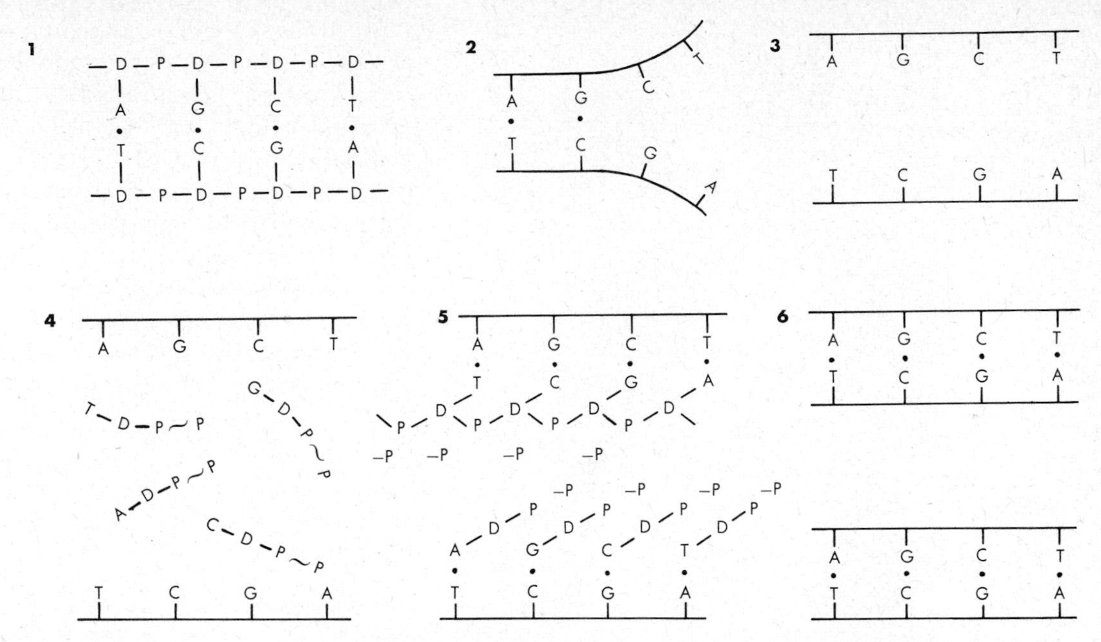

6.26. DNA duplication. *A, G, C, T,* nitrogen bases; *D,* deoxyribose; *P,* phosphate. If a double DNA chain (1) is to be duplicated, it separates (2) into two single chains (3). Nucleotide diphosphate raw materials (4) then become attached to each free nitrogen base (5, bottom), and the deoxyribose and phosphate parts of adjacent nucleotides become joined (5, top). Two DNA double chains thus result (6), identical to each other as well as to the original "mother" chain. Note that, in each newly formed double chain, one single chain preexisted and served as the code-specific template in the manufacture of the new single chain. In certain respects this code transcription from DNA to DNA resembles that of transcription from DNA to *m*RNA (cf. Fig. 6.18).

suggests that each newly manufactured DNA double chain contains one radio-labeled, newly produced nucleotide chain and one unlabeled, preexisting nucleotide chain. If now such an offspring cell is allowed to divide again (without further additions of radio-labeled raw materials), then only one of the resulting cells contains labeled chromosomes; the other contains unlabeled chromosomes. Such a result can be obtained only if DNA actually duplicates as described above (Fig. 6.27).

Reproduction of DNA characteristically takes place after more or less extended periods of cell growth. The genes of a cell normally transcribe their codes into newly forming *m*RNA, and abundant protein synthesis then occurs as a consequence. The structural proteins so formed increase cell size directly; the enzymatic proteins make possible the synthesis of other compounds, and these enlarge cell size in their turn. After a cell has grown in this manner for a given length of time, some—so far unknown—stimulus brings about a change in chromosome activity: the manufacture of *m*RNA stops for the time being and DNA produces new DNA instead. One set of genes so becomes two sets, and one chromosome set correspondingly becomes two sets. Chromosome reproduction then appears to be the trigger—again in an unknown manner—for cell division.

This event consists of at least two separate processes: cleavage of the cytoplasm into usually two parts, or *cytokinesis,* and reproduction of the nucleus, referred to as *karyokinesis* or *mitosis.* In animal cell division cytokinesis and mitosis normally occur together, and we may therefore speak of "mitotic division." After the chromosomes in the nucleus have duplicated as outlined above, a certain amount of time elapses before the visible events of division begin. These events consist

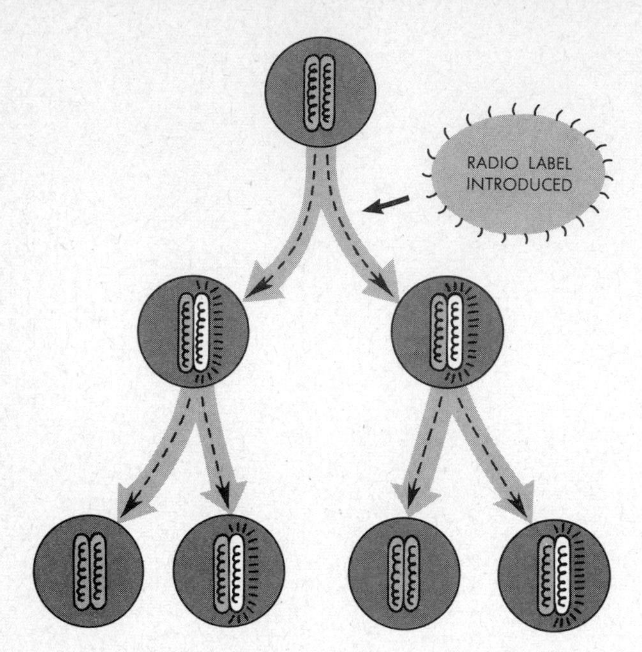

6.27. Experiment showing that newly produced DNA contains both preexisting and newly manufactured nucleotide chains. A cell about to divide is provided with radio-labeled raw materials for DNA synthesis (top). During division each chromosome duplicates, and both offspring chromosomes are then found to contain radio-labeled DNA (center). The diagram shows that, in *each* of these offspring chromosomes, only one of the two nucleotide chains is radio-labeled. This one has been newly manufactured; the other has preexisted. That such an interpretation is valid is proved if the offspring chromosomes are allowed to divide again (bottom). For when one of the radio-labeled chains now duplicates, it synthesizes a new, *unlabeled* chain as its partner. Hence in the DNA so formed, one chain (radio-labeled) has again preexisted; the other chain (unlabeled) has been newly formed. When an originally unlabeled chain duplicates, neither of the two resulting chains carries a label.

of four successive, arbitrarily defined stages: *prophase, metaphase, anaphase,* and *telophase.* One stage merges gradually into the next, and it is usually impossible to fix sharp lines of transition. Nuclear reproduction, i.e., mitosis proper, encompasses all four stages; cytokinesis occurs in the last stage.

One of the first events of *prophase* is the division of the centriole, just outside the nucleus (Fig. 6.28). As soon as daughter centrioles have formed, the two granules behave as if they repelled each other, i.e., they migrate toward opposite sides of the cell nucleus. Concurrently, portions of the cytoplasm transform into fine gel fibrils. Some of these radiate away from each centriole like the spokes of a wheel and form *asters.* Other gel fibrils develop between the two centrioles. Looping from one centriole to the other flat curves, these fibrils constitute a *spindle.* The centriole at each end marks a *spindle pole.* As the centrioles move farther and farther apart, the fibrils of the spindle and the asters lengthen and increase in number.

During these stages of prophase, the nuclear membrane dissolves, the nucleoli disintegrate, and nuclear and cytoplasmic substances mix freely. Moreover, the already duplicated chromosomes now become clearly visible (Fig. 6.29). Such

6.28. Early (A) and late (B) prophase in mitosis. In *A,* the nuclear membrane is just dissolving and chromosomes are already visible. To either side of the nuclear region is a darkly stained centriole area. These areas develop after a single centriole has divided and the two resulting daughter centrioles have migrated to opposite sides of the nucleus. From each centriole area fine fibrils are beginning to radiate out, i.e., asters are beginning to form. In *B,* asters are already conspicuous and spindle fibrils have formed between asters and chromosomes. The latter are migrating into a metaphase plate.

A

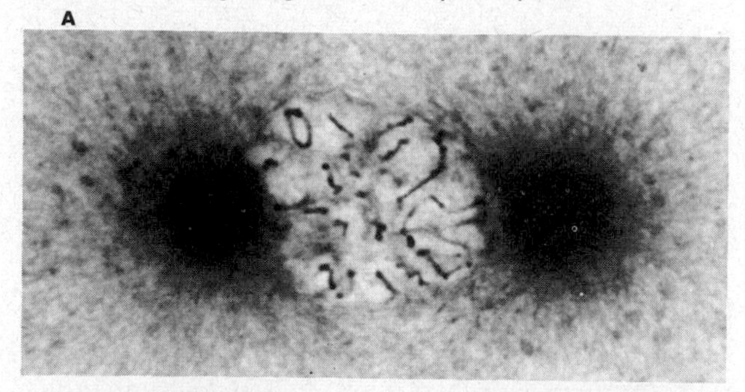

B

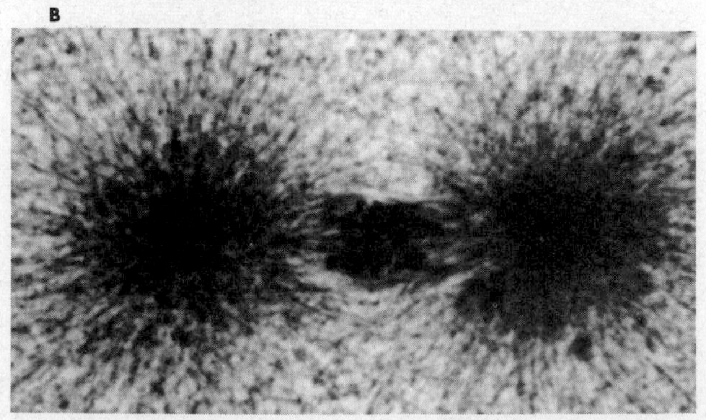

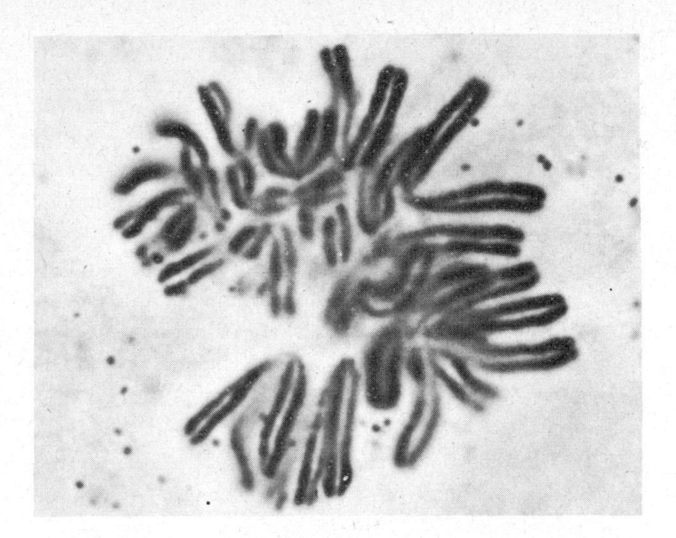

6.29. At the time of prophase, shown here, chromosomes have already duplicated, and doubled chromosomes, each known as a chromatid, are therefore present. The members of each pair of chromatids are still held together at one point, the centromere.

twin chromosomes lie closely parallel and are joined to each other only at a single point, the so-called *centromere*. The location of the centromere varies for different chromosome pairs. Two spindle fibrils become anchored to each centromere, one from each pole of the spindle. In this way, the chromosomes become linked to the spindle. At this general period, prophase comes to a close and *metaphase* begins.

Early during metaphase the chromosome pairs are still scattered randomly through the central portion of the cell, but later they begin to migrate. If we draw an imaginary line from one spindle pole to the other, we mark out a spindle axis. Chromosomes migrate into a plane set at right angles to the spindle axis, midway along it. Specifically, it is the centromere of each chromosome pair which comes to occupy a station precisely within this plane. During the migration, the chromosomes trail behind their centromeres like streamers. Lined up in one plane, the centromeres are said to form a *metaphase plate* (Fig. 6.30).

The lengthwise separation of the chromosome pairs now becomes complete. Each centromere divides and entirely independent chromosomes are produced in this manner. A small gel fibril arises at once between the centromeres of formerly joined chromosomes, and such chromosomes begin to move apart. Once they are completely separated, the members of a pair of chromosomes behave as if they repelled each other. Thus, one set of chromosomes migrates away from the metaphase plate toward one spindle pole, and an identical twin set migrates in the opposite direction, toward the other spindle pole. The centromeres again lead and the arms of the chromosomes trail. Also, the gel fibrils between twin centromeres lengthen, and fibrils between the centromeres and the spindle poles shorten. This period of poleward migration of chromosomes represents the *anaphase* of mitotic division (Fig. 6.31).

The beginning of *telophase* is marked by the appearance of a *cleavage furrow* in the plane of the earlier metaphase plate. At first the furrow is a shallow groove circling the surface of a cell. This groove then gradually deepens, cuts through the spindle fibrils, and eventually constricts the cell into two daughter cells. While such cytokinesis is in progress, the chromosomes within each prospective daughter cell aggregate near the spindle pole. Spindle fibrils subside; i.e., the gel composing them reverts to a sol state. A new nuclear membrane forms at each spindle pole, and this membrane surrounds the chromosomes. Concurrently, the chromosomes in each newly forming nucleus manufacture new nucleoli, in numbers characteristic of the particular cell type. These nuclear processes terminate roughly when cytoplasmic cleavage nears completion, and mitotic division then has reached its endpoint. In each of the newly formed daughter cells, the genes now re-

6.30. Metaphase in mitosis. Note asters, spindle, and the metaphase plate, halfway along and at right angles to the spindle axis. Note also the fibrils which join the chromosomes in the metaphase plate with the spindle poles.

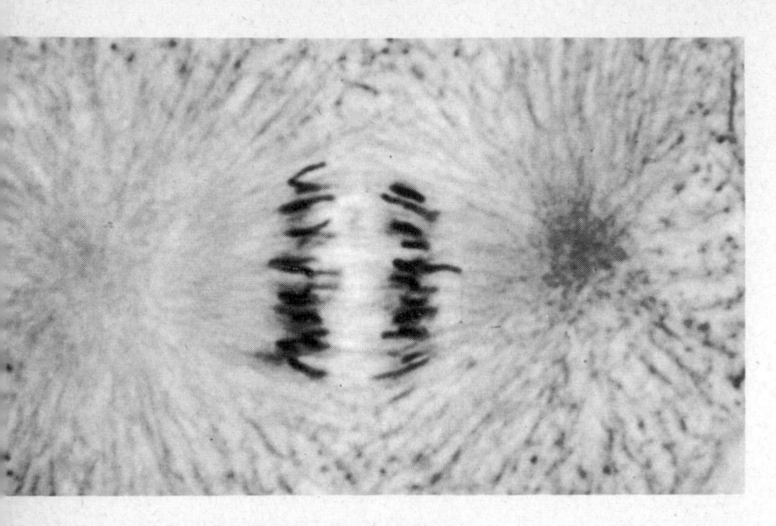

6.31. Mid-anaphase in mitosis. Chromosome sets are migrating toward the spindle poles.

sume control of *m*RNA manufacture, and a new growth cycle follows (Figs. 6.32 and 6.33).

The mechanical forces responsible for the chromosome movements in division cannot yet be identified precisely. It is clear, however, that these forces are brought into action ultimately by DNA and that they produce cells which contain exactly identical gene sets, incorporated in identical chromosome sets, and approximately equal quantities of all other cellular constituents. DNA thus is the key to cellular multiplication. It controls cell growth and, through its own periodic reproduction, triggers the reproduction of the whole cell. In the process, the all-important codes of DNA are transmitted exactly from cell generation to cell generation. This in effect constitutes *cellular heredity;* to the extent that genes themselves are stable, each new cell generation inherits the same genetic codes, hence the same structural and functional characteristics, that had been present in the previous cell generation.

DNA Reproduction: Adaptation and Evolution

On occasion, parts of the DNA codes actually do become altered. Such gene *mutations* may arise in several ways. For example, they probably occur as accidental errors during DNA duplication. Like any other process in nature, biological or otherwise, gene reproduction undoubtedly is not completely error-free. Even if a single nitrogen base of old DNA happens to be transcribed incorrectly into new DNA, one code triplet will become altered, and one of the protein types subsequently synthesized will contain a "wrong" amino acid at some point. Many mutations also arise through the effects of certain physical and chemical agents, e.g., X rays and other forms of high-energy radiation (both natural and manmade) or compounds such as nitrogen mustards. All such external agents in some way change either the position or the chemical nature of at least one nucleotide in DNA, resulting in an alteration of at least one code triplet. Such mutated triplets are just as stable as the unmutated ones, and the changed condition is inherited by successive cell generations. Every time a particular protein is then manufactured, it will contain a "wrong" amino acid. This may change or abolish entirely the specific enzymatic effect of the protein, and a given trait of a cell thus may become changed.

To be sure, if an amino acid is normally coded by more than one triplet, an error in one nitrogen base need not necessarily lead to a trait mutation. Partial degeneracy of the genetic code thus may protect against mutational consequences. A cell also has some additional protections. As already noted, for example, the nucleus itself is a protective device, which shields the genes from the destructive metabolism of the cytoplasm. Another important safeguard is *redundancy;* when one wishes to ensure that a message is not lost or altered, one makes it redundant, i.e., one repeats it several times. Just so, the genetic codes are stored in more than one place. First, an animal cell

6.32. Early telophase in mitosis. Asters are subsiding, nuclei are re-forming, chromosome threads have become indistinct, and cytoplasmic cleavage is under way, in the same plane as the earlier metaphase plate.

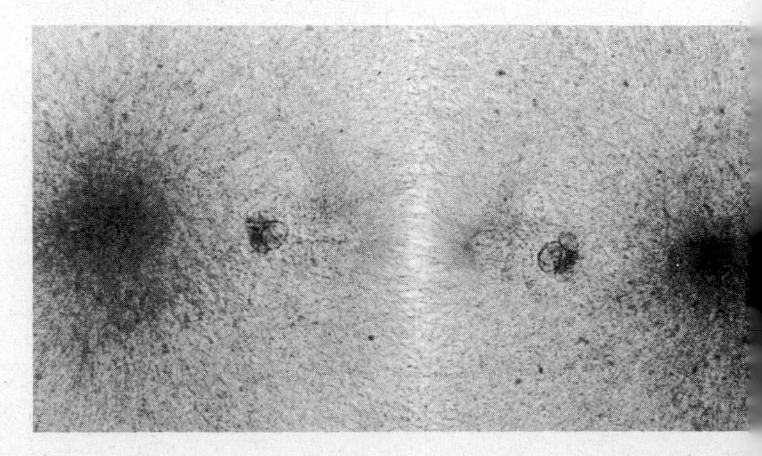

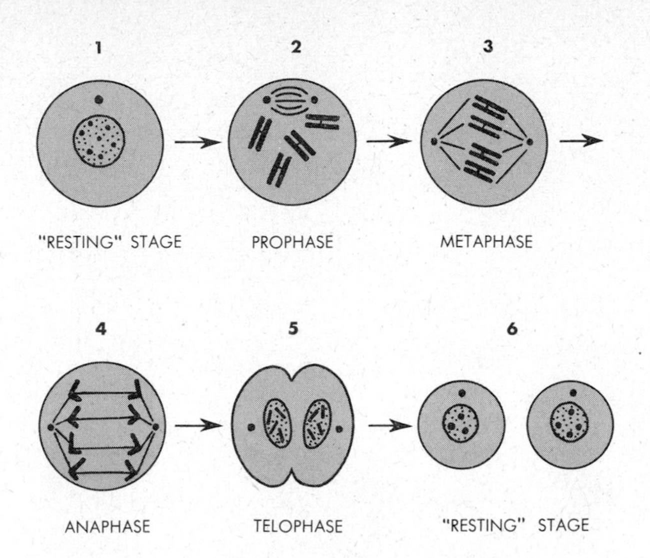

6.33. Mitosis, summary. The assumption here is that cytoplasmic cleavage accompanies mitosis. Note that a "resting" cell is resting only from the standpoint of reproductive activity; in all other respects it is exceedingly active.

ordinarily contains two complete sets of genes, one set having been inherited originally from the egg-producing parent and the other from the sperm-producing parent. Chromosomes and genes thus come in pairs, and even if one member of a gene pair mutates, the other member still carries the original code. Secondly, each type of cell is usually represented by many like cells. Even if some cells then die as a result of mutations, the genes of the remaining cells still possess the specific information characteristic of that cell type.

Despite such protection, mutations nevertheless do occur, with an estimated average frequency of about one per million cells (Fig. 6.34). Since many animals consist of trillions of cells, each such animal will carry several million mutated genes. Mutations appear to be completely random events. Any gene may mutate at any time, in unpredictable ways. A given gene may mutate several times in rapid succession, then not at all for considerable periods. It may mutate in one direction, then mutate back to its original state or in new directions. There is little question that every gene existing today is a *mutant* which has undergone many mutations during its past history.

The effect of a mutation on a cellular trait is equally unpredictable. Some are "large" mutations, i.e., they affect a major trait in a radical, drastic manner. Others are "small," with but little effect on a trait. Some mutations are said to be *dominant,* i.e., the mutated genes override the effects of their normal gene-partners, and thus they produce immediate alterations of traits. Other mutations are *recessive,* the unmutated gene-partners here masking the effects of the mutated genes and thus protecting the cell from actual trait changes. We shall return to this subject in Chap. 12.

A small percentage of mutants produces advantageous traits and new traits which are neither advantageous nor disadvantageous. But most mutations are disadvantageous. Indeed, inasmuch as a living cell is an exceedingly complex, very finely adjusted whole, it is to be expected that any permanent change in cellular properties would be more or less disruptive and harmful. In many cases, therefore, dominant mutations tend to be eliminated as soon as they arise, through death of the affected cell. In other cases, the effect of a dominant mutation, particularly a "small" dominant mutation, may become integrated successfully into cellular functions. Such a cell may then survive even though it exhibits an altered trait. By and large, however, recessive mutations are likely to persist more readily, since their effects may be masked by the normal

6.34. Mutant types in mice. **A,** the effects of the mutation "eyelessness;" **B,** the effects of the mutation "hairlessness." Each of such alterations in structure is correlated with a single mutant gene, and the alterations are stable and inheritable. Cf. also Fig. 12.25.

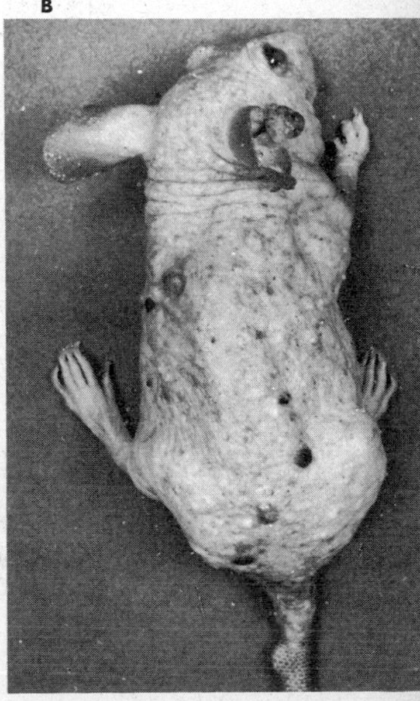

gene-partners. Accumulated evidence actually shows that surviving mutations are very largely recessive ones.

Gene changes which occur in body cells generally are known as *somatic mutations*. They affect the heredity of the cell progeny, i.e., a patch of tissue at most, and have little direct effect on the heredity of the whole animal. By contrast, the whole animal is affected by so-called *germ mutations*, stable genetic changes in reproductive cells. Such mutations are transmitted from reproductive cells to all cells which ultimately compose the offspring (Fig. 6.35). The latter may then come to differ from its parents in certain major or minor respects. If the difference happens to be advantageous in the environment in which the animal lives, the animal will be well *adapted* to its environment, possibly even better adapted than its forebears. Such an animal may then survive and produce many offspring in its own turn, passing on its well-adapted characteristics in the process. In this manner, an alteration of a single code triplet in a single cell may eventually come to affect a whole animal population. The final result, after numerous generations, may be a change which we recognize as *evolution*.

Thus, synthesis generally and DNA synthesis specifically actually do lie at the root of all cellular self-perpetuation and, through this, also at the root of animal self-perpetuation as a whole. Our survey of cellular operations is therefore substantially completed, for as a cell metabolizes and self-perpetuates, it lives. We are now ready to examine how large groups of living cells cooperate structurally and functionally, and indeed how they form whole animals and all higher levels of animal organization.

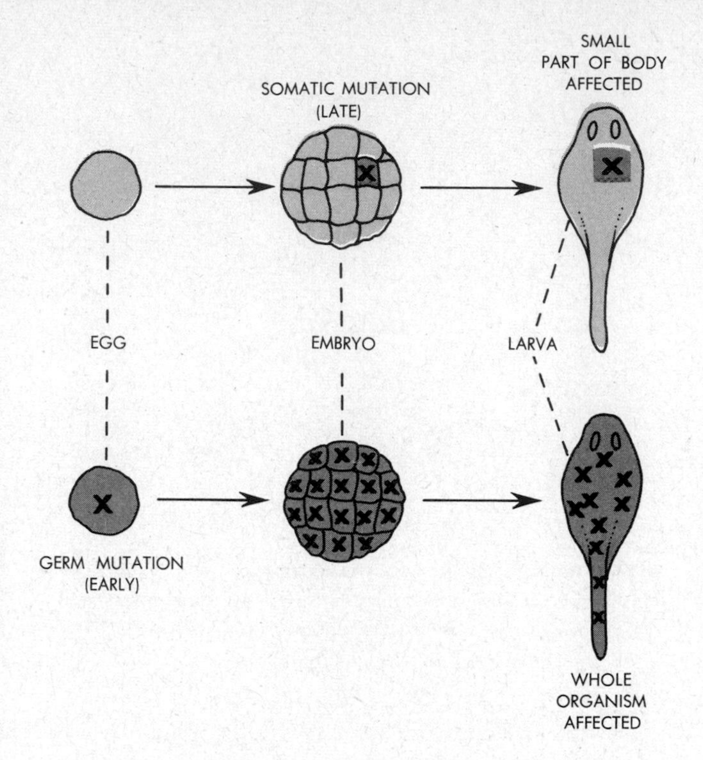

6.35. The effects of germ mutations and somatic mutations. If a mutation occurs late during development, in a somatic cell, then only the progeny of that cell will inherit the mutation and the total effect on the adult animal will be small. But if a mutation occurs early during development, e.g., in a reproductive cell, then all cells of the resulting adult will inherit such a germ mutation and all cells may feature altered traits.

REVIEW QUESTIONS

1. Describe the internal fine structure of a muscle. What and where is actomyosin? What are the roles of ATP in muscle? In what specific ways is the ATP supply maintained? Describe the energetic aspects of a unit cycle of muscle activity.

2. In what ways does an animal obtain and produce heat? What are the functions of heat in metabolism? How do animals produce bioluminescence? How do the properties of living light compare with those of nonliving light? How and by what animals is bioelectricity produced?

3. Review general and specific patterns of carbohydrate and fat synthesis. How can carbon chains be lengthened in animal cells? Shortened? What is the pathway of glycogen synthesis? How is a C_{18} fatty acid synthesized from acetyl CoA? How is a fat synthesized? Where in a cell do such syntheses take place?

4. Distinguish between maintenance and secretion syntheses. What are essential fatty acids and amino acids? What is a gland, and what general types of glands occur in animals? Name some of the secretions manufactured by different animal gland cells and state the function of each.

5. In what metabolic respects are animals dependent on plants, and what are the reasons for such dependence? What are the essential differences between autotrophic and heterotrophic patterns of nutrition?

6. Review the chemical structure of DNA and state the general way in which DNA is vital as the component of genes. Review the structure of RNA and distinguish between *m*RNA,

*r*RNA, and *t*RNA. Where in a cell do each of these occur and how do they operate?

7. Describe the mechanism of amino acid activation. What is the genetic code, what are code triplets, and on the basis of what reasoning has a triplet code been postulated for amino acid specification? What is a degenerate code?

8. How may amino acids lined up along *m*RNA become joined into polypeptides? Review the entire pattern of genetically controlled protein synthesis. Show how such synthesis is the key to all metabolism and self-perpetuation of a cell. How are the secondary, tertiary, and quaternary structures of a protein determined during synthesis?

9. What are chemical repression, induction, derepression, operator genes, regulator genes, structural genes? Review the operon hypothesis and show what role it plays in accounting for control of cellular operations. How might it account for developmental changes in a cell?

10. How does controlled DNA reproduction take place? How does a cell as a whole reproduce? Distinguish between cell division, cytokinesis, karyokinesis, mitosis. Describe the succession of events in cell division.

11. What are mutations and what are their characteristics? What type of change would constitute a mutation? Distinguish between small and large mutations and between somatic and germ mutations. What is the interrelation among mutation, adaptation, and evolution?

12. Review the various ways by which a cell obtains all the ingredients it requires for its survival. Describe the basic balance between synthesis and breakdown in animal cells and explain the meaning of metabolic turnover. Name some intracellular and extracellular functions of various compounds synthesized by an animal cell.

COLLATERAL READINGS

Excellent background material on many of the topics dealt with in this chapter will be found in the following articles:

Allfrey, V. G., and A. E. Mirsky: How Cells Make Molecules, *Sci. American,* Sept., 1961.

Crick, F. H. C.: The Structure of the Hereditary Material, *Sci. American,* Oct., 1954.

———: Nucleic Acids, *Sci. American,* Sept., 1957.

———: The Genetic Code, *Sci. American,* Oct., 1962.

Gamov, G.: Information Transfer in the Living Cell, *Sci. American,* Oct., 1955.

Green, D. E.: The Synthesis of Fat, *Sci. American,* Feb., 1960.

Hayashi, T.: How Cells Move, *Sci. American,* Sept., 1961.

Horowitz, N. H.: The Gene, *Sci. American,* Oct., 1956.

Hurwitz, J., and J. J. Furth: Messenger RNA, *Sci. American,* Feb., 1962.

Huxley, H. E.: The Contraction of Muscle, *Sci. American,* Nov., 1958.

Ingram, V. M.: How Do Genes Act? *Sci. American,* Jan., 1958.

Mazia, D.: Cell Division, *Sci. American,* Aug., 1953.

———:How Cells Divide, *Sci. American,* Sept., 1961.

McElroy, W. D., and H. H. Seliger: Biological Luminescence, *Sci. American,* Dec., 1962.

Mirsky, A. E.: The Chemistry of Heredity, *Sci. American,* Feb., 1953.

Nirenberg, M. W.: The Genetic Code, II, *Sci. American,* Mar., 1963.

Rich, A.: Polyribosomes, *Sci. American,* Dec., 1963.

Taylor, J. H.: The Duplication of Chromosomes, *Sci. American,* June, 1958.

Zamecnik, P. C.: The Microsome, *Sci. American,* Mar., 1958.

The following books and paperbacks include accounts of many of the subjects of this chapter:

Allen, J. M. (ed): "The Molecular Control of Cellular Activity," McGraw-Hill, New York, 1962.

McElroy, W. D.: "Cellular Physiology and Biochemistry," Foundations of Modern Biology Series, Prentice-Hall, Englewood Cliffs, N.J., 1961.

Loewy, A. G., and P. Siekevitz: "Cell Structure and Function," Modern Biology Series, Holt, New York, 1963.

the design of animals

PART 3

Animals are obviously put together in different ways; the anatomies of animals such as sponges, earthworms, clams, starfishes, insects, sharks, and men appear to have little in common. Yet we already know that, despite their highly varied gross organization, animals nevertheless are alike in fundamental respects; all are constructed out of cells, and in most cases these are further combined into tissues, organs, and a maximum of 10 organ systems: integumentary, skeletal, muscular, nervous, endocrine, circulatory, breathing, excretory, alimentary, and reproductive systems.

The three chapters of this part concentrate on the gross and fine aspects of animal design generally. First we examine the criteria of classification and body organization by which the gross designs of different animal types may be distinguished. Then we proceed with a careful study of the internal structures and functions of the 10 organ systems, the universal building blocks in animal design.

CHAPTER 7

animal types: body forms and body parts

All the components of an animal are arranged in a specific architectural pattern; the structure of each animal has a particular *form*. Such a structural makeup permits efficient performance of all internal functions and permits an animal to pursue a given way of life in a given external environment. Also, such a construction makes each animal type specifically different from every other.

Accordingly, the three sections of this chapter outline successively how different *animal types* are identified; how given animal types in turn are characterized by specific *body forms;* and how body forms in their own turn are attributes of particular body parts, specifically, of *tissues* and *organs*. The structures and functions of organ *systems* form the subject of the next two chapters.

ANIMAL TYPES

Taxonomic Classification

The classification of animal types is the special concern of a zoological subscience called *taxonomy* or *systematics*. A taxonomic system of classifying animal types was originated by Carolus Linnaeus, a Swedish naturalist of the early 18th century. This Linnaean system, now greatly elaborated and in universal use, is based on structural resemblances among animals; if certain animals can be shown to have similar body construction, they may be regarded as members of the same classification group. Superimposed on such a classification, moreover, evolutionary theory permits formulation of a historical inference: the more closely two animals resemble each other, the more closely are they likely to be related. Thus, taxonomy correlates the structural organization of animals directly and their evolutionary histories indirectly.

Within a given classification group, it is often possible to distinguish several subgroups, each containing animals characterized by even greater similarity of body structure and, by inference, evolutionary history. Each such subgroup may then be subclassified still further, and a whole *hierarchy* of classification groups so may be established. In this hierarchy, the progressively lower levels constitute so-called *taxonomic ranks*, or *categories*, and each is named; in succession from highest (most inclusive) to lowest (least inclusive), the main

7.1. The pattern of the hierarchy of taxonomic ranks.

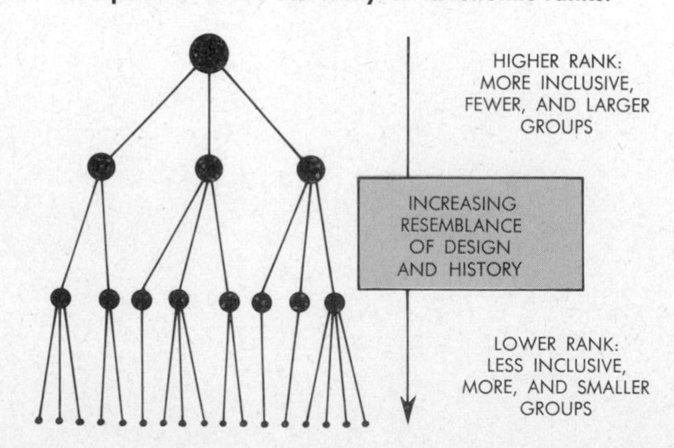

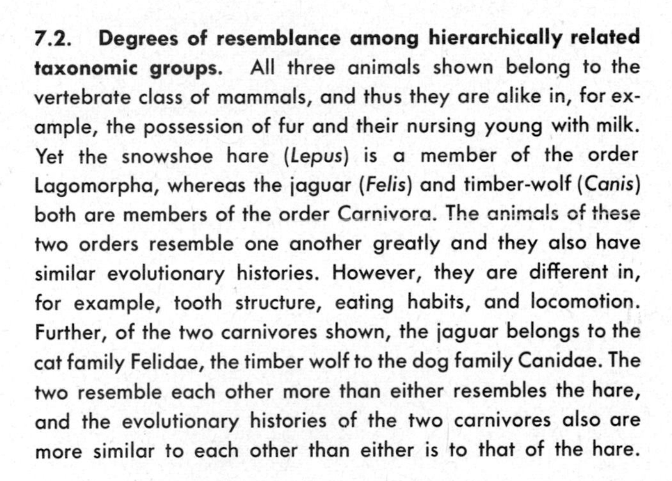

7.2. Degrees of resemblance among hierarchically related taxonomic groups. All three animals shown belong to the vertebrate class of mammals, and thus they are alike in, for example, the possession of fur and their nursing young with milk. Yet the snowshoe hare (*Lepus*) is a member of the order Lagomorpha, whereas the jaguar (*Felis*) and timber-wolf (*Canis*) both are members of the order Carnivora. The animals of these two orders resemble one another greatly and they also have similar evolutionary histories. However, they are different in, for example, tooth structure, eating habits, and locomotion. Further, of the two carnivores shown, the jaguar belongs to the cat family Felidae, the timber wolf to the dog family Canidae. The two resemble each other more than either resembles the hare, and the evolutionary histories of the two carnivores also are more similar to each other than either is to that of the hare.

categories are: *kingdom, branch, grade, phylum, class, order, family, genus,* and *species*. Additional ranks may be interpolated between any two main ranks. Such intermediate categories are identified by the prefixes *sub-* or *super-;* e.g., *subgrade, superclass, subgenus* (Fig. 7.1). The specific animal groups encompassed by a given category are often referred to as *taxa*. For example, the sponges constitute a taxon of phylum rank; mammals are a taxon at the class rank.

Note that, in the hierarchy as a whole, progressively lower ranks consist of progressively more but smaller groups. Thus, animals make up one kingdom, some two dozen phyla, and at least 1½ to 2 million species. Also, the progressively smaller groups at successively lower ranks are characterized by an increasing resemblance of body forms and an increasingly similar evolutionary history. For example, the members of a class resemble one another to a great extent, but the families within any of the orders of that class resemble one another to an even greater extent; and an analogous correlation holds for evolutionary histories (Fig. 7.2).

According to Linnaean tradition and internationally accepted rules, a species is always identified by two technical names. These names are in Latin or are latinized and are used uniformly all over the world. For example, the species of grass frogs is known technically as *Rana pipiens;* the species to which we belong is *Homo sapiens.* Such species names are always underlined or printed in italics, and the first name is capitalized. This first name always identifies the genus to which the species belongs. Thus, the human species belongs to the genus *Homo* and the grass-frog species to the genus *Rana. Homo sapiens* happens to be the only presently living species within the genus *Homo,* but the genus *Rana* contains *Rana pipiens* as well as many other frog species.

A complete classification of an animal tells a great deal about the nature of that animal. For example, if we knew nothing else about men except their taxonomic classification, then we would know that the design characteristics of men are as outlined in Table 6. Such data already represent a substantial detailing of the body structure. By implication, moreover, we would also know that the evolutionary history of men traces back to a common chordate ancestry.

That classification gives direct and inferential information about *two* kinds of animal attributes, i.e., body structure and evolutionary history, should be clearly kept in mind. Both attributes encompass not only the developmental and the

TABLE 6. A partial taxonomic classification of man

Rank	Name	Characteristics
phylum	Chordata	with notochord, dorsal hollow nerve cord, and gills in pharynx at some stage of life cycle
subphylum	Vertebrata	notochord supplemented or substituted by bony or cartilaginous vertebrae in adult; body with head, trunk, tail, basically segmented; skull enclosing brain.
superclass	Tetrapoda	terrestrial; trunk appendages are walking legs; gills embryonic only, breathing by lungs
(group)	Amniota	land-adapted eggs and embryos surrounded by water sac (amnion); heart 4-chambered; excretion by kidneys (metanephros); 12 pairs of cranial nerves
class	Mammalia	young nourished by milk glands; skin with hair; body cavity divided by diaphragm; aortic arch only on left; red corpuscles without nuclei; homoiothermic; 3 middle-ear bones; brain with well-developed cerebrum
subclass	Theria	eggs not laid, young born; teeth specialized (incisors, canines, premolars, molars)
infraclass	Eutheria	embryos attached to and nourished by maternal placenta; young born in fully developed condition; males with urethra and sperm ducts joining into single duct through penis; females with urethra and vagina opening at separate orifices
order	Primates	basically tree-dwelling; usually with fingers, flat nails; sense of smell reduced
suborder	Anthropoidea	flat face; eyes forward, with stereoscopic and color vision
superfamily	Hominoidea	arm freely movable in socket; hands and feet similarly specialized; tail internalized; menstrual cycles
family	Hominidae	upright, bipedal locomotion; living on ground; hands and feet differently specialized; family and tribal social organization
genus	*Homo*	large brain, speech; life span extended, with long youth
species	*Homo sapiens*	prominent chin, high forehead, thin skull bones; spine double-curved; body hair sparse

adult stages of animals but also all levels of structural organization, from the molecular to the organismic. Animals are characterized by other attributes as well, notably by functions and by ways of life. These, however, play only a limited role in defining animal types. Because basic functions are the same in all animals, such functions are not very useful as distinguishing traits. Moreover, both the ways of life and the detailed ways of performing functions can become modified greatly and can be adapted readily to fit particular environmental requirements. Such criteria therefore are usually less permanent than the architecture of the body and evolutionary history. Consequently, the Linnaean system incorporates essentially two kinds of data. And because it encodes a very considerable amount of information about animals, the system is far more than an empty naming scheme. That is why it is in universal use today and why other systems of classification, which *were* mere naming schemes, have not survived beyond the time of Linnaeus.

Another point requires special emphasis. As noted, the taxonomic system tacitly postulates or infers that a common body organization is correlated with a common ancestry. How firmly can such a causal interconnection actually be established? Can it be demonstrated that the body forms within a given taxonomic group resemble one another *because* the animals have all evolved from a common ancestry?

Structural resemblances of presently living animals can be studied readily, and indeed they are largely well known. Studies of ancestral fossil animals and deductions of historical interrelations represent an independent line of investigation, and the amount of information available here varies greatly. In general, evolutionary knowledge is progressively more precise for progressively lower taxonomic ranks. But above the phylum level, and in many cases at the phylum level, such knowledge is lacking almost entirely. It has been found that, for the lower ranks, a classification based on evolutionary history actually does dovetail well with an independently deduced classification based on living body forms. Such agreements justify and lend firm support to the generalization that "common body organization implies common ancestry."

However, where evolutionary information is incomplete or is lacking altogether, classification must be based almost wholly on studies of body organization. Conclusions about common ancestries then rest on very insecure foundations. It is well known, for example, that two structural patterns can have quite different ancestries yet be similar nevertheless, the animals having made similar adaptive responses to common survival requirements. We shall discuss such phenomena in Chap. 13. In short, unless structural information can actually be supported by independently obtained evolutionary information, inferences as to ancestries may be quite unwarranted. Such evolutionary inferences are being made nevertheless, and different zoologists frequently propose different ones from the same structural data. This explains why the classifications of the higher taxa often vary from text to text. Indeed, the groups above and in some cases also at the phylum rank are being reshuffled and reclassified more or less continually. This is as it should be, however, for as our knowledge of evolutionary histories improves, the taxonomic rankings must be adjusted accordingly.

We may note that, where structural and evolutionary data are in substantial agreement, classification is said to be *natural,* i.e., it reflects an actual interrelation of animals. The most natural taxonomic group is undoubtedly the *species,* which is defined as an *interbreeding* group; members of two different species normally do not interbreed (cf. Chap. 16). In this instance, therefore, a reproductive characteristic resulting from close similarity of structure and evolutionary history becomes a clearcut taxonomic criterion. On the other hand, where structural and evolutionary agreement is lacking, or where evolutionary histories are known incompletely or are only guessed at, classification remains far less natural. The best example of an "unnatural" taxon is the animal kingdom itself. By tradition going back to Linnaeus and even to earlier times, all living organisms are classified into two kingdoms, the plant kingdom and the animal kingdom. However, in view of vastly increased knowledge of both the structure and the evolution of organisms, it is highly questionable whether this 250-year-old tradition is still justifiable today, and there are good reasons to name the highest taxonomic ranks in a different way. We shall return to this point in Chap. 14.

In passing from the kingdom level to lower ranks, classifications begin to be more natural for phyla and at all events for classes. Thus, some notable exceptions notwithstanding (cf. Unit 2), we may generally define a phylum as the largest group of animals for which a common ancestry has been demonstrated reasonably well and which is characterized by a common, basically unique body organization. Table 2 (Chap. 3) indicates that most phyla represent fairly familiar type-categories of animals, usually named for one of their most distinctive design features. Subordinate ranks within phyla are identified by criteria which differ for different cases, but, as noted, the species level is defined rather uniformly by a common reproductive criterion, i.e., that of interbreeding.

Note, incidentally, that the hierarchy of taxonomic ranks intersects with that of organizational levels. The category of intersection may be taken as the *population*, a unit lower in rank than a species (e.g., in some cases a *subspecies*). Several populations (of the same kind) form a species in the taxonomic hierarchy; several populations (of different species) form a community in the organizational hierarchy (cf. Fig. 3.5.)

Taxonomic Comparisons

The taxonomic hierarchy suggests, and evolutionary studies fully confirm, that the pattern of animal evolution has the general form of a greatly branching bush (Fig. 7.3). All presently living animals and their body types are *contemporaneous*, appearing at the uppermost branch tips of the bush. Ancestral groups and body forms, mostly long extinct, appear lower on the bush, where branches join. Thus, a particular common ancestor may give rise to several different types of descendants, and a particular descendant living today may become a common ancestor of new and different types living tomorrow.

In a discussion of animal types, it is often desirable to make comparative statements about two or more body forms and/or evolutionary histories. The evolutionary-taxonomic bush then prescribes what kinds of comparisons are legitimately possible.

7.3. The bush pattern of evolution. The uppermost tips of the branches represent currently living forms and branches terminating below the top represent extinct forms. Fork points such as *B* and *C* are ancestral types. *B* is more ancient and of higher taxonomic rank than *C*. *A* represents the archancestor of all living types.

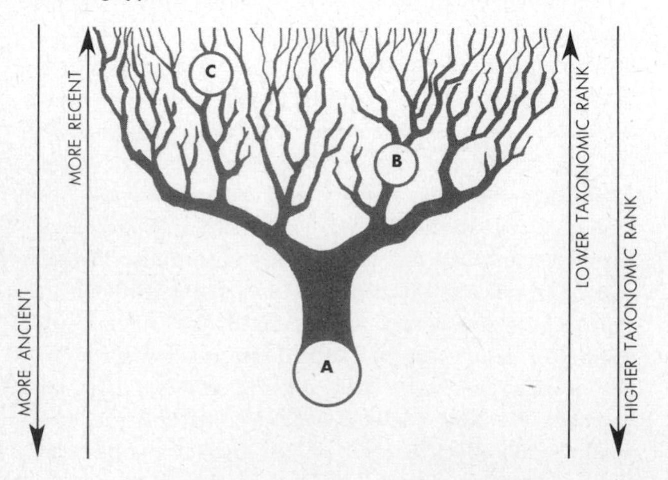

Several sets of contrasting terms are employed, and it is important to use them correctly.

Simple-Complex. These terms indicate comparative degrees of structural and/or functional elaboration of animals and their parts. Degrees of elaboration are judged by the *number* of components present and by the number of interrelations among the components (Fig. 7.4). The terms thus describe organizational and operational attributes of animals, *not* evolutionary attributes. If we say that one animal (e.g., a mammal) is more complex than another (e.g., an amoeba), we mean only that it possesses comparatively more cells, tissues, or other units, and comparatively more interrelations among these units; we do not make *any* evolutionary comparison. A higher level of organization, in the sense defined in Chap. 3, does signify "more complex." Within a given level of organization, comparative complexity must be judged on the basis of the number of parts and the abundance of interrelations. Note also that it is quite meaningless to speak of "simple" cells or animals; all units of life are exceedingly complex to begin with, and the terms "simple" and "complex" have zoological information content only if they are used in a *comparative manner.*

Primitive-Advanced. These adjectives are often—and wrongly—employed as equivalents for "simple" and "complex." Unlike the latter, "primitive" and "advanced" do have primarily historical, evolutionary connotations. If animals or their parts are structured according to ancient, ancestral patterns, we may say that such patterns are primitive; newer, more modern patterns superimposed on the ancient ones then are advanced, again by comparison only (cf. Fig. 7.4). Evolutionary "earliness" and "lateness" and similar time-contrasting terms are roughly equivalent to "primitive" and "advanced." Note that an animal or a part of it may be more primitive yet at the same time more complex than another (e.g., protozoan cells are more primitive and probably more complex than most mammalian cells); more advanced yet simpler than another (e.g., a human skull is more advanced but simpler—with far fewer bones—than a fish skull); and also more primitive as well as simpler, or more advanced as well as more complex (e.g., jellyfishes are more primitive and simpler than vertebrates).

Generalized-Specialized. These terms have both structural and evolutionary connotations. A pattern is generalized (=unspecialized) if it can be and actually has been further

A

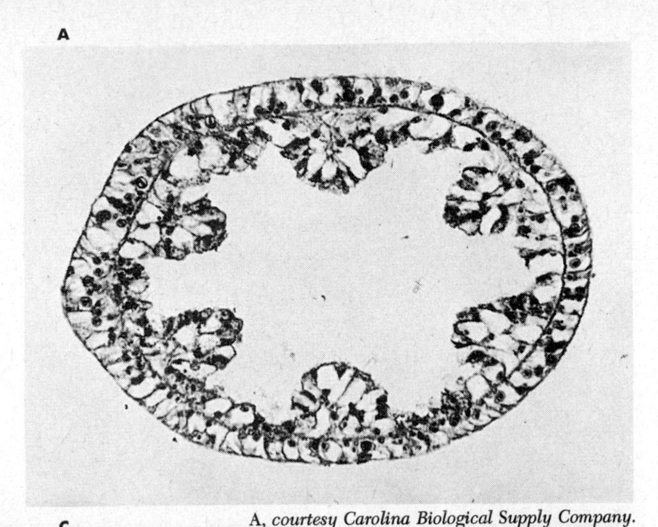

B

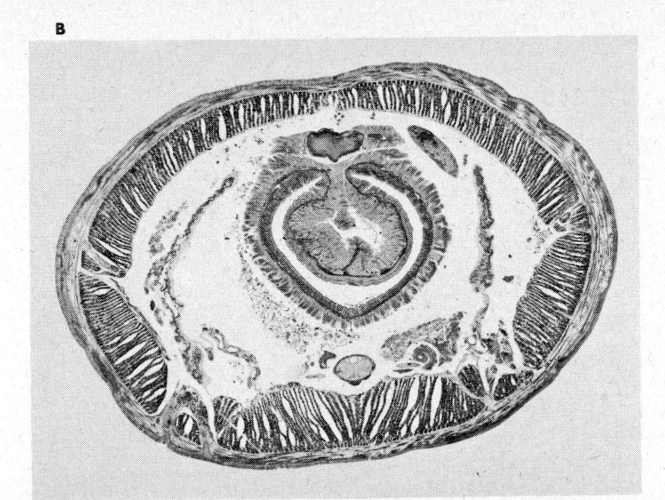

A, *courtesy Carolina Biological Supply Company.*

C

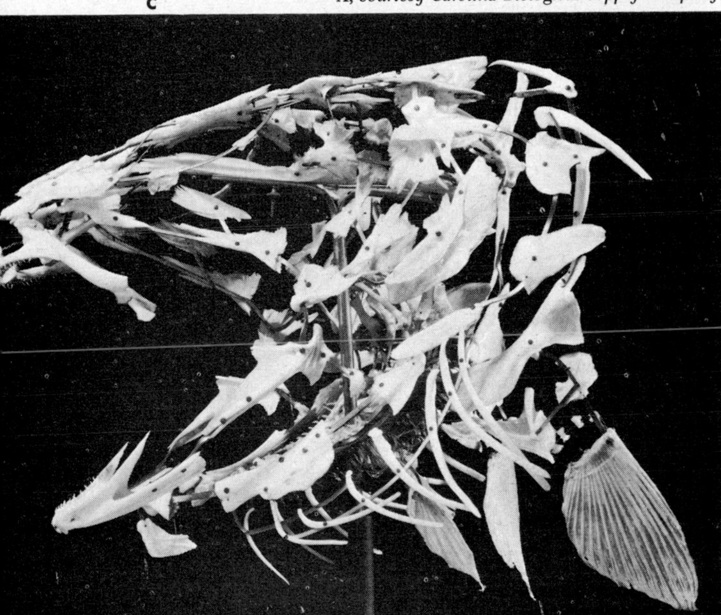

D

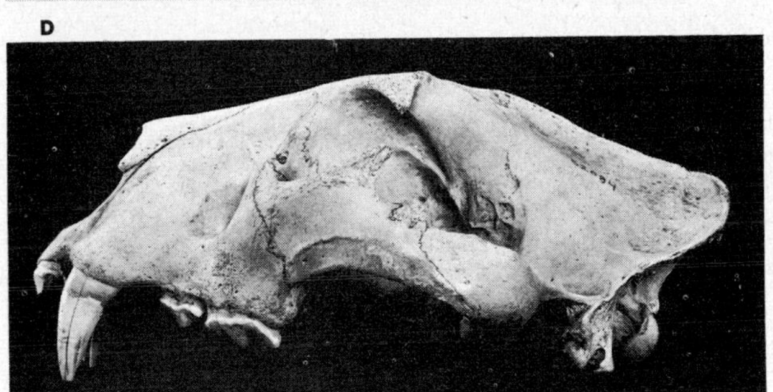

7.4. Structural and evolutionary comparisons. **A,** photomicrographic cross section through a hydra, a coelenterate; **B,** through an earthworm, an annelid. On the basis of the number of tissue layers present in each, the hydra can be considered to be simpler in structure than the earthworm. Hydra also happens to be more primitive, i.e., the evolutionary history of the coelenterate phylum goes back further than that of the annelid phylum. **C** shows the bones of a fish skull, and **D,** a lion skull. On the basis of the number of bones present in each, the fish skull is structurally more complex, yet it is more primitive from an evolutionary standpoint than the lion skull. Thus, a given degree of structural complexity does not necessarily or automatically signify a comparable degree of evolutionary advancedness.

modified, e.g., through simplification or complication; it is specialized if it is already modified. For example, the segmented body of an annelid, as exemplified by an earthworm, is generalized by comparison with the segmented body of an arthropod, as exemplified by an insect; the arthropod body is a modified annelid body (Fig. 7.5). Generalized structures originate earlier than and they later give rise to the specialized structures, both in evolutionary and in embryonic history. Hence generalized patterns are usually more primitive than specialized ones (but they are not always simpler).

Higher-Lower. These adjectives have a proper role only in reference to taxonomic and organizational levels and in a literal sense. Thus, a phylum is higher in the taxonomic

A

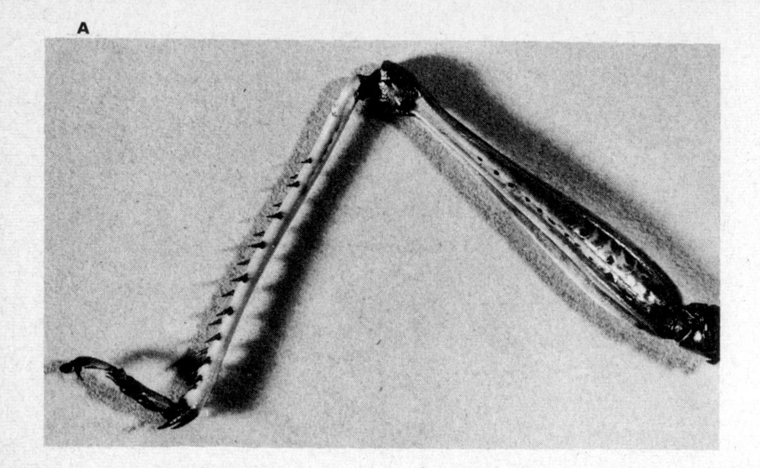

B

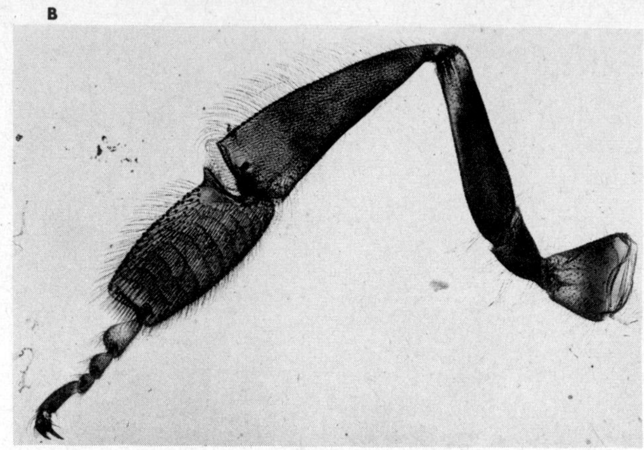

B, *courtesy Carolina Biological Supply Company.*

7.5. Generalized and specialized structures. A. The leg of a grasshopper, a rather generalized leg type within the broad scope of insect structure (even though such a leg is already specialized considerably, e.g., for jumping). **B.** The leg of a honeybee, a comparatively more clearly specialized type: a subterminal segment is enlarged into a bristly pollen basket, adapted specifically for pollen transport, and the leg as a whole is equipped abundantly with bristles, a feature which facilitates adherence of pollen.

hierarchy than a class, a tissue is higher in the organizational hierarchy than a cell, and a bird may be higher off the ground than a worm. In other zoological respects, the terms are either erroneous or meaningless if they are used in a structural or evolutionary sense—as they often are. Several decades ago it was believed that animal evolution occurred in a scalelike or ladderlike pattern, one animal type giving rise to another in serial fashion. The terms "lower" and "higher" were used to indicate "rung" positions on the evolutionary "ladder." Also, since "highest" generally meant "man," human vanity could be pleasantly satisfied by regarding all other animals as being "lower."

However, it is now clearly established that evolution is not ladderlike, but bushlike (cf. also Chap. 14); and that man stands only just as "high" on the bush as every other animal now living. The terms thus have lost any zoological significance, but old terminologies become habits which die slowly. Even professional zoologists still frequently speak of "higher" or "lower" animals, when the actually intended reference is to advanced and/or more complex ones or to primitive and/or simpler ones. Indeed, the terms are not only erroneous but also superfluous. For example, in a statement such as "man is a higher animal than a dog inasmuch as man has a more complex brain," the term "higher" is a man-centered value judgment without any zoological information content; for the zoologically significant information is fully expressed by the statement "man has a more complex brain than a dog." It might be noted that the sense of smell is developed far better in a dog than in a man, yet this does not make the dog "higher" either. The point is that the terms should be avoided if scientifically meaningful comparisons of body types are to be made.

Homologous-Analogous. As noted earlier, function and way of life play only relatively small roles as taxonomic criteria. It happens often, therefore, that the various subgroups within a taxonomic rank exhibit widely different methods of executing functions and highly different ways of life. Consider, for example, the different functions of the trunk appendages and the different ways of life of a fish and a man. Nevertheless, all members of the vertebrate subphylum use the same kinds of structures in solving the problems of their variously different ways of life. Thus, the lateral fins of a fish and the arms and legs of a man are basically the same kinds of structures, which have evolved from one common ancestral type of body appendage.

Whenever body parts within an animal or in different animals have evolved from a common ancestral starting point, as in the example just cited, and whenever they also develop embryologically in like fashion and thus have the same structure at least initially, then such components are said to be *homologous*. The forelimbs and hindlimbs of land vertebrates are homologous to one another and also to the lateral fins of fishes. Homology therefore connotes similarity of history and structure, without reference to function. By contrast, *analogy* connotes similarity of function or way of life, without reference to history or structure. For example, the wings of insects, bats, and birds are analogous. Bat and bird wings also happen to be homologous as well, but insect wings are not homologous to either bat wings or bird wings. We note that "homology" is essentially a single term for the two principal criteria on which taxonomy rests; a study of homologies is the basis of animal classification (Fig. 7.6).

7.6. Homology and analogy. The bird, bat, and insect wings diagrammed here are all analogous, since all serve the same function of flying. Bird and bat wings are also homologous, since they develop in similar fashion and have similar structure. But insect wings are not homologous to either of the other two.

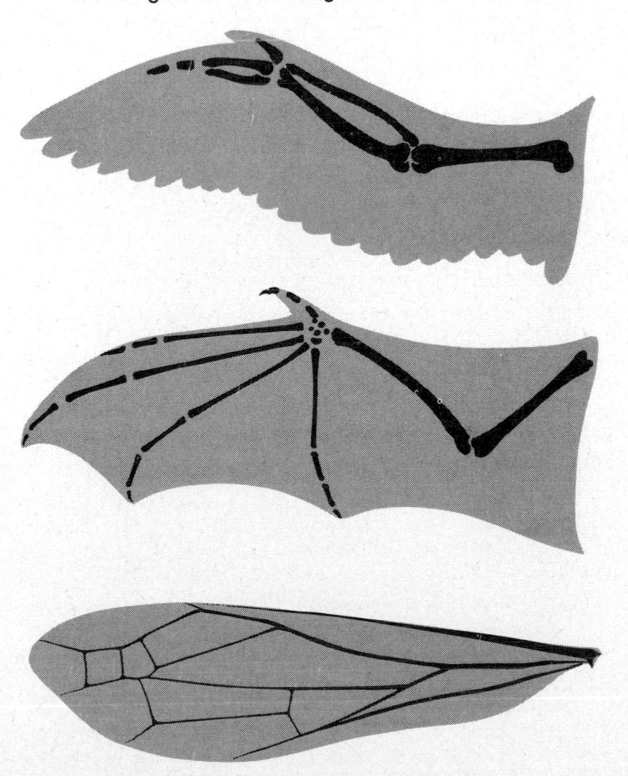

The above account has shown how animals and their body organizations are catalogued and compared, and we are now ready to examine the organizations themselves.

BODY FORMS

We know from the foregoing discussion that architectural patterns are the more generalized the higher the taxonomic rank, and the more specialized the lower the taxonomic rank. If we start with the animal kingdom as the highest taxon representing the most generalized architecture, what are the specific features of decreasing generality which characterize the body forms of progressively lower ranks? We may keep in mind here that, as noted, structure is often the only criterion for classifying animals down to about the phylum level, information on evolutionary histories being unavailable in many cases.

Only a single organizational feature can be considered to be shared in common by all animals: as pointed out in Chap. 3, all animals are constructed to pursue life on the basis of *heterotrophism* and *motility*. It is this most generalized identifying characteristic which has been used traditionally to establish animals as a separate kingdom. However, the adequacy of this tradition in the light of modern knowledge remains to be examined (Chap. 14).

Level of Organization

What is the next most generalized feature of animal body form? How can we even decide which of two features is more generalized than the other? We know that the choice must be made on the basis of animal structure, but we need not confine ourselves to adult structure; embryonic structure is part of the body organization too. Indeed, in assessing the comparative generality of traits we are often influenced strongly by the embryonic histories of animals. A century of embryological studies has shown that, of any two structural features, the one which appears earlier during embryonic development can usually be considered to be the more generalized. Some exceptions notwithstanding, embryos by and large do develop general traits first and more and more specific traits in later succession. For example, a human embryo develops chordate characteristics before vertebrate characteristics, mammalian characteristics before human characteristics; and only at the very last moment does the embryo acquire the specific personal traits which will distinguish the adult uniquely from all other human adults (Fig. 7.7). Analogous successions of progressively less general, hence more specific, developmental stages characterize the embryonic histories of all animals; and it is on such a basis that the degree of generality of a trait is often determined.

Accordingly, the most generalized trait should be the one which appears first in the embryo. That trait is the *level of organization* of animals, in this context also called the *grade of construction.* All animals typically begin life as single cells. Some never develop beyond this stage, others subsequently develop to supracellular levels. On this basis, the animal kingdom is traditionally subdivided into two *subkingdoms: Protozoa,* identified by a cellular grade of construction, and *Metazoa,* identified by various supracellular grades (Fig. 7.8).

Within the Metazoa, some animals are constructed almost entirely on the tissue level, whereas others pass beyond the tissue level in their embryonic development and become predominantly more complex. Accordingly, the metazoan subkingdom is considered to include two *branches.* In the branch *Parazoa,* the highest level of organization is the tissue. This branch encompasses just one phylum, the *sponges* (Porifera). The second metazoan branch comprises the *Eumetazoa,* and these animals are defined by the presence of permanent organs and particularly also of organ systems. The organ level is the highest principally in two phyla, the *coelenterates* and the *ctenophores.* All other Eumetazoa pass beyond the organ level in their embryonic development and come to exhibit a conspicuous system-level of construction (cf. Fig. 7.8). This structural difference between organ and organ-system levels offers a further criterion of taxonomic

7.7. The photos illustrate four successive stages in human development, viz., 25 days, 33 days, 6 weeks, and 8 weeks after fertilization, respectively. The series indicates that chordate features arise first (e.g., dorsal skeletal supports, gill pouches, as in **A**); that vertebrate features develop next (e.g., anteroposterior segments, paired limb buds, tail, as in **B**); that tetrapod and mammalian traits appear still later (e.g., four legs, umbilical cord, as in **C**); and that distinctly human traits appear last (e.g., arm-leg differences, flat face, individualized facial expression, as in **D**). Thus, earlier traits are more generalized and also taxonomically more inclusive.

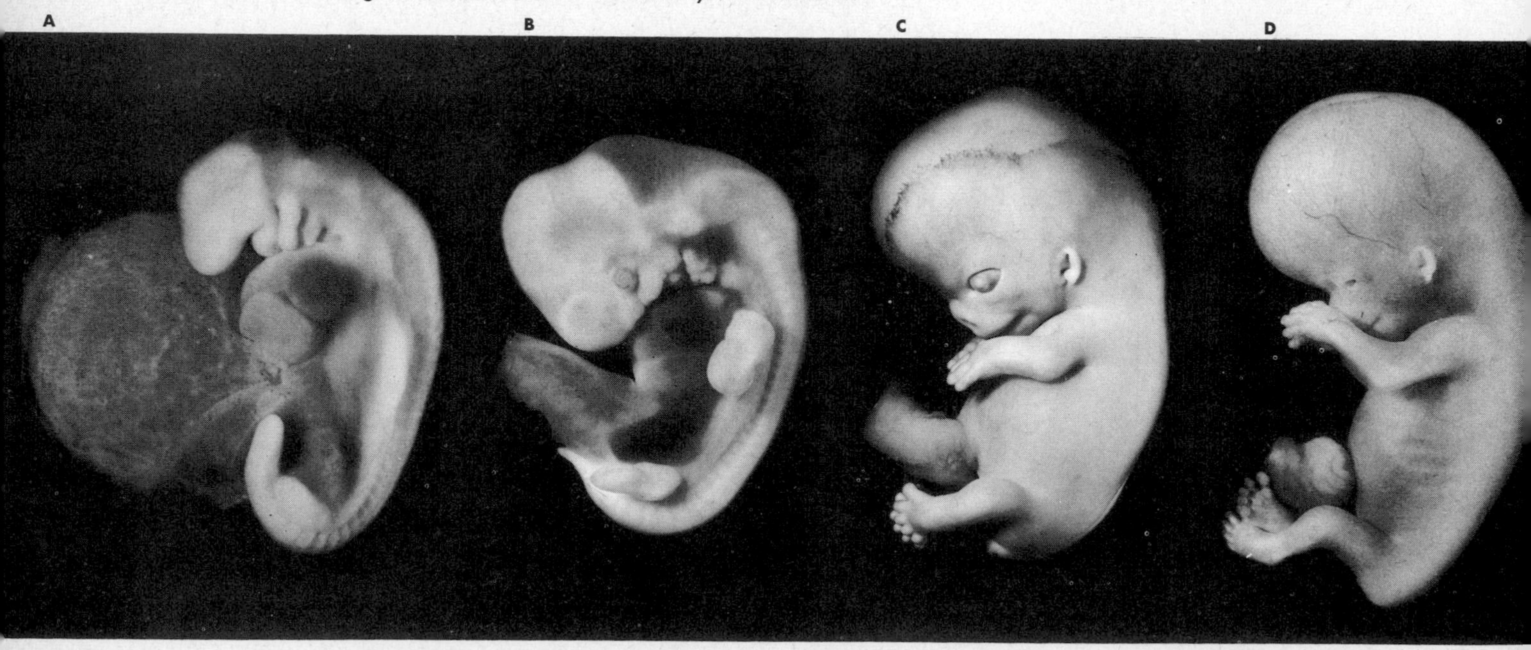

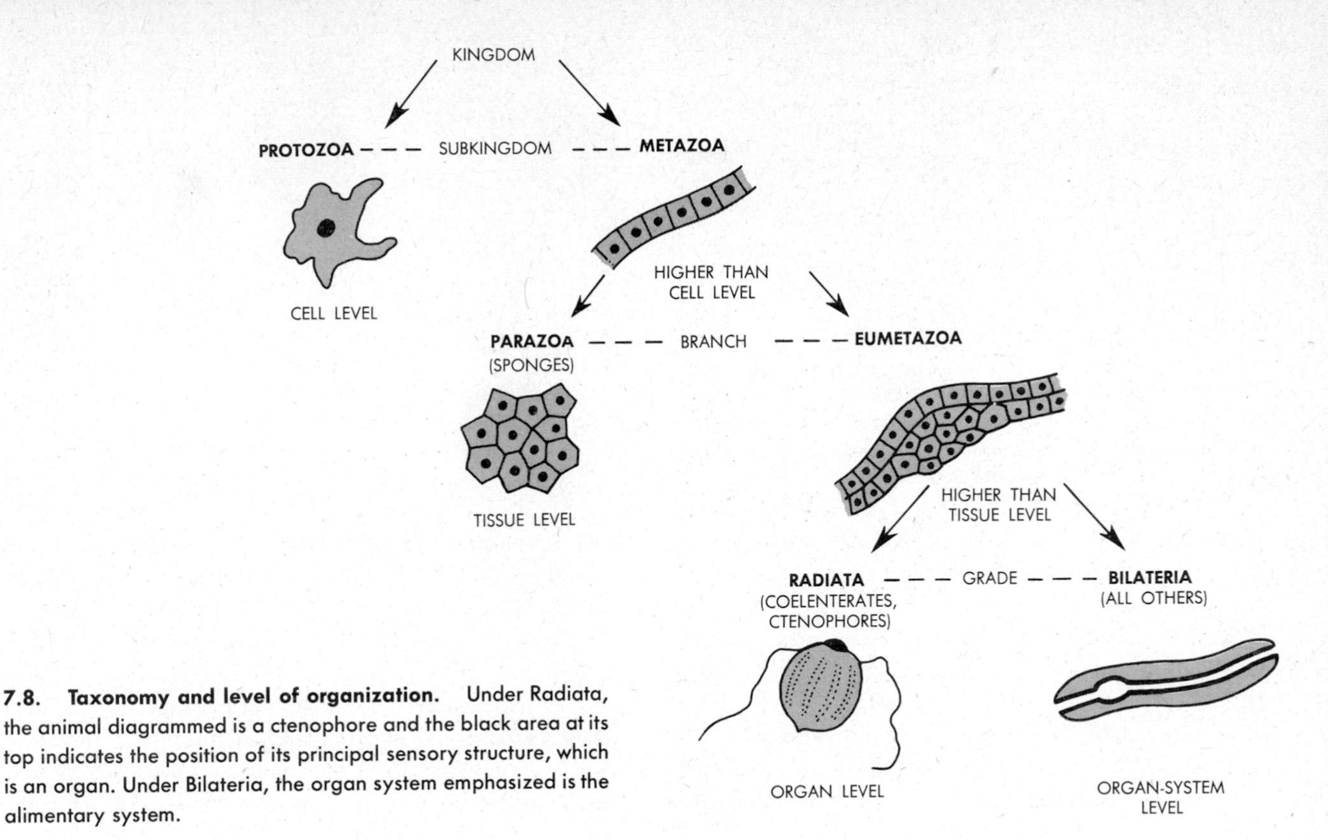

7.8. Taxonomy and level of organization. Under Radiata, the animal diagrammed is a ctenophore and the black area at its top indicates the position of its principal sensory structure, which is an organ. Under Bilateria, the organ system emphasized is the alimentary system.

distinction. Indeed, this distinction is reinforced by an additional architectural difference of lesser generality, viz., the basic symmetry of the animal body.

Symmetry

Four basic types of animal symmetries exist: *spherical, radial, bilateral,* and *asymmetrical* (Fig. 7.9). But note that geometric precision in animal structure is at best only approached and is hardly ever actually attained. Strictly speaking all animals are more or less asymmetrical. Any other label is to a greater or lesser extent superficial, though it may nevertheless be useful descriptively. Many an animal exhibits the same symmetry throughout its development and in the adult stage as well, but many another animal does not. In the latter case, adult symmetry is often greatly modified, usually as a specific adaptation to a particular mode of life. It is therefore important to distinguish between *primary* symmetry, as exhibited by the embryo and the larva, and *secondary* symmetry, as exhibited by the adult. Where the two are different, only the primary symmetry characterizes the original nature of the body form. Moreover, this characterization has a very high degree of generality: after levels of organization, symmetries are the first features to appear in animal development.

Protozoa are variously radial, bilateral, and asymmetrical, and a few also approach a more or less spherical design. All these protozoan symmetries are essentially secondary ones. More precisely, inasmuch as in these unicellular forms there is hardly any conspicuous difference between "developmental" and "adult" stages, it is impossible to decide to what extent a given symmetry is not merely an adaptation to a particular way of life. In effect, therefore, symmetry is a rather poor criterion on which to base further distinctions among protozoan body forms. This is also the case for sponges, but for a different reason: all these animals display a primary radial symmetry. The basic radiality is largely obscured in the adults, which are variously asymmetrical (cf. Fig. 7.9). The change from larval to adult symmetry here appears to be correlated with a change in way of life: sponge embryos and larvae are actively motile, the adults are exclusively sessile.

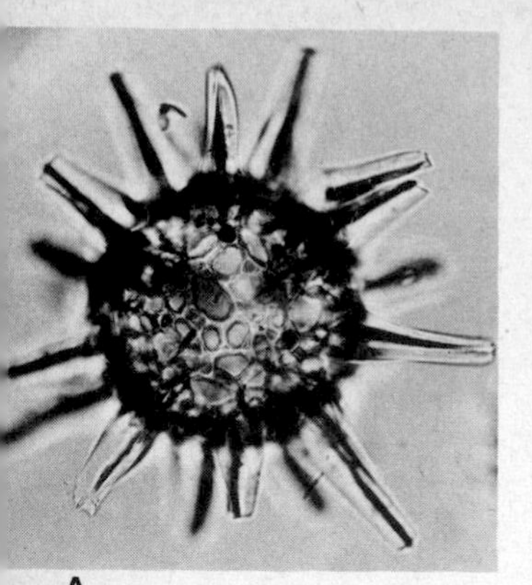
A

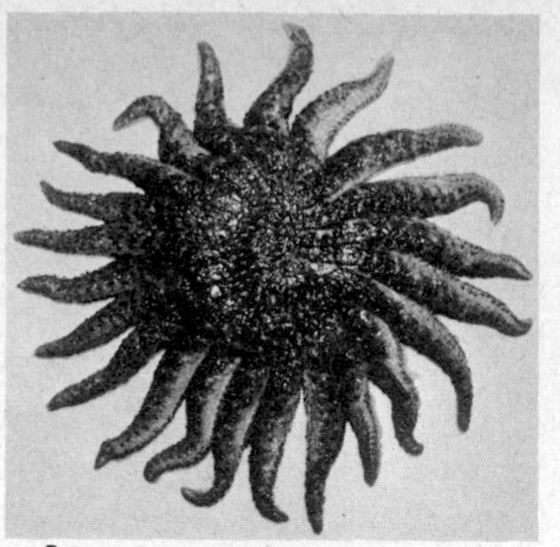
B

C

D

7.9. The four basic animal symmetries. **A,** a spherical radiolarian (protozoon); **B,** a radial starfish (echinoderm); **C,** a bilateral land crab (arthropod); **D,** an asymmetrical, irregularly shaped, branching sponge (poriferan).

In the Eumetazoa, however, the criterion of symmetry does serve as a basis for taxonomic classification. Some Eumetazoa develop a primary radial symmetry in the embryo, and this radiality then persists throughout embryonic and larval life and carries over more or less unchanged into the adult stage as well. Other Eumetazoa start life as radial early embryos, but this radiality rapidly changes to an embryonic bilaterality, and the remaining embryonic period as well as the larval and adult stages then are bilateral. On this basis, we may subclassify the Eumetazoa into two *grades* (Fig. 7.10). The grade *Radiata* includes the coelenterates and the ctenophores. These animals are identified both by an organ level of construction and by a pronouncedly radial body plan in the embryos, larvae, and adults.

Note that in some radiates, e.g., sea anemones, a measure of adult bilaterality is superimposed on the primary radiality. However, the adult symmetry of radiates cannot be correlated readily with way of life. For example, the superficial bilater-

7.10. Taxonomy and symmetry. The grade Radiata is identified by primary as well as secondary radial symmetry, regardless of whether the adults are sessile or motile. The grade Bilateria is identified by primary bilateral symmetry and by adults which are bilateral if motile or often secondarily radial if sessile.

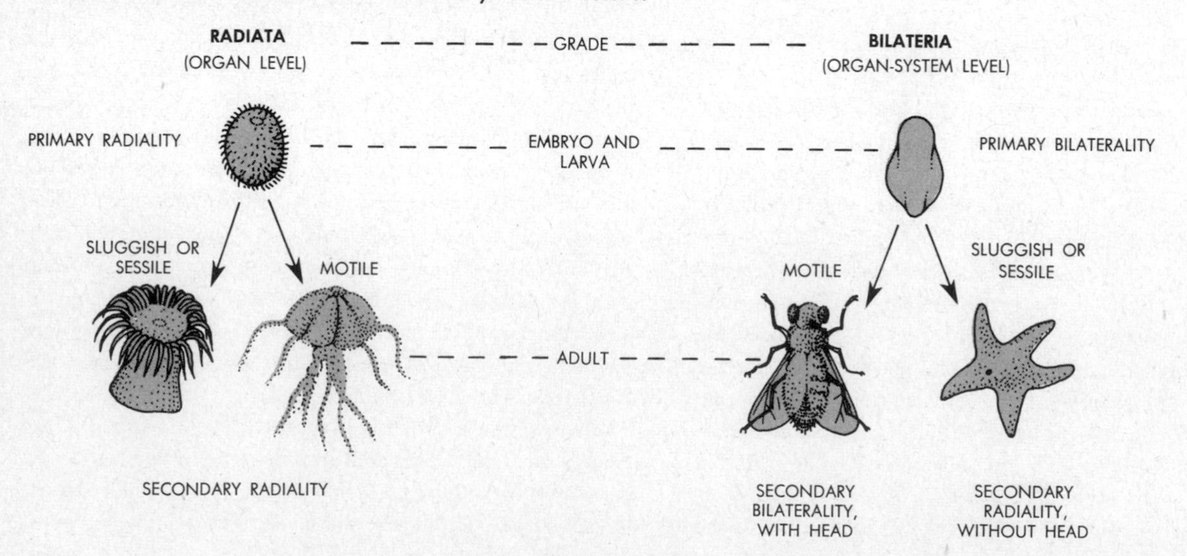

ality of sea anemones might suggest an adaptation for a motile way of life, yet sea anemones are virtually sessile. Analogously, the clearcut adult radiality of jellyfishes might be construed as an adaptation to sessilism, yet jellyfishes are highly motile. Thus, although adult radiates include sessile attached types as well as creeping and swimming motile types, all are basically radial. Evidently, broad generalizations correlating adult symmetry and way of life do not apply to the Radiata.

All other Eumetazoa are primarily bilateral, and they constitute the grade *Bilateria* (cf. Fig. 7.10). In most Bilateria not only the embryos and larvae but also the adults are bilateral. Such adults characteristically are motile, elongated, and they typically possess a head. In numerous Bilateria, however, the adult is specialized to pursue a sluggish or sessile way of life, and in most of these cases a secondary tendency toward radiality or some form of asymmetry is usually manifest. A head is then absent as well (e.g., starfishes and other echinoderms, numerous mollusks, and arthropods such as barnacles). Apparently, secondary symmetries and ways of life are correlated far more consistently among the Bilateria than among the Radiata (and we are led to conclude that a generalization applicable to one group of animals does not necessarily also apply to another group).

Alimentary Patterns

After symmetry, the next most generalized architectural feature is the form of the alimentary structures; once symmetry is established in the embryo, the pattern of the alimentary structures is among the first to become elaborated.

In the protozoa, alimentation is entirely an intracellular process and, like symmetry, it contributes little toward a distinction among protozoan body forms. The Metazoa, however, exhibit three different major alimentary patterns and these are useful in characterizing body forms (Fig. 7.11).

One alimentary architecture, unique to the sponges, may be described as exhibiting a *channel network* pattern. Alimentary channels ramify extensively throughout the body and communicate with the exterior via openings in the body wall. The channels are lined by flagellate digestive cells. The beat of the flagella draws food-bearing water into the channel system through most of the openings and expels water strained of food through one, usually larger opening. Microscopic food particles passing by the digestive cells are engulfed and are processed intracellularly, as in Protozoa. Unlike protozoan cells, however, the digestive cells of sponges function in cooperative groups, and they are specialized particularly for alimentation. Thus, as would be expected in a sponge, the alimentary apparatus exhibits a primitive tissue level of construction.

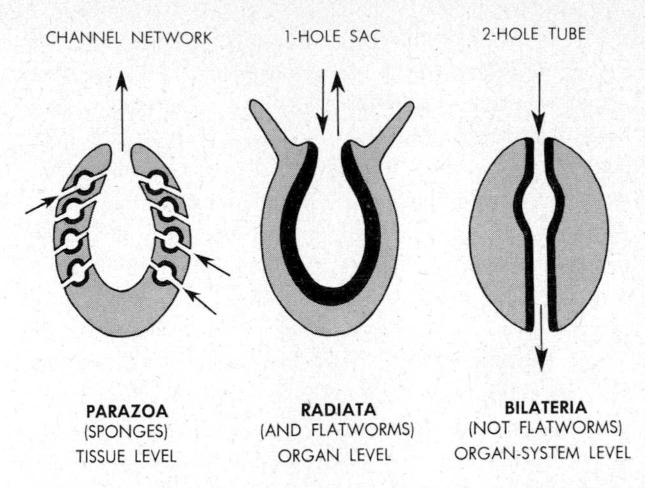

7.11. The alimentary patterns of Metazoa, diagrammatic. The positions of the alimentary cells are shown as dark layers.

The second type of alimentary architecture exhibits what may be called a *one-hole sac* pattern. A single opening in the sac functions as both mouth and anus, and the sac itself is formed by a single tissue layer which includes digestive cells (cf. Fig. 7.11). Food is brought into the sac partly by ingestive movements of the opening, partly by accessory ingestive organs such as tentacles or muscular sucking devices. The digestive cells again engulf microscopic food particles and process them intracellularly. Waste matter is expelled through the opening of the sac by contraction of the whole body. This alimentary pattern, which thus includes the organ grade of complexity, is encountered in all Radiata and in one phylum of the Bilateria, viz., the flatworms.

The third alimentary pattern characterizes all other Bilateria. Based on the organ-system level of construction, this pattern may be designated as forming a *two-hole tube* system (cf. Fig. 7.11). Developmentally, such a tube arises from a sac. The embryos of the Bilateria first develop a one-hole sac as do the embryos of the radiates and flatworms. But whereas the latter then retain the sac design permanently, the bilaterial embryos do not: a second opening soon develops in the sac, typically at the end of the body opposite to the first opening. One opening then specializes as a mouth, the other as an anus, and the tube interconnecting them becomes the alimentary tract in which food passes only one way, from mouth to anus. Moreover, the tract later specializes regionally as several different organs, e.g., pharynx, esophagus, stomach, intestine. Various additional organs such as liver, pancreas, and other glands contribute to the formation of a complex alimentary organ system (Fig. 7.12).

Of the two openings formed in the embryo, which becomes the mouth and which the anus? The answer varies for different bilaterial groups. In one group, the original opening of the one-hole sac forms the mouth, and the second opening developed later becomes the anus. In the other group the pattern is reversed, the first opening forming the anus, the second the mouth. We may use this distinction to define two general, descriptive categories among the Bilateria. In the *Protostomia,* the first opening is the mouth; and in the *Deuterostomia,* the second opening is the mouth. Protostomia include most of the phyla of worms, including segmented worms, as well as mollusks, arthropods, and numerous other phyla. There are fewer phyla among the Deuterostomia, their best-known members being the echinoderms and the chordates (cf. Fig. 7.12).

In bilaterally symmetrical adults of either category, the mouth is located at the head end, the anus at or near the hind end, and the interconnecting tract by and large follows the longitudinal axis of the body. In Bilateria in which embryonic

7.12. Taxonomy and alimentary structure. Among radiate animals, the single alimentary opening develops in the embryo and serves as both mouth and anus in the adult. Among Bilateria, the original alimentary opening of the embryo becomes the mouth in Protostomia, the anus in Deuterostomia. In each case a second alimentary opening develops later at the opposite end of the animal.

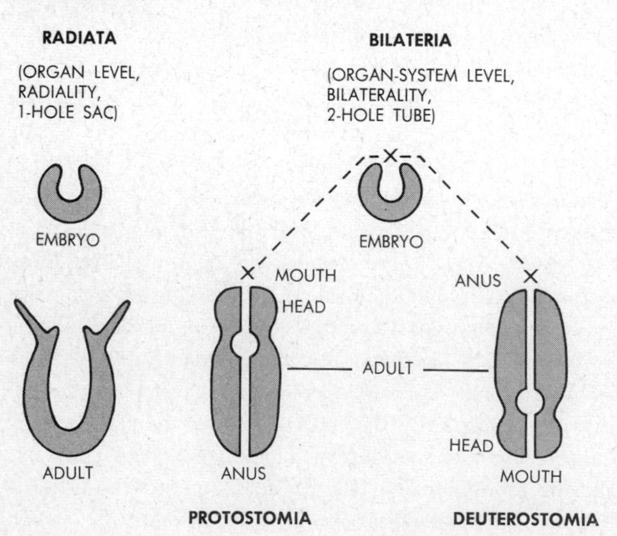

and larval bilaterality has given way to adult radiality or asymmetry, the course of the alimentary tract is usually modified as well. In many sessile, attached animals, for example, the tract has a U-shape, the bottom of the U marking the region where the animal is attached to the ground. Mouth and anus then lie close together and as far away from the ground as possible.

Taken together, the grade of construction, the symmetry, and the alimentary pattern provide a broad outline of the fundamental body form of any animal. Alimentary pattern and symmetry specify the basic interior and exterior architecture, respectively, and the organizational level specifies the complexity of the architectural building blocks. The information so available permits us to draw all animal body organizations in their most generalized form (Fig. 7.13). More detailed characterizations may now be obtained by studying the interior structure of animals, i.e., the regions between the alimentary and the external integumentary surfaces.

Mesoderm and Coelom

Fig. 7.13 indicates that all Metazoa may be considered to exhibit a basic three-ply construction; an outer layer is integumentary, an inner layer is alimentary, and a middle layer includes all other body parts. Such layering is comparatively indistinct in the sponges but is elaborated conspicuously in all Eumetazoa. In these animals the adult body is formed from well-defined embryonic *germ layers,* already referred to in Chap. 3. In the Radiata, the integumentary layer of the adult develops from an embryonic *ectoderm,* and the alimentary layer matures from an embryonic *endoderm.* Later, a middle *mesoderm* arises by the migration of a few cells from the ectoderm. Such cells are then called *mesenchyme* cells. This middle layer does not become elaborated much further even in the adults, though it may attain considerable bulk as a result of jelly secretion by the cells of the mesenchyme. In any event, the mesoderm of the Radiata adds little toward a taxonomic subclassification of the body forms.

In the Bilateria, however, the mesoderm does provide important further distinctions of body forms. In these animals mesoderm cells accumulate far more extensively than in the Radiata, and the cells also become organized into numerous tissues, organs, and systems. Indeed, the bulk of the adult body comes to have a mesodermal origin. We may note, moreover, that in different bilaterial embryos the mesoderm arises either from the ectoderm or from the endoderm or from both of these primary layers. Accordingly, we may distinguish be-

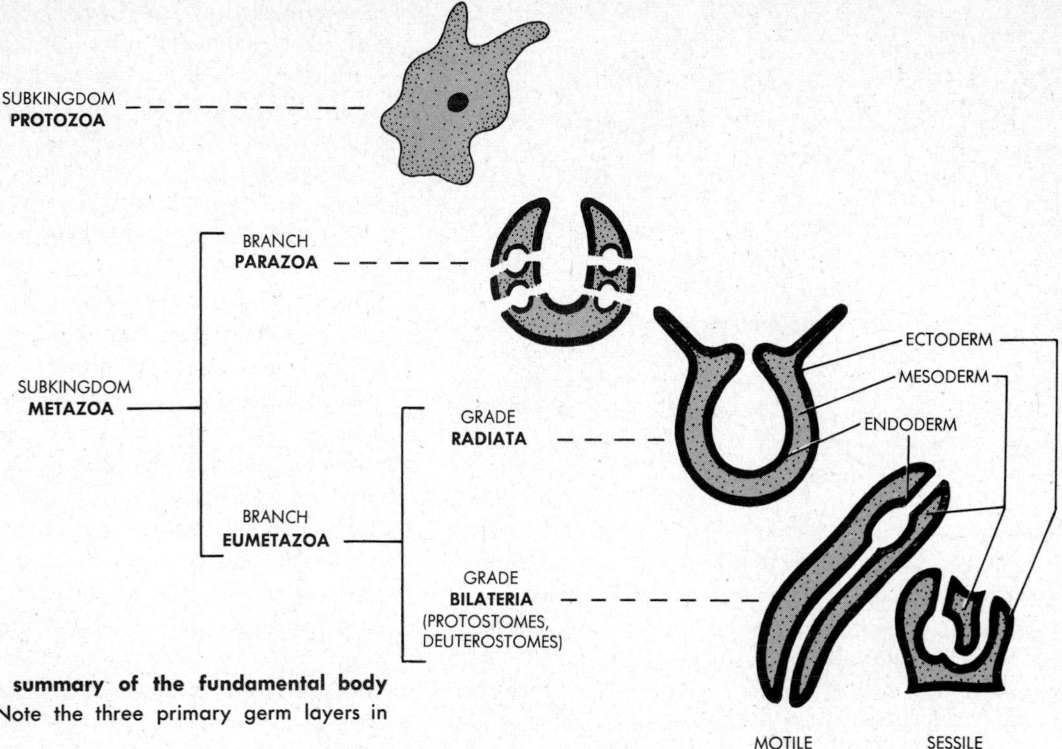

7.13. Diagrammatic summary of the fundamental body plans of animals. Note the three primary germ layers in eumetazoan animals.

tween *ectomesoderm* and *endomesoderm* (Fig. 7.14). Wherever ectomesoderm forms, it originates characteristically from a migration of loose cells from the ectoderm, as in the Radiata. However, such mesenchyme cells later multiply and become organized into cohesive tissues, organs, and systems. Wherever endomesoderm forms, it is usually a cohesive tissue from the start, and it too gives rise to organs and systems later. In only comparatively few cases does mesoderm arise purely from ectoderm or purely from endoderm; most Bilateria develop their middle layer from both of the other layers. Frequently, the mesoderm of the larva tends to be partly or largely ectomesodermal, and the mesoderm of the adult is then predominantly endomesodermal.

In some Bilateria, e.g., the flatworms, the mesoderm of the adult completely fills the region between the ectodermal integument and the endodermal alimentary system. In parts of this middle layer the mesoderm cells may remain loose, jelly-secreting mesenchyme, as in the Radiata, but in other parts the cells develop into the components of specialized organ systems. Animals so constructed may be considered to form a *subgrade* within the grade Bilateria, viz., the subgrade *Acoelomata.* The significance of this term will become apparent presently (Fig. 7.15).

In all other Bilateria, the middle body layer does not form a solid accumulation of body parts. Instead, as the mesoderm develops in the embryo, a more or less extensive free space is left between the ectoderm and the endoderm. This space later becomes the *principal body cavity* of the adult. Such a cavity is advantageous adaptively in several ways. For example, it makes the activities of the body wall mechanically independent of those of the alimentary tract. Moreover, the cavity permits an animal to attain considerable size, for fluid filling such a cavity may serve as a hydraulic "skeleton." In many cases, also, the fluid may aid in transporting food, wastes, and gases to and from deep-lying body parts, which in a large animal would otherwise be beyond the effective range of direct diffusion from integument and alimentary tract.

In many Bilateria, e.g., the rotifers, the roundworms, and the hair worms, the body cavity is bounded on the outside directly by the body-wall tissues and on the inside directly by the tissues of the alimentary system. The mesoderm in such cases is aggregated only in specific circumscribed regions

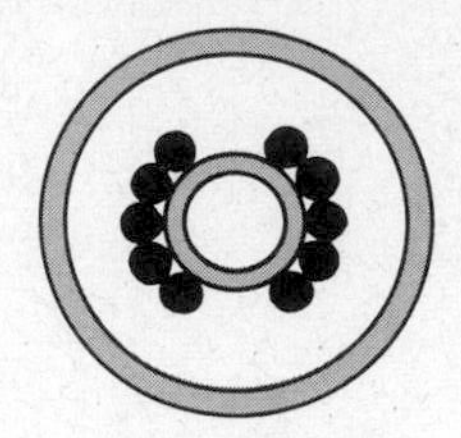

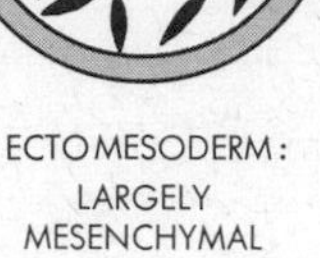

7.14. The light-gray layers represent embryonic ectoderm and endoderm, respectively; the dark cells are mesoderm. Mesenchyme ectomesoderm is characteristic of Radiata and many Bilateria, epithelial endomesoderm, of most Bilateria. Both kinds of mesoderm often form in bilaterial animals.

within the body cavity, and this cavity is not lined by any membranes. Animals so characterized constitute another subgrade of Bilateria, the *Pseudocoelomata* (cf. Fig. 7.15).

In all remaining Bilateria, the body cavity arises in a different way. During the embryonic development of mesoderm, part of it, the so-called *somatic* mesoderm, becomes applied against the inner surface of the ectoderm; and another part, the so-called *splanchnic* mesoderm, comes to surround the alimentary tract (cf. Fig. 7.15). In such animals, therefore, the body wall comprises both ectodermal and mesodermal layers, and the alimentary wall comprises both mesodermal and endodermal layers. The somatic and splanchnic parts of the mesoderm later give rise to various tissues and organs, and particularly also to a membrane which comes to enclose the free space between the two mesodermal parts. This mesodermal membrane is called the *peritoneum.* Vertical portions of it known as *mesenteries* suspend the alimentary tract from the body wall. The free space enclosed within the peritoneum represents the principal body cavity. We may now note that any cavity bounded completely by mesodermal components, especially by a peritoneal membrane, is known as a *coelom.* Accordingly, animals possessing a coelom may be said to constitute a bilaterial subgrade of *Coelomata.* The meaning of the terms "acoelomate" and "pseudocoelomate" then becomes clear. Acoelomates are animals without coelom and indeed without body cavity of any kind. Pseudocoelomates possess a *pseudocoel,* or "false" coelom, i.e., a body cavity lined by ectoderm, endoderm, perhaps partly by mesoderm, but in any event not by a peritoneum. The cavity thus resembles a true coelom superficially.

Among the coelomate Bilateria, further subgroups may be distinguished on the basis of how the coelomic cavities develop (Fig. 7.16). In one such subgroup, comprising among others the phyla of mollusks, annelids, and arthropods, the adult mesoderm arises in the embryo from two endoderm-derived cells, one on each side of the future gut. These so-called *teloblast* cells then proliferate into a pair of *teloblastic* or *mesoblastic* bands of tissue. At first the bands are solid cellular aggregates, but later each splits into somatic and splanchnic sublayers. Thus, because the coelom forms by a splitting

7.15. Diagrams of mesoderm and coelom localizations in the subgrades of the Bilateria.

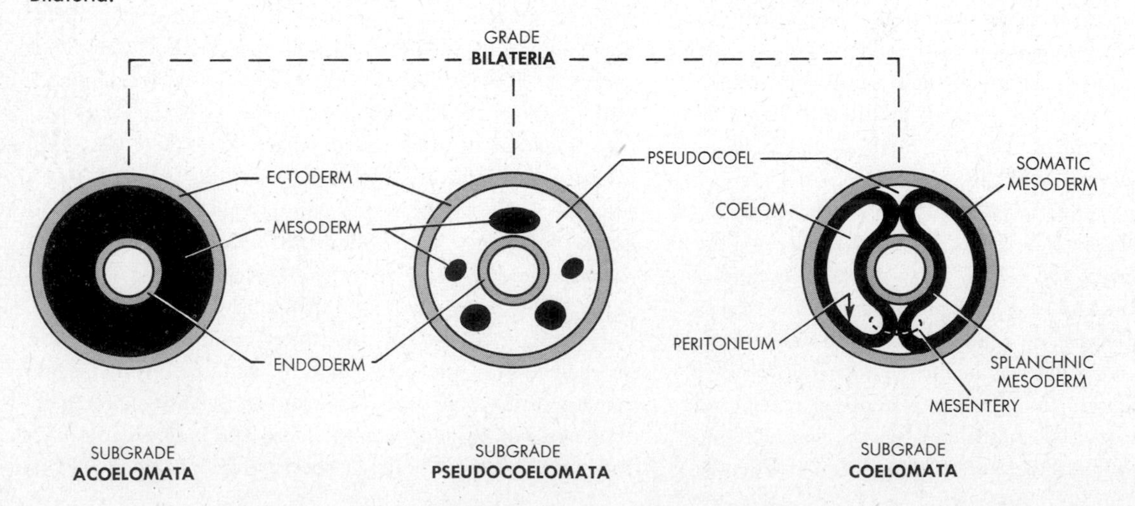

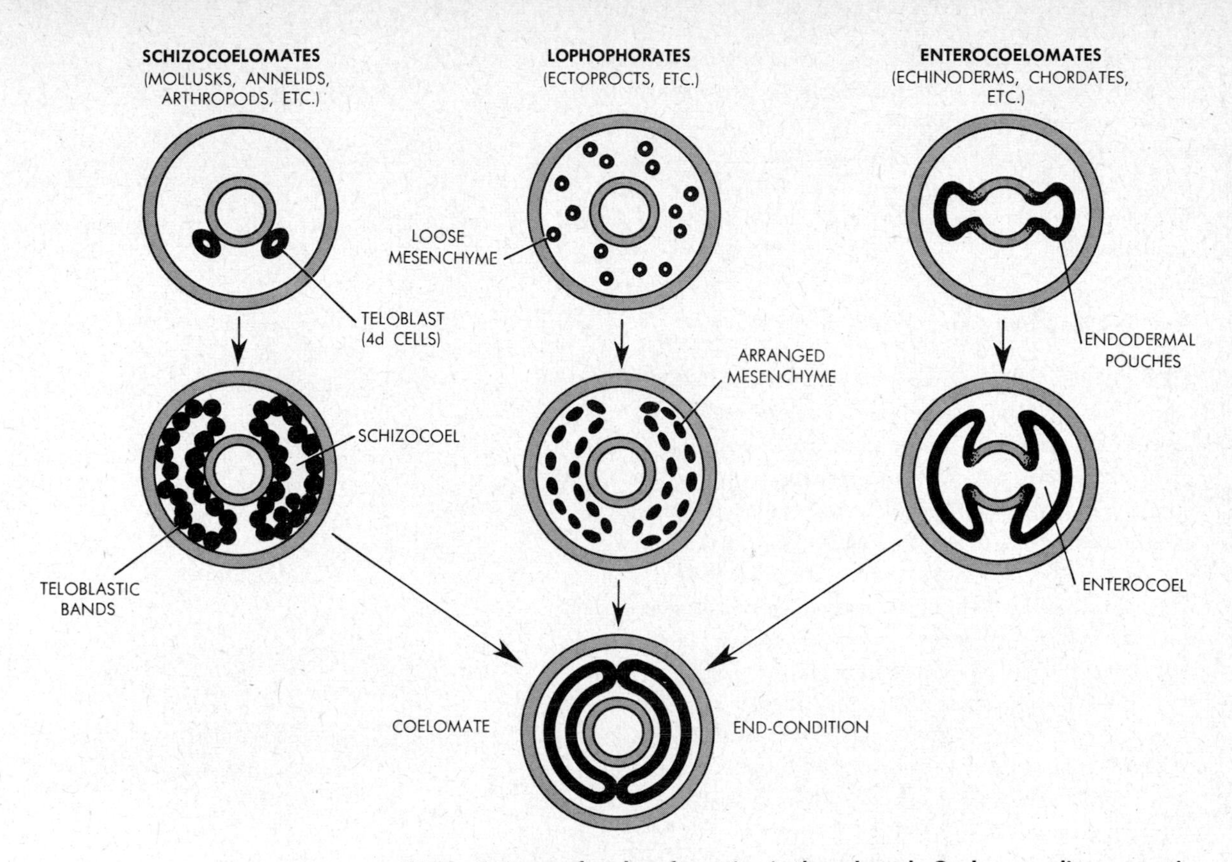

7.16. Patterns of coelom-formation in the subgrade Coelomata, diagrammatic. Note that, among lophophorates, the pattern shown is but one of several known to occur in that group.

of mesoderm, it is called a *schizocoel.* Animals characterized by body cavities of this type may be designated as *schizocoelomates.*

In another coelomate subgroup, represented principally by the echinoderms and the chordates, the mesoderm arises in the embryo as paired lateral pouches growing out from the endoderm (cf. Fig. 7.16). The pouches subsequently lose continuity with the endoderm, though their inner splanchnic portions remain applied against the developing alimentary system. Their outer somatic portions become applied against the developing body wall. The final condition is essentially quite similar to that in schizocoelomates. However, since the mesoderm and the coelom here are derivatives of the future gut, or *enteron,* the body cavity is called an *enterocoel.* Animals possessing such cavities are therefore known as *enterocoelomates.*

In a third subgroup, various other patterns of coelom formation are encountered. In one, for example, loose mesoderm cells of the embryo later migrate in amoeboid fashion and simply arrange themselves into a continuous peritoneal layer (cf. Fig. 7.16). Coeloms developed in this and various similar ways have not been given any special technical names. Animals in this subgroup also include types in which coeloms form in unique schizocoelic and enterocoelic fashion as well. The whole assemblage of animals may be referred to as the *lophophorates,* to which belong the phyla of phoronids, ectoprocts (moss animals), and brachiopods (lampshells). They are not particularly abundant today, but their ancestors may have been among the most ancient coelomate animals from which both the schizocoelomate and the enterocoelomate groups later evolved (cf. Chaps. 14, 24).

We may therefore characterize the Bilateria broadly both from the standpoint of mesoderm- and coelom-formation and from that of alimentary development, i.e., whether the mouth or the anus forms first in the embryo. The subgrades Acoelomata and Pseudocoelomata happen to be protostomial, the mouth here being formed first. Among the Coelomata, protostomial types include the lophophorate groups and the schizocoelomates, whereas deuterostomial types are represented by enterocoelomates such as echinoderms and chordates (Fig. 7.17).

Segmentation and Other Traits

An organizational feature of considerable anatomical and functional significance is the presence or absence of body segmentation. Encountered in some of the coelenterates and in many of the bilaterial groups, segmentation usually develops *after* the formation of the middle body layer. If it does, the characteristic thus is of lesser generality than the development or nondevelopment of body cavities. A segmented body is one which is marked by transverse constrictions into an anteroposterior succession of segments, or *metameres.* Two fundamentally different types of segmentation may be recognized. One type, called *superficial* segmentation, originates in the ectoderm and by and large affects only the body wall. The second type, called *metameric* segmentation, originates in the mesoderm, and it affects most mesodermal as well as most ectodermal derivatives (Fig. 7.18).

Superficial segmentation is exhibited in some body regions of certain coelenterates, and it occurs quite widely among various groups of acoelomates and pseudocoelomates. In several instances, the developing body elongates at one end in successive spurts, and in the pause following each spurt the surface cells secrete an inert external cuticle over the newly formed section of the body. In other cases, a cuticle develops over the entire body surface at more or less the same time but becomes thicker in alternate transverse sections of the body. Either method of formation produces a series of ringlike creases in the cuticle and the body wall, which may facilitate the bending or telescoping of the body. Superficial segmentation has little additional significance in animal organization, and its fairly haphazard occurrence also makes it unsuitable as a criterion for a taxonomic subclassification.

Of greater importance is metameric segmentation, encountered among some of the coelomate animals, both schizocoelomate and enterocoelomate. In all these cases the developing mesoderm of the embryo becomes constricted transversely into segmental portions, usually in anteroposte-

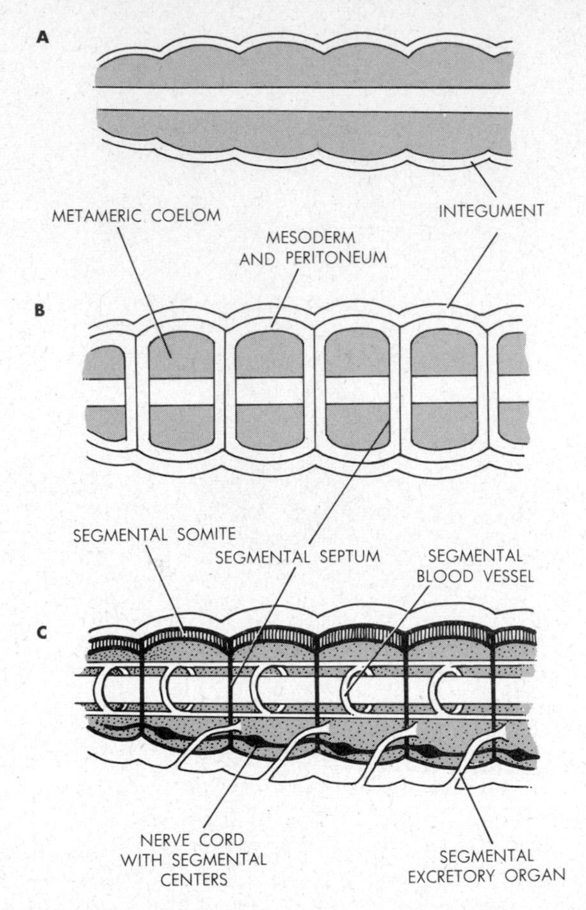

7.18. Segmentation. A. Superficial segmentation; only the integument exhibits ringlike creases. **B.** Metameric segmentation; both ectodermal and mesodermal derivatives are arranged segmentally, and the coelom is subdivided into metameric anteroposterior compartments. **C.** Diagram of the segmental arrangement of organ systems in metameric animals.

7.17. The interrelations of the major bilaterial groups.

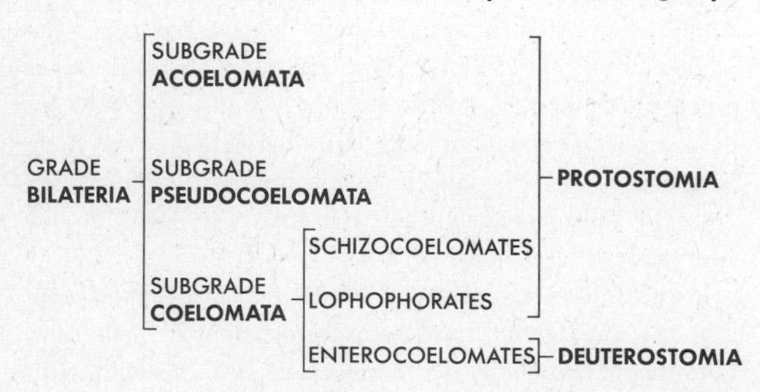

rior sequence. Each such portion then typically acquires its own coelomic body cavity and peritoneum. The most anterior segments thus are the oldest. Soon after such mesodermal segments have formed, the ectoderm overlying them comes to develop in a correspondingly segmental fashion. As a result, the whole thickness of the body with the exception of the endoderm becomes marked off into metameres. The segmental arrangement subsequently persists in later development, and most ectodermal and mesodermal organ systems form as segmented series of parts. For example, the muscular system arises in segmental blocks called *somites*. Each metamere may also have its own excretory, reproductive, skeletal, locomotor, and nervous organs. However, the metameres are not completely independent, for each organ system usually includes components which interconnect the organs in adjacent segments. For example, segmental nerve centers are joined by interconnecting nerve cords; segmental lateral blood vessels are joined by interconnecting longitudinal vessels (cf. Fig. 7.18). Only the alimentary system is not arranged metamerically to begin with.

Metamerically segmented groups include primarily the annelids, the arthropods, and the vertebrates. In the last two the metameric pattern is particularly evident in the embryos and larvae but becomes obscured to a greater or lesser extent in the adults. It is probably not an accident that arthropods and vertebrates also happen to be the most advanced and most successful animals, and it is fair to say that this success is in large part due to the segmental body design. For metamerism entails two basic adaptive advantages. First, the mere repetition of a pattern is itself often exceedingly useful, on the principle that there is safety in redundancy (cf. Chap. 6). In a sense, simple metameric segmentation as in annelids provides the same kinds of advantages on an organ-system level that simple multicellularity provides on a cellular level. Secondly, once a repeat pattern is available, the units of the pattern can become elaborated and specialized in different ways. This is exemplified in arthropods and vertebrates, which have become far more diversified structurally and functionally than any other animals primarily through diversification of their segments. Here again we find an organ-system level parallel to the different specializations possible among the cells of a multicellular aggregate. The recipe for zoological success thus appears to be described by the classic phrase "repeat, then vary."

Note, however, that metamerism is not particularly useful as a taxonomic criterion. Independent evidence does indicate that annelids and arthropods are closely related, and that both groups are related to unsegmented schizocoelomates such as mollusks. Yet annelids and arthropods are not closely related to the vertebrates. On their part, the vertebrates are related closely to the nonsegmented chordates (tunicates) and also reasonably closely to the other deuterostomial and nonsegmented enterocoelomates (e.g., echinoderms). It appears therefore that, among coelomate animals, metamerism evolved at least twice independently, once in the protostomial annelid-arthropod group and once in the deuterostomial chordate-vertebrate group. Consequently, it is not feasible to classify all metamerically segmented animals into one large taxonomic group and all nonsegmented ones into another. Nevertheless, despite this unsuitability of metamerism as a basic taxonomic criterion, the trait does constitute a major feature of the body organization wherever it is encountered.

At this point, our characterizations of body forms have reached a substantial level of detail; the next level is that of the phylum. Indeed, the phyla of protozoa and sponges are uniquely characterized already, by architectural details outlined above. To distinguish the body forms of the two radiate phyla, i.e., coelenterates and ctenophores, we must specify at least one more structural difference; and to distinguish among the several phyla within each subgrade of the Bilateria, we must likewise specify one and in most cases more than one additional architectural difference. It is one of the functions of Unit 2 to provide such detailed specifications. Here we may note generally that, at the phylum level of detail, criteria of distinction cease to be uniform. Thus, whereas groups of higher rank are distinguished by single features such as symmetry or coelom development, phylum groups are identified by many different features. For example, phyla may be defined by the nature of their locomotor structures (e.g., arthropods, echinoderms), or by the nature of their skeletal structures (e.g., chordates), or by the nature of their ingestive structures (e.g., coelenterates, rotifers). Analogously, various different features identify all other phyla. Such nonuniformity is largely due to the availability of a certain amount of evolutionary information. Many phyla are not defined by structural criteria alone, as noted, and the introduction of historical criteria usually necessitates some adjustment in classifications based purely on body forms.

Absence of uniformity in diagnostic features also characterizes the lower ranks within phyla. For example, the classes of sponges are distinguished on the basis of skeletal composition; as we shall see, one class is identified by calcareous supports, another by silica supports, a third by horny supports. On the other hand, the coelenterate classes are defined by differences in life cycle pattern and way of life, one class being predominantly sessile (sea anemones), a second predominantly motile (jellyfishes), and a third exhibiting both sessile and motile phases (hydroids). Among vertebrates, the class of birds is distinguished from that of mammals in part by the nature of the skin appendages (i.e., feathers vs. hair), and amphibia are set apart from reptiles in part by differences in reproductive pattern (i.e., aquatic vs. terrestrial eggs). Note incidentally that, as these examples indicate, way of life comes to play some role as a diagnostic criterion, increasingly so with the lower ranks; it is rarely the sole criterion, however. Note also that, as pointed out earlier, a single criterion is again used to define groups at the lowest main rank, i.e., the species.

Regardless of what the specific design features of a specific animal may be, each animal is composed of smaller constructional units which exhibit their own pattern of structure. The most fundamental of such units is the cell, and we have already discussed its structures and functions in earlier chapters. Now we may turn our attention to two types of units at higher levels of organization, viz., the tissue and the organ.

7.19. Fibroelastic connective tissue. Note the conspicuous fibers forming a meshwork and the cells (small dark dots) embedded in the meshwork.

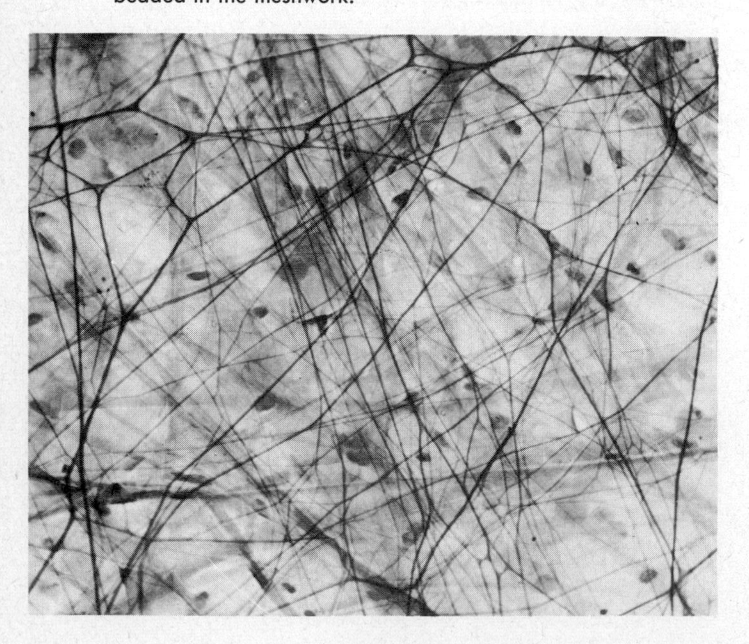

TISSUES AND ORGANS

As an animal matures, most of its cells specialize in given ways; they become more or less diversified in external appearance and internal structure, and they develop the capacity to perform some function or functions especially well. Such characteristics tend to become fixed and irreversible. Once a cell has become specialized—particularly highly specialized—in one way, it normally cannot change and respecialize in another way.

A *tissue* may be defined as an aggregation of cells in which each cooperates with all others in the performance of a given group function. Analogously, an *organ* is an aggregation of tissues all of which cooperate in the performance of a group function.

In a so-called *simple tissue,* all cells are of the same type. Two or more different cell types are present in a *composite* tissue. The cells of a tissue need not necessarily be in direct physical contact, but this is actually the case in many instances. Tissues may be highly or less highly specialized, according to the degree of specialization of the component cells.

Most tissues of most animals may be classified generally either as *connective tissues* or as *epithelia.* Not included are three specific tissues, viz., nerve, muscle, and blood, which will be discussed in the context of the organ systems of which they are a part. Also not included are occasional tissue types which are unique to given animal groups. Such tissues will be considered in later chapters, in the context of the specific animals in which they occur.

Connective Tissues

These tissues are identified by comparatively widely separated cells, the spaces between the cells being variously filled with fluid and solid materials. Another identifying characteristic is the relatively unspecialized nature of the cells. With appropriate stimulation, they may transform from one connective tissue cell type to another. Variants of connective tissues are distinguished on the basis of the types of intercellular deposits present and the relative abundance and the arrangements of the cells.

The most fundamental variant of the connective tissues may be considered to be *fibroelastic tissue* (Fig. 7.19). In it, the most conspicuous components are large numbers of threadlike fibers, some of them tough and strong (and made

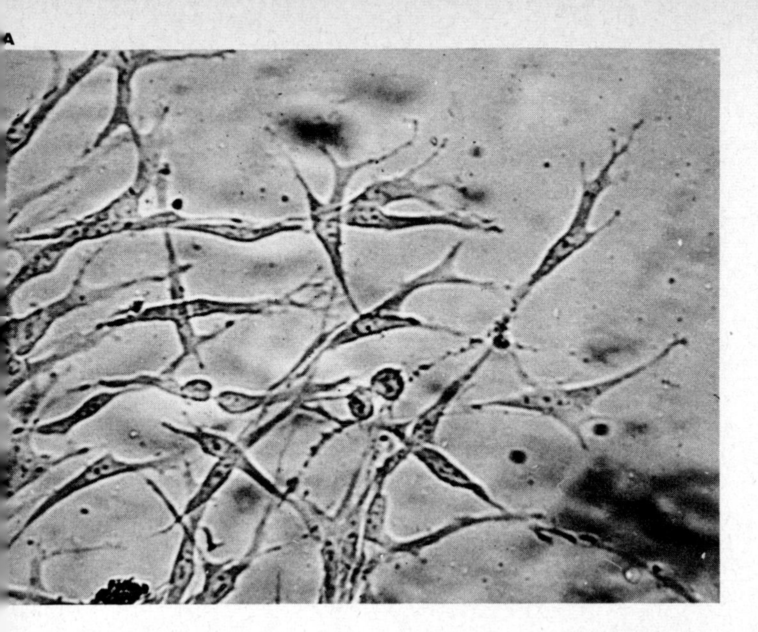

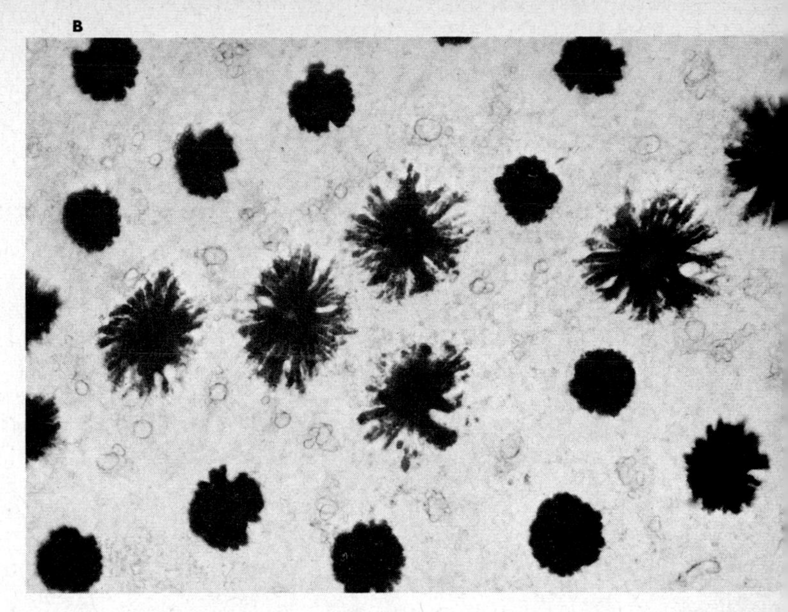

7.20. Some cellular components of fibroelastic connective tissue. **A.** Fibrocytes (growing in tissue culture). **B.** Pigment cells (melanophores) in various states of expansion (from the connective tissue in the skin of a flounder).

of the protein collagen), some of them elastic. These fibers are suspended in fluid and form an irregular, loosely arranged meshwork. The cells of the tissue are dispersed throughout the mesh. Materials secreted by the cells give rise to the fibers outside them.

The cells are of various types (Fig. 7.20). Many are so-called *fibrocytes,* generally spindle-shaped and believed to be the chief fiber-forming cells. Other cells, the *histiocytes,* are capable of amoeboid locomotion and of engulfing foreign bodies (e.g., bacteria in infected regions). Also present are *pigment cells (chromatophores), fat cells,* and, above all, *mesenchyme cells.* The latter are embryonic, undeveloped, and relatively quite unspecialized. The connective tissues of an embryo at first are entirely mesenchyme, and these cells later give rise to the various types of cells present in adult connective tissues. Such adult tissues usually still contain mesenchyme cells "left over" from embryonic stages, and the cells then serve as developmental reserves; they may develop into any of the cell types of connective tissues. Indeed, it is possible to define connective tissues as all those which have a mesenchyme origin in the embryo; and we may note that mesenchyme itself constitutes or arises from embryonic mesoderm. Hence all connective tissues are mesodermal (but not all mesoderm-derived tissues are connective tissues). In many embryonic or adult animals, mesenchyme cells play an important role in healing and regeneration. The cells may migrate to injured body regions and contribute to the redevelopment of lost body parts and to scar-tissue formation. Actually, most of the adult cell types of fibroelastic tissue may likewise transform into one another. For example, a fibrocyte might become a fat cell, then perhaps a histiocyte, then a fibrocyte again, and then a pigment cell. The specializations of any of these cells evidently are not fixed.

By virtue of its cellular components, fibroelastic tissue thus functions in food storage and in body defense against infection and injury; and by virtue of its fibers, the tissue is a major binding agent which holds one body part to another. For example, fibroelastic tissue connects skin to underlying muscle. The tough fibers provide connecting strength, yet the elastic fibers still permit the skin to slide over the muscle to some extent. The binding action of the tissue is also in evidence deep within an animal, where fibroelastic layers often form a tough but flexible link between different tissues within an organ or between different organs.

The relative quantities of both the cellular and the fibrous components may vary greatly, and on the basis of such variations one may distinguish other types of connective tissues (Fig. 7.21). For example, *tendons* are dense tissues containing only fibrocytes and tough collagen fibers, the fibers being arranged as closely packed parallel bundles. Tendons typically

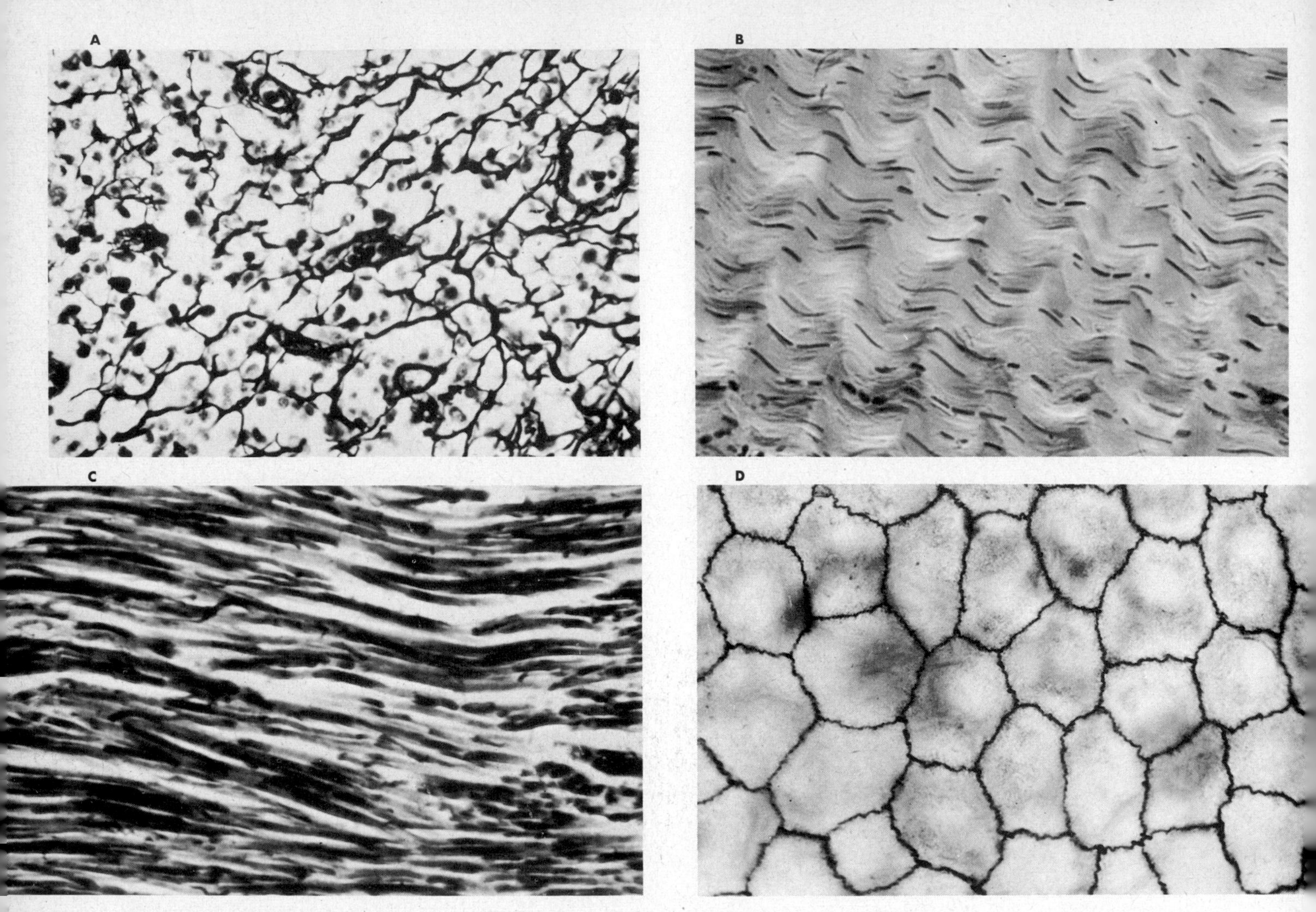

7.21. Some variants of fibroelastic connective tissue. A. Reticular tissue, composed of fibrous and cellular components. This tissue type is found in lymph nodes and lungs, and in preadult stages it is also a forerunner of mature fibroelastic tissue. **B,** tendon; **C,** ligament; **D,** mesothelium, the principal tissue of the peritoneal membrane.

connect muscles to parts of the skeleton. A *ligament* is similar to a tendon, except that both collagen and elastic fibers are present and that these are arranged in more or less irregular manner. Another variant of fibroelastic tissue is *adipose tissue,* of which fat cells are the most abundant components. Each fat cell contains a large fat droplet which fills almost the entire cellular space. A large collection of such cells has the external appearance of a continuous mass of fat. Still another variant of fibroelastic tissue occurs in several (but not all) types of tissue *membranes.* In these, the cellular components are abundant and packed together fairly densely, and the intercellular components are minimal. Various other types of loose connective tissues are known in addition. These arise in special locations at special times during animal development.

Many animals possess *jelly-secreting* connective tissues (cf. Fig. 7.21). The cells in such cases are mesenchyme, and they secrete gelatinous mucoid substances. As the latter accumulate, the cells become separated from one another more and more. Mesenchyme jelly tissue may become quite bulky, as in the middle body layer of the Radiata. Such tissues also occur abundantly in numerous other groups, either as parts of whole body layers (e.g., flatworms) or as parts of specific organs (e.g., around and in the eyes of vertebrates).

In some connective tissues, the cellular components secrete organic and, especially, inorganic materials, which form a solid precipitate around the cells. Thus the cells appear as islands embedded in hard intercellular deposits. The chief variants of this tissue type are *cartilage*, encountered in several invertebrate groups and in all vertebrates, and *bone*, characteristic of vertebrates. Both cartilage and bone arise from mesenchyme cells, and both function in support and protection (cf. Chap. 8 for details).

We may note that, as a group, the connective tissues serve largely in forming the structural scaffolding of the animal body. By contrast, the primarily functional parts of the body are formed chiefly by the epithelial tissues.

Epithelia

An epithelium is a tissue in which the cells are cemented directly to one another and so form single-layered sheets, multilayered sheets, or irregular, compact, three-dimensional aggregates (Fig. 7.22). The "cement" in such cases usually is complexly organic, one of the constituents often being *hyaluronic acid*, a carbohydrate derivative. Epithelia forming layered sheets generally rest on so-called *basement membranes*, flat networks of collagen-containing fibers secreted as supporting fabrics for the epithelial sheets. All three embryonic germ layers give rise to such tissues.

Sheets consisting of a single layer of cells are called *simple epithelia*. Distinctions among them are made principally on the basis of cell shape. If the cells are flattened and are joined along their edges, the tissue is known as a "pavement" or *squamous* epithelium. Many tissue membranes and the surface layer of the skin of many animals are of this type. If the cells have the shape of cubes, the tissue is a *cuboidal* epithelium. The walls of ducts and glands frequently consist of such tissues. Analogously, if the cells are prismatic and are joined along their long sides, the tissue is a *columnar* epithelium. In many animals, this type forms, for example, the innermost

7.22. Epithelia. **A.** Surface view of frog epidermis. Note the close packing of the cells and the angular outlines, produced by the pressure of cells against one another. **B.** Section through the lining of the human uterus. Note the progressive flattening of the closely packed cells toward the inner surface (top) of the organ.

A

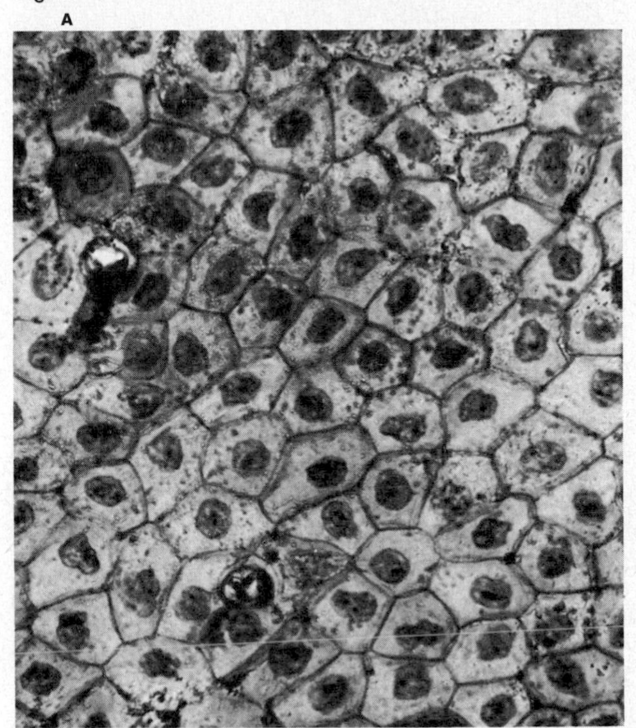

B

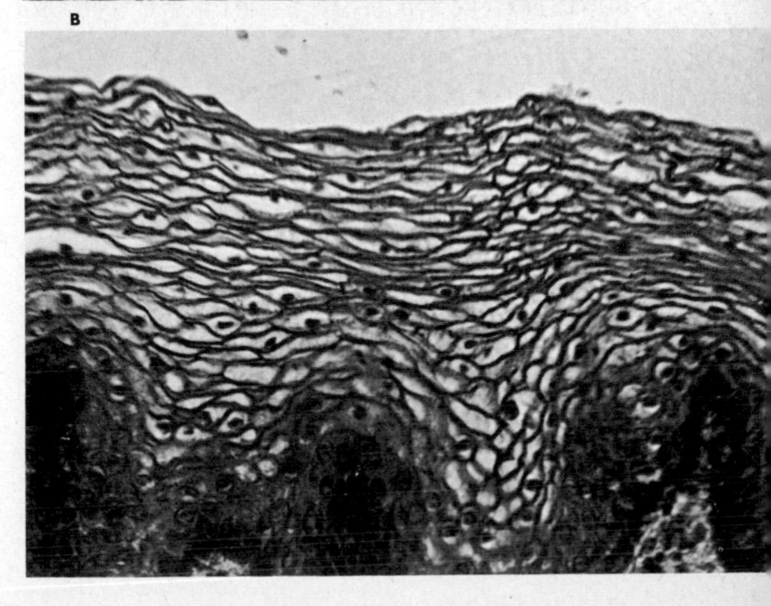

7.23. Simple epithelia. Diagrams: A, squamous epithelium, **B,** cuboidal epithelium; **C,** columnar epithelium. **Photos: A.** Simple columnar to cuboidal epithelium in lining of kidney tubule. **B.** Simple ciliated columnar epithelium in inner lining of frog gut wall.

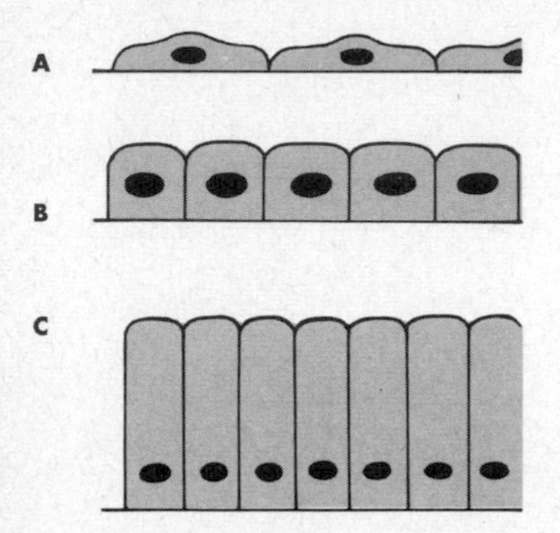

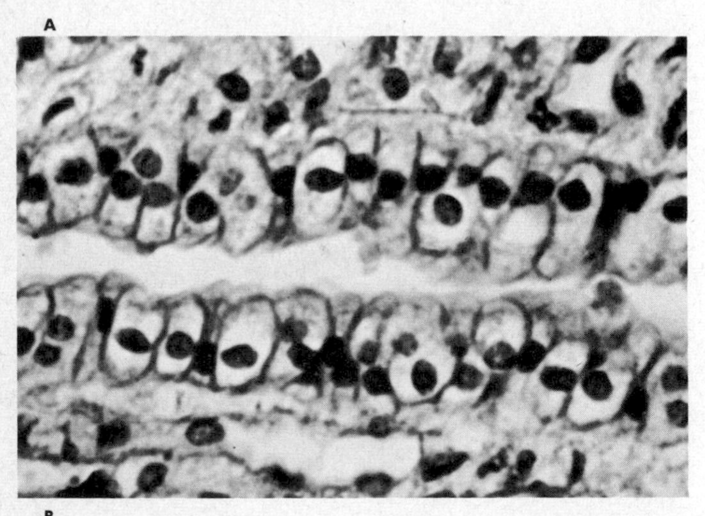

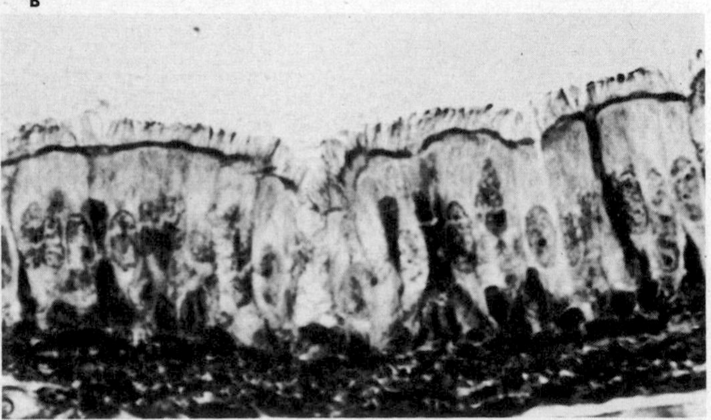

Courtesy Carolina Biological Supply Company.

(digestive-juice-secreting) layer of the intestine or the outermost (exoskeleton-secreting) layer of the skin (Fig. 7.23).

Intergradations between such cell shapes are very common, and many simple epithelia therefore cannot be classified into definite categories. If several epithelial layers of a given type are stacked into a multilayered sheet, the term *stratified epithelium* is often applied. Such complex epithelia thus may be of stratified squamous, stratified cuboidal, or of stratified columnar type. The mammalian epidermis, i.e., the outermost tissue of the skin, is a good example of a mixed stratified epithelium; the cells are squamous along the outer surface and become increasingly cuboidal with increasing distance from the surface (Fig. 7.24).

In contrast to the connective tissues, the epithelia are all fairly highly and permanently specialized. Once their cells are mature, they do not thereafter change in their basic structural characteristics. Also, by the time maturity is reached, the cells have acquired given fixed functions which are then performed throughout the life of the animal.

Some common animal tissues cannot be classified strictly as either connective tissues or epithelia, and they may share certain of the characteristics of both. Included here are particularly the *blood, muscle,* and *nerve* tissues. Blood is generally like a connective tissue in that it contains cellular components and intercellular deposits, fluid in this case; but although some of the blood cells have a mesenchyme origin, some do not. Muscle tissue has a mesenchyme origin, yet adult muscle resembles an epithelium more than a connective tissue. Nerve tissue has an epithelial origin, and it resembles an epithelium in some respects; but in others, e.g., its frequent netlike arrangement, it does not. All three tissues are therefore usually discussed as separate entities.

Organs

An organ typically consists of one or more epithelia and one or more connective tissues. The internal division of labor is such that the epithelia carry out the characteristic specialized functions of the organ, whereas the connective tissues serve in the necessary auxiliary roles. Thus, the connective tissues maintain the shape and the position of the organ as a whole, and they lead nerves, blood vessels, and other ducts to and from the epithelia. In sheetlike organs such as skin and in tubular organs such as intestine, epithelia and connective tissues form adjacent layers (cf. Figs. 7.23, 7.24). In compact three-dimensional organs such as liver, the connective tissues

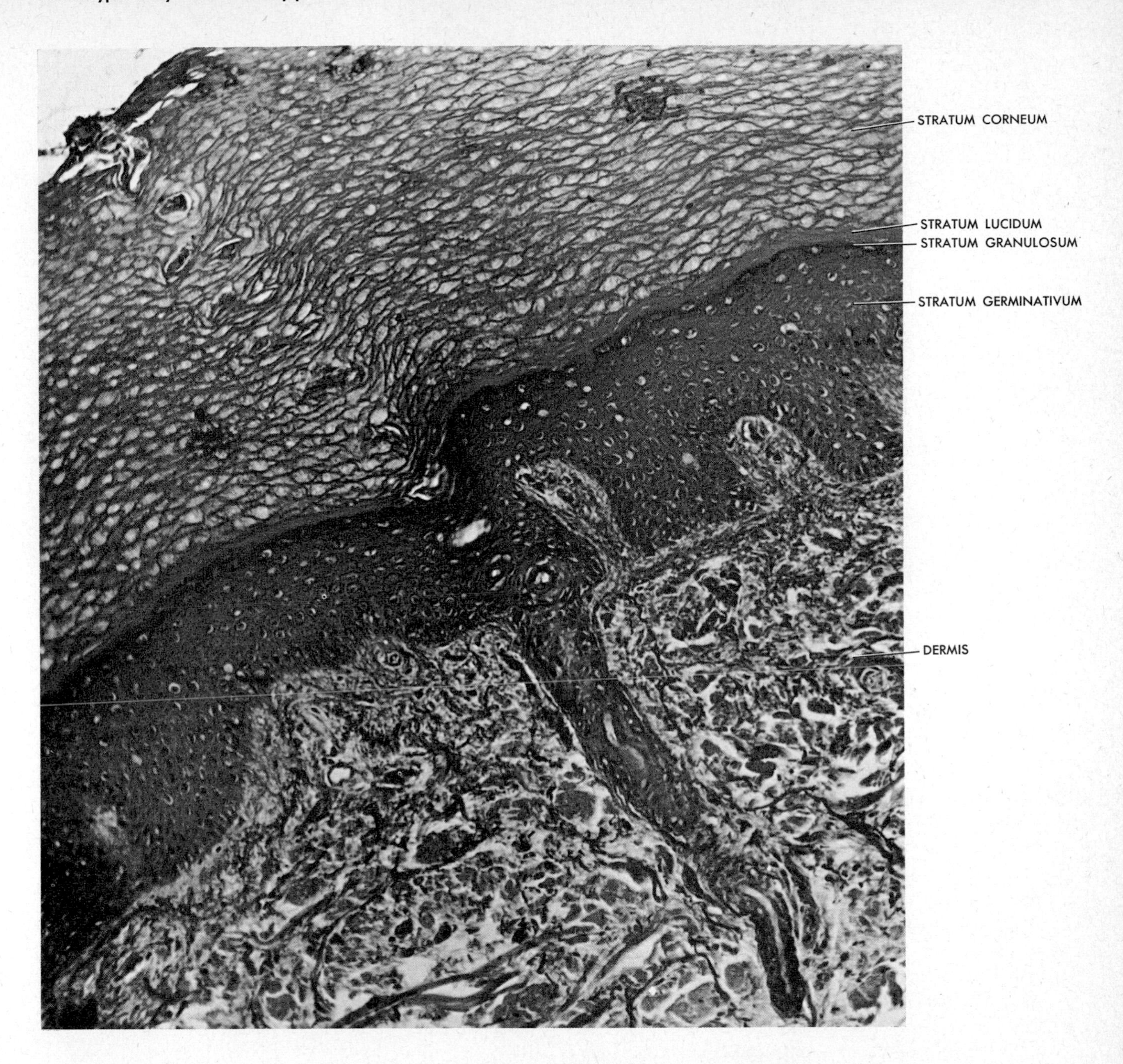

7.24. Histological section through mammalian skin. The epidermis consists of the four principal layers labeled near top of photo. Note the stratified epithelial nature of these layers. The underlying dermis consist largely of connective tissue. Parts of the duct of a sweat gland may be seen meandering from the dermis through the epidermis to the skin surface.

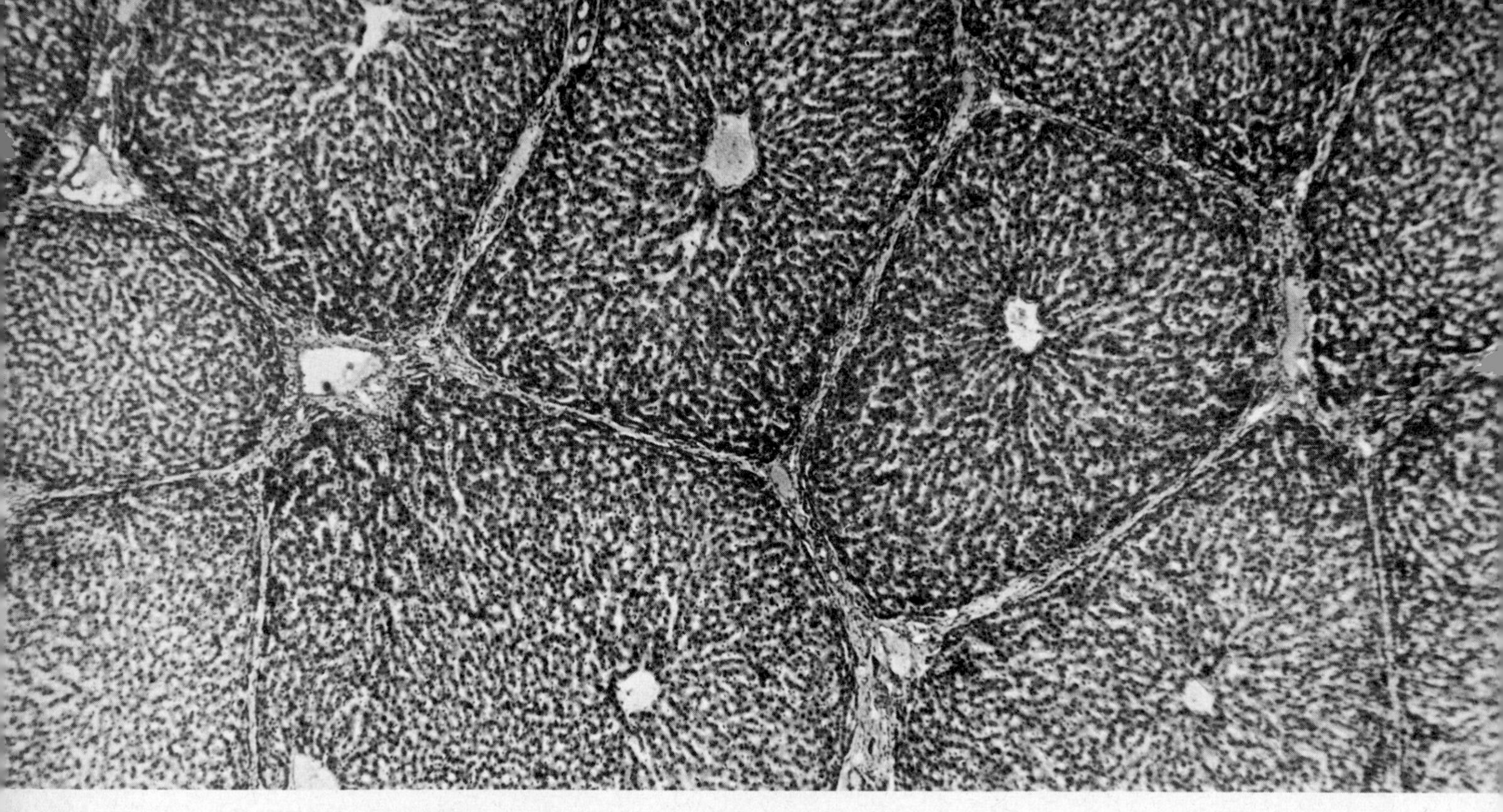

7.25. Histological section through a mammalian (pig) liver, showing a few of the epithelial lobules separated from one another by the stroma, i.e., layers of connective tissue. Branches of the hepatic portal vein in the stroma carry blood to a lobule. Blood then passes freely through the canallike spaces between the strands of lobule cells, and it eventually collects in and is carried off by a branch of the hepatic vein, seen as a large clear space in the center of a lobule. See also Fig. 9.24 for additional details on liver structure.

envelop the whole organ exteriorly and partition it extensively interiorly. The connective tissues here are said to form a *stroma,* a supporting framework which divides up the interior of the organ into islands of epithelial cells (Fig. 7.25). Such islands usually represent complete functional units of the organ, and the traffic of materials to and from the units is carried by the surrounding stroma. Small groups of such islands or each individual island may form a distinct *lobule,* and numerous lobules may be recognizable collectively as an anatomically distinct *lobe* of the organ.

In their turn, organs are linked together into organ systems, an organizational unit definable as a cooperating aggregation of organs. Certain organs within a system often merge into one another imperceptibly, without sharp microscopic lines of division. Nevertheless, grossly anatomical demarcations are usually recognizable even in such cases, and in others, microscopically visible partitions of connective tissue are present as well. Distinct layers of connective tissue also separate adjacent parts of different organ systems. The detailed organization and operation of the animal organ systems forms the subject of the two ensuing chapters.

REVIEW QUESTIONS

1. Review the structure of the Linnaean taxonomic system. What are the principal ranks? Name and define them and cite a taxon of each. How is this system related to the size and the numbers of groups at each rank? How is the system related to the hierarchy of organizational ranks?
2. What criteria of animal biology form the basis of the

taxonomic system? What are the advantages of these criteria, and why are other possible criteria not used? Why are animals not simply classified alphabetically or by some system equivalent to book-cataloguing in libraries?

3. What is meant by natural and unnatural taxonomic classifications? Give examples. What rules are in force in the naming of species? Review the taxonomic classification of man, with attention to the definition of each taxon.

4. Explain the proper zoological use of the following contrasting sets of terms: simple-complex; primitive-advanced; generalized-specialized. Criticize and correct the following statements: "Evolution consists of a progression from simple to more complex animal types." "In the evolutionary scale, higher animals such as vertebrates have descended from lower forms such as protozoa."

5. Define homology, analogy. Give specific examples of each. Why are homologies more important in taxonomy than analogies?

6. Name the subkingdoms, branches, grades, and subgrades of animals. What criteria define each of these ranks? What taxa of animals are included in each? List five or six of the most generalized architectural features of animals, and show on what basis they are considered to be more generalized than others.

7. Define Protostomia, Deuterostomia, Bilateria, Parazoa, coelom, schizocoel, enterocoel, mesoderm, ectomesoderm, splanchnic mesoderm, germ layer, peritoneum, mesentery, metamerism, somite, lophophorate, teloblast, mesenchyme.

8. Show how Bilateria are subdivided taxonomically on the basis of mesoderm- and coelom-formation, and show how coeloms arise in different ways in different groups. What is the adaptive advantage of a coelom? Does a pseudocoel offer analogous advantages?

9. Distinguish between and give the characteristics of superficial and metameric segmentation. In which animal groups is each encountered? Why is segmentation not particularly useful as a taxonomic criterion?

10. Draw up a comprehensive chart listing the classification of animals down to subgrade, and define each category so listed as fully as possible. Based on this chart, state the characteristics of the following animal groups: coelenterates, arthropods, echinoderms, flatworms, sponges, mollusks, ectoprocts, chordates.

11. What is a tissue? An organ? What is a connective tissue? Describe the makeup of several types of connective tissues and state their general functions in animal organization.

12. What is an epithelium? Describe the structure and function characteristic of an epithelium and list several variants of this type of tissue. Give specific examples of each. How are tissues joined into organs? What is a stroma and what is its role? Define organ system.

COLLATERAL READINGS

Readings on the principal features of animal design and on animal classification are included in the following books. The second entry refers to an outstanding five-volume series which represents the most comprehensive recent treatise in the English language on invertebrate zoology:

Griffin, D. R.: "Animal Structure and Function," Modern Biology Series, Holt, New York, 1962.

Hyman, L.: "The Invertebrates," vol. 1, chaps. 1 and 2; vol. 2, chap. 9 (p. 18 ff.), McGraw-Hill, New York, 1940 (1), 1951 (2).

Mayr, E.: "Systematics and the Origin of Species," Columbia, New York, 1942.

Mayr, E., E. G. Linsley, and R. L. Usinger: "Methods and Principles of Systematic Zoology," McGraw-Hill, New York, 1953.

Simpson, G. G.: The Principles of Classification and a Classification of Mammals, Bull. Amer. Mus. of Nat. Hist., vol. 85, New York, 1945.

Detailed accounts on tissues and organs are included in any text on histology, the zoological subscience dealing with tissue organization. Among such texts the following may be consulted:

Bloom, W., and D. W. Fawcett: "Textbook of Histology," Saunders, Philadelphia, 1962.

Greep, R. O. (ed): "Histology," McGraw-Hill, New York, 1954.

Maximov, A. A., and W. Bloom: "A Textbook of Histology," 7th ed., Saunders, Philadelphia, 1957.

CHAPTER 8

animal systems: support, motion, coordination

Every organ system of an animal contributes to the proper operation of every other, and all systems ultimately are equally essential in maintaining all metabolic and self-perpetuative processes. However, some systems contribute to given functions more directly than others, and on this basis we may conveniently divide the animal systems into two general groups. One group functions more directly than the other in relating an animal to its external environment. Such systems provide structural support, produce movement, and regulate animal activities in accordance with given environmental conditions. We shall examine these systems in this chapter. They are the *integumentary*, the *skeletal*, the *muscular*, the *nervous*, and the *endocrine* systems.

INTEGUMENTARY SYSTEMS

The first structure to form during the development of any metazoan, the integument represents the boundary between the interior of an animal and its physical environment. As such, the integument probably displays more variation among animals generally than any other system. Its different functions among different animals variously include protection, locomotion, support, breathing, sensory reception, internal steady-state control, nutrient uptake, and excretion. In addition, the integument often plays an important role in mate-selection, in expressing behavior, and in permitting recognition of other members of the species. In effect, the system performs a direct or indirect function in virtually every process of organismic metabolism and self-perpetuation.

This functional diversity of the integument is reflected in its structural variability. The basic organ of the system is the *skin*, which consists of one or of two main layers. An outer layer, the *epidermis*, arises from the embryonic ectoderm. In the Radiata and some of the noncoelomate Bilateria, the epidermis is the only skin layer present and usually constitutes the entire body wall. The skin of many other invertebrates similarly consists of epidermis only, but the body wall is more complex, the epidermis being underlain by one or more layers of muscle (derived from somatic mesoderm). In some invertebrates and all vertebrates, the skin is composed of two distinct layers, viz., an epidermis and an inner, mesoderm-derived *dermis*. The dermis in turn is underlain by muscle layers, skin and muscle forming a highly complex body wall (Fig. 8.1).

Epidermis

The epidermis is an epithelium, syncytial in some cases. Its basic function is to maintain the interior of the animal in optimal contact with the exterior environment. In invertebrates the epidermis generally is but one cell-layer thick. It is often ciliated, and it then usually represents the principal organ of locomotion. It may also function as the principal breathing, excretory, and osmoregulatory organ, exchange of respiratory gases, water, and mineral ions occurring directly through the skin. If the animal possesses a circulatory

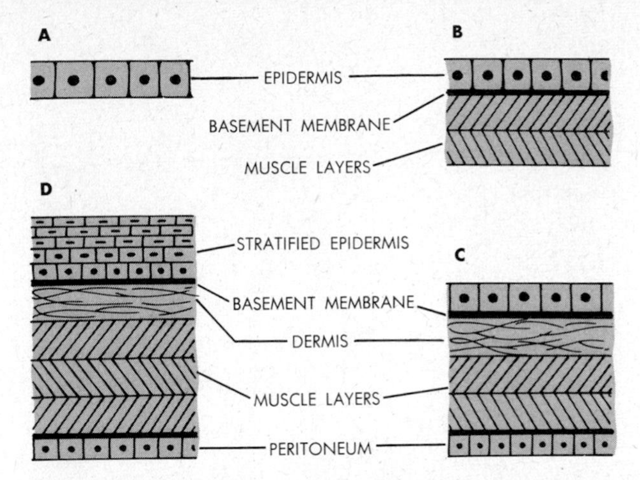

8.1. Skin and body-wall types, diagrammatic. A, as in radiates, some noncoelomates; **B,** as in numerous invertebrates, both noncoelomate and coelomate (in the latter case a peritoneum being present underneath the muscle layers); **C,** as in some coelomate invertebrates and also some chordates; **D,** as in most vertebrates. In all groups, the muscle layers are longitudinal or circular if only one is present, both longitudinal and circular if two are present, and any additional layers are variously diagonal. The outside-inside sequence of the layers varies considerably for different groups.

system, as in an earthworm, for example, blood coursing under the epidermis may play a major role in skin breathing and often also an accessory role in excretion. At various places on the body surface, single epidermal cells or small groups of them may be *glandular* and secrete mucous coats, cementing substances, or protective irritants and poisons. Other epidermal cells may be specialized as various *sensory* structures which connect with nerve fibers underneath the skin (Fig. 8.2).

In numerous invertebrates the epidermal cells are not ciliated. Instead, all are glandular in such cases and manufacture noncellular protective *cuticles.* Most commonly, horny *scleroprotein* is a principal constituent of such coverings. The cuticles may be thin, translucent, and pliable over the entire body surface; or they may have appreciable thickness in most body regions, with thin, motion-permitting areas then being confined to particular places. Thin cuticles do not interfere with molecular traffic through them. Thicker ones do, however, and in such cases the integument has only a reduced breathing and excretory function or none at all. In certain animals, notably the arthropods, cuticles containing *chitin* become extremely thick in most surface areas and acquire the functions of hard *exoskeletons.* An equally intimate interrelation between skin and exoskeleton exists in animals which manufacture calcareous *shells.* In such cases the epidermis again is the generating tissue of the exoskeletal secretion. In several groups of invertebrates, also, the epidermis produces skeletal secretions which do not remain in contact with the skin proper, but form a separate housing within which the animal may move freely. In various wormlike animals, for example, such housings are leathery or calcareous tubes, open at one or at both ends.

Among vertebrates, the epidermis of fishes is thin and without cuticles or cilia, but it contains numerous mucus-secreting glands which lubricate the body surface. Most other vertebrate groups are characterized by a more complexly structured epidermis; i.e., the tissue is a *stratified* epithelium, several to many cell layers thick (Figs. 8.1 and 8.3). The basal layers here continually produce new cells, and older cells thus become displaced outward. Such older cells typically manufacture *keratin,* an insoluble protein. As this protein accumulates it *cornifies* the cells; i.e., it makes them horny. The oldest cells, near and at the body surface, are completely cornified and are actually dead. By virtue of their keratin these layers are water-repelling; they prevent entry of exterior water and loss of interior water. The outer portions of the epidermis consequently form an excellent protective coat. This coat is also renewed continually; the outer layers wear

8.2. Some modifications of the epidermis.

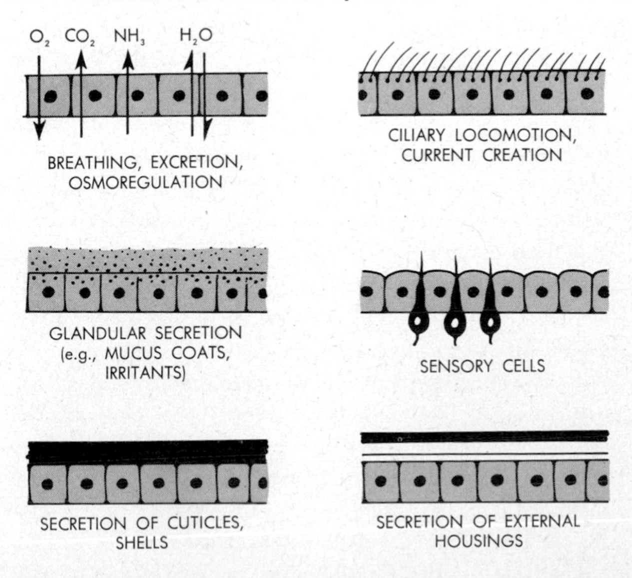

A

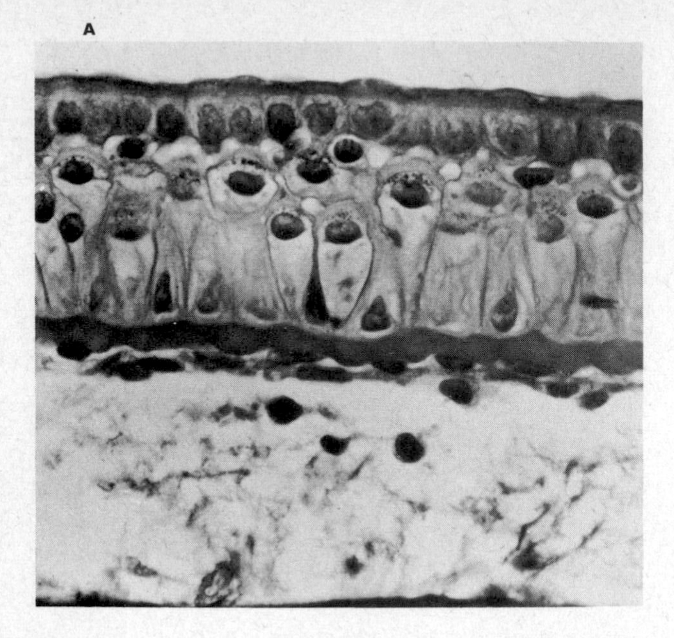

B

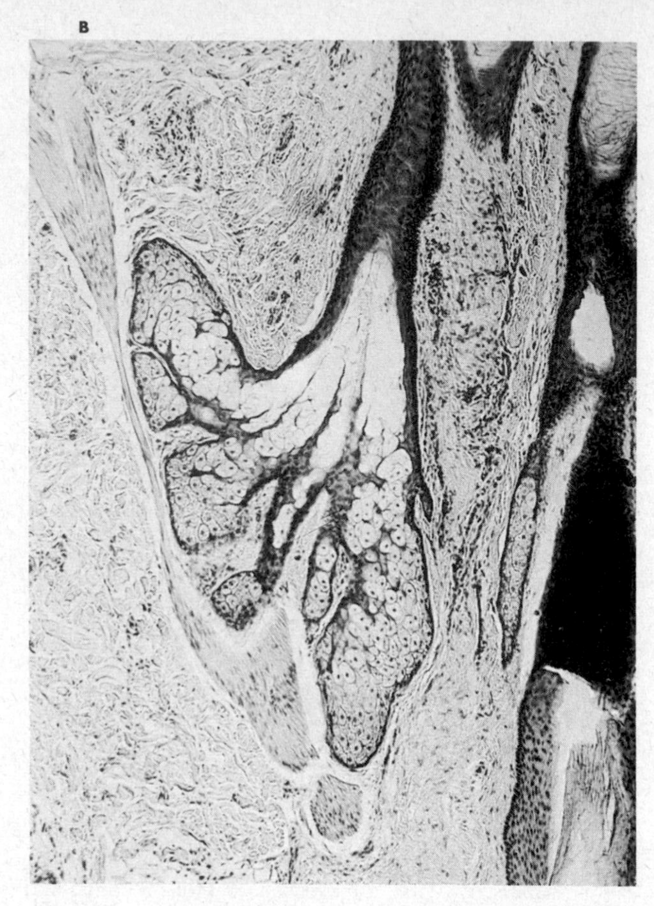

8.3. The skin of vertebrates. **A.** Section through skin of frog tadpole, showing stratified epidermis and connective tissue dermis underneath. Contrast epidermal structure here with that of mammalian skin, Fig. 7.24. Cf. also Fig. 7.22 for surface view of amphibian skin. **B.** Section through sebaceous gland of man. Such glands open along shafts of hairs and their oily secretions keep hair soft and pliable. Cf. Fig. 8.5 for hair structure.

off, but new layers produced in the basal regions of the epidermis move outward, cornify, and replace those lost.

Various ingrowths and outgrowths from the vertebrate epidermis give rise to numerous special integumentary structures. For example, ingrowths into the underlying dermis form a variety of glands, among them the *mucus-secreting* glands of frogs (animals in which the epidermis remains comparatively thin and uncornified), the excretory *sweat glands* of mammals, the oil-secreting *sebaceous glands* of mammals, and many others (cf. Fig. 8.3). Outgrowths of the epidermis may have the form of localized thickenings of the cornified layers, as in the *thumb pads* of male frogs and the *foot pads* of many mammals. More elaborately formed thickenings are represented by *beaks, claws, hoofs,* and *nails,* and by the outer horny layers of turtle *shells* and of cattle and sheep *horns.* Epidermal outgrowths also produce the horny teeth of the jawless fishes (e.g., lampreys), and the *scales* of snakes, lizards, and mammals such as anteaters and armadillos. Similarly, the *rattles* of rattlesnakes, the leg surfaces of birds, and the tail surfaces of rodents are formed by overlapping epidermal scales (Fig. 8.4).

The most conspicuous cornified epidermal outgrowths undoubtedly are *feathers* and *hairs,* which may be regarded as modified epidermal scales. Feathers in birds and hairs in mammals function primarily as insulators against heat loss, and thereby they contribute importantly to the maintenance of constant internal body temperatures. Both types of structures grow out from epidermal pits which penetrate deep into the dermis (Fig. 8.5). At the bottom of such a pit is a *follicle,* a structure consisting of a blood- and nerve-supplying *dermal papilla* and a surrounding jacket of epidermal cells. Feather or hair growth proceeds upward from the follicle, and the portions which come to project beyond the general body surface are highly cornified, dead cells. Most birds and mammals shed (*molt*) their feather and hairs from time to time. Reptiles too periodically molt the cornified layers of their epidermis.

The epidermis extends from the body surface into various

openings and cavities leading into the animal. For example, the integument not only lines the mouth cavity but also forms a variety of accessory ingestive structures in various animals, e.g., grinding, cutting, rasping, and other chewing organs, including toothlike formations. In arthropods, the chitin-covered skin even extends well beyond the mouth cavity and lines the anterior portion of the gut. Similarly, the posterior gut is lined by an interior extension of the integument. In vertebrates, the epidermis which lines the alimentary, excretory, reproductive, and other openings largely remains uncornified and thus differs in texture from external skin.

Dermis

The dermis is a fibroelastic connective tissue containing an abundance of tough and elastic fibers (cf. Fig. 8.3). Its primary function is to carry blood and nerves to and from the epidermis and to endow the whole skin with strength, toughness, and elasticity. Like the epidermis, the dermis similarly has appreciable thickness, and it too displays a wide range of structural modifications and specialized functions. For example, the cellular components of the connective tissue include pigment cells, or *chromatophores,* and thus it is the dermis which is chiefly responsible for skin coloration. Pigmentation patterns are important factors in camouflage, mate-selection, and in protection against the ultraviolet rays of sunlight. The dermis also contains the receptor organs of the cutaneous senses, e.g., touch, heat, cold, pressure, and pain. Nervous adjustment of the amount of blood flowing through the dermis governs the rate of heat radiation from the body surface; hence the dermis, like the epidermis, plays a major role in the regulation of body temperature. A localized permanent increase in the blood supply produces dermal modifications such as the combs and wattles of certain birds.

Most importantly, the dermis has a general tendency to produce calcareous deposits, e.g., the endoskeleton of echinoderms and parts of the body endoskeleton of vertebrates (Fig. 8.6). For example, the skull plates of all bone-possessing vertebrates are *dermal bones.* Groups of now extinct fishes (the jawless ostracoderms and the jawed placoderms) possessed armor plates of dermal bone over much of the body as well, and the present modern bony fishes still possess overlapping bony *scales.* These grow out from the dermis and are overlain by the epidermis on the outside. Sharks and related cartilage fishes possess, instead of scales, dermal *denticles,* composed of an outer layer of *enamel,* an inner layer of *dentine,* and a core of fibroelastic *pulp.* Such denticles are in fact structured like the *teeth* of most vertebrates, and we may regard teeth as essentially larger dermal variants of denticles. Evidently, teeth as well as scales occur in two structural varieties, according to whether they are epidermal or dermal in origin.

A

B

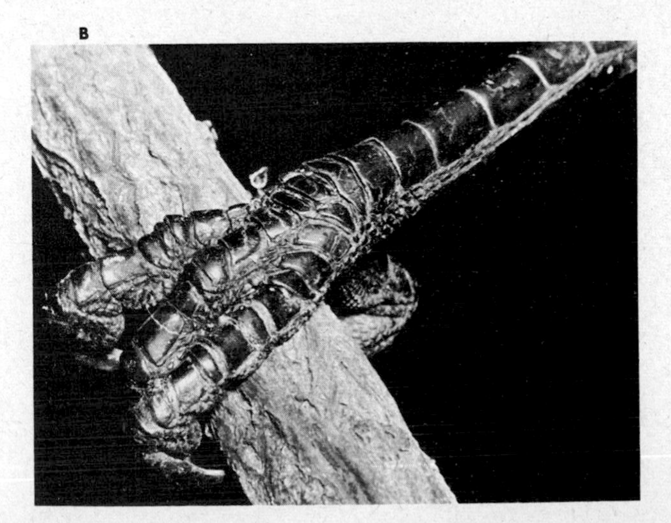

8.4. Two variants of vertebrate epidermis. A, the rattle of a rattle snake; **B,** scales and claws on a bird foot.

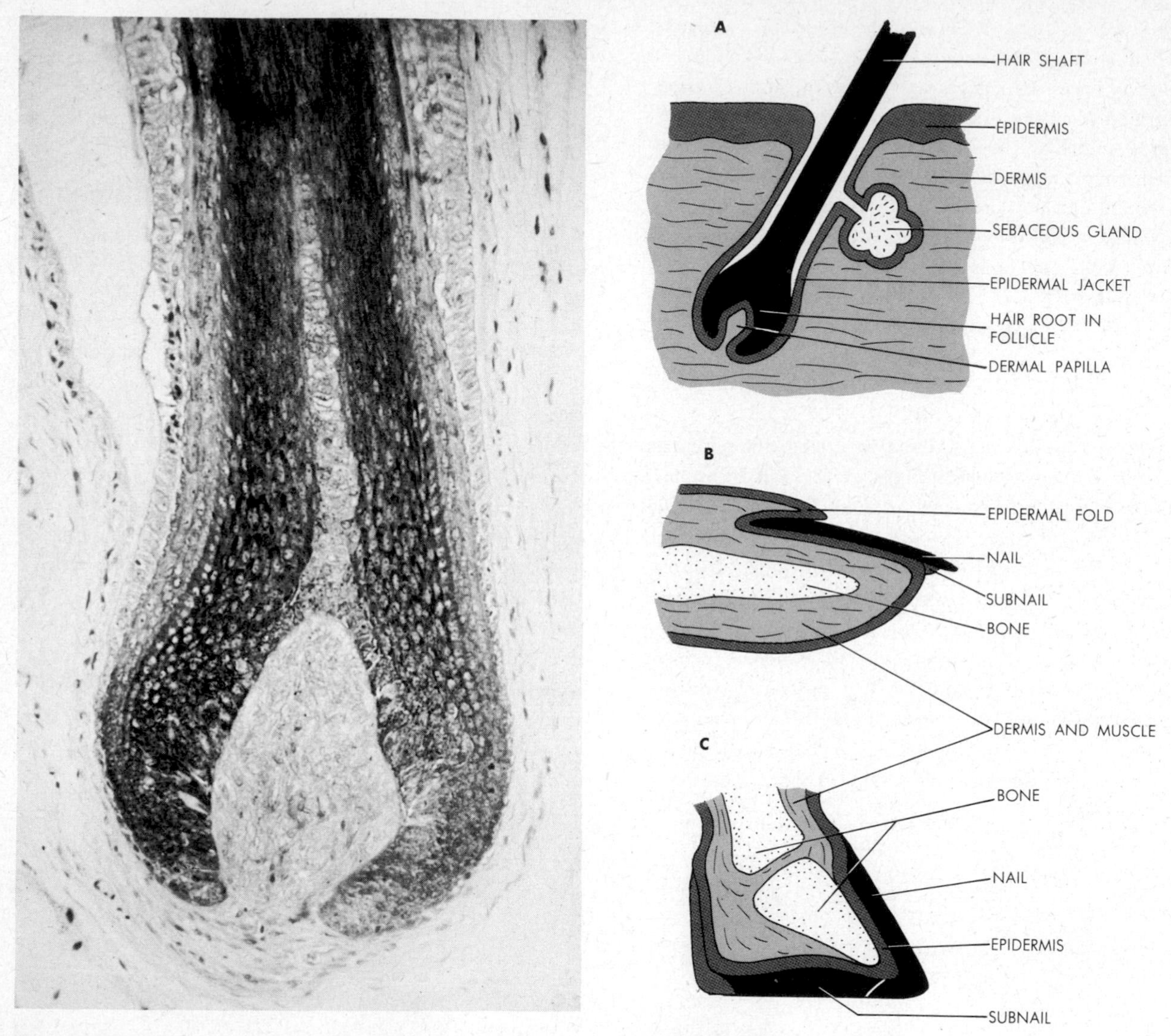

8.5. Photo: section through mammalian skin showing the part of a hair below the skin surface. Note the dermal papilla projecting into the expanded follicle at the hair base. **Diagrams: A.** Insertion of hair in skin. Note that the insertion and root structure of a feather are quite similar (cf. Fig. 32.19). **B.** Longitudinal section through a fingertip, to show nail; the tissue arrangement is similar in a claw. **C.** Longitudinal section through a hoof. (**B, C,** *after Boas.*)

In all animals, vertebrate or invertebrate, the epidermis may regenerate rather readily. The dermis in most cases regenerates less readily, and in some cases not at all. A deep wound affecting both skin layers usually becomes filled in by abnormally aligned dermal fibers, most of them tough and few of them elastic. A new epidermal layer grows over such a region by spreading and proliferation of cells from the surrounding intact epidermis. But because of the abnormal structure of the underlying dermis, the healed area differs from normal skin and remains recognizable permanently as a scar.

SKELETAL SYSTEMS

Invertebrates

Many animals, including even comparatively large ones such as giant earthworms, do not possess skeletons. Others are equipped with *exoskeletons,* which, as noted, are formed from epidermal secretions. Chemically, most exoskeletons consist either of calcium carbonate or of chitin (Fig. 8.7).

In sessile animals like corals and in slow-moving forms like snails and clams, the exoskeletons are entirely or almost entirely rigid. Made of thick calcareous deposits, the heavy shells afford excellent support and protection in an essentially stationary way of life. Locomotor and ingestive structures may be protruded from openings in the shells. These openings also allow for growth; as the animal enlarges beyond the confines of its old shell, a new and larger section of shell may be secreted in continuity with older portions. The chambered *Nautilus,* a mollusk related to squids, lives only in the newest, largest portions of its shell; earlier portions, each partitioned off from the others, are carried about empty (cf. Fig. 8.7). This also holds for some snails, though many of them, and most clams, occupy their whole shell.

Rapid locomotion becomes possible only if the animal possesses a highly mobile, flexible skeleton and one which can be carried about with a minimum of effort. Both conditions are realized in the chitinous exoskeleton of arthropods. Chitin is much lighter than lime, and substantial layers of chitin covered with a thin film of wax on the outside encase an arthropod completely. At specific regions the chitin shell remains thin and pliable; such *joints* permit free body movement. In insects, chitin also forms the wings (cf. Chap. 28).

However, an all-enveloping inert shell makes growth impossible. Indeed, arthropods must *molt* periodically and grow in spurts at such times. At each molt the exoskeleton loosens from the underlying epidermis and breaks open in one region, usually along the back. A thin new exoskeleton is secreted by the epidermis at the same time. The soft, defenseless animal then extricates itself from the old shell and enlarges rapidly during the next few days. The new and larger exoskeleton thickens in the course of this time, and when this new shell is

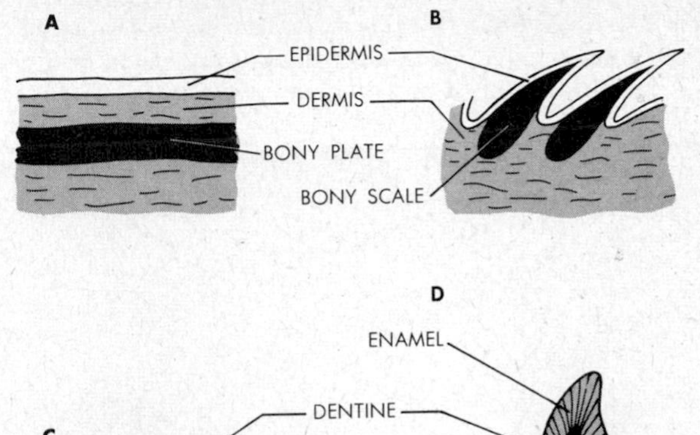

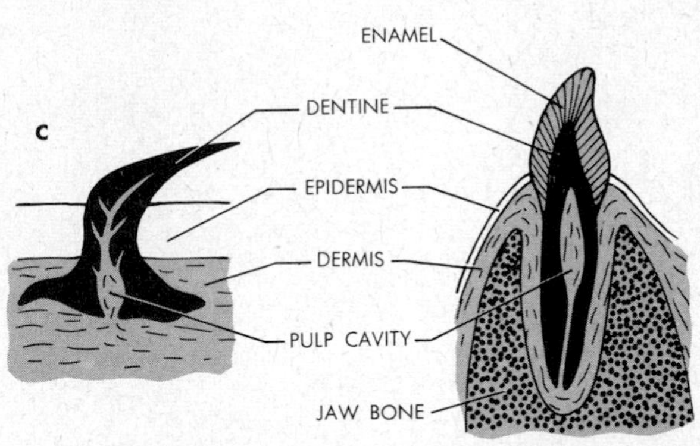

8.6. Some modifications of the dermis. **A,** dermal bony plate, as in echinoderms and vertebrates (e.g., skull bones); **B,** dermal scale, as in bony fishes; **C,** placoid scale, or denticle, as in shark skin; **D,** mammalian tooth. Note that **C** and **D** are fundamentally alike in structure. The pulp cavities arise from dermal papillae (as in hair) and are filled with connective tissue (in which are present blood vessels and nerve fibers).

fully hardened, growth ceases again. In Crustacea, periodic molting continues throughout life. Among insects, molts occur only during the larval period; adults neither molt nor grow in size (cf. Chap. 28, Fig. 28.33).

Affording structural support, protective armor, and freedom of movement, the chitinous exoskeleton is in large part responsible for the success of arthropods as a group. However, the very principle of exoskeletons entails severe inherent limitations. Most importantly, the larger the encased animal, the less support will be obtained by body parts deep in the interior. Indeed, animals with exoskeletons on the whole are comparatively small. Moreover, if the exoskeleton is calcareous, its proportionate weight (if not also its external position) restricts motion; and if the shell is chitinous, it is comparatively less sturdy, since calcium salts are the stronger material. Indeed, larger arthropods like lobsters impregnate their chitin shell with calcium salts. Furthermore, either an external

8.7. Exoskeletons. **A,** calcareous skeleton of a colonial coral, the sea fan *Eunicella;* **B,** chitinous exoskeleton of a bed bug. Cf. also Figs. 3.14, 4.34.

A

B

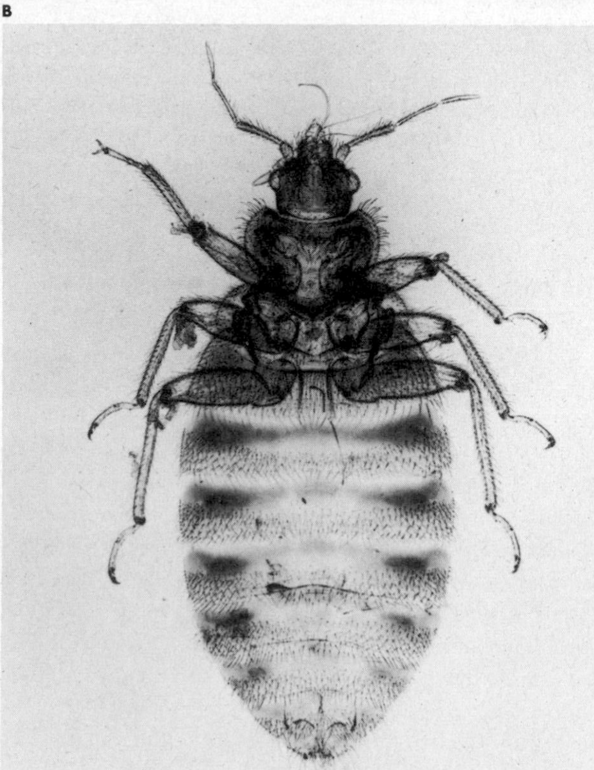

skeleton must be "open" to allow for growth, in which case support and protection are limited; or, if the shell is all-enveloping, the animal must molt at least during its early development, in which case support and protection are periodically lacking altogether.

Such disadvantages are absent from *endoskeletons,* which may support deep as well as superficial body parts. The animal consequently may attain considerable size and still be buttressed adequately. Large size also permits the development of proportionately large amounts of muscle, which may operate even a relatively heavy calcareous skeleton. The endoskeletal principle thus provides mobility, good internal support, and inherent strength, as well as the possibility of continuous growth.

Among several animal groups containing endoskeletal components, sponges possess *spicules,* needle-shaped deposits formed in the interior of certain specialized cells. Composed of calcium salts, silica, or horn (depending on the type of sponge), such spicules persist after the death of the animal in the form of densely intermeshed residues (Fig. 8.8). Squids possess a leaflike supporting blade, the *pen,* made of chitin in some types and of calcium carbonate in others. This single skeletal structure is located underneath the body wall and is thought to represent an evolutionary remnant equivalent to the exoskeletal shells of more primitive mollusks such as the chambered *Nautilus.* Pens are virtually absent in octopuses, but these animals as well as squids do possess true endoskeletal structures, in the form of mesoderm-derived *cartilage.* One fairly large plate of cartilage encases the brain, and additional cartilaginous supports are present near several other organs (cf. Chap. 25). In this possession of cartilage the squids and octopuses resemble the vertebrates; yet neither the pens nor the cartilages provide more than circumscribed localized support, and such isolated structures therefore cannot be considered to form a complete endoskeletal "system."

Echinoderms possess a well-developed endoskeleton consisting of many small calcareous plates formed in the dermis, typically around the whole body. Such plates are set together like tiles. These are held together by the dermal connective tissue and are moved by muscles in echinoderms such as starfishes, but the plates are fused to one another in forms such as sea urchins (cf. Fig. 8.8). The skeleton thus seems to form an external shell, yet it is covered by the epidermis and therefore is a true endoskeleton. Echinoderms move sluggishly or are sessile altogether, and the correlated presence of all-encasing supports again illustrates the suitability of shell-like skeletons in a slow way of life.

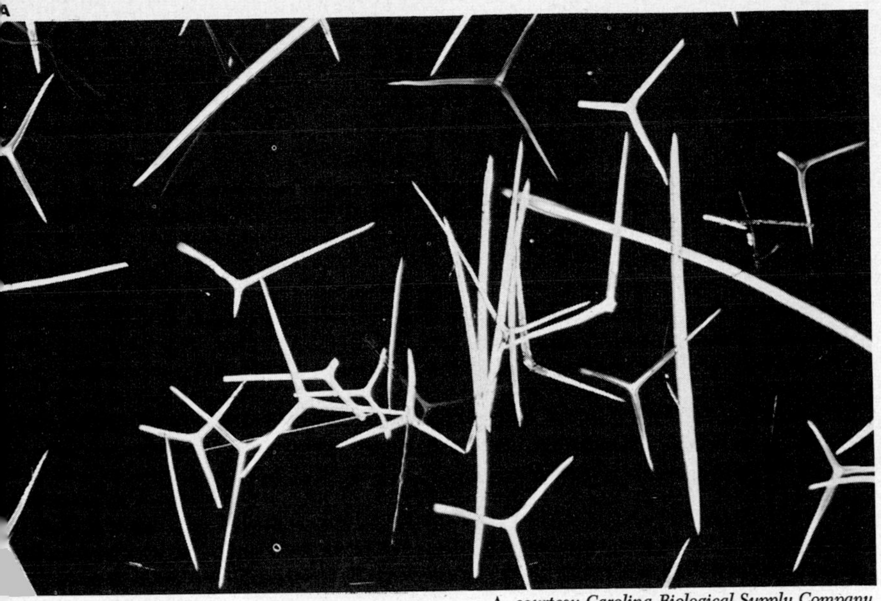

A, *courtesy Carolina Biological Supply Company.*

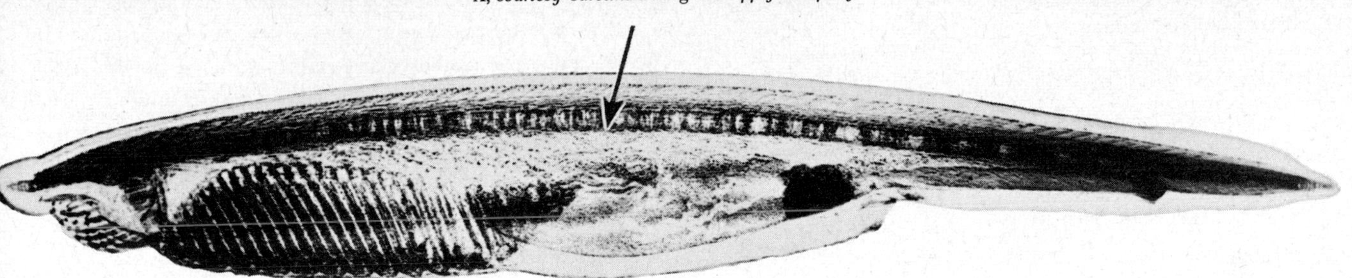

8.8. Endoskeletons. A, calcareous spicules of a sponge; **B,** the endoskeletal shell of a sea urchin (seen from the underside and showing the five teeth around the mouth); **C,** the notochord (arrow) in amphioxus, a primitive chordate. Cf. also Fig. 3.14.

Elaborate and unrestricting endoskeletons occur only in the chordates, vertebrates in particular. During the embryonic period all chordates possess a dorsal *notochord,* a stiff, elastic supporting rod situated along the midline of the body, between the spinal cord and the digestive tract. As noted earlier, the phylum name "Chordata" derives from this basic endoskeletal structure. In some chordates the notochord persists throughout life as the main or only skeletal support, but in most cases the notochord is supplanted in late embryos by a *vertebral column.* Additional supports develop as well, and the adult thus comes to possess a complex mesodermal skeletal system which buttresses all regions of the body (cf. Fig. 8.8).

Vertebrates

The vertebrate skeleton may be considered to consist of two parts, viz., an *axial* skeleton, comprising skull, vertebral column, and rib cage; and an *appendicular* skeleton, comprising the pectoral and pelvic girdles and the limb supports (Fig. 8.9). In jawless fishes (e.g., lampreys) and cartilaginous fishes (e.g., sharks), the whole skeleton is cartilaginous throughout life. In all other vertebrates the skeleton is very largely bony, bone developing in two ways. Most of the skull and part of the pectoral girdle, i.e., skeletal parts consisting of flat plates and lying close to the body surface, arise as *dermal bones,* directly from the connective tissue of the

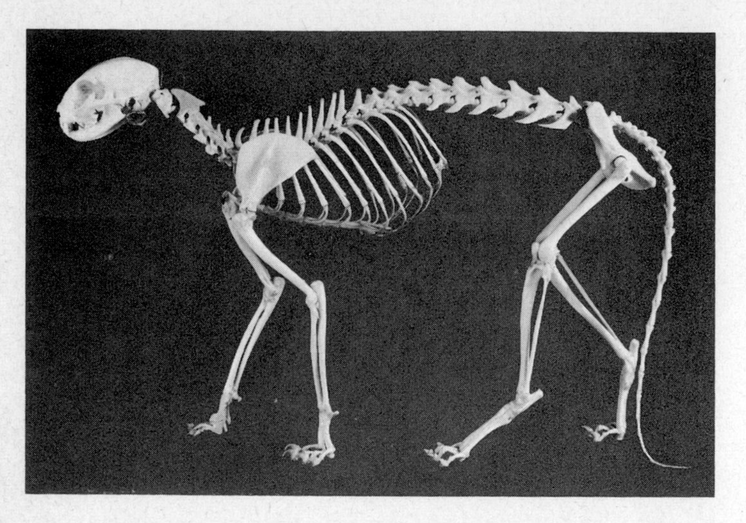

8.9. The skeleton of a cat, illustrating the general skeletal organization of vertebrates. Note skull, axial skeleton (vertebral column and tail), and appendicular skeleton (limbs and limb girdles). The suspension-bridge skeleton of four-footed vertebrates is turned upright in man.

dermis. All other skeletal parts, i.e., those which form deep-lying and predominantly elongated supports, develop as *replacement bones;* they are first laid down in cartilage, which is replaced subsequently by bone substance.

In the case of dermal bones, some of the mesenchyme cells in the dermis aggregate and form a tissue sheet, the *periosteum* (Fig. 8.10). If a bony plate is being formed, two parallel periosteal sheets usually develop, some distance apart. These periostea bud off new cells into the space between them; and the cells, known as *osteoblasts,* become specialized for the formation of bone substance. They produce a meshwork of fibers around themselves which later *ossifies,* i.e., becomes impregnated with hard mineral deposits. Such bone substance does not form as a solid mass but is honeycombed by channels; and in these are present not only the osteoblasts but also blood cells, fat cells, more mesenchyme cells—in short, the usual constituents of connective tissue. Also present are mesenchyme-derived *osteoclasts,* bone-destroying cells capable of resorbing the bone substance near them. Operating together, osteoblasts and osteoclasts may reshape and remold bone and alter its thickness and general configuration, in response to changes of stress and the growth of the animal as a whole.

Replacement bone similarly arises from mesenchyme cells in connective tissue (Figs. 8.11, 8.12). Such cells first form a *perichondrium,* a layer which buds off cartilage-forming *chondroblasts* and cartilage-destroying *chondroclasts.* Through their activity, cartilage in the shape of a future long bone is produced. Later the perichondrium specializes as a periosteum, and from then on bone formation occurs essentially as in dermal bones; as chondroclasts resorb cartilage, newly specialized osteoblasts secrete bone substance in substitution. Mineralization begins at three specific *ossification centers,* viz., in the *diaphysis,* or mid-region, of the cartilage rod, and at both ends, the *epiphyses.* Cartilage eventually persists only in two places: at the surfaces of the epiphyses, where cartilage provides smooth, friction-reducing joint pads; and at the juncture of epiphysis and diaphysis, where a layer of cartilage permits continuing bone growth in length. In the mature animal these growth centers ultimately ossify as well, and elongation of bone then ceases.

8.10. Development of dermal bone, diagrammatic.

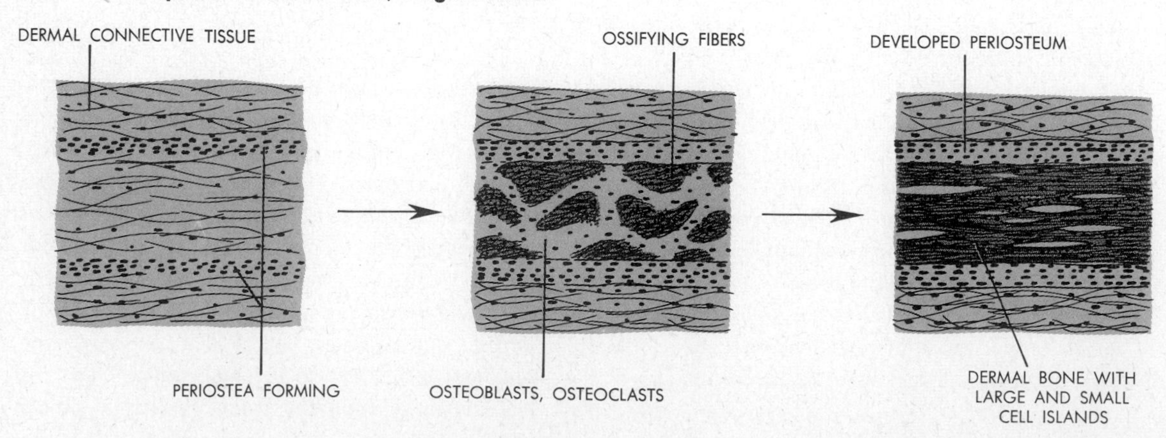

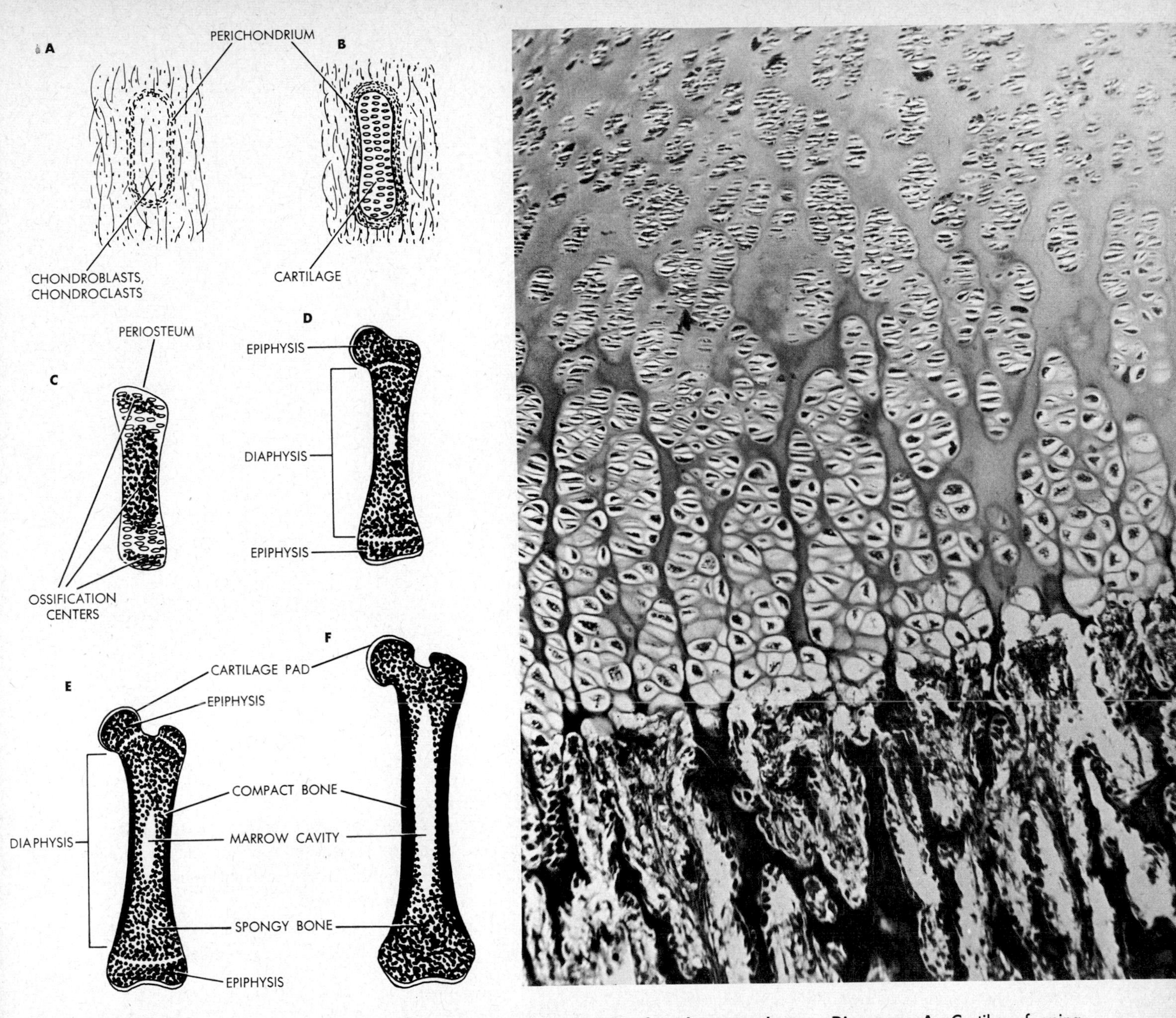

8.11. Development of replacement bone. Diagrams: A. Cartilage-forming cells aggregate in position of future bone. **B.** Formed cartilage rod. **C.** Perichondrium has become periosteum and bone-forming cells (osteoblasts) have begun secretion of bone substance at three centers. **D.** Spongy bone has replaced all cartilage except in regions near joints (white bands); compact bone begins to form at surface of shaft, and marrow cavity, in center of shaft. **E.** Compact bone (solid black) extensive, marrow cavity enlarging. **F.** Mature bone; cartilage layers near joints have been replaced by bone; in shaft, marrow cavity large and spongy bone largely replaced by compact bone. **Photo:** histological section through an ossification front, showing bone in process of formation (lower part of photo; note dark bone deposits, and cells in the spaces between) and cartilage in process of resorption (upper part of photo).

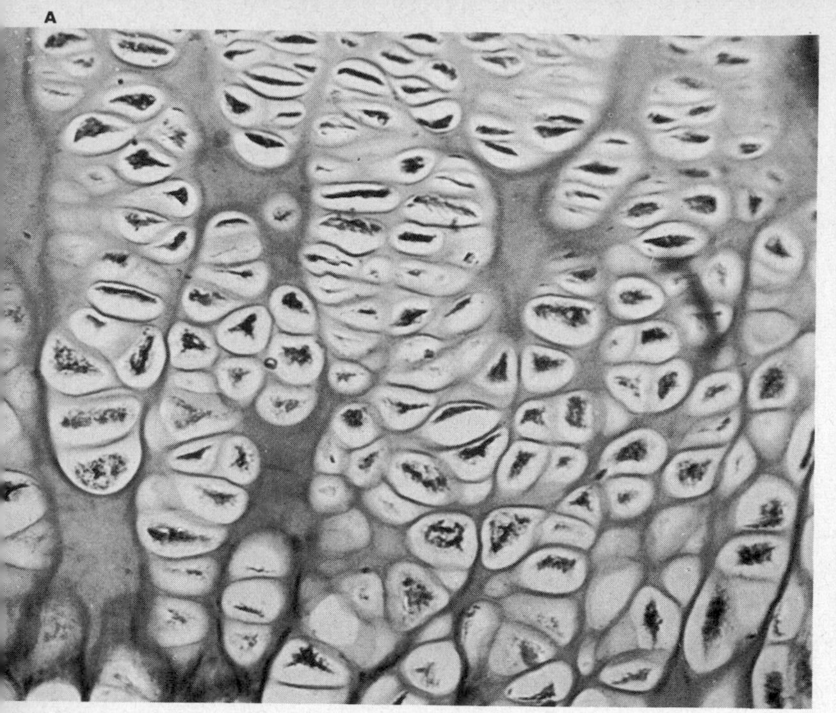

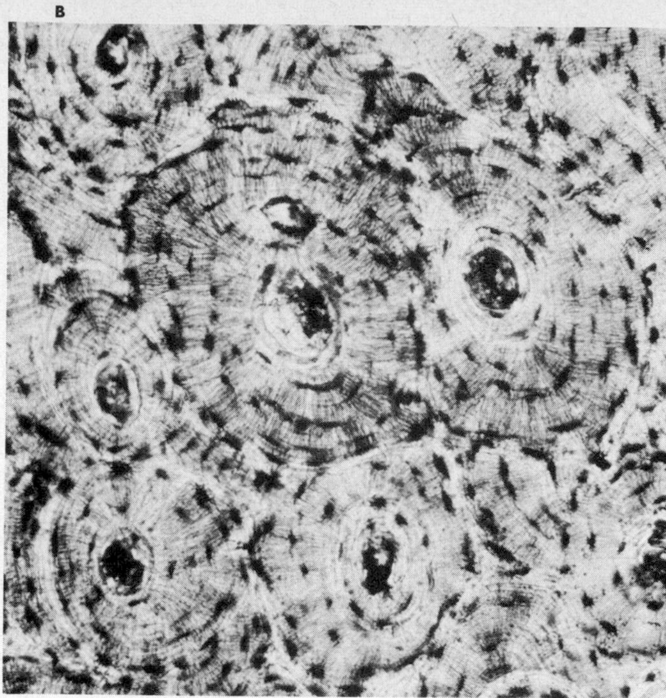

8.12. Cartilage and bone. **A.** Histological section through cartilage. Note the many cartilage-producing cells (chondroblasts), surrounded by their own secretions. **B.** Histological section through bone. Note the concentric layers of a Haversian system, with the bone-producing cells (osteoblasts) located in the dark patches. Hard bone substance, light in the photo, surrounds the cells. In the center of each such system is a wider Haversian canal.

Near the surface of a bone shaft, in the vicinity of the periosteum, the osteoblasts deposit bone substance in dense parallel layers. Some distance away from the surface, bone is laid down in dense concentric patterns, forming so-called *Haversian systems* (cf. Fig. 8.12). Present in such a system is a central canal and concentrically placed islands, filled with blood cells and the other cellular components of bone. Such densely organized *compact bone* constrasts with loosely organized *spongy bone,* formed in the core of the shaft and in the epiphyses (cf. Fig. 8.11). The bone substance here is traversed by comparatively wide, irregularly patterned channels, all again filled with soft tissue. Such tissue constitutes *marrow;* it is reddish in color if blood cells predominate in it, as in the epiphyses; and it is yellowish if fat cells predominate, as in the diaphysis. As a long bone matures, osteoclasts gradually resorb the spongy bone substance in the core of the shaft and a distinct *marrow cavity* appears. Like dermal bones, replacement bones too may be remolded and reshaped by the combined action of osteoblasts and osteoclasts.

Skull bones grow until they meet. Then they fuse along the edges into a continuous casing. Other bones articulate movably in joints. A joint is encapsulated by a strong ligament, and the interior of such a capsule is filled with a watery fluid which lubricates and facilitates movement.

In its gross anatomical construction, the vertebrate skeleton resembles a suspension bridge. The appendicular skeleton

8.13. The suspension-bridge principle of the mammalian skeleton. Cf. Fig. 8.9.

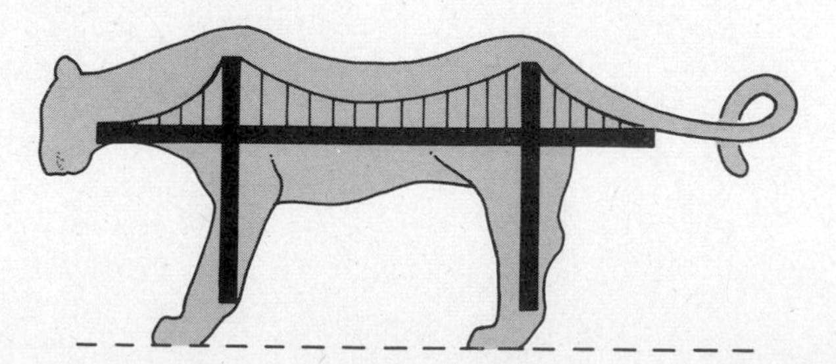

supplies the vertical supports, and the axial skeleton forms the horizontal parts (Fig. 8.13). In most modern bony fishes, the fins are supported by radial arrays of bone, so-called *fin rays*. In some primitive bony fishes and in all other bony vertebrates, the bones of the limbs exhibit a common structural plan (though not always the same shape). The upper part of a limb is supported by one bone, the lower part by two (Fig. 8.14). The extremities terminate with wrist and ankle bones and with the bones of the digits. Five digits per limb are characteristic generally, but the number is variously reduced in many specific vertebrates. Limbs as a whole connect to sockets formed by the appendicular girdles. In most vertebrates the pelvic girdle is fused directly to the vertebral column, whereas the pectoral girdle is held in place by ligaments and muscles (Fig. 8.15).

From head to tail, the vertebral column consists of five regions (cf. Fig. 8.9): *cervical* (neck), *thoracic* (chest), *lumbar* (lower back), *sacral* (hip), and *caudal* (tail). In each region the individual *vertebrae* are shaped in characteristic ways, but all vertebrae have a common basic structure (Fig. 8.16). The main portion is the *centrum*, and projecting from it are various bony outgrowths, called *processes*. Thus, vertebrae articulate with one another by means of *articular processes*. This jointing leaves room for *intervertebral disks*, pads of fibroelastic cartilage between adjacent vertebrae. Dorsal processes from the centrum form a *neural arch*, a bony enclosure for the spinal cord. Ventral processes, present primarily in bony fishes, form long, paired *ventral ribs* in the trunk region and *hemal arches* in the tail region. These arches enclose and protect the main artery and vein of the tail. In other bony vertebrates the ventral processes are not developed conspicuously, but lateral *transverse processes* are prominent instead. Attached to these in the thoracic region are *lateral ribs*. Such ribs extend around to the ventral side of the body,

8.14. Appendageal skeletons of vertebrates, diagrammatic. **A.** Fin of ray-finned bony fish; bony rays support the fin (cf. Chap. 31). **B.** Bony elements in fin of lungfish. **C.** Fin of lobe-finned fish, indicating the basic ancestral bony elements of all vertebrate walking limbs. **D.** Basic plan of bones in vertebrate limb; labels in brackets refer to names of bones in hind limb corresponding to those of forelimb. There are five distals, five metacarpals (metatarsals), corresponding to the five digits of the limb. In **C**, roman numerals in brackets refer to positions where the metacarpals and phalanges of **D** are added. Wrist bones collectively are carpals, ankle bones, tarsals. In the evolutionary transition from **C** to **D**, carpals and tarsals have become reduced in number. (**B, C,** *after Holmgren.*)

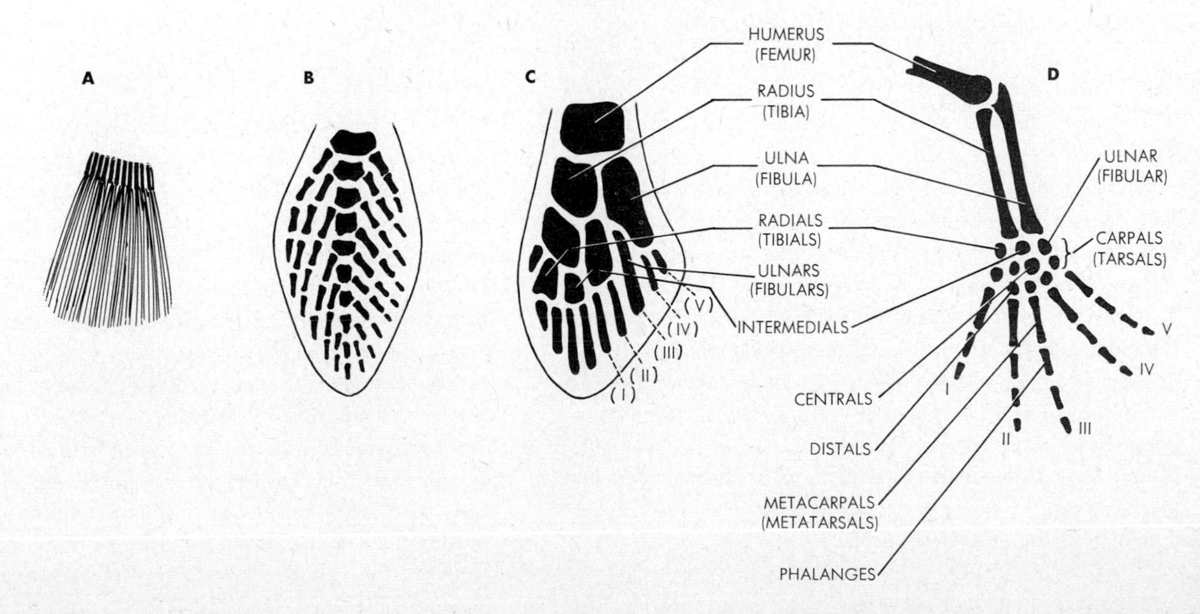

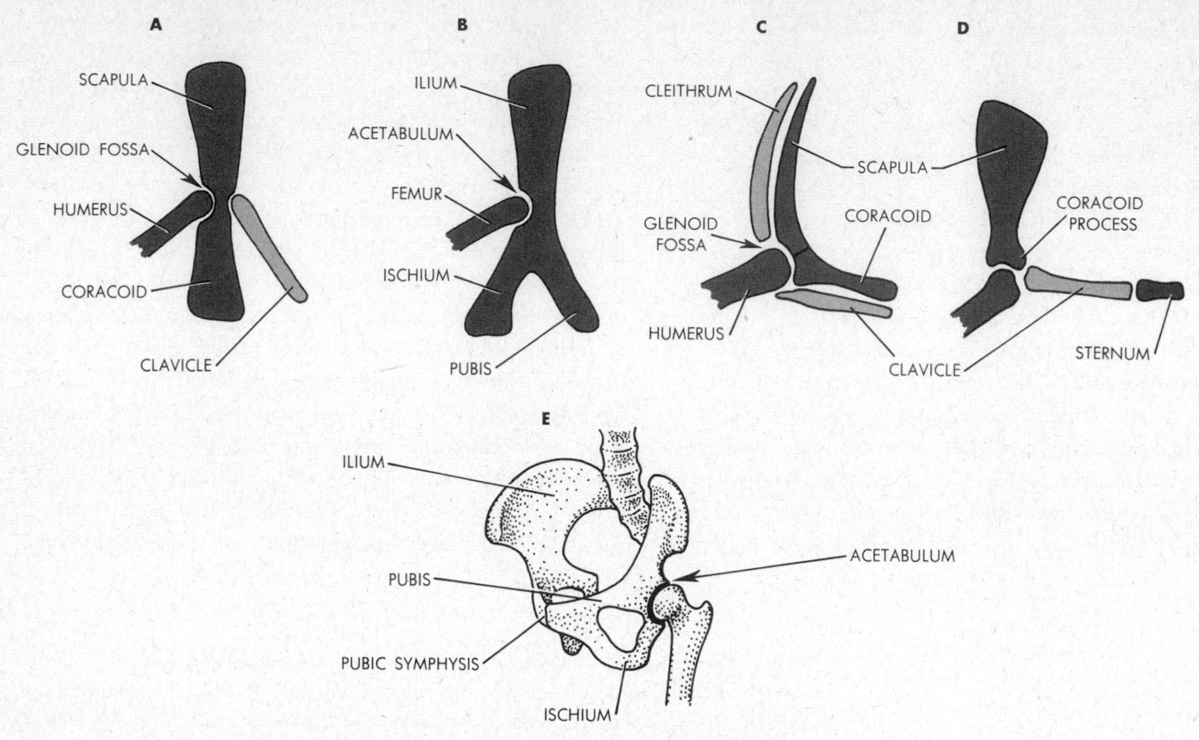

8.15. Limb girdles, diagrammatic. **A** and **B.** The basic plan of the vertebrate pectoral and pelvic girdles, respectively; note the fundamental similarity of design. The glenoid fossa is the socket for the humerus, corresponding to the acetabulum for the femur. **C.** Primitive vertebrate pectoral girdle as in fishes, showing dermal bones (light shading) and cartilage bones (dark shading). **D.** Pectoral girdle as in mammals; note reduction of coracoid to a bony process and disappearance of cleithrum. **E.** Oblique view of human pelvic girdle; the pubic symphysis is the median junction of the pubic bones. (*Partly adapted from Woodruff.*)

where they articulate with a *sternum*, or breastbone, and so form the rib cage.

The sturdiness of the whole skeleton is finely adjusted to body weight. Apart from the inherent strength of the construction material, the strength of a pillar is proportional to the square of its thickness. A column twice as thick as another is 2^2, or four times as strong. But body weight increases with the cube: an animal twice the diameter of another is roughly 2^3, or eight times as heavy. We find, indeed, that larger vertebrates possess more than just proportionately thicker legs in comparison with smaller vertebrates. Also, many light, slender-legged mammals are built to walk on their toes, for greater speed, whereas heavy mammals generally walk on the whole foot, for greater support.

In invertebrates generally, the hard skeleton-forming materials do not contain cells within them; they are inert deposits, most often on the body surface and secreted by layers of soft living tissues underneath. By contrast, the distinguishing feature of vertebrate endoskeletons is the presence of living tissue not only around but also directly within the hard supporting substance. The skeleton thus possesses active, plastic, living qualities which are largely absent from the solidly calcified or hornified skeletons in other animal groups. In this "aliveness" lies a major advantage: the skeleton may grow continuously along with the rest of the animal, and it may adjust to changing stresses and loads far better than a completely solidified, wholly passive support. Moreover, whereas a broken piece of clam shell, for example, must stay broken, a cracked piece of vertebrate bone may heal by renewed activity of the cells in its interior.

MUSCULAR SYSTEMS

Among protozoa, some groups possess in the cytoplasm contractile *myofibrils* which may produce changes in cell shape. Some cells of sponges are capable of slow contractions. The radiate phyla are characterized by the presence of *musculo-epithelial* cells in the epidermis and the alimentary wall. The basal portions of such cells contain bundles of myofibrils which may bring about contraction and bending of given body regions. These cells are not specialized exclusively for muscular activity; those in the integument also have protective functions, and those in the alimentary wall also serve in digestion (cf. Chap. 21, Fig. 21.4).

All other animals possess distinct mesoderm-derived muscle tissues specialized exclusively for the production of motion. Two basic types of such tissues may be distinguished, viz., *smooth* muscle and *striated* muscle (Fig. 8.17). In the smooth variety, the cells are elongated and spindle-shaped and each contains a single nucleus. Myofibrils are aligned longitudinally, and contraction shortens and thickens the cell. Striated muscle is made up of syncytial units. Each such unit, a *muscle fiber*, develops through repeated division of a single cell, and in this process the boundaries between daughter cells disappear. The resulting fiber therefore contains many nuclei embedded in a continuous mass of elongated muscular cytoplasm. As already described in Chap. 6 (Fig. 6.1), the longitudinally aligned myofibrils exhibit microscopic cross striations, i.e., alternate dark and light bands; hence the name of this type of muscle.

Smooth muscle is found in all animals which possess a musculature at all, and in most groups smooth muscle actually is the only type present. Characteristically, the cells are oriented in parallel and form muscular layers. Such layers are usually found, for example, underneath the integument, as part of the body wall. A single layer is present in roundworms. The cells here contract in the direction of the longitudinal axis of the body; hence the worms may bend sideways, but they cannot lengthen. In most other wormlike animals the body wall contains two layers of smooth muscle, with cells aligned longitudinally in one layer and circularly in the other (cf. Fig. 7.4). Such an arrangement permits lengthwise extension and contraction of the whole body as well as side-to-side

8.16. Vertebrae, diagrammatic. **A,** tail vertebra of bony fish; **B,** trunk vertebra of bony fish, showing ventral ribs; **C,** generalized trunk vertebra of a land vertebrate, showing dorsal ribs. The latter curve to ventral side of body and articulate with sternum. **D.** Side view of mammalian lumbar vertebrae; the spaces between the centra and the articular processes provide paths for the exit of spinal nerves. (*Partly after Kingsley.*)

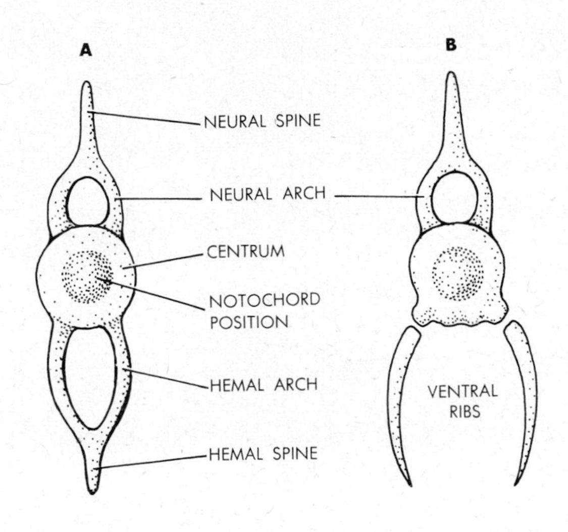

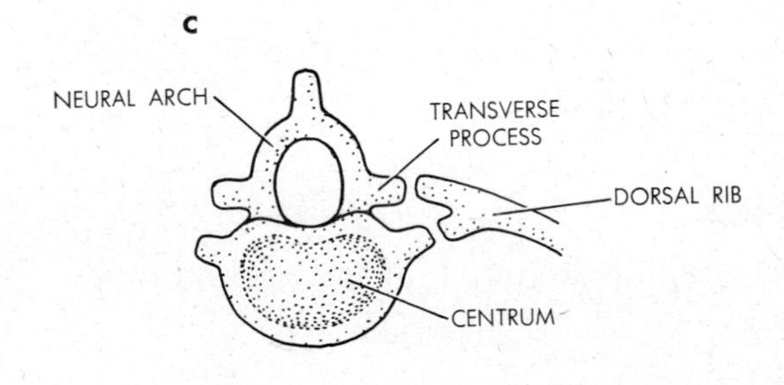

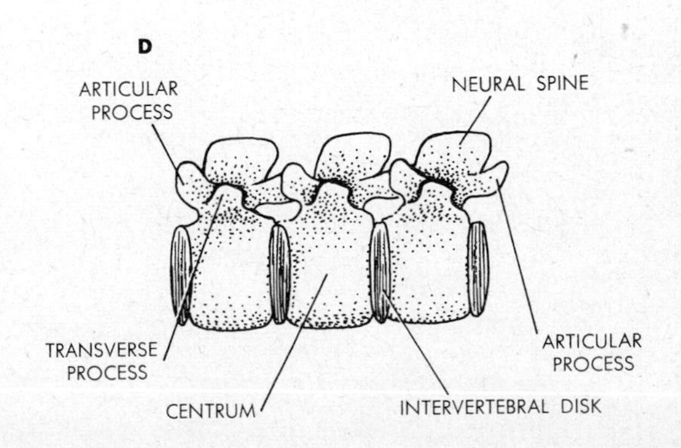

A

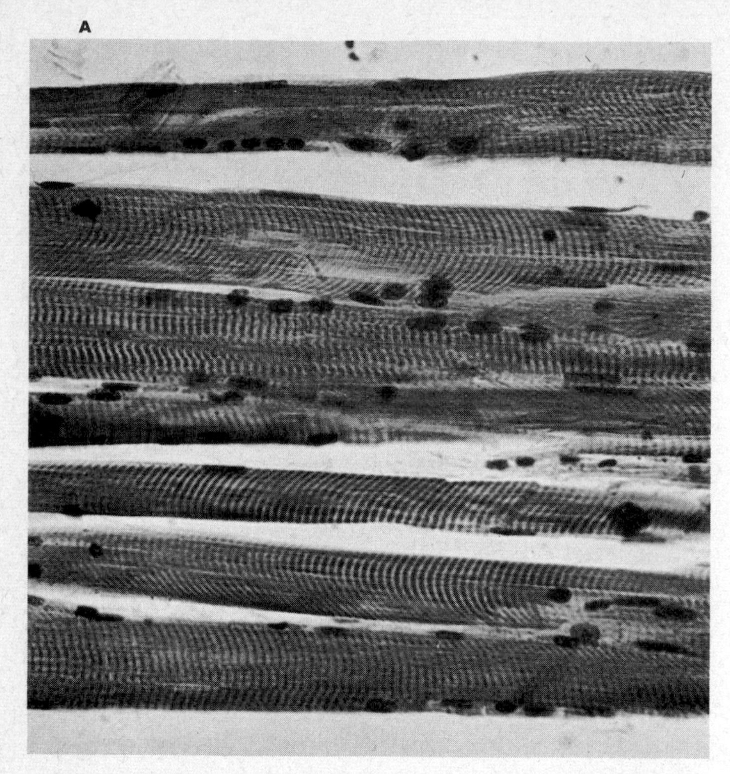

B

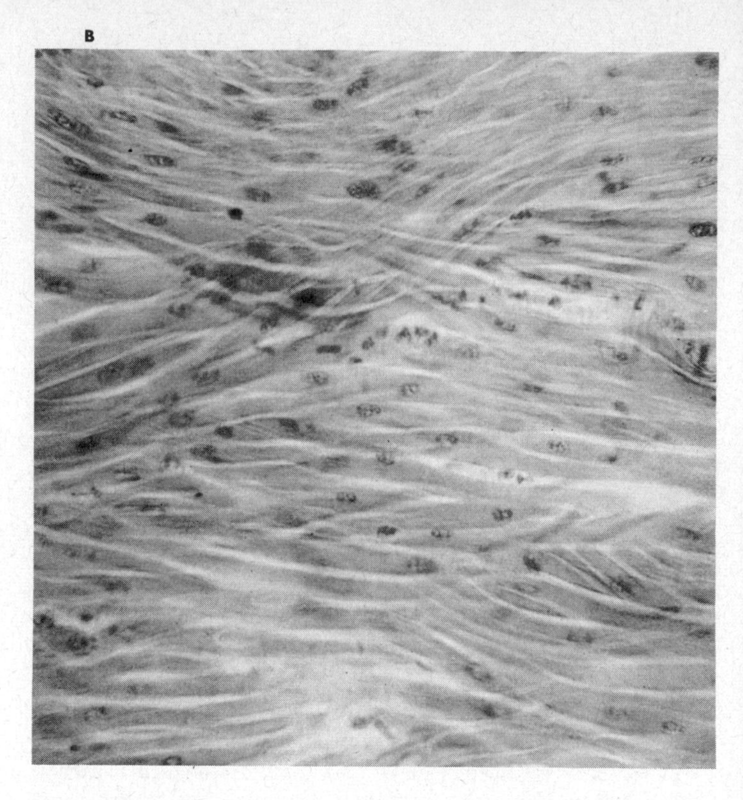

C

8.17. Muscle types. **A.** A few fibers of skeletal muscle. Note the cross striations and the many nuclei within each fiber. Cf. also Fig. 6.1 for structural details. **B.** Smooth muscle. Note the spindle-shaped cells. **C.** Cardiac muscle. Note the branching fibers, the nuclei, the faint longitudinal fibrils within each fiber, and the faint cross striations.

flexions and torsions. Many animals also possess smooth-muscle layers in the wall of the alimentary tract and in the walls of the larger blood vessels (cf. Chap. 9). In these instances, circularly and longitudinally aligned muscle cells are again common. Additional accumulations of smooth muscle are found in a large variety of other organs.

Striated muscle appears to occur specifically where great speed of motion is of advantage; it is encountered in addition to smooth muscle mainly in mollusks (particularly squids, octopuses, and clams such as scallops), in arthropods, and in vertebrates. The striated muscles here typically form a "skeletal" musculature; i.e., they connect to parts of the skeleton. Thus, striated muscles are fastened to the inner surfaces of the exoskeletons of arthropods and to the outer surfaces of the endoskeletons of vertebrates (Fig. 8.18). Most of such muscles connect one part of the skeleton to another, but some are joined to the skeleton only at one end, the other end inserting in another body part. In vertebrates, for example, several large muscles interconnect skeleton and skin and form part of the body wall.

A group of striated muscle fibers usually makes up a *muscle bundle,* which is enveloped by a layer of connective tissue. Several of such bundles form a *muscle,* an organ enclosed within a connective tissue sheath of its own. At either end this sheath may gradually become a *tendon,* attached to some part of the skeleton (cf. Fig. 8.18). In vertebrates, striated muscles are under voluntary nervous control, smooth muscles are not. Such a pattern probably holds also in other animals containing striated muscles. But in those that do not, smooth muscle evidently serves in both involuntary and voluntary functions. The hearts of vertebrates are composed of a specialized variant of striated muscle, viz., *cardiac muscle.* The fibers of cardiac muscle are syncytial, and they are fused to one another in intricate patterns. Thus, the whole vertebrate heart is a continuous multinucleate mass of contractile living matter. This muscle is not under voluntary control (cf. Fig. 8.17).

In vertebrates, all striated muscle fibers innervated by a single nerve fiber form a *motor unit.* Hundreds of such units may be present in a whole muscle (Fig. 8.19). Each motor unit operates in an *all-or-none* manner: either it contracts fully or it does not contract at all. To produce a contraction, the motor unit must be stimulated by a nerve impulse of minimum or *threshold* intensity. An adequate stimulus is followed in succession by a brief *latent period,* a *contraction period,* and a *relaxation period.* Such a unit cycle of activity may be referred to as a *simple twitch* (Fig. 8.20). If a second stimulus arrives during the latent period of a twitch, the stimulus will be without effect. But if a second stimulus arrives just after the latent period, e.g., during the contraction phase, then the motor unit may relax incompletely or not at all and begin a second twitch. If now a third stimulus again arrives during the contraction phase, a third contraction may follow without intervening relaxation. Through such *summation* of a rapid (but not too rapid) succession of stimuli, a sustained contraction may ensue. Known as *tetanus,* the sustained contraction lasts until the motor unit fatigues (cf. Fig. 8.20).

Most muscular activity in the living animal is tetanic in nature. Also, the motor units within a whole striated muscle work in relays, different ones contracting tetanically or being at rest at any given moment. Moreover, whereas a single motor unit operates in all-or-none fashion and thus cannot produce a graded response, a whole muscle may; its contraction will be the stronger the more motor units are active. Muscles actually

8.18. Diagrams: comparison of muscle-skeleton relation in endoskeleton (vertebrates, **A**) and exoskeleton (arthropods, **B**). Note tendon attachment of muscle to bone in **A. Photo:** transverse histological section through part of a striated muscle. The broad white spaces mark the position of layers of connective tissue which envelop and separate individual muscle bundles. Within each such bundle, narrower white spaces mark the connective tissue surrounding the individual muscle fibers. Black dots in the fibers are nuclei. Cf. also Fig. 6.1.

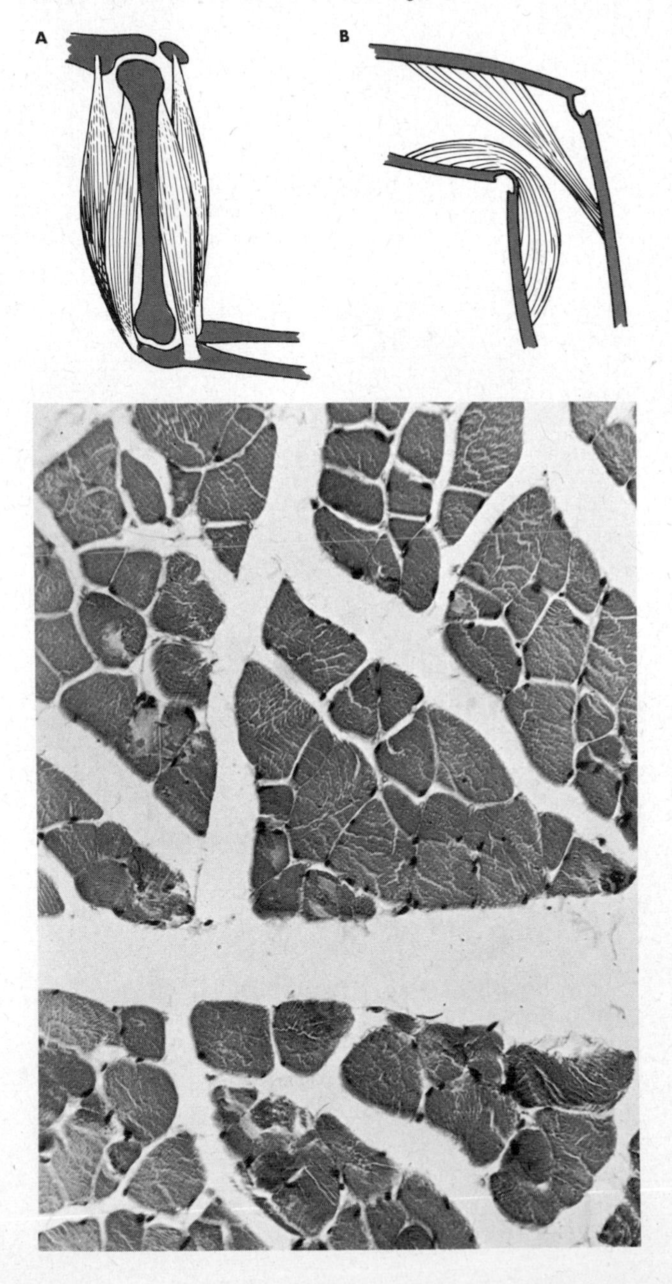

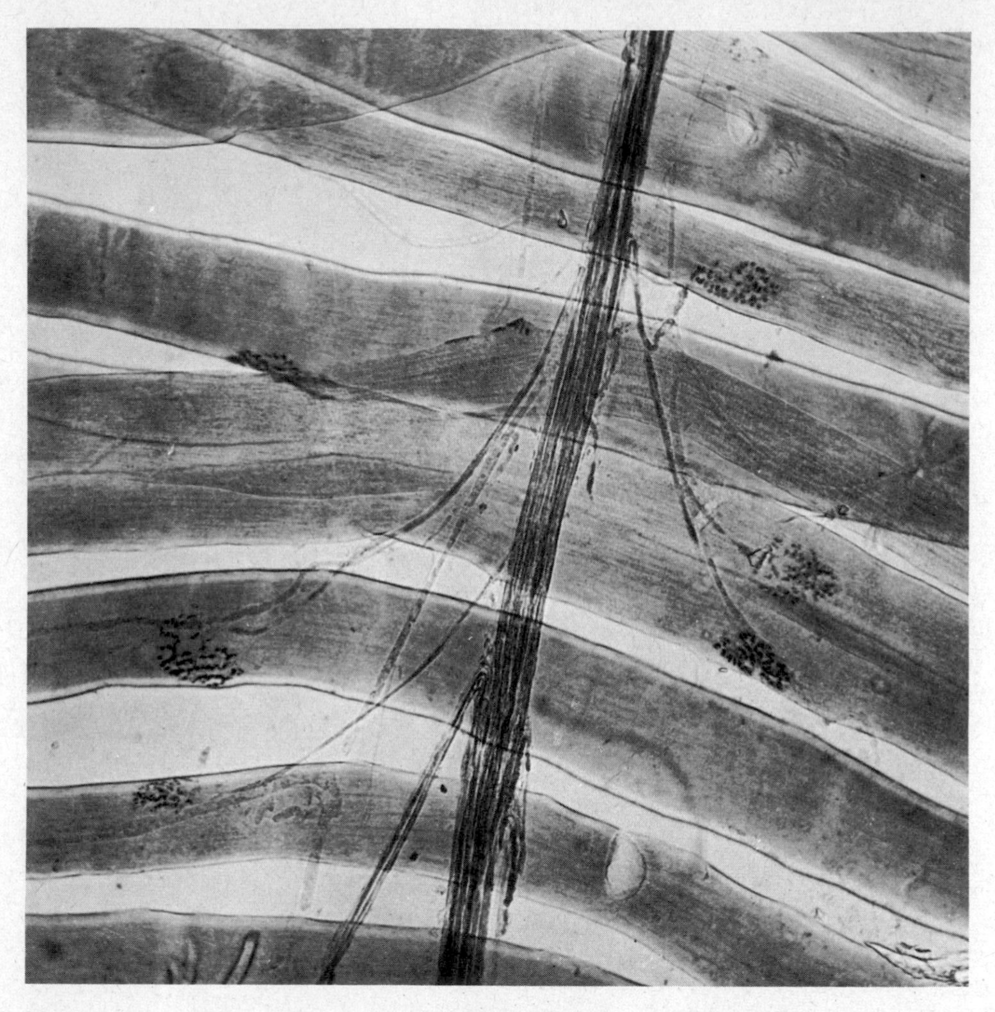

8.19. Motor end-plates, the knobbed terminals of the branches of a motor nerve on individual muscle fibers. The group of muscle fibers so innervated by a single nerve fiber and its terminal branches constitutes a motor unit, i.e., a set of muscle fibers functioning together.

are never relaxed completely. Even during periods of rest, muscles are held in a partially contracted state in which very little energy is expended. Such mild contractions constitute *tonus,* or muscle "tone," and it is through this that the muscular system preserves the shape of body parts, maintains posture, and provides mechanical support in general. Only stronger contractions, above and beyond tonic ones, result in outright movement of parts (and in pronounced energy expenditure).

Striated muscles operate far more rapidly than smooth muscles, and usually there are also many more motor units in a striated muscle than in a smooth muscle of equivalent size. Consequently, the striated musculature can produce faster, more abruptly alterable, and more finely adjustable motions. On the other hand, the smooth musculature requires comparatively less energy, and its slower, more sustained motions are well suited in steady, continuing activities; e.g., maintenance of blood pressure through vessel contractions or maintenance of digestion through gut-wall contractions.

Whereas many muscles function solely in producing internal movements, the bulk of the muscular system also contributes to the *locomotion* of the whole animal. All animal locomotion is based on the lever principle; a part of the body acts as a more or less rigid lever, and as muscles exert pull against it at one point, another point of the lever exerts push against the environmental medium (cf. Fig. 8.18). Where a skeleton is present, its components usually form part or all of the levers. Typically, the locomotor muscles are organized in opposing sets, one set producing a *flexion* or *adduction* of a body part, the other set producing an *extension* or *abduction*. Through such activities, animals propel themselves on solid surfaces by creeping, gliding, and walking; in water by paddling, lashing, and jetting; and in air by various forms of flying. Appropriate chapters of Unit 2 outline the characteristics of some of these locomotions. In view of the importance of locomotion in animal life generally, it is not surprising that, in forms such as vertebrates, the musculature has become the largest organ system of the body.

However, proper operation of the muscular system, as of any other, requires nervous control.

NERVOUS SYSTEMS

All nervous activity is based on *reflexes*, the operational units of nervous systems. Reflexes are routed through *reflex arcs*, each of which usually consists, like any control apparatus generally, of five components: *receptor, sensory pathway, modulator, motor pathway,* and *effector* (Fig. 8.21).

The neural receptors are specialized *sensory cells*, which may or may not be housed in elaborate sense organs. Receptors are sensitive to specific environmental changes, i.e., *stimuli;* and in a receptor an incoming stimulus initiates a *nerve impulse*. Such impulses are transmitted over *sensory nerve fibers* to the modulators. Neural modulators are specialized internal portions of a nervous system. Among many kinds of modulators, brains are the most familiar. It is the function of modulators to interpret incoming sensory impulses and in turn to initiate appropriate, steady-state-maintaining motor impulses. The latter are sent out over *motor nerve fibers* to the effectors, viz., *muscles* and *glands*. Such effectors then act on the motor impulses they receive by carrying out explicit *responses*. These usually bear a distinct relation to the original stimuli; i.e., they alter the internal or external activities of an animal in such a way that steady state can be preserved.

Neural Pathways

The structural units of reflex arcs, and thus of nervous systems as a whole, are nerve cells, or *neurons*. Each typically consists of a nucleus-containing cell body, the *cyton*, and of one or more *nerve fibers*, which are filamentous outgrowths extend-

8.20. Left: simple muscle twitch. *1*, stimulation applied; *2*, beginning of contraction (hence interval between *1* and *2* represents latent period); interval 2–3, contraction period; interval 3–4, relaxation period. **Right:** *1*, well-spaced stimuli are applied (as marked along bottom line), resulting in separate simple twitches; *2*, summation occurs when frequency of stimuli is increased; *3*, sustained tetanic contraction, resulting from exceedingly high stimulus frequency.

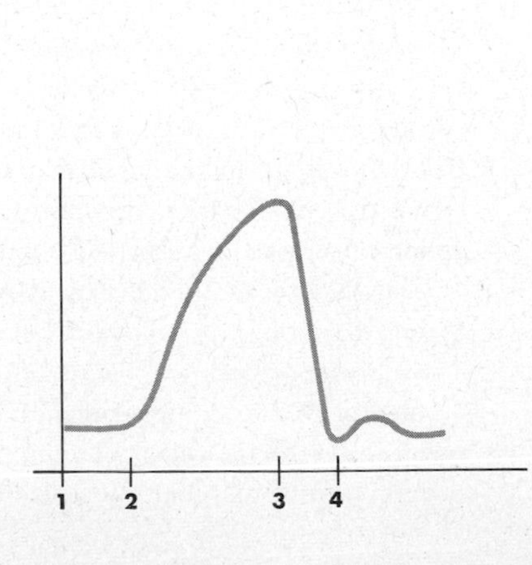

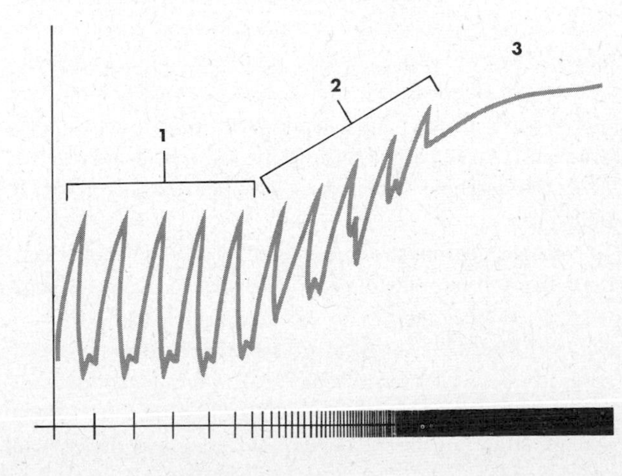

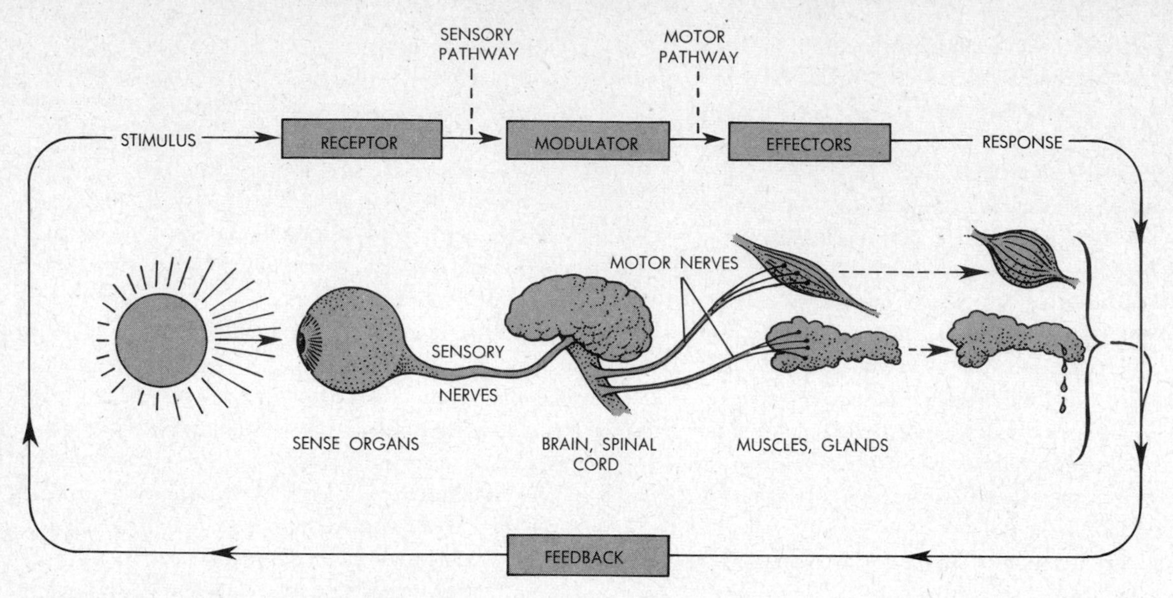

8.21. The general pattern of steady-state control and of reflex activity. The components of any kind of steady-state-maintaining device are indicated along the top row, those of a nervous reflex immediately underneath. A stimulus (such as light) will produce a response such as muscle contraction and/or glandular secretion. The response itself then becomes a feedback, i.e., a new stimulus, through which the modulator is informed whether or not the response produced an adequate reaction to the original stimulus. The feedback stimulus now may, or may not, initiate a new reflex.

ing away from the cell body (Fig. 8.22). Nerve impulses normally originate at the terminal of one of the fibers, travel toward the cell body, traverse it, then lead away from the cell body through another of its fibers. Nerve fibers in which impulses travel toward the cell body are termed *dendrites;* those carrying impulses away from the cell body are called *axons.* Dendrites and axons vary greatly in length, some being quite short, others often being as much as a yard or more long. Many nerve fibers branch along their course and at their terminals. On the basis of the number and arrangement of long fibers present, neurons may be said to be *unipolar, bipolar,* or *multipolar.*

Adjacent neurons are never fused to one another directly but remain structurally discrete. The fiber terminals are separated from one another by a microscopic space called a *synapse* (Fig. 8.23). Impulses are transmitted across such synapses by chemical means, from the axon terminal of one neuron to a dendrite terminal of an adjacent neuron (cf. below). Synaptic transmission usually imposes a one-way direction on impulse conduction as a whole. According to their position within a reflex arc, so-called *sensory,* or *afferent,* neurons transmit impulses from sense organ to modulator; and *motor,* or *efferent,* neurons transmit from modulator to effector. Neurons within a modulator, including those which transmit impulses from incoming sensory to outgoing motor fibers, are called *internuncial* neurons, or *interneurons.* Collectively, all modulator structures may be said to constitute the *central* portions of a nervous system; all sensory and motor components outside such central parts then form the *peripheral* portions. Groups of peripheral nerve fibers frequently traverse a body region as a single collective fiber bundle. Such a bundle, usually enveloped by a sheath of connective tissue, constitutes a *nerve.* Nerves are designated as being sensory, motor, or mixed, depending on whether they contain sensory fibers, motor fibers, or both (cf. Fig. 8.23).

Some protozoan groups contain impulse-conducting intracellular *neurofibrils* which coordinate the beat of flagella and cilia. Sponges are without specialized neural elements of any kind. All other animals possess distinct nerve cells. The most primitive type of neuron arrangement is a *nerve net*, or *plexus* (Fig. 8.24). In such an array, impulses may usually travel in either direction within a neuron, and an impulse originating at any given point in the net may spread diffusely to all other points. Synaptic gaps are known to separate the neurons. Nerve nets are characteristic particularly of the Radiata, flatworms, echinoderms, and hemichordates. In the first two of these groups such nets are the most elaborate neural structures present. But plexuses usually form at least part of the nervous systems of all other animals as well. For example, nets are common in the integument (as in earthworms and numerous other types of worms) and in the walls of the alimentary tract (as in animals generally, vertebrates included).

In addition to plexuses, most animals also possess more highly specialized neural aggregations. Among these are, for example, *nerve cords* and *ganglia* (Fig. 8.25). A nerve cord is a condensed, concentrated region of a plexus, forming a thickened bundle of nerve fibers which in effect resembles a nerve. A ganglion is a dense, localized accumulation of cell bodies and fiber terminals of neurons. Ganglia may be sensory, motor, or mixed, according to the kinds of fibers they connect with. The more complex ganglia contain meshworks of interneurons, organized into intricate internal pathways and neural circuits. In general, nerve cords and particularly

8.22. Diagrams: the labeled figure depicts the general structure of a neuron with a myelinated axon; a portion of a nonmyelinated fiber, possessing a Schwann sheath only, is sketched immediately below. At bottom are schematic drawings of different neuron types; axons only, not dendrites, are shown. The unipolar type corresponds to the labeled figure at top. **Photo:** micrograph of a stained (multipolar) motor neuron.

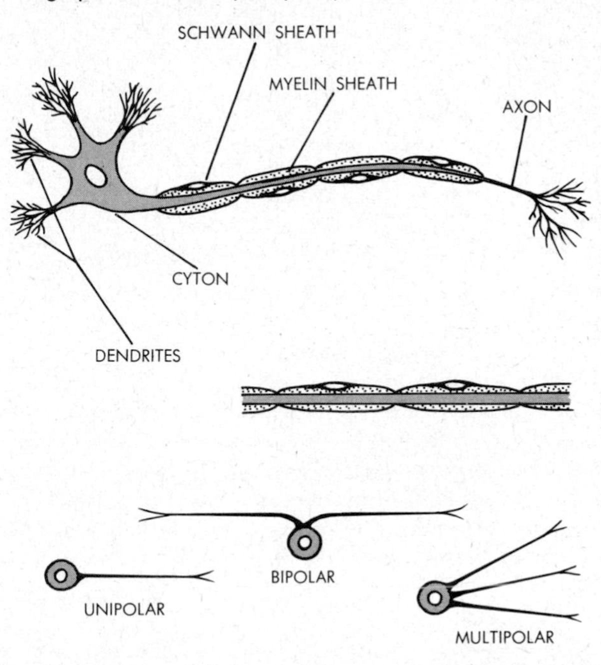

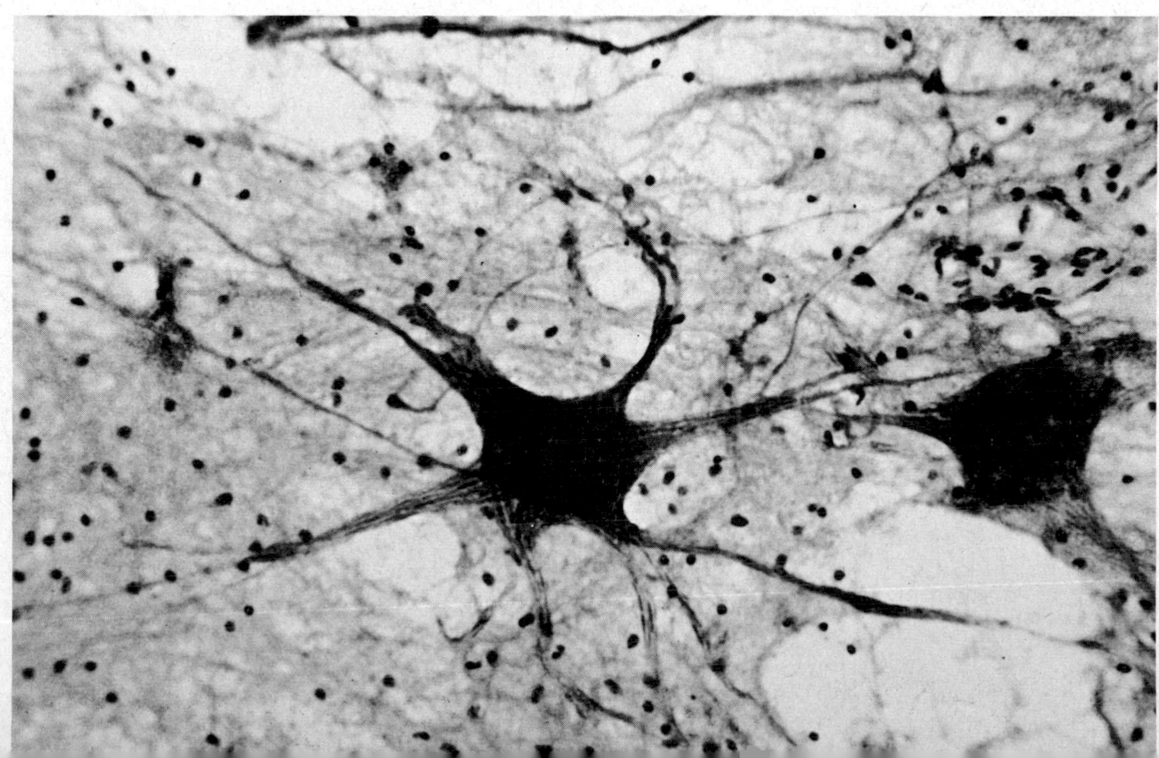

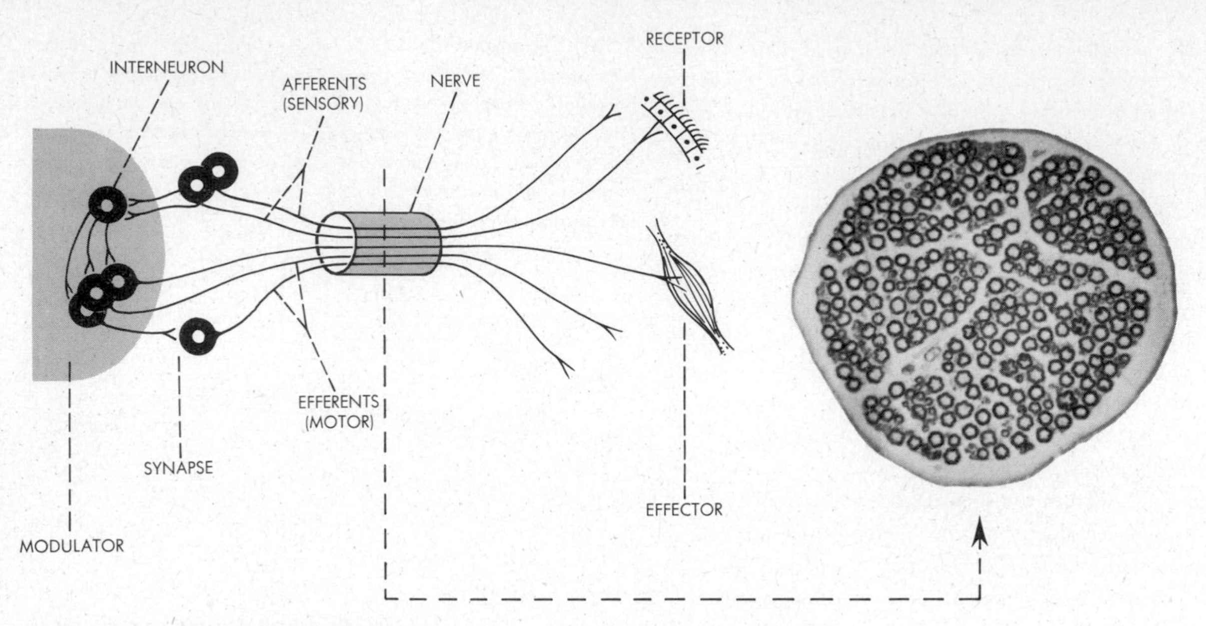

8.23. The pattern of neural pathways. Afferent fibers conduct impulses from receptors to modulators, where interneurons transmit the impulses to efferent fibers. The latter send impulses to effectors. Neurons interconnect functionally across synapses. Collected bundles of neuron fibers form nerves. The photo at right depicts a cross section through such a nerve, the dark circlets here representing the myelin sheaths of the individual nerve fibers.

ganglia function as *selective relays;* impulses coming into them may be suppressed or augmented, and new impulses may be sent out over specifically selected outgoing pathways. As we shall see, all modulator functions ultimately are based on such impulse modifications, channel selections, and switching activities. Within large ganglia there are usually functional subdivisions referred to as nerve *centers,* specialized groups of interneurons regulating specific activities. Thus, some centers are specialized to interpret incoming sensory information only, others transmit motor commands only; some centers control involuntary functions only, others are concerned with voluntary activities only. Moreover, the large ganglia may also store information as memory, may acquire and store new information by learning, and may control intricate patterns of behavior. Such ganglia in effect constitute *brains.*

As might be expected, the complexity of the nervous system of an animal matches the complexity of its locomotion and its behavior potential in general. In many groups, particularly in sessile and sluggish types, the central parts of the nervous system often consist of only a single main ganglion. The latter may form the hub of a peripheral array of sensory and motor pathways radiating to and from it. If several ganglia are present, they are usually interconnected by distinct nerves or by nerve cords. Among active, motile invertebrates, the central parts of the nervous system frequently have the form of a ladder (cf. Fig. 8.25). In its basic organization, this so-called *ladder-type* system consists of two parallel, ventrally located nerve cords, with ganglia along their course and transverse connecting cords between them. At the head end are several large, paired ganglia, usually arranged as a ring around the alimentary tract. The dorsal parts of this ring are the *brain ganglia,* or *cerebral ganglia,* which may be fused together partly or wholly. Numerous structural variants of the ladder-type system are known, and in many invertebrates the central neural components have their own unique anatomy. Nevertheless, as will be shown in appropriate later chapters, arrays of nerve cords and variously complex ganglia constitute the chief central portions in all such cases.

By far the most complex nervous systems are encountered in the vertebrates. The central portions here arise from a hollow dorsal nerve cord, or *neural tube,* formed in the embryo as an ingrowth from the ectoderm (cf. Chap. 11, Fig. 11.20). The anterior portion of the tube enlarges into a *brain;* the posterior portion becomes the *spinal cord.* The fluid-filled space within the tube persists as the brain *ventricles* anteriorly and the *spinal canal* posteriorly (consult Fig. 8.26). In the mature brain, three general regions may be distinguished, viz., *forebrain, midbrain,* and *hindbrain.* The originally evolved parts of these regions form a *brain stem.* Portions of this brain stem are enlarged in advanced vertebrates, and in such cases the original regions of the brain stem are hidden deep within the brain.

8.24. The configuration of a nerve net, or plexus. The coelenterate *Hydra* is diagrammed. Note that here, as in coelenterates generally, the density of the nerve plexus is greatest around the alimentary opening. Impulses may travel in any direction in such a net. The neurons interconnect functionally via synapses.

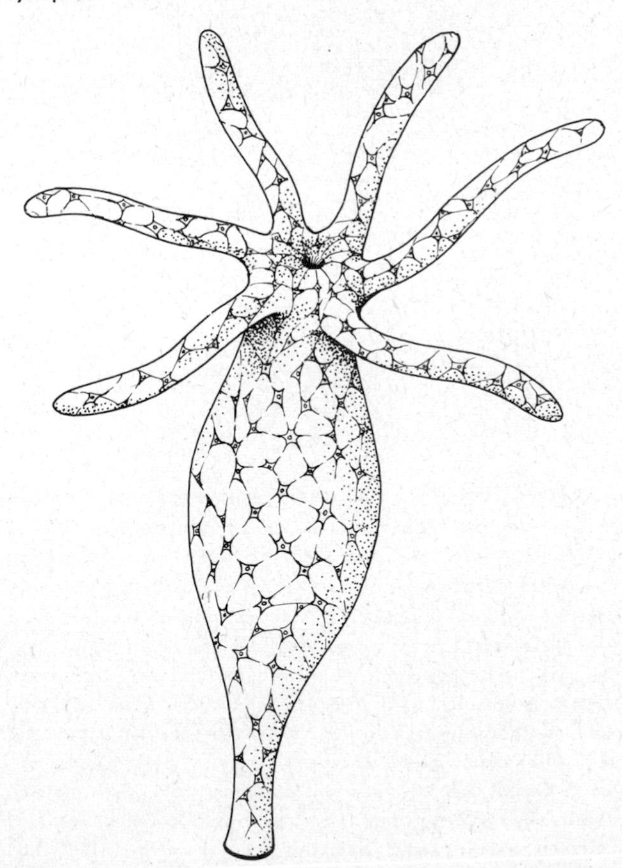

The forebrain consists of three subdivisions. Two of these form the *telencephalon* anteriorly, a left and right portion containing a brain ventricle each. In all vertebrates except the birds and mammals, the telencephalon remains comparatively small, and the most conspicuous components are the *olfactory lobes* extending forward. These contain the centers for the sense of smell. In birds and mammals, the telencephalon is enlarged greatly into two *cerebral hemispheres,* which cover virtually all of the rest of the brain stem dorsally, posteriorly, and laterally. The hemispheres contain the modulator centers for most sensory and motor activities; and they are also the seat of memory, intelligence, and all so-called "higher" mental functions. An extensive series of nerve tracts, called the *corpus callosum,* interconnects the two hemispheres.

Behind the telencephalon is the unpaired third part of the forebrain, the *diencephalon* (cf. Fig. 8.26). Its ventricle communicates with those anteriorly and those posteriorly. The principal modulator centers of the diencephalon are the *thalamus* and *hypothalamus,* lateral regions which control numerous involuntary muscular and glandular activities as well as consciousness, sleep, food intake, and the emotional state of the animal. On the ventral side of the diencephalon, the optic nerves from the eyes cross partially and lead into the brain via the *optic chiasma.* Just behind this region, the endocrine *pituitary gland* is attached to the brain. Nerves as well as blood vessels interconnect the hypothalamus and the pituitary gland, and it is through these pathways that the hypothalamus exercises major control over many glandular activities. Dorsally, the *pineal body* projects from the diencephalon. In lampreys and the reptile *Sphenodon,* the pineal body protrudes through the skull bones, and the integument overlying it is transparent. This structure in effect forms a third eye, located on top of the head. In other vertebrates such a pineal eye is not functional, and in birds and mammals the pineal body is covered over by the cerebral hemispheres.

The midbrain, or *mesencephalon,* contains a fourth brain ventricle and also dorsally located *optic lobes* (cf. Fig. 8.26). In all vertebrates except the mammals, these lobes represent the center of vision, and the nerve tracts from the eyes terminate there. In mammals, however, the optic nerve tracts are extended into newly evolved visual centers, located in the posterior regions of the cerebral hemispheres. The original optic lobes of the brain stem here are little more than relay stations for visual nerve impulses.

The hindbrain consists of two subdivisions, an anterior *metencephalon* and a posterior *myelencephalon,* or *medulla*

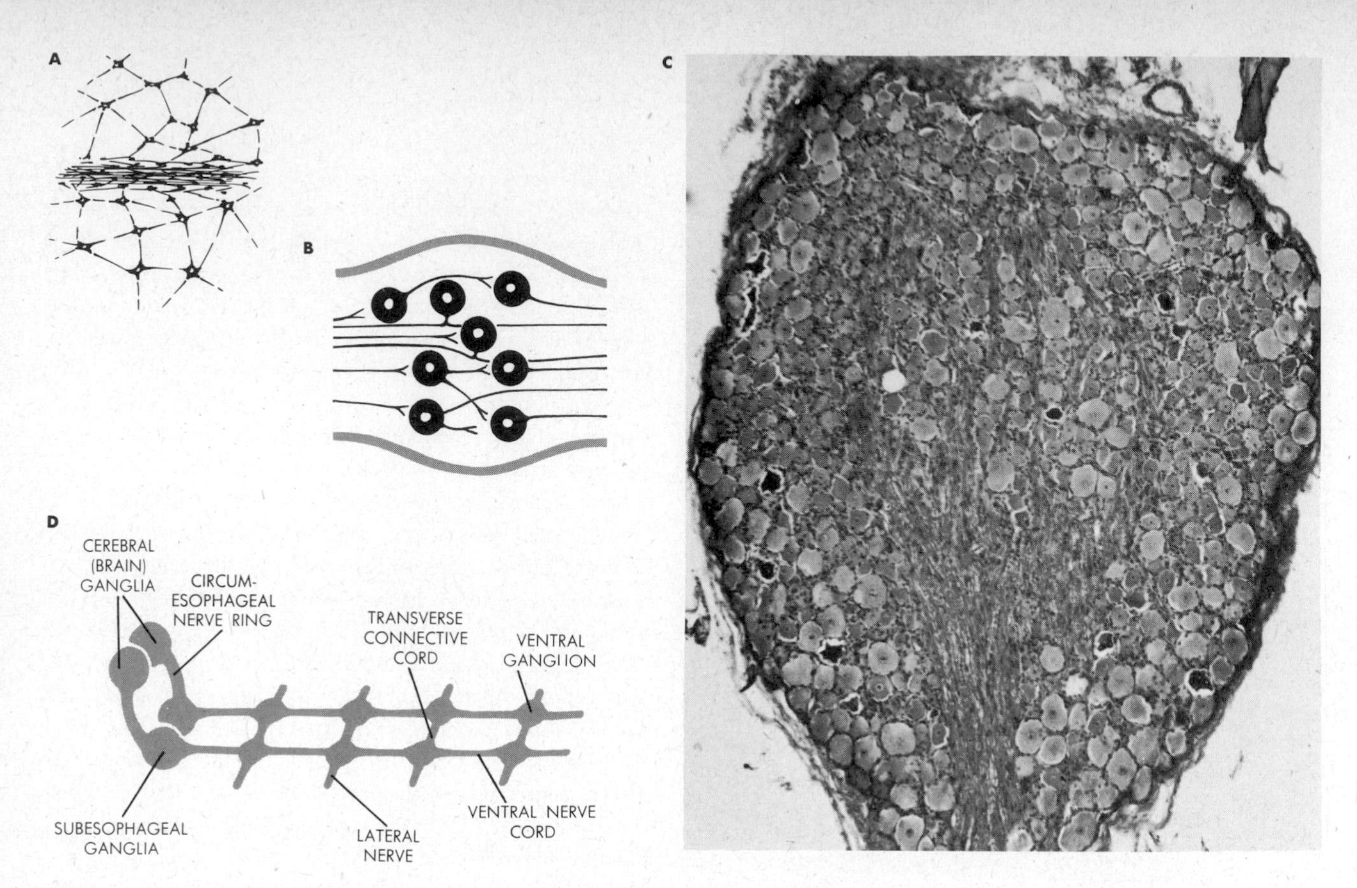

8.25. **A.** A portion of a nerve cord, indicating its origin as a dense concentration of parts of a nerve net. **B.** Schematic representation of a ganglion; a mixed ganglion is indicated here, with sensory and motor neurons as well as interneurons. **C.** Photographic section through a spinal (sensory) ganglion of a mammal; note the many cytons and also the nerve fibers, some seen in cross section, some in longitudinal section. **D.** Schematic representation of a ladder-type nervous system; most such systems possess the components shown, though not necessarily in exactly the indicated arrangement.

8.26. **The vertebrate brain** (right). **A.** Schematic representation of median section through primitive vertebrate brain (as in fishes), showing basic structural plan. Roman numerals refer to ventricles. Thalamus and hypothalamus regions are indicated in broken lines, since these brain parts lie on each side of the median plane (as do ventricles I and II). The hypophysis forms as an outpouching from the roof of the mouth cavity; it becomes the anterior lobe of the pituitary, the infundibulum becoming the posterior lobe. The two choroid plexi are membranous regions in which blood vessels are carried. **B, C, D, E,** dorsal views of brains of fish (shark), amphibia (frog), reptiles (alligator), and birds (goose), respectively. Note the proportionate progressive enlargement of the cerebrum in a posterior direction (and the parallel reduction of the olfactory lobes). **F.** Dorsal view of cerebrum of human brain. Note that the left cerebral hemisphere is slightly larger than the right, a usual condition in right-handed persons. **G.** Median section through human brain. The cerebrum here has become so large that, in dorsal view, it covers the cerebellum posteriorly.

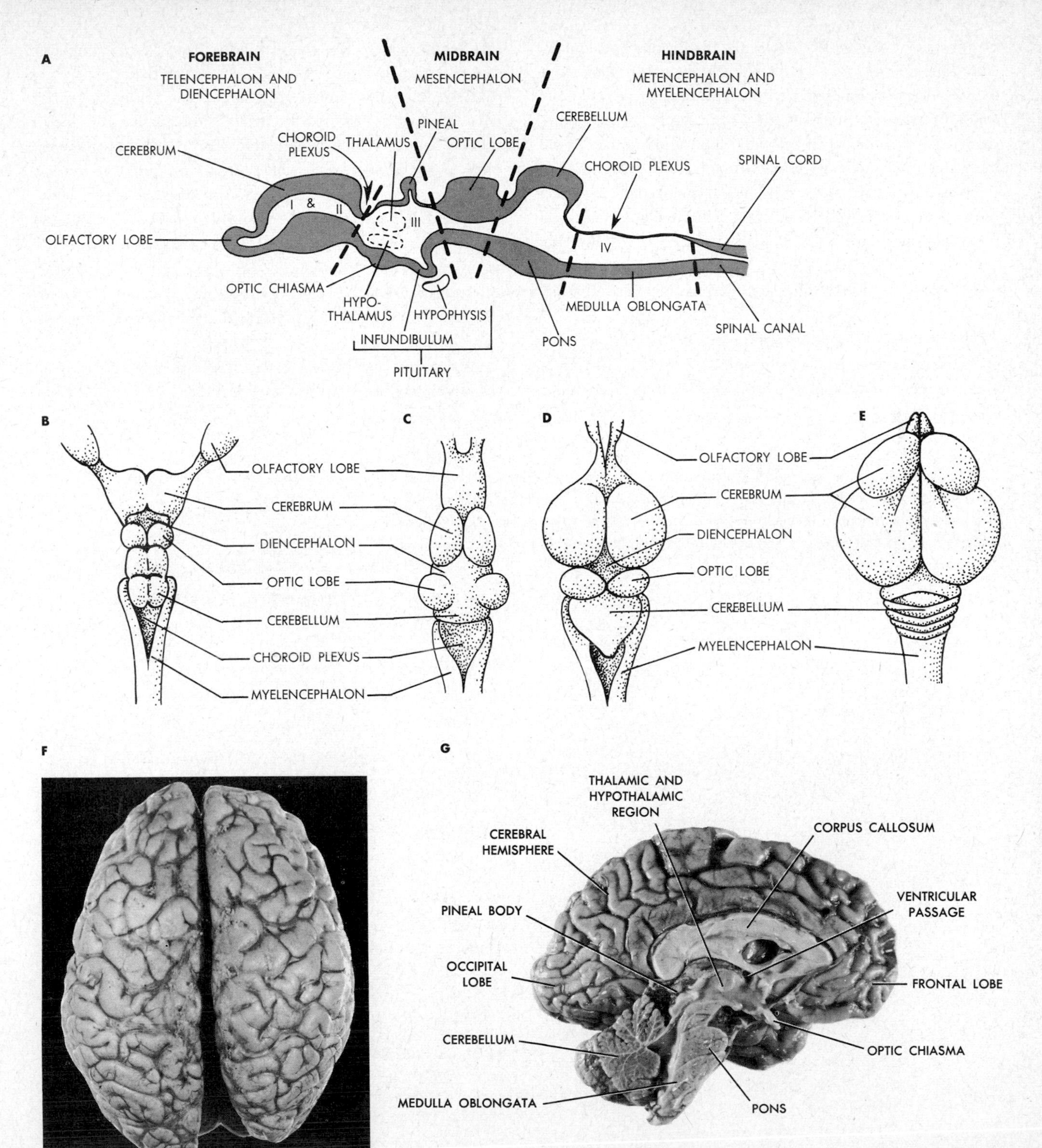

A
FOREBRAIN
TELENCEPHALON AND DIENCEPHALON
MIDBRAIN
MESENCEPHALON
HINDBRAIN
METENCEPHALON AND MYELENCEPHALON
CEREBRUM
CHOROID PLEXUS
THALAMUS
PINEAL
OPTIC LOBE
CEREBELLUM
CHOROID PLEXUS
SPINAL CORD
I & II
III
IV
OLFACTORY LOBE
OPTIC CHIASMA
HYPO-THALAMUS
HYPOPHYSIS
INFUNDIBULUM
PITUITARY
PONS
MEDULLA OBLONGATA
SPINAL CANAL
B
C
D
E
OLFACTORY LOBE
CEREBRUM
DIENCEPHALON
OPTIC LOBE
CEREBELLUM
CHOROID PLEXUS
MYELENCEPHALON
OLFACTORY LOBE
CEREBRUM
DIENCEPHALON
OPTIC LOBE
CEREBELLUM
MYELENCEPHALON
F
G
THALAMIC AND HYPOTHALAMIC REGION
CEREBRAL HEMISPHERE
CORPUS CALLOSUM
VENTRICULAR PASSAGE
PINEAL BODY
OCCIPITAL LOBE
FRONTAL LOBE
CEREBELLUM
OPTIC CHIASMA
MEDULLA OBLONGATA
PONS

oblongata (cf. Fig. 8.26). Dorsally, the metencephalon includes the *cerebellum*, a comparatively large lobe which coordinates all motor functions involving numerous muscles into smoothly integrated movements. For example, locomotion is regulated from this lobe. Ventrally, the metencephalon contains a conspicuous band of nerve tracts, the *pons*, in which the neural pathways between brain and spinal cord cross from the left side to the right side. Thus, the left side of the brain sends impulses to the right side of the spinal cord and thereby controls the activities on the right side of the body. Analogously, the right side of the brain controls the left side of the body.

The medulla oblongata is continuous posteriorly with the spinal cord. In the medulla are located, for example, the modulator centers controlling heartbeat and breathing. Ten pairs of *cranial nerves* (12 pairs in reptiles, birds, and mammals) emerge from the entire brain, and most of these lead away from the medulla oblongata (Fig. 8.27). The nerves are partly motor, partly sensory, partly mixed. Most fibers within these nerves control voluntary activities. The spinal cord connects with paired *spinal nerves*, which lead to the trunk and the appendages. Controlling voluntary activities in these body regions (e.g., skeletal muscles), the nerve pairs are arranged segmentally, and they are all mixed. In each such spinal nerve, sensory fibers enter the dorsal part of the spinal cord; motor fibers leave from the ventral part of the cord. The cytons of the sensory fibers lie just outside the spinal cord, in *spinal ganglia* (Fig. 8.28).

Vertebrates also possess a very well-developed *autonomic* subdivision of the nervous system. Controlling the involuntary

8.27. Diagram of the underside of brain and anterior part of spinal cord of man, showing the origin of the 12 pairs of cranial nerves and of a few of the spinal nerves. The names and functions of these nerves are given in the accompanying tabulation. Note that, in the spinal nerves, the dorsal roots (with ganglia) are sensory, the ventral roots, motor.

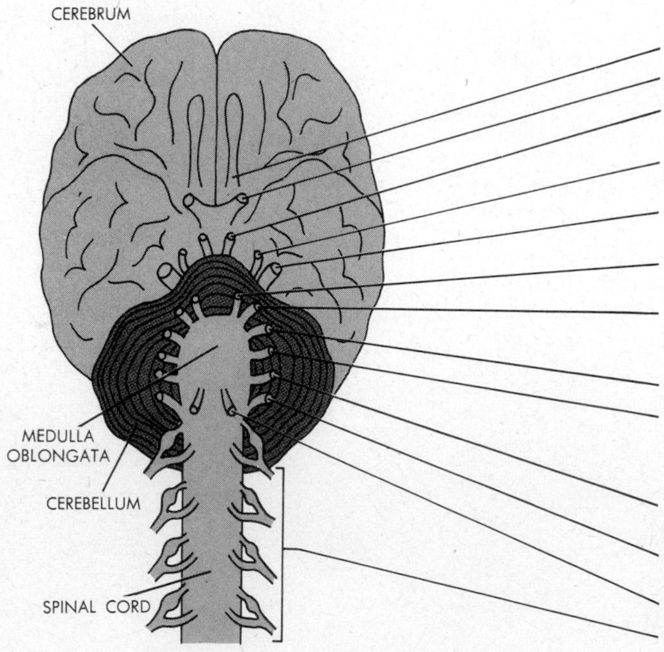

CEREBRUM
MEDULLA OBLONGATA
CEREBELLUM
SPINAL CORD

CRANIAL AND SPINAL NERVES IN MAMMALS

NAME	TYPE	INNERVATION
1. OLFACTORY	SENSORY	FROM NOSE
2. OPTIC	SENSORY	FROM EYE
3. OCULOMOTOR	MOTOR	TO MUSCLES OF EYEBALL
4. TROCHLEAR	MOTOR	TO MUSCLES OF EYEBALL
5. TRIGEMINAL	MIXED	FROM AND TO FACE, TEETH
6. ABDUCENS	MOTOR	TO MUSCLES OF EYEBALL
7. FACIAL	MIXED	FROM TASTE BUDS TO SALIVARY GLANDS AND FACIAL MUSCLES
8. AUDITORY	SENSORY	FROM EAR
9. GLOSSOPHARYNGEAL	MIXED	FROM AND TO PHARYNX, FROM TASTE BUDS TO SALIVARY GLANDS
10. VAGUS	MIXED	FROM AND TO CHEST AND ABDOMEN
11. SPINAL ACCESSORY	MOTOR	TO SHOULDER MUSCLES
12. HYPOGLOSSAL	MOTOR	TO TONGUE
SPINAL NERVES (31 PAIRS)	MIXED	FROM AND TO MUSCLES IN ARMS, LEGS, AND TRUNK

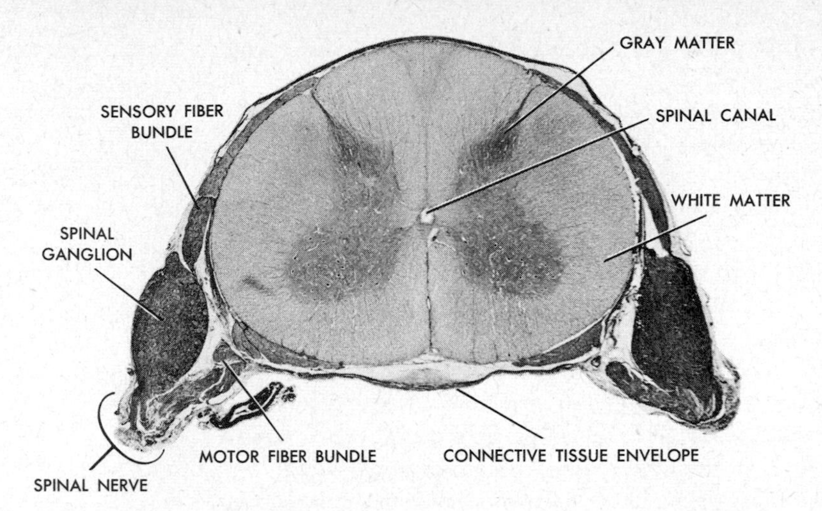

8.28. Cross section through mammalian spinal cord. Note the spinal nerves, each dividing into two fiber bundles. The motor bundle connects with the cord ventrally, and the sensory bundle passes through a spinal ganglion and connects with the cord dorsally. The spinal cord itself is a dense meshwork of neurons, the cytons of which are aggregated around the center and form so-called gray matter. The axons and dendrites of these neurons are collected around the gray matter, forming white matter. The central spinal canal contains lymphlike spinal fluid.

activities of smooth muscles, glands, and the heart, this autonomic system has its nerve centers located in the spinal cord and also partly in the brain stem, alongside the centers for voluntary functions. Two functionally different sets of *autonomic motor fibers* lead away from these centers (Fig. 8.29). Fibers from the brain stem and the most posterior part of the spinal cord constitute a so-called *parasympathetic* outflow of the autonomic system. Such fibers pass to all involuntarily controlled organs of the body, and they either stimulate (accelerate) or inhibit (decelerate) the functioning of such organs. Some of the parasympathetic fibers travel along with the fibers of certain cranial nerves; e.g., parasympathetic fibers from the medulla oblongata to the heart travel within the vagus nerve, the 10th cranial. From the middle portion of the spinal cord emanate motor fibers constituting a *sympathetic* outflow of the autonomic system. Sympathetic fibers likewise lead to all involuntarily controlled organs, and their effects are generally opposite to those of parasympathetic fibers. Thus, if parasympathetic fibers inhibit the functioning of a given organ, sympathetic fibers accelerate that organ; or vice versa. In effect, therefore, each organ functioning involuntarily is innervated by accelerating as well as by braking controls. On their way to particular organs, sympathetic fibers pass through *autonomic ganglia* located just outside the spinal cord (cf. Fig. 8.29). These ganglia are segmentally arranged, and they are interconnected to form *autonomic ganglion chains,* one on each side of the spinal cord. The anatomical relation between a spinal ganglion and an autonomic ganglion is shown in Fig. 8.30.

Apart from their particular anatomies, how do nervous systems work?

Neural Impulses

The precise nature of a nerve impulse is still unknown. We may say, in general, that an impulse is a sequence of reactions propagated along a nerve fiber. After an impulse has passed, the reaction balance returns to the original state, readying the fiber for a new impulse. These processes consume oxygen and energy.

Accompanying the chemical changes are electrical phenomena. Indeed, the intriguing resemblance of the nervous system to a meshwork of electrical wires conducting electric currents has been the basis of many attempts to explain nervous activity. Moreover, just as one can measure currents in wires by galvanometers, voltmeters, ammeters, and the like, so this same electrical equipment can be used on nerves. But nerve impulses are not simply electrical impulses. The latter travel some 100,000 miles per sec in a wire, the former at most about 100 yd per sec in a nerve fiber. Nerve impulses are neither purely electrical nor purely chemical, and at present they may best be described as *electrochemical* events.

Before a nerve fiber can transmit an impulse, it must be stimulated adequately. Whereas any environmental change may represent a stimulus, not every such change represents an adequate, i.e., effective, stimulus. To be effective, a stimulus must be at least of minimum strength, or *threshold intensity;* it must reach this threshold intensity fast enough, at an appropriately high *rate of change;* and it must last long enough and thus have appropriate *duration.* When a neuron is stimulated adequately, it will "fire," i.e., transmit an impulse from the point of stimulation over its fibers. Under normal conditions within the body, stimulation occurs at a dendrite terminal and an impulse then travels through or past the cell body to an axon terminal.

Whatever else an impulse may be, it is known that it is a *wave of electrical depolarization* sweeping along a nerve fiber. It can be shown that a resting, nonstimulated neuron is electrically positive along the outer side of its surface membrane and electrically negative along the inner side (Fig. 8.31). These electric charges are carried by mineral ions attached to the two sides of the neuron membrane. The outer, positive charges are due primarily to ions such as Na^+, the inner, negative ones, to ions such as Cl^-. As a result, an *electrical potential* is maintained across the cell membrane, and the membrane is said to be *polarized* electrically. In some ways this resembles the polarization believed to be maintained in the actomyosin units of muscles.

Polarization, as well as the integrity of a neuron membrane, appears to depend on *semipermeability.* The membrane is so constructed that it prevents the positive and negative ions from coming together. If the permeability state were altered, the membrane would depolarize; i.e., the positive and negative ions would join. Conversely, if depolarization were to occur, membrane semipermeability would be abolished. When a nerve impulse sweeps along a nerve fiber, local depolarization and simultaneous changes of permeability actually do occur at successive points of the fiber membrane. As this happens at any one point, an avenue is created through which positive and negative ions of an adjacent point may meet (Fig. 8.32). In other words, the impulse itself produces the necessary conditions which allow it to advance farther. In this manner, it travels wavelike along a fiber. Some short time after an impulse has passed a given point, the membrane at that point recovers; both the polarization and the original permeability state are restored.

These electrical aspects of transmission may be demon-

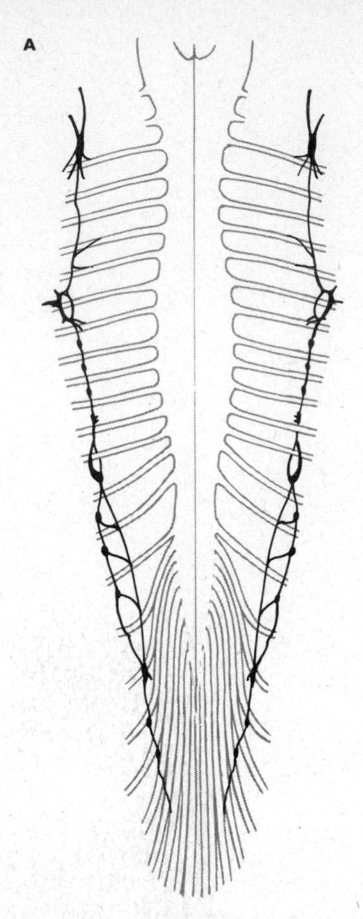

8.29. **A.** Sketch of mammalian spinal cord and autonomic chains to each side.

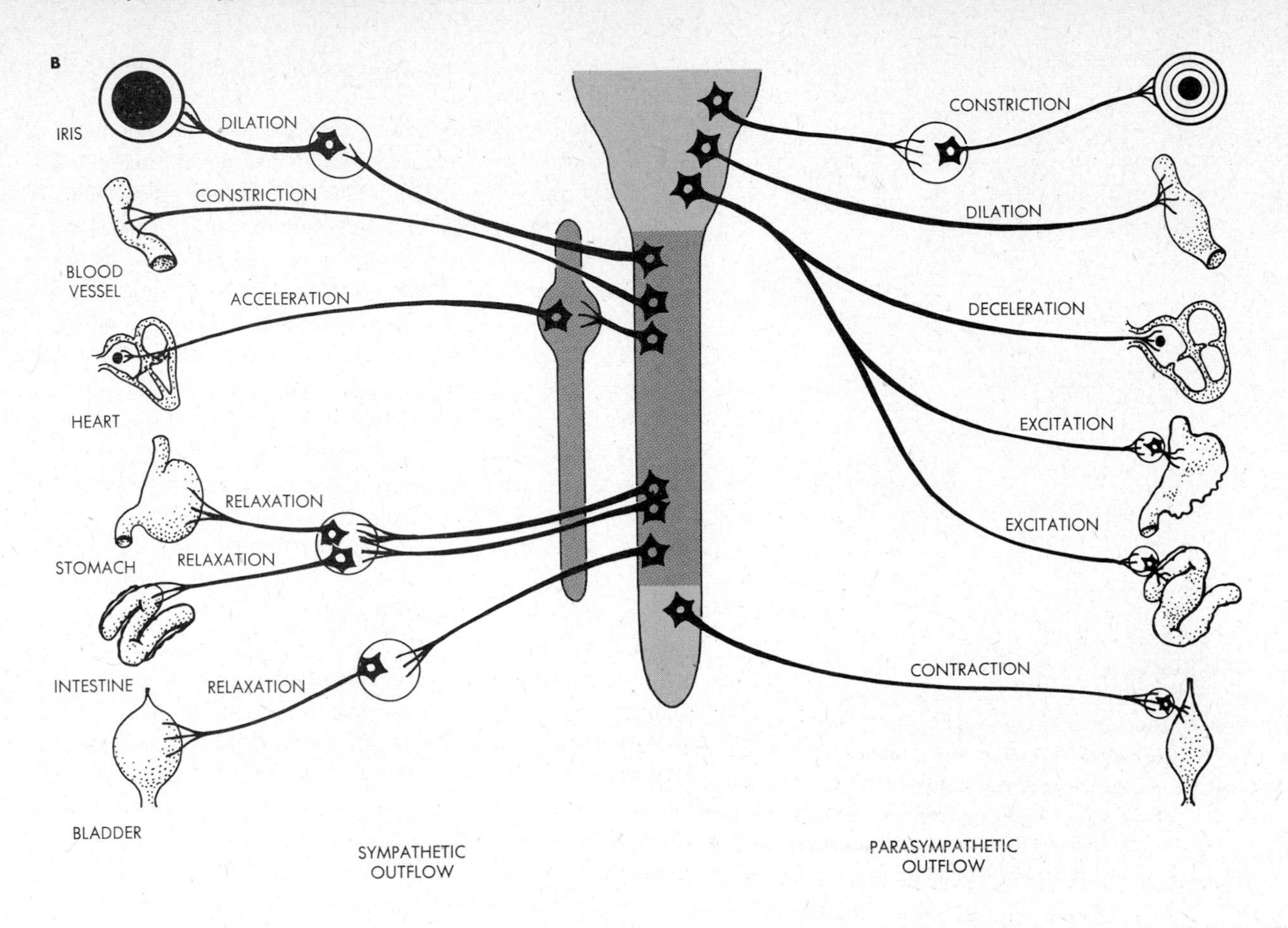

8.29. **B.** Schematic representation of some of the motor pathways of the autonomic nervous system. In spinal cord, parasympathetic centers shown light gray, sympathetic centers, dark gray. The column left of the spinal cord represents the sympathetic chain on that side. Each neural path shown occurs pairwise, one on the left and one on the right of the body. For simplicity, only one side is indicated in each case. Sympathetic motor paths are drawn on left, parasympathetic paths, on right. Note that each organ is innervated by both portions of the autonomic nervous system.

strated experimentally. For example, it is possible to expose a nerve of a test animal and to connect a galvanometer to this nerve with fine wires (Fig. 8.33). When the nerve is then stimulated, the needle of the galvanometer is deflected in one direction as the impulse passes the first wire contact and in the opposite direction as it passes the second. This indicates that electrical changes occur during impulse transmission. Moreover, the test reveals that the particular portion of the fiber which carries the impulse at any given moment is electrically less positive on the surface than the remainder of the fiber. This is what should be expected if an impulse were to cause local depolarization, i.e., reduction or removal of the outer positive charge. The current flow accompanying impulse transmission is called the action current, or the *action potential,* of a nerve fiber.

By measuring the action potentials of different nerves, one may determine many of the characteristics of the impulses these nerves carry, e.g., speeds, frequencies, and strengths. It has been found, for example, that impulse speeds tend to be directly proportional to the thickness of a nerve fiber. Moreover, speeds are also influenced by the presence or absence of so-called *myelin* sheaths. Myelin is a secreted fatty layer surrounding some nerve fibers, notably those which participate in the control of voluntary activities (cf. Fig. 8.23). In such myelinated fibers, impulse speeds can be up to 4 times as great as in nonmyelinated fibers.

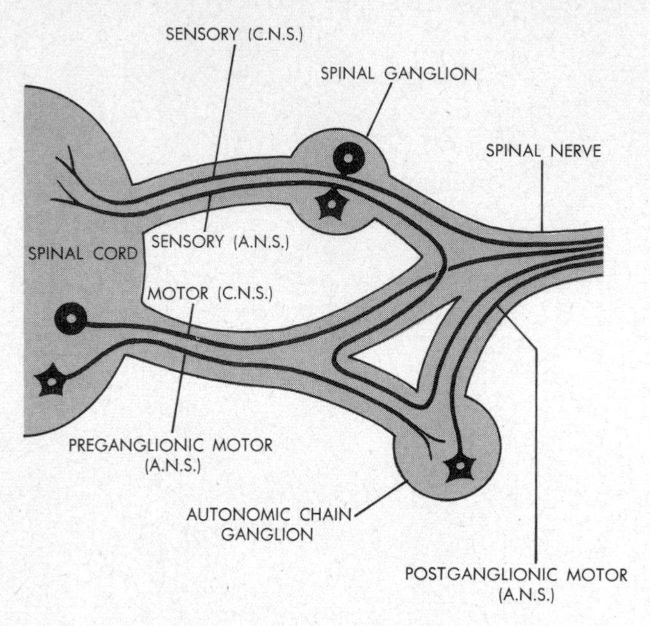

8.30. The anatomical interrelations of spinal C.N.S. and sympathetic A.N.S. fibers. C.N.S. neurons are shown with round cytons, A.N.S. neurons, with starshaped cytons. Note that a given fiber bundle or nerve connecting with the spinal cord contains sensory fibers of both C.N.S. and A.N.S., or motor fibers of both C.N.S. and A.N.S., or both sensory and motor fibers of both C.N.S. and A.N.S. Sensory fibers of all kinds always enter the spinal cord dorsally, motor fibers of all kinds always leave ventrally.

How does an impulse jump across the gap of a synapse? In certain cases it can be shown that, when an impulse reaches an axon terminal, the terminal acts like a miniature endocrine gland; it secretes minute amounts of a hormone. This hormone diffuses through the synaptic gap; some of it eventually reaches dendrite terminals of adjacent neurons, and the hormone there affects a dendrite in such a way that a new impulse is initiated in it (Fig. 8.34).

In vertebrates, four hormonal substances functioning in this manner have been identified. Called *neurohumors* collectively, they are *serotonin, acetylcholine, adrenaline,* and *noradrenaline,* the last two rather similar chemically. Serotonin or acetylcholine is secreted by some of the fiber terminals participating in the control of involuntary activities, and probably also by the terminals involved in the regulation of voluntary functions. Adrenaline or noradrenaline is produced by many of the fibers which innervate muscles and glands not under voluntary control. Analogous hormones probably occur among invertebrates.

Synaptic impulse transmission by chemicals has important consequences. For example, diffusion takes much longer than impulse conduction within a fiber. A complete reflex, which usually passes through many synapses, consequently lasts longer than would be expected on the basis of impulse speeds within fibers alone. Moreover, nerve fibers as such rarely fatigue but synapses get "tired" fairly easily. During intensive activity, axon terminals may temporarily exhaust their hormone-secreting capacity, and synaptic transmission then slows even more or stops altogether for the time being.

Note also that the synaptic hormones impose a one-way direction on neural pathways. A nerve fiber can be stimulated at either end or in the middle, and impulses then travel backward, forward, or in both directions. But only axon terminals are specialized to secrete hormones, and only dendrite terminals are sensitive to these hormones. As a result, impulse conduction is unidirectional.

The first nerve impulse in a reflex arc is generally produced by a receptor; a separate section below will show how such receptor structures operate. The last impulse in a reflex normally travels over a motor fiber to an effector, viz., a muscle or a gland. In the case of vertebrate muscle, the nerve fiber terminal branches into a *motor end-plate* (cf. Fig. 8.19). As noted earlier in this chapter, the muscle elements innervated by one motor end plate constitute a motor unit. Like neuron-neuron interconnections, nerve-muscle junctions analogously have synaptic gaps. These are bridged functionally as above, by neurohumors released from the motor end-plates. Nerve-gland junctions are organized similarly.

Apart from the first and last impulses in a reflex, most others are generated, transmitted, and delivered within the modu-

8.31. The polarization of an inactive resting neuron. The positive charges on the outside and the negative charges on the inside produce an electric potential across the cell membrane.

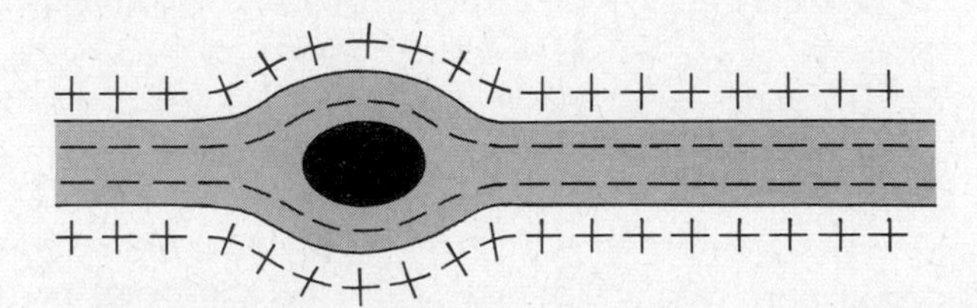

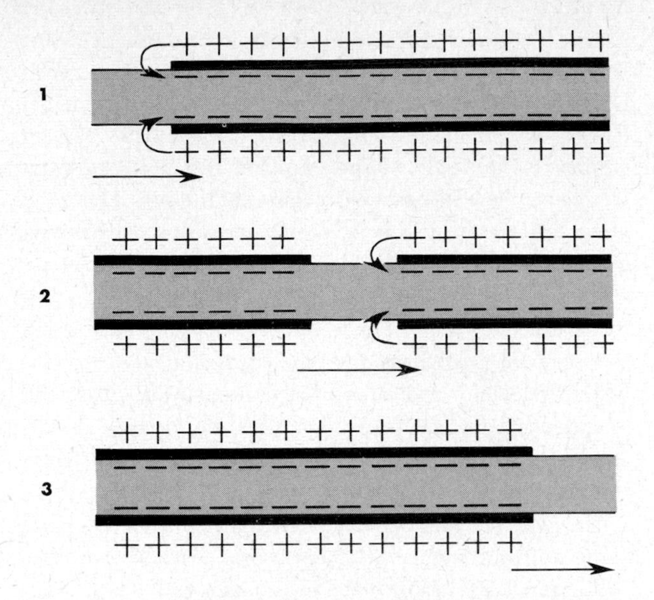

8.32. The passage of an impulse through a nerve fiber produces a local depolarization of the fiber membrane, propagated in wavelike manner through successive portions of the fiber. After an impulse has passed a given region, the original polarization reappears.

lators. How do these most critical components of nervous systems operate?

Neural Centers

The most basic type of modulator activity is probably a *reflex relay:* an impulse arriving in a ganglion or a brain over an incoming neural path is simply relayed via one or more synapses to an outgoing path (Fig. 8.35). However, even the least complexly organized modulators usually are capable also of *reflex modification.*

One important form of modification is *suppression* or *augmentation* of a reflex. Depending in part on the characteristics of the neurons involved and in part on the nature of the incoming impulses, outgoing impulses often may not be emitted at all, or may be emitted with greater intensities or frequencies. A suppression of impulses results from *inhibition* of given modulator neurons by others. Some neurons are so organized internally that, when their dendrites are stimulated by neurohumors, the production of impulses in these neurons becomes harder rather than easier. Inhibitory processes of this sort so provide "brakes" to impulse transmission. The opposite effect, augmentation, may result from a *summation* of impulses. In a synapse receiving many incoming impulses over numerous axon fibers, each arriving impulse may be individually too weak to produce an impulse in an outgoing fiber. However, the small quantities of neurohumors produced by the many incoming fibers may add together and may then become sufficiently powerful to initiate a strong impulse in an outgoing fiber. By virtue of its synapses, evidently, a modulator may carry out some degree of "interpretation" of the incoming signals and may exercise some measure of "decision" with regard to the outgoing signals (cf. Fig. 8.35).

Another, more complex form of modulator activity, based largely on summation and inhibition, is *channel selection.* Even a simply organized modulator contains the terminals of a large number of incoming neurons and the starting points of a large set of outgoing neurons. If impulses arrive via the first set, outgoing signals are not usually emitted over all outgoing paths present. Instead, again depending on the nature of the neurons and on the characteristics of the incoming sig-

8.33. The action potential of a nerve impulse. Left, resting fiber, connected to galvanometer by wire contacts. **Middle,** impulse passes first wire contact, which becomes electronegative relative to second contact. Hence current flows through galvanometer from right to left, as indicated by deflected galvanometer needle. **Right,** impulse passes second wire contact. Current now flows in opposite direction, the second contact being electronegative relative to the first. Current flow accompanying an impulse represents the action potential of that impulse.

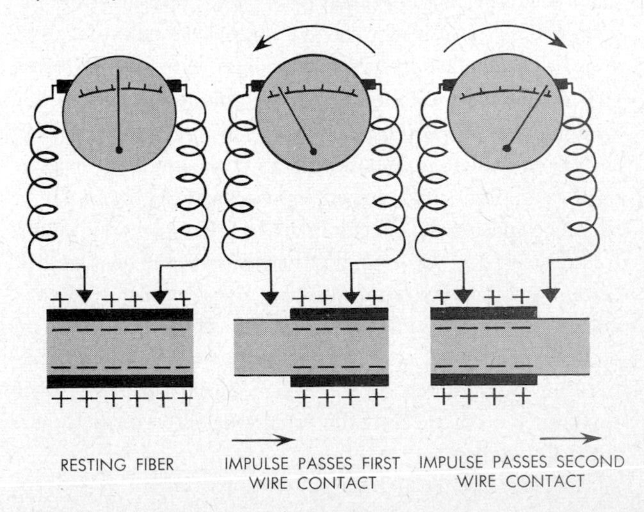

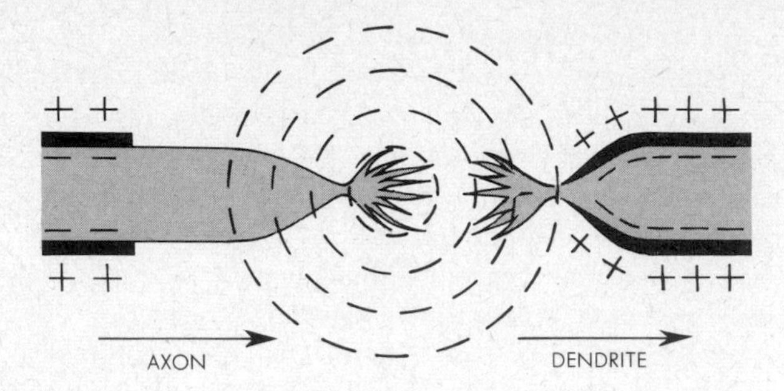

8.34. Diagram of a neural synapse, showing the release and local spreading of hormones from the axon terminal of one fiber to the dendrite terminal of another. Impulses are transmitted across synapses by such chemical means.

nals, modulator activity results in the *selection* among the hundreds or thousands of possible outgoing paths and the emission of signals only over some specifically chosen paths. The chosen paths are usually such that only *appropriate* effectors will receive motor commands. As a result, the effector response of an animal can be adaptively useful and can actually contribute to steady-state maintenance (cf. Fig. 8.35).

All modulator activity, no matter how complex, appears to be based ultimately on modification of impulses and on pathway selection. To be sure, little is known as yet about the internal mechanism by which some neural channels are selected in preference to others. In most cases, preferred circuits become established during the embryonic development of the nervous system. At any time thereafter, given sets of sensory impulses to a modulator result in more or less fixed, predictable sets of motor impulses to effectors. Such inherited circuit patterns govern most of the internal operations of most animals. In these operations, modulator activities are often considerably more complex than simple reflex relays. For example, sets of interneurons may be arranged as *oscillating circuits,* in which an impulse travels continuously over a circular route. Each time such an impulse passes a given synapse, a motor impulse to an effector might be initiated (cf. Fig. 8.35). It can be shown, for example, that the rythmic heartbeat of certain arthropods is controlled by nine modulator neurons embedded directly in the heart muscles. The neurons form an oscillator circuit which generates periodic motor commands resulting in rhythmic contraction of the heart. Many other rhythmic, automatic activities are governed by oscillator circuits of this type.

Still more complex circuit patterns, also inherited, are represented by *stored programs,* responsible for *instinctive* overt behavior. A given external stimulus here leads to the completion of several or many simultaneous reflex responses, all occurring as a single, *integrated* pattern of activity. A good example is the startle response in man: an unexpected blow directed at the head occasions a closing of the eyes, a lowering of the head and assumption of a crouching stance, and a raising of hands to the front of the face. These several dozen separate reflexes occur simultaneously, as a unified "program" of activity. The neural circuits are geared together and may be said to contain the program as *stored information.* In most animals, invertebrates in particular, virtually all behavior is based on inherited, instinctive, stored programs of this type. Fixed, automatic locomotor responses to particular stimuli are often referred to as animal *tropisms.*

The total number of circuit patterns a modulator may store varies with the number of interneurons available and thus with overall modulator size. In animals with small brain ganglia or brains, most available neurons are required for the establishment of inherited stored programs, and relatively few are left with which new circuit patterns could later be formed on the basis of experience. Such animals consequently are largely (though not entirely) incapable of converting experience into *learning* and *memory.* In comparatively large-brained animals, the majority of interneurons present again form inborn circuits, yet substantial numbers may remain for

8.35. Diagrammatic representation of various types of modulator activities. 1, simple reflex relay; **2,** suppression of impulse conduction; **3,** augmentation of outgoing impulse by summation of weak incoming impulses; **4,** channel selection; **5,** information storage by continued impulse conduction in oscillator (cyclical) circuit.

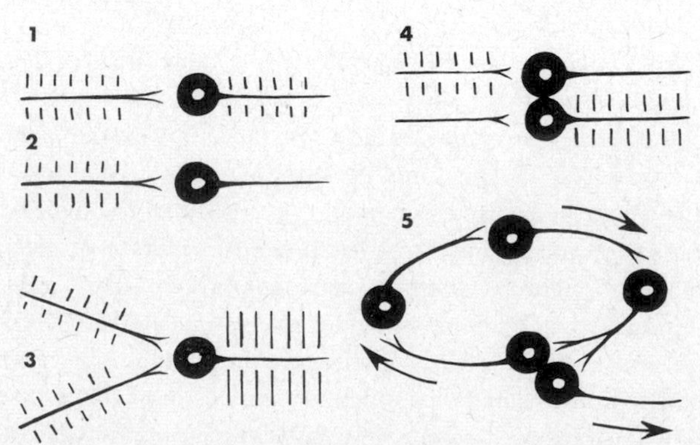

the later establishment of new circuits. Even so, learning and memory storage through experience are of more than incidental significance only in mammals.

It has been suggested that, where learning is possible, the establishment of new circuit patterns might occur through pathway *facilitation:* the more frequently impulses travel over a given neural circuit, the less resistance this circuit may offer to subsequent impulses. Wherever a choice of circuits exists, therefore, the often used, *facilitated* circuits may be selected in preference to the previously little used, unfacilitated ones. Conceivably, a circuit might become different in synaptic fine-structure and/or function the more often it is used. Learning might then occur by repetition and by trial and error. In a young animal, for example, few brain pathways are as yet firmly established by facilitation. Incoming impulses are transmitted more or less in all directions, and behavior is relatively uncoordinated and random. But among the random impulse paths, some will bring about advantageous effector results. The same pathway pattern may then be tried time and again, and a facilitated neural route may thus be established eventually. Training, habit formation, and memory accumulation are implicit in such learning by facilitation.

Facilitation may also account for learning by association, as in *conditioned reflexes.* Here two or more stimuli are presented to an animal simultaneously and repeatedly, until the animal learns to execute the same response to either stimulus. For example, if bright light is directed into the eyes of an animal, its pupils will contract by inherited reflex action. If the light stimulus is given repeatedly and is accompanied each time with the presentation of food, then pupillary contraction can eventually be initiated by food alone, without the stimulus of bright light. Evidently, the animal learns to associate food with light, and it has come to possess not only the inherited neural circuit to the pupillary muscles, but also an additional, facilitated circuit, acquired by learning through experience. Either one alone or both together may now produce the pupillary response. Conditioning of this sort plays a considerable role in the behavioral development of any animal capable of learning, vertebrate or invertebrate.

Undoubtedly the most complex modulator activities are those which produce consciousness, emotion, intelligence, personality, and ability to think abstractly. Such modulator functions are developed to any notable degree only in the most advanced mammalian brains. Regardless of the relative complexity of given modulator functions, however, the primary role of all modulators is to aid in steady-state maintenance by adjustment of muscular and glandular activity. And it should be clear that proper control over the effectors depends on adequate *information:* the modulators can only act on what the receptors tell them in the form of sensory impulses. The nature of animal receptors is examined in the following section.

Neural Receptors

Two basic varieties of receptor cells are known: *epitheliosensory* cells and *neurosensory* cells (Fig. 8.36). A receptor of the first type is a specialized, nonnervous epithelial cell which receives stimuli at one end and is innervated by a sensory nerve fiber at the other end. A receptor of the second type is a variously modified sensory neuron which at one end carries a dendritelike, usually short, stimulus-receiving extension and which at the other end carries an axon synapsing with other neurons. Thus, one type of stimulus receiver is part of the surface of a whole epithelial tissue cell, whereas the other type is essentially a free nerve ending. Both types of receptor cells may be aggregated into clusters and, together with accessory cells, form *sense organs.* All receptor cells of invertebrates are of the neurosensory type. Vertebrates possess both neurosensory and epitheliosensory receptors.

A light receptor is sensitive only to light and cannot be stimulated by sound waves, for example; all sensory structures are relatively *stimulus-specific.* If stimuli are to be effective, they must have certain threshold intensities, certain rates of change, and certain durations. Keep in mind also that sensory

8.36. The two kinds of receptor cells.

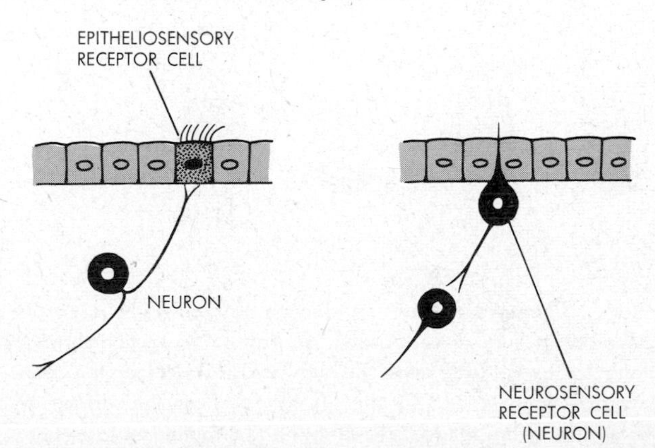

devices as such do not give an animal any perception of sensation. Receptors merely generate nerve impulses, and it is the modulator centers which must interpret the impulses as actual perceptions. In some instances such perceptions become conscious, but more often they do not, even in vertebrates. Thus, when nerve impulses reach the brain from a blood vessel, for example, sensing takes place, although without conscious awareness.

Receptors may be classified as *exteroceptors* and *interoceptors,* i.e., as structures receiving information either about the external or the internal environment of an animal. Receptors may also be classified according to the kinds of stimuli they are sensitive to, as in the following account.

Chemoreceptors. Some types of chemoreceptors give information about environmental *chemicals* generally; other types initiate sensations of *smell;* and still other types mediate the sense of *taste.* In most animals, the receptor structures are free sensory nerve endings which may be stimulated directly by given chemicals. In some cases the receptors are specialized ciliated *cells,* which receive chemical stimuli on the ciliated side and connect with a nerve fiber on the opposite side. An epithelial *tissue* of such cells forms the smell receptor of vertebrates. The taste buds of these animals are *organs* in which small clusters of ciliated receptor cells are embedded (Fig. 8.37).

The ability to sense common environmental chemicals has

A

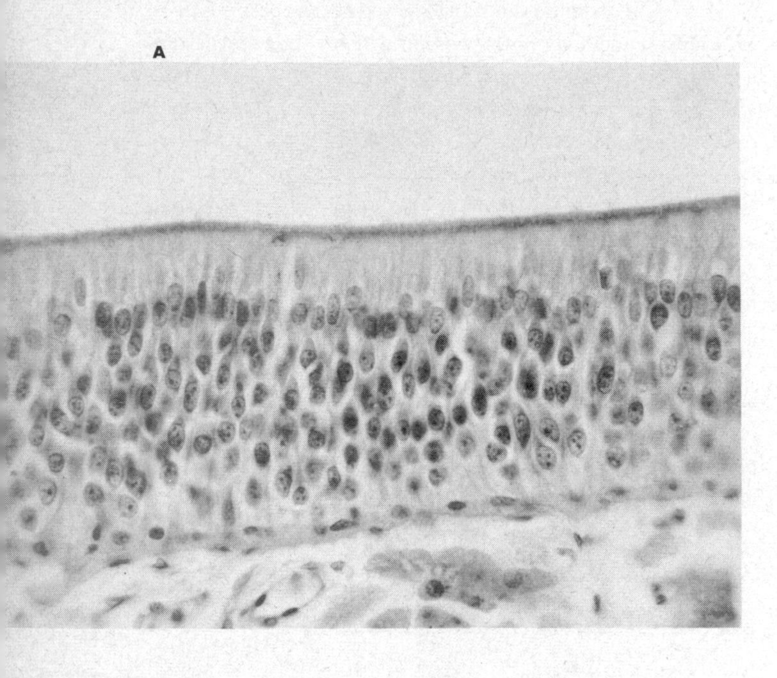

B

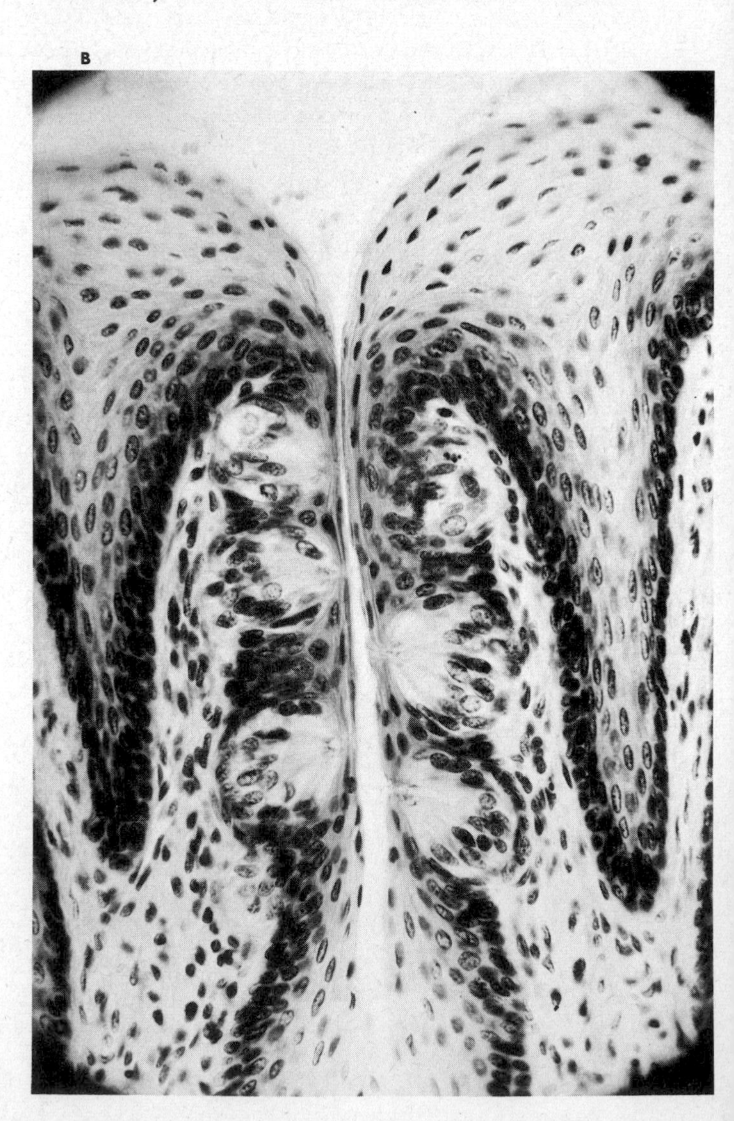

C

8.37. Chemoreceptors. **A.** Section through epitheliosensory nasal epithelium of vertebrates. **B.** Portion of section through tongue showing taste buds; the buds are located along the deep, narrow channels leading into the tongue from the surface. **C.** Diagrams of the tongue, showing the distribution of taste buds for the four taste sensations.

particular significance for many invertebrates, aquatic ones especially. This sense permits an animal to detect the presence of irritants or poisons and the chemical exudates of enemies, prey, food, and mates. The receptors usually are distributed abundantly over all parts of the integument, as might be expected, and they are believed to be distinct from those for smell and taste. These too are stimulated by direct contact with environmental chemicals. Smell is primarily a distance sense. Aquatic animals smell traces of chemicals in solution. Thus they may smell slight differences between different kinds of waters or the presence of plants, logs, and other objects. Terrestrial animals smell vaporized chemicals, of particular importance here being traces of chemicals which adhere to the ground; a worm or a four-footed mammal may receive as much information about its environment by sniffing the ground as an insect, a bird, or a man may receive by sight.

The sense of taste primarily conveys information about the general chemical nature of potential food substances. In most cases the receptors are localized in and around the mouth, but animals such as flies and moths taste with their legs; the receptors are at or near the distal ends of the legs. We may note here that tastes, and indeed chemical perceptions of all kinds, are subjective sensations, not objective properties of given chemicals. Sugar is not intrinsically sweet, nor is quinine intrinsically bitter. Sweetness, bitterness, and all similar sensory qualities are subjective interpretations made by the brain. This conclusion is clearly supported also by the observation that the same chemical which may be very pleasant to one animal may be distinctly less pleasant or actually repugnant to another.

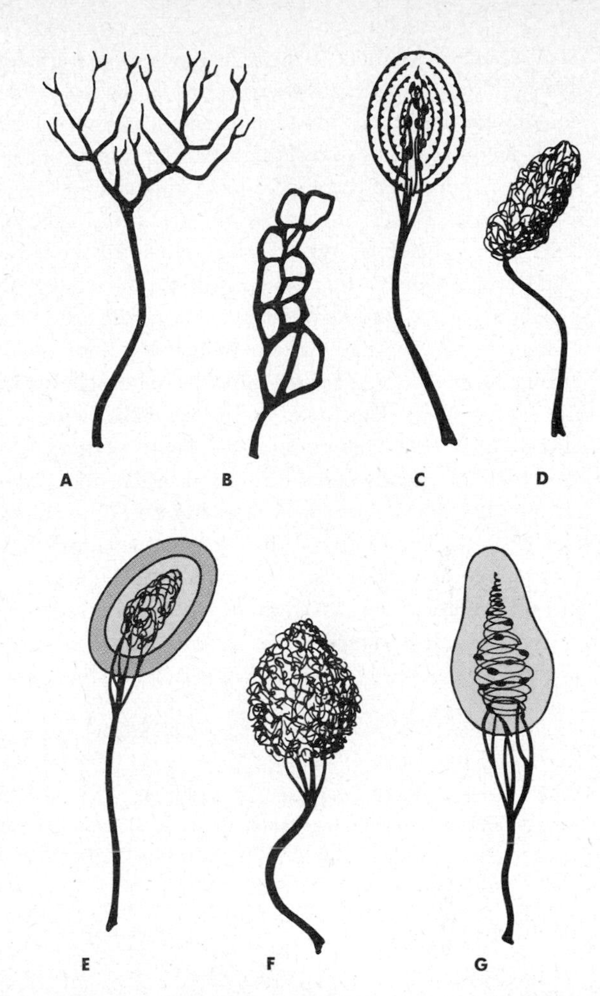

8.38. Some types of neural receptors in the mammalian skin. **A,** free nerve ending (pain); **B,** nerve net surrounding hair (touch); **C,** Pacinian corpuscle (pressure); **D,** organ of Ruffini (pressure); **E,** organ of Krause (cold); **F,** end organ of Ruffini (warmth); **G,** Meissner's corpuscle (touch). **B, C, D,** and **G** are tangoreceptors.

Tangoreceptors. These structures register stimuli of touch and of mechanical pressure generally. The receptors usually are free nerve endings, highly branched or elaborated into meshworks in many cases. Tangoreceptors occur abundantly in most animals, both in the integument and in the body interior (e.g., in muscles, tendons, and most connective tissues). The skin of vertebrates additionally contains minute touch-registering organs composed of nonnervous cells. Such organs connect to sensory nerve fibers (Fig. 8.38).

Tangoreceptors are stimulated by mechanical displacement or by changes in the mechanical stresses affecting surrounding parts. For example, the bending or the stretching of a part of the skin or of an internal organ is likely to result in receptor stimulation. An animal thereby receives information both about contacts with external objects and about movements of any body part. As might be expected, integumentary tangoreceptors are particularly abundant where contact with external objects occurs frequently—in and around the mouth, on appendages, and on the ventral body surfaces in creeping forms. In mammals, the base of each hair within the skin is surrounded by a meshlike terminal of a tangoreceptor fiber, and a nerve impulse is initiated if a hair is touched even lightly (cf. Fig. 8.38). Internal tangoreceptors are sometimes referred to as *stretch receptors* or *proprioceptors*. According to the impulses they transmit, the brain may initiate changes in the position and movement of body parts. Such receptors thus play an important role in maintaining animal posture and balance.

Statoreceptors. Present in many invertebrate groups and in all vertebrates, these organs are receptors for the sense of body equilibrium (Fig. 8.39). Most commonly, a statoreceptor is a small, fluid-filled sac, or *statocyst,* in which is present a cluster of ciliated *hair cells.* Attached to or resting against the hairs is a *statolith,* a grain of hard, usually calcareous material (a sand grain in crayfish). When an animal moves, the statolith shifts position under the influence of gravity, and it then presses against a somewhat different set of hair cells. Such a change in the pressure pattern produces a corresponding change in the pattern of nerve impulses traveling away from the sensory cells. The brain so receives information about altered body orientations. As noted, body position is also sensed independently by information received from tangoreceptors; but the statoreceptors must function if an animal is to carry out righting activities after being turned upside down, for example, and in general if an animal is to maintain a normal orientation in relation to gravity.

In most animals, statoreceptors are located in the head or in head appendages (e.g., antennae). In vertebrates they are part of the inner ear. Indeed, the vertebrate ear has evolved primarily as a statoreceptor and only secondarily as an organ of hearing. Hair cells with statoliths (here often called *otoliths* or *ear stones*) are located at several places along the walls of two inner-ear chambers, the *saccule* and the *utricle* (Fig. 8.40). These receptors give information primarily about position of the head; they may therefore be regarded as organs for static body balance.

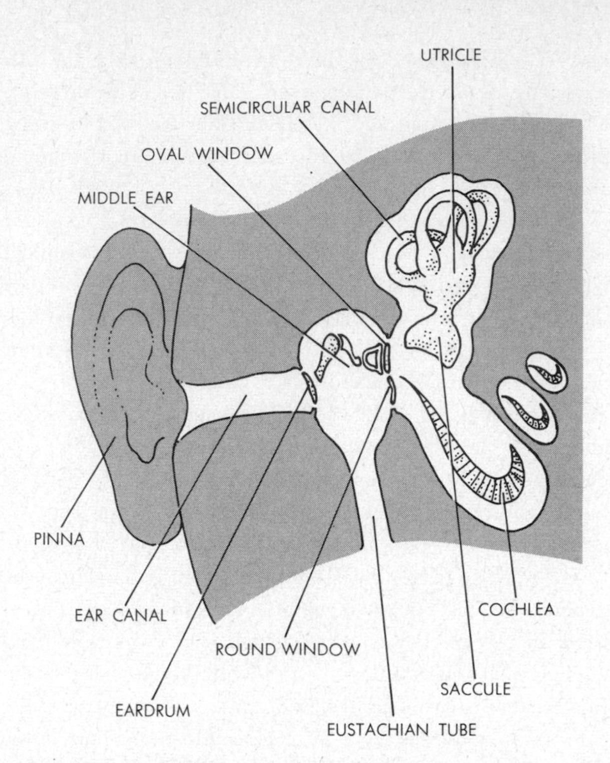

8.40. The gross structure of the ear of mammals. Note ear bones in the middle-ear cavity and attachment of semicircular canals to utricle. Statoreceptors are located in the utricle and the saccule.

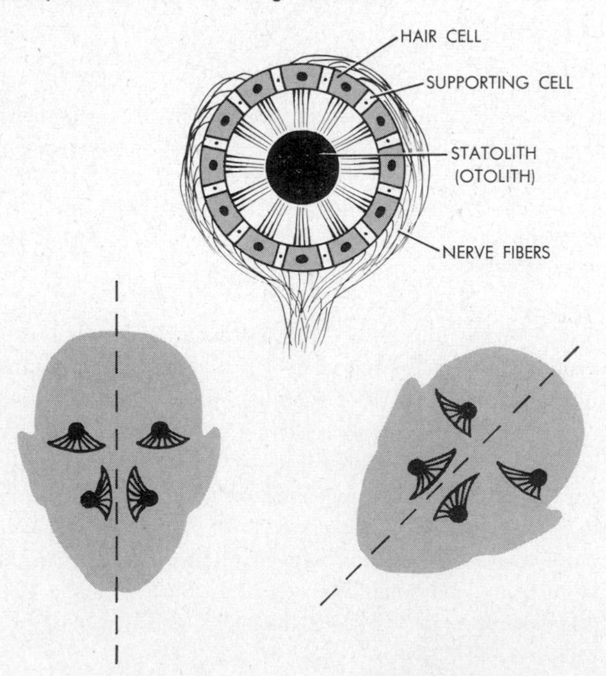

8.39. Upper figure, diagram of statocyst. The hair cells support an ear stone in the center of the cyst. (*After Hyman.*) Lower figures, position of the receptor organs in relation to the head, and the effect of tilting the head.

Present in addition in vertebrates are separate receptors, the *semicircular canals,* which inform about movement of the head and thus about the dynamic balance of the body. Three semicircular canals in each ear loop from the utricle back to the utricle. The canals are placed at right angles to one another in the three planes of space (Figs. 8.40 and 8.41). At one end of each canal is an enlarged portion, the *ampulla,* in which is found a cluster of hair cells. When the head is moved, the semicircular canals move with the head. But the fluid in the canals "stays behind" temporarily as a result of its inertia and "catches up" with the head only after the head has stopped moving. This delayed fluid motion bends the hairs of the receptor cells and produces nerve impulses. Different impulse patterns are transmitted to the brain according to the direction and intensity of fluid motion in the three pairs of canals.

Phonoreceptors. These organs are sensitive to pressure vibrations in displaceable media, e.g., water and air. Organs of hearing are particular kinds of phonoreceptors. Free nerve endings and hair cells in the integument function most commonly as the receptors for nonauditory vibrations. Many animals probably register pressure vibrations also by means of their tangoreceptors. Indeed, even where specialized phonoreceptors are absent, as in the skin of man, vibrations may still be felt. In such cases, direct conduction of pressure waves through skin, muscle, and bone probably leads to stimulation of tangoreceptors in various body parts. Most aquatic vertebrates possess highly developed, specialized phonoreceptors in the integument of the head and the trunk. Present there are so-called *lateral-line systems,* essentially series of water-filled canals which communicate via pores with the external medium (Fig. 8.42). Movement of the water in the canals initiates impulses in fiber terminals of nerves which travel alongside the canals. The lateral-line system probably registers vibratory currents produced both by moving objects in the water and by the movement of the fish itself. The system so

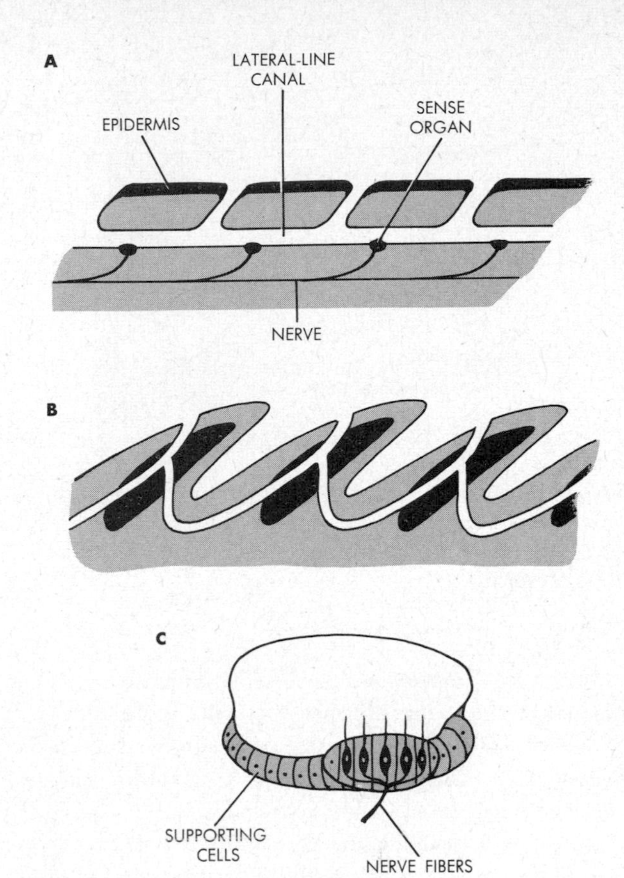

8.42. The lateral-line system. **A.** General plan of the system in relation to body surface. **B.** Relation of the system to the dermal scales of a bony fish; the lateral-line canal partly passes through the scales. **C.** Schematic cross section through a lateral-line canal, showing supporting cells and sense organ, the latter with phonoreceptive hair cells embedded among supporting cells. (After Goodrich.)

8.41. The semicircular canals of the left ear. Top of diagram is anterior, right side is toward median plane of head. The three canals are set at right angles to one another, hence only the horizontal canal reveals its curvature in this view. Both ends of each canal open into the utricle. The hair cells in the ampullae function as receptors for the sense of dynamic balance. When head is moved, fluid in the canals bends the hair cells, an action which initiates sensory nerve impulses.

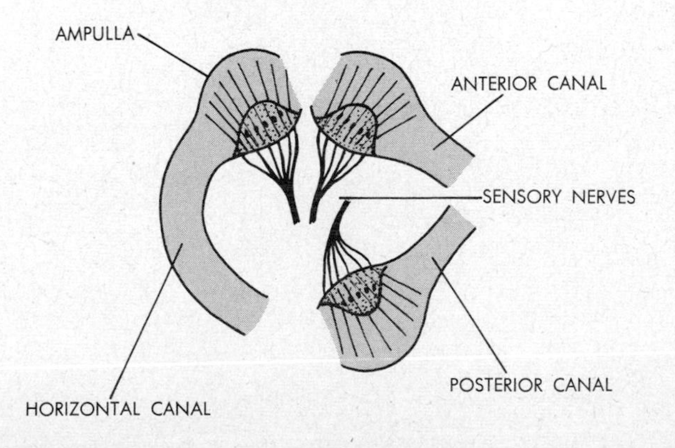

functions like the "listening" component of submarine sonar, and it probably serves not only to detect water turbulence as such but also to discriminate between different kinds of turbulence.

A distinct sense of hearing is restricted to some arthropods (certain crustacea, spiders, and insects) and to vertebrates; by and large, only animals that make sounds can also hear them. In arthropods, the receptor organs (Fig. 8.43) are located in various body regions, e.g., in the antennae (mosquitos), forelegs (crickets), thorax (cicadas), or abdomen (grasshoppers). In general, such "ears" of arthropods are sensitive to rather limited sound frequencies only. For example, a male mosquito

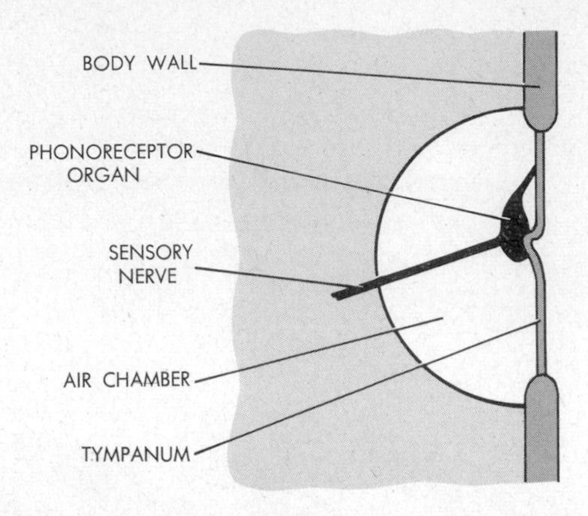

8.43. Diagram of an arthropod ear. An organ such as shown here occurs, for example, on grasshopper legs. The typanum corresponds functionally to the eardrum of mammals. (*After Wever.*)

can hear only sounds having the same frequency as those produced by the wings of female mosquitos in flight.

Among vertebrates, many fishes possess sound receptors in the inner ear (Fig. 8.44). In addition to the utricle and saccule, the inner ear of such fishes contains a third chamber equipped with a statolith, the *lagena.* This chamber connects with the swim bladder either directly or via a chain of small bones (*Weberian ossicles*). External sound waves produce resonating vibrations in the air within the swim bladder, and these vibrations are transmitted into the lagena and its statolith. Impulses are thereby initiated in auditory nerve fibers leading away from the hair cells.

Terrestrial vertebrate groups possess middle-ear cavities. These are closed off from the external environment by membranous *eardrums* and from the inner-ear cavities by other membranes. Middle-ear bones (one in amphibia, three in reptiles, birds, and mammals) transmit sound vibrations from the eardrum to the inner ear (cf. Fig. 8.40). A lagena is again the sound receptor in amphibia; but, in reptiles, birds, and mammals the lagena is developed as a coiled, highly elaborate receptor apparatus, the *cochlea* (Fig. 8.45). It contains a coiled *basilar* membrane with innervated clusters of hair cells, each cluster along the membrane being sensitive to a particular vibration frequency. External sound waves transmitted via the middle ear into the fluid of the cochlea produce fluid vibrations of given frequencies, and these set into vibratory motion the hair cells sensitive to such frequencies. The vibrating hair cells thus come into contact with an overhanging *tectorial membrane,* and these contacts initiate nerve impulses. Different patterns of impulses are interpreted in the brain as sounds of different pitch.

Mammals probably have the best sense of hearing. Their ears discriminate sounds of even slightly different intensity and pitch, and they are sensitive to a very wide range of sounds. For example, man may hear sounds ranging in frequency from 16 to 20,000 cycles per sec. Dogs are sensitive to frequencies up to 30,000 cycles per sec, and bats, to frequencies up to 100,000 cycles per sec. Sound reception in

8.44. Diagram of the left ear of a fish, showing the lagena and indicating the relation of the two ears to the Weberian ossicles and the swim bladder (the latter is in actuality proportionately much larger than drawn here). Note that utricle, saccule, and lagena contain ear-stone receptors and that the ampullae of the semicircular canals analogously contain hair-cell mechanisms (not shown). The mammalian cochlea has evolved as an extension of the lagena. (*After Wever.*)

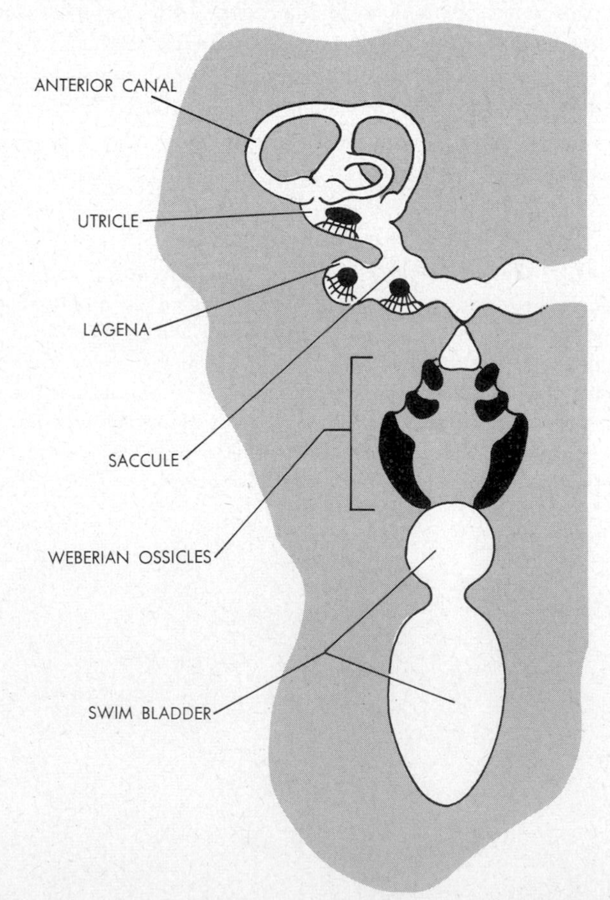

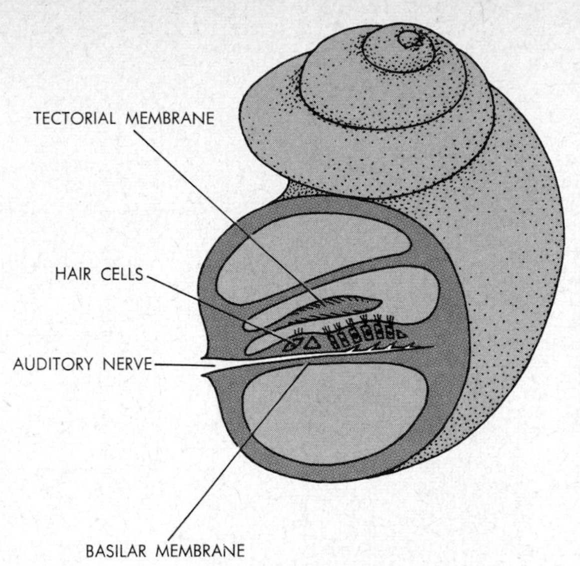

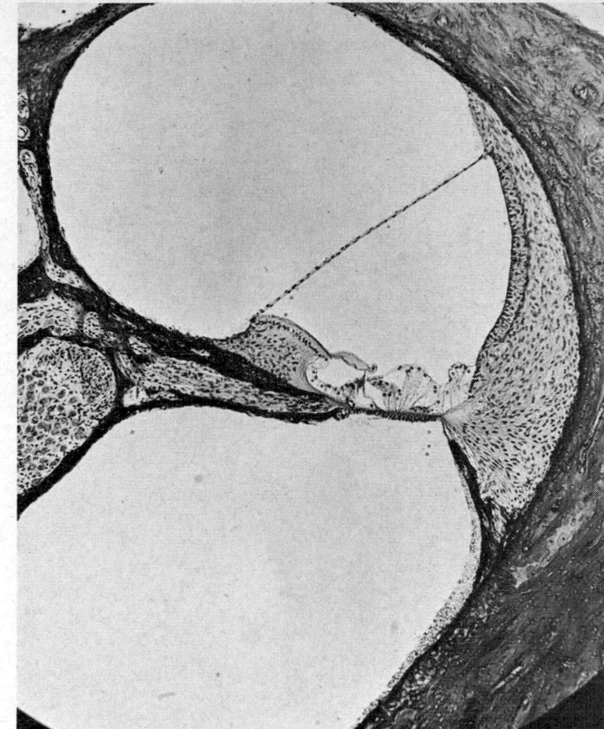

8.45. The cochlea. The diagram shows the coils of the cochlea and a cochlear cross section with the parts of the organ of Corti. A section through this organ is illustrated also in the photo. The tectorial and basilar membranes and the hair cells are clearly visible.

mammals is aided also by the presence of *outer ears,* composed of sound-collecting skin appendages and by external *ear canals* leading to the eardrums (cf. Fig. 8.40).

Photoreceptors. Notwithstanding their many differences on the organ level, all light-sensitive receptors of animals are alike on the molecular level: light-sensitive *photopigments* are present within specialized organelles of receptor cells. These pigments are unstable in light, and their breakdown in some way initiates an impulse in the sensory fiber innervating the receptor cell. Moreover, the photopigments appear to be chemically rather similar in all animals. Each consists of two joined molecular parts. One is a variant of *retinene,* an aldehyde derivative of vitamin A; the other is a variant of *opsin,* a protein to which retinene is linked. Light splits retinene away from opsin, resulting in a nerve impulse. In a subsequent series of ATP- and enzyme-requiring reactions, retinene is rejoined to opsin and the photopigment is thereby regenerated (Fig. 8.46).

Numerous invertebrate groups and all vertebrates possess elongated photoreceptor cells called *rods.* Their photopigment is "visual purple," or *rhodopsin.* With this pigment animals may detect different black-white intensities of light and changes in such intensities. Rods thus serve primarily as illumination- and motion-detectors. In addition to rods, a few animal groups, notably some insects, some reptiles, most birds,

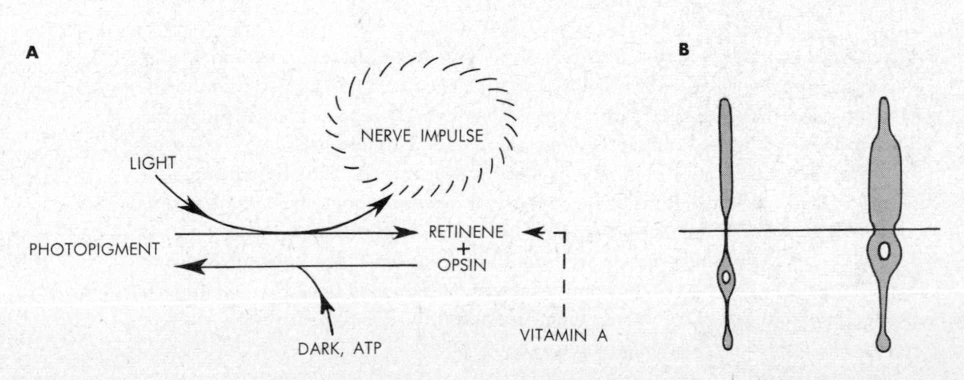

8.46. A. The pattern of the chemistry of vision. The photopigment is iodopsin in the case of cone cells, rhodopsin in that of rod cells. Note that vitamin A may replenish the supply of retinene. **B.** Schematic representation of an individual rod (left) and cone (right).

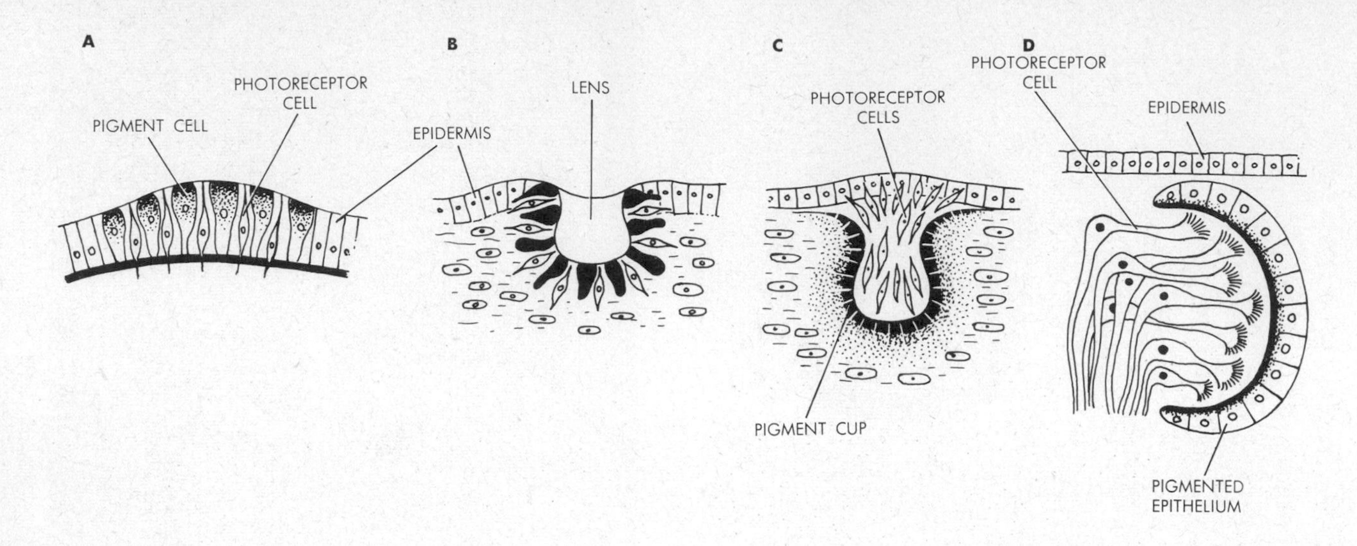

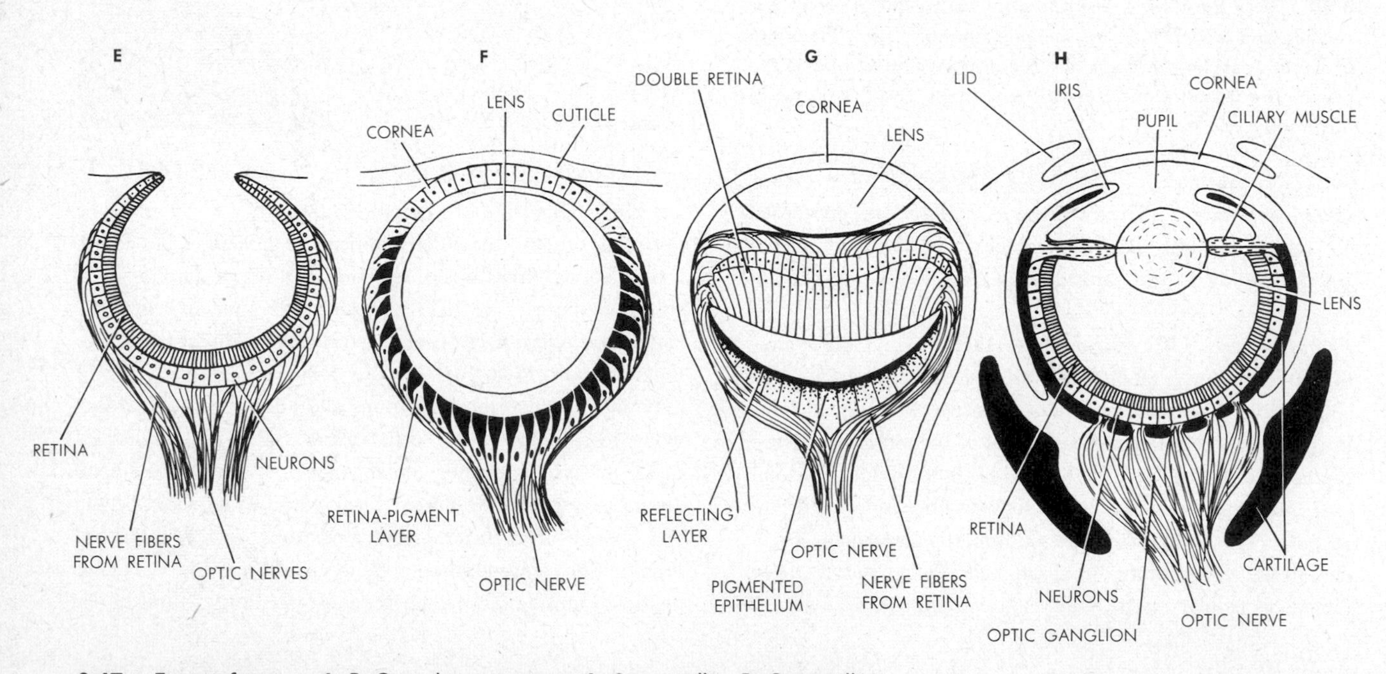

8.47. Types of eyes. **A, B, C,** coelenterate eyes. **A.** Spot ocellus. **B.** Cup ocellus with lens. **C.** Cup ocellus of inverted type, i.e., light must pass nerve fibers (not shown, part of epidermis) before reaching photosensitive cells. **D.** Cup ocellus of planarian flatworm, inverted type; light passes the neurons first before reaching the flared photosensitive ends of the neurons. **E.** Cup ocellus of the chambered *Nautilus,* converted type; light reaches the retina directly, the neurons being behind the photosensitive layer. **F.** Vesicular ocellus (eye) as in many snails and annelids, converted type. **G.** Eye of the scallop *Pecten,* with double retina; the outer retinal layer is inverted, the inner layer is converted and receives light bounced back from the reflecting layer. **H.** Eye of squid, converted type, structurally as complex as vertebrate eye. (*After Hesse, Borradaile, and other sources.*)

and monkeys, apes, and man, possess separate color detectors. The pigment in these is some form of *iodopsin*, a combination of variants of retinene and opsin different from those in rhodopsin. Animals able to see a complete color spectrum actually appear to possess at least three different kinds of iodopsin, sensitive respectively to red, blue, and yellow wavelengths of light. From various combinations of these three primary colors all other colors can probably be derived, as in color television. In the color-detecting vertebrates, iodopsin occurs in photoreceptor cells called *cones* (cf. Fig. 8.46).

Clusters or more extensive layers of photoreceptor cells form *retinas*, the principal components of seeing organs (Fig. 8.47). Some eyes (*ocelli*) are relatively simply constructed *eye spots* (*spot ocelli*), flush with the body surface. Others are *eye cups* (*cup ocelli*), and the most complex are more or less spherical *eyes* (*vesicular ocelli*). Most types of eyes are usually equipped with accessory structures such as transparent *lenses*. These are cellular in most animals, but are noncellular chitinous disks in arthropods, forms which possess two types of eyes: *simple eyes* and unique *compound eyes* (Fig. 8.48). Each compound eye consists of up to hundreds or thousands of complete visual units. Such a unit is an *ommatidium;* it contains its own lens and photoreceptor cells. The lenses of adjacent ommatidia form hexagonal *facets* on the eye surface.

The eyes of vertebrates (Fig. 8.49) and the remarkably similar eyes of certain squids (cf. Fig. 8.47) are the most complex photoreceptor organs. Apart from lenses, such eyes contain various additional accessory structures, among them curved transparent *corneas,* which aid in focusing; circularly and radially arranged *pupillary muscles,* which adjust the amount of light admitted into an eye; and a muscular *ciliary body* around the lens of each eye, which may vary the curvature of the lens surface and so contributes greatly to focusing. The retina of such an eye contains not only a layer of photoreceptor cells but also one or more layers of intersynapsing neurons. These are situated behind the photocells in mollusk eyes, and in front of them in vertebrate eyes (cf. Figs. 8.47 and 8.49).

Such retinal neuron layers have been shown to carry out a considerable amount of "data-processing" before impulses are sent via an optic nerve to the brain. It has been found, for example, that the retinal neurons of the frog eye arrange the impulses from the photoreceptors into four distinct sets of signals to the brain. One set provides the animal with a stationary black-and-white "outline drawing" of the visual field, a background picture of the world, as it were. A second set similarly provides an outline drawing, but only of illuminated objects which move across the visual field. A third set informs of objects which blot out illumination in part of the visual field and which become rapidly larger—undoubtedly indicating the approach of potentially dangerous animals. And a fourth pattern of impulses registers the presence of moving objects which blot out illumination but which remain small. This pattern probably serves as an insect-detector and thus may be of specific importance to a frog. The four sets of impulses become superimposed neatly in four adjacent neuron layers in the brain and appear to be interpreted as a single "picture." A frog evidently sees the world in its own unique and subjective manner. An analogous conclusion undoubtedly holds for light-sensitive animals of all kinds.

Other receptors. Numerous animal groups possess *temperature receptors* in the integument, in the form of either free nerve endings or minute organs. In arthropods, for example, such receptors are found particularly in the antennae and the mouth parts. Vertebrates possess distinct heat and cold receptors in all parts of the skin, specially dense concentrations usually being present around the mouth and within the mouth cavity (cf. Fig. 8.38). Especially sensitive heat receptors occur on the tongues of snakes, reptiles which may detect the body heat of other animals from appreciable distances.

In man and in vertebrates generally, distinct sets of free nerve endings in all parts of the body serve as *pain receptors* (cf. Fig. 8.38). Other animals may possess such receptors as well. However, inasmuch as feelings of pain must be communicated before we can be sure of their presence, it is actually quite difficult to determine how widespread a specialized sense of pain might be. Most animals do react sharply to stimuli which, by analogy with man, could be assumed to be painful. But such actions need not necessarily indicate pain as such; they could well result from stimulation of tangoreceptors or various other types of sense organs.

Apart from the senses here referred to, numerous others unquestionably exist. For example, man—and by inference presumably other animals as well—possesses senses of sexual excitation, hunger, thirst, and sleepiness; and man may also discriminate between sensations of burning, tickling, stinging, and limbs "falling asleep." Separate receptors need not necessarily be associated with each of these senses. It is known, for example, that some of the above sensations result when pressure, temperature, and pain receptors are stimulated simultaneously and in different combinations. In some instances, however, specialized receptors for so far unidenti-

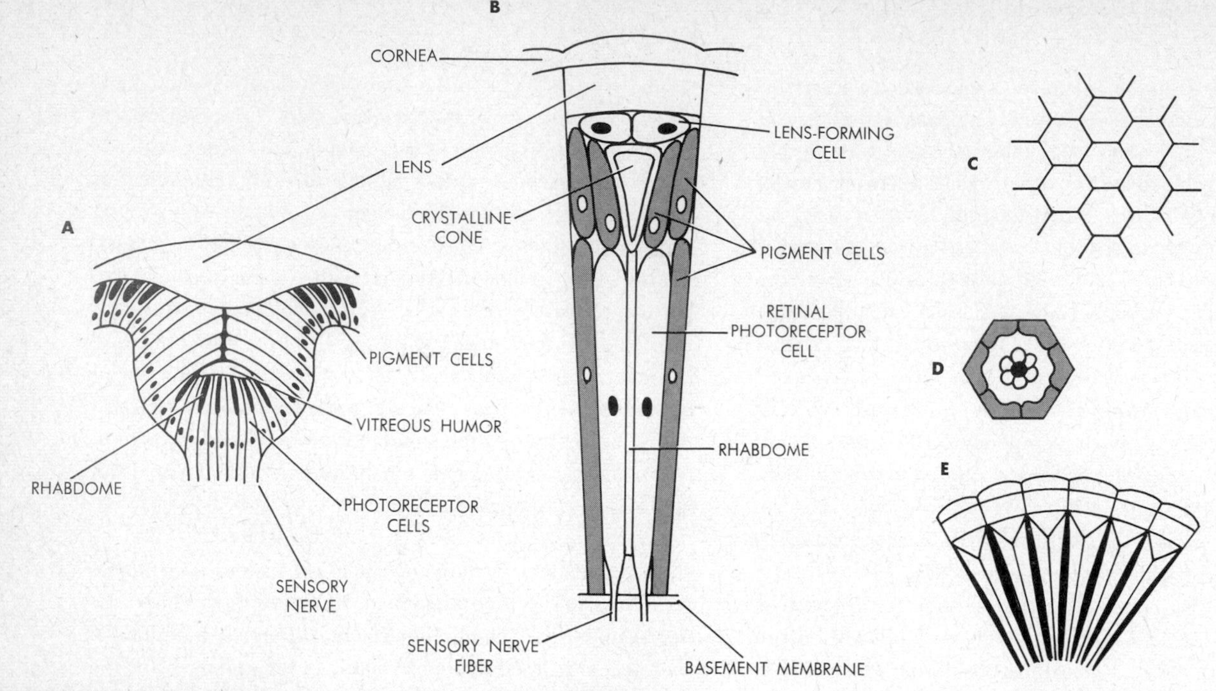

8.48. Arthropod eyes. Diagrams: A. Simple eye (ocellus), as in insects. **B.** Longitudinal section through a single visual unit (ommatidium) of a compound eye. **C.** Surface appearance of adjacent ommatidial corneas of a compound eye. **D.** Cross-sectional appearance of an ommatidium, cut at the level of the retinal cells; note the hexagonal arrangement of the pigment and retinal cells. **E.** Longitudinal section through several adjacent ommatidia. (*After Berlese, Snodgrass, and other sources.*) **Photo:** the compound eye of an insect.

fied senses undoubtedly do exist, particularly among invertebrates. Many of these animals possess organs which clearly reveal their sensory nature by their structure, but which so far have not yet been characterized functionally as registering this or that type of stimulus. The environments of many invertebrates often produce stimuli quite different from those we would expect in a terrestrial or human environment; and it is therefore not too surprising that invertebrate senses are as yet known far less completely than those of vertebrates.

ENDOCRINE SYSTEMS

It is fairly certain that all animals possess chemical agents which, after being produced in one body part, have specific regulatory or coordinating effects in other body parts. Such substances are usually manufactured in cells not specialized exclusively for regulatory functions, and they may be generally referred to as *humoral* agents. For example, CO_2 qualifies as a simple humoral agent in mammals; among other effects, it exerts a controlling role over breathing (cf. Chap. 9). Representing a class of highly diverse substances, humoral agents differ from other chemical controllers (such as vitamins and mineral ions) essentially in that they are manufactured within the body rather than being acquired from the outside.

In certain instances, humoral agents are produced in cells which *are* specialized particularly for such manufacturing functions. Agents of this sort are usually called *hormones*, and the cells producing them are *endocrine* cells. For example, inasmuch as neurons secrete neurohumors at their axon terminals, as outlined earlier, neurons qualify as endocrine cells. Indeed, in animals in which specialized endocrine cells have been found, such cells in numerous instances are *neurosecretory* cells: modified or unmodified nerve cells which secrete a variety of hormones having a variety of functions outside the nervous system. In other instances, endocrine cells are nonneural secreting cells. These are sometimes components within elaborate endocrine organs. Several of such organs within an animal, together with any neurosecretory cells present, then form an endocrine system. A distinguishing feature of endocrine glands of all types, both cellular and more complex ones, is that their hormonal secretions are not discharged into ducts; the hormones are secreted into and are carried by the body fluids. Glands which do discharge into ducts are known as *exocrine* glands (cf. Chap. 6), and the secretions here are not hormones.

Hormones vary greatly in chemical composition. Some are proteins, a few are amino acids, others are sterols, and the rest are various other simple or complex kinds of compounds. A few can be synthesized in the laboratory, a few have known chemical structure, and the remainder are known only through the effects of hormone deficiency (e.g., undersecretion, excision of the secreting cells) and of hormone excess (e.g., oversecretion, injection of hormone).

To date, the presence of endocrine cells has been demonstrated more or less definitely only in animals such as nemertine worms, certain segmented worms, mollusks, most arthropods, and chordates, including tunicates and all vertebrates. Cells of the neurosecretory type occur in all these

8.49. The vertebrate eye. **A,** general structure of mammalian eye; **B,** greatly simplified representation of a cross section through the retina: note that this is an inverted retina.

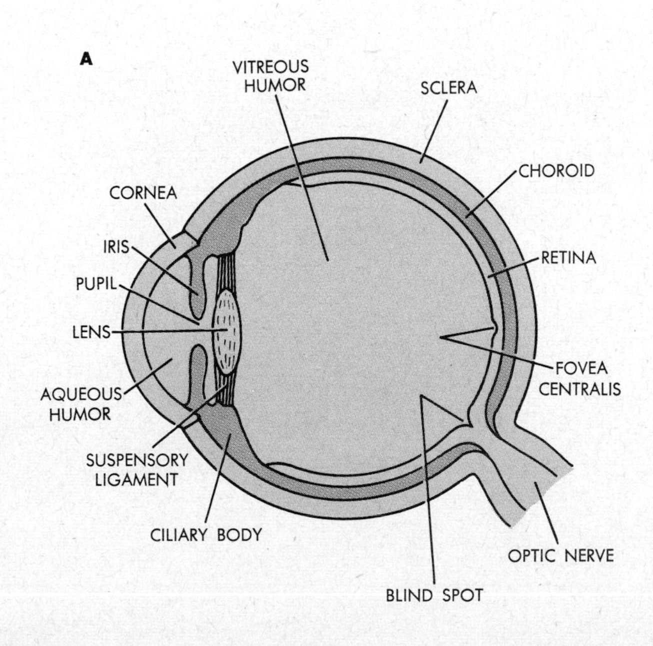

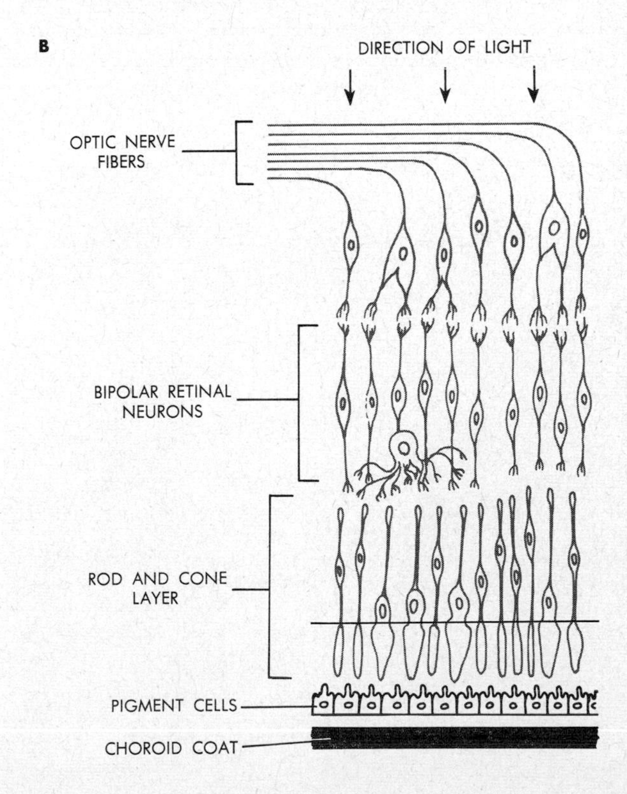

groups, but nonnervous endocrine cells are conspicuous only in arthropods and vertebrates. Also, only in these two are there distinct endocrine organ *systems*. The known functions of the endocrine secretions differ greatly among the groups named. In some worms the hormones play a role in growth and (regenerative) development; in mollusks such as squids and octopuses, neurosecretory hormones appear to control mainly the expansion and contraction of pigment-containing chromatophores in the integument. By such means, the animal may change its coloration in conformity with environmental backgrounds or in response to external stimuli of various kinds (e.g., color changes in an "excited" octopus). The external stimuli are primarily visual, and the neural centers then initiate neurosecretions.

Regulation of chromatophore activities also constitutes one function of arthropod hormones, particularly in crustaceans. Another appears to be control of breeding and mating behavior. Among insects, the best known effect of hormones is their growth-regulating role during development. Produced in part by neurosecretory cells in the brain ganglia, in part by nonnervous organs in the thorax, the hormones control the molting and the growth of larvae, the transformation of larvae into pupae, and the final transformation of either larvae or pupae into adults. These effects are discussed more fully in Chap. 28. But we may note here that, apart from influencing body tissues generally, the endocrines of crustaceans and insects also affect one another: a hormone synthesized by one gland exerts a specific excitatory or inhibitory effect on another gland. It is this which makes the endocrine glands an integrated system, even though they are structurally unconnected.

Such functional interdependences are particularly pronounced in the vertebrate endocrine systems, by far the most complex of all animals. First, the number of different glands is much greater than elsewhere (Fig. 8.50). Secondly, the organ systems of the vertebrate body perform hardly any function

8.50. The vertebrate endocrine system. Diagram: the locations of the principal component organs of the system. **Photo:** longitudinal section through a pituitary gland. The left side of the photo points in the direction of the face. Note the anterior lobe in the left part of the gland and the intermediate and posterior lobes in the right part. The posterior lobe continues dorsally as a stalk which joins the whole gland to the brain. The structure of the posterior lobe, i.e., a sac pouched out from the brain floor, is well visible here.

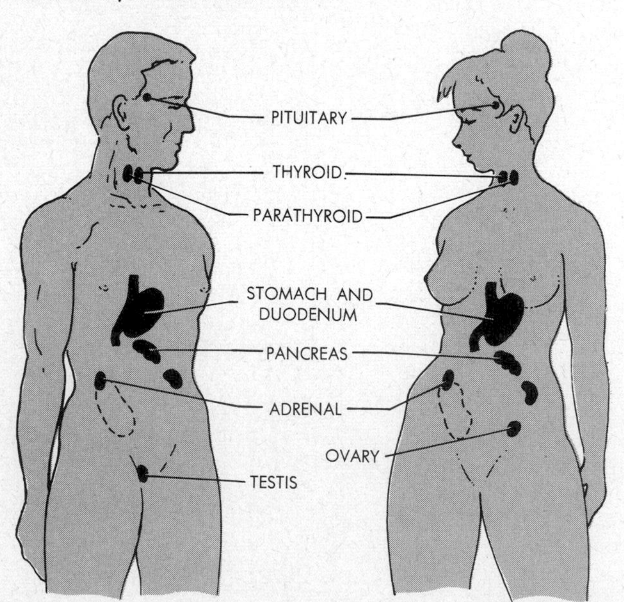

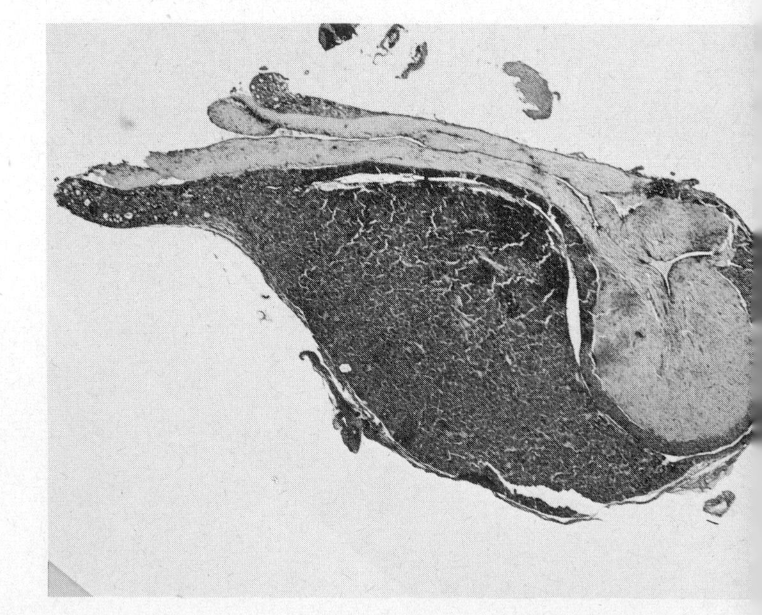

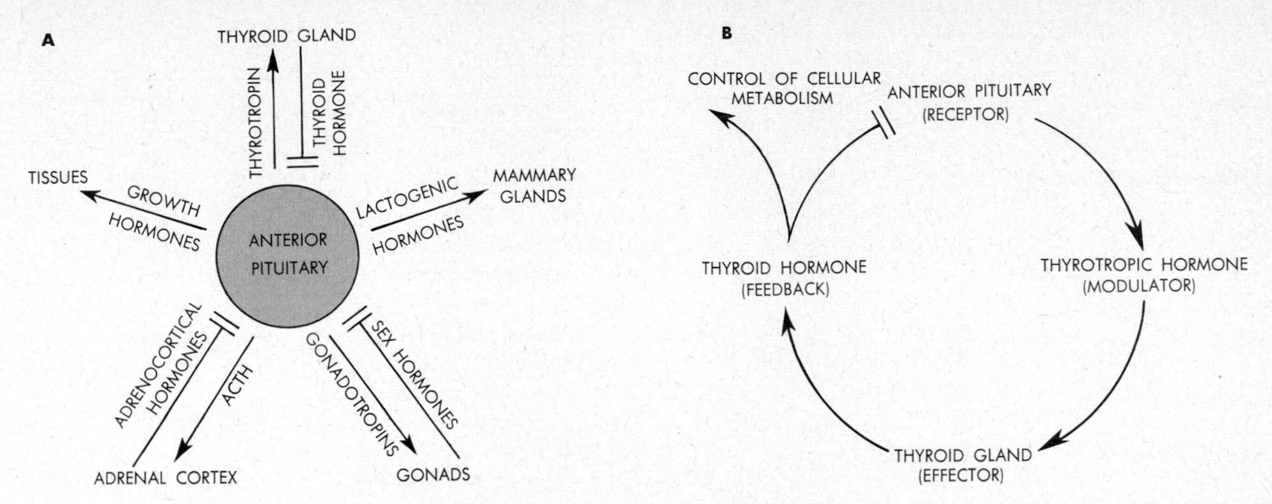

8.51. **Left,** summary of the secretions of the anterior lobe of the pituitary. Arrows tipped with transverse double bar symbolize known inhibitory feedbacks by which the secretion rate of pituitary tropic hormones is adjusted. **Right,** illustration of the specific stimulative effect of thyrotropic hormone on the thyroid gland and the inhibitive effect of thyroid hormone on the pituitary. Through such control cycles, the output of tropic pituitary hormones is automatically self-adjusting.

that is not influenced at least in part by hormones. Endocrine control usually operates in conjunction with nervous control, and in many instances the nervous system supplies information about the external environment while the endocrine system regulates the internal response to this information.

We may note also that, in vertebrates, *all* cells of the body probably require all hormones, just as all cells require all vitamins. Accordingly, terms such as "sex hormone," for example, are somewhat misleading. To be sure, sex hormones are manufactured in sex organs and the hormones contribute to the proper functioning of these organs. As we now know, however, sex hormones also contribute to the functioning of virtually every other organ in the vertebrate body. It happens that the effect of deficiency or excess of a given hormone may reveal itself first or most obviously in a particular body part. For convenience we may then name the hormone according to this body part, but we cannot conclude that the hormone functions only there.

Apart from their other controlling roles in cells, some vertebrate hormones perform an additional special function: they control the manufacture and secretion of one another. For example, many endocrine glands cannot secrete their hormones unless they are stimulated to do so by other hormones, secreted in other endocrine glands. As a group, such glands in effect function like a board of directors, in which the members hold one another in close mutual check. The output of each gland is controlled wholly or partially by the output of one or more other glands. As a result, the overall output by all glands is carefully balanced.

Of particular importance in this respect are the *tropic* hormones of the pituitary gland (Fig. 8.51). These stimulate the endocrine activity of the thyroid glands, the adrenal glands, and the sex organs. In their turn, the hormones of the latter inhibit the activity of the pituitary. Through such automatically self-adjusting feedback controls, the hormone concentrations are maintained at steady levels. For example, if the amount of thyroid hormone should rise unduly, the hormone would inhibit the pituitary, hence the output of thyrotropic hormone. The thyroid gland would then be stimulated less, and reduced secretion of thyroid hormone would follow.

The functions of the principal vertebrate endocrine glands are indicated in Table 7.

TABLE 7. The principal vertebrate endocrine glands and their hormones

Gland	Hormones	Chief functions	Effects of deficiency or excess
pituitary, anterior lobe	TTH (thyrotropic)	stimulates thyroid	
	ACTH (adrenocorticotropic)	stimulates adrenal cortex	
	prolactin (lactogenic)	stimulates milk secretion, parental behavior (nest building; care of young)	
	growth	promotes cell metabolism	dwarfism; gigantism
	FSH (follicle-stimulating)	stimulates ovary (follicle)	
	LH (luteinizing)	stimulates testes in male, corpus luteum in female	
pituitary, midlobe	intermedin	controls adjustable skin-pigment cells (e.g., frogs)	
pituitary, posterior	at least five distinct fractions	controls water metabolism, blood pressure, kidney function, smooth-muscle action	increased or reduced water excretion
thyroid	thyroxine	stimulates respiration; inhibits TTH secretion	goiter; cretinism; myxedema
parathyroid	parathormone	controls Ca metabolism	nerve, muscle abnormalities; bone thickening or weakening
adrenal cortex	cortisone, other steroid hormones	controls metabolism of water, minerals, carbohydrates; controls kidney function; inhibits ACTH secretion; duplicates sex-hormone functions	Addison's disease
adrenal medulla	adrenaline	alarm reaction, e.g., raises blood pressure, heart rate	inability to cope with stress
pancreas	insulin	glucose → glycogen conversion	diabetes
	glucagon	glycogen → glucose conversion	
testis	testosterone, other androgens	promote cell respiration, blood circulation; maintain primary and secondary sex characteristics, sex urge; inhibit FSH secretions (testis and ovary: follicle)	atrophy of reproductive system; decline of secondary sex characteristics
ovary: follicle	estradiol, other estrogens		
ovary: corpus luteum	progesterone	promotes secretions of oviduct, uterus growth in pregnancy; inhibits LH secretions	abortion during pregnancy

REVIEW QUESTIONS

1. Distinguish between epidermis and dermis and describe the basic structure of each. What derivatives does each layer produce in different animal groups? What are cornified cells? Name examples of tooth- and scale-types produced by (*a*) epidermis, (*b*) dermis. What is the structure of a hair? A mammalian tooth?

2. Which animal groups possess exoskeletons? Endoskeletons? What are the comparative advantages and disadvan-

tages of each? List structural variants of exoskeletons. What is a notochord? Distinguish between dermal and replacement bone, and describe the formation of each. What is the structure of cartilage?

3. Review the organization of the vertebrate skeleton. Name and describe the basic parts of the vertebral column, the limb girdles, and the appendages. Which vertebrates possess a cartilaginous skeleton? What are the parts of a vertebra?

4. What are the functions of a muscular system? Distinguish between smooth, striated, and cardiac muscle. In which animals and where in the body does each type occur? What is the anatomical relation among muscle, tendon, and skeleton? What is a motor unit? Describe the characteristics of a simple twitch. What are summation, tetanus, and tonus?

5. Name the components of a reflex arc and describe the course of a reflex in a vertebrate. Review the structure of a neuron. What are nerves, nerve nets, nerve cords, and ganglia? What is the difference between a ganglion and a brain? What is a ladder-type nervous system?

6. Describe the structure of the vertebrate brain. List the principal subdivisions, the principal components of each, and show how the elaboration of the subdivisions varies in different vertebrate groups. Distinguish between spinal and cranial nerves. List the mammalian cranial nerves and review the functions of each. Review the structure and functioning of the autonomic nervous system. Distinguish structurally and functionally between the sympathetic and parasympathetic subdivisions of that system.

7. What is a nerve impulse, and how is it transmitted (*a*) within a nerve fiber, (*b*) across a synapse? Describe simple and complex activities of neural centers. What are reflex modification, impulse summation, channel selection, oscillatory circuits, stored programs, tropisms? How are stored programs related to learning and memory? What are conditioned reflexes and what is their functional significance? What is meant by pathway facilitation?

8. Distinguish between neurosensory and epitheliosensory cells. Describe various kinds of chemoreceptors and their functions. What is the difference between smell and taste? How are these senses mediated in vertebrates? What are tangoreceptors, proprioreceptors? What are their functions, and where in an animal can such receptors be found?

9. Review the structure and function of statoreceptors generally. Then describe the organization of the mammalian ear and show where statoreceptors are located. Distinguish between static and dynamic body balance. What are the semicircular canals and how do they operate? What is a phonoreceptor?

10. Describe the structure and function of a lateral-line system. Which animals can hear, and by means of what kinds of receptors? Describe the structure of such receptors. What is a lagena, a cochlea, a Weberian ossicle? Show how hearing is accomplished in man and how pitch is discriminated.

11. Review the chemistry of vision. What are rods and cones? What is a retina? Which animals have color vision? Distinguish among spot, cup, and vesicular ocelli, and between simple and compound eyes. Describe the structure of such eyes. What is an ommatidium? Describe the eye structure of (*a*) various mollusks, (*b*) arthropods, (*c*) vertebrates.

12. What are humoral agents, hormones? Which animal groups possess (*a*) humoral agents, (*b*) endocrine cells, (*c*) endocrine systems? What are some of the known functions of hormones in different invertebrates? What are tropic hormones and what is their functional significance? Review the structure of the vertebrate endocrine system and describe the functions of various hormones.

COLLATERAL READINGS

On integumentary, skeletal, and muscular systems, the following books and articles may be consulted for additional information:

Benzinger, T. H.: The Human Thermostat, *Sci. American*, Jan., 1961.

Hocking, B.: Insect Flight, *Sci. American*, Dec., 1958.

Huxley, H. E.: The Contraction of Muscle, *Sci. American*, Nov., 1958.

McLean, F. C.: Bone, *Sci. American*, Feb., 1955.

Montagna, W.: "The Structure and Function of Skin," Academic, New York, 1956.

Prosser, C. L., and F. A. Brown, Jr.: "Comparative Animal Physiology," 2d ed., Saunders, Philadelphia, 1961.

The following are readings on endocrine secretions:

Beach, F. A.: "Hormones and Behavior," Hoeber, New York, 1947.

Constantinides, P. C., and N. Carey: The Alarm Reaction, *Sci. American*, Mar., 1949.

Fieser, L. F.: Steroids, *Sci. American*, Jan., 1955.

Funkenstern, D. H.: The Physiology of Fear and Anger, *Sci. American*, May, 1955.

Gorbman, A., and H. A. Bern: "Textbook of Comparative Endocrinology," Wiley, New York, 1962.
Gray, R. W.: Cortisone and ACTH, *Sci. American*, May, 1950.
Li, C. H.: The Pituitary, *Sci. American*, Oct., 1950.
———: The ACTH Molecule, *Sci. American*, July, 1963.
Rasmussen, H.: The Parathyroid Hormone, *Sci. American*, Apr., 1961.
Wilson, E. O.: Pheromones, *Sci. American*, May, 1963.

Senses and sense perception are the subject of the articles below:

Bekésy, G. von: The Ear, *Sci. American*, Aug., 1957.
Gibson, E. J., and R. D. Walk: The "Visual Cliff," *Sci. American*, Apr., 1960.
Griffin, D. R.: Sensory Physiology and the Orientation of Animals, *Amer. Scientist*, vol. 41, 1953.
Haagen-Smit, A. J.: Smell and Taste, *Sci. American*, Mar., 1952.
Kalmus, H.: Inherited Sense Defects, *Sci. American*, May, 1952.
Kennedy, D.: Inhibition in Visual Systems, *Sci. American*, July, 1963.
Lissman, H. W.: Electric Location by Fishes, *Sci. American*, Mar., 1963.
Melzack, R.: The Perception of Pain, *Sci. American*, Feb., 1961.
Miller, W. H., F. Ratcliff, and H. K. Hartline: How Cells Receive Stimuli, *Sci. American*, Sept., 1961.
Rushton, W. A. H.: Visual Pigments in Man, *Sci. American*, Nov., 1962.
Wald, G.: Eye and Camera, *Sci. American*, Aug., 1950.
———: The Molecular Basis of Visual Excitation, *Am. Scientist*, vol. 42, 1954.

Various components and activities of nervous systems are discussed in the following:

French, J. D.: The Reticular Formation, *Sci. American*, May, 1957.
Gerard, R. W.: What is Memory?, *Sci. American*, Sept., 1953.
Gray, G. W.: The Great Ravelled Knot, *Sci. American*, Oct., 1948.
Hubel, D. H.: The Visual Cortex of the Brain, *Sci. American*, Nov., 1963.
Hydén, H.: Satellite Cells in the Nervous System, *Sci. American*, Dec., 1961.
Katz, B.: The Nerve Impulse, *Sci. American*, Nov., 1952.
———: How Cells Communicate, *Sci. American*, Sept., 1961.
Keynes, R. D.: The Nerve Impulse and the Squid, *Sci. American*, Dec., 1958.
Liddell, H. S.: Conditioning and the Emotions, *Sci. American*, Jan., 1954.
Loewenstein, W. R.: Biological Transducers, *Sci. American*, Aug., 1960.
Olds, J.: Pleasure Centers in the Brain, *Sci. American*, Oct., 1956.
Snider, R. S.: The Cerebellum, *Sci. American*, Aug., 1958.
Sperry, R. W.: The Growth of Nerve Circuits, *Sci. American*, Nov., 1959.
Van der Kloot, W. G.: Brains and Cocoons, *Sci. American*, Apr., 1956.
Walter, W. G.: The Electrical Activity of the Brain, *Sci. American*, June, 1954.
Wooldridge, D. E.: "The Machinery of the Brain," McGraw-Hill, New York, 1963.

animal systems: supply, removal, transport

Of the five organ systems still to be discussed, four are considered in this chapter, viz., those of *circulation, alimentation, breathing,* and *excretion.* Reproductive systems are among the topics of the succeeding part.

The four systems to be examined here represent the internal "service" systems of cellular metabolism. Circulation plays the key role in this respect. As the system passes by all cells of the body, it supplies the cells with nutrients and oxygen and carries off their wastes; and as the system passes through the alimentary, breathing, and excretory systems, it delivers the wastes to all three of these and picks up fresh nutrient supplies from the first and oxygen supplies from the second. Because the service function of the circulation is so central, it is best to deal with this system first.

CIRCULATORY SYSTEMS

Pathway Patterns

Animals characteristically possess a *body fluid,* or *lymph,* which fills free spaces between cells and tissues. In ancestral animals, lymph was probably little different from the sea water in which these animals lived; modern animals have inherited some form of "sea water" as the universal internal medium of their bodies.

Lymph functions primarily in maintaining an adequate aqueous environment around all cells, i.e., in preserving *water, salt, pH,* and *osmotic* equilibria (and steady states) between the interior and the exterior of cells. Secondarily lymph also provides a medium for the diffusion and transport of foodstuffs, respiratory gases, waste materials, in some cases hormones, and any other substances transiting from one body region to another. In many animals, including particularly the sponges, radiates, and acoelomates, the free spaces between cells and tissues are minimal and the quantities of body fluid are comparatively small (Fig. 9.1). Also, the fluids here do not circulate in any specialized manner. Body movements redistribute lymph to some extent, and material exchanges between lymph, cells, and the external environment may contribute additionally to a certain amount of fluid motion. It should be noted that, in the animals just mentioned, most of the lymph occurs in the region between the ectodermal integument and the endodermal alimentary tract; i.e., lymph largely permeates the mesodermal tissues.

In pseudocoelomates, we recall, mesoderm is localized in limited regions only, and extensive pseudocoelic spaces are present between ectoderm and endoderm. Lymph here again permeates all mesodermal tissues, and it also fills the free pseudocoelic spaces. Such animals therefore possess substantial accumulations of internal fluid (cf. Fig. 9.1). By its very presence in the free spaces the lymph gives the animals a measure of hydrostatic "skeletal" support, and because of its bulk its movements are readily discernible: as the animal bends and twists, lymph ebbs and flows haphazardly but nevertheless quite conspicuously. Such motion has some resemblance to the movement of a blood, and lymph here may accordingly be called *hemolymph.* Thus, whereas lymph as such permeates the whole body, mesodermal tissues included, we may regard a hemolymph as that portion of the body fluids

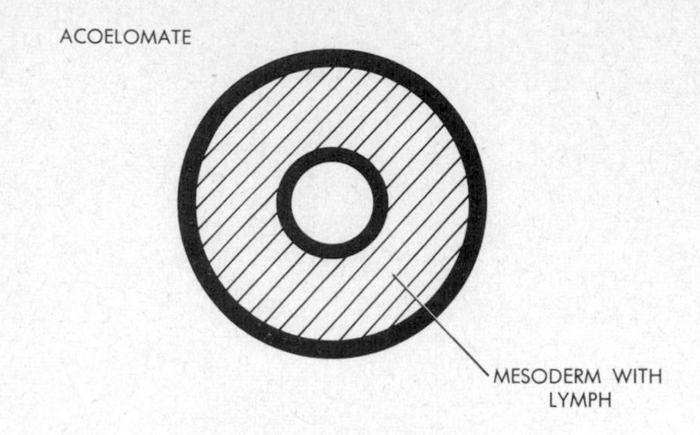

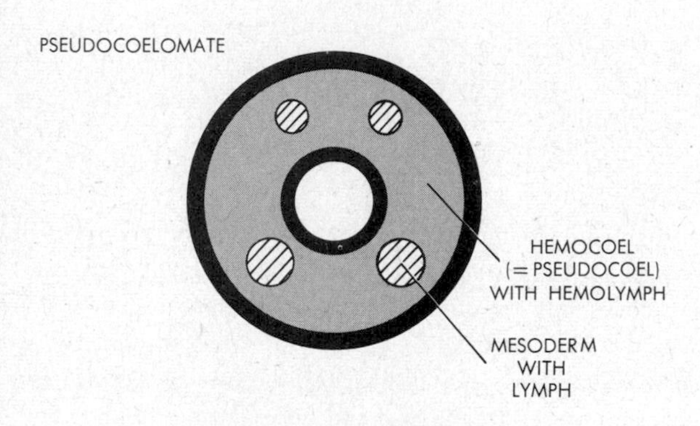

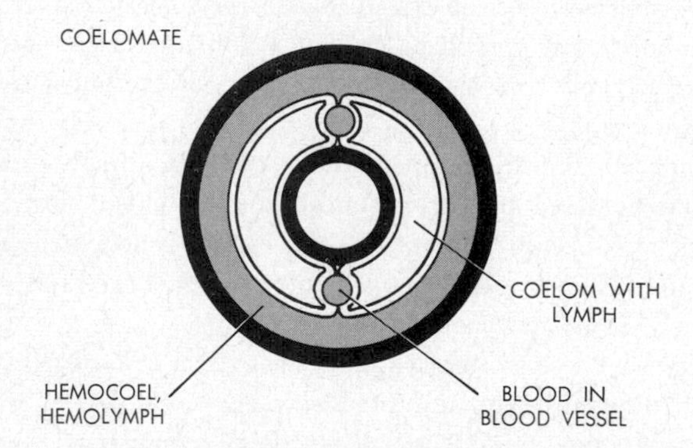

9.1. Lymph, hemolymph, blood, and their relation to the internal body spaces of animals.

which occupies a free pseudocoelic space; and a hemolymph-filled pseudocoelic space may itself be referred to as a *hemocoel.*

Inasmuch as movement of lymph generally and of hemolymph specifically is functionally useful in internal transport, regularization of such movements might be expected to be of considerable adaptive advantage. Indeed, many animals have evolved specialized devices for the regulation of fluid motion, namely, muscular pumping channels. These are derived from mesoderm, and they enclose part or all of the hemocoel and the hemolymph contained in it. We may now define any hemolymph confined partly or wholly within mesodermal channels as a *blood,* and the channels themselves as *blood vessels.* With certain exceptions among acoelomates and pseudocoelomates, vessel systems occur mainly in the coelomate animals (cf. Fig. 9.1). Groups without vessel systems may be said to be without blood, though such animals may contain hemolymph.

Where blood vessels are present, they form circulatory systems which are either *open* or *closed* (Fig. 9.2). In an open system, the vessels enclose a portion of the hemocoelic space and their ends are open to the remainder of this space. Blood therefore flows partly through vessels, partly through free hemocoelic spaces. The latter in such cases are often called *blood sinuses.* These may be large and extensive, the coelom then being crowded out and greatly reduced in size. Open systems of this type are encountered, for example, in mollusks (squids and octopuses excepted) and in arthropods.

In a closed system, the blood vessels form a complete, self-contained circuit. The only access to and exit from such a system is *through* the walls of the vessels. Free hemocoelomic

9.2. The principle of an open circulation (top, longitudinal and cross-sectional views) and a closed circulation (bottom, correspondingly). In animals with open systems, the coelom is small, the hemocoel large, and access to blood vessels is provided by the open ends of such vessels. In animals with closed systems, the coelom is large, the hemocoel is confined to the blood vessel spaces almost entirely, and access to the blood vessels is through the vessel walls only.

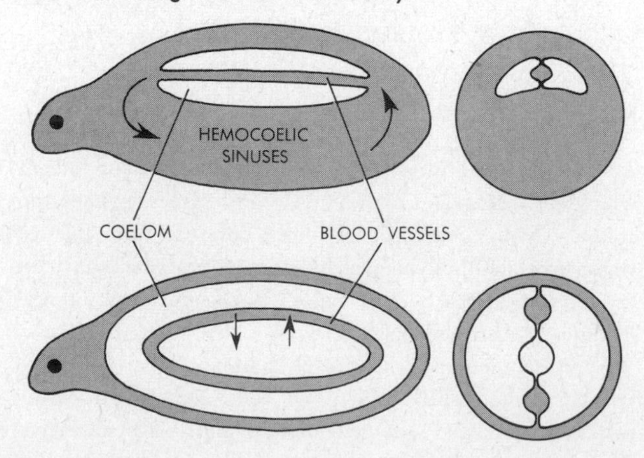

9.3. Circulatory vessels. A. Diagram illustrating the progressively greater thickness and tissue complexity of capillaries, veins, and arteries. The single cell-layer of squamous endothelium is continuous throughout a vessel system. Additional tissues do not necessarily occur in such neat layers as sketched here. **B.** Photomicrographic section through an artery and two veins. Note the thicker wall of the artery and the presence of many elastic fibers (dark wavy lines) in this wall. **C.** Longitudinal section through a lymph vessel, showing an internal valve. Such valves prevent backflow. Valves very much like this are present also in the larger veins.

A

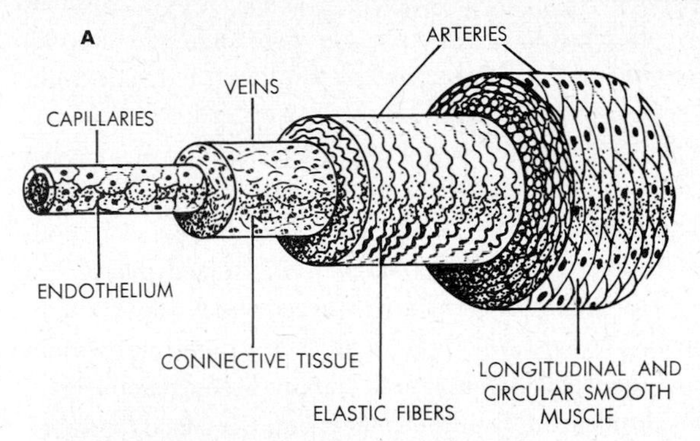

B

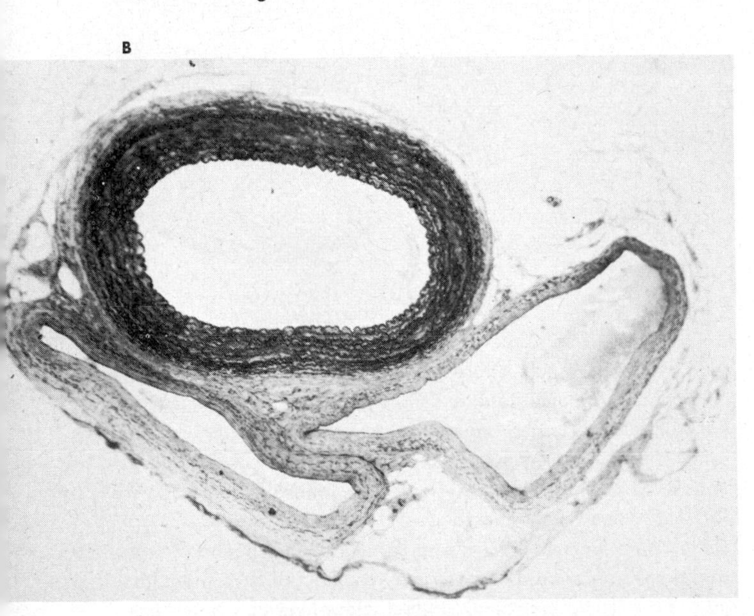

C

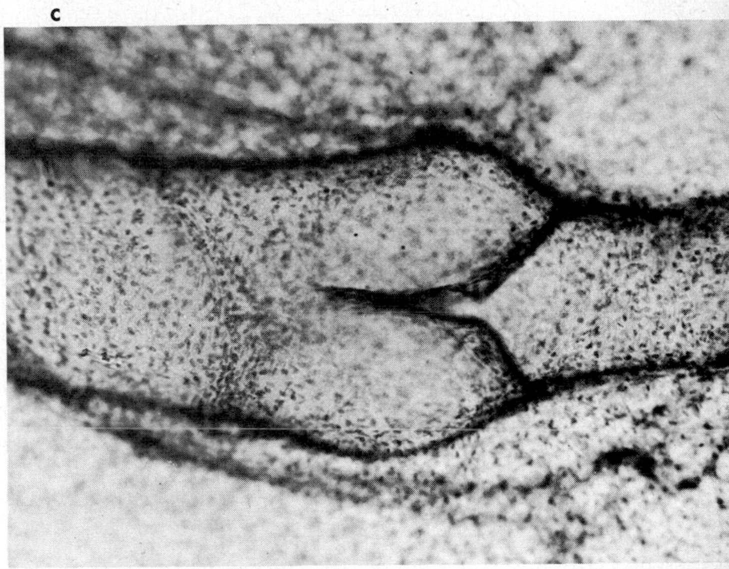

spaces outside the system are usually absent or greatly reduced in such cases; and if a coelom is present, it can therefore become large. Accordingly, the space within the vessel system is the main or only representative of the hemocoel, and the blood within the vessels is the main or only representative of the hemolymph. As in all other animals, however, the entire body is still permeated with lymph as such, and this fluid readily exchanges materials with blood through the vessel walls. Such closed systems are found, for example, in the acoelomate nemertine worms and in many coelomate animals, including annelids, squids and octopuses, and vertebrates.

The organs of a vessel system usually are *arteries, veins, hearts,* and, in most closed systems, also *capillaries* (Fig. 9.3). Arteries carry blood away from a heart, veins carry blood toward a heart. Capillaries are vessels of microscopic diameter interconnecting the narrowest arteries and veins. In closed systems the capillaries represent the regions where materials are exchanged between blood and lymph. The only tissue component of a capillary is a one-cell-thick *endothelium,* a mesoderm-derived simple squamous epithelium. This tissue is in direct contact with blood and is continuous throughout the entire circulatory system. Vessels with larger diameters possess additional tissues on the outside, principally layers of fibroelastic connective tissue and muscle. The wider a vessel, the thicker is its wall. Arteries, which carry blood under the greatest pressure, have thicker walls and more extensive layers of elastic fibers and muscles than veins. Hearts are characterized by exceptionally thick muscular walls, the muscle tissue being of the cardiac variety in vertebrates (cf.

Ch. 8). At intervals along the larger veins there are often internal *valves* which open toward the heart and so prevent backflow of blood (cf. Fig. 9.3).

In addition to blood vessel systems vertebrates also possess *lymph vessel systems* (Fig. 9.4). These are composed largely of *lymph capillaries* and *lymph veins*. The blind-ended capillaries receive body lymph, which is conducted into lymph veins. The latter become progressively wider and fewer in number. The widest lymph veins then empty into certain veins of the blood circulation. As lymph passes from the body tissues through the lymph channels into the blood, body tissues obtain new lymph by outflow of fluid from blood, through the walls of the blood capillaries. Lymph vessels are even thinner than blood veins, and they also contain a greater number of internal valves. Along the course of the lymph vessels there are glandular *lymph nodes*, which manufacture certain types of blood cells and perform several other functions to be described below. In frogs and some other vertebrates, certain regions of given lymph vessels are enlarged into pulsating *lymph hearts*. These aid in driving lymph into the blood circulation.

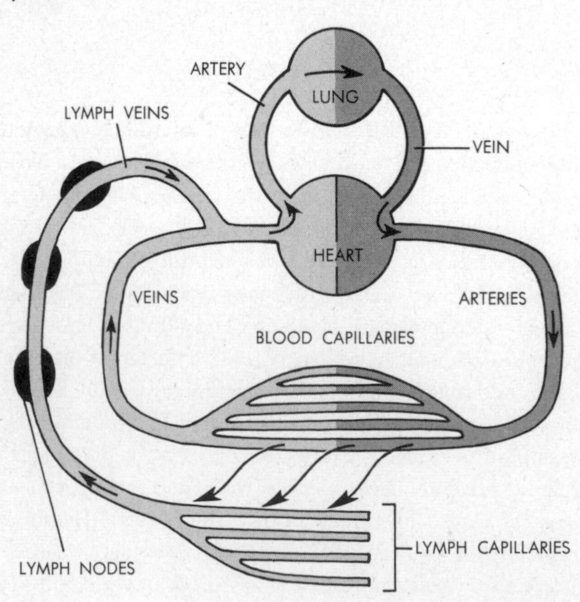

9.4. The general plan of the vertebrate blood and lymph circulations. Oxygenated blood, dark-shaded, venous blood, light-shaded. Oxygenation occurs in gills or lungs. Note that an artery carries blood away from the heart, a vein, toward the heart (regardless of the state of oxygenation of the blood carried). Fluid escaped from blood capillaries into surrounding tissues enters the lymph system through the walls of the lymph capillaries.

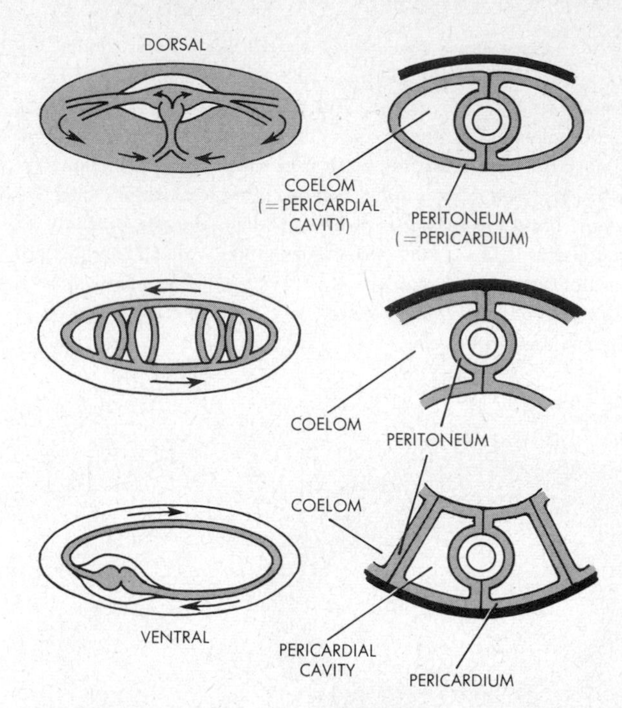

9.5. Left figures, directions of blood flow; right figures, pericardial and coelomic relations. Top, open systems (e.g., mollusks); the ventricle and one atrium are shown in left figure. The reduced coelom here becomes a pericardial space anatomically. Middle, closed invertebrate systems (e.g., annelids). Blood flows forward dorsally, the dorsal vessel and/or the anterolateral vessels here functioning as pumping hearts. The coelom is large and a distinct pericardial cavity may or may not be present (the latter case being illustrated). Bottom, closed vertebrate systems. The heart is ventral and blood flows backward dorsally. A portion of the peritoneum and coelom are partitioned off as a pericardium and pericardial cavity. Analogous partitions are present in some closed invertebrate systems (in dorsal locations, as in middle right figure).

As will become apparent in Unit 2, degrees of anatomical elaboration vary greatly among the circulatory systems of different animal groups. Closed systems include the simplest (in nemertine worms) as well as the most complex (in vertebrates). Open systems are believed to be specialized evolutionary derivatives of closed systems. Simple systems like those of nemertine worms are without heart or capillaries and consist of but a few interconnected main vessels (cf. Chap. 22,

Fig. 22.21). In more complex systems like those of annelids, capillaries and highly branched networks of larger vessels are usually present. Also, the widest, thickest vessels assume the functions of pumping "hearts" (Figs. 9.5 and 26.14). Very complex systems, both open and closed, contain hearts proper, anatomically distinct from vessels. Such hearts usually lie within a lymph-filled *pericardial cavity*. The latter is a portion of the coelom enclosed by a membrane, the *pericardium*, which is derived from the peritoneal lining of the coelom (cf. Fig. 9.5). The heart itself is either tubular, as in arthropods, or more compact, as in most other animals. In the latter case the organ generally contains two types of chambers, i.e., one or two relatively thin-walled *atria* (or *auricles*), which receive blood, and one or two thicker-walled *ventricles*, which pump blood out. Heart *valves* between atria and ventricles prevent backflow of blood (Fig. 9.6).

One atrium and one ventricle is present in the hearts of fishes and some mollusks, two atria and one ventricle in the hearts of amphibia and other mollusks. Reptiles, birds, and mammals possess four-chambered hearts, composed of two atria and two ventricles. The functional significance of this larger number of chambers will become clear in the section on breathing below. Some animals possess secondary hearts in addition to a main, or *systemic*, heart. For example, squids and octopuses have *gill hearts* which aid in driving blood through the gill capillaries (cf. Chap. 25, Fig. 25.30). Analogously, at the bases of the legs of certain insects are secondary hearts which aid in forcing blood through the narrow hemocoels of these appendages (cf. Chap. 28).

In invertebrates, hearts are located typically along the *dorsal* midline of the body. If the system is open, blood is pumped out dorsally, usually both toward the head end and the hind end of the body. Blood then flows back ventrally from both ends of the body and returns to the heart roughly at midbody. If the system is closed, blood by and large flows toward the head end dorsally and toward the hind end ventrally. In vertebrates, by contrast, the heart is situated on the *ventral* side, and blood is pumped toward the head end ventrally, toward the hind end dorsally (cf. Fig. 9.5).

The beat of a heart may originate *neurogenically* or *myogenically*. In a neurogenic beat, exemplified by the hearts of most arthropods, contraction is initiated by a neural ganglion situated in or close to the heart. Oscillator circuits within the ganglion deliver rhythmic impulses to the heart muscle. In a myogenic beat, contraction is initiated by a special node of modified or unmodified heart muscle, a so-called *pacemaker*,

9.6. Diagram: the course of the mammalian blood circulation. Arterial blood is in the left side of the circulatory system (right side of diagram), venous blood in the right side (left side of diagram). **Photo: the human heart, cut open to show the interior of the left ventricle.** Note the strands of tissue attached to the two flaps of the bicuspid valve. These strands prevent the valve from opening into the auricle (white area above ventricle).

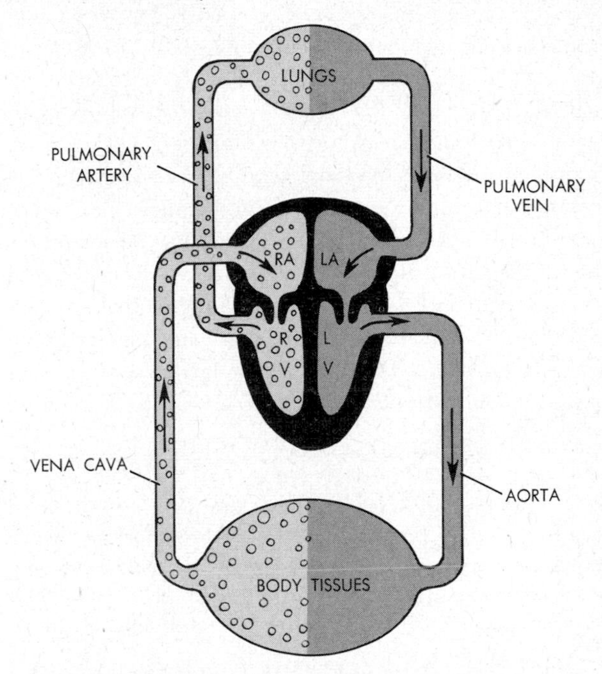

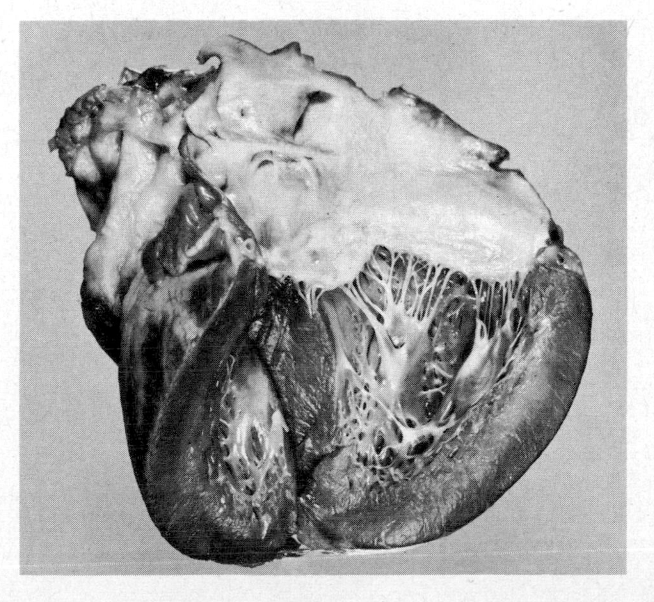

typically located in an atrium (e.g., the right atrium in mammals, Fig. 9.7). The pacemaker generates rhythmic excitatory signals which travel away through the whole heart muscle. In mammals, a bundle of specialized muscle fibers, the *bundle of His,* conducts the excitatory impulses through the heart. Thus, whether it contracts neurogenically or myogenically, the heart has an *intrinsic* beat which can be maintained even when the organ is isolated from the body. However, the intrinsic beat is subject to neural control. Both the heart ganglia and the pacemakers are innervated by accelerator and inhibitor nerves from the neural centers of an animal. Motor impulses through these nerves may override the intrinsic beat frequency and may adjust and vary the heart rate (Fig. 9.8). We may note here in passing that, in the myogenic hearts of tunicates, a chordate subphylum believed to be ancestral to that of the vertebrates, the location of the pacemaker alternates every minute or so from one end of the tubular heart to the other. The heart in effect produces an "alternating current"; blood is forced in one direction by one series of beats, in the opposite direction by the next series.

9.7. The motor innervation of the mammalian heart. Impulses through both inhibitor and accelerator nerves may affect the pacemaker. Impulses from there stimulate the atria, which contract as a result, then the A-V node, which in turn sends contraction signals to the ventricles through the bundle of His.

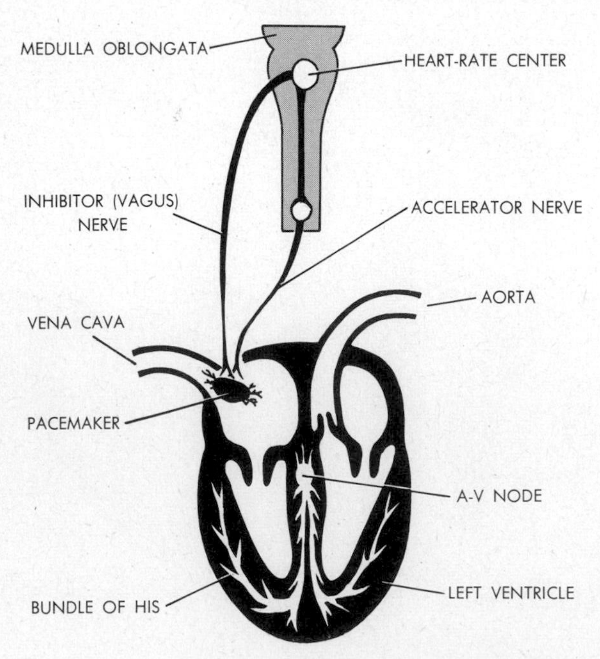

The force of the heart beat is a major determinant of *blood pressure.* In open systems, blood pressure diminishes greatly once blood has left the vessels and flows free in the hemocoelic sinuses. Return of blood to the heart is brought about in part by a "massaging" action resulting from body movements, i.e., the hemocoel is squeezed by the contraction of surrounding muscles. Also, blood is aspirated ("sucked back") to the heart as the latter empties with each beat.

In closed systems, blood pressure is determined not only by the force of the heart but also by the total volume of blood and by the total space available within the vessel system. This space may be varied nervously. So-called *vasomotor* nerves innervate the muscles in the walls of the larger vessels, particularly arteries. Contraction or relaxation of these muscles leads to *vasoconstriction* or *vasodilation,* i.e., a narrowing or widening of a vessel. Blood pressure so may increase or decrease, either locally or in the entire circulatory system (Fig. 9.9). Nevertheless, overall blood pressure falls gradually with increasing distance from the heart, for the total cross-sectional area of all capillaries is far greater than that of the arteries leaving the heart. The residual pressure after blood has passed through the capillaries is quite low, yet this *venous pressure* does contribute to returning blood to the heart. Other factors in venous return are again muscular squeezing of veins, aspiration of blood by the heart, and prevention of backflow by the valves in the veins.

In all its aspects, a vessel system is but a means to an end, namely, the movement of blood. Inasmuch as the basic function of blood is to service the body tissues in various ways, the most important parts of a circulation are the capillary beds in a closed system and the hemocoelic sinuses in an open system. It is in such regions that body tissues draw from and add to blood, and that blood in turn draws from and adds to the alimentary, breathing, and excretory systems.

Blood

Blood is a tissue consisting of two components: fluid *plasma,* in which are suspended loose blood *cells.*

Plasma. Dissolved in plasma are mineral ions as well as metabolites in transit from one body region to another. As already noted, such metabolites include waste materials, respiratory gases, and in some cases also hormones. To this extent, both the composition and the function of plasma and body lymph are indistinguishable.

The water component maintains an aqueous environment

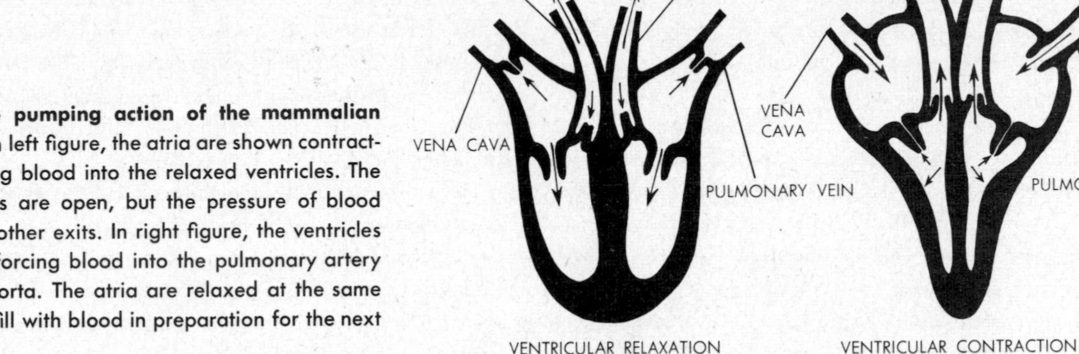

9.8. The pumping action of the mammalian heart. In left figure, the atria are shown contracting, forcing blood into the relaxed ventricles. The A-V valves are open, but the pressure of blood closes all other exits. In right figure, the ventricles contract, forcing blood into the pulmonary artery and the aorta. The atria are relaxed at the same time and fill with blood in preparation for the next beat.

around all tissues and, as pointed out above, its presence in a given volume also maintains a given blood pressure. The ionic mineral components contribute to *salt, pH,* and *osmotic* balance between plasma and lymph and between lymph and the interior of tissue cells. The most important minerals present are the positively charged sodium (Na^+), potassium (K^+), calcium (Ca^{++}), and magnesium (Mg^{++}) ions and the negatively charged phosphate ($H_2PO_4^-$, $HPO_4^=$, $PO_4^{\equiv}$), chlorine (Cl^-), bicarbonate (HCO_3^-), and sulfate ($SO_4^=$) ions. Over half of the total mineral content of plasma and lymph is sodium chloride, common table salt.

Ion pairs like $HPO_4^=/H_2PO_4^-$ are particularly significant as blood buffers. For example, if H^+ ions should accumulate in abnormally large quantities, then $HPO_4^=$ could combine with H^+, forming $H_2PO_4^-$. Concentrations of the former would thereby decrease, concentrations of the latter would increase, and the free H^+ ions would have been "taken out of circulation." The reverse reactions would occur if blood should tend to become too alkaline. Original ion balances may subsequently be restored by the excretory system, through removal of those ions which are present in relative excess.

Mineral ions exert important additional effects. For example, calcium ions regulate the sensitivity of nerves and muscles. Calcium also influences the sol-gel states within cells, and it is a component of the cementing substance between cells. Since all cells are in contact with blood or lymph, the concentration of blood calcium evidently is of enormous significance. Each of the other mineral components has its own specific biological effects. Water and mineral ions tend to be kept at constant concentrations, quantity control being exercised by the alimentary, breathing, and excretory systems. Some of the transiting foods, gases, and wastes in plasma are likewise held at constant levels, e.g. glucose in vertebrates (see below). The concentrations of most other metabolites usually vary, however, in accordance with the comparative rates of supply and utilization. Exchange of such components between plasma and lymph and between lymph and tissue cells is accomplished by diffusion and active transport.

Blood plasma differs importantly from lymph in that it contains plasma proteins.

9.9. The pattern of vasomotion. Vasoconstriction occurs when a blood vessel receives many constrictor impulses and few dilator impulses from the vasomotor center. Vasodilation occurs when constrictor impulses are few and dilator impulses are many.

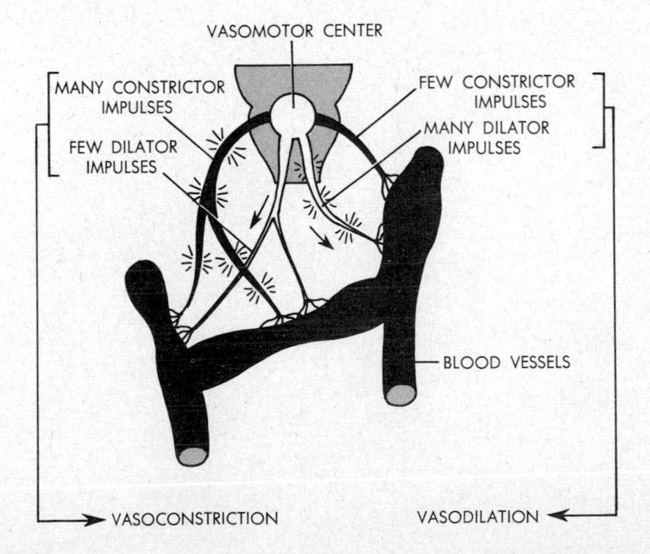

Plasma proteins. These constituents tend to remain at constant concentrations and, in closed systems, they do not leave the circulatory vessels. Most studied in vertebrates, the plasma proteins of these animals are manufactured principally in the liver, and they serve a variety of—nonnutritional—roles.

First, all these proteins contribute to the *osmotic pressure* of blood; and since most of them are ionized they also play a role in maintaining the *pH* of blood. Secondly, some of the proteins are ingredients in the *clotting* reaction; *fibrinogen* is one of these (see below). Thirdly, many of the proteins are active enzymatically. One of the catalytic agents is a clotting factor known as *prothrombin*. Others are enzymes such as are found in tissue cells generally and even in the intestine. Their catalytic functions in plasma are still largely obscure. Similarly obscure are the specific roles of plasma proteins called *albumins*, so named because of their chemical resemblance to egg white. Fourthly, proteins called *globulins* play a major role in *immunological* reactions. These include reactions resulting from *blood-type* differences and reactions involving *antigens* and *antibodies*.

That bloods of different animals differ in type has been known for some time. Where type differences exist, the blood cells of one animal act as foreign bodies when introduced into another animal, and the plasma globulins of the host may cause the foreign cells to clump. Clumped blood cells clog narrow blood vessels and may cause death. Type differences in man evidently are of major clinical significance, and before a blood transfusion is made, blood compatibility must be established. Zoologically, however, type differences are merely one further expression of the general phenomenon of protein specificity, i.e., of protein differences among all animals.

9.10. The action of specific antibodies. A foreign protein introduced into an animal is an antigen. It elicits the formation of antibodies, which "fit" precisely the surface configuration of the antigen. These specific antibodies may then combine with the antigens, making the latter harmless.

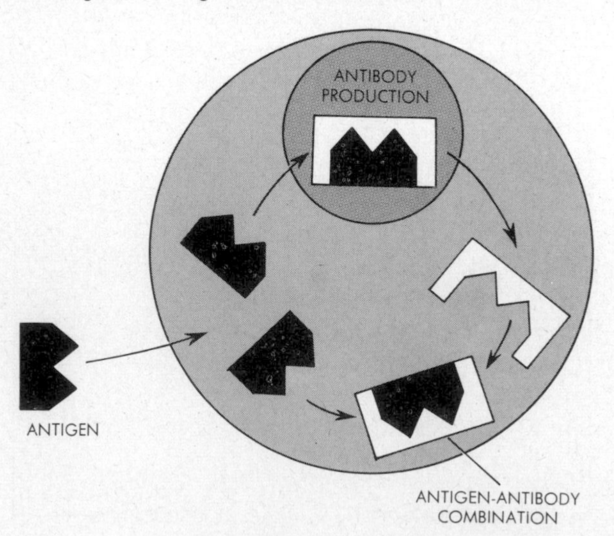

The antigen-antibody reactions referred to above likewise are an expression of protein specificity. An antigen is usually a foreign protein (in some cases it is a polysaccharide) which enters the body fluids of an animal as part of an infectious agent. For example, antigens may be introduced into an animal by viruses, bacteria, plant pollen, or any other protein-containing cell part or tissue part of a foreign organism. An antibody is a blood protein, specifically a globulin, which may make an infected animal immune to a foreign antigen. Thus, antibodies function importantly in body defense and in organismic steady-state control.

What does "making immune" mean? It means the combination of an antibody with an invading antigen. Such a union abolishes the free mobility of the antigen and thus prevents the antigen from acting in a damaging manner. The combining process is thought to be a "lock-and key" reaction, quite like the combining of an enzyme with a reactant. The molecular configuration of the antibody protein apparently is such that it "fits" into the configuration of the foreign antigen. In other words, antigen-antibody reactions are specific, and any given antigen requires a particular kind of antibody if *immunity* is to be established (Fig. 9.10).

However, animals do not normally possess preexisting antibodies for all types of antigens which might possibly invade the blood. This means that an appropriate type of antibody must be manufactured after a given antigen has invaded; and the more rapidly and abundantly such antibodies are formed, the less damage the animal will sustain. Several organs are believed to produce antibodies, the liver and the lymph nodes among them. The cells of such organs probably absorb samples of invading antigens and then synthesize large quantities of appropriate globulin antibodies. The immunity so established may protect an animal from particular diseases for long periods, often for life. (In man, such long-lasting *active* immunity may be induced by the injection of vaccines containing mild forms of disease-producing antigens. Short-term *passive* immunity is provided by injection of vaccines containing prefabricated antibodies obtained from other mammals.)

Apart from its protein content, plasma often differs from

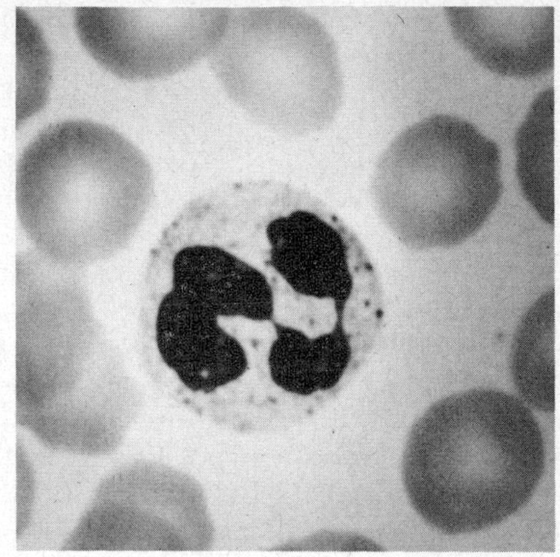
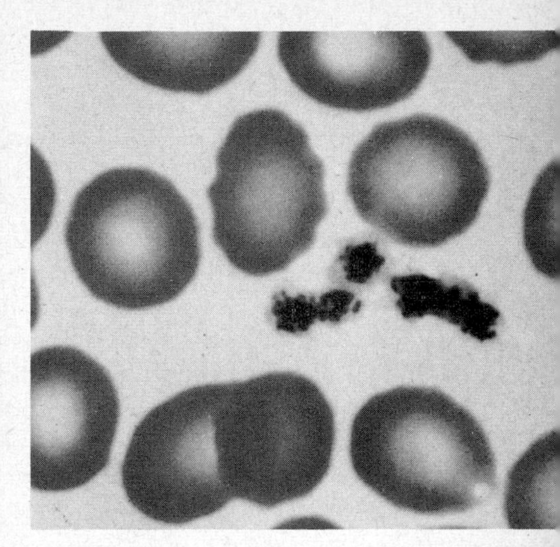

9.11. Human red corpuscles are shown in both photos. Note absence of nuclei. A white blood cell (nucleated) is shown in center of left photo, and a few blood platelets in center of right photo. See Fig. 4.26 for illustration of vertebrate red blood cells with nuclei.

lymph also in another respect, i.e., the presence of *respiratory pigments.* Inasmuch as in many animals such pigments occur not in plasma but in blood cells, the latter are best examined first.

Blood cells. Most bloods contain *nonpigmented* cells of various kinds, and many also contain *pigmented* cells. Both types are derived from mesoderm, and they are usually quite specialized; for example, normally they do not divide (Fig. 9.11).

To a greater or lesser extent, most of the nonpigmented cells are capable of amoeboid locomotion. Called *amoebocytes* generally and *white* cells in vertebrates, such cells may squeeze themselves between adjacent cells of a capillary vessel and may leave or reenter the circulatory system in this manner (Fig. 9.12). White cells are of two main types, *leucocytes* and *lymphocytes,* each with several subvarieties. In leucocytes the amoeboid habit is developed particularly well. Once leucocytes are out in the body tissues, they may migrate toward sites of infection and engulf the bacteria causing the infection. Leucocytes are probably guided to infected regions by their sensitivity to cellular disintegration products diffusing away from such regions. As noted in Chap. 4, amoeboid engulfment of foreign particles is called *phagocytosis;* leucocytes are said to be *phagocytes.* Phagocytic activity is generally characteristic also of the nonpigmented blood cells of invertebrates. An accumulation of phagocytes, bacteria, and cellular debris in wounded areas constitutes *pus.*

Lymphocytes too serve in body defense. These cells contribute importantly to scar-tissue formation after internal or external injury (via transformation into mesenchyme cells and fibrocytes), and in this manner they facilitate wound heal-

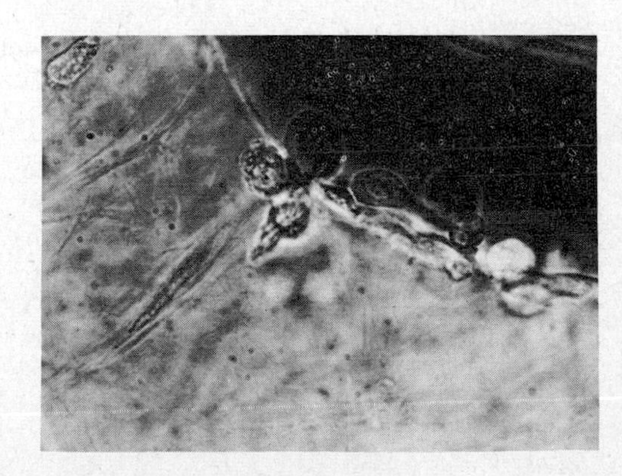
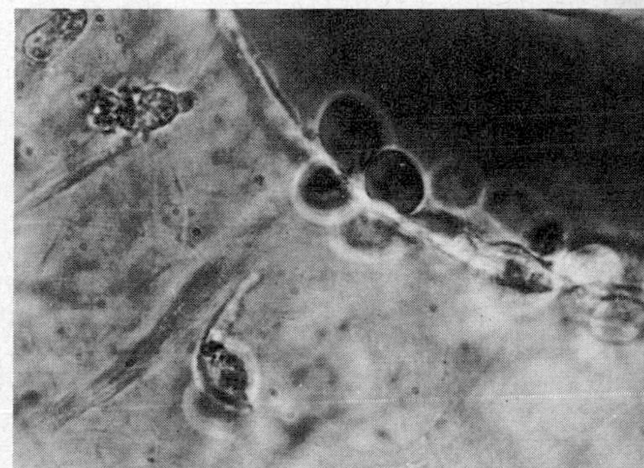

9.12. The migration of blood cells through capillary walls. In each photo, a blood-filled capillary is in upper right portion. In photo at left, two white blood cells have just penetrated through the capillary wall into surrounding tissues. In photo at right, the white cells have migrated farther into the tissue.

ing. Moreover, they aid in sealing off a surface wound against new infections. Numerous lymphocytes are normally present in the lymph nodes, where the cells act as lymph-purifying agents. For example, microscopic particles of dust, smoke, and other materials present in the atmosphere frequently get into lungs and may become embedded in lung tissue. Lymph then usually carries such particles to the lymph nodes. The lymphocytes there engulf the particles and retain them permanently.

Lymphocytes are manufactured primarily in the lymph nodes; leucocytes originate in the liver and spleen during embryonic stages but primarily in the red marrow of long bones during adult stages. Bone marrow is also believed to be the principal generating tissue of the blood *platelets* of vertebrates (cf. Fig. 9.11). These are not usually whole cells but cell fragments, often without nuclei. Platelets are the initiators of the *clotting reaction* in vertebrates. This self-sealing mechanism of the circulatory system is brought into action whenever platelets encounter obstructions and rupture. In most cases, such obstructions are the rough edges of torn blood vessels. External clotting then occurs. But air bubbles in blood (e.g., when dissolved gases effervesce) or roughness of the inner surfaces of blood vessels (e.g., as produced by solid deposits in hardened arteries) may suffice for the rupturing of platelets. An internal blood clot may then form.

Among the materials oozing out from ruptured platelets is an enzymatically active substance, *thrombokinase,* also called *thromboplastin* (Fig. 9.13). This substance interacts with two components of blood plasma, namely, calcium ions and the plasma protein *prothrombin.* The protein is an inactive precursor of the catalyst *thrombin.* In the presence of calcium ions and thrombokinase, prothrombin becomes converted to thrombin. Subsequently, thrombin reacts with fibrinogen, another of the plasma proteins. As a result of the reaction, fibrinogen becomes fibrin, an insoluble coagulated protein. Fibrin constitutes the blood clot. It is a yellowish-white meshwork of fibers in which pigmented blood cells are trapped, hence the color of the clot.

9.13. The main reactions of the clotting process.

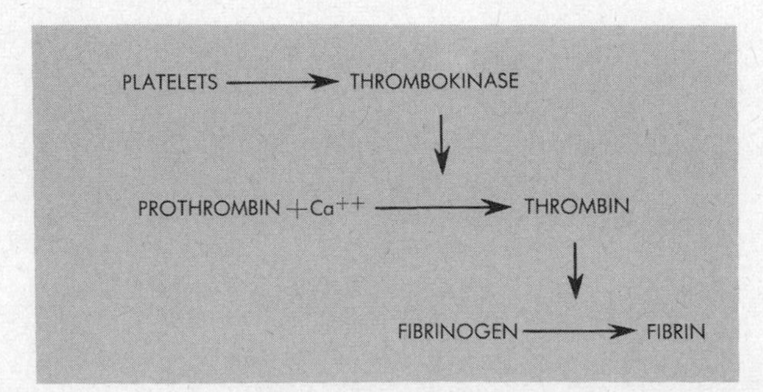

Invertebrate bloods too may clot. Although the detailed reactions are less well known, they also probably lead to the formation of coagulated protein meshworks. We may note that animals which feed on vertebrate blood, such as leeches, fleas, and bedbugs, secrete anticoagulants and mix it with ingested blood. For example, *hirudin* is a clotting inhibitor enabling leeches to store uncoagulated blood in their alimentary tracts for several months.

Where pigmented blood cells are present, they derive their color from *respiratory pigments* dissolved in their cytoplasms. The pigments serve principally in oxygen transport, and the cells as a whole are specialized for this function. Such cells usually contain nuclei, the only notable exception being the red blood cells of mammals. These cells are manufactured in the red bone marrow of adults (in liver and spleen of embryos), and in mammals the nuclei of the cells disintegrate as they mature. The resulting red *corpuscles* stay in the circulation for limited periods only, in numbers which are maintained constant: the liver of mammals and of vertebrates generally destroys red cells while bone marrow manufactures them. The rate-controlling factor is the amount of oxygen in the environment. A persistently high O_2 content in the environment (hence in blood) slows the production rate and increases the destruction rate of red cells; and a persistently low O_2 content in the environment (e.g., as at high altitudes) has the opposite effect. Red cells are by far the most abundant cellular components of vertebrate blood. For example, a cubic millimeter of human blood contains some 5 million red corpuscles, but only some 8,000 white cells and about 250,000 platelets.

Respiratory pigments. In all animals, the respiratory gases are *dissolved* in the water of plasma and lymph. In animals with blood, respiratory pigments provide an additional, far more efficient means of oxygen distribution. Carbon dioxide can be carried by some of the respiratory pigments, yet most CO_2 transport still depends on blood water; the gas is carried largely in the form of bicarbonate ions, according to the equation:

$$CO_2 + H_2O \underset{\text{in breathing system}}{\overset{\text{in body tissues}}{\rightleftharpoons}} H^+ + HCO_3^-$$

All respiratory pigments are linked to protein, the latter conferring reversibility on the reaction with oxygen:

$$\text{protein-pigment} + O_2 \underset{\textit{in body tissues}}{\overset{\textit{in breathing system}}{\rightleftharpoons}} \text{protein-pigment—}O_2$$

Without the protein, the pigments combine with O_2 irreversibly. Oxygen enters blood in the breathing system and leaves it in the body tissues; CO_2 takes the reverse path. Four different types of respiratory pigments are well known: *hemoglobin, chlorocruorin, hemerythrin,* and *hemocyanin.*

Hemoglobin (Hb) is by far the most widespread; its presence has been demonstrated in all phyla except the sponges and the radiates. However, the pigment does not always occur in all groups within a phylum, and it is not always a blood constituent. For example, Hb occurs in protozoan cells, in the eggs of many animals, and within vertebrate skeletal muscle (where it is responsible for the redness of flesh and the color of "dark meat" in cooked condition). In such noncirculatory locations, Hb serves as a supplier of oxygen, just as it does in blood. Blood hemoglobin occurs in the plasma of some animals, in the blood cells of others. For example, certain annelids, mollusks, and crustacea carry Hb in plasma. Other annelids, nemertine worms, sea cucumbers, and vertebrates are among animals in which Hb is in the blood cells. A few annelid worms are known in which the pigment occurs both in plasma and in blood cells. Virtually the only generalization possible here is that the distribution of Hb is quite haphazard and does not appear to follow any obvious taxonomic or evolutionary pattern.

Hemoglobin is purple-red without oxygen, orange-red when combined with oxygen. It consists of the carrier protein *globin* and the pigment *heme.* The latter is a tetrapyrrol containing a single iron atom, rather similar to the cytochromes (cf. Chap. 4, Fig. 4.21). Hb carries one molecule of oxygen per atom of iron. The nature of the globin fraction varies for different animal groups, and each such form of Hb endows blood with a unique oxygen-carrying capacity, as indicated in Table 8. This table shows that Hb is the most efficient of the respiratory pigments and that oxygen-carrying efficiency is correlated closely with the level of activity of which an animal is capable. Hb combines to some extent with CO_2, and it joins carbon monoxide (CO) in preference to O_2. Consequently, CO is an oxygen-displacing poison to Hb-possessing animals. In vertebrates, Hb within the red corpuscles is manufactured in bone marrow and destroyed in the liver. The iron fraction of destroyed heme is salvaged and becomes available again in marrow for the synthesis of new Hb. The tetrapyrrol part of destroyed heme becomes green *biliverdin* and red *bilirubin,* pigments which are excreted from the liver via bile and feces and via blood and urine; hence the characteristic colors of these elimination products. Some birds deposit the pigments in their egg shells, a disposal procedure which accounts, for example, for the blue color of robin's eggs.

Chlorocruorin is a green pigment with a reddish color in a concentrated state. Its composition is quite similar to that of Hb, i.e., it is an iron-containing heme protein. It also functions in essentially the same way as Hb (Table 8). Chlorocruorin is found only in certain sedentary marine annelids, in which it is dissolved in the plasma. The taxonomic distribution of this pigment similarly does not follow any obvious evolutionary pattern. For example, in one particular genus of annelids, one species is known to possess Hb and red blood; a second species possesses chlorocruorin and green blood; and a third species does not contain respiratory pigments at all and thus has colorless blood. In the annelid genus *Serpula,* moreover, both Hb and chlorocurorin are present (in plasma). This is

TABLE 8. Respiratory pigments and oxygen-carrying capacity of blood

Pigment	Animal group	cm^3 O_2 carried per 100 cm^3 blood	Occurrence in blood
hemocyanin (colorless to blue) Cu	some snails	2	plasma
	squids	8	plasma
	crustacea	3	plasma
hemerythrin (colorless to red) Fe	some annelids and related worms	2	cells
chlorocruorin (green to reddish) Fe	some annelids	9	plasma
hemoglobin (purple-red to orange-red) Fe	some mollusks	2	plasma
	some annelids	7	plasma or cells
	fishes	9	cells
	amphibia	11	cells
	reptiles	10	cells
	birds	18	cells
	mammals	25	cells

the only clearly known instance where two different blood pigments occur together.

Hemerythrin is a colorless pigment which turns red when combined with oxygen. Its chemistry is known only poorly, but it has been shown that the pigment is an iron compound containing two atoms of iron per molecule. Thus, two Fe atoms are required for the transport of one O_2 molecule. Hemerythrin too has a highly restricted distribution, being found only in one group of annelids, in some worms (related to annelids) and in one genus of brachiopods (*Lingula*). In these animals the pigment occurs exclusively in the blood cells.

Hemocyanin likewise is colorless, but it turns blue in combination with oxygen. It differs also from the pigments above in that it is a copper compound, two atoms of copper being required for the transport of one O_2 molecule. Unlike the other pigments, moreover, hemocyanin does not combine with CO. The pigment always occurs in plasma, and it is found in two phyla only, i.e., in mollusks (squids, octopuses, and certain snails) and in arthropods (crustacea most particularly). We may note that, in snails having hemocyanin in blood, hemoglobin usually occurs in the muscles.

An interesting group of blood pigments with a possible respiratory function is known to occur in tunicates. These marine chordates possess oxides of vanadium, combined with protein. The pigments are orange, green, and blue, and they occur in the blood cells. Hb is not present (and none, indeed, has as yet been demonstrated in any chordates other than the vertebrates). The vanadium pigments combine readily with O_2, and they release the gas in acid media. Since tunicate blood cells actually contain free sulfuric acid (!), a respiratory role of the pigments is probable though not fully proved. Evidently these presumable ancestors of the vertebrates constitute a most remarkable group. They are virtually unique among animals in possessing cellulose (in the external integumentary coats), and they are quite unique in having reversing hearts, free sulfuric acid within cells, and high concentrations of vanadium, an element so dilute in sea water that its presence there can barely be demonstrated.

We may conclude that lymph, blood, and circulatory systems as a whole serve in two principal capacities, viz., internal steady-state control and transport. The control functions include maintenance of internal ionic balances as well as internal body defense through clotting, immune reactions, and phagocytosis. The transport functions involve mainly foods, gases, and wastes, and in the final analysis such functions too are actually aspects of steady-state control.

ALIMENTARY SYSTEMS

If an animal could obtain all the nutrients it requires in the form of pure, immediately usable ions and molecules, it would not need an alimentary system. It could then simply acquire such nutrients from the environment by direct absorption through its cell surfaces. This is actually the nutritional pattern in many parasitic animals, in which alimentary structures are often highly degenerate or even absent altogether. But, apart from water and minerals dissolved in water, directly usable nutrients in ionic and molecular form are largely unavailable to free-living holotrophic animals. What a holotroph requires is plant or animal matter in bulk, living or dead. And it is the principal function of an alimentary system to permit the *ingestion* of such bulk nutrients, the *digestion* of these nutrients into ions and molecules directly absorbable by cells, and the *egestion* of all remaining materials.

The key process, digestion, is accomplished principally by chemical means, but is also aided in many animals by mechanical means. Mechanical digestion subdivides ingested food into fine, often colloidal particles suspended in water. Chemical digestion then reduces these particles to molecular dimensions. In the process, usable ions and molecules become separated out, and more complex molecules are broken up into smaller, usable ones.

Chemical dissolution of bulk foods depends on specialized *digestive enzymes*, secreted by various parts of the alimentary

9.14. Summary of the pattern of enzymatic digestion of the main classes of foods.

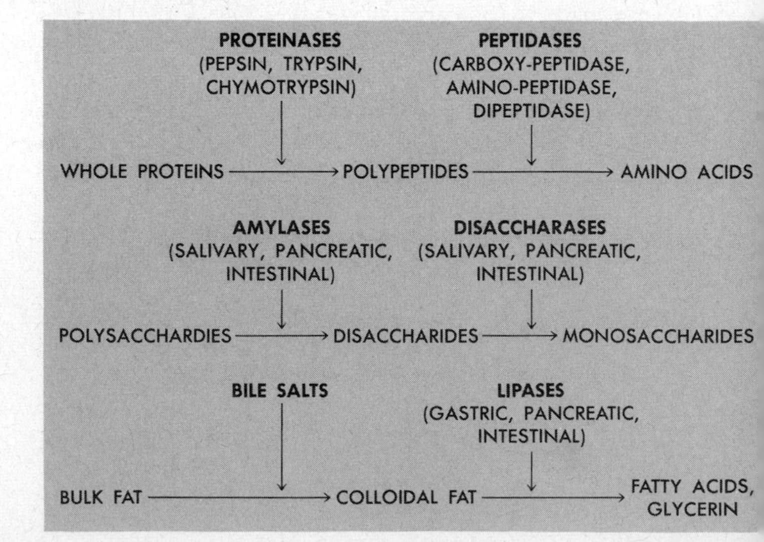

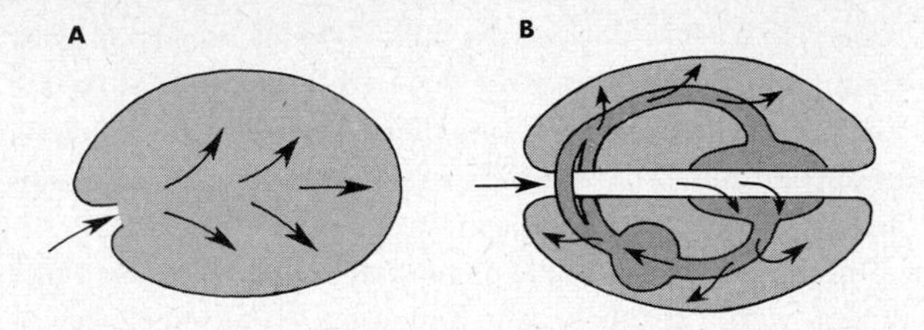

9.15. Short-distance food transfer as in A (path of arrows) takes place by direct diffusion and active transport of nutrient molecules from cell to adjacent cell. Long-distance transfer as in B occurs via the body circulation; i.e., nutrients are collected from the alimentary tract by blood and lymph (dark shading) and are transported to all body regions. In such locations, short-distance transfer then distributes food molecules from the circulation to nearby tissue cells.

apparatus. The enzymes principally catalyze the digestion of three classes of food compounds, viz., carbohydrates, fats, and proteins. The digestive reaction in all cases is a *hydrolysis*, i.e., a dissolution by water. A generalized digestive reaction may therefore be written as

$$\text{food} + \text{water} \xrightarrow{\text{enzyme}} \text{smaller nutrient components}$$

Complete hydrolysis of a given food compound usually occurs in a serial sequence of reaction steps, a given large molecule being decomposed into successively smaller ones. Each step is a hydrolysis, and each is catalyzed by a digestive enzyme specific for a particular type of chemical bond. A digestive lipase, for example, specifically breaks the link between the fatty acid and the glycerin component of a fat. Such an enzyme may therefore be effective in the digestion of virtually all kinds of fats. The general nature of the hydrolytic sequences and the enzymes required at each step are outlined in Fig. 9.14.

The combined result of mechanical and chemical digestion is an aqueous food solution. In it are present usable nutrients in the form of dissolved ions and molecules, as well as a variety of undigested and indigestible dissolved or suspended materials. The latter are eventually egested, but the usable components are absorbed from the digestive apparatus and are distributed to all interior parts of the animal body. In simply constructed animals, absorption and distribution of nutrients is accomplished by *short-distance transfer*, i.e., diffusion and active transport directly from cell to cell. In all more complexly constructed animals, *long-distance transfer* via lymph and the circulatory system occurs as well. In such cases, short-distance transport delivers nutrients from the digestive apparatus to the long-distance media and from the latter also to all cells of the body (Fig. 9.15).

The detailed patterns by which these alimentary processes take place vary greatly for different animals. Three main patterns may be distinguished, viz., *intracellular* alimentation, *extracellular* alimentation, and a combination of both. The first is characteristic mainly of protozoa and sponges. An individual cell here functions as a complete alimentary apparatus. The cell may ingest microscopic food particles by forming a *food vacuole*, either in amoeboid fashion or by first trapping food on the cell surface with the aid of current-producing cilia or flagella (Fig. 9.16). The cell cytoplasm then secretes digestive enzymes into the food vacuole and, after hydrolysis within the vacuole, the usable nutrients are absorbed into the cytoplasm. Indigestible remains are egested from the vacuole through a process which is essentially the reverse of ingestion. Protozoan cells live and feed as individual cells, and the digestive cells of a sponge form a tissuelike lining in the interior of the animal (cf. Chap. 20).

9.16. The pattern of intracellular alimentation. Ingestion by various means as shown leads to the formation of an intracellular food vacuole in which enzymatic digestion takes place. After usable nutrients are absorbed into the cell, the unusable remains of the original food are egested from the vacuole and the cell surface.

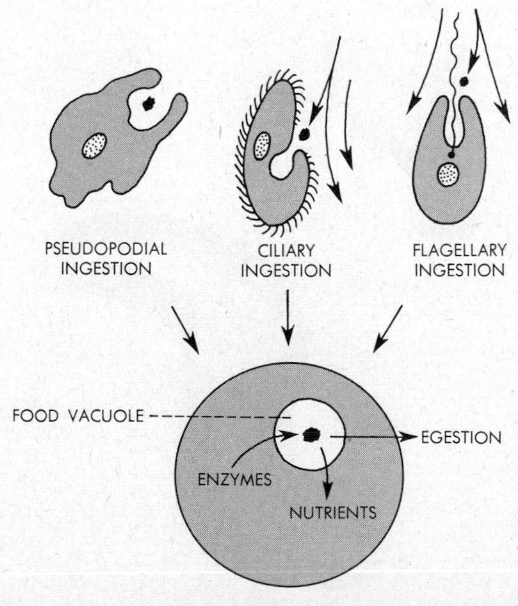

A combination of intracellular and extracellular alimentation occurs principally in the Radiata and the flatworms, i.e., largely in those groups in which the alimentary apparatus has a one-hole sac design (Fig. 9.17). Bulk food is brought into the alimentary cavity by various means, e.g., in coelenterates, by tentacles around the alimentary opening. Inside the alimentary cavity, the digestive cells lining the sac secrete enzymes which bring about a preliminary mechanical and chemical decomposition of ingested food. This constitutes the extracellular phase of digestion. The intracellular phase supervenes when microscopic bits of food, separated away from the main food mass, are ingested by the cells of the sac lining. Such food fragments are then processed intracellularly, essentially as in protozoa and sponges. All indigestible remains are eventually egested through the alimentary opening.

9.17. **Top,** extracellular and intracellular digestion in one-hole sac alimentary systems (radiates, flatworms). Enzymes secreted into the alimentary sac accomplish a preliminary extracellular phase of digestion, and small food particles engulfed into vacuoles of digestive cells are then digested intracellularly, as in protozoa. **Bottom,** extracellular digestion in two-hole tube alimentary systems (Bilateria). Enzyme-secreting digestive glands promote more or less purely extracellular digestion of foods. Nutrient molecules are then absorbed through the gut wall.

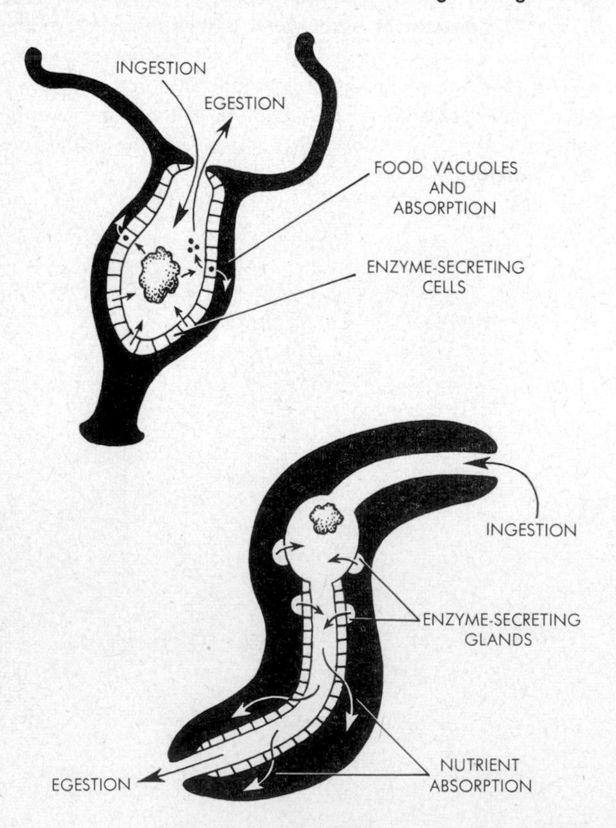

In animals in which the alimentary apparatus is an organ system with a two-hole tube design, purely extracellular alimentation is the rule in most cases: enzymes produced by specialized digestive glands are secreted into the cavity of the alimentary tube, and complete digestion takes place there, extracellularly (cf. Fig. 9.17).

The principal tissue component of the alimentary tube is the *mucosa*, an endodermal layer one cell thick, often ciliated over part or all of its surface. This layer bounds the alimentary cavity everywhere except near the mouth and the anus. At these ends the lining of the alimentary tube is ectodermal, the integument being tucked in for some distance. In some noncoelomate animals the mucosa represents the entire alimentary wall, but in most animals this layer is surrounded by various additional tissues. Derived largely from splanchnic mesoderm, these include layers of fibroelastic connective tissue, which carry networks of blood and lymph vessels; muscle layers, usually organized as circulatory and longitudinally arranged cells; and typically also nerve plexuses, which control the activities of the muscles and the digestive glands. In coelomate animals the outermost layer of the alimentary tract is a one-cell-thick *serosa*, part of the peritoneal lining of the coelom and continuous with the mesenteric portions of the peritoneum (Fig. 9.18).

Apart from the mouth opening itself, ingestive structures often include accessory organs present outside the mouth and also within the mouth cavity and in the pharynx behind it. Such organs are exceedingly varied, but they usually handle food in one of three general forms: fluids; solids of microscopic or near-microscopic dimensions; and solids of greater bulk (Fig. 9.19). Many animals subsist entirely on fluid foods such as plant juices, animal bloods, or liquefied animal matter. The ingestive organs in such cases are *sucking* devices of some kind, either tubular structures as in mosquitos and many other insects, or sucking disks equipped with cutting teeth, as in some leeches and other parasitic forms. In such animals suction is generally produced by a muscular widening of the mouth cavity, the pharynx, or the stomach. Numerous aquatic animals subsist on microscopic solid food such as minute plants and animals. Ingestion here is often accomplished by some form of *filter feeding*. In this process, ciliated ingestive struc-

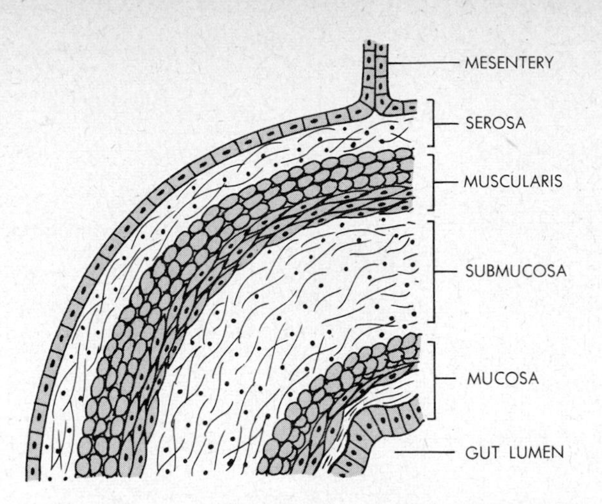

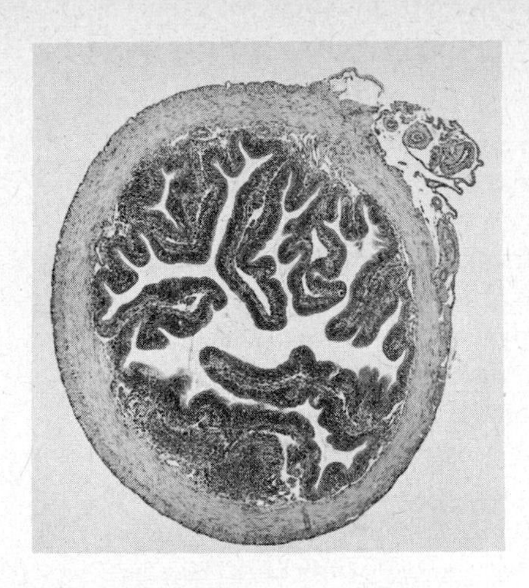

9.18. Diagram: cross section through a portion of the intestinal wall of mammals, to show the principal tissue layers. The mucosa, adjacent to the lumen (gut cavity), consists of columnar epithelium, connective tissue, and thin layers of muscle; the submucosa, of connective tissue, nerves, and blood vessels; the muscularis, of thick layers of inner circular and outer longitudinal muscles; and the serosa, of connective tissue and squamous epithelium, the latter continuous with the mesentery and representing part of the peritoneum. Note that the mucosa is usually extensively folded and that the gut lumen is far wider in proportion to the thickness of the gut wall than suggested in the diagram. **Photo: histological section through the intestine** (at the level of the duodenum), **showing the highly folded condition of the mucosa.**

9.19. Some forms of ingestion. **A** and **B.** Fluid ingestion via sucking pharynx (or stomach). Muscular dilation of the pharynx or stomach cavity provides suction for fluid intake through tubes (as in mosquitos, **A**), through saw-tooth-equipped suckers (as in leeches, **B**), and by many other means. **C.** Filter feeding via ciliated tentacles. The cilia create water currents, and the latter may carry microscopic particulate food. Such food is propelled into the mouth, while strained water passes out between the tentacles. **D.** Ingestion of minute particulate food by means of rasping organs, such as the mollusk radula diagrammed here. The rasper is a horny band with recurved teeth and is moved back and forth by muscles around a cartilaginous supporting prop (dark shading). In forms such an squids, sharp-edged biting jaws (shown below) may serve in ingestion as well. In addition to the ingestive patterns here illustrated, numerous others occur as well.

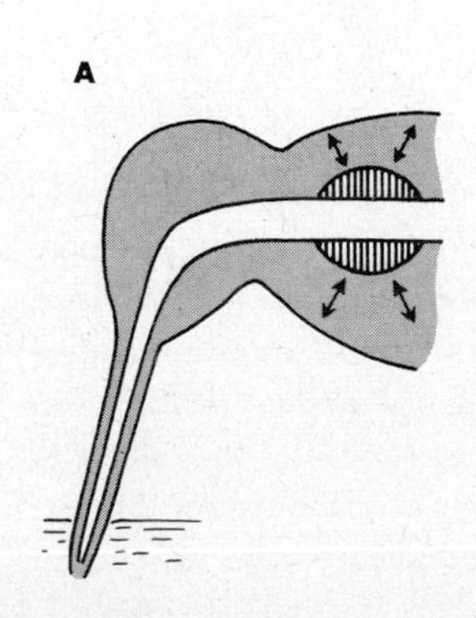

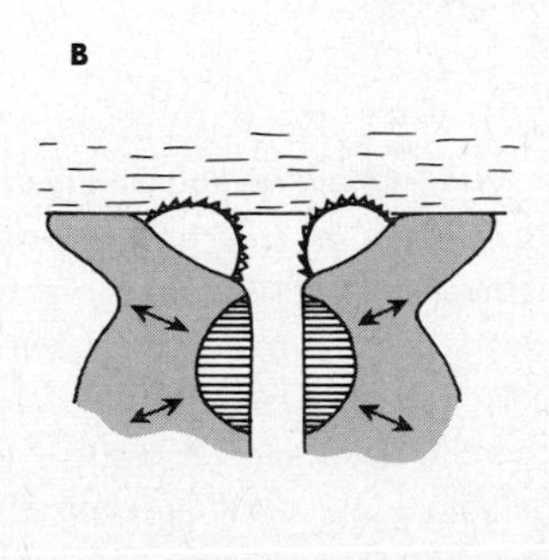

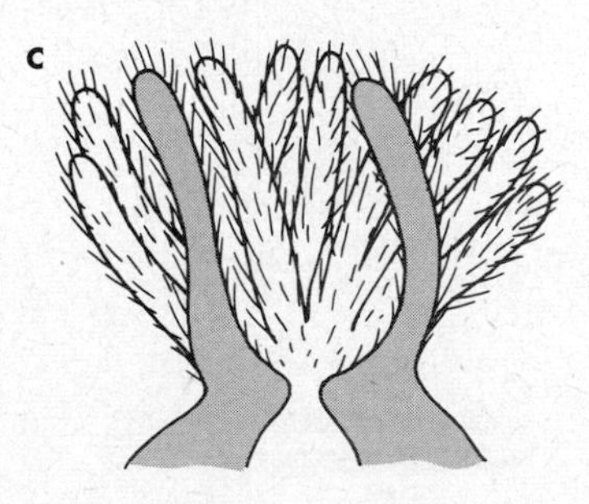

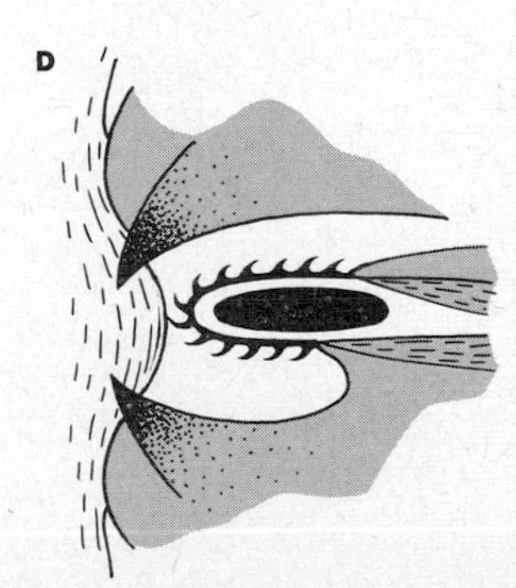

tures in or near the mouth serve to strain food particles from water flowing past. This type of ingestion is particularly widespread among burrowing and sessile animals. It occurs, for example, in mollusks such as clams and in chordates such as tunicates. Some animals, e.g., snails, obtain microscopic food with the aid of filelike *raspers,* which scrape off minute fragments from larger food masses (cf. Fig. 9.19).

The intake of bulkier food is usually accomplished by *biting,* an activity often preceded by some form of *tearing* and often followed by some form of *grinding* or *chewing.* Such processes serve not only in ingestion, but they also constitute a phase of mechanical digestion. The principal structures here are chitinous or calcareous *jaws* and *teeth.* Food intake by such means is perhaps the most widespread, and at all events the most familiar. Some animals obtain bulk food with the aid of the *pharynx,* just behind the mouth cavity. Such animals may, for example, possess thick muscular walls in the pharynx, and the wall may be either *protruded* or *everted* through the mouth. Food may then be gripped by the projecting muscular tube and pulled back (Fig. 9.20).

Once within the alimentary tract, food is propelled along by mucosal cilia and/or by *peristalsis,* i.e., wavelike sweeps of muscular contractions in the alimentary wall. In many animals the alimentary tract is without conspicuous modifications along its course, but in many others the tract is specialized regionally as a series of distinct organs (Fig. 9.21). Anteriorly, apart from mouth cavity and pharynx, such organs usually include *esophagus* and *stomach,* and in specific cases also organs such as *crop* and *gizzard.* Posteriorly, the principal organ is the *intestine.* In vertebrates the latter is subdivided into a long, comparatively narrow *small* intestine and a shorter, wider *large* intestine. All such alimentary subdivisions have the same general tissue composition, but they differ in the amounts and the specializations of these tissues. For example, a stomach is usually characterized by the presence of

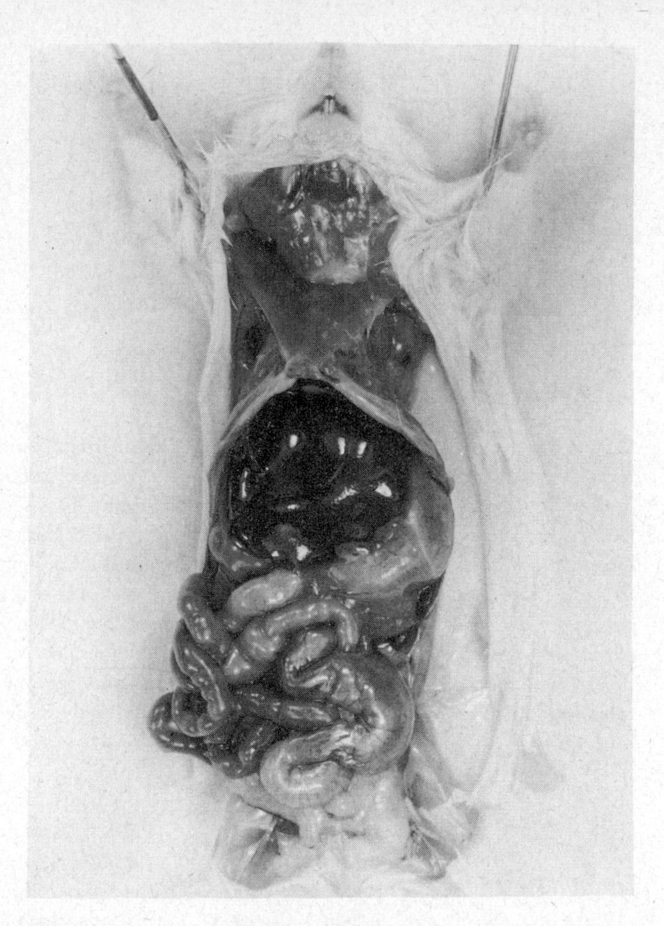

9.21. Some of the regionation of alimentary systems is illustrated in this abdominal dissection of a mouse. The most conspicuous organ of the digestive tract seen here is the small intestine. The dark organ just below the rib cage is the liver, and to its right (on the left side of the animal) lies the stomach. In the lower abdomen the small intestine connects with the wider large intestine, a portion of which is well visible in the photo. See later chapters for illustrations of regional alimentary organs in other animal types.

9.20. **Left:** rest (top) and extended (bottom) condition of eversible pharynx. **Right:** rest and extended condition of protrusible pharynx. The walls of the pharynx are shown in dark shading. In eversion, the whole pharynx slides out through the mouth; in protrusion, the pharynx narrows and elongates out of the mouth.

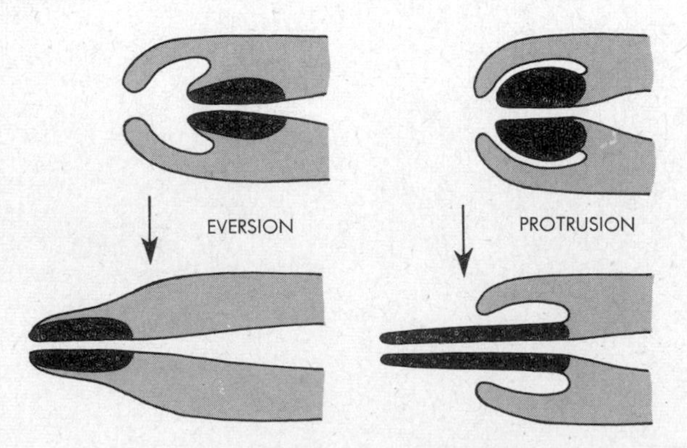

especially thick layers of muscle and of digestive glands of particular types. Most of the alimentary organs actually contribute to chemical digestion, digestive glands of various types being present almost all along the tract. In addition, the anterior organs serve in mechanical digestion and also in temporary food storage, and the posterior, intestinal regions are the chief absorptive structures.

Mechanical digestion, frequently already begun by jaws, teeth, and other ingestive organs in and around the mouth, may be continued by the grinding action of organs such as stomachs and gizzards, and also by specialized grinding organs located in the pharynx of some animals. In vertebrates, moreover, mechanical subdivision of food is accelerated greatly by strong HCl secreted from the stomach wall. This acid macerates food; i.e., it loosens tough fibrous components and erodes the cementing substance between cells. A food mass then literally falls apart.

Chemical digestion, which may begin in the mouth through saliva, accompanies the mechanical subdivision in the anterior portions of the tract and continues particularly as food reaches the intestine. Digestive juices containing lubricating mucus and hydrolytic enzymes are secreted in part by individual mucosal cells; in part by multicellular glands embedded within the alimentary wall; and in part by glands lying outside the tract but connecting to it via ducts. In vertebrates, such external glands are the *salivary glands*, the *pancreas*, and the *liver* (Fig. 9.22). Many invertebrates likewise possess digestive glands functionally comparable to the salivary and pancreatic organs of vertebrates, i.e., enzyme- and mucus-secreting glands which connect to mouth cavity and/or intestine. However, none of the invertebrates possesses an organ quite equivalent to the vertebrate liver. Certain invertebrate digestive glands are sometimes called "livers" (e.g., in lobsters), but such organs function either in the manner of a pancreas, or as absorptive structures, or in both capacities. On the other hand, the vertebrate liver is primarily a nutrient-processing, -storing, and -distributing organ. Moreover, it performs most of its functions after digestion and absorption have already taken place (see below). At least one liver function does contribute directly to digestion, i.e., the manufacture and secretion of *bile*, a digestive juice which may be stored temporarily in the *gallbladder*. Bile contains not enzymes but bile salts; and when these reach the intestine, they emulsify fats. This mechanical action subdivides fat into minute colloidal droplets which expose large surfaces to the chemical action of lipase.

Usable ionic and molecular nutrients are absorbed into the mucosa of the intestine and, in invertebrates, also into the interior cells of certain digestive glands ("livers") and digestive pouches connecting with the intestine. In many animals, the absorptive surfaces are enlarged greatly by the presence of intestinal coils, loops, or branches, and by numerous folds in the mucosal lining. In vertebrates, the mucosa additionally forms millions of microscopic fingerlike projections, so-called *villi*, which give the intestinal lining a fine carpetlike texture and enormous surface area (Fig. 9.23). In many animals, the anterior intestinal region (e.g., the small intestine of vertebrates) absorbs mostly organic nutrient molecules, whereas the posterior portions (such as the large intestine of vertebrates) absorb largely water and dissolved inorganic ions.

From the mucosa, absorbed nutrients are distributed to all parts of the body. In animals without circulatory systems and in those with open systems, long-distance distribution is accomplished by lymph, hemolymph, or blood in direct contact with the alimentary tissues. In animals with closed circulatory systems, the nutrients pass into blood vessels within the intestinal wall and are carried to all body regions within the vessels. Regardless of the specific transport pattern, distribution in invertebrates is always direct, from intestine to utilizing cells. In vertebrates, on the other hand, distribution is largely indirect, most foods first passing from the intestine to the liver. Some colloidal fats droplets usually escape hydrolysis in the intestine, and these, along with some water, are absorbed into intestinal lymph channels (cf. Fig. 9.23). The latter bypass the liver. However, all other usable nutrients are absorbed into blood vessels, and these are so arranged that they carry blood-borne foods to the liver (Fig. 9.24).

In effect, the alimentary tract of vertebrates merely makes foods available, but the liver determines what is to happen to them. The largest "gland" of the vertebrate body, the liver has been estimated to perform well over 100 different functions. Some of these are outlined in Fig. 9.25. It will be noted that liver cells act on all main classes of food compounds. Carbohydrates are released into the blood as glucose, at concentrations maintained constant by the liver. Excess quantities of carbohydrates are stored as glycogen or are converted into fats. Lipid nutrients from the intestine are partly stored, partly converted to carbohydrates, and partly released into the blood for utilization or for storage in adipose tissues elsewhere. Amino acids are partly released into the blood for distribution, partly transaminated and deaminated. Any resulting keto acids are converted into carbohydrates or fats,

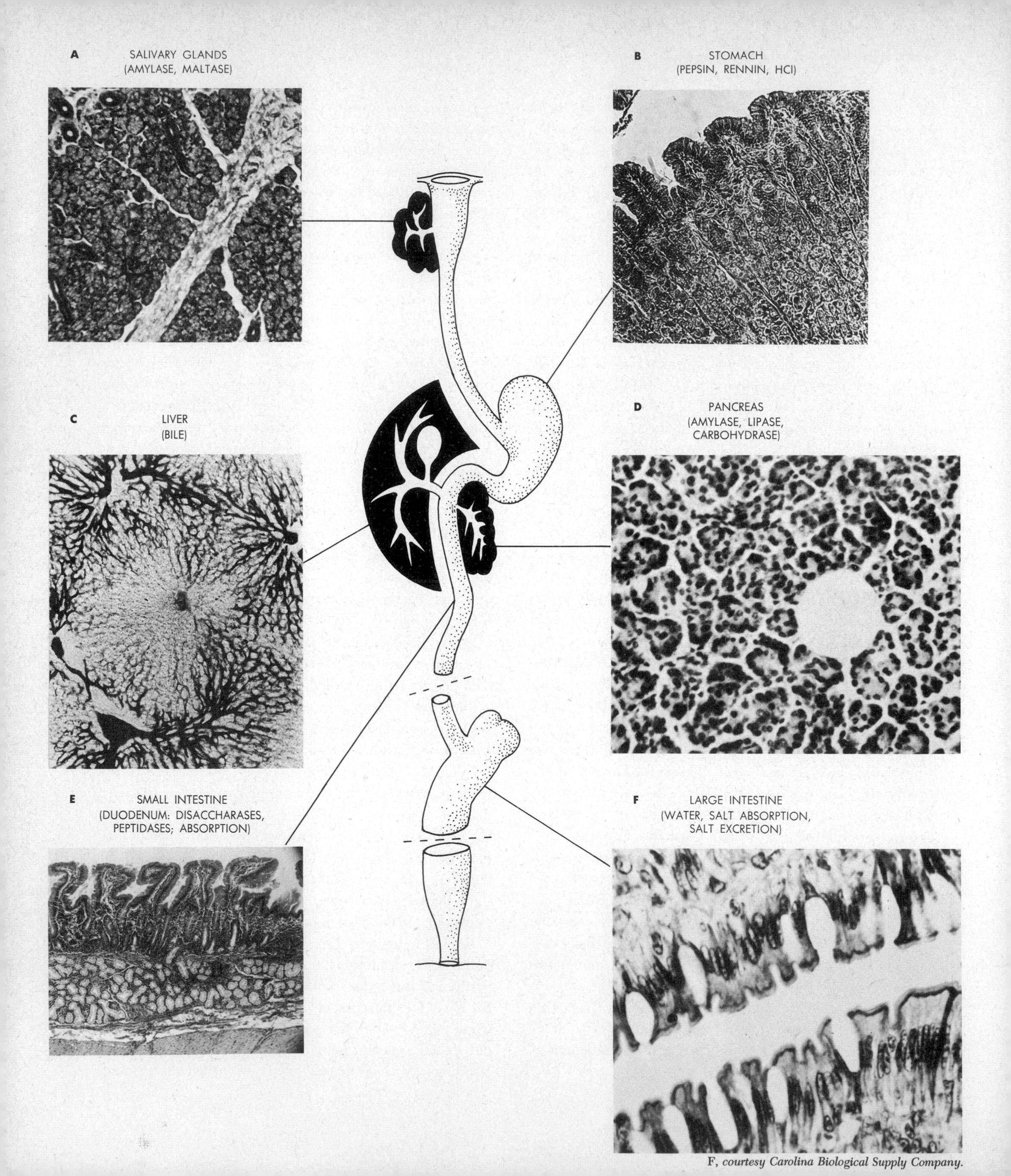

F, *courtesy Carolina Biological Supply Company.*

9.22. Schematic representation of the mammalian alimentary tract and photomicrographic sections through the principal organs (on facing page). The enzymes secreted and the other digestive roles of these organs are indicated above each photograph. In the stomach, secreted rennin is a milk-curdling agent. In the diagram of the liver, note the gall bladder and the common bile duct passing from both liver and gall bladder into the duodenum. Other representations of liver tissues are shown in Figs. 9.24 and 7.25. Photo insets: **A.** Section through a salivary gland. Note the connective tissue stroma (light areas in photo) traversing the gland and binding groups of gland cells together. Note also the several small salivary ducts (dark rings). **B.** Section through a portion of stomach wall. Note folded mucosa near top of photo. **C.** Section through liver, showing parts of a few lobules injected to reveal the blood channels (dark). Blood brought by the hepatic portal vein to a lobule passes to the hepatic vein in the center of the lobule (cf. Fig. 7.25). **D.** Section through pancreas. The large round space is a branch of the pancreatic duct (cf. also Fig. 8.50). **E.** Section through the wall of the duodenum. The cavity of the gut is toward the top. Underneath the folded inner surface tissues note the glandular layer. Its secretion is discharged into the gut cavity and contributes to the composition of intestinal juice. **F.** Section through a mucosal fold of the large intestine. Note the many so-called goblet cells, mucus-secreting cells in the mucosal lining.

and any resulting free amino groups appear as ammonia. This compound is excreted eventually, in many cases after prior conversion to urea or uric acid (Fig. 9.26). Liver cells also store fat-soluble vitamins and, as Fig. 9.25 indicates, they perform various functions not primarily nutritional in nature.

Liver cells thus regulate what kinds and what quantities of foods are sent out into body tissues. The adaptive advantage of the organ thus becomes clear. No matter when or at what regular or irregular intervals a vertebrate eats, the liver collects most of the food as it is absorbed from the gut and then releases it into the body at a pace adapted to the particular requirements of the moment. Therefore, whereas the metabolism of other animals reaches peaks just after food has been eaten, the metabolism of vertebrates may remain at a continuously steady level.

Materials which cannot or simply have not been digested move into the posterior portion of the tract. More water and inorganic ions are absorbed here, but such substances may also be *excreted* into the alimentary cavity if the body happens to contain them in excessive quantities (cf. Fig. 9.22). The lower tract usually harbors a rich permanent population of bacteria. These organisms live on the indigestible remains, and they may also excrete various substances useful to the animal host. For example, an animal frequently obtains some of the vitamins it requires from the byproducts of bacterial metabolism. Another result of bacterial activity is the *decay* of the intestinal remains; it is this decay which transforms the remains into *feces* ready to be egested.

In many animals, the intestine continues without further structural modifications directly to the anus. In other cases, an additional subdivision of the tract interconnects intestine and anus; i.e., a short elimination tube, called a *rectum,* may be present. In numerous animals the excretory and/or reproductive ducts open into the most posterior portion of the intestine, in which case this region is called a *cloaca.* The anus itself, like the mouth, is ringed by muscles which bring about the opening and closing of the aperture.

BREATHING SYSTEMS

The basic component of any breathing system is a *breathing surface,* i.e., a region exposed directly to the external environment where O_2 may diffuse in and CO_2 may diffuse out. In

9.23. Schematic representation of the blood and lymph circulation through the villi of the vertebrate intestine. Of the organic nutrients absorbed through the mucosa, colloidal whole fat enters the blind-ended lymphatic lacteals, passes from there into the lymph circulation of the body, and so bypasses the liver. Lymph ultimately drains into the blood circulation, hence whole fat eventually does come to be transported by blood. All other organic nutrients from the intestine are absorbed directly into the blood circulation of the villi, and from there these foods are transported to the liver.

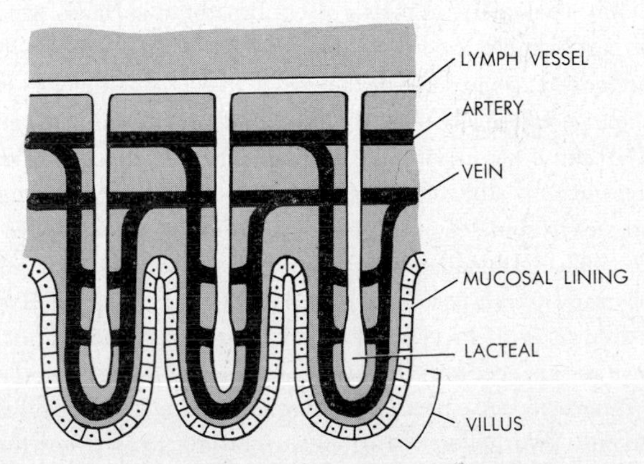

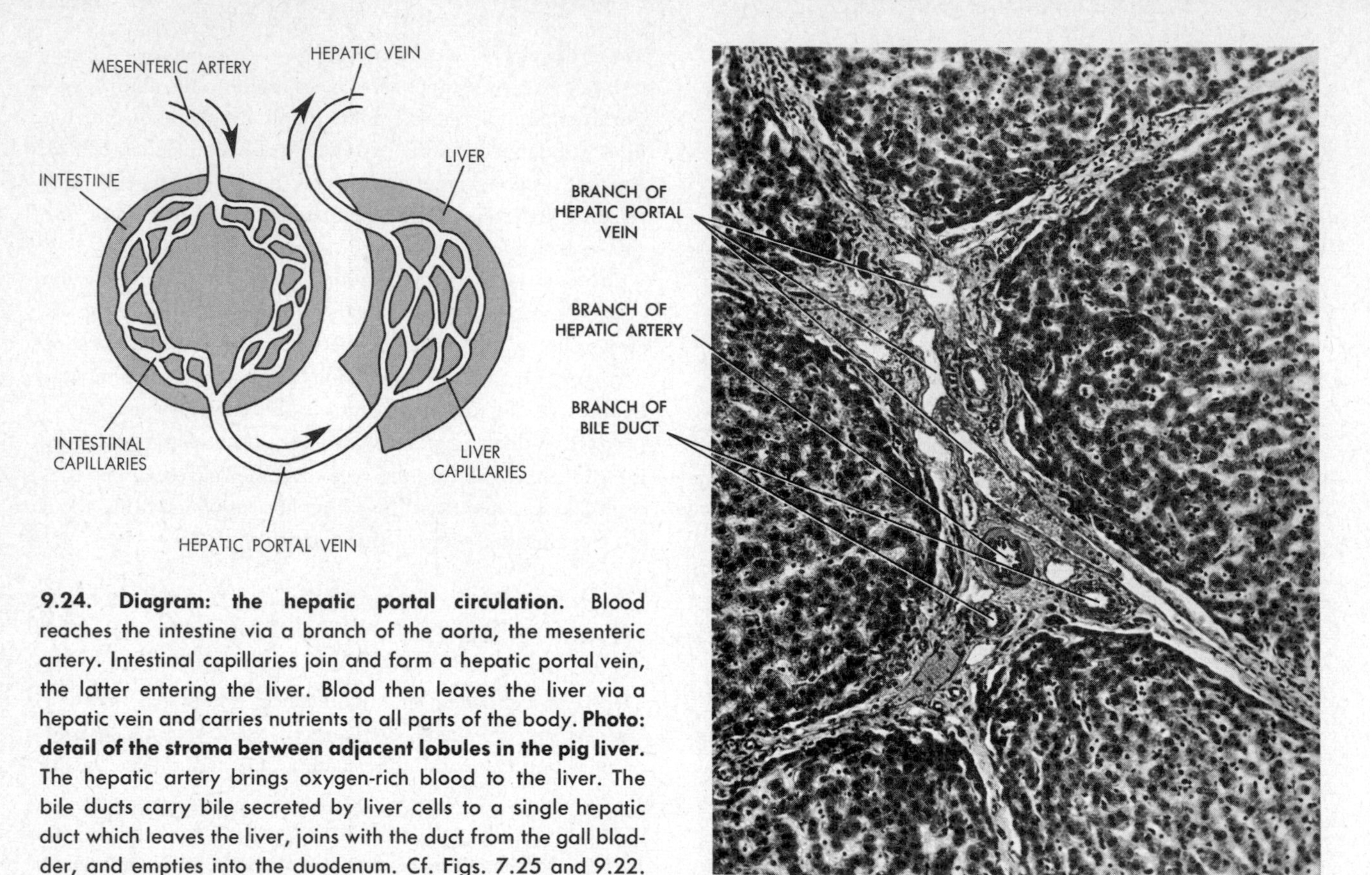

9.24. Diagram: the hepatic portal circulation. Blood reaches the intestine via a branch of the aorta, the mesenteric artery. Intestinal capillaries join and form a hepatic portal vein, the latter entering the liver. Blood then leaves the liver via a hepatic vein and carries nutrients to all parts of the body. **Photo: detail of the stroma between adjacent lobules in the pig liver.** The hepatic artery brings oxygen-rich blood to the liver. The bile ducts carry bile secreted by liver cells to a single hepatic duct which leaves the liver, joins with the duct from the gall bladder, and empties into the duodenum. Cf. Figs. 7.25 and 9.22.

all animals, terrestrial ones included, the actual gas exchange occurs in a liquid medium; the breathing surface is exposed to water if the animal is aquatic or is kept moist by the animal if the environment is air. Gases then must dissolve in a film of liquid covering the breathing surface before they can diffuse in or out. A dry breathing surface is a dead surface.

Only coelomate animals possess breathing surfaces which are part of specialized breathing systems. In all other animals, and indeed also in some of the coelomates (e.g., annelids), gas exchange essentially takes the form of *skin breathing:* O_2 and CO_2 diffuse across virtually all body surfaces exposed to the environment—the skin mainly, but also the alimentary surfaces to some extent. Internal distribution in such cases is accomplished adequately by direct diffusion between adjacent cells and by transport via lymph, hemolymph, or blood. Even in animals with specialized breathing systems, the skin may serve as an accessory breathing organ (e.g., frogs).

Where present, specialized breathing systems may be classified into two general categories according to whether they function in an *aquatic* or an *aerial* environment (Fig. 9.27). Among nonchordate invertebrates, both types of systems are *ectodermal* derivatives. Aquatic systems are *gills,* formed by outgrowths from the integument. Aerial systems are *lungs* or *tubes,* formed by ingrowths from the integument. Gills are usually feathery, leaflike, or filamentous; lungs are largely sacs. In both gills and lungs, the actual breathing surfaces are one-cell-thick epithelia, exposed to the environmental medium on one side and to the internal circulatory medium on the other. Where the circulatory system is open, hemocoelic sinuses usually extend right up to the breathing epithelum, and blood is in direct contact with it. In closed systems, dense networks of capillaries held within thin layers of connective tissue invest the inside surfaces of the breathing epithelium. Gases here must traverse two layers of cells intervening between environment and blood.

As later chapters will show, invertebrate gills are located in many different body regions. For example, the organs may be distributed over most of the body surface, as in starfishes; they

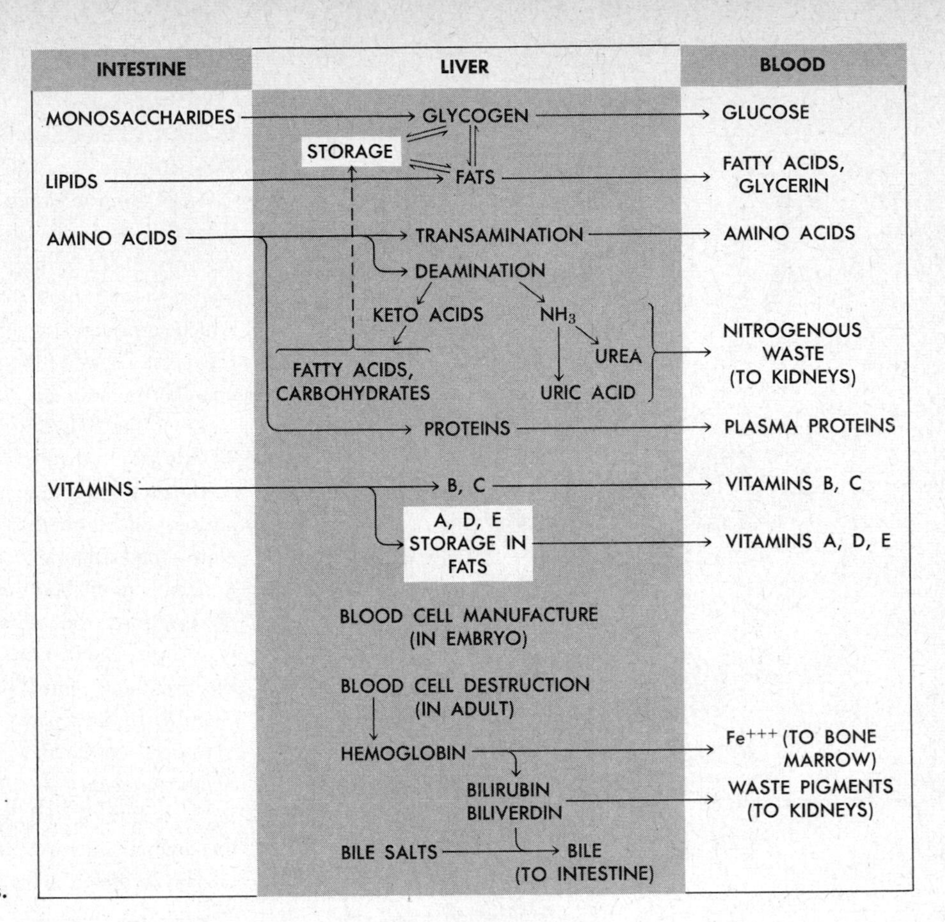

9.25. Summary of liver functions.

9.26. Synthesis patterns of nitrogenous excretion products. Urea production via the ornithine cycle in the vertebrate liver is outlined at left. Atoms indicate the fate of the NH_3 and CO_2 raw materials. The net input and output of the cycle is seen to be given by the equation $2\ NH_3 + CO_2 \longrightarrow NH_2CONH_2 + H_2O$. The pathways at right show that NH_3 may also participate in purine and pyrimidine synthesis (cf. also Fig. 6.14), and that stepwise degradation of these nitrogen bases leads to the formation of the excretion products indicated, urea included. Uric acid is a major excretory endproduct in, for example, insects, birds, and some reptiles; allantoin, in turtles and some mammals; allantoic acid, in some bony fishes. NH_3 as such is excreted in most invertebrate groups, and urea, in most vertebrate groups not named above. In actuality animals generally excrete mixtures of various endproducts, with one or the other substance predominating.

AMINO ACIDS
NH_3
CO_2
PURINES, PYRIMIDINES
XANTHINE
CITRULLINE
H_2O
NH_3
H_2O
ORNITHINE
URIC ACID
ARGININE
ALLANTOIN
H_2O
ALLANTOIC ACID
UREA

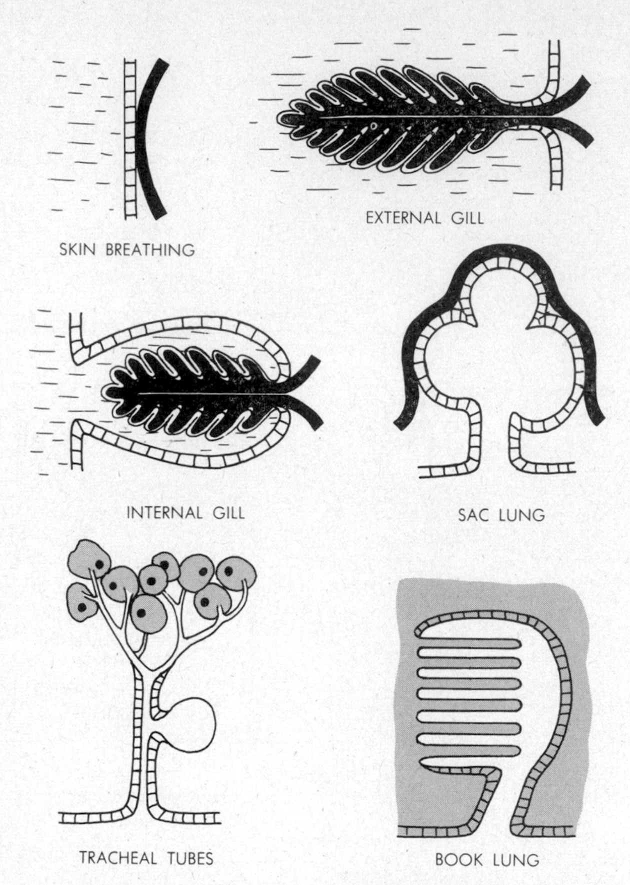

9.27. Patterns of nonchordate breathing. Epidermis shown as layer of cells, blood and vessels dark-shaded. As indicated, all such breathing organs are outfolded or infolded integumentary derivatives. External and internal gills differ in principle only in whether or not they are tucked into chambers. Tracheae occur in terrestrial arthropods, fine air tubes here reaching individual cells; muscle-enveloped sacs folded out from a main tube (as shown) often serve as air reservoirs and air pumps. Book lungs occur in most spiders, the gas-carrying medium here being hemolymph (shown dark-shaded). Free hemolymph rather than blood within vessels also transports gases in skin-, gill-, and lung-systems of animals in which vessel circulations are absent. Some aquatic arthropods possess book gills, constructed like book lungs but operating in water.

may be attached near the bases of the legs and other appendages, as in crustacea such as lobsters and crabs; they may be tucked into a chamber underneath the rim of the shell, as in snails; they may form flaps hanging underneath the shells, as in clams; or they may be tucked into the body and communicate with the cloaca, as in the sea cucumbers.

Apart from a variety of skin-breathing terrestrial worms, terrestrial invertebrates include only some snails and the majority of the arthropods. True lungs do not occur among any of these groups. In land snails, the space under the shell which in aquatic snails forms a gill chamber is used in terrestrial snails as a lung chamber (Fig. 9.28). The floor of this chamber is muscular and arches upward. When the muscles contract, the floor flattens out, the chamber thus enlarges, and air is sucked into it. Gas exchange then occurs through the epithelium lining the chamber. When the floor muscles relax, air is expelled from the chamber. Spiders and scorpions possess chitin-lined chambers containing *lung books.* Located on the underside of the body, each such chamber communicates with the exterior through a slitted opening, and hanging from the wall of the chamber are sets of leaflike plates. These represent the breathing epithelium. Air is sucked into and out of a chamber by muscular action.

Insects, centipedes, and millipedes breathe by means of unique *tracheal tubes,* or *tracheae.* These ducts originate at the body surface, and they branch extensively in the interior of the animal, microscopic branch terminals reaching into all tissues. In effect, air is piped from the outside to all interior cells, and cellular gas exchange then can take place even deeply within the animal. The tracheae are composed of chitin, and their openings to the exterior are equipped with valves called *spiracles.*

In chordates, gills and lungs are always *endodermal* derivatives. More specifically, they arise from the pharynx (cf. Fig. 9.29). In the embryo, a series of pouches grows out from each side of the pharynx, and a corresponding series of pouches grows in from the skin on each side of the head. The two sets of pouches on each side eventually meet and fuse, resulting in the formation of slitlike channels interconnecting the pharynx and the external environment. Lined with a ciliated breathing epithelium, these channels constitute the *pharyngeal gill slits.* The pharyngeal cilia draw water through the mouth, and they also strain food out of the water and pass it back into the esophagus. The water itself is diverted to the outside via the gill slits; and in its passage through the slits, gas exchange takes place.

The blood circulation is associated intimately with this breathing system (cf. Fig. 9.29). The heart is located ventrally, just behind the gills. A main artery, the *ventral aorta,* passes forward from the heart and splits into a left and right branch

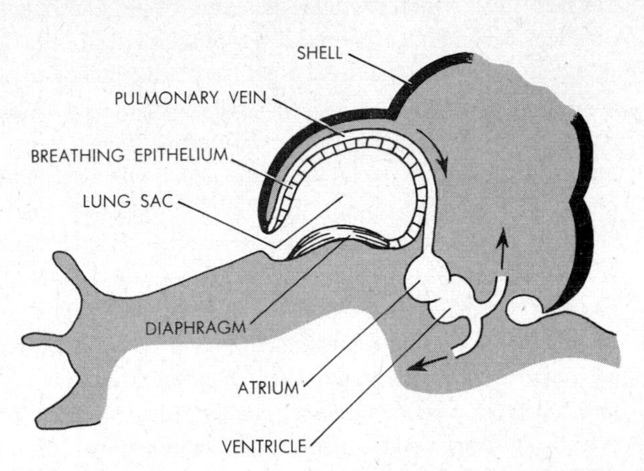

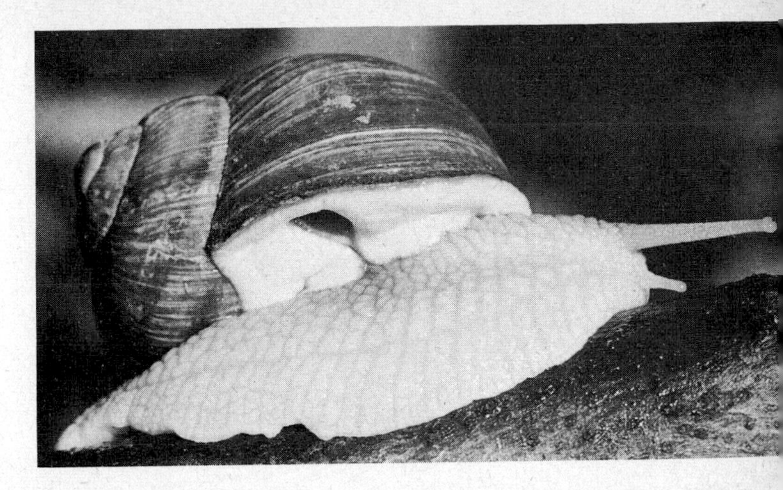

9.28. Diagram: schematic representation of the lung of the land snail *Helix*. Hemolymph in hemocoelic sinuses bathes the epithelium of the lung sac, and branches of a pulmonary vein then carry oxygenated blood to the heart. **Photo: side view of *Helix*, showing the opening to the lung chamber under the shell rim.**

lying ventral to the gills, one each side of the body. Each branch then rebranches into a series of *aortic arches* which pass dorsally through the tissue between the gill slits. Capillary beds from these aortic arches vascularize the breathing epithelia. On the dorsal side of the body the aortic arches all rejoin and form a median *dorsal aorta,* which carries oxygenated (arterial) blood posteriorly. All other body regions are supplied with arterial blood from this source. Veins from the body ultimately return oxygen-poor (venous) blood to the heart.

Notwithstanding many differences of detail, all aquatic chordates are characterized by a gill system and gill circulation of this type. Early vertebrate chordates also evolved the vertebrate lung. These animals are believed to have lived in fresh-water environments which dried out periodically. The adaptive response to such hazards was the evolution of a pouch which grew out ventrally from the pharynx and remained connected with the pharynx by a duct. This pouch was supplied with blood by a branch of the most posterior aortic arch, and a vein from the pouch returned blood to the heart. In a terrestrial environment, air could then be swallowed via the mouth into the pouch, and gas exchange could take place there. The pouch in effect was a lung, and the duct connecting it to the pharynx was a windpipe, or trachea (Fig. 9.30).

These early vertebrates gave rise to two main groups of descendants. In one group, the modern bony fishes, the animals still possess the air pouch, but they use it as a *swim bladder.* In some of these fishes the connecting duct to the pharynx actually still persists, but in other cases it has degenerated (cf. Chap. 31). The second group of descendants continued to use the air pouch as a lung, and from this group of early air-breathing fishes later arose the land vertebrates. The original single lung in time evolved into a paired sac, and an air passage from nose to pharynx became elaborated as well. Thus, air could pass from nose to lung *across* the food channel, and food could pass from mouth to esophagus across the air channel; the pharynx came to be a common segment in the food and the air stream, as in man. Amphibia still develop pharyngeal gills during their aquatic larval stages; but the gills later degenerate in most cases, and the adults acquire lungs instead. Reptiles, birds, and mammals possess lungs at all stages, and functional gills never form (cf. Fig. 9.30).

Nevertheless, the blood circulation between heart and lung in the land vertebrates still indicates its derivation from that of the original aortic gill arches. Fig. 9.30 outlines the circulation pattern in a lung-breathing fish. It will be noted that arterial blood returning from the lung to the two-chambered heart mixes with venous blood coming into the heart from the body. Thus, *mixed* arterial and venous blood is then pumped out from the heart, and most of the body actually never receives fully oxygenated blood. This is essentially still the pat-

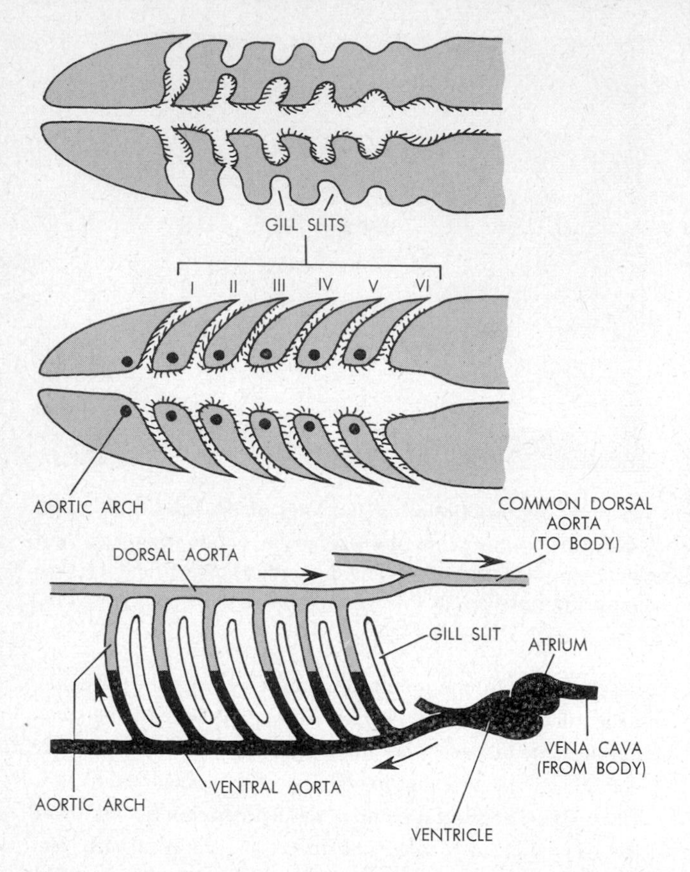

9.29. The endodermal gills of chordates and the basic vertebrate gill circulation. Top, diagram outlining the development of paired vertebrate gills as endodermal pouches from the pharynx meeting corresponding pouches from the integument. Successive stages are indicated in anteroposterior sequence. Middle, the six basic pairs of ciliated vertebrate gill slits. Food is strained into the esophagus (toward right), while water flows out through the slits and oxygenates blood in the aortic arches in the process. Bottom, side view of the basic gill circulation on the left side of the body. Venous blood, dark shading; oxygenated blood, light shading.

tern in adult amphibia, animals which possess a three-chambered heart. Clearly, however, breathing would be more efficient if a mixing of arterial and venous blood in the heart could be avoided. Such mixing has indeed been circumvented by the evolution of a four-chambered heart, as in reptiles, birds, and mammals. The four chambers here in effect represent a double heart, each unit composed of one atrium and one ventricle. A muscular partition separates the right and left unit; and the right side handles venous blood only, the left side arterial blood only (cf. also Fig. 9.6). Venous blood from the body passes through right atrium and right ventricle to the lungs, via *pulmonary arteries*. These are derived from the posterior pair of aortic gill arches which supplied blood to the ancestral vertebrate air pouch. Arterial blood from the lungs returns to the left heart via *pulmonary veins*, the latter again representing derivatives of the veins which returned blood from the air pouch to the heart. Unmixed, fully oxygenated blood then leaves the left ventricle via the aorta, a vessel which still arches from the ventral side dorsally and continues posteriorly as the dorsal aorta. In reptiles, the aorta again divides into a left and right branch after it leaves the heart, corresponding to one of the pairs of aortic gill arches (the fourth pair) in early vertebrates. The two branches then join dorsally and form a single median dorsal aorta. However, only the *right* aortic branch persists in birds, and only the *left* branch in mammals (cf. Fig. 9.30).

The breathing system of birds and mammals is characterized by the presence of voice-producing organs. In song birds, the organ is a *syrinx*, a cartilaginous chamber located at the end of the trachea, where the latter divides into a left and right *bronchus* to each lung. In mammals, an analogous structure, the *larynx*, is situated just behind the pharynx, at the origin of the trachea. Present within the larynx are the *vocal cords*, a pair of membranes with a slitted opening between them. The laryngeal cartilages may be moved by muscles,

9.30. Gills and hearts of vertebrates (on facing page). Venous blood, black; oxygenated blood, white; mixed blood, gray. The basic six pairs of aortic arches and gill slits are shown at top, middle; a side view is just below, indicating the nasal chamber and the primitive air pouch and duct to the pharynx. In sharks, the first gill slit has become reduced to a spiracle and the first aortic arch has degenerated. In bony fishes the first two aortic arches are absent. In lungfishes the air pouch functions as a lung, with a pulmonary artery branching to it from aortic arch VI (see also side view below main figure). Amphibian tadpoles essentially still exhibit the lungfish condition. But, with the disappearance of gills in adult amphibia, some of the aortic arches degenerate. Note also the two atria and, in side view, the posterior extension of the ventricle, which brings the atria anterior to the ventricle. The heart receives both venous blood from the body and oxygenated blood from the lungs (via pulmonary veins, not shown), hence blood pumped out by the heart is mixed (although the oxygen-richest blood does go to the head, the oxygen-poorest, to the lungs). In reptiles, birds, and mammals a four-chambered heart is present, with venous blood being confined to the right atrium and ventricle. Only the right aorta remains in birds, only the left in mammals.

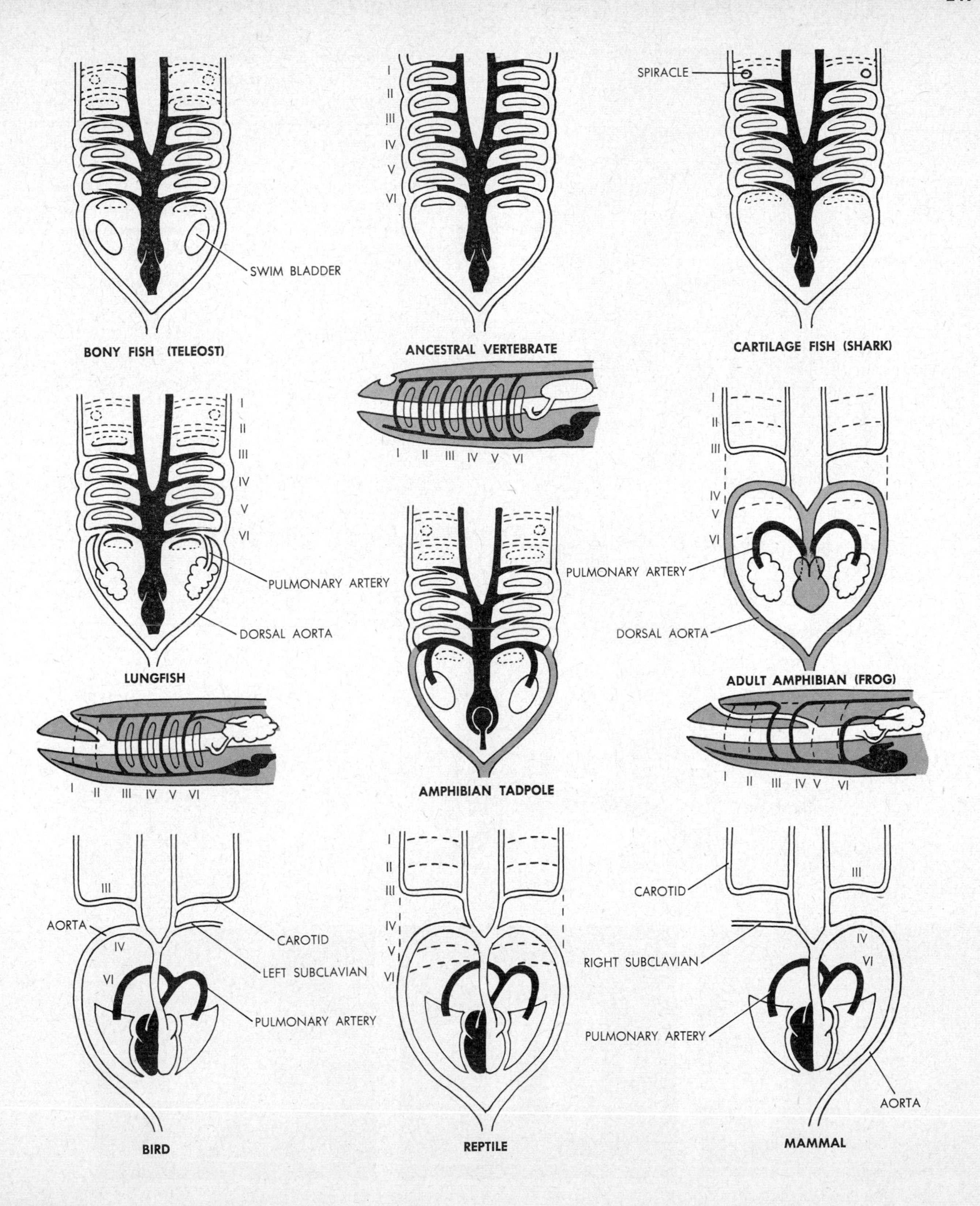
SPIRACLE
SWIM BLADDER
BONY FISH (TELEOST)
ANCESTRAL VERTEBRATE
CARTILAGE FISH (SHARK)
I
II
III
IV
V
VI
PULMONARY ARTERY
DORSAL AORTA
LUNGFISH
AMPHIBIAN TADPOLE
ADULT AMPHIBIAN (FROG)
AORTA
CAROTID
LEFT SUBCLAVIAN
RIGHT SUBCLAVIAN
BIRD
REPTILE
MAMMAL

A

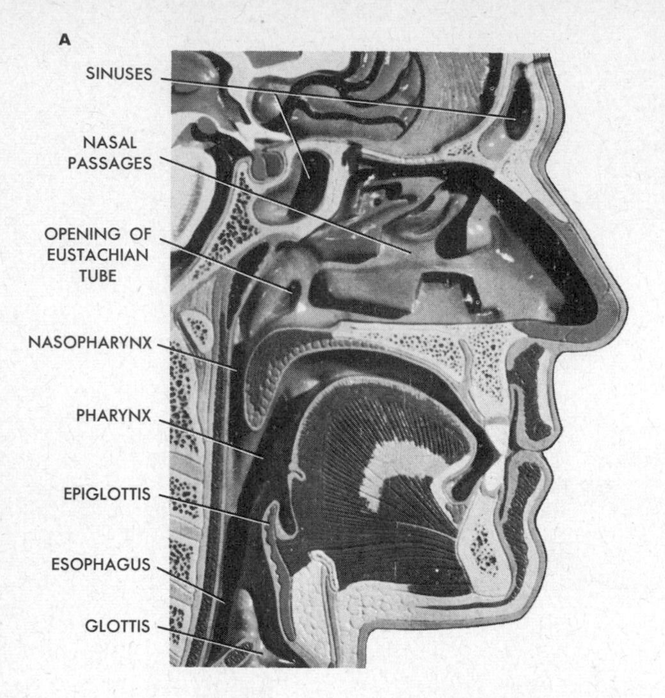

9.31. **A.** The nasal passages and the upper parts of the human breathing system. Note that in the pharynx the air path (from nasopharynx to glottis) crosses the food path (from mouth cavity to esophagus). **B.** The anatomical relations of the lower parts of the breathing system in birds (left) and mammals (right). Note the comparative position of syrinx and larynx. Note also that, in mammals, the intrathoracic cavity is sealed off. **C.** The vocal cords of man. The view is from above, looking into larynx and trachea. From left to right, sequence of vocal-cord positions during the transition from quiet breathing to voicing.

B

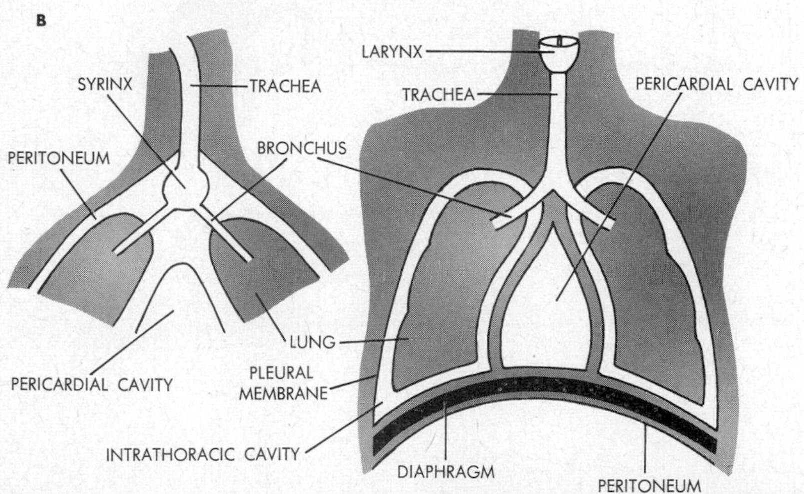

C

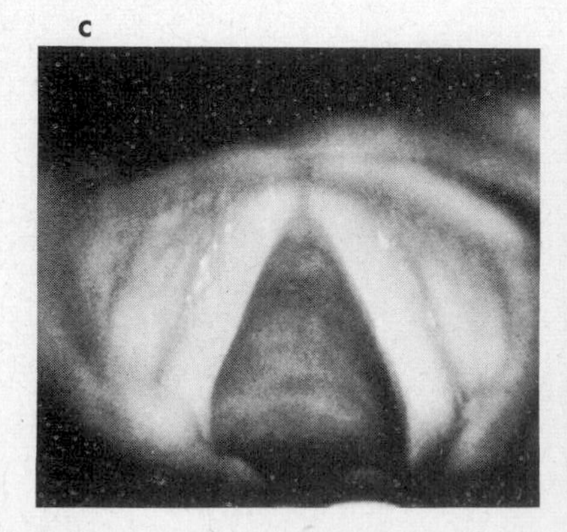
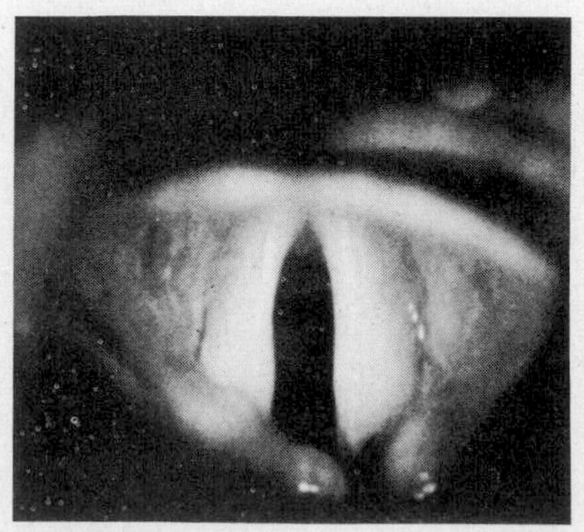
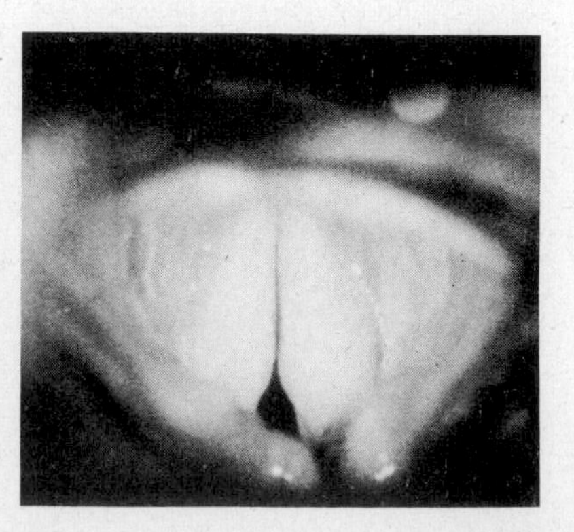

and the tension of the vocal cords may be altered in this manner. Sounds of different frequencies are produced when air is exhaled through the slit between the cords (Fig. 9.31).

The mammalian breathing system is also uniquely characterized by the presence of a *diaphragm*. This structure is an arched, muscular partition sealing off the chest cavity from the rest of the coelomic body cavity. When the diaphragm contracts, it flattens out, resulting in an enlargement of the chest cavity. The lungs are thereby "sucked" open, and air rushes into them. Relaxation of the diaphragm permits elastic recoil of the lungs and expulsion of air (cf. Fig. 9.31). Breathing is primarily under nervous control. A breathing center in the brain sends motor impulses via motor nerves to the diaphragm, which contracts as a result and produces an inhalation. Stretched lungs then generate sensory impulses in nerves leading to the breathing center, and such impulses inhibit the center from activating the diaphragm. The latter consequently relaxes and exhalation ensues. A new cycle is started when, with breathing now stopped, CO_2 concentrations in blood increase rapidly. High CO_2 concentrations in blood then override the inhibition of the breathing center, and the latter reactivates the diaphragm (Fig. 9.32).

The breathing systems of all animals not only supply oxygen, but they also excrete CO_2 in all cases, and water in numerous cases. Moreover, as will become apparent in the next section, the breathing structures of many animals also excrete other substances as well.

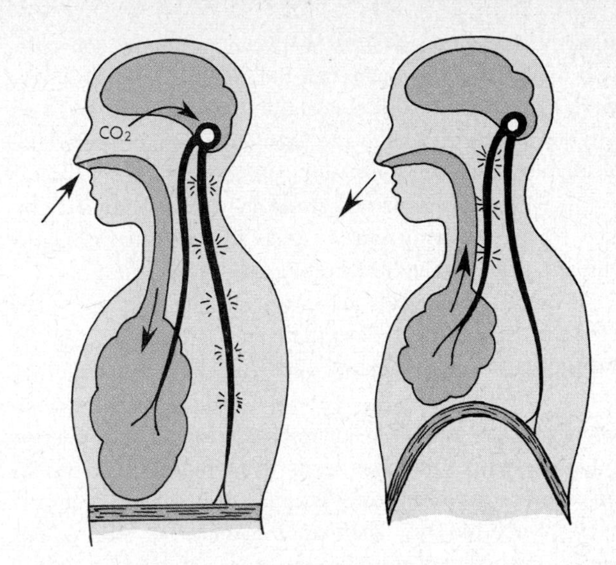

9.32. The control of inhalation (left) and of exhalation (right). Left, CO_2 in blood stimulates the breathing center to send impulses via the phrenic nerves to the diaphragm, leading to inhalation. Right, impulses from the inflated lung inhibit the breathing center, leading to exhalation.

EXCRETORY SYSTEMS

These systems are named somewhat inadequately, for a "casting out" of materials is only one of their functions; they also retain, supply, and adjust. Indeed, we may regard excretory systems as the ultimate controllers of chemical equilibria within the body and between the body and the environment; their primary role is to maintain optimum internal balances of water, salt, pH, and osmotic pressure, and to eliminate metabolic wastes. Actually, what is or is not "waste" at any given moment is precisely what the excretory system must determine.

Animal life originated in the sea, and the primitive members of most groups still live there. The internal inorganic composition of such animals generally matches that of the sea, i.e., the body fluids and the interior of the cells contain substantially the same kinds of salts as the sea, though usually in different concentrations. Even so, the interior of most marine invertebrates is in osmotic equilibrium with and is isotonic to sea water. When such an animal ingests food, it invariably acquires some sea water either as part of or along with the food. If an internal steady state is then to be maintained, this excess water and salt must be eliminated. Moreover, the food is metabolized, and we already know (Chap. 5) that the principal byproducts of metabolism are CO_2 and H_2O from respiration and NH_3 from interconversions among nitrogenous compounds. These byproducts too must be eliminated.

Ancestral marine animals have variously given rise to freshwater descendants. The latter must cope with a major additional problem of excretion. Concentrations within these animals are still essentially the same as those of their marine ancestors, i.e., roughly isotonic to sea water, yet the external environment now is far more dilute. Consequently, a permanent osmotic differential exists which draws water from the hypotonic to the hypertonic medium, i.e., from the freshwater environment into the animal. Such animals therefore are continuously in danger of internal flooding, which would overdilute all internal constituents and lead to death. To avoid this hazard the animals must excrete far more water, and far more continuously, than their marine relatives.

Thus, the basic problem of excretion is to eliminate NH_3, CO_2, excess salts, and excess water, the latter being always in great excess in freshwater forms. Any excretory structure then must be able to discriminate between waste and nonwaste, and remove the one but retain the other. To do so requires exquisite sensitivities to concentrations, for water and salts are both waste and nonwaste simultaneously, the difference being one of amount, not kind. Ammonia is a toxic alkaline compound which is usually waste in virtually any amount, and CO_2 is largely waste; as a gas it is excreted primarily through the breathing surfaces (Fig. 9.33).

As long as an animal is so constructed that all or most of its cells are exposed directly to the environment, each cell can function as its own excretory apparatus. The cell membrane is the excretory organelle, and waste substances can be eliminated into the external medium by diffusion and/or energy-requiring active transport across the cell boundary. Such a unicellular excretion pattern is characteristic of all protozoa, sponges, and radiates. The marine members of these groups largely are without excretory structures of any kind (other than the cell membranes themselves). The fresh-water members do possess specialized intracellular organelles for the continuous excretion of large quantities of water, viz., *contractile vacuoles.* Particularly well-seen in freshwater protozoa, the vacuoles are positioned near the cell surface, and they slowly enlarge as cytoplasmic channels empty water into them. At some critical stage a small region on the vacuole membrane apparently coalesces with the overlying cell membrane, a pore forms to the exterior, and the vacuole then shrinks rapidly and expels water to the outside. Subsequently, the process of vacuole growth begins again. A pulsation cycle of this sort may last some 30 to 60 sec, and it has been shown for some freshwater protozoa that they must excrete an amount of water equal to the volume of the whole cell every 4 min (Fig. 9.34).

9.33. The general pattern of excretion. The substances excreted primarily are ammonia, water, and carbon dioxide, or, secondarily, compounds derived from these (e.g., urea).

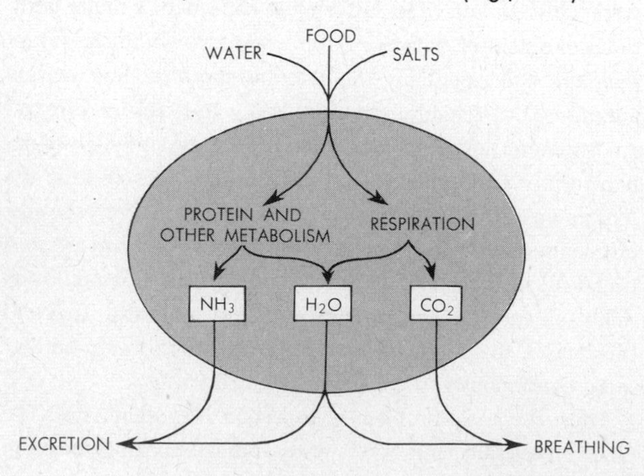

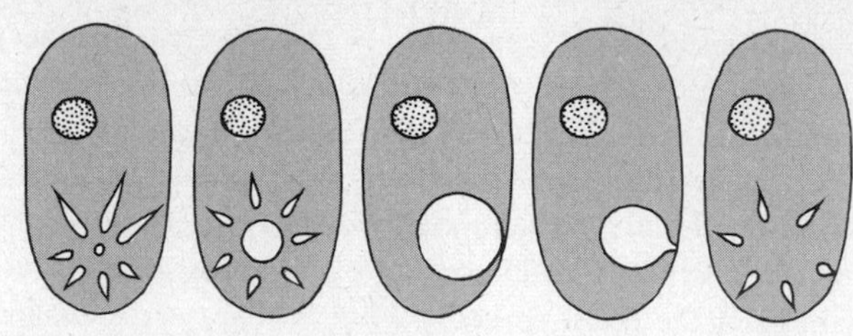

9.34. Schematic representation of the activity cycle of a contractile vacuole. From left to right, cytoplasmic water-channels empty into a growing vacuole, and the latter then expels water through a point on the cell surface. In the meantime water accumulates again in channels, and a new vacuole thus begins to form.

If all or most cells of an animal are not exposed directly to the environment, the interior cells can excrete only into lymph, hemolymph, or blood. Inasmuch as these internal fluids are without direct access to the exterior, excretory systems become a necessity. Such systems must operate in close association with the body fluids and act as screening devices, retaining valuable substances within the fluids and collecting only wastes for removal to the exterior. In all systems this excretory function appears to be performed through one or more of three basic processes: *filtration, reabsorption,* and *secretion* (Fig. 9.35).

Filtration takes place between the body fluids and the interior space of an excretory system, the filter usually being a surface layer of the system in contact with the body fluids. The pressure of the body fluids (e.g., blood pressure) supplies the force necessary for filtration. Cellular components and proteins in blood or hemolymph normally cannot pass through the filter, but most other components can; the filtration process is otherwise not too selective. Thus, the filtrate collecting within the space of the excretory structure is essentially lymph, which we may refer to here as *initial urine.* It still contains valuable substances as well as wastes. A separation of the two occurs when initial urine flows through another region of the excretory system on its way to the outside. Reabsorption then takes place, i.e., cells in contact with initial urine remove from it substances "judged" to be valuable, and they return such substances into the body fluids. We note that the reabsorbing cells represent the discriminators. They must expend a substantial amount of energy in carrying out these functions. If salts are being reabsorbed, the remaining urine will become more dilute, i.e., hypotonic to the body fluids; if water is reabsorbed, the remaining urine will become hypertonic and more concentrated than the body fluids.

The third process, secretion, may take place either in the same general region of the excretory system or in another body part altogether (e.g., gills). Secretion accomplishes the removal of waste or excess materials from the body fluids into either urine or the external environment directly (cf. Fig. 9.35). Secretion and reabsorption therefore appear to be essentially the same process but operating in opposite directions. However, the two do not always occur simultaneously, and the substances being reabsorbed may differ greatly from those being secreted. Thus, secretion provides an additional mechanism by which the composition of the internal body fluids can be regulated. The urine ultimately present may be called *final urine*. It is discharged either continuously or, after accumulating in a *bladder*, intermittently. In many cases the excretory system opens to the outside directly; in others it empties into the cloacal region of the alimentary tract or into the outgoing ducts of the reproductive system.

Among the simplest types of excretory systems are those containing *protonephridia*. A protonephridial system consists of a series of branching tubules which lead to the outside of the body via one or more *urinary pores*, also called *nephridiopores* (Fig. 9.36). The inside end of each branch tubule is closed off from the body fluids either by a *flame bulb* or by *solenocytes*. A flame bulb is a hollow, flask-shaped structure, often formed from a single cell. The neck of the flask is continuous with the tubule, and the outer surface of the flask is in contact with the body fluids. On the inner surface, the bottom of the flask carries a tuft of cilia whose beat suggests a flickering flame. A solenocyte is a variant form of a flame bulb; it too is a hollow tubular cell, but instead of cilia, the tubular portion contains a beating flagellum. Protonephridia of both types are known to be filtering structures primarily, and the cilia or flagella maintain the flow of urine to the outside. Reabsorption and secretion may take place along the ducts, but it is not known too well to what extent these processes actually occur. Systems with flame bulbs probably develop as ingrowths from the embryonic ectoderm, and systems with solenocytes are more definitely known to have an ectodermal origin. Flame bulb systems occur quite generally among the acoelomates and pseudocoelomates, though not in all groups of these animals. Solenocyte systems are found in the larvae of many pseudocoelomates and schizocoelomates, in some pseudocoelomate adults, in certain annelids, and in the cephalochordates, a chordate subphylum.

9.35. The basic action of an excretory system. A filtrate of blood or hemolymph forms initial urine, from which given materials are reabsorbed and to which other materials are secreted. These fundamental functions are performed by the lining cells of the system. Final urine is the ultimate product.

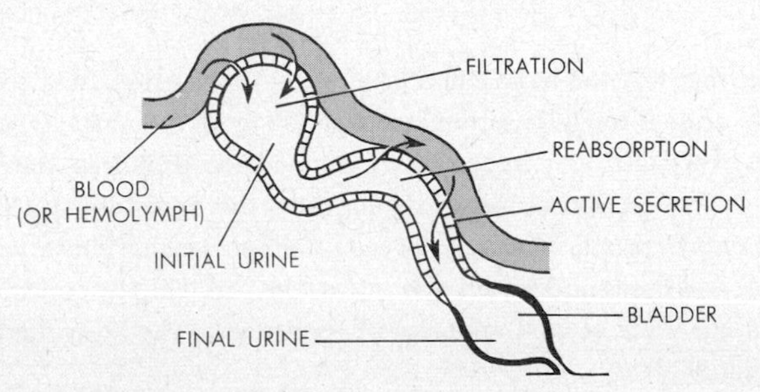

Among schizocoelomates with large coeloms (e.g., annelids), the adult excretory system is represented by *metanephridia*. These again are tubes emptying to the exterior via nephridiopores, but the inner ends of the tubes are open to the coelomic cavity. Such an opening, called a *nephrostome*, is funnel-shaped and ciliated, and the cilia propel lymph from the coelomic body cavity through the tubule of the metanephridium (cf. Fig. 9.36). Reabsorption and secretion take place along the course of the tubule. Filtration has occurred earlier, in the entire peritoneal lining of the coelomic cavity: as lymph passes from the body into the coelom, the fluid is filtered by the peritoneal cells. Near the nephridiopore a tubule may enlarge into a bladder, and tubules may open to the outside either individually or in groups sharing a single nephrodiopore. Most probably, metanephridia are in part mesodermal outgrowths from the coelomic lining, in part ectodermal ingrowths from the integument.

We recall that, in arthropods and most mollusks, the hemocoelic sinuses of the open circulation are large, whereas the coelomic body cavities are small. Metanephridia do not occur in such schizocoelomates with small coeloms. Instead, the adults possess *renal glands* of various types. In clams, for example, such glands excrete from the pericardial lymph, the latter representing the coelomic fluid in these animals. Filtration takes place earlier, from blood into the pericardial lymph through the wall of the heart. Arthropods have renal glands which filter from the hemocoelic sinuses and which also reabsorb and secrete. In crustacea these glands are located near the bases of the antennae and other appendages, and in insects the glands empty into the hindgut (cf. Fig. 9.36; also Chaps. 25 and 28).

Insects must cope with a problem common to all terrestrial

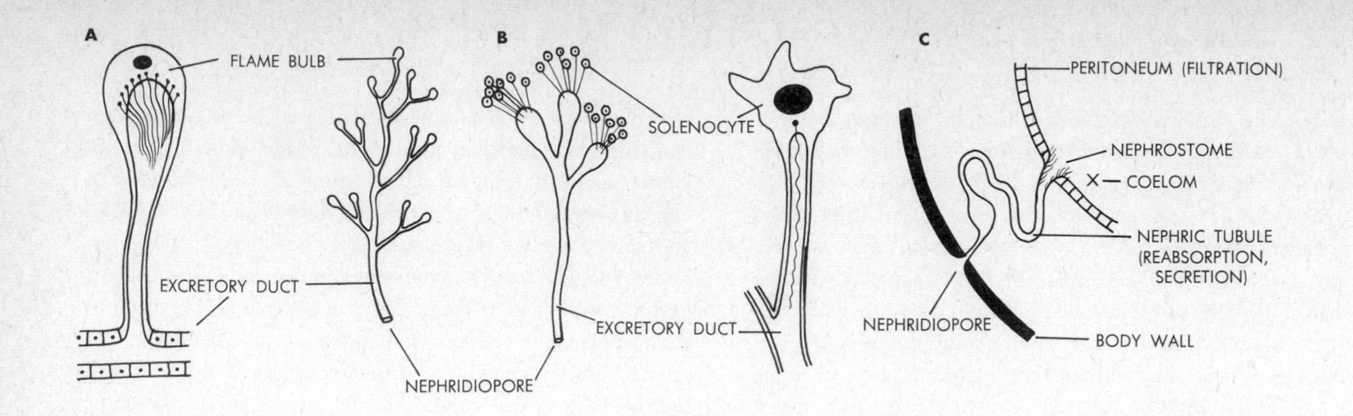

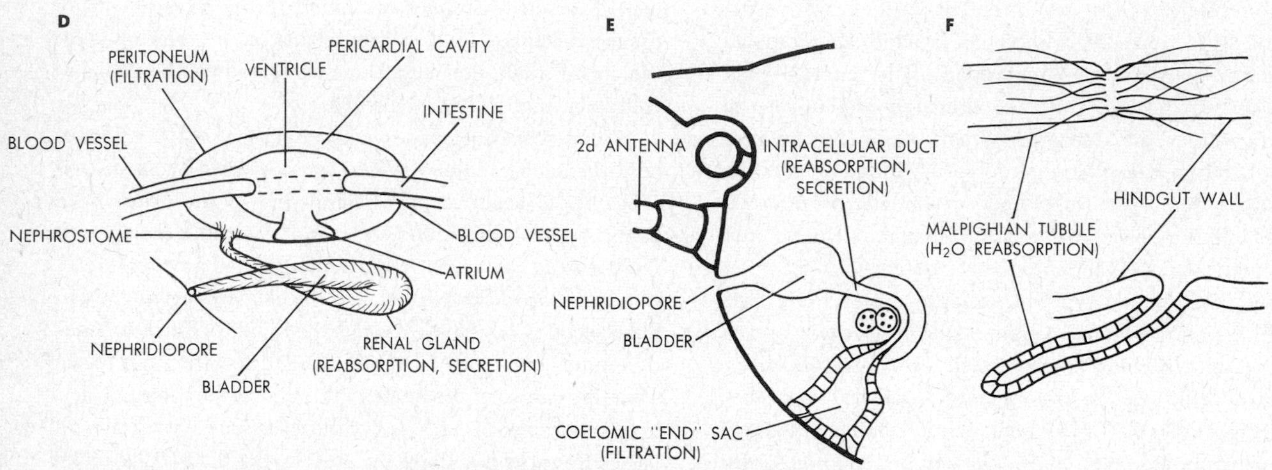

9.36. Excretory systems of invertebrates, diagrammatic. **A.** Protonephridia of the flame bulb type; single-celled flame bulb at left (note flamelike tuft of cilia), branch pattern of entire system at right. **B.** Protonephridia of the solenocytic type; single-celled solenocyte at right (note flagellum), branch pattern of entire system at left. **C.** The structural pattern of metanephridium; body fluids between the body wall and the peritoneum are filtered into the coelomic cavity. **D.** The excretory organ of a clam. The nephridiopore opens into a water-channel leading out of the animal (cf. Chap. 25). Note that the heart surrounds a portion of the intestine. **E.** The excretory organ of an aquatic arthropod (crustacean, lobster). Antennal, or "green," glands such as this often possess excretory giant cells (as shown), with two or more nuclei and cytoplasmic excretory channels. **F.** The excretory Malpighian tubules of terrestrial arthropods (e.g., insects). See Chap. 27 for other types of excretory systems in arthropods.

animals, viz., the reduced availability of external water. Thus, whereas a marine animal and even more so a freshwater animal is burdened with an internal water excess, a terrestrial animal has the very urgent opposite problem of conserving as much internal water as it can. The problem is solved in part by the development of water-retaining, evaporation-resistant integuments, in part by the excretory system. In an aquatic invertebrate the excretory system reabsorbs largely salts, leaving NH_3 and large amounts of water in the urine. In effect, the urine is roughly isotonic to the body fluids in marine types and very distinctly hypotonic to the body fluids in freshwater types. In terrestrial forms, by contrast, the excretory system reabsorbs primarily *water,* not salts, a process which conserves internal fluids and which also makes the urine highly concentrated; urine of land animals is hypertonic to the body fluids (Fig. 9.37).

However, a hypertonic urine creates a further problem, for NH_3 is toxic and can be tolerated only if it is diluted sufficiently in large amounts of water. The terrestrial animal cannot spare such water, and it relies instead on converting NH_3 into another, less toxic compound. One such compound is *urea,* another is *uric acid* (cf. Fig. 9.26). Both can be excreted in very much less water. Indeed, uric acid can be excreted in virtually solid form. Insects excrete uric acid, and their urine is almost solid; the hindgut reabsorbs practically all the water still left in the urine. The pattern is substantially the same in birds, whereas reptiles and mammals excrete a concentrated liquid urine containing mainly urea. To be sure, the vertebrate excretory system is structurally quite different from that of insects. Moreover, the vertebrate system serves aquatic animals as well as terrestrial ones.

The functional unit of the vertebrate *kidney* is a *nephron.* In its primitive form, still encountered in the larvae of all fishes and in the adults of a few, a nephron consists of three main parts (Fig. 9.38): a *glomerulus,* a tiny ball of capillaries receiving arterial blood from a branch of the dorsal aorta; a *nephric capsule,* a cup-shaped outgrowth from the dorsal part of the coelomic lining which partially envelops the glomerulus; and a *nephric tubule,* a relatively short duct which grows out and leads away from the nephric capsule. The opening from the capsule into the coelomic cavity is a *nephrostome.* Glomerulus and capsule together are referred to as a *renal corpuscle.* Nephrons of this sort develop on each side of the body, in a linear series along the roof of the coelom in the re-

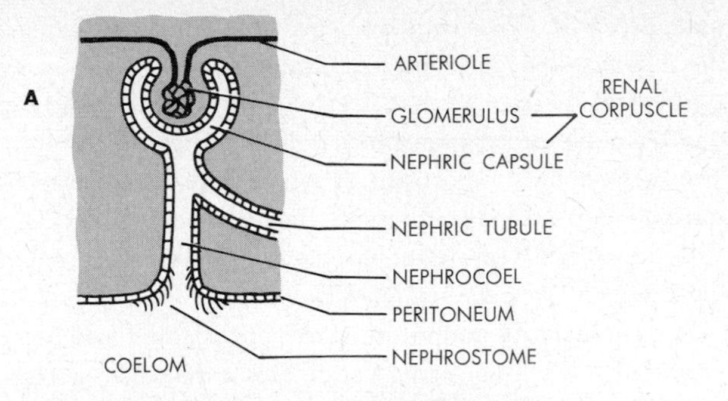

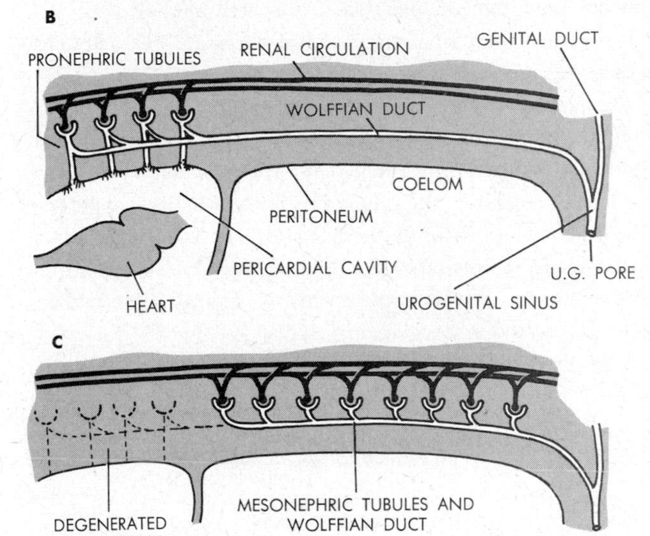

9.38. The primitive vertebrate kidney. A, a basic vertebrate nephron unit; **B,** diagram of a pronephric kidney; **C,** diagram of a mesonephric kidney (general labeling as in **B**). Note the nephrostomal openings into the pericardial (coelomic) cavity in **B,** absence of such openings in **C.** Many more nephron units than sketched here occur in an actual pronephros or mesonephros.

9.37. A freshwater animal takes up osmotic water and conserves internal salts; these two factors make the urine copious and dilute. By contrast, a land animal has a limited external water supply, and it conserves internal water; its urine is therefore limited in quantity and highly concentrated.

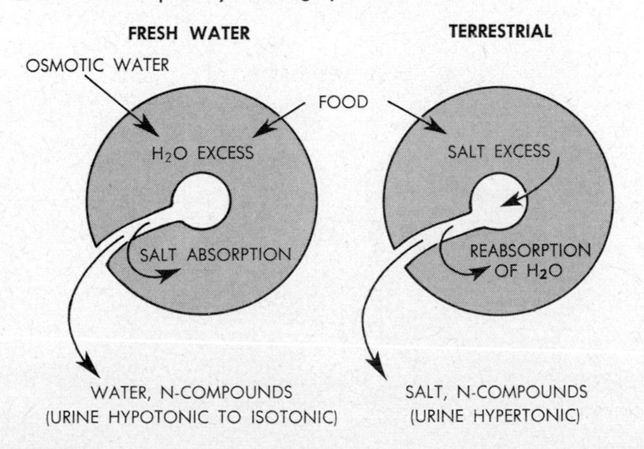

gion of the heart; the nephrostomes actually open into the pericardial portion of the coelom. The nephric tubules from all nephrons on each side of the body open into a common channel, the *Wolffian duct,* which proceeds posteriorly and empties into a *urogenital sinus.* The latter also receives the ducts of the reproductive system, and the sinus as a whole typically opens into the cloaca. Such an excretory system is known as a *pronephros,* or "head kidney." In it, the nephric capsules filter both the blood and the coelomic fluid (cf. Fig. 9.38).

In most fishes and in amphibia the pronephros is only a transient, embryonic kidney. As adulthood is attained, all parts except the Wolffian duct degenerate and a new kidney, the *mesonephros*, develops. This structure again forms along the roof of the coelom, but it extends from just behind the position of the earlier pronephros right to the level of the anus. The mesonephros too consists of sets of nephrons, but nephrostomes are no longer present; the nephric capsules filter only the blood. Nephric tubules again carry this filtrate into the common Wolffian duct (cf. Fig. 9.38).

Both pronephros and metanephros are adapted mainly for filtration, not reabsorption. Vertebrates evolved in a freshwater environment and, like freshwater animals generally, fishes in such an environment must eliminate the water that their gills and alimentary tracts take in by continuous osmosis. The pronephros and the metanephros provide the means for such water excretion; they largely filter, but they are not constructed to carry out much reabsorption of either water or salt. Consequently, the urine of freshwater fishes is hypotonic to (more dilute than) the body fluids. Salt loss through these kidneys is compensated for by absorption of salts from the environment, typically by means of special *salt-absorbing glands* located in the gills (Fig. 9.39).

Ancestral freshwater fishes have given rise to all the marine fishes now in existence. These live in an environment which is hypertonic to their body fluids; marine fishes tend to *lose* water osmotically, and they are in danger of dehydrating internally even though (because, really) a vast ocean is around them. In this tendency of dehydrating they actually resemble land animals but, unlike the latter, the marine fishes are unable to conserve water by producing a hypertonic urine: the fishes have inherited from their freshwater ancestors a mesonephric kidney designed basically to eliminate water, not to retain it.

Marine *bony* fishes solve this dilemma in two ways (cf. Fig. 9.39). First, they reabsorb as much water as their short nephric tubules will permit, and thus they make their urine as concentrated as possible, though it remains hypotonic. In some marine bony fishes, moreover, the glomeruli of the mesonephros degenerate altogether. This makes the kidney *aglomerular* and unable to filter much water from blood to begin with. A copious urine then cannot form, and the little that does form is essentially lymphlike, i.e., nearly isotonic to blood. Secondly, to compensate for the water lost by osmosis, marine bony fishes swallow sea water deliberately and retain the fluid component so acquired; but they excrete the salts of swallowed sea water through specialized *salt-excreting glands*

9.39. Water- and salt-balances in aquatic and terrestrial vertebrates.

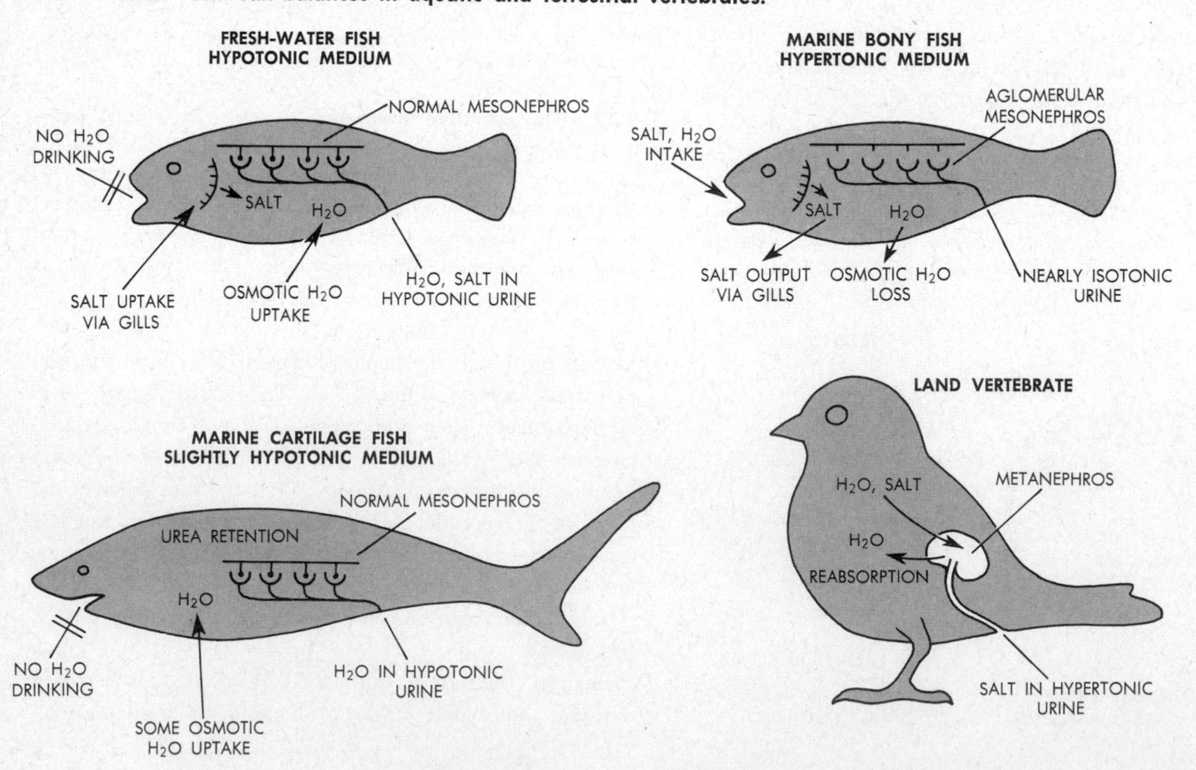

in the gills. A different solution to the osmotic problem was evolved by the marine *cartilaginous* fishes (e.g., sharks). These animals convert NH_3 into urea, and they retain this compound in all tissues and in blood, in high enough concentrations to make their interior slightly hypertonic to the external sea water. Thus they actually acquire some water osmotically, and their mesonephric kidneys may excrete hypotonic urine, like those of freshwater fishes (cf. Fig. 9.39).

As noted, land vertebrates produce a hypertonic urine. Inasmuch as a mesonephros cannot produce such a urine in marine fishes, we might also expect it to be unable to do so in land vertebrates. Indeed, reptiles, birds, and mammals develop not a mesonephros but a *metanephros.* This new type of kidney *is* capable of forming a hypertonic urine, and it is because a metanephros did evolve that some vertebrates could actually become efficient land animals (cf. Fig. 9.39). In the embryos of the land vertebrates, a pronephros and later a mesonephros develop only as transient structures. In a few primitive land vertebrates (some mammals included), the mesonephros persists till just after hatching or birth, but in most cases it degenerates much sooner. In place of it, the metanephros arises as the adult kidney.

This organ forms posterior to the mesonephros (Fig. 9.40). One portion of it develops as a branch duct growing out from posterior region of the Wolffian duct. This branch, which becomes the *ureter* of the adult system, divides repeatedly at its free end into a large set of tiny *collecting tubules.* A second portion develops around the terminals of these collecting tubules; it consists of numerous nephrons which establish connections with the collecting tubules. The nephrons differ from those in a mesonephros in one fundamental respect: their nephric tubules are far longer. Each nephric tubule contains two highly coiled, or *convoluted* regions, and an extensive loop between these two convoluted parts. This loop, *Henle's loop,* and the convoluted regions, are the essentially new features of the metanephric kidney; they function as a *water-reabsorbing* segment, and their presence permits the metanephros to produce a highly concentrated, hypertonic urine. The two kidneys of man, for example, contain some 2 million nephrons which filter the approximately 5 qt of water in blood once every 45 min, or about 32 times every 24 hr. Thus the kidneys filter the equivalent of about 160 qt of blood a day, but only about 1½ qt of fluid is actually released in the form of urine: some 99 per cent of the water in the blood filtrate is reabsorbed through the convoluted portions and the loops of Henle.

In reptiles and birds, the ureters from the kidneys typically still empty into a urogenital sinus, as do the reproductive ducts, and the sinus empties into the cloaca. In most male mammals, by contrast, the urogenital sinus has acquired an independent opening to the outside, and a cloaca is absent. The sinus here becomes a *urethra,* and it terminates at a *urogenital orifice.* In female mammals, furthermore, the excretory and reproductive ducts themselves open independently, and the urethra thus retains a urinary function only (cf. Fig. 9.40; also Chaps. 10 and 32). Many reptiles and virtually all mammals possess a *bladder* near the posterior end of the ureter, but such an organ is absent in some reptiles and in all birds (as would be expected, since the urine in these animals is semisolid and mixes with the feces in the cloaca).

Although a kidney is functionally the most essential component of an excretory system, it should be kept in mind that it is not the only component. As already noted, major excretory functions are also performed by breathing organs such as gills and lungs, and by alimentary organs such as the posterior portions of the intestine. Moreover, the total excretory process is aided materially by the skin of numerous animals, vertebrate as well as invertebrate. The sweat glands of mammals, for example, virtually represent tiny kidneys, and their water- and salt-excreting activities ease the work load of the main kidneys considerably. Additional excretory roles are played by structures such as salivary glands, nasal epithelium, tear glands, and liver. Indeed, we may conclude generally that any body part opening to the exterior either directly or indirectly contributes to at least some extent to excretion.

Considered on the level of organ systems, our examination of the main features of animal architecture is now substantially completed. However, still critically lacking is an appreciation of how animals perpetuate their architecture, i.e., how they maintain architectural *continuity* beyond their individual life spans. The next series of chapters considers this question.

REVIEW QUESTIONS

1. Define lymph, hemolymph, pseudocoel, hemocoel, blood, pericardium. What are the basic functions of lymph and blood? Distinguish between open and closed circulations and correlate each with coelom size. Which animals groups possess open and which closed circulations? Which are without blood circulations?

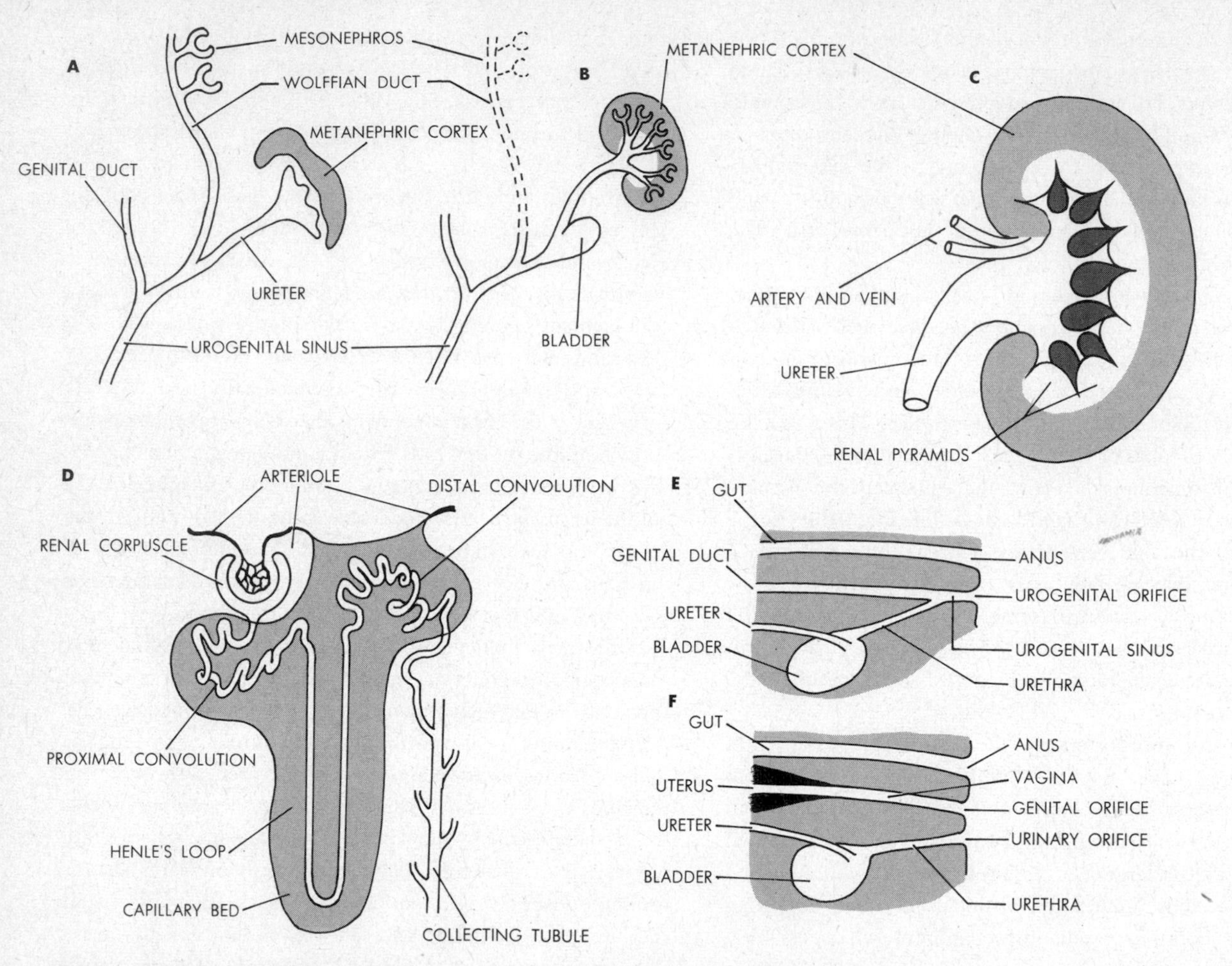

9.40. The metanephros. A and **B.** Stages in development of metanephros, schematic. In **B,** the mesonephros and most of the Wolffian duct have degenerated. **C.** The principal regions of the mature metanephric kidney; the renal pyramids are bundles of collecting tubules. **D.** A nephron unit, showing the convoluted portions of the tubule and Henle's loop. The capillary bed enveloping the coiled parts of a nephron is indicated as the shaded area. **E** and **F.** The anatomical relations of urinary and genital ducts in most mammals; **E,** pattern in most male mammals; **F,** pattern in most female mammals.

2. What organs are the principal components of a circulatory system? A lymph system? What is an endothelium? How do arteries differ from veins and capillaries (*a*) structurally, (*b*) functionally? What is the structure of different types of hearts? What are lymph hearts? Gill hearts? Lymph nodes? Review the course of blood and lymph circulation in a vertebrate.

3. How is the beat of a heart controlled? What is a pacemaker? How is blood pressure controlled? What is meant by vasomotion? Through what forces does blood return to the heart?

4. Describe the composition of blood plasma. Name some of the plasma proteins and state their functions. What is immunity and how is it produced? Name the different kinds of blood cells and state their functions. Review the clotting process. Name respiratory pigments, state in which animal groups each kind is encountered, and describe their functions. How is CO_2 transported in the body fluids?

5. Distinguish between short-distance and long-distance transport of nutrients, and show how each is accomplished. Describe the pattern of (*a*) intracellular, (*b*) extracellular alimentation. In which kinds of animals is each encountered?

What is the typical tissue structure of an alimentary tract? Describe some of the different ways in which (*a*) ingestion, (*b*) mechanical digestion take place.

6. What is peristalsis? An eversible pharynx? How and where is chemical digestion accomplished, and what are the specific chemical results? What digestive functions are performed by each of the different portions of an alimentary tract? How and where does nutrient absorption take place? What is the role of lymph in nutrient absorption?

7. Describe the structure and function of the vertebrate liver. Show how this organ processes each of the main classes of foods. How are urea and uric acid produced?

8. What different types of breathing processes and systems are encountered among invertebrates? How does each of these systems operate? How is a system operating in water different from one operating in air? What type of breathing system is characteristic of chordates and vertebrates? Review the pattern of the basic vertebrate gill circulation.

9. Show how the gill and lung circulations have changed in the course of vertebrate evolution. How did the vertebrate lung evole to begin with? What is the adaptive advantage of a four-chambered heart? Review the pattern of blood circulation between lung and heart. Describe the organization of the mammalian breathing system, and show how breathing movements are produced and controlled.

10. What is the basic structure and function of an excretory system? Show how a contractile vacuole operates. What does it excrete? Distinguish between excretory filtration, secretion, and reabsorption. Describe the structure and operation of flame-bulb protonephridia, solenocytic protonephridia, and metanephridia. What other types of excretory structures occur in invertebrates?

11. Show how the pattern of water and salt excretion varies according to whether an animal lives in a marine, a freshwater, or a terrestrial environment. Which animals produce hypotonic, isotonic, or hypertonic urine? Describe the structure of the vertebrate pronephros, mesonephros, and metanephros. How do these differ functionally? What is a nephron and how does it operate?

12. Describe the excretory patterns of different kinds of fishes and land vertebrates. Which groups of fishes do and which do not swallow water and why? What is an aglomerular kidney? Henle's loop? A ureter? A urethra? Which kinds of vertebrates have salt-absorbing and which have salt-excreting glands? What body parts other than the excretory system play a role in excretion?

COLLATERAL READINGS

Additional information on most of the topics covered in this chapter may be obtained from the following sources:

Baldwin, E.: "Introduction of Comparative Biochemistry," Cambridge, New York, 1949.

Carter, G. S.: Aquatic and Aerial Respiration in Animals, *Biol. Rev.*, vol. 6, 1931.

Griffin, D. R.: "Animal Structure and Function," Modern Biology Series, Holt, New York, 1962.

Hyman, L.: "The Invertebrates," vol. 2, chap. 9, p. 32 ff., McGraw-Hill, New York, 1951.

Pearse, A. S.: "The Migration of Animals from Sea to Land," Duke University Press, Durham, N.C., 1936.

Prosser, C. L., and F. A. Brown, Jr.: "Comparative Animal Physiology," 2d ed., Saunders, Philadelphia, 1961.

Smith, H.: "From Fish to Philosopher," Little, Brown, Boston, 1953.

———: "Evolution of Chordate Structure," Holt, New York, 1960.

Specific subjects touched on in this chapter are discussed in the articles below:

Burnet, M.: How Antibodies are Made, *Sci. American*, Nov., 1954.

———: The Mechanism of Immunity, *Sci. American*, Jan., 1961.

———: The Thymus Gland, *Sci. American*, Nov., 1962.

Clements, J. A.: Surface Tension in the Lungs, *Sci. American*, Dec., 1962.

Ebert, J. D.: The First Heartbeats, *Sci. American*, Mar., 1959.

Fox, H. M.: Blood Pigments, *Sci. American*, Mar., 1950.

Mayer, J.: Appetite and Obesity, *Sci. American*, Nov., 1956.

Mayerson, H. S.: The Lymphatic System, *Sci. American*, June, 1963.

McKusick, V. A.: Heart Sounds, *Sci. American*, May, 1956.

Ponder, E.: The Red Blood Cell, *Sci. American*, Jan., 1957.

Scholander: The Master Switch of Life, *Sci. American*, Dec., 1963.

Smith, H.: The Kidney, *Sci. American*, Jan., 1953.

Surgenor, D. M.: Blood, *Sci. American*, Feb., 1954.

Wiener, A. S.: Parentage and Blood Groups, *Sci. American*, July, 1954.

Wiggers, C. J.: The Heart, *Sci. American*, May, 1957.

Williams, C.: Insect Breathing, *Sci. American*, Feb., 1953.

Wood, W. B., Jr.: White Blood Cells vs. Bacteria, *Sci. American*, Feb., 1951.

Zweifach, B. J.: The Microcirculation of the Blood, *Sci. American*, Jan., 1959.

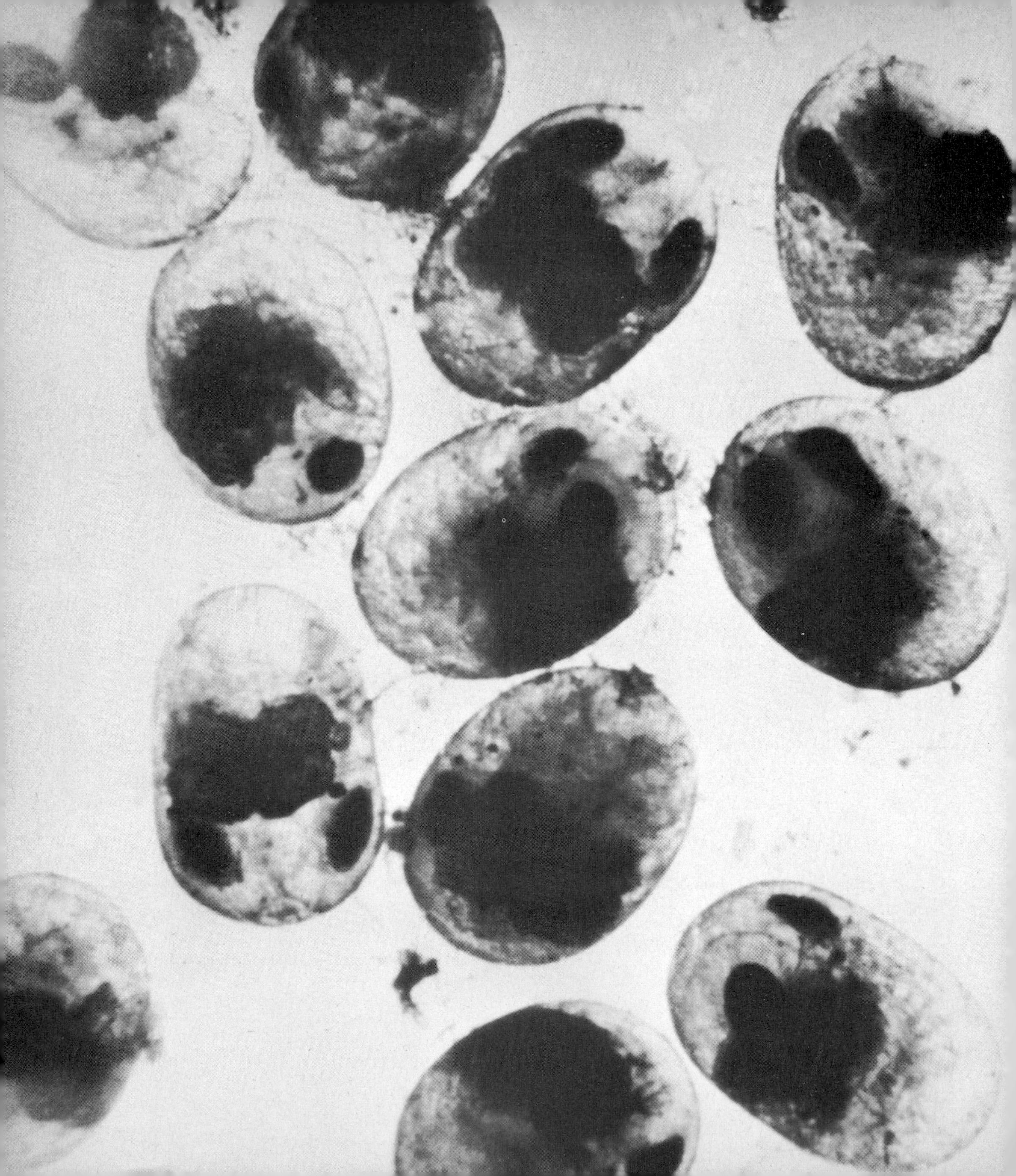

the continuity of animals

PART **4**

The ensuing chapters deal with the three kinds of processes which enable animals to persist on a long-term basis: *reproduction*, a mechanism of self-perpetuation which increases the number of animals and permits the formation of successive animal generations; *development*, the means by which new generations are actually molded from parts of the old; and *heredity*, a mechanism of adaptation which controls the forms and functions produced by development and plays a role in fitting a new generation to its environment. Reproduction implies and includes both development and heredity, yet all three processes contribute separately to the continuity and diversification of animals in time and to their expansion in space.

CHAPTER 10 reproduction

If we define reproduction broadly as extension of animal matter in space and in time, then its fundamental importance as a self-perpetuative device is readily apparent: the formation of new living units makes possible replacement and addition at every level of organization. Among molecules or cells, among whole animals or animal species, replacement offsets death from normal wear and tear and death from accident or disease. *Healing* and *regeneration* are two aspects of replacement. Above and beyond this purely restorative function of reproduction, addition of extra units at any level results in four-dimensional growth, i.e., increase in the net amount of existing animal matter.

Any new living unit resembles the old, and reproduction therefore implies reasonably exact duplication. To create new units, raw materials are required. Indeed, reproduction at any level depends on ample nutrition specifically and on properly controlled metabolism generally. It is also clear that duplication of a large unit implies prior or simultaneous duplication of all constituent smaller ones. Reproduction must therefore occur on the molecular level before it can occur on the cellular level, and on the cellular level before it can occur on the level of the whole animal.

FORMS OF REPRODUCTION

Animal propagation always includes at least two steps. First, a *reproductive unit* forms from the parent animal. Secondly, a duplicate animal arises from the reproductive unit through *development*.

On this basis, two major types of animal propagation may be recognized, viz., *vegetative reproduction* and *gametic reproduction*.

Vegetative Reproduction

In this form of multiplication a reproductive unit may consist of any portion of the parent animal; the unit is not specialized exclusively for reproduction and may range in size from the whole parent body to a minute fragment of that body. Moreover, the unit develops into an offspring directly, in a smooth sequence of events in which successive steps are not sharply marked off from one another. The principal variants of vegetative reproduction are *binary fission*, *multiple fission*, *fragmentation*, and *budding* (Fig. 10.1).

Binary fission is essentially mitotic cell division, i.e., the cleavage of one "mother" cell into two usually equally large "daughter" cells (cf. Chap. 6). These contain precisely identical genes sets incorporated in identical chromosome sets, and approximately equal quantities of all other cellular constituents. Consequently, the structural and functional potential of both daughter cells is the same as that of the original mother cell. In the unicellular protozoa, cell division is equivalent to reproduction of the whole individual. We may note that, in such types, binary fission actually is the *only* form of reproduction. Daughter protozoa generally separate, but in some cases they remain sticking together and form cell *colonies*.

In the multicellular Metazoa, binary fission either contributes to *cell replacement*, as in regeneration or wound healing, or adds to *cell number*. This leads to growth of tissues and organs. The growth of a metazoan thus may be a result either of molecular reproduction and increase in cell size or of cellular reproduction and increase in cell number, or of both. If the rate of cell division more than balances the rate of cell death, continued net growth occurs. In the opposite case, negative growth, or *degrowth*, will take place.

Rates of binary fission vary greatly. The most intense rates occur in embryonic stages, the least intense in old age. Cells which form membranous sheet-like tissues retain a fairly rapid but steadily decreasing rate of division throughout the life of an animal. By contrast, muscle, liver, or blood cells, for example, divide only rarely in the adult. And after being formed in the embryo, nerve cells do not divide at all. Nerve cells may grow, but only by increase in cell size. Destroyed neurons cannot be replaced. In general, the more highly specialized a cell, the less frequently it divides, and vice versa (cf. Chap. 3).

Why does tissue and organ growth slow down with increasing age? With few exceptions, cellular reproductive capacity in the adult remains potentially as great as in the embryo. This is shown, for example, by the high rates of cell division in wound healing, in regeneration, in cancers and other tumors, and in *tissue cultures*. Such cultures are prepared by separating groups of cells from an animal and growing them in artificial nutrient solutions. Under isolated conditions of this sort, cells are found to reproduce faster than if they had remained within an animal. Moreover, if newly formed cells in a tissue culture are cut away from time to time, the original bit of tissue may live almost indefinitely long, certainly far longer than it would have lived within the original donor animal. Through tissue culture, for example, a piece of the heart muscle of a chicken embryo has been maintained alive for more than 30 years, which exceeds the life span of the whole chicken several times.

10.1. The basic forms of vegetative reproduction.

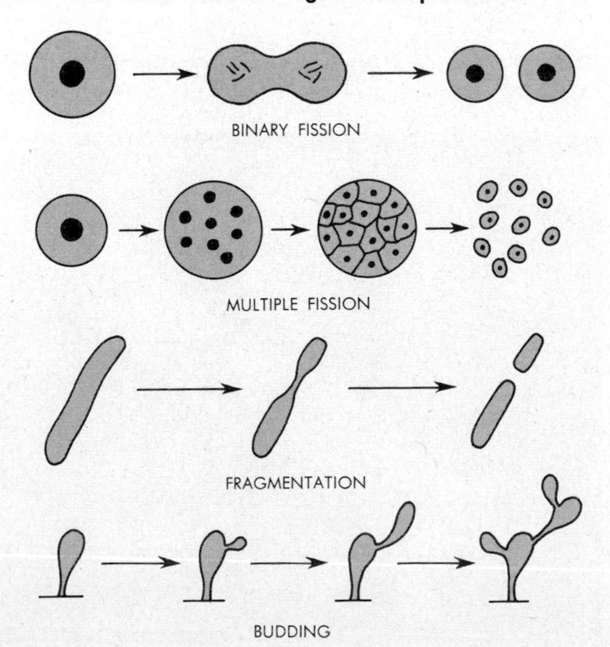

It is likely, therefore, that cell reproduction in intact animals may slow down mainly because the cells are not isolated as in a tissue culture. Instead, they are integrated very finely into a larger organization. Just as, in the human population, economic, social, and other checks hold the reproduction of the individual to less than maximum rates, so evidently is the rate of cell division held down by metabolic checks in the healthy cell population. Similarly, the slower expansion of an older, established society, compared with that of a new pioneer group, provides a close parallel to comparative growth rates in adult and embryo.

Although growth slows down with increasing age, it does not usually cease entirely; most animals continue to grow somewhat even in old age. The general range of body size is a genetically determined trait of the species, but within this range wide variations are possible. Net growth does stop in forms such as man after adulthood is attained, but this is not the case in fishes, for example. We may note here also that, in whales, elephants, and large animals in general, bulk arises primarily through increase in cell number, not increase in cell size.

Multiple fission is a variant form of cell division. In a preliminary process, the nucleus of a cell divides mitotically several times, but the cytoplasm remains undivided. Subsequently the cytoplasm does divide up into several smaller cells simultaneously, and each such cell incorporates at least one nucleus (cf. Fig. 10.1). Thus, numerous offspring cells form from one mother cell at the same time. Multiple fission occurs primarily in various groups of protozoa. In some of these the process is also known as *schizogony*, *sporulation*, or *spore formation* (cf. Chap. 19).

Note that, in both binary and multiple fission, the whole parent body constitutes the reproductive unit. By contrast, less than the whole body forms reproductive units in fragmentation. In this vegetative process, the parent animal spontaneously splits up into two or more fragments, each then regener-

10.2. Vegetative reproduction by fragmentation. The photo shows a sea anemone splitting lengthwise into two offspring animals.

ating the missing body parts. For example, certain sea anemones and flatworms occasionally fragment into two units. Other flatworms, and also certain hair worms, annelids, and echinoderms sometimes pinch off one or more smaller portions of their bodies, each such portion subsequently forming a whole new animal (Fig. 10.2).

Closely allied to this type of multiplication, and representing in fact a special form of vegetative propagation, is *regenerative reproduction*. Here the reproductive units arise fortuitously, as a result of injury to the parent by external agents. For example, many animals may be cut into several pieces, and each piece may then grow into a new, whole individual. A few segments of an earthworm, an arm of a starfish, a chunk of tissue from a hydra or a sponge—each is an effective reproductive unit. The parent animals losing such sections of their bodies usually regenerate the missing parts (Fig. 10.3).

In all these cases, the size and composition of the reproductive units obviously vary with the nature and extent of damage to the parent. It is clear, also, that regeneration may become reproduction only where the capacity of regeneration is extensive, as in the examples just cited. Regenerative capacity varies with the species, and in many animals it is highly limited. Salamanders may regenerate a whole limb, but a limb cannot regenerate a whole salamander (Fig. 10.4). In vertebrates generally, the regeneration potential is not even as great as in salamanders but, as in man, is limited to the healing of relatively small wounds.

Reproduction by budding involves the formation of reproductive units consisting initially of no more than perhaps a dozen cells each. Such *buds* are formed in different body regions in different animals. In many cases a bud retains a permanent anatomical connection with the parent, and after it has grown into an attached adult it may bud in its own turn and produce further generations of attached adults. Large colonies of joined individuals may form in this fashion. In other cases, a bud at first develops in anatomical continuity with the parent, but eventually it separates and becomes a new adult on its own. Budding is particularly widespread in sessile animals; it occurs, for example, among sponges, colenterates, flatworms, ectoprocts, and tunicates (Fig. 10.5).

The chief adaptive advantage of all forms of vegetative reproduction is that the process can be carried out by each animal without dependence on other animals. The only requirements are favorable environmental conditions, including ample food supplies. Thus, vegetative reproduction becomes particularly useful if the animal leads a sluggish or

10.3. Regenerative reproduction. An arm of a starfish regenerates all missing parts and becomes a whole animal.

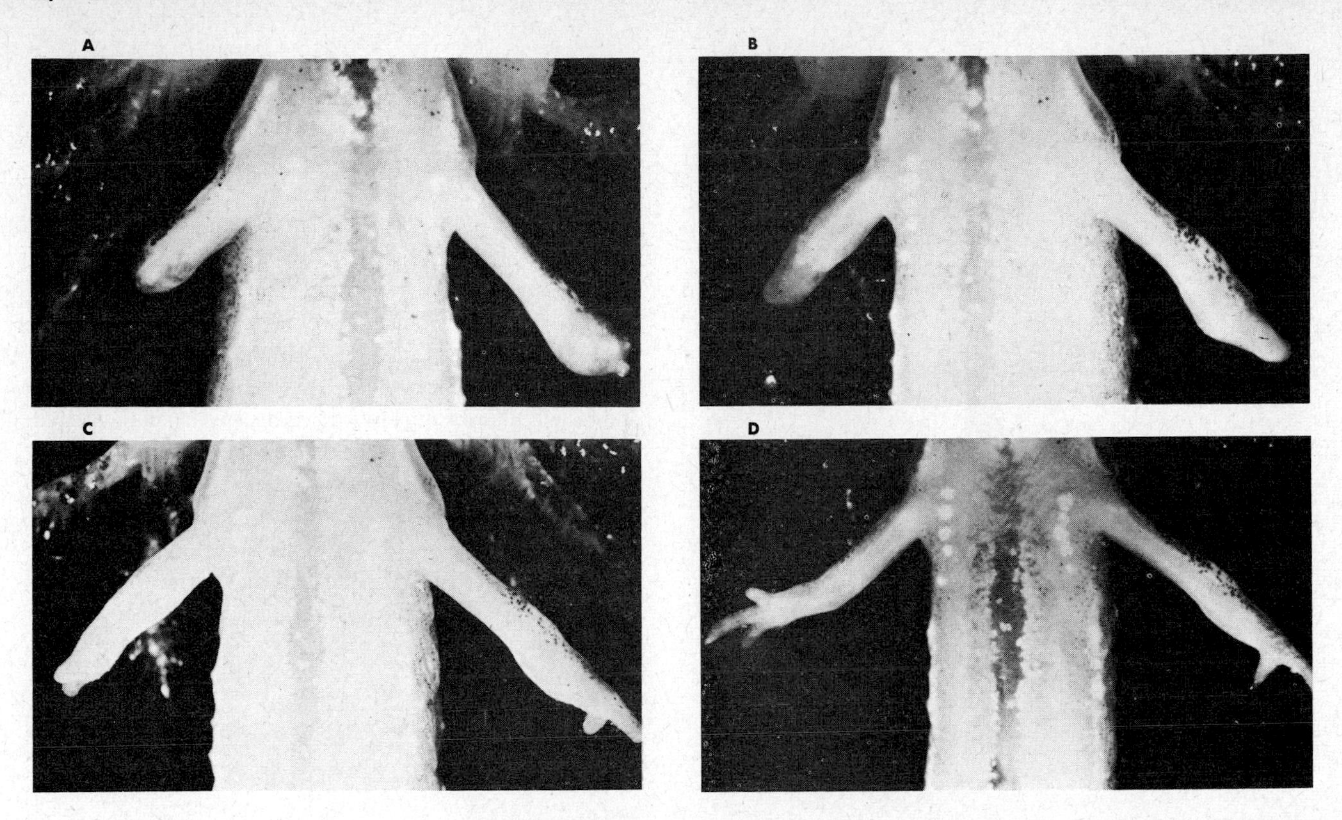

10.4. Regeneration. Both arms of a salamander larva were amputated, one above and the other below the elbow. The sequence of four photos (**A** to **D**) shows the degree of regeneration attained after 1, 14, 22, and 31 days, respectively. Note that, although a salamander can regenerate its limbs, an isolated limb cannot regenerate the missing salamander.

sessile way of life and is therefore relatively isolated from contact with other members of the species. Moreover, the method requires little tissue or cell specialization, and it also offers obvious advantages to an animal in the form of regenerative reproduction. As a result, vegetative multiplication has persisted as the only method of propagation in all protozoa and as a major subsidiary method in all but the most complex metazoans.

However, vegetative reproduction is characterized by a major limitation which can be circumvented only by the second basic method of multiplication, i.e., gametic reproduction.

Gametic Reproduction

On theoretical grounds, a reproductive unit of a metazoan should not have to be as large a portion of the parent animal as it is in vegetative reproduction. The smallest unit possessing the genetic information and the operating equipment representative of an entire multicellular animal is a single cell. Accordingly, the minimum unit for the construction of such an animal should be one cell. This is actually the universal case. Regardless of whether or not it may also reproduce vegetatively, every metazoan is capable of reproducing through single reproductive cells. All such cells are *specialized* for reproduction, and they are formed in specialized reproductive tissues or organs of the parent.

Moreover, the specialized reproductive cells typically can not develop directly. Instead, they usually must first undergo a *sexual process*, in which two reproductive cells fuse. The cells are therefore called *sex cells*, or *germ cells*, or *gametes*. Male gametes are *sperms;* female gametes are *eggs*. A mating process makes possible the pairwise fusion of gametes. This fusion is *fertilization*, and the fusion product is a *zygote*.

10.5. Vegetative reproduction by budding. A bud is developing on a parental hydra. When such buds are mature, they will separate from the parent and take up an independent existence.

Development of gametes into adults cannot occur until fertilization has taken place. In other words, if sex occurs, it is interpolated between the two basic phases of the reproductive sequence, namely, between the formation of reproductive cells and the development of these cells into adults (Fig. 10.6).

Note that, in a strict sense, the often-used terms "asexual reproduction" and "sexual reproduction" are meaningless as alternatives for vegetative and gametic reproduction, respectively. In all forms of multiplication, the essential "reproductive" event is the formation of reproductive units. The rest is development. And it is this developmental phase which may or may not require sexual triggering. Therefore, whereas development may be initiated sexually or asexually, reproduction as such, namely, the formation of reproductive units, is *always* "asexual."

Gametic reproduction entails a number of serious disadvantages. For example, the method depends on chance, for gametes must meet, and very often they simply do not. Much of the reproductive effort of the parent animals is then wasted. Gametic reproduction also depends on locomotion; for if gametes are to meet, they either must move themselves or must be brought together by locomotion of the parents. Yet eggs in virtually all cases and parent animals in many cases are incapable of locomotion. Above all, gametic reproduction invariably requires an aquatic medium. In air, gametes would dry out quickly unless they possessed evaporation-resistant outer shells. But if two cells were so encapsulated, they could then not fuse together. As we shall see below, terrestrial animals actually can circumvent this dilemma only be means of special adaptations.

However, all these various disadvantages are relatively minor compared to the one vital adantage offered by gametes. This advantage is sex. What is the crucial significance of this process?

SEXUALITY

The Role of Sex

In Metazoa, the sexual process always takes place in conjunction with a reproductive process. Consequently, the independent function of sex does not become readily apparent. In Protozoa, by contrast, sex and reproduction are not associated. The cells reproduce vegetatively whenever they can, and sexual processes occur separately at some stage of the life cycle. Reproduction takes place under optimal environmental conditions, when food is plentiful, when the animals are not too crowded together, and when the water has a proper temperature and pH. The sexual process typically takes place under the opposite conditions, when the environment is more or less unfavorable. For example, starvation or crowding or both usually lead to a sexual response. This response takes the form of *syngamy* in some protozoa and of *conjugation* in others. In syngamy, two whole individuals come to function as gametes and fuse into a zygote. In conjugation, two individuals fuse only partially, i.e., a cytoplasmic bridge forms between mating partners, or *conjugants*. Also, the nucleus of each conjugant gives rise to two *gamete nuclei*, and one of these migrates from each cell through the bridge into the other cell. After such a nuclear exchange, the two gamete

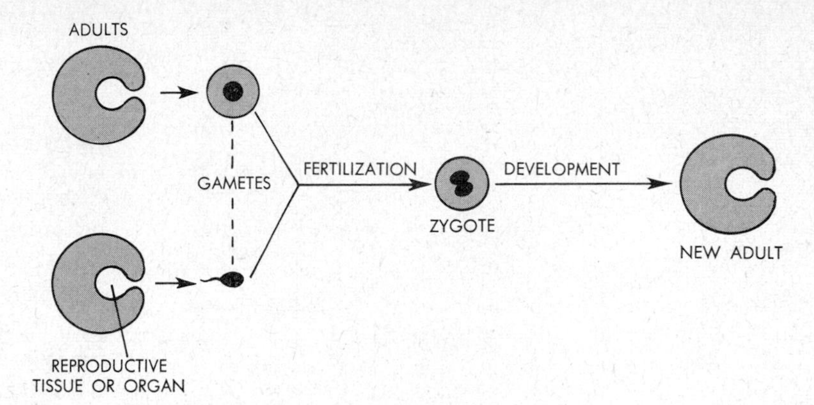

10.6. The pattern of gametic reproduction and of development in Metazoa. Gamete manufacture constitutes the actual reproductive phase.

nuclei now present in each conjugant fuse into a zygote nucleus, and the two mating partners then separate (Fig. 10.7).

Note, first, that the sexual process is fundamentally quite distinct from reproduction. Protozoa do not "multiply" through sex—if anything, quite the contrary. In syngamy two cells form one; in conjugation, two cells enter the process and two cells again emerge. Sex and reproduction are equally distinct in all Metazoa even though in these animals the two processes do occur together.

Note further that in protozoa and also in all other animals, man not excepted, sexual activity is particularly evident during periods of persistent environmental *stress.* Sexuality may be brought out or intensified by unfavorable climates, by widespread food shortages, by overpopulation, or by other

10.7. The pattern of syngamy (top) and conjugation (bottom) in protozoa. Note that, from comparison with Fig. 10.6, Metazoa are basically syngamous. A more detailed illustration of conjugation is given in Fig. 19.24.

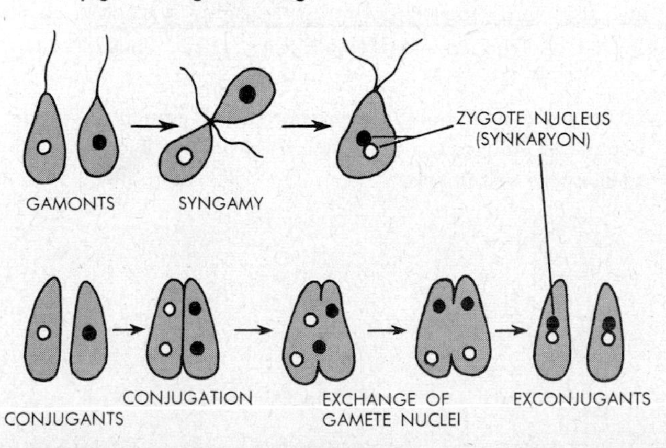

conditions which cannot be quickly responded to through steady-state control. Indeed, most animals living in temperate climates manifest sexual activity typically in the fall or in the spring. In each of these two seasons, sex is a response to the stress conditions of the preceding season, and it anticipates the stress conditions of the following season.

Just how is sexuality effective against conditions of stress? Events in protozoa supply the general answer: every cell resulting from the sexual process possesses the genes of *both* cells which entered the process. Stripped to its barest essentials, sex may be defined as the accumulation within a single cell of genes derived from two relatively unrelated cells. One method of achieving this is cell fusion, as in syngamous protozoa and all Metazoa; another method is exchange of duplicate nuclei, hence of duplicate gene sets, as in conjugating protozoa.

Sex therefore counteracts stress conditions on the principle of "two are better than one." If the self-perpetuating powers of two relatively unrelated parent animals are joined, through union of their genes, then the offspring animals produced later may acquire a survival potential which is greater than that of either parent alone. If parent A survives in environment a and parent B in environment b, then the offspring AB formed after sexual union may survive in environments a or b or in both (Fig. 10.8).

Moreover, a still poorly understood *rejuvenation,* on the biochemical, metabolic level, accompanies the sexual process. In certain protozoa, for example, if sex is experimentally prevented during an indefinite number of successive vegetative generations, then the vigor of the line eventually declines. The animals ultimately die, even under optimal environmental conditions. Internal stresses apparently appear

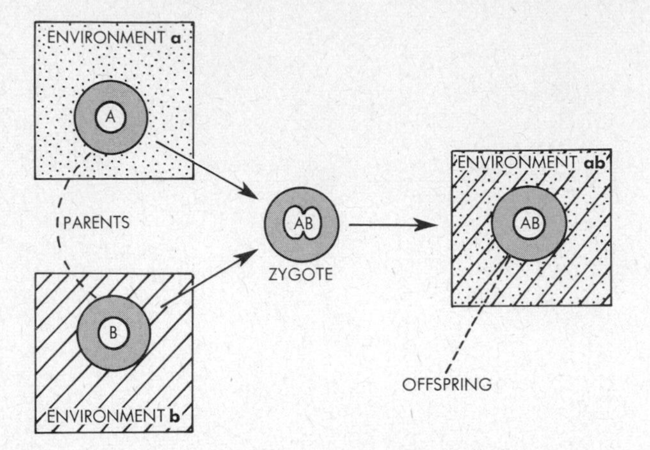

10.8. The role of sex in combating stress. *a* and *b* represent two different environments in which live two genetically different prospective parents; *A* and *B* symbolize their genes. Through sex, the offspring acquires the genes of both parents, hence also the ability to live in either environment *a* or *b*. Sex combines the adaptive potential of the parents and so endows the offspring with increased adaptation potentials.

in aged generations, and only a "rejuvenation" through sex may then save the reproductive succession and prevent the line from dying out (cf. also Chap. 19).

We may say that reproduction is a "conservative" process. Parental characteristics are passed on faithfully by reproduction from generation to generation, and so long as the external and internal environment remains favorable, succeeding generations survive as well as preceding ones. Sex, on the other hand, is a "liberalizing" process. It may offer survival under new or changed conditions. By combining the genes of two parents, sex introduces *genetic change* into the later offspring. And to the extent that such change may be advantageous for survival in new environments or under new conditions, sex has *adaptive* value. That is the key point; *sex is one of the chief processes of adaptation.* It is worth repeating that sex is *not* a process of reproduction.

Since the sexual process involves single cells, it must be carried out at a stage when an animal consists of but a single cell. Among the unicellular protozoa, therefore, sex may occur at any stage of the life cycle regardless of when reproduction occurs, and it may be dissociated completely from reproduction. By contrast, if in the life cycle of a multicellular animal sex is to take place at all, it must take place at a unicellular stage. Hence, "gametic reproduction" among Metazoa: sex occurs *after* the formation of gametes and *before* their development into multicellular adults.

Forms of Sexuality

Among many syngamous protozoa, the mating partners are not visibly distinguishable as being of two different types. Functional differences do exist, however, and it can be shown that mating cannot occur between just any two cells. Thus, even though the cells are all structurally alike, fertilization can only take place between two cells belonging to two functionally different varieties. Such structural likeness of two gamete types is referred to as *isogamy*, a condition encountered quite frequently among protozoa. In many other protozoa, however, the two mating cells do differ visibly. In numerous cases, for example, one mating cell is distinctly smaller than the other. Such forms are said to exhibit *anisogamy;* i.e., sexual processes involve gametes of unequal appearance. Still other protozoa, and all Metazoa, exhibit a special form of anisogamy, namely, *oögamy*. Here one gamete type is always flagellate and usually small, and the other gamete type is always nonmotile or amoeboid and usually large. The small type is called a *sperm*, the large type, an *egg* (Fig. 10.9).

In most animals, any one individual produces one gamete type only, and another individual produces the other type only. The sexes then are said to be *separate*. The mechanisms which determine whether a given animal will belong to one sex type or the other are discussed in Chap. 13. If separately sexed animals exhibit isogamy or anisogamy, the terms "male" and "female" are not strictly applicable. For example, there is little justification for regarding isogamous protozoa as either males or females. Instead, the two sex types, or *mating types*, are customarily identified by distinguishing symbols such as + and −. True male and female sexes are recognized only in

10.9. Size and motility characteristic of gametes. In the oögamous pattern, the large nonmotile gamete is an egg, the small motile one, a sperm.

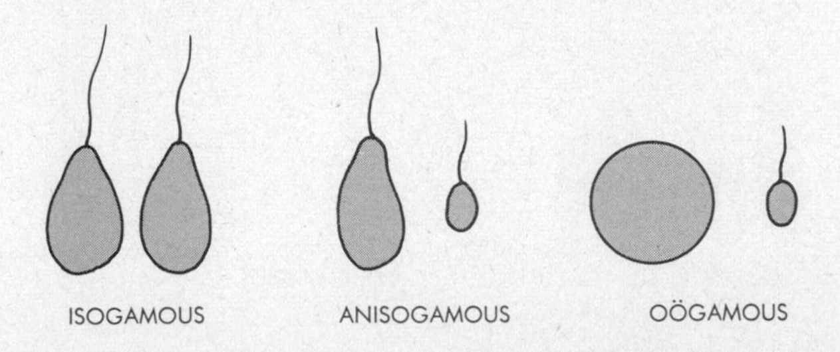

cases of oögamy, i.e., in Metazoa, in which distinct sperms and eggs are produced. In such animals males and females are often distinguished also in other ways. For example, sperms and eggs are produced in differently constructed sex organs. Sperms are formed in *testes*, eggs in *ovaries*. Collectively called *gonads*, these organs usually are components of somewhat differently constructed reproductive systems (cf. below). All such identifying traits of males and females constitute the so-called *primary sex characteristics*. In many animals males and females are otherwise quite alike, but in many other forms various *secondary sex characteristics* provide additional distinctions. Such secondary sex traits may appear in any body part other than the reproductive system. In man, for example, secondary sex characteristics include male-female differences in skeletal and muscular development, hair growth pattern, voice pitch, degree of fat deposition under the skin, mammary gland development, hormone secretions, metabolic rate levels, and other traits apart from primary reproductive ones by which males and females can be distinguished. We note that, where the sexes are separate, degrees of sex distinction may vary considerably. At one extreme are the isogamous protozoa, where visible differences between sex types are zero; and at the other extreme are advanced Metazoa, in which virtually every organ system may exhibit characteristics of maleness or femaleness.

In numerous animals, both gamete types are produced within the same individual. Known as *hermaphroditism*, this condition is believed to be more primitive than that of separate sexes; the latter condition may have evolved from hermaphroditism by suppression of either the male or the female potential in different individuals. For example, all vertebrates develop in the embryo with both potentials, but only one later becomes actual in a given individual (cf. also below). Most conjugating protozoa are hermaphroditic, each individual producing a gamete nucleus of each sex type. In Metazoa, hermaphroditism occurs in some groups of almost every phylum, and it is sometimes encountered as an abnormality in vertebrates, man included.

In most cases, hermaphroditism is a direct adaptation to either a parasitic or to a sluggish or sessile way of life. For example, since every hermaphrodite may function both as a "male" and a "female", a mating of two individuals may not even be required and *self-fertilization* may take place. In such instances, the gametes of one sex type are genetically compatible with the gametes of the other sex type produced in the same individual. Many of the conjugating protozoa are self-fertilizing hermaphrodites; exchange of gamete nuclei between two individuals need not occur, and the two gamete nuclei produced within one individual may fuse. Such self-fertilization is here called *autogamy* (cf. Fig. 19.24). In Metazoa, self-fertilization is comparatively rare. It occurs, for example, in parasitic forms such as tapeworms.

Most animal hermaphrodites must carry out *cross-fertilization;* a gamete of one sex type produced by one individual must fertilize a gamete of the other sex type produced by another individual (Fig. 10.10). Some kind of block against self-fertilization usually ensures cross-fertilization. For example, even if sperms and eggs of the same individual were to meet, a chemical or genetic incompatibility may make fertilization impossible. Most often, a meeting of sperms and eggs from the same individual is prevented by the anatomy of the body; i.e., the two reproductive systems open to the outside in different surface regions. In several hermaphroditic

10.10. The pattern of separate sexes and of hermaphroditism. The symbols ♀ and ♂ in the diagram identify female and male reproductive systems, respectively. These same symbols and also ⚥ outside the diagrams identify whole female, male, and hermaphroditic animals. The pattern of self-fertilizing hermaphroditism can be envisaged readily in the upper right figures. In protandry, below, the ♀ systems mature later than the ♂ systems, and after eggs are produced, already-stored sperms effect fertilization. In protogyny (not shown) the pattern is essentially reversed, i.e., eggs are already mature at a time when sperms first become available.

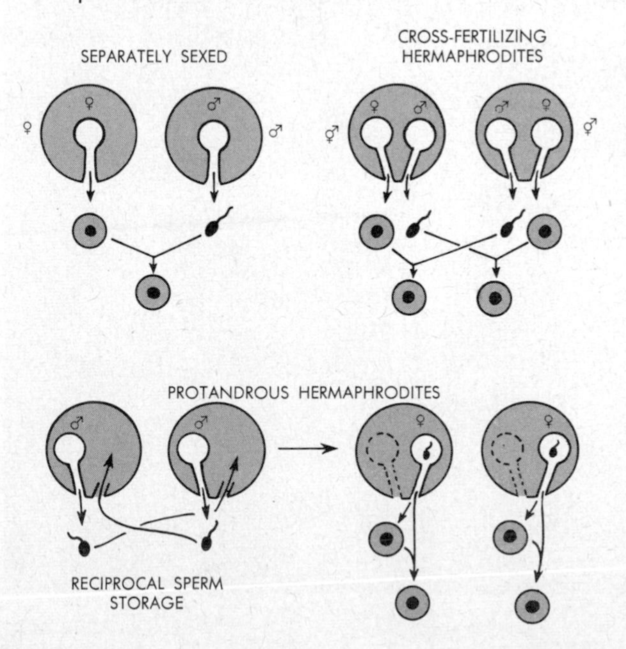

types, particularly among mollusks, for example, sperms and eggs are manufactured at different times during the breeding season. Some species produce ovaries first and testes next (the same gonad often switching function and producing both eggs and sperms in succession). Such animals are said to be *protogynous* hermaphrodites. Other species form testes first and ovaries later and are known as *protandrous* hermaphrodites (cf. Fig. 10.10).

Note that, in cross-fertilizing hermaphrodites, the mating pattern is essentially the same as in animals with separate sexes. Yet an adaptive advantage is apparent nevertheless; i.e., fewer reproductive cells are wasted than in species with separate sexes. Thus, if a given species is hermaphroditic and cross-fertilizing, sperms from one individual may meet eggs in any other individual, for every hermaphrodite does produce eggs. In species with separate sexes, by contrast, many sperms are wasted through chance misdistribution to the wrong sex. Similarly, if mutually cross-fertilizing hermaphrodites are capable of some locomotion, like earthworms, for example, then fertilization becomes possible whenever *any* two individuals meet. Since sluggish individuals are not likely to meet very frequently to begin with, and since every such meeting may result in fertilization, the adaptive value of hermaphroditism is clear (Fig. 10.11).

10.11. A copulating pair of the edible land snail *Helix pomatia*. These animals are cross-fertilizing hermaphrodites. Thus, whenever any two of them meet, each may be fertilized by the other.

In any animal, separately sexed or hermaphroditic, it happens often that given gametes fail to find partners of opposite type. Unsuccessful sperms always disintegrate very soon. This is also true of most unsuccessful eggs, but in exceptional cases unfertilized eggs do not disintegrate. On the contrary, the eggs may begin to develop as if they had been fertilized, and they often form normal adult animals. This phenomenon is known as *parthenogenesis*, "virginal development" of an egg *without* fertilization (Fig. 10.12). A *natural* form of parthenogenesis occurs as a normal reproductive event in, for example, rotifers, water fleas and other crustacea, and bees and other social insects. Natural parthenogenesis has also been encountered sporadically among birds, e.g., chickens and turkeys. Rotifers, water fleas, and other groups reproduce parthenogenetically during spring and summer, females giving rise to female offspring only; males are not formed. In the fall, however, males do develop from some of the eggs produced by the females, and many eggs then are fertilized normally. These winter over and develop into new females the following spring (cf. Chap. 23 for detailed account). Some species of rotifers appear to reproduce entirely by parthenogenetic means, males being completely unknown. In one genus of roundworms (*Rhabditis*), sperms enter the eggs but do not participate further in development. In effect, therefore, the eggs remain unfertilized and they develop parthenogenetically. Among bees, unfertilized eggs develop parthenogenetically into drone males, fertilized eggs become workers (cf. Chap. 16).

In many animals, *artificial* parthenogenesis may be induced by experimental means. For example, a frog egg can be made to develop before it has become fertilized by pricking its surface with a needle. The puncture simulates the entrance of a sperm and development then begins. But a sexual process has not taken place. By various experimental procedures, artificial parthenogenesis has also been achieved in the eggs of echinoderms, rabbits, and other animals.

One consequence of every normal fertilization is that a zygote formed from two gametes possesses twice the usual number of chromosomes. Animals counteract this increase by a series of special nuclear divisions known as *meiosis*. In most

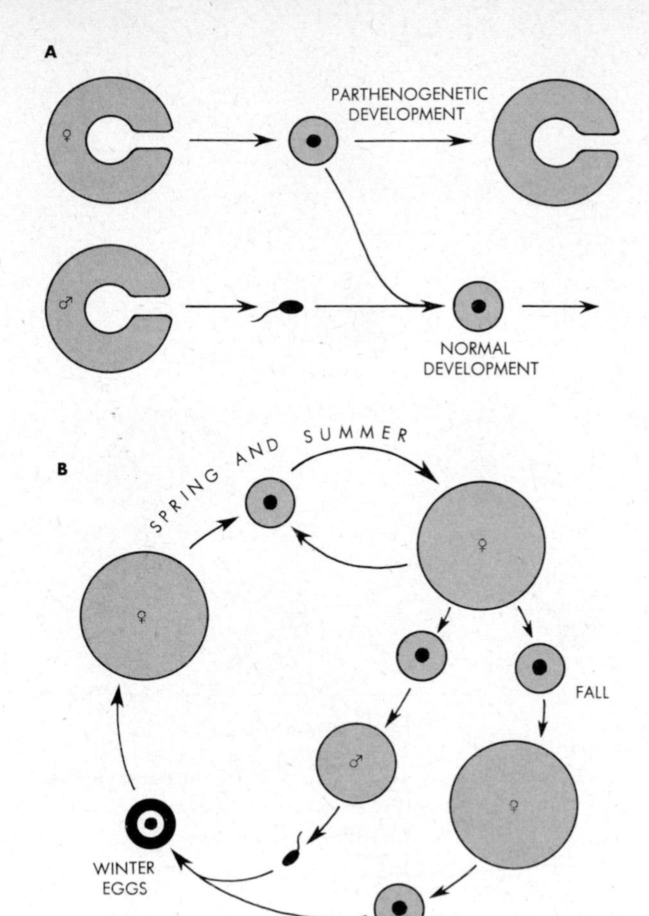

10.12. Parthenogenesis. **A.** The pattern of parthenogenetic development of unfertilized eggs. **B.** The general pattern of parthenogenetic life cycles (as in rotifers, water fleas). Adult female generations succeed one another in spring and summer, and only in fall do some of the eggs produced develop into males. Regularly fertilized eggs then winter over and develop into new females the following spring.

cases meiosis is accompanied by cytoplasmic divisions, and both events are then referred to collectively as meiotic cell divisions.

Meiosis and Life Cycle

As noted in Chap. 4, the number of chromosomes per cell is a fixed, genetically determined trait of every animal. However, whenever a sexual process takes place, the gene sets of two sex cells fuse together. In a zygote, therefore, the chromosome number is doubled. An adult animal developing from such a zygote would consist of cells which would all have a doubled chromosome number. If the next generation is again produced sexually, the chromosome number would then quadruple, and this process of progressive doubling would continue indefinitely through successive sexual generations.

This does not happen in actuality. Chromosome numbers do stay constant from one life cycle to the next, and the constancy is brought about by meiosis. It is the function of meiosis to counteract the chromosome-doubling effect of fertilization by reducing a doubled chromosome number to half. The doubled chromosome number, before meiosis, is called the *diploid* number, and it is symbolized as $2n$; the reduced number, after meiosis, is the *haploid* number, and it is symbolized as n (Fig. 10.13).

Meiosis occurs in every life cycle which includes a sexual process—in other words, more or less universally. Organisms differ according to when and where meiosis occurs in the life cycle. Excepting only a few protozoan groups, meiosis in all animals takes place just before fertilization, during the production of the gametes. The latter are haploid, and when two of them fuse, the resulting zygote is diploid. This diploid condition persists during the development of the zygote and into the adult stage. When the adult subsequently manufactures gametes of its own, the gamete-forming cells in the gonads undergo meiosis. Mature gametes are therefore haploid. A

10.13. The relation of meiosis and fertilization to chromosome numbers and life cycle.

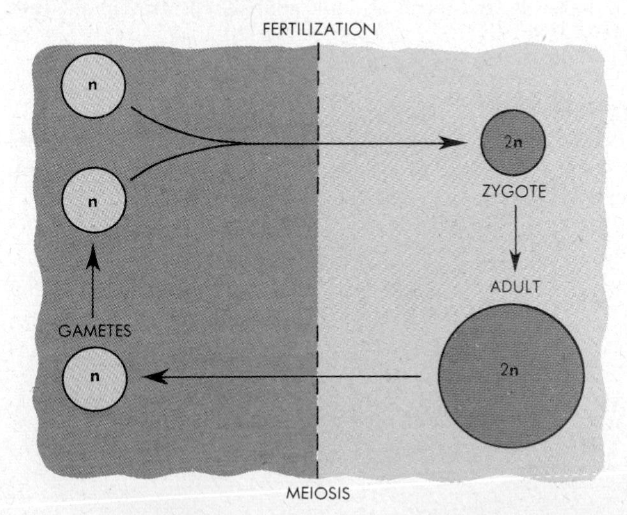

life cycle of this type, characterized by *diploid adults* and *gametic meiosis* is called a *diplontic* cycle; the only haploid stage in it is the stage of the mature sperm and egg (Fig. 10.14).

Note that, in such a cycle, the chromosome-doubling effect of fertilization is counteracted by meiosis only at the last possible moment, i.e., at the very end of the life cycle, just before a new generation is produced by the gametes. This late occurrence of meiosis gives the superficial impression that the response precedes the stimulus: fertilization is the chromosome-doubling stimulus to which meiosis is the chromosome-reducing response; but in animals, far less time elapses between meiosis and the ensuing fertilization than between fertilization and the ensuing meiosis. Nevertheless, a given meiosis is the response to the preceding fertilization. The fertilization following meiosis is a new stimulus having its own meiotic response at the end of that generation. Such a conclusion is suggested not only on logical grounds but also by the virtual certainty that the diplontic cycle of animals is derived from a more primitive cycle encountered in many nonanimal organisms and in certain parasitic protozoan groups. In these, fertilization is followed immediately by meiosis in the zygote, and the developing adults then are haploid. The ancestors of animals probably were characterized by such *haplontic* life cycles; and during the evolution of nonanimal ancestor to animal descendant meiosis apparently became *postponed*, from immediately after fertilization to the time of gamete formation in the adult. The response thus follows long after the stimulus, and the result is a diplontic cycle with diploid adults (cf. Fig. 10.14).

Such a postponement of meiosis is very advantageous adaptively. Inasmuch as adult animal cells are diploid, each

10.14. The haplontic life cycle at left has probably given rise, by postponement of meiosis, to the diplontic cycle at right.

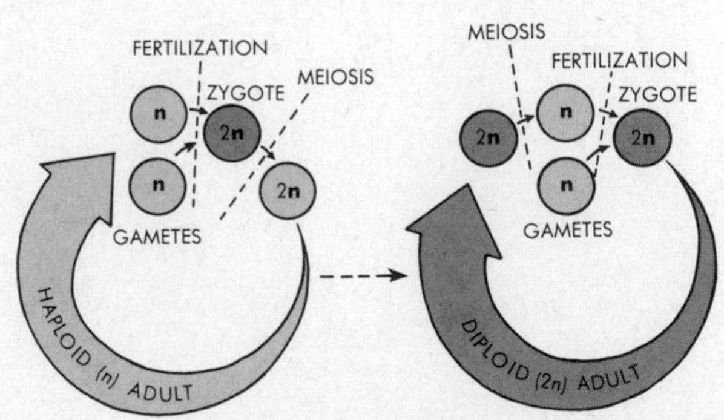

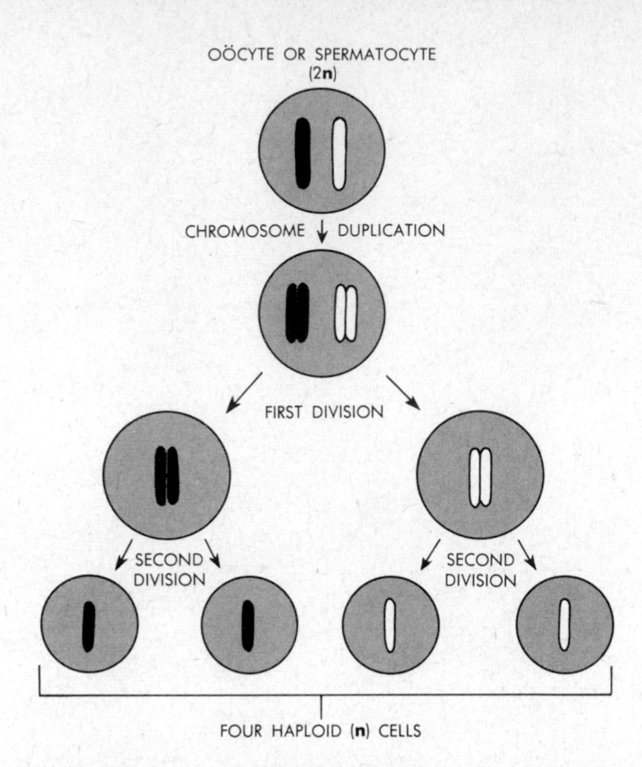

10.15. The general pattern of meiosis, on the assumption that $2n = 2$. During the meiotic divisions, each member of a chromosome pair is often referred to as a *chromatid.*

gene (and each chromosome) in each adult cell is represented twice rather than just once. One gene of a given pair has a maternal ancestry, the other a paternal ancestry. Consequently, one advantage of paired genes (and paired chromosomes) is that they increase the redundancy of the genetic information (cf. Chap. 6). Even if one gene of a pair changes in some way, e.g., by mutation, then the other gene still preserves the original message. In short, the diploid state increases the genetic stability of the individual. Moreover, the diploid state may also increase genetic variability, for occasionally a mutated gene might happen to produce a more advantageous trait than the original gene. Evidently, this possibility of permitting the stability or the variability to be exploited has been of sufficient adaptive significance to make diplontic life cycles nearly universal among animals.

Meiosis occurs through two *meiotic cell divisions.* The general pattern of these divisions is as follows. The gonads contain gamete-forming *generative cells* called *oöcytes* in ovaries and *spermatocytes* in testes. Both types of cells are diploid. When meiosis occurs, a generative cell undergoes two successive cytoplasmic cleavages, which transform the one original cell into four cells. During or before these cleavages, the chromosomes of the diploid cell duplicate once. As a result, $2n$ becomes $4n$. And of these $4n$ chromosomes, one n is incorporated into each of the four cells formed. In sum, *one diploid* cell becomes *four haploid* cells (Fig. 10.15).

In man, for example, a diploid generative cell in the gonads contains 46 chromosomes (or 23 pairs, one set of 23 being of maternal and the other set of 23 of paternal origin). During meiosis the chromosome number doubles once, to 92, and in the process the cell divides twice in succession. Four cells result, among which the 92 chromosomes are distributed equally. Hence each mature gamete contains 23 chromosomes, a complete haploid set.

The two meiotic divisions have many features in common with mitotic divisions. For example, each meiotic division passes through prophase, metaphase, anaphase, and telophase, as in mitosis. Moreover, spindles form and other nonchromosomal events are as in mitotic divisions.

10.16. A comparison of mitosis and meiosis, on the assumption that $2n = 6$. Note that the key difference between the two processes is the way the pairs of chromatids line up in metaphase.

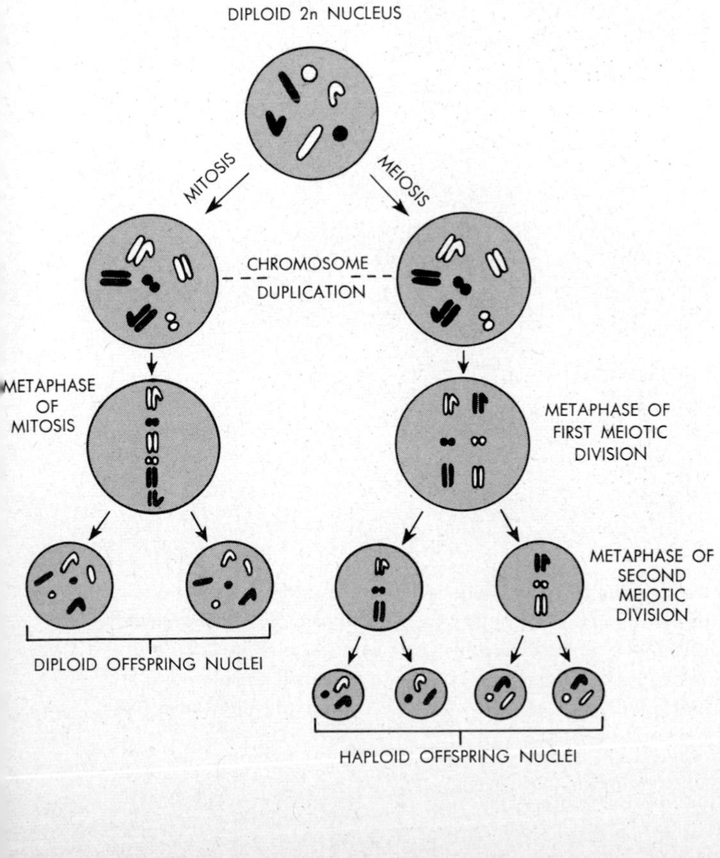

The critical difference between mitosis and the *first* meiotic division lies in their metaphases. In mitosis, we recall (cf. Chap. 6), all chromosomes, each of them already duplicated, migrate into the metaphase plate, where all the centromeres line up in the same plane. In the first meiotic division, the $2n$ chromosomes similarly duplicate during or before prophase. These $2n$ pairs, the members of each pair again joined at the centromere, also migrate into the metaphase plate. But now only n pairs assemble in one plane. The other n pairs migrate into a plane of their own, a plane which is closely parallel to the first. Moreover, every pair in one plane comes to lie next to the corresponding type of chromosome pair in the other plane. The metaphase plate is therefore made up of *paired chromosome pairs,* or *tetrads,* of like chromosomes lying side by side. And there are n of these tetrads in the whole plate (Fig. 10.16).

During the ensuing anaphase, two chromosomes of each tetrad migrate to one spindle pole, two to the other. At the end of the first meiotic division, therefore, there are two cells, each with n pairs of chromosomes. In the metaphase of the subsequent second meiotic division, the n pairs of chromosomes line up in the same plane, and n single chromosomes eventually migrate to each of the poles during anaphase. At the termination of meiosis as a whole, therefore, four cells are present, each with n single chromosomes, a complete haploid set.

Meiosis represents the *nuclear* phase of gamete maturation. In a male, a diploid spermatocyte in the gonads undergoes both meiotic divisions in fairly rapid succession, and all four resulting haploid cells constitute functional sperms. The pattern is somewhat different in females. An oöcyte in the ovary undergoes a first meiotic division and produces two cells. Of these, one is small and soon degenerates. Its remnants, now called the *first polar body,* remain attached to the other cell. This cell subsequently passes through the second meiotic division. Of the two cells produced here, one becomes the egg, and the other again is small and degenerates. Its remnants form the *second polar body,* which, like the first, remains attached to the egg. Each original oöcyte thus gives rise to only one functional egg (Fig. 10.17).

In some animals, e.g., coelenterates and echinoderms, oöcytes pass through both meiotic divisions in rapid sequence, as do the sperms of all animals. The eggs then formed are haploid and ready to be fertilized. In most other animals, however, vertebrates included, eggs are ready for fertilization

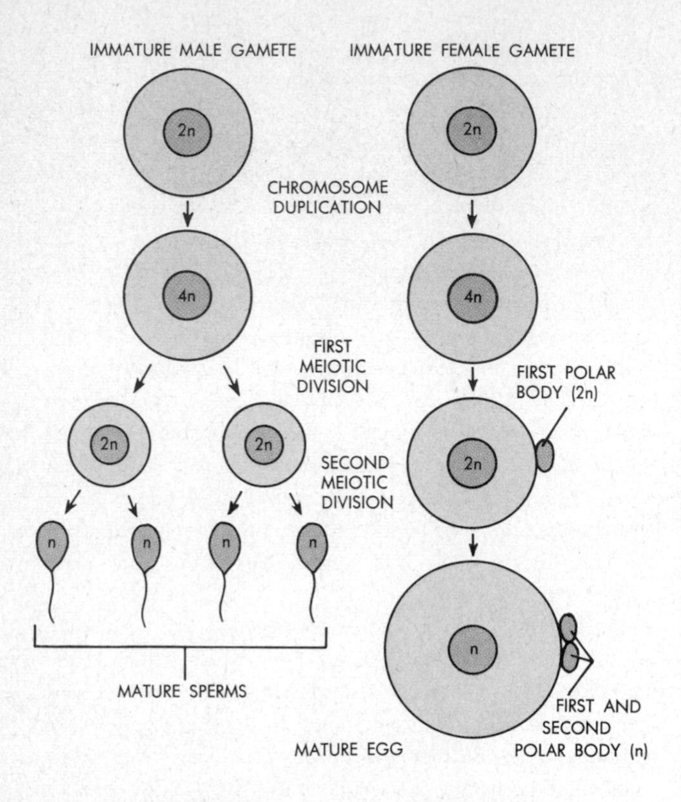

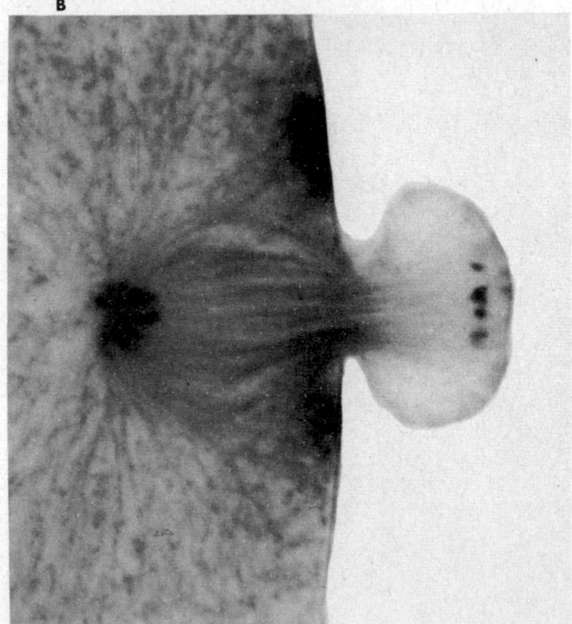

10.17. Diagram: meiosis in males and females. In males, all four haploid cells formed become functional sperms. In females, one cell formed by the first meiotic division is small and degenerates and becomes the first polar body. Similarly, one cell formed by the second meiotic division becomes the second polar body. Thus only one cell matures as a functional egg. **Photos: A and B. Polar-body formation. A.** Section through the edge of an immature whitefish egg, showing the extremely eccentric position of the spindle and the chromosomes during a meiotic division. The chromosomes are in anaphase, and cleavage, which will occur at right angles to the spindle axis, will therefore produce an extremely large and an extremely small cell. **B.** Cytoplasmic cleavage is under way. The small cell formed will degenerate and the remnants will persist as a polar body. **C, D,** and **E.** Meiosis in the egg maturation of the nematode *Ascaris,* in which $2n = 4$. **C.** First meiotic metaphase. Each of the two pairs of chromosomes has duplicated, and two tetrads are lined up in the metaphase plate. **D.** First telophase. One large and one small cell will be formed, each with four chromosomes. The small cell will degenerate and become the first polar body (as in **B**, above). The remaining large cell then undergoes the second meiotic division, the metaphase of which is shown in **E.** Of the two pairs of chromosomes here present, two (n) will go into each of the two cells yet to be formed. One of these will be the egg, the other will degenerate and become the second polar body. The first polar body may be seen as a dark spot at the top of the photo. Note also the dark central patch in **C, D,** and **E.** This is the sperm nucleus. When egg meiosis is completed, sperm and egg nuclei will fuse and fertilization will then have been accomplished.

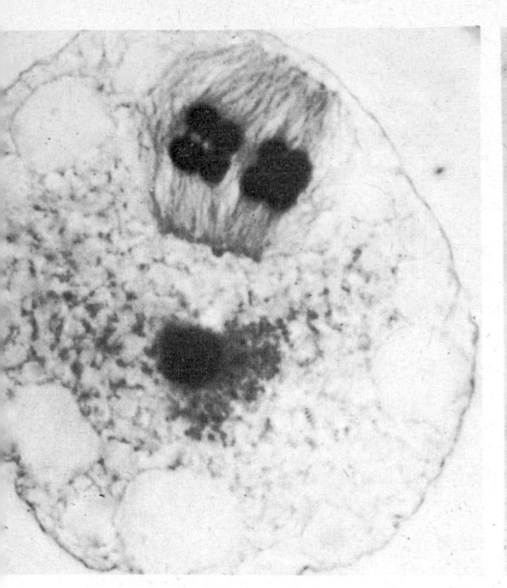

D

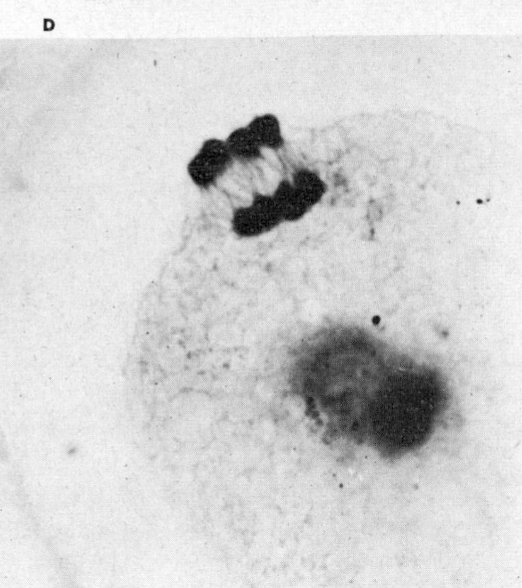

E

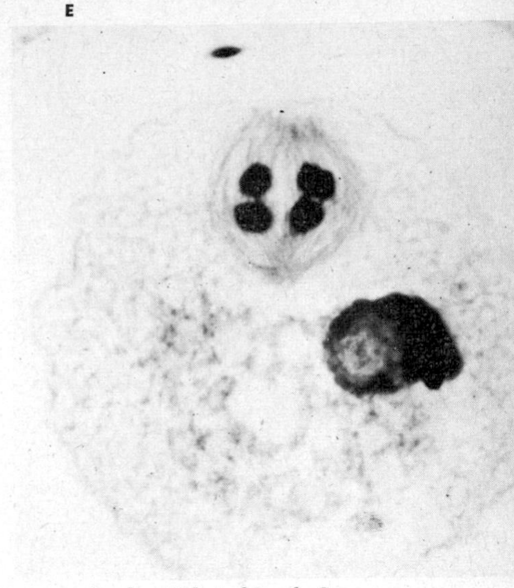

C,D,E, *courtesy Carolina Biological Supply Company.*

before meiosis is fully completed. Meiosis here proceeds part way, and the egg then remains in a state of meiotic arrest until fertilization occurs. At that time, the entrance of the sperm provides the stimulus for the completion of meiosis. In vertebrates, for example, oöcytes undergo the first meiotic division and produce the first polar body. The eggs are then fertilizable, and meiosis remains incomplete until fertilization actually occurs. When a sperm enters, the second meiotic division takes place and the second polar body is formed. Thereafter, the remaining haploid nucleus of the egg fuses with the haploid nucleus of the sperm, and this event completes fertilization.

In parallel with the nuclear maturation of gametes, cytoplasmic maturation takes place. The particular form varies for different animal groups. In the spermatocytes of most animals, much of the cytoplasm degenerates altogether. The nucleus enlarges into an oval *sperm head,* and the mature sperm retains only three structures having a cytoplasmic origin: a long posterior *sperm tail,* which serves as locomotor flagellum; a *middle piece,* which contains energy-supplying mitochondria and which joins the sperm tail with the sperm head; and an *acrosome,* a structure at the forward end of the sperm head, by means of which the sperm will make contact with an egg. As a result of losing all other cytoplasm, a mature sperm is among the smallest cells within the body (Fig. 10.18). Mature eggs, on the other hand, are among the largest cells; their cytoplasms have become specialized for the accumulation and storage of *yolk,* food reserves for the future embryos.

REPRODUCTIVE SYSTEMS

In sponges, individual mesenchyme cells undergo meiosis and either enlarge into amoeboid eggs or mature into small flagellate sperms. Distinct reproductive organs or systems are absent (cf. Chap. 20). The water current in the alimentary channels carries sperms and eggs to the outside. Eggs are usually liberated after they have been fertilized and are already developing. Mesenchyme cells also give rise to the gametes of radiate animals. Radiates typically possess gonads, and these discharge gametes into the alimentary cavity in most cases and through temporary ruptures in the epidermis in some cases.

The Bilateria all possess distinct reproductive *systems,* formed invariably from the embryonic mesoderm (Fig. 10.19). Gonads are usually globular organs containing gamete-forming *generative epithelia.* Mature gametes produced by these tissues may accumulate in chambers or spaces within the gonads. Note that, since the gonads are mesodermal, any such spaces are coelomic in nature, even in acoelomate or pseudocoelomate animals. Eggs within an ovary are often surrounded by nutritive *follicle* cells. These supply raw materials which

Courtesy Carolina Biological Supply Company.

10.18. Sperms of rats. In each, note sperm head (with acrosome faintly visible at forward end), sperm tail, and the middle piece (darkly stained).

accumulate in the eggs as yolk. Most animals typically possess a pair of gonads; but in many primitive members of certain groups, numerous gonads are present throughout the body. Some animals normally contain only a single gonad.

Apart from the gonads, a reproductive system contains *gonoducts* leading to the outside, viz., *sperm ducts* or *oviducts*. These are usually paired, and if the number of gonads exceeds two, smaller ducts from the gonads open into a main pair of gonoducts. Various accessory organs are generally associated with the gonoducts. For example, sperm ducts may pass through or past *sperm sacs*, chambers which may store sperms before discharge. Also, *prostate glands* may secrete *seminal fluids* which, together with the sperms, make up *semen*. Other glands may be present in addition. Many animals possess *copulatory organs* at or near the exterior termination of the sperm ducts. Such organs may be modified integumentary structures, e.g., anal fins or eversible regions in the wall of a cloaca. If the sperm duct passes through a copulatory organ, the latter is usually called a *penis*.

Analogously, along the course of oviducts may be present *seminal receptacles*, pouches which store sperms after mating and before fertilization, as well as *yolk glands*, *shell glands*, and other glands producing nutritive or protective layers around fertilized eggs. Before its exterior termination, an oviduct may be enlarged into a chamber, the *uterus*, in which egg development may occur. If the female mates by copulation, the last section of the oviduct receiving the copulatory organ during mating is usually called a *vagina* (cf. Fig. 10.19).

Gonoducts generally open to the outside via independent *gonopores*. In many cases, however, the gonoducts discharge into a cloaca, and in others they join with the excretory ducts and empty through a common *urogenital pore*. Hermaphroditic animals may possess separate gonopores for each of their two reproductive systems or a common single gonopore for both. Note that, in all copulating animals, fertilization takes place along the course of the oviduct, after sperms have been transferred from the male through the gonopore of the female.

In acoelomates and pseudocoelomates, the interior ends of the gonoducts connect directly with the gonads. By contrast, the gonoducts of coelomate animals typically connect interiorly with the lining of the coelom, at ciliated *gonostomes* (Fig. 10.20). The gonads here discharge gametes into the coelomic cavity, and the gonoducts then collect and transport them from the coelom to the exterior. Note that, in such cases, the gonoducts resemble metanephridia. Indeed, some coelomate animals are without gonoducts altogether, and their metanephridia serve both as excretory and as reproductive channels. In certain annelids with solenocytic protonephridia, the gonoducts open into the protonephridial tubules, and gametes as well as urine exit via the nephridiopores. Excretory and reproductive systems so joined are known as *nephromixia*. In a variation of this pattern, encountered in some of the annelids just referred to, the solenocytes degenerate, and excretory functions are taken over by the gonoducts. Such animals

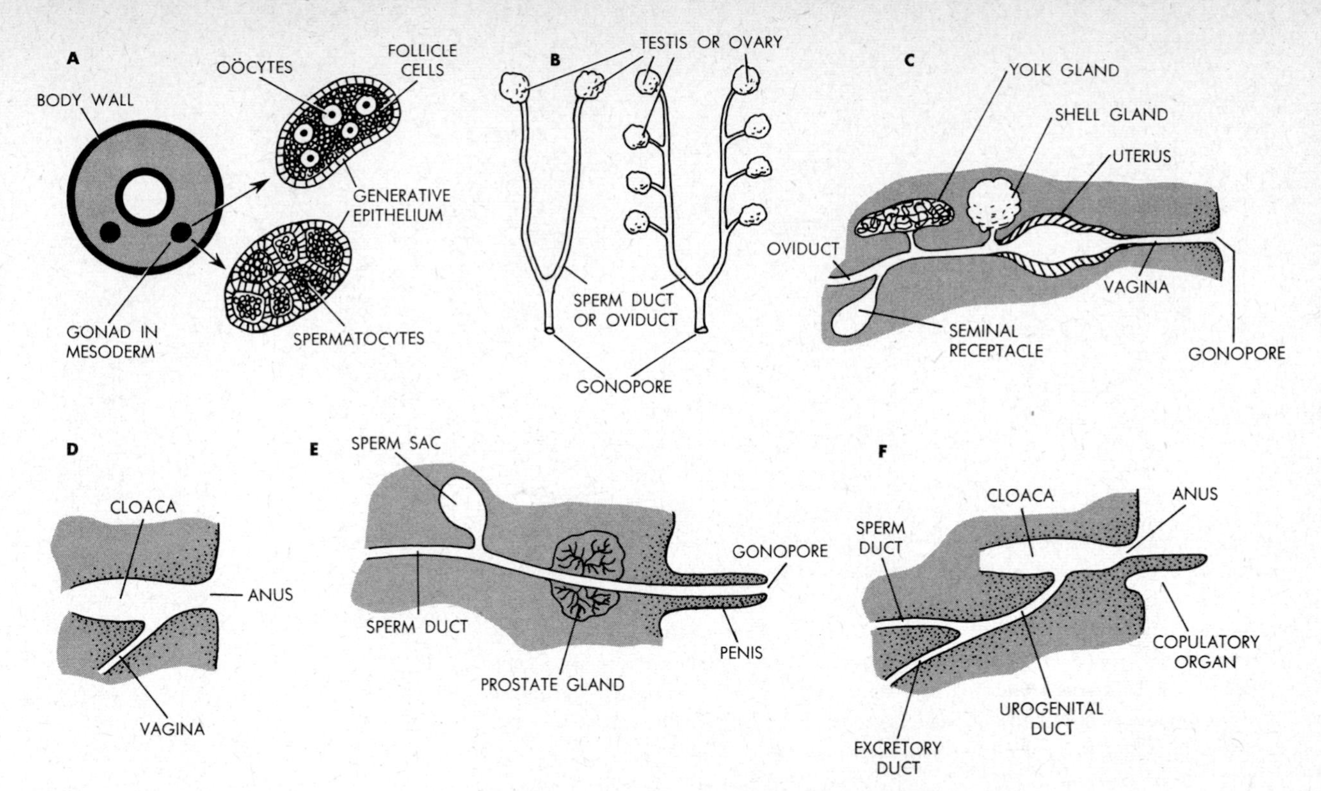

10.19. Gonads and reproductive systems, schematic. A. The mesodermal origin and the general structure of an ovary (top) and a testis (bottom). **B.** Reproductive systems with one (left) or multiple (right) pairs of gonads. **C** and **E.** Glands and appendages commonly associated with female and male reproductive systems, respectively; these are composite sketches, and not all such structures need necessarily occur in any one system. In **C,** the ♀ system opens via an independent gonopore; **D** shows the pattern if a cloaca is present; **E** and **F** illustrate corresponding exit modifications in male systems. Note penis with sperm-duct termination, copulatory organ without such termination.

therefore possess ducts which lead from the coelom to the exterior and are indistinguishable from true metanephridia. However, the interior parts of such ducts derive from the gonoducts, the terminal parts derive from the protonephridial tubules, and the ducts as a whole serve both excretory and reproductive functions. Note furthermore that, whereas gonoducts are always mesodermal, protonephridial ducts are ectodermal.

The gonoducts also have an intimate association with the excretory ducts among vertebrates. The gonads arise as paired pouches growing into the coelom from the dorsal coelomic lining, near the anterior portions of the mesonephros (Fig. 10.21). The epithelial tissue of such a pouch proliferates and forms the outer *cortex* of the developing gonad. The space within the pouch fills loosely with strands of mesonephric tissue (which to some extent resemble mesonephric tubules). This interior tissue represents the *medulla* of the developing gonad. The gamete-producing cells, called the *primordial germ cells,* arise not in the developing gonad itself, but in the head mesoderm of the embryo. From there the cells migrate into the embryonic gonad and disperse within it. At this stage the gonad is sexually still indifferent, i.e., it may develop in either a male or a female direction. The factors determining the actual direction will be examined in Chap. 13. If the gonad becomes a testis, the cortex develops very little more, but the medulla proliferates greatly and forms the bulk of the mature testis. The primordial germ cells then produce sperms after the animal is adult. By contrast, if the indifferent gonad transforms into an ovary, it is the medulla which undergoes little further development. The cortex on the other hand enlarges

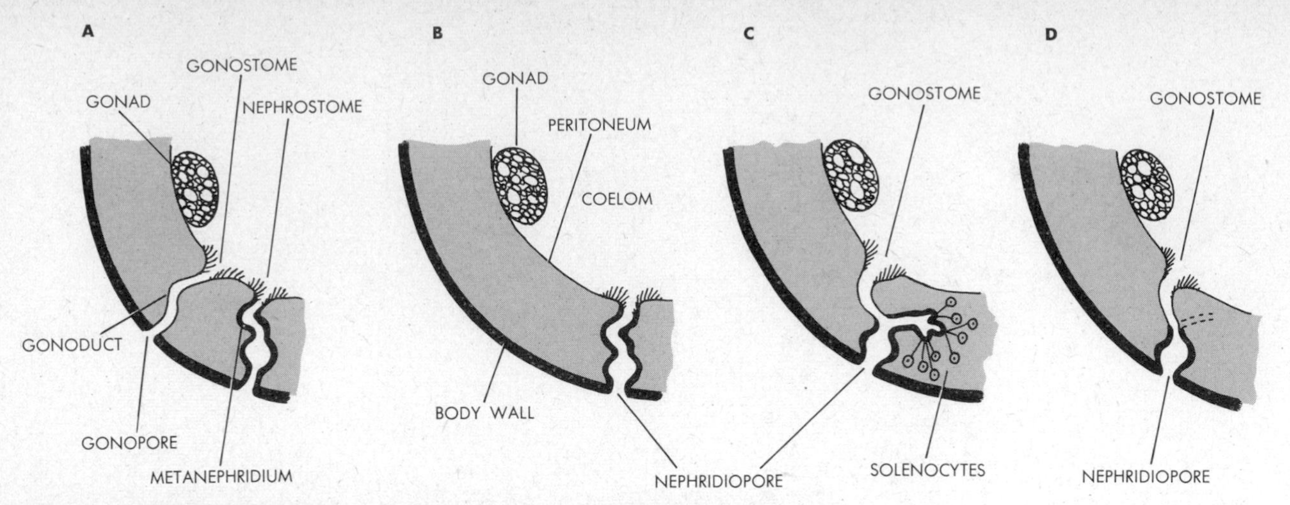

10.20. Relationships of reproductive and excretory systems in coelomate invertebrates. **A.** Basic condition of coelomic gonads and separate reproductive (fine line) and excretory (heavier line) systems. **B.** Gonoducts degenerated, metanephridium serving as both reproductive and excretory channels. **C.** Nephromixium (as in some annelids), with gonoduct joined to solenocytic protonephridium and both having common exit. **D.** Degeneration of solenocytes produces a joint system composed partly of gonoduct and partly of excretory duct and serving both reproductive and excretory (metanephridiumlike) functions, as in certain variants of **C.**

and gives rise to the bulk of the mature ovary (cf. Fig. 10.21). Each of the primordial germ cells present becomes surrounded by groups of nutritive follicle cells. After sexual maturity, the germ cells within the follicles develop into functional eggs, one or more of them at a time. An embryonic ovary of a human female contains some 400,000 primordial germ cells, but only about 400 of these will become functional eggs during adult life.

In the most primitive vertebrates (viz., jawless lampreys, hagfishes), the gametes of both sexes are discharged from the gonads directly into the coelom. Gametes then reach the outside via small pores in the posterior region of the Wolffian ducts (Fig. 10.22). In all other vertebrates, different exits for the gametes have evolved. During the larval stage, all these vertebrates develop a pair of *Muellerian ducts* running parallel to the Wolffian ducts. Anteriorly, the ducts open into

10.21. Development of the vertebrate gonad. **A.** Side view, indicating relation of the site of origin of the primordial germ cells and of gonad to mesonephros and coelom. **B.** Cross-sectional views of the early paired genital ridges of the indifferent gonads and the differential growth of either the medullary (mesonephros-derived) tissue, resulting in testes, or the cortical (peritoneum-derived) tissue, resulting in ovaries.

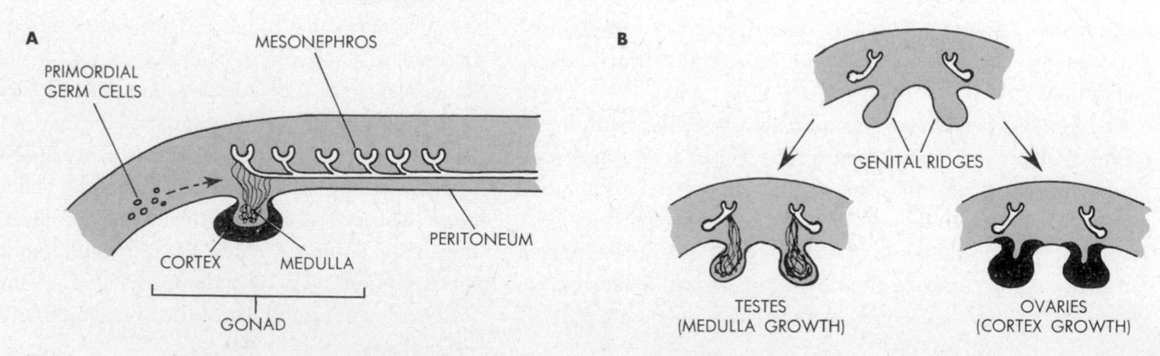

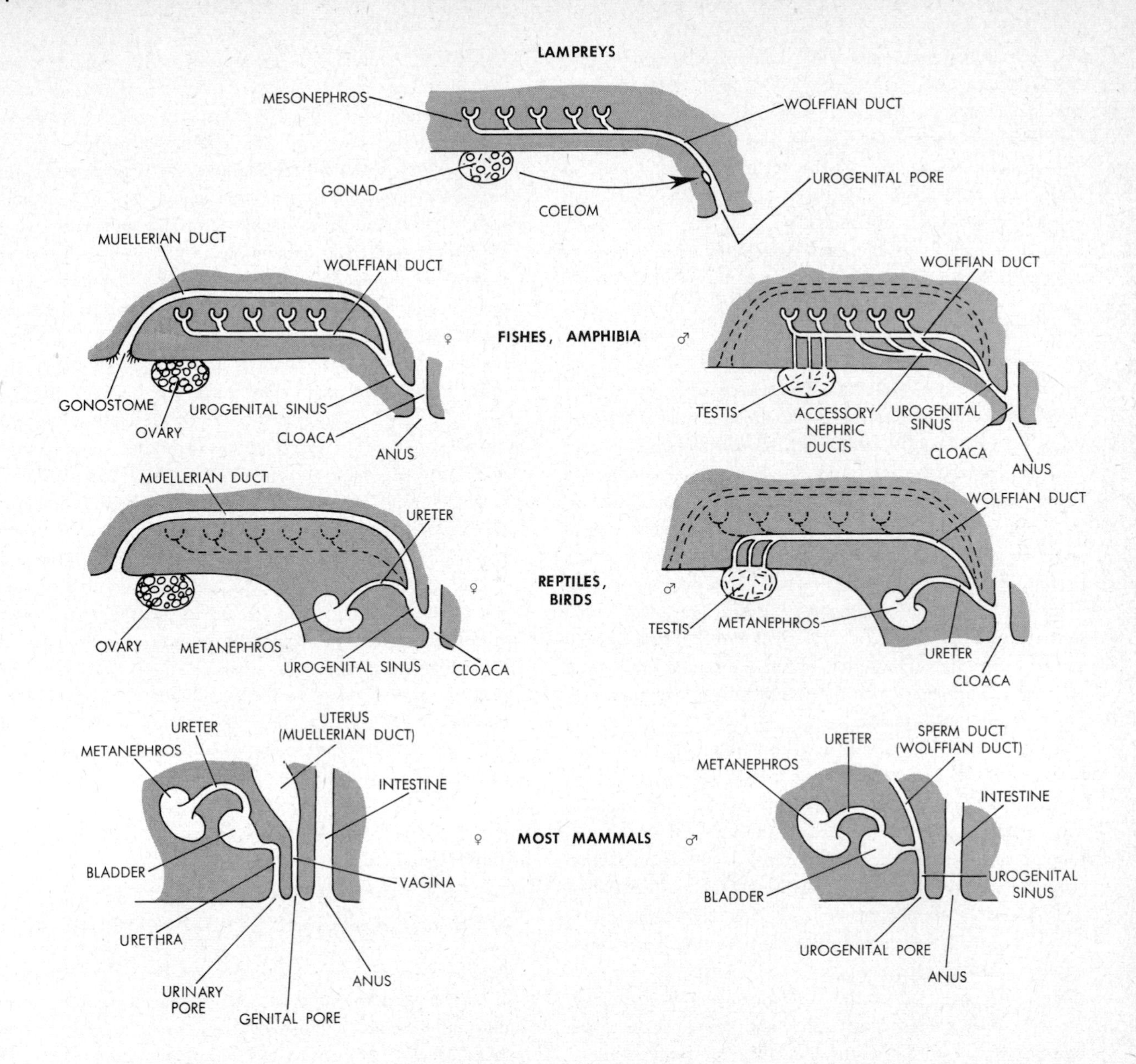

10.22. The reproductive ducts in vertebrates. Ducts of the left side of the body are diagrammed. In lampreys, note the pore in the Wolffian (mesonephric) duct for sperm or egg exit from the coelom. In fishes and amphibia, the Wolffian ducts carry urine in females but largely sperms only in males, urine here traveling mostly through accessory excretory ducts. In reptiles and birds, as also in mammals, the Wolffian ducts are sperm ducts exclusively, urine being carried through the ureters in males as well as females. Females have reproductive Muellerian ducts in all vertebrates except the lamprey group. The uterus and vagina of mammals are derived from the original Muellerian ducts. Such ducts form also in early embryos of males but degenerate soon after.

the coelom via gonostomes, and posteriorly they empty into the urogenital sinus along with the Wolffian ducts (and the sinus opens into the cloaca). It is probable that, in the ancestors of these vertebrates, the Muellerian ducts may have served as exit channels from the coelom for the gametes of both sexes. In modern vertebrates, however, the Muellerian exit path is used only in females.

In all female fishes and land vertebrates (man included), eggs are discharged from the ovary into the coelom and are carried from there to the outside typically via the Muellerian ducts. The latter thus represent oviducts. The females of fishes and amphibia also possess Wolffian ducts, and these serve as excretory channels. But the females of land vertebrates develop a metanephros and separate ureters, and the Wolffian ducts here degenerate along with the mesonephros. In reptiles and birds, the ureters and Muellerian ducts continue to empty into the cloaca via a urogenital sinus. In most mammalian females, however, the two types of ducts acquire separate openings to the outside; excretory ducts open via a urethra at a urinary pore, and reproductive ducts open via a vagina at a gonopore (cf. Fig. 10.22). Note incidentally that only the *left* ovary and oviduct persist in most birds; the right side of the reproductive system degenerates in the embryo, an adaptation correlated with the production of comparatively huge eggs in these animals.

The evolutionary pattern has been different in the male vertebrates (cf. Fig. 10.22). Here the Muellerian ducts degenerate in the embryo, and sperms in all cases are carried to the outside via the Wolffian ducts. In the embryos of fishes and amphibia, the developing testes retain a structural connection with the mesonephric tissue which contributes to testicular growth. Later, this connection becomes elaborated into a network of ducts which lead from the testes into the anterior nephric tubules of the mesonephros. Thus, sperms as well as urine are transported by the Wolffian ducts. We may note, however, that modern bony fishes and amphibia develop a number of accessory excretory ducts leading from the mesonephros to the urogenital sinus. In these animals, therefore, the Wolffian ducts come to function primarily as sperm ducts. Adult land vertebrates no longer possess a mesonephros, but the Wolffian ducts persist and function as sperm ducts exclusively. In reptiles and birds, the ducts still open into the urogenital sinus and cloaca along with the ureters, but in most male mammals the sinus is a urethra which opens independently of the alimentary tract. Mammalian males therefore possess a single urogenital orifice for both sperms and urine (cf. Fig. 10.22). Simultaneous ejection of these products is prevented by involuntary nervous reflexes and by constrictor muscles which close off either the sperm ducts or the ureters when the other ducts are discharging.

Many male mammals are characterized also by *descending* testes. In comparatively primitive mammals (e.g., opossums, bats, whales), the testes remain located permanently where ovaries are located in females, i.e., within the body, not far from the kidneys. In a second group of mammals, which includes elephants and many rodents, the testes are again found within the body for most of the year. But during the breeding season, when sperms are actually produced, the testes migrate into a *scrotum*, a skin sac between the hind legs. After the breeding season, the testes migrate back into the body, to their original positions. In a third group, exemplified by rodents such as mice and rats, the testes pass into a scrotum when the animals reach sexual maturity, and from then on the gonads remain there permanently. And in a fourth group, of which man is a member, the testes are internal only during embryonic stages; the organs migrate into a scrotum just before birth and then remain in this sac permanently.

It is known that the temperature in a scrotum is up to 7°C lower than within the body. It is also known that lower temperatures tend to promote sperm production and that higher temperatures tend to inhibit it. Temperature, testis location, and continuity of sperm manufacture therefore are probably correlated.

REPRODUCTIVE PATTERNS

Mating and Development

Regardless of whether an animal is aquatic, terrestrial, sessile, or motile, its sperms must be both aquatic and motile: sperms must move toward and fuse with eggs in an aqueous environment. This requirement has led to the elaboration of two basic mating patterns among animals. In *external fertilization*, mating partners are or come into more or less close proximity in natural bodies of water and both then simultaneously *spawn;* i.e., they release sperms and eggs directly into the water. Frequent chance collisions among the closely placed gametes then bring about many fertilizations. This pattern is characteristic of most aquatic animals, sessile as well as motile, and also of terrestrial animals such as certain insects and amphibia, which may migrate to permanent bodies of water for reproduction. The second pattern is *internal fertilization.* Mating partners here come into physical contact and *copulate,* i.e., by some means the male transfers sperms directly into the reproductive system of the female. Specialized copulatory organs may or may not be present in such cases. For example, birds are without copulatory organs (ostriches excepted), yet fertilization is internal; mating here requires the apposition of the cloaca of the male against that of the female. In all instances of internal fertilization, the internal tissues of the female provide moisture for the sperms and the need for external water is thereby circumvented. Internal fertilization is characteristic of most terrestrial animals, but the process occurs also in numerous aquatic groups (e.g., in many fishes).

Several variant forms of internal fertilization are known. For example, numerous groups of animals produce, not loose sperms in seminal fluid, but compact sperm packets, or *spermatophores.* These are transferred into females in a variety of different ways. Thus, squids and octopuses use their tentacles as transferring arms. The females of certain salamanders use cloacal lips to enfold spermatophores deposited on the ground by the males. Males of certain other amphibia transfer spermatophores by mouth. In certain spiders, sperms are placed on an anterior body appendage and the latter is then inserted directly into the female reproductive system. Some animals transfer sperms into females by a process akin to hypodermic injection, through any part of the skin. These and various other methods will be referred to specifically in Unit 2.

It should be readily apparent that, in both external and internal fertilization, mating is facilitated greatly by animal locomotion. All sessile animals are aquatic; sessile terrestrial animals could not survive because, in the absence of adult locomotion, sperms could not be brought to eggs, and sperms themselves cannot swim on land. Among the sessile aquatic animals, successful fertilization in large measure depends on the presence of reasonably dense populations of mating animals within circumscribed regions. Geographic species dispersal, accomplished among motile animals by larval and adult locomotion, is achieved among sessile animals mainly by the motile larvae. Keep in mind also that, regardless of the pattern of mating, *mutual* fertilizations usually take place if the mating partners are cross-fertilizing hermaphrodites.

Where fertilization is external, development of the zygotes into new adults takes place externally as well, in natural bodies of water. In many cases where fertilization is internal, the zygotes are released from the female parent, and zygote development then also occurs externally. All animals in which the eggs are shed to the outside, either in an unfertilized or a fertilized state, are said to be *oviparous.* Among vertebrates, for example, many fishes are oviparous and externally fertilizing, whereas all birds are oviparous and internally fertilizing. In all instances of oviparity, the eggs develop essentially on their own, food being supplied by yolk within each egg. Eventually the embryos *hatch* as larvae or as miniature, immature adults. Such a pattern of events is characteristic of most animals (Fig. 10.23).

If the development of oviparous animals takes place in water, the zygotes often have coats of jelly around them (e.g., frog eggs) but are otherwise protected very little. Coats of this sort are secreted by the oviducts before the eggs are laid. Zygotes developing on land possess more elaborate protection, particularly against evaporation. For example, earthworms,

10.23. Patterns of fertilization and offspring development in relation to the environment and the maternal body.

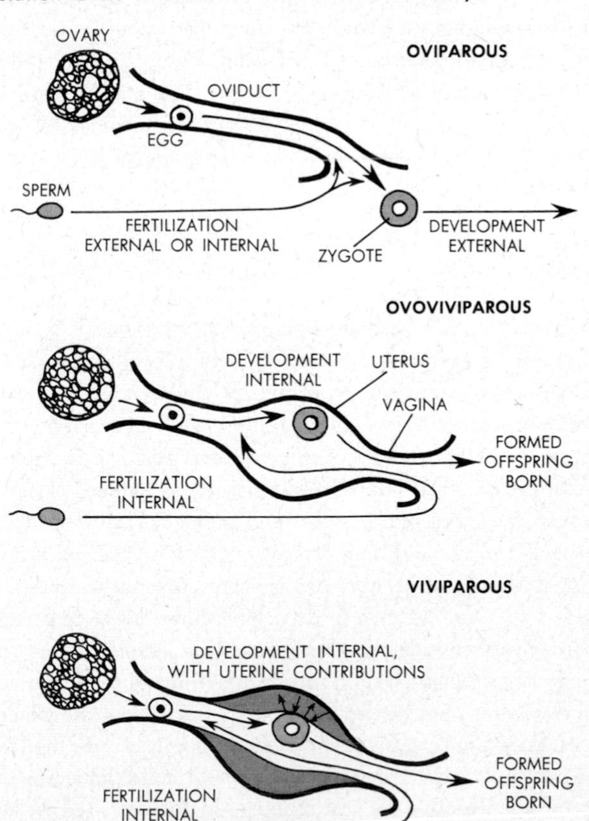

spiders, and insects such as grasshoppers and cockroaches form a cocoon or a hard casing around batches of just-laid fertilized eggs. Other insects and also reptiles and birds secrete shells around individual eggs after fertilization and before laying (cf. below).

Some animals are *ovoviviparous*. Fertilization in such cases is always internal, and the zygotes are then retained within the oviducts, specifically in uterine enlargements of these ducts. Development therefore occurs inside the female. However, beyond providing a substantial measure of protection, the female body does not otherwise contribute to zygote development; as in oviparous types, food is supplied by the yolk included within each egg. Ultimately the young are *born* rather than hatched, i.e., the females release fully formed animals, not eggs. Among vertebrates, some of the fishes, amphibia, and reptiles are ovoviviparous (cf. Fig. 10.23).

A third group of animals comprises *viviparous* types. In these, fertilization is again internal, zygotes are retained within a uterus, and the young are born as developed animals. However, the female body here influences the development of the young not merely by providing protection. It also supplies food and contributes to offspring metabolism generally in numerous and vital ways (cf. below). The females in such cases are *pregnant.* Viviparous vertebrates include, for example, some fishes, some snakes, and the majority of the mammals (cf. Fig. 10.23).

Breeding Cycles

Unlike vegetative reproduction, which may occur at any favorable time during the year, gametic reproduction takes place in most animals only during specific *breeding seasons.* Such seasons are largely annual, most of them occurring in spring or in fall. Among spring breeders (e.g., most fishes, birds, and mammals) the offspring are usually hatched or born in the same spring or in summer, and among fall breeders (e.g., deer, sheep) the offspring typically appear in the following spring. Many animals have two or more breeding seasons per year; e.g., there are two in dogs. The durations of breeding seasons vary considerably, being restricted in some cases to a single day (or night, as in clamworms and many other marine invertebrates), but extending in others throughout the whole year, as in monkeys, apes, and men. Breeding seasons tend to be continuous also where environmental conditions remain uniformly favorable the year round, as in domesticated cattle, chickens, and rabbits, and in laboratory mice and rats. Even in such cases, however, fertility is usually greatest during the spring (Fig. 10.24).

The external physical environment normally provides the stimulus for the onset of a breeding season. Particularly significant in this respect are light and temperature. For example, spring breeding in many, perhaps most, animals is induced by increasing temperatures and day lengths, fall breeding, by decreasing temperatures and day lengths. If habitual spring breeders or fall breeders are transported from the northern to the southern hemisphere or vice versa, their breeding time usually changes in line with the altered seasons in the new environment. In vertebrates, it can be shown that such external stimuli exert their effect via the sensory system, the brain, and the pituitary gland. Signals from the brain, including particularly blood-borne neurosecretory signals from the hypothalamus, activate the pituitary, and this gland then begins to secrete increasing amounts of gonadotropic hormones. As a result, all parts of the reproductive system increase in size considerably and become functional. Gonads may increase their weight up to 100 times. After a breeding season, the pituitary hormone output declines again, and the reproductive system becomes quiescent and reduced in size.

Initiation of a breeding season and actual mating thereafter often require biological, psychosomatic inducing stimuli. For example, female mammals such as mice can be induced to come into heat by the smell given off by the males. Many fishes and birds perform elaborate, instinctive courtship rituals, and the visual effect of one mate on the other brings about certain

10.24. Visualization of various breeding-season patterns. The monestrous condition characterizes the majority of mammals. In polyestrous types, fertilization of eggs in any given batch and onset of pregnancy stop further egg production during that breeding season.

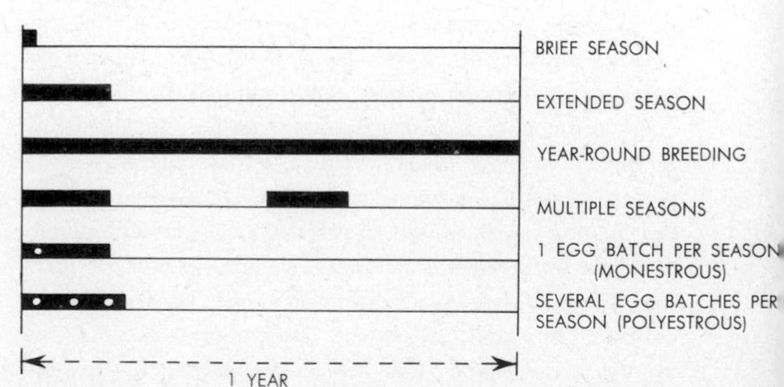

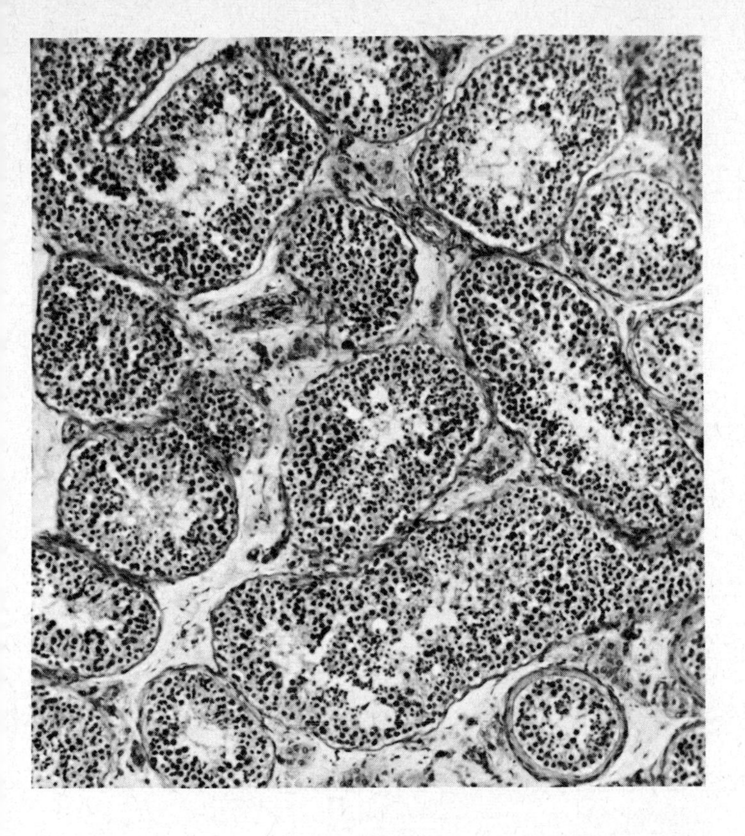

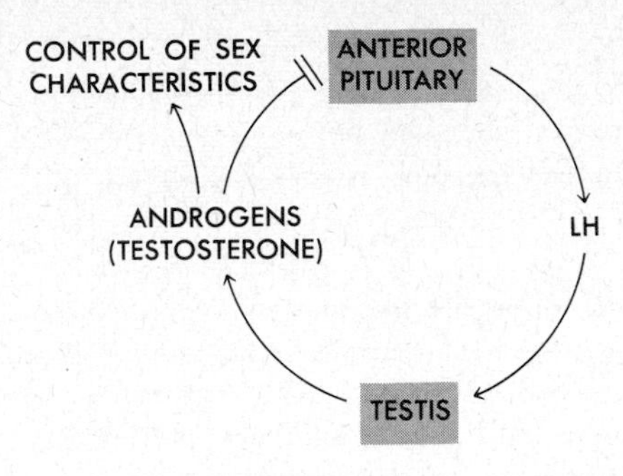

10.25. Photo: section through a mammalian testis. Note the tubular chambers in which sperms are produced. Mature sperms accumulate in the central spaces of the tubules. The tissue between the tubules contains the interstitial endocrine cells which manufacture androgens, the male sex hormones, under the stimulus of LH from the pituitary. **Diagram: the control of androgen secretion.** LH is one of the gonadotropic hormones of the pituitary. Arrow tipped with transverse double bar signifies inhibition.

hormone secretions internally and thus leads to reproductive readiness. In many instances, indeed, a fixed sequence of behavioral displays by one mate initiates a corresponding sequence of necessary internal hormonal events in the other.

Males produce sperms continuously during a breeding season and often also to a reduced extent between such seasons. In vertebrates, pituitary control is exercised mainly through *LH*, one of the three gonodotropic hormones (cf. Chap. 8, Table 7). Of the other two, *FSH* and *prolactin* (or lactogenic hormone), the function of FSH in males, if any, is still obscure. Prolactin likewise may play little or no role in sperm production as such, but the hormone is known to be responsible for any paternal behavior a male vertebrate may exhibit (e.g., protective attitudes toward mate and offspring, acquisition of nesting materials). Pituitary LH is the specific hormone which stimulates the testes to produce *androgens*, the male sex hormones. They are manufactured in cells located between the sperm chambers of a testis (Fig. 10.25). Of a series of androgens formed, *testosterone* is the most potent, and its concentration rises sharply at the start of a breeding season under the influence of LH. The entire reproductive system then becomes operational and sperms are actively produced. Also, sex urge increases and the secondary sex characteristics become pronounced (e.g., mating colors in plumage and the integument generally). If they are present in blood in excessive concentrations, androgens have an inhibitory effect on the pituitary. The latter produces less LH as a result, and androgen secretions consequently decline as well. Through such feedback control, the androgen concentration in males is maintained at a fairly steady level during the breeding season (cf. Fig. 10.25). In man and other mammals with a year-round breeding season, sperm production begins at puberty i.e., sexual maturity, and may continue for life.

Females do not produce eggs continuously during a breeding season. Rather, a given group of eggs matures together in a batch, and other eggs do not mature during the formation of such a batch. Most animals develop only a single batch of eggs during any one breeding season. This holds also for many mammals, animals in which the periods of egg maturation are called *estrus cycles*. Female dogs, for example, come into estrus just once during a breeding season. If the eggs produced in such a cycle are not fertilized, the animals remain infertile until the next breeding season. Other mammals (e.g., horses, sheep) are *polyestrous;* i.e., they may produce several successive batches of eggs per breeding season if they are not fertilized. If one batch does become fertilized, however, further egg manufacture then stops for the season (cf. Fig. 10.24).

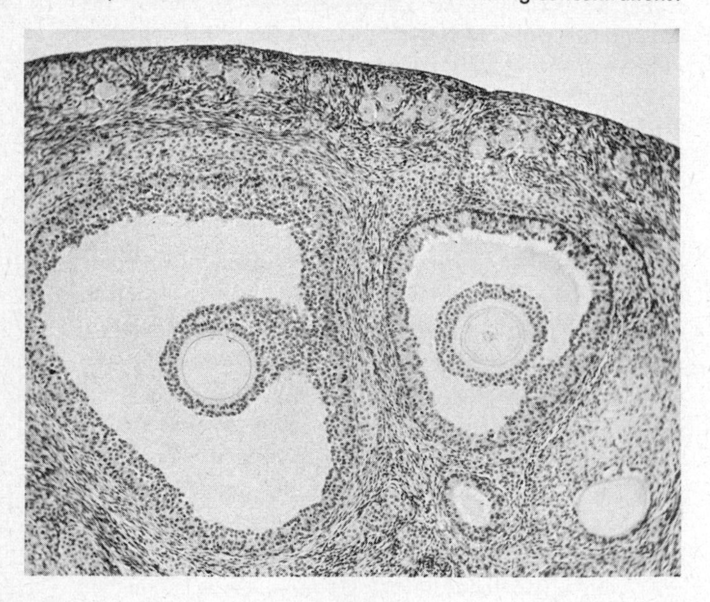

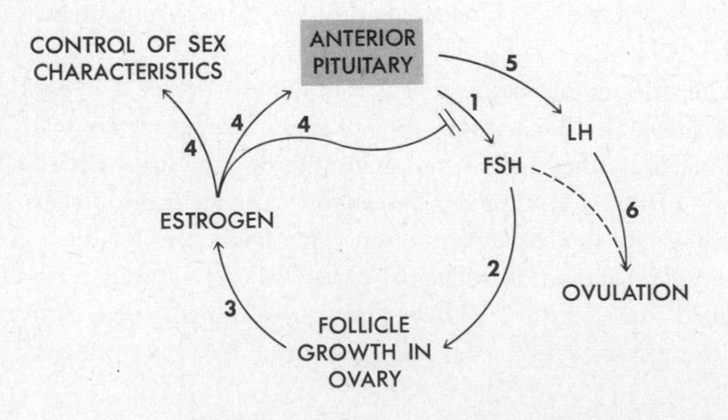

10.26. Photo: section through a mammalian ovary. Note the two large follicles, the follicular cavities, and the large egg cell in each follicle, embedded within a mass of cells along the follicular wall. Endocrine cells secrete estrogens into the follicular cavity. When the eggs are mature they will ovulate, i.e., escape by rupture of follicle and ovary walls. Near top of photo, along the edge of the ovary, note the relatively large cells. These are immature eggs which will become mature later, within follicles yet to be formed. **Diagram:** the feedback-control pattern of hormonal changes during the follicular phase of an egg-growth cycle, leading to ovulation. Numbers indicate sequence of steps. Regular arrows denote stimulation, arrows with double bar, inhibition. Broken line indicates decreasing concentrations.

The cycles of egg maturation have various durations. For example, year-round breeders such as domestic chickens may lay eggs as often as once a day. Man and apes produce eggs approximately once a month. Most animals *ovulate* spontaneously, i.e., the ovary releases eggs as soon as they are mature (Fig. 10.25). In some animals, however, ovulation is induced only by copulation. For example, rabbits, squirrels, and cats are among animals which come into estrus and then retain the mature eggs in the ovary. Depending on whether copulation subsequently does or does not take place, the eggs either ovulate into the oviduct and become fertilized or they degenerate within the ovary.

The hormonal controls of egg production in vertebrates parallel those of sperm production. In response to environmental stimuli at the onset of the breeding season, the pituitary secretes prolactin and FSH. Prolactin induces maternal behavior and broodiness, causing the female to contribute to nest building, to guard or sit on eggs after they are laid, or to care for the young after birth. FSH, the *f*ollicle-*s*timulating *h*ormone, influences ovarian activity. More specifically, the follicle cells around an egg, and perhaps also the ovarian cells between follicles, are stimulated to produce *estrogens,* the female sex hormones (cf. Fig. 10.26). These correspond functionally to the male hormones; i.e., they promote the growth of the reproductive system, follicles included, the pronounced development of secondary sex characteristics, and an increase in sex urge. As Fig. 10.26 indicates, a growing follicle acquires a central fluid-filled cavity, and the egg comes to be located excentrically, in a thickened region of the follicular wall. Developing follicles also migrate within the ovary, and at maturity they are stationed just underneath the ovary surface, where they may bulge outward pronouncedly.

When estrogen concentrations in blood exceed a certain threshold level, the hormones exert an inhibitory effect on FSH production in the pituitary. At the same time also, the estrogens stimulate the pituitary to produce another gonadotropic hormone, viz., LH. As a result, LH concentrations begin to rise just when FSH concentrations begin to fall. These shifting hormone balances are the specific stimulus for ovulation; the ovary surface and the follicle wall both rupture, and the mature egg escapes into the coelom and oviduct (Figs. 10.26, 10.27). In some animals, as noted above, ovulation also requires nervous triggering through copulation.

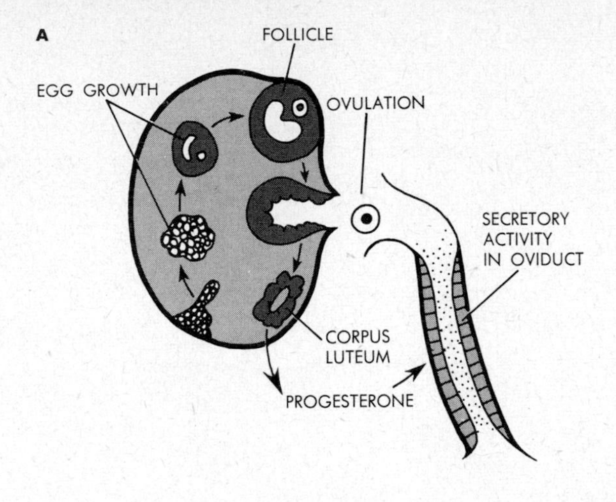

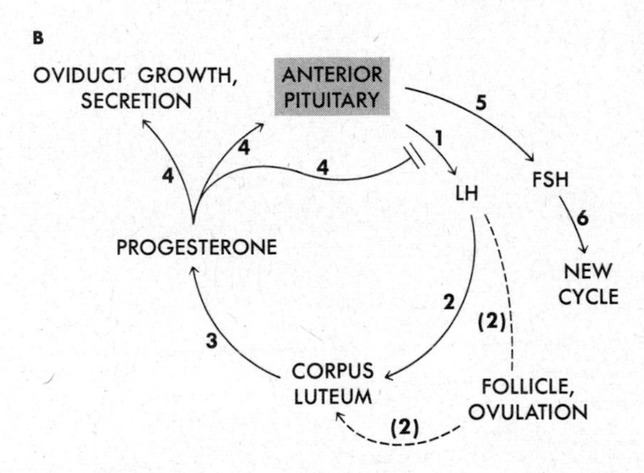

10.27. **A.** Ovarian events from egg growth to progesterone secretion in oviparous vertebrates, and the effect of progesterone on the oviduct; the secretions of the latter may be jelly coats around eggs or shell-forming substances. **B.** The hormonal changes during the luteal phase of an egg-growth cycle in oviparous vertebrates. Note that the feedback-control pattern is analogous to that in the follicular phase (Fig. 10.26). Numbers indicate sequence of steps.

An immediate consequence of ovulation is that the ruptured and eggless follicle remaining in the ovary loses its fluid and collapses. Another consequence is that, since FSH production by the pituitary has now ceased, the remnant of the follicle ceases to manufacture estrogen. Instead, under the specific influence of the LH produced in increasing quantities by the pituitary, the remnant of the follicle transforms into a yellowish body, the *corpus luteum.* The name "LH" stands for "*l*uteinizing *h*ormone." Under the continuing influence of this hormone, the corpus luteum begins to secrete a new hormone of its own, namely, *progesterone* (cf. Fig. 10.27).

This hormone is chemically very much like the estrogens and the male hormones (i.e., it is a steroid compound, cf. Fig. 4.22). In oviparous vertebrates, progesterone stimulates growth of the oviducts generally and the secreting activity of the ducts specifically. Thus, under the influence of this hormone, jelly coats or shells are secreted around eggs now passing through the oviduct. We may note that prolactin from the pituitary likewise stimulates the corpus luteum to produce progesterone. When the concentration of progesterone eventually exceeds a certain threshold level, the hormone inhibits LH production in the pituitary. The corpus luteum then ceases to manufacture progesterone, but by this time the eggs have already been shed. A new FSH-initiated egg-growth cycle may then begin if the breeding season does not come to a close.

The pattern of events is somewhat different in viviparous vertebrates, i.e., in those in which fertilized eggs are retained and developed in a uterus. In such cases protective layers are not secreted around an egg. Instead, the wall of the uterus thickens greatly and develops numerous glandular pockets and extra blood vessels (Fig. 10.28). Fertilized eggs become firmly embedded, or *implanted,* in this wall, and the eggs receive nourishment and oxygen from the maternal blood. In these animals, the function of progesterone is to stimulate and to maintain the growth of the uterus wall in preparation for this condition of pregnancy. However, if eggs are not fertilized on their way through the oviduct, they soon disintegrate, and the uterus then will have been made ready for nothing. In such an event, progesterone eventually inhibits pituitary LH production as above, and the corpus luteum then ceases to manufacture progesterone. Without the hormone, however, the ready condition of the uterus cannot be maintained, and the wall soon reverts to its normal thickness.

In Old World monkeys, apes, and men, the preparations for pregnancy in the uterus are so extensive that, if fertilization does not occur and progesterone production then ceases, the inner lining of the uterus actually disintegrates. Tissue fragments separate away, and some blood escapes from torn vessels. Over a period of a few days, all this debris is expelled through the vagina to the outside. This is *menstruation* (cf. Fig. 10.28). A *menstrual cycle* in such animals lasts about 28 days. Follicle maturation occurs during the first 10 to 14 days, a period terminating with ovulation. During the ensuing 14 to 18 days, the uterus grows in anticipation of pregnancy. If

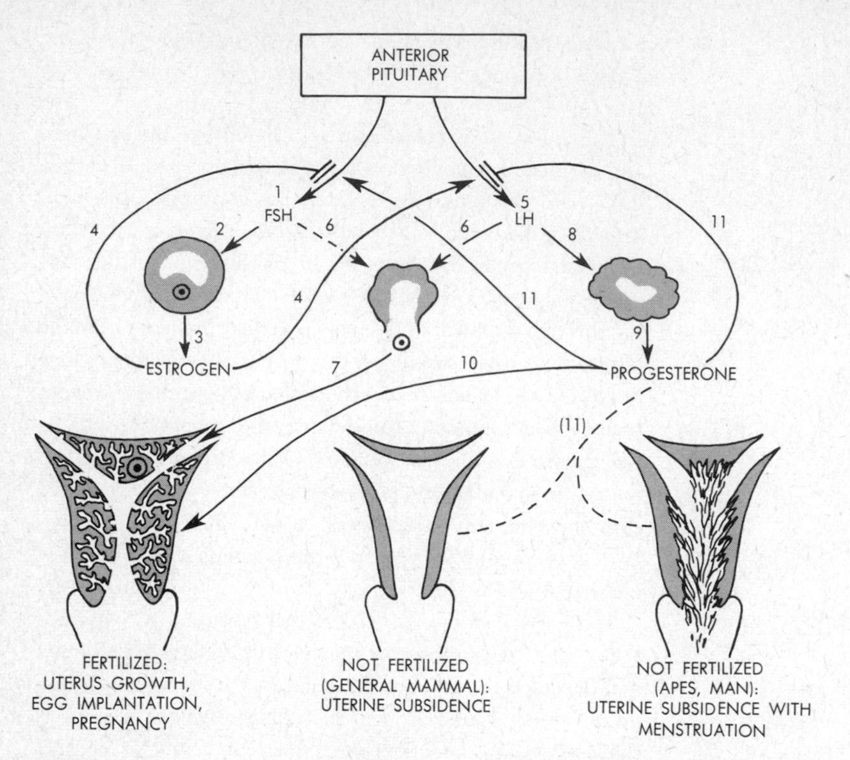

10.28. Egg-growth cycles in viviparous vertebrates. Steps 1 to 8 correspond generally to events in oviparous vertebrates (cf. two preceding figures). After progesterone production by the corpus luteum (step 9), events differ according to whether fertilization does or does not take place. If it does, pregnancy ensues (step 10); if it does not, progesterone production declines (via step 11), leading to tissue resorption in the uterus wall in most mammals and to menstruation in certain monkeys and in apes and man. The cycle then repeats (step 1).

pregnancy is then not initiated, menstruation takes place during the first few days of the ensuing menstrual cycle.

What happens in viviparous vertebrates if fertilization does occur?

Land Eggs and Pregnancy

In all viviparous vertebrates, fertilization in the oviduct and egg implantation in the uterus entail a suppression of further egg production during the ensuing pregnancy. Through still poorly identified nervous pathways, the presence of an embryo in the uterus wall is signaled to the hypothalamus and from there to the pituitary. This gland then continues to produce LH and the corpus luteum continues to secrete progesterone. Consequently, the thickened wall of the uterus can be maintained intact, and the developing embryo can remain implanted in it. Menstruations are likewise suppressed during pregnancy (Fig. 10.29).

The region where an egg implants in the uterus wall is called a *placenta*. In viviparous fishes, and indeed also in viviparous invertebrates, a fertilized egg contributes little to its own anchoring in the placental region. Such eggs receive raw materials partly from the yolk they contain, partly from maternal blood by direct diffusion and active transport in the placenta. Yolk also supplies many of the raw materials in the eggs of viviparous snakes, but here, and particularly also in all viviparous mammals, the developing eggs themselves contribute greatly to their firm implantation in the uterus; the eggs become anchored with the aid of so-called *extraembryonic membranes*. Evolved in ancestral reptiles, the first vertebrates to lay eggs on land, these membranes originally served as specific adaptations to egg development under terrestrial conditions. Early reptilian stocks subsequently gave rise to modern reptiles, to birds, and to mammals. All these groups inherited the membranes, and in the oviparous members they still serve in facilitating egg development on land. But in the viviparous forms the extraembryonic membranes came to function in new ways, and one of these is aiding in egg implantation in the uterus.

10.29. Photo: early monkey embryo just arrived in the uterus and beginning to implant. **Diagram:** the continuation of progesterone production during pregnancy. Nervous signals from uterus to hypothalamus and further signals from the latter to the pituitary lead to continuation of LH secretion, hence to continued progesterone manufacture and maintenance of the embryo implanted in the uterus. By this means, the usual inhibitory effect of progesterone on LH production is inhibited, i.e., the inhibition is removed and LH production can actually continue.

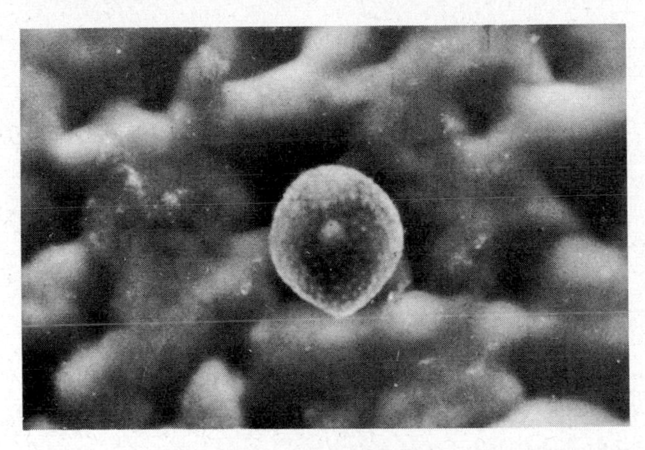

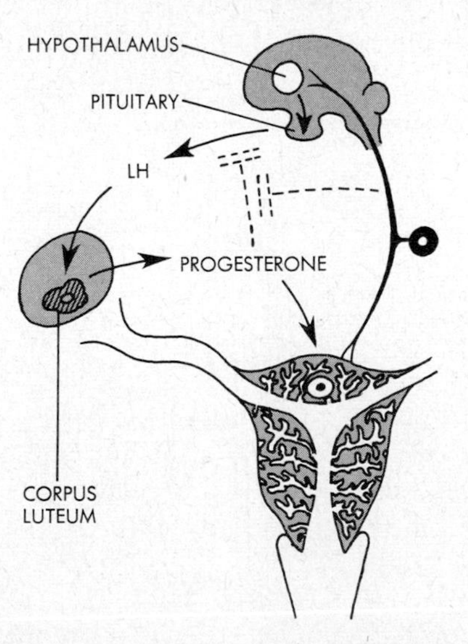

Most modern reptiles, all birds, and some primitive mammals are oviparous and lay shelled eggs on land. The shells are porous enough to permit aerial gas exchange, yet not porous enough to permit leakage of water. Just inside the egg shell and enclosing all interior structures lies one of the extraembryonic membranes, the *chorion* (Fig. 10.30). It prevents undue evaporation of water through the shell. A second membrane, the *amnion*, surrounds the developing embryo everywhere except on its ventral side. This membrane holds lymphlike fluid, the *amniotic fluid*, which bathes the embryo as in a "private pond." The fluid may be regarded as the equivalent of the freshwater ponds in which the aquatic ancestors of the land vertebrates developed.

Because reptiles, birds, and mammals all possess an amnion, these animals are called *amniotes* (and all other vertebrates are *anamniotes*). The two remaining extraembryonic membranes pouch out from the ventral side of the embryo, more specifically, from the alimentary tract. One of these is the *allantois*, which comes to lie against the egg shell just inside the chorion. Blood vessels ramify through the allantois, and this membrane is the breathing structure of the embryo; gas exchange occurs between it and the air outside the shell. Also, the allantois serves as an embryonic urinary bladder in which metabolic wastes are stored up to the time of hatching. The second membrane on the ventral side is the *yolk sac*, which contains the ample food stores for development and which gradually gets smaller as yolk is used up during the growth of the embryo.

The viviparous land vertebrates do not produce egg shells, but the four extraembryonic membranes are still in evidence in all cases (cf. Fig. 10.30). Thus, in mammals, the chorion again forms as an outer enclosure around the other membranes and the embryo; it is in direct contact with the tissue of the uterus. In one region the chorion develops numerous fingerlike outgrowths which branch extensively and erode paths through the thickened uterine wall. In this manner the tissues of the chorion and the uterus become attached to each other firmly. These interfingering and interlacing tissues represent the *placenta*. When fully developed, the placenta functions both as a mechanical and as a metabolic connection between the embryo and the maternal body.

The allantois in mammals still serves as in reptiles and birds

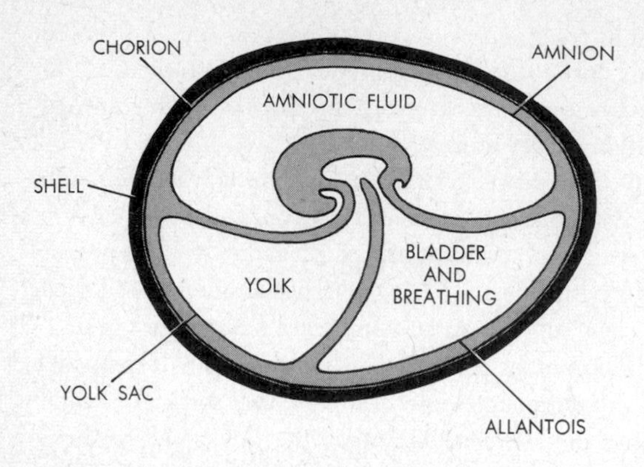

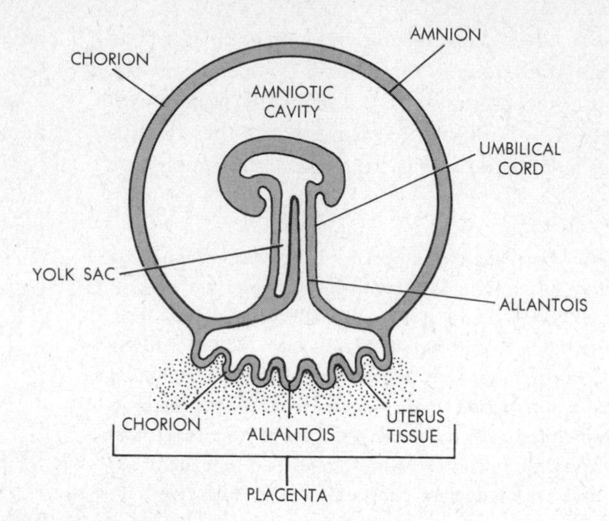

10.30. **Left:** the extraembryonic membranes in reptile and bird eggs. Note that yolk sac and allantois are large and functional. **Right:** the extraembryonic membranes in mammals and the placenta. Note that the yolk sac is rudimentary and collapsed, and that the allantois is no longer a bladder. However, it still functions as a breathing organ, via blood vessels it carries from placenta to embryo. In both figures, note that the amnion (and also the chorion) is ectodermal, that the yolk sac and allantois are endodermal, and that the gray-shaded areas correspond to mesodermal regions. More detailed illustrations of these germ-layer relationships are given in Figs. 11.18 and 11.19.

as an embryonic lung, except that now gas exchange occurs in the placenta, between the embryonic blood vessels of the allantois and the maternal blood vessels of the uterus (Fig. 10.31). However, the allantois has entirely lost its ancestral function as urinary bladder; embryonic wastes now are carried off by the maternal blood in the placenta. The allantois in mammals is actually a collapsed, empty sac. This is true also of the yolk sac, food being supplied by maternal blood, again through the placenta. On the other hand, the fluid of the amnion still functions as in reptiles and birds as a "private pond" and shock absorber. As more and more amniotic fluid accumulates during the course of pregnancy, the amnion distends greatly and the surrounding chorion and uterus are stretched correspondingly. This enlargement, more than growth of the embryo itself, eventually leads to the characteristic bulging out of the abdomen of the pregnant female.

In the mammalian placenta, the microscopic terminals of the chorionic outgrowths dip into pools of maternal blood which has accumulated within the placental spaces. The maternal blood circulates extensively through the maternal

10.31. Detail of mammalian placenta showing the embryonic and placental blood circulations. Note that embryonic and maternal bloods do not mix, being separated by the chorionic and allantoic membranes.

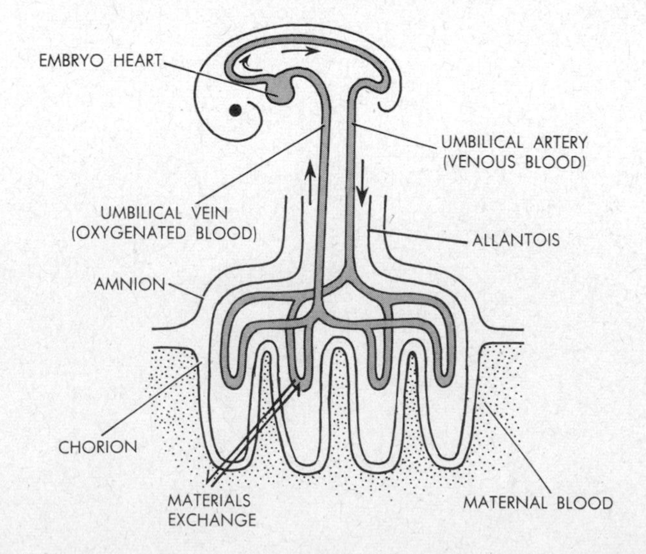

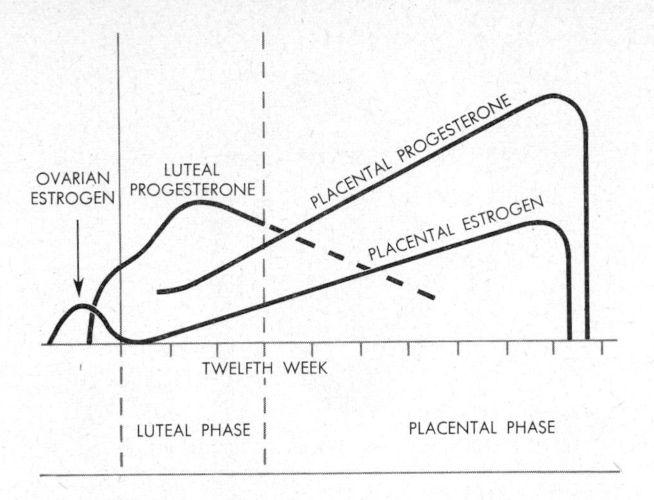

10.32. Curves indicating the amounts and sources of sex hormones present during pregnancy in man.

side of the placenta. On the embryonic side, an artery leaves the embryo proper, travels through the allantoic membrane, and capillarizes abundantly in the placenta, just underneath the chorion (cf. Fig. 10.31). The capillaries eventually join and form a large vein which leads back through the allantoic membrane to the embryo proper. The artery, the vein, the yolk sac, and the allantois become enveloped by connective tissue and skin, and the whole represents the *umbilical cord*. This is the lifeline between placenta and embryo; its point of origin in the embryo leaves a permanent mark in the later offspring in the form of the navel.

Note that maternal and embryonic bloods do not mix in the placenta. The two circulations approach each other closely, but the chorion always separates them. This membrane forms a selective boundary. Nutrients of all kinds and oxygen are passed across into the embryonic circulation, and metabolic wastes are passed in the opposite direction.

The placenta also specializes as an endocrine organ. As it grows and develops, it manufactures slowly increasing amounts of estrogen and progesterone. Indeed, the progesterone output eventually becomes far greater than that of the corpus luteum. The latter actually degenerates, and its hormone secretion subsides at some stage during pregnancy (roughly the twelfth week in man). From that time on, the placenta provides the main hormonal control of pregnancy; through its progesterone output it maintains its own existence (Fig. 10.32).

In parallel with the continuing growth and development of the mammalian embryo, the amnion gradually enlarges and the uterus stretches (Fig. 10.33). Also, the mammary glands enlarge markedly during the last months of pregnancy, under the stimulus of the increasing quantities of sex hormones from the placenta. Numerous ducts form in the interior of the glands, and in later stages the milk-secreting cells mature. Milk production is under the control of pituitary prolactin (called *lactogenic hormone* in mammals).

The process of birth normally begins when the chorion and the amnion rupture and when the amniotic fluid escapes to the outside. Labor contractions of the uterine muscles then occur with increasing frequency and strength, pressing against the offspring and pushing it out through the vagina. At the time of birth, the interlocked maternal and embryonic tissues which form the placenta loosen away from the wall of the uterus. The mechanical and metabolic connection between mother and offspring is thereby severed.

An important result of this is that CO_2 produced by the offspring must accumulate in his own circulation. Within seconds or minutes, the concentration of the gas then becomes high enough to stimulate the breathing center of the newborn. In correlation with this switchover from placental breathing to lung breathing, several structural changes occur

10.33. Photo of human embryo, about 8 weeks after fertilization, obtained after surgical removal of portions of the reproductive system of a female patient. The chorion is pushed to the side, revealing the amniotic sac. Note umbilical cord.

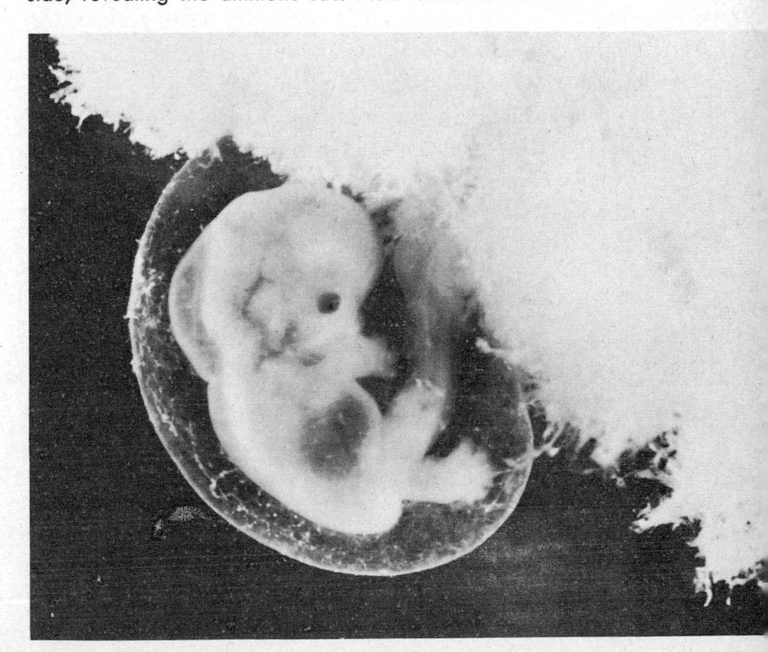

in the heart and in the large blood vessels around the heart. In the embryo before birth, the dividing wall between the right and left auricles is incomplete (Fig. 10.34). A movable flap of tissue provides an opening between these two chambers, and blood may pass freely from one chamber into the other. Once lung breathing is initiated at birth, the blood-pressure pattern within the heart changes, and the tissue flap is pressed over the opening interconnecting the auricles. The flap eventually grows into place, and the left and right sides of the heart so become separated permanently.

Another structural change involves an embryonic blood vessel, the *ductus arteriosus*, which before birth conducts blood from the pulmonary artery to the aorta. The ductus arteriosus shunts blood around the nonfunctional lung (and the vessel in fact corresponds to a part of the sixth aortic gill arch in fishes; cf. Fig. 9.30). At birth, a specially developed muscle in the ductus arteriosus constricts. This muscle never relaxes thereafter but degenerates into scar tissue. Blood is thus forced to pass through the lungs. The ductus arteriosus as a whole degenerates soon after birth.

The loosened placenta, still connected to the umbilical cord, is expelled to the outside as the *afterbirth* within an hour or so after the offspring is expelled. Mammalian mothers, modern human ones excepted, bite the umbilical cord off their young and then eat the cord and the placenta. Even herbivorous mothers do so, though they are vegetarians at all other times. The escaped amniotic fluid may also be lapped up. Indeed, among carnivorous mammals it is not uncommon that, in the course of drinking the fluid and eating the placenta, the just-born offspring is swallowed as well.

Up to this point, our attention has been focused primarily on the reproductive processes as such and on correlated changes taking place within the parental body. Our next concern is the fate of the reproductive unit itself, i.e., the development of a new animal.

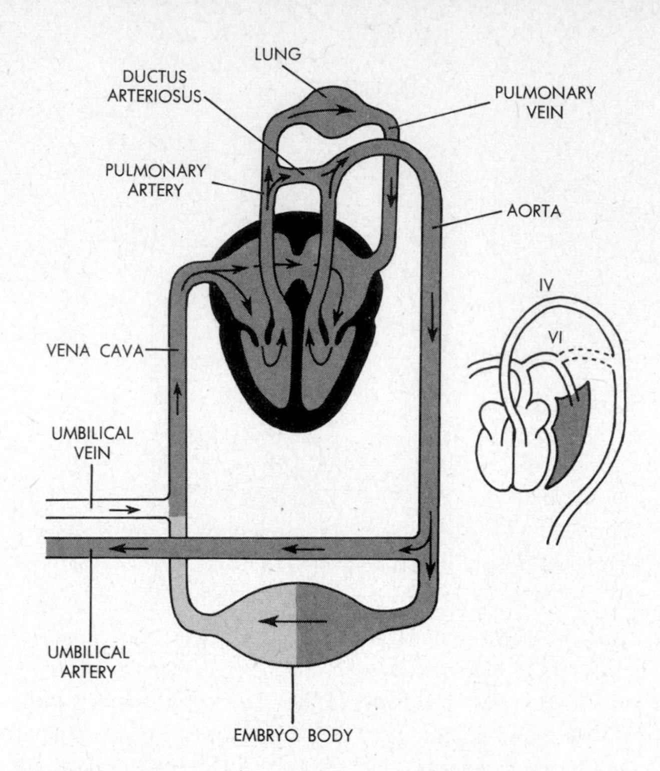

10.34. Diagram of the embryonic circulation in mammals. Oxygenated blood, white (only in umbilical vein, from placenta); venous blood, dark (from embryo body to vena cava); mixed blood, gray (in vena cava, heart, lung, and aorta). The embryo lung is nonfunctional as a breathing organ, and blood may pass directly from the pulmonary artery via the ductus arteriosus to the aorta. Note also the open passage between the two atria. The ductus arteriosus is a portion of the sixth aortic arch of the vertebrate gill circulation, as indicated in broken lines in the inset diagram. Compare the embryonic with the adult mammalian circulation (Fig. 9.30) and, in the diagram above, determine the effect on circulation if the umbilical vessels, the ductus, and the atrial passage all disappear and the lungs become functional.

REVIEW QUESTIONS

1. How does reproduction contribute to steady-state maintenance? To self-perpetuation in general? Define vegetative reproduction, binary fission, multiple fission, fragmentation, budding, regenerative reproduction. Give specific examples of each of these processes.

2. What is a tissue culture? What have experiments with tissue cultures shown about rates of cell division? When and where in an animal are division rates highest? Lowest? Distinguish between reproduction, development, and sex. What is gametic reproduction? What are the advantages and disadvantages of vegetative and gametic reproduction?

3. What basic processes occur in every sexual event? Under

what conditions does sex tend to take place? In what way is sex of adaptive value? Illustrate by means of examples among protozoa. Distinguish between syngamy and conjugation. Define mating, zygote, fertilization.

4. Define isogamy, anisogamy, oögamy, male and female, primary and secondary sex characteristics, natural and artificial parthenogenesis. What is hermaphroditism? What is its adaptive value? Distinguish between self-fertilization and cross-fertilization, and between protandry and protogyny. What is autogamy?

5. What is the basic function of meiosis, and what makes such a process necessary? Where does meiosis occur? Define haploid, diploid. How many pairs of chromosomes are found in a diploid cell? Of these, which and how many are maternal, and which and how many are paternal in origin?

6. How many chromosome duplications and how many cell duplications occur during meiosis? In what respects are mitosis and meiosis alike? What is the essential difference between the metaphase of mitosis and the metaphase of the first meiotic division? Describe the complete sequence of events during both divisions of meiosis.

7. Describe the nature of haplontic and diplontic life cycles. Which type appears to be primitive, and how has it given rise to the other? Which animal groups exhibit either type of cycle? What are the adaptive advantages of either type of cycle?

8. Describe the general structure of a male and a female reproductive system. What are follicle cells? How does the structure of the reproductive system generally vary for externally and internally fertilizing animals? Show how the reproductive and excretory systems of invertebrates are interrelated structurally and functionally. What is a gonostome? A nephromixium?

9. Describe the evolutionary development of the vertebrate reproductive system. What are primordial germ cells and where do they arise? Show how an indifferent gonad develops into either a testis or an ovary. What is the Muellerian duct? Trace the interrelations between the reproductive and excretory systems in vertebrate females and males. What is a urogenital sinus? What are descending testes? Spermatophores?

10. What are the first and second polar bodies? Are they found in males as well as females? Explain. What is the general structure of a mature sperm and of a mature egg? In which kinds of animals is fertilization (*a*) external, (*b*) internal? Define oviparity, ovoviviparity, viviparity. In which vertebrates does each occur?

11. How do breeding seasons vary with respect to time of occurrence, durations, and numbers of egg-growth cycles? What factors initiate breeding seasons? Describe the hormonal controls in sperm production of vertebrates. Describe analogous controls for egg production. What are monestrous and polyestrous mammals? What are the roles of prolactin in male and female vertebrates, and what are the roles of FSH and LH?

12. Relate ovulation, corpus luteum formation, progesterone production, and the role of the pituitary. Review the entire pattern of egg-growth cycles in viviparous mammals, and describe the special characteristics of menstrual cycles. Which mammalian groups exhibit menstrual cycles?

13. Describe the location and function of the extraembryonic membranes in reptiles and birds and in mammals. Relate the membranes to the formation and functioning of a placenta. Describe the pattern of the maternal and embryonic blood circulations through a placenta. What is the endocrine role of a placenta?

14. What events take place in the reproductive system of a pregnant mammal during birth of offspring? What changes take place in the blood circulation of the offspring at birth? How is milk production initiated and maintained?

COLLATERAL READINGS

The selections listed below cover various aspects of animal reproduction and sexuality.

Berrill, N. F.: "Sex and the Nature of Things," Dodd, Mead, New York, 1953.

Bishop, D. W.: Sperm Maturescence, *Sci. Monthly,* vol. 80, 1955.

Bullough, W. S.: "Hormones and Reproduction," Methuen, London, 1952.

———: "Vertebrate Reproductive Cycles," Wiley, New York, 1961.

Corner, G. W.: "The Hormones in Human Reproduction," Princeton, Princeton, N.J., 1942.

———: "Ourselves Unborn," Yale, New Haven, Conn., 1944.

Csapo, A.: Progesterone, *Sci. American*, Apr., 1958.
Farris, E. J.: Male Fertility, *Sci. American*, May, 1950.
Gray, G. W.: Human Growth, *Sci. American*, Oct., 1953.
Milne, L. J., and M. J. Milne: "The Mating Instinct," Little, Brown, Boston, 1950.
Monroy, A.: Fertilization of the Egg, *Sci. American*, July, 1950.
Nelsen, O. E.: "Comparative Embryology of the Vertebrates," Blakiston, New York, 1953.
Patten, B. M.: The First Heart Beats and the Beginning of the Embryonic Circulation, *Am. Scientist*, vol. 39, 1951.
Pincus, G.: Fertilization in Mammals, *Sci. American*, Mar., 1951.
Reynolds, S. R. M.: The Umbilical Cord, *Sci. American*, July, 1952.
———: Circulatory Adaptations at Birth, *Sci. Monthly*, vol. 77, 1953.
Stone, A.: The Control of Fertility, *Sci. American*, Apr., 1954.
Tyler, A.: Fertilization and Antibodies, *Sci. American*, June, 1954.
Wilson, E. B.: "The Cell in Development and Heredity," Macmillan, New York, 1928, reprinted 1953.

CHAPTER 11 development

Development is universal in scope among animals; any type of change, occurring on any level of organization and at any time in history, has developmental significance. Developmental changes can be structural or functional, quantitative or qualitative, progressive or regressive, normal or abnormal. Actually, development always involves all these simultaneously. But in given instances one or the other form of change may predominate or may be more readily apparent to the observer. Development is universal too with respect to the living unit in which change takes place and with respect to time. A molecule develops no less than a cell or a tissue, a whole animal no less than a whole species. And whether we measure it in microseconds as on the molecular level or in millions of years as on the species level, development occurs at every moment in animal history. The developmental domain, clearly, is as extended as that of zoology as a whole.

Within this domain, our present concern is chiefly the area of *embryology*, i.e., the study of those events which relate to the formation of individual animals and their parts. The problem is to describe, and where possible to explain, how single cells such as fertilized eggs are transformed into whole multicellular animals.

THE NATURE OF DEVELOPMENT

Morphogenesis

If single cells are to transform into whole animals, then a first obvious developmental requirement is increase in size, or *growth*. Overall growth may occur by either or all of three types of changes. Structural parts may increase in number, they may increase in size, or the spaces between the parts may enlarge.

Singly and in combination, all these alternatives actually occur. We already know, for example, that molecules increase in number either by being accumulated ready-made from the environment or by being newly synthesized within cells; that they increase in size by combining with other molecules; and that they increase in spatial distribution by dilution with water. Together, these ways of molecular growth constitute the means by which the size of cells increases. The number of cells increases by division; and the spacing increases by the accumulation between cells of water, cementing substance, or other secreted deposits. These ways of cellular growth in turn bring about increase in the size, the number, and the spacing of tissues and organs. The net result is overall growth of the animal. Note, however, that molecular growth is the fundamental prerequisite: the animal grows from its molecules up.

Growth introduces qualitative as well as quantitative changes. For example, certain types of molecules may be synthesized or accumulated at a greater rate or in greater amount than others. Indeed, some molecular types may disappear altogether, whereas others, not previously present, may appear for the first time. Similarly, the growth of cells, of tissues, or of organs may take place disproportionately in different parts of the developing animal. As a result of such *differential growth*, the structure and composition of the animal may be altered not only quantitatively but also qualitatively (Fig. 11.1).

Moreover, growth does not proceed randomly in all direc-

Courtesy Carolina Biological Supply Company.

11.1. Differential growth. The right claw of this crab has been lost and is regenerating. The new claw grows differentially at a far faster rate than the rest of the animal, for in the time the regenerate takes to reach the size of the left claw the rest of the body does not increase in size appreciably.

tions. How does it happen, for example, that developmental growth stops just when the nose, the brain, the heart and all other body parts are of the "right" proportional size and the "right" proportional shape? How does it happen that the different parts of the fully grown adult *retain* correct proportions and shapes? And how does it happen that, when the limb of a salamander is cut off, regenerative growth stops just when the newly developing limb has the size and the shape of the original one? In short, what determines the *form* of an animal, with respect to both size of parts and geometrical configuration of parts? Evidently, development of form, in addition to growth as such, is a second requirement if a reproductive unit is to be converted into a whole animal.

The basic aspects of form are *polarity* and *symmetry.* If *they* are given, a great deal about the general appearance of an object is already specified. The polarity of a structure indicates its orientation with reference to the three axes of space. A structure is polarized if one axis is in some way dominant. For example, the head-tail axis in most animals is longer than the other two. This axis is the principal guideline around which the whole animal is organized, and such an animal is polarized longitudinally. Symmetry indicates the degree of mirror-image regularity. A structure may be symmetrical in three, two, one, or in no dimensions; i.e., it may be spherical, radial, bilateral, or asymmetrical.

We already know from Chap. 7 that each animal exhibits a certain polarity and a certain symmetry, and we know also that these features are among the first and most permanent expressions of living form. Many traits of an animal can be changed by experimental means, but its original polarity and symmetry can hardly ever be changed. Millions of years later, long after the animal has become a fossil, polarity and symmetry may still be recognizable, even if all other signs of form have disappeared. It is a fairly general principle of development that the earlier a particular feature appears, the later it disappears.

Form is first blocked out in the rough, through establishment of polarity and symmetry, and then it becomes progressively more refined in regional detail. Whereas an animal grows from the molecule up, it forms from gross shape down. For example, the organ system is delineated ahead of its component organs. The tissue acquires definitive shape in advance of its component cells. And the molecules of the animal are last to assume final form. Evidently, form develops as in a sculpture, from the coarse to the fine, from the general to the specific. In both instances, this may be the only feasible way to ensure that the small remains appropriately subordinated to the large, structurally as well as functionally.

Specifically, establishment of form requires that cellular aggregates be molded into various configurations. Cells must become arranged and rearranged to produce regional enlargements and diminutions, to transform compact masses into sheets and vice versa, to produce channels, openings, cavities, and the like. Two general types of processes bring about such changes: *directed differential growth* and *form-regulating movements.*

For example, if differential growth proceeds differently in different parts of a developing animal, so that the amount and rate of growth vary for different directions of space, then regional enlargements and diminutions will be produced. Local elongations, thickenings, overgrowths, altered contours, layers, and other new shapes can arise in this manner. Also, a solid mass can become hollow if the outer layers of the mass grow faster than the core. And a hollow structure can become solid if the inner layers of the rind grow faster than the outside.

Form-regulating movements involve shifts and migrations of growing parts relative to one another. Directed migrations of parts can result in the piling up of material in one region and in attenuation in others. Sheets or compact masses can slide over one another, can fuse together, or can separate. Compact masses can spread out and become sheets or loose aggregations, or aggregations can condense and form larger masses. In short, if we add directed movements to directed growth, a sufficient machinery is available to translate the form of the reproductive unit into the specific form of the adult.

Growth, form, and all their qualitative and quantitative expressions together determine the architectural design of the animal. This architectural aspect of development is called *morphogenesis.* It is the first major component of the developmental process.

Differentiation

An animal develops not only architecturally but also operationally. Thus, growth of a zygote produces not simply an aggregate of many identical cells, but an aggregate of mutually different cells; for example, some become nerve cells, some liver cells, some skin cells, etc. How does a reproductive unit give rise to a multitude of differently specialized cells? Cell division as such certainly does not alter the characteristics of a cell. As already noted, daughter cells inherit the same set of genes and the same kinds of all other cellular components as are present in a mother cell. Cell division does copy faithfully, and a dividing reproductive cell therefore should give rise to many identical cells. Yet it does not; cell characteristics do change radically during development. As a result, every animal possesses structures and carries out functions which are not yet in existence at earlier developmental stages and which, indeed, may no longer be in existence at later stages.

Such dramatic changes of operational potential are brought about by the second and perhaps the most important major component of the developmental process, namely, by *differentiation.* A developing system need not necessarily grow, and it need not necessarily change form, but by the very meaning of development, it must differentiate. Through differentiation, structural units become functionally specialized in various ways. It is sometimes useful to distinguish between "chemodifferentiation," "cytodifferentiation," "histodifferentiation," "organ differentiation," etc., according to the level of organization at which operational change takes place.

The basis of differentiation, as of any living process, is *interaction.* In most interactions of animal parts with one another or with their physical environment, operational potentialities are not altered lastingly. But in some cases they are, and then the result is differentiation. For example, if some of the many interactions among molecules lead to the continuing production of novel categories of molecules, then these interactions contribute to chemodifferentiation. Or if a cell produces a hormone which, on reaching a second cell, causes that second cell to mature, to become abnormal, or to change operationally in some other lasting way, then this is an instance of cytodifferentiation. Or again, if in response to a persisting climatic change, animals transform into new types able to withstand the altered conditions, then this is a case of organismic differentiation, otherwise known as evolution (Fig. 11.2).

In short, to be differentiation, operational changes must have a certain degree of permanence. We may make an animal vitamin-deficient, for example, and many of its cells will then behave differently. But if we now add the missing vitamin to the diet, normal cellular operations will probably be resumed very promptly. Here cellular capacities have not been changed in any fundamental way. Only their expression has changed temporarily, in response to particular conditions. Such easily alterable, transient, reversible changes are spoken of as *modulations.* The concept of differentiation, on the contrary, implies a more or less fundamental, relatively lasting alteration of operational potentials. A vitamin-deficient cell which, after addition of the missing vitamin, *maintained* its altered characteristics would have differentiated (Fig. 11.3).

How does differentiation come about? On the organismic level, the process is understood comparatively well, and we shall discuss it in detail in the chapters on evolution. However, differentiation on the molecular and cellular levels is not fully understood as yet. Three general possibilities exist.

First, cell differentiation might be a result of progressive changes in gene action. Genes themselves probably do not change during development, for, as already noted, their stability is an essential requirement for the preservation of species characteristics. But the activity of different genes could vary with time. For example, in a given cell some genes might become active at certain developmental stages, whereas others might become inactive. The operon hypothesis (Chap. 6) now gives us some indication of how gene activity may become switched. Such differential activity patterns might occur differently in different cells, and this might contribute to differentiation.

Or, secondly, gene actions might remain the same, but the

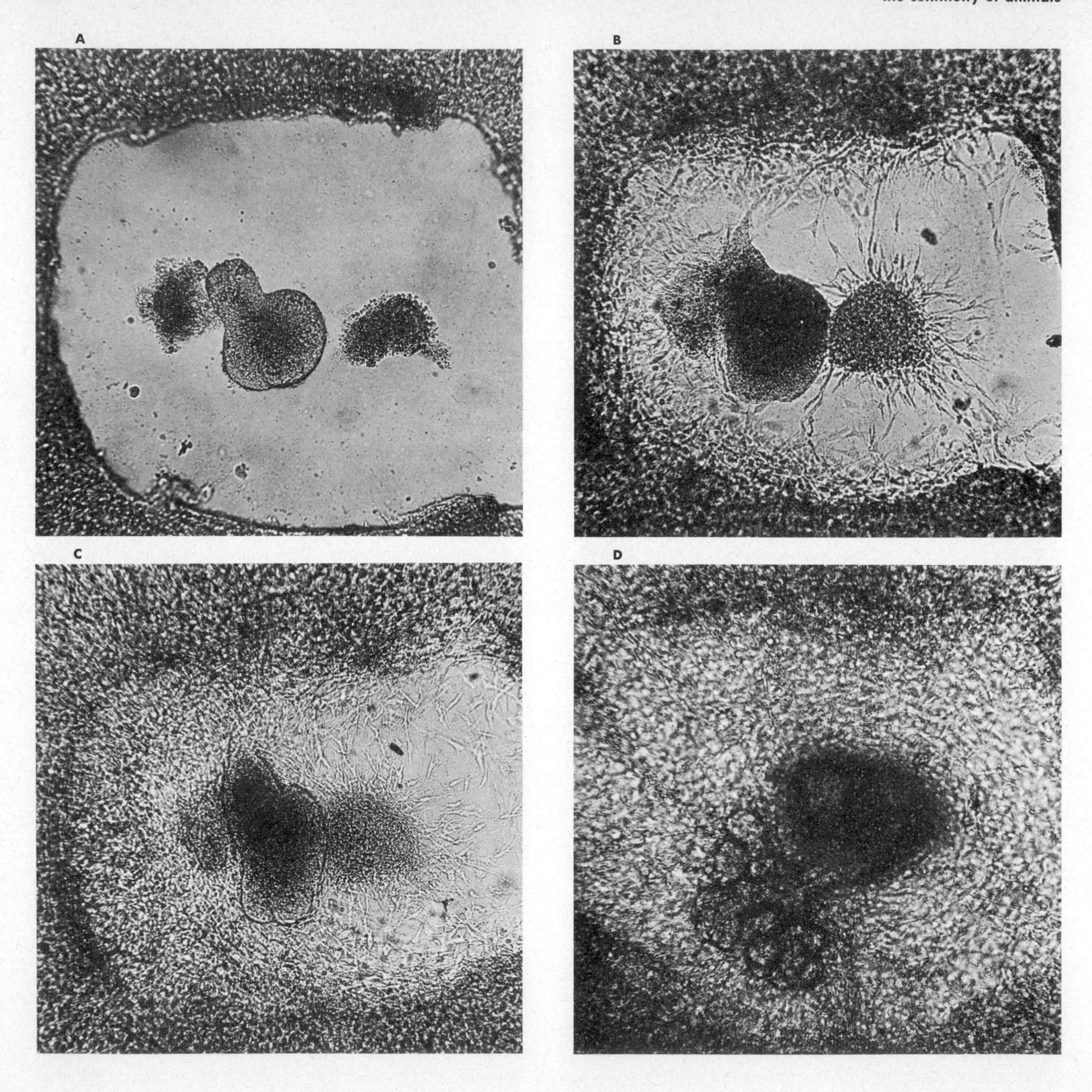

11.2. Salivary gland development in tissue culture, an example of differentiation by interaction. A piece of undifferentiated salivary ectoderm and two pieces of undifferentiated salivary mesoderm from a mouse embryo were put together into a culture **(A)**. These pieces grew and interacted **(B, C)** and eventually differentiated into secretion pockets and ducts characteristic of normal salivary glands **(D)**.

operations of the cytoplasm could become altered progressively. For example, one round of cytoplasmic reactions might use up a certain set of starting materials; and in the subsequent absence of these, similar reactions could then no longer take place. A next round of reactions would proceed with different starting materials and would therefore produce different endproducts. The net result could be progressive differentiation.

Or, thirdly, nuclear and cytoplasmic changes might both occur, in reciprocal fashion. This is probably the likeliest possibility, and much current research is devoted to a study of this very complex key problem.

Like growth, differentiation occurs from the molecule up. Just as a house cannot be any more serviceable than its component rooms will permit, so also the operational capacities of any living level are based on the capacities of subordinated levels. Chemodifferentiation therefore is the key to all differentiation. It is this which makes the problem of understanding so enormous. For if the process of differentiation is as complex as the totality of molecular interactions in cells, then it cannot be any less complex than the very process of life itself.

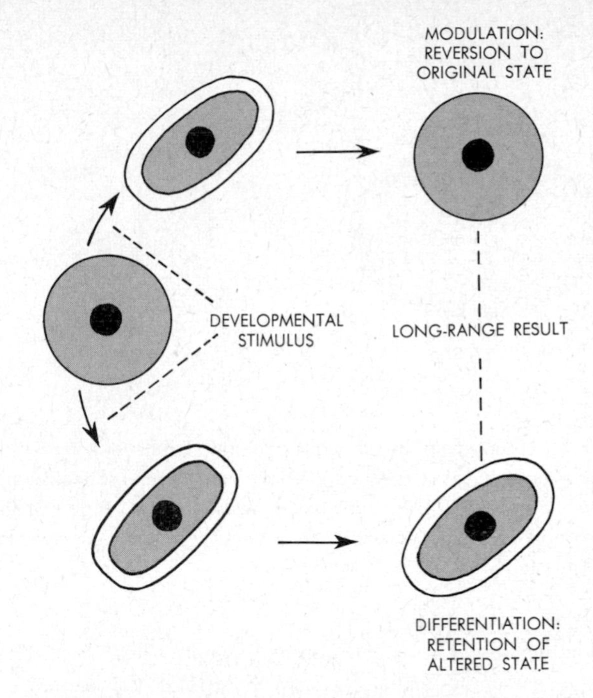

11.3. Differentiation versus modulation. In differentiation, the developmental change is permanent; in modulation, it is not.

Metabolism

Morphogenesis and differentiation are two of the forces which drive development processes. A third is metabolism. To be sure, metabolism is not a uniquely developmental requirement, but there could be no growth, no establishment of form, no differentiation, if energy were not available and if molecular syntheses did not occur. On the other hand, there could be no metabolism if morphogenesis and differentiation did not develop it.

Rates of metabolism are correlated with rates of development. At no point in the life cycle of any animal is metabolism more intensive and development more rapid than during the earliest stages. Both then decline in rate, until the zero point is reached at death; the metabolic clock is wound only once, at the beginning.

This circumstance introduces a number of major problems. Early in development, just when metabolic fires burn most fiercely, well-developed means of nutrition are not yet in existence. Neither the zygote nor in most cases the vegetative reproductive unit possesses a functioning food-procuring machinery. Three general solutions of this dilemma are possible; all three occur. First, enough food may be packed into the reproductive unit to last till it differentiates a functioning nutritional apparatus of its own. The yolk-filled egg provides the best example. Or, secondly, the developing unit may be fed more or less continuously by the parent, via a persisting functional connection between the two. This is well illustrated by the placental mechanism of mammals.

A third solution is frequently necessary in vegetative fragmentation and in regeneration, when the reproductive unit lacks a nutritional apparatus and internal food reserves are not available. Under such conditions, the regenerating unit may be able to draw foods from its own structural framework. One result of such partial self-destruction is decrease in size, or degrowth. Another is the mobilization of enough raw materials for effective redevelopment on a smaller scale. Mouthless fragments of many animals may degrow and regenerate with the foods so obtained (Fig. 11.4).

With fuel supplies assured, respiration and synthesis become possible. But initial dilemmas must be resolved here as well. Intensive respiration requires oxygen, and reproductive cells must exchange gases through their cell surfaces. But this requirement limits the size of a reproductive cell, for diffusion

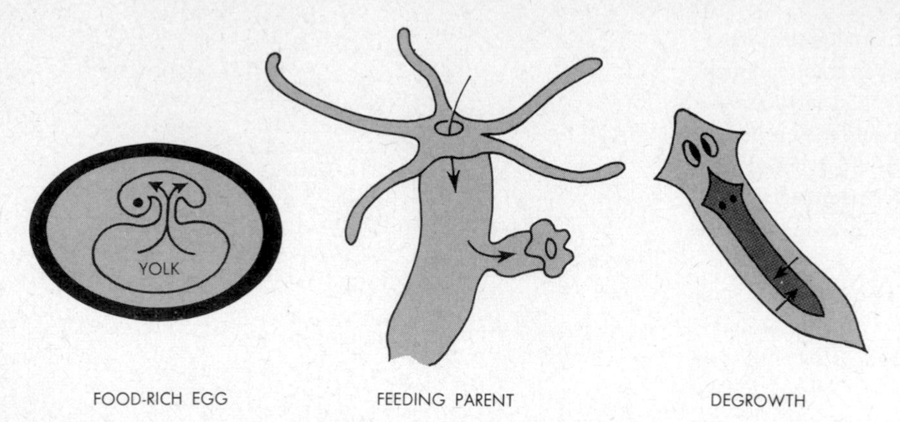

11.4. The three principal forms of nutrition in developing systems: inclusion of food in the embryo (as in yolky egg); attachment of embryo to parent and embryo nourishment by parent (as in hydra buds); degrowth, i.e., food obtained by partial breakdown of body and reorganization on smaller scale (as in planarians).

alone could not be effective in too large a cellular mass. The requirement of smallness, however, restricts the amount of food that can be stored in the reproductive cell, and this in turn places a time limit on the amount of development possible. Clearly, the developmental consequences of so "simple" a requirement as oxygen supply are quite far-reaching.

Once gas-supply problems are solved, respiration may proceed. The molecular equipment for energy production is inherited complete by all reproductive units and is more or less fully functional from the start. This is an absolute necessity for survival. But such is not the case for cellular syntheses; only relatively few kinds of synthetic reactions are possible initially. Most of the molecular equipment required for intricate developmental syntheses must itself first develop. End-products of a first round of synthesis must become the starting materials for a second, more complex round. In this manner, synthetic capacities must be increased and broadened progressively. Evidently, synthesis metabolism is as much a *result* of development as it is a prerequisite; it is one aspect of chemodifferentiation.

Morphogenesis, differentiation, and metabolism are three of the universal components of every developmental process. A fourth is perhaps the most essential.

Control

How does a reproductive unit happen to give rise to just the right kinds and right numbers of parts? For example, if the head of an earthworm is cut off, the worm develops a new head; not two or three heads or half a head, but one and only one; and not another tail, but another head. Even more strikingly, a zygote does not yet possess any of the features of the adult. How then does it happen to produce just one head and one tail, not two or more of each, but in a man, for example, two arms and two legs, not one of each? And why arms and legs at all—why not wings or fins?

Considerations such as these bring us to the most puzzling of all aspects of development. What integrates development? How do morphogenesis, differentiation, and metabolism mesh together to produce an elegant, sensibly functioning whole? By any standard, this smooth, seemingly unerring advance toward wholeness is probably the most remarkable property of development. The headless earthworm, for example, never ceases its quiet internal revolution till it has a new head. The transfigurations of the egg do not stop before the adult whole has come into being. Evidently, the healthy developing system behaves as if it "knew" its objectives precisely, and it proceeds without apparent trial and error. For normally there is no underdevelopment, no overdevelopment, and there are no probing excursions along the way. Development is directed straight toward wholeness.

Only one conclusion can be drawn: the course of development must somehow be rigorously controlled. However, recognition of the occurrence of control does not of itself provide an explanation of it. We know in general terms that the control systems must reside within a developing unit itself and that, like any other living process, development must be self-controlling. But the nature and operations of these built-in control systems have in most cases not yet been identified. Today it is fashionable to say that genes control development, as they control every other living process. This is unquestionably correct. But such an answer is not very informative and is actually little more than a restatement of the problem. *How* do genes control development? More specifically, how does a particular gene, through control over a particular enzyme or other protein, regulate a particular developmental occurrence? Answers to such small problems are just beginning to be obtained. The collective larger issue, i.e., the controlled, directed emergence of wholeness in an entire animal, remains a matter of future research. But even though we cannot now describe how the controls operate, we can nevertheless describe the results of such operations.

Developmental Patterns

The course of development varies considerably according to whether the starting unit is a zygote produced by a sexual process or an asexually developing vegetative body.

Zygotic development starts with *fertilization* and continues with the formation of an *embryo*. In oviparous animals the embryos develop outside the maternal body, in ovoviviparous and viviparous animals they develop within. During the embryonic period, all basic structures and functions of the future adult body are elaborated in at least rough detail. In oviparous animals, the embryonic phase typically terminates with a process of *hatching*, in which the embryo emerges from its original egg envelopes and becomes a free-living *larva*. A larval phase is characteristic of animals in virtually all phyla, but it is often absent in some of the more advanced subgroups within a phylum. Animals without larvae are usually (but not necessarily) ovoviviparous or viviparous.

In due course, larvae undergo *metamorphosis*, a more or less gradual but in many cases quite sudden transformation into the *adult* condition. In animals without larvae, the embryo becomes a young adult directly (the stage of transition being marked by birth in ovoviviparous and viviparous forms). Note that this last phase in the developmental history of an individual is not any more static than preceding phases. On the contrary, as shown in the chapters on metabolism, the components of the adult are steadily being demolished and redesigned or replaced. In this continuing turnover, internal as well as external features become altered. Youth and adolescence thus pass into maturity, marked by the onset of reproductive capacity. Maturity then gives way to senescence, and only death brings development to a halt (Fig. 11.5).

Thus, the typical developmental pattern following gametic reproduction is either fertilization ⟶ embryo ⟶ adult, as in some animals, or fertilization ⟶ embryo ⟶ larva ⟶ adult, as in most. In sharp contrast to this lengthy multistage course of sexual development, all forms of asexual devel-

11.5. These stages in the life cycle of a prawn (*Leander*) symbolize the main stages in the sexual development of animals generally. **A,** eggs carried by adult female; **B,** embryos, within egg membranes; **C,** free-living larva, which eventually develops into an adult as in **A,** completing the cycle. Cf. also Fig. 3.16 for analogous life-cycle stages in vertebrates.

B

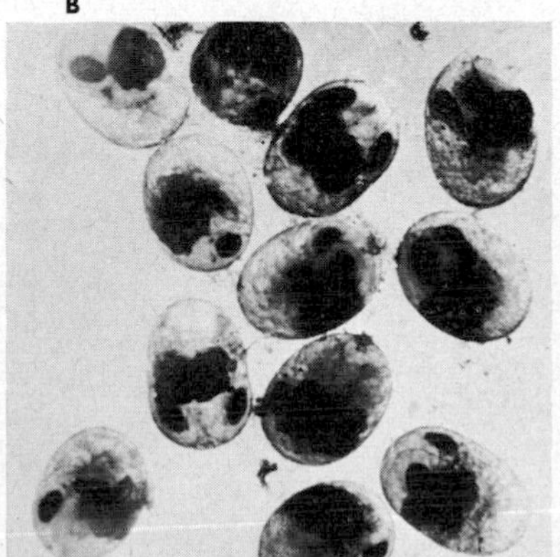

C

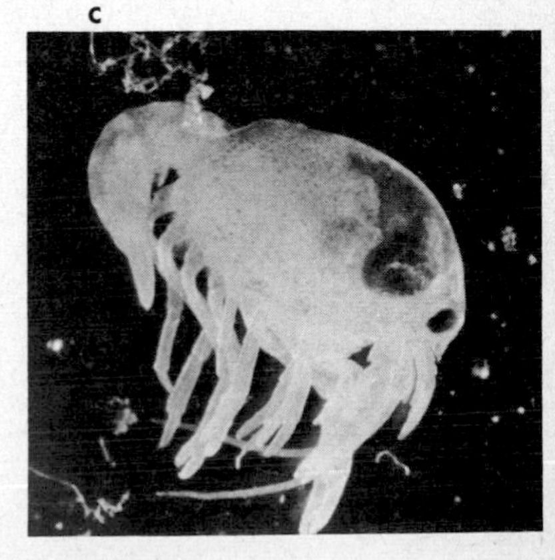

opment are exceedingly direct. In the development of vegetative units of any type there is no sex, hence no fertilization; there is no larva, hence also no metamorphosis. Instead, the reproductive unit becomes an adult in a smoothly continuous, single developmental step (Fig. 11.6).

Without doubt, this marked difference between sexual and asexual patterns of development must be due to the presence or absence of the sexual process itself. Unlike a vegetative body, an egg is *more* than simply a reproductive unit. As we have seen, it is also the agent for sex, i.e., it is an adaptive device. Through fertilization the egg acquires new genes, which may endow the future offspring with new, better adapted traits. However, before any new traits can actually be displayed, they must be *developed* during the transition from egg to adult. Embryonic and larval periods appear to be the result. These phases may be considered to provide the opportunity for translating the genetic instructions acquired sexually by the zygote into the adaptively improved structures and functions of the adult. Vegetative units do not acquire new genetic instructions through sex, and equivalent developmental processes for executing such instructions would then not be needed. Correspondingly, embryos and larvae are absent here.

11.6. Sexual vs. asexual development. In sexual development (top), new genetic instructions are introduced into the zygote via the gametes, and during subsequent embryonic and larval stages these instructions are elaborated explicitly. Hence the mature offspring may differ to greater or lesser extent from the parent. In asexual development (bottom), new genetic instructions are not introduced, and the offspring therefore resembles the parent fully.

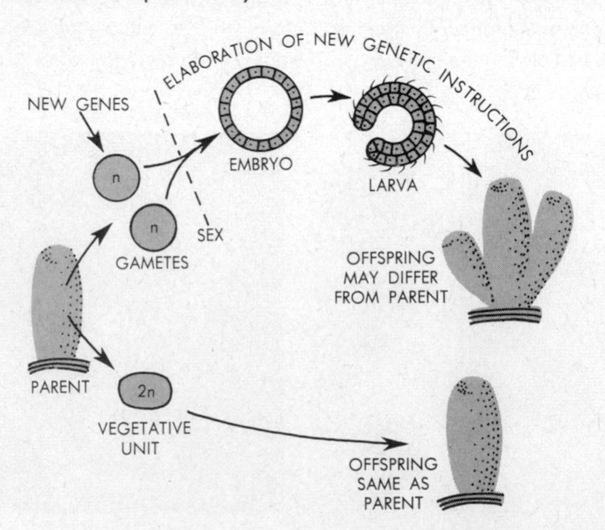

In the following sections, we shall examine the course of sexual development in greater detail.

EARLY DEVELOPMENT

Fertilization, Eggs, and Cleavage

The fertilization of an egg generally includes two steps, viz., *plasmogamy* and *karogamy*. Plasmogamy refers to the entrance of a sperm into an egg. If a sperm collides with an egg at an angle, it is likely to bounce off. By contrast, sperms hitting head on are likely to remain attached, for the acrosome at the sperm tip is specialized to adhere to the egg. One, and only one, sperm can normally enter any one egg. As soon as a first sperm makes contact, a *fertilization membrane* rises from the egg surface. This membrane has formed earlier, during egg maturation. On contact with a sperm, the egg rapidly secretes some water between its surface and the membrane. As a result, the membrane lifts off, the sperm which has made contact is trapped inside, and any other sperms are prevented from entering (Fig. 11.7). If by chance two or more sperms do enter an egg, either all but one of the sperms disintegrate within the egg, or the egg fails to develop normally and soon dies.

A successful sperm resting against the egg surface does not penetrate into the egg by boring in. Instead the egg engulfs the sperm; eggs are amoeboid at least to this extent, and in many animals they are very obviously amoeboid. During the entry of a sperm into the egg, the sperm tail drops off. At this point plasmogamy is completed, and the egg is *activated:* development of the egg has been triggered off. A mature egg is ready and able to develop, but this ability remains latent until a specific stimulus makes development start. Sperm penetration into the egg normally serves as this stimulus. Such an arrangement ensures that the sexual process occurs before development begins. As noted in Chap. 10, unfertilized eggs of many animals may become activated by natural or artificial parthenogenesis, without sperms. Such eggs remain haploid, and all the cells of the resulting embryos are correspondingly haploid.

Under normal conditions, a sperm nucleus which has activated an egg moves toward the egg nucleus, and the meeting of the two haploid nuclei then constitutes karyogamy, which completes fertilization. The membranes of the two nuclei dissolve, and a mitotic spindle forms. The chromosomes, now diploid in number, line up in a metaphase plate,

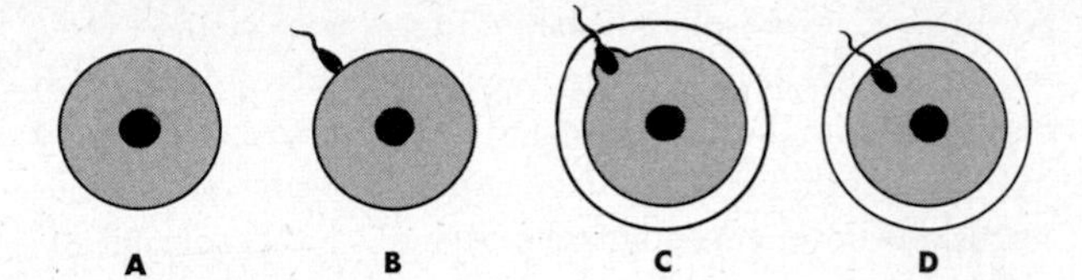

11.7. Diagrammatic representation of fertilization. A sperm enters an egg by being engulfed by the egg, through an egg cone which comes to surround the sperm **(C)**. A fertilization membrane lifts off the egg surface after a sperm has made contact, preventing additional sperms from being engulfed **(C, D)**. The sperm tail is left at the egg surface, and the sperm head (nucleus) alone migrates into the egg cytoplasm, where it fuses with the egg nucleus. An egg is fully fertilized only after sperm and egg nuclei have fused.

and the zygote undergoes its first *cleavage division*. Very shortly after fertilization, therefore, two cells are formed from the zygote. The two cells then divide again, and many successive mitotic cleavage divisions follow thereafter. The development of an animal is launched in this manner.

The cleavage divisions subdivide an egg into progressively smaller cells called *blastomeres*. As noted earlier, eggs are generally larger than the average adult cell size characteristic of the species. During cleavage, divisions occur so rapidly that the blastomeres have little opportunity to grow, hence the egg subdivides without enlarging as a whole. Cleavage divisions typically continue until adult cell sizes are attained. The early cleavages are unusual also in that they occur synchronously; all blastomeres present cleave simultaneously. However, such synchrony generally disappears after a few dozen blastomeres have formed.

Note that the function of cleavage is not merely to subdivide an egg into a random set of appropriately small cells. Sooner or later, different regions of an egg develop into qualitatively different *organ-forming zones*, i.e., areas which will later give rise to the various body parts of the adult. Cleavage *segregates* these zones into different cells and cell groups (Fig. 11.8). On the basis of how soon the zones become established, and thus how soon the later fate of the various egg parts becomes fixed, two categories of eggs may be distinguished: *mosaic* or *determined* eggs and *regulative* or *undetermined* eggs.

The first type is encountered generally among most protostomial animals. In these, the future developmental fate of every portion of the egg becomes fixed unalterably before or at the time of fertilization. The zygote therefore is already fully polarized and the head-tail, dorsal-ventral, and left-right axes are firmly established. Moreover, experiments show that each portion of the egg behaves as if it already "knew" what it is going to develop into. For example, after cleavage in such an egg has produced two or more cells, it is possible to separate these cells from one another. Each isolated blastomere then continues to develop and forms a *partial* embryo. More specifically, it produces the same portion of the embryo it would have produced if the cleaving egg had been left intact. In other words, the determined egg is like a quiltwork, a mosaic, in which each portion of the cytoplasm develops into a fixed, unalterable part of the whole embryo; and the nature of the mosaic is established before or, at the latest, during the time of fertilization (Fig. 11.9).

By contrast, the eggs of most deuterostomial animals, vertebrates included, are of the regulative variety. In these, the future fate of various egg portions similarly becomes unalterably fixed, but such fixing occurs comparatively much later, during the embryonic phase. Moreover, different features become determined at different times. For example, at fertilization only the main egg axis is determined. That is, the direction of "top" and "bottom" is already given, but other aspects of polarity and symmetry or of any other feature are not as yet fixed.

That this is so can be demonstrated very strikingly by experiment. If the two blastomeres formed by the first cleavage division are left as they are, then they will eventually form the left and right halves of the future animal. Further, cytoplasm in the center of the egg will develop into central internal

11.8. Generalized diagram of germ-layer- and organ-forming zones in egg and early cleavage stages. The principal egg axis is marked by so-called animal and vegetal poles at opposite points of the egg. Note that, as cleavage progresses, a given zone of the egg becomes segregated into progressively smaller but larger numbers of blastomeres.

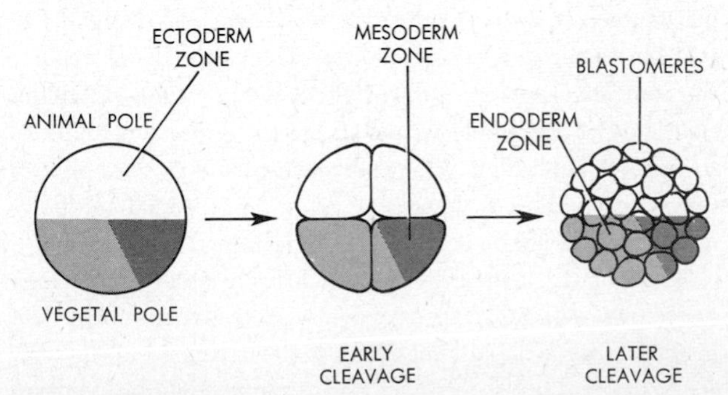

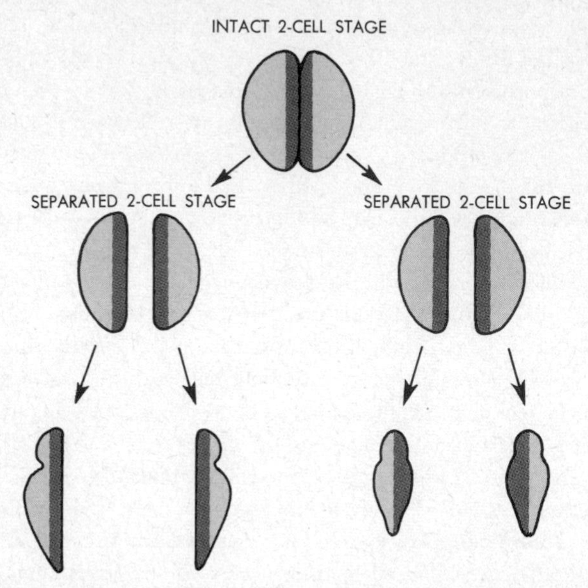

TWO HALF-EMBRYOS, NORMAL SIZE: MOSAIC OR DETERMINED DEVELOPMENT

TWO WHOLE EMBRYOS, HALF SIZE: REGULATIVE OR UNDETERMINED DEVELOPMENT

11.9. Mosaic vs. regulative development. If the cells of early cleavage stages of mosaic eggs are isolated experimentally (left), then each cell develops as it would have in any case. The inference is that the fates of cytoplasmic regions in such eggs (e.g., the dark-shaded central parts) are determined very early. But if the cells of early cleaving regulative eggs are isolated experimentally (right), then each cell develops into a smaller whole animal. The inference here is that the fate of the cytoplasm is still undetermined. Thus the dark-shaded central cytoplasm, which normally would form central body parts, actually forms left structures in one case, right structures in the other, if the two-cell-stage blastomeres are separated.

structures of the adult. But if the two blastomeres are separated from one another, then they do not develop into two half animals as would be the case in a mosaic egg. Instead, the two blastomeres develop into two whole animals. Moreover, the central cytoplasm of the original egg now gives rise to the right side of one whole animal and to the left side of another. Evidently, central material at the two-cell stage does not yet "know" whether to form left, right, or internal midbody structures. In short, it is not yet determined (cf. Fig. 11.9).

Analogously, if the cells of later cleavage stages are isolated and grown separately, then each may again give rise to a whole instead of a partial animal. But a limit is reached fairly soon. After the first three cleavages, for example, eight blastomeres are present. If these are separated from one another, eight whole animals usually cannot be obtained. Instead, each blastomere forms only one-eighth of an embryo, as it would have done if the eight-cell stage had been left intact. In other words, the developmental fate of the cells has become determined by now, and the embryo henceforth is like a mosaic.

We may conclude that mosaic and regulative eggs differ mainly in the timing of developmental determination. The early timing in mosaic eggs contrasts with the comparatively late timing in regulative eggs. During the undetermined phase in regulative eggs, any cell may substitute for any other cell and may develop into any structure, including a whole animal. Note here that developmental determination is a form of differentiation, particularly chemodifferentiation, and that the underlying mechanism is still completely unknown.

Note also that the formation of two or more whole animals from separated blastomeres is equivalent to the production of identical twins, triplets, quadruplets, etc. Natural *twinning* undoubtedly occurs through similar separations. However, the forces or accidents which actually isolate such blastomeres in nature are not understood. If the blastomeres are separated incompletely, Siamese twins result. This too can be demonstrated by laboratory experiments (Fig. 11.10). Twins are *identical* when they develop from a single fertilized egg, as above. They are *fraternal* when two or more whole eggs are fertilized separately at the same time. The offspring here may be of different sexes, and they need not resemble one another.

11.10. X-ray photo of Siamese twinning in fish. Abnormalities like these result from incomplete divisions of cells during early cleavage.

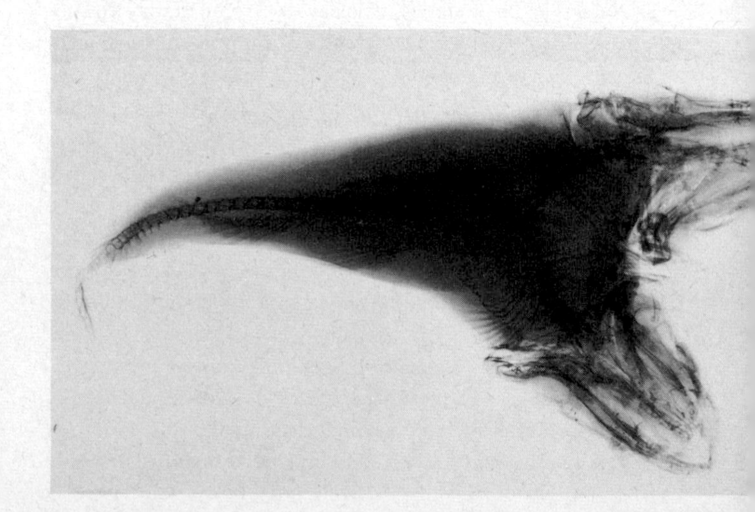

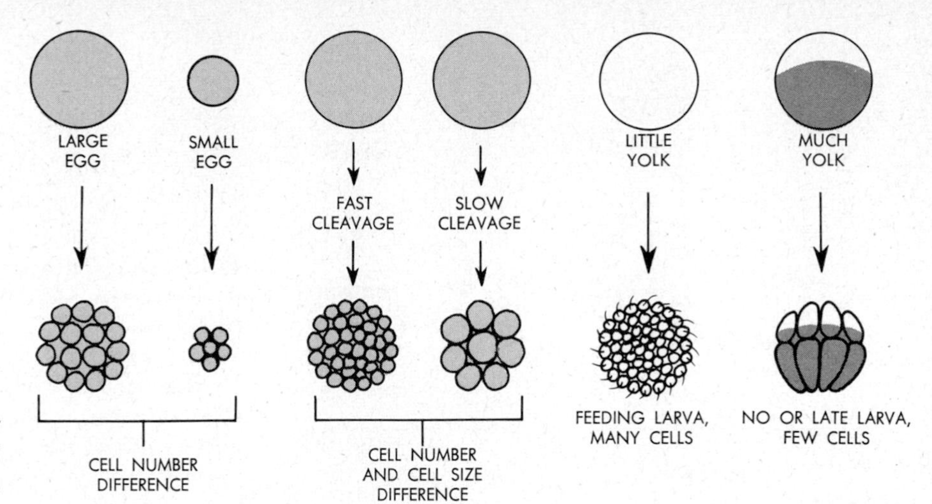

11.11. The effects on development of initial egg sizes, cleavage rates, and amounts and distributions of yolk.

By contrast, identical twins are of the same sex, and they do resemble one another. Indeed, they tend to be structural mirror images.

Apart from its mosaic or regulative properties, an egg also exhibits three additional properties which influence development greatly, viz., its original *size*, its *cleavage rate*, and the amount and distribution of its *yolk* (Fig. 11.11). If an egg is very much larger at the outset than the average adult cell size, cleavage will continue until many cells have formed. But if an egg is comparatively small initially, cleavage will result in correspondingly fewer cells. Consequently, the number of cells present at the end of cleavage will vary according to initial egg size, and we shall soon find that this cell number influences the architecture of the developing embryo in important ways.

The rate of cleavage likewise determines the number of cells available later. Assume that two eggs are equally large and that both undergo cleavage for an equal length of time. If one egg divides successively 7 times during this period, it will yield 128 cells. But if the other egg cleaves 8 times, it will yield 256 cells, fully 100 per cent more than the first egg. Clearly, even slight differences in cleavage rates can produce large differences in cell numbers; and this, like initial egg size, influences the nature of the embryo subsequently.

The quantity of yolk in an egg may be large or small and, other factors being equal, the egg with more yolk is generally larger and cleaves more slowly than the one with less yolk. Since the amount of yolk present determines how long an embryo can develop without external food sources, it follows that small eggs with little yolk must become independently feeding larvae sooner than large eggs with much yolk. Indeed, if the amount of yolk is exceptionally large, a feeding larvae may not be required at all, and the embryo may possess enough internal raw material to transform into a young adult directly. As a general rule, small, yolk-poor eggs actually do pass through the embryonic phase quickly and become feeding larvae very soon, whereas animals with large yolky eggs have long embryonic periods and late larvae or no larvae at all (cf. Fig. 11.11). Evidently, as will also become apparent later repeatedly, even small variations in initial properties such as egg size or yolk content can have profound developmental and indeed also evolutionary consequences.

The distribution of yolk is equally significant in early development. Four major egg types may be distinguished on the basis of yolk distribution. In *isolecithal* eggs, the yolk is dispersed rather evenly throughout the egg cytoplasm, and the amount of yolk present is comparatively small. Sponges and mammals are among animals with such eggs. In *centrolecithal* eggs, the amount of yolk is large and it is collected compactly in the center of the egg cell. Thus the egg possesses a clear, nonyolky layer of cytoplasm around the outside. Eggs of this type are characteristic of some coelenterates and most arthropods. In *telolecithal* eggs, the amount of yolk is again large, but it is collected excentrically in one region of the egg. Many annelids and mollusks and most amphibia have eggs of this type. In *discoidal* eggs, finally, the amount of yolk is so enormous that the nonyolky part of the cell forms a mere miscoscopic spot, or *blastodisc*, atop the yolk mass. Such eggs are encountered in squids and octopuses, in fishes, and in reptiles and birds (Fig. 11.12).

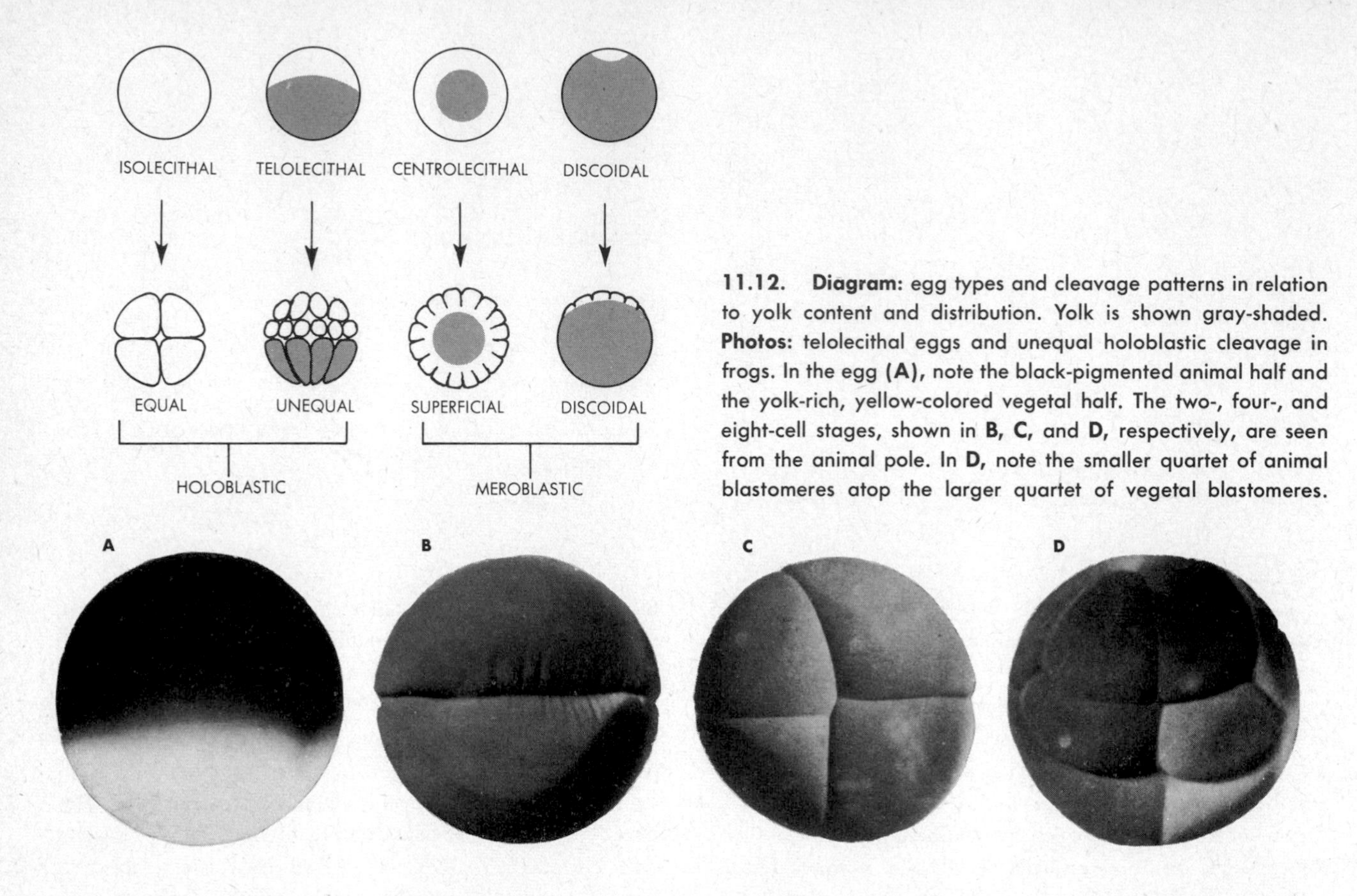

11.12. Diagram: egg types and cleavage patterns in relation to yolk content and distribution. Yolk is shown gray-shaded. **Photos:** telolecithal eggs and unequal holoblastic cleavage in frogs. In the egg **(A)**, note the black-pigmented animal half and the yolk-rich, yellow-colored vegetal half. The two-, four-, and eight-cell stages, shown in **B, C,** and **D,** respectively, are seen from the animal pole. In **D,** note the smaller quartet of animal blastomeres atop the larger quartet of vegetal blastomeres.

Yolk is a mechanically inert material which, when present in large quantities, may interfere physically with the subdivision of an egg during cleavage. Two major patterns of cleavage may actually be distinguished according to whether the blastomeres formed do or do not include most of the yolk. Isolecithal and telolecithal eggs undergo *holoblastic* cleavage, in which the cell membranes formed during cleavage cut completely through the egg, yolk included. In isolecithal eggs, the blastomeres formed are generally of roughly equal size, partly because the small amount of evenly distributed yolk interferes very little with the cleavage divisions. In telolecithal eggs, however, the blastomeres tend to be of various sizes, those containing most of the yolk being larger than the rest. Also, the larger blastomeres form more slowly than the smaller ones, yolk here delaying the completion of cytoplasmic cleavages. Cleavage is *meroblastic* in centrolecithal and discoidal eggs; i.e., such eggs contain so much yolk that most of it does not become included within the blastomeres. In the *superficial* meroblastic cleavage of centrolecithal eggs, the clear cytoplasm on the egg surface becomes multinucleate first, and then cell boundaries form and so produce a layer of cells which surrounds an undivided central yolk mass. Analogously, the blastodisc on top of discoidal eggs undergoes *discoidal* meroblastic cleavage, the resulting cells forming a layer on top of the undivided yolk (cf. Fig. 11.12).

Where yolk does not interfere with cleavage unduly, as in most holoblastically dividing eggs, cleavage occurs in a variety of inherent symmetry patterns. Initially, eggs are radially symmetrical around an axis passing from the so-called *animal pole* in the nonyolky half of a telolecithal egg to the *vegetal pole* in the yolky half. The animal-vegetal axis establishes the primary longitudinal polarity of an egg, and cleavages occur in patterns oriented with reference to this axis. Three such patterns are particularly common, viz., *radial, bilateral,* and *spiral* ones (Figs. 11.13 and 11.14). In all three, the first two cleavage divisions occur in essentially the same way, viz., the division planes pass through the animal-vegetal axis at right angles to each other. The result is a quartet of blastomeres, a tier of cells lying in the same plane.

In radial cleavage, the third division may again occur

longitudinally, yielding an octet of cells arranged as a single tier, or it may occur equatorially, forming two tiers of quartets one atop the other. In both cases, subsequent divisions then take place in such a way that a radial symmetry around the animal-vegetal axis is preserved. Such a cleavage pattern is encountered in sponges and some echinoderms, and variants of it are characteristic of the radiate animals. Thus, the cleavage symmetry is typically *tetraradial* in coelenterates and *biradial* (or *dissymmetrical*) in ctenophores (cf. Fig. 11.13).

Bilateral cleavage occurs, for example in most deuterostomial animals, including echinoderms and most chordates. After a first quartet of blastomeres has formed, the third division is equatorial, resulting in two tiers of quartets. Subsequent divisions establish a distinct bilateral symmetry (cf. Fig. 11.13). In the discoidal cleavages of fish, reptile, and bird eggs, a gross overall bilateral symmetry is evident, even though the detailed pattern of the early cleavages is modified by the yolk.

In the spiral pattern (Fig. 11.14) the quartet of blastomeres formed by the first two divisions are called *macromeres.* In subsequent equatorial cleavages, these bud off a succession of four or five quartets of *micromeres,* all quartets forming a series of tiers stacked one on top of the other. However, the first quartet of micromeres is cut off in such a way that it does not lie directly over the macromeres but is rotated 45° relative to the macromeres. Thus, the first micromeres formed lie over the valleys between the macromeres. Viewed from the animal pole, the positional shift is typically to the right. When the macromeres then cut off a second quartet of micromeres, the latter are rotated 45° to the left relative to the macromeres. The third quartet of micromeres rotates to the right again and thus comes to lie directly under the first micromeres.

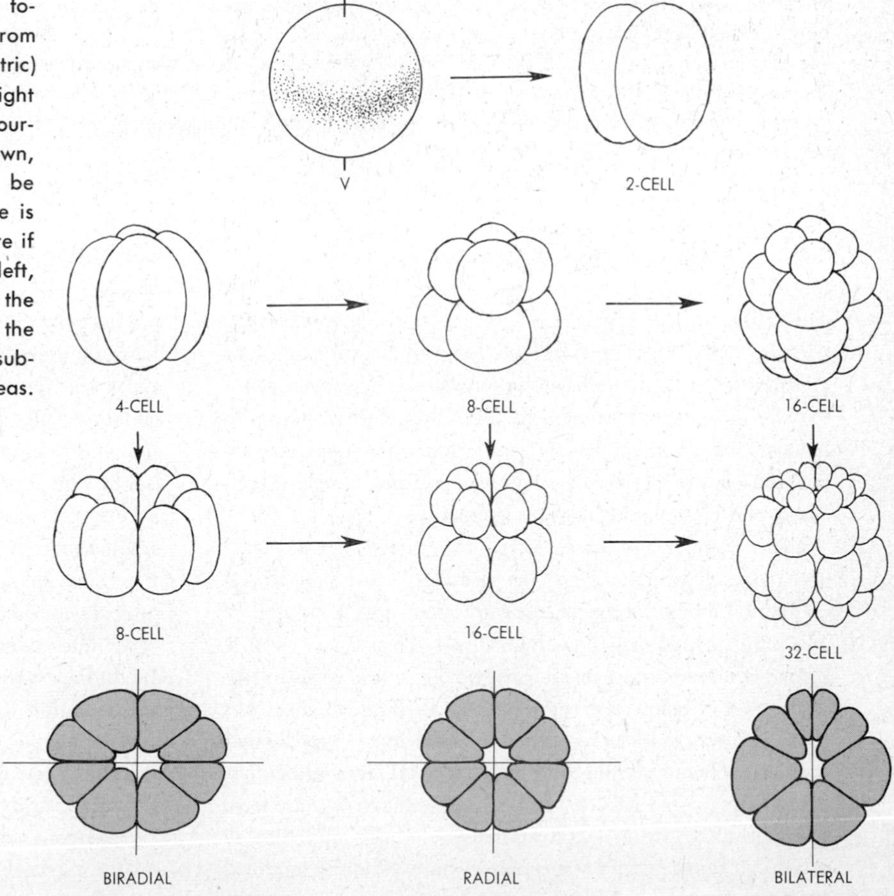

11.13. Cleavage patterns. In first three rows, animal half (A) toward top, vegetal half (V) toward bottom. Bottom row, sections as seen from animal pole. Left column, biradial (or dissymmetric) cleavage; middle column, radial cleavage; right column, bilateral cleavage. Note that the four-cell stage is similar for all three patterns shown, and that the eight- and 16-cell stages can be attained in different ways. The 32-cell stage is often more strongly bilateral than shown here if the pattern is a bilateral one. In the egg, top left, the animal region is prospective ectoderm, the vegetal region, prospective endoderm, and the mesoderm later arises from a belt of egg substances between the animal and vegetal areas.

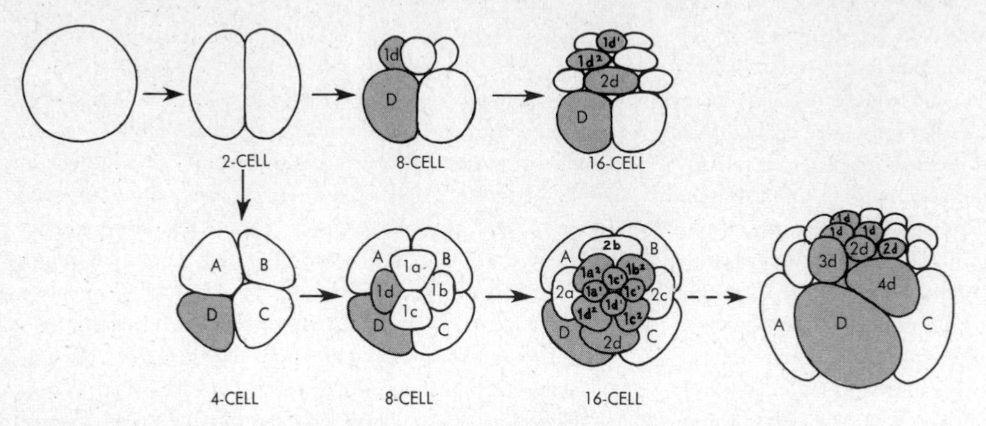

11.14. Spiral cleavage. Top row, side views; bottom row, view from animal pole for four-, eight-, and 16-cell stages, side view for diagram on right. Blastomeres in the *D* quadrant shaded gray, to indicate the alternating right and left spiralling of successive tiers of blastomeres. Capital letters identify macromeres, lower-case letters, micromeres. In the 16-cell stage, *D* has produced 2*d* (a micromere of the second quartet), while 1*d* (of the first quartet) has divided and has given rise to $1d^1$ and $1d^2$; analogously for the *A*, *B*, and *C* quadrants. Subsequent cleavages do not all occur synchronously. Note that the numbers before the letters indicate the quartet a blastomere belongs to; exponent numbers indicate sequence of blastomere origin. The diagram on lower right depicts an approximately 32-cell stage, with the 4*d* cell just cut off from *D*. This 4*d* micromere is the source of the later adult mesoderm. All other micromeres will develop most of the ectoderm, and the macromeres, most of the endoderm. (*Lower right, after Conklin.*)

The fourth tier of micromeres analogously comes to lie under the second, and this alternating positional shifting of the tiers constitutes the identifying characteristic of the spiral cleavage pattern. After the main tiers have formed, cleavage divisions continue in all blastomeres at various rates and, notwithstanding the initial spiral pattern, the resulting embryo comes to have an overall bilateral symmetry. Spiral cleavage characterizes the eggs of flatworms, nemertine worms, annelids, mollusks, and protostomial groups generally.

Many animals are characterized by other, *irregular* cleavage patterns which cannot be classified according to the categories above. Moreover, early divisions also do not follow any uniform patterns in meroblastically cleaving eggs. It is probable, however, that meroblastic eggs are evolutionary derivatives of holoblastic eggs, and that the large quantities of yolk have simply obscured any distinct cleavage pattern that may have been present originally. Indeed, it is altogether possible that isolecithal eggs and radial cleavages may have been primitive in ancestral animals generally, and that all other egg types and cleavage patterns are variously modified evolutionary derivatives. Note, however, that the distribution of egg types and cleavage patterns among animals is not always taxonomically tidy. Phyla believed to be related on the basis of other evidence often have distinctly different egg types and forms of cleavage, and such nonuniformity is encountered also within phyla. An egg may have a particular size and yolk content in the primitive members of a group, but in the advanced yet still closely related members the nature of the egg may have become modified drastically in adaptation to a different way of life. The cleavage pattern then is usually changed as well. For example, primitive marine tunicates possess small, isolecithal, mosaic eggs; but their vertebrate relatives, descended from the same common ancestors, typically possess large, discoidal, regulative eggs, in adaptation to life in fresh water (cf. Chap. 30). The land vertebrates have retained such eggs, but mammals have reverted to small isolecithal (yet still regulative) eggs, in correlation with the placental mode of development.

Analogous variations occur in almost all phyla and groups of related phyla. For this reason, considerable caution must be exercised in attempts to deduce evolutionary interrelations from the nature of the early embryos alone. Animal phyla with spirally cleaving eggs in particular have been regarded as being closely related, in part because of the common cleavage pattern. Thus, even though as adults annelids and mollusks are as radically different as animals can be, as spirally cleaving early embryos they are virtually identical cell for cell. We shall examine the legitimate contributions of embryology to evolutionary hypotheses in Chap. 14.

Inasmuch as in mosaic eggs the later fate of each blastomere is fixed from the start, it should be possible to trace the origin of each adult body part back to a particular blastomere in the early embryo. Such studies of *cell lineage* have been carried out in the bilaterally cleaving mosaic eggs of certain tunicates and the spirally cleaving mosaic eggs of certain annelids, mollusks, and other protostomes. In such studies, numbers and letters are used to identify successively cleaving blastomeres, and the developmental fate of each specific blastomere is followed into the later embryo. In the spiral mosaics, for example, it has been shown that, after four quartets of micromeres and one of macromeres are present, the macromeres will give rise later to most of the endoderm, whereas the micromeres will form ectoderm and mesoderm. Indeed, the entire adult mesoderm will arise from one particular blastomere in the fourth quartet of micromeres, viz., the so-called *4d teloblast* cell (cf. Fig. 11.14). Lineage studies must of necessity be less precise in regulative eggs, yet even here the later fate of groups of blastomeres can be followed after the first few cleavages, i.e., as soon as developmental determination has taken place. By such means, the prospective fate of each region of the early embryo has been mapped; and it has been found that, essentially as in mosaic eggs, the blastomeres in the animal half later usually form most of the ectoderm, those in the vegetal half form most of the endoderm, and the mesoderm arises from a band of blastomeres lying between the animal and vegetal parts of the early embryo (cf. Figs. 11.8 and 11.13).

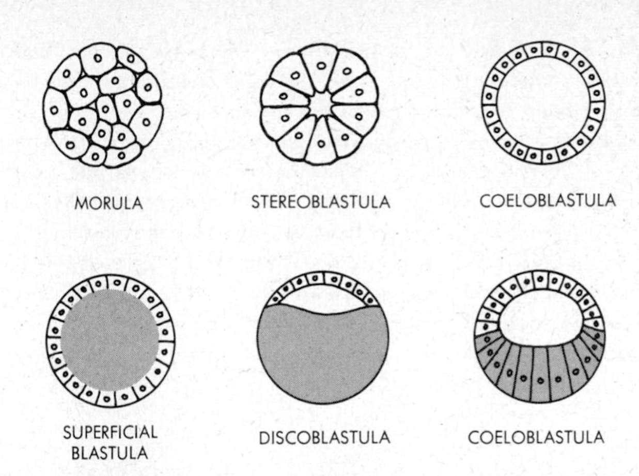

11.15. Blastula types. The diagrams represent sections through the animal-vegetal axis; yolk gray-shaded. The internal spaces in the two middle and two right blastulae are the blastocoels (= pseudocoels). Coeloblastulae may differ according to the amount and position of yolk in the original egg; e.g., a blastula such as at upper right might form from an isolecithal egg, and one such as at lower right, from a telolecithal egg.

Blastula, Gastrula, and Postgastrula

At the end of cleavage, the embryo consists of a ball of cells in holoblastic types and a layer of cells over yolk in meroblastic types. This developmental stage is referred to generally as a *blastula.*

In holoblastic embryos, the blastula is either solid or hollow (Fig. 11.15). If it is filled with irregularly arranged cells, it is called a *morula.* If it is hollow, the wall of the blastula may consist of several or of only one layer of cells. Also, the space within may be small or large, depending on the number and size of the cells present. If the cell number is small, as in cases in which a small egg undergoes few cleavage divisions or any egg cleaves at a slow rate, then the size of individual cells will be large relative to the size of the blastula as a whole. The free space within the blastula will then be correspondingly small or virtually absent. A blastula of this type is a *stereoblastula.* But if the number of cells present is large, as in cases in which large eggs undergo many cleavage divisions or any egg cleaves rapidly, then the size of individual cells will be small relative to the size of the whole blastula. The internal free space then will be large, and such a blastula is a *coeloblastula.* A whole range of intermediate conditions is known, however, and it is often difficult to distinguish between a morula and a stereoblastula, for example, or between a stereoblastula and a coeloblastula. Moreover, if much yolk is present, as in telolecithal eggs, the vegetal half of a blastula will consist of comparatively few, large cells, and the animal half, of many, small cells (cf. Fig. 11.15).

A space within a blastula represents a *blastocoel.* This embryonic body cavity originates during early cleavage; it may be identified in the eight-cell stage as the space enclosed by the two quartets of blastomeres. The cavity is the forerunner of the adult pseudocoel which, as we already know, becomes

11.16. Patterns of gastrulation. Diagrams: future and actual endoderm shaded gray. Top row, endoderm formation in a morula and formation of gut lumen by hollowing out, or cavitation. Second row, gastrulation by epiboly, or overgrowth of cells of the animal half over those of the vegetal half. Last two rows, four patterns of emboly. As the diagrams indicate, invaginative emboly is likely to occur only where the blastula is a coeloblastula with a large blastocoel. **Photos:** early development in starfish embryos. **A,** egg; **B,** two-cell stage; **C,** late cleavage; **D,** coeloblastula; **E,** embolic invagination, early gastrula; **F,** late gastrula, beginning of mesoderm formation; **G,** mesoderm formation under way.

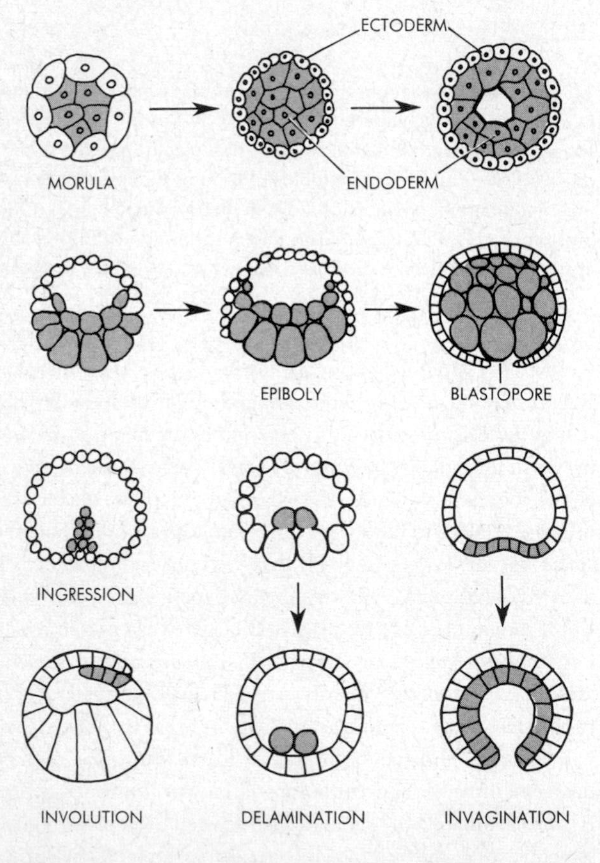

extensive in certain animals. A blastocoel is also present in meroblastically formed blastulae. In the superficially cleaved embryos, the blastocoel may be considered to be filled completely with yolk; and in the discoidal embryos the blastocoel forms by a lifting away of the blastodisc from the underlying yolk (cf. Fig. 11.15).

Once a blastula has formed, the next major developmental event is the establishment of the three embryonic germ layers in correct relative positions—ectoderm exteriorly, endoderm interiorly, and mesoderm between. The events which transform a blastula into such a developmental stage are collectively called *gastrulation*, and the resulting embryo itself is a *gastrula*.

There are almost as many different specific methods of gastrulation as there are animal types. In many radiate animals and some other groups, the blastula is a solid morula. Gastrulation here may consist of little more than the arrangement of the outer cells into a distinct ectoderm layer, the remaining inner cells then forming a solid endoderm. A space, constituting the forerunner of the alimentary cavity, usually develops later in the endoderm by *cavitation*. Still later, mesoderm forms by the inwandering of cells from the ectoderm (Fig. 11.16).

In the majority of animals, the blastula is a stereoblastula or a coeloblastula, or it is formed meroblastically. In all such cases, the regions which will give rise to the germ layers are already determined on the *surface* of the blastula and indeed even on the egg surface, as revealed by studies of organ-forming zones (cf. Fig. 11.8). Therefore, gastrulation here must involve an *interiorization* of the endoderm- and mesoderm-forming regions. Interiorization of the endoderm-forming regions usually takes place first, by either or both of two general processes: *epiboly*, overgrowth of the ectoderm-forming regions around the endoderm-forming regions; and *emboly*, ingrowth of the endoderm-forming regions under the ectoderm-forming regions (cf. Fig. 11.16). Epiboly is characterized by a rapid proliferation of cells in the animal half and

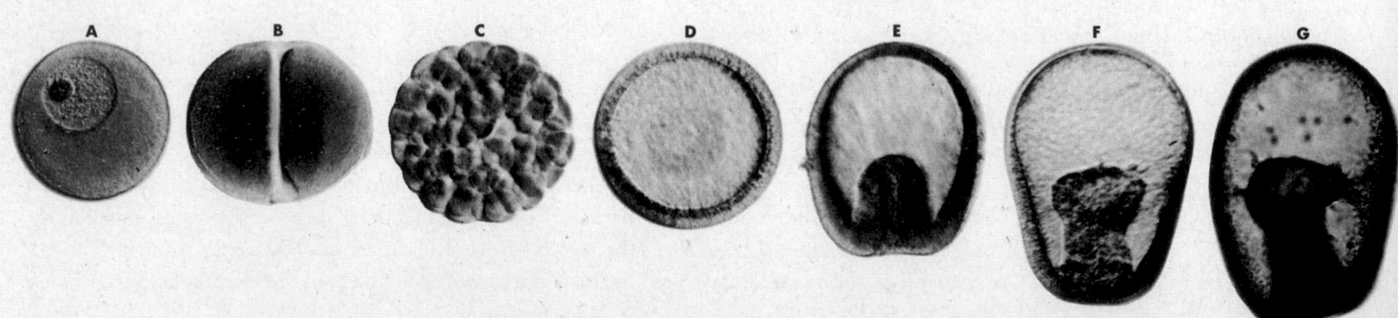

simultaneous spreading of this enlarging sheet of cells over the vegetal half. Emboly may occur in at least four different ways: by *ingression*, the inward migration of endoderm-forming cells from the vegetal parts of the blastula surface; by *delamination*, the budding off into the interior from the blastula surface of as much as a whole sheet of endodermal cells; by *involution*, the inrolling of a sheet of endoderm cells from the vegetal regions of the blastula surface and the subsequent interior spreading of this sheet underneath the surface layer; and by *invagination*, the indenting of the endoderm-forming part of the blastula surface, as when a balloon is pushed in with a fist in one region.

The original nature of the egg influences not only the pattern of cleavage but also that of gastrulation: whether epiboly or one or more variants of emboly will occur is determined in large part by the size and number of cells present, by the amount of yolk in the blastula, and by the extent of the blastocoel.

For example, emboly by invagination is mechanically possible only where the blastocoel is large and the cells of the blastula are small and numerous. In such cases, exemplified by the coeloblastulae of many echinoderms and nonvertebrate chordates, the endoderm-forming vegetal part of a blastula invaginates into the animal part, resulting in a two-layered cupshaped embryo. By contrast, invagination is mechanically difficult or impossible where the blastocoel is small and where vegetal cells are large, yolk-filled, and few in number. In such cases, gastrulation is achieved by combinations of epiboly and one or more forms of noninvaginative emboly. For example, stereoblastulae usually gastrulate by epiboly or involution or both. Many types of morulae gastrulate by epiboly, delamination, ingression, or by combinations of these processes. Meroblastic blastulae acquire interior layers by delamination or by involution. Regardless of the specific pattern, however, the ectoderm-forming portions of a blastula in all cases come to lie over and around the endoderm-forming portions; and the mesoderm-forming portions of the blastula then either remain on the outside as part of the gastrula ectoderm or become interiorized during gastrulation and come to form part of the endoderm (Fig. 11.17).

In holoblastically formed embryos, epibolic growth of ectoderm around endoderm usually results in the formation of an opening in the vegetal half where endoderm remains exposed to the last. In emboly, analogously, interiorization of endoderm usually occurs at a restricted vegetal region which has the form of an opening after gastrulation. Any such opening

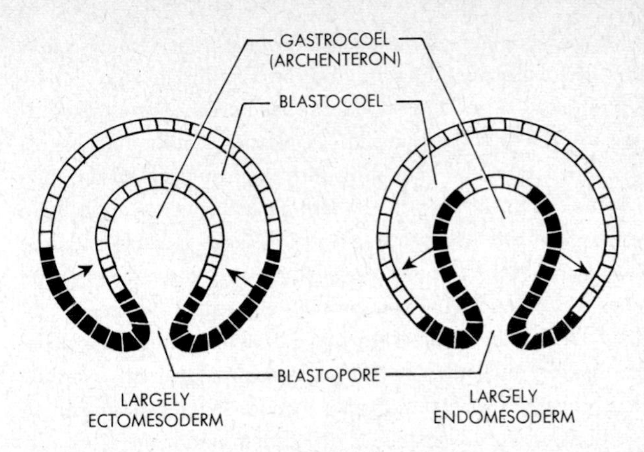

11.17. If little of the future mesoderm-forming zone is interiorized during gastrulation, as on left, then most (but not necessarily all) of the mesoderm eventually formed will be ectoderm-derived ectomesoderm (small arrows pointing into blastocoel). But if most of the mesoderm-forming zone is interiorized by gastrulation, as on right, then mesoderm will be largely (though not necessarily entirely) endoderm-derived endomesoderm. Although the diagram depicts gastrulation by invagination, the principle regarding the source of mesoderm applies to any gastrulation pattern. Primary (embryonic and larval) mesoderm is often ectomesodermal, adult mesoderm, usually endomesodermal. See also Fig. 7.16 for actual patterns of mesoderm formation among different animal groups.

is called a *blastopore*. It persists without change in many embryos, and in others it grows shut temporarily but then an opening re-forms at or near the original spot. In either event, this blastoporal perforation develops into the single alimentary opening of adult radiate animals and flatworms, into the adult mouth of Protostomia, and into the adult anus of Deuterostomia. In later embryonic stages, a second perforation breaks through near the animal pole, and this opening becomes the anus in the Protostomia and the mouth in Deuterostomia.

As a further result of the interiorization of endoderm, the blastocoel becomes obliterated partially and a new cavity appears: either the endoderm possesses an internal space from the outset or, if the interior is originally solid, a central space soon develops in it by cavitation. In both instances, the endodermal cavity is a *gastrocoel* or *archenteron*, the embryonic alimentary cavity. It communicates with the outside via the blastoporal opening (cf. Fig. 11.17). Note that this forerunner of the adult alimentary system is closely associated with food from the earliest moment: gastrula endoderm forms from the vegetal cells of the blastula, which contain the bulk of the yolk.

The beginning of mesoderm development usually marks the end of gastrulation. Formation of this third germ layer may be initiated as soon as endoderm has become interiorized. If most of the mesoderm-forming regions of the original blastula remain part of the ectoderm during early gastrulation, then the mesoderm arising in the late gastrula will be largely an *ectomesoderm*. As already noted in Chap. 7, such a middle germ layer forms characteristically by an inward migration (i.e., an *ingression*) of ectodermal cells which become established as mesenchyme. They may fill the remaining blastocoel completely and thus give rise to an acoelomate animal, or they may collect in specific regions only and then leave substantial portions of the blastocoel as an adult pseudocoel.

Alternatively, most of the mesoderm-forming regions of the blastula may have become interiorized as part of the gastrula endoderm, in which case the mesoderm arising later will be largely an *endomesoderm*. As also noted in Chap. 7, such a middle germ layer is generally epithelial from the start. It too fills larger or smaller portions of the blastocoel, and it may contribute to the formation of an acoelomate or pseudocoelomate body structure. Moreover, endomesoderm is also the main source of the adult mesoderm in almost all coelomate animals. In these, to be sure, some mesenchymal *primary mesoderm* in embryo and larva usually does form from ectomesoderm. However, most of the mesodermal components of the future adult, the epithelial *secondary mesoderm*, arises from endomesoderm. In the spirally cleaving schizocoelomates, the specific sources of the secondary, adult mesoderm are the 4d blastomeres (teloblasts) of early cleavage stages discussed above. As shown in Chap. 7, such cells proliferate teloblastic *mesoderm bands* on each side of the archenteron. Portions of these bands then form the splanchnic components of the middle body layer, other portions become the somatic components, and the coelomic body cavity remains between. In the enterocoelomates, mesoderm pouches out, i.e., *evaginates*, from both sides of the endoderm (cf. Fig. 7.16).

In meroblastically formed embryos, the sequence of gastrulation and postgastrulation events is substantially as above, but the architectural pattern is modified as a result of the amount and distribution of yolk. In the superficially cleaved eggs of arthropods, a blastula gastrulates mainly by a combination of ingressional and delaminative methods. The formation of mesoderm and the establishment of the body cavities are described in Chap. 27. In the discoidally cleaved eggs of, for example, reptiles and birds, gastrulation takes place primarily by involution of endoderm- and mesoderm-forming regions from the originally single-layered blastodisc (Fig. 11.18). After endoderm is interiorized as a layer underneath the ectoderm, mesoderm involutes at a highly elongated, slitlike blastopore, here called the *primitive streak*. Such mesoderm comes to lie in the blastocoel between ectoderm and endoderm, and it soon develops into coelom-enclosing splanchnic and somatic layers. This four-layered blastodisc represents the embryo proper. All four layers subsequently proliferate at the margins of the blastodisc and give rise to the extraembryonic membranes. More specifically, some portions of the germ layers spread by epiboly over and around the yolk. As a result, the yolk eventually comes to be enclosed completely by endoderm and splanchnic mesoderm, and the yolk may therefore be considered to fill a gastrocoelic space. The layers enclosing the yolk form the yolk sac. The sac shrinks progressively as yolk is used up, and it ultimately becomes the floor of the adult alimentary tract. At the head and tail regions of the blastodisc the ectoderm and the somatic mesoderm fold upward, and these folds come to fuse above the blastodisc. An ectoderm-lined amnion develops in this manner. Also, as shown in Fig. 11.18, the amnion, the blastodisc, and the yolk sac become enclosed by an ectodermal chorion. Similarly enclosed is the fourth extraembryonic membrane, the allantois, which later evaginates from the endoderm in the hindregion of the blastodisc.

A variant of this developmental pattern is encountered in placental mammals. The embryos of these animals form from small, isolectithal, almost yolkless eggs, yet the discoidal and involutionary pattern of gastrulation has been inherited from the reptilian ancestors. The blastula is a coeloblastula or a stereoblastula, in which a morulalike *inner cell mass* gives rise to the embryo proper (Fig. 11.19). The cell layer bounding the fluid-filled blastocoel, called the *trophoblast*, is ectodermal and is equivalent to the chorion. An amnion soon develops in the inner cell mass by cavitation, and some cells of this mass also proliferate and spread as a sheet around the inside of the trophoblast. The sheet is endodermal and represents the yolk sac; it converts the blastocoel into a yolkless gastrocoel. By this time the inner cell mass has become arranged into a double-layered disk, consisting of an ectodermal layer continuous with the amnion and an endodermal layer continuous with the yolk sac. This disk represents the embryo proper and is equivalent to the blastodisc of reptiles and birds. Other events occur as in the latter; i.e., mesoderm involutes via a primitive streak, and an empty allantois later evaginates from the posterior endoderm. The principal new feature is the proliferation of the chorionic trophoblast into the embryonic portion of the placenta.

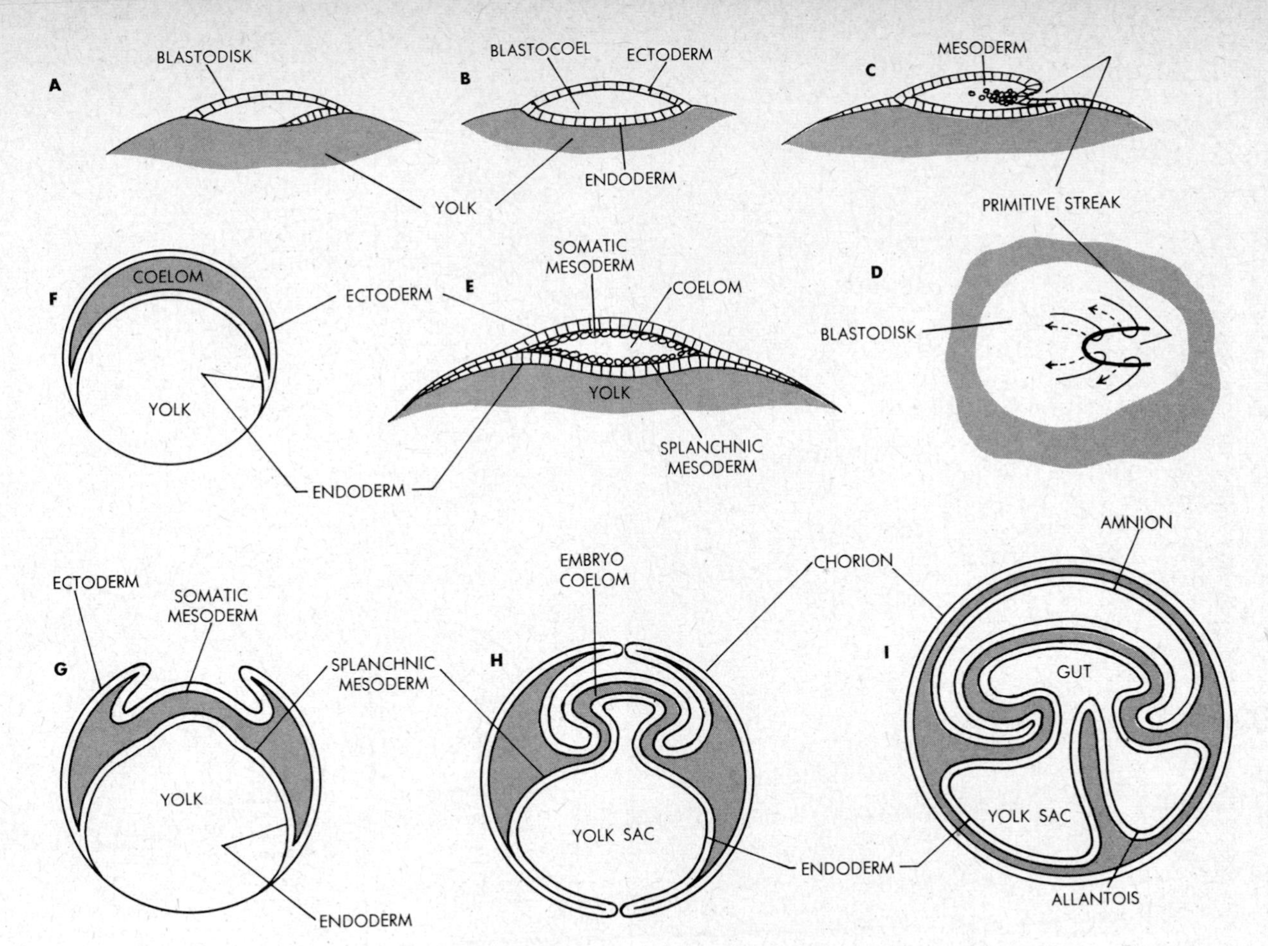

11.18. Early development in reptiles and birds. All figures except **D** represent longitudinal sections, with the head end toward left. **A.** Beginning of endoderm formation between outer blastodisc layer and yolk. **B.** Blastocoel established. **C.** Ingression and involution of mesoderm from primitive streak. **D.** Top view of **C;** arrows indicate the direction of tissue inrolling around the fold of the primitive streak (U-shaped line); the inrolled portions are shown as broken lines. **E.** Proliferation of mesoderm into somatic and splanchnic layers, and establishment of coelom. **F** through **I.** Progressive growth of ectoderm, somatic and splanchnic mesoderm, and endoderm. Each of these layers is shown as a single line, and the coelomic spaces are shaded gray. The embryo proper lies in the area directly above the yolk and becomes gradually covered over by amniotic-cavity-forming folds. Note the establishment of the four extraembryonic membranes, each accompanied by a mesodermal layer. The coelomic spaces outside the embryo proper form the extraembryonic coelom. Chorion and amnion are ectodermal, yolk sac and allantois are endodermal. See also Fig. 10.30.

Note that, in all animals, the late gastrula represents a key stage in embryonic development. At this stage the germ layers are established in proper positions, the prospective head and hind ends of the developing animal are marked, and in many cases the top-bottom and left-right axes are determined as well. Moreover, the extent and position of the pseudocoelomic or coelomic body cavities or both are already foreshadowed. In effect, the fundamental body plan of the future animal has become elaborated in rough outline. All later development consists essentially of a sculpturing of detail—formation of organ systems, then of organs within each system, then of tissues within organs, etc.

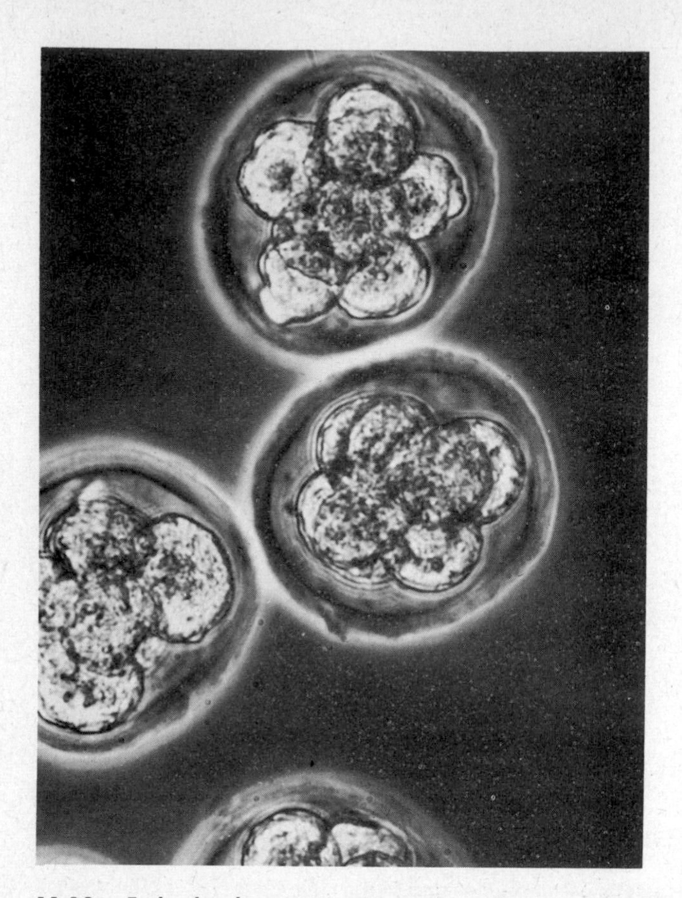

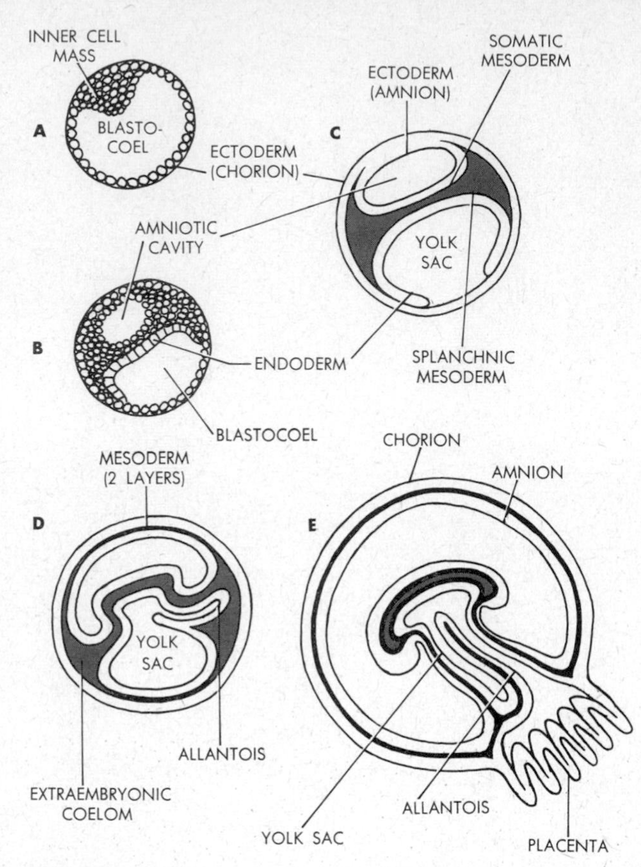

11.19. Early development in mammals. Photo: mouse embryos, at approximately eight-cell stage. **Diagrams: A.** The ectodermal trophoblast (later chorion) and the inner cell mass. **B.** Establishment of the amniotic space by cavitation in the inner cell mass and growth of endoderm layer from inner cell mass around blastocoel. **C.** Growth of mesoderm and coelom from inner cell mass and of amnion and empty yolk sac. The embryonic and extraembryonic coelom are shaded gray. **D.** Formation of allantois; this stage is quite comparable to diagram I of Fig. 11.18. **E.** Growth of allantois and chorion to form placenta, enlargement of amnion, and collapsed yolk sac (heavier lines between extraembryonic membranes symbolize double layer of mesoderm). See also Fig. 10.30.

Note, furthermore, that the broad sequence of the early developmental events is the same we have examined in Chap. 7, in a characterization of progressively less generalized features of animal body plans. Thus, eggs and cleavage establish the level of organization and the symmetry of an animal; gastrulation establishes alimentary design and the basic architecture of the mesoderm; and postgastrular development establishes the nature of the body cavities and traits such as segmentation and the specific makeup of organ systems.

How do organ systems actually arise in later embryos?

LATER DEVELOPMENT

Induction

The detailed patterns by which the germ layers are transformed into well-defined body parts differ vastly for different animals. Nevertheless, certain basic processes appear to be common to all such patterns. We may illustrate these by considering the development of the vertebrate nervous system (Fig. 11.20).

On the upper surface of the gastrula develop two ectodermal ridges, one along each side of the midline. These ridges grow upward and toward each other and soon meet along the midline. As their edges fuse, they form a tube of ectoderm which runs from front to back and is covered over by an outer ectoderm layer. This tube is the basis of the nervous system; it develops into brain in the front part of the embryo and into spinal cord in the hind part.

The essential event here is the outfolding of a tissue layer, followed by fusion of the fold edges. Almost all other formative processes of later embryonic development similarly consist of outfolding or infolding, outpouching or inpouching, of portions of the three germ layers of the gastrula. For example, limb buds arise by combined outpouchings from ectoderm and mesoderm. Lungs and digestive glands develop in part as outpouchings from various levels of the endoderm. The eye develops in part as an outpouching from the brain. All other body parts develop analogously. The ultimate result

11.20. The initial development of the vertebrate nervous system. **Top,** left to right, dorsal views of progressive stages in neural tube formation in frogs. The anterior ends of the embryos are toward the right. **Middle,** diagrammatic cross sections corresponding to the stages shown above. **Bottom,** left and right, cross sections corresponding to the first and last stages illustrated in top series. Note the large amounts of yolk. Middle, chicken embryo cross section corresponding to middle stage shown above. Note the mesoderm cells to either side of the developing neural tube and the notochord just below the tube.

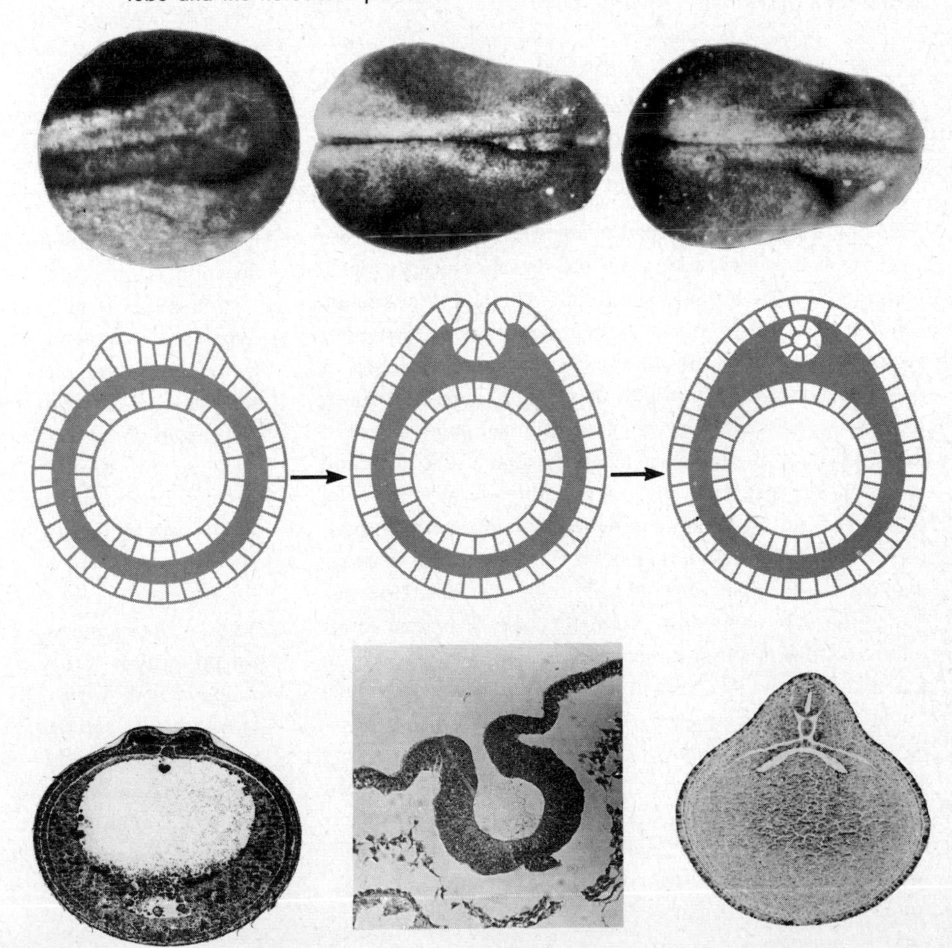

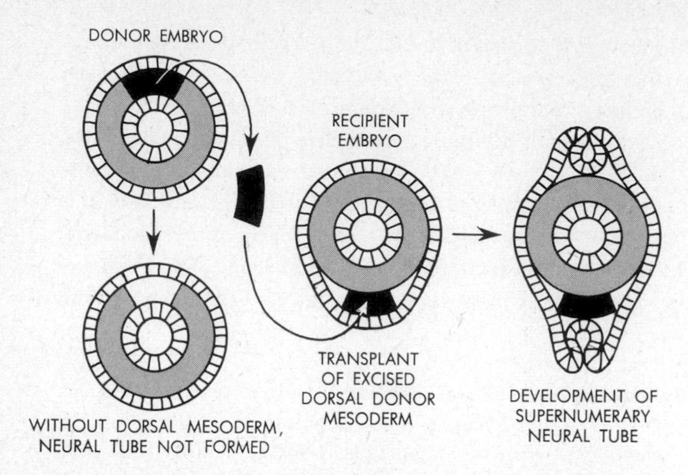

11.21. Neural induction. If the dorsal mesoderm of a donor embryo is transplanted under the belly ectoderm of a host embryo, then the transplant will induce the formation of an abnormally located neural tube in the host (and the host's own neural tube will develop as well).

of these processes of morphogenesis and differentiation is a fully formed embryo, clearly recognizable as a young stage of a particular species.

Experiments have shown how these orderly sequences of development may come about (Fig. 11.21). In amphibian embryos it is possible to cut out the dorsal ectoderm which, under normal circumstances, would fold up and form a neural tube. If this excised tissue is then transplanted to another region of the embryo, it will not form a neural tube and its cells will not differentiate into neurons. This result suggests that the dorsal mesoderm, which in an intact embryo lies just under the dorsal ectoderm, normally affects this ectoderm in such a way that it will fold out and differentiate into neural tissue. That this is actually so can be shown by another experiment. The dorsal mesoderm can be cut out and can be transplanted, for example, into the belly region of an embryo just under the belly ectoderm. Normally, belly ectoderm forms only skin. But if dorsal mesoderm lies under it, it will form a neural tube and its cells will differentiate into neural tissue.

The implications are clear. Somehow, the dorsal mesoderm of a normal embryo *induces* the outfolding of the overlying ectoderm, the formation of ridges, and the later differentiation of neural tissue. Such induction can actually be shown to occur whenever outfoldings or infoldings and outgrowths or ingrowths develop in the embryo. The formation of the vertebrate eye provides a particularly striking example.

Eye development (Fig. 11.22) begins with the evagination of a pocket from the side of the future brain. This pocket is narrow at the base and bulbous at the tip. Soon the bulbous portion invaginates from the forward end and a double-layered cup is formed. The cup represents the future eyeball. As it grows outward from the brain, its rim comes into contact with the outer ectoderm layer which overlies the whole nervous system and which represents the future skin. Just where the eyecup rests against it, the ectoderm layer now begins to thicken. This thickening eventually grows into a ball of cells, which is nipped off toward the inside. It fits neatly into the mouth of the eyecup and represents the future lens. The cells of this ball and the ectoderm overlying them later become transparent. The basic structure of the eye is then established.

The following type of experiment has shown dramatically how these developmental processes are controlled. It is possible to cut off the eyecup and its stalk before they have grown very far. Eyecup and stalk may then be transplanted. For example, they may be inserted into a region just under the belly ectoderm of an embryo. Under such conditions, the patch of belly ectoderm overlying the eyecup soon thickens, a ball of cells is nipped off toward the inside, and a lens differentiates. Moreover, lens and overlying skin become transparent. In effect, the transplanted structures have caused the formation of a structurally normal eye in a highly abnormal location (Fig. 11.23).

A common conclusion emerges from this and many similar types of experiments. One embryonic tissue layer interacts with an adjacent one, and the latter is thereby induced to differentiate, to grow, to develop in a particular way. This developed tissue then interacts with another one in turn and induces it to develop. In short, sequential induction must occur if progressive development is to take place. As in the induction sequence: dorsal mesoderm ⟶ neural tube ⟶ eyecup ⟶ lens, so also generally; one tissue provides the stimulus for the development of the next. The phenomenon of *embryonic induction* consequently may account well for the orderly, properly timed, and properly spaced elaboration of body parts.

Although inductive processes among embryonic tissues may be identified and described, the nature of such interactions in terms of reactions within and among cells is still obscure in many respects. Even so, the ultimate result of these various occurrences is a fully formed embryo, which may later hatch and become a larva or may develop into an adult directly.

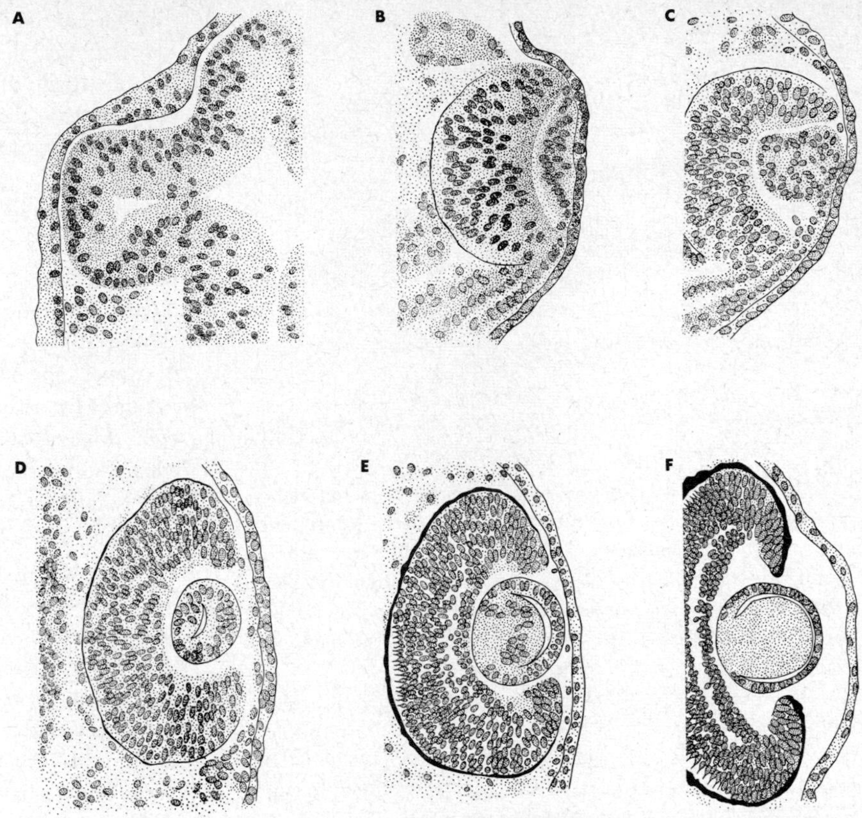

Courtesy Dr. Dietrich Bodenstein, from figs. 2 and 3, J. Exp. Zool., *vol. 108, pp. 96, 97.*

11.22. Development of the vertebrate eye (amphibian), semidiagrammatic. This series of successive stages shows the outgrowth of a pocket from the brain, contact of this pocket with the outer body ectoderm, formation of an eyecup, gradual formation of a lens from the outer ectoderm, and development of the pigmented and other tissue layers of the eyeball.

Larvae

In oviparous animals, a larval period begins when the embryo escapes from its enclosing membranes and protective coats and, after such a process of hatching, assumes a free-living existence. As an immature, preadult stage of development, a larva is usually without reproductive capacity. However, some types of larvae do reproduce vegetatively and others even form gametes (cf. below).

Larvae play a genetic and evolutionary as well as a functional role. As noted earlier, animals pass through embryonic and larval phases only if development is initiated by a sexual process. Such developmental phases then provide the means and the necessary time for the translation of genetic instructions present in the egg into actual structures and functions. To the extent that the genetic instructions may be different from those received by the ancestors, embryos and larvae permit the introduction of evolutionary changes. We shall discuss this important evolutionary role of the preadult stages in Chap. 14.

Functionally, larvae serve mainly in one or both of two capacities. In many animals, sessile ones in particular, the larvae are the chief agents for geographic distribution of the species. In some cases, indeed, larvae are still without fully formed alimentary systems; the mouth may not be functional as yet, and the larvae then serve exclusively as nonfeeding

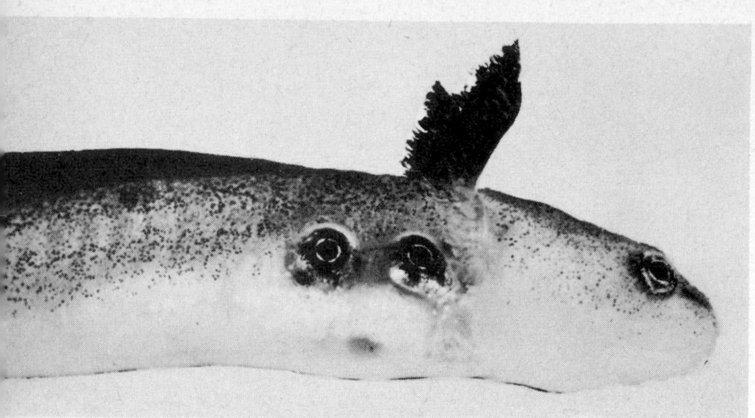

Courtesy Drs. S. R. Detwiler and R. H. Van Dyke, from fig. 16, J. Exp. Zool., *vol. 69, p. 157.*

11.23. Experiments in eye transplantation. If an embryonic eyecup is excised from a donor embryo and is transplanted into an abnormal location in a host embryo, then a structurally perfect eye may develop at that abnormal location. The photo shows a larva of a salamander (*Amblystoma*) with two supernumerary eyes grafted into abnormal locations. The photo was taken 43 days after the transplant operation.

dispersal devices (e.g., tunicate tadpoles). As might be expected, locomotor structures are well developed in such species-distributing larvae. For example, chordate tadpoles possess strongly muscled tails, and the swimming larvae of many invertebrates have greatly enlarged, variously folded and ciliated epidermal surfaces, with bands and tufts of extra-long cilia in given regions (Fig. 11.24). In many other animals, the larvae can disperse geographically far less well than the adults (e.g., insect caterpillars). The primary function of larvae here appears to be nutritional. Embryonic development alone evidently provides only enough building materials for the construction of a transitional feeding apparatus in the form of a larva. The latter then may accumulate the necessary food reserves for the ultimate formation of a more complex adult body. Larvae of this type have particularly well developed alimentary systems, whereas the ectodermal components remain comparatively immature. Note that the larvae of numerous animals function both as dispersal and as feeding machines. And note also that, in many cases, feeding larvae have nutritional requirements differing from those of the adults; hence, the existence of two distinct stages may permit more complete utilization of foods available in the environment and may reduce competition for food within a population.

What are the developmental characteristics of larvae? We may note, first, that development via larval stages is characteristic of virtually all metazoan phyla. Within a phylum or class, however, the relative importance of larvae in the overall life cycle usually varies greatly. Typically, some members of a phylum or a class have short embryonic and long larval periods; others have long embryonic and short larval periods; and still others are without larvae altogether. Animals with larvae are said to develop *indirectly,* those without larvae, *directly.* In most cases, the indirect developers tend to be the comparatively primitive members of the group, the direct developers, the more advanced or more specialized members.

Where larvae do occur, some resemble the adults substantially, whereas others are structurally quite distinct from the adults. For example, the appearance of fish larvae leaves little doubt that these animals will become fishes, but it is not nearly so obvious that tadpoles will become frogs or that caterpillars will become butterflies. Furthermore, where the resemblance is great, metamorphosis into the adult is usually a very gradual, barely perceptible process. But where a resemblance is lacking, metamorphosis generally involves a drastic and abrupt reorganization of the whole body. What developmental factors account for such differences in the durations, forms, and terminal events of larval life?

The answer has three parts. One part is illustrated particularly clearly by the larvae of crustacea (Fig. 11.25). Like arthropods generally, crustacea develop after germ-layer formation by elongating at the posterior end. In the process, the body segments are laid down in anteroposterior succession and paired appendages grow from each segment (cf. also Chap. 28). A sequence of developmental stages may therefore be recognized, each stage characterized by the presence of several more segments than in the preceding stage and several fewer segments than in the succeeding stage. These stages have been named; in sequence they are the *nauplius, metanauplius, protozoaea, zoaea, mysis,* and the *adult.* The nauplius stage possesses only the most anterior head segments in reasonably developed condition, and the preadult mysis stage possesses all but the most posterior abdominal segments.

The important consideration here is that different crustacea become larvae at different points in the above developmental series. For example, the brine shrimp *Artemia* develops as an embryo up to the nauplius stage, and then it hatches as a nauplius larva. In its subsequent development, the animal passes through a succession of larval stages, each separated from the next by a molt. Thus, the nauplius molts into a metanauplius

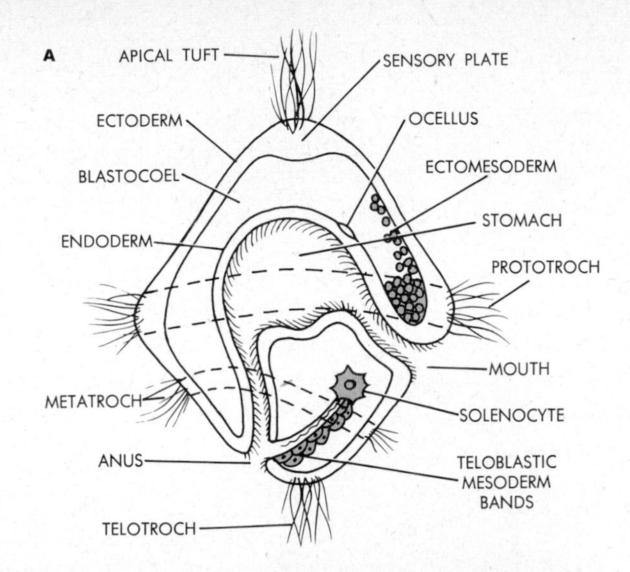

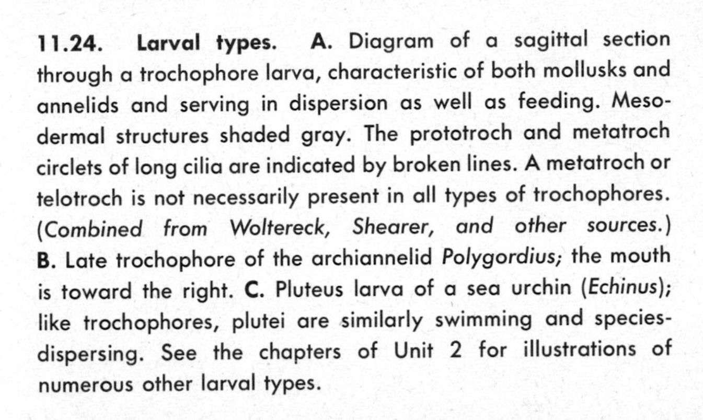

11.24. Larval types. A. Diagram of a sagittal section through a trochophore larva, characteristic of both mollusks and annelids and serving in dispersion as well as feeding. Mesodermal structures shaded gray. The prototroch and metatroch circlets of long cilia are indicated by broken lines. A metatroch or telotroch is not necessarily present in all types of trochophores. (*Combined from Woltereck, Shearer, and other sources.*) **B.** Late trochophore of the archiannelid *Polygordius;* the mouth is toward the right. **C.** Pluteus larva of a sea urchin (*Echinus*); like trochophores, plutei are similarly swimming and species-dispersing. See the chapters of Unit 2 for illustrations of numerous other larval types.

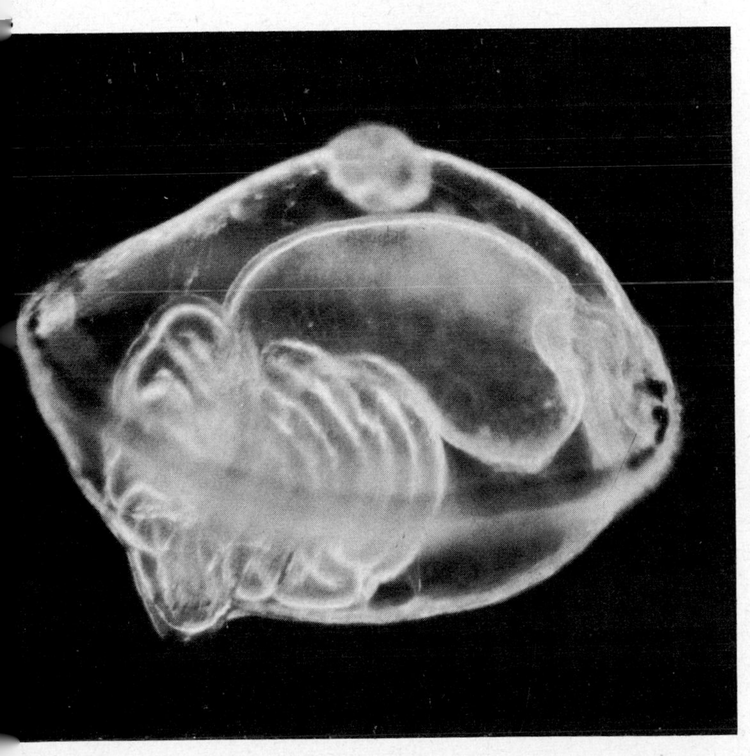

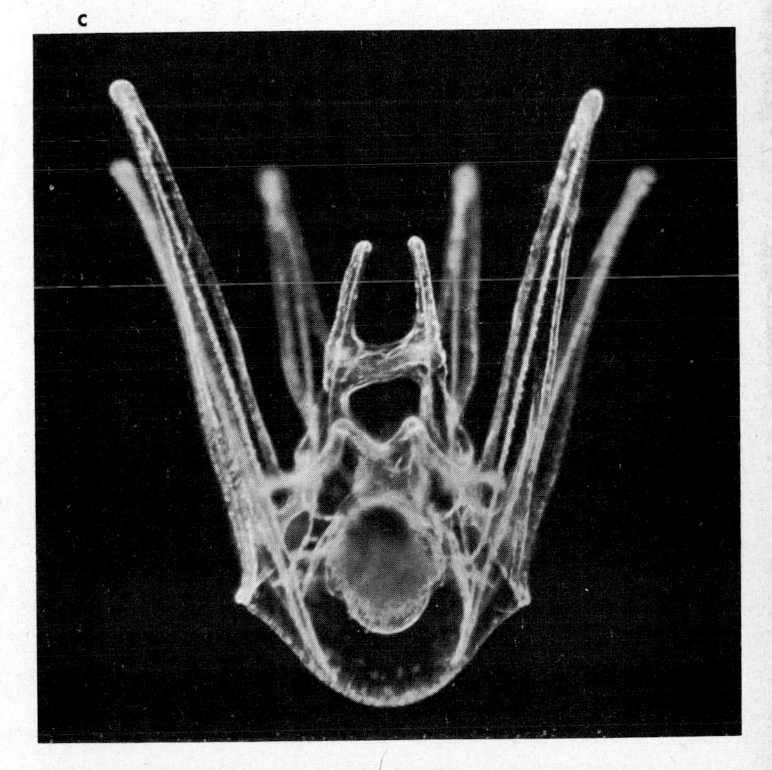

larva, and subsequent molts then give rise to a sequence of larvae corresponding to the series above. By contrast, the related freshwater fairy shrimp *Branchipus* passes the nauplius stage during the embryonic phase and hatches only later as a metanauplius larva. The mantis shrimp *Squilla* passes both the nauplius and metanauplius stages during the embryonic phase and hatches only as a protozoaea larva. Still later hatching occurs in many crabs, which emerge as zoaea larvae; in lobsters, which do not become larvae until the mysis stage; and in many shrimps and prawns related to lobsters, which hatch only as fully formed adults. The complete series of developmental stages does not necessarily occur in all crustacea;

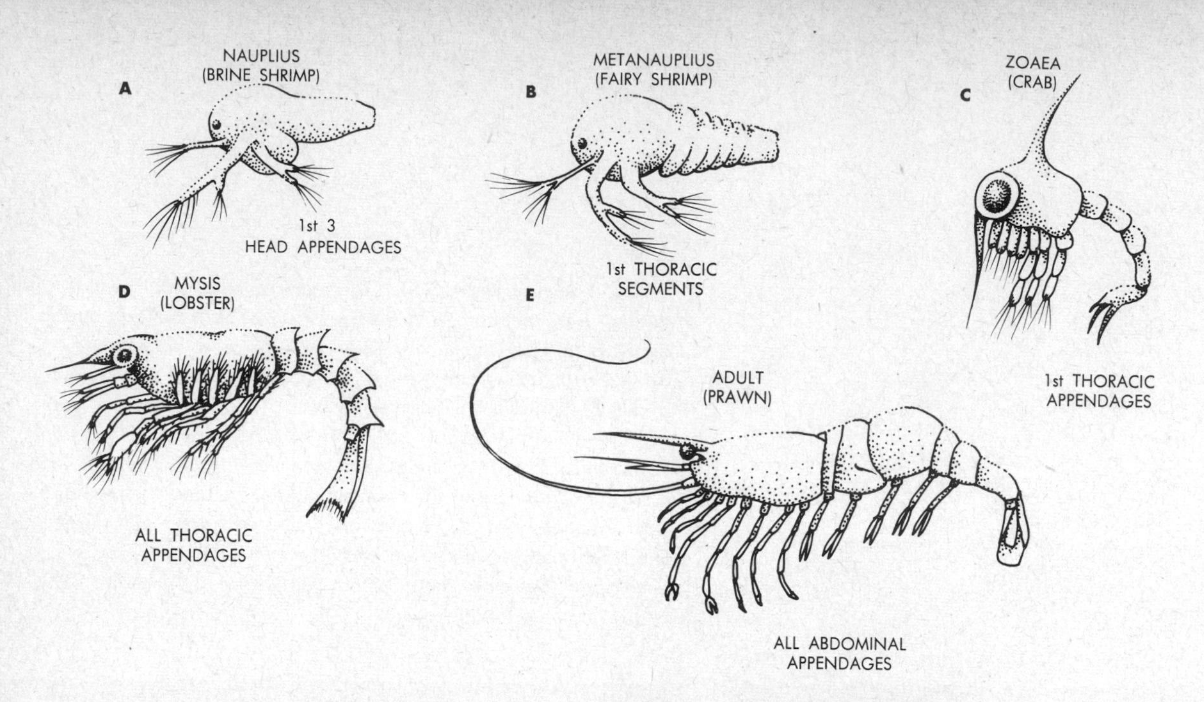

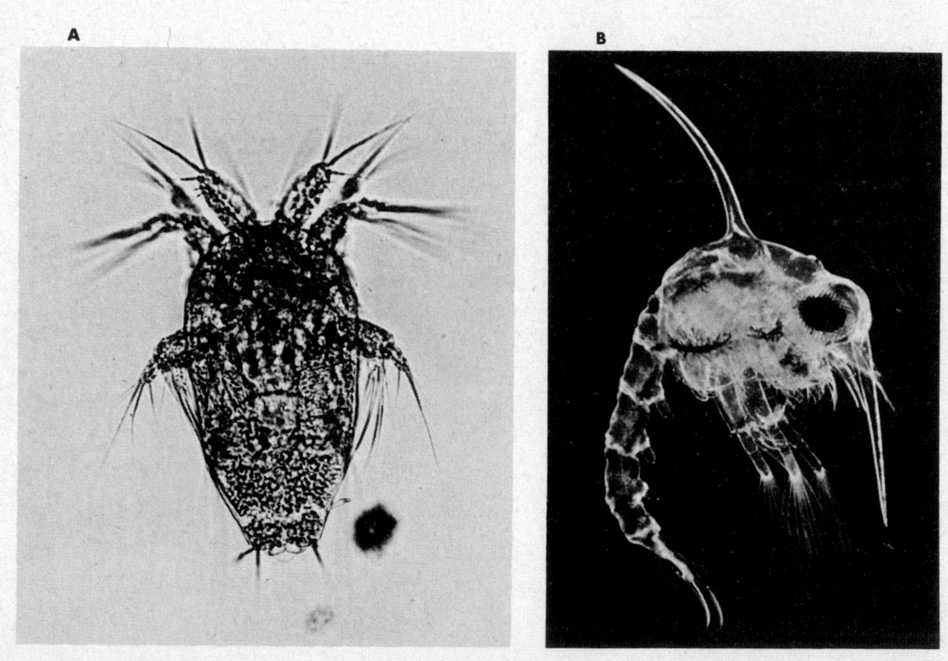

11.25. Crustacean larvae. Diagrams: A through **E,** series of larvae hatched at successively later developmental stages (segmental stage attained is indicated below each figure). Thus, a just-hatched brine shrimp nauplius still must develop a great deal before the adult condition is attained, but a just-hatched mysis of a lobster already resembles the adult considerably. If hatching occurs very late, as in some shrimps and prawns, a larval phase is absent altogether. (*After Claus, Faxon, Herrick, Ortman, Weisz.*) **Photos:** two larval forms corresponding to stages illustrated in the diagrams. **A,** nauplius of the copepod *Cyclops;* **B,** late (third) zoaea of the shore crab *Carcinus*. Additional crustacean larvae are illustrated in Chap. 28.

i.e., in given species two or more of the stages may be condensed into a single one, either during the embryonic or during the larval phase. Nevertheless, different crustacea do become larvae at successively later developmental stages.

The inference is clear. The earlier the time of hatching, the more larval development must take place before the adult condition is attained; and the less therefore does the young larva resemble the adult. Conversely, the later hatching occurs, the less larval development is necessary to produce the adult and the more will larva and adult resemble each other. Moreover, if hatching takes place very late, a larval phase may be absent altogether and the emerging animal will be an adult.

These correlations hold not only for crustacea but for other animals as well. Among insects, for example, some are early hatchers and the form of the larvae then is more embryonic than adult. All caterpillar-producing types belong to this category. Other insects, e.g., grasshoppers, pass through the wormlike caterpillar stage as embryos, and when they then hatch comparatively late they already resemble the adults substantially; the growth of wings constitutes one of the principal remaining developmental requirements. Still other insects, exemplified by the flightless "silverfish," hatch comparatively even later, and their "larvae" are virtually young adults.

We are led to the general principle that the durations and forms of larval periods vary according to the developmental stage the animals have attained at the time they *hatch*. The principle answers in part why some types of larvae resemble the adults more than other types and why larvae do not necessarily occur in all members of a group.

Another part of the answer to the general question above is revealed very clearly in the development of brittle stars, a class of echinoderms. Among these animals, the eggs vary considerably in size and yolk content. At one extreme, eggs are small and isolecithal; at the other, they are large and highly telolecithal (Fig. 11.26). The small eggs undergo equal cleavage and form single-layered coeloblastulae. The latter then gastrulate embolically by invagination, and the mesoderm forms in a manner typical for echinoderms, i.e., by enterocoelous outpouching from the endoderm. Hatching occurs soon thereafter and a long larval period ensues. Called *ophiopluteus* or simply *pluteus,* the larva functions primarily as a dispersal device; it develops a large, ciliated, ectodermal swimming surface folded into conspicuous arms. In the course of larval life, the number of such arms increases from two to four to six and finally to eight. After a long free-swimming period, the pluteus metamorphoses into a small pentagonal adult.

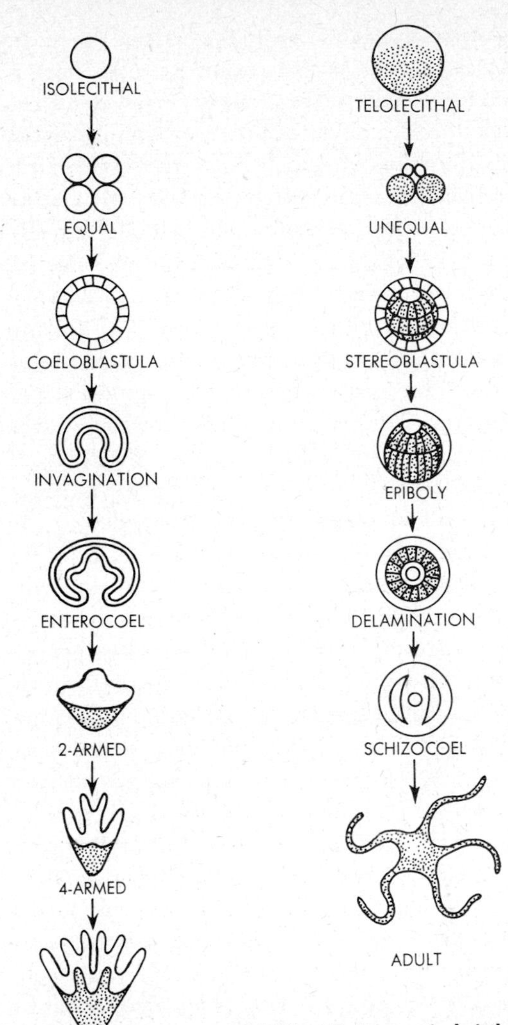

11.26. Extremes in brittle star development. Left column, pattern starting with small yolkless egg leads through many-armed larval stages. Right column, pattern starting with large telolecithal egg leads to absence of larval stages and viviparous adults. (*Adapted from Fell.*)

A quite different pattern of development characterizes brittle stars with larger eggs. The yolkier the egg, the fewer arms develop in the larva; some plutei form only six arms, others only two, still others none at all. Moreover, the smaller the number of arms, the shorter is the larval period and the earlier does metamorphosis occur. In the extreme condition, development as a whole comes to follow a course which is quite atypical for echinoderms generally (cf. Fig. 11.26). Thus, a highly telolecithal egg produces a many-layered stereoblastula; gastrulation occurs by epiboly, not emboly; mesoderm arises by delamination, not as enterocoelous pouches; and the coelom forms as in schizocoelomates by splitting of mesoderm, i.e., by cavitation. Indeed, even the gastrocoel arises by cavitation. In short, the large initial amounts of yolk change the entire developmental pattern. And what is particularly significant in the present context, a larva does not develop at all in such cases: very soon after germ-layer formation, the embryo transforms directly into a small pentagonal adult. Brittle stars of this type are actually viviparous. Their eggs develop within the maternal adult, attached to placentalike regions; and the adult releases not eggs, but young adults.

Several interrelated points are illustrated here. First, we note again that the nature of the egg influences the pattern of all further development. Secondly, small eggs are correlated with long larval periods, yolky eggs with short larval periods or with absence of larvae. Thirdly, prolonged larval periods are adaptively useful in geographic dispersal, and they are correlated with a precocious "ectodermization" of the larva, i.e., an overgrowth of ectoderm and formation of extensive swimming surfaces. Conversely, a short larval period or an absence of a larva is correlated with abundant food supplies in the form of yolk and a precocious "endodermization" of the larva, i.e., the endoderm is developed more extensively than the ectoderm. Finally, it is clear that the later metamorphosis occurs, the more structures will develop with purely larval significance (like the arms of a pluteus, which are larval only and do not become part of the adult); and that the earlier metamorphosis occurs, the fewer body parts of larval significance will form. If metamorphosis takes place exceedingly early, a larva will in effect be suppressed altogether and the embryo will become an adult directly.

These correlations again are applicable quite generally. Among schizocoelomates such as mollusks and annelids, for example, many groups develop from small, somewhat telolecithal eggs. The larvae are precociously ectodermized *trochophores* which, like ophioplutei, function in geographic dispersal; they have prolonged free-swimming periods and metamorphosis occurs comparatively late in development (cf. Fig. 11.24). By contrast, mollusks such as squids and octopuses develop from large, yolk-laden, discoidal eggs, and annelids such as earthworms analogously develop from large, highly telolecithal eggs. In these groups metamorphosis occurs so early that in effect it does not occur at all: the stages corresponding to the trochophore are suppressed and the embryos become adults directly (though not usually viviparously in these particular instances; cf. Chaps. 25, 26).

We are led to a second general principle: the durations and forms of larval periods vary according to the developmental stage given animals have attained at the time they *metamorphose*. The principle shows why larval periods have different durations in different cases and again why larvae do not necessarily occur in all members of a group. Note that if a larval period is long, for example, it is so either because hatching occurs early or because metamorphosis occurs late or because of both these reasons. And note also that absence of a larva may be due either to a forward extension of the embryonic phase up to the adult or to a backward extension of the adult

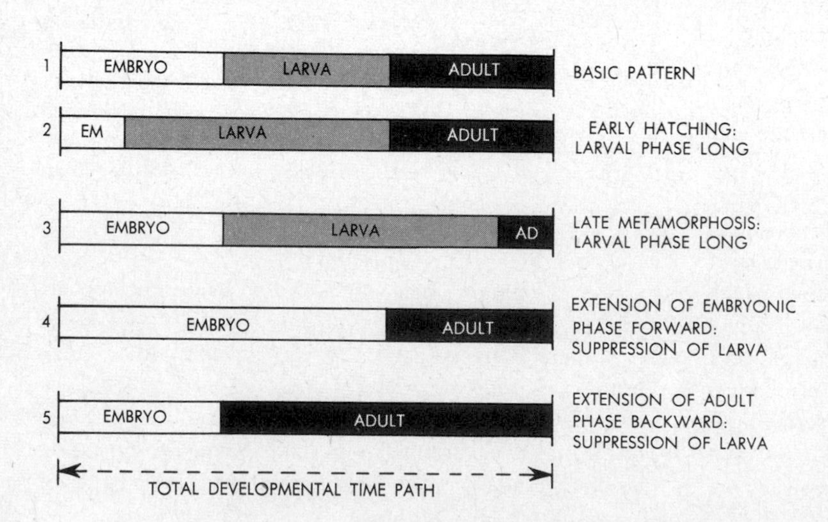

11.27. The general nature of larvae as determined by the time relations between the main developmental periods. Larval differences due to variations in hatching time (as in pattern 2) are illustrated by crustacea; those due to variations in time of metamorphosis (pattern 3), by brittle stars. Early hatching and late metamorphosis yield the same result, namely, prolonged larval periods. Pattern 4 results from extremely late hatching (as in some crustacea), pattern 5, from extremely early metamorphosis (as in some brittle stars); both again produce the same result, i.e., complete suppression of larval periods.

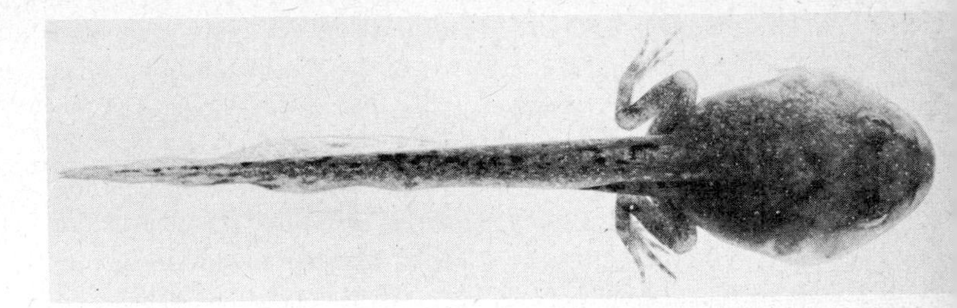

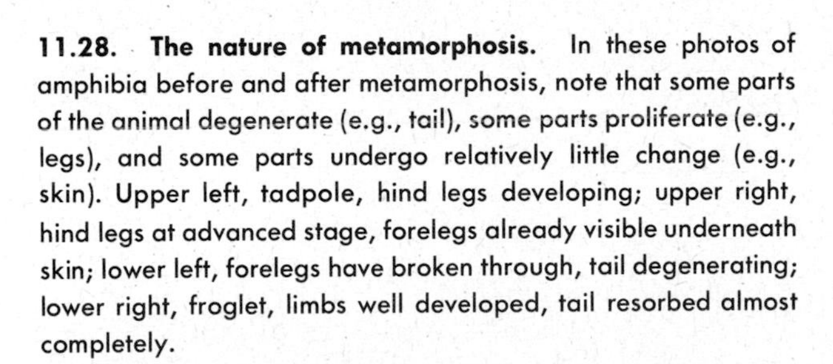

11.28. The nature of metamorphosis. In these photos of amphibia before and after metamorphosis, note that some parts of the animal degenerate (e.g., tail), some parts proliferate (e.g., legs), and some parts undergo relatively little change (e.g., skin). Upper left, tadpole, hind legs developing; upper right, hind legs at advanced stage, forelegs already visible underneath skin; lower left, forelegs have broken through, tail degenerating; lower right, froglet, limbs well developed, tail resorbed almost completely.

phase down to the embryo. Although the result in both cases is the same, i.e., a maximum compression, suppression, or abbreviation of the larval period, the mechanism of achieving such a result is not the same (Fig. 11.27).

The relative timing of metamorphosis also influences the general nature of that transformation. As just noted above, the later metamorphosis occurs, the more purely larval body parts will have developed. In other words, all larvae are *composite* organisms to a greater or lesser extent: they contain parts serving more or less exclusively during larval life only, other parts serving during adult life only, and still other parts which serve both in the larva and in the adult. For example, the ectodermal locomotor systems of plutei and trochophores are purely larval structures, but the precursor tissues of the reproductive systems will be purely adult; and the alimentary apparatus serves both the larva and the adult. Analogously, a frog tadpole possesses gills and a tail for larval life, the beginnings of lungs and limbs for adult life, and nervous, alimentary, and other parts functioning in the larval as well as the adult phase.

Accordingly, metamorphosis usually includes three general kinds of processes: resorption or disintegration of purely larval components, rapid proliferation and differentiation of purely adult components, and continuation of growth of all other components (Fig. 11.28). It follows also that metamorphosis will entail more radical reorganization of the body, the more purely larval and the fewer purely adult parts there are present. Clearly, if a pluteus is without arms to begin with, metamorphosis of the epidermis will require far less tissue resorption than if the larva possessed eight arms. Two factors determine to what extent the larval and the adult tissues have developed. One is the timing of metamorphosis as such, as outlined above. Another is the *rate* of development of the various body parts.

For example, assume an extreme condition in which the developmental rates of the larval and adult components are nearly equal. All parts of a larva will then mature simultaneously, and it will actually be impossible to distinguish between "larval" and "adult" components; the whole larva will become the adult. In such instances, therefore, metamorphosis will not include drastic tissue destruction and the transformation of larva into adult will tend to occur smoothly and gradually. Metamorphoses of this kind, called *gradual* or *incomplete* metamorphoses, are encountered, for example, among the crustacea (cf. above), the flightless silverfish insects (which hatch late and thus in effect are without larval periods to begin with), and the comparatively simply constructured *planula* larvae of coelenterates (Fig. 11.29). Such larvae consist of a ciliated layer of ectoderm and an interior mass of endoderm which is often solid originally but which later acquires a gastrocoel by cavitation. At metamorphosis, a planula settles and, as in other larvae of this category, all tissues present become part of the adult.

Assume next an intermediate condition in which purely

11.29. Types of metamorphosis. Graphs on left indicate the comparative durations and degrees of body reorganization during metamorphosis (met), and also the relative amounts of transient larval and prospective adult tissues (shaded areas) composing a larva and later adult. Specific examples are sketched at right. **A.** If the larva consists entirely of tissues which are essentially part of the later adult, then metamorphosis will be so gradual and prolonged as to be almost unnoticeable. **B.** If the larva consists partly of purely larval and partly of future adult tissues, then metamorphosis will be somewhat less gradual than in **A** and will also involve a certain amount of body reorganization (dissolution of larval tissues, proliferation of adult tissues). **C.** If the larva consists very largely of purely larval tissues, metamorphosis will be a very abrupt and drastic transformation. **D.** If the larva is entirely without prospective adult tissues (opposite case of **A**), then the larva in effect will become permanent and the "adult" state requires only the later development of reproductive organs (neoteny).

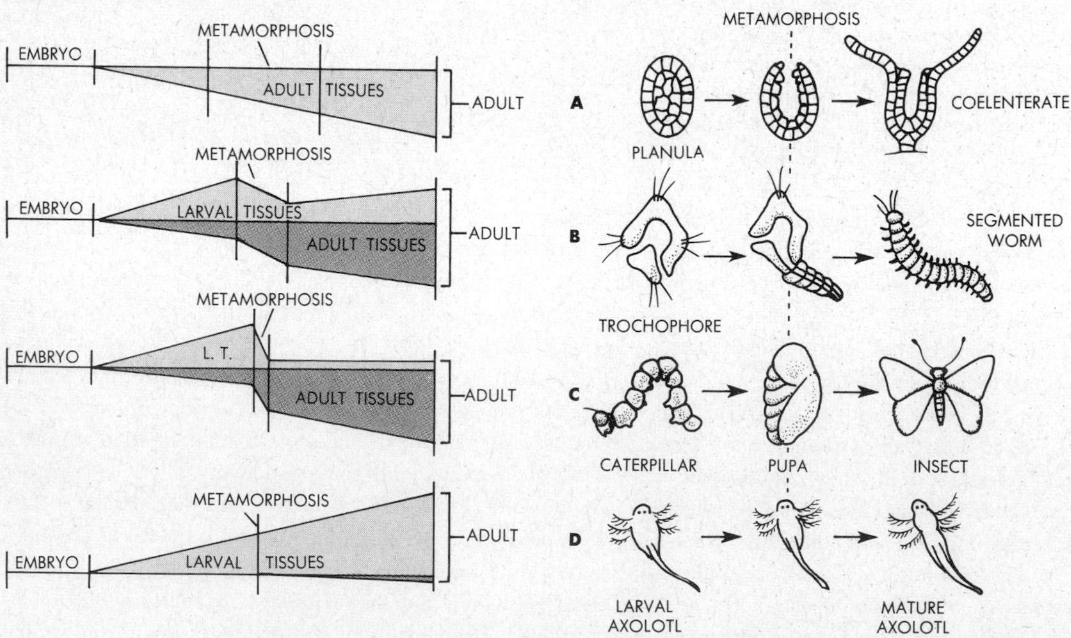

larval components develop at a substantially faster rate than future adult components. To this category belong most larval types, including, for example, the tadpoles of tunicates and frogs, the plutei of brittle stars, and the trochophores. In all of these, the purely larval tissues develop so rapidly that they are already senile when the purely adult tissues just begin to mature. Because they are senile, the larval tissues then die off during metamorphosis, whereas the young, immature adult tissues differentiate rapidly; and other tissues become part of the adult without major changes in growth rates (cf. Fig. 11.29). Thus, frog tadpoles become adults when senile gill and tail tissues (among others) disintegrate, when immature lung and limb tissues (among others) proliferate greatly, and when mesonephric tissues (again among others) continue to develop at substantially the same pace as before. Analogously, the metamorphosis of trochophores includes a disintegration of many ectodermal and most ectomesodermal components and a rapid development of the endomesodermal teloblasts. We note that metamorphoses of this type are partly smooth transitions, partly drastic reorganizations; they represent another variety of gradual or incomplete metamorphoses.

A far more drastic metamorphosis occurs if the developmental rate of the adult components is exceedingly slow relative to that of the larval components. In such cases, almost the whole larva consists of purely larval tissues; and at a time when all larval parts are already senile, the adult parts may not even have begun to develop at all. Metamorphosis then will entail an almost total dissolution of the larva (cf. Fig. 11.29). Highly precipitous transformations of this kind are exemplified well in, for example, the caterpillars of insects. These are composed almost entirely of purely larval tissues. Future adult tissues are represented only by small, undeveloped plaques of cells, so-called *imaginal disks,* located in various regions of the larval body. At metamorphosis a caterpillar forms a cocoon around itself, and in this *pupa* virtually the whole caterpillar disintegrates. The imaginal disks thus come to lie in a more or less loose aggregation of larval debris. With the raw materials supplied by the debris, the imaginal disks then proliferate rapidly and develop into the adult flying insect. The latter eventually extricates itself from the pupal cocoon (cf. also Fig. 28.33). Larvae which become adults by drastic and abrupt reorganizations as in caterpillars are said to undergo *complete* metamorphosis.

The most extreme result would be obtained if the developmental rate of the adult tissues were comparatively so slow that it would in effect be zero. Such tissues then would not mature at all, metamorphosis would not occur, and the larval condition would become permanent: the adult stage would remain suppressed (cf. Fig. 11.29). This direct opposite to the total suppression of a larva is encountered, for example, in salamanders such as axolotls. In amphibia generally, growth of larval tissues and metamorphosis are controlled by the hormone of the thyroid gland. However, in axolotls the thyroid mechanism does not become operational, and the animals therefore remain aquatic larvae permanently, with gills but without lungs (cf. Chap. 32). One adult component later does escape the growth suppression, viz., the reproductive system. Thus, the animals may breed despite their otherwise larval state. That axolotls are indeed *larval* animals can be proved in one species by the injection of thyroid hormone. Such individuals metamorphose immediately; i.e., they lose their gills and become lung-breathers instead. Permanent retention of larval traits specifically and of youthful traits generally is called *neoteny.* The axolotl example above suggests that neoteny might provide a means of creating essentially new types of animals. Indeed, as will be shown in Chap. 14, neoteny is believed to have been an exceptionally important mechanism in the evolution of major new categories of animals.

We may now formulate a third principle; viz., the forms and characteristics of larvae and of metamorphoses differ according to differences in the relative developmental rates of larval and adult body parts. Note again that, as already pointed out in Chap. 8, larval growth and metamorphosis are known to be under specific hormonal control not only in vertebrates but also in insects and in arthropods generally (cf. Chap. 28). Equivalent controls in other invertebrates are as yet far less clearly identified. In some cases, chemical stresses such as copper ions or acids (CO_2) in sea water can be shown to promote disintegration of senile larval tissues.

We may thus conclude quite generally that the form and nature of animal development as whole is determined by relatively few basic factors. Looming large among them are egg size, yolk content and yolk distribution, cleavage rate, and developmental rate (i.e., the rates of differentiation and morphogenesis). As these vary even slightly, the characteristics of embryos and of larvae may often come to vary greatly. Such developmental processes on the anatomic level undoubtedly are reflections of more fundamental events on the molecular level. It is known, for example, that many specific structural changes in embryos and larvae are accompanied by—and presumably are a result of—chemical changes such as appearance of new types of proteins, activation of new enzyme systems, or establishment of new metabolic pathways. In due time probably all gross developmental events will be shown to be based on fine changes in biochemical patterns. And there can be little question even now that all the molecular determining factors are in turn controlled by the genetic complement of the fertilized egg. This consideration leads us to the last and undoubtedly most fundamental ingredient of animal continuity, namely, the mechanism of genetic inheritance.

REVIEW QUESTIONS

1. Define morphogenesis, differential growth, form-regulating movements, polarity. Through what types of growth processes does an animal enlarge in size? Explain the meaning of the phrase "Animals grow from their molecules up."
2. What different types of symmetries are exhibited by living units? In what ways do polarity and symmetry circumscribe the form of an animal? Specify the polarity and symmetry of man. What is the role of differential growth in the development of form?
3. Define and distinguish between differentiation and modulation. What is the relation between differentiation and specialization? Cite examples of differentiative changes on the level of (*a*) molecules, (*b*) cells, (*c*) whole animals, and (*d*) societies. What kinds of changes within cells might bring about cytodifferentiation?
4. What role does metabolism play in development? How does metabolic rate vary during the developmental history of an animal? In what different ways may an incompletely developed reproductive unit acquire (*a*) nutrients and (*b*) respiratory gases? What cellular metabolic capacities (*a*) are and (*b*) are not in existence in a zygote?
5. Describe and define the principal developmental phases in the life history of an animal if this history (*a*) includes and (*b*) does not include a sexual process. What is the significance of the greater number of phases under condition *a*?
6. Define karyogamy, plasmogamy, blastomere, organ-forming zones. Distinguish between mosaic and regulative eggs. How can it be established by experiment whether a given egg is mosaic or regulative? In which animals do each of these egg types occur? How are twins formed? Distinguish between identical and fraternal twinning. Can identical twinning take place in mosaic eggs?
7. Show how development is influenced by egg size, cleavage rate, yolk content, yolk distribution. What events occur during the cleavage of an egg? What types of holoblastic and meroblastic cleavage patterns can be distinguished? Which animal groups exemplify each type? Describe the patterns of radial, bilateral, and spiral cleavage. What is a 4d cell? A macromere? A micromere?
8. What basic types of blastulae may be distinguished? In what general ways does gastrulation occur? What are blastocoels and gastrocoels? Review the processes of mesoderm- and coelom-formation in different animal groups. How do the characteristics of an egg influence the subsequent process of gastrulation in a mechanical sense? Review the pattern of gastrulation in amniote vertebrates and show how the extraembryonic membranes form.
9. By what general processes of morphogenesis do the primary germ layers develop into adult structures? Illustrate this in the development of the vertebrate nervous system and the eye. What differentiative role does induction play in such transformations? Again illustrate in the development of the nervous system and the eye and describe supporting experiments.
10. What are the general functional roles of larvae? Name and describe some of the larval types encountered among animals. What is direct and indirect development? List and describe the larval types of crustacea, and show how the relative time of hatching influences the nature of a larva. Show analogously how the relative time of metamorphosis influences the nature of a larva. What developmental factors determine whether metamorphosis will be complete or incomplete? What is neoteny?

COLLATERAL READINGS

The following sources may be consulted for additional information on topics discussed in this chapter:

Balinsky, B. I.: "An Introduction to Embryology," Saunders, Philadelphia, 1960.

Barth, L. J.: "Development," Addison-Wesley, Reading, Mass., 1964.

Berrill, N. J.: "Growth, Development, and Pattern," Freeman, San Francisco, 1961.

Brachet, J.: "Chemical Embryology," Interscience, New York, 1950.

Ebert, J. D.: "Interacting Systems in Development," Modern Biology Series, Holt, New York, 1965.

Gabriel, M. L., and S. Fogel: "Great Experiments in Biology," section on Embryonic Differentiation, Prentice-Hall, Englewood Cliffs, N.J., 1955.

Hyman, L.: "The Invertebrates," vol. 1, pp. 255–263, McGraw-Hill, New York, 1940.

McElroy, W. D., and B. Glass: "The Chemical Basis of Development," Johns Hopkins, Baltimore, 1958.

Nelsen, O. E.: "Comparative Embryology of the Vertebrates," Blakiston, New York, 1953.

Raven, C. P.: "An Outline of Developmental Physiology," McGraw-Hill, New York, 1954.

Sussman, M.: "Animal Growth and Development," Foundations of Modern Biology Series, Prentice-Hall, Englewood Cliffs, N.J., 1960.

Waddington, C. H.: "Principles of Embryology," G. Allen, London, 1956.

Willier, B. H., P. A. Weiss, and V. Hamburger: "Analysis of Development," Saunders, Philadelphia, 1955.

The books and articles listed below give popular accounts of development and related topics:

Bonner, J. T.: "Morphogenesis," Princeton, Princeton, N.J., 1952.

Dahlberg, G.: An Explanation of Twins, *Sci. American,* Jan., 1951.

Fischberg, M., and A. W. Blackler: How Cells Specialize, *Sci. American,* Sept., 1961.

Frieden, E.: The Chemistry of Amphibian Metamorphosis, *Sci. American,* Nov., 1963.

Gray, G. W.: The Organizer, *Sci. American,* Nov., 1957.

Moog, F.: Up from the Embryo, *Sci. American,* Feb., 1950.

Moscona, A. A.: How Cells Associate, *Sci. American,* Sept., 1961.

Puck, T. T.: Single Human Cells in Vitro, *Sci. American,* Aug., 1957.

Singer, M.: The Regeneration of Body Parts, *Sci. American,* Oct., 1958.

Tyler, A.: Fertilization and Antibodies, *Sci. American,* June, 1954.

Waddington, C. H.: How Do Cells Differentiate?, *Sci. American,* Sept., 1953.

Wigglesworth, V. B.: Metamorphosis and Differentiation, *Sci. American,* Feb., 1959.

CHAPTER 12 heredity

Like sex, heredity has adaptive value, for what an animal inherits will determine its survival potential in large measure. It should be clear here that animals do *not* inherit strong muscles, sensitive hearing, red blood, or any other trait. Animals inherit *genes*, not traits. Moreover, they do not inherit genes only, but all the contents of reproductive units, i.e., whole *cells*. Visible traits then *develop* in an offspring, under the control of inherited genes and within the limitations imposed by given intracellular and extracellular environments.

The key problem in studies of heredity is to explain the inheritance of *likeness* and of *variation:* how an offspring usually comes to resemble its parents in certain major respects but differs from the parents in many minor respects. Are such hereditary patterns in any way regular and predictable, and if so, what are the underlying principles? Answers can be found by examining the traits of successive generations of animals and inferring from the visible likenesses and variations what the inheritance of the genes has been. The first important studies of this sort were made in the last half of the nineteenth century by the Austrian monk Gregor Mendel. He discovered two basic rules of inheritance which laid the foundation for all later advances in understanding of processes of heredity.

Accordingly, this discussion of heredity will include an examination of the general relationship between *genes* and *traits*, an account of the rules of *Mendelian inheritance*, and a survey of the main aspects of *non-Mendelian inheritance* brought to light since the time of Mendel.

GENES AND TRAITS

The pattern of inheritance varies according to whether reproduction is *uniparental* or *biparental.* If an animal is produced by a single parent, as in vegetative reproduction, the genes of the parent are passed on unchanged to the offspring. In uniparental reproduction, therefore, offspring and parent are genetically identical and usually display the same visible traits. The only source of genetic variation in such cases is *mutation*. For example, if some gene in a cell forming a bud undergoes a mutational change, then, and only then, may the offspring become genetically different from the parent. Trait variations may then also be displayed.

By contrast, *two* sources of genetic variation exist in cases of biparental, gametic reproduction. One is again mutation, in this instance germ mutation in one or both gametes (cf. Chap. 6). The other is a direct result of sex; two sets of genes are pooled in the zygote. The genetic endowment of the offspring consequently may differ from that of either parent. Through such *sexual recombination* of genes, the offspring may become unlike the parents (Fig. 12.1).

We conclude, for both uniparental and biparental reproduction, that likeness to parents will be inherited to the extent that the genes of the offspring are the same as those of the parents; and that variation will be inherited to the extent that mutation, recombination, or both, have changed the genes of the offspring (Fig. 12.2).

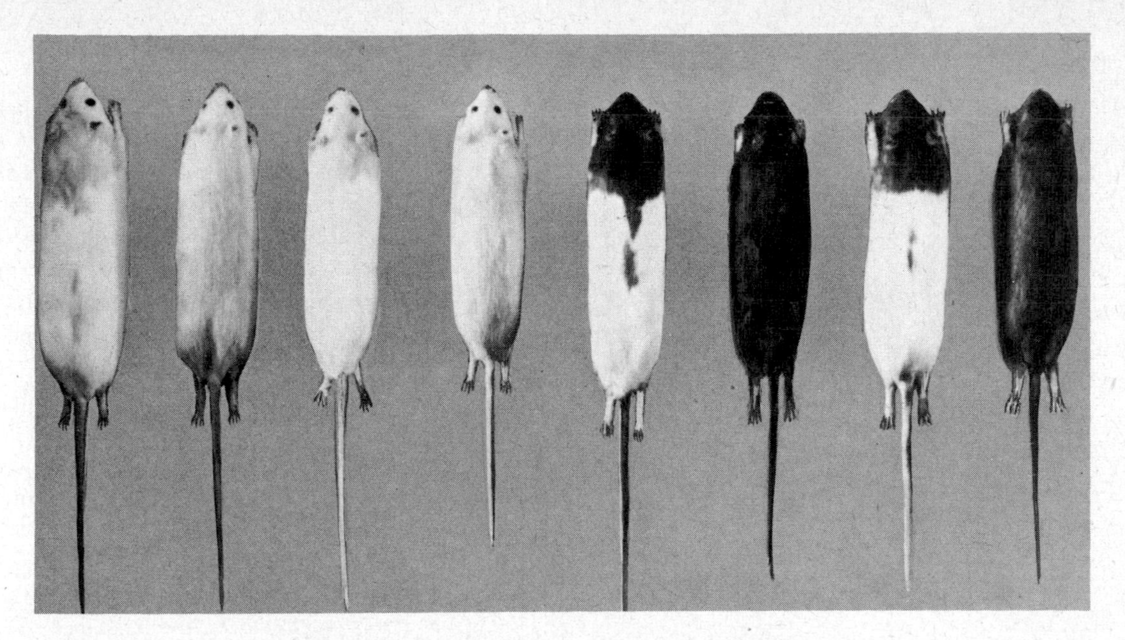

12.1. Inheritable, gene-controlled trait variations. These are litter-mate rats, produced by the same two parents. Considerable variation in coat color is evident. Such differences among offspring are due to the gene-shuffling effects of sex, and as a consequence even brothers and sisters of the same family may be different genetically. Variant traits of this type are superimposed on common, more basic traits; e.g., despite the color differences all the litter mates above are distinctly rats and, indeed, members of a single breeding line with specific fine characteristics.

But inheriting a certain gene is not automatically equivalent to developing a certain trait; the development of traits is affected by the environment. Genes supply a reasonable promise, as it were, and the total environment of the genes subsequently permits or does not permit the translation of promise into reality.

The environment of genes includes, first of all, other genes. Indeed, gene interactions are exceedingly common (cf. below). The functional integration of genes in a cell is actually so intimate and so complex that it becomes relatively meaningless to speak of "a" gene as if it were an independently acting particle. Only the interacting totality of genes in a cell, called the *genome*, has functional reality.

The environment of genes also includes the cell cytoplasm, and it too influences the development of traits in major ways. We know that, in the cytoplasm, genes indirectly exercise their basic function, i.e., they control protein synthesis. We may regard proteins as the *primary* traits of a cell, and indeed we may formulate an acceptable definition of "gene" on this basis: a gene is that minimum section of a chromosome which controls the synthesis of a single type of protein molecule. Genes thus may be considered to be *units of biochemical action*. In the form of enzymes or structural components, the proteins manufactured under gene control then bring about the development of various *secondary*, often visible traits. But such traits usually differ in different cell types even though the genes are the same in all. At a given time, for example, different genes may be active in different cells, the result then being different cytoplasmic traits. Or again, all cells of man possess eye-color genes, for example, but only iris cells actually develop the color. Evidently, the cytoplasms of different cells may react differently to the genes they contain, and trait expression will then differ correspondingly. The various developmental differentiations and specializations of cells and larger body parts are the result.

We may therefore distinguish between *inherited* traits, con-

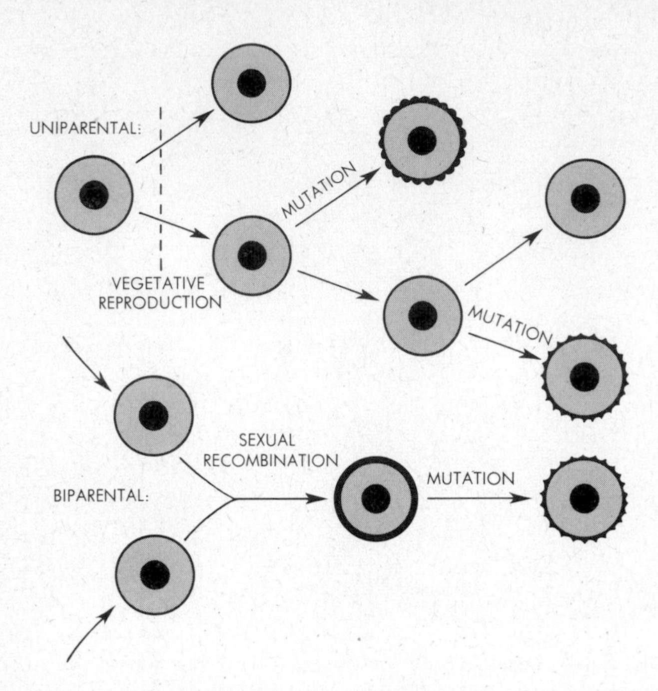

12.2. The sources of genetic variations. In uniparental inheritance the only source is mutation. In biparental inheritance genetic variations may arise both by mutation and by sexual recombination of genes.

trolled by genes, and noninherited *acquired* traits, superimposed on the inherited ones and produced by nongenetic environmental and developmental effects. And we are led to the fundamental conclusion that visible traits are always a product of inherited genes *and* of the nongenetic environment. To the extent that variations of traits may be advantageous to an animal in its way of life, heredity, like sex, has adaptive value.

Note here that, unless we refer to a particular cellular protein synthesized under gene control, we must exercise caution in designating a given characteristic of an animal as "a" trait. For example, the hereditary characteristic of disease resistance is not really "a" trait. Inasmuch as it is a functional property of a whole animal, it is a composite property of millions of cooperating cells. Each of these contributes some particular function to the total trait, and specific genes in each of these cells control each of these functions. Therefore, disease resistance must be a combination of perhaps millions of different cellular activities controlled by a large, equally unknown number of genes.

In very many instances, what is normally regarded as one trait is of such composite nature. Body size, the fine structure and the functional capacities of organs, general vigor, intelligence, fertility, and many others—all are interaction products of several dozens or hundreds or thousands of different genes. Indeed, there is reason to believe that most of such highly composite traits are controlled by the collective action of possibly all genes of an animal, each contributing a tiny effect to the total trait. Such traits may then be expressed in a correspondingly great variety of ways. As is well known, for example, the expression of traits like body size or intelligence in different individuals may range from one extreme to another, through enormously varied series of intergradations.

But if all genes contribute to the control of a trait like body size, for example, and if all genes also control disease resistance and other highly composite traits, then any one gene clearly must contribute to the control of more than one trait. We are led to the generalization that *one composite trait may be controlled by many genes, and one gene may contribute to the control of many composite traits* (Fig. 12.3).

In some instances, the functional relation between genes and traits is comparatively less complex. We know from Chap. 6 that the general pattern of gene action within a cell may be symbolized by the sequence: gene ⟶ enzyme protein ⟶ reaction ⟶ reaction product. Sometimes such a reaction product does not participate in the elaboration of a more complex trait but constitutes a final trait itself. For example, a pigment produced within a cell is a gene-controlled reaction product, and it is often a visible *end*product, a final trait. In cases of this sort, a readily specifiable unit trait is correlated directly with one particular gene (Fig. 12.4). From the pattern in which such a trait is expressed visibly in successive generations, one may readily infer the pattern of gene inherit-

12.3. The principle that one trait (for example, *D*) is controlled by many genes and that one gene (for example, *B*) controls many traits.

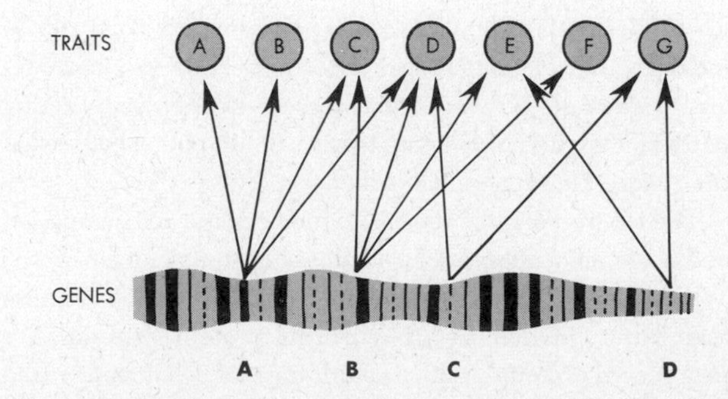

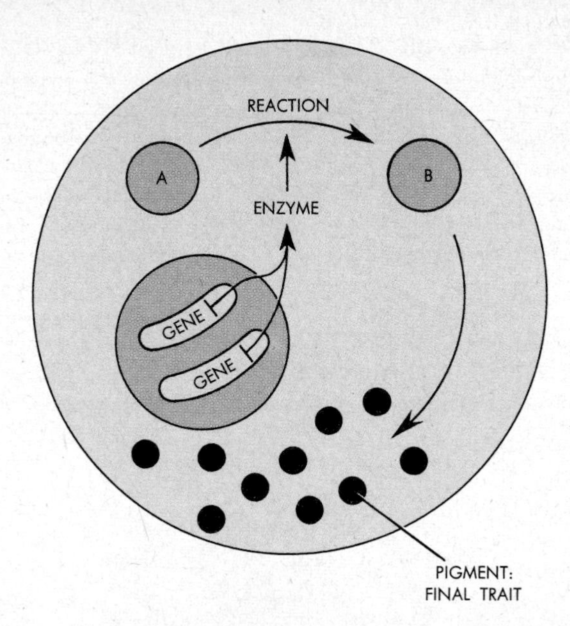

12.4. In certain special cases, one gene is known to control just one trait. A cytoplasmic pigment may be final trait, and a single gene may govern the manufacture of the enzyme which in turn controls the metabolic reaction producing such a pigment.

ance. It was from studies of just such color traits (in plants) that Mendel deduced his two rules of heredity. If he had happened to investigate, instead, any of the numerous composite traits of organisms, then regularities in hereditary patterns would not have been clearly apparent and his name might not be immortal today.

MENDELIAN INHERITANCE

The Chromosome Theory

The fruit fly *Drosophila* has long been among the most widely used animals in genetic research. The flies can be bred easily; offspring are sexually mature after 3 weeks; the chromosome number is desirably small ($2n = 8$); and the larval salivary glands contain giant chromosomes large enough to permit detailed microscopic study of chromosome structure (Fig. 12.5). Like Mendel's plants, moreover, the flies exhibit numerous variant forms of given unit traits. By mating animals displaying such trait variations it is possible to investigate the patterns of gene inheritance. For example, the trait of body pigmentation is expressed in at least two alternative forms. In one, the general coloration of the animal is gray and the abdomen bears thin transverse bands of black melanin pigment. A gray body constitutes the *wild type,* i.e., the predominant type of body color in nature. By contrast, some flies are pigmented black uniformly all over the integument, a variant form of body coloration referred to as the *ebony* trait (cf. Fig. 12.5).

If two gray-bodied wild type flies are mated, all offspring produced are also gray-bodied. Indeed, all later generations again develop only wild type colorations. Analogously, a mating of two ebony flies yields ebony offspring in all subsequent generations. Gray and ebony body colors here are said to be *true-breeding* traits (Fig. 12.6).

In Mendel's time, it was generally supposed that if alternative forms of a trait are cross-bred, a *blending* of the traits would result. Thus, if gray and black were mixed together, like paints, a dark-gray color should be produced. And if blending really occurred, dark-gray should subsequently be true-breeding as well; for mixed traits, like mixed paints, should be incapable of "unblending." In reality, however, the results of cross-breeding are strikingly different. When a wild type and an ebony fly are mated (parental generation, P), all offspring (first filial generation, F_1) are gray-bodied, exactly like the wild type parent. Furthermore, when two of such gray-bodied F_1 flies are subsequently mated in their turn, some of the offspring obtained are gray-bodied, others are ebony; color mixtures do not occur. Numerically, some 75 per cent of the second (F_2) generation are gray-bodied, like their parents and one of their grandparents, and the remaining 25 per cent are ebony, unlike their parents but like the other grandparent (Fig. 12.7).

Evidently, cross-bred color traits do *not* blend or mix, and the traits of the offspring do not breed true: from gray-bodied flies in the F_1 can arise ebony flies in the F_2. Large numbers of tests of this kind have clearly established that, quite generally for any trait, blending inheritance does not occur and that traits remain distinct and intact. They may become joined together in one generation, but they may then again become separated, or *segregated,* from one another in a following generation. Mendel was the first to reach such a conclusion from his studies on plants, and this denial of blending was his most significant contribution. It ultimately reoriented the thinking about heredity completely and paved the way for all modern insights. Mendel himself supplied the first of such insights, for he not only negated the old interpretation but also postulated a new one.

12.5. Chromosomes. Photos: A. Portion of the larval salivary gland of the midge *Chironomus,* prepared to show the cell nuclei with their giant chromosomes. **B.** Isolated chromosomes corresponding to those in **A.** Chromosomal cross-banding apparent in such preparations is encountered also in the giant salivary chromosomes of other insects, e.g., fruit flies, and indeed is characteristic of all kinds of chromosomes generally. **C** and **D.** Isolated chromosomes from cells of a human male and a human female, respectively. In each case the photo shows 46 chromosomes (or 23 pairs), each already duplicated as a preliminary to cell division. **Diagrams: A.** The chromosomes of a diploid ($2n = 8$) *Drosophila* cell (female). **B.** The phenotypic appearance of a female gray-bodied and an ebony-bodied fruit fly.

A

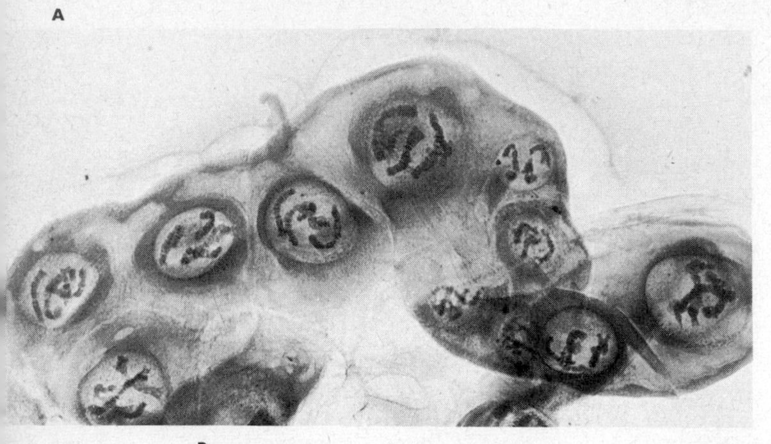

B

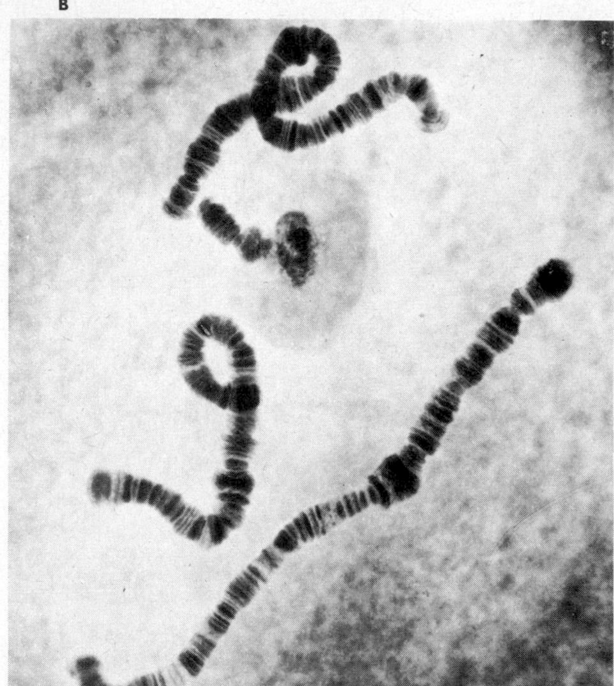

B, *courtesy Drs. D. F. Poulson and C. W. Metz, from fig. 1,* J. Morph., *vol. 63, p. 366.*

C

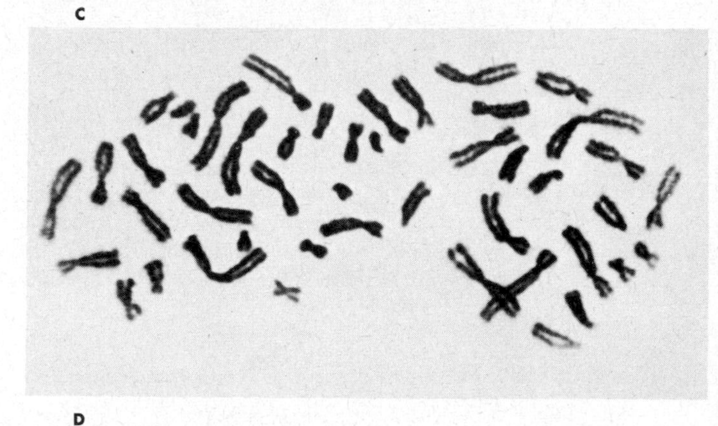

D

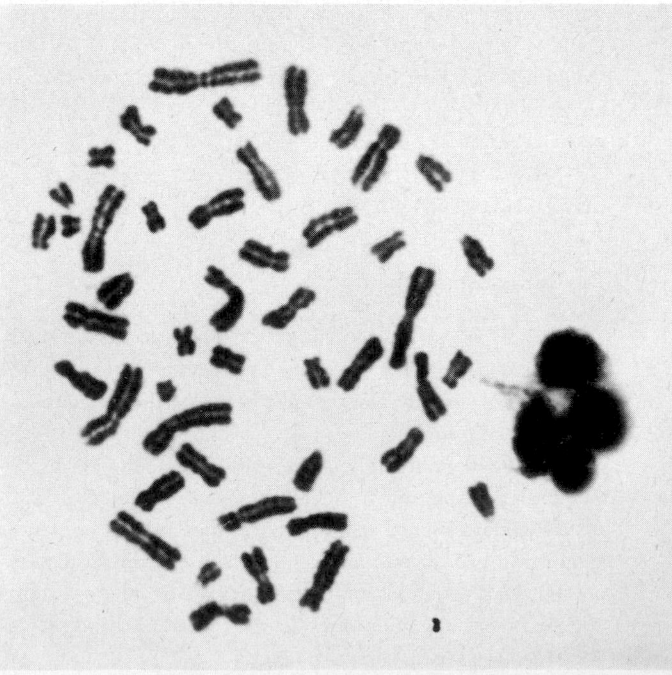

C,D, *courtesy Carolina Biological Supply Company.*

A

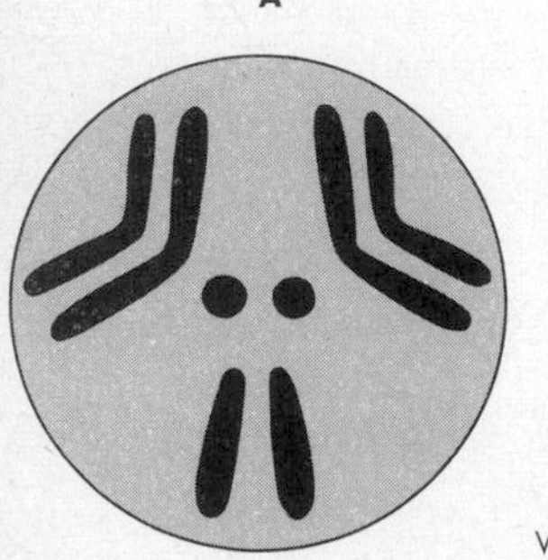

B

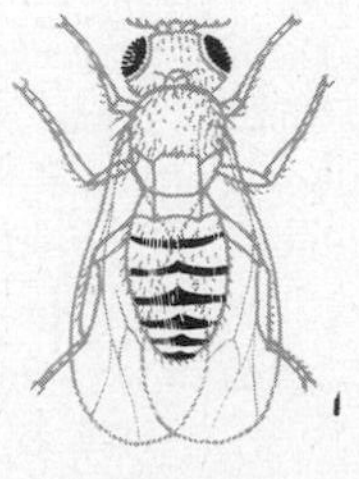

WILD-TYPE GRAY BODY (E)

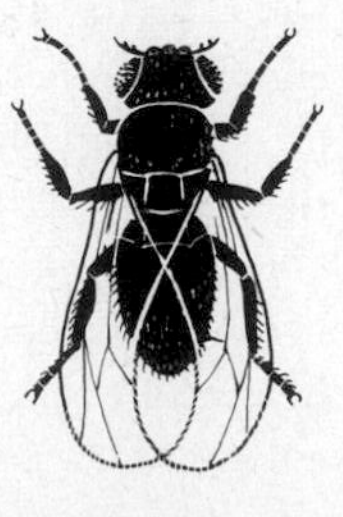

EBONY BODY (ee)

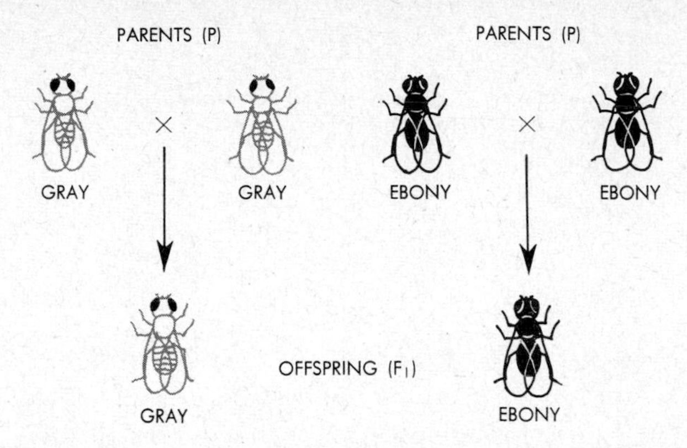

12.6. True-breeding in *Drosophila*. If two gray-bodied (wild type) flies are mated, all offspring will be wild type (left); and if two ebony flies are mated, all offspring will be ebony (right).

He realized that traits trace back to the sperm and the egg which produce an organism, and he suspected that some specific components within the gametes controlled the later development of traits. Mendel called these hypothetical components "factors." For any given trait, he argued, an organism must inherit at least one factor from the sperm and one from the egg. Therefore, the offspring must possess at least two factors for each trait. When that offspring becomes an adult and produces gametes in its own turn, each gamete must similarly contribute one factor to the next generation. Hence, at some point before gamete production, two factors must be reduced to one. Mendel consequently postulated the existence of a factor-reducing process.

With this he in effect predicted the occurrence of meiosis. When near the end of the nineteenth century meiosis was actually discovered, it was recognized that the reduction of chromosomes at some point before fertilization matched precisely the postulated reduction of Mendel's factors. Chromosomes then came to be regarded as the carriers of the factors, and the *chromosome theory of heredity* thus emerged. This theory has since received complete confirmation, and Mendel's factors became the genes of today.

The Law of Segregation

On the basis of the chromosome theory, we may interpret the fruit fly data above as follows. A true-breeding wild type fly possesses a pair of gray-color–producing genes in each cell. These genes, which may be symbolized by the letters *EE* (or more usually as $++$) are located on a given pair of chromosomes, one of which is maternal and one paternal in origin. We say that the *genotype*, or gene content, of the fly is *EE* (or $++$) and that the *phenotype*, or visible appearance, is gray. Before such an animal produces gametes, meiosis occurs. Mature gametes therefore contain only one of the two chromosomes, hence only one of the two genes (Fig. 12.8).

Note that it is entirely a matter of chance which of the two adult chromosomes will become incorporated into a given gamete. Since both adult chromosomes here carry the same color gene, all gametes will be genetically alike in this respect. We may understand now why *EE* animals are true-breeding, i.e., why a mating of $EE \times EE$ will produce only gray-bodied offspring (Fig. 12.9).

In precisely analogous manner, we may symbolize the genotype of a true-breeding ebony fly as *ee*. A mating of two such flies will yield only black-bodied offspring (Fig. 12.10).

If we now mate a wild type and an ebony fly, we already know that all offspring will be gray-bodied (Fig. 12.11). In such offspring, the *E* and *e* genes are present together, yet the effect of the *e* gene evidently is suppressed completely. In other words, the single gene *E* by itself exerts the same effect as two *E* genes; the trait produced by *E* and *e* is indistinguish-

12.7. If a gray-bodied fly is mated with an ebony fly (P generation), all offspring will be gray-bodied (F_1 generation). And if two of the F_1 flies are then mated in turn, the offspring will be gray-bodied and ebony in the ratio shown.

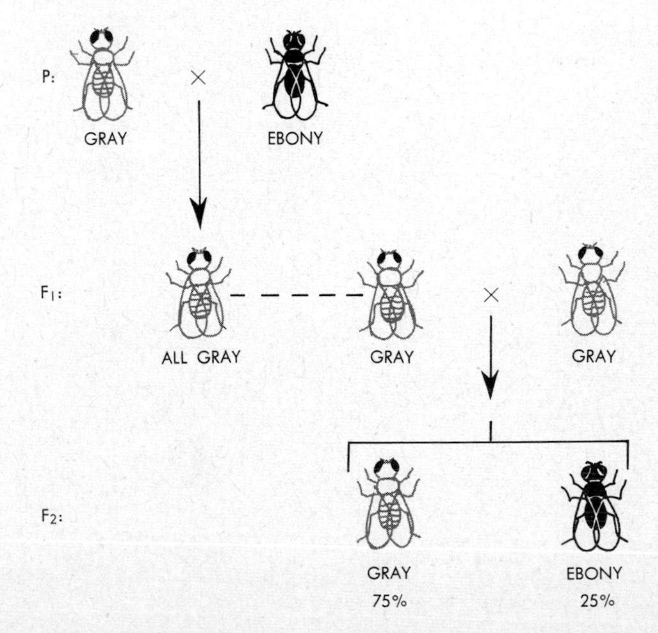

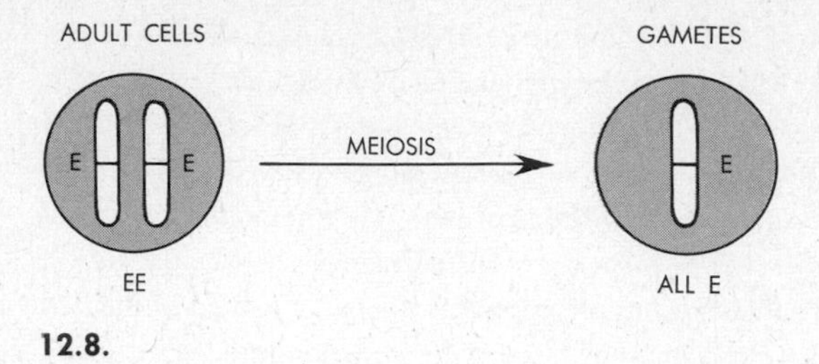

12.8.

able from that produced by *EE*. By contrast, the single gene *e* by itself does not exert the same effect as that produced by *ee*. We may say that *E* has a maximum effect even in single dose, but that *e* has an effect only in double dose, i.e., in the absence of *E*. Genes like *E*, which exert a maximum effect in single dose are said to produce *dominant* traits. Such genes mask more or less completely the effect of correlated genes such as *e*, which are said to produce *recessive* traits.

Genes affecting the same trait in different ways and occurring at equivalent locations in a chromosome pair are called *allelic* genes, or *alleles*. Genes such as *E* and *e* are alleles, and pairs such as *EE*, *ee*, and *Ee* are different allelic pairs. If both alleles of a pair are the same, as in *EE* or *ee*, the combination is said to be *homozygous*. The combination *EE* is *homozygous dominant*, the combination *ee*, *homozygous recessive*. A *heterozygous* combination is one such as *Ee*, in which one allele produces a dominant and the other a recessive trait.

12.9. **12.10.**

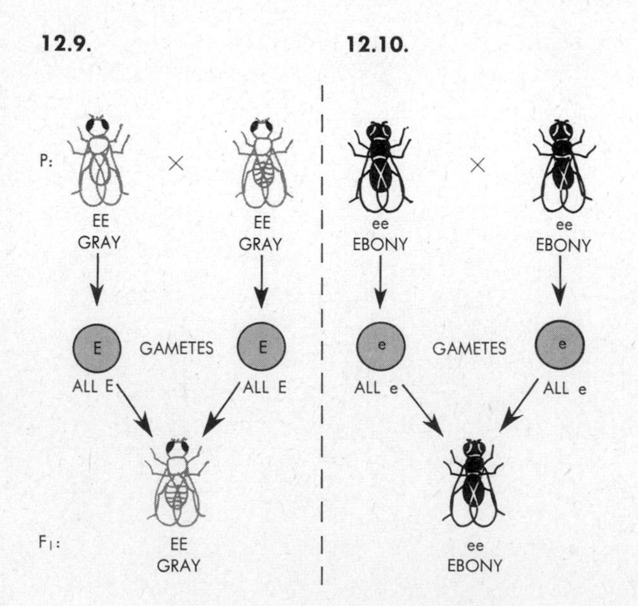

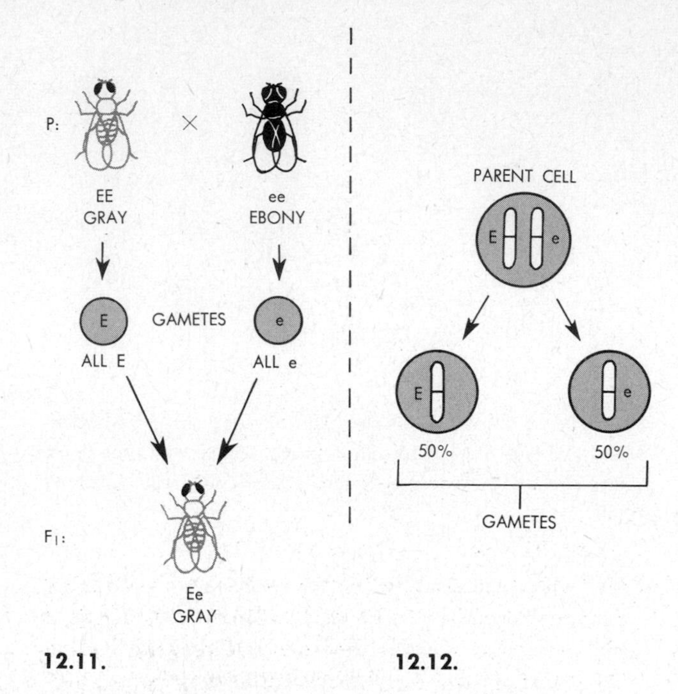

12.11. **12.12.**

Thus, the F_1 resulting from a mating of a wild type and an ebony fly as above is heterozygous, and this F_1 reveals that the wild type trait is dominant over the ebony trait. That the heterozygous F_1 condition is not a true-breeding blend is now shown if two F_1 flies are mated. After meiosis, each fly will give rise to two types of gametes. Given the genes *Ee*, either the *E* gene or the *e* gene could become incorporated into any given gamete. What actually happens in each specific case is determined by chance. Hence if, as is usually the case, large numbers of gametes are produced, each possibility will be realized with roughly equal frequency. Consequently, approximately 50 per cent of the gametes will carry the *E* gene, the other 50 per cent the *e* gene. We may write: *Ee* parent $\longrightarrow$ 50 per cent *E* gametes, 50 per cent *e* gametes (Fig. 12.12).

Now fertilization occurs. There are two genetically different sperm types and two genetically different egg types, and it is wholly a matter of chance which of the two sperm types fertilizes which of the two egg types. If many fertilizations occur simultaneously, as is usually the case, then all possibilities will be realized with appropriate frequency (Fig. 12.13).

We note that three-quarters of the offspring are gray-bodied and resemble their parents in this respect. One quarter is ebony and these offspring resemble one of their grandparents. We may conclude that the visible results can be explained adequately on the basis of nonblending, freely segregating genes and the operations of chance. Offspring in ratios of ¾:¼ are usually characteristic for matings of heterozygous animals as above.

Not all genes occur in sharply dominant and sharply recessive alternative forms. It has been found that many allelic genes are neither dominant nor recessive. In such cases, *each* allele in a heterozygous combination such as *Aa* may exert a definite effect. The combined effects of an *A* and an *a* gene then usually result in a trait intermediate between those produced by *AA* and *aa* combinations. However, the intermediate result again is not an instance of blending, for the *Aa* condition is not true-breeding; a mating of *Aa* × *Aa* segregates *AA*, *Aa*, and *aa* offspring in a characteristic phenotype ratio of ¼:½:¼. The inheritance pattern of the genotypes in such cases is precisely the same as where genes are sharply dominant and recessive, and only the phenotype ratios are different (Fig. 12.14).

In modern terminology, Mendel's first law, the *law of segregation*, may now be stated as follows: *Genes do not blend, but behave as independent units. They pass intact from one generation to the next, where they may or may not produce visible traits, depending on their dominance characteristics. And genes segregate at random, thereby producing predictable ratios of traits in the offspring.* Implied in this law are chromosome reduction by meiosis and the operation of chance in the transmission of genes.

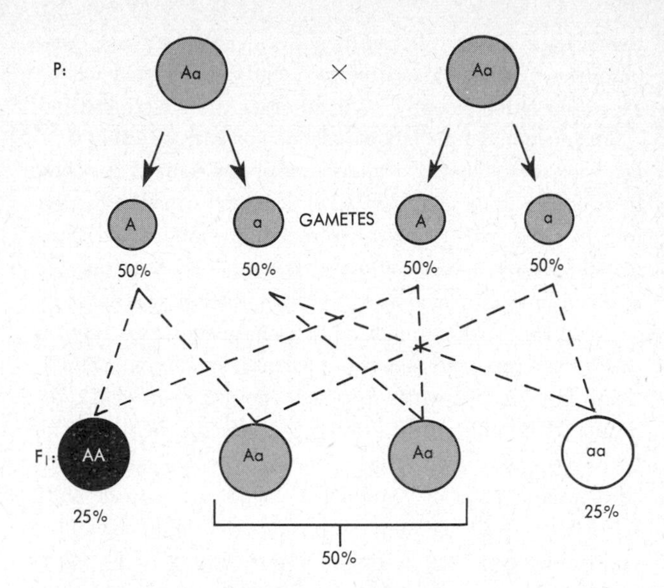

12.14. Partial dominance. If genes A and a each have their own definite effect on phenotype, i.e., if neither is completely dominant or recessive, then a mating of two heterozygous adults will produce an F_1 with phenotypes and genotypes in the ratios shown. Note that, in this F_1, only 50 per cent of the offspring are like the parents (the other 50 per cent resembling two of their grandparents; verify this).

12.13.

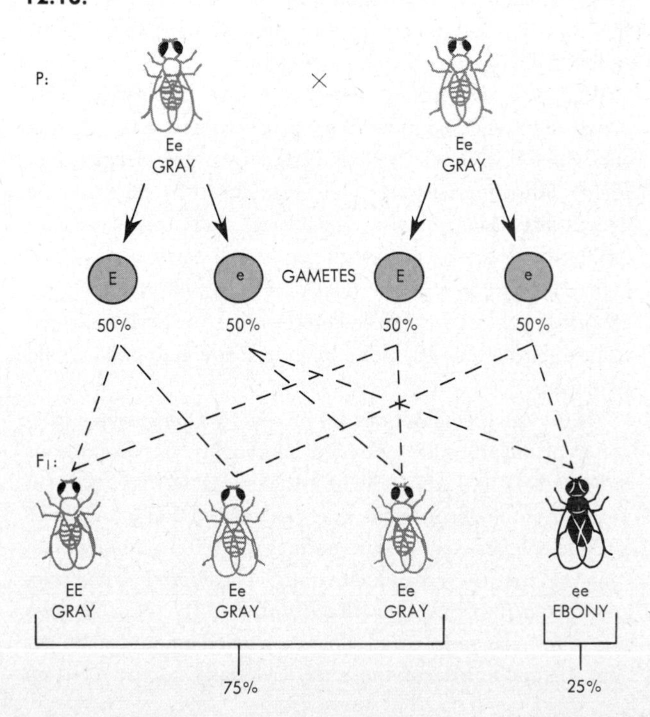

The Law of Independent Assortment

Animals do not express traits one at a time, but exhibit all their traits simultaneously. Analogously, genes are not inherited one at a time, but all of them are inherited together. Therefore, given certain parents, what will the offspring be like with respect to two or more simultaneous traits?

Mendel discovered a fundamental rule here. Phrased in modern terms, this *law of independent assortment* states: *The inheritance of a gene pair located on a given chromosome pair is unaffected by the simultaneous inheritance of other gene pairs located on other chromosome pairs.* In other words, two or more traits produced by genes located on two or more chromosome pairs "assort independently"; each trait will be expressed independently, as if no other traits were present.

The meaning of the law may be demonstrated readily if we analyze the simultaneous inheritance of two traits of fruit flies, *body color* and *wing shape*. We already know that body color may be either wild type gray or recessive ebony. Wing shape may be either normal or *vestigial*, i.e., the wings may be reduced in size to such an extent that the animal cannot fly (Fig. 12.15). Such stunted wings can be shown to develop whenever a recessive gene *vg* is present in homozygous condition, *vgvg*. Normal wings represent the dominant wild type, produced by either *VgVg* or *Vgvg* gene combinations. The body color and wing shape genes are located on different chromosome pairs of *Drosophila*. As in the case of body coloration, a mating of a normal-winged *VgVg* fly and a vestigial-winged *vgvg* fly results in a heterozygous F_1, *Vgvg*, which displays normal wings. A mating of two such F_1 flies then produces an F_2 composed of roughly 75 per cent normal-

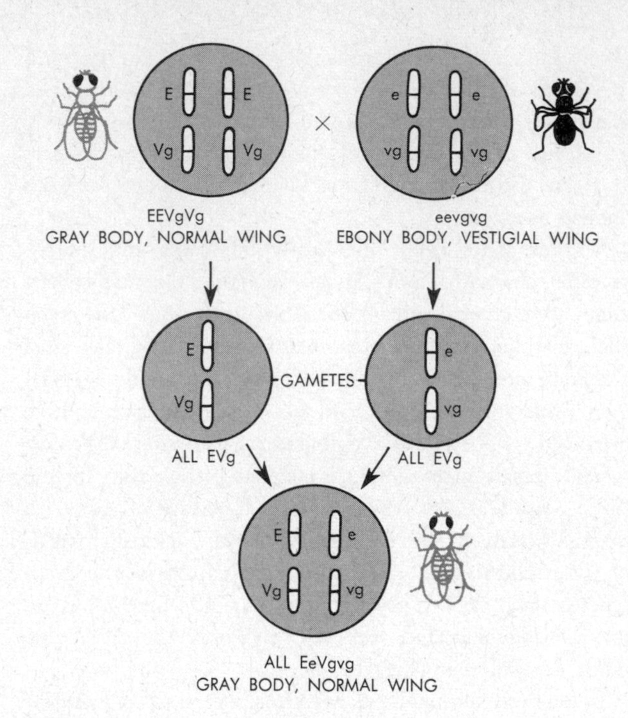

12.16.

12.15. Vestigial wings in fruit flies, as at upper right, are due to the recessive gene vg in homozygous condition. A mating as shown (P) thus yields a normal-winged F_1 (in which the vg gene is present, however). In the F_2, vg is segregated out in homozygous condition 25 per cent of the time, hence one-quarter of these offspring possess vestigial wings.

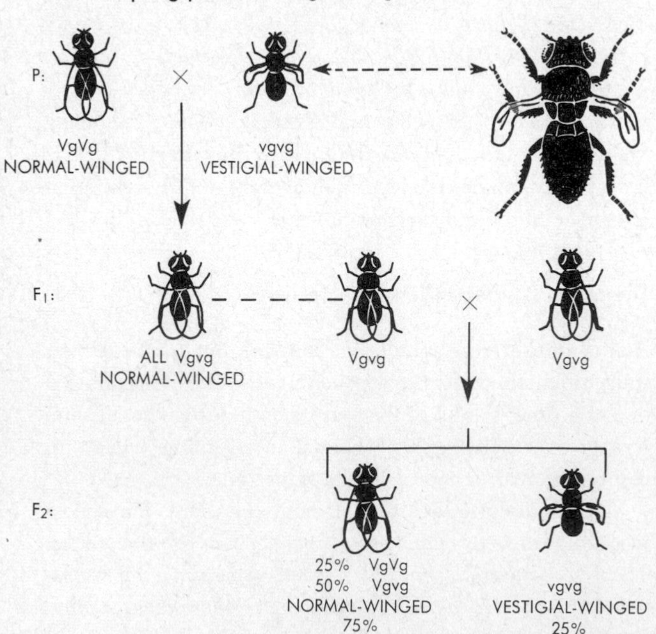

winged and 25 per cent vestigial-winged flies (cf. Fig. 12.15). In short, both the body color and the wing-shape traits individually obey the law of segregation.

What will now be the result of a mating between a fly which is gray-bodied as well as normal-winged, *EEVgVg*, and one which is ebony as well as vestigial-winged, *eevgvg* (Fig. 12.16)? After meiosis, all gametes produced by the wildtype fly will carry one gene of each pair, viz., *EVg*; and all gametes produced by the partner analogously will carry one gene of each pair, viz., *evg*. All fertilizations thus will yield *EeVgvg* offspring, i.e., the F_1 will be heterozygous for both traits and will exhibit both body color and wing shape in dominant wild type form.

If two such F_1 flies then are mated, four phenotypic categories of offspring are obtained, in roughly the proportions indicated in Fig. 12.17. Note here that a total of 75 (56 plus 19) per cent of the offspring are gray-bodied and that a total of 75 (56 plus 19) per cent are also normal-winged. In other words, each of the two dominant traits, considered *separately*, amounts to 75 per cent, or three-fourths, of the total. The two recessive traits, considered separately, each amount to 25 per cent, or one-fourth, of the total. Evidently, as expected on the basis of the law of segregation, each dominant and its correlated recessive appear in a ratio of ¾ : ¼; i.e., dominants are three times as abundant as recessives.

Moreover, the two dominants are also three times as abundant *even if they are considered together*. That is, among the 75 per cent total of gray-bodied offspring, 56 per cent, or very nearly three-fourths of 75, are at the same time also normal-winged. And among the 75 per cent total of normal-winged offspring, 56 per cent, or again nearly three-fourths of 75, are at the same time also gray-bodied. In other words, the 56 per cent gray-bodied and normal-winged offspring amount to *three-fourths of three-fourths*, or nine-sixteenths, of the total. The overall ratio thus is very nearly 9/16 : 3/16 : 3/16 : 1/16. Such a ratio can be obtained only if each trait obeys the law of segregation and is therefore expressed *independently* of other traits. This consideration underscores the essence of the law of independent assortment.

The validity of the law emerges clearly if we consider chromosomes, meiosis, and gametes. In the mating above, the cells of the F_1 parents are as shown in Fig. 12.18.

After meiosis, each gamete will contain only *one* color gene and only *one* wing gene. But which of each pair? The dominant or the recessive gene? This is a matter of chance. There are four possibilities. A gamete might contain the genes *E* and *Vg*, or *E* and *vg*, or *e* and *Vg*, or *e* and *vg*. Many gametes are produced and all four combinations will therefore occur with roughly equal frequency (cf. Fig. 12.18).

Fertilization is also governed by chance. Consequently, *any* one of the four sperm types might fertilize *any* one of the four egg types. Hence there are 16 different ways in which fertilization can occur. If large numbers of fertilizations take place

12.17.

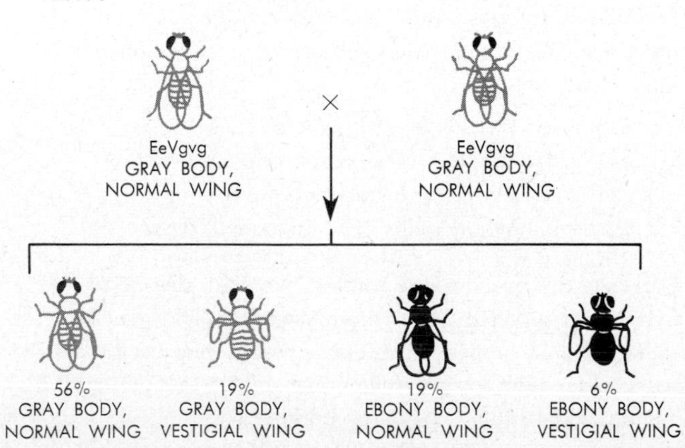

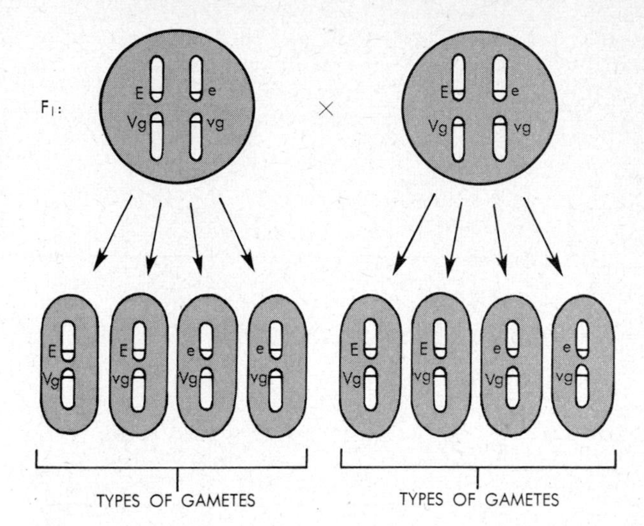

12.18.

simultaneously, all 16 ways will be realized with roughly equal frequency. We may determine these 16 ways by using a grid in which the gametes of one parent are put along a horizontal edge and the gametes of the other parent along a vertical edge (Fig. 12.19).

Among the 16 offspring types now formed, we find some individuals which contain *both* dominant genes at least once, some which contain one *or* the other of the dominant genes at least once, and some which contain none of the dominant genes. A count reveals gray-normal, gray-vestigial, ebony-normal, and ebony-vestigial to be present in a ratio of 9 : 3 : 3 : 1. This is equivalent to the 56 : 19 : 19 : 6 per cent ratio actually obtained above.

Mendel's second law applies specifically to gene pairs located on different chromosome pairs. The law will therefore hold for as many different gene pairs as there are chromosome pairs in each cell of a given animal. Suppose we considered the inheritance of three different gene pairs, each located on a different chromosome pair. For example, what would be the offspring of a mating of two triple heterozygotes, such as *AaBbCc* × *AaBbCc*?

We have found above that a double heterozygote *AaBb* produces *four* different gamete types. It should not be too difficult to verify that a triple heterozygote produces *eight* different gamete types, namely, *ABC, ABc, AbC, Abc, aBC, aBc, abC,* and *abc*. To determine all possible genotypes of the offspring, we may make a grid eight squares by eight squares and place the eight gamete types of each parent along the

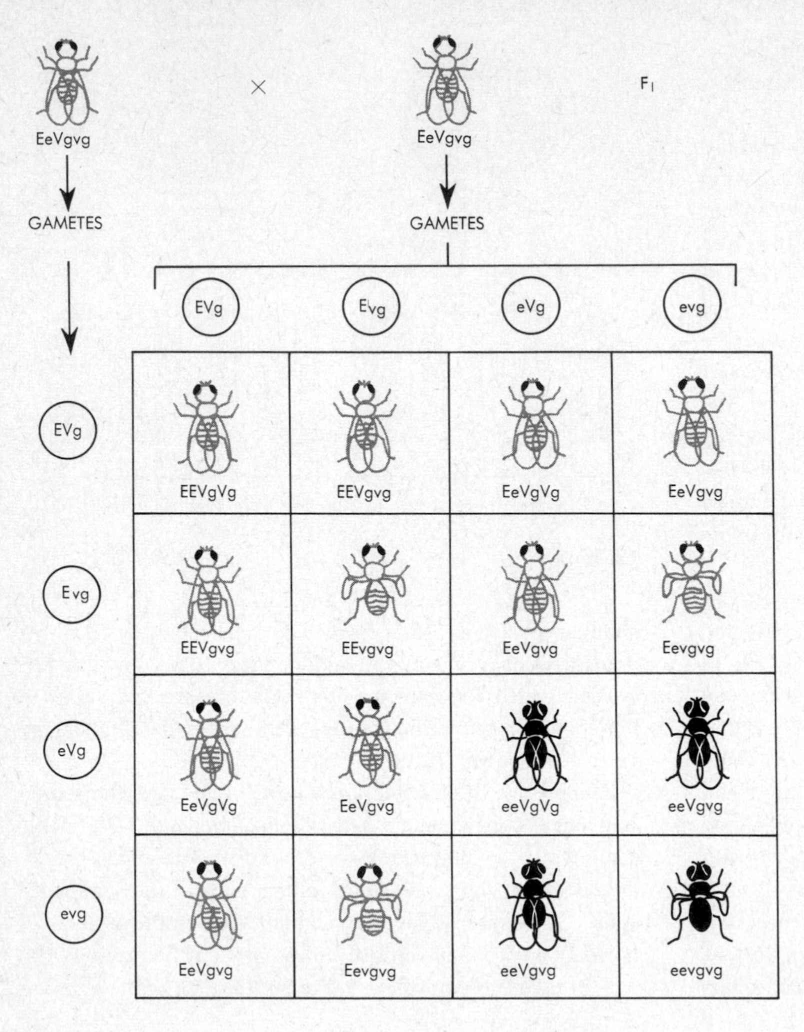

12.19.

sides of the grid, as above; and 64 offspring types will then result. Of these, 27 will express all three traits in dominant form. The complete phenotype ratio may easily be verified as 27:9:9:9:3:3:3:1.

Two quadruple heterozygotes, *AaBbCcDd*, would manufacture 16 gamete types each, and we would need a grid 16 by 16 to represent the 256 different genotype combinations. Evidently, the possibilities rapidly become astronomical once we consider more than a few traits simultaneously.

Animals which are heterozygous for a large number of traits are known as *hybrids*. *Aa* types are sometimes referred to as *monohybrids*, *AaBb* types as *dihybrids*, *AaBbCc* types as *trihybrids*. In man there are 23 pairs of chromosomes per cell. Consequently, Mendel's second law will apply to any 23 different traits controlled by genes located on different chromosome pairs. We might then study a mating of, for example, two 23-fold hybrids: *AaBb* . . . *Ww* × *AaBb* . . . *Ww*. How many gamete types would each such hybrid produce? We know that:

a monohybrid yields $2^1 = 2$ gamete types
a dihybrid yields $2^2 = 4$ gamete types
a trihybrid yields $2^3 = 8$ gamete types
a quadruple hybrid yields $2^4 = 16$ gamete types

Carrying this progression further, we find that a 23-fold hybrid will produce 2^{23} or more than 8 million genetically different gamete types. Therefore, in considering just 23 traits, we would require a grid 8 million by 8 million to represent the more than 64 trillion possible genotypes.

A particular individual then inherits just one of these genotypes. Of all the possible genotypes, a few millions or billions will produce resemblance to parents and another few millions or billions to grandparents or earlier ancestors. But there are bound to be a good many million or billion genotypes which have never yet become expressed during the entire history of man. Accordingly, there is a very excellent chance that every newly born human being differs from every other one, past or present, in at least some genes controlling just 23 traits. And the genetic differences for all traits must be enormous indeed. Here is one major reason for individual variations and a genetic basis for the universal generalization that no two animals are precisely identical.

Any given chromosome contains not just one gene but anywhere from a few hundred to a few thousand genes. What is the inheritance pattern of two or more gene pairs located on the same chromosome pair? This question leads us beyond Mendel's two laws.

The Law of Linear Order

Genes located within the same chromosome are said to be *linked:* as the chromosome is inherited, so are all its genes inherited. Such genes clearly do not assort independently, but are transmitted together in a block. The traits controlled by linked genes are similarly expressed in a block. For example, assume that in the heterozygote *AaBb* the two gene pairs are linked. When such a dihybrid produces gametes, only *two* different gamete types are expected, 50 per cent of each (Fig. 12.20). We recall that if the gene pairs *Aa* and *Bb* were not linked, we should expect four gamete types through independent assortment, namely, *AB*, *ab*, *Ab*, and *aB*, 25 per cent of each.

Linkage studies were first undertaken by T. H. Morgan, a renowned American biologist of the early twentieth century. The first to study genetics by using fruit flies, Morgan discovered a curious phenomenon. When genes were linked, the expected result of two gametes types in a 50:50 ratio was obtained relatively rarely. Instead, there were usually somewhat fewer than 50 per cent of each gamete type, and there were correspondingly small percentages of two additional, completely unexpected gamete types.

In fruit flies, for example, the same chromosome pair which carries the wing shape genes also carries one of many known pairs of eye color genes: a dominant allele *Pr* produces red, wild type eyes, and a recessive allele *pr* in homozygous condition produces distinctly purple eyes. If now a normal-winged, red-eyed heterozygous fly *VgvgPrpr* produces gametes, only

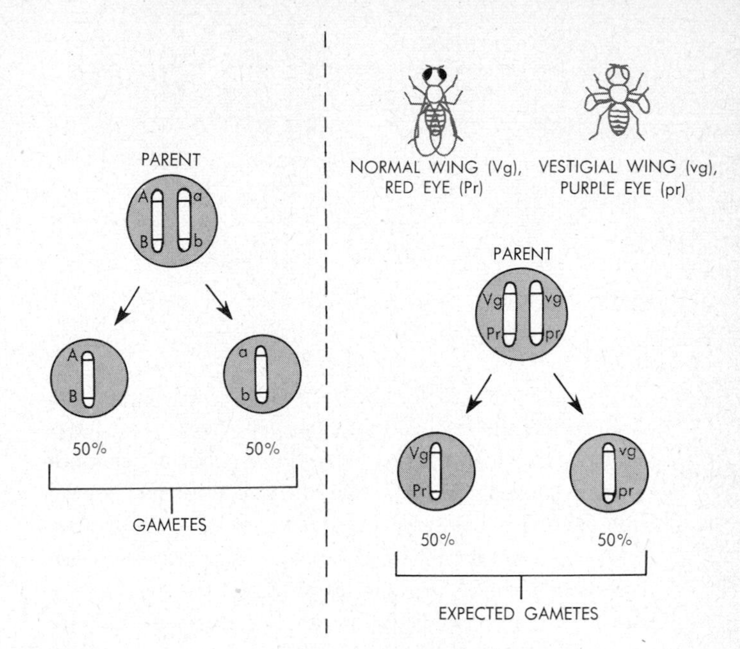

12.20.

12.21. Top, the phenotypes of two traits (wing shape and eye color) of *Drosophila*, controlled by linked genes. Bottom, the expected gametes of a heterozygous fly, *VgvgPrpr*.

two types should be expected, namely *VgPr* and *vgpr* (Fig. 12.21). However, in actuality one consistently obtains four gamete types, and their proportions might be as in Fig. 12.22.

If these four types would form to an extent of about 25 per cent each, the result could be regarded simply as a case without linkage, governed by Mendel's second law. But the actual results include significantly *more* than 25 per cent each of the expected gamete types and significantly *fewer* than 25 per cent of the unexpected types.

To explain odd results of this sort, Morgan proposed a new hypothesis. He postulated that, during meiosis, paired chromosomes in some cases might twist around each other and might break where they are twisted. The broken pieces might then fuse again in the "wrong" order (Fig. 12.23).

This would account for the large percentage of expected and the small percentage of unexpected gamete types. To test the validity of this hypothesis, cells undergoing meiosis were examined carefully under the microscope: could chromosome twists and breaks actually be seen? They could indeed, and the phenomenon of *crossing over* was so proved.

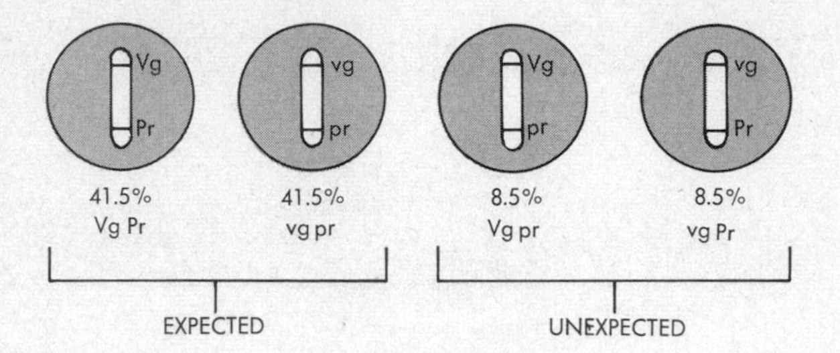

12.22.

The implications of this discovery were far-reaching. It was reasoned that the frequency of crossovers should be an index of the *distance* between two genes. If two genes on a chromosome are located near each other, the chances should be relatively small that a twist will occur between these close points. But if two genes are relatively far apart, then twists between these points should be rather frequent. In general, the frequency of crossovers should be proportional to the distance between two genes (Fig. 12.24).

Inasmuch as the crossover percentage of two genes could be determined by breeding experiments, it became possible to construct *gene maps* showing the actual location of given genes on a chromosome. Since Morgan's time, the exact positions of few hundred genes have been mapped in the fruit fly. Smaller numbers of genes have similarly been located in mice and in various other organisms. Many of these determinations have been corroborated by X-ray work. When irradiated, a chromosome may break into pieces and a small piece of this sort may be lost from a gamete. Offspring resulting from such deficient gametes will be abnormal in certain traits. In many cases, microscopic examination can show where a chromosome piece is missing and a trait thus can be correlated with a particular spot on a chromosome.

12.23.

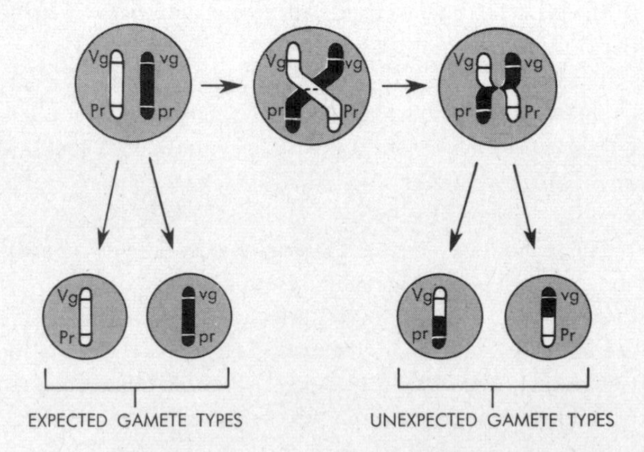

A second implication of crossing over is that genes on a chromosome must be lined up single-file. Only if this is the case can linkage and crossing over occur as it actually does occur. This generalization has become known as the *law of the linear order of genes*. It constitutes the third major rule which governs Mendelian inheritance.

Thirdly, crossing over has provided another good definition of "gene": A gene is the smallest section of a chromosome within which crossovers do not take place. The assumption here is that the minimum chromosome unit able to cross over is one whole gene, not a fractional part of one gene. Evidently, genes may be regarded either as units of biochemical action or as *crossover units*. A third possible definition will emerge below. It follows that a gene is not a singularly definable object; what we mean by "a gene" depends entirely on the techniques we use to study it. Put another way, genes are *operational concepts*.

Lastly, a further implication of crossing over during meiosis is that meiosis is a *source of genetic variations*. For example, when a diploid cell in a testis undergoes meiosis and produces four haploid sperms, these four do not contain merely the same whole chromosomes as the original cell, even though redistributed. For if the original diploid cell possesses a chromosome pair M′ and M″, then a given sperm will not simply receive either the M′ or the M″ chromosome. Instead, as a result of crossing over, it will receive a quiltwork chromosome composed of various joined *pieces* of *both* M′ and M″. In each set of four sperms, actually, two will contain like chromosomes composed of one set of M′ and M″ pieces, and two will contain like chromosomes composed of the remaining, complementary M′ and M″ pieces. Moreover, the original diploid cell possesses not just a single chromosome pair but several pairs, and each such pair is likely to be subject to crossing over in an unpredictable fashion. Chances are therefore excellent that the four sperms will be genetically different from one another as well as from the original diploid cell. Further, even two genetically identical diploid cells are not likely to give rise to genetically identical sets of sperms. In general, the gene-shuffling effect of crossing over is in evidence wherever or whenever meiosis occurs. Genetic variations consequently are produced by both phases of sex, i.e., by chromosome doubling through fertilization as well as by chromosome reduction through meiosis.

The three rules of heredity here outlined describe and predict the consequences of sexual recombination, i.e., the vari-

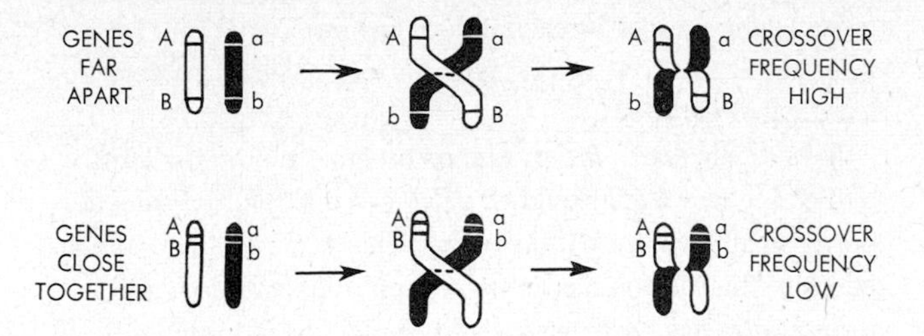

12.24. Crossover frequency in relation to gene distances. If two genes are far apart, crossing over between them is likely to occur rather frequently (top). But if genes are close together, crossing over between them is less likely. In general, the farther apart given genes are on a chromosome, the more frequent crossing over will be.

ous results that can be obtained when different sets of genes become joined through fertilization and are pooled in the zygote. In other words, sexual recombination of genes leads to Mendelian inheritance. However, a great many hereditary events have been found which do not obey the three basic rules. Some of these non-Mendelian processes will be the subject of the following section.

NON-MENDELIAN INHERITANCE

Mutation

Mutations are among the most important types of non-Mendelian variations, of universal significance in all animals. Any stable, inheritable change in the genetic material present in a cell constitutes a mutation. For example, the accidental doubling, tripling, etc., of the normal chromosome number represents a stable, transmissible change. This is a *chromosome mutation.* Accidental loss or addition of a whole chromosome, loss of a chromosome piece, fusion of such a piece with another chromosome or fusion with the original chromosome in inverted position—these also are chromosome mutations. But by far the most common type of mutation is a *point mutation,* a stable change of one gene. A third definition of "gene" is that it is a *unit of mutation,* i.e., that minimum part of a chromosome which, after becoming altered in a stable manner, changes just one trait of a cell (Fig. 12.25).

It has been known for many years that mutational changes can be induced by high-energy radiation such as X rays. The frequency of mutation has been found to be directly proportional to the amount of radiation a cell receives. Are naturally occurring mutations similarly produced by radiation, e.g., by cosmic rays and other space radiation, or by radioactive elements in the earth? Probably not entirely; it can be shown that the unavoidable natural radiation which affects all living organisms is not sufficiently intense to account for the mutation frequency characteristic of genes generally. This frequency has been estimated as about one mutation per million cells, on the average. However, natural "background" radiation does produce some mutations. Most others probably represent errors in gene reproduction (cf. Chap. 6). And still others are undoubtedly caused by man-made radiation, which adds to and so increases the natural background radiation. Mutations can also be produced experimentally by physical agents other than radiations and by various chemical agents.

Apart from limitations as to numbers, types, and range of

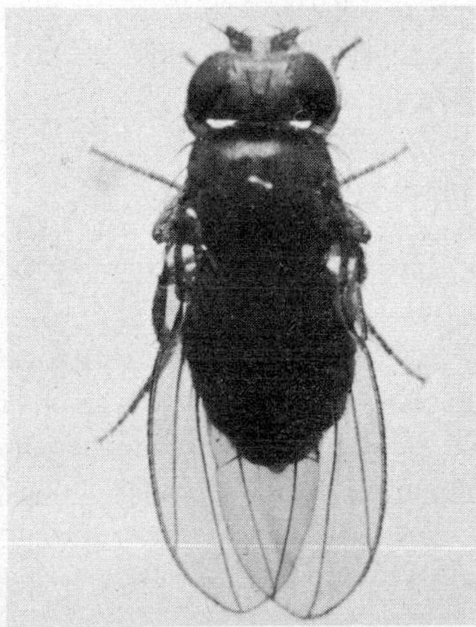

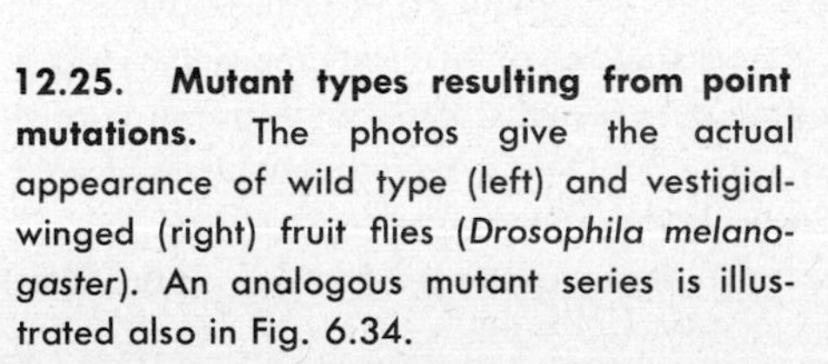

12.25. Mutant types resulting from point mutations. The photos give the actual appearance of wild type (left) and vestigial-winged (right) fruit flies (*Drosophila melanogaster*). An analogous mutant series is illustrated also in Fig. 6.34.

possible effects, mutations occur entirely at random and most of them are recessive and disadvantageous (cf. Chap. 6). Most mutated genes therefore accumulate in a cell without immediate visible effect, their action being suppressed by the dominant wild type alleles. Traits will become visibly altered only if a mutation has a dominant effect from the outset or if a mutation with recessive effect becomes homozygous, e.g., by the segregation of two mutant genes of the same kind into one offspring. Either event is relatively rare in nature, and when it does occur the trait alteration must be "small" if the cell is to survive (cf. Chap. 6). Furthermore, even a persisting small mutation will affect subsequent generations of animals only if it is a *germ mutation*, i.e., a genetic change in the reproductive cells. However, to the extent that mutations do persist into successive animal generations, they may affect the inheritance and adaptation of individuals in non-Mendelian fashion just as sexual recombination does in Mendelian fashion.

Mutons, Recons, Cistrons

We already know from Chap. 6 that a mutation need not change the structure of an entire gene; an alteration of a single nitrogen base in a code triplet of DNA probably suffices to produce an effective mutational change. If so, should a whole gene be regarded as the "minimum" unit of heredity? If a gene represents a series of joined nucleotides in a DNA chain, could not the internal parts of such a section of DNA represent functional *subunits* of a gene?

That a gene actually might contain functional subunits can be inferred from the phenomenon of *multiple alleles*. Many genes are known to occur in not just two but in several dozens of alternative allelic forms, each allele having a specifically different effect on the expression of the same trait. Thus, all allelic forms of a gene might affect eye color, but the actual color produced by one of the alleles might differ slightly from that produced by any of the other alleles. Inasmuch as allelic forms do affect the same trait, it is quite unlikely that their DNAs are totally different. It is far more likely that all alleles of a gene have essentially the same basic sequence of joined nucleotides, and that the functional differences between alleles are due to relatively slight chemical differences. For example, one or perhaps a few of the nucleotides present at a given point along the DNA chain of one allele might differ from the nucleotides present at the corresponding point of another allele. In terms of molecular structure, therefore, any given allelic state of a gene might differ from any other by not more than one or at most a few nucleotides along otherwise identical DNA chains.

New allelic states are produced by gene mutation. Thus, a detectable new allele could be produced if a mutation affected only a small segment within a gene. The creation of new allelic genes in this fashion has actually been demonstrated, and the term *muton* has been proposed to designate the smallest segment within a gene which by mutation can produce an altered trait. A muton has been estimated to consist of perhaps just one and probably no more than three joined nucleotides, and a whole gene may be envisaged to consist of a linear array of very many mutons.

If consecutive segments within a single gene may mutate independently, a gene should include a linear series of functionally different regions. That this is actually the case has been shown in recent studies on gene-controlled enzyme synthesis (principally in bacteria). These studies indicate that small, functionally distinguishable segments of genes can become transferred to other chromosomes. Such transfers are not crossovers. As outlined earlier, crossing over takes place between whole genes, and we recall that a possible definition of a whole gene is that it is the smallest chromosome unit capable of crossing over. Crossing over is also a reciprocal process. For example, suppose that genes A and B are linked on one chromosome and that the corresponding alleles a and b are linked on the other chromosome of a pair. If crossing over occurs between A and B, the result will be the combination Ab as well as the reciprocal combination aB.

By contrast, transfers of small segments within a gene are occasionally nonreciprocal. Suppose that a gene R contains the subunits R_1 and R_2 and that an allele r on the other chromosome of a pair contains the corresponding subunits r_1 and r_2:

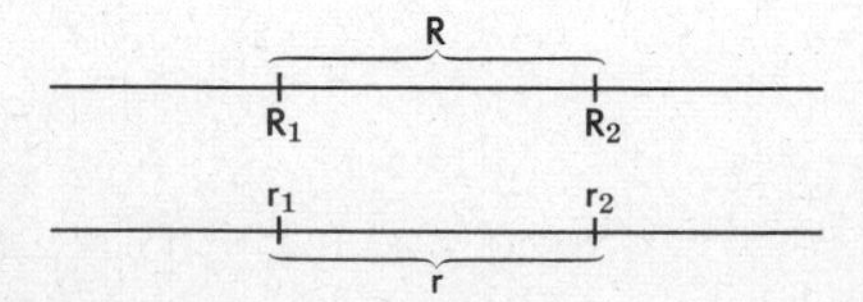

When the genes R and r reproduce and form duplicates, it happens in some cases that one of the genes appears to make a reproduction error. For example, R_1R_2 may form a normal duplicate but r_1r_2 may not. The r_1 region may be duplicated correctly, but instead of also forming a copy of the r_2 region, the gene copies the R_2 region of the nearby allele on the other chromosome. The result is a new gene which contains the subunits r_1R_2:

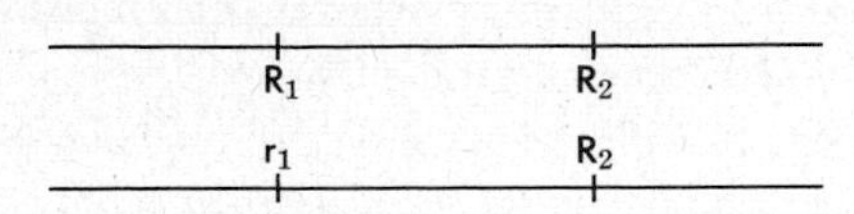

The combination r_1R_2 is clearly nonreciprocal, for R_1r_2 is not formed.

Evidently, the r_2 region behaves as if it could make a *copy choice* during reproduction; it may produce either another r_2 or an R_2 region. Recombinations of subunits within genes generally appear to result from "wrong" copy choices. The smallest segment within a gene capable of forming new recombinations as just described has been called a *recon*. Such a unit may consist of perhaps not more than a single nucleotide. It is therefore conceivable that, in some cases at least, a recon may be identical with a muton.

Can a genetic function continue to be performed if the mutons and recons of a gene are not united together on the same chromosome? For example, suppose that a given gene A controls the synthesis of a particular enzyme protein and that this gene consists of a series of consecutive mutons. Suppose also that the mutons form two adjacent groups A_1 and A_2, and that an allelic recessive gene a contains the corresponding muton groups a_1 and a_2:

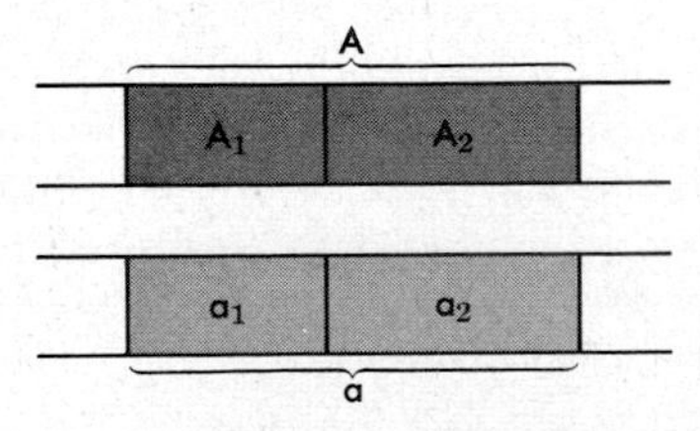

So arranged on the same chromosome, the muton groups A_1 and A_2 are said to be in a *cis* position. Under such conditions the whole gene A will function normally and will control the synthesis of the enzyme, as assumed.

What is the effect on genetic function if the muton groups A_1 and A_2 are not within the same chromosome? For example, A_1 and A_2 may come to be located on different chromosomes and may be arranged in a *trans* position:

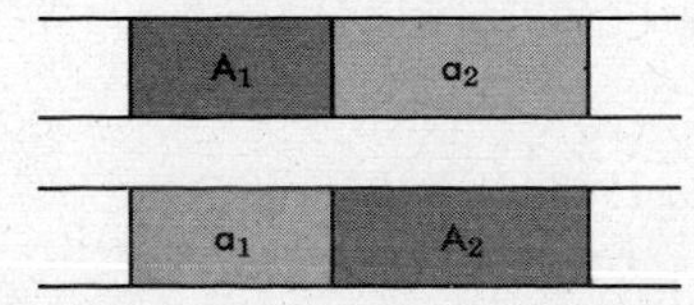

In such a *trans* arrrangement, control of enzyme synthesis may or may not continue normally, depending on the specific nature of the gene and its muton groups A_1 and A_2. If the genetic function of A does continue normally, the muton groups A_1 and A_2 are said to *complement* each other. In this case, evidently, the functional effectiveness of the gene is not interfered with even though the gene is subdivided structurally and each muton group is located on a different chromosome. In other cases, however, a *trans* arrangement of muton groups does lead to cessation of enzyme synthesis; complementation does not occur and genetic function then stops.

Such results can be explained if it is assumed that given numbers of consecutive mutons (or recons) within a gene form larger functional sets. For example, if a gene consisted of 50 mutons, 30 of these might form an integrated block which might have to stay intact structurally if function is to be maintained. Thus, if this block of 30 were on one chromosome and the remaining 20 were in *trans* position on another chromosome, then complementation could occur and genetic function could continue. But if only 29 mutons were on one chromosome and the remaining 21 were on the other, then complementation could not occur since an intact block of 30 mutons would no longer be present. The actual existence of such integrated blocks of mutons within genes has been verified. The term *cistron* has been coined for the smallest set of mutons or recons of a gene which must remain together on one chromosome if genetic activity is to be preserved. The mutons within a cistron undoubtedly interact with one another in some intimate way, and if the interaction is prevented by a subdivision of the cistron, then the whole cistron will become nonfunctional. Tests show that a single gene consists of relatively few cistrons, one adjacent to the next and each composed of specific numbers of mutons (or recons).

We may conclude, therefore, that for each operational definition of a whole gene we now also have a corresponding operational definition of a genic subunit. A whole gene is a unit of mutation, or a unit of recombination by crossing over, or a unit of biochemical action. Within a gene, analogously, the mutational unit is the muton, the recombinational unit is the recon, and the functional unit is the cistron. Both sets of definitions are operational ones reflecting the nature and refinement of the experimental methods we use. If we regard a whole protein molecule as the smallest genetically controlled trait, then our experimental methods will lead us correspondingly to a smallest unit of heredity; we have called such a unit a gene. Until recently, the best conclusion actually permitted by available methods was that a whole gene controlled the synthesis of a whole protein molecule, more specifically, an enzyme. However, we know that a protein molecule has subtraits, namely, amino acids and polypeptide chains. And it is now possible to study the inheritance of such subtraits by means of refined experimental methods. The results have shown that a whole Mendelian gene is a relatively crude hereditary unit within which finer subunits may be identified —mutons, recons, and cistrons. Conceivably, mutons and recons may control the triplet coding of individual amino acids, and cistrons may control the synthesis of polypeptide chains. In any event, a gene must be regarded as a complex chromosome region composed of many interacting functional parts. The latter appear to have a definite though not yet fully specified relation to the structural units of the DNA chain.

Interaction is known to occur not only among the parts within a gene but also among whole genes; as pointed out earlier, the activity of a gene is influenced by its environment and this environment includes other genes. Like the interactions within genes, those among genes again tend to produce non-Mendelian results, i.e., results which cannot be predicted by the three rules of Mendelian inheritance.

Gene-Gene Interactions

Groups of genes within a cell often cooperate in controlling a highly composite trait. One of the best illustrations is the trait of sexuality, which in certain animals is controlled not by individual genes acting separately but by whole chromosomes acting as functionally integrated units.

Sex determination. Each animal is believed to possess genes promoting the development of male traits as well as genes promoting the development of female traits. Such genes are not usually specialized "sex genes," but they tend to be of a type which, among other effects, also happen to influence the sexual development of an animal. Fundamentally, therefore, all animals are potentially *bisexual;* each possesses a genetic potential for *both* maleness and femaleness. Animals may be classified into two categories according to how this genetic potential is translated into actual sexual traits.

In one category, comprising the vast majority of animal groups, the masculinizing genes are equal in effect to the feminizing genes. The two genetic influences are matched exactly in "strength," and in the absence of other influences an animal will then develop as a hermaphrodite. In many cases, however, other sex-determining factors do exert an effect. Such factors are *nongenetic* and *environmental;* different conditions in the external or internal environment affect an animal in such a way that it develops either as a male or as a female. The genetic characteristics of the species determine whether nongenetic influences will or will not play a role and thus whether the species will be separately sexed or hermaphroditic (Fig. 12.26).

Also, the genetic nature of the species determines *when* during the life cycle the sex of an animal may become fixed by nongenetic means. In hydras and other groups, for example, sex determination does not occur until the animals are fully adult and ready to produce gametes. Up to that time, an individual is sexually indifferent, i.e., all cells retain a bisexual potential. Later, certain environmental influences affect the gamete-producing cells in such a way that they mature either as sperms or as eggs. The precise nature of these influences has in most cases not yet been identified. In hermaphroditic species, the gamete-producing cells remain unaffected and both sperms and eggs are formed.

In other instances, nongenetic sex determination takes place earlier in the life cycle, e.g., during the larval phase.

12.26. Nongenetic sex determination. Left, *M* and *F* in zygote and adults indicate that male- and female-determining genetic factors are equally balanced. The actual maleness, femaleness, or hermaphroditic condition of the adult will depend on external nongenetic influences acting at some point during development. Right, such influences act at different developmental stages in different types of animals. Thus, separate sexuality may already become manifest in the embryo or, by contrast, not until the adult produces gametes. Up to such an externally influenced determination stage, the animal retains a bisexual (hermaphroditic) potential.

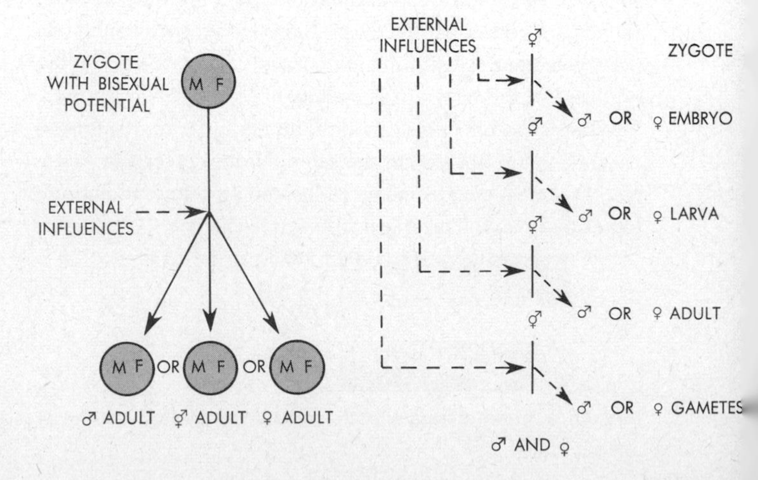

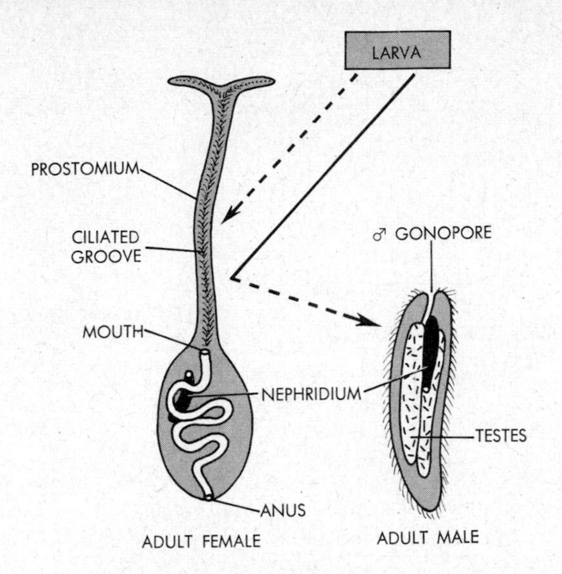

12.27. Nongenetic sex determination in the larvae of the echiuroid Bonellia. A free larva develops into a female adult, as at left. A larva which makes contact with a female develops into a reduced male adult, as at right. This male is parasitic, and small enough to live within the nephridium of the female.

After metamorphosis, therefore, such animals are no longer bisexual but are already determined as males, females, or hermaphrodites. In at least one case, viz., the schizocoelomate worm *Bonellia,* the sex-determining factor is known to be environmental CO_2. The relatively low concentrations of CO_2 in sea water affect free-swimming larvae of this worm to develop as females. But if a sexually still indifferent larva happens to come into contact with an adult female or with a larva already determined as a female, then the added respiratory CO_2 produced by the latter causes the indifferent larva to be determined as a male; it becomes a small, sperm-producing, and structurally quite distinct parasitic animal, permanently attached to the body of the female (Fig. 12.27). In still other animal groups, nongenetic sex determination occurs even earlier during the life cycle, e.g., after cleavage or within the fertilized egg itself. The sexually indifferent period then is reduced correspondingly.

An altogether different pattern of sex determination characterizes a second category of animals, represented chiefly by the arthropods (insects particularly) and the vertebrates. In these, the masculinizing genes are *not* equal in effect to the feminizing genes, and the primary determination of sex has a purely *genetic* basis. Any individual becomes either a male or a female; hermaphroditism does not occur except as an abnormality. Also, the determination of sex always takes place at the time of fertilization, and every cell of the embryo, the larva, and the adult is therefore genetically male or female (Fig. 12.28).

Animals in this category possess a pair of special *sex chromosomes,* different in size and shape from all other chromosomes. The latter are referred to as *autosomes.* The sex chromosomes are of two types, *X* and *Y*, and the pair of sex chromosomes may consist either of one *X* and one *Y* chromosome or of two *X* chromosomes. For example, in mammals and flies (fruit flies included), all cells of adult *males* contain an *XY* pair, all cells of *females,* an *XX* pair. By contrast, in butterflies, most moths, some fishes, and in birds, the cells of females contain an *XY* pair and the cells of males, an *XX* pair (cf. Fig. 12.28). In either case, the *Y* chromosomes are inert from the standpoint of sex determination; they may be lost from cells without appreciable effects. But the *X* chromosomes and the autosomes do

12.28. Patterns of genetic sex determination. **1.** In the zygote, either the male-determining genes outweigh the effect of the female-determining genes (*M* over *f*), or vice versa (*F* over *m*); all subsequent stages are then separately sexed, and hermaphroditic or bisexual conditions do not (normally) occur. **2.** If *X* chromosomes determine femaleness and autosomes (*A*) maleness, then the cells of the two sexes will have the chromosome balances shown (many pairs of *A* are assumed to be present here). This pattern occurs, for example, in insects such as fruit flies and in mammals; actual chromosome balances for man are indicated in **3.** The reverse of **2** is shown in **4,** *X* here determining maleness, *A*, femaleness. This pattern is encountered, for example, in butterflies, fishes, and birds. The *Y* chromosomes in **2, 3,** and **4** are genetically inert.

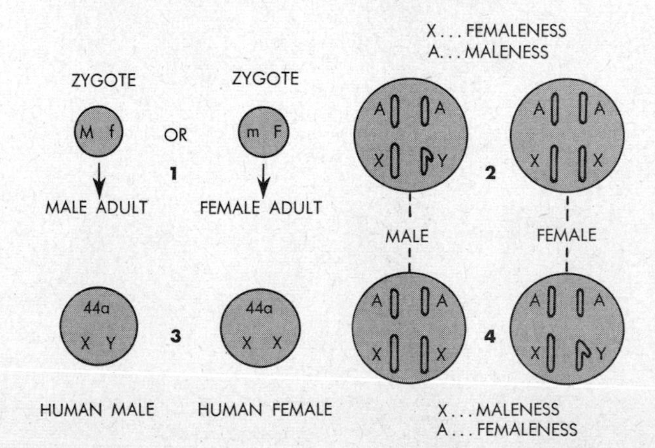

determine sexual traits. Where the males are of the XY type, the X chromosomes promote the development of femaleness and the autosomes promote the development of maleness. Conversely, where the males are of the XX type, the X chromosomes promote male traits, the autosomes, female traits.

In man, for example, each adult cell contains 22 pairs of autosomes plus either an XY or an XX pair. The chromosome complement of male cells may be symbolized as $44A + XY$, that of female cells, as $44A + XX$ (cf. Fig. 12.28). In view of the inertness of the Y chromosome, the cells of human females thus possess 46 functional chromosomes but male cells possess only 45. This difference of one whole X chromosome, with its hundreds of genes, lies at the root of the sexual differences between males and females. More specifically, in a $44A + XX$ cell, the total feminizing influence of the two X chromosomes outweighs the total masculinizing effect of the 44 autosomes, and such cells are therefore female. But in a $44A + XY$ cell, the total masculinizing effect of the 44 autosomes outweighs the feminizing influence of the single X chromosome and such cells are male.

That the sexual nature of an individual does indeed depend

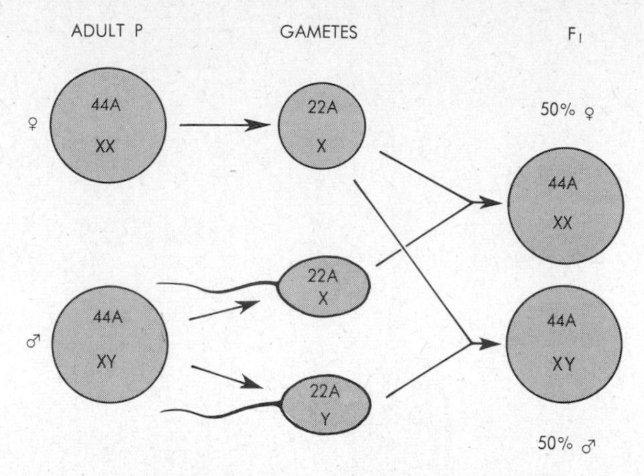

12.30. Sex determination in man. Males produce two genetically different types of sperms, roughly 50 per cent of each. Offspring will then be male and female in a 1 : 1 ratio.

on a particular balance of X chromosomes and autosomes has been proved by experiments in fruit flies. In these animals, it is possible by certain laboratory procedures to vary the numbers of X chromosomes and autosomes normally present in the sperms and eggs. One may then obtain offspring characterized by normal paired sets of autosomes, but by three X chromosomes instead of two. Such individuals grow into so-called *superfemales:* all sexual traits are generally accentuated in the direction of femaleness. *Supermales* and *intersexes* may be produced analogously. In intersexes, sexual traits are intermediate between those of males and females. The chromosome balances are shown in Fig. 12.29. Paradoxically, supersexes and also intersexes are generally sterile; as a result of the abnormal chromosome numbers, meiosis occurs abnormally and the sperms and eggs then produced are defective.

In the light of such balances, we may appreciate readily how the sex of an offspring is normally determined at the time of fertilization. For example, human females, $44A + XX$, produce eggs of which each contains $22A + X$ after meiosis. Males, $44A + XY$, produce two kinds of sperms, namely, $22A + X$ and $22A + Y$, in roughly equal numbers. Fertilization now occurs at random, i.e., a sperm of either type may unite with an egg. Therefore, in about 50 per cent of the cases the result will be $(22A + X) + (22A + X)$, or $44A + XX$, or zygotes developing into females. In the remaining 50 per cent of the cases, the zygotes will be $(22A + X) + (22A + Y)$, or $44A + XY$, or prospectively male (Fig. 12.30).

Note that it is the paternal parent who, at the moment of fertilization, determines the sex of the offspring. When only

12.29. Top, the chromosomes of the fruit fly Drosophila. In each cell, $2n = 8$. Note the differences in the sex chromosomes of males and females. **Bottom, sex and chromosome balances in the fruit fly.** The sexual character of an individual is determined by the specific balances of autosomes and X chromosomes.

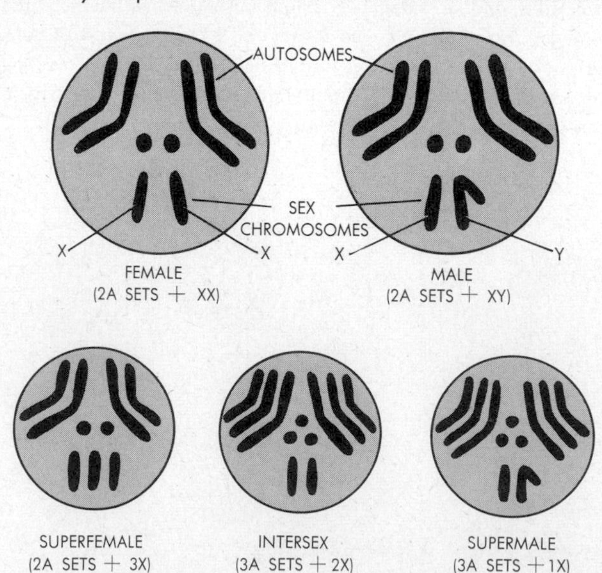

a single offspring is produced, there exists a 50:50 chance of its being a son or a daughter. When many offspring are produced, the number of males will generally equal the number of females.

Note also that the absence of a functional mate to the X chromosome in males has other genetic consequences. In females, the effect of a recessive gene located on one X chromosome may be masked by the effect of a dominant located on the other X chromosome. But in males, recessive genes on the X chromosome may exert their effect, since another X chromosome with masking dominants is never present. Genes located on X chromosomes are called *sex-linked* genes.

Because males possess only a single X chromosome, such genes are inherited according to a characteristic pattern. For example, red-green color blindness in man is traceable to a sex-linked recessive gene c. Suppose that a color-blind male, X_cY, marries a normal female, XX. In this symbolization, an X chromosome without the subscript c is tacitly assumed to contain the dominant gene C, which prevents the expression of color blindness. The offspring of such a mating, shown in Fig. 12.31, include sons and daughters in equal numbers. The daughters carry recessive gene c, but *all* offspring have normal vision.

Suppose now that one of these daughters marries a normal male, as shown in Fig. 12.32. All daughters resulting from such a mating are normal, but half the sons are color-blind. Thus the trait has been transmitted from color-blind grandfather via normal mother to color-blind son. Such a zigzag pattern of inheritance is characteristic of all recessive sex-linked traits. Males typically exhibit the trait; females merely transmit it. The second X chromosome in females prevents expression of recessive sex-linked traits.

The discussion above applies specifically to animals in which the males are of the XY type. Where the females are of the XY type, all generalizations apply equally if every reference to "male" above is changed to "female" and vice versa.

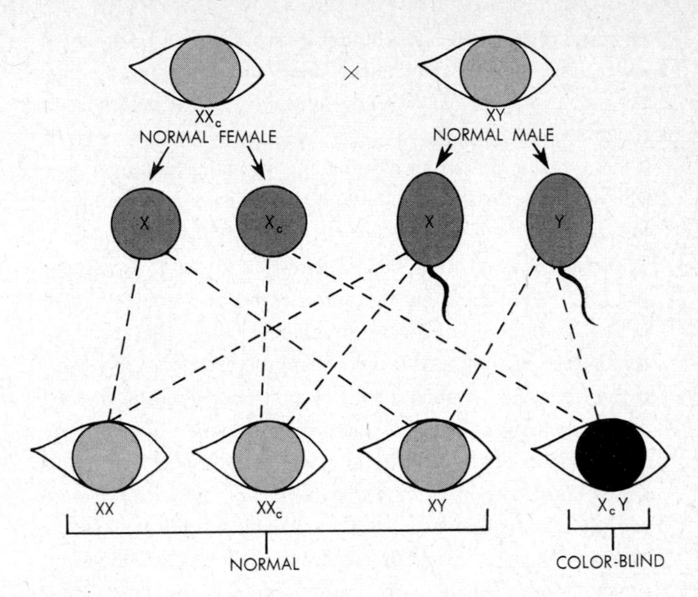

12.32.

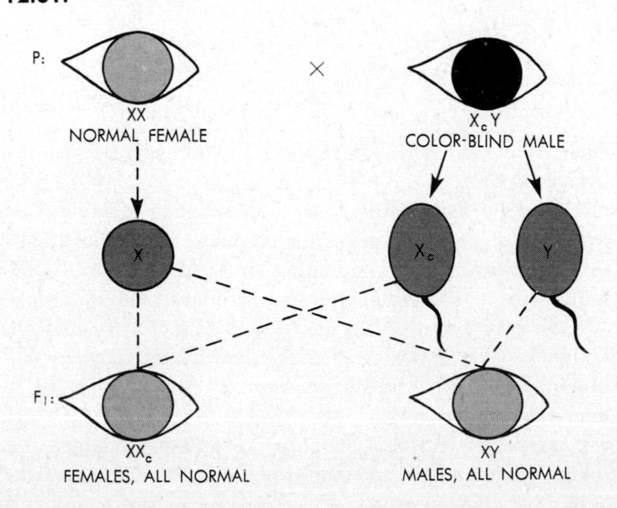

12.31.

Genetic systems. The examples of genetic sex determination just described show clearly that genes of one or more chromosomes may act in concert and control one highly composite trait. The implication is that genes are not merely independent "beads on a string," lined up haphazardly on given numbers of chromosomes. On the contrary, the genes in every chromosome appear to interact in very specific ways, and the expression of traits is influenced by such interactions.

Many other illustrations of this principle are known. For example, if genes were simply independently functioning units, then it should not matter if the position of genes relative to one another were rearranged. But experiment shows that such rearrangement actually does matter. It is possible to change the position of given sections of a chromosome. A piece lost by one chromosome may become attached to another, or it may become reattached to the same chromosome but in inverted position or at the other end. Genes here are neither removed from nor added to a cell; only their position relative to one another is rearranged. Under such conditions the cell may nevertheless develop altered traits, a clear indication that genes normally interact with their neighbors.

The phenomenon of dominance provides another good illustration of the interdependence of genes. A dominant gene acts as it does not only because of its inherent characteristics, but also because other genes *permit* it to act in dominant fashion. If the functional characteristics of the recessive allele of a given dominant were to change, then the status of dominance of that gene would change correspondingly. And if the functional characteristics of any other genes in the cell were to change, then the status of dominance of that gene would again change. It is now well established that given genes boost, suppress, partially inhibit, or otherwise change the effects of other genes. For example, *modifier* genes are known, which intensify or minimize the traits produced by other genes. Analogously, *suppressor* genes completely prevent traits produced by other genes from becoming expressed explicitly. Some genes affect not the traits produced by other genes but the other genes themselves. Among them are the *operator* and *regulator* genes we have discussed in Chap. 6, in connection with the operon hypothesis. As in that earlier context, we now may therefore conclude again that genes are of two general types, viz., "structural" genes, effective in producing cellular traits, and "coordinating" genes of various kinds, effective in influencing the activity of the structural genes and their products. Indeed it is quite possible that *every* gene in a cell might have a trait-producing role in some hereditary functions and a coordinating role in others.

Thus, whereas the pre-Mendelians thought that *traits* were inherited, and whereas the Mendelian era advanced to the concept that individual factors, or *genes,* were inherited, the modern post-Mendelian era recognizes that actually neither traits nor genes nor even subunits of genes are inherited. Instead, what are inherited are whole chromosome *sets,* coordinated complexes of genes, subtly integrated and interacting *genetic systems.* Moreover, even genetic systems are not inherited by themselves but are transmitted within whole cells. The functional integration between genetic system and cell system is never lost, and it is zoologically almost meaningless to consider one without the other. Actually, the cell cytoplasm is a carrier of heredity too. For example, cell membranes, mitochondria, kinetosomes, nuclear membranes, and many other intracellular organelles are inherited not through genes but through "self-duplication" of preexisting parental structures. The genetic system of a cell undoubtedly controls such cytoplasmic self-duplication, but the cytoplasmic system certainly also controls the self-duplication of the genetic system. Ultimately, therefore, the smallest real unit of inheritance is one whole cell.

In this chapter we have found that the inheritance of genetic systems and of the traits they control is governed by the biological nature of sex, by various probabilistic Mendelian rules, by random mutational changes, by interactions within genes and between genes, and by the effects of the environment on gene-trait relationships. To the extent that this interplay between sex, heredity, and environment remains the same in the course of successive generations, it maintains animal continuity and given degrees of individual adaptation. And to the extent that the interplay changes as the reproductive succession continues, it produces animal *evolution* and altered adaptations.

REVIEW QUESTIONS

1. What are the sources of genetic variations in (*a*) uniparental, (*b*) biparental inheritance? Distinguish between inherited and acquired variations. What contributions are made to the expression of traits by (*a*) genes, (*b*) the environment? What is an "inherited disease"?

2. What was meant by "blending inheritance"? Describe breeding experiments showing that blending does not occur. What hypothesis did Mendel substitute for the blending concept? State the chromosome theory of heredity. What is the evidence that genes are actually contained within chromosomes?

3. Define: genome, true-breeding, phenotype, genotype, allele, dominant gene, recessive gene, homozygous, heterozygous, hybrid.

4. Review the experiments on inheritance of body color in fruit flies in terms of genes and chromosomes. What are the quantitative results of the mating $Aa \times Aa$ if (*a*) A is dominant over a, (*b*) neither gene is dominant over the other?

5. In your own words, state the law of segregation. If A is dominant over a, what phenotype ratios of offspring are obtained from the following matings: (*a*) $Aa \times aa$, (*b*) $AA \times aa$, (*c*) $Aa \times Aa$, (*d*) $Aa \times AA$?

6. In your own words, state the law of independent assortment. What kinds of breeding experiments and what reason-

ing substantiate this law? Interpret the law in terms of genes, meiosis, and gametes. How many genetically different gamete types will be produced by an animal heterozygous in 10 gene pairs? If two such animals were mated, how many genetically different offspring types could result?

7. Define linkage. Why does inheritance of linked genes not obey Mendel's second law? What are the quantitative and qualitative differences here? What were Morgan's observations which led him to the hypothesis of crossing over? Describe this hypothesis. How do crossover data permit the construction of gene maps? State the law of the linear order of genes. What definition of gene is based on the phenomenon of crossing over? Review other definitions.

8. Distinguish between chromosome mutations and point mutations and between somatic mutations and germ mutations. What is the relation between mutation frequency and radiation intensity? What are the characteristics of mutations from the standpoint of (*a*) predictability, (*b*) functional relation to normal alleles, (*c*) effects on traits, and (*d*) relative advantage to the animal?

9. What are mutons, recons, and cistrons? Review the patterns of sex determination based on (*a*) nongenetic, (*b*) genetic mechanisms. What is the significance of a given numerical balance between autosomes and sex chromosomes? What are supersexes and intersexes?

10. What are sex-linked genes? Describe the inheritance pattern of the sex-linked recessive hemophilia gene *h*, assuming that a hemophilic male mates with a normal female. What are modifier genes? Describe specific instances of gene-gene interactions.

COLLATERAL READINGS

The original work of the founder of modern genetics will always be of special interest:

Mendel, G.: "Experiments in Plant Hybridization," translation of the 1865 original, Harvard, Cambridge, Mass., 1941.

The following may be consulted for additional information on genetics:

Bonner, D. M.: "Heredity," Foundations of Modern Biology Series, Prentice-Hall, Englewood Cliffs, N.J., 1961.
King, R. C.: "Genetics," Oxford, New York, 1962.
Moore, T. A.: "Heredity and Development," Oxford, New York, 1963.
Sinnott, E. W., L. C. Dunn, and T. Dobzhansky: "Principles of Genetics," McGraw-Hill, New York, 1958.
Srb, A., and R. D. Owen: "General Genetics," Freeman, San Francisco, 1955.

Specific genetic topics are discussed in the books and articles below:

Beadle, G. W.: The Genes of Men and Molds, *Sci. American*, Sept., 1948.
Bearn, A. G., and J. L. German: Chromosomes and Disease, *Sci. American*, Nov., 1961.
Benzer, S.: The Fine Structure of the Gene, *Sci. American*, Jan., 1962.
Deering, R. A.: Ultraviolet Radiation and Nucleic Acid, *Sci. American*, Dec., 1962.
Dobzhansky, T.: "Evolution, Genetics, and Man," Wiley, New York, 1955.
Dunn, L. C., and T. Dobzhansky: "Heredity, Race, and Society," Penguin, Baltimore, 1946.
Goldschmidt, R. B.: "Understanding Heredity," Wiley, New York, 1952.
Hollaender, A., and G. E. Stapleton: Ionizing Radiation and the Cell, *Sci. American*, Sept., 1959.
Knight, C. A., and D. Fraser: The Mutation of Viruses, *Sci. American*, July, 1955.
Mittwoch, U.: Sex Differences in Cells, *Sci. American*, July, 1963.
Müller, H. J.: Radiation Damage to the Genetic Material, *Am. Scientist*, vol. 38, 1950.
———: Radiation and Human Mutation, *Sci. American*, Nov., 1955.
———, C. C. Little, and L. H. Snyder: "Genetics, Medicine, and Man," Cornell, Ithaca, N.Y., 1947.
Snyder, L. H.: Human Heredity and Its Modern Applications, *Am. Scientist*, vol. 43, 1955.
Sonneborn, T. M.: Partner of the Genes, *Sci. American*, Nov., 1950.
Stern, C.: Two or Three Bristles, *Am. Scientist*, vol. 42, 1954.

history of animals

PART 5

On the molecular as on the organismic level, in structure as in function, every animal is *adapted* to its environment. For example, among thousands of shapes that a fish might possess, it actually possesses one which is well suited for rapid locomotion in water. A bird is cast in a form eminently suited for aerial life, yet its ancestry traces to fish. Over long periods of time, clearly, animals may change their particular adaptations in response to new environments.

Based on steady-state control and reproduction, adaptation is brought about by sex, by heredity, and by evolution. In preceding chapters we have discussed the first two of these processes and in this series of chapters we concentrate on the third. We begin in the first chapter with an account of the *mechanism* of evolution, i.e., the basic forces which produce and direct evolutionary change; and we continue in the second and third chapters with an examination of the results of this mechanism.

The results encompass a time span of about 2 to 3 billion years, a period which for convenience we may subdivide into four major segments: the origin of life on earth; the subsequent origin of the first animals; the later diversification of the ancestral animals into phyla; and the history of the phyla up to the present time. This last segment is documented to some extent by fossils, but the first three segments are not. Consequently, knowledge about the early history of life can be obtained only from a study of presently living forms, supplemented to a considerable degree by guesswork and speculation. The second chapter of this series deals with this early history, with particular emphasis on the presumed *interrelations* of animals and other types of organisms. The third chapter then describes the later history of animals, i.e., the *descent* of phyla as indicated by the fossil record.

CHAPTER 13

animal evolution: mechanisms

No zoologist today seriously questions the principle that species arise from preexisting species. Evolution on a small scale can actually be brought about in the laboratory, and the forces which drive and guide evolutionary processes are understood quite thoroughly.

That evolution really occurs did not become definitely established till the nineteenth century. For long ages man was unaware of the process, but he did wonder about the origin of his kind and of other living creatures. Indeed, he developed a succession of simple and rather crude theories about evolution. Unsupported by real evidence, these were ultimately proved untenable one by one. Yet the early ideas occasionally still color the views of those who are unacquainted with the modern knowledge.

It is advisable, therefore, that we begin this chapter with a brief survey of the historical *background* of evolutionary thought. Based on such a perspective, we may then discuss the *forces of evolution*, as these are understood today, and follow with an analysis of the *nature of evolution*, as determined by the underlying forces.

BACKGROUND

Early Notions

The earliest theory of organic creation is contained in the Old Testament: God made the world and its living inhabitants in six days, man coming last. Later ideas included those of *spontaneous generation* and of *immutability of species*, which largely held sway until the eighteenth and nineteenth centuries. Each species was considered to have been created spontaneously, completely developed, from dust, dirt, and other nonliving sources. And once created, a species was held to be fixed and immutable, unable to change its characteristics.

In the sixth to fourth centuries B.C., Anaximander, Empedocles, and Aristotle independently considered the possibility that living forms might represent a *succession* rather than unrelated, randomly created types. However, the succession was thought of in an essentially philosophical way, as a progression from "less nearly perfect" to "more nearly perfect" forms. The historical nature of succession and the continuity of life were not yet recognized. Nor was the notion of continuous succession exploited further in later centuries, for clerical dogma by and large discouraged thinking along such lines.

Francesco Redi, an Italian physician of the seventeenth century, was the first to obtain evidence against the idea of spontaneous generation, by showing experimentally that animals could not arise from nonliving sources. Contrary to notions held at the time and earlier, Redi demonstrated that maggots would never form spontaneously in meat if flies were prevented from laying their eggs on the meat. But old beliefs die slowly, and it was not until the nineteenth century, chiefly through the work of Louis Pasteur on bacteria, that the notion of spontaneous generation finally ceased to be influential.

By this time, the alternative to spontaneous generation, namely, the idea of continuity and historical succession, or *evolution*, had occurred to a number of thinkers. Some of them recognized that any concept of evolution demanded an earth of sufficiently great age, and they set out to estimate that age. Newton's law of gravitation provided the tool with which to calculate the weight of the earth. One could then bring a small weighed ball of earth to white heat and measure its rate of cooling. From such measurements, one could calculate how long it must have taken the whole earth to cool to its present state. Determinations of this sort provided the many millions of years required to fit evolution into, and this time span gradually lengthened as techniques of clocking improved. As a result of these efforts, the notion of evolution was clearly in the air when the nineteenth century began. In 1809, the first major theory of evolution was actually published. This was the theory of the French naturalist Lamarck.

Lamarck

Lamarck considered the reality of evolution as established. He believed, correctly, that to explain how evolution occurred was equivalent to explaining *adaptation*—how individual variations arise among animals and how such variations lead to the emergence of different species suited to different environments and ways of life. To account for such evolution, Lamarck proposed the two ideas of *use and disuse of parts* and of *inheritance of acquired characteristics*. He had observed that if a part of an animal was used extensively, such a part would enlarge and become more efficient, and that if a structure was not fully employed, it would degenerate and atrophy. Therefore, by differential use and disuse of various parts during its lifetime, an animal would change to some extent and would acquire individual variations. Lamarck then thought that such acquired variations were inheritable and could be transmitted to offspring.

Evolution, according to the Lamarckian scheme, would come about somewhat as follows. Suppose a given short-necked ancestral animal feeds on tree leaves. As it clears off the lower levels of a tree, it stretches its neck to reach farther up. During a lifetime of stretching, the neck becomes a little longer, and a slightly longer neck is then inherited by the offspring. These in turn feed on tree leaves and keep on stretching their necks; and so on, for many generations. Each generation acquires the gains of previous generations and itself adds a little to neck length. In time, a very long-necked animal is formed, something like a modern giraffe.

This theory was exceedingly successful and did much to spread the idea of evolution. But Lamarck's views ultimately proved to be untenable. That use and disuse do lead to acquired variations is quite correct. For example, it is common knowledge that much exercise builds powerful muscles. However, Lamarck was mistaken in assuming that such (nongenetic) acquired variations were inheritable. We may say categorically that *acquired characteristics are not inheritable.* They are effects produced by environment and development, not by genes. Only *genetic* characteristics are inheritable, and then only if such characteristics are controlled by the genes of the reproductive cells. What happens to cells other than reproductive cells through use and disuse, or in any other way for that matter, does not affect the genes of the gametes. Accordingly, although Lamarck observed some of the effects of use and disuse correctly in some cases, such effects cannot play a role in evolution.

Darwin and Wallace

The year in which Lamarck published his theory was also the year in which Charles Darwin was born. During his early life, Darwin undertook a 5-year-long circumglobal voyage as the biologist on the naval expeditionary ship *H.M.S. Beagle*. He made innumerable observations and collected a large number of different plants and animals in many parts of the world. Returning home, he spent nearly twenty years sifting and studying the collected data. In the course of this work, he found evidence for certain generalizations. Another biologist, Alfred R. Wallace, had been led independently to substantially the same generalizations, which he communicated to Darwin. Darwin and Wallace together then announced a new theory on the mechanism of evolution, which was to supplant that of Lamarck. Darwin subsequently elaborated the new theory into book form. This famous work, entitled "On the Origin of Species by Means of Natural Selection, or the Preservation of Favored Races in the Struggle for Life," was published in 1859.

In essence, the Darwin-Wallace *theory of natural selection* is based on three observations and on two conclusions drawn from these observations.

Observation. Without environmental pressures, every species tends to multiply in geometric progression.

In other words, a population doubling its number in a first year possesses a sufficient reproductive potential to quadruple its number in a second year, to increase eightfold in a third year, etc.

Observation. But under field conditions, although fluctuations occur frequently, the size of a population remains remarkably constant over long periods of time.

We shall come to appreciate the validity of this point in Chap. 16.

Conclusion. Evidently, not all eggs and sperms will become zygotes; not all zygotes will become adults; and not all adults will survive and reproduce. Consequently, there must be a "struggle for existence."

Observation. Not all members of a species are alike; i.e., there exists considerable individual variation.

Conclusion. In the struggle for existence, therefore, individuals featuring favorable variations will enjoy a competitive advantage over others. They will survive in proportionately greater numbers and will produce offspring in proportionately greater numbers.

Darwin and Wallace thus identified the *environment* as the principal cause of natural selection. Through the processes above, the environment would gradually weed out animals with unfavorable variations but preserve those with favorable variations. Over a long succession of generations and under the continued selective influence of the environment, a group of animals would eventually have accumulated so many new, favorable variations that a new species would in effect have arisen from the ancestral stock.

Laymen today often are under the impression that Darwin and Wallace's theory is *the* modern theory of evolution. This is not the case. Indeed, Darwinism was challenged even during Darwin's lifetime. What, it was asked, is the source of the all-important individual variations? How do individual variations arise? Here Darwin actually could do no better than fall back on the Lamarckian idea of inheritance of acquired characteristics. Ironically, the correct answer regarding variations began to be formulated just six years after Darwin published his theory, when a monk named Mendel announced certain rules of inheritance. But Mendel's work remained unappreciated for more than thirty years and progress in understanding evolutionary mechanisms was retarded correspondingly.

Another objection to Darwinism concerned natural selection itself. If this process simply preserves or weeds out what already exists, it was asked, how can it ever create anything new? As we shall see, natural selection actually does create novelty. The earlier criticism arose in part because the meaning of Darwin's theory was—and still is—widely misinterpreted. Social philosophers of the time and other "press agents" and disseminators of "news," not biologists, thought that the essence of natural selection was described by the phrase "struggle for existence." They then coined alternative slogans like "survival of the fittest" and "elimination of the unfit." Natural selection so came to be conceived almost exclusively as a negative, destructive force. This had two unfortunate results. First, a major implication of Darwin's theory, namely, the creative role of natural selection, was missed, and, secondly, the wrong emphasis was often accepted in popular thinking as the last and final word concerning evolution.

Such thinking proceeded in high gear even in Darwin's day. Many still did not accept the reality of evolution and were prompted variously to debate, to scorn, and to ridicule the merits of the evidence. It was felt also that evolution implied "man descended from the apes," and man's sense of superiority was duly outraged. Moreover, because evolutionary views denied the special creation of man, they were widely held to be antireligious. In actuality, the idea of evolution is not any more or less antireligious than the idea of spontaneous generation. Neither really strengthens, weakens, or otherwise affects belief in God. To the religious person, only the way God operates, not God as such, is in question.

But many were properly convinced by the evidence for evolution. However, under the banner of phrases like "survival of the fittest," evolution was interpreted to prove an essential cruelty of nature; and human behavior, personal and national, often came to be guided by the ethic of "jungle law," "might is right," "every man for himself." Only in that way, it was thought, could the "fittest" prevail. Even today, unfortunately, the mechanism of evolution is still commonly—and erroneously—thought to be a matter of "survival of the fittest."

By now, a full century after Darwin and Wallace, the emotion-charged atmosphere has cleared, and the impact of their theory may be assessed calmly. That Darwin made the greater contribution cannot be questioned. In voluminous writings, he, far more than Wallace, marshaled the evidence for the occurrence of evolution so extensively, and so well, that the reality of the process has never been in doubt since. Moreover, the theory of natural selection was the most convincing explanation of the evolutionary mechanism offered up to that time. Indeed, carrying new meaning today, it still forms an essential part of the modern theory of evolution. As now understood, however, natural selection has very little to do with "struggle," "weeding out," or "the fittest." Also, we know that Darwin and Wallace, like Lamarck, were unsuccessful in identifying the actual sources of individual variations. In short, the explanation supplied by Darwin and Wallace was incomplete, but as far as it went, theirs was the first to point in the right direction.

The modern theory of evolution is not the work of any one man and it did not arise by "spontaneous generation," fully developed. Rather, it evolved slowly during the first half of the current century, many biologists of various specializations contributing to it. The theory is the spiritual offspring of Mendel and of Darwin, but the family resemblance, though present, may not be immediately evident. We shall be concerned with this modern theory in what follows.

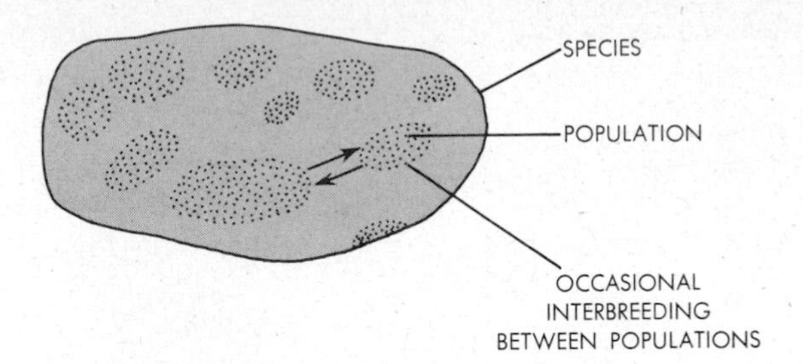

13.1. The concept of a gene pool. In a species, gene flow occurs within and between populations. The total gene content of the species thus represents a gene pool to which all members of the species have access. Gene flow normally cannot occur between the gene pools of two different species.

THE FORCES OF EVOLUTION

The Evolutionary Process

The medium of evolution is the *population*, i.e., a geographically localized aggregation of members of a given species. The raw materials of the evolutionary process are the *inheritable variations* which appear among the individuals of such a population. And the mechanism of evolution may be described as *natural selection acting on the inheritable variations of a population.*

In a population, the members interbreed preferentially with one another and they also interbreed occasionally with members of neighboring sister populations. The result of the close sexual communication within a population is a *free flow of genes*. Hereditary material present in a part of a population may in time spread to the whole population, through the gene-pooling and gene-combining effect of sex. Therefore, in the course of successive sexual generations, the total genetic content of a population may become shuffled and reshuffled thoroughly. We may say that a population possesses a given *gene pool* and that the interbreeding members of the population have free access to all components of that pool. Moreover, inasmuch as sister populations are in occasional reproductive contact, the gene pool of one population is connected also to the gene pools of sister populations. In this way, the total genetic content of an entire species continues to be shuffled about among the member animals (Fig. 13.1).

Evolution operates via the gene pools of populations. We already know from Chap. 12 how changes in genetic systems, hence inheritable variations, may arise: by sexual recombination and by mutation. In each generation, some individuals may appear featuring new trait variations, as a result of either recombinational or mutational processes (e.g., cf. Fig. 12.1). If these variant animals survive and have offspring of their own, then their particular genetic innovations will persist in the gene pool of the population. In the course of successive generations, the genetic novelty may spread to many or all members of the population.

Whether or not such spreading actually takes place depends on natural selection. This term is synonymous with *differential reproduction*. Either "natural selection" or "differential reproduction" means simply that some individuals of a population have more offspring than others. Clearly, those which leave more offspring will contribute a proportionately greater percentage of genes to the gene pool of the next generation than those which leave fewer offspring. If, therefore, differential reproduction continues in the same manner over many generations, the abundant reproducers will contribute a progressively larger number of individuals to the whole population. As a result, *their* genes will become preponderant in the gene pool of the population (Fig. 13.2).

Which individuals leave more offspring than others? Usually, but by no means necessarily, those that are *best adapted* to the environment. Being well adapted, such individuals on the whole are healthier and better fed, may find mates more readily, and may care for their offspring appropriately. However, circumstances may on occasion be such that comparatively poorly adapted individuals have the most offspring. Instances of this are sometimes encountered in human populations, for example. In any event, what counts most in evolution is not how well or how poorly an animal copes with its environment, but how many offspring it manages to leave. The more there are, the greater a role will the parental genes play in the total genetic content of the population. By and

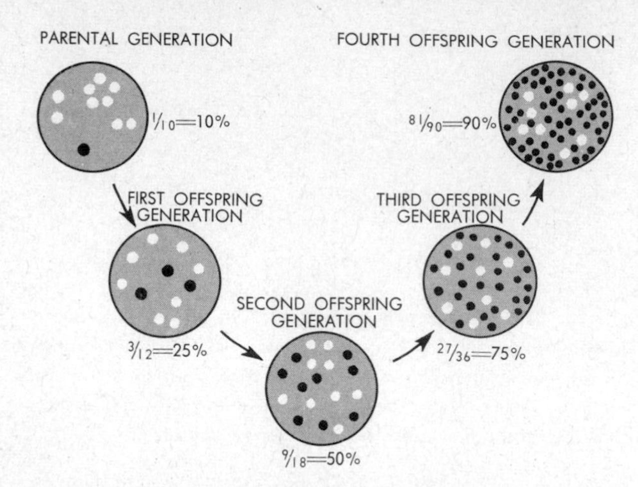

13.2. The effect of differential reproduction, or natural selection. Assume that a variation arises in one individual of a parental generation (black dot) and that the variant animal is able to leave three offspring. Each nonvariant animal (white dots) on the other hand only manages to leave one offspring. The complexion of the population will then change as shown during subsequent generations; i.e., the variant type will represent a progressively larger fraction of the numerical total. Such spreading of variations, brought about by differential reproduction, constitutes natural selection.

large, the well-adapted animal contributes most to the gene pool.

Therefore, if an inheritable variation appears in an animal and if, through differential reproduction in successive generations, the progeny of that animal becomes numerically more and more abundant, then a given genetic novelty will spread rapidly throughout the population. As a result, a trait variation originating in one animal will have become a standard feature of the population as a whole.

This is the unit of evolutionary change. Many such unit changes must accumulate in a population before the animals are sufficiently altered in structure or function to be established as a new species. All evolution operates through the basic process just described. In brief, it consists of:

1. appearance of inheritable variations by sexual recombination and mutation
2. spreading of these variations through a population, by differential reproduction in successive generations

Inasmuch as inheritable variations originate at random, evolutionary innovations similarly appear at random. And inasmuch as the best reproducers are generally the best adapted, evolution as a whole is directed by adaptation and is oriented toward continued or improved adaptation; thus it is not a random process.

Note that, in this modern view of evolution, natural selection is fundamentally a creative force; for its important effect is to spread genetic novelty, hence new traits, through a population. It is also a peaceful force, involving reproduction, not "struggle for existence" or "survival of the fittest." Animals actually struggle rather rarely. Indeed they try to avoid struggle and attempt to pursue life as inconspicuously as possible, eating when they can, reproducing when they can. Moreover, natural selection does not "eliminate the unfit." The "fit" may be the mightiest and grandest animal in the population, but it might happen to be sterile. And the "unfit" could be a sickly weakling, yet have numerous offspring. The point is that neither "survival" nor "elimination" is actually at issue. The only issue of consequence here is comparative reproductive success. Indirectly, to be sure, health, fitness, and even actual physical struggles may affect the reproductive success of animals. To that extent, such factors can have evolutionary consequences. But what in Darwin's day was regarded as the whole of natural selection is now clearly recognized to have only a limited, indirect effect on evolution. The whole of natural selection, directly and indirectly, undoubtedly is differential reproduction.

The Genetic Basis

The Hardy-Weinberg law. From the preceding, we may describe evolution as *a progressive change of gene frequencies.* This means that, in the course of successive generations, the proportion of some genes in the population increases and the portion of others decreases. For example, a mutation may at first be represented by a single gene, but if by natural selection this mutation spreads to more and more individuals, then its frequency increases whereas the frequency of the original unmutated gene decreases. Clearly, the rates with which gene frequencies change will be a measure of the speed of evolution. What determines such rates?

Suppose we consider a large population in which two alleles, *A* and *a*, occur in certain frequencies. In such a population, three kinds of individuals will be found, namely, *AA*, *Aa*, and *aa*. Let us assume that the numerical proportions happen to be

AA	*Aa*	*aa*
36%	48%	16%

Assuming further that the choice of sexual mates is entirely random, that all individuals produce roughly equal numbers of gametes, and that the genes *A* and *a* do not mutate, we may then ask how the frequency of the genes *A* and *a* will change from one generation to the next.

Since *AA* individuals make up 36 per cent of the total population, they will contribute approximately 36 per cent of all the gametes formed in the population. These gametes will all contain one *A* gene. Similarly, *aa* individuals will produce 16 per cent of all gametes in the population, and each will contain one *a* gene. The gametes of *Aa* individuals will be of two types, *A* and *a*, in equal numbers. Since their total amounts to 48 per cent, 24 per cent will be *A* and 24 per cent will be *a*. The overall gamete output of the population will therefore be

parents		*gametes*	*parents*		*gametes*
36% *AA*	⟶	36% *A*	16% *aa*	⟶	16% *a*
48% *Aa*	⟶	24% *A*	48% *Aa*	⟶	24% *a*
		60% *A*			40% *a*

Fertilization now occurs in four possible ways: two *A* gametes join; two *a* gametes join; an *A* sperm joins an *a* egg; and an *a* sperm joins an *A* egg. Each of these possibilities will occur with a frequency dictated by the relative abundance of the *A* and *a* gametes. There are 60 per cent *A* gametes. Accordingly, *A* will join *A* in 60 per cent of 60 per cent of the cases, i.e., in 60 × 60, or 36 per cent of the time. Similarly, *A* sperms will join *a* eggs in 60 × 40, or 24 per cent of the cases. The total result:

sperms		*eggs*				*offspring*
A	+	*A*	⟶	60 × 60	⟶	36% *AA*
A	+	*a*	⟶	60 × 40	⟶	24% *Aa*
a	+	*A*	⟶	40 × 60	⟶	24% *Aa*
a	+	*a*	⟶	40 × 40	⟶	16% *aa*

We note that the new generation in our example population will consist of 36 per cent *AA*, 48 per cent *Aa*, and 16 per cent *aa* individuals. These are precisely the same proportions we started with originally. Evidently, *gene frequencies have not changed.*

It can be shown that such a result is obtained regardless of the number and the types of gene pairs considered simultaneously. The important conclusion is that, *if mating is random, if mutations do not occur, and if the population is large, then gene frequencies in a population remain constant from generation to generation.* This generalization is known as the *Hardy-Weinberg law.* It has somewhat the same central significance to the theory of evolution as Mendel's laws have to the theory of heredity (Fig. 13.3).

The Hardy-Weinberg law indicates that, when a population is in genetic equilibrium, i.e., when gene frequencies do not change, the rate of evolution is zero. That is, genes continue to be reshuffled by sexual recombination and, as a result, individual variations continue to originate from this source. But the overall gene frequencies do not change. Of themselves, therefore, the variations are *not* being propagated differentially. Evolution consequently does not occur.

What does make evolution occur are deviations from the "ifs" specified in the Hardy-Weinberg law. Thus, mating is decidedly not random whenever natural selection takes place; genes actually do mutate; and populations are not always large. Singly and in combination, these three factors may disturb the genetic equilibrium of a population and may produce evolutionary change.

The effect of nonrandom mating. This effect may be appreciated readily if we assume that, in our example above, *AA*, *Aa*, and *aa* individuals are not adapted equally well. Suppose that the *A* gene in double dose, as in the *AA* combination, has a particular metabolic effect, such that death in embryonic stages will occur in one-third of the individuals possessing these genes. Under these conditions, 36 per cent *AA* individuals will be produced as zygotes, but only two-thirds of their number will reach reproductive age. Consequently, the *Aa* and *aa* individuals will constitute a proportionately larger fraction of the reproducing population and will contribute proportionately more to the total gamete output. The ultimate result over successive generations will be a progressive decrease in the frequency of the *A* gene and a progressive increase of the *a* gene. A certain intensity of natural selection, or *selection pressure,* here operates against the *A* gene and for the *a* gene (Fig. 13.4). Whenever such selection pressures exert an effect, Hardy-Weinberg equilibria are not maintained. Instead, as gene frequencies become altered more or less rapidly, given traits spread or disappear, and this represents evolutionary change. In nature, most traits are steadily being selected for or selected against. In the course of many generations, even a very slight selection pressure affects the genetic makeup of a population substantially.

The effect of mutations. Inasmuch as mutations do occur in populations, Hardy-Weinberg equilibria change for this reason also. Depending on whether a mutation has a beneficial or harmful effect on a trait, selection will be made either for or against the mutated gene. In either case gene frequencies will change, for the mutated gene will either increase or decrease in abundance.

The evolutionary effect of mutations varies according to whether the trait changes produced are dominant or recessive. A newly originated mutation with dominant effect will influence traits immediately, and selection for or against the mutation will take place at once. But if a mutation has a recessive effect, it does not influence traits immediately. Natural selection therefore does not influence the mutation immediately either. This is the case with most mutations, since, as noted in Chap. 12, most actual mutations produce recessive effects.

Nevertheless, such mutations may spread through a population. For example, an animal may carry a mutant gene a' having a recessive effect, and it may also carry a linked gene B which produces an adaptively very desirable dominant trait. Natural selection could then operate *for* the gene B; i.e., the

13.3. The Hardy-Weinberg law. If mating is random, if mutations do not occur, and if the population is large, then gene frequencies do not change from one generation to the next.

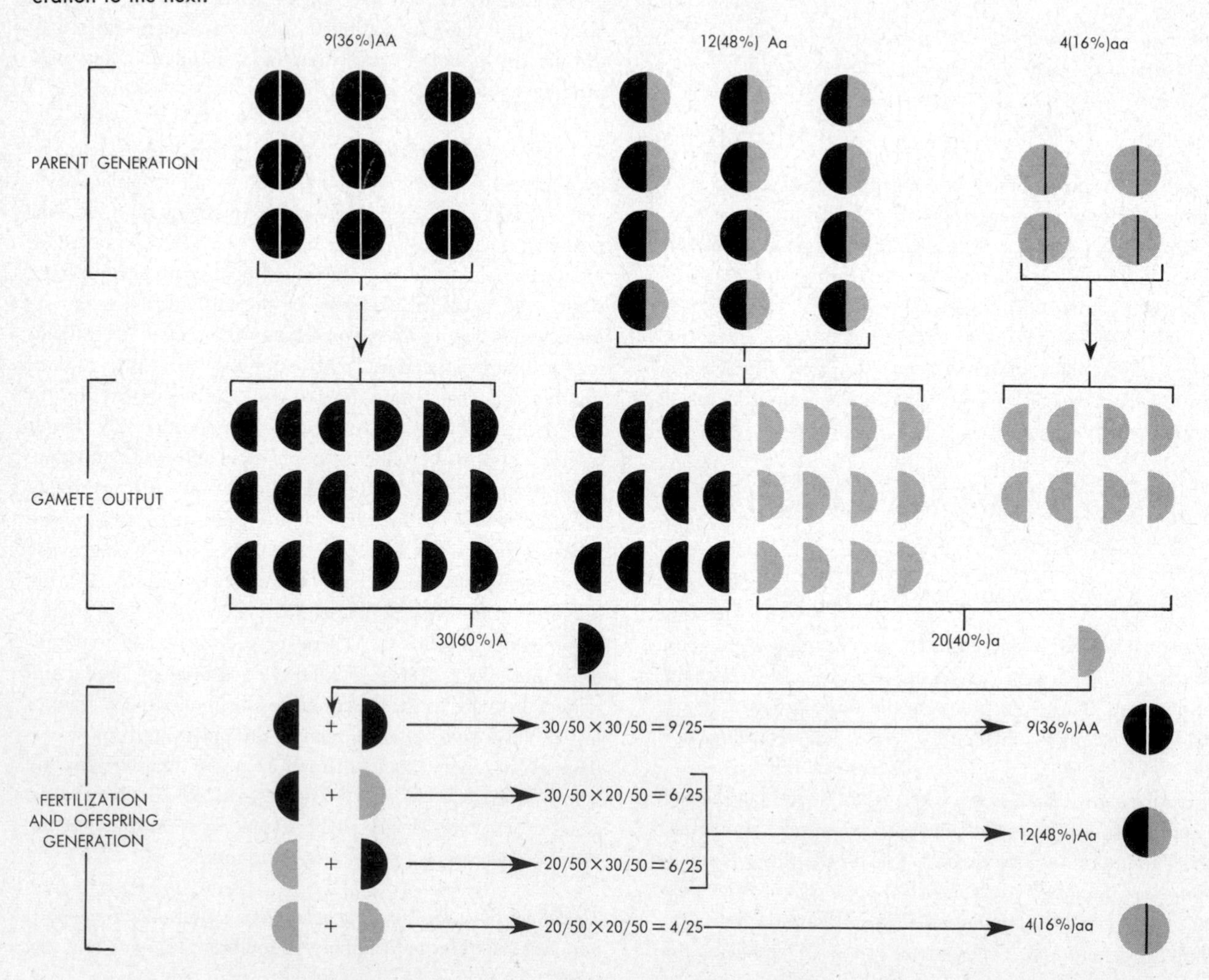

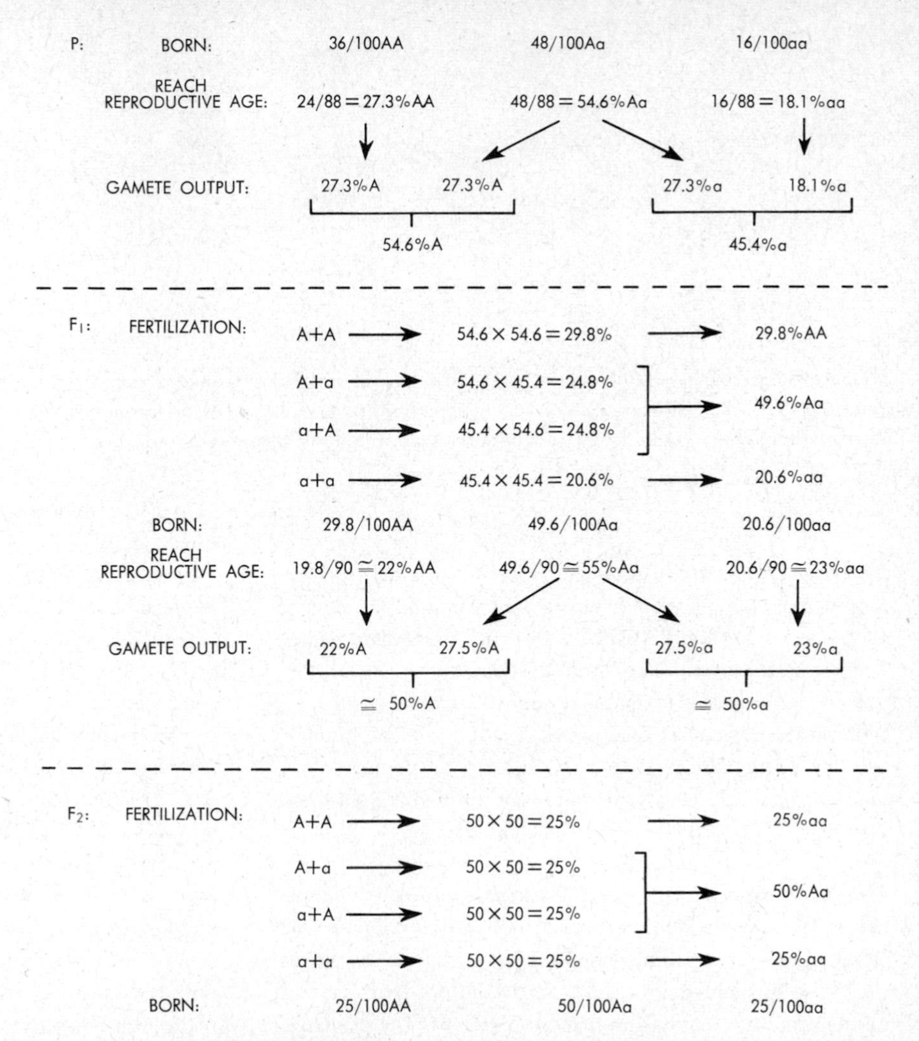

13.4. The effect of nonrandom mating. If only two-thirds of all *AA* individuals reach reproductive age, then in the course of two generations the frequency of the *A* gene will decrease and the frequency of the *a* gene will increase, as shown in the calculation.

animal possessing *B* might reproduce abundantly, and its genes would spread through the population. As a result, the linked mutant gene *a′* would be spread at the same time. Many mutations with recessive effects actually do propagate in this way, by being inherited along with other, adaptively useful genes having dominant effects. Note, moreover, that many genes are only partially recessive and that these will be subject to natural selection even more directly.

Completely recessive mutants simply accumulate in the gene pool without visible expression. However, if two individuals carrying the same mutation happen to mate, then one-fourth of their offspring will be homozygous recessive: $Aa' \times Aa' \longrightarrow$ 25 per cent $a'a'$. These offspring will exhibit altered visible traits, and natural selection will then affect the mutation directly (Fig. 13.5).

Mutational effects in evolution also vary according to how greatly a given mutation influences a given trait. A "large" mutation which affects a vital trait in major ways is likely to

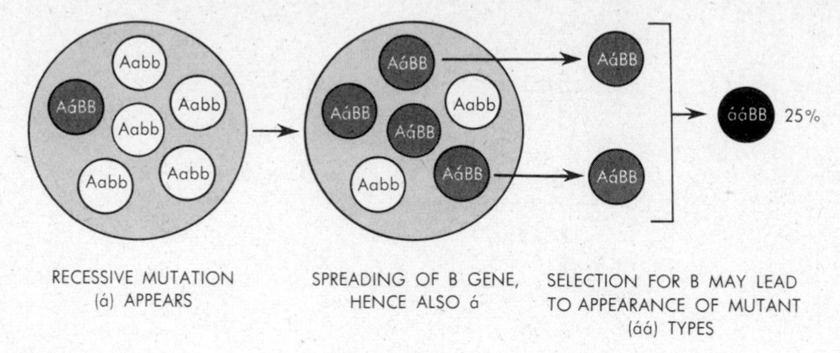

13.5. If a recessive mutation *a*′ appears in an animal and if that animal also carries a gene *B* which is strongly selected for, then both *B* and *a*′ may spread through a population. The appearance of mutant phenotypes *a*′*a*′ then becomes rather likely.

be exceedingly harmful and will usually be lethal. For example, *any* change in the principal structure and function of the heart of an animal is likely to cause immediate death. Indeed, large variations are usually eliminated as soon as they arise. By contrast, an animal may survive far more readily if a mutation is "small." Evolutionary alterations of animals actually occur almost exclusively through the accumulation of *many, small* changes in traits, not through single large changes.

The effect of population size. The third condition affecting Hardy-Weinberg equilibria is population size. If a population is large, any regional imbalances of gene frequencies which may arise by chance are quickly smoothed out by the many random matings among the many individuals. The principle underlying this holds in statistical systems generally. In a coin-flipping experiment, for example, heads and tails will each come up 50 per cent of the time, but only if the number of throws is large. If only three or four throws are made, it is quite possible that all will come up heads, by chance alone. Analogously, gene combinations attain Hardy-Weinberg equilibria only if a population is large. In small groups, chance alone may produce major deviations.

Assume, for example, that *AA*, *Aa*, and *aa* individuals are expected in a certain ratio, in accordance with existing gene frequencies. If the population contains many hundreds of individuals, this ratio will actually materialize. But if the population consists of a few individuals only, all these might by chance turn out to be of the same genotype, rather than of the three expected genotypes. We say that, in small populations, chance leads to *genetic drift,* i.e., to the random establishment of genetic types which numerically are not in accordance with Hardy-Weinberg equilibria (Fig. 13.6).

This effect resembles that of natural selection; if several genotypes are possible, a particular one would likewise come to predominate if there were a selection pressure for it. But whereas natural selection normally propagates the adaptive trait, genetic drift is governed primarily by chance and is initially not oriented by adaptation. The result is that, in small populations, nonadaptive and often bizarre traits may become established. These may actually be harmful to the population and may promote its getting even smaller. On the other hand, genetic drift may happen to adapt a given small population rather well to a given environment, and such a population might subsequently evolve selectively and eventually give rise to a new species. Genetic drift is often observed among animals on islands and in other small, reproductively isolated groups.

By way of summary, the forces of evolution may now be described as follows. First, recombinational or mutational genetic novelty originates at random among certain individuals of a population. If this novelty happens to be adaptively advantageous in a given environment and if the population is large, then greater or lesser selection pressure for the novelty will disturb the equilibria of existing gene frequencies. Consequently, this pressure of natural selection, operating through differential reproduction, will bring about a correspondingly rapid or slow propagation of the genetic innovation throughout the population. The final result will be the establishment of new adaptive traits.

13.6. Genetic drift. Given a population as at top of figure, the genetic constitution of offspring populations is influenced by population size. In large populations (left), gene combinations produced in different territories will average out to form a total offspring population in which gene frequencies are as in the parent generation. But in small populations (right), chance alone may produce significant deviations from Hardy-Weinberg expectations.

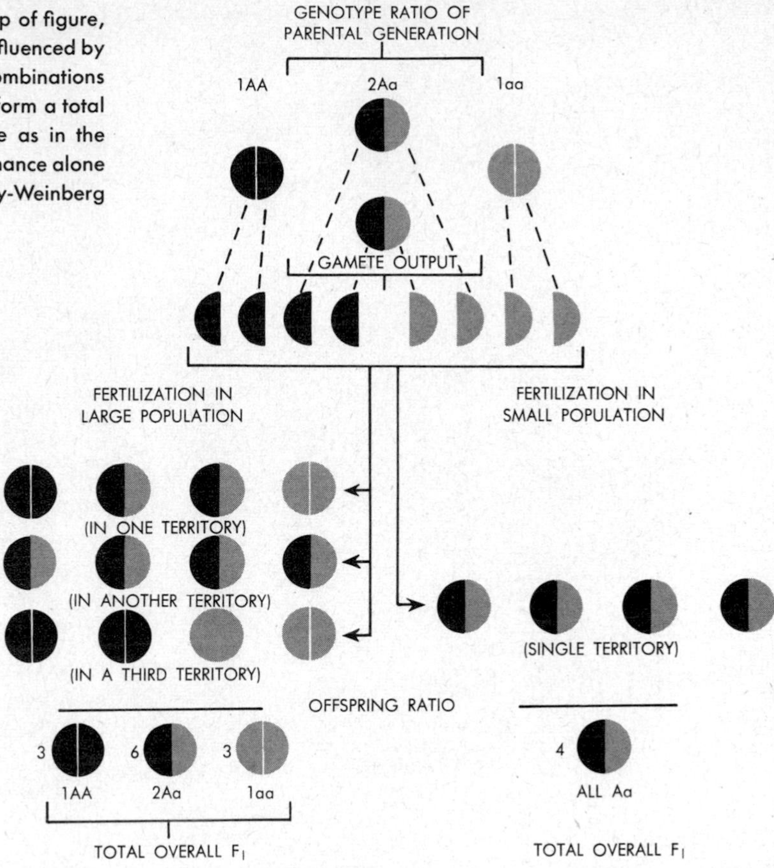

Evolution as it actually occurs must be interpreted in terms of this mechanism. That it in fact can be interpreted on this basis will become clear in the following section.

THE NATURE OF EVOLUTION

Speciation

The key process to be explained is how unit evolutionary changes in a population eventually culminate in the establishment of new species and higher taxonomic categories. As noted in Chap. 7, a species may be defined as a collection of populations within which reproductive communication is maintained by interbreeding. We may now define a species alternatively as a group of populations sharing the same gene pool (cf. Fig. 13.1). Within the pool a free flow of genes is maintained, but genetic flow between two such pools does not occur; a reproductive barrier isolates one species from another. The problem of speciation, therefore, is to show how reproductive barriers arise.

Geographic barriers between sister populations usually develop before biological reproductive barriers come into existence. Among geographic barriers, *distance* is probably the most effective. Suppose that, in the course of many generations, the populations of a given species grow in size and number and that, as a result of the increasing population pressure, the animals radiate into a progressively larger territory. In time, two populations *A* and Z at opposite ends of the territory may be too far apart to permit direct interbreeding of their members. Although gene flow still takes place via the interconnecting populations between *A* and Z, individuals of *A* and Z no longer come into reproductive contact directly (Fig. 13.7).

It is then almost certain that, by chance, different genetic innovations arise in *A* and Z and that different ones will be

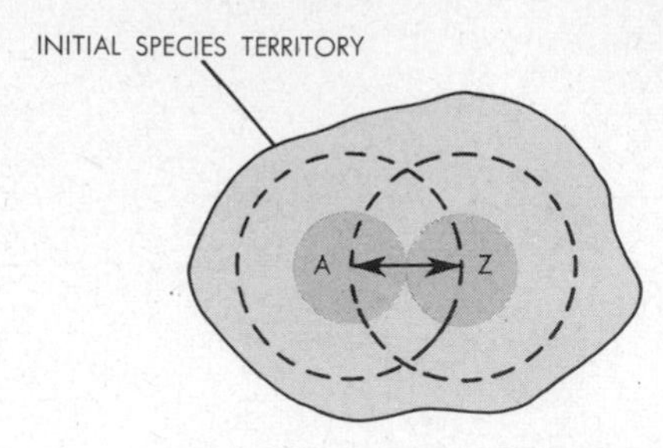

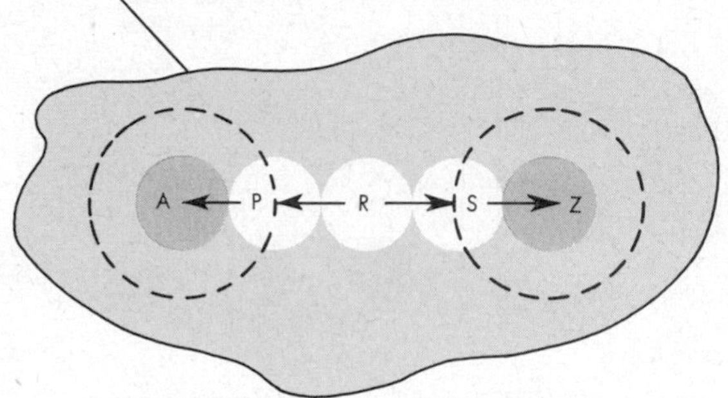

13.7. Species populations may in time become separated by distance, with the result of comparative reproductive isolation, hence a reduction of gene flow.

propagated within A and Z by natural selection. Such an effect will be particularly pronounced if the environments of A and Z are or become more or less different. If now the evolutionary changes *within* A and *within* Z occur faster than the speed of genetic flow *between* A and Z, then A and Z will actually become progressively different in structure or function. These two populations thus may come to represent two distinct *subspecies* (Fig. 13.8).

Geographic isolation here has set the stage for the development of initial differences between members of A and Z. If the differences accumulate, they may eventually become so great that gene flow between A and Z will stop altogether. For example, population A (or Z) may undergo a change in the reproductive organs such that mating with neighboring populations becomes mechanically impossible. Or the protein specificities of A may so change that the gametes become incompatible with those of neighboring populations. Or the time of the annual breeding season in A may become advanced or delayed relative to that of neighboring populations. Or the individuals of A may become changed psychologically, so that they no longer accept mates from neighboring populations. *Biological* barriers of this sort will interrupt all gene flow between A and Z. These subspecies, isolated reproductively, then in effect will have become two different *species* (Fig. 13.9).

Although an initial isolation due to distance is probably the most common kind, other forms of geographic isolation are encountered also. The development of terrestrial islands surrounded by water or of aquatic islands surrounded by land, the interposition of a forest belt across a prairie or of a prairie belt across a forest, the appearance of mountain barriers, river barriers, temperature barriers, or of many another physical barrier, each may result in geographic isolation. With reproductive contact then lost between two populations, evolution in each may henceforth follow entirely different courses. In effect, the parental species will be split into two new ones (cf. Figs. 13.8 and 13.9). At first, the descendant species will still be rather similar structurally and functionally. In time, however, evolutionary changes are likely to introduce progressively pronounced differences, including biological barriers to interbreeding. These add to and reinforce the environmental ones already in existence.

13.8. Different populations of a species may, by selective spreading of variant types, develop into different subspecies. Cf. also Fig. 16.2.

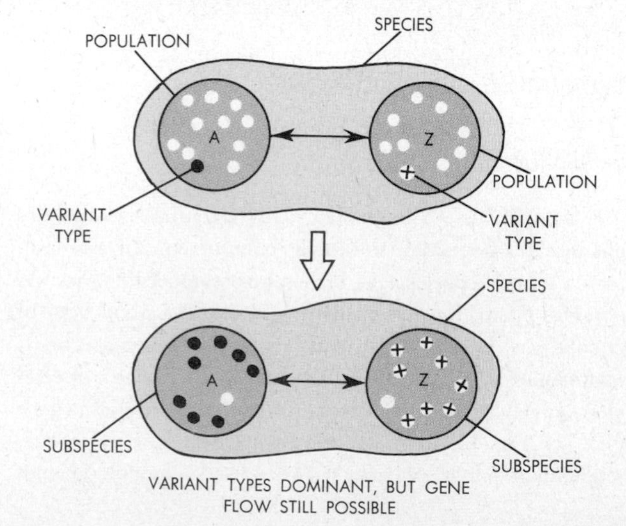

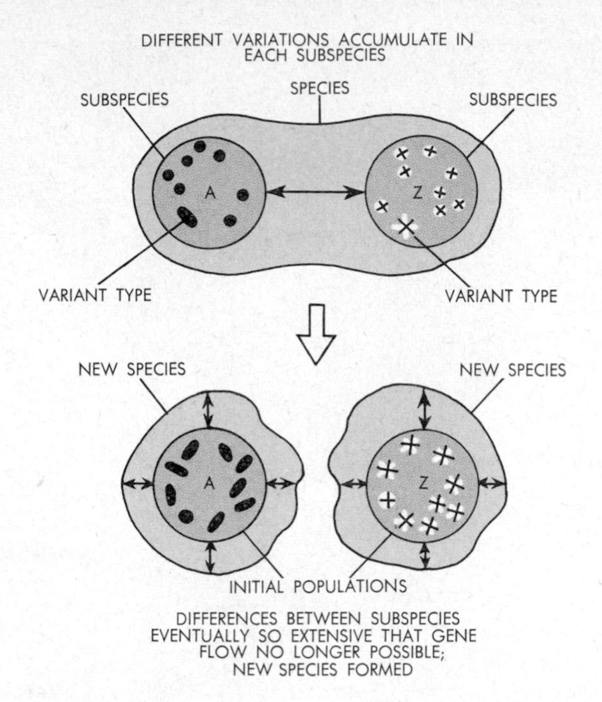

13.9. The origin of two new species from two subspecies of a single ancestral species.

In two just-formed sister species, interbreeding often may still take place if the isolating condition is removed, but in nature such removals do not normally occur. Therefore, when two different species *do not* interbreed in nature, this does not always mean that they *cannot* interbreed. In many cases, members of different species may be brought together in the laboratory and there they interbreed perfectly well. For example, swordtails and platys, two species of tropical fish (Fig. 13.10), may under certain conditions have offspring in the laboratory. But in nature they almost never do because they are isolated reproductively; i.e., although they live together in the same rivers, biological barriers discourage crossbreeding. And after two sister species have been separated for long periods, interbreeding will no longer be possible even if members of the two are brought together artificially; the biological differences sooner or later become sufficiently pronounced to preclude such interbreeding.

Speciation by this means is the principal way in which new species evolve. Such a process takes, on an average, about 1 million years. Consciously or unconsciously making use of this principle of reproductive isolation, man has been and is now contributing to the evolution of many other organisms. Here may be found direct proof that evolution actually occurs and, indeed, that it operates according to the mechanism described above.

The most ancient evolution-directing effort of man is his successful *domestication* of various plants and animals. Darwin was the first to recognize the theoretical significance of domestication, and it was this, actually, which led him to his concept of natural selection. He reasoned that if man, by *artificial selection* and isolation, can transform wild varieties of given plants and animals into domesticated varieties, then perhaps natural selection and isolation, acting for far longer periods, can bring about even greater evolutionary transformations in nature. We know now that the domesticating process in fact does involve all the elements of natural evolution: first, deliberate physical, hence reproductive and genetic, isolation of a wild population by man; and secondly, long-continued, carefully controlled, differential reproduction of individuals "adapted" to human desires, i.e., of individuals exhibiting traits considered desirable by man. The result is the creation of new strains, races, subspecies, and even species (Fig. 13.11).

13.10. Platyfish female at top, swordtail male at bottom. These animals belong to different species, and in nature they do not interbreed. But they can and do interbreed in the laboratory.

A

B

13.11. **A.** Wild boar; ancestral forms of animals such as this have also given rise to domesticated pigs, as a result of man-directed artificial selection. **B.** Red jungle fowl, another example of a wild animal from which man has bred domesticated varieties.

Furthermore, during the last few decades, rather rapid, man-directed evolution has taken place among certain viruses, bacteria, insects, various parasites, and other pests. These live now in an environment in which antibiotics and numerous pest-killing drugs have become distinct hazards. And the organisms have evolved and are still evolving increasing resistance to such drugs. Indeed, the very rapid evolution of viruses and bacteria becomes a problem in research; laboratory populations of microorganisms may evolve resistance to a drug even while the drug is being tested. Because microorganisms have exceedingly short generation times, because their populations are physically small, compact, and easily reared, and because high mutation rates may be induced readily by X rays, they have become favorite test objects in evolution experiments.

Clearly, then, small-scale evolution unquestionably occurs and is observable directly. Moreover, it may be made to occur under conditions based on the postulated modern mechanism of evolution. That this mechanism actually operates as implied by theory is therefore no longer in doubt.

Does the same mechanism operate in large-scale evolution, i.e., in the formation of higher taxonomic groups? In the recent past, a few zoologists have expressed the belief that small-scale and large-scale evolution are not governed by the same forces. Differences between orders, classes, and phyla are far too great, it has been argued, to be accounted for by a gradual accumulation of many small, minor variations among animals. Whereas the origin of species and even of genera can be explained on this basis, it has been maintained that for higher taxonomic categories a different machinery may be required.

A mechanism involving "large" mutations has therefore been postulated. According to this hypothesis, a major mutation affecting many vital traits simultaneously transforms an animal suddenly, in one jump, into a completely new type which represents a new, high-ranking taxonomic category. In most cases such an animal could not survive, for it would undoubtedly be entirely unsuited to the local environment. But it is assumed that, in extremely rare cases, such "*hopeful monsters*" might have arisen by freak chance in environments in which they could survive. Only a few successes of this sort would be needed to account for the existing major taxonomic variants among animals. Evolution by jumps would also explain why transitional fossil forms between various phyla are rare, whereas transitional fossils between different species and different genera, created by gradual evolution, are extremely common.

Few zoologists today accept the hypothesis of jump evolution. In studies of natural and experimental mutations over many years, it has always been found that sudden genetic changes with major effects are immediately lethal. This is the case not only because the external environment is unsuitable, but also, and perhaps mainly, because the internal metabolic upheaval caused by a major mutation is far too drastic to permit continued survival. Indeed, large mutations lead to death well before hatching or birth. But even supposing that a hopeful monster could develop beyond birth, it would by definition be so different from the other individuals of the population that it certainly could not find a mate. Also, although transitional fossils between major taxonomic categories are rare, they are by no means nonexistent. On the whole, therefore, it is far more consistent with available evidence to explain the evolution of high taxonomic groups on a basis other than that of fortuitous hopeful monsters.

The almost universally accepted view is that large-scale evolution is governed by the very same mechanism as small-scale evolution. Thus, the origin of high-ranking taxonomic categories is again envisaged to involve isolation and accumulation of small trait variations, only more of them than in the case of a species and accumulating for a longer period of time. Although the differences between phyla and other major categories are great, they are not so great that one such category could not have evolved gradually from another category. Indeed, the evidence from fossils and embryos shows reasonably well how such derivations might have been achieved (cf. Chap. 14). Moreover, as the next section will show, important aspects of the evolutionary process cannot be explained in terms of jump evolution but can be explained rather well in terms of gradual evolution.

Characteristics of Evolution

Rates of change. Even on the species level, evolution is an exceedingly slow process. As noted, a very large number of very small variations of traits must accumulate, bit by bit over many generations, before a significant structural or functional alteration of organisms is in evidence. Moreover, genetic innovations occur at random, whereas natural selection is directed by adaptation. Therefore, if a substantial environmental change necessitates a correspondingly substantial adaptive change in a group of animals, then the animals must *await* the random appearance of appropriate genetic innovations. If useful innovations do not happen to arise by chance, then the animals will not be able to readapt and will die out. Yet even if useful genetic novelty does arise in a given generation, there is no guarantee that more novelty of similar usefulness will originate in the next generation. In short, even though evolution may occur, it could occur too slowly to permit successful adaptation to changed environments.

The actual speeds of past evolution, though slow in all instances, have varied considerably for different types of animals, differently at different times. As a rule, the more stable a given environment has been, the slower has been the evolution of the animals living in it. Thus, terrestrial types by and large have evolved faster than marine types. Also, during periods of major geologic upheavals, e.g., in times of glaciation or of mountain building (cf. Chap. 15), evolution has been fairly rapid generally. On the other hand, in a few existing types of animals the rate of evolution has been practically zero for hundreds of millions of years. Horseshoe crabs, certain lampshells, and some of the radiolarian protozoa are among the oldest of such "living fossils" (cf. Chaps. 24, 27). In these and similar cases, the specific environment of the animals has been stable enough to make the ancient way of life still possible. Given the general evolutionary mechanism of small random variations acted on by adaptively oriented natural selection, it is not surprising that speeds of evolution should have varied in step with environmental changes.

Adaptive radiation. A general feature of evolution is the phenomenon of adaptive radiation. We have seen how, in speciation, one original parent species gives rise simultaneously to two or more descendant species. A similar pattern of *branching* descent characterizes evolution on all levels. A new type evolves, and it then becomes a potential ancestor for many different, simultaneous descendant lines. For example, the ancestral mammalian type has given rise simultaneously to several lines of grazing plains animals (e.g., horses, cattle, goats), to burrowing animals (e.g., moles), to flying animals (e.g., bats), to several lines of aquatic animals (e.g., whales, seals, sea cows), to animals living in trees (e.g., monkeys), to carnivorous predators (e.g., dogs, cats), and to many others. Evidently, the original mammalian type branched out and exploited many different available environments and ways of life. Each descendant line thereby became adaptively specialized in a particular way. The sum of the various lines, all leading away from the common ancestral type, formed an "adaptive radiation."

Within each such line, furthermore, adaptive radiations of smaller scope can take place. For example, the line of tree-living mammals in time evolved several simultaneous sublines, and each of these in turn gave rise to subsublines, etc. The specific results today are animals as varied as monkeys, lemurs, tarsiers, apes, and men. Evidently, man did not descend from the apes, if by "apes" we mean modern ones such as gorillas or gibbons. Rather, living apes and man have had a common ancestor, and they are contemporary members of the same adaptive radiation.

The important implication here is that evolution is *not* a "ladder" or a "scale." As also pointed out in Chap. 7, the pattern is more nearly that of a greatly branching bush, where the tips of all uppermost branches represent currently living species (cf. Fig. 7.3). Of these, none is "higher" or "lower" than any other. Instead, they are simply contemporary groups of different structure, function, and history. Thus, the all-too-frequent picture of evolution as a "progression from amoeba to man" is and always has been utterly without foundation. Leading down from the branch tips to progressively thicker branches, the evolutionary bush goes backward in time. Junctions of branches represent common ancestors, and these are the higher in taxonomic rank the more closely the main stem is approached.

Extinction. Not all the branches on a bush ramify right to the top, but some terminate abruptly at various intermediate points. In evolution, similarly, *extinction* has been a general feature. In many actual cases of extinction, the specific causes may never be known. But the general cause of all extinctions emerges from the nature of the evolutionary mechanism. That cause is change in environment without rapid enough re-adaptation of animals to the change. Evidently, unlike death, which is inherent in the life history of every individual, extinction is not a foregone conclusion inherent in the evolutionary history of every group. Rather, extinction occurs only if and when the group cannot make adaptive adjustments to environmental change (Fig. 13.12).

Such change need not necessarily be physical. For example, biological competition between two different types occupying the same territory often has led to the extinction of one. However, note that competition most often does not involve direct combat or "struggle." Characteristically, the competition is usually quite indirect, as when two different types of herbivores draw on the same limited supply of grass. The more narrowly specialized type here usually prevails over the more generalized type. For example, a herbivore like a rabbit is specialized to feed on vegetables. It is therefore likely to have the competitive advantage over an omnivore like a man or a bear if that omnivore happens by circumstance to be forced to eat only vegetables. The rabbit will be able to find vegetables more easily and to make more efficient use of them. On the other hand, if vegetables should disappear locally, the specialized herbivore would quickly become extinct, whereas the generalized omnivore might find other food and survive. Clearly, specialization and adaptive flexibility each has certain evolutionary advantages and certain disadvantages. The issue of survival or extinction depends on a fine balance between the two.

This probably accounts for the observation that extinction is the more common the lower the taxonomic category. Extinction of species and even of genera has been a nearly universal occurrence, but relatively few orders and still fewer classes have become extinct. And virtually all phyla that ever originated continue to be in existence today. The phylum evidently includes so broad and so far-flung an assemblage of different adaptive types that at least some of them have always persisted, regardless of how evironments have changed. Species, on the other hand, are usually adapted rather narrowly to limited, circumscribed environments. Given these rigid conditions, the chances for extinction are therefore greater.

Replacement. In conjunction with extinction, *replacement* has been another common occurrence in evolution. As noted, competition may be a direct cause for the replacement of one group in a given environment by another. For example, pouched marsupial mammals were very abundant in the Americas a few million years ago, but with the exception of forms like the opossum, they were replaced in the Western Hemisphere by the competing placental mammals. Competition is not a necessary prerequisite for replacement, however. A group may become extinct for some other reason and another group may then evolve into the vacated environment and way of life. A good example of this is provided by the *ammonites,* fossil mollusks related to the living chambered nautilus (cf. Chap. 15). Some 200 million years ago, ammonites were represented by about a dozen families. All but one of these later became extinct, and the surviving group rapidly evolved into some two dozen new families of ammonites. The

13.12. Restorations of animals which have become extinct relatively recently. **A**, dodo; **B**, Irish elk; **C**, sabertooth cat; **D**, woolly mammoths. Of these, the dodo survived the longest (till just a few hundred years ago), mammoths and sabertooths became extinct some 20,000 years ago, and Irish elks, some thousands of years before that.

latter then exploited the adaptive niche vacated by the earlier ammonites.

Replacement in this case was more or less immediate. On occasion, however, many millions of years may elapse before a new group evolves into a previously occupied environmental niche. *Delayed replacement* of this sort took place, for example, in the case of the ichthyosaurs. These large, marine, fishlike reptiles became extinct some 100 million years ago,

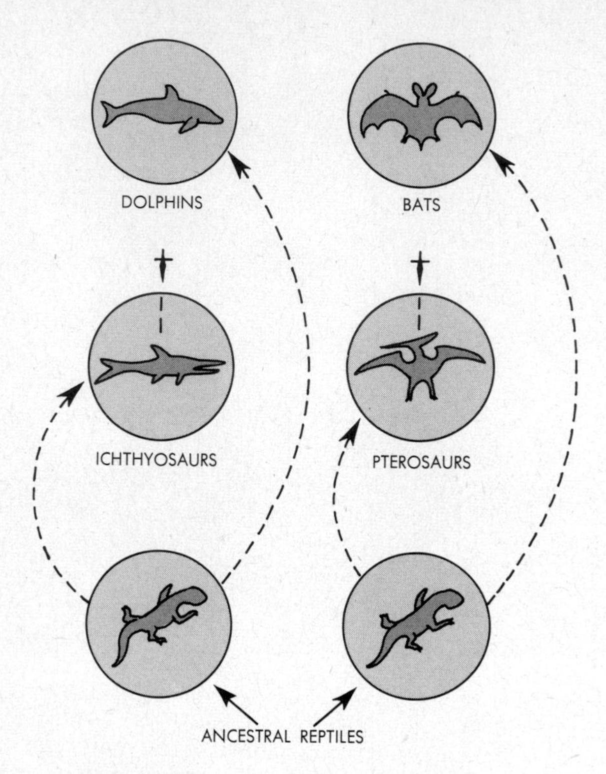

13.13. Diagrams of evolutionary replacement.

and their particular niche subsequently remained unoccupied for about 40 million years. Dolphins and porpoises evolved then, and these mammals replaced the ichthyosaurs. Similarly delayed replacement occurred between the flying reptilian pterosaurs and the later mammalian bats (Fig. 13.13).

Convergence and divergence. The phenomenon of replacement is often accompanied by that of convergence, a frequent feature in evolution generally. We have seen how, in an adaptive radiation, a common ancestral type gives rise to two or more descendant lines, all adapted in different ways to different environments. Such development of dissimilar characteristics in closely related groups is often called evolutionary *divergence*. By contrast, when two or more unrelated groups adapt to the same type of environment, then their evolution is oriented in the same direction. Such animals may come to resemble one another in one or more ways. Evolution of a common set of characteristics in groups of different ancestry is called *convergence* (Fig. 13.14).

For example, the development of wings in both pterosaurs and bats or of finlike appendages in both ichthyosaurs and dolphins illustrates evolutionary convergence in replacing forms. Inasmuch as the replacing type occupies a similar adaptive niche as the type which is being replaced, the ap-

13.14. In evolutionary divergence **(A)**, a common ancestor gives rise to different descendant lines. In evolutionary convergence, **(B, C)**, relatively unrelated ancestors give rise to rather similar lines.

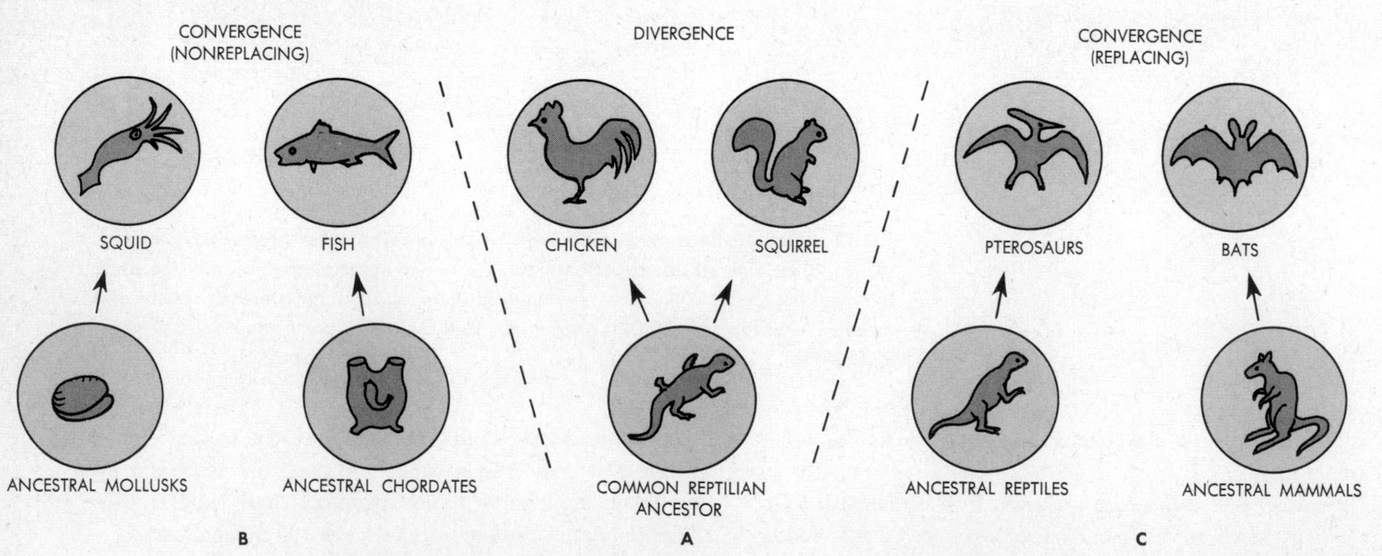

pearance of convergent features is not surprising. But convergence is also encountered in nonreplacing forms. For example, the eyes of squids and of fish are remarkably alike (cf. Chap. 8). Squids and fish are not related directly and neither replaces the other. However, both groups comprise large, fast swimmers, and good eyes of a particular construction are a distinct advantage in the ways of life of both. Selection actually has promoted variations which have led to eyes of similar structure, and the observed convergence is the result.

Opportunism. We may note that although the eyes of squids and fish are strikingly alike, they are by no means identical. Similarly, although the wings of pterosaurs and bats or of insects and birds are convergent, in the sense that all carry out the same functions of flying, the various wing types are quite different structurally and operate in different ways. Convergence leads to *similarity*, never to identity. Moreover, neither squids nor fish possess a theoretically "best" eye structure for fast swimmers, and none of the flying groups possesses a theoretically "best" wing design. Actually, the design of an organ or of a whole animal need not be theoretically "best" or "most efficient." The design only needs to be practically workable and just efficient enough for a necessary function. In a way of life based on flying, wings of *some* sort are clearly essential. But virtually all requirements for living can have *multiple* solutions, and so long as a given solution works at all it does not matter how the solution is arrived at. The various animal wings do represent multiple solutions of the same problem, each evolved from a different starting point and each functioning in a different way. All other instances of evolutionary convergence are similar in these respects.

We are led to one of the most important and most universal characteristics of evolution, that of *random opportunism*. Evolution has produced not what is theoretically desirable or best, but what is practically possible. There has been no predetermined plan, no striving for set "goals," but only the exploitation of actually available opportunities offered by selection among random hereditary changes. For example, it might have been adaptively exceedingly useful for terrestrial animals to grow wheels. But such a development did not occur, because it could not occur; the ancestors simply did not possess the necessary structural and functional potential. However, they did possess the potential to evolve adequate, workable, alternative solutions. Among vertebrates, for example, already existing fins could be reshaped into walking legs.

Clearly, evolution can only remodel and build on what

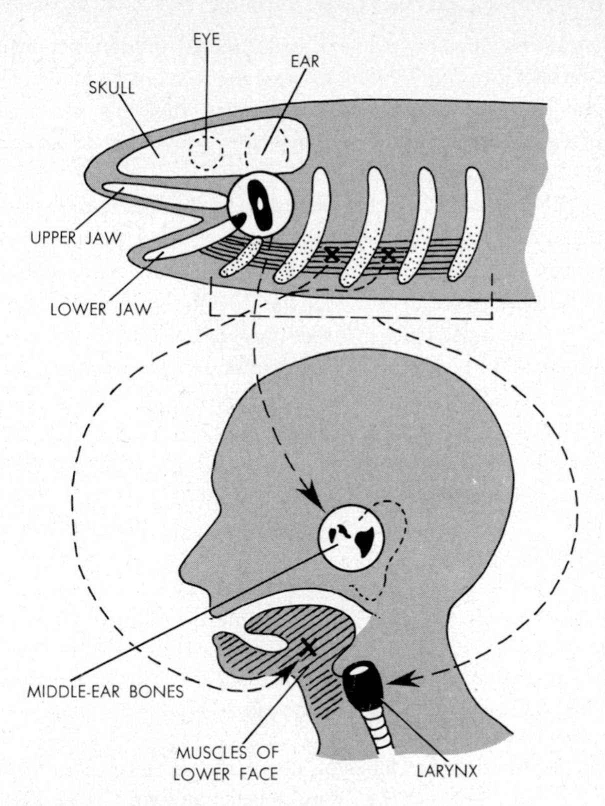

13.15. Evolutionary opportunism. Parts of the lower jaw and the upper bone of the next bony gill arch of ancestral fishes have been the evolutionary sources of the middle-ear bones of man. Analogously, parts of the other gill supports of fishes have evolved into the cartilages of the mammalian larynx, and the gill musculature of fishes has contributed the muscles of the lower face of mammals.

already exists, in small, successive steps. Since, given a long enough time span, every feature of every animal undergoes random variations in many different directions, opportunities for diverse evolutionary changes have been and still are very numerous. Some of these opportunities have been and are actually exploited.

Therefore, every animal, man not excepted, is a patchwork of good opportunities seized by selection at the right time. In man, for example, the bones of the middle ear have arisen opportunistically from pieces of earlier vertebrate jawbones. The musculature of the lower face has evolved from the gill muscles of ancestral fish. The larynx has developed from the skeletal gill supports of ancient fish (Fig. 13.15 and cf. Chaps. 31 and 32). Such instances of evolutionary opportunism are legion. We consequently conclude that specific animals are not the result of any planned, goal-directed, or predetermined course of creation. Instead, they are the result of a cumulative, opportunistic process of piece-by-piece building, based on preexisting animals and governed entirely by natural selection acting on random variations.

The above outlines the general characteristics of the evolutionary process. We have found that, in the past, evolution has proceeded at various rates through successive adaptive radiations, has led to extinction here and to replacement there, to further divergence in some instances, to convergence in others, and to opportunistic exploitation of possibilities in all. As an overall result, the living mass on earth has been increasing fairly steadily in individual numbers and types and has seeped into practically all possible environments. Indeed, it has created new environments in the process. For example, the evolution of trees has created new possibilities of life in the treetops, exploited later by very many new animals, including our own ancestors. The evolution of warm-blooded birds and mammals has created a new environment in the blood of these animals, exploited later by many new parasites. The evolution of man has created numerous new environments in human installations, and these have been exploited by a large variety of new animals.

We recognize here yet another general characteristic of evolution: a progressive, creative *expansiveness,* as regards both living mass and ways of life. The expansion is still under way, faster in some cases than in others, and the end cannot be predicted as yet.

REVIEW QUESTIONS

1. Describe the essential points of the evolutionary theories of (*a*) Lamarck, (*b*) Darwin and Wallace. How could the evolution of giraffes from short-necked ancestors be explained in terms of each of these two theories? What were the weaknesses of each theory?

2. What different kinds of inheritable variations may arise in animals? Do such variations appear randomly or are they oriented toward usefulness? How do noninheritable variations arise and what role do they play in evolution?

3. Define the modern meaning of natural selection. Show how natural selection has little to do with "survival of the fittest" or "struggle" or "weeding out" and how it is both a peaceful and a creative force. How does it happen that natural selection is oriented toward improved adaptation?

4. State the Hardy-Weinberg law. If a population consists of 49 per cent *AA*, 42 per cent *Aa*, and 9 per cent *aa* individuals, show by calculation how the law applies. If a Hardy-Weinberg equilibrium exists in a population, what are the rate and amount of evolution?

5. What three conditions disturb Hardy-Weinberg equilibria? For each condition, show in what way such equilibria are disturbed and how evolution is therefore affected. How do recessive genes spread through a population? What is genetic drift and where is it encountered?

6. Define "species" in genetic terms. Describe the process of speciation. What are some common geographic isolating conditions and what is their effect on gene pools? What is a subspecies? How do reproductive barriers arise between populations?

7. Review some actual evidence for past and present evolution. Describe the hypothesis of jump evolution. What are its weaknesses and what is the commonly accepted alternative hypothesis?

8. How have rates of evolution varied in the past? What is an adaptive radiation? Illustrate in the case of mammals. What are the general causes of extinction? What has been the pattern of extinction on different taxonomic levels?

9. How do narrow specialization and broad adaptability contribute to either extinction or survival? What is evolutionary replacement? Distinguish between immediate and delayed replacement and give examples. Distinguish between evolutionary divergence and convergence and give examples.

10. In what important way is evolution randomly opportunistic? List 10 structural and functional features of man and show for each (*a*) how it has evolved opportunistically, and (*b*) that it cannot be labeled as being "theoretically best." What has been the general evolutionary trend regarding the total quantity of life on earth? Show how evolution has created new environments, hence new opportunities for evolution.

COLLATERAL READINGS

This group of references is of outstanding historical importance:

Darwin, C.: "The Origin of Species & the Descent of Man," Modern Library, New York, 1948.

——— and A. R. Wallace: On the Tendency of Species to Form Varieties; and of the Perpetuation of Varieties and Species by Natural Means of Selection, original 1858 statement of theory of natural selection, reprinted in M. L. Gabriel and S. Fogel, "Great Experiments in Biology," Prentice-Hall, Englewood Cliffs, N.J., 1955.

Lamarck, de, J. P. P. A.: Evolution through Environmentally Produced Modifications, translation of 1809 original, in T. S. Hall, "A Source Book in Animal Biology," McGraw-Hill, New York, 1951.

Pasteur, L.: Examination of the Doctrine of Spontaneous Generation, translation of 1862 original, in Gabriel and Fogel.

Redi, F.: Experiments on the Generation of Insects, translation of 1688 original, in Gabriel and Fogel.

Any of the following books, especially the one by Mayr and the first of the two by Simpson, is recommended for background reading on evolutionary theory as a whole:

Blum, H.: "Time's Arrow and Evolution," Princeton, Princeton, N.J., 1951.

Dobzhansky, T.: "Genetics and the Origin of Species," 3d ed., Columbia, New York, 1951.

Dodson, E. O.: "Evolution: Process and Product," Reinhold, New York, 1960.

Ehrlich, P. R., and R. W. Holm: "The Process of Evolution," McGraw-Hill, New York, 1963.

Huxley, J. S.: "Evolution: The Modern Synthesis," Harper, New York, 1943.

Mayr, E.: "Animal Species and Evolution," Harvard, Cambridge, Mass., 1963.

Ross, H. H.: "A Synthesis of Evolutionary Theory," Prentice-Hall, Englewood Cliffs, N.J., 1962.

Savage, J. M.: "Evolution," Modern Biology Series, Holt, 1963.

Sheppard, P. M.: "Natural Selection and Heredity," Harper, New York, 1960.

Simpson, G. G.: "The Meaning of Evolution," Yale, New Haven, Conn., 1949; or Mentor Books, M66, New York, 1951.

———: "The Major Features of Evolution," Columbia, New York, 1953.

The articles and books below discuss various specific topics in the field of evolution:

Crow, J. F.: Ionizing Radiation and Evolution, *Sci. American*, Sept., 1959.

Deevey, E. S.: The End of the Moas, *Sci. American*, Feb., 1954.

Dobzhansky, T.: The Genetic Basis of Evolution, *Sci. American*, Jan., 1950.

Dunn, L. C.: Genetic Monsters, *Sci. American*, June, 1950.

Eiseley, L. C.: Charles Darwin, *Sci. American*, Feb., 1956.

———: Charles Lyell, *Sci. American*, Aug., 1959.

Kettlewell, H. B. D.: Darwin's Missing Evidence, *Sci. American*, Mar., 1959.

Lack, D.: Darwin's Finches, *Sci. American*, Apr., 1953.

Metcalf, R. L.: Insects vs. Insecticides, *Sci. American*, Oct., 1952.

Ryan, F. J.: Evolution Observed, *Sci. American*, Oct., 1953.

Stebbins, G. L.: Cataclysmic Evolution, *Sci. American*, Apr., 1951.

CHAPTER 14

animal origins: phylogeny

One of the main lines of investigation revealing the time course of past evolution is *paleontology*, the study of fossils. Representing the remains of formerly living individuals, fossils provide the most direct evidence of the kinds of animals in existence at various earlier times. Unfortunately, the animal fossil record does not go back more than 500 million years, a span of time representing only the last quarter or so of living history. Moreover, the fossil record shows that all animal phyla known today were already in existence 500 million years ago, when fossils themselves were first being formed. Consequently, animal history prior to the 500-million-year mark can only be inferred indirectly. This early history includes most particularly the *origin of life* in the form of the first cells, the *origin of early groups* from which modern ones later evolved, and the *origin of animals* as a diversified assemblage of organisms exhibiting a variety of body plans.

Indirect inference about these events must be based chiefly on *comparative morphology*, the study of the structure of presently living animals and other organisms. Being the products of past organisms, modern ones reflect in their design the evolutionary history of their antecedents; from molecules to organ systems, all body parts of organisms existing today embody the record of past evolution. A first problem is to detect and to interpret the historical clues contained in this living record. And a second is to construct from such clues a consistent picture of *phylogeny*, i.e., of evolutionary interrelations among organisms. Since independent evidence from fossils is unobtainable, phylogeny is necessarily rich in hypothesis and speculation, and divergent views are correspondingly abundant. One of the functions of this chapter is to present some of the chief phylogenetic hypotheses now current.

THE ORIGIN OF LIFE

It is thought today that life originated through a progressive series of synthesis reactions; atoms combined into simple compounds, these combined into more complex ones, and the most complex compounds formed eventually became organized into living cells.

The details of these processes are at present known only partly. Some of the existing knowledge results from a backward projection of cellular types and activities encountered today. For example, one may deduce from viruses, bacteria, and other primitive existing forms what the earliest living forms might have been like. Other clues come from astronomy, physics, and geology, sciences which contribute information about the probable physical character of the ancient earth. Important data are also provided by ingenious chemical experiments designed to duplicate in the laboratory some of the steps which many millennia ago may have led to the beginning of life.

What has been learned in this way indicates that living creatures on earth are a direct product of the earth. Moreover, there is every reason to believe that living things owe their origin entirely to certain physical and chemical properties of the ancient earth. Nothing supernatural appears to be involved—only time and natural physical and chemical laws operating within the peculiarly suitable earthly environment. Given such an environment, life probably *had* to happen. Put another way, once the earth had originated in its ancient form, with particular chemical and physical properties, it was then virtually *inevitable* that life would later originate on it also. The chemical and physical properties of the earth permitted certain chemical and physical reactions to occur, and one result of these reactions was something living. We may infer, moreover, that if other solar systems possess planets where chemical and physical conditions resemble those of the ancient earth, then life would originate on these other planets as well. Indeed, it is now believed strongly that life occurs not only on this earth but probably widely throughout the universe as well.

The life-producing chemical and physical properties of the early earth were a result of the way the earth and our solar system as a whole came into being to begin with. Available evidence indicates that the solar system is anywhere from 5 to 10 billion years old. Several hypotheses have been proposed to explain how the sun and the planets were formed. According to one, now widely accepted, the whole solar system started out as a hot, rapidly rotating ball of gas. This gas was made up of free atoms. Hydrogen atoms probably were the most abundant, and other, heavier kinds were present in lesser quantities. The sun was formed when most of this atomic gas, hence most of the hydrogen, gravitated toward the center of the ball. Even today, the sun is composed largely of hydrogen atoms. A swirling belt of gas remained outside the new sun. Eddies formed in this belt and in time it broke up into a few smaller gas clouds. These spinning spheres of fiery matter were the early planets.

The earth thus probably began as a glowing mass of free hydrogen and other types of atoms. These eventually became sorted out according to weight. Heavy ones, such as iron and nickel, sank toward the center of the earth, where they are still present today. Lighter atoms, such as silicon and aluminum, formed a middle shell. The very lightest, such as hydrogen, nitrogen, oxygen, and carbon, collected in the outermost layers (Fig. 14.1).

At first, temperatures were probably too high for the formation of compounds—bonds would have been broken as fast as they might have formed. But under the influence of the cold of cosmic space, the earth began to cool down gradually. In time, temperatures became low enough to permit the formation of relatively stable bonds between atoms. Compounds then appeared in profusion and free atoms largely disappeared. With this we reach the beginning of the chemical history of the earth, which henceforth will accompany the physical history. As far as they are known or suspected, what were the life-producing chemical reactions?

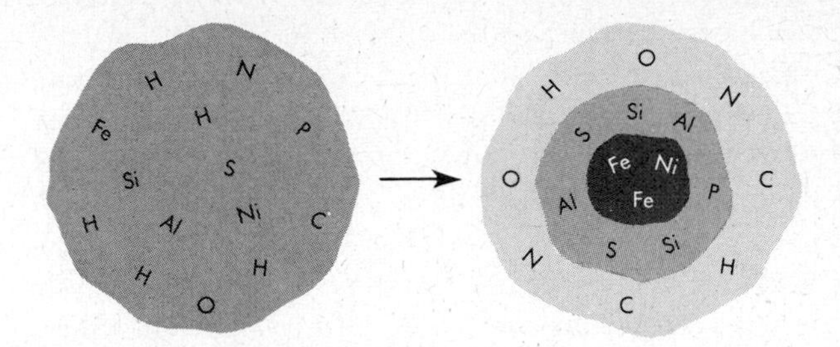

14.1. The elements composing the early earth became sorted out according to weight. Heavy elements like iron sank to the center; lighter ones like silicon formed a middle shell; and very light elements like hydrogen collected into an outer mantle.

Chemical Evolution

Among the lightest and most abundant materials in the surface gas of the early earth were, as noted above, atoms of hydrogen, oxygen, carbon, and nitrogen. Consequently, when temperatures became low enough to allow formation of compounds, the atoms of these four elements must have played a conspicuous role. (And it is therefore not a coincidence that, even today, some 95 per cent or so of the substance of every living creature consists of just these four elements.)

What simple compounds could have formed from the four elements? On the basis of their known chemical properties and their presumed relative abundance on the early earth, H, C, O, and N should have joined into some half-dozen different combinations: water (H_2O); methane (CH_4); ammonia (NH_3); carbon dioxide (CO_2); hydrogen cyanide (HCN); and hydrogen molecules (H_2). We have evidence that at least the first three of these compounds actually came into being not only on the early earth but on other planets as well. For example, on the cold, distant planet Jupiter, water, methane, and ammonia are present today in the form of thick layers of permanently frozen solids. Apparently, these compounds must have formed there as on earth, but at that great distance from the sun the surface of the planet probably froze before much additional chemical change could occur. On the hot earth, by contrast, the early compounds could interact further and give rise to new compounds in the later course of time.

Moreover, there is every reason to believe that, under the

conditions prevailing on the early earth, simple compounds which theoretically could have appeared actually did appear. On this basis, the formation of H_2, H_2O, CH_4, NH_3, CO_2, and HCN must have been a strong probability. Practically all of the hydrogen molecules (and any free hydrogen atoms still present) must soon have boiled off the surface layers of the earth, for the gravitational attraction of the comparatively small planet could not have been great enough to hold these extremely light substances. The other compounds remained and constituted a hot, gaseous *atmosphere.*

In time, as the gas ball which was the earth continued to cool, temperatures became low enough to allow some of the gases to liquefy and some of the liquids in turn to solidify. Heavy substances near the center of the earth probably tended to liquefy and solidify first. But the heat of the materials prevented complete solidification, and to this day the earth contains a hot, thickly flowing, deformable center. On the other hand, the middle shell of lighter substances did congeal, and a solid, gradually thickening crust developed. As the crust thickened and cooled, it wrinkled and folded and gave rise to the first mountain ranges. Overlying this crust was the outer atmospheric mantle, which at temperatures then prevailing still remained gaseous.

Then the rains started. All the water on earth up to this stage was in the atmosphere, forming clouds probably hundreds of miles thick. The solidifying crust underneath at first was sufficiently hot that any liquid water would boil away instantly. But eventually the crust became cool enough to hold water in liquid form. Then rain began falling in unceasing, centuries-long downpours. Basins and shallows filled up and torrential rivers tore down from the mountains. The oceans formed in this way.

Dissolved in these seas were quantities of the atmospheric CH_4, NH_3, CO_2, and HCN, compounds which persist as gases at temperatures at which water is liquid. Also accumulating in the ocean were salts and minerals. At first there were none, but as the rivers eroded the mountainsides and dissolved them away and as violent tides battered the shores and reduced them to powder, salts and minerals came to be added to the ocean in increasing quantities. Moreover, massive submarine bursts of molten lava probably erupted frequently through the earth's crust, and they too added their substance to the mineral content of the world's waters. Thus the oceans acquired their saltiness relatively early and to a small extent they became saltier still during subsequent ages (Fig. 14.2).

The formation of large bodies of liquid water containing the early atmospheric gases and many minerals in solution was the key event which made the later origin of life possible. Water was and is now the most essential single component of living matter. This fundamental role of water traces primarily to two of its properties.

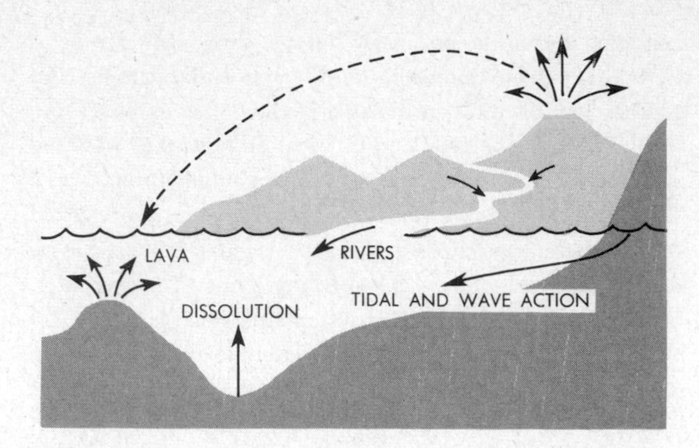

14.2. The original sources of oceanic salt. Some was derived from volcanoes, both submarine and terrestrial; some was dissolved out of the sea bottom; a third source was tidal action, which crumbled and dissolved the shore lines; and a good deal of salt also came from the land surface, leached out by rain and rivers.

First, water is virtually the best of all possible solvents. It dissolves a greater variety of substances and greater quantities of each substance than practically any other liquid. This means that it is an excellent medium for chemical reactions. Chemical processes also occur in gases and solids, but many more can occur in liquids and much more readily. Since living processes are based on chemical processes, the abundant supply of liquid water on the early earth was a promising circumstance.

Secondly, water was originally the only good source of hydrogen and oxygen. Both elements have exceedingly useful properties, and the construction of a living system on a chemical basis virtually demands their availability. But, as noted, free hydrogen and free oxygen became unavailable soon after the origin of the earth. Water molecules then came to serve as the principal suppliers. Water remains today virtually the only usable source of hydrogen and one of the important sources of oxygen.

Thus, oceanic water set the stage for the formation of living matter. The actors on this stage were the various gases and minerals dissolved in water, plus water itself. And the title role was played by the carbon atoms present in gases such as methane; for from methane could be formed numerous kinds of *organic* compounds, i.e., compounds containing *linked* carbon atoms and comparatively large amounts of chemical energy. The formation of organic compounds would have required external energy sources, but at least two such sources must actually have been available.

One of these was the sun. Although the dense cloud layers of water vapor at first must have prevented sunlight from reaching the earth's surface (which must have made the earth quite dark for long ages), the ultraviolet rays, X rays, and other high-energy radiations of the sun must have penetrated the clouds well. Some of the radiation could have provided the necessary energy for reactions among methane, ammonia, hydrogen cyanide, and water. Solar radiation certainly is known to support various chemical reactions today.

Moreover, a second energy source must have been the powerful electric discharges in lightning, which must have occurred almost continuously in the early cloud-laden, storm-lashed atmosphere. Like solar radiation, lightning is capable of activating and sustaining chemical reactions. Either lightning or solar energy (or also radioactive materials in the earth's crust) could have acted directly on the gas molecules of the atmosphere, as still happens today to some extent. The resulting aerial chemicals could then have been washed down into the seas by rain. Alternatively, reactions could have taken place directly in the waters of the ocean, where methane and all other necessary ingredients were dissolved.

That simple organic materials can indeed be created in such fashion was demonstrated in the early 1950s through dramatic and now classic laboratory experiments. In these experiments, the presumable environment of the early earth was duplicated in miniature. Into a flask were put inorganic mixtures containing water, methane gas, and ammonia gas, and electricity was discharged through these mixtures for several days to simulate the lightning discharges of the early earth. When the contents of the flask were then examined, many amino acids, fatty acids, and other simple organic compounds were actually found to be present.

Thus there is excellent reason to think that, under the impact of early energy sources, simple gases and other inorganic materials not only could but probably did react with one another and gave rise to a variety of simple organic compounds which accumulated in the ancient seas. These organic substances then constituted the biological "staples" out of which more complex organic materials could become synthesized (Fig. 14.3).

It now appears quite plausible from a chemical standpoint that proteins and nitrogen bases could have formed on the early earth from simpler biological staples already present. For example, recent experiments show that if concentrated mixtures of amino acids are heated under nearly dry conditions, proteinlike complexes are formed. Similarly, mixtures of other appropriate starting compounds heated under almost dry conditions can yield products having the characteristics of nitrogen bases (Fig. 14.4). Proteins would then have provided enzymatically active materials, and the chemical tempo on earth would henceforth have quickened substantially. Among their other catalytic effects, the proteins might have promoted the combination of nitrogen bases, simple sugars and phosphates into nucleotides; and the latter are only a simple chemical step away from energy carriers such as ATP. With enzymes and ATP then being available, a few further chemical steps lead to hydrogen carriers such as coenzymes and to nucleic acids such as DNA and RNA (cf. Chap. 6). To be sure, direct laboratory demonstration of these last steps of synthesis is still lacking. Even so, there does not appear to be any basic chemical problem in envisaging that such syntheses can, and probably did, take place.

With the formation of proteins and nucleic acids, all the essential ingredients for the eventual origin of a living entity must have been in existence. Events up to this point may be referred to collectively as *chemical evolution:* the production and gradual accumulation in the early seas of all the various compounds which later came to function as components of living matter. The most fundamental of these compounds must have included at least seven categories of substances, namely, inorganic materials such as water and dissolved mineral substances, and organic materials such as adenosine phosphates, carbohydrates, fats, proteins, and nucleic acids.

The events which followed subsequently may be described as *biological evolution:* the actual putting together of the chemical components into the first living units, viz., cells. Note here that chemical evolution did not simply stop at one point and that biological evolution then took over. On the contrary, chemical evolution continued and indeed goes on even now. Rather, at some stage of chemical evolution an *additional* kind of creation must have taken hold. Molecules no longer gave rise just to new molecules only, but some of the molecules also produced something entirely new, something hitherto completely nonexistent, namely, fully living cells. These in turn then produced more cells through processes of multiplication still going on today. In other words, the new dimension of biological evolution must have become *superimposed* on the still continuing older dimension of chemical evolution.

How did the first cell arise?

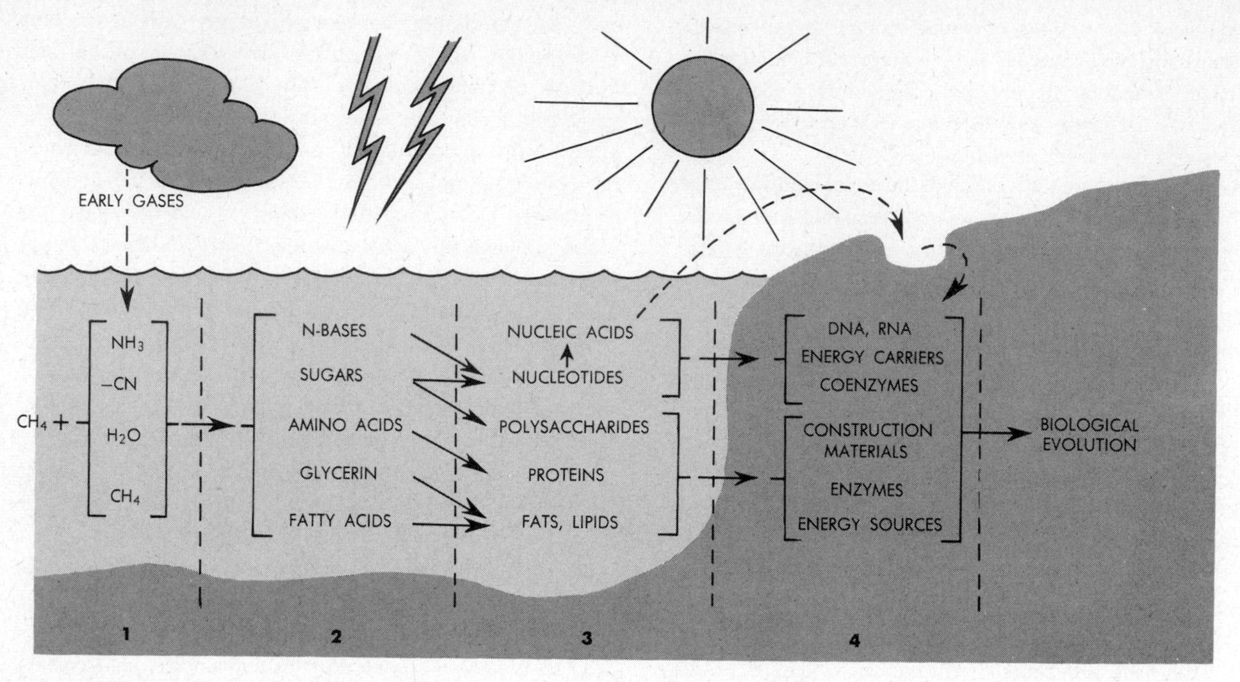

14.3. Summary of probable early synthesis reactions constituting chemical evolution. At least three or four successive phases may be envisaged to have occurred. The original gaseous raw materials came from the early atmosphere, and with the aid of energy from lightning and the sun, key biological compounds were progressively synthesized in the ocean. The later, more complex synthesis reactions perhaps took place in sand pockets along the shore, where required ingredients could become highly concentrated by evaporation of water. After the origin of the first living units, the latter presumably were washed out to sea and further biological evolution then must have occurred in the sea.

Biological Evolution

The nature of the events which resulted in cell formation can be deduced in very general outline only; the details are considerably less clear than even those of chemical evolution. Conceivably, the same end result may have been achieved in several different ways. However, we might assume that, by one means or another, sets of all the key compounds already present in the early ocean must have collected together in tiny spaces; and each set of materials so accumulated must have remained aggregated in a cohesive drop. By virtue of the various properties of the aggregated materials, the drop would have been alive.

If this describes the overall process, we may ask first whether or not such aggregates could actually have formed by known physical and chemical means. Certain possibilities which might appear reasonable theoretically must probably be regarded as unlikely in practice. For example, it is probably unlikely that aggregation of the necessary compounds could have occurred directly in the open ocean. The concentration of compounds must have been quite low, and the open water must therefore have been too dilute a solution to provide a reasonable chance for repeated aggregation of just the right sets of materials. Moreover, even if such aggregation could have occurred, it is difficult to imagine how the aggregated materials could have been kept together in open water for any length of time. After making chance contacts, the compounds most likely would have dispersed again.

It is physically and chemically more plausible to assume that the critical aggregations took place along the shores of the ocean. The solid ground available there would have provided appropriate surfaces to which oceanic molecules could have adhered. Many organic compounds are known to be readily *adsorbed* to various surfaces, a property which we often recognize as stickiness. For example, sugars, fats, and proteins stick very readily to many kinds of surfaces, and nucleic acids are extraordinarily adsorbable. Also, finely divided sand and clay particles are excellent adsorbing materials and such particles must have been abundant along the ocean shore. Accordingly, it is reasonable to think that some of the organic compounds which had formed in the ocean were washed to the shore, where some of them became adsorbed more or less at random to various surfaces. This process might have occurred progressively. The adsorbing surfaces could have trapped those molecules which did happen to make contact, and other molecules of the same or of different types might or might not have become added later. The concentrations of the molecules so would have increased slowly, a process which would have been reinforced by considerable evaporation of water in the tide zone.

Moreover, it is not necessary to assume that nucleic acids, proteins, complex polysaccharides, and large molecules generally first had to be formed in the open ocean and then had to become aggregated along the shore. On the contrary, it is physically and chemically more plausible to think that the first aggregations involved only the relatively simple organic compounds. Large molecules later could have become synthesized directly within, and indeed as a consequence of, the simple accumulations. Thus, if microscopic organic pockets could become established along the shore, and if these accumulated high enough concentrations of biological staples, then such systems probably would have been adequate for the production of larger molecules within them. In line with the heating experiments cited earlier (cf. Fig. 14.4), evaporation of water and solar heat would actually have promoted the synthesis of more complex substances.

It is unlikely, therefore, that compounds as complex as proteins and nucleic acids could have formed by the chance accumulation of enough ATP, amino acids, and nucleotides in the open ocean. It is more likely that proteins and nucleic acids arose *after* the precursor compounds had already become collected together in a small enough space and in high enough concentrations to provide an adequate and perhaps fairly dry chemical environment for synthesis. And after nucleic acids had become synthesized, they themselves could then have directed the synthesis of specific, no longer randomly formed proteins, again right within the aggregations already present. Some of these proteins would subsequently have acted as specific enzymes and so would have facilitated the rapid formation of a wealth of new compounds, e.g., polysaccharides out of sugar raw materials.

Moreover, some of the proteins and also some of the fats and carbohydrates would have represented building materials. These could have become organized into a structural framework which ramified through and around the aggregated drop. Many proteins are known to precipitate out of solution and to form solid granules or threadlike fibrils, for example. Also, as has been shown in Chap. 4, mixtures containing proteins or protein-fat complexes may, as a consequence of their physical state, form membranous surface films (e.g., like the "skins" on custards). And the heating experiments described above similarly show that proteinaceous aggregates form surface membranes quite readily. Moreover, polysaccharides like cellulose similarly may form fibrils or films. Thus the aggregated drops on the ocean shore could well have developed external boundary membranes and some measure of internal scaffolding. They would henceforth have been distinctly individual units marked off from the surrounding ocean water, and they would have remained individualized even if they absorbed more water and were later washed back into the open ocean. Indeed, such units would have been primitive cells (Fig. 14.5).

14.4. Summary of thermal-synthesis experiments in the test tube. Such experiments indicate that, even under essentially simple conditions, amino acids can give rise to proteinlike complexes having a primitive gross structure (top), and that nucleotide precursors such as ureidosuccinic acid can arise from relatively simple starting compounds (bottom). Consult Fig. 6.14 and verify that the basic pattern of thermal synthesis in the test tube, as outlined in the bottom series, actually does parallel that of biological synthesis in the living cell.

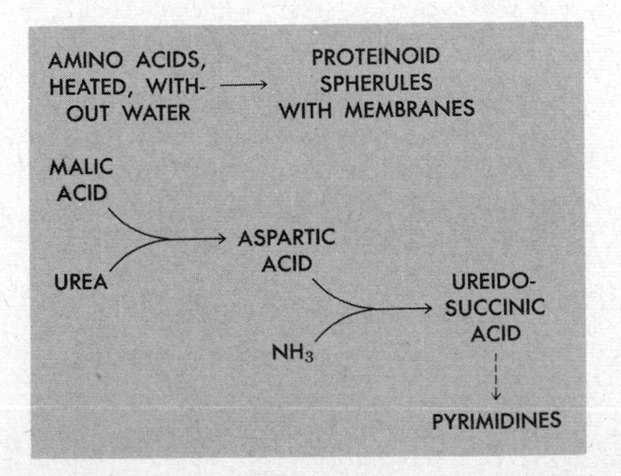

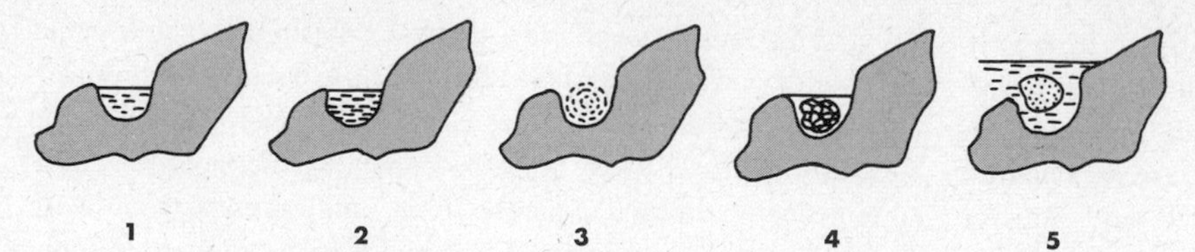

14.5. The possible origin of the first cells. Appropriate chemical ingredients might have accumulated by adsorption in microscopic pockets along the seashore **(1)** and these ingredients could have become concentrated progressively **(2)**. Under relatively dry conditions and perhaps with the aid of ATP, which might have been present, nucleic acids and proteins could have become synthesized **(3)**. The proteins then would have permitted the occurrence of enzymatically accelerated reactions and the formation of structural membranes and internal fibrils **(4)**. Finally, primitive cellular compartments might have been washed out to sea **(5)**.

Undoubtedly, numerous trials and errors must have occurred before the first actual cells left their places of birth. Many and perhaps most of the aggregations on the shore probably were "unsuccessful." In given cases, for example, the right kinds of starting ingredients might never have come together; or the right amounts of ingredients might never have accumulated; or the mixture might have dried up completely; or it might have been washed out to sea prematurely and dispersed. Clearly, numerous hazards must have led to many false starts and to many incomplete endings. Yet when an appropriate constellation of materials did form a persisting aggregate, the formation of a cell would have been a likely result.

A basic problem in tracing the possible origin of cells (and of later groups of organisms generally) is to decide whether they evolved *monophyletically*, from a single common ancestral stock, or *polyphyletically*, from several separate ancestral stocks (Fig. 14.6). Did early cells arise from a single first cell, an archancestor of all life, or did numerous "first" cells originate independently? The answer to this question (and to similar ones relating to many later groups) is simply not known. Nevertheless, biologists often try to arrive at a tentative answer on the basis of generalized arguments. For example, proponents of monophyletism invoke a *probability* argument: cells are so complex that multiple origins of such complexities are statistically quite unlikely. The argument leads logically to the view that all life started in a single place at a single time, from a single cell which formed by a "lucky accident."

Proponents of polyphyletism counter this line of reasoning with a probability argument of their own. They point out, first, that compared to their later complexities, the original cells could not really have been so complex. Initially they probably were little more than loose aggregates of chemicals surrounded by membranes, and structural complexity must have evolved subsequently, in gradual steps. Secondly, if the required conditions for original cell formation could actually have developed in one place at one time, it is statistically most likely that such conditions also developed in other places and at other times. In one sense, therefore, chance undoubtedly did play a role in cell formation: many aggregations never became cells, and those that did formed by the chance accumulation of the right ingredients. But in another sense, cell formation was not simply an enormously "lucky accident," a one-time occurrence of very remote probability. On the contrary, given an early earth so constituted that certain compounds could form and given these compounds and their special properties, then cell formation *had* to take place sooner or later, inevitably and *repeatedly*. The only element of chance here was time; the uncertainty was not in the nature of "if," but in the nature of "when" and "how often." Accord-

14.6. Monophyletism vs. polyphyletism.

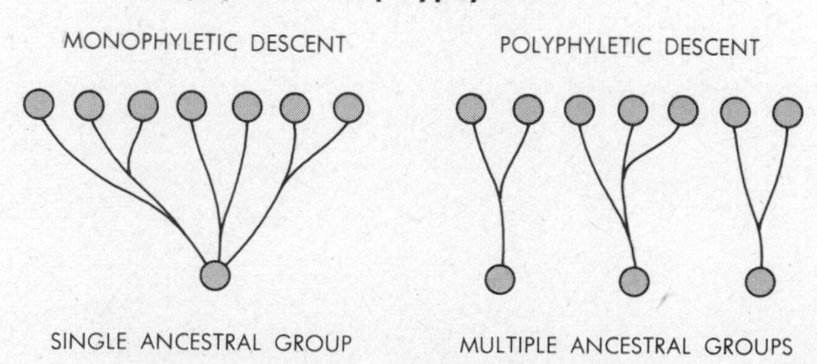

ingly, the origin of cells was as little an accident as is the eventual appearance of sevens and elevens in a succession of dice throws.

Monophyletists also invoke a *homology* argument: all cells are fundamentally so remarkably alike in composition, in chemical and microscopic structure and in function, that a common origin from a single ancestry seems clearly indicated. Polyphyletists here counter with an *analogy* argument: all aggregates which actually became living cells must *necessarily* have shared the same broad structures and functions; for those aggregates that did not simply would never have remained "alive," by definition of this term. Thus, all living cells must have contained at least water, various mineral substances, adenosine phosphates, carbohydrates, fats, proteins, and nucleic acids. Only aggregates possessing these substances would have been able to metabolize and to self-perpetuate. Biological staples still in the ocean would have provided nutrient raw materials; internal decomposition reactions with the aid of enzyme proteins, nucleotide coenzymes, and ATP would have permitted respiration; and nucleic acids would have constituted genes which controlled synthesis, growth, and cell reproduction. Moreover, such genes would have mutated occasionally, would have been inherited by offspring cells, and would have become exchanged sexually if two cells happened to meet and then fused temporarily or permanently. Therefore, even the earliest cells must have been capable of evolving, and indeed they must have exhibited all the properties by which we define "life."

Actual evidence is unavailable, and at present it is very difficult to decide whether a monophyletic or a polyphyletic origin is more plausible. As we shall soon see, the problem of monophyletism vs. polyphyletism recurs many times in phylogenetic considerations, and favoring one view in one instance does not require that the same view be favored in all other instances. Regardless of problems of this nature, best estimates at present suggest that the first cell or cells probably arose some 2 billion years after the origin of the earth, i.e., perhaps 3 billion years ago.

It is not necessary to assume that cellular life originated exclusively by the processes just described or even that the above outline corresponds in detail to the actual events of the distant past. The important consideration at this stage of knowledge is mainly that we *can* envisage processes of cell formation which are plausible within the limits of the physics and chemistry of the earth. This in itself represents a major advance over knowledge available just two or three decades ago. In another two or three decades we are quite likely to have far surer knowledge, plus, perhaps, the ability to duplicate in the laboratory some of the key steps of these early events.

THE ORIGIN OF EARLY GROUPS

Cellular Evolution

In the earliest cells, gene-forming nucleic acids probably were suspended free within the cell substance or, judging from primitive cells today, were aggregated into tiny nucleic acid clumps. It must have happened on occasion that such clumps, along with other cell components, broke free from a cell into the open ocean. This could have occurred, for example, after an accidental temporary rupture in the boundary membrane of a cell or after a cell died and disintegrated. In this free state, a nucleic acid clump would have been simply a lifeless and inert chemical aggregate, for it lacked its normal cellular "housing." But it must have happened often that such inert aggregates by accident met up with other early cells and entered them. Within these host cells, the inert aggregates could become active again; i.e., the living machinery of the host could again provide the means for nucleic acid control activities and reproduction.

Such nucleic acid clumps, escaped from one cell, existing free for a time in an inert state, and then reentering and being reactivated by another cell, may have been the ancestors of the modern *viruses*. Viruses today behave exactly that way. They consist mainly of nucleic acids, the only other structural component being an external mantle of protein (Fig. 14.7). Note therefore that viruses are *not* cells and *not* organisms; they are considerably less than cells or organisms. We know also that at least some viruses arise as fragments broken off from the nucleic acid components of a given cell. Such fragments then direct the cell to manufacture protein mantles around them. The so-formed viruses subsequently escape from the host, often disintegrating the host cell in the process. Then they exist free in air or water. They are quite inert in the free state and they become reactivated if, and only if, they enter some new cell. In such *infections*, a virus becomes attached to the surface of the new host and then the nucleic acid mass of the virus is squeezed out from the protein mantle, through the cell surface, into the cell interior. The empty mantle itself does not enter the host. Within the host cell, the

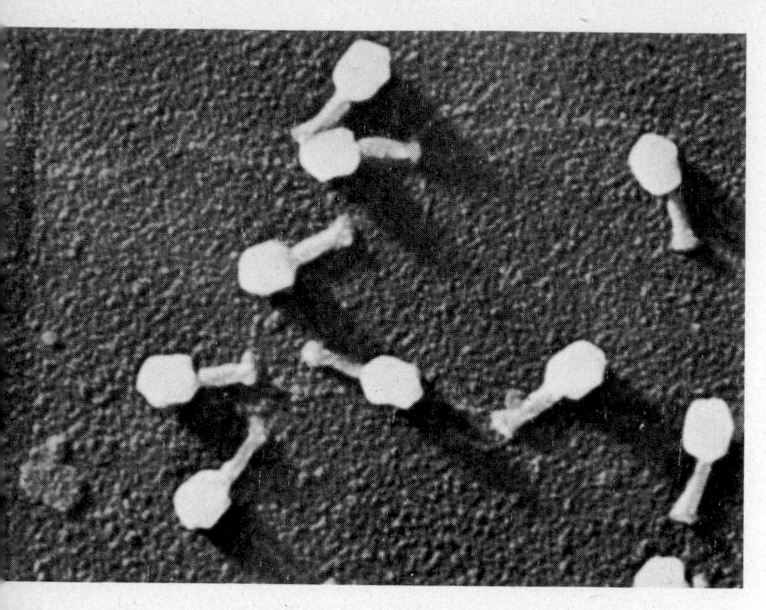

14.7. The shape and structure of modern viruses. Virus types of many other shapes are known. The diagram shows the typical composition of viruses, with nucleic acid in the center and a protein mantle on the outside. The photo is an electron micrograph.

viral nucleic acids now may use the living apparatus of the cell for renewed virus formation; the nucleic acids reproduce and new protein mantles are manufactured around them.

Viruses can therefore be classified as infective *parasites*. Modern viruses are known to be highly complex in structure and behavior, and it is likely that their distant ancestors were far less complex. Indeed, such ancestors may have been naked nucleic acids. If so, formation of viruslike nucleic acid units and transfer from cell to cell could have occurred as soon as cells themselves were in existence.

Such transfers must have had important consequences. For example, some cells would have lost certain properties and other cells would have gained them; the transferred nucleic acids were genes, and as they became shuffled among cells, so did the activities which these genes controlled. Therefore, in addition to sex and mutation, transfer of viruslike nucleic acids would have constituted a third way in which the gene content of cells could have become altered. This process similarly must have promoted cellular evolution and must have contributed to the emergence of a great variety of different cell types. Certain kinds of modern viruses still transfer genes from one cell to another, a phenomenon known as *transduction*.

By hindsight, we know that among the early single-celled organisms two main structural types came to have particular significance in later evolution. As already noted briefly, the first cells probably possessed freely suspended gene-forming nucleic acids. In some of the descendants of the early cells, the genes apparently aggregated together into threadlike filaments which formed loose clumps. In each cell, such a clump then remained embedded within the cell substance and in direct contact with it.

The group of organisms characterized by this type of internal cellular arrangement may be referred to collectively as the *Monera* (or *Procaryota*, i.e., organisms preceding the evolution of cell nuclei). Representatives of this group are still in existence today; the most familiar are the *bacteria* (Fig. 14.8). The exact ancestry of modern bacteria is uncertain. However, in structure as well as function, bacteria now living are very close to our conception of what the first Monera might have been like. Conceivably, therefore, the latter may have been the ancestors of modern bacteria. Another group now living and probably descended from the first Monera are the *blue-*

14.8. The dispersed nucleic acid of bacteria. Some of the cells of this bacterial type, *Escherichia coli*, occur as single individuals, others are joined into chains. Staining makes the nucleic acids appear as dark bodies. Note the dispersion of these bodies throughout the bacterial cytoplasm.

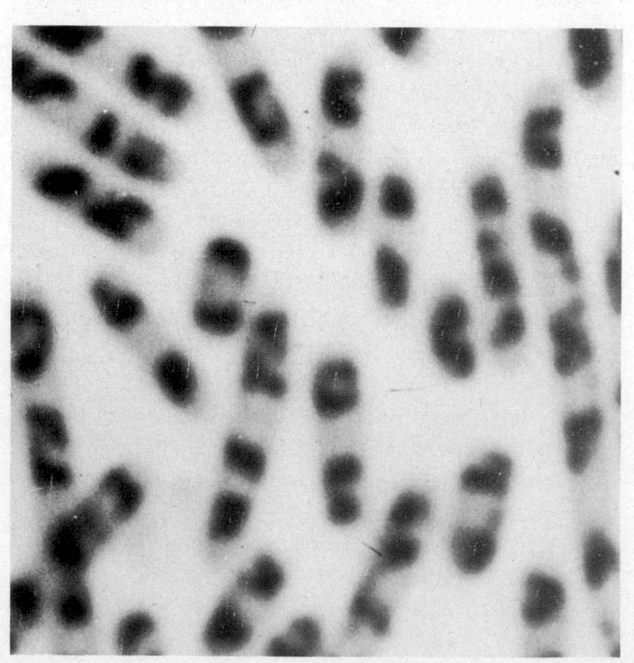

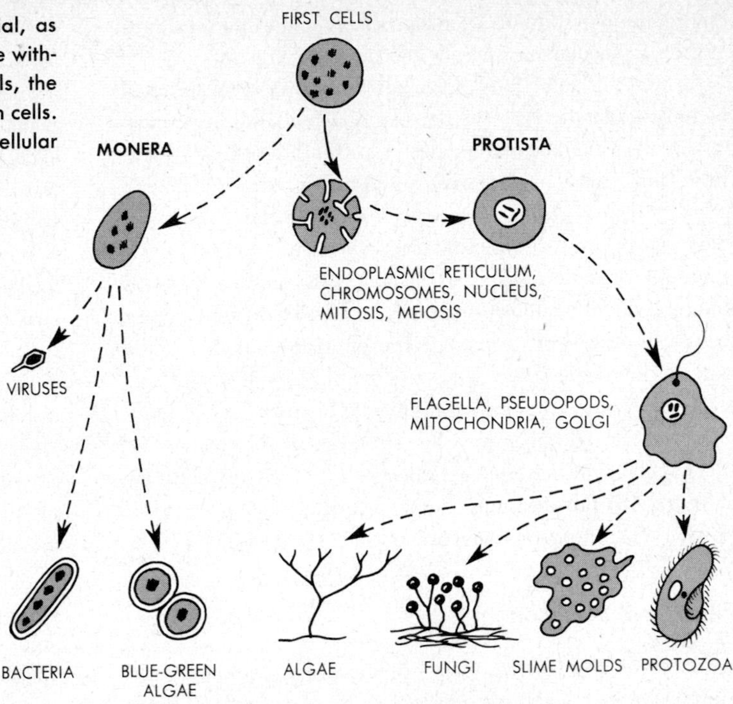

14.9. The earliest cells (with dispersed genetic material, as indicated) probably gave rise to two structural types, one without nuclear membrane and referred to as moneran cells, the other with nuclear membrane and referred to as protistan cells. The presumable later evolutionary history of these two cellular stocks is sketched in the diagram.

green algae. These primitive cellular organisms resemble bacteria in many ways, including the way in which the genes are arranged within their cells. We may say, therefore, that the first cells on earth probably gave rise to an early group of Monera and that these in turn were the ancestors of the modern Monera, represented today chiefly by the bacteria and the blue-green algae.

In a second major cellular type descended from the first cells, the genes became associated prominently with proteins, and these *nucleoproteins* became organized into microscopically distinct, filamentous *chromosomes.* In addition, a fine membrane formed around the chromosomes, probably by inward growth of part of the cell membrane. A *nucleus* so became established, and the genes were no longer in direct contact with the cytoplasm. As pointed out in Chap. 6, the adaptive advantage of such an arrangement appears to be protection, i.e., the nuclear membrane shields genes from respiratory decomposition. Furthermore, ancestral cells of this type sooner or later also evolved kinetosomes, flagella, amoeboid locomotor capacity, centrioles, mitotic and meiotic mechanisms, as well as membranous organelles such as endoplasmic reticula, mitochondria, Golgi bodies, and vacuoles—in short, all the cellular structures and processes we have already studied in earlier contexts. Ancestral organisms characterized by such an internal makeup, sharply different from the moneran pattern, may be referred to collectively as the early *Protista* (or *Eucaryota,* i.e., organisms with distinct nuclei). Four of their descendant groups form a major part of the living world today, viz., the *algae* (other than the blue-greens), the *fungi,* the *slime molds,* and the *protozoa* (Fig. 14.9).

There is reasonably good general agreement among biologists that Monera and Protista probably have had separate evolutionary histories going back right to the earliest cells. It is unknown whether Monera and Protista arose monophyletically from just one successful original group of cells or polyphyletically from two or more such groups. The many structural differences between the moneran and protistan cell types do suggest separate polyphyletic origins, but a monophyletic beginning is not necessarily ruled out. Indeed, the extensive chemical correspondences of the two groups suggest that the ancestor of the Protista must at least have been exceedingly similar to the early Monera.

Within both the early moneran and protistan groups, abundant diversification must have occurred and many new types of organisms must have arisen through evolution. We may surmise, moreover, that this evolutionary branching out must have been promoted and oriented by a powerful environmental stimulus, namely, the gradual disappearance of free molecular foods from the ocean.

As more and more food molecules were withdrawn and used by more and more cells, the rate of global food utilization must eventually have become greater than the rate of food formation from methane, ammonia, and the other atmospheric constituents. In time, therefore, free molecular foods must have disappeared completely from the ocean, and that environment then became as exclusively inorganic as it still is today. Evidently, almost as soon as it originated, living matter began to affect and to change the physical character of the earth. Thus, unless new ways of procuring foods could have evolved, the ever-increasing multitude of reproducing, food-using cells would soon have nourished itself into extinction. Early cells evidently did not succumb, but on the contrary gave rise to the far-flung, richly diversified living world of today. What nutritional inventions made this possible?

Nutritional Evolution

One of the first evolutionary responses to dwindling food supplies probably was the development of *parasitism*. If foods could not be obtained from the open ocean, they still could be obtained within the bodies of living cells. As already noted, a virus, for example, could penetrate right into a cell and use the foods accumulated in such a host. Similarly, a small cell could solve its food supply problem if it could manage to invade a larger cell. Methods of infecting cellular hosts undoubtedly evolved early, and today all viruses as well as many of the descendants of the first Monera and Protista are infective and parasitic.

For many of the early organisms, parasitism undoubtedly was an effective new way of life. Another new way which required relatively little evolutionary adjustment was *saprotrophism*. Here an organism drew food molecules not from the decreasing supply in the ocean but from the bodies of dead cells or disintegrated cellular material. Many early bacterial and fungal groups probably adopted this comparatively easy method of getting food and became the ancestors of the modern saprotrophic bacteria and fungi. Note that organic *decay* is a result of the nutrient-gathering activities of saprotrophic organisms. Before the evolution of saprotrophism, decay was unknown on earth. Today, saprotrophic types—especially bacterial and fungal saprotrophs—are so abundant that virtually any substance begins to decay almost immediately after exposure to air or water.

A third new process which permitted survival despite dwindling food supplies was *holotrophism*, i.e., the process of eating other living cells whole. This became possible through evolution of cellular mouths or amoeboid pseudopodia and through flagellary or amoeboid locomotion. Many algal and protozoan protists adopted this method of nutrition. Note, incidentally, that the difference between cellular eating and cellular parasitism is largely defined by the final result. In both cases, one cell gets inside another, but in one instance the larger host lives off the guest and in the other instance the guest lives off the host (Fig. 14.10).

But all three of these new food-gathering procedures were ultimately self-limiting. Parasitism, saprotrophism, and holotrophism, collectively constituting *heterotrophic* forms of nutrition, merely changed the distribution of already existing organic matter; they did not add any new food to the global supply. Clearly, if totally new food sources had not become available, life would have had to cease sooner or later.

What was needed, fundamentally, was a new way of making organic substances, preferably right within cells. The original way, in which sun and lightning formed food compounds out of materials such as methane, ammonia, and water, was no longer adequate, if it occurred at all at that late period. But the raw materials for a new process were still available in abundance. Water was in inexhaustible supply and, in addition to methane or hydrogen cyanide, there now existed, directly within cells, an even better source of carbon: carbon dioxide, byproduct of respiration. Given CO_2 and water, organic molecules could be manufactured in cells, provided that new external sources of energy could be found. Internal energy in the form of ATP was still available, to be sure; but ATP was itself an organic compound and was therefore among

14.10. The three noncreative, heterotrophic methods of obtaining food. In parasitism, one organism obtains food from another living one (a small parasitic cell is shown inside a larger host cell). In saprotrophism, food is obtained from dead organisms or from organic derivatives of other organisms. And in holotrophism, one organism eats another, in whole or in part, and obtains food in this manner. These three methods are noncreative because they merely redistribute already existing foods and do not create new supplies. See also Fig. 3.13.

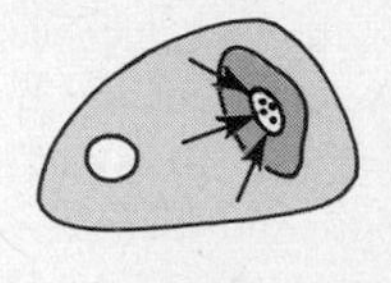

PARASITISM

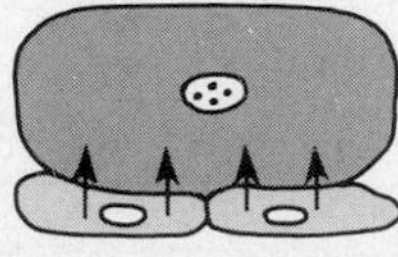

SAPROTROPHISM

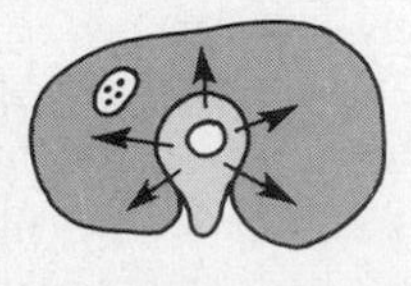

HOLOTROPHISM

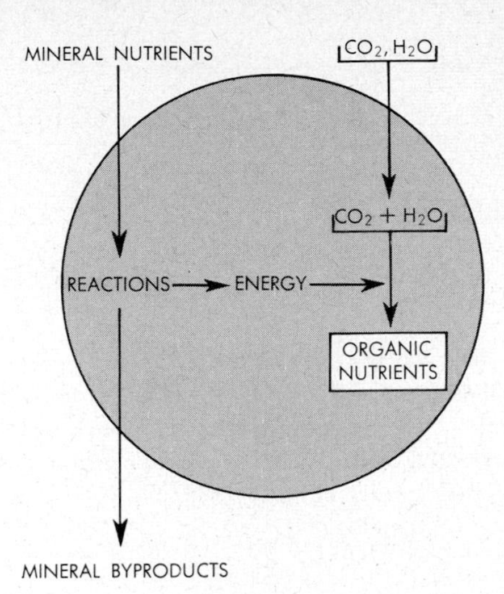

14.11. The general pattern of chemosynthesis. With energy obtained from inorganic nutrients, the organism creates new organic nutrients out of carbon dioxide and water.

the very substances which would have disappeared if new external energy had not made possible their continued manufacture. Organisms which evolved means of utilizing external energy in the production of organic compounds are collectively known as *autotrophs*. Two broad categories of them came into existence.

Some of the early Monera, particularly certain bacterial types, found new external sources of energy in sulfur, in iron, in nitrogen, and in a number of other metallic and nonmetallic materials obtainable from the environment. Several groups of the early bacteria must have evolved in such a way that they could absorb various inorganic molecules into their cell substances and there make them undergo various exergonic reactions. Chemical energy so obtained was then used within the cells to combine CO_2 and water into food molecules, specifically, carbohydrates. The whole process is called *chemosynthesis* (Fig. 14.11). Certain bacteria living today still manufacture foods in this manner.

Judging from the results some 2 billion years later, early chemosynthesis apparently was only a limited solution of the energy- and food-supply problem. Possibly it depended too much on particular inorganic materials available only in particular localities. A more generally useful solution required a steady, more nearly universal external energy source. Such a source was the sun. High-energy solar radiations such as ultraviolet and X rays no longer were a sufficient energy source for the amount of food production required. Similarly, the earlier permanent cloud cover had disappeared by now and lightning too became inadequate as an energy source. But solar radiation of lower energy content, especially *light*, now beamed down to earth as predictably and dependably as could be desired. If sunlight could be used, the energy problem, hence the food problem, would be solved. Sunlight actually did become the ultimate energy supplier for the vast majority of organisms, and it has played that role ever since.

Utilization of light energy within cells requires a cellular light-trapping device. Many kinds of photosensitive compounds are known to be able to absorb light and to trap more or less of its energy. By chance reactions, such compounds may have formed very early in the open ocean, along with all the others we have discussed. And it is likely that some of these substances were among the many materials which collected together and formed cells. Alternatively, light-trapping compounds might have been manufactured directly within cells already in existence, as one of the new materials produced by cellular synthesis. In some such way, some of the early cellular organisms came to possess substances which were more or less efficient in trapping the energy of sunlight. This energy could be then used to transform CO_2 and water into organic food compounds.

One of the early light-trapping substances has been perpetuated to the very present. It is green, and we call it *chlorophyll*. The new process, in which sunlight and chlorophyll promote the transformation of CO_2 and water into foods, is called *photosynthesis* (Fig. 14.12).

14.12. The general pattern of photosynthesis. With energy obtained from the sun and by means of energy-trapping molecules such as chlorophyll, the organism creates organic nutrients out of carbon dioxide and water.

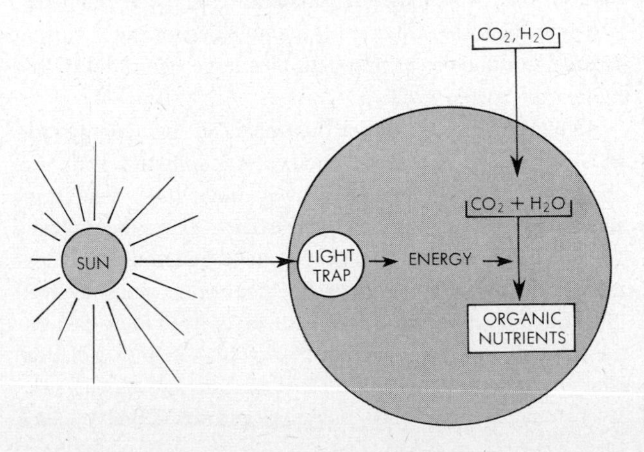

With this new source of organic compounds assured, it did not matter that free molecular foods in the ocean finally disappeared. Photosynthesizing cells could make foods for themselves; holotrophic organisms could eat such cells and then each other; parasites could invade photosynthesizers or eaters; and saprotrophs in turn could find foods in the dead bodies of any of these. Consequently, excepting only the chemosynthesizers, which made their own foods, all other organisms were saved from premature extinction by photosynthesis. Today, photosynthesis still supports all living creatures except the chemosynthesizers.

We note that, sooner or later after the appearance of the first cells, five kinds of food-getting methods were evolved: parasitism, saprotrophism, holotrophism, chemosynthesis, and photosynthesis. Only the last two added to the net global supply of foods. The Monera apparently adopted all methods except eating, and the Protista, all methods except chemosynthesis.

As photosynthesis occurred to an ever-increasing extent, it brought about far-reaching changes in the physical environment. A byproduct of photosynthesis is free molecular oxygen (O_2), a highly reactive gas which combines readily with other substances. Before the advent of photosynthesis, free oxygen had not existed since the early days of the earth, when oxygen atoms were still uncombined. Later, such small quantities of free oxygen as might occasionally have formed would have combined quickly with materials in the vicinity. Now, increasingly large amounts of free oxygen escaped from photosynthesizing cells into the ocean and from there into the atmosphere. The gas must have reacted promptly with everything it could, and this probably initiated a slow, profound "oxygen revolution" on earth (Fig. 14.13). This revolution ultimately transformed the ancient atmosphere into the modern one, which no longer contains methane, ammonia, and cyanide. Instead, it consists mainly of water vapor, carbon dioxide, and molecular nitrogen, plus large quantities of free molecular oxygen itself.

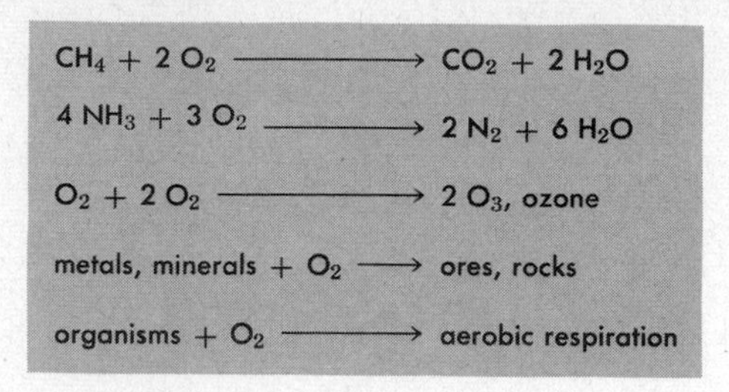

14.13. The oxygen revolution. Oxygen resulting from photosynthesis reacted with other materials as shown and brought about the changes indicated. A major result was the establishment of a new, modern atmosphere, containing N_2, CO_2, and H_2O in addition to O_2 itself.

At higher altitudes, under the impact of high-energy radiation from space, oxygen molecules combined with one another. The result was a layer of *ozone* (O_3). This layer, several miles up, has been in existence ever since. Ozone formed an excellent screen against deep penetration of high-energy radiation. Consequently, organisms which evolved after the establishment of the ozone layer lived in an environment more or less completely free of high-energy radiation. This is why modern advanced plants and animals are comparatively unadapted to such radiation and are killed by even small doses of it. By contrast, the earliest organisms had evolved before the large-scale formation of ozone and had become more or less well adapted to space radiation. Some of their modern relatives among the Monera and Protista still display this radiation resistance. They now can withstand exposures to X rays and similar radiation that would kill an army of men.

Free oxygen also reacted with the solid crust of the earth and converted most pure metals and mineral substances into *oxides*—the familiar ores and rocks of which much of the land surface is now made. A few relatively unreactive metals like gold resisted the action of oxygen, but others could not. And if today we wish to obtain pure iron or aluminum, for example, we must smelt or otherwise process appropriate ores to separate out the firmly bound oxygen.

Free oxygen, finally, made possible a new, much more efficient form of respiration. The earliest cells respired anaerobically (and to some bacteria now living oxygen is actually a poison). But when free environmental oxygen began to accumulate in quantity, newly evolving organisms developed means of *aerobic* respiration and this method soon became the standard way of extracting energy from foods.

We note that the effects and activities of the early organisms greatly altered the physical character of the earth and also the biological character of the organisms themselves. So it has been ever since, even if never again so dramatically and incisively: the physical earth creates and influences the development of the biological earth, and the biological earth then reciprocates by influencing the development of the physical earth.

Protistan Evolution

Of the four protistan groups now in existence, viz., algae, fungi, slime molds, and protozoa, each encompasses a wide range of different structural types and each is generally believed to be highly polyphyletic. The algae probably include the most ancient protists. Indeed, the characteristics of the primitive types among algae are thought to be rather like those the ancestral protists must have exhibited. Judging from primitive living algae, the early protists were unicellular, structurally different from the Monera, and the single-celled individuals functioned as reproductive units directly, both vegetatively and gametically. The offspring cells were new adults immediately and embryonic development was absent.

In addition, again judging from primitive algal types today, early protists must have had the capacity of obtaining food by two or more methods simultaneously (Fig. 14.14). As noted above, we know from their living representatives that Protista as a whole had evolved four methods of nutrition, viz., photosynthesis, saprotrophism, parasitism, and holotrophism. Early protists appear to have had the potential of *multiple* means of nutrition. Thus, it must have been quite common for a protist to ingest or absorb food ready-made from the external environment *as well as* to photosynthesize it internally. The adaptive advantage of such multiple nutrition can be appreciated readily. For example, at night or at the dimly lit bottoms of natural waters, an ancestral protist could hunt for food holotrophically, but in the presence of ample light it could save energy by photosynthesizing. In this respect, evidently, the ancient protists probably were both plantlike and animallike simultaneously, like some of their algal descendants today.

14.14. Ancestral protists (top) presumably possessed chloroplasts as well as gullets and motility, hence their nutrition may have been both autotrophic and heterotrophic. From such stocks with joint plantlike and animallike traits probably evolved purely autotrophic and purely heterotrophic types, represented today by the groups indicated. Primitive groups which retained, and still retain, mixed traits may have contributed even later to the evolution of more nearly plantlike and animallike protists.

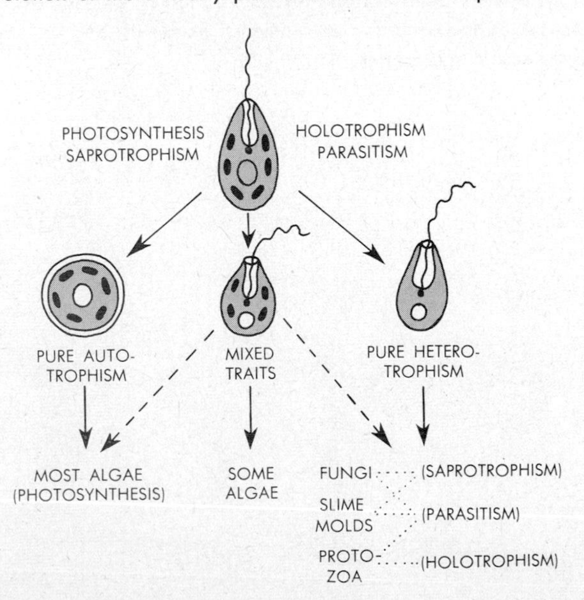

From such a joint plant-animal starting point later evolved a purely photosynthetic, plantlike branch of protistan life and a purely heterotrophic, animal-like branch. Among algae, most groups today no longer possess multiple methods of nutrition. Instead they have perfected photosynthetic nutrition, have lost the heterotrophic potential, and are distinctly plantlike. We can be fairly sure that such differential losses of nutritional capacities resulted from mutations, for it is possible today to induce just such loss-mutations in the laboratory. Thus, primitive algae with multiple nutrition may be converted experimentally to either purely plantlike or purely animallike organisms. Such experiments undoubtedly duplicate the ancient natural process of nutritional evolution among protists.

Furthermore, there is little doubt that early groups of colorless protists came to be the ancestors of the modern fungi, slime molds, and protozoa. These present Protista all are exclusively heterotrophic. Fungi are partly saprotrophic, partly parasitic; slime molds are saprotrophic for the most part; and protozoa are largely holotrophic, though many are parasitic and a few are saprotrophic.

Another basic characteristic of ancient protists must have been their capacity of existing in at least four alternative states, two of them motile and two nonmotile (Fig. 14.15). Motility is displayed in *flagellate* and *amoeboid* states. In a free-swimming flagellate condition, a parent cell can give rise by successive divisions to many, similarly flagellate offspring. The amoeboid state arises when a flagellate cell casts off its flagellum (but retains its kinetosome). Such a cell can then move along a surface and feed by means of pseudopodia. At some later time, even after offspring cells are produced by division, new flagella can grow from the kinetosomes, amoeboid activity may cease, and the free-swimming flagellate state may be resumed.

The two nonmotile states analogously develop by loss of flagella. One is a special type of multinucleate state which

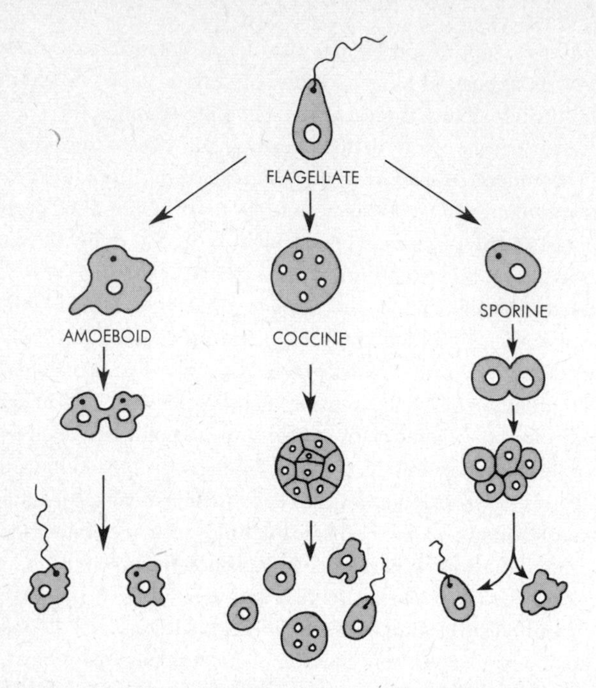

14.15. The four basic vegetative states of protistan existence. A flagellate cell may become amoeboid, and later that same cell or any of its offspring may revert to a flagellate state. Loss of flagella and development of multinuclearity leads to the nonmotile coccine state, in which, after cell-boundary formation during reproduction, the offspring cells may remain coccine or resume flagellate or amoeboid existence. Loss of flagella and successive vegetative divisions produce the sporine state. The resulting cells may separate or stay together as colonies, and any of the cells may also assume the flagellate or the amoeboid condition.

for present purposes we may refer to as the *coccine* condition. Here the cytoplasm of the nonmotile cell does not divide vegetatively but the nucleus does continue to divide. The result is a progressively more multinucleate but still unicellular organism. At the time of reproduction, multiple cytoplasmic fission takes place: the organism becomes partitioned into numerous cells simultaneously, each cell containing at least one nucleus. Such offspring cells may disperse passively or develop flagella and disperse actively. In either case they soon settle and grow into new nonmotile multinucleates. The second nonmotile condition we shall call the *sporine* state. A cell here remains uninucleate and it does divide vegetatively. Since the cells are nonmotile, daughter cells formed by division tend to remain in close proximity, a condition which facilitates the development of multicellular aggregates. In these, individual cells may later redevelop flagella or become amoeboid and disengage from the aggregates.

Of these four different states of existence, the flagellate condition is believed to be basic and primitive; all others may be derived from it. The adaptive advantage of such multiple alternative states must have been exceedingly important, particularly in conjunction with the multiple means of nutrition. Thus, the motile states undoubtedly facilitated heterotrophic nutrition generally and made possible holotrophic nutrition specifically; and the sessile states became particularly economical when photosynthesis was under way. In the majority of modern Protista, to be sure, given individuals no longer retain all four alternative states but exhibit a particular one more or less permanently. We may infer, therefore, that ancestral types with the potential of multiple states gave rise to separate flagellate, amoeboid, coccine, and sporine lines of descent. We know also that each such line produced not only unicellular members but multicellular ones as well; Protista today include motile *flagellate colonies,* motile *amoeboid colonies,* sessile *coccine colonies,* and sessile *sporine colonies* (Fig. 14.16).

Among algae particularly, many phyla independently include three or all four of these lines of descent in both uni-

14.16. The multicellular (or multinucleate) derivatives of the four states of protistan existence. Note that multinucleate and filamentous coccine types (as exemplified by most fungi and some algae) are also called *siphonaceous* forms. Multicellular sporine types occur not only as simple filaments as shown, but also as complexly branched filaments, as sheets, and as three-dimensional aggregates.

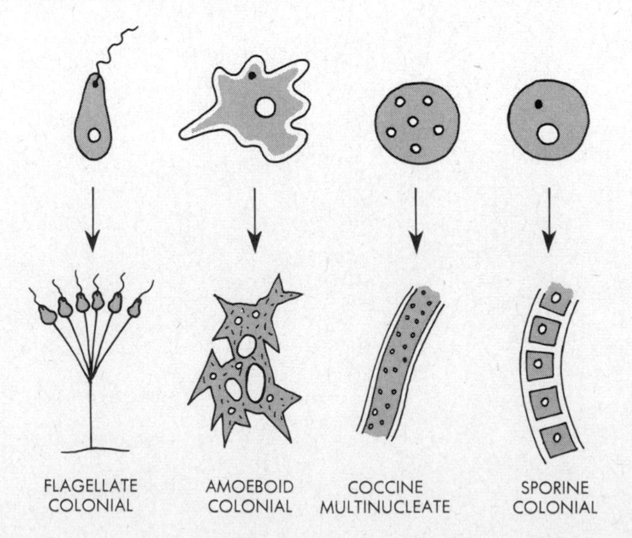

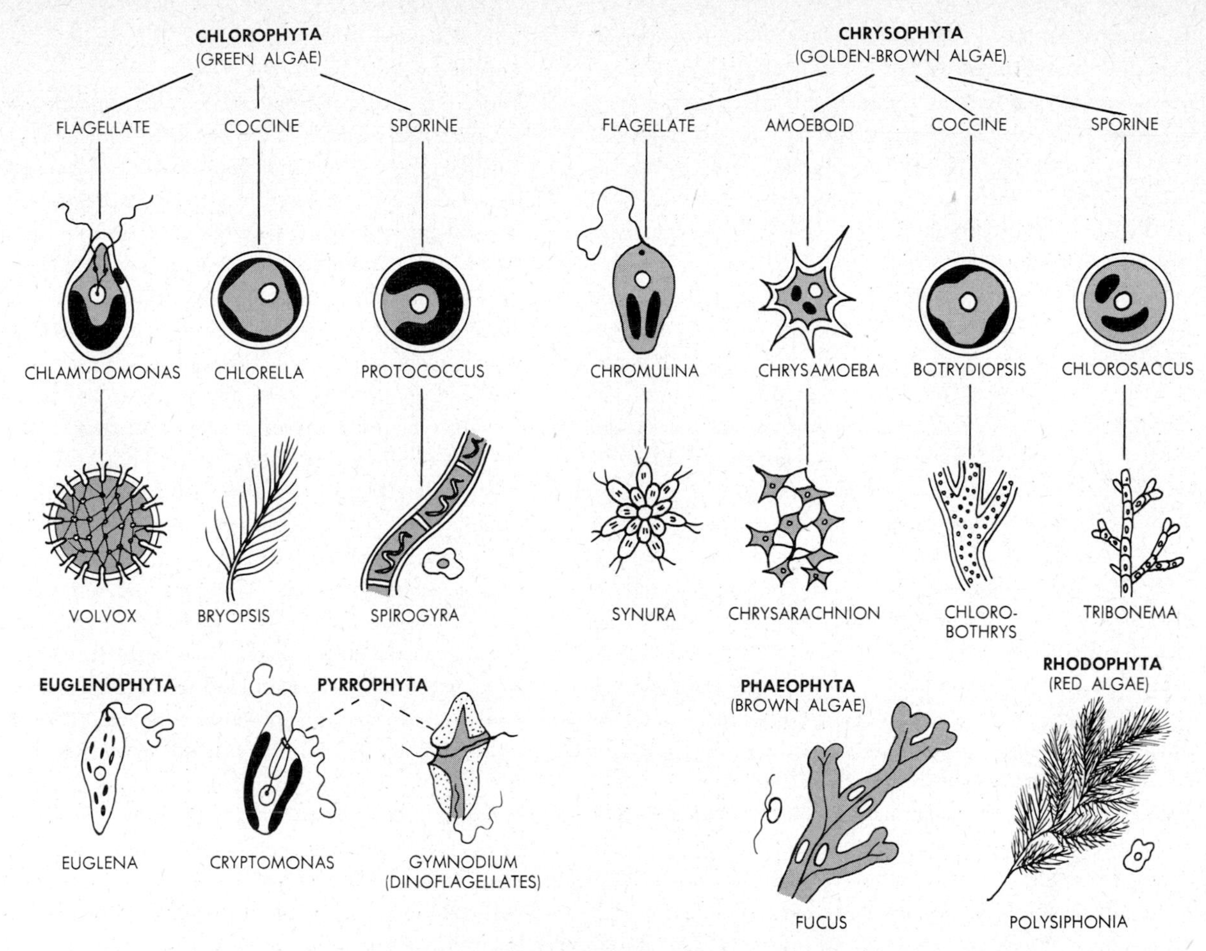

14.17. Diagrams of representatives of the various photosynthetic (i.e., algal) lines of protistan evolution. Unicellular members of green and golden-brown algae are sketched in top row, multicellular members underneath. Euglenophytes and pyrrophytes occur virtually only as unicellular flagellate forms, phaeophytes and rhodophytes virtually only as multicellular sporine forms. Next to *Spirogyra* and *Polysiphonia* are diagrammed the amoeboid gametes of these protists, next to *Fucus* (and brown algae generally), their flagellate gametes. Each major algal group is a taxonomic "division," a taxon corresponding to an animal phylum.

cellular and multicellular form (Fig. 14.17). In green algae, for example, three lines are now in existence. Among the representatives of the flagellate line are the unicellular *Chlamydomonas* and the colonial *Volvox*. In the coccine line are the initially uninucleate *Chlorella* and the permanently multinucleate *Bryopsis*. The sporine line includes the unicellular *Protococcus* and the multicellular *Spirogyra*. And an amoeboid line either never evolved or has since become extinct. Among golden-brown algae, analogously, one of several flagellate lines is exemplified by the unicellular *Chromulina* and the colonial *Synura;* one of several amoeboid lines, by the unicellular *Chrysamoeba* and the multicellular *Chrysarachnion;* one of several coccine lines, by the unicellular *Botrydiopsis* and the colonial *Chlorobotrys;* and one of several sporine lines, by the unicellular *Chlorosaccus* and the multicellular *Tribonema.*

An analogous multiplicity of lines is encountered in many other algal phyla. However, some have specialized in only one state of existence; other states may never have evolved or may since have become extinct. For example, euglenoid algae occur virtually entirely as flagellate unicells, *Euglena* being a familiar example. By contrast, brown and red algae exist today almost entirely as highly multicellular sporine organisms (cf. Fig. 14.17). Early sporine colonies must have displayed particularly rich evolutionary potentialities. Since they were sessile and their cells were joined directly and tightly, such colonies could hold together well and could grow to extremely large size, well beyond the microscopic range of other types of colonies. Moreover, depending on the planes of cell division, they could form one-dimensional filaments, or two-dimensional disks and sheets, or three-dimensional compact masses. Such structurally elaborate sporine aggregates are quite indistinguishable from, and are actually identical with, true tissues. Complex sporine protistan types now occur conspicuously among the advanced green, brown, and red algae.

In the colorless Protista, analogously, the several states of existence are well represented. However, as in some of the algal forms, each heterotrophic group has specialized in just one or two of the states (Fig. 14.18). Thus, fungi may be regarded as the culmination of the colorless coccine state. All fungi are or become multinucleate organisms which partition off distinct uninucleate cells only at the time of reproduction. Slime molds represent a culmination of the amoeboid state. These organisms are large, creeping, multicellular or multinucleate amoeboid masses without permanent shape. Protozoa are basically flagellate and amoeboid, and in these organisms the multicellular condition has developed to only a limited degree. Instead, protozoa appear to have exploited the unicellular way of life perhaps more fully than any other group of protists. An early flagellate stock probably gave rise to the ciliate protozoa, and from both flagellate and amoeboid stocks appear to have evolved the spore-forming and rather coccinelike protozoa. We may note here that if a unicellular algal flagellate or amoeba loses its green color, it becomes virtually indistinguishable from a protozoan flagellate or amoeba. It is very likely, therefore, that protozoa are highly polyphyletic as a group, and that their membership has been and perhaps still is being augmented by continual addition of colorless types newly formed among algae.

We may note also that any of the Protista which have specialized in just one of the states of existence often reveal, at the time of reproduction, the ancestral potential of developing other states. Thus, even if the adults are sessile, the reproductive cells are still flagellate or amoeboid. For example, most of the sessile aquatic algae and fungi produce flagellate reproductive cells. Some, like *Spirogyra* and red algae, produce amoeboid ones. Among slime molds, some groups form flagellate, others amoeboid reproductive cells. Moreover, numerous algae and fungi form both kinds at the same time: flagellate cells become sperms, amoeboid cells become eggs.

The evolution of Protista as a whole thus may be envisaged broadly as an adaptive radiation which began with unicellular ancestral stocks possessing both plantlike and animallike characteristics simultaneously (Fig. 14.19). From such stocks then descended at least three kinds of groups. One remained relatively unchanged, and its algal representatives today still display joint plant and animal traits. A second group became exclusively photosynthetic, and it is represented today by the

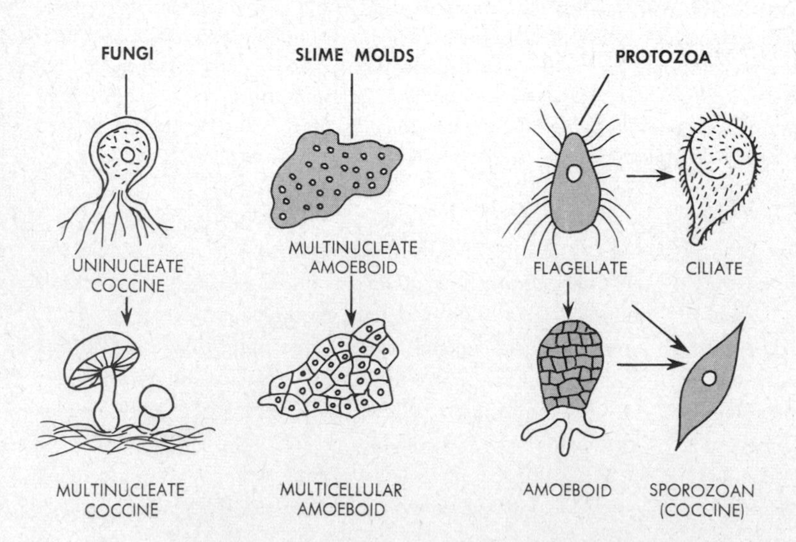

14.18. Diagram of the nonphotosynthetic lines of protistan evolution. All three of the major groups (corresponding taxonomically to phyla) probably arose directly or indirectly from protistan ancestors which originally may have possessed photosynthesizing capacity but then lost it.

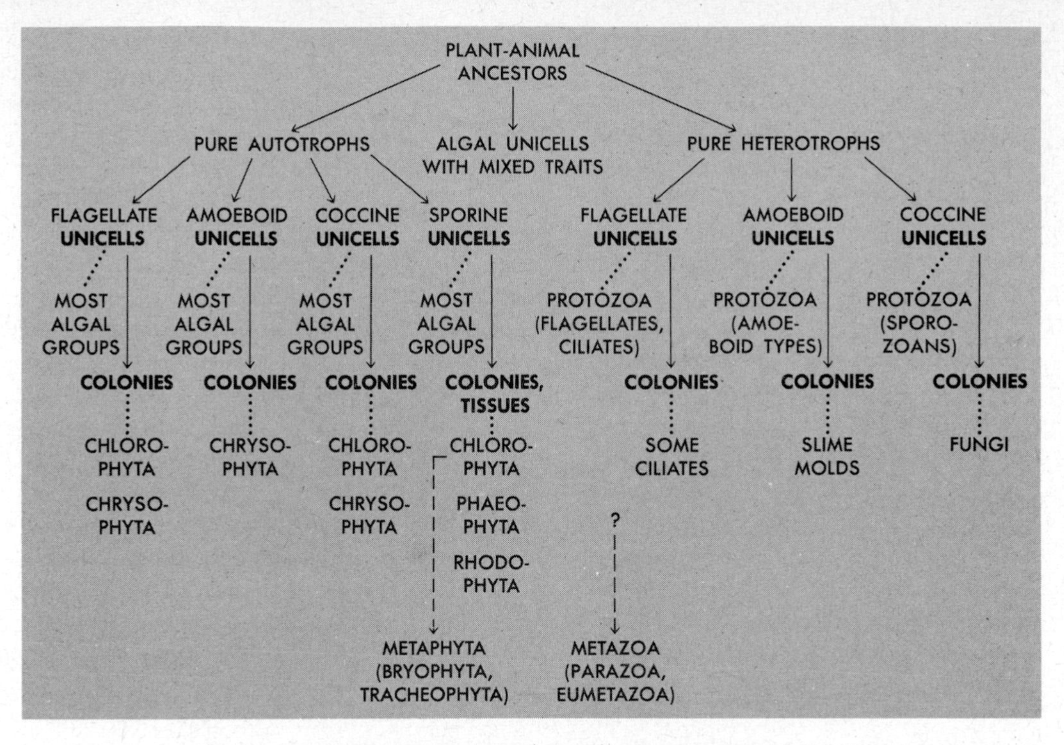

14.19. Summary of the main aspects of protistan evolution. Note that the heterotrophic branch of evolution does not include sporine protists.

flagellate, amoeboid, coccine and sporine algae. The third group became exclusively heterotrophic, and its present membership includes the primarily flagellate and amoeboid protozoa, the amoeboid slime molds, and the coccine fungi. And in all of these groups and subgroups, multicellularity developed to different degrees of complexity.

At some points during their evolutionary history, the Protista unquestionably also gave rise to the two largest groups of organisms now in existence, the *Metaphyta* and the *Metazoa.* Metaphytes encompass the terrestrial green plants, i.e., the moss plants (*bryophytes*) and the vascular plants (*tracheophytes*). Metazoa encompass the multicellular animals, as we already know. There can be little doubt that both Metaphyta and Metazoa have had protistan, not moneran origins. The cell structure is typically protistan, and flagellate, amoeboid, and sporine states of existence are common throughout these groups. Sperms and many integumentary and sensory cells are flagellate (or ciliate); eggs, mesenchyme cells, and blood and connective tissue cells are actually or potentially amoeboid; and the vast bulk of the metaphytan and metazoan body is sporine. Metaphyta and Metazoa are distinguished from Protista in possessing *multicellular* reproductive structures, i.e., at least reproductive *tissues* and in most cases *organs;* in being exclusively *oögamous,* i.e., the gametes are always differentiated into distinct sperms and eggs; and in developing via distinct *embryonic* stages. The two groups are distinguished from each other in that Metaphyta do not but Metazoa typically do possess larval stages. Also, Metaphyta are photosynthetic and sessile, whereas Metazoa are heterotrophic and motile (either in the larval or the adult stage or in both).

In rather general terms, the probable ancestry of the Metaphyta can be pinpointed comparatively well. Metaphyta appear to be descended from early sporine green algae which also gave rise to the complex sporine green algae of today (cf. Fig. 14.19). The tissue-level construction of these algae is repeated in the Metaphyta, and the latter subsequently have added the organ and organ-system levels as well. Moreover, such algae and all metaphytes possess identical varieties of chlorophyll and other pigments, and their biochemical traits as a whole are virtually identical. Metaphyta only arose some 350 million years ago, i.e., about 150 million years after the beginning of the fossil record. As it happens, the early history of the metaphytes is documented reasonably well in this record.

By contrast, metazoan beginnings must be ascertained without fossil evidence and without the aid of built-in biochemical clues such as chlorophyll. Thus, although we can be reasonably sure that Metazoa did originate among the Protista, it is at present quite impossible to decide which of the Protistan groups might actually have been ancestral. Nevertheless, numerous hypotheses have been proposed and we shall examine the chief ones in the next section.

THE ORIGIN OF ANIMALS

The First Animals

As indicated firmly by their name, Protozoa have been and still are widely regarded as the "first animals." Moreover, they are traditionally considered to be the ancestors of the Metazoa. Before any group can be labeled as animal, first or otherwise, it is of some importance to be clear about the intended meaning of such a label. As noted in Chap. 7, Protozoa and Metazoa traditionally form the Linnaean *animal kingdom*. All other living groups, including the Metaphyta, all Protista other than the Protozoa, and in most classifications also all Monera, form the traditional *plant kingdom*.

In such a Linnaean sense, plants and animals are completely undefinable by biological criteria. Every trait usually regarded as characteristic of one kingdom occurs also in the other. For example, if plants are defined as sessile photosynthesizers and animals as motile heterotrophs, as is customary, then some of the primitive algae and all fungi and slime molds would have to be animals, for all these organisms are colorless heterotrophs and a good many of them are motile as well, at least in the reproductive stage. Yet sponges, corals, barnacles, tunicates, and other metazoan groups could not be strictly labeled as animals since they are sessile as adults, just as many algae are sessile in the mature state. Moreover, certain primitive algal types and carnivorous green plants (such as Venus fly traps) would be both plant and animal at once, since they are both photosynthetic and heterotrophic. Indeed, algae of this type can also be alternately sessile and motile, as noted earlier.

Traits other than nutrition or motility fail similarly as distinguishing criteria. If animals are defined as organisms possessing neural and muscular structures, or at least equivalent conductile and contractile elements, then many photosynthetic algae would have to be animals; and sponges, which are without such structures, would have to be plants. If plants are defined as organisms possessing carbohydrates such as cellulose, then tunicates would have to be plants and a number of entire phyla of photosynthetic algae without cellulose would have to be animals. In effect, there does not appear to be a single characteristic which would distinguish the traditional plant and animal kingdoms uniquely.

The necessary conclusion is that the Linnaean kingdoms are not natural classification groups. Instead, they represent a man-made, arbitrary division of the living world, and this division evidently is no longer meaningful in a strict taxonomic sense. The kingdoms lack such meaning today because we now know that plants and animals, as defined according to our traditional preconceptions, were not in existence right from the beginning. As noted previously, separate plant and animal traits *evolved gradually* from a common protistan base in which both types of traits were present jointly. The Linnaean kingdoms do not take into account this gradual evolutionary development, and they do not provide any place for primitive organisms like certain Protista and also many Monera which are still both "plant" *and* "animal." Hence the difficulty of fitting the whole living world into just two major categories.

This taxonomic difficulty is circumvented if we divide the living world into not two but four highest taxa, viz., Monera, Protista, Metaphyta, and Metazoa, and if we also restrict the applicability of the terms "plant" and "animal." Thus, if "plant" designates *only* the Metaphyta and "animal" *only* the Metazoa, then the terms reacquire exact meaning, for Metaphyta and Metazoa are sharply definable. On such a basis, protozoa are then neither animals nor first animals; the *first* animals would have been the first Metazoa. This is probably as it should be, for protozoa are protists above all, and as such they actually have far more in common with all other motile, heterotrophic protists than with the Metazoa.

However, regardless of whether we call protozoa first animals or late protists, they are still regarded traditionally as the ancestors of the Metazoa. Most zoologists agree that the Metazoa are probably not a monophyletic but at least a diphyletic group: sponges are believed to represent one separate line of descent, all Eumetazoa, another. Most zoologists also agree that sponges probably did evolve from protozoa, and that they represent complexly elaborated flagellate colonies, just barely at a tissue level of construction. Several cogent reasons are advanced in support of this view. As already pointed out in Chap. 7, for example, the alimentary architecture of sponges is quite unique, and these animals also possess unique digestive collar cells very much like the protozoan collar flagellates (cf. Chap. 19). Unlike the Eumetazoa, also, sponges are without neural and muscular elements and, as we shall see in Chap. 20, the embryonic development of the animals differs in important respects from the patterns typical among Eumetazoa. In effect, therefore, the question to be decided is, not whether protozoa are the ancestors of the Metazoa, but whether protozoa are the ancestors of the Eumetazoa. And most traditional hypotheses answer in the affirmative (Fig. 14.20).

In one group of hypotheses, some of them rather recent, the specific ancestors of Eumetazoa are envisaged to have been *ciliate* protozoa. Modern ciliates are multinucleate, and some of their early multinucleate ancestors are postulated to have become multicellular by the development of internal cell boundaries, rather like coccine algae and fungi. The first Eumetazoa so formed are considered to have been flatworms, more specifically, types resembling the bilateral *acoel* flatworms, a group now quite generally believed to represent the most primitive Bilateria. In the acoel worms a ciliated epidermis covers the outside, and the interior is filled solidly with alimentary and reproductive cells. The latter represent comparatively poorly elaborated endoderm and mesoderm. A single alimentary opening is present midventrally and digestion is carried out intracellularly by individual interior cells (cf. Fig. 14.20).

The "ciliate-flatworm hypothesis" thus includes two basic assumptions, namely, that ciliate protozoa became Eumetazoa by internal coccinelike cell formation, or *septation*, and that the first Eumetazoa were acoel flatworms. The septational part of this view is totally without evidence, even of a suggestive kind. Moreover, the notion of an origin of multicellularity by septation is completely at variance with the actual development of multicellularity in any organism, protist, metaphytan, or metazoan, acoels included. Even in the coccine algae and fungi, the cells formed after internal partitioning usually do not stay together as multicellular aggregates. On the contrary, the cells are reproductive cells (spores) which disperse geographically; and septation is primarily an adaptive device to make just such dispersal possible, not to produce multicellularity. Furthermore, in the comparatively few instances where the cells do stay together and form coccine colonies, the level of structural complexity attained always remains exceedingly low; coccine types of any known kind apparently do not possess the evolutionary potential to reach even the level of tissues. But an acoel or any other eumetazoan does possess tissues. Because of this complete lack of support from living forms, septational views have found comparatively few adherents. The same also holds for the acoel part of the hypothesis, for reasons to be outlined below.

Far more widely accepted are hypotheses postulating the eumetazoan ancestors to have been early *flagellate* protozoa,

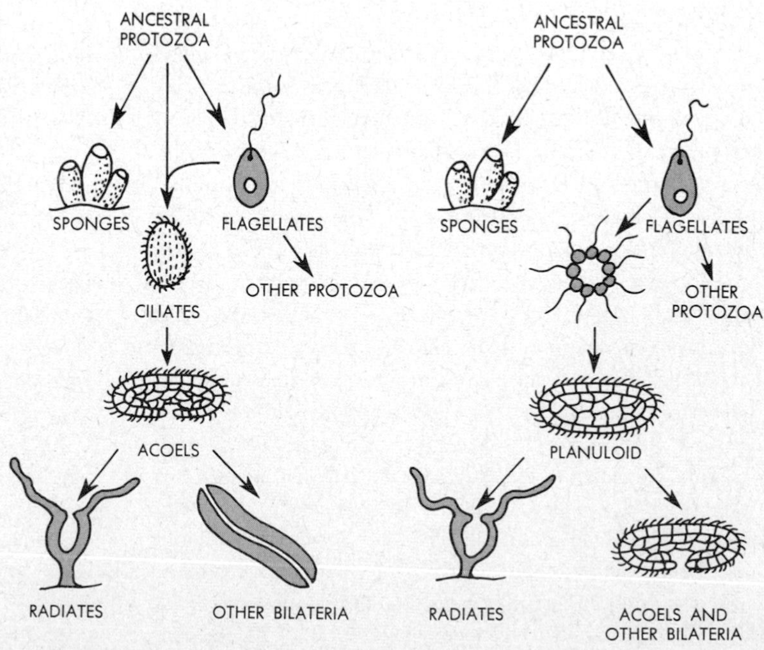

14.20. A comparison of the "ciliate-flatworm" (left) and "flagellate-planuloid" (right) hypotheses of eumetazoan origins. At left, multicellularity is assumed to have arisen by septation of a ciliate into an acoel; at right, by aggregation of flagellates into a blastulalike form. At left, bilaterality is assumed to be basic via the acoels, and radiates are taken to be secondarily radial. At right, bilaterality is assumed to have arisen during the evolution of radial planuloids into acoels, which thus would make the radiates primitively radial (cf. also below and Fig. 14.25).

particularly types which, like some of the primitive flagellates today, also exhibited conspicuous amoeboid tendencies. In support of such views it is pointed out that all Metazoa still develop flagellate and amoeboid gametes as well as other, similarly motile cell types; and that flagellate-amoeboid protists have a known evolutionary capacity to become multicellular by colony formation, or *aggregation*, just as metazoan eggs become multicellular embryos in similar fashion. Indeed, in such hypotheses it is generally postulated that the first Eumetazoa might have resembled either hollow flagellate colonies such as the algal *Volvox* or solid flagellate colonies such as the algal *Synura*. In their turn, such colonies would have resembled the *planula* larvae of coelenterates, and it is generally held today that coelenterates actually were the first Eumetazoa.

Like the flatworm hypothesis, this "coelenterate hypothesis" consists of two parts, i.e., the flagellate origin and the planulalike, or *planuloid*, result. The first part is subject to an objection similar to one raised above. Judging from living Protista, flagellate-amoeboid organisms have exhibited only quite limited evolutionary resourcefulness; like the coccine forms, none of these types has attained a tissue level of complexity. Protozoan flagellates have remained unicellular almost entirely, and among algal flagellates the most complex structures actually evolved have been types such as *Volvox* and *Synura*. It is therefore unlikely that flagellate ancestors could have evolved much beyond a colonial level.

On the other hand, we do know that another state of existence, viz., the sporine condition, has exhibited a pronounced capacity to evolve beyond the colonial level, as in the complex algae and the Metaphyta. Moreover, Eumetazoa do represent predominantly sporine types of organisms. Accordingly, it may be suggested that Eumetazoa originated from sporine, not flagellate, ancestors. Since sporine types are not included among protozoa or any other colorless protists (cf. Fig. 14.19), such ancestors would have had to be early, *preprotozoan* protists (Fig. 14.21). Colorless unicellular forms of this kind could have given rise to sporine, *nonmotile* colonies, which might have floated passively in shallow water and perhaps settled to the bottom occasionally. Inasmuch as photosynthesis would not have occurred, each cell would not have required direct exposure to sunlight. Consequently, such colonies need not have been hollow spheres or filaments or sheets, and they need not have resembled sporine algal colonies. Instead, they could have been solid, morulalike aggregates without locomotor organelles, exactly like many early eumetazoan embryos today. Nutrition could have been saprotrophic or holotrophic

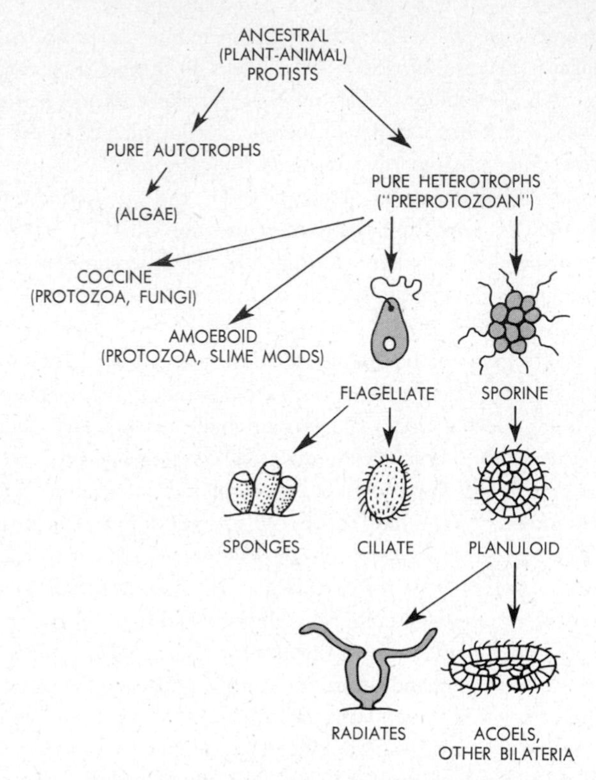

14.21. A hypothesis of eumetazoan origins alternative to those outlined in Fig. 14.20. Multicellular Metazoa here are postulated to have arisen as a preprotozoan *sporine* line of evolution, not present otherwise among heterotrophic Protista (cf. Fig. 14.19). Thus, Protozoa and Metazoa are assumed to represent parallel, not sequential, groups. Other aspects of this hypothesis are analogous to the flagellate-planuloid postulates (Fig. 14.20).

or both, and digestion could have been carried out intracellularly by all cells. At some later stage of evolution, some or all of the exterior cells of such colonies might have redeveloped flagella, just as given cells of sporine algae today may become flagellate. In this manner, the exterior cells could have assumed primarily locomotor functions, and alimentary and reproductive functions could have become the specialized responsibility of the interior cells.

In the end, therefore, primitive Eumetazoa would again have been exteriorly flagellate balls of cells, but note that the original evolutionary starting point would have been a sporine and not a flagellate state. The difference is important, for, as noted, sporine types do and flagellate types do not have a known potential to develop beyond the colony level. Moreover, a sporine origin would move the start of eumetazoan history desirably far back into protistan history, to a time when early colorless protists still had most of their evolutionary potentialities before them and when presumably none of the modern Protista were as yet established. It would thus become possible to derive Eumetazoa from generalized, primitive ancestors, as is proper, not from already specialized ancestors such as protozoa. On this basis, Eumetazoa and Protozoa would have evolved not one after the other but in parallel, as separate lines of an adaptive radiation emanating from early colorless protists. And if protozoa subsequently gave rise to sponges, the latter might therefore not have come into existence until *after* eumetazoan history was already well launched (cf. Fig. 14.21).

Notwithstanding the variety of opinions concerning the exact protistan starting point of the Eumetazoa, most of the current hypotheses do envisage the early representatives of these animals to have been multicellular aggregates with exterior locomotor cells. The view that such aggregates were hollow, like *Volvox*, is now less prevalent than the view that they were solid; and the belief that they represented coelenterates is accepted widely. The hollow-colony concept and the coelerate concept as a whole received their strongest original impetus in a famous hypothesis proposed by E. Haeckel, a German zoologist of the late nineteenth century. This hypothesis dealt not only with the problem of metazoan origins but also with several other phylogenetic questions. Most of Haeckel's ideas are now discredited, but they were once so influential that they still persist today under various guises. For this reason, even though the issues are quite dated, it may nevertheless be of some value to review them briefly.

Addition and Divergence

Haeckel recognized, as did others before him, that the early embryonic development of all Eumetazoa passed through certain common stages. Thus, development starts with the unicellular zygote, proceeds by cleavage to a blastula stage, then to a gastrula stage, and continues later with mesoderm formation. Haeckel then thought that each such stage corresponded to an adult form of an ancestral type.

Accordingly, the succession of embryonic stages would mirror a succession of past evolutionary stages (Fig. 14.22). The zygote would represent the unicellular protistan stage of evolution. The blastula would correspond to an evolutionary stage when animals were, according to Haeckel, hollow one-layered spheres. Haeckel coined the term *blastea* for such hypothetical adult animals, and he thought that his ancestral blasteas may have been quite similar to currently living forms such as *Volvox*. Analogously, the gastrula stage in animal development would correspond to a hypothetical ancestral adult type which

14.22. Haeckel's "gastrea" hypothesis, exemplifying the recapitulatory view that successive developmental stages of animals repeat the successive evolutionary stages of their ancestors. Based primarily on coelenterate development, the hypothesis is at variance with the actual developmental patterns typical of most coelenterates, as sketched in right column. Thus (notwithstanding some exceptions) most coelenterates do not gastrulate by invagination, and mesoderm usually arises by delaminative and ingressional methods, a gastrocoel forming by central cavitation. The presence of mesoderm actually makes not only the Bilateria but also the Radiata (i.e., *all* Eumetazoa) "triploblastic," or formed from three primary germ layers, contrary to Haeckel's hypothesis.

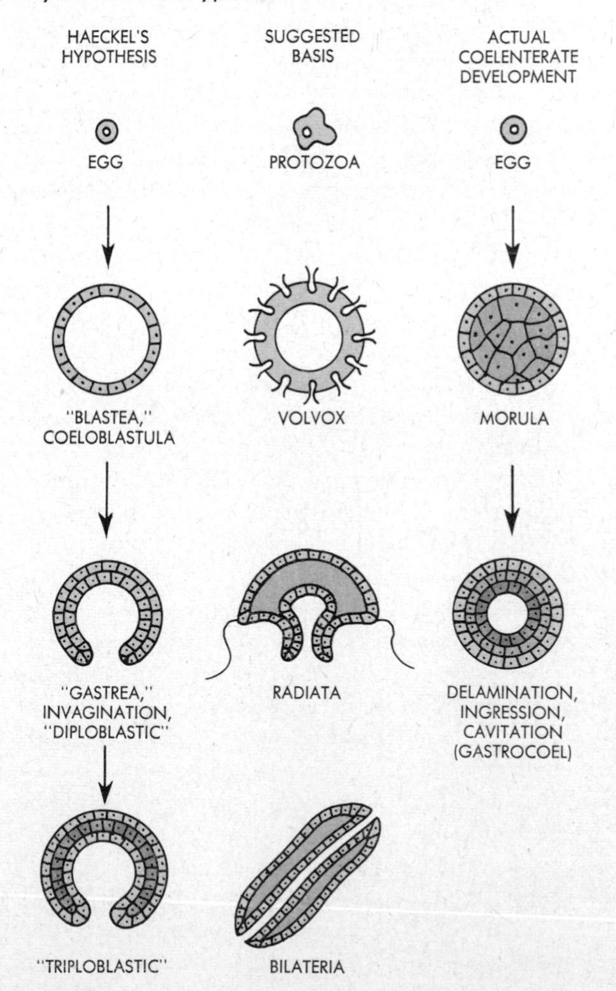

he called a *gastrea.* He believed that gastreas arose from blasteas by embolic invagination, and that such animals are still represented today by the living coelenterates. In some respects the latter do resemble early embolic gastrulae; e.g., the body contains an ectodermal and an endodermal epithelial layer and the single alimentary opening is reminiscent of the blastopore of a gastrula.

On such grounds, Haeckel considered embolic gastrulation to be primitive and he regarded the gastrea to have been the common ancestor of all Eumetazoa. Moreover, he assumed the two-layered gastrealike condition to represent a *diploblastic* stage in animal evolution, attained by the coelenterates and the ctenophores. Further evolution then added mesoderm to the gastrealike radiates and so produced a three-layered *triploblastic* condition, as in flatworms and all other Bilateria. Thus, he considered coelenterates to represent a stage in flatworm evolution. By extension, this Haeckelian hypothesis later also came to imply, for example, that a caterpillar larva represented an annelid stage in insect evolution, that a frog tadpole represented a fish stage in frog evolution, and that a human embryo, which exhibits rudimentary gill structures at certain periods, represented a fish stage in man's evolution.

Haeckel condensed his views into a *law of recapitulation,* the essence of which is described by two phrases, viz., "ontogeny recapitulates phylogeny" and "phylogeny causes ontogeny." The first statement means that the embryonic development of an egg (ontogeny) repeats the evolutionary development of the phyla (phylogeny); and the second, that "because" animals have evolved one phylum after another, their embryos still pass through this same succession of evolutionary stages.

Therefore, if one wishes to determine the course of animal evolution, he need only study the course of embryonic development. For, according to the law, evolution occurs by the addition of extra embryonic stages to the end of a given sequence of development. If to a protozoon is added cleavage, the protozoon becomes a zygote and the new adult is a blastea. If to a blastea is added the process of gastrulation, then the blastea becomes a blastula and the new adult is a gastrea. Similarly, if to a fish are added lungs and four legs, then the fish represents a tadpole and the new adult is a frog. And if to such an amphibian are added a four-chambered heart, a diaphragm, a larger brain, an upright posture, and a few other features, then the frog is a human embryo and the new adult is a man.

We can attribute to the lingering influence of Haeckel, not to Darwin, this erroneous idea of an evolutionary "ladder" or "scale," proceeding from "simple amoeba" to "complex man," with more and more rungs being added on top of the ladder as time proceeds. All such notions are invalid because Haeckel's basic thesis is invalid. Indeed, Haeckel's arguments were shown to be unsound even in his own day, but his generalizations were so neat and they seemed to explain so much so simply that the fundamental difficulties were ignored by many.

For example, it was already well known in Haeckel's time that the coelenterates, which should develop most nearly like the postulated gastrea, in actuality develop quite differently (cf. Fig. 14.22). Most form not hollow coeloblastulae but solid morulae or stereoblastulae; and gastrulation occurs only very rarely by embolic invagination but most often by delamination and ingression. Moreover, apart from exceptional forms like the hydras, the radiate animals do not really have two-layered bodies but distinctly three-layered ones, even though the middle mesoderm remains mesenchymal and poorly developed. Thus, two-layered animals virtually do not exist and a distinction between diploblastic and triploblastic types does not correspond to reality. In effect, the central conceptual foundation on which the recapitulatory law was based was never valid.

We know also that the evolution of new types generally does not occur through addition of extra stages to ancestral adults. Instead, new evolution occurs for the most part by *developmental divergence:* a new path of embryonic or larval development branches away from some point along a preexisting developmental path (Fig. 14.23). The best example is evolution by larval neoteny, a mechanism which probably has been responsible for a great deal of past evolution. In such cases, a new group arises from the larva of an old group. The existing larva does not metamorphose into the customary adult, but instead develops sex organs precociously and becomes established in this larval form as a new type of adult animal. The evolution of tunicates into vertebrates provides a good illustration (Fig. 14.24). Tunicates develop via tadpole larvae into sessile adults; and it is now virtually certain that vertebrates represent neotenous tunicate tadpoles. The tadpoles of certain tunicate ancestors appear to have retained their tails, with notochord and dorsal hollow nerve cord intact, and to have become reproductively mature in such a larval condition. These permanent larvae then came to represent a new chordate subphylum, viz., the vertebrates (cf. also Chaps. 30 and 31).

Analogously, it can be shown that man probably represents a neotenous product derived from late apelike embryos (cf.

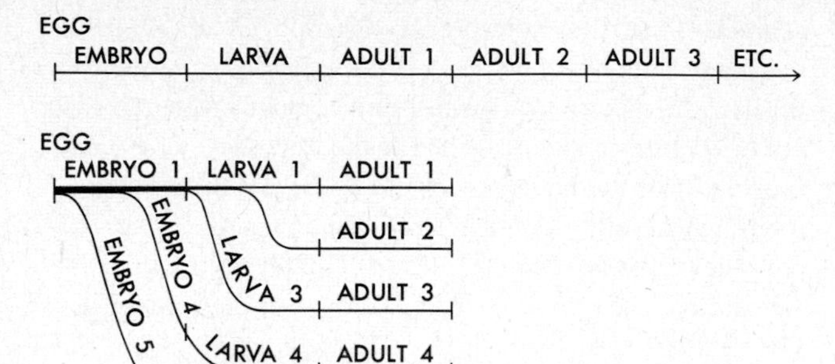

14.23. Top, the pattern of evolution according to Haeckel, by addition of extra stages to a preexisting ancestral path of development. Bottom, the actual pattern of evolution, by divergence of new developmental paths (2, 3, 4, 5) from various points of a preexisting ancestral path (1). Thus, if divergence occurs comparatively late, as in adult 2, its embryonic and most of its larval development may be very similar to that of adult 1. But if divergence takes place early, as in adult 5, then almost the whole developmental pattern may be quite dissimilar from that of adult 1. Adults 1 and 5 then might represent different phyla, whereas adults 1 and 2 might belong to the same class or order.

Chap. 15). Numerous other instances of evolution by neoteny are known, and some will be encountered in later chapters. In all of them, the new developmental path branches away rather late along the course of the old path. But note that the point of developmental divergence often occurs earlier, e.g., in the embryo, or even as early as in the egg itself. And it should also be clear that the sooner two such paths do diverge, the more dissimilar will be the two subsequent developmental sequences and the two types of resulting adults. For example, the embryos of man and of apes resemble each other till relatively late in development, and the developmental paths diverge only then. The embryos of man and of tunicates are similar for considerably shorter periods; here the developmental paths diverge much sooner (cf. Fig. 14.24).

Such developmental correlations were clearly recognized before Haeckel and even before Darwin. We know today that they have evolutionary meaning, but not in the Haeckelian sense. It is quite natural that related animals descended from a common ancestor should resemble one another in some of their adult as well as some of their embryonic features. Such resemblances may or may not be pronounced, depending entirely on how widely the evolutionary paths have diverged. Accordingly, human embryos resemble those of fish and frogs in certain respects not because an egg of man becomes a fish embryo first, changes to a frog embryo next, and transforms into a human embryo last. The similarities arise, rather, from the common ancestry of all these three animal types, including their common developmental histories up to certain stages. Beyond such stages, each type has modified its developmental processes in its own specific way. Note in this connection that, the sooner two developmental paths diverge, the greater will be the taxonomic gulf between the resulting types. Thus, tunicate and vertebrate development diverge at an early stage, and the results then represent two different subphyla. But ape and human development diverge comparatively much later, at a stage equivalent to very late larvae, and the results then differ only to the extent of taxonomic families (cf. Fig. 14.24).

The mechanism of developmental divergence is consistent

14.24. A specific set of examples of evolution by developmental divergence. Tailed, tadpolelike early stages are characteristic of all chordates, and the adult and taxonomic results are the more different the earlier the developmental paths diverge from the ancestral one. Keep in mind, however, that chordate eggs, for example, are not all the "same"; they are merely similar. Analogously, "common" developmental stages of two animals are merely similar.

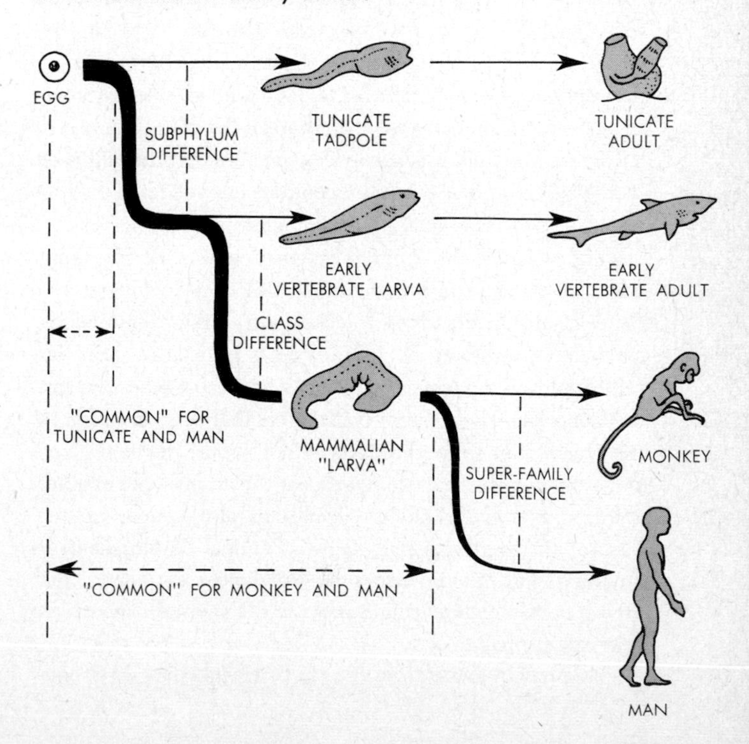

also with the basic *genetic* role of development. Eggs, embryos, and larvae are precisely the stages still undifferentiated enough to permit actual execution of new developmental programs, as coded by the genes of the zygote. On the other hand, the adults are already fully specialized in all their parts and are no longer plastic developmentally. Contrary to the Haeckelian postulate, therefore, the potential of developing extra stages then no longer exists. Moreover, we know from available fossil data and from our present understanding of evolutionary mechanisms that evolution must have occurred in bushlike fashion; and a branching pattern of developmental divergence is fully consistent with this knowledge, but ladderlike end-addition is not.

Furthermore, detailed examinations of embryos have shown clearly that the "common" developmental stages of animals are not nearly so alike as Haeckel's law postulates. For example, although both tunicates and vertebrates possess similarly *named* stages such as blastulae and gastrulae, the actual structure of these stages differs greatly in the two cases; e.g., coeloblastulae are largely typical of tunicates, discoidal blastulae of vertebrates. In effect, each animal is a unique construction which must be developed in a unique way. This uniqueness starts directly with the egg and becomes progressively more pronounced as development proceeds. The *whole* life cycle is therefore species-specific, all embryonic, larval, and adult stages included. And if we cast aside theoretical preconceptions, it should be very obvious that mammalian development, for example, does not really represent a successive transformation of a protozoon into a coelenterate, a flatworm, a tunicate, a fish, etc. Actually it is hardly conceivable that the billion or more years of animal evolution could be crowded into the few weeks or months of animal development.

Thus, whereas the common stages in animal development may give evidence of general similarities, they do not give evidence of specific identities. Haeckel ignored these embryological realities. Instead, he formulated a pure abstraction far removed from the actual evidence. For these various reasons, the conclusion is certainly warranted that recapitulation in the Haeckelian sense simply does not occur; the embryonic stages of given animals do *not* repeat the adult stages of other animals. Moreover, phylum evolution also cannot be the "cause" of the progressive stages in animal development. If anything, just the reverse probably holds; as pointed out above, developmental stages provide the sources from which new groups may evolve.

What now remains from the era of Haeckel are three conceptions. First, the three primary germ layers are probably homologous in all Eumetazoa. Often called the "germ-layer theory," this generalization did not originate with Haeckel but he certainly popularized it. The theory is acceptable today provided that the homology of the layers is not extended to imply an analogy. In other words, although the layers are equivalent developmentally and structurally in the different Eumetazoa, they do not always serve equivalent functions. For example, in many cases of reproduction by budding, the entire animal may develop from ectoderm alone. Analogously, although muscles are usually mesodermal, the pupillary muscles of the vertebrate eye and some of the body wall muscles of certain invertebrates are derived from ectoderm.

Secondly, similarities among paths of development do exist, and inasmuch as they are consequences of common ancestries they may indicate degrees of evolutionary interrelations among animals. This hypothesis likewise predates Haeckel, and he then converted or subverted it into the recapitulation law. In the latter sense the hypothesis is no longer acceptable, but in the former sense it is still valid: the development of two individuals or groups on the whole does tend to be similar if the animals are related, and the similarity lasts longer the closer the relationship. Indeed, as outlined in Chap. 7, phyla are grouped into subgrades for the most part on the basis of common body plans, and in such classifications developmental resemblances do play a substantial contributing role.

And thirdly, coelenterates can be hypothesized to be the most ancient Eumetazoa. However, those who accept this view today, i.e., the majority of zoologists, do so largely on a modified non-Haeckelian or neo-Haeckelian basis. As outlined, the first Eumetazoa are now usually envisaged to have been solid, not hollow, colonies. In such a form, a "coelenterate" hypothesis is more nearly in line with the solid morulae and the general absence of invaginative gastrulation in most living coelenterates.

The Later Animals

Both the coelenterate and flatworm hypotheses consider the Eumetazoa to be a monophyletic group; the first members are assumed to have given rise to all others. Neither hypothesis succeeds in giving a fully satisfactory picture of eumetazoan phylogeny, and it is possible that this failure might be due at least in part to the basic assumption of eumetazoan monophyletism.

According to the most common current form of the coelenterate hypothesis, the planulalike first Eumetazoa gave rise to two groups, viz., the modern radiate phyla and the acoel flatworms (Fig. 14.25). The interior of the ancestral planuloid would have acquired an alimentary cavity by cavitation, and a single opening to the exterior would have developed in the blastopore region. Such processes actually occur in the embryonic development of most modern radiates. Furthermore, a planula is structurally quite similar to an acoel worm, and the evolutionary transition from one to the other fundamentally would have required only one major change, namely, transformation of a radial to a bilateral symmetry. This is believed to have occurred via a change in way of life. Planulae are free-swimming forms, acoels are creeping. Thus, if a planuloid ancestor assumed a creeping mode of existence, with the alimentary opening directed toward the ground, then it could readily have evolved into a flattened, conspicuously elongated, and bilateral worm. Such acoelous animals subsequently would have given rise to all other flatworms and eventually also to all other acoelomates.

According to the flatworm hypothesis, the above events are postulated to have taken place in the reverse order (cf. Fig. 14.25). Disregarding the implausible origin of acoels by septation of ciliates and assuming an origin from sporine protists, the possibility remains that acoels in turn could have given rise to two groups, viz., the modern acoelomates and the radiate animals. On this basis, Eumetazoa would have been bilateral from the start and the radiality of coelenterates would be a secondarily derived condition. Proponents of this view also point out that coelenterates such as sea anemones today actually display bilaterality (cf. Chap. 7), and they maintain that these animals therefore represent the most primitive radiates; their bilaterality can be construed as a residual inheritance from the bilateral acoel ancestor.

On the contrary, proponents of the coelenterate hypothesis maintain that the fully radial coelenterates are primitive and that those with superficial bilaterality are the more advanced members of the group. In support of this view it can be argued that if bilaterality really were primitive, then the embryos of coelenterates and certainly of acoels should show signs of this symmetry. Yet all these embryos are radial and all the larvae of coelenterates are radial as well (acoels are without distinct larvae). Most zoologists accept the postulate that radiality evolved into bilaterality and that radiate ancestors gave rise to flatworms. A third and equally plausible possibility is not generally considered, namely, that early radial coelenterates and early bilateral flatworms arose not one from the other but each independently from even earlier protistan ancestors (Fig. 14.26).

Although opinion regarding the origin of the first Bilateria is thus divided, it is agreed quite widely that the acoelomates generally and the acoel flatworms specifically produced all other Bilateria. Moreover, most theorists also agree that the Protostomia, i.e., acoelomates, pseudocoelomates and protostomial coelomates, are related more or less closely. In all probability, the acoelomates are ancestral to the other Protostomia (Fig. 14.27). Several observations support such conclusions. All these groups possess predominantly mosaic eggs, and their development is generally of the determined type. Cleavage is variously spiral, radial, and bilateral in pseudocoelomates, and it is entirely spiral in both acoelomates and protostomial coelomates. All three groups are protostomial; i.e., the mouth develops at or near the blastopore. The mesoderm typically arises in part as a mesenchymal ectomesoderm, in part as an endomesoderm, the latter in many cases from teloblastic 4d cells. Moreover, although the adults are clearly different in structure, in many instances drastically so, they are nevertheless not so different that similar design characteristics are not present. For example, adult Protostomia possess broadly common features such as ventral ladder-type nervous systems, protonephridial and metanephridial excretory systems, and main blood vessels or hearts in dorsal locations. Among the coelomate protostomes, the schizocoelomates additionally have in common a developmental pattern which typically includes trochophore larvae.

It is more difficult to assess the relation between Protostomia and Deuterostomia. Deuterostomes develop predominantly from regulative eggs via bilateral cleavage, the blastopore becomes the anus, the principal mesoderm arises enterocoelously from endoderm, and the animals are coelomate. Moreover, most deuterostome groups (i.e., all except the chaetognaths, or arrow worms, cf. Fig. 14.27) have been postulated to be derived from a hypothetical *dipleurula* ancestor, a type corresponding to a common structural stage encountered during the early larval development of these forms (cf. Chap. 29). Hypotheses of an earlier day have linked the enterocoelomates with schizocoelomate ancestors. More specifically, annelids were postulated to have given rise to the chordates. The principal supporting arguments were that both groups possess true coeloms, that metameric segmentation occurs in both, and that solenocytic protonephridia are found in certain primitive annelids as well as in certain primitive chordates (*amphioxus*). Also, if the body of an annelid is

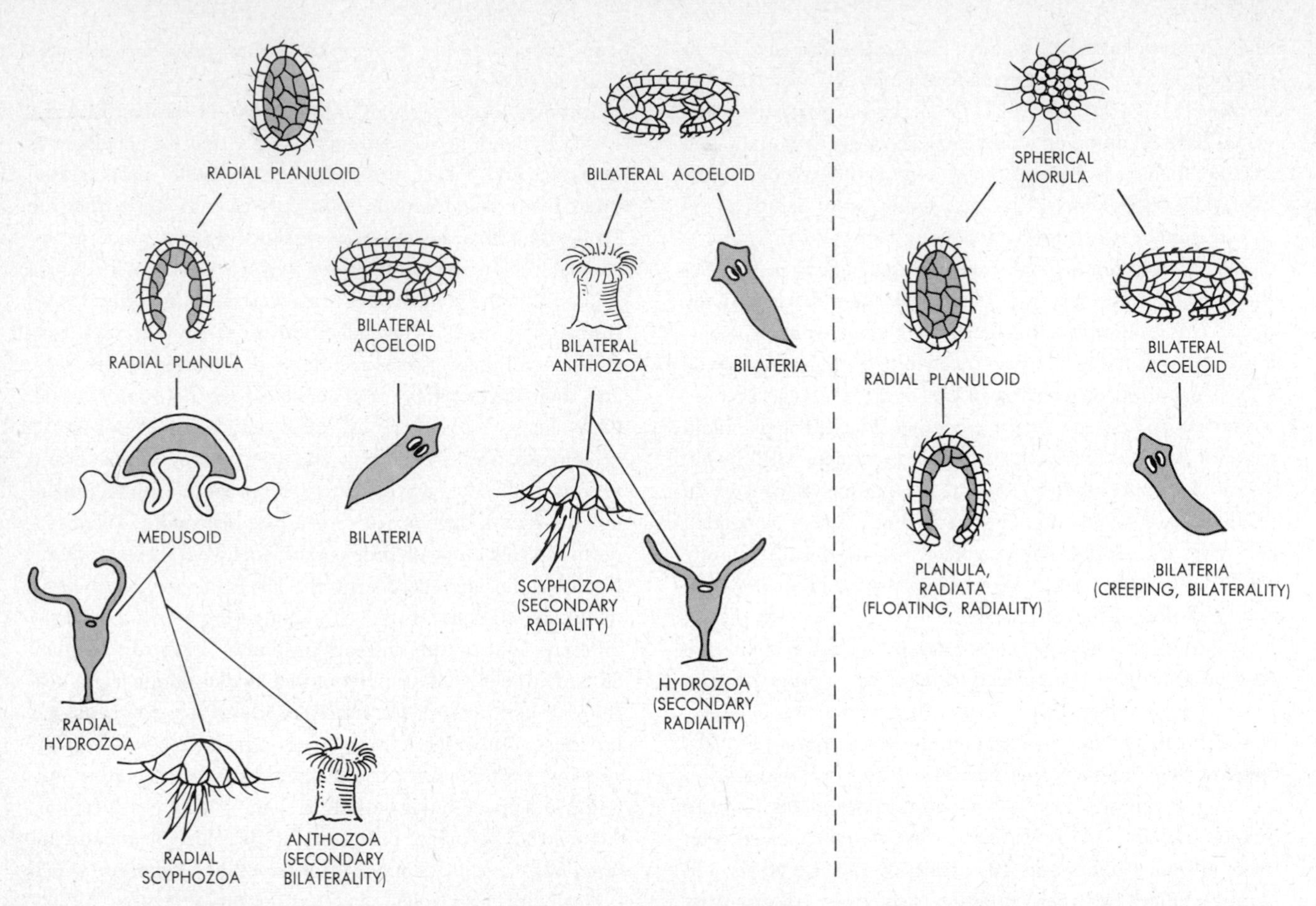

14.25. **Left,** the widely accepted "coelenterate" hypothesis of bilateral origins. Radial coelenterate ancestors (planuloids) are postulated to have given rise to bilateral acoels and Bilateria and to primitively radial coelenterates. Hydrozoa then would be the most primitive living coelenterates, Anthozoa (sea anemones), the most advanced (with secondary bilaterality in some of their features). **Right,** the disputed "flatworm" hypothesis. Already bilateral acoel ancestors are here postulated to have given rise to Bilateria and to basically bilateral coelenterates, Anthozoa being regarded as most primitive. Radial features of coelenterates then would have evolved secondarily, Hydrozoa becoming the most advanced group. See text for critiques.

14.26. A hypothesis alternative to those outlined in Fig. 14.25. The first eumetazoan here is postulated to have been a morulalike, spherical, sporine form. It then is assumed to have given rise, on the one hand, to the radiate groups, which were medusalike floaters primitively and possessed a basic radiality in adaptation to this mode of life; and on the other, to the bilateral groups, which adopted a creeping mode of life and developed bilaterality in conjunction.

turned upside down, so that dorsal becomes ventral, then the nervous system would come to lie dorsally and the principal blood vessel ventrally, as in the body plan of vertebrates (Fig. 14.28). However, such views left all other enterocoelomates unaccounted for, including the nonsegmented chordate tunicates. In addition, the hypothesis did not reconcile the many differences of development between annelids and vertebrates: mosaic vs. regulative eggs, spiral vs. bilateral cleavage, protostomial vs. deuterostomial mouths, teloblastic vs. enterocoelous mesoderm, and trochophore vs. tadpole larvae.

The original "annelid" hypothesis has since been abandoned, but one basically similar to it has recently been advanced by the proponents of the ciliate-flatworm hypothesis. According to this newer scheme, early segmented schizocoelo-

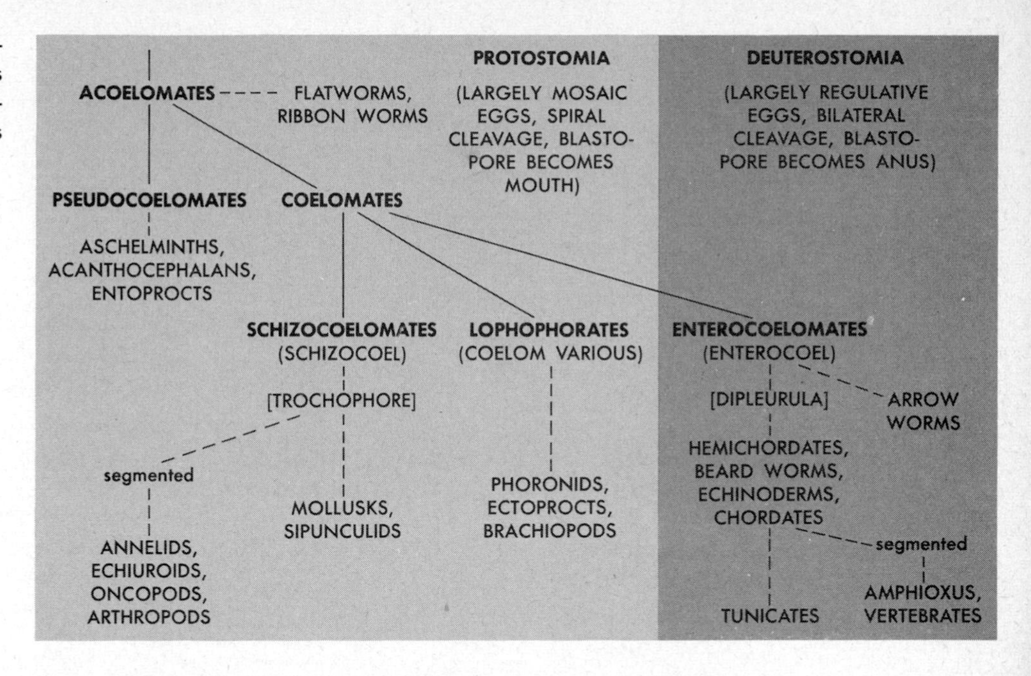

14.27. The probable evolutionary interrelations of the protostomial groups and of the Bilateria as a whole. Diagrams of postulated "dipleurula" stages will be found in Chap. 29.

mate protostomes are postulated to have given rise to the deuterostomes and the developmental differences are regarded as fundamentally unimportant; although the two types of embryonic development differ, they are regarded as representing merely variants of the same basic pattern. Thus, mosaic and regulative eggs differ only in the timing of developmental determination; cleavage patterns are probably related in any case and are without phylogenetic significance; the fate of the blastopore might be expected to change if the body plan is turned upside down; the primary mesoderm is mesenchymal ectomesoderm in both groups, just as the secondary mesoderm is endomesodermal; and we know that the size of the blastocoel determines whether the coelom is formed enterocoelously from endodermal pouches or schizocoelously from teloblast cells. Moreover, free-living larvae must be adapted to different specific environments, and their structure therefore cannot be relied upon as a basis for phylogenetic deductions. Most of these points do have considerable general validity, but a basic difficulty remains nevertheless: if enterocoelomates are derived from segmented schizocoelomates, *all* deuterostomes would have to be regarded as fundamentally segmented groups. Such a view is

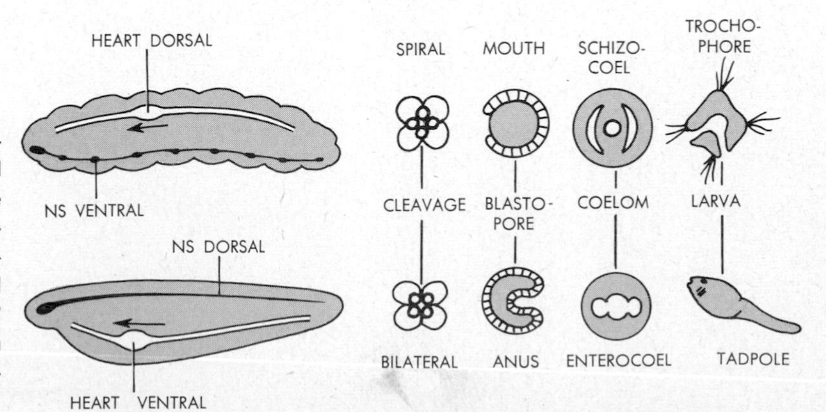

14.28. The old "annelid" hypothesis of vertebrate origins (left part of figure), which postulated that the vertebrate body plan arose when the annelid body plan somehow became turned upside down. However, the basic developmental differences, as sketched at right would then still remain, and it will be noted also that a mere 180° rotation of the dorsoventral axis does not change a protostomial mouth-anus relation to a deuterostomial one.

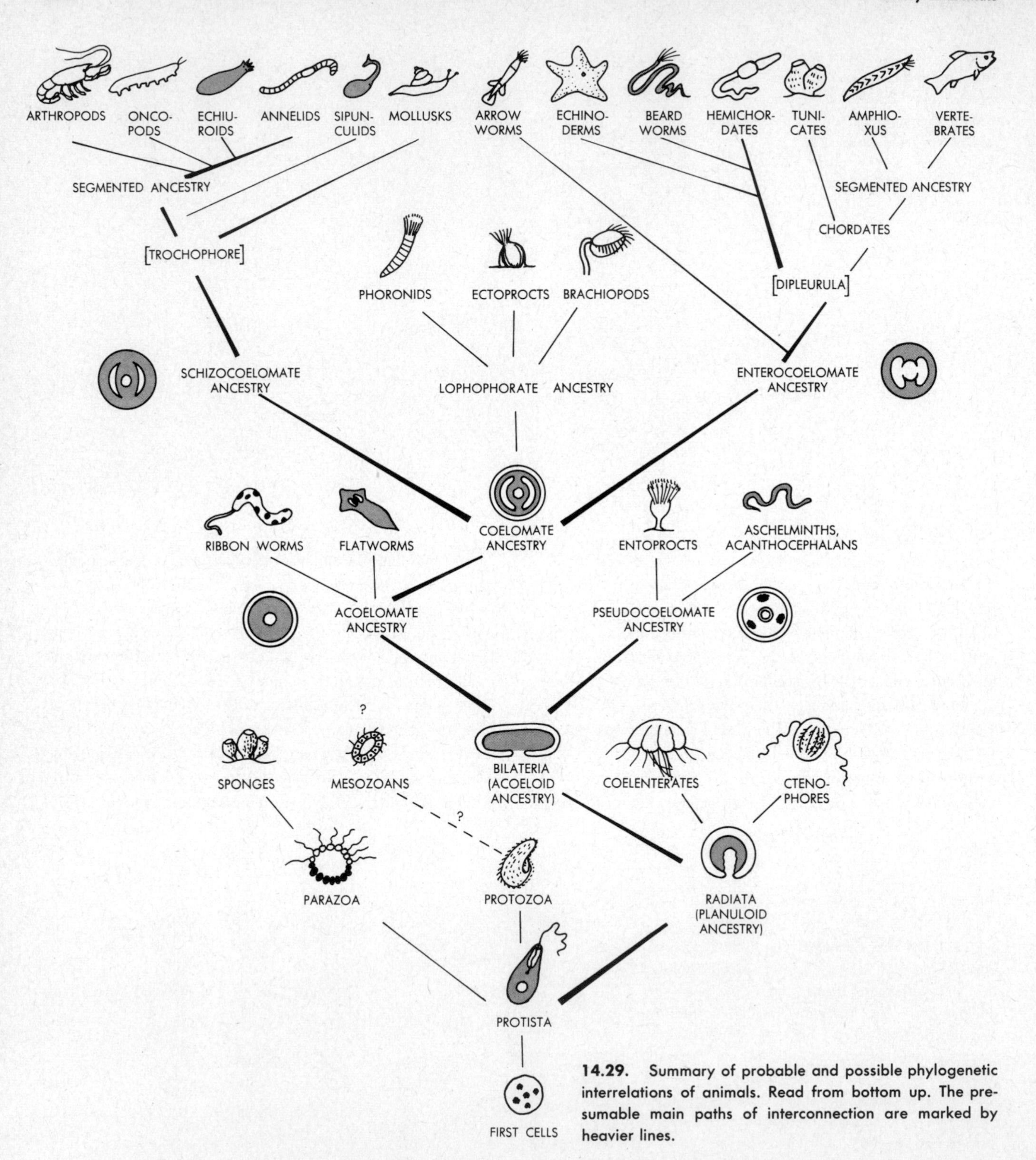

14.29. Summary of probable and possible phylogenetic interrelations of animals. Read from bottom up. The presumable main paths of interconnection are marked by heavier lines.

quite at variance with the actual evidence, for among deuterostomes only some of the chordates are segmented.

Most zoologists adhere to the view that deuterostomes evolved from early coelomate protostomes, possibly protostomes of a type not too different from some of the ancestors of the lophophorates (cf. Fig. 14.27). As already noted in Chap. 7, the methods of mesoderm- and coelom-formation are not at all uniform among the lophophorate phyla; some develop a coelom schizocoelously, others enterocoelously, still others by various unique processes. Such nonuniformity suggests that, among early coelomate ancestors, methods of coelom formation may have been subjected to considerable evolutionary "experimenting." The schizocoelous methods then would have become standard in some of the descendant lines, the enterocoelous methods in the others. It is less clear how such enterocoelous groups might also have become deuterostomial, but a reversal of anteroposterior polarity unquestionably did take place. Thus, lophophorates might represent the surviving remnants of a possible evolutionary connection (though not a particularly direct one) between the schizocoelous protostomes and the enterocoelous deuterostomes. On this basis a comprehensive phylogenetic scheme such as outlined in Fig. 14.29 appears to be most consistent with the incomplete evolutionary evidence presently available.

Of numerous other hypotheses proposed in the course of the last century, most have not come widely into vogue either because, like some we have dealt with in this chapter, they were at variance with the evidence from living animals, or because they entailed conceptual consequences later found to be untenable, or because they lacked direct or even indirect support altogether. A few of the more prominent and to some extent still current earlier postulates will be mentioned in the chapters of Unit 2. In these chapters we shall also examine the possible or probable evolutionary interrelations among individual phyla and of groups within phyla.

For the present we can really be sure only that the various animal groups did evolve in *some* phylogenetic pattern, and that the period of these events extended from roughly 2 billion to about 500 million years ago. At the latter date all phyla were already well established, and from then on our knowledge of evolutionary histories moves to surer ground by virtue of fossils.

REVIEW QUESTIONS

1. Which gases in the early atmosphere of the earth may have contributed to the formation of organic compounds? What were some of these compounds and what evidence do we have that they could actually have formed? Review the role of (*a*) temperature (*b*) water, (*c*) organic compounds, and (*d*) enzymes in the origin of life.

2. Review the synthesis reactions through which compounds required for the origin of living systems might have occurred. Describe experiments duplicating some of these reactions. How are the first cells believed to have evolved? Distinguish between chemical and biological evolution.

3. What was the physical character of the earth (*a*) at the time it formed, (*b*) before living systems originated, and (*c*) after living systems originated? Review the principal events of the oxygen revolution and describe the consequence of this revolution. How many years ago did (*a*) the earth form, (*b*) life originate?

4. Distinguish between monophyletism and polyphyletism and cite arguments for and against each view as related to the origin of cells. How are moneran and protistan cell types distinguished and how might both have arisen from the first cells? What living groups belong to the Monera and the Protista? What are viruses and how are they believed to have originated?

5. Through what processes of evolution may moneran and protistan nutritional patterns have arisen? What processes probably necessitated and promoted such nutritional evolution? Describe the principal chemical events of chemosynthesis and photosynthesis. In what groups of organisms do these occur?

6. On what basis may it be assumed that ancestral protists probably could obtain nutrients by more than one method? What states of existence could early protists probably exhibit? Define each of these states and give specific examples of each.

7. List specific examples of unicellular and multicellular algae exhibiting each of the states of existence. Which protistan group probably gave rise to the Metaphyta? What is the basis for such an assumption? What states of existence are exhibited by (*a*) fungi, (*b*) slime molds, (*c*) protozoa? How are Metaphyta and Metazoa distinguished from each other and how are both distinguished from Protista?

8. Why are the terms "animal" and "plant" inadequate in a Linnaean taxonomic sense? Show how different taxonomic definitions could make these terms precise. Review traditional

hypotheses of metazoan origins. Describe the ciliate-flatworm and the flagellate-coelenterate hypotheses of eumetazoan origins. What are the weaknesses of each? What alternative hypotheses might circumvent such weaknesses?

9. Review the recapitulatory hypotheses of Haeckel and show why they are not tenable. What Haeckelian views are still valid today? Review hypotheses proposed to account for the origin of the first Bilateria. What are the presumable phylogenetic interrelations between acoelomates, pseudocoelomates, and coelomates? On what evidence are such assumptions based?

10. What are the likely phylogenetic interrelations between schizocoelomates, lophophorates, and enterocoelomates? On the basis of current phylogenetic thinking, how many times would (*a*) the coelom, (*b*) segmentation have evolved independently? Which groups exhibit metameric segmentation? Draw up a comprehensive diagram indicating the presumable phylogenetic interrelations of all major animal groups.

COLLATERAL READINGS

Supplementary information relating to topics in this chapter may be found in the following sources. The articles by Miller and Fox deal with some of the experiments on chemical evolution referred to above, and the readings in Kerkut and Hyman are recommended as background to various hypotheses on animal phylogeny:

Anfinson, C. B.: "The Molecular Basis of Evolution," Wiley, New York, 1959.

Bonner, J. T.: "The Evolution of Development," Cambridge, Cambridge, England, 1958.

Brown, H.: The Age of the Solar System, *Sci. American,* Apr., 1957.

De Beer, G. R.: "Embryos and Ancestors," Clarendon, Oxford, 1940.

Fox, S. W.: The Evolution of Protein Molecules and Thermal Synthesis of Biochemical Substances, *Am. Scientist,* vol. 44, 1956.

Gamow, G.: The Origin and Evolution of the Universe, *Am. Scientist,* vol. 39, 1951.

Horowitz, N. H.: On the Evolution of Biochemical Syntheses, in M. L. Gabriel and S. Fogel, "Great Experiments in Biology," Prentice-Hall, Englewood Cliffs, N.J., 1955.

Hyman, L.: "The Invertebrates," vol. 1, chap. 5; vol. 2, chap. 9, McGraw-Hill, New York, 1940, 1951.

Kerkut, G. A.: "The Implications of Evolution," Pergamon Press, New York, 1960.

Landsberg, H. E.: The Origin of the Atmosphere, *Sci. American,* Aug., 1953.

Marcus, E.: On the Evolution of Animal Phyla, *Quart. Rev. Biol.,* vol. 33, 1958.

Miller, S. L.: The Origin of Life, in W. H. Johnson and W. C. Steere, "This is Life," Holt, New York, 1962.

Oparin, A.: "The Origin of Life," 3d ed., Dover, New York, 1945.

Urey, H.: The Origin of the Earth, *Sci. American,* Oct., 1952.

Wald, G.: The Origin of Life, *Sci. American,* Aug., 1954.

———: Innovation in Biology, *Sci. American,* Sept., 1958.

———: Life and Light, *Sci. American,* Oct., 1959.

CHAPTER 15

animal descent: paleontology

Inasmuch as the animal phyla had already originated before the beginning of the fossil record, paleontology can only provide data relating to the evolutionary histories *within* individual phyla. A further limitation is that, to the extent it is known, the fossil record is scanty or incomplete for many animal groups. Nevertheless, where fossils are available, they contribute a body of evidence equal in importance to that obtained from the embryology and adult anatomy of the living animals. In most instances, the two kinds of data have proved to be mutually consistent and have formed a reasonably firm base for phylogenetic deductions. In some instances, however, the two lines of evidence have led to flatly contradictory conclusions. It is then sometimes possible to decide whether more weight should be given to the fossil or to the embryological evidence, but such decisions often cannot be reached. Unresolved differences of opinion then simply persist.

In this chapter, we begin with a survey of the geologic record as a whole and continue with a more detailed survey of the fossil record of animals. For the most part, the account emphasizes broad historical trends in successive segments of time; where warranted, references to specific fossil groups of given phyla are also included in the chapters of Unit 2.

THE GEOLOGIC RECORD

Fossils

Any long-preserved remains of animals are fossils. They may be skeletons or shells, perhaps recrystallized under heat and pressure and infiltrated with mineral deposits from surrounding rock. They may be footprints later petrified or the remnants of animals trapped in arctic ice, amber, quicksand, gravel pits, tar pits, and swamps. Or they may be imprints of carbon black on rock, left when the soft parts of animals vaporized under heat and pressure. Whenever a buried animal or any part of it becomes preserved in some way before it decays, it will be a fossil.

Fossils formed in the past are embedded in earth layers of different ages. In a geologically undisturbed section of the earth's crust, the deeper layers are the older layers. Material eroded from high-lying land gradually piles up on low land and on the sea bottom. A deep layer today therefore was on the surface in past ages and the earth's surface today will be a deep layer in the future. Fossils embedded in successive layers so provide a time picture of evolution. To be sure, deep-lying fossils are normally not accessible. But on occasion a canyon-cutting river, an earthquake fracture, or an upbuckling and consequent breaking of the earth's crust may expose a cross-section through the rock stata. Moreover, erosion gradually wears away top layers, exposing deeper rock. Geologic changes of this sort have been sufficiently abundant to expose layers of all different ages in various parts of the world (Fig. 15.1).

How is the actual age of a rock layer determined? Very excellent clocks are built right into the earth's crust: radioactive substances. The disintegration rate of these substances is known accurately, as are the endproducts of disintegration. For example, a given quantity of radium is known to "decay" into lead in a certain span of time. When radium and lead are found together in one mass within a rock, the whole mass presumably had been radium originally, when the rock was formed. From the relative quantities of radium and lead

15.1. Rock layers of different ages are often exposed to view. Generally speaking, the deeper a layer in the earth's crust, the older it is.

present today, one can then calculate the time required for that much lead to form. This dates the rock, with an error of about 10 per cent of its total age.

An analogous principle underlies age determinations by potassium-argon dating and by radiocarbon dating. In the potassium-argon process, one measures how much of the unstable isotope potassium 40 has decayed into the isotope argon 40. Radiocarbon dating involves measurements of carbon 14, an isotope of "natural" carbon 12. Whereas the potassium-argon method can be used for dating fossils many millions of years old, the carbon 14 method is accurate only for fossils formed within the last 50,000 years. Fossils themselves often help in fixing the age of a rock layer. If such a layer contains a fossil which on the basis of other evidence is known to be of a definite age (*index fossil*), then the whole layer, including all other fossils in it, is likely to be of the same general age.

Based on data obtained from radioactive and fossil clocks, geologists have constructed a *geologic time table* which indicates the age of successive earth layers and so provides a calendar of the earth's past history. This calendar consists of five successive main divisions, or *eras*. The last three of these are subdivided in turn into a number of successive *periods*. The names of the eras and periods and their approximate durations are indicated in Table 9.

The beginning and terminal dates of the eras and periods have not been chosen arbitrarily but have been made to coincide with major geologic events known to have occurred at those times. The transitions between eras in particular were times of great upheaval, characterized by mountain building and by severely fluctuating climates. For example, the transition from the Paleozoic to the Mesozoic dates the *Appalachian revolution*, during which the mountain range of that name was built up. By now, these mountains are already greatly reduced by erosion. Similarly, the transition between the Mesozoic and the Cenozoic was marked by the *Laramide revolution*, which produced the high mountain ranges of today: the Himalayas, the Rockies, the Andes, and the Alps. As we shall see, these major geologic events led to major zoological ones, marked by evolutionary crises and large-scale replacement of types.

The Precambrian Era

The first geologic era, the immensely long Azoic, spans the period from the origin of the earth to the origin of life. Living history begins with the next era, the Precambrian.

Fossils are not lacking altogether from these distant Precambrian ages. But the record is exceedingly fragmentary and it shows mainly that life, simple cellular life at least, already existed about 1 billion years ago. This must mean that the actual origin of life must have occurred earlier; we place it at about 2 billion years ago, at the start of the Precambrian. We also know how far evolution must have proceeded by the end of the Precambrian, for from that time on we have a continuous and abundant fossil record.

It is a very curious circumstance that rocks older than about 500 million years are so barren of fossils whereas rocks younger

TABLE 9. The geologic time table*

Era	Period	Duration		Beginning date
Cenozoic ("new life")	Quaternary	75	1	1
	Tertiary		74	75
Mesozoic ("middle life")	Cretaceous	130	60	135
	Jurassic		30	165
	Triassic		40	205
Paleozoic ("ancient life")	Permian	300	25	230
	Carboniferous		50	280
	Devonian		45	325
	Silurian		35	360
	Ordovician		65	425
	Cambrian		80	505
Precambrian		1,500		2,000
Azoic ("without life")		3,000		5,000

*All numbers refer to millions of years; older ages are toward bottom of table, younger ages toward top.

than that are comparatively rich in them. Many hypotheses have been proposed to account for this, but to date a satisfactory explanation has not been found. Did the Precambrian environment somehow preclude the formation of fossils? Were fossils destroyed in some way before the Paleozoic? Or is the Precambrian fossil record so scanty because the animals then were still too insubstantial to leave fossilizable remains? We simply cannot be sure.

But we *are* reasonably sure that Precambrian evolution must have brought about not only the origin of life and the origin of cells but also the origin of three of the four present main groups of organisms, namely, the Monera, the Protista, and the Metazoa. Moreover, virtually all phyla within these three groups were in existence by the end of the Precambrian. To be sure, the organisms then representing these phyla were not the same kinds as those living today; extinction and replacement by new types were still to occur many times. But the ancient types nevertheless did belong to the same phyla we recognize now.

Evidently, the long Precambrian spanned not only three-quarters of evolutionary time but also three-quarters of evolutionary substance. Nearly all the organisms in existence at the end of the Precambrian were aquatic. With the probable exception of some of the bacteria and some of the Protista, the land apparently had not been invaded as yet. The ensuing last quarter of evolution brought about principally a rich and extensive further diversification within the existing phyla. This process resulted in a replacement of ancient forms by new ones, including in each of the three main groups the evolution of types which could live on land. And among the land-adapted descendants of the Protista, more specifically the green algae, there were organisms which established a new main group, namely, the Metaphyta. Their fossil history begins in the Silurian, and they appear to have been the last to evolve among the four main categories now living. Very soon thereafter, still during the Silurian, some of the Metazoa began to follow the plants to the land.

Starting with the Cambrian period of the Paleozoic era, the course of evolution is documented fairly amply by fossils. These show that every animal phylum in existence in the Cambrian has persisted to the present. However, not a single species has persisted. So far as is known, only a single genus has survived from the Ordovician, the period after the Cambrian. This genus is *Lingula,* of the phylum Brachiopoda (cf. Chap. 24). Apart from this 400-million-year-old relic, all ancient genera have become extinct as well. Indeed, the dominant theme of the animal fossil record as a whole is very extensive and repeated replacement within major groups and relatively few additions of new major groups.

THE PALEOZOIC

Cambrian and Ordovician

During these first two periods of the Paleozoic, the land remained free of animals but life in the sea was already abundant. As might be expected, the most conspicuous forms known from these periods were types which could fossilize readily, i.e., groups with prominent exoskeletons and endoskeletons (Fig. 15.2). For example, protozoa are represented

15.2. Seascapes of the early Paleozoic, restorations. A. Cambrian seas. Various algae, trilobites (in center foreground), eurypterids (in center background), sponges, jellyfishes, brachiopods, and different types of worms are the most prominent organisms shown. **B.** Ordovician seas. The large animal in foreground is a straight-shelled nautiloid.

A

B

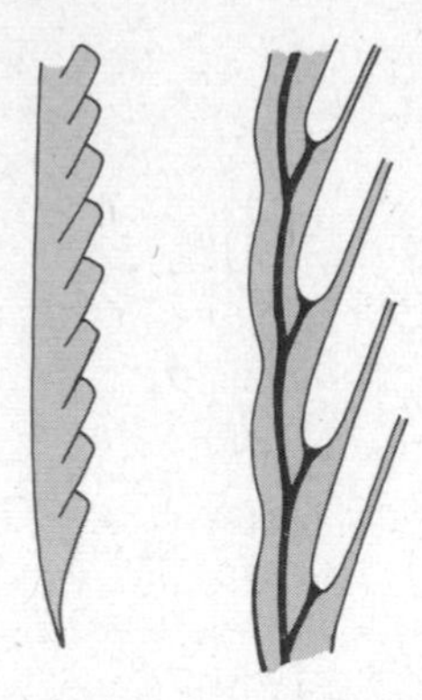

15.3. **Left,** a portion of a graptolite colony; **right,** detail of the exoskeleton of a few individuals (zooids) in a colony, showing their interconnecting *black stolon* (cf. also Chap. 29). (*Adapted from Kozlowsky.*)

by fossils of shelled amoeboid *foraminiferans* and *radiolarians*, some of the latter remarkably like certain Radiolaria today (cf. Chap. 19). Sponge fossils are known from earliest Cambrian times onward, and the fossil history of coelenterates begins with Cambrian jellyfishes and Ordovician corals and sea anemones. A variety of tube-forming worms is known from these and even earlier periods, and stalked brachiopods already existed in the Cambrian; as noted, *Lingula* has persisted since the Ordovician. The latter period also marks the beginning of a rich record of endoprocts and ectoprocts (*bryozoa*), tiny colonial animals whose skeletons form extensive crusts on rocks and algae even today.

Similarly rock-encrusting and colonial were the *graptolites*, abundant especially during the first half of the Paleozoic (Fig. 15.3). The nature of the graptolites cannot be fully determined from their skeletal remains alone. At first the animals were thought to be related to the coelenterates, but they are now believed to represent a group possibly allied to primitive hemichordate stock; they may have been a class of Hemichordata or possibly a separate related phylum. In either case, graptolites became extinct by the end of the Paleozoic and it is unknown if they left any descendants in the form of animals now living.

Echinoderms were amply represented by six archaic groups (Fig. 15.4). Three of these, the *cystoids, carpoids,* and *blastoids* have since become extinct, but the stalked sessile *crinoids* and the ancient *asteroids* and *echinoids* subsequently gave rise to the present-day echinoderms (cf. below). Mollusks too were prominent. Archaic clams and snails were exceedingly abundant, as were the *nautiloids*, a group related closely to modern squids and octopuses and still represented today

15.4. **Fossil echinoderms. A,** a cystoid; **B,** a blastoid; **C,** a crinoid.

A

B

C

15.5. Paleozoic arthropods. **A,** trilobite fossils; **B,** eurypterid fossil.

by the chambered nautilus. The early nautiloids included types with coiled and uncoiled shells, and the uncoiled forms may have been the largest animals of the time; some of their shells reached lengths of 5 to 6 yd.

Of three groups of ancient arthropods, one was that of the *ostracod* crustacea, descendants of which are still abundant today. A second group comprised the *trilobites,* believed to have been the most primitive of all arthropod types (Fig. 15.5). Trilobites were segmented, they possessed segmental body appendages, and their body was marked into three lobes by two longitudinal furrows; hence the name of the group. The animals resembled crustacea and they may have been their ancestors. Naupliuslike fossil larvae of trilobites have been found (cf. also Chap. 27). Trilobites were already exceedingly abundant when the Cambrian began, and they are among the most plentiful of all fossil forms. It has been suggested that these animals perhaps might not actually have been so abundant as the fossil record appears to indicate, since, like arthropods generally, trilobites must have molted; and many of the fossils may merely represent the remains of molted exoskeletons. The third group of arthropods was that of the *eurypterids,* large animals which resembled the scorpions of today. Indeed, the first fossil scorpions were marine animals of the early Silurian. Scorpions are related to spiders and horseshoe crabs, and it appears likely that eurypterids may have been ancestral to this whole present assemblage of *arachnids.* We may note here in passing that *Limulus,* a surviving genus of horseshoe crabs, has existed unchanged for the last 200 million years (cf. Chap. 27).

Vertebrate history begins in the late Ordovician. The marine tunicate ancestors probably were already present at or near the start of the Paleozoic, and the vertebrates evolved from them as freshwater forms. As pointed out in Chap. 14, the transition from the sea to fresh water was probably achieved by an evolutionary derivative of the tunicate tadpole. In this new environment it became necessary to adapt to the force of river currents and to the lower salt concentrations. Currents can be overcome by muscle-powered swimming, but even larvae must be able to swim from the moment they hatch. The tailed tadpole satisfied this requirement. Indeed, the power of the tail was reinforced by a result of an evolutionary change in egg size, from the small, primarily isolecithal eggs of early chordates to the large, highly telolecithal eggs typical of vertebrates. Well supplied with yolk, the embryos could develop for comparatively long periods before hatching, which allowed ample time for the elaboration of the whole body, the muscular system included. At hatching, therefore, the tail of the tadpole could already be well muscled and could

15.6. Restoration of two ostracoderms. See also Fig. 31.2.

serve efficiently in maintaining the animal against river currents. We note that the large vertebrate egg is a direct adaptation to the environment in which these animals evolved.

Salt and water balances in fresh water were regulated by the newly evolved pronephric and mesonephric kidneys, as outlined in Chap. 9. Moreover, much of the vertebrate endocrine system probably developed primarily as an adaptation to freshwater conditions. With fewer salts available in the environment, tight control over salt management undoubtedly became important, both qualitatively and quantitatively. Endocrine organs such as the thyroid, parathyroid, and adrenal glands came to serve in these salt control functions, as they still do today. However, as also pointed out in Chap. 9, the vertebrate kidney evolved primarily as a *water*-excreting system, and control over salt excretion at first was probably quite inefficient. Excretion of calcium appears to have been especially difficult originally, for the early vertebrates all displayed a tendency to retain excess calcium in their bodies and to deposit it as *bone*, particularly external dermal bone.

Thus, the first fossil vertebrates are named the *ostracoderms* because they were "shell-skinned," i.e., covered with bony armor plates (Fig. 15.6). These animals were members of the class Agnatha, the *jawless fishes*. From their ancestral freshwater home some of them invaded the ocean, and the whole group flourished until the end of the Devonian. Most of them became extinct then, and their only surviving descendants today are the lampreys and the hagfishes. In these modern Agnatha, the cartilage skeleton formed in the embryo remains cartilaginous and all traces of external bone have been lost.

The Cambrian and Ordovician periods lasted 150 million years, a span of time encompassing nearly a third of the entire fossil record. Such a span also represents fully half of the duration of the whole Paleozoic, more than the whole of the Mesozoic, and twice the time of the whole Cenozoic. Note therefore that what might be dismissed as merely two of many geologic periods is actually an immensely long stretch of time, in which an enormous amount of evolution took place. In a comparative amount of time, man could have evolved from an apelike starting point five times over. A summary of the Cambrian-Ordovician record is included in Fig. 15.7.

Silurian and Devonian

The Silurian was the period during which the first Metaphyta evolved. Providing a food source for animals, these land plants were soon followed by land animals; late Silurian land scorpions, probably evolved from earlier sea scorpions, are the earliest known terrestrial animals. Additional groups of arthropods invaded the land during the latter part of the Silurian and the beginning of the Devonian. The first mites and centipedes and the first insects appeared during these times. Also, the first spiders evolved, probably from marine eurypterid stocks. In the sea, the nautiloids gave rise to a new molluscan group, the shelled *ammonites,* which were to flourish for long ages (Fig. 15.8). Echinoderms underwent major evolutionary changes, i.e., cystoids and carpoids became extinct near the end of the Devonian and the ancient asteroids branched into two descendant groups, the brittle stars and the starfishes.

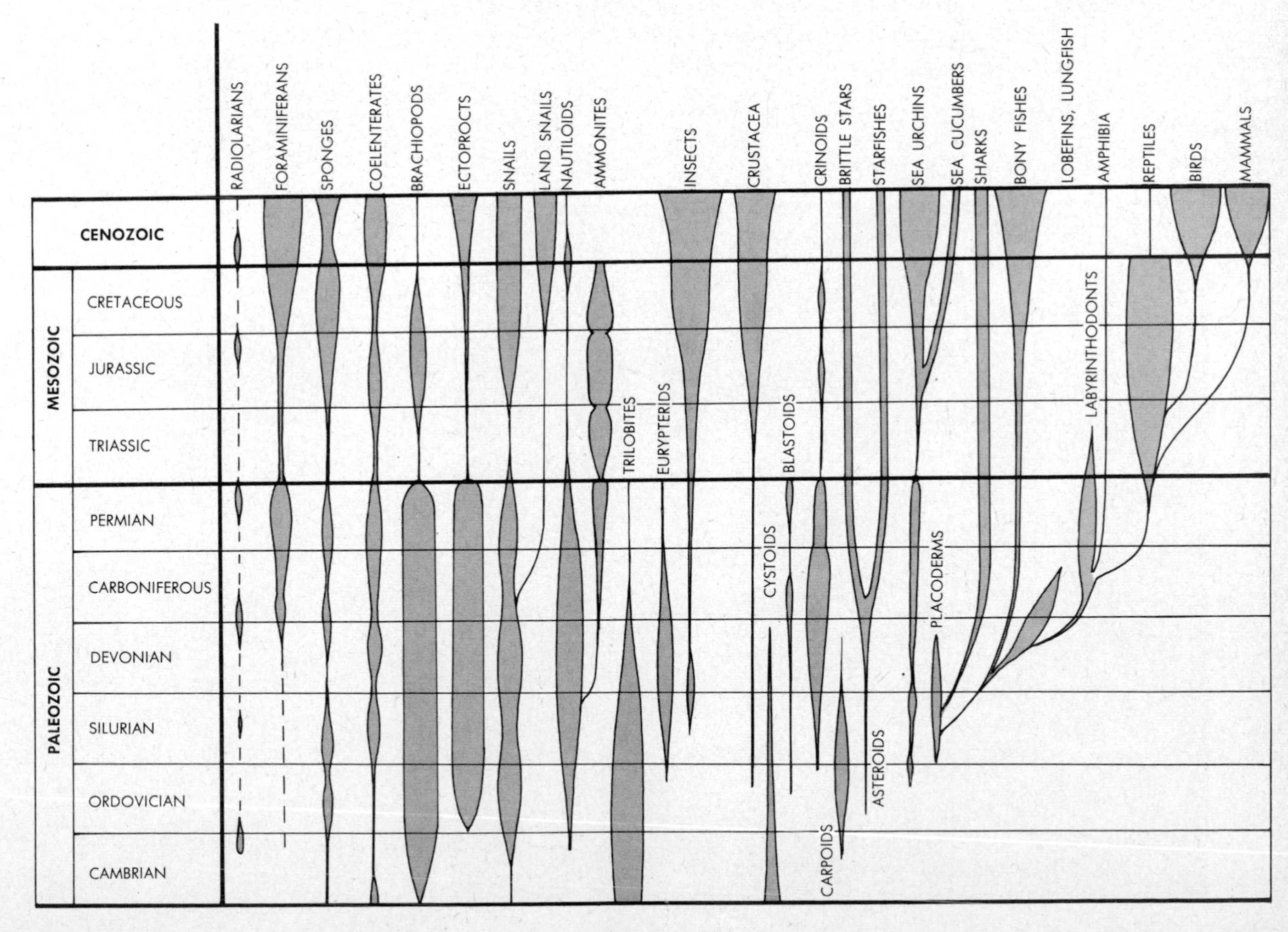

15.7. Summary of the fossil record of various animal groups. The varying widths of each vertical graph indicate roughly the time changes in the abundance of a given animal group, as judged by the numbers of fossil species found. Such widths are directly comparable only for different times within a group, not between groups. For example, the absolute Cenozoic abundance of insects would equal or exceed that of all other groups combined. The graphs clearly show the major decline which took place at the end of each era, particularly so during the Permo-Triassic transition, less so during the Mesozoic-Cenozoic transition. (*Adapted from Dunbar, Romer, and other sources.*)

Moreover, the vertebrates produced a major adaptive radiation during the Silurian and Devonian (Fig. 15.9).

The Devonian as a whole is often called the "age of fishes." During the early Silurian, some of the jawless ostracoderms had given rise to a new line, the *placoderms*, or *jawed fishes*. The name of this separate class of vertebrates again refers to the bony armor plates with which the skins of these fishes were equipped (cf. Fig. 3.4). Probably evolved in fresh water, the placoderms became abundant when the Devonian began and some stocks then spread into the ocean. Thus the placoderms replaced their own ancestors, the ostracoderms, more or less completely in all aquatic environments. Some of the placoderms were small, but others reached lengths of 12 yd or more. Most exploited the possession of jaws by adopting a fiercely carnivorous way of life.

The placoderms remained dominant for some 70 million years. During the later Silurian, ancestral placoderms gave rise to two new lines of fishes which slowly came to replace the later placoderms. By the end of the Devonian placoderms had become reduced greatly in numbers, and by the end of the Paleozoic they had disappeared completely. We may note that this is the only vertebrate class which has become extinct.

The two new types of fishes evolved from placoderms during the late Silurian were the *cartilage fishes* and the *bony fishes*, each representing a separate class. Both groups arose in fresh water, but the cartilage fishes rapidly adopted a marine habit and their present membership of sharks, skates, and rays is almost exclusively marine. These animals also lost the bone-forming potential of their placoderm ancestors; the fishes acquire a cartilaginous endoskeleton during embryonic stages and they then retain it as such, without later replacement by bone. Replacement does occur in the bony fishes, however, and in all later vertebrates as well.

The bony fishes at first remained in fresh water, where they soon radiated into three main subgroups: the so-called *paleoniscoid fishes*, the *lungfishes*, and the *lobe-finned fishes*. All three groups possessed pharyngeal air sacs. The paleoniscoids used them primarily as swim bladders, the other two

15.8. Restoration model of an ammonite.

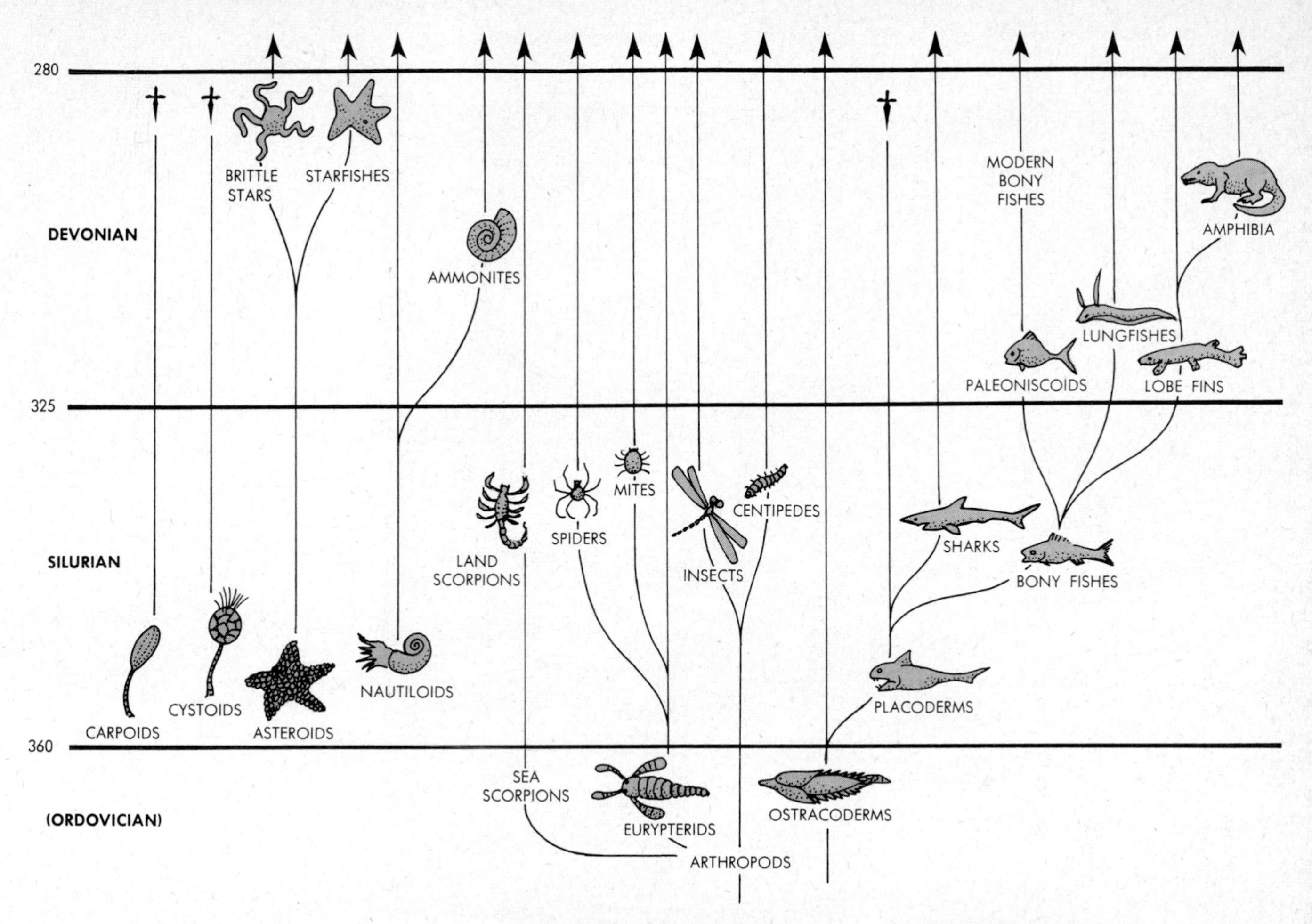

15.9. Summary of Silurian-Devonian events. The chart shows only newly evolved groups and those which became extinct during the period under consideration. Groups already established during earlier periods are not otherwise indicated. The level at which a given group is placed here corresponds roughly to the time of its first appearance in the fossil record. Daggers signify extinction, at the appropriate time level.

groups, mainly as lungs (cf. Chap. 9). The paleoniscoids later spread into the ocean and became the ancestors of virtually all bony fishes in existence today, both freshwater and marine. The lungfishes were common in Devonian and later Paleozoic times, but thereafter they declined and today they are represented by only three surviving genera. The lobe-fins similarly are almost extinct today (Fig. 15.10).

But the Devonian representatives of the lobe-fins included the ancestors of the first land vertebrates, the *amphibia.* As indicated by their name, the lobe-fins had fleshy appendages, usable to some extent as burrowing organs or perhaps even walking legs. These fishes probably lived in fresh waters which dried out periodically, and their lungs and fins may have enabled them to crawl overland to other bodies of water or to embed themselves in mud and to breathe air through the mouth. We may conclude therefore that terrestrial vertebrates probably arose not because certain fish preferred the land, but because they had to become terrestrial if they were to survive as fish.

Thus, when the Devonian came to a close, sharks dominated in the ocean and bony fishes in fresh water. On land, terrestrial arthropods had become abundant and the first

15.10. **A,** restoration of fossil lobe-finned fishes; **B,** a modern lungfish (*Protopterus*) from West Africa.

A

B

amphibia had made their appearance. Many of the land animals could shelter in the stands of primitive trees already established at that time. Fig. 15.7 includes a summary of Silurian-Devonian events.

Carboniferous and Permian

The Carboniferous is often regarded as representing two distinct periods, an earlier *Mississippian* and a later *Pennsylvanian* (Fig. 15.11). The Carboniferous is so called because during its Pennsylvanian phase many tracts of land became so wet that they were transformed into vast swamps and marshes. In these, many woody plants died and later geologic changes then converted the bodies of the plants into coal. For example, the coal beds of Pennsylvania and West Virginia came into being in this manner.

During these later Paleozoic times, the character of aquatic life did not change in major ways. The ancient coelenterates experienced a Mississippian decline, and in the Pennsylvanian the first advanced crustacea appeared in the form of crab- and crayfishlike animals. On land the changes were more pronounced. Additional terrestrial groups evolved from aquatic ancestors, and some other groups, already terrestrial, began to diversify. More specifically, land snails appeared for the first time during the Mississippian, and spiders and scorpions became more abundant. Above all, insects produced extensive adaptive radiations during the Pennsylvanian and Permian. Some of these ancient insect types reached sizes well

15.11. Some of the evolutionary events of the Carboniferous-Permian and the Permo-Triassic crisis. Placement of groups corresponds roughly with the time of their first appearance in the fossil record. Only some of the newly evolved groups and those which became extinct are indicated.

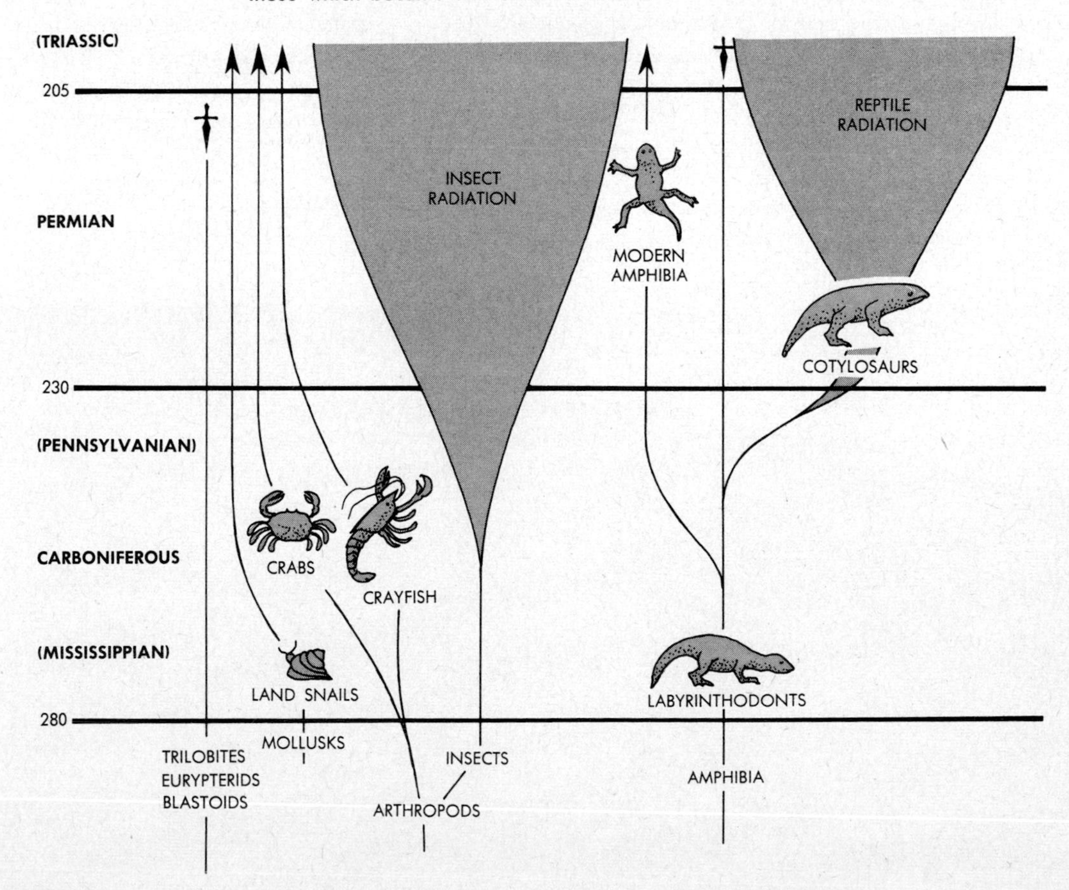

above the modern maximum. A Permian dragonfly, for example, is known to have had a wingspread of close to a yard (Fig. 15.12).

Among vertebrates, the early amphibia gave rise to a large variety of more or less clumsy, often bizarre forms, the *labyrinthodonts* (Fig. 15.13). These arose during the Mississippian, and during the Pennsylvanian they became the ancestors of two groups, viz., types which later evolved into the modern amphibia and types which represented a new vertebrate class, the *reptiles.* The latter were represented at first by one main group, the stem reptiles, or *cotylosaurs* (Fig. 15.14). They were characterized by hard-shelled land eggs with extraembryonic membranes. Thus, the stem reptiles were the first amniotes and the first completely terrestrial vertebrates. They produced a major reptilian radiation during the Permian, which brought about the decline of the labyrinthodonts and also set the stage for a subsequent "age of reptiles" during the Mesozoic era.

As noted earlier, the Paleozoic era terminated with a major *Permo-Triassic crisis* precipitated by the Appalachian revolution. This unstable time of transition was characterized by widespread extinction of archaic animal forms, by replacement and rapid evolution of new groups, and by a general, temporary decrease in the total amount of life (cf. Figs. 15.7 and 15.11). In the sea, the protozoan Foraminifera declined in numbers. The brachiopods and ectoprocts, abundant since Cambrian times, became almost extinct. All mollusks passed through a major decline, nautiloids becoming virtually extinct and ammonites being reduced to a small group. The trilobites, eurypterids, and blastoids disappeared altogether, and the crinoids were left with only a few surviving types. Similarly, only a single echinoid type survived into the Triassic. Placoderms died out, and extensive intragroup replacement occurred among the cartilage and bony fishes. Land animals were less affected on the whole, though their numbers did decline temporarily. Labyrinthodonts lingered on into the Triassic, but soon they too became extinct. However, the reptiles survived the crisis well, and when the new Mesozoic era opened they were already dominant.

THE MESOZOIC

The era as a whole was characterized by a reexpansion of virtually all groups which survived the Permo-Triassic crisis and by extensive intragroup replacements. Thus, protozoa, sponges, and coelenterates underwent major adaptive radiations during the Jurassic, with the result that these animals

15.12. The wing of a Permian insect, actual size. Insects larger than this existed in the Permian, but even the owner of the wing shown was far larger than any insect today.

15.13. **A.** Reconstruction of *Diplovertebron,* a Permian labyrinthodont amphibian. **B.** Reconstruction of *Seymouria,* a transitional amphibian type probably related to the stock from which reptiles appear to have evolved.

exist today in greater numbers than ever before. Ectoprocts and brachiopods similarly expanded; but whereas the former still appear to be gaining today, the latter became virtually extinct again at the end of the Mesozoic. The brachiopods are still a very reduced group today (cf. Fig. 15.7).

Among mollusks, clams and snails diversified greatly, and land snails in particular became abundant. The nautiloids did not regain their Paleozoic importance, however, and the chambered nautilus is their sole present survivor (Fig. 15.15). The ammonites began the Mesozoic with 12 families, and 11 of these became extinct during the Triassic. The one remaining family then evolved into more than 20 new groups during the Jurassic and the whole stock of ammonites reexpanded during the later Mesozoic. Yet at the end of the era all ammonites became extinct. In the echinoderm group, the crinoids managed to linger on as relics, and the more abundant brittle stars and starfishes held their own. But the single echinoid group which survived from the Paleozoic underwent an explosive expansion during the late Mesozoic. In the course of it the modern sea urchins and sea cucumbers evolved. Sea urchins today are more abundant than all other echinoderms combined. The crustacea gained slowly and steadily in numbers and types. Insects reradiated enormously, and their present importance traces to this Mesozoic expansion. This rise of insects paralleled the evolution of flowering plants. The latter appeared during the Jurassic and then they underwent an explosive expansion which established them as the dominant land plants from then on.

15.14. Reconstruction of *Labidosaurus,* one of the cotylosaurian stem reptiles.

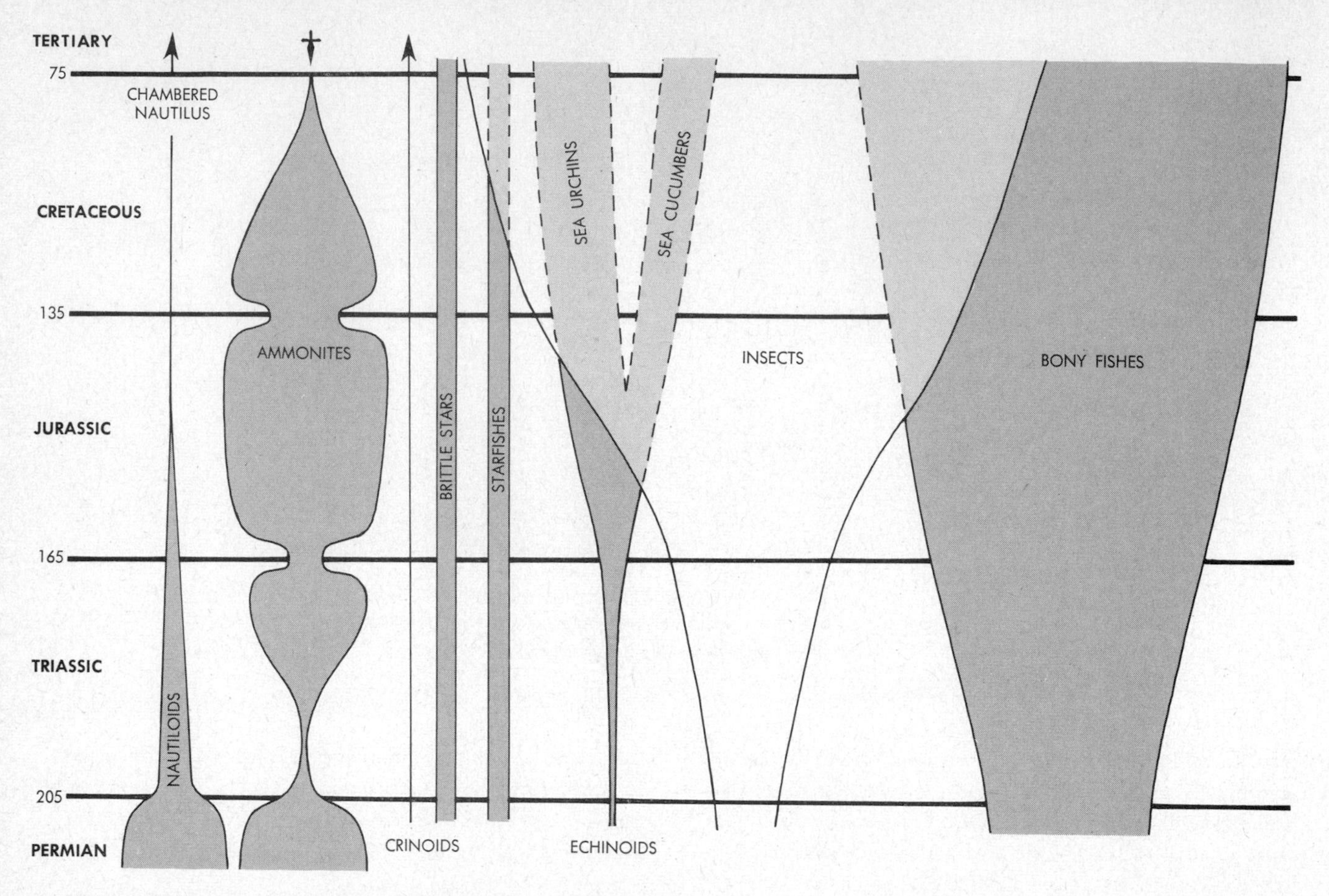

15.15. Summary of nonreptilian events during the Mesozoic.

An extensive radiation also occurred among the bony fishes. During the Cretaceous, the paleoniscoid fishes gave rise to a multitude of new freshwater and marine types, the modern bony fishes. These became the dominant animals of the aquatic environment, a status they still retain today (cf. Fig. 15.15).

However, the most spectacular Mesozoic event was the expansion of the reptiles. These animals not only evolved many different terrestrial ways of life but also invaded the water and the air. As a group they reigned supreme on earth for 130 million years, longer than any other animals to date. When their dominance was eventually broken, they were replaced by two new groups they themselves had given rise to, the birds and the mammals.

At the beginning of the Mesozoic, five major reptilian stocks were in existence, all evolved during the Permian from the cotylosaurian stem reptiles (Fig. 15.16). One group, the so-called *thecodonts,* reradiated extensively during the Triassic and in turn gave rise to the following types: the ancestral *birds;* the ancestors of the modern *crocodiles, lizards,* and *snakes;* the flying *pterosaurs;* and two other groups, referred to collectively as *dinosaurs.* A second reptilian stock was ancestral to the modern *turtles.* A third and fourth produced two kinds of marine reptiles, the porpoiselike *ichthyosaurs* and the unique, long-necked *plesiosaurs.* The fifth stock comprised the so-called *therapsids,* reptiles which included the ancestors of the mammals (Fig. 15.17).

These various reptilian types did not all flourish at the same time. The Triassic was dominated largely by the ancestral thecodonts and the therapsids. The former were rather bird-like in appearance. They possessed large hind limbs for walking, an enormous supporting tail, and diminutive forelimbs, often not even long enough to shovel food into the mouth. Therapsids were four-footed walkers. Mammals evolved from some of them during the late Triassic or early Jurassic (cf. below).

During the Jurassic, ichthyosaurs became abundant in the ocean and one of the thecodont groups evolved into birds. This transition is documented beautifully by a famous fossil called *Archeopteryx* (Fig. 15.18). The animal possessed teeth and a lizardlike tail, two features which are distinctly reptilian. But it also possessed feathers and wings, and presumably it flew like a bird. Like the previously evolved mammals, birds remained inconspicuous during the whole remaining Mesozoic. They were overshadowed particularly by their thecodont kin, the pterosaurs. These flying reptiles had their heyday during the Cretaceous, the period when reptiles as a whole attained their greatest abundance and variety. Plesiosaurs then were common in the ocean, and the dinosaurs came into undisputed dominance on land (Fig. 15.19).

The two dinosaurian groups belong to two taxonomic orders called the *Ornithischia* and the *Saurischia*. Both evolved from the thecodonts. Not all dinosaurs were large, but some of the group were enormous. The saurischian *Brontosaurus* was the largest land animal of all time, exceeded in size only by the modern blue whale. This dinosaur was herbivorous and it probably lived in swamps or lagoons, where it could support its 20- to 30-ton bulk in water. Another saurischian, the giant *Tyrannosaurus*, probably was the fiercest land carnivore of all time. Among its victims undoubtedly were animals like *Ankylosaurus* and *Triceratops*, herbivorous and heavily armored ornithischian giants (Fig. 15.20).

15.16. The great reptilian radiation of the Mesozoic. Placement of groups corresponds roughly with the time of their greatest abundance.

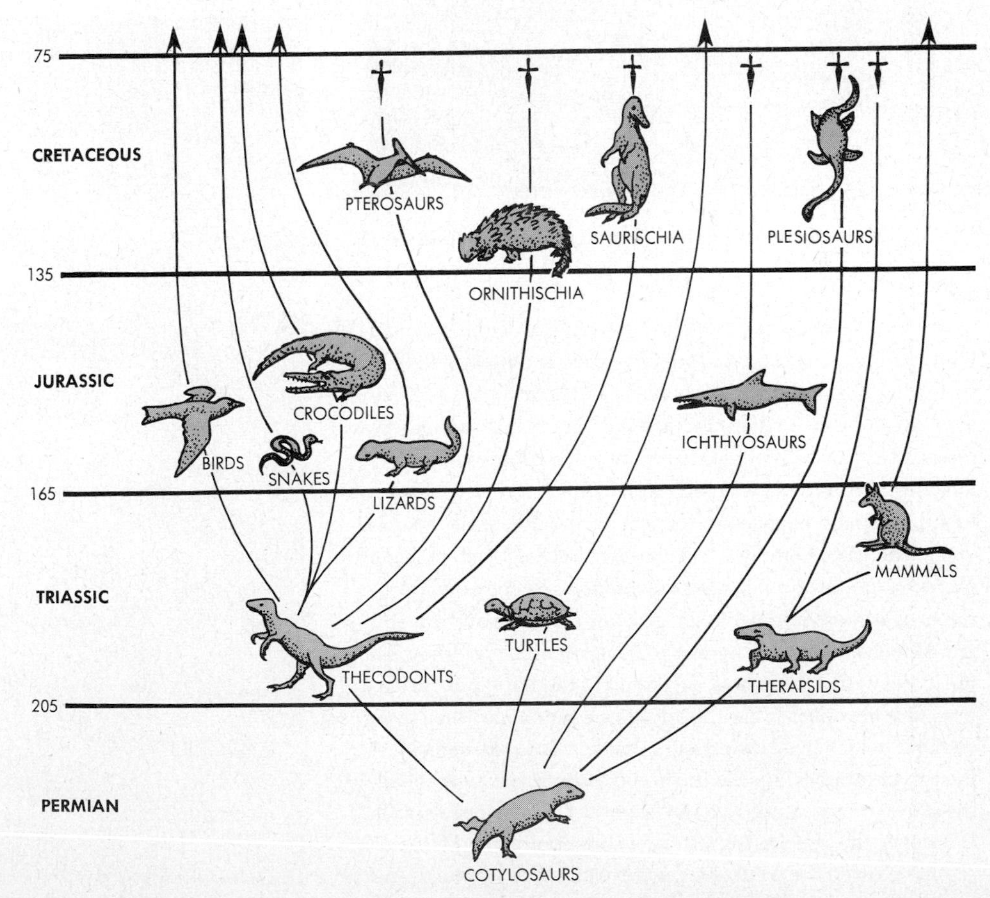

15.17. A therapsid, a mammallike reptile of the Triassic.

As the Cretaceous came to a close, most of the reptilian multitude became extinct. Today the class is represented only by turtles, crocodiles, lizards, snakes, and *Sphenodon*, a lizardlike "living fossil" in New Zealand (cf. Chap. 32). The specific reasons for this large-scale dying out have been sought for a long time, but fully satisfactory explanations have not yet been found. Climatic changes at the end of the Mesozoic, coincident with the Laramide revolution, are believed to have played a decisive role. Mesozoic reptiles were adapted to rather warm environments, as their modern descendants still are. However, climates appear to have become cooler toward the close of the Cretaceous, as a result of the Laramide revolution. Much tropical and subtropical vegetation may then have died out, which must have meant that herbivorous reptiles lost their food supplies. And as the herbivorous stocks so declined, the carnivorous reptiles would have had to die out as well. Whatever the precise causes, the extinction of the Mesozoic reptiles cleared the way for a great expansion of mammals and birds.

Some of the factors which must have promoted the evolution of mammals can be deduced from a comparison of their structure with that of reptiles. We know that the earliest mammals were small, in the size range of mice. Their ancestors among therapsid reptiles probably were comparably small, and this size disadvantage must have entailed perpetual danger in a world dominated by far larger animals; survival must have depended on ability to escape danger, primarily by running. But the running capacity of early reptiles probably was as limited as in their modern descendants: the metabolic machinery of reptiles is keyed to maintain only a relatively low level of bodily activity. The air-holding capacity of reptilian lungs is moderate, and breathing movements are fairly shallow. Also, the oxygen-holding capacity of blood is only about

15.18. Plaster cast of *Archeopteryx*. Note feathered tail, wings. The head is bent back, and the tooth-bearing mouth is not easily visible here.

15.19. Reconstruction of plesiosaurs (left) and ichthyosaurs (right).

9 to 10 cm^3 of O_2 per 100 ml of blood, roughly comparable to that of amphibian and fish blood (cf. Table 8, Chap. 9). Such oxygen potentials are far greater than those of most invertebrates (e.g., five times greater than in earthworms), and they permit fishes to be the most active aquatic animals. On land, however, the same potentials are too low to support sustained high levels of activity.

The evolutionary transition from reptiles to mammals was actually characterized by a pronounced elevation of metabolic capacities. A newly evolved diaphragm increased breathing efficiency; nonnucleated red blood corpuscles developed, maximally specialized for oxygen transport; hemoglobin changed into chemical variants which permitted 100 ml of blood to carry 25 cm^3 of O_2, nearly three times as

15.20. Restoration of extinct saurians. Left, *Triceratops*, a Cretaceous horned herbivore; **right,** *Tyrannosaurus*, a giant Cretaceous carnivore.

much as before; and along with an evolution from reptilian scales to an insulating coat of mammalian hair, a temperature-control mechanism came into being which permitted maintenance of a constant, uniformly high body temperature. The whole *rate* of metabolism thus intensified and could be kept at an increased level. As a result, mammals became perhaps the most active animals of all time. In part for similar reasons of escaping danger, in part because of the energy-requirements of flight, birds became nearly as active. However, they did not evolve a diaphragm, the oxygen capacity of their blood is only twice that of reptiles, and their red cells remain nucleated. Yet body temperature is maintained at an even higher level than in mammals.

As early mammals adapted to a life of running, they also sought safety in the forests; they became primarily nocturnal and furtive, with a distinct preference for hiding in darkness. Their sense of sight was of limited value in such an existence and, like virtually all mammals today, they were color-blind. Their sense of smell became dominant instead, and their basic orientation to the environment came to depend on odor cues from the forest floor. Also, in an active life geared to running and hiding, it is unsafe merely to lay eggs and leave them, in reptilian fashion. The first mammals undoubtedly did just that and a few primitive *egg-laying mammals* actually still exist. Yet most evolved means to carry the fertilized eggs with them, within the body of the females; i.e., they became viviparous. The pouched *marsupial* mammals and the *placental* mammals arose in this manner.

After mammals originated during the late Tertiary, their evolutionary potentialities remained latent for the rest of the Mesozoic, i.e., for a period of about 80 million years. But when at the end of the Mesozoic the dominance of the reptiles ceased, mammals, and also birds, produced explosive adaptive radiations which established them rapidly as the new dominant forms.

THE CENOZOIC

Just as each geologic era may be subdivided into periods, so each period in turn may be subdivided into *epochs*. The periods and epochs of the Cenozoic era are shown in Table 10. As noted earlier, the era as a whole began with the great upheavals of the Laramide revolution, which produced the present high mountain ranges and led to progressively cooler climates during the Tertiary period. These climatic changes

TABLE 10. The epochs and periods of the Cenozoic era*

Period	Epoch	Duration	Beginning date
Quaternary	Recent	20,000 years	20,000 B.C.
	Pleistocene	1	1
Tertiary	Pliocene	11	12
	Miocene	16	28
	Oligocene	11	39
	Eocene	19	58
	Paleocene	17	75

*Unless otherwise stated, all figures refer to millions of years.

culminated in the ice ages of the last 600,000 years, the latter part of the Pleistocene epoch.

Ice ages had probably occurred before during the earth's history, and in the Pleistocene there were four. In each, ice sheets spread from the North Pole southward, covered much of the land and sea of the Northern Hemisphere, then receded. The Southern Hemisphere was subjected to analogous advances of ice. Warm interglacial periods intervened between successive glaciations. The last recession began some 20,000 years ago, at the beginning of the Recent epoch, and it is still in progress: polar regions are still covered with ice. The biological importance of Cenozoic climates in general and of Pleistocene ice in particular is great, for these environmental conditions materially influenced the evolution of all organisms, plant or animal, man not excepted. Man in a sense is one of the products of the ice ages.

The radiation of mammals and birds came to be the main feature of animal evolution during the Cenozoic era. Terrestrial mammals replaced the dinosaurs; aquatic mammals eventually took the place of the former ichthyosaurs and plesiosaurs; and bats, but more especially birds, gained the air left free by the pterosaurs. The Cenozoic is often designated as the "age of mammals"; it might equally well be called the "age of birds."

The Mammalian Radiation

When the Cenozoic began, the great mammalian radiation was just getting under way. A total of some two dozen independent lines came into existence, each ranked as an order and belonging to one of three larger taxonomic units, viz.,

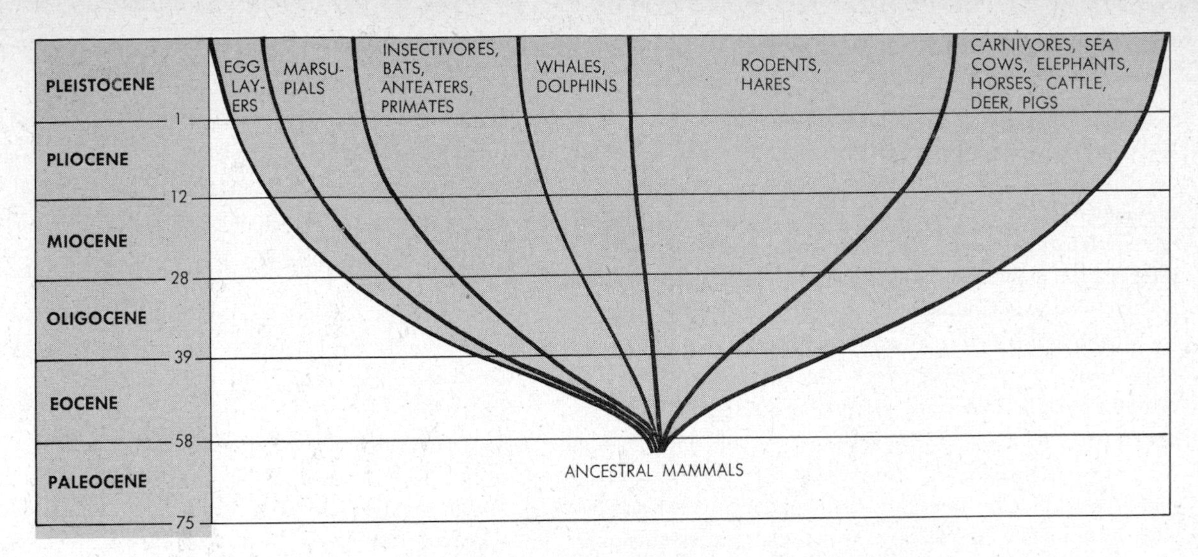

15.21. Some of the main features of the mammalian radiation during the Cenozoic. The animal types are grouped according to their evolutionary affinities (cf. Chap. 32); but note that extinct groups are not shown and that the groups now living have by no means undergone so steady and progressive a quantitative increase as this simplified diagram might tend to suggest. It is true, however, that the rodent-hare group today is the most abundant in terms of species number.

egg-laying, marsupial, or placental mammals (cf. Chap. 32). The placental mammals represent the most abundant and most familiar group; it includes cats, dogs, seals, and walruses; rodents; whales and dolphins; bats; moles and shrews; cattle, sheep, pigs, and camels; horses and zebras; elephants and tapirs; monkeys and men; and many others (Fig. 15.21).

The fossil record of this mammalian radiation is fairly extensive for most types and extremely good for a few, such as horses and elephants (Fig. 15.22). In the majority of cases, the record documents a general erratic increase in body size, a pattern which appears to be almost universal in animal evolution; ancestral primitive types are small, later advanced types

15.22. Reconstruction of evolution of the horse during the Cenozoic. The evolutionary sequence begins at left, with the fossil horse *Eohippus,* and proceeds via *Mesohippus, Hypohippus,* and *Neohipparion* to *Equus,* the modern horse at right. The drawings are to scale and show how the average sizes and shapes of horses have changed. Progressive reduction in the number of toes took place, as well as changes in dentition. Note, however, that the animals shown represent a highly selected series, and it should not be inferred that horse evolution followed a straight-line pattern. Here, as elsewhere, a bush pattern is actually in evidence.

often tend to be larger. Each mammalian line descended from the common ancestral stock exploited a particular way of life available at the time. The animals came to occupy either a new environmental niche or one left free after the extinction of the Mesozoic reptiles. One mammalian line is of particular interest, for it eventually led to man. This line exploited a relatively new environmental possibility. Its members took to the trees and adapted to an *arboreal* life.

Soon after such a stock of arboreal mammals had evolved during the early Paleocene, it must have reradiated and produced two major sublines, the order *Insectivora* and the order *Primates* (cf. Fig. 15.21). Most modern insect-eating mammals, particularly the moles and the hedgehogs, are clearly distinct from modern primates, of which man is a late member. But some of the shrews now living are exceedingly like insectivores on the one hand and like primitive living primates on the other. Indeed, one group of shrews is actually classified with the Insectivora and another with the Primates. Fossil data similarly support the view that insectivorous mammals and primates are very closely related, through a common, shrewlike, arboreal, insect-eating ancestor (Fig. 15.23).

15.23. An arboreal squirrel shrew, order Insectivora.

The Primate Radiation

The first distinct primates evolved from this insect-eating ancestor during the Paleocene may be referred to as the *early prosimians*. They were still small, shrewlike in appearance, with a fairly long snout and a long bushy tail. They probably had poor sight and good smelling ability, like the primitive first mammals; and although they lived in trees their close evolutionary connection with ancestral ground animals is still apparent from their fossils. Of the many sublines which radiated from the early prosimians during the Paleocene, five major ones survive today: the *lemuroids*, the *tarsoids*, the *ceboids*, the *cercopithecoids*, and the *hominoids* (Fig. 15.24). Each of these five has adapted to an arboreal existence and, roughly in the order listed, the extent of this adaptation has become progressively greater.

The lemuroids include the *lemurs* and *aye-ayes*, found today largely on the island of Madagascar (Fig. 15.25). These animals are a little larger than the ancestral primates, and they still possess long snouts and tails. However, instead of claws and paws they possess strong flat nails, a general characteristic of all modern primates. Long nails are probably more useful than claws in anchoring the body on a tree branch. In a tree, moreover, smelling is less important than seeing, and lemuroids have made a beginning toward improved vision. For example, each of their eyes can be directed forward independently, permitting better perception of branch configurations than if the eyes were fixed on the side of the head.

The tarsoids, represented today by the *tarsiers* of Southern Asia and Indonesia, have evolved several additional improvements. In these animals the ancestral "smell brain" has become a "sight brain"; i.e., the olfactory lobes have become small but the optic lobes have increased in size. In parallel with this reduction of the olfactory lobes, the snout has receded and a fairly well-defined face has appeared (cf. Fig. 15.25). Indeed, the eyes have moved into the face and, though they are still movable independently, both eyes can be focused on the same point. As a result, tarsiers are endowed with stereoscopic vision and efficient depth perception, traits which all other primates also share. Such traits evidently are of considerable adaptive value if balance is to be maintained in a tree. (The only other mammals with eyes directed toward the front are the carnivores; in hunting for prey, they too must

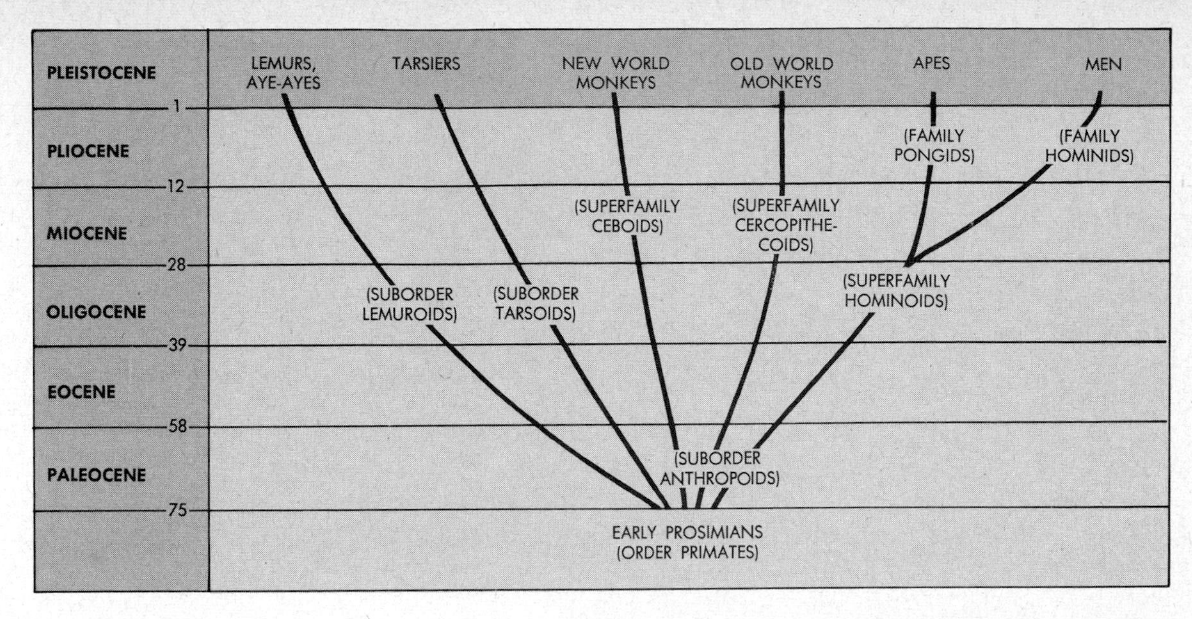

15.24. The main lines of the primate radiation.

gauge jumping distances. In a herbivorous mammal, by contrast, the eyes are lateral, where they enable the animal to scout the open environment even while grazing.) Tarsiers have also adapted to arboreal life by developing independently movable fingers and toes, with a branch-gripping pad at the end of each. Moreover, the several offspring in a typically mammalian litter have been reduced to a single offspring, clearly a safer reproductive pattern among the branches of a tree.

The ceboids comprise the *New World monkeys,* confined to South and Central America. These animals are characterized by long strong tails used as fifth limbs. The cercopithecoids are the *Old World monkeys,* found in Africa and Asia. They are identified by tails which are not used as limbs. Both groups of

15.25. *A,* a modern lemur from Madagascar; *B,* a modern tarsier from Indonesia.

15.26. *A*, howler monkey, a ceboid (New World) type; *B*, rhesus monkey, a cercopithecoid (Old World) type.

A

B

monkeys attained their present diversity during the Oligocene and Miocene, and note that monkey status evolved twice independently; the two groups represent separate, though remarkably similar, evolutionary developments. In both groups, adaptations to arboreal life have evolved a good deal farther than in the primates above (Fig. 15.26).

If tarsoids possess a "sight brain," monkeys may be said to possess a "space brain." The two eyes are synchronized and can be focused in different directions without movement of the head as a whole. Each eye also possesses a *fovea centralis*, a retinal area of most acute vision (cf. Chap. 8). Above all, monkeys are endowed with cone cells and *color* vision, the only mammals other than the hominoids so characterized. And colors are actually there to be seen: flowers, foliage, sky, and sun provide an arboreal environment of light and space, far different from the dark forest floor which forced early mammals literally to keep the nose to the ground. The smelling capacity of monkeys has actually become as poor as their sight has become excellent. Correspondingly, the cerebrum has enlarged greatly and its posterior regions contain not only extensive vision-control centers but also vision-memory areas; monkeys store visual memories of shape and color as we do, for reference in future activities.

The evolution of an enlarged brain has been correlated additionally with the adoption of a predominantly *sitting* way of life. Sitting on branches and in branch forks was probably not only safer than lying, but this posture also relieved the forelimbs of locomotor functions and resulted in a new freedom to use hands for touch exploration. The consequences were many, and monkeys today still exhibit them. The ability to *feel* out the environment and one's own body led to a new self-awareness and to curiosity. The ability to touch offspring and fellow inhabitants of the tree led to new patterns of communication and social life, reinforced greatly by good vision, by voice, and by varied facial expressions. Touch-control areas of the brain increased apace, centers controlling hand movements became extensive, and the brain as a whole so enlarged in parallel with the new patterns of living. The mind quickened as a result, and the level of intelligence increased well above the earlier primate average. Thus, whereas a ground mammal such as a dog sniffs its environment, the later arboreal primates began to explore it by sight and by touch. And we may note that evolution of intelligence has been correlated particularly with the improvement of coordination between eye and hand. Fundamentally, therefore, primate intelligence too is an adaptation to the arboreal way of life.

We may note also that tree life became a basically secure

way of life. Actually only two types of situations constituted significant dangers, viz., the hazard of falling off a tree and the hazard of encountering snakes. The first was minimized by the evolution of opposable thumbs and of precocious gripping ability in general (as displayed also in newborn human babies). The second was countered by the strength of the body, which increased as the evolution of larger bodies continued. As a result of this increase in body size, and perhaps aided also by the emancipation from perpetual fear, the inherent life span lengthened considerably. Monkeys may become up to four decades old compared with a life expectancy of a single year in shrews. Moreover, absence of danger and continuously warm climates led to a breeding season spanning the whole year (and in Old World monkeys the year-round breeding potential came to be accompanied by menstrual cycles).

Trends of the same kind, but developed very much farther than in monkeys, are apparent also in the hominoids. During the early Miocene, some 30 million years ago, the hominoid line branched into two main sublines. One of these led to the *pongids*, or apes, the other to the *hominids*, the family of man. Before this branching took place, the whole hominoid stock evolved important skeletal modifications over and above those attained by the monkeys. Early hominoids developed a fully upright posture, and by hand-over-hand locomotion between two levels of branches they became tree-*walkers* more than tree-sitters. Also, limbs lengthened generally, and universal limb sockets evolved as well. As a result, only an ape or a man can rotate the arm completely within its socket, and even a monkey can match neither the acrobatics of an ape in a tree nor that of a man on a trapeze. Moreover, only apes and men are capable of swiveling their hips, and only they have broad chests with lengthened and strengthened collar bones. Tree-walking was also facilitated by a reduction of the tail and its interiorization between the hind limbs. The tail skeleton shortened and broadened into an internal pelvic support, and the tail muscles came to form a pelvic floor. In this position the tail helped to counteract the internal sag produced by gravity acting on an upright body. In response to the effect of gravity, furthermore, the lumbar portion of the vertebral column became thickened and shortened. In addition, far more so than in horizontally placed mammals, exceedingly firm placental attachments came to be established between embryo and uterus.

After the hominoid stock split into a pongid and a hominid line, each of these gave rise to an adaptive radiation of its own. Apes today are represented by four genera, viz., gibbons, orangutans, chimpanzees, and gorillas (Fig. 15.27). Gibbons are survivors of an early branch of ape evolution, characterized by comparatively light bodies and retention of a fully arboreal way of life. Indeed, the gibbon is undoubtedly the most perfectly adapted arboreal primate. But the other three types of apes, representing later and heavier pongid lines, have abandoned life in the trees to a greater or lesser extent. Orangutans and especially chimpanzees can be quite at home out of trees, and gorillas are ground animals as fully as we are. It appears that these later apes ceased to be completely arboreal when their bodies had become so large and heavy that trees could no longer support them aloft. The feet of the apes give ample evidence of the weight they have to support; foot bones are highly foreshortened and stubby, as if crushed by heavy loads. Correspondingly, the agile grace of the arboreal gibbon is not preserved in the ground ape. Large apes cannot and do not walk, but they *scamper* along in a crouching shuffle-gait.

15.27. A gibbon.

So far as is known, the hominid line left the trees completely and almost as soon as it split away from the common hominoid stock, i.e., just when the adaptations to tree life had finally become perfected. Was this descent prompted by great body weight, too, as in the case of the large apes? Probably not, or else the hominid foot should resemble that of a ground ape, and man should scamper rather than walk. In actuality the hominid foot is very much like that of a gibbon, and as a gibbon swings in a tree so a man literally swings on the ground; the human bipedal gait is quite unique among ground forms. It is likely, therefore, that the hominid line left the trees when its evolution had progressed to a level comparable to that of the early gibbonlike apes, i.e., when the body was still comparatively small and light. On the ground the foot could then remain efficient and not overloaded unduly, and the grace won in the trees could persist. Thus, the early hominids may have been small creatures originally and, in contrast to the later apes, their size may have increased only *after* they had come out of the trees (Fig. 15.28).

But if not body weight, what other conditions could have forced hominids to the ground? It is pertinent to note that a similar descent was undertaken by late evolutionary members of the monkeys, as represented by the baboons today. Weight probably was not a factor here either. Instead, the chief cause appears to have been a change in climate. One of the long-range results of the Laramide revolution was a progressive lowering of average annual temperatures during the Tertiary. At first the temperatures were high and even the year round, as indicated by the absence of annual rings in the wood of fossil trees of the time. The poles of the earth were ice-free and much of the earth was subtropical and tropical. It was this luxuriant tree growth which promoted the evolution of the arboreal primates. But climates gradually became cooler during the later Cenozoic and the earth surface became differentiated into distinct polar, temperate, and tropic zones. Plants and animals already living in the tropics could then remain tropical, but organisms elsewhere could not. Three choices were open to them. They could migrate to the tropics; or they could readapt right where they lived to the succession of winter and summer; or they could die and become extinct. All three possibilities were actually realized to different degrees in different groups.

15.28. One group of apes probably developed to a large size in the trees, and their modern descendants such as the gorilla therefore were already heavy when they adopted life on the ground. Another group, exemplified by the gibbon, remained light and arboreal. Early hominids likewise probably remained light and small. They presumably left the trees as small types and their evolutionary size increase then occurred on the ground.

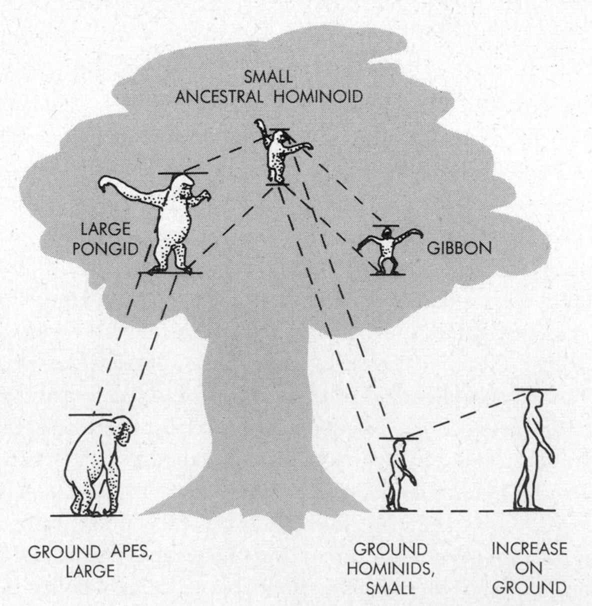

In the hominid line, as in the ancestors of baboons, descent to the ground probably became necessary when, as a result of cooler Tertiary climates, forests thinned out in many regions. Continuous overhead canopies of branches and foliage disappeared, and our prehuman ancestors would have had to travel on the ground if they wished to move from one stand of trees to another. Such forced excursions may well have been fraught with considerable danger, for saber-toothed carnivores and other large mammals dominated the ground at those times (cf. Fig. 13.12). Consequently, ability to dash quickly across open spaces may have had great selective value, and strong muscles would be required to move the hind limbs in new ways. Indeed, a unique trait of the hominid line is the possession of such muscles in the form of curving calves, thighs, and buttocks. In this respect, man is also different from apes.

These and other features which now distinguish men and apes came to be superimposed on the characteristics of pre-Miocene arboreal primates. We recognize, therefore, that the modern human type could not have evolved if the ancestral type had not first been specialized for life in trees.

The Hominid Radiation

After the hominid stock had originated in the Miocene, each of its evolving main lines must have produced various sublines and subsublines in turn. The detailed pattern of this hominid radiation is unknown, but that it occurred can be inferred from available fossil evidence. To be sure, this evidence is tantalizingly scanty; we can trace the recent evolution of almost any mammal far better than our own. Nevertheless, such fossils of hominid types as have been found show clearly that a substantial number of separate lines of descent must have arisen. Consequently, the members of the hominid radiation known to date, including ourselves, do not appear to be related directly. They probably trace back independently and through an unknown pattern of branching to various earlier and equally unknown common ancestors, and ultimately to the original hominid stock of the early Miocene. The path of descent of our own species therefore remains undiscovered as yet; other known hominids are related to us somewhat as uncles or cousins (Fig. 15.29).

With the exception of the line leading to ourselves, all other lines of the hominid radiation have become extinct at various periods during the last 30 million years. Exact times of extinction are as uncertain as times of origin, for whereas a fossil find indicates when a given hominid was alive, it does not indicate when it originated or died out. It is not necessary to find whole fossilized skeletons to reconstruct the probable appearance of their-once-living owners. The proportions of body parts to one another may be deduced with reasonable accuracy from living man, from apes, and from such whole skeletons of hominid types as have been found. For example, a tooth, a jawbone, a skullcap, or a leg bone may not only be identified from its shape as belonging to a particular hominid, but may also give important clues about the missing remainder. By and large, the skull gives the greatest amount of information. Thick or thin bones, prominent or reduced eyebrow ridges, receding or vertical forehead, small or large brain case, poorly or well-defined chin, all indicate fairly well whether or not a given fossil is a primitive or an advanced hominid.

Much may also be learned from various signs of cultural activity often associated with a fossil find. For example, the type of tool, the type of camp site, the type of weapon found with a fossil, each may reveal a great deal about the evolutionary status of the hominid in question. Note, incidentally, that apart from their zoological distinctions, hominids are regarded as being prehuman or truly human on the basis of cultural achievements. Any hominid which *made* tools in addition to using them can be called a "man." If a hominid only used stones or sticks found ready-made in his environment, he is considered prehuman; if he deliberately fashioned natural objects into patterned tools, no matter how crude, he is considered human. By this criterion, quite a few hominid types were men.

The early parts of hominid history are almost completely unknown. Some clues about the common ancestor from which both the pongids and the hominids have arisen are provided by 25-million-year-old fossil apes of the genus *Proconsul*, found in East Africa. *Proconsul* clearly belongs to the pongid radiation, but certain features of the skull, particularly the teeth, suggest that the base of the hominid radiation could well have been represented by an animal similar to this ape. Also more nearly allied to the pongid rather than the hominid radiation is *Oreopithecus*, the "mountain ape," whose remains were found in northern Italy. This primate dates back some 10 million years, to the early Pliocene. The teeth and the organization of the jaw of *Oreopithecus* were rather hominid, but in other respects the animal was still apelike.

The hominid record of the Tertiary notably includes *Zinjanthropus*, the "East Africa man," discovered in Tanganyika in 1959 (Fig. 15.30). Potassium-argon dating has shown that *Zinjanthropus* lived 1¾ million years ago, in the late Pliocene. This hominid made tools and thus was a true man. The tools included wooden clubs and stone hammers with which *Zinjanthropus* killed small animals and broke open their bones. The diet was mainly coarse vegetation, however, as the large molars clearly indicate. Bone structure in the skull reveals that the head was held very erect and that jaw muscles were attached as in modern man, suggesting that *Zinjanthropus* probably knew speech. On the other hand, a forehead was almost absent; the volume of the brain could not have been larger than 600 cm^3, comparable to the brain volume of a modern gorilla. Also, a low bony ridge was present on top of the skull, another apelike characteristic. *Zinjanthropus* is not the most ancient man known, for the location where he was found gives evidence of another, even more ancient tool-making hominid called *Homo habilis*. As the genus name indicates, this man appears to have been fairly close to our own line of descent.

All other known hominid fossils are of Pleistocene origin, i.e., not older than 1 million years. The oldest of these form a group of several genera of which *Australopithecus*, the "southern ape," is representative (Fig. 15.31). The remains of this australopithecine group have been discovered in South Africa and have been shown to date back roughly 1 million years or somewhat less. Although the australopithecines thus lived much later than *Zinjanthropus*, they probably were not so far advanced. For example, they apparently did not make tools. Their brain volume averaged 600 cm^3, and they prob-

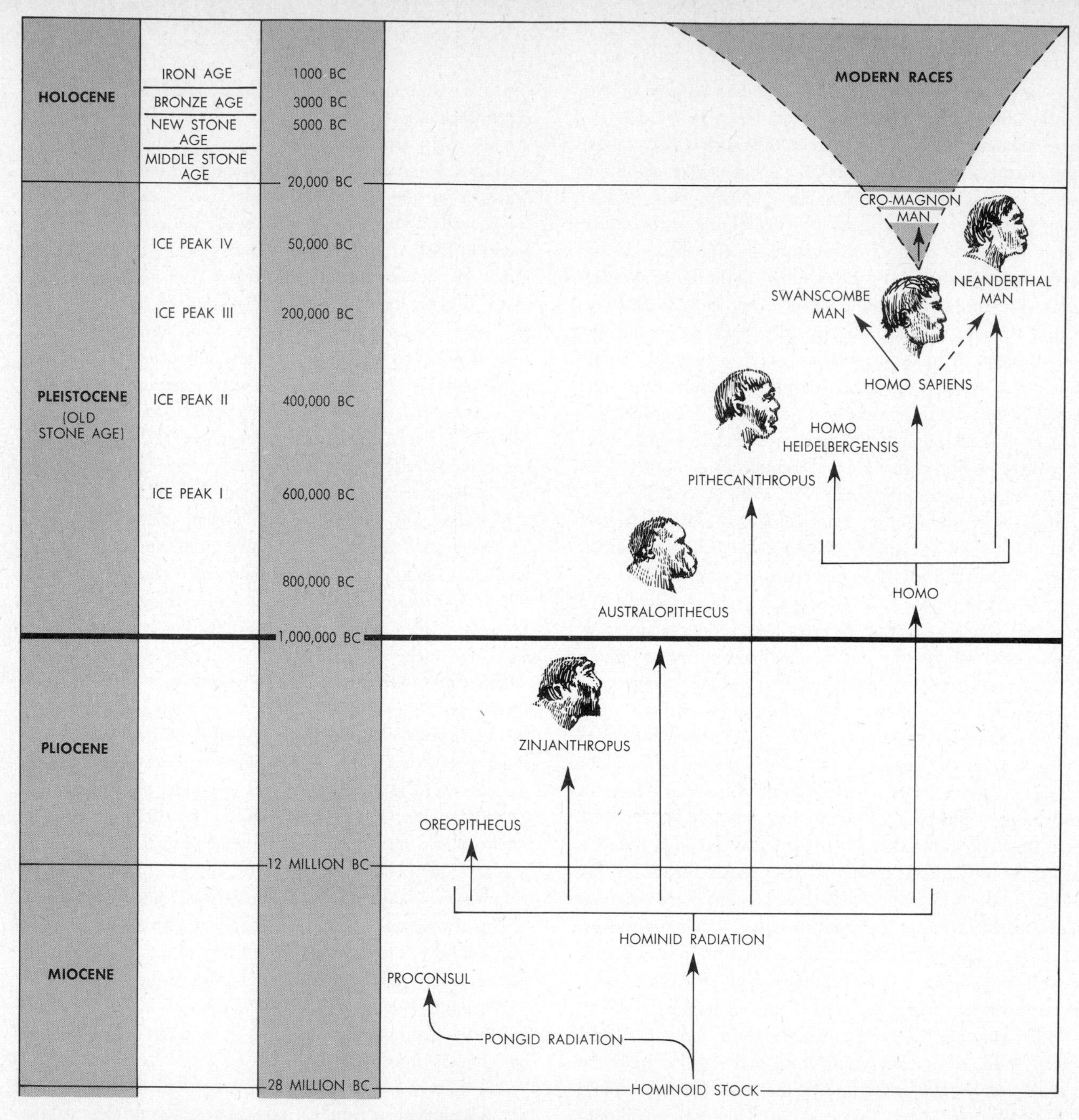

15.29. The hominid radiation and some of its principal members. Each hominid type is shown roughly at a time level at which that type is known to have existed. The detailed interrelations of the various hominid lines are unknown.

ably walked erect or almost erect. Skulls and skeletons reveal a mixture of apelike and manlike traits, but the latter predominate and it is clear that the australopithecines are within the hominid family. Indeed, some investigators regard these near-men to be fairly closely related to the line which gave rise to modern man.

A more recent and comparatively much better known hominid is *Pithecanthropus erectus,* also called *Homo erectus,* a true man who made tools of stone and bone and used fire for cooking (Fig. 15.32). Remains of several species of *Pithecanthropus* were found in Java and China and were shown to be about 500,000 years old. The brain volume of this "erect ape-man" averaged 900 to 1,000 cm^3. The skull had a flat, sloping forehead and thick eyebrow ridges, and the massive protruding jaw was virtually chinless. Like several other hominids, *Pithecanthropus* probably practiced cannibalism. His fossil remains include separate skullcaps detached cleanly from the rest of the skeleton; sheer accident does not appear to have caused such neat separations.

Homo heidelbergensis, the Heidelberg man, appears like *Pithecanthropus* to have lived about 500,000 years ago. Unfortunately, Heidelberg man is known only from one fossil jaw, and his status therefore cannot be fully assessed. Far more complete information is available about another representative of the genus *Homo*, namely, *Homo sapiens neanderthalis*, a subspecies of our own species and the best known of all prehistoric men (Figs. 15.32 and 15.33). Neanderthal man probably arose some 150,000 years ago, flourished during the period of the last ice age, and became extinct only about 25,000 years ago, when the ice sheets began to retreat. The brain of the Neanderthalers had a volume of 1,450 cm^3, which compares with a volume of only 1,350 cm^3 for modern man. The Neanderthal brain was also proportioned differently; the skull jutted out in back, where we are relatively rounded, and the forehead was low and receding. Heavy brow ridges were still present, and the jaw was massive and again almost without chin.

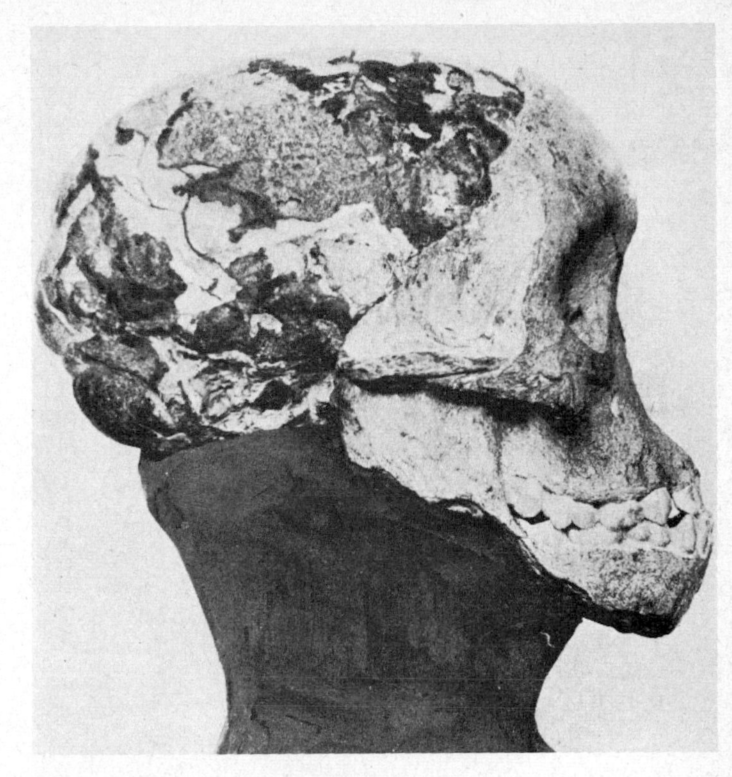

15.31. Skull of *Australopithecus*.

15.30. Reconstruction of *Zinjanthropus*, the East Africa man. Note the exceptionally low forehead and small brain case.

Adapted from painting by Peter Bianchi, © National Geographic Society, 1961.

Culturally, the Neanderthalers were Stone Age cavemen. All Pleistocene hominids are generally regarded as belonging to the *Old Stone Age*. But whereas earlier hominids made only crude stone implements, Neanderthal man fashioned a variety of weapons, tools, hunting axes and clubs, and household equipment. Yet he was still a nomad living from hand to mouth and he had neither agriculture nor domesticated animals. He did not make pottery and did not leave any art. His territory covered most of Europe, with fringe populations along the African and Asian coasts of the Mediterranean. He was a contemporary of modern man, and it appears that the Neanderthal type may have become integrated genetically with modern man.

Homo Sapiens Sapiens

The time of origin of our own subspecies, *H. sapiens sapiens*, cannot be pinpointed very precisely. The oldest representative appears to be *Swanscombe man*, known only through a few skull bones. These are from 500,000 to 250,000 years old. Early groups of modern man thus may have been contemporaries of *Pithecanthropus*.

Later groups include *Cro-Magnon man* (Figs. 15.32, 15.34), who lived from about 50,000 to 20,000 years ago and who may have caused the extinction of Neanderthal man in Europe. Cro-Magnon was 6 ft tall on the average, with a brain volume of about 1,700 cm^3. His culture still belongs to the Old Stone Age. In addition to stone implements, Cro-Magnon used bone needles with which he may have sewn animal skins into crude garments. The dog became his companion, but he still did not domesticate food animals and he did not practice agriculture; Cro-Magnon was a cave-dwelling hunter. He developed a remarkable art, however, as his murals on cave walls indicate.

15.32. Reconstruction of *Pithecanthropus*.

Cro-Magnon man was a contemporary of other groups of *Homo sapiens* living in different parts of the world. The racial division of modern man into *caucasoids, negroids,* and *mongoloids* may have taken place then. However, any original racial traits became diluted or obliterated fairly rapidly, through interbreeding among the extensively migrating human populations. None of the present types of man (or any other animal for that matter) represents a "pure" race.

By the time the Pleistocene came to a close, some 20,000 or 25,000 years ago, all human species and subspecies other than our own had become extinct. The ice started to retreat, milder climates gradually supervened, and eventually man no longer needed to shelter in caves. For the next 15,000 years he produced what is known as the *Middle Stone Age* culture. It was characterized chiefly by great improvements in stone tools. Man was still a nomadic hunter.

The *New Stone Age* began about 5,000 B.C., about the time Abraham settled in Canaan (cf. Fig. 15.29). A great cultural revolution took place then. Man learned to fashion pottery; he developed agriculture; and he was able to domesticate animals. From that period on, modern civilization moved on with rapid strides. By 3,000 B.C. man had entered the *Bronze Age*. Some 2,000 years later the *Iron Age* began. And not very long afterward man discovered steam, electricity, and now the atom and outer space. Measured by geologic standards, the hairy beast which lumbered down from the trees 30 million years ago turned into college professor in a flash.

That modern man has evolved through the operation of the same forces which produced all other creatures is clear. And it should also be clear that this creature is by far the most remarkable product of evolution. Man is sometimes described rather offhandedly as being "just" another animal. Often, on the contrary, he is considered to be so radically distinct that the appellation "animal" assumes the character of an insult. Neither view is justified.

Man certainly *is* an animal, but an animal with many unique attributes. Structurally, man is fully erect and possesses a double-curved spine, a prominent chin, and walking feet with arches. He is a fairly generalized type in most respects, being

15.33. Reconstruction of Neanderthal man.

not particularly specialized for either speed, strength, agility, or rigidly fixed environments. However, he possesses a brain proportionately far larger and functionally far more elaborated than any other animal. And most of the uniquely human traits have their basis in man's brain.

Man acquired his brain by an exceedingly rapid, explosive process of evolution. Judging from Miocene fossils and living apes, the hominoids 30 million years ago may have had a brain volume of about 300 to 400 cm^3, comparable to that of a newborn human baby today. From then to the beginning of the Pleistocene, brain volume increased to an average of 600 cm^3, as in *Zinjanthropus* and the australopithecines. In some of the latter the brain was actually somewhat larger, i.e., up to 800 cm^3. Thus, in a span of about 29 million years, brain size roughly doubled. But during the first half of the Pleistocene, in only ½ million years, brain size more than doubled again, from the 600 to 800 cm^3 range to the 1,400 to 1,700 cm^3 range of *Homo sapiens* (Fig. 15.35).

Evidently, the human brain has become considerably larger than might be expected on the basis of general increase in body size alone. Moreover, the increase is not due simply to a proportional enlargement of all brain parts; certain parts have grown far more than others. The greatest growth has occurred in the temporal lobes, which participate in the control of speech, and especially in the frontal lobes, which control abstract thought. As a result, a basic *qualitative* difference between man and an ape or any other primate is that man has the capacity to think in a new time dimension, viz., the *future*. An ape or any other mammal has a mind which may grapple well with problems of the present and to a certain extent also with those of the past; the hind portions of the cerebrum are well developed. But such an animal at best has only a rudimentary conception of future time. It does not possess elaborate control centers for this dimension of existence, viz., frontal lobes. Man does, and he alone therefore, far more so than even a genius chimpanzee, is able to plan, to reason out the consequences of future actions not yet performed, to choose by deliberation, and to have aims and purposes. Also,

15.34. Reconstruction of Cro-Magnon man.

only man is able to any appreciable extent to think in symbolic terms, to generalize, and to envision beauty and to weep and to laugh. Directly or indirectly, all such unique attributes of human mentality are based on ability to deal with the non-concrete, the nonspecific, and on abstract forward projections of consequences. Such attributes in effect are fundamentally future-directed. Man has rightly been called a "time-binder," and he may be regarded also as the only philosophical animal. What has been the original adaptive value of these remarkable capacities and what accounted for their evolution? The answer may lie in food, family, and neoteny.

Living as omnivorous ground animals, our ancestors subsisted partly on food plants obtained relatively easily, and more especially on food animals which they had to hunt down. Since they were far less fleet of foot and also far less strong than almost any desirable food animal, *group* hunting undoubtedly became a necessity. Clearly, however, a group would be the more successful the better it could preformulate its hunting plans, i.e., the more it could project into a future situation. Consequently, natural selection would have operated in favor of enlarged frontal lobes. Furthermore, hunting probably was, as it still is, a differential function of the male. But like most mammals, early men were probably polygamous, each male exerting his dominance over as many females as he could and aggressively warding off any likely competitors. Yet such a social system is decidedly incompatible with group hunting, since going out to hunt means leaving the females unguarded and subject to the attentions of other males. Here too the survival potential would have been increased by foresight and reasoning capacity, for it must have become apparent that undue concern over numerous females ultimately proves to be less important than cooperative hunting; hunting leads to food and survival but females do not, and a single female still suffices for reproduction. Thus, in the conflict between emotional preoccupation with females and realistic preoccupation with food, natural selection would again have favored reason and food, hence larger forebrains. And as a further result, the monogamous human family would have come into existence, a social institution effective *both* as a food-procuring and reproducing unit.

If food and family may have promoted the evolution of larger brains, neoteny in any event provided the mechanism. Embryonic development in man, as indeed in all animals, begins sooner in the head region than in the more posterior parts of the body; the tail region differentiates last. Head development later also slows down sooner and the posterior body regions then catch up. For example, when a mammalian embryo is about the size of a sand grain, three-quarters of it may consist of head structures. But at birth, the head may take up no more than one-sixth to one-eight of the total length. If now the rate of embryonic development were to slow down, one consequence would be that the stage of birth would be reached when the embryo is still relatively undeveloped, i.e., when the head is still comparatively large. If growth subsequently continues at an even rate in all body parts, the large head size would be maintained and this neotenous condition would become permanent (Fig. 15.36).

Such a slowing of developmental rates actually occurred in the course of human evolution and large heads with large brains were the result. An additional result was the presence of other neotenous features, among them, for example, the absence of furry skin. In mammals generally, growth of a fur coat is one of the last developmental events before birth. But if birth occurs at a relatively earlier developmental stage, as in man, fur will not yet have developed. The hair cover of man is therefore sparse, and we note that a large brain and lack of fur go together. Other neotenous features of man include a smaller jaw with fewer teeth than in apes, absence of prominent eyebrow ridges, and thinner, more delicate skulls and bones generally.

Above all, the deceleration of early development resulted in a very substantial lengthening of the whole human life cycle. Man so became almost the longest-lived of all animals. All phases of the life cycle stretched out, including the period of postnatal growth and adolescence. In this manner another uniquely human characteristic emerged, namely, a proportionately very long *youth*. An ape such as a chimpanzee is mature at an age of about 5 or 6 and is senile at age 30, when

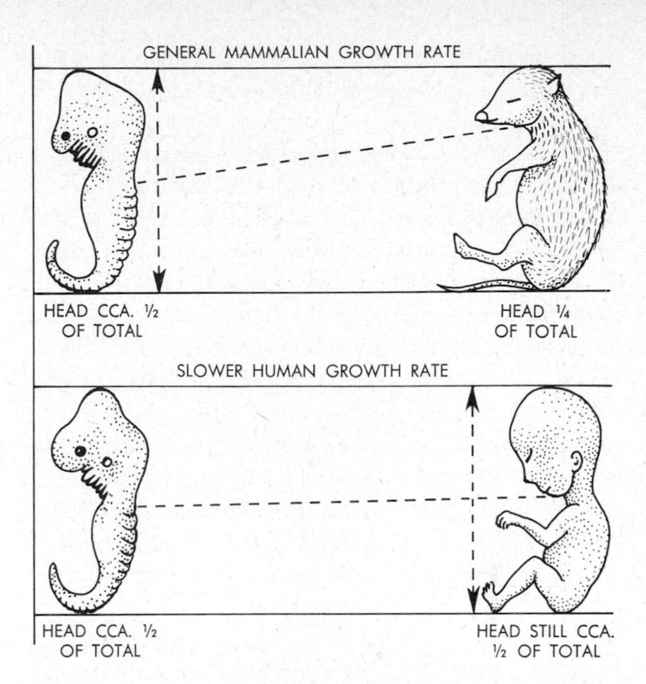

15.36. These diagrams show two successive developmental stages of pig and man, drawn to the same scale. The openings behind the eyes at left mark the ears; note also the rudimentary gill arches, tissues of which contribute to the development of the lower parts of the face. In the time the pig embryo has grown to a stage at which its head is about one-quarter of its total length (right), the human embryo has grown only to a stage at which its head is still roughly one-half of its total length. The slopes of the dashed lines indicate this comparatively slower human growth, and the vertical dotted lines mark the time difference of attaining equivalent body proportions. The result is that man is in a neotenous, i.e., a comparatively more embryonic, state at birth than mammals generally. A proportionately larger brain, thin, delicate bones, lack of body fur, and a prolonged life span are among the consequences of this slower rate of early human development.

15.35. In the course of the time periods indicated, the hominid brain not only doubled in overall size twice in succession, but it also changed in the relative proportion of parts. The frontal and temporal lobes particularly increased more than proportionately, indicated also by the changes in the contours of the skull.

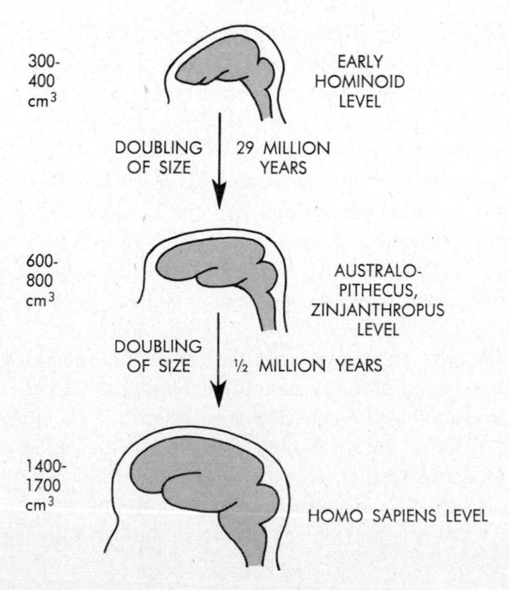

man is just attaining adulthood. Man therefore has *time* to learn and to gather experiences, and in the extent of his learning capacity man is also unique.

Learning was, as it still is, associated closely with family life, and in the monogamous human family the male adult actually came to assume a new function; in addition to being merely the biological parent he also became a social parent, i.e., a *father* who taught hunting skills to the young. And the longer the learning period of the young could be extended, the greater a survival potential the young would acquire. Speech here came to play an increasingly important role. Indeed, enlargement and loosening of the tongue was accompanied by the evolution of a prominent chin, another uniquely human trait. Furthermore, learning and speaking led to *traditions* and to the *accumulation* of knowledge over successive generations. This transmission of knowledge occurred by nonbiological means, and we actually deal here with a new kind of evolution. The old is biological evolution, its vehicle being the gene. The new is social evolution, its vehicle being spoken and later written *speech*. Man is quite unique in having evolved and in continuing to evolve through inherited traditions passed on not only by genes, but also by words.

Conceivably, this changeover from the merely biological to the human may have as much future significance as the earlier changeover, 2 billion years ago, from the inorganic to the biological. The first transition gave rise to totally new opportunities through which matter became organized into a wealth of previously nonexistent arrangements. The recent transition may create new possibilities of like scope. But the realization of this potential is now in the hands of man, for with the coming of man, the chance operations of nature have begun to be modified and manipulated by human purpose. The activities of man block chance increasingly, and man's fate therefore will be decided by man's purpose.

REVIEW QUESTIONS

1. What is a fossil? How can the age of a fossil be determined? Review the names and dates of the geologic eras and periods. What were the Appalachian and Laramide revolutions? List the major groups of organisms generally and animals specifically not yet in existence 500 million years ago.

2. Describe the key events of animal evolution during the Cambrian and Ordovician periods. What were trilobites, eurypterids, blastoids, and nautiloids, and what phyla did they represent? When did they become extinct? What events of vertebrate evolution took place during the Cambrian-Ordovician phase?

3. Describe the key events of animal evolution during the Silurian and Devonian periods. What were the first land animals? What group did the placoderms arise from and what groups did they in turn give rise to? What are lobe-finned fishes? Ammonites?

4. Describe the key events of animal evolution during the Carboniferous and Permian periods. What were the labyrinthodonts and how did they evolve? What group represented the first reptiles? Which animal types became extinct at the end of the Permian and which declined in abundance? Describe the causes and events of the Permo-Triassic crisis. At what date did it take place?

5. Review the evolutionary happenings during the Mesozoic among groups other than reptiles. Make an analogous review for reptile evolution. Which reptilian groups exist today and what ancient groups did they derive from? What group was ancestral to mammals? What conditions promoted the evolution of mammals and what mammalian characteristics today are consequences of these conditions?

6. Name the time and the events of the Laramide revolution. What were the long-range climatic consequences of that revolution? When did the last ice ages occur? Review the evolutionary events of the whole Cenozoic era, with special attention to the mammalian radiation. What factors delayed this radiation until Cenozoic times?

7. Describe the origin of ancestral primates. Then describe the major features and the time pattern of the primate radiation and name living animals representing each of the main lines. When and from where did the line leading to man branch off?

8. Describe the various adaptations of each of the primate stocks to arboreal life. Which structural, functional, and behavioristic features of man trace back specifically to the arboreal way of life of his ancestors? How does the hominoid line differ from other descendants of early prosimians? How does the hominid line differ from the pongid line?

9. Describe the known members of the hominid radiation. When was each of them probably in existence? What culture was associated with each? Roughly when did *Homo sapiens* arise? Review the biological characteristics which *Homo sapiens* shares with (*a*) all other hominids, (*b*) all other hominoids, (*c*) all other primates, (*d*) all other mammals.

10. What was the course of brain evolution in the hominid

radiation? By what mechanism did man acquire a large brain? What was the adaptive advantage of a large brain to early man? What conditions may have promoted the evolution of human familial and social institutions?

COLLATERAL READINGS

The following, some of which are textbooks and some of which are popular books and articles, supplement the topics dealt with in this chapter. The first book by Berrill is recommended particularly as a masterful account of human evolution.

Ableson, P. H.: Paleobiochemistry, *Sci. American,* July, 1956.

Berrill, N. J.: "Man's Emerging Mind," Dodd, Mead, New York, 1955; or Premier Books, Fawcett World Library (pocket books), New York, 1957.

———: "The Origin of Vertebrates," Oxford, New York, 1955.

Bogert, C. M.: The Tuatara: Why Is It a Lone Survivor? *Sci. Monthly,* vol. 76, 1953.

Broom, R.: The Ape-Men, *Sci. American,* Nov., 1949.

Clark, J. D.: Early Man in Africa, *Sci. American,* July, 1958.

Colbert, E. H.: The Ancestors of Mammals, *Sci. American,* Mar., 1949.

———: "The Dinosaur Book," American Museum of Natural History, 1951.

———: "Evolution of the Vertebrates," Wiley, New York, 1955.

Coon, C. S.: "The Origin of Races," Knopf, New York, 1962.

Deevey, E. S.: Living Records of the Ice Age, *Sci. American,* May, 1949.

———: Radiocarbon Dating, *Sci. American,* Feb., 1952.

Dobzhansky, T.: The Present Evolution of Man, *Sci. American,* Sept., 1960.

Eiseley, L.: "Firmament of Time," Atheneum, New York, 1960.

Ericson, D. B., and G. Wollin: Micropaleontology, *Sci. American,* July, 1962.

Glaessner, M. F.: Pre-Cambrian Animals, *Sci. American,* Mar., 1961.

Hockett, C. F.: The Origin of Speech, *Sci. American,* Sept., 1960.

Jarvik, E.: The Oldest Tetrapods and Their Forerunners, *Sci. Monthly,* vol. 80, 1955.

Leakey, L. S. B.: Finding the World's Earliest Man, *Nat. Geographic,* Sept., 1960.

———: Exploring 1,750,000 Years into Man's Past, *Nat. Geographic,* Sept., 1961.

Millot, J.: The Coelacanth, *Sci. American,* Dec., 1955.

Moore, R.: "Man, Time, and Fossils," Knopf, New York, 1953.

Moore, R. C., C. G. Lalicker, and A. G. Fisher: "Invertebrate Fossils," McGraw-Hill, New York, 1952.

Napier, J.: The Evolution of the Hand, *Sci. American,* Dec., 1962.

Newell, N. D.: Crises in the History of Life, *Sci. American,* Feb., 1963.

Romer, A. S.: "Vertebrate Paleontology," The University of Chicago Press, Chicago, 1945.

Sahlins, M. D.: The Origin of Society, *Sci. American,* Sept., 1960.

Shrock, R. R., and W. H. Twenhofel: "Principles of Invertebrate Paleontology," McGraw-Hill, New York, 1953.

Simpson, C. G.: "Life of the Past," Yale, New Haven, Conn., 1953.

———: "The Major Features of Evolution," Columbia, New York, 1953.

Tax, S. (ed.): "Evolution after Darwin: The Evolution of Man," The University of Chicago Press, Chicago, 1960.

Washburn, S. L.: Tools and Human Evolution, *Sci. American,* Sept., 1960.

Weckler, J. E.: Neanderthal Man, *Sci. American,* Dec., 1957.

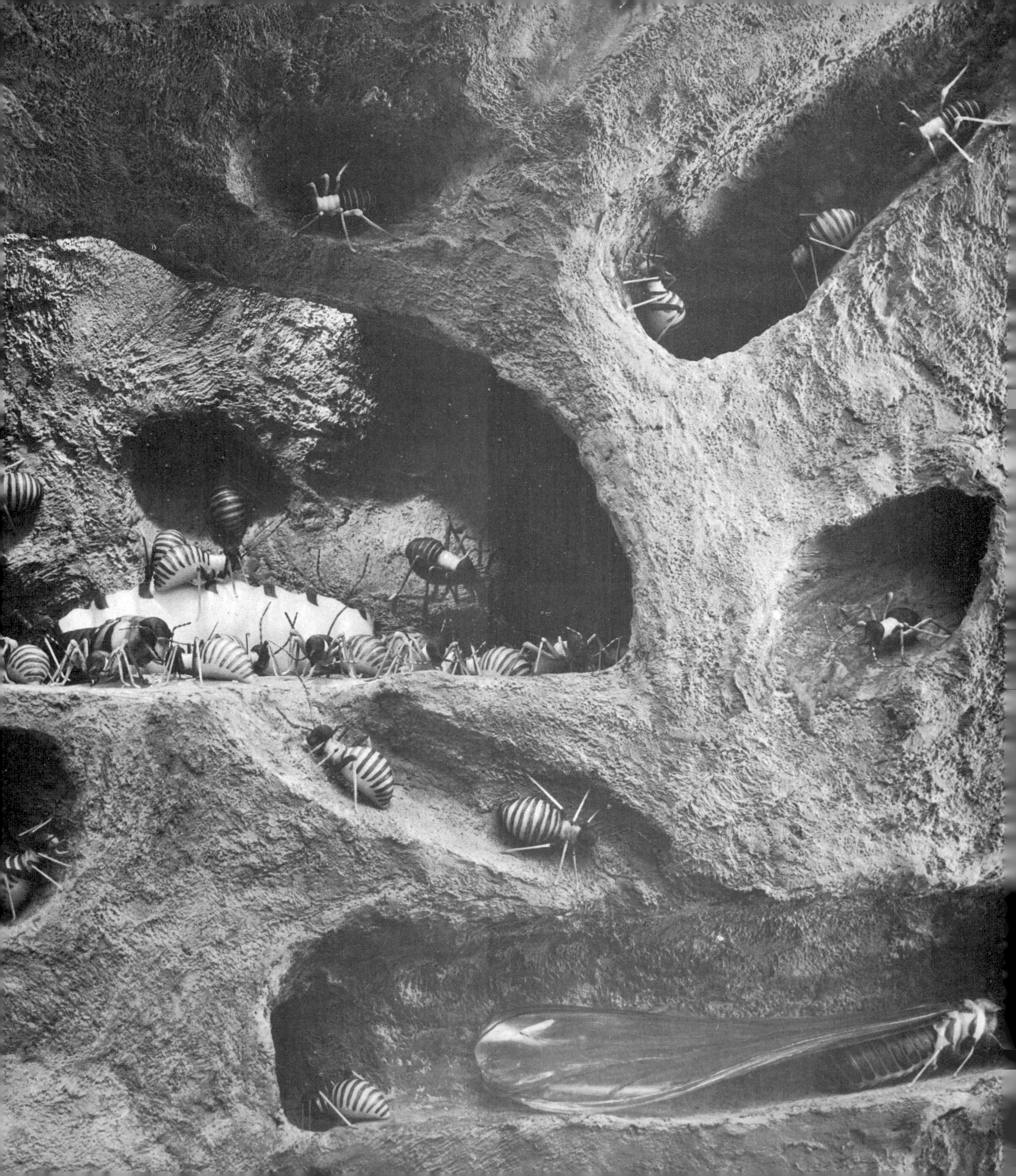

the associations of animals

PART 6

Inasmuch as every animal is specialized to greater or lesser degree, it invariably depends on other animals, on other organisms, and on the environment generally for some essential product or process; survival requires *group* association. Indeed, interacting and interdependent natural groupings of animals are as ancient as the animals themselves, and as the animals evolved, so did their groupings.

We already know that a geographically localized group of individuals of the same species represents a *population*, and that a whole *species* normally consists of many such populations. Each population lives in a particular territory in which are usually present additional populations of other, different species. Such localized groups of associated populations form a *community*. The physical living space of a community together with the community itself is referred to as an *ecosystem*. All ecosystems on earth collectively form the *biosphere*. This most inclusive association therefore encompasses the entire inhabited part of the globe and comprises all nonliving and living components.

All parts of the biosphere are linked together by evolutionary, developmental, sometimes structural, but particularly by functional bonds; any living unit, large or small, interacts with all other living units and with the physical environment. The biological subscience dealing with the biosphere as a whole and the associations within it is called *ecology*. In the chapters to follow we shall discuss the various ecological aspects of nature.

CHAPTER 16 species and population

Every species has an internal structure, a unique functional interrelation with the environment, and a particular developmental history. By virtue of its structure, function, and history, each species is able to live in a given region of the world and thus exhibits certain *zoogeographic* characteristics. The first part of this chapter examines these properties of species.

Representing the main subordinate associations within species, populations have structural, functional, and developmental characteristics of their own. For example, the individuals within populations display variously intimate forms of group *behavior*. We refer to the most intimate of these as *social* behavior. The properties of populations generally and of social populations specifically are outlined in the second part of this chapter.

THE SPECIES

Structure and Function

Several types of bonds unify the members of a species into an associated *natural* grouping of animals. First and foremost, each species is an *evolutionary* unit: all member animals share the same gene pool and thus are more closely related to one another than to members of any other species. As a result, the members of a species have in common a basic set of structural and functional traits.

However, no two animals are exactly alike, and despite their close relation the members of a species actually differ from one another quite considerably; superimposed on the common traits, *individual variations* are characteristic within each species (cf. Fig. 12.1). Indeed, the range of such variations within one species may be directly continuous with the range within a closely related species. It may happen, therefore, that two animals belonging to two different species might not differ too much more in structure and function than two animals belonging to the same species.

We already know from earlier chapters that variations can be acquired or genetically controlled, i.e., they can be noninheritable or inheritable. Most of the acquired variations encountered within a species actually represent genetically controlled traits which have been modified in noninheritable fashion. For example, body weight is a general, inherited species characteristic. But what the actual weight of an individual will be depends partly on his eating habits. Similarly, a generalized level of intelligence is characteristic of the human species and is inherited, but actual mental capacities depend greatly on the training of each individual and on other noninherited factors.

Inherited variations are often correlated with the environmental characteristics of the different geographic regions a species inhabits. In warm climates, for example, individuals of many animal species tend to have smaller body sizes, darker colors, and longer ears, tails, and other protrusions than fellow members of the species living in cold climates (Fig. 16.1). Such structural variations are *adaptive;* i.e., they are advantageous to the individuals in the different environments. Smaller bodies and longer ears, for example, make for a large

16.1. Evaporation surfaces tend to be larger in warm-climate than in corresponding cool-climate animals. For example, the Arabian desert goat illustrated here possesses external ears very much longer than those of related types in temperate regions.

body surface relative to the body volume. Under such conditions evaporation from the skin surface is rapid and the cooling effect of this enhanced evaporation is of considerable benefit in a warm climate. The converse holds in a cool climate. In many instances it may be very difficult to recognize the adaptive value of a variation. And some variations conceivably may be *nonadaptive,* without inherent advantage to the possessors. Bizarre traits produced by genetic drift in small isolated species generally tend to be nonadaptive (cf. Chap. 13).

Still other kinds of variations are adaptive not primarily in an environmental but in a developmental or functional sense. Thus, a species always includes temporarily variant individuals in the form of embryos, larvae, and young and mature adults. Also present in most species are permanently variant individuals in the form of males and females. Moreover, a species usually encompasses subordinate classification groups such as subspecies (or *varieties*), subsubspecies (or *races*), and strains, and these too are distinguished by particular, unique variations (Fig. 16.2).

By virtue of the structural variations of their members, all species may be said to exhibit a greater or lesser degree of *polymorphism,* i.e., to be composed of individuals of "many

16.2. Individual variation. These two umbrella birds belong to the same species, viz., *Cephalopterus ornatus.* But they are members of different populations, and the structural differences between the birds are quite pronounced. Technically these birds are said to belong to different subspecies (also referred to as *varieties*) of the same species.

shapes." The structural differences between males and females are instances of *dimorphism*, a form of polymorphism (Fig. 16.3). But polymorphism may be far more pronounced. Two individuals of the same species may be so different structurally that their common traits become evident only through the most careful study. For example, many coelenterate species are highly polymorphic. These animals occur in two structurally quite different forms, viz., as free-swimming

16.3. Sexual dimorphism, a form of polymorphism: swordtails, male on top, female below. Note sex differences in size and structure.

medusae like jellyfish and as sessile *polyps* like sea anemones. One entire class of coelenterates consists of species in which the life cycle of the individuals typically includes *both* polyp and medusa stages (cf. Chap. 21).

Moreover, the polyp and medusa forms of a coelenterate may themselves be highly polymorphic. For example, *Physalia,* the Portuguese man-of-war, is a colonial coelenterate common on the surface of warm seas (Fig. 16.4). A colony is made up of several dozens or hundreds of individuals and several classes of polymorphs are present: *feeding* individuals, with tentacles and mouths adapted for the ingestion and digestion of small fish; *protective* individuals, whose long trailing tentacles are equipped with batteries of sting cells which paralyze prey and ward off predators; individuals modified into the air-filled *float,* which buoys the whole colony; and *reproductive* individuals, which manufacture gametes and propagate the species.

Many species of several other phyla exhibit analogously pronounced forms of polymorphism. The most familiar probably are the social insects, a given species of which includes queens, drones, soldiers, workers, and others, all structurally quite dissimilar (cf. Fig. 16.17).

All such instances of polymorphism are expressions of structural *specialization.* And where animals exhibit great polymorphic diversity, a high degree of functional interdependence necessarily follows as well. In *Physalia,* for example, only the feeding individuals can feed and the whole colony depends on that. Only the protective individuals can protect and all other polymorphs depend on that also. And only the whole, tightly integrated colony is a self-sufficient unit. Indeed, such units display all the attributes of a society. We note that structural variation generally and polymorphism specifically produce functional bonds which link the members of a species into a necessarily cooperating association of animals.

However, regardless of the scope of their structural variations, the members of a species usually *must* cooperate functionally in any case; for by virtue of being an evolutionary unit, a species is also a *reproductive unit.* Males and females are interdependent reproductively, and interbreeding thus links the members of a species into a functionally cohesive association. We already know that such reproductive bonds are the customary criterion used to define a species as a taxonomic unit: a species is a group of individuals which shares the same gene pool through interbreeding and which is separated from the gene pools of other species by reproductive barriers (Chaps. 7, 13).

Not all species can actually be defined uniquely by this

16.4. Model of *Physalia,* the Portuguese man-of-war. Each tentacle suspended from the gas-filled float represents a portion of a single coelenterate individual. The several different types of tentacles here indicate the high degree of polymorphism encountered in such colonies.

criterion of interbreeding. As already pointed out in Chap. 13, *interspecific fertility* is a well-known phenomenon; sister species recently originated from the same parent species (or sister subspecies about to become separate species) usually are interfertile at least potentially, even though in nature they are normally isolated from one another by geographic barriers. Moreover, even more distantly related species are known to be interfertile in some cases. For example, matings between male lions and female tigers may produce *ligers;* the reverse crosses may yield *tiglons;* and horses and asses have long been used for the production of *mules.* These and many other known interspecific hybrids are themselves usually sterile, but they do illustrate the occasional possibility of interbreeding between members of two different species.

In some cases, furthermore, interbreeding is *not* possible between two members of the *same* species. Such *intraspecific infertility* is encountered among protozoa, for example. Certain protozoan species are split up functionally into several mating types such that one type can mate only with a particular one among all the other types. The best illustration is the ciliate protozoan species *Paramecium aurelia,* which consists of 34 hereditary mating types forming 16 distinct mating groups, or *syngens.* Of these, 15 syngens contain two mating types each and one syngen contains four mating types. Mating within a syngen is possible, but mating between syngens is not. Intraspecific infertility is even more pronounced in all vegetative and parthenogenetic species among animals, in which fertilization does not occur at all and in which interbreeding therefore is completely absent (Fig. 16.5).

Evidently, common reproductive bonds characterize more than a single species in some cases (or what appears to be more than a single species) and less than a whole species in others. The definition of a species as a reproductive unit consequently does not always hold, and the derived taxonomic definition similarly is not applicable in all instances. It remains true, nevertheless, that reproduction does represent an important unifying link among the members of *most* species.

Another type of link is probably even more important and universal. By virtue of being evolutionary units, all species are also *ecological units.* Each species is defined by its *ecological niche,* i.e., its place in nature: it inhabits a certain territorial range, it uses up particular raw materials in that range, and it produces particular byproducts and endproducts. For example, the biosphere offers numerous opportunities for carnivorous modes of existence, all differing from one another in hundreds of fine details. Such opportunities represent ecological

16.5. Interspecific fertility is illustrated also by matings between tigers and lions, two species of the same genus (*Felis leo, Felis tigris*). Middle diagram, mating types I and II of *Paramecium aurelia* belong to syngen (or "variety") 1, within which mating may occur. Individuals of types VII and VIII belong to syngen 4, and mating within this group may occur similarly. But interbreeding between syngens may not occur, even though such individuals may live in close proximity and belong to the same species. Right, exclusively vegetative reproducers such as *Amoeba proteus* are by definition infertile, hence also intraspecifically infertile.

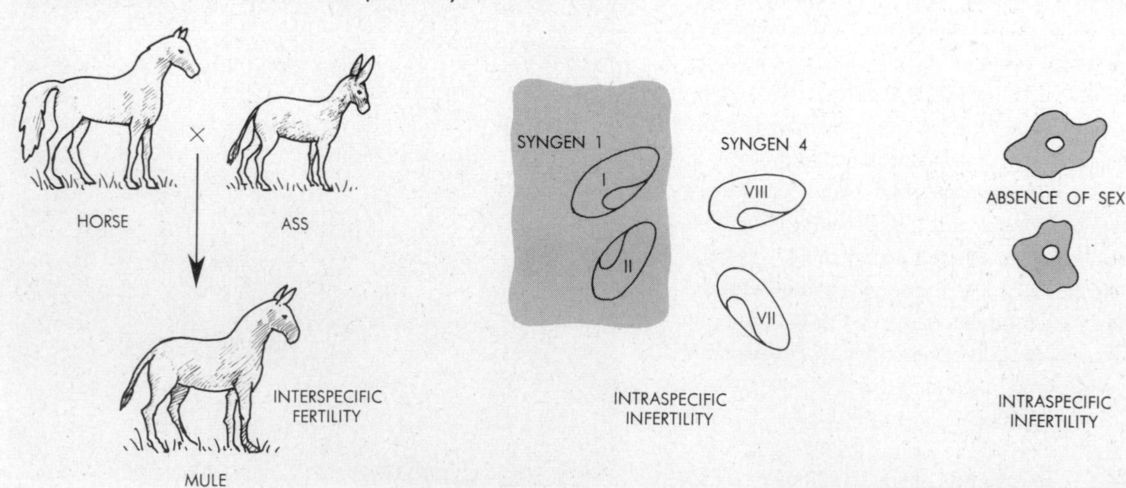

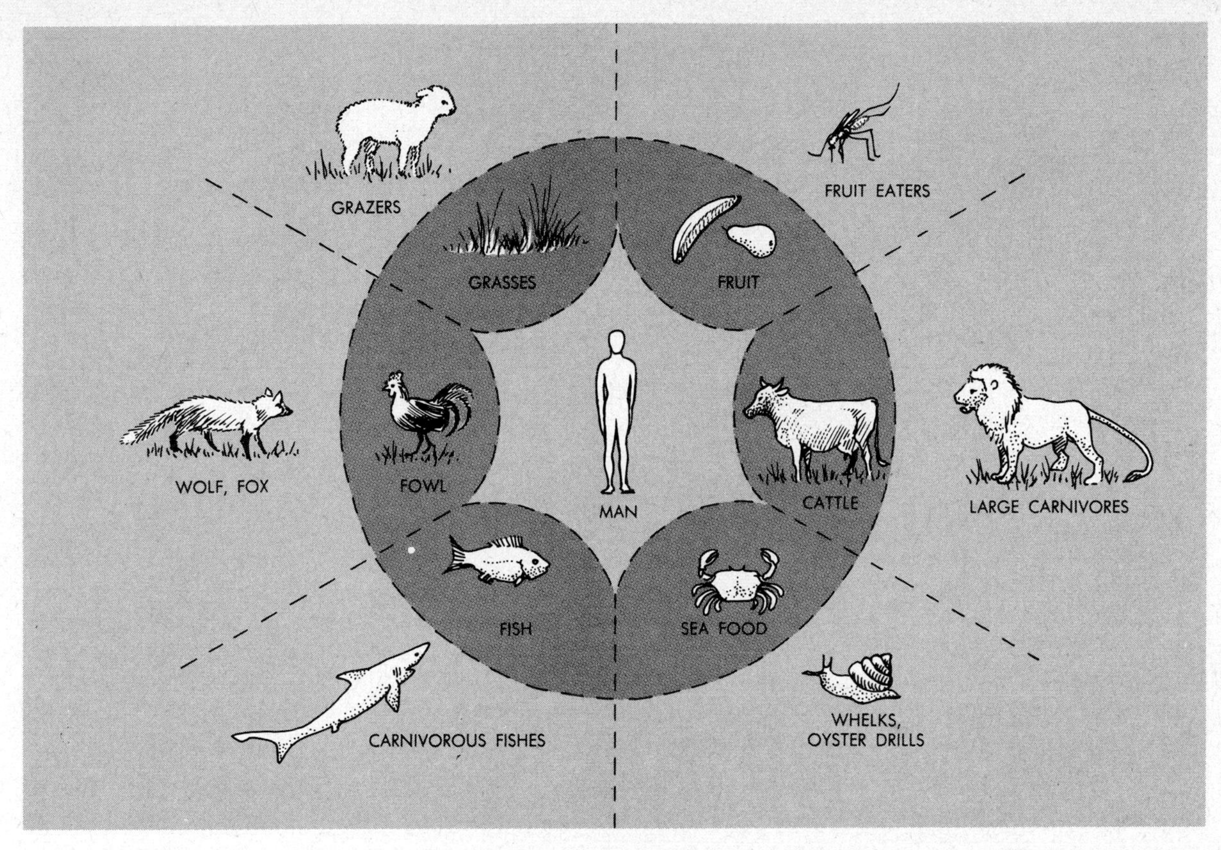

16.6. The ecological niche of *Homo sapiens* is characterized in part by man's requirement of foods such as are shown within the circle around man. These food organisms also happen to form parts of the ecological niches of other animals, as indicated outside the circle. However, the latter animals also use foods not used by man. Thus, although these various ecological niches and that of man overlap partially, each niche is nevertheless distinct in its entirety and characteristic of a particular species only. Note that ecological niches are defined not only by the food organisms but also by the geographic territory, the waste products, the structures, and other attributes of given species.

niches, and given species occupy them. Thus, species can subsist carnivorously by being terrestrial, aquatic, or aerial; sessile or motile; cold-, temperate-, or warm-climate types; daytime or nighttime hunters; small-prey or large-prey specialists. And each such coarse category additionally encompasses innumerable finer categories of possible carnivorous ways of life. Analogously, all other opportunities of pursuing specific ways of life in the biosphere constitute ecological niches which are occupied by particular species. Consequently, just as each species is identified by a basic internal anatomy and body design, so it is also identified by its ecological niche (Fig. 16.6).

Note that such niches are strongly correlated with geography: in similar kinds of environments, even if widely separated, species with similar ways of life will be found. For example, prairies offer opportunities for grazing animals, and each prairie region of the world actually has its own species filling available grazing niches; e.g., antelopes in Africa, bisons in North America, kangaroos in Australia. Analogously, species on different high mountains occupy similar ecological niches. At the same time, *several* similar niches may be available in a single territory. Thus, the Central African plains support not only numerous types of antelopes but also zebras, giraffes, and other grazing species. The ways of life of such species overlap in many respects, but they are not precisely identical in all details; each species normally fills a unique ecological niche.

By being adapted to the same niche, the members of a species are linked together through powerful ecological bonds. These have cooperative as well as competitive aspects. For example, *intraspecific cooperation* is often necessary to execute the way of life of a species; hunting may have to be done in cooperative packs, migrations may have to be undertaken in the comparative safety of herds or flocks, and groups may be required for the construction of complex nests, hives, or dams. At the same time, inasmuch as the members of a species must share the living space and the food sources within their common ecological niche, *intraspecific competition* occurs as well. However, such competition is generally quite indirect. Although direct competition by combat does occur, in most cases one member of the species affects another via the environment only and physical or even visual contact need not be involved (Fig. 16.7). We may note here incidentally that the reproductive bonds within a species have cooperative and competitive aspects too. Thus, males and females must cooperate to achieve fertilization, yet the members of one sex compete among one another for acceptance by members of the other sex.

Development and Zoogeography

Each species occupies an ecological niche for the duration of its evolutionary life, i.e., in the order of a million years on the average. In the course of this time, the history of a species typically follows a characteristic pattern. A species originates in a small home territory, spreads from there over as wide a geographic range as it can, then dies out in various localities of that range and eventually becomes extinct altogether (Fig. 16.8). As noted in Chap. 13, species extinction is not a theoret-

16.7. Intraspecific cooperation is usually pronounced in herding animals, as in this walrus herd. But note that, since the members of a herd use the same restricted space and limited food sources, a measure of intraspecific competition is usually likely to be manifested as well.

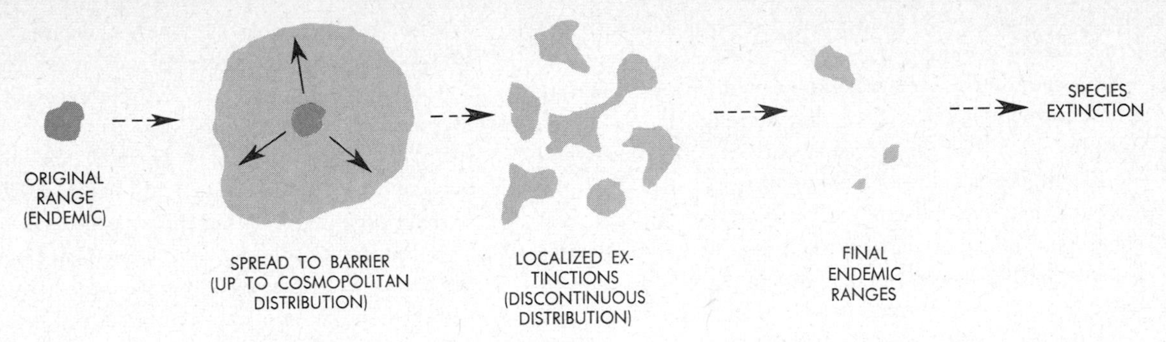

16.8. The general life history of a species.

ically inherent necessity, but all species known to date actually did become extinct sooner or later. What factors govern this pattern of species development?

Speciation occurs through events already described in Chap. 13: in a parent species, a population becomes isolated reproductively from neighboring sister populations and so forms a new descendant species. In the process of being "born," a new species also acquires its own ecological niche. It may do so, for example, by occupying a niche that had been free in the territory. Or if a change in the physical environment has created a new niche, the new species may come to occupy it. Or the new species may itself create its own niche, by evolving a way of life which overlaps partially with the ways of life of several similar species living in the same area. Or the new species may simply encroach on a niche that another species in the area already occupies. The result then is *interspecific competition,* with consequences to be described below.

During its early development, a new species is said to be *endemic:* its members may form just one or at most a few populations, all localized in a single small territory. Further development depends on the geographic extent of the ecological niche of the species. In general, a species tends to expand in widening circles around the home territory until the boundaries of its niche are reached; i.e., until ecological barriers impose conditions to which the species is not adapted. Such barriers are biological as well as physical. For example, spreading in a given direction may be blocked by the presence of competing species or by the absence of appropriate food sources in the path of expansion. Physical barriers can be crudely geographic, e.g., water, deserts, or mountains may stop the spread of a terrestrial species, and land may halt an aquatic species. More subtle physical barriers exist as well. On land, for example, the area of an ecological niche is usually delimited by temperature ranges, by the extent of seasonal and daily climatic changes, by precipitation and soil characteristics, by proximity of water, and by numerous other environmental factors. In the ocean, species may be confined to given temperature zones, salinity zones, pressure zones, oxygen zones, illumination zones, current zones, and many more. Analogously, dispersal opportunities in fresh water may depend on the strength of currents, amount of silting and pollution, oxygen content, bottom conformation, and a wide variety of other conditions.

Thus, if the boundaries of its ecological niche are close to the original home territory, a species cannot spread significantly and it must then remain endemic. Such species usually become extinct fairly rapidly, i.e., as soon as changes in the physical or biological environment make survival impossible. However, most species can find the living conditions they require in more extensive areas and they usually succeed in spreading quite widely. Many cover all or major parts of a continent or an ocean, and many others expand even farther and become *cosmopolitan,* i.e., distributed around the world. In numerous instances, a species may expand to a physical barrier and may subsequently cross that barrier and reexpand on the other side of it. Barrier crossing may occur in one of three general ways, viz., by *sweepstake bridges,* by *filter bridges,* and by *corridor bridges.*

A sweepstake bridge provides variously fortuitous or accidental means of species expansion. For example, terrestrial species may cross even extensive water barriers by floating across on uprooted trees or on driftwood. Also, eggs and microscopic animals may be carried across water in mud clinging to the feet of birds. Moreover, if a water barrier is not too wide, large animals may swim across or may be forced across by currents or storms. Most small oceanic islands were probably populated via sweepstake bridges of this sort. Islands contain comparatively few species, terrestrial mammals and amphibia are generally absent (unless introduced by man), and the species that are present often are quite unique and are found nowhere else. However, they do resemble related species present on the nearest mainland. Such a pattern of island life is fully consistent with the hypothesis (first proposed by Darwin) that islands have been colonized from the closest continent. The pattern also fits the further hypothesis that the colonization has been effected by sweepstake routes: relatively few terrestrial species would be able to reach outlying islands even by chance; terrestrial amphibia and mammals could neither swim nor be carried across hundreds of miles of ocean; and such species that could reach islands would be subject to little competition and could then evolve into unique new species by genetic drift (Fig. 16.9).

Whereas a sweepstake bridge normally operates in one direction only, a filter bridge accommodates traffic in both directions. Such a bridge is a narrow land connection between two continents and it usually exists for only brief geologic periods. It is a filter inasmuch as it prevents many species from crossing over it. A good example of a filter bridge is the land connection between North and South America (Fig. 16.10).

16.9. Sweepstake bridges operate unidirectionally from mainland to islands via accidental and fortuitous means of transport, as indicated. The ancestral finches of the Galapagos islands probably came there from South America via sweepstake routes, as Darwin had surmised, and then evolved on each island into unique types by genetic drift. The heads of five such finches are shown below; counterclockwise from the top: *Geospiza difficilis*, a ground finch; *Camarhynchus cassirostris*, a vegetarian tree finch; *C. psittacula*, an insect eater; *Pinarolaxias inornata*, a warbling finch; and *C. heliobates*, another insect eater. (*Birds after Lack.*)

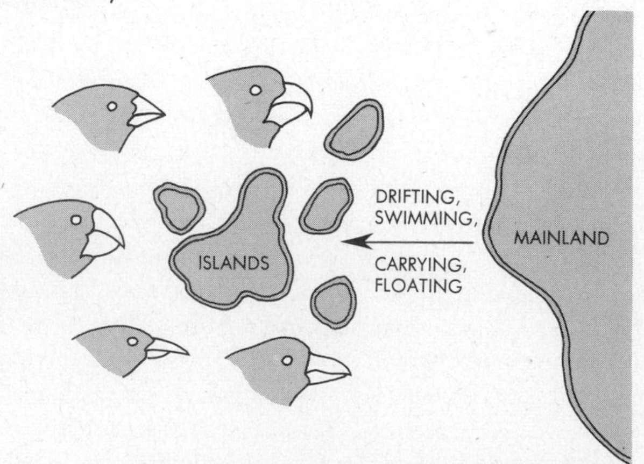

During the late Mesozoic, this connection permitted the spread of marsupial mammals northward (and the North American oppossum of today is a relic species tracing back to Mesozoic ancestors in South America). Concurrently, primitive placental mammals such as sloths, anteaters, and armadillos crossed from North to South America. The land bridge became submerged during the Paleocene but it reemerged later and formed a continuous connection from the Pliocene on. During the Pliocene, only some 5 per cent of all mammalian types in the Americas occurred in common on the northern and southern continent. Today, nearly 50 per cent occur in common. Evidently, the bridge allowed many species to cross but it filtered out many others.

Another major filter bridge was the land link between Asia and Australia. This link remained open till late Mesozoic times and permitted the spread of marsupial mammals from South Asia into Australia. When the bridge later became submerged (except for island chains still present today), placental mammals evolving at that time in Asia could no longer reach Australia. That continent thus became an exclusive preserve for marsupials and, in the absence of placental competitors, these animals then produced an extensive adaptive radiation. Only the recent introduction of placental mammals in the form of man and animals associated with him has seriously jeopardized the survival of the Australian marsupial fauna.

Probably the most studied filter bridge is the link between Alaska and Siberia (cf. Fig. 16.10). At present this bridge is submerged under the shallow waters of the Bering Straits, but an intermittent land connection existed during the ice ages of the Pleistocene. As water became locked into continental ice the global ocean level sank enough to expose the Bering land bridge, and as ice melted (as it does now) the bridge became submerged again. During its intermittent existence, the link permitted the crossing of successive waves of species. For example, among the mammals spreading from Asia to North America are believed to have been bisons, bears, deer, rhinos, mammoths, and cats; and a reverse crossing was probably achieved by horses, dogs, and camels (the latter then spreading throughout Asia into Africa). The bridge acted as a filter, since its far northern location permitted only cold-adapted types to make the crossing.

Corridor bridges are far rarer than either filter or sweepstake routes. A corridor is a broad land connection between continents which persists for long geologic periods and allows free and substantial exchange of species. The best known corridor of the past was the land link between North America and Asia, the same which much later became reduced to the Ber-

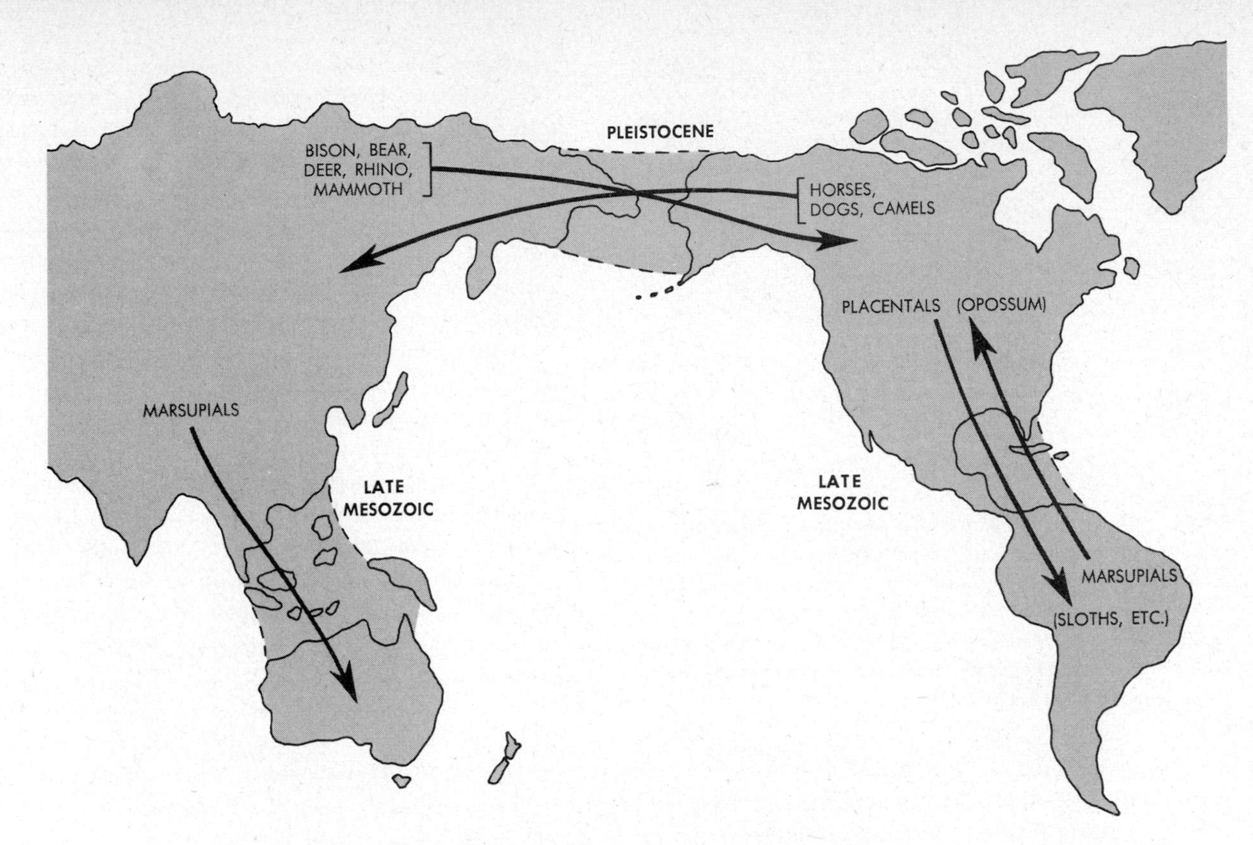

16.10. The three best-known filter bridges. An intermittently open Pleistocene land bridge between Siberia and Alaska permitted exchange of animals such as those named above. A land connection open till near the end of the Mesozoic permitted marsupials, but not the Cenozoic placental mammals, to spread from Asia into Australia. And a Caribbean land bridge during the late Mesozoic and early Cenozoic allowed marsupials to migrate northward, placentals such as sloths, armadillos, and anteaters to migrate southward.

ing filter bridge. During most of the earth's history, including virtually the whole Mesozoic, the land connection between Asia and North America was very extensive (and these two land masses may actually have been part of a single continent; cf. below). Mesozoic reptiles could cross quite freely, and this probably accounts for the present distribution of, for example, alligators. Today these animals are found only in China and North America, but in Mesozoic times their forebears probably ranged throughout Asia and North America. The type probably became extinct in all but its present localities when climates became cooler during the Cenozoic and when the Pacific Ocean came to extend northward and submerged the corridor between the two continents (Fig. 16.11). Corridors undoubtedly exist today, but their location will become apparent only in future ages. If some continuous continental land mass of today should later become subdivided into two subcontinents, then future zoologists may discover that the subcontinents were joined by a connecting corridor in the present age.

The example of alligators above illustrates a very general phenomenon in the life history of a species. After a species becomes cosmopolitan or at least manages to become distributed widely, its populations in numerous localities cease to exist sooner or later. Such local disappearances are brought about in three ways, viz., a population either dies out, or it emigrates, or it evolves into a new species. The underlying causes can be physical, biological, or both. Thus, if a locality undergoes a change in physical characteristics, a population

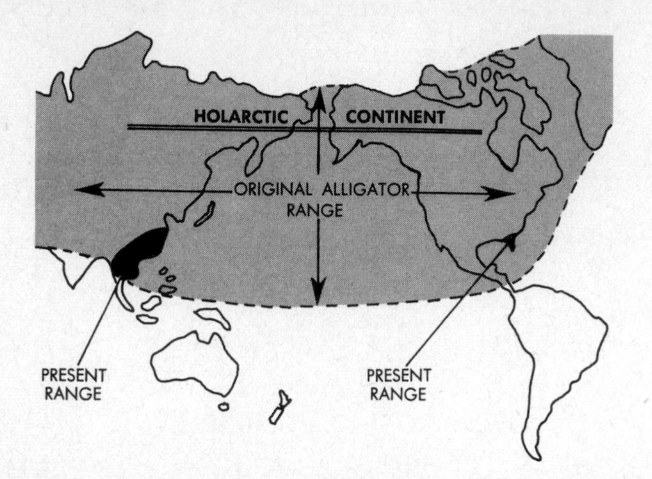

16.11. The light-shaded area represents part of the postulated original Holarctic land mass, which later became separated into Asia and North America by sinking of land in between and by northward extension of the Pacific Ocean. Alligators presumably ranged throughout Holarctica originally, that continent forming a corridor bridge. The present restricted distribution of alligators is indicated in dark shading.

living there must respond to the change in one of the three ways above. Evidently, alligators responded by dying out in all regions except those which remained suitable. Tapirs provide a similar example. These animals are found today only in South America and in the Malaysian Archipelago. Presumably they originated somewhere in Central Asia and spread to Southern Asia and across the Bering bridge into the Americas. Then they became extinct in all intermediate regions as a result of the cold climates of late Cenozoic times (Fig. 16.12).

Biological change in a locality is brought about by other species living in the same locality. For example, if two populations of different species have a predator-prey relationship, then disappearance of the prey for any reason will constitute a biological change to which the predator must respond in one of the three ways above. Another frequent cause of biological change is the invasion of a region by members of other species which occupy an ecological niche very similar to that occupied by the earlier inhabitants. Such a condition leads to *interspecific competition* for the same living space or the same food sources or both. Coexistence may be possible at first, though each competing species is usually hampered in some way by the others. In time, however, one population usually becomes dominant and those of other species then must emigrate, evolve into new species, or die out.

In widely distributed and cosmopolitan species, therefore, population gaps are likely to appear in various parts of the territorial range. Over an extended period of time, such gaps tend to become progressively more numerous, and what may have been a fairly continuously distributed species earlier may eventually become a discontinuous array of more or less isolated populations. Literally hundreds of species are known, those of alligators and tapirs among them, which consist of widely separated, isolated populations in various parts of the globe. They represent remnant populations in what originally had been continuous species ranges. Ultimately, a species may again become endemic and may consist of but a few localized populations or even just a single one. Total extinction then is usually not far away (cf. Fig. 16.8).

Note that a species may be endemic either because it is young or because it is old. Old endemic species are more numerous today (and probably at any time) than young endemic ones, presumably because young endemic species become cosmopolitan more rapidly than old endemic ones become extinct. A progressive increase in the total number of existing species is the result.

About 1½ million animal species are known and identified today, and possibly ten or more times that many may actually exist. Some 10,000 new ones are described each year. Apart from cosmopolitan species, characteristic groups of terrestrial species are found on each of the continental land areas of the world. Based especially on the types of birds and mammals present, six major *zoogeographic regions* can be identified. The locations and a few of the most typical mammals of these regions are shown in Fig. 16.13. This figure indicates, for example, that the widely separated faunas of Alaska and Northern Mexico by and large resemble each other more than the

16.12. The present distribution of tapirs through Malaya, Sumatra, Borneo, and most of central and northern South America.

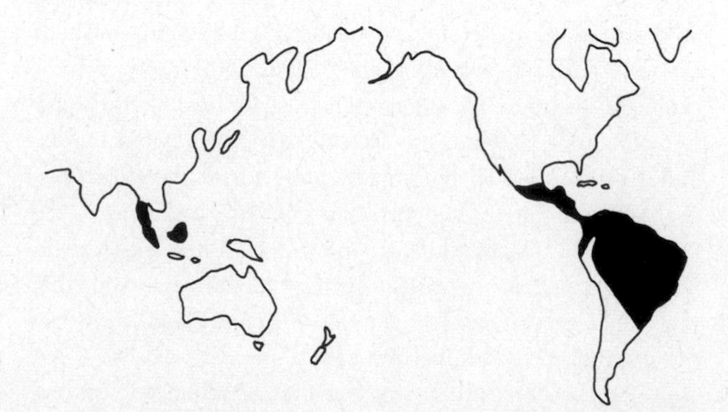

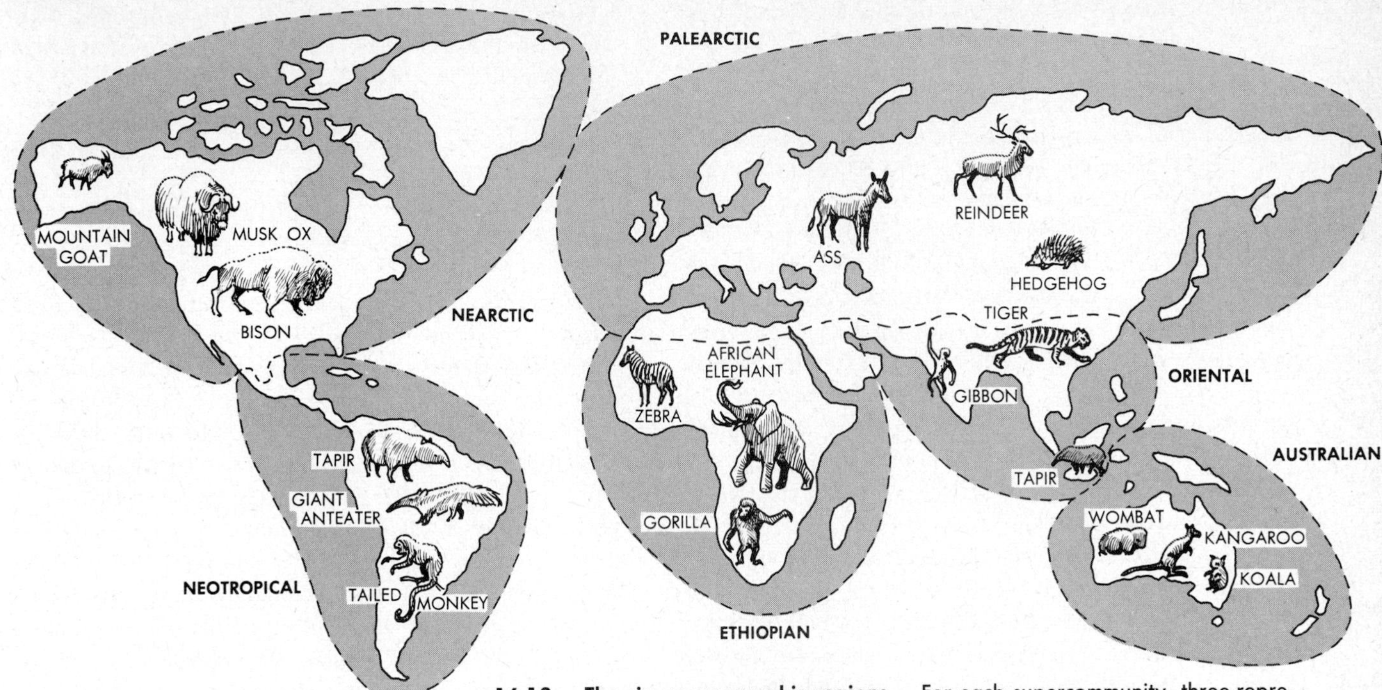

16.13. The six zoogeographic regions. For each supercommunity, three representative and characteristic mammals are indicated. Note that, within each supercommunity, the positioning of the animals is dictated largely by the drawing space available, not by geographic considerations. Thus the positioning is precise only for the zoogeographic regions regarded as wholes, but not necessarily for the actual habitats of given animals.

directly adjacent faunas of Northern Mexico and Southern Mexico.

If species are considered singly, the pattern of their life histories in most cases accounts adequately for their present global distribution. However, the distribution of groups of species is not always explainable so readily. As noted earlier, for example, we know that during late Mesozoic and early Cenozoic times groups of marsupial mammals existed in South America and in South Asia, from where they later spread to North America and Australia, respectively (cf. Fig. 16.10). If marsupials arose in some original Asian or American territory and then expanded to South Asia and South America, the migration route would have had to pass over the Bering bridge. Yet fossil evidence does not support a hypothesis of a Bering crossing. Indeed, the nonoccurrence of such a crossing might actually be expected since marsupials are predominantly temperate- and warm-climate types. How then can the early Cenozoic and the present distribution of marsupials be explained? Did these pouched mammals migrate from Southern Asia to South America or vice versa via sweepstake routes over South Pacific islands and Antarctica? Possibly. Or did they evolve twice independently in two different regions of the world? Again, possibly. Another explanation has also been advanced, viz., the hypothesis of *continental drift*.

According to the most common form of this hypothesis (first proposed in the early part of this century), all the land mass of the globe formed a single continent, *Pangaea,* up to roughly Carboniferous times (Fig. 16.14). Pangaea then is assumed to have split into a northern *Laurasia* (or *Holarctica*) and a southern *Gondwana,* with a *Tethys* sea between them. During the late Mesozoic, each of these land masses is postulated to have split further, Laurasia into what later was to become North America and Eurasia, Gondwana into the later Antarctica, Africa, India, Australia, most Pacific islands, and South America. All these fragments subsequently are thought to have drifted to their present locations and to have become joined here and there into the familiar land masses of today.

The hypothesis was suggested originally by the contours of the present continents, some of which seem to fit together like pieces of a jigsaw puzzle (e.g., South America and Africa). Moreover, continental drift could account for numerous other geographic and geologic features of the earth. Above all, the drift hypothesis was clearly and neatly consistent with the known zoogeographic attributes of the world, as outlined above by Fig. 16.13. Indeed, it was possible to explain distributions such as those of the marsupials. In line with the drift concept, marsupials would have originated on Gondwana, and when this continent later foundered, the animals would have become split into South Asian and South American groups. A drift hypothesis could also explain curious migration patterns such as those of eels, animals which travel to common North Atlantic spawning grounds from both American and European rivers (cf. also below). North America and Europe were part of the postulated Holarctic land mass, and the ancestral eels once may have spawned in a common central Holarctic river system. Correspondingly, their present descendants still are assumed to travel from both west and east to the same ancestral spawning area, which now however happens to lie in the middle of an ocean (cf. Fig. 16.14).

Geologic evidence for continental drift has been generally absent, and during the last 50 years the idea has been alternately abandoned and resurrected. Various other hypotheses have been proposed instead, most of them based on the assumption of permanently stationary continents. Very recently, the notion of continental drift has again come into prominence, mainly because for the first time good—though still indirect—evidence for it has become available through modern geologic studies. Indeed, it appears that the evidence is becoming more and more convincing. If the hypothesis now does attain the status of a well-supported theory, the task of the zoogeographer will be clarified considerably.

THE POPULATION

Internal Organization

A population, often also called a *species-population,* is a relatively stable, geographically localized association of members of the same species. All species are substructured into population units. Such units are exemplified by the dragon flies in a field, the minnows in a pond, the earthworms in a plot of ground, or the people in a village. Individual animals multiply and die, emigrate or immigrate, but collectively a population persists. It may split into subpopulations or fuse with adjacent sister populations, yet the basic characteristics of the group as a whole do not thereby change. The fundamental unifying links within a population are that all members share the same ecological niche and the same local territory, and particularly also that they interbreed more or less preferentially with one another. However, fairly frequent interbreeding with members of sister populations does occur in addition. A population thus is a reproductively cohesive unit, integrated more loosely with other such units.

The significant structural characteristics of populations are their *dispersion* and their *growth* patterns. Dispersion is a measure of the density of the animals present and also of their distribution pattern within the population territory. A *random* distribution is typical in populations of solitary ani-

16.14. The basic elements of the continental drift hypothesis. Continental areas at right are identified by abbreviations; they are assumed to have drifted to their present positions since the late Mesozoic. Dark-shaded areas in Holarctica and at right indicate how the origin and later distribution of eels might be accounted for; dark-shaded areas in Gondwana and at right, analogously for origin and later distribution of marsupials.

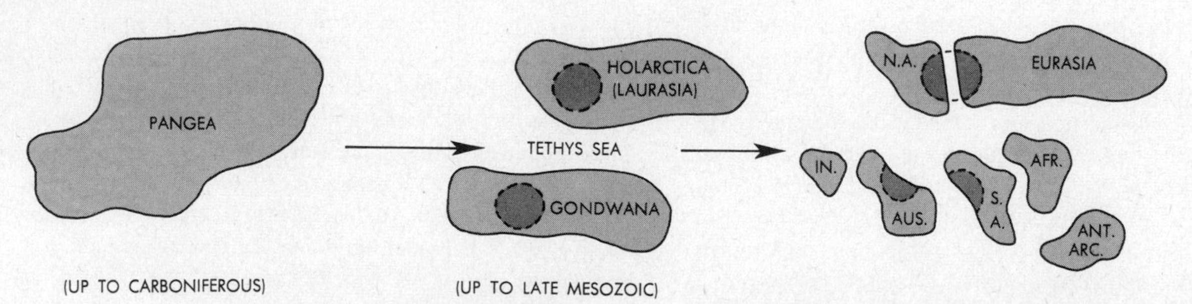

16.15. Patterns of population dispersion. This photo of nesting cormorants illustrates uniform distribution, the birds occupying nesting mounds more or less exactly equidistant from one another. Moving herds usually exhibit clumped distributions, i.e., distances between animals vary considerably and in shifting patterns.

mals, e.g., earthworms, spiders. Distribution tends to be more or less *uniform* in sessile populations, e.g., permanently sessile groups such as colonial corals or temporarily sessile groups such as flocks of birds nesting on an island. In highly motile populations, e.g., in birds or insects in flight or in moving herds, the distribution is usually *clumped* in shifting patterns of densities (Fig. 16.15).

The growth characteristics of a population are determined by the rates of reproduction, mortality, and mobility (i.e., emigration and immigration), by age distribution, and by the effects of population density on growth. Such effects can be of three kinds. In some species, the growth rate of a population is inversely proportional to the density of the member animals. Thus, population growth may occur more rapidly if the density is low. In such cases lessened intraspecific competition promotes faster propagation of the fewer individuals present. In other species, population growth tends to increase in geometric proportions until the limits of the available living space and food supply are reached. And in still other species, population growth is most intense at intermediate densities, i.e., both undercrowding and overcrowding depress growth. Each of these expansion patterns is described by a characteristic growth curve (Fig. 16.16).

In all patterns, we may note, population growth is restricted by *limiting factors.* These include not only population density but also the territory available and the abundance of the food supply, i.e., any conditions contributing to intraspecific competition. Moreover, growth is limited also by interspecific

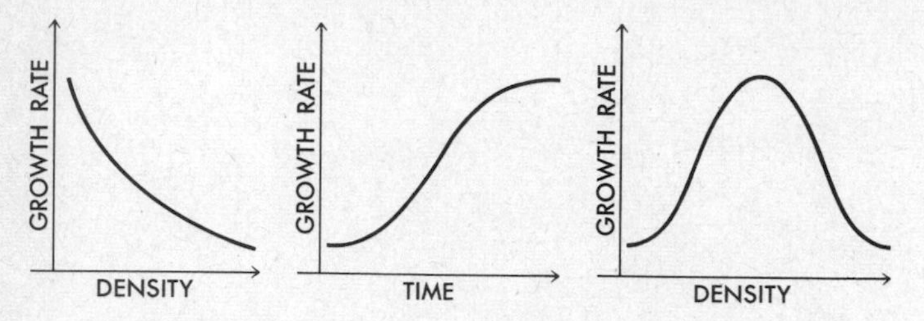

16.16. Patterns of population growth. Left, growth rates are higher at lower population densities. Middle, growth rates are independent of densities and reach a plateau after a population reaches food and territorial limits. Right, growth rates are highest at intermediate population densities, both undercrowding and overcrowding having a growth-depressing effect.

competition and by mortality resulting from predation, parasites, and epidemic diseases. Such factors are usually linked to population density; both predation and epidemics have more pronounced effects if a population is denser.

The important net result of all such variables is that, over appreciable periods of time, *natural populations tend to remain constant in size*. The conditions which promote and those which limit growth normally attain a balanced steady state such that the numbers of individuals in a population do not change significantly on a long-term basis. We shall see in the next chapter how such steady states are automatically self-adjusting. Here we may reiterate that Darwin was already aware of the long-range numerical constancy of populations, despite their inherent potential of growing reproductively in geometric progression (cf. Chap. 13).

The most significant functional attribute of populations is that they *interact* with all components of their surroundings. The members of a population interact with their physical environment, with members of other species-populations (interspecific interaction), and with one another (intraspecific interaction).

Reference to interactions with the physical world has already been made. For example, the external environment provides inorganic nutrients, a living space with certain physical characteristics, and it orients the life history of a population by remaining stable or by changing in various ways. In turn, a population also affects its physical surroundings through the sheer chemical and mechanical impact of the living mass on the nonliving environment. We shall discuss such effects more fully in Chap. 18. Furthermore, the physical environment is often responsible for certain forms of population behavior, especially *rhythmic behavior*. For example, the populations of virtually all species, those of man included, respond behaviorally to the seasonal and daily motions of the sun. They do so differently in different parts of the world, as if populations possessed internal clocks set to local time. Thus, as already noted in Chap. 10, breeding seasons are set to local time. So are daily feeding rhythms, seasonal migration rhythms, and activity rhythms in general. The moon exerts its own effects, particularly on aquatic populations living in the tide zone. If members of two widely separated tidal populations of the same species are brought together in the laboratory, each group continues to behave in accordance with the tidal rhythm of its own locality, as if it responded to built-in tide tables computed for local latitude, longitude, and season. Numerous other environmentally determined types of population behavior are known.

Interspecific interaction of a population with those of other species living in the same area produces *community* life, a subject treated in detail in the next chapter. Of more immediate concern are the intraspecific interactions of the members within a single population. In the majority of animal species, the members of a population live as solitary individuals and, as pointed out earlier, interactions other than reproductive ones are largely indirect, mediated by the environment. By contrast, some species are characterized by extensive direct interactions within populations. The animals in such cases live not as solitary individuals but as more or less closely knit *social* populations. In all of these, cooperative interdependence is correlated with various functional specializations of the member animals, and in many instances structural specializations in the form of polymorphism are encountered as well. As already noted, societies based on functional and pronounced structural specializations are exemplified best by insects, those based primarily on functional specializations, by vertebrates.

Social Populations: Insects

Highly developed societies occur among termites, ants, bees, and wasps. Each member of such populations is adapted structurally from the outset to carry out specific functions in the society. Insect societies, organized somewhat differently in each of the four groups just named, operate in fixed, stereotyped, largely unlearned behavior patterns. In its rigid, inflexible ways, the insect society resembles a human dictatorship, except that among insects there is no dictator, no rule by force. Each member is guided by inherited, instinctive reactions and is unable to carry out any functions other than those for which built-in instincts exist. Insects *can* learn, though only to a limited extent. For example, a bee may be taught to respond differently to different colors and scents, and it may learn a new route to its hive if the hive has been moved.

All social insects build variously intricate *nests,* and their societies are stratified into *structurally* distinct castes. In each of the four groups, different species form populations of different degrees of social complexity. We may profitably examine the organization of a few of the more complex associations.

Honeybees. A population of honeybees is made up of three social ranks: a *queen,* tens or hundreds of male *drones,* and from 20,000 to 80,000 *workers* (Fig. 16.17). The queen and the stingless drones are fertile, and their main functions are reproductive. The smaller-bodied workers are all sterile females. They build the hive, ward off enemies, collect food, feed the queen and the drones, and nurse the young.

When a hive becomes overcrowded, the queen together with some drones and several thousand workers secedes from the colony. The emigrants swarm out and settle temporarily in a tree or other suitable place until a new hive is found (Fig. 16.18). In the old hive, meanwhile, the workers which remain behind raise a small batch of the old queen's eggs in large,

16.18. A swarm of honeybees, emigrated from a parental hive and searching for a new hive.

16.17. A female worker bee. A queen bee is illustrated in Fig. 16.19.

specially built honeycomb cells. These eggs develop into new queens. The first one to emerge from its cell immediately searches out the other queen cells and stings their occupants to death. If two new queens happen to emerge at the same time, they at once engage in mortal combat until one remains victorious. The young queen, her succession now undisputed, soon mates with one of the drones. In a nuptial flight high into the air, she receives millions of sperms which are stored in a receptacle in her abdomen. The sperms from this single mating last through the entire egg-laying career of the queen.

Among the eggs laid individually into honeycomb cells (Fig. 16.19), some escape fertilization, even in a young queen. None is fertilized in an older queen once her sperm store is exhausted. Unfertilized eggs develop into drones. Fatherless development of this sort, an instance of *natural parthenogenesis,* is widespread among social insects generally. Fertilized eggs develop into larvae and these either into queens or into workers, depending on the type of food the larvae receive from their worker nurses. Larvae to be raised into workers are fed a "regular" diet of plant pollen and honey. Queens

A

B

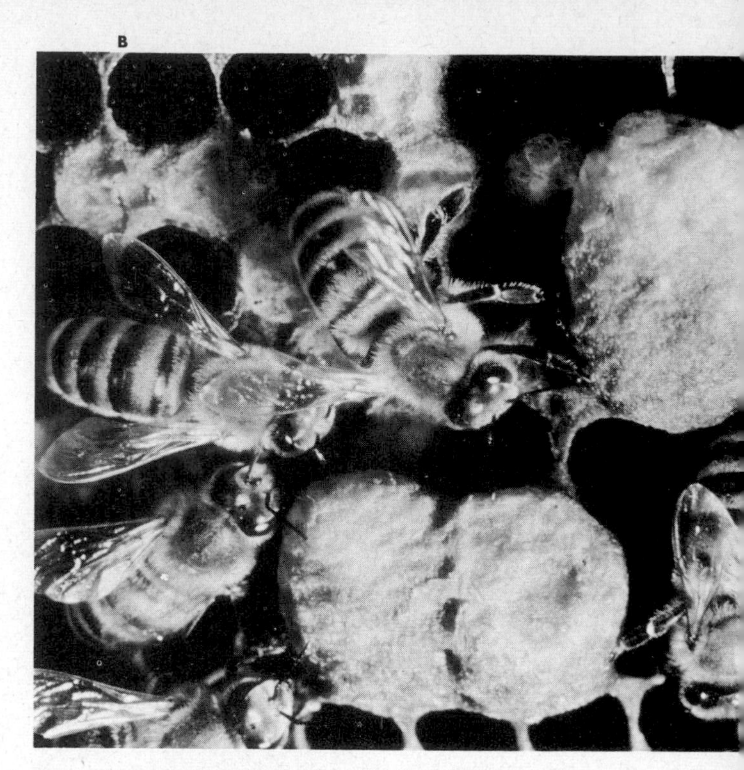

C

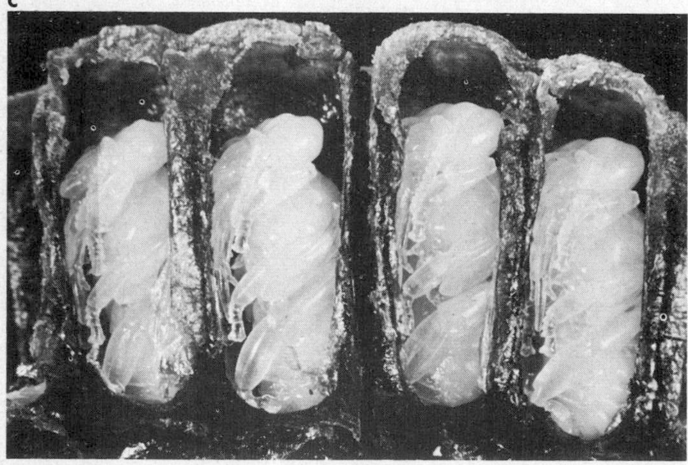

16.19. **A.** Egg-laying queen bee in center, tended by surrounding workers. **B.** Workers finish sealing off enlarged brood cells containing eggs and royal jelly; the eggs will develop into new queens. Note normal-sized honey cells at upper left of photo. **C.** Brood cells cut away, exposing worker larvae within.

form when the larvae receive an especially rich royal jelly, containing pollen, honey, and comparatively huge amounts of certain vitamins (particularly pantothenic acid). But new queens are not raised while the original queen remains in the hive, healthy and fertile. If the queen produces eggs faster than honeycomb cells can be built, she receives less food from her attendants. Egg production then slows down. Conversely, if she is behind in her egg laying, she is fed more intensively.

In the six weeks or so of its life, a worker bee does not perform the same duties continuously. The age of a bee determines what work it can do; housekeeping tasks are performed by young bees, food-collecting trips are made by older ones. On a food-collecting trip, the bee gathers pollen, rich in protein, and nectar, a dilute sugar solution. Pollen is carried home in *pollen baskets* on the hind legs. Nectar is swallowed into the *honey crop*, a specialized part of the alimentary tract, where saliva partially digests the sugar of nectar. On arriving at the hive, the bee first passes a security check on the way in, then unloads its pollen into one cell and regurgitates its nectar into another. Other bees which happen by pack the pollen tight and start converting nectar into honey. They rapidly beat their wings close to a nectar-filled cell, a process which is continued until most of the water has evaporated. Every now and then a bee samples the product (probably more a matter of hunger than of professional pride in the work). And when the honey is just right (or when all the bees standing by have had their fill?) the cell is sealed up with wax. This is the principal food store for the winter. Pollen is unobtainable at that time and, being perishable, cannot be stored so readily.

Bees and other social insects possess remarkable powers of orientation and communication. On food-collecting trips, bees have been shown to navigate by the sun. They are able to relate the position of their hive with the direction of polarized light coming from the sun; hence they may steer a beeline course home from any compass point. On arrival in its hive, a scouting bee which has found a food-yielding field of flowers communicates with its fellow workers by means of an *abdominal dance,* a side-to-side wiggle of the hind portion of the bee's body. The violence of the dance gives information about the richness of the food source. Flight distance is indicated by the duration of the dance, and flight direction, by the specific body orientation the dancing bee assumes on the honeycomb surface.

In winter, bees cling together in compact masses. Animals in the center always work their way out; those near the surface work their way in. A clump of bees thereby withstands freezing, even when exposed to very low temperatures. Smoke calms bees, as is well known. The animals react to smoke by rushing to their food stores and gorging themselves with honey. They are too busy at that time to sting an intruder. This is probably an inherited adaptive response to fire. Smoke might indicate a burning tree, and it is of obvious advantage if the bees are well fed when they are forced to abandon their nest. Similarly adaptive is the expulsion of all drones from the colony at the approach of winter. Not contributing to the well-being of the population, males merely use up food, which is at a premium in the cold season. Reactions such as these might appear to be thought out. Yet bees probably do not "reason" at all.

Other insect societies. Polymorphic castes and functional division of labor are in evidence among other social insects also. Many species of ants and termites include, in addition to sterile wingless workers, sterile wingless *soldiers.* These are strong-jawed, heavily armored individuals which accompany work crews outside and keep order within the nest. Soldiers in many cases cannot feed themselves and are cared for by workers. Besides a winged fertile queen (Fig. 16.20) and one or several winged fertile males (kings), ant and termite societies may maintain structurally distinct lesser "royalty," probably developed by over-feeding larvae; not enough to produce queens, but more than enough to produce workers (Fig. 16.21).

16.20. The nest of a wasp colony.

Agricultural societies occur among both termites and ants. Populations of certain termite species make little garden plots of wood, excrement, and dead termites. There they plant and rear fungi for food. *Leaf-cutting ants,* similarly, prepare pieces of leaves on which fungi are grown. The fungi are systematically pruned and cared for by gardening crews. *Magnetic termites* of Australia are of interest because they appear to be sensitive to the magnetic field of the earth. They build their hive in such a way that its long axis points precisely north and south, its short axis, exactly east and west. The adaptive significance of such hive orientation is not known.

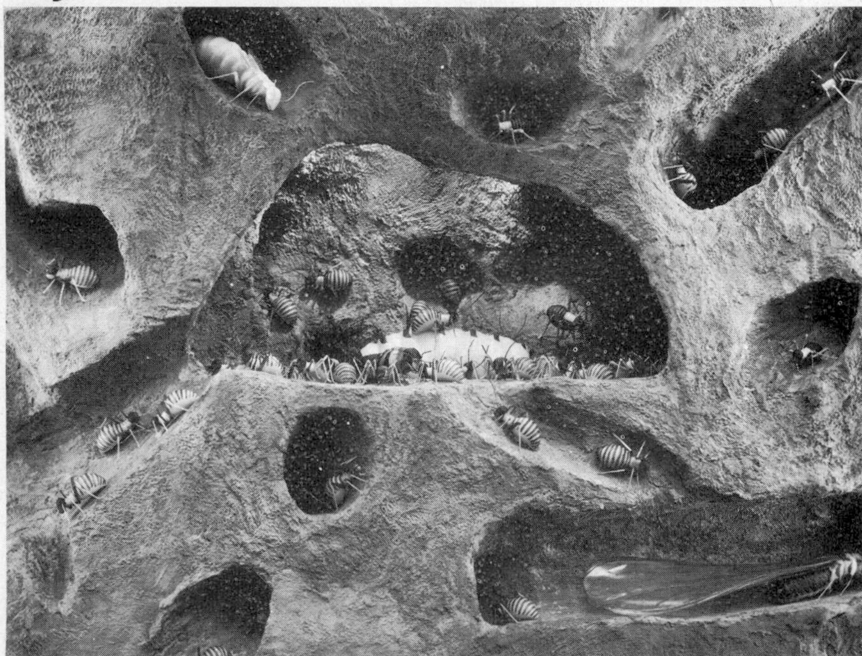

16.21. Social insects: termites. **A,** worker; **B,** soldier; **C,** winged king; **D,** portion of nest. In central chamber note queen, swollen with eggs and being cared for by workers. Winged king in lower right corner, larval queen in upper left corner.

Dairy ants exist which keep aphids, tiny green insects (plant lice), as food suppliers. The aphids secrete honeydew, a sugar- and protein-containing mixture, on which the ants depend. A common species of garden ant, for example, places "domesticated" plant lice on the roots of corn. The aphids feed there, and the ants thereafter milk these "ant cows" by gently stroking them. At the approach of winter, the aphids are carried into the ant nest and are put back on corn roots the following spring (cf. Fig. 17.10).

Slave-making ants exist which can neither build nests, feed themselves, nor care for their larvae. They form workerless soldier societies capable only of making raids on populations of other ant species. These victims are robbed of their pupae. The captive pupae mature, and the emerging slaves then care for their masters, performing all the functions they would have carried out in their own nest.

Tropical *army ants* (also called *driver* or *legionary* ants) march across country in raiding expeditions. They travel in columns, and larger-bodied "officer" ants march alongside. Everything living in path of such columns is devoured, even large animals, including man, if they should be unable to move away. The instinctive, unreasoned nature of insect behavior is shown particularly well in these ants. If a column of army ants is made to travel in a circle, so that the first animals of the column come to march right behind the last, then these ants will continue to circle endlessly. And unless they are diverted by an outside agency they may march themselves to death. Each ant evidently is so completely "disciplined" that it is incapable of thinking itself out of even a slightly changed situation (Fig. 16.22).

Certain desert ants, called *honeypot ants,* collect nectar from flowers and feed it to some of their fellow workers which are kept within the nest. These "living bottles" become greatly distended and serve as bacteria-free storage bins; during the dry season they dispense drops of honey to their thirsty mates.

Among insect societies generally, the fixed specialization of each individual constitutes a potential long-range disadvantage. Death of a queen bee and the destruction of

16.22. A. A marching column of army ants. If such a column is made to travel in a circle, as in the photo, then these ants will circle endlessly. Unless they are diverted by an outside force, they may march themselves to death. Each ant evidently is governed by inherited instinct so completely that it is capable only of following the ant before it and is unable to think itself out of an even slightly altered situation. **B.** Nests of Australian magnetic ants. The edges of the anthills are oriented in precise compass directions.

A

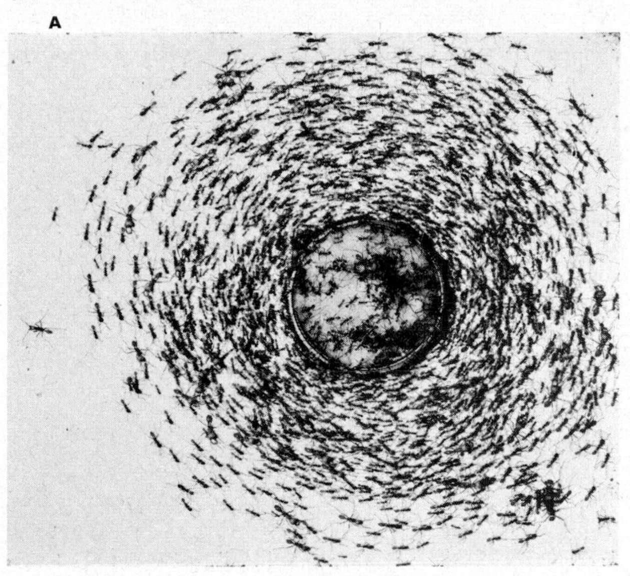

B

honeycomb cells which contain larvae still young enough to be reared into queens usually spell the end of a bee population; new workers are not produced, and old ones die out. Local eradication of the fungus on which agricultural termites depend and the destruction of their gardens spell the end of the termite population, its members not being equipped to grow any other food. The victims of slave-making ants could better preserve their population if all the workers could be mobilized into defending soldiers at the moment of attack.

Among insects, destructive social crises of this sort are offset by the establishment of numerous populations and by enormously high reproduction rates. The safety of the population lies in the number of its individuals. We recognize, however, that it would be immediately advantageous if, in addition to safety through numbers, the society were organized more flexibly; if each member could perform the functions of every other member and if the population as a whole could learn to adopt new ways of life in the face of changed environmental conditions. Flexible social organization is actually in evidence to greater or less degree among vertebrate populations.

Social Populations: Vertebrates

In contrast to insects and apart from the sexual dimorphism of males and females, the members of vertebrate populations are structurally more or less alike, at least during the early stages of adult life. If specialization does become pronounced later, it is predominantly functional and behavioral and is based on variations in physical strength, developed skills, mental acuity, and in man also on social tradition. Group behavior may be minimal or well developed. In either case the main determinant of behavior is inherited instinct, tempered here and there with a more or less thin veneer of *learning*. Learning goes hand in hand with *training*, and both are made possible largely by *family* groupings. Populations substructured into families are particularly characteristic of vertebrates.

Group behavior is quite distinct in populations such as schools of fish, packs of wolves, flocks of birds, or herds of deer. In most of these, functional specialization is not particularly pronounced. In travel, the individual which happens to be in the lead position, usually a male, guides the population temporarily. Other males, often stationed along the outskirts of the group, may take the lead in frequent rotation. The advantages of such associations are largely protective. Many eyes see more than two; a closely huddled herd stays warm; a group is more effective in attack and in defense. Family life within such populations may or may not be evident. There is hardly any in schools of fish. But a duck or a doe trains its young.

Closely associated with family life specifically and reproductive behavior generally is the phenomenon of *territoriality*, i.e., the tendency of each family or mating unit to maintain its own physical space within the population territory. For example, in a herd of seals resting on an island, males take up stations at more or less regular intervals, and each male gathers his family around him. The individual patriarch jealously guards his territory, driving off bachelor males and keeping a sharp eye on his females (Fig. 16.23). A strongly developed territorial sense is in evidence also among birds, where the males of a population stake out the borders of their domains by flying around them. Song birds then protect their claim vocally and they later also use song to attract females into their domain. A domain is usually just large enough to provide sufficient nesting materials and adequate food for the prospective family. Not only terrestrial but also numerous aquatic animals display territory-claiming behavior. Indeed, such behavior is more or less characteristic of all animals, vertebrate as well as invertebrate, for all have distinct territorial requirements for reproduction and for survival generally. The result of territoriality may be reduced intraspecific competition within a population (and often also interspecific competition with members of other species claiming the same territory).

Group behavior becomes particularly pronounced during seasonal *migrations* of populations. Not all species are migratory. Those that are undertake migrations in search of richer or safer pastures, in response to seasonal changes in climate, or to reach geographically fixed breeding grounds. Eels, seals, salmon, and many types of birds are among familiar migrants. When not migrating, solitary individuals or families of these animals may be dispersed widely over a considerable territory. At specific times, as if on cue, individuals draw together from far and near to a common jumping-off point, and then they travel to their destination together.

How do these animals know where to gather before the journey? How do they time their arrival there, often exactly to the day? And what leads them unerringly to their destination, thousands of miles away in many cases? The navigation problem is sufficiently puzzling among types which make the same trip every year, like seals and birds. If nothing else, a remarkable memory for landmarks, prevailing winds, or ocean

16.23. Family organization and territoriality. The photo shows a family of elks, consisting of a bull, a cow (right foreground), and their calves. Family groups such as this occupy and control a particular territory and the animals tend to ward off any fellow members of the species, or other potential competitors, which might invade this territory accidentally or deliberately.

currents may be indicated. The problem becomes even more puzzling, however, when none of the migrating animals has ever been at its destination before. This is the case among eels.

The spawning grounds of both European and American eels are situated in the deep waters of the Sargasso Sea, southeast of Bermuda and northeast of Puerto Rico. The eggs hatch there, and the near-microscopic larvae, or *elvers*, then travel toward the coasts; larvae of American eels turn west, those of the European eel turn east. The spawning beds of the American type lie farther west. Differences in the direction of ocean currents probably contribute to the initial separation of the two species.

Elvers of the American species travel for about a year before they reach continental waters—and maturation of the larvae requires just 1 year. The voyage of European elvers lasts 3 years— and their maturation requires precisely 3 years.

In coastal estuaries the elvers change into adults. The glassy transparency of the larval body changes to an opaque brown-gray, and the fishlike larval shape changes to the characteristic elongated form of the adult (Fig. 16.24). Adult males remain in estuaries. Females ascend rivers and settle in headwaters and in lakes. Some 7 to 15 years now pass. Then the females migrate back to the estuaries, rejoin the males, and all head out into the Atlantic. Reproductive organs mature during this migration, and upon arrival in the Sargasso the females spawn and the males fertilize the eggs. The adults then die.

How do the adults find their breeding grounds? It is hardly conceivable that they memorized the route in reverse when they made the trip as immature larvae, a decade earlier. And how do the larvae find coastal waters from which to ascend rivers? In line with the hypothesis of continental drift (cf. above), it might be postulated that the Atlantic spawning grounds corresponded to the location of ancient Holarctic river systems. If so, however, then it is still not clear how a "geographic memory" could survive countless generations of fishes as well as geographic changes themselves stretching over some 70 million years. Nor is it clear how, in eels and migrating populations generally, migration behavior is regulated in terms of neural, sensory, or other internal control mechanisms. On the other hand, it is rather obvious that banding together during travel has great adaptive value.

In migrating as well as nonmigrating social species, the behavioral structure of a population is influenced greatly by the mating structure. In many populations mating occurs promiscuously, as in numerous rodent and bird groups, and a distinct familial organization is then usually absent. In other types of social populations family units are distinguishable, such units being polygamous in some instances, monogamous in others. For example, herds of seals, deer, and flocks of many types of birds consist of polygamous families, whereas packs of wolves and flocks of other types of birds are made up of monogamous units. Families are not always associated as herds or herdlike groups but may also live as solitary units. For example, chickens form solitary polygamous family units and sticklebacks, parrots, bears, and wolverines form solitary monogamous units. In some cases, monogamy persists for a single breeding season only, in others it lasts for life. Bears and wolverines are among animals which usually mate for life.

16.24. A mature European freshwater eel (*Anguilla anguilla*). Such an adult will eventually migrate to sea and spawn.

A solitary polygamous family, like that of chickens, for example, is generally made up of a single dominant male, a series of females, their young, and sometimes a few unrelated young bachelor males. The rule of the dominant male is frequently challenged by the bachelors. If one of these succeeds in defeating his opponent in battle, the loyalty of the females is transferred to the winner. In this way the group is assured of continuously fit, healthy leadership.

An interesting social organization exists among the females of a polygamous family. In a flock of chickens, for example, hens are ranked according to a definite *peck order*. A given hen may peck without danger all hens below her in social rank but may be pecked in turn by all hens above her in the scale. If a new hen is introduced into the flock, she undertakes or is made to undertake a pecking contest with each fellow hen. Winning here and losing there, she soon finds her level in the society. A high ranking carries with it certain advantages, such as getting first to the food trough and obtaining a position of prestige on the perch. Very-high-ranking birds often are so aggressive that they persistently reject the attentions of the rooster. More submissive hens then produce most of the offspring. Social rankings of a similar nature are found also among female elephants as well as in most other polygamous families.

The success of vertebrate societies as a whole lies primarily in the functional versatility of the individual. In the insect society, as we have noted, reproduction of the majority is suppressed, and reproduction of the minority serves not only toward the new formation of individuals, but also toward the new formation of the whole society. Thus, among insects, the fate of the society hinges on the fate of a single female, and her genes alone provide continuity from one social generation to the next. By contrast, virtually all members of a vertebrate society are reproducers. Social continuity consequently is the responsibility of many, and reproduction of any one individual is less vital for the propagation of the society.

The phenomenon of "society" as a whole appears to be bound up with advanced evolutionary status; both insects and vertebrates are elaborately evolved groups. And although societies have evolved independently in these two groups and have different detailed organization, remarkably similar patterns of social behavior are in evidence nevertheless. Ants and man are unique among animals in making deliberate war, in practicing slavery, in pursuing agriculture, and in domesticating other organisms.

Social populations of all kinds, like populations generally, are components of species. At the same time, populations of social and nonsocial animals are also units within communities, associations which will occupy our attention next.

REVIEW QUESTIONS

1. What are individual variations? Distinguish between inheritable and noninheritable variations and give examples of each. Why are noninheritable variations without direct importance in species evolution? What are adaptive variations?
2. Define ecology, ecosystem, biosphere, species, polymorphism, dimorphism, and give an example of each. What characteristics does a species possess as an evolutionary unit? As a reproductive unit?
3. Distinguish between interspecific fertility and intraspecific infertility. What is a syngen? Is the usual definition of "species" universally applicable? What is an ecological niche? Give examples. Distinguish between intraspecific cooperation and intraspecific competition and again give examples.
4. Through what evolutionary processes does a new species originate? Describe the typical life cycle of a species. What are endemic and cosmopolitan species? What effect does interspecific competition have on species survival?
5. Define and give specific examples of sweepstake bridges, filter bridges, and corridor bridges. What are the zoological characteristics of small oceanic islands and how can such characteristics be accounted for? Describe the geographic and zoological characteristics of the six zoogeographic zones.
6. How can widely discontinuous distributions of given animal types be explained? Describe the hypothesis of continental drift and show how it may contribute to explaining animal distribution patterns.
7. What is a population and what is its relation to a species? What are the structural characteristics of a population with reference to dispersion and growth pattern? What are some of the physical and biological limiting factors of population growth? How does a population interact with its physical environment? Its biological environment?
8. Distinguish between interspecific and intraspecific interactions of populations. What is a social population? Review the structural and functional organization of honeybee societies.
9. What forms of polymorphism are encountered in honeybee and other insect societies? Describe the characteristic structure of vertebrate societies and contrast it with that of insects. What are the adaptive advantages and disadvantages of each? What is meant by territoriality of animals? What is the adaptive advantage of this behavior trait?
10. What is the significance of animal migrations? Which animal groups are migratory? How does the mating structure of different vertebrate populations vary? What is a peck order and what is its significance?

COLLATERAL READINGS

The following books and articles include discussions of species, populations, and ecological topics generally (see also listing at end of Chap. 17):

Allee, W. C.: "Animal Aggregations," The University of Chicago Press, Chicago, 1931.

Cole, L. C.: The Ecosphere, *Sci. American,* Apr., 1958.

Dodson, E. O.: "Evolution: Process and Product," Reinhold, New York, 1960.

Elton, C.: "Animal Ecology," Macmillan, New York, 1937.

Hesse, R., W. C. Allee, and K. P. Schmidt: "Ecological Animal Geography," 2d ed., Wiley, New York, 1951.

Odum, E. P.: "Ecology," Modern Biology Series, Holt, New York, 1963.

Ecological effects of geography and the physical environment generally are treated in the following articles:

Brown, F. A.: Biological Clocks and the Fiddler Crab, *Sci. American,* Apr., 1954.

Johnson, C. G.: The Aerial Migration of Insects, *Sci. American*, Dec., 1963.
Hasler, A. D., and J. A. Larson: The Homing Salmon, *Sci. American*, Aug., 1955.
Howells, W. W.: The Distribution of Man, *Sci. American*, Sept., 1960.
Kay, M.: The Origin of Continents, *Sci. American*, Sept., 1955.
Lack, D.: Darwin's Finches, *Sci. American*, Apr., 1953.
Sauer, E. G. F.: Celestial Navigation by Birds, *Sci. American*, Aug., 1958.
Wilson, J. T.: Continental Drift, *Sci. American*, Apr., 1963.

Among many sources on social behavior and societies, the following popular readings may be consulted:

Eibl-Eibesfeldt, I.: The Fighting Behavior of Animals, *Sci. American*, Dec., 1961.
Guhl, A. M.: The Social Order of Chickens, *Sci. American*, Feb., 1956.
Hess, E. H.: "Imprinting" in Animals, *Sci. American*, Mar., 1958.
Imms, A. D.: "Social Behavior in Insects," Methuen, London, 1947.
Krough, A.: The Language of the Bees, *Sci. American*, Aug., 1948.
Lorenz, K. Z.: The Evolution of Behavior, *Sci. American*, Dec., 1958.
Schneirla, T. C., and G. Piel: The Army Ant, *Sci. American*, June, 1948.
Shaw, E.: The Schooling of Fishes, *Sci. American*, June, 1962.
Thorpe, W. H.: The Language of Birds, *Sci. American*, Oct., 1956.
———: "Learning and Instinct in Animals," 2d ed., Methuen, London, 1963.
Tinbergen, N.: The Curious Behavior of the Stickleback, *Sci. American*, Dec., 1952.
———: "Social Behavior in Animals," Wiley, New York, 1953.
———: The Evolution of Behavior in Gulls, *Sci. American*, Dec., 1960.
Von Frisch, K.: "Bees, Their Vision, Chemical Senses, and Language," Cornell, Ithaca, N.Y., 1950.
———: Dialects in the Language of Bees, *Sci. American*, Aug., 1962.
Washburn, S. L., and I. DeVore: The Social Life of Baboons, *Sci. American*, June, 1961.
Wilson, E. O.: Pheromones, *Sci. American*, May, 1963.

CHAPTER 17 community and symbiosis

A community, technically often referred to as a *biota* or a *biotic* community, is a localized association of several populations of *different* species. Almost always, a community contains representatives of Monera, Protista, Metaphyta, and Metazoa, all being required for group survival. Moreover, the populations of a community consist only partly of free-living organisms; also present are *parasites* and other types which live together in so-called *symbiotic* associations.

Whole communities together with the physical environments in which they live represent *ecosystems,* the basic units of the biosphere (Fig. 17.1). Examples are a pond, a forest, a meadow, a section of ocean shore, a portion of the open water of the sea, a coral reef, or a village with its soil, grasses, trees, people, bacteria, cats, dogs, and other living and nonliving contents. We shall examine the properties of ecosystems generally in the first part of this chapter and those of the symbiotic components specifically in the second.

THE ECOSYSTEM

Structure and Growth

The living portion of an ecosystem, i.e., a community, exhibits a characteristic species structure: a few species are represented by large populations and many species are represented by small populations. For example, large populations of just two or three kinds of monkeys may form the bulk of a jungle community. The remainder may consist of small populations of numerous kinds of birds, bats, snakes, and other animals. Analogously, several dozen species of trees may be present, but populations of just two or three may constitute some 70 to 80 per cent of the total number of trees. In general, species diversity in an ecosystem is inversely proportional to the sizes of organisms. Thus, a forest is likely to contain more species of insects than of birds and more species of birds than of large mammals. The reasons for such correlations will become apparent presently.

Species diversity probably contributes to the stability of a community; the presence of many different kinds of organisms

17.1. When the territorial ranges of several species overlap, populations of these different species may coexist within a given limited area as a community. The organisms in such a community plus the territory they occupy represent an ecosystem.

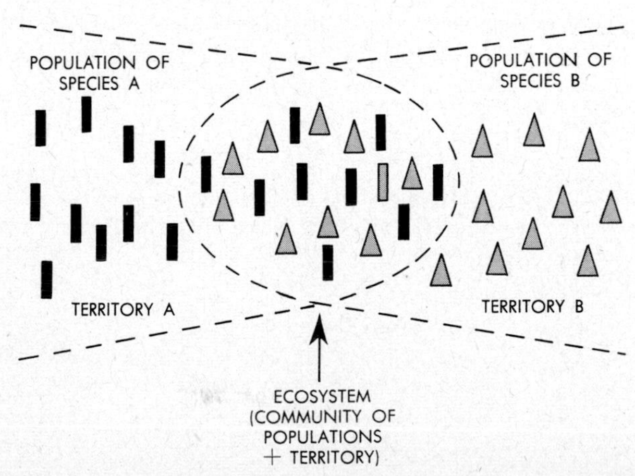

may provide a reservoir of adaptive types able to withstand many changes in the physical environment. Ecosystems are actually highly stable. In some cases, their stability has been estimated to be up to five times greater than that of the physical parts of the systems alone. Yet even such a high degree of stability is only relative for, like other living entities, communities grow, develop, pass through mature phases, reproduce, and ultimately die. To be sure, the time scale is in hundreds and thousands of years.

Such life cycles result from an interplay between the living and the nonliving components of an ecosystem. Being specialized to occupy particular ecological niches, different species must live in different types of environments. Consequently, the physical characteristics of a given region determine what types of organisms can settle there originally. Temperature, winds, amount of rainfall, the chemical composition of the surroundings, latitude and altitude, soil conditions, and other similar factors decisively influence what kinds of plants will be able to survive in a given locale. Vegetation in turn, as well as the physical character of the locale, has a selective effect on the types of animals that may successfully settle in the region.

By its very presence, however, a particular set of organisms gradually alters local conditions. Raw materials are withdrawn from the environment in large quantities and metabolic wastes are returned. To the extent that these wastes differ from the original raw materials, the environment becomes altered. Moreover, the components of dead organisms also return to the environment, but not necessarily in the same place or necessarily in the same form in which they were obtained. In time, therefore, communities bring about profound redistributions and alterations of vast quantities of the earth's substance.

This means that later generations of the original community may find the changed local environment no longer suitable. The populations of the community must then resettle elsewhere or readapt or die out, and the result is the gradual development of population gaps within species, as already noted in the last chapter. A new community of different plants and animals thus may come to occupy the original territory, and as this community now alters the area according to its own specializations, type replacement, or *ecological succession*, may eventually follow once more. We note how closely the nonliving and the living components of an ecosystem are interlinked; change in one produces change in the other, and the passage of time alone is bound to initiate change.

Continued ecological succession ultimately leads to the establishment of a *climax community:* a set of populations which alters the local environment in such a way that the original conditions are repeatedly re-created more or less exactly. The North American prairie and forest belts are good examples of climax communities; so are the communities in large lakes and in the ocean. Climax communities represent ecological steady states, and they are perpetuated within a territory as long as local physical conditions are not altered drastically by climatic or geologic upheavals. If that happens, communal death usually follows. Development of new communities by immigration or major evolutionary adjustment of the remnants of the old community may then occur.

In an ecological succession culminating in a climax community, each developmental stage is called a *sere* (Fig. 17.2). A sequence of seres is characterized not only by changes in the sets of populations present, but also by a progressive increase in the diversity of species and the total quantity of the living mass. The sequence of seres for a given region is often fully predictable, both with respect to the general types of populations expected at each sere and to seral durations. On land, for example, a climax stage is often represented by a forest community. If the original physical environment is sand or equivalent virgin territory, then the succession of seres generally follows the pattern: soil-forming organisms (bacteria, lichens, mosses) ⟶ annual grasses ⟶ perennial grasses ⟶ shrubs ⟶ trees. Characteristic animal populations

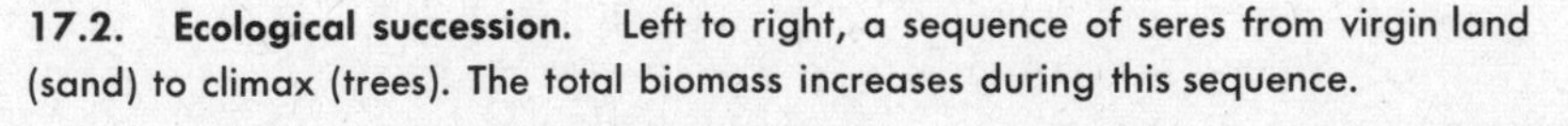

17.2. Ecological succession. Left to right, a sequence of seres from virgin land (sand) to climax (trees). The total biomass increases during this sequence.

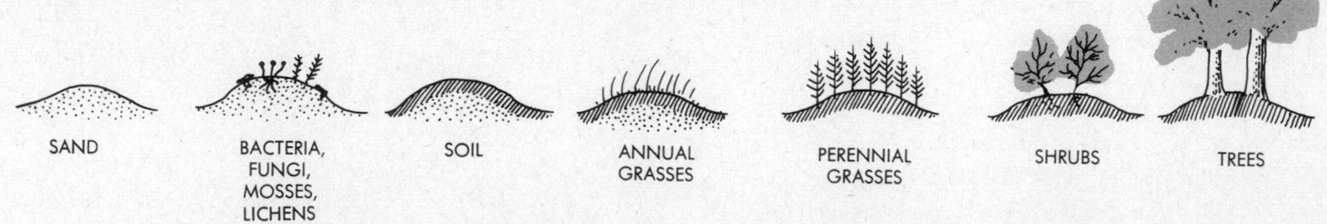

are associated with each of such seral plant populations. It can be shown that, from sand to forest climax, in the order of 1,000 years may have to elapse. If denuded but previously cultivated land is the starting condition, i.e., if soil is already present, then a grass climax may be attained in about 50 years, a forest climax is about 200 years. We may note here also that successive generations of the same plant population gradually exhaust the soil in specific ways. Moreover, the dead bodies of the plants add materials to the soil which often prove to be toxic to subsequent generations of the same plants. Thus, both specific exhaustion of soil and specific additions to it promote seral succession, not only among the plants but also among the animal populations dependent on them.

Clearly, *turnover* occurs on the level of the ecosystem just as it does on all other levels of living organization. Population flux continues even after a climax has been attained, e.g., given populations emigrate or die out and the ecological niches so emptied come to be reoccupied by other similar populations. The important point is that, after a climax has been reached, such population flux is automatically self-adjusting and the community remains *balanced* internally with respect to types of populations present. Furthermore, the community also exhibits a numerical steady state: in all populations present, the numbers of individuals remain relatively constant on a long-term basis. In a large, permanent pond, for example, the numbers of algae, frogs, minnows and other organisms stay more or less the same from decade to decade. Annual fluctuations are common, but over longer periods of time constancies of numbers are characteristic in most natural communities.

Three main factors create and control these striking numerical balances: *food, reproduction,* and *protection*. They are the principal links which make the populations of a community interdependent.

Links and Balances

From the standpoint of its nutritional structure, a stable ecosystem generally consists of four parts (Fig. 17.3). The *abiotics* are the nonliving physical components of the environment on which the living community, or *biomass,* ultimately depends. The *autotrophs* are the photosynthetic organisms, the *producers* which subsist entirely on the abiotic portion of the ecosystem. The *heterotrophs* are the *consumers,* largely animals. Herbivorous consumers subsist on the producers, and carnivorous consumers consume one another as well as the herbivores. Lastly, the *saprotrophs* are the *decomposers,* i.e., bacteria and fungi. These subsist on the excretion products and the dead bodies of all producers and consumers, and thus they bring about decay and the return of raw materials to the abiotic part of the ecosystem. Evidently, the physical substance of an ecosystem *circulates,* from the abiotic part through the biomass back to the abiotic part. The circulation is maintained by the continuous influx of energy from the sun.

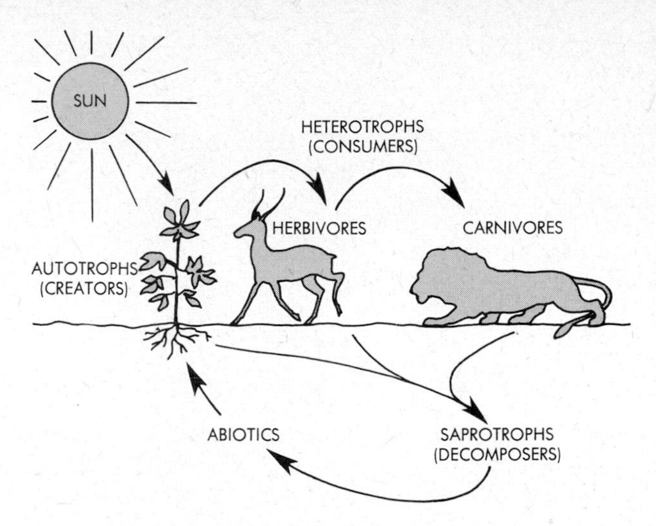

17.3. The nutritional structure of an ecosystem. Abiotics represent the physical, nonbiological components of the system. Arrows indicate flow of energy and materials.

However, energy dissipates in the course of such cycles, and only a small fraction of a given amount of solar energy is recoverable in the form of new living plant matter. Similarly, a pound of plant food cannot make a pound of new living animal matter; for much of what a plant consists of, cellulose, for example, cannot be digested or used otherwise by animals. Moreover, even the usable portions of plants yield energy which dissipates as heat and thus cannot become useful in animal metabolism. Therefore, as raw materials are transferred from soil to plants and from plants to animals, these transfers are not 100 per cent efficient. More than a pound of soil is needed to make a pound of plant matter, and more than a pound of plant matter is needed to make a pound of animal matter. Similarly, more than 300 lb of antelope meat or even lion meat is required to produce a 300-lb lion.

This inescapable condition leads to the establishment of *food pyramids* in the community (Fig. 17.4). So many tons of soil can support only so many *fewer* tons of grass. Grass in turn supports herbivores which together weigh less than the grass. And only a relatively small weight of carnivores can

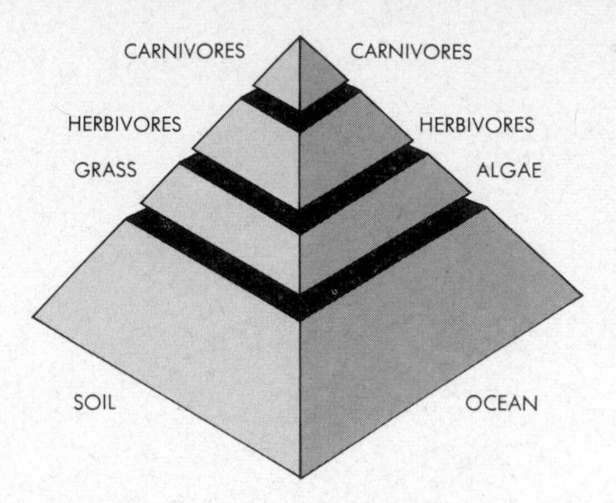

17.4. The general pattern of food pyramids. Soil and ocean support plant life; herbivorous animals subsist on the plants; and carnivorous animals subsist on the herbivores.

find sustenance in such a community. Several acres of ground thus might just suffice to support a 150-lb man. Such a pyramid of total weights and relative productivity also delineates a pyramid of individual numbers and individual sizes, for prey is generally smaller than predator; hence the balanced community may contain millions of individual grasses, but only one man.

Pyramids of this sort are one of the most potent factors in balancing communal populations; significant variations of numbers at any level of a pyramid entail automatic adjustments at every other level. For example, overgrowth of land by plants soon results in nutritional depletion of soil, since more raw materials are withdrawn from the soil. This depletion eventually leads to a starvation of the plants and decimation of their numbers. But the bodies of the dead plants now enrich the soil again, and the fewer plants which still live make less total demand on the raw materials once more present in soil. These living plants therefore can become well nourished. Hence they may reproduce relatively rapidly, and this circumstance increases their numbers again. The cycle then is repeated (Fig. 17.5).

Analogous cycles probably occur among animals. For example, overpopulation of carnivores might result in the depletion of herbivores, since a greater number of herbivores is eaten. This depletion might lead to starvation of carnivores, hence to a reduction of their numbers. Underpopulation of carnivores then could result in overpopulation of herbivores, since fewer herbivores are eaten. But the fewer carnivores could be well fed. They might therefore reproduce relatively rapidly, and this would increase their numbers again (cf. Fig. 17.5). As a general result, although the numbers of all kinds

17.5. Left, population balance in plants. A large population on land reduces the soil nutrients available (top). This eventually leads to starvation of plants and reduction of their numbers (lower right). But the dead plants now enrich the soil (lower left), which again permits an increase in the numbers of living plants. **Right, population balance in animals.** A large carnivore population reduces the herbivore population by predation (top). This eventually decreases the food supply of carnivores and leads to decrease of their population (bottom right). This in turn then permits the herbivore population to flourish again (bottom left), which also permits an increase in the numbers of carnivores. Through continued repetitions of such cycles, the sizes of populations, both plant and animal, are maintained fairly constant over the long term.

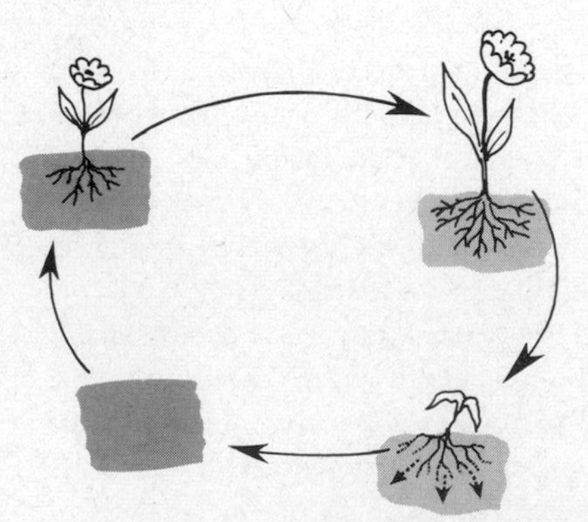

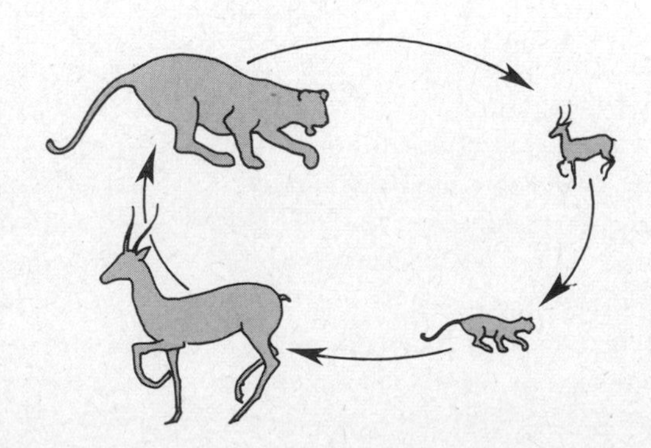

of organisms would undergo short-term fluctuations, the total quantities could remain relatively constant over the long term.

The territory of an ecosystem usually supports more than one food pyramid. Each of these is characterized by a different *food chain,* culminating in a different carnivore. For example, a lion would find it extremely expensive in terms of locomotor energy to live on insects, worms, or even lizards and mice. Bigger prey, like antelope or zebra, is obviously more appropriate. On the other hand, insects and worms are suitable food for small birds; and small birds, lizards, and mice provide adequate diet for larger predator birds. In this example, two food pyramids are based on the same plot of land. The pattern is generally much more complex. Different types of plants in one territory may sustain many different herbivores. These may form the basis of different, intricately interlocking animal food chains. As in the case of elephants, a herbivore may itself represent the peak of a food pyramid (Fig. 17.6).

It may be noted also that, in a balanced ecosystem, the total biomass yields just enough dead matter and other raw materials to replenish the soil or the ocean. This permits the continued existence of the various food pyramids above ground or in water. In such delicate nutritional interdependencies, minor fluctuations are rebalanced fairly rapidly. But serious interference, by disease, by man, or by physical factors, is likely to topple the whole pyramid. If that happens, the entire community may cease to exist.

A second link between the members of a community is reproductive interdependence. A familiar and most important example of this is the pollinating activity of insects. In some well-known cases of remarkable specialization, a given insect visits only one or a few specific flower types for pollen and nectar. The flowers (e.g., snapdragons) in turn are structurally adapted to facilitate entry of the insect. Such intimate reciprocity testifies to a closely correlated evolutionary development of animal and plant. It is fairly obvious how such interdependence contributes to population balance: reduction of the insect population entails reproductive restriction of the plant, and vice versa. Similarly significant in balancing the reproductive growth of plant populations is the seed-dispersing activity of birds and mammals, man in particular.

Other examples of reproductive dependence are many. Birds such as cuckoos lay eggs in nests of other birds. Insects such as gall wasps embed their eggs deep in the tissues of particular plants, where the hatching larvae find food and protection. Other insects deposit eggs on or under the skin of various animals. Certain wasps, for example, kill tarantulas and lay their eggs in them (cf. Fig. 17.18).

17.6. Several different food chains, each culminating in a different animal, may be supported by a single plot of ground.

Reproductive growth and geographic expansion of a community are intimately correlated with nutritional balances. In new territory, a pioneer association of populations will first form a small food pyramid, occupying perhaps only part of the available territory. The pyramid may still be too "low" to support any big herbivores or carnivores. Abundance of food and absence of competition promote high reproduction rates and a rapid increase of numbers at all levels of the pyramid. The base of the pyramid therefore widens, and a larger area of the territory will be occupied. Sizable herbivores and even a few larger carnivores may gradually be assimilated into the community. As a result, the rate of predation will increase, which in turn will slowly decrease the net reproductive population gain. A turning point will be reached eventually. Prior to it the community grows at an increasing rate; after it the community still grows, but at a decreasing rate. Net expansion finally comes to a standstill, and from then on the pyramid retains relatively stable proportions (Fig. 17.7).

In such a growth pattern, it is assumed that territorial and numerical expansion can follow its inherent trend without external restriction. Yet geographic and biological barriers often delimit an area. A small forest may be surrounded by water or by land on which trees cannot grow; a meadow may be ringed in by forest, or a valley by high mountains. In such cases, growth of a community is stopped before its inherent potential is fully expressed. The food pyramid on such a limited territory may never become high enough to support large herbivorous or carnivorous members. One searches in vain for stag in a tiny forest, for large fish in a small pond. But the one is likely to abound in worms, mice, and small birds; the other in algae, protozoa, and frogs.

In a community incapable of further expansion, steady reproduction may produce a centrifugal *population pressure.* This condition may be relieved by emigration of the overflow population. If emigration is not possible or if it is not sufficiently effective, numbers will be decimated by starvation or even sooner by epidemic diseases. The latter spread rapidly through an overpopulated, undernourished, spatially delimited community. Even if disease affects only one of the component populations, the whole communal web is likely to be disrupted.

The third main link among the populations of a community is protective interdependence. Plants in forest and grassland usually protect animals by providing shelter against enemies and adverse weather. If the opportunities for protection are reduced, both the animal and the plant population may suffer. For example, if an overpopulation of insects makes available plant shelter inadequate, the insects will become easier prey for birds and bats. But this circumstance may also decimate the plant populations, for their major pollinating agents may no longer be sufficiently effective.

Given animals in many cases are protected from other animals by *camouflage.* Such a protective device may involve body color or body shape or both (cf. Fig. 3.11). Probably the most remarkable instance of color camouflage is the phenomenon of *mimicry,* widespread particularly among butterflies and moths. In certain of these animals, pigmentation patterns

17.7. Population growth and geographic expansion. As the number of individuals increases, more territory will be occupied and the food pyramid will become wider and higher (left part of diagram). Rates of population increase are indicated in the curve at right; increasing rates are in evidence prior to a turning point (arrow), decreasing rates thereafter.

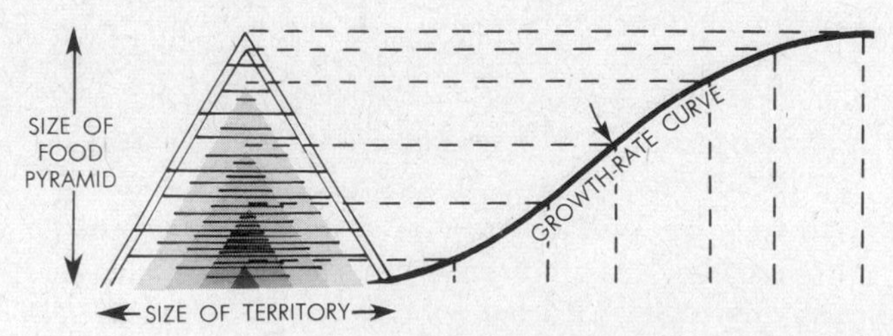

exist which are virtually indistinguishable from those of other, unrelated species. Usually those species are mimicked which are strong or fast and have few natural enemies. The advantage is that an animal resembling even superficially another more powerful one will be protected too, by scaring off potential predators. Insects also display a variety of structural camouflages. For example, the individuals of certain species possess the detailed shape of leaves, of branches, or of thorns. This serves not only defensively but also as a disguise against potential victims (Fig. 17.8).

Other protective devices vary widely in type. Various birds and some mammals mimic the song and voice of other species, either defensively or as an aggressive lure. The hermit crab protects its soft abdomen in an empty snail shell of appropriate size. Schools of small pilot fish scout ahead of large sharks, leading their protectors to likely prey. Significant protection is also afforded by man, through domestication, game laws, parks, and sanctuaries.

These various examples illustrate how the member populations of a community are specialized nutritionally, reproductively, and protectively. Carnivorous populations cannot sustain themselves on plant food and not even on every kind of animal food. Herbivorous populations require plants and are incapable of hunting for animals. The populations of green plants depend on soil or ocean, and the populations of saprotrophs cannot do without dead organisms. These are profound specializations in structure and function, and they imply loss of individual self-sufficiency as well as a need for cooperative interdependence. Communal associations of populations evidently are a necessity. They are but extensions, on a higher biological level, of the necessarily cooperative aggregation of cells into tissues, organs, and organ systems.

Indeed, the development of "community" appears to be as integral a part of organic evolution as the development of individual organisms. Events were probably not such that a particular organism first evolved structurally and functionally in a certain way and then happened to find the right community into which it could fit. Rather, the community probably existed from the very beginning, and all its member populations evolved together; the community itself evolved. The histories of the bumblebee and the snapdragon are linked as intimately as the histories of every man's hand and foot.

A community within a given territory includes not only free-living organisms in loose cooperative association but also organisms which live together in more or less permanent *physical* contact. Two individuals of different species may be joined so intimately that one lives right *within* the other. All

17.8. Animal disguise. Great resemblance to the stems and branches of a bush is exhibited by the body shapes of stick insects (**A**) and South African mantises (**B**). The forms and colors of foliage are in evidence in the bodies of dead-leaf butterflies (**C**, top of photo).

such instances of physically intimate living together of members of different species are instances of *symbiosis*, a special form of communal life.

SYMBIOSIS

The Pattern

A free association in which an animal habitually shelters under a plant might, in a relatively simple evolutionary step, become an association in which the animal and the plant have entered a more permanent protective union. A plant which depends on some animal for seed dispersal might advantageously live in, or on, the animal altogether, not only at the time of seed production but throughout life. A soil bacterium or a scavenging protozoon living on the undigested elimination products of larger forms might find a surer food supply if it could adapt to an existence right in the gut cavity of its supplier.

Among ancestral populations of free-living forms, ample opportunity existed for the development of such symbiotic relationships. These opportunities were exploited to the full, and many associations arose in which two organisms of different species came to live together in intimate, lasting physical contact. Today there is no major group of animals which does not include symbiotic species, and there is probably no individual animal which does not play *host* to at least one *symbiont*.

The phenomenon of symbiosis is expressed in two basic patterns. In *facultative* associations, two different animals "have the faculty" of entering a more or less intimate symbiotic relationship. But they need not necessarily do so, being able to survive as free-living forms. In *obligatory* associations, on the other hand, one animal *must* unite symbiotically with another animal or plant, usually a specific one, if it is to survive. The ancestors of obligatory symbionts have invariably been free-living animals which in the course of history have lost the power of living on their own. Before becoming obligatory symbionts, they formed facultative associations with organisms on which they came to depend more and more (Fig. 17.9).

Symbionts affect each other in different ways. Thus, *mutualism* describes a relationship in which both associated partners derive some benefit, often a vital one, from living together. *Commensalism* benefits one of the partners, and the other is neither helped nor harmed by the association. *Parasitism* is of advantage to the parasite but is detrimental to the host to greater or lesser extent. These categories intergrade imperceptibly, and in many boundary cases clearcut distinctions cannot be made.

Mutualism

An example of a loose mutualistic association is the tickbird-rhinoceros relationship. The tickbird feeds on skin parasites of the rhinoceros, and in return the latter is relieved of irritation and obtains warning of danger when the sharp-eyed bird flies off temporarily to the security of the nearest tree. Another

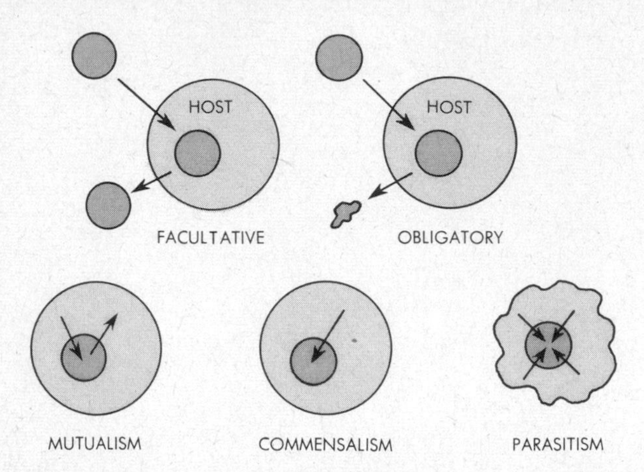

17.9. Forms of symbiosis. Symbionts are represented by small dark circles, hosts by larger, lighter ones. Top, a facultative symbiont may live apart from a host, but an obligatory symbiont dies under such conditions. Bottom, both symbiont and host benefit in mutualism, only the symbiont benefits in commensalism and parasitism. In parasitism, indeed, the effect of the symbiont is damaging to the host.

example is the relationship between dairy ants and aphids, already cited in Chap. 16 in the section on insect societies. The ant obtains food from the aphid, and the aphid in turn secures protection, food, and care from the ant. These two examples also illustrate the difference between facultative and obligatory mutualism. Both tickbird and rhinoceros can get along without each other if necessary; but the ant cannot do without its aphids and the aphid cannot do without its ants (Fig. 17.10).

A somewhat greater degree of physical intimacy is exhibited in the mutualistic symbiosis of sea anemones and hermit crabs. Sea anemones attach themselves to empty snail shells, and hermit crabs use these shells as protective housings. The sea anemone, an exceedingly slow mover by itself, is thus carried about on the shell of the hermit crab—an obvious advantage to the anemone in its search for food and in geographic dispersal. The hermit crab in turn benefits from the disguise. Moreover, since the anemone is not a dainty eater, scraps of food become available to the crab when the anemone catches prey. This is a facultative association; sea anemones and hermit crabs may and largely do live on their own.

An example of rather more intimate mutualism is provided by *lichens*, grayish and yellowish incrustations commonly found on rock surfaces and on tree bark (Fig. 17.11). These crusts are associations of photosynthesizing single-celled algae and saprotrophic threadlike fungi. The meshes of the fungal threads support the algae, and they also hold rain water like a sponge. The algae produce food for themselves and for the fungus. The fungus in turn contributes water, nitrogenous wastes, and respiratory carbon dioxide, substances which allow for continued photosynthesis and food production. Lichens may consequently survive in relatively dry terrestrial environments. The fungus may live alone in a water-sugar medium, and the alga may persist by itself in mineral-containing water. Separately, they are merely two types of organisms not particularly different from many others like them. But together they become a combination of considerable evolutionary importance. Lichens were among the first organisms capable of eking out a terrestrial existence. Contributing to the crumbling of rock and the formation of soil, they paved the way for a larger-scale colonization of the land.

The most intimate forms of mutualism involve organisms which live directly within other organisms. For example, the roots of certain vascular plants, particularly legumes like soybeans, clover, and peas, form important mutualistic associations with so-called *nitrogen-fixing bacteria*. The bacteria invade the roots of the hosts and the infected root cells respond

17.10. Mutualistic symbiosis: carpenter ant protecting a larva of a tree hopper insect. The larva benefits from the protection and in return secretes sugary honeydew, which is licked off by the ant and serves as its food.

B

17.11. Mutualism. **A.** Lichens, representing mutualistic associations of algae and fungi. **B.** Root nodules, representing mutualistic associations of nitrogen-fixing bacteria and roots of pea plants, as shown, or other legumes.

by increasing in size and number. The result is the development of *root nodules* (cf. Fig. 17.11). In them, the host provides nutrients for the bacteria and the bacteria in turn fix atmospheric nitrogen; i.e., they make this essential element chemically usable for both themselves and the host plant. As we shall see in the next chapter, this is a major source through which usable nitrogen becomes available to animals.

Analogously invasive forms of mutualism occur in associations between algae and various other organisms. For example, many free-living protozoa, (e.g., a species of *Paramecium*) and coelenterates (e.g., several species of *Hydra*) harbor green single-celled algae within their translucent bodies. Known as *zoochlorellae*, the algae supply food and oxygen, as in lichens, and in turn receive protection, water, and other materials essential for continued photosynthesis.

In the gut of termites live flagellate protozoa which secrete an enzyme capable of digesting the cellulose of wood. Termites chew and swallow wood, the intestinal flagellates then digest it, and both organisms share the resulting carbohydrates. Thus, to the detriment of man, termites may exploit unlimited food opportunities open to very few other animals. And the protozoa receive protection and are assured of a steady food supply.

Virtually every animal which possesses an alimentary canal houses billions of intestinal bacteria, particularly in the lower gut. These bacteria draw freely on materials not digested or not digestible by the host, and as a result of their activities, they initiate fecal decay (cf. Chap. 9). The host generally benefits from the auxiliary digestion carried out by the bacteria and in many instances is also dependent on certain of the bacterial byproducts. For example, man and other mammals obtain many vitamins in the form of "waste" materials released by the bacterial symbionts of the gut.

Parasitic associations may sometimes develop into mutualistic ones. In the course of successive generations, a relationship in which one partner originally lives at the expense of the other may change gradually into an association beneficial to both. Thus, mutualistic intestinal bacteria might have evolved from originally parasitic ancestors. As we shall see shortly, it is adaptively advantageous to a parasite not to jeopardize its own survival opportunity, and this factor often tends to change a parasitic relation into either a commensalistic or a mutualistic one in the course of evolution.

Commensalism

Just as the chance association of two free-living animals may develop into mutualism, so an analogous chance association may develop into commensalism. As far as can be demonstrated, the commensal neither harms nor helps its host, and the host appears neither to resist nor to foster the relationship in any way.

Commensalism is illustrated, for example, by a species of small tropical fish. Individuals of this species find shelter in the cloacas of sea cucumbers. The fish darts out for food and returns, to the utter indifference of the host. The so-called sharksucker, or *remora*, provides another example. This fish (Fig. 17.12) possesses a dorsal fin which is modified into a holdfast device. By means of it, the fish attaches to the underside of sharks and thereby secures scraps of food, wide geographic dispersal, and protection. The shark neither benefits nor suffers in any respect. In still another example, barnacles may attach to the skin of whales, an association which secures geographic distribution and wider feeding opportunities for the sessile crustaceans. In this instance, a trend toward parasitism is in evidence; in some cases the barnacles send rootlike processes into the whale, outgrowths which eat away bits of host tissue.

These and most other existing commensalistic unions tend to be facultative; for a symbiont is not likely to be allowed to impose on a host in intimate, obligatory fashion unless the host derives at least some benefits from such an imposition and therefore fosters the association, or unless the symbiont has overcome the host's defenses and is frankly parasitic. Consequently, although obligatory commensalistic associations may have evolved quite often, most of them have probably been unstable; they would soon have changed either into mutualism or into parasitism.

Parasitism

Parasitic ways of life. It has probably become apparent in the above that symbiosis revolves largely, though not exclusively, around the problem of food. We might suspect, therefore, that symbiosis in general and parasitism in particular would be most prevalent among organisms in which competition for food is most intense. This is actually the case. Parasitism flourishes primarily among organisms which must obtain food from others, i.e., in heterotrophs: in viruses, in bacteria, in fungi, and in animals.

All viruses are parasitic. Of the bacteria, those which are not photosynthetic or saprotrophic are parasitic. Among fungi, some are saprotrophic, the rest are parasitic. And in animals, many phyla and classes are wholly parasitic; virtually all others include important parasitic subgroups.

As we have seen in Chap. 14, parasitism is almost as old as life itself. So advantageous and economical is the parasitic mode of living that many parasites may be infested with

17.12. Commensalism. A. Shark with three remoras, or suckerfish, attached to underside. **B.** The dorsal sucker of a remora, a modified dorsal fin.

A

B

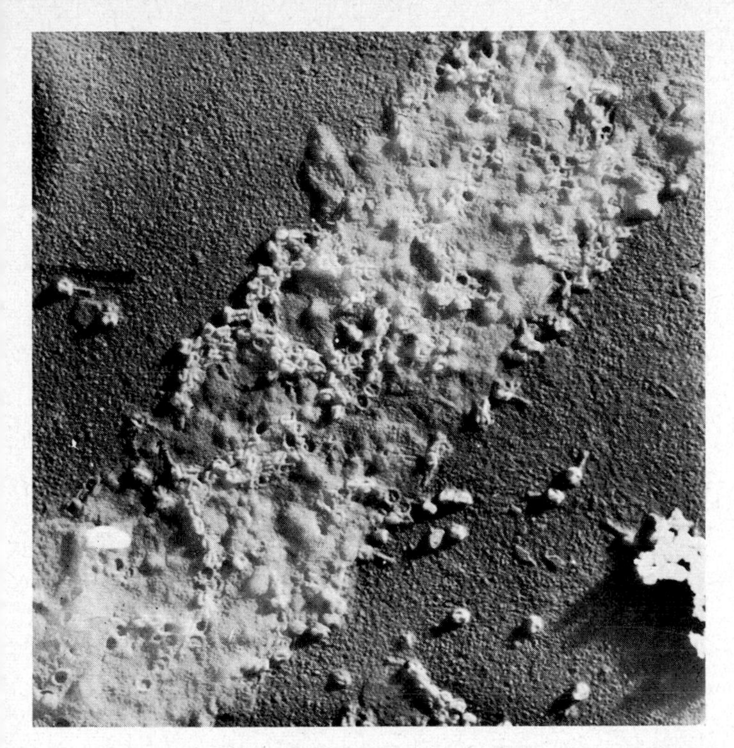

17.13. Electron micrograph of the remnants of a bacterium after attack by bacteriophage viruses. The virus parasites are the small rodlets with knobbed ends. Cf. Fig. 14.7.

smaller parasites of their own and these in turn may support still smaller ones. For example, a mammal may harbor parasitic worms; these may be invaded by parasitic bacteria; and the bacteria may be infected by *bacteriophages,* i.e., viruses which parasitize bacteria (Fig. 17.13). *Hyperparasitism* of this sort, i.e., one parasite inside another, is very common. It represents a natural exploitation of the very condition of parasitism. Inasmuch as the parasite is generally smaller than the host and inasmuch as one host may support many parasites, parasitic and hyperparasitic relationships form inverted food pyramids contained within the pyramids of the larger community.

The first problem a potential parasite faces is the defense mobilized by a potential host. Attachment to the outer body surface can be prevented only with difficulty, particularly if the host does not possess limbs. Numerous *ectoparasites* exploit this possibility. Equipped with suckers, clamps, or adhesive surfaces, they hold onto skin or hair, and with the aid of cutting, biting, or sucking mouth parts, or with rootlike outgrowths, they feed on the body fluids of the host. Examples among animals are leeches, lice, ticks, mites, and lampreys.

Endoparasites, within the body of the host, must breach more formidable defenses. Cellular enzymes of a host, digestive juices and strong acids in the alimentary tract, antibodies in the blood, white blood cells and other cells which engulf foreign bodies in amoeboid fashion (e.g., histiocytes), these are among the defensive agents which guard against the invader. Overcoming such defenses means specialization: development of resistant outer cuticles, as in most parasitic worms; development of cyst walls and calcareous capsules; development of hooks or clamps with which to hold onto the gut wall; development of enzymes which, when secreted, erode a path through host tissues (Fig. 17.14).

Specialization of the parasite also involves the selection of

17.14. Parasite attachment. The photo shows a section through the anterior part of a hookworm (a nematode), clamped to the mucosal tissue in the intestine of the host.

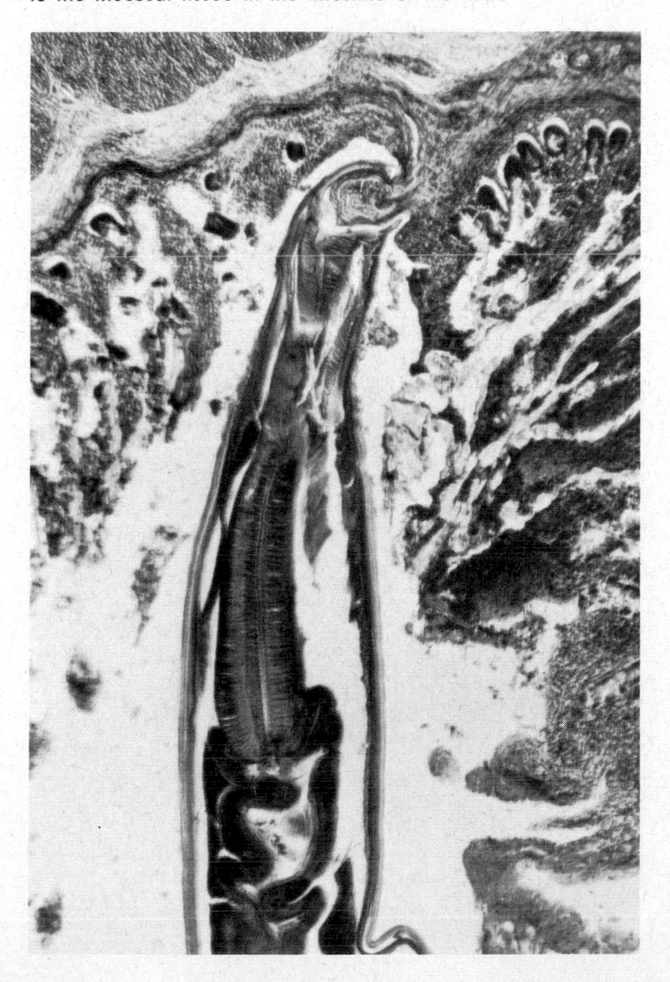

specific hosts. Highly advanced parasites cannot pick a host at random, even if many similar ones offer the same type of nutrients. During the evolution of a parasite, structural and functional specializations have developed in adaptation to particular hosts only. Thus, most parasites enter a host's body by fixed routes, then settle in fixed regions, as if in the course of time they had learned to channel their attack through points of weakness characteristic of particular hosts.

Breaching the host's defenses is a perennial problem to the parasite. No sooner has it developed an avenue to a comfortable existence than the infected individual is discriminated against in his environment; healthy hosts which have evolved a resistance to the parasite have a better chance of surviving. For example, large-scale infection of a population with parasites will lead to the preferential survival of those hosts which, through random mutations, develop specific means of combating the infectious agents. Hence if a parasite is to prevail against host defenses continually improved by evolution, it too must readjust and evolve. Through its own random mutations, it must develop new means of attack.

We recognize that parasite and host evolve *together*, first the parasite, then the host being one jump ahead. The very fact that free-living animals exist at all today signifies that they are resistant to a good many potential parasites by which they are constantly besieged. The very fact that parasites continue to exist signifies that free-living animals are not completely resistant—and they probably can never be, in view of the evolutionary inventiveness of the parasites.

It may be noted in this connection that it is to the obvious advantage of the parasite to keep the host alive. We find, indeed, that the virulence of a parasite often decreases with time. When a parasite-host relationship is first established, the invader is likely to be *pathogenic*, i.e., disease-producing. Two parallel evolutionary trends tend to reduce this pathogenicity. One is natural discrimination against infected hosts, as indicated above; the least resistant will be eliminated through plagues and epidemics. At the same time, less virulent populations of a given parasite will be favored; for when a parasite kills a host, the killer is generally killed as well. Therefore, the more harmful the parasite, the more difficult is its perpetuation. Many parasites are only mildly pathogenic, or not at all, often indicating long association with a particular host. In time the parasitic relation may actually become commensalistic or mutualistic.

Parasitic simplification. Once established in the body of a

17.15. Tapeworms. **A**, head; **B**, segmental sections near middle of body; **C**, segmental sections near hind end of body. Tree-shaped structures in **B** and **C** are reproductive organs. Note testes filling segments in **B** and genital pores opening on the sides of the segments. In **C**, the uterus filled with eggs is conspicuous. A digestive system is absent in these worms. Cf. Fig. 22.17.

A

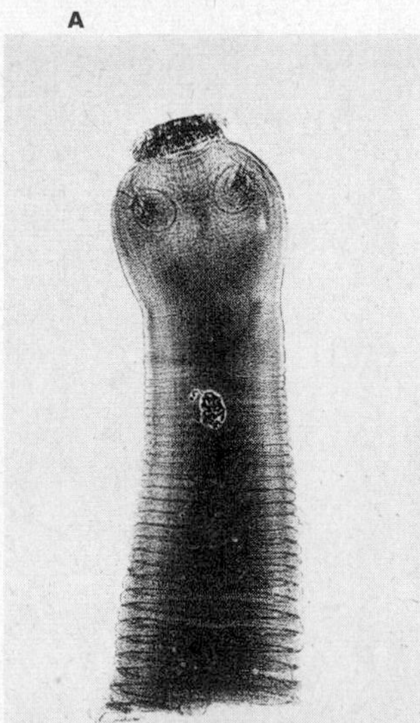

B

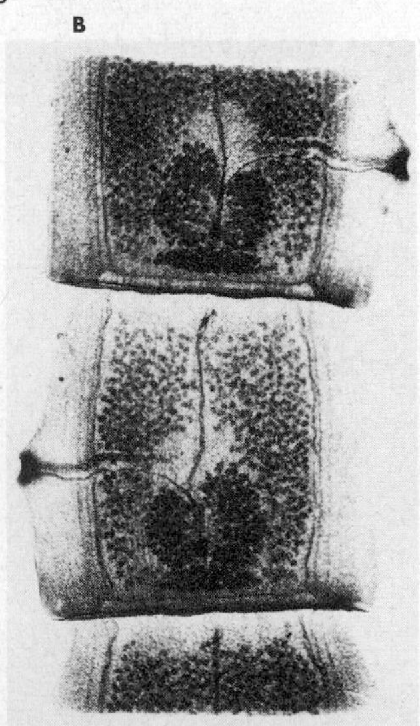

C

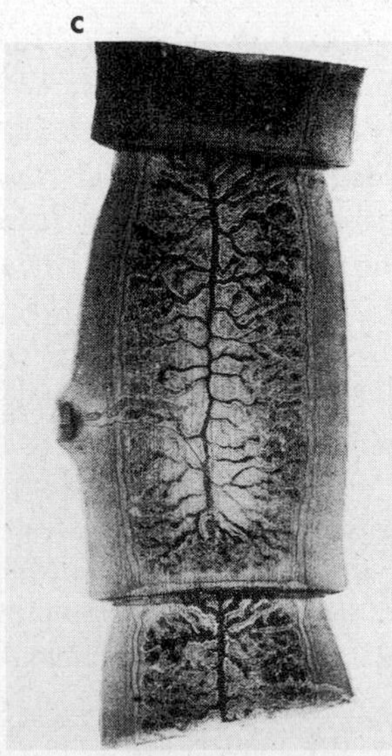

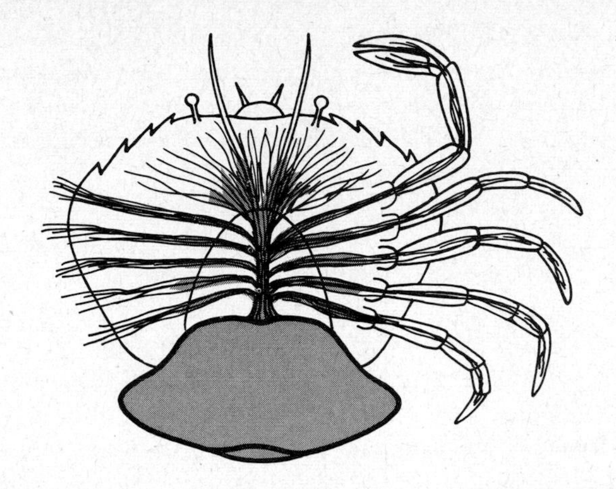

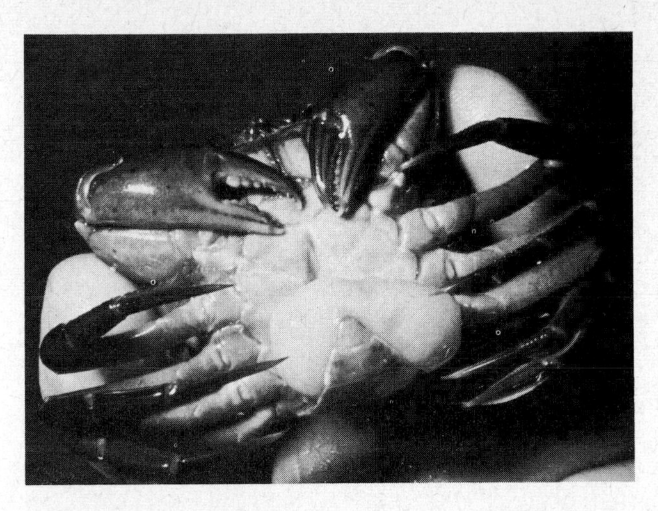

17.16. Diagram and photo of the crustacean parasite *Sacculina*, attached to the abdomen of a crab. The exteriorly formless adult parasite, related to barnacles, spreads tissue outgrowths into the host. (*Diagram adapted from Calman.*)

host, the parasite may pursue a life of comparative ease. Embedded in food, it needs no locomotor equipment, few sense organs, no fast nervous reflexes. Indeed, structural and functional *simplification* is a nearly universal characteristic of parasites. Here we encounter the ultimate expression of the principle that loss of self-sufficiency tends to be proportional to the degree of interdependence of animals.

Structural simplification is pronounced in tapeworms, for example (Fig. 17.15). These parasitic flatworms possess only a highly reduced nervous system, a greatly reduced muscular system, and not even a vestige of a digestive system. Almost like blotting paper, the worms soak up through their body walls the food juices in the host gut. Even more simplified is the adult of *Sacculina*, a crustacean which parasitizes its not too remote relatives, crabs. The parasitic adult is little more than a formless, semifluid mass of cells which spreads through a crab like a malignant tumor. The invader later produces sperms and eggs, and fertilized eggs then develop into recognizably typical, free-swimming crustacean larvae. These attach to crabs, enter them, and change into the simplified adults (Fig. 17.16).

Simplification also extends to metabolic activities. In particular, the synthetic capacities of a parasite are almost invariably restricted. For example, in the presence of nitrogen sources and simple carbohydrates like glucose, a *free-living* soil bacterium or fungus may synthesize amino acids, proteins, vitamins, numerous antibiotics useful to man, in short, all the complex compounds which make up the living matter of the organism. By contrast, an obligatorily *parasitic* bacterium or fungus promptly dies when given nitrogenous and simple organic substances alone. It has reduced synthesizing capacities and has become dependent on its host to supply it with most of the components of its living substance in prefabricated form.

In this respect, the modern virus is the most simplified. It cannot metabolize or self-perpetuate at all except within living host cells. Removed from cells, it becomes a lifeless crystal of complex chemicals; and it may resume metabolic and self-perpetuative activities only when it is reintroduced into living host cells (Fig. 17.17). Other parasites may be free-living at least at some stage of their life cycle, but viruses are never free-living. Their parasitism is total, complete, obligatory. It should be noted here, however, that viruses actually cannot be considered to be in the same category as other parasites; for, as shown in Chap. 14, viruses are chemical complexes and not cellular organisms like other parasites. Moreover, the ancestors of all other parasites were free-living organisms, whereas the ancestors of viruses probably were the genetic substances of bacteria. These substances were never free-living to begin with.

In parasitic animals simplification is probably an adaptive advantage, for the reduced condition may be more economical than the fully developed condition of the free-living ancestor. A tapeworm, for example, being structurally reduced, may concentrate all its resources into parasitizing the host; it need not divert energy and materials into maintaining elaborate

17.17. Crystal of a virus. In this state viruses are nonliving. They exhibit living properties only when they are present within host cells.

nervous, muscular, or digestive systems, which are unnecessary anyway in this parasitic way of life.

Parasitic reproduction. In one respect parasites are far from simplified: reproduction. In this function they are as prolific as the most prolific free-living forms. The practical necessity of an enormous reproductive potential is correlated with a major problem confronting the parasite, particularly the endoparasite, namely, how to get from one host to another. The problem is severely compounded by the requirement that not any new host will do. Another individual of the same host species must be found.

Parasites succeed in two ways, both of which involve reproduction: *active transfer* and *passive transfer*. In the former, one stage of the life cycle of the parasite is free-living *and* motile; i.e., this stage transfers from one host to another through its own powers of locomotion. For example, the adult phase may be parasitic and the free-living embryo or larva may be capable of locomotion, as in *Sacculina*. Or the larval phase may be the parasite, the adult then being free-living and capable of locomotion. This is the case in a number of parasitic insects which deposit their eggs within or on individuals of other species (Fig. 17.18).

Passive transfer is encountered among parasites which in *no* phase of the life cycle is capable of locomotion. Propagation here is accomplished by wind, by water, or by *intermediate hosts*. The latter offer a means of transfer which is not quite as chancy as random distribution by wind or water. What is involved here is well illustrated in the propagation of tapeworms (Fig. 17.19).

These parasites of man and other mammals, like numerous others, exploit one of the easiest routes into and out of the host, namely, the alimentary tract. Entering through the

17.18. A caterpillar of a sphinx moth, carrying the pupal cocoons of a parasitic insect (*Microgaster*).

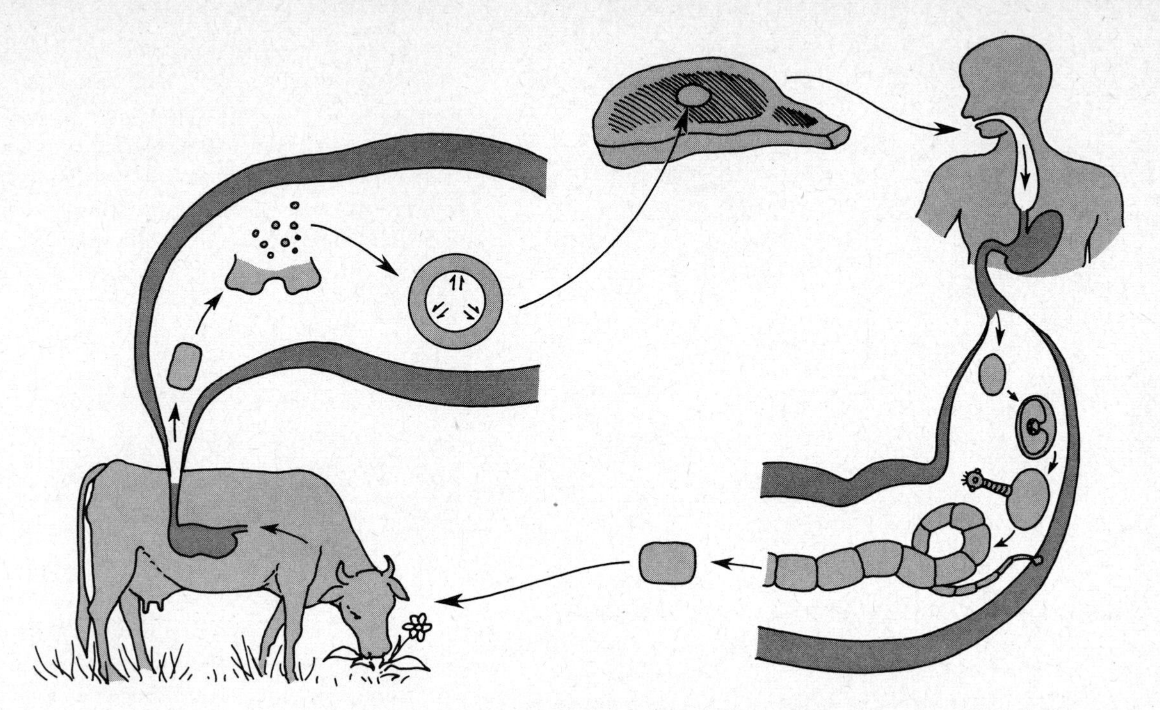

17.19. The life-cycle pattern of a beef tapeworm. Ripe sections of the worm pass with the feces from the human gut. Eggs are released from these sections in the gut of cattle. Walled, hook-bearing tapeworm embryos then encapsulate in beef muscle, and the embryos become adults in the intestine of man. The head (scolex) of the worm is invaginated at first, but it soon everts and with the hooks and newly developed suckers it attaches to intestinal tissues.

host's mouth by way of eaten food and leaving through the anus by way of feces, tapeworms spend their adult life directly in the gut cavity of the host. Other endoparasites utilize the gut as a springboard from which to invade interior tissues. The problem is to transfer offspring from one host, e.g., a man, to another by passive means. Tapeworms accomplish a first phase of this readily; namely, mature eggs are released to the outside with the host's feces.

Since man does not eat feces, the eggs evidently cannot reach new human hosts directly. However, tapeworms ingeniously take advantage of the food pyramids of which man is a member; man eats beef, and cattle eat grass. A ready-made pathway from grass to man thus exists, and the transfer chain becomes complete if, as happens on occasion, human feces are deposited on grass. Tapeworm eggs clinging to such vegetation may then be eaten by cattle.

In the intestine of a cow, a tapeworm egg develops into an embryo and such an embryo bores a path through the gut wall into the cow's bloodstream. From there the embryo is carried into beef muscle, where it encapsulates and matures. If man then eats raw or partially cooked beef, the capsule surrounding the young tapeworm is digested away in the human gut and the free worm now hooks on to the intestinal wall of its new host (cf. Fig. 17.19).

This history illustrates a very widely occurring phenomenon. Many kinds of parasites utilize well-established food pyramids in transferring to new hosts. Often there is more than one intermediate host, as in the life cycle of the Chinese liver fluke (Fig. 17.20). The adults of this parasitic flatworm infest the liver of man. Fertilized eggs are released via the bile duct into the gut of the host and pass to the outside with the feces. If the feces get into ponds or rivers, as happens frequently, the eggs develop into so-called *miracidium* larvae.

Such a larva must then enter a snail, and in the tissues of this animal the miracidium develops into another larval type, called a *sporocyst*. The latter subsequently gives rise to many

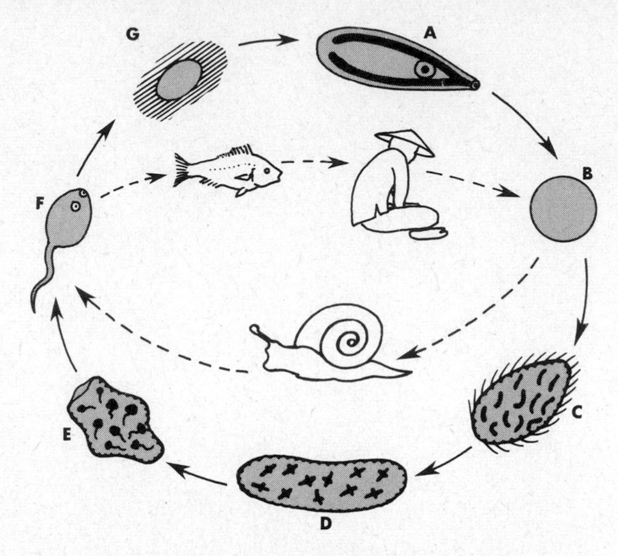

17.20. The general life-cycle pattern of a Chinese liver fluke. The detailed structure of the adult and the complete outline of the life cycle are examined in Chap. 22 (cf. Fig. 22.15). **A,** adult in liver of man; **B,** egg, passing out with feces and eaten by snail; **C,** miracidium larva in snail; **D,** sporocyst; **E,** one of many rediae, formed from sporocyst; **F,** cercaria; **G,** cercariae escape from snail and encapsulate in fish muscle.

redia larvae, which feed on snail tissue and grow. Then each of the rediae produces yet another set of many larvae, called *cercariae* (cf. Chap. 22 for details). These fourth-generation larvae escape from the snail and swim about freely. If within a short time they happen to find a fish, they bore into it and encapsulate in muscular tissue. And if man subsequently eats raw or incompletely cooked fish, the young adult flukes find their way from the human gut into the liver.

Note that this cycle involves two intermediate hosts, the snail and the fish. Transfer is partly passive (man to snail, fish to man), partly active (snail to fish). Note particularly the multistage, larva-within-a-larva type of development. Characteristic of flukes generally, larval polymorphism of this sort constitutes a highly efficient method of enormously increasing the number of reproductive units. A single fluke egg is estimated to yield a final total of some 10,000 cercariae—and a single adult fluke may produce many tens of thousands of eggs. Hence the chances become fairly good that at least some of the millions or billions of larvae will reach final hosts.

Through active locomotion, through physical agents such as air and water, and through routes involving food pyramids and intermediate hosts, parasites have solved their transfer problems most successfully. So successfully, indeed, that there are many more individual parasites in existence than free-living animals.

A community consists of various kinds of free-living and various kinds of symbiotic populations. Which particular ones of each type actually compose a given community, hence the nature of a whole ecosystem, is determined largely by the nature of the *physical environment.* We shall examine this abiotic part of the biosphere in the following chapter.

REVIEW QUESTIONS

1. Define ecosystem, biomass, biota, community. What is the species structure of a community? What factors produce and maintain such a structure?

2. Describe the life cycle of a community. What are ecological succession, climax communities, seres? Describe the nutritional structure of a community. What maintains such a structure?

3. What are food pyramids and food chains, and what factors produce and maintain them? Show how nutritional factors contribute to the long-range numerical constancies within communities.

4. Describe reproductive links which make the components of communities interdependent. What are the growth characteristics of communities? What is population pressure?

5. What kinds of protective links unite the members of communities? Show how such links contribute to numerical population balances. Give examples of mimicry and other forms of camouflage in animals.

6. What are the various forms of symbiosis and how are they defined? Give specific examples of each. What are lichens, root nodules, zoochlorellae? To what other organisms is man a host in mutualistic or commensalistic relationships?

7. What general structural and functional characteristics distinguish parasites from free-living animals? Distinguish between ectoparasites and endoparasites and give examples.

8. What is hyperparasitism? In what ways are parasites adapted to ectoparasitic or endoparasitic modes of life? What is the adaptive advantage of parasitic simplification? In what ways are parasites simplified?

9. Distinguish between active and passive transfers in

parasite life cycles. What is the role of food pyramids in parasite transfers? What are intermediate hosts?

10. Review the life cycles of tapeworms and liver flukes and show what general principles of parasite transfer are illustrated by these cycles.

COLLATERAL READINGS

The books and other readings on ecology cited at the end of Chap. 16 may be profitably consulted also for additional information on the subject of ecosystems and communities. Further pertinent discussions, including some on symbiosis and parasitism, may be found in the following:

Bigger, J. W.: "Man against Microbe," Macmillan, New York, 1939.

Burnet, F. M.: Viruses, *Sci. American,* May, 1951.

Clarke, G. L.: "Elements of Ecology," Wiley, New York, 1954.

Cleveland, L. R.: An Ideal Partnership, *Sci. Monthly,* vol. 67, 1948.

Deevey, E. S.: The Human Crop, *Sci. American,* Apr., 1956.

———: Bogs, *Sci. American,* Oct., 1958.

———: The Human Population, *Sci. American,* Sept., 1960.

Dunbar, M. J.: The Evolution of Stability in Marine Environments. Natural Selection at the Level of the Ecosystem, *Am. Naturalist,* vol. 94, 1960.

Harzen, W. E.: "Readings in Population and Community Ecology," Saunders, Philadelphia, 1964.

Kendeigh, S. C.: "Animal Ecology," Prentice-Hall, Englewood Cliffs, N.J., 1961.

Lack, D.: "The Natural Regulation of Animal Numbers," Clarendon, Oxford, 1954.

Lamb, I. M.: Lichens, *Sci. American,* Oct., 1959.

Limbaugh, C.: Cleaning Symbiosis, *Sci. American,* Aug., 1961.

Luria, S. E.: The T2 Mystery, *Sci. American,* Apr., 1955.

Lwoff, A.: The Life Cycle of a Virus, *Sci. American,* Mar., 1954.

Odum, E. P.: "Fundamentals of Ecology," 2d ed., Saunders, Philadelphia, 1959.

Pequegnat, W. E.: Whales, Plankton, and Man, *Sci. American,* Jan., 1958.

Rogers, W. P.: "The Nature of Parasitism," Academic, New York, 1962.

Slobodkin, L. B.: "Growth and Regulation of Animal Populations," Holt, New York, 1961.

Zinsser, H.: "Rats, Lice, and History," Little, Brown, Boston, 1935.

CHAPTER 18 biosphere and habitat

All individuals, populations, and communities depend directly on their physical *environment;* their evolution is oriented by it and their continued existence is made possible by it. In the first section below we shall examine this interrelation between the nonliving and the living components of the biosphere. In the second section, our attention will be focused on the different types of homes, or *habitats,* actually provided by the environment for living communities.

THE ENVIRONMENT

All living functions begin with raw materials, and all basic raw materials ultimately come from the physical environment of the earth. The environment also influences animals in other major ways. For example, every cellular reaction, hence every animal as a whole, is affected greatly by environmental temperature, pressure, the nature of the surrounding medium, in short, by geography and weather generally. In this respect, forces of global dimensions have a direct bearing on forces of molecular dimensions. We may note, therefore, that the environment sets the stage for life principally in two ways: the environment is the ultimate supplier of all raw materials, and it provides the physical and chemical background against which living processes must be carried out. How does the environment function in these roles?

The most important general observation we can make about the environment is that it is forever changing, on every scale from the submicroscopic to the global. The physical world is subjected unceasingly to various astrophysical, meteorologic, geologic, and geochemical forces which alter every component of the earth sooner or later, very rapidly in some cases, rather slowly in others. Being part of the earth's substance, living matter too is subjected to these forces and it therefore undergoes unceasing change. As we have seen, the very origin of living matter on earth was itself a result of environmental change. Animals and organisms generally then became a powerful cause of continued change.

The fundamental reason for uninterrupted environmental change is that the earth as a whole, hence also living matter and every other component, is an *open system.* Such systems exchange materials, energy, or both with their surroundings. By contrast, a *closed system* exchanges nothing with its surroundings (cf. Chap. 2 and Fig. 18.1). On earth, to be sure, the amounts of material entering from space or leaving into

18.1. In a closed system, nothing enters or leaves. In an open system, materials or energy or both may enter or leave.

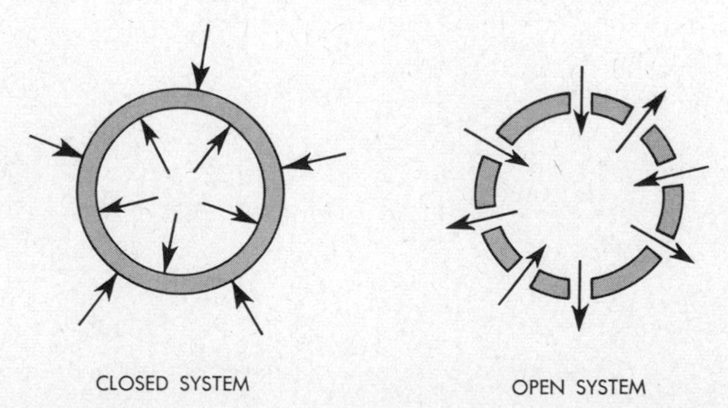

space are negligible. However, *energy* both enters and leaves, and this makes the earth an open system. Most importantly, various forms of solar energy—heat, light, X rays, ultraviolet rays, electric waves and many others—beam to earth uninterruptedly; and enormous amounts of energy radiate out, principally in the form of heat. As a result, the earth's material substance can never attain static equilibrium; so long as the sun shines and the earth spins, energy flux creates balance-upsetting disturbances. Every imbalance creates new imbalances of its own, and, as a general consequence, the earth's environment is forever changing.

Such changes, being produced primarily by sun and planetary motion, occur predominantly in rhythmic, patterned *cycles*. Daily and seasonal climatic cycles are familiar examples. Other environmental cycles may be less readily discernible, particularly if their scale is too vast or too minute or if they occur too fast or too slowly for direct observation. Animals are interposed into these cycles; and as the earth's components circulate, some of these components become raw materials in living processes.

The physical environment is the ultimate source which supplies plants and animals with all required *inorganic* metabolites. From some of these, the required *organic* metabolites must then be manufactured and distributed within the living world itself. Thus all organisms build up their bodies at the ultimate expense of inorganic materials withdrawn directly from the physical environment. Excretion products formed within organisms return to the environment largely while the organisms live. And when they die, all other materials of their bodies return to the environment as well. As we shall see, *decay* caused by saprotrophic bacteria and fungi gradually retransforms all the returned substances into the same kinds of inorganic materials which were withdrawn from the environment originally (Fig. 18.2).

Living organisms may therefore be envisaged as transient constructions built out of materials "borrowed" temporarily from the environment. One important corollary of this is that, despite its material contribution to the formation of organisms, the physical earth *conserves* all its raw materials on a long-term basis; and this makes possible an indefinitely continued, repeated re-creation of living matter. Therefore, the continuity of life depends on the parallel continuity of death.

A second corollary is that, because of their life and their death, organisms contribute in major ways to the movement of earth substances in cycles. Billions of tons of materials are withdrawn from the environment into billions of organisms all over the world, are made components of living matter, are redistributed among and between organisms, and are finally put back into the environment as they were obtained. Such life-involving cycles become part of and often reinforce and contribute to the physical cycles on earth. As in the purely physical cycles, those in which living matter participates run on energy supplied by the sun. Solar energy is trapped by living organisms via photosynthesis, and some of this energy is later spent by organisms in moving parts of the earth through their bodies.

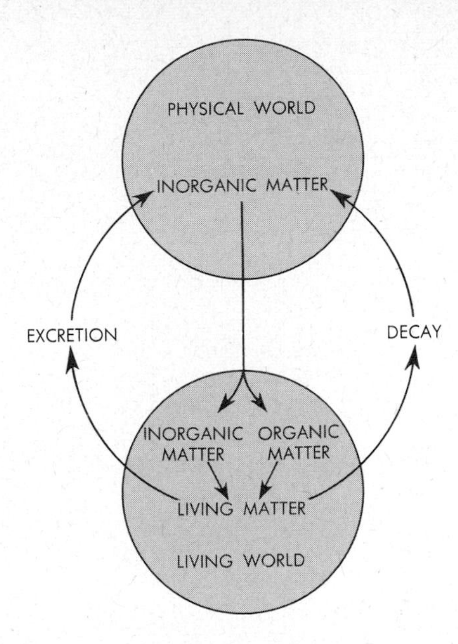

18.2. The cyclical interaction between the physical environment and living matter. Raw materials taken by living matter from the physical environment return to the latter through excretion and decay.

The global environment consists of three main subdivisions. The *hydrosphere* includes all liquid components, i.e., the water in oceans, lakes, rivers, and on land. The *lithosphere* comprises the solid components, i.e., the rocky substance of the continents. And the *atmosphere* is the gaseous mantle which envelops the hydrosphere and the lithosphere. Living organisms require inorganic metabolites from each of these subdivisions. The hydrosphere supplies liquid *water;* the lithosphere supplies all other *minerals;* and the atmosphere supplies *oxygen, nitrogen,* and *carbon dioxide*. Together, these inorganic materials provide all the chemical elements needed in the construction and maintenance of living matter (Fig. 18.3). In addition to being sources of supply, the three subdivisions of the environment also affect animals in various

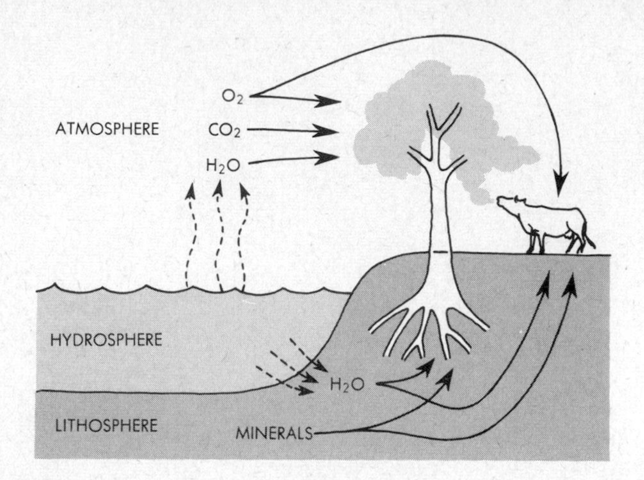

18.3. The material contributions of each of the three subdivisions of the environment to the maintenance of living matter.

other specific ways. We shall examine the contributions of each of the subdivisions in turn.

The Hydrosphere

Water is the most abundant mineral of the planet. It covers some 73 per cent of the earth's surface entirely and it is a major constituent of the lithosphere and the atmosphere. As shown in Chap. 4, water is also the most abundant component of living matter. It is not surprising, therefore, that it is the major inorganic nutrient required by all living organisms. In metabolism, water is the exclusive source of the element hydrogen and one of several sources of oxygen.

The basic water cycle which moves and conserves water in the environment is quite familiar. Solar energy evaporates water from the hydrosphere into the atmosphere. Subsequent cooling and condensation of the vapor at higher altitudes produces clouds, and precipitation as rain or snow then returns the water to the hydrosphere. This is the most massive process of any kind on earth, consuming more energy and moving more material than any other (Fig. 18.4).

In using water as metabolic raw material, animals withdraw it principally from the hydrospheric segment of the global cycle. Aquatic animals absorb water directly from their liquid environment; they excrete some of it back while they live; and after death the remainder, still in the form of liquid water, is returned through decay. Terrestrial animals and also plants are interposed more extensively in the global cycle, and indeed they contribute substantially to its continuance. These organisms absorb liquid water from the reservoir present in soil and in bodies of fresh water. Plants and animals move such water through their bodies and in the process they retain required quantities. The remainder is excreted, partly as liquid water but more particularly as water vapor which raises the moisture content of the atmosphere. We may note that a given quantity of environmental water is moved from hydrosphere to atmosphere far faster through the metabolic agency of living organisms than if that water were simply allowed to evaporate directly from the hydrosphere. In other words, the metabolism of terrestrial organisms actively accelerates the global water cycle. Sometimes this may have an effect on the climate. For example, the trees of tropical jungles release so much water vapor that the air over vast areas remains permanently saturated with moisture, cloudbursts occurring almost every evening. After terrestrial organisms die, any liquid water in their bodies again returns to the hydrosphere through decay.

Water influences animals not only through its function as a prime nutrient but also through its effect on almost all aspects of climate and weather, both in the sea and on land. Very largely, these effects are immediate or distant consequences of the interplay between solar energy, the rotation and revolution of the earth, and the hydrosphere—the same interplay which also produces the global water cycle.

In the ocean, water warmed in the tropics becomes light and rises to the surface, whereas cool polar water sinks. These up-down displacements bring about massive horizontal shifts of water between equator and pole. The rotation of the earth introduces east-west displacements. These effects, reinforced substantially by similarly patterned wind-producing air

18.4. The global water cycle. Evaporated water eventually returns to earth through precipitation.

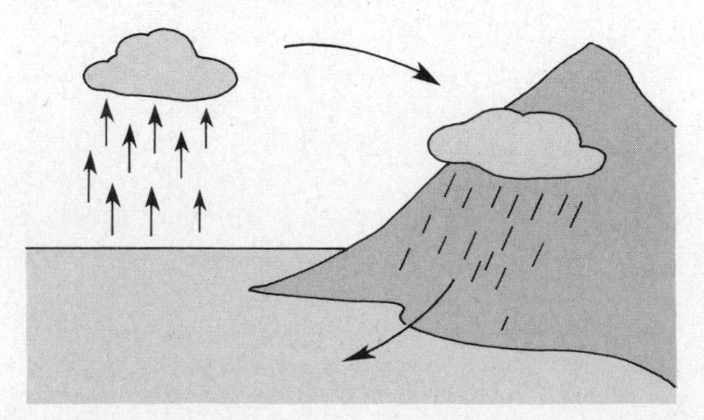

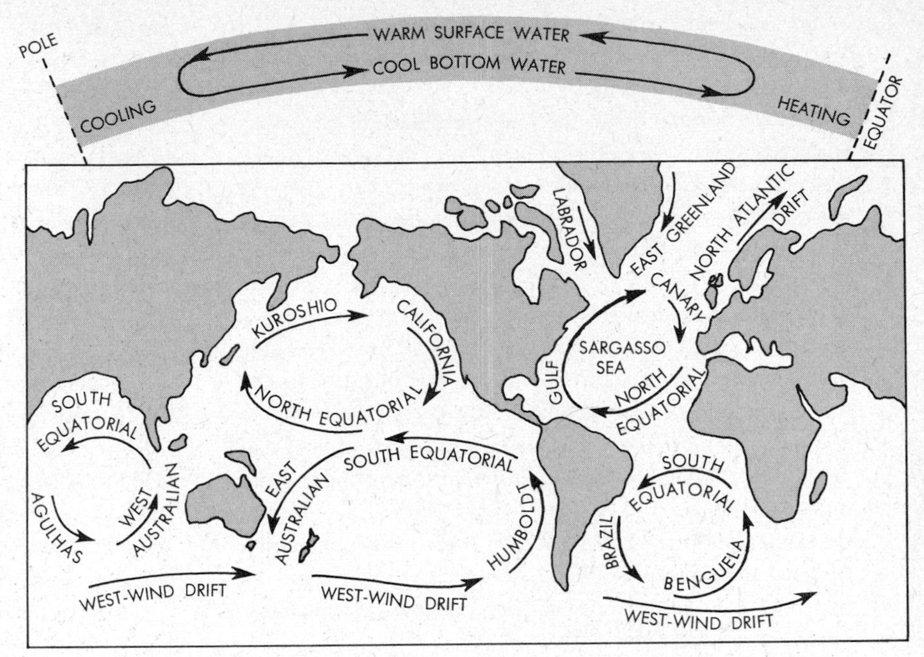

18.5. Top, the depth circulation of ocean water. A cross-sectional ocean profile between equator and pole is shown. Water warmed in equatorial regions rises and water cooled in polar regions sinks. This produces north-south and up-down circulation as indicated. However, this basic equator-to-pole movement is modified by the rotation of the earth, by winds, and by the position of the continents. **Bottom,** the actual circulation so produced and the names and flow directions of the chief currents shown in surface view.

movements, result in *oceanic currents*. The latter influence climatic conditions not only within the seas, but also in the air and on land (Fig. 18.5).

Another climatic effect is a result of the thermal properties of water. Of all liquids, water is one of the slowest to heat or cool, and it stores a very large amount of thermal energy. The oceans thus become huge reservoirs of solar heat. The result is that sea air chilled by night becomes less cold because of *heat radiation* from water warmed by day. Conversely, sea air warmed by day becomes less hot because of *heat absorption* by water cooled by night. Warm or cool onshore winds then moderate the inland climate in daily patterns. Analogous but more profound effects are produced by heat radiation and absorption in seasonal summer and winter patterns.

Thirdly, global climate over long periods of time is determined by the relative amount of water locked into *polar ice*. Temperature variations averaging only a few degrees over the years, produced by still poorly understood geophysical changes, suffice for major advance or retreat of polar ice. As noted earlier, ice ages have developed and waned rhythmically during the last million years, and warm *interglacial* periods, characterized by ice-free poles, have intervened between successive advances of ice. At the present time, the earth is slowly emerging from the last ice age, which reached its peak some 50,000 to 20,000 years ago (cf. Fig. 15.29). As polar ice is melting, water levels are now rising and coast lines are gradually being submerged. If trends during the past 50 years are reliable indications, the earth appears to be warming up generally. Deserts are presently expanding; snow lines on mountains are receding to higher altitudes; in given localities, more days of the year are snow-free; and the flora and fauna native to given latitudes are slowly spreading poleward. It is difficult to be sure whether these changes are merely part of a short warm cycle or are really indicative of a long-range trend.

All these various cyclic changes in the hydrosphere have profound impact on animals. By influencing temperature, humidity, amount of precipitation, winds, waves, currents, and indeed the very presence or absence of water in given localities, they play a major role in determining what kinds and amounts of metabolism are possible in such localities, hence what kinds and amounts of animals may live there.

The Lithosphere

This subdivision of the environment plays two vital roles. First, as already pointed out above, it is the exclusive source of most *mineral* metabolites for all animals, terrestrial as well as aquatic; and, secondly, it forms the bulk component of *soil*, required specifically by terrestrial plants and by numerous subterranean animals.

Like the world's water, the rocky substance of the earth's surface moves in a gigantic cycle, but here the rate of circulation is measured in thousands and millions of years. One segment of this global mineral cycle is *diastrophism*, the vertical uprising of large tracts of the earth's crust. Major parts of continents or indeed whole continents may undergo such diastrophic movements. They occur when a land mass is pushed up from below or is subjected to great lateral pressure, generated in adjacent portions of the earth's crust. Uplifting or upbuckling then follows. Changes of this sort take place exceedingly slowly. The most striking instance of diastrophism is *mountain building*. Presently the youngest and highest mountain ranges are the Himalayas, the Rockies, the Andes, and the Alps. All of them were thrown up during the Laramide revolution, some 70 million years ago, and we may note that the earth crust in these regions is not completely settled even now.

Quite apart from the tremendous upheaval caused by mountain formation itself, such an event has long-lasting effects on climate, hence on animals. A high, massive mountain barrier is likely to interfere drastically with continental air circulation. For example, moisture-laden ocean winds may no longer be able to pass across the barrier. Continual rain will therefore fall on the near side and the region may become lush and fertile. By contrast, the far side will be arid and desert conditions are likely to develop (Fig. 18.6). The following are two good examples: fertile California on the ocean side of the Sierras and the deserts of Arizona and New Mexico east of the Rockies; fertile India on the ocean side of the Himalayas and the belt of deserts north of them. Animals living on either side of a newly formed mountain range must adapt to the new environmental conditions by evolution. As pointed out in Chap. 15, periods of extensive mountain building have always been followed by major evolutionary turnover among animals.

18.6. The effect of a mountain on climate. A mountain deflects moisture-rich ocean winds upward and causes rain to remain confined to the slope facing the ocean. That slope will therefore be fertile, but the far slope will become a desert.

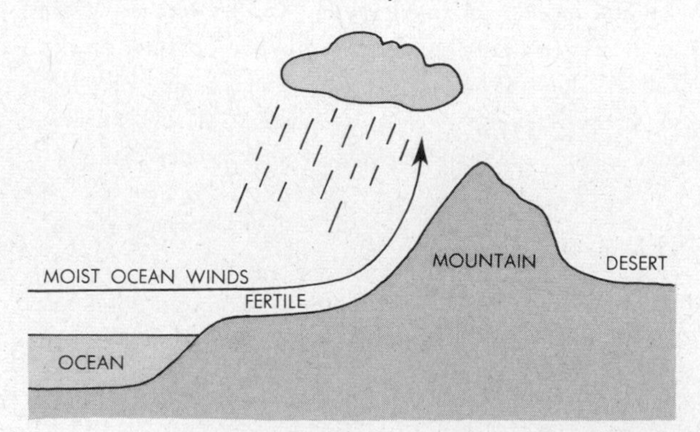

The second segment of the global lithospheric cycle involves *gradation*, the lowering of high land and the leveling of mountains. These changes are brought about in part by actual geologic sinking of land and in part by actions of the hydrosphere and the atmosphere. These actions usually take the form of *erosion* and *dissolution* of rock. Many erosive processes are quite familiar. For example, water and gravity produce shearing, canyon-cutting rivers (Fig. 18.7). Water and high temperatures produce corrosive humidity. Water and low temperatures produce grinding, rock-pulverizing glaciers. And as freezing water expands in rocky crevices, it carves boulders and stones off the face of a mountain. Water, wind, and sun in time so reduce mountain to hill and hill eventually to plain. Together with geologic sinking, these processes often may make land lie so low that substantial parts of it become overrun by the ocean.

Accompanying the physical forces of gradation are chemical forces, and these are of particular direct and indirect importance to animals. First, the chemical action of water, and also chemical processes which accompany the decay of dead terrestrial organisms, are major erosive factors. They contribute to breaking large stones into smaller ones and small pebbles into tiny sand grains and microscopic rock fragments. Gradation thus plays a principal role in the formation of the rocky components of *soil*.

Secondly, whenever water is in contact with rock, it dissolves small quantities of it and so acquires a *mineral* content. The dissolved minerals are carried largely in the form of ions. Accordingly, as rain water runs off high land, it becomes progressively laden with minerals. Streams and rivers form, and these irrigate adjacent areas and contribute to the water content of soil. Rain adds to soil water directly. In soil, the water leaches more minerals out of the many rock fragments present and the total supply then serves as the mineral source for terrestrial animals. After the animals die and decay, the mineral ions of their bodies return to the soil.

Dissolved soil minerals eventually drain back into rivers, and rivers drain into the ocean. Therefore, as the lithosphere is slowly being denuded of mineral compounds, the hydrosphere fills with them. It was partly by this means that the

18.7. The cutting, erosive effect of a river. The canyon was channeled out by the stream flowing through it.

early seas on earth acquired their original saltiness, and as the global water cycle now continues, it makes the oceans even saltier. Animals in the sea freely use the mineral ions as metabolites.

The death of marine organisms subsequently helps to complete the global mineral cycle. Many plants and animals use mineral nutrients in the construction of protective shells and supporting bones. After these organisms die, their bodies sink down toward the sea floor, in a slow, steady rain (cf. also below). All organic and some of the inorganic matter dissolves during the descent, but much of the mineral substance persists in solid form and reaches the sea bottom. So abundant and uninterrupted is this rain that it forms gradually thickening layers of ooze over huge tracts of the sea floor. Various types of Protista contribute particularly to these mineral accumulations. In the course of millennia, the older, deeper layers of the ooze may compress into rock. The global lithospheric cycle then becomes complete when a section of sea bottom or low-lying land generally is subjected to new diastrophic forces. High ground or mountains are thereby regenerated, and such parts as were sea floor originally may be thrust up as new land in the process (Fig. 18.8).

The lithospheric cycle supplies many more *types* of minerals than organisms normally require. From the standpoint of quantities, the relatively *least* plentiful mineral circulating in the environment will determine the maximum quantity of living matter that can be supported on earth. Despite the dense cover of life now carpeting the earth, available quantities of most minerals are still well in excess of currently required amounts. However, because of the uninterrupted, global growth of organisms for millions of years, a few key minerals now tend to be in relatively short supply. In particular, phosphates and nitrates have become significant limiting factors. For example, agriculturally used soils may often be burdened with so much vegetation that they may become exhausted unless artificially enriched with fertilizers—largely phosphates and nitrates. Similarly, the amount of microscopic life sustainable in the ocean is now determined principally by the amounts of available phosphates and nitrates. As we shall see presently, however, nitrates are supplied not only by the lithospheric cycle but by an atmospheric cycle as well.

The Atmosphere

Like the hydrosphere, the atmosphere as a whole is subjected to physical cycles by the sun and the spin of the earth.

18.8. The global mineral cycle. Minerals absorbed by terrestrial plants and animals return to soil by excretion and death. Rivers carry soil minerals into the ocean, where some of them are deposited at the bottom. Portions of sea bottom then may be uplifted geologically, and the new land so formed reintroduces minerals to a global cycle.

Warmed equatorial air rises and cooled polar air sinks, and the axial rotation of the earth shifts air masses laterally. The resulting global air currents basically have the same general pattern as the ocean currents, and the winds of the former strongly reinforce the latter. Also, the motions of air substantially influence climatic conditions.

Equally significant to animals are the chemical cycles of the atmosphere. Air consists mainly of oxygen, O_2 (about 20 per cent); carbon dioxide, CO_2 (about 0.03 per cent); nitrogen, N_2 (about 79 per cent); water (in varying amounts, depending on conditions); and minute traces of inert gases (helium, neon, krypton, argon, xenon). Except for the inert gases, all these components of air serve as metabolites and each circulates in a global cycle in which organisms play a conspicuous role. Also, all the gases are dissolved in natural waters, and in this respect the hydrosphere is in equilibrium with the atmosphere. Regardless of whether they are aquatic or terrestrial, therefore, all organisms have access to the aerial gases.

The role of the water vapor in air has already been discussed, for this vapor represents the atmospheric segment of the global water cycle. The cycles of the other gases are as follows.

The oxygen cycle. Atmospheric oxygen enters the living world as a gas required in respiration (Fig. 18.9). As shown in Chap. 5, the function of oxygen in respiration is to collect hydrogen, resulting in the formation of water. This water joins all other water present in organisms and as such it may undergo three possible fates. Some of it may be excreted immediately and so add to the water content of the environment. Another fraction may be used as a building material in the construction of more living matter, water here being the

18.9. The oxygen cycle. Components of the cycle within organisms are inside the shaded rectangle.

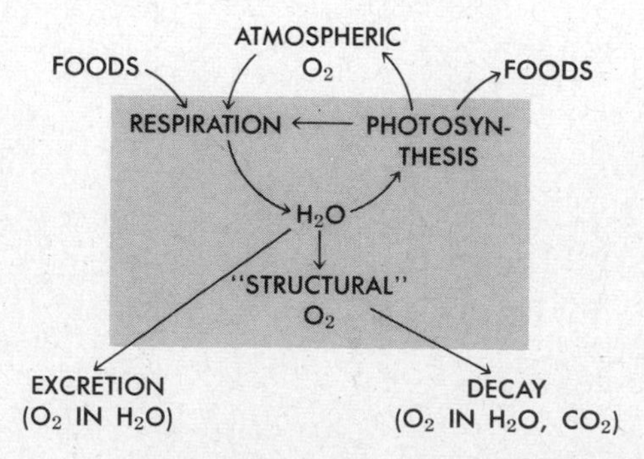

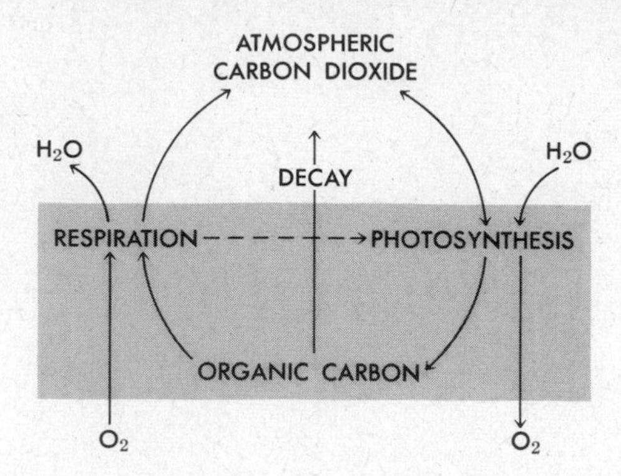

18.10. The carbon cycle. Components of the cycle within organisms are inside the shaded rectangle.

source of the elements hydrogen and oxygen. Such structural oxygen remains within an organism until death, and subsequent decay returns it to the environment. However, the return is not usually in the form of free atmospheric oxygen, but is either in the form of water or in the form of carbon dioxide. Evidently, the global oxygen cycle is closely interlinked with the global water and CO_2 cycles. A third possible fate of water within organisms is its utilization as a fundamental raw material in photosynthesis. In this process, water is split apart into hydrogen and oxygen, the hydrogen then being used in food manufacture. The oxygen is a byproduct. Such free oxygen may now again be used in respiration, or it may be returned to the environment as molecular atmospheric oxygen, completing the cycle.

In sum, atmospheric molecular oxygen *enters* organisms only through respiration and *leaves* organisms only through photosynthesis. In intervening steps, the oxygen is incorporated into water and in this form it may interlink with the water cycle or, indirectly, with the carbon cycle.

We may note here that atmospheric oxygen is the source of the ozone (O_3) layer which envelops the earth at an altitude of some 10 miles. This layer prevents a great deal of the high-energy radiation of the sun (ultraviolet rays, X rays) from reaching the earth's surface, and so it affects organisms indirectly, by shielding out potentially lethal rays.

The carbon cycle. Atmospheric carbon dioxide is virtually the exclusive carbon source and, with water, one of the two major oxygen sources for the construction of living matter. The gas enters the living world through photosynthesis, in which it is a fundamental raw material (Fig. 18.10). Photosynthesis incorporates CO_2 into organic substances, which serve two principal functions. One fraction is used in the construction of more living matter. The carbon and oxygen so supplied by CO_2 remain in living matter until death. Decay subsequently returns CO_2 to the atmosphere, and this completes one possible carbon cycle. Another fraction of the organic substances is used as fuel in respiration. This process releases CO_2 as a byproduct. Such carbon dioxide may now be used in photosynthesis again, or it may return to the environment and complete a second possible carbon cycle. The interrelations of the CO_2, O_2, and H_2O cycles are outlined in Fig. 18.11.

The carbon dioxide content of the atmosphere is replenished not only through biological combustion, i.e., respiration, but also through nonliving combustion, i.e., real fires. For example, forest fires and burning of industrial fuels release CO_2 into the air. Such events represent a long-delayed completion of the carbon cycle, for wood, coal, oil, and natural gas all contain combustible organic substances which were manufactured through photosynthesis, in many cases millions of years ago. Aerial CO_2 was then used up, and the gas is returned to the atmosphere only now. The rapid, very voluminous release of CO_2 by man-made combustions today may increasingly affect global climates, for atmospheric CO_2 acts as a heat screen. That is, it permits various solar energies other than heat to reach the earth's surface readily, where some of them are transformed into heat, but it retards the radiation of earth heat into space. Carbon dioxide therefore has a "greenhouse" effect, which in some measure probably contributes to the present warming up of the earth. Note, finally, that occasional net additions of CO_2 to the atmosphere are brought about by volcanic eruptions.

18.11. The interrelations of the oxygen, water, and carbon cycles.

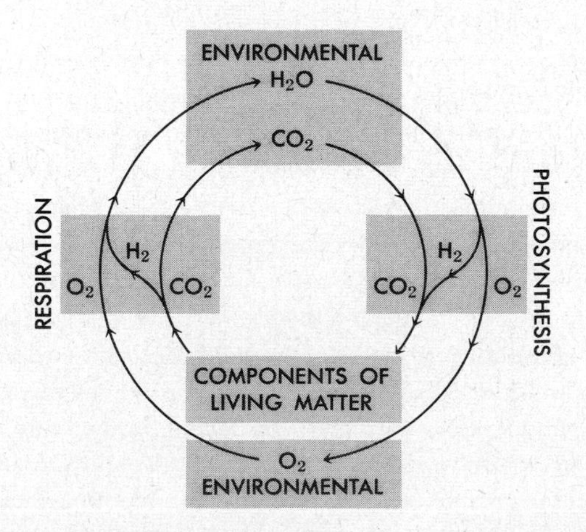

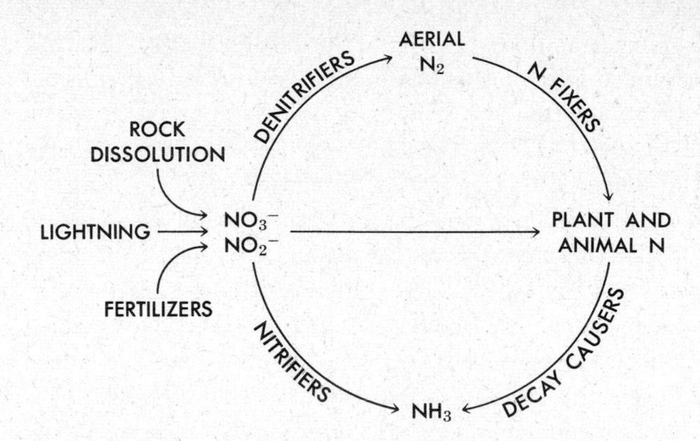

18.12. The basic pattern of the nitrogen cycle (simplified diagram). Participating bacterial types are shown over the curved arrows.

The nitrogen cycle. Nitrogen is required by all organisms in the construction of proteins, nucleic acids, and other nitrogenous compounds. Atmospheric nitrogen serves as the ultimate source (Fig. 18.12). But aerial N_2, the most abundant component of air, is rather inert chemically and it actually cannot be used as such by the majority of organisms. The most common *usable* nitrogen source frequently is the *nitrate* ion, NO_3^-. This ion may be absorbed by plants as a mineral metabolite from the environment and may be converted by them (but not by animals) into amino groups ($-NH_2$) and other nitrogen-containing components of living matter. Some of the environmental sources of nitrate have already been referred to above; e.g., the mineral accumulates through dissolution of rock by water or through addition of soil-fertilizers by man. Small quantities also form in the air when the energy of lightning bolts combines aerial nitrogen and oxygen. Rain then carries such nitrates to the ground. As we shall see presently, another major nitrate source is the global nitrogen cycle.

Nitrogen incorporated into the organic structure of plants stays there until death. Animals must obtain their usable nitrogen from one another and ultimately by eating plants. Eventually animals die as well. Through subsequent decay, all organic nitrogen of dead plants and animals is ultimately converted into *ammonia*, NH_3. This substance becomes available in the environment, where it forms a nutrient for so-called *nitrifying bacteria*. These are of two types. One type absorbs ammonia and converts it into *nitrite* ions, NO_2^-, which are excreted into the environment. The second type absorbs the nitrite and converts it into *nitrate* ions, NO_3^-, which are similarly excreted. Thus, the combined metabolic activities of the nitrifying bacteria provide environmental nitrates. This is a major source from which plants obtain their nitrogen supplies.

Environmental nitrates are also acted upon by *denitrifying bacteria.* These are of various kinds, and the net result of their combined metabolic activities is that nitrate is converted into molecular, atmospheric nitrogen, N_2. They therefore reduce the available supply of environmental nitrates and increase the nitrogen content of the air. However, still another set of bacteria indirectly compensates for this loss of nitrates. Atmospheric N_2 can be used directly by so-called *nitrogen-fixing* organisms, namely, certain bacteria and blue-green algae which live in water and soil. These absorb aerial nitrogen as a metabolite and are able to incorporate it into their amino acids and proteins. Some nitrogen-fixing bacteria are free-living soil saprotrophs. When they die, the nitrogenous materials of their bodies decay and yield ammonia to the soil, which is then transformed into nitrates by the nitrifiers. Other nitrogen-fixing bacteria are mutualistic symbionts on the roots of leguminous plants. The bacteria here bring about the formation of characteristic root nodules (cf. Chap. 17), and the nitrogen fixed by the bacteria becomes available to the legumes as *usable* nitrogen. Legumes actually may in some situations acquire most or all of their nitrogen from the nitrogen-fixing bacteria in the nodules.

Thus, by converting aerial nitrogen into usable nitrogen, the nitrogen-fixing bacteria complete the global nitrogen cycle. We note that this cycle depends on at least four different sets of bacteria: the decay causers, the nitrifiers, the denitrifiers, and the nitrogen fixers. We may emphasize also that these bacteria act as they do, not because they are aware of the grand plan of the global nitrogen cycle, but because they derive immediate metabolic benefits from their action. For example, the decay causers are saprotrophs which require dead organisms as their source of food. Ammonia happens to be the excretory byproduct of their gathering and processing activities involving nitrogen-containing foods. Analogously, the other bacterial types above utilize one kind of nitrogenous compound as nutrient and eliminate an altered kind as excretion product.

So long as the global environment makes available water, mineral ions, and atmospheric gases, and so long as it provides suitable operating conditions generally, the living components of the biosphere may obtain not only their required basic nutrients but also their necessary *living space:* the places, or habitats, in which animal communities must pursue life are portions of the hydrosphere, the lithosphere, or the atmosphere.

HABITATS

With the possible exception of the most arid deserts, the high, frozen mountain peaks, and the perpetually icebound polar regions, probably no place on earth is devoid of life. The subdivisions of the planetary environment represent *habitats* in which communities live. The two principal habitats are the *aquatic* and the *terrestrial.* Both range from equator to pole and from a few thousand feet below to a few thousand feet above sea level. *Ocean* and *fresh water* are the principal components of the aquatic habitat and *air* and *soil* of the terrestrial.

The Ocean

The ocean basin. Even the land dweller will appreciate readily that the sea is not a single, unified environment. Indeed, an examination of its structure and of its content of living matter shows clearly that this birthplace of life comprises nearly as many distinct habitats as the land.

The most conspicuous attribute of sea water is its high mineral content. The *proportions* of the different types of salt present in sea water are almost the same all over the globe, as a result of thorough mixing of all waters by currents. More than four-fifths of the total mineral content consists of the ions of table salt; 55 per cent of all ions present are chlorine, 30 per cent are sodium. However, although the proportions are invariant, the total salt concentration, or *salinity,* varies greatly from region to region. The highest salinities are encountered in tropic waters, where high temperatures and extensive evaporation concentrate the quantities of oceanic salts. In the Red Sea, for example, ocean water is a 4 per cent solution; every 100 parts of sea water include 4 parts of minerals (and 55 per cent of these 4 parts are chlorine, as just noted above). By contrast, sea water in higher latitudes, north or south, evaporates less and is therefore less salty. In such latitudes, moreover, the fresh water melt from the polar ice caps dilutes the oceanic salts considerably. As might be expected, furthermore, salinities are lower for often several hundred miles around the mouths of great rivers. The lowest known salinity occurs in the Baltic Sea, where ocean water is a 0.7 per cent solution (comparable to the total salinity of the blood of frogs and somewhat less than that of human blood). Salinity determines the buoyancy (density) of ocean water, buoyancy being the greater the higher the salinity. Both salinity and buoyancy are of considerable significance to all marine life (cf. below).

An ocean basin has the general form of an inverted hat (Fig. 18.13). A gently sloping *continental shelf* stretches away from the coast line for an average distance of about 100 miles (discounting often extreme deviations from this average). The angle of descent then changes more or less abruptly and the shelf grades over into a steep *continental slope.* Characteristically, this slope is scored deeply by gorges and canyons, carved out by slow rivers of mud and sand discharging from estuaries. Several thousand feet down, the continental slope levels off into the ocean floor, a more or less horizontal expanse known as the *abyssal plain.* Mountains rise from it in places, with peaks sometimes so high that they rear up above sea level as islands. Elsewhere the plain may be scarred by deep rifts, e.g., the Japan and Philippine Deeps along the western edge of the Pacific. These plunge 35,000 ft down and are the lowest parts of the earth's crust.

Three major habitats may be distinguished in such a basin. The sea floor from the shore out to the edge of the continental shelf forms the *littoral* zone. Its most important subenvironment is the narrow *intertidal* belt, between the high- and low-tide lines. Beyond the littoral, the sea floor along the continental slope and the abyssal plain constitutes the *benthonic* habitat. The third principal habitat is the *pelagic*—the water itself which fills the ocean basin. This environment includes a *neritic* subdivision over the littoral zone and an *oceanic* subdivision over the benthonic zone.

A most important vertical subdivision of the pelagic habitat is brought about by the sun. Acting directly or via the overlying medium of air, the sun produces "weather" in the surface layers of the sea: waves, currents, storms, evaporation, seasons, daily climatic rhythms, and other changes. Deep water is not so affected. Moreover, sunlight penetrates into water only to an average depth of about 250 ft and to at most 600 ft in certain seas. Within this sunlit layer, called the *photic zone,* light dims progressively to zero with increasing distance from the surface. The most significant consequence of this circumstance is that photosynthesizing vegetation can exist only in the uppermost layers of the sea. Animal life directly dependent on plant foods therefore must similarly remain near the surface. As a result, the top 250 ft or so of the oceans contains a concentration of living matter as dense as any on earth. In sharp contrast, the *aphotic zone,* i.e., the dark region underneath the photic zone, is completely free of photosynthetic organisms and contains only animals, bacteria, and possibly fungi.

On the basis of its relationship to these various habitats, marine life has been classified into three general categories: *plankton, nekton,* and *benthos.* Plankton includes all passively drifting or floating forms. Most of them are microscopic and are found largely in the surface waters of the sea, i.e., in the photic zone. Even though some of these forms possess loco-

18.13. The structure of an ocean basin. The littoral zone is not labeled; it is that part of the benthonic zone which forms the floor of the continental shelf (and is covered by the waters of the neritic zone).

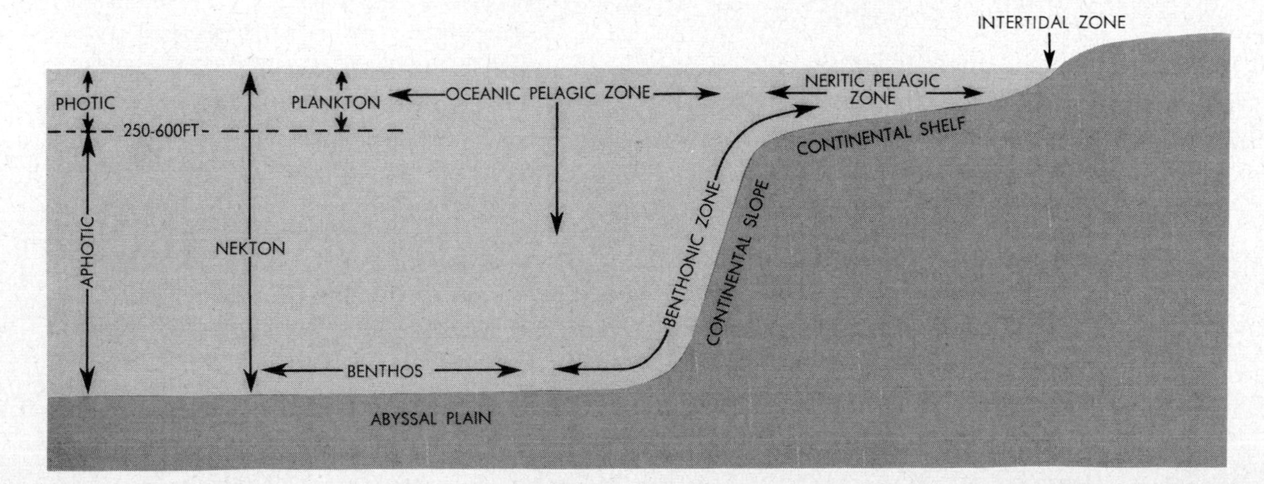

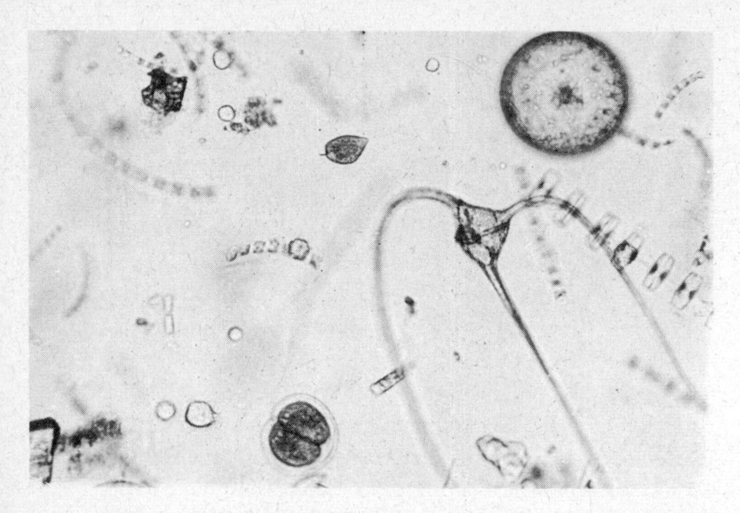

18.14. Marine plankton, including diatoms, dinoflagellates, protozoa, animal eggs, and other organisms.

motor systems, they are nevertheless too weak or too small to counteract currents and movements of water. Nekton comprises the active swimmers, capable of changing stations at will. All nektonic types are therefore animals, and they are found in all waters, along the surface as well as in the sea depths. The benthos consists of crawling, creeping, and sessile animals along the sides and the bottom of the ocean basin.

The photic zone. Since photosynthetic organisms do not possess powerful locomotor systems such as muscles, such organisms in open water can stay within the range of sunlight only if they float. And since living material is slightly heavier than water, passive floating is possible only if an organism possesses a special floating device or if it is small enough to be buoyed up by the salt water.

Thus, the predominant marine photosynthesizers are planktonic. They include teeming trillions of algae which, as a group, probably photosynthesize more food than all land plants combined. Collectively called *phytoplankton*, this oceanic vegetation represents the richest pasture on earth; directly or indirectly, it forms the nutritional basis of all marine life (Fig. 18.14).

Most of the algal types included in this "grass of the sea" are microscopic. Unquestionably the most abundant are the *diatoms*. Each of these single-celled protists is enclosed within a delicate, intricately sculptured, silicon-containing shell. Reddish *dinoflagellates* also abound in surface waters, sometimes in populations so dense that they tint acre upon acre of ocean with a coppery hue (e.g., "red" tides). Other marine algae include many types of variously pigmented forms, and some of these, as well as countless numbers of marine bacteria, are bioluminescent. They emit flashes of cold light, which dot the night seascape with a billion pin points of greenish fire.

Surrounded on all sides by raw materials and bathed in sunlight, the passively drifting phytoplankton community inhabits a highly favorable, chemically rather stable habitat. The death rate resulting from animal feeding is high, but rapid reproduction sufficiently offsets it. Physical and climatic changes do not affect an algal cell too greatly. In winter, the temperature of surface waters may fall below the freezing point, but the salts of the ocean prevent actual freezing. Cold merely reduces the rate of metabolic processes and algal life continues at a slower pace.

Indeed, low temperatures promote algal growth. When surface temperatures are high, as in tropical waters throughout the year and in northern and southern waters in summer, pronounced *temperature layering* of water prevents much vertical mixing. A warm-water layer is less dense and thus lighter than a colder layer below it; it "swims" on top of the colder layer without mixing. Under such conditions, organisms in the warm layer deplete the surface waters of mineral raw materials and at death these materials sink down without being brought back to the surface by vertical mixing. As a result, the amount of surface life is limited, and warm seas are actually relatively barren (Fig. 18.15).

18.15. Temperature layering of surface water, as in summer, leads to the formation of a thermocline, a temperature barrier responsible for poor vertical circulation of water (left). In early spring, late fall, and winter, surface waters do acquire the same temperature as deeper layers; the thermocline then disappears, and vertical mixing does become possible (right).

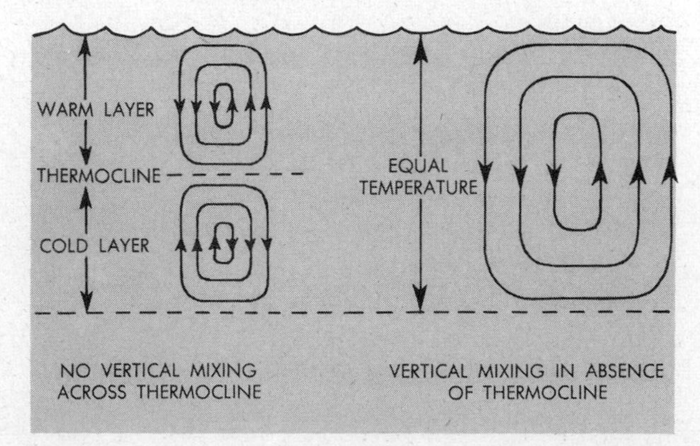

18.16. Diurnal migrations of plankton organisms. At night (left), plankton is distributed throughout the dark surface water. In daytime (right), plankton migrates away from the surface and accumulates in darker, deeper waters.

By contrast, when surface and deeper waters have roughly the same low temperature, vertical mixing becomes possible and minerals are recirculated more rapidly. Surface life may therefore be more abundant. The perennially cold arctic, antarctic, and subpolar waters actually support huge permanent populations of algae. And, as is well known, the best commercial fishing grounds are in the high north and south, not in the tropics, and the best fishing seasons are spring and fall, not summer.

Warm and cold waters differ not only in the total amount of life but also in its diversity. Although the biomass in tropic seas is comparatively smaller than in temperate and polar seas, its species diversity is far greater. Thus, whereas warm oceans sustain small populations of many species, cold oceans harbor large populations of few species. The reason is that higher temperatures promote all reactions, including those leading to evolution. Warm-climate life will therefore tend to become more diverse than cold-climate life. However, the limited mineral content of warm seas will keep individual numbers low. Note also that, in all oceans, the sun produces *diurnal migrations* of most planktonic organisms. During the night the organisms are distributed vertically throughout the surface layers of the sea, but during the day most of the plankton shuns the bright light and moves down into the lower strata of the photic zone. Larger animals feeding on plankton migrate up and down correspondingly. As a result, even richly populated seas are quite barren on the surface during the daytime, and it is well known that surface fishing is most fruitful at night (Fig. 18.16).

In certain circumscribed regions, phytoplankton also includes larger, multicellular algae: flat, sheetlike seaweeds, often equipped with specialized air bladders which aid in keeping the organisms afloat. Such seaweeds may sometimes aggregate in considerable numbers over wide areas, particularly if a region is ringed in by ocean currents and therefore remains relatively isolated and stagnant. The Sargasso Sea in the mid-Atlantic is a good example. This sea has figured prominently in marine lore. For example, stories are told of ships trapped in "floating jungles," rapidly overgrown by plants, and sunk without a trace. Such accounts are wholly legendary, since the organisms are nowhere dense enough to prevent a ship's passage. Yet the Sargasso *is* unique from a botanical and also a zoological standpoint. The comparative isolation of the region has led to the evolution of distinct plants not found elsewhere on earth, and an equally distinct fauna finds shelter and food in this vegetation.

Living side by side with the photosynthetic phytoplankton in the open waters of the photic zone are the small nonphotosynthetic forms. These include bacteria and members of the *zooplankton:* protozoa, eggs, larvae, tiny shrimp (krill) and other crustacea (particularly *copepods*), and countless other small animals carried along by surface drift (Fig. 18.17). They feed directly on the microscopic vegetation; hence as the

18.17. Freshwater plankton. The algal components here are largely *Volvox,* and the animals are crustacean water fleas (*Daphnia*) and copepods (*Cyclops*).

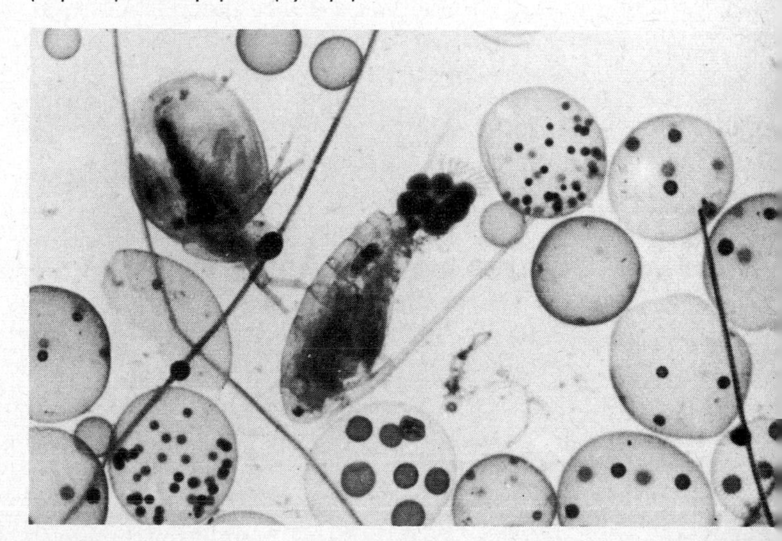

phytoplankton waxes and wanes, so does the zooplankton. A good part of the nekton, largely fishes and marine mammals, comes into these waters to feed either on zooplankton or on phytoplankton directly.

Nearer to shore, in the neritic waters above the littoral zone, the problem of remaining afloat is not so critical for a photosynthesizer as in open water, for here even a bottom dweller is likely to be within the range of sunlight. The problem, rather, is to remain attached to solid ground, for close to shore the force of waves and of ground swells is considerable. In the intertidal belt, moreover, an even more profound problem is the ebbing of water twice daily and the consequent rhythmic alternation between aquatic and essentially terrestrial conditions. Also, in waters in and for miles beyond estuaries, fresh water discharging from rivers mixes with ocean water, a circumstance introducing additional environmental inconstancies. Being the meeting ground of water, land, and air, the intertidal belt is actually among the most violently changing habitats on earth.

Vegetation here and in the littoral and the overlying neritic region as a whole is again largely algal. In addition to the single-celled and small planktonic types, attached multicellular forms abound. Most of these are equipped with specialized holdfasts which anchor the organisms to underlying ground. Green, brown, and red algae are particularly common. For example, the soft, slippery mats of vegetation encrusting rocks along the shore are familiar to many, as is *Fucus*, a common leathery brown alga found in dense populations on coastal rock.

Animals in coastal waters include representatives of almost all major groups. In addition to the abundant planktonic types, sessile and creeping animals occur which are variously adapted to rocky, muddy, or sandy bottoms. The animals make use of all conceivable dwelling sites—for example, tide pools left on rock by ebbing water, crevices and hollows in and under rock, burrows in sand or mud, the sheltered water among vegetation and among sessile animal growths, empty shells and other skeletons of dead animals, and flotsam and jetsam along the shore and in deeper water. Among the very abundant nektonic animals in these regions, largely fish, many normally do not stray very far from a particular home, even though an efficient locomotor system would permit them to do so. They foray into surrounding waters for food and mates, but they always tend to return to the same base of operations. However, another group of nektonic animals consists of perpetual wanderers without permanent homes.

The aphotic zone. The contrast between the surface habitats within reach of the sun and those underneath is dramatic. As the ones are forever fluctuating, so the others are perennially steady and relatively unchanging. The deep ocean is still little explored, and, for many, this "last frontier" has acquired a romance and mystery all its own. Several unique physical conditions characterize this world of the sea depths.

First, the region is one of eternal night. In the total absence of sunlight, the waters are pervaded with a perpetual blackness of a kind found nowhere else on earth. Secondly, seasons and changing weather are practically absent. Localized climatic changes do occur as a result of occasional submarine volcanic activity or, more regularly, through deep-sea currents. These produce large-scale shifts of water masses and, incidentally, bring oxygen to even the deepest parts of the ocean. Being beyond the influence of the sun, the deep waters are cold, unchangingly so. Temperatures range from about 10°C at the top of the dark zone to about 1°C along the abyssal plain.

Thirdly, water pressure increases steadily from the surface down, 1 atmosphere (atm) for every 33 ft of descent. Thus, in the deepest trenches of the ocean, the pressure is about a thousand times as great as at sea level. And fourthly, a continuous slow rain of the dead remnants of surface organisms drifts down toward the sea bottom. Much of this material, particularly the organic fraction, dissolves completely during the descent. But much microscopic mineral matter reaches the abyssal plain, where it forms ever-thickening layers of ooze. Accumulating over the millennia, the older layers eventually compress into rock. Vertical bore samples of such rock have revealed a great deal of the past history of the oceans and their once-living surface inhabitants.

Contrary to early beliefs that life should be impossible in such an environment, a surprisingly rich diversity of organisms has been found to exist virtually everywhere in the free water and along the floor of the deep sea. Apart from containing bacteria and perhaps fungi, the community is characteristically *animal*—photosynthesizing organisms are confined to the sunlit surface. Virtually all animal groups are represented, many by—to us—strange and bizarre types uniquely adapted to the locale (Fig. 18.18).

If a deep-sea animal is to avoid death from explosion or implosion, its internal pressure must equal the external pressure of the water. A few of the nektonic animals, toothed whales, for example, are adapted to resist the harmful effects of rapid changes of external pressure. These animals are cap-

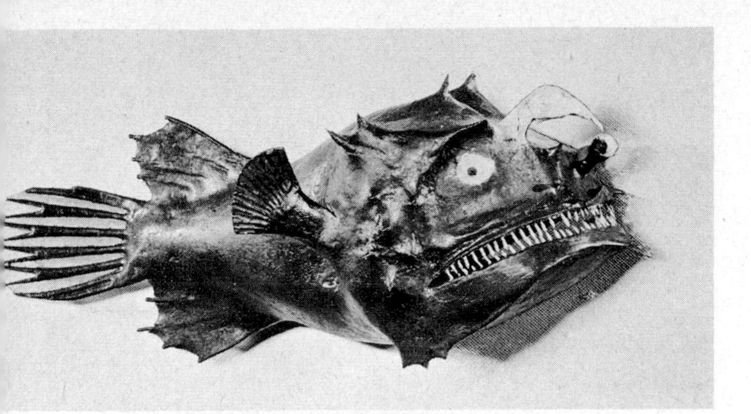

18.18. The deep-sea angler fish shown here is a female. The structure above the eye is a parasitic male, which is carried about permanently attached. This neatly solves the problem of finding a mating partner in the dark. Many of these large-mouthed, dagger-toothed fishes are surprisingly small; e.g., the fish above fits comfortably into a person's palm.

able of traversing the whole ocean from bottom to surface. They may therefore feed directly on the rich food supplies in surface waters. But the bulk of the deep-sea nekton is adapted to particular water pressures only, and given animals are rigidly confined to limited pressure zones at given depths. Such animals must therefore obtain food either from the dead matter drifting down from the surface—a meager source, particularly in deeper water—or from within the nekton itself.

This last condition makes the deep sea the most fiercely competitive habitat on earth. The very structure of the animals underscores their violently carnivorous, "eat-or-be-eaten" mode of existence. For example, most of the fishes have enormous mouths equipped with long, razor-sharp teeth, and many can swallow fish larger than themselves.

Since the environment is pitch-black, one of the critical problems for these animals is to *find* food to begin with. A highly developed pressure sense provides one solution. Turbulence in the water created by nearby animals can be recognized and, depending on the nature of the turbulence, may be acted upon either by flight or by approach.

Another important adaptation to the dark is bioluminescence. Many of the deep-sea animals possess light-producing organs on the body surface, of different shapes, sizes, and distributions in different species. The light patterns emitted may include a variety of colors and probably serve partly in species recognition. Identification of a suitable mate, for example, must be a serious problem in an environment where everything appears equally black. Another function of the light undoubtedly is to warn or to lure. Some of the bioluminescent lures have evolved to a high degree of perfection. Certain fish, for example, carry a "lantern" on a stalk protruding from the snout (Fig. 18.19). An inquisitive animal attracted to the light of the lantern will discover too late that it has headed straight into powerful jaws.

The Fresh Water

Physically and biologically, the link between ocean and land is the fresh water. Rivers and lakes were the original invasion routes over which some of the descendants of ancestral marine animals reached land and, in the process, evolved into terrestrial forms. Certain of the migrant types never completed the transition but settled along the way, in fresh water.

Among such animals, some adapted to the brackish water in estuaries and river mouths or to a life spent partly in the ocean, partly in fresh water (e.g., salmon, eels). Very many types could leave the ocean entirely and adapt to an exclusively freshwater existence. The descendants of these animals in-

18.19. A deep-sea angler fish with a stalked, luminescent "lantern" over the mouth. Note the vertical position of the mouth, a feature which facilitates catching prey lured to the light of the lantern.

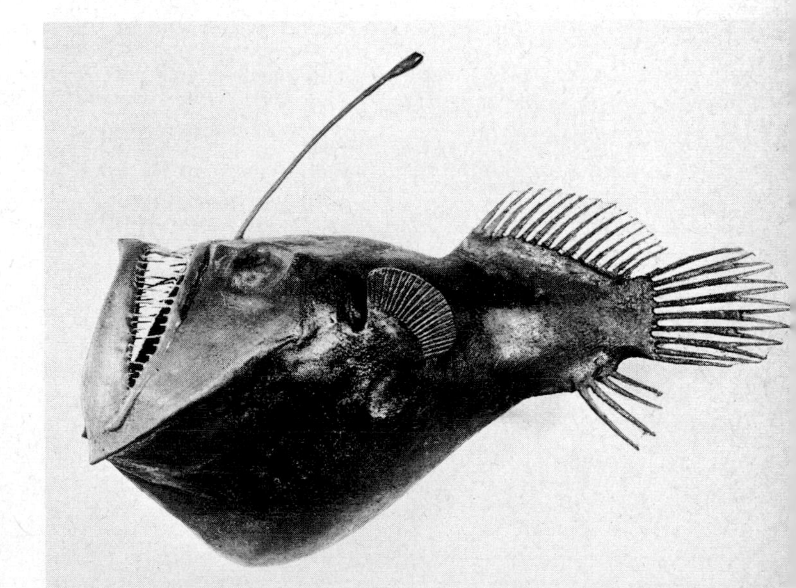

clude representatives of most of the groups present in the ocean. Certain of the freshwater types later managed to gain a foothold on land. Of these, some continued to spend part of their lives in or near fresh water (e.g., frogs), but more became wholly terrestrial. And among the terrestrial forms, some subsequently returned to water and adapted secondarily to an aquatic existence (e.g., some snails, many insects). Thus, animals which inhabit the fresh water today constitute a rich and major subdivision of the living world.

In addition to rivers, freshwater habitats also include lakes, ponds, and marshes. Contrary to what might be expected, each category of these environments harbors communities which are remarkably similar throughout the world. Thus, lakes of a given size in any part of the world are likely to contain roughly identical types of communities. Such similarities are probably due to several factors. First, if a given marine ancestor could give rise to a particular freshwater descendant in one part of the world, any such descendant in another part would probably have been quite similar, since the freshwater environments are themselves similar. Secondly, freshwater systems within a continent are often interconnected, and direct migration from one part of such a system to a far distant part is possible in many cases. Thirdly, many freshwater animals are able through their own locomotion to travel overland for longer or shorter distances; and small animals also are carried overland accidentally, by sweepstake routes which often involve birds as carrying agents (e.g., animal-containing mud clinging to the feet of birds). Whether or not such explanations are fully adequate, it is true nevertheless that freshwater communities are far more similar on a global basis than either marine or terrestrial communities.

Three main conditions distinguish the freshwater habitat from the ocean. First, the salinity is substantially lower. In various earlier contexts we have already discussed the osmotic consequences of this difference and the excretory adaptions evolved by freshwater animals. A second condition characterizing much of the freshwater environment is the presence of strong, swift currents. Where these occur, passively floating life so typical of the ocean surface is not likely to be encountered. On the contrary, the premium will be either on maintaining firm anchorage along the shores and bottoms of rivers or on ability to resist and to overcome the force of currents by muscle power.

Indeed, the vegetation found in swift rivers consists almost entirely of plants possessing rootlike holdfasts or actual roots. Since true roots are characteristic only of terrestrial plants, the presence of such plants in rivers means that some ancestral land plants have adapted secondarily to a life in water. Pertinent examples are reed grasses, water foxtails, wild rice, and watercress. By contrast, where fresh water is not flowing strongly, as in lakes, ponds, bogs, and marshes, not only rooted but also floating planktonic vegetation may be exceedingly abundant. In stagnant or near-stagnant water, algal communities forming continuous layers of green surface scum are particularly conspicuous.

Among animals, analogously, those in quiet fresh waters

18.20. Amphibian eggs, each surrounded by a sticky jelly coat which attaches such eggs to vegetation or other objects in the water.

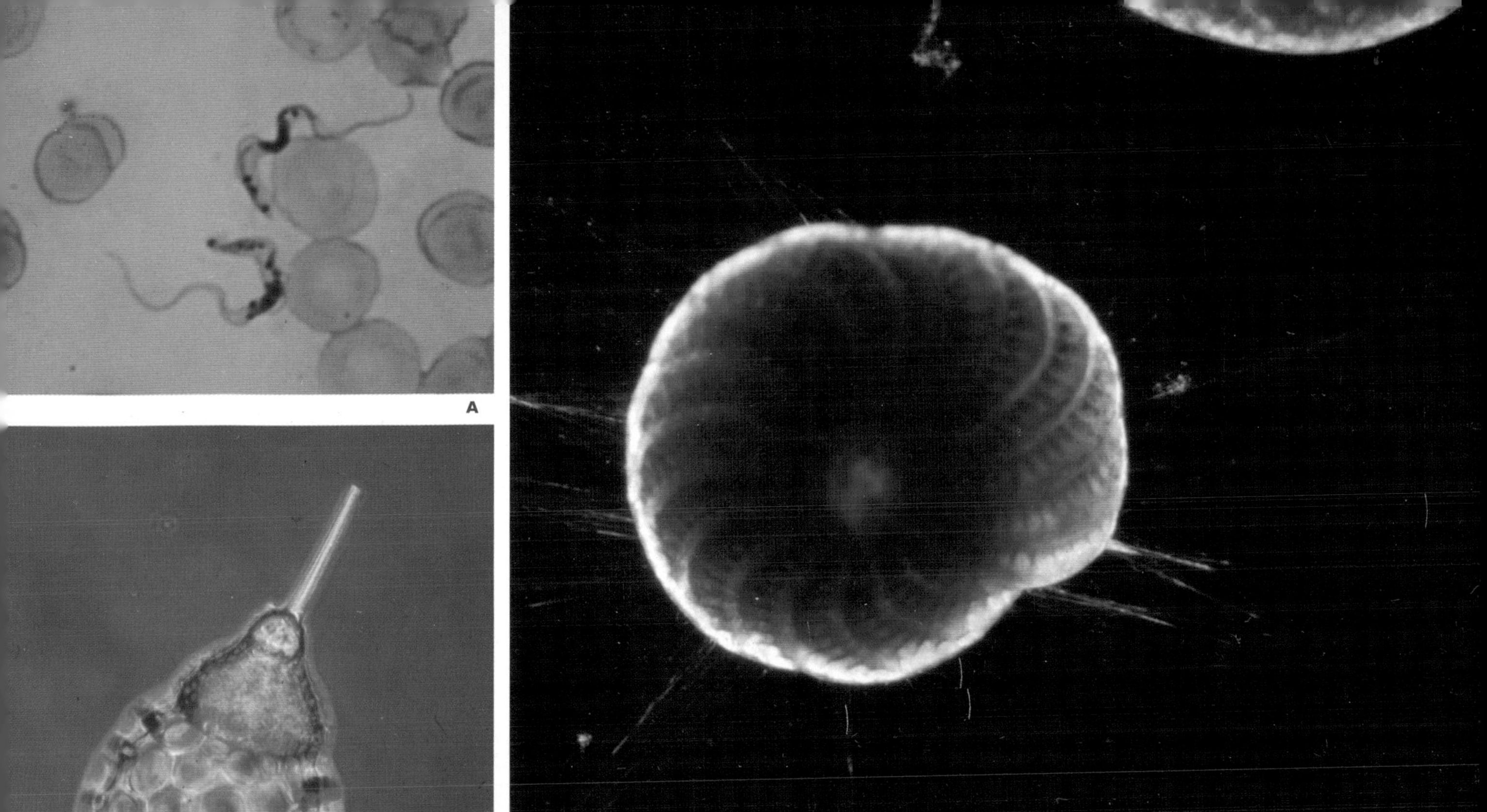

PLATE I. Flagellate and amoeboid protozoa. A, *Trypanosoma,* a flagellate parasite in blood. **B,** *Elphidium* (= *Polystomella*), a foraminiferan. **C,** *Podocystis,* and **D,** *Saturnulus,* two radiolarians.

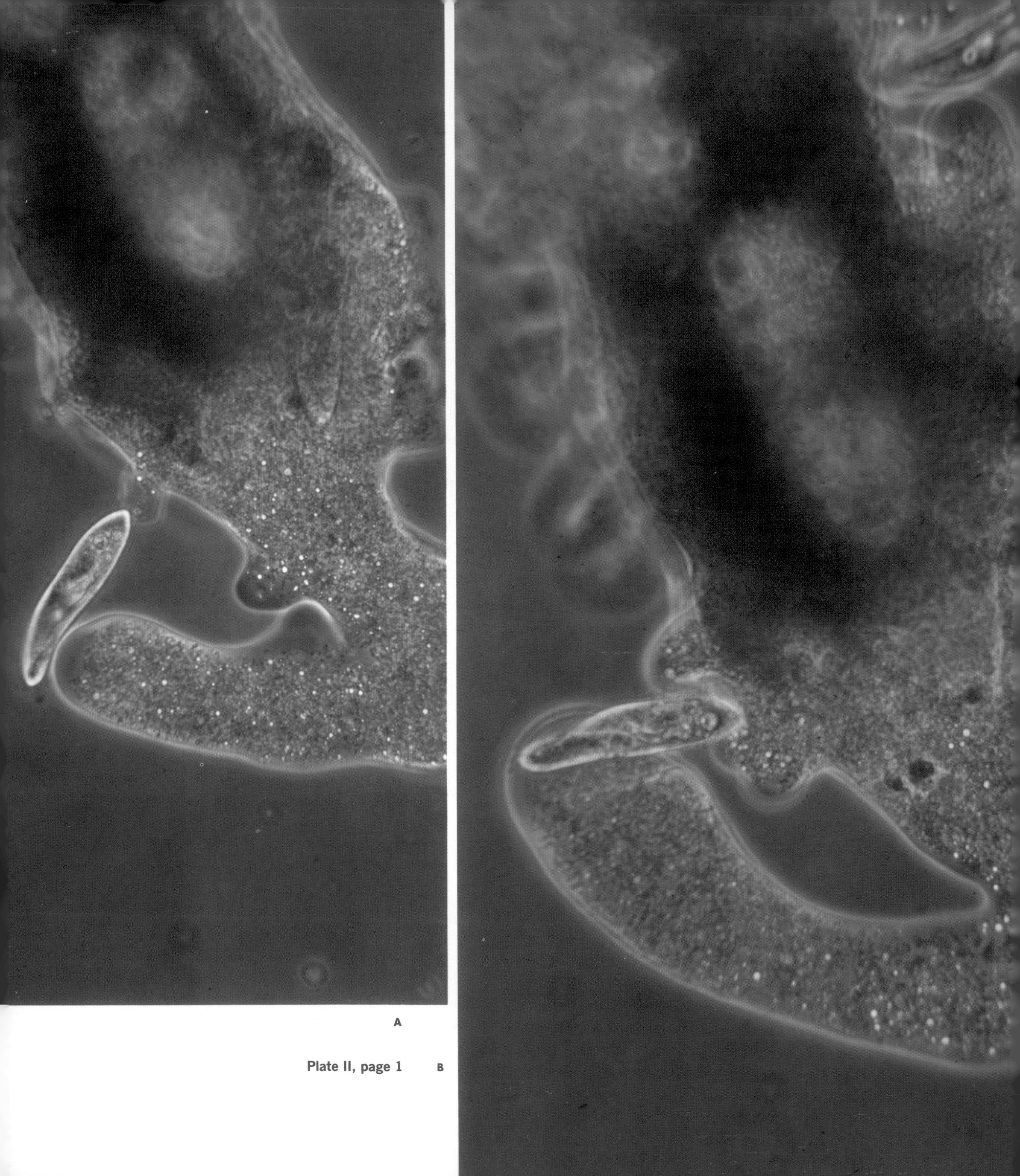

Plate II, page 1

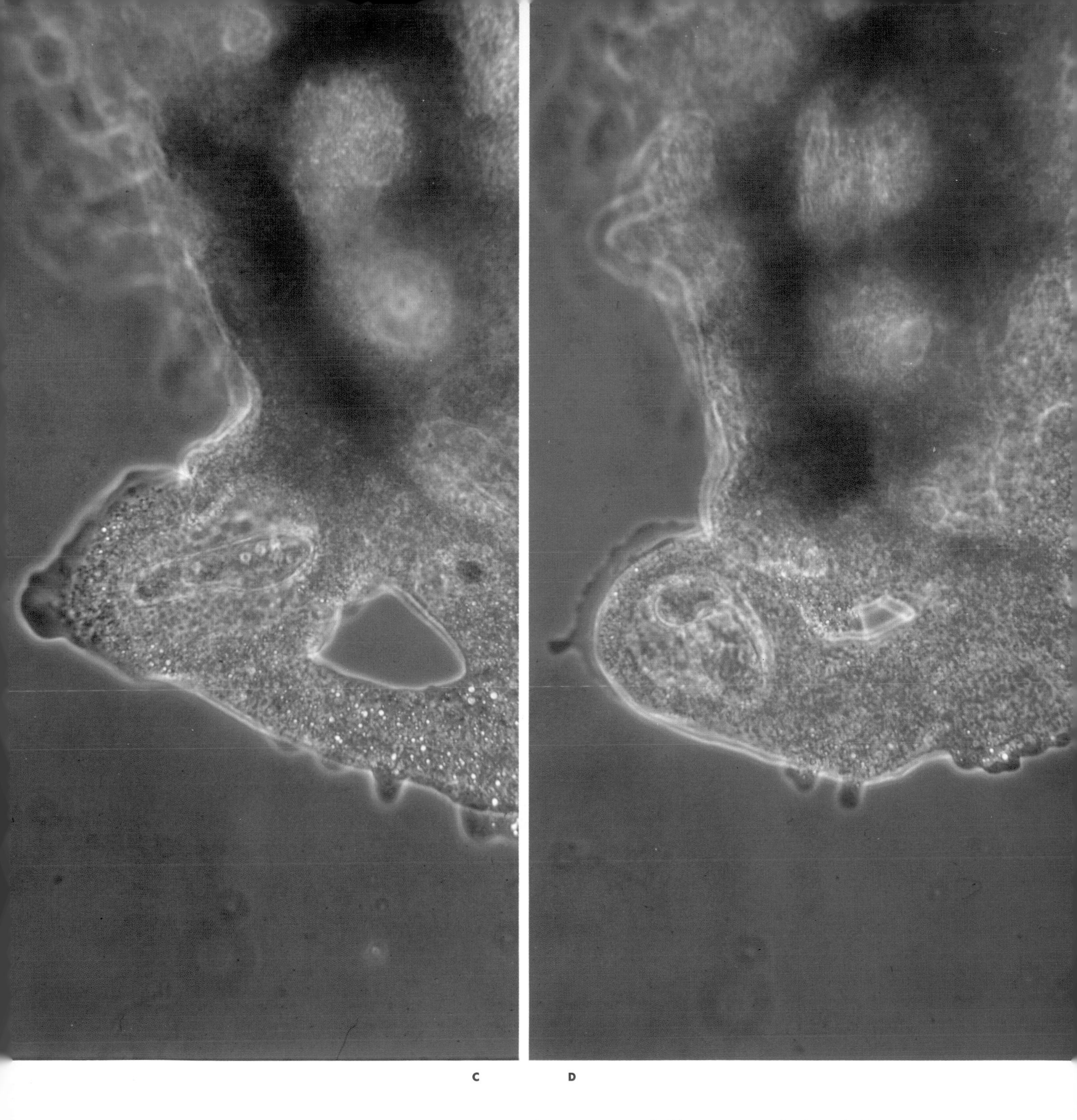

C D

Plate II, page 2

PLATE II. Amoeboid protozoa. A through **D,** sequence showing an amoeba catching and engulfing a paramecium.

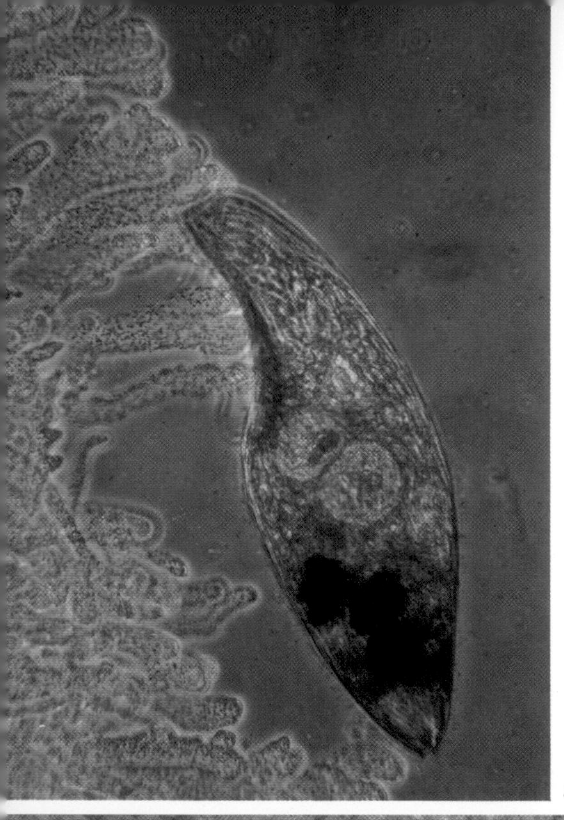

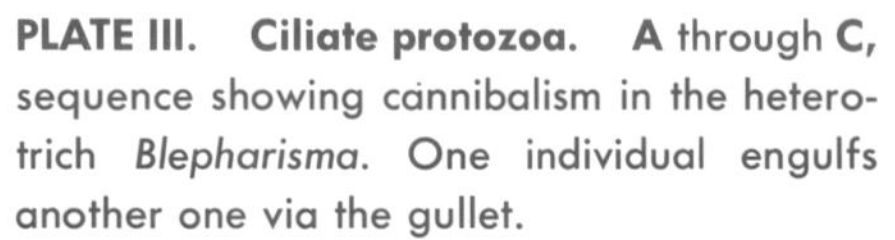

PLATE III. Ciliate protozoa. A through **C,** sequence showing cannibalism in the heterotrich *Blepharisma*. One individual engulfs another one via the gullet.

Plate III, page 1

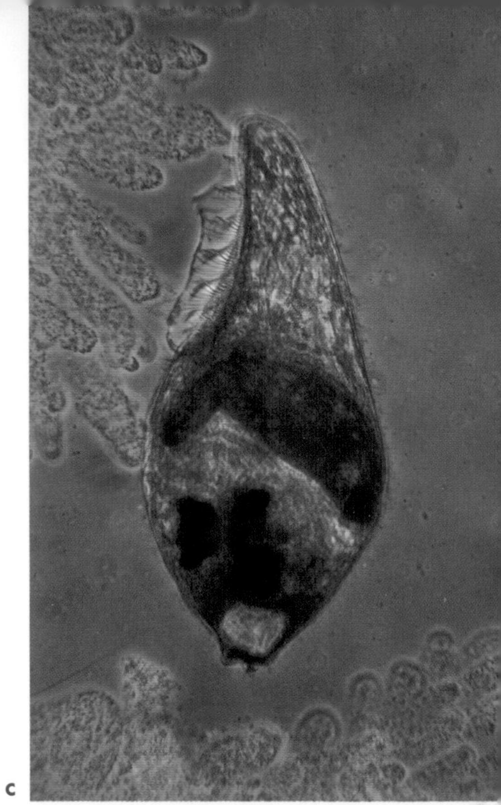

A B C

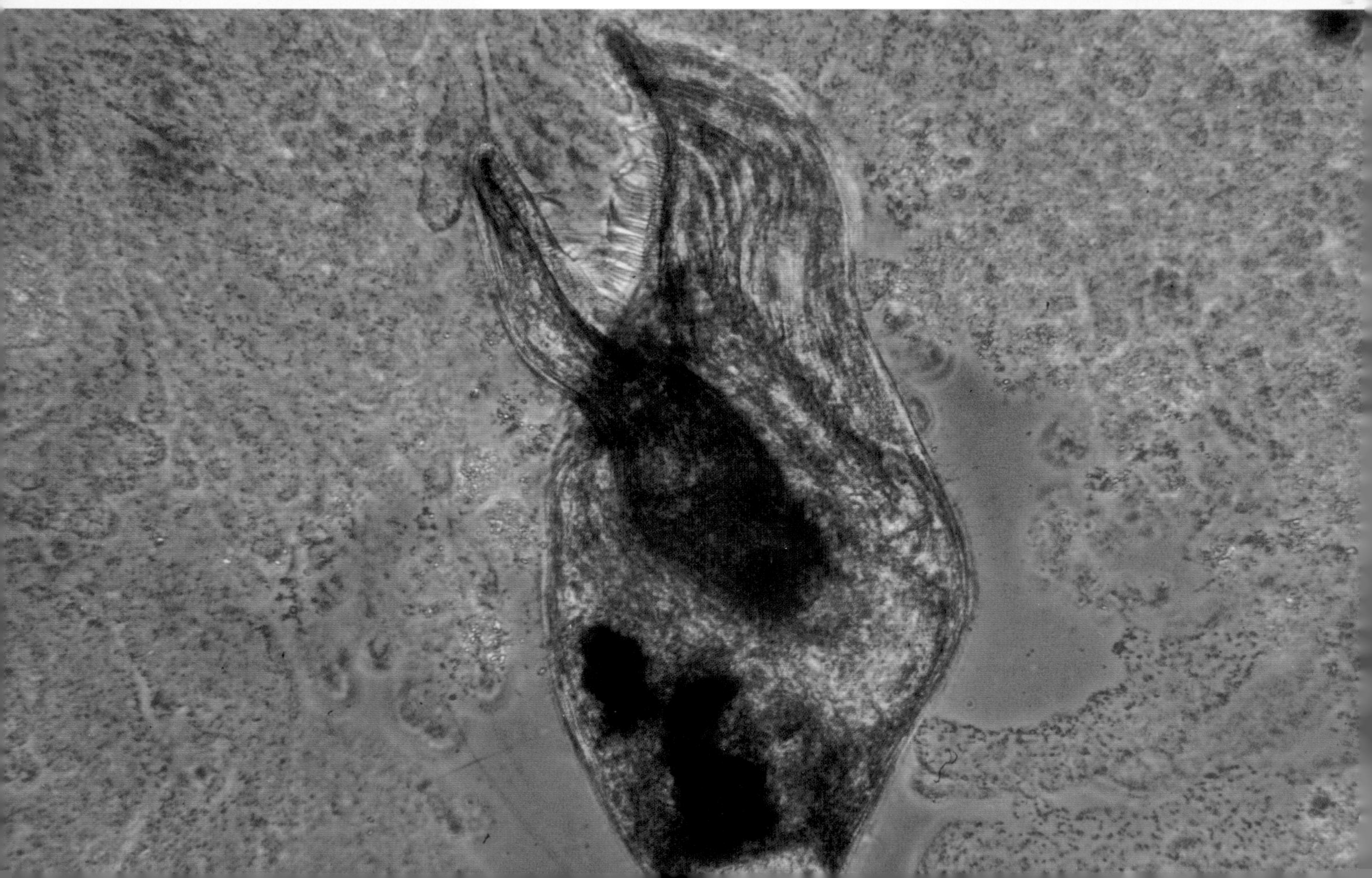

include both planktonic and nektonic types, but those in swiftly flowing water are either attached and sessile or nektonic and swimming. The eggs of nektonic animals cannot swim, to be sure, but they are enveloped by sticky jelly coats which adhere firmly to plants or other objects in the water (Fig. 18.20). And the young are strongly muscled from the moment they hatch (e.g., freshwater vertebrates).

A third major distinction between fresh water and ocean is that the former, with the exception of only the very large lakes, is affected much more by climate and weather than any part of the latter. Bodies of fresh water often freeze over in winter and may dry up completely in summer. Water temperatures change not only seasonally but also daily, frequently to a considerable extent. Gales or flood conditions may bring bottom mud and silt to the surface and upset the freshwater habitat in major ways. A large number of factors may alter flow conditions and produce, for example, stagnant water or significantly altered chemical content or situations facilitating infectious epidemics. We note that the fresh water shares the environmental inconstancies of the land in very large measure. Notwithstanding the aquatic nature of the freshwater habitat, its living component reflects the ebb and flow of land life as much as that of ocean life.

The Land

That land habitats differ vastly in their characteristics is eminently clear to a land dweller as efficient and far-ranging as man. It should also be clear that, regardless of which particular subdivision of the terrestrial environment one considers, the sustaining foundations of all land life are air and, directly or indirectly, also soil. Air and soil are to the terrestrial habitat what the surface waters of the ocean are to the marine.

Like air, soil is itself a terrestrial home, providing a habitat for a vast array of subsurface organisms. And by creating the conditions necessary for the survival of all other terrestrial organisms, soil becomes a major agency which transforms terrestrial environments into life-sustaining "habitats." Two other agencies play a vital role here: annual *temperature* and *rainfall*. As these vary with geographic latitude and altitude, they divide the soil-covered land surface into a number of distinct habitat zones, or *biomes: desert, grassland, rain forest, deciduous forest, taiga*, and *tundra*.

In the tropics are found representatives of the first three of the six biomes just named. They are characterized here by comparatively high annual temperatures and by daily temperature variations which are greater than the seasonal variations. Differences in the amount of precipitation largely account for the different nature of these habitats.

A *desert* (Fig. 18.21) usually has less than 10 in. of rain per year, concentrated largely in a few heavy cloudbursts. Desert

18.21. The desert habitat. The bobcat and kangaroo rat shown are characteristic members of the fauna of many deserts.

A

B

life is well adapted to this. Plants, for example, grow, bloom, are fertilized, and produce seeds, all within a matter of days after a rain. Since the growing season is thus greatly restricted, such plants stay relatively small. Leaf surfaces are often reduced to spines and thorns (as in cacti), minimizing water loss by evaporation. Desert animals too are generally small, and they include many burrowing forms which may escape the direct rays of the sun under the ground surface. In most deserts, the homoiothermic mammals and birds are comparatively rare or are absent altogether; maintenance of constant body temperature is difficult or impossible under conditions of great heat and practically no water. By contrast, animals which match their internal temperature to the external can get by much more easily.

Grassland is not an exclusively tropical habitat but extends into much of the temperate zone as well (Fig. 18.22). The more or less synonymous terms "prairie," "pampas," "steppe," "puszta," and many other regional designations underscore the wide distribution of this biome. The common feature of all grasslands is intermittent, erratic rainfall, amounting to about 10 to 40 in. annually. Grasses of various kinds, from short buffalo grass to tall elephant grass and thickets of bamboo, are particularly adapted to irregularly alternating periods of precipitation and dryness. Grassland probably supports more species of animals than any other terrestrial habitat. Different kinds of mammals are particularly conspicuous.

In those tropical and subtropical regions where torrential rains fall practically every day and where a well-defined rainy season characterizes the winter, plant growth continues the year round. *Rain forests* have developed here (Fig. 18.23), typified particularly by the communal coexistence of up to several hundred different species of trees. Rain forests are the "jungles" of the adventure tale. They cover much of central

18.22. Two mammals characteristic of African grasslands. A, oryx; **B,** wild buffalo.

A

B

18.23. The habitat of the rain forest. Many dozens of different plant types, coexisting in dense formations, are generally characteristic of it. Note moss plants hanging from many tree branches.

Africa, south and southeast Asia, Central America, and the Amazon basin of South America. Trees in such forests are normally so crowded together that they form a continuous overhead canopy of branches and foliage, which cuts off practically all the sunlight, much of the rain water, and a good deal of the wind. As a result, the forest floor is exceedingly humid and quite dark, and it is populated by plants requiring only a minimum of light. Animal communities too are stratified vertically, according to the several very different habitats offered between canopy and ground. The tropical rain forest is singularly quiet during the day, but it erupts into a cacophony of sound at night, when the largely nocturnal fauna becomes active.

In the temperate zone, apart from extensive grasslands and occasional deserts, the most characteristic biome is the *deciduous forest* (Fig. 18.24). The fundamental climatic conditions here are cold winters, warm summers, and well-spaced

18.24. Two mammals characteristic of many deciduous forests. **A,** Virginia deer; **B,** mountain lion.

A

B

rains bringing some 30 to 40 in. of precipitation per year. The habitat is characterized also by seasonal temperature variations which are greater than the daily variations. Winter makes the growing season discontinuous, and the flora is adapted to this. Trees are largely deciduous, i.e., they shed their leaves and hibernate; and small annual plants produce seeds which withstand the cold weather. A deciduous forest differs from a rain forest in that trees are spaced farther apart and in that far fewer species are represented. Compared with the hundreds of tree types in the one, there may be only some ten or twenty in the other. The many familiar animal types in this biome include deer, boars, raccoons, foxes, squirrels, and, characteristically, woodpeckers.

18.25. Two mammals characteristic of the habitat of the taiga. **A,** moose; **B,** North American wolverine.

A

B

North of the deciduous forests and the grasslands, across Canada, northern Europe, and Siberia, stretches the *taiga* (Fig. 18.25). This is a biome of long, severe winters and of growing seasons limited largely to the few months of summer. Hardy conifers, spruce in particular, are most representative of the flora, and moose, wolves, and bears of the fauna. The taiga is preeminently a zone of forests. These differ from other types of forests in that they usually consist of a single species of tree. Thus, over a large area, spruce, for example, may be the only kind of tree present. Another conifer species might be found in an adjacent, equally large area. Occasional stands of hardy deciduous trees are often intermingled with conifers. An accident of geography makes the taiga a habitat characteristic of the northern hemisphere only: little land exists in corresponding latitudes of the southern hemisphere.

The same circumstance makes the *tundra,* most polar of terrestrial biomes, a predominantly northern phenomenon (Fig. 18.26). Much of the tundra lies within the Arctic Circle. Hence its climate is cold and there may be continuous night during the winter season and continuous daylight, of comparatively low intensity, during the summer. Some distance below the surface, the ground is permanently frozen (*permafrost*). Above ground, frost can form even during the summer—plants often freeze solid and remain dormant until they thaw out again. The growing season is very brief, as in the desert, but in the tundra the limiting factor is temperature, not water supply. Plants are low, ground-hugging forms, and trees are absent. Lichens, mosses, coniferous and other shrubby growths, and herbs with brilliantly colored flowers, all blooming simultaneously during the growing season, are characteristic of the habitat. Conspicuous among the animals are hordes of insects, particularly flies, and a considerable variety of mammals: caribou, arctic hares, lemmings, foxes, musk oxen, and polar bears. Birds are largely migratory, leaving for more southern latitudes with the coming of winter.

Life does not end at the northern margin of the tundra but extends farther into the ice and bleak rock of the soilless polar region. Polar life is almost exclusively animal. And it is not really terrestrial anyway but is based on the sea (e.g., walrus, seals, penguins; Fig. 18.27).

The horizontal sequence of biomes between equator and

18.26. Two animals characteristic of the habitat of the tundra. A, musk ox; **B,** ptarmigan.

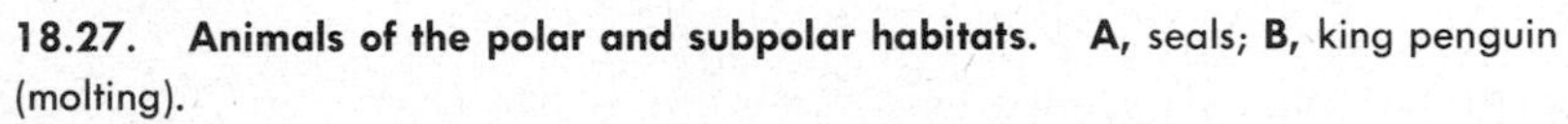

18.27. Animals of the polar and subpolar habitats. A, seals; **B,** king penguin (molting).

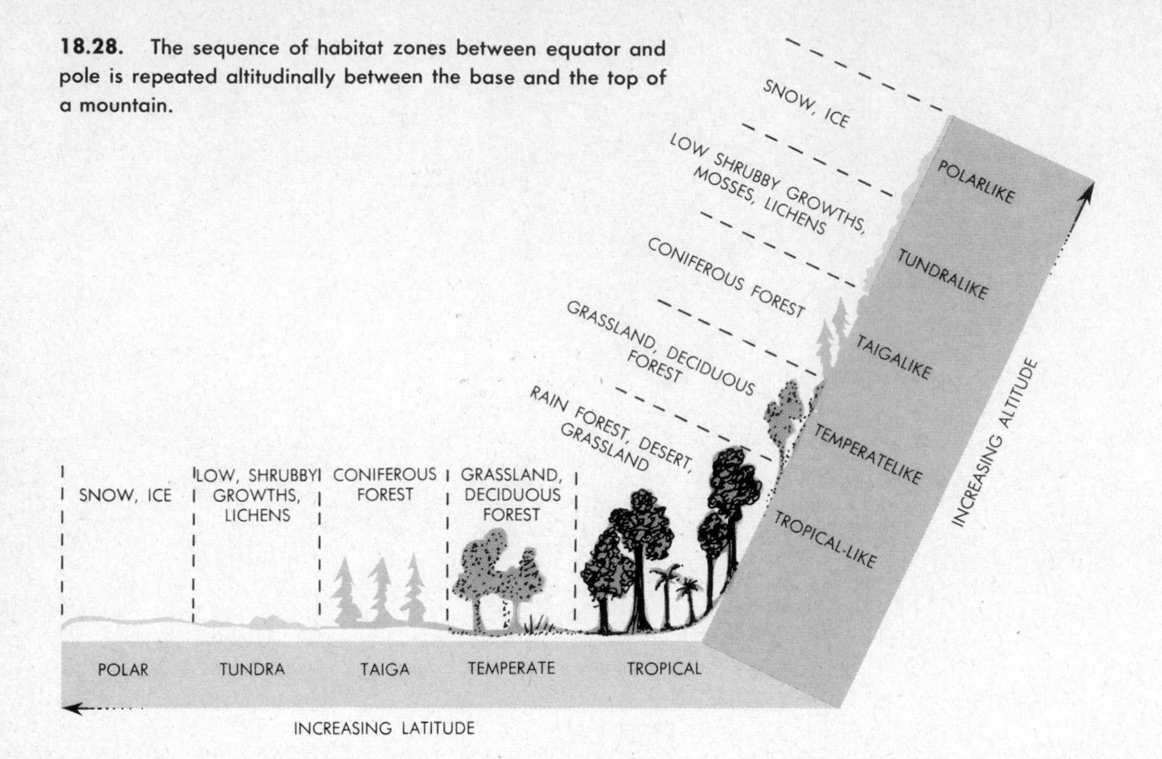

18.28. The sequence of habitat zones between equator and pole is repeated altitudinally between the base and the top of a mountain.

pole is repeated more or less exactly in a vertical direction, along the slopes of mountains (Fig. 18.28). Here too temperature and precipitation are the decisive variables. On a high mountain in the tropics, for example, the succession of biomes from mountain base to snow line is tropical rain forest, deciduous forest, coniferous forest, and lastly, low shrubby growths and lichens. The farther north a mountain is situated, the more northern a biome covers its base and the fewer biomes cover its slopes. In the taiga, for example, the foot of a mountain is coniferous forest and the only other biome higher along the slopes is the zone of low shrubby plants. Thus, habitat zones which are spread over thousands of miles latitudinally are telescoped altitudinally into a few thousand feet.

The foregoing should make it clear that the nature of any kind of habitat, terrestrial, freshwater, or marine, is determined by a few persistently recurring physical variables. Among them are solar light, solar heat, geographic latitude, vertical depth and altitude, rainfall, wind and water currents, and the chemical composition of the locale. As such environmental variables differ in different geographic regions of the world, the habitats present there, hence the ecosystems as a whole, differ accordingly.

This concludes our characterization of the biosphere and the living associations within it. In subsequent chapters we shall systematically examine the phyla of animals as such.

REVIEW QUESTIONS

1. Why is the global environment always changing and never stable? What are open and closed physical systems? How do animals and organisms generally contribute to cyclic environmental change?

2. Review the global water cycle. What forces maintain it? How do animals participate in this cycle? In what different ways does the world's water influence climates?

3. What are diastrophism and gradation? How does the formation of mountains influence climates? Cite examples. In what ways is the global lithospheric cycle of nutritional importance?

4. Review the general pattern of mineral cycles. On the basis of this, construct diagrams showing the pattern of global phosphate and calcium cycles.

5. What is the chemical composition of the atmosphere? Which of these components do not play a role in the maintenance of animals?

6. Review (*a*) the oxygen cycle, (*b*) the carbon cycle. Show how these cycles are interlinked with each other and with the global water cycle.

7. Review the global nitrogen cycle. How many different groups of bacteria aid in the maintenance of this cycle? What is the role of decay in the atmospheric, lithospheric, and hydrospheric cycles?

8. What is the structure of an ocean basin? What are the major habitats in such a basin? What role does the sun play in creating subdivisions in these habitats? What physical conditions characterize the various subdivisions?

9. Define plankton, nekton, and benthos. Give specific examples of each. Where in the ocean are each of these types of organisms found? Why is life in tropical waters generally less abundant but more diverse than in temperate and subpolar waters?

10. What physical and biological conditions characterize the sea depths? What are diurnal migrations? What are the patterns of ocean currents? How do different oceans vary in (*a*) density, (*b*) salinity of water? What are the proportions of oceanic salts?

11. Review the essential physical differences between oceanic and freshwater habitats. What major types of animals occur in fresh water, and in what general ways are they adapted to this habitat?

12. What are the main terrestrial habitats and what physical and biological conditions characterize each of them? In what way are latitudinal terrestrial habitats related to altitudinal habitats?

COLLATERAL READINGS

The various physical components of the biosphere are discussed in the following books and articles:

Bailey, H. S.: The Voyage of the "Challenger," *Sci. American*, May, 1953.

Bascom, W.: Ocean Waves, *Sci. American*, Aug., 1959.

Carson, R.: "The Sea around Us," Oxford, London, 1951.

Coker, R. C.: "This Great and Wide Sea," 2d ed., University of North Carolina Press, Chapel Hill, N.C., 1949.

Dietz, R. S.: The Sea's Deep Scattering Layers, *Sci. American*, Aug., 1962.

Ellison, W. D.: Erosion by Raindrop, *Sci. American*, Nov., 1948.

Fairbridge, R. W.: The Changing Level of the Sea, *Sci. American*, May, 1960.

Henderson, L. J.: "The Fitness of the Environment," Macmillan, New York, 1913.

Kellogg, C. E.: Soil, *Sci. American*, July, 1950.

Kuenen, P. H.: Sand, *Sci. American*, Apr., 1960.

Munk, W.: The Circulation of the Oceans, *Sci. American*, Sept., 1955.

Ommanney, F. D.: "The Oceans," Oxford, London, 1949.

Opik, E. J.: Climate and the Changing Sun, *Sci. American*, June, 1958.

Plass, G. N.: Carbon Dioxide and Climate, *Sci. American*, July, 1959.

Stetson, H. C.: The Continental Shelf, *Sci. American*, Mar., 1955.

Stommel, H.: The Anatomy of the Atlantic, *Sci. American*, Jan., 1955.

Wexler, H.: Volcanoes and World Climate, *Sci. American*, Apr., 1952.

The readings below deal with the relation of the living to the nonliving components of the biosphere and with the nature of various habitats:

Beebe, W.: "Edge of the Jungle," Little, Brown, Boston, 1950.

Berrill, N. J.: "The Living Tide," Dodd, Mead, New York, 1951.

Deevey, E. S.: Life in the Depths of a Pond, *Sci. American*, Apr., 1951.

Ingle, R. M.: The Life of an Estuary, *Sci. American*, May, 1954.

Nicholas, G.: Life in Caves, *Sci. American*, May, 1955.

Pequegnat, W. E.: Whales, Plankton, and Man, *Sci. American*, Jan., 1958.

Ryther, F. H.: The Sargasso Sea, *Sci. American*, Jan., 1956.

Tiffany, L. H.: "Algae, the Grass of Many Waters," 2d ed., Charles C Thomas, Springfield, Ill., 1958.

Vevers, H. G.: Animals of the Bottom, *Sci. American*, July, 1952.

Walford, L. A.: The Deep-sea Layers of Life, *Sci. American*, Aug., 1951.

UNIT **2**

the kinds of animals

In the parts and chapters of the preceding unit, we have concentrated primarily on the various main aspects of animal nature; the emphasis has been not so much on particular animals as on particular attributes typical or characteristic of animals generally. Thus, we have dealt with structures and body designs, with functions at different levels of organization, with developmental and evolutionary histories, and with ecological associations. Now we are ready to consider all these attributes jointly and to concentrate on the animals themselves: in what forms and combinations are these different characteristics actually exhibited by specific animals?

The four parts of this unit are organized along taxonomic and evolutionary lines. The phylum is the basis of discussion, superphyla or correlated groups of phyla form broader frames of reference; and the general objective is to identify and to characterize major animal types, in most cases at least down to the level of taxonomic orders. Pursuit of this objective not only acquaints us with the vast diversity of the animal creatures on earth but, through this, we may also come to appreciate the innumerable ways in which the problems of living have found "animal" solutions.

early groups

PART 7

Subkingdom PROTOZOA

Protists of the cellular grade of construction; heterotrophic; predominantly motile and unicellular; cells in colonies are alike and may survive as isolated individuals; flagellate-ciliate, amoeboid, and coccine states of existence. *PROTOZOA.*

Subkingdom METAZOA

Multicellular animals; oögamous, sexuality by syngamy; life cycle diplontic; development via embryos and, typically, larvae.

Branch MESOZOA: animals at tissue grade of construction; adult body a stereoblastula. *MESOZOA.*

Branch PARAZOA: animals at tissue grade of construction; adult body more complex than a blastula. *PORIFERA* (sponges).

Branch EUMETAZOA: animals above tissue grade of construction.

Grade RADIATA: animals attaining organ level of complexity; with primary and typically also secondary radial symmetry. *CNIDARIA* (coelenterates); *CTENOPHORA* (comb jellies).

Grade BILATERIA: animals attaining organ-system level of complexity; with primary and typically also secondary bilateral symmetry; all other animals.

The three chapters of this part deal with the protozoa, mesozoa, sponges, coelenterates, and comb jellies. These five phyla represent an evolutionary radiation of at least three separate main lines of descent, protozoa exemplifying one, sponges another, and the radiate groups a third. The mesozoa are a comparatively obscure group which may or may not constitute a fourth original line of descent. Apart from a body design based on cells and variously organized cellular aggregations, these phyla have one other general attribute in common, viz., they are living representatives of what undoubtedly were among the most ancient animals and animallike forms on earth. As such, they constitute the most direct clues we have as to what the base of the bush of animal evolution might have been like.

CHAPTER 19 protozoa

We recall that protozoa are *protists,* with structural and evolutionary affinities to slime molds, fungi, and algae. Protozoa are therefore not "animals" in the same sense that Metazoa are animals (cf. discussion in Chap. 14). Nevertheless, they are the most animal*like* of all protists. As a group, they have exploited the unicellular way of life more diversely than any other organisms, and by any criterion they are certainly among the most successful of all living creatures. Indeed, major segments of both the physical and biological parts of the biosphere owe their development and present organization to the protozoa.

GENERAL CHARACTERISTICS

Some zoologists regard protozoa not as cellular but as "noncellular" or "acellular" types, i.e., organisms "not substructured into cells." Apart from purely semantic considerations, such a view implies, first, that the cell theory is invalid, since "true" cells would exist only in multicellular organisms; secondly, that *all* organisms other than the multicellular ones are acellular, and that real cells therefore could have been invented only *after* acellular organisms had evolved; and thirdly, that the cells of *all* multicellular organisms could have originated only by septation (cf. Chap. 14), for if they had originated by aggregation the units which aggregated would have been "cells" already.

These implications make the acellular conception untenable. For example, an egg is universally admitted to be a "true" cell; yet numerous protists, postulated to be "acellular," may function as eggs directly. How then could an "acellular" unit be a "cellular" unit at the same time; and how could eggs have existed before cellular units are supposed to have existed? It is also completely unwarranted to suppose that multicellularity in *all* cases could have arisen *only* by septation, a position one is forced into by the "acellular" view. Indeed, even the most ardent septationists are not willing to go quite that far (though it is true that most advocates of the "acellular" idea are zoologists who champion the ciliate protozoa as the ancestors of the Eumetazoa). Other zoologists who adhere to the "acellular" view do so mainly because they are impressed with the great complexity and diversity of protozoa, and they assume that by attaching the label "acellular" they somehow elevate the organisms to some status more elaborate than "just" cells.

Actually, the cell concept encompasses even the most complex protozoa quite adequately. The essence of that concept is that the functional units of living matter are differentiated internally into nuclear or predominantly genetic and cytoplasmic or predominantly nongenetic structures. Consequently, "acellular" or "noncellular" would mean that a nucleocytoplasmic differentiation does not exist, hence in effect that life is not exhibited. Protozoa certainly are alive, and they display a very obvious nucleocytoplasmic differentiation. They are therefore clearly "cellular"; and any other designation not only denies their membership in the living fabric of the earth but also introduces more problems, unnecessarily so, than it intends to solve. Accordingly, we shall here consider nonmulticellular types, most protozoa included, to be *unicellular,* and to be homologous to any individual cell of a multicellular organism. Unlike the acellular view, the unicellular one does not automatically prejudge the method of origin of multicellularity, and it is also semantically quite consistent with the observation that a single cell is "not substructured into cells."

It has long been customary to include among protozoa numerous photosynthetic organisms that botanists claim to be algae. Thus, in zoological tradition virtually all flagellate groups in all algal phyla are lumped with the protozoa, mainly on the ground that such organisms are motile and that motility is an animal trait. By the same criterion, numerous bacterial types would have to be included among the protozoa (yet other bacterial types would have to be excluded). The issue relates directly to the inadequate Linnaean practice of rigidly labelling all organisms as either "plants" or "animals." In line with the critical discussion of this point in Chap. 14, we shall for present purposes consider motility *and* heterotrophism combined to be more characteristically animallike traits than motility alone. Accordingly, we shall as in earlier chapters regard protozoa to contain nonphotosynthetic, heterotrophic types *only* and shall consider forms such as *Euglena* or *Volvox* to be algae, not protozoa.

Four main groups of protozoa are recognized, viz., flagellate, amoeboid, spore-forming, and ciliate protozoa. There is fairly general agreement that the flagellate and amoeboid groups are interrelated closely, that flagellate ancestors probably have also given rise to the ciliate group, and that the origin of the spore-forming group is obscure (Fig. 19.1). Opinion is less uniform as to how the four groups should be classified, both internally and in relation to one another. Each has been considered traditionally to represent a taxonomic class and in more recent practice a subphylum. All four together have been and generally still are regarded as a single phylum. Although such a single-phylum designation cannot be objected to by overriding arguments against it, its present adequacy may perhaps be questioned. Botanists have long classified the algae as several separate phyla, partly because the traits of different algal groups are sufficiently distinct, partly because the organisms give strong internal evidence of great evolutionary diversity. Being far more heterogeneous in their traits than algae, and indicating an at least equally great evolutionary diversity, protozoa probably should be classified analogously into several phyla. On the cellular level of organization, the differences between, say, an amoeba and a paramecium may well be considered to be comparatively just as profound as the higher-level differences between an earthworm and a caterpillar, for example; and if the latter belong to different phyla, as they do, so probably should the former. However, it does not actually matter too much whether protozoa are regarded as a single phylum or as several, for in traditional practice the whole group represents a subkingdom in any case, distinctly set off from and taxonomically equivalent to the subkingdom Metazoa. A similarly high rank is implied if protozoa are considered to be a main subgroup within a category Protista.

19.1. Possible interrelations among protozoan groups. Contributions to protozoan evolution by algae, slime molds, and fungi are suggested by dashed lines.

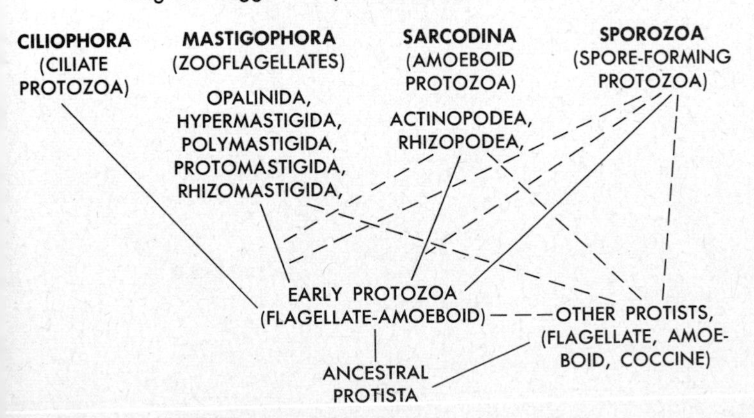

Protozoa exhibit all known forms of heterotrophism. Most species are free-living and holotrophic, the organisms subsisting on particulate food. So-called "carnivorous" types feed on microscopic Metazoa or on other protozoa (including fellow members of the same species in cannibalistic forms); "herbivorous" types feed largely on bacteria and microscopic algae; and omnivorous types subsist on any microscopic food. In their turn, protozoa are a food source for many Metazoa. Some protozoan species are saprotrophic, and many are symbiotic—mutualistic, commensalistic, and particularly parasitic. The spore-forming protozoa are exclusively parasitic, and all other groups include some parasitic subgroups. Protozoa themselves are hosts to various parasites, mainly bacteria, fungi, and other protozoa. Also, many of the organisms are hosts to photosynthetic algae, so-called *zoochlorellae* and *zooxanthellae,* which live in protozoan cytoplasm as mutualistic symbionts.

The organisms are components of all ecosystems in all aquatic environments, in soils, and generally in any environments containing some moisture. Indeed, protozoa occupy almost as many different ecological niches on the cellular, microscopic level as Metazoa occupy on higher levels; and by virtue of their microscopic organization they may also subsist in numerous niches not open to larger organisms. This tremendous ecological diversity is reflected in the number of protozoan species. Figures often quoted are in the order of 15,000, but there are known to be more than that many foraminiferan species alone. Moreover, very many types of Metazoa (but very few Metaphyta) are hosts to at least one unique type of protozoan parasite, which means that protozoan species could well number in the hundreds of thousands. As a conservative estimate, at least 100,000 species of protozoa may probably be assumed to exist at present; and some 10,000 extinct species have been described to date as well.

As already pointed out in Chap. 14, protozoa are diversified primarily in flagellate and amoeboid directions. Coccine potentials are exhibited by the spore-formers, but sporine types are absent. The protozoan cell (Fig. 19.2) is either naked and plastic or is surrounded by a flexible to rigid cuticle, the *pellicle*, composed of a variety of organic and horny substances. Most species have asymmetrically shaped bodies, but some approach regular symmetries to greater or lesser degree. Thus, sessile types often tend toward radiality, free-floating forms towards sphericity. Bilaterality is rare; actively motile, swimming types typically exhibit a spiral asymmetry, correlated with a corkscrew pattern of locomotion. Many species secrete shells, or *tests*, as permanent external covers, and some manufacture *loricae*, secreted housings within which the organisms may move. Numerous protozoa may also secrete temporary shells, or *cysts*. In a *protective* cyst, viable for up to 5 years in some cases, a protozoon may remain dormant and withstand unfavorable environmental conditions such as droughts; in a *reproductive* cyst, division or sexual processes may take place. Encysted states are expressions of protozoan dimorphism or polymorphism. Other such expressions are common. For example, sexual partners or gametes may differ structurally; young and adult stages may differ; and flagellate states may become nonflagellate or amoeboid and vice versa. Cell sizes among protozoa range from about 2 or 3 microns to several millimeters and more. Some of the shelled forms, foraminifera particularly, attain diameters of up to 10 or more centimeters.

Flagella and cilia are attached to kinetosomes underneath the cell surface. Numerous modifications of flagella and cilia are known, and in many cases the kinetosomes too give rise to a variety of other protozoan organelles (cf. below). Also

19.2. A composite of some of the organelles found in various protozoa. No single protozoan possesses all the organelles shown here, and most protozoa possess others, not shown here. See also subsequent illustrations.

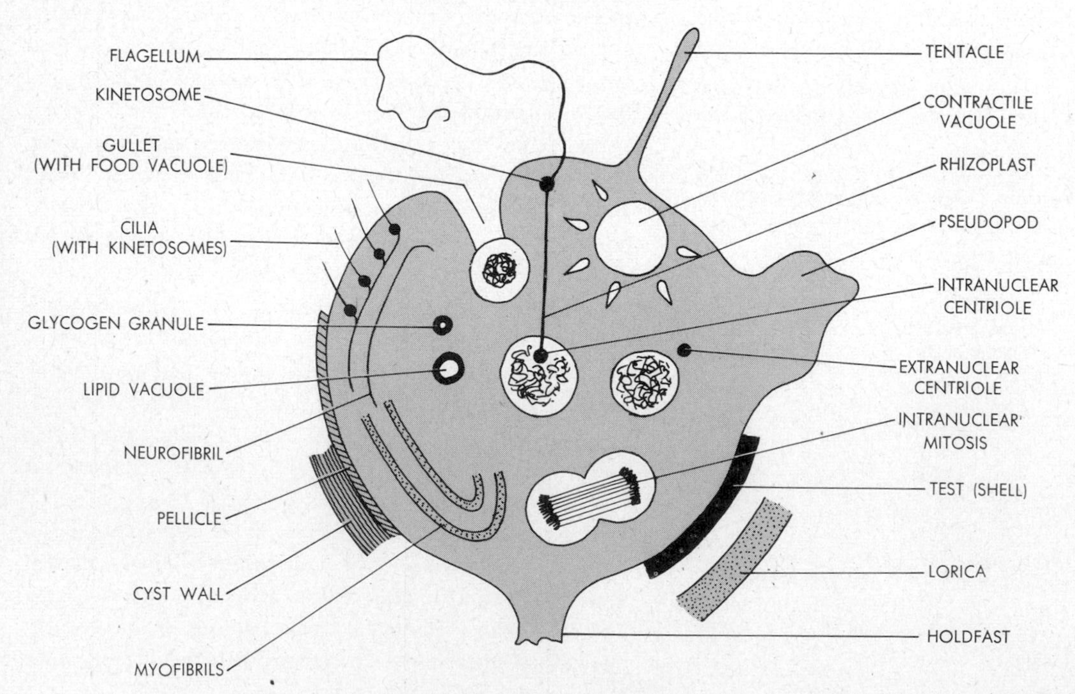

present underneath the cell surface in a number of protozoan groups are conductile neurofibrils, contractile myofibrils, and contractile vacuoles. The latter occur in virtually all fresh-water forms as well as in a few marine and parasitic types. Ingestive structures in holotrophic protozoa are pseudopodia, gullets, and in certain cases tentacles. As in Metazoa, the food of protozoa must contain organic carbon and nitrogen, some 10 or more amino acids, and several vitamins. Digestion takes place intracellularly within food vacuoles, and reserve food is stored principally in the form of granules of glycogen or related polysaccharides and as lipid droplets or vacuoles. The latter also tend to increase the buoyancy of the organisms. Internal transportation of metabolites is achieved by diffusion and cyclosis, and gas exchange and excretion take place directly through the cell surface. If contractile vacuoles are present, they are the principal osmoregulating and water-balancing organelles.

Though definite sensory structures can in most cases not be identified, protozoa are nevertheless exquisitely sensitive to their surroundings. They may "taste" food and refuse to ingest unsuitable materials; they give distinct avoidance responses to undue temperatures, light, electric charges, pH, mechanical stimuli, and chemicals in the water; they seek out optimum environments by trial-and-error behavior; and many of them have been trained through conditioning and have been made to give "learned" responses to specific stimuli.

Protozoa are uninucleate as well as multinucleate. In ciliates two types of nuclei are present, viz., *micronuclei* and *macronuclei*, often more than one of either or both (cf. below). Mitotic division is *intranuclear* in the vast majority of species, i.e., two sets of chromosomes are formed within the original nucleus, which then merely constricts into two (cf. Fig. 19.2). Centrioles are not present in some protozoa; in other cases such granules are either intranuclear or, less often, extra-nuclear. As noted in Chap. 4, an extranuclear centriole may be connected with a kinetosome by a fine rhizoplast.

Mitotic nuclear division precedes or accompanies the basic forms of protozoan reproduction, i.e., equal binary fission, unequal binary fission (*budding*), and multiple fission (*sporulation*, or spore-formation; Fig. 19.3). Equal binary fission is by far the most common form. In this process, flagella, cilia, gullets, and certain other types of organelles may be resorbed in the mother cell and new sets of organelles may then grow in each offspring; or the existing set in the mother cell may become part of one offspring, the other then developing a new set; or the existing set in the mother cell may be duplicated prior to fission, each offspring then acquiring one of these

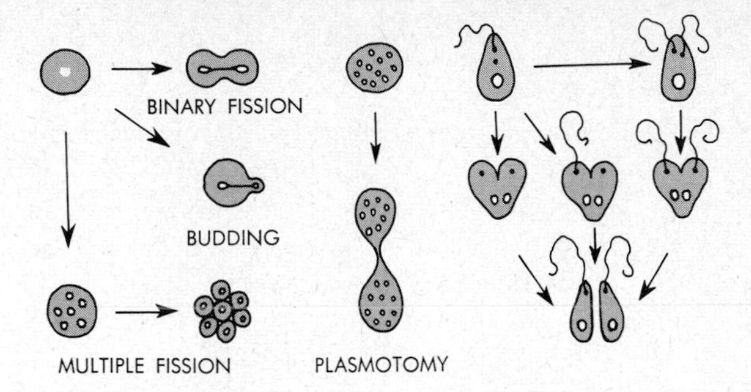

19.3. Protozoan cell divisions and reproductive patterns. Diagram at right shows that, during division, neither offspring or one offspring or each offspring may inherit flagella (or other formed organelles) directly from the parental cell.

sets during fission. Cell division occurs longitudinally in many protozoa, transversely in many others (cf. below). A few multinucleate protozoan types reproduce vegetatively by fragmentation (*plasmotomy*), a process in which the organism divides into two or more parts without nuclear divisions, the nuclei merely becoming distributed among the offspring (cf. Fig. 19.3). Regeneration potentials are high in protozoa generally, and many multinucleate types may be cut into virtually as many pieces as there are nuclei, each piece then redeveloping into a complete organism.

Sex occurs through syngamy or through conjugation. Hermaphroditism is common, and such species are either cross-fertilizing, self-fertilizing (autogamous), or both. With some exceptions to be referred to below, protozoan life cycles are generally diplontic, i.e., meiosis takes place during gamete formation. In parasitic types the life cycles are usually exceedingly complex and are geared to the ways of life of the specific hosts.

SUBPHYLUM MASTIGOPHORA

Flagellate protozoa; one or more flagella present throughout life or at given stages; predominantly uninucleate; reproduction by longitudinal binary fission; sexuality by syngamy.

Class ZOOFLAGELLATA

Order Rhizomastigida: one to many flagella; permanently amoeboid. *Mastigamoeba, Mastigina, Mastigella, Multicilia*

Order Protomastigida: one to three flagella; often amoeboid; cells naked, without gullet. *Bodo, Codosiga, Trypanosoma*

Order Polymastigida: two to many flagella; organelles often in duplicate or multiple sets. *Giardia*

Order Trichomonadida: three to five flagella, plus one forming edge of undulating membrane. *Trichomonas*

Order Hypermastigida: numerous flagella; body complex. *Trichonympha*

Order Opalinida: numerous flagella, shortened to cilia, arranged in rows. *Opalina*

With the exception of the order listed last, the others represent a series ranging from predominantly free-living, simply constructed types to predominantly symbiotic, complexly constructed types. The former have probably given rise to the latter. The body of the organisms is elongated in the direction of motion, and where flagella are few in number they originate anteriorly (Fig. 19.4). A flagellum is believed to propel an organism on the principle of the screw; flagellate protozoa typically spiral through the water. Each flagellum consists of an outer sheath continuous with the cell surface and, as noted in Chap. 4, an inner bundle of 11 fibrils. Referred to collectively as the *axoneme,* these fibrils are produced by and attached to a basal granule, the kinetosome. In most species with multiple flagella, each flagellum originates from a separate kinetosome. But in some forms, notably the Hypermastigida (cf. Fig. 19.6), a single kinetosome may be greatly

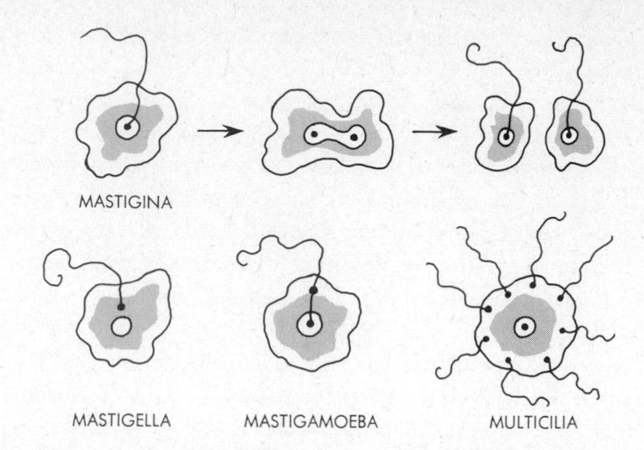

19.5. Diagrams of Rhizomastigida, permanently amoeboid as well as flagellate. In *Mastigina* and *Mastigella,* the basal granule serves joint centriole and kinetosome functions; the division pattern in *Mastigina* is sketched at top. In the other two genera shown, the centriole and kinetosome functions are performed by separate granules.

19.4. Composite diagram of some mastigophoran organelles associated with the kinetosome. The flagellum base is drawn in exaggerated size relative to the cell as a whole. The outer flagellum sheath covers the 11 fibrils which form the axoneme attached to the kinetosome.

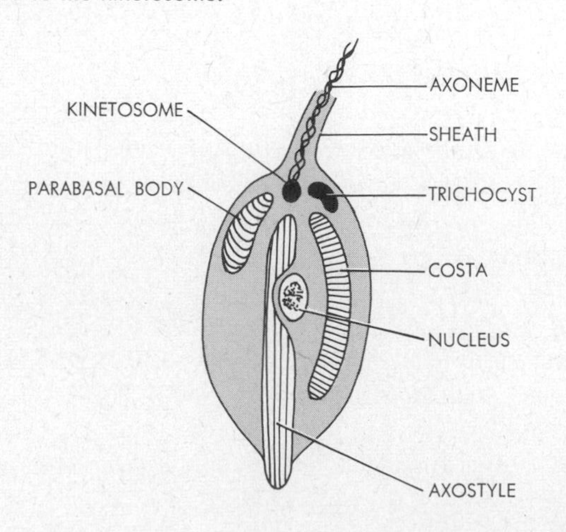

enlarged into a spiral ribbon or other complexly shaped body and all flagella then emanate from it.

In Rhizomastigida and Opalinida the flagellum is the principal product of the kinetosome, but in the other orders the basal granule also gives rise to a variety of other cytoplasmic organelles (cf. Fig. 19.4). Thus, most groups possess a *parabasal body,* which exhibits a Golgi-like structure under the electron microscope. An *axostyle* is often present, composed of bundles of fibrils attached to the kinetosome and possibly serving a supporting function. Similarly supporting may be the *costa,* which under the electron microscope displays a cross-banded structure like that of the tough collagen fibers of metazoan connective tissues. Other kinetosomal products include *trichocysts,* organelles which may be discharged. Their functions are uncertain. In some instances they are believed to paralyze potential prey; in others they may aid in anchoring a cell during feeding.

The Rhizomastigida are excellent examples of organisms with permanently combined flagellate and amoeboid traits (Fig. 19.5). In some of these amoeboid flagellates, the flagella attach to granules which are either intranuclear (e.g., *Mastigina*) or extranuclear (e.g., *Mastigella*) and which serve *both* as kinetosomes and as centrioles. Such a dual role of the granules unquestionably represents a primitive condition. At cell division the flagellum is usually resorbed, the kinetosome-centriole divides, and each daughter granule then produces a new flagellum in each offspring cell. Fission occurs

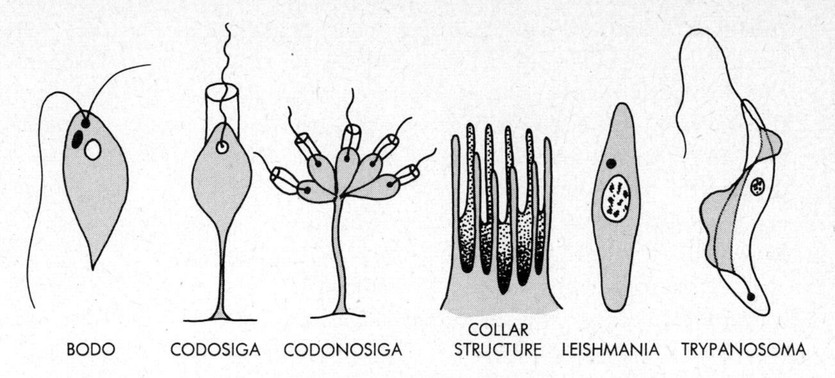

19.6. Various Protomastigida. In *Bodo,* a trichocyst is indicated. The electron microscope shows that the collar of collar flagellates is a circlet of long cytoplasmic extensions, as indicated here. In *Trypanosoma,* note the posterior kinetosome and the undulating membrane edged by the flagellum. *(Adapted from Minchin, Kudo, Lapage.)* See also color plate I.

longitudinally, as in flagellates generally. In most members of the other flagellate orders, centrioles and kinetosomes typically are separate organelles; centrioles may again be intranuclear or extranuclear, but kinetosomes are always extranuclear and they are generally located just underneath the cell surface. Forms like *Multicilia,* for example, possess numerous flagella over the entire body surface, each flagellum emanating from a separate subpellicular kinetosome.

The Protomastigida include free-living saprotrophic types like *Bodo,* as well as the holotrophic *choanoflagellates,* or collar flagellates, so named because of their membranous sleeve around the single flagellum (Fig. 19.6). The electron microscope reveals that such a collar consists of a circlet of fingerlike cytoplasmic extensions from the main part of the cell. *Codosiga* is a stalked, sessile, solitary collar flagellate, and other genera (e.g., *Codonosiga*) form stalked sessile colonies. *Proterospongia* was long thought to belong to this group of colonial flagellates, but these "organisms" have since been shown to be fragments of sponges, animals in which flagellate collar cells are conspicuous. Among numerous symbiotic and parasitic Protomastigida, the trypanosomes are the most noteworthy. These organisms are intracellular parasites in lymph and blood cells of various animals, vertebrates and mammals particularly. For example, a species of *Leishmania* is the causative agent of the serious oriental disease kala azar, which affects lymphoid tissues; and various species of *Trypanosoma* are blood parasites which produce sleeping sickness in man and other mammals.

Polymastigida are predominantly intestinal symbionts in insects and vertebrates. Many, like *Giardia,* are interestingly bilateral, with symmetric duplicate sets of nuclei, flagella, axostyles, and other organelles (Fig. 19.7). Certain species of both *Giardia* and *Trichomonas* occur in man, as free-swimming holotrophic or saprotrophic commensals. They are transmitted in encysted condition via the feces. The Hypermastigida are the most complex flagellates (cf. Fig. 19.7). All are symbiotic in the intestines of cockroaches and termites. In these insects the flagellates digest the wood eaten by their hosts, for the benefit of both hosts and symbionts. It has been shown that the timing of the sexual process of the flagellates is geared to the molting time of the hosts. Undoubtedly, the evolution of numerous flagella in these symbiotic flagellates is

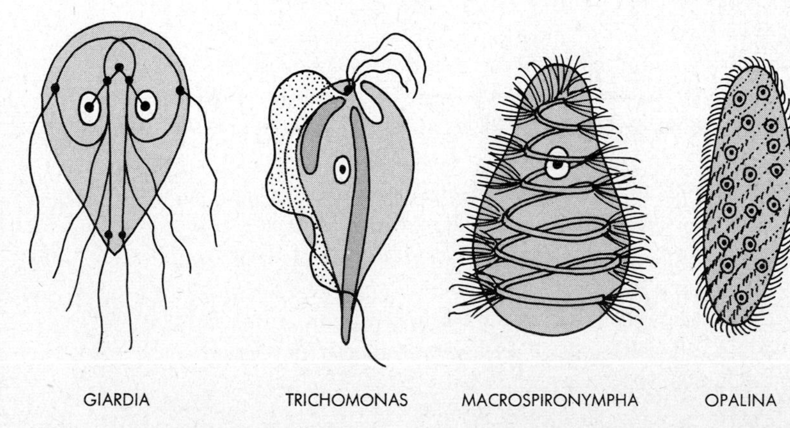

19.7. In the polymastigid *Gardia,* note the two nuclei (with centrioles) and the symmetrically paired kinetosomes, rhizoplasts, and flagella. In *Trichomonas,* an axostyle, a parabasal body, a gullet, and an undulating membrane are present, in addition to a nucleus with intranuclear centriole and a kinetosome with additional flagella. The hypermastigid *Macrospironympha* possesses a spiral, ribbonlike kinetosome and hundreds of flagella emanating from it. *Trichonympha* is similarly hyperflagellate. *Opalina* is characterized by diagonally spiralling rows of kinetosomes, each such granule bearing a short flagellum. The nucleus is without centriole, as in ciliate protozoa.

correlated adaptively with their life spent in the thick, viscous contents of the host gut; numerous locomotor organelles are clearly advantageous in such an environment. Note that an almost identical evolutionary development has taken place in the sperms of cycads and ginkgoes, coniferous trees in which the sperms must swim through the viscous cytoplasm of pollen tubes on their way toward eggs. The sperms here are likewise hyperflagellate and, like *Trichonympha,* they too contain spiralled, ribbon-shaped kinetosomes.

The Opalinida live as saprotrophic commensals in the gut of amphibia. These flagellates are of considerable interest from an evolutionary standpoint, for their body is covered uniformly with neat rows of shortened flagella which are indistinguishable from cilia (cf. Fig. 19.7). Each cilium originates from a separate kinetosome. Because of this ciliation pattern and also in view of the absence of centrioles, opalinids actually have been classified until recently as a primitive group of ciliate protozoa. However, their basic characteristics are now recognized to be distinctly flagellate. More specifically, fission is longitudinal, as in flagellates but not in ciliates; opalinids possess a single type of nucleus, like flagellates but unlike ciliates; and opalinids mate by syngamy, again like flagellates and unlike ciliates. Conceivably, the opalinids represent still-living descendants of an evolutionary transition stage between flagellates and ciliates, a possibility strengthened further by the resemblance of opalinids to forms such as *Multicilia.* Thus, ancestral flagellates may have given rise to two principal evolutionary lines, one leading via Rhizomastigida and Protomastigida to the Hypermastigida, the other to the ciliate protozoa via opalinid-like stocks (cf. Fig. 19.1). In some classifications the Opalinida are ranked as a superorder of flagellates.

SUBPHYLUM SARCODINA

Amoeboid protozoa; feeding and locomotion by pseudopodia; uni- and multinucleate; reproduction by binary and multiple fission and by plasmotomy; sexuality by syngamy.

Class ACTINOPODEA: pseudopodia in the form of axopodia
- Order Helioflagellida: flagella present. *Dimorpha*
- Order Heliozoida: radial to spherical, with vacuolated outer cytoplasm. *Actinophrys, Actinosphaerium, Camptonema*
- Order Radiolarida: radial to spherical, with central capsule and with skeletal shell, predominantly of silica. *Heliosphaera*

Class RHIZOPODEA: pseudopodia not in the form of axopodia
- Order Amoebida: without shells. *Naegleria, Amoeba, Pelomyxa*
- Order Testacida: with noncalcareous shells. *Arcella, Difflugia*
- Order Foraminiferida: with calcareous shells. *Lagena, Globigerina*

Pseudopodia are of several types (Fig. 19.8). An *axopodium* is a cytoplasmic extension supported internally by a stiff, nonmotile axoneme, i.e., a fibril bundle of the kind also present within a flagellum (though motile in that case). In an axopodium, the axoneme originates from a granule which is either an extranuclear kinetosome or a kinetosome-centriole located extranuclearly or intranuclearly. Axopodia occur in the class Actinopodea. Pseudopods without axonemes, as in the class Rhizopodea, are of three principal forms. Slender, terminally branching pseudopods are termed *filopodia;* broad, rounded ones are *lobopodia;* and filamentous, netlike ones are *rhizopodia.*

The Helioflagellida undoubtedly are a primitive group, with close affinities to the flagellates. For example, *Dimorpha*

19.8. Types of pseudopodia. In an axopodium, the axoneme is a stiff, internal supporting rod, structured like a flagellum and emanating from an extranuclear or intranuclear kintosome-centriole. Axonemes are absent in the other pseudopodial types.

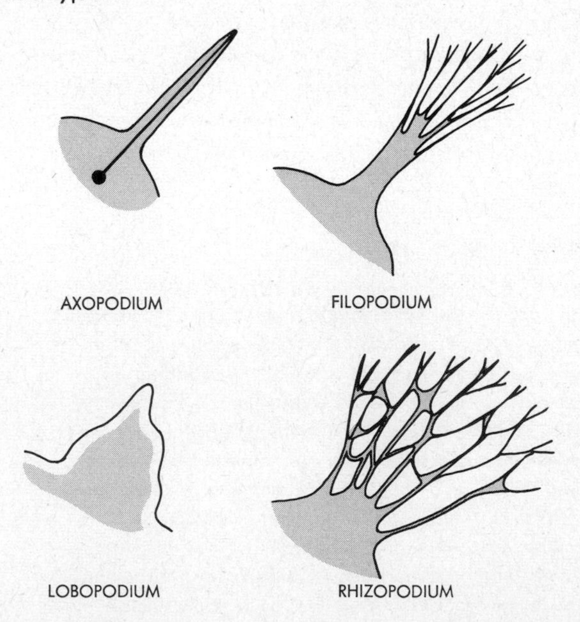

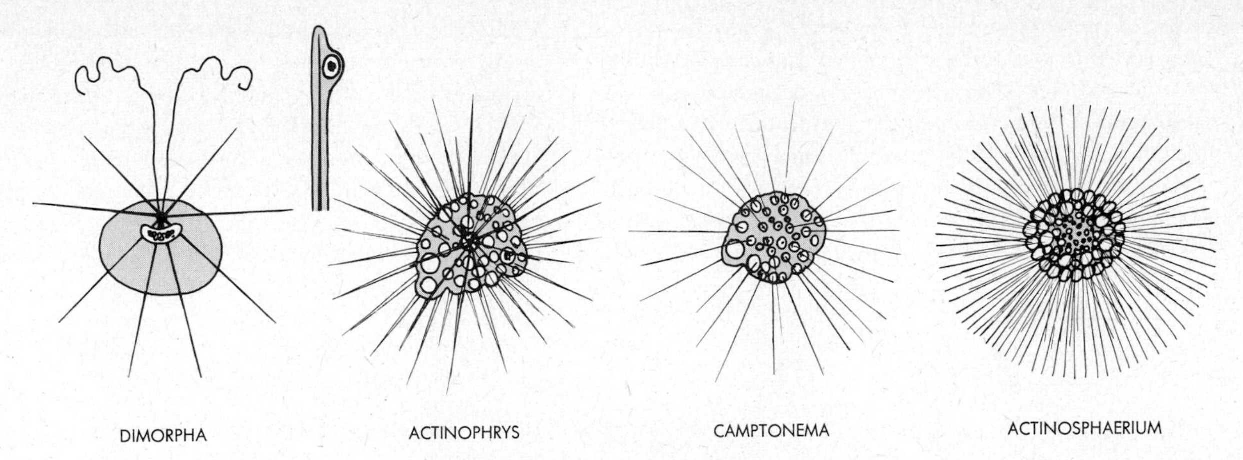

19.9. Actinopodea. In *Dimorpha* (a helioflagellid), note the motile anterior flagella and the stiff axopodia, all radiating from an extranuclear centriole. Next to this organism is a sketch of the tip of an axopodium, showing the axoneme and a food vacuole in the surrounding cytoplasm. In *Actinophrys* (a heliozoid), note the foamy cortex, with contractile and food vacuoles, and the denser medulla with a nucleus. Axonemes radiate out from an intranuclear centriole. The heliozoids *Camptonema* and *Actinosphaerium* are multinucleate, the former with intranuclear and the latter with extranuclear centrioles.

possesses a single extranuclear kinetosome-centriole, from which radiate two anteriorly directed motile flagella as well as numerous axopodia with axonemes (Fig. 19.9). As in Actinopodea generally, the naked exterior cytoplasm of an axopodium is sticky and traps food organisms coming into contact with it. A food vacuole then forms at the point of contact and the vacuole subsequently may flow into the main portion of the protozoon.

In the Heliozoida, the "sun animalcules" or *Heliozoa,* motile flagella are not present (cf. Fig. 19.9). *Actinophrys* is uninucleate, and an intranuclear kinetosome-centriole produces the stiff axonemes of the axopodia. Other genera, e.g., *Camptonema,* are multinucleate, each nucleus possessing a kinetosome-centriole internally and giving rise to one axoneme. *Actinosphaerium* is also multinucleate, but here the kinetosome-centriole granules are extranuclear. In all these forms, the axopodia may be temporarily shortened or resorbed completely, as during reproductive and sexual stages and during a rolling form of locomotion on solid surfaces. The main part of the body consists of a highly vacuolated *cortex* on the outside and a denser, nucleus-containing *medulla* in the center. Heliozoa are largely freshwater types; their contractile vacuoles are located in the cortex. Some Heliozoa possess latticelike silica skeletons around the surface, the axopodia protruding through the spaces in the lattices. Reproduction occurs by binary fission, by budding, and by plasmotomy. Sexual processes are known for *Actinophrys* and *Actinosphaerium,* where they take the form of autogamous self-fertilization. An individual resorbs its axopodia, encysts, and divides by binary fission. Each sister cell then undergoes two meiotic divisions with polar-body formation and in this manner gives rise to one haploid gamete cell. The two sister gametes in a cyst then fuse into a diploid zygotic individual, and the latter eventually encysts and resumes a vegetative existence (Fig. 19.10).

The Radiolarida, or *Radiolaria,* are marine and planktonic. Many species have diameters of several centimeters. The body

19.10. Sex by autogamous self-fertilization in the heliozoan *Actinophrys.* Inside a cyst wall the cell divides, each daughter undergoes meiosis (with polar-body formation), and the resulting single gamete cells then fuse and form a zygotic diploid adult.

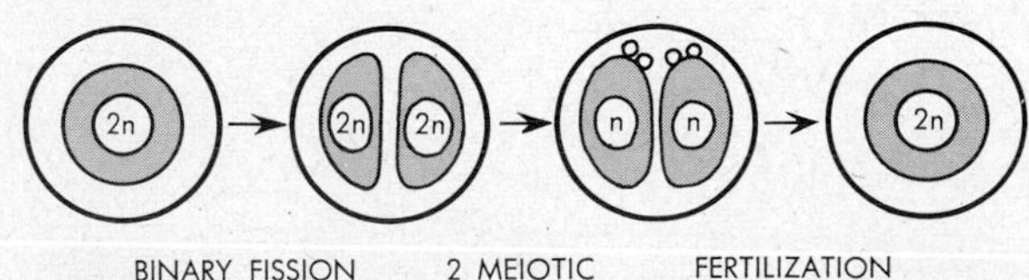

of these organisms is subdivided into an inner medulla and an outer cortex by a *capsule*, a membrane composed of chitinlike material (Fig. 19.11). Numerous perforations in the capsule permit the cortical and medullary cytoplasm to remain in continuity. The medullary portion contains one to many nuclei, and the cortical portion is highly vacuolated and foamy in appearance. Radiolaria are noted for their exceedingly complex, beautifully and intricately sculptured skeletons. Most of these are composed of glassy silica; in one group they consist of strontium sulfate. The skeletons are spherical and perforated in a very large variety of patterns. In many cases a whole series of latticelike skeletal spheres is present, one inside the other. The outermost sphere is usually ornamented with spines, needles, and hooklike extensions. In addition, needleshaped *spicules* lie free in the cytoplasm of many species. Protruding through the perforated skeletons are the

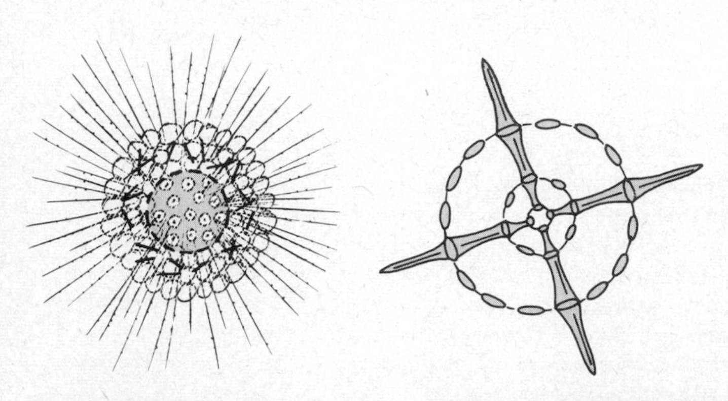

19.11. Radiolarida. Diagrams: left, the general structure of a radiolarian, showing vacuolated cortex with free spicules, the perforated capsule, and the interior medulla with nuclei; right, section through the silica skeleton of a particular radiolarian, showing three spherical lattices one inside the other, the outermost with projecting spines. The three lattices are kept in position by interconnecting struts. (*Adapted from Bütschli.*) **Photo:** a variety of radiolarian shells. See also color plate I.

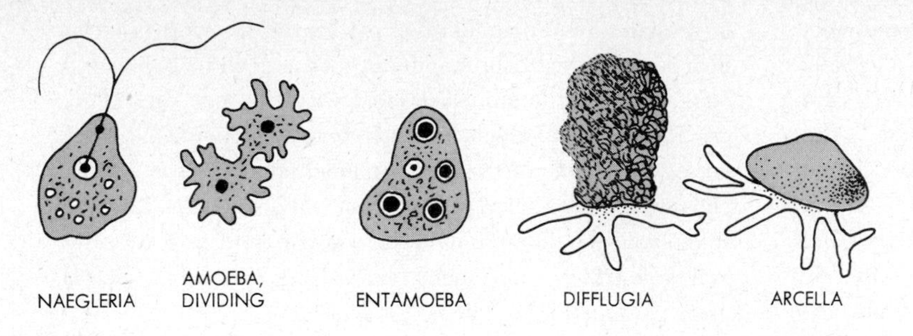

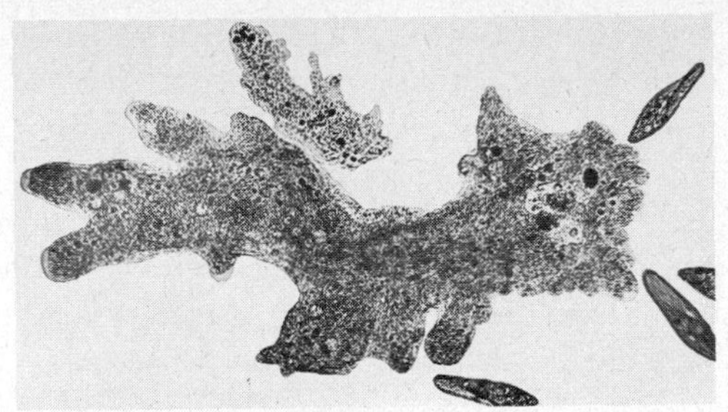

Courtesy Carolina Biological Supply Company.

19.12. Rhizopodea. *Entamoeba histolytica* is diagrammed with ingested blood corpuscles in food vacuoles. The photo shows *Pelomyxa,* one of the naked, multinucleate Amoebida (with paramecia around it and *Amoeba proteus* near top, for comparison). An amoeba is also illustrated in Fig. 3.8. Dividing amoebae characteristically produce numerous pseudopods, as shown above. *Difflugia* and *Arcella* are shelled rhizopods, of the order Testacida. See also color plate II.

pseudopodia, which are axopodia in most cases but filopodia without axonemes in some. In certain forms, reproduction by binary fission includes a division of the skeleton into two halves, each daughter cell regenerating the missing half of the skeleton. In most species one daughter escapes from the skeletal enclosure of the parent and develops a whole new skeleton on its own; the other daughter inherits the parental skeleton. Some Radiolaria are or become multinucleate and then subdivide by multiple fission into biflagellate spore cells, which may function as gametes.

Buoyed up by the vacuolated cortical cytoplasm and the floating surfaces formed by pseudopodia and skeleton, the Radiolaria inhabit both the surface and the deeper waters of all oceans. In colder latitudes they constitute an important food source for plankton feeders. During their slow fall into the ocean depths, the silica skeletons of dead Radiolaria resist dissolution well. Such skeletons consequently become a major component of the bottom deposit of deep oceans. Some 5 per cent of the world's ocean floor actually consists of "radiolarian ooze." Compressed into rock, these silicaceous deposits form *flint* and *chert*. Radiolaria are among the few fossil organisms known from Precambrian times, and some fossil types from the earliest Cenozoic are virtually identical with forms living today.

In the class Rhizopodea, the order Amoebida includes flagellate amoebae like *Naegleria,* which occupy an evolutionary position equivalent to that of the helioflagellates among the Actinopodea. Thus, *Naegleria* is permanently amoeboid but under certain conditions it may develop from one to three flagella (Fig. 19.12). These are attached to an extranuclear kinetosome, which in turn connects via a rhizoplast to an intranuclear centriole. *Naegleria* does not possess axopodia, however, and its pseudopods are lobopodial. Lobopodia are also generally characteristic of the other members of the order. The "common" amoeba is *Amoeba,* and the most familiar species, one of the best known and popularly most misunderstood organisms of all kinds, is *Amoeba proteus*. Kinetosomes are absent here and the flagellary potential has been lost. The related genus *Entamoeba* is symbiotic in the intestine of many animals. One parasitic species, *E. histolytica,* causes amoebic dysentery in man. *Pelomyxa* is a large, multinucleate genus which reproduces by plasmotomy. The many nuclei in this form, as indeed in most multinucleate protozoa, undergo mitosis synchronously, suggesting the existence of a common trigger mechanism for the simultaneous initiation of division in all nuclei present. Most members of the Amoebida are freshwater and marine bottom creepers, largely omnivorous and subsisting on bacteria and other protozoa (e.g., amoebae feed on ciliates such as paramecia). The Testacida occupy similar ecological niches. The shells of these protozoa usually have one opening from which lobopodial or filopodial pseudopods are protruded. *Arcella* secretes a chitinlike test; *Difflugia* manufactures organic secretions to which it cements sand grains as a protective cover (cf. Fig. 19.12).

The Foraminiferida, or *Foraminifera,* correspond among the Rhizopodea to the Radiolaria among the Actinopodea. The organisms are marine, largely deep-water forms and bottom creepers, often very large, and exceedingly abundant. *Globigerina* is among several planktonic surface floaters. Foraminifera are equipped with tests or shells which are of two general types (Fig. 19.13). One type is single-chambered, with a single opening, as in *Lagena*. Such tests are usually composed of organic materials, often chitinlike, to which sand grains or other foreign bodies may become attached. The tests expand as the organisms grow. The other type is many-chambered and heavily impregnated with secreted calcium carbonate. The organism first produces a single chamber and as it grows it manufactures additional chambers in continuity with the first. The chambers are interconnected by internal pores and the protozoon occupies all chambers present. One, and most often many, exterior perforations permit the extrusion of pseudopods, which are sticky rhizopodial networks in all Foraminifera. The skeletal perforations give the group its name, viz., "hole bearers."

The shells are formed in a large variety of species-specific patterns (cf. Fig. 19.13). Most species produce spiralled shells, either flat like coiled rope or conical as in snails (e.g., *Globigerina*). In other cases the chambers of the shells are arranged linearly, or in concentric cyclic patterns, or in two or three alternating braidlike rows, or in variously irregular ways. When the organisms die, their shells sink toward the sea floor but they dissolve before they reach very great depths. At intermediate depths, however, they accumulate in fantastically large quantities. Some 30 to 40 per cent of the world's ocean floor is covered with foraminiferan ooze (and with *Globigerina* ooze specifically in numerous localities). *Limestone* and *chalk* are the rocky products of such bottom deposits; uplifted geologically, they form limestone mountain ranges and chalk cliffs such as those along the English Channel coasts. Like the Radiolaria, the Foraminifera are known from at least early Paleozoic times, and many early Cenozoic fossil species greatly resemble those living today.

Foraminifera with many-chambered shells exhibit life cycles encompassing two generations of adults (Fig. 19.14). A young, small-shelled organism is uninucleate at the start but becomes multinucleate by mitotic nuclear divisions. A process of multiple fission then takes place which includes meiosis. The resulting haploid offspring cells are *spores* which escape from the parent shell. They subsequently develop into large-shelled haploid individuals. These too become multinucleate eventually and they then undergo a process of *mitotic* multiple fission. The products so formed are haploid gametes, amoeboid in some species, biflagellate in others. Syngamy occurs, and the resulting diploid zygotes mature into small-shelled individuals which start a new life cycle. Evidently, the foraminiferan cycle differs from that of all

19.13. Foraminiferida. Diagrams: shell types. 1. Single-chambered shell, as in *Lagena*. 2 and 6. Flat coils; 2. Spiral type, as in majority of genera; 6. Cycloid type, as in *Discospirulina* and extinct nummulites. 3. Helical coil, as in *Globigerina* (pseudopodia projecting from shell). 4. Linear type, as in *Nodosaria*. 5. Braided type, as in *Textularia*. **Photo:** *Globigerina*. See also color plate I.

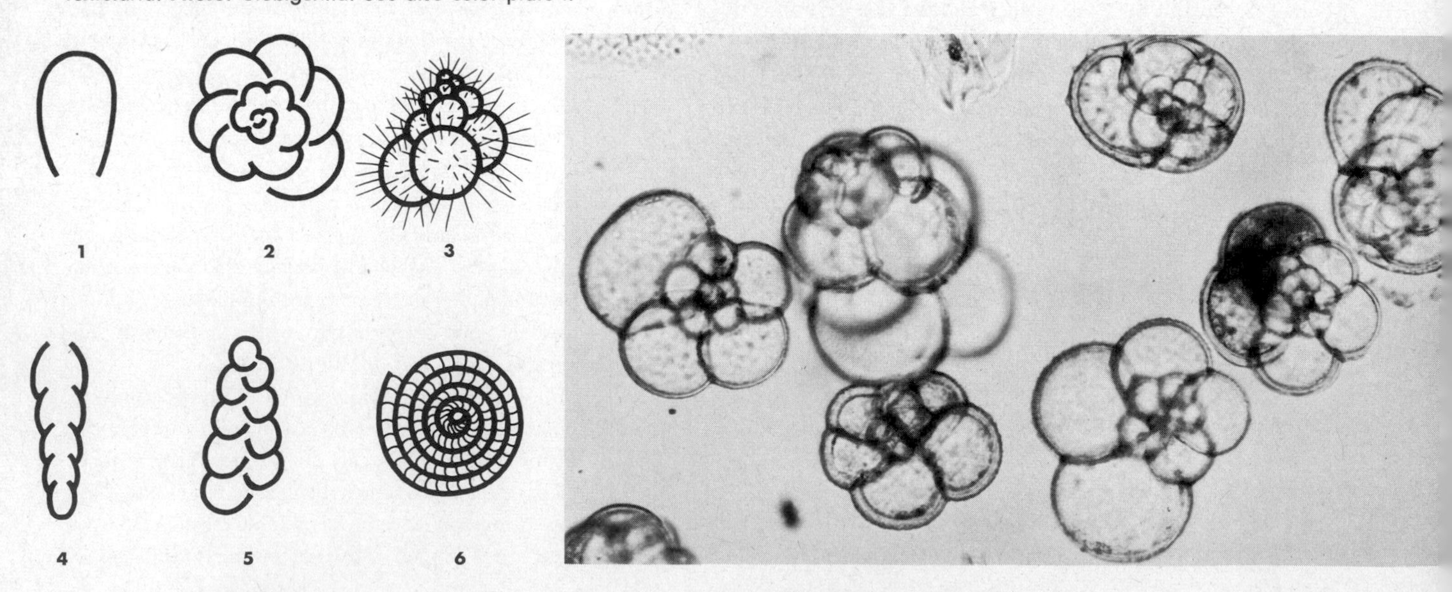

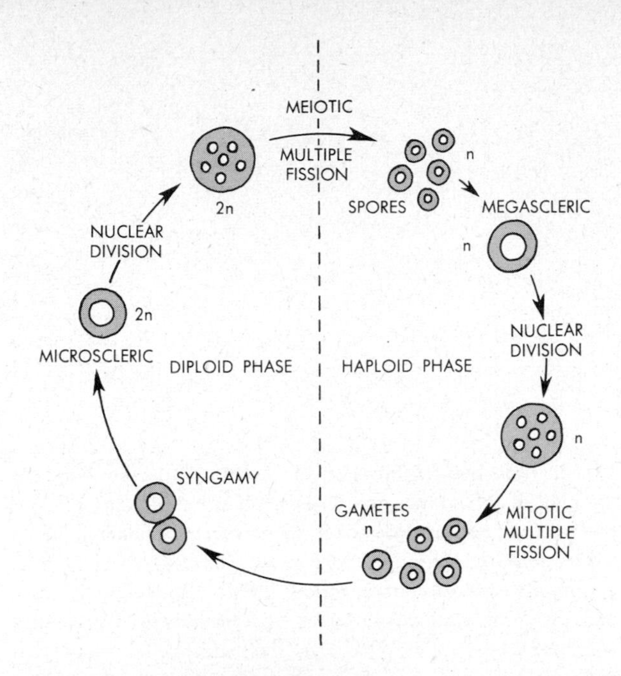

19.14. The foraminiferan life cycle. In this alternation of generations (diplohaplontic cycle), the diploid microscleric adult is small-shelled, the haploid macroscleric adult, large-shelled.

other protozoa. A diploid, small-shelled (*microscleric*) phase produces haploid spores meiotically and the latter give rise to a haploid, large-shelled (*megascleric*) phase. Haploid gametes are then formed mitotically, from which arises a new diploid microscleric phase. Such a cycle, characterized by meiosis during spore-formation, may be referred to as a *diplohaplontic* cycle. It involves an alternation of two dimorphic generations of adults, and it corresponds to the cycle encountered in many algal groups and in all plants (Metaphyta).

It is generally agreed that the subphylum Sarcodina as a whole represents a highly polyphyletic assemblage of types. A number of different algal flagellate stocks may have given rise to the amoeboid flagellates and to the flagellate amoebae. Moreover, as already noted in Chap. 14, the nonflagellate amoebae may have evolved by loss of chlorophyll from numerous algal amoeboid types. Furthermore, the Radiolaria and Foraminifera are similar in many respects to certain shelled flagellate groups among the golden-brown algae, and the nature of the life cycles in these protozoan orders additionally suggests algal affinities. It is therefore probable that Sarcodina may have descended from a number of diverse sources among early flagellate and amoeboid algal protists. Furthermore, Mastigophora such as the Rhizomastigida may have contributed to sarcodine evolution (cf. Fig. 19.1). Hence it is rather pointless to speculate which of the sarcodine classes might be more primitive than the other, or whether or not Sarcodina as a whole are more primitive or less primitive than the Mastigophora.

SUBPHYLUM SPOROZOA

Spore-forming coccinelike protozoa; exclusively parasitic; cells without contractile vacuoles; uni- and multinucleate; reproduction by multiple fission; sexuality by syngamy; life cycle typically haplontic.

Class TELOSPORIDIA: spores naked or encapsulated; polar capsules absent.

- Subclass Gregarinida: extracellular parasites, often with myonemes; spores encapsulated
  - Order Eugregarinida: schizogony absent. *Gregarina, Monocystis*
  - Order Schizogregarinida: schizogony present, alternating with sporogony. *Schizocystis, Ophryocystis*
- Subclass Coccidia: intracellular epithelial parasites; spores encapsulated; alternation of sporogony and schizogony. *Eimeria, Hepatozoon, Adelea*
- Subclass Hemosporidia: intracellular blood parasites; spores naked; alternation of sporogony and schizogony. *Plasmodium*

Class CNIDOSPORIDIA: sporogony absent; polar capsules present. *Nosema, Myxobolus*

Class ACNIDOSPORIDIA: sporogony absent; polar capsules absent. *Sarcocystis, Haplosporidium*

The key to an appreciation of the Sporozoa is their life cycle. A complete cycle incorporating all stages has the following general pattern (Fig. 19.15).

Living inside or outside a cell of a host, a vegetative haploid sporozoan parasite is called a *trophozoite*. It is uninucleate at first and it later develops into a multinucleate reproductive individual referred to as a *schizont*, or *agamont*. This individual then undergoes multiple fission, i.e., *schizogony*, or *agamogony*. The resulting cells are spores termed *merozoites*. The latter either may develop into new vegetative trophozoites, which reinfect other cells or body parts of the host; or they may grow into sexual individuals called *gamonts*. These undergo a process of multiple fission, *gamogony*, resulting in

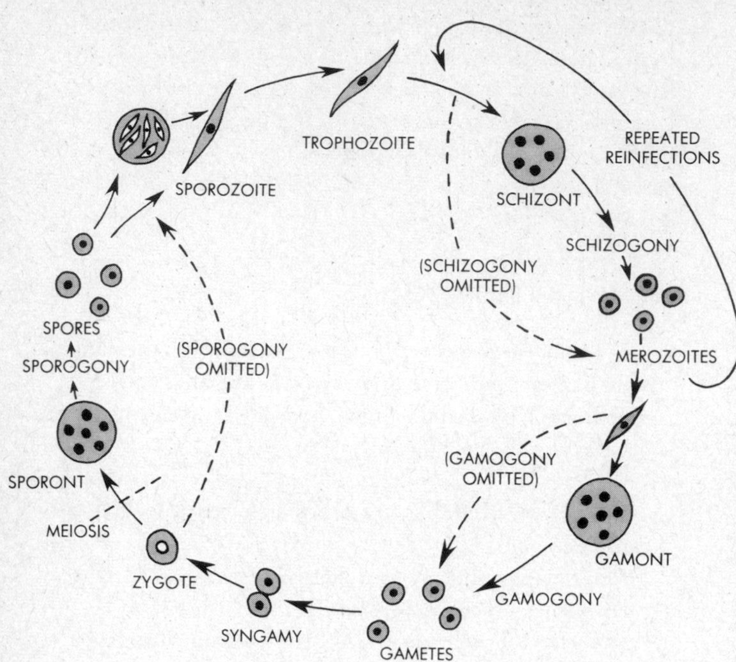

19.15. The basic sporozoan life cycle. In it, the zygote is the only diploid stage, and three haploid generations succeed one another, each characterized by a process of multiple fission (schizogony, gamogony, sporogony). The first is omitted in eugregarines, the second in various different sporozoan groups, and the third in cnidosporidians and acnidosporidians. After sporogony, spores may become free sporozoites directly or encapsulated sporozoites first.

gametes. The latter join pair-wise in syngamy, and the diploid zygotes so formed then undergo still another process of multiple fission which now includes meiosis. Such meiotic fission is *sporogony.* The cells produced are all haploid, and they are either naked *sporozoites* from the start or encapsulated spores. In the latter case, a spore divides within its capsule into a number of sporozoites, which subsequently escape from the capsule. Free sporozoites finally become vegetative trophozoites and a new cycle is thereby initiated. In effect, the characteristic sporozoan life cycle is haplontic, the zygote representing the only diploid stage. Moreover, three successive haploid generations may be present: one which reproduces by schizogony and forms vegetative spores; a second one which arises from the vegetative spores, reproduces by gamogony, and forms gametes; and a third one which arises from the gametes after syngamy, reproduces by meiotic sporogony, and forms sporozoites either directly or via walled spores.

Variations of this basic pattern result from an absence of any of the three phases. If schizogony is omitted, as in the Eugregarinida, the trophozoites become gamonts directly. If gamogony is omitted, as in numerous types of several groups, the gamonts become gametes directly. And if sporogony is omitted, as in Cnidosporidia and Acnidosporidia, the zygotes become sporozoites and thus trophozoites directly. Further variations result from the presence or absence of intermediate hosts in given life cycles.

As a group, the Sporozoa parasitize all phyla, including other protozoa and even other Sporozoa. Most require specific hosts. In the class Telosporidia, the gregarines are found in numerous invertebrate phyla but apparently not in vertebrates. Adult trophozoites live extracellularly in the intestine and in body cavities, where they adhere to tissues by means of various holdfast devices. For example, the eugregarine *Monocystis* is parasitic in the seminal vesicles and sperm ducts of earthworms (Fig. 19.16). When trophozoites are mature they join into chains of two individuals (more than two in other gregarines), a formation known as *syzygy.* One individual is male, the other female (and males and females alternate where chains of more than two members are formed). The sexual partners in syzygy (gamonts) then encyst within a common wall, the *gamocyst,* undergo gamogony, and form numerous gametes. Syngamy follows, and each resulting zygote then encysts individually within the gamocyst. The walled zygotes subsequently leave the earthworm via the sperm ducts, or the host is eaten by a bird which releases the zygotes unchanged with its feces. If such cells are later swallowed by another earthworm, they undergo meiosis and sporogony and form eight sporozoites. These escape from the spore wall when the latter is digested in the host gut. The sporozoites then migrate through the intestinal tissues and they eventually reach and enter sperm-forming cells of the earthworm. There the sporozoites mature into trophozoites which soon become extracellular and attach to seminal tissues. *Gregarina*, parasitic in the intestine of insects, has a similar history.

The eugregarines are relatively harmless, for in the absence of schizogony their numbers in any one host individual stay low. However, schizogony does occur in the schizogregarines, and repeated vegetative reinfection cycles here produce numerous parasitic individuals causing considerably more damage. Representative examples are *Schizocystis*, an intestinal parasite of larval flies, and *Ophryocystis*, a parasite in the Malpighian excretory tubules of beetles (Fig. 19.17).

Members of the subclass Coccidia occur in most coelomate animals, vertebrates included. The parasites live within the epithelial cells of the intestine, liver, spleen, and other organs. Syzygy occurs in some species but not in others, and in most cases the gamonts are dimorphic, males being structurally

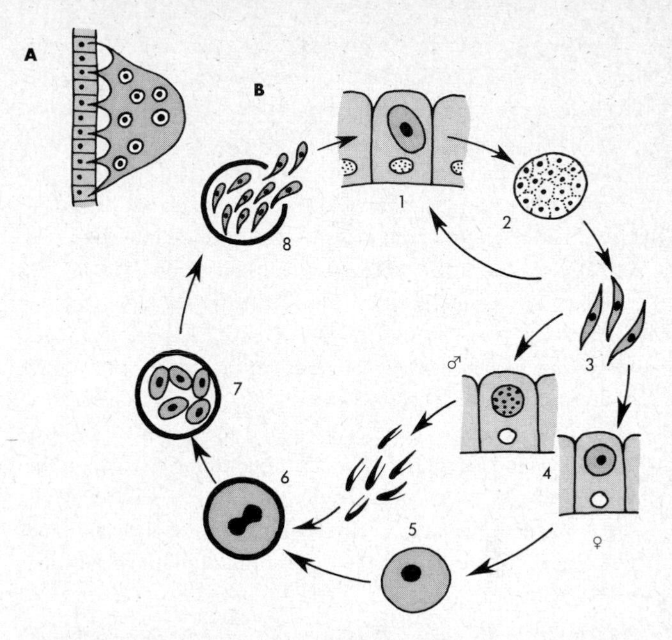

19.17. **A.** a schizont of the schizogregarine *Ophryocystis*, attached to the lining epithelium of a Malpighian tubule. (*After Leger.*) **B.** The life cycle of a species of the coccidian *Eimeria* (*after Borradaile*). 1, trophozoite in cell of mucosal lining of host; 2, schizogony; 3, merozoites, which may reinfect other mucosal cells; 4, merozoites become male and female gamonts in mucosal cells and undergo gamogony; 5, sperms and egg; 6, zygote encysts after fertilization and divides, forming encysted sporoblasts; 7, each sporoblast divides into two sporozoites; 8, the whole cyst with sporozoites leaves the host, is swallowed by another, and sporozoites escape and infect mucosal cells.

19.16. The life cycle of the eugregarine *Monocystis*, diagrammatic. (*After Borradaile.*) 1. Adult trophozoite in seminal vesicle of earthworm host. 2. A syzygy chain of two gamonts encapsulated in gamocyst. 3. Gamogony, resulting in gametes. 4. Pairwise syngamy, resulting in zygotes. 5. Individual zygote, encysted, nucleus undergoing two meiotic and one mitotic divisions. 6. Sporogony, resulting in formation of eight sporozoites. 7. Escape of sporozoites from original zygote cyst (and gamocyst) in new host.

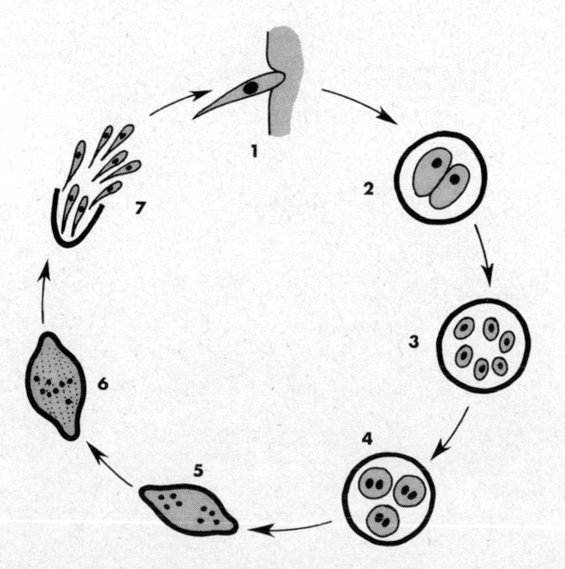

different from females. As in *Eimeria*, moreover, the male gametes are small biflagellate sperms, the female gametes are distinctly amoeboid eggs (cf. Fig. 19.17). Again because of the occurrence of schizogony, the Coccidia are relatively harmful. *Eimeria*, for example, is the causative agent of coccidiosis, a disease common in domesticated and other mammals and characterized by severe digestive disturbances.

The Hemosporidia include the best known and most studied sporozoan genus, *Plasmodium*, various species of which cause malaria in mammals, birds, and occasionally in lizards. Man is subject to infection by three species, each responsible for a different type of malaria. Repeated cycles of schizogony and release of merozoites from red blood corpuscles result in successive attacks of fever, the time interval between attacks being a main diagnostic feature of each of the three types of malaria. Completion of the plasmodial life cycle requires a specific blood-sucking intermediate host, viz., the *Anopheles* mosquito in man, the *Culex* mosquito in birds. In the intestinal tissues of such insects gamete formation and sporogony take place, without apparent harm to the hosts. Naked sporozoites then invade the salivary glands of the insects, and from there the parasites are injected into the blood streams of new main hosts through mosquito bites (Fig. 19.18).

The class Cnidosporidia (Fig. 19.19) is characterized by walled spores which contain *polar capsules,* oval organelles containing coiled, hollow filaments. In the intestine of a host, the spore walls are digested and the threads of the polar capsules are discharged, i.e., everted like the finger of a glove. Such a thread may pierce the intestinal lining and fasten the spore cell to the gut wall. Trophozoites are amoeboid. They become multinucleate, eventually undergo schizogony, and the amoeboid merozoites then mature into multinucleate and in some cases multicellular gamonts. The latter give rise to sporelike gametes with polar capsules, and after syngamy a new generation of amoeboid trophozoites is formed. These life cycles are exceedingly complex and many details are not yet known. The class is divided into several orders on the basis of differences in spore structure. The members of the class are parasitic in annelids, arthropods (e.g., *Nosema*), and fishes (e.g., *Myxobolus*).

The Acnidosporidia are similarly amoeboid, but their spores are without polar capsules. Some of these Sporozoa inhabit the muscles and connective tissues of various vertebrates, including particularly mammals and occasionally man (e.g., *Sarcocystis*). Others parasitize various tissues of aquatic invertebrates such as annelids (e.g., *Haplosporidium*). As in the preceding class, many details of the life histories are still unknown.

Like the Sarcodina, the Sporozoa represent a diverse as-

19.18. The life cycle of the malarial parasite *Plasmodium vivax*. Trophozoites invading human red blood corpuscles (top left) undergo schizogony. Production of merozoites destroys a corpuscle and free merozoites (which at this point occasion an attack of fever) then reinfect new corpuscles and lead to repetition of a 48-hr fever cycle. Merozoites entering red corpuscles also function as gamonts, and if human blood is sucked up by an *Anopheles* mosquito, the gamonts break free in the mosquito stomach, transform into male and female gametes, and effect fertilization. The zygote then encysts in the stomach wall and undergoes meiotic sporogony. The resulting free sporozoites subsequently migrate through the body cavities and organs of the insect, including the salivary glands. From there the sporozoites are injected into the human circulation when the mosquito bites a man. Sporozoites mature in human liver cells into trophozoites, and the latter begin a new life cycle.

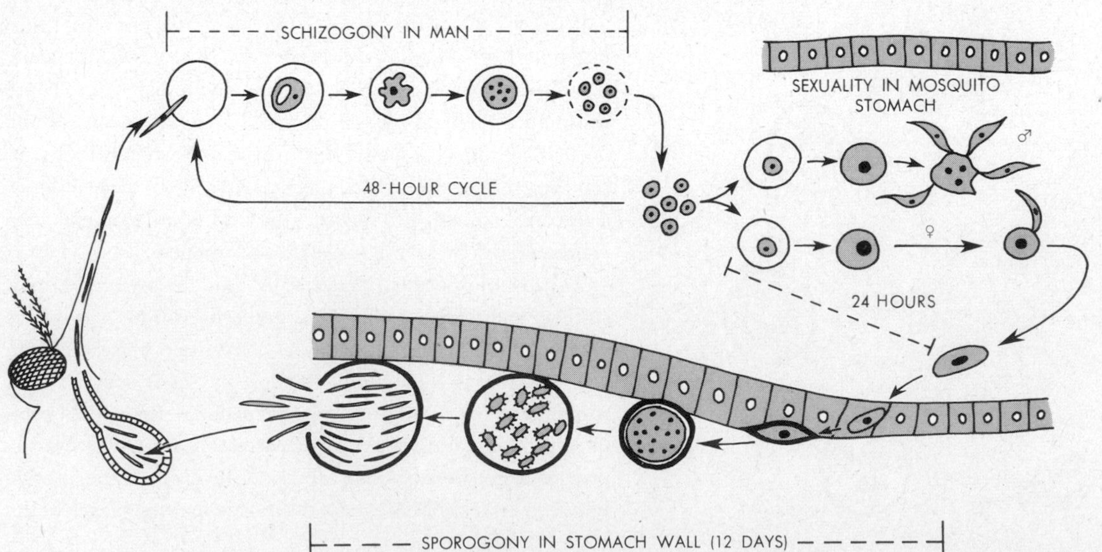

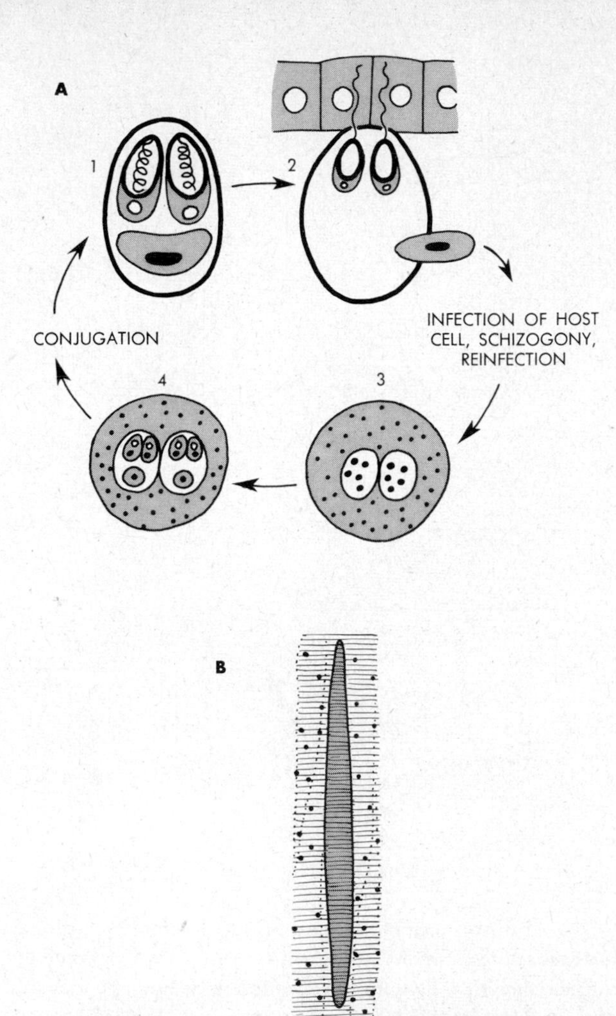

19.19. **A.** Stages in the life cycle of a myxosporidian (Cnidosporidia). 1, mature spore after conjugation, with zygote (bottom) and two polar-capsule-containing cells (top); 2, discharge of polar capsules attaches the spore to the intestinal lining of host, and zygote (trophozoite) is set free; 3, after schizogony, merozoites grow into syncytial complexes in which multinucleate spore-forming cells appear; 4, the spore-forming cells develop polar capsules and spore cells; conjugation of latter completes the cycle. **B.** A walled, spore-containing individual of the acnidosporidian *Sarcocystis*, embedded in vertebrate straited muscle.

semblage of groups. The Telosporidia appear to constitute a related set of organisms, with possible affinities to the flagellate-amoeboid stock of free-living protists. However, spore-formation and the haplontic life cycles are particularly reminiscent of coccine protists, both algal and fungal. Assessment of the evolutionary status of the Telosporidia and of Sporozoa generally is exceedingly difficult in view of the parasitic modes of life; one cannot be sure whether any given trait is original and primitive or is a secondarily specialized adaptation to parasitism. The classes Cnidosporidia and Acnidosporidia do not appear to be related too closely to the Telosporidia. The amoeboid trophozoites and the protractedly multinucleate and multicellular stages vaguely suggest affinities to slime molds, but before any meaningful interpretations can be made the organisms will have to be known far more completely. Polar capsules appear to be original evolutionary adaptations to parasitism, and it is possible that Acnidosporidia might have evolved from Cnidosporidia by loss of the polar capsules. On balance, the Sporozoa may constitute at least three and possibly more separate groupings of protists, all or some of which may or may not have any direct evolutionary relation to the other protozoa (cf. Fig. 19.1).

SUBPHYLUM CILIOPHORA

Ciliate protozoa; cilia present throughout life or at young stages, generally arranged into rows; body with micronuclei and macronuclei, often numerous; centrioles absent; reproduction by binary fission, largely transverse; sexuality by conjugation.

Class CILIATA

- Subclass Holotrichida: ciliation uniform over entire body surface
  - Order Gymnostomatida: gullet simple, with trichites. *Prorodon, Didinium*
  - Order Trichostomatida: gullet with vestibule. *Colpoda, Balantidium*
  - Order Chonotrichida: ciliation reduced; vaseshaped body; ectocommensalistic on crustacea. *Spirochona*
  - Order Apostomatida: rosette around gullet; ectocommensalistic on marine crustacea. *Foettingeria*
  - Order Astomatida: gullet absent; parasitic in annelid gut. *Anoplophyra*
  - Order Suctorida: stalked; feeding by tentacles; reproduction by budding. *Podophrya, Acineta*
  - Order Hymenostomatida: gullet ciliation elaborate. *Paramecium, Tetrahymena*
  - Order Thigmotrichida: gullet absent or located posteriorly. *Boveria*
  - Order Peritrichida: ciliation transverse, fission longi-

tudinal; many forms stalked, also colonial. *Vorticella, Epistylis*

Subclass Spirotricha: ciliation nonuniform, body ciliation largely reduced, compound ciliary organelles conspicuous

Order Heterotrichida: adoral area elaborate, body ciliation uniform. *Stentor, Blepharisma, Spirostomum*

Order Oligotrichida: adoral area elaborate, body ciliation absent. *Halteria*

Order Tintinnida: adoral area elaborate; body in shell or lorica. *Favella*

Order Odontostomatida: little ciliation, body compressed laterally. *Saprodinium*

Order Entodiniomorphida: body ciliation absent, internal structure complex; endocommensalistic in herbivorous mammals; *Ophryoscolex, Cycloposthium*

Order Hypotrichida: body ciliation absent, cirri present. *Euplotes, Stylonychia*

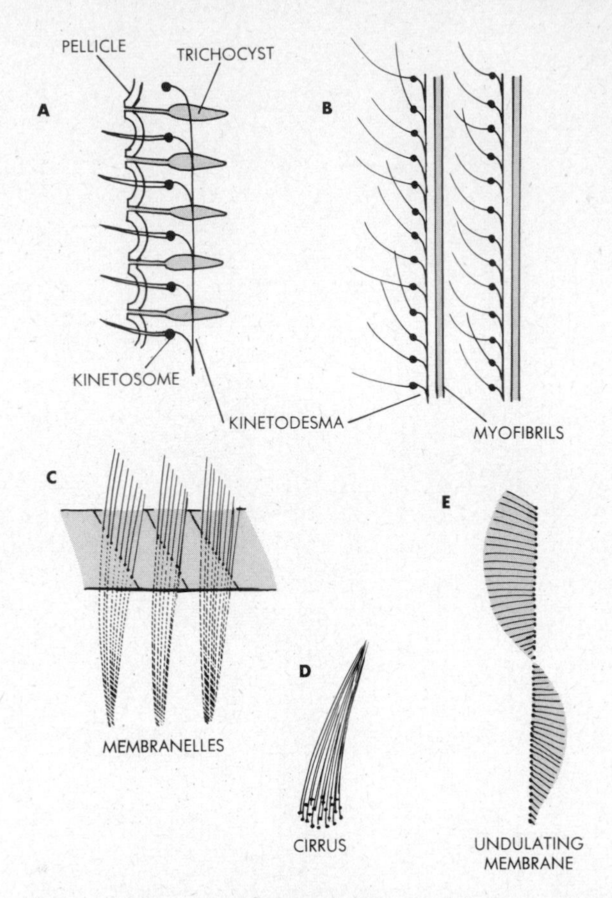

19.20. Surface organelles and ciliary variants in ciliate protozoa, diagrammatic. A. Cross section through portion of *Paramecium* surface; in surface view the pellicle has a hexagonal pattern, with a depression in each hexagon. **B.** Surface view of *Stentor,* showing portions of two adjacent ciliary rows, each accompanied by a myofibril. The cilia are drawn to illustrate their wavelike, metachronous beat. In **C,** the triangular basal portion of each membranelle is an anchor lying in the surface cytoplasm of the organism. **D** and **E** show the organelle portions projecting beyond the cell surfaces.

The ciliates form a more nearly homogeneous evolutionary group than the other protozoan subphyla. They are also the most complexly elaborated protozoa, and they represent the most diversely specialized of all known cell types. The organisms probably constitute a monophyletic assemblage, with an ancestry tracing back presumably to early zooflagellates, possibly to opalinidlike stocks. Ciliates include some 5,000 described species. Most of them are free-living and may be found in all aquatic environments. A substantial number of species is commensalistic in various animal groups, and comparatively few are parasitic. The majority are solitary and motile organisms, some are sessile, and of the latter a few form branching, treelike colonies.

The structural characteristics of the group are most distinctive. The body is generally plastic but has a species-specific shape and in most cases also a distinct longitudinal axis. A pellicle covers the outside, and protruding through it in all or certain circumscribed body regions are the cilia (Fig. 19.20). Typically, cilia are arranged in orderly rows and they beat in a coordinated, *metachronous* rhythm, i.e., in wavelike sequence. The rows of cilia are attached internally to corresponding subpellicular rows of kinetosomes, the latter being joined to one another by complex systems of fibrils (*kinetodesmas*). Also present subpellicularly is a system of conductile neurofibrils. These presumably control the ciliary beat, probably in conjunction with the kinetosomal fibrils. In many species, contractile myofibrils, or *myonemes*, parallel the rows of kinetosomes. Rows of dischargeable *trichocysts* may also be found in given species. Cilia often occur as modified compound organelles. For example, if cilia in a row are fused into a sheet, an *undulating membrane* is formed. Such a membrane may serve in locomotion or in producing food-bearing currents. Tapered tufts of fused cilia are *cirri,* strong bristlelike organelles functioning as locomotor legs. Fused cilia from several rows form *membranelles,* tiny paddles which create an extra-strong beat (cf. Fig. 19.20).

Most ciliates possess a permanent gullet, or *cytostome* (Fig. 19.21). The simplest type consists merely of a *mouth* opening on the body surface. In more complex gullets, the mouth is situated at the bottom of a *vestibule,* a funnelshaped depression in the body surface. In the most complex gullets, a vestibule leads into a short canal, the *cytopharynx,* at the bottom of which is the ingestive area proper. Ciliates feed either by using the gullet apparatus to catch prey, often larger than the ciliate itself, or by creating water currents which bring microscopic food to the gullet. In the first case the mouth area or vestibule may be strengthened by a circular set of stiff rods called *trichites.* In the second case a specialized *adoral zone,* or *peristome,* is present near or around the gullet area and produces the food-bearing currents. The adoral zone is usually equipped with membranelles, and additional membranelles as well as undulating membranes may be present in the cytopharynx. Food vacuoles migrate over a more or less definite path within the ciliate body, and digestive remains are egested at a fixed point, the *cytopyge,* often located in or near the cytopharynx. Contractile vacuoles occur at fixed positions near the body surface, and in many cases definite cytoplasmic channels form an internal drainage system leading to the contractile vacuoles (cf. Fig. 19.21).

19.21. Mouth structure in ciliates. **A.** Simple mouth opening on cell surface, food vacuole formed directly underneath. **B.** Mouth located at bottom of vestibular pit. **C.** Like **B** but vestibule supported by circlet of rodlike trichites. **D.** Vestibular depression and peristome area (adoral zone) circled by ring of adoral membranelles; vestibule leads into funnelshaped cytopharynx, with mouth and food vacuole at bottom. **E.** Sectional view of **D;** dashed line indicates path of food vacuole, arrow points to egestive cytopyge region, here located in cytopharynx. Water-drainage channel and contractile vacuole sketched on right side of organism.

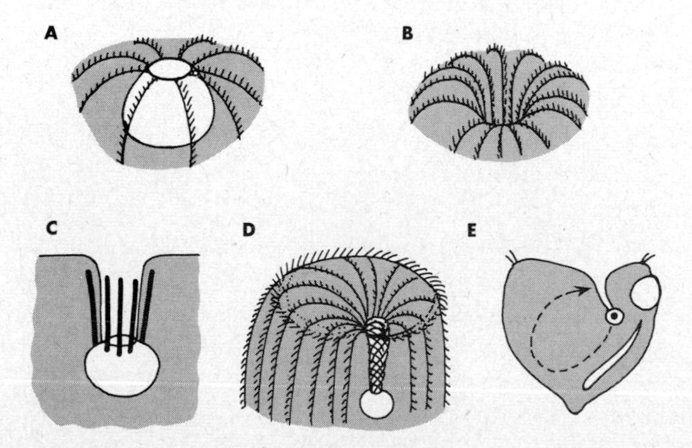

Ciliates are always multinucleate. They possess at least one and often many (up to several hundred) *micronuclei,* and at least one or many (up to several dozen) *macronuclei* (Fig. 19.22). The diploid micronuclei contain typical chromosomes but the macronuclei do not, at least not in the usual identifiable form. Instead, a macronucleus is believed to consist of numerous genetic units or "subnuclei," each containing a diploid set of genes which is not organized into the typical chromosomal arrangement. In the ordinary microscope this genetic substance of a macronucleus appears as a homogeneous clump of material. The structural dimorphism of the nuclei is accompanied by important functional differences. Micronuclei produce and exert long-range control over the macronuclei, and they are also the principal controllers of sexual processes. Macronuclei govern all metabolic and developmental functions, and they are directly responsible for the maintenance of the visible traits of the organism. Micronuclei may be lost or removed and a ciliate may survive without difficulty as a vegetative individual, which may continue to reproduce normally by fission. However, if the macronuclei are removed or lost, even if the micronuclei are still present, all structures such as gullet, membranelles, cirri, contractile vacuoles, and body cilia degenerate and the organism dies. Even so, it can be shown that all macronuclear functions are ultimately determined by the genes in the micronuclear chromosomes, and only the latter are transmitted from one sexual generation to the next; the macronuclei disintegrate at the end of one sexual generation and form anew from the micronuclei at the beginning of the next (cf. below).

This specialization of the genetic material of ciliates into two types of nuclei is correlated directly with the high degree of cytoplasmic specialization. Maintenance of the whole complex surface apparatus of a ciliate is controlled by the macronuclei; and it appears that the evolution of such a level of cell specialization was possible only through the parallel evolution of a separate nuclear machinery serving specifically in the control and maintenance of the surface elaboration. Thus, complex ciliation and macronuclei go together, and apparently there cannot be a ciliate level of structure without a macronucleus. This is an added reason, incidentally, why a ciliate hypothesis of eumetazoan origins is quite unsatisfactory. Such hypotheses usually postulate a ciliate ancestor *without* macronucleus. A creature of this kind remains completely imaginary, for actual ciliates certainly do not provide a model for it. And if a ciliate ancestor *with* macronucleus is postu-

A

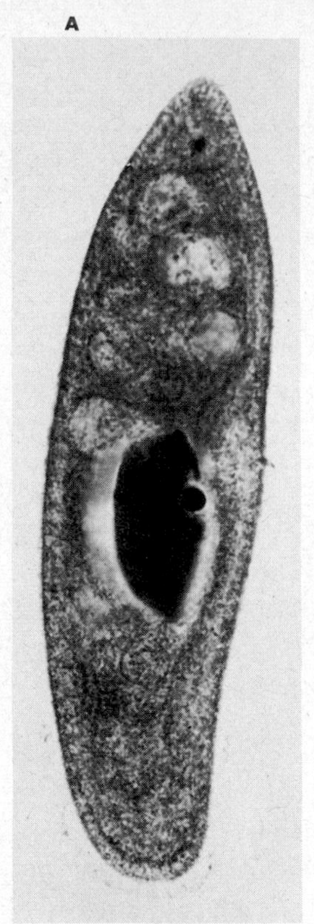

A, *courtesy Carolina Biological Supply Company.*

B

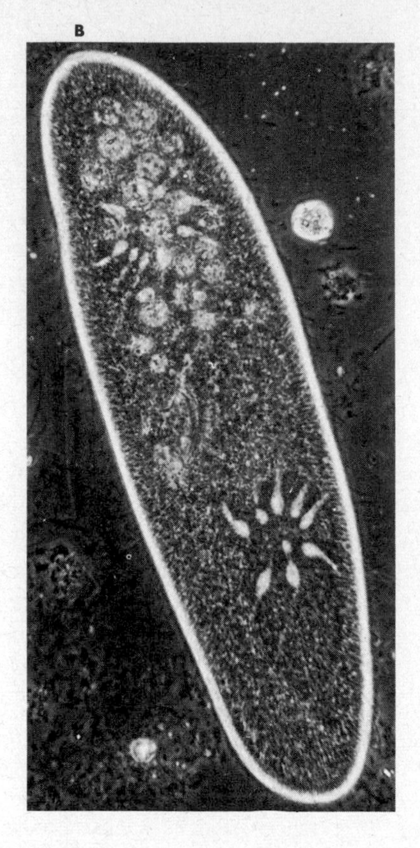

C

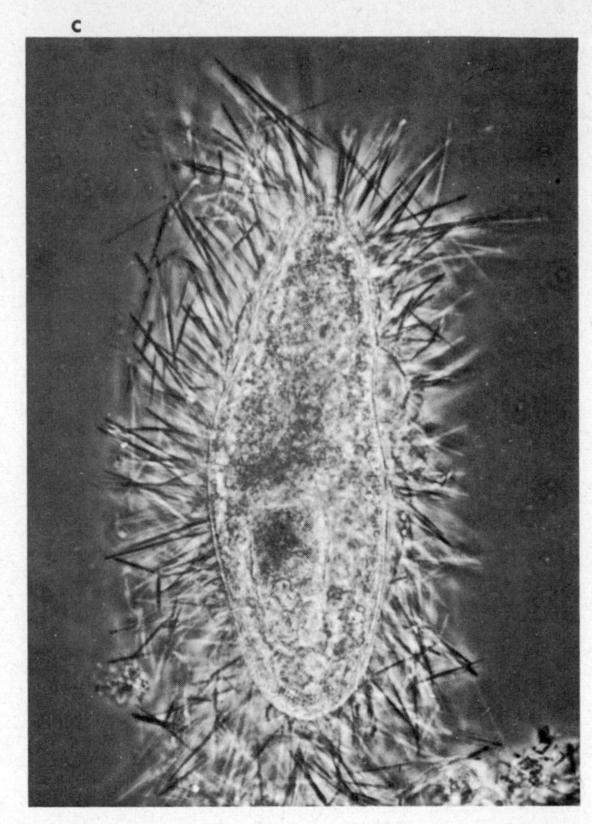

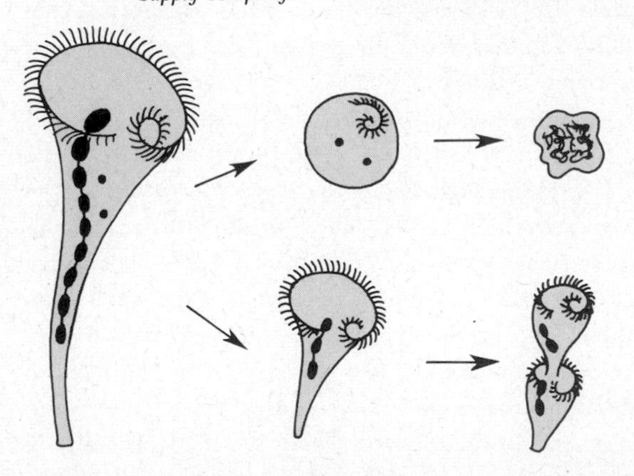

19.22. Photos: A. *Paramecium caudatum,* stained to reveal the macronucleus (large dark central body) and the micronucleus (small dark body partly overlapping the macronucleus on one side). **B.** *Paramecium multimicronucleatum,* from life, showing contractile vacuoles. **C.** *Paramecium caudatum,* discharging trichocysts. **Diagram:** *Stentor,* with macronucleus and two of numerous micronuclei shown. If the macronucleus is removed, the organism dedifferentiates and dies (top figures); but if the micronuclei are removed (and even if only part of the macronucleus is left), the organism continues to be normal and may even divide (bottom figures).

lated, what would have been the fate of this organelle in the supposed metazoan descendants? The latter are totally without any signs or remnants of it, and the single type of nucleus of their cells controls *all* activities, as in all cells other than those of ciliate protozoa. What the ciliates actually do suggest is that they represent a highly specialized, separate line of cellular evolution. This line has fully exploited the possibilities of cytoplasmic specialization based on special nuclear machinery, and presumably has given rise to nothing but more ciliates.

A capacity of encystment is in evidence throughout the ciliate subphylum. Reproduction occurs by transverse binary fission in virtually all cases, though longitudinal fission is the rule in one group (order Peritrichida) and reproduction by budding in another (order Suctorida). In fission (Fig. 19.23), the micronuclei divide mitotically and intranuclearly; centri-

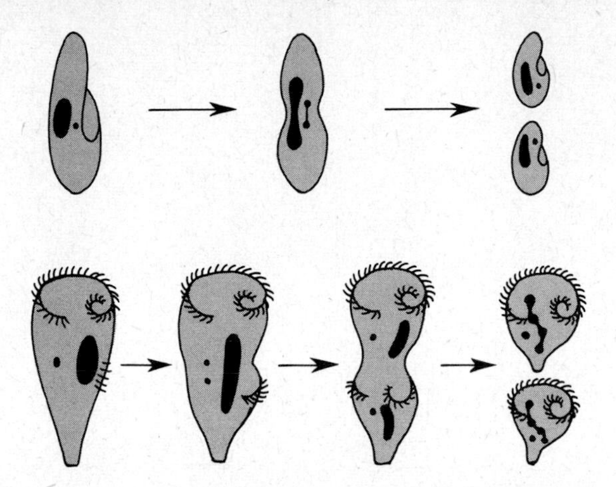

19.23. Ciliate reproduction. Top, fission in *Paramecium*. The micronucleus undergoes mitosis and the macronucleus elongates and constricts into two parts. Also, the original mouth organelles are resorbed and each offspring develops a new set. **Bottom,** fission in *Stentor*. The nodulated macronucleus (as in Fig. 19.22) condenses into a compact mass and then re-elongates as each micronucleus divides mitotically. Concurrently, the future anterior offspring inherits the original set of mouth organelles, while the future posterior one develops a new set. After the two offspring are constricted apart, the macronuclear portion inherited by each renodulates.

oles are absent. The macronuclei divide *amitotically*, i.e., they split into approximately equal halves and each daughter nucleus so receives roughly equal numbers of genetic subnuclei. In many species, the gullet, membranelles, cirri, and other specialized surface structures of a parent individual are resorbed at the onset of fission, each daughter cell then developing a new set. In many other species, one daughter, usually the anterior one, inherits the parental organelles and the other daughter develops a new set. Ciliates have a highly developed regeneration potential. In numerous instances, almost any cell fragment containing at least one macronucleus may regenerate all the missing organelles characteristic of a normal adult (cf. Fig. 19.23).

Virtually all ciliates are hermaphroditic and the sexual process is conjugative. A substantial number of species is known to include two or more mating types, conjugation being possible only between members of two specific different types (cf. also discussion of *syngens,* Chap. 16). Conjugation involves the temporary partial fusion of two individuals, usually near their gullet areas. Concurrently, the macronuclei degenerate and the micronuclei undergo a series of divisions. In *Paramecium,* for example, some species normally contain only one (diploid) micronucleus and one macronucleus (Fig. 19.24). At the beginning of conjugation, the micronucleus passes through two meiotic divisions, and of the four resulting haploid nuclei three degenerate. In species with many micronuclei, all undergo meiosis, and all but a single resulting haploid nucleus degenerate. The remaining haploid nucleus divides mitotically, and one product represents the "male" or *migratory* gamete nucleus, the other the "female" or *stationary* gamete nucleus. Nuclear exchange between the conjugated partners occurs next, and the two gamete nuclei then present in each conjugant fuse into a diploid zygote nucleus. The latter subsequently undergoes a series of mitotic divisions and gives rise to a number of new micronuclei and new macronuclei. In certain species of *Paramecium,* for example, the zygote produces eight nuclei in three divisions; four of these eight become micronuclei, the other four, macronuclei. The mating partners then separate and each such *exconjugant* undergoes two successive fissions. In the process, each of the four cells formed acquires one micronucleus and one macronucleus, the adult complement in such cases. In other ciliates the exconjugants divide analogously until each resulting cell contains the normal number of nuclei (cf. Fig. 19.24).

Some ciliates (e.g., the peritrich *Vorticella*) exhibit mating dimorphism. One mating partner is a large *macroconjugant,* the other a smaller *microconjugant.* After conjugation only the macroconjugant survives and among its later offspring new microconjugants may develop. In *Paramecium* and several other genera, it has been shown that the successive generations of offspring descended from a single exconjugant pass through a kind of "super-life-cycle," also called a *Maupasian* or "physiological" cycle. Early vegetative generations after conjugation are juvenile and sexually immature, i.e., vegetative reproduction may occur but conjugation may not. A mature phase then follows, during which conjugation does become possible. If sex actually takes place, the participating organisms become "rejuvenated" and their immediate descendants are again juvenile. But if conjugation does not occur (e.g., if individuals are isolated), then a senile phase eventually follows, characterized by nuclear and genetic abnormalities and other signs of aging. Such a line sooner or later becomes incapable of undergoing conjugation and it ultimately dies out. In many cases, genetic or actual death of a line is avoided even in the absence of conjugation, by self-fertilization in the form of autogamy (cf. Fig. 19.24). Though conjugative nuclear exchange here does not take place, fusion of two sister gamete nuclei within a single individual has the same rejuvenating

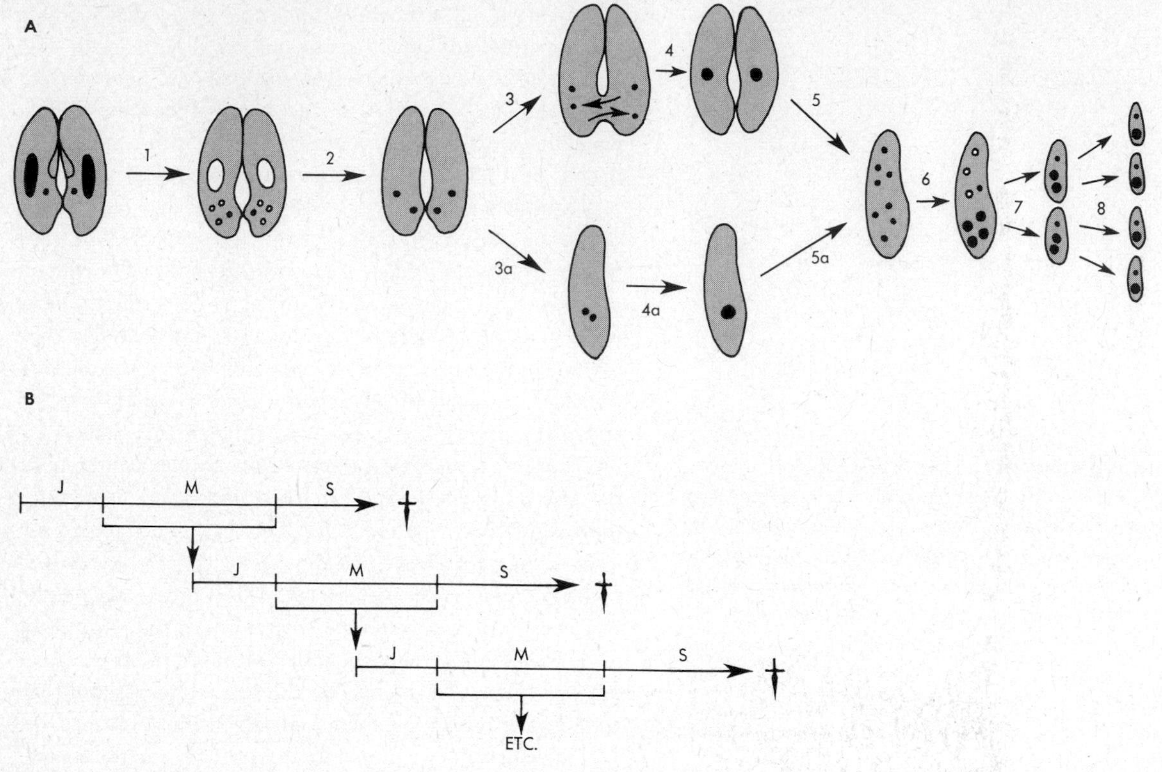

19.24. Sexuality in *Paramecium*. **A.** The sexual process 1. Meiosis in each conjugating partner, followed by degeneration of macronucleus and of three haploid micronuclei resulting from meiosis. 2. Mitotic division of remaining haploid nucleus, yielding one stationary and one migratory gamete nucleus (the latter situated near surface in paroral cone). 3 and 4. Nuclear exchange and formation of synkaryon (diploid zygote nucleus). 3a and 4a. Autogamy, or self-fertilization, i.e., fusion of haploid gamete nuclei within same individual. 5 and 5a. Three mitotic nuclear divisions, resulting in eight diploid nuclei. 6. Four of the eight become macronuclei, three degenerate, and one becomes a micronucleus. 7 and 8. Cell and micronuclear divisions and parceling out of macronuclei, till each exconjugant possesses normal nuclear complement. **B.** The Maupasion life cycle. *J, M, S,* juvenile, mature, and senile phases, respectively. If sex occurs it takes place during the *M* phase, and the resulting exconjugants then are juveniles starting a new cycle. If sex does not occur, it no longer can occur during the *S* phase and the line eventually dies out. See also color plate IV.

effect as conjugation; the descendants after autogamy are juveniles.

As given above, the modern classification of ciliates differs in substantial respects from earlier schemes. The single class within the subphylum consists of two subclasses, of which the Holotrichida are undoubtedly more primitive than the Spirotrichida. Representative types of each order are shown in Figs. 19.25 and 19.26, and the presumed evolutionary interrelations of the orders within the subphylum are depicted in Fig. 19.27. The main evolutionary path is believed to have led from Gymnostomatida to Hymenostomatida to Heterotrichida to Hypotrichida, with side branches at each step. The main path is characterized by a progressive increase in the complexity of the ciliary apparatus and the gullet structures. Thus, primitive Gymnostomatida such as *Prorodon* possess only a simple mouth and they are without specialized com-

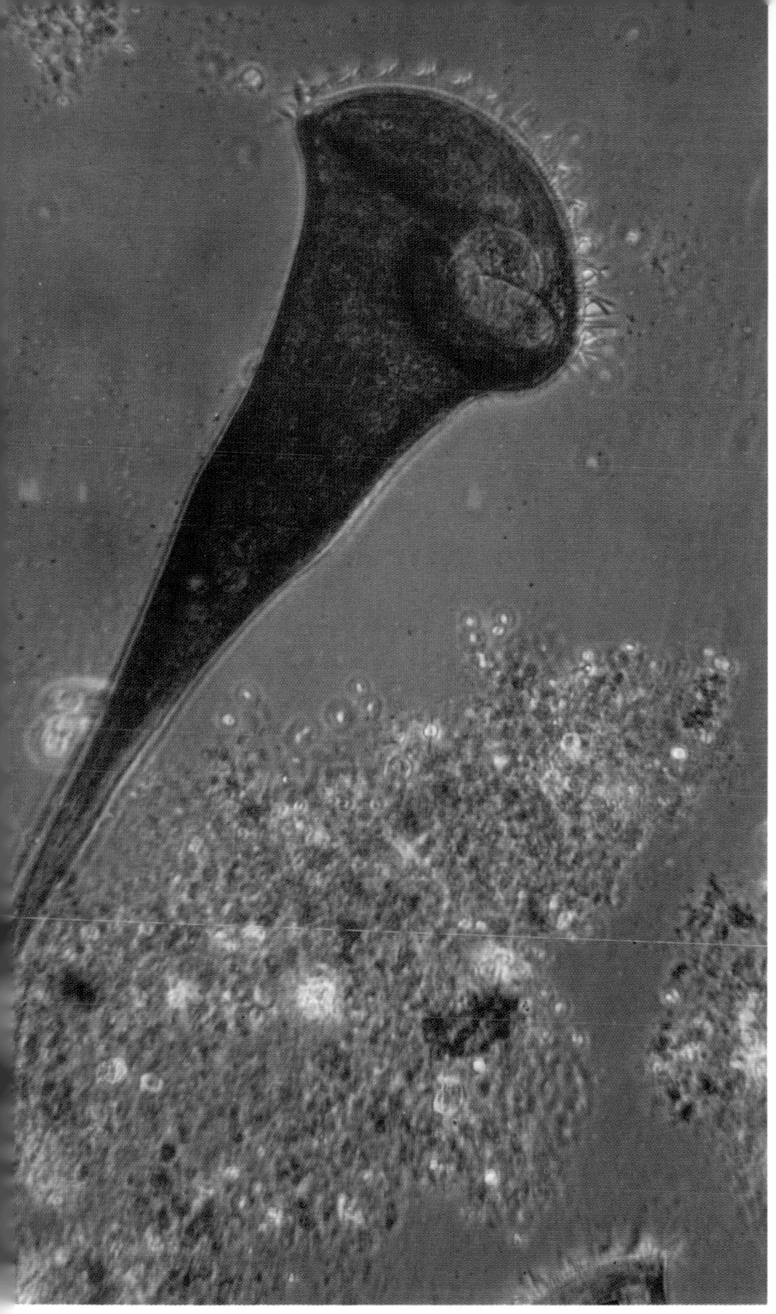

PLATE IV. Ciliate protozoa. **A, B,** the heterotrich *Stentor coeruleus*. **A,** individual attached by holdfast and body extended; **B,** same but body contracted after surface stimulation. **C,** the heterotrich *Spirostomum*. **D,** conjugating pair of *Blepharisma;* note temporary fusion of mating partners near their gullet regions. **E,** the hymenostome *Paramecium bursaria;* the green coloration is due to the presence of zoochlorellae in the protozoan cytoplasm. **F,** the hypotrich *Stylonychia*.

Plate IV, page 1

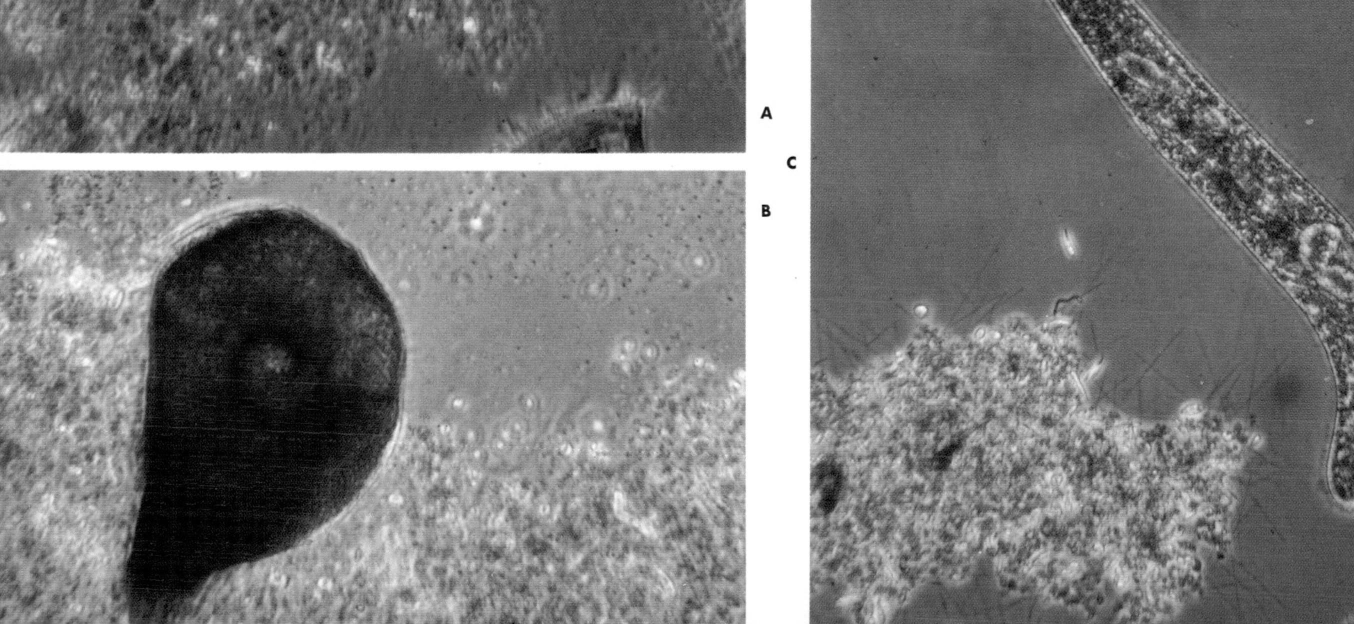

A

C

B

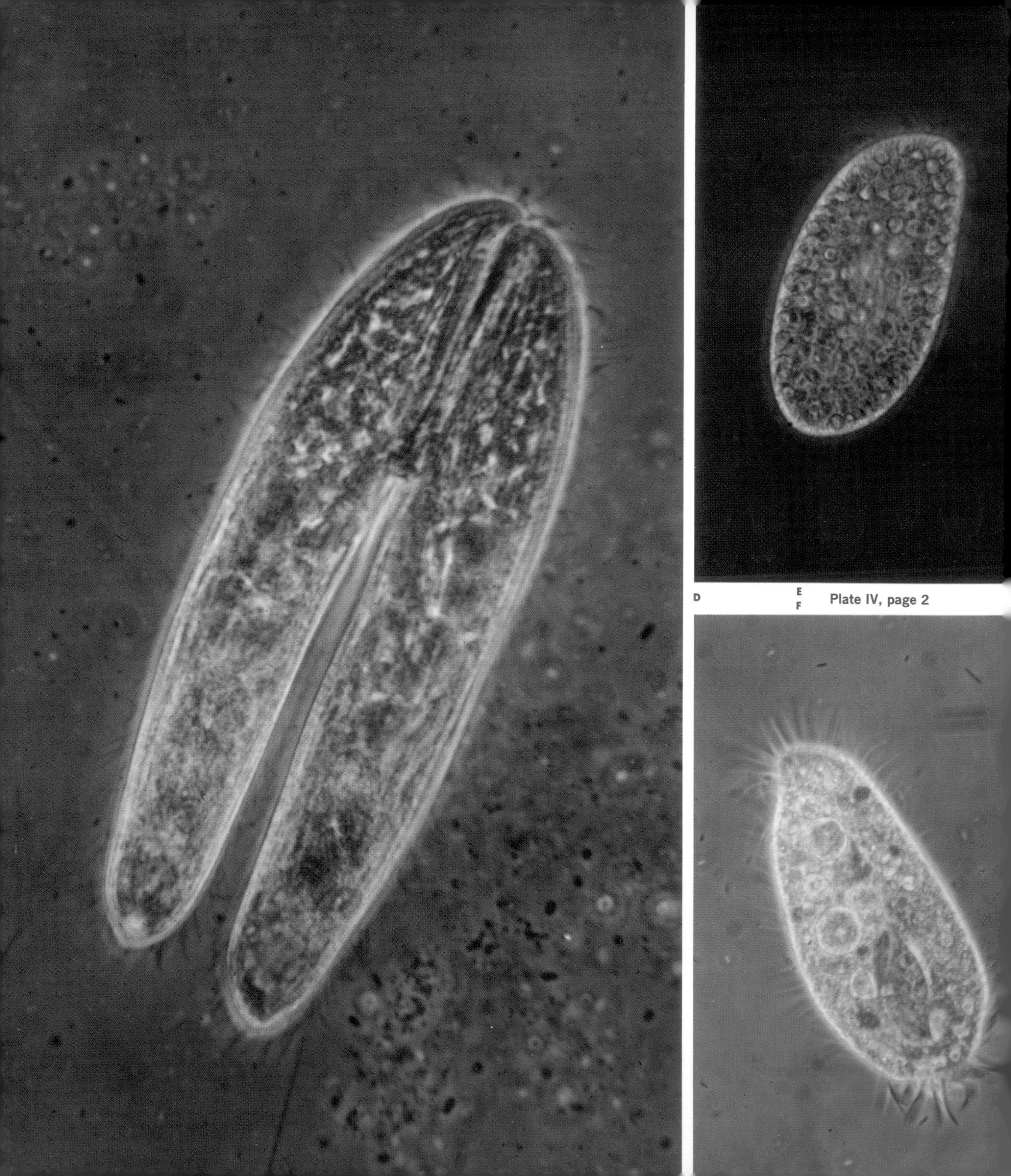

Plate IV, page 2

pound ciliation. Hymenostomatida like *Paramecium* or *Tetrahymena* possess a gullet with vestibule and cytopharynx, the latter also containing a four-part, or "tetrahymenal," arrangement of three membranelles and one undulating membrane. Heterotrichida such as *Stentor* are further equipped with a complex peristome, ringed by sets of adoral membranelles. And Hypotrichida like *Euplotes* possess the most complex sets of membranelles, cirri, and undulating membranes.

Gymnostomes are saltwater and freshwater forms. The order includes *Didinium,* a carnivore with a protrusible proboscis by means of which it may engulf a whole paramecium at one time. The four orders Trichostomatida, Chonotrichida, Apostomatida, and Astomatida consist largely of commensalistic and parasitic types, *Balantidium* representing the only known ciliate to be parasitic in man. Suctorida are stalked, sessile types, formerly classified as a separate class or subclass within the subphylum. The suctorians trap prey with tentacles, and by some still poorly understood process they suck the contents of the prey through the tentacles into their bodies. Adult suctorians are without body cilia; cilia are required neither for feeding nor for locomotion in these sessile organisms. Nevertheless, a complete set of kinetosomes is present underneath the cell surface. Also, cilia do appear in the motile offspring. Such offspring are produced by internal or external budding. In internal budding, a small portion of the parent is divided off into a chamber formed within the adult cell. Whether it grows internally or externally, a bud develops cilia, escapes from the parent, and after swimming about for some time it settles and transforms into a sessile adult without cilia (Fig. 19.28).

The hymenostomes occur in all aquatic environments. *Tetrahymena* is of considerable importance in nutritional and other biological research, for this ciliate can be grown in *axenic* cultures, i.e., in chemically completely defined nutrient media and in the absence of any food organisms. Thigmotrichs

19.25. Holotrichous ciliates, representative genera of each order. The lines over the body surface of many of the organisms represent rows of kinetosomes; they carry cilia, not shown here. The positions of macronuclei, micronuclei, and contractile vacuoles are indicated in many instances. See also color plate IV. (*Adapted from Corliss, Hyman, Lwoff, and other sources and from life.*)

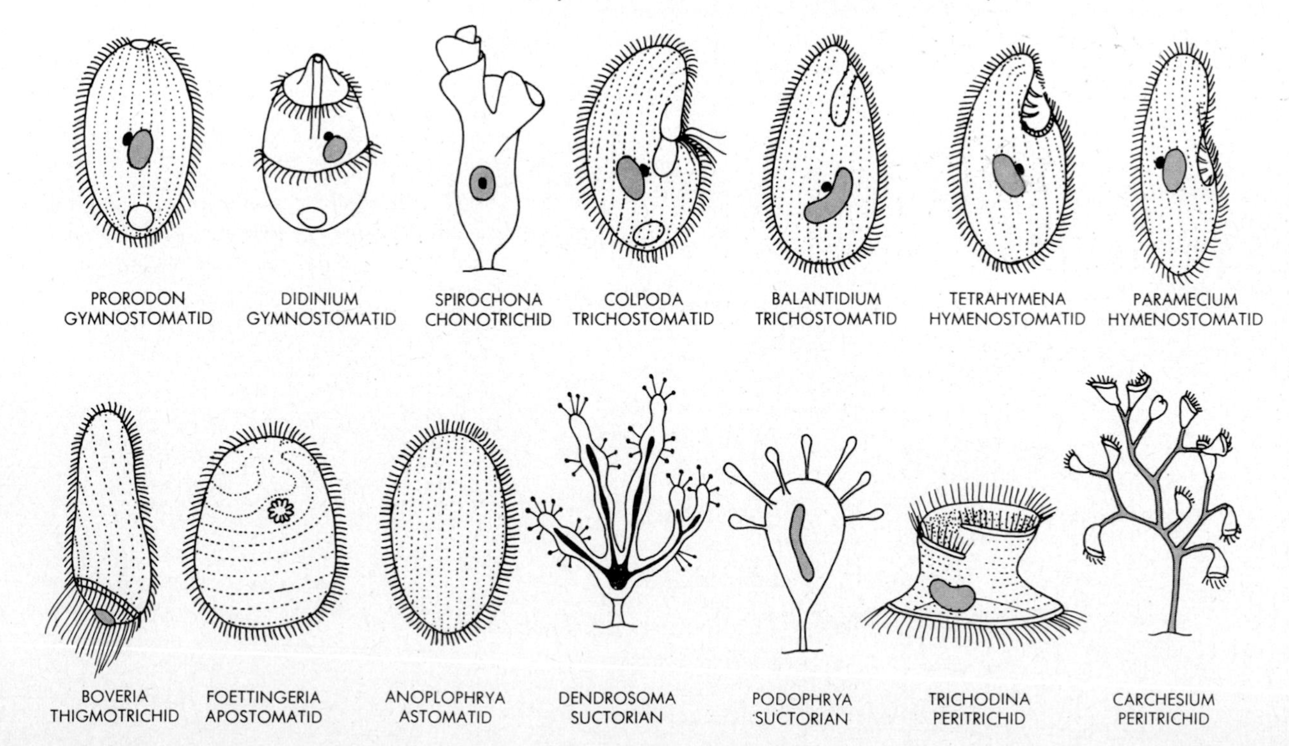

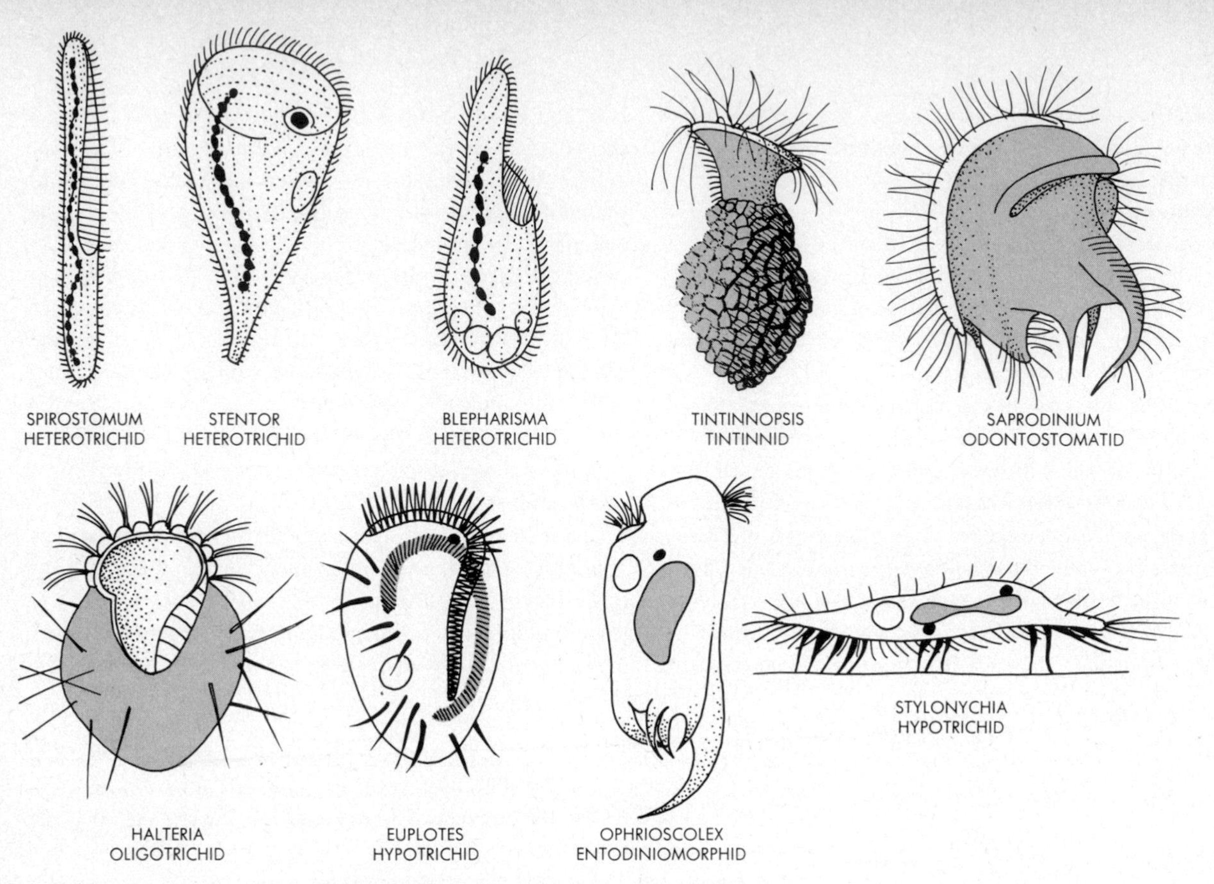

19.26. Spirotrichous ciliates, representative genera of each order. *Stylonychia* is shown in side view. Otherwise as in Fig. 19.25. See also color plate IV. (*Adapted from Corliss, Bütschli, and other sources and from life.*)

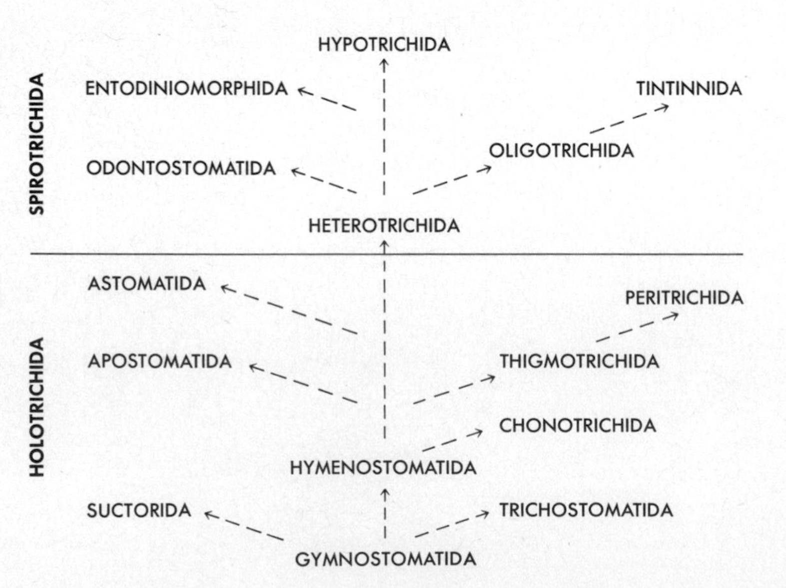

19.27. Presumed evolutionary interrelations among the orders of ciliates. All connecting paths should be regarded as tentative. (*Adapted from Corliss.*)

are mouthless in some cases; these ciliates occur as commensals in freshwater and saltwater mollusks. In the peritrichs, the body surface (but not the peristome) is without cilia. Kinetosomes are present, however, and they are arranged in transverse rows. Fission is longitudinal. Sessile adults include stalked solitary forms such as *Vorticella* (which possesses a myoneme-like contractile organelle in the stalk), and branched colonial types such as *Epistylis*. When fission occurs, one daughter develops a circlet of body cilia, separates from the other daughter which remains stalked, and swims

19.28. A. Internal (left) and external (right) budding in Suctorida, diagrammatic. **B.** Daughter individuals of *Vorticella* just after fission. One daughter inherits the stalk of the parent (with spiral myoneme); the other develops a posterior ciliary girdle and will migrate away and settle via a newly formed stalk elsewhere. Note the transverse rows of kinetosomes. **C.** The entodiniomorph *Cycloposthium*. (*Schematic, after Dogiel.*) The whole structural complexity of this ciliate is not apparent in a sketch such as this. Photos: **A,** contracted, and **B,** expanded individual of *Vorticella*. **C.** A suctorian. See also color plate IV.

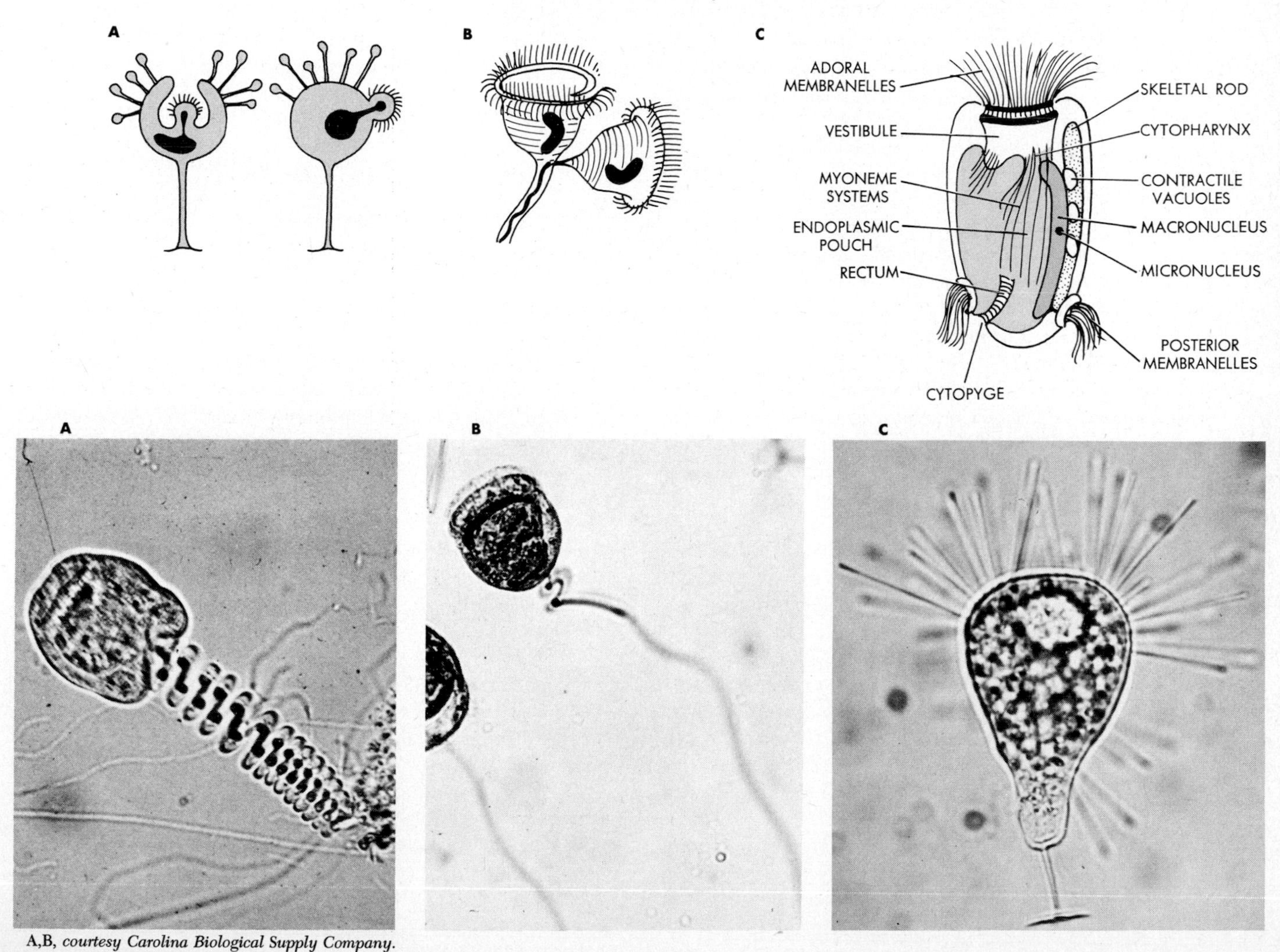

A,B, *courtesy Carolina Biological Supply Company.*

about for a time. Then it acquires a stalk of its own, settles, and loses its body cilia (cf. Fig. 19.28).

The Spirotrichida are generally larger organisms than the Holotrichida. In the order of heterotrichs a uniform body ciliation is still present, but the peristome is quite elaborate, as noted above. Many heterotrichs are beautifully pigmented, in green, blue, pink, orange, and other colors. Most are free-living, but some are commensals. Among freshwater types, *Spirostomum* and some species of *Stentor* are large enough to be visible with the naked eye. The oligotrichs include *Halteria,* a type without general ciliation but which possesses bristles used as tiny stilts in a jumping form for locomotion. The tintinnids are marine and pelagic, without cilia except in the peristome area. Their secreted housings have been preserved as fossils, the only ciliate fossils known. The odontostomes live in sewage and in natural waters in which the oxygen content is low. Their bizarrely shaped bodies are compressed laterally, and the ciliation is highly reduced. The remaining two orders, Entodiniomorphida and Hypotrichida, include the most complex of all the ciliates. For example, *Cycloposthium,* an intestinal commensal of herbivorous mammals, contains skeletal supports, virtually a complete alimentary "system" with a permanent rectal tube, a "muscular" system composed of numerous sets of myonemes and also neurofibrils, systems of contractile vacuoles, and complex groups of membranelles in various regions (cf. Fig. 19.28). General ciliation is absent. This is true also in the hypotrichs, ciliates which are flattened dorsoventrally and in which ventral cirri serve as walking legs. Hypotrichs are largely free-living in both salt and fresh water.

Even this brief account probably shows clearly that the protozoa are far from "simple." They are small, to be sure, but in this very smallness lies perhaps their most remarkable property; namely, that despite being limited to the confines of single cells they can be as diverse, varied, and complex in the microsphere of life as only very few groups of far larger organisms can be in the macrosphere.

REVIEW QUESTIONS

1. Define protozoa taxonomically and give analogous definitions for each of the four main subgroups. Distinguish protozoa and (*a*) other protists, (*b*) Metazoa. What undesirable conceptual consequences follow if protozoa are labelled "acellular" organisms?

2. Review the possible evolutionary relations of protozoan groups to one another and to other organisms. Describe the general structural characteristics of protozoan cells. What cytoplasmic and nuclear organelles are typical of such cells? How does reproduction occur? Describe the life cycle of the majority of protozoa.

3. State the characteristics of Mastigophora and of each zooflagellate order. How are these orders probably related? Describe the structure of a zooflagellate cell. What are axonemes, parabasal bodies, axostyles, costae, trichocysts?

4. In what way do the Rhizomastigida suggest a close interrelation of Mastigophora and Sarcodina? Name a representative genus of each mastigophoran order and describe its (*a*) structural, (*b*) ecological characteristics.

5. What characteristics do the genera *Mastigamoeba, Dimorpha,* and *Naegleria* have in common? Why are the first and last of these not classified as members of the same taxonomic unit (e.g., subphylum)? Name the classes and orders of Sarcodina and a representative genus of each. Describe the structure of various types of pseudopodia and indicate in which group each occurs.

6. What is the structure of Heliozoa? How do these organisms feed? Reproduce? Undergo sexual processes? Describe the structure of Radiolaria. Where do these organisms live?

7. What are Foraminifera and where do they live? Describe the structure of their interior and their shells. Review the life cycles of these organisms and indicate why such cycles are called diplohaplontic ones.

8. Describe the basic life cycle of Sporozoa and distinguish between schizogony, sporogony, and gamogony. Review the classification of the subphylum and indicate the type of life cycle encountered in each major taxonomic group. What is syzygy? A polar capsule? In which groups are polar capsules found? Describe the life cycle of the malarial parasite *Plasmodium.*

9. What is the general structure of a ciliate? What are cirri, membranelles, and undulating membranes? How does the gullet structure vary in different ciliate groups? Distinguish micronuclei and macronuclei (*a*) structurally, (*b*) functionally.

10. Show how a ciliate divides. What is the detailed sequence of events in ciliate conjugation? What is autogamy? A Maupasian life cycle? Name various ciliate orders, indicate their characteristics, and review their postulated evolutionary interrelations.

COLLATERAL READINGS

The reference works below cover most major aspects of protozoan biology:

American Association for the Advancement of Science: "Sex in Microorganisms," AAAS Symposium, Washington, D.C., 1954.

Calkins, G. N, and F. M. Summers (eds.): "Protozoa in Biological Research," Columbia, New York, 1941.

Hall, R. P.: "Protozoology," Prentice-Hall, Englewood Cliffs, N.J., 1953.

Hutner, S., and A. Lwoff (eds.): "Biochemistry and Physiology of the Protozoa," vols. 1 and 2, Academic, New York, 1951, 1955.

Hyman, L. H.: "The Invertebrates," vol. 1, chap. 3, McGraw-Hill, New York, 1940.

Jahn, T. L., and F. F. Jahn: "How to Know the Protozoa," William C. Brown, Dubuque, Iowa, 1949.

Lwoff, A.: "Problems of Morphogenesis in Protozoa," Wiley, New York, 1950.

Rivera, J. A.: "Cilia, Ciliated Epithelium, and Ciliary Activity," Pergamon Press, New York, 1962.

Sonneborn, T. M.: Breeding Systems, Reproductive Methods, and Species Problems in Protozoa, in "The Species Problem," American Association for the Advancement of Science, Washington, D.C., 1957.

Tartar, V.: "The Biology of *Stentor*," Pergamon Press, New York, 1961.

Weisz, P. B.: Morphogenesis in Protozoa, *Quart. Rev. Biol.*, vol. 29, 1954.

Wichterman, R.: "The Biology of *Paramecium*," McGraw-Hill, New York, 1953.

CHAPTER 20 mesozoa, parazoa

With this chapter we begin a study of the Metazoa, or animals proper. Each of the two groups considered here represents a separate taxonomic branch within the subkingdom Metazoa. As it happens, each branch encompasses only a single phylum, i.e., the phylum Mesozoa in one case, the phylum Porifera in the other. We shall be concerned mainly with the Porifera, or sponges, and we may recall in this connection that many characteristics of these animals have already been outlined in earlier contexts, e.g., in Chaps. 7, 8, and 9.

PHYLUM MESOZOA

Mesozoans; parasitic; body consisting of a single layer of ciliated surface cells serving digestive and locomotor functions and of one or a few interior cells serving reproductive functions; cell numbers and arrangements constant for any given species; life cycles complex, incompletely known, with vegetative and sexual generations.

Order Dicyemida: endoparasitic in mollusks, mostly squids.
Order Orthonectida: endoparasitic in various worms and other invertebrates.

This small group of minute animals (Fig. 20.1) constitutes a taxonomic and evolutionary puzzle, for its affinities are altogether obscure. Many zoologists, possibly the majority, regard the animals as degenerate offshoots of flatworms, more specifically, of trematodes (cf. Chap. 22). The principal basis for such a view is the mesozoan life cycle which, like that of certain of the trematodes, includes vegetative and sexual generations. The first take place in some still unidentified intermediate hosts, the last in the molluscan main hosts. A taxonomic rank on approximately the level of a class within the phylum of flatworms is therefore proposed. Other authorities consider mesozoan life cycles to be simply adaptations to parasitism, and they point out that Mesozoa have so simple a structure (including a ciliated cellular surface rather than a cuticle as in trematodes) that a flatworm affinity is unlikely. Instead, the animals are regarded to be on a level between protozoa and coelenterates and their life cycles are considered to be no more complex than those of the Sporozoa. A protozoan derivation of Mesozoa is therefore suggested. On such a view the group is assigned the rank of a phylum and the animals are proposed to represent a branch within the Metazoa, equivalent to the branches Parazoa and Eumetazoa.

The issue remains obscure not only because of the parasitism of the animals, but also because the body layers of Mesozoa are not homologous to the eumetazoan ectoderm and endoderm. Thus, mesozoan "ectoderm" is digestive and "endoderm" is purely reproductive, never digestive. All zoologists agree that, notwithstanding differences of opinion regarding possible affinities, the actual status of the Mesozoa remains an enigma at present.

PHYLUM PORIFERA

Sponges, "pore-bearing" animals; marine and freshwater; adults sessile, often in colonies; body radial to asymmetrical; alimentation via channel system with choanocytes; development regulative; larvae flagellated and free-swimming.

Class CALCAREA (chalk sponges): spicules calcareous with

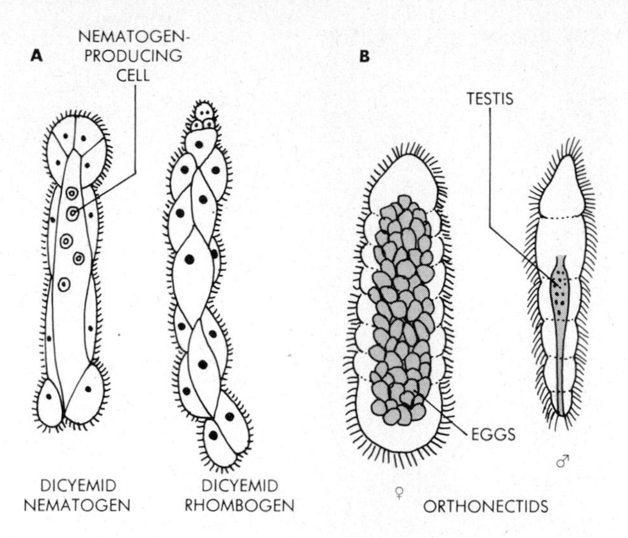

20.1. Mesozoa. (*After Hyman.*) **A. Dicyemids.** Left, section through *nematogen,* the mature parasitic stage in cephalopod hosts. By nuclear divisions in the single elongated interior cell, more cells are formed which escape the nematogen and produce new generations of nematogens. Upon sexual maturity of the host, interior cells of a nematogen give rise to a *rhombogen* stage, shown in surface view at right. The internal cell of such a stage produces larva-forming cells. Larvae leave the host and presumably enter intermediate hosts in which the sexual process of dicyemids takes place. **B. Orthonectids.** Left, mature female with eggs; right, mature male with testis. The exterior tissues of these animals consist of numerous small cells. Fertilized eggs develop within the female, and the larvae then escape and infect hosts where orthonectids have the form of multinucleate amoeboid masses. From the latter, males and females as shown above develop and escape from the hosts.

one, three, or four rays; larvae amphiblastulae or stereoblastulae. *Leucosolenia, Sycon, Leuconia*

Class HEXACTINELLIDA (glass sponges): spicules siliceous with six rays, often fused into continuous networks; surface epithelium absent; choanocyte chambers in layer. *Hexactinella, Euplectella, Farrea*

Class DEMOSPONGIAE (horny sponges): spicules of spongin, silica, or both, sometimes none; if present, silica spicules with one to four rays; body of leuconoid type; larvae amphiblastulae or stereoblastulae. *Halisarca, Oscarella, Spongia*

Of the approximately 5,000 species of sponges described to date, most are marine. Some 150 species live in fresh water and all of these are members of the class of horny sponges. Among the marine types, the chalk sponges are largely shallow-water forms, glass sponges inhabit deep waters predominantly, and horny sponges are found in all regions, from the tide zone to depths of 3 to 4 miles. Many of the glass sponges are anchored to the sea floor by means of ropelike tufts of skeletal materials (Fig. 20.2). As a group, sponges provide a home for numerous other animals. Shrimps, crabs, copepods, sea anemones, octopuses, barnacles, and brittle stars are among many types sheltering in the spaces within sponge colonies. Such colonies must often compete for living space with other sessile colonial forms, e.g., corals, tunicate ascidians, and bryozoan ectoprocts. Snails and chitons prey on sponges, and some crabs give themselves a protective disguise by attaching pieces of sponges to their backs.

Most sponges are asymmetrical, with flattened, globular, lobed, and variously branched shapes (cf. Fig. 20.2). Horny sponges often occur as encrustations on objects in water. However, a good many sponges tend toward radiality, with vaselike or cupshaped forms. Some types are brightly pigmented, the predominant colors in such cases being red, yellow, blue, and black.

The body of a chalk sponge is made up of an external epithelium, an internal epithelium, and an intermediate *mesogloea* (Fig. 20.3). The first is composed of *pinacocytes,* flat polygonal cells which form an epidermal surface. Many of the pinacocytes contain myofibrils, and such *myocytes* are contractile. The interior epithelium is formed from *choanocytes,* flagellated collar cells which maintain a flow of food-bearing water through the sponge. The mesogloea consists of jelly in which are embedded amoeboid mesenchyme cells of various types, collectively called *amoebocytes.* Certain amoebocytes known as *archeocytes* give rise to the gametes and function also as important digestive cells; they engulf food particles too large for choanocytes and also particles passed to them from the choanocytes. Other amoebocytes serve as skeleton-forming cells, as food-storing cells, as nurse cells for eggs, as pigment cells, as cells harboring symbiotic zoochlorellae, and as jelly-secreting cells. The nuclei of pinacocytes and choanocytes generally have small nucleoli or none; amoebocytes and most of their variants, archeocytes particularly, are characterized by nuclei with large, prominent nucleoli. Many of the cell types of a sponge are not rigidly specialized. For example, some of the mesenchyme cells may transform from one variety into most of the others.

Chalk sponges exhibit three degrees of architectural com-

20.2. Sponges. **A,** a calcareous type; **B,** the silicaceous sponge *Hyalonema;* the ropy tuft anchors the sponge to the sea floor; **C,** the silicaceous Venus flower basket *Euplectella;* **D,** the horny type *Stelospongia*. See also Figs. 7.9 and 8.8.

plexity (Fig. 20.4). The simplest types, exemplified by *Leucosolenia,* have a so-called *asconoid* structure. Such a sponge is essentially a straight-walled sac, the wall being formed by an epithelial pinacocyte layer exteriorly, a layer of mesogloea underneath, and an epithelial choanocyte layer interiorly. The choanocyte layer lines a *spongocoel,* the central cavity of the sac. Water enters the spongocoel via *porocytes,* pore cells scattered in the body wall. Probably derived from amoebocytes, porocytes are tubular cells each pierced by a canal so situated that water may flow from the exterior through the body wall into the spongocoel. Porocytes are contractile and may close their canals. Water leaves the sponge via an *osculum,* a wide exit from the spongocoel.

More complexly constructed are the *syconoid* sponges, e.g., *Sycon.* In these, the body wall is deeply pitted or folded along its interior surface (cf. Fig. 20.4). The spaces between the folds are lateral extensions of the spongocoel called *radial canals.* The choanocyte layer here is not continuous but lines the radial canals only. The rest of the spongocoel is covered by pinacocytes. Porocytes are absent and water enters the sponge via numerous *ostia,* porelike spaces between the external pinacocytes. Intercellular spaces within the mesenchyme form *incurrent canals* which lead from the ostia into the radial canals. Water leaves a sponge as above, via spongocoel and osculum.

The *leuconoid* sponges, exemplified by *Leuconia,* are architecturally the most complex. Here the choanocytes are confined to extensive systems of small interconnected chambers (cf. Fig. 20.4). Incurrent canals lead into such a chamber via several *prosopyles,* tiny spaces between the choanocytes. Water exits from a chamber by a single opening, the *apopyle,* which leads into an *excurrent canal.* The excurrent canals from numerous chambers join into progressively larger canals, and the largest ultimately lead to the exterior via one or more oscula. In such sponges, spongocoel and radial canals are reduced or are absent. Note that the structural series from asconoid to syconoid to leuconoid types is characterized by a progressively greater subdivision of the choanocyte-lined spaces and by a correlated complication of the channel system to and from these spaces. The adaptive advantage of increased structural complexity undoubtedly lies in increased operational efficiency.

In any sponge, the diameter of an osculum is greater than the total diameters of all the ostia or porocyte canals combined. Water therefore enters a sponge at a greater velocity and at a lower pressure than it leaves. In effect, the flow maintained by the choanocytes is such that water is actively sucked into the sponge and is actively pushed out of it. Food in the water consists largely of bacteria, microscopic algae, and organic debris. The food-bearing water may be considered to pass through a succession of progressively finer screens. Thus, ostia average 50 microns in diameter, prosopyles are about 5 microns wide, and the slits between the cytoplasmic extensions in the choanocyte collars have a width of about 0.10 microns. Hence only the smallest food particles can be engulfed by choanocytes. Larger food is taken up by archeocytes stationed along the incurrent canals. Such cells are also believed to engulf small food particles which have been taken up by choanocytes and are then released by them into the mesenchyme. After they digest food intracellularly in protozoan fashion, archeocytes migrate to the external body surface or to the excurrent canals and discharge indigestible matter at such locations.

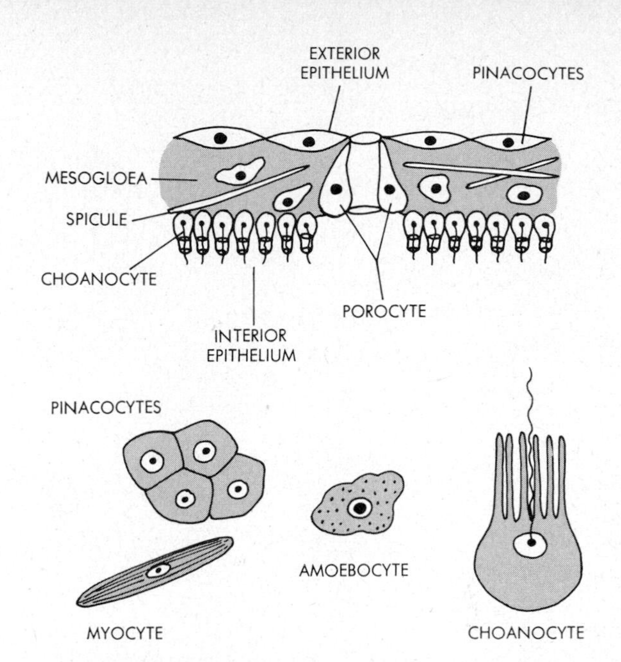

20.3. Tissues and cells of a sponge, schematic. Top, the three basic tissue layers. In the mesogloea, free skeletal spicules and amoebocytes are indicated. A porocyte is a tubular cell with a water canal into the sponge (cf. Fig. 20.4). **Bottom,** some of the cell types. Note the larger nucleolus in an amoebocyte. Myocytes are modified pinacocytes with contractile myofibrils. The choanocyte collar consists of a circlet of cytoplasmic extensions, as shown (cf. Fig. 19.6). Variants of amoebocytes form numerous other cell types with different functions.

Virtually all horny sponges (e.g., the bath sponge *Spongia*)

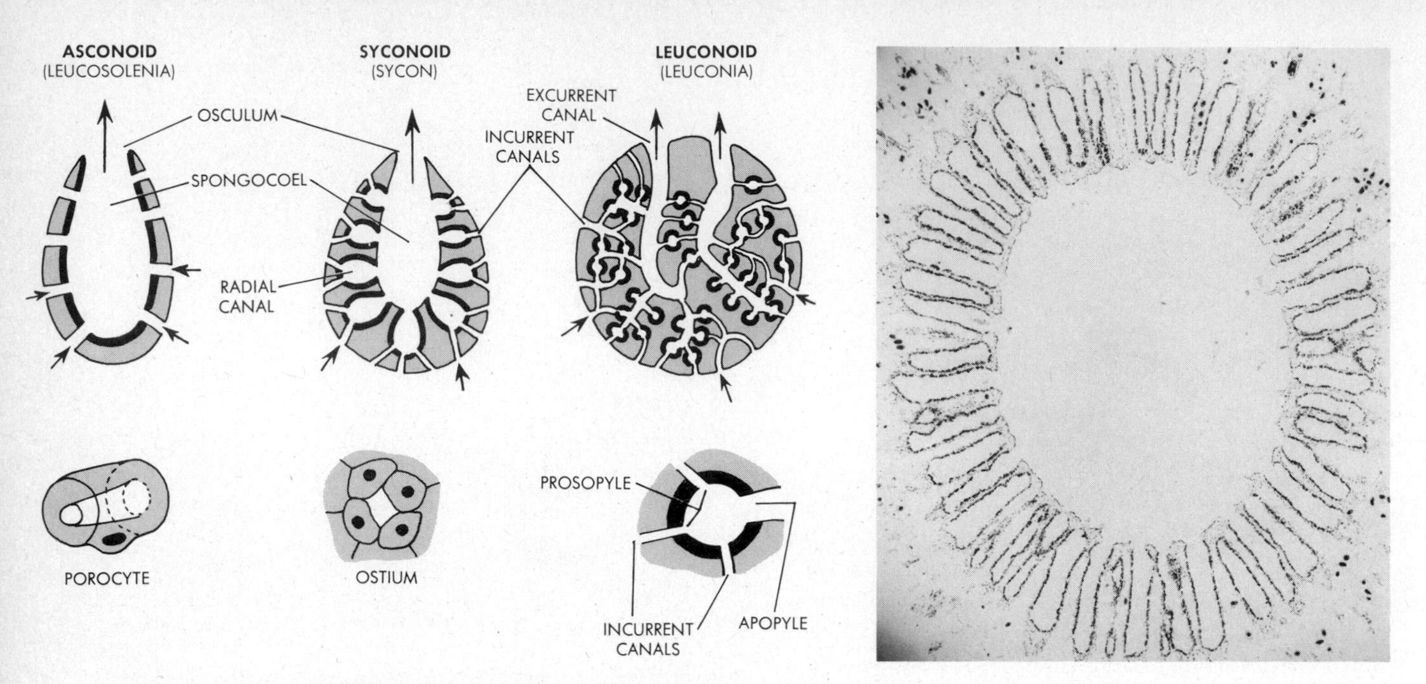

20.4. Diagram: the three levels of structural complexity in calcareous sponges. Dark strips indicate position of choanocyte layers. Lower figures outline the structures of the incurrent water pores. **Photo:** cross section through a syconoid sponge, showing the choanocyte-lined radial canals (dark layers) opening into the large central spongocoel.

are of the leuconoid type, i.e., they correspond structurally to the leuconoid chalk sponges. Glass sponges have a somewhat different architecture (Fig. 20.5). The body of these animals is basically syncytial. Mesenchyme cells are joined together into widemeshed networks of cellular strands, and individual cells may detach from such strands, migrate about, then rejoin the syncytium. A definite epidermal epithelium is absent. The choanocytes are in chambers with wide apopyles, hence the chambers have a thimblelike appearance. The basal portions of the choanocytes are fused to one another, but this syncytium remains netlike, with tiny prosopylar spaces between the cellular units. The chambers themselves are arranged into a layer which traverses the mesenchyme between the body surface and the boundary of the spongocoel. This layer is folded in many cases, the folds delimiting lateral extensions of the spongocoel like the radial canals of syconoid chalk sponges.

Although claims to the contrary have been made, sponges do not appear to possess nerve cells. However, the animals do respond to external stimuli to some extent. For example, bright light or mechanical irritation may initiate a slow contraction, beginning at the point of stimulation and progressing very gradually over the rest of the sponge. Such responses are undoubtedly brought about by the myofibril-containing pinacocytes. Also, myocytes located around ostia and oscula may produce a widening or narrowing of these openings. To a certain extent a sponge may therefore regulate the velocities and pressures of the water flowing through it. None of the behavioral reactions of sponges is rapid, however, nor are such reactions particularly diverse; any one of many different types of stimuli produces roughly the same type of response, viz., a slow regional contraction which may or may not spread to other parts of the body.

Contraction responses could not be too varied in any event, for the rigid skeletal elements of sponges preclude any substantial amount of body deformation. The skeletal units are called *spicules*. Each consists of up to six needlelike rays (Fig. 20.6). Thus, spicules are said to be monactinic, diactinic, triactinic, tetractinic, pentactinic, or hexactinic. They may also be classified according to how many axes the rays mark out, e.g., the spicules may be monaxons, triaxons, or tetraxons. Spicules are manufactured by variants of amoebocytes called *calcoblasts, silicoblasts,* and *spongioblasts.* Their skeletal products are composed of, respectively, calcium carbonate, silica, and *spongin,* a horny organic material (scleroprotein).

Chalk sponges possess calcoblasts. A single cell of this type manufactures a monactinic ray within its cytoplasm, the calcium carbonate being laid down around an organic axial fibril. As the ray grows the cell divides and leaves most of the ray exposed. One daughter cell (the "founder") continues to establish the shape of the ray, and the other cell (the "thickener") secretes more skeletal substance around it. In the production of complex spicules, groups of cells cooperate. The cells of a group are aggregated in such a way that, after each manufactures a single ray, the several rays can become fused together in a given geometric pattern. Only one-, three-, and four-rayed spicules occur in chalk sponges (cf. Fig. 20.6).

Hexactinic silica spicules are characteristic of glass sponges. The process of spicule formation here is similar to that in the chalk sponges, i.e., single silicoblasts form monactinic rays and several silicoblasts cooperate in the production of more complex spicules. In many of these sponges spicules project from the body surface, and in forms like *Euplectella,* the Venus flower basket, the skeletal elements are fused to one another and form a continuous glassy latticework (cf. Fig. 20.2). A sieve of interlaced spicules may also cover the osculum. In the horny sponges, spongin spicules are produced by spongioblasts grouped into linear rows. Each cell manufactures a section of the spicule and the several sections then fuse together into a rod, the usual form of a spongin element. In all sponges, the producing cells usually migrate away or degenerate after a spicule has been formed. Spicules are then left as free skeletal elements lying in mesenchyme. A few genera of primitive horny sponges are without skeletal parts of any kind (e.g., *Halisarca, Oscarella*).

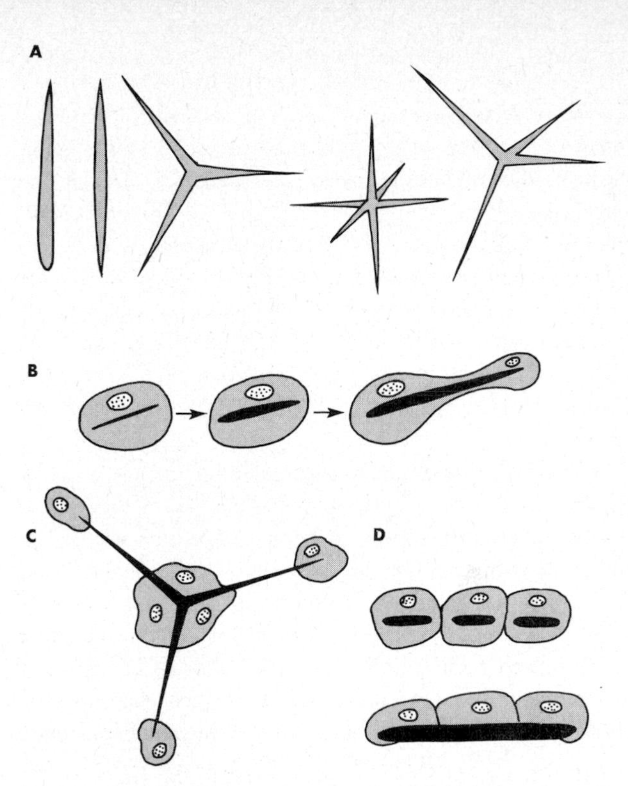

20.6. Sponge spicules. Cf. also Fig. 8.8. **A.** Some of the principal types of spicules. **B.** Formation of a spicule. A single cell (far left) manufactures a monactinic spicule around an organic fibril. Division of the cell results in a founder at the base of the developing spicule and a thickener at the tip. The founder shapes the spicule. **C.** Groups of cells cooperate in the formation of many-rayed spicules. **D.** Formation of a horny spicule by groups of spongioblasts. (*After Tuzet.*)

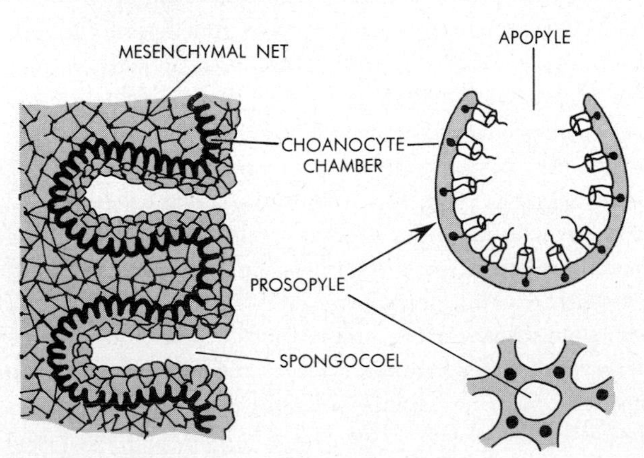

20.5. The syncytial structure of glass sponges, schematic. Left, section through such a sponge, showing absence of definite epidermal layer, the mesenchymal net, and the layer of thimble-shaped choanocyte chambers. (*From Hyman after Schulze.*) A single chamber is sketched at right (top) the wide opening forming the apopyle. The bases of adjacent choanocytes are fused syncytially, leaving prosopylar openings between the nucleated portions (bottom). (*After Ijima.*)

Sponges exhibit remarkably extensive capacities of vegetative reproduction. The animals may reproduce by *budding*, a process in which small aggregates of amoebocytes and archeocytes collect at some region of the body surface of a parent sponge (Fig. 20.7). Such aggregates then may fall off and develop on their own into new sponges. Alternatively, buds may retain continuity with the parent sponge, and in this fashion extensive colonies of sponges may be formed. Sponges may also be fragmented into pieces, and each piece usually reconstitutes a whole individual. Indeed, it has long been known that a sponge may be dissociated completely into a suspension of loose cells (e.g., by pressing the animal through fine-meshed cheesecloth). The separated cells then migrate in amoeboid fashion into an aggregated mound and reconstitute an intact, normally structured whole. If cell suspensions of two different species are mixed together, the cells "recognize" which species they belong to; two separate mounds will form, each growing into an intact whole.

Sponges subjected to unfavorable conditions in nature often form *reduction bodies*, i.e., small globules composed of a pinacocyte layer exteriorly and packed amoebocytes interiorly. The parent sponge may die off but the reduction bodies

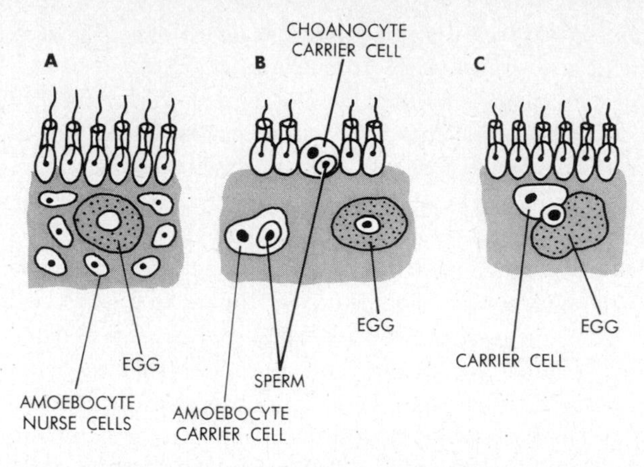

20.8. Gametes and fertilization, schematic. A. An egg, surrounded by amoebocytic nurse cells underneath the choanocyte layer of the parent. **B.** Formation of sperm-carrying cells, either from choanocytes or from amoebocytes. **C.** Migration of a carrier cell and transfer of sperm to egg, resulting in fertilization.

persist and may later develop into new individuals. All freshwater and some marine sponges produce *gemmules* under conditions of drought or low temperature. A gemmule is a reduction body containing an external layer of columnar cells with spicules. The internal cells include a variety of amoebocytes, particularly archeocytes and cells rich in protein and lipid food reserves (cf. Fig. 20.7). Note that buds, reduction bodies, and vegetative reproductive units in general do not normally contain choanocytes as such; collar cells develop later from some of the amoebocytes. In a normal sponge, also, a choanocyte may resorb its flagellum and collar and become a migratory amoebocyte.

Most sponges are hermaphroditic. Self-fertilization is rare, for a given sponge does not form sperms and eggs at the same time. The gametes arise from archeocytes, and after meiosis the eggs are amoeboid, the sperms flagellate. Eggs come to lie in the mesenchyme directly underneath the choanocyte layer, where they may receive food materials from amoebocytic nurse cells (Fig. 20.8). Sperms leave the parent via spongocoel and osculum and enter another sponge with the incoming water current. A sperm then becomes trapped either by an amoebocyte in the mesenchyme or by a choanocyte. A sperm-containing choanocyte resorbs its flagellum and collar and, like a sperm-containing amoebocyte, assumes the function of an amoeboid "carrier" cell: it migrates to an egg and makes intimate contact with it. The sperm is then transferred from the carrier cell into the egg, resulting in fertilization.

20.7. Vegetative reproduction in sponges. In budding, a bud may form either a separate (left) or an attached (right) individual. In a reduction body, spicules and pinacocytes form exterior layers, amoebocytes fill the interior. The structure of a gemmule is similar, except that the exterior cells are columnar and the spicules are elongated. Bottom, right, dissociated sponge cells migrate and reaggregate into an intact sponge.

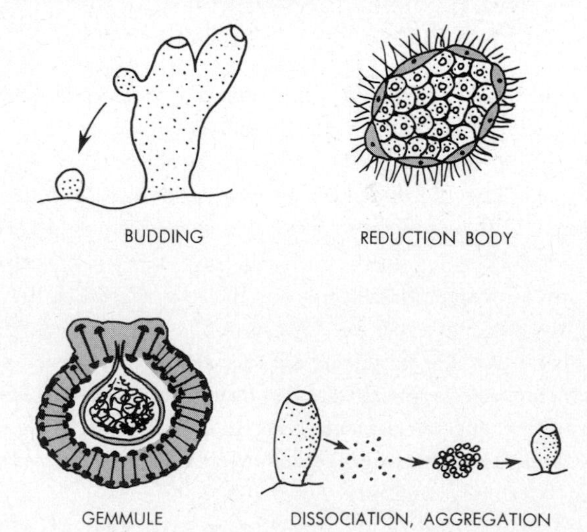

In the chalk sponges, embryonic development occurs according to two main patterns. One is exemplified by *Sycon* (Fig. 20.9). The zygote undergoes four successive cleavage divisions and produces a hollow blastula of 16 cells. One set of eight cells lies next to the parental choanocyte layer; these cells are without flagella and will eventually form the epidermal epithelium of the new sponge. The other eight cells lie on the side away from the parental choanocyte layer. These cells develop flagella directed into the space within the blastula, i.e., into the blastocoel. The cells will give rise to the choanocytes of the new sponge. During subsequent cleavages an opening develops among the cells in the nonflagellated half of the blastula. The whole blastula then turns inside out through this opening. As a result, the flagella of the flagellated cells come to be directed outward. Such a curious process of embryonic *inversion* is encountered not only in sponges but also in the development of spherical colonial algae, e.g., *Volvox*. This similarity has sometimes been postulated to suggest an evolutionary affinity between sponges and *Volvox*, but inversion in the two cases probably represents little more than parallel evolution in similar embryonic situations.

After inversion, the sponge embryo is a hollow *amphiblastula*, with an anterior flagellated hemisphere and a posterior nonflagellated hemisphere. At this stage the embryo escapes as an amphiblastula larva. After it swims about for a time, it settles at the anterior, flagellated end. The flagellated hemisphere then invaginates embolically into the nonflagellated hemisphere. A basic asconoid structure is thereby established, with the nonflagellated cells on the outside and the flagellated cells on the inside. The former give rise to pinacocytes, porocytes, and skeleton-forming calcoblasts. The latter develop into choanocytes and all other mesenchyme cells.

In the second pattern of development, exemplified by *Leucosolenia*, the zygote produces a many-celled hollow blastula, in which all but a few cells at the posterior end are flagellate (Fig. 20.10). These posterior cells, together with some of the surrounding cells (which lose their flagella), then migrate into the blastocoel, where they proliferate and fill up the entire space. A *stereoblastula* larva so forms, which subsequently escapes, swims about, and settles. Inversion then takes place. More specifically, the cells of the exterior layer lose their flagella and migrate into the interior, where they form choanocytes. Concurrently, the interior cells migrate to the exterior, where they develop into pinacocytes and mesenchyme. A spongocoel becomes hollowed out subsequently and an asconoid structure so results. *Leucosolenia* remains

20.9. Development of calcareous sponges, *Sycon* pattern. (*Adapted from Hartmann and Brien.*) 1, coeloblastulalike embryo underneath maternal choanocyte layer; 2, inversion, i.e., embryo turns inside out through opening in nonflagellated half, resulting in exteriorized flagella; 3, the amphiblastula larva; 4, invagination of flagellated half into nonflagellated half; 5, larva settles with "blastopore" end toward ground; 6, blastopore end closes over and an osculum breaks through at free side, establishing asconoid structure.

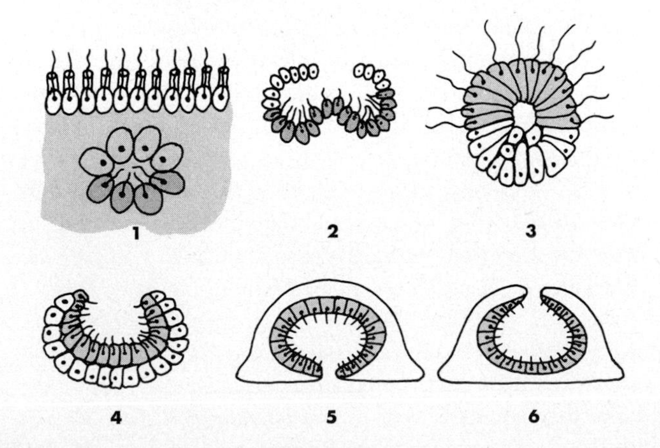

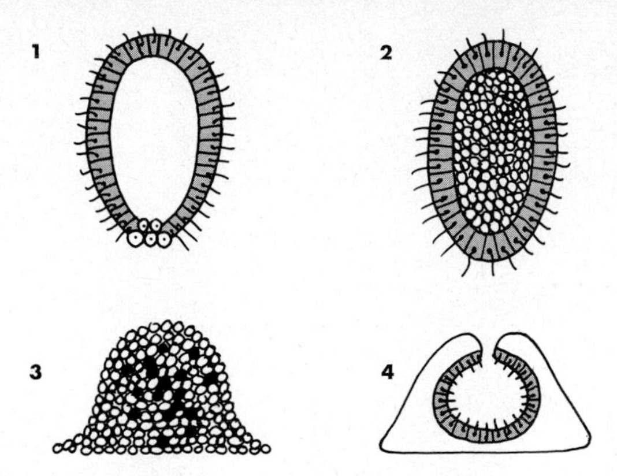

20.10. Development of calcareous sponges, *Leucosolenia* pattern. (*Adapted from Hartmann, Metschnikoff.*) 1, coeloblastula with posterior nonflagellated cells; 2, stereoblastula larva formed by ingression and proliferation of nonflagellated cells; 3, settling of larva and inversion, i.e., migration of flagellated cells (dark) to center, nonflagellated cells (light) to outside; 4, arrangement of flagellated cells into choanocyte layer and formation of osculum, establishing asconoid structure. The latter may subsequently develop further and acquire a syconoid or leuconoid architecture, depending on the genus.

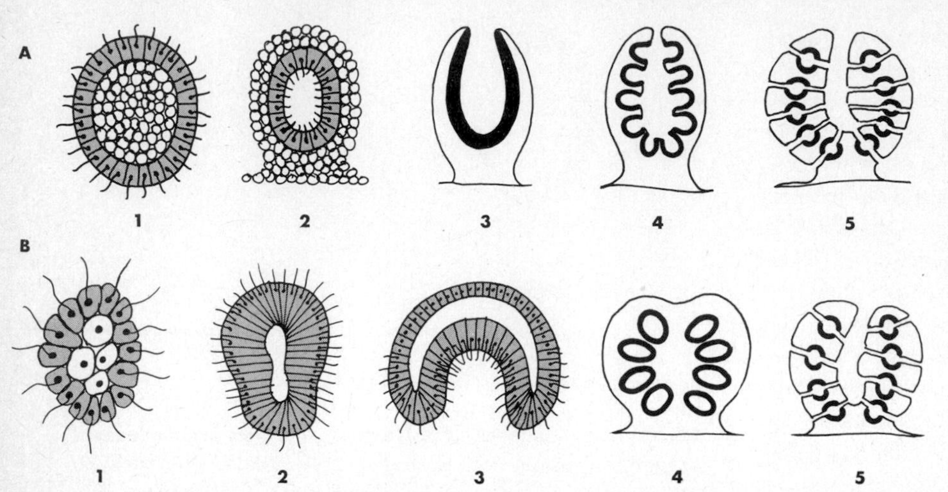

20.11. Development of horny sponges. (*Adapted from Lévi.*) **A.** *Halisarca.* 1, stereoblastula larva; 2, settling, inversion, and hollowing out of spongocoel; 3, osculum formation, asconoid stage; 4, proliferation and folding of choanocyte layer, syconoid stage; 5, pinching off of choanocyte chambers, young adult leuconoid stage. **B.** *Oscarella.* 1, morula embryo, after inversion; 2, amphiblastula larva, center degenerated and hollow; 3, invagination of strongly flagellated half and settling at blastopore end; 4, direct pinching off of choanocyte chambers, osculum beginning to form; 5, osculum established, young adult leuconoid stage achieved directly.

asconoid even as an adult; but in other chalk sponges, an early asconoid stage reached by either of the two patterns above later undergoes further structural transformations into syconoid or leuconid adults.

The embryonic development of the glass sponges has been studied in only a few cases. Genera such as *Farrea* produce stereoblastula larvae, with a tightly packed layer of cells on the outside and amoeboid cells embedded in jelly on the inside. Some of the amoeboid cells migrate to form an interior layer around a central spongocoelic cavity, and such cells later give rise to the choanocytes. Spicules soon begin to form in the interior and come to project through the surface of the larva. After the larva settles, an osculum breaks through at the nonattached side and the basic architecture of the sponge is then established.

In the horny sponges, development occurs either via stereoblastulae or via amphiblastulae (Fig. 20.11). In *Halisarca,* for example, the embryo and later larva is a solid stereoblastula, with flagellated cells on the outside and various types of amoebocytes in the interior. The outer cells lose their flagella, migrate into the interior, and develop into choanocytes. These come to surround an enlarging cavity, the later spongocoel. After the larva settles, an osculum breaks through and an asconoid stage of development is so reached. Later, folding of the choanocyte layer produces a transitional syconoid stage, and a leuconoid adult eventually develops when choanocyte chambers become pinched off. In *Oscarella,* the embryo is a morula which subsequently becomes hollow by the degeneration of the interior cells. An amphiblastula larva forms in this fashion. The larva later settles at the anterior pole and, as in the amphiblastulae of the calcareous sponges, the flagellated hemisphere invaginates. However, the cells of this hemisphere then become arranged into choanocyte chambers directly. Osculum and spongocoel ultimately come to connect with them. A leuconoid architecture thus arises right at the start, without asconoid or syconoid transitional stages.

All the various developmental patterns of sponges have one characteristic in common, viz., they do *not* correspond to the patterns encountered in Eumetazoa. More specifically, the exterior cells of sponge embryos are not "ectoderm," for they eventually come to form some of the mesenchyme as well as the *interior* choanocyte layer; and the interior cells of sponge embryos are not "endoderm," for they later come to form the remainder of the mesenchyme as well as the *exterior* epidermal layer. In effect, the embryonic layers of sponges are not homologous to those of other animals.

This absence of embryonic homology is one of the principal reasons for regarding sponges as an independent branch of metazoan evolution. Other such reasons include the presence of choanocytes (and, indeed, of unique porocytes); the pattern of alimentation based on the water-channel system; and the construction of the body generally, which constitutes a loosely integrated aggregation of cells at an advanced colony-level or a very primitive tissue-level of organization. Because of their choanocytes, sponges are traditionally considered to have eveolved from zooflagellates such as the Protomastigida. The latter do display a potential of forming sessile colonies of collar cells (cf. Chap. 19). However, sponge choanocytes are secondarily developed adult cells not present in embryos; and choanocytes are not the predominant sponge cells in any case, since the majority of cells exhibit an *amoeboid* state of existence. Thus, it is quite possible that choanocytes in protozoa and sponges are independent evolutionary developments; and that sponges may trace their ancestry back to stocks which were primarily amoeboid but also possessed flagellate potentials. Such stocks need not have been protozoan, moreover, but could have been more broadly protistan in nature.

Within the phylum of sponges, the three classes presumably represent separate lines of an adaptive radiation which may have emanated from a common ancestor. Early members in each line may have been asconoid types without skeletons. Later members then may have evolved variously syconoid and leuconoid levels of complexity, with increasingly elaborate skeletons. Fossil sponges with complex skeletons are known from earliest Cambrian times.

REVIEW QUESTIONS

1. Review the status of the Mesozoa. What are the structural, developmental, and ecological characteristics of these animals?

2. Define the phylum Porifera taxonomically and give analogous definitions for each class. Name representative genera in each class. What is the general ecology of sponges and what are the adult shapes and growth characteristics of these animals?

3. Describe the cell types and principal tissues of sponges. What structures does each cell type produce? Describe the three architectural variants of sponges and indicate groups and genera in which each variant is encountered.

4. Outline the fine structure of glass sponges and show how their architecture differs from that of chalk or horny sponges. What are porocytes, prosopyles, apopyles, spongocoels, ostia, oscula, radial canals, and incurrent and excurrent canals?

5. What is the course of water through a sponge in each of the classes? How is the water flow maintained and adjusted and what are the hydrodynamic characteristics of this flow? Show how a sponge feeds. Which cell types participate in ingestion, digestion, and egestion?

6. What responses to external stimuli may a sponge make? Describe the structure of spicules and show how such supports are manufactured. How are spicule-forming cells named and what cell types are they derived from? What is spongin?

7. Describe the various forms of vegetative reproduction in sponges. Which of these forms do not generally occur in marine species? What adaptive advantages can you suggest for the occurrence of special reproductive methods in freshwater species? Can sponges regenerate?

8. Describe the process of fertilization in sponges. What cell types give rise to the gametes? Show how embryonic and larval development take place in (*a*) *Sycon*, (*b*) *Leucosolenia*. What is embryonic inversion and in what variant forms does it occur in (*a*) and (*b*)?

9. Describe the embryonic and larval development of horny sponges as exemplified by *Halisarca* and *Oscarella*. Name the adult structures produced by the exterior and interior cells of an amphiblastula. Why are the tissue layers of Mesozoa and Porifera not homologous to the germ layers of Eumetazoa?

10. Review the evidence suggesting that Mesozoa, Parazoa, and Eumetazoa represent separate and independent branches of animal evolution. Review arguments for and against the traditional hypothesis that sponges have evolved from flagellate protozoa.

COLLATERAL READINGS

Most books on marine life and texts on zoology include accounts on sponges. Specifically recommended for further reading are the following:

Hyman, L.: "The Invertebrates," vol. 1, chaps. 4 and 6, McGraw-Hill, New York, 1940.

Wilson, H. V., and J. T. Penney: Regeneration of Sponges from Dissociated Cells, *J. Exp. Zool.*, vol. 56, 1930.

CHAPTER 21

radiata

Within the Metazoa, the principal taxonomic branch is that of the Eumetazoa, animals characterized by an organizational complexity above that of the tissue level. In these animals, also, the embryonic germ layers of given groups are homologous to those of other groups (though they are not necessarily analogous, as pointed out in Chap. 14). Two eumetazoan grades are distinguished, the Radiata and the Bilateria. The radiate grade, subject of the present chapter, is identified by a primitive organ level of construction and by a basic radial symmetry. Two phyla are included, the Cnidaria and the Ctenophora. Note here again that many of the traits of these animals have already been referred to earlier, e.g., in Chaps. 7, 8, and 9.

PHYLUM CNIDARIA

Coelenterates; largely marine; tentacle-bearing radiates with *cnidoblasts* containing *nematocysts;* gastrovascular cavity (*gastrocoel, coelenteron*) with single opening; polymorphic, with medusae and/or polyps; solitary or colonial; development regulative, via *planula* larvae.

Class HYDROZOA: both medusae and polyps; some medusae as floating colonies, some polyps as colonial corals; hydrozoans.

Class SCYPHOZOA: medusae dominant; jellyfishes.

Class ANTHOZOA: polyps only; corals, sea anemones, sea fans, sea pens, sea feathers, sea pansies.

General Characteristics

The approximately 10,000 species of coelenterates form a phylum of first importance both historically and ecologically. Historically, the animals are generally believed to represent the most primitive living Eumetazoa. Thus, the phylum is of special interest as a group at or close to the base of the animal evolutionary bush. Ecologically, coelenterates are possibly the most common macroscopic animals of the oceans, and they have shaped substantial parts of the earth surface as the principal builders of coral reefs and coral islands. Though some groups are benthonic in deep waters, most coelenterates inhabit the shores and the shallow sunlit zones of warm, tropical and subtropical seas. Dead, discolored laboratory specimens usually reveal very little of the delicate watery transparency of most of these animals in the living state; nor do such specimens indicate their myriads of living colors and often flowerlike shapes or their eerie bioluminescent beauty at night.

Coelenterates occur both as sessile and motile forms and both as solitary animals and colonies. Individuals as well as colonies can be near-microscopic or exceedingly large. The largest solitary coelenterate is a species of jellyfish which may attain a diameter of up to 12 ft and may have trailing tentacles up to 100 ft long. Coelenterates are exclusively carnivorous. They eat any animals their tentacles can hold and paralyze, and their highly distensible bodies can admit prey far larger than themselves. Only very few species are parasitic, but many animals of other phyla live symbiotically with coelenterates and even more make their home in and among coelenterate growths.

The name "coelenterate" originally designated not only the phylum here under discussion but also the sponges and the ctenophores. Today the term "coelenterate" is still used descriptively, but it is restricted to the phylum known technically as the Cnidaria. This name derives from the most characteristic single trait of the group, viz., the presence of *cnidoblasts* (Fig. 21.1). These are stinging cells found singly or in grouped "batteries" around the mouth, on the tentacles, and elsewhere on and within the body. Each cnidoblast contains a horny *nematocyst*, a stinging capsule similar in some respects to the polar capsules of certain sporozoan groups. Within a nematocyst is a coiled, hollow thread, continuous at one end with the nematocyst wall and either closed or open at the free end. From the cnidoblast projects a pointed spike, the *cnidocil*. Appropriate stimulation of it leads to explosive discharge of the nematocyst thread, a process in which the thread turns inside out. Threads with closed ends (*volvants*) are primarily adhesive; they are used in trapping and holding prey. Those with open ends (*penetrants*) pierce through the body surface of prey (even chitinous armor) and are used to paralyze; a toxic secretion is released through the thread. A massive injection of coelenterate toxin is powerful enough to kill a man. Some varieties of penetrants are armed with barbs and hooks (located on the inside of the thread in the undischarged condition). Discharge of nematocysts appears to be brought about mainly by mechanical and chemical stimuli, possibly by nervous ones as well. Well-fed coelenterates in process of digesting prey do not discharge their nematocysts even if more of the same type of prey touches the cnidocils. Many animals live commensally on coelenterates and others feed on coelenterates, yet none of these animals causes nematocyst discharge. In some animals, indeed, cnidoblasts from eaten coelenterates migrate from the gut to the body surface of the eaters, where these cells then become part of the epidermal tissues and serve as they do in coelenterates.

21.1. The two basic types of cnidoblasts, in undischarged and discharged condition (schematic). Numerous variants of each type are known.

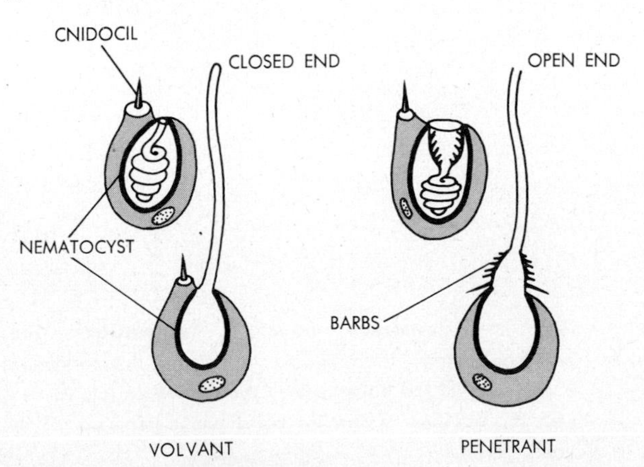

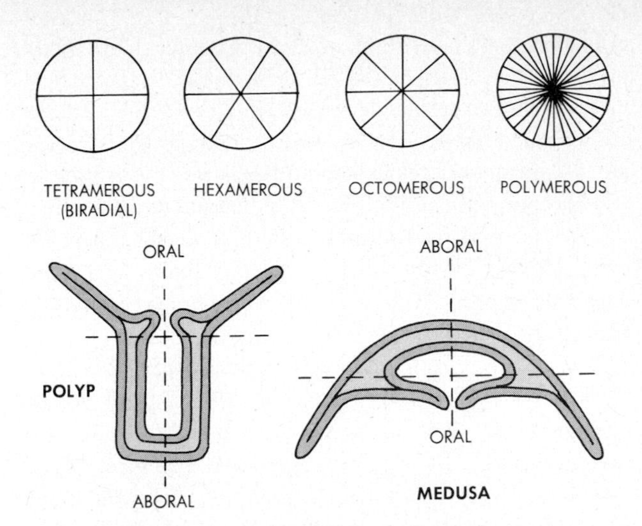

21.2. Coelenterate symmetries and body forms. Different types of coelenterate radialities are schematized at top. Bottom figures indicate that the oral-aboral axis is long compared to the body diameter in a polyp, whereas the reverse holds true in a medusa. Both polyp and medusa are variants of a single common body design.

The Cnidaria are organized radially around an *oral-aboral axis*, i.e., a main axis passing from the alimentary opening to the opposite end of the animal (Fig. 21.2). In many cases two planes passing through this axis at right angles to each other mark a coelenterate into four equal quadrants, a form of radiality known as *biradial* or *tetramerous* symmetry. Other coelenterates are *hexamerous, octomerous,* or *polymerous,* i.e., they are hexagonally, octagonally, or polygonally radial. In the latter case, almost any plane through the main axis marks out two mirror-image halves.

Coelenterates are basically dimorphic, the two structural forms being the *polyp* and the *medusa* (cf. Fig. 21.2). Both are derivable from a single form. If the oral-aboral axis is long relative to the diameter of the animal, the design is that of a cylindrical polyp; and if the main axis is short relative to the diameter, the design is that of an umbrella- or bellshaped medusa. Polyps are sessile vegetative individuals; medusae are pelagic, free-swimming, sexual individuals. In many groups, a zygote develops into a planula larva which settles and becomes a polyp. The latter then buds off medusae, these produce gametes, and the zygote formed begins a new life cycle. In other groups, either the polyp or the medusa phase may be more or less reduced or suppressed or may be omitted altogether (cf. below).

The bodies of both polyp and medusa contain a central *gastrovascular cavity*, or *coelenteron*, with a single gastric opening serving both as mouth and anus (Fig. 21.3). This cavity is bounded by a body wall consisting fundamentally of three tissue layers, derived from the three embryonic germ layers characteristic of Eumetazoa. The principal derivative of the ectoderm is the epithelial (occasionally syncytial) *epidermis*. The principal derivative of the endoderm is the epithelial *gastrodermis*. The epidermis and often also the gastrodermis are underlain by a thin layer of neural tissue and, in the majority of coelenterates, by a thin layer of muscular tissue as well. The principal derivative of the mesoderm is the *mesogloea*, a poorly to well developed connective tissue. The mouth areas of polyps and medusae and also the bell margins of many medusae bear sets of *tentacles*. These may be hollow or solid. If they are hollow, they are formed from all three body layers and the internal spaces are extensions of the coelenteron; if they are solid, the gastrodermis is not part of their structure.

In the many species in which the polyps form an exoskeleton, such supports are secreted by the epidermal cells at the base and the sides of the body (cf. Fig. 21.3). For example, most hydrozoan polyps secrete a *perisarc*, a tube of transparent chitin, and numerous corals embed their bases and sides in massive calcareous secretions (which in many cases accumulate as the main constituents of coral reefs). Any epidermal cells of polyps not covered by an exoskeleton and the epidermal cells of medusae variously have a smooth surface (Hydrozoa), a flagellated surface (Scyphozoa), or a ciliated surface (Anthozoa). In the Hydrozoa such epidermal cells are epitheliomuscular, i.e., the bases of the cells are expanded and contain longitudinally oriented myofibrils (Fig. 21.4). In the two other classes the epidermal cells are without myofibrils and the inner ends of the cells are narrow and pseudopodial. Such animals possess a layer of distinct muscle cells underneath the epidermis. Subepidermal muscles are

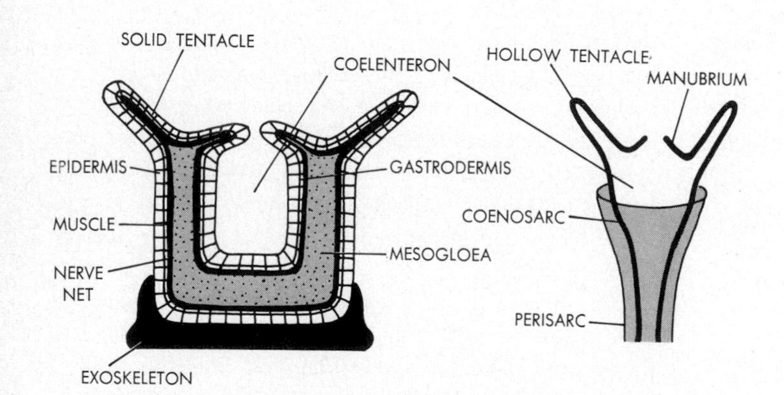

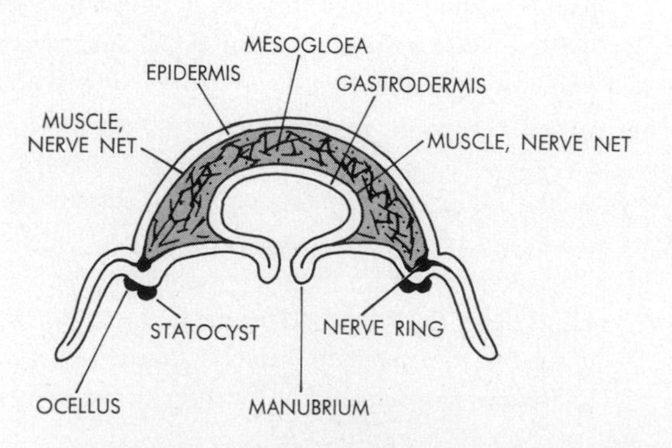

21.3. Coelenterate structure, schematic. Top, polyps; **bottom,** medusa. At **top right,** the coenosarc consists of all the body tissues shown individually at left, i.e., the cavities of the tentacles are extensions of the coelenteron.

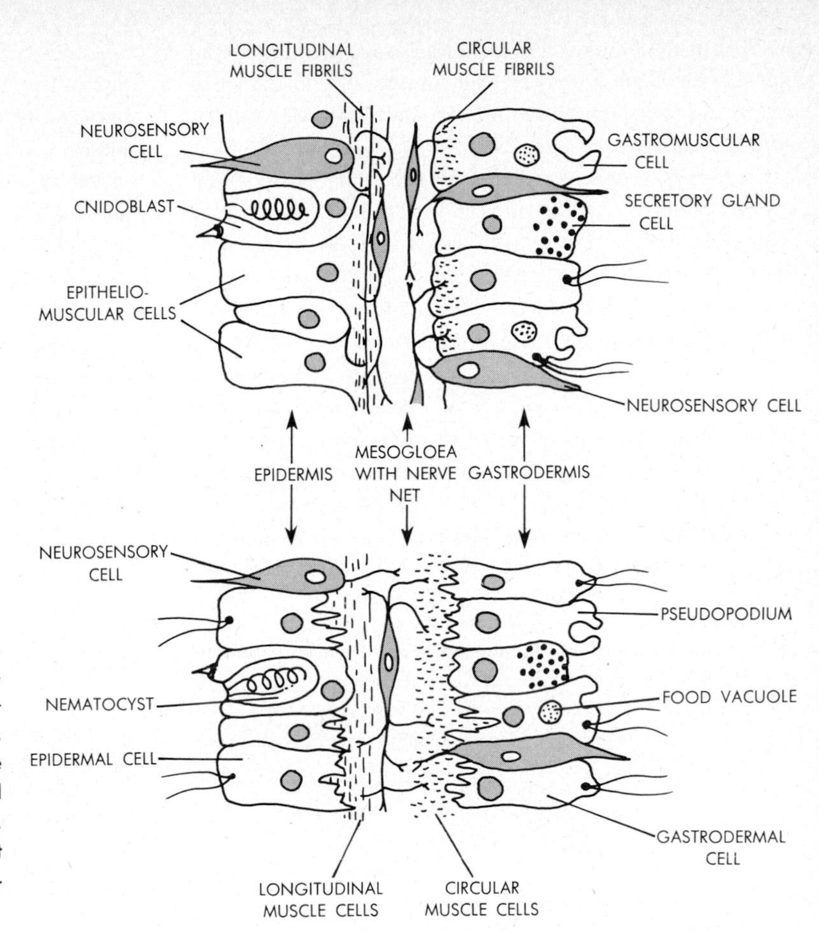

21.4. The epidermis and gastrodermis. Top, Hydrozoa; note muscle fibrils within gastromuscular and T-shaped epitheliomuscular cells. **Bottom,** Scyphozoa and Anthozoa; separate muscle cells are present here, and the epidermal and gastrodermal cells have pseudopodial bases. The nerve net is double or single in different coelenterates, and in either case the neurons innervate the muscular elements.

developed particularly well in Scyphozoan medusae. Scattered among the epidermal cells in all classes are neurosensory cells which probably serve as receptors for thermal, mechanical, and chemical stimuli; and cnidoblasts, concentrated particularly in the oral region and the tentacles. The epidermis of medusae usually also contains either or both of two kinds of true organs, viz., sense organs for vision and balance. The former are *ocelli,* i.e., either eye cups with or without chitinous lenses or simple eyespots. The balance organs are *statocysts* with statoliths (cf. Figs. 8.39, 8.47). Such organs are usually located along the bell margins, particularly at the bases of the tentacles.

The gastrodermis consists largely of ingestive flagellate-amoeboid cells (cf. Fig. 21.4). Each such cell normally bears two flagella on a kinetosome. But the flagella may be resorbed, and the cell may engulf small food particles by means of pseudopodia. Intracellular digestion in protozoan fashion then follows. In the Hydrozoa, the expanded bases of these cells again contain myofibrils, here aligned circularly. The gastrodermal cells of Scyphozoa and Anthozoa are without such myofibrils; a separate muscle layer may be present underneath the gastrodermis. Anthozoa particularly possess well-developed subgastrodermal muscles. Apart from flagellate-amoeboid cells, the gastrodermis also contains large numbers of gland cells. Most of these participate in the extracellular phase of digestion by secreting digestive enzymes into the coelenteron (cf. Chap. 9). Mucus-secreting gastrodermal cells occur near the mouth. Neurosensory cells are again present, as are cnidoblasts in certain gastrodermal regions of Scyphozoa and Anthozoa. In many species, the cells of the gastrodermis harbor symbiotic zoochorellae or zooxanthellae.

Underlying the epidermis and often also the gastrodermis are *nerve nets* (cf. Fig. 21.4 and Fig. 8.24). Neuron fibers from these nets connect with the myofibril-containing cells, the muscle cells, and the neurosensory cells. The greatest concentration of neurons occurs in the mouth and tentacle regions, and in a medusa the net also includes a *nerve ring* around the bell margin. Such a ring receives impulses from the sense organs and emits impulses to the myofibrils or muscles which produce contraction of the bell.

Nerve nets, muscles, and the epidermal and gastrodermal epithelia abut against the mesogloea, the middle body layer (Fig. 21.5). The most primitive form of this layer is probably found in the hydrozoan medusae, some of which are believed to represent the most primitive coelenterates. The mesogloea of these animals consists largely of substantial quantities of jelly in which are embedded comparatively small numbers of mesenchyme cells and connective tissue fibers. The jelly contains some 95 or more per cent water and makes up the bulk of such a medusa. From this condition, one line of evolution has apparently led to an even less elaborate mesogloea, another line to a more elaborate one. A less complex mesogloea is encountered in the majority of hydrozoan polyps, where the jelly forms a thin layer containing neither cells nor fibers. The extreme stage of simplification has been attained by some of the hydrozoan polyps, e.g., *Hydra*, in which a mesogloea is virtually absent. A more complex level of mesogloeal development is exhibited by the Scyphozoa. These animals possess a jelly containing substantial numbers of amoeboid mesenchyme cells as well as numerous fibers. The most complex degree of elaboration is reached in the Anthozoa, where the mesogloea is a true connective tissue, with large numbers of various types of mesenchyme cells and fibers embedded in small amounts of jelly.

21.5. The structure of the mesogloea. From a probably basic condition in hydrozoan medusae developed, on the one hand, reduced mesogloeae in hydrozoan polyps, and, on the other, more elaborate mesogloeae in Scyphozoa and Anthozoa.

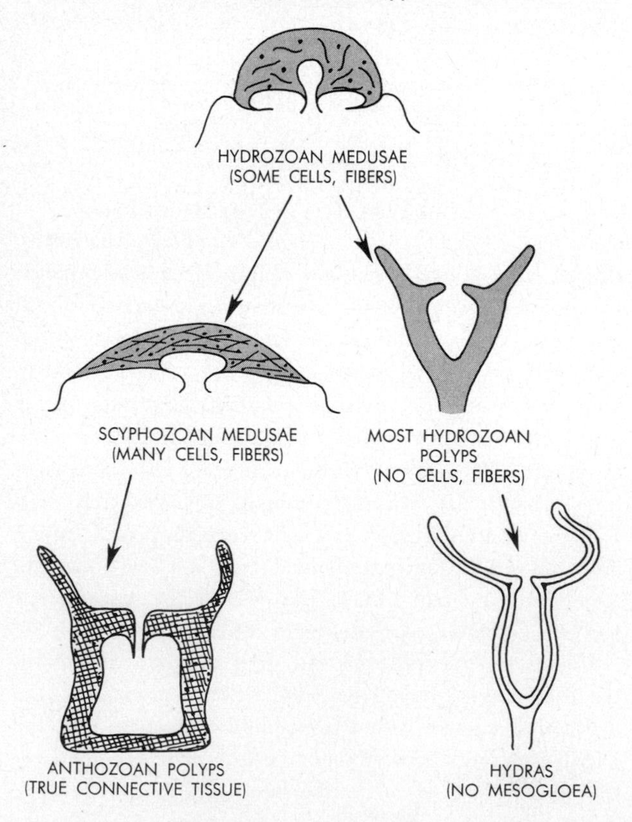

A mesogloea is the more resilient the more fibers it contains, and despite its high water content this middle layer is actually quite firm and elastic. In the locomotion of a medusa, for example, the bell margin contracts muscularly and pushes water out from underneath the bell, producing propulsion in the opposite direction. Muscular relaxation is then followed by elastic recoil of the mesogloea. Repeated contraction-recoil cycles of this sort lead to locomotion in rhythmic spurts. The fibers of the mesogloea are presumably secreted by the mesenchyme cells. Most of the latter are amoeboid, and some of them actually wander into and become part of the epidermal and gastrodermal epithelia. These so-called *interstitial cells*, believed to be equivalent to the archeocytes of sponges, have been considered to be embryonic reserve cells. They are also thought to be the source of the sex cells. Either interstitial cells or other mesenchyme cells are probably the source of muscle cells and of cnidoblasts. The latter are known to originate and to differentiate in regions far removed from the locations in which they will eventually become incorporated into the epithelia. Embryologically, the mesogloea is largely of ectodermal origin; coelenterates may therefore be considered to possess an ectomesoderm.

The principal adult tissue types of coelenterates thus are basically of four kinds, viz., epithelial, nervous, muscular, and mesenchyme. But note that the cells are not yet as firmly or rigidly integrated into tissues as they are in other Eumetazoa. Hydrozoan polyps such as *Hydra* may be turned inside out experimentally, and in a small percentage of successful cases all cells migrate back to their normal positions: exteriorized gastrodermal cells glide back into the interior, past epidermal cells on their way back to the exterior. A normal polyp reforms in this manner. As in sponges, moreover, coelenterates such as hydras may be dissociated into loose suspensions of cells and such cells may then migrate and reorganize into whole intact animals.

Coelenterates reproduce by budding, by fragmentation,

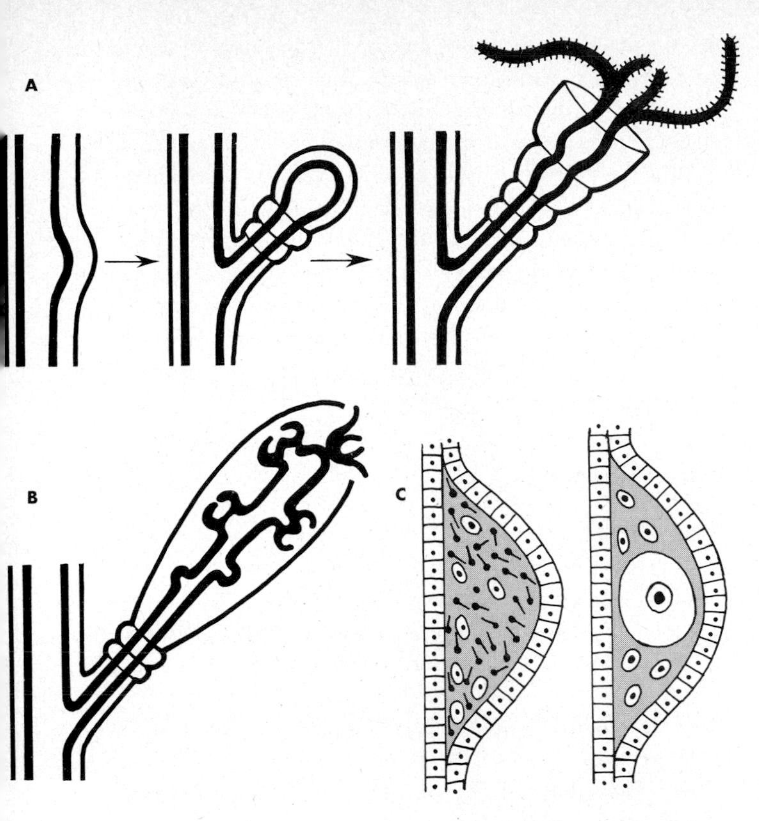

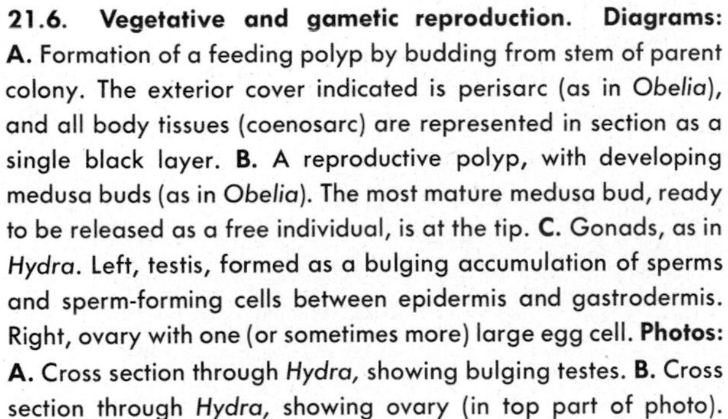

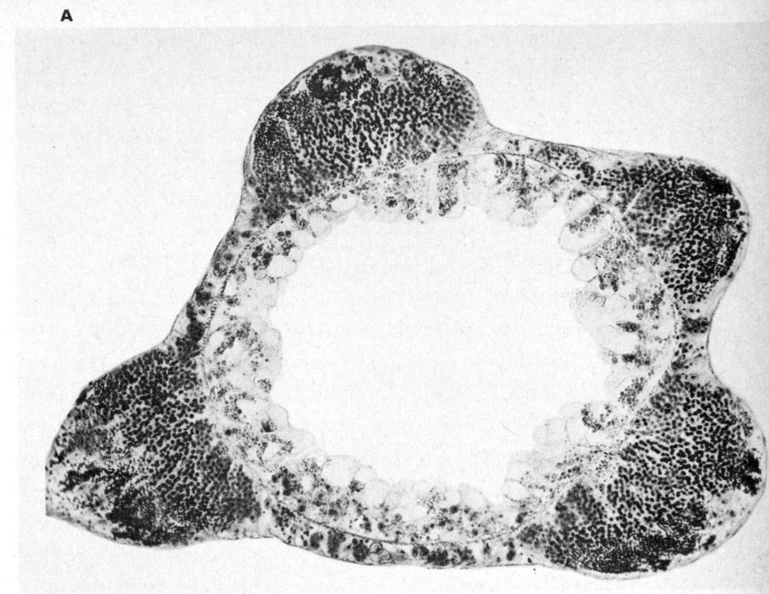

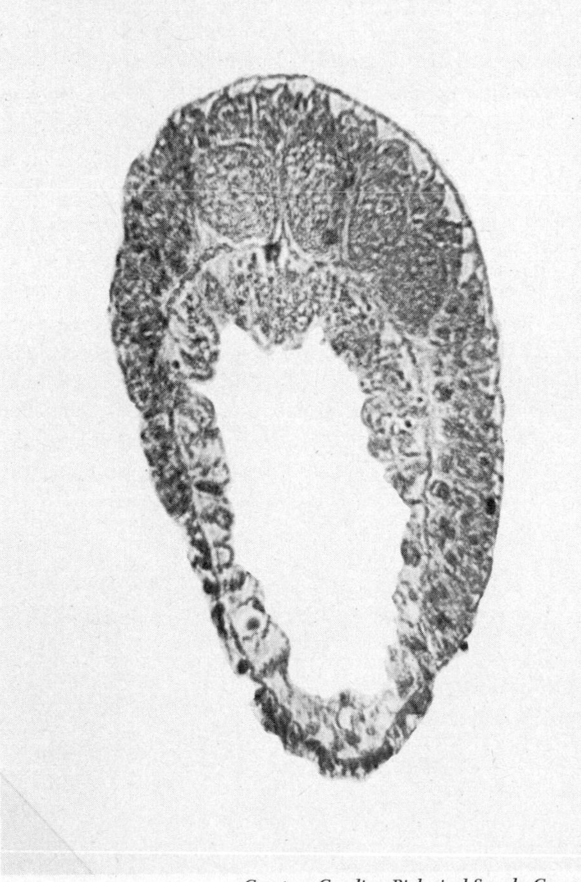

Courtesy Carolina Biological Supply Company.

21.6. Vegetative and gametic reproduction. Diagrams: A. Formation of a feeding polyp by budding from stem of parent colony. The exterior cover indicated is perisarc (as in *Obelia*), and all body tissues (coenosarc) are represented in section as a single black layer. **B.** A reproductive polyp, with developing medusa buds (as in *Obelia*). The most mature medusa bud, ready to be released as a free individual, is at the tip. **C.** Gonads, as in *Hydra*. Left, testis, formed as a bulging accumulation of sperms and sperm-forming cells between epidermis and gastrodermis. Right, ovary with one (or sometimes more) large egg cell. **Photos: A.** Cross section through *Hydra*, showing bulging testes. **B.** Cross section through *Hydra*, showing ovary (in top part of photo).

and via gametes. Budding capacity is particularly well developed; in many cases even the larvae may bud. Solitary polyps form other solitary polyps by budding, and this process also produces extensive interconnected colonies of polyps. A bud typically arises as a fingerlike evagination from the body wall of a polyp; all three body layers usually participate (Fig. 21.6). The protruberance then elongates, and mouth and tentacles develop anteriorly. In a polyp colony all individuals share a common, continuous coelenteron. Apart from producing other polyps, polyps may also form buds which develop into medusae. The latter may bud in their turn, usually in the mouth region or at the bell margin. But whereas polyps may bud off either more polyps or medusae, medusae bud off only more medusae, never polyps. Reproduction by fragmentation may occur by spontaneous splitting, as in sea anemones (cf. Fig. 10.2), or as regenerative reproduction following injury. The regeneration potential is particularly high in polyps.

Most coelenterates are separately sexed. Gametes arise seasonally, probably from interstitial cells located either in the epidermis or the gastrodermis. Some of these cells migrate into specific regions of the body, accumulate there, and mature into sperms or eggs (cf. Fig. 21.6). Mesenchyme nurse cells often supply growing eggs with yolk and in some cases an egg is formed by the fusion of several cells. Collections of sex cells represent the gonads, i.e., testes or ovaries. They are found in different locations in different coelenterate groups (cf. below). In many cases the gametes are shed; in others the eggs are fertilized and undergo part or all of their early development while remaining attached to the parental body. Cleavage is superficial in some coelenterates, radial and holoblastic in most (Fig. 21.7). In the holoblastic cleavages, the blastomeres at first adhere to one another only loosely and they are usually not arranged in any fixed pattern. The result of cleavage in the majority of cases is an externally ciliated morula or stereoblastula. The external cells subsequently become organized as a definite ectoderm layer and the interior cells come to represent the endoderm. The embryo so forms a stereogastrula. In some cases cleavage produces a hollow blastula. Gastrulation then occurs by ingression of cells from one or several surface regions or, less often, by delamination. A stereogastrula forms again in either event. At this point the developing coelenterate has reached the *planula* stage, characteristic of the entire phylum. Depending on how long a parent retains its eggs, if at all, the embryo may be set free as a larva at any time before or not until the planula has developed.

21.7. Early development of coelenterates. Irregularly cleaved blastomeres (left) become a morula or stereoblastula in most cases (top middle), a coeloblastula in some cases (bottom middle). In all cases a stereogastrula (right) forms ultimately. The endoderm is already foreshadowed in a morula or it develops by ingression in a coeloblastula. The stereogastrula represents the common *planula* stage of coelenterate development.

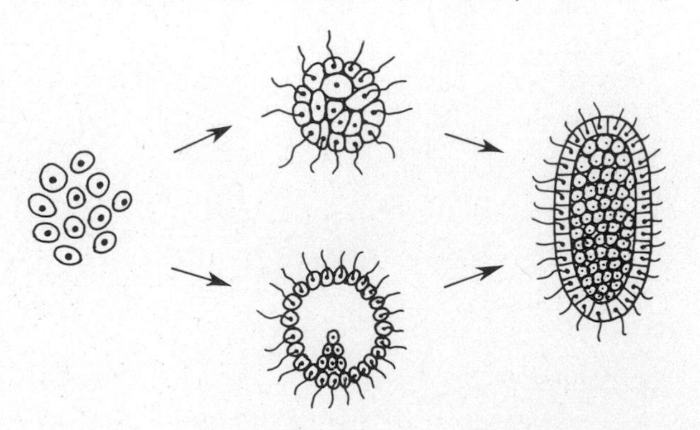

The further fate of the planula larva and also the particular characteristics of adult coelenterates are best discussed separately for each class.

Class Hydrozoa

Body tetramerous or polymerous; polyp and medusa phases alternating or either phase reduced or omitted; medusa usually with *velum* and with ocelli and/or statocysts; coelenteron smooth-walled; gonads usually epidermal; solitary and colonial; mostly marine, some in fresh water.

Order Hydroida (hydroids): polyp and medusa phases usually alternating. *Obelia, Tubularia, Hydractinia, Hydra*

Order Milleporina (millepore corals): colonies on extensive calcareous exoskeletons; polyps dimorphic, medusae reduced. *Millepora*

Order Stylasterina: like millepores in most respects. *Stylaster*

Order Trachylina: sessile polyp phase absent. *Gonionemus*

Order Siphonophora: floating composite colonies of polyps and medusae. *Physalia, Velella*

Hydroid planulae typically develop into polyp colonies (Fig. 21.8). The endoderm of a planula hollows out into an early coelenteron, a thin mesogloea begins to form, and the free-swimming larva eventually settles at its anterior end (which corresponds to the vegetal pole of the egg and the aboral end of the future adult). Elongation into a polyp *stem* follows, and a *hydranth* develops at the free end. A hydranth is a bellshaped expanded body part ringed by whorls of tentacles. In the center of the tentacle wreath is the mouth, located at the tip of a conical *manubrium* (Fig. 21.9). Many hydroids produce *stolons,* i.e., stalks growing along the ground, from which upright stems may then bud. Branch stems may also grow out laterally from an upright stalk. A hydranth develops at the end of each branch. As noted earlier, the epidermis of hydroid polyps usually secretes a chitinous perisarc; the body tissues within the perisarc tube are then referred to as the *coenosarc.* In some forms, e.g., *Obelia,* the perisarc widens into a cup around each hydranth. In other cases, e.g., *Tubularia,* the perisarc terminates below a hydranth and the latter remains exposed.

Hydranths are feeding individuals, or *gastrozooids.* Polyp colonies also produce several types of polymorphic variants of hydranths. For example, one variant type is a *dactylozooid,* a protective individual without mouth, tentacles, or coelenteron, but richly studded with batteries of sting cells. Dactylozooids may or may not form in given species, but virtually all hydroid species give rise to reproductive, medusa-forming polyps. In many cases such polyps are *gonozooids,* stalked, perisarc-enclosed individuals as in *Obelia* (cf. Fig. 21.9). A gonozooid buds off *gonophores* in anteroposterior succession, and each gonophore develops into a medusa. When such a medusa is mature it detaches from the gonozooid and becomes a free-swimming individual.

On the underside of the bell, a hydrozoan medusa possesses

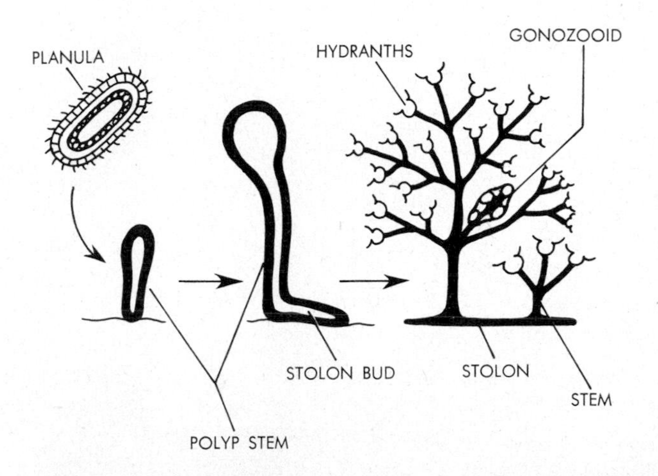

21.8. Diagrams: development of hydrozoan polyps. The planula larva acquires a thin mesogloea and a hollowed-out center which represents the early coelenteron. The larva eventually settles at its anterior end and proliferates into the stem of the first polyp. Branching growth from both upright stems and stems along the ground (stolons) then gives rise to a colony of polyps. Most of the latter are tentacled feeding individuals, i.e., gastrozooids, or hydranths; some are medusa-forming gonozooids. **Photos: hydrozoan polyp colonies. A.** *Sertularia;* note clubshaped gonozooids along hydranth-bearing branches; **B,** *Clava* (on seaweed); note unbranched stems of hydranth-bearing polyps. In **A,** the perisarc extends over and protects much of each hydranth; in **B,** the perisarc terminates below a hydranth. Cf. also color plate V.

B

21.9. Diagrams: polymorphism of polyps (schematic). **A.** Gastrozooid (hydranth) of *Obelia* type, with mouth at tip of conical manubrium and cuplike perisarc over lower part of hydranth. **B.** Gastrozooid of *Tubularia* type, with double wreath of tentacles and perisarc terminating below hydranth. **C.** Dactylozooid, with batteries of stinging cells and without mouth or tentacles. **D.** Gonozooid, studded with medusa-producing gonophores. In types such as *Obelia,* a perisarc envelops the gonozooid. **Photos: A.** Close-up view of two hydranths of *Obelia.* **B.** Close-up view of portion of *Campanularia* colony, showing feeding hydranths (gastrozooids) and gonozooids. In the latter, note medusa buds. In both photos, note perisarc extending cuplike over the hydranths.

a short, four-cornered manubrium, often equipped with long tentaclelike trailing arms or actual tentacles (Fig. 21.10). The manubrium leads into a coelenteron from which four *radial canals,* 90° apart from one another, pass toward the bell margin. There each canal opens into a common circular *ring canal.* Near this juncture, a small egestive pore leads from the radial canal to the exterior. These channels of the gastrovascular system undoubtedly facilitate circulation of food to outlying parts of the bell. A nerve ring parallels the ring canal. The bell margin of hydrozoan medusae generally is not scalloped but is extended into a *velum,* a contractile, shelflike membrane projecting inward under the bell. The velum aids in locomotion by increasing the force of bell contraction and by reducing the opening through which water is expelled from the underside of the bell; the water jet thus acquires a greater velocity. *Obelia* and a few other hydrozoan medusae are without a velum. Tentacles emanate from the bell margin in tetramerous patterns. The gonads are epidermal folds, usually located either along the side of the manubrium or underneath

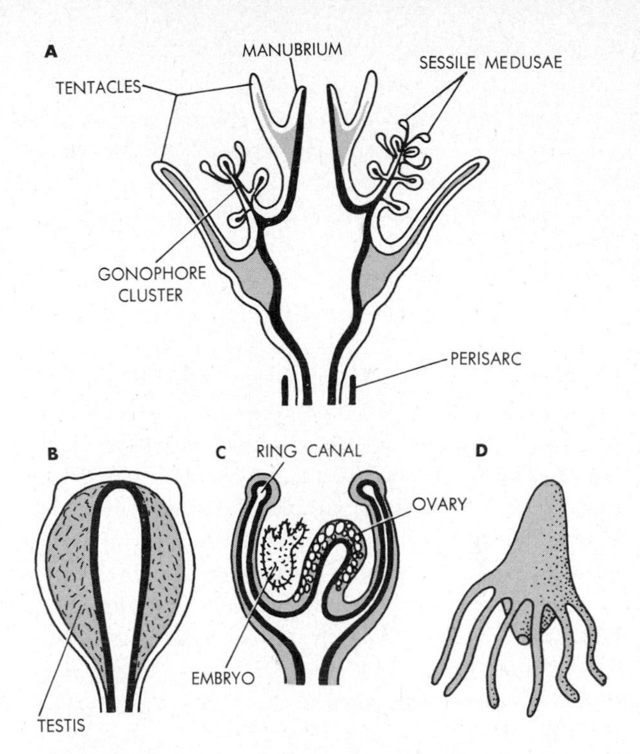

21.11. The life cycle of *Tubularia*. A. Section through a hydranth, with gonophore-bearing gonozooids on manubrium. Epidermis white, gastrodermis black, gastrodermal tentacle-supporting cells gray. **B.** Immature male gonophore (which eventually acquires a medusoid structure resembling **C**). **C.** Mature female gonophore, representing a sessile medusa as indicated in **A.** The embryo is a late planula, just developing tentacles. **D.** Young actinula larva, developed from planula and set free by parent medusoid. The larva eventually settles at the aboral end and grows into a polyp colony with hydranths as in **A.**

21.10. Hydrozoan medusae. Diagrams: top, sectional view; bottom; bottom view. **Photos: A.** *Obelia* medusa, top view; note central manubrium, sex organs along radial canals, marginal tentacles, and absence of velum in this genus. **B.** *Lizzia* medusa, side view; note new medusa budding off manubrium in center of bell, and faintly visible velum projecting inward around tentacled margin of bell. Cf. also color plate V.

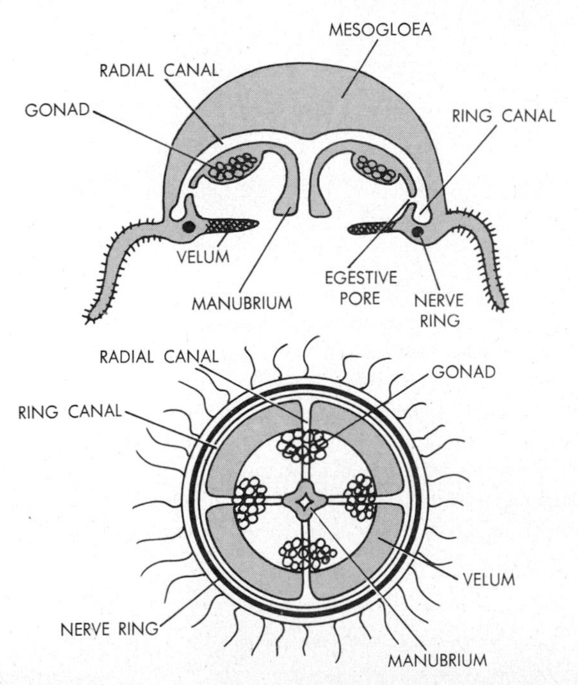

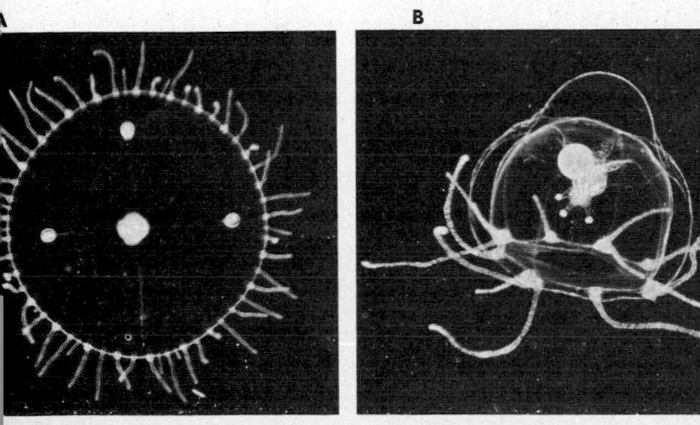

the radial canals. As pointed out earlier, medusae may reproduce vegetatively by budding off other medusae.

In many hydroids, the medusa phase is suppressed to a greater or lesser extent. Polyp colonies in such cases typically give rise to variant individuals called *medusoids.* These are stalked buds which may form almost anywhere on a colony. In *Tubularia,* for example, medusoids arise on the manubria of the hydranths (Fig. 21.11). A medusoid produces gonophores and these generally develop into *sessile* medusae, i.e., individuals which in most instances are not set free. Such medusae are also structurally reduced; locomotor or other body parts not actually required for gamete production are not developed. In *Tubularia,* eggs are fertilized and then retained in the sessile medusae. After a planula stage has developed, the embryo acquires tentacles, a mouth, and a rudimentary coelenteron. Only when such a polyplike *actinula* stage has been reached is the embryo liberated as a larva. The actinula larva leaves the sessile medusa and creeps about with the aid of its tentacles. It later settles and gives rise to a new polyp colony.

Suppression of the medusa phase has become complete in hydroids such as *Hydra*, one of the few freshwater coelenterates. In these animals, gonads develop in the epidermis of the solitary polyp (Fig. 21.12). Ovaries typically form near the base, testes at higher body levels. Mature, yolky eggs protrude through the epidermis and after fertilization they are retained in such positions. An encysted stereogastrula forms, which then falls off the parent and remains dormant for some weeks. During winter or under conditions of drought, the dormancy period may last several months. When development is resumed a new polyp arises directly. Note, incidentally, that none of the freshwater coelenterates possesses free-swimming planula larvae, undoubtedly an adaptation to their particular environment.

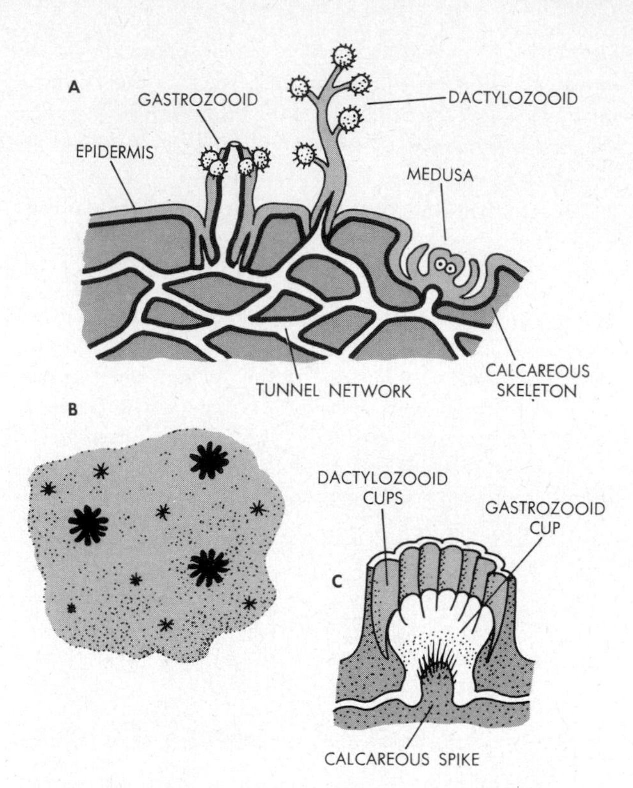

21.13. **A.** Schematic section through a surface region of *Millepora*, showing continuous surface epidermis, the subsurface canal system, and the pits for the dimorphic polyps. In the reduced medusa, note the eggs (two are shown). **B.** Surface view of a portion of *Millepora*; the larger starshaped figures represent the protrusion pores of gastrozooids, the smaller ones, of dactylozooids. **C.** Schematic cutaway sketch of the skeletal supports in *Stylaster*; dactylozooids ring in a larger gastrozooid which sits on a calcareous spike. (*Adapted from Hickson, Moseley.*)

21.12. Gametic reproduction in *Hydra*. **A** and **B.** Adults, showing position of testes and ovaries, respectively (internal gonad structure is illustrated in Fig. 21.6). **C.** Embryos develop within cysts protruding from the body wall of the parent. Such cysts will eventually drop off and independent offspring hydras will emerge directly.

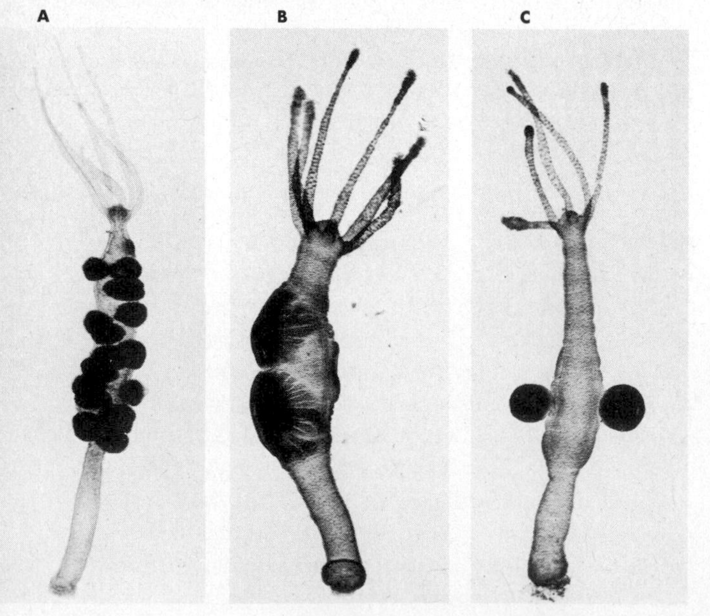

An abbreviated though not completely suppressed medusa phase also characterizes the life cycle of *Millepora*, the only genus of the order Milleporina. Millepore corals live in shallow tropical seas where they produce colonies up to 2 ft in size (Fig. 21.13). Sometimes such colonies are upright branching growths, at other times they occur as encrusting formations. The living portion of a colony is a thin film covering the surface of a massive calcareous exoskeleton. Individual polyps sit in small pits in the surface of the exoskeleton and a layer of epidermis covers the skeletal mass between adjacent polyps. The polyps may protrude their tentacles and mouths from the pits. Just under the surface of the exoskeleton, the pits are interconnected by networks of tunnels. The animals are dimorphic; some are squat cylindrical gastrozooids, others are thinner, taller dactylozooids, with tentacles reduced to cnidoblast-studded knobs. Gonophores are budded into some of the tunnels within the exoskeleton. Small medusae form which are structurally incomplete; they are without mouth or alimentary system and without tentacles or velum. Such medusae are set free but they live only for a few hours, just long enough to produce and release gametes.

Stylasters are quite similar to the millipores. A diagnostic difference is the presence of a calcareous spike in the bottom of each pit containing a polyp. The spike pushes the base of the polyp well into the coelenteron.

The Trachylina (Fig. 21.14) are of considerable theoretical interest, for they are widely believed to represent the most primitive coelenterates and thus the most primitive living Eumetazoa. Trachylines are medusae; their life cycle is without polyps. In all but a very few cases, the planula larva does not become sessile but develops instead into a motile actinula, i.e., a second larval stage, with mouth and tentacles. The actinula then becomes an adult medusa directly. It is hypothesized that the original coelenterate stock was medusoid and free-swimming, that its life cycle included planula and actinula stages, and that the actinula larva of some of these primitive forms subsequently became established as a prolonged, sessile phase of the life cycle. Thus, the polyps of the present coelenterates would represent long-lived larvae. Their budding potential could be considered to be derived from that of the actinula, which is known to be able to bud off other actinulae. Early trachylines thus could have been directly ancestral to all other Hydrozoa and to the Scyphozoa. Such a pattern of descent would account also for the occurrence of actinulae in hydroids like *Tubularia.* Moreover, neotenous acquisition of gonads in early actinulae may have led to a complete loss of the medusa phase, hence to the evolution of exclusively polyp-forming coelenterates such as the hydras and all Anthozoa. In effect, this view postulates that radially symmetrical free-swimming medusae were the first coelenterates and that the bilateral sessile polyps of Anthozoa were the last and most specialized. The point is discussed further below (and cf. Fig. 21.24).

If trachylines may be primitive coelenterates, the Siphonophora are certainly among the most advanced. These floating social colonies are made up of various types of both polypoid and medusoid individuals, many of them representing polymorphic variants not encountered in other Hydrozoa. Polypoid polymorphs include gastrozooids, two or more kinds of dactylozooids, and gonozooids. Medusoid polymorphs include attached medusa bells, gonophores, and *pneumatophores,* i.e., floats which buoy up the whole colony. The Portuguese Man-of-War *Physalia* possesses an air-filled baglike float (cf. Fig. 16.4); the float of *Velella* is diskshaped, with an upright

21.14. Trachylina. A. Actinula larva of *Gonionemus,* budding off another actinula. This stage has arisen from a planula. **B.** Adult medusa of *Gonionemus,* with gonads along radial canals. Note velum. Fertilized eggs become planulae which form actinulae directly, without sessile polyp stages. **C.** Tentacle tip of *Gonionemus,* showing spiral band of sting-cell batteries and an adhesive pad. **D.** Photo of *Gonionemus,* from life.

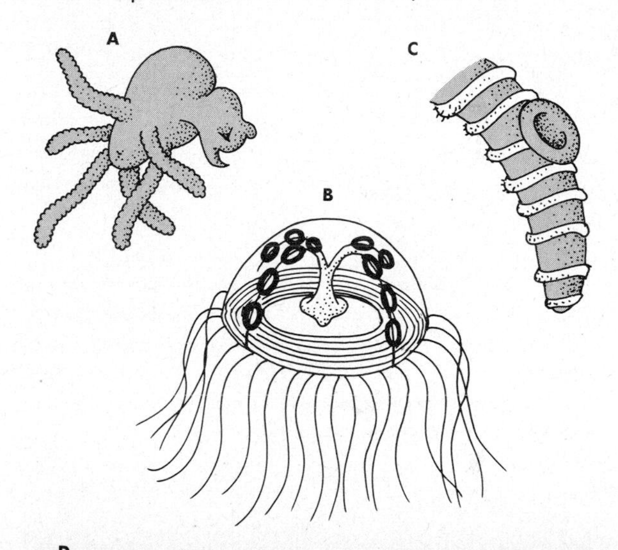

"sail" catching the ocean winds (Fig. 21.15). On the underside of *Velella,* the margin of the float is ringed by dactylozooids, the center bears a single large gastrozooid, and the remaining space is studded with gonozooids. In *Physalia* the arrangement of the individuals has a less precise pattern, and there are also more kinds of polymorphs in the colony. The medusae of the Siphonophora become free in only a few genera, and then they are highly reduced in structure. The poisons released by their stinging batteries make the members of this order exceedingly dangerous even out of water.

Class Scyphozoa

Jellyfishes; all marine; tetramerous symmetry; pelagic medusa stage dominant, without velum; epidermis with flagellated cells and with cnidoblasts in all body regions; subepidermal muscle cells; mesogloea with abundant cells and fibers; coelenteron with gastric filaments, typically partitioned into four gastric pouches by septa; gonads endodermal; development via planula, *scyphistoma,* and *ephyra* larvae. *Aurelia, Pelagia, Cynaea*

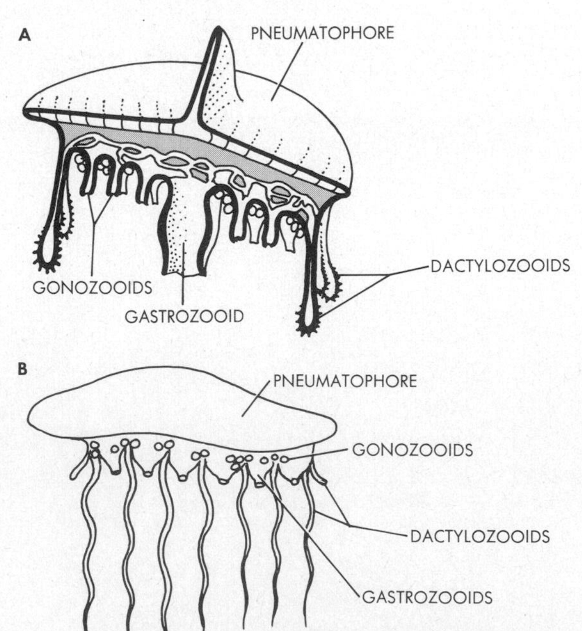

21.15. Siphonophora. (*After Haeckel.*) **A.** Section through *Velella.* Note the gastrodermal channels, which provide continuity between the coelenterons of the gastrozooid and the gonozooids; on the latter note the medusa buds. **B.** Arrangement of polymorphs in *Physalia* (schematic). *Physalia* is illustrated also in Fig. 16.14 and color plate VI.

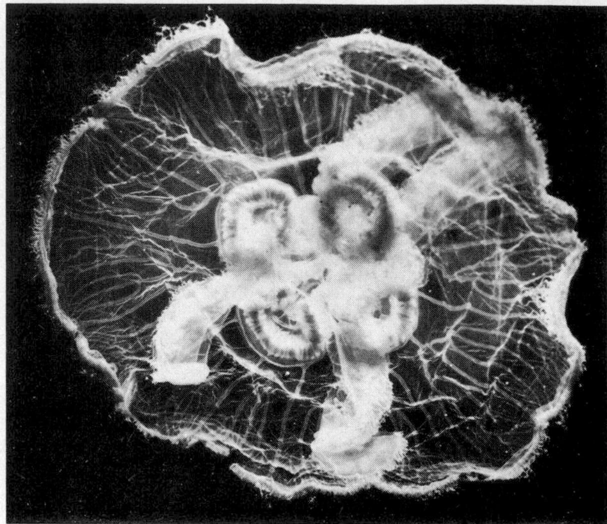

Courtesy Carolina Biological Supply Company.

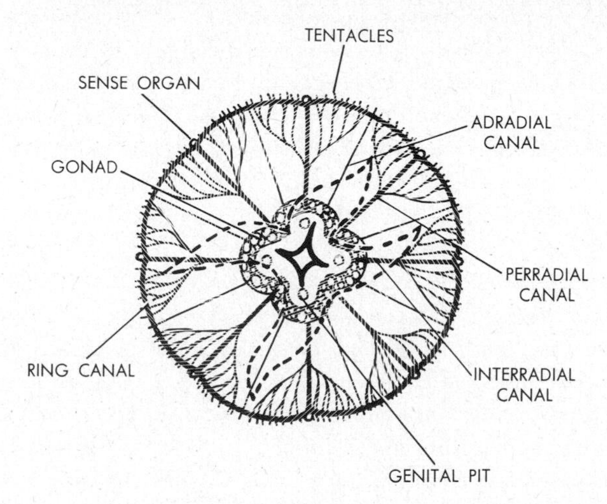

21.16. Photo: *Aurelia,* the common jellyfish. Note the four gonads near center. **Diagram:** the canal system and symmetry in *Aurelia.* Interradial and periradial canals are branched, adradials are unbranched. Dashed lines mark the positions of the tentacular arms of the manubrium. A ring canal, present in *Aurelia,* is not found in most other scyphomedusae.

Scyphomedusae can be distinguished readily from hydromedusae by the absence of a velum and the presence of scalloped *lappets* along the bell margin (Fig. 21.16). The number of lappets is usually divisible by four and the body as a whole is strongly tetramerous. The four-cornered manubrium under the umbrella bears long, trailing *oral arms,* richly supplied with cnidoblasts. In some Scyphozoa, e.g., *Aurelia,* the coelenteron leads into radial canals. Four main canals are present, each with numerous branches. Such canals continue into the tentacles along the bell margin. Ring canals are generally absent, though not in *Aurelia.* In the majority of Scyphomedusae, however, canal systems are not formed. Instead, the coelenteron is partitioned into four radially arranged *gastric pouches,* formed by outfoldings of the gastrodermis and underlying mesogloea into the coelenteron space (Fig. 21.17). Each such outfolding, or *gastric septum,* is pierced by a perforation which provides communication between adjacent gastric pouches. The free edges of the septa bear fringes of *gastric filaments,* short enzyme-secreting tentacles. Such filaments occur in all Scyphozoa, even those without gastric pouches.

21.17. The structure of most scyphozoan medusae (schematic). **A,** horizontal section; **B,** cross section. The gonads lie on the floor of the gastric pouches, in the gastrodermis.

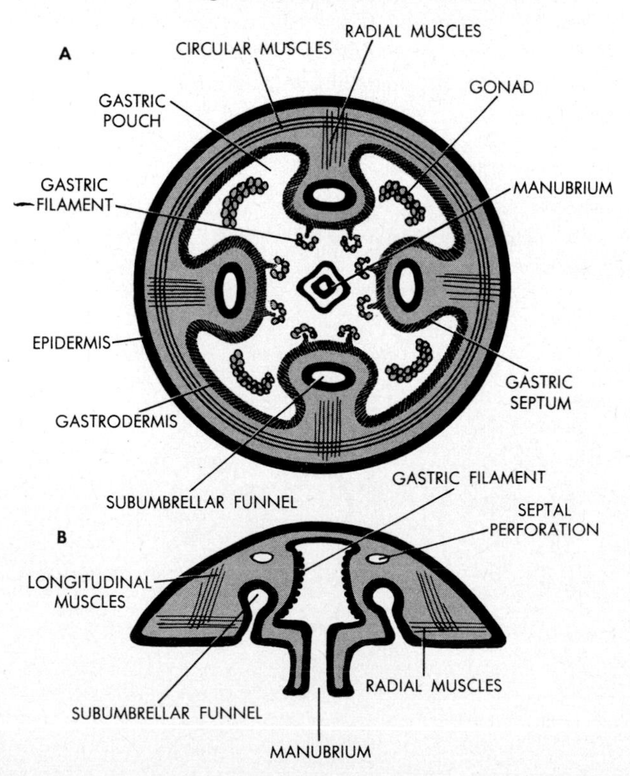

Each of the four gastric septa contains a mesogloeal cavity opening to the exterior via a pore on the underside of the umbrella. Called *subumbrellar funnels,* these cavities fill with sea water and they presumably facilitate breathing (cf. Fig. 21.17). As pointed out earlier, Scyphozoa possess distinct muscle cells. They are ectomesodermal in origin and form a layer underneath the epidermis. Particularly prominent are a circular muscle band around the bell margin, radial bands to the bell margin, and a longitudinal band traversing the base of each gastric septum. Subgastrodermal muscles are not well developed. The margin of the umbrella bears sense organs and short, often structurally reduced tentacles. With some exceptions, most scyphomedusae are without marginal nerve rings.

The animals characteristically possess four gonads, located in the gastrodermis along the floor of the gastric pouches (cf. Figs. 21.16 and 21.17). The sexes are usually separate. Planula larvae settle and develop into a polypoid larval form, the *scyphistoma* (Fig. 21.18). This tentacled larva lives as a solitary individual, in some cases for years. Ultimately it transforms either into a single free-swimming larval medusa, the *ephyra,* or into several such ephyrae. A single ephyra is formed in *Pelagia,* for example. *Aurelia* exemplifies the group in which one scyphostoma larva produces numerous ephyrae. Such young medusae arise through a budding process called *strobilation.* In it, ephyra buds develop like body segments in oral-aboral succession; and when the topmost bud is mature, it is cut off transversely from the scyphistoma and set free. Strobilation is the chief form of vegetative reproduction among Scyphozoa; the medusae reproduce exclusively via gametes.

Class Anthozoa

Polyps only, all marine; externally radial but internally more or less bilateral; with oral disk, pharynx, and siphonoglyphs; epidermis with ciliated cells; subgastrodermal muscles well developed; mesogloea a fibrous connective tissue; coelenteron always partitioned by septa; gonads endodermal; development via planula.

Subclass Octocorallia: octomerous; one ventral siphonoglyph; always colonial; usually with endoskeleton. *Tubipora, Heliopora, Gorgonia, Pennulata*

Subclass Hexacorallia: hexamerous or polymerous; two or more siphonoglyphs; solitary or colonial. *Metridium, Adamsia, Meandra, Fungia*

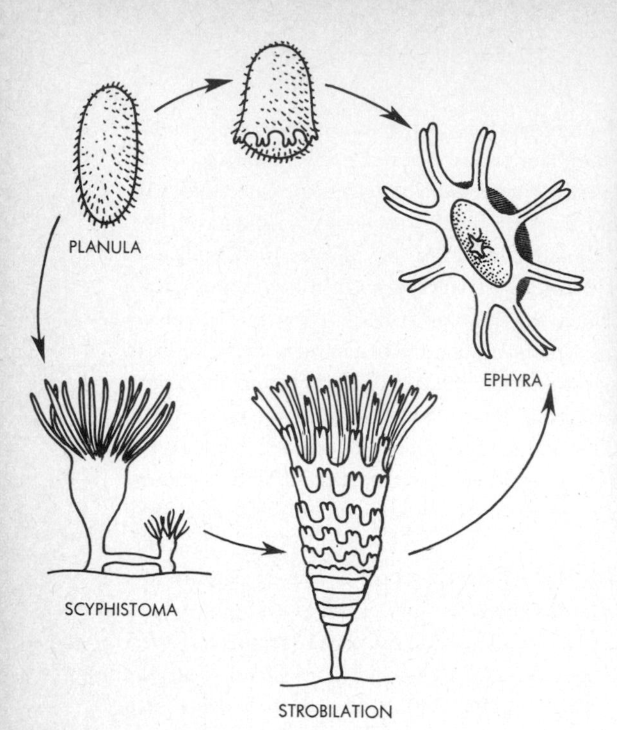

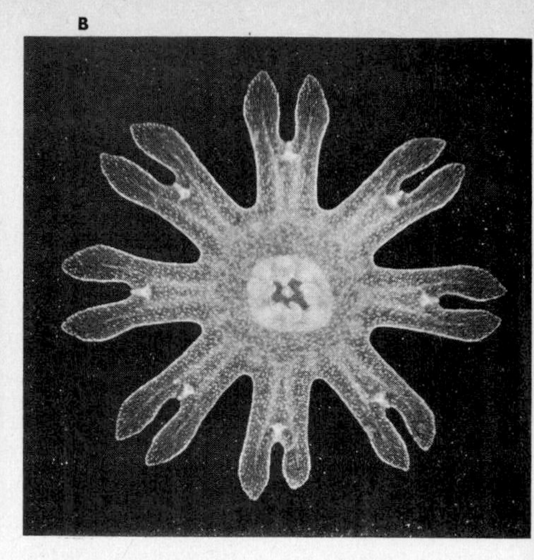

21.18. Development of Scyphozoa. Diagrams: the planula may transform into an *ephyra* larva directly, as in *Pelagia* (top), or may settle and become a *scyphistoma* larva (bottom). The latter may bud via stolons and eventually via strobilation, a segmentation process which results in free ephyrae. **Photos: A.** Scyphistoma larvae of *Aurelia*, hanging from underside of rock; young larva at right, strobilating older larvae in middle and at left. Note ephyrae being cut off at left. **B.** Individual free ephyra larva of *Aurelia*, seen from underside.

The oral end of anthozoan polyps is a flat *oral disk*, ringed by tentacles (Fig. 21.19). In the center of the disk the epidermis continues past the mouth opening as a short internal tube, the *pharynx*. The mouth is often oval or in the form of a slit, one expression of anthozoan bilaterality. A plane through the slit marks the animal into right and left halves. One end of the slit is said to be ventral, the other dorsal. The end referred to as ventral in Octocorallia is identified by a pharyngeal band of flagellated cells, the *siphonoglyph*. More than one such structure is usually present in the Hexacorallia. Siphonoglyphs aid in breathing by maintaining a flow of water into the coelenteron.

21.19. Anthozoan structure. A. Cutaway section through a sea anemone, to show the gastric partitions. **B** and **C.** The symmetries and ground plans of Octocorallia and Hexacorallia, with longitudinal muscles on the ventral surfaces of the gastric septa and with one or two siphonoglyphs.

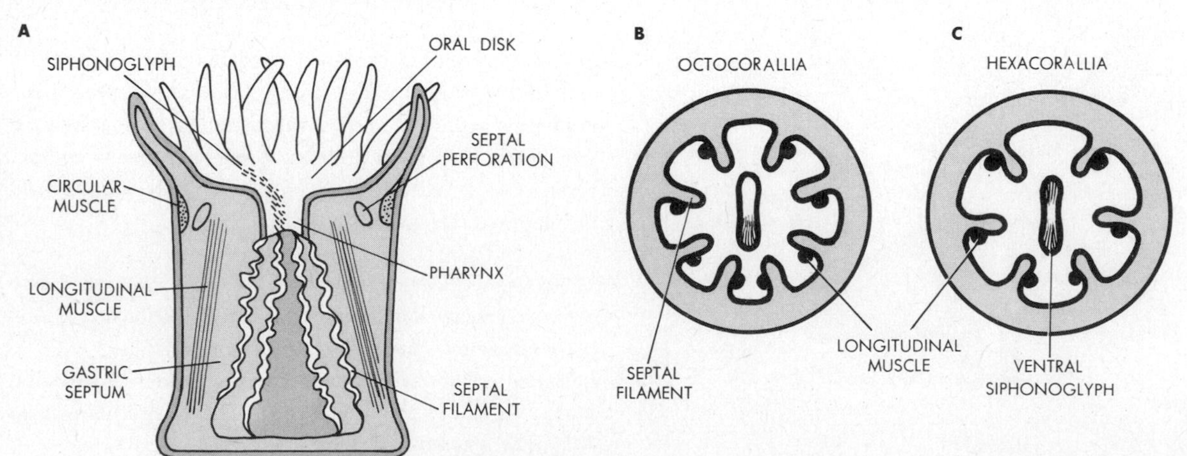

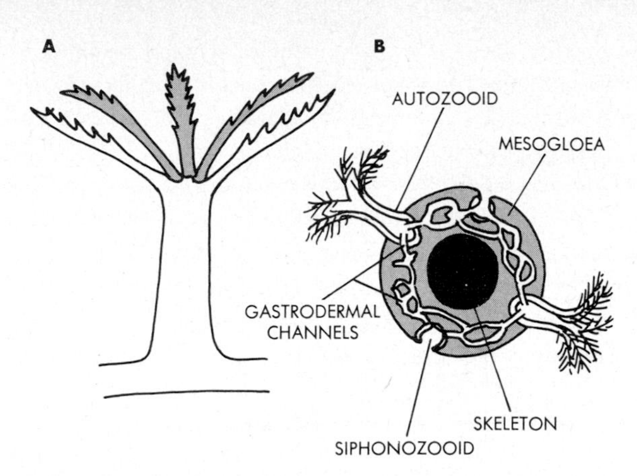

21.20. Octocorallia. Diagrams: A. Outline of single octocorallian polyp, with eight hollow, feathery tentacles. **B.** Section through a branch of the precious red coral *Corallium,* showing the skeletal core and the living surrounding tissue (coenosarc). Autozooids are tentacled feeding individuals, siphonozooids are individuals without tentacles which aid in circulating water through the colony. **Photos: A.** Close-up of portion of gorgonian sea fan *Eunicella.* Note expanded polyps. A whole colony is shown in Fig. 8.7. **B.** Close-up of portion of soft coral *Alcyonium.* Polyps are largely retracted; note the eight tentacles on each, the characteristic octocorallian condition. Cf. color plate VIII.

The coelenteron is partitioned conspicuously by radial septa, i.e., outfoldings of the gastrodermis and mesogloea into the coelenteron space. At the oral end the septa connect to the oral disk and the pharynx; they become narrower toward the base of the animal. Eight such septa characterize the Octocorallia, six or multiples of six often occur in the Hexacorallia. The free edge of each septum forms a thickened *septal filament;* it is glandular and studded with cnidoblasts. A strong longitudinal subgastrodermal muscle lies along the ventral surface of each septum, another expression of internal bilaterality: the muscles on the two most ventral septa face each other, those on the two most dorsal septa face away from each other. Sets of circular muscles ring the body of the polyp. The gastric septa also bear the gonads located under the gastrodermis. Sexes are most often separate. Eggs are shed in some forms, retained up to the planula stage in others. Planulae settle and develop directly into polyps.

Anthozoa comprise some two-thirds of all the species of coelenterates. Octocorallia (Figs. 21.20, 21.21) are exclusively colonial; their branching, often delicately plantlike and multicolored growths are among the most beautiful objects in the sea. Each individual polyp is characterized by eight hollow,

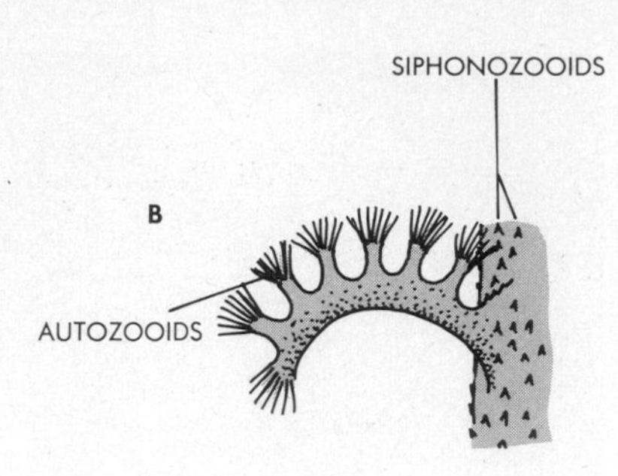

21.21. Sea pens. Diagrams: A, whole view of the sea pen *Pennulata,* showing stem (axial polyp) with lateral "leaves"; **B,** detail of a leaf, with autozooids along it. Siphonozooids are tentacleless polyps along the stem. **Photo:** portion of the sea pen *Acanthoptilum,* showing autozooids along lateral leaves. Cf. color plate VIII.

feathery tentacles, the space within a tentacle being an extension of one of the eight gastric pouches. Colonies are supported by endoskeletons which are either calcareous, horny, or both. Such skeletons arise as spicules, formed as in sponges by mesenchyme cells. In many cases the spicules are fused together into a continuous framework supporting the entire colony. Horny skeletons most often consist of rods passing through the cores of stems and branches; the polyps are arranged around these rods. Animals in this subclass include the purple organ-pipe corals (*Tubipora*), the blue corals (*Heliopora*), the soft corals (*Alcyonium*), and the red, or precious, corals (*Corallium*). The latter are members of an order of so-called "gorgonian" corals, named after *Gorgonia,* the common sea fan. Also in this order are the sea whips and sea feathers. Most of these types have horny skeletons, not calcareous ones like *Corallium.* The Octocorallia additionally include the sea pens (e.g., *Pennulata*), each of which consists of a long upright axial polyp to which are attached numerous smaller lateral polyps. These latter are dimorphic; *autozooids* are feeding individuals, and *siphonozooids* are structurally reduced individuals functioning primarily in maintaining water circulation through the colony.

The Hexacorallia (Figs. 21.22 and 21.23) comprise the sea anemones and the stony corals. Sea anemones (e.g., *Metrid-*

21.22. Hexacorallia. True anemones are illustrated in color plate VII. The photo below depicts *Cerianthus,* an anemonelike type representing a separate hexacorallian order.

A

Plate V, page 1

PLATE V. Coelenterates: Hydrozoa. A, polyp colony of *Plumularia*. **B,** polyp colony of *Tubularia*. **C,** a hydrozoan medusa. **D,** the hydrozoan coral *Allopora,* order Stylasterina; in foreground note nudibranch snail.

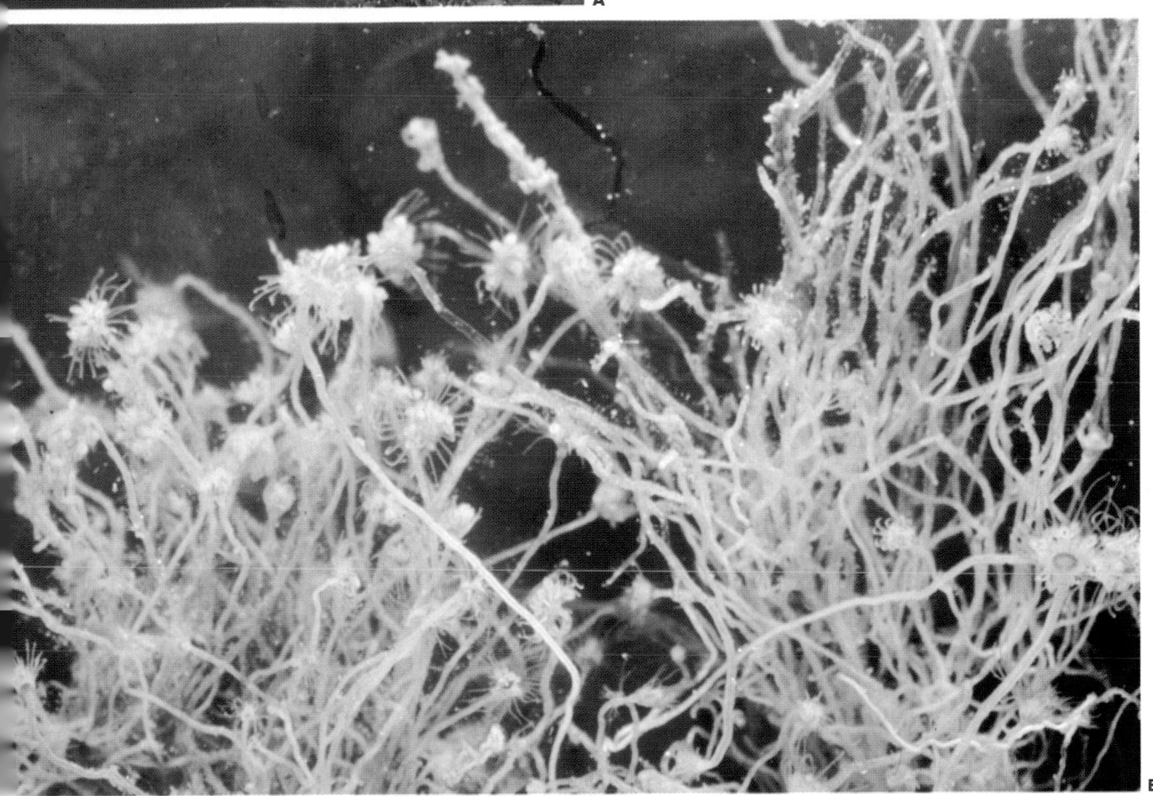

B

C D

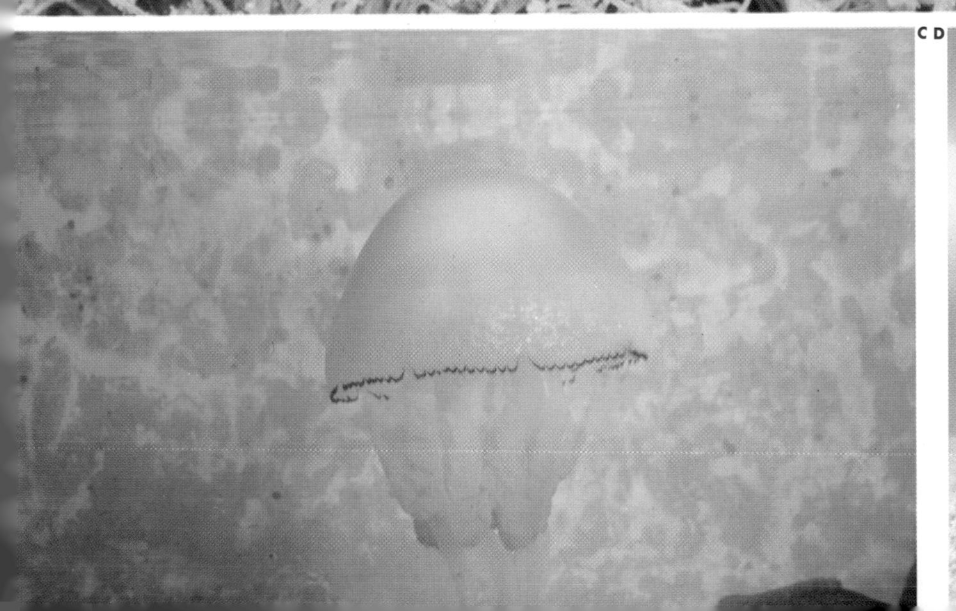

A

PLATE VI. **A variety of coelenterate types.** **A, B,** whole view and close-up of the siphonophoran *Physalia,* the Portuguese man-of-war. **C,** *Cyanea,* a scyphozoan jellyfish.

B

c Plate VI, page 2

Plate VII, page 1

PLATE VII. Hexacorallian Anthozoa. A, the greenish *Anemonia.* **B,** a species of *Actinea.* **C,** a species of *Calliactis.* **D,** *Boloceroides,* an anemone in which offspring may be budded off from the bases of the tentacles. **E,** *Porites,* a hexacorallian stony, or madreporian, coral.

Plate VII, page 2

D

E

A

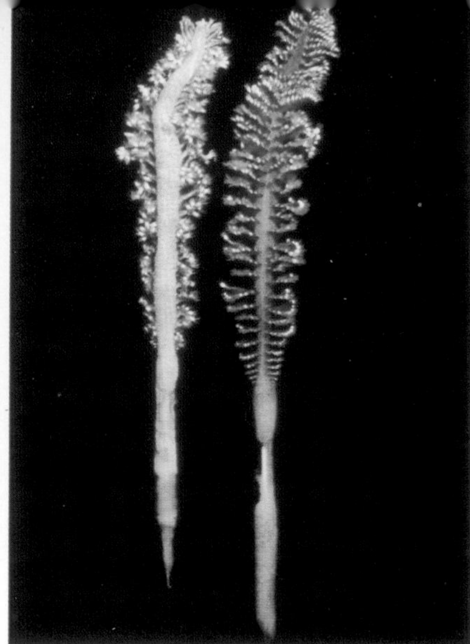

B

Plate VIII, page 1

PLATE VIII. Octocorallian Anthozoa. A, *Alcyonum,* an octocorallian "soft" coral. **B,** a species of the sea pen *Pennulata.* **C, D,** whole view and close-up of the sea fan *Gorgonia,* a horny, or gorgonian, coral.

C

D

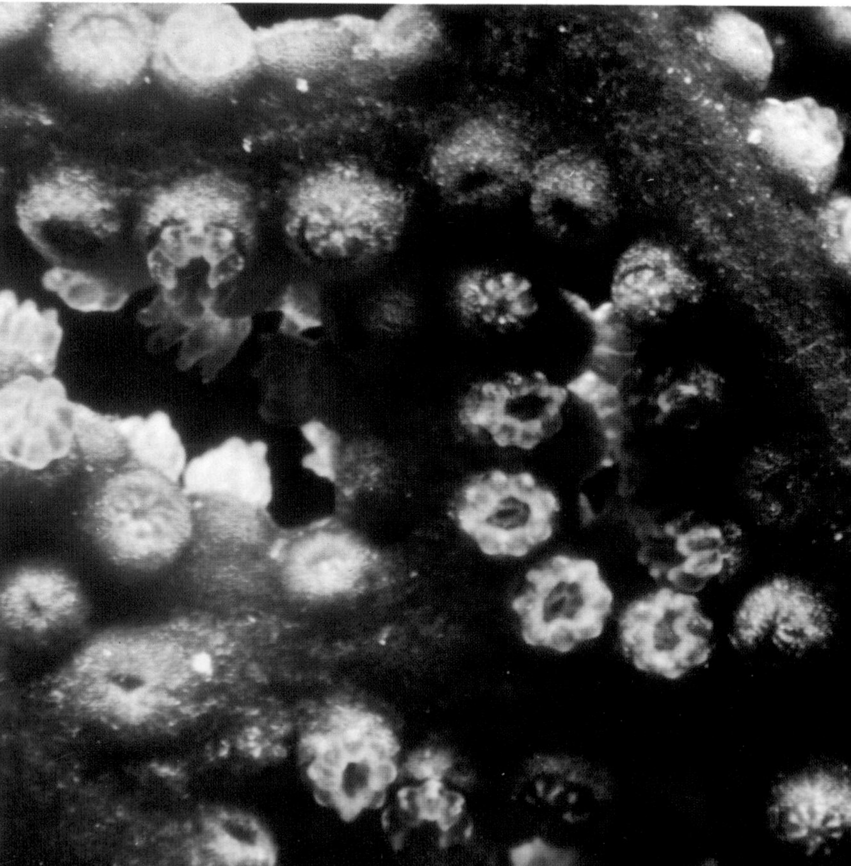

ium) are solitary and they are attached firmly to rock like suction cups: the animals lift the central portions of their basal disks away from the stone surface. Occasionally an anemone will relinquish its hold and may creep slowly to another location. Anemones such as *Adamsia* live mutualistically with hermit crabs, attached to the snail shells the crabs inhabit (cf. Chap. 17). The stony corals are the principal reef-formers. Colonies of such animals secrete massive globular exoskeletons up to several yards in width. Individual polyps live in tiny pits along the surface, essentially like the millipore corals. The various genera are identified by the characteristic surface configurations of their exoskeletons. For example, the skeletons of brain corals (*Meandra*) suggest cerebral hemispheres; those of fungus corals (*Fungia*) are reminiscent of the gill plates on the underside of a mushroom.

Though stony corals contribute most to reef formation, they are not the only organisms to do so. Among the many others are, for example, the coralline (red) algae, the foraminiferan protozoa, and corals such as *Millipora* and *Tubipora*. Reef formations were far more extensive in earlier geologic ages than at present, and it is possible that, as a group, coral animals are on the decline. Climate is probably a major limiting factor. Coral-forming coelenterates require warm, shallow seas, and reefs today are confined to a geographic belt bounded approximately by the Tropic of Cancer and the Tropic of Capricorn.

Anthozoa most particularly but also coelenterates of the other classes are known as fossil forms from early Paleozoic times. As noted in Chap. 15, the graptolites may conceivably be coelenterate fossils, though they are now generally thought to constitute a separate phylum with other affinities. The earliest known coelenterate fossil is a jellyfish from the very late Precambrian, and the circumstance that this ancient coelenterate was a scyphozoan may prove to be of considerable phylogenetic significance. The fossil suggests that early coelenterates were *medusae*, and thus it tends to support the generally held view that medusae like the trachylines were the first coelenterates. As pointed out in Chap. 14, a minority opinion holds that sessile bilateral Anthozoa are primitive and that the Hydrozoa are the most advanced class. The issue transcends coelenterates as such, for it relates to the problem of the ancestors of the Bilateria. If Hydrozoa are primitive, then radial animals have given rise to bilateral ones; but if Anthozoa are primitive, then Eumetazoa could have been bilateral from the start, a basic postulate of the ciliate-flatworm hypothesis of eumetazoan origins. Various arguments against this latter hypothesis have already been cited in previous contexts, and the comparative structure of the

21.23. Hexacorallia: corals and reef formers. A, *Caryophyllia,* a solitary stony coral of the order Madreporia; **B,** skeleton of a brain coral, another madreporian type; **C,** a black coral, member of the separate hexacorallian order Antipatharia. Cf. color plate VII.

A

B

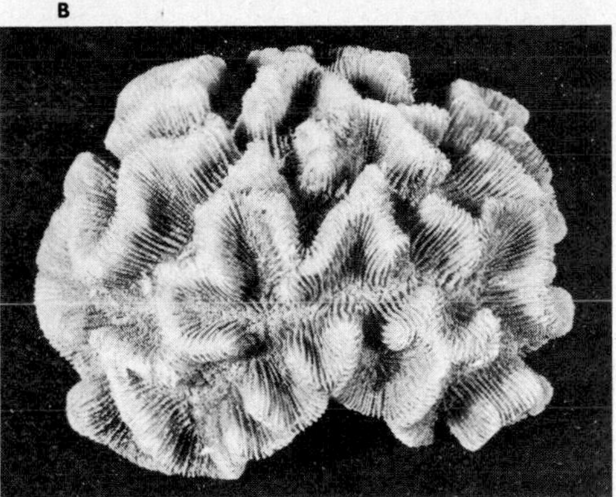

B, *courtesy Carolina Biological Supply Company.*

C

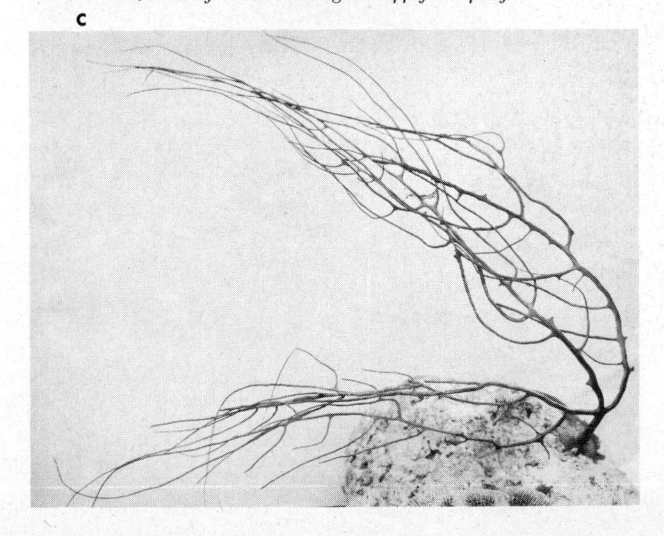

coelenterates probably provides an additional one. The assumption of anthozoan primitiveness would imply that true muscle cells must have evolved first, and that such cells later must have degenerated or have somehow become fused with epithelial cells to form the epitheliomuscular cells of Hydrozoa. Such a possibility is hardly credible, especially since the reverse hypothesis does happen to be entirely plausible. Thus, epitheliomuscular cells are at substantially the same level of evolutionary development as the myofibril-containing pinacocytes of sponges and the myoneme-containing cells of protozoa. Epitheliomuscular cells later could then have readily specialized in two different directions, one leading to primarily epithelial cells, the other to primarily contractile cells, as in the Scyphozoa and Anthozoa (and also the primitive flatworms, cf. Chap. 22). On this view, Anthozoa could not have been ancestral to Hydrozoa, but Hydrozoa could well have been ancestral to Anthozoa. If so, the bilaterality of Anthozoa would represent a uniquely anthozoan adaptation, and the bilaterality of the later Eumetazoa would have been derived independently from radial Hydrozoa.

At this stage neither view can be proved, but as a working hypothesis the concept of hydrozoan primitiveness receives far more support from the internal evidence of the actual animals than the concept of anthozoan primitiveness. In the next chapter we shall proceed on the assumption that Bilateria arose from radial hydrozoan coelenterates (Fig. 21.24).

PHYLUM CTENOPHORA

Comb jellies (sea gooseberries, sea walnuts); all marine and hermaphroditic; biradial symmetry; mesogloea with muscle cells; locomotion by eight meridional comb plates; gonads endodermal; development mosaic, via *cydippid* larvae.

Class TENTACULATA: with tentacles containing colloblasts. *Pleurobrachia, Cestum, Coeloplana*

Class NUDA: without tentacles; gastrovascular cavity with large pharynx. *Beroë*

The 80 species of this phylum have worldwide distribution in all oceans. The animals are mostly small, pelagic, and bioluminescent. They form part of the zooplankton and they feed on plankton themselves. Polymorphic variants do not occur,

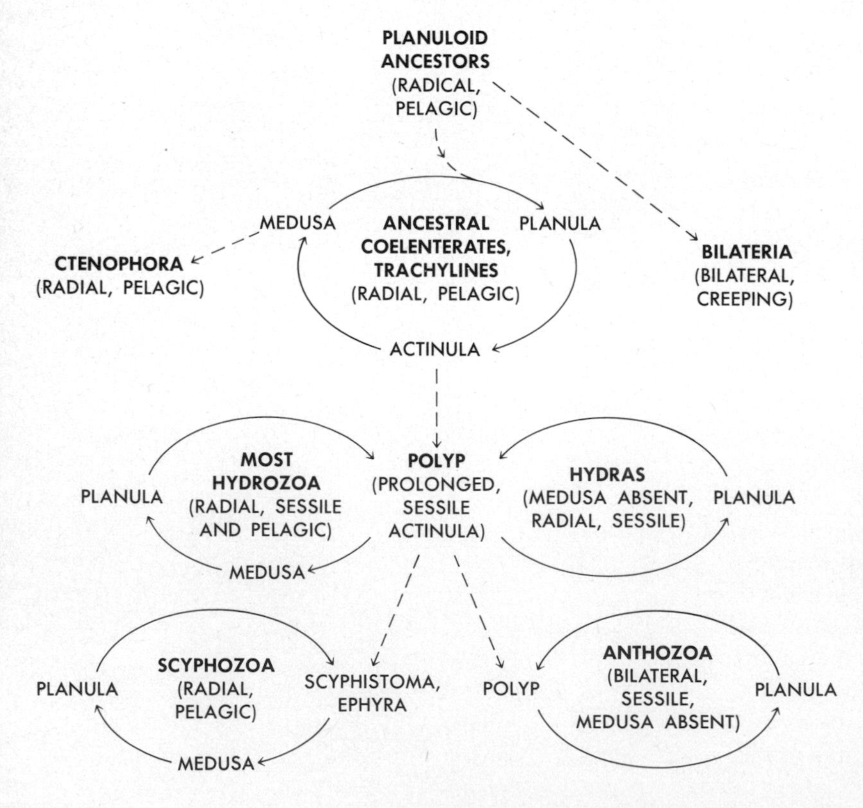

21.24. Coelenterate life cycles and their probable evolutionary interrelations. Actual life cycles indicated by solid-line arrows, presumed evolutionary interconnections, by broken-line arrows. The ancestral (possibly trachyline) life cycle at top is assumed to have become a typical hydrozoan cycle by prolongation of the actinula phase into a sessile polyp. In most Hydrozoa the medusa phase was retained (middle left), but in some the polyp phase became dominant by neoteny and a medusa phase was suppressed (hydras, middle right). Emphasis on the medusa phase is also hypothesized to have produced the scyphozoan life cycle (bottom left), and neotenous emphasis on the polyp phase, like that for the hydras, the anthozoan cycle (bottom right). Accompanying changes of symmetry and ways of life were probably ecologically adaptive and of little evolutionary significance by themselves.

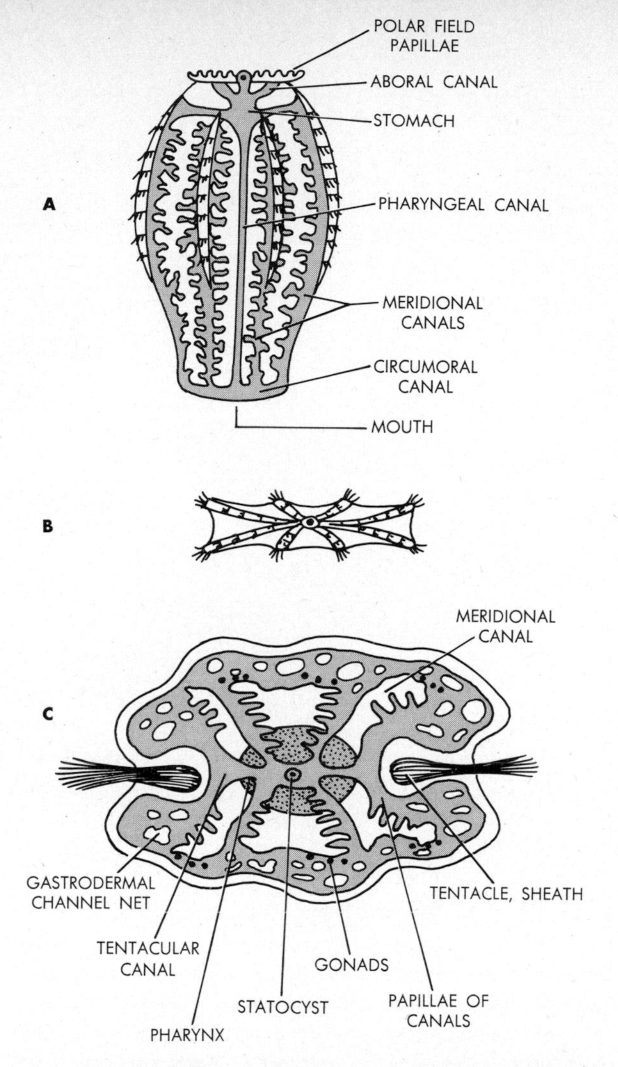

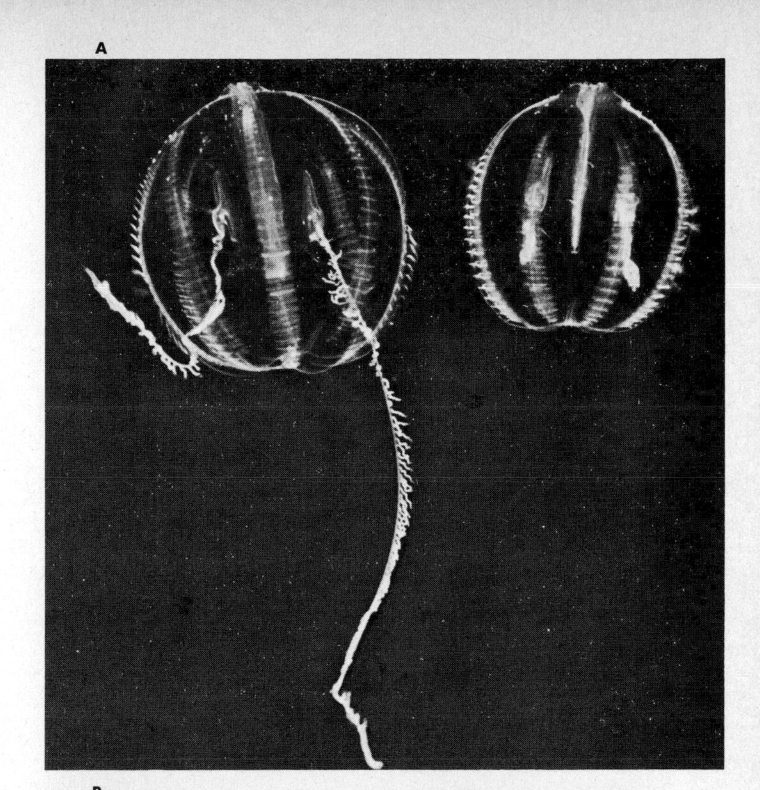

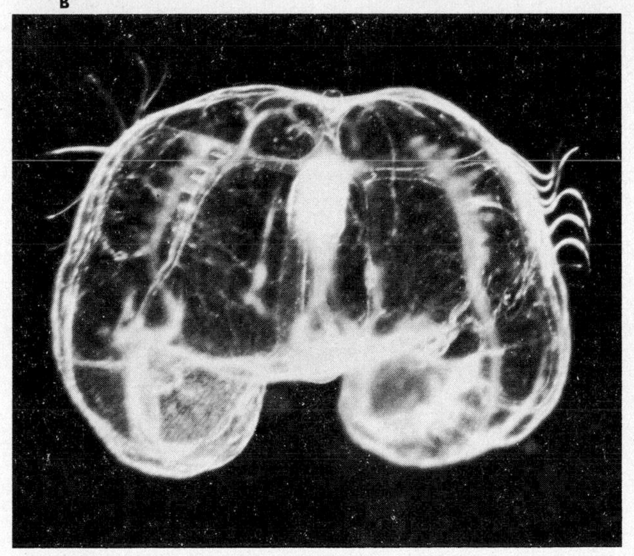

21.25. Ctenophores. Diagrams: A. Side view of *Beroë*. Gastrodermal channels are shaded; note the highly branched meridional canals. **B.** *Beroë* seen from aboral pole, indicating the greatly flattened design of the animal. **C.** Top view of *Coeloplana*, a creeping ctenophore flattened in the oral-aboral direction; gastrodermal channels shaded. (*Adapted from Mayer, Dawydoff.*) **Photos: A,** *Pleurobrachia*, and **B,** *Bolinopsis*, both from life.

nor do colonial forms. The bodies of comb jellies are glassily transparent, generally globular or bellshaped (Fig. 21.25). Some types are compressed laterally, e.g. the Venus girdle *Cestum*, which has the form of a gelatinous band. This animal may be from 1 to 2 yd long. Other types are flattened in an oral-aboral direction; they are adapted secondarily to a creeping mode of life (e.g., *Coeloplana*).

In their basic design, comb jellies rather resemble coelenterate medusae (Figs. 21.26, 21.27). The body consists of epidermis, subepidermal nerve net, bulky mesogloea with mesenchyme and muscle cells, gastrodermis, and a gastrovascular cavity with one principal opening. The symmetry is always biradial. The epidermis bears eight meridional rows of *comb plates*, two rows per quadrant. A comb plate is essen-

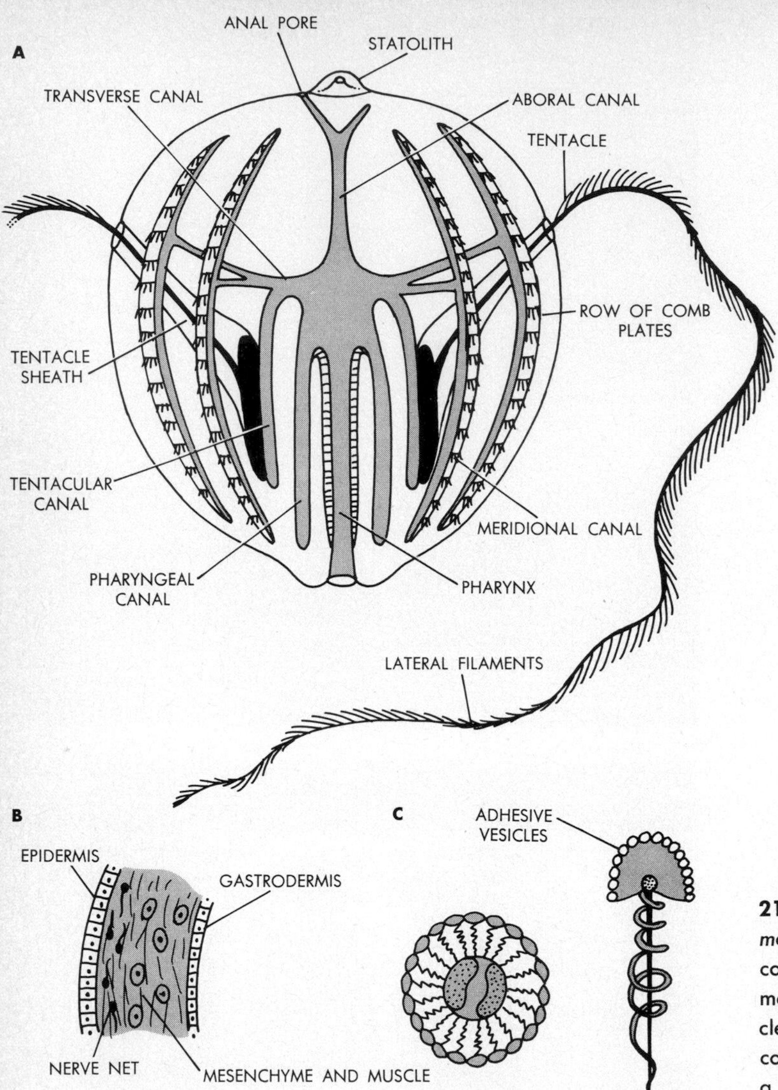

21.26. The structure of ctenophores. (*Pleurobrachia, largely modified from Hyman.*) **A.** Arrangement of gastric canals and comb rows on one side. **B.** The tissue layers of the body, schematic. **C.** Left, cross section through a lateral filament of a tentacle, with colloblasts forming an external layer; right, single colloblast cell with adhesive vesicles on outside and a spiral and a straight filament extending from the nucleated portion.

tially a membranelle, i.e., a short band of cilia. Rows of such plates beat in metachronous rhythm. They have a locomotor function and they represent the principal diagnostic feature of the phylum.

The main axis of the body passes through the mouth orally and a statocyst aborally. The latter is more complexly organized than the equivalent sense organ of coelenterates. In a statocyst of ctenophores, a statolith is supported by four stiff bristles, so-called *balancers*. Fused cilia form a dome over the statolith. From each balancer two ciliary tracts lead over the epidermal surface to the two comb rows of that body quadrant. On opposite sides of the statocyst are two ciliated areas, the *polar fields,* which are believed to serve as chemoreceptors.

Comb jellies of the class Tentaculata typically possess two tentacles, one on each side of the body between two comb rows. The tentacles lie in one of the two main symmetry planes, the so-called transverse plane. Each tentacle arises from and is anchored at the bottom of a sheath, a deep epidermis-lined pit in the body. The tentacles contain longitudinal muscles and may be withdrawn completely into their sheaths. Numerous *lateral filaments* give the tentacles a feathery appearance. Cnidoblasts are entirely absent (one aberrant species probably excepted), but *colloblasts* occur abundantly on the lateral filaments. Colloblasts are specialized epidermis-derived adhesive cells; they serve in trapping plankton organisms, which the tentacles then convey to the mouth. Tentacles are highly reduced in some of the Tentaculata, e.g., in *Cestum;* tentacles are absent altogether in the members of the class Nuda.

The mouth leads into a tubular, epidermis-lined *pharynx,* a ciliated canal which is flattened in the sagittal plane, the symmetry plane lying at right angles to the transverse plane. Extracellular digestion occurs mostly in the pharynx. From this tube food passes into a gastrodermis-lined *stomach,* a central gastrovascular cavity. Series of canals lead away from the stomach. An *aboral canal* passes toward the statocyst, where it branches into four small *anal canals.* Two of these have closed ends just under the epidermis, the other two conduct indigestible remains to the exterior via *anal pores.* Two blind-ended *pharyngeal canals* lead from the stomach in an oral direction, and two oppositely directed *transverse canals* project laterally. Each of the latter branches into a *tentacular canal* terminating at the tentacle base and into four *meridional canals* lying under the comb rows. The gastrodermal lining of all these canals is flagellated or ciliated, and the cells carry out intracellular digestion.

21.27. A, *Pleurobrachia* **seen from aboral side; B,** structure of the aboral sense organ; left, top view, right, side view.

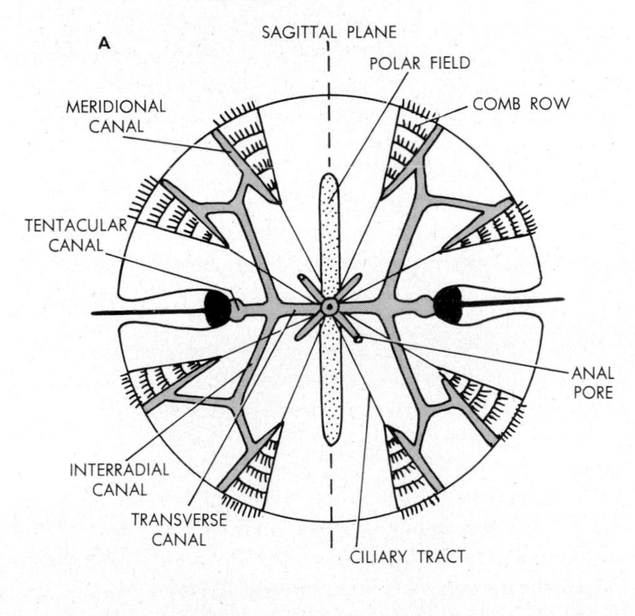

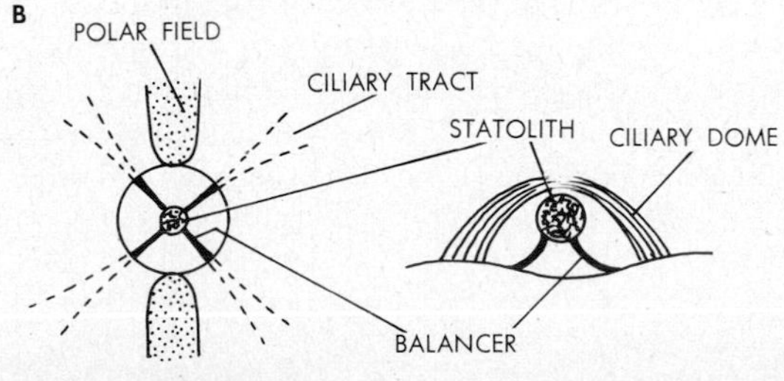

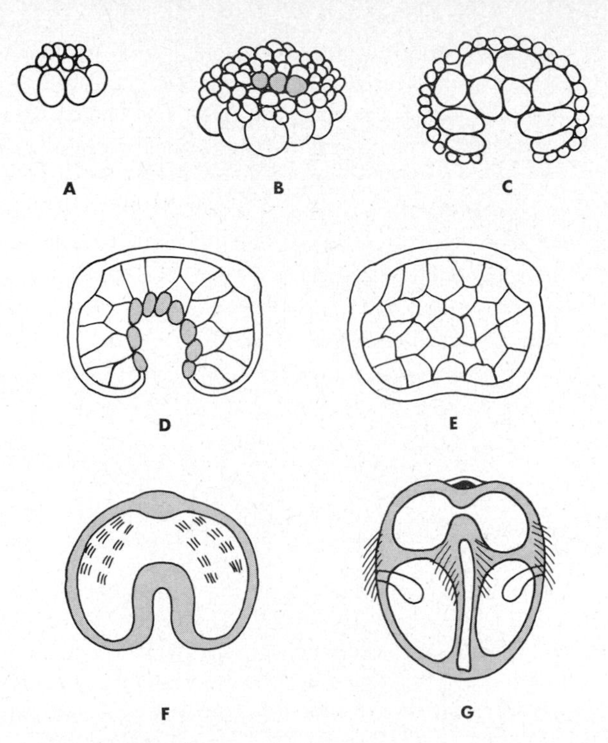

21.28. Ctenophore development. A. Sixteen-cell stage, with a tier of eight macromeres and one of eight micromeres; note biradial (disymmetric) symmetry. **B.** Late cleavage, micromeres beginning to move by epiboly over macromeres. **C.** Section to show epibolic gastrulation. **D.** Formation of mesoderm (shaded cells.) **E.** Ectodermal thickenings for comb rows (top) and mouth (bottom). **F.** Invagination of pharynx and early gastrocoel; early comb rows have formed and aboral sense organ begins to differentiate. **G.** Section through young cydippid larva, showing stomach, pharynx, tissue of future transverse canals, and beginnings of aboral canal system; positions of comb rows and tentacle sheaths are indicated.

Vegetative reproduction is unknown in ctenophores. All these animals are hermaphroditic. Their endodermal gonads are located along the meridional canals, and gametes are shed via the gastrovascular channels. Cleavage follows a unique and exact biradial pattern, the planes of symmetry here corresponding to those of the adult (Fig. 21.28). In sharp contrast to the regulative development of coelenterates, ctenophore development is mosaic. At the 16-cell stage an embryo consists of a tier of eight micromeres and a tier of eight macromeres.

The micromeres become ectoderm, i.e., they proliferate epibolically over and around the macromeres. Gastrulation in this fashion results in a stereogastrula. The middle body layer later arises as an ectomesoderm. The larva characteristic of the whole phylum is the *cydippid*, a globular stage which develops comb rows and tentacles. In forms like *Pleurobrachia*, larva and adult are quite similar and in such cases only few metamorphic changes take place. In other instances, e.g., in *Cestum* or *Coeloplana*, formation of the adult requires considerable additional development.

As a group, ctenophores display fundamental affinities to the coelenterates, though many of their features are more specialized and advanced than equivalent ones of coelenterates. It is now generally believed that ctenophores arose from primitive trachyline medusoid stocks, as an independent line of an adaptive radiation which also led to the various coelenterate lines. The ctenophore branch then must have evolved on its own, to a level of somewhat greater structural elaboration and specialization than the coelenterate branches. Based mainly on the exceptional creeping ctenophores such as *Coeloplana*, earlier hypotheses have postulated a direct evolutionary connection between ctenophores and flatworms. Such concepts have proved to be largely untenable and are now generally discounted.

REVIEW QUESTIONS

1. Give taxonomic definitions of the branch Eumetazoa, the grade Radiata, and the phyla Cnidaria and Ctenophora. Describe the ecology of coelenterates and the body symmetry of these animals. What general structural features distinguish polyps and medusae? Describe the structural and functional characteristics of cnidoblasts.

2. What cell types occur in the coelenterate epidermis, gastrodermis, and mesogloea? How does the organization of these tissues vary in the three coelenterate classes? What is the structure of a tentacle?

3. Review the characteristics of the skeletal elements formed by coelenterates. What is the organization of nervous and muscular elements? How does a coelenterate ingest, digest, and egest? What is the role of the mesogloea in locomotion? How does medusa locomotion take place? What is the embryonic origin of the mesogloea and the gametes?

4. Describe the reproductive processes of coelenterates. How and where does budding occur in polyps and medusae? How and where do gonads form? Describe the development of coelenterates up to the planula stage. Define Hydrozoa taxonomically and name representative genera of each order.

5. Describe the structure of a hydrozoan polyp colony. What are perisarc, coenosarc, stolon, hydranth, manubrium? What are the characteristics of polymorphic variants of polyps? How do polyps of (*a*) *Obelia*, (*b*) *Tubularia* form medusae?

6. Describe the structure of a hydrozoan medusa. What is an actinula? In what respects is *Hydra* an atypical hydrozoan? Review the organization of millepore corals, siphonophorans, and trachyline medusae. What is the possible evolutionary significance of the last-named? What is a pneumatophore?

7. Define Scyphozoa taxonomically and describe the structure of a medusa in that class. How does *Aurelia* differ in organization from related medusae? What are gastric filaments, subumbrellar funnels?

8. Describe the life cycle of (*a*) various Hydrozoa, (*b*) Scyphozoa. What are the larval stages of Scyphozoa and what are their characteristics? Define Anthozoa taxonomically and distinguish between the subclasses. Describe the structure of a sea anemone. In what ways do its tissues differ from equivalent ones of other classes?

9. Define septal filament, siphonoglyph. In what respects are Anthozoa bilateral? Describe the life cycle of an anthozoan. What kinds of coelenterates occur among (*a*) Octocorallia, (*b*) Hexacorallia? Which coelenterates contribute greatly to formation of coral reefs?

10. Review the possible evolutionary interrelations of the coelenterate classes and of coelenterates as a whole and the ctenophores and Bilateria. Review hypotheses on the possible ancestors of coelenterates and discuss the relative merits of various views.

11. Characterize each of the ctenophore classes taxonomically. Name representative genera in each class. Describe the exterior structure of ctenophores, with particular attention to locomotor, sensory, and ingestive organs.

12. Describe the gastrovascular system of ctenophores and contrast with that of coelenterates. Review the development of ctenophores and show in what ways it differs from that of coelenterates. How many described species are known in ctenophores, coelenterates, sponges, and protozoa?

COLLATERAL READINGS

As in the case of sponges, most books on aquatic life contain discussions about various radiate animals. The texts and general readings below may be consulted specifically:

Berrill, N. J.: "Growth, Development, and Pattern," chaps. 8–11, Freeman, San Francisco, 1961.

Brien, P.: The Fresh-Water Hydra, *Amer. Scientist,* vol. 48, 1960.

Darwin, C. R.: "The Structure and Distribution of Coral Reefs," 3d ed., Appleton, New York, 1896.

Hyman, L.: "The Invertebrates," vol. 1, chaps. 7 and 8, McGraw-Hill, New York, 1940.

Lane, C. E.: Man-of-War, the Deadly Fisher, *Nat. Geographic,* Mar., 1963.

Lenhoff, H. M., and W. F. Loomis (eds.): "Symposium on the Physiology and Ultrastructure of *Hydra* and Some Other Coelenterates," University of Miami Press, Coral Gables, Fla., 1961.

Wiens, H. J.: "Atoll Environment and Ecology," Yale, New Haven, 1962.

Yonge, C. M.: "A Year on the Great Barrier Reef," Putnam, New York, 1930.

Zahl, P. A.: Glass Menageries of the Sea, *Nat. Geographic,* June, 1955.

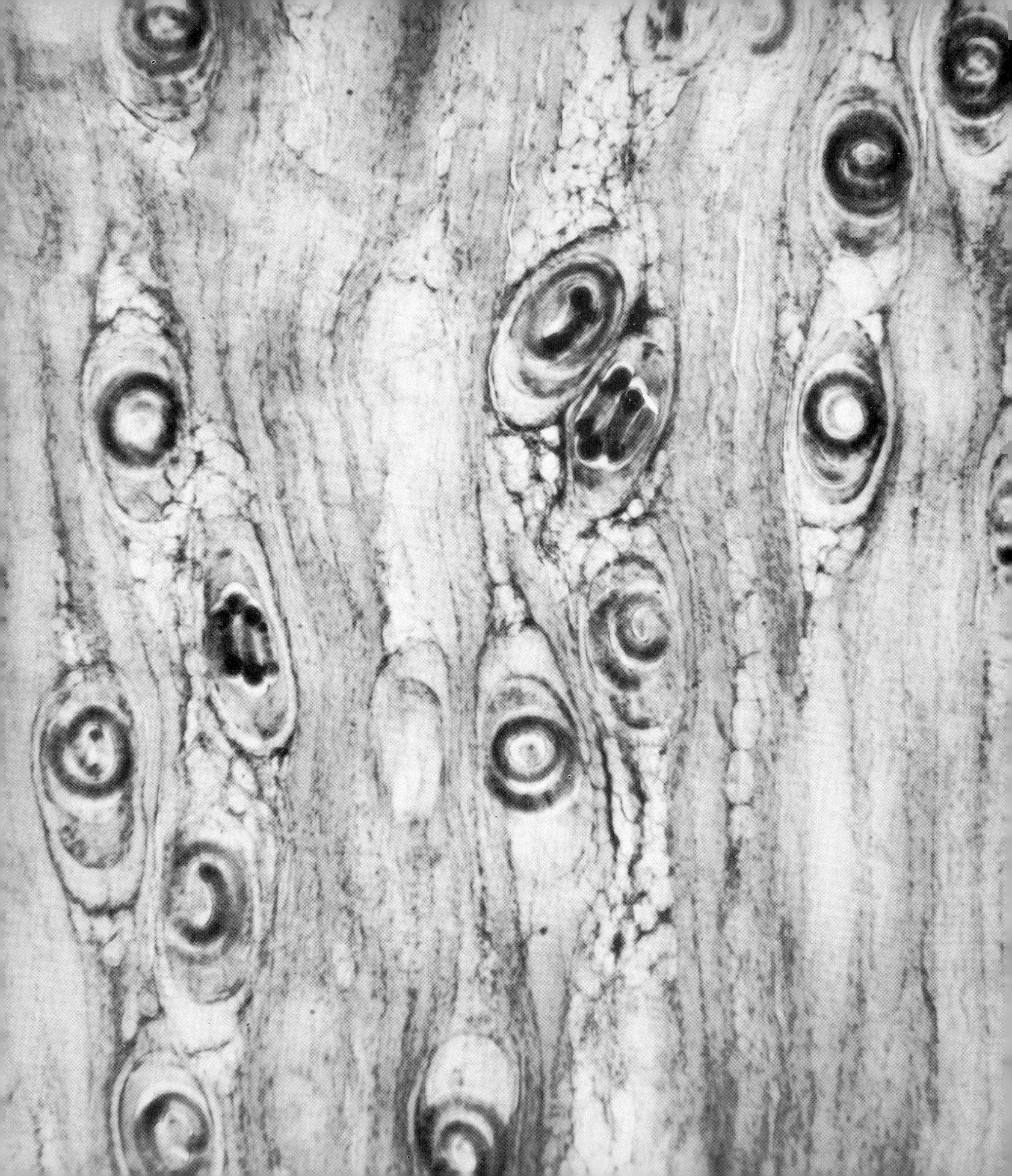

noncoelomate bilateria

PART 8

Grade BILATERIA: animals at organ-system level of complexity; with primary and typically also secondary bilateral symmetry.

Subgrade ACOELOMATA: development mosaic, typically with spiral cleavage; adult mouth formed at or near embryonic blastopore; mesoderm from ectoderm and endoderm, filling space between body surface and alimentary system, hence internal body cavities not present; with and without larvae; skeletal or breathing systems absent; nervous system basically a nerve net, with localized ganglionic thickenings; excretion by protonephridial flame-bulb system. *PLATYHELMINTHES, NEMERTINA*

Subgrade PSEUDOCOELOMATA: development mosaic, cleavage typically in spiral or bilaterally modified pattern; adult mouth formed at or near embryonic blastopore; mesoderm from ectoderm and endoderm, aggregated locally between body surface and alimentary system, hence leaving pseudocoelic, blastocoel-derived body cavity lined by ectoderm and endoderm; with and without larvae; nervous system not a nerve net; skeletal, circulatory, or breathing systems absent; excretion largely by protonephridial flame-bulb system. *ASCHELMINTHES, ACANTHOCEPHALA, ENTOPROCTA*

This and all subsequent parts deal with the fundamentally bilateral animals, collectively constituting a taxonomic grade within the Eumetazoa. Within the Bilateria in turn, three subgrades may be distinguished, the acoelomates, the pseudocoelomates, and the coelomates. The first two of these are discussed in the two chapters of this part. Both groups are characterized by the absence of a true coelomic body cavity, the acoelomates lacking a body cavity altogether and the pseudocoelomates possessing a pseudocoelic cavity. Both groups are also protostomial, i.e. the mouth of the adult arises from or near the blastopore of the embryo.

CHAPTER 22 acoelomates

Of the two phyla in this group, the Platyhelminthes, or flatworms, are generally considered to include the most primitive representatives of all bilateral animals. These worms therefore have special theoretical significance not only as the presumable ancestors of the Nemertina, the other acoelomate phylum, but also as the presumable direct or indirect evolutionary source of all other animal phyla as well.

PHYLUM PLATYHELMINTHES

Flatworms; body flattened dorsoventrally, alimentary system without anus; without circulatory system; largely hermaphroditic. *Turbellaria,* free-living flatworms; *Trematoda,* flukes, parasitic; *Cestoidea,* tapeworms, parasitic.

The flatworms comprise some 10,000 species, two-thirds of them parasitic in invertebrates and vertebrates. A few symbiotic types are included also among the *Turbellaria,* but most members of this class live free in fresh water, in the littoral regions of the oceans, and in some cases in moist terrestrial environments. These free-living forms are carnivorous (and occasionally cannibalistic). Flatworms are generally small animals, ranging from microscopic sizes to lengths of 2 to 3 inches. However, some of the tapeworms can be up to 15 to 20 yd long. Most flatworms are colorless or display the colors of the symbiotic algae living within them, and a few are pigmented in a variety of patterns. The worms are generally elongated, conspicuously flattened, and a definite head region is present. The latter is not often marked off externally from the rest of the body, but it is always characterized by the presence of the chief nervous ganglia and the main sense organs. The mouth is on the underside of the body, commonly in a midventral location, in other cases situated more anteriorly. There can be little doubt that free-living flatworms are the primitive members of the phylum and that the parasitic members are evolutionary derivatives.

Class Turbellaria

Free-living flatworms; epidermis cellular or syncytial, without cuticle.

Order Acoela: marine; without intestine. *Convoluta, Childia*

Order Rhabdocoela: marine and freshwater; intestine straight and saclike. *Stenostomum, Mesostoma*

Order Alloeocoela: marine, freshwater, and terrestrial; intestine straight, often with lateral pouches. *Monocoelis, Plagiostomum*

Order Tricladida: marine, freshwater, and terrestrial; intestine with three branches. *Dugesia, Bipalium, Bdelloura*

Order Polycladida: marine; intestine with many branches. *Cestoplana, Stylochus*

The Acoela are the primitive members of the class and thus in effect the most primitive of all Bilateria. Inasmuch as they are believed to have evolved from hydrozoan coelenterates, their characteristics are of considerable significance.

Acoels on the whole resemble planula larvae (Fig. 22.1). They are small, often microscopic, sometimes up to ½ in. long. The body is flattened down and elongated. A ciliated epidermis, often syncytial, covers the exterior, and the interior is filled with digestive-mesenchyme cells. A simple mouth opening is present midventrally, but other formed alimentary structures are lacking; food passess through the mouth directly into the interior cells. In some acoels (e.g., *Childia*), the epidermal components are epitheliomuscular, as in Hydrozoa. In other forms, separate muscle cells may underlie the epidermis. The nervous system is a nerve net of the coelenterate type. The net is concentrated anteriorly into a poorly defined nerve ring or brain ganglion, and the remainder forms numerous longitudinally concentrated neuron strands, or nerve cords. A coelenteratelike statocyst is present anteriorly, and some acoels also possess ocelli. Excretory systems are absent. Like flatworms generally, acoels are hermaphroditic.

22.1. Frontal view **(A)** and side view **(B)** of an acoel flatworm (schematic). The epidermis here is drawn as a syncytium, though in many cases it is epithelial and ciliated. The nervous system, emphasized particularly in **B**, is part of the epidermal tissue layer.

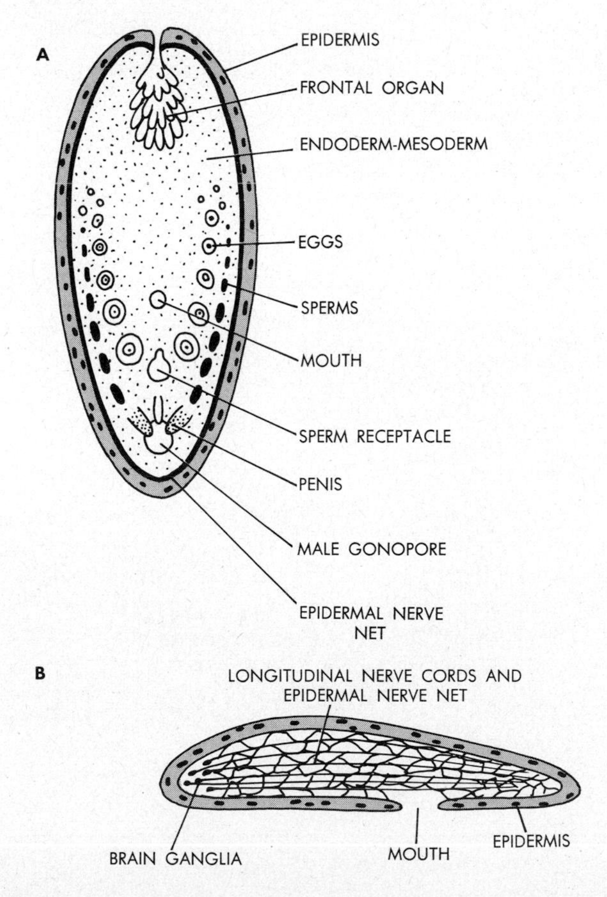

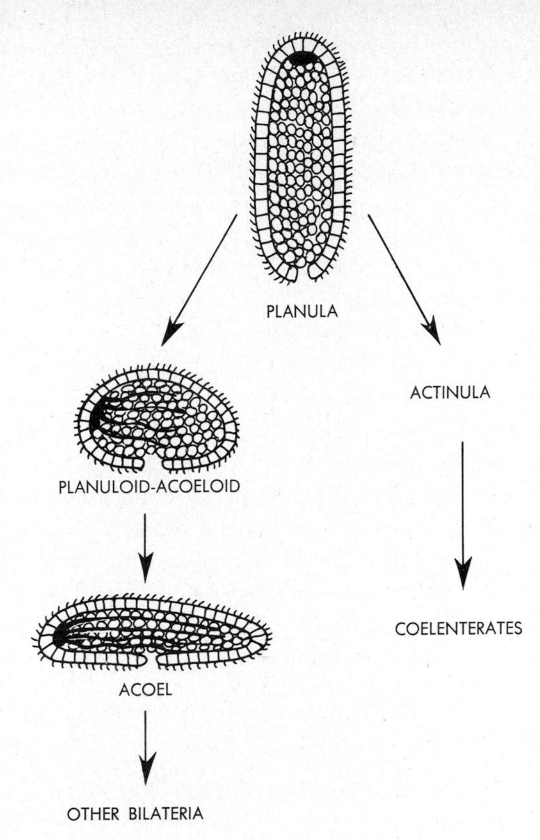

22.2. The possible derivation of acoels and flatworms generally from a coelenterate planuloid ancestor. (*Adapted from Hyman.*) In a radial, pelagic planula, sensory structures are situated aborally, opposite the blastopore-mouth. If the oral-aboral axis became shortened and if, in correlation with a creeping mode of life, the sensory center also shifted in the direction of forward propulsion, then a bilateral acoel structure can be derived.

Gametes form in clusters directly from mesenchyme cells. Primitive sperm ducts develop seasonally, but duct systems are not produced for the eggs; the latter are shed through the mouth or through ruptures in the body surface. Zygotes cleave spirally (cf. also below), gastrulation is epibolic, and the blastopore of the stereogastrula becomes the adult mouth. Indeed, the stereogastrula is virtually an adult already.

As pointed out in Chaps. 14 and 21, acoels and flatworms generally are thought to have arisen from radial, hydrozoan planulalike stocks (Fig. 22.2). Some of these would have become actinulae with tentacles and evolved into coelenterates. Others would have adopted a creeping mode of life and so would have become bilateral; the planula would have flattened down and become elongated in the direction of motion, a process which would have left the blastopore-mouth in a ventral position but would have shifted the main neural concentration anteriorly. Primitive acoels would have evolved in this manner. They would have inherited epitheliomuscular cells as well as the potential of developing a saclike gastrovascular cavity, a potential which would have become actual in the flatworms later derived from the acoels. All other major flatworm characteristics would have been independent original inventions of these later descendants.

On this view, it is therefore not surprising that modern Turbellaria other than the Acoela still exhibit numerous coelenteratelike traits. The alimentary system consists of a mouth, a well-developed and usually eversible pharynx, and a saclike intestine (Fig. 22.3). The cells forming the intestine are ciliated and phagocytic; digestion occurs both extracellularly and intracellularly, as in coelenterates. The middle body layer is a mesenchyme without body cavities, often called *parenchyma* in the acoelomate animals. This layer is a connective tissue, composed largely of a syncytium which also contains free amoeboid cells.

The exterior is covered with a wholly or partly ciliated epidermis, still syncytial in the primitive orders, but cellular in the more advanced ones. The epidermis rests on a basement membrane, and in many cases the cells are drawn through the membrane so that the inner, nucleus-containing portions come to lie deep in the mesenchyme. Such sunken cells include particularly the many mucus-secreting gland cells with which the worms are equipped. Many of the animals secrete slime tracks on which they then move by ciliary action. Glandular cells are often aggregated into multicellular glands, and these may function as adhesive organs or may aid in trapping prey. Certain of the epidermal gland cells manufacture *rhabdites*, intracellular rodlike bodies specifically characteristic of flatworms. The organelles contain a viscous substance, they are dischargeable, and they may play a role in the formation of secreted mucus; their exact function is unknown, however. Many flatworms feed on hydras and other coelenterates, and in such cases the epidermis may contain functional cnidoblasts which have migrated from the intestine to the surface.

The epidermis is underlain by two or more layers of muscle

22.3. Turbellarian structure. A. Cross section through anterior portion of a turbellarian. The epidermis is wholly ciliated or only partly, as shown here. Mesenchyme constitutes the parenchyma. **B.** Detail of a section of ventral epidermis, to show the sunken nucleated portions of the epidermal cells. **C.** Schematic diagram of a triclad planarian, with the alimentary system shaded. Note the one anterior and two posterior pouches, each with branch pouches, or diverticula. The pharynx is of the protrusible type (cf. Fig. 9.20).

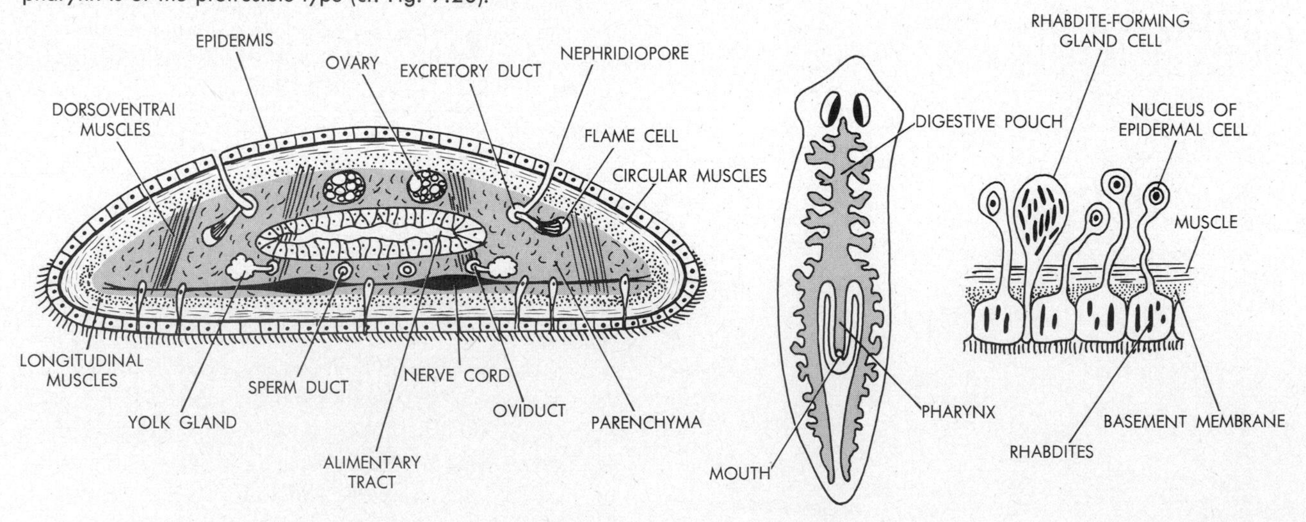

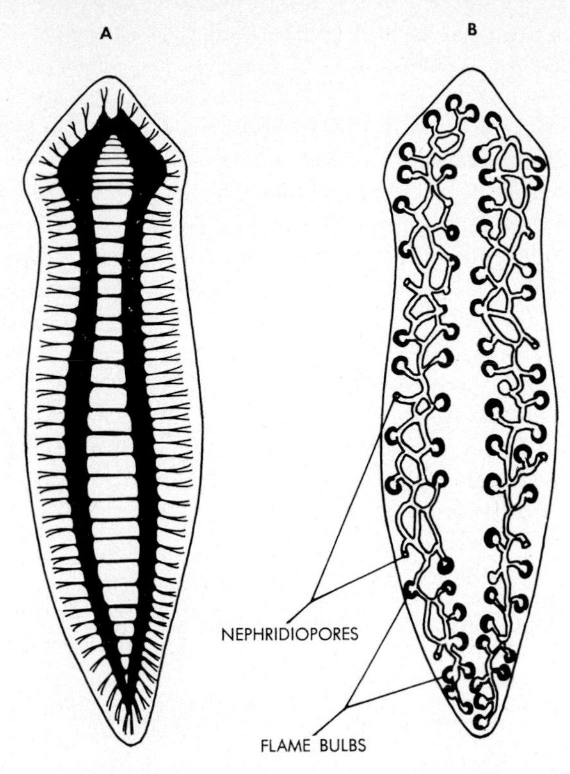

22.4. Turbellarian structure. Triclad planarian nervous system **(A)** and protonephridial (flame-bulb type) excretory system **(B)**.

tissue. If two layers are present, the cells of one are aligned longitudinally, those of the other circularly. If a third layer is present, its cells are usually aligned diagonally. Other sets of muscles traverse the parenchyma, and still others operate the pharynx and the copulatory organs. Muscles also occur in the suckers and other types of muscular adhesive organs found in some of the animals. The body musculature permits flatworms to swim or creep by sinuous movements; as noted above, another method of locomotion is ciliary gliding on slime tracks.

The nervous system is fundamentally a nerve net, with ganglia or rings anteriorly and sets of longitudinal cords passing posteriorly. In the more advanced flatworms, such a system has become elaborated further in two ways. First, two of the ventrally located nerve cords and their cross-connections are developed more than the others, resulting in a basic ladder-type system (Fig. 22.4). Secondly, the whole net with its cords and ganglia has sunk deeper into the body, coming to lie in or under the muscle layers or even within the mesenchyme. However, numerous neurosensory cells have been "left behind" in the epidermis. Thus, tangoreceptors and chemoreceptors occur abundantly all over the body surface, and in the head region more elaborate sensory areas have evolved. Statocysts are still common, particularly in marine types, and, excepting only the cave-dwelling land flatworms, virtually all others possess one or more pairs of ocelli, largely of the eyecup type (cf. also Fig. 8.47).

Excretion is accomplished by a protonephridial flame-bulb system. It is embedded in mesenchyme and opens to the exterior via several nephridiopores in various body regions (cf. Fig. 22.4). Also in the mesenchyme is the reproductive system, which has become the most complex part of the flatworm body (Fig. 22.5). Both male and female systems develop

22.5. Turbellarian structure. The reproductive system, as in a triclad planarian. **A,** top view of whole system; **B,** side view of terminal portions.

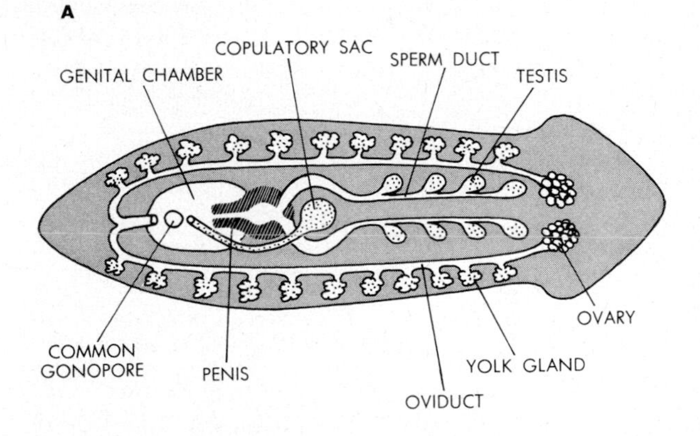

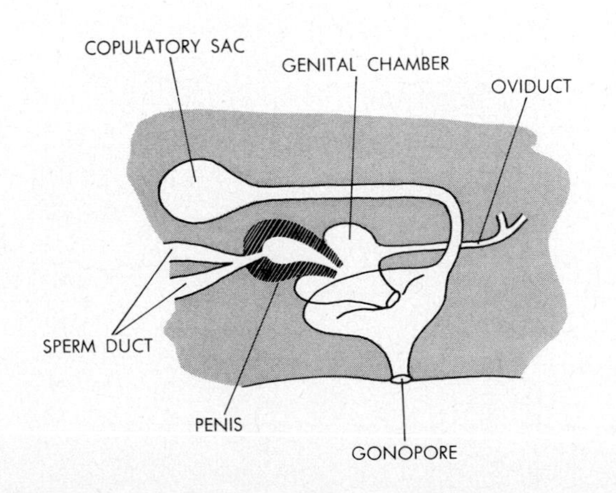

seasonally in each individual. The male system typically contains numerous testes arranged as anteroposterior pairs. From them lead tubules which open into a pair of common sperm ducts. The latter terminate in a single, muscular, midventrally located tube, the copulatory organ. This organ is often armed, i.e., equipped with hooks. The organ projects into a genital chamber, which in turn opens to the outside via a gonopore behind the mouth. In many worms the gonopore opens into the cavity of the pharynx. The female system consists of one to many pairs of ovaries connecting via short ducts with a pair of oviducts. In many cases, numerous yolk glands lie alongside and open into the oviducts. In other cases yolk glands are part of the ovaries. The oviducts join into a single canal which may lead into the genital chamber where the male system also terminates. Alternatively, the oviducts may open into a genital chamber of their own, the latter then exiting to the outside through a separate gonogore. A *copulatory sac* projects into the common genital chamber or into the

22.6. Turbellarian development. A. Left, egg cyst with ectolecithal eggs of directly developing turbellarians; right, later stage, blastomeres more numerous, embedded in yolk from yolk cells. **B.** Spiral cleavage in Acoela, spiralling already beginning at the second cleavage, not the third; the last stage shown is a stereoblastula; after blastopore-mouth formation, the adult condition will be essentially established. **C.** Development of polyclad turbellarians. 1. Postgastrula stage, attained after regular spiral cleavage and epibolic gastrulation; the mouth opening has formed and the alimentary pouch is organizing. 2. Later stage, larval alimentary system fully formed. 3. Mueller's larva, side view, showing the eight ectodermal lobes and the position of internal organs. 4. Mueller's larva, ventral view.

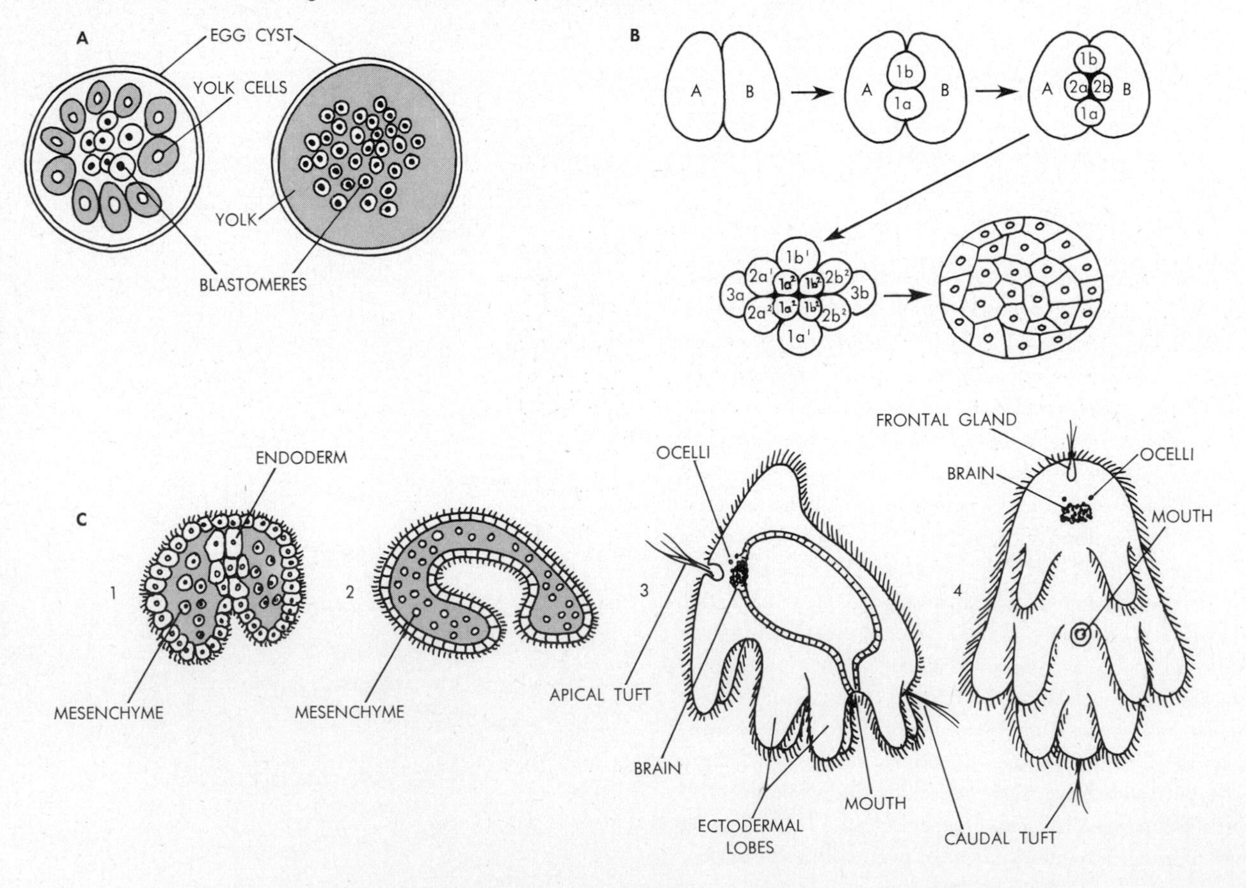

separate chamber of the female system. This sac receives the copulatory organ of the mating partner.

Sex cells arise in the mesenchyme and migrate into the gonads, where they become mature gametes. The eggs of all but the acoels and polyclads are distinguished by being *ectolecithal,* i.e., the yolk is outside an egg proper, within surrounding *yolk cells.* Such cells may be evolutionary derivatives of true egg cells. Yolk cells are formed either in the ovaries or in yolk glands; the latter invest the eggs with yolk cells as the eggs pass from the ovaries through the oviducts. Many flatworms can be self-fertilizing, but mutual internal cross-fertilization takes place in most cases. In some worms mating occurs by hypodermic impregnation, a process in which the armed copulatory organ of one worm hooks anywhere into the epidermis of another and ejects sperms into the parenchyma of this partner. The sperms then migrate toward the eggs into the oviducts. In polyclads and acoels, fertilized eggs (with yolk inside the zygotes) are shed into sea water and they develop there into swimming larvae. In the other orders, and particularly in freshwater and terrestrial types, the oviduct secretes a wall around a zygote and its surrounding yolk cells (Fig. 22.6). Such encysted eggs are released by the worms, and development is then direct; larvae are not formed and young adults eventually hatch from the cysts. Some groups (e.g., the freshwater rhabdocoel *Mesostoma*) are ovoviviparous; zygotes are retained within the female system and the offspring are born as young adults.

In these encysted, directly developing types, the embryonic processes are greatly modified owing to the presence of yolk cells. Blastomeres at first lie free within the cysts as separate cells. Later they aggregate more or less directly into the parts of a future worm. By contrast, well-defined developmental patterns are in evidence among polyclads. Cleavage here is spiral, as in acoels. We may note in this connection that, in polyclads, the spiral shifting of blastomeres begins at the third cleavage, i.e., after a tier of four cells has formed, as is usual for this pattern (cf. Fig. 11.14). In acoels, by contrast, spiralling already begins at the second cleavage (cf. Fig. 22.6). The result is an early embryo composed of several two-celled tiers, not four-celled ones. Gastrulation in polyclads is epibolic, as in acoels, the blastopore becoming the mouth of the adult. The interior of the polyclad gastrula then differentiates into distinct endoderm and mesoderm layers. Mesoderm arises in part from the 4d blastomere formed at earlier stages, in part from ectodermal cells which migrate to the interior and become mesenchyme. The middle body layer later forms the parenchyma and the muscular, excretory, and reproductive systems. The larva of polyclads is a so-called *Mueller's larva,* a swimming, externally ciliated form with eight epidermal lobes (cf. Fig. 22.6). Internally the larva is essentially like the later adult. Metamorphosis involves principally the resorption of the exterior lobes.

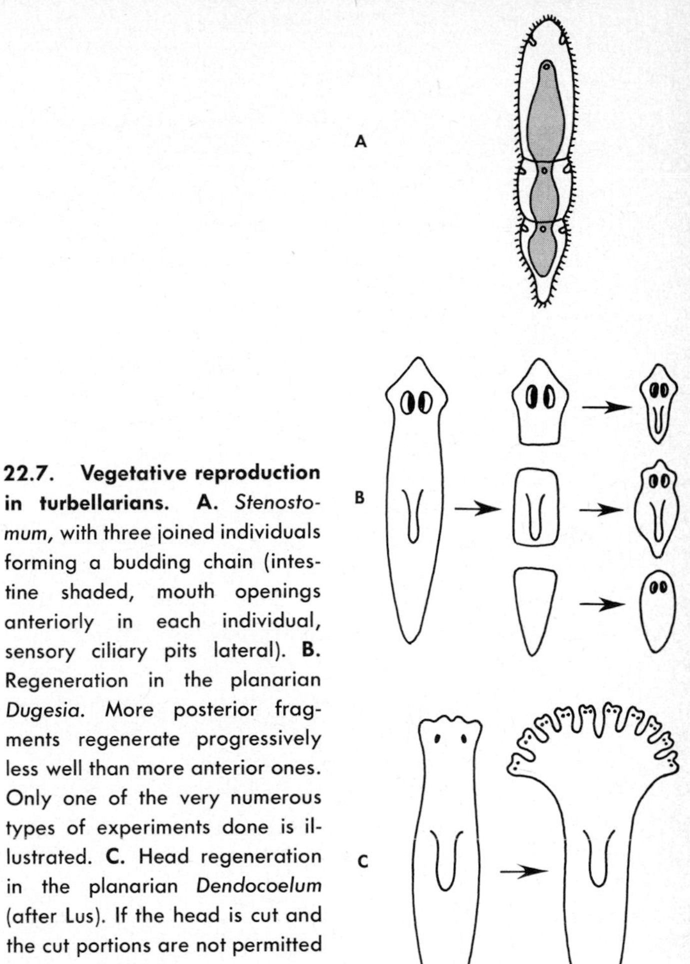

22.7. Vegetative reproduction in turbellarians. A. *Stenostomum,* with three joined individuals forming a budding chain (intestine shaded, mouth openings anteriorly in each individual, sensory ciliary pits lateral). **B.** Regeneration in the planarian *Dugesia.* More posterior fragments regenerate progressively less well than more anterior ones. Only one of the very numerous types of experiments done is illustrated. **C.** Head regeneration in the planarian *Dendocoelum* (after Lus). If the head is cut and the cut portions are not permitted to heal together, each cut part becomes a complete head.

Flatworms are noted for their extensive powers of vegetative reproduction (Fig. 22.7). Many of the animals may fragment spontaneously into an anterior and a posterior part, each part then regenerating the missing portions. The rhabdocoel *Stenostomum,* one of the most common freshwater flatworms, may form several temporary anteroposterior segments, each developing into a complete worm and then separating from the rest. Flatworms such as the triclad planaria *Dugesia* have long been among the most widely used animals in regenera-

PLATYHELMINTHES
TURBELLARIA
ACOELS POLY-CLADS TRI-CLADS ALLOEO-COELS RHABDO-COELS FLUKES TAPE-WORMS
NEMERTINA
PSEUDOCOELOMATE GROUPS
ACOELOID ANCESTORS

22.8. The likely evolutionary interrelations of the turbellarian orders and of acoelomate animals as a whole. Read from bottom up. Cf. Fig. 23.20.

tion research. It has also been found that if such animals are subjected to prolonged starvation they undergo a process of degrowth, i.e., a partial self-destruction. They then use the raw materials so gained to maintain themselves as intact worms, though on a progressively smaller scale. We may note, incidentally, than planarians such as *Dugesia* have been used not only in regeneration studies but in recent years also in learning and behavioral research. It has been found, for example, that a worm can be trained by conditioning to respond in a given manner to a particular stimulus. Such a worm may then be cut into several pieces, and after the pieces have regenerated into whole new worms, it is found that each of the latter still "remembers" the training of the original worm. RNA has been implicated in these knowledge-storage and knowledge-transfer processes.

Among the descendant groups of the early acoels, one presumably gave rise to the Rhabdocoela (Fig. 22.8). Undoubtedly endowed with an extensive adaptive potential, the original rhabdocoels must have elaborated all the characteristic features of flatworms and must have become highly diversified in the process. Indeed, modern rhabdocoels are often classified as three separate orders. The primitive members still contain four pairs of longitudinal nerve cords, but in more advanced types (Fig. 22.9) a ventral pair is already formed into a well-developed ladder-type system. In many rhabdocoels ocelli are lacking, and as in the acoels the female reproductive system is still without ducts. Forms such as *Gyratrix* have acquired a highly differentiated pharynx, viz., an armed, protrusible proboscis of a type very similar to that encountered in the phylum Nemertina. The rhabdocoel flatworms actually are believed to have been ancestral to the nemertines. Moreover, rhabdocoels have probably also been the evolutionary source of all the parasitic flatworms, i.e., the flukes and tapeworms. Some important differences notwithstanding, certain flukes, for example, may be regarded as "parasitic rhabdocoels" (cf. below).

A second group of early acoels probably gave rise independ-

22.9. First three diagrams on left, representative rhabdocoel genera; two on right, representative alloeocoel genera (*Adapted from Hyman.*) Note the straight anteroposterior digestive pouches, those of alloeocoels such as *Bothrioplana* with lateral diverticula. In *Prorhynchus*, a single common anterior channel serves for feeding and sperm release.

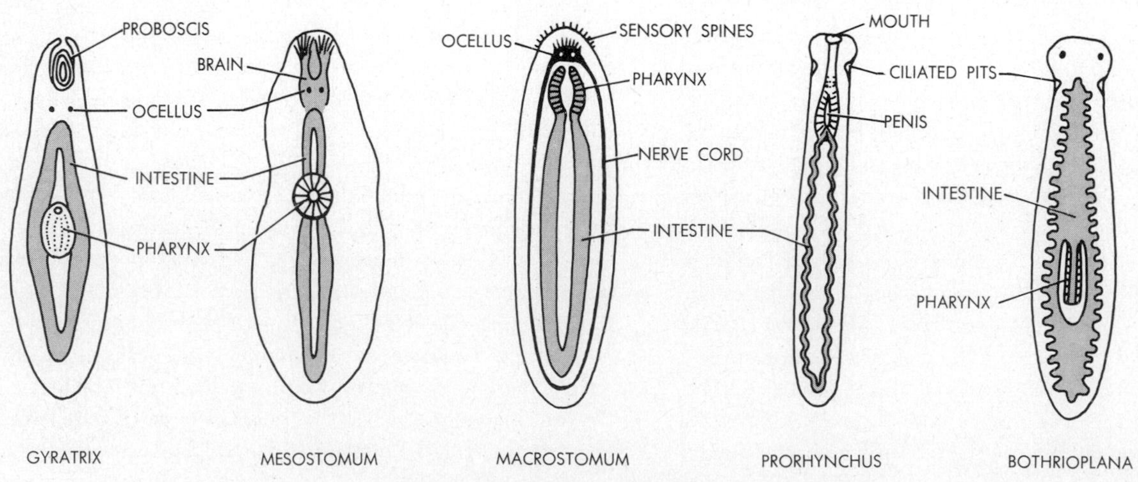

A

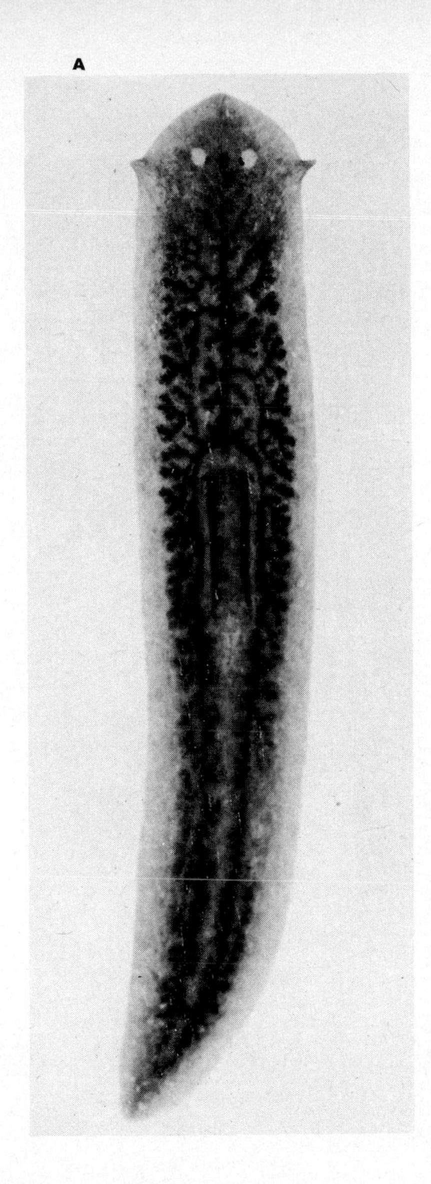

B

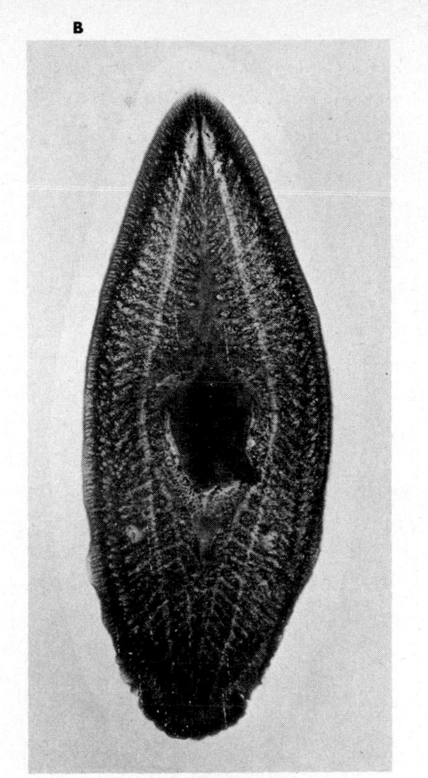

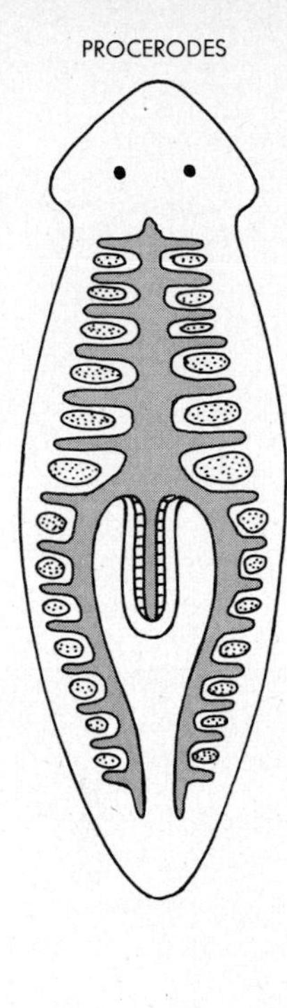

22.10. Triclads. Photos: A. The free-living planarian *Dugesia.* Note eyes, pointed lateral lobes (auricles) at level of eyes, pharynx in middle of underside, and darkly stained branches of digestive tract. **B.** *Bdelloura,* a commensalistic marine triclad. Note nerve cords (light), pharynx, and faintly visible digestive tract with diverticula. The posterior part of the worm is an adhesive pad. **Diagram:** *Procerodes,* a triclad with regularly alternating pairs of digestive diverticula and gonads. The most anterior pair of gonads represents ovaries, all others, testes. (*After Hyman.*)

ently to the Alloeocoela. These flatworms are distinguished by a straight intestine which is usually extended into numerous lateral pouches. Called *diverticula,* the pouches increase the effective digestive and absorptive area. Diverticula also characterize the intestine of the Tricladida, an order presumably derived directly from the Alloeocoela. In the triclads, the intestine divides into three main branches, one directed anteriorly, two posteriorly (Fig. 22.10). The most familiar triclads—and the most familiar flatworms generally—are the freshwater *planarians,* of which *Dugesia* (= *Euplanaria*) is the best known genus. The many species of these animals have a pair of well-developed eyes and, laterally at the level of the eyes, pointed sensory lobes called *auricles.* Triclads also include marine types such as *Procerodes,* a genus characterized by a very regular anteroposterior alignment of gonads and intestinal diverticula, in an arrangement suggestive of segmentation. Another marine form is *Bdelloura,* a commensal on horseshoe crabs. The terrestrial triclads are on the whole larger than their aquatic relatives. For example, *Bipalium* may reach lengths of about 1 ft.

Acoels most probably gave rise directly to the marine Polycladida (Fig. 22.11). These are comparatively large, oval, platelike animals in which the intestine is divided into numerous branches. Exteriorly, the worms usually possess a pair

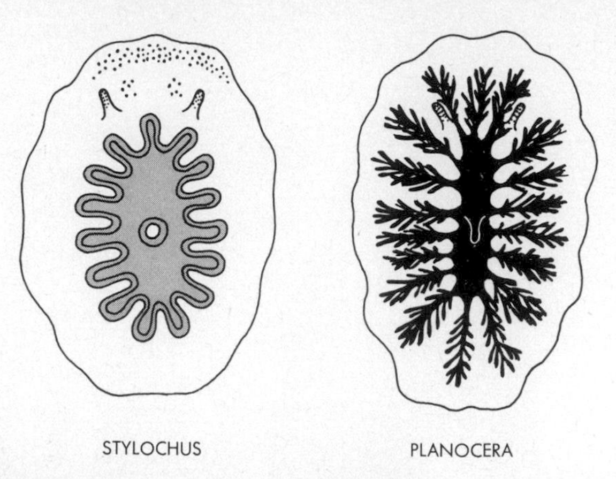

22.11. Polyclads. Diagrams: two widely distributed and common polyclads. The many-branched digestive sacs are shaded. The mouth is at midbody. Note the tentacles anteriorly. In *Stylochus,* the dots on the tentacles and the front region of the body represent clusters of ocelli. (*Adapted from Hyman.*) **Photos: A,** *Prosteceraeus;* the anterior end of this animal is near top of photo; **B,** Mueller's larva, characteristic in polyclad development.

A

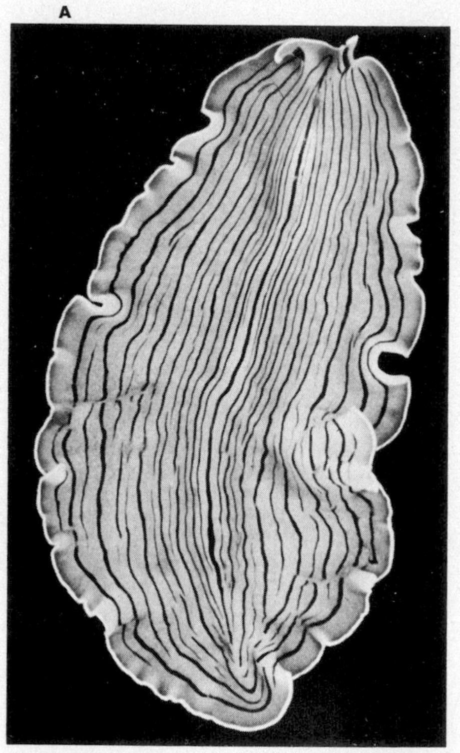

B

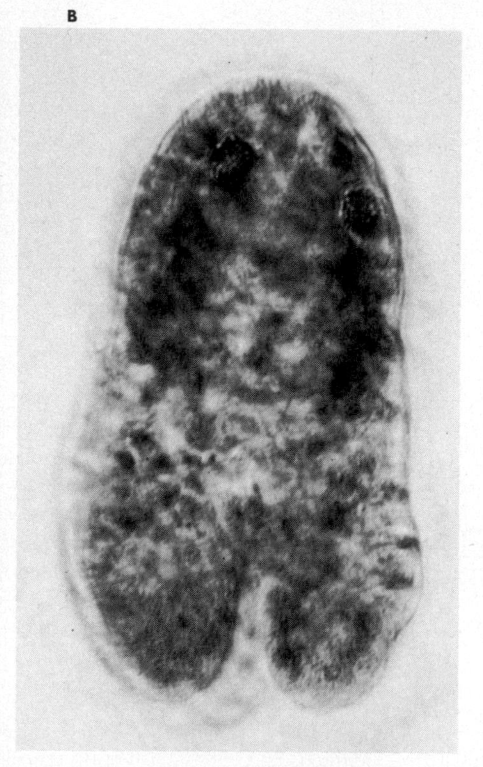

of short tentacles equipped with ocelli. Additional groups of ocelli may be present in other body regions as well. Posterior and ventral adhesive organs are also common.

Class Trematoda

Flukes; holotrophic parasites; epidermis absent; exterior with cuticle and one or more suckers.

Subclass Digenea: endoparasitic, with one or more (up to three) intermediate hosts. *Dicrocoelium, Schistosoma, Clonorchis*

Subclass Aspidogastrea: mainly endoparasitic, with and without intermediate hosts. *Aspidogaster*

Subclass Monogenea: mainly ectoparasitic, without intermediate hosts. *Gyrodactylus*

The most notable structural differences between the flukes and the rhabdocoels are epidermal. First, flukes lack an epidermis entirely. Present instead is a tough *cuticle,* secreted by cells in the mesenchyme (Fig. 22.12). The cuticle is horny, of a scleroprotein composition. Secondly, sense organs and sensory cells are highly reduced in number, and eyes or statocysts are largely absent. Thirdly, flukes possess one or more well-developed muscular *suckers* or equivalent adhesive organs. In the Digenea, one sucker is usually anterior, near or surrounding the mouth, the other is generally midventral. In the Monogenea, the main sucker (or group of suckers) is located posteriorly and is often equipped with hooks or claws. Anteriorly, adhesive disks or pits may be present but not usually a sucker. The Aspidogastrea possess a ventral plate which is subdivided into rows of suckers. In all groups, the suckers operate on the principle of suction cups. Note that the cuticle, the lack of sensory receptors, and the conspicuous adhesive devices are all specific adaptations to the parasitic mode of life.

22.12. Flukes. A. Diagram of a surface section of a fluke, showing the tissues and the absence of an epidermal layer. **B.** Outline representations of the basic structure of each of the three orders of trematodes; Digenea represented by *Dicrocoelium,* Monogenea by *Polystomoides,* Aspidogastrea by *Cotylogaster.* Cf. also Figs. 22.13 through 22.16.

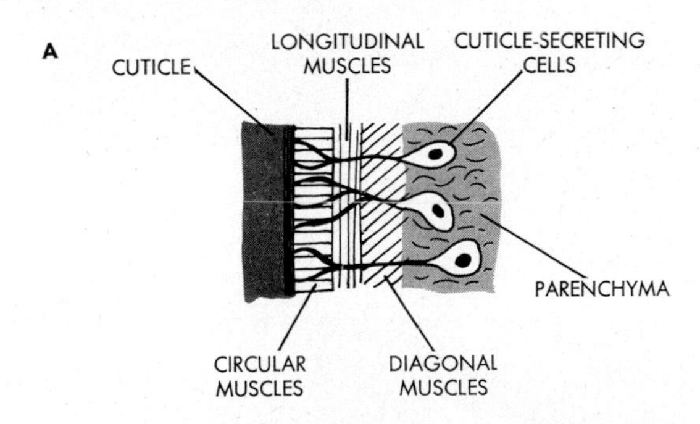

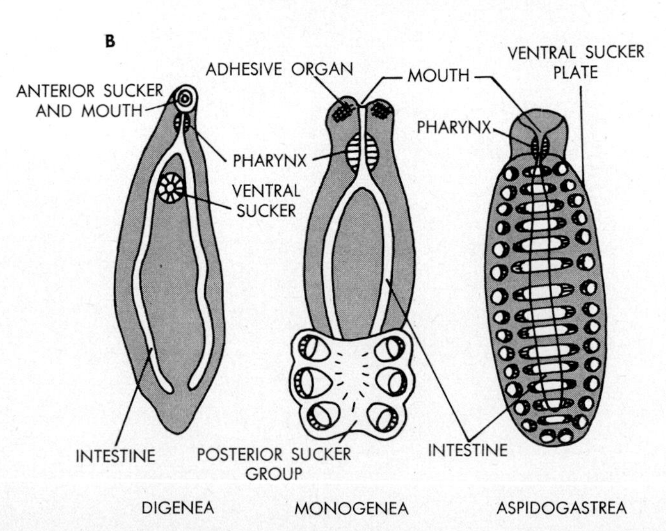

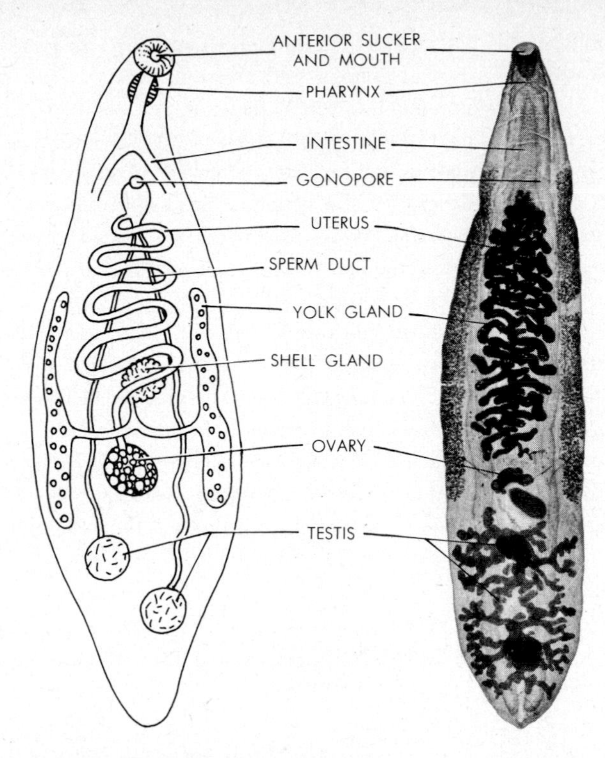

22.13. The trematode reproductive system. The diagram gives a simplified outline representation. Note that in many cases a shell gland is not present and that the testes may be large and branched, as in the photo, rather than globular as in the diagram.

Interiorly, the structure of flukes is essentially rhabdocoel-like. The intestine usually divides into two blind-ended branches (though in some forms each branch leads to the exterior via a posterior anal pore). The nervous, muscular, and excretory systems correspond to those of the free-living flatworms. Many flukes possess networks of channels in the parenchyma; called *lymphatic channels,* they are thought to facilitate the circulation of the body fluids. The reproductive system likewise has a rhabdocoel pattern (Fig. 22.13). A pair of posteriorly located testes connects with sperm ducts which lead via copulatory organ and genital chamber to a gonopore. The latter is often situated behind the ventral sucker. The female system typically contains a single ovary, numerous yolk glands, and an oviduct which is a wide, much-coiled tube functioning as a *uterus;* eggs may be retained in it for considerable periods. The uterus exits through either a separate gonopore or one shared in common with the male system.

With some exceptions in which the sexes are separate (notably the blood flukes of the genus *Schistosoma*), flukes are typically hermaphroditic. Self-fertilization may occur on occasion, but cross-fertilization is the general rule. Copulation takes place as in Turbellaria, and fertilized eggs surrounded by yolk cells are encysted in the uterus. The subsequent history of the eggs differs for the three subclasses.

In the Digenea, the so-called "digenetic" trematodes, the life cycle typically includes three or four successive larval stages and up to four different hosts. The main or final hosts are vertebrates, and the parasites live in the intestine, in blood, and in the body cavities. For example, *Schistosoma japonicum* inhabits the intestinal blood vessels of man. Released egg cysts accumulate till the blood vessels rupture, and the cysts then pass into the cavity of the gut from which they are discharged with the feces. The egg cysts of *Clonorchis sinensis*, the Chinese liver fluke, reach the exterior via bile duct and intestine (Fig. 22.14; cf. also Fig. 17.20).

Egg cysts released by adult flukes develop into free-swimming, ciliated *miracidium* larvae (Fig. 22.15). These are miniature flatworms with nervous, muscular, and excretory systems. They also possess balls of *propagatory cells* in the interior which will give rise to the next larval stage. A miracidium must find an intermediate host, namely, a snail, within several hours. If it does, it enters the body of this host, sheds its cilia, and transforms into a *sporocyst*. The propagatory balls within it then proliferate and produce either *redia* larvae or daughter sporocysts. Rediae arise in *Clonorchis*, for example. Such larvae feed on snail tissue, and they subsequently produce either *cercaria* larvae or another generation of rediae,

22.14. Photos: A, redia larva of a fluke, with cercaria larvae present inside it (dark bodies in curved end); **B,** adult male blood fluke *Schistosoma japonicum*. **Diagram:** *Schistosoma*, male and female copulating. (*After Looss.*)

A

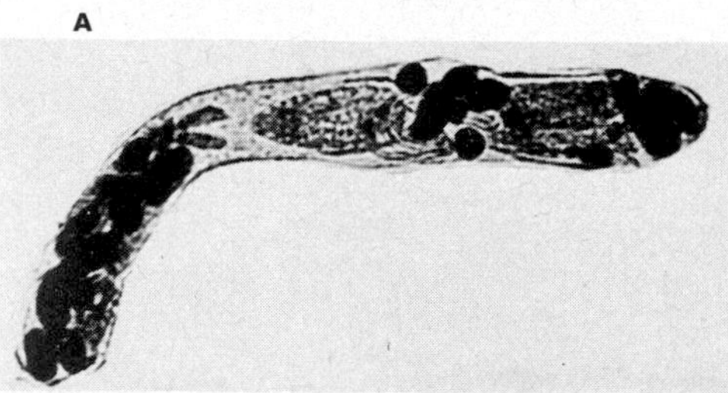

A, *courtesy Carolina Biological Supply Company.*

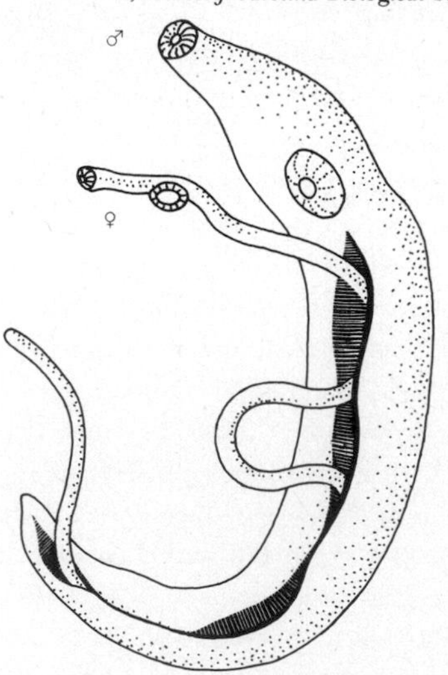

B

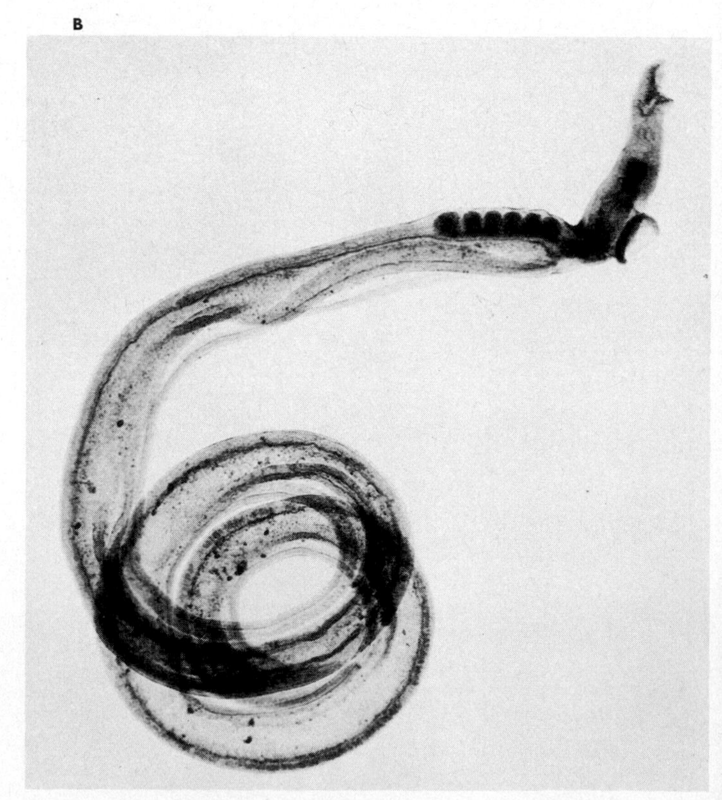

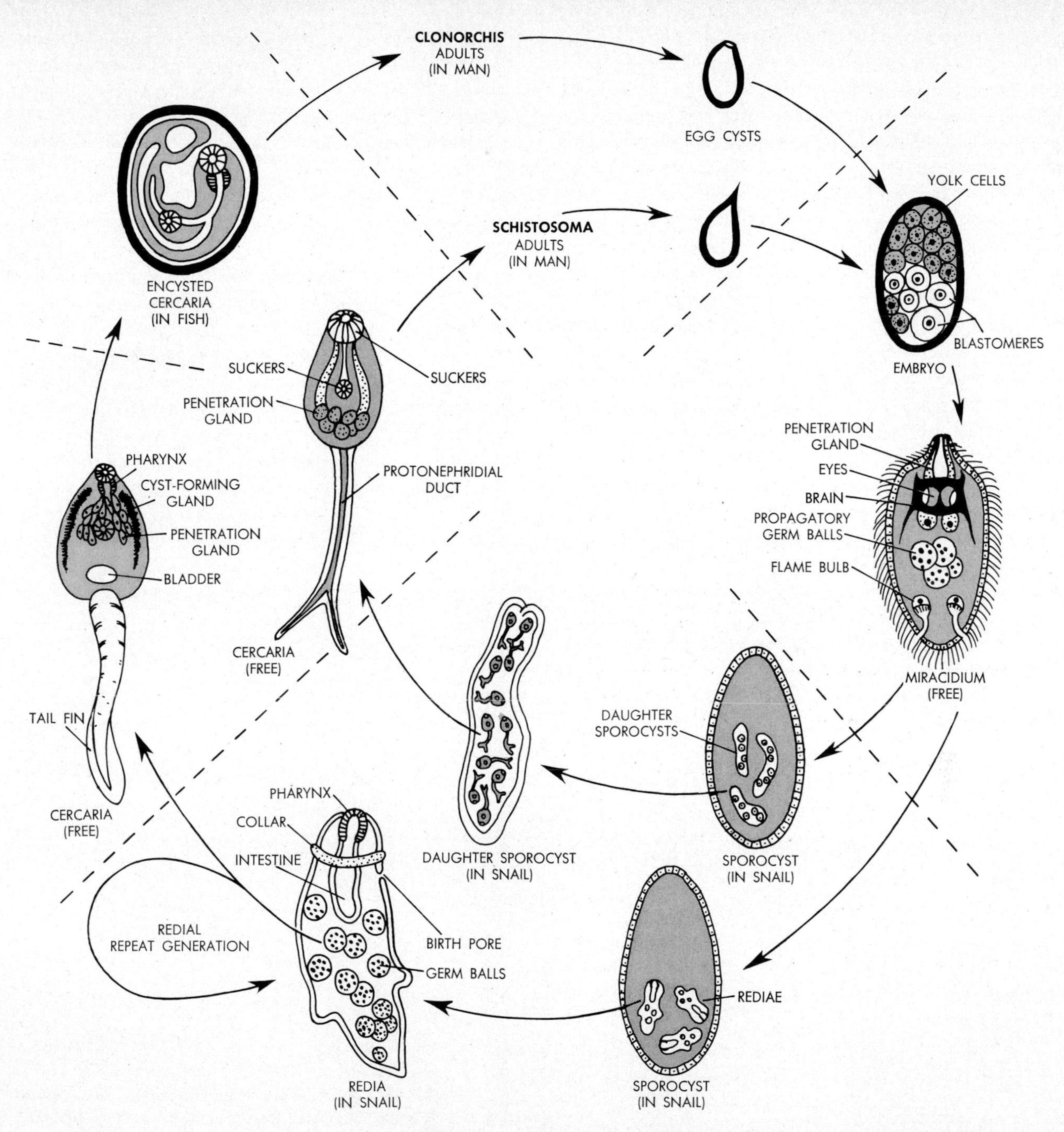

22.15. The life cycles of the liver fluke *Clonorchis sinensis* (outer pathway) and *Schistosoma japonicum* (inner pathway). Adults are illustrated in Fig. 22.14. Broken lines mark the extent of developmental phases passed in given environments, as indicated.

which then form cercariae. Rediae do not develop in *Schistosoma*. Instead, the sporocysts give rise to daughter sporocysts and these produce cercariae directly. Thus, an original sporocyst produces either numerous rediae each of which then forms numerous cercariae, or numerous sporocysts which give rise to cercariae without intervening redial stages. In all cases, cercariae are the final larval forms in snails.

Cercariae are tailed, with typical trematode structure internally. After leaving a host snail, a cercaria swims about as a free-living larva. It must soon enter either a final host or another intermediate host. *Schistosoma* cercariae enter their final host, man, through his skin, migrate via the circulation to intestinal blood vessels, and become adults there. *Clonorchis* cercariae enter second intermediate hosts, viz., fishes, and encyst in fish muscle. Raw or partially cooked fish must then be eaten by man before the encysted flukes become adults (cf. Chap. 17).

The Aspidogastrea parasitize aquatic invertebrates such as clams and snails and vertebrates such as fish. Egg cysts are swallowed by mollusks and develop there into larvae resembling adults. Eventually the larvae transform into actual adults right within their molluscan hosts (Fig. 22.16). In such life cycles, therefore, intermediate hosts are not required. In some cases, however, the mollusks do come to serve accidentally as intermediate hosts: the infected mollusks may be eaten by fish and the latter then acquire the parasites and become final hosts in this manner.

The life cycles of the Monogenea, or "monogenetic" trematodes, are always without intermediate hosts. These worms are largely ectoparasites, usually on aquatic and amphibious vertebrates. For example, *Gyrodactylus* lives on fish gills. The eggs of this fluke become larvae and later adults which attach directly to new fish hosts (cf. Fig. 22.16). An interesting phenomenon in this genus is the continuous development of embryos: an early embryo forms a second embryo inside it, the second one forms a third one inside it, and often a fourth generation embryo may form inside the third. In effect, therefore, any given adult carries within it several successive offspring generations.

Because of their generally greater harmfulness to their hosts, their endoparasitic nature, and their complicated life cycles, the Digenea are usually considered to represent the most primitive flukes. Note that the multiple larval stages formed by vegetative reproduction are instances of *internal budding* and are direct adaptations to the parasitic way of life. Recall also the often presumed evolutionary connection between trematodes, Digenea particularly, and the Mesozoa (cf. Chap. 20).

Class Cestoidea

Tapeworms; fluid-feeding parasites; epidermis absent, exterior with cuticle; without alimentary system.

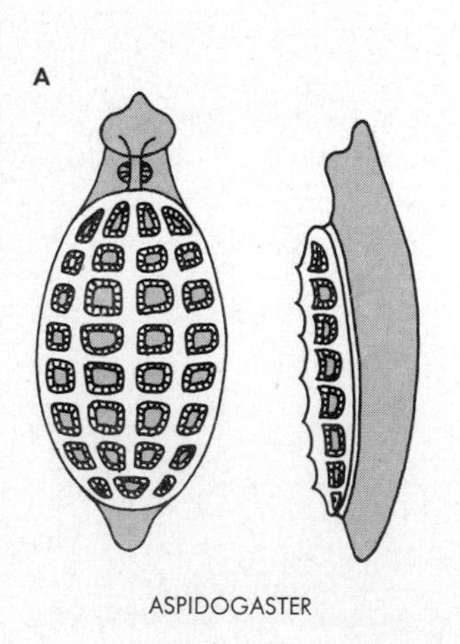

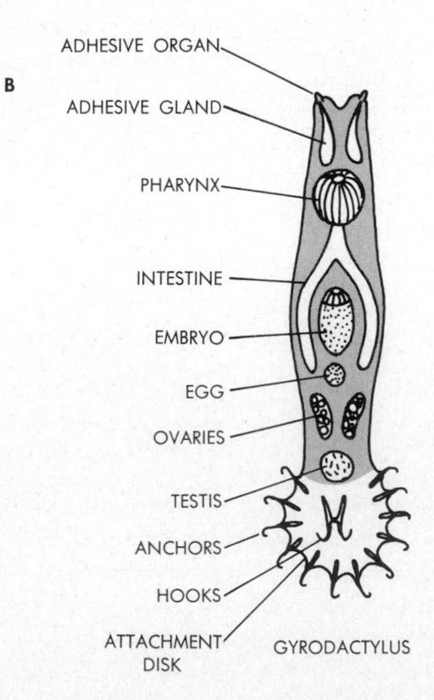

22.16. A. *Aspidogaster*, representing the Aspidogastrea, possesses a ventral sucker plate subdivided into four longitudinal columns and several transverse rows of individual suckers; ventral view on left, side view on right. (*Adapted from Monticelli.*) **B.** *Gyrodactylus*, a monogenetic trematode. (*Adapted from Mueller, VanCleave.*) Note the developing embryo within the adult.

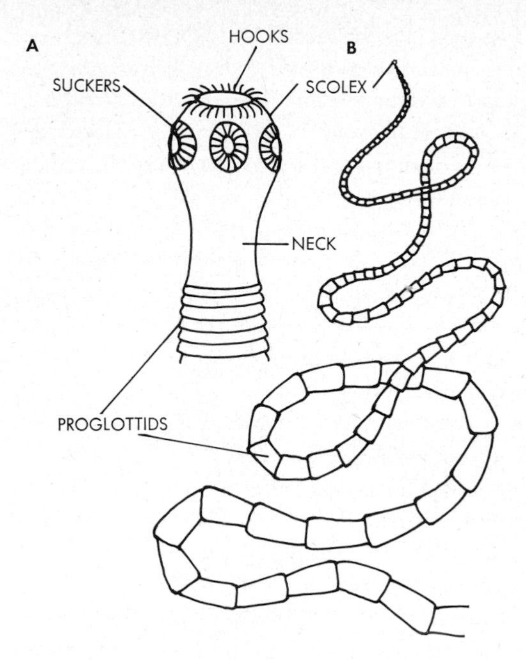

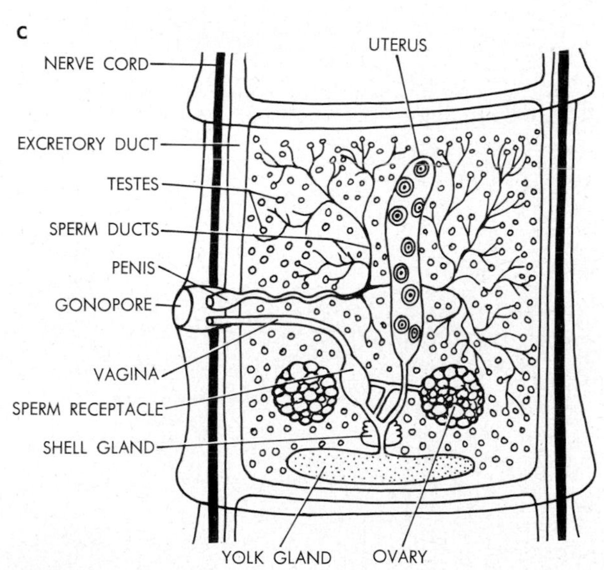

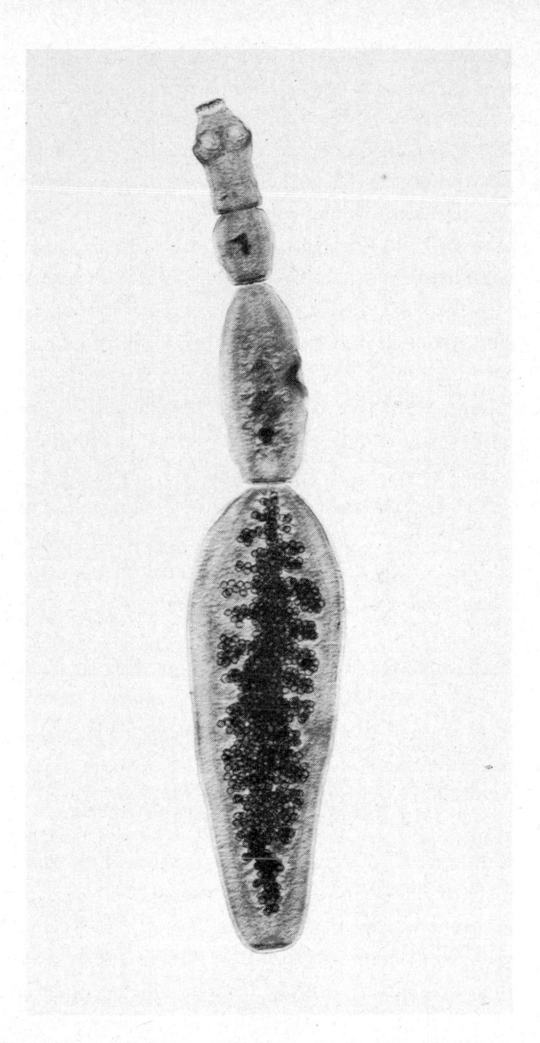

22.17. Tapeworm structure. Diagrams: A. Head (scolex) of adult. **B.** Whole worm, showing change of proportions in length and breadth of proglottids with increasing distance from scolex. **C.** Organs of reproductive systems in a proglottid. Cf. Fig. 17.15. **Photo:** the minute tapeworm *Echinococcus*, parasitic in dogs. The worms consist of only a few proglottids. In the most posterior one, note ripe uterus (dark).

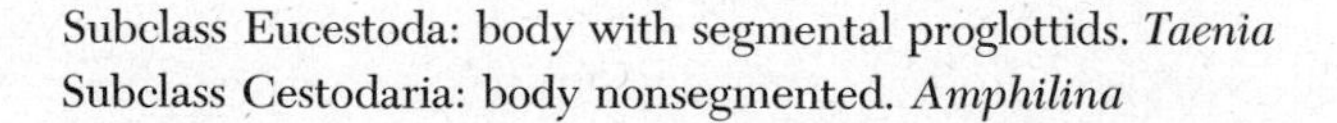

Subclass Eucestoda: body with segmental proglottids. *Taenia*
Subclass Cestodaria: body nonsegmented. *Amphilina*

Tapeworms are believed to have evolved from early rhabdocoel stocks as a specialized line of fish parasites. Later the group evidently diversified and came to infect all other vertebrates as well. The cestodes are intestinal parasites completely lacking an alimentary system; they absorb molecular nutrients directly through the body surface. This surface is characterized as in flukes by a secreted cuticle, underlain by muscles. The interior is filled solidly with parenchyma. At the head end are reduced brain ganglia, and a longitudinal nerve cord runs along each edge of the body. A simplified protonephridial system is present, with longitudinal ducts paralleling the nerve cords (Fig. 22.17).

The majority of tapeworms are members of the subclass

Eucestoda. In these animals the body consists of a head, or *scolex,* equipped with six attaching hooks and several suckers; a short *neck;* and a series of segments called *proglottids.* The latter are formed continuously behind the neck, older proglottids thereby being pushed back to progressively more posterior positions. Young segments are broader than long, older ones become gradually longer than broad. In each, a cross-connection joins the lateral nerve cord and a cross-canal unites the excretory ducts; the longitudinal cords and ducts are continuous throughout the worm. Each proglottid contains a complete male and female reproductive system. Testes may number from 1 to 1,000, all discharging into a common sperm duct which exits via copulatory organ, genital chamber, and gonopore. The latter is situated laterally along a proglottid. A pair of ovaries leads via an oviduct and a vagina to the genital chamber and the gonopore which are also part of the male system. Along its course the oviduct connects with the ducts of a large yolk gland, a shell gland, and a blind-ended pouch, the uterus (cf. Fig. 22.17).

The male system matures earlier than the female system. Consequently, proglottids near the middle of a worm contain ripe sperms, those closer to the posterior end contain ripe eggs. Fertilization is internal and may be accomplished in four different ways: by hypodermic impregnation (in some cases); by self-fertilization within a single proglottid; by cross-fertilization between two proglottids of the same worm (the most common process); or by cross-fertilization between two different worms. Zygotes together with yolk cells become encysted by secretions from the shell gland, and the egg cysts then accumulate within the uterus. The latter enlarges and branches extensively as it fills with more and more eggs. Concurrently, all other parts of the reproductive systems degenerate. Ripe proglottids of this type occur at the most posterior regions of the tapeworm. Such segments break off and are discharged with the feces of the host.

The subsequent life history of one of the tapeworms of man, the beef tapeworm *Taenia saginata,* has been outlined in Chap. 17 (cf. Fig. 17.19). In this case cattle serves as intermediate host. Another species, *T. solium,* has a similar life cycle, but its intermediate host is the pig; man is infected with the encysted embryos by eating raw or undercooked pork. These species are among many tapeworms with only one intermediate host, but many others have two. For example, the fish tapeworm of man, *Diphyllobothrium latum* (over 50 ft long and by far the most harmful of the tapeworms of man) must have its ripe proglottids deposited in water, where the embryos hatch into larvae. These must then be eaten by copepods. To complete the life cycle the infected copepods subsequently must be eaten by fish, and raw or undercooked fish finally must be eaten by man. Some tapeworms use given mammals as intermediate hosts and other mammals as final hosts. The encysted embryos of such worms occasionally enter man by accident, when man eats the undercooked meat of the intermediate hosts. The embryos then may cause serious damage, particularly if they lodge in vital organs other than muscles. It happens that such tapeworm embryos never complete their life cycle, for man is not normally eaten by the usual final host; if he were, he would long since have become a regular secondary intermediate host. Multiple intermediate-host stages may actually have evolved from such originally accidental infections of given animals.

Note that proglottid manufacture is a form of vegetative reproduction, similar in most respects to the strobilation process characteristic of the scyphistoma larvae of scyphozoan coelenterates.

The Cestodaria are believed to be neotenous offshoots of the Eucestoda, i.e., animals which retain larval characteristics but become sexually mature (Fig. 22.18). Cestodaria retain an

22.18. Cestodaria. A. *Amphilina. (Adapted from Hein.)* **B.** *Gyrocotyle. (Adapted from Lynch.)*

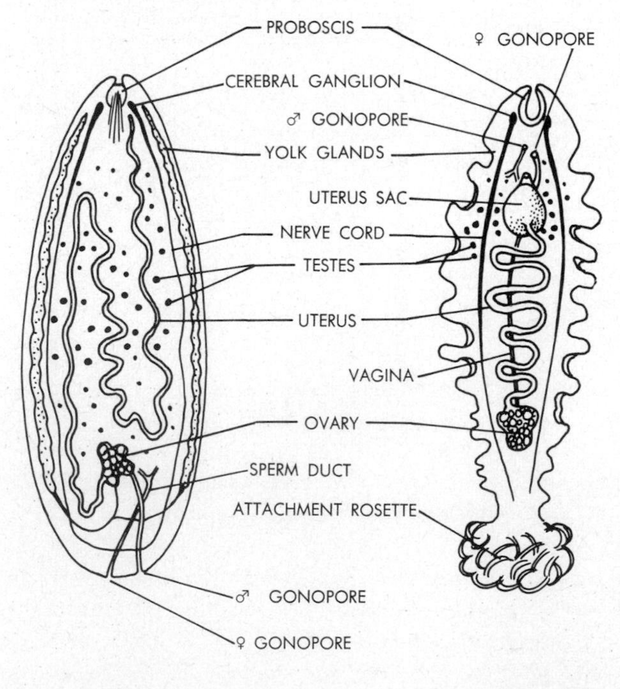

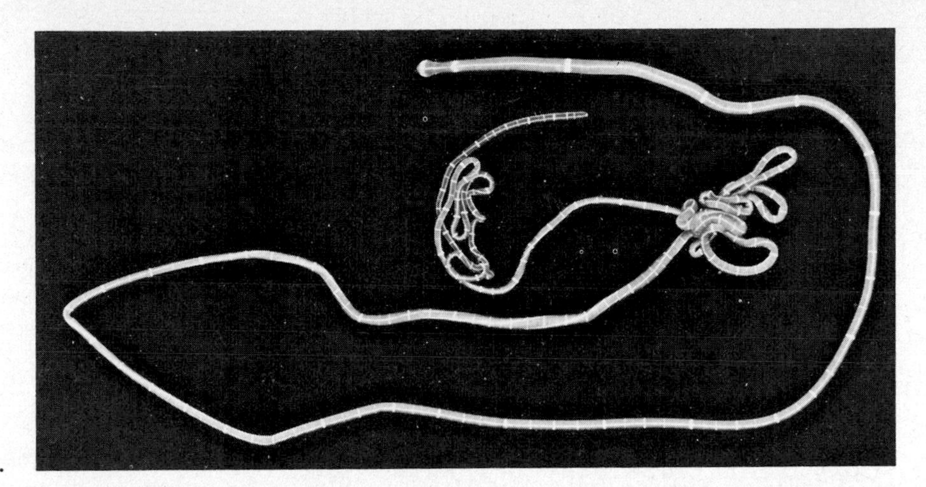

22.19. The nemertine *Tubulanus*.

unsegmented body, like eucestode larvae. These worms infect the intestine and also the coelomic cavities of primitive fishes, attaching themselves by means of posterior holdfast organs. Life histories are known only for a few members of the group. In these, the larvae mature in invertebrate intermediate hosts. The latter are then eaten by fishes and the larvae become adults in these final hosts.

PHYLUM NEMERTINA (RHYNCHOCOELA)

Ribbon worms, proboscis worms; alimentary system with separate mouth and anus; circulatory system present; eversible proboscis within rhynchocoel; sexes separate.

Class ANOPLA: mouth posterior to brain; nervous system subepidermal. *Tubulanus, Cerebratulus, Lineus*

Class ENOPLA: mouth anterior to brain; nervous system submuscular. *Prostoma, Malacobdella*

The ribbon worms encompass some 600 species of largely marine animals. A few live in fresh water (e.g., *Prostoma*), and some are terrestrial. Though certain of the marine forms are pelagic, most are bottom dwellers in sand, mud, or under stones along the coasts of North Temperate regions. None of the nemertines is parasitic, but some are commensals. The worms range in length from less than an inch to under 2 ft. However, one species of *Lineus* may attain a length of some 100 ft. Many types are brightly pigmented, red, green, brown, and other colors often forming striped patterns (Fig. 22.19).

The body of the worms is flattened dorsoventrally to some extent and the exterior is covered with a ciliated, highly glandular epidermis. As in flatworms, many of the epidermal cells contain rhabdites. Some nemertines secrete mucus tubes around themselves and live or copulate within them. The epidermis is underlain by a *dermis,* which may be gelatinous and contain few fibers (e.g., *Tubulanus*) or may be a fibrous connective tissue (e.g., *Cerebratulus*). Epidermis and dermis together form a well-defined skin. A definite body-wall musculature is present underneath, composed of two but more often of three or more muscle layers, with cells aligned circularly, diagonally, and longitudinally. These muscles may extend inward right to the alimentary tract or they may be separated from the tract by a thin layer of mesenchyme. The latter is a connective tissue which fills all spaces not occupied by other tissues and organs. Such spaces are not very extensive; nemertines are compactly and sturdily built animals (Fig. 22.20).

This phylum is the first in which we encounter an alimentary system exhibiting a one-way tube design (Fig. 22.21). Mouth, foregut, intestine, and anus are the principal components. The tract is straight and equipped with diverticula in types such as *Prostoma;* it forms many loops and is without diverticula in *Malacobdella.* The wall of the system is a single layer of ciliated cells. Digestion occurs extracellularly, and food is moved through the tract by the muscular contractions of the body wall. The nemertines are also the first phylum in which we encounter a circulatory system with blood. As pointed out in Chap. 9, the system of these worms is closed. It lies in the mesenchyme and consists basically of two longitudinal lateral vessels joined anteriorly and posteriorly (cf. Fig. 22.21). In many cases a longitudinal dorsal vessel or cross-

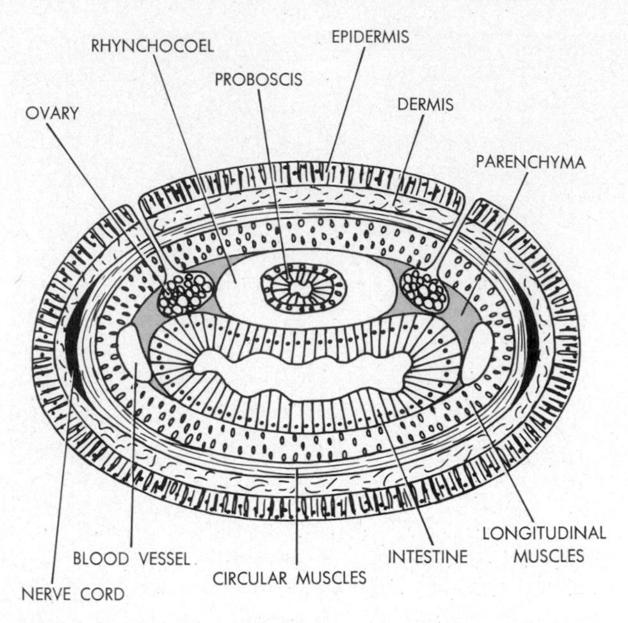

22.20. Cross section through the nemertine body. Note the sparseness of parenchymal spaces and the glandular epidermis. The proboscis is lined by an epithelium which overlies several radial and longitudinal muscle layers.

connecting transverse vessels or both are also present. The blood of most nemertines is colorless but in some forms it is pigmented (red, orange, yellow, or green); the red pigment is hemoglobin, contained within the blood cells.

Excretion is accomplished by a protonephridial flame-bulb system. Its principal ducts lie roughly parallel to the lateral blood vessels. The flame bulbs tend to abut against the walls of the vessels, an arrangement which undoubtedly facilitates nephric filtration (cf. Figs. 22.20 and 22.21). One or more pairs of nephridiopores open to the outside along the lateral margins of the body. The nervous system consists principally of a nerve net, thickened anteriorly into a pair of brain ganglia and in the rest of the body into a pair of longitudinal lateral nerve cords. Numerous accessory thickenings form ganglia and secondary cords connected with the main cords. Tango-receptor and chemoreceptor cells may occur all over the body, but specialized sense organs are found chiefly in the anterior regions. Thus, many nemertines possess paired anterior grooves, pits, or canals (*frontal organs*), lined with ciliated cells and probably serving chemoreceptor functions. Ocelli of the eyecup type may be limited to a single pair in front of the brain ganglia. In some cases, however, up to 200 or more ocelli may be distributed in clusters over the front end of the body. Some nemertines possess paired statocysts directly dorsal to the brain ganglia. A notable phylum characteristic is a pair of *cerebral organs*, anterolateral flaskshaped invaginations lined with glandular cells. Surrounded by neurons which lie close to or are part of the brain ganglia, the organs are believed to serve chemoreceptor and possibly also endocrine functions (cf. Fig. 22.21).

The most distinctive trait of a ribbon worm is its *proboscis*, an anterior muscular tube which in the rest state is invaginated into the body dorsal to the intestine (Fig. 22.22). Except for its basal portion, the proboscis lies free in a closed, fluid-filled cavity, the *rhynchocoel*. The latter may be short or may extend posteriorly almost to the anus. Within it, the proboscis is often looped and this organ may therefore be longer than the worm itself. The blind-ended tip of the proboscis is at-

22.21. Nemertine structure, schematic. A, alimentary, circulatory, and excretory systems; **B,** nervous, sensory, reproductive, and proboscis systems.

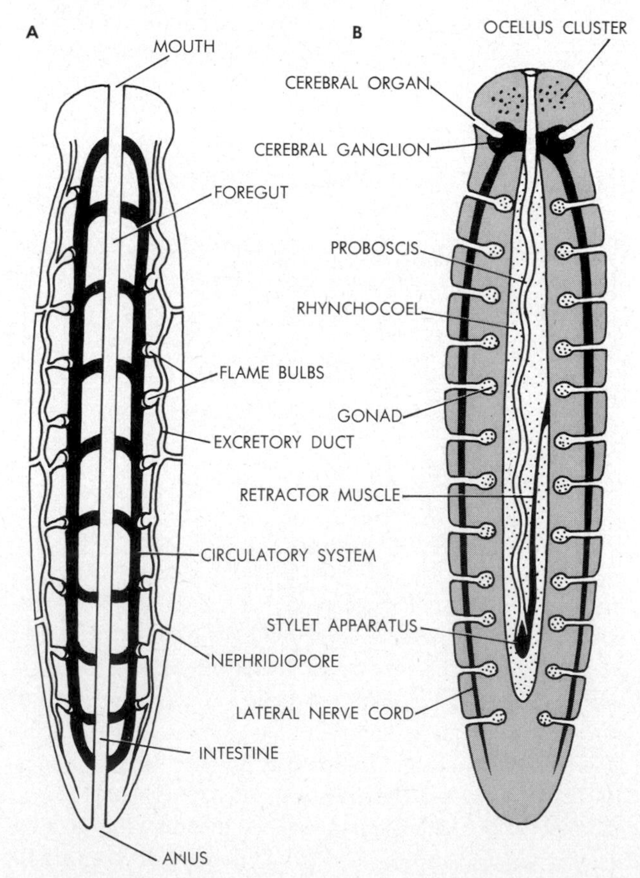

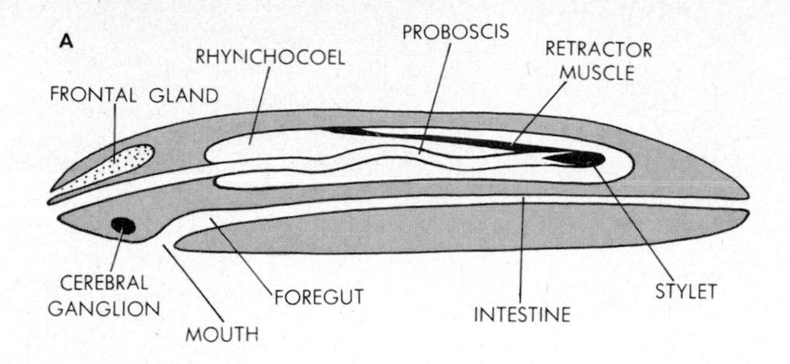

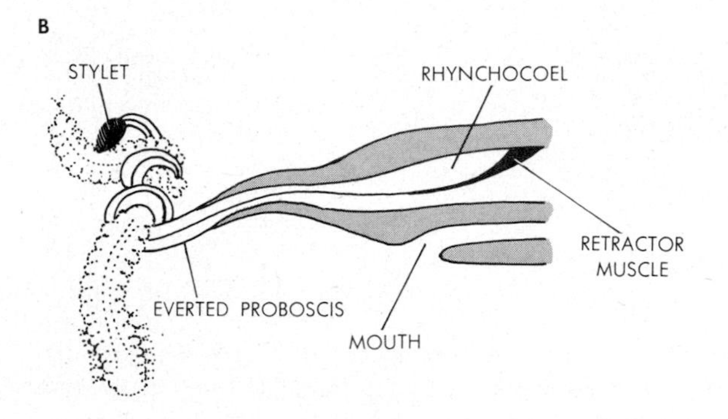

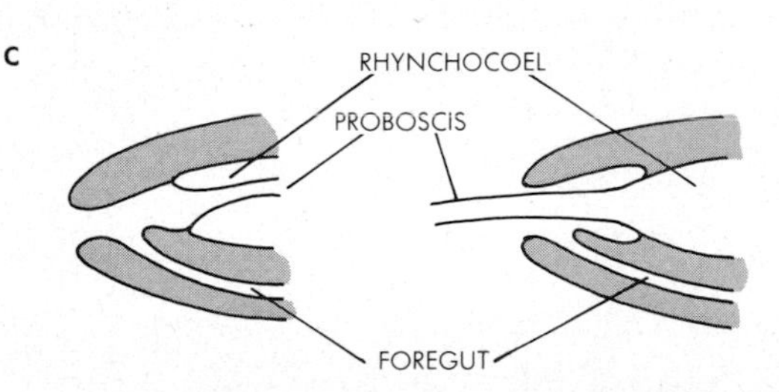

22.22. The proboscis apparatus, schematic. **A,** position of rhynchocoel and proboscis in rest condition; **B,** everted proboscis, entangling and capturing a clam worm; **C,** rest and everted positions of the proboscis of forms such as *Malacobdella,* in which the foregut opens into the base of the proboscis tube and a separate mouth is absent.

tached by a retractor muscle to the wall of the rhynchocoel. Muscular contraction of this wall exerts pressure on the fluid within the cavity, resulting in an explosive eversion of the proboscis to the exterior. The animal uses this organ in locomotion, in burrowing, and principally in trapping annelids and other worms, food animals which become encoiled by the proboscis. The surface of the organ also secretes mucus, an additional aid in catching and holding prey. In many cases, moreover, the tip of the proboscis is armed with a *stylet,* a sharp-pointed spike, and with glands which secrete poison into a wound made by the stylet. One nemertine is known in which the proboscis branches in bushlike fashion from the base forward, ending in up to 32 separate tips. In some ribbon worms (e.g., *Malacobdella*) a primary mouth is absent. Instead, the foregut and the proboscis base open anteriorly through a common single canal (cf. Fig. 22.22). The whole proboscis apparatus and nemertine structure generally give evidence of a fairly close evolutionary relation between ribbon worms and rhabdocoel flatworms.

Like flatworms, also, the nemertines are capable of reproducing vegetatively by fragmentation. When irritated unduly, forms like *Lineus* are known to fragment spontaneously into a dozen or more pieces. Each piece then regenerates into a new, smaller worm, like the multiplying brooms of the sorcerer's apprentice. Some nemertines may shoot out their probosces so forcefully that these organs break off, yet new ones regenerate subsequently.

Most ribbon worms are separately sexed, although hermaphroditism is common in freshwater and terrestrial groups. Gonads arise in mesenchyme as anteroposterior pairs (cf. Fig. 22.21). During the breeding season a short duct develops from each gonad and leads to separate lateral gonopore. Most species are externally fertilizing. In some fertilization is internal, but in most of these the zygotes nevertheless develop externally. In a few groups (exemplified by *Prostoma*) development occurs internally and directly; larvae are absent and young adults form ovoviviparously.

The cleavage pattern is typically spiral, the blastomeres are mosaic, and the blastula is a hollow coeloblastula (Fig. 22.23). Gastrulation takes place by embolic invagination in most cases, by ingression or by epiboly in the others. The mesoderm is derived partly from ectoderm, partly from endoderm, as in flatworms.

Larvae are of two types. The more common *pilidium* larva (e.g., as in *Cerebratulus*) is a helmet-shaped ciliated form, with an aboral *apical tuft* of long cilia and two epidermal *oral lobes* on each side of the mouth. The alimentary system terminates interiorly as a blind pouch, the *stomach.* The *Desor's larva* (e.g., as in *Lineus*) remains within its egg membrane and does not develop an apical tuft or oral lobes. In other respects, however, this larval type is similar to the pilidium and it also develops similarly. Note that the oral lobes and particularly also the saclike alimentary systems of the larvae are reminiscent of the flatworm level of evolution. Metamorphosis is a drastic transformation. The larval ectoderm invaginates in a number of places and buds off small

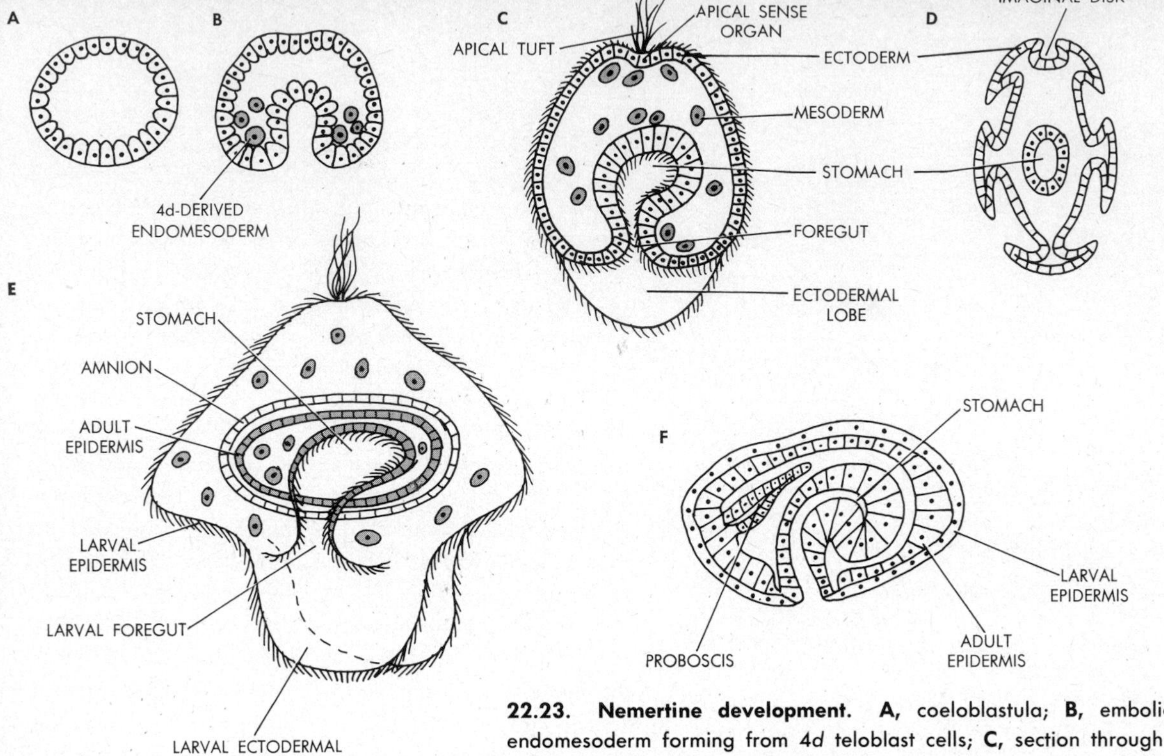

22.23. Nemertine development. **A,** coeloblastula; **B,** embolic gastrulation, endomesoderm forming from *4d* teloblast cells; **C,** section through early pilidium larva, as in *Cerebratulus* (*adapted from* Coe). **D.** Frontal section of pilidium, showing the five invaginations from the larval epidermis forming imaginal disks. Fusion of the disks with one another will give rise to a continuous amnion layer. **E.** Late pilidium, beginning of metamorphosis; the amnion is complete and has given rise to the adult epidermis. Structures outside the amnion will not become part of the adult. **F.** Desor's larva, as in *Lineus*. (*After Arnold.*) In early stages, the structure of the larva corresponds substantially to that of an early pilidium.

imaginal disks into the interior. These disks proliferate and eventually form a continuous layer of ectoderm around the larval stomach. The layer and its contents then elongate and develop into the future worm. All external parts degenerate and are cast off (cf. Fig. 22.23). In effect, therefore, virtually the whole larval ectoderm and much of the larval mesoderm are lost, the adult arising from small plaques of larval ectoderm, the remaining larval mesoderm, and the larval stomach. Note incidentally that the nemertine embryo, being mosaic, is without regenerative capacity, whereas the adult regenerates readily. We shall encounter an analogous disparity in the entoprocts and the chordate tunicates.

REVIEW QUESTIONS

1. Give taxonomic definitions of the Bilateria, the acoelomates, and the phyla of flatworms and ribbon worms. Analogously define the classes of flatworms and characterize each turbellarian order. Name representative genera of each such order.

2. Describe the structure of the Acoela and contrast it with that of a planula. How do Acoela develop?

3. Describe the tissue and organ-system organization of turbellarians generally. What are rhabdites? What is the specific organization of the reproductive system of planarians and how does fertilization take place?

4. Describe the development of (*a*) turbellarians with ectolecithal eggs, (*b*) polyclads. What larval types occur in (*b*)? What are the probable evolutionary interrelations of the turbellarian orders and the other flatworm classes? What is the basis for such views?

5. In what structural respects do flukes differ from turbellarians? Describe the internal structure of flukes, with particular attention to the reproductive system. What are Digenea? What is the life cycle of (*a*) *Schistosoma,* (*b*) *Clonorchis*? Describe the larval stages of these worms in detail. Review the life cycles of Aspidogastrea and Monogenea.

6. Define and distinguish the subclasses of tapeworms and name representative genera of each. Describe the structure of *Taenia.* What is the organization of the reproductive system in a proglottid? In which different ways may fertilization occur? Review the life cycle of tapeworms.

7. Define the nemertine classes taxonomically and name representative genera of each. Describe the structure and function of the proboscis apparatus. Review the sequence and structure of the body tissues from the epidermis inward. By what means is food (*a*) moved through the gut, (*b*) digested?

8. What is the organization of the nemertine (*a*) nervous system, (*b*) circulatory system, (*c*) excretory system? What are the functional characteristics of the blood? What kinds of sensory structures does a nemertine possess?

9. Describe the development of nemertines. What are structural and functional differences between pilidium and Desor's larvae? Describe the structure of a pilidium and give an account of the events of metamorphosis. Are acoelomate animals (*a*) derived from mosaic or regulative eggs, (*b*) protostomial or deuterostomial?

10. What environments do (*a*) flatworms, (*b*) nemertine worms occur in? Be specific and name actual groups or genera. How abundant are both phyla in terms of species numbers? Can these animals regenerate and does vegetative reproduction occur? Discuss.

COLLATERAL READINGS

The books on parasitism and disease below include accounts on flukes and tapeworms. The article by Best reviews conditioning and learning experiments in flatworms.

Beauchamp, R. S. A.: The Rate of Movement of Planaria, *J. Exp. Biol.,* vol. 12, 1935.

Best, J. B.: Protopsychology, *Sci. American,* Feb., 1963.

Bronstedt, H. V.: Planarian Regeneration, *Biol. Rev.,* vol. 30, 1955.

Cameron, T. W. M.: "Parasites and Parasitism," Wiley, New York, 1956.

Caullery, M.: "Parasitism and Symbiosis," Sidgwick and Jackson, Ltd., London, 1952.

Coe, W. R.: Analysis of the Regenerative Process in Nemerteans, *Biol. Bulletin,* vol. 66, 1934.

Faust, E. C.: "Animal Agents and Vectors of Human Disease," Lea & Febiger, Philadelphia, 1955.

Hyman, L.: "The Invertebrates," vol. 2, chaps. 10 and 11, McGraw-Hill, New York, 1951.

Jenkins, M. M., and H. P. Brown: Sexual Activities and Behavior in the planarian *Dugesia, Am. Zoologist,* vol. 2, 1963.

CHAPTER 23

pseudocoelomates

This grade consists of a heterogeneous assemblage of animals currently often classified into three phyla, viz., Aschelminthes, Acanthocephala, and Entoprocta. The first two of these cannot yet be considered as taxonomically stabilized units, for the acanthocephalans could be regarded with some justification as one of the classes within the aschelminths. It has also been suggested that the animals now classified as aschelminths and acanthocephalans be regrouped into two differently constituted and differently named phyla. Alternatively, the various aschelminth classes could be—and indeed they often are—treated as separate phyla altogether. Whatever taxonomic system may finally prove to be most appropriate for these pseudocoelomate groups, there is little doubt that they are set off fairly distinctly from the entoprocts, animals which do form a clearly separate phylum unit.

PHYLUM ASCHELMINTHES

Wormlike animals with cuticle, often segmented superficially; cells or nuclei constant in number and arrangement for each species; musculature typically not as circular and longitudinal layers; pharynx usually highly differentiated; with and without excretory system; reproductive system simple; sexes usually separate. *Rotifera, Gastrotricha, Kinorhyncha, Priapulida, Nematoda, Nematomorpha*

The phylum as a whole is probably derivable from flatworms, particularly rhabdocoel flatworms. Within the phylum, the interrelations of the principal groups are not always obvious. These groups are here treated as classes, a practice which appears to be gaining in popularity. But we may reiterate that each group can also be considered to represent a separate phylum, a procedure which would tend to underscore the present uncertainties concerning the interrelations of the groups. Aschelminths include some of the rarest as well as some of the most abundant of all animals; priapulids encompass but five known species, nematodes perhaps 500,000.

Class Rotifera

Microscopic aquatic animals with anterior wheel organ; pharynx with jaws; flame-bulb protonephridia. *Philodina, Trochosphaera*

Most of the approximately 1,500 species of rotifers are freshwater types; the rest are marine. The animals live as swimmers, creepers, floaters, and sessile forms, and their shapes reflect their modes of life. Floaters tend to be globular and saclike, sessile species are vaselike and usually enveloped by a lorica or a cuticular envelope, and creepers and swimmers are elongate and roughly wormshaped (Fig. 23.1). The animals are about as large as ciliate protozoa, and their *corona*, or wheel organ, actually gives them the superficial appearance of ciliates. Representing the major distinguishing feature of the class, the corona is an anterior wreath of cilia, beating metachronously. The organ is used in swimming locomotion and in the creation of food-bearing water currents.

Rotifers are syncytial animals, and each member of a given species is constructed from exactly the same number of embryonic cells. After these become syncytial in the adult, each individual still retains the exact same number of nuclei,

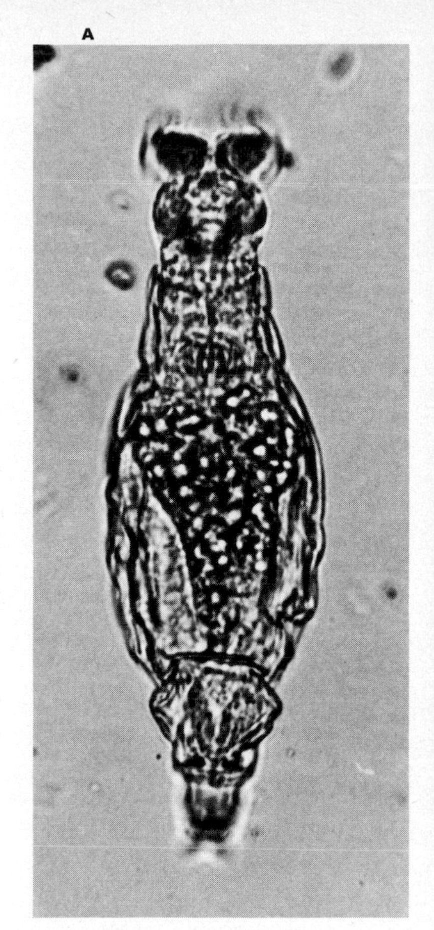

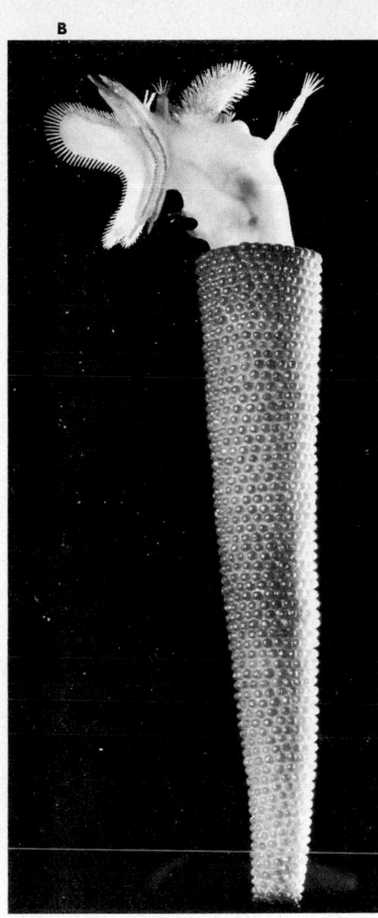

23.1. Rotifers. A, *Philodina,* from life; **B,** glass model of *Melicerta,* a sessile type which constructs its housing from fecal pellets cemented together.

positioned at exactly fixed locations. Thus, any two members of a species are structurally identical and the body design of each species can be mapped out precisely, cell for cell and nucleus for nucleus. Such cell and nuclear constancies also characterize virtually all other groups of the Aschelminthes. By actual count, a rotifer consists of 1,000 to 2,000 cells or nuclei, a particular number in that range identifying each species.

The body of a rotifer is usually organized into a *head,* a *trunk,* and a tapered *foot* (Fig. 23.2). The end of the foot is typically bifurcated into two *toes,* each containing a *cement organ.* Its secretion permits the animal to attach itself temporarily to a solid object. The toes are also used in a creeping, caterpillarlike locomotion. The exterior of the body is a tough scleroprotein cuticle, secreted by the thin epidermis underneath. Transverse ringlike grooves scored into the cuticle give the animal a segmented appearance. This superficial segmentation permits a rotifer to telescope and reextend its body, and to flex and twist it in highly mobile fashion. The epidermis bounds the lymph-filled pseudocoel. A small number of ectomesodermal amoeboid mesenchyme cells is present in the lymph. Muscles are not arranged as distinct layers. Instead, individual muscle cells crisscross in specific patterns throughout the pseudocoel. The alimentary tract is a straight, syncytial tube. It begins at the ciliated mouth ventral to the corona and leads into a pharynx. Here called *mastax,* the pharynx is a complex muscular chewing organ containing cuticular jaws studded with teeth and receiving secretions of salivary glands. The detailed construction of the mastax differs for different species; rotifers are distinguished taxonomically on this basis. Behind the mastax, the remainder of the alimentary tract has a ciliated inner surface. The tract terminates at the anus, located in the region between the toes of the foot.

From a brain ganglion dorsal to the pharynx lead two main ventral nerve cords (cf. Fig. 23.2). Secondary ganglia are present in other body regions. In the head are found a variety of sensory cells and pits. Just behind the brain ganglion lies a *cerebral organ,* a glandular invagination believed to be

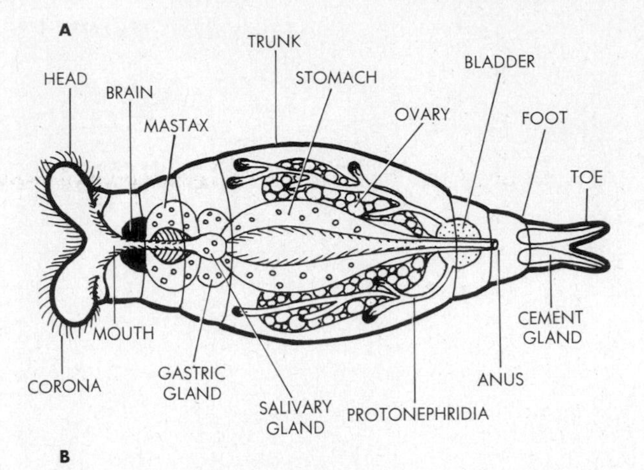

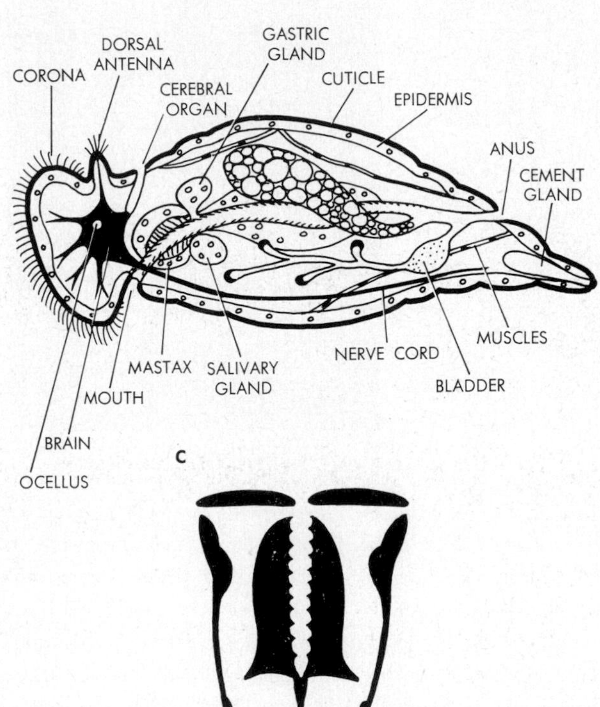

23.2. Rotifer structure (*Philodina*), schematic. A, ventral view; **B,** side view; **C,** the horny elements of one type of mastax (*after Beauchamp*).

homologous to the frontal organs of acoelomates. Most rotifers possess an *ocellus* directly on the dorsal surface of the brain ganglion. Particularly characteristic of these animals is a *dorsal antenna,* a small surface projection dorsal to the brain ganglion. The antenna is innervated and carries a tuft of cilia; its precise function is obscure. The excretory system consists of a single pair of protonephridial tubules branching into terminal flame bulbs in the pseudocoel. At their posterior ends the tubules form a urinary bladder which opens into the hind part of the intestine. Rotifers may therefore be considered to possess a cloaca.

The sexes of rotifers are separate, and a pronounced sexual dimorphism is in evidence; males are small, structurally reduced individuals (Fig. 23.3). In one group, exemplified by the common freshwater genus *Philodina,* males are absent altogether, the females reproducing parthenogenetically only. Where males do occur, such an individual possesses one or a pair of testes. These lead via a sperm duct to a posterior gonopore. Some species mate with the aid of a male copulatory organ, but in most rotifers fertilization takes place by hypodermic impregnation. The injected sperms then migrate through the pseudocoel to the ovaries. The latter are single or paired, and they open into the cloaca. Like other body parts, an ovary is syncytial; it may contain some 10 to 50 nuclei. In the lifetime of a female, the ovary can bud off only as many eggs as there are nuclei.

In exclusively parthenogenetic types like *Philodina,* unfertilized eggs are shed one at a time and develop into new females. In bisexual forms, the eggs are of two kinds (Fig. 23.4). So-called *amictic* eggs are parthenogenetic and cannot be fertilized. As such eggs mature from diploid ovary nuclei, they undergo only one meiotic division and form only one polar body. As a result such eggs remain *diploid,* and they develop parthenogenetically into females. If environmental conditions remain stable and favorable, as is usual in a pond during the spring and summer months, female rotifers produce amictic eggs only. The population of females grows in this manner. But if the animals are subjected to more or less sudden environmental changes, as during the fall months, then *mictic* eggs are formed. These are smaller because during their maturation they undergo two normal meiotic divisions. Mictic eggs thus are *haploid.* If they are not fertilized, as would necessarily be the case in the first mictic eggs of the

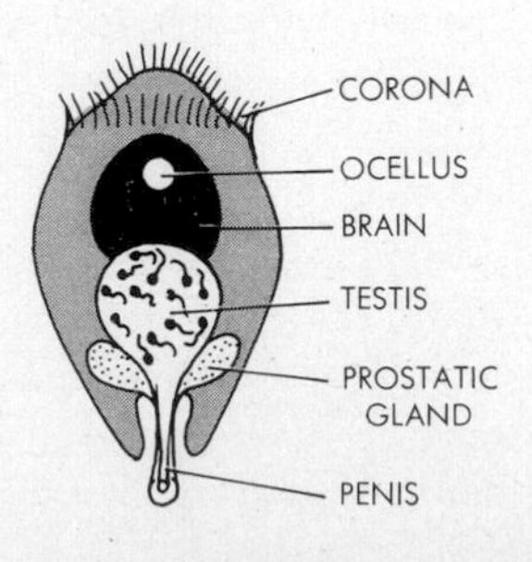

23.3. A male rotifer (*Trichocerca*), structurally reduced. (*Adapted from Wesenberg-Lund.*)

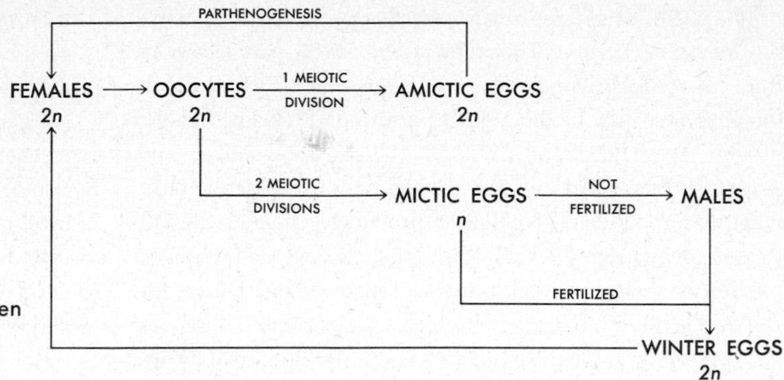

23.4. The rotiferan life cycle and the difference between mictic and amictic eggs.

population, they develop parthenogenetically into males. But if they are fertilized by some of the males now appearing in the population, then the mictic eggs acquire a heavy cyst wall in the oviduct and they are shed as resistant *winter eggs*. The adult rotifers all die during the winter, and in the following spring the winter eggs develop into amictic females.

The pattern of embryonic development is typically spiral in some cases and uniquely bilateral in others (Fig. 23.5). In the latter, cleavage results in tiers of blastomeres which are not shifted spirally relative to one another. A stereoblastula forms, composed of only a few, large cells. Gastrulation takes place by emboly, though only a single large cell invaginates into the blastocoel. This cell is the source of the future gonads. The mesoderm is almost entirely ectomesodermal. Development is mosaic, and even adult rotifers cannot regenerate. Moreover, other forms of vegetative reproduction also do not occur. However, freshwater rotifers are remarkable in being able to withstand drying out. Slow desiccation shrivels such an animal into a cystlike body, and in this condition it may remain dormant but viable for years. When water becomes available again, the animal "inflates" back to normal and resumes an active life.

Experiments have shown that the longevity of a rotifer is correlated with how old its mother had been at the time the rotifer was produced. The first egg formed by a given female develops into a longer-lived offspring than the second, third, or any later egg. In the course of its reproductive life, therefore, a female produces a succession of progressively less long-lived offspring. And when a longer-lived offspring produces eggs in its own turn, the first of these becomes a still longer-lived offspring. Thus, by continued selection of first eggs in successive generations, a line of progressively longer-lived generations may be raised (until a limit to increase in longevity is reached). It is possible that this phenomenon is correlated adaptively with the seasonal life of these animals. Inasmuch as first-formed individuals live longer, they may have a better chance to survive into the fall season, when the production of winter eggs will ensure species survival into the following year.

Class Gastrotricha

Microscopic aquatic animals with cilia and cuticular spines.
Chaetonotus

23.5. Development of a rotifer (*Asplanchna*). (*Adapted from Tannreuther, Nachtwey.*) **A,** four-cell stage; **B,** 16-cell stage (the bottom cell will invaginate); **C,** stereoblastula, beginning of gastrulation; **D,** sagittal section through later embryo. The large invaginated cell gives rise to the ovary and the reproductive system. Mesoderm arises from ectoderm.

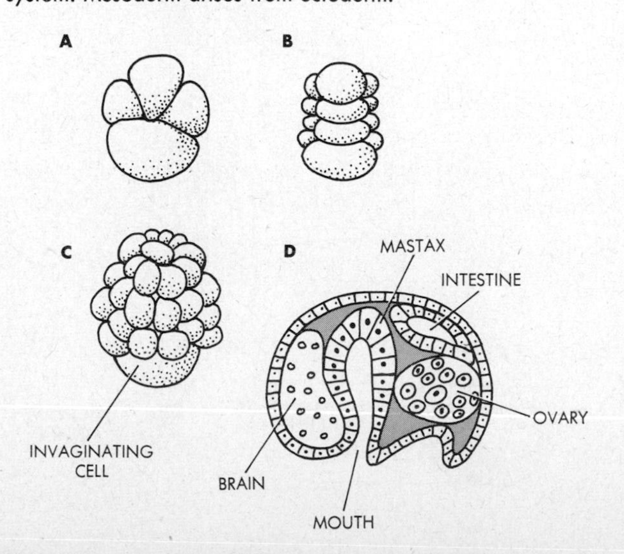

The 200 or so species of gastrotrichs have the same general size range as rotifers. Their characteristics in part resemble those of rotifers, in part those of nematodes (Fig. 23.6). In the elongate body of the animals, a rounded lobelike head is usually set off from the trunk by a slight constriction. The ventral surface is flat and it is ciliated either completely or in longitudinal bands. This ciliation produces a smooth, gliding type of locomotion. Dorsally the surface is covered by spines, bristles, or scales, an identifying trait of the class. The posterior end is often bifurcated into toes, as in rotifers. These toes contain adhesive glands, and in some cases such glands are also present along the lateral margins of the body. The epidermis, musculature, and nervous systems are essentially as in rotifers. Lymph without cells fills a not very extensive pseudocoel. Some gastrotrichs possess epidermis-derived noncellular membranes which divide the pseudocoel partially into compartments.

The pharynx is quite unlike that of rotifers but resembles that of nematodes to a remarkable degree. It is a tubular structure, roughly as long as one-third of the whole body, and it may be expanded into one or more bulbous enlargements along its length. The wall contains columnar epithelial cells and closely associated, radially aligned muscle fibers. The interior canal of the pharynx has a roughly triangular cross-sectional shape (cf. Fig. 23.6). An excretory system is present in only one group, exemplified by *Chaetonotus*. In these ani-

23.6. Gastrotrich structure. Diagrams: *Chaetonotus.* (*Adapted from Zelinka.*) **A,** external side view; **B,** cross-sectional outline to show ventral ciliary bands; **C,** internal organs, top view (uterus not shown); **D,** cross section of pharynx, showing radially aligned muscles and triangular outline of pharyngeal canal. **Photo:** external view of the gastrotrich *Lepidodermella,* from life.

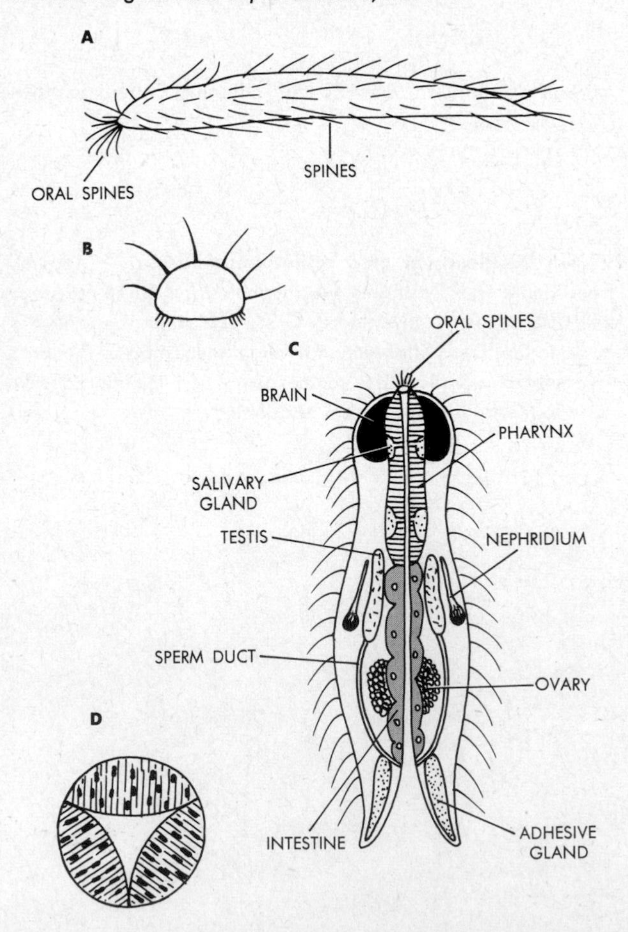

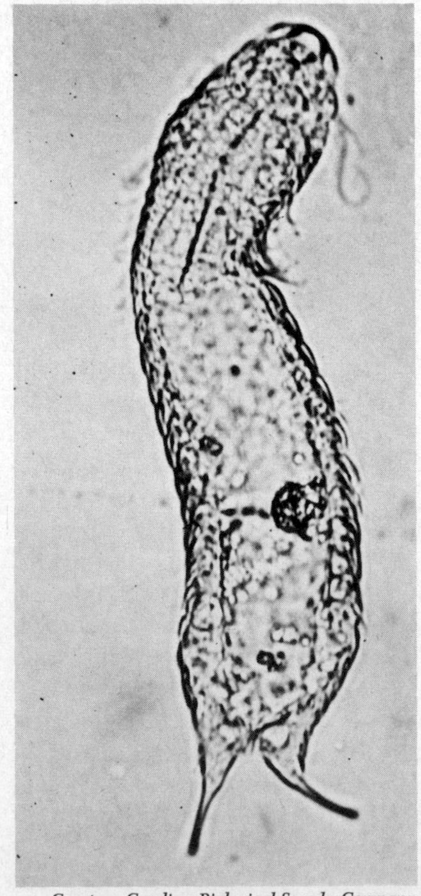

Courtesy Carolina Biological Supply Company.

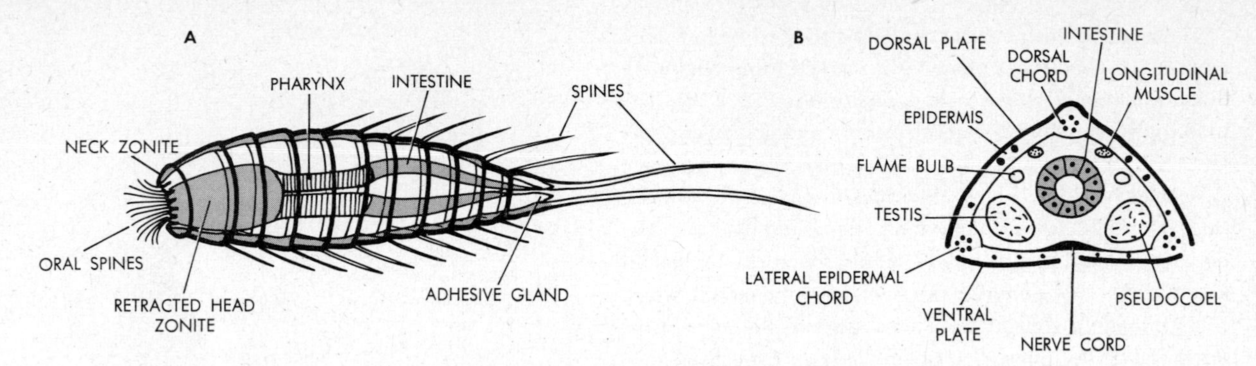

23.7. Kinorhyncha. **A,** side view of *Echinoderella;* **B,** cross section through tenth zonite of *Pycnophyes.* (*Adapted from Zelinka.*)

mals the system consists of a single pair of anterior, elongated flame bulbs, which continue as highly coiled tubules and terminate at midventral nephridiopores. The structure of the reproductive systems is not fully known. However, it is known that gastrotrichs are basically hermaphroditic and that in some groups the male system has apparently degenerated, the animals reproducing exclusively as parthenogenetic females. Development is mosaic. The first three cleavages tend to be spiral, but subsequent divisions occur bilaterally. Cleavage results in a coeloblastula, and two vegetal cells then invaginate embolically.

Because gastrotrichs resemble nematodes in cuticular specializations and in pharyngeal structure, the animals are believed to represent an evolutionary link between rotifers and nematodes.

Class Kinorhyncha

Microscopic marine animals without cilia; body segmented conspicuously. *Echinoderella*

The approximately 100 species of kinorhynchs (also called *Echinodera*) occur abundantly on and in the shallow mud bottoms along ocean shores. They are minute animals, with a body divided into 13 or 14 segments (Fig. 23.7). The latter are far less superficial than in other Aschelminthes; they are called *zonites* in distinction to the mesodermal, metameric segments of coelomate groups. The first zonite constitutes the *head,* which bears the mouth and a circlet of *oral spines.* The head zonite may be retracted completely into the second, or *neck,* zonite. All other zonites form the *trunk.* Each such zonite is covered by one dorsal and two ventral cuticular plates, which lap posteriorly over the plates of the adjacent zonite. The cuticle bears sets of spines and bristles, and at the posterior end there are a pair of adhesive glands, as in gastrotrichs. The syncytial epidermis is thickened dorsally and laterally into a series of *longitudinal chords,* a feature which relates the animals to the nematodes. Indeed, the structure of the pharynx is very similar to that of nematodes and of gastrotrichs, and a gastrotrich-rotiferlike musculature is present. However, the muscle fibers are arranged on a segmental basis. Analogously segmental is the nervous system, which is so intimately joined to the epidermal layer that the two tissues appear to be one. The pseudocoel is spacious, and numerous cells are present in the lymph. One pair of flame bulbs is situated in the tenth zonite.

Sexes are separate. The pattern of embryonic development is unknown, but it is known that the early larvae are quite unlike the adults and that they pass through a succession of larval stages. At the end of each such stage the cuticle is molted and a new, larger one then forms. The phenomenon of molting and the segmental design of the adult have at one time been interpreted to indicate an evolutionary connection between kinorhynchs and arthropods. However, it is now evident that these similarities are incidental and little more than instances of parallel evolution; larval molting is known to occur also in the nematodes (see below), and although the segmentation of kinorhynchs is pronounced, it is nevertheless superficial as in other Aschelminthes and is not metameric. There can be little doubt that, in view of their pseudocoelomate structure generally and their particular traits specifically, the kinorhynchs are related to nematodes and gastrotrichs.

Class Priapulida

Marine animals up to 3 or 4 in. long; body with invertible prosoma and superficially segmented trunk; with solenocytic protonephridia. *Priapulus, Halicryptus*

These rare animals live in the mud bottoms of shallow shore regions. The body of a priapulid is marked externally into a bulbous *prosoma* and a cylindrical *trunk* (Fig. 23.8). The prosoma is ridged with wartlike sensory and glandular papillae, and it may be invaginated and retracted into the trunk. A cellular, cuticle-covered epidermis is underlain by a distinct outer circular and inner longitudinal muscle layer. The nervous system is structurally almost identical to that of kinorhynchs. Anteriorly the prosoma bears the mouth, which leads through a straight alimentary tract to the anus at the other end of the animal. Sets of circularly and radially aligned muscles are present in the pharynx, which is lined with a toothed cuticle, an interior continuation of the cuticle covering the body. The spacious body cavity is partitioned partially by noncellular membranes, as in gastrotrichs. In this cavity lie extensive clusters of solenocytic nephridia, priapulids being the only Aschelminthes with protonephridia of this type. The solenocytes and the gonad on each side of the body open into a common urogenital duct which exits via a pore near the anus. Sexes are separate. The developmental pattern has not been fully elucidated as yet, but cleavage appears to be radial. The larva is similar to the adults. Unless more complete future data about these animals should indicate otherwise, the priapulids appear to be part of the same adaptive radiation which has also given rise to the gastrotrichs, kinorhynchs, and nematodes.

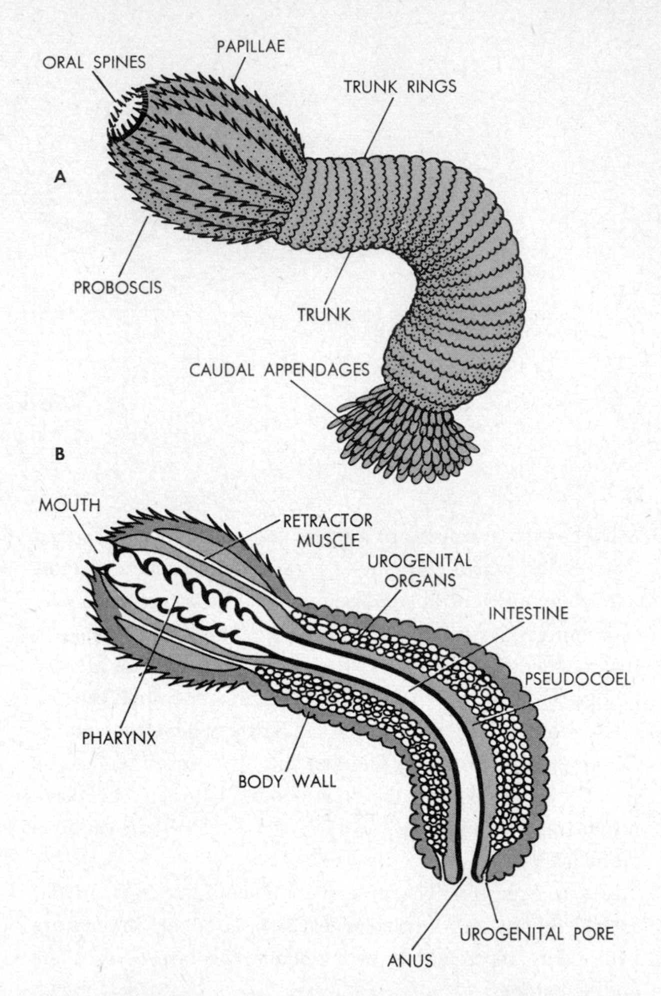

23.8. Priapulid structure. (*Adapted from Theel.*) **A,** external view; **B,** sectional view. The retractors in the proboscis (shown only in part) attach to the inner surface of the trunk body wall.

Class Nematoda

Roundworms; cylindrical, elongated worms without cilia; excretory system not protonephridial; free-living and parasitic. *Ascaris, Rhabditis, Trichinella, Necator, Wuchereria, Dracunculus*

After the insects, roundworms are probably the most abundant of all animals, both in numbers of individuals and numbers of species. Named species total only about 12,000 at present, but new ones are being described at an average rate of one per day. Each vertebrate species is known to harbor at least one and usually more than one type of parasitic nematode; hence there should exist a minimum of 50,000 different kinds of the worms. Moreover, vertebrates are by no means the only nematode hosts among animals, and roundworms additionally parasitize plants. Further, it is estimated that the free-living species of roundworms are even more numerous than the parasitic ones. Informed guesses therefore place the number of existing nematode species at about ½ million. Free-living nematodes occur in the sea, in fresh water, and in soil. Parasitic types are exclusively endoparasitic, infecting virtually any tissue of plants or animals. Many of the animal nematodes live in the intestines of their hosts and they are usually implicated when an animal is said to suffer from "worms." Man alone harbors some 50 species. Most of these are relatively harmless, but some cause serious diseases.

Most nematodes, free-living ones in particular, are under 2 in. long, though some parasitic types have lengths of well over a yard. The worms are cylindrical in cross-section and at the tapered ends are the mouth and anus, respectively (Fig. 23.9). A complex three-layered cuticle covers the outside. The outer layer is keratinlike, the middle layer consists of a net of spongy fibers, and the inner layer is a network of collagen fibers. Externally the cuticle may be smooth or may bear bristles, scales, or a variety of wartlike papillae. A basement membrane separates the cuticle from the cellular or syncytial epidermis. The latter is thickened on the inner surface into four *longitudinal chords*, one dorsally, one ventrally, and one on each side of the body. The lateral chords are visible externally as faint lines. If the epidermis is syncytial, its nuclei are situated only in the chords. In each body quadrant between the chords the epidermis is underlain by a layer of body wall muscles, the cells of which are aligned longitudinally only; the worm can bend but it cannot lengthen. Cells and fibers are present in the pseudocoelic lymph.

The nervous system consists of a nerve ring around the pharynx, a dorsal and ventral nerve cord, and from one to three pairs of lateral cords. Ganglia are associated with both ring and cords. Free-living worms contain numerous sensory bristles and papillae on the body surface, and some aquatic types also possess an ocellus with a cuticular lens on each side of the anterior body region. Sensory receptors are reduced in parasitic species. Accessory ingestive structures such as cuticular jaws or teeth may be present around the mouth. The well-developed pharynx is a tube of epithelial, glandular, and muscular tissue, organized as in gastrotrichs. Also, the pharyngeal canal is three-cornered in cross-section. The excretory system is unique. It consists of a pair of large cells under the pharynx, the so-called *ventral glands*. The necks of these cells lead forward to anteriorly located nephridiopores. Such a system is evidently not protonephridial, nor does it resemble any other excretory system of animals.

Many species are hermaphroditic, and in some of these the same gonad produces sperms first, which are stored, then eggs which are fertilized by the stored sperms. Some nematodes are exclusively parthenogenetic. In the vast majority of nematodes, however, the sexes are separate. Males are smaller than females and they are identifiable also by their curled posterior ends (Fig. 23.10). The tubular gonads may be single or paired. Testes lead via sperm ducts into the hind portion of the intestine. The sperms of nematodes are unique in that they are amoeboid, not flagellate (and we may note that nematodes as a whole are unique in that flagella or cilia are totally absent). Ovaries lie anteriorly in the pseudocoel, and they pass eggs through a uterus and a vagina to a separate gonopore located ventrally in the posterior region of the body. Fertilization occurs internally. The males of many species possess a copulatory spicule near the anus, used to widen the female gonopore during the introduction of sperms. The latter enter eggs near the ovary, and the zygotes become encysted and begin to develop during their passage through uterus and vagina. In

23.9. Nematode structure. **A,** ventral view of organs in anterior portion of a nematode (*adapted from Chitwood*); **B,** cross section at level of excretory glands. The contractile portions of the longitudinal body-wall muscles (cross-barred) are adjacent to the epidermis, the noncontractile portions curve toward the nearest longitudinal nerve cord. In some nematodes a longitudinal excretory canal runs within each lateral epidermal chord and connects with the excretory glands. The lateral chords mark the position of the externally visible lateral lines. **C,** the layers of the cuticle secreted by the epidermis; **D,** cross section through the pharynx, showing triangular channel, radial muscles, and gland cells.

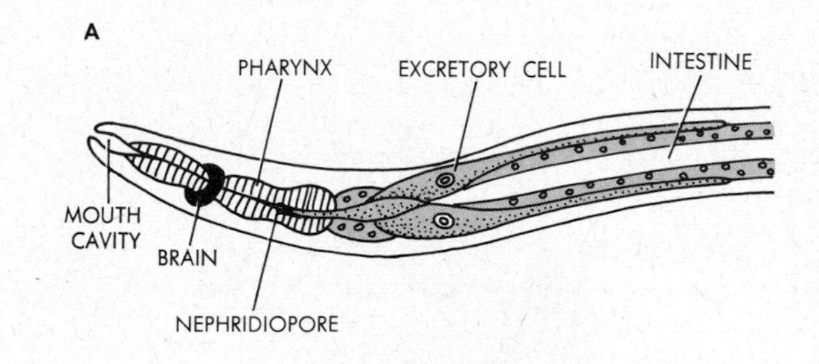

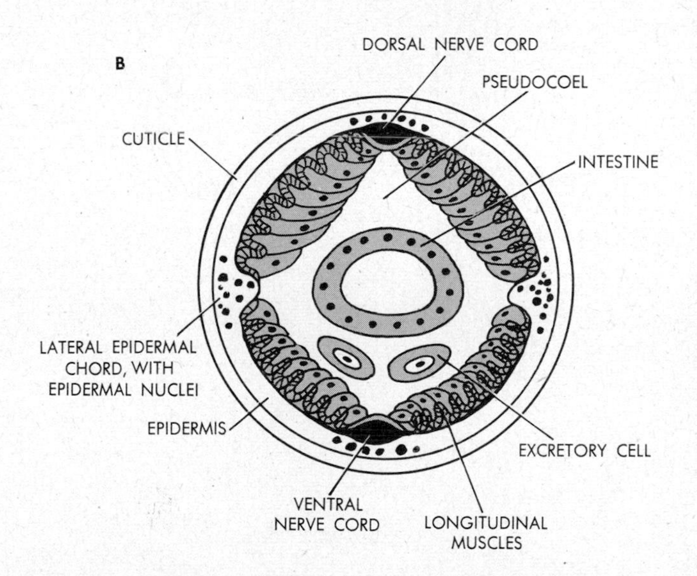

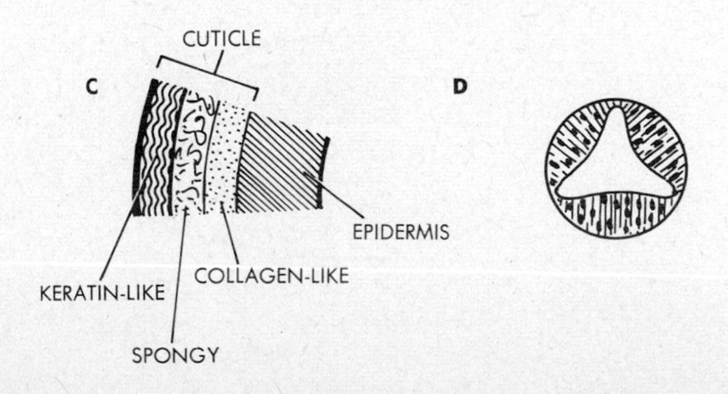

23.10. Reproductive systems in nematodes (*Ascaris*). A, female worm, **B,** male worm, side views, each showing position of reproductive system. Only the left gonads are shown; the terminal portions of the systems are unpaired and are situated along the ventral midline. Note the smaller size and the curled posterior end of the male. In nematodes other than *Ascaris,* the female gonopore may be located in the middle or the posterior portion of the body. **C.** Sagittal section through posterior end of male *Ascaris,* to show relation of reproductive organs to alimentary terminus.

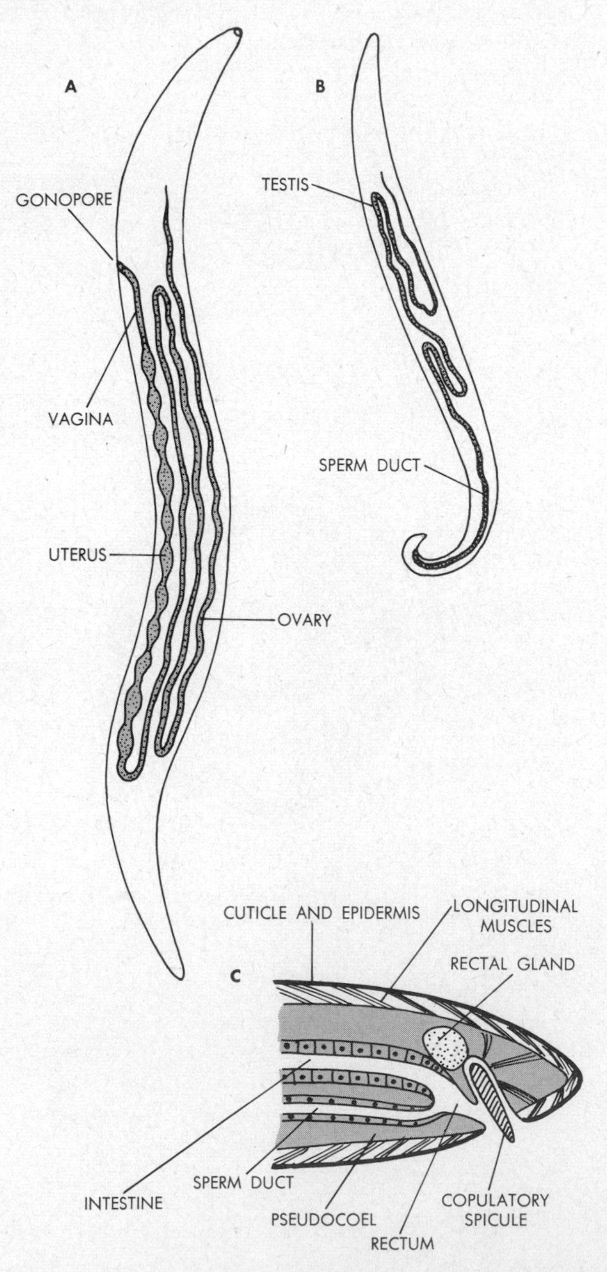

Ascaris, for example, the entrance of the sperm is the initiating stimulus for the completion of meiosis in the egg (cf. Chap. 10). By the time such an egg reaches the female gonopore, it is already an embryo of several cells.

Development is mosaic. Cleavage follows a distinctive pattern, undoubtedly a modification of the spiral pattern of flatworms (Fig. 23.11). The two-cell stage consists of a so-called S1 and a *P*1 cell. Each divides, the S1 cell giving rise to an *A* and a *B* cell, the *P*1 cell to an S2 and a *P*2 cell. These at first form a T-shaped four-cell stage, but the cells soon shift into a rhomboidal arrangement. In it, the *A* cell is anterior, the *B* cell is dorsal, the S2 cell is ventral, and the *P*2 cell is posterior. The progeny of the *A* and *B* cells will form almost all of the ectoderm and the excretory glands; the *P*2 cell will give rise to the remainder of the ectoderm and to the reproductive system. The S2 cell produces an *StM* cell and an *E* cell, and the *StM* cell then divides to form a separate *St* cell and an *M* cell. The *St* cell will develop into the pharynx, the *M* cell into the mesoderm (musculature and pseudocoel cells), and the *E* cell into the endoderm (midgut). Continued cleavages produce a bilateral coeloblastula which gastrulates by epiboly; the progeny of the *St*, *M*, and *E* cells as well as some of the progeny of the *P*2 cell become overgrown by the progeny of the *A* and *B* cells. The mesoderm is therefore largely endomesodermal.

Nematode cleavages are particularly noteworthy for the very early differentiation of the cellular source of the later reproductive cells. At the two-cell stage, only the *P*1 cell retains normal chromosomes; those of the S1 cell become reduced to small granular bodies, a process known as *chromosome diminution.* When the *P*1 cell then divides into the S2 and *P*2 cells, S2 again undergoes chromosome diminution and only *P*2 retains normal chromosomes. Analogous diminutions occur in the subsequent progeny of *P*2; only the cells which will form the reproductive system and the gametes retain normal chromosomes. Also, only such cells continue to divide in later embryonic stages. Divisions soon stop in all cells with diminished chromosomes, and the cell numbers present at that time will remain constant from then on throughout the life of the worm. It appears therefore that the adult characteristics of all but the reproductive cells become fixed permanently and very early by *chromosomal* specializations. At the same time, the cellular derivation of the reproductive system is marked precisely, in the form of normal chromosomes, right from the zygote stage on.

During their larval development, nematodes typically molt their cuticles, usually four times in succession; enlargement of the larvae takes place at these molting stages. The life cycles of the worms are more varied than those of almost any other

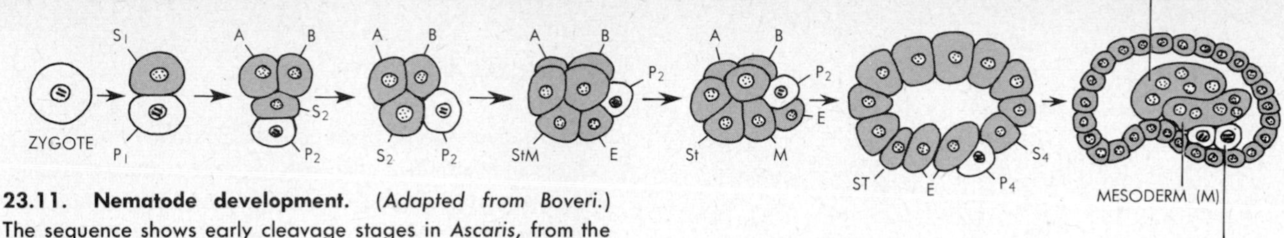

23.11. Nematode development. *(Adapted from Boveri.)* The sequence shows early cleavage stages in *Ascaris*, from the zygote to the postgastrula. The anterior end is toward the left. Blastomeres in which chromosome diminution occurred are shaded and the nuclei are dotted. The next-to-last stage shown is a sagittal section through the coeloblastula (the *M* cell and its progeny are lateral and thus not included in the section). The last stage is a section after epibolic gastrulation has occurred, the *A* and *B* progeny having overgrown the *E*, *M*, and *P* progeny. A mouth invagination is beginning to form from *St* progeny in the blastopore region. Cells in the interior are becoming syncytial.

animals. In the free-living types, the larvae develop directly into new free-living adults. In parasitic types, an infective stage is reached at a given point in development. Up to that point the worms are free-living; and when they become infective, they must enter a specific plant or animal host within a short time or perish. The infective stage may be attained as soon as the earliest larval form has developed or at almost any time thereafter. In many cases the worms become adults and live free for substantially long periods before they become infective. Numerous nematodes require one or more intermediate hosts for the completion of their life cycles.

Among the serious nematode pests of man (Fig. 23.12) are the *trichina* worms (*Trichinella spiralis*), introduced into the human body via insufficiently cooked pork; the *hookworms* (e.g., *Necator americanus*), which live in soil and infect man by boring through his skin; the *guinea* worms (e.g., *Dracunculus medinensis*), which develop as larvae in copepods, which enter man via copepod-containing drinking water, and which form ulcerating blisters in the skin of man from which the larvae of the next generation are released; and the *filaria* worms (e.g., *Wuchereria bancrofti*), transmitted by mosquitos and causing blocks in human lymph vessels. The disease resulting from filarial infections is characterized by immense swellings and is known as *elephantiasis*. One of the relatively less dangerous roundworms is *Ascaris lumbricoides*, a species which lives in the intestine of man and other mammals.

All nematodes, parasitic or free-living, probably have evolved from free-living marine nematode ancestors. In their turn, the latter have undoubtedly been derived from the same stocks which also have given rise to the priapulids, kinorhynchs, and gastrotrichs.

Class Nematomorpha

Hairworms; highly elongate animals without epidermal chords or excretory systems; alimentary system reduced; parasitic in arthropods during young stages. *Gordius*, *Paragordius*

The 80 species of these worms are readily identified by their very great lengths relative to their diameters (Fig. 23.13). Ranging in size from less than ¼ in. to nearly 2 yd, the animals

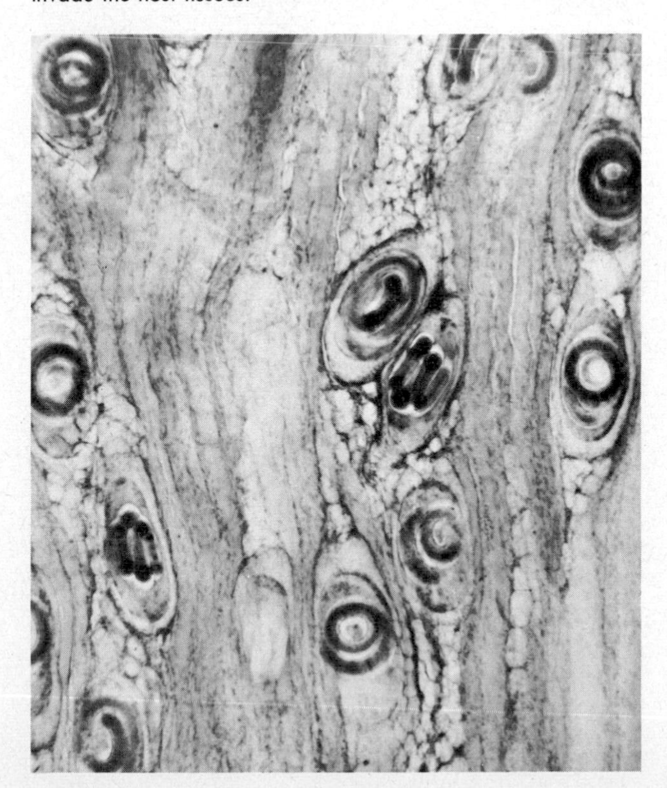

23.12. Larvae of trichina worms, encapsulated in pig muscle. If infected pork is cooked improperly, the larvae are digested out in the intestine of the host and the worms then invade the host tissues.

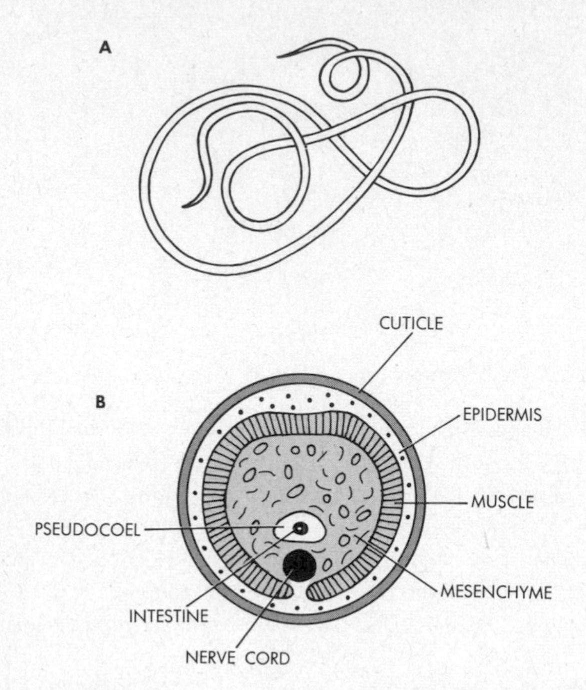

23.13. Hairworms. **A,** outline sketch of *Paragordius;* **B,** cross section through a nematomorph. (*Adapted from Montgomery.*) Note absence of epidermal chords and excretory organs.

resemble roundworms in general internal structure. However, epidermal chords and excretory systems are lacking, and the alimentary tract may remain nonfunctional and permanently undeveloped at either the mouth end, the anus end, or both. The pseudocoel is partitioned by membranes as in gastrotrichs and priapulids, and mesenchyme cells are abundant in it. The larval stages are parasitic in arthropods such as grasshoppers and beetles. Larvae are equipped with a retractible proboscis apparatus. This structure is lost at metamorphosis, when the worms leave their hosts and pursue a free-living adult existence in bodies of water. The sexes are separate. Cleavage is rhomboidal and later bilateral, as in nematodes, and similarly results in a coeloblastula. Embolic gastrulation ensues, the ectoderm later producing most of the mesoderm. Larvae can remain free-living for only a short time and must soon enter insect hosts.

The hairworms are clearly related to nematodes. We may note therefore that, among Aschelminthes as a whole, the rotifers provide a link, tenuous though it may be, with the flatworms on the one hand and with the gastrotrichs and thus all other aschelminth classes on the other (cf. Fig. 23.19).

PHYLUM ACANTHOCEPHALA

Spiny-headed worms; parasitic with arthropod intermediate hosts and vertebrate final hosts; nuclear constancy in adult; with armed proboscis; alimentary system absent; excretory system protonephridial if present. *Moniliformis, Corynosoma*

This group comprises some 600 species of intestinal parasites of vertebrates. The worms are similar in some respects to the parasitic flatworms, and they have often been classified with the acoelomates. However, the design of the worms is clearly pseudocoelomate, and their structure actually relates them to the Aschelminthes; the worms could well be regarded as a class within the Aschelminthes (as indeed they have been by some zoologists).

Most of the acanthocephalans are quite small, though some may attain lengths of about 1 ft. The larger species generally infect birds and mammals, the smaller ones, predominantly fishes, particularly bony fishes. Spiny-headed worms are identified by an anterior, comparatively short *proboscis,* armed with numerous recurved spines or hooks (Fig. 23.14). This organ of attachment can be withdrawn into the anterior portion of the spiny or warty *trunk.* Withdrawal is effected by means of internal retractor muscles which connect the tip of the proboscis with the wall of the trunk. Like most of the aschelminths, the Acanthocephala exhibit a syncytial structure as adults and the nuclear number is constant for each species.

In many cases the external cuticle is superficially segmented, very conspicuously so in types like *Moniliformis.* In the epidermis, a unique system of fluid-filled channels is formed by unlined spaces in the epidermal cytoplasm. This *lacunar system* consists of two longitudinal main channels, running either laterally or dorsally and ventrally in the trunk of the worm, and of numerous transverse cross-connecting channels. The system is closed and does not communicate with any other parts or cavities of the body. It is believed to function in food transport; the worms are fluid-feeders without alimentary systems, and food absorbed through the body surface is thought to be distributed by the lacunar system. At the juncture of proboscis and trunk, the epidermis is extended into the pseudocoel in the form of two elongated bodies, the *lemnisci,* which are thought to serve as reservoirs for the lacunar fluid when the proboscis is retracted.

The epidermis is underlain by an outer circular and an inner longitudinal layer of syncytial muscle fibers. The nervous system consists of an anterior brain ganglion in the proboscis

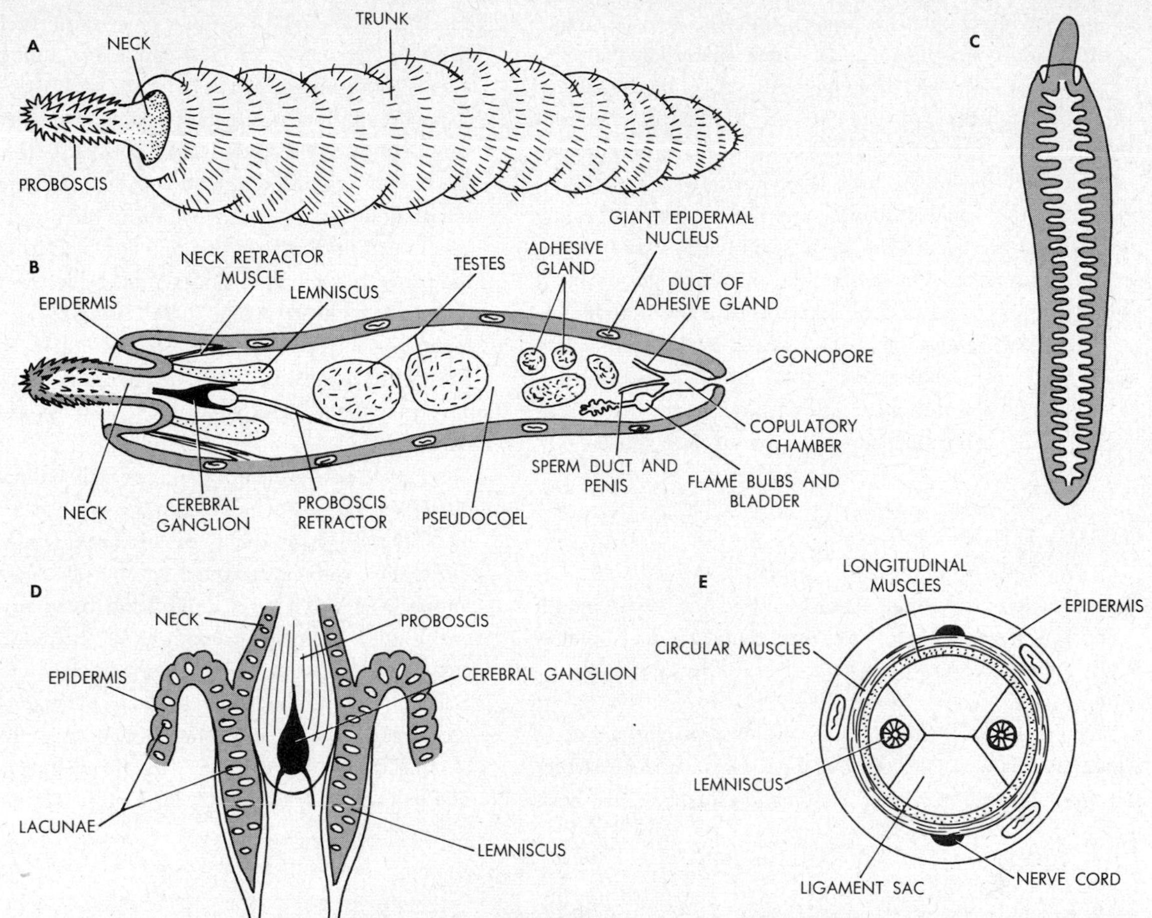

23.14. Acanthocephala. A. External appearance of an acanthocephalan, proboscis extended. **B.** Longitudinal section through **A,** proboscis retracted, body wall muscles and female reproductive system not shown. (*Adapted from Yamaguti.*) Note the giant nuclei of the epidermis. **C.** The lacunar system in the epidermis of *Moniliformis,* dorsal view. (*After Meyer.*) Transverse branch channels lead from a longitudinal main channel. **D.** Section through the neck region of *Acanthocephalus,* showing lacunar channels in epidermis and lemnisci; body wall muscles not indicated. (*Adapted from Hamann.*) **E.** Cross section through lemniscal region, showing ligament sacs. (*After Meyer.*) More posteriorly these sacs envelop the gonads. In some acanthocephalans the nerve cords are lateral and the ligament sacs are reduced and incomplete.

and two longitudinal lateral cords. As in parasitic worms generally, sensory receptors are greatly reduced. Most acanthocephalans are without excretory system. In worms with such a system, the latter consists of clusters of flame bulbs without nuclei. These clusters empty into a common (nucleus-containing) excretory duct which opens into either the sperm duct or the uterus.

The sexes are separate, and female worms typically are larger than male worms. A pair of reproductive organs lies in *ligament sacs,* longitudinal membranous (but noncellular) chambers within the pseudocoel. The sacs are probably homologous to the pseudocoelic chambers encountered in various aschelminths. A single ligament sac or a dorsal and a ventral one may be present. Fertilization is internal. Males are equipped with copulatory organs, and the females retain zygotes within their ligament sacs until larvae have formed. Development is mosaic. The encysted zygotes tend to cleave spirally, though the cyst walls prevent the blastomeres from becoming arranged in typical spiral patterns. The blastomeres soon fuse and the embryo becomes syncytial very early. Consequently, the usual gastrulation- and germ-layer-forming processes do not occur. The larva is solid at first, but it later acquires a pseudocoel by the establishment of a space around the developing ligament sacs. Upon release from the female parent, the larvae must enter the body cavity of intermediate hosts, which are always insects. If an infected insect is then eaten by a final vertebrate host, the larvae attach to the intestinal lining of the vertebrate and mature into adults.

PHYLUM ENTOPROCTA

Adults stalked and sessile, solitary and colonial; with mouth and anus inside a circlet of ciliated tentacles; development via unique larvae. *Loxosoma, Pedicellina, Urnatella*

The 60 species of entoprocts are the only pseudocoelomates which are not wormlike. All but the freshwater genus *Urnatella* are marine. Types like *Loxosoma* are solitary, and others, e.g., *Pedicellina*, form colonies encrusting rocks, shells, and algae in shallow water. The individual animals are largely microscopic; none exceeds ¼ in. in size.

Each entoproct is attached to a solid surface by a stalk, a continuation of the body wall (Fig. 23.15). The remainder of the body is organized around a U-shaped alimentary system, mouth and anus opening away from the attachment surface. The mouth-anus side of the animal is said to be ventral, and the plane through these alimentary openings is the sagittal, or principal, symmetry plane. The ventral surface is ringed by tentacles which, unlike those of coelenterates, are ciliated. Entoprocts are *filter-feeders,* i.e., microscopic food organisms are strained out by the tentacles from water currents created by the cilia.

A cuticle covers the body and the stalk but not the tentacles or the ventral side. The epidermis is cellular and is underlain by a layer of longitudinal muscle fibers. These continue into the tentacles and permit the latter to be curved inward. A muscle ring also circles around the bases of the tentacles, enabling the animal to contact its ventral side with its tentacles tucked in. A gelatinous, mesenchyme-containing pseudocoel extends into the hollow tentacles. The alimentary system is ciliated internally and widens into a stomach at the bottom of the U. In the curve of this U lies the main nervous ganglion. The pairs of nerves from it lead to subsidiary ganglia near the

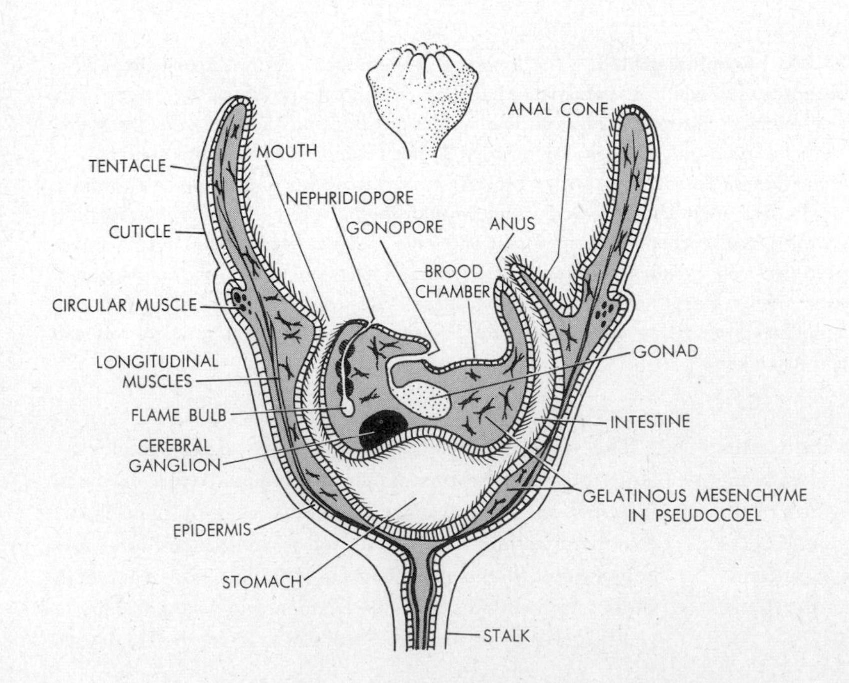

23.15. Sagittal section through an entoproct, schematic. Anterior is toward left, dorsal, toward bottom. Inset diagram shows individual with tentacles tucked in.

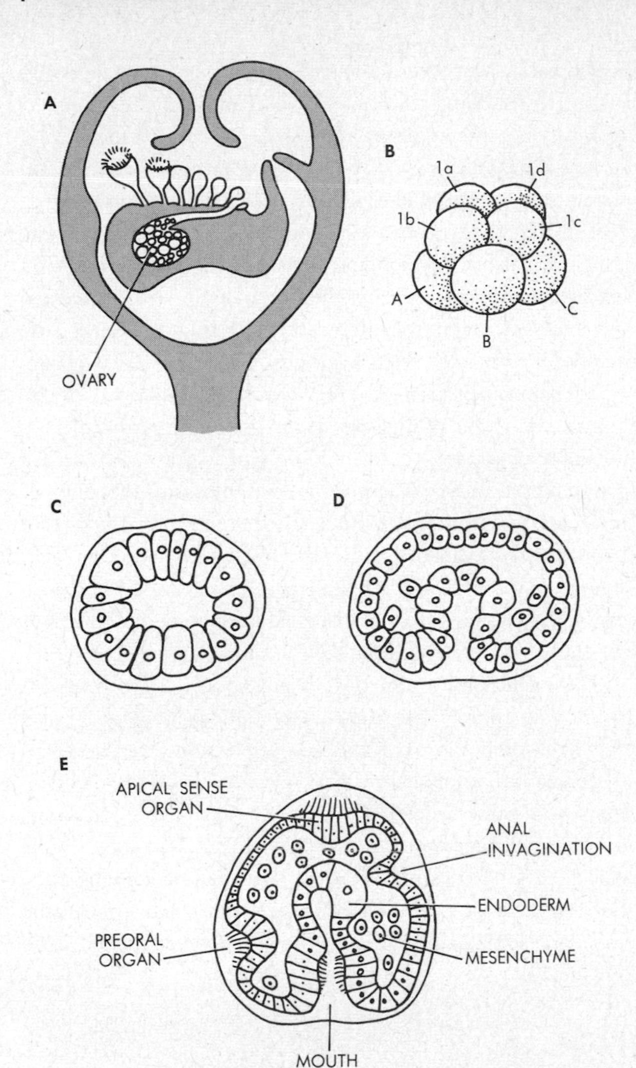

23.16. Early development of entoprocts. (*Adapted from Marcus.*) **A.** Section through adult (with incurled tentacles), showing series of embryos at different stages in brood pouch. **B.** Eight-cell embryo showing spiral cleavage pattern. The *D* macromere is hidden in this side view. **C.** Section through coeloblastula (with external embryo membrane). **D.** Embolic gastrulation, mesenchyme arising from ectoderm. **E.** Section through late embryo, alimentary system nearly complete, mesenchyme arising from both ectoderm and endoderm. A free larva is released soon after this stage.

tentacle crown. The excretory system consists of a single pair of flame bulbs. These lead via ducts to a common nephridiopore located on the ventral side, between mouth and anus. One pair of saclike gonads lies near the main ganglion in the stomach region; ducts lead to a single gonopore between the nephridiopore and the anus.

Some entoprocts are hermaphroditic, but in most the sexes are separate. Fertilization is internal. Eggs develop in a "brood chamber," a depression on the ventral side of the female, between the gonopore and the anus (Fig. 23.16). Early embryos just entering the chamber push aside older embryos already there. A brood chamber is therefore likely to contain a series of embryos at all stages of development. Cleavage occurs in a modified spiral pattern which in some respects is similar to that of rotifers. The blastomeres are mosaic; they form a coeloblastula which gastrulates by emboly. Mesoderm arises partly from the interiorized 4d blastomere, partly from ectoderm. A free-swimming larva results which is often said to be "trochophorelike" but which actually is very unlike a trochophore and is quite unique to entoprocts (Fig. 23.17). It possesses an apical tuft, a girdle of cilia around the blastopore-mouth region, and a sensory *preoral organ* laterally near

23.17. Later development of entoprocts. A, section through free larva, **B,** metamorphosis; diagrammatic sequence shows settling at oral side, internal rotation of alimentary system, growth of stalk and tentacles.

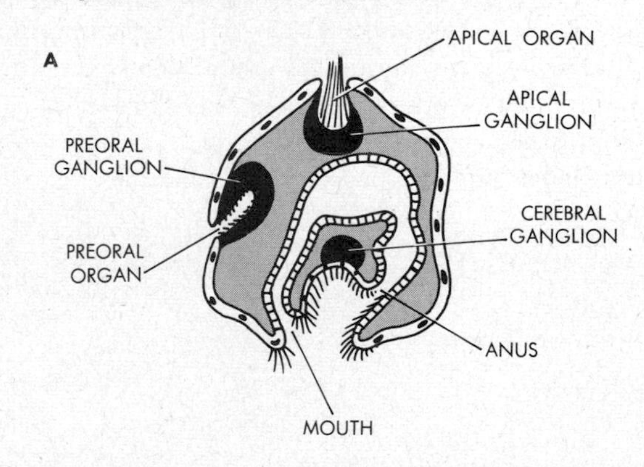

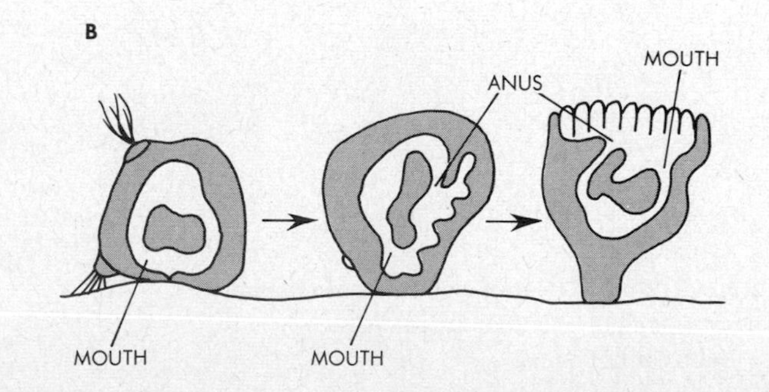

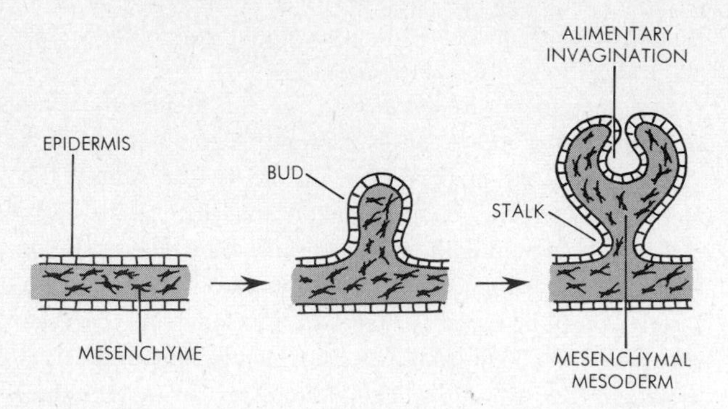

23.18. Entoproct budding from horizontal stolon, diagrammatic. The ectodermal alimentary invagination will later differentiate into an entire alimentary system, which thus arises from parental ectoderm. Parental mesenchyme gives rise to the mesodermal components of the budded individual.

the mouth. Later larvae acquire a complete U-shaped alimentary tract with an anus near the mouth. A remarkable metamorphosis eventually ensues. The larva settles at the rim carrying the ciliary girdle. This rim then grows inward, closing over the mouth and the anus, and elongates into a stalk. Concurrently, the whole alimentary system undergoes an internal rotation of 180°, which brings the mouth and anus away from the attached side and toward the ventral side of the adult. A crown of tentacles then develops around this ventral side.

Notwithstanding the mosaic nature of the embryos, the adults exhibit extensive regeneration capacities, like the acoelomates. Moreover, entoprocts reproduce abundantly by vegetative budding and they form colonies in this manner. Upright individuals arise from stolons creeping over solid surfaces, and buds may also form from upright stalks. A bud starts as a localized evagination from the epidermis (Fig. 23.18). Included within the young bud is some mesenchyme, which will later develop into mesodermal body parts. The epidermal exterior then invaginates and nips off an internal vesicle which will become the alimentary tract. Evidently, the ectoderm gives rise here to a structure, the alimentary system, which is normally derived from endoderm. Analogous development of given body parts from "wrong" germ layers is generally characteristic in budding and will be encountered again in other phyla. The phenomenon indicates that animals may develop their parts from any parental structures that happen to be available, regardless of the original germ-layer source of these structures; and that the germ layer theory evidently does not imply rigidly fixed developmental functions for the three embryonic layers (cf. also Chap. 14).

The evolutionary affinities of entoprocts are obscure. A relation to rotifers has been postulated on occasion, but a cogent argument for such an assumption cannot be made. For the present, entoprocts must be regarded as an independent branch of pseudocoelomate animals, presumably stemming from very distant acoelomate stocks which also produced the coelomates. Thus, on the basis of their major characteristics, the tentative interrelations of the various acoelomate and pseudocoelomate groups may therefore be envisaged as in Fig. 23.19.

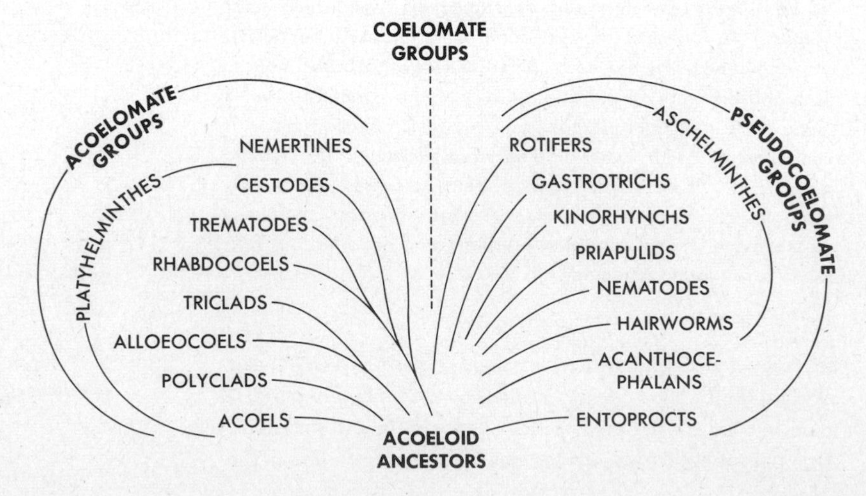

23.19. The likely evolutionary interrelations of acoelomate and pseudocoelomate animals. Read from bottom up.

REVIEW QUESTIONS

1. Give taxonomic definitions of pseudocoelomates, aschelminths, acanthocephalans, and entoprocts. Name the classes of aschelminths and indicate the characteristics which justify inclusion of these animals within a single phylum. On what grounds could Acanthocephala be regarded as an aschelminth class?

2. Describe the structure of a rotifer. What is the phenomenon of cell constancy and in which animal groups is it encountered? Distinguish between amictic and mictic eggs and show by what processes each kind is formed. Describe the life cycle of bisexual rotifers. How do rotifers develop?

3. Describe the structure of a gastrotrich. In what respects is the pharynx similar to that of nematodes? Name representative genera of all aschelminth classes. How does a gastrotrich develop?

4. What is the structure of a kinorhynch? Why are the segments called zonites? What traits of kinorhynchs are reminiscent of arthropods? Describe the excretory system of all aschelminth classes. Which aschelminth groups are hermaphroditic?

5. Describe the structure of a priapulid. Where do these animals live? In what environments may members of other aschelminth classes be found? Describe the organization of the nervous system in all aschelminth classes. How does a priapulid develop?

6. What is the structure of a nematode? In what external traits are males and females different? What are longitudinal chords? Describe the cleavage pattern and the later development of a nematode. What is chromosome diminution and what is its developmental significance? Name nematodes found in man.

7. How many species of (*a*) hairworms, (*b*) other aschelminth classes, have been described? In what structural respects does a hairworm differ from a nematode? Describe the life cycle of a hairworm. Review the structure of the body wall in all aschelminth classes.

8. Describe the structure of a spiny-headed worm. What are lemnisci, the lacunar system, ligament sacs? How does the proboscis of an acanthocephalan differ from that of a nemertine? How does mating take place in (*a*) acanthocephalans, (*b*) all other aschelminth classes?

9. Describe the structure of an entoproct. How does such an animal feed? Mate? Where and according to what pattern does development take place? Describe the organization of the larva and the process of metamorphosis. In what respects are entoprocts different from all other pseudocoelomates?

10. Which pseudocoelomate groups exhibit vegetative reproduction and in what forms? Which of these animals (*a*) can, (*b*) cannot regenerate? Make a diagram indicating possible evolutionary interrelations among the various pseudocoelomate and acoelomate groups. On what morphological evidence can you base such a diagram?

COLLATERAL READINGS

The article by Lansing below reviews the work on longevity in rotifers referred to in this chapter. Accounts of nematode and other pseudocoelomate parasites are included in the texts listed:

Berrill, N. J.: "Growth, Development, and Pattern," chap. 7, Freeman, San Francisco, 1961.

Cameron, T. W. M.: "Parasites and Parasitism," Wiley, New York, 1956.

Caullery, M.: "Parasitism and Symbiosis," Sidgwick and Jackson, Ltd., London, 1952.

Chitwood, B. G.: Nematoda, in "McGraw-Hill Encyclopedia of Science and Technology," vol. 9, 1960.

Hyman, L.: "The Invertebrates," vol. 3, chaps. 12, 13, and 14, McGraw-Hill, New York, 1951.

Lansing, A.: Experiments in Aging, *Sci. American*, Apr., 1953.

Moore, D. V.: Acanthocephala, in "McGraw-Hill Encyclopedia of Science and Technology," vol. 1, McGraw-Hill, New York, 1960.

coelomate bilateria: protostomes

PART 9

[Grade BILATERIA]

Subgrade COELOMATA: mesoderm in part from ectoderm but primarily from endoderm; with true coelom, formed in various ways as a body cavity lined entirely by mesoderm.

PROTOSTOMIAL COELOMATES: development mosaic; cleavage typically in spiral or radially modified pattern; adult mouth formed at or near embryonic blastopore; coelom formed by schizocoelous, enterocoelous, or unique methods; larvae if present various, many with trochophores; primitive groups unsegmented, advanced groups segmented. *Phoronida, Ectoprocta, Branchiopoda, Mollusca, Sipunculida, Annelida, Echiuroida, Oncopoda, Arthropoda*

[*DEUTEROSTOMIAL COELOMATES:* all remaining animals]

As pointed out in Chap. 14, a basic stock of coelom-possessing animals is generally assumed to have evolved from (unspecifiable) acoelomate ancestors. In this original coelomate stock, one major group evidently retained the protostomial tradition of the acoelomate ancestors; development remained mosaic, cleavage typically occurred in a spiral pattern, and the blastopore formed the mouth. Such early coelomate protostomes then appear to have produced an extensive adaptive radiation, and the nine phyla listed above are the present result. The last six of these may be referred to as the *schizocoelomate* animals; they are characterized fairly uniformly by mesoderms formed from 4d-derived teloblast cells, by coeloms developed through splitting of the mesoderm layers, and primitively by aquatic trochophore larvae. The last four of the schizocoelomate phyla comprise segmented animals.

The first three phyla listed all possess a similar food-catching apparatus, the *lophophore;* hence these animals are often referred to as *lophophorates.* They differ from schizocoelomates in that their coeloms form in a variety of unique ways and that their larvae are not trochophores. It appears likely that the lophophorates represent very early offshoots of the main line of descent which led from the original coelomate protostomes to the schizocoelomates (cf. Fig. 14.27). The phoronids appear to have diverged least from the ancient ancestral type, the brachiopods most. Indeed, the brachiopods exhibit certain features which seem to link these animals with the deuterostomial coelomates. As a group, therefore, lophophorates may exemplify a primitive, transitional level of evolution at which an original coelomate stock produced both protostomial and deuterostomial branch lines. The protostomial lines, viz., the lophophorates themselves and the schizocoelomates, form the subject of the five chapters of this part.

CHAPTER 24 lophophorates

The three lophophorate phyla, i.e., Phoronida, Ectoprocta, and Brachiopoda, all comprise sessile or sedentary *filter feeders:* animals which strain microscopic food organisms from their aquatic environment by means of ciliated tentacles. The latter are part of a *lophophore,* a body part in which the mouth is located. The alimentary tract is U-shaped, as is common in sedentary animals, and the anus opens outside the lophophore region. None of these animals possess a head or a clearly distinct head region. The body is simply marked into a lophophore-bearing forepart and a trunk. Each of these two parts encloses a portion of the coelom. The lophophore-coelom is the *mesocoel,* the trunk-coelom is the *metacoel,* and these two cavities are separated completely or partially by a transverse peritoneal *septum.* Lophophorates are without breathing systems, and all are endowed with a potential of forming an exoskeleton, expressed differently in the three phyla.

Lophophorates typically develop via free-swimming larvae. Although such larvae are frequently described as "trochophorelike" or as "modified trochophores," they are actually different from trochophores and quite unique to each of the three phyla.

PHYLUM PHORONIDA

Marine, wormlike, tubedwelling; with horseshoe-shaped lophophore; excretion through metanephridia; with closed circulatory system; development mosaic, cleavage spiral to irregular; *actinotroch* larvae with solenocytic protonephridia. *Phoronis, Phoronopsis*

The only two existing genera of this phylum encompass 16 described species. The animals live in shallow water along sandy or muddy shores, either as solitary individuals in upright tubes or in tangled aggregations of numerous tubes. Representing the exoskeleton of phoronids, the tubes are epidermal, parchmentlike secretions in which the animals may move freely. Chitin is the main constituent of the tubes and sand grains or other particles adhere to them externally. The animals within them range in size from about ½ in. to roughly ½ ft. (Fig. 24.1).

From the upper end of the tube a phoronid may project its lophophore, which consists of a double row of ciliated tentacles set on a double ridge of the body wall. The mouth is a slitlike, crescentshaped funnel between the two ridges, centered in the middle between the left and right *arms* of the lophophore. The outer, convex ridge is considered to be on the ventral side of the animal, and the lophophore arms coil in a dorsal direction. A plane through the mouth and a dorsal, anus-containing elevation marks the main symmetry plane. To either side of the anus is a nephridiopore, and in some species a surface groove leads from a nephridiopore to a glandular depression, the *lophophoral organ,* embraced by the lophophore coil of that side. The space surrounded by the lophophore arms serves as a brood chamber in several species. Overhanging the mouth from the dorsal lophophore ridge is

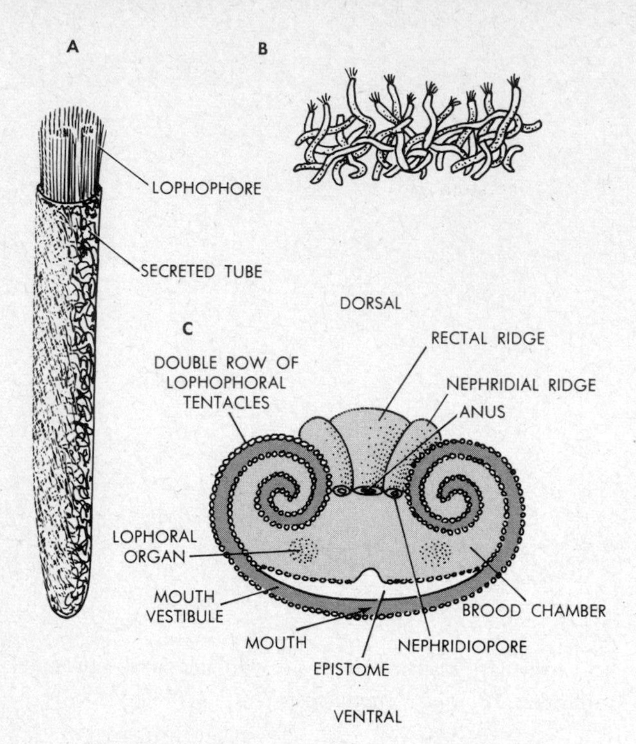

24.1. Phoronids, external views. **A,** general appearance of a phoronid in straight tube; **B,** tangled aggregation of phoronids (*adapted from Shipley*); **C,** the anterior, lophophoral end of a phoronid (*adapted from Benham*).

a flap of tissue, the *epistome,* which may close off the mouth. This midregion of the lophophore is also the place where new tentacles are developed; the oldest tentacles are at the outer extremities of the lophophore arms.

Inasmuch as they are extensions of the trunk, the lophophore and its tentacles have the same tissue structure as the body wall of the trunk (Fig. 24.2). Externally, the epidermis contains many glandular and neurosecretory cells, and the tentacular epidermis is also covered with a thin cuticle. The epidermal tissue rests on a basement membrane which is underlain first by a layer of circular muscles and then by one of longitudinal muscles. In the midtrunk region, the longitudinal muscles usually form radial, ribbonlike bundles projecting into the coelom. The innermost layer of the body wall, lining the coelom, is a syncytial peritoneum. This membrane also traverses the coelom as mesenteries and is applied against the outer surface of the U-shaped alimentary tract. In a phoronid there are typically four mesenteries, and Fig. 24.2 shows that they are double peritoneal membranes; they divide the metacoel, or trunk coelom, into four longitudinal compartments. The metacoel is also separated by a transverse peritoneal septum from the mesocoel, or lophophore-coelom. The latter extends into all tentacles and is lined analogously by a peritoneal membrane of its own.

The nervous system of phoronids is an integral part of the epidermis; neural elements are said to be *intraepidermal* (Fig. 24.3). The epidermis therefore probably qualifies as a neuroepithelial tissue, undoubtedly a primitive condition not too different from that in primitive flatworms. However, the phoronid system is not quite as diffuse. The nervous elements

24.2. **A,** tissues of the phoronid body wall, diagrammatic. **B,** cross section through phoronid trunk. (*Adapted from Pixell.*) Note ridges of longitudinal body-wall muscles, overlain by peritoneum.

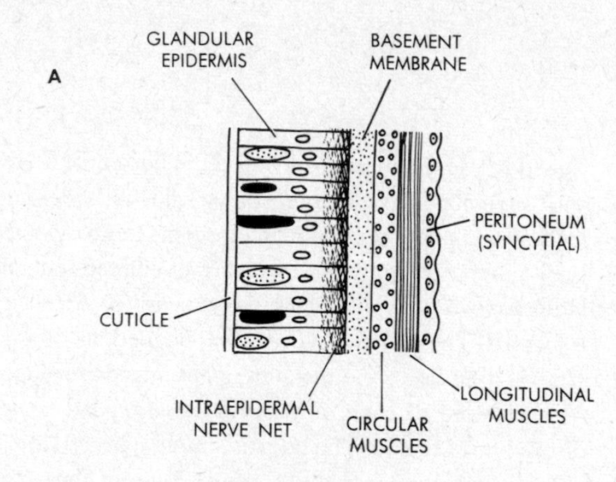

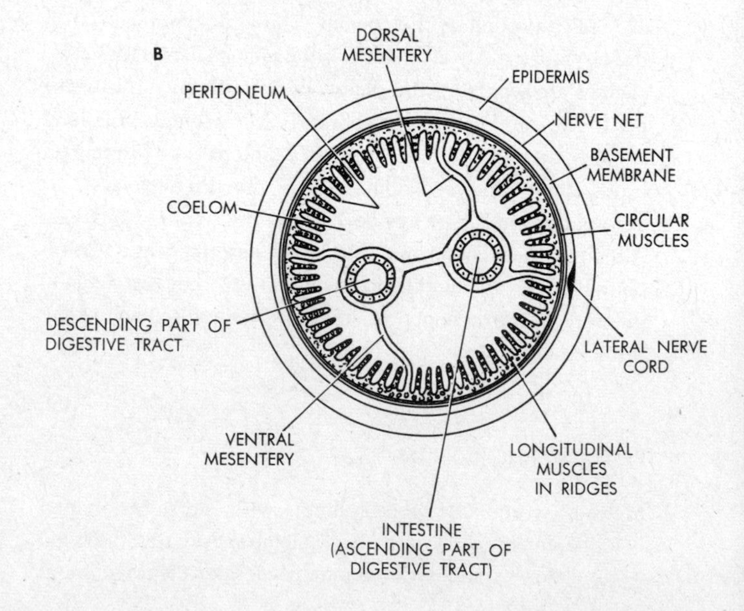

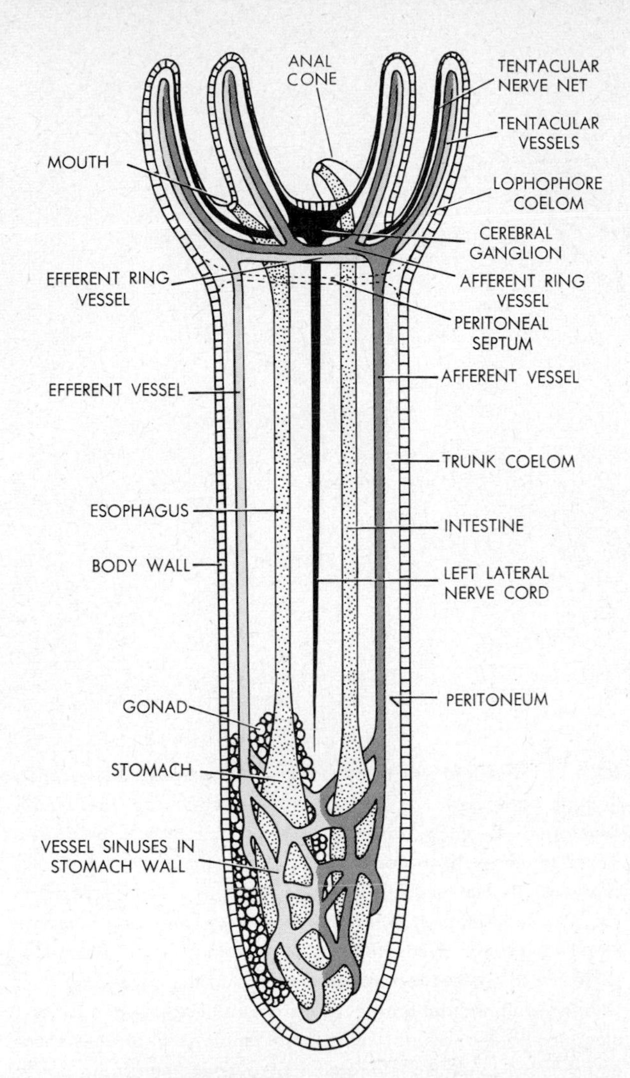

24.3. The alimentary, nervous, circulatory, and reproductive systems of a phoronid, schematic. Peritoneal membranes enveloping alimentary tract and blood vessels not shown.

are thickened into an epidermal nerve ring along the outer edge of the lophophore, and a median ganglionic enlargement in this ring represents the main neural center. Branch nerves from the ring pass into the epidermis of each tentacle, and a single longitudinal cord runs through the trunk epidermis along the left side of the body. In some cases a smaller right lateral cord is also present. The remainder of the body is innervated by transverse branch nerves emanating from the left longitudinal cord.

The alimentary tract is regionated into esophagus, stomach, and intestine, and the inner mucosal lining is ciliated throughout. Muscle and connective tissue layers form the wall of the tract between mucosa and peritoneum. Food consists largely of planktonic microorganisms, predominantly diatoms. Digestion is mainly intracellular, another primitive trait. The circulatory system (cf. Fig. 24.3) is composed of two longitudinal trunk vessels interconnected posteriorly and opening anteriorly into a pair of ring vessels in the lophophore base. From each ring vessel emanate branches into the tentacles, the two branches to each tentacle joining into a single blind vessel within the tentacle. An extensive blood sinus in the stomach wall is part of the longitudinal trunk circulation. The blood contains hemoglobin-carrying red corpuscles. Such cells are present also in the coelomic fluid, in which are found amoeboid white cells as well. The metanephridia are a pair of ciliated tubules located anteriorly in the trunk and held in place by the mesenteries. These tubules open into the coelom through ciliated nephrostome funnels, and they communicate with the outside via the nephridiopores next to the anus. Metanephridia provide the only exit from the coelom; they serve also as gonoducts.

Most phoronids are hermaphrodites, though some species are separately sexed. The gonads are loose masses in the peritoneum, near the blood sinuses of the stomach wall. Gametes are shed into the coelom and from there they reach the exterior via the metanephridia. In some phoronids fertilization takes place internally right within the coelom; in other cases the zygotes form externally. Development may occur in the open sea or, more usually, in the brood space between the lophophore arms. By producing adhesive secretions the lophophoral organs are believed to assist in retaining the embryos at such brood sites.

Cleavage patterns apparently vary; some are typically spiral, others are more nearly radial, still others are irregular. In either case a coeloblastula with an apical tuft is formed, and gastrulation occurs by embolic invagination (Fig. 24.4). The mesoderm arises predominantly by ingrowth of cells from the endoderm. A larval and later adult mouth is formed at the blastopore, and the blind inner end of the endodermal invagination comes to extend in an L-shaped curve until it meets the ectoderm on one side; the anus then breaks through at that point. A ring of long cilia, the *telotroch,* soon develops around the anal region, and another ciliary band, the *postoral girdle,* forms in an oblique plane between the level of the mouth and that of the anus. This girdle proliferates into a wreath of larval tentacles. Extensive growth of the ectoderm between the mouth and the apical tuft leads to the establishment of a *preoral lobe,* a prominent hood over the mouth. The

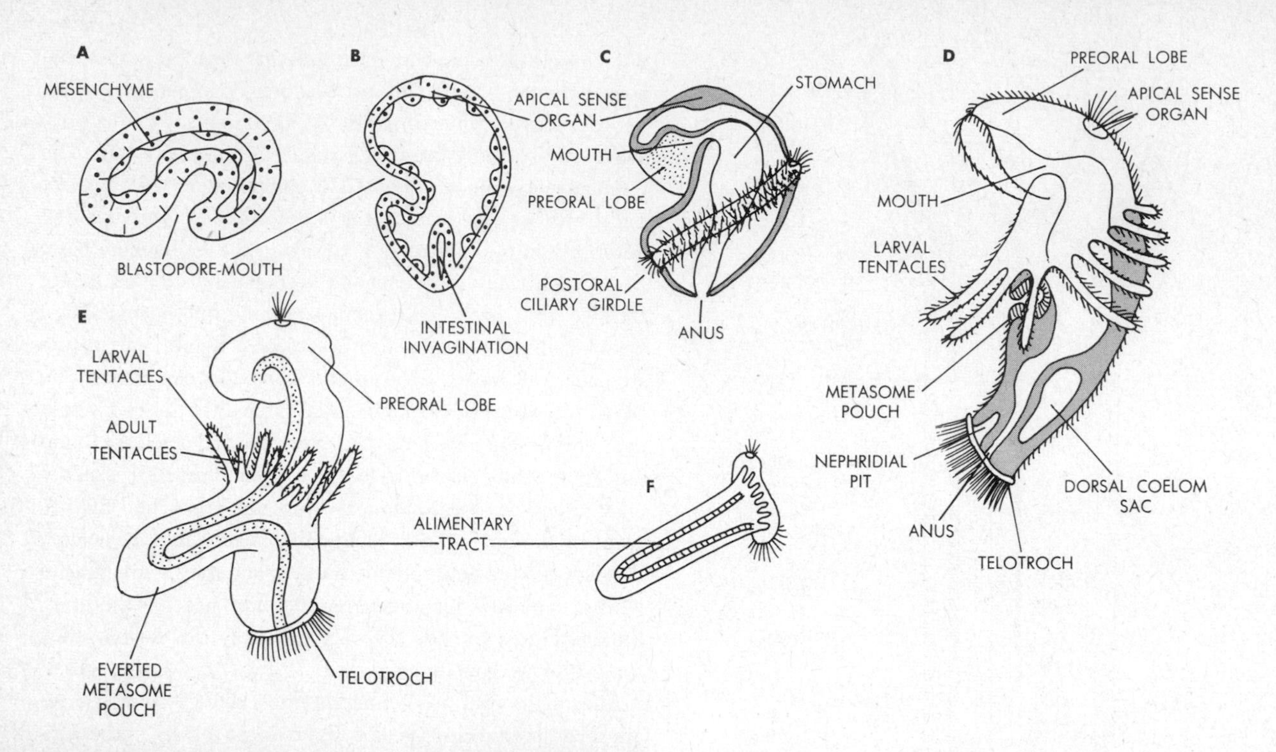

24.4. Phoronid development. Diagrams: A, section through early gastrula; **B,** postgastrula, alimentary tract nearly developed; **C,** early larva; **D,** mature actinotroch larva. The larval tentacles have formed as outgrowths from the earlier ciliary girdle. Note invagination of metasome pouch, and coelom sac. The nephridial-pit tissue will extend interiorly and bifurcate into two groups of solenocytes. **E.** Metamorphosis. The metasome pouch has everted, carrying a loop of the alimentary tract with it. Adult lophophoral tentacles have already begun to form along the band of degenerating larval tentacles. **F.** Establishment of adult by continued elongation of everted metasome pouch and degeneration of other larval parts. **Photo:** actinotroch larva, from life. Note larval tentacles, digestive tract, metasome pouch at midbody, and posterior ciliary girdle.

coelom arises near the anus from mesenchyme cells, which migrate and arrange themselves into a hollow dorsal sac. Note that this pattern of coelom formation is neither schizocoelous nor enterocoelous. The coelomic sac later enlarges and grows around the intestine, forming a ventral mesentery where the advancing parts meet and a peritoneal septum anteriorly. At this stage the embryo usually escapes from the brood chamber as a young *actinotroch* larva.

This larva is about ¼ in. long and it leads a free-swimming existence for several weeks. In the course of this time it develops complete nervous, muscular, and circulatory systems, as well as solenocytic protonephridia. These form as a bifurcating invagination from a *nephridial pit* near the anus. Also, the posterior region of the larva elongates greatly and the alimentary tract lengthens considerably. Moreover, a progressively deepening invagination of the body wall develops in the midregion of the larva, between mouth and anus (cf. Fig. 24.4). This *metasome pouch* plays a key role in the ensuing metamorphosis. The basic event of metamorphosis is completed in the span of a few minutes. The larva undergoes spasmic contractions which result in the sudden eversion of the metasome pouch. Body-wall muscles and a loop of the alimentary tract are carried along into the everted part, which now constitutes the adult trunk. Most of the other larval components then degenerate and are cast off. The adult lophophore grows in the region of the larval tentacles, and it later acquires its own mesocoel from mesenchyme cells. The solenocytes degenerate, but the protonephridial tubules of the larva become part of the adult metanephridia.

Note that the larval solenocytes represent the only fundamental feature which actinotrochs share in common with trochophores. To be sure, the two larval types correspond also in possessing apical tufts and ciliary girdles, but neither of these is particularly unique in pelagic, free-swimming larvae; nor, indeed, are solenocytes particularly uncommon at these evolutionary levels. On the other hand, actinotrochs and trochophores do differ very significantly in methods of mesoderm- and coelom-formation and in patterns of metamorphosis. For these reasons it is erroneous to regard actinotrophs merely as "modified trochophores," as is often the case. The two larval types are clearly and basically different, and they cannot be adduced as evidence to prove any close phylogenetic affinities of the animals they represent.

Phoronids regenerate well as adults. Cut sections of the trunk may develop into whole individuals and damaged or lost tentacle crowns are regenerated readily. Like flatworms, moreover, phoronids may reproduce vegetatively by spontaneous transverse fragmentation, a normal process in species which form intertwined aggregations of individuals.

PHYLUM ECTOPROCTA

Moss animals (also *Bryozoa* or *Polyzoa*); always sessile in colonies, formed by budding from single microscopic zooids; with lophophore and U-shaped alimentary tract, the anus opening outside the tentacle crown; without circulatory or excretory systems; development mosaic, cleavage largely radial; with *cyphonautes* or other larvae.

Class GYMNOLAEMATA: marine; lophophore circular, without epistome; body-wall musculature absent; without coelomic connection between zooids; exoskeleton calcified or membranous; often polymorphic. *Bugula, Electra, Vesicularia, Membranipora*

Class PHYLACTOLAEMATA: fresh water; lophophore horseshoe-shaped, with epistome; body-wall musculature present; with coelomic connection between zooids; exoskeleton gelatinous; not polymorphic. *Plumatella, Cristatella, Pectinatella*

Ectoprocts are among the most ancient fossil animals known, the earliest dating back to Cambrian and possibly pre-Cambrian times. Some 15,000 extinct species have been recognized, and living species are variously stated to number between 5,000 and 10,000. The animals have a remarkably close though only superficial resemblance to the Entoprocta. Indeed, both groups are often—and unjustifiably—classified even now as subphyla within a single phylum Bryozoa (or Polyzoa in British tradition). Entoprocts are pseudocoelomates, ectoprocts are true coelomates, and this very basic difference alone requires that the two groups be regarded as separate phyla. As primitive coelomates, to be sure, the ectoprocts may well be very distantly related to the entoprocts; but the ectoprocts appear to be more nearly allied to the phoronids.

The individual ectoproct is a zooid of microscopic dimensions, encased in a secreted exoskeleton which is part of its body wall. In the marine forms, the exoskeleton is essentially a box open at the unattached side, where the lophophore may be protruded. In simply constructed types (e.g., *Vesicularia*), the exoskeleton is a vaselike chitinous membrane fused to and secreted by the epidermis. In most other genera, a substantial layer of calcareous deposits lies between the chitinous exterior and the epidermis (Fig. 24.5). The peritoneum directly lines the epidermis interiorly in marine ectoprocts, a body-wall musculature being lacking in these animals. It is customary to refer to the fixed body wall of ectoprocts as the *cystid* and to the movable interior parts as the *polypid*. The latter consists of the lophophore apparatus, the alimentary system, the internal musculature, and the reproductive organs.

The unattached side of the exoskeletal enclosure is often partially covered by a *frontal membrane*, a chitinous shield continuous with the lateral body wall (Fig. 24.6). This mem-

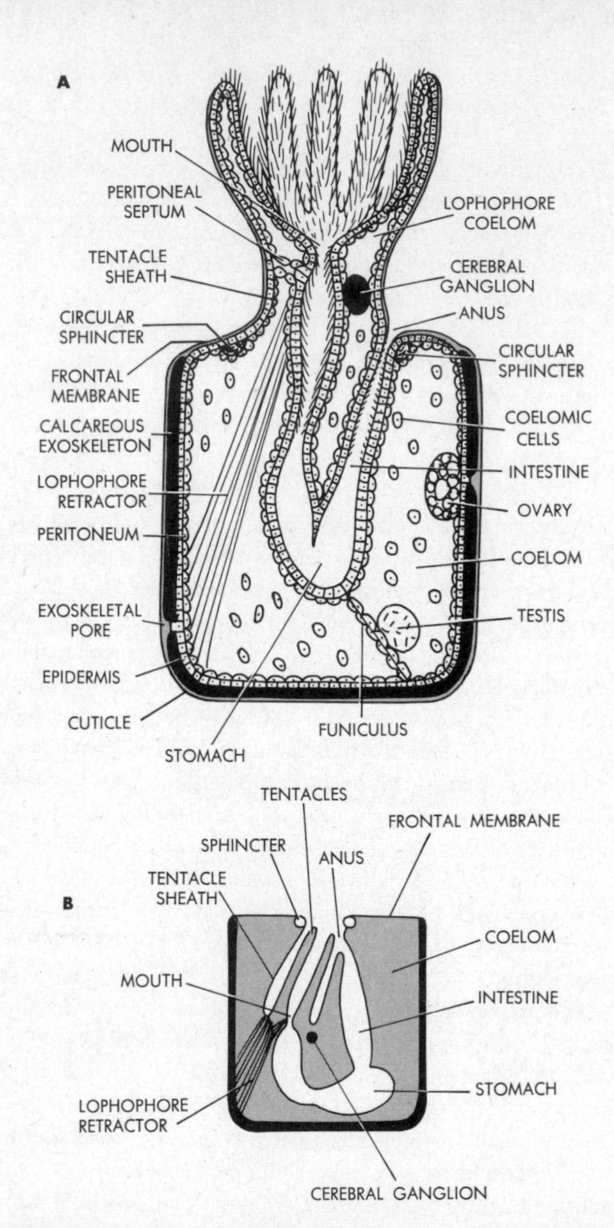

24.5. Ectoproct structure schematic. *(Adapted from Marcus.)* **A,** section through a marine ectoproct, lophophore extended; **B,** diagram of position of lophophore in retracted condition.

brane may or may not be calcified, and it leaves an opening just large enough for protrusion of the lophophore. In many cases this opening is protected by an *operculum,* a movable cover, and often also by spines arching over the operculum. The lophophore is structured essentially as in phoronids. A single row of tentacles is set on a *tentacle sheath,* which inverts when the lophophore is withdrawn and which then forms a tube enclosing the tentacles. Withdrawal is accomplished by prominent lophophore retractor muscles. Other muscles operate the operculum, if present. Protrusion of the lophophore is accomplished by hydrostatic pressure exerted on the fluid interior of the animal. In some cases such pressure is generated by the frontal membrane, which may be depressed by muscles and which thus may function like a diaphragm. In other cases the animal possesses an air sac with a separate opening to the outside. This sac may be pumped full with sea water, and the enlarging sac then squeezes the rest of the interior and forces the lophophore out (cf. Fig. 24.6).

Circular and radial muscles surround the mouth, which leads into a ciliated pharynx. The remainder of the alimentary tract consists of esophagus, stomach, and intestine, the latter passing to the anus located on the tentacle sheath, i.e., outside the lophophore (cf. Fig. 24.5). The esophagus is lined with epitheliomuscular cells, of a type similar to those of Hydrozoa. The whole alimentary tract is covered externally by the peritoneum. As in phoronids, the coelom is subdivided into a mesocoel and a metacoel. Neither cavity opens to the exterior; nephridia are absent. The nervous system consists of a main ganglion between mouth and anus, a nerve ring around the lophophore base, and a plexus in the body wall. The coelomic fluid contains free amoeboid cells and connective tissue fibers.

Gonads without ducts are attached to the peritoneum. Most ectoprocts are hermaphroditic. Fertilization occurs internally in the coelom, and self-fertilization may be the rule. It is not clear in any event how reproductive cells get into or out of the coelom; temporary openings in the lophophore region are undoubtedly formed. Many ectoprocts brood their eggs, either externally in the lophophore region like phoronids, or in invaginated epidermal sacs formed around the tentacle sheath, or directly within the coelom. Some species exhibit *polyembryony,* equivalent to identical twinning; a zygote divides a number of times and each resulting separate cell gives rise to a complete individual. Cleavage is radial or biradial, and a coeloblastula gastrulates by delamination at the vegetal pole; four vegetal cells elongate into the blastocoel and their inner ends then constrict off as endoderm. The larval mesoderm arises from these cells (Fig. 24.7).

In marine ectoprocts two kinds of larvae develop. Nonbrooding species (e.g., *Electra*) produce a *cyphonautes,* a conical ciliated larva with apical sensory bristles and a prominent cilary girdle around its base. This girdle rings the larval

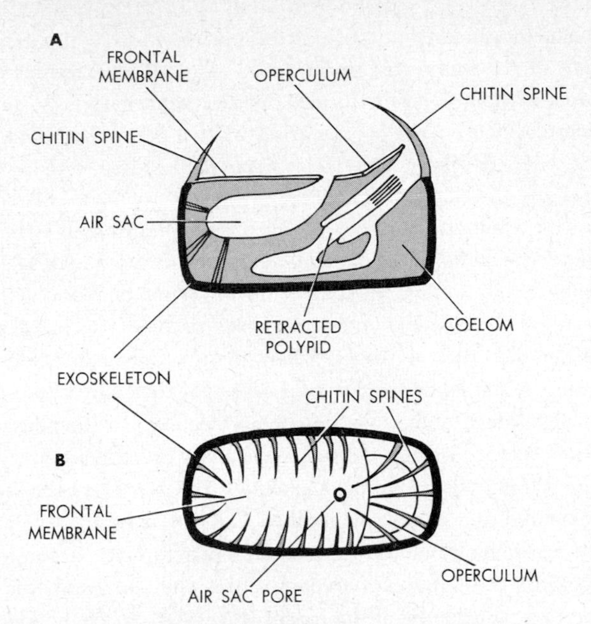

24.6. Marine ectoprocts. Diagrams: A, sectional side view of ectoproct showing air sac, operculum and protecting spines; **B,** top view of **A. Photos: A,** colonies of *Bugula,* suspended from rock overhang; **B,** colony of *Electra,* on red algae.

mouth and anus. Between these two openings is an invaginated ectodermal *adhesive sac,* and another ectodermal invagination near the mouth forms a glandular *pyriform organ.* A cyphonautes lives free for about 2 months, during which time it also secretes two shell plates, or *valves,* on its surface. Brooding species form larvae of a variety of shapes. They are generally without alimentary systems and they live free for a short time only. Such larvae have not been specially named.

Metamorphosis involves very drastic transformations which occur similarly in the various larval types (cf. Fig. 24.8). The adhesive sac of a larva everts and attaches the animal to a solid surface. The whole larva then disintegrates. It becomes little more than a mound of loose cells covered by a layer of larval ectoderm. The latter subsequently invaginates and forms a vesicle, and a second vesicle later constricts off the first. Both then produce a second cell layer around them. The outer layer so formed represents the adult mesoderm; it gives rise to the peritoneum, the musculature, and the gonads. The inner layer of the first vesicle later develops into the lophophore and the pharynx, and the inner layer of the second vesicle represents the endoderm, which will form the remainder of the adult alimentary system.

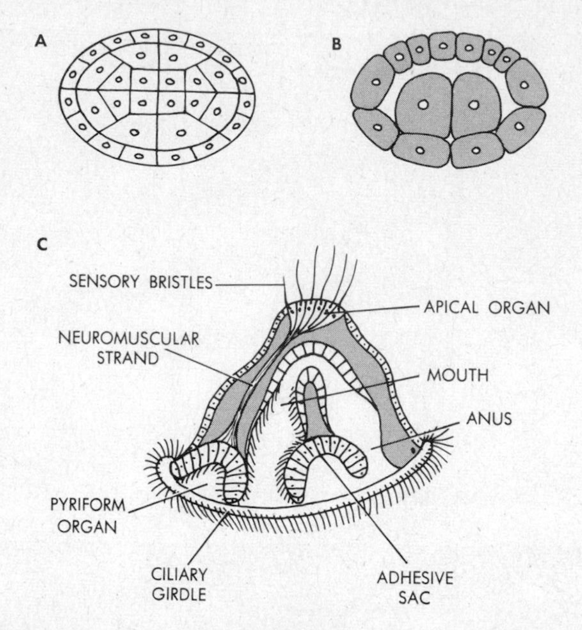

24.7. Development of marine ectoprocts. (*After Marcus.*) **A.** Top half of 64-cell embryo seen from animal pole, showing biradial cleavage pattern. **B.** Section through early gastrula, showing delamination of endoderm cells at vegetal pole. **C.** Sectional view of cyphonautes larva.

Three points are noteworthy here. First, with the exception of the exterior ectoderm, the structure of the larva is in no way related to the structure of the adult. Secondly, *all* adult body parts arise from larval ectoderm, the alimentary system included—another instance of the diverse functional potentials of the embryonic germ layers. And thirdly, the ectoproct coelom is again formed neither schizocoelously nor enterocoelously. The coelom arises when a peritoneal boundary layer develops directly between the exterior ectoderm and the developing alimentary system.

The resulting ectoproct is called an *ancestrula;* it is the source of the adult colony. All ectoprocts are colonial, such colonies often attaining sizes of up to a yard or more, with millions of individual zooids. Colonies may be attached flat to surfaces such as rocks, wharf pilings, seaweeds, shells, sponges, and other objects in the water, or they may be more or less erect and have branching, leaflike, or numerous other shapes. In many instances, stolons or rootlike downgrowths provide additional anchorage. The zooids are usually cemented directly to one another. Colonies arise by vegetative budding from the ancestrula. Buds form in a parent zooid by an ingrowth of epidermis from the body wall, leading to a partitioning of the parent (Fig. 24.9). In the budded compartment all structures then develop directly from the epidermis, as in the original formation of the ancestrula from larval ectoderm. In the marine ectoprocts adjacent zooids are not interconnected, though the developing exoskeleton leaves pores in places. However, such pores are covered by the epidermis of each zooid and the two coeloms therefore remain separate (cf. Fig. 24.5).

Many marine ectoprocts are polymorphic. In addition to the feeding *autozooids* a variety of *heterozooids* is formed (Fig. 24.10). All these are modified autozooids in which the

24.8. Metamorphosis in marine ectoprocts. A. Cyphonautes with everted adhesive sac, just before settling. **B.** Attached larva with disintegrating interior tissues. **C.** Formation of first ectodermal vesicle. **D.** Secondary vesicle formed from larval ectoderm, giving rise to adult endoderm. Note mesoderm cells becoming arranged into a peritoneum. **E.** Adult body plan established in essential features.

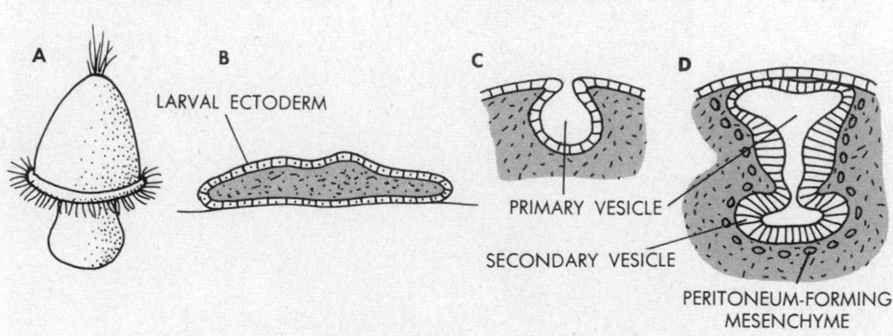

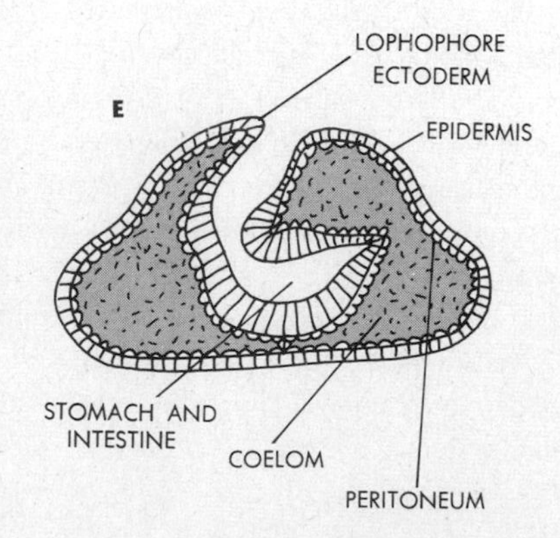

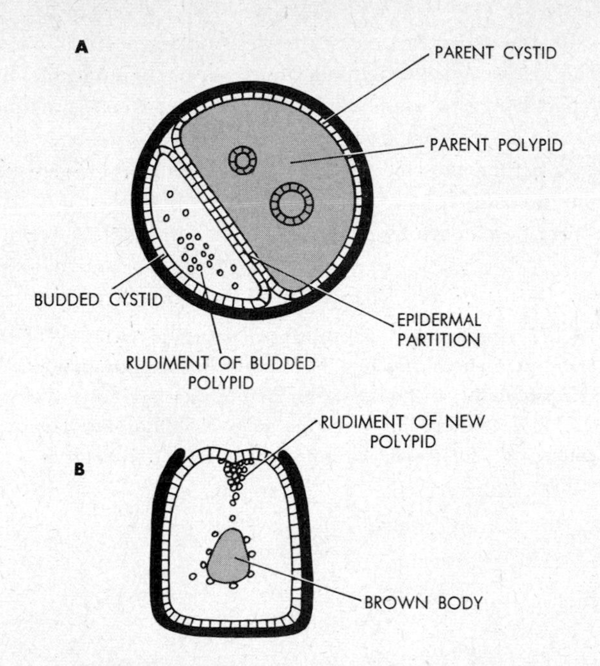

24.9. Ectoproct budding and regeneration. **A.** Formation of bud by partitioning of parent cystid, cross-sectional view. **B.** Sectional longitudinal view of cystid, with old polypid degenerated into brown body and new polypid beginning to regenerate.

cystid portions are overdeveloped and the polypid portions remain greatly underdeveloped. A common type of heterozooid is an *avicularia* (e.g., as in *Bugula*), in which the operculum is enlarged into the form of a bird's beak. Interiorly a coelom is present, as are highly developed muscle systems which may open and close the opercular beak. The polypid is either absent altogether or is reduced to a tiny *setiferous organ.* Avicularia snap their beaks and prevent larvae or other organisms from settling on an ectoproct colony. Another type of heterozooid is a *vibracula,* in which the operculum has become a long bristle. The latter sweeps back and forth over the colony surface. Stolons, stalks, and rootlike anchors in a colony represent heterozooids called *kenozooids.* In some species, given individuals are specialized to form brood chambers and they then are called *gonozooids.*

As might be expected from their budding potential, ectoprocts exhibit marked powers of regeneration. Indeed, most ectoprocts regularly degenerate their polypids; the polypid of a given zooid becomes reduced to a compact *brown body* lying in the cystid (cf. Fig. 24.9). A new polypid subsequently regenerates from the cystid, and the brown body is digested in the newly developing stomach. Brown-body formation is believed to represent a form of excretion. Colonies are often zoned according to the degeneration-regeneration cycles of their zooids. Active young zooids are usually present along the peripheral zone of a colony. Farther inward is a zone of degeneration and brown-body formation, and still farther inward lies a regeneration zone, where active individuals are again present. The oldest zooid, the ancestrula, never develops any gonads. Sexual activity occurs only in zooids farther out, i.e., only after colonies have attained substantial size and age.

Freshwater ectoprocts number only about 50 species. Their

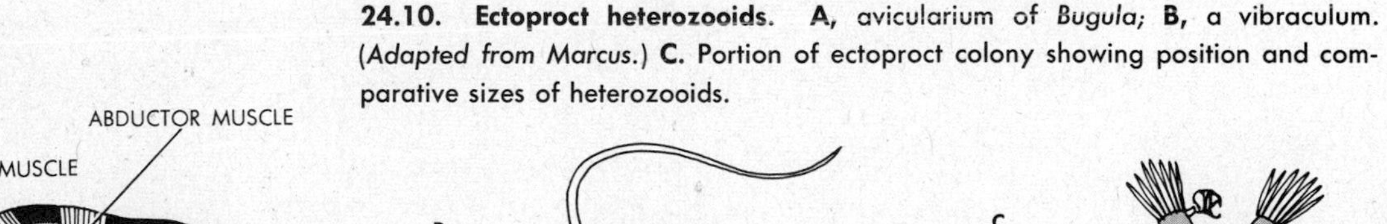

24.10. Ectoproct heterozooids. **A,** avicularium of *Bugula;* **B,** a vibraculum. (*Adapted from Marcus.*) **C.** Portion of ectoproct colony showing position and comparative sizes of heterozooids.

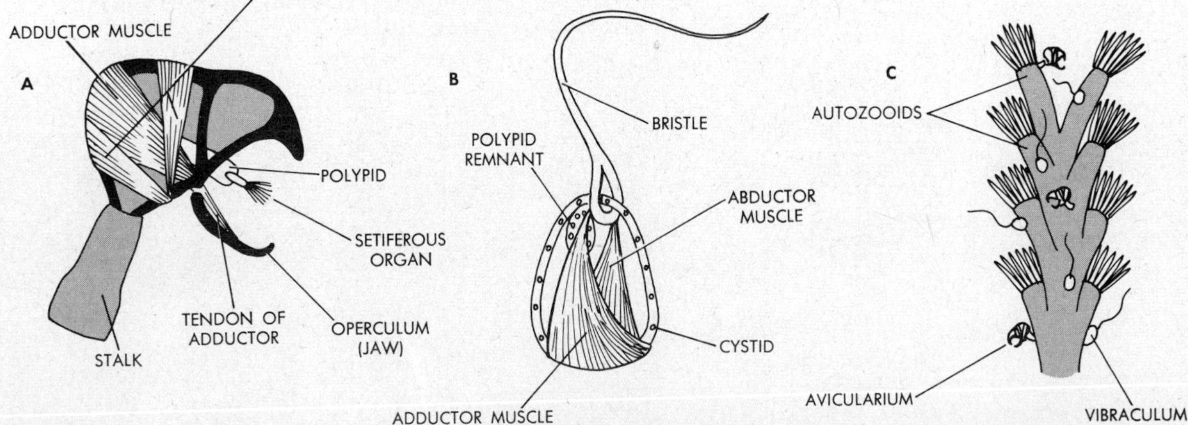

structure is basically the same as that of their marine relatives (Fig. 24.11). However, the exoskeleton is never calcified but forms large gelatinous masses (which often clog drain pipes and spillways). Opercula are absent. The zooids are interconnected, all individuals of a colony sharing a common, continuous cystid wall. The layers of this wall are, from the outside inward, chitin and gelatinous exoskeleton, epidermis, ectoderm-derived circular muscles, basement membrane, mesoderm-derived longitudinal muscles, and peritoneum. The lophophore is horseshoe-shaped, not circular as in marine types. The stomach is connected to the base of the body wall by a peritoneum-covered and muscle-containing tissue cord, the *funiculus*.

The developmental pattern is modified in adaptation to the

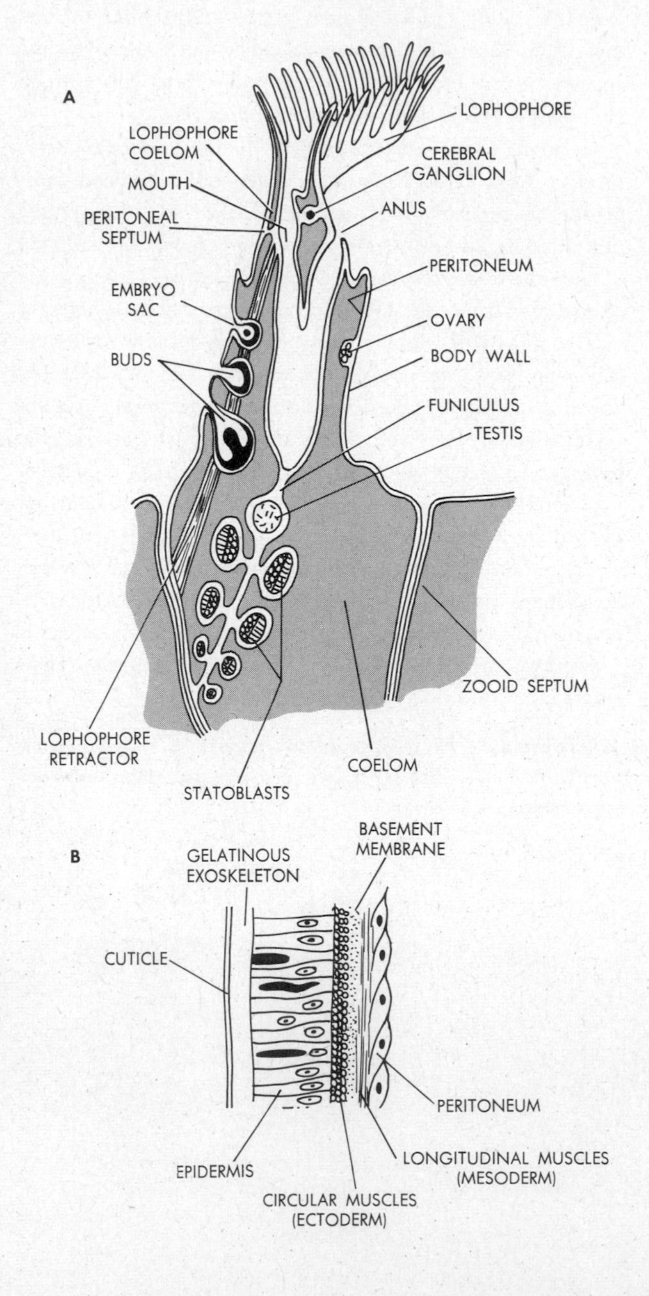

24.11. Freshwater ectoprocts. Diagrams (*Adapted from Brien*): **A.** General structure. Note horseshoe-shaped lophophore and statoblasts in funiculus; for details on statoblasts, cf. Fig. 24.12. **B.** The tissue layers of the body wall. **Photo:** portion of a colony of *Plumatella*; a statoblast-bearing funiculus is well visible.

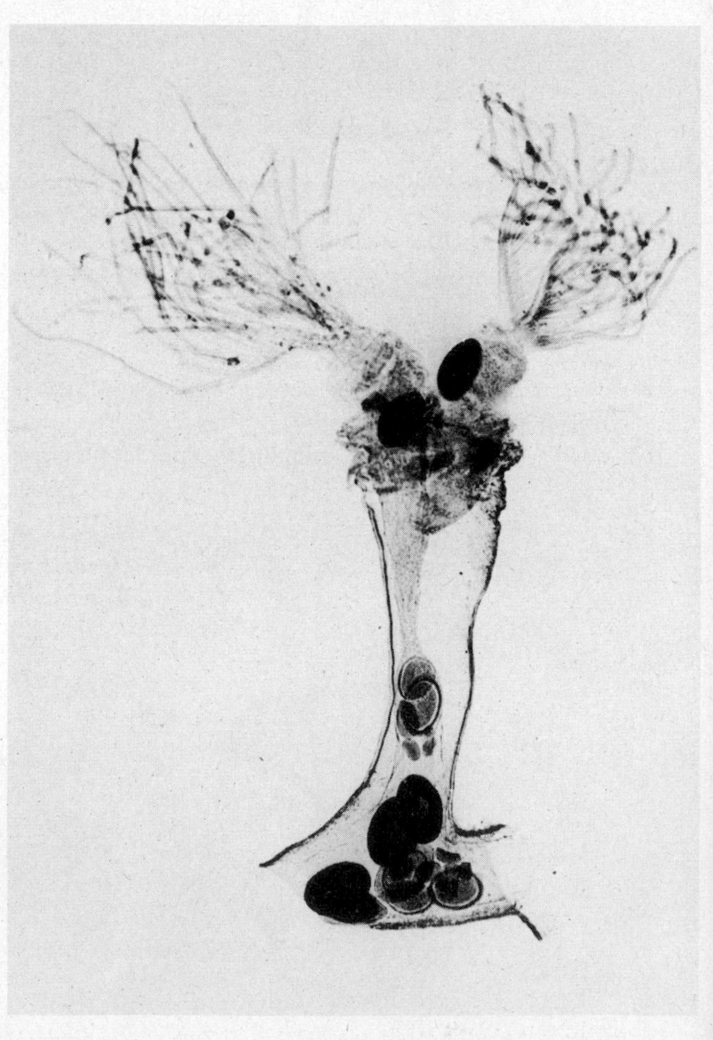

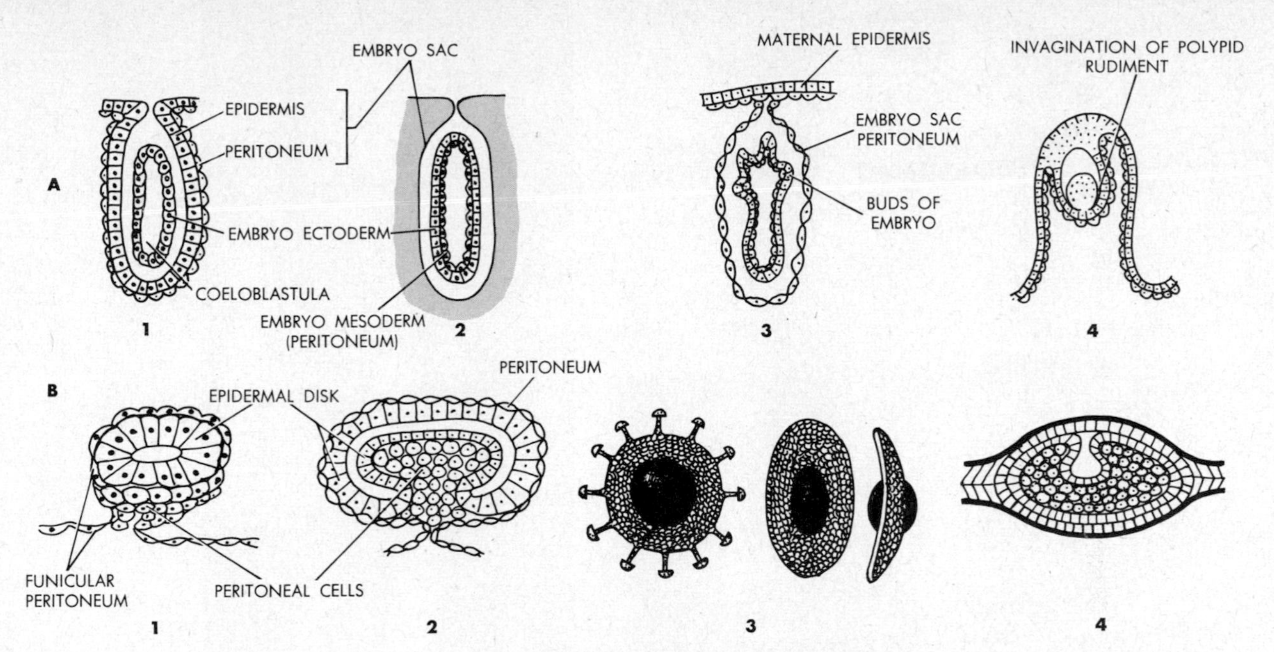

24.12. Development of freshwater ectoprocts. **A.** embryonic development. 1. Cleavage and formation of coeloblastula in parental embryo sac. 2. The embryo has become two-layered. 3. Additional cystid buds develop at one end of embryo, forming embryonic colony; note that epidermis of original embryo sac largely degenerates. 4. Enlarged view of single cystid bud on embryo colony, showing invagination which will develop into polypid portion of individual zooid. At this or a somewhat later stage the whole embryo colony escapes as a larval complex. **B.** Development from statoblasts. 1. Internal tissues of statoblast formed in parental funiculus. 2. Later stage, epidermal disk growing as a double layer around peritoneal inner cell mass; the outer epidermal layer forms the shell. 3. External views of mature statoblast shells. Hooked type is of *Pectinatella*. Statoblasts of *Cristatella* are similarly hook-equipped. Other sketches are side and top views of statoblasts of *Hyalinella*. 4. Statoblast germination; zooid-forming invagination develops from inner epidermal layer.

freshwater environment. All freshwater ectoprocts are hermaphroditic, and all brood their eggs in sacs invaginated from the body wall. After cleavage, the coeloblastula elongates and becomes two-layered, the inner layer representing the peritoneum (Fig. 24.12). The embryo is now in effect the cystid of an ancestrula. It produces several additional cystid buds at one end, and such a larval colony then escapes from the brood sac. The colony soon settles and continues to bud. Later, alimentary systems develop in each cystid from invaginated ectoderm vesicles, as in the marine types.

Freshwater ectoprocts also reproduce vegetatively by means of *statoblasts,* formed in the funiculus of an adult (cf. Figs. 24.11 and 24.12). A statoblast consists of an internal mass of epidermal and peritoneal cells and an exterior protective shell of species-specific construction. Functioning like the gemmules of freshwater sponges, statoblasts are formed in huge numbers, usually in the fall, when the adult colonies disintegrate. The reproductive bodies withstand dry conditions and low winter temperatures, and they germinate in the spring by forming two-layered vesicles as in budding.

Freshwater ectoprocts are generally considered to be more primitive than their marine relatives. This assumption is based on the primitive phoronidlike horseshoe shape of the lophophore, the presence of a body-wall musculature, and the absence of specialized features such as complex exoskeletons, opercula, and polymorphism. The freshwater ectoprocts are

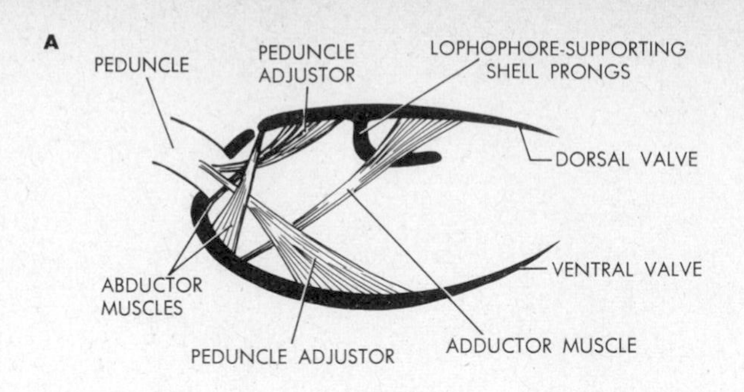

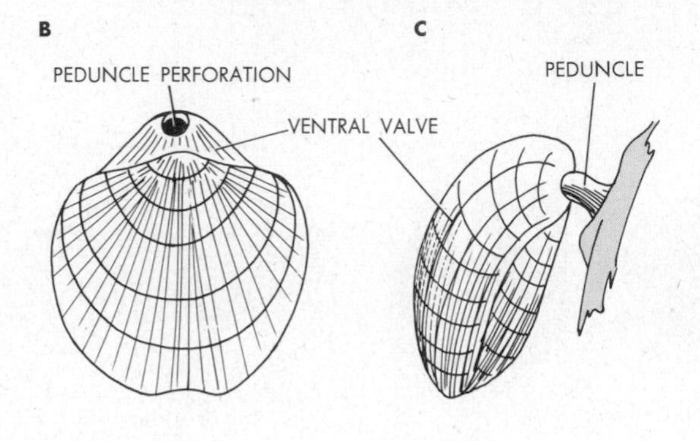

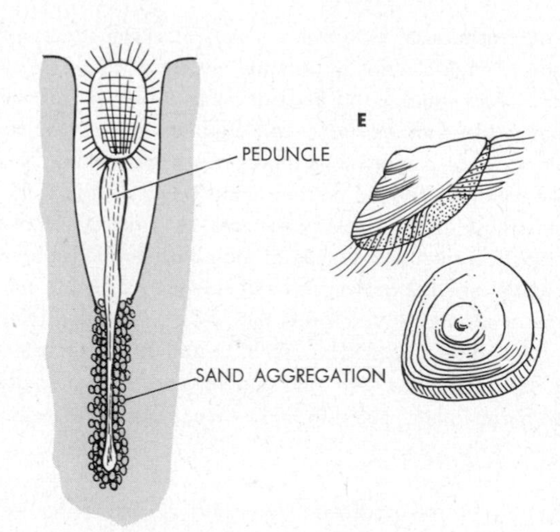

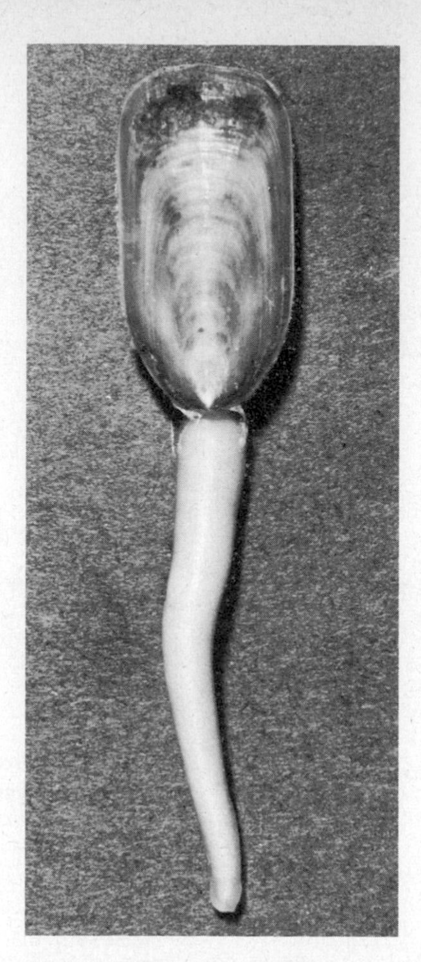

24.13. Brachipods. Diagrams: A. Sagittal section showing some of the valve muscles. Shell prongs from dorsal valve occur in Articulata only. **B.** View of *Terebratella* from dorsal side. **C.** Side view of a brachiopod, attached to a rock by the peduncle. **D.** *Lingula* in its tubular burrow; note sand adhering to terminal portion of peduncle. **E.** Views of *Crania*, one with extended lophophore, showing dorsal valve and circular growth lines; the ventral valve is cemented directly to stone. **Photo:** *Lingula*. Cf. color plate IX.

believed to have evolved from independent marine ancestors, i.e., not from the same stocks which presumably gave rise to the present marine forms.

PHYLUM BRACHIOPODA

Lamp shells; marine lophophorates with dorsal and ventral bivalve shells, often with stalk; shells lined by mantle lobes of body wall; with open circulatory system and metanephridia; development mosaic, cleavage radial; coelom formed as schizocoel or enterocoel; with free-swimming unique larvae.

Class INARTICULATA (ECARDINES): valves held by muscles only; lopohophore without skeletal support; anus present. *Lingula, Crania*

Class ARTICULATA (TESTICARDINES): valves interlocked; lophophore with skeletal support; anus absent. *Terebratula*

As fossil types, the lamp shells are even older and more

abundant than the ectoprocts; some 30,000 extinct species are known. *Lingula* and *Crania* have the distinction of being the most ancient of all fossil genera, these animals going back to at least Ordovician times (though the present *species* within the genera are not very ancient). Today only some 300 species of brachiopods are known to exist. Extinct lamp shells attained sizes of more than 1 ft, but living species generally do not exceed lengths of about 3 in.

By virtue of their two-shelled or *bivalve* exoskeletons, brachiopods resemble clams superficially (Fig. 24.13). However, whereas the valves are lateral in clams, they are dorsal and ventral in brachiopods. The ventral valve (often borne uppermost in the living animal) is usually the larger one. Leading away from its inner surface is the stalk, or *peduncle*, by which many brachiopods are anchored to a solid surface. The peduncle passes through a perforation in the ventral valve (hence the name "lamp shell," from the resemblance of the empty exoskeleton to an ancient oil lamp). Most stalked brachiopods are attached permanently, but *Lingula* is not. Its peduncle merely sticks in the bottom of a vertical mud burrow along tidal flats, and the animal may change its location. It uses its peduncle in locomotion, and it may also snap its valves and propel itself to some extent in that manner. Some specialized brachiopods, e.g., *Crania*, are without peduncles, their ventral valves being cemented directly to a solid surface.

In the Articulata, the valves are hinged together posteriorly, by means of teeth in one valve and corresponding pits in the other. A valve consists of an exterior chitinous cuticle and a thicker layer of calcium carbonate or calcium phosphate underneath (Fig. 24.14). The inner surface is clothed by the body wall. The animal occupies only the posterior space between the valves, the anterior portion being filled by the large lophophore. In this anterior *mantle cavity*, the inner lining of

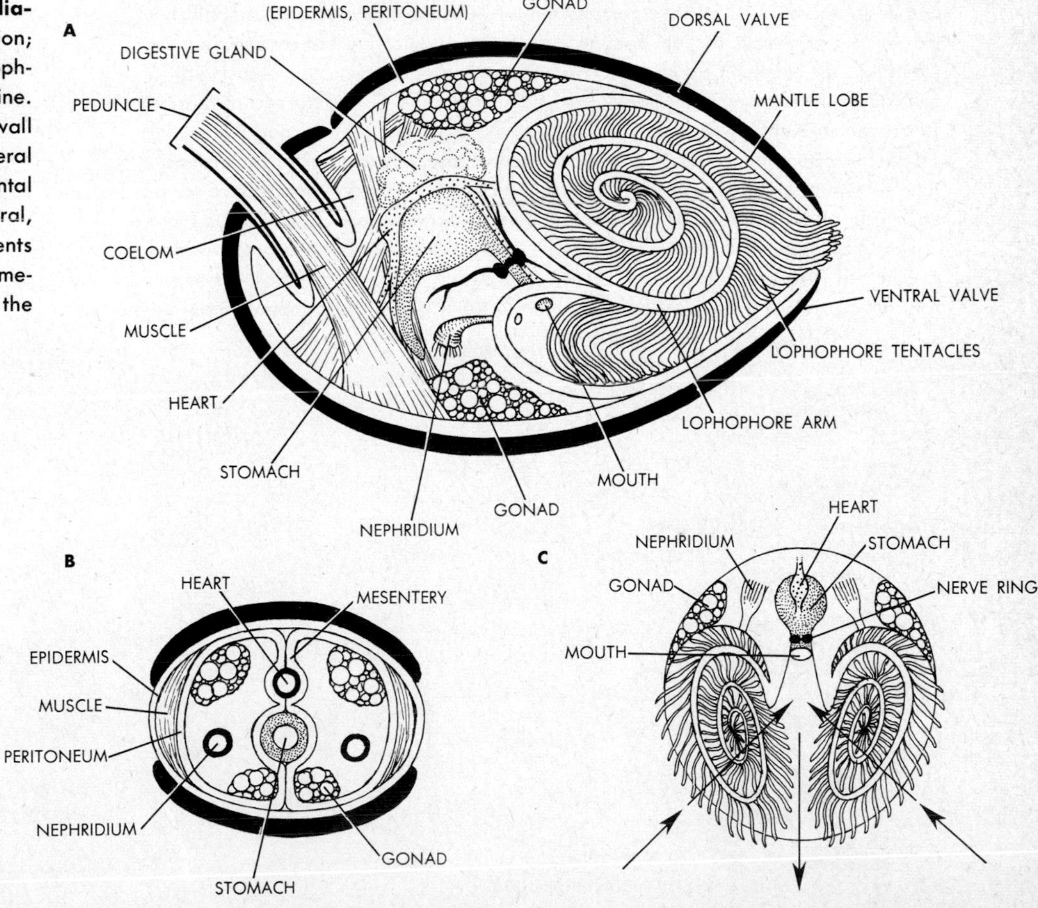

24.14. Brachiopod structure, diagrammatic. A. Longitudinal section; note the nerve ring around the esophagus and the blind-ended intestine. **B.** Cross section; note that body-wall muscles are present only in the lateral parts of the body wall. **C.** Horizontal section; the arrows indicate lateral, lophophore-produced water currents passing toward the mouth and a median current passing away from the mouth.

a valve is a double-layered flap of body wall, forming a *mantle lobe*. The body wall consists of epidermis, mesenchyme (and often cartilagelike) connective tissue, and peritoneum. Between the last two layers, muscles are present in the lateral body wall, i.e., that portion of the wall which is not in contact with the valves. The peduncle is an extension of the ventral body wall and has the same tissue construction. In some cases (e.g., *Lingula*), the coelom extends into the peduncle, but in others the coelomic space of the stalk is filled secondarily with muscles. Attached to the inner surfaces of the valves are *adductor* and *abductor* muscles which close and open the valves, respectively.

The lophophores of brachiopods vary in shape considerably. Tentacles are quite long and can be protruded beyond the shell margins. A single row of tentacles emanates from the lophophore arms. The latter are coiled and, in Articulata, are supported internally by a pair of skeletal prongs growing out from the dorsal valve. The mouth is a transverse slit in the center of the lophophore base. This opening leads into an internally ciliated alimentary system, which is complete only in the Inarticulata. In these animals the alimentary tract is U-shaped, the anus opening into the mantle cavity. In the Articulata the intestine is a blind-ended tube. The stomach in all brachiopods connects with a conspicuous digestive gland ("liver"), shaped like bunches of grapes. A nerve ring around the esophagus is thickened dorsally and ventrally into ganglia, and nerves radiate away from the ring both into the lophophore and into posterior body regions. A neural plexus lies in the stomach wall. Brachiopods possess only a poorly developed sensory system.

The alimentary system is supported by a dorsal and a ventral vertical mesentery, and over the stomach in the dorsal mesentery lies a simply constructed contractile vesicle, or *heart* (cf. Fig. 24.14). Leading away from it in both directions

24.15. Brachiopod development, schematic. (*Adapted from Conklin, Yatsu, and other sources.*) **A.** Embolic gastrulation (after radial cleavage and coeloblastula stage). **B.** Transverse section through *Lingula* larva, showing schizocoelic coelom-formation by development of cavity within endomesodermal mass on each side of endoderm; the mantle fold has adult (forward) orientation. **C.** Sagittal section through *Terebratulina* embryo, showing development of dorsal endodermal partition cutting off coelomic sac from posterior portion of archenteron. **D.** Longitudinal (anteroposterior) section at stage later than in **C,** showing anterior extension of coelomic sac on each side of archenteron. **E.** Longitudinal section through early *Terebratulina* larva; coelom has expanded greatly and mantle fold has developed over posterior end. **F.** External view of mature larva. **G.** Metamorphosis; after larva settles, the mantle fold turns forward over anterior lobe, and the latter gives rise to lophophore.

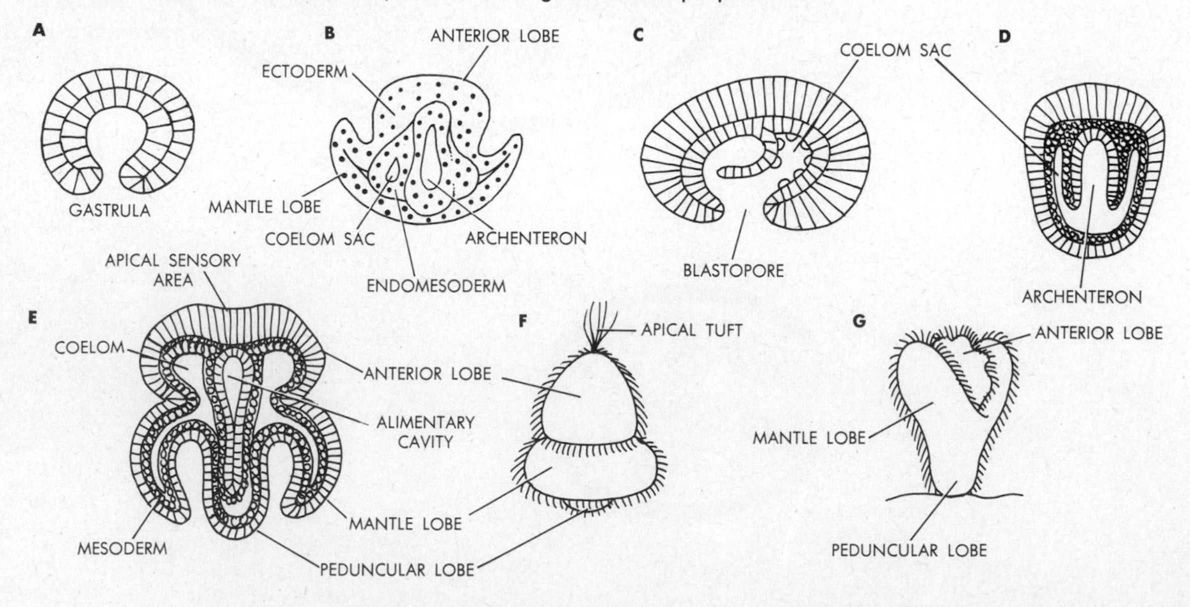

are branched dorsal vessels. In *Crania,* a well-developed transverse septum separates a metacoel from a mesocoel. In other genera, however, such a septum is less fully formed. Free cells are present in the coelomic fluid. From the trunk coelom, funnellike nephrostomes lead into metanephridia which open into the mantle cavity. These excretory tubules also serve as gonoducts. Gonads (usually four) without ducts are attached to the peritoneum of the body wall or to the mesenteries.

Sexes are separate in almost all cases. Fertilization occurs in the coelom or when the gametes exit via the nephridiopores, and the zygotes are generally brooded in the mantle cavity. Radial cleavage produces a coeloblastula which gastrulates by emboly in the majority of cases (Fig. 24.15). In *Lingula,* mesoderm proliferates as a solid cell mass from each side of the endodermal invagination. These masses then split internally, and the cavities so formed represent the coelom. In this genus, therefore, and presumably in Inarticulata generally, the coelom is a schizocoel. In Articulata, by contrast, an enterocoelous coelom develops. Thus, in *Terebratella* and *Terebratulina,* a partition grows down from the roof of the archenteron and separates a single dorsal coelomic sac from the endoderm. Note that, although the coeloms of these brachiopods are technically true schizocoels and enterocoels, both types are formed here in special ways, different from those characteristic of other protostomial schizocoelomates and deuterostomial enterocoelomates (cf. below and Chaps. 7 and 11). Like the development of other lophophorates, evidently, that of brachiopods is also unique. Nevertheless, the very occurrence of schizocoels and enterocoels does suggest that the group may possibly represent a living remnant of an ancient evolutionary link between schizocoelomates and enterocoelomates.

The ciliated, globular larva of brachiopods possesses an apical sensory area and one pair of statocysts (cf. Fig. 24.15). An equatorial ectodermal *mantle fold* grows like a collar over the aboral hemisphere of the larva. This fold subsequently turns forward over the mouth-containing hemisphere and from it later develop the adult mantle lobes. Valves are beginning to be secreted even during the larval stages. A lophophore then becomes elaborated around the mouth, and the adult thus emerges through a very gradual metamorphosis. Like other lophophorate larvae, those of brachiopods evidently are equally quite unlike trochophores.

Inarticulata are probably the primitive brachiopods. Such a conclusion is supported by the characteristics of exceedingly ancient forms such as *Crania* and *Lingula,* in which hinged valves are absent and in which complete, U-shaped alimentary tracts are present. Like other lophophorates, brachiopods as a whole possess a transversely divided coelom, as noted, a trait encountered also in deuterostomial animals (cf. Chap. 29). In addition, as we have seen, some brachiopods actually acquire coeloms in enterocoelic fashion. Brachiopods thus exhibit more deuterostomial features than other lophophorates, but phoronids exhibit more protostomial features; e.g., spiral cleavage in some cases and solenocytic protonephridia in the larvae. The propensity of all lophophorates to form chitinous covers also is reminiscent of schizocoelomate protostomes generally. Despite these vague indications of possible affinities, however, the most essential traits of lophophorates establish these animals as a distinct group, not obviously allied to any other existing coelomates; and each lophophorate phylum analogously appears to be a distinct group in its own right.

REVIEW QUESTIONS

1. Give taxonomic definitions of coelomates, protostomial coelomates, lophophorates, and schizocoelomates. Which phyla are included in each of these groups? What is a lophophore, a mesocoel, a metacoel, and a peritoneal septum? How does a lophophorate animal feed?
2. In what environments does each of the lophophorate groups occur? How many species are known in each group? What is their possible evolutionary relation to one another and to deuterostomes and other protostomes? Which of the lophophorates are most like (*a*) deuterostomes, (*b*) other protostomes, and in what respects?
3. Describe the general structure of a phoronid. What is the epistome? Describe the organization of the body wall and the subdivisions of the coelom. What is the organization of the (*a*) nervous system, (*b*) circulatory system, (*c*) excretory system?
4. Show how fertilization takes place in phoronids and describe the pattern of development. How does an actinotroch larva form and how does it metamorphose? What is the metasome pouch and what is its role in metamorphosis? In what respects are actinotrochs different from trochophores?
5. Name and define the classes of ectoprocts. Describe the

structure of an ectoproct and distinguish between cystids and polypids. What is the structural interrelation between adjacent zooids? How does an ectoproct withdraw and extend its lophophore?

6. Describe the structure of the nervous and alimentary systems of ectoprocts. How do ectoprocts differ from entoprocts and why must these two groups be classified as separate phyla? Through what developmental processes does a cyphonautes form? How does the coelom arise?

7. Describe the metamorphosis of ectoprocts and the processes by which adults arise. How do (*a*) budding, (*b*) brown-body formation take place? Describe the structure of an ectoproct colony and show how it grows.

8. What kinds of polymorphs occur among ectoprocts and what is their structure and function? In what (*a*) structural, (*b*) reproductive respects are marine and freshwater ectoprocts different? What are statoblasts? Name representative genera of all lophophorate phyla and classes.

9. Define the classes of brachiopods taxonomically. Describe the structure of a brachiopod and show how this structure differs in the different classes. How does brachiopod symmetry differ from that of a clam? What is the organization of the nervous, circulatory, alimentary, and excretory systems?

10. Describe the development and metamorphosis of brachiopods. What does the developmental pattern suggest with regard to the evolutionary position of brachiopods? Which lophophorate groups are hermaphroditic and which are separately sexed? Which groups brood their young and where?

COLLATERAL READINGS

Additional references to the lophophorate phyla may be found in the following sources. All but the text by Hyman are articles in the technical literature:

Hyman, L.: The Occurrence of Chitin in Lophophorates, *Biol. Bulletin,* vol. 114, 1958.

———: "The Invertebrates," vol. 5, chaps. 19, 20, and 21, McGraw-Hill, New York, 1959.

Lynch, W.: Factors Influencing Metamorphosis of *Bugula* Larvae, *Biol. Bulletin,* vol. 103, 1952.

Marsden, J.: Regeneration in *Phoronis, J. Morphol.,* vol. 101, 1957.

Morse, W.: The Chemical Constitution of *Pectinatella, Science,* vol. 71, 1930.

Rattenbury, J.: The Embryology of *Phoronopsis viridis, J. Morphol.,* vol. 95, 1954.

Rogick, M., and H. Croasdale: Bryozoa Associated with Algae, *Biol. Bulletin,* vol. 96, 1949.

Stehli, F.: Evolution of the Loop and Lophophore in Terebratuloid Brachiopods, *Evolution,* vol. 10, 1956.

Weiss, C.: The Seasonal Occurrence of Sedentary Marine Organisms in Biscayne Bay, Florida, *Ecology,* vol. 29, 1948.

Williams, S.: Larval Colonies of *Pectinatella, Ohio J. Sci.,* vol. 21, 1921.

PLATE IX. Lampshells and mollusks. **A,** *Terebratulina,* lampshells of the phylum Brachiopoda. **B,** underside of the chiton *Katharina;* note gills in posterior part of mantle cavity. **C,** the pectinibranch cowry *Cypraea,* a prosobranch snail. **D,** a neogastropod (prosobranch) "finger shell" of Guam, turning over in sand.

Plate IX, page 1

A

B

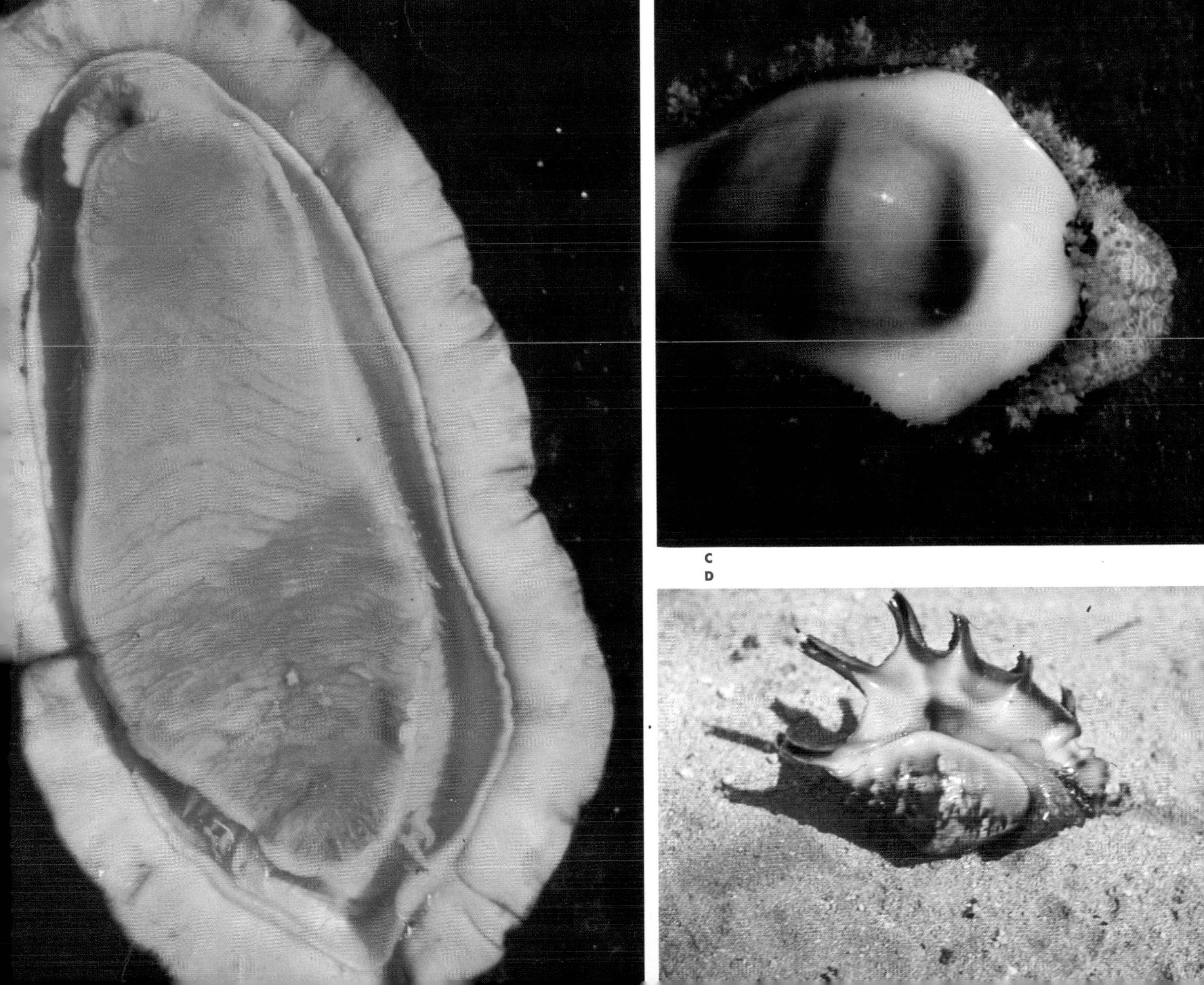

C

D

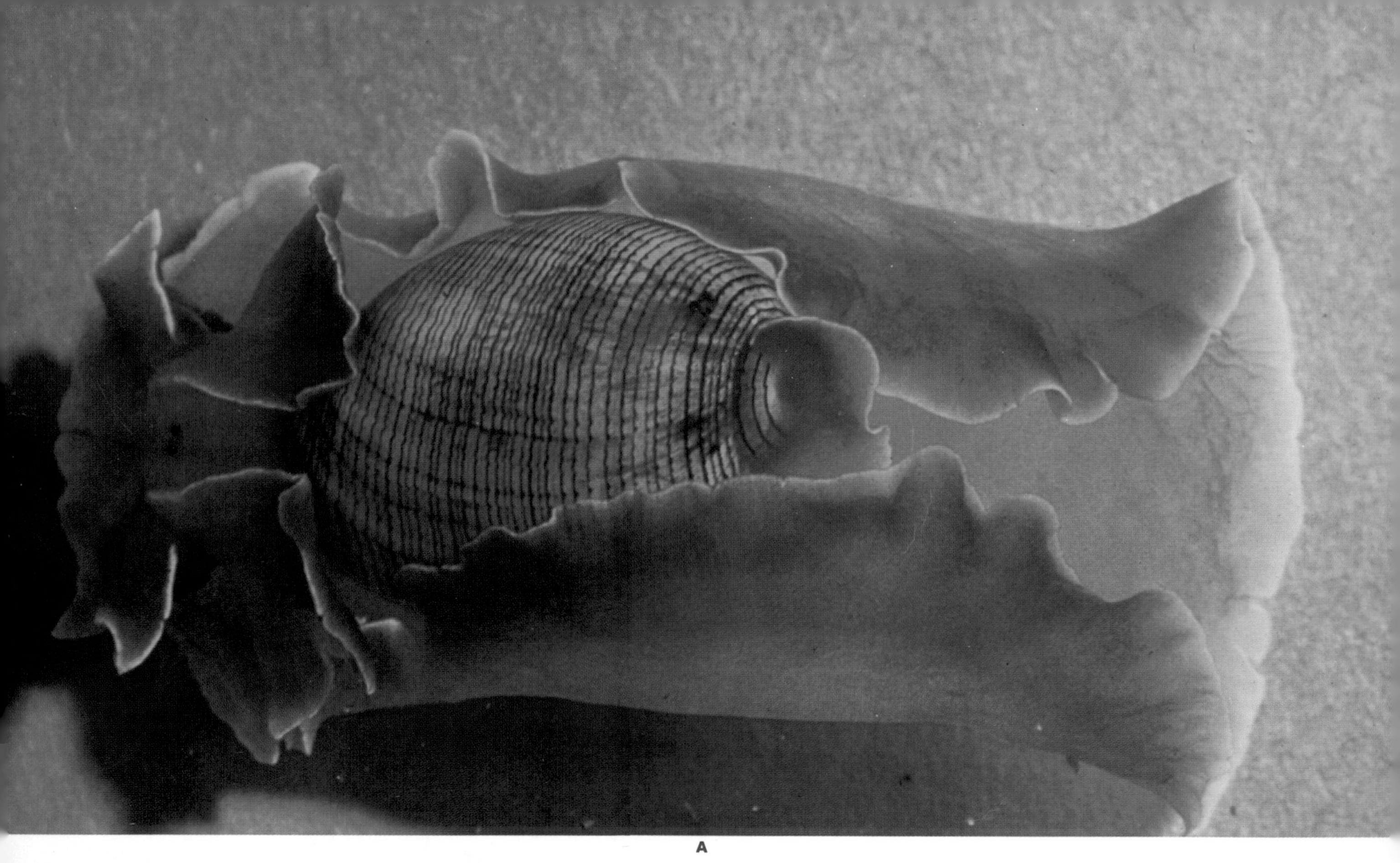

A

B

PLATE X. Opisthobranch snails. A, *Bulina,* a seahare-like tectibranch. **B,** the sea lemon *Doris;* note tentacles toward left foreground, gill toward right background. **C,** the nudibranch *Melibe.* **D, E,** adult and egg ribbon of the nudibranch *Hexabranchus.*

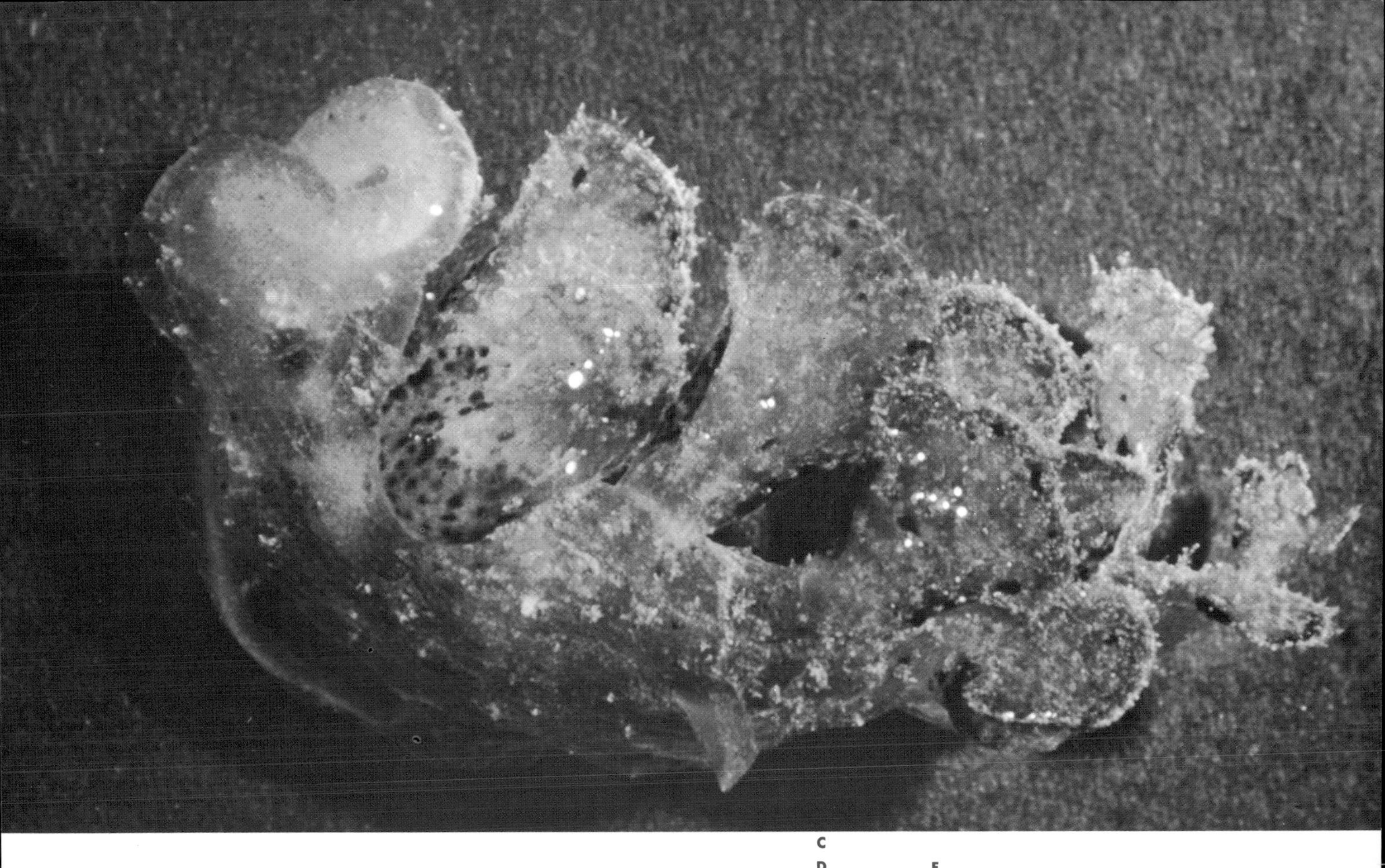

C

D

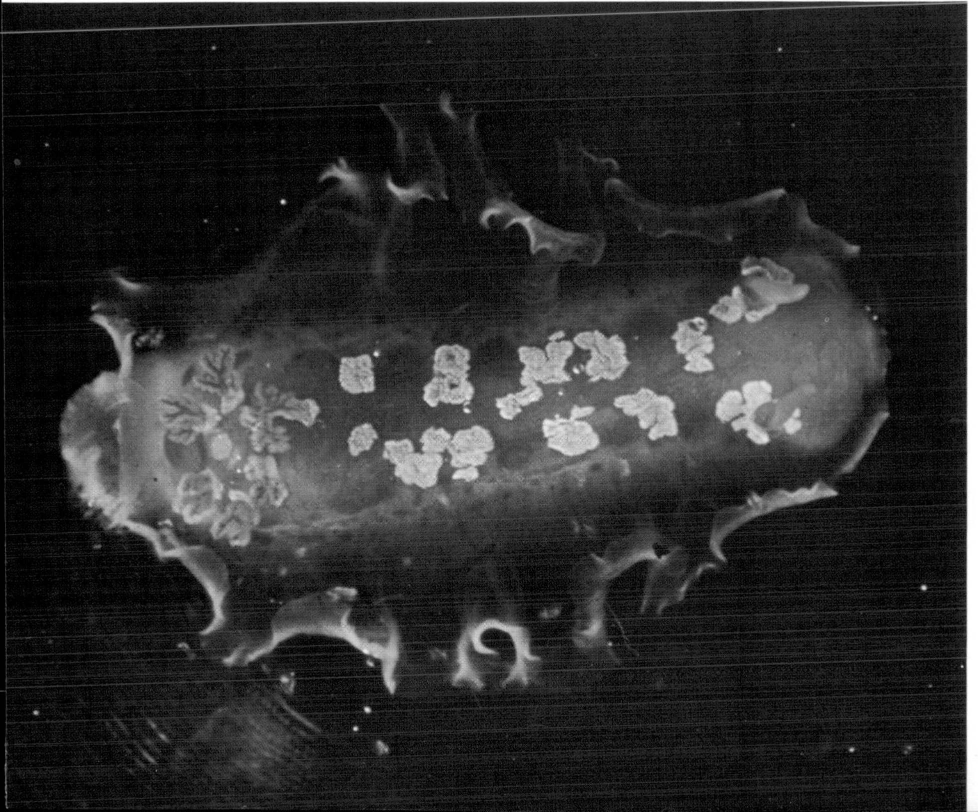

E

Plate X, page 2

A
B
C

PLATE XI. Pelecypod mollusks. **A,** *Mytilus,* the filibranch mussel. **B,** *Gryphea,* the Portuguese oyster. **C,** *Chlamys,* a scalloplike filibranch clam; note the small light-blue ocelli on the fringed mantle along the shell edge. **D,** close-up of mantle edge of the scallop *Pecten;* note dark ocelli. **E,** the rock-boring eulamellibranch clam *Pholadidea,* positioned in rock.

Plate XI, page 2

D

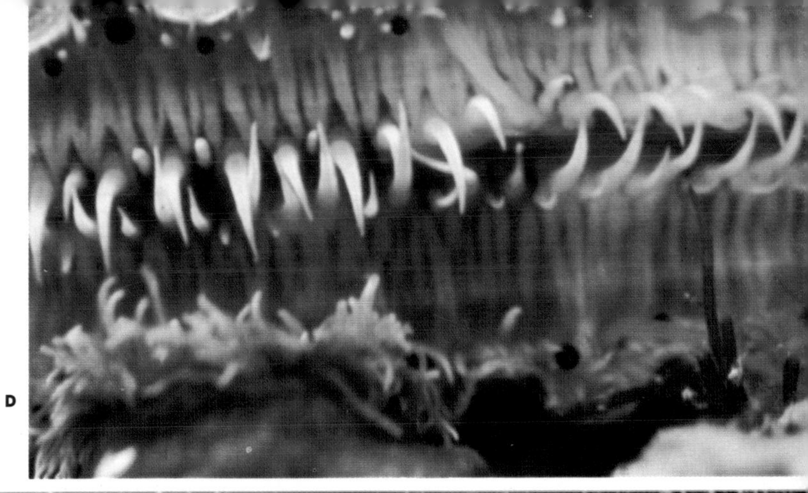

E

Plate XII, page 1

PLATE XII. Annelids and other worms. A, *Phascolosoma,* worms of the phylum Sipunculida. **B,** anterior part of the palolo worm *Leodice,* a polychaete annelid. **C,** tube and tentacular crown of the feather duster *Hydroides,* a sabellid polychaete. **D,** the rhynchobdellid leech *Placobdella.*

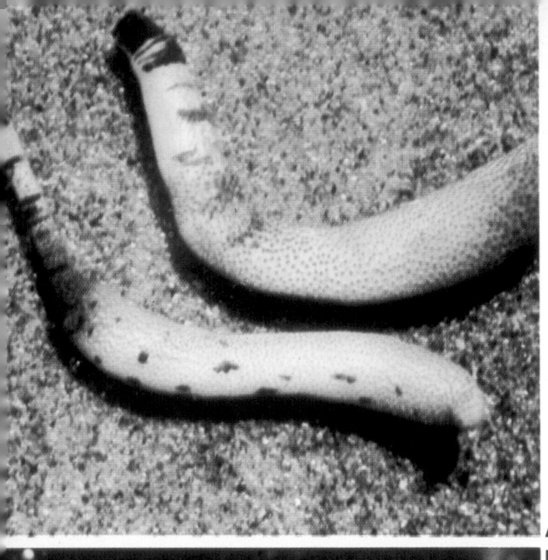

A

B

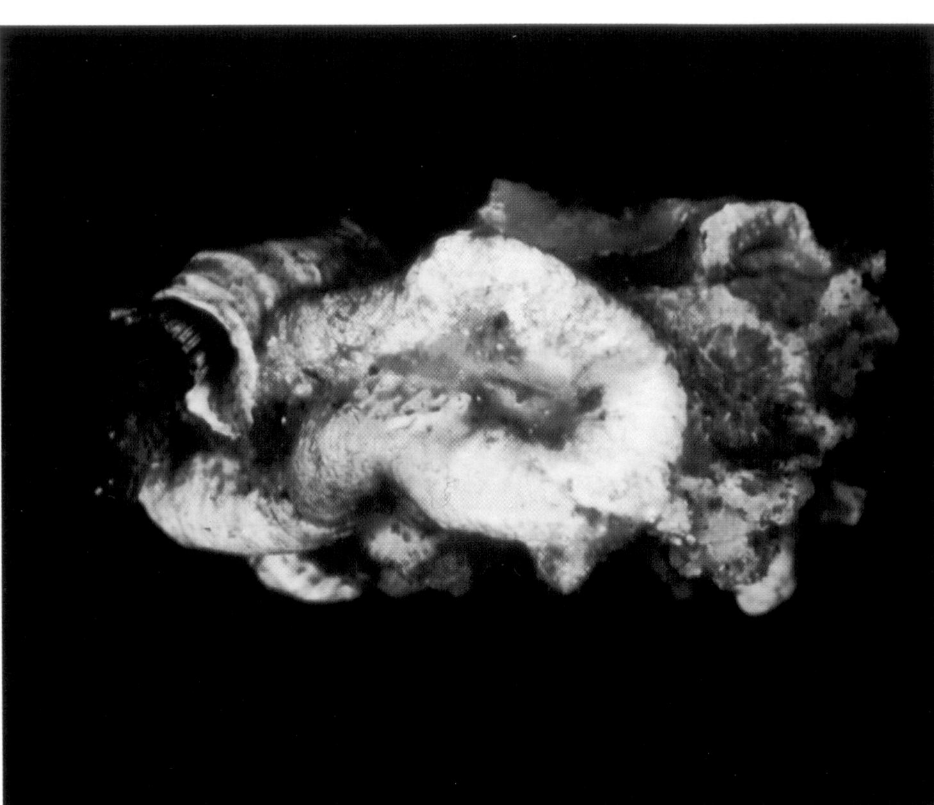

C

D

CHAPTER 25

mollusks

This chapter begins an examination of the schizocoelomate protostomes. As pointed out earlier, the schizocoelomate phyla as a whole are characterized by almost identical patterns of early embryonic development and by the occurrence of almost identical trochophore larvae in many different groups. Yet, the adults include types as diverse as clams, earthworms, and houseflies. Ancestral protostomial coelomate stocks probably gave rise to two main series of schizocoelomate animals, viz., an earlier one which remained unsegmented and a later one in which metameric segmentation became highly elaborated. Both series proved to be exceedingly successful; the first is represented today chiefly by the mollusks, the second, by the arthropods.

Phylum Mollusca: mollusks, soft-bodied animals; body bilateral, usually composed of *head*, ventral *foot*, and dorsal *visceral hump*, the latter covered by *mantle* which typically secretes an *exoskeleton;* alimentary system with *radula* and *hepatopancreas;* breathing via gills; circulatory system open in most cases, with chambered heart and extensive hemocoel; excretion through renal organs; sexes separate or hermaphroditic; development mosaic, cleavage spiral, mouth at or near blastopore, coelom schizocoelic; larvae *trochophores* followed in many cases by *veliger* stages. *Amphineura, Gastropoda, Scaphopoda, Pelecypoda, Cephalopoda*

GENERAL CHARACTERISTICS

Among Metazoa the mollusks constitute the second-largest phylum; over 100,000 species have been described to date. Only the arthropods include more known species (and nematodes and protozoa probably include more existing but so far undescribed species). Like large, successful phyla generally, mollusks are adapted to practically all available living conditions. The animals occur in the sea, at all depths from the deepest trenches to the surface, in all kinds of fresh waters, and on land, to altitudes as high as the snowline. Most mollusks have a length in the order of an inch or two; but the phylum also includes the largest and most highly elaborated of all invertebrates, viz., the squids, some of which are 50 to 60 ft long. Mollusks are known from Cambrian times on. The nautiloids were prominent inhabitants of Paleozoic seas, and ancient snails were among the first of all terrestrial animals. Today mollusks are even more abundant than ever before (and man also uses them as food more widely and in greater quantities than any other invertebrate). Very few members of the phylum are symbiotic, though the animals are themselves hosts to a wide variety of commensals and parasites.

As noted, mollusks are evolutionary derivatives of early schizocoelomate stocks which did not evolve body segmentation. The molluscan line produced a very extensive adaptive

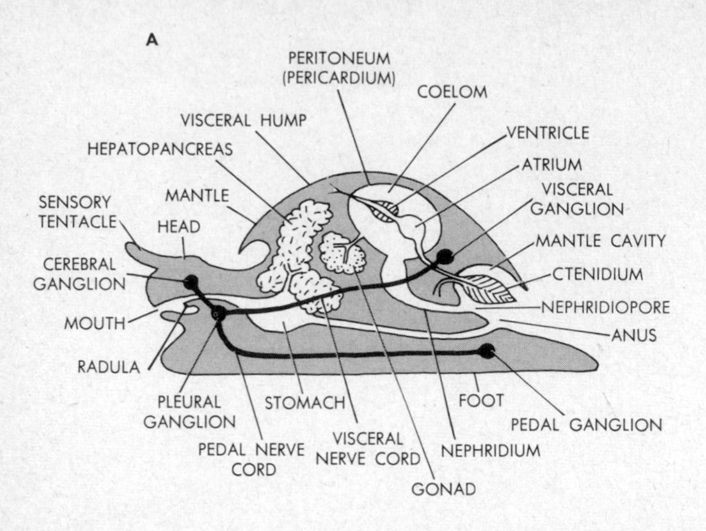

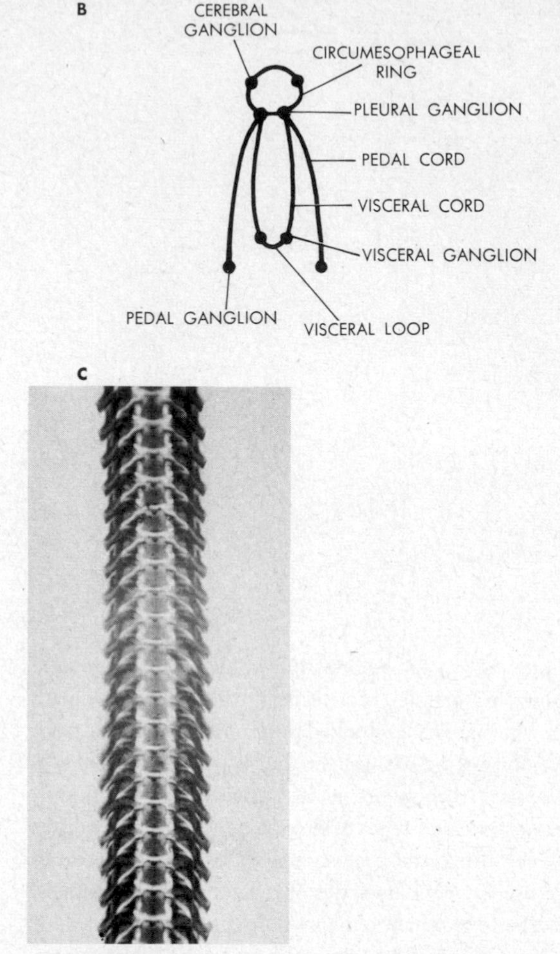

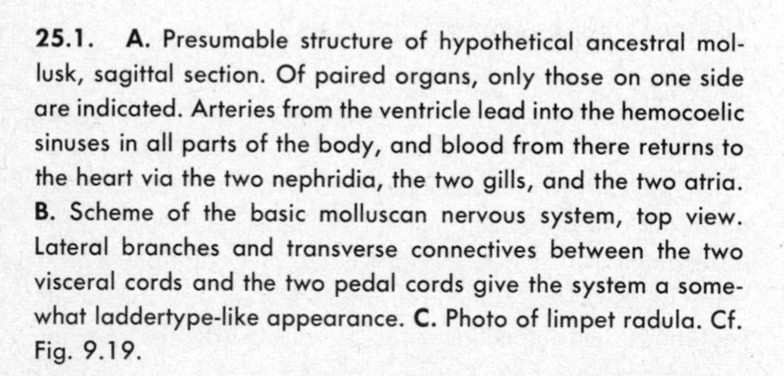
25.1. **A.** Presumable structure of hypothetical ancestral mollusk, sagittal section. Of paired organs, only those on one side are indicated. Arteries from the ventricle lead into the hemocoelic sinuses in all parts of the body, and blood from there returns to the heart via the two nephridia, the two gills, and the two atria. **B.** Scheme of the basic molluscan nervous system, top view. Lateral branches and transverse connectives between the two visceral cords and the two pedal cords give the system a somewhat laddertype-like appearance. **C.** Photo of limpet radula. Cf. Fig. 9.19.

radiation of numerous sublines, all based on a common ancestral body plan. Thus, although the molluscan phylum includes such diverse and seemingly unrelated animals as snails, clams, and squids, the body plans of all of them nevertheless are readily derivable from a single ancestral pattern. The original mollusks which exhibited this pattern can be hypothesized to have had the following characteristics (Fig. 25.1).

The body consisted of a *head*, which may have borne a pair of sensory tentacles; a broad, flat, muscular *foot*, ventrally located and serving in a creeping form of locomotion; and a domeshaped, dorsal *visceral hump*, which contained the principal organ systems. The body wall of the dome was a *mantle* which secreted calcareous spicules in the epidermal layer and which was extended into an overhanging rim around the sides of the body, particularly at the posterior end. The space under this posterior rim represented a *mantle cavity*. Projecting into it were a pair of feathery or leaflike gills, the *ctenidia*. The mouth was located in the head, and it led into a pharynx equipped with a *radula*, a rasping organ characteristic of mollusks generally. A radula is a horny band studded with recurved teeth arranged in a variety of patterns. Muscles move this band back and forth over a cartilaginous supporting rod in the ventral wall of the pharynx. Protruded through the mouth, the radular apparatus serves in rasping pieces of tissue from plant or animal food organisms. From the pharynx, food passed through an esophagus into a stomach, which connected with a conspicuous *hepatopancreas*, or "liver." In the latter, a substantial amount of digestion occurred intracellularly. Extracellular digestion took place in the intestine, which opened posteriorly into the mantle cavity.

The nervous system consisted of a nerve ring around the esophagus, thickened dorsally into a pair of *brain ganglia* and ventrally into a pair of *pleural ganglia*. From the latter emanated two pairs of longitudinal nerve cords; the *pedal cords* traveled posteriorly into the foot, where they thickened into a pair of *pedal ganglia*, and the *visceral cords* passed posteriorly into the dorsal hump, where they formed a pair of *visceral ganglia*. A transverse connection between the visceral ganglia established a *visceral loop*. The ancestral mollusk also possessed an open circulatory system. It contained a heart with one ventricle and two posterior atria, as well as systems of arterial and venous vessels from and to the gills. Additional vessels passed into all other body regions and opened into extensive hemocoelic blood sinuses which permeated all organs. The coelom was comparatively reduced. Its principal component was a *pericardial cavity* around the heart. Leading into this coelomic space were the interior cavities and the ducts of the gonads, located anteriorly, and passing posteriorly were a pair of excretory tubules. The latter were essentially metanephridial; i.e., they connected the pericardial coelom with the exterior, via nephridiopores in the mantle cavity near the anus. Gametes were shed into the pericardial coelom and they then exited through the excretory ducts. Development was mosaic, with spiral cleavage, 4d-derived mesoderm, teloblastic bands, schizocoelic coelom, and free-swimming trochophore larvae (cf. Figs. 7.16, 11.14, and 11.24).

In its essential features, such an ancestral design is still in evidence in the most primitive mollusks today, i.e., the Amphineura (Fig. 25.2). The most familiar members of this class are the chitons, in which the head has become reduced and the exoskeleton has become more specialized. In virtually all other respects, however, the amphineurans still display the basic ancestral body structure. In snails and allied forms of the class Gastropoda, the ancestral structure is again largely preserved, but the early evolution of these mollusks has included a pronounced dorsal growth of the visceral hump. This growth has occurred unequally on the left and right sides, resulting in a spiral coiling of the hump (and the shell covering it). More-

25.2. Molluscan body plans. Shells drawn in heavy lines, alimentary systems, stippled, foot shaded, other body divisions labeled. Note that, among cephalopods, the squid is drawn in a position comparable to that of other groups, i.e., the tentacle-bearing head-foot end is ventral, the mantle cavity is posterior. The chambered *Nautilus* is drawn in swimming position, with the buoyant shell uppermost.

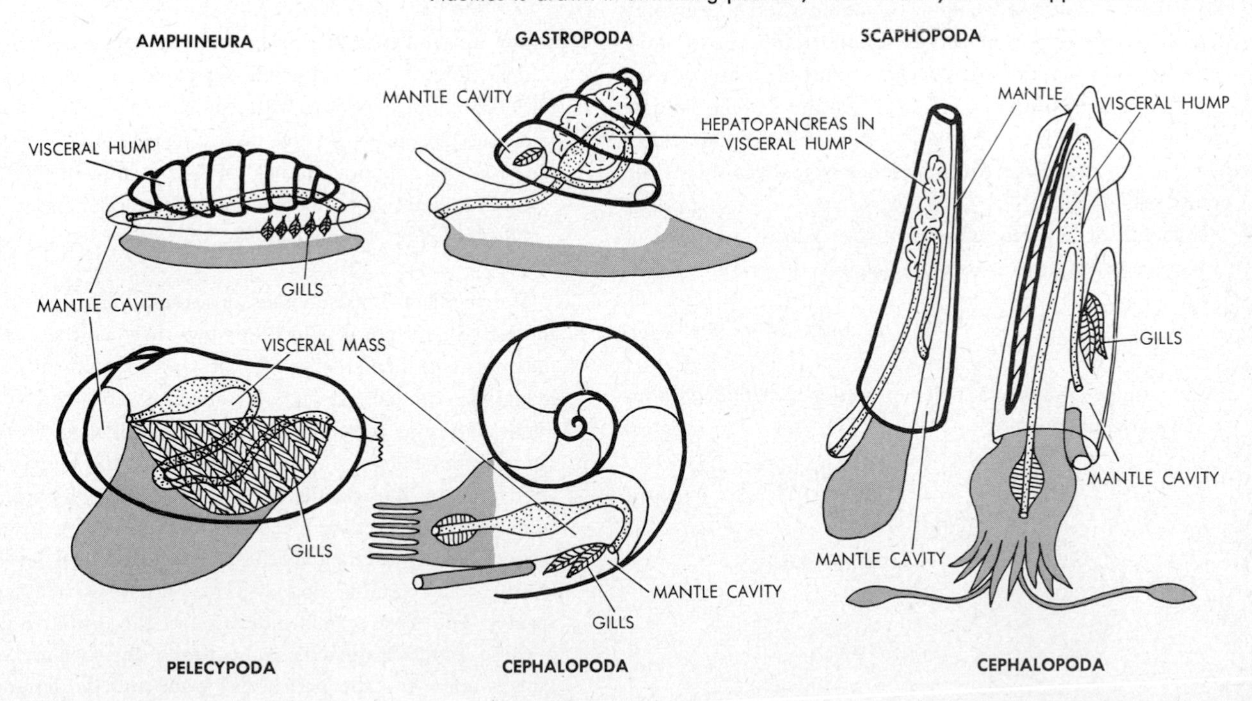

over, the whole coiled hump has also become rotated by 180° relative to the head and the foot, a process which has brought the mantle cavity to an anterior position and has occasioned corresponding positional shifts of the internal organs. The Scaphopoda, or tusk shells, similarly have elongated greatly in a dorsal direction, but a coiling has here not taken place. Pelecypoda, or clams, have adapted to a sedentary filter-feeding existence. In these animals the ancestral body has become flattened from side to side, the head has disappeared, and the gills have expanded into ciliary food-collecting organs (though they retain their breathing function as well).

Finally, the Cephalopoda have evolved the most pronounced departures from the ancestral pattern. As in snails, early cephalopods elongated the visceral hump in a dorsal direction. But growth remained equal on the left and right sides; hence the hump came to form a flat coil. Also, the covering shell became partitioned into progressively larger compartments as the animal grew, and only the last, largest compartment was occupied by the animal. All earlier compartments were filled with air; this gave the early cephalopods considerable buoyancy and permitted them to adopt a free-swimming existence. In correlation with this newly developed pelagic mode of life, the foot became modified partly into muscular prehensile tentacles, which also equipped the animals as predatory, carnivorous types. The nautiloids of the Paleozoic and the chambered nautilus today exemplify this early stage of cephalopod evolution. In other lines of the cephalopod group, the shells became reduced greatly or were lost altogether, and the nervous and sensory systems became highly developed in conjunction with rapid, swimming locomotion. The result was the emergence of modern squids and octopuses.

Evidently, the basic ancestral body pattern admitted of numerous and highly varied modifications; the present success of mollusks unquestionably is a consequence of this original adaptive potential.

25.3. Aplacophora. Schematic cross section through body to show foot and mantle.

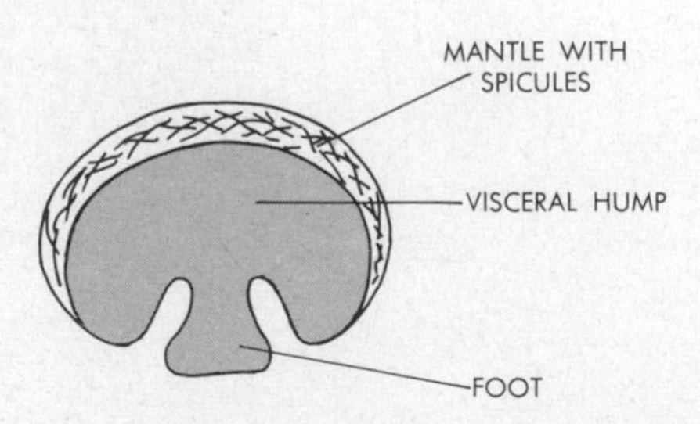

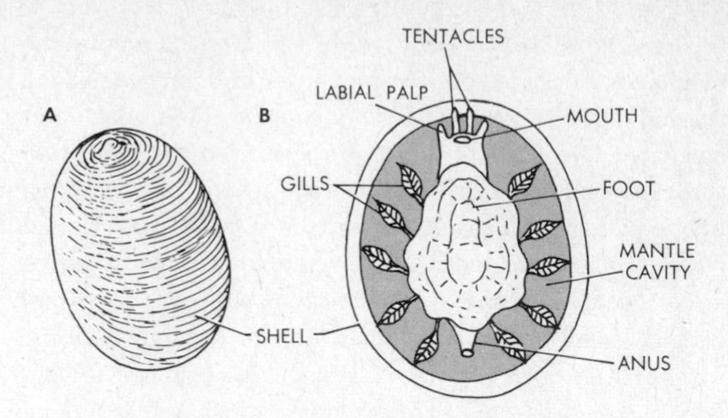

25.4. Monoplacophora (*Neopilina*). (*Adapted from Moore.*) **A,** dorsal view of the single shell plate; **B,** ventral view of body. Note one pair of tentacles and of labial palps on head and five pairs of ctenidia in mantle cavity.

CLASS AMPHINEURA

Chitons and allied types; marine; bilateral symmetry; head reduced, without eyes or tentacles; nervous system without ganglia; hermaphroditic or separately sexed.

Order Aplacophora: wormlike; with calcareous spicules but shell absent; foot reduced. *Chaetoderma, Neomenia*

Order Monoplacophora: with caplike shell; most internal organs present as several pairs. *Neopilina*

Order Polyplacophora (chitons): shell consisting of eight overlapping plates; gonoducts with separate openings. *Chiton, Chaetopleura, Craspedochilus*

The shell-less amphineurans, Aplacophora, are atypical of the class in certain respects, but they may be the primitive members and thus possibly the most primitive mollusks now living. The animals possess a rudimentary foot consisting of a muscular ridge projecting from a longitudinal groove on the ventral side (Fig. 25.3). The remainder of the body surface is covered by the mantle. In it are spicules, which provide some support, but they are not abundant enough to form a distinct shell. A radula is absent in some cases, well developed in others, and ctenidia too may or may not be present in given species. The most certain indication of the primitive nature of these animals is the ancestral nature of the coelom; i.e., the gonads open into the pericardial space and the latter communicates with the outside via nephridial ducts. Apart from the specific characteristics here noted, all other aplacophoran features are as in chitons (cf. below).

Represented by the single living genus *Neopilina,* the Monoplacophora are sometimes classified as a separate class of mollusks, sometimes also as a group within the class Gastropoda. The group has come under investigation only recently and the results have proved to be extremely interesting. As indicated by the name of the order, these animals are covered dorsally by a single shell plate, formed by an accumulation of calcareous spicules secreted by the underlying mantle (Fig. 25.4). Underneath the rim of the shell is a mantle cavity, circling the animal as in chitons. The reduced head bears a pair of sensory tufts and a pair of tentacles, and the mouth is equipped with a pair of *labial palps,* flaps of ciliated tissue which direct microscopic food toward the mouth. Five pairs of ctenidia project into the posterior part of the mantle cavity. The internal organs analogously occur as multiple pairs, e.g., several paired atria, six pairs of excretory tubules, two pairs of gonads, 10 pairs of transverse neural connections. These organs are arranged in regularly spaced anteroposterior arrays, a design which has made some investigators speculate about possible segmented ancestors of mollusks. Such speculations are unwarranted, however. Signs of segmentation are totally absent in molluscan embryos and larvae, contrary to what would be expected if the ancestors had been segmented; and serial repetitions of organs are by no means unusual in nonsegmented groups (e.g., flatworms such as *Procerodes,* cf. Chap. 22). Indeed, organs such as gills often occur in numerous multiple pairs also in chitons, animals which are certainly

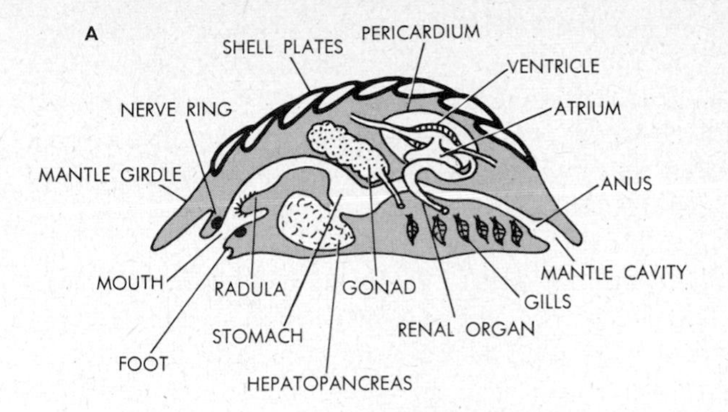

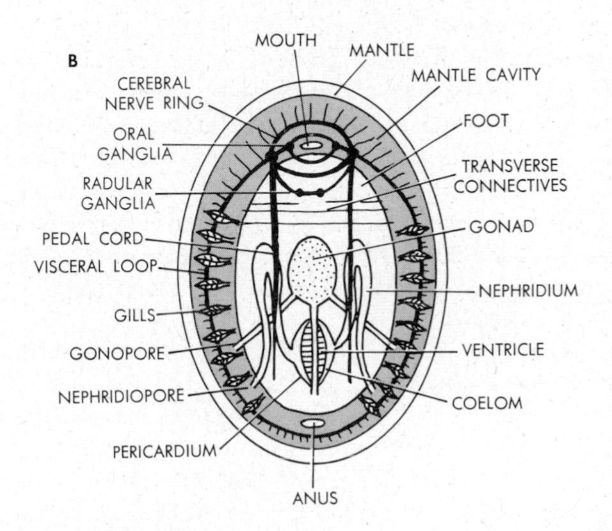

25.6. Internal structure of chitons, schematic. **A,** sagittal section; **B,** organ systems as seen from ventral side. The transverse nerve connections between the pedal cords occur in midbody and posterior regions as well, but they are not drawn in to preserve the clarity of the other structures.

25.5. A chiton, member of the amphineuran order Polyplacophora. The animal is seen from the dorsal side. Note the eight shell plates set into the mantle. Cf. color plate IX.

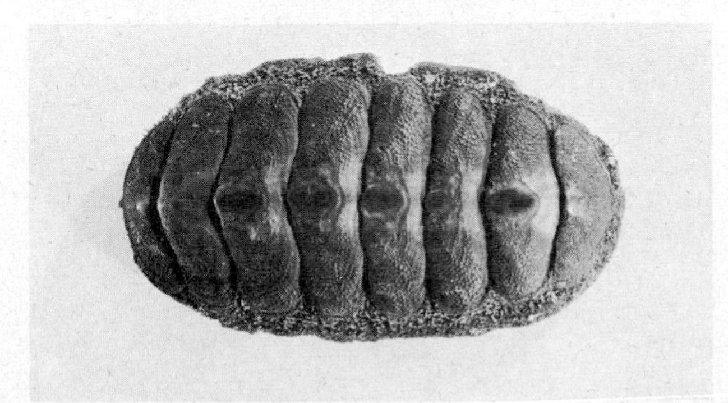

nonsegmented. It appears, therefore, that the monoplacophorans exemplify little more than a transitional evolutionary stage in the line leading to the Polyplacophora, the principal group of amphineurans.

Chitons with eight shell plates are known from Ordovician times on. Each shell plate consists of two calcareous layers. These plates do not cover the mantle entirely, but a *mantle girdle* rings the shells along the free edges (Figs. 25.5 and 25.6). The mantle cavity is a narrow groove circling the entire animal between the edge of the mantle girdle and the flat, oval foot. The anus opens posteriorly into the mantle cavity, and on either side of this opening is an excretory pore and a separate gonopore. Laterally, some six to 80 pairs of small ctenidia project into the mantle cavity. The mouth is anterior, and a well-developed radula may be protruded through it. The head as a whole is quite reduced and is not marked off in any way from the trunk. Also, tentacles and eyes are absent. In these respects chitons are specialized and divergent from the probable ancestral molluscan condition.

Internally a chiton exhibits the basic molluscan design except that, in correlation with the reduction of the head, the nervous system is simplified correspondingly. Thus, definite ganglionic thickenings are usually absent. A circumesophageal nerve ring merely extends posteriorly as two pairs of cords. The visceral cords typically are joined posteriorly and form a loop, and all cords bear lateral branches. The only other principal deviation from the ancestral molluscan pattern is the presence of separate gonoducts to the outside, a condition characteristic also of all other mollusks. A pair of tubular excretory ducts still emanates from the pericardial coelom, but the simple nephridial nature of the ducts no longer persists. Instead, the ducts have become complex *renal organs* with distinct filtering and reabsorbing portions, again as in all other mollusks.

Chiton larvae are typical free-swimming trochophores with solenocytic protonephridia (Fig. 25.7). A foot begins to develop in the region between the larval mouth and anus. The larva eventually settles wih its apical tuft directed forward and the foot directed downward. Shell plates begin to be secreted at the free dorsal side. Chitons inhabit the rocky sea shores, where they cling to rocks so tightly that they cannot be pried off without serious injury to the animals. However, a chiton may release its hold on its own by pressing its whole foot against the rock and thereby abolishing the suction which kept it attached. The animal may then creep over the rock surface, secreting a slime track from the foot in the process. Algal vegetation is the principal food.

CLASS GASTROPODA

Snails and allied types; marine, freshwater, and terrestrial; visceral hump typically coiled, with torsion or various degrees of detorsion; usually with shell; head with eyes and one or two pairs of tentacles; trochophore and veliger larvae or direct development.

Subclass Prosobranchia: mostly marine, some freshwater and terrestrial; 180° torsion, visceral loop in figure-of-eight; shell caplike or coiled; one pair of tentacles; foot with or without operculum; sexes usually separate.

Order Aspidobranchia: shell caplike or moderately coiled; operculum absent; nervous system not concentrated; internal organs paired. *Patella,* limpets; *Crepidula,* slipper limpets; *Haliotis,* abalones.

Order Pectinibranchia: shell well humped, usually coiled; operculum present; nervous system not concentrated; internal organs single. *Littorina,* periwinkles; *Paludina,* freshwater snails; also cowries.

Order Neogastropoda: shell coiled; operculum present; nervous system concentrated; internal organs single. *Buccinum, Busycon,* whelks; *Murex,* rock snails; *Strombus,* conches.

Subclass Opisthobranchia: mostly marine; 90 or 180° detorsion, visceral loop partly or wholly uncoiled; shells reduced or absent; two pairs of tentacles; foot usually without operculum; almost all hermaphroditic.

25.7. Chiton development beyond the trochophore stage. *(Adapted from Heath.)*

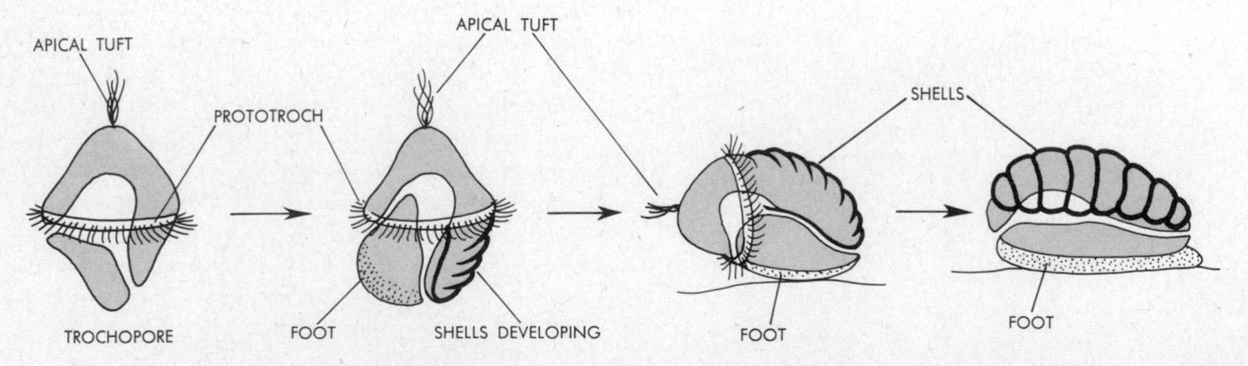

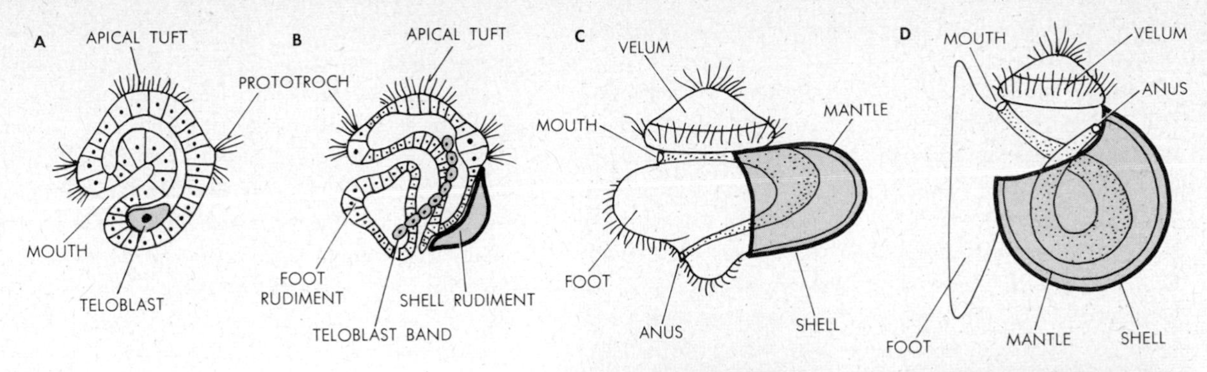

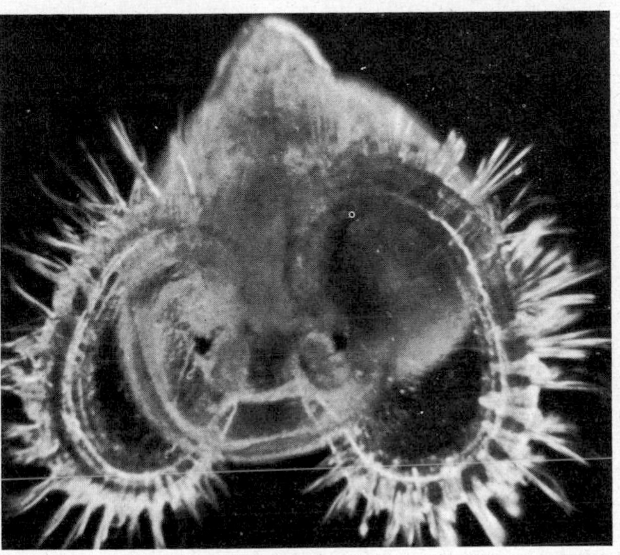

25.8. Gastropod development. Diagrams: ventral side in all figures toward left, dorsal side toward right. **A.** Sagittal section through trochophore. **B.** Early veliger larva. Foot and shell rudiments are present. The prototroch and the anterior larval portion become the velum. **C.** Mature veliger, before torsion. The dorsal visceral hump has enlarged greatly, producing a U-loop in the alimentary tract. **D.** Veliger after torsion. The alimentary tract is now coiled, with the anus shifted to same (anterior) side as mouth. **Photo:** veliger larva of the marine snail *Rissoa*, seen from the ventral side. The ciliated velum is nearest the viewer, the visceral hump with shell faces away from the viewer.

Order Tectibranchia: shell usually present; with mantle cavity and gill. *Aplysia*, sea hares; *Cavolinia*, pteropods, or sea butterflies.

Order Nudibranchia: shell, mantle cavity, and gill absent. *Eolis*, sea slugs.

Subclass Pulmonata: terrestrial and fresh water; modified torsion, with or without shells; two pairs of tentacles; without operculum; gills absent, air breathing by means of pulmonary sac; all hermaphroditic; development direct, without larvae.

Order Basommatophora: eyes at base of posterior tentacles. *Limnaea*, water snails.

Order Stylommatophora: eyes at tip of posterior tentacles. *Helix*, land snails; *Limax*, land slugs; *Physa*, *Helisoma*, water snails.

In this largest molluscan class, the most characteristic group-identifying events take place during the larval stages. A trochophore develops in the usual manner and this larva subsequently becomes a *veliger*, a larval form typical of mollusks generally (Fig. 25.8). It is characterized particularly by a *velum*, a ciliary swimming organ elaborated and enlarged from the prototrochal ciliary girdle of the trochophore. In the veliger, also, a foot begins to develop as in chitons, between mouth and anus, and the dorsal ectoderm becomes a humped, shell-secreting mantle. The progressive dorsal growth of the mantle has the effect of bending the alimentary tract more and more into a U-shape, until the anus comes to lie quite close to the mouth. The hepatopancreas arises dorsally from the larval stomach, and dorsal growth of this gland pushes the developing visceral hump still higher. Growth is somewhat or greatly unequal on the left and right sides, resulting in a certain degree of coiling of both the visceral hump and the shell on the outside.

A distinct and separate event then takes place as well; referred to as *torsion*, it is achieved in the course of a few minutes: the whole visceral hump of the veliger rotates 180° relative to the rest of the body, usually in a counterclockwise direction. The results of torsion are (Fig. 25.9) that the mantle cavity comes to lie anteriorly, above the head; the gills are anterior; the anus, excretory pores, and gonopores all are anterior, in the mantle cavity; the alimentary tract is twisted from a U-shape into a loop; the visceral nerve loop becomes twisted into a figure-of-eight, with the left cord coming to lie under the alimentary tube and the right cord over it; and the atria come to lie anterior to the ventricle, i.e., the heart is turned around. The evolution of torsion in the veliger is undoubtedly correlated with the way of life of the adults. Snails protect themselves by withdrawing into the shell, and inasmuch as the shell has only one opening, it is clearly advantageous that both the mouth and the anus should open through the shell aperture. Also, it is perhaps of some advantage in a creeping way of life that the mouth should remain closer to the ground than the anus. Torsion produces these basic characteristics of gastropod structure. Most gastropods pass the trochophore stage as embryos within the egg membranes, and only the veliger becomes a free larva. In freshwater and terrestrial forms even the veliger stage remains embryonic; development is then direct, the young adults representing the first free stage.

25.9. Scheme of adult gastropod structure resulting from larval torsion (as exemplified in some primitive Prosobranchia, i.e., aspidobranchs; *modified from Naef*). **A.** Sectional side view. Note anterior position of anus, gills, atria, and nephridiopores (contrast with Fig. 25.1). The gonads open into the renal organs. **B.** Top view. Note figure-of-eight formed by visceral nerve loop, the left visceral cord passing under the alimentary tract, the right cord over it. Paired organs are still of equal size in primitive prosobranchs, as shown below.

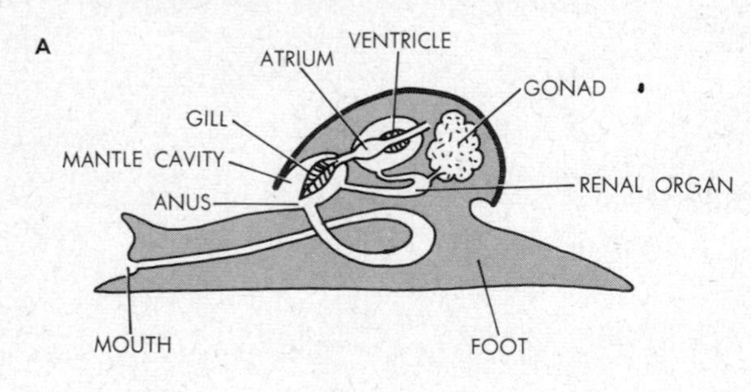

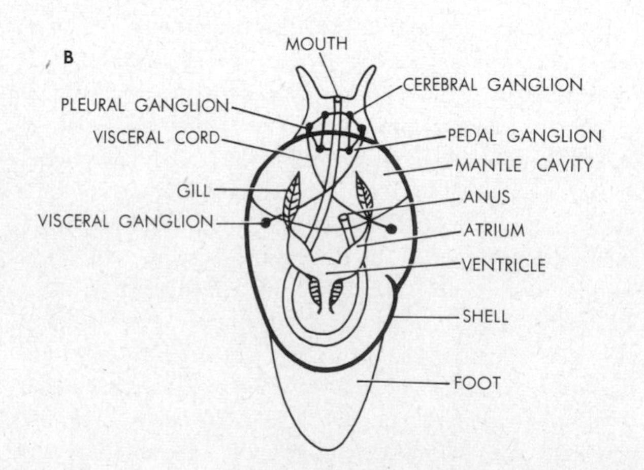

The largest group of gastropods is the subclass Prosobranchia, in which full 180° torsion is in evidence. In the Aspidobranchia (Fig. 25.10), the most primitive order, the left and right sides of the trochophore and veliger stages grow at more or less equal rates before torsion, and after torsion the visceral hump is coiled only minimally. As a result, aspidobranchs such as the limpet *Patella* possess a caplike, uncoiled, symmetrical shell in the adult condition (though some measure of coiling is in evidence in the larva). In *Crepidula* the shell is slightly more asymmetrical, and in the abalone *Haliotis* spiral coiling is already fairly pronounced.

In some respects these gastropods resemble chitons, for the anterior mantle cavity is reduced and a secondary mantle cavity tends to circle almost the entire animal. In contrast to chitons, however, a head with sensory tentacles is well developed. Ctenidia are reduced or absent in some cases, and anterior folds of the mantle under the rim of the shell acquire the functions of *mantle gills.* Aspidobranchs exhibit a primitive design of the nervous system and of the internal body cavities, i.e., the gonads open into the right kidney ducts which serve as gonoducts. Limpets are very common along stony beaches, attached to or creeping on rocks between tidewater marks. Abalones attain lengths of up to 10 ins. The shell of such a mollusk bears a row of perforations through which water is expelled from the mantle cavity.

In pectinibranchs and neogastropods, a very pronounced coiling of the visceral hump and the shell results from highly unequal growth on the left and right sides. Indeed, many of the organs on one side actually fail to develop. In most instances it is the original left side which remains suppressed before torsion. For example, a periwinkle or a whelk possesses only one gill, one atrium, and one kidney, all developed on

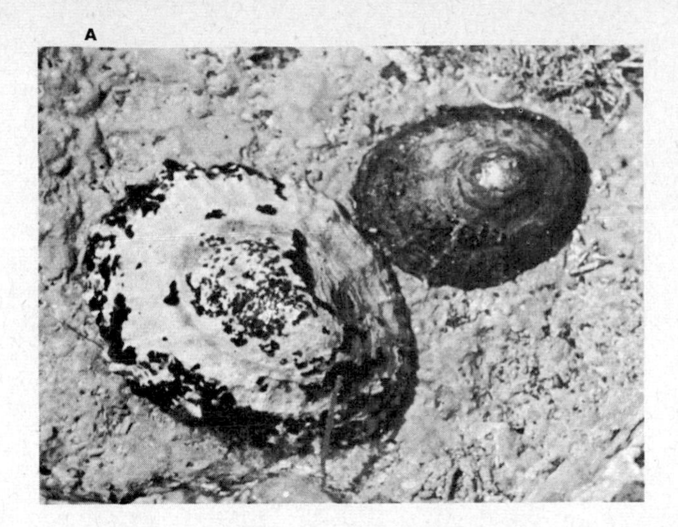

25.10. Aspidobranch snails. **A,** limpets (*Patella*); **B,** abalone (*Haliotis*); note row of shell perforations for excurrent water.

the right side before torsion and lying on the left side after torsion (Fig. 25.11). These animals also possess gonads which open via separate gonoducts into the mantle cavity. The nervous system still has an ancestral design in the pectinibranchs; i.e., the four pairs of ganglia are well separated and are interconnected by nerve cords twisted after torsion. In neogastropods, by contrast, the ganglia are aggregated close together around the anterior nerve ring, and series of nerves lead away from the ring into the body.

25.11. Scheme of organs in adult pectinibranchs and neogastropods (after torsion). Note that only the right members of paired organs are developed; these organs lie on the left side of the adult. A concentrated nerve ring as sketched is characteristic of neogastropods.

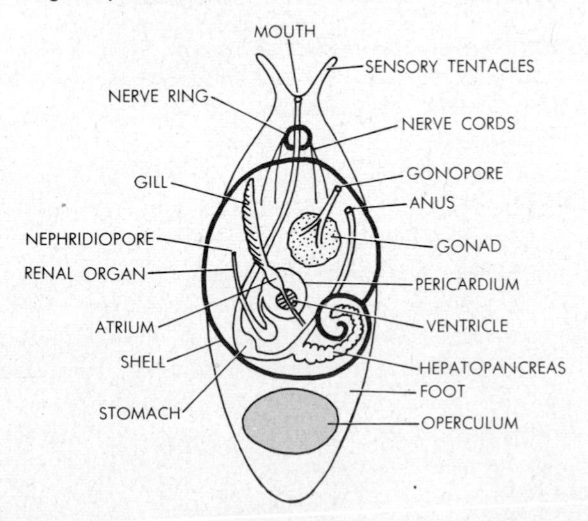

Both pectinibranchs and neogastropods possess an *operculum*, a horny plate on the dorsal posterior part of the foot (Figs. 25.12 and 25.13). When the animal withdraws into its shell, the operculum covers the shell opening. *Littorina*, the exceedingly common periwinkles found between high- and low-water marks along most sea shores, are actually amphibious in habit. When the tide ebbs, these snails withdraw into their shells and secrete a mucus seal around the opercular cover. In this state the animals await the return of the tide, several hours later. Pectinibranchs also include the *cowries*, snails with brilliantly pigmented and highly polished shells greatly prized by collectors. Freshwater pectinibranchs like *Paludina* are ovoviviparous, directly developing types.

Neogastropods include the most advanced prosobranchs. These animals typically possess a *siphon*, a tubular extension of the body wall from the mantle cavity, used in directing incoming water drawn by the gill toward the latter. Most of the neogastropods are scavenging and predatory carnivores. They clasp lobsters, crabs, or other mollusks with the foot and bore a hole through the protective armor of the prey by means of the radula (hence the designation "oyster drills" for many

A

B

25.12. Pectinibranch snails. **A.** *Littorina,* the common tidal periwinkles. **B.** The freshwater type *Paludina,* withdrawing into its shell and closing the operculum. Cf. color plate IX.

of these snails). Indeed, whelks are equipped with a protrusible proboscis bearing the mouth at the tip. The radula may be protruded through the mouth and prey may be attacked in this fashion. In some of the neogastropods the salivary glands also secrete poison into wounds made by the radula. Male whelks possess a penis protrusible from the mantle cavity, a copulatory organ containing the terminal part of the sperm duct. Female whelks lay eggs in batches of several hundred, each batch enclosed in a capsule. Many of such capsules are joined to one another, the whole forming a spongy mass.

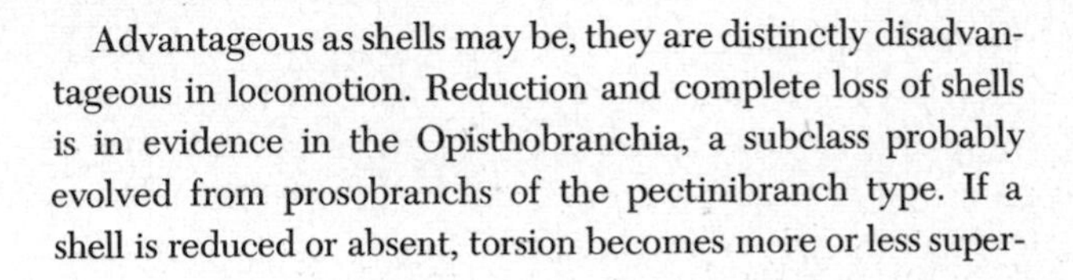

25.13. Neogastropod snails. A. The whelk *Buccinum,* seen from above. The long tube projecting anteriorly is the siphon, which directs water toward the gill. The pair of smaller projections are sensory tentacles on the head. **B.** Egg capsules of *Buccinum.* Cf. color plate IX.

Advantageous as shells may be, they are distinctly disadvantageous in locomotion. Reduction and complete loss of shells is in evidence in the Opisthobranchia, a subclass probably evolved from prosobranchs of the pectinibranch type. If a shell is reduced or absent, torsion becomes more or less superfluous. We actually find that opisthobranchs exhibit a greater or lesser reversal of torsion, or *detorsion.* Thus, veligers first undergo torsion in typical prosobranch fashion, but during later development the visceral hump rotates back partly or wholly to its original position (Fig. 25.14). Some opistho-

25.14. Torsion and detorsion, schematic. A. Original condition, paired organs present, anus posterior. **B.** After torsion, as in most prosobranchs (aspidobranch pattern, cf. Fig. 25.9; pectinibranch and neogastropod pattern, cf. 25.11). Only right members of paired organs present, lying on left side, anus anterior, visceral cords form figure-of-eight. **C.** 90° detorsion, as in some opisthobranchs. Mantle cavity, gill, and anus lie on right side, heart lies transversely, visceral cords partially untwisted. **D.** Complete 180° detorsion, as in other opisthobranchs (e.g., nudibranchs). Pattern resembles **A,** but only right members of paired organs present.

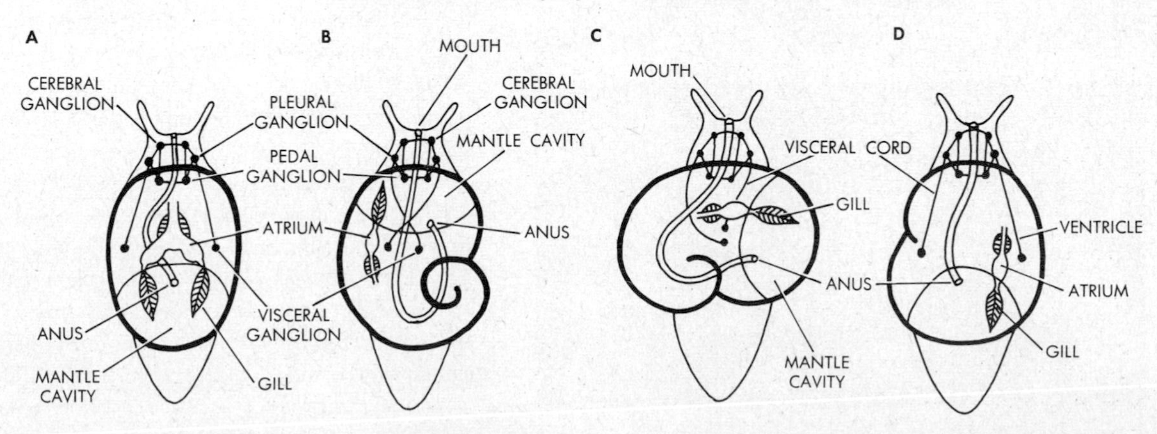

25.15. Tectibranchs. The diagram of the sea hare represents *Aplysia,* that of the sea butterfly, *Cavolinia.* **Photos: A,** the sea-hare *Aplysia,* in eelgrass; **B,** a sea butterfly, from life. Cf. color plate X.

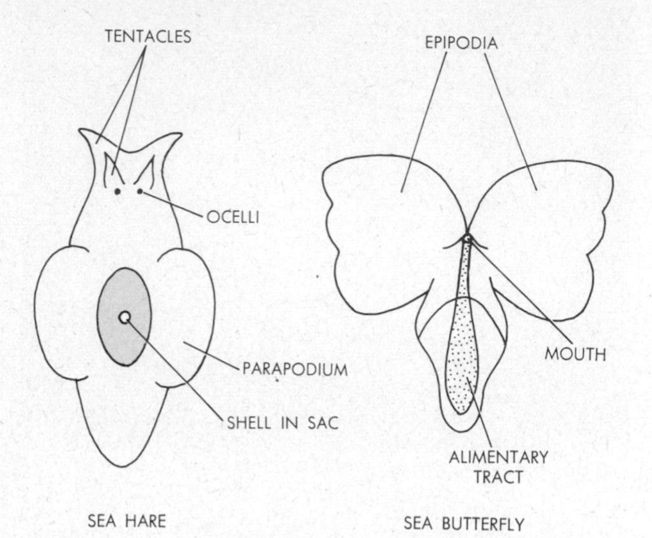

A

B

branchs undergo a 90° clockwise detorsion, a process which brings the anus and the mantle cavity to the right side. Other members of the subclass, especially the shell-less nudibranchs, undergo a complete 180° detorsion. Here the anus is posterior and the visceral nerve loop no longer has the form of a figure-of-eight. Nevertheless, before the original torsion occurs in the veliger, the organs of the left side again fail to develop, as in pectinibranchs. Thus, despite the later detorsion, adult opisthobranchs still possess only a single atrium, gill, and kidney. These organs develop on the right side before the original torsion, come to lie on the left after torsion, and end up on the right side again after detorsion. We note that although the adults may be perfectly bilateral externally, their internal structure and curious development clearly testify to their derivation from asymmetrical ancestors. Opisthobranchs also illustrate the general evolutionary principle that organs once lost are never regained (though other organs with analogous functions may well evolve).

Tectibranchs such as *Aplysia* undergo full detorsion. These animals possess two pairs of tentacles, the anterior pair ear-shaped (hence the name "sea hares"), the posterior pair each with an eye at the base (Fig. 25.15). The shell is small and partly interiorized; i.e., it is covered over much of its surface by turned-up portions of the mantle. In the region of the shell, the foot extends on each side as an upward-projecting flap. Called *parapodia,* these flaps make swimming possible. Sea hares feed on seaweeds; the esophagus is expanded into a crop lined with horny plates, where food is ground up. Tectibranchs also include the *pteropods,* or sea butterflies, adapted even more than the sea hares to an actively swimming, pelagic life. The shells of these animals are generally transparent and either coiled (e.g., *Limacina*) or uncoiled and vaseshaped (e.g., *Cavolinia*). Extending from the shell opening is the foot, which is expanded into a pair of conspicuous finlike paddles, the *epipodia.* Cilia on these direct planktonic food organisms toward the mouth.

The Nudibranchia encompass some of the most beautifully pigmented of all sea animals (Fig. 25.16). Because they are without shells or mantle cavities, these opisthobranchs are referred to generally as *slugs.* Detorsion is often complete, as in *Eolis,* and in such cases the anus is posterior. Gills are usually absent, the mantle surface functioning in breathing. In *Eolis* and many other nudibranchs, gas exchange is assisted further by *cerata,* conspicuous tentaclelike projections from the sides of the animal which hide much of the dorsal body surface. In the interior of a ceratum is a diverticulum from the alimentary tract. Such pouches in some species do and in

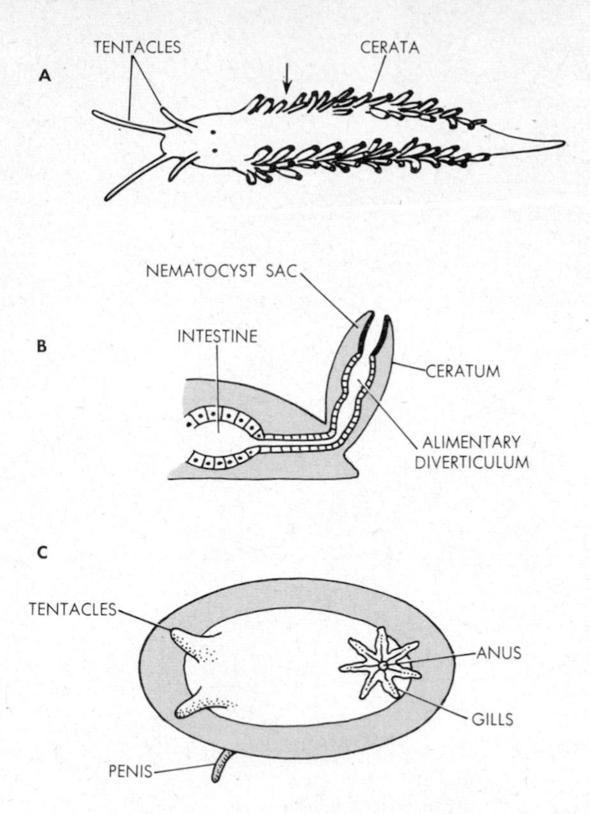

25.16. Nudibranchs. Diagrams: A. Outline sketch of *Eolis*, showing position of cerata; arrow points to anus, on side (90° detorsion). **B.** Section through a ceratum of *Eolis*, showing terminal opening and alimentary channel and sac in which coelenterate stinging cells accumulate. **C.** Outline sketch of the sea lemon *Doris*, a nudibranch without cerata. **Photo:** *Hermissenda*, on hydroid colony serving as food for the nudibranchs. Note prominent cerata. Cf. color plate X.

others do not open to the exterior at the tips of the cerata. *Eolis* feeds on coelenterates, and the cnidoblasts of the latter become collected in the cerata. The nudibranch then uses the sting cells defensively, just as the original coelenterate would have done. Not all nudibranchs possess cerata, however. For example, the sponge-eating sea lemon *Doris* is without such projections.

Prosobranchs of the pectinibranch type are believed to have given rise also to the subclass Pulmonata. These most-specialized gastropods exhibit torsion and they possess a single atrium and kidney, but the alimentary tract is U-shaped without being looped. In the most primitive pulmonates the visceral nerve loop again forms a figure-of-eight. In more advanced members of the subclass the longitudinal cords are shortened, and in the most specialized types all ganglia become part of the circumesophageal nerve ring. A concentrated, symmetrical, untwisted nervous system thus results, symmetry evidently being achieved here by an extreme shortening of the nerve cords, not by a detorsion. Ctenidia are absent in all pulmonates. Instead, the dorsal part of the mantle cavity is highly vascularized and the cavity as a whole functions as an air-breathing lung (cf. Fig. 9.28). It is likely that pulmonates evolved as terrestrial forms directly from marine pectinibranch ancestors. Some pulmonates have adapted secondarily to aquatic life (e.g., the freshwater snail *Limnaea*), but these are still lungbreathers which must surface periodically for air.

Terrestrial pulmonates include shell-less land slugs like

A

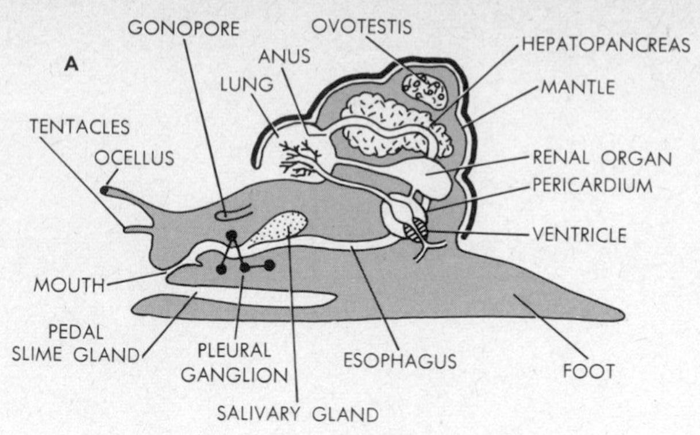

B

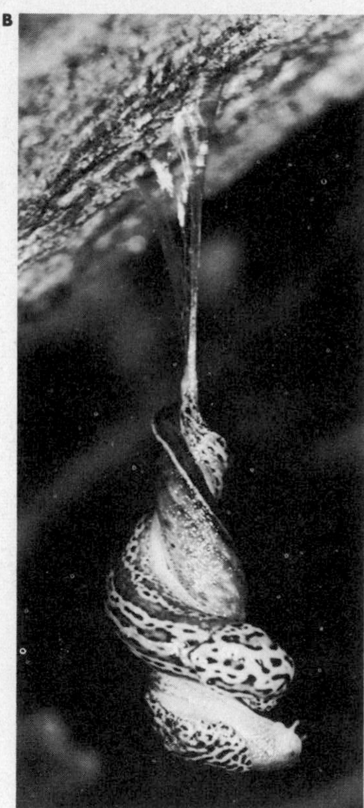

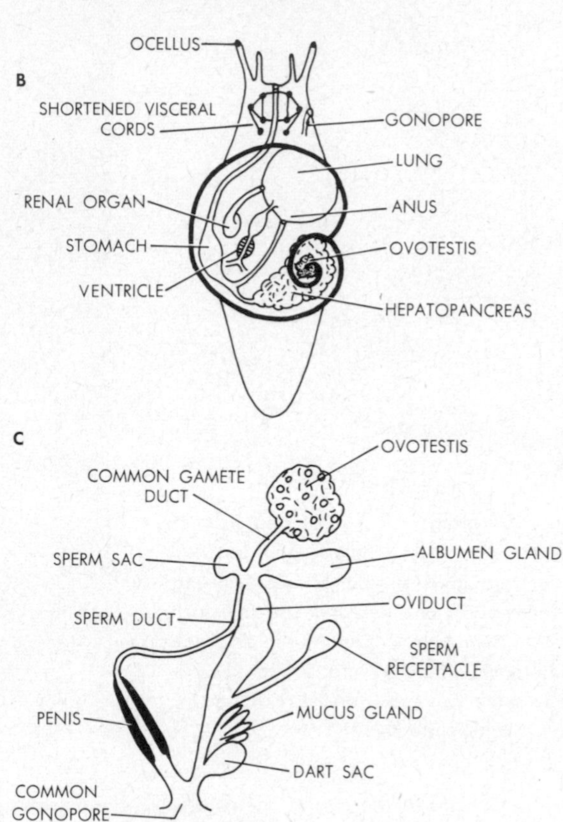

25.17. Pulmonates. Photos (cf. also Figs. 9.28 and 10.11): **A.** *Planorbis,* a fresh-water pulmonate with a shell coiled in one plane. **B.** A mating pair of land slugs, suspended from a tree by a mucus cord. **Diagrams: A.** Schematic sectional side view of *Helix,* showing position of some of the internal organs. The gonopore opens on left side of anterior region, near head (as indicated in **B**). The ovotestis and the gonopore are connected by a duct system (as shown in **C**). For illustration of breathing system, cf. Fig. 9.28. To preserve clarity of other parts, the alimentary system is not drawn in completely. **B.** Sectional top view of *Helix,* correlated with **A.** Note anteriorly concentrated ganglia with shortened cords, hence absence of figure-of-eight loop. **C.** The hermaphroditic reproductive system of *Helix,* simplified. The sperm sac stores sperms produced by the snail itself, whereas the sperm receptacle stores sperms produced in the mating partner and introduced into the snail during mating.

Limax and the familiar shelled land snails of the genus *Helix* (Fig. 25.17). One species of the latter, *H. pomatia,* is edible. All pulmonates, and *Helix* in particular, are noted for their exceedingly complex reproductive systems. Like opisthobranchs but unlike prosobranchs, pulmonates are hermaphroditic. *Helix* possesses a single gonad which produces *both* sperms and eggs (*ovotestis*). Sperms are formed first and are stored, and eggs developed later are fertilized by sperms from another snail. A sperm duct leads to the outside via a penis, and an oviduct exits via a vagina. Sperms are packed together into a *spermatophore,* or sperm package, before transfer during copulation. The gonopores lie anteriorly on the right side behind the head. Mating is preceded by a remarkable preparatory process. Two snails approach each other, and from the female gonopore of each a calcareous *dart* is shot out into the body of the partner. Coming to rest among the internal organs of the snails, such darts appear to stimulate the animals to undertake copulation. The penis of each snail is inserted into the vagina of the other and the spermatophores are transferred. Fertilized eggs become enveloped by bulky layers of albumen and by external layers of calcium deposits secreted around them (comparable to events in bird eggs). Some pulmonates are ovoviviparous and these retain their eggs. Others, like *Helix,* lay the eggs in batches of several dozens or hundreds under leaves or in holes in soil. Development then takes place directly.

Pulmonates winter over by hibernating. They creep into soil burrows under leaves, withdraw into their shells, and secrete one or more protective calcareous membranes over the shell opening.

CLASS SCAPHOPODA

Tusk shells; marine; shell tubular, open at both ends; foot a burrowing organ; ctenidia absent; circulatory system reduced; development via trochophore and veliger stages. *Dentalium*

This small class comprises some 350 species of sand-burrowing mollusks. The animals are elongated in a dorsal direction and they are covered by tapered, tubular shells. These give them the appearance of canine teeth or tusks (Fig. 25.18). From the wider, ventral end of a shell projects the muscular, conical foot, which serves as a digging organ. Also protruding from the ventral end is a reduced, proboscislike head, to which are attached numerous prehensile tentacles, the *captacula,* each with an expanded, suckerlike tip. Between the visceral hump and the posterior portion of the shell passes a channel-like mantle cavity, formed in the larva by the fusion of lateral mantle lobes (as in some clams). Water circulates in and out of the mantle cavity via the dorsal shell opening, which projects beyond the burrow of the animal into clear water. The mantle tissues function in breathing; gills are absent. A radula can be protruded through the mouth, and the anus opens posteriorly into the mantle cavity. The nervous system is symmetrical and of primitive structure, with four pairs of ganglia and connective cords. A heart is absent and the system of blood vessels is reduced. Kidneys are paired, without opening into the coelom, and the right kidney also serves as gonoduct, as in aspidobranch gastropods.

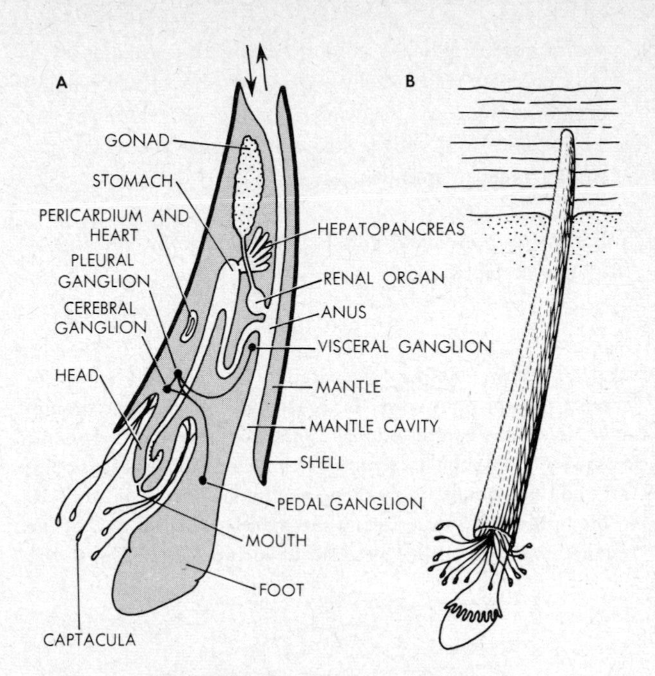

25.18. Scaphopoda. A. Sagittal section through *Dentalium* (schematic, *adapted from Naef*). **B.** Body proportions and position of animal in its sand burrow.

Scaphopods appear to represent a separate line of molluscan evolution, on a level of specialization intermediate between that of gastropods and that of pelecypods.

CLASS PELECYPODA

Clams and allied bivalve mollusks; aquatic, mostly marine; bilateral, laterally compressed, with dorsally hinged valves; head rudimentary, without tentacles; foot usually tongue-shaped and used in burrowing; mouth with labial palps;

radula absent, stomach in most cases with crystalline style; gills complex and usually expanded into ciliary feeding organs; sexes nearly always separate; development via trochophore and veliger stages in marine types, via *glochidia* stages in freshwater types.

Subclass Protobranchia: gills for breathing only, feeding by means of labial palps. *Nucula*

Subclass Filibranchia: gill filaments usually connected by ciliary junctions; gills primarily for feeding. *Mytilus*, mussels; *Pecten*, scallops.

Subclass Eulamellibranchia: gill filaments connected by tissue junctions; gills for feeding primarily. *Ostrea*, oysters; *Mya*, soft-shelled mud clams; *Mercenaria*, hard-shelled quahog clams; *Tridacna*, giant clams; *Teredo*, shipworms.

Subclass Septibranchia: gills organized as muscular septum with pores. *Cuspidaria*

25.19. Clam shells. **A.** Exterior view of left valve. **B.** Interior view of right valve. **C.** Schematic cross section through one valve near ventral edge, to indicate layers. **D.** Schematic cross section through both valves, to indicate manner of fusion of left and right mantle layers along ventral midline, as in scallops. In the latter, ocelli occur along the exposed ventral part of the mantle, as indicated (for structure of such ocelli, cf. Fig. 8.47).

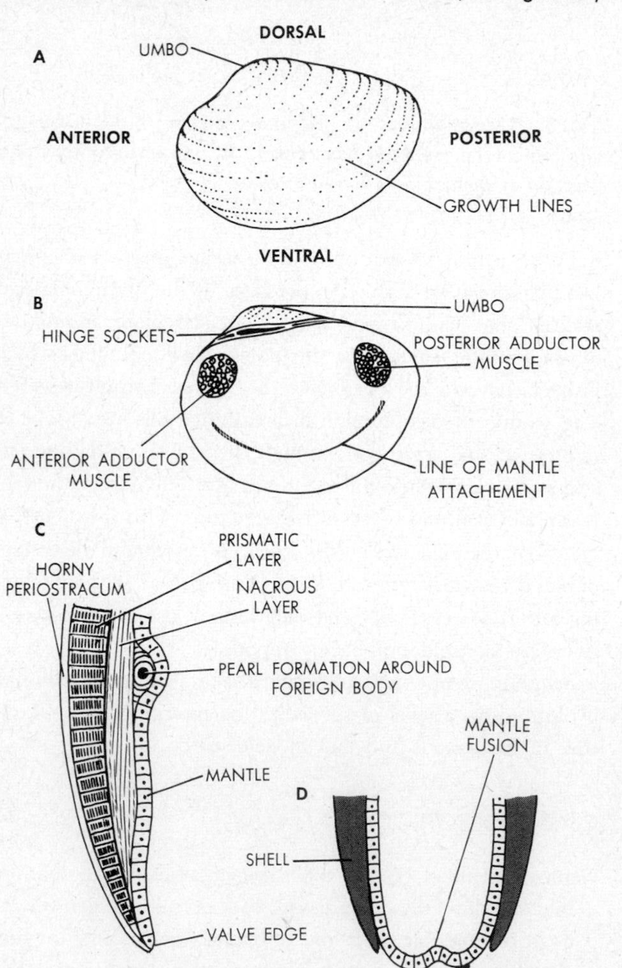

In contrast to brachiopods, the mantle lobes of a clam consist of a single layer of the body wall. The ventral edges of the lobes may hang free or may be fused together along the ventral midline and enclose the mantle cavity more or less completely (Fig. 25.19). In the latter case the shell valves may gape open only to a very limited extent. The dorsal hinge of the valves is of the tooth-and-pit type, and a horny elastic ligament keeps the valves joined along the hinge and also presses them open. The valves are closed by usually two powerful adductor muscles, one anteriorly and one posteriorly between the two valves.

A valve typically has the shape of a flattened and distorted cone. The apex of the cone, or *umbo*, lies toward the anterior end of the animal. The umbo is the oldest part of a valve, growth proceeding around it in concentric ellipsoid rings. The outermost layer of a valve is a horny *periostracum*, an acid-(CO_2)-resistant cuticle which protects the underlying portions from dissolution by sea water. This periostracal layer is secreted by the whole mantle epithelium in the larva and by the mantle edge in the adult. Under the periostracum is the *prismatic layer*, a calcareous shell plate in which $CaCO_3$ crystals are aligned perpendicularly to the mantle surface. This layer too is formed only by the mantle edge in the adult. The innermost part of a valve is the pearly *nacrous layer*, in which $CaCO_3$ crystals are aligned parallel to the mantle surface. Nacre is laid down by the whole mantle epithelium even in the adult; hence this layer is oldest and thickest near the umbo, youngest and thinnest along the mantle edge. Note that a valve thus grows in thickness in all parts but in area only along the mantle edge. Pearls are formed when a foreign object (even a fluke larva or other parasite in some cases) lodges between the valve and the mantle. The object then becomes enveloped by successive layers of nacre secreted by the mantle.

The internal structure of pelecypods generally is illustrated by the anatomy of a clam (subclass Eulamellibranchia). The visceral mass is suspended from the dorsal midline (Fig.

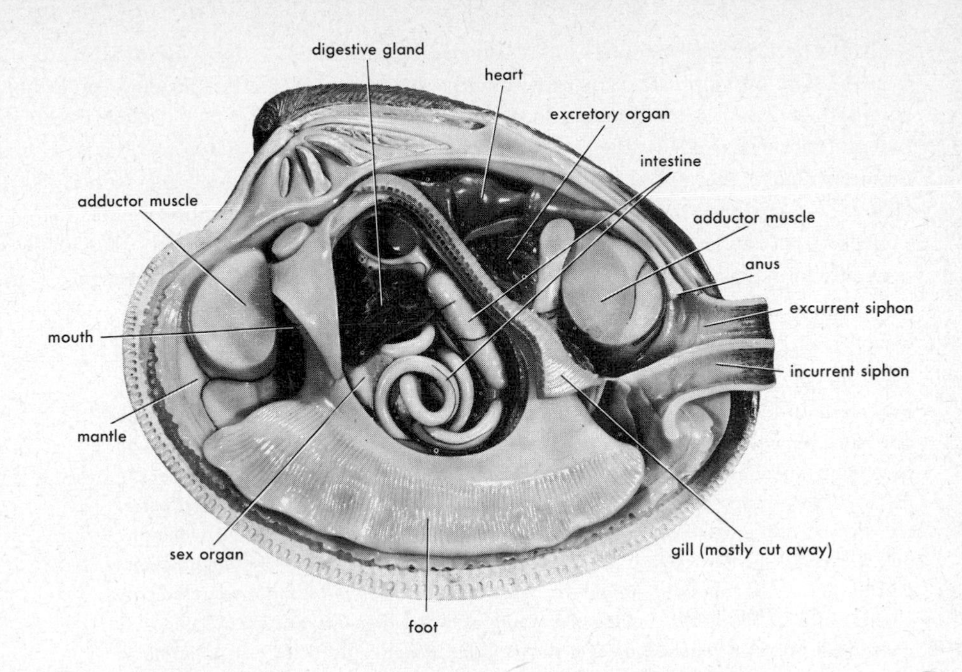

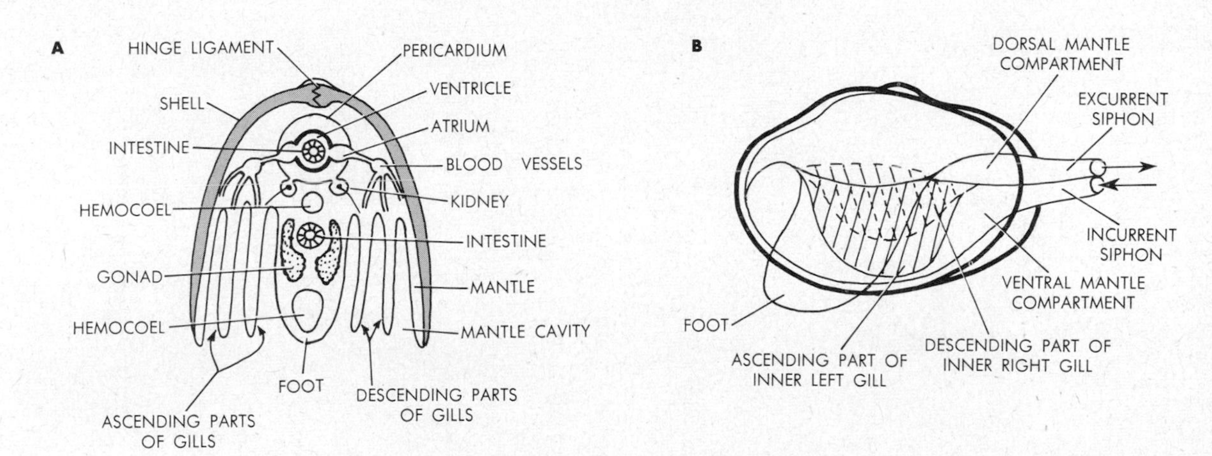

25.20. Clam structure. Photo: model indicating general internal anatomy. **Diagrams: A.** Transverse section. A portion of the intestine passes through the ventricle of the heart. Note that the hinge ligament is a thickened dorsal part of the periostracum. **B.** View of left side, indicating anatomical interrelations of foot, gills, mantle compartments, and siphons.

25.20). Continuous ventrally with the visceral mass is the foot, which in many cases may be protruded between the valves. Two pairs of ctenidia project into the mantle cavity, one member of each pair on the left side of the foot, the other member on the right. The ctenidia are suspended dorsally from the mantle and they form large, conspicuous *gill plates.* Each is actually a double plate; a descending part passes ventrally from its dorsal attachment, then folds up into an ascending part which continues back to the dorsal attachment. On each side of the foot, the ascending part of the inner gill lies directly against the foot, and behind the foot these two ascending parts on the left and right fuse to each other and continue posteriorly as a horizontal partition. The latter divides the posterior mantle cavity into a dorsal compartment and a ventral main compartment. Both compartments open to the outside along the posterior valve edges, the dorsal chamber forming an *excurrent siphon*, the ventral one an *incurrent siphon*. In some clams these siphons are extended into long retractile tubes projecting beyond the valves. The two siphons in such cases generally form a single tubular projection covered by a common sheath of mantle tissue. In some cases the siphons form two separate projections.

A gill plate is composed of numerous fused, parallel *gill filaments* (Fig. 25.21), each consisting of a descending and an ascending section like the gill plate as a whole. These two sections are held together in places by transverse tissue struts. A blood vessel with descending and ascending portions passes through the whole filament. Adjacent filaments are fused to

25.21. Clam structure: food and water paths. **A.** Cutaway diagram of ventral edge of gill, showing ascending and descending gill plates, and tissue junctions between individual gill filaments. Afferent vessels carry blood down through tissue partitions between water tubes, and efferent vessels then conduct blood up through ascending and descending portions of a gill. Microscopic food particles collect in mucus strands in the ventral food groove, heavy particles fall off the gill and out of the animal, and water passes through the gill pores into the water tubes within the gill. **B.** Frontal section through single gill plate, to indicate ciliation of gill epithelium. **C.** The gills of the left side, the outer gill shown as a cutaway. Solid arrows indicate ciliary paths of food particles toward mouth and of water into excurrent siphon via dorsal gill passage; dashed arrows show paths of heavy particles not adhering to gill.

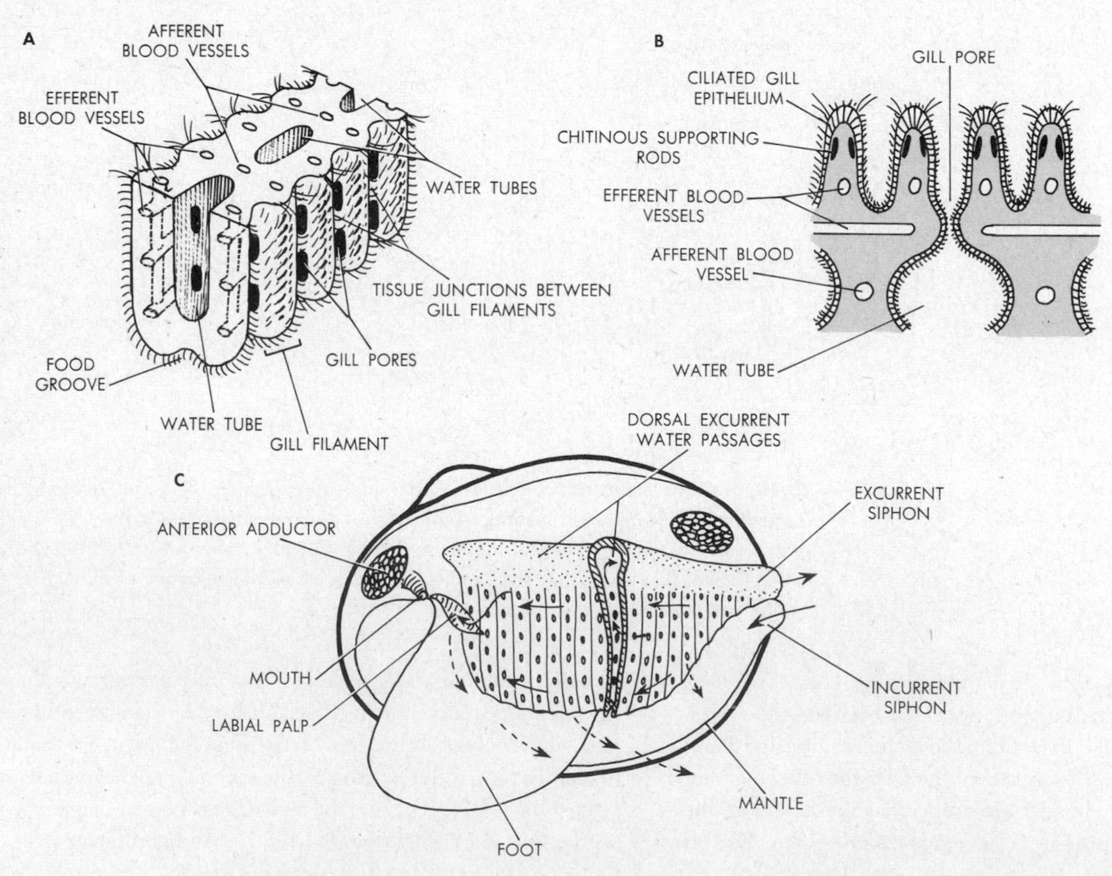

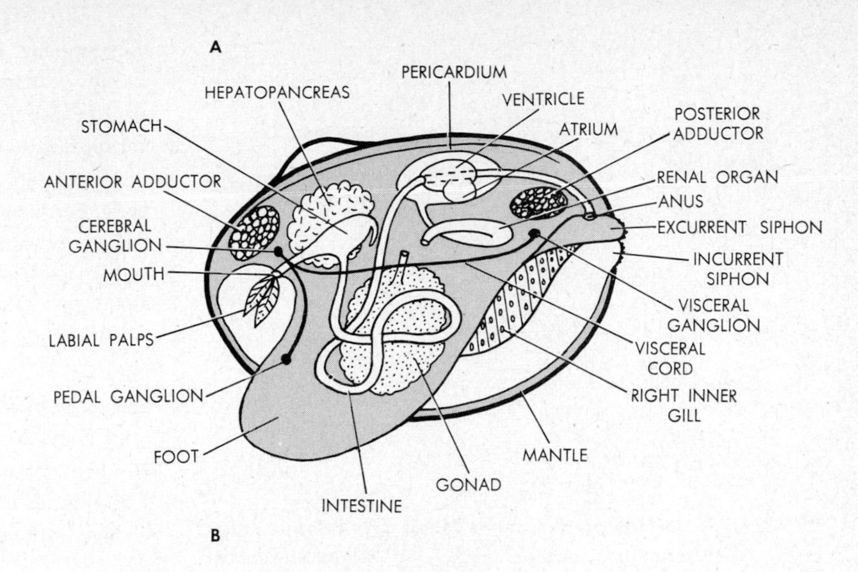

25.22. Clam structure: internal organs. A. Sectional sagittal view. The nephridiopores and gonopores open into the excurrent water passages. **B.** Sectional view through stomach, showing crystalline style and gastric shield (*adapted from Barrois*).

one another by tissue junctions, which leave *gill pores* through which water may pass. The whole outer gill surface is covered by cilia. These beat in such a way that food-bearing water is drawn into the mantle cavity via the incurrent (ventral) siphon. Water then passes through the gill pores into the *water tubes* present between the descending and ascending parts of a gill plate. From there the water is propelled dorsally and eventually out of the clam via the excurrent (dorsal) siphon. Food particles are strained out on the gill surface, where they become caught up in secreted mucus. Strands of such mucus are propelled by the cilia first to the ventral gill edges, then anteriorly toward the *labial palp* on that side (cf. Fig. 25.21). The palps are ciliated tissue flaps which conduct food-containing mucus toward the mouth. A considerable amount of sorting takes place before food is actually ingested. For example, sand grains and larger food organisms drop off the gills and usually do not reach the mouth.

Clams (and pelecypods generally) are without a radula, a superfluous structure in a filter-feeding animal. However, a well-developed hepatopancreas is present, as is typically a *crystalline style* (Fig. 25.22). This is a rodshaped gelatinous body lying in and being secreted by a diverticulum of the stomach. Ciliated cells in this gastric pouch keep the style rotating and pressed against a horny *gastric shield* along the inner surface of the stomach. As the style rotates against the shield, tiny particles of style material wear off and mix with food. This style material contains an amylase, the only extracellular digestive enzyme produced by clams. Proteins and lipids are digested intracellularly within the hepatopancreas. A coiled intestine leads from the stomach through the pericardial coelom to the anus, and the latter opens ino the dorsal siphon.

In correlation with the reduction of the head, the nervous system is to some extent reduced. A pair of fused cerebral-pleural ganglia lies anteriorly in a nerve ring around the esophagus. Nerve cords from there connect with a pair of

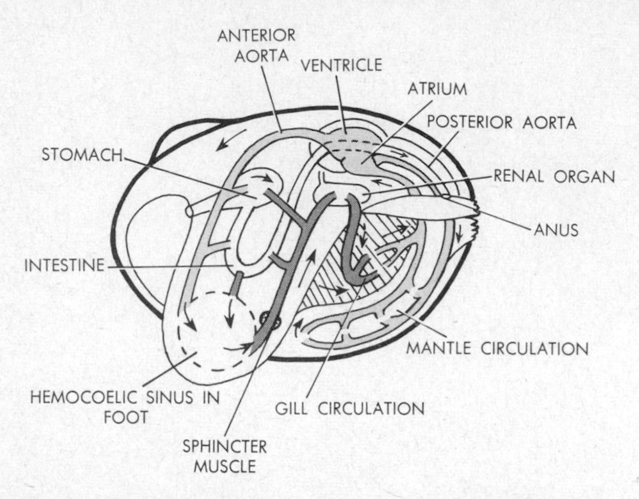

25.23. The general pattern of the blood circulation in most clams, diagrammatic. Arterial blood, light-shaded; venous blood, dark-shaded. Oxygenation of blood takes place in the gills and, most particularly, in the mantle. Arrows indicate direction of flow. The sphincter around the vein in the foot accomplishes an accumulation of blood in the foot sinuses and so produces enlargement and stiffening of the foot during locomotion.

pedal ganglia in the foot and a pair of visceral ganglia near the anus and the posterior adductor muscle. The renal organs are paired. They lead from the pericardial coelom to an internally ciliated bladder which opens into a dorsal water passage on that side (cf. Figs. 25.22 and 9.36). Urine flow is maintained by the bladder ciliation. The renal organs are associated intimately with the circulatory system, which is complex and open (Fig. 25.23). A ventricle in the pericardial coelom surrounds the intestine, and arteries from the ventricle pass both anteriorly into the hemocoelic sinuses of the visceral mass and the foot and posteriorly into the mantle lobes. The latter are important breathing organs in clams; veins from the mantle lobes carry oxygenated blood directly back to the heart, via a pair of atria. Blood from the visceral and pedal sinuses travels via veins to the kidneys. The main vein from the foot is equipped with a sphincter muscle, contraction of which brings about an accumulation of blood in the foot and thus a distension of that organ (cf. also above). After its passage through the kidneys, venous blood is channeled into the gill circulation, and oxygenated blood from there joins that from the mantle lobes in the atria. The blood is colorless in most cases; amoeboid white cells are present in it.

In most clams the sexes are separate. Gonads are formless masses around the intestinal coils. Gonoducts discharge into the outgoing water current. In marine forms fertilization usually occurs externally, and the trochophores and veligers are free-swimming. Freshwater clams typically exhibit internal fertilization and abbreviated development. More specifically, eggs are discharged into the water tubes within a gill plate. Fertilization occurs there, the sperms entering with the incurrent water. Zygotes develop in the water tubes into highly modified veligers, so-called *glochidia* larvae, each equipped with two valves (Fig. 25.24). These larvae are expelled through the excurrent siphon; and if they are not to perish, they must attach themselves within a short time to the gills or fins of fishes. At such sites the glochidia live parasitically until they are mature. Then they drop off and become independent adults.

Clams lead a semisedentary, burrowing life. They may change station by digging the foot into sand or mud, distend-

25.24. Development of freshwater clams. Diagrams (*adapted from Lefevre and Curtis*): **A,** glochidia larva, seen from anterior end; the hooks clamp on to gill tissue of fish host; **B,** young free clam, after detaching from fish gills. **Photo:** glochidia larvae of a freshwater mussel.

A HOOK VALVE ADDUCTOR MUSCLE B FOOT

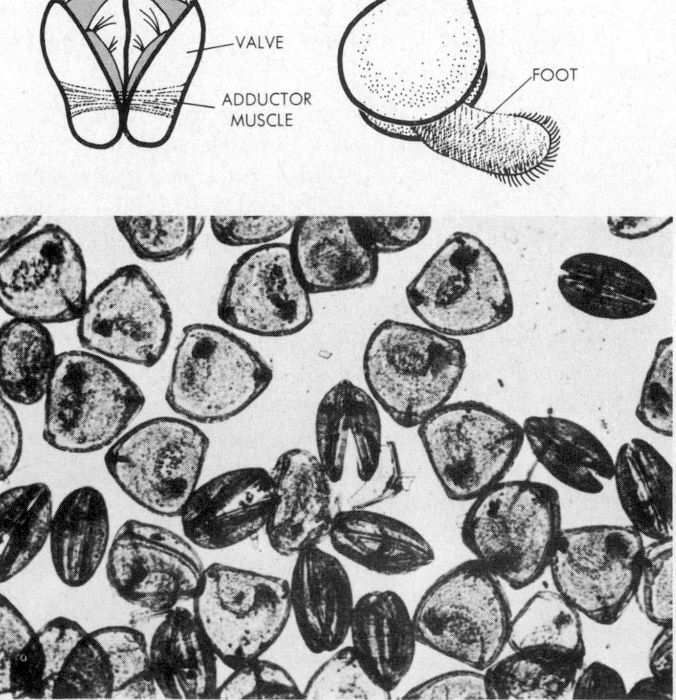

Courtesy Carolina Biological Supply Company.

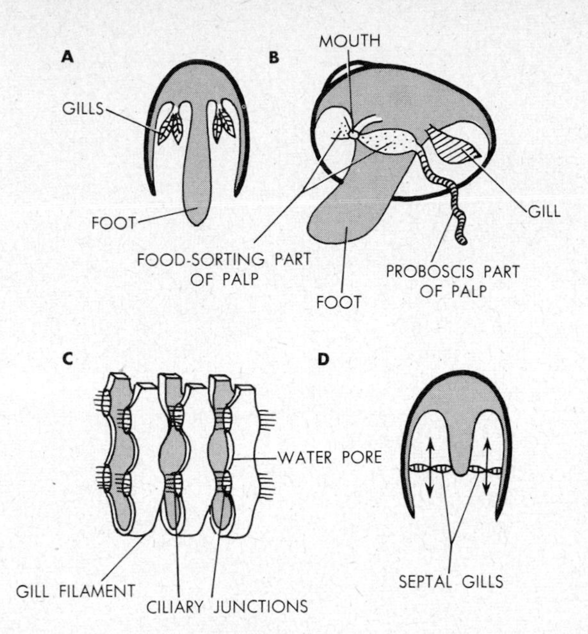

25.25. A and B, Protobranchia (*Nucula*). **A,** schematic cross section showing small breathing ctenidia; **B,** diagrammatic sectional side view showing food-collecting apparatus and separate breathing gill. **C.** Filibranchia, portion of ventral part of gill, showing ciliary junctions between gill filaments (contrast with eulamellibranch condition, Fig. 25.21.) **D,** Septibranchia, diagrammatic cross section showing perforated partition formed by ctenidia on each side. (*Adapted from Naef, Sedgwick, and other sources.*)

ing it with blood to form an anchor, and then pulling the rest of the body after it. Otherwise clams remain partly or wholly buried, with the siphon protruding up into clear water. Some clams, notably oysters, are permanently attached, their larger left valve being cemented to the ground.

The most primitive pelecypods are the Protobranchia, in which the ctenidia are small and, as in other molluscan groups, function only in breathing (Fig. 25.25). Feeding is accomplished entirely by highly developed labial palps. These consist of an elongated, proboscislike, food-collecting part and of two platelike, ciliated, food-sorting parts. The nervous system of these animals still preserves the ancestral design and contains four separate pairs of ganglia. Another primitive trait is the continuity of the gonad with the kidney.

In the Filibranchia (now often classified as two separate subclasses), adjacent gill filaments are held together not by tissue connections as in clams, but only by ciliary junctions (cf. Fig. 25.25). The common mussel *Mytilus* secretes numerous mucous *byssus threads,* which harden on contact with sea water and which attach these animals to rocks and to one another (Fig. 25.26). The secretion is formed in a glandular *byssus pit* at the posterior part of the foot. A ventral groove from the pit along the foot permits the secretion to flow as a thread. Mussels are characterized also by the location of the gonads within the tissue of the mantle lobe. The scallop *Pecten* is a highly specialized filibranch. It possesses but a single adductor muscle (the posterior one), and this muscle consists of two parts, one with smooth fibers which close the valves, the other with striated fibers which produce a rapid clapping of the valves. Such clapping enables a scallop to propel itself in jetlike fashion for short distances. The foot of scallops is highly reduced and is used mainly to clean the labial palps and the gills. Numerous stalked blue eyes, each with a lens, are present along each mantle edge (cf. Figs. 8.47 and 25.26). Scallops are among the few pelecypod hermaphrodites; ovaries are pink, testes are cream-colored.

25.26. Filibranchs (cf. color plate XI). **A.** Some of the organs of the scallop *Pecten.* (*Modified from Dakin.*) **B.** Section through valve edge of *Mytilus,* showing eggs in mantle tissue. **C.** The byssus apparatus of the mussel *Mytlius,* schematic.

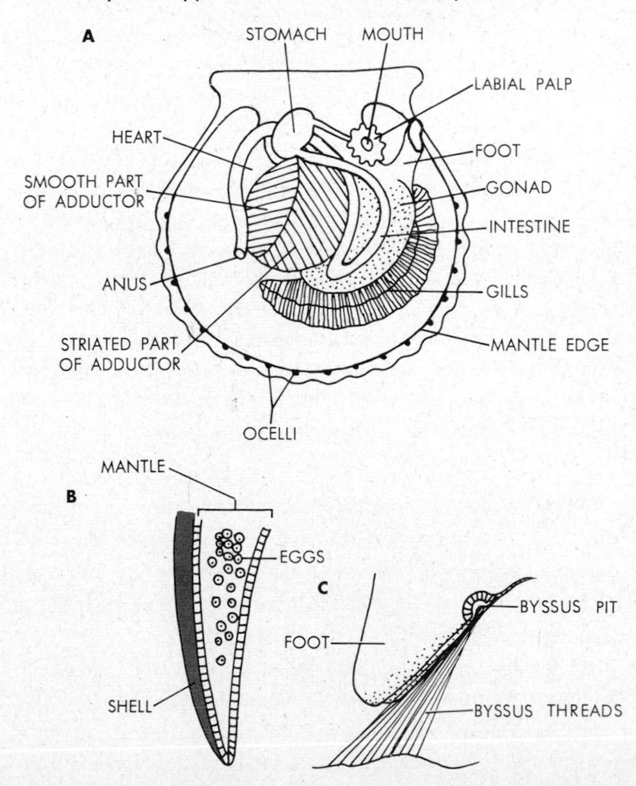

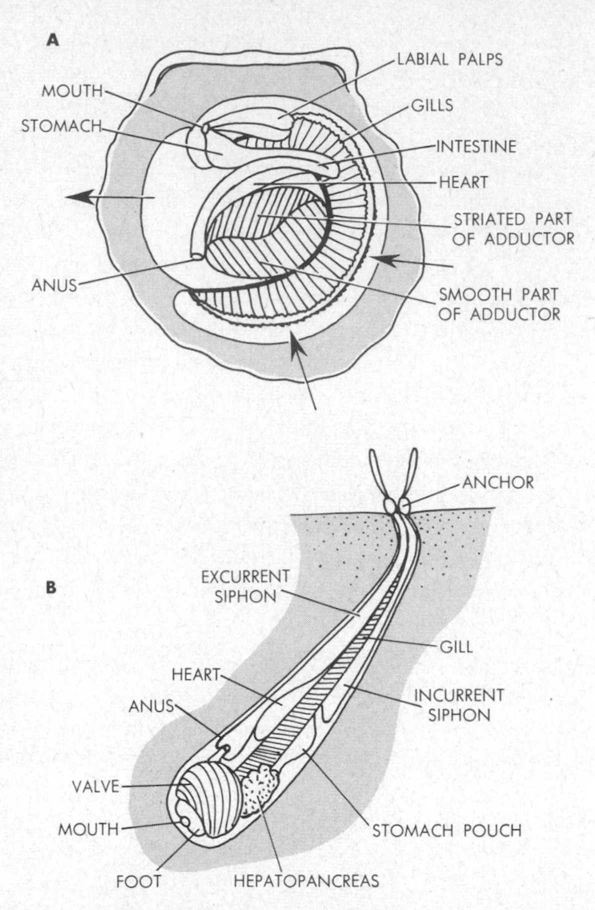

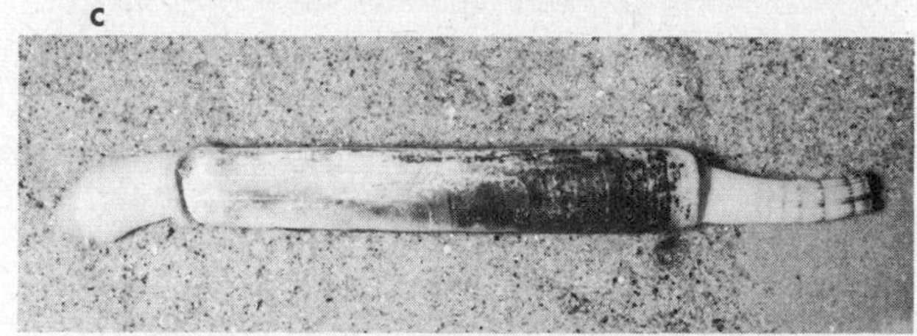

25.27. Eulamellibranchs. Diagrams: A. Some of the organs of the oyster *Ostrea.* Arrows indicate incurrent and excurrent pathways of food and water (*modified from Yonge*). **B.** Sectional side view of the shipworm *Teredo* in wood burrow. Sawdust produced by rotational motion of the valves passes into the mouth and may be stored in the stomach pouch. (*Modified from Borradaile.*) **Photos: A.** The marine horse clam *Schizothaerus,* with long extensible siphon. **B.** The freshwater clam *Anodonta;* note conspicuous burrowing foot. **C.** The razor clam *Solen.* **D.** Shell of the giant clam *Tridacna.* Cf. color plate XI.

Septibranchia are a small group in which the gills on each side have become modified into a horizontal muscular "diaphragm," equipped with pores. As such a septum moves up and down, it circulates water between the dorsal and ventral chambers of the mantle cavity (cf. Fig. 25.25).

By far the most abundant pelecypods are the eulamellibranchs, identified by fused gill filaments as described above (Fig. 25.27). In this group are the oysters, which in some respects resemble scallops: a foot is absent, only a single adductor muscle is present, and the animals are hermaphrodites. Indeed, the single gonad here is an ovotestis functioning alternately as an ovary and a testis, as in the land snail *Helix*. Of interest for sheer size are the giant clams such as *Tridacna*, which may be 2 yd wide and weigh ¼ ton. Rock-boring and wood-boring clams are also noteworthy, the latter mainly because of their destructiveness. For example, the "shipworm" *Teredo* uses its reduced, roughened valves to bore tunnels into ship bottoms and wharf pilings, reducing them to sawdust. The clams eat the sawdust and digest it. Evidently these forms secrete a cellulase, a unique digestive enzyme among animals (even termites do not digest wood on their own, but depend on flagellate protozoa; cf. Chap. 19). As the tunnel bored by *Teredo* lengthens, the clam extends its long siphon and keeps the tip anchored at the tunnel opening.

CLASS CEPHALOPODA

Squids and allied types; marine; bilateral, elongated in a dorsal direction; head with tentacles; foot formed into a funnel; shell either external and chambered, or internal and reduced, or absent altogether; nervous system exceedingly well developed; with cartilaginous endoskeleton; circulatory system usually closed; eggs telolecithal, cleavage discoidal; development usually direct, without larvae.

Subclass Tetrabranchiata: with two pairs of ctenidia; shell external; ink sac absent; circulatory system open.
- Order Nautiloidea; extinct nautiloids; *Nautilus*, chambered nautilus, living.
- Order Ammonoidea: ammonites, all extinct.

Subclass Dibranchiata: with one pair of ctenidia; shell not external; with ink sac; circulatory system closed.
- Order Decapoda: with 10 tentacles; coelom well developed; shell internal; pelagic. *Sepia*, cuttlefish: *Loligo*, *Ommatostrephes*, *Vampyroteuthis*, *Spirula*, squids; *Architeuthis*, giant squids.
- Order Octopoda: with eight tentacles; coelom reduced; shell highly reduced or absent; semisedentary. *Octopus*, *Argonauta*

Cephalopods are the most advanced and specialized mollusks; they rank among the most highly evolved animals generally. The name of the subclass, "head-footed" animals, refers to the continuity of the head with the surrounding tentacles, which may represent part of the molluscan foot. This head-foot is ventral and the elongated visceral mass is dorsal. However, inasmuch as cephalopods swim in a horizontal position, we shall here refer to the head end as anterior, the other end as posterior, the upper body surface as dorsal (instead of anterior), and the lower surface as ventral (instead of posterior).

Squids illustrate the structure of cephalopods generally. A thick muscular mantle surrounds the visceral mass (Fig. 25.28). Posteriorly the mantle comes to a point, laterally it is extended into a pair of horizontal fins, and anteriorly it terminates at a free edge, the *collar*. Underneath the mantle on the dorsal side lies the remnant of the shell, which in squids is a horny, leafshaped plate called the *pen* (cf. Fig. 25.33). Cartilaginous ridges on the inner surface of the collar fit into corresponding grooves of a midventral muscular tube, the *funnel* (or *siphon*). This structure definitely corresponds to the molluscan foot (whereas the tentacles are regarded by some as part of the head rather than the foot). The channel within the funnel leads into the mantle cavity. When the mantle musculature is relaxed, water passes between the collar and the funnel into and out of the mantle cavity. On contraction of the mantle muscles, the collar clamps tightly around the funnel and water is forced out through the funnel tube. The funnel may be bent backward and in other directions, enabling a squid to change swimming direction. In ordinary cruising, a squid swims mainly by means of its fins and with the head directed forward; in pursuit of prey or in flight, the animal uses its funnel mechanism and jet-propels itself "backward." Squids are among the fastest swimmers, and they readily match the maneuverability of fishes.

A squid possesses 10 tentacles, or *arms*. The first three and

25.28. The general structure of a squid (*Loligo*). Tentacles shown on left side only. Arrows indicate path of water into and out of mantle cavity, resulting in jet propulsion in posterior (dorsal) direction.

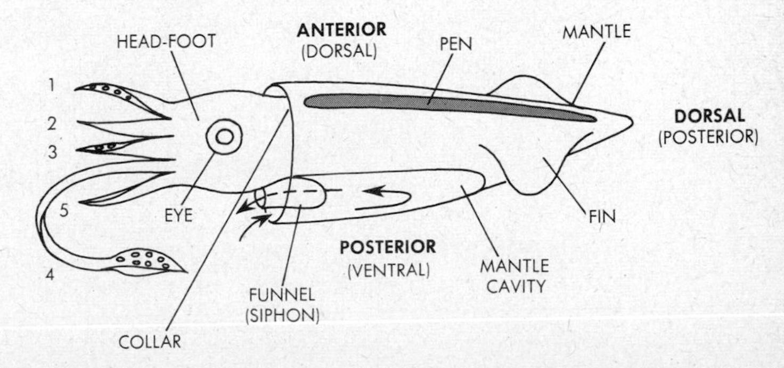

the fifth pairs counted from the dorsal side are short and are studded on the inner surface with stalked, cupshaped suckers (cf. Fig. 25.31). The fourth pair is long (and retractile in some cephalopods), with suckers only at the expanded tip. These long arms catch prey (e.g., various crustacea or fishes) and bring it to the short arms. The latter then hold the prey while the powerful horny beaked *jaws* in the mouth bite chunks out of it. A radula is present but is probably used little, food being swallowed rapidly. Cephalopods possess two pairs of salivary glands, the posterior pair being poison-secreting in many cases (e.g., *Octopus, Sepia*).

The alimentary tract is U-shaped (Fig. 25.29). The stomach connects with a large digestive gland consisting of two distinct parts: a spongy pancreatic part which secretes enzymes and is also excretory to some extent; and a large, elongated, solid liver, which absorbs food molecules from blood and stores them. The stomach communicates through a sphincter-equipped aperture with a long *caecum* extending to almost the posterior tip of the animal. This pouch separates out solid food particles and absorbs only liquid food. The anus opens anteriorly into the mantle cavity, on an *anal papilla*. An *ink sac* discharges into the mantle cavity via rectum and anus. Expelled through the funnel, a cloud of ink probably distracts an enemy while the squid makes its escape. The ink contains finely dispersed melanin granules. We may note, incidentally, that squids and octopuses may change body color when they are excited or disturbed. Pigment-containing cells, *chromatophores,* in the mantle contract and expand and so lighten or darken the animals according to backgrounds or according to their emotional states.

A highly developed nervous system is particularly characteristic of squids and of cephalopods generally. Several fused pairs of ganglia form a complex brain, which is surrounded by a cartilage capsule. Located between the eyes, the brain represents the dorsal part of a nerve ring which circles the esophagus. From the ring elaborate tracts of nerves ramify throughout the body. Dorsally, giant nerve fibers innervate the mantle musculature and control the rapid locomotor activities of the animals. Because of their large size, these giant neurons are employed frequently in research. The eyes of squids are as complex as those of vertebrates; their structure has been discussed in Chap. 8 (cf. Fig. 8.47).

Breathing is accomplished by a pair of feathery ctenidia structured as in mollusks generally. These gills project forward into the mantle cavity (Fig. 25.30). At the base of each is a *gill heart,* which pumps venous blood into the gill. Oxygenated blood from the gills is sucked into the atria and then into a median ventricle. From there blood is carried by arteries into all parts of the body. A median ventral *vena cava* returns blood from the body tissues. This vessel splits into two *branchial*

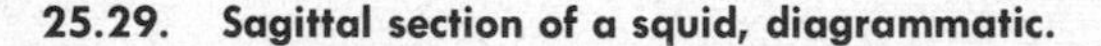

25.29. Sagittal section of a squid, diagrammatic.

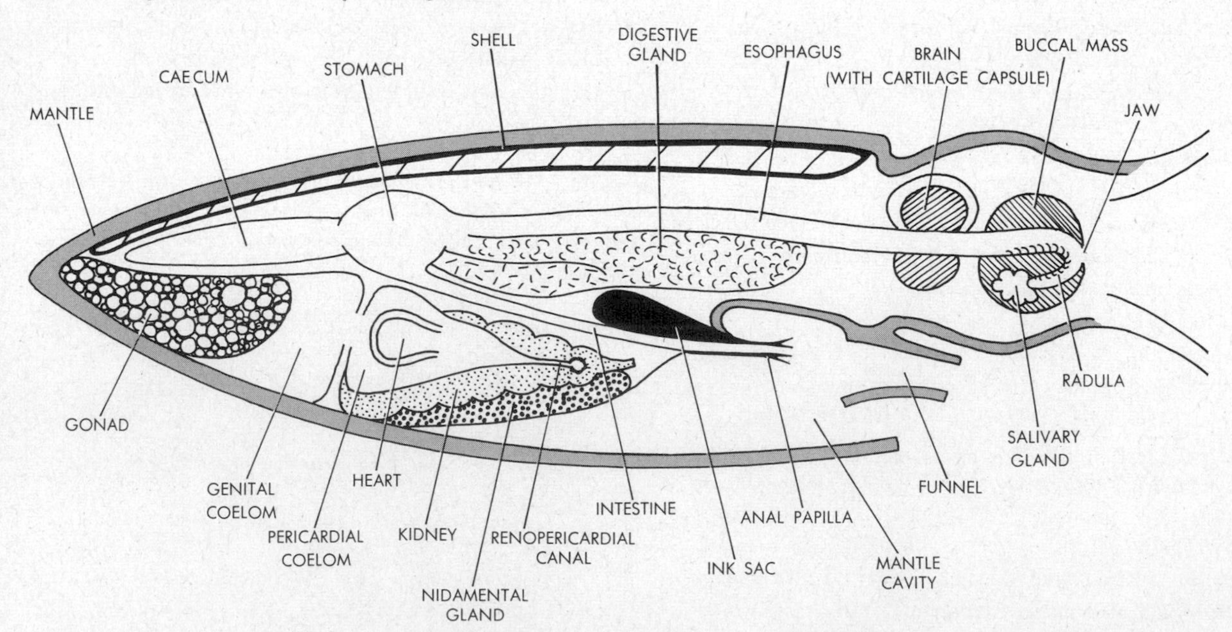

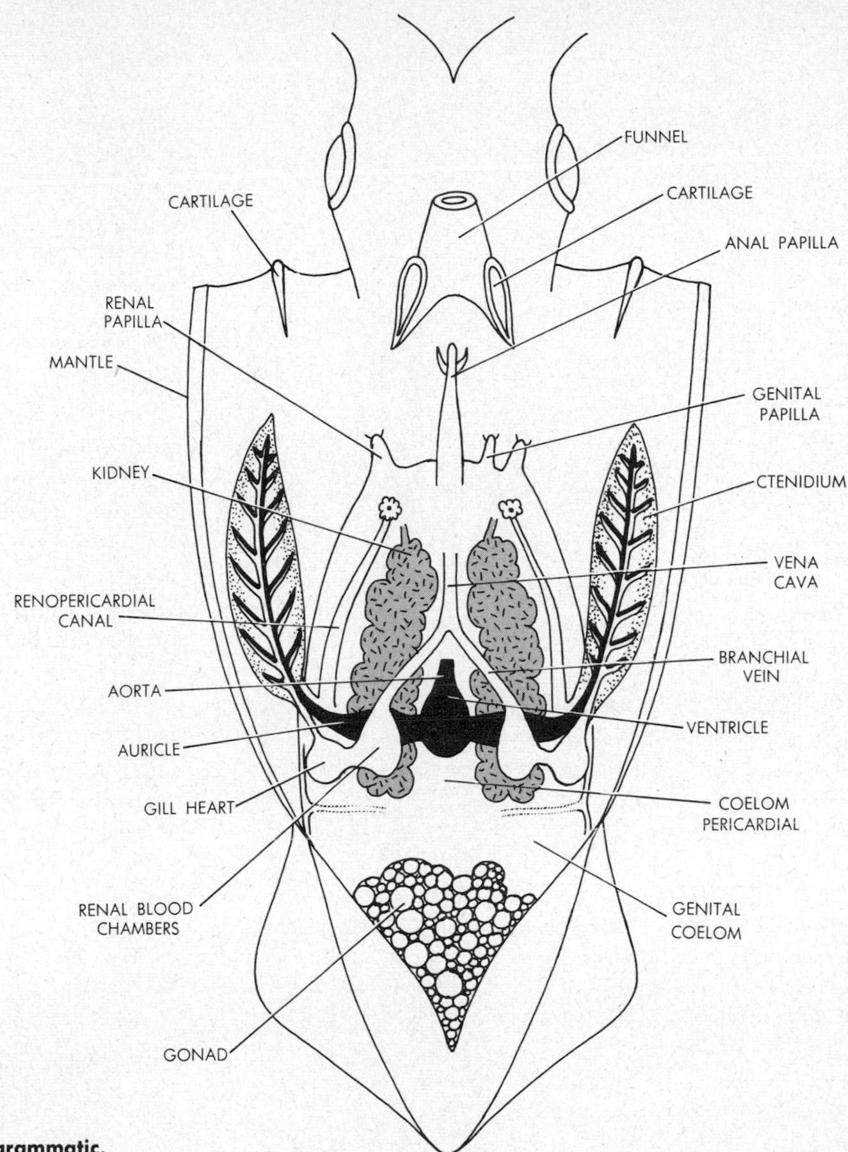

25.30. Ventral dissection of a squid, diagrammatic.

veins near the heart, and each branchial vein passes into a gill heart. Note that the circulation of squids is a *closed* system, the connection between arteries and veins being formed not by hemocoelic blood sinuses but by *capillary* vessels. The blood contains the blue respiratory pigment hemocyanin.

The coelom is spacious and occupies the posterior part of the visceral mass (cf. Fig. 25.30). Most posteriorly, along the wall of the coelom, lies the single gonad. A single gonoduct from this genital part of the coelom leads into the mantle cavity via a gonopore, located on a *genital papilla* to the left of the anal papilla. Anteriorly, the genital coelom is partially separated by incomplete septa from a pericardial coelom. A pair of kidneys containing spongy excretory tissue lies just anterior to the pericardial cavity. Excretory ducts open into the mantle cavity through *renal papillae,* one on each side of the anal papilla. From the pericardial coelom leads a pair of ducts, the *renopericardial canals,* which traverse the substance of the kidneys and open just behind the renal papillae. The coelom thus communicates with the outside via the excretory pores. The kidneys are associated intimately with

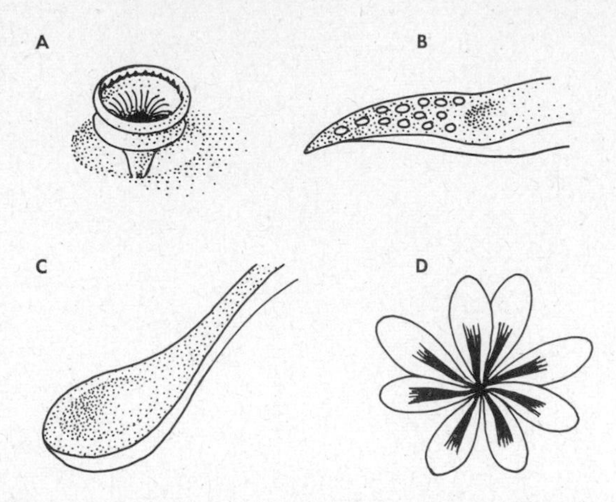

25.31. **A.** Stalked sucker as on squid tentacles. **B.** Diagram of squid hectocotylus, showing suckerless depression for spermatophore transfer. In *Loligo* the third right tentacle and in *Sepia* the fifth left tentacle are of this type **C.** The spoonshaped hectocotylus of *Octopus*. **D.** A cluster of *Octopus* egg cases, with directly developing embryos within.

blood chambers along the branchial veins; blood is filtered before it passes into the gills.

The sexes are separate. In their passage through the gonoduct, sperms are compacted and become enveloped by an elastic tube, the whole forming a *spermatophore*. This packet passes via mantle cavity and funnel to one of the tentacular arms, the *hectocotylus*, which is modified for spermatophore transfer to the female. The hectocotylus is a different arm in different kinds of cephalopods, and its modifications vary in type (Fig. 25.31). For example, the fifth left arm and the third right arm serve as hectocotyli in *Sepia* and *Loligo*, respectively. These arms merely bear suckerless depressions where the spermatophores are held. By contrast, *Octopus* possesses a spoonshaped hectocotylus. Spermatophores are introduced into the mantle cavity of the female, and the male of *Argonauta* actually detaches the whole arm in the process and leaves it in the female.

Eggs pass via genital coelom and gonoduct into the mantle cavity. After they are fertilized, the eggs become encased in a protective capsule, secreted by *nidamental glands* lying ventral to the kidneys. Zygotes either are shed into the open sea (e.g., *Vampyroteuthis*) or, more commonly, become attached in clusters to rocks or other objects in water. Females of *Octopus* brood over the eggs. Development is basically schizocoelic, but the pattern is highly modified owing to the presence of very large amounts of yolk. The telolecithal eggs cleave superficially and a discoblastula results (cf. Figs. 11.12 and 11.15). Larvae are generally not formed, and miniature adults eventually emerge from the egg capsules.

The chambered nautilus is the only living survivor of the originally very large subclass of Tetrabranchiata. Nautiloids

25.32. **Nautiluses.** **Diagram** (*modified from Naef*): sectional longitudinal view through chambered nautilus, showing position of some of the organs. Note double circlet of tentacles, absence of ink sac. Gills shown on one side only. **Photo:** section through shell of *Nautilus*. Note septa, siphuncle.

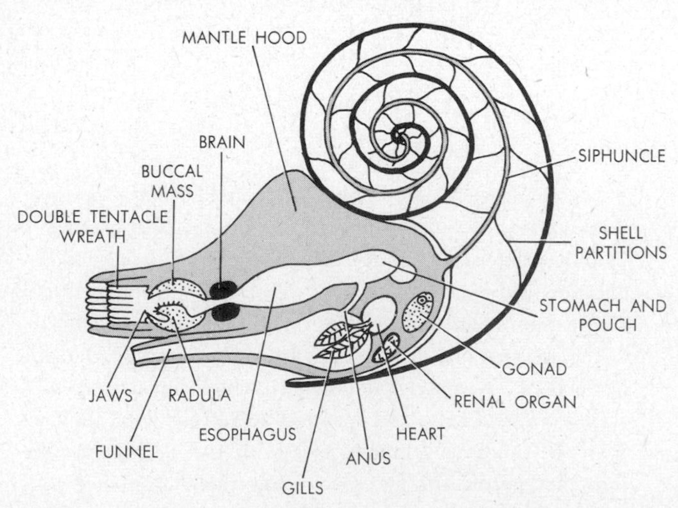

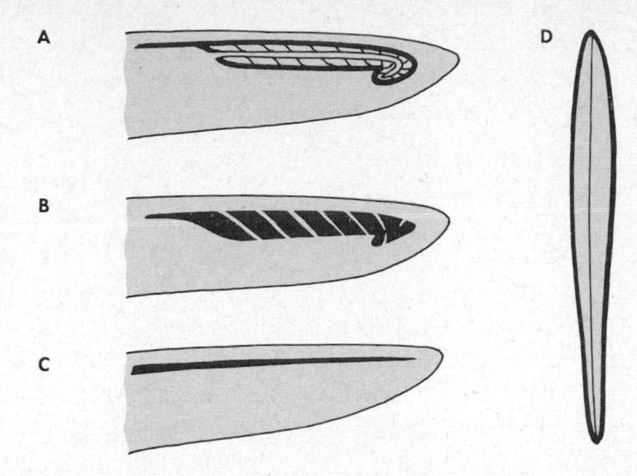

25.33. Shell reduction in cephalopods. *(Partly adapted from Naef.)* **A.** Internalized reduced shell in *Spirula;* note presence of siphuncle. **B.** Calcified shell of *Sepia,* corresponding to upper part in **A** (i.e., part above siphuncle); part below siphuncle is highly reduced, hence siphuncle absent as such. **C** and **D.** Highly reduced horny pen of *Loligo* (**D,** surface view), corresponding to uppermost shell part in **A** (i.e., top part without partitions).

differ from squids in several respects. The tentacles are numerous and are arranged in a double wreath around the head. They are retractile; and although they are without suckers, they are adhesive. The funnel consists of two unfused lobes, as in the embryos of other cephalopods. Four gills project into the mantle cavity (hence the designation "tetrabranchs"). Two pairs of kidneys are present. An ink sac is absent, and the eyes are simple ocelli of the cup type (cf. Fig. 8.47). The circulatory system is open, as in the other molluscan classes. Thus, nautiloids undoubtedly represent the primitive cephalopods. A further indication of their primitive nature is the external, calcareous shell, which is a flat, chambered coil (Fig. 25.32). The posterior part of the mantle secretes the curved partitions in the shell. A strand of mantle tissue, the *siphuncle,* perforates all partitions and passes right back to the innermost, youngest chambers. Blood vessels in the siphuncle are believed to secrete the gas which fills the unoccupied chambers. The gas is airlike, but it contains less oxygen than ordinary air.

A progressive internalization and reduction of the shell is illustrated in a series of genera of the Dibranchiata (Fig. 25.33). In *Spirula,* the shell is still coiled and chambered, but it is internalized and reduced in size. The cuttlefish *Sepia* has a calcareous internal shell which is solid, yet the evolutionary derivation from a chambered condition is indicated by oblique, more or less regularly spaced growth planes within such a cuttlebone. In *Loligo* the shell is a horny internal pen, as noted; in *Octopus,* only vestigial remnants of a shell are present; and in forms like *Ideosepius* all traces of a shell have disappeared. As a group, dibranchs are identified by two ctenidia, closed circulatory systems, and ink sacs. Squidlike eyes are present in the cuttlefish, but giant squids of the genus *Architeuthis* have eyes without corneas or lenses. Giant neurons are also absent. These large squids inhabit the deep ocean, where they are preyed on by large fishes and particularly by toothed killer and sperm whales. All decapod cephalopods are more or less streamlined and adapted to a pelagic life (Fig. 25.34). Most of them are bioluminescent. Squids like *Loligo* typically travel in schools. Some of the decapods are cannibalistic. Octopods on the contrary are of rounded, saclike shapes and are adapted to sedentary life. Most octopods are quite small, some measuring no more than

A

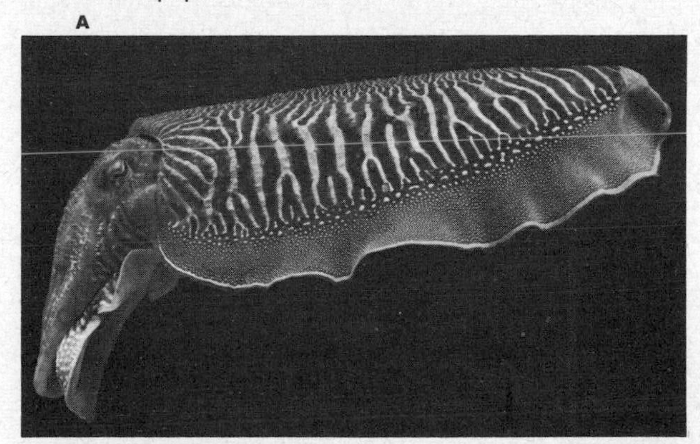

B

25.34. Squids and octopuses. A, *Sepia officinalis,* the cuttlefish; **B,** shell of *Argonauta,* an octopus despite the common name "paper nautilus."

2 in. in length. However, the largest known specimens of *Octopus* have an arm spread of about 30 ft.

Cephalopods as a whole appear to be on the decline; they are not nearly so abundant at present as in Paleozoic times. On the other hand, gastropods and pelecypods are far more abundant than ever before. These two groups make up the bulk of the molluscan phylum today.

REVIEW QUESTIONS

1. Give taxonomic definitions of the phylum of mollusks and of each of the molluscan classes. Describe the structural traits of the hypothetical molluscan ancestor and show in what ways living mollusks still exhibit the ancestral body structure to greater or lesser degree.
2. Describe the structure of chitons. Which features of these animals appear to be specialized traits and which ancestral ones? What is the organization of the excretory and breathing systems? Name the orders of Amphineura, their characteristics, and representative genera of each.
3. Define the subclasses and orders of gastropods. Describe the pattern of (*a*) embryonic, (*b*) larval development of gastropods. What is torsion, when and in what groups does it occur, and what are its structural and functional consequences? Distinguish between torsion and coiling.
4. What is the adaptive advantage of torsion? What is detorsion, when and in what groups does it occur, and what are its consequences? What is the adaptive advantage of detorsion? Which gastropod groups possess opercula? Which are hermaphroditic? How many atria and kidneys are present in a (*a*) nudibranch, (*b*) whelk, (*c*) pulmonate?
5. What are cerata, parapodia, epipodia? Do pulmonates exhibit torsion and detorsion? Describe the structure of scaphopods. Name and define the subclasses of pelecypods and name representative genera of each.
6. Describe the structure and growth of pelecypod valves. What are the functions of clam gills and what is their structure? How do they execute their functions? What are the siphons of clams and how are they formed?
7. Describe the organization of the alimentary system of a clam and the structure and function of the crystalline style. Describe the organization of the nervous, circulatory, and excretory systems of a clam. What are glochidia? How do (*a*) marine, (*b*) freshwater clams develop?
8. Define the subclasses and orders of cephalopods and name representative genera. Describe the structure of a squid. What is the organization of the alimentary, nervous, and sensory systems? Describe the organization of the coelom and of the breathing, circulatory, and excretory systems.
9. What skeletal supports are present in a squid? What is a renopericardial canal? A spermatophore? A hectocotylus? How does mating take place in cephalopods? Describe the pattern of cephalopod development and contrast it with that of other mollusks.
10. How do tetrabranch and dibranch cephalopods differ structurally? Describe the structure of a nautiloid. Show how the skeleton of cephalopods has become reduced during the evolution of these mollusks. What is the adaptive significance of this reduction?
11. What types of mollusks may be found in (*a*) deep oceans, (*b*) the photic pelagic zone of the ocean, (*c*) the intertidal zone, (*d*) the fresh water, (*e*) land environments? Which molluscan groups are (*a*) permanently sessile, (*b*) sedentary and semisedentary, (*c*) predominantly motile? How many species of mollusks are known, and which are the most abundant classes?
12. What is the presumable evolutionary relation of the molluscan classes and orders to one another? How are mollusks as a whole presumably related to (*a*) lophophorates, (*b*) segmented protostomial coelomates? Make a diagram indicating these various interrelations and review the evidence on which you base them.

COLLATERAL READINGS

Various aspects of molluscan biology are discussed in the following books:

Abbott, R. T.: "American Sea Shells," Van Nostrand, New York, 1954.

Morton, J. E.: "Mollusks," Hutchinson University Library, London, 1958.

Mozley, A.: "An Introduction to Molluscan Ecology," H. K. Lewin, London, 1954.

Nichols, J. T., and P. Bartsch: "Fishes and Shells of the Pacific World," Macmillan, New York, 1945.

Tressler, D. K.: "Marine Products of Commerce," 2d ed., Reinhold, New York, 1951.

Webb, W. F.: "A Handbook for Shell Collectors," Rochester, N.Y., 1935.

CHAPTER 

annelids and allied groups

The evolutionary change from unsegmented to wormlike, segmented schizocoelomates (Fig. 26.1) appears to have left living evidence of at least one intermediate transitional stage, in the form of the small phylum *Sipunculida*. These animals are generally annelidlike, but they are unsegmented and they probably never were segmented during their past history. Later, after the fully segmented *Annelida* had become established, an offshoot of this phylum apparently evolved in such a way that the young stages still retained a segmented condition, whereas the adults reverted to an unsegmented state. This offshoot line is represented today by the small phylum *Echiuroida*. By contrast, other offshoots of annelid stocks not only retained the adult segmented state but actually elaborated it further, and from such ancestors the arthropods arose eventually. This transition too appears to have left living evidence, viz., various annelid- and arthropodlike animals now provisionally grouped into the phylum *Oncopoda*.

The characteristics of annelids and these more or less annelid-related forms are described in this chapter.

PHYLUM SIPUNCULIDA

Peanut worms; marine; body with introvert and trunk; alimentary tract recurved and coiled, anus anterodorsal; circulation open and rudimentary or absent; excretion metane-

26.1. The likely evolutionary interrelations among schizocoelomates.

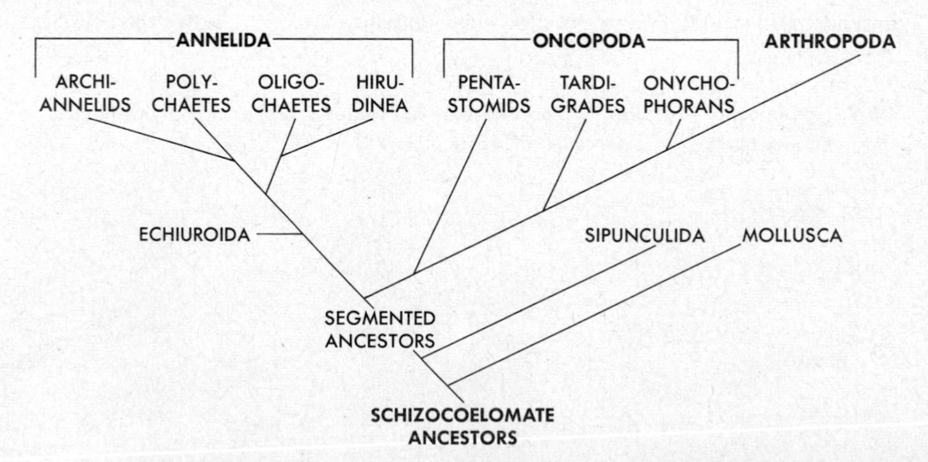

phridial; sexes separate; development mosaic, with spiral cleavage, schizocoelic coelom, trochophore larvae. *Sipunculus, Phascolosoma, Phascolion*

The 250 or so species of these worms lead a sedentary, burrowing life in sandy and muddy tidal flats. Ranging in length from less than ¼ in. to 2 ft, the animals are readily identified by their long, slender *introvert,* a proboscislike anterior tube which may be invaginated and retracted into the plump *trunk* (Fig. 26.2). At the tip of the introvert is an *oral disk* with a mouth and a wreath of tentacular outgrowths. Both introvert and trunk may be covered with papillae and spines, and in some species a shield of calcareous plates protects the anterior region of the trunk, where the introvert joins. The worms may crawl to new locations and they may also use the introvert in a lashing type of swimming. A strong retractor muscle pulls the introvert into the trunk, and the tube may be forced out again by contractions of the body wall of the trunk.

This wall consists of an external nonchitinous cuticle; an epidermis with glandular and sensory cells; a connective tissue dermis traversed by coelomic canals in some species; an outer circular and an inner longitudinal layer of muscles; and a peritoneum (Fig. 26.3). Mesenteries in the large, spacious coelom are poorly developed, and the alimentary tract is held in place largely by muscles. This tract passes from the mouth through the introvert into the intestine, which then recurves in the posterior part of the trunk and coils forward to a rectum and anus. The latter lies dorsally, in the front part of the trunk or the hind part of the introvert. Most sipunculids are detritus feeders. The intestinal mucosa is ciliated and thrown into longitudinal folds, and a conspicuous ciliated groove runs throughout the length of the intestine to the rectum.

The nervous center is a circumesophageal nerve ring in the anterior portion of the introvert, thickened dorsally into a brain ganglion. A ventral cord with lateral branches passes from the ring posteriorly. As will be seen below, the anatomy of this system is essentially annelidlike. A sensory *nuchal organ,* consisting of a pad of ciliated cells and believed to be chemoreceptive, lies in the oral disk. Also in the oral disk are the openings of a pair of *cephalic tubes,* which lead backward to a pair of ocelli of the cup type embedded in the brain ganglion. Most sipunculids are without a circulatory system, though some possess a dorsal and ventral open-ended blood vessel and a connecting blood sinus in the anterior part of the introvert. In addition to mesenchyme cells, the coelomic fluid also contains red blood corpuscles with hemoerythrin pigments. Further present in the fluid are interesting microscopic cellular aggregates called *urns.* Some of these are fixed to the peritoneal lining; others swim free in the coelomic fluid by means of their cilia. The function of these bodies is obscure. Bacteria, cellular debris, and disintegration products of blood cells can be observed to adhere to the urns; possibly their main role is to collect such debris. Urns are not obviously excretory, though they might serve in this capacity indirectly. Sipunculids possess one or a pair of metanephridia. Located anteriorly in the trunk, these organs open to the outside via nephridiopores.

The sexes are separate. Gonads form along the peritoneum and the gametes are shed into the coelom. They mature there and exit subsequently via the metanephridia. Development follows the typical schizocoelomate pattern. Spiral cleavage produces a coeloblastula which gastrulates partly by epiboly, partly by emboly. Endodermal 4d cells proliferate into teloblastic mesoderm bands; these split and form lateral schizocoelic sacs, and a *trochophore* larva eventually results (cf. Figs. 7.16 and 11.14). This larva later elongates greatly in the region between the mouth and the anus, a process which leaves the anus at a progressively more anterodorsal position

26.2. Sipunculid structure. Composite external features are illustrated in the diagram. The photo is of *Sipunculus.* Cf. color plate XII.

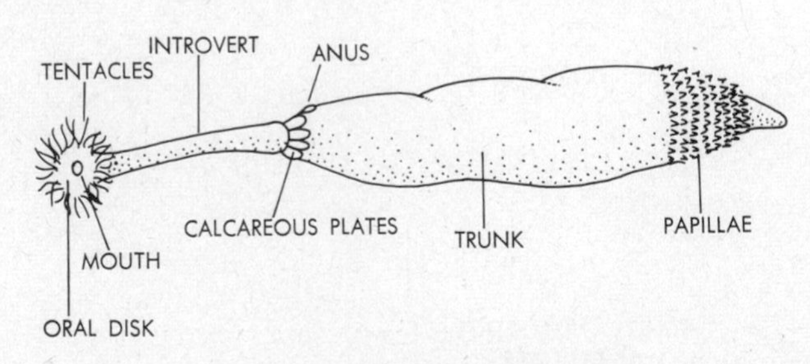

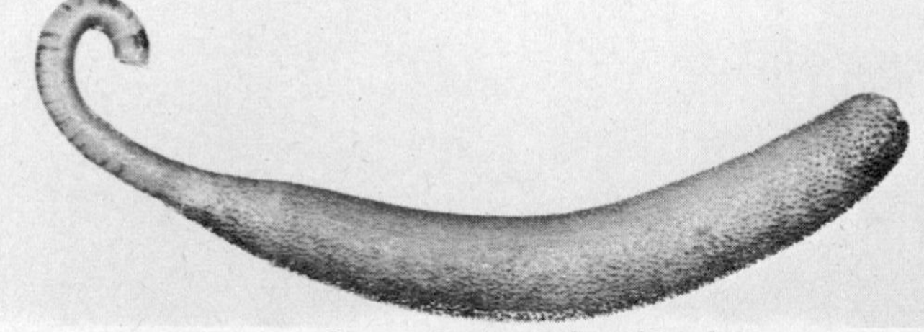

Courtesy Carolina Biological Supply Company.

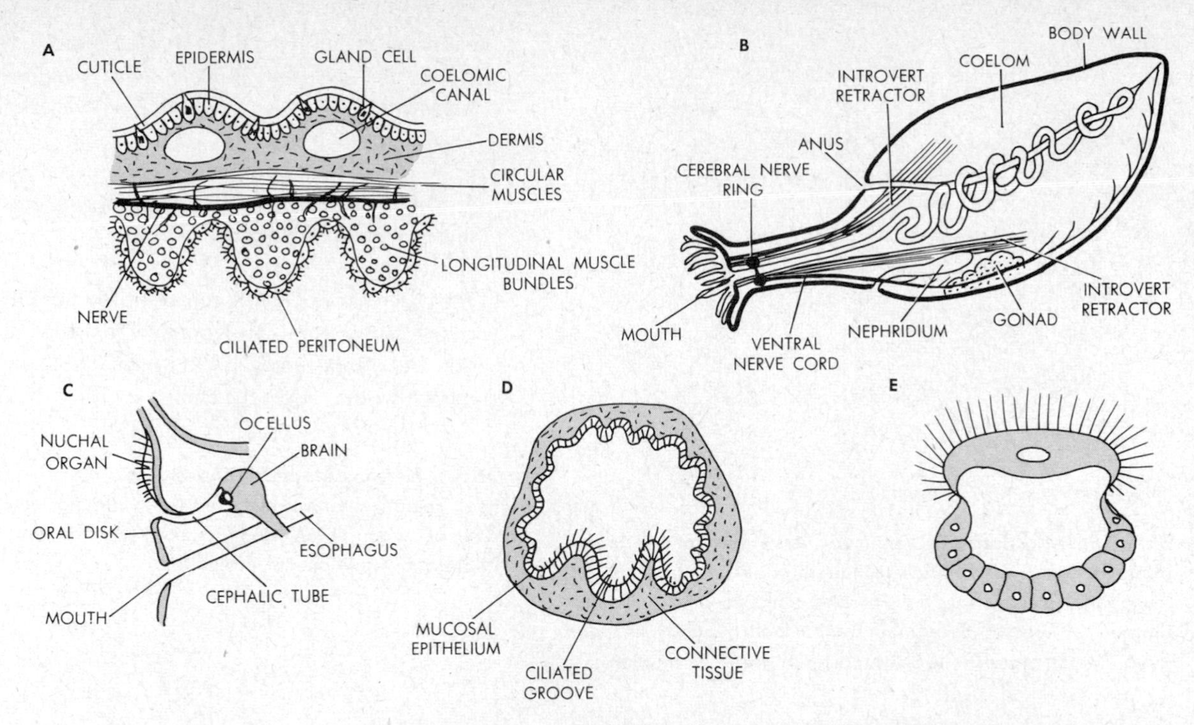

26.3. Sipunculid structure. **A.** Transverse section through body wall. **B.** Sagittal section through body, showing position of various organs. **C.** Sagittal section through dorsal part of oral disk, indicating relation of cephalic tube to ocellus and brain. The nuchal organ is a ciliated pad of tissue. **D.** Schematic cross section through intestine, to show the ventral ciliated mucosal groove. **E.** Section through a free coelomic urn, with a large ciliated cell forming a "lid" and a layer of smaller cells forming a cup-shaped vesicle.

(Fig. 26.4). The adult thus arises by gradual metamorphosis. Adult worms are endowed with an extensive capacity of regeneration.

Although sipunculids play a minor ecological role, they are interesting and important from a theoretical standpoint. The worms are clearly related to annelids, as indicated by their entire embryonic and larval development, their annelidlike nervous systems, and their wormlike and coelomate structure generally. However, sipunculids are completely unsegmented as adults, nor does the embryo or the larva show any signs of segmentation. The worms therefore appear to represent a transitional stage on the path of schizocoelomate evolution which eventually led to segmentation. Thus, primitive schizocoelic ancestors may have given rise to two major descendant lines, one culminating in the mollusks, the other later splitting into two sublines; one of these would have given rise to unsegmented worms such as sipunculids, the other to fully segmented worms such as annelids (cf. Fig. 26.1).

PHYLUM ANNELIDA

Metamerically segmented worms, typically with chitinous setae; coelom septate; organ systems arranged on segmental basis; circulatory system usually closed; development mosaic, with spiral cleavage, teloblastic mesoderm, and schizocoelic coelom; larvae trochophores or absent. *Polychaeta, Archiannelida, Oligochaeta, Hirudinea*

Numbering some 15,000 species, annelids constitute one of the important phyla of animals. In these worms metamerism has become evolved in fully elaborated form, presumably for

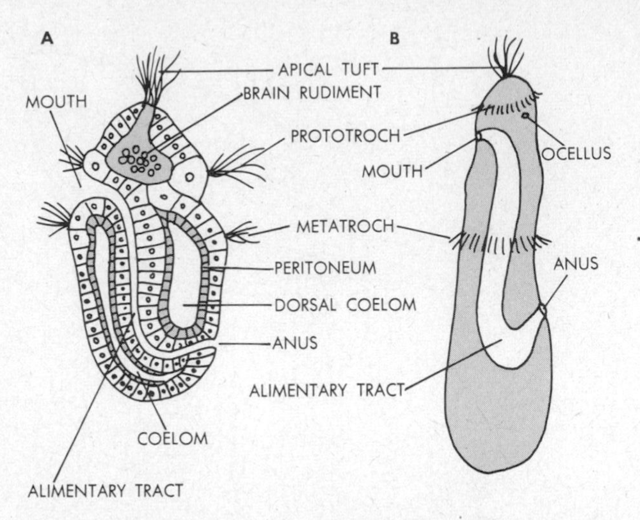

26.4. Post-trochophoral sipunculid development. (*Adapted from Gerould.*) **A.** Sagittal section through post-trochophoral stage. The coelomic sacs have formed schizocoelically, in mesodermal cell masses developed from teloblast bands. **B.** Later larva. Metamorphosis involves posterior trunk elongation.

the first time among protostomial coelomates. Characteristically, the first anterior segment forms the head, and all others except the last, anus-bearing segment are alike developmentally. In given species up to 800 trunk segments may be produced. They often remain more or less alike even in the adult worm, but in numerous instances considerable segmental specialization is in evidence, particularly in external features. As noted in Chap. 7, it is just this possibility of intersegmental modification that represents the principal adaptive advantage afforded by metamerism.

The phylum encompasses four classes. Some two-thirds of the species belong to the class of polychaetes, a group which may serve to illustrate the major characteristics of the phylum as a whole.

Class Polychaeta

Largely marine; with segmental parapodia and numerous setae; sexes separate; oviparous, with free-swimming trochophore larvae.

Order Errantia: pelagic worms; actively motile, often in temporary tubes or burrows; largely herbivorous; pharynx armed, eversible; body modified in breeding season. *Eunice, Leodice,* palolo worms; *Nereis,* clamworms; *Aphrodite,* scale worms; *Syllis, Eulalia,* leaf worms.

Order Sedentaria: *a.* tubiculous, or tube-forming worms, permanently in secreted tubes; filter-feeding; pharynx unarmed, not eversible. *Chaetopterus,* parchment worms; *Amphitrite,* sea mice; *Sabella, Serpula,* feather-duster worms.

b. burrowing worms, permanently in sand or mud borrows; detritus-feeding; pharynx unarmed, eversible. *Arenicola,* lugworms.

Polychaetes undoubtedly represent the primitive annelids. Very few of the worms are symbiotic, most living free in the neritic and littoral zones of the ocean. The basic taxonomic group within the class is the family, and there has as yet been

26.5. *Nereis,* external features. Diagrams: A, top view of head region, pharynx retracted (and thus not visible); **B,** side view of head region, pharynx everted, with left jaw visible. **Photo:** the whole animal, from life.

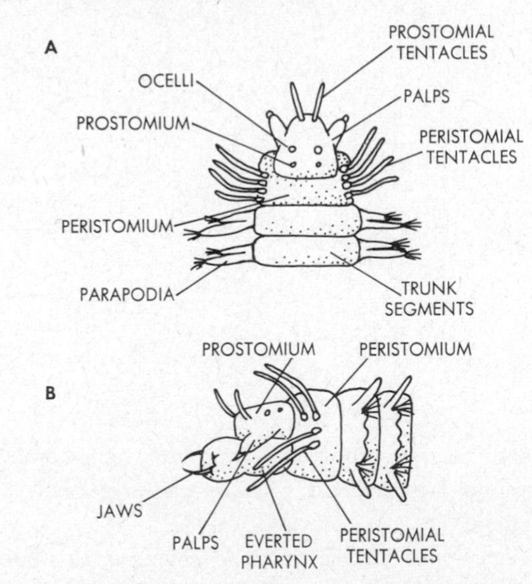

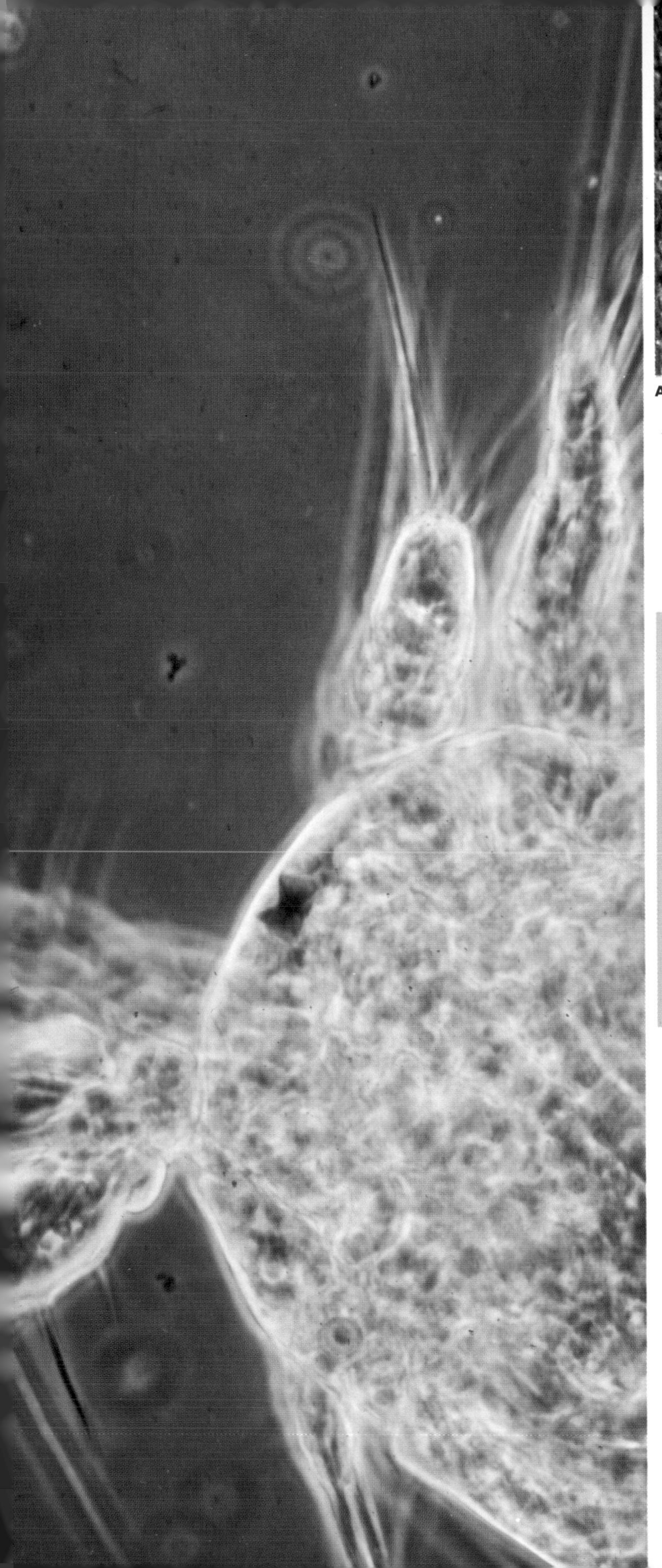

A B

PLATE XIII. Crustacea. **A,** anterior part of nauplius larva of the copepod *Cyclops;* note median ocellus, paired appendages. **B,** the decapod shrimp *Bandana.* **C, D,** crabs (**C,** *Portunus;* **D,** *Phynodius*). **E,** the fiddler crab *Uca,* exhibiting the daytime dark color phase at top, the nighttime light color phase at bottom.

C

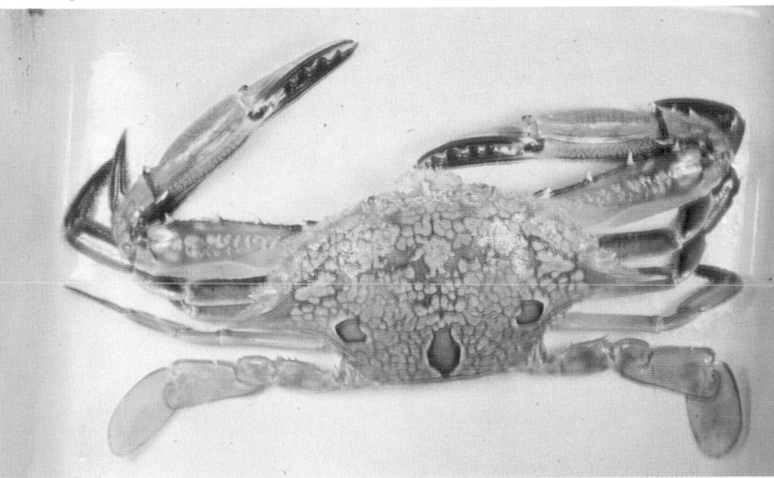

Plate XIII, page 1

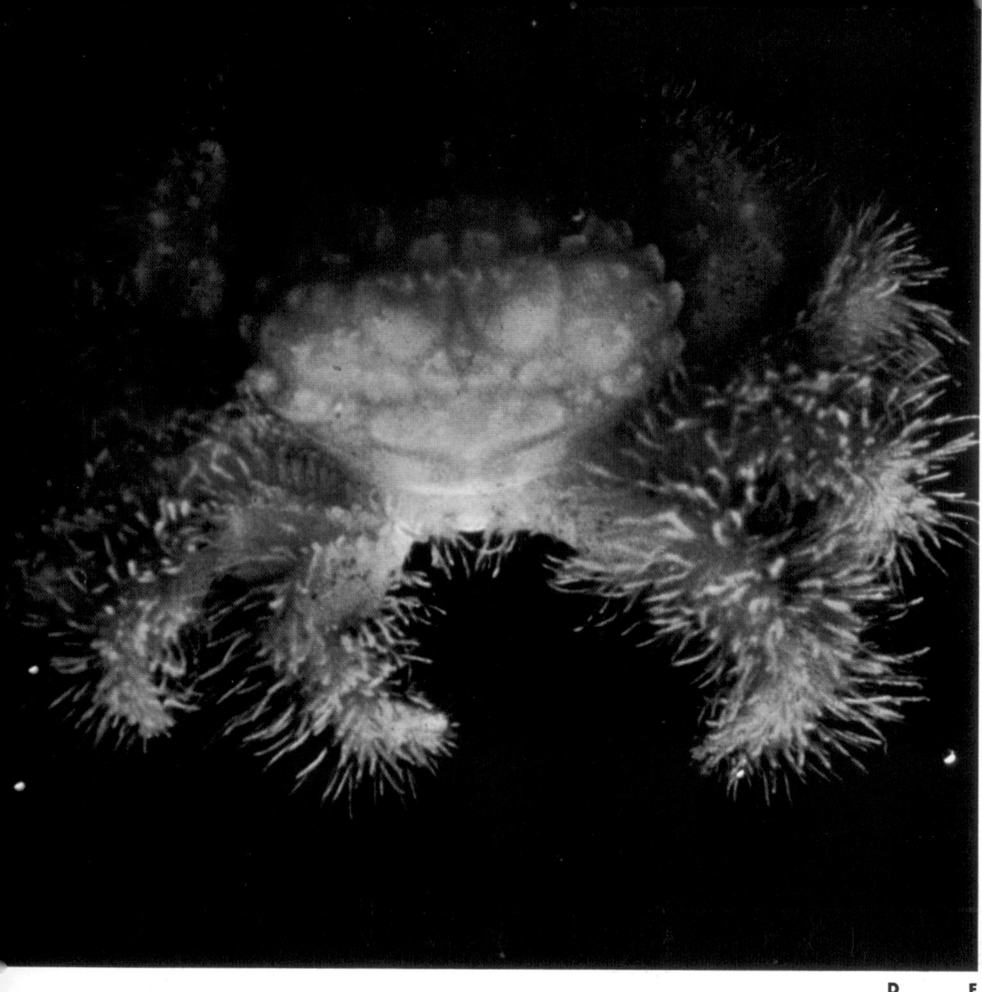

Plate XIII, page 2

D E

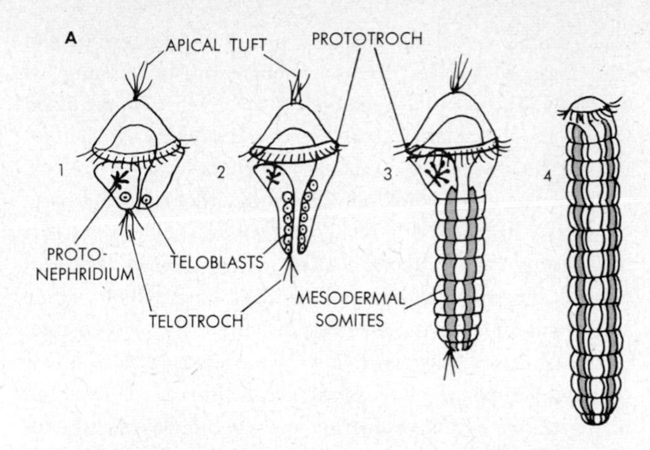

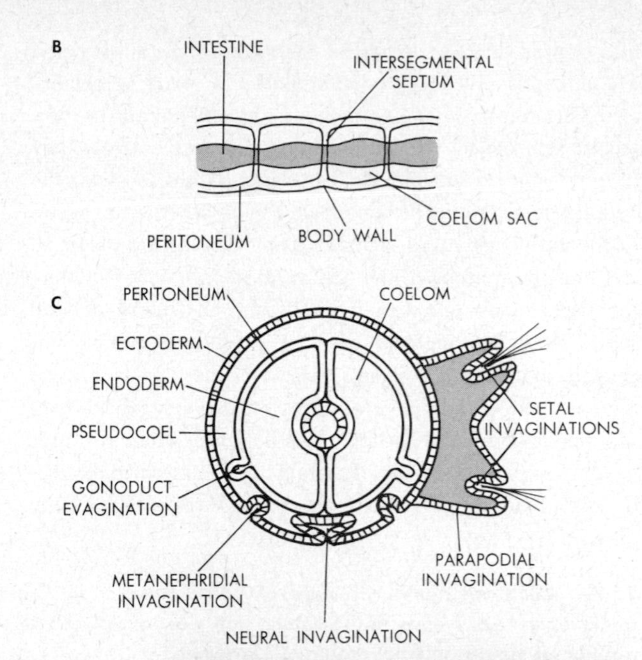

26.6. The segmental development of annelids. Diagrams: A. Somite formation (*modified from Woltereck*). 1, trochophore; 2, posterior elongation, proliferation of teloblastic mesoderm bands; 3, mesodermal somites forming; 4, later stage in segmental development, reduction of larval head structures. **B.** Side view of early segments, showing segmental coelomic sacs and intersegmental peritoneal septa. **C.** Diagrammatic cross section through developing segment, showing formation of ventral, lateral, and ventrolateral organs. **Photo:** posttrochophoral larva of the marine polychaete *Phyllodoce*.

little agreement on how the families should be arranged into orders. Thus, one speaks of "nereids," "syllids," "sabellids," "serpulids," and similar divisions, each referring to a family of polychaetes (the name being based on that of the representative genus). However, the families are customarily grouped into two orders and three general categories listed above, according to characteristic modes of life and certain structural traits.

All polychaetes possess a preoral head segment, or *prostomium,* equipped with sensory structures (Fig. 26.5). In the common clamworm *Nereis,* for example, the prostomium bears one or two pairs of dorsal, lens-containing *eyes,* a pair of anterodorsal sensory *tentacles,* and a pair of anteroventral sensory *palps*. Behind the head is a *peristomium,* in which the mouth is located anteroventrally. Sensory extensions are present on the peristomium as well; that of *Nereis* bears four pairs of tentacular palps. As will become apparent below, however, the numbers and types of prostomial and peristomial appendages vary greatly in different polychaete groups. The mouth leads into a pharynx, which in the pelagic polychaetes is eversible and armed; chitinous jaws as in *Nereis* or teeth as in *Syllis* are used in biting off pieces of algal vegetation. Posteriorly, the trunk is marked conspicuously into segments, and the only internal system not designed metamerically is the alimentary tract. The latter consists, behind the pharynx, of an esophagus with digestive glands and a straight intestine which terminates at the anus in the last segment.

Segmental development begins with a pronounced posterior elongation of the trochophore, very different from events in molluscan trochophores (Fig. 26.6). This posterior elongation is accompanied by a forward proliferation of the mesodermal teloblast bands on each side of the larval body. These bands give rise to anteroposterior pairs of cellular aggregations, the *somites*, which later hollow out and so form paired schizocoelic coelom sacs. As the sacs enlarge, they become applied against the ectoderm on the outside, the endoderm on the inside, and against each other anteroposteriorly and dorsally and ventrally along the sagittal plane. In this manner are formed the peritoneal membranes, including the dorsal and ventral mesenteries and the intersegmental septa. Note that segments so arise in anteroposterior succession, the anterior segments being the oldest. In each larval segment, the ectoderm invaginates on the ventral side and produces a segmental portion of the nervous system as well as a pair of nephridial tubules. Laterally the ectoderm and mesoderm fold out on each side into a *parapodium*, a flap of body wall. Within epidermal pits of a parapodium later develop stiff chitinous bristles, the *setae* characteristic of annelids generally. The peritoneal mesoderm will give rise to the blood vessels, the muscular system, the gonads, and, by evagination toward the ventral ectoderm, also the gonoducts.

Accordingly, an adult polychaete has the following basic segmental structure (Fig. 26.7). The body wall consists of a thin, horny (nonchitinous) cuticle, an epidermal layer, an outer (segmental) circular and an inner longitudinal muscle layer, and the somatic peritoneum. Laterally, a segmental parapodium is either *uniramous*, i.e., a single flap, or *biramous*, composed of a dorsal *notopodium* and a ventral *neuropodium*. Set into epidermal pits are bundles of chitinous setae, two such bundles being present in biramous parapodia. The setae can be retracted by special parapodial muscles.

26.7. Adult segmental structure of polychaetes. **A.** Cross section through trunk segment of *Nereis*. **B.** Schematic side view of polychaete segments showing position of various internal organs. **C.** Schematic cross section of *Nereis* trunk segment, showing pattern of circulation. Blood flows from dorsal vessels to ventral ones in the anterior body segments. **D.** Scheme of anterior portion of nervous system, as in *Nereis*.

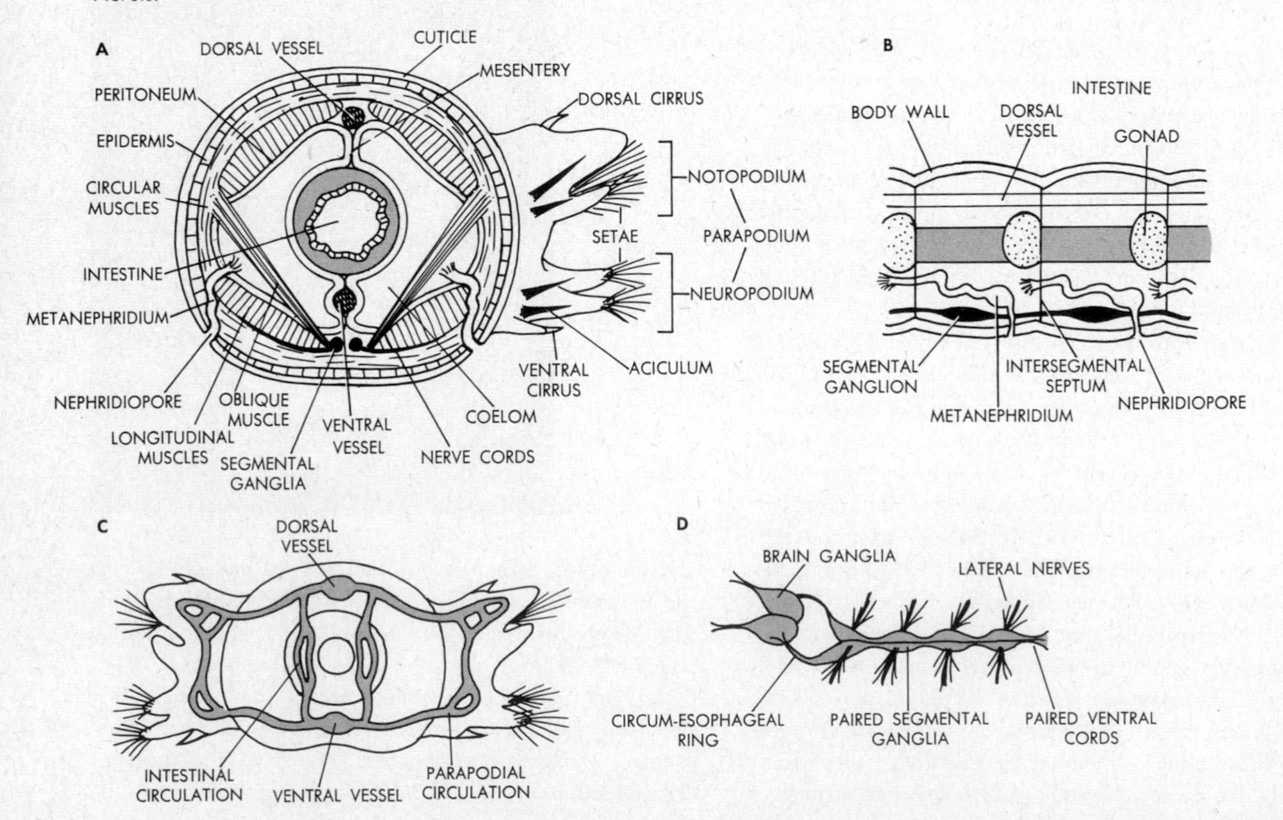

These bristles grow continuously and they can also be replaced by regeneration. They serve partly in protection, but more particularly as locomotor levers and holdfast spikes. The parapodium as a whole is often stiffened internally by embedded, nonprojecting setae called *acicula*. Each parapodium typically also bears a dorsal and a ventral projecting *cirrus*. Parapodia are absent from the prostomium and the peristomium, but the four pairs of peristomical palps of *Nereis* are homologous to the parapodial appendages of the trunk.

Subepidermally on the ventral side, each trunk segment contains a pair of ganglia with transverse connectives and lateral branches, as well as longitudinal cords passing into adjacent segments. The paired ganglia and cords may lie close together or may be more or less fused along the midline, approximating the appearance of an unpaired system. In the peristomium the cords form a ring around the pharynx. Dorsally this ring contains a pair of large brain ganglia, located in the prostomium. The circulatory system consists of a longitudinal dorsal and ventral trunk vessel, supported in each segment by the dorsal and ventral mesenteries, respectively. These vessels are interconnected in each segment by transverse systems which include capillary nets in the alimentary tract, in the parapodia, and in the other segmental organs. Blood flows forward dorsally by peristaltic contractions of the dorsal vessel, backward ventrally. In certain cases, some of the anterior segmental connecting vessels are similarly contractile. The circulatory system is thus typically closed, but in many of the pelagic polychaetes an open system without capillaries has developed secondarily. Some of the small pelagic polychaetes are without circulatory system altogether. The blood may be colorless or colored; if present, respiratory pigments are usually dissolved in plasma. Most polychaetes possess hemoglobin, but green chlorocruorin is common in the tubiculous types and other pigments occur as well (cf. Chap. 9).

Breathing is accomplished principally through the epidermis, particularly in the parapodia, which wave back and forth and thereby circulate water around the body. In numerous worms the notopodia of some or many of the segments are extended into branched *gills*, usually without setae in such cases. A substantial amount of breathing also occurs through the wall of the alimentary tract. Nephridiopores are located ventrally, near the base of each parapodium. A few polychaetes, among them the pelagic *Eulalia* and the burrowing *Glycera*, possess solenocytic protonephridia, like their trochophore larvae. Such nephridia do not open into the coelom. Some polychaetes, e.g., the pelagic *Vanadia*, are characterized by nephromixia, each with solenocytes as well as a separate gonostome into the coelom (cf. Chap. 10). In the majority of polychaetes, however, typical metanephridia are present. The nephrostomes in such cases are supported by the intersegmental septa, the funnels opening into the coeloms of the anteriorly adjacent segments (cf. Fig. 26.7). Gonads are peritoneal sacs, usually attached to the septa. In most polychaetes gonoducts are absent. Gametes exit via the metanephridial channels or, particularly in the pelagic types, they simply accumulate in the coelom until a segment bursts open (cf. below).

Various polychaete groups differ from one another most in their external features. As in *Nereis*, pelagic polychaetes generally are characterized by a well-developed head. The elaboration of head structures with sense organs (*cephalization*) is adaptively advantageous in a pelagic existence and is exemplified in many cases by the presence of a peristomium consisting of more than one segment. In *Nereis*, for example, the four pairs of peristomial palps correspond to the dorsal and ventral pairs of cirri of two segments, indicating that the peristomium of the worm represents two fused segments. Three (unfused) segments form the peristomium in *Eulalia*.

Pelagic polychaetes swim by sinuous body movements, the parapodia aiding materially as paddles. Neural control of muscles is metameric in annelids generally; segmental reflexes leading to contraction of longitudinal muscles in one segment lead simultaneously to relaxation of the circular muscles of that same segment and to contraction of the longitudinal muscles in adjacent segments. The ventral ganglia exercise the chief local control, and the brain ganglia serve importantly as locomotor *inhibitors*. For example, a decapitated worm tends to move more violently and more continuously, hence with less control and less purposefully, than an intact worm.

Many pelagic polychaetes swim freely only when locomotion is actually unavoidable, the worms preferring to spend most of their time in burrows. *Nereis* is among numerous types which dig simple burrows in sand. Others line their burrows with secreted tubes. In such cases, the setal pits in the notopodia produce plastic threads which are then matted together by the setae into a tubular fabric. *Aphrodite* is characterized by *elytra*, paired platelike outgrowths from the dorsal epidermis of each segment (Fig. 26.8). The notopodia secrete a woven mat of fibers over the elytra, and respiratory water is flushed through the space between the mat and the elytra. The tube-formers among the pelagic polychaetes differ from the true tubiculuous polychaetes in that they readily leave their burrows and tubes for feeding and to migrate to

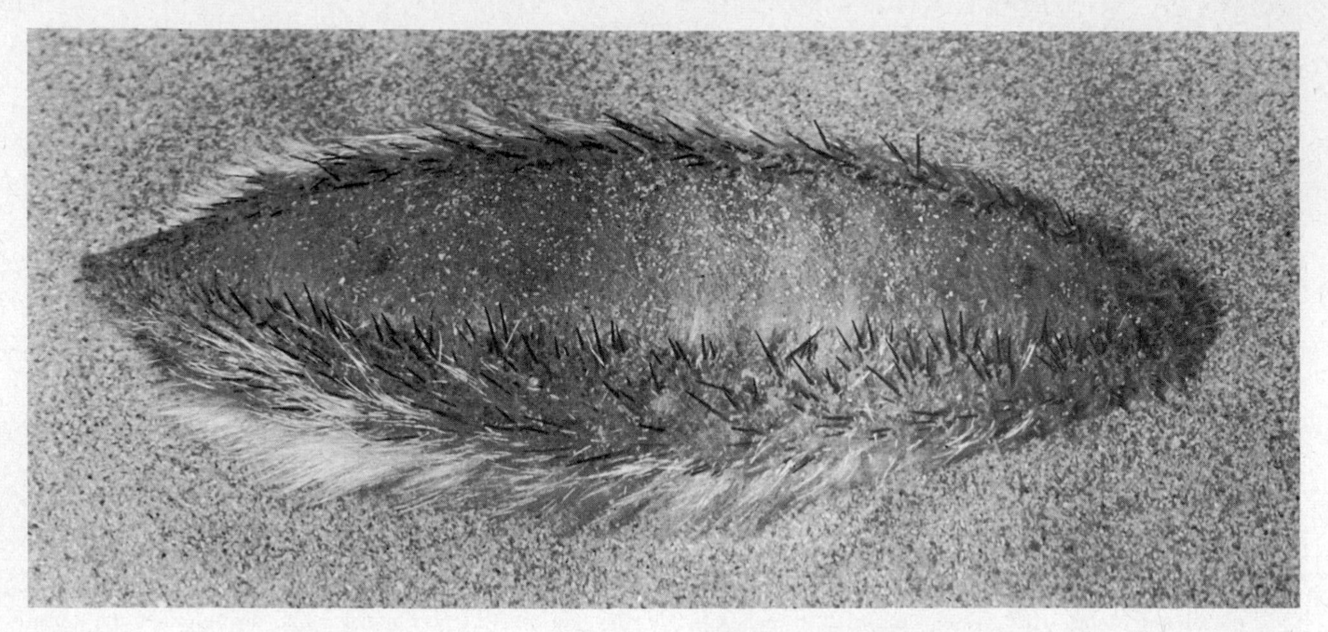

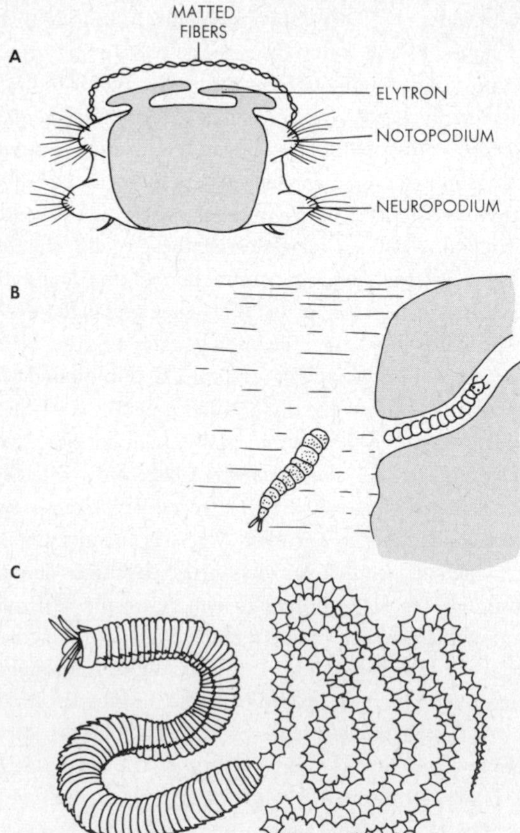

26.8. Errant polychaetes. Photo: the sea mouse *Aphrodite;* note notopodial setae. **Diagrams: A.** Transverse section through *Aphrodite,* showing dorsal elytra and matted notopodial fibers (composed of chitin, like setae). **B.** Swarming palolo worms. The anterior portion of a worm is shown in its burrow. The posterior, gamete-filled portion is detached and will swarm to the water surface, where it will burst open and release the gametes. **C.** The palolo *Eunice.* (*Adapted from Woodworth.*)

other locations (hence the designation of the animals as *Errantia,* i.e., errant, migratory worms).

Errantia are characterized further by interesting breeding habits. Like polychaetes generally, the pelagic types are separately sexed. Breeding occurs seasonally and is accompanied by often pronounced structural changes in all or part of the worms. For example, body colors and textures may become altered, the parapodia may enlarge, and long swimming setae may be developed. In *Nereis,* such dimorphic changes occur in each sex in the posterior half of the body, where the segmental coeloms concurrently fill with gametes (cf. Fig. 26.8). At certain fixed nights, determined in part by environmental factors such as the amount of moonlight in a particular month of the year, the posterior sexual parts of the worms become detached and swarm separately to the surface of the sea. Called *heteronereis,* these worm sections then burst and disintegrate, and the released gametes effect fertilization. The nonswarming body parts subsequently regenerate the lost heteronereis regions. These regenerated sections fill up again with gametes by the following year, becoming ripe in the same calendar period.

In the reef-dwelling palolo worms, swarming takes place precisely at one or two predictable nights of the year, e.g., during the last quarter of the October-November moon in a South Pacific type (*Eunice*) and during the third quarter of the June-July moon in a West Indies type (*Leodice*). These nights of swarming are anticipated eagerly by the local islanders, who gather up the swarming palolos in buckets, broil them, and eat them as rare delicacies. It should be clear that if all worms in a local population swarm at the same, precisely fixed time, the chances of successful encounters between gametes are increased greatly. The two sexes of *Odontosyllis* enchance the chances of meeting still further by exchanging flashes of bioluminescent light just before swarming.

In many of the syllid polychaetes, the sexual body parts are not merely posterior sections but whole, posteriorly budded worms complete with heads. Some genera develop a single posterior worm bud, the head in such cases usually lacking a pharynx and, consequently, feeding capacity.

26.9. Budding in syllids. Sexual parts shaded and stippled. **A.** *Syllis,* budding of single posterior sexual individual. **B.** *Autolytus,* budding of posterior chain of sexual individuals. **C.** *Trypanosyllis,* formation of posterior lateral sexual buds. The alimentary tract is continuous into the posterior bud. Note that the most posterior individuals are the oldest and thus the most developed.

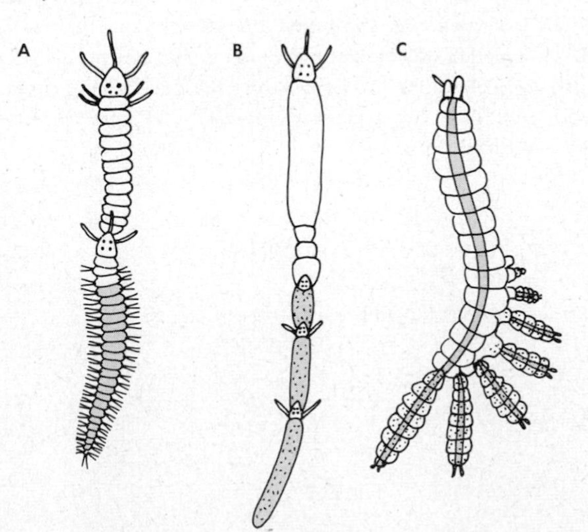

Others, like *Autolytus,* bud off a long string of posterior sexual worms in succession, the most posterior here being the oldest and ripest. Budded sexual worms develop as a succession of lateral buds in *Trypanosyllis,* and numerous other variations on this general theme are known to occur (Fig. 26.9).

The tubiculous polychaetes, belonging to the *Sedentaria,* live in secreted tubes which are left only very rarely, if at all. Composed of a variety of materials in different genera, such tubes either may be straight and positioned vertically in sand or mud, or may be curved into U-shapes as in *Chaetopterus* (cf. Fig. 26.11). Tubiculuous worms are adapted structurally to a ciliary plankton-feeding existence. The head segment is greatly reduced and the pharynx is unarmed and not eversible. However, the appendages of the prostomia and peristomia are highly elaborated and form a wide variety of different plankton-catching divices. Among these are the conspicuous, feathery, retractile tentacles of sabellids and serpulids (hence the designation of such worms as "feather dusters"; Fig. 26.10). Protruded beyond the tube opening, the tentacles trap food organisms, which are then swept over ciliary grooves toward the mouth. Longitudinal ciliary grooves along the body of such worms propel feces from the posterior, closed end of the tube toward the open anterior end. In some serpulids one of the oral tentacles is often modified into a stopper-like *operculum* which, on retraction into the mouth of the tube, closes the tube opening. In many terebellids and other tubiculous worms, some of the anterior parapodia are modified into gills. Parapodial movement and undulatory movements of the whole body maintain a respiratory water flow into and out of the tubes.

A one-way flow of water is maintained through the U-shaped parchmentlike tube of *Chaetopterus,* one of the most bizarrely modified tubiculous polychaetes (Fig. 26.11). The worm lies at the bottom of the U with its ventral side downward. The peristomium is enlarged as a tubular cuff around the prostomium, and the notopodia of the anterior segments form food-collecting flaps. The neuropodia of anterior segments are short and stumpy, and they probably aid in maintaining the position of the worm in the tube. In the midregion of the body, notopodia are fused into large *fans,* flaps which maintain water circulation through the tube. Plankton carried in with the water is trapped along the ciliated surfaces of the collar and the anterior notopodia. Food particles are then propelled over ciliated grooves to an organ in front of the fans, where compact food balls entangled in mucus are formed. Such balls are subsequently conducted forward dorsally,

26.10. Sedentary tubiculous polychaetes. A. Anterior end of sabellid *Eudistylia,* showing parchmentlike tube and crown of "feather duster" tentacles. **B.** Anterior end of *Serpula,* showing feather duster tentacles and the retractible operculum which closes the tube. Cf. color plate XII.

toward the mouth. Tubiculous polychaetes do not swarm in the breeding season. Instead, the gametes are shed via metanephridia (usually) and fertilization occurs in free water.

The burrowing polychaetes, exemplified by the lugworm *Arenicola* (Fig. 26.12), to some extent resemble earthworms in their feeding habits. These marine animals dig tunnels in mud or sand along the shore, swallow some of the mud, and live on the organic detritus contained in it. Residual mud is expelled

26.11. The parchment worm *Chaetopterus*. Photo: section through the tube of the worm. The head of the animal is at left. Note the greatly elaborated parapodia. Between the arms of the U tube is a sipunculid. **Diagrams:** side view **(A)** and dorsal view **(B)** of anterior region of *Chaetopterus.* Food in mucus passes posteriorly along the food grooves to the ball-forming cup, and food balls then are propelled forward and into the mouth. (*Adapted from Borradaile.*)

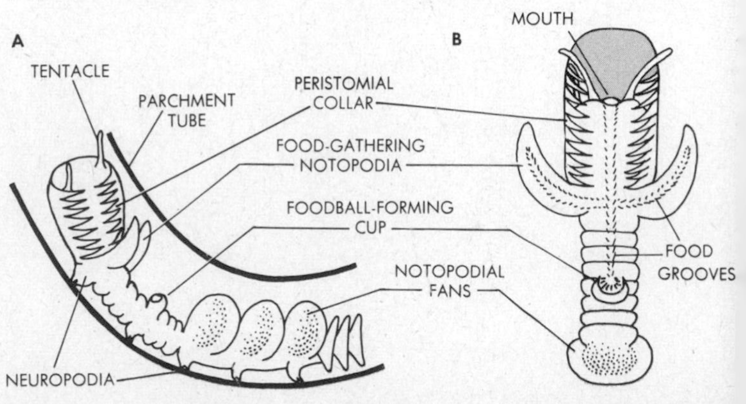

26.12. Burrowing polychaetes. The lugworm *Arenicola* is shown.

through the anus. The coiled earth-castings of lugworms are familiar sights along beaches. *Arenicola* moves peristaltically, like an earthworm, not by sinuous side-to-side undulations like other polychaetes. Also, the head of these burrowing types is highly reduced, appendages or eyes being absent. The unarmed pharynx is eversible, is studded with papillae, and forms an efficient digging organ. The first few segments behind the head are without setae. In midbody setae are present, as are branched segmental gills on the parapodia. The posterior section of the worm lacks gills, parapodia, and setae. Internally, *Arenicola* is virtually without intersegmental septa, and nephridia and gonads are formed only in the midsection of the body. Fertilization occurs as in tubiculous worms.

We note that the basic segmental structure of polychaetes permits different body regions to become specialized in different ways, enabling the animals to be adapted to a wide variety of living conditions. Polychaetes as a whole include comparatively few symbiotic types. Among the pelagic worms only a few genera are commensalistic (e.g., in sponges), and some syllids and serpulids are parasitic in crabs, echinoderms, or other polychaetes. We may note also that the sabellids include freshwater species which live in lakes. Like annelids generally, polychaetes have a highly developed regeneration capacity; a few segments normally suffice to produce a whole new worm. Such regenerative reproduction is undoubtedly correlated with the metameric design, i.e., even a single segment represents virtually a complete "subindividual" containing a more or less complete set of internal organs.

Class Archiannelida

Marine, some parasitic; structurally simplified; without external segmentation; parapodia and setae generally absent; with epidermal ciliation and nerve cords within epidermis; sexes separate. *Polygordius, Dinophilus, Saccocirrus*

This small group of annelids probably evolved from pelagic polychaete ancestors by neoteny (Fig. 26.13). Permanently retained larval trochophoral traits include the absence of external segmentation, the presence of ciliation on the ventral side of the body, the persistence of the nervous system as part of the epidermis, the frequent development of solenocytic protonephridia, and the general absence of parapodia and setae (though setae do occur in *Saccocirrus*). Internal segmentation is well developed, and intersegmental septa are present. In *Polygordius* the longitudinal muscles of the body wall are arranged in four quadrants, and circular muscles are usually lacking. The animal possesses a well-developed prostomium, with a pair of tentacles and a pair of eyes.

26.13. Archiannelids. A. *Dinophilus,* male. Note five pairs of nephridia and five ciliary rings, one each per segment, and note trochophorelike head. **B.** *Polygordius,* anterior region. Again note trochophore remnants. **C.** *Saccocirrus,* side view of anterior region.

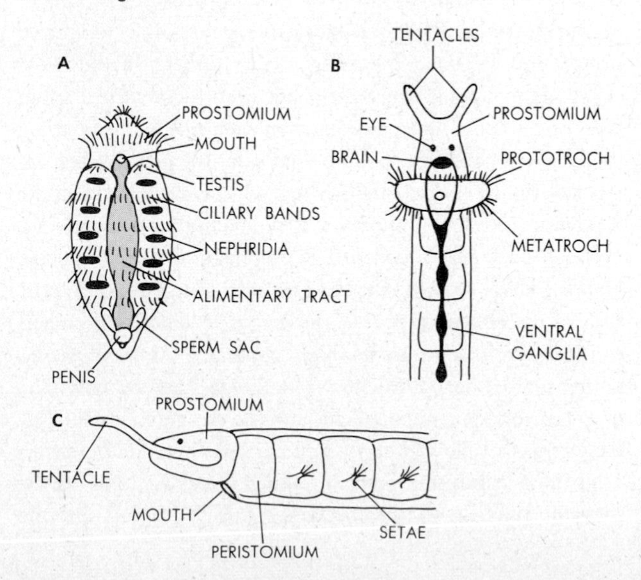

Dinophilus contains only five trunk segments, with solenocytic nephridia in each. The coelom of this worm is reduced, and its eggs are of two types. Small ones develop into males, larger ones into females. Development passes through typical free-living trochophore stages.

Archiannelida were once thought to represent the most primitive annelids (hence the name, "ancient" annelids). It is now recognized, however, that these worms are probably specialized, i.e., secondarily simplified from a polychaete level of organization. As noted, neoteny appears to have been the evolutionary mechanism by which the structural simplification has been achieved.

Class Oligochaeta

Earthworms and allied annelids; mostly terrestrial and freshwater; head reduced, without appendages; parapodia absent, setae few; hermaphroditic, gonads in specific segments only; fertilization internal; development direct, without larvae. *Lumbricus, Eisenia, Megascolides,* terrestrial; *Lumbriculus, Tubifex, Aeolosoma,* aquatic.

Like oligochaetes generally, the familiar earthworms of the species *Lumbricus terrestris* are detritus-feeders. The animals dig extensive tunnel systems in earth by swallowing soil as they burrow along, making use of any food present in it and egesting the remaining soil. Such castings form mounds commonly seen along the ground. These burrowing activities of earthworms probably aid plant growth, inasmuch as soil becomes aerated and plowed over on a fine scale.

Earthworms and oligochaetes as a whole possess a reduced prostomium without appendages (Fig. 26.14). In the trunk segments parapodia are absent, but each segment bears four pairs of retractile setae, two pairs on each side, corresponding to the notopodial and neuropodial setae of polychaetes. Internally the pharynx is neither armed nor eversible, and the remainder of alimentary tract is regionated into esophagus, crop, gizzard, and intestine. The esophagus bears paired *lime glands,* which secrete calcium ions into the gut and thereby reduce the concentration of these ions in blood. The blood calcium levels tend to be high generally, for earthworms absorb a great deal of calcium from the soil they swallow. The crop is a storage compartment, and the strongly muscled gizzard grinds swallowed earth. Soil particles here aid in macerating food organisms present. Along the dorsal side of the intestine runs a fold, the *typhlosole,* which projects into the intestinal cavity and increases the absorptive area. The nervous system is similar to that of polychaetes, except that a pair of subesophageal ganglia is present and that the nerve cords and segmental ganglia are fused into an unpaired ventral tract.

In addition to a longitudinal dorsal and ventral blood vessel, earthworms also possess a longitudinal subneural vessel, ventrally in the midline between the body wall and the nerve tract. The dorsal vessel is equipped with internal valve flaps which prevent blood from flowing backward. This vessel is not contractile. Circulation is maintained instead by five pairs of contractile "hearts," connective vessels joining the dorsal and ventral main channels in segments 7 to 11. The metanephridia are as in polychaetes, and oligochaetes in addition possess excretory *chloragogue* cells in the peritoneum surrounding the gut and the main blood vessels. These cells absorb wastes and then detach and float free in the coelomic fluid. Nephridial elimination may follow, or amoebocytes in the coelom may engulf the chloragogues, migrate into the body wall, and there deposit the excretion products as pigments. In some oligochaetes, the nephridia open not to the outside directly but into the alimentary tract, either anteriorly into the pharynx or posteriorly into a rectum. Moreover, in some oligochaetes more than one pair of nephridia are present in each segment. For example, *Megascolides* possesses up to 2,000 or more "micronephridia" per segment.

The most pronounced differences between polychaetes and oligochaetes are reproductive. Oligochaetes are hermaphrodites (and many develop parthenogenetically). The reproductive systems are developed in specific segments (Fig. 26.15). In segments 10 and 11 of *Lumbricus* are paired testes, each pair enclosed within a testis sac. Pouched extensions from the latter serve as sperm sacs in which gametes released from the testes mature. Sperm ducts from the testis sacs lead to a pair of male gonopores located ventrally in segment 15. In some oligochaetes, but not in *Lumbricus,* the terminal parts of the sperm duct pass through an eversible penis. Present in segments 9 and 10 of *Lumbricus* are paired *spermathecae,* receptacles in which sperms from another worm will be stored after mating. The female system consists of a pair of ovaries in segment 13 and a pair of egg-sac-containing oviducts leading from segment 13 to female gonopores in segment 14. An important reproductive role is played by a conspicuous *clitellum,* a thickened glandular band around the epidermis in which the segmental divisions are obscured. Formed only during the breeding season in many oligochaetes

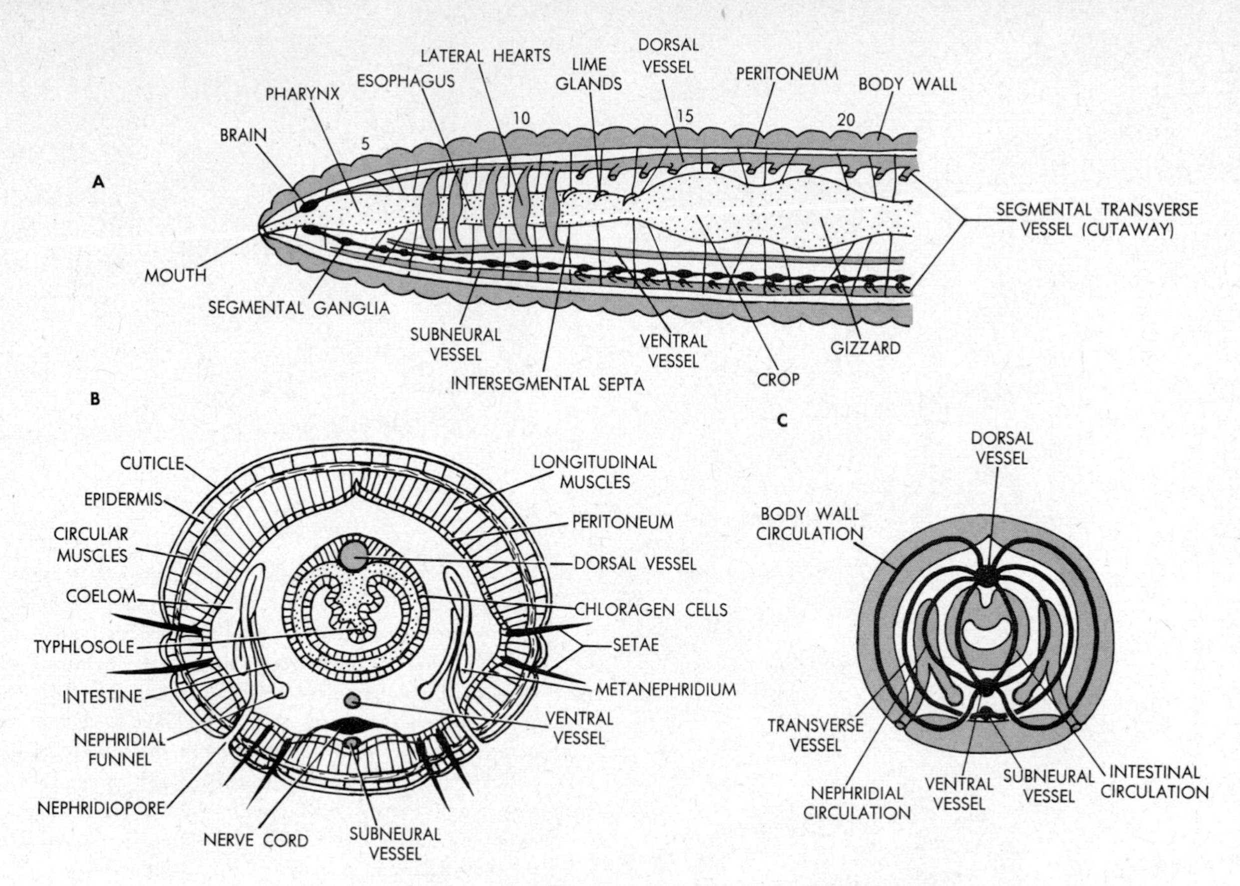

26.14. Earthworm structure. **A.** Cutaway diagram of anterior body region, segments numbered. Segment 12 and all segments behind it contain, on each side, a transverse vessel (shown in A in cutaway view) interconnecting the dorsal and subneural longitudinal blood vessels. **B.** Cross section through intestinal region, diagrammatic. **C.** Scheme of dorsoventral blood circulation pattern in each trunk segment. Blood vessels form extensive capillary beds (not shown) in the body wall, the nephridia, and the gut wall.

but present premanently in earthworms, the clitellum likewise occupies a fixed position; in an earthworm it extends over segments 31 to 37. A pair of ventral mucus-forming grooves connects the male gonopores in segment 15 with the clitellum.

In mating, two worms come into contact with their heads pointing in opposite directions (cf. Fig. 10.11). The region of closest contact ranges from segment 9 to the posterior end of the clitellum, and in this region each worm secretes a mucus sheath around itself. Sperms are then released by both worms, and these gametes are propelled in the mucus grooves from segment 15 of one worm to segments 9 and 10 of the other. There the sperms enter the spermathecae and are stored, whereupon the worms separate. Subsequently, the clitellum of each worm secretes a mucus sheath which is propelled anteriorly by peristaltic movements of the body wall. As this sheath passes segment 14, several eggs are shed into it; and as it continues past segments 9 and 10, sperms from the other worm are deposited into it. Fertilization then takes place within the mucus sheath, and after the sheath slips over the head of the worm the open ends seal up and an egg *cocoon* is so formed. Development follows the typical spiral annelid

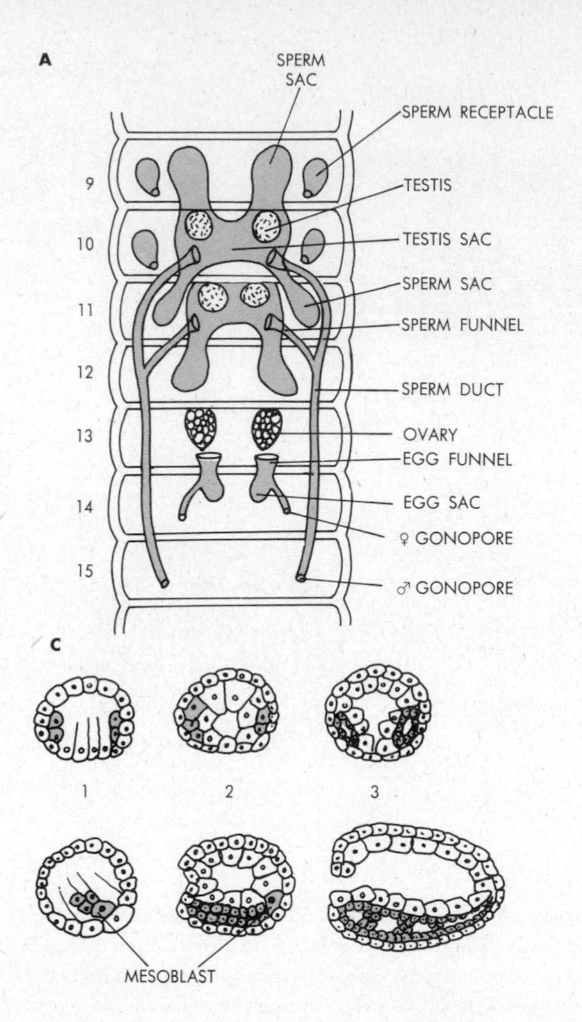

26.15. Earthworm reproduction and development. A. The reproductive system of *Lumbricus terrestris,* ventral view, diagrammatic; segments numbered. (*Adapted from Hesse.*) **B.** External view of whole worm, showing position and extent of clitellum. **C.** Stages in early development. Top row, transverse cross sections; bottom row, longitudinal sagittal sections, anterior ends toward left; mesoderm shaded. 1. Postgastrula, teloblastic bands forming. 2. Later stage, mouth invagination beginning (at former blastopore); teloblast bands (double row of mesoblast derivatives) well developed and proliferated in anterior direction. 3. Early embryo, mouth present, schizocoelic coelom sacs formed ventrally in anteroposterior succession. These sacs will later expand dorsally on each side of the alimentary tract.

pattern, but the trochophore stage is greatly abbreviated and a free larval phase is absent.

Lumbricus is a large oligochaete, though the Australian *Megascolides* attains lengths of 10 ft and diameters of an inch or more. By contrast, the aquatic *Aeolosoma* measures less than 1 mm. Aquatic oligochaetes are quite small generally and they also appear to be the more primitive types. Thus, *Aeolosoma* still retains a ciliated ventral side and related forms possess prostomial eyes. *Tubifex* lives at the mud bottoms of lakes, ponds, and stagnant waters, with its head buried in mud and its posterior end waving incessantly in the open water. Such movements aid in breathing. Many of the freshwater oligochaetes reproduce by spontaneous fragmentation. Moreover, forms like *Chaetogaster* and *Aeolosoma* produce anteroposterior strings of individuals. When mature, each such individual constricts off the string and becomes an independent worm.

Class Hirudinea

Leeches; mostly freshwater parasites, some carnivorous; with terminal suckers; parapodia and setae absent; number of segments fixed throughout life; coelom packed with mesenchyme; hermaphroditic, with clitellum; copulation and development as in oligochaetes. *Acanthobdella, Placobdella, Macrobdella, Hirudo*

There is little doubt that leeches represent evolutionary offshoots of oligochaetes which have become ectoparasitic. Thus, transitional types such as the salmon leech *Acantho-*

bdella still possess two pairs of setae per segment. *Acanthobdella* also contains fewer segments than is usual for leeches generally, and a sucker is present only at the posterior end. In other Hirudinea setae are entirely absent, the number of internal segments is fixed at 34, and suckers are typically present at both ends of the body (Figs. 26.16 and 26.17). The head region, usually comprising the first six segments, bears a pair of eyes and the anterior sucker. The most posterior seven segments form the other sucker. A "segment" in a leech refers to an *internal* metamere only; each such segment is annulated externally into two to 16 superficial cuticular rings, the exact number being fixed for each species. Thus there are always more external "segments" than true internal ones.

Leeches are classified into three orders. One is exemplified by *Acanthobdella,* animals in which the pharynx is unarmed and not eversible. In a second order, the *Rhynchobdellida,* the pharynx is an eversible unarmed proboscis which may be protruded through the mouth and may be forced into soft tissues of host animals, e.g., the gills of fishes. *Placobdella* is a member of this group. In the third, most specialized order, the *Gnathobdellida,* the pharynx is not eversible but is armed with three sawlike chitinous jaws. With these a leech may cut a Y-shaped incision through even tough skin or armor such as fish scales.

Most of these leeches are blood suckers. Adaptations to such a mode of life include salivary glands which secrete the anticoagulant *hirudin* into a wound cut by the pharyngeal saws. Also present are a crop and numerous intestinal diverticula, which may hold up to 10 times as much blood as the weight of the leech itself and which enable the animal to survive as long as 9 months from a single feeding. The "medicinal" leech, *Hirudo medicinalis,* was widely used in earlier times for "bloodletting," then thought to be beneficial generally and in a large number of human diseases specifically. Today leeches still occasionally obtain human blood in freshwater environments, but nonhuman vertebrates are the more

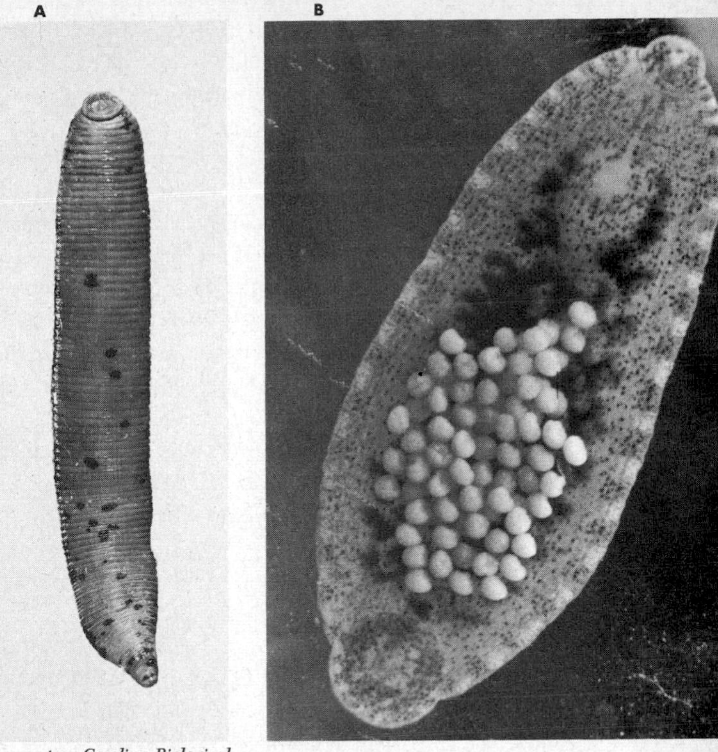

A, *courtesy Carolina Biological Supply Company.*

26.17. Leeches. **A,** specimen showing exterior annulation; **B,** ventral view of a leech of turtles, with developing eggs attached to underside of body. Cf. color plate XII.

26.16. Diagram of a leech (*Hirudo*) showing some of the internal organs. The transverse exterior creases, or annuli, are not true metameric segments but superficial folds. Two or more such annuli correspond to an interior metameric segment. The numbers below refer to such segments. The ingestive saw apparatus of bloodsucking leeches is illustrated in Fig. 9.19.

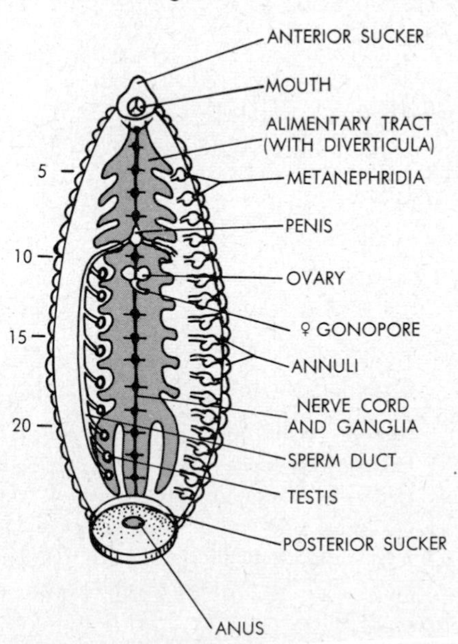

common hosts. Moreover, many of the leeches are not obligatory blood-feeders; they normally survive as carnivores, catching worms and other invertebrates (e.g., *Macrobdella*). Some leeches are actually nonparasitic altogether.

Leeches reproduce like oligochaetes. Several pairs of testes and one pair of ovaries are present along the sides of the gut. A single male gonopore opens midventrally in the anterior region, just in front of a single female gonopore. Some leeches possess an eversible penis. Others form *spermatophores*, which are attached by one worm to the body of a mating partner. The sperms then migrate through the body tissues to the ovaries. Most leeches copulate like oligochaetes, however, and a clitellum performs a similar cocoon-forming function.

Note that, structurally and reproductively, leeches superficially resemble flukes to a certain extent, an instance of parallel evolution under similar conditions of life.

PHYLUM ECHIUROIDA

Spoon worms; marine; somites in larva but unsegmented as adults; with proboscislike prostomium; one pair of anteroventral setae; coelom nonseptate; one to three pairs of metanephridia; sexes separate, dimorphic in some cases; development schizocoelic, with trochophore larva. *Echiurus, Urechis, Bonellia*

This small phylum of about 60 species comprises marine plankton-feeders which live in sand or mud burrows in coastal regions (Fig. 26.18). Attaining lengths of some 3 in., the plump trunk of such a worm bears an extended, tubular prostomium anteriorly. The tip of this prostomium is roughly spoonshaped in forms like *Echiurus* and *Urechis* but is bifurcated into two tissue flaps in *Bonellia*. A highly mobile organ, the prostomium may be contracted, but it does not introvert or retract into the trunk. At the base of the prostomium lies the mouth.

The body wall is thick, fleshy, and studded with glandular papillae. A single pair of setae is set into the ventral surface, and *Echiurus* also possesses a circlet of setae at the posterior end of the trunk, around the anus. The alimentary tract is tubular and highly coiled. It is supported largely by muscles which traverse the coelom. A pair of mucus-secreting *anal vesicles* connects with the rectum. The nervous system is simplified. It consists of ventral and lateral cords and a nerve ring around the rim of the prostomium; ganglia are absent. Dorsal and ventral blood vessels interconnected anteriorly constitute the principal parts of the circulatory system. The coelom is large but without mesenteries or septa. Metanephridia open anteroventrally on the trunk, usually behind the setae. The excretory tubules also serve as gonoducts. A single posterior gonad discharges gametes into the coelom. Sexes are separate and interestingly dimorphic in *Bonellia*, the dwarfed males being parasitic in the females (cf. Fig. 12.27). Such dimorphism is not in evidence in other genera.

The early echiuroid larva is a typical trochophore. In later stages it also exhibits rudimentary segmentation: 15 mesodermal somites arise, and the developing ventral nerve cord acquires a corresponding number of segmental swellings and

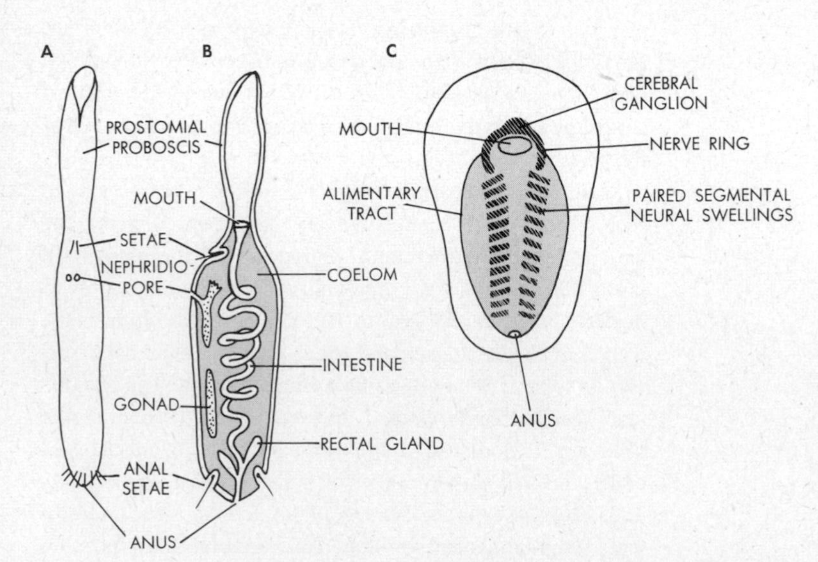

26.18. Echiuroids. A. External features of *Urechis*, ventral view. **B.** Sagittal section of *Urechis*, schematic. (*Adapted from Fisher.*) **C.** Ventral view of *Echiurus* larva, showing abortive neural segments, diagrammatic. (*Adapted from Baltzer.*) See Fig. 12.27 for illustration of *Bonellia;* also color plate XII.

paired lateral branches. But these beginnings of metamerism never become further elaborated. Indeed they become obscured later, and they eventually disappear altogether as the adult stage is reached.

In the past, echiuroids have been loosely grouped with the sipunculids, which they resemble superficially (and even with the priapulids, which they do not resemble). However, in view of the larval metamerism and the sparse setae of the adults, it is now clearly recognized that echiuroids form a separate and distinct group. Whereas sipunculids probably exemplify an unsegmented *pre*annelid stage of schizocoelomate evolution, echiuroids appear to represent a *post*-annelid stage in which segmentation and other annelid traits have been reduced or lost (cf. Fig. 26.1).

PHYLUM ONCOPODA

Claw-bearing animals, fundamentally segmented; usually with unjointed legs; sexes separate, development various. *Onychophora, Tardigrada, Pentastomida*

As presently constituted, the Oncopoda represent a "phylum of convenience." The assemblage includes three small groups which may perhaps be interrelated very distantly and which have two features in common: they possess claws of a type characteristic of arthropods, and their other traits are variously intermediate between those of polychaetes and those of arthropods. However, at least for the Pentastomida, it is not clear whether these intermediate traits are primitive and original or the results of secondary adaptations. Each of the three groups has been and still is classified in widely different ways, either as a separate phylum or as a subphylum or class with arthropod affinities. The present grouping of all three into one phylum follows a recent proposal and appears to be generally justified on at least structural grounds. We may therefore tentatively consider the Oncopoda to represent surviving remnants of three separate and independant lines, which appear to have been offshoots of a main evolutionary line leading from annelid ancestors to arthropod descendants (cf. Fig. 26.1).

Subphylum Onychophora

Terrestrial, wormlike; with many pairs of claw-bearing legs and corresponding internal segments; external segmentation absent; annelid and arthropod traits mixed; sexes separate; mostly ovoviviparous, development direct. *Peripatus, Eoperipatus, Paraperipatus, Peripatopsis*

Of all the Oncopoda, the Onychophora are most clearly intermediate between annelids and arthropods, and their very existence provides one of the best proofs that annelids and arthropods are closely related. *Peripatus* and its allies comprise some 70 species, distributed discontinuously but widely in tropical and subtropical regions of the world. The animals are about 2 to 3 in. long, they live in damp, leafy places on the ground, and they feed mainly on insects (Fig. 26.19). A rather caterpillarlike body is covered with a thin, nonchitinous, velvet-textured cuticle, and under the epidermis are two muscle layers as in annelids. The head consists of three segments (in contrast to the single head segment in annelids and the six in most arthropods). The first bears a pair of annelidlike eyes and a pair of nonretractile tentacles, or *preantennae.* A pair of *oral papillae* on the second segment contains the openings of elongated *slime glands.* The secretion of these glands may be shot out explosively and is used to entangle prey or enemy. The third segment contains the mouth, equipped with a pair of horny biting *jaws.* Along the trunk, the stumpy paired legs each carry two recurved claws. Legs and claws are undoubtedly homologous to the parapodia and setae of polychaetes. The legs mark the position of internal segments and of segmentally arranged organs. The nervous system is annelidlike, though distinct ganglia are absent. Also annelidlike are the multiple pairs of metanephridia which open ventrally at the bases of the legs. However, the nephridial tubules only drain small coelomic sacs—the coeloms are greatly reduced in size and segmental septa are absent, as in arthropods. The hemocoelic blood spaces are correspondingly large and, again as in arthropods, the blood circulation is open; the only vessel present is a single dorsal *heart* with lateral segmental openings, or *ostia.* Another very arthropodlike feature is the breathing system, which consists of *tracheal tubes.* But the onychophoran system is unique in that the tracheal openings are not arranged segmentally. Instead, they lead into the body from randomly scattered surface pores. Also, these pores are without closing mechanisms and the tracheae are unbranched interiorly.

The sexes are separate. Gonads are paired, with a median gonoduct opening just in front of the anus. The gonoducts are ciliated internally, as in annelids and in contrast to arthropods, in which cilia are absent altogether. The males of some species produce spermatophores and in all species fertilization occurs internally. In some cases development is external, but

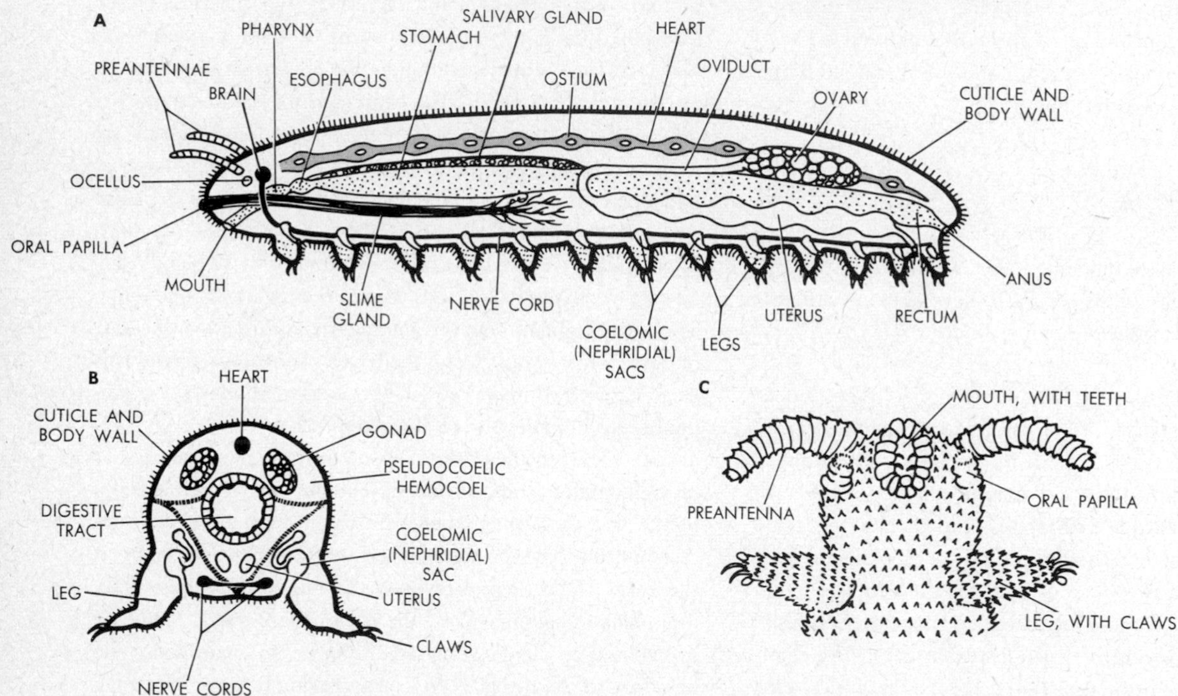

26.19. Peripatus. Photo: external dorsal view, head at left. **Diagrams:** sagittal section **(A)** and cross section **(B)** of female, showing the position of the principal organs. **C.** Ventral view of head region.

most onychophorans develop internally and ovoviviparously, eggs being retained in uterine enlargements of the oviducts. A true viviparous condition is actually in evidence in certain instances; i.e., the maternal body contributes to the nutrition of the embryo. As in arthropods, the eggs are yolky and centrolecithal and they cleave superficially (cf. Fig. 11.12). After gastrulation by delamination the blastopore elongates and closes midventrally, and the persisting anterior opening becomes the mouth, the posterior one the anus (Fig. 26.20). Teloblastic mesoderm bands then form and proliferate into an anteroposterior series of paired somites. In the latter the coeloms arise in schizocoelic fashion. The dorsal coelomic portions subsequently separate from the ventral ones, the former giving rise to the heart anteriorly and the reproductive system posteriorly, the latter to the nephridia. Development is thus rather arthropodlike, with residual polychaetelike processes producing the annelidlike traits of the adult. The whole animal actually represents as perfect a "missing link" between two other animal types as is known in zoology.

Subphylum Tardigrada

Water bears; microscopic, mostly terrestrial; with six segments and four pairs of claw-bearing legs; cell numbers constant; without circulatory, excretory, or breathing systems; sexes separate; enterocoelic coelom; development oviparous and largely direct, without larvae. *Milnesium, Macrobiotus*

This interesting group of some 350 species has a worldwide distribution. The animals are barrelshaped and they do not exceed 1 mm in length. Most of them are terrestrial and herbivorous; they live among lichens, mosses, or liverworts and eat the cellular interiors of such plants. Some species are marine and some occur in freshwater. Such aquatic types are

26.20. *Peripatus* development. (*Adapted from Sedgwick.*) **A.** Early gastrula seen from vegetal pole, showing elongating blastopore closing along midregion and thus defining mouth and anus. **B.** Side view of early embryo with anteroposterior series of coelomic sacs. **C.** Cross section of stage as in **B. D.** Cross section of late embryo, many organ systems already delineated (cf. Fig. 26.19).

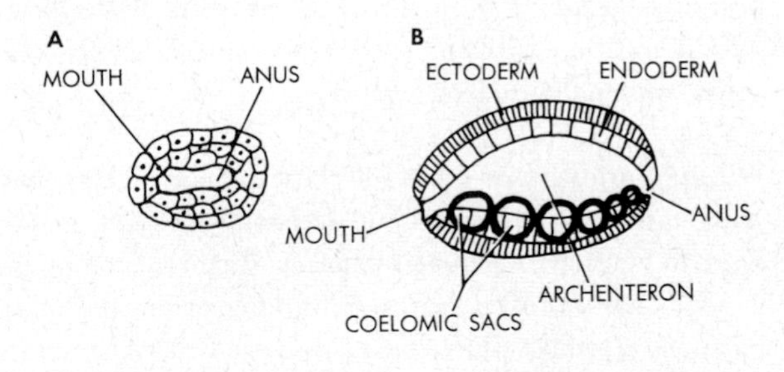

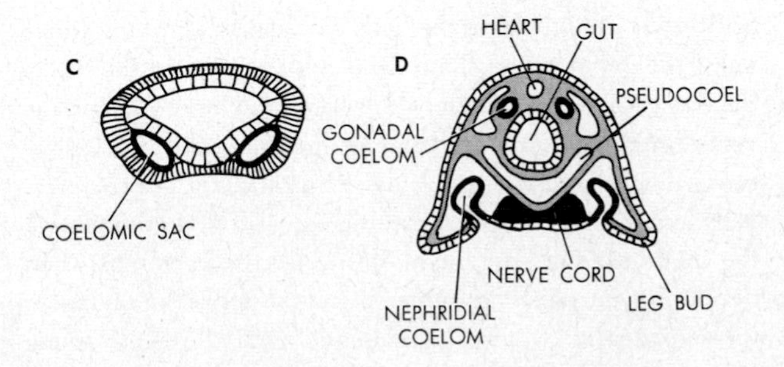

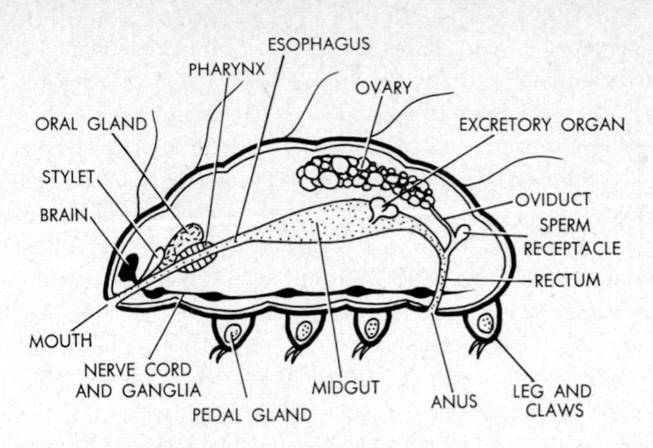

26.21. Tardigrades. Sagittal section of *Macrobiotus* showing principal organs. The oral gland and stylet are each paired, one on each side of the pharynx, but only the right members are shown here. (*Adapted from Marcus.*)

often carnivorous, i.e., they pierce nematodes, rotifers, or other tardigrades and suck up their fluid contents.

The body consists of a prostomium and five trunk segments (Fig. 26.21). The first four trunk segments bear short, stumpy legs with claws. On the head segment are the openings of a pair of *oral glands,* and also horny, partly chitinous *stylets,* used in piercing the walls of food organisms. The mouth leads into a straight alimentary tract with diverticula, and the tract terminates at the anus in the last segment. The body is covered with a thin, pliable, nonchitinous cuticle which is tucked in at the mouth and anus and lines the foregut and hindgut, respectively. Spines and analogous cuticular outgrowths are usually present on the head and elsewhere on the outside. A segmentally arranged body musculature underlies the epidermis, the longitudinal muscles being organized in quadrants. The nervous system is annelidlike, with supra- and subesophageal ganglia and a ventral cord. The latter contains four segmental ganglia. A pair of eyecup ocelli is embedded on each side of the brain ganglion. The body cavity is nonseptate and in the adult a peritoneum is not present, the cavity being hemocoelic as in arthropods. Food-storing amoebocytes float in the body lymph. Excretion takes place through the oral glands, the alimentary tract, and the body surface. The epidermis also serves as the principal breathing structure.

Like rotifers, tardigrades exhibit cell constancy, a phenomenon which appears to be common generally in very small but comparatively complexly constructed animals. Tardigrades molt their cuticles several times during their adult life. The animals are separately sexed, a single gonad lying in a dorsal sac. From it two gonoducts lead either into the hindgut or to a separate posterior gonopore. Fertilization is external and the eggs also develop externally, directly in most cases. Cleavage follows an irregular pattern, a coeloblastula forms, and gastrulation occurs by delamination of endoderm cells at the vegetal pole. The nervous system then delaminates ventrally from the ectoderm. After an embryonic gut has developed, five pairs of coelom sacs form from it by *enterocoelic* lateral evaginations (Fig. 26.22). The first four pairs of sacs give rise to the musculature, the fifth pair to the reproductive system. By the time the adult stage is attained, few remnants of the coelom sacs are left. The body cavity thus remains a pseudocoelic hemocoel.

Again like rotifers, tardigrades have the capacity to dehydrate down to cystlike bodies. In such a dormant state an animal may remain viable for up to 5 years. Lack of oxygen, food, or water will induce the dormant condition. It has been shown that a tardigrade can be revived some 14 successive times from a dormant condition before death supervenes. Hence such animals might have a theoretical life span of at least 70 years.

Tardigrade development and adult structure indicate that the animals combine certain polychaete features, a substantial number of arthropod features, and quite a few features peculiar only to tardigrades. Most unexpected is the method of coelom-formation, which is unequivocally enterocoelic (through the pattern differs from that encountered in deuterostomial enterocoelomates). It might perhaps be hypothesized that the small number and comparatively large size of the blastomeres leave not enough room within the embryo for the accumulation of teloblastic somites; direct formation of coelom sacs from the endoderm evidently represents an available shortcut. In view of their adult features, tardigrades can justifiably be regarded as representatives of a transitional evolutionary level on the main path from annelids to arthropods. Yet this level appears to be a different one from that of the Onychophora (cf. Fig. 26.1).

26.22. Cross-sectional view **(A)** and sagittal view **(B)** of tardigrade embryo, showing enterocoelous formation of coelomic sacs from early alimentary tract. (*Adapted from Marcus.*)

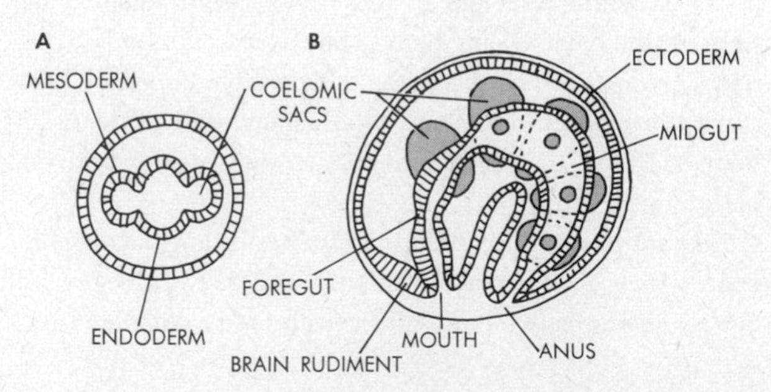

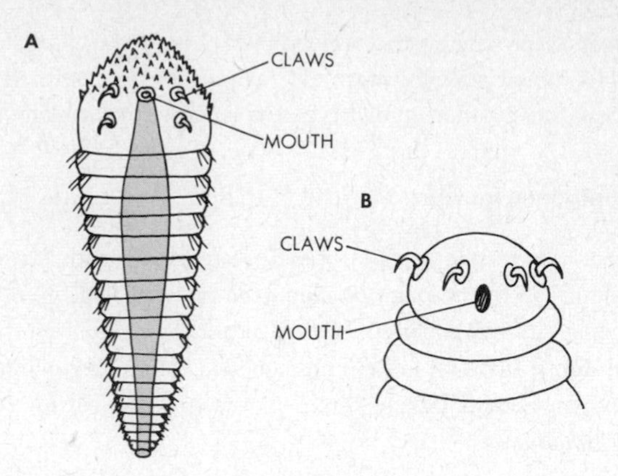

26.23. Pentastomids. A. *Linguatula,* from ventral side. (*Adapted from Leuckart.*) **B.** *Porocephalus,* head region. (*Adapted from Hill.*)

Subphylum Pentastomida

Blood-sucking endoparasites in vertebrates; larvae with claw-bearing legs; adults wormlike, with two pairs of claws on side of mouth; circulatory, excretory, breathing systems absent. *Linguatula, Porocephalus*

The approximately 50 species of these animals (also known as *Linguatulida*) live primarily in the nasal passages, windpipes, and lungs of amphibia, reptiles, and other vertebrates. The adults are wormlike, up to 5 in. long, with superficial annulations externally (Fig. 26.23). Recurved claws lateral to the mouth enable them to remain attached to host tissues. Scattered randomly over the body surface are air pores. Internally, the nervous system is of the ladder type and the reproductive system is elaborate. Gonopores and anus open on the ventral side. The sexes are separate and dimorphic, the female being much larger than the male. Fertilization occurs internally and the zygotes reach external aquatic environments via the mouth of the host. In most species the eggs must then develop in intermediate hosts such as fishes or amphibia. If the latter drink water contaminated with the eggs, young larvae emerge in the host gut, migrate to internal organs, and encyst there. The arthropod affinities of the group are suggested principally by these larvae. They have the general appearance of mites, and they are equipped with four to six pairs of stumpy legs, each with two terminal claws. Moreover, the larvae also molt their cuticles at intervals. If the intermediate host is later eaten by a final host, the larvae excyst in the gut and find their way into the breathing system of the host, where they become adults. Some pentastomids complete their life cycles without intermediate hosts.

Evidently, the phylogenetic connection between pentastomids and arthropods is distant and tenuous, and the parasitic nature of the animals obscures any affinities still further. That the worms are allied more nearly to arthropods than to annelids is probably indisputable; but it is not clear whether they represent degenerate mites or a basically primitive, prearthropod line of evolution which later became parasitic. The claw-bearing legs of the larvae, the larval molts, and also the air pores of the adults constitute the main justification for including these animals among the Oncopoda. As members of this phylum, pentastomids are certainly much less close to the tardigrades and the onychophorans than the latter two groups are to each other.

REVIEW QUESTIONS

1. Define Sipunculida taxonomically and describe the structure of these animals. Where do sipunculids live and how do they feed? Move? What are urns and what is their possible function? How do sipunculids develop? On what basis and how may they be regarded to be related to annelids?
2. Give taxonomic definitions of annelids and of each of the annelid classes. Describe the development of a segmented body organization from a trochophore larva. Then describe the segmental structure of an adult polychaete. What is the organization of a parapodium? What are the basic adaptive advantages of metamerism?
3. Describe the nervous, circulatory, and excretory systems of *Nereis*. How is breathing accomplished? Where do the gonads form and how does fertilization occur? What is heteronereis? In what different ways do polychaetes form swarming body parts?
4. Distinguish between errant, tubiculous, and burrowing polychaetes. Name representative genera of each group. Show how feather-duster and parchment worms are adapted to tubiculous ways of life. Describe the external appearance of a lugworm and show how it is adapted to a burrowing existence.
5. In what respects do (*a*) archiannelids, (*b*) oligochaetes differ from polychaetes? Describe the external and internal structure of an earthworm and contrast the body design with that of *Nereis*. What are typhlosole, crop, and lime glands and what are their functions?
6. Describe the structure of the reproductive systems of an earthworm and show how mating and fertilization occur. How does earthworm development take place? Name aquatic and terrestrial genera of oligochaetes.
7. Describe the structure of a leech and contrast it with that of an earthworm. In what different ways do leeches feed? What adaptations to ectoparasitism are in evidence? How do leeches mate and develop?
8. Define echiuroids taxonomically and describe the structure of such a worm. In what respects is the structure different from that of (*a*) an annelid, (*b*) a sipunculid? How does an echiuroid develop and in what sense do these animals represent a postannelid stage of evolution? Review the development of sexual dimorphism in *Bonellia*.
9. Define Oncopoda and each subphylum taxonomically. Describe the organization of *Peripatus* and list features which are (*a*) annelidlike, (*b*) arthropodlike, (*c*) unique to onychophorans. How does *Peripatus* develop and where and how does it live?
10. Describe the structure and development of tardigrades and pentastomids. In what respects are tardigrades rotiferlike and pentastomids mitelike? Make a diagram indicating presumable evolutionary interrelations of all phyla and major subgroups within the phyla of schizocoelomate protostomes. Relate the whole assemblage to lophophorates and to ancestral coelomates.

COLLATERAL READINGS

The references cited deal with most of the groups studied in this chapter:

Allen, E. J.: Fragmentation in the Genus *Autolytus* and in other Syllids, *J. U.K. Marine Biol. Assoc.*, vol. 14, 1927.

Bell, A. W.: The Earthworm Circulatory System, *Turtox News*, vol. 25, 1947.

Berrill, N. J.: "Growth, Development, and Pattern," chaps. 12, 13, Freeman, San Francisco, 1961.

——— and D. Mees: Reorganization and Regeneration in *Sabella*, *J. Exp. Zool.*, vol. 73, 1936.

Gates, G. E.: On Segment Formation in Normal and Regenerative Growth of Earthworms, *Growth*, vol. 12, 1948.

Grove, A. J.: On the Reproductive Processes of the Earthworm, *Quart. Rev. Microsc. Sci.*, vol. 69, 1925.

Hyman, L.: "The Invertebrates," vol. 5, chap. 22 (sipunculids), McGraw-Hill, New York, 1959.

Marcus, E.: Tardigrada, in "McGraw-Hill Encyclopedia of Science and Technology," vol. 13, McGraw-Hill, New York, 1960.

Moment, G. B.: A Study of Growth Limitation in Earthworms, *J. Exp. Zool.*, vol. 103, 1946.

Morgan, T. H.: A Study of Metamerism, *Quart. Rev. Microsc. Sci.*, 1895.

Shipley, A. E.: Tardigrada and Pentastomida, in S. F. Harmer and A. E. Shipley (eds.), "The Cambridge Natural History," vol. 4, Macmillan, London, 1920.

Stephenson, J.: "The Oligochaeta," Clarendon Press, Oxford, 1930.

CHAPTER 27

arthropods: chelicerates

The Arthropoda, or jointed-legged animals, represent the culmination of schizocoelomate evolution. The general characteristics of the whole phylum and the specific features of two of the subphyla are discussed in this chapter; the following chapter deals with the remaining subphylum.

Phylum ARTHROPODA: metamerically segmented; with jointed legs and chitinous exoskeleton; characteristically with compound eyes; sexes largely separate; eggs centrolecithal, cleavage superficial; development mosaic, with various types of larvae or direct.

† Subphylum TRILOBITA: trilobites, extinct
Subphylum CHELICERATA: without jaws or antennae
  Class XIPHOSURIDA: horseshoe crabs
  † Class EURYPTERIDA: eurypterids, extinct
  Class PANTOPODA: sea spiders
  Class ARACHNIDA: scorpions, spiders, ticks, mites
Subphylum MANDIBULATA: with jaws and antennae
  Class CRUSTACEA: crustaceans
  Class CHILOPODA: centipede myriapods
  Class DIPLOPODA: millipede myriapods
  Class PAUROPODA: myriapods
  Class SYMPHYLA: myriapods
  Class HEXAPODA: insects

GENERAL CHARACTERISTICS

This is the largest phylum not only among animals but among all living organisms. Some 1 million species have been described already, a number amounting to about 75 to 80 per cent of all animal species and to more than 50 per cent of all species. Thus there are more arthropod types already known than all other living types combined, plants included; and three or four out of every five animals are arthropods. About 80 per cent of the described species comprise the insects, but according to one estimate some 10 million insect species may actually exist. One of the orders of insects, the beetles and weevils, includes 300,000 species alone, which makes this order larger than any other *phylum* of organisms.

Arthropods do not exhibit particularly large sizes, microscopic mites representing one extreme, 5-ft giant crabs with 12-ft leg spans the other. But the animals are more widely and more densely distributed throughout the world than any others, and they occur in all environments and ecological niches. And inasmuch as they include the insects, most of which fly, they are also able to pursue ways of life not available to other forms. Moreover, arthropods represent the principal group that other animals must compete with, man not excepted. The chief competitors of arthropods are arthropods themselves, bacteria, and vertebrates; the latter as a whole probably consume more arthropods as food than any other animals, both in the sea and on land. In their turn, and notwithstanding the vast numbers of carnivorous, omnivorous, and symbiotic types, arthropods depend most on algae if they live in water, on plants if they live on land.

Such unrivalled success is a consequence of the basic arthropod design, the essential features of which are the metameric structure and the exoskeletal armor. We do not have a record of the original schizocoelomate ancestor from which arthropods evolved, but we are well justified to assume that this

27.1. Lateral view of a grasshopper larva, showing segmental structure of arthropods generally. Head, at right, is externally unsegmented and bears antennae, eyes, and mouth parts. The thorax, consisting of three segments in insects, bears three pairs of legs (one per segment) as well as two pairs of wings in adults (on the second and third thoracic segments). The abdomen in insects typically consists of 11 segments and is without appendages. Numbers of segments and types of appendages vary considerably for different arthropod groups, but sets of mutually different segments are present in all.

ancestor possessed metameric segments with paired segmental appendages. One descendant line then must have given rise to polychaetes with parapodia and thus annelids generally, and another line, to arthropods. The polychaetelike early stock presumably ceased to pursue a swimming mode of life and adopted a crawling existence, the parapodial appendages serving as legs. Later, the epidermal potential of forming chitin, restricted at first (and in annelids even today) to the production of setae, evidently became general and led to development of a chitinous cover over the whole body. Such an exoskeleton not only protected but also strengthened and supported mechanically; it must have permitted the legs to become more elongated and jointed into segments, hence more efficient.

Furthermore, far more so than in the annelid group, the parapodial appendages could become structurally and functionally different in different segments, for the chitinous cover could become molded and elaborated into diverse permanent shapes. As a result, the segmental appendages of the evolving arthropods became not only walking legs, but also structures adapted to biting, cutting, sucking, piercing, cleaning, grasping, carrying, breathing, swimming, flying, egg-laying, sperm-transferring, sensing, and even silk-spinning. In effect, the possibilities of diversification inherent in a segmental repetition of parts became exploited in an almost infinite variety of ways. Moreover, a chitin cover also served well in the elaboration of complex sensory receptors, including most particularly the unique compound eyes. We may note that arthropods actually differ from one another far more in their exterior, chitin-covered parts than in their interior design; and it is just this exterior variability which has made arthropods so hugely diversified a group.

As already pointed out in Chap. 8, an exterior chitin envelope puts constraints on body size and on growth. Further, the presence of an inert cover in early stages rules out surface ciliation, hence also pelagic, feeding trochophore larvae. Eggs must therefore be large, with enough yolk to suffice for the elaboration of either a nonciliated, nontrochophore larva or a small adult. Indeed, arthropod development has deviated drastically from the ancestral pattern of spiral cleavage and trochophores, and although the eggs have remained mosaic, they have become centrolecithal and they cleave superficially (cf. Figs. 11.12 and 11.15). The blastulae typically gastrulate by delamination, ingression, or invagination of cells from the ectoderm. Interiorized cells then proliferate both into an endodermal midgut around the yolk and into paired bands of mesodermal somites. Each of the latter subsequently acquires a schizocoelic cavity, but such coelom sacs remain small. By the time the mesodermal organs have developed from the somites, the coeloms have actually disappeared as closed sacs and the body cavity has become a hemocoel. Segmental appendages develop as buds evaginated from the ectoderm and the somatic mesoderm. Larvae may or may not form and, as outlined in Chap. 11, if they do form they metamorphose gradually or abruptly, depending on the specific arthropod group (cf. also below and Chap. 28).

The fundamental body divisions of an adult arthropod are *head, thorax,* and *abdomen* (Fig. 27.1). These may or may not be marked off sharply, and in many cases the head and thorax are fused together into a *cephalothorax* (also called *prosoma,* the abdomen then being referred to as *opisthosoma*). Any of the body divisions may be unsegmented externally as a result of fusion of segments during embryonic stages. Fused or unfused segments of head, thorax, or cephalothorax are often covered by a *carapace,* a continuous exoskeletal shield. The head often arises from six embryonic segments which in almost all cases later fuse together. Nevertheless, the head appendages and the ganglia (generally concentrated together to form the brain) still reveal the segmental origin of this body division. Head appendages vary greatly in different groups; their functions are largely sensory and ingestive. The mouth arises in the embryo just behind the first segment, but in later stages it comes to lie more posteriorly. In different groups of adults, therefore, the mouth is generally found various segmental distances behind the front end of the animal, on the ventral side. Group variation also characterizes the number of segments in thorax and abdomen (cf. Table 11 below and Table 12, Chap. 28). The appendages of these divisions are to some extent ingestive, reproductive, and respiratory, and to a large and conspicuous extent locomotor. Typically, the most posterior part of the abdomen is a *telson;* it is essentially a terminal body section behind the trunk segments and it is not counted as a last segment (Table 11).

Segments and appendages are covered by a continuous chitinous cuticle, often richly studded with bristly "hairs" or spines. This cuticle is secreted by the underlying epidermis (here also called *hypodermis*), and it is basically thin and pliable. It forms the joint membranes and the breathing surfaces, and it is tucked in at both ends of the alimentary tract as a lining of the foregut and the hindgut. In the regions between the segmental and appendageal joints, the cuticle is thickened into a hard, exoskeletal cover. Hardness results not so much from chitin itself as from secreted impregnating materials such as horny scleroproteins and, in aquatic forms, also calcium salts. Around a segment this hard cover occurs fundamentally in the form of four plates, or *sclerites:* a dorsal *tergum,* a ventral *sternum,* and two *pleura,* i.e., one pleuron on each side (Fig. 27.2). These and the sclerites of adjacent segments are articulated by interior muscles which lie in the hemocoel and do not form a continuous body-wall musculature. The muscles are organized segmentally.

An appendage basically consists of several joints, or *endites,* arranged either as a linear set (uniramous) or as a bifurcated set (biramous). A uniramous appendage is composed of three main parts: *coxopodite, basipodite,* and *endopodite* (cf. Fig. 27.2). The coxopodite is basal. It may or may not bear a medial protruberance studded with chitinous teeth; if such a protruberance is present, it is called a *gnathobase.* In many cases, also, the coxopodite bears a lateral endite, the *epipodite* (which may be a gill in aquatic arthropods). The main endite attached to the coxipodite is the basipodite, which in turn carries the endopodite. The latter may itself consist of several

TABLE 11. Segments and appendages in trilobites and chelicerate arthropods

	Trilobite			Horseshoe crab	Eurypterid	Scorpion	Spider	Pantopod
head	embryo only	1	*cephalothorax* (append. typic. uniramous)	embryo only	embryo only	embryo only	embryo only	embryo only
	antennae (uniramous)	2		chelicerae (chelate)	chelicerae (chelate)	chelicerae (chelate, small)	chelicerae (fangs)	chelicerae (chelate)
	walking legs (biramous)	3		pedipalps (chelate)	pedipalps	pedipalps (chelate, large)	pedipalps (tactile)	pedipalps (tactile)
	walking legs (biramous)	4		walking legs (chelate)	walking legs	walking legs (clawed)	walking legs (clawed)	ovigerous legs
	walking legs (biramous)	5		walking legs (chelate)	walking legs	walking legs (clawed)	walking legs (clawed)	walking legs (clawed)
	walking legs (biramous)	6		walking legs (chelate)	walking legs	walking legs (clawed)	walking legs (clawed)	walking legs (clawed)
pygidium, thorax	walking legs (biramous)	7		walking legs (chelate)	walking legs	walking legs (clawed)	walking legs (clawed)	walking legs (clawed)
	walking legs (biramous)	8		chilaria	embryo only	embryo only	embryo only	walking legs (clawed)
	walking legs (biramous)	9	*abdomen* (append. typic. biramous)	operculum, gonopores	operculum, gonopores	operculum, gonopores	embryo only	reduced abdomen
	walking legs (biramous)	10		gill books	gill books	pectines	lung books/tracheae	
	walking legs (biramous)	11		gill books	gill books	lung books	lung books/tracheae	
	walking legs (biramous)	12		gill books	gill books	lung books	spinnerets	
	walking legs (biramous)	13		gill books	gill books	lung books	spinnerets	
	walking legs (biramous)	14		gill books	gill books	lung books	14–18 embryo only	
				telson	15–20 without appendages	15–20 "tail" without appendages		
	telson				telson	telson		

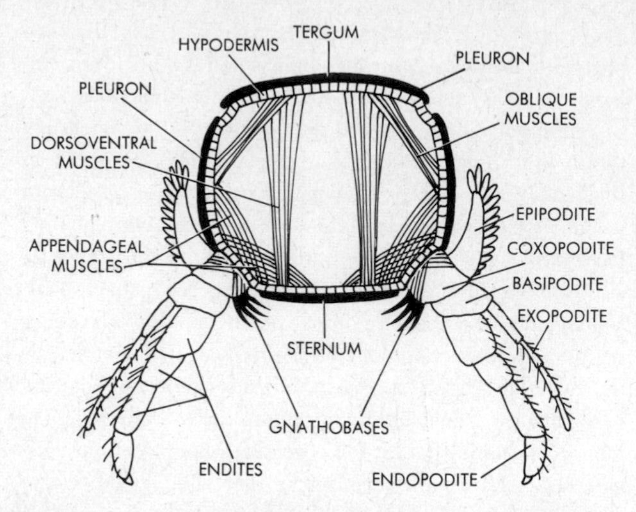

27.2. Segmental exoskeletal parts and basic appendage structure in arthropods. Epipodites typically are gills. Endopodites typically terminate in claws or pincers. Only a few of the trunk and appendageal muscles are shown.

endites, and it forms the extremity of the uniramous appendage. If the appendage is biramous, the basipodite carries not only an endopodite but also an *exopodite,* again composed of one or more endites.

All appendages of arthropods are modifications of this basic structure. Given endites may be suppressed entirely or may be fused with others or may be enlarged or transformed in shape. The results then are appendages as diverse as jaws, gills, poison fangs, egg depositors, antennae, pinching claws, adhesive pads, walking legs, and numerous others. Note, however, that eyes and wings are not homologous to these appendages; the latter are ventral and ventrolateral outgrowths, whereas eyes and wings are dorsal and dorsolateral outgrowths.

Internally, the alimentary tract consists of foregut, midgut, and hindgut (Fig. 27.3). The first and last portions are lined with chitin, as noted, and they are derived from ectoderm; only the midgut is endodermal. Salivary glands are typically present. They open into the mouth or the anterior part of the foregut. The latter is often regionated into pharynx, esophagus, and stomach. Digestive glands, in many cases large and conspicuous, connect with the stomach of the intestinal midgut. The hindgut, which may or may not include a rectal section, terminates at the anus, usually located ventrally in the telson. Breathing is accomplished in exceptional cases entirely through the body surface. However, most arthropods are equipped with specialized organs such as gills, gill books, lung books, or tracheal systems, all of these representing chitin-lined and metamerically arranged outgrowths or ingrowths on the body segments or the appendages (cf. Chap. 9 and below).

The circulatory system is open. A dorsal heart forms the principal vessel, though others are often present. Excretion takes place principally through two kinds of organs, one kind opening at the bases of given body appendages, the other attached to and opening into the hindgut (cf. Fig. 27.3). Organs of the first type are variously called *coxal glands, renal glands, green glands,* and other names; they are characteristic primarily of aquatic arthropod groups, e.g., horseshoe crabs and crustacea. Such excretory structures are probably

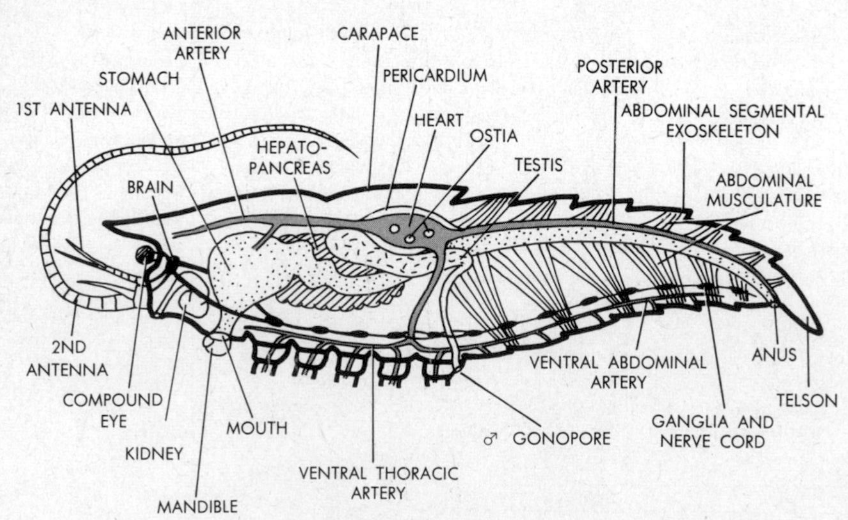

27.3. The basic internal structure of arthropods, exemplified by a lobster. For additional details see illustrations in Chap. 28.

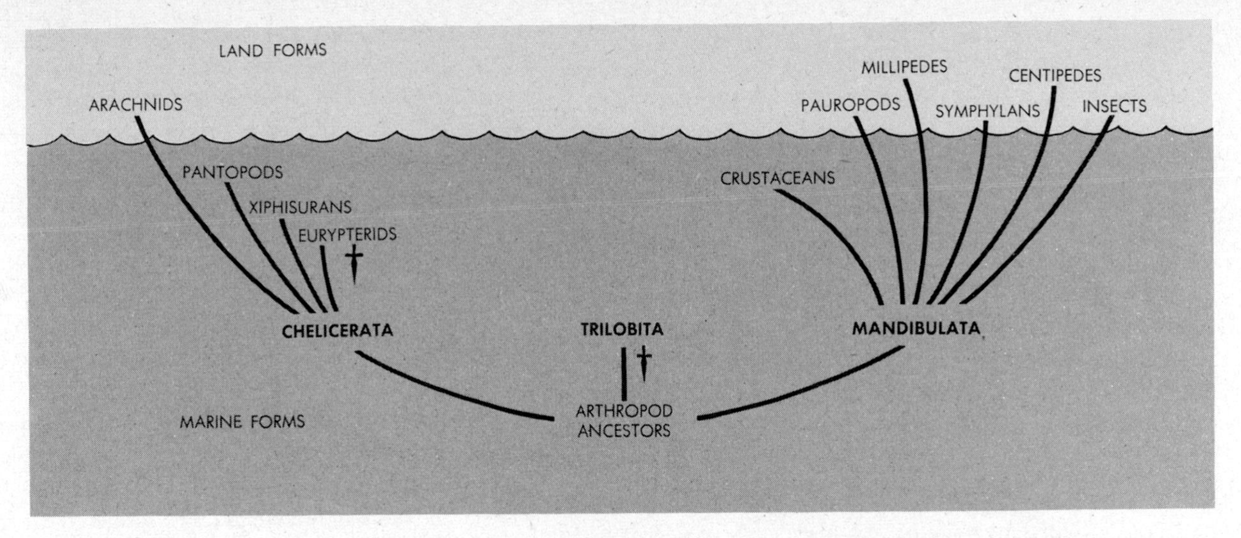

27.4. The presumable evolutionary interrelations of the principal arthropod groups. Primarily aquatic groups are indicated below the water surface, primarily terrestrial groups, above.

homologous to metanephridia, and they may represent remnants of an originally complete metameric series of ancestral metanephridia. Organs of the second type are known as *Malpighian tubules;* they are characteristic mainly of terrestrial arthropods, e.g., arachnids and insects. In such animals the principal nitrogenous waste is uric acid and the urine is semisolid (cf. Chap. 9).

Arthropods typically also possess endocrine systems. These consist of neurosecretory cells in the brain and of a variety of glands in the thorax and at the bases or in the stalks of the compound eyes. Hormone secretions play a major role in the control of molting and in reproduction and development, particularly well studied in crustacea and insects (cf. Chap. 28). In many crustacea, as we shall find, the hormones also govern the activities of pigment-containing chromatophores, which adjust the body coloration according to external backgrounds. In the vast majority of arthropods, as in most active, free-living animals, the sexes are separate. Hermaphroditic types do occur, however, and we may here recall also that parthenogenetic groups are fairly common. Paired gonads are located dorsally in the thorax or the abdomen. They lead to gonopores located in different specific segments in different groups. All patterns of fertilization and egg management are encountered; the principal ones are described below.

Arthropods are largely free-living, but symbiotic types of all categories are abundant as well. In their turn, arthropods are favorite hosts of many other symbiotic animals, a predictable circumstance in view of the sheer numbers of potentially available arthropod hosts. Arthropods also exhibit a wide variety of unique behavioral and social traits, as outlined in Chap. 16; and the organizational complexity of their constructed home sites is matched only by the installation of modern man. In effect, there is hardly a single phase or aspect of animal biology in which arthropods are not of foremost significance.

Evolution within the phylum appears to have proceeded along three separate major lines (Fig. 27.4). One gave rise to the subphylum Trilobita, extinct since the Permo-Triassic crisis and probably representing the most ancient group. The second line now constitutes the subphylum Chelicerata, and the third, the subphylum Mandibulata. Trilobites were entirely aquatic, and the early members of each of the other two subphyla were and still are aquatic. Thus, the extinct eurypterids and the living horseshoe crabs and sea spiders represent the aquatic chelicerates, and the crustacea represent the aquatic mandibulates. Later branch lines of both subphyla became terrestrial, culminating in arachnids and insects, respectively. Evidently, each of the three subphyla produced a highly diversified adaptive radiation. Each line within these radiations then gave rise to a huge subradiation in its turn, and vast arrays of subsubradiations evolved as well. As a result,

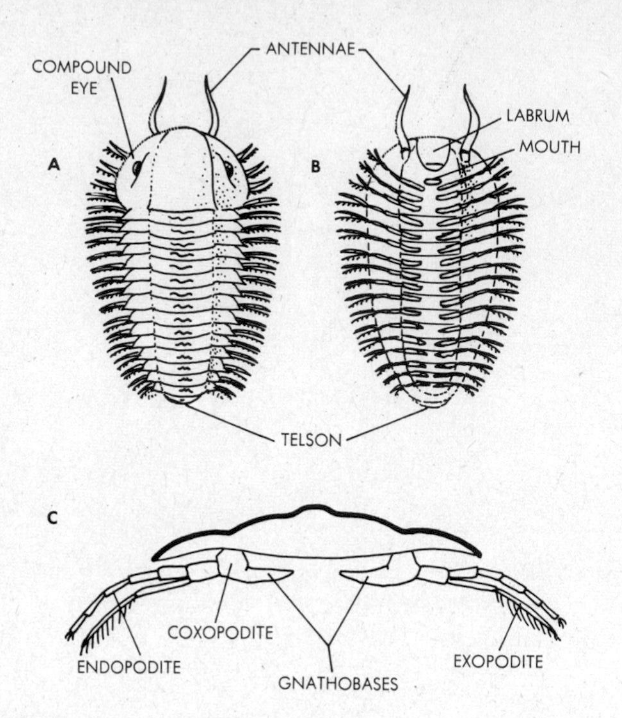

27.5. Trilobite structure. (*After Beecher.*) *Triarthus* is illustrated in dorsal **(A)**, ventral **(B)**, and cross-sectional **(C)** views. The last few segments in front of the telson represent the pygidium, the generating region (itself segmented in the adult) which in the larva has given rise to all the more anterior segments (cf. Fig. 27.6). Photos of trilobites are shown in Fig. 15.5.

arthropods encompass more classes, orders, and families than any other phylum.

SUBPHYLUM TRILOBITA

Extinct; marine; body marked dorsally into three lobes by two longitudinal furrows; paired identical ventral legs on all segments except first two and last; compound eyes present; development via larvae. *Triarthus*

Some 2,500 of these animals are known as fossil species (cf. Fig. 15.5). They possessed a *head* with fused segments and a trunk divided into a conspicuously segmented *thorax* and a fused posterior *pygidium*. The latter bore a terminal *telson* (Fig. 27.5). Dorsally the head contained a pair of compound eyes. Ventrally the first head segment was without appendages, to the second segment were attached a pair of uniramous antennae, the third segment contained the mouth, and the third as well as every other more posterior segment of the entire body (telson excepted) bore a pair of identical legs (cf. Table 11). These were biramous, with gnathobases and epipodite gills. The presence of a complete series of undiversified appendages, reminiscent of annelids, underscores the primitive nature of the animals and also suggests that the appendages of early arthropods were predominantly locomotor. Judging from living arthropods, one may surmise that the gnathobases probably functioned in chewing food and in passing it anteriorly along the ventral midline toward the mouth.

Trilobites developed via larvae shaped like circular disks in the early stages (Fig. 27.6). Almost the whole disk represented the head, and a small telson was present posteriorly. Between the head and the telson then formed the pygidium. The anterior boundary of the pygidium later generated thoracic segments in anteroposterior succession, while at the same time the posterior boundary developed a corresponding number of new segmental sections in front of the telson. Thus the pygidium retained constant length but the thorax gradually increased its number of segments. Segmental growth occurs in precisely this fashion in the nauplius larvae of living crustacea (cf. Chap. 28). Arthropod development such as this, in which the just-hatched animal possesses only an incomplete number of segments, is generally said to be *anamorphic*.

SUBPHYLUM CHELICERATA

Adult body usually comprising cephalothorax, generally unsegmented, and abdomen, either segmented or not; cephalothorax typically with six pairs of appendages: first pair

27.6. Trilobite larvae. (*After Barrande.*) Three successive stages are shown. Development here is anamorphic, i.e., segments continue to be formed after hatching, throughout the larval phase. The pygidium is the segment-generating body part.

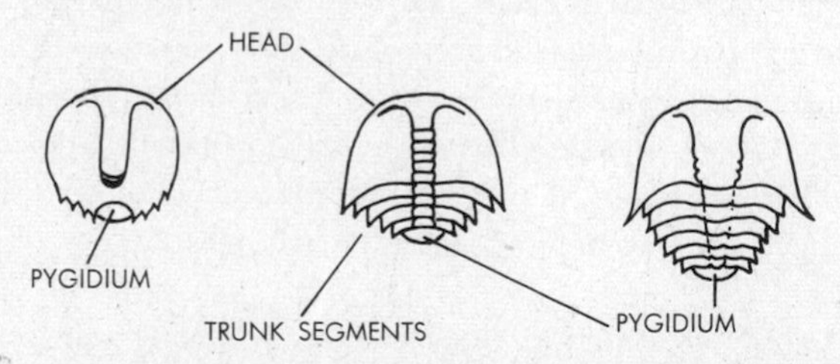

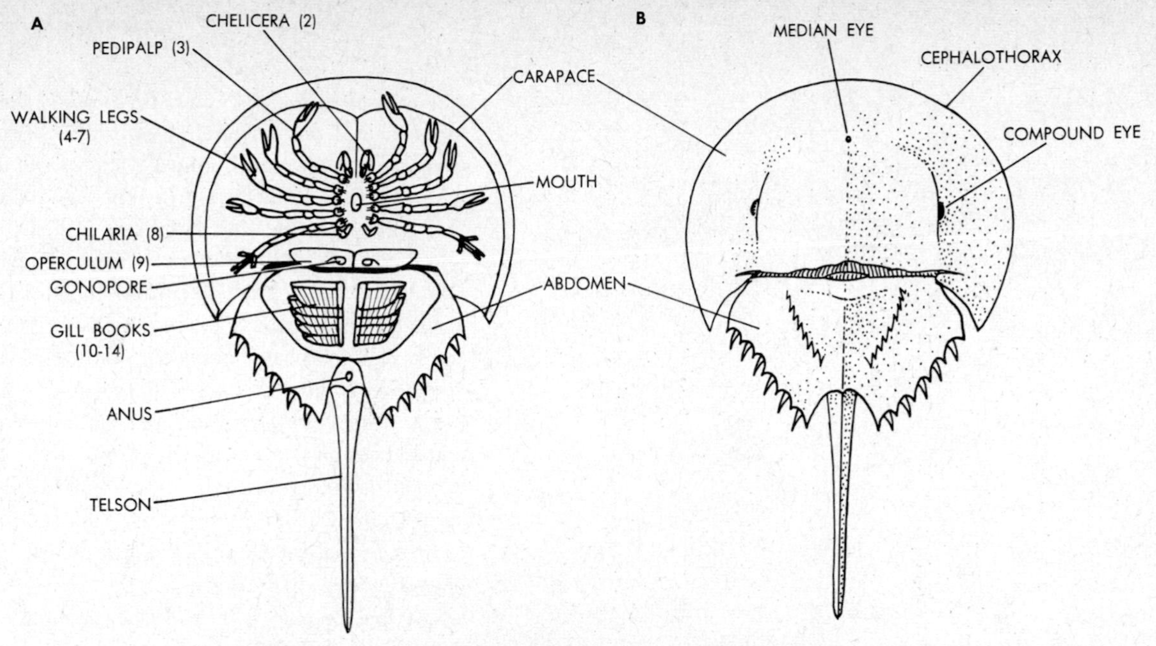

27.7. *Limulus*, external features. Diagrams: A. Ventral view, with appendages. Segments numbered according to sequence in embryo. The operculum is drawn turned forward to show gonopore on underside and to expose the gill books. **B.** Dorsal view. **Photo:** living animals on tidal flat.

chelicerae, typically three-jointed; second pair *pedipalps,* typically six-jointed; last four pairs, *walking legs;* jaws or antennae never present; abdominal appendages not primarily locomotor or absent. *Xiphosurida, Eurypterida, Pantopoda, Arachnida*

Class Xiphosurida

Horseshoe crabs; marine; cephalothorax with carapace and hinged to fused abdomen; telson a spine; compound eyes present; development via "trilobite" larvae. *Limulus*

The horseshoe crab *Limulus* is the most familiar of the few present-day survivors of this class (four species in two genera). Representing the nearest existing relatives of trilobites, these "living fossils" are found along sandy shores, where they lead a burrowing, semisedentary existence (Figs. 27.7 and 27.8). The body plan of a horseshoe crab exemplifies the segmental design of chelicerates generally.

The cephalothorax consists of eight embryonic segments (cf. Table 11). These later fuse and develop a conspicuous, horseshoe-shaped carapace dorsally. The carapace bears two small anterior ocelli and two lateral compound eyes which

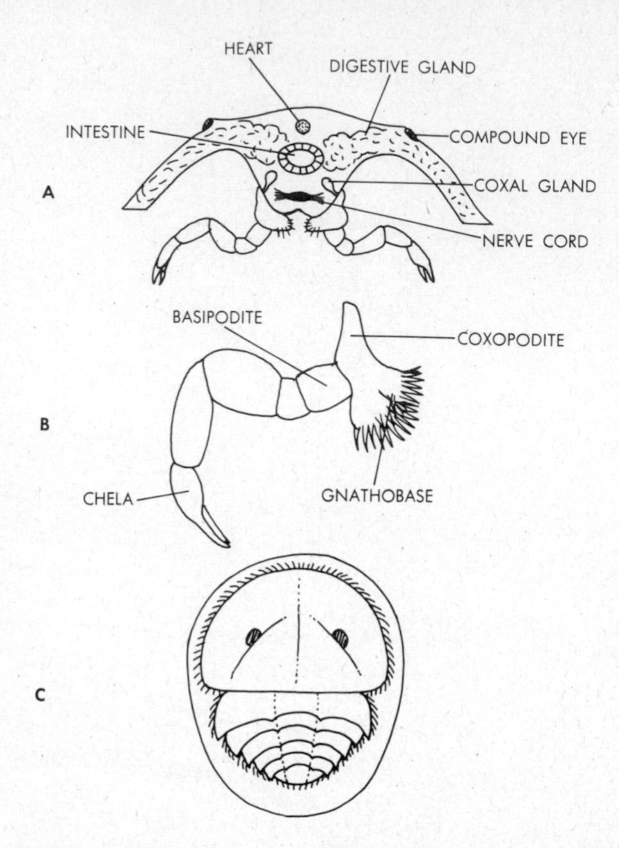

27.8. ***Limulus.*** **A,** cross section through cephalothorax showing a few of the organs; **B,** detail of a walking leg showing gnathobase on coxopodite; **C,** the ciliated "trilobite" larva of *Limulus,* dorsal view, still in its egg membrane; it will later swim free (cf. Fig. 27.6).

have been shown to be sensitive to both unpolarized and polarized light. Ventrally, the first segment is embryonic only and does not develop appendages. The remaining seven cephalothoracic segments each possess a pair of uniramous appendages. The first pair are the three-jointed *chelicerae,* little legs with terminal pincers. The pair of *pedipalps* behind them is indistinguishable from the next three pairs of *walking legs.* All these are six-jointed; they are pincer-equipped (*chelate*) in females and young males, but they terminate in claws in adult males. The last pair of legs is larger than the preceding but otherwise similar in structure. Each of these two legs also carries four subterminal spines which aid in pushing sand behind when the animal burrows into the ground. All legs and the chelicerae are equipped with prominent gnathobases; legs are positioned in such a way that the gnathobases surround the mouth, located relatively far back in the adult. Horseshoe crabs feed on clamworms, soft-shelled clams, and other small animals. Such food is crushed and minced by the gnathobases when the legs are moved, and the food pulp, including sand and pieces of shells, is then pushed into the mouth by the chelicerae. These animals are the only chelicerates eating solid food; all others subsist on liquid food, a practice conditioned by the absence of jaws.

The eighth cephalothoracic segment bears a seventh pair of appendages, the *chilaria,* small projections without known function; they may represent vestigial remnants of an additional ancestral pair of legs. In other chelicerates this eighth segment is embryonic only and without appendages of any kind. The cephalothorax as a whole is hinged to the abdomen, composed of a broad anterior region of fused segments and, hinged to them posteriorly, a telson spine. Apart from the telson, the abdomen is formed from six embryonic segments. Each of these develops a pair of biramous appendages. The first pair is fused along the midline and forms a transverse plate, the *operculum,* which covers and protects all posterior appendages. On the back face of the operculum are the gonopores. The next five pairs of appendages are *gill books.* Each such appendage develops from a narrow endopodite to which is fused laterally a broad, platelike exopodite. On the back face of such a plate are numerous gill plates. By flapping their gill books horseshoe crabs may not only breathe but also swim to some extent. If the telson spine is disregarded, *Limulus* thus is constructed out of 14 embryonic segments, eight in the cephalothorax, six in the abdomen. The first of these and the telson are without appendages.

Internally, the alimentary tract contains a chitin-lined gizzard and a large, highly lobulated digestive gland which occupies much of the space under the carapace. The anus opens at the base of the telson. The ventral ganglia of the cephalothorax are concentrated into an anterior fusion ganglion, and the nervous system as a whole is of interest in that the ganglia and the principal cords are sheathed by arterial blood vessels. The blood pigment is hemocyanin, dissolved in plasma. Excretion occurs via a pair of coxal glands opening at the bases of the last pair of legs. Male horseshoe crabs are somewhat smaller than females and they possess a more strongly arched carapace. Also, the males possess hard, conical, gonopore-bearing papillae on the operculum. Females lay large, yolky eggs in sand burrows and males fertilize the spawn externally. The larvae resemble trilobites and also the adult horseshoe crabs, though a telson is absent. Maturation lasts up to 10 or 11 years, and several (usually annual) larval molts occur during this time. Adults do not molt. As pointed out in Chap. 15, the genus *Limulus* is at least 200 million years old.

Class Eurypterida

Extinct, marine; cephalothorax unsegmented, abdomen segmented, with telson; compound eyes present. *Eurypterus*

The chelicerae of these animals (Figs. 15.5 and 27.9) were long and equipped with powerful pincers, the pedipalps were legs, and of the following four pairs of legs the last were larger than the rest and paddleshaped. Twelve segments and a telson formed the abdomen. The first abdominal segment bore an operculum and gonopores, and the fossils suggest that the next five segments carried gill books. All other segments were without appendages. Thus the appendageal complement and the general segmental structure of eurypterids appears to have been almost identical with that of horseshoe crabs (and the two groups are actually often classified as subclasses within a single class *Merostomata;* cf. Table 11). However, the shape of eurypterids resembled that of scorpions, and some types actually had narrowed posterior abdominal segments reminiscent of scorpion tails.

27.9. The segmental structure of eurpyterids. *(Adapted from Schmidt.)* Diagram of dorsal view, only cephalothoracic appendages shown. Segmental numbers take account of embryonic segments as well. A photo of a eurypterid is in Fig. 15.5.

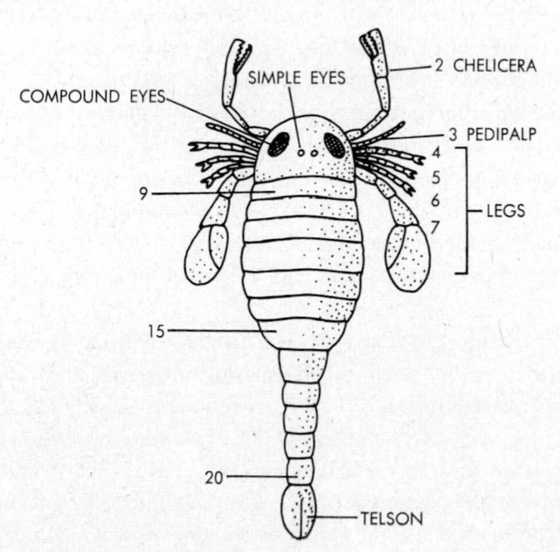

Class Pantopoda

Sea spiders; head and thorax with long, thin legs; abdomen vestigial; mouth on proboscis; compound eyes absent; egg-carrying *ovigerous* legs on head of both sexes; development via larvae or direct. *Nymphon, Pycnogonum*

The 500 species of these spiderlike animals (also known as *Pycnogonida*) occur at different ocean depths from the surface to the sea bottom. The bodies of pantopods are usually minute, but their legs are exceedingly long and thin and they move rather slowly over the ground (Fig. 27.10). The head is equipped with four eyes and a sucking proboscis (pharynx), at the tip of which is the mouth. Pantopods feed mainly on the tentacles of sea anemones, which they tear open with the pincers of their chelicerae. Then they insert the proboscis and suck up the juices. In addition to the pair of chelicerae, the head also bears a pair of tactile pedipalps, a pair of 10-jointed *ovigerous* legs, and a first pair of eight-jointed, claw-bearing walking legs. The ovigerous legs are generally smaller than the others; on these legs the *males* carry the developing eggs. Three segments behind the head bear the remaining three pairs of walking legs (cf. Table 11). The abdomen is reduced to a tiny posterior protrusion. All legs contain pouches of the alimentary tract as well as portions of the gonads. The latter may open on the last pair of legs, as in *Pycnogonum*, or on all legs, as in other genera. The animals breathe through the body surface. Larvae are four- or six-legged.

Class Arachnida

Arachnids; terrestrial, some secondarily aquatic; usually carnivorous, predatory; cephalothorax typically unsegmented, abdomen segmented or unsegmented; compound eyes absent; breathing by lung books, tracheae, or both; development largely direct.

Order Scorpionida (scorpions): chelicerae small, pedipalps large, both with pincers (chelate); abdomen segmented, second segment with tactile *pectines;* last six segments elongate, tail-like, terminating in poison sting (telson); ovoviviparous. *Scorpio, Centrurus*

Order Pedipalpi (whip scorpions): small chelicerae and large

27.10. **Pantopods.** **A,** diagram (*after Mobius*), and **B,** photo, of the sea spider *Nymphon.*

A

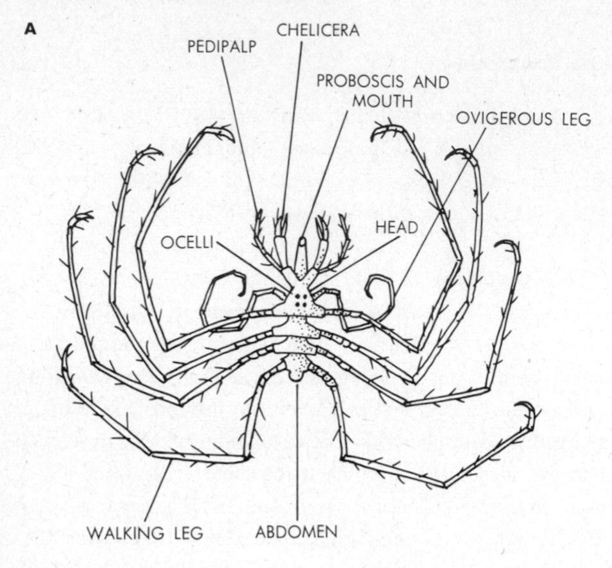

B

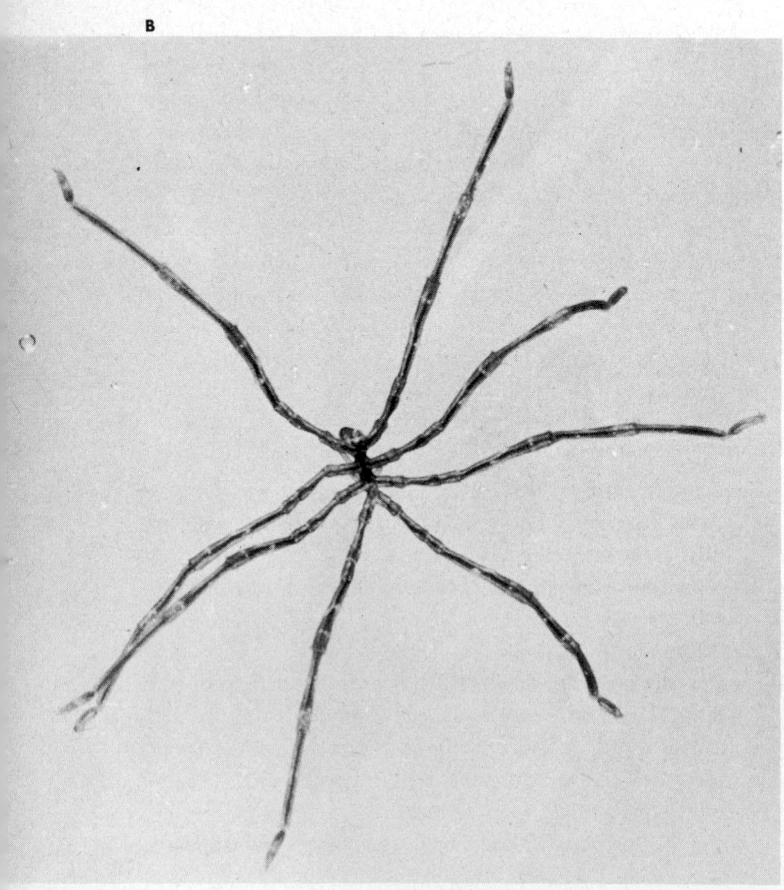

pedipalps, both chelate; first pair of legs tactile; four pairs of ocelli; abdomen flattened, segmented; with or without whiplike tail formed by appendage of last segment. *Mastigoproctus*

Order Palpigradi: small, scorpionlike; chelicerae strong, pedipalps leglike; without eyes; abdomen segmented, last 15 segments form narrow scorpionlike tail. *Prokoenenia*

Order Araneida (Araneae, spiders): cephalothorax and abdomen unsegmented, joined by narrow waist; chelicerae fangs, pedipalps leglike; up to four pairs of eyes; abdomen with spinnerets and lungbooks, some with tracheae; development usually external, direct. *Theridion,* house spiders; *Agelenopsis,* grass spiders; *Pirata,* wolf spiders; *Salticus,* jumping spiders; *Bothryocyrtum,* trap door spiders; *Aphonopelma,* tarantulas; *Latrodectus,* black widow spiders.

Order Solpugida (sunspiders): short pincered chelicerae, pedipalps leglike; two pairs of eyes; first pair of legs sensory; junction of cephalothorax and abdomen not narrow; abdomen segmented, without spinnerets. *Galeodes,* Egyptian "tarantulas."

Order Pseudoscorpionida: scorpionlike; chelicerae spin silk from anterior glands; pedipalps chelate, large; abdomen segmented, rounded, without tail or sting. *Chelifer,* book "scorpions."

Order Ricinulei: spiderlike, small; without eyes; anterior abdominal segments from waist. *Cryptocellus*

Order Phalangida (harvestmen, "daddy longlegs"): body small, cephalothorax and abdomen fused, abdomen segmented; chelicerae short and chelate, pedipalps leglike; one pair of eyes; legs long and thin; without spinnerets or lungbooks, breathing by tracheae. *Phalangium*

Order Acarina (mites and ticks): free-living and ectoparasitic; body fused, unsegmented; breathing by tracheae; with larvae. *Sarcoptes,* itch mites; *Tyroglyphus, Acarus,* cheese mites; *Trombicula,* chigger mites; *Dermacentor,* dog ticks and spotted-fever ticks; *Boöphilus,* cattle-fever ticks; *Ornithodorus,* ear ticks and relapsing-fever ticks.

Arachnids include well over 30,000 species. Of these, spiders number about 20,000 species, mites and ticks about 6,000, harvestmen about 2,500, pseudoscorpions about 1,000, and the true scorpions roughly 700. The other orders are less abundant and also less familiar (Fig. 27.11). Scorpions are probably the primitive members of the class, as indicated by the presence of a complete series of segments and appendages characteristic of chelicerates generally (recall also the finds of

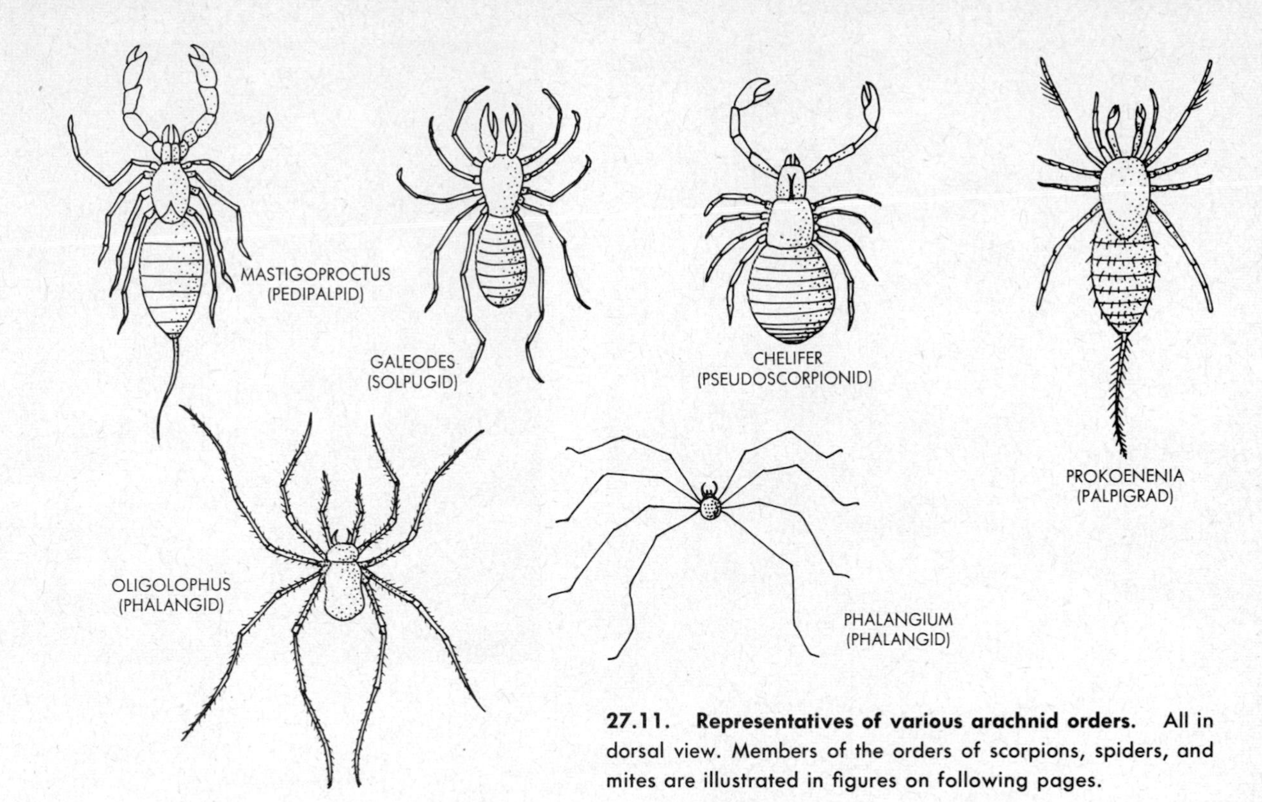

27.11. Representatives of various arachnid orders. All in dorsal view. Members of the orders of scorpions, spiders, and mites are illustrated in figures on following pages.

fossilized Paleozoic sea scorpions; Chap. 15). The other orders are listed above roughly in a sequence of increasing levels of structural specialization, characterized by a progressively greater fusion of segments and body divisions. Thus, spiders exemplify an intermediate level of structural specialization and mites and ticks are the most specialized arachnids.

Scorpions. A scorpion has a segmental structure similar to that of a horseshoe crab (Fig. 27.12, Table 11). The cephalothorax is formed from eight embryonic segments and is covered dorsally by a carapace. On the carapace are a pair of median ocelli and from two to five lateral ocelli on each side. Some scorpions are without eyes, however. Ventrally, the second segment bears short, three-jointed, pincered chelicerae; the third carries the large, six-jointed pedipalps, also chelate; and on the next four segments are the six-jointed walking legs, each terminating in a pair of claws. The pedipalps and the first two pairs of legs have gnathobases with a chewing function, as in *Limulus*. Scorpions feed mostly on insects and spiders, which are crushed by the gnathobases. The juices are then sucked into the mouth, located between the second and third head segments. The eighth segment, like the first, is embryonic only (and corresponds to the chilarial segment of *Limulus*).

The abdomen consists of 12 segments and a telson. On the first is a small operculum containing the gonopores, and the second bears a pair of *pectines*, fringed, comblike flaps. These structures are tactile; they brush over the ground when a scorpion walks and they represent important thigmoreceptors in these nocturnal animals. On each of the next four abdominal segments is a pair of ventrolateral *spiracles*, slits leading into chambers containing the *lung books*. Both the pectines and the lung books are undoubtedly homologous to the gill books of horseshoe crabs. The seventh abdominal segment is without appendages. This is true also for the remaining five segments, which are narrowed into a "tail." The telson contains a poison gland with a sharp, terminal sting. Scorpions carry their tail arched up and forward, and they use the poison sting to paralyze and kill prey that has been grasped with the pedipalps.

The mouth leads into a sucking pharynx which siphons in fluid food when its diameter is enlarged by muscular action

A

B

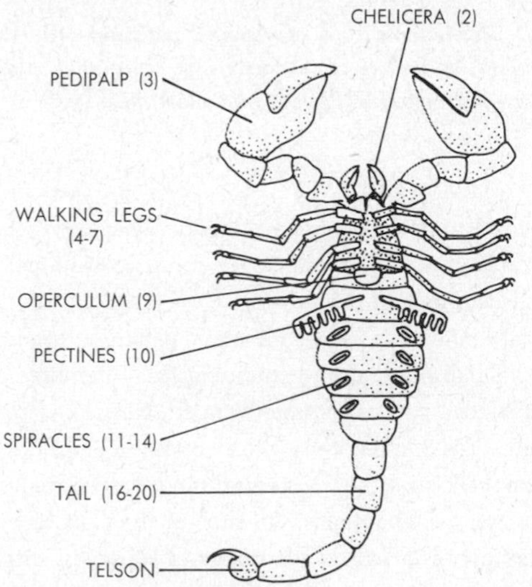

27.12. Scorpions, external features. Photos: A, *Androctonus,* from life; **B,** ventral view showing pectines. **Diagram:** ventral view of a scorpion showing segmental structure.

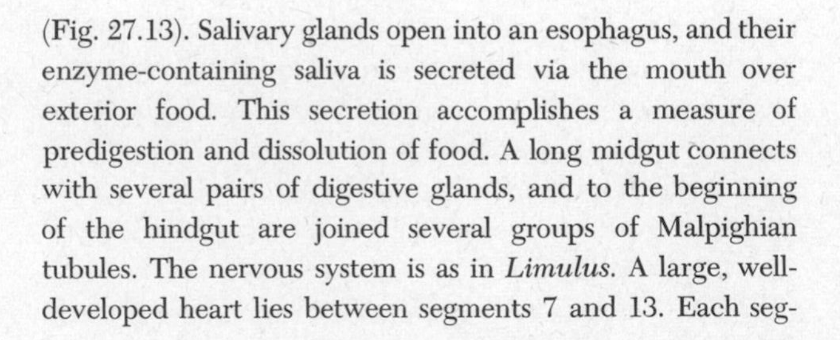

(Fig. 27.13). Salivary glands open into an esophagus, and their enzyme-containing saliva is secreted via the mouth over exterior food. This secretion accomplishes a measure of predigestion and dissolution of food. A long midgut connects with several pairs of digestive glands, and to the beginning of the hindgut are joined several groups of Malpighian tubules. The nervous system is as in *Limulus.* A large, well-developed heart lies between segments 7 and 13. Each segmental section of the heart bears a pair of lateral openings, or *ostia,* as well as a pair of ventrolateral arteries. Anterior and posterior arteries are present as well. The blood contains hemocyanin in plasma. Gonads are diffuse, with gonoducts particularly well developed in females. Fertilization is internal and occurs after elaborate courtship dances. Internal development of the zygotes follows. Some types have highly yolked eggs and such scorpions are ovoviviparous. Others,

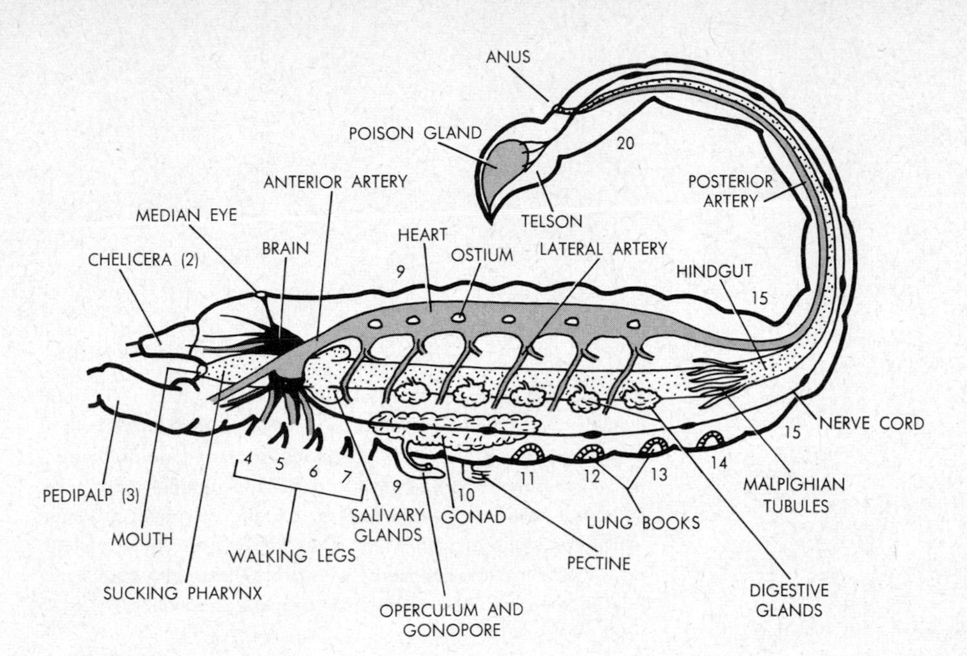

27.13. The interior structure of a scorpion. Segments numbered according to embryonic count. (*Adapted from Leuckart.*)

like *Scorpio*, have small, isolecithal eggs; these develop viviparously, placental connections being formed in pouches of the uterus.

Spiders. Although a spider is quite dissimilar in appearance to a scorpion, the basic segmental structure of the two is nevertheless almost identical, at least in the embryos. In a spider embryo are formed 18 anteroposterior coelomic sacs, each corresponding to an embryonic segment. The first eight give rise to the adult cephalothorax, the last 10 to the abdomen (cf. Table 11). In the cephalothorax, the first embryonic segment remains without appendages, the second and third form chelicerae and pedipalps, respectively, the next four develop legs, and the eighth is again without appendages. In the abdomen, the first segment disappears, the second and third each develop a pair of lung book chambers (or tracheal bundles, cf. below), and the fourth and fifth form the silk-spinning *spinnerets.* All remaining embryonic abdominal segments disappear in later development. In the adult (Fig. 27.14), the segments of both the cephalothorax and the abdomen are fused, and these two body parts are joined by a narrow *peduncle,* the characteristic "waist" of a spider. Up to eight eyes are borne on the cephalothorax; cave-dwelling spiders are blind.

Chelicerae are only two-jointed; a coxopodite contains a poison-secreting gland and the rest of the appendage forms a fang, with a poison duct within it opening at the tip. In tarantulas the two fangs point downward; in all other spiders they point toward each other. The six-jointed pedipalps are without pincers. They are tactile over most of their length, and in males they contain a specialized receptacle for the transfer of sperms during copulation. Above all, the gnathobases on the coxopodites of the pedipalps are the principal chewing devices of spiders. Food consists chiefly of insects, though *Mygale* attacks animals as large as birds, and tarantulas readily subdue mice or small lizards. Prey is killed with the poison fangs, is chewed and torn with the gnathobases, and is predigested for an hour or more by proteinases secreted from salivary glands opening into the mouth. The food juices are then sucked up, suction being produced by the stomach.

Walking legs often end in claws. The lungbooks are in chambers as in scorpions, and the same embryonic segments which form these breathing organs also give rise to tracheae in some spiders. Such tracheae are bundles of unbranched tubules emanating from the walls of the lung chambers (Fig. 27.15). Certain groups of spiders possess two pairs of lungbooks only. Some possess two pairs of tracheal bundles only. And in the majority there is an anterior pair of lung books and

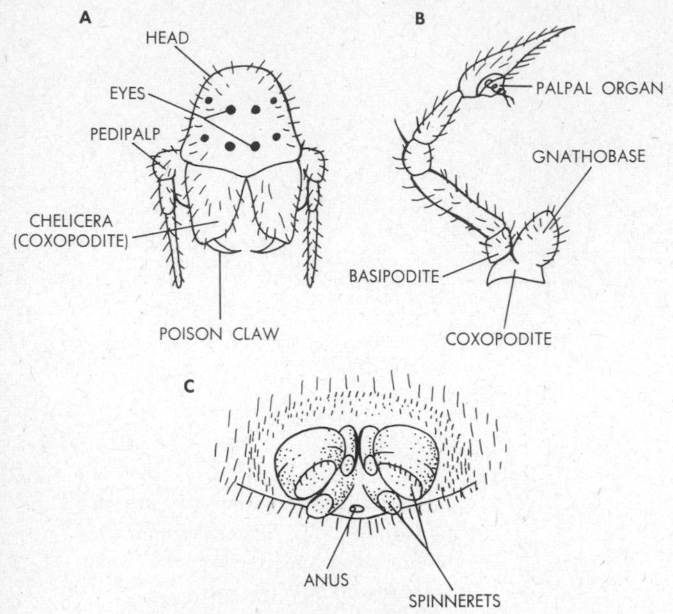

27.14. External features of spiders. Diagrams: A. Head from the front, showing chelicerae with fangs. **B.** The pedipalp of a male house spider, showing gnathobase and palpal organ, in which sperms are transferred to the female gonopore. (*Adapted from Shipley and McBride.*) **C.** Tip of spider abdomen with three pairs of spinnerets. **Photos: A.** Front view of head of wolf spider. Note chelicerae with fangs, large and small eyes. **B.** Abdomen of a house spider, in process of spinning silk from spinnerets.

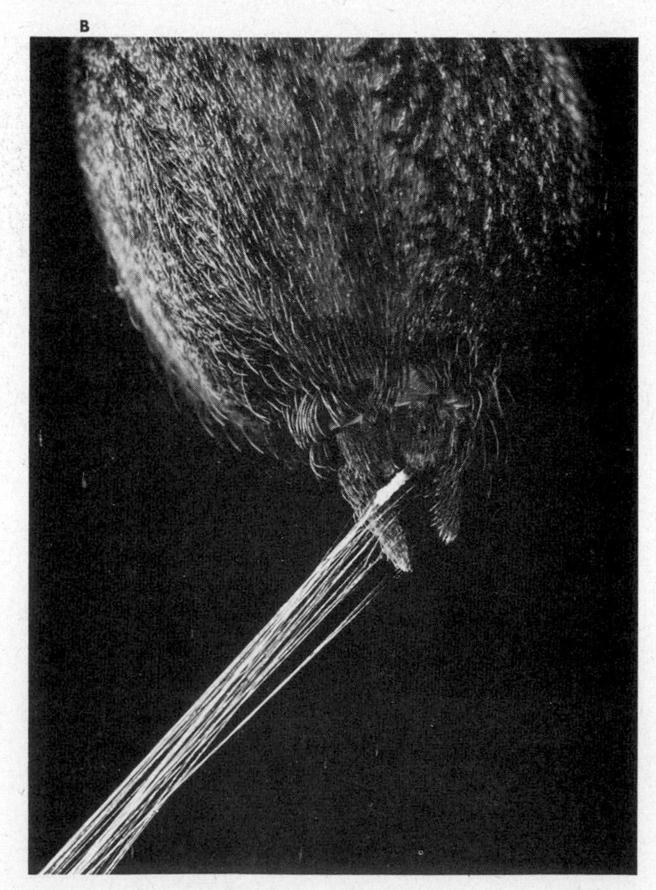

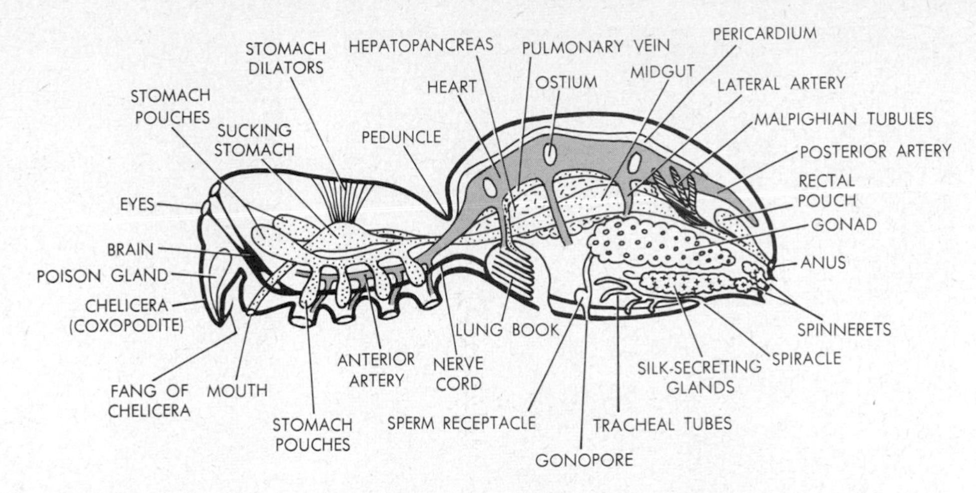

27.15. The interior structure of a spider. The peduncle, or "waist", connects the cephalothorax with the abdomen. (*Adapted from Warburton.*)

a posterior pair of tracheal bundles, the latter either fully developed or rudimentary. Behind the spiracle openings of the breathing chambers are the gonopores. Behind them are two or more pairs of *spinnerets*, modified abdominal appendages which provide outlets for the *silk glands* in the abdomen.

Spider silk is a scleroprotein which hardens on contact with air. All spiders produce silk, but not all construct webs. Those that do form them in different patterns, e.g., sheet webs, funnel webs, netted webs, and the beautiful orb webs (Fig. 27.16). Some species construct only horizontal webs, others only vertical ones, still others form their webs only in corners, tree forks, or in other specific locations. In addition to fashioning webs, spiders use silk in numerous other ways. For example, the animals spin egg cocoons, sperm webs, molting sheets, attachment disks, burrow linings, trapdoors with hinges, binding threads for prey, drag lines, and also gossamer for "ballooning," i.e., riding air currents at the end of a free-floating thread.

Internally, the midgut pouches into a pair of sacs which pass forward and send branches into the legs (cf. Fig. 27.15).

27.16. Spider webs. **A,** an orb web; **B,** hover fly caught in the orb web of a garden spider.

A

B

A

B

C

27.17. Three spider types. **A,** aquatic diving spider, with web and spun air bell; **B,** trapdoor spider in sectioned burrow; **C,** tarantula; note vertical fangs.

The midgut also connects via several ducts with a large, highly lobulated digestive gland located dorsally on each side of the heart. Malpighian tubules open into the hindgut. The heart has three pairs of ostia and it gives off an anterior and posterior artery as well as three pairs of lateral vessels. In the nervous system, all ventral ganglia are concentrated anteriorly.

Fertilization occurs internally. The male sheds sperms either on a leaf or on a specially spun sperm web, then dips a pedipalp into the sperms and so fills the receptacle on that appendage. The tip of the pedipalp is then inserted into the gonopore of the female. Highly elaborate courtship dances usually precede copulation. In some species the females kill and eat the males after sperm transfer. Sperms are stored within the females temporarily, and batches of fertilized eggs are usually laid into spun cocoons. Development is direct.

Among the spiders which do not trap prey by means of webs, the wolf spiders hunt and run after prey, and the jumping spiders ambush prey and pounce on it. Trap-door spiders live in burrows covered with hinged lids spun from silk. The animals make brief excursions to the outside for food. The largest spiders are the Amazonian tarantulas, which have a body 3 in. long and a leg span of 10 in. The most dangerous spiders are the black widows, whose poison can be fatal to man (Fig. 27.17).

Mites and ticks. The acarines are mostly minute animals in in which all body divisions are fused together (Fig. 27.18). Free-living types are terrestrial and freshwater predators and scavengers; parasitic types live both on plants and animals and many are blood suckers. Mites differ from ticks in that, as a group, they are the smaller animals. The two groups represent different suborders within the acarine order. In the free-living types, the chelicerae typically are chelate and the pedipalps are tactile and leglike. In parasitic forms, the chelicerae are usually modified into cutting devices and also into elon-

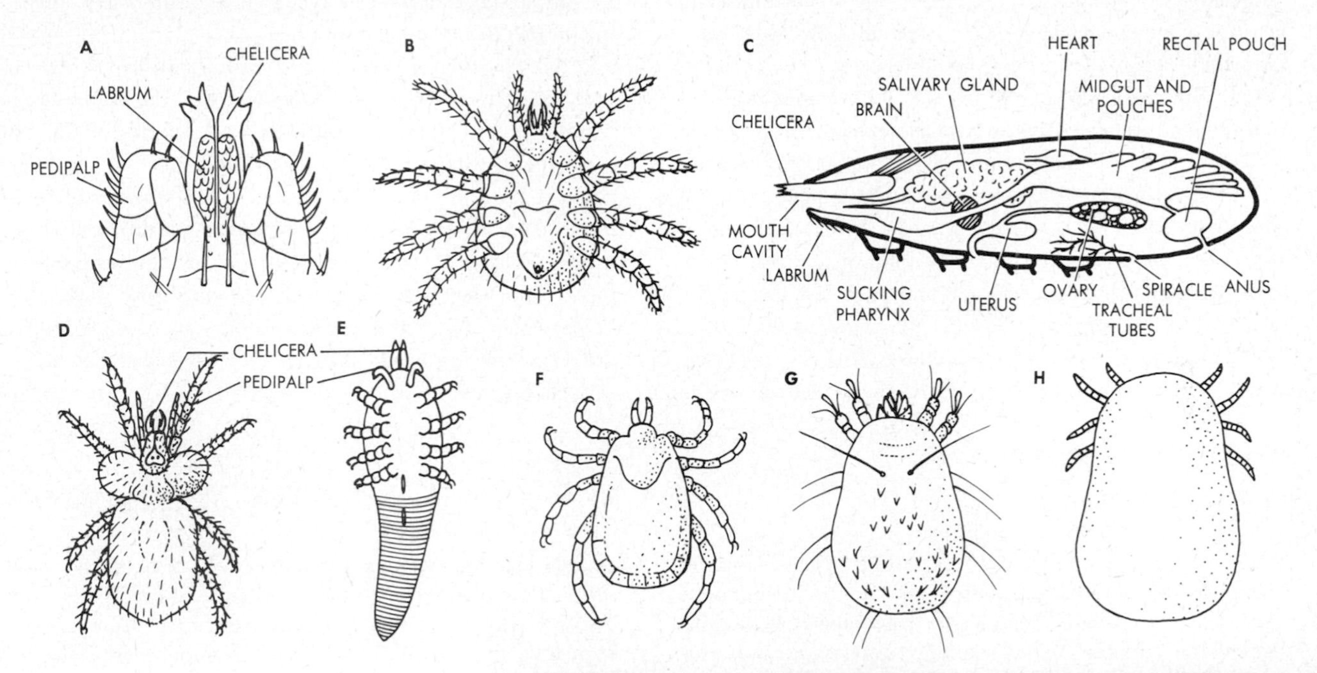

27.18. Mites and ticks. Diagrams: A, ventral view of acarine head region, showing cutting chelicerae and hooked four-jointed pedipalps; **B,** generalized ventral view of acarine body, showing relatively wide spacing of leg bases; **C,** sagittal section through *Dermacentor;* **D,** the chigger mite *Trombicula,* dorsal view; **E,** the follicle mite *Demodex,* ventral view; **F,** *Dermacentor,* dorsal view; **G,** the itch mite *Sarcoptes,* dorsal view; **H,** the Texas cattle tick *Boöphilus,* filled with blood. **Photo:** wood tick in fur of a deer.

gated, fused, sucking channels. Pedipalps in such cases may again be sensory or may form hooklike attaching organs. The four pairs of legs are spaced apart rather widely, a general acarine trait. A sucking pharynx is connected with salivary glands which, in the blood-sucking forms, secrete an anticoagulant as in leeches. Some acarines digest blood intracellularly, in amoeboid lining cells of the stomach. Most mites are without hearts, but ticks generally possess a simple dorsal vessel. Breathing occurs through well-developed tracheal tubes, which open on the abdominal segments in primitive acarines and at the bases of any of the appendages in more advanced forms. Some of the blood-sucking parasites complete their entire life cycle on one host, relinquishing their attachment only temporarily at the time of molting. Other forms fall off the host at each molt and must seek a new host thereafter. A succession of hosts is also required during the larval development of some of the ticks. In these, six-legged larvae feed on one host, drop off, molt and become eight-legged larvae, feed on another host, drop off again and molt, and eventually become adults which must find new hosts once more. Each larval stage may survive for many months on a single blood meal, and adults often may live for years from one feeding.

Acarines are notorious carriers of disease organisms affecting vertebrates, mammals and man particularly. Thus, mites and ticks are the exclusive carriers of the viruslike *rickettsial* parasites, responsible for many diseases; e.g., Rocky Mountain spotted fever of man (transmitted by a species of *Dermacentor*), and many other rickettsial fevers and "poxes." The acarines also transmit diseases caused by other kinds of organisms, e.g., a species of *Boöphilus* carries a sporozoan which causes Texas cattle fever. In addition, the acarines produce diseases on their own. For example, numerous gall mites damage plants, and the mite *Sarcoptes* and other related types are responsible for mange disease in domestic birds and mammals and for itch in man.

REVIEW QUESTIONS

1. Characterize the phylum Arthropoda taxonomically and name the subphyla and classes. How are these subgroups presumably interrelated historically? How have the chitinous exoskeleton and the metameric body design probably promoted the broad diversification of the phylum?
2. Describe the basic exterior structure of the arthropod body. What are the skeletal parts of a segment and the endites of a segmental appendage? Describe the basic arthropod organization of each of the 10 animal organ systems.
3. Review the structural characteristics of trilobites and show how these animals developed segmentally. Define the chelicerate subphylum taxonomically and similarly define each of the classes.
4. Review the segmental structure of the chelicerate cephalothorax and the structure and function of the appendages of that body division. Describe the structure of a horseshoe crab. How does this arthropod feed, move, breathe, and mate? Where does it live? What is the organization of its excretory system? How does the animal develop?
5. Review the structure of a compound eye and contrast it with that of an arthropod ocellus. Which chelicerate groups possess either, neither, or both types of eyes? Describe the segmental structure of (*a*) a eurypterid, (*b*) a sea spider, and show how their appendages are specifically different. What are ovigerous legs?
6. Name several of the arachnid orders and describe the general characteristics of the animals they encompass. Describe the segmental and internal structure of a scorpion. How does a scorpion obtain, ingest, and digest food? Compare the segmental structure of the abdomen of all chelicerate classes. What are chilaria? Pectines?
7. What are chelate legs? How many embryonic segments form the body of a spider? Describe the structure of the chelicerae of spiders. What is the meaning of the following sentence: "In Araneida, the gnathobases on the coxopodites of the pedipalps are masticatory."
8. How does a spider breathe? Where are the breathing and the excretory organs located? Describe the segmental and the internal structure of a spider. Describe and contrast the circulatory systems in all chelicerate classes. Do analogously for the nervous systems.
9. What is spider silk chemically and where is it produced? Describe the structure and function of various types of webs. What other uses do spiders make of spun silk? Name various types of spiders and describe their general characteristics. How does a spider mate and develop?
10. Distinguish between mites and ticks and describe the characteristics of these arthropods. What is their external structure and what are their feeding methods. Name some of the diseases for which acarines are responsible directly or indirectly. When does molting take place in the life histories of the chelicerate arthropods?

COLLATERAL READINGS

Among many readings on chelicerate arthropods, the following selections provide further information on most subjects dealt with in this chapter:

Baker, E. W., and G. W. Wharton: "An Introduction to Acarology," Macmillan, New York, 1952.

Comstock, J. H., and W. J. Gertsch: "The Spider Book," 2d ed., Doubleday, New York, 1940.

Gertsch, W. J.: "American Spiders," Van Nostrand, New York, 1949.

Kaston, B. J., and E. Kaston: "How to Know the Spiders," William C. Brown, Dubuque, Iowa, 1953.

Savory, T. H.: "The Arachnida," Edward Arnold, Ltd., London, 1935.

———: Daddy Longlegs, *Sci. American,* Oct., 1962.

Snodgrass, R. E.: "A Textbook of Arthropod Anatomy," Cornell, Ithaca, N.Y., 1952.

———: Arthropoda, in "McGraw-Hill Encyclopedia of Science and Technology," vol. 1, McGraw-Hill, New York, 1960.

Waterman, T. H.: A Light Polarization Analyzer in the Compound Eye of *Limulus, Science,* vol. 111, 1950.

CHAPTER 28

arthropods: mandibulates

Subphylum MANDIBULATA: body comprising cephalothorax and abdomen; or head, thorax, and abdomen; or head and trunk; cephalothorax or head unsegmented externally; thorax, abdomen, or trunk segmented; head segments typically with one or two pairs of *antennae,* one pair of *mandibles,* two pairs of *maxillae;* thoracic appendages two or more pairs; with or without abdominal appendages. *Crustacea, Myriapoda, Insecta*

CLASS CRUSTACEA

Mainly aquatic, both marine and freshwater, some terrestrial; free-living and parasitic; head and thorax or cephalothorax, often with carapace; thorax and abdomen usually segmented; with telson; two pairs of antennae; compound eyes typically present; appendages largely biramous; breathing via body surface or gills; excretion through antennal or maxillary glands; early development usually in or on female; typically with *nauplius* or other free larval stages.

Subclass Cephalocarida: marine; 20 trunk segments; appendages jointed, not present on abdomen. *Hutchinsoniella*

Subclass Branchiopoda: largely freshwater; mostly with carapace, often bivalved; trunk legs phyllopodial in all but last order.

Order Anostraca: without carapace. *Artemia,* brine shrimps; *Branchipus,* fairy shrimps.

Order Notostraca: carapace flat, platelike. *Apus, Lepidurus*

Order Conchostraca: carapace bivalved, enclosing whole body. *Lynceus, Estheria*

Order Cladocera: carapace tentlike, enclosing trunk only; appendages jointed. *Daphnia,* water fleas; *Leptodora*

Subclass Ostracoda: marine and freshwater; carapace bivalved, with adductor; two pairs of trunk appendages. *Cypridina, Cypris*

Subclass Mystacocarida: marine; five pairs of thoracic appendages, last four reduced; six abdominal segments, without appendages; compound eyes absent. *Derocheilocarus*

Subclass Copepoda: marine and freshwater; many parasitic; five or six thoracic segments with appendages, abdomen four segments without appendages; with ocellus, compound eyes absent. *Calanus, Cyclops*

Subclass Branchiura: marine and freshwater; body flat, discoid; second maxillae are suckers; ectoparasitic on fish. *Argulus*

Subclass Cirripedia (barnacles): marine; adults sessile, some parasitic; carapace forms mantle enclosing body, often calcareous; antennae and compound eyes absent in adult; abdomen usually vestigial; hermaphroditic; free-swimming larvae. *Balanus,* acorn barnacles; *Lepas,* goose barnacles; *Sacculina.*

Subclass Malacostraca: marine and freshwater; typically eight thoracic segments, six abdominal segments, all with appendages; usually with carapace; female gonopore on sixth, male gonopore on eighth thoracic segment; compound eyes mostly stalked.

Infrasubclass Leptostraca: seven abdominal segments, last without appendages; bivalve carapace with adductor, not fused to thorax, marine. *Nebalia*

Infrasubclass Eumalacostraca: six abdominal segments; carapace if present not bivalved, fused variously to thorax.

Superorder Syncarida: without carapace.

Order Anaspidacea: freshwater; primitive "living fossil" structure. *Anaspides*

Order Bathynellacea: in subterranean wells; eyeless. *Bathynella*

Superorder Pancarida: brood pouch dorsal, formed by carapace.

Order Thermosbaenacea: in hot springs and subterranean lakes; eyeless. *Thermosbaena, Monodella*

Superorder Peracarida: carapace fused with up to four thoracic segments or absent; first thoracic segment fused with head; brood pouch on thoracic appendages.

Order Spelaeogriphacea: one thoracic segment fused with carapace; in caves. *Spelaeogriphus*

Order Tanaidacea: one or two thoracic segments fused with carapace; marine, in mud. *Tanais, Apseudes*

Order Mysidacea (opossum shrimps): one to four thoracic segments fused with carapace. *Mysis*

Order Cumacea: three to four thoracic segments fused with carapace; marine, burrowing. *Diastylis*

Order Isopoda (fish lice, wood lice, slaters, sow bugs, pill bugs): marine, on shore, freshwater, and terrestrial, some parasitic; without carapace; eyes sessile; breathing via abdominal appendages. *Ligia, Asellus,* slaters; *Porcellio,* sow bugs; *Armadillium,* wood lice and pill bugs.

Order Amphipoda: marine, on shore; without carapace; eyes sessile; epipodial gills on thoracic appendages. *Orchestia,* beach fleas; *Gammarus,* sand hoppers.

Superorder Hoplocarida: carapace fused with three thoracic segments

Order Stomatopoda (mantis shrimps): first five thoracic appendages grasping; first five abdominal appendages with gills; marine, on bottom. *Squilla*

Superorder Eucarida: carapace fused with whole thorax; compound eyes stalked; without brood pouch.

Order Euphausiacea (krill): pelagic; first three thoracic appendages not maxillipeds; gills exposed; without statocysts. *Euphausia, Nyctiphanes*

Order Decapoda: marine and freshwater; first three thoracic appendages are maxillipeds; five pairs of legs, first pair chelate, not locomotor; with pleural gills in carapace chamber; statocyst in first antennae.

Macrura: abdomen long. *Crago,* shrimps; *Penaeus, Leander,* prawns; *Homarus,* American lobsters; *Nephrops,* Norway lobsters; *Palinurus,* spiny lobsters; *Cambarus Astacus,* crayfishes

Brachyura: abdomen short. *Cancer,* rock crabs; *Callinectes,* blue crabs; *Uca,* fiddler crabs; *Pagurus,* hermit crabs; *Libinia, Maia,* spider crabs; *Machrocheira,* giant crabs

The roughly 30,000 described species of crustacea, animals with "crust"-like shells, are among the most ubiquitous living organisms. They range from all depths of the ocean to freshwater lakes over 10,000 feet high, from 50°C waters of hot springs to glacial waters at the freezing point, from briny salt lakes to subterranean cave waters (where the animals are blind); and the sow bugs, pill bugs, and wood lice of the order Isopoda live on land. Most members of the class are free-living, but commensalistic and parasitic types are quite common. The free-living crustacea generally are filter-feeders on plankton; paddling locomotor movements by the legs create currents from which the bristles of the legs strain food particles. These are then passed forward toward the mouth, in a stream along the ventral midline between the rows of legs on each side. Other crustacaea are scavengers, and among the advanced forms are many carnivorous predators. However, the majority of crustacea are themselves part of the plankton. Indeed, copepods and krill occur so abundantly that they form the animal base of most aquatic food pyramids. Though most crustacea thus are quite small, the group also includes the largest arthropods (*Macrocheira*) as well as the heaviest—the lobster *Homarus* can attain weights of well over 40 lb.

The primitive crustacean ancestor probably was one of three main offshoots of an early mandibulate stock; the other two offshoots culminated in the myriapods and the insects, respectively (cf. Fig. 27.4). The crustacean ancestor may be presumed to have possessed a head with fused segments and numerous similar trunk segments. All of the latter bore similar biramous appendages which served simultaneously in breathing, locomotion, and filter feeding through locomotion. From such a beginning an adaptive radiation of three major lines later evolved (Fig. 28.1). In one of these, represented by the very recently discovered *Cephalocarida*, the ancestral structure has largely persisted; the only principal change has been the loss of abdominal appendages (Fig. 28.2).

The second major line is represented by the subclasses *Branchiopoda, Ostracoda, Mystacocarida, Copepoda, Branchiura*, and *Cirripedia*. Of these, the branchiopods include the most primitive groups (fairy shrimps, order Anostraca), which still retain many of the ancestral features, particularly the presence of numerous like segments with like appendages serving in locomotion, feeding, and breathing. The other branchiopod orders and the other subclasses have become specialized in various different directions, though in most cases the changes have followed a fairly general common pattern: (1) shortening of the trunk and reduction of the number of segments and of trunk appendages, paralleled by an enlargement of the head appendages, which then assume the main feeding and locomotor functions; (2) development of a carapace which covers progressively more of the body; and (3) adoption of parasitic or sessile ways of life (cf. Fig. 28.2).

The third major crustacean line is represented by the subclass *Malacostraca.* This line too has specialized in three general directions: (1) retention of all trunk segments and appendages, including abdominal ones, but increasing diversification of these appendages in different body regions; (2) as above, development of a progressively larger carapace which eventually fuses the entire thorax to the head; and (3) substitution of filter feeding by predatory and scavenging means of nutrition in the advanced groups. The primitive members of this line are represented by the Leptostraca and the Syncarida, the most advanced ones by the Eucarida, particularly the order Decapoda, which includes the shrimps, lobsters, and crabs (cf. Figs. 28.1 and 28.2).

The segmental structure of a decapod is outlined in Table 12 and Fig. 28.3; note that the head segments are typically the same in all crustacea. A median ocellus is already present in the nauplius larvae (cf. Fig. 28.7) and persists into the adult in most cases (however, it is vestigial in adult Malacostraca). Lateral compound eyes develop in later larval stages; such

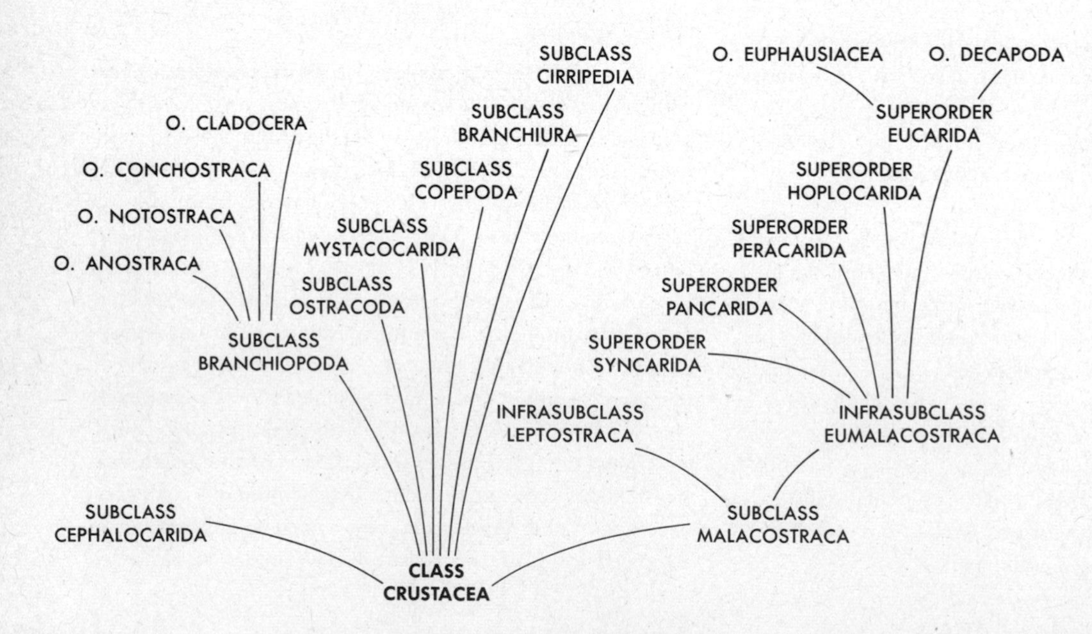

28.1. The main taxonomic groups among crustacea and their probable evolutionary interrelations.

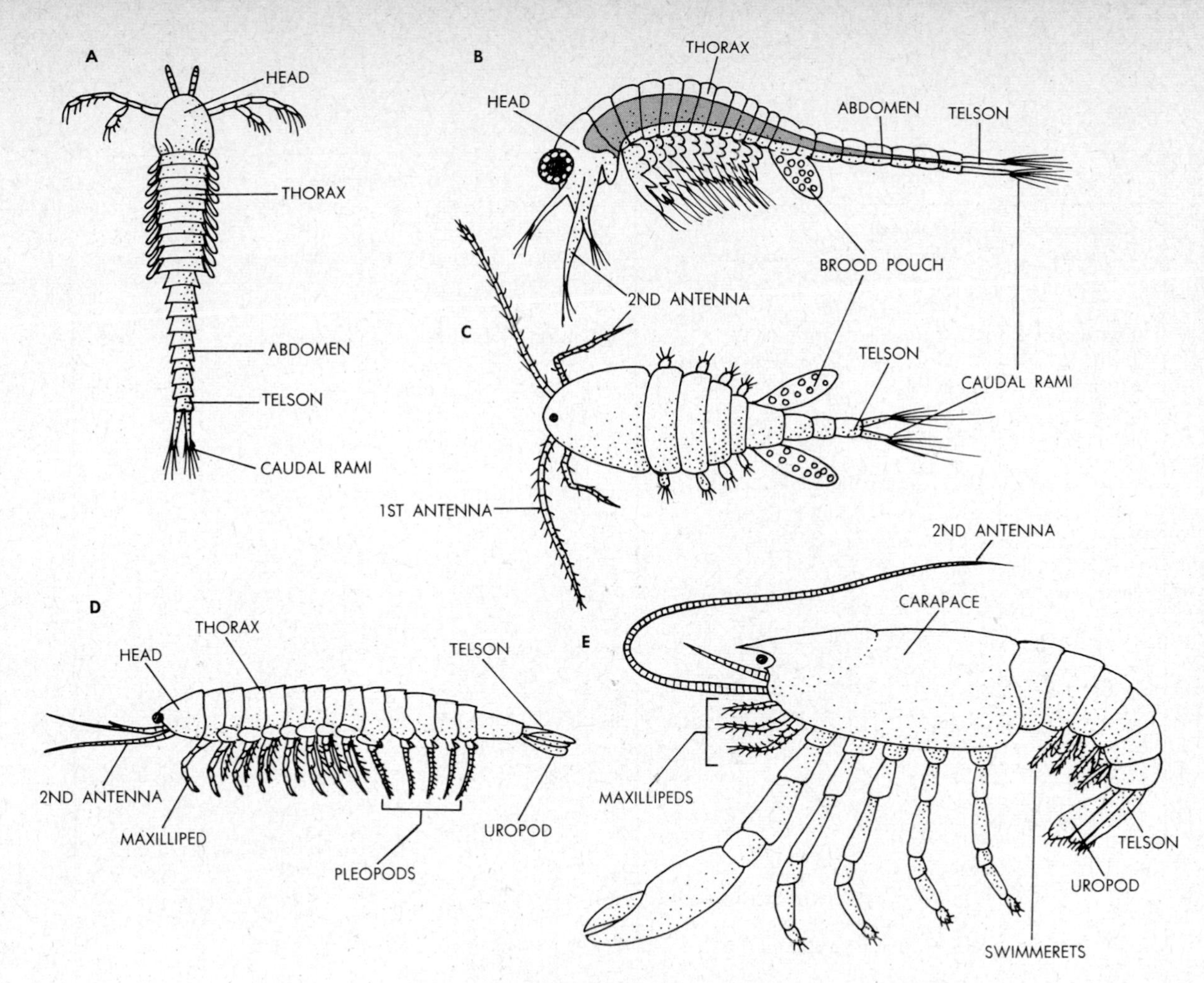

28.2. Evolutionary trends in Crustacea. **A.** The cephalocarid line of evolution, as in *Hutchinsoniella,* showing primitive condition of large numbers of similar segments. (*Adapted from Sanders.*) **B** and **C.** A second line, exemplified by the branchiopod brine shrimp *Artemia* (**A**) and the copepod *Cyclops* (**B,** *after Hartog*). **A** displays the primitive condition of numerous unfused segments, **B,** the advanced condition in which cephalothorax incorporates some of the thoracic segments and in which the numbers of paired appendages are reduced. **D** and **E.** The malacostracan line of evolution, exemplified by the syncarid *Anaspides* (**D,** after Snodgrass) and the eucarid crayfish *Astacus* (**E**). **D** is primitive, with comparatively little appendageal diversification (pleopod is a general term for an abdominal appendage). **E** is advanced, with the same number of appendages as **D** but with considerable appendageal diversification in structure and function (cf. Fig. 28.3).

eyes may be sessile, i.e., set directly into the head, or stalked. Also present in the nauplius stage are the first four head segments with their three pairs of appendages; the maxillae develop as part of the head later. The first antennae are basically uniramous (though in the lobster they secondarily acquire a second branch later). All other appendages of crustacea typically are biramous, at least in the larva and largely also in the adult. In the mandibles, or jaws, the chewing parts are the gnathobases of the coxopodites. The mouth lies above the mandibles; it is often guarded by an upper lip, or *labrum,* and an underlip, or *metastoma.* These two flaplike structures are accessory extensions of the body wall, not segmental append-

TABLE 12. Segments and appendages in mandibulate arthropods

	Crustacea			Myriapoda				Hexapoda
	Cephalocarida	Brine shrimp (*Artemia*)	Decapods	Centipede (*Lithobius*)	Symphyla	Millipede	Pauropod	Insects
1	—	—	—	—	—	—	—	—
2	antennae	antennae	antennae	antennae	antennae	antennae	antennae	antennae
3	antennae	antennae	antennae	—	—	—	—	—
4	mandibles	mandibles	mandibles	mandibles	mandibles	mandibles	mandibles	mandibles
5	maxillae	maxillae	maxillae	maxillae	maxillae	—	—	maxillae
6	maxillae	maxillae	maxillae	labium (maxillae)	labium (maxillae)	maxillae	maxillae	labium (maxillae)
7	legs	legs	maxillipeds	prehensors	legs	—	—	legs
8	legs	legs	maxillipeds	legs	legs	legs (1 pr)	legs	legs, wings, (spiracles)
9	legs	legs	maxillipeds	legs	legs	legs (1 pr), (gonopores)	legs	legs, wings, (spiracles)
10	legs	legs	chelate legs	legs	legs	legs (1 pr)	legs	(spiracles)
11	legs	legs	legs	legs	legs	↑	legs	(spiracles)
12	legs	legs	legs (♀ gonopore)	legs	legs		legs	(spiracles)
13	legs	legs	legs	legs	legs			
14	legs	legs	legs (♂ gonopore)	legs	legs		legs	(spiracles)
						variable number of double segments, 2 pr of legs each	legs	(spiracles)
15	legs	legs	—	legs	legs		legs	(spiracles)
16	legs	legs	swimmerets	legs	legs		legs	(spiracles)
17	↑	legs	swimmerets	legs	legs		—	(spiracles), ♀ ovipositor, ♂ copulatory organ
18		brood pouch	swimmerets	legs	legs, spinnerets	↓	—	
19		penis, (♀ ♂ gonopores)	swimmerets	legs		telson		(anus)
20	without abdominal appendages	↑	uropods	legs				(anal cerci)
21			telson					
22		without abdominal appendages		legs				
23				legs (gonopores)				
24								
25	↓	↓		telson				
	telson	telson						

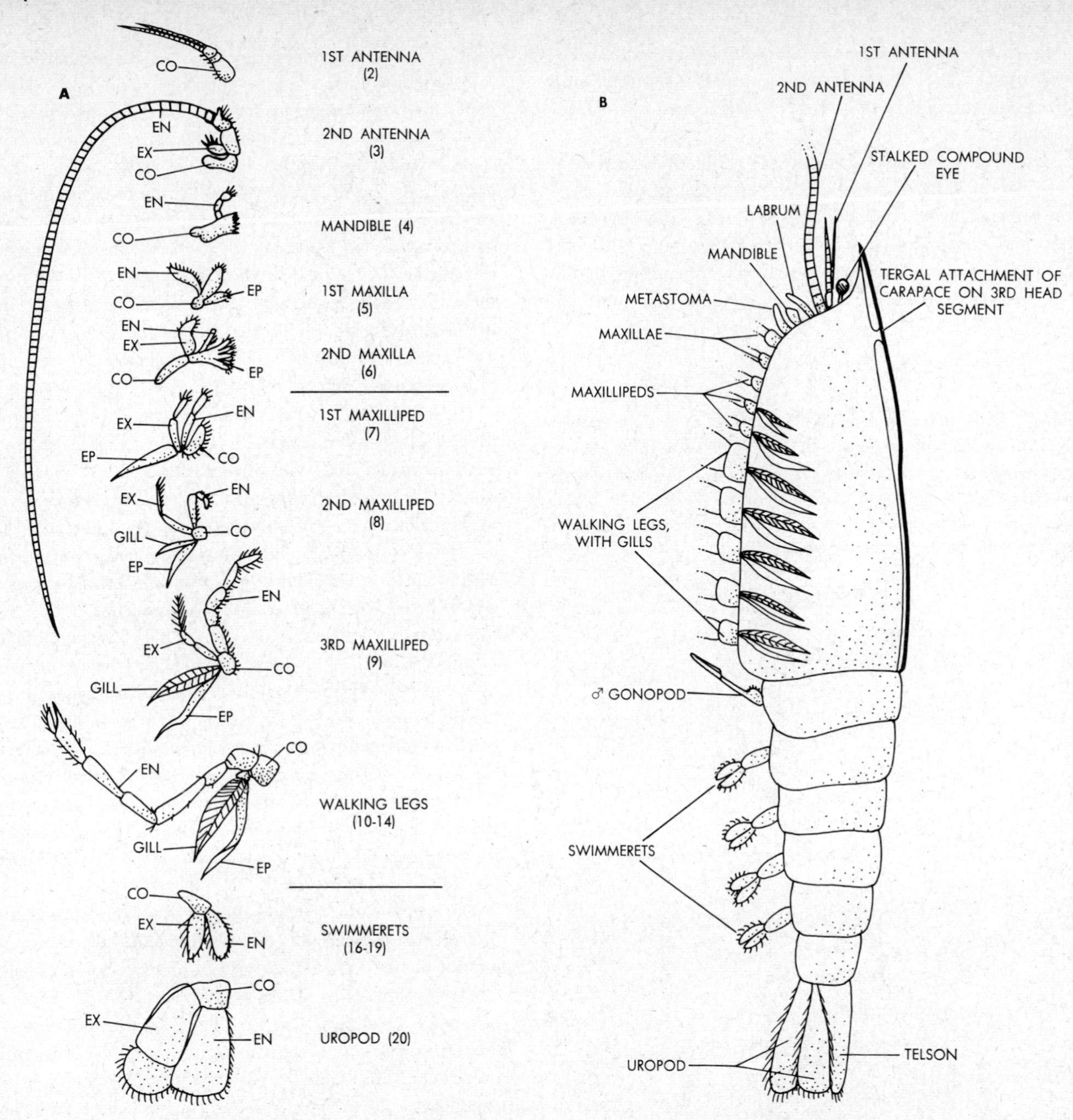

28.3. The segmental structure of the (decapod) lobster *Homarus*. A. Segmental appendages of left side. co, coxopodite; *ex*, exopodite; *en*, endopodite; *ep*, epipodite. Segmental numbers counted from anterior end and indicated in parentheses. Note that some appendages are without exopodites. Gills are part of the epipodites. The first abdominal segment (no. 15) bears a copulatory appendage in males (gonopod). **B.** Diagram of position of appendages in whole animal. The labrum and metastoma, which guard the mouth, and the eyes are extensions of the body wall, not metameric appendages.

ages. In cases where a carapace develops over part or all of the trunk, such a shield always represents a posterior (and often also anterior) extension of the tergum of the third head segment.

The thorax and abdomen include a different number of segments in different crustacea. In primitive groups, where the appendages of the trunk tend to be alike and serve feeding and locomotor functions, breathing occurs directly through the unmodified body wall of the appendages (and to some extent through the body wall generally). In more advanced forms, the appendages are variously differentiated into food-handling, sensory, grasping, swimming, walking, copulatory, sperm-transferring, egg-carrying, and other organs. Breathing is then accomplished by feathery gills, developed from the epipodites of some or most of the thoracic appendages (Fig. 28.4). In such animals, moreover, the dorsal carapace is often extended down on each side into an overhang over the leg bases, leaving a lateral chamber in which the gills lie. The epipodial gill of a segment then may be attached not to the coxopodite of the appendage itself but may be fused to the pleuron on that side of the segment. *Pleural gills* of this type occur in lobsters, for example, and are typical of decapods generally.

28.4. Gills in crustaceans (diagrammatic). A, epipodial gills, as in majority of groups; **B,** pleural gills and blood circulation, as in decapods (such as lobsters). In **B,** the sinuses refer to hemocoelic blood spaces. Arrows indicate direction of blood flow. Blood in the ventral sinus and in parts of the gills is venous, blood in the other spaces shown is arterial.

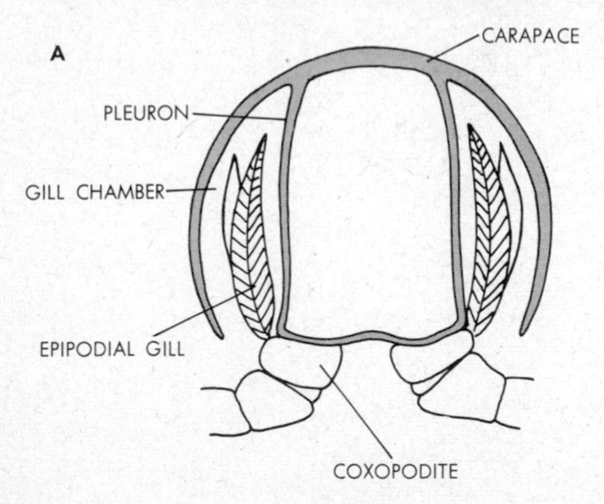

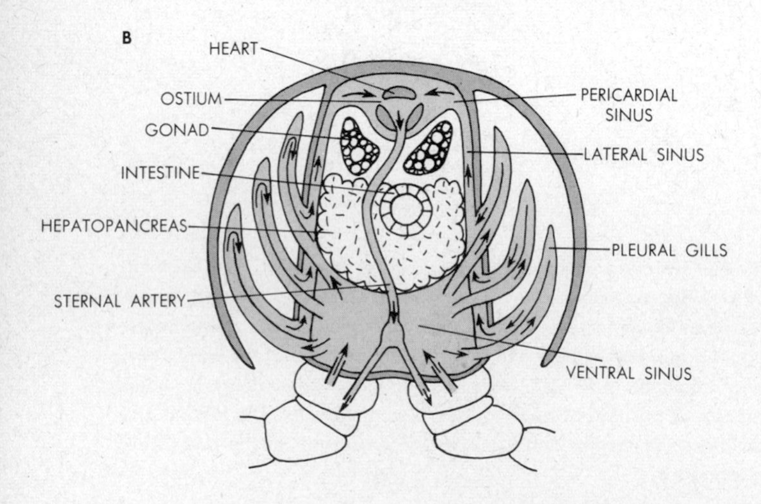

The last thoracic segment bears the gonopore in males and in most cases also in females; the female gonopore of Cladocera, Cirripedia, and Malacostraca is located on a more anterior segment. In the first or first two abdominal segments of most crustacea, the appendages are modified into structures associated with reproduction. Thus, in a male, these appendages often form copulatory or sperm-transferring organs; in a female, they are fashioned into brood pouches or into structures to which batches of developing eggs remain attached. Lobsters possess egg-holding appendages, the *swimmerets,* on the second to fifth abdominal segments. In other crustacea, the posterior abdominal appendages, if present at all, serve chiefly in locomotor or breathing functions. The last abdominal appendages in many cases are flat, platelike *uropods,* which together with the telson may form a tailfan as in lobsters. In all crustacea except the Isopoda and Eumalacostraca, the telson typically bears a pair of *caudal rami,* tapered posterior extensions (cf. Fig. 28.3).

The exoskeleton of the larger crustacea is impregnated heavily with calcium salts and is extended in places into the interior of the body in the form of ingrown folds, so-called *apodemes* (Fig. 28.5). These provide surfaces for muscle attachments. In decapods, the apodemes on the ventral side of the thorax are united internally into a continuous supporting framework, the *endophragmal skeleton.* Apodemes in the appendages also provide breakage planes where limbs may be readily snapped off by voluntary muscular action of the animal. Decapods are particularly noted for this capacity of self-amputation, or *autotomy;* the process usually takes place when a limb becomes injured or trapped. Little blood loss is sustained in such autotomies, for the apodeme at the breakage plane already covers the limb stump almost entirely; clotted blood soon seals it completely. Lost limbs then regenerate readily, though not usually back to the original size.

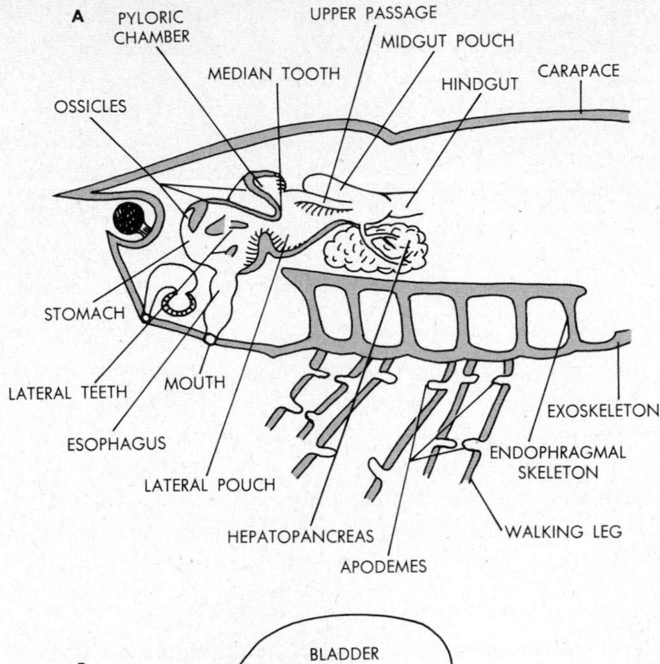

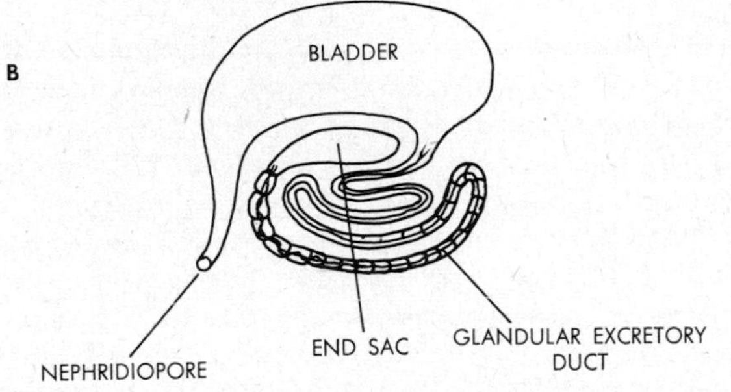

28.5. Decapods, internal structure. A. Sectional side view of cephalothorax, showing endophragmal skeleton, apodemes in appendages, and some structural details of the gastric mill. The ossicles (shaded areas) are calcifications in the stomach wall, and they and the stomach teeth constitute the gastric mill (i.e., they serve in macerating food into fine particles). A lateral pouch on each side of the pyloric chamber then functions as a fine-filter; contractions of each pouch press liquefied food into the upper passage and from there into the very short midgut, which communicates with the hepatopancreas via a pair of ducts. Food residues pass through a valve from the midgut into the hindgut. **B.** Section through an antennal (green) gland of a crayfish (cf. also Fig. 9.36).

Internally, crustacea exhibit typical arthropod structure (Figs. 28.5 and 28.6). In the Malacostraca the stomach has two compartments. An anterior *gastric mill* equipped with chitinous teeth and other hard outgrowths grinds coarse food. And a posterior filter compartment sorts food: coarse particles are returned into the gastric mill; fine particles pass into the intestine; and liquefied food is taken up into the capacious *hepatopancreas* ("liver") of the midgut, where the bulk of digestion and absorption occur. The nervous system ranges from a typical ladder-type design in many groups to a highly concentrated ventral ganglionic mass in crabs. Sensory structures include, apart from eyes and antennae, hundreds of tactile and chemoreceptive bristles all over the body. Moreover, statocysts are located at the bases of the first antennae in decapods or in other appendages in other groups. The heart

28.6. Decapods, internal structure. A. General plan of the nervous system. The brain supplies nerves to both pairs of antennae, and the first ventral ganglion innervates the mouth, the maxillae, and the maxillipeds. The second ventral ganglion innervates the large chelate legs, and the remaining thoracic ganglia supply nerves to the last four pairs of legs. **B.** Basal piece of first antenna, showing statocyst, schematic. **C.** Longitudinal section through stalked compound eye, showing location of endocrine sinus gland, schematic. See Fig. 27.2 for other anatomical features of decapods.

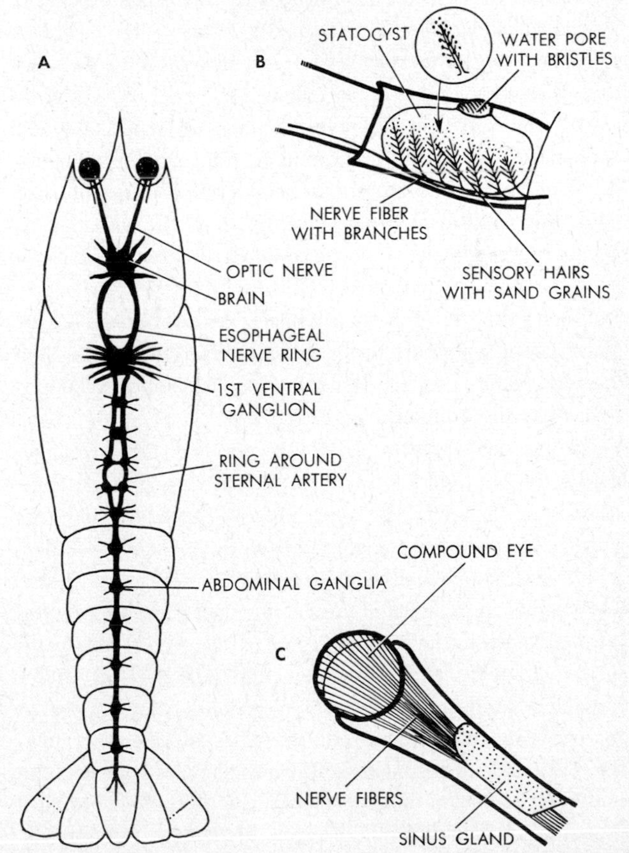

is primitively elongated, with a pair of ostia in each segmental portion. In advanced groups the heart is shortened and possesses fewer ostia. For example there are three pairs of ostia in lobster hearts, one pair in the hearts of water fleas. Many ostracods, copepods, and barnacles are without hearts altogether, and the parasitic copepod *Lernanthropus* possesses a *closed* vessel system without heart. The blood pigment in that genus is hemoglobin, but the blood of most crustacea is either colorless or, as in Malacostraca, hemocyanin-containing. Excretion occurs through saclike glands with ducts opening at the bases of the second antennae or the maxillae (cf. Fig. 9.36). Nauplii generally develop antennal glands first, then these degenerate and maxillary glands take over in later larval and adult stages. Some groups, among them the Decapoda, retain antennal glands throughout life and do not acquire maxillary glands. The primitive leptostracan *Nebalia* possesses a renal gland at the base of each thoracic appendage; this metameric array of eight pairs is an interesting holdover of a presumably ancestral condition.

Crustacea possess a well-developed endocrine system, most studied in Malacostraca and particularly in the malacostracan order Decapoda. Important elements of this system include a so-called *X-organ*, consisting of several groups of neurosecretory cells located in each eyestalk near the eye, and of a *sinus gland*, also located in each eyestalk, near its base. This gland appears to be primarily a temporary storage site for hormones produced in the X-organ. Another endocrine component is the nonneural, epithelial *Y-organ*, situated in the antennal or maxillary segments of the head. Behind the head is present a *postcommissural organ*, near the heart lies a *pericardial organ*, and along the sperm ducts of male crabs are *androgenic organs*. The first two are probably endocrine storage sites like the sinus gland, and the last are epithelial, nonneural endocrines like the Y-organ.

The hormones released by these various glands have been shown to play a variety of roles. At least one group of substances produced in the X-organs and stored in the sinus glands controls the chromatophores in the crustacean epidermis. These pigment cells contain melanin as well as red, green, and yellow components of varied chemical composition. The hormones govern the relative degree of dispersion of such pigments within the chromatophores, and thus they may make the animal lighter or darker and, in general, color-adapted to the predominant background hues of the environment. Relative dispersion of pigments in the eyes likewise appears to be under the control of these hormones. Another set of hormones regulates the molting of crustacea. Hormones from the X-organ and distributed via the sinus gland can be shown to inhibit the Y-organ in some crustacea, to stimulate it in others. The Y-organ then does not or does produce a hormone of its own, and whenever such a hormone is produced it initiates a molt. Depending on the species, evidently, the sequence

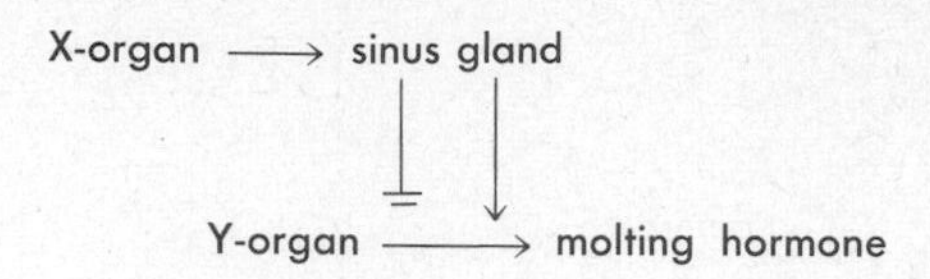

may effect either molting inhibition or molting acceleration. Crustacea molt frequently in larval stages, progressively less frequently as adults. Another Y-organ effect is gonadotropic, i.e., the organ accelerates growth of the reproductive organs in both sexes. The androgenic organ of male crabs functions additionally in such a gonadotropic role. It is known also that a hormone from the pericardial organ accelerates heart beat, and it is very likely that still other endocrine effects will be

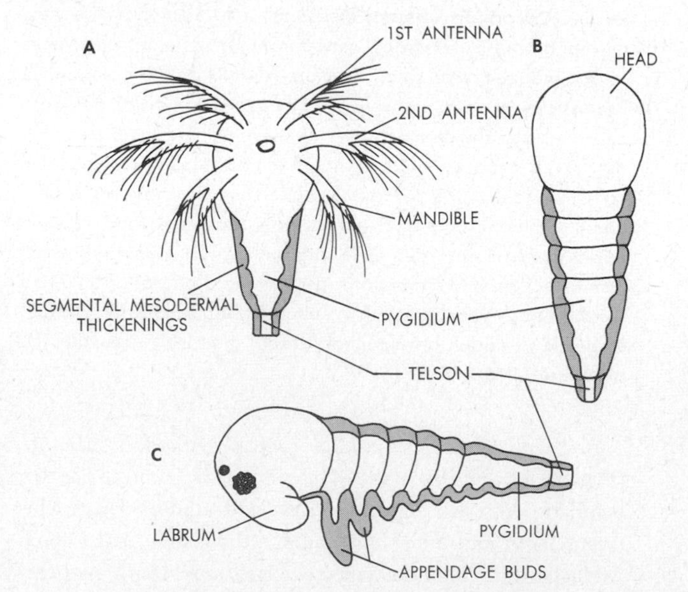

28.7. Anamorphic development in crustacea. (*After Weisz.*) **A.** Nauplius larva, as in *Artemia,* with first three pairs of head appendages and thoracic segments foreshadowed in pygidium as ringlike mesodermal thickenings. **B.** Later stage, with first three thoracic segments cut off from pygidium anteriorly, the pygidium having elongated posteriorly by an equivalent amount. **C.** Side view of still later stage, showing anteroposterior sequence of appendage buds. After all thoracic segments are laid down, the pygidium itself becomes subdivided into the abdominal segments.

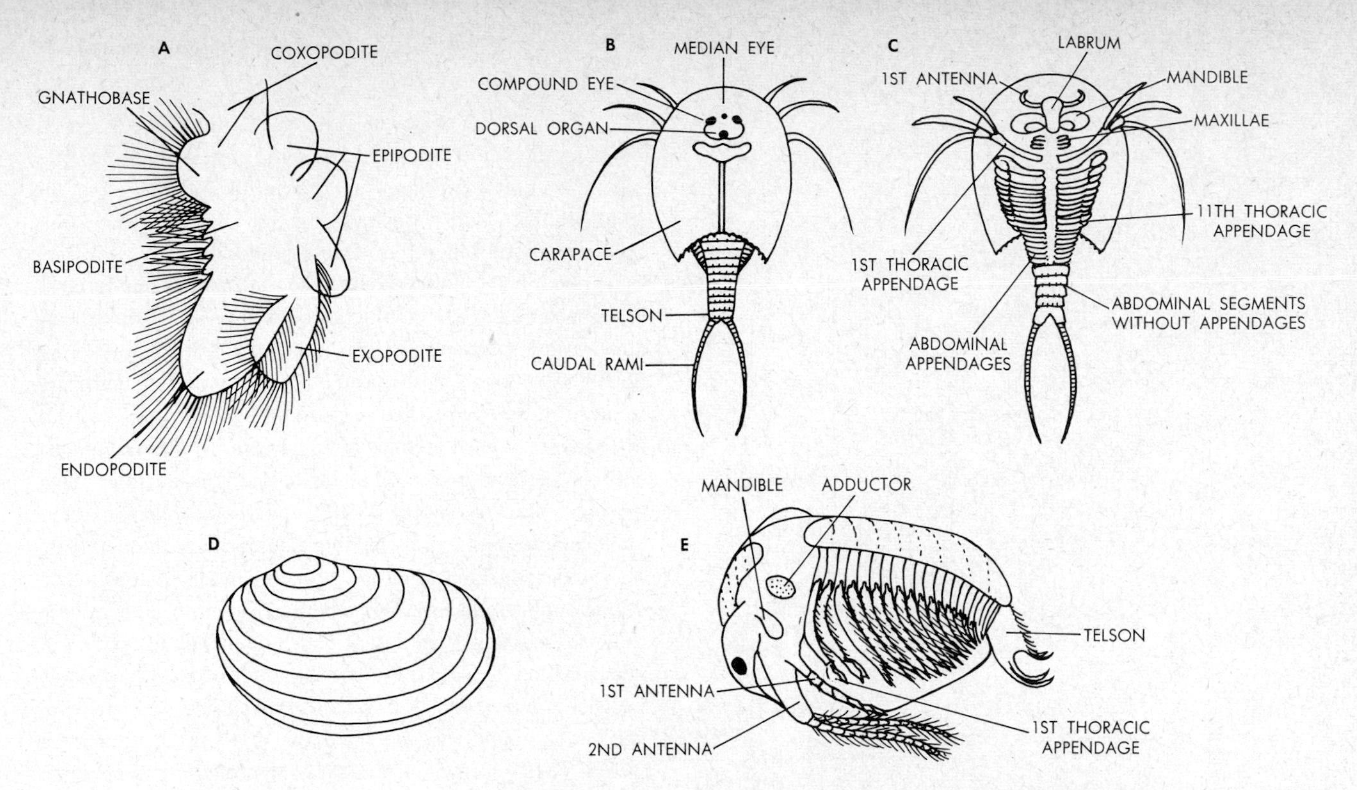

28.8. Branchiopods (fairy shrimps). **A.** A phyllopodial leg from right side of thorax of an anostracan. For anstracan structure cf. *Artemia,* Fig. 28.2. **B** and **C.** The notostracan *Apus,* dorsal and ventral views. (*After Borradaile.*) **D.** Left valve of carapace of conchostracan *Estheria.* (*After Sars.*) **E.** Left side view of *Estheria,* left valve of carapace removed. (*After Sars.*)

discovered in time. It is already quite clear in any event that the crustacean endocrine system has a level of complexity comparable to that of insects and vertebrates.

Parthenogenesis is common in water fleas and ostracods, and the (sessile) barnacles and some of the parasitic isopods are hermaphroditic. In some crustacea fertilization and development occur externally, but in the majority fertilization is internal, often via spermatophores. Eggs develop on or in the females, held in brood pouches or on abdominal appendages. As outlined in Chap. 11 (cf. Fig. 11.25), development typically passes through a series of free larval stages, including in some cases a complete series of nauplius, metanauplius, one or more zoaeae, and mysis larvae (*anamorphic* development). As also noted, however, some or all of these stages may be passed in the embryo, leading to an abbreviation or absence of free larval forms and an ovoviviparous condition. The hatched young then possess a complete number of segments from the start, a form of development said to be of the *epimorphic* type. During larval development, segments arise in anteroposterior succession by the appearance of metameric constrictions at the anterior end of a larval *pygidium.* As in trilobites, the pygidium is located in front of the telson and in an early nauplius it constitutes the entire trunk (Fig. 28.7). As the pygidium then generates segments, it retains its length by posterior growth. After a certain fixed number of segments has formed, the pygidium ceases to extend posteriorly and becomes segmented itself. These somites then constitute the last segments of the adult abdomen.

Primitive Branchiopoda are exemplified by the remarkable brine shrimp *Artemia,* found naturally in salt lakes (cf. Fig. 28.2). However, the animal may survive in waters of virtually any salinity, from almost distilled water to concentrated brine. Like Anostraca generally, *Artemia* is without carapace and its legs (thoracic only) are *phyllopodial,* i.e., leaflike and without joints. Nevertheless, the parts of such an appendage correspond precisely to those of a jointed leg. Phyllopod appendages are specializations to a swimming life; they occur also in the orders Notostraca and Conchostraca (Fig. 28.8). Notostraca have a flat, rounded carapace which does not overhang the sides of the body and thus does not interfere with leg

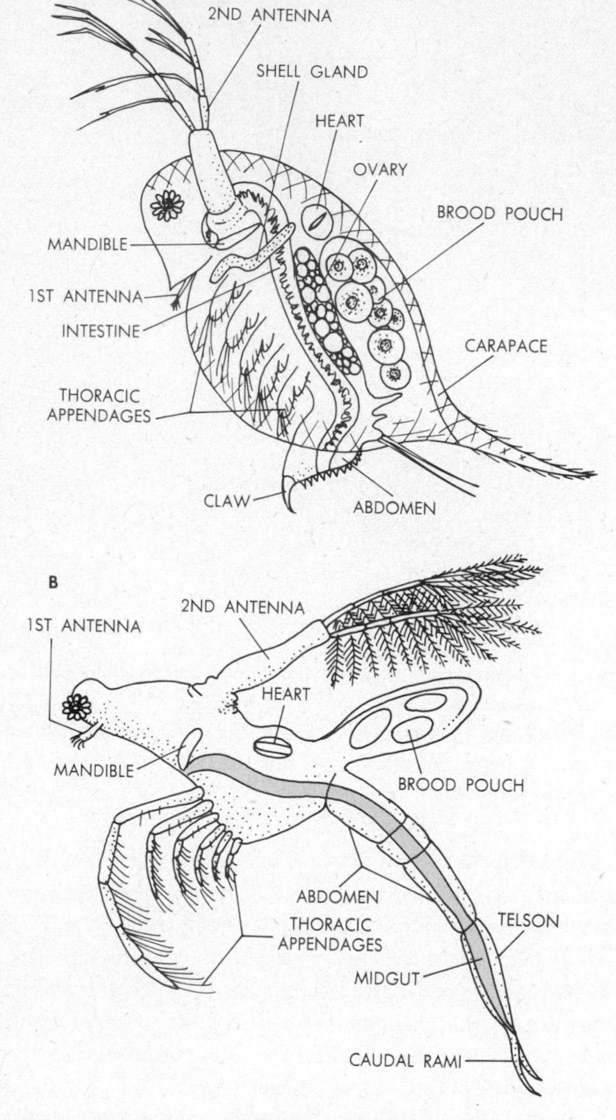

28.9. Cladoceran branchiopods (water fleas). A. *Daphnia. (After Sars.)* **B.** *Leptodora. (After Lilljeborg.)*

mobility. But the Conchostraca already exhibit the trend of specialization referred to above: the carapace is bivalved (held by a transverse adductor muscle), and it covers almost the whole laterally compressed body. At the same time, the second antennae are enlarged and form the main locomotor organs. A similar trend is in evidence in the Cladocera, among which the daphnias are common and familiar representatives (Fig. 28.9). These water fleas have an unhinged carapace covering the whole trunk; an adductor muscle is absent. The second antennae are large and powerful and they serve in locomotion. In both Conchostraca and Cladocera the trunk appendages still retain the feeding function.

In the subclass Ostracoda, however, feeding has in most cases become the special function of the well-developed head appendages, particularly the antennae and maxillae (Fig. 28.10). The trunk is shortened greatly, and the limbs are reduced to two pairs; they serve mainly in cleaning the head appendages. These animals in effect are composed almost entirely of head parts. The whole body is enclosed in a bivalved carapace equipped with an adductor muscle. Antennal locomotor and in part feeding functions are characteristic also of the Mystacocarida and the Copepoda, groups which are without carapace. In the copepods, however, one or more thoracic segments are fused with the head, a feature of cephalization similar to the development of a cephalothorax in other groups (Figs. 28.2 and 28.11). *Calanus* is a main component of animal marine plankton; *Cyclops* lives in fresh water. These animals are readily identified by their large median ocellus and by the conspicuous pair of lateral brood pouches in females. Numerous copepods have become para-

28.10. A. Ostracods; side view of *Cypris*, left half of carapace removed. *(Adapted from Zenker.)* **B.** Mystacocarids; side view of *Derocheilocarus*. *(After Pennak.)*

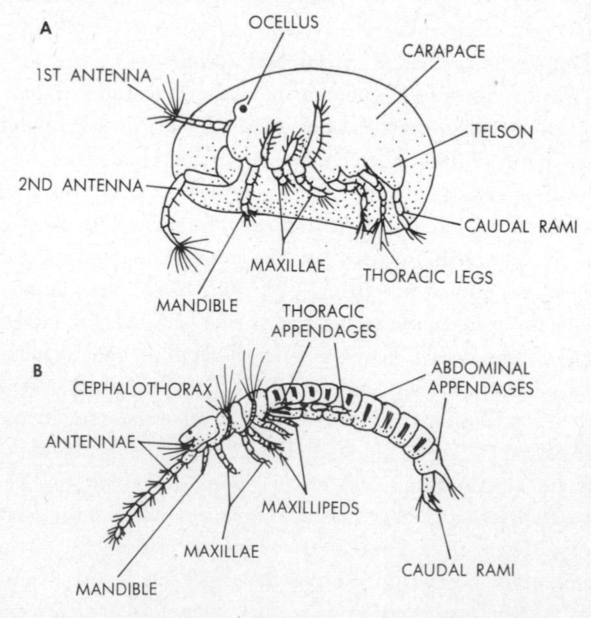

A

B

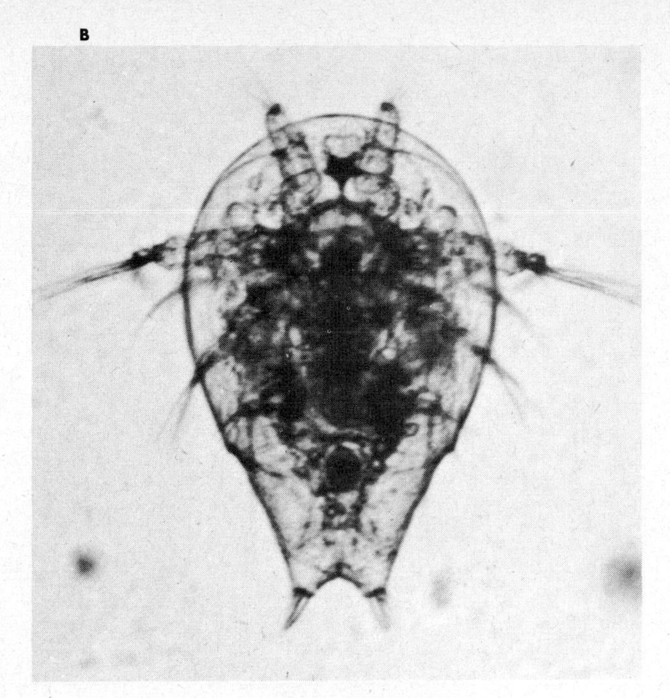

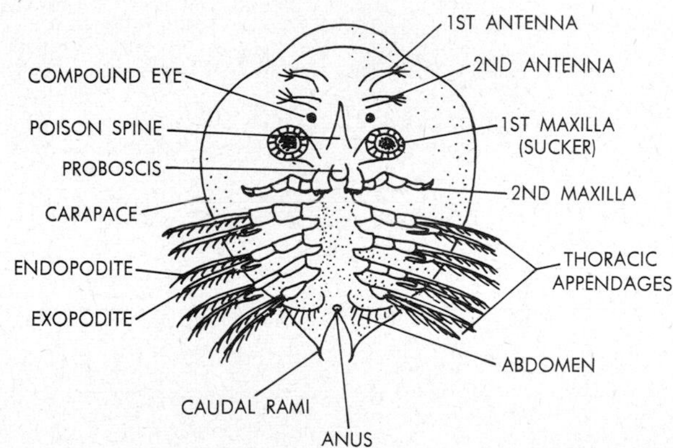

28.11. Photos: copepods. **A,** the fresh-water type *Cyclops,* with brood pouches; **B,** a nauplius larva hatched from such pouches. **Diagram:** ventral view of the branchiuran *Argulus.* (*After Wilson.*) Note the two lobes of limbless abdomen.

sitic. Parasitic types are included also in the small subclass Branchiura (cf. Fig. 28.11).

Cirripedia had originally been classified as mollusks. However, the crustacean nature of barnacles is quite evident from their larvae and also from their modified adult structure (Fig. 28.12). Free-swimming larval barnacles (i.e., zoaea-like *cypris* larvae) settle with their head end to the ground, a permanent attachment being formed by cement organs opening on the first antennae. The antennae and the compound eyes then degenerate, and the carapace comes to form a *mantle* around almost the whole animal. Calcareous plates are secreted on the mantle in the familiar acorn barnacles *Balanus,* a genus found on rocks and wharf pilings along most sea shores (and even on the backs of large lobsters). The thoracic limbs become highly developed; they may be protruded from the mantle and are used in food-gathering. Thus, a barnacle is a crustacean which stands on its head and kicks food into its mouth with its hind legs. The abdomen is usually reduced or absent. Forms such as *Lepas* are stalked, the stalk representing an enlarged anterior region of the head. Quite a number of barnacles are parasitic. The extreme parasitism of *Sacculina* has already been noted in Chap. 17 (cf. Fig. 17.16). Barnacles are hermaphroditic, an adaptation to sessilism and also to parasitism.

The largest crustacean group is the subclass Malacostraca

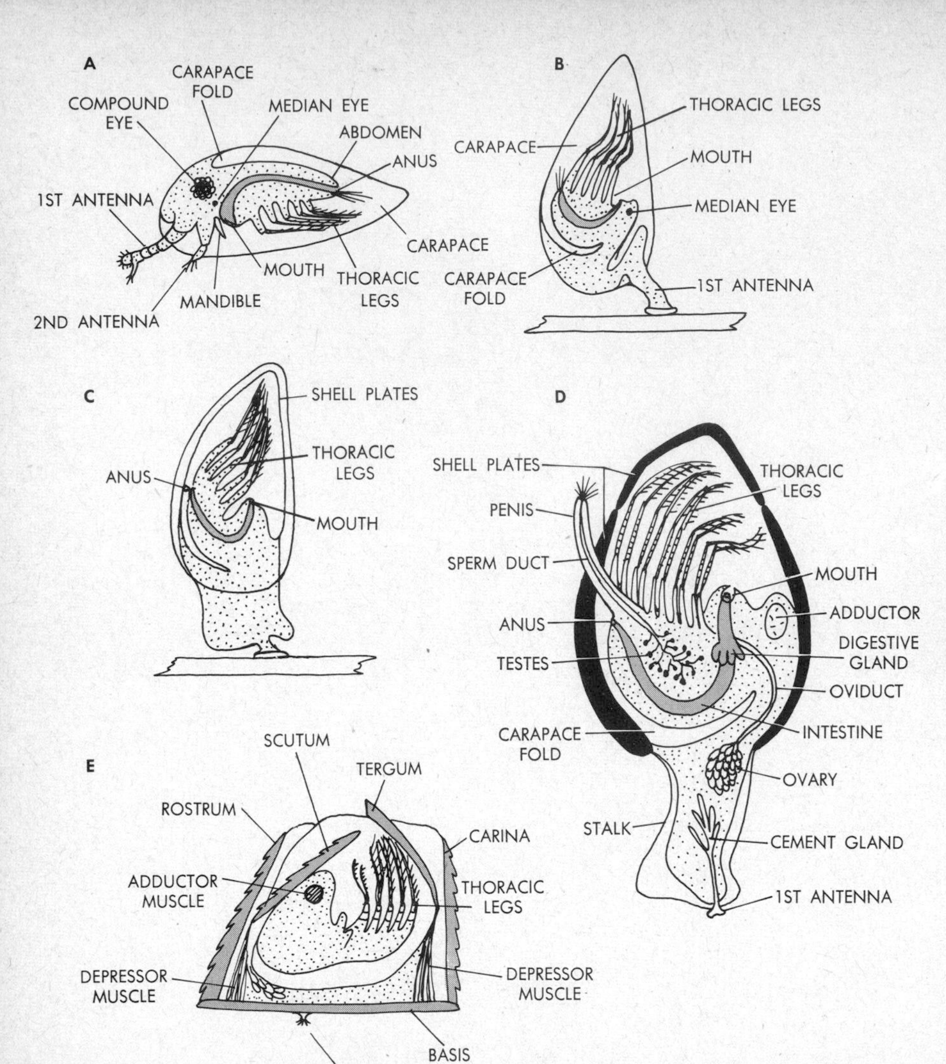

28.12. Cirripeds (barnacles). Diagrams: A. Free cypris larva of goose barnacle *Lepas*. **B.** Attachment of larva at head end. Note that ingrowth of a dorsal skin fold from the carapace brings about a rotation of the internal organs. **C.** Later stage of metamorphosis; stalk already well formed, shell plates developing. **D.** Section through adult. Mandibles, maxillae, and labrum (not shown) guard the mouth. The carapace fold of larval stages is still present. **E.** Section through rock barnacle *Balanus*, showing position of body and principal shell plates. (**A** to **D**, *after Korschelt and Heider, and modified from Leuckart;* **E**, *after Henry*). **Photos: A, B** and **C.** Nauplius larva, cypris larva, and feeding adult of *Balanus*. Cf. Fig. 17.16 for illustration of *Sacculina*.

A

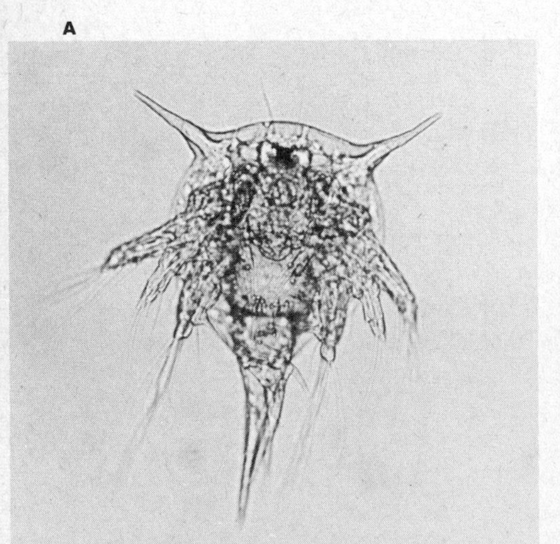

B

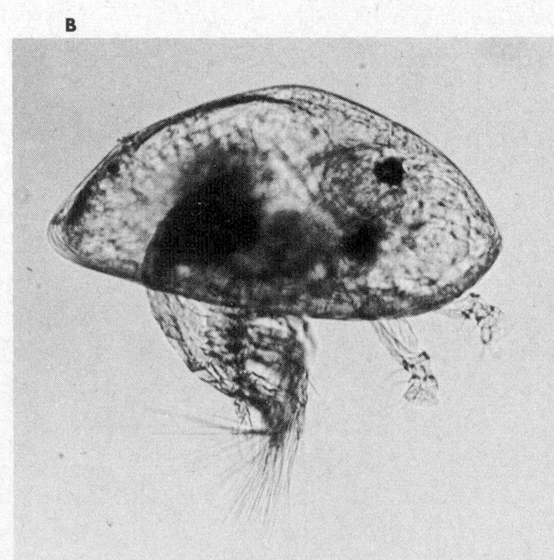

C

(Figs. 28.13, 28.14, and 28.15). As indicated in the taxonomic listing, the subclass includes two evolutionary series, the Leptostraca and the Eumalacostraca. In the latter, *Anaspides* of South Australia represents the primitive type, and the main groups are the two superorders Peracarida and Eucarida. Of more than passing interest are the several forms which have adapted to unusual ecological niches, e.g., subterranean waters, hot springs, and terrestrial environments. The terrestrial isopods breathe through their abdominal appendages. In some cases, the latter are equipped with water-retaining recesses, in others the appendages contain tracheal structures remarkably similar to those of insects. Clearly the most familiar crustacea, and among the most familiar invertebrates generally, are the decapods. They have largely adopted scavenging and carnivorous ways of life, and the crabs are the most highly organized among them.

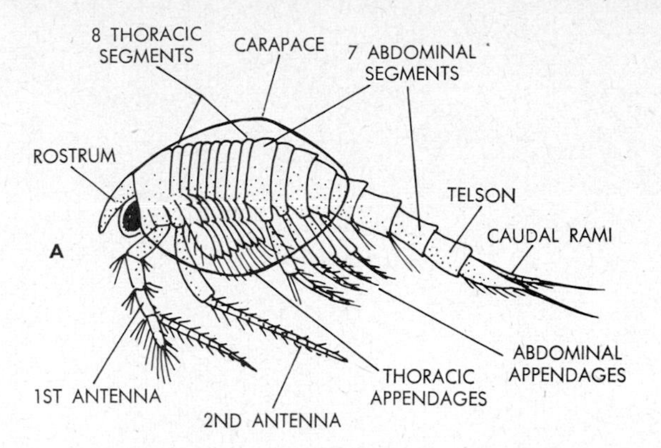

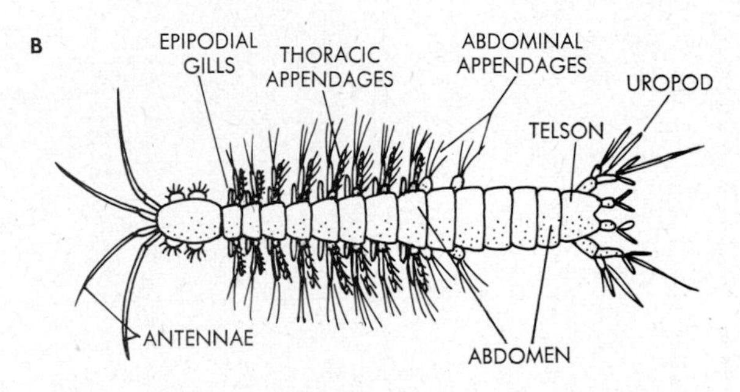

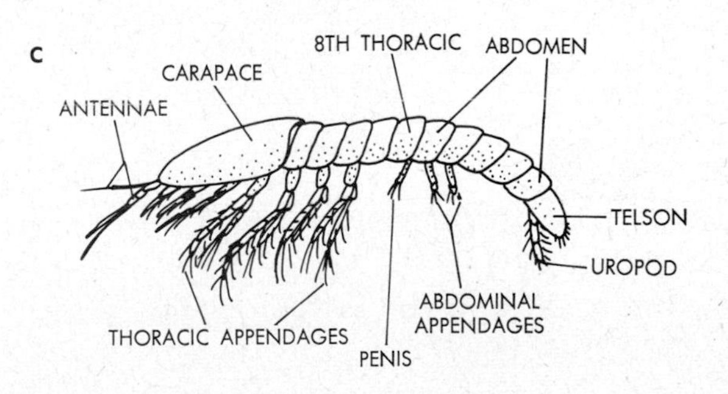

28.13. Malacostraca. A. The leptostracan *Nebalia.* (*After Claus.*) **B.** The syncarid *Bathynella.* (*After Jacobi.*) The syncarid *Anaspides* is shown in Fig. 28.2. **C.** The pancarid *Thermosbaena.* (*After Brunn.*)

MYRIAPODA

Class CHILOPODA (centipedes): carnivorous, predatory; head with one pair of antennae, one pair of mandibles, two pairs of maxillae; trunk with one pair of poison-claw-containing prehensors and 15 or more pairs of walking legs; eyes compound or simple or absent; spiracles dorsal or lateral, tracheae branched; gonopores at posterior end of trunk; young hatch with seven or all pairs of legs. *Scutigera,* house centipedes; *Lithobius, Scolopendra*

Class SYMPHYLA: centipedelike, in soil, debris, humid environments; head as in centipedes, but with one pair of spiracles; trunk with 12 pairs of legs, 15 or more tergum plates, and one pair of terminal spinnerets; gonopores at anterior end of trunk; young hatch with six or seven pairs of legs. *Scutigerella*

Class DIPLOPODA (millipedes): herbivorous, scavenging; head with one pair each of antennae, mandibles, and maxillae; first four trunk segments single, rest double and fused; with ocelli; tracheae unbranched; gonopores on third trunk segment; young hatch with three pairs of legs. *Iulus, Glomeris*

Class PAUROPODA: somewhat centipedelike; head as in millipedes, antennae branched; 12 trunk segments, nine pairs of legs; trunk terga fused in pairs; without eyes, breathing, or circulatory systems; gonopores at anterior end of trunk; young hatch with three pairs of legs. *Pauropus*

The term "Myriapoda" is now used descriptively but not taxonomically for the above four classes of superficially similar animals. The likeness of the many trunk segments of these forms suggests that the four classes represent (separate) descendant lines of quite primitive mandibulate stocks of arthropods. The Chilopoda and in particular the Symphyla appear to be close to the mandibulate branch which led to the insects (cf. Fig. 27.4). All four classes are terrestrial. The animals are worldwide in distribution, but they reach their greatest abundance in humid, tropical and subtropical environments.

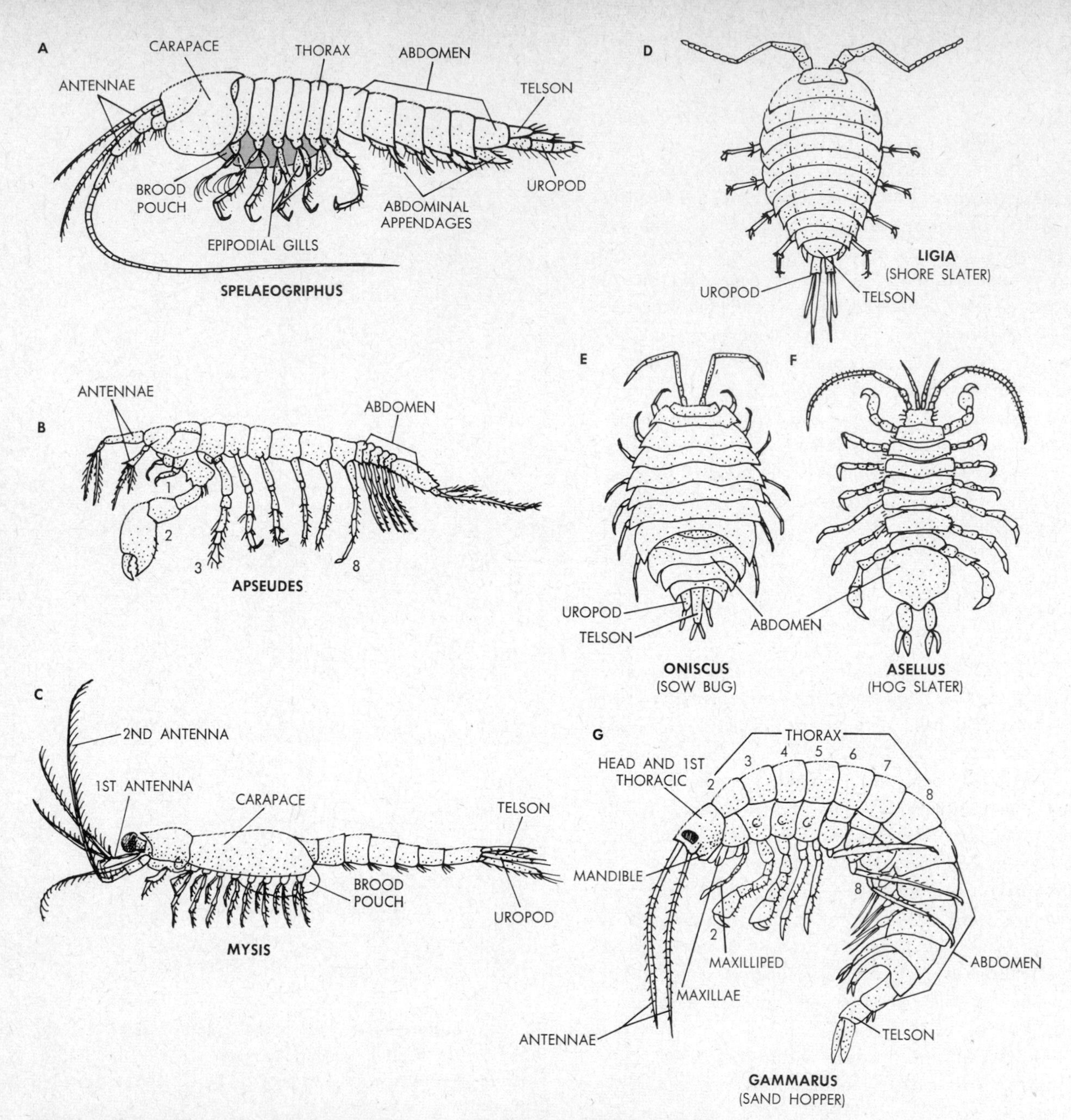

28.14. Peracarida. A. Order Spelaeogriphacea, female specimen. (*After Gordon,* © *I. Gordon and British Museum of Natural History.*) **B.** Order Tanaidacea. (*After Sars.*) Numbers refer to appendages on thoracic segments. **C.** Order Mysidacea. (*After Sars.*) **D, E** and **F.** Order Isopoda. **D.** Marine, tidal genus. **E.** Terrestrial genus. (*After Palmer.*) **F.** Freshwater genus. (*After Smith.*) **G.** Order Amphipoda. (*After Leuckart.*) Numbers refer to thoracic segments. Of the seven pairs of thoracic appendages, the first two (on segments 2 and 3) are semichelate, the third and fourth help in feeding, and the fifth, sixth, and seventh are turned backward and serve when the animal crawls on its side. Modified parts of the large second, third, and fourth thoracic coxopodites serve as egg pouches (labeled c). Of the five pairs of abdominal appendages, the first three serve in swimming, the last two, in jumping.

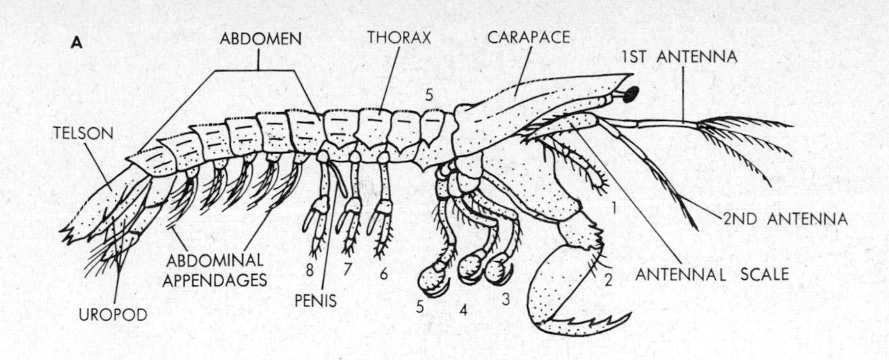

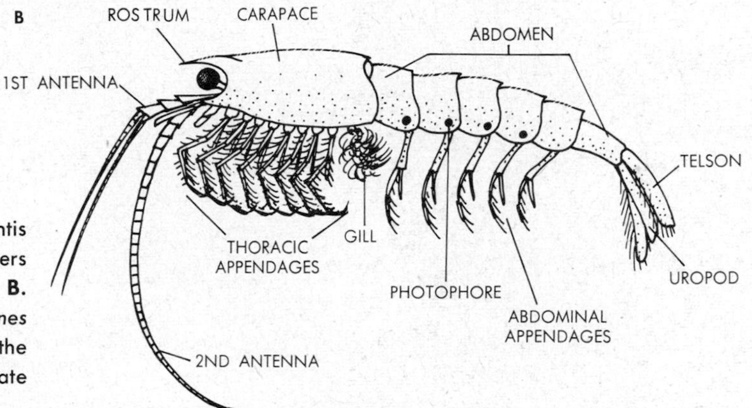

28.15. **A.** Hoplocarida. The (stomatopod) mantis shrimp *Squilla* is shown. (*After Calman.*) Numbers refer to thoracic segments and appendages. **B.** Eucarida, order Euphausiacea. The krill *Nyctiphanes* is shown. (*After Boden, Watasé.*) Photophores on the abdomen are bioluminescent organs. Cf. color plate XIII.

The head structure of a centipede is like that of an insect (Fig. 28.16, Table 12). The first segment is embryonic; the second bears antennae; the third begins to develop antennae as in crustacea but these appendages never mature and are absent in the adult; the fourth is mandibulate, i.e., it possesses jaws, and fifth and sixth each bears a pair of maxillae, the second pair being fused into a *labium*. In the trunk of forms like *Scutigera* and *Lithobius*, 18 embryonic segments arise. The first is equipped with fanglike *prehensors*, which contain poison glands and ducts opening at the tips. The next 15 segments bear a pair of seven-jointed legs each. On the seventeenth segment are the gonopores, and the eighteenth is a reduced telson segment with anus. Other groups of centipedes are similar except that the number of leg-bearing segments may be greater; types with up to 183 pairs of legs are known. In *Scutigera*, the 15 leg-bearing segments are covered dorsally with only seven tergum plates. In other forms the number of terga corresponds to the number of segments. Each trunk segment contains a pair of dorsal (*Scutigera*) or lateral (*Lithobius*) spiracles which, as in insects, lead into branching and intersegmentally connected tracheal tubes. The nervous system is of the primitive ladder-type pattern, and the heart extends throughout the trunk and contains in each segmental portion a pair of ostia and lateral arteries. Excretion occurs through Malpighian tubules. *Scutigera* possesses compound eyes, *Lithobius* only simple ocelli, and many other centipedes are blind. Fertilization is internal, often via spermatophores, and the fertilized eggs are laid into holes in the ground. Forms such as *Scutigera* and *Lithobius* are anamorphic; they hatch with an incomplete number of segments, the complete number developing gradually in a series of molting steps. Other groups are epimorphic, the hatched young having a complete set of segments from the start. In the latter case the adult females usually brood the eggs and guard the young for a time. Centipedes molt throughout life. The animals typically are rapid runners, in correlation with their predatory mode of existence. They shun exposure, however, and during the day they tend to hide under stones, leaves, and in crevices and crannies. At night they emerge in search of food, which consists of earthworms, insects, and snails. Some of the relatives of *Scolopendra* attain lengths of 1 ft, but most members of the class are far smaller. The class includes some 2,500 species.

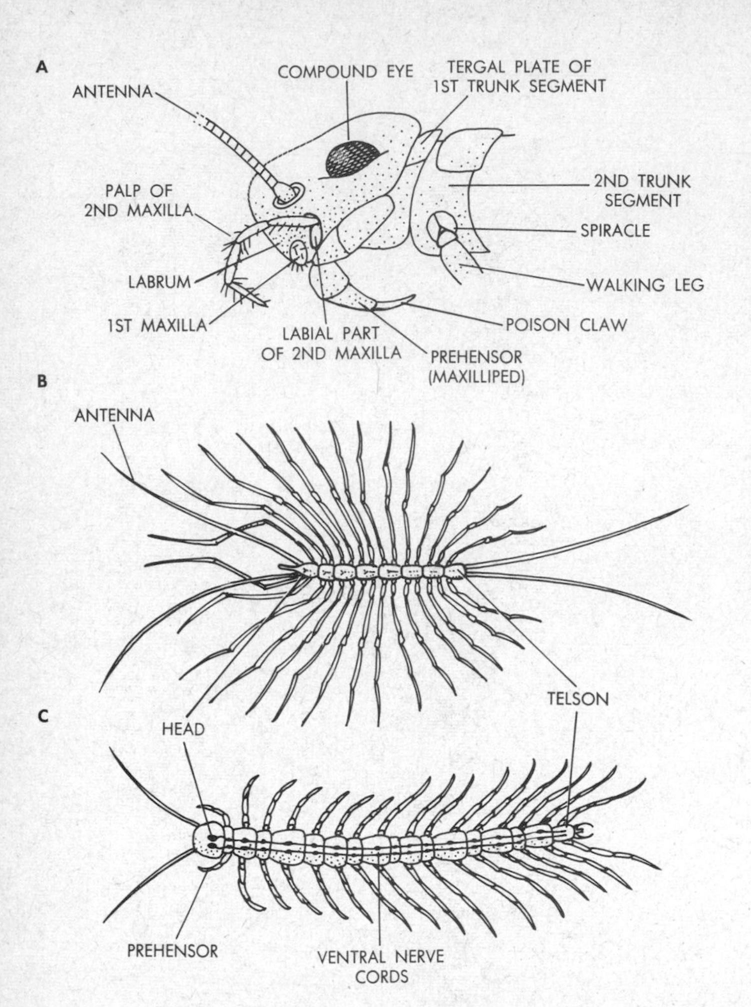

28.16. Centipedes. Diagrams: A. Head end of *Scutigera*. (*Modified from Snodgrass.*) Mouth and mandibles are hidden behind the labrum and the maxillae. Note that the prehensors on the first trunk segment are homologous to crustacean maxillipeds. **B.** Dorsal view of *Scutigera*. (*After Snodgrass.*) **C.** Dorsal view of *Lithobius*, with primitive ladder-type nerve cords drawn in. (*Adapted from Shipley and McBride.*) **Photo:** *Lithobius*, from life.

Symphyla are a group of only 60 or so species. They are generally centipedelike in appearance and habits, and their head structure is quite similar as well. But they differ from centipedes in the segmental structure of the trunk, in the anterior location of the gonopores, and in several other respects. Pauropods too number only about 60 species. These animals have anterior gonopores also, but their heads are like those of millipedes. Both the symphylans and pauropods develop anamorphically. The biology of both groups is known only poorly as yet (Fig. 28.17).

The millipede head bears but a single pair of maxillae, on the sixth head segment; the appendages on the fifth fail to mature in the embryo and are absent in the adult (Fig. 28.18). In the first four trunk segments, the second, third, and fourth each bears a single pair of legs, and the gonopores open on the third. In this anterior position of the genital openings the millipedes thus are like pauropods and symphylans but unlike centipedes. The subsequent trunk segments of millipedes are fused in pairs. Each such double segment arises from two pairs of coelomic sacs and possesses two pairs of ganglia, two pairs of ostia in the heart, two pairs of spiracles, and two pairs of legs. The number of such double segments varies considerably in different groups. Spiracles lead into bundles of unbranched tracheae, another difference from centipedes. Millipedes possess simple ocelli and excretion takes place as in centipedes via Malpighian tubules. Fertilization is internal, in some cases by means of spermatophores which are transferred to the female in the mouth of the male. Eggs are laid and often brooded, and development is anamorphic. Millipedes are poor runners despite their large number of legs; the herbivorous life of the animals does not require rapid locomotion. The animals are retiring in habit, and when they are exposed they may roll up into a ball. Their exoskeleton is usually impregnated heavily with calcium carbonate. More than 8,000 species have been described, but estimates place the number of actually existing species at about 25,000.

CLASS INSECTA (HEXAPODA)

Terrestrial, some secondarily freshwater, exceptional cases marine; head six segments, typically with antennae, mandibles, maxillae, labium; thorax three segments, typically with three pairs of legs, two pairs of wings, two pairs of spiracles; abdomen typically 11 segments, without locomotor appendages, with eight pairs of spiracles; compound eyes and ocelli present; excretion via Malpighian tubules; fertilization internal, development external; larvae with gradual (direct, incomplete) or abrupt (indirect, complete) metamorphosis.

Subclass Apterygota (Ametabola): Primitively wingless; without metamorphosis

Order Protrura: piercing mouth parts; without antennae, eyeless; with and without tracheae; abdomen 12 segments, first three with small appendages; development anamorphic, young hatching with nine abdominal segments only; in moist places, e.g., under leaves, twigs; 100 species. *Acerentulus*

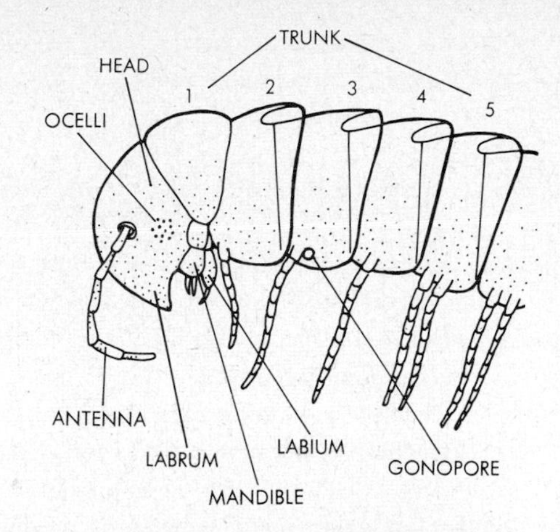

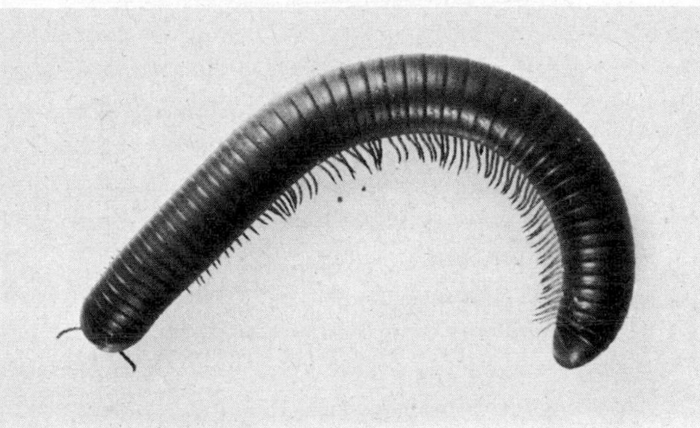

Courtesy Carolina Biological Supply Company.

28.18. Millipedes. Diagram: head region of *Iulus*. (*Modified from Borradaile.*) Numbers refer to trunk segments. Note dorsally overlapping plates of segmental exoskeleton. **Photo:** note two pairs of legs in most trunk segments of this animal.

28.17. A, the symphylan *Scutigerella,* dorsal view, and **B,** the pauropod *Pauropus,* side view. (*After Snodgrass.*) Numbers refer to trunk segments.

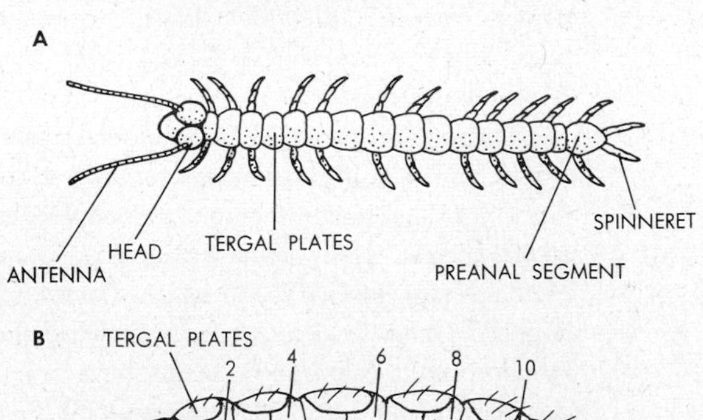

Order Diplura: biting mouth parts, deep in head; with long antennae; eyeless; with tracheae, without Malpighian tubules; abdomen 11 segments, with cerci or forceps; in moist, rotting places, e.g., under leaves, bark; 400 species. *Machilis, Campodea, Japyx*

Order Thysanura (bristle tails): biting mouth parts; antennae long; 11 abdominal segments and two or three cerci; in buildings, nooks and crannies, book bindings; 750 species. *Lepisma,* silver fish.

Order Collembola (spring tails): biting mouth parts; without compound eyes; six abdominal segments with three pairs of appendages (not legs), modified for jumping and adhesion; without tracheae or Malpigian tubules; in moist, rotting places; 2,000 species. *Axelsonia*

Subclass Pterygota (Metabola): with wings, with metamorphosis.

Superorder Exopterygota (Hemimetabola): wing growth external on larval body; metamorphosis gradual; larvae are *nymphs* if terrestrial, *naiads* if aquatic; compound eyes already present in larvae.

Order Orthoptera (grasshoppers, locusts, crickets, cockroaches): biting mouth parts; forewings narrow, leathery; hind wings membranous, folding under forewings; many wingless; typically with cerci; herbivorous; 25,000 species. *Romalea, Melanoplus,* grasshoppers; *Locusta,* migratory locusts; *Gryllus,* field and house crickets; *Gryllotalpa,* burrowing mole crickets; *Stagomantis,* praying mantis; *Microcentrum,* katydids; *Anisomorpha,* walking-stick insects; *Periplaneta, Blatta,* cockroaches.

Order Dermaptera (earwigs): biting mouth parts; forewings short, leathery, hind wings large, membranous; many wingless; cerci form terminal forceps; in dark places, nocturnal; 1,200 species. *Forficula*

Order Plectoptera (stone flies): mouth parts biting or absent; wings pleated, hind wings larger and hidden under forewings; weak fliers; cerci prominent; naiad larvae, with tracheal gill tufts behind each leg; 1,500 species. *Perla*

Order Embioptera (embiids): body flattened, elongate; in males wings similar, females wingless; cerci two-jointed, asymmetrical in males; colonial, in tunnel networks of secreted silk; mainly tropical; 1,000 species. *Embia,* webspinners.

Order Isoptera (termites): biting mouth parts; wings similar, membranous; social and polymorphic, largely tropical. 2,000 species. *Termes, Neotermes, Calotermes, Macrotermes* (with 30-ft nests).

Order Psocoptera (book lice): biting mouth parts; wings without cross-veins or absent. 1,300 species. *Atropus,* book lice; *Peripsocus,* bark lice.

Order Zoraptera: biting mouth parts; often wingless, eyeless, in warm, dark, rotting places; 21 species. *Zorotypus*

Order Odonata (dragonflies, damsel flies): biting mouth parts; wings similar, net-veined; compound eyes conspicuous; abdomen elongate and slender; predatory, legs catch insects in flight; naiad young with tracheal gills; 5,000 species. *Aeschna, Anax,* dragonflies, wings do not fold back; *Argia,* damsel flies, wings fold up at rest.

Order Anopleura (sucking lice): piercing-sucking mouth parts, retractable; body flattened; wingless; thorax fused; without eyes; one claw per leg for clinging; ectoparasitic blood-suckers on hair of mammals, transmitters of typhus fever, trench fever, and other diseases; 500 species. *Pediculus,* human body lice.

Order Mallophaga (biting lice): biting mouth parts; body flattened; wingless; eyes reduced or absent; two claws per leg for clinging; ectoparasitic on birds, some mammals, feeding on skin, feather, and hair fragments; 3,000 species. *Menopon*

Order Ephemeroptera (May flies): biting mouth parts vestigial; wings membranous, folded up at rest, hind wings small; long cerci; naiad young with multiple paired tracheal gills, biting mouth parts, long-lived; adults nonfeeding, live less than 24 hr; 1,500 species. *Ephemera*

Order Hemiptera (bugs): piercing-sucking mouth parts; forewings thickened basally, membranous terminally, crossed flat on body at rest; hind wings membranous, folded under forewings; herbivorous, predatory, and ectoparasitic; some viviparous; 40,000 species. *Corixa,* waterboatmen; *Nepa,* water scorpions; *Belostoma,* water bugs; *Notonecta,* back swimmers; *Helobates,* water striders (marine); *Rhodnius,* bugs; *Cimex,* bed bugs; *Lygus,* plant bugs; *Perilus,* stink bugs; *Blissus,* cinch bugs.

Order Homoptera (plant lice, scale insects): piercing-sucking mouth parts; wings tentlike at rest, forewings thickened or membranous; sapfeeders; 30,000 species. *Aphis,* aphids, plant lice; *Cicada,* cicadas; *Magicicada,* 17-year locust; *Ceresa,* tree hoppers; *Circulifer,* leaf hoppers; *Lepidosaphes, Pseudococcus,* scale insects, mealy bugs.

Order Thysanoptera (thrips): sucking mouth parts; wings similar, veins few or absent; some parthenogenetic; mostly herbivorous on fruit and grasses; 3,500 species. *Thrips, Taeniothrips*

Superorder Endopterygota (Holometabola): wing growth internal in larval body; metamorphosis abrupt; larval stages are caterpillars and pupae; compound eyes not yet present in larvae.

Order Mecoptera (scorpion flies): biting mouth parts on turned-down beak; wings similar, with rhomboid venation, roofed over body at rest; end of abdomen arched up and forward. 500 species. *Panorpa*

Order Neuroptera (dobson flies, lace wings): biting mouth parts; wings similar, roofed over body at rest; larvae with biting or sucking mouth parts, abdominal gills, carnivorous insect- and mite-feeders. 5,000 species. *Sialis,* alder flies, *Chrysopa,* lace wings; *Cordalis,* dobson flies; *Myrmeleon,* ant lions.

Order Trichoptera (caddis flies): sucking mouth parts; wings roofed over body at rest; wings and body hairy or scaly; aquatic; larvae aquatic and carnivorous, with biting mouth parts; 7,000 species. *Rhyacophila*

Order Strepsiptera: biting mouth parts reduced or absent; in males, forewings reduced to halteres, hind wings membranous; females larvalike, wingless, eyeless, legless; larvae and females parasitic in insects, permanently within body of host. 300 species. *Stylops*

Order Lepidoptera (moths, butterflies): sucking mouth parts, with proboscis; wings membranous, scaly; larvae with biting mouth parts; moths, wings horizontal at rest, antennae feathery; butterflies, wings folded up at rest, antennae filamentous; 125,000 species. *Sphinx,* hawk moths; *Lymantria,* gypsy moths; *Actias,* luna moths; *Tinea,* clothes moths; *Ephestia,* flour moths; *Samia,* cecropia silk moths; *Bombyx,* silk worms; *Papilio,* swallowtail butterflies; *Danaus,* monarch butterflies.

Order Diptera (flies, gnats, mosquitos): piercing-sucking-biting mouth parts in proboscis; hind wings reduced to halteres; 100,000 species. *Musca,* houseflies; *Simulium,* black flies; *Sarcophaga,* flesh flies; *Tabanus,* horseflies; *Gastrophilus,* bot flies; *Drosophila,* fruit flies; *Glossina,* tsetse flies; *Chironomus,* midges; *Culex, Anopheles, Aedes,* mosquitos.

Order Siphonaptera (fleas): piercing-sucking mouth parts; wingless; laterally compressed; ectoparasitic blood suckers; 1,500 species. *Pulex,* human fleas; *Ctenocephalides,* dog and cat fleas; *Xenopsylla,* rat fleas, transmitting bubonic plague.

Order Coleoptera (beetles, weevils): biting mouth parts; forewings horny *elytra,* hind wings folded under forewings. 300,000 species. *Popillia,* Japanese beetles; *Gyrinus,* whirligig beetles; *Lucanus,* stag beetles; *Scarabaeus,* dung beetles, scarabs; *Necrophorus,* carrion beetles; *Calosoma,* ground beetles; *Cicindela,* tiger beetles; *Tribolium,* flour beetles; *Anthrenus,* carpet beetles; *Coccinella,* ladybird beetles; *Tenebrio,* meal worms; *Lampyris, Photinus,* fireflies, glowworms; *Anthonomus,* cotton-boll weevils.

Order Hymenoptera (bees, ants, wasps, sawflies): biting, sucking, and lapping mouth parts; hind wings smaller than forewings; first abdominal segment fused to thorax, waist behind it; ovipositor for piercing, stinging, or sawing, often long and looped forward and downward; social and polymorphic, parthenogenesis common; pupae typically in cocoons; 110,000 species. *Vespa,* yellowjacket wasps, hornets; *Andricus,* gall wasps; *Sirex,* wood wasps; *Therion,* ichneumon flies; *Cimbex,* saw flies; *Formica,* ants; *Polyergus,* slavemaking ants; *Monomorium,* black ants; *Atta,* fungus ants; *Mutilla,* velvet ants; *Solenopsis,* fire ants; *Apus,* honeybees; *Bombus,* bumblebees; *Megachile,* leaf-cutter bees.

An assemblage constituting the largest group of animals, with more than ¾ million described species and possibly 10 times that many actually in existence, is bound to be replete with more superlative attributes than any other. Within the membership of the Coleoptera, the largest order, some 40,000 species comprise the family of the weevils, the largest single family in the animal world. Three other orders encompass more than 100,000 species each, viz., the Lepidoptera, including butterflies and moths, the Hymenoptera, including bees and ants, and the Diptera, including flies and mosquitos. Next in abundance are the Hemiptera, or bugs, the Homoptera, or plant lice, and the Orthoptera, or grasshoppers.

Insect evolution has produced three levels of specialization. The most primitive level is represented by the wingless Apterygota (Fig. 28.19), in which the hatched young are miniature adults and in which therefore larvae and metamorphosis are essentially absent (hence the alternative name Ametabola). Such insects have appeared in the fossil record first, viz., in Devonian times. We may note here also that, because the order Protrura exhibits anamorphic development, uniquely distinct from the epimorphic pattern in other apterygotes and also other insects, some zoologists do not even consider Protrura to be insects but a separate class of arthropods.

28.19. Apterygote insects. **A.** The protruran *Acerentulus*. (*After Ewing.*) Note absence of antennae. **B.** The dipleuran *Anajapyx*. (*After Essig.*) **C.** The thysanuran silverfish *Lepisma*. (*After Lubbock.*) **D.** The collembolan spring tail *Axelsonia*. (*After Folsom.*) Note forked spring on fourth abdominal segment, engaged in hooklike appendage of third abdominal segment. When released, the spring flies backward and catapults the insect forward. An adhesive ventral tube projects from the first abdominal segment.

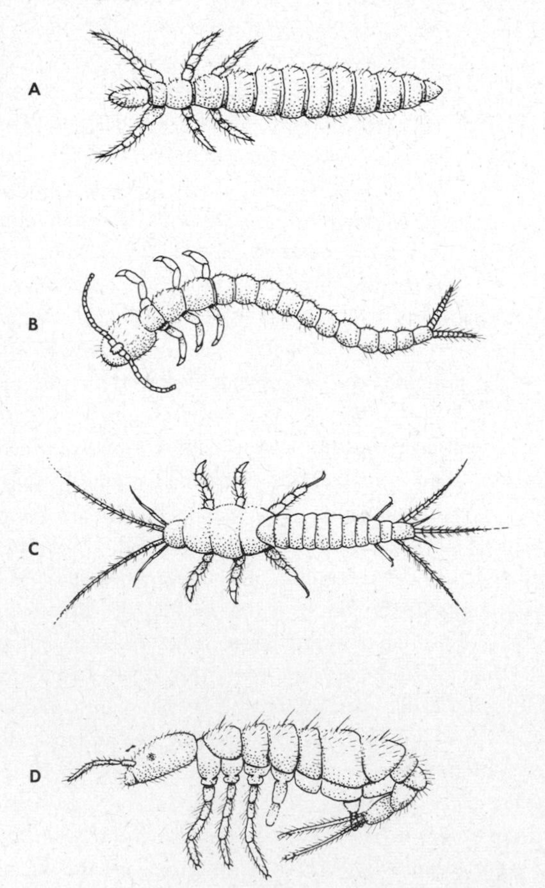

28.20. Exopterygote insects. **A,** cockroach (Orthoptera; cf. also Figs. 17.8 and 27.1); **B,** earwigs (Dermaptera), male at left, female at right; **C,** damsel fly (Odonata); **D,** human body louse (Anopleura).

A

B

C

D

A second level of specialization is displayed by the winged Exopterygota, also called Hemimetabola (Figs. 28.20 and 28.21); in these insects the hatched young are larvae, or *nymphs* (*naiads* if they are aquatic). They are wingless initially, but they metamorphose gradually into winged adults via a series of molting steps called *instars*. In the process, wing buds on the outside of the body grow progressively larger (cf. below).

The third and most specialized level is represented by the Endopterygota, or Holometabola (Figs. 28.22, 28.23). Here the larvae are caterpillars or caterpillarlike. They undergo a series of molts and the last of these transforms the caterpillar into a *pupa*. During the pupal phase a drastic metamorphosis takes place, in the course of which wings and other adult structures arise from internal buds called *imaginal disks*. The adult, or *imago*, then extricates itself from the pupal envelopes (cf. below). Endopterygotes did not appear in the fossil record till Permian times. Beetles arose only in the Triassic, and the origin of the most advanced endopterygotes, viz., bees, butterflies, and flies, coincided with the Jurassic and Cretaceous expansion of the flowering plants.

The head of an insect consists of six fused segments (Fig.

28.21. Exopterygote insects. A, mayfly (Ephemoptera); **B,** bedbug (Hemiptera); **C,** 17-year locust (Homoptera).

A

B

C

28.22. Endopterygote insects. A, dobson fly (Neuroptera); **B,** flea (Siphonaptera); **C,** Saturnia moth; and **D,** peacock butterfly (Lepidoptera; cf. Fig. 28.24); **E,** several larvae and one pupa of the mosquito *Culex* (Diptera), suspended from water surface.

A

B

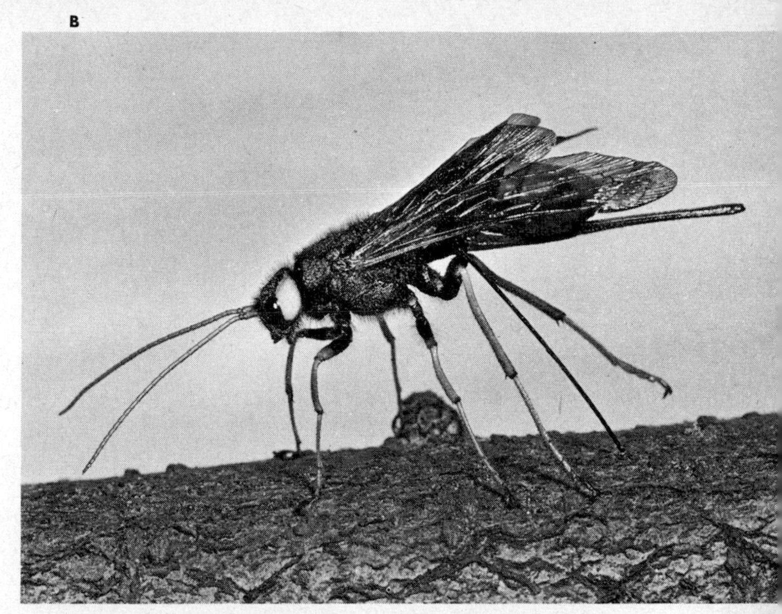

C

28.23. Endopterygote insects. **A,** hercules beetle (Coleoptera); **B,** wood wasp (Hymenoptera; note ovipositor serving as wood borer); **C,** tropical fire ant (Hymenoptera; other members of this order are illustrated in Chap. 16, section on social insects).

28.24 and cf. Table 12). The first is embryonic and without appendages; the second bears antennae; the third is embryonic; the fourth develops mandibles (coxopodites only); the fifth bears maxillae; and the sixth similarly grows a pair of maxillae, but these fuse along the midline and form an underlip, the *labium.* The mandibles and first maxillae lie lateral to the mouth, and an upper lip, the *labrum,* protects the mouth anteriorly. Primitively, labrum, mandibles, maxillae, and labium form *biting-and-chewing* mouth parts. Most exopterygote orders possess oral structures of this type.

In more advanced forms, however, these basic oral appendages are modified into *sucking, licking,* or into *piercing-and-sucking* structures (Fig. 28.25). In a butterfly, for example, the maxillae are drawn out into an elongated proboscis, each maxilla forming a complete sucking tube. The two tubes are interlocked, can be extended deep into the nectar-containing region of a flower, and can be rolled up toward the head when not in use. In a bee, the median parts of the labium form a tubular sucking proboscis, the lateral parts form a pair of elongated protective sheaths, and the maxillae are developed into an additional external jacket. In aphids, the mandibles and maxillae form an elongated retractable tube, which pierces plant cells and carries the juices up through the hollow interior. The extended and jointed labium here provides a dorsally grooved guiding structure for the piercing-sucking apparatus. In a female mosquito, the labium forms an elongated protective proboscis with an anterior groove. In this groove lie rapierlike mandibles and maxillae, which are the skin-piercing structures. Into a wound are then inserted two sucking tubes, an *epipharynx* formed from the labrum and the mouth roof, and a *hypopharynx*, formed from the floor of the mouth. Muscular sucking action by the pharynx behind the mouth draws up blood through the hypopharynx, and a duct within that tube also carries saliva into the wound. In a housefly, the mandibles are absent and the maxillae are reduced, but the epipharynx and hypopharynx tubes are present. Also, the labium is a foldable proboscis expanded terminally into a conspicuous pad. The latter contains trachea-

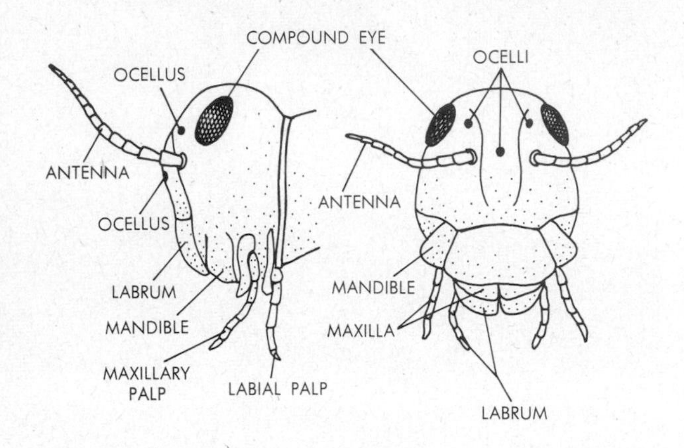

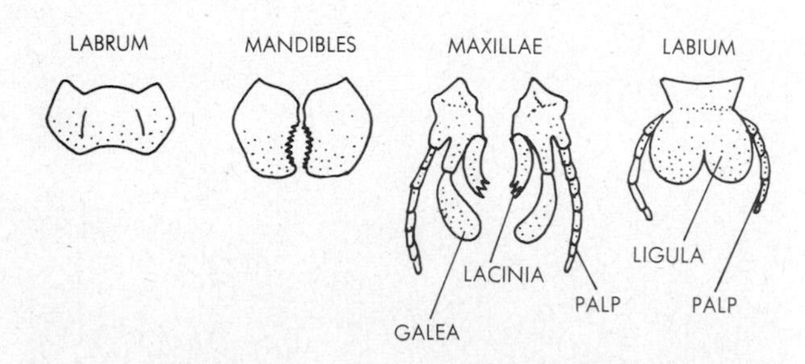

28.24. **Diagrams:** external head structure of insects (grasshopper). (*Adapted from Snodgrass.*) Side view and front view are shown, as well as diagrammatic outlines of the individual components of biting mouth parts. **Photos: A,** side view of head and thorax of locust (*Locusta*); **B,** front view of head of a moth, showing feathery antennae (a diagnostic distinction from butterflies, in which the antennae are unbranched).

A

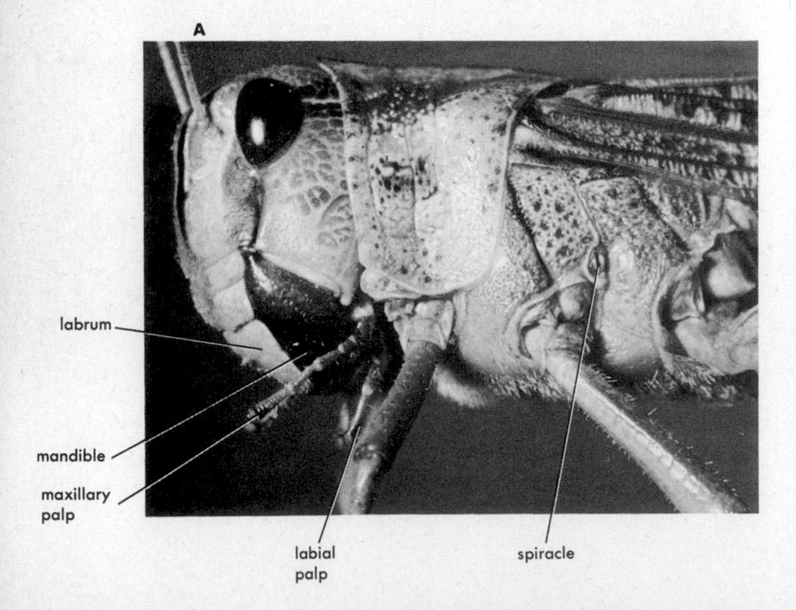

B

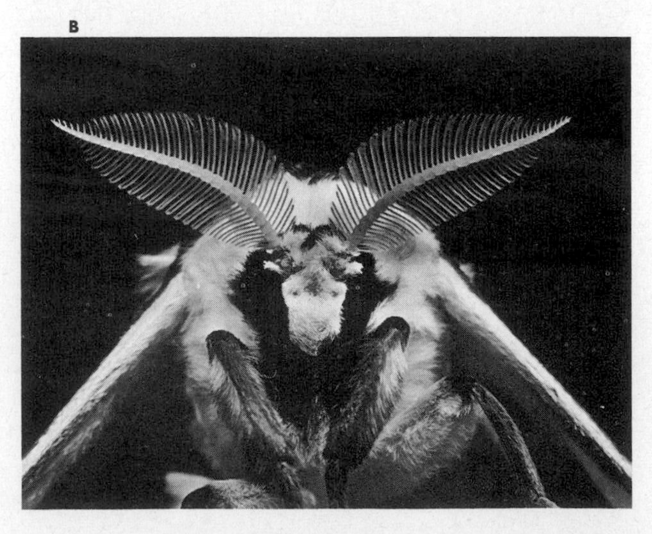

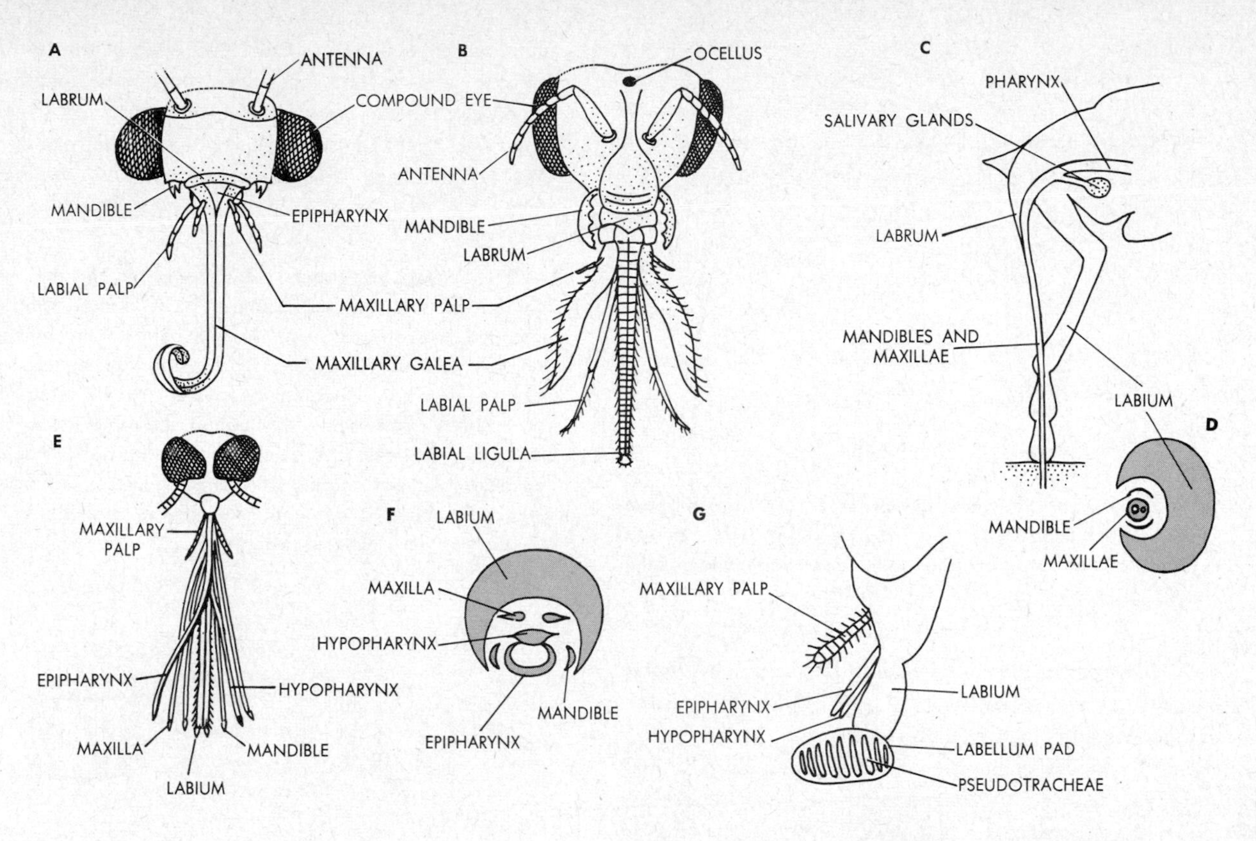

28.25. Mouth parts of insects. Diagrams: A. Front view of moth-butterfly head, showing sucking proboscis formed from maxillary components. (*After Metcalf and Flint.*) **B.** Front view of bee head, showing sucking proboscis formed from labial components. (*After Cheshire.*) **C.** Side view of hemipteran head, showing mandibular-maxillary sucking stylet and grooved labial guide. (*After Weber.*) **D.** Cross section through hemipteran stylet; in the maxilla, the forward tube conducts food up, the hind tube, saliva down. (*After Imms.*) **E.** Front view of mosquito head. The epipharynx is an extension of the labrum, the hypopharynx, an extension of the floor of the mouth; together they form a food-conducting channel, as shown in cross section in **F.** (*After Patton and Cragg.*) **G.** Side view of proboscis apparatus of house fly. (*Adapted from Borradaile.*) **Photos: A** and **B.** Head of cabbage butterfly (*Pieris*), showing proboscis in rolled-up and extended position.

like tubules and fine chitinous rasping teeth. This proboscis shields and guides the sucking tubes. Numerous other modifications of mouth parts are encountered in different insect groups.

Embryos develop lateral ocelli which become substituted in the nymphs and adults of exopteryogotes and in the adults of endopterygotes by compound eyes. Adult median ocelli develop in addition. The head joins the thorax via a narrow

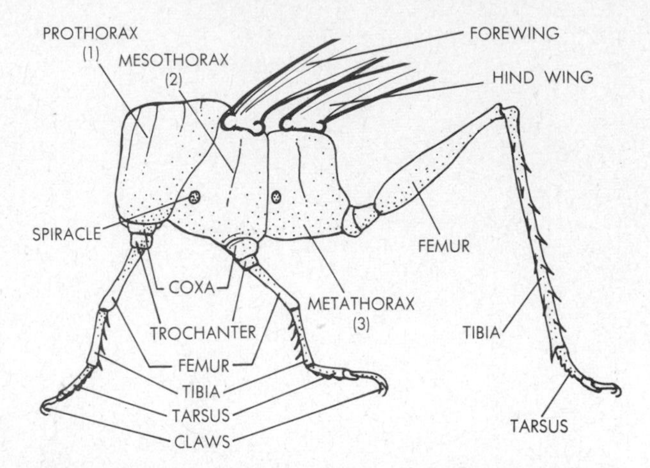

28.26. External thorax structure of insects (grasshopper, *adapted from Snodgrass*). The coxa of a leg is homologous to the coxopodite, the trochanter to the basipodite, and all other endites to the endipodite. Cf. Fig. 7.5.

neck. Of the three thoracic segments (*prothorax, mesothorax, metathorax*), each carries a pair of legs (Fig. 28.26). These are uniramous and consist of coxopodite (*coxa*), basipodite (*trochanter*), and endopodite (*femur, tibia, tarsus, claws*). Femur and tibia usually are the longest joints. Legs too are variously modified for specialized functions. For example, they may be adapted to grasping, as in a praying mantis, to jumping, as in a grasshopper, to digging, as in a mole cricket, to sound production, as in numerous Orthoptera, to food collection, as in the pollen baskets of bees, to adhesion, as in the adhesive pads between the claws of flies, and to swimming, as in the diving beetles.

Each of the second and third thoracic segments typically also bears a pair of wings. Their form and venation constitute an important taxonomic criterion. A wing is a flattened fold of the body wall, extending out dorsolaterally between the tergum and the pleuron (Fig. 28.27). Air-filled tubular spaces within the wing form the supporting veins; the chitin cuticle is thicker around these tubules than elsewhere. Muscles attaching directly to a wing are used primarily to move the appendage into a flying or a rest position. An indirect musculature between tergum and sternum brings about the flying motions themselves. Contraction of these muscles flattens the normally arched tergum and moves the wing up; relaxation permits the elastic tergum to recoil to its arched position and this brings the wing down. In butterflies and other groups, both wings on one side function in unison. Such unity is achieved by bristles or hooks which form a locking or coupling device; chitinous projections from the hind edge of the front wing lap over the back wing, and similar projections from the leading edge of the back wing lap over the front wing. In general, wings beat faster the smaller the insect. Either pair of wings may be modified. Thus, the hind wings are reduced to

28.27. Wings of insects. Diagrams: A, B, the indirect, flight-producing wing musculature. Note that a wing is an extended fold of body wall. Arrows point to the fulcrums of wings, at the dorsal edges of the pleura. In **A,** tergosternal muscles relaxed, longitudinal muscles contracted, wings moved down. **B** depicts the converse situation, the wings having moved up. **C,** diagrammatic cross section through part of a wing, showing tracheal tubes forming thickened veins. **D,** scheme of a knobbed haltere, a reduced hindwing as in Diptera. **Photo:** Venation in the wing of a dragonfly.

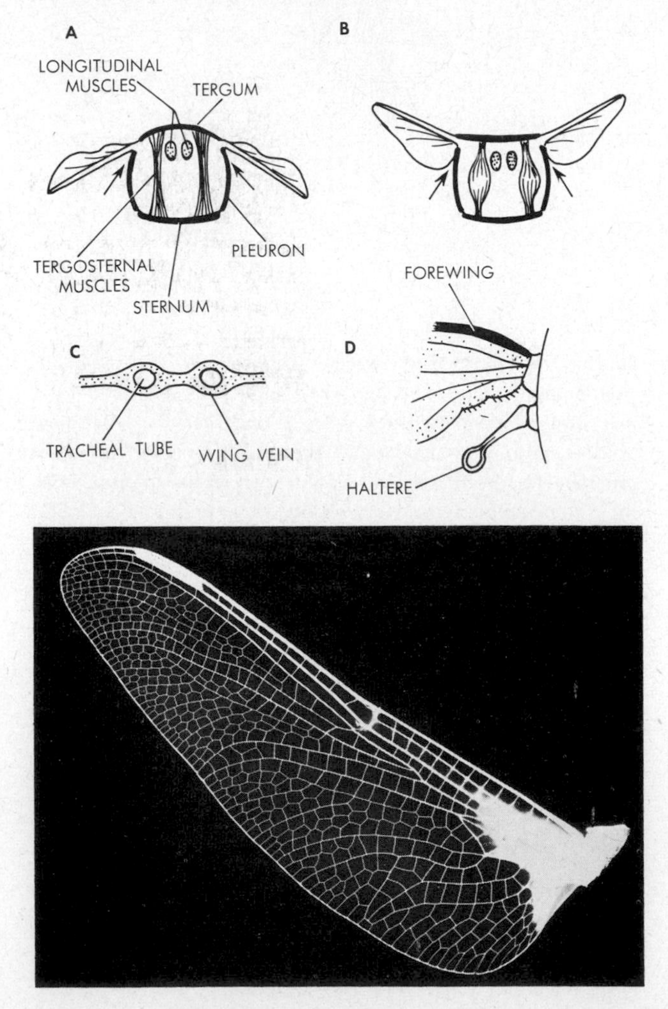

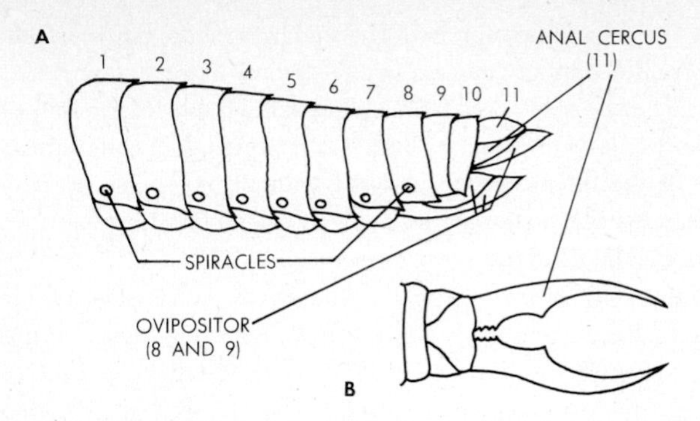

28.28. External abdomen structure of insects. **A,** side view of abdomen as in grasshopper; numbers refer to segments. **B,** dorsal view of hind end of earwig *Forficula* (order Dermaptera), showing anal cerci elongated into horny forceps.

knobbed stumps, the *halteres,* in Diptera (and the vibrations of the halteres during flight produce the buzzing sounds made by flies). In Strepsiptera it is the front wings which form halteres. The forewings are hardened into protective covers in Orthoptera, Dermaptera, and Coleoptera, and in the latter order these covers are horny *elytra.* In fleas and many members of other orders, both pairs of wings are secondarily absent.

The abdomen (Fig. 28.28) typically consists of 11 segments, though the number is often reduced secondarily. In the first seven segments, appendages begin to form in the embryo but they never mature (except in some of the Apterygota) and are absent in the adult. The appendages of the eighth and ninth segments in females become *ovipositors,* accessory egg-laying structures in most cases but formed into stings, saws, and piercers in bees and other Hymenoptera. The appendages of the ninth segment in males form copulatory organs. The anus is on the tenth segment in most cases. The eleventh, if present, is often extended as a pair of variously shaped posterior projections, the *anal cerci* (probably homologous to the caudal rami of crustacea).

The last two thoracic and the first eight abdominal segments typically bear a pair of *spiracles* each, equipped with muscle-operated valves (Fig. 28.29). These air pores lead into a highly developed tracheal system, which includes interconnecting longitudinal tubes, air sacs, and branching tracheae. Contraction of muscles attached to the terga and sterna of the segments bring about exhalation, and relaxation of these muscles leads to recoil of the skeletal plates and thus to inhalation. Such breathing motions of the abdomen can be observed readily in a quiescent fly. It has been shown that, in grasshoppers and probably many other insects, the first four pairs of spiracles are open and inhalatory while the remaining six pairs are closed; and that the first four pairs are closed while the posterior six pairs are open and exhalatory. Primitive insects such as Collembola are skin breathers without tracheal systems. The aquatic larvae of the midge *Chironomus* also breathe through the skin, the tracheal system here being undeveloped as yet. Most aquatic larvae possess *tracheal gills,*

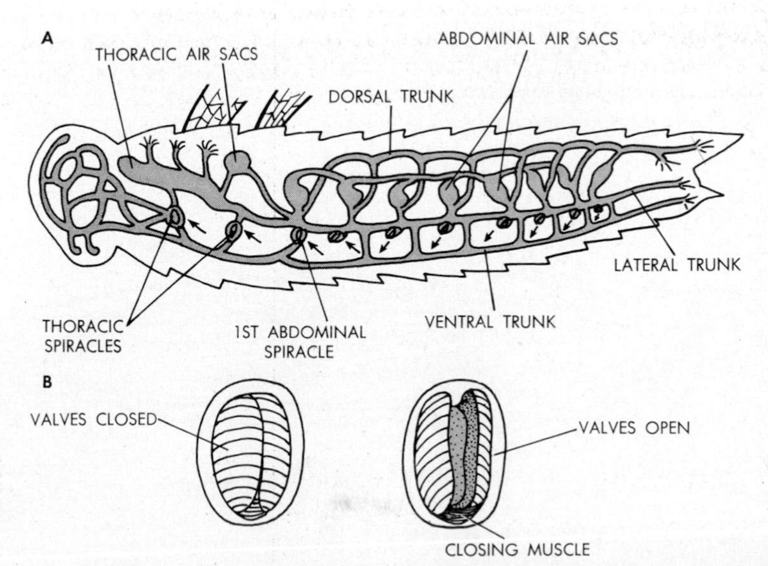

28.29. The tracheal system of insects. A. Pattern of main tracheal tubes and air sacs as in grasshopper. *(Modified from Vinal.)* Arrows show air movement through spiracles. **B.** The lidlike valves of a spiracle, in closed and open position. *(Partly after Snodgrass.)* Cf. Fig. 28.24.

expanded, hollow tufts of body wall in various segments. However, mosquito larvae breathe air via spiracles, the larvae hanging suspended from the surface film of the water in which they live; and the diving water beetle *Dytiscus* periodically carries a fresh air bubble down into the water and breathes air in that fashion.

The nervous system of insects is of the ladder type, various numbers of ventral ganglia often being concentrated into single units, as in other arthropods. Insects also possess a well-developed autonomic, sympathetic system, which innervates the alimentary tract and the musculature of the spiracles and the breathing apparatus. Sensory devices include, apart from the eyes, chemoreceptive and tactile "hairs" over most of the body and, in many cases, phonoreceptors located at various surface regions (cf. Chap. 8). In addition, insects are able to taste and to smell. The antennae are the principal olfactory organs; in many cases a male may detect the presence of an even very distant female, by antennal reception of odors secreted in scent-producing glands of a female.

The circulatory system consists of an elongated heart, typically with 13 but often with fewer pairs of ostia (Fig. 28.30). The heart lies dorsal to a usually well-developed pericardial membrane. In aphids and some other forms, accessory hearts are located at the bases of the legs (cf. Chap. 9). Blood is generally colorless, but a few types (e.g., *Chironomous*) possess hemoglobin. Excretion occurs through Malpighian tubules opening into the hindgut.

The alimentary system is fairly typically arthropodlike in structural respects but is often highly specialized functionally. Salivary glands open into the floor of the mouth cavity, and their secretions often contain proteinases and other enzymes which accomplish a measure of external digestion (e.g., in bees). The pharynx is adapted in many groups to sucking fluid foods, and a crop and a gizzard may form other subdivisions of the chitin-lined foregut. The midgut is short but is usually extended laterally into numerous surface-increasing pouches. The enzymes secreted by it are geared to the feeding habits of the animal. For example, insects subsisting largely on proteins (e.g., the blood suckers) may be without lipases or carbohydrases or may manufacture such enzymes in limited quantities only. The clothes moth *Tinea*, similarly, eats the protein keratin on sheep hair (wool), and its digestive enzymes are specialized accordingly. Analogous digestive adaptations are in evidence in most insects which subsist on unusual foods, e.g., the wood-eating and cellulase-possessing woodboring beetles, the paste- and glue-eating book lice, the silk-eating museum beetles, and others. The chitin-lined hindgut of most insects is equipped with *rectal glands*, which absorb water from the feces and so assist water conservation.

With very few exceptions, insects are separately sexed and fertilization occurs internally, often by means of spermatophore-transfer from males. Gonads are paired, and their ducts lead to a single gonopore located ventrally between the ninth and tenth abdominal segments in males, on the eighth, ninth, or tenth in females. Parthenogenesis is common in aphids and in the social insects (cf. Chap. 16). Tsetse flies and a few other types are viviparous, but most insects are oviparous and the ovipositors assist in egg-laying. The eggs are generally elliptical, large, and centrolecithal. Many nuclei are formed by division of the zygote nucleus and they migrate toward the egg surface, where they become the centers of the superficially cleaved blastomeres (Fig. 28.31 and cf. Fig. 11.12). In a *germ band* on the ventral side gastrulation then takes place, by means of either invagination, delamination, or ingression of a layer of cells. This layer is *mesoderm*, which later buds into somites; embryonic (schizo-)coelomic sacs then arise in the latter. Strictly speaking, therefore, insects possess an ecto-mesoderm. Anteriorly and posteriorly, the ectoderm also

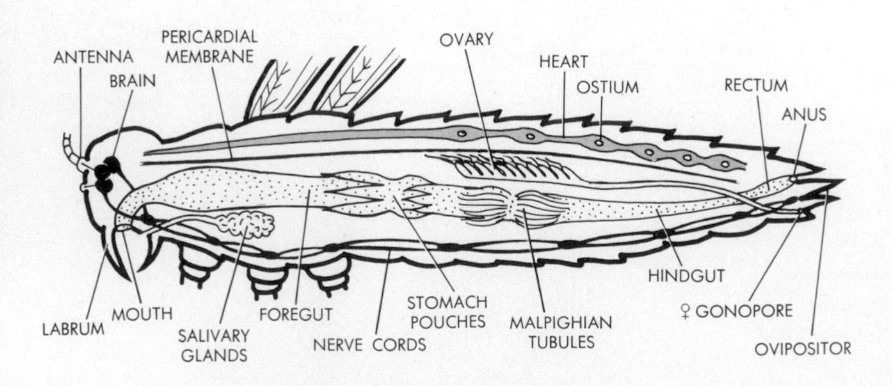

28.30. The internal structure of an insect (grasshopper), schematic.

28.31. Early insect development. (*Adapted from Eastham, Graber.*) **A** to **E.** Sagittal sections of progressively later embryos, head regions toward left. **A,** superficial cleavage of centrolecithal egg and blastula (cf. Fig. 11.12); **B,** late blastula, germ band thickened in ventral ectoderm; **C,** segmental accumulation of mesoderm, formed by various gastrulation methods; **D,** mouth and anus invaginations, and ingrowth from both of these of an endoderm layer between mesoderm and yolk; **E,** appearance of schizocoelic coelom sacs, and separation of endoderm from mesoderm, leaving hemocoel. **F.** Cross section at stage after **E.** The heart rudiments are mesodermal and will later fuse along the dorsal midline. Yolk mass disappearing. **G** and **H.** Ventral views of later embryos, showing gradual anteroposterior development of segments and segmental appendages. Note movement of mouth backward.

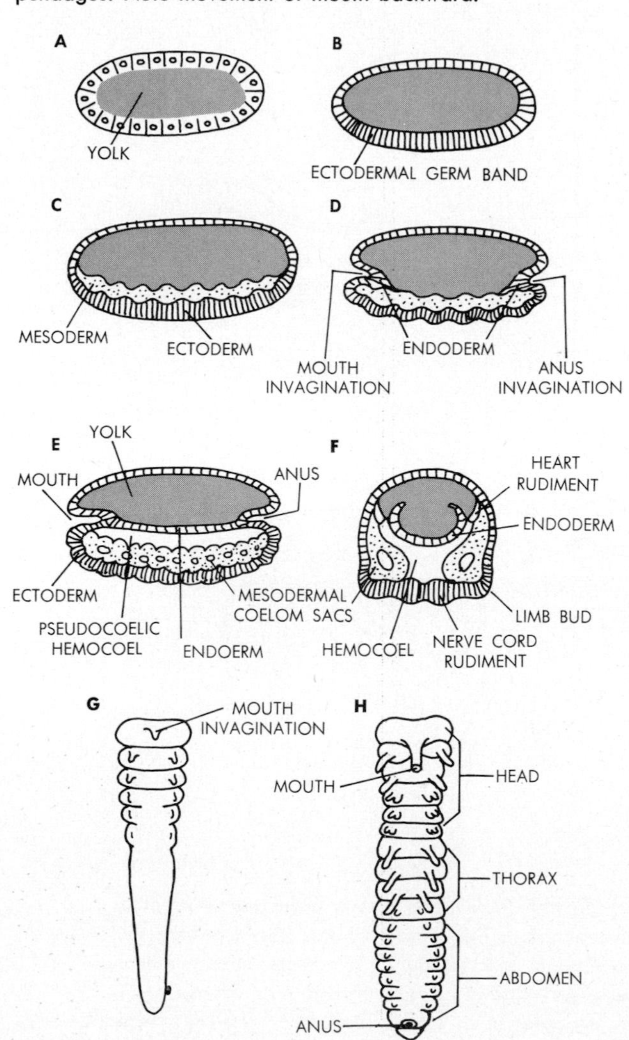

invaginates to form the later foregut and hindgut, respectively. Ingrowths from these invaginations then give rise to an endodermal, midgut-forming layer ventral to the yolk.

Appendage buds arise in anteroposterior succession on the ventral side. In Apterygota and Exopterygota, such buds develop into adult appendages. All these are typically formed in the embryonic phase (Protrura excepted). Thus, when apterygotes hatch, they are essentially miniature adults; and when exopterygotes hatch, they resemble adults greatly though the wings are still lacking. The wings then grow during the nymphal instars, as exterior folds on the second and third thoracic segments (Fig. 28.32). In Endopterygota, however, the embryos hatch in a less fully developed condition, i.e., all appendages are not yet laid down. The free caterpillar larvae therefore retain an embryonic character, and they correspond to the later *embryonic* (prenymphal) stages of exopterygotes. In the wormlike, annelidlike caterpillars, invaginated pockets arise in the body wall, and at the bottom of such pockets the imaginal disks begin to develop. These disks will give rise to the adult appendages, e.g., mouth parts, legs, and wings. Note therefore that the larva possesses mouth parts, for example, developed originally in the embryo, whereas the adult possesses a new set of mouth parts developed later from imaginal disks. In some groups it happens that the caterpillar is equipped with biting-chewing parts but that the imago comes to possess piercing-sucking parts. Transformation from the embryonic caterpillar to the imago occurs in the pupa (which consequently corresponds to the nymphal phase of exopterygotes). In a pupa, the pockets containing the imaginal disks open out and the developing appendages become exteriorized. The wings, for example, so make their first external appearance (Fig. 28.33).

The nymphal molts in exopterygotes and the larval and pupal molts in endopterygotes are under precise hormonal control. Three types of glands have been shown to participate in these metamorphoses: certain specialized *neurosecretory cells* (NS) in the brain ganglia; a pair of *corpora allata* (CA) located in the mandibular segment dorsal to the pharynx; and a pair of *prothoracic glands* (PT) in the first thoracic segment (Fig. 28.34). The NS cells rhythmically secrete a hormone which stimulates the PT glands, and the latter then secrete a hormone in their turn which acts on the body tissues and induces molting. The nature of the molt is governed by whether or not hormones are secreted by the CA glands. These glands are active during nymphal and larval stages; and under the influence of CA hormones, the PT-induced molt will produce

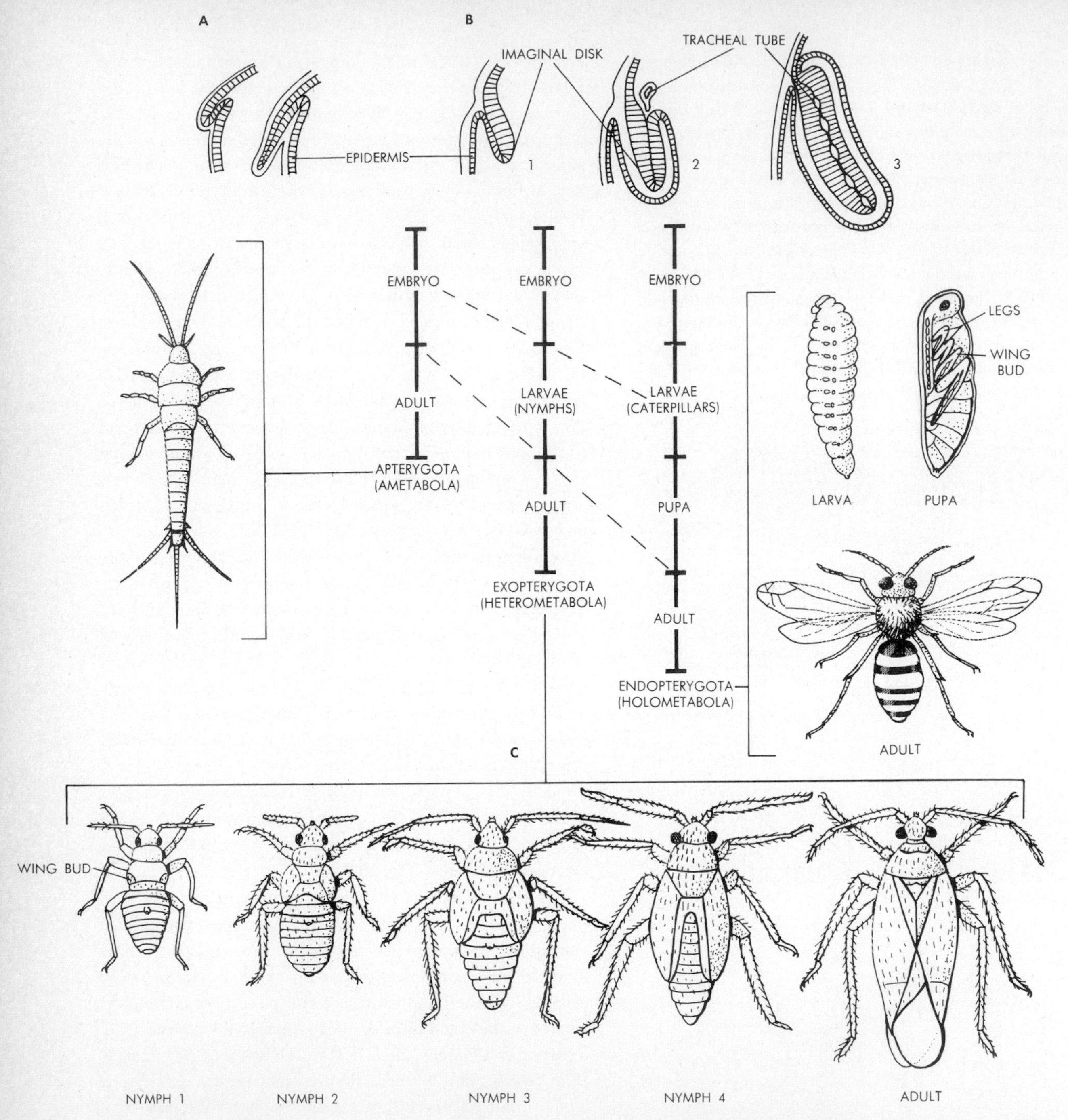

28.32. Later insect development. A. growth of a wing as an exterior epidermal fold on the body, as in exopterygotes. **B.** growth of a wing as an imaginal disk in an interior epidermal pocket, as in endopteryogotes. **C.** The main stages of development in the three large insect groups are indicated in the chart; the dashed lines suggest that embryonic stages of apterygotes correspond developmentally to exopterygote larvae, which in turn correspond developmentally to endopterygote pupae. These points are illustrated in the drawings. The apterygote shown is a silverfish, the larva here being equivalent to the adult (except for reproductive maturity). The exopteryogotes shown are the successive larval stages and the adult of a hemipteran bug, indicating progressive external wing growth, with little change otherwise. The endopterygote stages are of a hornet (order Hymenoptera). The caterpillar here is legless, and the pupa, which possesses young though adult appendages (including wing buds), corresponds to the nymphs of the exopteryogote bug.

28.33. Insect development: stages in the development of the silkworm moth *Bombyx mori.* **A,** caterpillar; **B,** spinning of pupation cocoon; **C** and **D,** cocoon and pupa within; **E,** emergence of adult, wings still uninflated; **F,** some minutes later, wings attaining mature size.

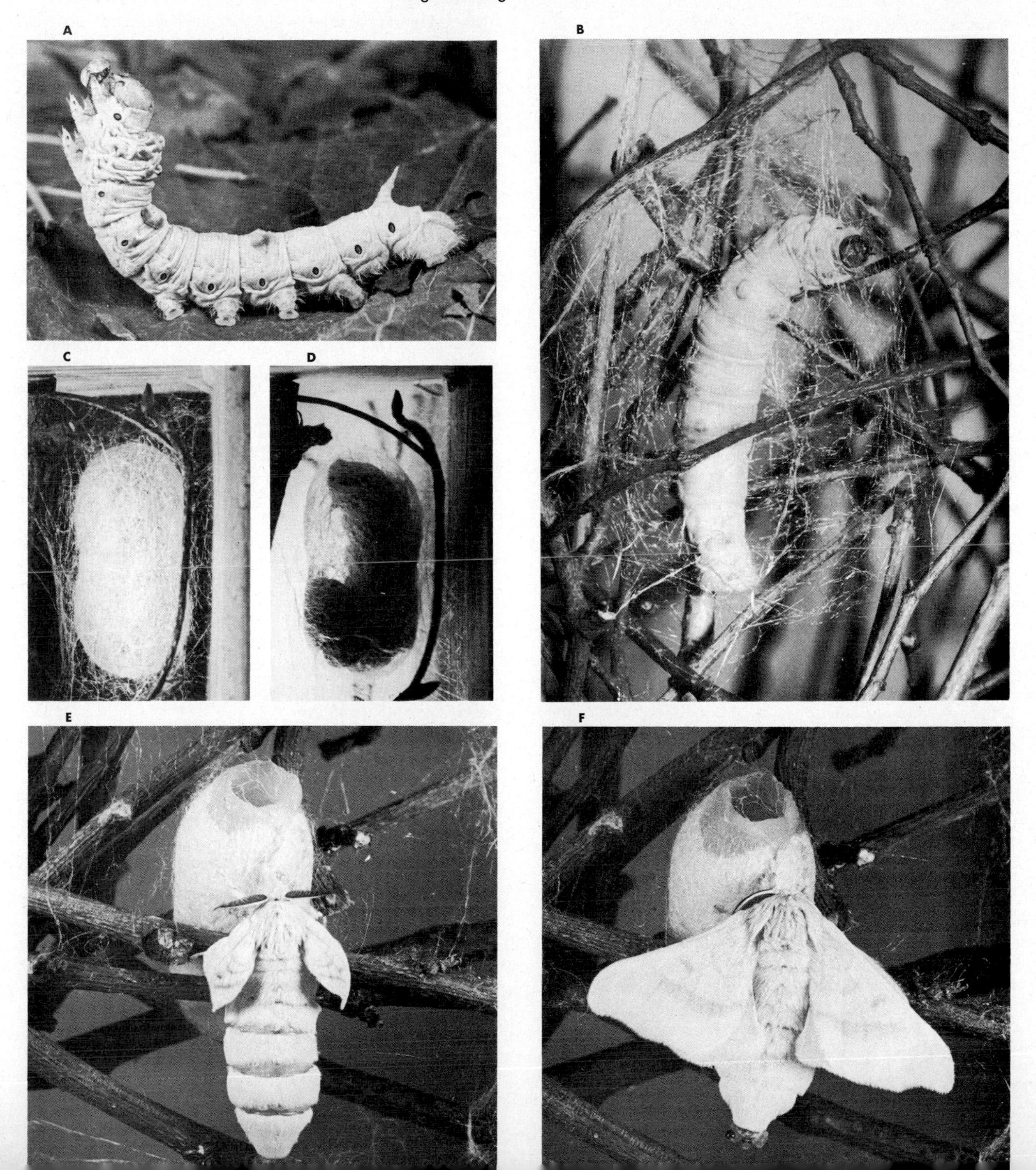

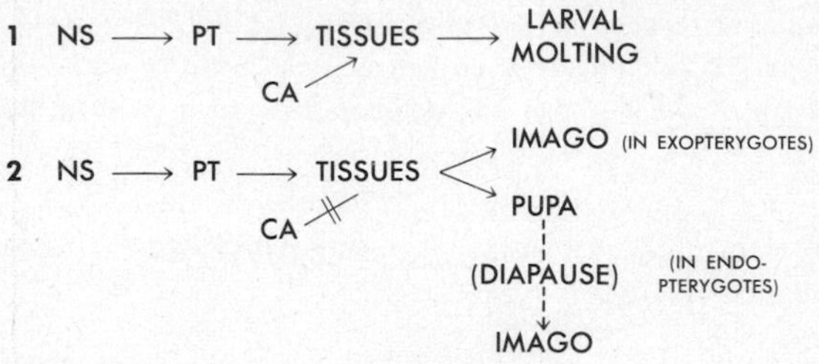

1 NS ⟶ PT ⟶ TISSUES ⟶ LARVAL MOLTING
CA ↗ (TISSUES)

2 NS ⟶ PT ⟶ TISSUES ⟶ IMAGO (IN EXOPTERYGOTES)
CA ↗̸ (TISSUES) ⟶ PUPA ⋮ (DIAPAUSE) ⋮ IMAGO (IN ENDOPTERYGOTES)

28.34. Hormonal controls of insect development. Neurosecretory cells from the brain (*NS*) stimulate production of prothoracic gland hormone (*PT*) which, in conjunction with corpora allata hormones (*CA*), determine larval molting (step 1). After CA ceases to become available (step 2), the next *PT*-induced molt produces the adult in exopterygotes, the pupa in endopterygotes. The latter usually passes through a dormant phase, the diapause, and the following *PT*-induced molt then yields the adult.

the next nymphal or larval stage. At a certain point in development, however, the CA glands cease their hormone secretion (and it is not yet known how this cessation is regulated). When that happens the next PT-induced molt will form an adult in exopterygotes and a pupa in endopterygotes. And when a pupa later undergoes the next PT-induced molt, it too will become an imago. In effect, therefore, PT hormone acting alone transforms a preadult stage into an adult, either in one molting step or in two; but PT hormone acting in conjunction with CA hormone transforms a preadult stage into another preadult stage. The CA glands thus maintain a juvenile condition. That this is so can be shown by experimentally grafting a nymphal head and thorax of one insect to an adult abdomen of another. Normally, the adult abdomen would not molt at all or, under the influence of PT hormone alone, would molt but remain adult. However, under the influence of the CA hormone in the experimental graft, the abdomen molts and reverts at least partially to a "rejuvenated" larval stage.

We may note that after endopterygotes attain the pupal stage, the pupa normally passes through a *diapause*, a dormant phase of various durations in different groups. During a diapause the NS cells and the PT glands are inactive and molting does not occur. In some cases at least, emergence from the diapausal condition is apparently triggered by low temperatures. The NS cells become reactivated by cold, secretion of PT hormone then follows, and the adult-producing imaginal molt eventually supervenes. It is possible that, in nature, the low temperatures of the winter season prepare the way for an imaginal molt in the spring.

REVIEW QUESTIONS

1. Give taxonomic definitions of the mandibulate subphylum and the crustacean class. Name each of the crustacean subclasses and the superorders of Malacostraca. Name representative orders and genera in as many of these groups as you can.

2. Review the probable evolutionary interrelations of the crustacean groups. What evolutionary trends are in evidence in each broad line of descent? How do crustacea typically feed? What are phyllopodial legs and in which groups do they occur?

3. Describe the segmental structure of the crustacean head, thorax, and abdomen. Name and state the functions of all appendages of a lobster. Show how appendageal structures and functions vary among crustacea. What are apodemes? Where is the sinus gland located and what is its function?

4. Describe the internal anatomy of a lobster. In the process review the organization of every organ system. What are pleural and epipodial gills?

5. Distinguish between anamorphic and epimorphic development. Show how crustacea develop segmentally through activities of a pygidium. Review the nature of the larval stages of crustacea.

6. Describe the general characteristics and state the taxonomic position of (*a*) water fleas, (*b*) copepods, (*c*) barnacles. Describe the development and general adult structure of a barnacle. How do crustacea mate? Review the ecology of crustacea and show in what ways certain crustacea (which?) are land-adapted.

7. Give taxonomic definitions of each of the myriapod classes. Describe the segmental head structure of myriapods. Review the trunk structure and contrast the external anatomy of a centipede and a millipede. Which features of which myriapods are similar to those of insects? Which myriapods develop anamorphically? What are the ways of life of different myriapods?

8. Give taxonomic definitions of the class of insects and of each subclass and superorder. Within each such subgroup name as many orders and representative genera as you can.

In any event characterize the orders Orthoptera, Homoptera, Hemiptera, Lepidoptera, Diptera, Coleoptera, and Hymenoptera. In what respects are Protrura not insectlike?

9. Describe the segmental structure of an insect and contrast it with that of (*a*) a myriapod, (*b*) a crustacean. Review the internal anatomy of an insect and in the process describe the organization of every organ system.

10. Contrast the structure of biting, sucking, piercing, and other types of mouth parts and give specific examples of each. What is the structure of an insect wing and how is flight motion produced? Show how (*a*) wings, (*b*) legs are modified in different insect groups.

11. How do aquatic insects breathe? What alimentary specializations occur among insects in conjunction with particular modes of life? How do insects mate? In what environments are insects not found? Compare the species abundance of crustacea, myriapods, and insects.

12. Describe the early development of an insect and show how later development differs according to the subclass or superorder. Describe the hormonal controls of insect molting and development. What are trochanters, elytra, ovipositors, naiads, corpora allata, diapause?

COLLATERAL READINGS

Most of the following articles are popularly written:

Carpenter, F. M.: The Geological History and Evolution of Insects, *Amer. Scientist*, vol. 41, 1953.

Davidson, T.: Rose Aphids, *National Geographic*, June, 1961.

Fraenkel, G.: The Function of the Halteres of Flies, *Proc. Zool. Soc. London* (A), vol. 109, 1939.

——— and J. W. S. Pringle: Halteres of Flies as Gyroscopic Organs of Equilibrium, *Nature*, vol. 141, 1938.

Hocking, B.: Insect Flight, *Sci. American*, Dec., 1958.

Johnson, C. G.: The Aerial Migration of Insects, *Sci. American*, Dec., 1963.

Murphy, R. C.: The Oceanic Life of the Antarctic, *Sci. American*, Sept., 1962.

Pringle, J. W. S.: The Excitation and Contraction of the Flight Muscles of Insects, *J. Physiol.*, vol. 108, 1949.

Thorpe, W. H.: Orientation and Methods of Communication of the Honeybee and Its Sensitivity to the Polarization of Light, *Nature*, Vol. 164, 1949.

Van der Kloot, W. G.: Brains and Cocoons, *Sci. American*, Apr., 1956.

Waterman, T. H.: Flight Instruments in Insects, *Amer. Scientist*, vol. 38, 1950.

Wigglesworth, V. B.: Metamorphosis and Differentiation, *Sci. American*, Feb., 1959.

Williams, C. M.: The Metamorphosis of Insects, *Sci. American*, Apr., 1950.

———: Insect Breathing, *Sci. American*, Feb., 1953.

Zahl, P. A.: Mystery of the Monarch Butterfly, *Nat. Geographic*, Apr., 1963.

Insects and mandibulates generally are the subjects of the books below:

Chu, H. F.: "How to Know the Immature Insects," William C. Brown, Dubuque, Iowa, 1949.

Green, J.: "A Biology of Crustacea," Quadrangle Books, Chicago, 1961.

Imms, A. D.: "A General Textbook of Entomology," 9th ed., Methuen, London, 1947.

Jacques, H. E.: "How to Know the Insects," 2d ed., William C. Brown, Dubuque, Iowa, 1947.

Johannsen, O. A., and F. H. Butt: "The Embryology of Insects and Myriapods," McGraw-Hill, New York, 1941.

Metcalf, C. L., and W. P. Flint: "Destructive and Useful Insects," 3d ed., McGraw-Hill, New York, 1951.

Roeder, K. D. (ed.): "Insect Physiology," Wiley, New York, 1953.

Ross, H. H.: "A Textbook of Entomology," 2d ed., Wiley, New York, 1956.

Snodgrass, R. E.: "Textbook of Arthropod Anatomy," Cornell, Ithaca, N.Y., 1952.

Wigglesworth, V. B.: "The Principles of Insect Physiology," Methuen, London, 1950.

coelomate bilateria: deuterostomes

PART 10

[Grade BILATERIA]

[Subgrade COELOMATA]

DEUTEROSTOMIAL COELOMATES: development typically regulative; cleavage radial or bilateral; adult anus formed at or near embryonic blastopore; coelom arises enterocoelically; larvae if present not trochophores; most groups unsegmented, some advanced groups segmented. *Chaetognatha, Branchiata, Hemichordata, Echinodermata, Chordata*

After primitive coelomate animals had arisen from acoelomate ancestors, (cf. Fig. 14.29), one series of lines evolved into the protostomial animals dealt with in preceding chapters. A second major series concurrently must have produced deuterostomial types, characterized by regulative eggs, radial and bilateral (i.e., not spiral) cleavage, anus-formation at or near the blastopore, and coelom-formation in enterocoelic fashion, i.e., by separation from the embryonic gut. Moreover, where larvae developed, these were neither trochophores nor like any of the larval types characteristic of the protostomes.

Like the protostome radiation, that of the deuterostomes analogously gave rise to unsegmented types first. These are represented today by the first four phyla listed above and by the most primitive group of the fifth. The remaining groups of the fifth phylum, notably the vertebrates, comprise segmented animals. The nonchordate phyla form the subject of the first chapter of this part and the three following chapters are devoted to the chordates.

CHAPTER 29 echinoderms and allied groups

Of the four phyla to be discussed below, the chaetognaths do not appear to be related very closely to the other deuterostomes; they probably exemplify an independent, early branch of deuterostome evolution. The remaining phyla, i.e., branchiates, hemichordates, and echinoderms, and indeed also the chordates, probably do represent a broadly interrelated assemblage. Hemichordates may be closest to an ancestral deuterostome line, and branchiates, echinoderms, and chordates may have arisen from it as independent branch lines. However, we may note here generally that precise phylogenetic interrelations of deuterostome groups can be determined even far less definitely than those of protostomes.

PHYLUM CHAETOGNATHA

Arrowworms; marine, mainly planktonic; head with grasping spines and hood; trunk with lateral and tail fins; coelom subdivided into three compartments; without circulatory, breathing, or excretory systems; hermaphroditic, oviparous, eggs regulative; development without larvae. *Sagitta, Spadella*

Chaetognaths are small, torpedoshaped animals, most of them well under 2 in. long (Fig. 29.1). Although they number only about 50 species, some of them, particularly *Sagitta,* are so abundant that they often are among the principal constituents of zooplankton. Arrowworms are carnivorous, feeding on minute planktonic animals such as copepods, and they are themselves eaten by fishes and plankton-feeders generally.

The body of an arrowworm is marked into head, trunk, and tail, and the internal coelomic cavity is subdivided analogously by partitions. A transverse head-trunk septum lies just behind the head, and a vertical longitudinal septum divides the trunk cavity into two lateral compartments. These are the primary coelomic divisions of the animal. Secondarily, another transverse septum partitions the trunk cavity into an anterior main compartment and a posterior tail compartment. The body wall consists of (nonchitinous) cuticle; epidermis (without cuticle in glandular and sensory regions); a conspicuous basement membrane; and a layer of body-wall muscles arranged in the form of six longitudinal bands (Fig. 29.2). Circular muscles and a peritoneum are absent. The septa in the body cavity represent interior extensions of the basement membrane. This membrane is thickened in parts of the body wall into strengthening "skeletal" plates and is also extended exteriorly into *fin rays,* i.e., supporting strands in the cuticular fins. One or two pairs of trunk fins and a posterior rounded tail fin are present. All are positioned horizontally and they are unmuscled; they are not used in locomotion but probably serve primarily as stabilizers and buoyancy-promoting devices.

The head bears one or two pairs of rows of tiny, anterior teeth; a pair of dorsolateral eyes (each consisting of a cluster of cup-ocelli); and a row of prominent, chitinous *grasping*

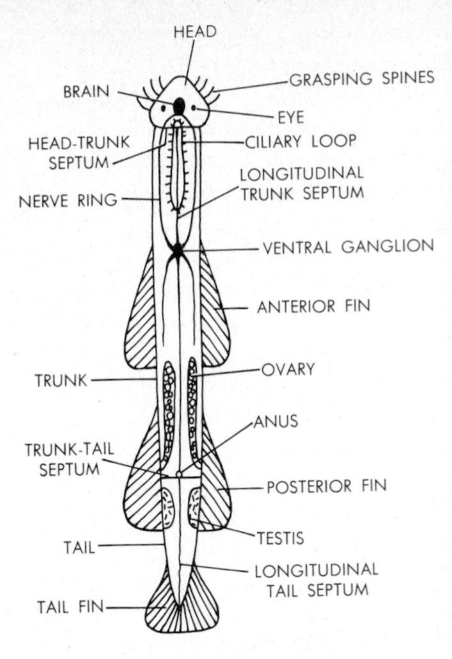

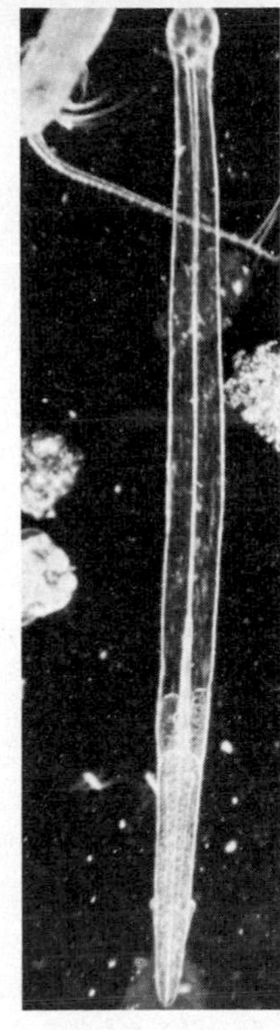

29.1. Chaetognaths. Diagram and photo of *Sagitta,* showing general body design.

spines on each side of a ventral mouth vestibule (cf. Fig. 29.2). The teeth and the spines are moved by a powerful head musculature, and the spines are the principal food-catching structures; they can be closed in toward the mouth, forming a tiny cage in which food animals may become trapped. Head muscles also operate a *hood,* a fold grown out from the body wall at the head-trunk juncture (and containing part of the coelom). This hood can be extended forward to cover almost the whole head, spines included; it probably streamlines the head when the animal darts about rapidly. All the head- and body-wall muscles are striated.

The nervous system comprises a dorsal brain ganglion, a ventral trunk ganglion, a connective cord on each side between these (located between epidermis and basement membrane), and posterior cords and plexi emanating from the trunk ganglion. Partially embedded in the brain is a pair of dorsal *retrocerebral organs,* sacs which are possibly glandular and which open dorsally and immediately behind the brain through a retrocerebral pore. Just posterior to this pore the epidermis is modified into a *ciliary loop,* a patch of tissue composed of glandular cells and ringed by flagellated cells. It has been suggested that the loop might serve sensory and/or excretory functions. The alimentary tract is straight and is supported in the trunk within the vertical mesenteric septum. The esophagus is expanded posteriorly into a bulbous enlargement, and the intestine extends into a pair of anteriorly pointing digestive pouches. The anus opens ventrally, just before the trunk-tail septum. Fluid fills the coelomic cavities. Specialized circulatory, breathing, excretory systems are absent.

All arrowworms are hermaphrodites. Paired testes are situated just behind the trunk-tail septum, paired ovaries just in front of it. Gonopores open at corresponding locations, though male openings apparently are not permanent, being formed only during the breeding season. An oviduct has an-

29.2. Chaetognath structure, diagrammatic. (*Based on Ritter-Zahony.*) **A,** cross section through trunk of *Sagitta;* **B,** anterior end of *Sagitta,* seen from ventral side, hood retracted.

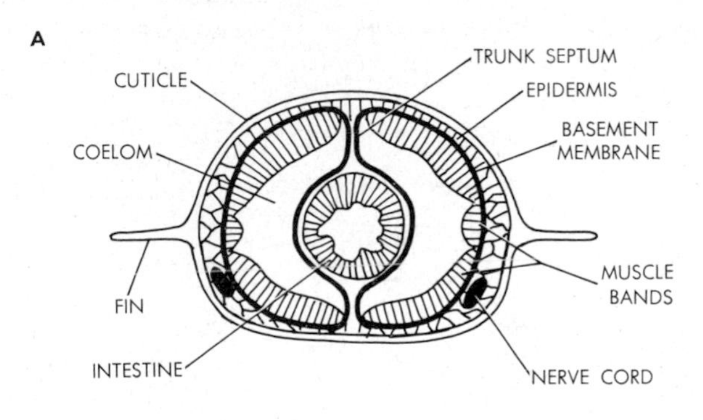

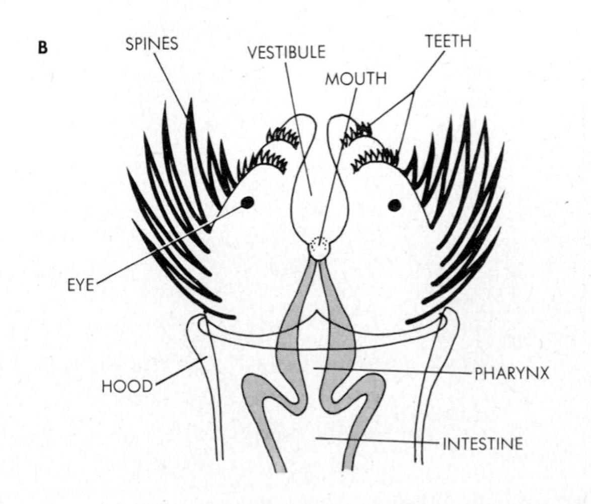

other syncytial tube within it, and this inner tube leads to the female gonopore, where it provides an entry path for sperms (Fig. 29.3). Eggs are shed into the space between the outer and inner tube, and sperms somehow reach this location and effect fertilization there. Zogotes are then shed via temporary openings through oviduct and body wall, not through the female gonopore.

Eggs develop externally, in some cases attached in batches to stones or other objects in water. The eggs are isolecithal, cleavage is radial and regulative, and a blastula with a large blastocoel is formed (cf. Fig. 29.3). The ensuing gastrulation constitutes a textbook example of embolic invagination and produces a perfectly two-layered embryo. The posterior blastopore soon closes over but the anus will later form at that region. Anteriorly, a pair of large, subsequently gamete-forming cells is budded into the archenteron. Also, two invaginated endodermal folds begin to grow from the anterior end backward, until they meet and fuse with the posterior wall of the endoderm. Two coelomic sacs are cut off in this manner, one on each side of the future alimentary tract. The gamete-forming cells come to lie in these sacs. The sacs then also partition transversely into head and trunk coeloms, and mouth and anus break through at the ends of the alimentary tube. All these coelomic and alimentary cavities then fill up temporarily with cells, and the embryos hatch at this stage as entirely solid constructions. Later the internal spaces reappear, but it is not quite clear if these adult body cavities correspond exactly to the embryonic cavities. The temporarily solid phase also explains the absence of a peritoneum in the adult; the embryonic peritoneum produces the adult mesodermal structures and it disappears as a distinct membrane during the solid condition of the embryo. After hatching, the animals resemble miniature adults, and they gradually become mature adults without special larval stages. Like unsegmented deuterostomes generally, chaetognaths regenerate well.

29.3. Chaetognath development. *(Adapted from Burfield.)* **A.** Cross section through female reproductive system. Sperms pass from the inner tube into the oviduct channel, where eggs are present and fertilization occurs. **B.** Sequence of embryonic stages. 3 and 4 show development of enterocoelic mesoderm and coeloms.

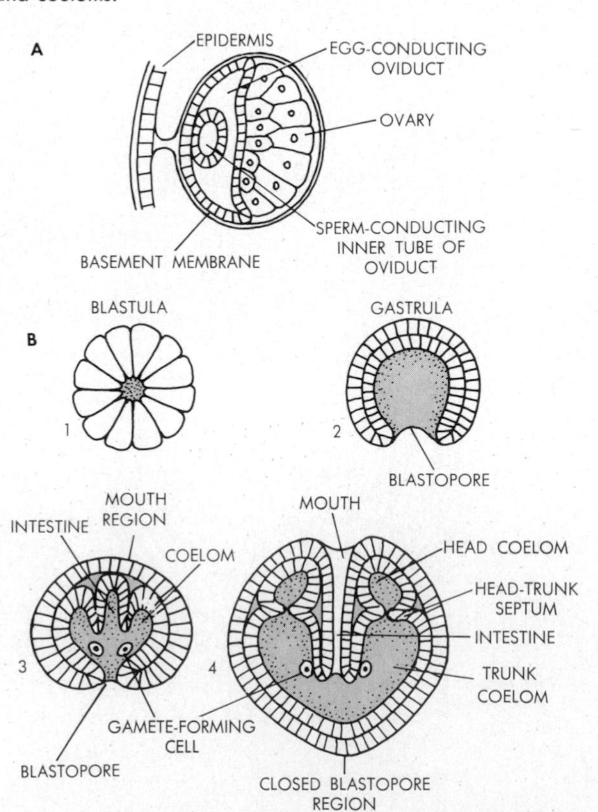

The specific evolutionary affinities of chaetognaths are obscure. Certain traits of the adults are superficially reminiscent of the pseudocoelomate groups, e.g., the internal membranous septa, the lack of a peritoneum, the longitudinal bands of body-wall muscles. However, the developmental pattern of arrowworms is unequivocally deuterostomial and enterocoelic. Yet it must be noted that this pattern differs from those of the other deuterostomial groups. Indeed, the manner of coelom-formation by a partitioning off of sacs from the archenteron closely resembles that of some of the lophophorate brachiopods. Also, the presence of chitin in the grasping spines is a trait quite atypical of deuterostomes generally, all of which are otherwise entirely without chitin. On balance, therefore, chaetognaths may exemplify a very early offshoot of deuterostomial evolution; the line presumably branched away from a main deuterostomial stock very soon after such a stock itself became separated from the protostomial groups (cf. Figs. 14.29 and 29.15). The residual resemblances of chaetognath embryos to those of pranchiopods, and of chaetognath adults to those of pseudocoelomates, might be explained on this basis.

PHYLUM BRANCHIATA (POGONOPHORA)

Beard worms; marine, tube-dwelling, abyssal; with anterior tentacles; alimentary system entirely absent; circulatory

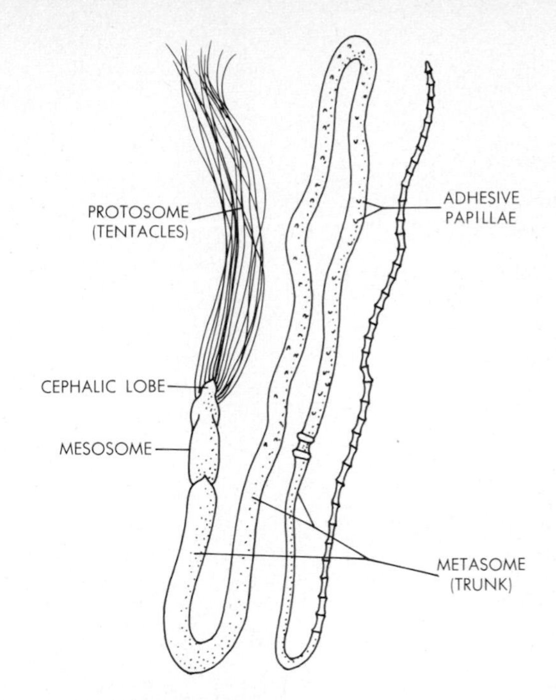

29.4. Pogonophora. The general features of the beard worm *Lamellisabella* are diagrammed. (*After Ivanov.*)

system closed; without breathing system; coelom divided into *protocoel, mesocoel,* and *metacoel;* sexes separate. *Polybranchia, Lamellisabella*

This interesting group, presently encompassing 25 species, was first discovered in 1933 in material dredged up from deep sea bottoms. The animals range in length from about 4 in. to 1 ft, but they are exceedingly thin, with a diameter of not more than about 1 mm. They live in close-fitting secreted tubes which may be up to 5 ft long and which probably stick vertically in the ooze of the ocean floor. Such tubes are composed of polysaccharide-containing *tunicin,* the same celluloselike material which also forms the test of the chordate tunicates. Beard worms are without mouth, anus, digestive tract, or any other trace of an alimentary system—the only free-living animals so characterized (Fig. 29.4).

The body of a beard worm is marked by transverse internal septa into an anterior *protosome,* a middle *mesosome,* and posterior *metasome.* The coelomic cavities corresponding to these divisions are the *protocoel,* the *mesocoel,* and the *metacoel,* respectively. As we shall see, such divisions are generally characteristic also of all other deuterostome phyla. The protocoel is probably homologous to the head coelom of chaetognaths, and the mesocoel and metacoel together correspond to the trunk coelom of chaetognaths. This internal division of beard worms is not clearly marked externally, but the protosome bears the tentacles, the mesosome forms a short, slightly thickened collarlike section, and the metasome represents the entire remainder of the body. The body wall consists of a polysaccharide-containing cuticle, a glandular epidermis, and a thin circular and a thick longitudinal muscle layer (Fig. 29.5). A distinct peritoneum is lacking in the metacoel but is present in the protocoel and the tentacles. The nervous system is a primitive intraepidermal plexus, thickened into a longitudinal cord which enlarges anteriorly into a ganglionic center. The latter is located in a *cephalic lobe,* at one side of the tentacle bases. Inasmuch as a mouth is absent, the cephalic lobe and the nerve cord have been taken arbitrarily to define a "dorsal" side.

The protocoel leads to the exterior via pair of ducts, presumed to be nephridial in function. Also, the protocoel extends into the tentacles, which are long, fingerlike outgrowths from the body wall. The tentacles bear ciliary tracts as well as numerous tiny lateral extensions, the *pinnules.* Each of these is an elongated epidermal cell. A blood-vessel loop passes through the peritoneum of each tentacle, with branches into the pinnules. These tentacular vessels connect with a main longitudinal dorsal and a similar ventral trunk vessel, in which blood flows *backward* dorsally and *forward* ventrally (as dorsal and ventral are here defined). The system is closed, and the trunk vessels are supported in a vertical mesentery which divides both the mesocoel and the metacoel into lateral compartments. The mesocoel is without ducts to the exterior, but the metacoel communicates with the outside via gonoducts. In males, one pair of gonopores opens ventrally just behind the mesocoel-metacoel septum, and in females the gonopores lie near the middle of the trunk.

Beard worms are believed to feed by arranging their tentacles into a tube, the pinnules serving as interlocking devices. The ciliary tracts on the tentacles are assumed to draw food-bearing currents into the tube. Trapped inside, food organisms would be digested extracellularly, and the resulting nutrients would be absorbed directly into the tentacular blood vessels. We may note, however, that actual feeding has so far remained unobserved. Moreover, the tentacular epidermis is not glandular, contrary to what should be expected if it were to secrete enzymes. Thus, the nutrition of beard worms still remains somewhat of an enigma. Better known is the pattern of embryonic development (Fig. 29.6). Cleavage is radial or bilateral, a stereoblastula is formed, and

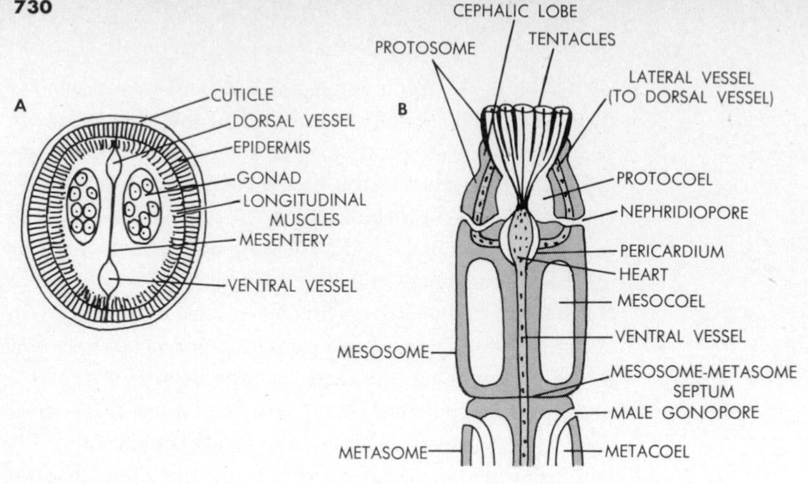

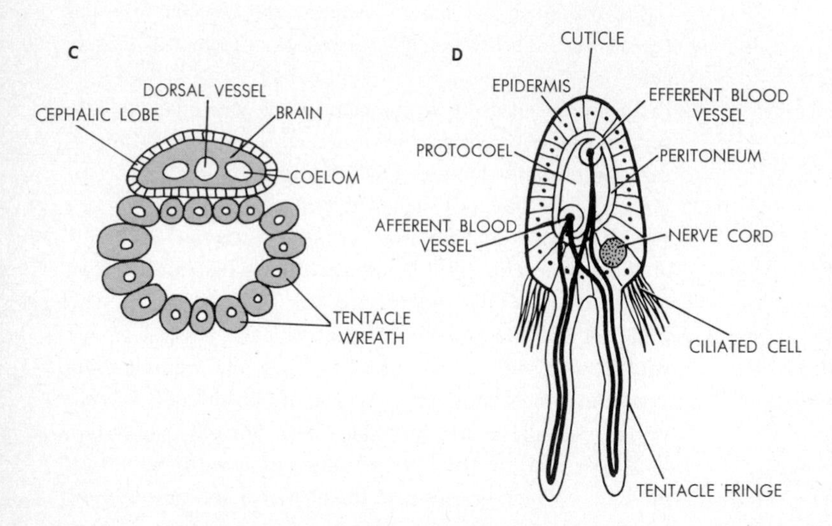

29.5. Pogonophoran structure, diagrammatic. A. Cross section through trunk (metasome). **B.** Schematic sectional ventral view of anterior end of body. The lateral blood vessels join dorsal to the heart into the dorsal longitudinal vessel (hidden in this view). **C.** Cross section near tentacle bases, at level of cephalic lobe, showing cerebral ganglionic mass. **D.** Cross section through a tentacle. (**A, B, D,** *after Ivanov;* **C,** *after Johansson.*)

gastrulation takes place by delamination. A blastopore never arises, but the region where it would be expected to form becomes the posterior end of the animal. A pair of coelomic mesodermal sacs cuts off anteriorly from the endoderm. These two sacs enlarge in a posterior direction and then constrict into proto-, meso-, and metacoelic portions. The endoderm never acquires a cavity, and an alimentary system consequently does not develop even in rudimentary form. Embryos are apparently brooded within the tube of the female parent, and in some genera elongated, wormlike larvae resembling the adults have been observed.

The pogonophorans evidently are enterocoelomates, and their embryology and adult coelomic structure relate them clearly to the other deuterostomes, hemichordates in particular.

PHYLUM HEMICHORDATA

Marine; wormlike, colonial in secreted housings or solitary in sand burrows; body marked prominently into protosome, mesosome, metasome, and coelom divided into protocoel, mesocoel, metacoel; mostly with gill slits; sexes largely separate; development typically via larvae.

Class PTEROBRANCHIA: colonial, zooids microscopic;

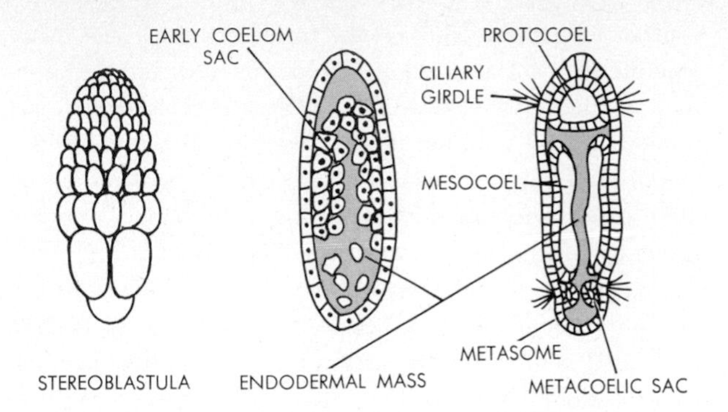

29.6. Stages in pogonophoran development. (*After Ivanov.*) The stereoblastula is an external view, the other two stages are sections. Note that the metasome at first is restricted to the posterior portion of the embryo. The three coelomic body divisions arise from the early coelomic sacs. Inasmuch as this mesoderm develops from the endodermal mass, mesoderm and coelom may be considered to form enterocoelically.

with or without gill slits; alimentary tract U-shaped; circulation open; with tentacles on mesosome. *Cephalodiscus, Rhabdopleura, Atubaria*

Class ENTEROPNEUSTA (acorn worms): solitary; with many gill slits; alimentary tract straight; circulation closed in some; without tentacles; often with *tornaria* larvae. *Protoglossus, Balanoglossus, Saccoglossus*

The approximately 100 species of these animals share one important trait with the chordates, viz., gill slits in the pharynx. In the past, the animals were also believed to possess a notochord and were actually classified as a subphylum within the chordate phylum. However, it is now clear that the supposed "notochord" of hemichordates is little more than a pouch from the mouth cavity, and that the group as a whole is more closely related to echinoderms than to chordates. It is therefore properly recognized as a separate deuterostome phylum, allied with beard worms and echinoderms and only distantly with chordates. Of the two hemichordate classes, the pterobranchs are probably the more primitive group; but the enteropneusts are the more familiar and their larvae may be of great theoretical significance.

Enteropneusts live singly in sand burrows along shallow water shores (Fig. 29.7). They feed on the microscopic food in sand. From 1 in. to 2 yd long, the body of an acorn worm consists of a short or long *proboscis* (protosome), a short *collar* (mesosome), and a long *trunk* (metasome). The proboscis is a muscular burrowing organ. Also, copious mucus secretion from its surface entangles sand and food which pass into the mouth. This opening lies ventrally at the base of the proboscis, under a forward flange of the collar. On the trunk are a prominent dorsal and a similar ventral longitudinal ridge, within which lie nerve cords and blood vessels. Anteriorly to each side of the dorsal ridge are a series of gill slits, usually U-shaped. Laterally, the body wall may be extended into prominent folds which may curve up and cover the gill slits partially. The gonads lie within these folds, which are therefore called

29.7. Model of *Dolichoglossus*, an acorn worm; dorsal view. Note proboscis at right, conspicuous collar, and row of paired gill slits along anterior portion of trunk. The lateral edge of this region of the trunk is folded up on each side, forming the genital ridges.

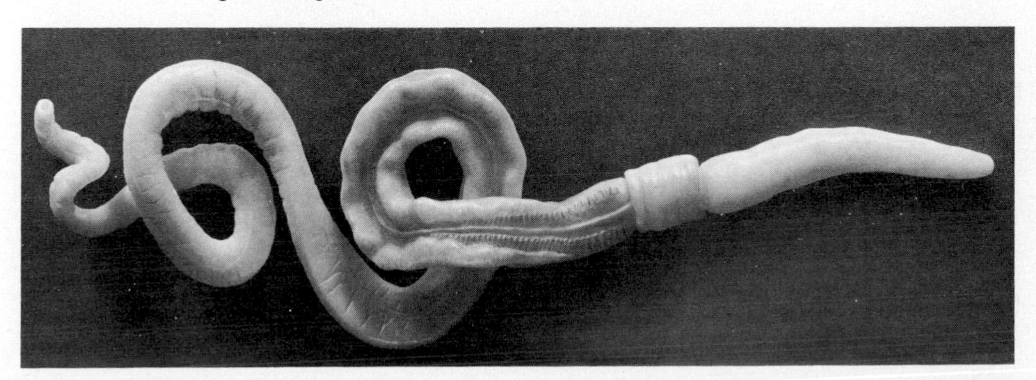

genital ridges. The posterior portions of the trunk may be variously ridged transversely; at the end of the trunk is the anus.

The body wall (Fig. 29.8) consists of a glandular ciliated epidermis, mucus-secreting in most regions; a basement membrane, thickened in parts of the protosome into a proboscis "skeleton"; and an outer circular and an inner longitudinal muscle layer in the proboscis but only a longitudinal layer in most of the collar and in the trunk. In *Protoglossus* a peritoneum is present but in other enteropneusts such a membrane is generally absent. Much of the coelom of these animals is filled with connective tissue and muscle, formed directly from the peritoneum in the embryo and thus accounting for the lack of this membrane in most adults. The nervous system is a primitive intraepidermal plexus, thickened into a ventral and dorsal cord in the trunk. At the collar-trunk juncture the ventral cord joins the dorsal cord, and the latter then continues forward as the *collar cord.* This section lies inside the body wall, not in the epidermis. It represents the neural center of the animal. In many cases it contains one or more small cavities, and in a few cases a continuous cavity is present throughout the collar cord. Where such hollow cords occur, one or two *neuropores* provide communication between the neural cavity and the exterior of the body. Hollow cords and neuropores are reminiscent of chordate nervous systems. Note that the systems of hemichordates are without definite ganglia, undoubtedly a primitive condition.

The proboscis is largely filled with muscle tissue, but a small protocoel persists and communicates with the outside via a dorsal *proboscis pore.* The mesocoel in the collar is similarly

29.8. Hemichordate structure, schematic. **A.** Sagittal section of anterior region of *Balanoglossus.* The parabranchial ridge is a fold of endoderm separating the pharynx into an upper breathing and a lower alimentary chamber. **B.** Cross section through *Balanoglossus* at level just behind gill slit region.

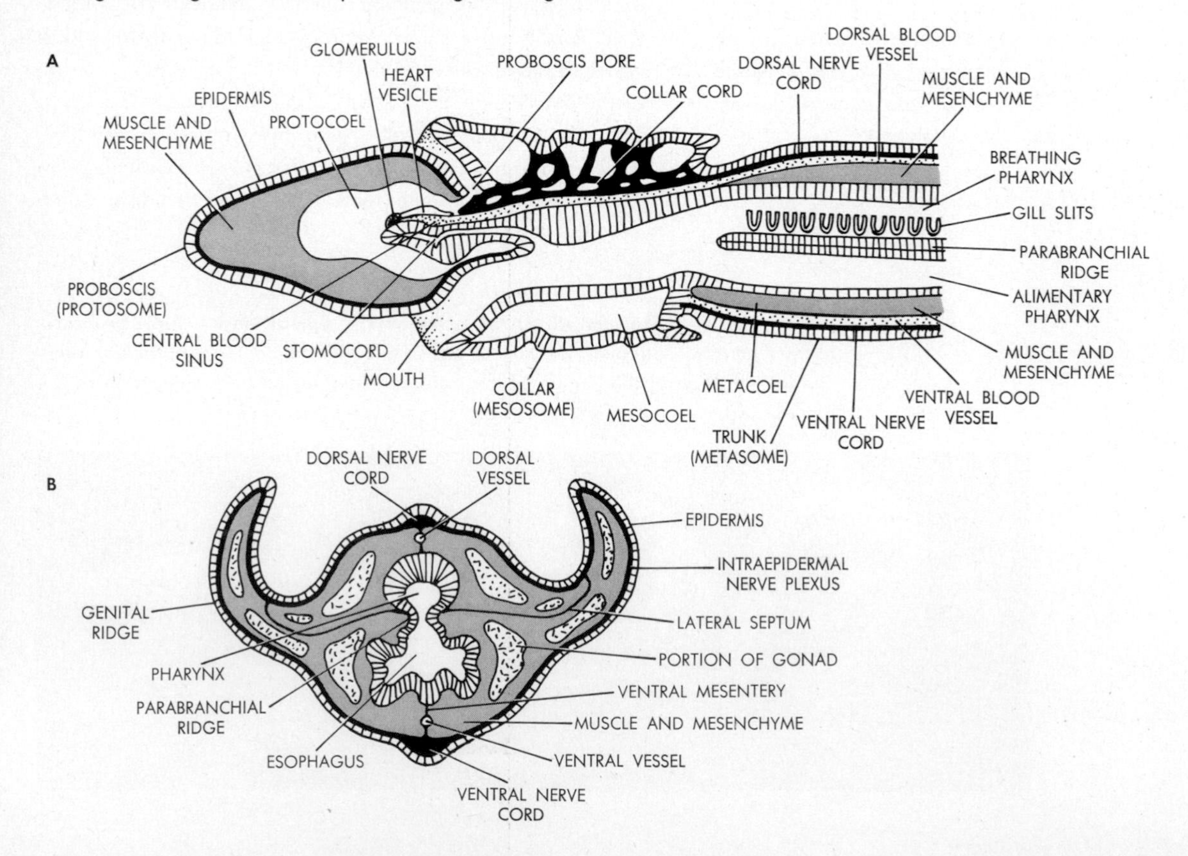

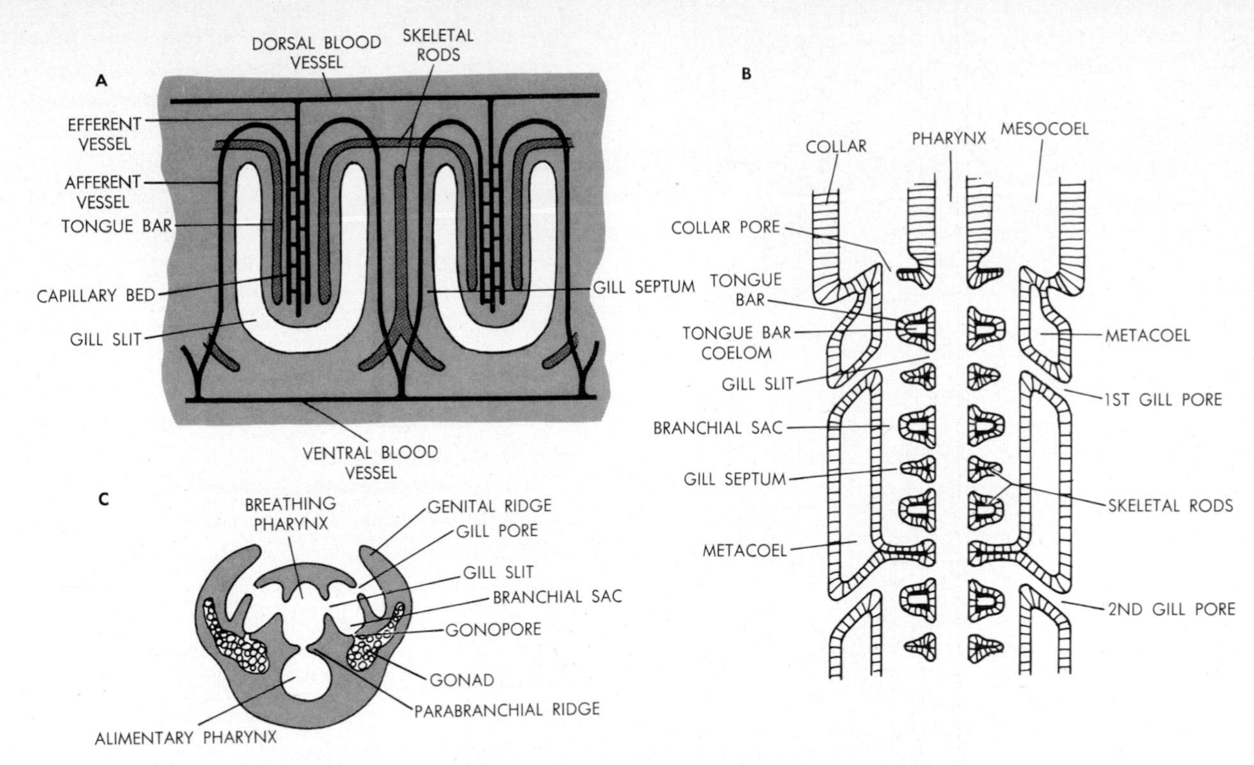

29.9. The gill region of hemichordates, schematic. (*After Van der Horst and other sources.*) **A.** Side view of two gill slits, showing skeletal supports and pattern of blood circulation. **B.** Frontal section through anterior gill region of *Balanoglossus*, showing relations between mesocoel, branchial sacs, pharynx, and gill pores. **C.** Cross section through gill region.

reduced, and it opens posteriorly into the first gill chamber (cf. below). Both the proboscis and collar coeloms thus have access to sea water and are probably filled with it. By contrast, the metacoel of the trunk is closed and contains coelomic fluid with amoeboid cells. A vertical mesentery in both mesocoel and metacoel supports the alimentary tract. This tract leads from the mouth into a mouth cavity which is extended anteriorly into a hollow pouch, the *stomocord.* It is this pouched structure which had been mistaken originally for a notochord. In the anterior portion of the trunk the alimentary tract contains a long pharynx, perforated dorsally by the paired gill slits (Fig. 29.9). Each such slit is originally oval, but a *tongue bar* later grows down from the dorsal edge of the oval and produces the characteristic U-shape of a slit. Basically the slits are food strainers: water passes through them to the outside, while food and sand are retained and fall to the ventral gutter of the pharynx. Secondarily such slits also serve in breathing, a specialized function developed most clearly in chordates. Gill slits open into deep, communicating pockets invaginated from the body wall, the *branchial sacs,* or gill chambers. These in turn open to the outside via *gill pores.* The collar coelom on each side of the body communicates with the most anterior branchial sac on that side. Behind the pharynx the alimentary tract continues as an esophagus and an intestine.

The circulatory system is probably closed, but in parts of it blood passes through sinuses without vessel walls. A dorsal and ventral longitudinal trunk vessel is supported by the mesentery, and blood flows forward in the dorsal vessel, backward in the ventral one (Fig. 29.10). Anteriorly, just above the stomocord at the proboscis base, the dorsal vessel widens into a noncontractile "heart." Above the latter in turn is a contractile *heart vesicle,* a muscular coelom sac homologous to a pericardial sac (and probably derived in the embryo from the right protocoel). Pulsations of this sac press rhythmically against the heart and maintain blood circulation through it.

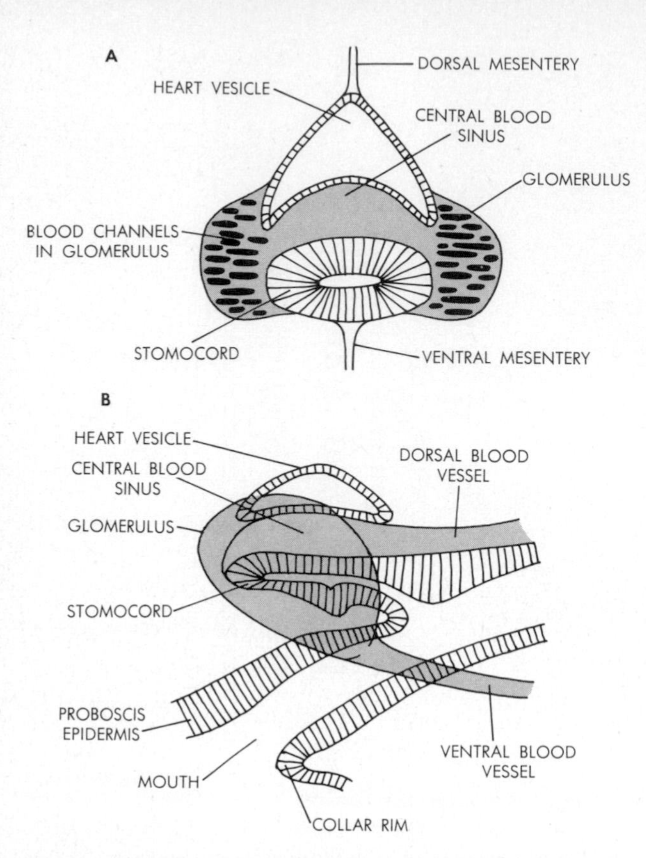

29.10. The organs in the proboscis base of hemichordates. (*Based on Hill, Van der Horst.*) **A,** cross section at level of stomocord; **B,** sectional side view of proboscis base.

On each side of the heart is a *glomerulus,* an accumulation of mesodermal tissue believed to have an excretory function. Blood is forced to pass from the heart through the glomeruli and is presumably filtered there. Excretion products probably leave the body via the protocoel or mesocoel or both. In the pharyngeal region the ventral trunk vessel sends branches into the gill-slit region, and another set of branch vessels leaves that region and enters the dorsal trunk vessel. Oxygenated blood thus passes anteriorly. The blood is colorless.

The gonads lie lateral to the pharynx, as noted, and they open into the branchial sacs. Fertilization takes place externally in sea water, during a spawning period. Eggs are isolecithal or telolecithal, and they develop into two different kinds of larvae according to the amount of yolk they contain. In both cases cleavage is radial, and a resulting coeloblastula gastrulates by embolic invagination (Fig. 29.11). The blastopore closes over and the anterior part of the archenteron becomes cut off as the embryonic protocoel. This coelomic pouch then grows back along each side of the archenteron and constricts off mesocoelic and metacoelic portions; it also develops an opening to the outside, the later proboscis pore. The anterior part of the archenteron proliferates to the ventral side, where a mouth then breaks through. The anus analogously breaks through at the point of the earlier blastopore. If development started from a yolky egg, as in *Saccoglossus,* for example, the larva now established is ellipsoid, with an anterior ciliated apical tuft and a posterior ciliated girdle (telotroch). But if the egg is isolecithal, as in *Balanoglossus,* the resulting larva is a *tornaria,* with apical tuft, telotroch, and a conspicuous ciliary band which winds sinuously over the larval surface.

The tornaria is remarkably like the larva of starfishes. This similarity constitutes a main argument for regarding hemichordates and echinoderms as being closely related. Indeed, it is widely believed that both phyla originated from a hypothetical *dipleurula* ancestor, a form postulated to have resembled the tornaria and the starfish-type of larva (cf. also below and Fig. 29.16).

Both the oval and the tornaria larvae of enteropneusts eventually cease their pelagic life and metamorphose gradually into young worms. In the process the larval pharynx evaginates lateral sacs, and corresponding invaginations from the body wall grow inward. Gill slits form in this manner, in anteroposterior sequence. Adult enteropneusts may readily regenerate a new proboscis, and forms such as *Balanoglossus* and others also are known to bud off new individuals at the posterior end of the trunk.

The whole class Pterobranchia consists of just the three genera listed above. In their basic structure these animals are similar to the enteropneusts, but they differ in a number of conspicuous ways (Fig. 29.12). Thus, they are individually of microscopic dimensions and they are colonial. All zooids of a colony arise by budding from a single starting ancestrula, as in ectoprocts, and a whole colony shares a common secreted housing. The latter may have tubular, branching, globular, or other forms. In *Cephalodiscus* the zooids of a colony are separate, but in *Rhabdopleura* they are interconnected by a dark cord, the *black stolon.* Each zooid consists of a plump metacoel with a long slender stalk at the posterior end, a mesosome collar which bears a set of tentacles anterodorsally, and a shieldlike protosome which tilts down over the mouth area like a lid. Internally, only a single pair of gill passages is present and the circulatory system is distinctly open, formed vessels being entirely absent. Sexes are separate, but hermaph-

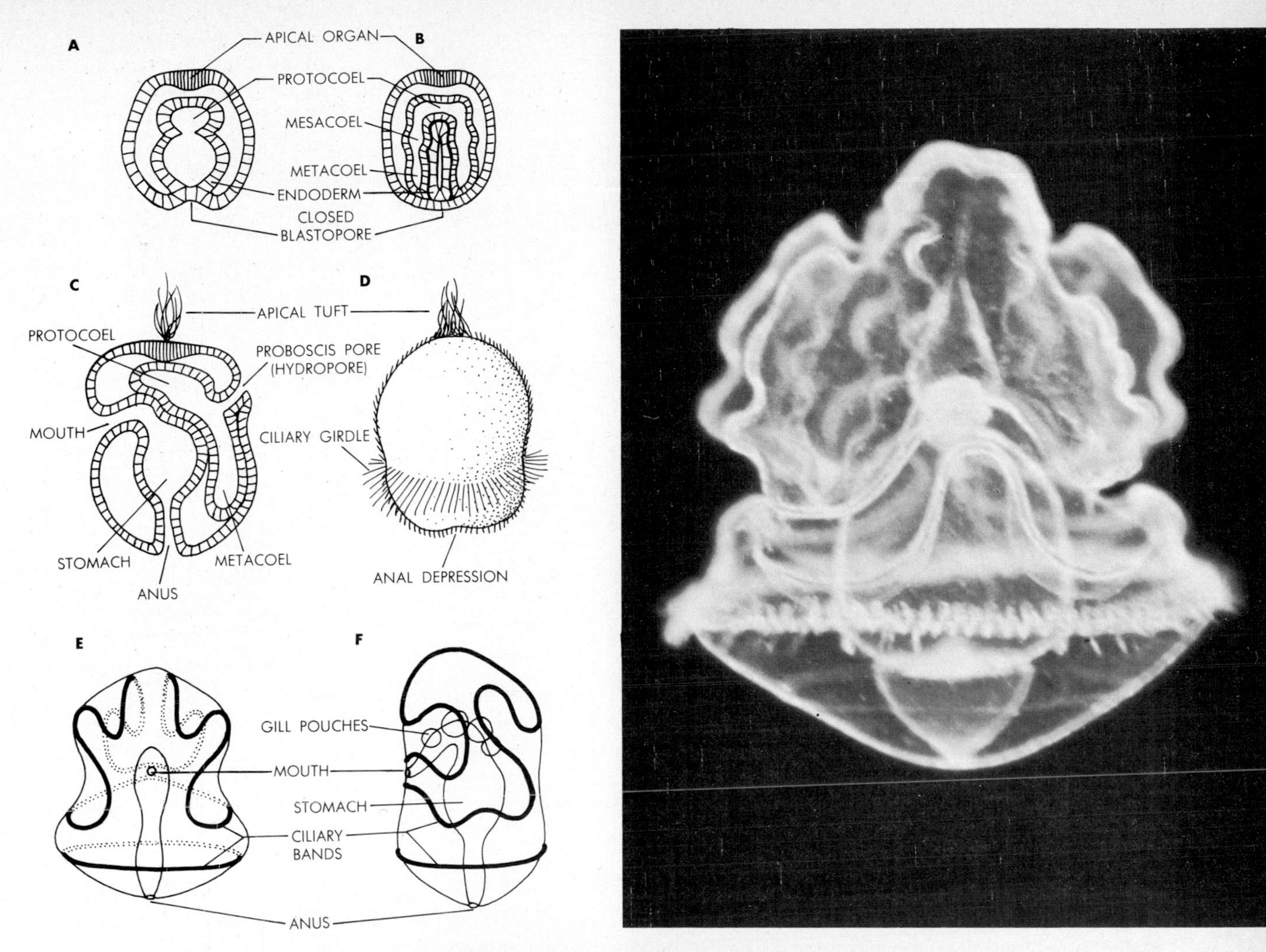

29.11. Hemichordate development. Diagrams (*based on Heider, Davis, Morgan, and other sources*): **A.** Frontal section of postgastrula, showing protocoel being cut off from archenteron. **B.** Frontal section of later stage, showing posterior enlargement of coelom on each side of archenteron. **C.** Sagittal section through late embryo. Alimentary tract is complete and protocoel communicates through dorsal pore with exterior. **D.** Type of larva formed from telolecithal egg, as in *Saccoglossus*. **E** and **F.** Dorsal and side views of tornaria larva formed from isolecithal egg, as in *Balanoglossus*. The gill pouches evaginate from the pharynx and become gill slits after acquiring openings through corresponding invaginations from the epidermis. **Photo:** tornaria larva of *Glossobalanus*, from life.

roditism is exhibited in some species. The eggs are yolky and they generally develop via ciliated larvae with apical tufts. At the posterior end of the metasome stalk is a budding zone where numerous new individuals may be formed. *Atubaria* does not secrete a housing.

The housings of *Rhabdopleura* closely resemble those of the fossil *graptolites*, and the latter clearly show that their once-living inhabitants were likewise joined by interconnecting stolons (cf. Fig. 15.3). These similarities constitute the basis for considering the graptolites as relatives or even one of the

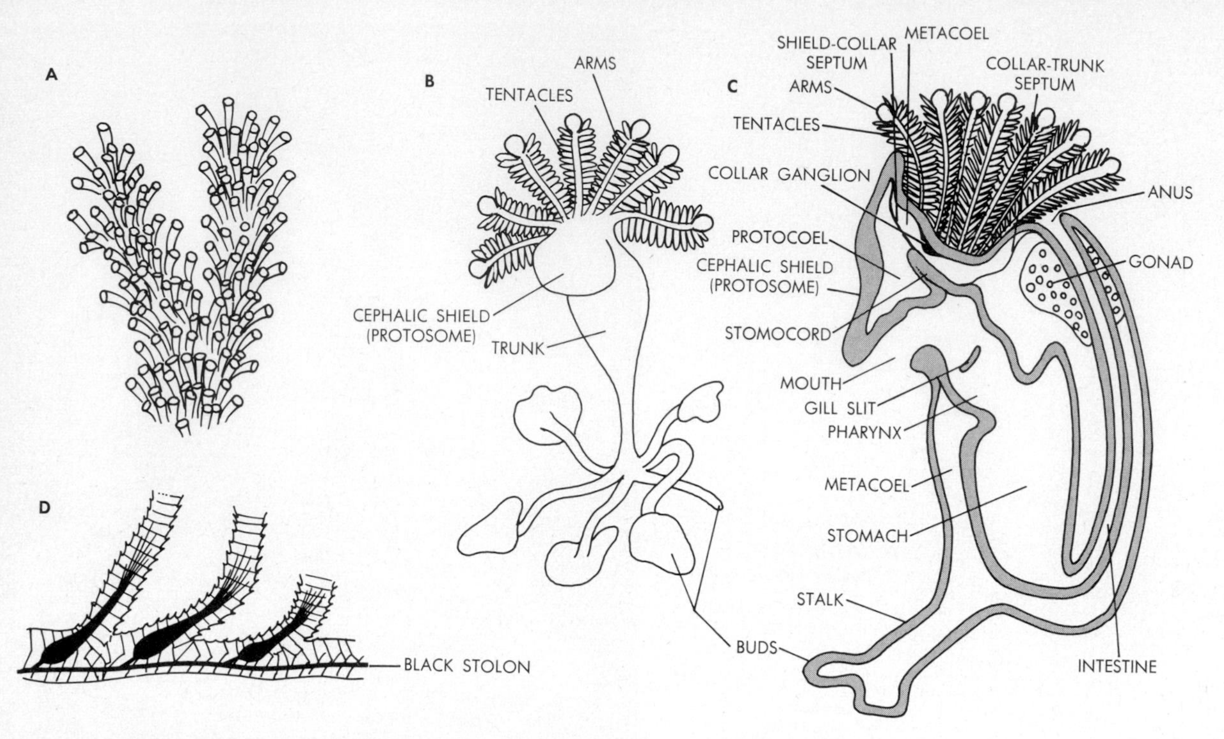

29.12. Pterobranch hemichordates. **A.** Schematic diagram of a colony of *Cephalodiscus.* (*Based on Ridewood.*) **B.** An individual zooid of **A.** (*Adapted from Andersson.*) **C.** Sagittal section through a *Cephalodiscus* zooid. (*After Schepotieff.*) **D.** portion of a colony of *Rhabdopleura,* showing creeping and erect skeletal tubes, position of zooids within, and interconnecting black stolon. (*After Schepotieff.*)

classes of hemichordates. However, some zoologists hold such similarities to be unconvincing, and they consider the problem of graptolite affinities to be unsettled as yet.

Inasmuch as hemichordates as a whole suggest both echinodermlike traits in some of their larvae and chordatelike traits in their pharynx, quite aside from exhibiting embryological similarities to both these phyla, they presumably represent an evolutionary link between echinoderms and chordates. However, the relationship of hemichordates to echinoderms appears to be much closer than to the chordates. Hence it is possible that hemichordates, echinoderms, and pogonophora all may have arisen from a more or less common deuterostomial ancestry. The chordate group then may have branched off later from the line leading to hemichordates (cf. Fig. 29.16). None of these likely interrelations appears to be very direct, close, or certain; the supporting evidence from the living animals here is far more tenuous than that available in support of postulated protostomial interrelations. Most particularly, the ancestry of chordates is comparatively obscure.

PHYLUM ECHINODERMATA

Spiny-skinned animals; exclusively marine; larvae bilateral, with protocoel, mesocoel, and metacoel in early stages; adults pentaradial; with calcareous endoskeleton and coelomic water-vascular system; mostly separately sexed; development regulative, typically oviparous with larvae, some ovoviviparous and viviparous without larvae.

Subphylum PELMATOZOA: typically attached, oral surface facing upward, aboral surface with or without stalk; ambulacral grooves ingestive, tube feet food-catching; both mouth and anus on oral side; main nervous system aboral.

†Class CARPOIDEA: extinct; with horizontal stalk, body bilaterally symmetrical and flattened side to side, typically without arms. *Trochocystites*

†Class CYSTIDEA: extinct; with or without upright stalk, body oval to spherical, with arms; pinnules absent. *Caryocrinites*

†Class BLASTOIDEA: extinct; with or without upright stalk; body rounded; with arms, pinnules often present. *Pentremites*

†Class EDRIOASTEROIDEA: extinct; body diskshaped, free or attached, without stalk or arms; ambulacral grooves often sinuous; endoskeletal plates with pores for tube feet. *Edrioaster*

Class CRINOIDEA: body cupshaped, free or attached; endoskeleton limited to aboral side, oral side membranous; arms branched, usually with pinnules; development via *doliolaria* larvae or direct. *Cenocrinus, Ptilocrinus,* sea lilies, with stalk; *Antedon, Neometra,* feather stars, without stalk.

Subphylum ELEUTHEROZOA: not attached, oral surface facing downward or to side; ambulacral grooves not ingestive; tube feet locomotor, with ampullae; mouth oral, anus aboral or absent; main nervous system oral.

†Class OPHIOCISTIOIDEA: extinct; body diskshaped, without arms; up to six pairs of large tube feet per ambulacrum. *Volchovia*

Class HOLOTHUROIDEA (sea cucumbers): secondarily bilateral; mouth region with tentacles; ambulacral grooves closed; endoskeleton reduced to ossicles; oral-aboral axis horizontal; development via *auricularia* larvae or direct. *Cucumaria, Holothuria, Thyone, Synapta*

Class ASTEROIDEA (sea stars, starfishes): starshaped, with arms; open ambulacral grooves on oral side; with digestive glands; tube feet project between endoskeletal plates; development via *bipinnaria* larvae or direct. *Asterias, Asterina, Odontaster, Asterodiscus, Solaster, Heliaster*

Class OPHIUROIDEA (brittle stars, serpent stars, basket stars): starshaped, with long, highly flexible arms; ambulacral grooves closed, tube feet reduced; madreporite on oral side; without intestine or anus; development via *ophiopluteus* larvae or direct. *Ophiura, Ophiocoma, Ophiothrix, Orchasterias, Gorgonocephalus*

Class ECHINOIDEA (sea urchins, sand dollars): spherical to diskshaped, without arms; ambulacral grooves closed; endoskeleton fused, nonflexible, with pores for tube feet and with movable spines; development via *echinopluteus* larvae or direct. *Arbacia, Echinus, Paracentrotus, Strongylocentrotus, Clypeaster, Echinarachnius*

Evolutionary Considerations

This important phylum today includes some 6,000 living species representing five classes, viz., crinoids, holothuroids, asteroids, ophiuroids, and echinoids. Echinoderms generally develop via bilateral free-swimming larvae which later metamorphose into sessile or sluggish adults. The latter are organized pentaradially around a comparatively short oral-aboral axis. In crinoids and Pelmatozoa generally, the oral side is directed upward and both mouth and anus are present on that side. In sea cucumbers the oral-aboral axis is horizontal, the mouth marking one end of the axis and the anus the other. In all other groups the oral side is directed downward, with the mouth in the center (Fig. 29.13). Two fundamental features characterize all echinoderms: an *endoskeleton* produced in the dermis and consisting of separate or fused calcareous plates overlain by epidermis; and, most particularly, a unique *water-vascular system* consisting of a series of coelomic tubes filled with sea water. The exterior parts of this system are numerous hollow, muscular tube feet, or *podia,* which starfishes, for example, use as little legs.

Nothing like a water-vascular system is encountered in any other animals. How did it evolve and how, therefore, did echinoderms evolve? It is quite clear that the original function of the water system could not have been locomotor, for the ancient fossil echinoderms were, and many of the primitive living crinoids still are, sessile types, attached aborally. Crinoids possess highly branched arms, and the tube feet present on the oral side of the arms serve entirely in feeding; they trap small organisms which are then passed along an arm toward the mouth, in a ciliated *ambulacral groove.* Primitively, therefore, the tube feet are food-catching *tentacles,* and the system of arms as a whole represents a tentacular apparatus. Further, we know that the earliest known fossil echinoderms, the carpoids, were bilateral animals, not radial ones (cf. Fig. 15.4 and below). Also, the larvae of all echinoderms still are bilateral, and we know that during crinoid metamorphosis the larval anterior end becomes the attached, aboral end of the adult. Accordingly, we may assume that echinoderms evolved from bilateral, free-swimming ancestors whose descendants

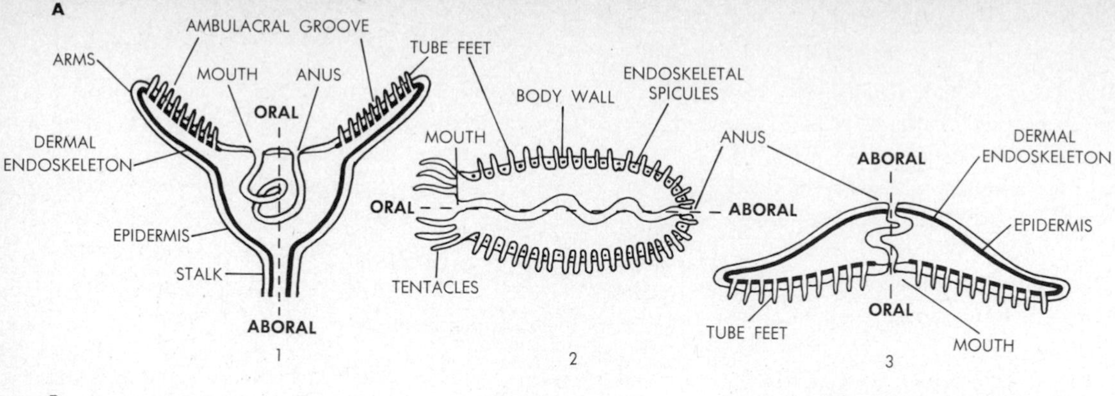

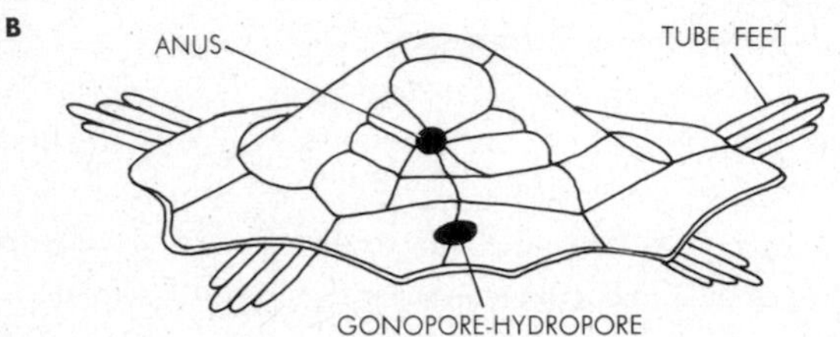

29.13. Echinoderm body plans. **A.** Schematic diagrams of general body designs. 1, the crinoid pattern, with vertical oral-aboral axis, mouth and tube feet pointing up; 2, the holothuroid pattern, with horizontal axis; 3, the asteroid-ophiuroid-echinoid pattern, with vertical axis, mouth and tubefeet pointing down. **B.** Restoration of the extinct ophiocystioid *Volvochia*. (*Redrawn from Hecker.*)

later adopted a sessile way of life through anterior attachment.

Three consequences ensued, all still in evidence in the larval development of crinoids (Fig. 29.14). First, the anterior organs of the larva near the attached end undergo an internal rotation, a process which brings the mouth to the same upper side as the anus and thus establishes the adult oral side. Concurrently a radial symmetry develops by a suppression of growth on the right side, the left side twisting from a vertical alignment to a horizontal one and then proliferating horizontally in a pentaradial manner. Secondly, as is common in sessile animals, a protective skeleton develops, of dermal origin in this case. And thirdly, as is again quite common in sessile forms, arms develop as feeding tentacles. However, the tentacles themselves become clothed with endoskeletal plates, a circumstance which aids in protection but which severely limits the mobility and thus the food-trapping capacity of the arms. This restriction is then circumvented by the formation of small mobile branch tentacles on the arms, i.e., tube feet, as well as food passages to the mouth in the form of ambulacral grooves. We may therefore interpret the water-vascular system as a modified tentacular system, evolved in primitive echinoderms in specific correlation with sessilism and endoskeleton-induced body rigidity.

Once a tube-foot system had evolved, it could be adapted secondarily to functions other than feeding, most particularly, to locomotion. Thus, the later echinoderm groups relinquished the stalked, attached mode of life, and the Eleutherozoa became motile, even though sluggishly so, with the oral side directed downward. Tube feet, and in some cases also endoskeletal spines and the arms themselves, could function in propulsion. The animals so could move to food actively, and no longer needed to depend on what the arms could strain out of the water. More and more, therefore, the mouth came to participate in feeding directly, and the ambulacral grooves largely ceased to be of importance. Indeed, in all Eleutherozoa except the asteroids, these grooves closed over by growth of folds of the body wall and the ancestral grooves are still represented in the structure of these animals by so-called *epineural canals,* just under the body wall (Fig. 29.15 and below). Moreover, inasmuch as the mouth then no longer needed to lie on the same side as the anus, the sharp internal rotation of the larval organs no longer needed to take place and the alimentary tract could assume forms other than U shapes. Even so, the organs of all echinoderm larvae still undergo the internal twisting from a vertical alignment to a horizontal one, and an original bilaterality changes over to radiality via suppression of growth on the right side. Also, the left side of the larva still becomes the oral side of the adult. Above all, in the course of developing the adult water-vascular system, echinoderm larvae still pass through an at least rudimentary tentaclelike stage.

These common features in echinoderm development are widely postulated to represent evolutionary reminders of a hypothetical bilateral ancestor, the *dipleurula* (Fig. 29.16). As noted earlier, a dipleurula is assumed to have been a common evolutionary starting point of hemichordates as well as of echinoderms, and it is true that the embryos and early larvae

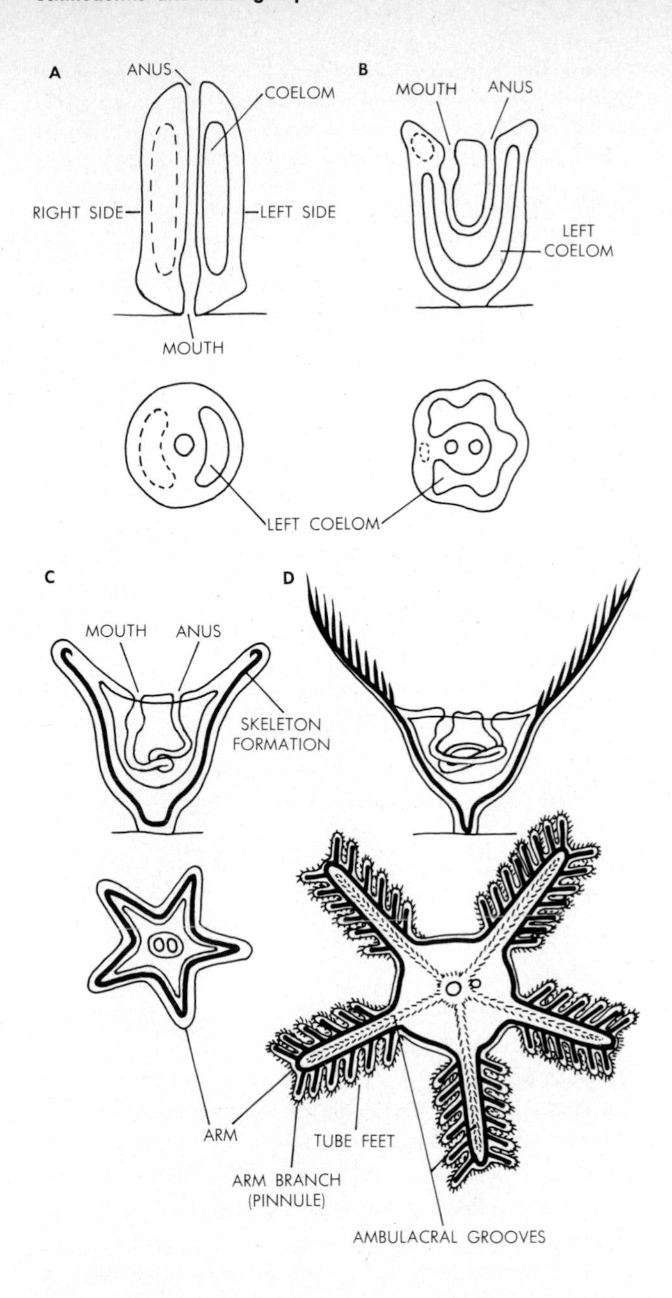

29.14. The general pattern of development of echinoderm radiality (as in crinoids), schematic. Top figures, sagittal sections; bottom figures, cross sections. **A.** Oral settling and attachment of bilateral larva, growth suppression on right side. **B.** Proliferation of left side, rotation of mouth to new oral position away from attached end, beginning of horizontal radiality. **C.** Horizontal pentaradial growth of left coelomic derivatives and body as a whole, formation of horizontal arms, development of enveloping and motion-restricting endoskeleton. **D.** Development of feathery arms, with branch arms and tube feet for food gathering, and growth of ciliated ambulacral grooves for food conduction toward mouth.

of both groups are remarkably alike. Most particularly, as we shall see, echinoderms develop protocoel, mesocoel, and metacoel cavities just as do hemichordates. Moreover, the tentaclelike stage in echinoderm development is highly reminiscent of pterobranchs, and these primitive hemichordates actually are believed to be closest to the primitive echinoderm stock. Finally, as also noted, the tornaria larvae of some hemichordates are quite similar to the bipinnaria larvae of asteroids. It is therefore plausible to envisage that a dipleurulalike ancestor may have given rise, on the one hand, to bilateral pterobranchs and hemichordates generally, and on the other, to secondarily radial echinoderms, which still exhibit pterobranchlike features during development. It is only on such a basis, actually, that the exceedingly complex developmental processes and the adult structures of echinoderms are at all meaningful.

General Characteristics

Although each echinoderm class is characterized in its own specific way, the basic traits of echinoderms as a whole are exemplified well in the asteroids, animals which combine some of the primitive as well as some of the advanced traits of the

29.15. Diagrams of cross sections through echinoderm arms, illustrating open ambulacral groove as in starfish and crinoids (left) and closed ambulacral groove as in other living echinoderms. The epineural canal is the result of groove closure, the latter brought about by body-wall folds which grow over and fuse around the open groove.

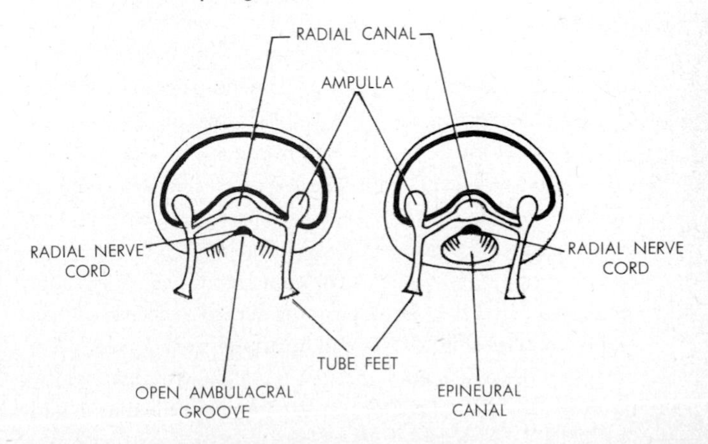

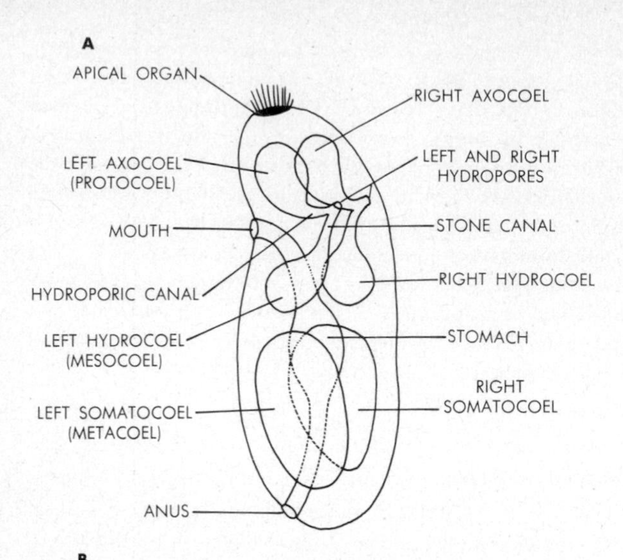

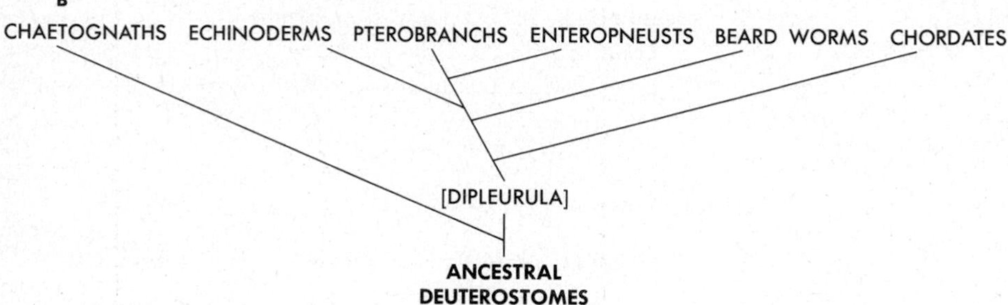

29.16. Deuterostome evolution. A. The structure of the hypothetical dipleurula ancestor of deuterostomes (*After Bather*). Note the resemblance to hemichordate larvae and also to embryonic stages in echinoderm development. **B.** The possible (tentative) evolutionary interrelations among deuterostome groups.

phylum. A starfish of the genus *Asterias*, a common and familiar seashore inhabitant, may serve as illustrative example.

The body of such an animal consists of a central *disk* and, typically, five *arms* (Fig. 29.17). In the center of the disk on the underside, or oral surface, is the mouth. Leading to the mouth along the oral side of each arm is an *ambulacral groove*, bordered on each side by a row of hollow, muscular tube feet, or *podia*, the terminal parts of the water-vascular system. On the upper, or aboral, side, a tiny anal opening lies in the center of the disk, and near the angle between two of the arms is a reddish *madreporite*, a sieve plate representing the entrance to the water-vascular system. The body wall contains a cuticle, an epidermis, and a well-developed underlying dermis. The latter consists of connective tissue which secretes the endoskeleton. In a starfish, such a skeleton is made up of separate calcareous plates held together by the dermal connective tissue and by muscles (Fig. 29.18). On their outer surfaces the plates bear wartlike thickenings and low spinelike extensions which produce an uneven exterior texture. In places the dermis and the overlying epidermis are evaginated into microscopic *pedicellariae*, muscle-operated pincers which protect the epidermal surface of a starfish from small animals. Tiny calcareous bodies, or *ossicles*, stiffen the stem and the jaws of a pedicellaria internally. Underneath the dermis lie an outer circular and an inner longitudinal muscle layer, well developed only on the oral side, particularly along the ambulacral grooves. A flagellated peritoneum lines the body wall internally. Studding the exterior surface are numerous microscopic fingerlike projections, the *skin gills*. Made up of all layers of the body wall, peritoneum included, the gills are hollow; their internal spaces are extensions of the coelom which project between adjacent skeletal plates. The pedicellariae protect the skin gills particularly.

Sea water communicates with the water-vascular system via the madreporite, which leads into a small bulbous cavity, the *madreporic ampulla* (Fig. 29.19). From there a calcified *stone canal* conducts water into a circular channel ringing the esophagus of the animal. Along this *ring canal* are usually present one or more sacs, the *polian vesicles,* which are probably water reservoirs (not present in *Asterias*), and five pairs of *Tiedemann's bodies,* whose specific function is unknown. From the ring canal also emanate five *radial canals,* one into each arm. There each radial canal gives off short lateral branches, and each of the latter bifurcates at its end into a saclike *podial ampulla,* located inside the arm, and a tube foot protruding from the oral surface of the arm between the skeletal plates. Stiffened by water pressed into them from the ampullae, tube feet can be used as tiny walking legs. A podium also serves as a suction disk when its end is placed against a solid surface and the inner parts of the end plate are then lifted away. By such means the tube feet of a starfish may exert enough steady pull on the shells of a clam to tire the adductor muscles of the clam and to force the shells open.

Clams and oysters constitute the main diet of starfishes. The mouth leads through a short esophagus into a spacious stomach (Fig. 29.20), which may be everted through the mouth into the soft tissues of a clam. Stomach tissues may even be squeezed through minute spaces usually present between the *closed* shells of a clam. Small food particles and fluid foods pass into a short intestine, and from there into five pairs of large digestive glands, the *pyloric caeca,* one pair occupying most of the free space within each arm. Each caecum is suspended from the aboral body wall by a double mesentery (cf. Fig. 29.18). Near the anus the intestine connects with two *rectal pouches* which probably have an excretory function. The anus itself lies somewhat excentrically on the aboral surface of the disk.

As in echinoderms generally, the nervous system consists of three subsystems, each in the form of a plexus thickened regionally into cords. The oral, or *ectoneural,* system contains

A

29.17. Starfish structure, external features. A. The oral side of *Ophidiaster.* The mouth is at the center of the disk, and a tube-foot–lined ambulacral groove passes along the oral side of each of the five arms. **B.** The aboral side of *Asterias.* An anus lies at the exact center of the disk but is too small to be visible. The buttonlike madreporite is apparent excentrically on the disk, between the uppermost and the upper right arm.

B

B, courtesy Carolina Biological Supply Company.

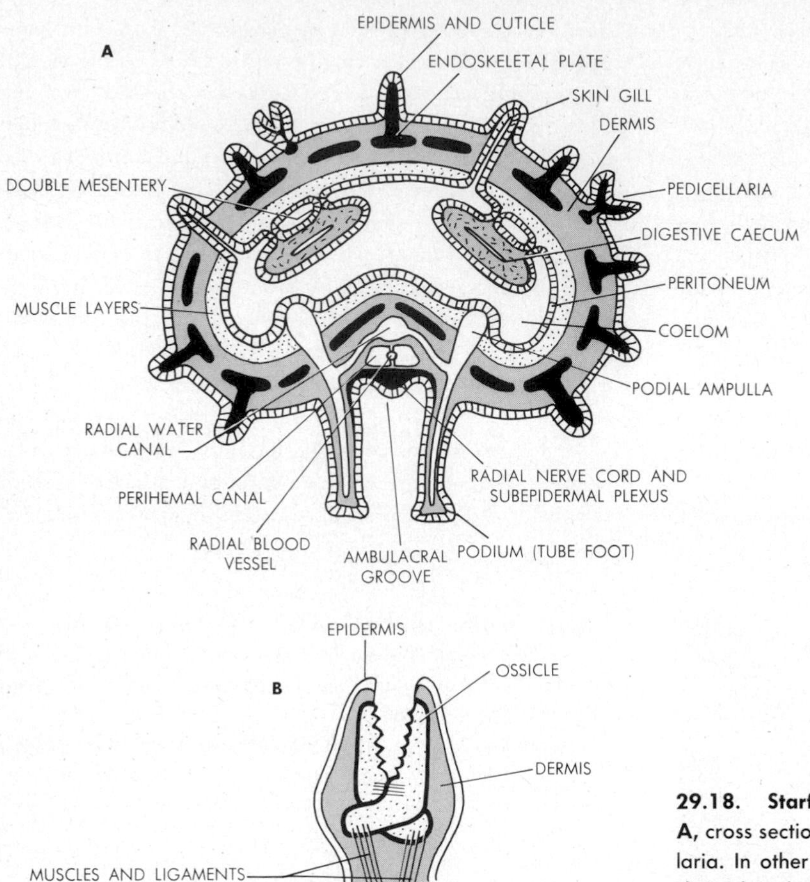

29.18. Starfish structure, diagrammatic. (*Based on Smith.*) **A**, cross section through an arm; **B**, detail of one type of pedicellaria. In other types the endoskeletal elements, or ossicles, are shaped and arranged differently.

29.19. Plan of the water-vascular system. Note that polian vesicles are absent in *Asterias.*

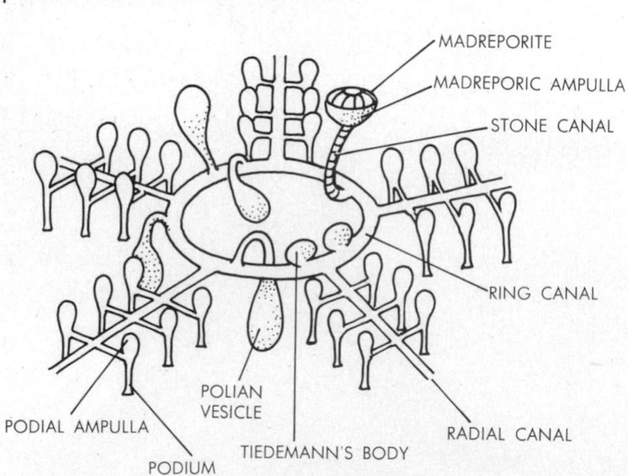

a *nerve ring* under the body wall on the oral side, near the ring canal of the water-vascular system (cf. Fig. 29.20). From this nerve ring emanate five *radial nerves*, one into each arm. These nerves run along the bottom of the ambulacral grooves, just underneath the epidermis, and they are part of a subepidermal neural plexus. At the tip of each arm a radial nerve terminates at an ocellus. The ectoneural system is the main system in starfishes and in all other members of the subphylum Eleutherozoa as well. A second, less developed subsystem is the aboral, or *endoneural* system, and a third is the *hyponeural* system. Both of these are patterned like the ectoneural system; i.e., they each contain a nerve ring, radial nerves, and a plexus. The nerve ring of the hyponeural system lies near that of the ectoneural system, and the ring of the endoneural system is located under the aboral body wall.

The circulatory system consists of a series of blood channels

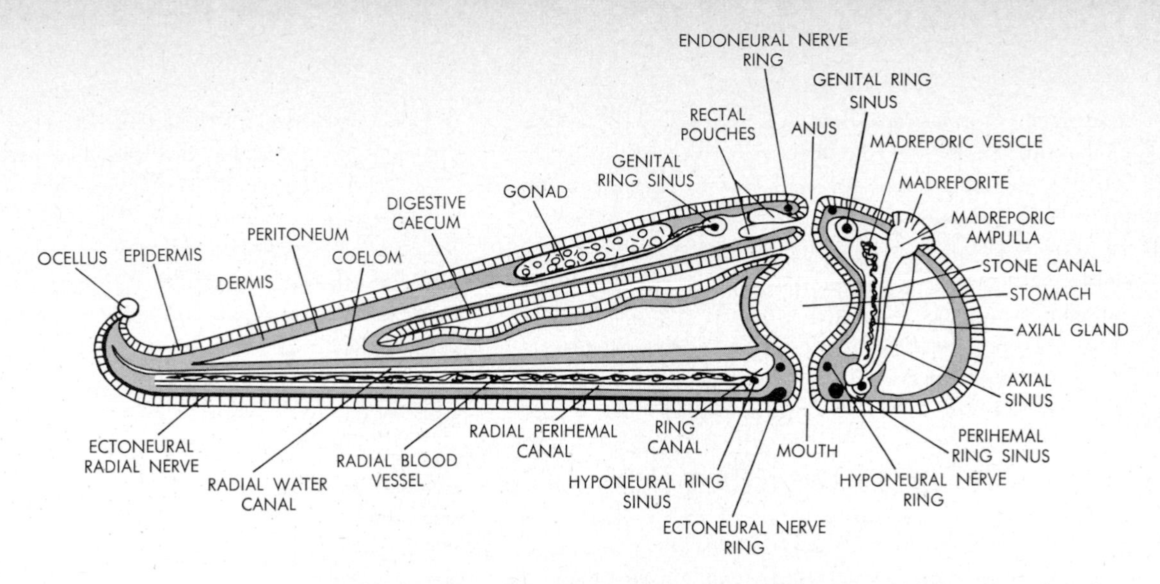

29.20. Section through the disk and one arm of a starfish. (*Based on Borradaile.*) For a more detailed illustration of the various sinuses and coelomic channels see Fig. 29.21.

which are without walls but which lie within coelomic ducts. The center of the system may be considered to be the *axial gland,* a meshwork of contractile channels lying next to the stone canal (Fig. 29.21). It is likely that this gland may in part have an excretory function. Both the stone canal and the axial gland traverse a coelomic duct called the *axial sinus.* Aborally the axial gland projects into a *madreporic vesicle,* a contractile coelomic sac near the madreporic ampulla. Pulsations of this vesicle and the axial gland presumably maintain a circulation of blood. The axial gland also continues into blood spaces within an aboral *genital ring sinus,* a circular coelomic channel from which emanate paired branch channels to the gonads in each arm. At its oral end the axial gland connects with blood spaces within another circular channel, the *hyponeural ring sinus,* located near the rings of the nervous system. From this sinus branches a *radial sinus,* containing a radial blood channel, into each arm. The blood is colorless and differs little from coelomic fluid. All the blood-carrying coelomic channels lie in the spacious coelomic body cavity, which contains numerous amoeboid cells. The latter are partly excretory; they may absorb excretory waste and carry it to the exterior via the skin gills. Apart from the rectal pouches, the axial gland, and the coelomic amoebocytes, other excretory structures are absent.

Starfishes possess five pairs of gonads, one pair per arm, lying laterally near the arm bases. Each gonad is enveloped by a coelomic membrane continuous with one of the branch channels from the genital ring sinus. Each gonad also opens

29.21. The plan of the circulatory and coelomic channels in starfishes. (*After Hyman.*) Only one of the five pairs of gonads is shown. All blood channels are surrounded by a perihemal membrane which encloses a coelomic space, i.e., a perihemal channel.

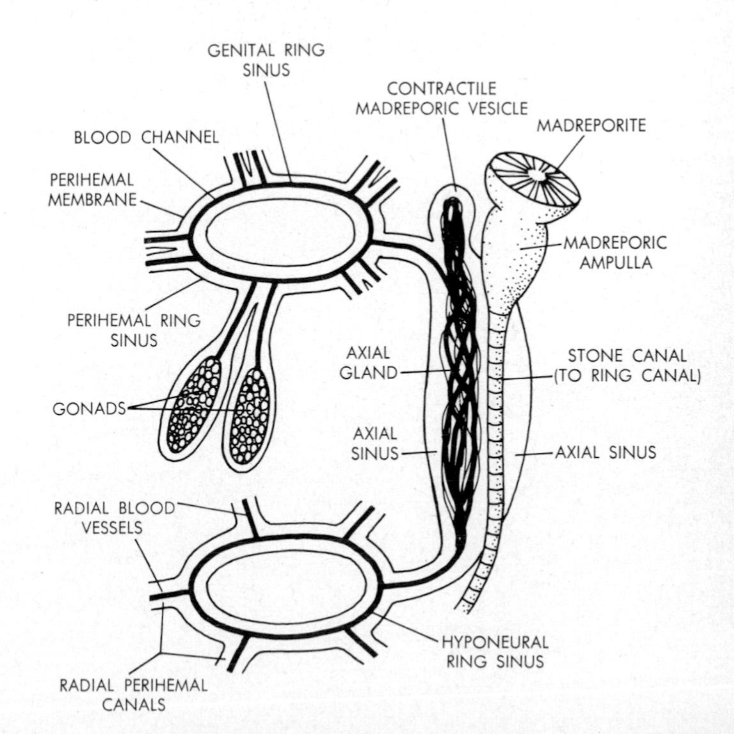

individually via a short gonoduct and a gonopore near the base of the arm (cf. Fig. 29.18). The sexes are characteristically separate, though asteroids do include some hermaphroditic types. A few asteroids mate by copulation, some brood their offspring in chambers on the oral side of the disk, but most spawn, fertilization and development occurring externally. Brooding types have yolky eggs which develop directly, but spawning types shed isolecithal eggs which develop indirectly via free-swimming larvae. The larvae differ in their external features from those formed in other echinoderm classes; but in fundamental respects the internal structure and the developmental pattern are the same for all echinoderms.

An isolecithal echinoderm egg typically cleaves radially and regulatively into a coeloblastula (Fig. 29.22). Such a blastula may become a free-swimming early larva, which then gastrulates by embolic invagination. The archenteron produces much primary mesoderm in the form of loose mesenchymal cells. The epithelial main mesoderm soon arises either as a pair of lateral enterocoelic pouches evaginated from the anterior part of the archenteron or as a single pouch which sub-

29.22. Starfish development. Early stages are illustrated in Fig. 11.16. **Diagrams: A.** Postgastrula, formation of protocoelomic sacs. **B.** Subdivision of coelom into three sacs on each side; the later fate of each sac is indicated in parentheses. **C.** Early bipinnaria larva, seen from left side; the left hydrocoel is already characterized by the five lobes which foreshadow the five radial canals of the adult. **D.** Later bipinnaria, ventral view. The hydrocoel is in process of growing around the esophagus, foreshadowing the ring canal of the adult. (**C** and **D** *based on Hörstadius.*) **Photo:** ventral view of bipinnaria larva of *Asterias*, from life. Alimentary tract and epidermal ciliary bands are well visible.

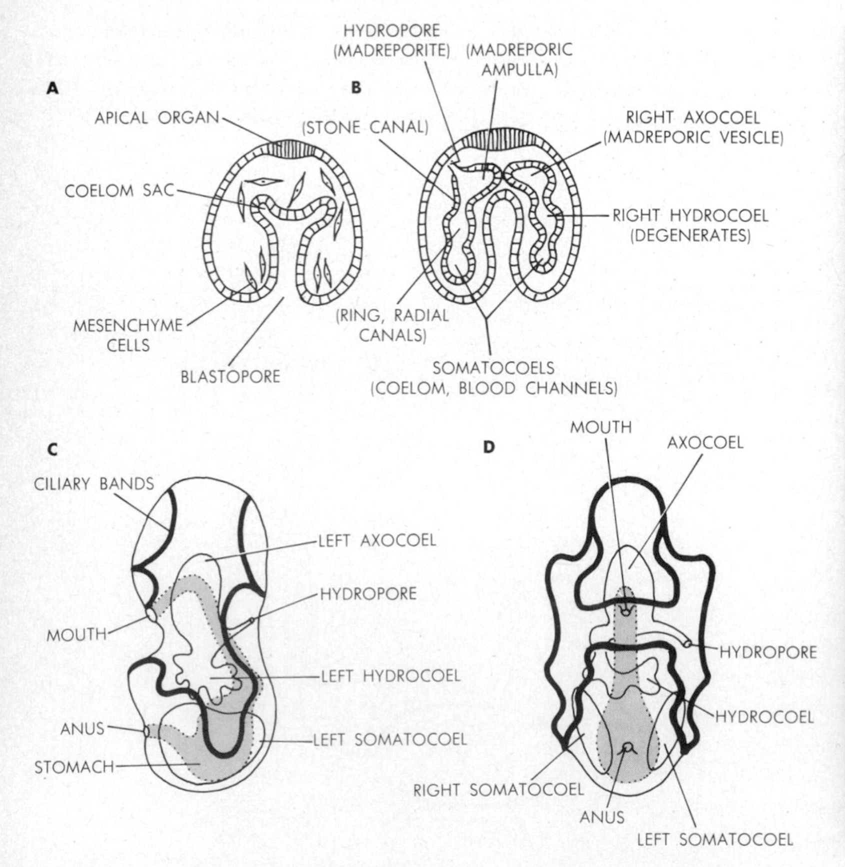

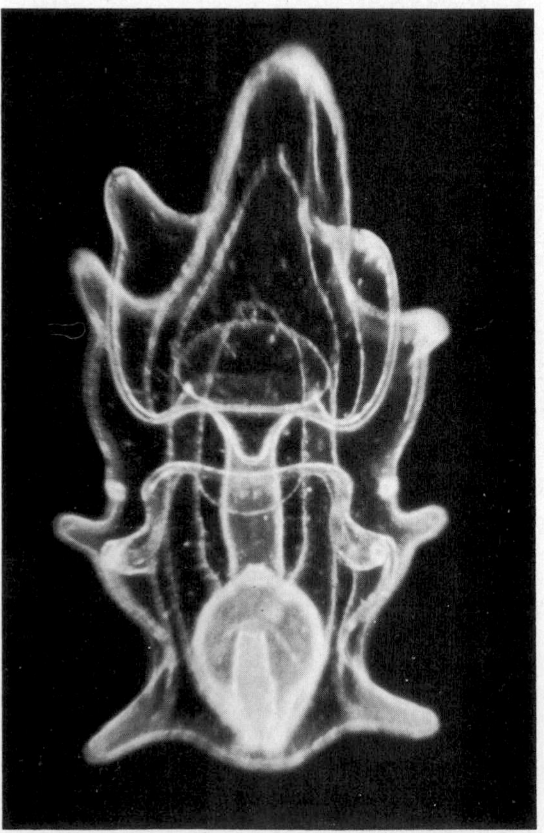

sequently divides into a left and a right portion. These coelomic sacs then enlarge and typically split into three sacs on each side, viz., the *axocoel* (protocoel) anteriorly, the *hydrocoel* (mesocoel) behind, and the *somatocoel* (metacoel) posteriorly. In most cases the axocoel and hydrocoel actually do not separate completely but remain linked by a tubular connection. The left axocoel then puts out a duct which breaks through the overlying ectoderm and establishes an opening to the exterior, the *hydropore*. It will be noted that events up to this point are essentially identical to the corresponding processes in hemichordates, the hydropore in one case being equivalent to the proboscis pore in the other. Indeed, as pointed out earlier, a hypothetical dipleurula is postulated to have been the common ancestor of hemichordates and echinoderms. Such a dipleurula would have been characterized by three pairs of coelomic sacs, and the three sacs on each side may originally have developed in bilaterally symmetrical fashion. Thus, both the left and right axocoels would have formed a hydropore (cf. Fig. 29.16). In echinoderms, however, the adult radial condition has become correlated with a suppression of larval growth on one side, viz., the right side. Consequently, only the left axocoel develops a hydropore.

At this stage, the later fates of the coelomic sacs are clearly fixed. The hydropore represents the later madreporite; the left axocoel becomes the madreporic ampulla; the tubular connection between the left axocoel and the left hydrocoel forms the stone canal; and the left hydrocoel gives rise to the ring and radial canals of the water-vascular system. The left somatocoel will form part of the adult coelum as well as the blood channels and the coelomic tubes surrounding them. On the right side, the right axocoel persists as the madreporic vesicle, believed to be homologous to the heart vesicle of hemichordates. The right hydrocoel degenerates altogether, and the right somatocoel forms another portion of the adult coelom.

The later starfish larva is a free-swimming *bipinnaria* (cf. Fig. 29.22). It is characterized by a lobed, buoyancy-increasing ectoderm and by ciliary surface bands which serve in locomotion and feeding. A telotroch is absent, but in other respects the bipinnaria is very similar to a hemichordate tornaria, as noted earlier. Internally, the left and right somatocoels enlarge and come to suspend the archenteron in a vertical mesentery. The anterior part of the archenteron proliferates toward the ventral side and eventually breaks through the ectoderm as the larval mouth. From the left hydrocoel evaginate five fingerlike pouches, and the main part of the hydrocoel itself proliferates in such a way that it eventually forms a closed ring. This ring represents the later ring canal, and the five pouches become the radial canals. They eventually push out of the left side of the larva as five fingers covered by ectoderm externally. It is at this general stage that the bipinnaria is reminiscent of a tentacled, pterobranchlike ancestral form. The early radial canals will later extend in length and, by lateral branching, give rise to the system of tube feet.

In the course of its free life a bipinnaria acquires additional ectodermal lobes and this later larva is known as a *brachiolaria* (Fig. 29.23). At its anterior end the brachiolaria develops a special attachment disk by which it settles in preparation for metamorphosis. Of all Eleutherozoa, only the asteroids still attach during metamorphosis, probably a reminder of a sessile ancestral condition. The whole attachment apparatus and most of the anterior larval ectoderm will be discarded and left behind after metamorphosis; the young adult star forms primarily from the posterior part of the larva. In the process of metamorphosis, the larval mouth and anus close over, and most of the larval esophagus and intestine degenerate. From the persisting stomach a new adult esophagus evaginates toward the left side and then grows through the hydrocoel ring, establishing a new adult mouth on the left, the adult oral side. Analogously, another stomach evagination grows toward the right and forms an adult anus on what so becomes the adult aboral side. Evidently, the positions of the adult mouth and anus do not bear any direct relation to the alimentary openings of the larva. Note also that starfish development does not include any pronounced rotation of larval organs, such nonrotation being correlated with the motile, free existence of the adults.

Echinoderm Types

Pelmatozoa: crinoids. The stalks of Pelmatozoa are hollow aboral extensions of the body wall, covered by calcareous rings (which, like all skeletal parts of echinoderms, are overlain by epidermis). Rootlike extensions or attachment disks form the holdfast devices (Fig. 29.24). At intervals along the stalks are present whorls of movable *cirri*, again hollow and protected by skeletal plates. Such cirri are grasping organs which aid in maintaining a hold on rocks or seaweeds. In the extinct pelmotozoans, the main portion of the body was clothed entirely by endoskeletal plates. In many of the cystoids, for

29.23. Starfish metamorphosis. A. Brachiolaria larva seen from left side. (*After Mead.*) Only the hydrocoelic portion of the coelom is indicated. The posterior part of this larval stage will become the adult, the anterior part contributes to the formation of a temporary attachment apparatus and then degenerates. **B.** Photo of brachiolaria larva of *Asterias*, from life. At top, note the three fringe-ended arms at which attachment will occur. At bottom, the definitive star is developing. **C.** Larvae of *Leptasterias*, developing without typical bipinnaria or brachiolaria stages in the brood chamber of female parent (cf. Fig. 29.30). When set free, such larvae attach by means of their knobbed anterior disks and the adult star arises from the saclike posterior part. **D.** Young star at metamorphosis. The hydrocoelic lobes form the radial canals, and lateral branches give rise to the first tube feet. **E.** Photo of a young star, just metamorphosed.

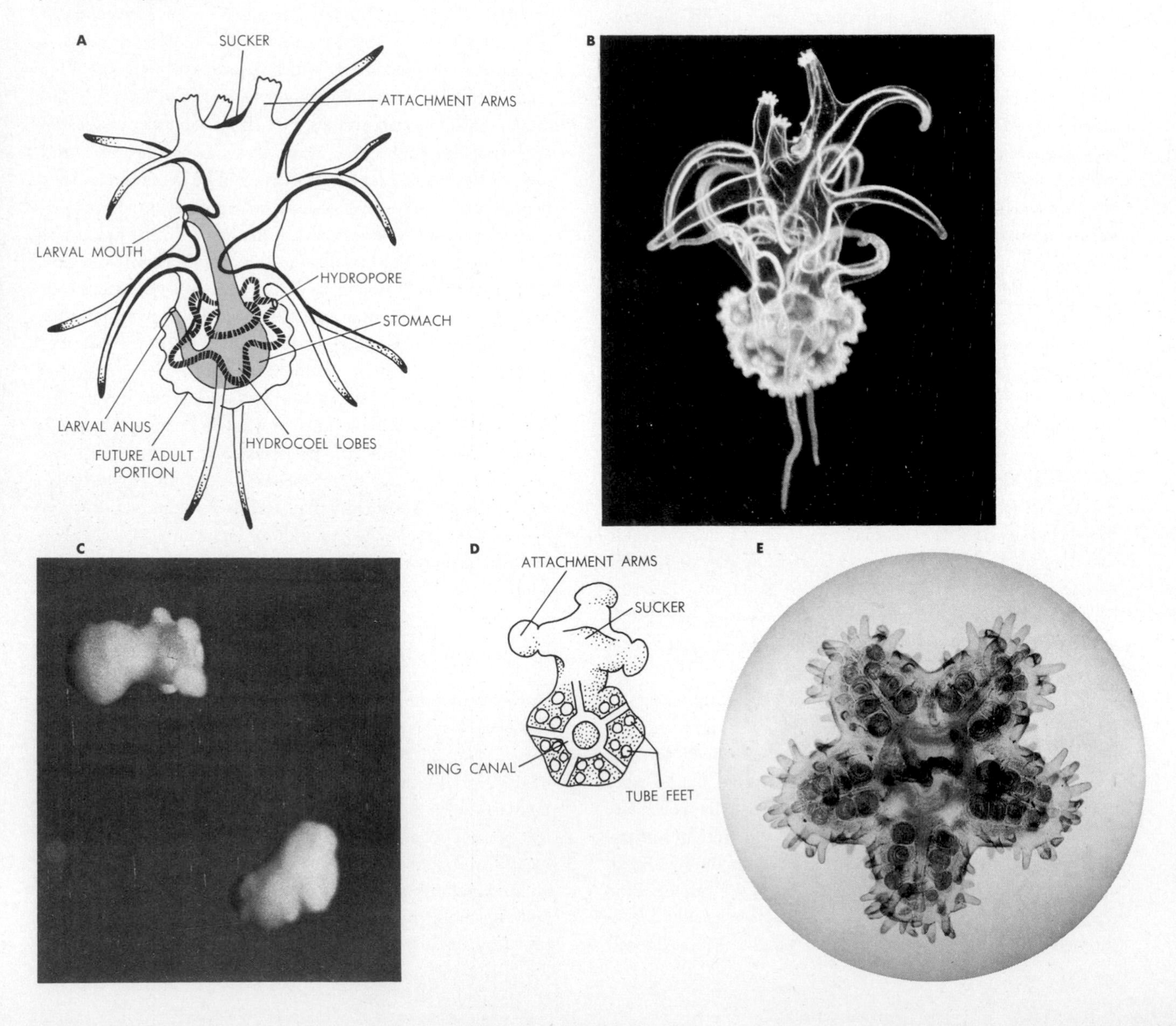

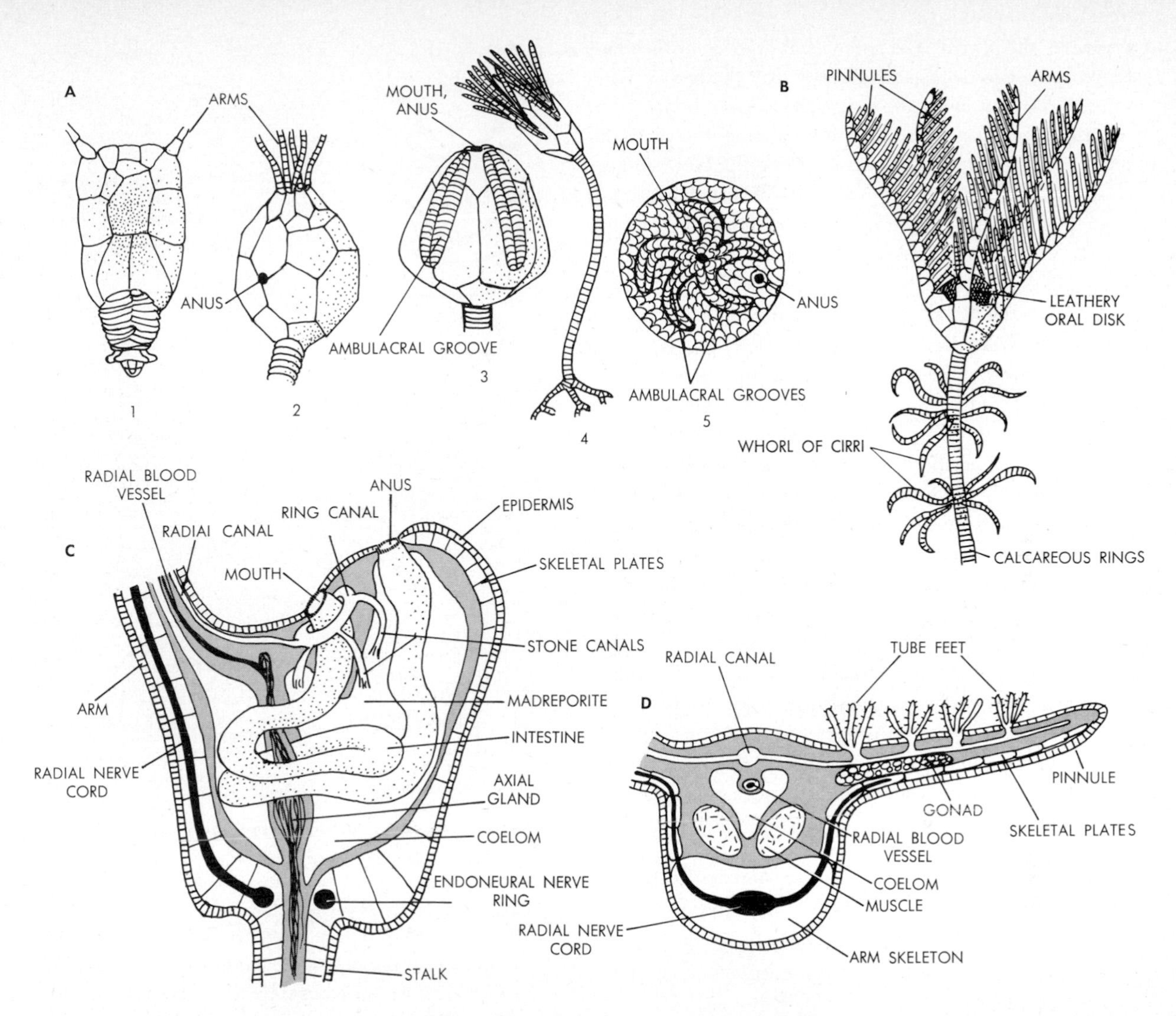

29.24. Pelmatozoa. **A.** Extinct pelmatozoans. 1. Concave side of the roughly spoonshaped skeleton of the carpoid *Enoploura.* (*After Caster.*) 2. Skeleton of the cystoid *Echinoencrinus.* (*After Moore.*) 3, the body skeleton, and 4, view of whole animal, of the blastoid *Pentremites.* (*After Moore.*) 5. Oral side of the diskshaped edrioasteroid *Carneyella.* (*After Bassler.*) **B.** Plan of the external body design of a living stalked crinoid. (*After Clark, Carpenter.*) **C.** Sagittal section through a crinoid, schematic. (*Adapted from Reichensperger.*) Only one set of structures leading into an arm is indicated. **D.** Section through an arm and one of its pinnules. (*Adapted from Hamann, Chadwick.*) Note that one side branch of the water canal in a pinnule leads into a set of three tube feet.

example, the arms projected from a small oral area on the upper surface of the more or less spherical body, and the ambulacral grooves were on the oral side of the arms. Not all members of this class were as yet fully radial. Complete pentaradiality characterized the blastoids, which often possessed arms with side branches, or *pinnules.* Both cystoids and blastoids also included types in which the arms were located not at extreme oral positions but farther down along the body, toward the aboral end. The ambulacral grooves in such animals passed for some distance directly over the body surface and continued out along the arms. This trend culminated in the edrioasteroids, in which arms were absent altogether, the ambulacral grooves running entirely along the body surface.

Living crinoids differ from the extinct pelmatozoans in that the endoskeleton of the body has the form of a cup, the oral side being protected by a leathery body wall without calcareous plates. The arms, usually branched and with pinnules, emanate from the rim of the cup, and the ambulacral grooves pass from the oral side of the arms across the oral part of the body wall toward the mouth. Sea lilies are stalked, sessile types, the stalks attaining lengths up to 2 ft (compared with the sometimes 70-ft-long stalks of extinct sea lilies). Some 5,000 extinct crinoid species are known, and of the 700 or so living species only about 100 are sea lilies; this group appears to be on the road toward ultimate extinction. By contrast, feather stars may be on the increase today. These animals are unstalked, but they often possess very long cirri with which they may hold on temporarily to rocks (Fig. 29.25). Feather stars can creep and swim gracefully by means of their arms, and we may also note that these crinoids are beautifully colored in a large variety of pastel shades. Like echinoderms generally, crinoids have a well-developed regeneration potential. The animals actually exhibit the capacity of autotomy; i.e., they may spontaneously cast off one or more arms if such arms should become trapped.

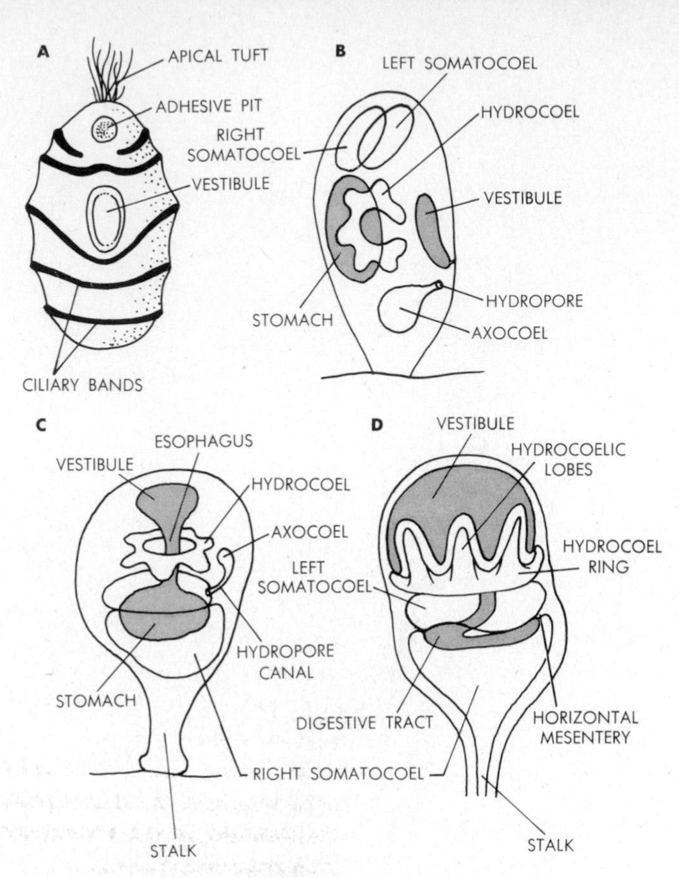

29.26. Crinoid development and metamorphosis. A. doliolaria larva, external ventral view. Attachment will occur at the adhesive pit. The vestibule is an invaginated pouch, at the bottom of which the mouth will form in the adult. **B.** Doliolaria after anterior attachment, seen from left side (schematic, ventral is toward right of page). Rotation of internal organs has not yet occurred and the hydrocoel ring lies in a vertical plane. **C.** Later stage, after rotation and horizontal shifting of organs. The vestibule has made connection with the stomach via the esophagus, the latter now encircled by the horizontally lying hydrocoel ring. Rotation has also brought the left somatocoel over the right one, and a horizontal mesentery between these two sacs comes to support the horizontally developing digestive tract. **D.** Preadult stage. Formation of arms, each with a hydrocoelic lobe within it, has begun. The vestibule will soon break open and an anus will develop. (**A,** *after Bury;* **B, C,** and **D,** *combined and greatly modified from Barrois, Bury, and Seeliger.*)

29.25. Oral view of the feather star *Antedon*, a crinoid. Cf. color plate XIV.

The internal structure of crinoids corresponds substantially to that of asteroids. However, the water-vascular system is without connection to the exterior. Instead, the ring canal ex-

PLATE XIV. Crinoids and holothuroids. A, B, two species of the crinoid featherstar *Antedon*. **C,** the sea cucumber *Holothuria*. **D,** the holothuroid *Opheodesoma,* a type in which tubefeet are totally lacking (as in the related *Synapta,* Fig. 29.27).

Plate XIV, page 1

A

B

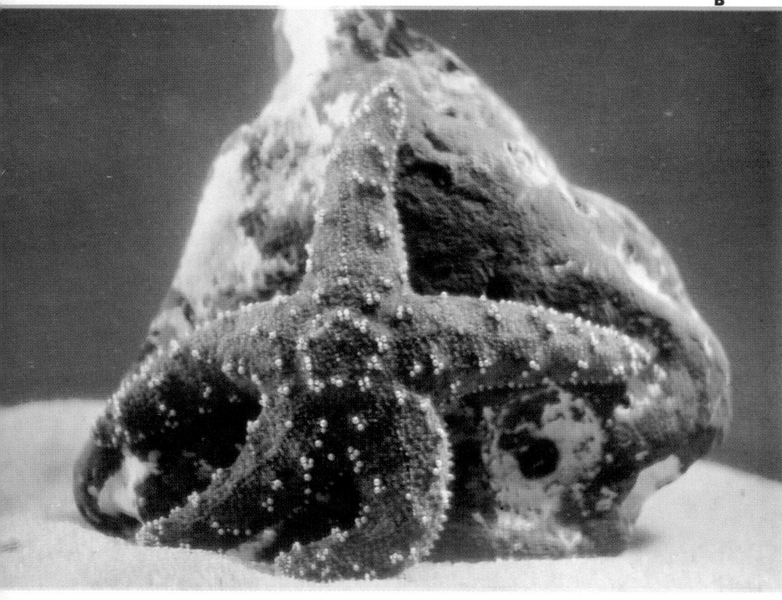

PLATE XV. Asteroids. A, *Asterias,* regenerating an arm. **B,** *Pisaster,* and **C,** *Acanthaster,* two types of starfishes. **D, E,** oral and aboral view of the cushion star *Culeita.*

C

Plate XV, page 2

D

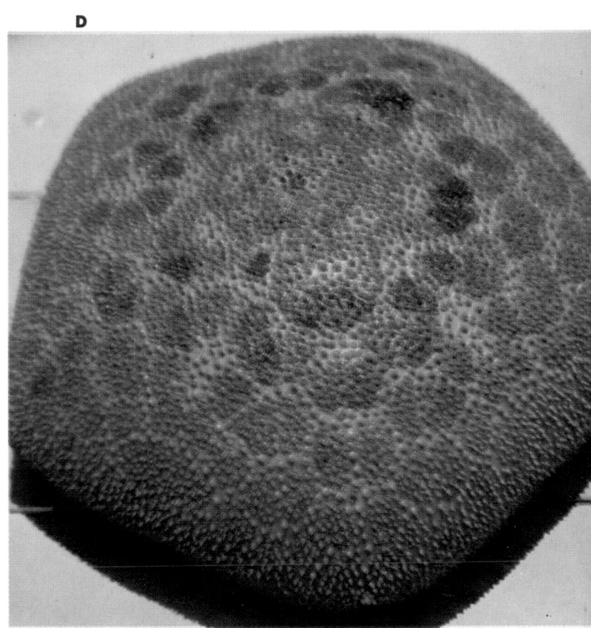

E

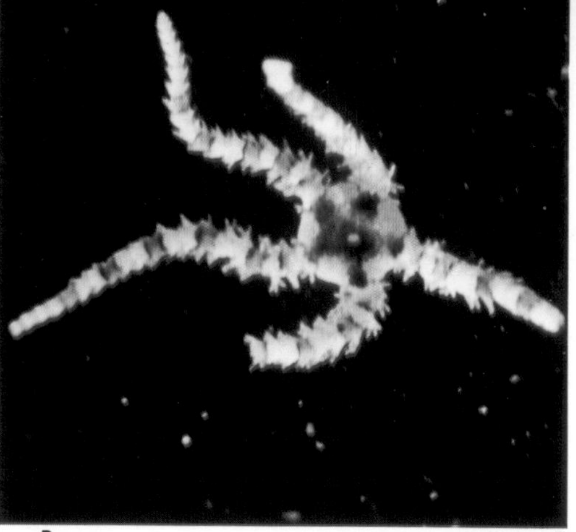

PLATE XVI. Ophiuroids and echinoids. A, *Ophiothrix,* and **B,** *Ophiocoma,* two types of brittle stars. **C,** the ophiuroid basket star *Gorgonocephalus.* **D,** the echinoid *Heterocentrotus.* **E,** a sea urchin with venomous spines.

A B D

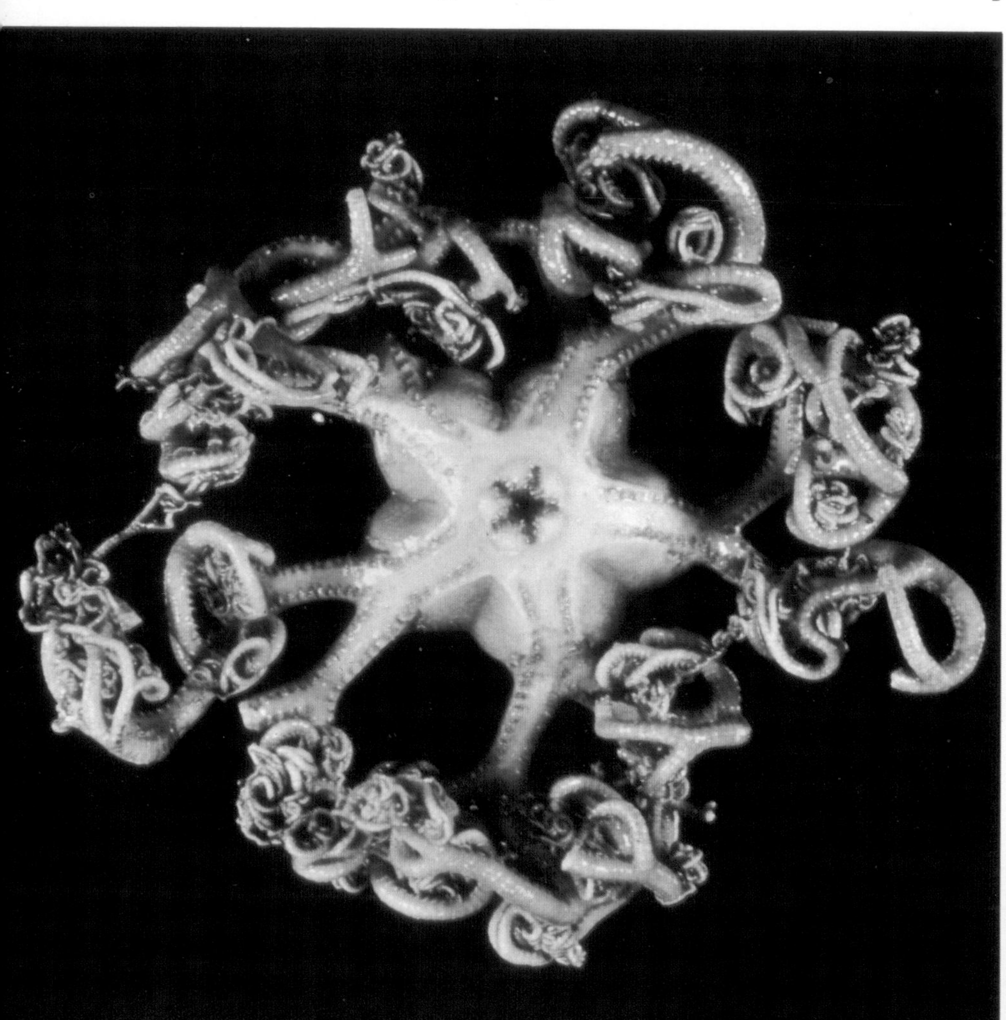

C E

tends into numerous stone canals which terminate within the coelom and which communicate directly with that cavity. Another difference is that the most developed part of the nervous system is the aboral endoneural system. Further, the gonads are located in the pinnules, which burst open when the gametes are ripe. As in each of the other living echinoderm classes, some crinoids develop directly from yolky eggs, and larval stages are then abbreviated or omitted. Typically, however, crinoids develop via a *doliolaria*, a bilateral barrelshaped larval form equipped with an apical tuft and four or five separate transverse bands of cilia (Fig. 29.26).

As pointed out above, larval development in crinoids passes through many presumably primitive processes. Most particularly, a pronounced rotation of the internal organs occurs, such that the larval mouth shifts to the side near the anus. Also, the left coelomic sacs twist toward this newly established oral side, the right sacs toward the aboral side. As a result, the mesentery and the alimentary tract come to lie in a transverse plane, at right angles to the oral-aboral axis. These movements also account for the highly coiled condition of the greatly elongated adult intestine. In later larval stages the left hydrocoel grows in a circle around the larval esophagus and evaginates the five fingers which will form the radial canals around the mouth. In contrast to asteroids, crinoids retain the larval alimentary openings as the adult mouth and anus. (In crinoids, therefore, the esophagus maintains its original position and the left hydrocoel shifts toward it; but in asteroids it is the left hydrocoel which maintains its original position and a new esophagus then grows toward it.)

Inasmuch as crinoids as a whole represent descendants of comparatively primitive echinoderm stocks, it could be assumed that doliolaria larvae likewise may be closer to ancestral forms than bipinnarias or other echinoderm larvae. If so, the phylogenetic significance of bipinnaria-tornaria resemblances becomes questionable; it is entirely possible that these resemblances are little more than instances of parallel evolution. Even in such a case, however, the dipleurula hypothesis could remain tenable, for the early embryonic developments of hemichordates and all echinoderm groups do follow a remarkably common pattern.

Holothuroids. The horizonal position of the oral-aboral axis in sea cucumbers introduces a secondary bilateral symmetry, often reinforced by the elaboration of a distinct ventral *sole* on which the animals lie (Fig. 29.27). Thus, whereas *Cucumaria* possesses five double rows of tube feet positioned in

29.27. Sea cucumbers. **A.** *Cucumaria.* **B.** *Synapta* (model), a type in which the body is highly elongate and the podial apparatus is reduced to warty knobs on the body surface. **C.** Dermal ossicles in the skin of *Leptosynapta.* Cf. color plate XIV.

A

Courtesy Carolina Biological Supply Company.

B

C

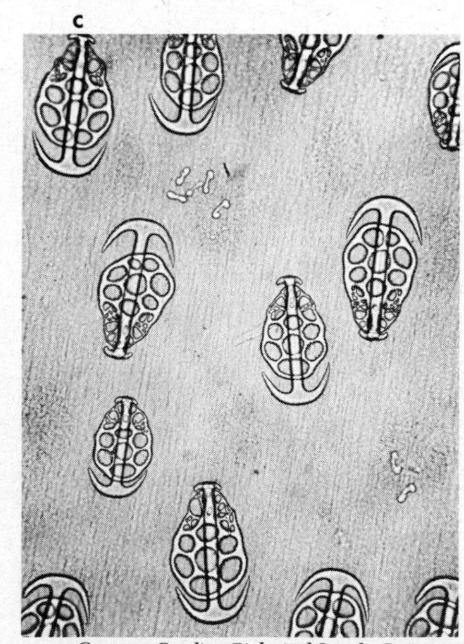

Courtesy Carolina Biological Supply Company.

perfect pentaradial symmetry, *Holothuria* is characterized by a ventral sole which alone bears tube feet; on the rest of the body surface the tube feet are reduced to warty papillae. In forms like *Thyone* tube feet are distributed more or less randomly over the entire body, and in types such as *Synapta* all tube feet are reduced to warty knobs. The endoskeleton of sea cucumbers is reduced greatly. The dermis manufactures only microscopic *ossicles,* scattered randomly throughout the body wall. Sea cucumbers consequently have a soft exterior. The epidermis is glandular, however, and secretes large amounts of mucus.

Holothurians are sand-burrowers and plankton-feeders. The principal food-gathering organs are the retractile, usually highly branched *tentacles* around the mouth (Fig. 29.28). They are evaginated extensions of the body wall and they contain canals of the water-vascular system branched off from the five radial canals. The latter emanate from the ring canal around the pharynx, give off the branches into the tentacles and then continue posteriorly into the body, underneath the rows of actual or reduced tube feet. Between each radial canal and the body surface lies an ectoderm-lined *epineural canal,* the remnant of the ambulacral groove which in these animals has ceased to be functionally significant and has closed over by growth of the body wall. At the base of the tentacles on the dorsal side many sea cucumbers possess a hydropore, the equivalent of a madreporite. In many cases, however, the stone canal terminates directly in the coelom, as in crinoids, and a hydropore is then absent in the adult. The main nervous system is the oral ectoneural system, as in Eleutherozoa generally, and an endoneural system is actually lacking in sea cucumbers. The animals are also without axial glands, but around the pharynx they do possess an oral ring with blood spaces. This ring communicates with two very well developed blood sinuses along the alimentary tract. The sinuses are dorsal and ventral, and the dorsal one is contractile and pumps blood in an oral direction.

Around the pharynx lies a *calcareous ring,* which provides attachment for the retractor muscles of the tentacles. This ring probably corresponds to the calcareous chewing apparatus of echinoids (cf. below). The alimentary tract is long and looped and is supported in a vertical mesentery. Near the anus are a pair of large, extensively branched evaginations from the cloaca. These are the *respiratory trees,* which serve in breathing and also in excretion. Close to the juncture of the trees

29.28. The internal structure of a sea cucumber, schematic.

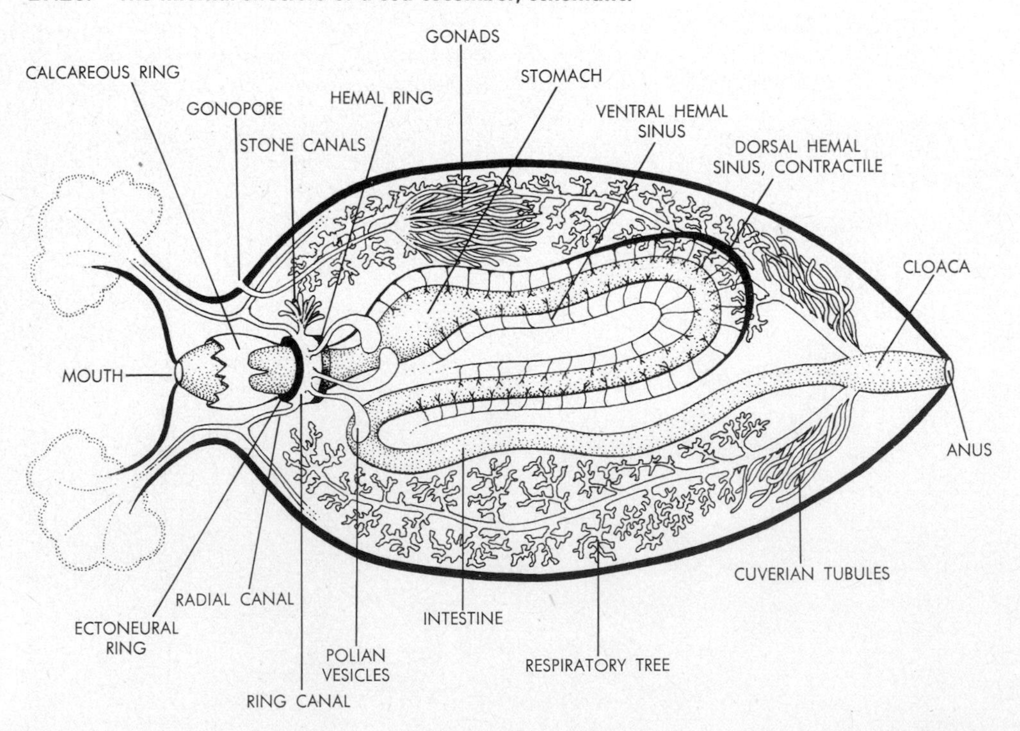

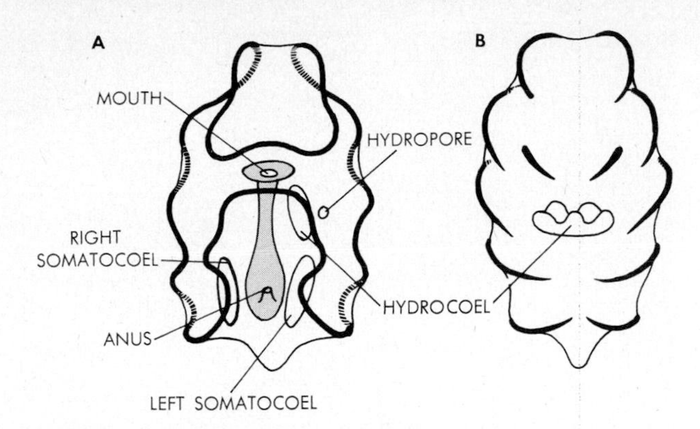

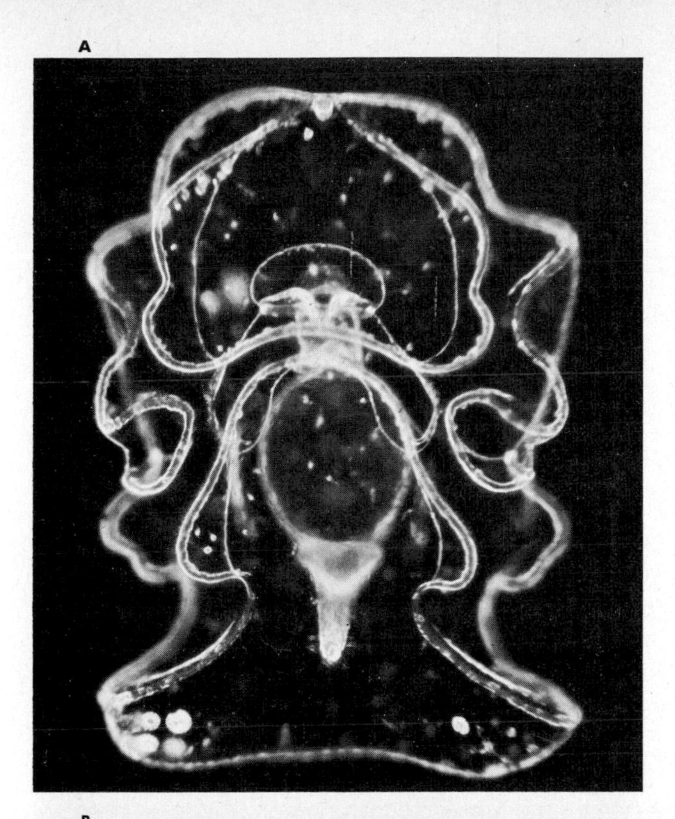

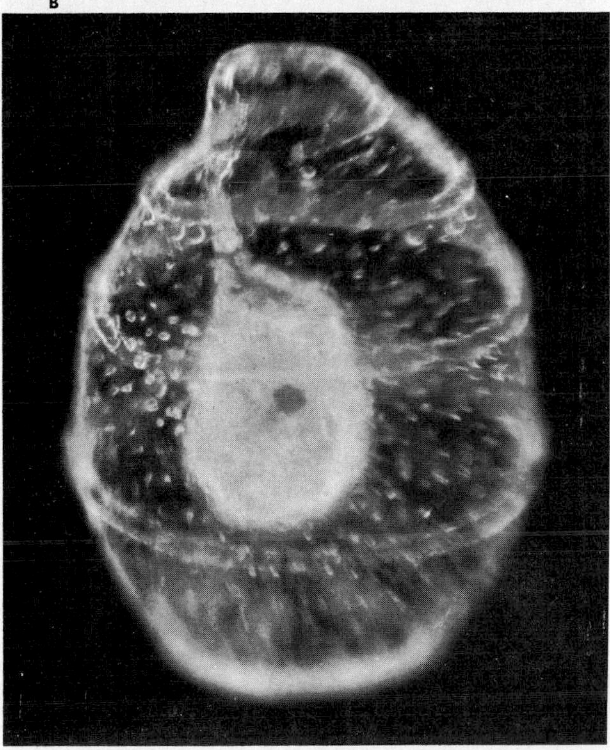

29.29. Holothuroid larvae. Diagrams: A. An auricularia, ventral view. Note resemblances to bipinnarias of starfishes. **B.** The later doliolarialike stage in holothuroid development, ventral view. (*After Mortensen.*) **Photos: A.** auricularia of *Labidoplax,* ventral view. **B.** Later doliolaria stage, both from life.

with the cloaca are bundles of *Cuverian tubules.* When a sea cucumber is irritated it may expel these tubules through the anus; in some unknown way they apparently distract or scare off potential enemies. We may also note that under unfavorable conditions the animals may rupture spontaneously at or near the anus and expel almost all their internal organs. A new set of organs then regenerates from the eviscerated remaining body.

Like primitive Pelmatozoa, holothurians possess but a single gonad. Located anterodorsally, it discharges via a gonopore near the hydropore. Species with yolky eggs are without larvae, and they brood their young either externally on the female or in invaginated pockets of the body wall or even directly within the coelom. Most holothurians spawn, however, and the fertilized eggs then develop externally via *auricularia* larvae. These resemble bipinnarias in most respects. Also, their internal development is essentially like that of asteroids. For this reason, holothuroids and asteroids are believed by some zoologists to represent closely related echinoderm classes. In later stages, an auricularia transforms into a crinoidlike doliolaria larva, a process involving mainly a partial degeneration of the ciliary bands on the surface. Accordingly, holothurians are also thought to represent a likely evolutionary link to the crinoids and the Pelmatozoa generally (Fig. 29.29).

Asteroids. Little needs to be added here to the account on starfishes above. The class comprises over 2,000 living species, and most are characterized by arms numbering five or multiples of five. In some, however, there are six, seven, or eight arms, as in sunstars such as *Solaster;* numerous arms, up to 20 or more, identify sunstars such as *Heliaster* (Fig. 29.30). In addition to its other specific features, the class as a whole is defined by the possession of open ambulacral grooves and by the large digestive glands, traits which are unique among Eleutherozoa. The extensive regeneration capacity of these animals has already been noted in Chap. 10 (cf. Fig. 10.3), and we may note that the animals occasionally also reproduce vegetatively by spontaneous fragmentation.

Ophiuroids. In their adult traits, the 2,000 or so species of brittle stars may be regarded as highly specialized variants of starfishes. Ophiuroids too are starlike, but their arms are marked off sharply from the central disk (Fig. 29.31). Moreover, the arms are long and highly mobile in sinuous fashion (hence the name "serpent stars"). Some species can move the arms only from side to side, but others may extend them in all directions and use them as grasping organs. Many of the animals have branched arms, e.g., the basketstar *Gorgonocephalus.* Mobility of the arms results from a fairly wide spacing of the endoskeletal plates. Inasmuch as the arms serve very adequately in locomotion, tube feet are reduced to warty knobs and podial ampullae are lacking internally (Fig. 29.32). However, ophiuroids do possess a stone and a ring canal, the latter bearing organs corresponding to Tiedemann's bodies. Radial canals are present as well. The ambulacral grooves are closed over as in holothurians, and the madreporite lies on the oral side of the disk.

29.31. Close-up of disk of the brittle star *Ophiothrix.* Cf. color plate XVI.

29.30. Asteroids. The photo shows the underside of *Leptasterias,* with embryos in brood pouch (cf. larvae in Fig. 29.23). The many-armed starfish *Pycnopodia* is shown in Fig. 7.9. Cf. also color plate XV.

Ophiuroids also differ from asteroids in the absence of pedicellariae and skin gills. Breathing is accomplished instead by five pairs of *bursae,* specialized pouches located on the oral side of disk near the bases of the arms. Each such bursa communicates with the outside via a *bursal slit.* Further, brittle stars are without intestine or anus, the stomach ending as a blind sac. The mouth is armed with five muscle-operated calcareous teeth. Ophiuroids are primarily carnivorous, but they usually will eat any kind of food found along the sea bottom. The circulatory and nervous systems are generally as in asteroids. Gonads are attached to the bursae, and in spawning species the gametes escape via the bursal slits. Ovoviviparous

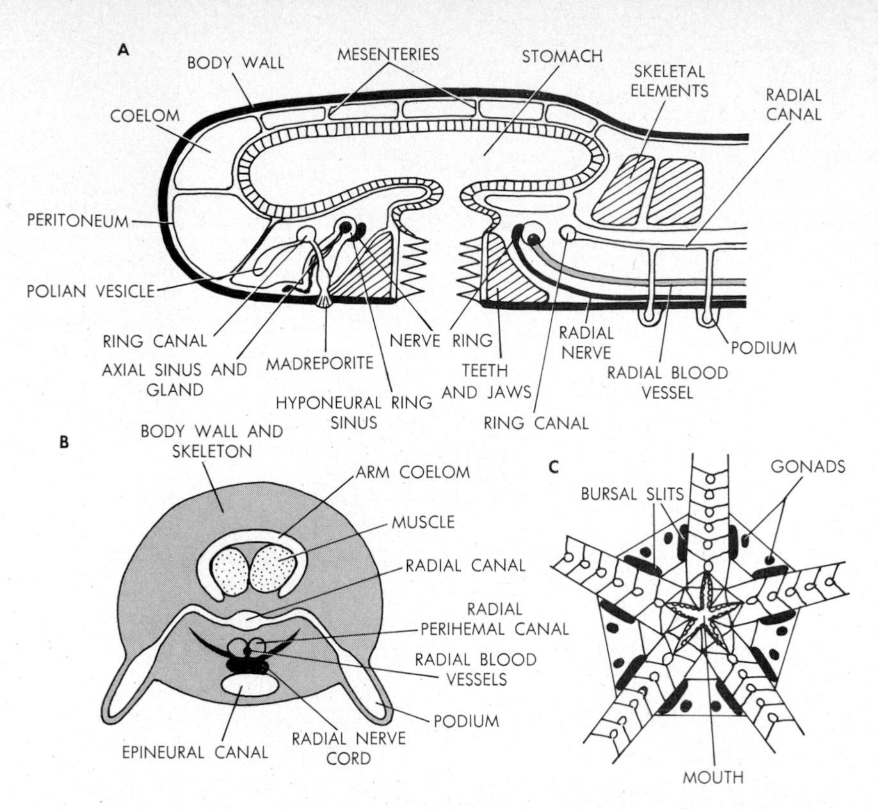

29.32. Ophiuroid structure. **A.** Longitudinal section through the disk and the base of one arm, schematic. (*Adapted from Ludwig.*) **B.** Cross section through an arm, showing epineural canal, schematic. (*Based on Hamann, Cuenot.*) **C.** Plan of the oral disk, to show position of bursal slits and gonads. (*Based on Reichensperger.*)

and viviparous forms retain their yolky eggs within the bursae, where they develop directly (cf. Fig. 11.26).

Indirectly and externally developing eggs pass through substantially the same stages as asteroids, but the larvae formed are *plutei.* In the formation of a pluteus, the gastrula acquires a conical shape and its left (future oral) side flattens (Fig. 29.33). Anterior and posterior ectodermal outgrowths from this flat side then produce the characteristic arms of a pluteus, each arm being stiffened internally by a calcarous spicule. Such arms form in successive pairs; as many as four pairs are present in older larvae. The pluteus arms increase buoyancy like the ectodermal lobes of other echinoderm larvae, and the cilia on the surface of the larva are the means of locomotion. During metamorphosis all arms and their spicules are shed. Also, the larval anus closes over and the larval intestine degenerates; adult intestine or anus are then never formed. Ophiuroids regenerate well and their arms are autotomized readily (hence the name "brittle stars").

Echinoids. Sea urchins are readily identified by their globular shapes and their movable spines (Fig. 29.34). Often up to a foot long, the spines are outgrowths from the endoskeletal plates and they are covered over their whole surface with epidermis. In some cases the epidermis at the tip of the spines is elaborated into poison glands. Spines are protective as well as locomotor. The skeletal plates are fused together into a rigid shell, a trait unique among living Eleutherozoa. Ten pairs of meridional rows of plates are present, and in every other pair the plates carry pores through which the tube feet protrude. Thus there are five pairs of podial rows, as in other echinoderms. (Note, however, that these rows extend from near the mouth right to the apex of the aboral side, whereas in ophiuroids and asteroids the tube feet occur only on the oral side.) The ambulacral grooves are closed in under the podial plates. Five pairs of skin gills are present along the rim of the small, leathery oral disk; on other parts of the surface gills are lacking. However, the tube feet play a substantial role in

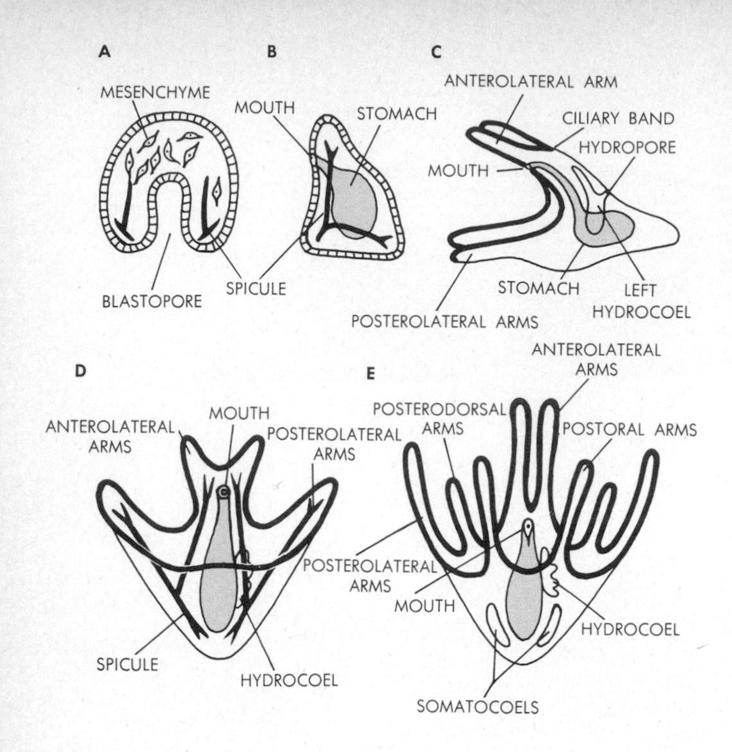

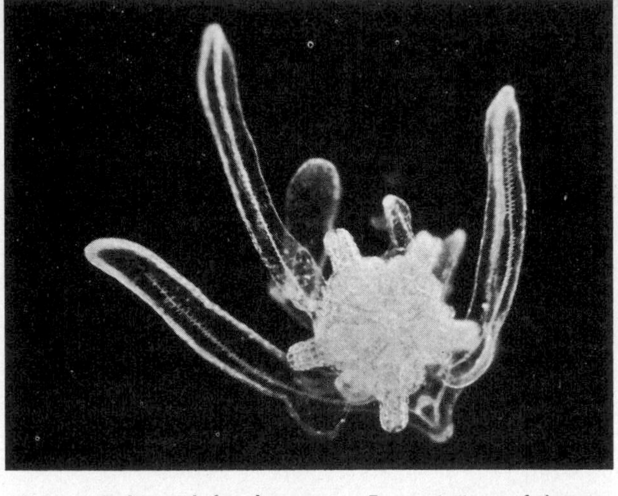

29.33. Ophiuroid development. For variations of the pattern see Fig. 11.26. **Diagrams: A,** gastrula; **B,** later stage, from left side; **C,** early, four-armed ophiopluteus, from left side (note absence of anus); **D,** four-armed pluteus in posteroventral view; **E,** eight-armed pluteus in posteroventral view. **Photo:** metamorphosing ophiopluteus of *Ophiura,* from life.

breathing. Pedicellariae, which represent modified spines, occur abundantly on the body surface. In many sea urchins the anus lies in the center of the aboral side, but in others this opening has become shifted to an excentric position.

In many respects the internal structure of sea urchins is like that of holothurians, particularly in the region of the pharynx. Thus, corresponding to the calcareous ring of sea cucumbers, sea urchins possess an exceptionally well developed chewing organ, the *lantern of Aristotle* (Fig. 29.35). It consists of 40 separate calcareous ossicles, including five

29.34. Sea urchins. A, spines and tube feet; **B,** the shell of a sea urchin, showing fused endoskeletal plates (cf. Fig. 8.8). A pluteus larva of *Echinus* is illustrated in Fig. 11.24. Cf. also color plate XVI.

A

B

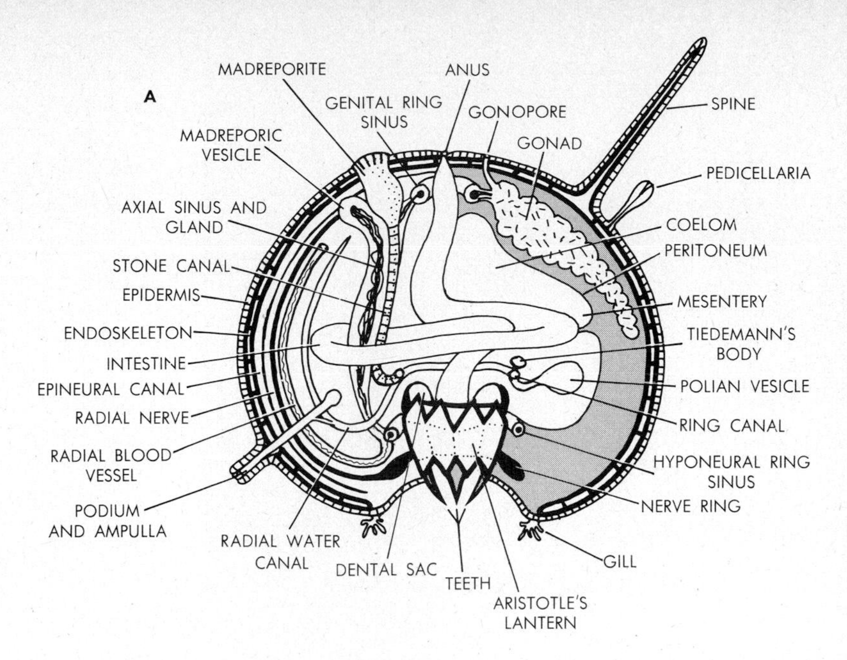

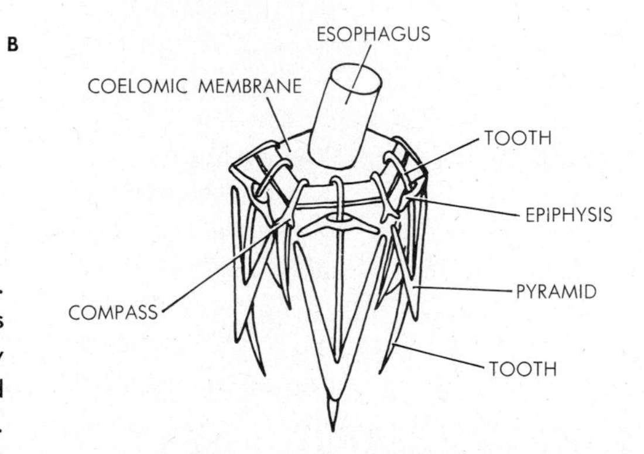

29.35. Echinoid structure (*combined from various sources*). **A.** Schematic section through a sea urchin. For clarity, many parts are here exaggerated or reduced in relative size. Also, only single representatives of organ systems such as tube feet and spines are shown. **B.** Some of the ossicles of Aristotle's lantern.

sharp-pointed teeth which converge toward the center of the mouth. Interiorly these teeth lie in and grow continuously from five *dental sacs,* evaginated from the pharynx. Muscles move the teeth and all other hard parts of the lantern apparatus. Sea urchins are omnivorous and scavenging, mostly along rocky or stony sea bottoms; they can readily bite through the armor of a dead arthropod. Indeed, some sea urchins may literally bite through rocks, in which they tunnel burrows. *Paracentrotus* is one of the rock-boring echinoids. The nervous, circulatory, and water-vascular systems follow the typical echinoderm pattern.

Sand dollars (e.g., *Clypeaster, Echinarachnius*) are highly flattened in an oral-aboral direction, and their movable locomotor spines are generally quite short (Fig. 29.36). The podial rows occur in an oral and an aboral set, those on the aboral side usually being arranged like the outline of flower petals. None of the tube feet are locomotor here, breathing being their main function. The anus lies at a point along the edge of the disk. As their name suggests, sand dollars are sand-burrowers. Some possess a broad and flattened lantern of Aristotle, but in many such a structure is absent.

Echinoids as a whole typically possess five gonads, with five aboral gonopores. Most members of the class spawn and the eggs develop externally, but brooding species are again known. The larvae are *plutei,* hardly distinguishable from those of ophiuroids. Accordingly, many zoologists have considered echinoids and ophiuroids to be closely related, just as, again on the basis of larval similarities, asteroids and holo-

29.36. Aboral and oral views of a sand dollar.

thuroids have been assumed to be closely related (Fig. 29.37). It is clear, however, that on the basis of *adult* structure asteroids resemble ophiuroids most, whereas echinoids have many traits in common with holothuroids. The adult structure of extinct species and the fossil record likewise suggest such interrelations (cf. also Chap. 15). Thus, the echinoderm classes provide an excellent example of a direct conflict between phylogenetic relations suggested by embryology and those suggested by adult morphology and paleontology. If the embryological data are accorded more weight, then starfishes and sea cucumbers would represent one branch of Eleutherozoan evolution, characterized by bipinnaria or similar larvae. Brittle stars and sea urchins would represent another branch, characterized by pluteus larvae. The adult differences between starfishes and sea cucumbers and between brittle stars and sea urchins would then have to be interpreted as instances of adaptive divergence.

On the other hand, if the paleontological data are given more weight, then the resemblances between adult starfishes and brittle stars would be due to derivation of these animals

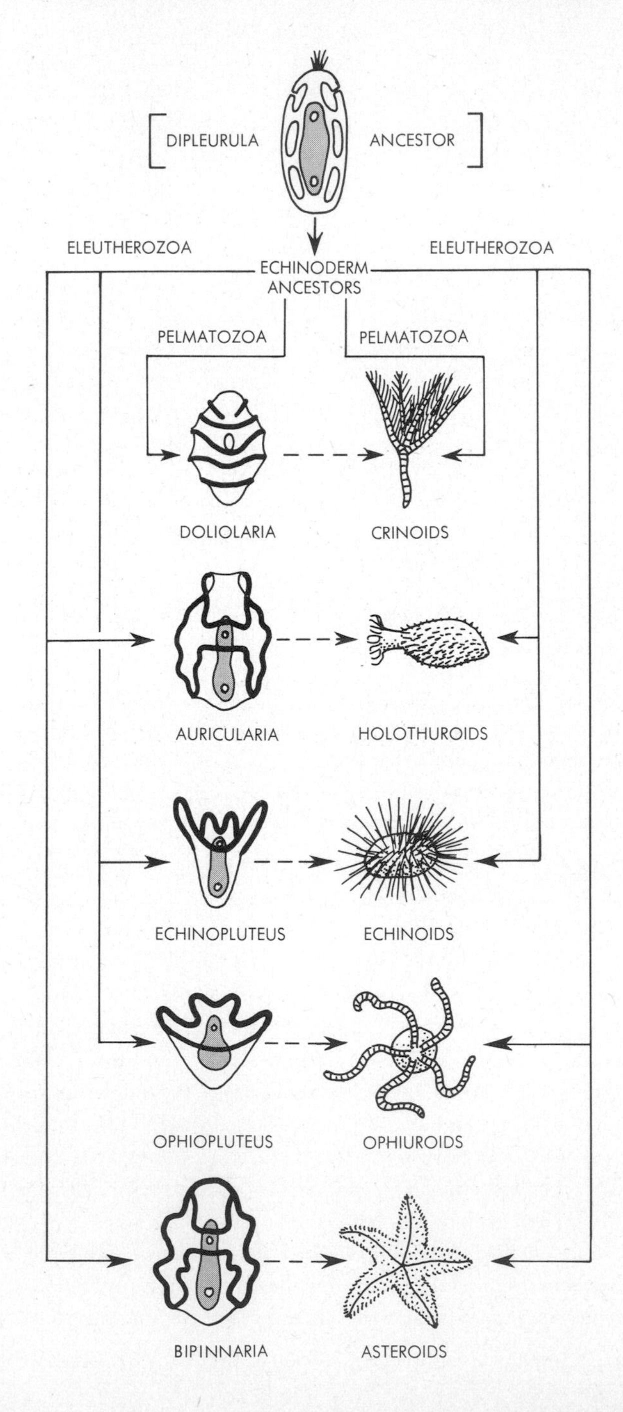

29.37. The conflicting data on evolutionary relations among echinoderms. Left side, summary of embryological conclusions, indicating evolution of two eleutherozoan lines, one leading to echinoids and ophiuroids via similar plutei, the other to holothuroids and asteroids via similar bipinnarialike larvae. Right side, summary of adult morphological and paleontological conclusions, indicating again evolution of two eleutherozoan lines, but one leading to the structurally similar holothuroids and echinoids, the other to the similar ophiuroids and asteroids. In either case, Pelmatozoa probably represent a separate branch of echinoderm evolution, and all echinoderms may be related to dipleurulalike deuterostome ancestors.

from a common stock of ancestral forms. In that case, the differences between their larvae would have to be ascribed to adaptive larval divergence; and the similarities between ophiopluteus and echinopluteus larvae would be due not to a close ancestral connection but to a secondary larval convergence. At present the issue cannot be resolved, for there are roughly as many cogent arguments in support of one view as in support of the other. In recent years more zoologists have tended to lend increasing weight to the paleontological evidence, on the ground that resemblances or differences between larval forms are too often misleading and have proved to be without much phylogenetic significance in many cases. Thus, where two groups of animals are clearly *known* to be closely related on the basis of other evidence, dissimilar larvae may nevertheless characterize them (e.g., the tornaria and other larval types among enteropneusts, the cyphonautes and other larval types among ectoprocts, the nymphal and caterpillar larvae among insects, and others). Moreover, larvae are well known to become modified greatly or even to be suppressed altogether simply by changes in the yolk content of the eggs. However, other zoologists argue *for* the great phylogenetic significance of larvae; they cite examples such as the trochophores or the tadpoles, on the basis of which even adults as dissimilar as mollusks and annelids or tunicates and frogs are generally conceded to be closely related.

In effect, therefore, the pros and cons concerning possible affinities among echinoderm groups are about evenly matched and opinion regarding echinoderm evolution remains divided. And if to this dilemma is added the uncertainty as to whether crinoid doliolaria larvae are primitive or not, we are left with a series of still unsolved problems concerning the internal evolutionary pattern of the echinoderm phylum. The only point on which there appears to be fairly general agreement is the dipleurula postulate, i.e., the assumption that echinoderms, hemichordates, and also pogonophora have evolved from a common bilateral ancestor with three pairs of coelomic sacs.

REVIEW QUESTIONS

1. Give taxonomic definitions of deuterostomial coelomates as a whole and of each of the phyla. Discuss the possible evolutionary interrelations of the phyla and the evidence from which such possibilities are deduced. What is the dipleurula hypothesis and what are its merits and weaknesses? Describe the structure of the hypothetical dipleurula.

2. Describe the structure of a chaetognath and review the organization of every organ system. Show how the coelom is subdivided and in what respects these divisions (*a*) correspond, (*b*) do not correspond to coelomic subdivisions of other deuterostomes. Where do arrowworms live and how and on what do they feed? What are the principal sense organs of the animal? How does a chaetognath develop?

3. Describe the structure of a pogonophoran and review the organization of every organ system. What are the external and internal subdivisions of the body? How does a beard worm feed and where does such an animal live? How does it develop?

4. Define and distinguish between the hemichordate classes and name representative genera. Describe the exterior and interior structure of an acorn worm and review the organization of every organ system. What structural features do such worms have in common with chordates? Show how acorn worms develop and in what respects this pattern is typical among deuterostomes generally.

5. What is the possible evolutionary significance of a tornaria? Contrast the structure of an enteropneust with that of a pterobranch. What is the basis for the assumption that pterobranchs and graptolites might be related? On what grounds could pterobranchs be considered to be more primitive than enteropneusts?

6. Define the subphyla and living classes of echinoderms. What are the presumable evolutionary interrelations of the subphyla and the possible interrelations of the classes? In what respects is the evidence bearing on such problems conflicting? What evolutionary and adaptive factors have probably led to the development of echinoderms and their water-vascular systems?

7. Describe the symmetries of various echinoderm body plans. Define podium, ambulacral groove, epineural canal. Describe the structure of a skeletal element and of the whole skeleton of echinoderms. What variant skeletal organizations are encountered in different classes?

8. Describe the external and internal structure of a starfish and review the organization of every organ system. What is the function of the water-vascular system and how are such functions performed? How and on what does a starfish feed? Show how the functional importance of the three neural subdivisions varies for different classes?

9. Distinguish between (*a*) madreporic ampulla and vesicle, (*b*) genital and hyponeural ring sinus, (*c*) radial and axial sinus, (*d*) axial sinus and gland. Describe the complete

development of a starfish, including dipleurulalike, bipinnaria, and brachiolaria stages. Show in detail how the water-vascular system develops.

10. To what hemichordate structures do axocoels, hydrocoels, and somatocoels correspond? What are the developmental fates of these cavities? How does metamorphosis take place? How does (*a*) early, (*b*) larval development differ for different echinoderm classes? Name the various echinoderm larvae and contrast their structure.

11. Describe the structure of a crinoid and distinguish structurally between a sea lily and a feather star. What are cirri? Pinnules? Can any of the echinoderm groups regenerate? Reproduce vegetatively? How many species are known in each living class?

12. Describe the structure of a sea cucumber and show to what extent this structure (*a*) corresponds, (*b*) differs from that of a starfish or a sea urchin. Which eleutherozoan groups have closed ambulacral grooves? How and where does a holothurian live? What are Cuverian tubules and what are their functions?

13. Contrast the structure of a starfish and a brittle star. Contrast the development of (*a*) directly, (*b*) indirectly developing brittle stars. In which echinoderm groups is ovoviviparity and viviparity encountered and, in the latter case, where in the adult body does offspring development take place?

14. Describe the structure of a sea urchin and show how it differs from that of other echinoderms. What are the unique features of the skeleton? How and on what does a sea urchin feed? Distinguish sea urchins and sand dollars structurally. How do these animals breathe?

15. Name and characterize the extinct classes of echinoderms. Which of these are believed to be the most primitive and why? Review the paleontological history of echinoderms. Consult Chap. 15 for dates if necessary.

COLLATERAL READINGS

Additional readings on the phyla discussed in this chapter may be found in the following:

Fell, H. B.: Echinoderm Embryology and the Origin of Chordates, *Biol. Reviews*, vol. 23, 1948.

———: Echinodermata, in "McGraw-Hill Encyclopedia of Science and Technology," vol. 4, McGraw-Hill, New York, 1960.

Hyman, L. H.: "The Invertebrates," vol. 4 (echinoderms); vol. 5, chaps. 16 (chaetognaths), 17 (hemichordates), and 18 (pogonophora), McGraw-Hill, New York, 1955, 1959.

Ivanov, A. V.: Pogonophora, *Systemat. Zool.*, vol. 4, 1955.

———: On the Systematic Position of Pogonophora, *Systemat. Zool.*, vol. 5, 1956.

Manton, S. M.: Embryology of Pogonophora and Classification of Animals, *Nature*, vol. 181, 1958.

Moore, R. C.: Echinodermata Fossils, in "McGraw-Hill Encyclopedia of Science and Technology," vol. 4, McGraw-Hill, New York, 1960.

———, C. G. Lalicker, and A. G. Fischer: "Invertebrate Fossils," McGraw-Hill, New York, 1952.

Morgan, T. H.: The Development of *Balanoglossus*, *J. Morphol.*, vol. 9, 1894.

Pierce, E. L.: The Chaetognatha over the Continental Shelf of North Carolina, *J. Marine Research.*, vol, 12, 1953.

Shrock, R. R., and W. H. Twenhofel: "Principles of Invertebrate Paleontology," 2d ed., McGraw-Hill, New York, 1953.

CHAPTER 30

chordata: protochordates

Phylum CHORDATA: notochord, pharyngeal gill slits, and dorsal hollow nerve cord present in preadult stages only or throughout life; coelom enterocoelic; development via tailed larvae (*tadpoles*) or direct.

PROTOCHORDATES (ACRANIA): without head. *Urochordata*, tunicates; *Cephalochordata*, amphioxus
VERTEBRATES (CRANIATA): with head. *Vertebrata*

Because this phylum includes man and the animals most directly important to man, it is unquestionably the most interesting from almost any standpoint. The phylum also has a special evolutionary significance, for it comprises what is by far the most progressive group of animals. Other phyla may include more species and may be more diversified; but only among chordates do we encounter, within one and the same phylum, types as primitively organized as tunicates and types as complexly organized as men. Evidently, the ancestral chordate body plan was richer in major evolutionary potentialities than the body plans of other phyla.

The specific ancestors of chordates are unknown, but a distant and probably indirect affinity to hemichordates, pterobranchs particularly, is suggested by chordate morphology. Thus, both hemichordates and chordates exhibit a basically similar pattern of embryonic development, and both groups possess a pharynx with gill slits used primitively in ciliary plankton feeding. A dorsal hollow nerve cord also qualifies as a common trait. Two additional major chordate features, viz., notochord and metameric segmentation, evolved as original inventions directly within the phylum, as we shall see.

Chordates encompass some 50,000 species classified into three subphyla: the headless and unsegmented *urochordates*, or *tunicates;* the headless and segmented *cephalochordates*, or *amphioxus;* and the head-possessing and segmented *vertebrates*, or *craniates*. Tunicates generally and ascidian tunicates specifically are undoubtedly the primitive members of the phylum. Ancestral stocks of ascidians appear to have given rise to all present tunicates, independently to the amphioxus group, and, again independently, to the vertebrates (Fig. 30.1). Much of this evolutionary diversification has probably been achieved through the mechanism of neoteny. Tunciates are marine, amphioxus lives along shores, and vertebrates primitively are freshwater animals. We note that early evolution within the chordates appears to have been oriented by shifts in habitat. Such shifts have certainly also played a major role later, during the diversification of vertebrates.

The Acrania, encompassing the urochordates and cephalochordates, are often also referred to as *protochordates*. We shall discuss these animals in this chapter; the vertebrates form the subject of the two following chapters.

SUBPHYLUM UROCHORDATA

Tunicates; marine, sessile or pelagic; with secreted external envelope (*tunic*); unsegmented; coelom not clearly elaborated; pharynx with endostyle and gill slits, used for breathing and ciliary filter feeding; circulatory system open, heart with reversing beat; often colonial through budding;

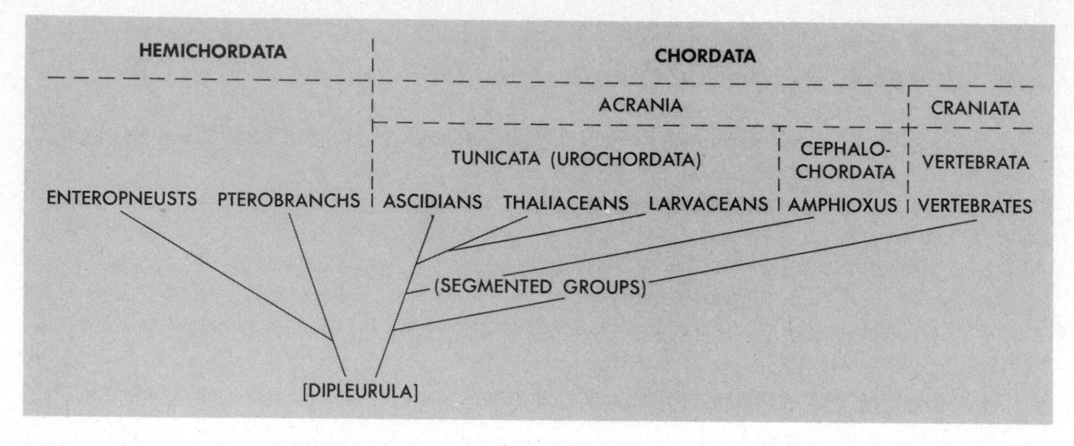

30.1. The probable evolutionary and the taxonomic interrelations among chordates.

hermaphroditic; development mosaic, indirect or direct. *Ascidiacea, Thaliacea, Larvacea*

Class Ascidiacea

Ascidians, sea squirts; larvae if present free-swimming tadpoles, nonfeeding, with notochord and dorsal nerve cord in tail; adults without tail, sessile, colonial through budding or solitary; gill slits numerous; alimentary tract U-shaped. *Ascidia, Ciona, Styela, Botryllus, Molgula*

Most of the approximately 2,000 species in the tunicate subphylum belong to the sessile ascidians. These animals live in sand, mud, or attached to rocks, and many of them form flat, budding colonies. An individual sea squirt or "sea potato" is covered externally by a tunic, or *test,* composed of *tunicin,* largely cellulose (Fig. 30.2). Secreted by the underlying epidermis, the test may or may not have appreciable thickness and it may be transparent or opaque, leathery or gelatinous, and in some cases also stiff and cartilaginous. It sloughs off along the surface but is continuously re-formed from the epidermis underneath. On the upper, nonattached side the animal has two openings, an incurrent *branchial siphon,* through which water and food enter, and an excurrent *atrial siphon,* through which water, elimination products, and reproductive cells leave. A plane through the two siphons defines dorsal and ventral aspects, the branchial siphon being ventral. Each siphon, equipped with ocelli along its rim, may be closed by circular sphincter muscles.

The atrial siphon provides exit from a large cavity which lies dorsally and along each side of the animal. This *peribranchial cavity,* or *atrium,* arises in the larva by invagination from the dorsal ectoderm, hence the atrial lining is ectodermal; it forms the innermost layer of the body wall in much of the animal. Between the epidermis and the atrial lining are connective tissue and circular and longitudinal muscle strands. The branchial siphon leads to the mouth, ringed with short tentacles, and then into a large pharynx, or *branchial basket.* This chamber is perforated by numerous gill slits, arranged in several transverse (dorsoventral) rows. Blood vessels traverse the pharyngeal wall between the slits. Each such slit is elongated anteroposteriorly and is just wide enough to accommodate the cilia set around its rim (Fig. 30.3). Cilia also line the interior surface of the pharynx. Along the ventral gutter of the chamber is a band of specialized tissue called the *endostyle.* In it, a median strip of cells bears long flagella and strips adjacent along each side are ciliated and glandular. Note that an endostyle is already foreshadowed in the ventral pharyngeal gutter of hemichordates (and that this organ is also the evolutionary source of the vertebrate thyroid gland). The endostyle secretes mucus continuously, and the flagella and all the pharyngeal cilia distribute this mucus over the inner surface of the branchial basket. The cilia, particularly those around the gill slits, also draw in a continuous food-bearing water current through the branchial siphon and the mouth. Food particles become entangled in the mucus and water passes through the gill slits into the atrium, oxygenating blood in the process. From the atrium, water is expelled in a forceful excurrent stream through the atrial siphon. Concurrently, food in mucus collects along the dorsal wall of the

30.2. Ascidian adult structure. Photo: cutaway model of an adult ascidian. Food-bearing water is drawn into the pharynx through the incurrent siphon. Food passes into the U-shaped alimentary tract, and water emerges through the gill slits and via atrium and excurrent siphon to the outside. **Diagrams: A.** schematic section as in photo, showing anatomical orientations and the ectodermal nature of the lining of the branchial sac. The ectoderm layer is drawn with cross-lines suggesting cells, the endoderm layer is without such lines. Internal organs such as heart and gonads lie in the gray-shaded regions between the epidermis and the branchial lining. **B.** Schematic dorsoventral (cross) section at a level just below the excurrent siphon, showing how the branchial sac envelops the pharynx; layers and regions drawn as in **A. C.** section through ganglion region, showing neural gland and duct to pharynx (*after Roule*).

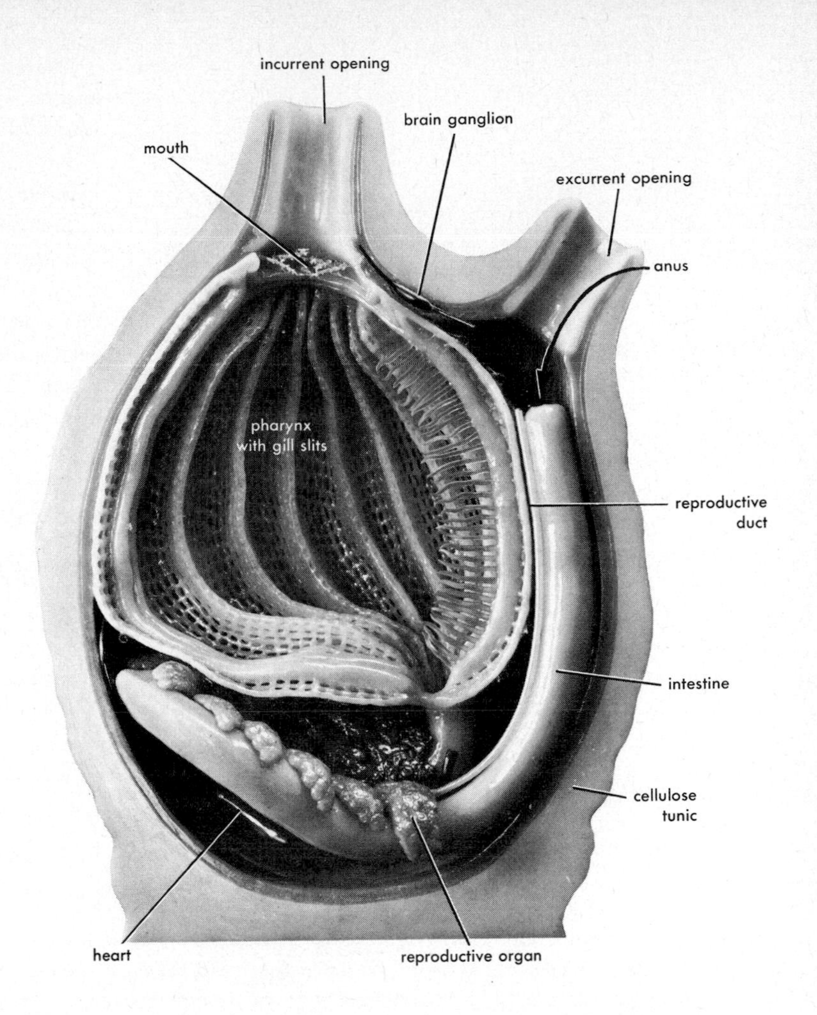

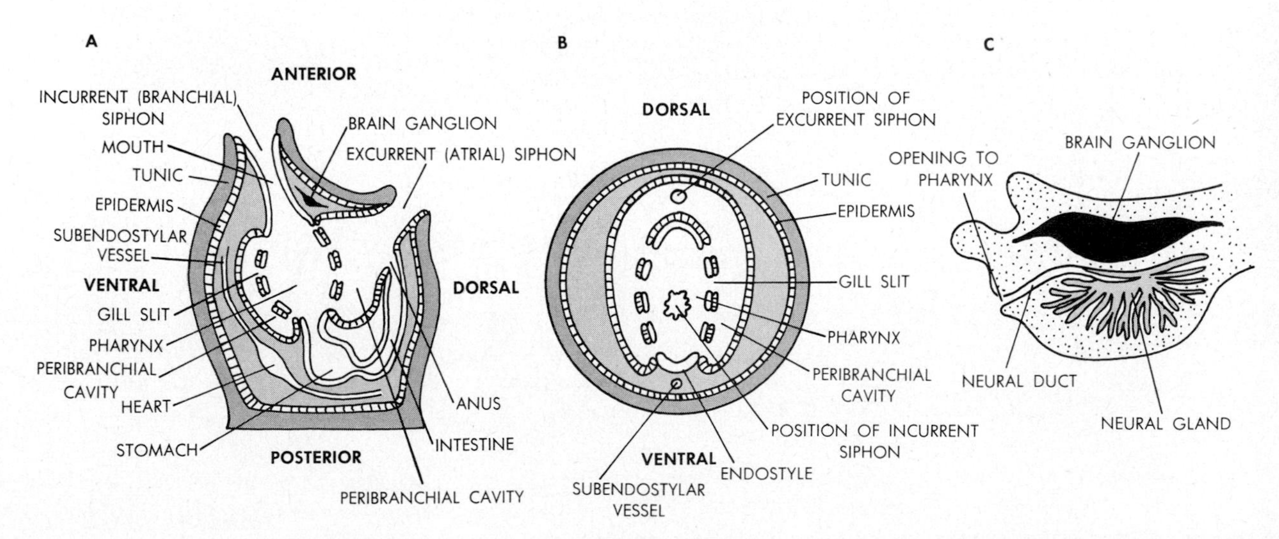

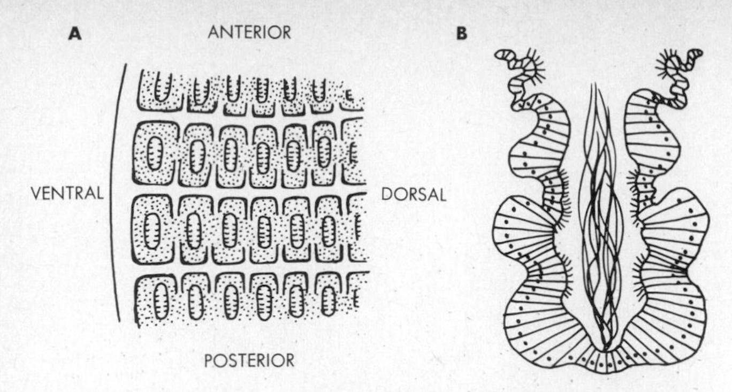

30.3. A. Diagram of portion of pharyngeal wall, showing dorsoventral rows of gill slits, each elongated anteroposteriorly and rimmed with cilia. Transverse (dorsoventral) blood vessels between rows branch open-endedly into the tissue spaces between adjacent slits of a given row. **B.** Cross section through the endostyle, showing the three pairs of glandular and three pairs of ciliated cells on each side, as well as the flagellated cells in the ventral gutter. (*After Sokolska.*)

branchial basket and is propelled by ciliary action into the esophagus, continuous posteriorly with the pharynx. The esophagus leads into a stomach at the bottom of the alimentary U, and an intestine terminates at the anus, which opens into the atrial cavity.

The nervous system consists of a single neural ganglion located anteriorly between the two siphons (cf. Fig. 30.2). A *neural gland*, ventral or dorsal to the ganglion, leads through a duct into the pharynx just behind the mouth. This gland may be absorptive or may possibly function as a smell- or taste-receptor or both; it has on occasion been homologized with the vertebrate pituitary. The circulatory system includes a ventral heart as well as blood vessels passing anteriorly into the pharynx and posteriorly into the stomach region. These vessels are open-ended; capillaries are absent, and blood flows mostly through hemocoelic spaces. The heart arises as an invaginated fold in a pericardial tube, the latter itself forming from a mesodermal tissue mass between the ventral ectoderm and endoderm (Fig. 30.4). The pericardial space thus surrounding the heart is the only cavity qualifying as a coelom. The several peculiarities of ascidian blood and the heart have already been discussed in Chap. 9.

Many sea squirts possess an *epicardium* close to or around the heart (cf. Fig. 30.4). This structure arises as a pair of evaginations folded out from the posterior part of the pharynx. In *Ciona* the folds enlarge and envelop all the organs in the posterior part of the body, like peritoneum and mesentery. In other cases the folds fuse into one, and in forms such as

30.4. Heart and body cavities. A through **D** represent a sequence of cross sections through the ventral body region near the posterior end of the pharynx, showing the development of the heart, the epicardium, and the renal organ. Note that the heart arises as an invaginated tube from the pericardial coelom; that the epicardia are a pair of folds evaginated from the posterior end of the pharynx which eventually envelop all ventral and posterior organs; and that the renal organ, formed from the epicardia, accumulates a solid excretory concretion in its interior.

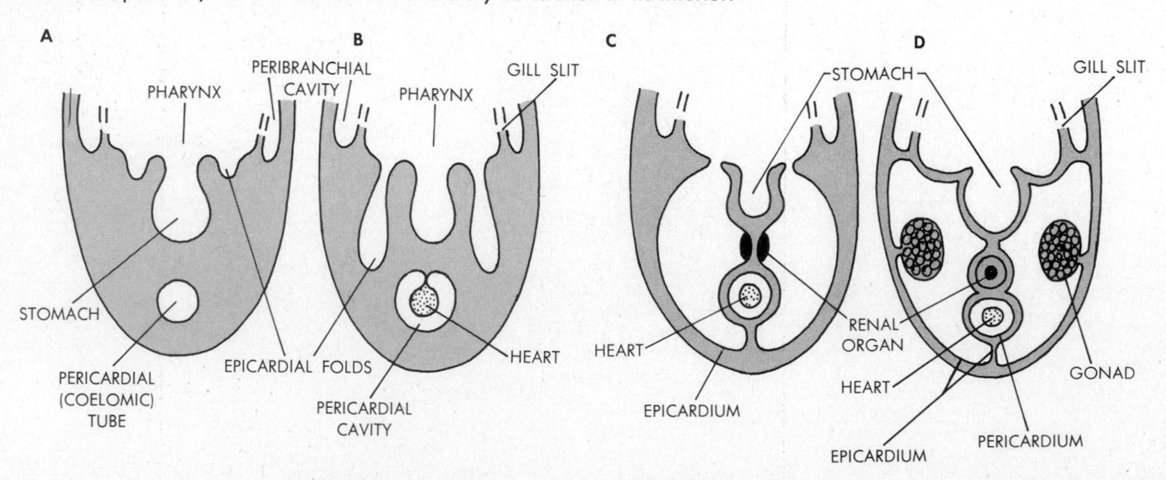

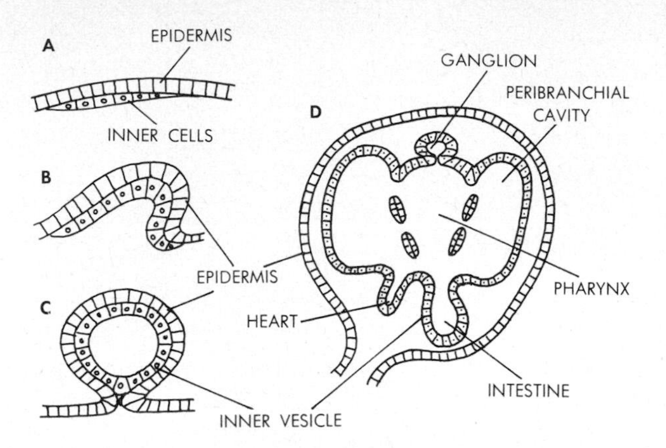

30.5. Ascidian budding. *(Slightly modified from Berrill.)* The sequence sketches bud growth in *Botryllus,* in which buds arise from the atrial body wall. In other cases the inner cells do not always form a neat layer from the start. In all cases, however, the inner cells (shown here with nuclei) eventually do form a vesicle from which all endodermal and mesodermal body parts arise by invagination and evagination.

Molgula they condense into a *renal organ* which comes to lie alongside the heart. This organ stores excretory wastes in the form of uric acid crystals. The latter simply continue to accumulate in the renal organ throughout the life of the animal. In many cases the epicardial membranes play a role in budding (cf. below). Inasmuch as they are derived from endodermal tissues, the epicardial folds may represent homologues of true coelomic membranes. To be sure, they are formed later than in early embryonic stages, when coeloms normally develop.

Buds arise in various specific ways in different ascidians. In all cases, however, budding involves a pinching off from the parent animal of a small epidermal sphere in which are included pieces of some interior tissue (Fig. 30.5). In different ascidians such buds can form almost anywhere on the epidermis, and the bud interior can derive from almost any relatively unspecialized adult tissue. In some instances, e.g., in *Botryllus,* buds arise in the body wall of the atrial region, mesenchyme from the body wall supplying the internal cells of the bud. In other instances buds form more posteriorly and the epicardium then usually contributes the interior bud cells.

The epidermal exterior of a bud produces only the epidermal components of the new adult. The interior cells develop the mesodermal as well as the endodermal components, regardless of the original germ-layer derivation of these cells. If many buds originate more or less simultaneously in a compact group, they may form a colony surrounded by a common tunic and even a common atrial siphon, as in the rosette colonies of *Botryllus* (Fig. 30.6). A single bud may also elongate into a stalk, or stolon, along which more or less separate zooids may develop at intervals. Remaining interconnected by the stolon, such individuals may come to form a branching colony, as in *Perophora.* The capacity of budding is believed to be primitive among ascidians; solitary forms such as *Ciona* or *Styela* are considered to have lost the budding potential. As is characteristic of budding animals generally, ascidians regenerate readily.

Sea squirts are hermaphroditic. Male and female gonads lie in the loop of the alimentary tract in one group (cf. Fig. 30.2), in the atrial body wall in another. Gonoducts open into the atrium. A number of species is viviparous, and in these the female gonoduct is enlarged into a brood pouch, or uterus. The eggs of such animals are large and yolky. Most ascidians spawn, however, and their smaller, less yolky eggs develop externally. Cleavage is bilateral and interestingly mosaic, a unique and exceptional condition among deuterostomes (Fig. 30.7). The fertilized eggs of several ascidian types possess distinct, differently pigmented inclusions which become aggregated in definite, fixed areas of the egg; and because cleavage proceeds in a rigidly patterned mosaic fashion, the colored regions become part of specific blastomeres. These have rigidly fixed, predictable fates; hence the precise developmental history of each cell of the early embryo can be traced readily by the pigmentation.

The blastomeres form a coeloblastula which gastrulates (in *Styela,* between the sixth and seventh cleavage) by embolic invagination. The future dorsal and lateral parts of the embryo wall near the blastopore then proliferate posteriorly, establishing the rudiments of the tail. A specific group of cells invaginated during gastrulation and lying in the roof of the archenteron participates in this posterior proliferation; the cells come to form a single longitudinal row in the tail and they constitute the future *notochord.* Concurrently with these events, a solid band of mesoderm cells is budded off from each side of the archenteron. Similarly extending back into the tail, these bands later develop into the tail musculature. The orderly, linear alignment of such muscle-forming cells may represent a presegmented condition from which a segmented musculature may subsequently have evolved in other chordate groups. Note that the archenteron here does not produce coelomic sacs but solid mesoderm bands only; embryonic coelomic sacs do not develop.

A nerve cord arises as in vertebrates by the infolding of a

A

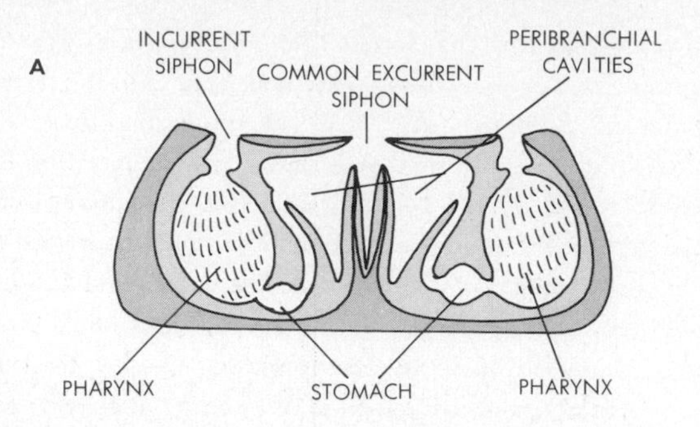

B

C

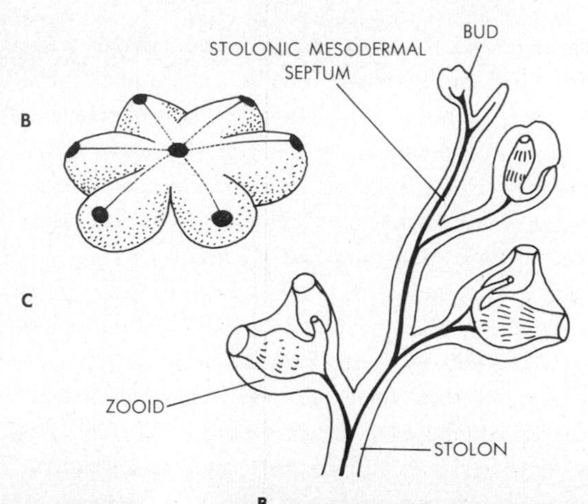

30.6. Solitary and colonial ascidians. Diagrams: A. Section through a rosette colony of *Botryllus*, showing common atrial siphon. (*Based on Delage and Herouard.*) **B.** Exterior view of **A,** attached to rocks. (*After Edwards.*) **C.** Diagram of stolonic colony of *Perophora*, growing on sea weeds. (*After Listen.*) **Photos: A.** *Ciona*, a solitary type; the incurrent siphon is at top, fully open. **B.** Colony of *Ascidia*. **C.** Rosette colonies of *Botryllus*, attached to twigs, rocks, or other objects in water.

A

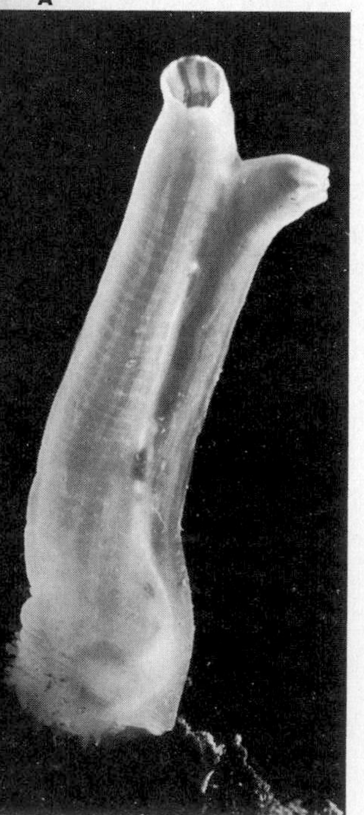

B

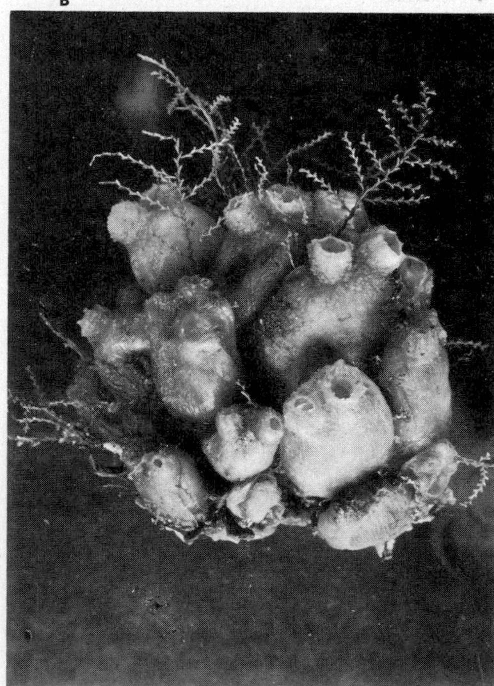

C

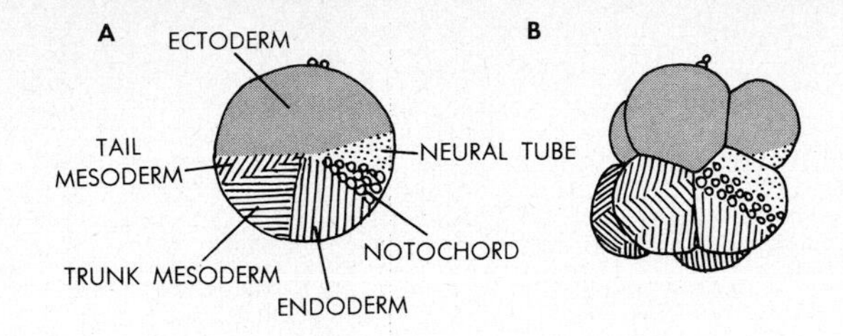

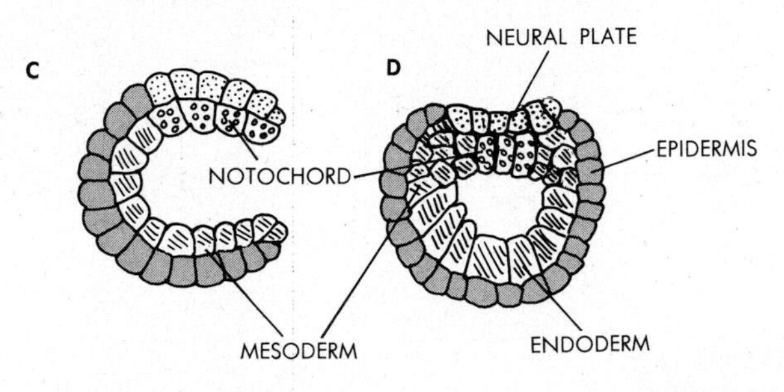

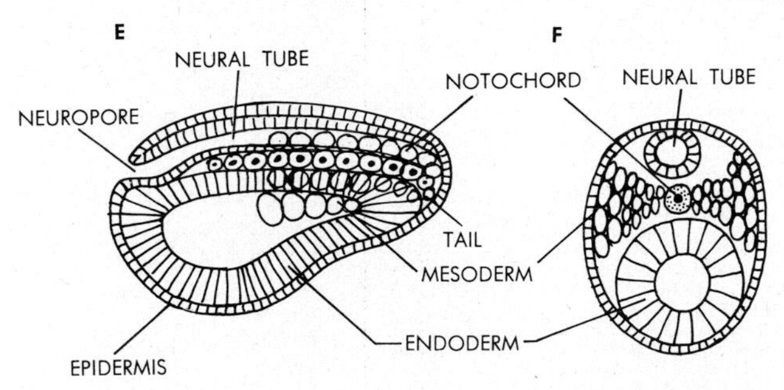

30.7. Early ascidian development. *(After Conklin, Van Beneden.)* **A.** side view of mosaic egg, showing location of prospective organ-forming regions (areas are marked similarly in **B** to **D**). **B.** Side view of eight-cell stage. **C.** Sagittal section through gastrula. **D.** Cross section through gastrula. **E.** Side view of tail bud stage, notochord cells shaded and with black nuclei, lateral mesoderm bands superimposed. **F.** Cross section through tail bud stage as in **E.**

neural tube from the dorsal ectoderm (cf. Fig. 11.20). The anterior part of the tube enlarges into a brain vesicle. At the forward end of the vesicle a neuropore comes to open into the developing pharynx (Fig. 30.8). This pore is homologous to the neuropore of hemichordates, and it represents the forerunner of the pharyngeal opening of the adult neural gland. Inside the brain vesicle, an ocellus develops as an inward projection from the roof of the brain wall and a statocyst forms on the floor. Just behind the level of the brain, the dorsolateral ectoderm on each side of the larva invaginates and produces a pair of atrial pouches; the two openings to the outside will later fuse and form the single atrial siphon of the adult. The archenteron in the meantime has differentiated into the parts of the alimentary tract. An endostyle arises in the pharynx, and the first three pairs of gill slits break through and establish communication between the pharyngeal cavity and the atrial chambers. Additional gill slits form later, by subdivision of the first ones.

Anteriorly the tadpole possesses *adhesive papillae* by which it will eventually become attached. Everywhere else the surface of the larva is enveloped in the rudiments of the tunic, which covers the mouth and so makes this opening nonfunctional. The tadpole swims about for only a short period—from minutes to a few hours at most; as pointed out in Chap. 11, it serves primarily as a site-selection device. After anterior attachment, metamorphosis involves degeneration of the tail, including notochord, nerve cord, and the tail muscles. Also, a rapid proliferation of the body wall between the attachment area and the mouth shifts the mouth toward the free end of the animal and produces the U shape of the alimentary tract. Parts of the brain vesicle persist as the neural ganglion and neural gland. The tunic is resorbed in the region over the mouth and a feeding adult so becomes established.

In earlier theories, the ascidian tadpole was considered to represent the living reminder of a primitive, original chordate ancestor, whereas the ascidian adult was regarded to represent a degenerate form, adapted secondarily to a sessile mode of life. Today, however, it is appreciated that the sessile ascidian adults are basically primitive types, and that they probably represent direct descendants of possibly hemichordatelike, filter-feeding, chordate ancestors. The main dispersal mechanism of such ancestral chordates would have been budding, and the tadpole larva is envisaged to constitute an original evolutionary invention of this early chordate stock. Tadpoles would have been adaptively highly advantageous to the sessile adults, as efficient dispersal and site-selecting agents. Moreover, a tailed, well-muscled tadpole could become adapted readily to a prolonged pelagic existence; and if the larva also developed gonads precociously, a neotenous new type of adult could be evolved. It is believed that from such a free-swimming larval starting point have arisen the pelagic tunicates, viz., Thaliacea and Larvacea, as well as all other chordates.

Class Thaliacea

Chain tunicates; larvae if present free-swimming, with notochord and dorsal nerve cord in tail; adults without tail,

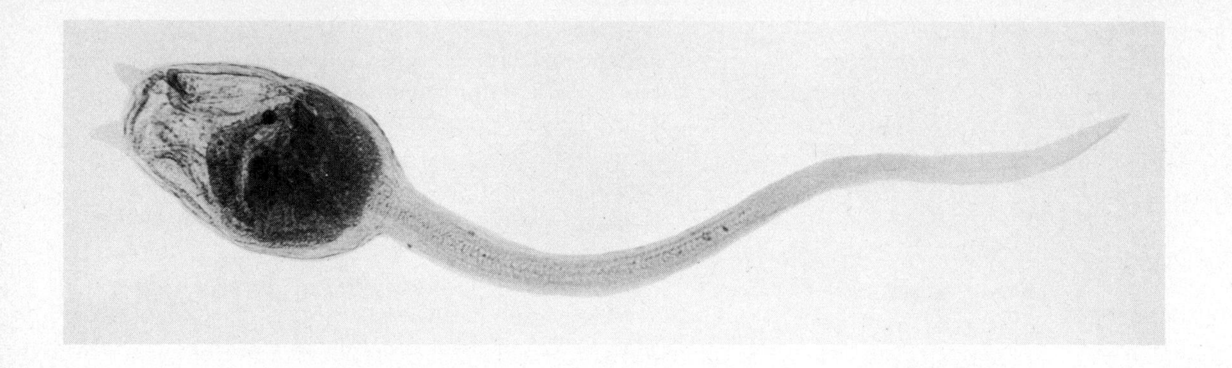

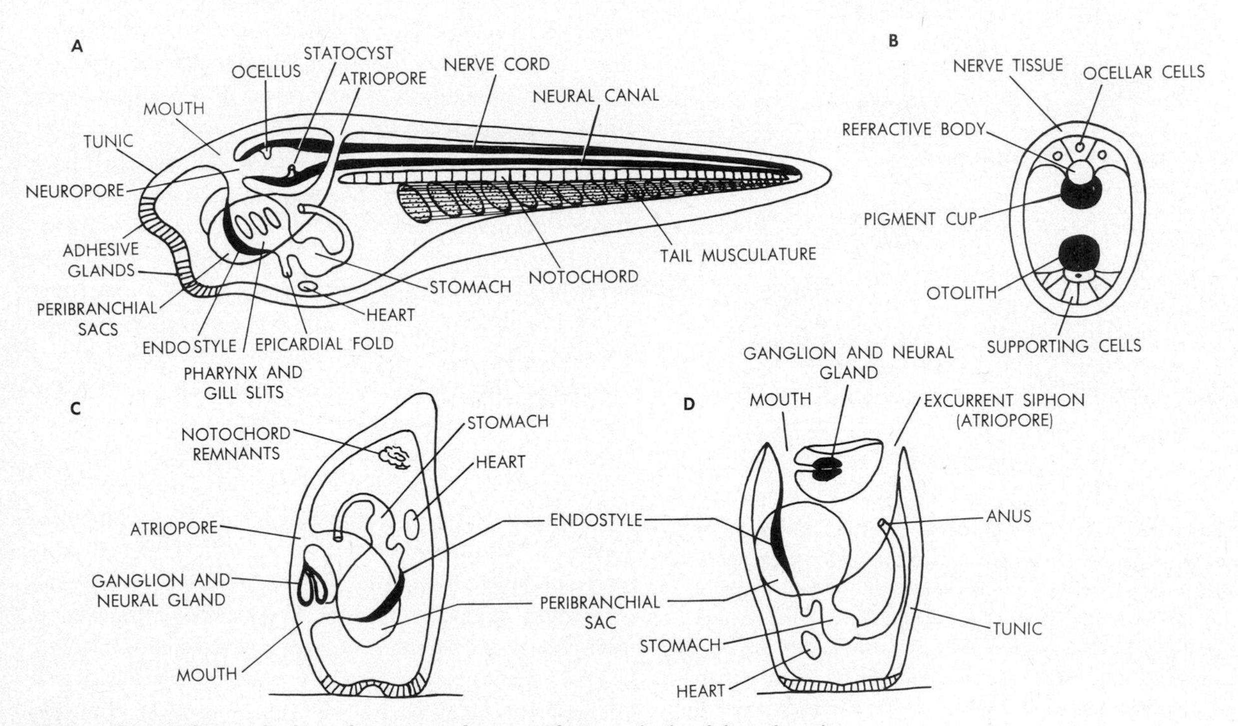

30.8. The ascidian tadpole and metamorphosis. Photo: tadpole of the colonial ascidian *Aplidium* (=*Amaroucium*). **Diagrams,** schematic: **A.** Sagittal section through a tadpole. **B.** Cross section through brain vesicle, showing three-celled ocellus and one-celled statocyst. **C.** Early metamorphosis; tail, notochord, and most of nerve cord resorbed, neuropore becoming opening from neural gland duct into pharynx. **D.** Late metamorphosis; rotation of internal organs completed and siphons functional.

pelagic and free-swimming, locomotion by "jet" propulsion; body wall with transverse muscle hoops; gill slits few to numerous; polymorphic, with colonial stages budded in chains from stolons. *Pyrosoma, Doliolum, Salpa*

The members of this and the following class include some of the most fascinating of all animals. Individual zooids are minute, barrelshaped or cylindrical, with conspicuous bands of circular body-wall muscles arranged in the form of transverse hoops (Fig. 30.9). There are eight such hoops in *Doliolum* and seven in *Salpa*. In the latter genus the hoops are incomplete ventrally and in *Pyrosoma* they are poorly developed

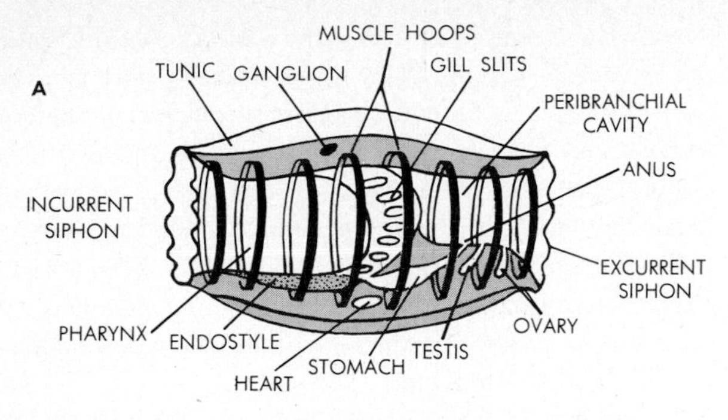

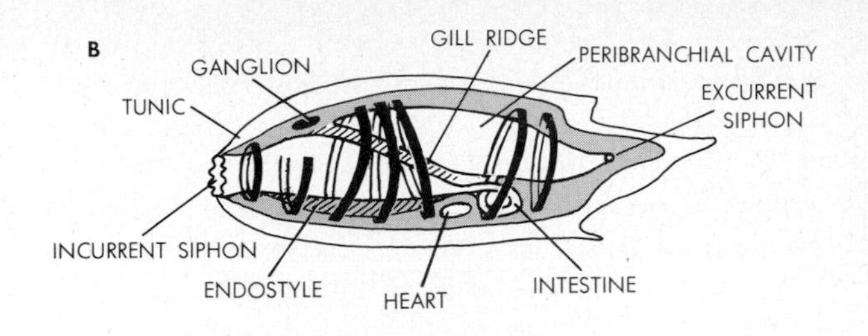

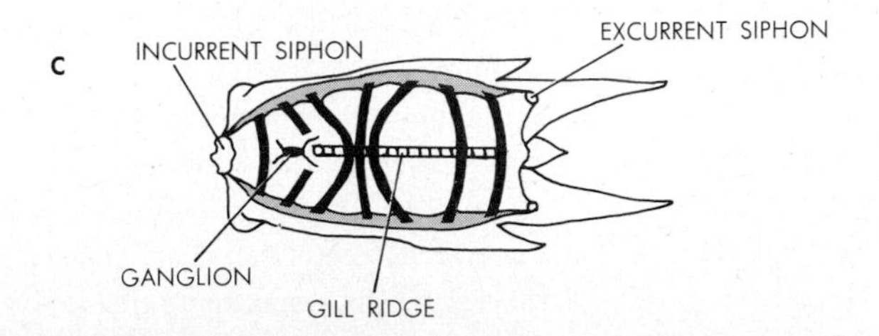

30.9. Thaliacea. Diagrams: A. Adult (gonozooid) of *Doliolum,* from left side. Note the eight muscle hoops. (*After Uljanian, Berrill.*) **B.** Adult (gonozooid) of *Salpa,* from left side. Note the seven muscle hoops. (*After Metcalf.*) **C.** Dorsal view of *Salpa.* Note the incompleteness of the muscle hoops and the paired openings from the peribranchial cavity. (*After Metcalf.*) **Photo:** *Doliolum,* in plankton, from life.

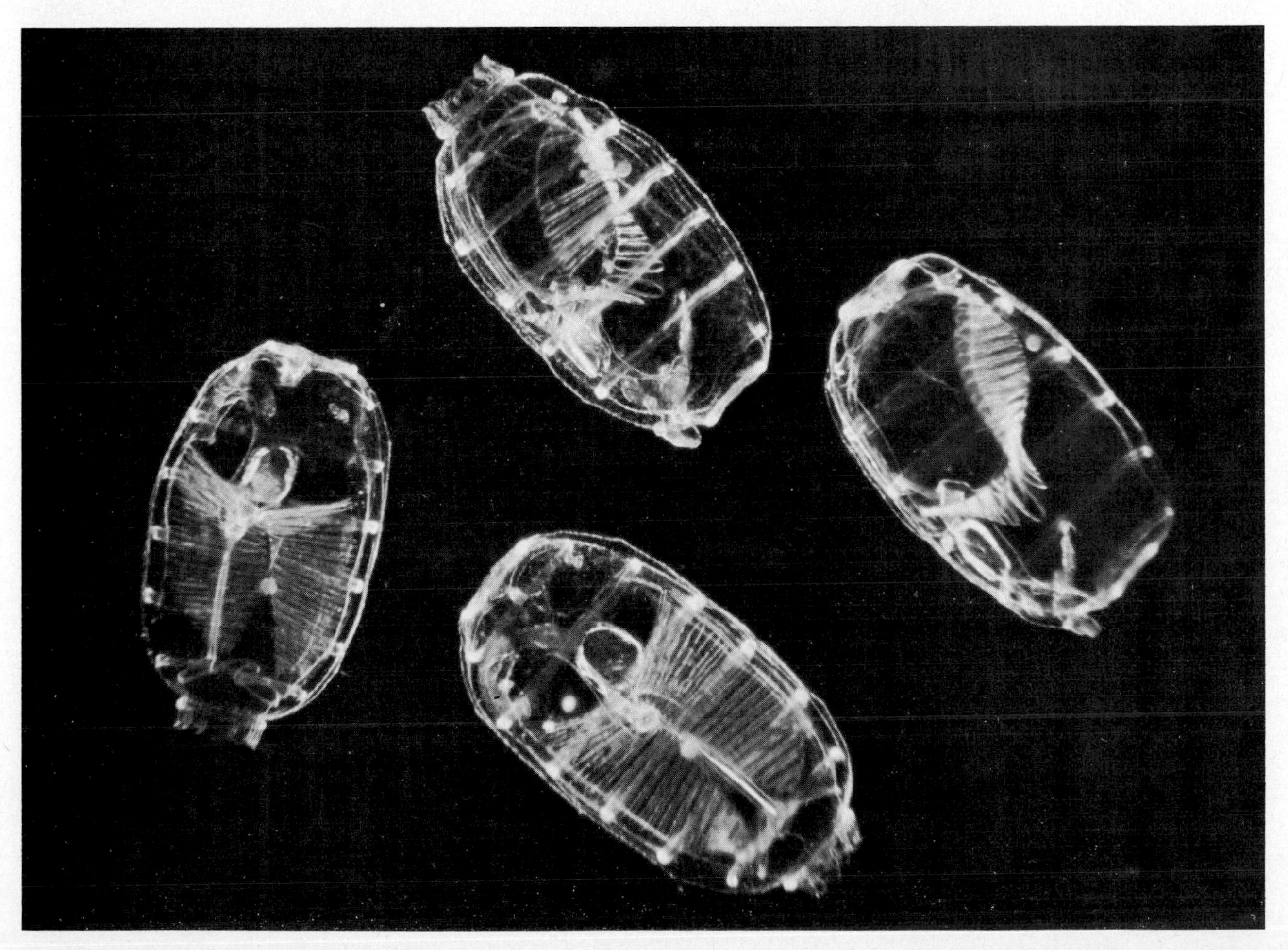

altogether. The siphons are at opposite ends of the animal, and the hoop muscles at these ends serve as sphincters; when one contracts the other relaxes, actions resulting in a more or less rhythmic expulsion of water from the posterior atrial siphon and a jetlike propulsion of the animal. Pharyngeal cilia maintain the water flow, as in ascidians. Other internal structures too are substantially as in the sea squirts, though the numbers of gill slits vary with the group and the positioning of the organs is somewhat modified in correlation with the linear organization of the body.

The main interest lies in the colonial structure and the life cycles of these animals. *Pyrosoma*, brightly bioluminescent (hence its name), represents the only genus of one thaliacean order. The animal forms gelatinous tubular colonies (Fig. 30.10). Such a tube is closed at one end, open at the other, and the zooids lie within the wall of the tube. Their branchial siphons face outward, their atrial siphons face into the cavity of the tube. All excurrent water streams thus are expelled into this cavity, and the common jet emerging from the open end of the tube propels the colony in the opposite direction. Fertilization takes place internally. The highly yolky eggs, retained in the parent colony, cleave superficially and develop directly into incompletely elaborated *oözooids*, i.e., individuals produced from eggs. From the region near the heart, i.e., near the posterior end of the endostyle, an oözooid then proliferates a stolon which projects out of the body. On this stolon develops a connected chain of four *blastozooids*, i.e., individuals formed by budding. These four migrate around the oözooid surface and become stationed on it in a circlet. Such a young colony now escapes from the cavity of the parent colony and gives rise to a new offspring colony by continued budding from the four blastozooids.

The doliolids, representing another thaliacean order, undergo one of the most remarkable life cycles. The mature sexual adult is a *gonozooid* which produces externally developing eggs (Fig. 30.11). Each egg grows up into a torpedoshaped tailed tadpole enveloped in a thick tunic. The tail is later resorbed, and the adult so formed is an *oözooid*, or "nurse," distinguished (among other traits) by a median dorsal projection, or *spur*, at the posterior end. Within the tunic of this oözooid, a ventral stolon arises behind the endostyle. From the end of the stolon are constricted off a succession of *blastozooid* buds. These migrate to the dorsal spur, where they become attached in a median and two lateral rows. As more

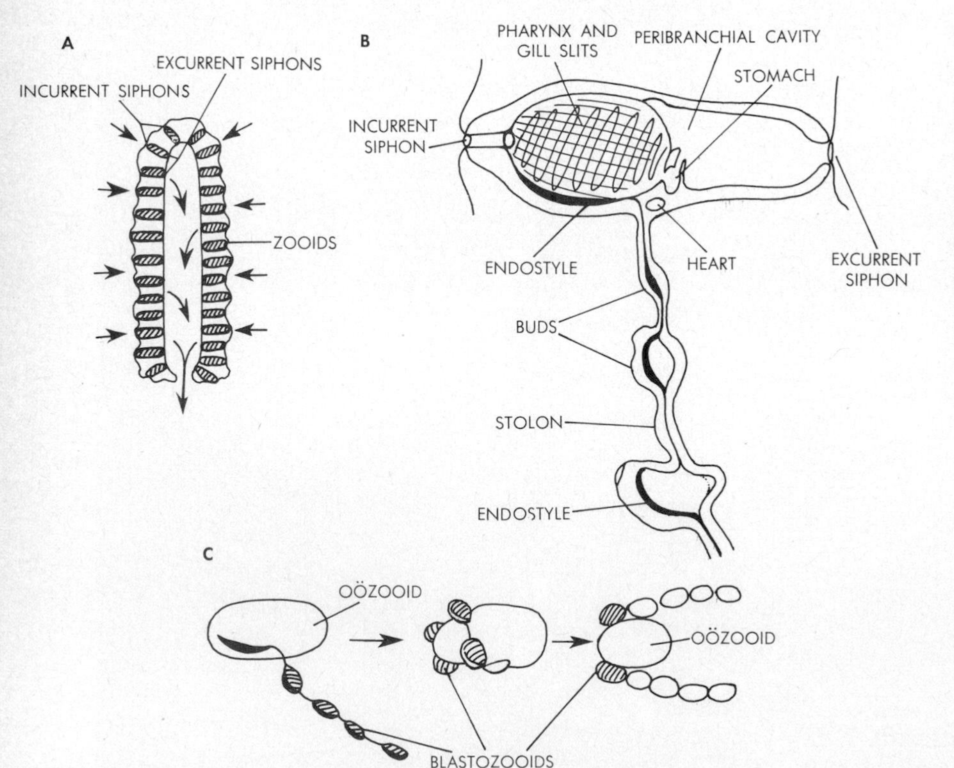

30.10. ***Pyrosoma.*** **A.** Section through colony. (*After Metcalf, Berrill.*) Note orientation of zooids in tubular colony and flow of water (arrows). **B.** Diagram of individual zooid, with stolon growing from endostyle and chain of buds formed along stolon. Note that endostylar stolon of one individual forms next individual along chain, including endostylar stolon of that individual. (*After Metcalf, Neumann.*) **C.** Four blastozooid buds as in **B** become arranged as a circlet around the oözooid and by continued budding give rise to a new colony.

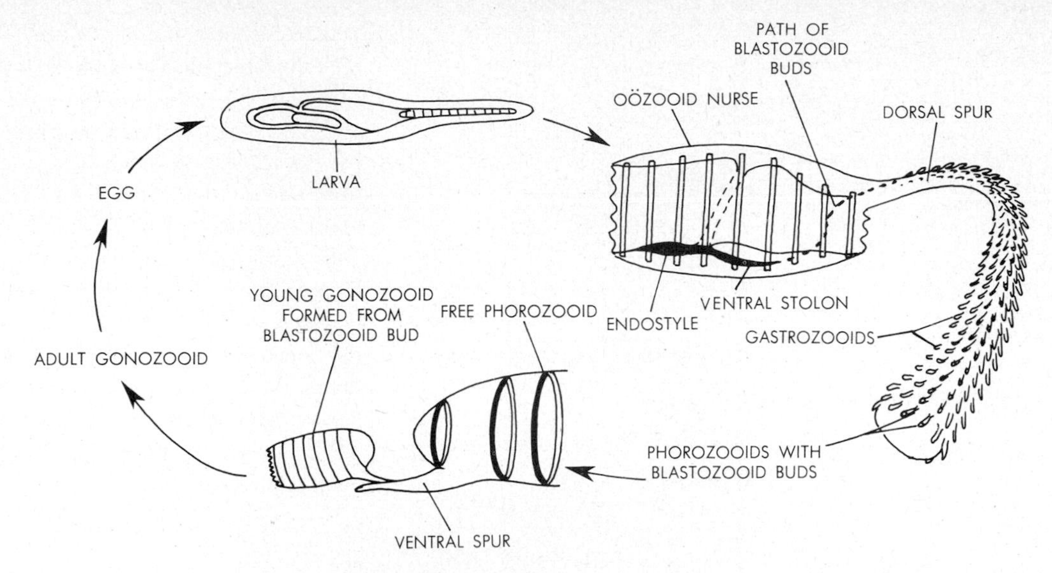

30.11. The life cycle of doliolids. (*Adapted from Neumann, Berrill.*) An adult gonozooid is illustrated in Fig. 30.9. The oözooid is shown in side view; note the nine muscle hoops. On the dorsal spur, note the paired lateral rows of gastrozooids and the paired median row of phorozooids, each of the latter with a wandering blastozooid migrated to that position from the ventral stolon. In the free phorozoid, only a side view of the posterior end is shown.

and more of such wandering buds settle there the spur extends, often up to a length of ½ yd. The blastozooids settled in the lateral rows are stalked and without gonads; they function as a colony of *gastrozooids* which, curiously enough, feed the nurse. Blastozooids in the median row become *phorozooids,* also without gonads but each equipped with a *ventral spur.* Some of the wandering blastozooids continuously budded from the stolon of the nurse then come to settle on the ventral spurs of the phorozooids. The latter subsequently break free from the nurse, and each bud carried on their ventral spurs then gives rise to a young gonozooid. Such individuals eventually detach and develop into new adults, completing the cycle. Evidently, the life history represents a multi-stage succession of budded polymorphic generations, one kind of bud (phorozooid) functioning as transport vehicle for others (wandering buds), which then grow and re-create gonozooids.

Analogously polymorphic are the *salps,* a third order of thaliaceans. Solitary individuals of these animals are oözooids produced from eggs (Fig. 30.12). They themselves are without gonads, but they produce a chain-budding stolon like the oözooids of doliolids. In *Salpa* the buds in a stolonic chain are arranged in a linear array, but in *Cyclosalpa* the older, more posterior buds become organized into wheellike complexes. All such stolonic individuals are blastozooids. They do not develop stolons of their own, but they do acquire gonads. Their eggs develop viviparously, without larval stages, in pouches of the oviduct. The young eventually set free represent a new generation of oözooids. Thus, salps undergo an alternation of generations: oözooids are sexually produced and they propagate vegetatively by forming blastozooids; the latter then propagate by sexual means into new oözooids.

Pyrosomas appear to be thaliaceans most closely related to ascidians. Doliolids may have evolved from ancestral pyrosoma stock, and salps from ancestral doliolid stock. The latter also may have given rise to the Larvacea.

Class Larvacea

Appendicularians; tail with notochord and nerve cord permanent throughout life; adults larvalike, pelagic; tunic formed into complex housing used with tail in feeding; one pair of gill slits. *Appendicularia, Oikopleura, Fritillaria*

These animals are the most specialized tunicates; they are neotenous, permanently tailed forms which have adapted their

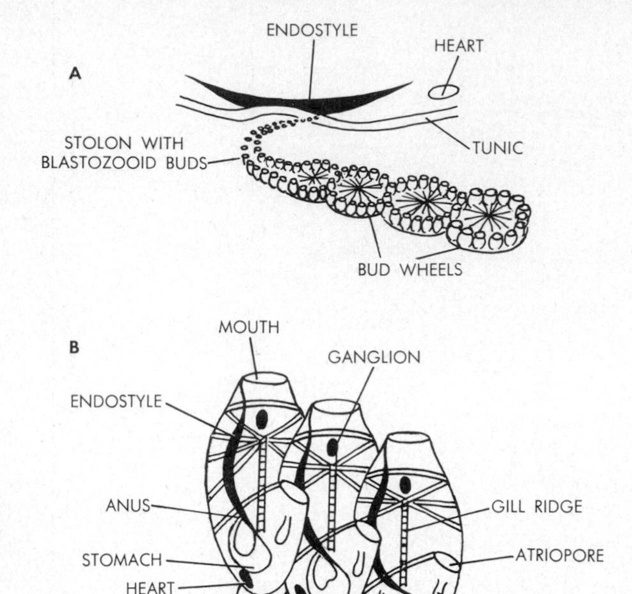

30.12. The life cycle of salps. **A.** Ventral region of parent oözooid, showing endostylar stolon carrying chains and wheel arrays of sexual blastozooids. (*After Ritter and Johnson, Berrill.*) Terminal bud wheels of this type form in *Cyclosalpa*, straight chains in *Salpa*. **B.** Detail of stolonic chain showing a few paired blastozooids. (*After Claus.*) Eggs from such blastozooids give rise to new adult gonozooids (as in Fig. 30.9).

tail and their highly elaborated tunic into a remarkable feeding apparatus. The body of an appendicularian is minute, structurally rather like an adult thaliacean with a larval tail (Fig. 30.13). This tail is from three to seven times as long as the trunk. Also, it is twisted 90° on its axis from a vertical into a horizontal plane and is bent forward under the trunk. The whole body, tail included, is enclosed within a roughly tubular tunic which forms a spacious "house" well separated from the surface of the body. Dorsolaterally the house is equipped with a pair of windows, covered with a straining grid of fine fibers. Food-bearing water enters through the windows into the house, and only the finest microscopic particles can enter. The water stream is then directed through a filtering system in the forward part of the house, where food particles are separated from the water and are sucked into the mouth of the animal. Filtered water is then expelled forcefully through a forward door in the house, the outgoing jet propelling the animal in the opposite direction. The entire flow of water is maintained and controlled by undulating movements of the tail, which lies along the floor of the house. A posterior ventral trap door serves as an excape hatch; after a given period of occupancy in the house, the animal leaves through the hatch and sets about secreting a new domicile.

Larvacea are hermaphrodites whose eggs escape through temporary ruptures in the body wall. Gastrulation occurs one cleavage earlier than in *Styela,* hence at any given later developmental stage larvacean embryos consist of half the number of cells of *Styela* embryos; differentiation evidently occurs faster relative to cleavage, as would be expected in a neotenous type (cf. Chap. 11). Development is direct, i.e., the embryos hatch as miniature adults.

SUBPHYLUM CEPHALOCHORDATA

Amphioxus, lancelets; marine, in sand; notochord and dorsal nerve cord throughout life; head or brain absent; coelom well developed, metamerically segmented; filter-feeding, with atrium, numerous pharyngeal gill slits, and endostyle; circulation open, without heart; excretion via solenocytic protonephridia; sexes separate, fertilization and development external; eggs regulative; with asymmetrical larvae. *Branchiostoma, Asymmetron*

The whole subphylum consists of about 30 species included in the two genera above. Amphioxus is a slender, laterally compressed animal, 2 to 3 in. long, and pointed at both ends (hence its name; Fig. 30.14). It lies covered in sand of shallow coastal waters, only its anterior end sticking out of its burrow. From time to time it swims to a new location by sinuous lateral undulations of the body.

Externally, a median dorsal fin extends over most of the length of the animal, a median ventral fin occurs in roughly the posterior third, and a tail fin connects the dorsal and ventral fins around the hind end. The dorsal and ventral fins are strengthened by boxlike *fin rays,* composed of connective tissue and lying underneath the epidermis. On the left side, at the junction between the ventral fin and the tail fin, lies the anus. Anteriorly the ventral fin terminates at the *atriopore,* the exit from the atrium. The alimentary system begins anteroventrally at an *oral hood,* a funnelshaped epidermal fold fringed with fingerlike extensions, or *cirri.* These form

30.13. Larvacea. Diagrams (*After Lohmann, Berrill*): **A,** side view, and **B,** dorsal view, of *Oikopleura*. Note that the tail is far longer than the body. The flow of water and food is indicated by arrows. **C.** The position of some of the organs in *Oikopleura*. **D.** Diagram of section through *Oikopleura,* to show horizontally twisted tail and tail structure. **Photo:** *Oikopleura,* in plankton, from life.

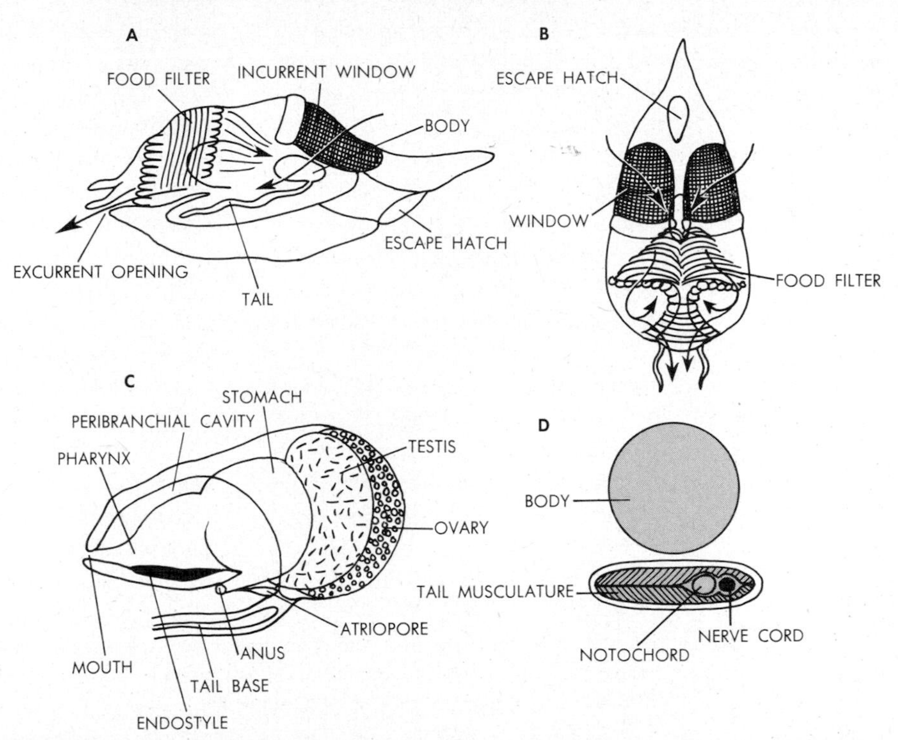

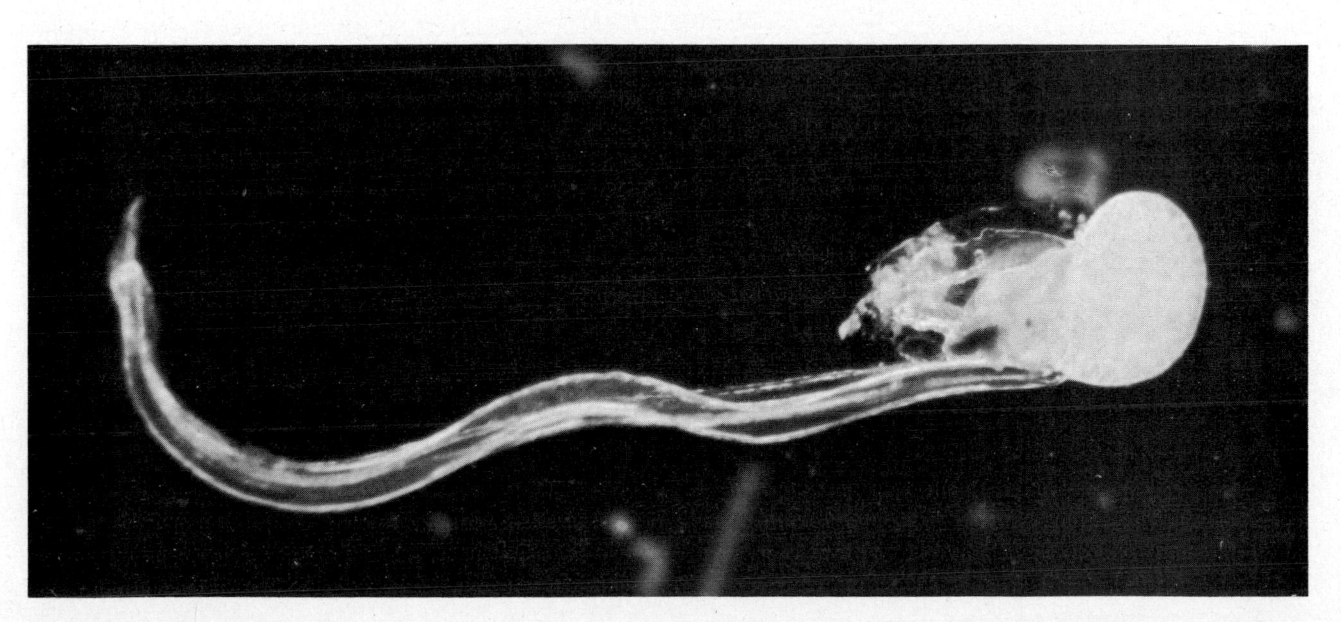

a coarse screen when the animal is feeding. On the inner surface of the oral hood are projections with complex bands of cilia, the whole forming a *wheel organ.* It produces a spiraling ingoing water current. The longest projection on the dorsal side contains a groove and a pit, called *Hatschek's groove* and *pit,* respectively (Fig. 30.15). These structures are remnants of an embryonic duct and an opening developed from the first coelomic sac on the left. The groove and pit are therefore believed to be homologous with the protocoelic duct and the hydropore of hemichordates, the duct of the neural gland of ascidians, and the pituitary of vertebrates. The cavity of the oral hood extends posteriorly to a transverse membrane, the *velum,* in the center of which is the mouth. This opening is fringed by *velar tentacles,* corresponding to the oral tentacles of ascidians (and the mesosomal tentacles of hemichordates).

Behind the mouth, the long pharynx contains 60 to more than 100 pairs of lateral sloping gill slits (cf. Figs. 30.14 and 30.15). The pharyngeal wall between two adjacent slits is known as a *gill bar;* it is stiffened internally by a skeletal rod of tissue. Each slit is subdivided by a *tongue bar,* here fused ventrally with the rim of the slit. As in ascidians, cilia are present around the gill slits and on the inner pharyngeal lining. Moreover, an endostyle forms the ventral pharyngeal gutter and an analogous strip of ciliated tissue lies along the roof of the pharynx. The whole branchial basket operates as in ascidians. Gill slits open into the atrium, formed by a pair of lateral folds of body wall grown down over the sides of the body and fused midventrally (except for the atriopore). Body-wall muscles on the floor of the atrium can contract and force water out through the atriopore. The ventrolateral edge of the atrium projects down on each side as a finlike *metapleural fold.* Note that the atrium has a ventral position in amphioxus, whereas it lies dorsally in ascidians. Behind the pharynx, a straight intestine receives enzymatic secretions from a digestive gland, a hollow elongated sac extending forward from its ventral intestinal attachment along the right side of the pharynx. The gland is often called "liver," but it actually appears to be more nearly equivalent functionally to the vertebrate pancreas.

The notochord passes through the entire length of the body. Directly over it lies the hollow nerve cord. Anteriorly the cord tapers to a point and contains a brain vesicle only slightly larger than the neural canal. A pigment spot and an olfactory pit are the only sensory structures present. The absence of a distinct brain is probably not primitive but a secondarily

30.14. Amphioxus, gross anatomical features. The many pharyngeal gill slits are very prominent just behind the mouth. Note also the notochord, i.e., the very dark rod just above the gill slits running from front to back. Cf. also Fig. 8.8.

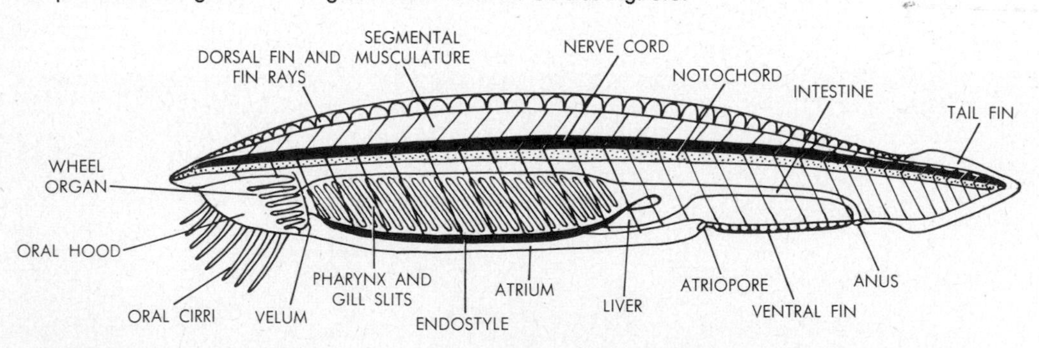

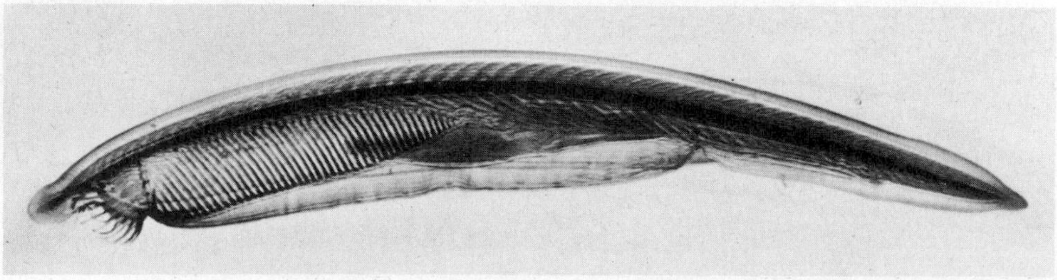

Courtesy Carolina Biological Supply Company.

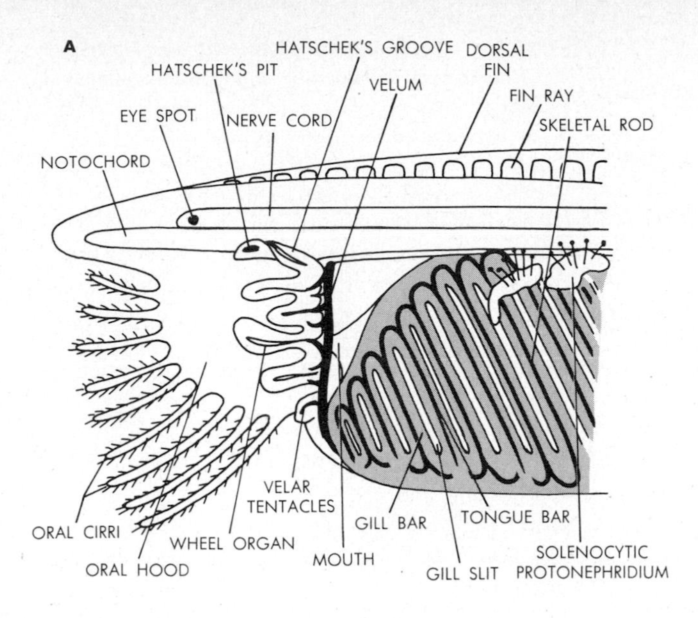

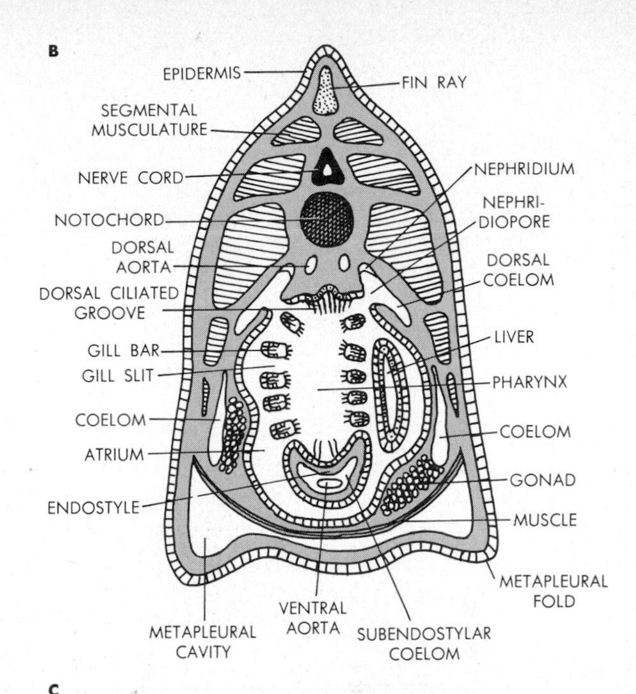

C

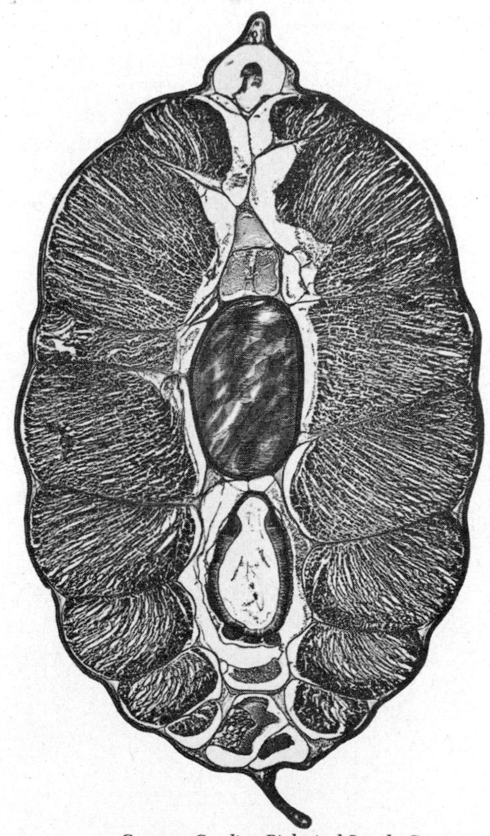

Courtesy Carolina Biological Supply Company.

30.15. The structure of amphioxus. A. side view of anterior end; **B,** cross section through pharyngeal region; **C,** photo of cross section through intestinal region.

simplified condition correlated with the predominantly sedentary life of the animal. On each side the nerve cord gives off paired segmental nerves, the dorsal members of a pair being largely sensory, the ventral ones largely motor. This arrangement corresponds to the vertebrate pattern (though the dorsal and ventral fibers here are not combined into single trunks like the spinal nerves of vertebrates). Most of the motor nerves innervate the lateral segmental musculature. Each muscle segment, or *myotome*, is V-shaped, and those on one side of the body are not in line with those on the other side but alternate with them in a staggered pattern.

The circulation is open; capillaries are lacking, but arteries and veins are numerous. Disregarding the absence of a heart, the vessels are arranged in substantially vertebratelike fashion (Fig. 30.16). A contractile ventral aorta below the pharynx pumps blood forward and then up through aortic arches in the gill bars. At the ventral end of each aortic arch is a small contractile bulb which aids in driving blood through the arch. Blood collects above the pharynx in a pair of dorsal aortae. These two vessels join behind the pharynx into a single systemic aorta which gives off branches to the tissues. From an intestinal plexus of spaces, blood drains as in vertebrates

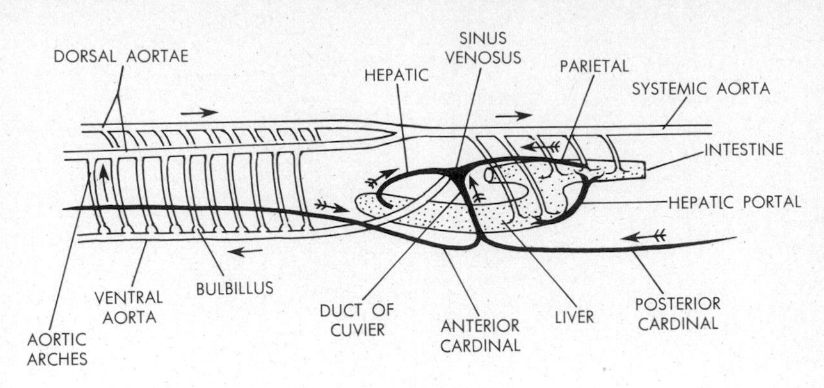

30.16. The basic plan of the circulatory system of amphioxus. Veins dark-shaded, arteries light-shaded. Direction of venous blood flow indicated by feathered arrows, that of arterial blood flow, by plain arrows. A bulbillus is a contractile vesicle. The sinus venosus is formed by the confluence of the main veins of the body and the exit of the ventral aorta; it corresponds in position to (and forms part of) the heart of vertebrates.

into a hepatic portal vein which passes into the liver. A hepatic vein from there, and other veins from blood spaces in various body parts, conduct blood to a *sinus venosus*, a chamber which continues anteriorly as the ventral aorta. The blood is colorless and without cells. Oxygenation is probably achieved primarily in the atrial walls and only secondarily in the gill bars.

In the pharyngeal region the coelom is small and consists only of a pair of flattened sacs dorsal to the pharynx (cf. Fig. 30.15). Posteriorly the coelom is represented by the space

30.17. Amphioxus development. (*After Hatschek.*) **A, B,** and **C.** Cross sections through sequence of embryos. **D** and **E.** Sagittal sections, position of coelom sacs superimposed. In **E,** the anterior outgrowth of the first coelom sac on the left may correspond to the hydroporic canal of ancestral deuterostomes, and from it arises Hatschek's pit and groove. **F.** Cross section through late embryo, at time of subdivision of coelom into dorsal and ventral sacs. In the myocoel, the inner cells (shown with nuclei) give rise to the segmental muscles. **G.** Larva seen from left side, showing asymmetric position of mouth and early gill slits on left.

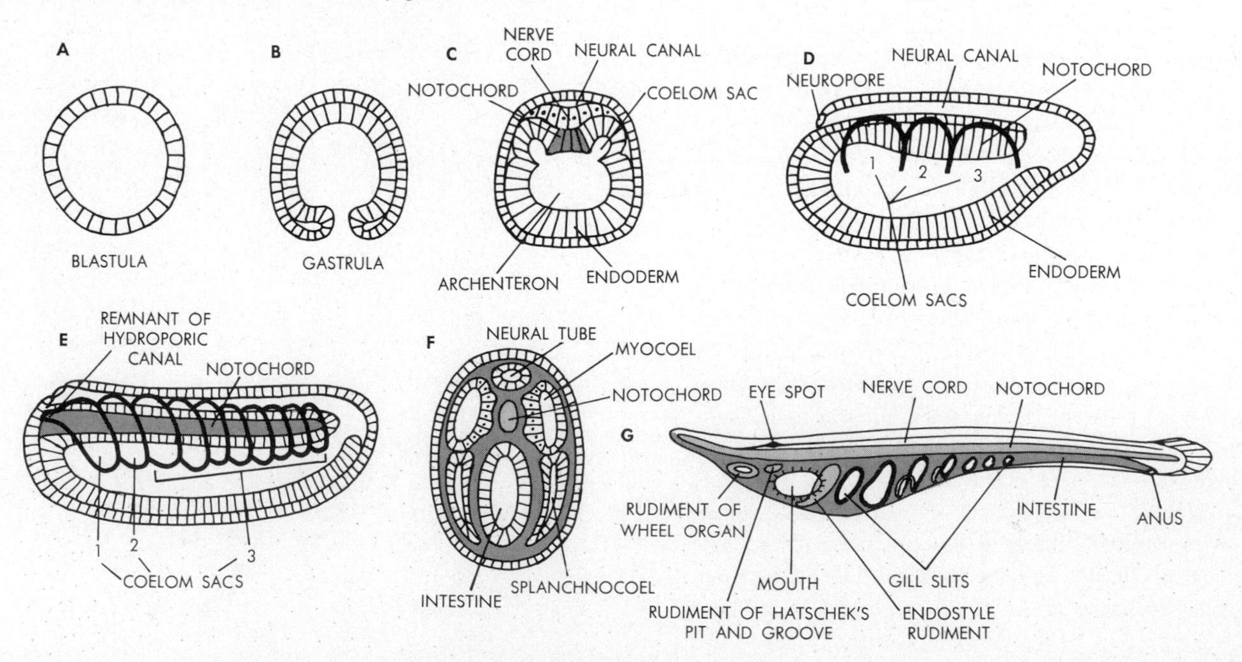

surrounding the alimentary tract. As will become apparent presently, the coelom arises from segmental coelomic sacs in the embryo. The excretory system consists of ectoderm-derived solenocytic protonephridia, an inexplicable feature and a source of considerable puzzlement to phylogenists. Such nephridia occur nowhere else among deuterostomes, and indeed they are otherwise encountered mainly in polychaete annelids. Most zoologists consider the nephridia of these two groups to represent a very remarkable instance of parallel evolution. In amphioxus one such organ is located near the dorsal end of each gill bar. The solenocytes abut against blood vessels in the coelom lining, and the collecting tubule of each nephridium opens into the atrium (cf. Fig. 30.15).

Gonads are paired in *Branchiostoma*, unpaired and present on the right side only in *Asymmetron*. They form an anteroposterior succession of saclike organs in the lateral body wall of the atrium. Gametes escape through ruptures into the atrium and out via the atriopore. The sexes are separate, and fertilization takes place externally. Eggs cleave bilaterally into a coeloblastula (Fig. 30.17). Gastrulation is embolic. The notochord forms from the roof of the archenteron, the nerve cord by invagination of a dorsal neural tube; an anterior neuropore is present temporarily. The anterior end of the archenteron soon buds off three pairs of coelomic sacs in succession. In the first pair, the right sac becomes the coelom of the most anterior region. The left sac comes to communicate with the exterior through a hydroporic connection which persists in the adult only as Hatschek's pit and groove. The second pair of sacs represents the first pair of segmental coeloms, and the third pair will later give rise, by subdivision in anteroposterior succession, to the more posterior coelomic segments. During this formation of the coeloms, the whole embryo elongates posteriorly. Each coelomic sac becomes subdivided horizontally into a dorsal *myocoel* and a ventral *splanchnocoel*. The myocoels subsequently fill up with cells which form the segmental musculature, but the splanchnocoels remain hollow; by fusion they give rise to the continuous coelomic cavity of the adult.

The embryo hatches after the first few muscle segments have formed. A mouth has broken through and now comes to be located on the left side of the young larva. Gill slits begin to form in pairs, but all of these lie on the right side of the body. Evidently, the larva is quite asymmetrical. Note also that at this stage the gill slits open directly to the exterior. The later larva does acquire adult bilateral symmetry; the mouth moves into the center anteriorly, and one set of gill slits becomes shifted from the right to the left side. Additional pairs of slits then continue to form symmetrically. Metapleural folds eventually appear and grow over the gill slits, forming the atrium. In effect, the larva undergoes a fairly gradual metamorphosis.

The asymmetry of the amphioxus larva undoubtedly represents a specialized adaptation to its particular type of filter-feeding existence. The adult too is quite specialized in many respects. Nevertheless, the animal displays numerous vertebrate features in very primitive form and thus it may suggest in a rather distant way what ancient vertebrates might have been like. Most probably, amphioxus exemplifies a transitional stage in the evolution of ancestral ascidian stocks into the first vertebrate stocks. The ascidian ancestors probably were neotenous, tadpolelike, and pelagic. In such a group, segmentation may have developed originally in the tail musculature and its nerve supply, an advantageous step which would have increased the locomotor power and the fine control of the tail. Such added efficiency would have been particularly important if the early chordate shifted its habitat from the open sea to shallow shore waters, especially around river mouths, where influx of minerals promotes rich plankton growth. Waves, surf, undertow, and also the currents of discharging river water would have made efficient locomotor control exceedingly important. Segmentation thus may have evolved in adaptation to such conditions.

Through segmented and finely controlled tails, furthermore, the basic means would have become available for an ascent into the rivers. However, actual entry into rivers would have required at least two additional evolutionary changes, and these the ancestors of amphioxus apparently did not perfect. Instead, they merely appear to have adopted a sedentary existence in shallow coastal regions, and modern amphioxus still exhibits this mode of life. By contrast, related ancestral types certainly did succeed in evolving the necessary evolutionary adjustments, and they became the first vertebrates. The required adjustments were larvae with more than just the first few segments developed upon hatching, i.e., strongly muscled larvae which could surmount the force of river currents as soon as they hatched; and excretory systems which could cope with the continuous osmotic inflow of water into the body, unavoidable in a freshwater environment. The first requirement was met by the evolution of a large, very yolky egg, which could develop without additional food to an advanced, fully segmented stage even before hatching; and the second was met by the development of a pronephric and metanephric kidney, specialized primarily for the elimination of osmotic water.

REVIEW QUESTIONS

1. Give taxonomic definitions of the phylum Chordata and the acraniate subphyla. Analogously define the tunicate classes. Review the presumable evolutionary interrelations of these various groups. What role has neoteny probably played in chordate evolution?

2. Describe the structure of an adult ascidian and review the organization of every organ system. Which structures are specifically homologous to equivalent ones of hemichordates? Describe the feeding process of a sea squirt. What are endostyle, epicardial folds, neural glands? Review the unique attributes of ascidian hearts and of blood.

3. How do buds form and develop in ascidians and what types of colonies may arise? Review the embryonic development of ascidians and describe the prospective fates of the blastomeres. In what respects are ascidian and annelid eggs and cleavage patterns (*a*) similar, (*b*) different?

4. Show how the ascidian tadpole develops from the embryo, how mesoderm arises, and how the adult ascidian structure emerges during larval development and metamorphosis. What is the (*a*) ecological, (*b*) probable evolutionary significance of the ascidian tadpole?

5. Describe the structure of an individual and of a colony of *Pyrosoma*. Show how locomotion occurs. By what processes does a colony develop? Distinguish between oözooids and blastozooids.

6. Describe the structure of a doliolid gonozooid and review the life cycle of these animals. What are the structural and functional differences between oözooids, blastozooids, gastrozooids, and phorozooids? Which thaliacean groups are hermaphroditic?

7. Describe the structure of an oözooid salp and outline the life cycle. Contrast this cycle with those of other thaliaceans. Similarly describe the structure of a larvacean and review the feeding and locomotor processes of the animal. What role does the tail play here?

8. Describe the complete structure of amphioxus and outline the organization of every organ system. What features appear to be specifically homologous to those of tunicates and hemichordates? How and where does amphioxus feed? Is the persistence of the notochord and nerve cord in amphioxus a neotenous trait?

9. Show how cephalochordates develop and how the coelom forms. Describe the general structure of the early and later larva and outline the development of the adult atrium. Are cephalochordates hermaphroditic? Oviparous?

10. What ecological factors might have played a role in the evolution of cephalochordate segmentation? What factors probably contributed to keeping cephalochordates in a marine environment whereas vertebrates could become freshwater forms?

COLLATERAL READINGS

Of the selections below, the writings by Berrill are recommended most particularly:

Berrill, N. J.: Metamorphosis in Ascidians, *J. Morphol.*, vol. 81, 1947.

———: "The Tunicata," Ray Society, London, 1950.

———: Budding in *Pyrosoma, J. Morphol.*, vol. 87, 1950.

———: Budding and Development in *Salpa, J. Morphol.*, vol. 87, 1950.

———: Regeneration and Budding in Tunicates, *Biol. Reviews*, vol. 26, 1951.

———: "The Origin of the Vertebrates," Oxford, London, 1955.

———: "Growth, Development, and Pattern," chaps. 13, 14, 18, and 19, Freeman, San Francisco, 1961.

Conklin, E. G.: The Embryology of Amphioxus, *J. Morphol.*, vol. 54, 1932.

Hyman, L. H.: "Comparative Vertebrate Anatomy," 2d ed., The University of Chicago Press, Chicago, 1942.

Watkins, M. J.: Regeneration of Buds in *Botryllus, Biol. Bulletin*, vol. 115, 1958.

Young, J. Z.: "The Life of Vertebrates," 2d ed., chaps. 2 and 3, Oxford, London, 1962.

CHAPTER 31

vertebrates: fishes

Subphylum VERTEBRATA: vertebrates; segmented, with head, trunk, and tail; cranium (skull) enclosing brain; dermal bone typically present; embryo with notochord; adult with notochord and/or vertebral column of cartilage or replacement bone; typically with two pairs of trunk appendages; with pharyngeal gills or lungs; coelom well developed; circulation closed, heart with two, three, or four chambers; excretion pronephric, mesonephric, or metanephric; endocrine system elaborate; sexes usually separate, fertilization various; eggs regulative, development various. *Agnatha,* jawless fishes; *Placodermi,* extinct armored jawed fishes; *Chondrichthyes,* cartilage fishes; *Osteichthyes,* bony fishes; *Amphibia,* amphibians; *Reptilia,* reptiles; *Aves,* birds; *Mammalia,* mammals.

GENERAL CONSIDERATIONS

The overriding orienting factor in vertebrate evolution was fresh water. As pointed out in the preceding chapter, ancestral vertebrates probably invaded the rivers as segmented, tailed, neotenous derivatives of ascidian stocks. Their larvae were well developed at hatching, and their adults were pharyngeal filter feeders like their marine forebears. Also, they possessed kidneys capable of forming a copious watery urine.

An early fundamental consequence of such an excretory pattern was that whereas the kidneys could eliminate water well, they were not as yet adapted to eliminate salt. Retention of minerals was undoubtedly counteracted by export via the gills, as still happens today in aquatic vertebrates (cf. Chap. 9), yet it is possible that in early stages of vertebrate evolution, minerals may have tended to accumulate internally. Many of them apparently were disposed of primitively by deposition in the skin; dermal bone in the form of heavy plates so probably evolved as a basic vertebrate trait. The earliest known fossil vertebrates were jawless fishes which were armored in bone, and most other vertebrates still possess dermal bone in their skulls. Calcium deposits later also came to form replacement bone in the axial and appendicular skeletons, where cartilage was present primitively as a strengthening material around the notochord.

This conversion of an original mineral liability into an adaptive asset may have been facilitated by the concurrent evolution of an endocrine system which could control mineral metabolism. The glands of the system would necessarily have had to be sensitive to salt levels coming into the body via food and leaving the body via excretion. It is therefore probably not a coincidence that, although other organs contribute as well, mineral-regulating functions in vertebrates are carried out to some extent by the anterior pituitary, perhaps evolved from the protocoel canal and the hydropore of ancient deuterostomes; indirectly by the thyroids and directly by the parathyroids, evolved from the endostyle in the floor of the chordate pharynx; and to a major extent by the adrenal cortex, situated along the kidney.

Like their ancestors, the most primitive vertebrates of record still are pharyngeal filter feeders: the larvae and adults of Agnatha are jawless, and in the larvae the ciliation of the pharynx sucks in water with small food particles through a round, open mouth. Such a pattern holds in principle even for the living ectoparasitic adult lampreys, whose food consists of body fluids and tiny tissue fragments rasped off the host. A considerable evolutionary gap then separates the agnaths from all other vertebrates. In the latter, the principal food-collecting organ is no longer the pharynx but the mouth itself, which has become equipped with jaws and true teeth. The nature of the food has changed concurrently, to bulk nutrients, and the ancestral method of filter feeding has largely ceased. This change has left the branchial basket primarily as a breathing organ; and since therefore tiny particles no longer needed to be strained out of water, the number of gill slits could become reduced. Indeed, such a reduction has proceeded in parallel with jaw development, for the skeletal supports in the most anterior gill bars came to be remodeled into jaw supports (a condition persisting even in mammals and man).

But even though fewer gill slits now remained, these could become more efficient oxygenators; for the circulation had become closed, extensive capillary nets were developed, and an originally two-chambered heart moved blood more rapidly than in any of the chordate ancestors. Such increased circulatory efficiency actually constituted a necessary corollary to the active, food-hunting mode of existence, and also to the still-persisting requirement of excreting osmotic water rapidly. A further corollary to the development of a food-hunting way of life was a new burst of neural evolution, resulting in the elaborate brains and sense organs of vertebrates.

The freshwater environment continued to orient also the later stages of vertebrate evolution. Pharyngeal air sacs appear to have been a primitive, original trait of bony fishes, in adaptation to occasional periods of drought. Such sacs then evolved into swim bladders in some of these fishes, into lungs in others (cf. Fig. 9.30 and below). Other adaptations to at first temporary and later permanent terrestrial life developed as well: fleshy fins, elongated subsequently into legs; yolky eggs and aquatic tadpoles, followed by even yolkier land eggs with shells, amnion and allantois, and direct development; three and then four chambers in the heart, permitting efficient separation of arterial and venous blood and thus providing enough oxygen for the increased energy requirements on land; mesonephric and later metanephric kidneys, with reversal from an original water-excreting to a new water-retaining function; a distinct energy-saving neck, with a swivel joint for the head; controls for maintenance of constant body temperature, including insulating surface layers of fat, feathers, and fur; improved breathing and circulatory machinery in mammals, including a diaphragm and highly efficient hemoglobin; and a placental reproductive mechanism, with viviparity and with milk for the young.

The basic structural and functional characteristics of vertebrate animals have already been discussed fully in the chapters on animal systems and in the various other contexts of Unit 1; a systematic review at this juncture is recommended. The developmental aspects have similarly been outlined, but an additional note concerning the manner of coelom-formation is appropriate here. Inasmuch as the vertebrate egg is basically large and yolky, the fundamental chordate pattern of enterocoelic development is to some extent obscured, as in other cases where yolky eggs mask a primitive original pattern. Thus, gastrulation occurs not by emboly as in amphioxus, but by various combinations of epiboly, involution, and ingression. The pattern is modified particularly greatly in the groups with discoidal eggs and superficial cleavage, i.e., fishes and birds (cf. Figs. 11.15 and 11.18). After gastrulation, a sheet of cells in the roof of the archenteron, called the *chorda-mesoderm*, gives rise to the notochord along the midline and the mesoderm along each side of it (Fig. 31.1). The lateral mesoderm sheets then thicken in segmental fashion and each segmental portion becomes subdivided into three parts. The dorsal part, closest to the notochord, represents the *myotome*, which will form the segmental muscles. The middle part, lateral (ventral) to the myotome, constitutes the *nephrotome*, which will contribute to the segmental development of the mesonephric excretory system. And the outermost part is the *lateral plate*, which develops in *non*segmental fashion and in an essentially schizocoelic manner: a split arises within it and the cavity so formed represents the body coelom. The lateral plate on one side eventually meets and fuses ventrally with the one on the other side, resulting in the formation of the mesentery. The coelom on each side then enlarges and ultimately becomes quite spacious.

The eight classes of vertebrates can be ranked into larger units on the basis of a variety of criteria. For example, one such grouping recognizes *Agnatha* and *Gnathostomata*, the jawless fishes constituting the first unit, all remaining seven classes of jaw-possessing vertebrates the other. A group *Anamniota* can be set off from a group *Amniota*, the latter embracing the amnion-possessing reptiles, birds, and mammals. Birds and mammals are *homoiothermic*, with blood at

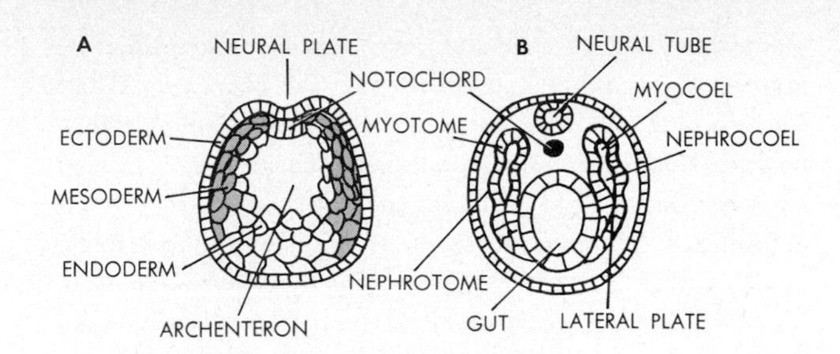

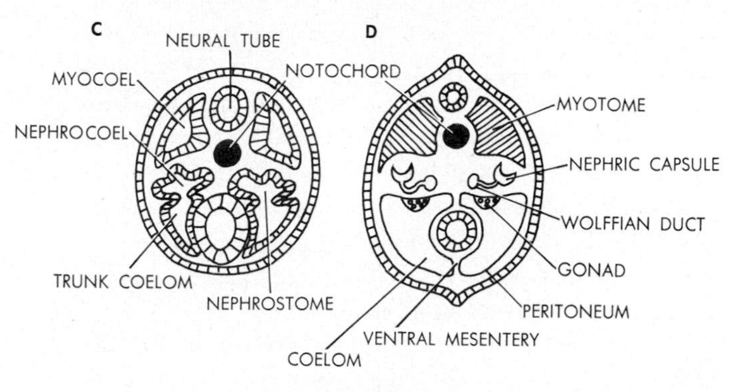

31.1. The coelom in vertebrates. A through D, cross sections through successive embryonic stages. A, Neural plate stage, mesoderm formed from endoderm as anteroposterior segmental cell masses. B. Neural tube formed, mesoderm subdivided into three portions on each side, with coelomic cavities present in upper two. Note that the term "myotome" refers to the cellular portion, the term "myocoel" to the cavity, of the coelomic compartment; analogously for "nephrotome" and "nephrocoel." C. Coelomic space formed in lateral plate, and nephrocoel communicates with this coelom via a ciliated nephrostome. D. Left and right trunk coeloms envelop intestine and form vertical mesentery, and nephrotome differentiated into segmental pronephric and mesonephric components; gonad rudiment formed in dorsal coelomic lining (cf. Fig. 10.21).

constant temperature, and all other classes are *poikilothermic*, or "cold"-blooded. A superclass *Pisces* includes the four classes of fishes, and a correlated superclass *Tetrapoda* comprises the four classes of four-legged vertebrates. We shall discuss the first of these superclasses in this chapter, the second, in the next.

With at least 25,000 species, the bony fishes represent the largest class, roughly half of all vertebrates. Birds constitute the next largest, with about 10,000 species. Next in order are reptiles (6,000 species), mammals (5,000 species), amphibia (3,000 species), cartilage fishes (600 species), and jawless fishes (50 species).

CLASS AGNATHA

Jawless fishes; with notochord throughout life; internal skeleton cartilaginous, dermal bony armor absent in living forms; without true (dermal) teeth; typically with single nostril; pineal eye present.

† Subclass Cephalaspidomorphi: extinct ostracoderms; single nostril with external opening; gill pores numerous; tail heterocercal or hypocercal.

Order Osteostraci: dorsoventrally flattened; bony armor as single shield; with pectoral fins; eyes on top of head; tail heterocercal. *Cephalaspis*

Order Anaspida: fishlike; with small, thick scales; tail hypocercal. *Pterolepis*

† Subclass Pteraspidomorphi: extinct ostracoderms; nostril without external epening; one pair of gill openings; tail hypocercal.

Order Heterostraci: bony armor of several plates; without fins but with lateral keels. *Pteraspis*

Order Coelolepida: with small scales. *Thelodus*

Subclass Cyclostomata: without bony armor or scales; paired fins absent; sucking mouth; nostril single; heart two-chambered; excretion pronephric and mesonephric; gonad single, without duct.

Order Petromyzontiformes (lampreys): seven pairs of gills, each opening separately; without internal nares; two pairs of semicircular canals; oviparous, with *ammocoete* larvae. *Petromyzon*, sea lampreys; *Lampetra*, brook lampreys.

Order Myxiniformes (hagfishes): five to 15 pairs of gills, one pair of external openings; with internal nares; one pair of semicircular canals; oviparous, development direct. *Myxine, Bdellostoma*

The extinct agnaths, often classified as a single subclass *Ostracodermi*, were comparatively small, 2 in. to 2 ft long, and primarily freshwater types (Fig. 31.2). Most of them were bottom inhabitants, possibly weighed down by the often massive dermal armor. When present, such armor covered the head and part of the trunk, and the rest of the body was often scaled. Osteostracs possessed a *heterocercal* tail; i.e., the upper tail lobe was larger than the lower one and the notochord extended into it. In the other groups the tail was *hypocercal*, the

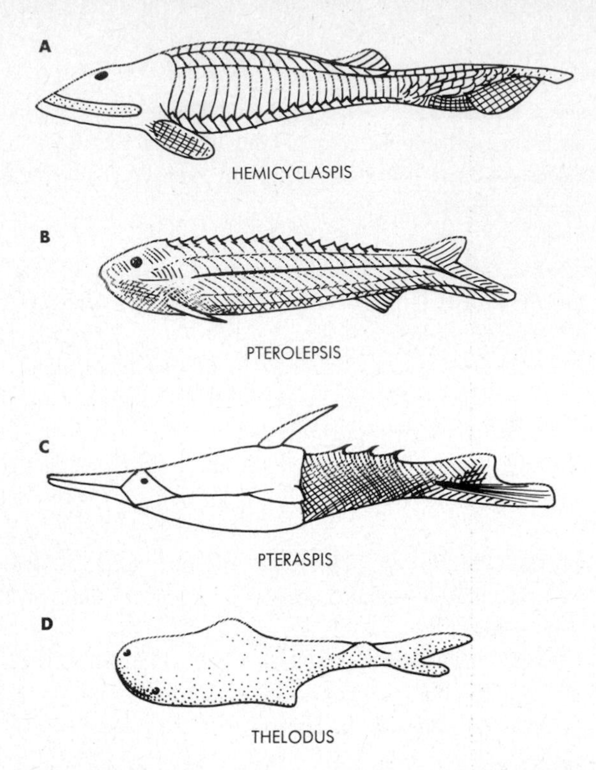

31.2. Ostracoderms. **A.** An osteostracan. (*After Stensiö.*) Note heterocercal tail and presumed electric organ along front edge of head. **B.** An anaspid. (*After Stensiö.*) **C.** A heterostracan. (*After White.*) **D.** A coelolepid. (*After Traquair.*)

lower lobe being the larger and containing the main skeletal support. This hypocercal condition is believed to have facilitated upward planing during swimming (and a heterocercal tail probably facilitates downward planing). Paired pectoral fins appear to have been present in osteostracs. The other groups probably were without such appendages, though lateral keels or flaps often did occur. A single external nasal opening identified the cephalaspidomorphs, the subclass from which the living agnaths probably descended.

The cyclostomes of today exhibit many secondarily reduced traits as well as numerous others which represent special adaptations of the group. Dermal armor, scales, and all bone have been lost; the skin is soft, glandular, and slimy (Fig. 31.3). Paired fins are absent, and the low median dorsal and caudal fins on the eel-like body are supported by cartilaginous fin rays. The single lobe of the caudal fin corresponds to the upper lobe of the ancestral heterocercal condition. Internally, the persisting notochord is stiffened by incompletely developed segmental cartilage supports, and cartilage also forms the brain case and additionally surrounds the heart (Fig. 31.4). The round mouth lies in the center of a shallow funnel which bears cornified horny epidermal teeth. This funnel is fringed by short tentacles in lampreys and by long fleshy *barbels* in hagfishes. A tongue, similarly studded with horny teeth, can be protruded through the mouth. In lampreys, a single dorsal nostril leads into a closed nasal sac, but in hagfishes the nasal passage opens into the posterior part of the mouth cavity (*internal nares;* cf. below and Fig. 9.30). Just behind the nostril lies a functional *pineal eye* (cf. Figs. 31.4 and 8.26). Only one or two semicircular canals are present in the ears, hence the animals presumably sense horizontal, two-dimensional motion rather well but up-down motion probably far less well. Cyclostomes possess a lateral-line system and a pituitary gland, and the structure of the nervous system conforms to the typical pattern of primitive vertebrates (cf. Chap. 8). An autonomic system is absent, however.

The gill chamber is an elongated sac pouched out from the pharynx and lying ventral to the esophagus. In lampreys the seven pairs of gill slits open directly to the exterior, but in hagfishes water from the gills exits via a single pair of gill pores. A stomach is absent, the esophagus continuing directly into the intestine. The circulation is patterned according to the characteristic vertebrate plan. A single atrium and a single ventricle compose the heart, as in all fishes. The blood too is typically vertebrate, with hemoglobin-containing red corpuscles and leucocytes. A lymph system is also present. The adult excretory system is mesonephric, the coelom is well developed, and a single gonad discharges into the coelom. Gametes reach the outside via a genital pore along the course of the mesonephric duct. The latter opens just behind the anus (cf. Fig. 10.22).

Sea lampreys are blood-sucking ectoparasites which attach to fish by their mouths and rasp through the skin of the hosts. While attached, the animals pump water both in and out of the gills directly, the mouth not being free for the breathing function. Some lampreys are separately sexed, others are hermaphrodites; all spawn in rivers, whether or not they also live in fresh water permanently. An egg develops into an eyeless *ammocoete* larva. The latter possesses an endostyle and it filter-feeds in upstream river bottoms in the ancestral manner (Fig. 31.5). Metamorphosis does not occur for 3 to 7 years and then the young adults migrate downstream. Sea lampreys head into the open sea where they lead a parasitic mature life for usually not more than a single season. Brook lampreys are

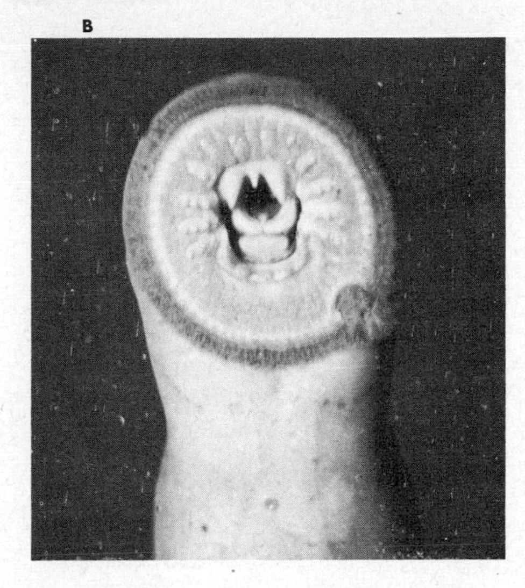

31.3. Agnatha. A. The lamprey *Petromyzon.* Note gill slits. **B.** View into the sucker mouth of a lamprey, showing epidermal teeth.

permanently freshwater forms. The adults here are not parasitic. Indeed they do not feed at all, but spawn within a few days and die.

Hagfishes are marine cyclostomes. All are hermaphroditic and oviparous; they develop directly, without larval stages.

CLASS PLACODERMI

Placoderms; extinct jawed fishes; true teeth present; typically with dermal armor over head and front of trunk; bone in internal skeleton, notochord persisting and partly ossified;

31.4. Lamprey structure. (*After Goodrich.*) A sagittal section of the anterior region of *Petromyzon* is shown. Cf. Figs. 9.38 and 10.22.

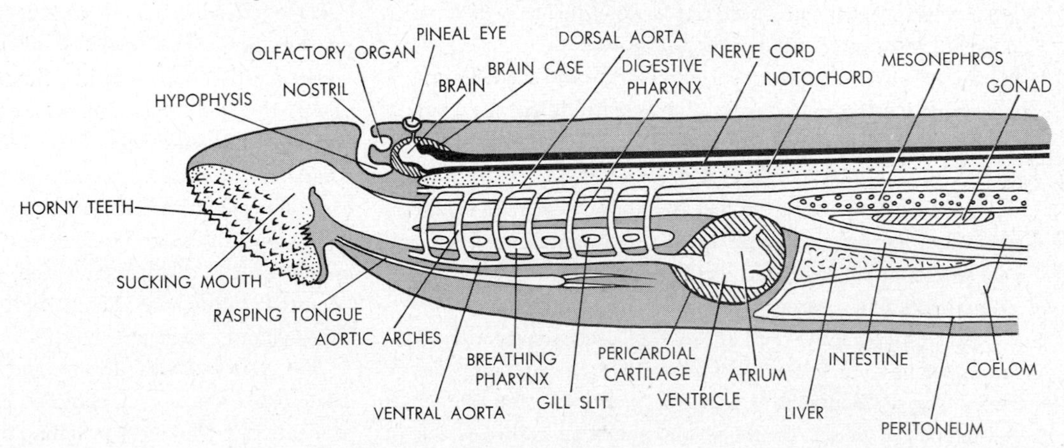

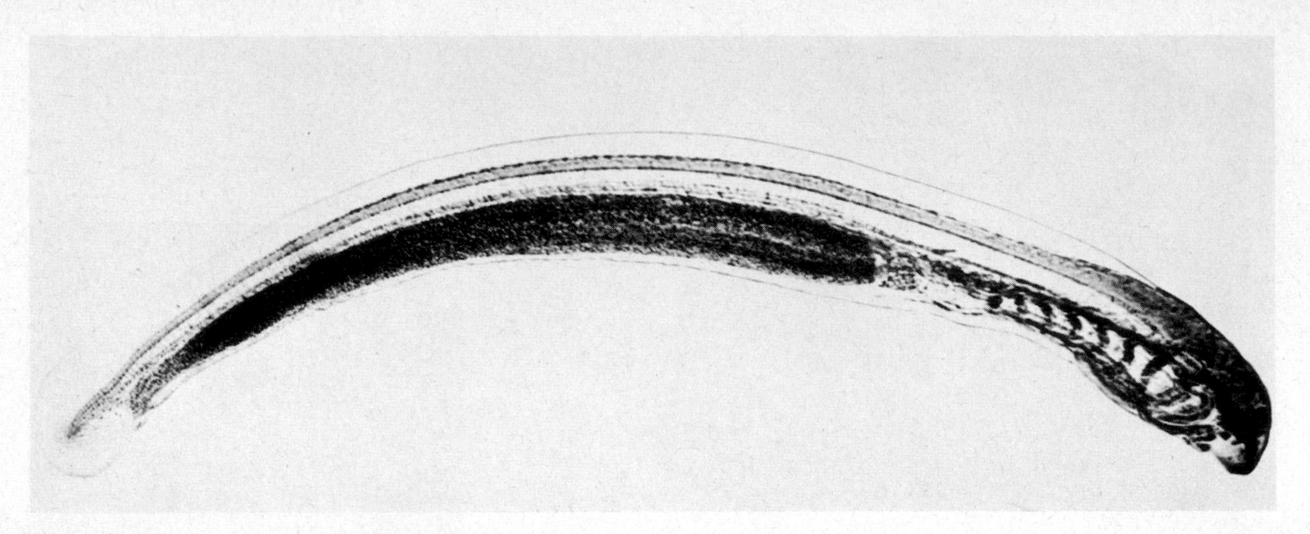

31.5. Ammocoete larva; cleared specimen. Note gill slits, nerve cord and brain, and position of notochord just underneath nerve cord.

first pair of gill slits not reduced, second pair of gill arches unaffected by jaws; with paired fins, plus one or two dorsal, one caudal, and one anal median fins; tail heterocercal.

† Order Acanthodii (spiny sharks): head with armor plates, trunk and tail with small diamond-shaped enamel scales; dermal spine at front edge of each fin except caudal; freshwater. *Climatius*

† Order Antiarchi: head with armor, trunk and tail scaled or naked; turtlelike appearance; pectoral fins flippershaped and armored, pelvic fins probably small or absent; probably with lungs; freshwater, bottom forms. *Pterichthyodes*

† Order Arthrodira: head armor movable on trunk armor by ball-and-socket joints; posterior body fishlike, scaled or naked; freshwater and marine. *Coccosteus*

† Order Stegoselachii (armored sharks): ray- and skatelike appearance, flattened body, eyes at top of head; body covered with small dermal plates; braincase ossified. *Gemuendina*

The placoderms were the first to exhibit two traits which since have become universal among vertebrates, viz., jaws, and paired appendages on pectoral and pelvic girdles.

As pointed out earlier, jaws probably evolved from the skeletal supports of the most anterior gill arch. In placoderms, the upper jaw has become jointed with the brain case (Fig. 31.6). This development of jaws apparently affected neither the first pair of gill slits nor the second pair of gill arches, which in placoderms remained as large as all the others. In later fishes, however, these gill structures did become changed in correlation with the posterior expansion of the jaws (cf. below). The evolution of paired fins is more difficult to interpret. According to a so-called "fin fold" hypothesis, ancestral vertebrates are assumed to have possessed a continuous ventral pair of fins, just as amphioxus and lampreys still possess a continuous dorsal fin. In the descendants of this presumed ancestor, all ventral fin parts except those which were to become the pectorals and pelvics then would have disappeared. This hypothesis probably also accounts for the origin of the median fins of fishes, i.e., the dorsals, the caudal fin, and the ventral anal fin behind the anus (cf. Fig. 31.6).

Placoderms generally were small freshwater bottom dwellers (Fig. 31.7). The head and anterior part of the trunk was covered with a dermal armor of bony plates, and the remainder of the body was either scaled or naked. Considerable ossification of the internal skeleton was already manifest. Thus, the brain case was bony at least in part and in some cases almost entirely; the jaw supports were partially ossified; and segmental bony elements (including neural and hemal arches, cf. Fig. 8.16) surrounded the notochord, which still persisted throughout life. The gill skeleton was largely cartilaginous, however.

The class includes diverse and probably not too closely related subgroups, each classified either as an order or often as a subclass. The most primitive placoderms appear to have been the acanthodians. These animals possessed three to five pairs of gills, covered externally with a lateral armor plate similar in some respects to the operculum of bony fishes. Internally, the gill skeleton of each arch was extended into a *gill raker,* a screening device which probably prevented food from entering the gill passages (cf. Fig. 31.10). Antiarchs are of great interest inasmuch as they appeared to have possessed

31.6. **A.** The evolution of jaws. (*After Romer.*) 1. Scheme of gill slits, skeletal gill arches, and brain case in ancestral jawless vertebrates. 2. Scheme in jaw-possessing vertebrates. The principal bones of the gill arches are named. The first gill arch has given rise to the main elements of the upper and lower jaw. The upper component of the second gill arch (hyomandibular) may aid in suspending the gill skeleton from the brain case. Also, backgrowth of the jaw has reduced the first gill slit to a spiracular opening. **B.** Fin evolution according to the fin-fold hypothesis. (*After Wiedersheim.*) 1. Scheme of continuous fin fold assumed to have characterized body of ancestral vertebrates. 2. Later evolutionary degeneration of parts of the fin fold is postulated to have left only the fins shown.

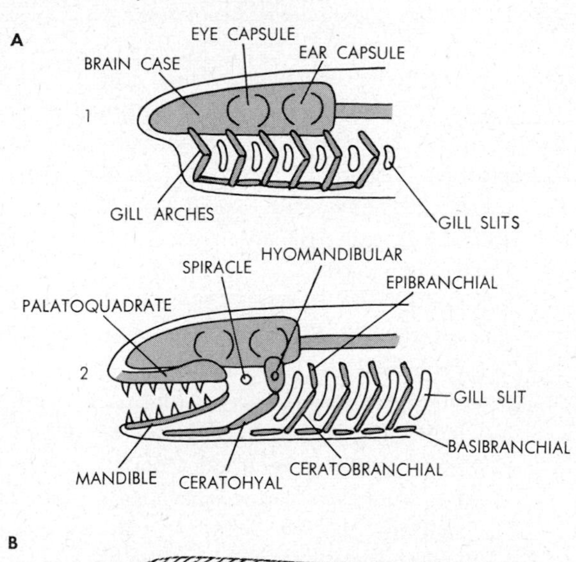

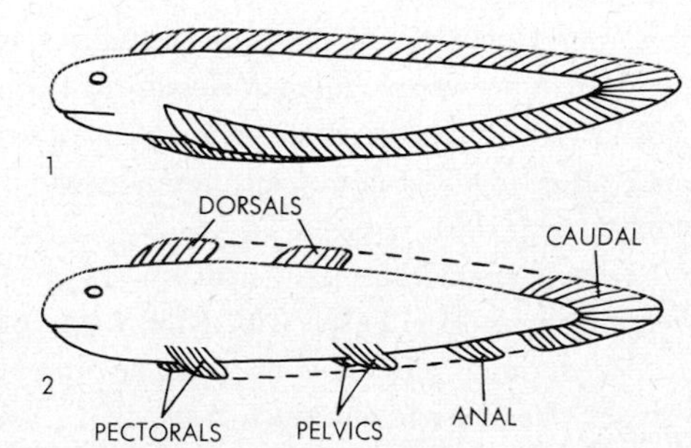

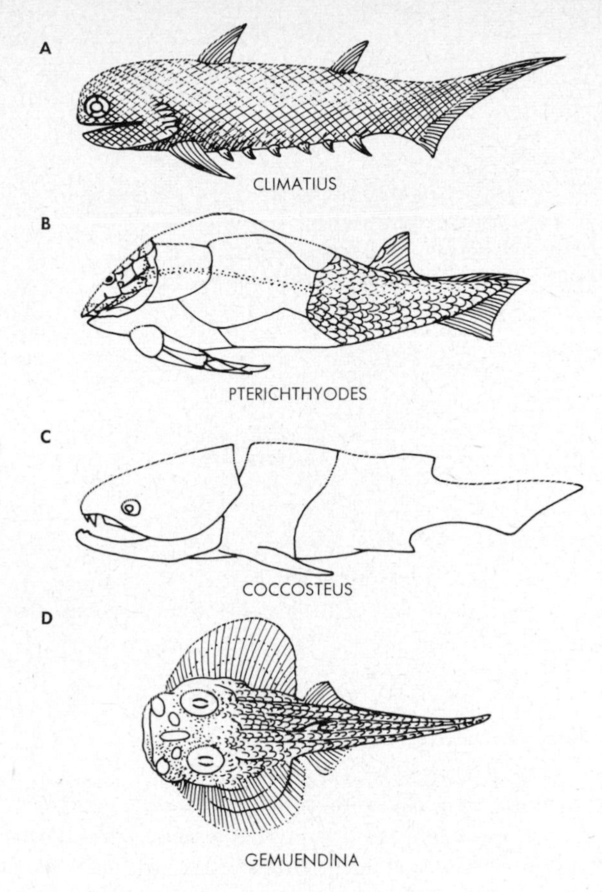

31.7. **Placoderms.** **A.** An acanthodian. (*After Traquair and Watson.*) **B.** An antiarch. (*After Traquair.*) **C.** An arthrodire. (*After Heintz.*) **D.** A stegoselachian. (*After Broili.*)

lungs. These animals had flipperlike armored pectoral fins and a humped dorsal armor, features which gave them a superficially turtlelike appearance. In arthrodires, the head and trunk armor was jointed together in such a way that the animal could move its head up but not sideways or down. These placoderms appear to have been the most diversified members of the class. The majority was of moderate size, but some arthrodires attained lengths of 6 yd and represented the giants of the vertebrate world of the day. A few were marine, others were without dermal armor, and still others had a skate- or raylike appearance. Very distinctly skatelike and flattened dorsoventrally were the stegoselachians, a group which was entirely marine.

The specific placoderm ancestors of later fishes are unknown; it is possible that bony fishes may have evolved from

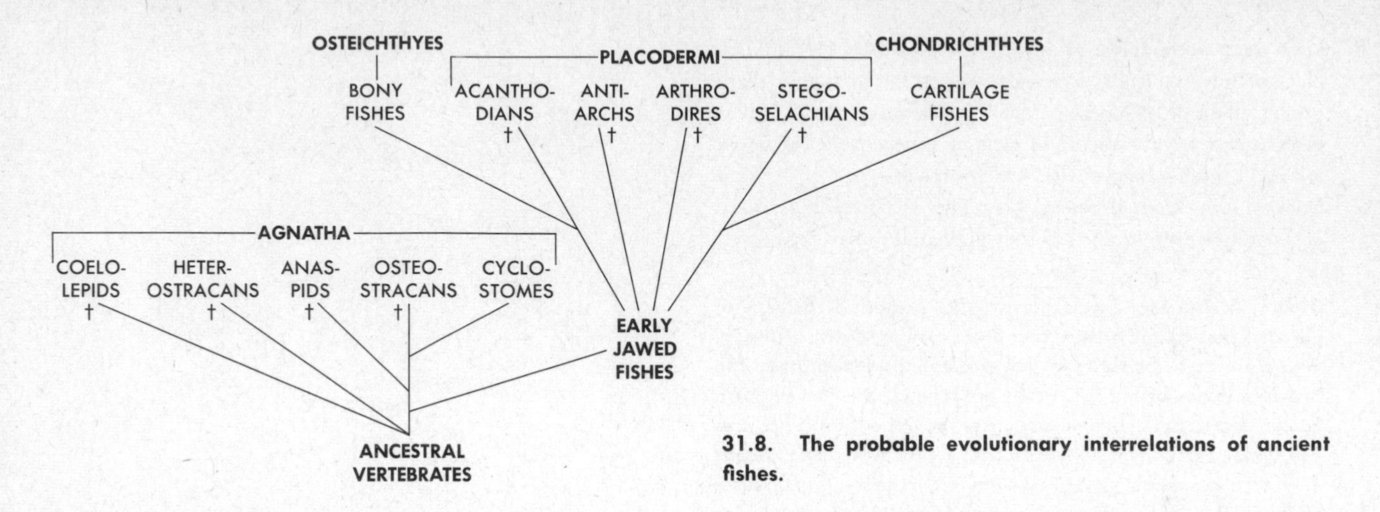

31.8. The probable evolutionary interrelations of ancient fishes.

acanthodian stocks and cartilage fishes from stegoselachian stocks (Fig. 31.8).

CLASS OSTEICHTHYES

Bony fishes; typically with scales, of ganoid, cycloid, or ctenoid type; spiracles primitively present; with opercula, hinged on the hyoid arch; nostrils paired, with or without internal nares; lung or swim bladder present, the latter with or without duct; lateral-line system present; sexes separate; mostly oviparous with external fertilization, some ovoviviparous or viviparous.

Subclass Sarcopterygii (flesh-finned fishes): paired fins with internal bony skeleton and fleshy exterior; notochord persisting; two dorsal fins; often with internal nares; air sac functions as lung.

Superorder Crossopterygii (lobe-finned fishes): teeth not fused into plates.

†Order Osteolepiformes: extinct; with pineal eye; tail heterocercal; ancestral to amphibia; freshwater. *Osteolepsis*

Order Coelacanthiformes (coelacanths): skull largely cartilaginous; tail diphycercal, with three lobes; without internal nares; extinct types in freshwater, living forms marine. *Latimeria*

Superorder Dipnoi (lung fishes): with three pairs of fused tooth plates.

Order Dipteriformes: tail diphycercal in living forms; ossification reduced; freshwater. *Protopterus, Neoceradotus, Lepidosiren*

Subclass Actinopterygii (ray-finned fishes): paired fins with soft or hard fin rays; single dorsal fin; without internal nares; air sac usually functions as swim bladder.

Superorder Chondrostei: notochord persisting; spiracles present; tail heterocercal; paired fins with broad-base attachment; scales ganoid or reduced; lung or swim bladder with duct.

Order Polypteriformes (bichirs): skeleton ossified; with paired ventral air sac lunglike; freshwater. *Polypterus*

Order Acipenseriformes (sturgeons): skeleton largely cartilage; scales reduced, skin often with bony plates; swim bladder dorsal, with duct; freshwater and marine, spawning in freshwater. *Acipenser*

Superorder Holostei: notochord persisting, replaced in some by bony vertebrae; spiracles present or absent; tail shortened heterocercal; paired fins with narrow-base attachment; scales usually ganoid; swim bladder dorsal, with duct.

Order Semionotiformes (garpikes): jaws elongated; swim bladder dorsal, with duct. *Lepidosteus*

Order Amiiformes (bowfins): spiracles absent; swim bladder dorsal, with duct. *Amia*

Superorder Teleostei: notochord replaced by bony

vertebrae; spiracles absent; tail homocercal; paired fins with narrow-base attachment; pectorals often high lateral, with pelvics anterior to pectorals; scales cycloid, ctenoid, or secondarily reduced; swim bladder with or without duct.

Order Clupeiformes (herrings, sardines, anchovies, tarpons, salmon, trout, whitefishes, pikes): soft-rayed, without fin spines; pelvics abdominal; scales cycloid or reduced; swim bladder with duct; Weberian ossicles absent; mostly marine.

Order Myctophiformes (lantern fishes, lizard fishes): pelvics abdominal; swim bladder with duct or absent; Weberian ossicles absent; deep sea.

Order Saccopharyngiformes (gulpers): without opercula, swim bladder, pelvics, caudals, scales, or ribs; gill openings minute; mouth tremendous, pharynx highly distensible; eyes tiny; tail slender; deep sea.

Order Cypriniformes (carps, goldfishes, minnows, catfishes, neon fishes, electric "eels"): like clupeiform fishes, but with Weberian ossicles (cf. Fig. 8.44); most important freshwater order.

Order Anguilliformes (true eels): pectorals often absent, pelvics absent; with or without scales; swim bladder with duct; gill apertures reduced; freshwater and marine, all spawning in sea.

Order Notacanthiformes (spiny eels): pectoral girdle loosely suspended, pelvics abdominal; anal fin long, caudal absent, tail tapered; swim bladder without duct; deep sea.

Order Beloniformes ("flying" fishes, needle fishes): pectorals large, high lateral, used in gliding; pelvics abdominal; swim bladder without duct; lateral-line system ventrolateral; mostly marine.

Order Cyprinodontiformes (guppies, swordtails, platyfishes, killifishes): pectorals high lateral, pelvics abdominal or absent; swim bladder without duct; mostly marine.

Order Gasterosteiformes (sea horses, snipefishes, pipefishes, trumpetfishes, sticklebacks): snout long, mouth at end of tube; with or without fin spines; swim bladder without duct; mostly marine.

Order Gadiformes (cod, haddock, pollock, hake, grenadiers): pelvics jugular; swim bladder without duct; dorsal often subdivided into three parts, anal into two parts; without fin spines; marine.

Order *Lampridiformes* (ribbonfishes, oarfishes): pelvics thoracic; dorsal with one or two fin spines, rest soft-rayed; swim bladder without duct; body large, serpentine; marine.

Order Percopsiformes (sand rollers): pelvics thoracic; dorsal with one to four fin spines, rest soft-rayed; scales ctenoid; swim bladder without duct; in sluggish freshwater.

Order Beryciformes (squirrel fishes): pelvics thoracic; with fin spines but soft-rayed; scales ctenoid; swim bladder without duct; marine.

Order Zeiformes (John Dories): pelvics with fin spines but soft-rayed; marine.

Order Perciformes (perches, tunas, mackerels, basses, rockfishes, marlins, sailfishes, bluefishes, jacks, sunfishes): pectorals high lateral, pelvics thoracic or jugular; fins with spines, hard-rayed; scales ctenoid; swim bladder without duct; most important marine order.

Order Pegasiformes (sea moths, sea dragons): external bony casing, bony snout; without teeth; pectorals horizontal, winglike, pelvics abdominal; swim bladder absent; marine.

Order Pleuronectiformes (halibut, flounders, plaice, sole): flatfishes, on left or right side, with loss of bilaterality; dorsals and anals long; marine, on bottom.

Order Echeneiformes (remoras, shark suckers): dorsal modified into oval adhesive disk; scales cycloid; pectorals high lateral, pelvics thoracic.

Order Tetraodontiformes (puffers, triggerfishes, trunk fishes, porcupine fishes): armored with bony plates and spines, prickly or naked; fin spines present; mostly marine.

Order Gobiesociformes (cling fishes): without fin spines, scales, ribs, or swim bladder; both girdles and pelvics modified into thoracic sucking disk for attachment to bottom; mostly marine.

Order Batrachoidiformes (toadfishes): first vertebra fused with flattened cranium; ribs absent; three pairs of gill arches; pelvics on throat; marine.

Order Lophiiformes (angler fishes): first ray of dorsal modified as angler with luminescent lure; pelvics thoracic or absent; ribs absent; gill aperture small, far back; males sometimes dwarfed, on female; deep sea.

Order Mastacembeliformes: eel-like; median fins

continuous, with fin spines; snout long, with tentacles; with or without scales; pelvics absent; swim bladder without duct; fresh water.

Order Synbranchiformes: eel-like; small gill apertures joined across throat; gills reduced, breathing partly by mouth-pharynx pouches; pelvics on throat, small; without pectorals, fin spines, or swim bladder; with or without scales; in swamps, caves, sluggish fresh water.

Bony fishes arose from placoderm ancestors approximately at the same time (or possibly even before) the cartilage fishes had evolved. As their name suggests, these fishes are characterized by a basically bony skeleton, though in many primitive forms the degree of ossification is reduced. Dermal armor is no longer present, but dermal bones form most of the skull, parts of the girdles, and all of the scales. In the primitive groups these scales are most often *ganoid,* i.e., diamond-shaped and covered externally with a layer of hard, glossy enamel (the whole scale being homologous to the denticles of sharks and the teeth of mammals). In more advanced fishes the scales are *cycloid* if their free posterior parts are smooth, *ctenoid* if the free parts are rough-textured or spiny (Fig. 31.9). The tail of bony fishes is primitively heterocercal but in most cases it is symmetrical. In some symmetrically tailed forms the main skeletal support extends to the tip of the tail (*diphycercal* condition), in others the skeletal support terminates at the tail base (*homocercal* condition). The median fins, i.e., dorsals, anals, and caudals, are supported by elongated bony fin rays. The support of the paired fins differs sharply in the two great subclasses of bony fishes. In the flesh-finned Sarcopterygii, the pectorals and pelvics contain internal bony skeletons, the parts corresponding substantially to the bones of tetrapod limbs. In the ray-finned Actinopterygii, by contrast, the paired fins are supported like the median fins by bony rays (cf. Figs. 31.9 and 8.14).

The skull is composed of some 60 bones, a number which has become reduced in later vertebrate classes (e.g., to about 20 in mammals). In primitive bony fishes the notochord often still persists into the adult stage, but in most groups it is replaced partly or (more usually) wholly by bony vertebrae which are preformed in cartilage. The vertebrae characteristically bear neural and hemal arches as well as ventral ribs (cf. Fig. 8.16). The gill region has become modified partly by

31.9. Scales and tails in bony fishes. The three principal scale types are illustrated in the photos: **A,** ganoid scale; **B,** cycloid scale; **C,** ctenoid scale. In the latter, the smooth portion is overlapped by the scale before it and the spiny part is exposed. The diagrams show the three principal tail types. The characteristics of flesh-finned and ray-finned pectoral and pelvic paired fins are diagrammed in Fig. 8.14.

A

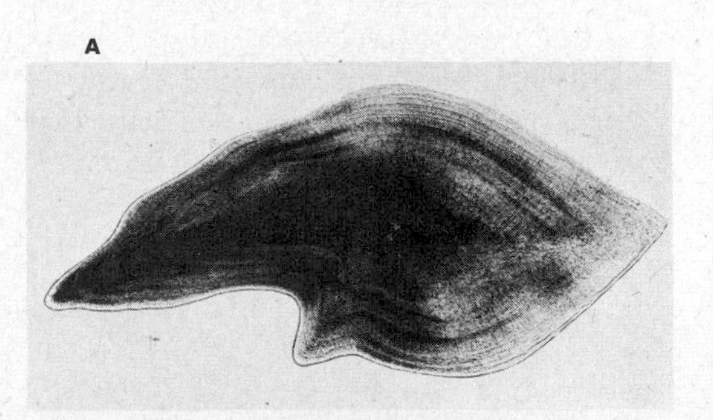

B

C

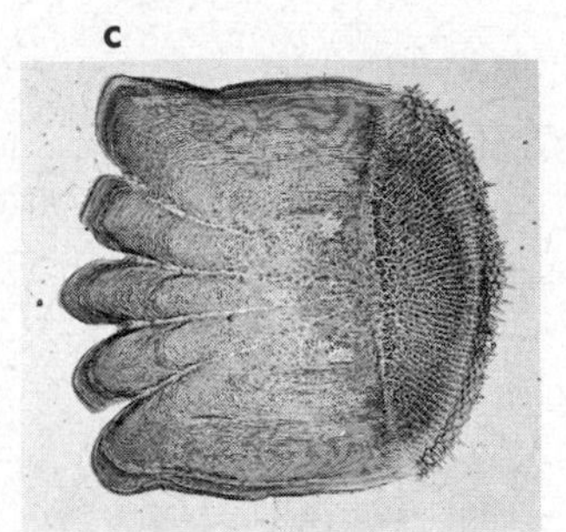

A,B,C, *courtesy Carolina Biological Supply Company.*

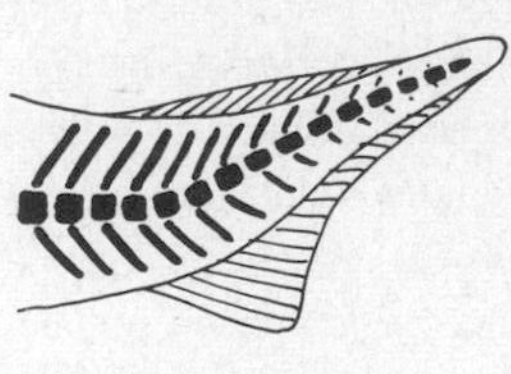

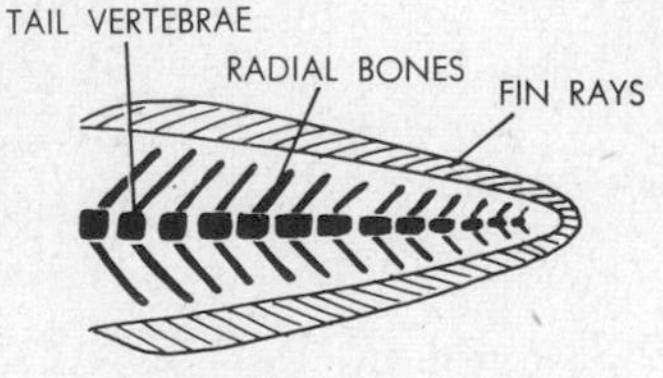

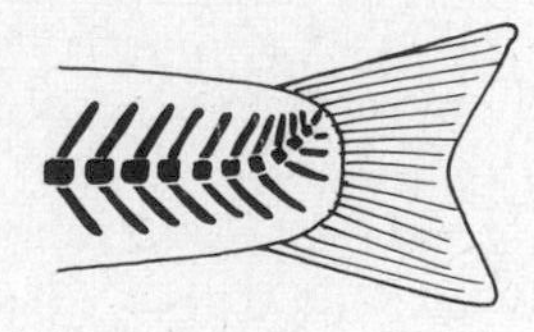

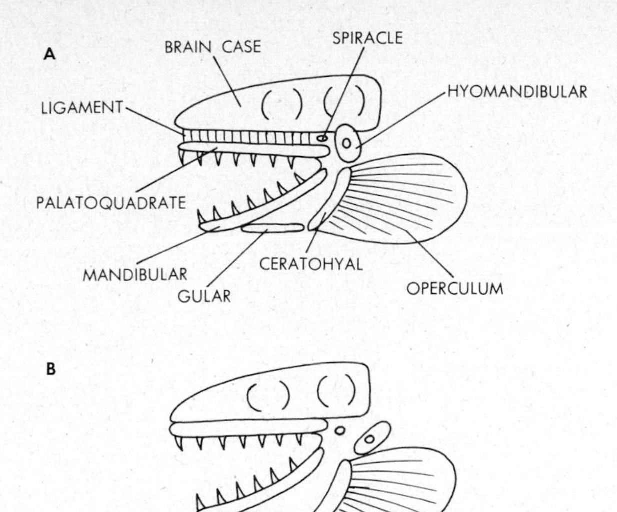

31.10. **Jaws in bony fishes.** **A.** Condition as in most such fishes, principal bony prop between skull and upper jaw being the hyomandibular (though ligaments aid in supporting the palatoquadrate). Position of spiracles is indicated, though in most instances this remnant of the first ancestral gill slit is absent altogether. **B.** Condition as in primitive sarcopterygians (and amphibia), the upper jaw here being fused directly to the brain case; the hyomandibular is free and without supporting function.

a further elaboration of the jaws (Fig. 31.10). The jaws themselves represent the *mandibular arch,* the first of the ancestral skeletal gill arches. Backgrowth of the bones of this arch on each side of the body has crowded the first gill slit and the second gill arch. As a result, the first gill slit has become reduced primitively to a small tubular opening, the *spiracle.* In most bony fishes, actually, not even such a spiracle is present and the first gill slit has disappeared completely. The second skeletal gill arch, called the *hyoid arch,* has acquired two new functions. The lower bone of the arch bears a hinged posterior bony plate, the *operculum,* which forms a cover over the gills behind; and the upper bone, called the *hyomandibular,* in most cases serves to hold the upper jaw to the brain case. In some primitive bony fishes the upper jaw has become fused directly with the brain case, leaving the hyomandibular without function (the ancestors of amphibia were of this type, and the unused hyomandibular subsequently acquired a new function as the stirrup bone of the middle ear).

The nostrils of bony fishes are paired, and they lead either into closed nasal sacs or into the mouth cavity via internal nares (cf. Fig. 9.30). In the pharynx, gills usually number four and never more than five pairs. Each gill is a double plate composed of fine *gill filaments,* epithelial and highly vascularized (Fig. 31.11). A skeletal gill arch supports the gill tissue. The inner edges of the gills are often expanded into *gill rakers,* which prevent food from passing through the gill slits. An air sac pouches out from the pharynx. The sac serves as a lung primitively and as a swim bladder in more advanced forms (cf. Fig. 9.30). The remainder of the alimentary system is typically vertebrate, with esophagus, stomach, and intestine. Liver and pancreas are present, as is a spleen. Bony fishes do not possess a cloaca, i.e., the intestine opens separately at the anus. Excretory and reproductive ducts join into a urogenital sinus which exits via its own pore (cf. Chaps. 8, 9, and 10 for details on these and also the nervous, circulatory, and other systems).

31.11. **The gills of bony fishes.** (*Partly after Goodrich.*) **A.** Diagram of a left gill in side view, showing double row of gill filaments pointing away from pharynx and gill rakers pointing into pharynx. **B.** Horizontal section through a gill filament; the afferent vessel is the part of the aortic arch coming from the ventral aorta, the efferent vessels, the parts leading to the dorsal aorta. **C.** Horizontal section through fish head, showing position of the four pairs of gills.

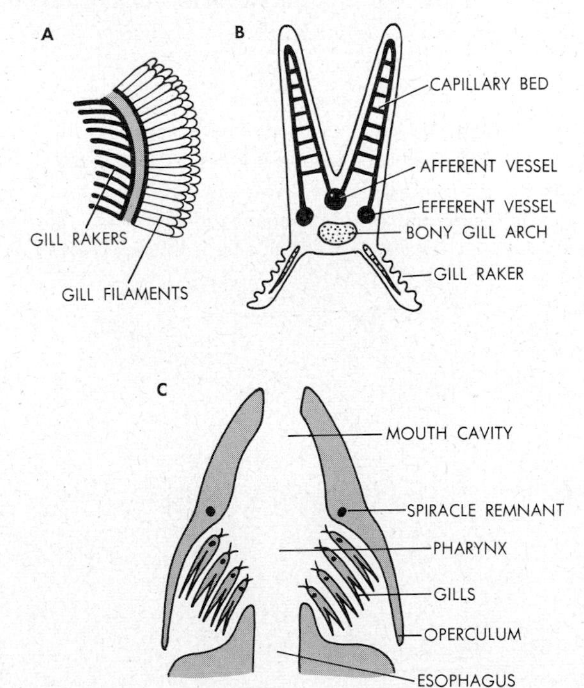

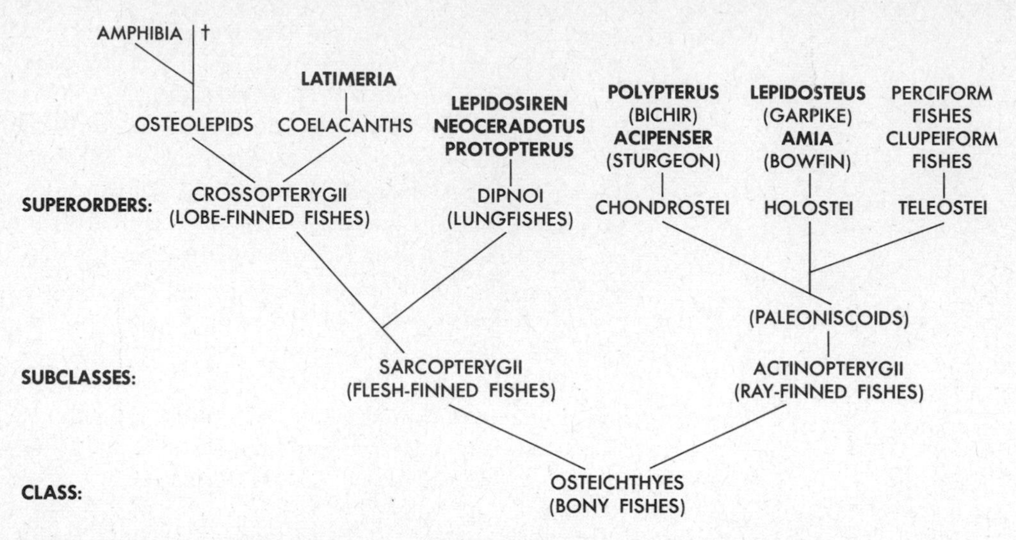

31.12. Taxonomic and probable evolutionary interrelations of major groups within the bony fishes. Presently living genera of all but the teleosts are indicated above each superorder.

Both the flesh-finned and the ray-finned bony fishes were primitively freshwater forms, and both presumably used their air sacs originally as lungs. The flesh-finned groups then continued and perfected the air-breathing habit, but the ray-finned types soon came to use the air sac purely as a swim bladder. Flesh-finned fishes also used their skeleton-supported pectorals and pelvics as primitive walking organs. One group among them, the lobe-finned osteolepids, probably gave rise to the amphibia (Figs. 31.12 and 31.13). In the course of this transition the fin skeleton elongated into the typical tetrapod

31.13. Sarcopterygii. Crossopterygian lobefins are illustrated in diagram **A**, representing the extinct *Osteolepis* (*after Traquair*), and in the photo, representing the living coelacanth *Latimeria*. Dipnoan lungfishes are shown in diagram **B**, depicting the living *Neoceradotus* (*after Norman*), in diagram **C**, showing the living *Lepidosiren* (*after Norman*), and in Fig. 15.30, which contains a photo of *Protopterus*.

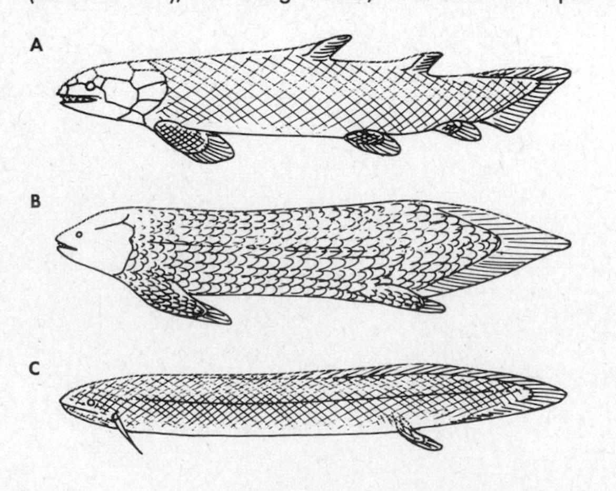

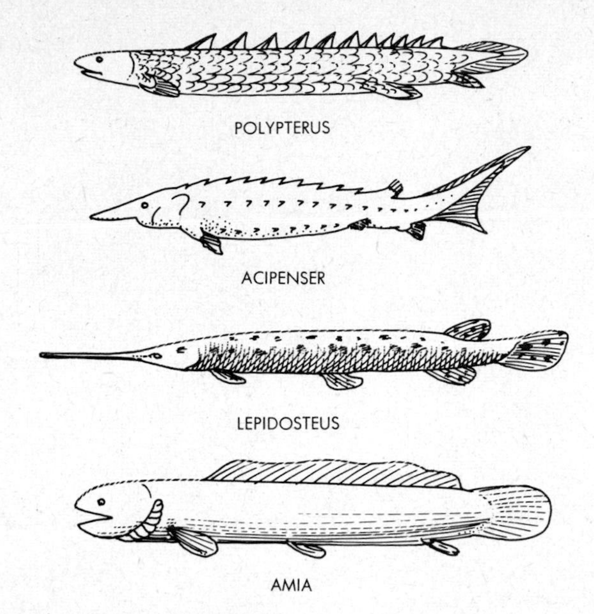

31.14. The bichir *Polypterus* and the sturgeon *Acipenser* are representatives of the superorder Chondrostei. Note the broad-base attachment of the paired fins and the diphycercal or heterocercal tail. The garpike *Lepidosteus* and the bowfin Amia are members of the superorder Holostei. Fins here are attached by narrow bases and tails are homocercal.

leg (cf. Fig. 8.14). Amphibia and all later vertebrates also inherited the internal nares from these ancestors, as well as an upper jaw fused directly with the skull.

Osteolepids are now extinct, and the related lobe-finned order of coelacanths similarly was believed to have been extinct since the Cretaceous. However, a marine coelacanth was found off Madagascar in 1938, and since this dramatic and completely unexpected discovery, additional specimens have been caught. All these coelacanths are members of the single known surviving lobe-finned genus *Latimeria.* The allied lungfishes, analogously flesh-finned types, are represented today by three surviving genera, *Protopterus* in Africa, *Lepidosiren* in South America, and *Neoceradotus* in southern Australia. These fishes live in muddy fresh waters. When water dries up seasonally, the animals wrap themselves in layers of mud and breathe air until the rains restore an aquatic environment. The whole subclass Sarcopterygii today thus consists of only four genera.

By contrast, the subclass of ray-fins now constitutes the most abundant group of vertebrates. These fishes are without internal nares, they possess a single dorsal fin, and, as noted, their paired fins are supported by fin rays only. The two primitive superorders, Chondrostei and Holostei (Fig. 31.14) are alike in that their scales are generally ganoid, the tail is basically heterocercal, and the notochord tends to persist. Also, a secondary reduction of ossification is quite common. Both of these superorders are represented today by only about 40 surviving species. All are either freshwater forms or, like some of the sturgeons, they live in the sea but return to rivers for spawning. *Polypterus,* found in Africa, is still a lung-breather (and has formerly been classified, erroneously so, with the lungfishes). In the other genera the air sac is a swim bladder primarily and the duct to the pharynx persists. *Amia,* represented by a single species, is an interesting "living fossil" found exclusively in the larger rivers of North America.

All remaining bony fishes are members of the large and highly diversified superorder Teleostei. In this group, the skeleton is fully ossified and the tail is homocercal. Two broad general patterns of teleost structure may be distinguished (Fig. 31.15). In one, the *clupeiform* pattern, fin rays are soft, scales are cycloid, the swim bladder has a persisting duct, and

31.15. The principal characteristics of the clupeiform and perciform patterns of teleost structure. Note that many teleosts possess a mixture of clupeiform and perciform traits. Note also that all teleosts possess homocercal tails and narrow-base attachments of the paired fins.

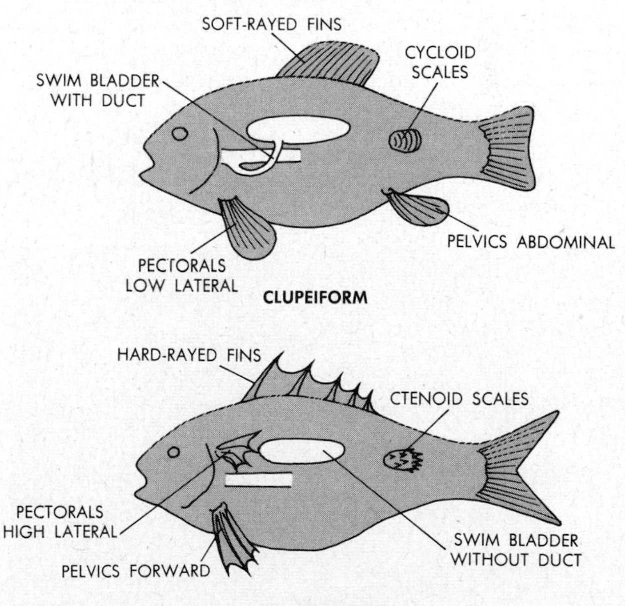

the pelvic fins are in a primitive abdominal, i.e., posterior, position. In the other, *perciform* pattern, fin rays are hard and spiny, scales are ctenoid, the swim bladder is without duct, the pectoral fins have moved high up on the sides of the body, and the pelvic fins have moved far forward, either into a thoracic position or even as far anteriorly as the head region (throat or jugular position). Thus, the pelvic fins lie in *front* of the pectorals. These two patterns are not always sharply distinct, and many teleost groups exhibit mixed and intergrading characteristics. For example, the fin rays may be soft generally, but the first or the first few anterior rays of a fin may be hard and spiny; cycloid scales may occur together with ductless swim bladders, or with pelvics in thoracic position.

The clupeiform pattern takes its name from the Clupeiformes, the third largest order of teleosts (1,000 species). The pattern occurs also in pure form in the *cyprinoid* fishes, of the order *Cypriniformes.* The latter constitutes the second largest teleost group (5,000 species) and the most important among freshwater fishes. Perciform types take their name from the order Perciformes, the largest order not only among fishes but among vertebrates as a whole (8,000 species). These fishes constitute one of the most important marine groups. Note that, in the tabulation above, the sequence of orders by and large forms an intergrading series from more or less pure clupeiform to more or less pure perciform types. Thus, all orders down to but not including the Notacanthiforms possess swim bladders with ducts; all orders down to but not including the Percopsiformes are characterized by cycloid scales; and orders down to but not including Perciformes have basically soft-rayed fins, though hard anterior fin spines are present in many cases. Teleost classification has not yet been completely stabilized; the one given above follows a recent tabulation. Representatives of various orders are shown in Figs. 31.16 and 31.17.

31.16. A pike (*Esox;* order Clupeiformes), illustrating the clupeiform teleost type.

31.17. A perch (*Perca;* order Perciformes), illustrating the perciform teleost type.

CLASS CHONDRICHTHYES

Cartilage fishes; mostly marine; bone entirely absent; notochord reduced but persisting in adult; scales placoid; tail heterocercal; nostrils paired, internal nares absent; without air sac; lateral-line system present; sexes separate; fertilization internal, development direct; oviparous, ovoviviparous, or viviparous.

Subclass Elasmobranchii (Selachii): with spiracles; five to seven pairs of gills, operculum absent; teeth numerous; upper jaw not fused with brain case; dorsal fin nonerectile.

Order Squaliformes (sharks): gill slits lateral; pectorals free anteriorly; swimming by tail; males with pelvic fin-claspers. *Squalus,* spiny dogfishes; *Mustelus,* smooth dogfishes; *Rhineodon,* whale sharks; *Cetorhinus,* basking sharks; *Sphyrna,* hammerhead sharks.

Order Rajiformes (rays, skates): gill slits ventral; pectorals attached to side of head anteriorly; swimming by pectorals. *Raja,* rays and skates; *Pristis,* sawfishes;

Rhinobatus, guitar fishes; *Torpedo,* electric "eels"; *Manta,* devil fishes, manta rays; *Dasyatis,* sting rays.

Subclass Holocephali (chimeras): without spiracles; four pairs of gills, operculum present; teeth fused into six pairs of plates; upper jaw fused to brain case; dorsal fin erectile; scales absent in adult; tail whiplike; without ribs or cloaca; males with claspers. *Hydrolagus*

At one time it was thought that cartilage fishes represented a more primitive group than the bony fishes. Such an assumption was generally based on the lack of bone, the presence of spiracles in most cases, and the persistence of parts of the notochord even in the modern members of the class. As noted above, however, it is now clear that cartilage fishes did not appear any sooner during vertebrate evolution than the bony fishes and that they are a highly specialized group. Although spiracles and notochord are indeed primitive traits (just as they are in early bony fishes), the lack of bone is probably not. Other indications of the specialized nature of these fishes are their almost exclusively marine occurrence, their excretory adaptations correlated with this habitat (cf. Chap. 9 and Fig. 9.39), and their large, yolky eggs and exclusively direct development.

In a shark, the skin is studded with tiny *placoid* scales, a modified ganoid type with enamel on the outside and dentine on the inside. The scale is extended into a pointed spine, and the whole *denticle* resembles and is homologous to a mammalian tooth (Fig. 31.18 and cf. Fig. 8.6). The notochord, though present, no longer forms a continuous rod. It persists

31.18. External features of sharks. Note fins, gill slits, teeth.

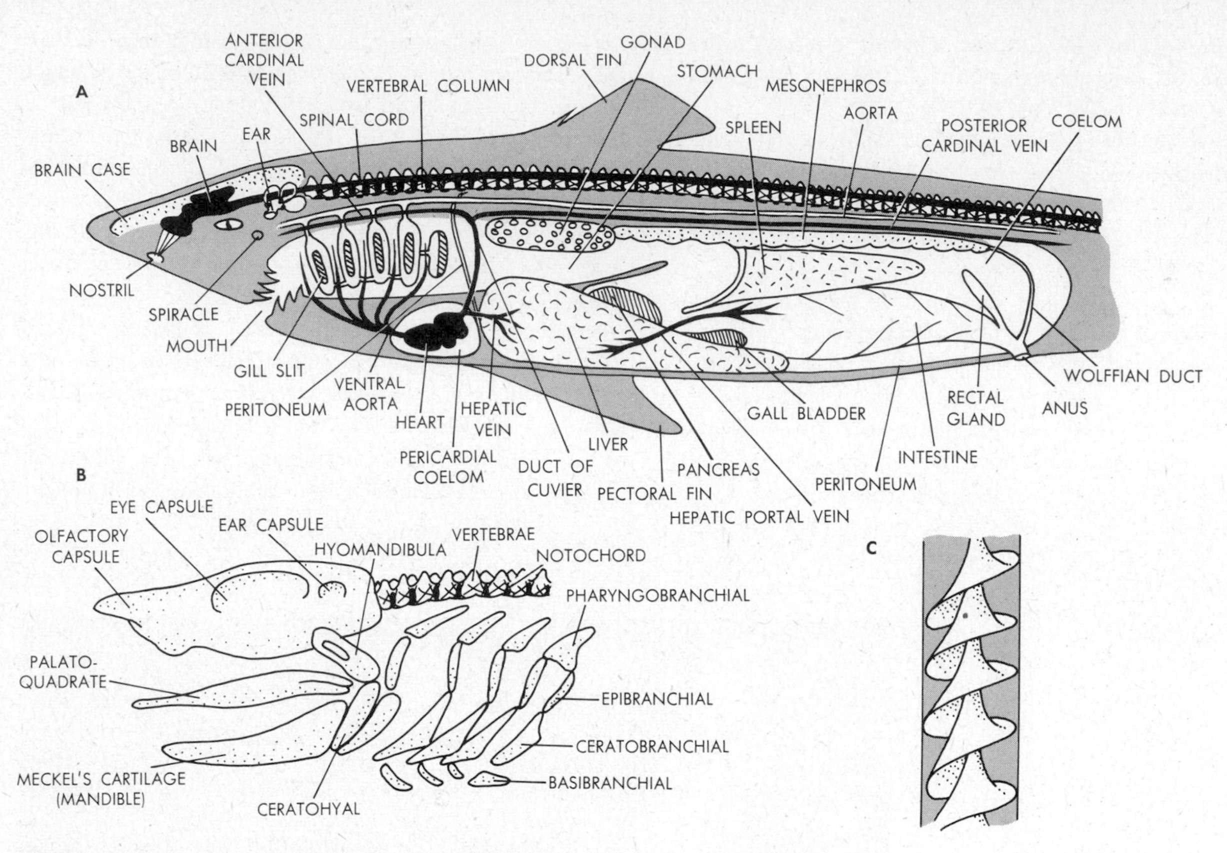

31.19. The internal structure of sharks. A. Sagittal section of the spiny dogfish *Squalus*. Note that in a female a uterus (Muellerian duct) would be present in addition to the Wolffian duct (cf. Fig. 10.22). Note also that the dorsal aorta gives off arteries (not shown) to stomach, intestine, liver, kidney, and the other internal organs. **B.** The anterior skeleton of a shark, all elements being composed of cartilage. The principal jaw support is the hyomandibular, and the notochord still persists between the vertebrae. (*After Goodrich.*) **C.** Diagram of the spiral valve in the intestine. (*After James.*)

only between adjacent vertebrae, which are cartilaginous but well developed. Anteriorly the skull is a cartilaginous box, with lateral capsules for the eyes and ears. In extinct elasmobranchs the upper jaws were joined to the skull partly directly, partly by means of the hyomandibular supports. The presently living forms retain only the hyomandibular prop, probably a specialization which permits the mouth to gape open more widely (Fig. 31.19). In the chimeras the upper jaws are fused directly to the skull, indicating an evolutionary trend parallel to that in the flesh-finned bony fishes. Elasmobranchs possess conspicuous spiracles, but these fishes are without opercula; the reverse holds for the chimeras. The median fins of the cartilage fishes are supported by fin rays, the paired fins by small jointed cartilages. Pectoral as well as pelvic girdles are present.

Internally, the mouth is rimmed by several rows of pointed teeth, homologous to the placoid denticles. Anchored in flesh, the anterior teeth often break loose but they are replaced by new ones formed behind. The alimentary tract includes a stomach and, opening into the intestine, a large liver and a separate pancreas. In the intestine is present a conspicuous *spiral valve,* a tissue fold which increases the absorptive surface. Cartilage fishes possess a cloaca, i.e., the reproductive and excretory ducts open into the terminal part of the hindgut. A salt-secreting *rectal gland* is also attached to the cloaca. In the many ovoviviparous and viviparous types, the female gonoducts (Müllerian ducts, cf. Fig. 10.22) are enlarged into uteri. Oviparous forms shed eggs protected in rectangular horny cases (commonly seen washed up and dried out along beaches; Fig. 31.20). The males typically possess a pair of copulatory organs, the *fin-claspers*, between the pelvic fins.

Chimeras, mainly deep-water forms, are represented by

A

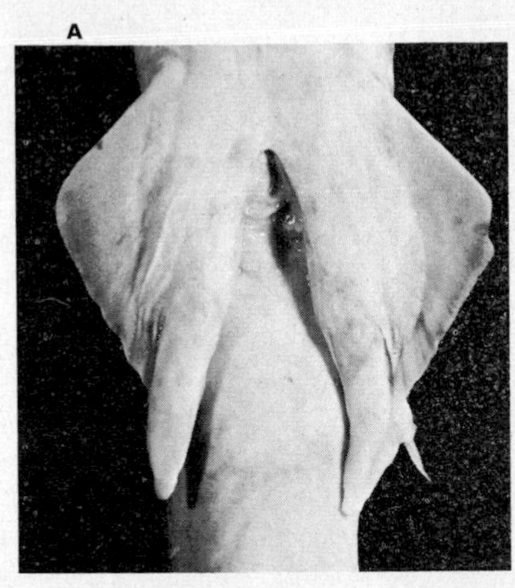

B

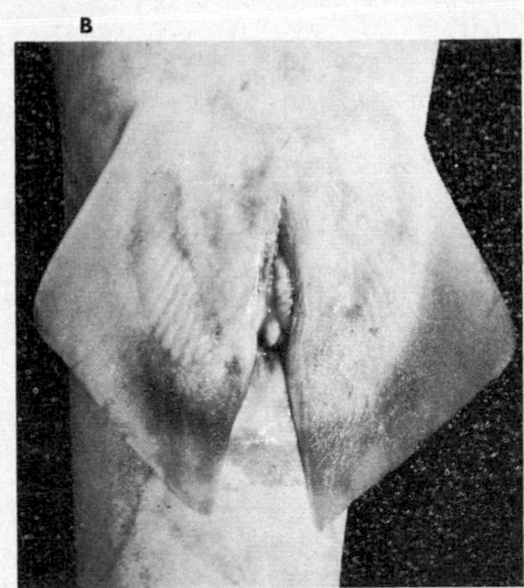

A,B, *courtesy Carolina Biological Supply Company.*

31.20. A and **B.** Ventral views of cloacal region of male and female shark, respectively, showing presence of fin claspers in male and their relation to the pelvic fins. **C.** Horny egg cases of an oviparous shark. The four filamentous extensions from each case anchor it on objects in water. Note developing embryos in some of the cases.

C

31.21. ***Scylliorhinus*, a dogfishlike shark.**

only some two dozen species; all other cartilage fishes are elasmobranchs, i.e., sharks and rays (Figs. 31.21, 31.22). Most sharks are fiercely carnivorous, but the largest, viz., whale and basking sharks, are plankton-feeders. Basking sharks are the largest of all fishes, attaining lengths of over 50 ft. Many rays and skates exhibit interesting offensive and protective adaptations, among them snouts extended into "saws" in the sawfishes, poison spines at the end of the tail in sting rays, and electric organs lateral to the eyes in the torpedo rays. Most rays and skates are bottom inhabitants, as would be expected from the flattened form of these fishes. Giant mantas become up to 20 ft long.

31.22. **Ventral view of the ray *Raia*.**

REVIEW QUESTIONS

1. Give taxonomic definitions of the subphylum Vertebrata and of each of the classes in the superclass Pisces. Show how most of the basic vertebrate traits are direct adaptations to a life in fresh water. Describe the pattern of mesoderm- and coelom-formation in vertebrates.

2. Define the class Agnatha and the subclass Cyclostomata. Characterize the extinct ostracoderms and show in what respects modern agnaths are structurally different. What different types of tail fins occur among Pisces and which type occurs in which groups? What is a pineal eye? How many species of each class of Pisces are known?

3. Describe the external and internal structure of a lamprey, with attention to all organ systems. Review pertinent data of Chaps. 8, 9, and 10. In what ways are lampreys adapted to an ectoparasitic existence? What are ammocoetes? Which cyclostomes (*a*) live permanently, (*b*) spawn, in fresh water?

4. Define the class Placodermi and characterize each order. How did jaws and paired and median fins presumably evolve? What changes in vertebrate anatomy were correlated with the change from ancestral filter feeding to bulk feeding by mouth? Review the pattern of embryonic development among fishes.

5. Give taxonomic definitions of all subclasses and superorders of bony fishes. Which group was probably ancestral to amphibia? What are the genera of living flesh-finned types? Name and characterize representative types of as many different orders of ray-finned fishes as you can.

6. What different scale types and tail types occur among bony fishes? In which groups is ossification reduced or incomplete? What changes in vertebrate anatomy were correlated with jaw elaboration as in bony fishes? Which groups possess spiracles? How is the upper jaw suspended in (*a*) placoderms, (*b*) bony fishes, (*c*) cartilage fishes?

7. Which groups of bony fishes possess (*a*) connecting ducts between air sac and pharynx, (*b*) internal nares, (*c*) air-breathing lungs? Describe the gill structure of a bony fish. Review the organization of every organ system of a bony fish (consult Chaps. 8, 9, and 10, if necessary).

8. Which groups of bony fishes (*a*) can hear, (*b*) possess lateral-line systems? Distinguish between clupeiform and perciform fishes and name orders of each type. How do fishes mate? Does viviparity occur among (*a*) cyclostomes, (*b*) bony fishes, (*c*) cartilage fishes?

9. Define the subclasses and orders of chondrychthians. Why is the cartilage skeleton now thought to represent a specialized rather than a primitive trait? Do (*a*) chondrychthians, (*b*) any other classes of fishes possess a complete or partial notochord as adults?

10. Contrast the structure of the gill region in sharks and chimeras. What are fin-claspers? How do sharks differ from rays? Review the adaptations of sharks and rays to their respective ways of life.

COLLATERAL READINGS

Popular and technical references on fishes are mixed in the listing below:

Applegate, V. C., and J. W. Moffett: The Sea Lamprey, *Sci. American*, Apr., 1955.

Colbert, E. H.: "Evolution of the Vertebrates," Wiley, New York, 1955.

Daniel, J. F.: "The Elasmobranch Fishes," University of California Press, Berkeley, Calif., 1934.

Gilbert, P. W.: The Behavior of Sharks, *Sci. American*, July, 1962.

Goodrich, E. S.: "Studies on the Structure and Development of Vertebrates," Macmillan, London, 1930.

Hyman, L. H.: "Comparative Vertebrate Anatomy," 2d ed., The University of Chicago Press, Chicago, 1942.

Lennon, R. E.: The Feeding Mechanism of the Sea Lamprey and its Effect on Host Fishes, *U.S. Fish and Wildlife Service Fisheries Bulletin*, vol. 56, 1954.

Millot, J.: The Coelacanth, *Sci. American*, Dec., 1955.

Romer, A. S.: "Man and the Vertebrates," The University of Chicago Press, Chicago, 1941.

———: "Vertebrate Paleontology," 2d ed., The University of Chicago Press, Chicago, 1945.

———: "The Vertebrate Body," Saunders, Philadelphia, 1950.

Schultz, L. P., and E. M. Stern: "The Ways of Fishes," Van Nostrand, New York, 1948.

Young, J. Z.: "The Life of Vertebrates," 2d ed., Oxford, London, 1962.

CHAPTER 32

vertebrates: tetrapods

CLASS AMPHIBIA

Amphibians; freshwater and terrestrial; paired appendages are legs; skin without scales (most living species); with internal nares; upper jaws fused to skull; heart three-chambered; breathing via gills, lungs, skin, and mouth cavity; 10 pairs of cranial nerves; sexes separate; fertilization mostly external, development via tadpole larvae; some with internal development, ovoviviparous or viviparous.

†Subclass Stegocephalia: extinct amphibians; head covered with bony plates, skin often with scales.
- Order Lepospondyli: *Diplocaulus*
- Order Phyllospondyli: *Branchiosaurus*
- Order Labyrinthodonti: *Eryops, Seymouria*

Subclass Gymnophiona (Apoda) (cecilians): wormlike; up to 200 vertebrae; without limbs or girdles; skull topped with bone; ribs long; skin smooth, often with embedded dermal scales; eyeless; small tentacles in front of nostrils; tail short; fertilization internal, male with copulatory organ; *Typhonectes, Ichthyosis*

Subclass Urodela (Caudata) (salamanders, newts): body with head, trunk, tail; aquatic larvae resemble adults; larvae and some adults with teeth on upper and lower jaws; eyelids typically present in terrestrial adults, absent in aquatic adults.
- Order Proteida: permanently aquatic; with permanent gills and lungs; tail with fin. *Necturus,* mudpuppies.
- Order Meantes: permanently aquatic; with permanent gills; jaws with horny cover; without pelvic limbs. *Siren,* mud eels.
- Order Mutabilia: adults usually without gills, with lungs, air-breathing, though habitat often aquatic. *Cryptobranchus,* hellbenders; *Amphiuma,* congo eels; *Amblystoma,* axolotls; *Triturus,* newts, efts; *Salamandra, Pseudotriton,* salamanders. *Megalobatrachus,* giant salamanders.

Subclass Anura (Salientia) (toads, frogs): tailless, without neck; hind limbs usually long, feet webbed; 10 vertebrae; with eyelids; adults mostly terrestrial.
- Order Amphicoela: fertilization internal. *Liopelma,* bell toads.
- Order Opisthocoela: fertilization external; many aquatic, without eyelids. *Xenopus,* South African clawed toads; *Pipa,* Surinam toads; *Alytes,* European midwife toads.
- Order Anomocoela: hind feet with horny style. *Pelobates,* spade-foot toads.
- Order Procoela (true toads): without teeth. *Bufo,* common toads; *Hyla,* tree toads; *Gastrotheca,* pouched toads.
- Order Diplasiocoela (true frogs): teeth on upper jaw;

tongue forked. *Rana,* common frogs; *Polypedates,* tree frogs.

Amphibia mark the transition from aquatic to terrestrial existence in vertebrate evolution. Accordingly, the most notable traits of amphibians are those which adapt the animals to an at least partly terrestrial life; and we may note that most of these traits trace their origin to the ancestral lobe-finned fishes. Thus, the paired appendages have elongated to distinct walking legs, each with five (or fewer) toes. Though primitive amphibia still live entirely in water, the majority of these possesses lungs which serve actively in air breathing. Gills occur in the larvae and in exceptional cases also in some of the adults (where lungs may or may not be present in addition). Correlated with the air-breathing habit, the internal nares inherited from the lobe-fins provide an important adaptation, for they make breathing possible while the mouth remains closed. As in ancestral lobe-fins, moreover, the upper jaw is fused to the skull. The free hyomandibular has become a *columella,* a middle-ear bone. One end of it is attached to an external eardrum flush with the body surface, the other end to the inner ear. This bone transmits aerial sound waves into the ear, as in a frog, clearly adaptive in terrestrial living (Fig. 32.1).

A further adaptation of presently living amphibia is that the skin has become thin and scaleless, providing a breathing organ auxiliary to gills, lungs, and the lining of the mouth. At the same time the skin has become highly glandular and mucus-secreting, a necessary protection against desiccation. Also adaptive in air breathing is the three-chambered heart, capable to some extent of preventing the mixing of oxygenated blood returning from the lungs and venous blood returning from the body: arterial blood enters one atrium, venous blood the other. The two bloods then mix in the single ventricle, to be sure, but not so much as they do in a two-chambered heart. Where gills persist they are reduced to three pairs, and three corresponding pairs of aortic arches are present in such cases; where gills do not persist, however, the aortic arches have become reduced to a single pair (cf. Fig. 9.30).

Thus, from a lobe-finned starting point, the amphibian descendant has become able to walk and to breathe air on a permanent basis, even though these activities are still carried out in an aquatic environment in some instances. Actually, even the most land-adapted amphibian requires a moist environment at all times. We may note, indeed, that although amphibia have become land-adapted in locomotor and respiratory respects, they have not become so in most other respects. For example, the nervous system is still substantially fishlike, as is the mesonephric excretory system, which remains adapted to eliminate water primarily but not salt. We may appreciate therefore why amphibia are unable to subsist in a marine or otherwise salty environment, and why these animals have generally been incapable of colonizing oceanic islands by their own efforts. The skeleton is still basically as in fishes, particularly so in the urodeles, even though the appendages have become modified and the number of skull bones has become reduced. Other specialized skeletal alterations have occurred in forms such as frogs (Fig. 32.2).

Above all, reproductive and developmental processes still retain ancestral fishlike characteristics. Thus, eggs must typically be laid into water, and fertilization and development take place externally, as in bony fishes. Some amphibian eggs develop directly and without larvae, in a number of cases within brood pouches on the backs of the adults (cf. Figs. 32.6 and 32.7). In most cases, however, development includes aquatic tadpole larvae and these are thoroughly fishlike, regardless of whether or not the adults are also aquatic. The tadpole is characterized by a finned tail; by a lateral-line system; by three pairs of gills and three pairs of aortic arches, but in early stages not by lungs; by a two-chambered heart; by a hyomandibular piece which is not yet a middle-ear bone;

32.1. Sections through the hind region of the head, illustrating the transition from fish ear to amphibian ear. (*After Romer.*) **A.** Condition in primitive bony fishes, with free hyomandibular (and upper jaw fused to skull), the ear consisting only of the canals and sacs within the bony ear capsule. **B.** Amphibian condition, with the hyomandibular (also called columella in amphibia) functioning as sound transmitter between tympanum and inner ear. The spiracular duct to the pharynx (remnant of the original first gill slit) has become the middle ear cavity and the eustachian tube.

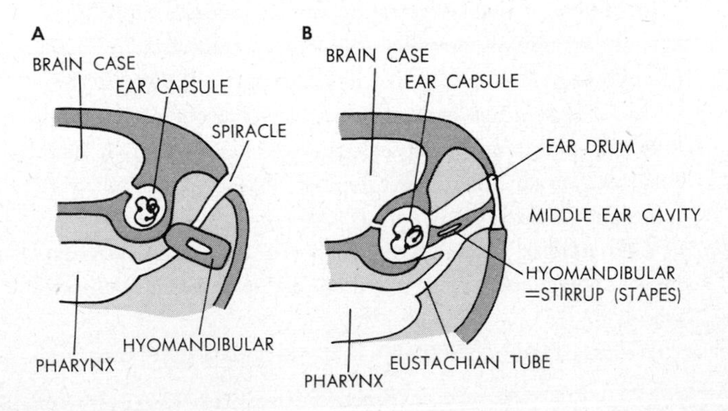

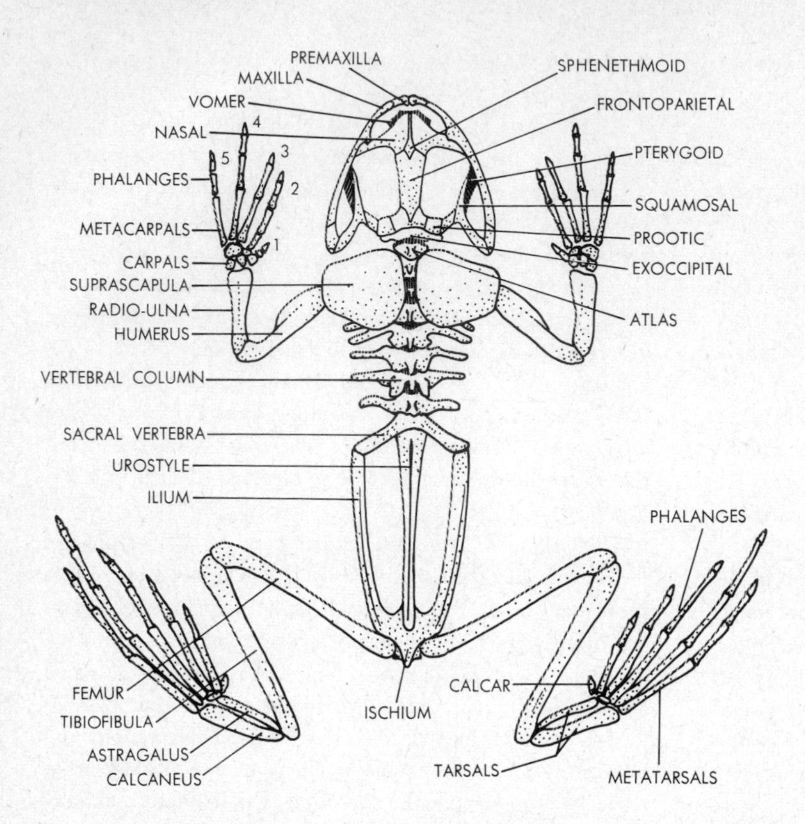

32.2. The skeleton of a frog. (*After Marshall.*)

and in early stages by an absence of limbs (Fig. 32.3). In effect, therefore, any terrestrial adaptations exhibited by adult amphibia are acquired anew in each amphibian life cycle.

Amphibians are distinguished in large measure on the basis of skeletal differences, particularly variations in the structure of the vertebrae, the limbs, and the presence or absence of a tail in the adult. Of the three living subclasses, the cecilians are the smallest and least familiar group. They probably represent a very primitive amphibian stock, though the animals today are secondarily highly specialized. Encompassing about 70 mostly tropical species, these amphibians are blind, limbless soil-burrowers. They lead a life somewhat like earthworms. Their fertilization is internal, the males possessing a penis. Eggs are large and yolky and they cleave superficially, as in fishes. Though some species develop via aquatic larvae, most are direct developers and of these many are viviparous. Judging from their skeletal traits, cecilians appear to be most nearly related to the extinct Lepospondyli and the living urodeles. The animals exhibit one major ancestral trait, i.e., dermal scales embedded in the skin.

The 250 or so species of the subclass Urodela include a wide range of types, from fully aquatic ones to those in which terrestrial habits are developed to various degrees (Figs. 32.4 and 32.5). Two of the three orders, represented by mud puppies and mud eels, are strictly aquatic, with permanent gills in both and with lungs in addition in mud puppies. The gills are external in these animals; i.e., they protrude beyond the general body surface and are not covered by opercular folds. Fully aquatic types are included also in the order of newts and salamanders, e.g., the congo eel *Amphiuma,* which possesses permanent gills as well as lungs, and the hellbender *Cryptobranchus,* in which only lungs are present (but in which a spiracle is open and functional). Newts such as *Triturus* and salamanders such as *Salamandra* similarly possess only lungs, and these animals actually are largely terrestrial in habit (though they can usually live in water as well). *Pseudotriton* exemplifies a group of largely terrestrial urodeles in which both gills and lungs are absent, the skin here being the main breathing organ. The most nearly terrestrial urodeles are internally fertilizing, viviparous forms such as certain species of *Salamandra.* Although these do require a moist environment, they can otherwise be independent of natural bodies of water.

We may note, incidentally, that the aquatic gill-possessing

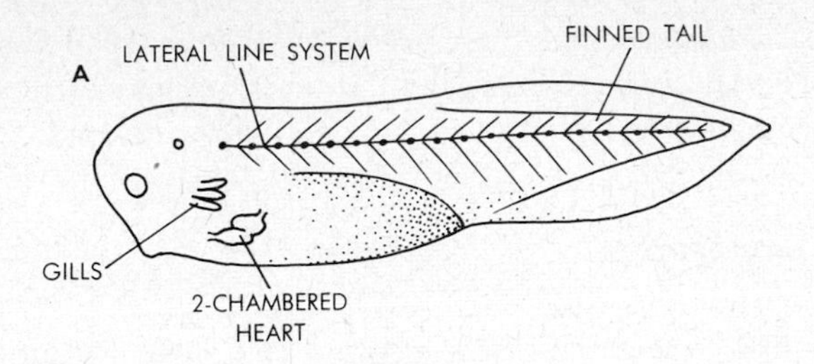

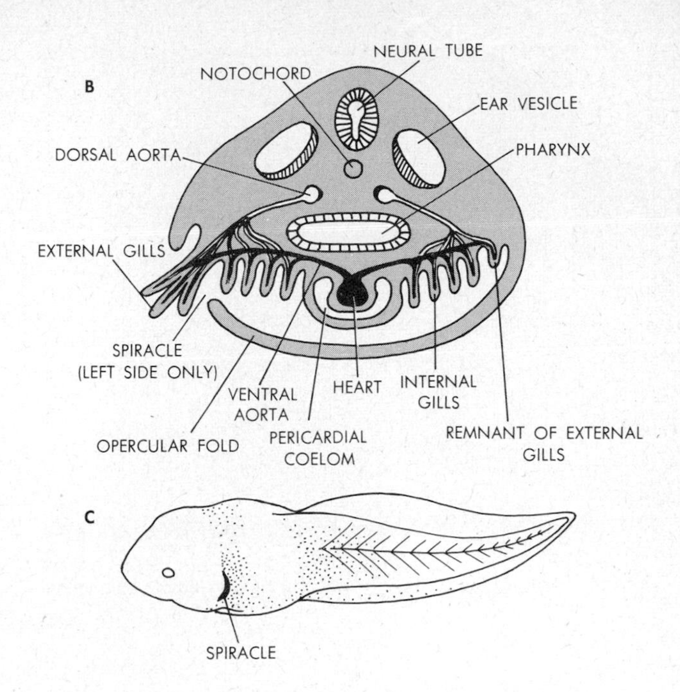

32.3. Traits of the amphibian larva. A. Fully formed tadpole, showing characteristics particularly typical also of fishes. **B.** Cross section through gill region of later larval stage, at time of transition from external to internal gills. (*Adapted from Goodrich.*) On right side, external gills have already degenerated and the newly developed internal gills are covered by the opercular fold. On left side, external gills still protrude for a time through the spiracle, but ultimately they too degenerate. From then until the lungs become functional, breathing is accomplished through the internal gills. **C.** Later tadpole showing position of the spiracular opening on left side (tadpoles and larval circulation are illustrated also in Figs 3.16 and 9.30).

urodeles represent neotenous types—animals which have retained the larval state but have acquired reproductive organs. That this is so can be shown dramatically in the axolotl *Amblystoma mexicanum,* a species which normally remains in a gill-possessing larval condition, develops sex organs, and breeds. However, if such an animal is injected with thyroid hormone it metamorphoses at once, loses its gills, and acquires lungs. Other species of axolotls, and gill-possessing urodeles generally, are *genetically* neotenous, i.e., they retain the larval state no matter how much thyroxine might be injected into them; their tissues are incapable of undergoing metamorphic change (cf. also Chap. 11).

Whereas urodeles are on the whole more aquatic than terrestrial, the reverse holds for the Anura, toads and frogs.

32.4. The congo eel *Amphiuma,* a urodele. Note reduced appendages.

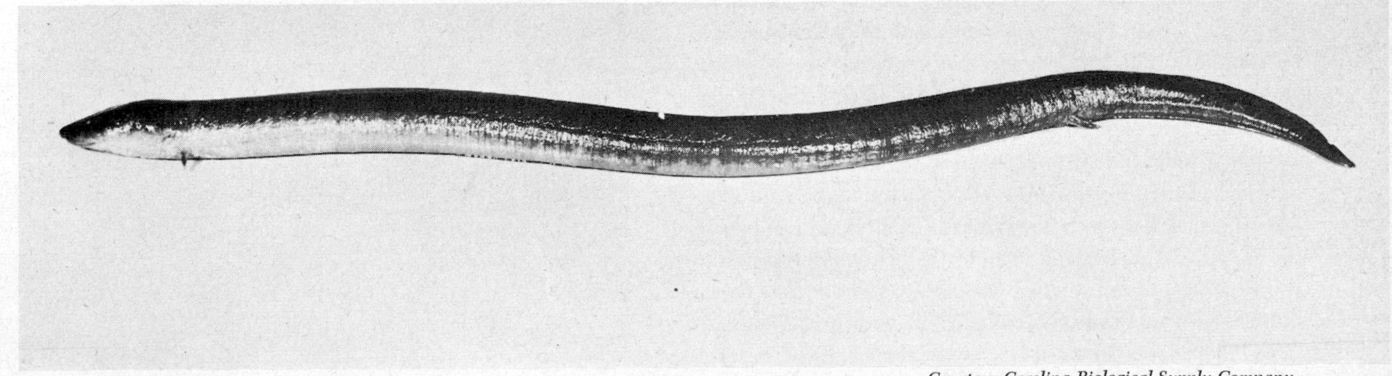

Courtesy Carolina Biological Supply Company.

32.5. The neotenous Mexican axolotl *Amblystoma mexicanum.*

These animals make up the vast bulk of the amphibian class (Figs. 32.6 and 32.7). Most anurans have eyelids, an adaptation to their life in an aerial environment. Also, they are often highly specialized as jumpers, with hind legs far longer and stronger than the forelegs. Correlated with this locomotor specialization the tail is absent and the head is joined directly to the trunk, without neck. Indeed, the whole body has become squat and foreshortened, and the number of vertebrae is reduced to 10 (cf. Fig. 32.2). Jumping aids in feeding, i.e., insect-catching. Frogs are equipped with a highly extensible, sticky tongue, attached at its anterior end to the lower jaw. The tongue can be hurled out of the mouth at an insect, posterior end first.

Toads are toothless, frogs possess teeth on the upper jaw and the roof of the mouth. None of the anurans possesses gills as adults and all possess lungs. A few toads, e.g., *Xenopus,* are permanently aquatic though air-breathing forms. Most anurans require an outright aquatic environment only for egg-laying and for tadpole development. However, a good many are water-independent even in these processes, for the eggs develop directly. For example, toads such as *Alytes* and *Pipa* incubate their eggs in brood pouches or in epidermal pits on the back. It is interesting to note that parallel evolution in toads and frogs has extended not only to reproductive habits but also to ways of life generally. Thus, both frogs and toads have independently evolved tree-living forms, e.g., tree toads such as *Hyla* and tree frogs such as *Polypedates.* Apart from the direct developers among them, all others still must lay their eggs into water.

CLASS REPTILIA

Reptiles; epidermis with cornified scales; limbs five-toed, clawed; breathing by lungs; heart four-chambered, ventricular separation usually incomplete; one pair of aortic arches; 12 pairs of cranial nerves; excretion metanephric; fertilization internal; eggs with extraembryonic membranes and shells; development direct; oviparous, ovoviviparous, or viviparous.

Subclass Anapsida

extinct: cotylosaurian stem reptiles

living: Order Chelonia (turtles, tortoises, terrapins): body encased in bony shell and epidermal cover; teeth absent in living types; single penis in male; oviparous. *Chrysemys,* common painted turtles; *Chelydra,* snapping turtles; *Caretta,* loggerhead turtles; *Chelonia,* green turtles; *Terrapene,* box turtles; *Dermochelys,* leatherback turtles; *Malaclemys,* diamondback terrapins; *Testudo,* Galapagos giant tortoises.

32.6. Egg-carrying male of the midwife toad *Alytes obstetricians.*

32.7. *Hyla, a tree frog.*

†Subclass Synapsida
extinct: therapsids, mammallike reptiles, and mammalian ancestors
living: none
†Subclass Ichthyopterygia
extinct: ichthyosaurs
living: none
†Subclass Synaptosauria
extinct: plesiosaurs
living: none
Subclass Lepidosauria
extinct: lizardlike reptiles
living: Order Rhynchocephalia (tuataras): males without penis; pineal eye functional; oviparous. *Sphenodon punctatus*
Order Squamata (lizards, snakes): skull bones reduced; epidermal scales horny; males with paired penis; oviparous, ovoviviparous, or viviparous.
Suborder Sauria (lizards): eyelids movable; usually with external ear opening; pectoral girdle present, limbs usually present; halves of lower jaw articulated; urinary bladder generally present. *Phyllodactylus,* geckos; *Iguana,* iguana lizards; *Phrynosoma,* horned "toads"; *Draco,* "flying" dragons; *Lacerta,* common lizards; *Chamaeleon,* chameleons; *Eumeces,* skinks; *Varanus,* monitor lizards; *Heloderma,* Gila monsters.
Suborder Serpentes (snakes); eyelids not movable; without external ear openings; pectoral girdle absent, limbs absent; halves of lower jaw not articulated, united by ligament; urinary bladder absent. *Typhlops,* worm snakes; *Eunectes,* anacondas; *Python,* phytons, boas; *Thamnophis,* garter snakes; *Natrix,* water snakes; *Hydrus,* sea snakes; *Naja,* cobras; *Bitis,* puff adders; *Vipera,* vipers; *Lachesis,* bushmasters; *Bothrops,* fer-de-lance; *Crotalus,* rattlesnakes.
Subclass Archisauria
extinct: dinosaurs, flying pterosaurs, ancestors of birds.
living: Order Crocodilia (alligators, crocodiles, caimans, gavials): heavy-bodied, semiaquatic; epidermal scales with dermal reinforcements; webbed feet; teeth in sockets; urinary bladder absent; males with single penis; oviparous. *Alligator, Crocodylus, Gavialis, Caiman*

As the first fully terrestrial vertebrates, reptiles exhibit many characteristics encountered also in the later mammals and birds. The reptilian skin is dry and protected from desiccation typically by cornified epidermal scales, homologous to fur and feathers. Limbs are perfected as walking legs, and they raise the body off the ground more than the amphibian limbs. Reptiles are strictly air breathers, with a breathing system constructed like that of mammals and birds, i.e., tracheal and bronchial tubes pipe air to the lungs. The heart is four-chambered, completely so in crocodiles, nearly so in the other orders; it keeps arterial separated from venous blood, essentially as in mammals and birds. However, a single pair of aortic arches is still present, as in the ancestral (labyrinthodont) amphibia (cf. Fig. 9.20). The adult excretory system has become metanephric, as in mammals and birds, and is capable of producing a highly hypertonic, water-conserving urine. Secondarily aquatic turtles and alligators excrete urea, like mammals; all terrestrial reptiles excrete a semisolid urine containing uric acid, like birds (cf. Fig. 10.22).

In the nervous system, 12 pairs of cranial nerves emanate from the brain, as in birds and mammals, in contrast to the 10 pairs in amphibia. Reptilian ears still retain an essentially amphibian structure, with a single middle-ear bone and an external eardrum. However, a notable advance is in evidence in the eyes which, in lizards and turtles at least, are capable of recording color. The single most essential adaptation to land life is the shelled, amniote egg, as already discussed in Chap. 11 (cf. Fig. 11.18). The shells are leathery in some cases, calcareous in others. As in birds, the hatching young possesses a temporary *egg tooth* on its upper jaw, with which it cuts itself out of its shell. *All* reptiles lay eggs on land, typically in burrows on the ground. Secondarily aquatic species likewise deposit their eggs on the shore, though mating itself may occur in water. Fertilization is always internal, as in mammals and birds; a penis in males may be absent (tuatara), single (turtles, crocodiles), or double (lizards, snakes). These organs are located in or near the cloaca and they are protruded during copulation.

32.8. **A.** reptilian skull types. (*After Romer.*) p, parietal; o, postorbital; s, squamosal; j, jugal. In the anapsid type, openings in the skull are not present behind the eye. In the synapsid skull, a single opening is bounded by the postorbital, squamosal, and jugal bones. In the parapsid skull, a single opening is bounded by the postorbital, squamosal, and parietal bones. And in the diapsid type, both of the above openings are present. **B.** Tooth attachments, illustrated in front and side views. The acrodont type is encountered in sharks, for example, a tooth here being held in place only by connective tissue. In a pleurodont attachment, as in lizards, a tooth is joined to bone on one side only. And in the thecodont pattern, as in alligators and mammals, a tooth is imbedded by its root within a bony socket.

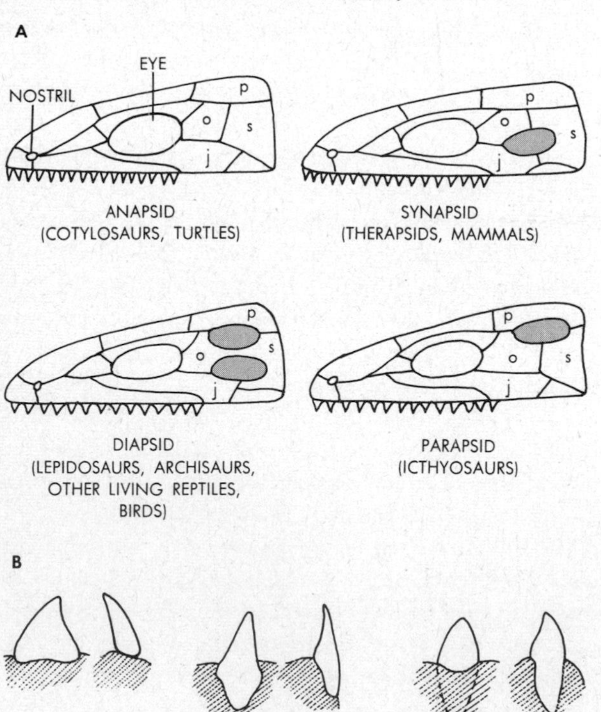

Reptiles are distinguished taxonomically mainly on the basis of skull structure. Of particular usefulness is the number and precise location of lateral windows in the skull bones behind the eyes (Fig. 32.8). A so-called *anapsid* skull, without lateral windows, is primitive. This type persists today in turtles, a highly specialized group but one which exemplifies the primitive ancestral reptiles. In a *diapsid* skull two pairs of lateral windows are present, a skull type characteristic of the lepidosaurian and archisaurian subclasses, hence of all other living reptiles. Skull types with variously placed single pairs of lateral windows are no longer in existence—except, to be sure, the *synapsid* type, preserved in modified form in the mammals. Among other diagnostic criteria used in reptile classification are the nature of the tooth insertion (which is *thecodont*, i.e., in jaw sockets, in Archisauria and in the synapsid line leading to mammals), the structure of the pelvis, and other skeletal traits.

With their unique protective body covering, turtles are among the most readily recognizable animals (Fig. 32.9). The shell consists of a dorsal *carapace* and a ventral *plastron*, the two parts being joined along the sides either by bony or by membranous bridges (Fig. 32.10). The carapace is a dome of bone, formed from fused vertebrae and from broadened and fused ribs. Analogously, the bony plate of the plastron is constructed from flattened and fused parts of the pectoral and pelvic girdles and the sternum. The pectoral girdle is actually wholly inside the shell of the rib cage, a unique condition encountered nowhere else. The bony parts of the shell are overlain by skin, which may be soft but tough, as in the leatherbacks, or hardened into cornified epidermal horn, as in painted and numerous other types of turtles.

In conjunction with this extensively armored condition the trunk musculature has become highly reduced, and since the rigid shell precludes breathing movements of the body wall, the lungs must be ventilated by special internal muscles. Many

32.9. The green sea turtle *Chelonia.*

turtles cannot withdraw their heads into the shells. Whereas extinct chelonians usually possessed teeth, all living species have horny beaks. In marine turtles the forelimbs are modified into flippers (and we may note here that the term "turtle" properly applies only to the aquatic chelonians; terrestrial types are "tortoises"). The majority of chelonians are aquatic or semiaquatic. In these, mating usually takes place in water. Males often have a concave plastron, a feature which makes mounting the back of females easier during copulation. The largest turtles are the marine leatherbacks, which are up to 8 ft long and weigh nearly a ton. The giant Galapagos tortoises attain lengths of about 4 ft (and recorded life spans of often well over 60 years). Chelonians are generally omnivorous (and *Chelonia* in turn is prized by man as contributor of the essential ingredient in turtle soup).

Some 95 per cent of all living reptile species embrace the lizards and snakes, order Squamata. Lizards (Fig. 32.11), characterized like snakes by horny epidermal scales, are predominantly insectivorous. Good examples of such insect-eating types are the chameleons, which have a long, explosively protrusible tongue (and a well-known ability to change color according to the background foliage of their arboreal environment). Some lizards are herbivorous, and still others, particularly the larger forms, subsist on mice, frogs, or bulkier vertebrates. The largest lizard is the Komodo dragon *Varanus*, up to 10 ft long. Two species, both of the genus *Heloderma*, are venomous; the Gila monster *H. suspectum* is one, the bearded lizard *H. horridum* is the other. Some lizards (e.g., "glass snakes") are limbless and these are soil animals which propel themselves in snakelike fashion. Lizards generally are noted for their capacity of autotomizing the tail. A cast-off tail usually lashes and wriggles about for some time and it

32.10. The turtle skeleton. Diagrams: outlines of dorsal carapace **(A)** and of ventral plastron **(B)**. Heavy lines mark extent of external horny plates (scutes), light lines mark the bony plates underneath the horny layer. Note that the midrow of bone represents portions of the vertebral column, the lateral rows, expanded ribs. Note also that the suture lines of bone and of horn in most cases alternate in position, a feature which strengthens the shell. **Photo:** the whole skeleton. The plastron is cut away and hinged back, showing the limb girdles underneath the carapace formed by the rib cage.

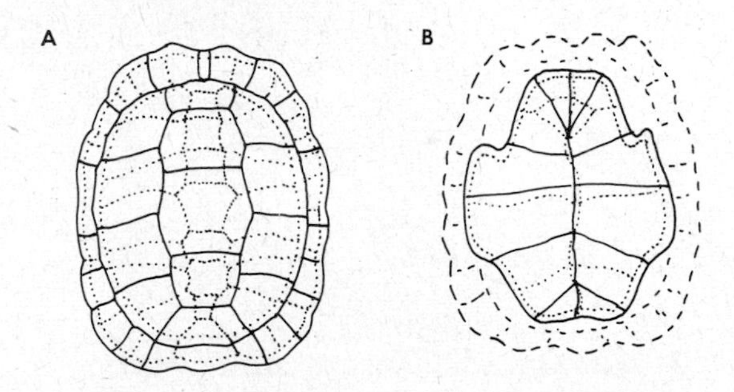

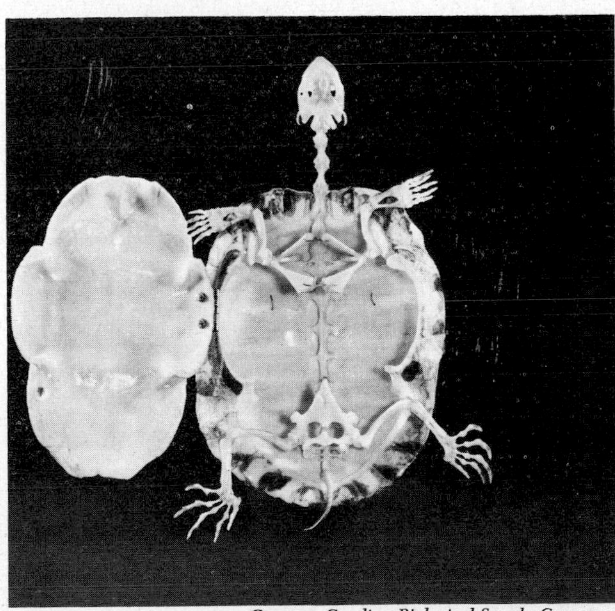

Courtesy Carolina Biological Supply Company.

32.11. The lizard *Chamaeleon,* catching grasshopper with tongue.

often distracts an enemy. A new tail regenerates without bony support. Like many snakes, many lizards are ovoviviparous or fully viviparous. Most are oviparous, however.

Snakes may be characterized as limbless lizards, though they differ from lizards in many specialized respects. Nevertheless, there is little doubt that snakes have evolved from lizards, probably several times independently. In snakes, the left and right halves of the lower jaws are not articulated but are held together by ligaments, and the upper and lower jaws are often similarly joined by ligaments only. Consequently, the mouth can be distended greatly and a snake may ingest an animal several times wider than its own diameter (Fig. 32.12). In the course of such ingestion, which here cannot be assisted by limbs, a snake works the two halves of the jaws forward in alternation, resulting in a "climbing" type of food engulfment. In venomous snakes, poison fangs may be formed by teeth in the rear of the upper jaw (e.g., certain tree snakes), or by fixed, immobile front teeth (e.g., cobras), or by hinged front teeth which can be folded back along with the upper jaw bone when not in use (e.g., vipers, rattlers). In some cases poison from the salivary glands flows along an anterior groove on a fang, in others, through a duct within the fang.

Snakes are without external ear openings but they "hear" vibrations transmitted from the ground via the skeleton. Eyes are lidless and covered by a transparent membrane. Vertebrae may number up to 400. A sternum is absent, hence ribs can spread apart readily. Limb girdles and limbs are generally absent as well, though vestigial remains of pelvic girdles persist in pythons and related types. Snakes usually employ the free posterior edges of their large, transverse belly scales to obtain leverage on the ground. Locomotion is achieved in four different ways in different groups, viz., by "snakelike" lateral undulations of the body, by straight-line rippling undulations, by accordion-like folding and extension of the body, and by sidewinding (as in the horned rattlesnake, in which the body is thrown almost at right angles to the path of progression; Fig. 32.13). Anacondas and pythons are the largest snakes; they reach lengths of 30 to 35 ft. The largest poisonous snake is the king cobra *Ophiophagus*, with a length of about 18 ft. Snakes are strictly carnivorous, and as a group they occur in

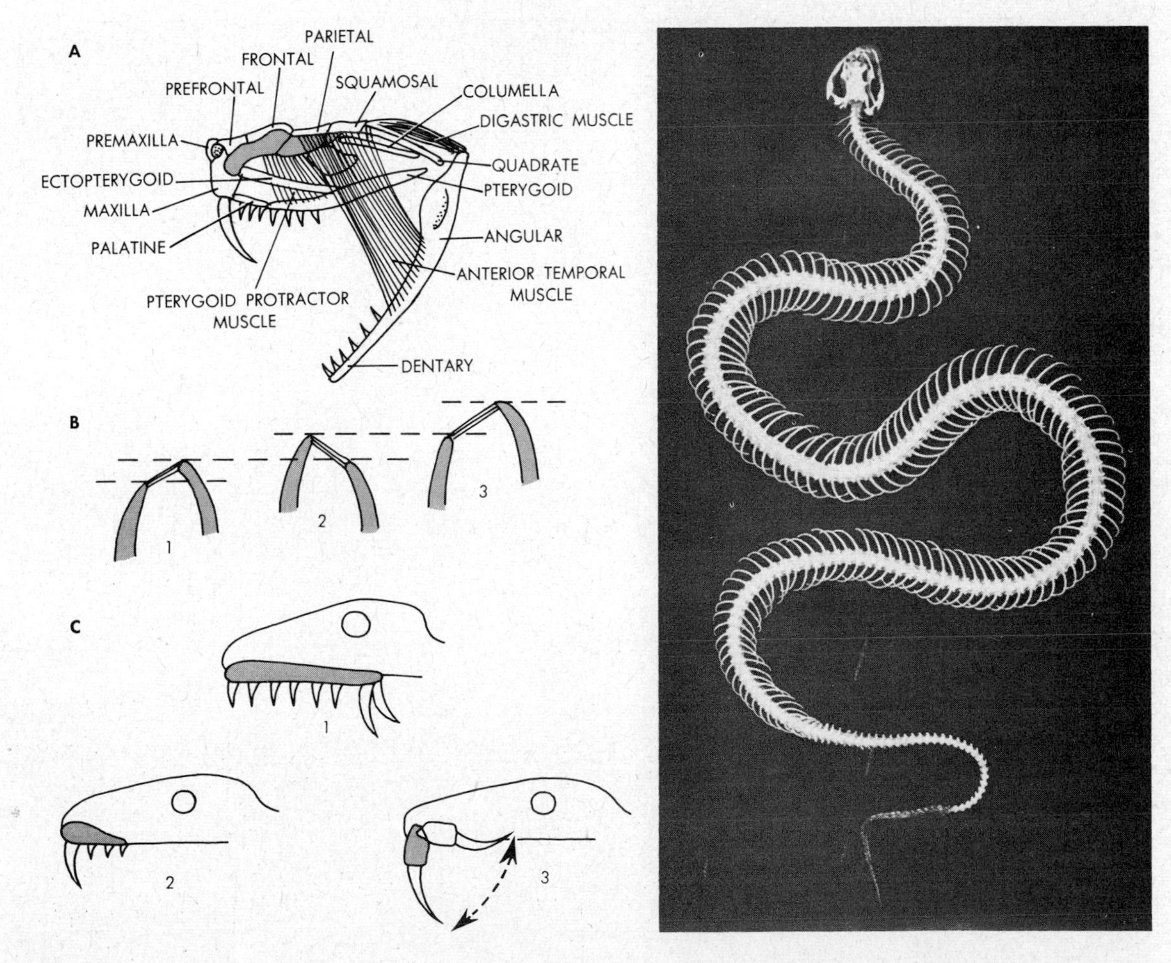

32.12. Skeletal adaptations in snakes. Diagrams: A. The skull of a rattlesnake, showing the principal bones, reduced in number in this group of animals. (*Modified from Gadow.*) Of the three muscles indicated, the digastric opens the jaw, the anterior temporal closes it, and the pterygoid protractor pushes the palatine-pterygoid bony rod forward, an action which in turn rotates the maxilla and erects its fang to a striking position (as shown). **B.** sketches of the two halves of the lower jaw, joined by an elastic ligament only, and the alternating forward movement of each half during ingestion. Numbers indicate sequence of movements of jaw halves. **C.** The three fang positions in venomous snakes. 1, fangs are fixed rigidly at rear of elongated maxillae; 2, fangs are fixed rigidly on front rim of maxillae; 3, a maxilla and its fang is hinged and can be rotated forward and backward (through muscle referred in **A**). **Photo:** the whole skeleton. Note the numerous vertebrae and the absence of a sternum.

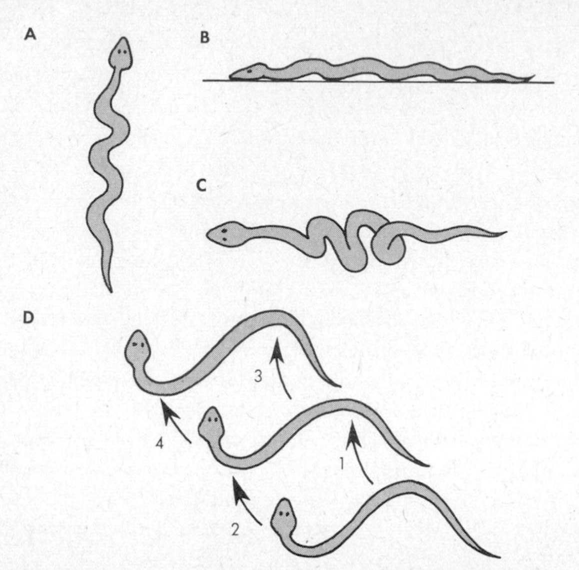

32.13. Locomotion patterns in snakes. Diagrams: A, lateral undulations; **B,** straight-line rippling; **C,** accordionlike folding; **D,** side-winding (successive positions are illustrated). **Photo:** a sidewinder rattler (*Crotalus cerastes*) and its characteristic track in sand.

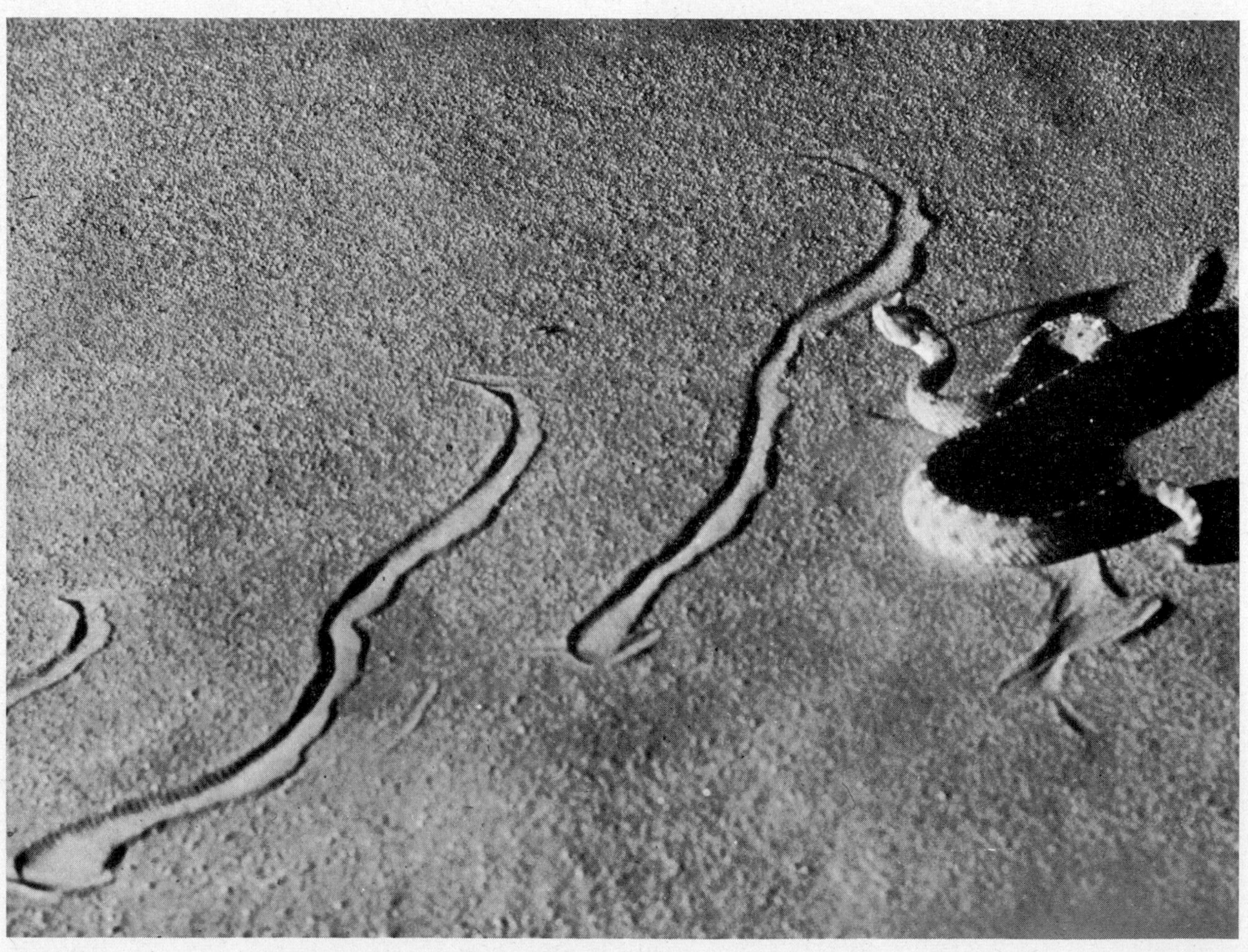

32.14. A python.

a wide variety of habitats, including trees, the ground, caves, and fresh and salt water (Fig. 32.14).

Members of the order Crocodilia are the closest living relatives of dinosaurs on the one hand and of birds on the other. Also, these reptiles exhibit three characteristics which, in different form and for different adaptive reasons, occur also in mammals: a four-chambered heart with fully separated ventricles; a palate extended far back, separating the nasal passages from the mouth cavity right to the pharynx (Fig. 32.15); and a peritoneal but nonmuscular septum which separates the chest cavity completely from the abdominal cavity. These traits have evolved independently in crocodilians as specific adaptations to a secondarily aquatic existence; all three improve breathing efficiency, especially when the mouth is submerged or full with food or both. Other adaptations to aquatic life include closable nostrils; recessed eardrums which can be covered by a flap of skin under water; webbed feet; and a powerful, laterally flattened tail which serves in swimming as well as in offense and defense.

Crocodiles and gavials have narrow jaws, and the fourth pair of teeth on the lower jaw remains exposed when the mouth is closed. Alligator and caiman jaws are broader and rounded anteriorly, and none of the teeth is exposed after mouth closure. The skin is armored with epidermal horny scales, reinforced along the back by rows of underlying dermal bony plates (Fig. 32.16). Crocodilians are strictly carnivorous and all are oviparous. Gavials live primarily in India, caimans in South America, and crocodiles and alligators are distributed more widely in tropical and semitropical regions (cf. also Chap. 16 and Fig. 16.11). The largest members of the order are the American crocodiles and alligators, which attain lengths of about 20 ft, and a marine South Pacific crocodile (*C. porosus*), which can be 25 ft long. The whole order includes only about two dozen species.

Sphenodon punctatus, the tuatara of New Zealand, is the

32.15. Characteristics of Crocodilia. A, some of the adaptations in head structure to a semiaquatic existence; **B,** identifying differences in head structure among Crocodilia.

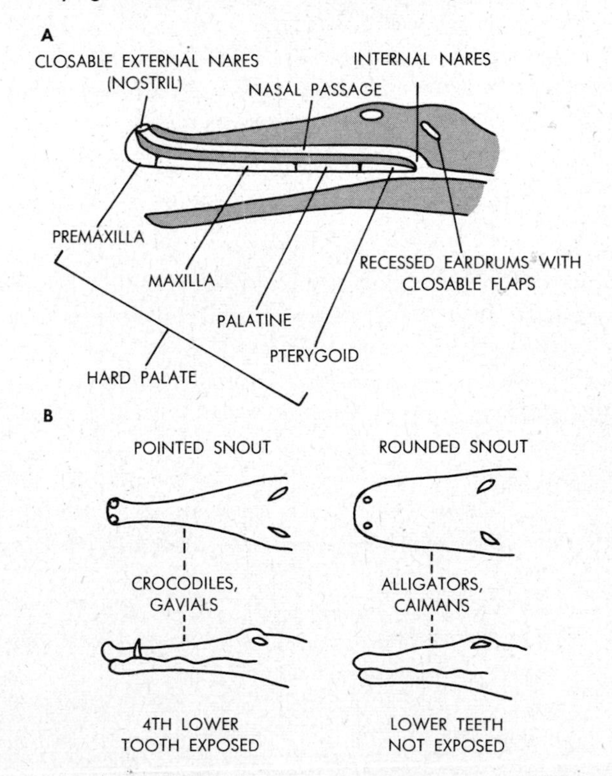

32.16. A young caiman.

sole surviving species of the order Rhynchocephalia. The animal is generally lizardlike and it possesses a functional pineal eye (Fig. 32.17). This reptile is a good example of a living fossil, but in a sense all living reptiles (lizards and snakes perhaps excepted) can be considered as such.

CLASS AVES

Birds; skin with feathers; forelimbs wings, with three fingers in hand; hind limbs legs, each with four or fewer toes; living types with horny beak, teeth absent; heart four-chambered, single aortic arch on right; lungs with extended air pouches; syrinx at base of trachea; 12 pairs of cranial nerves; pelvis fused to sacrum, pubic bones typically not fused ventrally; external ear lobes, external genitalia, and urinary bladder absent; urine semisolid; homoiothermic; fertilization internal; eggs amniote; oviparous.

†Subclass Archaeornithes (early birds): extinct; with thecodont teeth; three fingers not fused, clawed; tail long. *Archeopteryx* (cf. Fig. 15.18).

Subclass Neornithes (modern birds): fingers fused; tail shortened.

†Superorder Odontognathae: extinct; with teeth. *Hesperornis*

†Superorder Ichthyornithes: extinct; without teeth, billed. *Ichthyornis*

Superorder Palaeognathae (walking birds, or "ratites"): without teeth; generally flightless, sternum typically without keel.

Order Struthioniformes (ostriches): two-toed; pubic bones fused ventrally; feathers without aftershafts; males incubate eggs; omnivorous. *Struthio*, largest living bird, 200 lb, 7 ft tall.

Order Casuariiformes (emus, cassowaris): wings vestigial; feathers with long aftershafts; three-toed; males incubate eggs; 5 ft tall. *Dromaeus*, emus; *Casuarius*, cassowaris.

Order Apterygiformes (kiwis): long-billed, nostrils at tip; wings vestigial; feathers without aftershafts; four-toed; nocturnal, omnivorous. *Apteryx*, in New Zealand.

†Order Aepyornithiformes ("rocs," elephant birds): wings rudimentary. *Aepyornis*, in Madagascar, extinct for a few centuries only, the largest birds (900 lb; 10 ft tall; eggs 13 in., 20 lb, largest eggs of all animals).

†Order Diornithiformes (moas): feathers with long

32.17. *Sphenodon*, the lizardlike tuatara of New Zealand, sole surviving member of the reptilian order Rhynchocephalia.

aftershafts; four toed; *Diornis*, extinct for a few centuries only; 8 ft tall.

Order Rheiformes (rheas): three-toed; feathers without aftershafts; males incubate eggs. *Rhea*, 4 ft tall.

Order Tinamiformes (tinamous): weak fliers, keel present; good runners; males incubate eggs. *Tinamus*

Superorder Neognathae (flying birds): without teeth; generally capable of flight, sternum typically with keel.

Order Sphenisciformes (penguins): flying in water, not air, forelimbs are water wings; four-toed, feet webbed; feathers scaly; marine. *Aptenodytes*, emperor penguins; *Spheniscus*, Galapagos penguins; *Pygoscelis*, Adelie penguins.

Order Procellariformes (petrels, fulmars, albatrosses): wings narrow, long; excellent fliers, poor walkers; feet webbed, hind toes absent; nostrils with tubular openings; marine. *Fulmarus*, *Oceanodroma*

Order Gaviiformes (loons): diving birds; bill pointed, large; legs short, set far back; toes webbed; mostly marine. *Gavia*

Order Podicipitiformes (grebes): diving birds, freshwater; toes lobed; tail vestigial. *Podiceps*

Order Pelecaniformes (gannets, frigate birds, cormorants, boobies, pelicans): nostrils reduced or absent; throat pouch present; all four toes in foot web; marine and freshwater. *Sula*, *Morus*, gannets, boobies; *Pelecanus*, pelicans; *Phalacrocorax*, cormorants.

Order Ciconiiformes (herons, storks, ibises, spoonbills, flamingos): wading birds, long-necked, long-legged; feet webbed in some; aquatic or semiaquatic, near marshes. *Ciconia*, storks; *Ardea*, herons; *Casmerodius*, egrets; *Phoenicopterus*, flamingos.

Order Anseriformes (swans, ducks, geese): water fowl; broadbilled, with hard cap at end; feet webbed; tail short. *Anas*, mallard ducks; *Aythya*, canvasbacks; *Somateria*, eider ducks; *Anser*, common geese; *Branta*, Canada geese; *Cygnus*, swans.

Order Falconiformes (eagles, vultures, hawks, falcons, kites, ospreys): bill hooked, sharp-edged; grasping feet with talons; predaceous. *Falco*, falcons; *Buteo*, hawks; *Neophron*, *Coragyps*, vultures; *Vultur*, *Gymnogyps*, condors; *Cathartes*, buzzards; *Aquila*, *Haliaetus*, eagles (*H. leucocephalus* is bald eagle, on U.S. official seal).

Order Galliformes (chickens, turkeys, quail, grouse): game birds; poor fliers; feathers with aftershafts; feet for walking, scratching. *Gallus*, chickens; *Meleagris*, turkeys; *Opisthocomus*, hoatzins; *Pavo*, peacocks; *Colinus*, quails; *Lagopus*, ptarmigans; *Perdix*, partridges; *Phasianus*, pheasants; *Bonasa*, grouse.

Order Gruiformes (cranes, coots): long-necked; feathers with aftershafts; marsh birds. *Grus*, cranes; *Fulica*, mud hens, coots; *Rallus*, rails, some flightless.

Order Charadriiformes (gulls, terns, snipes, sandpipers, plovers): wading shore birds, with webbed feet, dense plumage. *Larus*, gulls; *Gallinago*, *Capella*, snipes; *Uria*, murres; *Fratercula*, puffins; *Pluvialis*, plovers; *Sterna*, terns; *Erolia*, *Calidris*, sandpipers; *Plautus*, little auks; *Pinguinus*, flightless great auks, extinct for 100 years.

Order Columbiformes (pigeons): with large crop secreting "pigeon milk" for young. *Columba*, common pigeons; *Goura*, crowned pigeons; *Zenaidura*, doves; *Ectopistes*, passenger pigeons, extinct since 1900; *Raphus*, dodo, extinct since seventeenth century.

Order Cuculiformes (cuckoos, turacos): two toes in front, two in back, outer back toe reversible; tail long. *Cuculus*, *Coccyzus*, cuckoos; *Centropus*, coucals.

Order Psittaciformes (parrots): two toes in front, two in back, outer back toe not reversible; grasping feet; beak hooked, sharp-edged, upper beak hinged on facial bones; plumage brilliantly colored. *Ara*, macaws; *Myopsitta*, parakeets; *Agapornis*, love birds; *Psittacus*, parrots.

Order Strigiformes (owls): head large, eyes directed forward; ear openings often with long feathers; beak short, hooked; grasping feet with sharp claws; predaceous, nocturnal. *Tyto*, barn owls; *Bubo*, horned owls; *Otus*, screech owls; *Nyctea*, snowy owls.

Order Caprimulgiformes (nighthawks, whippoorwills): mouth wide, edged with bristly feathers;

beak small, legs short; insect-eating, nocturnal. *Caprimulgus*, nightjars; *Chordeiles*, nighthawks; *Antrostomus*, whippoorwills.

Order Apodiformes (hummingbirds, swifts): wings pointed, most perfectly aerial birds. *Archilochus*, *Trochilus*, *Mellisuga*, hummingbirds, with tubular beak and tongue (lastnamed genus includes smallest of all birds); *Apus*, *Chaetura*, swifts, small-beaked, insectivorous.

Order Coliiformes (colies, mousebirds): beaks used in climbing; tail long; all four toes can be directed forward; arboreal, fruit-eating. *Colius*

Order Trogoniformes (trogons): most brilliantly plumed birds; bill short; feet small, first and second toes directed backward; tropical. *Pharomacrus*, quetzals; *Trogon*, trogons.

Order Coraciiformes (kingfishers, hornbills, motmots, ground rollers, bee-eaters): brilliantly plumed; third and fourth toes fused basally; mostly tropical. *Alcedo*, *Megaceryle*, kingfishers.

Order Piciformes (woodpeckers, toucans): beak long, pointed; tongue protrusible; tail feathers pointed; mainly insectivorous. *Sphyrapicus*, sapsuckers; *Picus*, *Dendrocopus*, *Melanerpes*, woodpeckers.

Order Passeriformes (perching birds): four-toed, three toes in front; order encompasses over half of all bird species, including all song birds; familiar representatives are: crows, larks, jays, magpies, nuthatches, swallows, nightingales, chicadees, titmice, wrens, mockingbirds, robins, bluebirds, thrushes, pipits, shrikes, starlings, sparrows, warblers, orioles, blackbirds, tanagers, finches, grosbeaks.

Because flying requires a structural design which cannot deviate too greatly from fixed aerodynamic specifications, birds are more like one another than the members of most other animal groups. Flying birds have sizes within a fairly narrow range, neither overly large and heavy nor overly small and weak; it is only in the flightless birds that very large sizes are in evidence. In consequence of the weight limitation, flying birds cannot store up much food reserve in the form of body fat, hence they must eat more or less continuously. Also, a considerable total quantity of food is required to provide the energy, the high operating temperature, and the high metabolic level generally essential for efficient flight. Indeed, birds are homoiothermic, with a constant temperature at about 41°C, higher than in mammals.

Feathers are the important heat regulators and also the means of flight. But note that different sets of feathers serve these two different functions. All types of feathers are horny outgrowths from the epidermis (Fig. 32.18). Flight feathers, also called *contour* feathers, are located on the wings and usually also on the tail. Each consists of a long *shaft* with primary branches (*barbs*), secondary branches (*barbules*), and tertiary hooklike branches (*barbicels*). These parts are interlocked into a closely woven surface, and orderly series of such feathers are attached to the forearms and hands. Muscles in the arm maintain an overlap between adjacent feathers when the wing moves down; and during the upstroke the feathers are canted like Venetian blinds, letting air pass through with but little drag. Heat-regulating body feathers and *down* feathers are shorter, with long, flexible, and not closely woven barbs. In many birds, moreover, an *aftershaft* with barbs is joined to the base of the primary shaft of such feathers. Down feathers provide a light mat which efficiently retains a layer of insulating dead air between the skin and the environment. The same smooth muscles in the skin which produce "gooseflesh" in man may erect the down feathers of birds and permit cooling of the body. Several structurally modified feather types, adapted to special functions, occur in a number of bird groups.

Birds typically possess a *uropygial gland* dorsally at the base of the tail. With their bills the animals spread the oil secreted by this gland over the feathers during preening; the oil keeps the plumage water-repellent and pliable. Like mammals, birds molt their skin cover. This is a gradual process, enough feathers usually being left at any given time to maintain flying and thermoregulator capacity.

The skeleton of birds is thoroughly flight-adapted (Fig. 32.19). Bones are light and delicate, with a minimum of spongy bone in the interior and correspondingly larger free spaces. Skull bones are fused, and the whole head is rounded and streamlined. Jaw bones are extended into a toothless bill or beak and are covered with a layer of epidermal horn. The eyes are rimmed with a circlet of *sclerotic plates*, which minimize the effect of wind pressure on the eyes during flight. As in reptiles, the eye can also be covered by a transparent *nictitating membrane*, drawn over the cornea from the outer corner of the eye. Eardrums are recessed deeply, and the bony canal leading to them is usually without external lobes or other

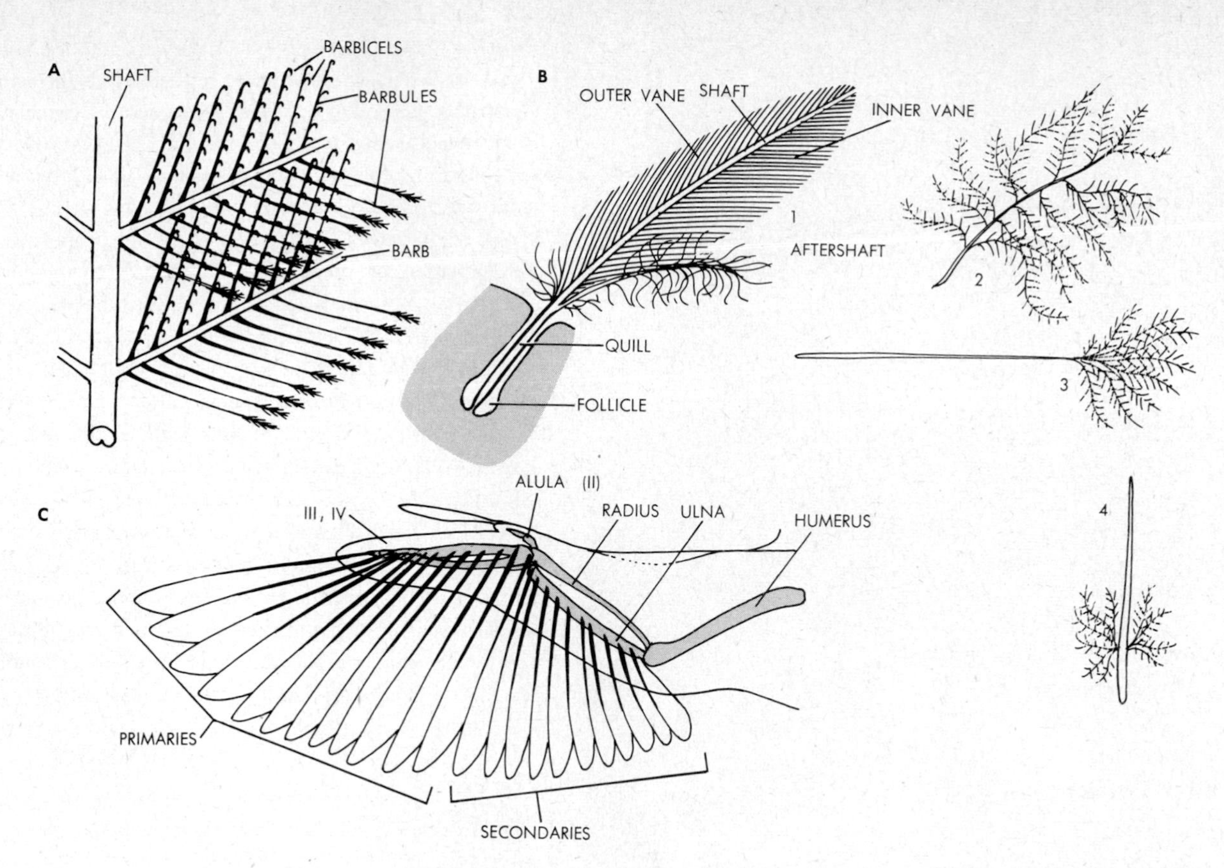

32.18. Feathers. **A,** diagram of part of a contour feather showing interlocking of the barbicels of adjacent barbules. (*After Mascha.*) **B.** 1, Whole view of contour feather, with aftershaft and pattern of insertion in skin; 2, down feather; 3, filoplume; 4, bristle. **C.** Arm skeleton with position of flight feathers superimposed. Roman numerals refer to digits of hand. (*After Pycraft.*)

projections which might disrupt smooth air flow over the head. The neck vertebrae provide extreme head mobility, but these vertebrae lock to one another firmly during flight. Flying birds possess a large sternum with a prominent keel, the latter serving as the attachment surface for the powerful flight muscles. Of these, the *pectoralis major* connects to the underside of the humerus and pulls the wing down. The *pectoralis minor* loops to the upper surface of the humerus, through a canal formed by the bones of the pectoral girdle near the arm socket (Fig. 32.20). Three digits remain in the hand, a small anterior and separate *alula,* and two posterior digits which are elongated and fused to one another (cf. Fig. 32.18). These three digits correspond to the index, middle, and ring fingers of man. The alula is lifted up during landing maneuvers, and it disrupts the smooth, lift-producing air stream over the wing at that time.

The pelvic girdle is fused to the sacral vertebrae and, except in ostriches, the pubic bones are not joined midventrally but leave a wide open passage (for the large eggs). This "bird-hipped" condition had already existed also in one group of archisaurian dinosaurs. Legs are typically four-toed, one toe being directed backward in most cases. In perching birds, systems of tendons permit the toes to curl tightly over a tree branch. Body weight alone ensures maintenance of a tight hold, and the energy expenditure and mental concentration required for perching are minimal (cf. Fig. 32.20). The tail vertebrae are reduced in number and the tail feathers function as a rudder.

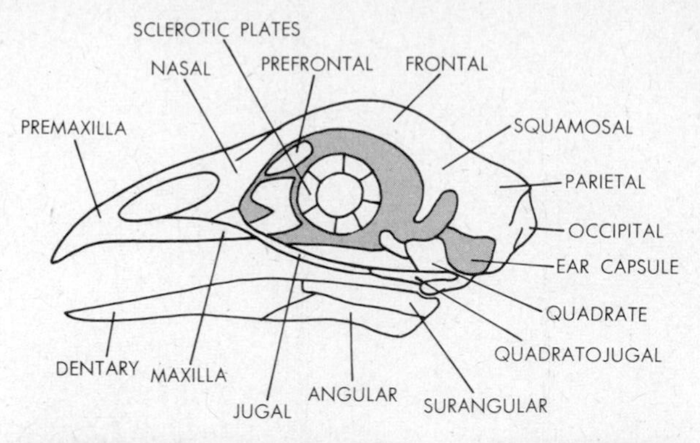

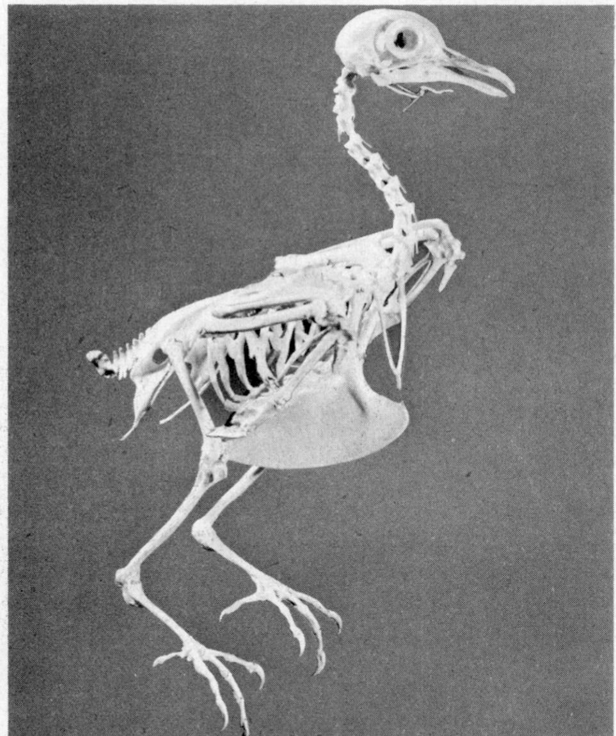

32.19. The skeleton of birds. Diagram: the structure of the head, and the ring of sclerotic bones around the eye. Note that many of the skull bones are fused, and that the labels refer to the general area where the individual bones had developed originally. Note also that almost the whole upper beak consists of the premaxilla. Photo: the whole skeleton of a pigeon.

A remarkable breathing system provides the large quantities of oxygen needed in flying (Fig. 32.21). Lungs are actually rather small, but from them are pouched out several large air sacs which occupy much of the space between the internal organs. Some of the sacs extend forward into the neck, where they may play an added role in distending the neck during courtship and other activities. In some birds one sac on each side even passes into the internal cavity of the humerus, through an opening near the base of that bone. These sacs not only provide buoyancy but they also permit air to flow *past* the breathing surfaces of the lungs proper. As a result, oxygenation of pulmonary blood may occur during both inhalation and exhalation. The heart is completely four-chambered, as in mammals, and the left aortic arch is absent in the adult bird (cf. Fig. 9.30).

Excretion and gamete production occur as in reptiles, but note that the reproductive pattern too is adjusted to the condition of flying. First, external genitalia are absent (except in the flightless ostriches, in which the males possess a penis). This lack, perhaps another adaptation toward maintenance of smooth body contours, necessitates strong cooperation during mating, for the cloaca of the male must be apposed precisely against that of the female. Such cooperation in turn requires behavioral adaptations in the form of more or less elaborate courtship activities. The latter in their own turn

32.20. Flying and perching. A. The arrangement of the flight muscles in relation to bones of shoulder girdle and arm. The pectoralis minor pulls the arm up, the pectoralis major pulls it down. **B.** The tendon system in the foot does not stretch. When the leg is straight, as in walking, the tendon permits the toes to move readily. But when the leg and the toes are bent, as in perching, the tendon becomes taut as a result of the flexed heel and holds the toes tightly curled around the perching branch. (*Based on Wolcott.*)

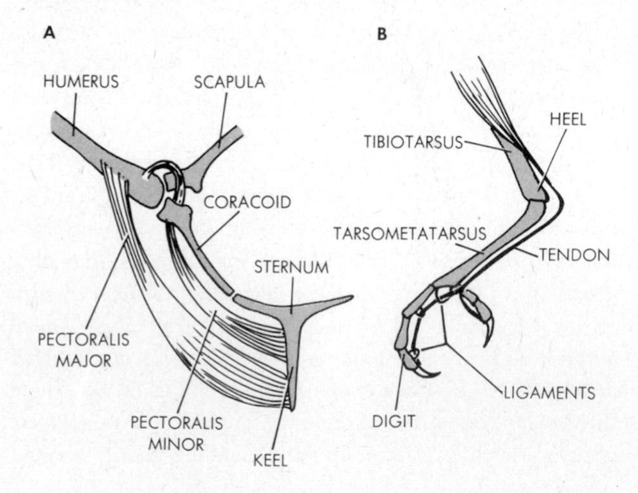

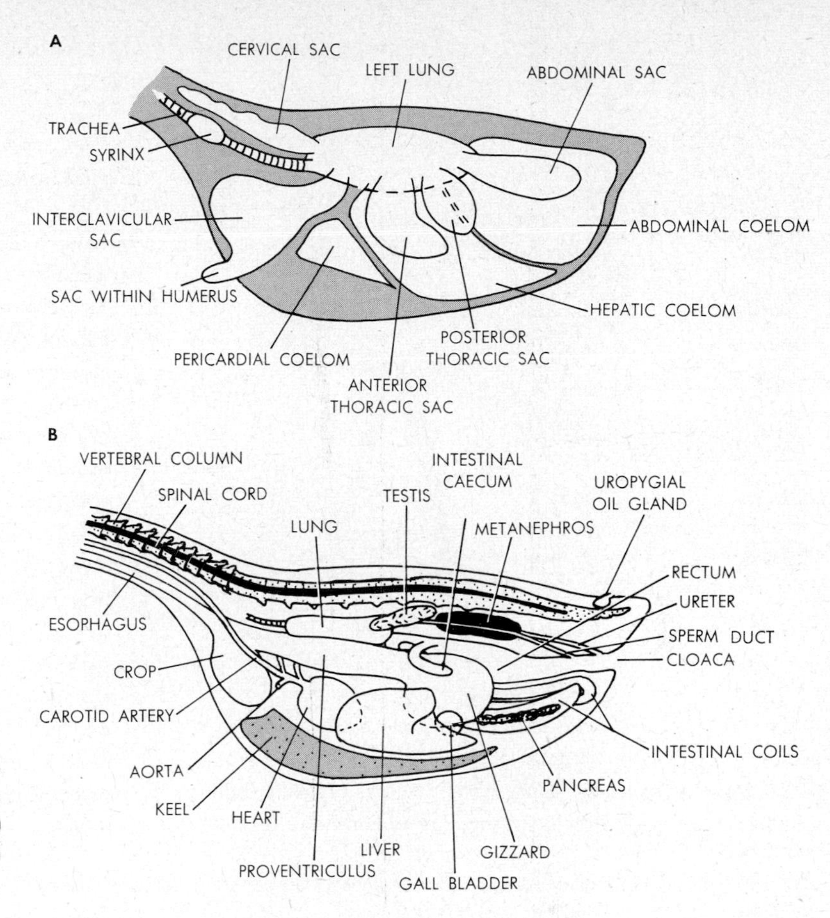

32.21. The internal structure of birds. A. Position of lung, air sacs, and coelomic spaces. (*Adapted from Goodrich.*) Note that five (paired) air sacs pouch out from the lungs. **B.** The position of some of the internal organs.

are correlated with the evolution of colorful display plumage and the persistence of the color vision inherited from reptiles. Thus, it is possible that birds see color at least partly because the males are without a penis.

Furthermore, oviparity becomes adaptively most advantageous, for an internally carried offspring would increase the flying weight unduly. But since the young are flightless at first and cannot be made too fat, they must be fed rather continually and in small doses, essentially like the parents themselves. Consequently, the young must be cared for by the parents for extended posthatching periods. Nesting behavior is one adaptive solution here; another is the presence of a storage crop in the alimentary tract, from which food can be dispensed intermittently. We note that the whole biological nature of birds, structural, functional, and behavioral, is subordinated to and explainable in terms of the basic activity of flying. Such subordination clearly has proved to be exceedingly worthwhile, for, judged by their evolutionary diversity and species number, birds today are the most successful terrestrial vertebrates.

The very earliest birds already flew and had feathers; flightless birds living today are descendants of flying ancestors. The superorder of flightless *ratites* includes several different and independent lines of evolution, and these birds probably arose through the mechanism of neoteny; the flightless condition of the young has been retained into the adult stage. Such birds therefore are not primitive (and they do not exemplify an early flightless stage of bird evolution, as had been supposed formerly). Ratites tend to large sizes. Their wings are generally vestigial, a keel is absent, but their legs are long and strong and the animals are good runners. Interestingly, the eggs of ratites typically are incubated by the males (Fig. 32.22).

Flying birds have adapted to a very wide range of habitats and modes of existence. They are among the best known and most familiar of all animals, and the domesticated groups among them are of considerable economic importance. More

32.22. An emu, a flightless ratite bird of the order Casuariiformes.

than 5,000 species, over half of the whole class, belong to a single order, the passerine, or perching birds. This group includes the so-called song birds, which typically build elaborate nests and in which song and brilliant mating plumage are usually characteristic of the males only (Figs. 32.23 and 32.24).

CLASS MAMMALIA

Mammals; skin with hair and various types of glands; teeth thecodont; seven neck vertebrae; ears with three middle-ear bones; limbs typically five-toed, with various terminal differentiations; heart four-chambered, single aortic arch on left; red corpuscles nonnucleated; coelom divided by muscular diaphragm; larynx at upper end of trachea; pelvis fused to sacrum, pubic bones fused ventrally; with urinary bladder, urine liquid; 12 pairs of cranial nerves, brain comparatively elaborate; homoiothermic; external genitalia present; fertilization internal, eggs amniote; some oviparous, mostly viviparous; young nourished by milk from mammary glands.

Subclass Prototheria (oviparous mammals): teeth present or absent in young, adults with horny bill; ears without external lobes; testes in abdomen; nipples absent, numerous ducts of mammary glands open individually; with cloaca; without uterus or vagina; egg-laying, eggs with pliable shells.

32.23. **A,** shoebill stork (order Ciconiiformes); **B,** peacock (order Galliformes).

A

B

32.24. **A,** hornbill (order Coraciiformes); **B,** Australian lyre bird (order Passeriformes).

Order Monotremata (monotremes): cervical vertebrae with ribs; in females, urogenital canal leads into cloaca; in males, urogenital canal forks terminally into separate urinary canal to cloaca, genital duct through penis. *Ornithorhynchus,* duck-billed platypus, aquatic, in burrows, with webbed feet, clawed toes, flattened tail, eggs in nest within burrow; *Tachyglossus, Zaglossus,* spiny anteaters (*echidnas*), terrestrial, noctural, with tubular beak, protrusible tongue, coarse hair and spines, clawed toes (Fig. 32.25).

Subclass Theria (viviparous mammals): teeth in young and adult; ears with external lobes; testes in abdomen or scrotum; nipples present; with or without cloaca; with uterus and vagina; young born in immature or mature condition.

Infraclass Metatheria (pouched mammals): cloaca present, shallow; ureters enter bladder, urethra from bladder opens into urogenital canal leading into cloaca in both sexes; penis protruded through cloaca, carries both urine and sperms; young born in immature condition.

Order Marsupialia (marsupials): nipples in ventral abdominal pouch, or *marsupium;* immature young complete development in pouch, attached to nipples; uterus and vagina double, vaginae opening separately into urogenital canal; penis often forked terminally. *Didelphis,* opossums; *Macropus,* kangaroos; *Vombatus,* wombats; *Phascolarctus,* koala bears; *Thylacinus,* Tasmanian wolves; *Sarcophilus,* Tasmanian devils (Fig. 32.26).

Infraclass Eutheria (placental mammals): cloaca absent; ureters enter bladder; in male, urethra from bladder and sperm ducts join into single duct through penis opening at urogenital orifice; in females, urethra and vagina open at separate urinary and reproductive orifices (cf. Fig. 10.22); uterus double or single, vagina always single; young develop in uterus throughout, attached and nourished via chorionic placenta, born in mature condition.

Order Insectivora (shrews, moles, hedgehogs): snout long: clawed toes; arboreal, terrestrial, subterranean (cf. Chap. 15). *Sorex*, shrews; *Scalopus*, moles; *Erinaceus*, hedgehogs.

Order Dermoptera (colugos): aerial gliding; skin web between limbs and tail form parachute. *Galeopithecus*, "flying lemurs."

Order Chiroptera (bats): forelimbs and four digits elongated, support flying web with hind limbs (and with tail in some); first digit of forelimbs and all of hind limbs clawed; true flight; nocturnal. *Desmodus*, vampire bats; *Pteropus*, fruit bats, "flying foxes" (include largest bats); *Myotis*, whiskered, brown bats.

Order Edentata (sloths, anteaters, armadillos):

32.25. Monotremes (Prototheria), the oviparous mammals. Photos: A, *Ornithorhynchus,* the duck-billed platypus; **B,** nest and eggs of a platypus; **C,** *Tachyglossus,* the echidna or spiny anteater. **Diagrams:** the urogenital ducts of monotremes. **A,** male; **B,** female. (*After Ihle.*)

A

B

C

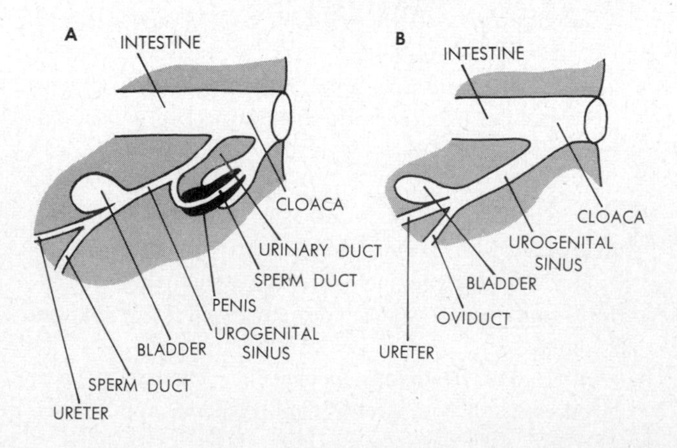

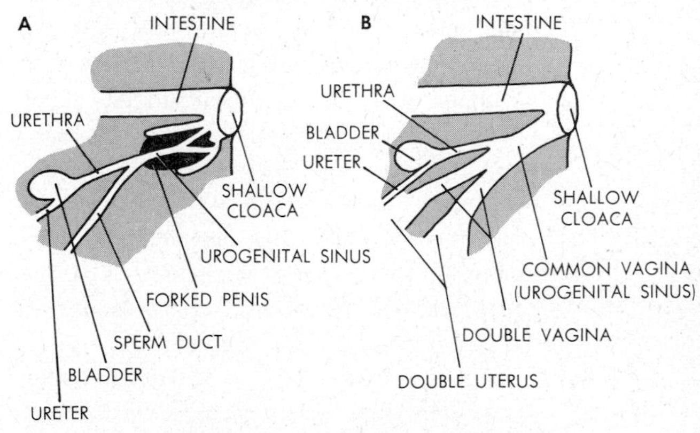

32.26. Marsupials (Metatheria), the pouched mammals. Photos: A, Australian cuscus, a member of the opossum family; **B,** crest-tailed marsupial mouse; **C,** Tasmanian devil. **Diagrams:** the urogenital ducts of marsupials. **A,** male; **B,** female.

teeth absent or only molars, without enamel. *Bradypus,* three-toed sloths, claws long, arboreal, suspended from branches; *Myrmecophagus,* anteaters, with tubular snout, protrusible long tongue; *Dasypus,* armadillos, with nonoverlapping epidermal horny plates.

Order Pholidota (scaly anteaters): teeth absent; tongue elongate; with overlapping epidermal horny plates. *Manis,* pangolins.

Order Primates (primates): cf. Chap. 15 and Table 6, Chap. 7.

Suborder Lemuroidea: *Lemur*

Suborder Tarsioidea: *Tarsius*
Suborder Anthropoidea
Superfamily Ceboidea: *Cebus,* capuchins; *Hapale,* marmosets.
Superfamily Cercopithecoidea: *Macacus,* rhesus; *Cynocephalus,* baboons.
Superfamily Hominoidea
Family Pongidae (apes): *Hylobates,* gibbons; *Pongo,* orangutangs; *Pan,* chimpanzees; *Gorilla*
Family Hominidae (men): *Homo*

Order Lagomorpha (rabbits): incisors without roots, growing continuously, two pairs in upper jaw, one pair in lower, enamel on front and back surfaces; canines absent; jaw motion laterally only; elbow nonrotatable. *Lepus,* hares, rabbits; *Sylvilagus,* cottontails; *Ochotona,* conies.

Order Rodentia (rodents): incisors one pair per jaw, without roots, growing continuously, enamel on front surface only; canines absent; jaw motion back and forth as well as laterally; elbow rotatable; toes clawed; most numerous mammalian order. *Mus, Peromyscus,* mice; *Rattus,* rats; *Castor,* beavers; *Cavia,* guinea pigs; *Hystrix,* porcupines; *Tamias,* chipmunks; *Sciurus,* squirrels; *Geomys,* gophers.

Order Cetacea (whales, dolphins, porpoises): neck absent; forelimbs flippers, digits within; hindlimbs absent; transverse tail flukes, with median notch; teeth without enamel or absent; ears small, nostrils dorsal; hairs sparse, around mouth. Toothed whales: *Delphinus,* dolphins; *Phocaena,* porpoises; *Physeter,* sperm whales; *Orcinus,* killer whales; *Monodon,* narwhals. Whalebone whales: *Balaena,* right whales; *Balaenoptera,* blue (sulfulbottom) whales; *Megaptera,* humpback whales.

Order Carnivora (carnivores): canines long; teeth pointed (carnassial); clawed toes.
Suborder Fissipedia (toe-footed carnivores): *Canis,* dogs, wolves, jackals; *Vulpes,* foxes; *Procyon,* racoons; *Ailurus,* pandas; *Ursus,* bears; *Gulo,* wolverines; *Mustela,* minks, weasels, ferrets; *Lutra,* otters; *Mephitis,* skunks; *Meles,* badgers; *Herpestes,* mongoose; *Hyaena,* hyenas; *Vivera,* civets; *Lynx,* lynxes; *Felis,* all "cats," including house cats, lions, tigers, jaguars, ocelots, leopards, pumas, cheetahs.
Suborder Pinnipedia (fin-footed carnivores): *Odobenus,* walruses; *Phoca,* seals; *Eumetopias,* sea lions.

Order Tubulidentata (aardvarks): ears long; snout long, tubular, with protrusible tongue; teeth without enamel; strong claws; insect-eating. *Orycteropus*

Order Hyracoidea (coneys): forelimbs four-toed, hind limbs three-toed; ears and tail short. *Procavia*

Order Proboscoidea (elephants): skin thick, neck short, hairs sparse; nose and upper lip form proboscis; incisors are tusks; molars with transverse enamel ridges, one set of molars present at any time; feet with elastic bottom pads, toes with naillike small hooves. *Elephas,* Indian elephant; *Loxodonta,* African elephant.

Order Sirenia (sea cows, manatees): forelimbs are paddles; hindlimbs absent; transverse tail flukes without notch; snout blunt, lips fleshy; external ears absent; teeth with enamel; hairs sparse; coastal, estuarine, herbivorous. *Trichechus,* manatees; *Halicore,* dugongs; *Hydrodamalis,* sea cow extinct for 100 years.

Order Perissodactyla (odd-toed ungulates): one- or three-toed, toes terminating in hoofs. *Equus,* horses, asses, zebras; *Tapirus,* tapirs; *Rhinoceros*

Order Artiodactyla (even-toed ungulates): two- or four-toed, toes terminating in hoofs.
Suborder Bunodontia: nonruminant, stomach simple. *Sus,* pigs, boars; *Phacochoerus,* warthogs; *Pecari,* peccaries; *Hippopotamus*
Suborder Pecora: ruminant (cud-chewing), stomach four-chambered. *Camelus,* bactrian camel (two-humped), dromedary (1-humped); *Auchenia,* alpacas, llamas; *Tragulus,* chevrotains; *Rangifer,* reindeer, caribous; *Alces,* moose; *Cervus,* deer elk; *Giraffa,* giraffes; *Ocapia,* okapis; *Antilocapra,* pronghorns; *Bos,* cattle, yaks; *Gazella,* gazelles; *Taurotragus,* elands; *Aepyceros,* impala; *Bison,* buffalo; *Bubalus,* water buffalo; *Ovibos,* musk oxen; *Capra,* goats; *Ovis,* sheep.

The probable forces which oriented the evolution of reptiles into mammals have already been described in Chap. 15. The basic structural results of that evolutionary transition are outlined in substantial detail throughout the chapters of Unit 1. Chap. 15 also deals fully with the special evolutionary and organizational history of the order Primates, with particular attention to man. The present discussion may therefore be devoted to a brief survey of the adaptive characteristics of mammals as a whole.

Fossil evidence and the structure of living mammals indicate that the mammalian orders form not simply a single series of heterogeneous groups but at least six parallel series of groups (Fig. 32.27). The egg-laying monotremes represent one; the pouched marsupials represent another. And, in the tabulation of the living placental orders above, one series includes the first six orders listed, from insectivores to primates; a second series is formed by the rodents and lagomorphs; a third, by the cetaceans; and a forth, by all the remaining orders, from carnivores to artiodactylans. Each of these series appears to have followed its own evolutionary paths from a more or less central ancestral stock of mammals. The interesting point is that, in each of these series independently, the same five general adaptive trends are in evidence. All five relate more or less directly to feeding and locomotion.

First, body size has tended to increase, an advantageous change within limits, helpful both in searching for food and in avoiding becoming food. Secondly, the number of teeth has tended to decrease; and, instead of all teeth remaining alike as in the reptilian ancestors, the fewer teeth have become differentially specialized, in parallel with a specialization in the types of food eaten. Thirdly, the legs have tended to become longer and stronger, the body and often also the heels being lifted off the ground more and more. The result has been an evident improvement in locomotor efficiency and, indeed, a significant diversification in the types of locomotion, an essential correlate to the alimentary specializations. Fourthly, in conjunction with newly developed modes of locomotion, a radiation into diverse habitat zones has occurred. And lastly, partly as a result of the increase in body size, partly in conjunction with the improved motility, the size of the brain has increased; in the course of evolutionary time, more powerful intellects have simply been acquired by mammals of all kinds.

That these five adaptive trends have become elaborated independently in the different evolutionary series of mammals is exemplified best among the placentals. Thus, as already shown in Chap. 15, the evolutionary progression from insectivore to primate has indeed been accompanied by size increase; by brain increase; by a locomotor change from quadrupedal to bipedal methods and to flying; and by a habitat diversification from an original arboreal to later terrestrial, aerial, and subterranean environments (Fig. 32.28). With regard to teeth, a mole, for example, possesses the basic eutherian dentition, viz., 44 teeth, each quadrant of the mouth containing three incisors, one canine, four premolars, and three molars. These figures can be written as a tooth formula, $\frac{3\ 1\ 4\ 3}{3\ 1\ 4\ 3} = 44$, where the upper and lower rows denote the number of teeth in one-half of the upper and lower jaws, respectively, in sequence for incisors, canines, premolars, and molars. We may also note here that incisors are frontal, sharp-

32.27. The orders of mammals, arranged according to the six main evolutionary groups; cf. Fig. 15.21.

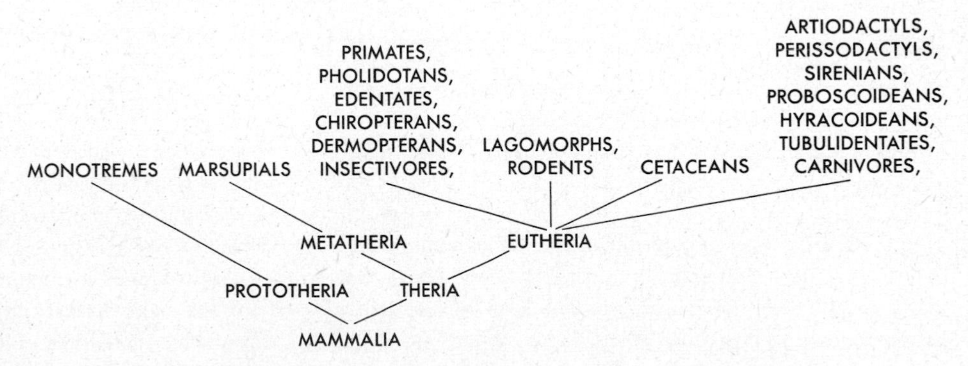

A

B

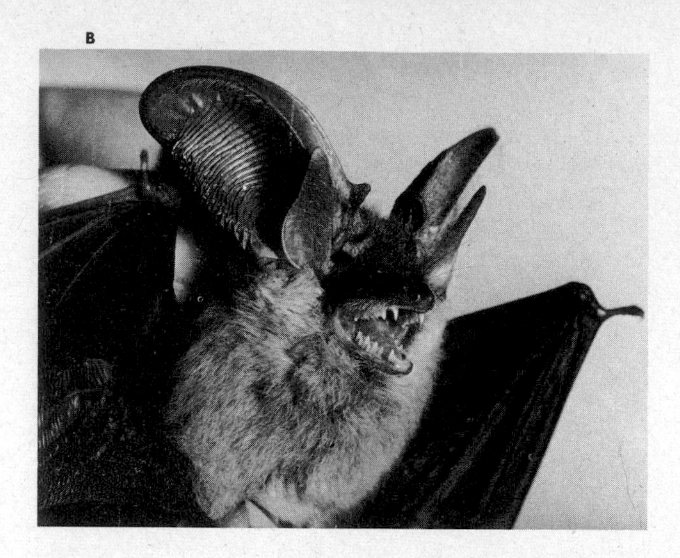

C

32.28. Representatives of the insectivore-primate series of mammals. **A.** Hedgehog (order Insectivora). **B.** *Plecotus* (order Chiroptera), a type with highly elaborated ears for improved phonoreception. **C.** An Indian pangolin, or scaly anteater (order Pholidota). Primates are illustrated in Chap. 15.

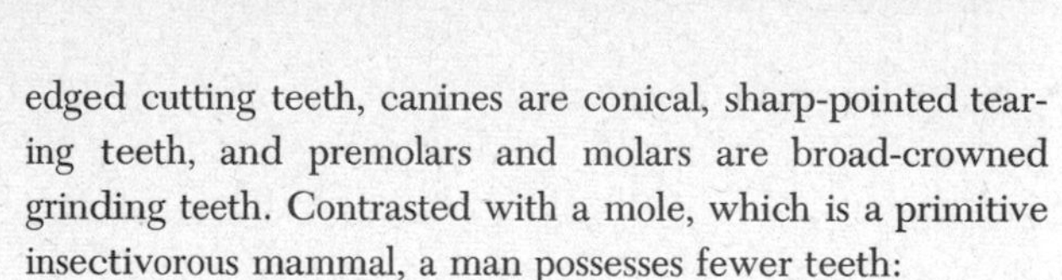

edged cutting teeth, canines are conical, sharp-pointed tearing teeth, and premolars and molars are broad-crowned grinding teeth. Contrasted with a mole, which is a primitive insectivorous mammal, a man possesses fewer teeth:

$$\frac{2\ 1\ 2\ 3}{2\ 1\ 2\ 3} = 32.$$

However, even a mole already has a reduced number of teeth as compared with a marsupial, for example. Thus, the tooth formula for an opossum, a fairly primitive pouched mammal, is $\frac{5\ 1\ 3\ 4}{4\ 1\ 3\ 4} = 50$ (Fig. 32.29).

In the second series of placental mammals, lagomorphs and rodents, adaptive change in brain size has not been particularly spectacular (though on a rodent level rats are fairly brainy animals), but the four other adaptive trends are amply evident. Fossils show clearly that these groups have become larger-bodied; lagomorphs exhibit a very obvious locomotor specialization, i.e. long hind legs adapted to leaping; and dental specialization involves not only a well-known modification of the incisors, but also a wide variation in tooth numbers (e.g., conies, $\frac{3\ 1\ 4\ 4}{3\ 1\ 4\ 4} = 48$; rabbits, $\frac{2\ 0\ 3\ 3}{2\ 0\ 3\ 3} = 28$; squirrels, $\frac{1\ 0\ 2\ 3}{1\ 0\ 1\ 3} = 22$; rats, $\frac{1\ 0\ 0\ 3}{1\ 0\ 0\ 3} = 16$). All these changes are correlated with a vast habitat diversification; rodents comprise nearly half the number of all mammalian species, and we already know that each species represents a distinct ecological niche (cf. Chap. 16). In the whale series, analogous trends are manifest. No group has increased in body size more dramatically or has altered its habitat and its locomotion more drastically. Tooth reduction has become maximal, i.e., vertical baleen plates replace teeth altogether in the whalebone whales. Also, the intelligence of the whale group generally is well known, porpoises particularly being noted for their mental prowess (Fig. 32.30).

Finally, in the carnivore-ungulate series, most of the groups have exploited the possibilities of a running life on the plains.

32.29. Mammalian teeth. **A.** The dentition of a dog in one half of the head. These teeth are carnassial, i.e., pointed, as in carnivores generally. **B.** The incisors of lagomorphs number four and two in upper and lower jaw respectively (1) and enamel covers both front and back surfaces (2). In rodents, two incisors are present in each jaw (3) and enamel covers only the front surfaces (4). **C.** In teeth with more than one root, as in premolars and molars (right), the pulp cavity passes through each root. Incisors, canines, and the anterior premolars are single-rooted (left, cf. Fig. 8.6).

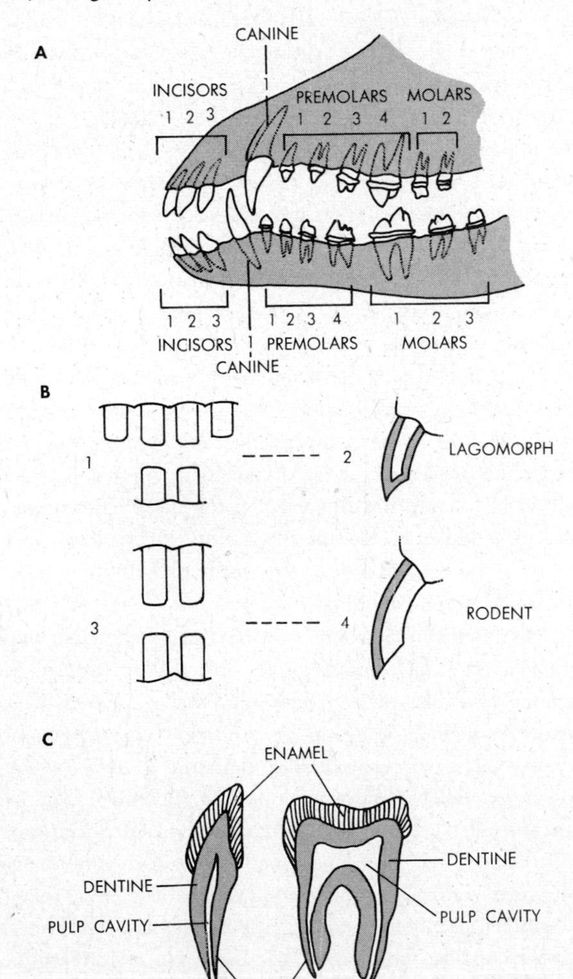

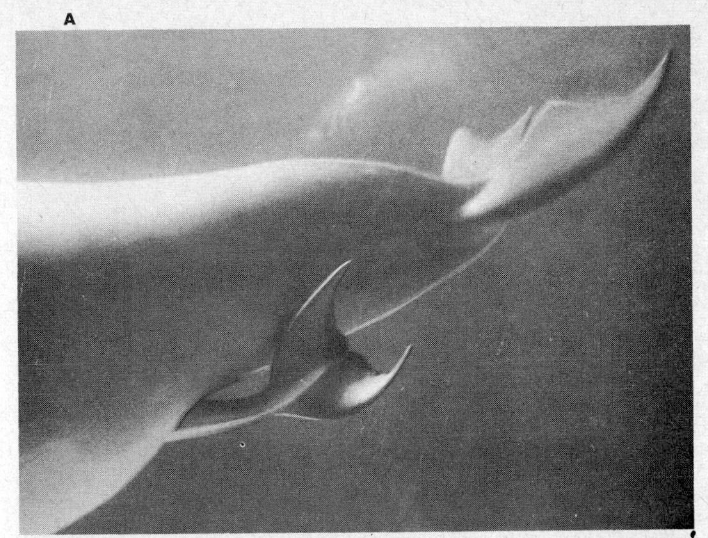

32.30. Some representatives of the whale series of mammals. **A.** Birth of a dolphin takes place tail first, for during this up to 2-hr long process the offspring still must obtain its oxygen supply from its mother and would drown if its head were in open water. As soon as the head is free, the offspring swims to the surface and begins breathing on its own. **B.** Dolphin female and young.

Leg and foot structure should therefore show major adaptive modifications, and that is the case (Fig. 32.31). In several subseries within the group, the heel has become lifted off the ground, the foot has become elongated, and the animal has come to walk on its toes, as in cats or gazelles, for example. Concurrently, toes have become increasingly *ungulate*, i.e., hoofed, and the number of toes has become reduced. Thus, carnivores are five-toed and still nonungulate; coneys exhibit a reduction in toe number; elephants have small naillike hoofs on each toe; and a fully ungulate condition with varying degrees of reduced toe numbers is in evidence in the horse group

32.31. Foot structure in the carnivore-ungulate group of mammals. Note claws and absence of hoofs in carnivores (**A**, cat); presence of small naillike hoofs in proboscoideans (**B**, elephant); and fully ungulate condition in perissodactylans (**C**, three-toed rhinoceros, **D**, one-toed horse) and artiodactylans (**E**, four-toed pig, **F**, two-toed cow). Note also that elephants, like men, are plantigrade, i.e., they walk on the whole foot, whereas all other mammals referred to in this illustration are digitigrade, i.e., they walk on the toes.

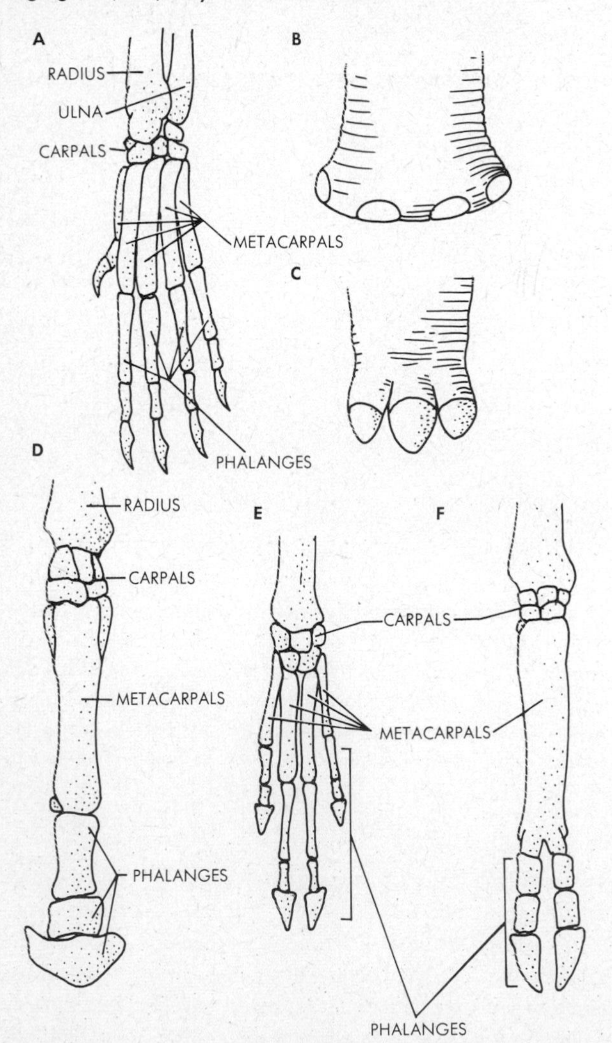

and the deer-cattle group. The dentition has undergone concomitant modifications. Although elephants exhibit a sharp reduction of tooth numbers, the other groups have changed less in that respect (e.g., the tooth formula for both the horse and the pig is the same as that for the mole, and the formula for the dog is quite similar, viz., $\frac{3\ 1\ 4\ 2}{3\ 1\ 4\ 3} = 42$). However, major changes have occurred in differential tooth specializations, in line with the different eating habits and the diversified ecological niches of these mammals. Thus, in carnivores all teeth are *carnassial*, i.e., pointed and adapted for tearing, and the canines form particularly long fangs. By contrast, the herbivorous ungulates have incisors adapted for cutting, molars adapted for grinding, and canines are often absent altogether. Body size has increased independently in the several subseries of the group; e.g., bears, horses, elephants, and bisons all have evolved from far smaller ancestors. And intelligence likewise has reached higher levels several times; bears, cats, dogs, wolverines, horses, and elephants all are well known as "smart" animals (Fig. 32.32).

Though similar adaptive trends characterize all these mammalian series, each series has had its own area of special success. The insectivore-primate group has come to excel in aerial and bipedal locomotion and in mental capacities; the rodent group, in habitat diversification; the whale group, in aquatic locomotion and body size; and the carnivore-ungulate group, in ground locomotion and speed. Note also that a good deal of convergent evolution has taken place. For example, aquatic types with flippers and finlike tails occur in the cetacean series as well as twice in the carnivore-ungulate series (pinnipeds, manatees); ant-eating types with long snouts and protrusible tongues have evolved in the insectivore-primate series (pholidotans, edentates), in the carnivore-ungulate series (aardvarks), and indeed also in the primitive egg-laying monotreme group (echidnas).

In this connection we may note that the pouched marsupial mammals duplicate almost all the adaptive trends of the placental mammals, in a remarkably parallel manner. Thus, marsupials include a group of arboreal types which are variously insectivorous, slothlike, or aerial gliders ("flying" opossums); molelike burrowing types; rodentlike pouched mice and rats; leaping harelike kangaroos; ant-eating types with tubular snouts and long tongues; and carnivorous types which are variously wolflike, doglike, catlike, and, though now extinct, also lionlike (cf. Fig. 32.26). Such parallel results offer perhaps the best proof that the same adaptive forces operate universally for all animals, and that, if the environ-

A

B

C

32.32. Representatives of the carnivore-ungulate series of mammals. **A,** sea lion, a pinniped carnivore; **B,** manatees, sirenian mammals; **C,** llamas, ruminant even-toed ungulates. See Chap. 18 for illustrations of other mammalian types.

mental opportunities are similar, remarkably similar animals may result several times independently. Furthermore, it should be clear that the primary adaptive forces which have shaped mammalian nature ultimately reduce to just two, feeding and locomotion, the very same which have also shaped the nature of all other animals since the first primaeval cell types appeared in the ocean.

Mammals, vertebrates, and chordates as a whole also illustrate a final point exceedingly well, namely, that significant evolutionary advance proceeds from the primitive type, not the specialized one. Thus, as these chapters have shown, chordates probably trace back to *primitive* pterobranchlike hemichordates, not to advanced forms such as acorn worms; vertebrates probably originated from the *primitive* sessile ascidians, not from the specialized pelagic tunicates; amphibia evolved from early *primitive* lungbreathing fishes, not from the later bony teleosts; reptiles have their ancestry among the *primitive* labyrinthodonts, not the later frogs or toads; mammals derive from the *primitive* thecodont reptiles, not any of the subsequent forms; primates evolved from *primitive* insectivorous stocks, not from modern mammals; and man arose from *primitive* primates, not from the later monkeys or apes. We note again that, in an adaptive radiation branching away from an ancestor, the evolutionary source of new types usually is to be found along the primitive branch lines, not along the lines which specialize more rapidly and more spectacularly.

REVIEW QUESTIONS

1. Give taxonomic definitions of amphibia and of each amphibian class. Name representative genera of each class. Which amphibian features and adaptations trace their origin directly to ancestral fishes? Which group of fishes was ancestral to amphibia?

2. Review the structural changes in the breathing and circulatory systems during amphibian metamorphosis. Review the structural organization of every amphibian organ system. In what respects are amphibia (*a*) water-adapted, (*b*) land-adapted? Characterize the nature of cecilians. Name urodeles possessing (*a*) adult gills only, (*b*) adult gills and lungs, (*c*) adult lungs only.

3. Name urodeles which are predominantly (*a*) aquatic, (*b*) terrestrial. In what respects are urodeles neotenous? Describe the skeletal structure of an anuran and show how it differs from that of a urodele. Distinguish between frogs and toads. What instances of parallel evolution are in evidence among anurans?

4. Define the class Reptilia and each of its living orders. Review the characteristics of the extinct groups. What adaptations to terrestrial life are in evidence among reptiles? Describe the structure of all organ systems of reptiles. Show which of the ancestral skull types are represented among reptiles today.

5. Characterize the skeletal structure of Chelonia. In what environments do the chelonians live and how do they mate? How are lizards different from snakes? Describe the structural and functional adaptations of snakes to their particular mode of life. How do snakes move? In which reptilian groups is viviparity encountered?

6. Describe the anatomical traits of crocodiles. Which of these traits are similar to those of mammals? How are crocodiles and alligators distinguished? How are reptiles and amphibia distinguished? How many species of each tetrapod class are known?

7. Define Aves taxonomically and distinguish between palaeognathous and neognathous birds. Name as many different orders of birds as you can and list representative types or genera of each. Describe the structure of feathers and discuss the role of feathers in heat regulation and flying. How is bird flight produced? Contrast with insect flight. What skeletal adaptations to flight are in evidence among birds?

8. Describe the organization of all organ systems of birds and show particularly which traits are adaptations to flying. Which behavioral, reproductive, and ecological traits are consequences of or adaptations to a flying life? What are ratite birds? How do birds manage to sleep on a tree branch without falling off? How do birds produce song and what is the function of song?

9. Define the class of mammals and each subclass and order. Name representative types or genera of each order. Review the different ways in which mammalian embryos and young depend structurally and functionally on the maternal parents.

10. Show which mammalian orders form various supergroups and describe the similar evolutionary trends in evidence in all these supergroups. What are the adaptive advantages of these trends? Indicate variations in (*a*) dentition, (*b*) foot structure, (*c*) locomotion, (*d*) food sources among mammals. Describe the structure and function of every organ system of a mammal, drawing on all information accumulated during the study of the whole book. Review the detailed taxonomic classification of man.

COLLATERAL READINGS

The following selections from the enormous literature on tetrapod vertebrates provide background reading to the topics dealt with in this chapter:

Audubon, J. J.: "The Birds of America," reprint, Macmillan, New York, 1937.

Barbour, T.: "Reptiles and Amphibians, Their Habits and Adaptations," 2d ed., Houghton-Mifflin, Boston, 1934.

Bourliere, F.: "The Natural History of Mammals," Knopf, New York, 1954.

Bradley, O. C.: "The Structure of the Fowl," Lippincott, Philadelphia, 1950.

Butler, R. A.: Curiosity in Monkeys, *Sci. American,* Feb., 1954.

Colbert, E. H.: "Evolution of the Vertebrates," Wiley, New York, 1955.

Ditmars, R. L.: "Reptiles of the World," Macmillan, New York, 1926.

———: "The Reptiles of North America," Doubleday, New York, 1936.

Frieden, L.: The Chemistry of Amphibian Metamorphosis, *Sci. American,* Nov., 1963.

Goodrich, E. S.: "Studies on the Structure and Development of Vertebrates," Macmillan, London, 1930.

Heilman, G.: "The Origin of Birds," Witherby, London, 1926.

Henderson, J., and E. L. Craig: "Economic Mammalogy," Charles C Thomas, Springfield, Ill., 1932.

Hyman, L. H.: "Comparative Vertebrate Anatomy," 2d ed., University of Chicago Press, Chicago, 1942.

Klauber, L. M.: "Rattlesnakes: Their Habits, Life History, and Influence on Mankind," University of California Press, Berkeley, Calif. 1956.

Kortlandt, A.: Chimpanzees in the Wild, *Sci. American,* May, 1962.

Lincoln, F. C.: "Migration of Birds," Doubleday, New York, 1952.

Noble, G. K.: "Biology of the Amphibia," McGraw-Hill, New York, 1931.

Oliver, J. A.: "The Natural History of North American Amphibia and Reptiles," Van Nostrand, New York, 1955.

Pope, C. H.: "Snakes Alive and How they Live," Viking, New York, 1939.

Romer, A. S.: "Man and the Vertebrates," The University of Chicago Press, Chicago, 1941.

———: "Vertebrate Paleontology," 2d ed., The University of Chicago Press, Chicago, 1945.

———: "The Vertebrate Body," Saunders, Philadelphia, 1950.

Stebbins, R. C.: "Amphibia and Reptiles of Western North America," McGraw-Hill, New York, 1954.

Sturkie, P. D.: "Avian Physiology," Cornell, Ithaca, N.Y., 1954.

Wallace, G. J: "An Introduction to Ornithology," Macmillan, New York, 1955.

Warden, C. J.: Animal Intelligence, *Sci. American,* June, 1951.

Young, J. Z.: "The Life of Vertebrates," 2d ed., Oxford, London, 1962.

glossary

SECTION A

Common Prefixes, Suffixes, and Anatomical Terms in Zoological Usage*

Note: The meaning of many terms not specifically listed in Part B, below, can be ascertained from Part A. For example, the text chapter on fishes contains the term *Actinopterygii.* The parts of this term are the prefix *actino-* and the suffix *-pterygii;* the list below indicates the meanings of these word parts as "ray" and "fin," respectively. Hence the whole term denotes "ray-finned." In general, a very large number of technical designations may be translated into English equivalents by separating the words into parts and consulting this first section of the glossary.

a- [Gr. not]: negates succeeding part of word: e.g., *acoel,* without coelom; *Acrania,* without head.

ab- [L. away, off]: opposite of *ad,* e.g., *aboral,* away from mouth: *abductor,* muscle which draws away from given position.

acro- (ăk′rō) [Gr. *akros,* outermost]: e.g., *acrodont, acromegaly, acrosome.*

actino- (ăk′tĭ·nō) [Gr. *aktis,* ray]: e.g., *actinopodial,* ray-footed; *actinotroch,* ray-wheel.

ad- [L. toward, to]: opposite of *ab-;* e.g., *adrenal,* at (near) the kidney; *adductor,* muscle which draws toward given position.

afferent (ăf′ẽr·ĕnt) [L. *ad* + *ferre,* to carry]: to lead or carry toward given position, opposite of efferent; e.g., afferent nerve, afferent blood vessel.

amphi- (ăm′fĭ) [Gr. on both sides]: e.g., *amphioxus,* pointed at both ends; *amphipod,* possessing both thoracic and abdominal types of legs.

*The system of indicating pronunciation is used by permission of the publishers of Webster's New Collegiate Dictionary. Copyright 1949, 1951, 1953, 1956, 1958 by G. C. Merriam Co.

an- [Gr. not]: like *a-,* used before vowel or "h"; e.g., *anamniote, anhydride.*

ana- [Gr. up, throughout, again, back]: e.g., *analogy,* likeness, resemblance; *analysis,* thorough separation; *anatomy,* cutting up.

andro- (ăn′drō) [Gr. *aner,* man, male]: e.g., *androgen,* male-producing hormone.

anterior, antero- (ăntĕr′ĭ·ẽr, ăn′tĕr·ō) [L. *ante,* before, in front of]: at, near, or toward front end.

antho- (ăn′thō) [Gr. *anthos,* flower]: e.g., *Anthozoa,* flower-like coelenterate animals.

anthropo- (ăn′thrō·pō) [Gr. *anthrōpos,* man, human]: e.g., *anthropoid,* manlike ape.

apical (ăp′ĭ·kăl) [L. *apex,* tip]: belonging to an apex, being at or near the tip.

apo- (ăp′ō) [Gr. away, off, from]: comp. to L. *ab-;* e.g., *apopyle,* far-lying opening.

-apsid (ăp′sĭd) [Gr. *apsis,* loop, arch]: e.g., *diapsid,* having two arches or windows in skull.

arch-, archeo- (ärch, är′kē·ō) [Gr. *archos,* chief]: first, main, earliest; e.g., *archenteron,* first embryonic gut; *archeocyte,* chief cell; *archiannelid,* primitive annelid.

arthro- (är′thrô) [Gr. *arthron,* joint]: e.g., *arthropod,* jointed-legged; *arthritis,* joint inflammation.

-aspid, aspido- (ăs′pĭd, ăs′pĭ·dô) [Gr. *aspis,* shield]: e.g., *cephalaspid,* head-armored; *anaspid,* without shield.

aster-, -aster (ăs′tẽr) [Gr. star]: e.g., *asteroid,* starshaped animals (starfish).

auto- (ô′tô) [Gr. same, self]: e.g., *autogamy,* self-fertilization; *autozooid,* "standard" animal type.

axo- (ăk′sô) [Gr. *axine,* axis]: pertaining to an axis; e.g., *axoneme,* axial filament.

bi- (bī) [L. *bis,* twice, double, two]: e.g., *bilateral; bicuspid,* having two points.

bio- (bī′ô) [Gr. *bios,* life]: pertaining to life; e.g., *biology, biomass, biosphere.*

-blast, blast-, blasto- (blăst, blăst′ô) [Gr. *blastos,* embryo]: pertaining to embryo; e.g., *blastocoel,* embryonic cavity; *mesoblast,* midlayer of embryo.

brachio- (brā′kĭ·ô) [L. *brachium,* arm]: e.g., *brachiopod,* arm-footed animal; *brachiolaria,* arm-possessing larva.

-branch, branchio- (brăng′kĭ·ô) [Gr. *branchia,* gills]: e.g., *lamellibranch,* having platelike gills; *branchiopod,* animal with gills on feet; *branchial* sac, breathing sac.

cardio- (kär′dĭ·ô) [Gr. *kardia,* heart]: e.g., *pericardial,* around the heart.

caudal (kô′dăl) [L. *caudo,* tail]: at, near, or toward the tail.

cephalo- (sĕf′á·lô) [Gr. *kephalē,* head]: e.g., *cephalothorax; cephalopod,* head-footed animal.

cerci, cercal, cerco- (sûr′kĭ, sûr′kăl, sûr·kô) [Gr. *kērkos,* tail]: pertaining to tail; e.g., *cercaria,* tailed fluke larva; *cercopithecoid,* tailed monkey; *anal cerci.*

cervical (sûr′vĭ·kăl) [L. *cervix,* neck]: at, near, or toward the neck region.

chaeto- (kē′tô) [Gr. *chaitē,* bristle, hair]: e.g., *chaetognath,* bristle-jawed animal.

chilo- (kī′lô) [Gr. *chilos,* lip]: e.g., *chilopod,* "lip"-footed.

chloro- (klō′rô) [Gr. *chloros,* green]: e.g., *chlorophyte,* green alga; *chlorocruorin,* green blood pigment.

choano- (kō′ă·nô) [Gr. *choanē,* funnel]: e.g., *choanocyte; Choanichthyes,* subclass of teleosts, comprising bony fishes with internal nares.

chondro- (kŏn′drô) [Gr. *chondros,* cartilage]: e.g., *chondrocranium,* cartilage skull; *Chondrostei,* cartilage-possessing fish group.

-chord, chorda- (kôrd, kôr′dă) [L. *chorda,* cord, string]: e.g. *notochord,* cord along back; *Chordate,* notochord-possessing animal.

-chrome, chromo-, chroma- (krōm, krō′mô) [Gr. *chroma,* color]: e.g., *cytochrome,* cell pigment; *chromatophore,* pigment-carrier cell.

-clad, clado- (klăd, klă′dô) [Gr. klādos, branch, sprout]: e.g., *triclad,* three-branched; *Cladocera,* branch-"horned" animals.

-coel, coela-, coelo- (sēl) [Gr. *koilos,* hollow, cavity]: e.g., *pseudocoel,* false coelomic cavity; *coelenterate,* possessing "gut cavity."

cranio- (krā′nĭ·ō) [Gr. *krānion,* skull]: e.g., *craniate,* skull-possessing.

crypto- (krĭp′tô) [Gr. *kryptōs,* hidden]: e.g., *cryptobranch,* hidden-gilled.

cten-, cteno- (tĕn) [Gr. *kteis,* comb]: e.g., *ctenoid,* comblike scale or gill structure; *Ctenophora,* comb-bearing animals.

-cyst (sĭst) [Gr. *kystis,* bladder, pouch, sac]: e.g., *sporocyst,* spore-containing cyst.

-cyte, cyto- (sīt) [Gr. *kytos,* vessel, container]: pertaining to cell; e.g., *cytoplasm,* cell substance; *fibrocyte,* fiber-forming cell.

-dactyl, dactylo- (dăk′tĭl) [Gr. *daktylos,* finger, toe]: e.g., *pterodactyl,* wing-fingered; *dactylozooid,* fingerlike individual.

de- (dē) [L. away, from, off]: like Gr. *apo-;* e.g., *dehydration,* removal of water.

-dent, denti- (dĕnt) [L. *dens,* tooth]: like Gr. *-dont;* e.g., *denticle,* little tooth.

dermis, -derm (dûr′mĭs) [Gr. *derma,* skin]: e.g., *ectoderm,* outer skin tissue; *epidermis,* over-skin (i.e., exterior layer).

di- (dī) [Gr. twice, double, two]: like L. *bi-,* e.g., *disect,* to cut in two.

dia- (dī·a) [Gr. through, across, thorough]: e.g., *diaphragm,* across the midriff; *dialysis,* separation (dissolution) through a membrane.

diphy- (dĭ′fĭ) [Gr. *diphyses,* double]: e.g., *diphycercal,* having tail with equal shape above and below vertebral support.

diplo- (dĭ′plô) [Gr. *diploos,* twofold]: e.g., *diploid,* with two chromosome sets; *Diplopoda,* animals with double set of legs; *diplontic,* having two life-cycle phases.

dis- [L. apart, away]: e.g., *dissect,* cut apart (distinct from *di-,* cf. above).

distal (dĭs′tăl): situated away from or far from point of reference (usually the main part of body); opposite to *proximal.*

-dont (dŏnt) [Gr. *odontos,* tooth]: e.g., *thecodont,* having encased (socketed) teeth.

dorsal (dôr′săl) [L. *dorsum,* back]: at, near, or toward the back; opposite to *ventral.*

dys- (dĭs) [Gr. hard, bad]: e.g., *dysfunction,* malfunction; *dysentery,* "bad intestine."

echino- (ê·kī′nô) [Gr. *echinos,* spiny, bristly]: e.g., *echinoderm* spiny-skinned; *echiuroid,* having spines at hind end.

eco- (ē′kŏ) [Gr. *oikos,* house, home]: e.g., *ecology,* study of organism-habitat (home territory) relationships.

ecto- (ĕk′tô) [Gr. *ektos,* outside]: e.g., *ectoproct,* outside-anus (i.e., exterior to tentacle ring).

-ectomy (ĕk′tômĭ) [Gr. *ek,* out of, + *tomein,* to cut]: excision; e.g., *thyroidectomy,* partial or complete excision of the thyroid gland.

efferent (ĕf′ēr·ĕnt) [L. *ex,* out, away + *ferre,* to carry]: e.g., to lead or carry away from given position; opposite of *afferent;* e.g., efferent nerve, efferent blood vessel.

endo- (ĕn′dô) [Gr. *endon,* within]: e.g., *endoderm,* inner skin (tissue) layer; *endothermic,* requiring heat (i.e., within reacting system).

entero-, -enteron (ĕn′tērô) [Gr. *enteron,* intestine]: e.g., *enterocoel,* coelom formed from intestine; *archenteron,* first intestine.

ento- (ĕn′tô) [Gr. var. of *endo-,* within]: e.g., *entoproct,* inside-anus (i.e., within tentacle ring); *entoderm,* endoderm.

entomo- (ĕn′to·mô) [Gr. *entomon,* insect]: e.g., *entomology,* study of insects.

epi- (ĕp′ĭ) [Gr. to, on, over, against]: e.g., *epistome,* over the mouth; *epicardial,* over the heart.

erythro- (ê rĭth′rô) [Gr. *erythros,* red]: e.g., *erythrocyte,* red (blood) cell.

eu- (ū) [Gr. good, well, proper]: e.g., *Eumetazoa,* metazoa proper.

ex-, exo-, extero- (ĕks, ĕk′sô, ĕks′tēr·ô) [L. out, from, exterior]: e.g., *exopterygote,* exterior-winged; *exothermic,* releasing heat (i.e., out of reacting system).

-fer, -fera (fēr, fē′rȧ) [L. *ferre,* to carry]: like Gr. *-phore;* e.g., *foraminifer,* hole-carrier; *rotifer,* wheel-carrier; *Porifera,* pore-carrying animals (sponges).

-form, -formes (fôrm, fôr′mēz) [L. *-formis,* having the form of]: e.g., *perciform,* perchlike; *Galliformes,* order of chickenlike birds.

frontal [L. *front, frons,* forehead]: in a horizontal plane separating dorsal half from ventral half.

gamo-, -gamy (găm′ô) [Gr. *gamein,* to marry]: pertaining to gametes or fertilization; e.g., *gamocyst,* capsule around zygote; *autogamy,* self-fertilization.

gastro- (găs′trô) [Gr. *gaster,* stomach]: e.g., *gastrozooid,* feeding individual; *gastrocoel,* stomach cavity.

-gen, -genic, geno- [Gr. *genēs,* born, created]: e.g., *hydrogen,* water-producing; *myogenic,* produced by muscles; *genotypic,* according to genetic constitution (gene content).

-gest, gest- (jĕst) [L. *gestare,* to carry]: e.g., *ingest,* to carry in (food); *gestation,* period of fetus-carrying within pregnant female.

glosso-, -glossus (glŏs′ô) [Gr. *glossa,* tongue]: e.g., *glossopharyngeal,* pertaining to tongue and pharynx; *Saccoglossus,* sac-tongue–possessing worm (hemichordate).

gnatho-, -gnath (nă′thô) [Gr. *gnathos,* jaw]: e.g., *gnathobase,* jawlike base; *chaetognath,* bristle-jawed; *agnath,* jawless.

gono- (gŏn′ô) [Gr. *gonos,* seed, generation]: pertaining to reproduction; e.g., *gonozooid,* reproductive individual; *gonopore,* reproductive opening; *gonoduct,* reproductive duct.

gymno- (jĭm′nô) [Gr. *gymnos,* naked]: e.g., *gymnostome,* naked mouth.

gyn-, gyno- (jĭn′nô) [Gr. *gynē,* woman, female]: opposite of *andro-;* e.g., *gynogenic,* female-producing.

haplo- (hăp′lô) [Gr. *haploos,* single]: e.g., *haploid,* with one chromosome set; *haplontic,* having single life-cycle phase.

-helminth (hĕl′mĭnth) [Gr. *helminthos,* worm]: e.g., *platyhelminth,* flatworm; *aschelminth,* sacworm.

hem-, hemo-, hemato- (hēm, hē′mô, hĕm′ȧ tô) [Gr. *haima,* blood]: var. of *haem-;* e.g., *hemoglobin,* red blood pigment.

hemi- (hĕm′ĭ) [Gr., half]: like *semi-;* e.g., *hemichordate,* similar to chordate.

hetero- (hĕt′ēr·ô) [Gr. *heteros,* other, different]: opposite to *homo-;* e.g., *heterotrophic,* feeding on other living things.

hex-, hexa- (hĕks, hĕk′sȧ) [Gr. six]: e.g., *hexapod,* six-legged; *hexose,* six-carbon sugar.

holo- (hŏl′ô) [Gr. *holos,* whole, entire]: e.g., *holotrophic,* feeding on whole animals; *holoblastic,* cleavage into whole blastomeres in embryo.

homeo-, homoio- (hō′mê·ô, hô·moi′ô) [Gr., similar]: e.g., *homeostatic,* remaining similar in state; *homoiothermic,* possessing constant temperature.

hydro (hī′drô) [Gr. *hydōr,* water]: e.g., *hydrolysis,* dissolution by water.

hyper- (hī′pēr) [Gr. above, over]: opposite of *hypo-;* e.g., *hypertrophy,* overgrowth.

hypo- (hī′pō) [Gr. under, less]: opposite of *hyper-;* e.g., *hypodermis,* "underskin."

ichthyo- (ĭk′thĭ·ô) [Gr. *ichthyos,* fish]: e.g., *Osteichthyes,* bony fishes; *ichthyosaur,* fishlike reptile.

inter- (ĭn′tûr) [L. between, among]: e.g., *intercellular,* between cells.

intra- (ĭn′trȧ) [L. within]: e.g., *intracellular,* within cells.

iso- (ī′sô) [Gr. *isos,* equal]: like *homo-;* e.g., *isolecithal,* having evenly distributed yolk.

-lecithal (lĕs′ĭ·thăl) [Gr. *lekithos,* egg yolk]: e.g., *lecithin,* any of several complex nitrogenous substances found especially in the brain and nerve tissue and in egg yolk.

-logy (lō′jĭ) [Gr. *logos,* discourse, study]: e.g., *Zoology,* animal studies.

lumbar (lŭm bēr) [L. *lumbus,* loin]: at, near, or toward loin region.

-lysis, -lytic, -lyte [Gr. *lysis,* a loosening]: pertaining to dissolving; e.g., *electrolytic,* dissolution by electricity; *autolysis,* self-dissolution.

macro- (mă′krō) [Gr. *makros*, long]: opposite of *micro-;* e.g., *macromere*, large embryo cell.
mastigo- (măs′tĭ·gô) [Gr. *mastix*, whip]: e.g., *Mastigophora*, flagellum-bearing (protozoa); *mastigoneme*, whip-hair.
mero-, -mere, -mer (mē′rô, mēr, mĕr) [Gr. *meros*, part]: e.g., *blastomere*, embryo part (cell); *polymer*, chemical of many (similar) parts; *meroblastic*, cleavage into partial blastomeres in embryo; *merozoite*, partly matured individual.
meso- (mĕs′ô) [Gr. *mesos*, middle]: e.g., *mesoderm*, middle "skin"; *mesosome*, middle body region.
meta- (mĕt′ȧ) [Gr. after, behind]: e.g., *metacoel*, hind cavity; *Metazoa*, later (advanced) animals.
micro- (mī′krō) [Gr. *mikros*, small]: e.g., *micromere*, small embryo cell.
mono- (mŏn′ô) [Gr. *monos*, single]: e.g., *monosaccharide*, single sugar (unit).
-morph, morpho- (mŏrf, mŏr′fô) [Gr. *morphē*, form]: e.g., *morphology*, study of form (structure); *metamorphosis*, process of acquiring later (i.e., adult) structure.
myo- (mī′ô) [Gr. *mys*, muscle]: e.g., *myocyte*, muscle cell; *myoneme*, muscle (-like) hair; *myocoel*, muscle-associated coelom.
-neme, nemato- (nē′mĕ, nĕm′ȧ·tô) [Gr. *nema*, thread]: e.g., *axoneme*, axial filament; *nematocyst*, thread-containing capsule; *nematode*, threadlike worm.
neuro- (nū′rô) [Gr. *neuron*, nerve]: e.g., *neuropore*, opening from nervous system.
noto- (nō′tô) [Gr. *nōton*, the back]: e.g., *notopodium*, foot along back; *notochord*, cord along back.
octo- [Gr. *okto*, eight]: e.g., *octopus*, eight-"legged" animal.
-oid, -oida, -oidea (oid, oid′ȧ, oi′dê·ȧ) [Gr. *eidos*, form]: having the form of; like L.*-form;* e.g., *echinoid*, like *Echinus*.
oligo- (ŏl′ĭ·gô) [Gr. *oligos*, few, small]: e.g., *oligochaete*, having few bristles.
onto- (ŏn′tŏ) [Gr. *on*, being]: e.g., *ontogeny*, production of being (= development).
oö- (ō′ô) [Gr. *ōion*, egg]: e.g., *oöcyte*, egg cell; *oözooid*, individual formed from egg.
opistho- (ŏp′ĭs·thô) [Gr. *opisthen*, *opithen*, rear, hind]: e.g., *opisthobranchia*, rear-gilled animals (snails); *opisthosoma*, hind body (= abdomen).
oral (ō′răl) [L. *or-*, *os*, mouth]: at, near, or toward the mouth.
osteo- (ŏs′tê·ô) [Gr. *osteon*, bone]: e.g., *osteoblast*, bone-forming cell; *osteology*, study of bones; *periosteum*, tissue layer covering a bone.
ostraco- (ŏs′trȧ·kô) [Gr. *ostrakon*, shell]: pertaining to a skeletal (calcareous) cover or shield; e.g., *ostracoderm*, armor-skinned.
oto-, otic (ō′tô, ŏ′tĭk) [Gr. *ous*, ear]: e.g., *otic* vesicle; *otolith*, earstone.
ovi- ovo- (ō′vĭ, ō′vô) [L. *ovum*, egg]: e.g., *oviduct; ovotestis*, egg-and-sperm-producing organ.
para- (păr′ȧ) [Gr., beside]: e.g., *parapodium*, side foot; *Parazoa*, animals on side branch of evolution (sponges).
pectin- (pĕk′tĭn) [L. *pecten*, comb]: e.g., *pectinibranch*, having comblike gills; *pectine*, comblike organ.
pectoral (pĕk′tô·răl) [L. *pectorale*, breastplate]: at, near, or toward chest or shoulder region.
-ped, -pedia, pedi- (pĕd, pĕd′ĭ·ȧ, pĕd′ĭ) [L. *pes*, foot]: like Gr. *-pod;* e.g., *bipedal*, two-footed; *pedipalp*, leg-like appendage.
pelvic (pĕl′vĭk) [L. *pelvis*, basin]: at, near, or toward the hip region.
pent-, penta- (pĕnt, pĕn′tȧ) [Gr. *pente*, five]: e.g., *pentose*, five-carbon sugar; *pentaradial*.
peri- (pĕr′ĭ) [Gr. around]: e.g., *peristomial*, around the mouth; *perisarc*, around the flesh; *peristalsis*, wavelike compression around tubular organ (e.g., gut).
phago-, -phage (făg′ô, fāj) [Gr. eating]: e.g., *phagocyte*, cell eater; *bacteriophage*, bacterium eater (virus).
phono-, -phone (fōn′ô, fōn) [Gr. *phonē*, sound]: e.g., *phonoreceptor*, sound-sensitive sense organ.
phoro-, -phore (fōr′ô, fōr) [Gr. *phoros*, bearing, carrying]: like L. *-fer;* e.g., *phorozooid*, carrying individual; *trochophore*, "wheel"-bearing (larva).
photo-, photic (fō′tô, fō′tĭk) [Gr. *photos*, light]: e.g., *photophore*, light-possessing part or organ; *photosynthesis*, synthesis with aid of light.
-phragm (frăm) [Gr. barrier]: e.g., *diaphragm; endophragm*, internal skeletal plates.
phyllo-, -phyll (fĭl′ô, fĭl) [Gr. *phyllon*, leaf]: e.g., *phyllopodium*, leafshaped foot; *chlorophyll*, green pigment in leaf.
phyto-, -phyte (fī′tô, fīt) [Gr. *phyton*, plant]: e.g., *phytoplankton*, plankton consisting of plant life; *Metaphyta*, later (advanced) plants.
-pithecus (pĭ′thê·kŭs) [Gr. *pithēkos*, ape]: e.g., *Australopithecus*, southern ape; *Pithecanthropus*, manlike ape.
placo- (plă′kô) [Gr. *plax*, tablet, plate]: e.g., *placoderm*, plate-skinned; *Monoplacophora*, single-plate–bearing animals; *placoid*, platelike.
-plasm, plasmo-, -plast (plăz′m, plăz′mô, plăst) [Gr. *plasma*, form, mold]: e.g., *protoplasm*, first-molded (living matter); *plasmotomy*, cutting (division) of protoplasm; *chloroplast*, green-formed (body).
-pleur, pleuro- (plo͝or, plo͝or′ô) [Gr. *pleuron*, side, rib]: e.g., *pleuron*, lateral exoskeletal plate; *pleura*, membrane lining rib cage.
-ploid [Gr. *-ploos*, -fold]: number of chromosome sets per cell; e.g., *haploid, diploid, polyploid*.
poly- (pôl′ĭ) [Gr. *polys*, many]: e.g., *polymorphic*, many-shaped; *polychaete*, many-bristled animal.

poro- (pŏ′rō) [Gr. *poros,* pore]: e.g., *porous,* full of pores.

post-, postero-, posterior (pōst, pŏs′tēr·ō) [L. behind, after]: opposite of *pre-, antero-;* at, near, or toward hind end or part.

pre- (prē) [L. before, in front of]: opposite of *post-;* e.g., *preoral,* in front of mouth.

pro- (prō) [Gr. before, in front of]: like L. *pre-;* e.g., *prostomial,* in front of mouth.

-proct, procto- (prŏkt, prŏk′tō) [Gr. *proctos,* anus]: e.g., *ectoproct,* having anus outside of ring of tentacles; *proctodeal,* anal.

proto- (prō′tō) [Gr. *prōtos,* first]: e.g., *protozoa,* first animals; *protostome,* first mouth; *protocoel,* first coelom.

proximal (prŏk′sĭ·măl) [L. *proximus,* near]: situated near to point of reference (usually the main part of body); opposite of distal.

pseudo- (sū′dō) [Gr. *pseudēs,* false]: e.g., *pseudocoel,* false coelom; *pseudopodium,* false foot.

ptero-, -ptera, -ptery (tĕr′ō, tĕr′ȧ, tĕr′ĭ) [Gr. *pteron,* wing, fin]: e.g., *Aptera,* wingless animals (insects); *pterodactyl,* wing-fingered; *exopterygote,* exterior-winged.

-pyge, pyg- (pī′jī, pĭj) [Gr. *pygē,* rump]: e.g., *cytopyge,* cellular elimination pore in protozoa; *pygidium,* posterior abdominal region (in arthropods); *uropygial gland,* oil-secreting gland near tail base in birds.

rami-, -ramous (răm′ĭ, rā′mŭs) [L. *ramus,* branch]: e.g., *biramous,* two-branched; *ramified,* branched.

retro- (rĕt′rō) [L. backward, behind]: e.g., *retrocerebral,* behind the cerebrum (or brain).

rhabdo- (răb′dō) [Gr. *rhabdos,* rod]: e.g., *rhabdite,* rod-shaped organelle; *rhabdocoel,* flatworm possessing straight (rodlike) intestine.

rhizo- (rī′zō) [Gr. *rhiza,* root]: e.g., *rhizopod,* having rootlike feet; *rhizoplast,* rootlike organelle.

-rhynch, rhyncho- (rĭngk, rĭng′kō) [Gr. *rhynchos,* snout]: e.g., *kinorhynch,* having movable snout; *rhynchocoel,* snout (proboscis) cavity.

sagittal (săj′ĭ·tăl) [L. *sagitta,* arrow]: at, near, or toward plane bisecting left and right halves; in median plane.

sarco-, -sarc (sär′kō, särk) [Gr. *sarkos,* flesh]: e.g., *coenosarc,* common flesh (living portions); *Sarcopterygii,* flesh-finned (fishes).

saur-, -saur (sôr) [Gr. *sauros,* lizard]: e.g., *saurian,* pertaining to lizards (and reptiles generally, including extinct); *pterosaur,* flying reptile.

schizo- (skĭz′ō) [Gr. *schizein,* to split, part]: e.g., *schizocoel,* coelom formed by splitting of tissue layer; *schizogamy,* reproduction by splitting.

sclero- (sklēr′ō) [Gr. *sklēros,* hard]: e.g., *scleroprotein,* hard (horny) protein; *sclera,* hard layer.

scypho- (sī′fō) [Gr. *skyphos,* cup]: e.g., *Scyphozoa,* cupshaped animals (jellyfish); *scyphistoma,* cupped-mouth–possessing larva.

seti-, seta (sĕt′ĭ, sĕt′ȧ) [L. *seta,* bristle]: like Gr. *chaeto-;* *setiferous,* bristle-bearing.

-soma, -some, somato- (sō′mȧ, sōm, sō′mȧ·tō) [Gr. *sōma,* body]: e.g., *chromosome,* pigmented body; *somatocoel,* body coelom; *somatic mesoderm,* outer mesoderm.

spermo-, sperma-, spermato- (spûr′mō, spûr′mȧ, spûr·mȧ′tō) [Gr. *sperma,* seed]: e.g., *spermatophore,* sperm (-bearing) capsule.

splanchno- (splăngk′nō) [Gr. *splanchnon,* entrails]: e.g., *splanchnic mesoderm,* inner mesoderm.

sporo- (spō′rō) [Gr. *sporā,* seed]: e.g., *Sporozoa,* spore-forming (protozoa); *sporogony,* reproduction by spores.

stato- (stăt′ō) [Gr. *statos,* standing stationary, positioned]: e.g., *statolith,* position (-indicating) stone; *statoblast,* stationary embryo.

stereo- (stĕr′ē·ō) [Gr. *stereos,* solid]: e.g., *stereoblastula,* solid blastula.

-stome, -stoma, -stomato-, (stōm, stōm′ȧ, stō·mă′tō) [Gr. *stoma,* mouth]: *peristomial,* around the mouth; *stomatogenesis,* mouth-formation.

sym-, syn-, (sĭm, sĭn) [Gr. *syn,* together, with]: like L. *con-;* e.g., *syngamy,* coming together of gametes; *synapse,* looping together (of neurons); *synthesis,* construction, putting together; *symbiosis,* living together.

tango- (tăng′gō) [L. *tangere,* to touch]: e.g., *tangoreceptor,* touch-receptive sense organ.

taxo-, taxi-, -taxis (tăksō, tăksĭ, tăks′ĭs) [Gr. *taxis,* arrangement]: e.g., *taxonomy,* "arrangement" laws; *taxidermy,* skin arrangement; *chemotaxis,* arrangement (movement) produced by chemicals.

tel-, tele-, teleo- (tĕl, tĕl′ē, tĕl′ē·ō) [Gr. *telos,* end]: e.g., *telophase,* end phase; *telotroch,* preanal tuft of cilia in a trochophore larva; *teleost,* (fish with) bony end (adult) state; *teleology,* knowledge of end conditions.

tetra- (tĕt′rȧ) [Gr. four]: e.g., *tetrabranch,* four-gilled; *tetrapod,* four-footed.

theco-, -theca (thē′kō, thē′kȧ) [Gr. *thēkē,* case, capsule]: e.g., *thecodont,* having socketed teeth; *spermatheca,* sperm receptacle.

thigmo- (thĭg′mō) [Gr. *thigma,* touch]: like L. *tango-;* e.g., *thigmotropy,* movement due to touch.

thoracic (thō·răs′ĭk) [L. *thorax,* chest]: at, near, or toward chest region, or region between head and abdomen.

-tome, -tomy (tōm, tō′mĭ) [Gr. *tomē,* section, a cutting apart]: e.g., *autotomy,* self-dismemberment (of appendage); *myotome,* muscle-producing part.

transverse (trăns′vûrs) [L. *transversare,* to cross]: at, near, or toward plane separating anterior and posterior; cross-sectional.

tri- (trī) [L. *tria,* three]: e.g., *triclad,* three-branched (digestive tract).

-trich, tricho- (trĭk, trĭk′ō) [Gr. *trichos,* hair]: e.g., *hetero-*

trich, (ciliate) possessing different types of hairs (cilia); *trichocyst,* hair-containing sac.

-troch, trocho- (trŏk, trŏk'ō) [Gr. *trochos,* wheel]: e.g., *prototroch,* first (i.e., anterior) wheel (of cilia); *trochophore,* (larva) bearing wheel (of cilia).

-troph, tropho- (trŏf, trŏ'fō) [Gr. *trophos,* feeder]: e.g., *autotrophic,* self-nourishing; *trophozoite,* feeding (i.e., mature) animal; *trophoblast,* feeding embryonic part.

uro-, ura (ū'rō, ū'rȧ) [Gr. *oura,* tail]: e.g., *uropod,* tail-foot; *urochordate,* tailed chordate; *Branchiura,* gill-tailed animals (crustacea): *Anura,* tailess animals (amphibia).

ventral (vĕn'trăl) [L. *venter,* belly]: opposite to *dorsal;* at, near, or toward the belly or underside.

zoo-, -zoa, -zoon (zō'ō, zō'ȧ, zō'ŏn) [Gr. *zōion,* animal]: e.g., *zooplankton,* animal life of the plankton; *protozoon; zooid,* individual animal (usually in a colony).

SECTION B

General Listing of Technical Terms

Note: In most cases in this section where derivations of particular word parts are not given, such parts and their derivations may be found in section A, above.

abdomen (ab'dŏ·mĕn) [L.]: region of animal body posterior to thorax.

Acanthocephala (ă·kăn'thō·sĕf'ă·lȧ) [Gr. *akantho,* thorn]: spiny-headed worms, a phylum of pseudocoelomate parasites.

Acarina, acarine (ăkă'rī·nȧ, ăkă'rĭn) [Gr. *akar,* mite]: (1) the arachnid order of mites and ticks; (2) pertaining to mites and ticks.

acentric (â·sĕn'trĭk): without center; applied specifically to type of mitosis in which a centriole is absent.

acicula (ȧ·sĭk'ū·lȧ) [L. dim. of *acus,* needle]: needlelike bristle embedded within polychaete parapodium.

acid (ăs'ĭd) [L. *acidus,* sour]: a substance which releases hydrogen ions in water; having a pH of less than 7.

acrosome (ăk'rō·sōm): structure at the tip of the sperm head (nucleus) which makes contact with the egg during fertilization.

adenine (ăd'ē·nēn): a purine component of nucleotides and nucleic acids.

adenosine (di-, tri-) phosphate (*ADP, ATP*) (ȧ·dĕn'ō·sēn): phosphorylated organic compounds functioning in energy transfers within cells.

adipose (ăd'ĭ·pōs) [L. *adipis,* fat]: fat, fatty; fat-storing tissue.

adrenal, adrenaline (ăd·rē'năl, ăd·rĕn'ăl·ĭn) [L. *renalis,* kidney]: (1) endocrine gland; (2) the hormone produced by the adrenal medulla.

adrenergic (ăd'rĕn·ûr'jĭk): applied to nerve fibers which release an adrenalinelike substance from their axon terminals when impulses are transmitted across synapses.

aerobe, aerobic (ā'ĕr·ōb, -ō'bĭk) [Gr. *aeros,* air]: (1) oxygen-requiring organism; (2) pertaining to oxygen-dependent form of respiration.

aldehyde (ăl'dĕ·hīd) [L. abbrv. for *alcohol dehydrogenatum,* dehydrogenated alcohol]: organic compound possessing a —CHO grouping.

aldose (ăl'dōs): one of a series of sugars possessing a terminal aldehyde grouping.

alga (ăl'gȧ), pl. *algae* (-jē): any member of a largely photosynthetic superphylum of protists; blue-green, green, golden-brown, brown, red algae.

alkaline (ăl'kä·lĭn): pertaining to substances which release hydroxyl ions in water; having a pH greater than 7.

allantois (ă·lăn'tō·ĭs) [Gr. *allantoeides,* sausage-shaped]: one of the extraembryonic membranes in reptiles, birds, and mammals; functions as embryonic urinary bladder or as carrier of blood vessels to and from placenta.

allele (ă·lēl') [Gr. *allēlōn,* of one another]: one of a group of alternative genes which may occupy a given locus on a chromosome; a dominant and its correlated recessive are allelic genes.

alula (ăl'ū·lȧ) [L. dim. of *ala,* wing]: the first digit (thumb) of a bird wing; reduced in comparative size.

alveolus (ăl·vē'ō·lŭs), pl. *alveoli* (-lī) [L. dim. of *alveus,* a hollow]: a small cavity or pit, e.g., a microscopic air sac of lungs.

ambulacrum, ambulacral (ăm'bū·lă'krŭm, -ăl) [L. walk, avenue]: (1) tube-feet–lined ciliated groove leading over arm to mouth in certain echinoderms; conducts food to mouth; (2) adjective.

amino, amino acid, amination (ă·mē'nō, ă·mĭnā'shŭn): (1) —NH_2 group; (2) acid containing amino group, constituent of protein; (3) addition of amino group to other compound.

ammocoete (ăm'mō·sēt) [Gr. *ammos,* sand]: lamprey larva.

amnion, amniote, amniotic (ăm'nĭ·ŏn) [Gr. dim. of *amnos,* lamb]: (1) one of the extraembryonic membranes in reptiles, birds, and mammals, forming a sac around the embryo; (2) any reptile, bird, or mammal, i.e., any animal possessing an amnion during the embryonic state; (3) pertaining to the amnion, as in *amniotic fluid.*

amphiblastula (ăm′fĭ·blăst′ū·lȧ): larval stage in certain sponges.

ampulla (ăm·pŭl′ȧ) [L. vessel]: enlarged saclike portion of a duct, as in ampullae of semicircular canals in mammalian ear, or in ampullae of echinoderm tube feet.

amylase (ăm′ĭ·lās) [L. *amylum,* starch]: an enzyme promoting the decomposition of polysaccharides into smaller carbohydrate units.

anaerobe, anaerobic (ăn·ā′ēr·ōb, -ō′bĭk): (1) an oxygen-independent organism; (2) pertaining to an oxygen-independent form of respiration.

anaphase (ăn′ȧ·fāz): a stage in mitotic division, characterized by the migration of chromosome sets toward the spindle poles.

anisogamy (ăn·ī′sŏg′ȧm·ī): sexual fusion in which the gametes of opposite sex types are unequal in size.

Annelida (ăn′ĕ·lĭd·ȧ) [L. *anellus,* a ring]: the phylum of segmented worms.

annulus (ăn′ū·lŭs) [L. *ring*]: a ringlike structure.

antibody (ăn′tĭ·bŏd′ĭ): a substance, produced within an animal, which opposes the action of another substance; in specific usage, an antibody is a globulin type of protein which combines and renders harmless an antigen, i.e., a foreign protein introduced into an animal by infectious processes.

antigen (ăn′tĭ·jĕn): a foreign substance, usually protein in nature, which elicits the formation of specific antibodies within an animal.

Anura (ă·nū′rȧ): order of tailless amphibia, including frogs and toads.

apodeme (ăp′ō·dēm) [Gr. *demos,* district]: breakage plane in crustacean appendage where autotomy may occur readily.

apopyle (ăp′ō·pīl) [Gr. *pylē,* gate]: the excurrent opening of a choanocyte chamber in sponges.

Arachnida (ȧ·răk′nĭd·ȧ) [Gr. *arachnē,* spider]: class of chelicerate arthropods including spiders, scorpions, mites, ticks, and other orders.

Arthropoda (är·thrô′/ŏd·ȧ): the phylum of jointed-legged invertebrates.

Artiodactyla (ăr′tĭ·ō·dăk′tĭl·ȧ) [Gr. *artios,* even]: order of even-toed ungulate mammals; includes cattle, swine, deer, camels.

Aschelminthes (ăs·kĕl·mĭn′thēs) [Gr. *askos,* bladder, sac]: bladder-like or bladder-forming worms; a pseudocoelomate phylum including rotifers, roundworms, and other groups.

asconoid (ă′skŏn·oid) [Gr. *askos,* bladder, sac]: saclike; refers specifically to a type of sponge architecture.

atom (ăt′ŭm) [Gr. *atomos,* indivisible]: the smallest whole unit of a chemical element; composed of given numbers of protons, neutrons, and other particles which form an atomic nucleus, and of given numbers of electrons, which orbit around the nucleus.

atrium, atrial (ā′trĭ·ŭm, -ăl) [L. yard, court, hall]: entrance or exit cavity, e.g., entrance chamber to heart, exit chamber from chordate gill region.

auricle (ô′rĭ·k′l) [L. dim. of *auris,* ear]: earshaped structure or lobelike appendage; e.g., atrium in mammalian heart, lateral flap in vertebrate hind brain, lateral flap near eyes in planarian worms.

auricularia (ô·rĭk′ū·lă′rĭa) [L. *lar,* larva]: larva of holothuroid echinoderms, with ear-lobe–like ciliated bands.

autosome (ô′tō·sōm): a chromosome which is not a sex chromosome.

avicularium (ā·vĭk′ū·lā′rĭ·ŭm) [L. dim. of *avis,* bird]: a specially differentiated polymorphic individual in a colony of ectoprocts, shaped like a bird's head, serving a protective function.

axenic (ā·zĕn′ĭk) [Gr. *xenos,* foreigner]: pertaining to a culture medium in which the available food sources are completely and specifically identified.

axon (ăk′sŏn): an outgrowth of a nerve cell, conducting impulses away from the cell body; a type of nerve fiber.

bacterium (băk·tēr′ĭ ŭm) [Gr. dim of *baktron,* a staff]: a small, typically unicellular (moneran) organism characterized by the absence of a formed nucleus; genetic material is dispersed in clumps through the cytoplasm.

benthos, benthonic (bĕn′thŏs) [Gr., depth of the sea]: (1) collective term for organisms living along the bottoms of oceans and lakes; (2) adjective.

beriberi (bĕr′ĭ·bĕr′ĭ) [Singhalese *beri,* weakness]: disease produced by deficiency of vitamin B_1 (thiamine).

bioluminescence (bī′ō·lū′mĭ·nĕs′ĕns) [L. *lumen,* light]: emission of light by living organisms.

biome (bī′ōm): habitat zone, e.g., desert, grassland, tundra.

biota, biotic (bī·ō′tȧ, -ŏt′ĭk): (1) the community of organisms of a given region; (2) adjective.

bipinnaria (bī′pĭn·âr′ĭ·ȧ) [L. *pinna,* feather, fin]: larva of asteroid echinoderm, with ciliated bands suggesting two wings.

blastula (blăs′tū·lȧ): stage in early animal development, when embryo is a hollow or solid sphere of cells.

blepharoplast (blĕf′ȧ·rō·plăst′) [Gr. *blepharon,* eyelid]: the basal granule of a flagellum or cilium; equivalent to *kinetosome.*

bronchus, bronchiole (brŏng′kŭs, brŏng′kĭ·ōl) [Gr. *bronchos,* windpipe]: (1) a main branch of the trachea in air-breathing vertebrates; (2) a smaller branch of a bronchus.

buffer (bŭf′ēr): a substance which prevents appreciable changes of pH in solutions to which small amounts of acids or bases are added.

bursa (bûr′sȧ) [L. sac, bag]: saclike cavity.

byssus (bĭs′ŭs) [Gr. *byssos,* flax, linen]: silky threads secreted by mussels for attachment to rocks.

caecum (sē′kŭm) [L. *caecus,* blind]: cavity open at one end, e.g., the blind pouch at the beginning of the vertebrate large intestine, connecting at one side with the small intestine.

calorie (kăl′ô·rĭ) [L. *calor,* heat]: unit of heat, defined as the amount of heat required to raise the temperature of 1 g of water by 1° C.

captaculum (kap·ta′kū·lŭm) [L. *captare,* to capture]: tentaclelike outgrowth from head of tooth-shell mollusks, serving in food capture.

carapace (kăr′ȧ·pās) [fr. Sp. *carapacho*]: a hard case or shield covering the back of certain animals, e.g., many crustacea.

carbohydrate, carbohydrase (kär′bô·hī′drāt): (1) an organic compound consisting of a chain of carbon atoms to which hydrogen and oxygen, present in a 2:1 ratio, are attached; (2) an enzyme promoting the synthesis or decomposition of a carbohydrate.

carnassial (kär·năs′ĭ·ăl) [Fr. *carnassier,* flesh-eating]: pertaining to mammalian tooth type adapted for flesh-eating.

carnivore, Carnivora (kär′nĭv′ō·rȧ) [L. *carnivorus,* flesh-eating]: (1) any holotrophic animal subsisting on other animals or parts of animals; (2) an order of mammals; includes cats, dogs, seals, walruses.

carotene, carotenoids (kăr′ô·tēn, kȧ·rŏt′ê·noid) [L. *carota,* carrot]: (1) pigment producing cream-yellow to carrot-orange colors; precursor of vitamin A; (2) a class of pigments of which carotene is one.

catalysis, catalyst, catalytic (kȧ·tăl′ĭ·sĭs) [Gr. *katalysis,* dissolution]: (1) acceleration of a chemical reaction by a substance which does not become part of the end-product; (2) a substance which accelerates a reaction as above; (3) adjective.

ceboid (sē′boid): a New World monkey; uses its tail as a fifth limb.

Cenozoic (sē′nô·zō′ĭk) [Gr. *kainos,* recent + *zōē,* life]: geologic era after the Mesozoic, dating approximately from 75 million years ago to present.

centriole (sĕn′trĭ·ōl): cytoplasmic organelle forming spindle pole during mitosis and meiosis.

centrolecithal (sĕn′trô·lĕs′ĭ·thăl): pertaining to eggs with yolk accumulated in center of cell, e.g., in arthropods.

centromere (sĕn′trô·mēr): region on chromosome at which spindle fibril is attached during mitosis and meiosis.

cercopithecoid (sûr′kô·pĭ·thē′koid): an Old World monkey; possesses tail which is not used as limb.

cerebellum (sĕr′ê·bĕl′ŭm) [L. dim. of *cerebrum*]: a part of the vertebrate brain controlling muscular coordination.

cerebrum (sĕr′ê·brŭm) [L. brain]: a part of the vertebrate brain, especially large in mammals; controls many voluntary functions and is seat of higher mental capacities.

chelate (kē′lāt) [Gr. *chēlē,* claw]: claw-possessing, esp. a limb or appendage.

chelicera (kĕ·lĭ′sĕr·ȧ): a pincerlike appendage in a subphylum of arthropods (chelicerates).

chemosynthesis (kĕm′ô·sĭn′thê·sĭs): a form of autotrophic nutrition in certain bacteria, in which energy for the manufacture of carbohydrates is obtained from inorganic raw materials.

chitin (kī′tĭn): a horny organic substance forming the exoskeleton of arthropods, the epidermal cuticle or other surface structures of many other invertebrates, and the cell walls of Protista such as certain fungi.

chloragogue (klō′rȧ·gŏg) [Gr. *agōgos,* leader]: excretory cell in annelids and some other invertebrates, leading wastes from body fluids to epidermis.

chlorocruorin (klō′rō·krū·ôr·ĭn) [L. *cruor,* blood, gore]: green blood pigment in the plasma of certain annelids.

cholinergic (kō′lĭn·ûrjik): refers to a type of nerve fiber which releases acetycholine from the axon terminal when impulses are transmitted across synapses.

Chondrichthyes (kŏn·drĭk′thĭ·ēz): fishes with cartilage skeleton, a class of vertebrates comprising sharks, skates, rays, and related types.

Chordata (kôr·dā′tȧ): animal phylum in which all members possess notochord, dorsal nerve cord, and pharyngeal gill slits at least at some stage of life cycle; three subphyla, the Urochordata, Cephalochordata, and Vertebrata.

chorion (kō′rĭ·ŏn) [Gr.]: one of the extraembryonic membranes in reptiles, birds, and mammals; forms outer cover around embryo and all other membranes, and in mammals contributes to structure of placenta.

choroid (kō′roid): mid-layer in wall of vertebrate eyeball, between retina and sclera; carries blood supply to eye and contains light-absorbing black pigment.

Chrysophyta (krĭs′ô·fĭt·*a*) [Gr. *chrysos,* gold]: the phylum of golden-brown algae; includes diatoms.

Ciliophora (sĭl′ĭ·ŏ′fôr·ȧ) [L. *cilium,* eyelid]: a protozoan subphylum; the ciliate protozoa.

cilium (sĭl′ĭ ŭm): microscopic bristlelike variant of a flagellum, present on surfaces of many cell types and capable of vibratory motion; functions in cellular locomotion and in creation of currents in water.

cirrus (sĭr′ŭs) [L. tuft, fringe]: a movable tuft or fingerlike projection from a cell or a body surface.

cloaca (klō ā′ kȧ) [L. sewer]: exit chamber from alimentary system; also serves as exit for excretory and/or reproductive system.

Cnidaria (nī·dă′rĭ·ȧ) [Gr. *knidē,* nettle]: coelenterates; the phylum of cnidoblast-possessing animals.

cnidoblast (nī′dō·blăst): stinging cell characteristic of coelenterates; contains nematocyst.
cnidocil (nī′dō·sĭl): spike or hair-trigger on cnidoblast serving in nematocyst discharge.
coccine (kŏk′sēn) [Gr. *kokkos,* grain]: pertaining to sessile protistan state of existence in which reproduction does not take place during vegetative condition.
cochlea (kŏk′lē·ȧ) [Gr. *kochlias,* snail]: part of the inner ear of mammals, coiled like a snail shell; houses the organ of Corti.
coelom (sē′lŏm): body cavity lined entirely by mesoderm, especially by peritoneum.
coenosarc (sē′nō·särk) [Gr. *koinos,* common]: the living parts of a coelenterate hydroid colony, as distinguished from external secreted perisarc.
coenzyme (kō·ĕn′zīm): a substance, usually organic, required if a given enzyme is to be active.
colloblast (cŏl′ō·blăst) [Gr. *kolla,* glue]: adhesive cell type in tentacles of ctenophores.
colloid (kŏl′ oid): a substance divided into fine particles, where each particle is larger than one of a true solution but smaller than one in a coarse suspension; a colloidal system contains particles of appropriate size and a medium in which the particles are dispersed.
colon (kō′lŏn): the large intestine of mammals; portion of alimentary tract between caecum and rectum.
columella (kŏl′ū·mĕl′ȧ) [L. little column]: an elongated skeletal or supporting structure; e.g., hyomandibular bone in amphibian ear.
commensal, commensalism (kŏ·mĕn′ săl, -ĭz′m) [L. *cum,* with + *mensa,* table]: (1) an animal living symbiotically with a host, where the host neither benefits nor suffers from the association; (2) noun.
compound (kŏm′pound) [L. *componere,* to put together]: a combination of atoms or ions in definite ratios, held together by chemical bonds.
conjugation (kŏn·jo͝o·gā′shŭn) [L. *conjugare,* to unite]: a mating process characterized by the temporary fusion of the mating partners; occurs in ciliate protozoa and other Protista. (2) alternation of double and single bonds in a chemical compound.
convergence (kŏn·vûr′jĕns) [L. *convergere,* to turn together]: the evolution of similar characteristics in animals of widely different ancestry.
Copepoda (kō′pē·pŏd·ȧ) [Gr. *kope,* oar]: a subclass of crustaceans.
corona (kō·rō′nȧ) [L. garland, crown]: any wreath or circlet of cilia or tentacles.
corpus allatum (kôr′pŭs ă·lā′tŭm) pl. *corpora allata* [L. added body]: endocrine gland in insect head, behind brain, secreting hormone inducing larval molt.
corpus callosum (kôr′pŭs kă·lō′sŭm) pl. *corpora callosa* [L. hard body]: broad tract of transverse nerve fibers uniting cerebral hemispheres in mammals.
corpus luteum (kôr′pŭs lū′tē·ŭm) pl. *corpora lutea* [L. yellow body]: progesterone-secreting bodies in vertebrate ovaries, formed from remnants of follicles after ovulation.
corpuscle (kôr′pŭs′l) [L. dim. of *corpus,* body]: a small, rounded structure, cell, or body; e.g., blood corpuscle, renal corpuscle.
cortex (kôr′tĕks), pl. *cortices* [L. bark]: the outer layers of an organ or body part, e.g., adrenal cortex, cerebral cortex.
costa (kŏs′tȧ) [L. rib, side]: a rib or riblike supporting structure.
cotylosaur (kŏt′ĭ·lō sôr′) [Gr. *kotylē,* something hollow]: a member of a group of Permian fossil reptiles, evolved from labyrinthodont amphibian stock and ancestral to all other reptiles.
coxopodite, coxal (kŏks·ŏ′pō·dīt, kŏk′săl) [L. *coxa,* hip]: (1) the first, most basal joint of a segmental appendage of arthropods; (2) adjective.
cretinism (krē′tĭn·ĭz′m) [fr. L. *christianus,* a Christian]: an abnormal condition resulting from underactivity of the thyroid in the young mammal, specifically man.
crinoid (krī′noid) [Gr. *krinoeides,* lilylike]: a member of a class of echinoderms; a sea lily or feather star; also used as adjective.
Crossopterygii (krŏ′sŏp·tē′rĭ·jē) [Gr. *krossoi,* tassels]: a superorder of bony fishes, within the subclass of Sarcopterygii; the lobe-finned fishes.
Crustacea (krŭs·tā′shē ȧ) [L. *crusta,* shell, rind]: a class of mandibulate arthropods; crustaceans.
crystalloid (krĭs′tăl·oid) [Gr. *krystallos,* ice]: a system of particles within a medium, able to form crystals under appropriate conditions; a true solution.
Ctenophora (tē·nŏf′ō·rȧ): a phylum of radiate animals characterized by comb plates; the comb jellies.
cutaneous (kū·tā′nē·ŭs) [L. *cutis,* skin]: pertaining to the skin; e.g., cutaneous sense organ.
Cyanophyta (sī′·ă·nôf′·ĭtȧ) [Gr. *kyanos,* dark-blue]: the moneran phylum of blue-green algae.
cyclosis (sī·klō′sĭs) [Gr. *kyklos,* circle]: circular streaming and eddying of cytoplasm.
cydippid (sī′dĭp·ĭd): a larva of ctenophores.
cyphonautes (sī′fō·nôt′ēs) [Gr. *kyphos,* crooked, + *nautēs,* sailor]: a larva of ectoprocts.
cytochrome (sī′tō·krōm): one of a group of iron-containing hydrogen carriers in aerobic respiration.
cytolysis (sī·tŏl′ĭ·sĭs): dissolution or disintegration of a cell.
cyton (sī′ton): the nucleus-containing main portion (cell body) of a neuron.
cytoplasm (sī′tō plăz′m): the living matter of a cell between cell membrane and nucleus.

cytosine (sī'tō·sēn): a nitrogen base present in nucleotides and nucleic acids.

decapod (dĕk'ȧ·pŏd) [Gr. *deka,* ten]: ten-footed animal, specifically decapod (malacostracan) crustacean (e.g., lobster), decapod (cephalopod) mollusk (e.g., squid); order Decapoda; also used as adjective.

deciduous (dê·sĭd'ū ŭs) [L. *decidere,* to fall off]: to fall off at maturity, as in trees which shed foliage during the autumn.

dedifferentiation (dē·dĭf'ẽr·ĕn'shĭ·ā'shŭn): a regressive change toward a more primitive, embryonic, or earlier state; e.g., a process changing a highly specialized cell to a less specialized cell.

degrowth (dē'grōth): negative growth; becoming smaller.

dehydrogenase (dê·hī·drŏ'·jĕn·ās): an enzyme promoting dehydrogenation.

denaturation (dē·nă'tûr·ā'shŭn): disruption of the tertiary or secondary structure of a protein molecule.

dendrite (dĕn'drīt) [Gr. *dendron,* tree]: filamentous outgrowth of a nerve cell, conducting nerve impulses from its free end toward the cell body.

denitrify, denitrification (dē·nī'trĭ·fī): (1) to convert nitrates to ammonia and molecular nitrogen, as by denitrifying bacteria; (2) noun.

deoxyribose (dē·ŏk'sĭ·rī'bōs): a 5-carbon sugar having one oxygen atom less than parent-sugar ribose; component of deoxyribose nucleic acid (DNA).

Deuterostomia (dū'tẽr·ô·stō'mē·ȧ) [Gr. *deuteros,* second]: animals in which blastopore becomes anus and mouth forms as second embryonic opening opposite blastopore.

diabetes (dī'ȧ·bē'têz) [Gr. *diabainein,* to pass through]: abnormal condition marked by insufficiency of insulin, sugar excretion in urine, high blood-glucose levels.

diastole (dī·ăs'tô·lē) [Gr. *diastolē,* moved apart]: phase of relaxation of atria or ventricles, during which they fill with blood; preceded and succeeded by systole, i.e., contraction.

diastrophism (dī·ăs'trô·fĭz'm) [Gr. *diastrophē,* distortion]: geologic deformation of the earth's crust, leading to rise of land masses.

dichotomy (dī·kŏt'ô mĭ) [Gr. *dicha,* in two + *temnein,* to cut]: a repeatedly bifurcating pattern of branching.

diencephalon (dī'ĕn·sĕf'*a* lŏn) [Gr. *enkephalos,* brain]: hind-portion of the vertebrate forebrain.

differentiation (dĭf'ẽr·ĕn'shĭ·ā'shŭn): a progressive change toward a permanently more mature or advanced state; e.g., a process changing a relatively unspecialized cell to a more specialized cell.

diffusion (dĭ·fū'zhŭn) [L. *diffundere,* to pour out]: migration of particles from a more concentrated to a less concentrated region; the process tends to equalize concentrations throughout a system.

dimorphism (dī·môr'fĭz'm): difference of form between two members of a species, e.g., as between males and females; a special instance of polymorphism.

dipleurula (dī'ploor·ū·lȧ): hypothetical ancestral form of most deuterostomial animals, resembling developmental stage of hemichordates and echinoderms.

diplohaplontic (dĭp'lô·hăp·lŏn'tĭk): designating a life cycle with alternation of diploid and haploid generations.

disaccharide (dī·săk'ȧ rīd) [Gr. *sakcharon,* sugar]: a sugar composed of two monosaccharides; usually refers to 12-carbon sugars.

dissociation (dĭ·sō'sĭ·ā'shŭn) [L. *dissociare,* to dissociate]: the breakup of an electrolyte in water, resulting in the formation of free ions.

diurnal (dī·ûr'năl) [L. *diurnalis,* daily]: e.g., as in daily up-and-down migration of plankton in response to absence or presence of sunlight.

divergence (dī·vûr'jĕns) [L. *divergere,* to incline apart]: evolutionary development of dissimilar characteristics in two or more lines descended from the same ancestral stock.

diverticulum (dī'vẽr·tĭk'ū lŭm) [L. byway]: branch or sac off a canal or tube; e.g., digestive diverticulum.

DNA: abbreviation of deoxyribose nucleic acid.

doliolaria (dŏ·li·ô·lā'ri·ă) [L. *dolium,* small cask]: yolky barrel-shaped larva of crinoid echinoderms and transient larval stage in holothuroid development.

dominance: a functional attribute of genes; a dominant gene exerts its full effect regardless of the effect of its allelic partner.

DPN: abbreviation of diphosphopyridine nucleotide; a hydrogen carrier in respiration.

ductus arteriosus (dŭk'tŭs är·tē'rĭ·ō'sŭs): an artery present in the embryo and fetus of mammals, which conducts blood from the pulmonary artery to the aorta; shrivels at birth, when the lungs become functional.

duodenum (dū'ô·dē'nŭm) [L. *duodeni,* twelve each]: most anterior portion of the small intestine of vertebrates; continuation of the stomach, bile duct and pancreatic duct open into it.

Echinodermata (ê·kī'nô·dûrm·ā'tă): the phylum of spiny-skinned animals; includes starfishes, sea urchins.

Echiuroida (ê·k'ĭ·ûr·oi'dȧ): a phylum of wormlike, schizocoelomate animals, characterized by spines at hind end.

Ectoprocta (ĕk'tô·prŏk·tȧ): a phylum of sessile coelomate animals, in which the intestine is U-shaped, the mouth is surrounded by a lophophore with ciliated tentacles, and the anus opens outside this lophophore.

egestion (ê·jĕs'chŭn) [L. *egerere,* to discharge]: the elimination from the alimentary system of unusable and undigested material.

elasmobranch (ē·lăs′mô·brăngk) [Gr. *elasmos,* plate]: a member of a subclass of cartilage fishes (sharks and rays); also used as adjective.

electrolyte (ē·lĕk′trô·līt) [Gr. *ēlektron,* amber + *lytos,* soluble]: a substance which dissociates into ions in aqueous solution and so makes possible the conduction of electric current through the solution.

element (ĕl′ē·mĕnt): one of about 100 distinct natural or man-made types of matter, which, singly or in combination, compose all materials of the universe; an atom is the smallest representative unit of an element.

Eleutherozoa (ĕlū·thĕ·rô·zō′*ȧ*) [Gr. *eleutheros,* free]: a subphylum of echinoderms comprising all living classes except the crinoids.

elytron (ĕl′ĭ·trŏn) pl. *elytra* [Gr. cover, sheath]: e.g., the hardened forewings of beetles, the dorsal notopodial sheath of certain polychaete annelids.

embolus (ĕm′bô·lŭs) [Gr. *embolos,* peg, stopper]: blood clot formed within a blood vessel.

emboly (ĕm′bô·lĭ): invaginative gastrulation.

embryo (ĕm′brĭ·ō) [Gr. *en* in, + *bryein,* to swell]: an early developmental stage of an animal produced from a fertilized egg.

emulsion (ē·mŭl′shŭn) [L. *emulgere,* to milk out]: a colloidal system in which both the dispersed and the continuous phase are liquid.

endemic (ĕn·dĕm′ĭk) [Gr. belonging to a district]: pertaining to a limited locality; ecologically, occurring in a particular region only; opposite of cosmopolitan.

endergonic (ĕn′dẽr·gŏ·nĭk): energy-requiring, as in a chemical reaction.

endocrine (ĕn′·dô·krĭn) [Gr. *krinein,* to separate]: applied to type of gland which releases secretion not through a duct but directly into blood or lymph; functionally equivalent to hormone-producing.

endoplasm, endoplasmic (ĕn′dô·plăz′m): the inner portion of the cytoplasm of a cell, i.e., the portion immediately surrounding the nucleus; contrasts with ectoplasm or cortex, i.e., the portion of cytoplasm immediately under the cell surface.

energy (ĕn′ẽr·jĭ) [Gr. *energos,* active]: capacity to do work; the time rate of doing work is called power.

enterokinase (ĕn′tẽr·ô·kī′nās) [Gr. *kinētos,* moving]: an enzyme present in intestinal juice of vertebrates; it converts trypsinogen into trypsin.

enteropneust (ĕn′tẽr·ô·nūst) [Gr. *pnein,* to breathe]: a member of a class of hemichordates; an acorn worm.

enthalpy (ĕn′thăl·pĭ) [Gr. *enthalpein,* to warm in]: a measure of the amount of energy in a reacting system.

Entoprocta (ĕn′tô·prŏk′t*ȧ*): a phylum of sessile, pseudocoelomate animals, possessing a U-shaped alimentary tract, a mouth surrounded by a ring of ciliated tentacles, and an anus opening within this ring.

entropy (ĕn′trô·pĭ) [Gr. *entropia,* transformation]: a measure of the distribution of energy in a reacting system.

enzyme (ĕn′zīm) [Gr. *en,* in + *zymē,* leaven]: a protein produced within an organism, capable of accelerating a particular chemical reaction; a type of catalyst.

ephyra (ĕf′ĭ·r*ȧ*) [L. *ephyra,* a nymph]: free-swimming larval stage in scyphozoan coelenterates; larval jellyfish.

epiboly (ē·pĭb′ô·lĭ) [Gr. *epibolē,* throwing over]: gastrulation by overgrowth of animal region over vegetal region of embryo.

epididymis (ĕp′ĭ·dĭd′ĭ mĭs) [Gr. *didymos,* testicle]: the greatly coiled portion of the sperm duct adjacent to the mammalian testis.

epiglottis (ĕp′ĭ·glŏt′ĭs) [Gr. *glōssa,* tongue]: a flap of tissue above the mammalian glottis; contains elastic cartilage, and in swallowing folds back over the glottis, so closing the air passage to the lungs.

epithelium (ĕp′ĭ·thē′lĭ·ŭm) [Gr. *thēlē,* nipple]: tissue type in which the cells are packed tightly together, leaving little intercellular space.

esophagus (ē·sŏf′*ȧ*·gŭs) [Gr. *oisō,* I shall carry]: part of alimentary tract connecting pharynx and stomach.

estrogen (ĕs′trô·jĕn) [Gr. *oistros,* frenzy]: one of a group of female sex hormones of vertebrates.

estrus (ĕs′trŭs) [L. *oestrus,* gadfly]: egg production and fertilizability in mammals; e.g., estrus cycle, monestrous, polyestrous.

eurypterid (ū·rĭp′tẽr·ĭd) [Gr. *eurys,* wide]: extinct Paleozoic chelicerate arthropod.

eustachian (ū·stā′kĭ·*ȧ*n): applied to canal connecting middle-ear cavity with the nasopharynx of mammals.

exergonic (ĕk′sẽr·gŏ·nĭk): energy-yielding, as in a chemical reaction.

exocrine (ĕk′sô·krĭn): applied to type of gland which releases secretion through a duct.

exteroceptor (ĕk′stẽr·ô·sĕp′tẽr): a sense organ receptive to stimuli from external environment.

feces (fē′sēz) [L. *faeces,* dregs]: waste matter discharged from the alimentary system.

femur, femoral (fē′mẽr, fĕm′ô·r*ă*l) [L. thigh]: (1) thighbone of vertebrates, between pelvis and knee; (2) adjective.

fermentation (fûr′mĕn·tā′shŭn): synonym for anaerobic respiration, i.e., fuel combustion in the absence of oxygen.

fetus (fē′tŭs) [L. offspring]: prenatal stage of development in man and other mammals, following the embryonic stage; roughly from third month of pregnancy to birth.

fiber (fī′bẽr) [L. *fibra,* thread]: a strand or filament produced by cells but located outside cells.

fibril (fī′brĭl) [L. dim. of *fibra*]: a strand or filament produced by cells and located within cells.

fibrin, fibrinogen (fī′brĭn, fī·brĭn′ô·jĕn): (1) coagulated blood protein forming the bulk of a blood clot in vertebrates; (2) a protein present in blood which upon coagulation forms a clot.

fibula (fĭb′ū·lă) [L. buckle]: the usually thinner of the two bones between knee and ankle in the vertebrate hind limb.

filopodium (fī·lô·pō′dĭ·ŭm) [L. *filum*, thread]: a filamentous type of pseudopodium in sarcodine protozoa.

flagellate, flagellum (flăj′ĕ·lāt, -ŭm) [L. whip]: (1) equipped with one or more flagella; an organism possessing flagella; (2) a microscopic, whiplike filament serving as locomotor structure in flagellate cells.

follicle (fŏl′ĭ·k′l) [L. *folliculus*, small ball]: ball of cells; as in egg-containing balls within ovaries of many animals, or cellular balls at base of hair or feather.

Foraminiferida, foraminifera (fō·râmĭ·nĭ′fĕr′ĭ·dă, -nĭf′ĕr·ă) [L. *foramen*, hole]: an order of sarcodine protozoa, characterized by delicate calcareous shells with holes, through which pseudopods are extruded.

fovea centralis (fō′vê·ȧ sĕn·trā′lĭs) [L. central pit]: small area in optic center of mammalian retina; only cone cells are present here and stimulation leads to most acute vision.

fundus (fŭn′dŭs) [L., bottom]: the bottom or base of a hollow structure, i.e., the fundus of the mammalian stomach, farthest away from the intestinal opening.

funiculus (fū·nĭk′ū·lŭs) [L. dim of *funis*, rope]: tissue strand, as in attachment of stomach to body wall of ectoprocts.

gamete (găm′ēt): reproductive cell which must fuse with another before it can develop; sex cell.

ganglion (găng′glĭ·ŭn) [Gr. a swelling]: an aggregated collection of cell bodies of neurons typically less complex than a brain.

ganoid (găn′oid) [Gr. *ganos*, brightness]: pertaining to shiny, enamel-covered type of fish scale.

gastrin (găs′trĭn): a hormone produced by the stomach wall of mammals when food makes contact with the wall; stimulates other parts of the wall to secrete gastric juice.

Gastropoda (găs·trŏp′ô·dȧ): a class of mollusks; comprises snails and slugs.

Gastrotricha (găs′trŏt′rĭ·kȧ): class of minute, aquatic, pseudocoelomate animals, possessing cilialike bristles on the ventral side and often elsewhere; members of the phylum Aschelminthes.

gastrula, gastrulation (găs′troo·lȧ, -lā′shŭn): (1) a two-layered and later three-layered stage in the embryonic development of animals; (2) the process of gastrula formation.

gel (jĕl) [L. *gelare*, to freeze]: quasi-solid state of a colloidal system, where the solid particles form the continuous phase and the liquid forms the discontinuous phase.

gemmule (jĕm′ūl) [L. *gemma*, bud]: vegetative, multicellular bud of (largely freshwater) sponges.

gene (jēn): a segment of a chromosome, definable in operational terms: repository of a unit of genetic information.

genome (jēn′ōm): the totality of genes in a haploid set of chromosomes, hence the sum of all different genes in a cell.

genus (jē′nŭs) [L. race]: a rank category in taxonomic classification between species and family; a group of very closely related species.

globulin (glŏb′ū·lĭn): one of a class of proteins present in blood plasma of vertebrates; may function as antibody.

glochidia (glô·kĭd′ĭȧ) [Gr. *glochis*, arrow point]: pincer-equipped bivalve larvae of freshwater clams, parasitic on fish.

glomerulus (glô·mĕr′ū·lŭs) [L. dim. of *glomus*, ball]: small meshwork of blood capillaries or channels; e.g., in the nephron of vertebrates, the renal organ of hemichordates.

glottis (glŏt′ĭs) [Gr. *glōssa*, tongue]: slitlike opening in the mammalian larynx, formed by the vocal cords.

glucogenic (glōō′kô·jĕn′ĭk) [Gr. *gleukos*, sweet wine]: glucose-producing, especially amino acids which, after deamination, metabolize like carbohydrates.

glucose (glōō′kōs): a 6-carbon sugar; principal form in which carbohydrates are transported from cell to cell.

glycerin (glĭs′ĕr·ĭn) [Gr. *glykeros*, sweet]: an organic compound possessing a 3-carbon skeleton; may unite with fatty acids and form a fat.

glycogen (glī′kô·jĕn): a polysaccharide consisting of joined glucose units; a principal storage form of carbohydrates.

glycolysis (glī·kŏl′·ĭ·sĭs): respiratory breakdown of glucose (or starch or glycogen) to pyruvic acid; anaerobic respiration of carbohydrates.

goiter (goi′tĕr) [L. *guttur*, throat]: an enlargement of the thyroid gland; may be an overgrowth resulting in excessive secretion of thyroid hormone or may be a compensatory overgrowth occasioned by undersecretion of thyroid hormone.

Golgi body (gôl′jê): a cytoplasmic organelle playing a role in the manufacture of certain cell secretions.

gonad (gōn′ăd) [Gr. *gonē*, generator]: reproductive organ; collective term for testes and ovaries.

gradation (gra·dā′shŭn) [L. *gradus*, step]: leveling of land by the geologic effects of erosion.

graptolite (grăp′tô·līt) [Gr. *graptos*, written]: one of a phylum group of fossil exoskeletons of uncertain affinities; may be related to coelenterates or pterobranch hemichordates.

hectocotylus (hĕk′tô kŏt′ĭ·lŭs) [Gr. *hekto-*, hundred, + kōtylē, cup]: modified arm of male cephalopod mollusks serving in sperm-transfer to female.

helix (hē′liks) [L. a spiral]: spiral shape; e.g., polypeptide chain, snail shell.

heme (hēm): an iron-containing, red, cyclic pyrrol pigment.

hemoglobin (hē′mô·glō′bĭn) [L. *globus,* globe]: oxygen-carrying constituent of blood; consists of red pigment heme and protein globin.

hemophilia (hē′mô fĭl′ĭȧ) [Gr. *philos,* loving]: a hereditary disease in man characterized by excessive bleeding from even minor wounds; clotting mechanism is impaired by failure of blood platelets to rupture after contact with torn edges of blood vessels.

Hemichordata (hĕm′ĭ·kŏr dā′tȧ): a phylum of deuterostomial, enterocoelomate animals.

hepatic (hê·păt′ĭk) [Gr. *hēpar,* liver]: pertaining to the liver; as in hepatic vein, hepatic portal vein.

herbivore (hûr′bĭ·vōr) [L. *herba,* herb + *vorare,* to devour]: a plant-eating animal.

hermaphrodite (hûr·măf′rô·dīt) [fr. Gr. *Hermes* + *Aphrodite*]: an organism possessing both male and female reproductive structures.

heterozygote, heterozygous (hĕt′ĕr·ô·zī′gōt) [Gr. *zygōtos,* yoked]: (1) an animal in which a pair of alleles for a given trait consists of different (e.g., dominant and recessive) kinds of genes; (2) adjective.

holothuroid (hŏl·ô·thū′·roid) [L. *holothuria,* water polyp]: a member of a class of echinoderms; a sea cucumber; also used as adjective.

hominid (hŏm′ĭ·nĭd) [L. *homo,* man]: a living or extinct man or manlike type; the family of man or pertaining to this family.

hominoid (hŏm′ĭ·noid): a superfamily including hominids, the family of man, and pongids, the family of apes.

homozygote, homozygous (hō′mô·zī′gōt) [Gr. *zygōtos,* yoked]: (1) an animal in which a pair of alleles for a given trait consists of the same (e.g., either dominant or recessive, but not both) kinds of genes; (2) adjective.

hormone (hôr′mōn) [Gr. *hormaein,* to excite]: a secretion produced by an endocrine gland within an animal and affecting another part of that animal.

humerus (hū′mēr·ŭs) [L. shoulder]: the bone of the vertebrate upper forelimb, between shoulder and elbow.

humoral (hū′mēr·ăl) [L. *humor,* moisture, liquid]: pertaining to body fluids, esp. biologically active chemical agents carried in body fluids; e.g., hormones or similar substances.

humus (hū′mŭs) [L. soil]: the organic portion of soil.

hybrid (hī′brĭd) [L. *hibrida,* offspring of tame sow and wild boar]: an animal heterozygous for one or more (usually many) gene pairs.

hydranth (hī′drănth) [Gr. *anthos,* flower]: flowerlike terminal part of hydroid polyp, containing mouth and tentacles; a feeding polyp.

hypertonic, hypertonicity (hī′pēr·tŏn′ĭk): (1) exerting greater osmotic pull than the medium on the other side of a semipermeable membrane, hence possessing a greater concentration of particles and acquiring water during osmosis; (2) noun.

hypothalamus (hī′pô·thăl′ȧ mŭs) [Gr. *thalamos,* chamber]: a region of the forebrain, containing various centers of the autonomic nervous system.

hypothesis (hī·poth′ê sĭs) [Gr. *tithenai,* to put]: a guessed solution of a scientific problem; must be tested by experimentation and, if not validated, must then be discarded.

hypotonic, hypotonicity (hī′pô·tŏn′ĭk): (1) exerting lesser osmotic pull than the medium on the other side of a semipermeable membrane; hence possessing a lesser concentration of particles and losing water during osmosis; (2) noun.

imago, imaginal (ĭ·mā′gō, ĭ·măj′ĭ·nȧl) [L., image]: (1) an adult insect; (2) adjective.

induction, inductor (ĭn·dŭk′shŭn) [L. *inducere,* to induce]: (1) process in embryo in which one tissue or body part causes the differentiation of another tissue or body part; (2) an embryonic tissue which causes the differentiation of another.

instar (ĭn′stär) [L. likeness, form]: period between consecutive molts in insect development.

insulin (ĭn′sū·lĭn) [L. *insula,* island]: a hormone produced by the islets of Langerhans in the vertebrate pancreas; promotes the conversion of blood glucose into tissue glycogen.

integument (ĭn·tĕg′ū·mĕnt) [L. *integere,* to cover]: covering; external coat, skin.

intermedin (ĭn·tēr·mē′dĭn): hormone produced by the midportion of the vertebrate pituitary gland; adjusts degree of extension of pigment cells in skin of certain vertebrates, e.g., frogs.

interoceptor (ĭn′tēr·ô·sĕp′tēr): a sense organ receptive to stimuli generated in the interior of an animal.

invagination (ĭn·văj′ĭ·nā′shŭn) [L. *in,* in + *vagina,* sheath]: local infolding of a layer of tissue, leading to the formation of a pouch or sac; as in invagination during a type of embolic gastrulation.

involution (ĭn′vô·lū′shŭn) [L. *involutio,* a rolling or folding in]: inrolling of a tissue layer underneath another one, as in a type of embolic gastrulation.

ion, ionization (ī′ŏn, -ĭ·zā′shŭn) [Gr. *ienai,* to go]: (1) an electrically charged atom or group of atoms; (2) addition or removal of electrons from atoms.

isogamy (ī·sŏg′ȧ·mĭ): sexual fusion in which the gametes of opposite sex types are structurally alike.

isolecithal (ī′sô·lĕs′ĭ·thăl): pertaining to eggs with yolk evenly distributed throughout egg cytoplasm.

isotonic (ī′sô·tŏn′ĭk): exerting same osmotic pull as medium on other side of a semipermeable membrane, hence possessing the same concentration of particles; net gain or loss of water during osmosis is zero.

isotope (ī′sô·tōp) [Gr. *topos,* place]: one of several possible forms of a chemical element, differing from other forms in atomic weight but not in chemical properties.

karyogamy (kăr′ĭ ŏg′·*ȧ*·mĭ) [Gr. *karyon,* nut]: fusion of nuclei in process of fertilization.

kenozooid (kē′nô·zō′oid) [Gr. *koinos,* common, communal]: stalk- and stolon-forming polymorphic individuals of an ectoproct colony.

keratin (kĕr′*ȧ*·tĭn) [Gr. *keratos,* horn]: a protein formed by certain epidermal tissues, e.g., those of vertebrate skin.

ketogenic, ketone, ketose (kê′tô·jĕn′ĭk, -tōn, -tōs): (1) keto-acid-producing, especially amino acids which after deamination metabolize like fatty acids; (2) organic compound possessing a —CO— grouping; (3) one of a series of sugars characterized by a (nonterminal) ketone group.

kinetosome (kĭ·nĕt′ô·sōm) [Gr. *kinētos,* moving]: granule at base of flagellum, presumably motion-controlling.

Kinorhyncha (kĭn′ô·rĭng′k*ȧ*): a class of pseudocoelomate animals; members of the phylum Aschelminthes.

labium, labial (lā′bĭ·ŭm, -ăl) [L. lip]: (1) any liplike structure; especially underlip, posterior to mouth, in insect head; (2) adjective.

labrum (lā′brŭm) [L. lip]: a liplike structure; especially upper lip, anterior to mouth, in arthropod head.

labyrinthodont (lăb′ĭ rĭn′thô·dŏnt) [Gr. *labyrinthos,* labyrinth]: extinct, late-Paleozoic fossil amphibian.

lacteal (lăk′tê·*ă*l) [L. *lactis,* milk]: lymph vessel in a villus of intestinal wall of mammals.

lactogenic (lăk′tô jĕn′ĭk): milk-producing; as in lactogenic hormone, secreted by vertebrate pituitary.

lacuna (lä·kū′n*ȧ*) [L. gap]: a cavity or channel; e.g., in bone or cartilage, where cells are embedded in solid substance.

lagena (l*ȧ*·jēn′*ȧ*) [L. large flask]: portion of the primitive vertebrate (fish) ear in which sound is translated into nerve impulses; evolutionary forerunner of cochlea.

larva (lär′v*ȧ*), pl. *larvae* (-vē) [L. mask]: period in developmental history of animals between embryo and adult; the larval period begins at hatching and terminates at metamorphosis.

larynx (lăr′ĭngks) [Gr.]: voice box; sound-producing organ in mammals.

lemniscus (lĕm·nĭs′·kŭs) [L. a ribbon hanging down]: an elongated, paired, interior extension of the body wall in the anterior region of acanthocephalan worms.

leucocyte (lū′kō·sīt) [Gr. *leukos,* white]: a type of white blood cell in vertebrates characterized by a beaded, elongated nucleus.

leukemia (lū·kē′mĭ·*ȧ*): a cancerous condition of blood, characterized by overproduction of leucocytes.

lipase (lĭ′pās) [Gr. *lipos,* fat]: an enzyme promoting the conversion of fat into fatty acids and glycerin, or the reverse.

lipid, lipoid (lĭp′ĭd): (1) fat, fatty, pertaining to fat; (2) fatlike.

lithosphere (lĭth′ô·sfēr) [Gr. *lithos,* stone]: collective term for the solid, rocky component of the earth's surface layers.

littoral (lĭt′ô·răl) [L. *litus,* seashore]: the sea floor from the shore to the edge of the continental shelf.

lophophore (lō′fô·fōr) [Gr. *lophos,* crest]: tentacle-bearing arm in anterior region of certain coelomates (lophophorate animals); serves in food-trapping.

lorica (lô·rī′k*ȧ*) [L. sheath, cover]: secreted protective covering, e.g., as in some ciliate protozoa.

luciferase, luciferin (lū·sĭf′ēr·ās, -ĭn) [L. *lux,* light]: (1) enzyme contributing to the production of light in animals; (2) a group of various substances essential in the production of bioluminescence.

lymph (lĭmf) [L. *lympha,* goddess of moisture]: the body fluid outside the blood circulation.

lymphocyte (lĭm′fô·sīt): a type of white blood cell of vertebrates characterized by a rounded or kidney-shaped nucleus.

macronucleus (măk′rô·nū′klê·ŭs): a large type of nucleus found in ciliate protozoa; controls all but reproductive functions in these organisms.

madreporite (măd′rê pô rīt) [It. *madre,* mother, + *poro,* passage]: a sievelike opening on the surface of echinoderms, connecting the water-vascular system with the outside.

Malacostraca (măl′*ȧ*·kŏs′·trä·kă) [Gr. *malakostraka,* soft-shelled]: a subclass of crustaceans.

maltose (môl′tōs): a 12-carbon sugar formed by the union of two glucose units.

mandible (măn′dĭ·b′l) [L. *mandibula,* jaw]: in arthropods, one of a pair of mouth appendages, basically biting jaws; in vertebrates, the principal bone or cartilage of the lower jaw.

marsupial (mär·sū′pĭ·ăl) [Gr. *marsypion,* little bag]: a pouched mammal, member of the mammalian infraclass Metatheria.

mastax (măs′t*ȧ*ks) [L. *masticare,* to chew]: horny, toothed chewing apparatus in pharynx of rotifers.

Mastigophora (măs′tĭ gô′fōr*ȧ*): a subphylum of primarily unicellular flagellate protozoa; zooflagellates.

maxilla (măk·sĭl′*ȧ*) [L.]: in arthropods, one of the head appendages; in vertebrates, one of the upper jawbones.

maxilliped (măk·sĭl′ĭ·pĕd): one of three pairs of segmental appendages in decapod crustacea, located posterior to the maxillae.

medulla (mē·dŭl′ȧ) [L.]: the inner layers of an organ or body part, e.g., adrenal medulla; the medulla oblongata is a region of the vertebrate hindbrain which connects with the spinal cord.

medusa (mē·dū′sȧ): the free-swimming stage in the life cycle of coelenterates; a jellyfish.

meiosis (mī·ō′sĭs) [Gr. *meioun*, to make smaller]: process occurring during gamete maturation in Metazoa, in which the chromosome number is reduced by half; compensates for the chromosome-doubling effect of fertilization.

melanin (mĕl′ȧ·nĭn) [Gr. *melas*, black]: black animal pigment, in cytoplasmic granules of chromatophore cells known as melanocytes.

menopause (mĕn′ō·pôz) [Gr. *menos*, month + *pauein*, to cause to cease]: the time at the end of the reproductive period of (human) females when menstrual cycles cease to occur.

menstruation (mĕn′stro͞o·ā′shŭn) [L. *mensis*, month]: the discharge of uterine tissue and blood from the vagina in man and apes at the end of a menstrual cycle in which fertilization has not occurred.

mesencephalon (mĕs′ĕn·sĕf′ȧ·lŏn) [Gr. *enkephalos*, brain]: the vertebrate midbrain.

mesenchyme (mĕs′ĕng·kīm) [Gr. *enchyma*, infusion]: non-epithelial mesoderm, especially abundant in embryos and in primitive adult animals; often jelly-secreting.

mesogloea (mĕs′ō·glē′ȧ) [Gr. *gloios*, glutinous substance]: the often jelly-containing layer between the ectoderm and endoderm of coelenterates and comb jellies.

metabolism (mĕ·tăb′ō·lĭz′m) [Gr. *metabolē*, change]: a group of life-sustaining processes including principally nutrition, production of energy in usable form (respiration), and synthesis of more living substance.

metabolite (mĕ·tăb′ō·līt): any chemical participating in metabolism; a nutrient.

metachronous (mĕt′ȧ·crō′nŭs) [Gr. *chronos*, time]: pertaining to successive beating of adjacent cilia in a row, resulting in wavelike progression of beat.

metamorphosis (mĕt′ȧ môr′fō·sĭs) [Gr. *metamorphoun*, to transform]: transformation of a larva into an adult.

metaphase (mĕt′ȧ·fāz): a stage during mitotic division in which the chromosomes line up in a plane at right angles to the spindle axis.

metencephalon (mĕt′ĕn·sĕf′ȧ·lŏn) [Gr. *enkephalos*, brain]: anterior portion of vertebrate hindbrain.

micron (mī′krŏn), pl. *microns, micra:* one-thousandth part of a millimeter, a unit of microscopic length.

mictic (mĭk′tĭk) [Gr. *mixis*, act of mixing]: pertaining to fall and winter eggs of rotifers, which if fertilized produce males and if not fertilized produce females.

mimicry (mĭm′ĭk·rĭ) [Gr. *mimos*, mime]: the superficial resemblance of certain animals, particularly insects, to other more powerful or more protected ones, resulting in a measure of protection for the mimics.

mineral (mĭn′ēr·ăl) [L. *minera*, ore]: a compound or substance of the inorganic world; an inorganic material.

miracidium (mī′ră sĭd′ĭ·ŭm): a larval stage in the life cycle of flukes; develops from an egg and gives rise in turn to a sporocyst larva.

mitochondrion (mī′tō·kŏn′drĭ·ŏn) [Gr. *mitos*, thread, + *chondros*, grain]: a cytoplasmic organelle serving as site of respiration.

mitosis (mī·tō′sĭs): a form of nuclear division characterized by complex chromosome movements and exact chromosome duplication.

molecule (mŏl′ō·kūl) [L. *moles*, mass]: a compound in which the atoms are held together by covalent bonds.

Mollusca, mollusk (mŏ·lŭs′kȧ, mŏl′ŭsk) [L. *molluscus*, soft]: (1) a phylum of nonsegmented schizocoelomate animals; (2) a member of the phylum Mollusca.

Monera (mŏn·ē′rȧ) [Gr. *monos*, alone]: a major category of living organisms comprising the bacteria and the blue-green algae; characterized in part by absence of true nuclei or chromosomes.

monestrous (mŏn·ĕs′trŭs) [Gr. *oistros*, frenzy]: having a single estrus (egg-producing) cycle during a given breeding season.

monophyletic (mŏn′ō·fī·lĕt′ĭk) [Gr. *phylon*, tribe]: developed from a single ancestral type; contrasts with polyphyletic.

monosaccharide (mŏn′ō·săk′ȧ·rīd) [Gr. *sakcharon*, sugar]: a simple sugar such as 5- and 6-carbon sugars.

morphogenesis (môr′fō·jĕn′ē·sĭs): development of size, form, and other architectural features of animals.

morula (mŏr′ū·lȧ) [L. little mulberry]: solid ball of cells resulting from cleavage of egg; a solid blastula.

mucosa (mū·kō′sȧ) [L. *mucosus*, mucus]: a mucus-secreting membrane, e.g., the inner lining of the intestine.

mutation (mū·tā′shŭn) [L. *mutare*, to change]: a stable change of a gene, such that the changed condition is inherited by offspring cells.

Mycophyta (mī′kō·fī′tȧ) [Gr. *mykēs*, mushroom]: the phylum comprising the fungi.

myelencephalon (mī′ĕ·lĕn·sĕf′ȧ lŏn) [Gr. *myelos*, marrow]: the most posterior part of the vertebrate hindbrain, confluent with the spinal cord; the medulla oblongata.

myelin (mī′ĕ·lĭn): a fatty material surrounding the axons of nerve cells in the central nervous system of vertebrates.

myofibril (mīō·fī′brĭl): a contractile filament within a cell, especially in a muscle cell or muscle fiber.

myosin (mī′ō·sĭn): a protein which can be isolated from muscle; forms an integral component of the contraction machinery of muscle.

Myriapoda (mĭ′·rĭ·ȧ′pŏ·dȧ) [Gr. *myrias,* ten thousand]: collective descriptive term for the arthropod classes Chilopoda, Diplopoda, Symphyla, and Pauropoda.

myxedema (mĭk′sē·dē′mȧ) [Gr. *myxa,* slime + *oidēma,* a swelling]: a disease resulting from thyroid deficiency in the adult characterized by local swellings in and under the skin.

Myxophyta (mĭks·ŏf′ĭ tȧ): the protistan phylum of slime molds.

nacre, nacreous (nā′kẽr, -krê·ŭs) [Pers. *nakdra,* pearl oyster]: (1) mother-of-pearl; (2) adjective.

nauplius (nô′plĭ·ŭs) [L. shellfish]: first in a series of larval phases in crustacea.

nekton (nĕk′tŏn) [Gr. *nēktos,* swimming]: collective term for the actively swimming animals in the ocean.

Nematoda (nĕm′ȧ tōd·ȧ): the class of roundworms, of the pseudocoelomate phylum Aschelminthes.

Nematomorpha (nĕm′ȧ·tō·môr′fȧ): the class of hairworms, of the pseudocoelomate phylum Aschelminthes.

Nemertina (nĕm·ẽr tĭn′ȧ): ribbon or proboscis worms, an acoelomate phylum (also called *Rhynchocoela*).

neoteny (nē′ō·tēnē) [Gr. *neo,* new + *teinein,* extend]: a permanent, sexually mature, larval state.

nephric, nephron (nĕf′rĭk, -rŏn): (1) pertaining to a nephron or excretory system generally; (2) a functional unit of the vertebrate kidney.

nephromixium (nĕf′rō·mĭks′ĭ·ŭm): joint excretory and reproductive unit in certain polychaete annelids, composed of solenocytic nephridia and a gonoduct from the coelom which opens into the nephridial tubule.

neritic (nê·rĭt′ĭk) [fr. Gr. *Nereus,* a sea god]: oceanic habitat zone, subdivision of the pelagic zone, comprising the open water above the continental shelf, i.e., above the littoral.

neuron (nū′rŏn) [Gr. nerve]: nerve cell, including cyton, dendrites, and axons.

nictitating (nĭk′tĭ·tāt′ĭng) [L. *nictare,* wink]: pertaining to thin transparent eyelid-like membrane in many vertebrates, which opens and closes laterally across cornea.

nidamental (nĭd′ȧ·mĕnt·ăl) [L. *nidus,* nest]: pertaining to gland in female cephalopod mollusks which secretes protective capsule around eggs.

nitrify, nitrification (nī′trĭ·fī, -fĭ·kā′shŭn): (1) to convert ammonia and nitrites to nitrates, as by nitrifying bacteria; (2) noun.

nuchal (nū′kăl) [L. *nucha,* nape of neck]: pertaining to nape of neck; e.g., nuchal ligament in tetrapods, aiding in holding head in horizontal position.

nucleic acid (nū·klē′ĭk): one of a class of molecules composed of joined nucleotide complexes; the principal types are deoxyribose nucleic acid (DNA) and ribose nucleic acid (RNA).

nucleolus (nū·klē′ō·lŭs): an RNA-containing body within the nucleus of a cell; a derivative of chromosomes.

nucleoprotein (nū′klê·ō-): a molecular complex composed of nucleic acid and protein.

nucleotide (nū′klê·ō·tīd): a molecule consisting of joined phosphate, 5-carbon sugar (either ribose or deoxyribose), and a purine or a pyrimidine (adenine, guanine, uracil, thymine, or cytosine).

nucleus (nū′klê·ŭs) [L. a kernel]: a body present in all cell types except those of the Monera, and consisting of external nuclear membrane, interior nuclear sap, and chromosomes and nucleoli suspended in the sap.

nutrient (nū′trĭ·ĕnt) [L. *nutrire,* to nourish]: a substance usable in metabolism; a metabolite; includes inorganic materials and organic materials (foods).

ocellus (ō·sĕl′ŭs) [L. dim of *oculus,* eye]: eye or eyespot, of various degrees of structural and functional complexity; in arthropods, a simple eye, as distinct from a compound eye.

olfaction, olfactory (ŏl·făk′shŭn, -tō·rĭ) [L. *olfacere,* to smell]: (1) the process of smelling; (2) pertaining to smell.

ommatidium (ŏm′ȧ tĭd′ĭ·ŭm) [Gr. *omma,* eye]: single visual unit in compound eye of arthropods.

omnivore (ŏm′nĭ·vōr) [L. *omnis,* all]: an animal which may subsist on plant foods, animal foods, or both.

Oncopoda (ŏn·kŏ′pô·dȧ) [Gr. *onkos,* bulk]: tentative phylum comprising three small groups related to arthropods, viz., Onychophora, Tardigrada, Pentastomida.

Onychophora (ŏnĭ·kŏ′fŏr·ȧ) [Gr. *onych,* claw]: a subphylum of Oncopoda, comprising *Peripatus* and related types.

oögamy (ō·ŏg′ȧ·mĭ): sexual fusion in which the gametes of opposite sex type are unequal, the female gamete being an egg, i.e., nonmotile, the male gamete being a sperm, i.e., motile.

operculum (ō·pûr′kū·lŭm) [L. a lid]: a lidlike structure.

ophiuroid (ō·ē′roid) [Gr. *ophis,* snake]: a member of a class of echinoderms; a brittle star; also used as adjective.

organ (ôr′găn) [fr. Gr. *organon,* tool, instrument]: a group of different tissues joined structurally and cooperating functionally to perform a composite task.

organic (ôr·găn′ĭk): pertaining to organisms or living things generally; chemically, compounds of carbon of nonmineral origin.

organelle (ôr·găn·el′): a formed body in the cytoplasm of a cell; a cytoplasmic structure.

organism (ôr′găn·ĭz′m): an individual living creature, either unicellular or multicellular.

ornithine (ôr′nĭ·thēn) [Gr. *ornithos,* bird]: an amino acid which, in the liver of vertebrates, contributes to the conversion of ammonia and carbon dioxide into urea.

osmosis (ŏs·mō′sĭs) [Gr. *ōsmos,* impulse]: the process in which water migrates through a semipermeable membrane, from a side containing a lesser concentration of particles to the side containing a greater concentration; migration continues until particle concentrations are equal on both sides.

ossicle (ŏs′ĭ·k'l) [L. dim. of *ossis*, bone]: a small bone or hard bonelike supporting structure.

Osteichthyes (ŏs·tē·ĭk′thĭ·ēz): a class of vertebrates, comprising the bony fishes.

ostium (ŏs′tĭ·ŭm) [L. door]: orifice or small opening; e.g., one of several pairs of lateral pores in arthropod heart, pore for entry of water in certain sponges.

ovary (ō′vȧ·rĭ): egg-producing organ of animals.

oviparity, oviparous (ō′vĭ·păr′ĭ·tĭ, ō·vĭp′ȧrŭs) [L. *parere*, to bring forth]: (1) reproductive pattern in which eggs are released by the female and offspring development occurs outside the maternal body; (2) adjective.

ovoviviparity, ovoviviparous (ō′vô·vĭv′ĭ·păr′ĭ·tĭ, ō′vô·vī vĭp′ȧ·rŭs): (1) reproductive pattern in which eggs develop within the maternal body, but without nutritive or other metabolic aid by the female parent; offspring are born as miniature adults; (2) adjective.

ovulation (ō′vŭ·lā′shŭn): expulsion of an egg from the ovary and deposition of egg into the oviduct.

oxidation (ŏk′sĭ·dā′shŭn): one half of an oxidation-reduction (redox) process at the end of which the free energy change is negative, i.e., the process is exergonic and the endproducts are more stable than the starting materials; often takes the form of removal of hydrogen or electrons from a compound, as in respiration.

paleoniscoid (pā′lē·ô·nĭs′koid): extinct Devonian bony fish, ancestral to modern bony fishes, lungfishes, and lobe-fin fishes.

paleontology (pā′lē·ŏn·tŏl′ô·gĭ) [Gr. *palaios*, old]: study of past geologic times, principally by means of fossils.

Paleozoic (pā′lē·ô·zō′ĭk) geologic era between the Precambrian and the Mesozoic, dating approximately from 500 to 200 million years ago.

palp (pălp) [L. *palpus*, feeler]: a feelerlike appendage; e.g., labial palp in clams, pedipalp in chelicerate arthropods.

Pantopoda (păn·tŏ′pôdȧ) [Gr. *pantos*, all]: a subphylum of Oncopoda, comprising Linguatula and related types.

papilla (pȧ·pĭl′ȧ) [L. nipple]: any small nipplelike projection.

parasite (păr′ȧ·sīt) [Gr. *sitos*, food]: an organism living symbiotically on or within a host organism, more or less detrimental to the host.

parasympathetic (păr′ȧ·sĭm′pȧ·thĕt′ĭk): applied to a subdivision of the autonomic nervous system of vertebrates; centers are located in brain and most anterior part of spinal cord.

parathyroid (păr′ȧ·thī′roid): an endocrine gland of vertebrates, usually paired, located near or within the thyroid; secretes parathormone, which controls calcium metabolism.

parenchyma (pȧ·rĕng′kĭ mȧ) [Gr. *para* + *en*, in + *chein*, to pour]: name applied to mesenchyme tissues of acoelomate animals.

parthenogenesis (pär′thē·nô·jĕn′ē·sĭs) [Gr. *parthenos*, virgin]: development of an egg without fertilization; occurs naturally in some animals (e.g., rotifers) and may be induced artificially in others (e.g., frogs).

pathogenic (păth′ô jĕn′ĭk) [Gr. *pathos*, suffering]: disease-producing.

Pauropoda (pô·rŏ′pô dȧ) [Gr. *pauros*, small]: a small class of myriapod mandibulate arthropods, probably related to millipedes.

pedicellaria (pĕd′ĭ·sĕl′ā′rĭ·ȧ) [L. *pedicellus*, little stalk]: a pincerlike structure on the surface of echinoderms; protects skin gills.

pedipalp (pĕd′ĭ·pălp): one of a pair of head appendages in chelicerate arthropods.

peduncle (pē·dŭng′k'l) [L. *pedunculus*, little foot]: a stalk or stemlike part.

pelagic (pē·lăj′ĭk) [Gr. *pelagos*, ocean]: oceanic habitat zone, comprising the open water of an ocean basin; subdivided into the neritic zone and the oceanic zone.

Pelecypoda (pĕ′lē·sĭp′ô·dȧ): a class of the phylum Mollusca, comprising clams, mussels, oysters.

pellicle (pĕl′ĭ·k'l) [L. dim. of *pellis*, skin]: a thin, membranous surface "skin", as on the exterior of many protozoa.

Pelmatozoa (pĕl·mă′tô·zō′ȧ) [Gr. *pelma*, sole of foot]: a subphylum of echinoderms, comprising only the crinoids among living groups.

penis (pē′nĭs) [L.]: a copulatory organ containing the terminal portions of the sperm duct.

pepsin (pĕp′sĭn) [Gr. *peptein*, to digest]: a protein-digesting enzyme present in gastric juice of vertebrates.

peptidase (pĕp′tĭ·dās): an enzyme promoting the liberation of individual amino acids from a peptide, i.e., an amino acid complex smaller than a whole protein.

peptide (pĕp′tīd): the type of bond formed when two amino acid units are joined end to end; the resulting double unit is a dipeptide, and a joining of many amino acid units into a chain results in a *polypeptide*, the basic structural component of a protein molecule.

Perissodactyla (pē·rĭs′ô·dăk′tĭ·lȧ) [Gr. *perissos*, odd]: the order of odd-toed ungulate mammals; includes horses, tapirs, rhinoceroses.

peritoneum (pē·rĭ′tô·nē′ŭm) [Gr. *peritonos*, stretched over]: a mesodermal epithelial membrane lining the coelomic body cavity in coelomate animals.

pH: a symbol denoting the relative concentration of hydrogen ions in a solution; pH values run from 0 to 14, and the lower the value, the more acid is a solution, i.e., the more hydrogen ions it contains.

Phaeophyta (fē′·ō·fī′t·ȧ): the phylum of brown algae.

pharynx (făr′ĭngks) [Gr.]: the part of the alimentary tract between mouth cavity and esophagus; in vertebrates it is also part of the air channel from nose to larynx.

Phoronida (fō′rŏn′ĭ·dȧ): a phylum of wormlike, marine, tube-dwelling, coelomate animals, characterized by a lophophore.

phosphagen (fŏs′fă·jĕn): collective term for compounds such as creatine-phosphate and arginine-phosphate, which store and may be sources of high-energy phosphates.

phosphorylation (fŏs′fō·rĭ·lā′shŭn): the addition of a phosphate group (for example, $-H_2PO_3$) to a compound.

phrenic (frĕn′ik) [Gr. *phrēn,* diaphragm]: pertaining to the diaphragm, e.g., phrenic nerve, innervating the diaphragm.

phylogeny (fī·lŏj′ē·nĭ) [Gr. *phylon,* race, tribe]: the study of evolutionary descent and interrelations of animal groups.

phylum (fī′lŭm), pl. *phyla:* a category of taxonomic classification, ranked above class.

physiology (fĭz′ĭ·ŏl′ō·jĭ) [Gr. *physis,* nature]: study of living processes, activities, and functions generally; contrasts with morphology, the study of structure.

pilidium (pī·lĭd′ĭ·ŭm) [L. *pilus,* hair]: a larval type characteristic of many nemertine worms.

pinacocyte (pĭn′ȧ·kō·sīt′) [Gr. *pinax,* tablet]: an epithelial cell type on the body surface of sponges.

pineal (pĭn′ē·ăl) [L. *pinea,* pine cone]: a structure in the brain of vertebrates; functions as a median dorsal eye in a few (e.g., lampreys), but does not have a demonstrable function in most.

pinocytosis (pĭ′nō·sī·tō′sĭs) [Gr. *pinein,* to drink]: cell "drinking"; the intake of fluid droplets into a cell.

pinna, pinnule (pĭn′ȧ, pĭn′ūl) [L. feather, fin]: a featherlike structure; e.g., the pinnules on the arms of crinoids; cf. also Pinnipedia, fin-possessing members of the mammalian order Carnivora.

pituitary (pĭ·tū′ĭ·tĕrĭ) [L. *pituita,* phlegm]: a composite endocrine gland in vertebrates, attached ventrally to the brain; composed of anterior, intermediate, and posterior lobes, each representing a functionally separate gland.

placenta (plȧ·sĕn′tȧ) [L. cake]: a tissue complex formed in part from the inner lining of the uterus and in part from the chorion of the embryo; develops in most mammals and serves as mechanical, metabolic, and endocrine connection between the adult female and the embryo during pregnancy.

planarian (plȧ·nâr′ĭ·ăn) [L. *planarius,* level]: any member of the class of free-living flatworms.

plankton (plăngk′tŏn) [Gr. *planktos,* wandering]: collective term for the passively floating or drifting flora and fauna of a body of water; consists largely of microscopic organisms.

planula (plăn′ū·lȧ) [L. dim. of *planus,* flat]: basic larval form characteristic of coelenterates.

plastron (plăs′trŏn) [fr. It. *piastrone,* breastplate]: the ventral shell part, as in turtles.

Platyhelminthes (plăt′ĭ·hĕl·mĭn′thēz) [Gr. *platys,* flat]: flatworms, a phylum of acoelomate animals; comprises planarians, flukes, and tapeworms.

plesiosaur (plē′sĭ·ō·sôr) [Gr. *plesios,* near]: a long-necked, marine, extinct Mesozoic reptile.

plexus (plĕk′sŭs) [L. braid]: a network, especially of nerves or of blood vessels.

pluteus (plo͞ot′ē·ŭs) [Gr. *plein,* to sail, float, flow]: the larva of echinoids and ophiuroids; also called echinopluteus and ophiopluteus, respectively.

pneumatophore (nū·mȧ′tō·fōr) [Gr. *pneumatos,* air, wind]: the air-filled float of siphonophoran hydroid coelenterates.

Pogonophora (pō·gŏ′nō·fōr′ȧ) [Gr. *pōgōn,* beard]: beard worms, a phylum of deuterostomial deep-sea animals.

polyclad (pŏl′ĭ·klăd): a member of an order of free-living flatworms, characterized by a digestive cavity with many branch-pouches.

poikilothermic (poi′kĭ·lō·thûr′mĭk) [Gr. *poikilos,* multicolored]: pertaining to animals without internal temperature controls; "cold-blooded."

polymer, polymerization (pŏl′ĭ·mĕr, -ĭ·zā′shŭn): (1) a large molecule composed of many like molecular subunits; (2) process of polymer formation.

polymorphism (pŏl′ĭ·môr′fĭz′m): differences of form among the members of a species; individual variations affecting form and structure.

polyp (pŏl′ĭp) [L. *polypus,* many-footed]: the sessile stage in the life cycle of coelenterates.

polyphyletic (pŏl′ĭ·fī·lĕt′ĭk) [Gr. *phylon,* tribe]: derived from more than one ancestral type; contrasts with monophyletic.

polysaccharide (pŏl′ĭ·săk′a·rīd): a carbohydrate composed of many joined monosaccharide units, e.g., glycogen, starch, cellulose, all formed out of glucose units.

Porifera (pō·rĭf′ĕr·ȧ): the phylum of sponges.

porocyte (pō′rō·sīt): a cell type in certain sponges characterized by a pore or canal passing through it; serves in water intake.

Priapulida (prī′ă·pû′·lĭ·dȧ): a small class of pseudocoelomate animals in the phylum Aschelminthes.

primordium (prī·môr′dĭ·ŭm) [L. beginning]: the earliest

developmental stage in the formation of an organ or body part.

proboscis (prô·bŏs'ĭs) [L.]: any tubular process or prolongation of the head or snout.

progesterone (prô·jĕs'tēr·ōn): hormone secreted by the vertebrate corpus luteum and the mammalian placenta; functions as "pregnancy hormone" in mammals.

proglottid (prô·glŏt'ĭd): a segment of a tapeworm.

prophase (prō'fāz'): a stage during mitotic division in which the chromosomes become distinct and a spindle forms.

proprioceptor (prō·'prĭ·ô·sĕp'·tēr) [L. *proprius,* one's own]: sensory receptor of stimuli originating in internal organs; e.g., as for muscle sense.

prosimian (prô·sĭm'ĭ·ăn) [L. *simia,* ape]: an ancestral primate.

prosopyle (prŏ'·sô·pīl) [Gr. *proso,* forward]: one of several incurrent openings of a choanocyte chamber in sponges.

protein (prō'tê·ĭn) [Gr. *prōteios,* primary]: one of a class of organic compounds composed of many joined amino acids.

proteinase (prō'tê·ĭn·ās): an enzyme promoting the conversion of a protein into smaller units, e.g., amino acids, or the reverse; also called protease.

prothrombin (prô·thrŏm'bĭn) [Gr. *thrombos,* clot]: a constituent of vertebrate blood plasma; converted to thrombin by thrombokinase in the presence of calcium ions, and so contributes to blood clotting.

Protista (prô·tĭs'tȧ) [Gr. *prōtistos,* first]: a major category of organisms, including all groups of algae, slime molds, protozoa, and fungi; characterized by usually unicellular reproductive structures, true nuclei, and chromosomes.

pseudopodium (sū'dô·pō'dĭ·ŭm): a temporary cytoplasmic protrusion from an amoeboid cell; functions in locomotion and feeding.

pupa (pū'pȧ) [L. doll]: a developmental stage, usually encapsulated or in cocoon, between the larva and the adult in holometabolous insects.

pylorus (pī·lō'rŭs) [Gr. *pylōros,* gatekeeper]: the opening from stomach to intestine.

pyriform (pĭr'ĭ·fôrm) [L. *pirum,* pear]: pertaining to a pear-shaped structure; e.g., pyriform organ.

Radiata (rā·dĭ·ă'tȧ) [L. *radius,* ray]: a taxonomic grade within the Eumetazoa, comprising animals with primary and secondary radial symmetry; viz. coelenterates and ctenophores.

Radiolarida, radiolaria (rā'dĭ·ô·lâr'ĭ·dȧ, -ȧ): an order of sarcodine protozoa, characterized by silicon-containing shells.

radula (răd'ū·lȧ) [L. *radere,* to scrape]: the horny rasping organ in the alimentary tract of chitons, snails, squids, and other mollusks.

recessive (rê·sĕs'ĭv) [L. *recedere,* to recede]: a functional attribute of genes; the effect of a recessive gene is masked if the allelic gene is dominant.

rectum (rĕk'tŭm) [L. *rectus,* straight]: a terminal nonabsorptive portion of the alimentary tract in many animals; opens via the anus.

redia (rē'dĭ·ȧ): a larval stage in the life cycle of flukes; produced by a sporocyst larva and in turn gives rise to many cercariae.

reduction (rê·dŭk'shŭn) [L. *reducere,* to lead back]: one half of an oxidation-reduction (redox) process; the phase of such a process which yields the net energy gain; often takes the form of addition of hydrogen or electrons to a compound, as in respiration.

reflex (rē'flĕks) [L. *reflectere,* to bend back]: the unit action of the nervous system; consists of stimulation of a sense receptor, interpretation and emission of nerve impulses by a neural center, and execution of a response by an effector organ.

renal (rē'năl) [L. *renes,* kidneys]: pertaining to the kidney.

rennin (rĕn'ĭn) [Middle Engl. *rennen,* to run]: an enzyme present in mammalian gastric juice; promotes the coagulation of milk.

respiration (rĕs'pĭ·rā'shŭn) [L. *respirare,* to breathe]: the liberation of metabolically useful energy from fuel molecules within cells; may occur anaerobically or aerobically.

reticulum, reticulate (rê·tĭk'ū·lŭm, -lāt) [L. *reticulum,* little net]: (1) a network or mesh of fibrils, fibers, or filaments, as in *endoplasmic reticulum* within cytoplasm; (2) netlike.

retina (rĕt'ĭ·nȧ) [L. *rete,* a net]: the innermost tissue layer of the eyeball; contains the receptor cells sensitive to light.

Rhodophyta (rô'dŏf'ĭ·tȧ) [Gr. *rhodon,* red]: the phylum of red algae.

ribosome (rī'bô·sōm): a cytoplasmic organelle which contains RNA and is site of protein synthesis.

rickettsia (rĭk·ĕt'sĭ·ȧ) [after H. T. Ricketts, American pathologist]: a type of microorganism intermediate in nature between a virus and a bacterium, parasitic within cells of insects and ticks.

Rotifera (rō·tĭf'ĕrȧ) [L. *rota,* wheel]: a class of microscopic pseudocoelomate animals within the phylum Aschelminthes.

rudimentary (roo'dĭ·mĕn'tȧ·rĭ) [L. *rudis,* unformed]: pertaining to an incompletely developed body part.

saccule (săk'ūl) [L. *sacculus,* little sac]: portion of the inner ear of vertebrates containing the receptors for the sense of static balance.

saprotroph (săp'rô·trōf) [Gr. *sapros,* rotten]: an organism subsisting on dead or decaying matter.

Sarcodina (sär′kō·dīnȧ): a subphylum of protozoa; amoeboid protozoa.

Sarcopterygii (sär′kō·tĕr′ĭ·jē): a subclass of bony fishes; flesh-finned fishes comprising the lobe-finned fishes and the lungfishes.

Scaphopoda (skä·fŏp′ō·dȧ) [Gr. *skaphē*, boat]: tooth shells, a class of the phylum Mollusca.

Schizophyta (skĭ·zŏf′ĭ·tȧ): the phylum of bacteria.

scolex (skō′lĕks) [L. worm, grub]: the head of a tapeworm.

scrotum (skrō′tŭm) [L]: external skin pouch containing the testes in most mammals.

sebaceous (sē·bā′shŭs) [L. *sebum*, tallow, grease]: pertaining to sebum, an oil secreted from skin glands near the hair bases of mammals.

seminal (sĕm′ĭ·nȧl) [L. *semen*, seed]: pertaining to semen, or sperm-carrying fluid.

septum, septate (sĕp′tŭm, -tāt) [L. enclosure]: (1) a complete or incomplete partition; (2) adjective.

sere (sēr) [fr. L. *series*, series]: a stage in an ecological succession of communities, from the virginal condition to a stable climax community.

serum (sē′rŭm) [L.]: the fluid remaining after removal of fibrinogen from vertebrate blood plasma.

simian (sĭm′ĭ ăn) [L. *simia*, an ape]: pertaining to monkeys; also used as noun.

sinus (sī′nŭs) [L. a curve]: a cavity, recess, space or depression; e.g., blood sinus, bone sinus.

siphon (sī′fŏn) [Gr. *siphōn*, a pipe]: tubular structure for drawing in or ejecting fluids, as in mollusks, tunicates.

siphonoglyph (sī·fŏn′ō·glĭf) [Gr. *glyphein*, to carve]: flagellated groove in pharynx of sea anemones; creates water current into gastrovascular cavity.

siphonozooid (sī·fŏn′ō·zō′oid): specialized polymorphic individual in colonies of certain coelenterates; produces water currents aiding in breathing of colony.

siphuncle (sī′fŭngk′l): gas-filled canal passing through coiled shell of chambered nautilus.

Sipunculida (sī·pŭng′kū·lī′dȧ): a phylum of wormlike schizocoelomate animals.

sol (sŏl): quasi-liquid state of a colloidal system, where water forms the continuous phase and solid particles the dispersed phase.

solenocyte (sō·lēn′ō·sīt) [Gr. *sōlēn*, channel, pipe]: excretory cell in a type of protonephridium.

somite (sō′mīt): one of the longitudinal series of segments in segmented animals; especially an incompletely developed embryonic segment or a part thereof.

species (spē shĭz), pl. *species* (spē′shēz) [L. kind, sort]: a category of taxonomic classification, below genus rank, defined by breeding potential or gene flow; interbreeding and gene flow occur among the members of a species but not usually between members of different species.

spermatogenous (spûr′mȧ·tŏj′ē·nŭs): sperm-producing.

sphincter (sfĭngk′tēr) [Gr. *sphingein*, to bind tight]: a ring-shaped muscle capable of closing a tubular opening by constriction, e.g., pyloric sphincter, which closes the opening between stomach and intestine.

spicule (spĭk′ūl) [L. *spiculum*, little dart]: a slender, pointed, often needleshaped secretion; usually a skeletal support.

spiracle (spi′rȧ·k′l) [L. *spirare*, to breathe]: reduced evolutionary remnant of first gill slit in fishes.

spore (spōr): a reproductive cell capable of developing into an adult directly.

sporocyst (spō′rō·sĭst): a larval stage in the life cycle of flukes; produced by a miracidium larva and in turn gives rise to many rediae.

sporine (spō′rin): pertaining to a sessile state of protistan existence in which cell division may take place during the vegetative condition.

sternum (stûr′nŭm) [Gr. *sternon*, chest]: in arthropods, ventral exoskeletal plate of a body segment; in vertebrates, breast bone, articulating with ventral ends of ribs on each side.

sterol, steroid (stĕr′ōl, stĕr′oid): one of a class of organic compounds containing a molecular skeleton of four fused carbon rings; includes cholesterol, sex hormones, adrenocortical hormones, and vitamin D.

stimulus (stĭm′ū·lŭs) [L. goad, incentive]: any internal or external environmental change which activates a receptor structure.

stolon (stō′lŏn) [L. *stolo*, shoot, branch]: in colonial animals, a (usually) horizontal branch or runner from which upright individuals may bud.

strobilation (strō·bĭ·lā′shŭn) [L. *strobilus*, pine cone]: process of segmentlike budding in sessile scyphistoma larvae of scyphozoan coelenterates, resulting in cutting off of successive free-swimming ephyra larvae.

stroma (strō′mȧ) [Gr. couch, bed]: the connective tissue network supporting the epithelial portions of an organ.

style, stylet (stīl, stī′lĕt) [Gr. *stylos*, pillar]: a stalklike or elongated body part, often pointed at one end.

substrate (sŭb′strāt) [L. *substratus*, strewn under]: a substance which is acted upon by an enzyme.

symbiont, symbiosis (sĭm′bĭ·ŏnt, sĭm′bĭ·ō′sĭs): (1) an organism living in symbiotic association with another; (2) the intimate living together of two organisms of different species, for mutual or one-sided benefit; the principal variants are mutualism, commensalism, and parasitism.

sympathetic (sĭm′pȧ·thĕt′ĭk): applied to a subdivision of the autonomic nervous system; centers are located in the midportion of the spinal cord.

synapse (sĭ·năps′): the microscopic space between the axon terminal of one neuron and the dendrite terminal of another, adjacent neuron.

syncytium (sĭn·sĭ′shĭ·ŭm): a multinucleate animal tissue without internal cell boundaries.

syngen (sĭn′jĕn): a mating group (or variety) within a protozoan species; mating may occur within a syngen but not usually between syngens; a functional (as distinct from taxonomic) "species".

synthesis (sĭn′thē·sĭs) [Gr. *tithenai,* to place]: the joining of two or more molecules resulting in a single larger molecule.

syrinx (sĭr′ĭngks) [Gr. a pipe]: the vocal organ of birds, located where the trachea branches into the bronchi.

systole (sĭs′tô·lē) [Gr. *stellein,* to place]: the contraction of atria or ventricles of a heart.

syzygy (sĭz′ĭ·jĭ) [Gr. *syzygia,* conjunction]: sex cells of Sporozoa united into a chain of alternating male and female cells.

taiga (tī′gȧ) [Russ.]: terrestrial habitat zone characterized by large tracts of coniferous forests, long, cold winters, and short summers; bounded in the north by tundra; found particularly in Canada, northern Europe, and Siberia.

tardigrade (tär′dĭ·grād) [L. *tardigradus,* a slow-stepper]: a member of a subphylum of Oncopoda; water bears.

tarsus (tär′sŭs) [Gr. *tarsos,* sole]: in insects, the terminal parts of a leg; in vertebrates, the ankle.

taxon (tăks′ŏn) pl. *taxa:* a rank category in animal classification.

taxonomy (tăks·ŏn′ô·mĭ) [Gr. *nomos,* law]: classification of organisms, based as far as possible on natural relationships.

tectorial membrane (tĕk·tō′rĭ·ȧl) [fr. L. cover, covering]: component of the organ of Corti in cochlea of ear.

telencephalon (tĕl′ĕn·sĕf′ȧ·lŏn) [Gr. *enkephalos,* brain]: the vertebrate forebrain.

telolecithal (tĕl′ô·lĕs′ĭ·thăl): pertaining to eggs with large amounts of yolk accummulated in the vegetal half of the cell; e.g., as in frog eggs.

telophase (tĕl′ô·fāz): a stage in mitotic division during which two nuclei form; usually accompanied by partitioning of cytoplasm.

telson (tĕl′sŭn) [Gr. boundary, limit]: terminal segmentlike body part of an arthropod (not generally counted as a segment).

template (tĕm′plĕt): a pattern or mold guiding the formation of a duplicate; term applied especially to gene duplication, which is explained in terms of a template hypothesis.

temporal lobe (tĕm′pô rȧl) [L. *tempora,* the temples]: a part of the vertebrate cerebrum; contains neural centers for speech and hearing.

tergum (tûr′gŭm) [L. the back]: the dorsal exoskeletal plate of a body segment in arthropods.

testis, pl. *testes* (tĕs′tĭs, tĕs′tēz) [L.]: sperm-producing organ.

tetrad (tĕt′răd): a pair of chromosome pairs present during the first metaphase of meiosis.

tetrapyrrol (tĕt′rȧ·pĭ′rŏl): a molecule consisting of four united pyrrol units, each of the latter being a five-membered ring of carbon and nitrogen; the four pyrrols may be joined linearly or as a larger ring;. tetrapyrrols include pigments such as heme.

thalamus (thăl′ȧ·mŭs) [Gr. *thalamos,* chamber]: a lateral region of the diencephalic portion of the vertebrate forebrain.

theory (thē′ô rĭ) [Gr. *theōrein,* to look at]: a scientific statement based on experiments which verify a hypothesis; the last step of the scientific method.

therapsid (thê·răp′sĭd) [Gr. *thērion,* beast]: extinct Mesozoic mammallike reptile; true mammals evolved from same group.

Theria (thēr′ĭ·ȧ) [Gr. *thērion,* beast]: a subclass of mammals comprising marsupials and placentals; the viviparous mammals.

thrombin (thrŏm′bĭn) [Gr. *thrombos,* clot]: a substance participating in vertebrate blood clotting; formed from prothrombin and in turn converts fibrinogen into fibrin.

thrombokinase (thrŏm′bô·kĭn′ās): enzyme released from vertebrate blood platelets during clotting; transforms prothrombin into thrombin in presence of calcium ions; also called *thromboplastin.*

thrombus (thrŏm′bŭs): a blood clot within the circulatory system.

thymus (thī′mŭs) [fr. Gr.]: a lymphoid gland in most young and many adult vertebrates; disappears in man at puberty; located in lower throat and upper part of thorax.

thyroxine (thī·rŏk′sĭn): the hormone secreted by the thyroid gland.

tibia (tĭb′ĭ·ȧ) [L.]: in vertebrates, usually the larger shin bone of the hind limb between knee and ankle; in insects, the leg portion between femur and tarsus, part of the arthropod endopodite.

tissue (tĭsh′ū) [L. *texere,* to weave]: an aggregate of cells of similar structure performing similar functions.

tornaria (tôr·nā′rĭ·ȧ) [L. *tornus,* lathe, chisel]: the larva of certain enteropneust hemichordates.

trachea, tracheal (trā′kê·ȧ) [Gr. *trachys,* rough]: (1) air-conducting tube, as in windpipe of mammals and breathing system of insects; (2) adjective.

transduction (trăns·dŭk'shŭn): transfer of genetic material from one bacterium to another through the agency of a virus.

trilobite (trī'lô·bīt): an extinct, marine, Paleozoic arthropod, marked by two dorsal longitudinal furrows into three parts or lobes.

trochanter (trô·kăn'tẽr) [Gr. *trechein*, to run]: in insects, the part of a leg adjoining the coxa, equivalent to the basipodite of arthropods.

trochophore (trŏk'ô fōr): a free-swimming ciliated marine larva, characteristic of schizocoelomate animals.

tropic, tropism (trŏp'ĭk) [Gr. *tropē*, a turning]: (1) pertaining to behavior or action brought about by specific stimuli, e.g., phototropic (light-oriented) motion gonadotropic (stimulating the gonads); (2) noun.

trypsin (trĭp'sĭn) [Gr. *tryein*, to wear down]: vertebrate enzyme promoting protein digestion; acts in small intestine, but produced in pancreas as inactive trypsinogen.

tundra (to͝on'drȧ) [Russ.]: terrestrial habitat zone, between taiga in south and polar region in north, characterized by absence of trees, short growing season, and frozen ground during much of the year.

Turbellaria (tûr'bĕ·lar'ĭ·ȧ) [L. *turba*, disturbance]: the class of free-living flatworms; planarians.

turgor (tûr'gŏr) [L. *turgere*, to swell]: the distention of a cell by its fluid content.

typhlosole (tĭf'lô·sōl) [Gr. *typhlos*, blind]: dorsal fold of intestinal wall projecting into the gut cavity in oligochaete annelids such as earthworms.

umbilicus (ŭm·bĭl'ĭ·kŭs) [L.]: the navel in mammals; during pregnancy, an umbilical cord connects the placenta with the offspring, and the point of connection with the offspring later becomes the navel.

umbo (ŭm'bō) [L. boss of shield]: rounded prominence near hinge of clam valve; oldest part of shell.

ungulate (ŭng'gû·lāt) [L. *ungula*, hoof]: hoofed, as in certain orders of mammals.

urea (û·rē'ȧ) [Gr. *ouron*, urine]: an organic compound formed in the vertebrate liver out of ammonia and carbon dioxide and excreted by the kidneys; represents principal means of ammonia disposal in mammals and some other animal groups.

ureter (û·rē'tẽr) [fr. Gr.]: duct carrying urine from a metanephric kidney to the urinary bladder.

urethra (û·rē'thrȧ) [fr. Gr.]: duct carrying urine from the urinary bladder to the outside of the amniote body; in the males of most mammals, the urethra also leads sperms to the outside during copulation.

Urodela (û'·rŏ·dĕ'lȧ) [Gr. *dēlos*, visible]: a subclass of tailed amphibia, comprising newts and salamanders.

uterus (ū'tẽr·ŭs) [L. womb]: enlarged region of a female reproductive duct in which embryo undergoes all or part of its development.

utricle (ū'trĭ·k'l) [L. *utriculus*, little bag]: portion of the vertebrate inner ear containing the receptors for dynamic body balance; the semicircular canals lead from and to the utricle.

vacuole (văk'û ōl) [L. *vacuus*, empty]: a small, usually spherical space within a cell, bounded by a membrane and containing fluid, solid matter, or both.

vagina (vȧ·jī'nȧ) [L. sheath]: the terminal, penis-receiving portion of a female reproductive system in animals in which fertilization is internal.

vagus (vā'gŭs) [L. wandering]: the tenth cranial nerve in vertebrates; it is a mixed nerve, innervating many organs in the chest and the abdomen.

valence (vā'lĕns) [L. *valere*, to have power]: a measure of the bonding capacity of an atom; an atom may be monovalent, divalent, trivalent, etc., indicating the number of other atoms it may bond with; bonds may be electrovalent, formed through electron transfer, or covalent, formed through electron sharing.

vasomotion (văs'ô·mō'shŭn) [L. *vasum*, vessel]: collective term for the constriction (vasoconstriction) and dilation (vasodilation) of blood vessels.

veliger (vĕl'ĭ·jẽr) [fr. L. *velum*, veil]: post-trochophoral larval stage in many mollusks.

velum (vē'lŭm): a membranous curtainlike band of tissue, as on underside of many hydroid medusae.

venous (vē'nŭs) [L. *vena*, vein]: pertaining to veins; also applied to oxygen-poor, carbon-dioxide-rich blood.

ventricle (vĕn'trĭ·k'l) [L. *ventriculus*, the stomach]: a chamber of a heart which receives blood from an atrium and pumps out blood from the heart.

vestigial (vĕs·tĭj'ĭ·ȧl) [L. *vestigium*, footprint]: degenerate or incompletely developed, but more fully developed at an earlier stage or during the evolutionary past.

villus (vĭl'ŭs) *pl. villi* [L. a tuft of hair]: a tiny fingerlike process projecting from the intestinal lining into the cavity of the mammalian gut; contains blood and lymph capillaries and is bounded by the intestinal mucosa.

virus (vī'rŭs) [L. slimy liquid, poison]: a submicroscopic noncellular particle, composed of a nucleic acid core and a protein shell; parasitic, and within a host cell it may reproduce and mutate.

viscera (vĭs'ẽr·ȧ), sing. *viscus* [L.]: collective term for the internal organs of an animal.

vitamin (vī'tȧ·mĭn) [L. *vita*, life]: one of a class of organic growth factors contributing to the formation or action of cellular enzymes.

vitreous (vĭt′rē·ŭs) [L. *vitrum,* glass]: glassy; as in vitreous humor, the clear transparent jelly which fills the posterior part of the vertebrate eyeball.

viviparity, viviparous (vĭv′ĭ·păr′ĭ·tĭ, vī·vĭp′*à*·rŭs) [L. *vivus,* living + *parere,* to bring forth]: (1) reproductive pattern in which eggs develop within female body with nutritional and other metabolic aid of maternal parent; offspring are born as miniature adults; (2) adjective.

volvant (vŏl′vănt) [L. *volvere,* to roll, twist]: closed-ended adhesive thread in a type of nematocyst of coelenterates.

Xiphosura (zĭ·fōs′ū·r*à*) [Gr. *xiphos,* sword]: a class of chelicerate arthropods; the horseshoe crabs (*Limulus*).

zoaea (zō·ē′à) pl. *zoaeae:* a larval form of crustaceans.

zooid (zō′oid): an individual animal in a colonial aggregation; often physically joined with fellow zooids and may be a polymorphic variant.

zygote (zī′gōt) [Gr. *zygōtos,* yoked]: the cell resulting from the sexual fusion of two gametes; a fertilized egg.

index